BIOLOGY

A Journey Into Life

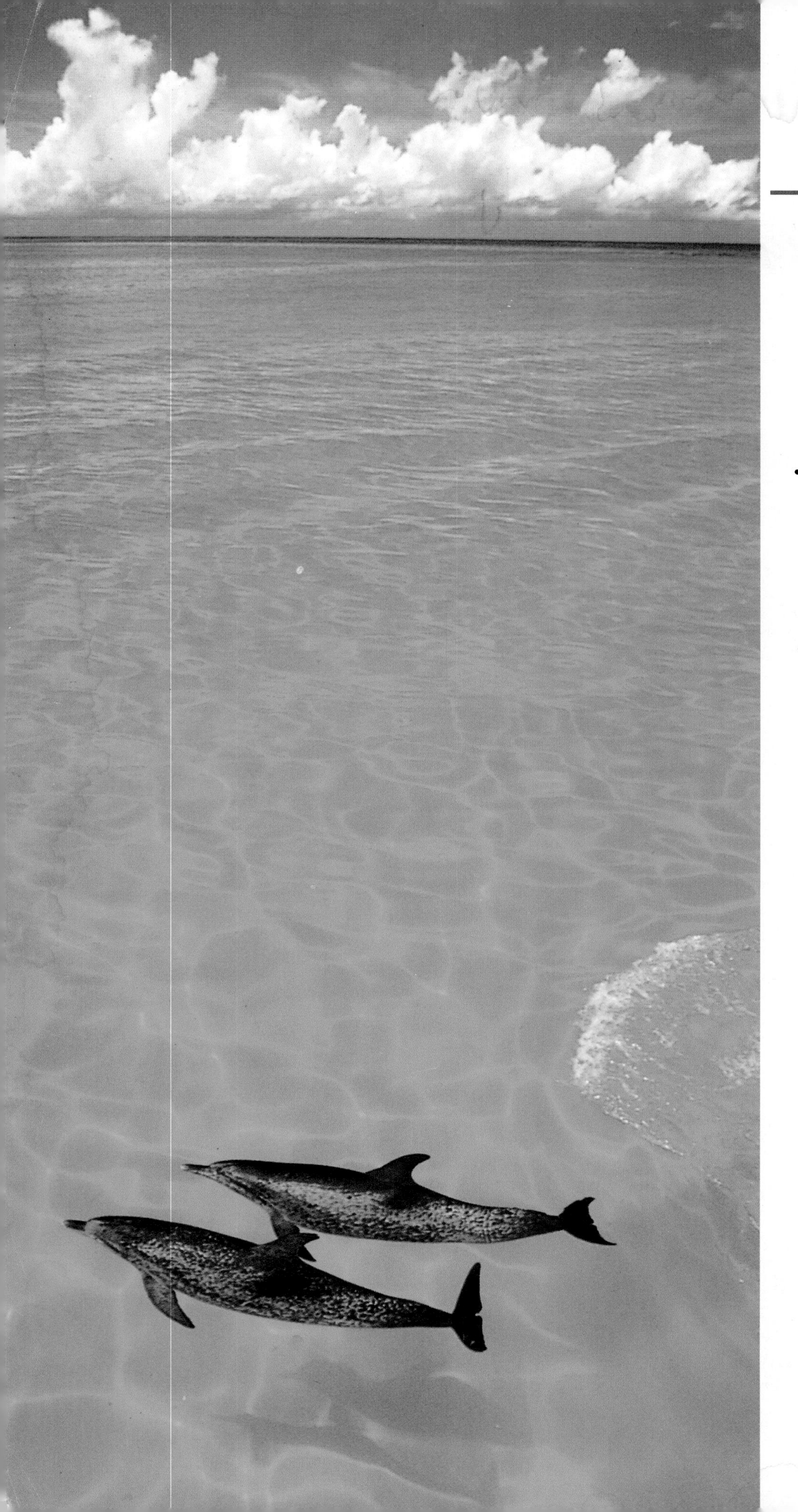

Karen Arms

Pamela S. Camp

Janann V. Jenner
Talladega College

Edward J. Zalisko
Blackburn College

BIOLOGY

THIRD EDITION

A Journey Into Life

SAUNDERS COLLEGE PUBLISHING
HARCOURT BRACE COLLEGE PUBLISHERS

Fort Worth Philadelphia San Diego New York Orlando
San Antonio Toronto Montreal London Sydney Tokyo

Text Typeface: Garamond
Compositor: York Graphic Services
Acquisitions Editor: Julie Levin Alexander
Developmental Editor: Gabrielle Goodman
Managing Editor: Carol Field
Project Editor: Margaret Mary Anderson
Copy Editor: John Beasley
Manager of Art and Design: Carol Bleistine
Text Designer: Tracy Baldwin
Cover Designer: Lawrence R. Didona
Art Development/Text Artwork: J/B Woolsey Associates
Photo Research: Laurel Anderson/Photosynthesis, Inc.
Layout Artist: Rebecca Lloyd Lemna
Director of EDP: Tim Frelick
Production Manager: Joanne Cassetti
Director of Marketing: Marjorie Waldron
Marketing Manager: Sue Westmoreland

Cover Credit: ©Norbert Wu (inset)
©COMSTOCK, Inc. (background)

Printed in the United States of America

BIOLOGY: A JOURNEY INTO LIFE, Third Edition

0-03-0098796-2

Library of Congress Catalog Card Number: 93-085642

4567-032-987654321

To Richard, Sarah, and Patrick

K.A.

To Walter

P.C.

For all my students who, try as they might, could never see anything through the microscope but the reflection of their own eyelashes. . .

J.J.

To my late grandfather, John Kozlowsky, who shared his love of life and learning, and to my parents, Elizabeth and Miles, for nurturing those pursuits.

E.Z.

About the Author Team

Jan Jenner lived in Manhattan's Greenwich Village for many years, and most recently has been transplanted to the Alabama pine woods. She received her doctorate in biology from New York University, specializing in herpetology, and taught introductory biology there for more years than she cares to mention. Although snakes are a particular interest, Jenner enjoys all aspects of whole animal studies, especially those that involve field work. She currently teaches introductory biology, comparative vertebrate zoology, and environmental science at Talladega College (AL) and supervises undergraduate researchers as the Director of Research at TC's Holtzclaw Wildlife Preserve. Jenner spends free time writing works of fiction and non-fiction (the novel *Sandeagozu,* as well as *A Bird Watcher's Companion* and *Backyard Birds*), birding, studying insects, painting, and tending her pet tortoises.

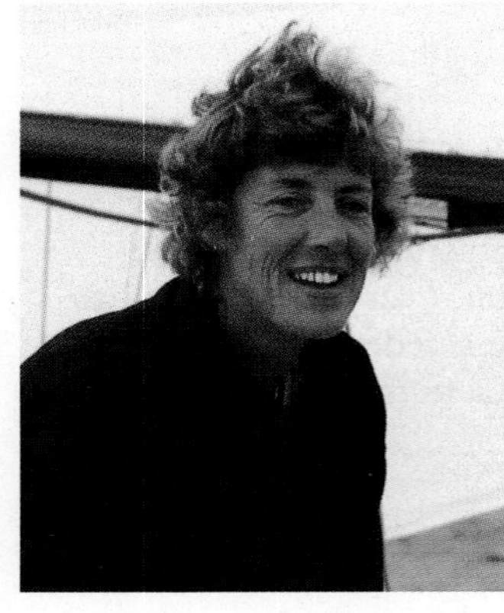

Karen Arms grew up in Oxford, England and received her doctorate in molecular embryology at Oxford University. After relocating to the United States, she taught a wide range of courses at both Stanford and Cornell Universities and began to shift the emphasis of her work from teaching to writing, in response to the needs of introductory biology students at Cornell University. Arms now lives in Savannah, Georgia, where she sails, gardens, appears as guest lecturer at nearby schools, works as an editor and teacher with the University of Georgia Marine Extension Service, is President of Halfmoon Publishing, is an active participant of the 1996 Olympic sailboat committee, and more than anything else, writes, writes, and writes about biology, the subject she adores. Arms is married to an artist and has three children in college.

Pam Camp grew up in suburban Washington, D.C. and attended Allegheny College in Meadville, Pennsylvania and Cornell University. Subsequently she became involved in the development of the introductory biology course at Cornell with Karen Arms. The first edition of their book for majors, *Biology,* was written from the Cornell course. The highly successful author team of Arms and Camp has produced seven editions of three biology textbooks and Camp now devotes all of her time to writing. When she's not at the computer, pressured by a deadline, she's most likely to be found behind the lens of her camera, swimming, out on a long walk, gardening, biking, traveling, or blissfully engrossed in a mystery or romance novel with one of her two cats purring at her side. Camp lives in Ithaca, New York.

Ed Zalisko was raised in East Peoria, Illinois. At Washington State University he earned a doctorate in zoology as well as a state teaching certificate in biology and chemistry. His research projects have included numerous anatomical and ecological studies of salamanders and snakes. His interests in biology stem from many outdoor vacations and a grandfather who stimulated his natural curiosity. Zalisko uses his enthusiasm, sense of humor, and analogies to help motivate the students in the seven courses he teaches as Professor of Biology at Blackburn College in Carlinville, IL. A specialist in the methodology of science teaching, he regards his biology classes as "a chance to tell some great stories." When not in class Zalisko can be found presenting results of his research at scientific meetings, conducting research with undergraduates, or visiting local elementary schools to show off "beasts." In his spare time Zalisko reads, plays with his two small children, Benjamin and Sarah, and enjoys small town life with his wife, Amy.

PREFACE

The concepts of biology are threads in the fabric of daily life.

For example,

Is it safe to eat genetically engineered tomatoes?
What is a hiccup?
Is it necessary to take megadoses of vitamins?
Will your unborn child inherit your curly, red hair?
Can scientists actually recover dinosaur DNA and bring *Tyrannosaurus rex* back to life?
What are the side effects of spraying insecticides around your home and garden?
Why should you follow your medical doctor's instructions when taking antibiotics?
Which came first, the chicken or the egg?

Biological questions run the gamut from the silly to the sublime, but biology is the most visible, and arguably the most important science in the news. Every day each person makes many decisions centered around biological concepts. The philosophical basis of this revision of *Biology: A Journey Into Life* revolves around the goal of making the critical concepts of biology more accessible, thus enabling students, as educated citizens, to better understand the issues that confront them, both in the headlines, and in their own lives.

During the entire revision process we have consciously kept the nonmajor student, or the student with little or no knowledge of biology, in mind. We have tried to make this new edition inviting, clear, and easy to understand, without sacrificing the excellent science of the second edition. To facilitate a student's grasp of the critical concepts of biology, we have developed an array of artistic and pedagogical features designed to highlight biological concepts, enhance learning, increase comprehension, and develop critical thinking skills. In addition, we have placed an even greater emphasis, integrated throughout the text, on the environmental issues that confront us—issues that every well-educated citizen should understand from a biological perspective.

Updating the text and underscoring environmental issues were relatively easy goals to achieve; re-concepting the art and developing the pedagogical features involved integrating more ideas and thus were more difficult, but after a year and a half of intensive work that has involved scores of people, the process has finally resulted in the book you now hold in your hands. This text is appropriate for courses taught to nonmajors, courses with a mix of majors and nonmajors, and even for courses for some majors.

Improved Illustrations

Because we know that so much of human learning is visual, a special Art Focus Group of active educators was convened to explore how students and teachers use an art program. As a result, every illustration has been scrutinized, reconsidered, redeveloped, and revised by a team of biology teachers, writers, and artists. Many of the ideas for the illustrations are completely new; all of the line drawings are new; and many new art features have been added that will facilitate learning. Context has been emphasized in all of our newly designed artwork to provide the student with a frame of reference for each figure. Context is also provided by the use of standardized icons throughout the text, such as those of the human body, and of an idealized plant and animal cell.

Biology is full of complicated processes that challenge the skills of learners and teachers alike. Respiration, photosynthesis, evolution, DNA replication, protein synthesis, mitosis, and meiosis are only a few examples. In each chapter these important pieces of the biological puzzle have been made more accessible through the use of a new illustration feature, **"A Journey Through,"** which combines multi-part artwork and photos with a unique labeling system that guides students through the figure. Examples include "A Journey Through Photosynthesis," "A Journey Through Meiosis, "A Journey Through An Animal Cell," etc.

New and Improved Pedagogical Features

The following new and/or improved pedagogical features have been added:

1. **Study Techniques for Biology Classes.** Biology demands an array of study skills that are quite different from those needed in other college courses. Written by one of the text authors with 20 years of experience in teaching introductory biology, "Study Techniques for Biology Classes" gives careful, time-tested, practical advice for the beginning student on how to maximize efficiency when studying for biology.
2. **Curiosity Questions.** Each chapter begins with a few attention-grabbing questions that articulate things that students may have wondered about, or

things that they might have asked after giving the subject some thought. Curiosity Questions encourage students to dive into the chapter in search of answers. The questions are referenced to icon-marked text or specific figures within the chapter where the answer is integrated into the text or figure legend. Examples include,

1? Why aren't there as many bald women as bald men?

2? How can parents have a child who grows taller than either one of them?

3? Are the cells of an elephant bigger than the cells of a mouse?

3. **Chapter Opening Photograph.** Rather than using chapter opening photos as window dressing, a single chapter opening photograph illustrates an important point within the chapter and piques the curiosity of the reader, encouraging him or her to read further. After briefly explaining the figure and the concept it illustrates, the figure legend ends by directing students to the text section where they can "read more about this topic."
4. **Concept Guides.** These frame the objectives for each chapter as directives that have been rewritten at higher cognitive levels to encourage critical thinking skills. We consider objectives such as "list" or "name" too basic to develop critical thinking skills. Instead, Concept Guides ask the student to "explain," "describe," "compare," etc.
5. **Key Concepts.** Brief statements of the central concepts of each chapter set the stage for the development of the ideas within each chapter.
6. **Concept Capsules.** Concept Capsules are a new feature of this edition. They appear between two red lines at the end of each main section, and summarize and reinforce main concepts before the student reads on. To encourage students to read entire sections instead of just the summaries, Concept Capsules use terms presented in the section to discuss the key concepts without repeating their definitions.
7. **Bio-Bits.** Interesting, topical biological facts directly tied to the text appear in each chapter to emphasize important points and highlight biological issues that are newsworthy and/or relevant to students' lives. Examples include,

   "In tundra permafrost entire, intact mammoths have been found with edible meat still clinging to their skeletons."

   "It is likely that within your lifetime, nearly one-quarter of all of the species of plants, animals, and microorganisms on Earth will become extinct due to destruction of tropical rain forests. The vast majority of the species lost will never even have been studied."

   "It is currently estimated that skin cancer rates increase by 6% for every 1% loss of the ozone layer. The recent thinning of the ozone layer by 3% near middle latitudes has therefore resulted in a nearly 20% increase in skin cancer."

8. **Tool Boxes.** Specially designed to look like windows on a computer screen, these are short boxes that explain some of the technicalities which are necessary for a complete understanding of the main ideas of a chapter. For example, moles as units of measurement, and surface area-to-volume ratios, are explained in Tool Boxes.
9. **Journey Boxes.** These boxed essays are aimed at providing deeper knowledge of a topic that is ancillary to material presented in a chapter. Boxes fall into one of four categories: **"A Journey into Science in Process," "A Journey into Evolution," "A Journey into Healthy Living,"** or **"A Journey into the Environment."** More than one-third of the boxes are completely new, and all have been updated where necessary. In particular, "A Journey into the Environment" plays an important role in integrating environmental issues throughout the entire text.
10. **Chapter Summary.** To facilitate comprehension, the chapter summaries have been revised and changed from a paragraph format to a numbered list of the main points.
11. **Self-Quiz.** This is a series of straightforward questions that test the basic comprehension of the ideas presented in the chapter. The answers to the Self Quiz are provided at the end of the book.
12. **Thinking Critically.** These are probing questions that require additional thought and consideration. They can form the basis of class discussions, be the springboard for lectures, or be the basis for writing exercises.
13. **Selected Key Terms.** This list of key terms is new to this edition. One of the most daunting aspects of learning biology is mastering the terminology. For quick review this feature lists the chapter's new and important terms alphabetically with page references.
14. **Suggested Readings.** The list of books and articles at the end of each chapter has been edited and updated.
15. **The Educated Citizen.** These excerpts of recent *Discover* magazine articles, chosen by the authors, appear at the end of each of the book's six parts. The articles explore topical issues that are controversial and often have many sides to them. At the end of each Educated Citizen article are several questions, called **Connecting the Concepts.** These questions may be used as either essay or discussion topics and link the issues discussed in the articles with the concepts that have been

introduced and explained in the preceding Part. They are tied to the specially edited *Infinite Voyage* videos that are available as part of the supplement package.

Organization and Content Changes

A new Chapter 1 sets the tone for our revision. Emphasizing our Journey Into Life theme, this chapter gives the student an overview of the intellectual terrain that lies ahead and emphasizes the central biological concepts that each student will encounter. Our environmental emphasis is introduced and highlighted in a new Table on the "Top Ten" global environmental challenges, which will be underscored throughout the text. In Part 1, The Unity of Life: Cells, the entirely rewritten chapters on chemistry (2 and 3) make the bane of most freshman biology classes much less threatening and much easier to understand. The excellent science of the second edition's cell membrane and cell structure and function chapters (4 and 5) has been retained, updated where necessary, and now is further clarified by improved artwork and accompanying photos. A rewritten chapter on energy (6) uses everyday analogies to explain energy transformations and the importance of energy intermediates. The chapters on cellular respiration (7) and photosynthesis (8) have been completely rewritten to present the complexities of these often daunting topics in the clearest manner. The accompanying "A Journey Through" features walk the reader through these processes in a careful, stepwise fashion.

In Part 2, The Unity of Life: Genetic Information and Its Expression, the focus of the revision is on the artwork. In Chapter 9 we highlight the experimental work that established DNA as the genetic material, explain the structure of the DNA molecule, and build a chromosome from a strand of DNA. Much of Chapter 10 has been rewritten, including a new Journey box, entitled, "Smart Genes." The artwork for "A Journey Through Protein Synthesis" clarifies this challenging topic. In Chapter 11 two new two-page spreads, "A Journey Through Mitosis" and "A Journey Through Meiosis" detail these processes in both the plant and animal cell, and match explanatory artwork with photographs similar to microscope slides the student may encounter in laboratory exercises. The artwork in Chapters 12 and 13 brings new clarity to Punnett squares. Not only are the results of the various genetic crosses clearly shown, but each Punnett square figure reminds students of the context with a small, accompanying icon.

In Part 3, Evolution and the Diversity of Life, the introductory and initial sections of Chapter 14, Evolution and Natural Selection, have been rewritten, with the comments of many veteran teachers in mind. The unique Table 14-1 directly addresses and dispels many of the major misconceptions students have about the theory of evolution. The art of Chapter 15 has also been carefully revised. Concepts such as Hardy-Weinberg equilibrium and the founder effect were especially challenging design problems, and we believe our illustrations are exceptionally clear. Reflecting the environmental emphasis of this revision, Chapter 16 has an especially important new Journey box on overpopulation and Chapter 17 has a new Journey box on ozone depletion. New students are often mystified by the emphasis that biology places on correct scientific names. In a new Journey box "The Name Game" (Chapter 18), we explore the utility of scientific names. Chapters 19–21 have been updated and trimmed and in all of these chapters, new artwork displays the diversity and beauty of living organisms.

The emphasis of the revision in Part 4, Animal Biology was twofold: improved, more inviting artwork and reworking, clarifying, and tightening of the text. Special attention has been paid to Chapter 25, Defenses Against Disease and in Chapter 27 a new section on Fertilization and Implantation clarifies these topics. New findings, such as those in the treatment of cystic fibrosis and PKU have been incorporated into the text. In each chapter in Part 4 (Chapters 22–31) a special, large-format Journey Through figure integrates details of human anatomy with details of physiology, presenting the student with a new visual summary of the workings of each of the organ systems.

The emphasis in the revision of Part 5, Plant Biology (Chapters 32–35), was on the development of the artwork. A new Journey box "Tropical Rain Forest Canopy" continues the environmental theme, highlighting the importance and uniqueness of this new biological frontier.

In Part 6, The World of Life: Ecology, the fundamental concepts of ecology, such as energy flow, biogeochemical cycles, niche, community structure, biomes, are explained and developed. New illustrations and new photographs make these important chapters even more vivid. The content of Chapter 39, Human Impact On The Environment, has been completely revised, and much of it newly written to reflect our environmental concerns and suggest appropriate responses to the environmental crises that confront us.

Supplements

We are pleased to be able to offer more ancillary teaching aids with this edition of *Journey.*

Six **videos** in the critically acclaimed *The Infinite Voyage* series have been chosen by the authors and specially edited to augment the text and the Educated Citizen articles in the book.

***Infinite Voyage* Videodisc Series:** The Year in Review examines various biological topics in a collection of eight 1-hour video programs, including “Secrets from a Frozen World” and “Insects: The Ruling Class.”

Saunders General Biology Videodisc offers live action and still images from six Saunders biology textbooks, including *Biology: A Journey Into Life,* Third Edition. Features include random access, light scanner/barcode access, and extensive lecture presentation capabilities through LectureActive™ Software (see description below). A Directory with descriptions, barcode labels, and reference numbers for each image accompanies the Videodisc.

LectureActive™ Software contains all video clip and still images from Saunders Videodisc, allowing instructors to create custom lectures quickly and easily. Lectures can be read from the computer screen or printed with accompanying barcodes for all Videodisc instructions.

Study Guide written by author Ed Zalisko includes questions keyed to text sections, Concept Guides and “Journey” boxes. In addition, each chapter offers specific study advice, additional challenging questions, a key to word roots, and a crossword puzzle which includes key terms and word roots.

Overhead Transparency Acetates feature 250 pieces of lustrous art from the text, using labels with large type for easy classroom viewing.

Slides feature the same 250 images in a 35-mm format.

Sequence Overhead Transparencies are a separate set of 25 sequential overhead transparencies, containing topics displayed in a series of stages of layers. Each overhead transparency includes four to six overlays held together with removable plastic pins.

Bio-Art reproduces 150 selected pieces of art from the text as black-and-white unlabeled line drawings, encouraging students to learn the labeling process and take notes. Bio-Art also provides a handy study tool or can be used as a test item.

Instructor’s Manual contains key words, an overview, lecture outline, suggested readings, and teaching suggestions.

Lecture Outline on Disk, an ASCII version of the lecture outline, is available for IBM and Macintosh computers. Instructors are able to edit, expand the outline, and create study guideline handouts for students.

Test Bank provides 3,000 printed test questions in several formats and levels of difficulty. Instructors can add to, alter, or edit to develop their own tests.

ExaMaster+™ Computerized Test Bank offers instructors a software version of the printed Test Bank, complete with the same capabilities.

BIO-XL, a unique, computer-assisted tutorial software package, is available in two modes. Test Mode quizzes students about chapter material and assesses students’ knowledge; Tutor Mode adds pedagogical support through immediate feedback to responses. Corresponds to specific page references in the text.

SimLife™, an advanced biological simulation, enables students to design creatures and plants from a genetic level and manipulate the environment in which they live, testing their ability to survive. The software is based on the latest advances in Artificial Life research, and is accompanied by a Teacher’s Guide and Sample Lab Book.

Acknowledgments

We thank the artists at **J/B Woolsey Associates** for their superb efforts in revising the illustrations for this new edition. The entire art team has been seemingly tireless in the production and execution of our art and we thank them especially for giving a 110 percent effort. Patrick Lane, John Woolsey, and Dave McShane were responsible for the concepting and development of illustrations, while Patrick Lane, Regina Hollister, Greg Gambino, Mark Desman, Tom Sincak, Dawn Derosa and Todd Smith executed the illustrations. We think they are beautiful as well as functional and feel confident that our readers will concur.

Laurel Anderson and the staff at **Photosynthesis, Inc.** were our photo researchers on this revision. Laurel and her staff were able to find photos to match nearly all of our excruciatingly obscure requests and we thank Photosynthesis for their superb attention to detail and eye for beauty.

The entire staff at **Saunders College Publishing** has cooperated to make the revision process as amiable and pleasant as possible. We thank them for their "can-do" attitude and the cheerful goodwill they have extended to this author team. **Elizabeth Widdicombe,** Publisher, was always supportive of our efforts. She and **Julie Levin Alexander** brought the team together, and got the process started. Many of the original ideas for this revision were Julie's and her synthetic perspective, unfailing support, and guidance have been greatly appreciated. As Developmental Editor, **Gabe Goodman** was the mastermind behind this revision. Gabe's impeccable ear for language and dogged attention to detail enabled her to coordinate efforts of authors, artists, and reviewers, and prepare the manuscript for production. In addition to her professional tasks, Gabe provided much-needed reassurance to the authors when the pressures of deadlines seemed unbearable. The authors could not have asked for a better editor than Gabe Goodman. **Laura Coaty,** Field Product Manager, was one of our essential links to the needs of teachers and students. Laura was closely involved in the initial planning stages of this revision, conducting the Art Focus Group, preparing invaluable reviewer data bases, and relaying her knowledge to the revision team. We greatly appreciate her buoyant dedication to making this the best of all possible books. **Christine Rickoff,** Associate Developmental Editor, coordinated the development of the ancillary package and Study Guide. As Project Editor, **Margaret Mary Anderson** shepherded the manuscript through all the various incarnations that have resulted in this finished textbook. We thank her for tying up all of our loose ends and for her careful coordination of the work of copy editor, author team, and artists. **Carol Bleistine,** Manager of Art and Design, supervised both the creation of the design of our book and the completion of the beautiful artwork. We thank her for making *Journey,* Third Edition a stunning example of book design. **Joanne Cassetti,** Production Manager, worked with an impossible schedule, and like everyone else associated with this revision, did the impossible, and got our book off the presses on time.

Reviewers

This book's official reviewers are teachers. They have taken time from their busy schedules to give us their suggestions for improvement, read the manuscript, and suggest numerous changes. For your many thoughtful contributions, our thanks to:

Frances Abbott, *Broward Community College*

Richard Alford, *Jefferson State Community College*

Jane Aloi, *Saddleback College*

Mark Armstrong, *Blackburn College*

Charlotte Bacon, *University of Hartford*

Susan Brawley, *University of Maine, Orono*

Linda Butler, *University of Texas at Austin*

Sharon Clark, *Golden West College*

William Coleman, *University of Hartford*

Don Collins, *Orange Coast College*

Don Defler, *Portland Community College*

Jean De Saix, *University of North Carolina, Chapel Hill*

Larry Drummond, *Talladega College*

Bob Eagan, *Gold West College*

Stephen Hedman, *University of Minnesota - Duluth*

Julius Ikenga, *Mississippi Valley State University*

Alice Jacklet, *State University of New York, Albany*

Andrew Langford, *Community College of Denver*

Donny Lawson, *Copiah-Lincoln Community College*

Charles Leavell, *California State University - Fullerton*

Om P. Madhok, *Minot State University*

Phil Mathis, *Middle Tennessee State University*

Neil Miller, *Memphis State University*

Thomas Milton, *Richard Bland College*

Roy Olson, *Southwestern College*

Mike Palmer, *Oklahoma State University*

Rudolph Prins, *Western Kentucky University*

Doug Reynolds, *Eastern Kentucky University*

Franklin Roberts, *University of Maine, Orono*

Doug Schecnayder, *Copiah-Lincoln Community College*

David Shannon, *Hiwassee College*

May Shumakihuro, *Morehead State University*

Frank Sivik, *Broward Community College*

Dan Tallman, *Northwestern State University*

Gary Tallman, *Pepperdine University*

David Thorndill, *Essex Community College*

John Thornton, *Oklahoma State University*

Bob Turner, *Western Oregon State College*

Jack Turner, *Sam Houston State College*

M. R. Uddin, *Rust College*

Jack Waber, *West Chester University*

Mark Wallert, *Morehead State University*

Patricia Walsh, *University of Delaware*

Ted Weinheimer, *California State University - Bakersfield*

Albert Will, *Broward Community College*

John B. Williams, *South Carolina State University*

Ron Williamson, *Copiah-Lincoln Community College*

Daniel Wivagg, *Baylor University*

Paul Wright, *Western Carolina University*

Calvin Young, *California State University, Fullerton*

Anne Zayaitz, *Kutztown University*

Art Focus Group Participants

Kristen Bender, *California State University at Fullerton*

Christine Collins, *Orange Coast College and California State University at Fullerton*

Don Collins, *Orange Coast College*

Bob Eagan, *Golden West College*

Tina Hartney, *California State University at Fullerton*

Charles Leavell, *Fullerton College*

Calvin Young, *California State University at Fullerton*

Many others who were not official reviewers for *Journey* were patient and helped in various ways in the development of this book. We would like to thank:

Valerie Antoine *U.S. Metric Association*

Richard Crowell, *Blackburn College*

R. K. Dillon, *Magic Fingers, Inc.*

Herndon G. Dowling, *New York University*

L. K. McAndrews

Dilbagh Singh, *Blackburn College*

Marvett Sharpe, *Talladega College*

Contents Overview

Study Techniques for Biology Classes xxvii

1 Introduction 1

PART 1
The Unity of Life: Cells 21

2 Some Basic Chemistry 22
3 Biological Chemistry 40
4 Cells and Their Membranes 64
5 Cell Structure and Function 84
6 Energy and Living Cells 110
7 Food as a Fuel: Cellular Respiration and Fermentation 124
8 Photosynthesis 141

PART 2
The Unity of Life: Genetic Information and Its Expression 163

9 DNA and Genetic Information 164
10 RNA and Protein Synthesis 178
11 Reproduction of Eukaryotic Cells 202
12 Mendelian Genetics 226
13 Inheritance Patterns and Gene Expression 246

PART 3
Evolution and the Diversity of Life 275

14 Evolution and Natural Selection 276
15 Population Genetics and Speciation 300
16 Evolution and Reproduction 322
17 Origin of Life 340
18 Classification of Organisms and the Problem of Viruses 356
19 Bacteria, Protists, and Fungi 372
20 The Plant Kingdom 404
21 The Animal Kingdom 430

PART 4
Animal Biology 471

22 Animal Nutrition and Digestion 472
23 Gas Exchange in Animals 490
24 Internal Transport 506
25 Defenses Against Disease 524
26 Excretion 544
27 Sexual Reproduction and Embryonic Development 560
28 The Nervous System and Sense Organs 582
29 Muscles and Skeletons 610
30 Animal Hormones and Chemical Regulation 628
31 Behavior 644

PART 5
Plant Biology 667

32 Plant Structure and Growth 668
33 Nutrition and Transport in Vascular Plants 686
34 Regulation and Response in Plants 710
35 Reproduction in Flowering Plants 726

PART 6
The World of Life: Ecology 747

36 Distribution of Organisms 748
37 Ecosystems and Communities 770
38 Populations 788
39 Human Impact on the Environment 810

Contents

STUDY TECHNIQUES FOR BIOLOGY CLASSES xxvii

CHAPTER 1 Introduction 1

1-A Biology: A More Complete Picture 2

1-B Biology: Saving Your Life and Preserving the Earth 4

1-C Life: Only On Earth 15

Summary 19

A Journey into Science in Process: *Scientific Method* 6

A Journey into Evolution: *Natural Selection* 10

PART 1

THE UNITY OF LIFE: CELLS 21

CHAPTER 2 Some Basic Chemistry 22

2-A Chemical Elements and Atoms 24

2-B Bonds Between Atoms 26

2-C Molecules and Compounds 28

2-D Movement of Molecules 29

2-E Chemical Reactions 30

2-F Water 31

2-G Dissociation and the pH Scale 33

Summary 37

A Journey into the Environment: *Acid Precipitation* 34

CHAPTER 3 Biological Chemistry 40

3-A Structure of Organic Molecules 42

 Carbon Skeletons 42

 Functional Groups 42

 Building Biological Polymers 42

3-B Lipids 45

 Fatty Acids 46

 Fats and Oils 47

 Phospholipids 47

 Steroids 47

3-C Carbohydrates 48

3-D Proteins 51

 Protein Structure 53

3-E Enzymes 56

 Enzyme-Substrate Complexes 56

 Factors That Affect Enzyme Activity 57

3-F Nucleic Acids and Nucleotides 58

3-G Metabolism 58

Summary 61

A Journey into Science in Process: *Kitchen Chemistry* 55

CHAPTER 4 Cells and Their Membranes 64

4-A Structure of Biological Membranes 66

4-B Roles of the Plasma Membrane 67

4-C How Small Uncharged Molecules Cross Membranes 71

 Diffusion 71

 Osmosis 72

 Cells as Osmotic Systems 73

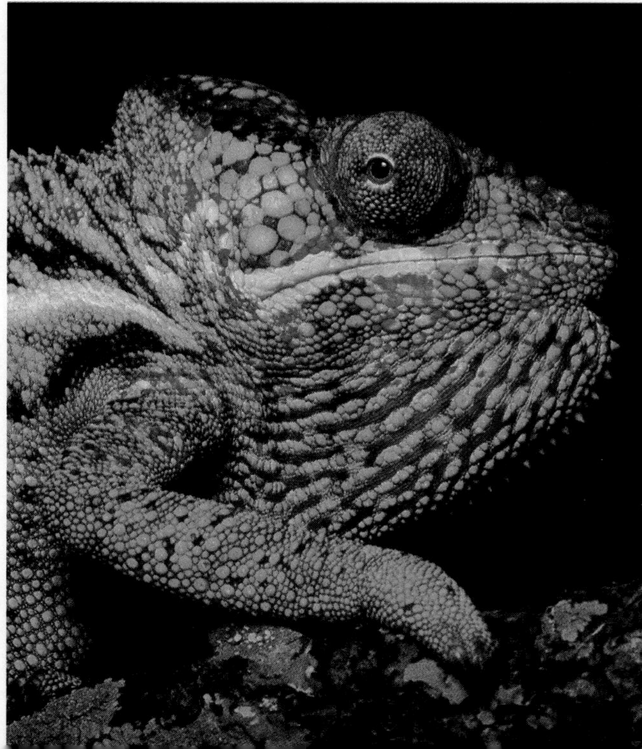

4-D Transport By Membrane Proteins 74
Passive Transport Using Channel Proteins 75
Passive Transport Using Carrier Proteins 76
Active Transport Using Carrier Proteins 76
4-E Membrane Transport of Large Particles 77
4-F Membrane Attachments Between Cells 78
4-G Communication Between Cells 79
Summary 81
A Journey into Science In Process: *Looking at Cells: Microscopy* 68

CHAPTER 5 **Cell Structure And Function** 84
5-A Prokaryotic Cells: Cells Without Nuclei 86
5-B Eukaryotic Cells: The Cells of Higher Organisms 87
5-C The Nucleus: Genetic Message Center 89
5-D Ribosomes: Workbenches for Protein Assembly 93
5-E Endoplasmic Reticulum 93
5-F Golgi Complex: Molecular Finishing and Sorting Area 94
5-G Lysosomes: Sacs of Hydrolytic Enzymes 95
5-H Mitochondria: Mills to Make ATP 96
5-I Plastids: Food Factories and Storehouses 96
5-J Cell Walls: Protection and Support 97
5-K Vacuoles: Sacs Full of Fluid 99
5-L The Cytoskeleton: Cell Shape and Movement 99
Microtubules 102
Microfilaments 103
Organization of the Cytoskeleton 104
5-M Tissues and Organs: Cells Organized Into Work Parties 104
Animal Tissues 104
Plant Tissues 106
Summary 107
A Journey into Evolution: *How Are Prokaryotes and Eukaryotes Related?* 100

CHAPTER 6 **Energy And Living Cells** 110
6-A Energy Transformations 112
6-B Chemical Reactions and Energy 112
6-C Photosynthesis and Respiration 116
6-D Oxidation-Reduction Reactions 117
6-E Energy Intermediates 118
Summary 123
A Journey into the Environment: *Global Warming* 120

CHAPTER 7 **Food As A Fuel: Cellular Respiration And Fermentation** 124
7-A Respiration—In a Nutshell: It's *Why* You Breathe, But It's *Not* Breathing 126
7-B Glycolysis 129
7-C Mitochondria: ATP Factories 129
7-D Respiration 129
Preparation for Respiration 129
Citric Acid Cycle 130
Electron Transport and Chemiosmotic ATP Synthesis 132
Role of Oxygen in Cellular Respiration 134
The Energy Yield of Glucose 134
7-E Fermentation 135
7-F Alternative Food Molecules 137
Summary 139

CHAPTER 8 **Photosynthesis** 141
8-A Light 144
8-B Trapping Light Energy: Photosynthetic Pigments 145
8-C Tour of a Leaf 147
Overview: What Happens In Photosynthesis? 149
8-D Photosynthesis in Detail 149
Energy Capture 149
Carbon Fixation 152
8-E What Controls the Rate of Photosynthesis 153
8-F Ecological Aspects of Photosynthesis 154
Summary 158
A Journey into Evolution: *Coevolution of Plants and Herbivores* 156
The Educated Citizen: *No Longer Human* 160

PART 2

THE UNITY OF LIFE: GENETIC INFORMATION AND ITS EXPRESSION 163

CHAPTER 9 DNA and Genetic Information 164

9–A Evidence that DNA is the Genetic Material 166
 The Riddle of Bacterial Transformation 166
 Bacteriophages 166
9–B The Structure of DNA 168
9–C DNA Replication 170
9–D DNA Repair 171
9–E Mutations 171
9–F Structure of Eukaryotic Chromosomes 172
9–G Organization of the Genome 172
 Jumping Genes 172
 Repetitive DNA 176

Summary 176

A Journey into Science In Process: *Genetic Engineering* 174

CHAPTER 10 RNA and Protein Synthesis 178

10–A RNA 180
 Transcription of DNA Into RNA 181
10–B The Genetic Code 182

Overview of Protein Synthesis 184

10–C Kinds of RNA 185
 Messenger RNA 185
 Ribosomal RNA and Ribosomes 186
 Transfer RNA 186
10–D Protein Synthesis 187
 Initiation 187

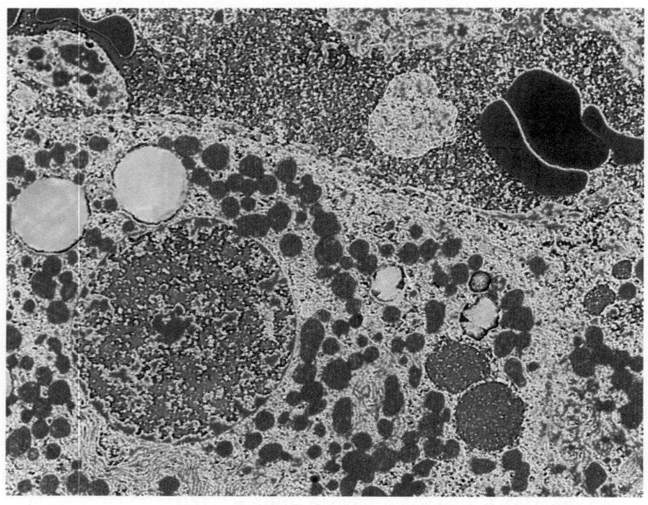

 Peptide Chain Formation 187
 Termination 187
10–E Control of Protein Synthesis 190
 Control in Prokaryotes 190
 Control in Eukaryotes 193
10–F Control of Gene Activity During Development 194
 Metamorphosis 194
10–G Cancer 196

Summary 200

A Journey into Science in Process: *Smart Genes* 192

A Journey into Science in Process: *Gene Therapy* 199

CHAPTER 11 Reproduction of Eukaryotic Cells 202

11–A Mitosis and Meiosis: An Overview 204
11–B Eukaryotic Chromosomes 204
 Haploid and Diploid Chromosome Numbers 204
11–C The Cell Cycle 210
11–D Mitosis 211
11–E Cytokinesis 214
11–F Meiosis 215
11–G Genetic Reassortment 218
11–H Gamete Formation in Animals 219

Summary 223

A Journey into Science in Process: *The Human Genome Project* 207

CHAPTER 12 Mendelian Genetics 226

12–A A Simple Breeding Experiment 228
 Gene Pairs 229
 Dominant and Recessive Alleles 229
 Genotype and Phenotype 231
 Law of Segregation and Meiosis 231
 Monohybrid Cross 232
12–B Predicting the Outcome of a Genetic Cross 233
12–C Test Cross 233
12–D The Dihybrid Cross: Independent Assortment of Genes 234
12–E Incomplete Dominance and Codominance 236
12–F Linkage Groups 237
12–G Crossing Over 239

Summary 243

A Journey into Science in Process: *Secrets of Mendel's Success* 240

CHAPTER 13 **Inheritance Patterns And Gene Expression** 246

13–A Phenotypic Expression of Mutations 248

13–B Lethal Alleles 248

13–C Inborn Errors of Metabolism 252

13–D Multiple Alleles 254

13–E Polygenic Characters 255

13–F Sex Determination 256

13–G Sex Linkage 259

13–H Sex-Influenced Genes 262

13–I Some Factors That Affect Gene Expression 263

13–J Nondisjunction and Translocation 265

Summary 268

A Journey into Healthy Living: *Genetic Counseling and Fetal Testing* 266

The Educated Citizen: *Gay Genes* 272

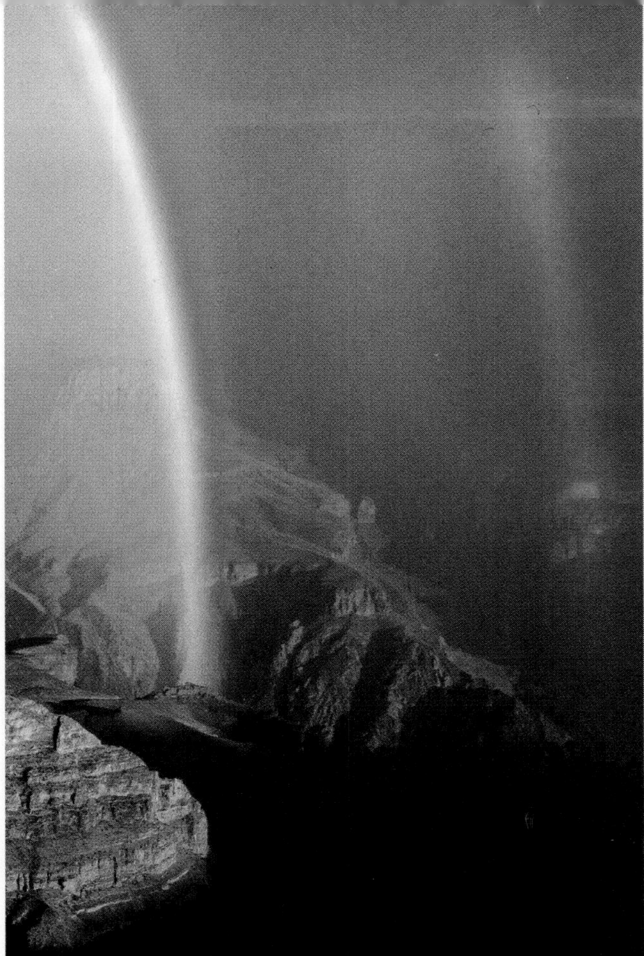

PART 3

EVOLUTION AND THE DIVERSITY OF LIFE 275

CHAPTER 14 **Evolution And Natural Selection** 276

14–A Overview: The Theory of Evolution 278

14–B History of the Theory of Evolution 280

- Lamarckism 280
- Darwin and Wallace 280

14–C The Evidence for Evolution 283

- The Evidence from Artificial Selection 283
- The Evidence from the Fossil Record 283
- The Evidence from Comparative Anatomy 285
- The Evidence from Comparative Biochemistry 288
- The Evidence from Developmental Biology 288
- The Evidence from Biogeography 289

14–D Evolution by Means of Natural Selection 290

- The Peppered Moth 291

14–E Genetic Contribution to Future Generations 294

14–F Adaptations 295

- Resistance to Pesticides and Antibiotics 295

Summary 297

A Journey into the Environment: *Going. . . . going* 292

CHAPTER 15 **Population Genetics And Speciation** 300

15–A The Hardy-Weinberg Principle 302

15–B Causes of Evolution 302

- Natural Selection 302
- Mating Preferences 305
- Gene Flow 305
- Genetic Drift 307

15–C What Promotes and Maintains Variability in Populations? 309

- Heterozygote Advantage and Hybrid Vigor 310

15–D What Is a Species? 312

15–E Speciation 314

- Allopatric Speciation 314
- Pleistocene Glaciations 315
- Sympatric Speciation 316
- Selection Against Hybrids 317

15–F How Quickly Do New Species Form? 318

Summary 319

A Journey into Evolution: *Double Trouble for Cheetahs* 308

CHAPTER 16 **Evolution and Reproduction** 322

16–A Is Sex Necessary? 324

16–B Evolution of Sexual Reproduction 326

16–C Evolution of Mechanisms that Ensure Fertilization 328

16–D Evolutionary Roles of Male and Female 328

Sexual Differences 328

Mating Systems 330

Polygyny 330

Polyandry 331

Monogamy 332

Ecology and Mating Systems 332

16–E Selfishness and Altruism 333

The Social Insects 336

Summary 338

A Journey into the Environment: *Overpopulation* 334

CHAPTER 17 **Origin Of Life** 340

17–A Conditions for the Origin of Life 342

17–B The Prebiotic Earth 343

17–C Production of Organic Monomers 344

17–D Formation of Polymers 346

17–E Formation of Aggregates 346

17–F Beginnings of Metabolism 347

Origin of Energy Metabolism 347

17–G The Beginnings of Biological Information 348

17–H Heterotrophs and Autotrophs 349

17–I Respiration 350

17–J Origin of Eukaryotes 350

17–K Early Fossils 354

Summary 354

A Journey into the Environment: *Ozone Depletion* 352

CHAPTER 18 **Classification Of Organisms And The Problem Of Viruses** 356

18–A Binomial Nomenclature 358

18–B Taxonomy 358

18–C Interpreting the Characteristics of Organisms 359

Monophyletic or Polyphyletic? 361

18–D The Five Kingdoms 362

Kingdom Prokaryotae 362

Kingdom Protista 362

Kingdom Fungi 364

Kingdom Plantae 364

Kingdom Animalia 364

Difficulties With the Five-Kingdom System 364

18–E The Problem of Viruses 364

18–F Viral Reproduction 366

18–G Viruses and Evolution 370

Summary 370

A Journey into Science in Process: *The Name Game* 368

CHAPTER 19 **Bacteria, Protists, and Fungi** 372

Kingdom Prokaryotae: The Bacteria 374

19–A Reproduction and Evolution in Prokaryotes 374

19–B Bacterial Metabolism and Ways of Life 375

Autotrophic Bacteria 376

Nitrogen-Fixing Bacteria 377

Heterotrophic Bacteria 378

19–C Classification of Prokaryotes 378

Archaeobacteria 379

Eubacteria 379

19–D Symbiotic Bacteria 380

Origin of Eukaryotes and the Kingdom Protista 381

19–E Kingdom Protista: Unicellular Algae and Protozoa 381

Physiology of Protists 381

19–F Photosynthetic Protists 383

Phylum Pyrrophyta (Dinoflagellates) 384

Phylum Euglenida 384

Phylum Chrysophyta (Diatoms and Golden Algae) 385

19–G Protozoa (Heterotrophic Protists) 386

Phylum Zoomastigina (Zooflagellates) 386

Phylum Sarcodina 387

Phylum Apicomplexa 387

Phylum Ciliophora (Ciliates) 388

Slime Molds: Protists or Fungi? 389

Phylum Myxomycota 389
Phylum Acrasiomycota 390
Phylum Oomycota 390
19-H On To Multicellularity! 390
19-I Kingdom Fungi 392
Body Plan 393
Reproduction 393
19-J Classification of Fungi 394
Division Zygomycota 395
Division Ascomycota 395
Division Basidiomycota 395
Division Deuteromycota 397
19-K Symbiotic Relationships of Fungi 397
Mycorrhizae 397
Lichens 398
19-L Bacteria and Fungi: Friends and Foes 399
Diseases 399
Bacteria, Fungi, and Food 399
Summary 400

CHAPTER 20 **The Plant Kingdom** 404

The Multicellular Algae 407
20-A Division Rhodophyta: Red Algae 407
20-B Division Phaeophyta: Brown Algae 408
20-C Division Chlorophyta: Green Algae 409
20-D Life Histories 411
Land Plants 411
Reproduction on Land 414
20-E Division Bryophyta: Mosses and Liverworts 415
20-F Vascular Plants 416
20-G Lower Vascular Plants 418
Division Lycophyta: Club Mosses and Ground Pines 418
Division Spenophyta: Horsetails or Scouring Rushes 418
Division Pterophyta: Ferns 419
20-H Gymnosperms 421
Division Cycadophyta 423
Division Coniferophyta 423
Division Ginkgophyta 423
20-I Division Anthophyta: Angiosperms or Flowering Plants 425
Summary 427
A Journey into Healthy Living: *A Tree That Changed History* 424

CHAPTER 21 **The Animal Kingdom** 430

21-A Animals and Their Environments 433
21-B Animal Structure 434
Body Symmetry 434
Body Layers 435
The Coelom 435
Invertebrates 436
21-C Phylum Porifera: Sponges 436
21-D Phylum Cnidaria: Jellyfish, Corals, Sea Anemones 436
21-E Phylum Platyhelminthes: Flatworms 438
21-F Phylum Nematoda: Roundworms 439
21-G Phylum Annelida: Segmented Worms 442
21-H Phylum Mollusca 442
21-I Phylum Arthropoda 445
Class Arachnida 445
Class Crustacea: Lobsters, Crabs, Wood Lice, and Their Relatives 446
Class Insecta 447
21-J Phylum Echinodermata 449
21-K Phylum Chordata 449
Chordate Subphylum Urochordata: Tunicates 451
Chordate Subphylum Cephalochordata 451

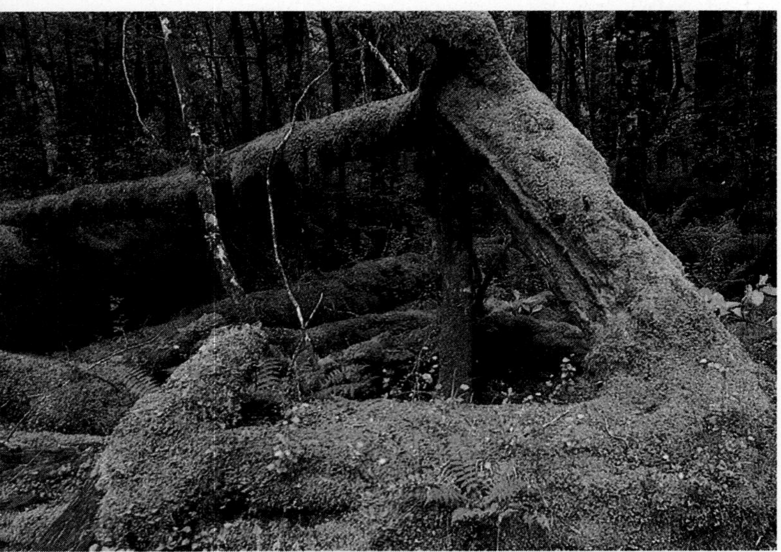

Vertebrates 451

21-L Three Classes of Fishes 453

Class Agnatha: Jawless Fishes 453

Evolution of Fishes 454

Class Chondrichthyes: Cartilaginous Fishes (Sharks and Rays) 454

Class Osteichthyes: Bony Fishes 455

21-M The Move to Land: Tetrapods 455

Class Amphibia: Frogs, Salamanders, Toads, Newts 457

Class Reptilia: Lizards, Snakes, Turtles, Crocodiles 458

Class Aves: The Birds 460

Class Mammalia 462

Summary 465

A Journey into Evolution: *Parasitism* 440

A Journey into Evolution: *Insects in Our Environment* 448

The Educated Citizen: *Can We Wipe Out Disease?* 468

PART 4

ANIMAL BIOLOGY 471

CHAPTER 22 **Animal Nutrition and Digestion** 472

22-A Nutrients 474

Macronutrients 474

Micronutrients 474

22-B Digestive Systems 478

22-C Human Digestion 480

The Human Digestive Tract 480

Human Digestive Enzymes 482

Absorption of Nutrients 483

22-D Feeding and Digestion in Herbivores 485

22-E Adaptations of Mammalian Carnivores 486

22-F Feeding in Birds 487

22-G Functions of the Mammalian Liver 488

22-H Stored Food and Its Uses 488

Summary 488

A Journey into Healthy Living: *Diet and Cardiovascular Disease* 476

CHAPTER 23 **Gas Exchange In Animals** 490

23-A Supplying Oxygen 492

23-B Factors That Influence Gas Exchange 493

23-C Respiratory Surfaces and Ventilation 494

The Body Surface 494

Gills 494

Lungs 496

Tracheal Systems 500

23-D Respiratory Pigments 501

23-E Carbon Dioxide Transport 502

23-F Regulation of Ventilation 502

Summary 503

A Journey into Healthy Living: *Air Pollution and Smoking* 498

CHAPTER 24 **Internal Transport** 506

24-A Transport in Invertebrates 508

Cnidaria 508

Planaria 508

Annelida 508

Insects 510

24-B Circulation in the Vertebrates 510

Fishes 510

Amphibians and Reptiles 510

Mammals and Birds 512

24-C The Mammalian Circulatory System 512

Blood Vessels 512

The Pathway of Blood Flow in the Body 514

The Heart Cycle 514

Blood Pressure and Circulation 514

The Circulatory System's Adjustment to Exercise 516

Diseases of the Circulatory System 516

24-D Blood 518

Blood Clotting 518

24-E The Lymphatic System 519

Summary 522

A Journey into Evolution: *Emperor Penguins: Unanswered Questions* 520

CHAPTER 25 **Defenses Against Disease** 524

25-A Nonspecific Defenses 526

External Barriers 526

Internal Defenses 526

25-B Specific Defenses: An Overview of Immune Responses 528

25-C The Immune System 529

Cells of the Immune System 530

25-D Recognizing Foreign Antigens: Receptors and Antibodies 531

The Genetics of Antibodies and T Cell Receptors 532

25-E Attacking Pathogens 533

Cellular Immunity 533

Humoral Immunity 533

25-F Immunological Memory: Primary and Secondary Responses 535

25-G Medical Aspects of Immune Responses 536

Vaccination 536

Passive Immunity 537

Tissue and Organ Transplantation 537

25-H Malfunctions of the Immune System 537

Autoimmune Diseases 537

Allergies and Anaphylaxis 540

Summary 541

A Journey into Healthy Living: *Understanding AIDS* 538

CHAPTER 26 **Excretion** 544

26-A Substances Excreted 546

Nitrogenous Wastes 546

26-B Osmoregulation in Different Environments 548

Overview: How Excretory Systems Work 550

26-C Excretory Organs of Invertebrates 550

26-D The Vertebrate Kidney 551

Functions of the Nephron 553

Concentration of the Urine 553

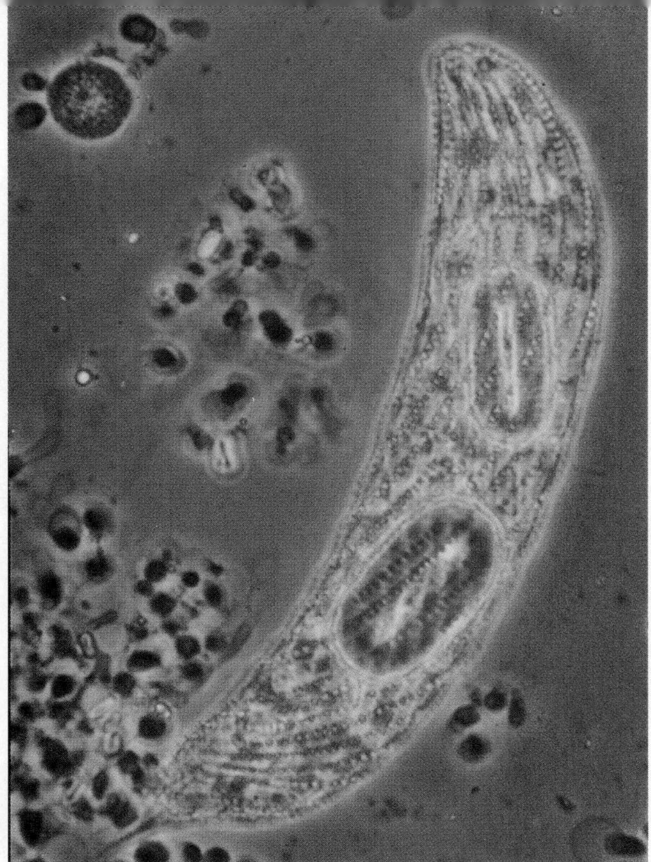

26-E Regulation of Kidney Function 555

Vasopressin (ADH) 555

Aldosterone, Renin, and Angiotensin 555

Summary 557jxxii

A Journey into Evolution: *Adaptations of Mammals to Sodium-Deficient Environments* 554

CHAPTER 27 **Sexual Reproduction And Embryonic Development** 560

27-A Reproductive Patterns 562

Human Reproduction 563

27-B Human Reproductive Organs 563

Female Reproductive Organs 563

Male Reproductive Organs 563

27-C Physiology of Sexual Intercourse 563

27-D Hormones and Reproduction 566

Male Hormones 566

Female Hormones 566

27-E Birth Control 570

Vasectomy and Tubal Ligation 572

Abortion 572

27-F Fertilization and Implantation 572
27-G Stages of Embryonic Development 574
Cleavage 574
Gastrulation 574
Neurulation 576
Organogenesis 576
27-H Birth 577
27-I Maturation, Aging, and Death 578
Summary 579
A Journey into Healthy Living: *Sexually Transmitted Diseases* 571
A Journey into Science in Process: *Test Tube Babies and Surrogate Mothers* 578

CHAPTER 28 The Nervous System And Sense Organs 582

Neurons 584

Overview: How Neurons Work 584
28-A Electrical Properties of Neurons 584
The Resting Potential 584
Local Potential and Action Potentials 586
Nerve Impulses 587
28-B Synaptic Transmission 589
28-C Neurotransmitters 592
28-D Organization of Neurons Into Nervous Systems 593

The Vertebrate Nervous System 594
28-E The Vertebrate Brain 596
Hindbrain 596
Midbrain 597
Forebrain 597
Motor Pathways 600
28-F The Spinal Cord 600
Reflex Arcs 600
28-G Cranial and Spinal Nerves 602
28-H The Autonomic Nervous System 602
28-I Sense Organs and Their Functions 602
Summary 607
A Journey into Healthy Living: *This Is Your Brain On Drugs* 596

CHAPTER 29 Muscles and Skeletons 610
29-A Muscle Tissue 612
Smooth Muscle 612
Cardiac Muscle 612
Skeletal Muscle 614
29-B Muscle Contraction 615
Control of Contraction 615
Tetanization and Fatigue 615
Graded Response of an Intact Muscle 618
29-C How Muscles and Skeletons Interact 618
Antagonistic Muscles 618
29-D The Vertebrate Skeleton 620
Joints in the Vertebrate Skeleton 620
29-E Connective Tissue 622
Cartilage 622
Bone 622
Summary 625
A Journey into Healthy Living: *Preventing Osteoporosis* 625

CHAPTER 30 Animal Hormones And Chemical Regulation 628

Chemical Messengers: An Overview 630
30-A Hormones 630
Feedback Control of Secretion 632

30-B How Chemical Messengers Affect Cells 633
Steroid and Thyroid Hormones 633
Water-Soluble Hormones 633
30-C Hormonal and Nervous Control 634
Fight or Flight 634
The Hypothalamus-Pituitary Connection 635
30-D Local Chemical Messengers 636
Prostaglandins 637
Neurotransmitters as Local Messengers 637
Growth Factors 637
30-E Hormones and Seasonal Changes 637
Environmental Control of Reproduction 637
30-F Biological Rhythms 638
Circadian Rhythms 638
Annual Rhythms 638
Biological Clocks 638
30-G Pheromones 641
Summary 642
A Journey into Healthy Living: *Our Daily Spread* 640

CHAPTER 31 Behavior 644

31-A Short- and Long-term Causes of Behavior 646
31-B Genes and Environment 646
31-C Development of Behavior 647
31-D Instinct Versus Learning 648
Adaptive Value of Learned and Innate Behaviors 649
31-E The Neural Basis of Behavior 649
Stereotyped Behavior 650
Sign Stimuli 651

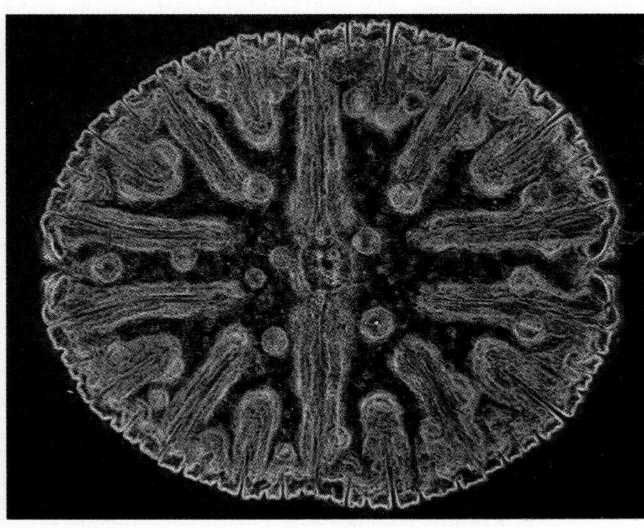

Drive and Motivation 652
31-F Kinds of Learning 652
31-G Territorial Behavior 653
31-H Conflict and Courtship 654
31-I Migration and Homing 655
31-J Social Behavior (Sociobiology) 657
Communication 658
Honeybee Societies 658
Vertebrate Societies 659
Summary 662
A Journey into Evolution: *Chimpanzee Societies* 660
The Educated Citizen: *The Transplant Gap* 664

PART 5
PLANT BIOLOGY 667

CHAPTER 32 Plant Structure And Growth 668

32-A The Bean Seed 670
32-B The Root System 671
Primary Growth of Roots: Growth in Length 671
Primary Structure of Roots 672
Primary Growth of Roots: Production of Laterals 672
Functions of Roots 674
32-C Stems 674
Primary Growth of Stems: Growth in Length 674
Primary Structure of Stems 674
Primary Growth of Stems: Production of Laterals 675
Functions of Stems 675
32-D Leaves 676
Structure of Leaves 676
32-E Secondary Growth 678
32-F A Comparison of Monocotyledons and Dicotyledons 681
Summary 684
A Journey into Evolution: *Leaves to the Defense* 680

CHAPTER 33 Nutrition And Transport In Vascular Plants 686

33-A Nutritional Requirements of Plants 688
33-B Soil 688
The Soil Solution 690

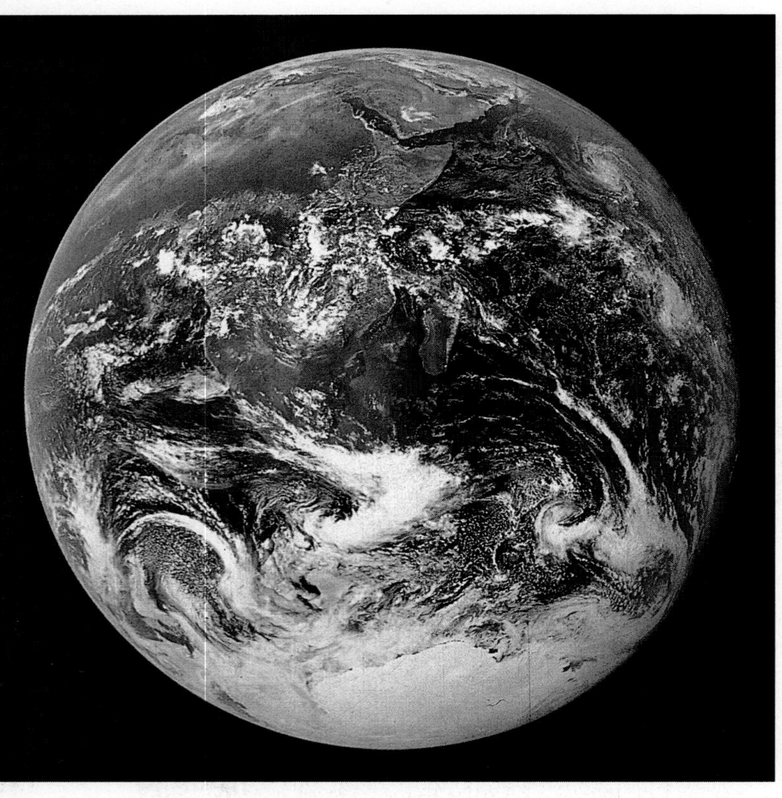

33-C Absorption by the Roots 692
33-D Functions of Xylem and Phloem 694
33-E Structure of Xylem 694
Changes in the Xylem of Woody Plants 695
33-F Transport in the Xylem 696
Root Pressure 698
Transpiration Pull 699
33-G Phloem Structure 701
33-H Transport in Phloem 702
33-I Distribution of Substances 705
Practical Applications 706
Summary 707
A Journey into Science in Process: *How to Care for Cut Flowers* 705

CHAPTER 34 Regulation And Response In Plants 710
34-A Plant Hormones 712
Auxin 712
Gibberellins 713
Cytokinins 714
Abscisic Acid 715
Ethylene 715
34-B Apical Dominance 716
34-C Responses to the Environment 717
Tropisms: Growth in Response to Gradients 718
34-D Flowering 720
Photoperiodism: Responses to the Length of Night and Day 720
34-E Senescence 721
Summary 724
A Journey into the Environment: *Tropical Rain Forest Canopy—Where the Action Is* 722

CHAPTER 35 Reproduction In Flowering Plants 726
35-A Flowers 728
35-B Pollen, Pollination, and Ovule Preparation 729
Pollination 730
Pollen Maturation 730
Preparation of the Ovule 730
35-C Fertilization 731
35-D Development of the Seed and Fruit 731
35-E Dispersal of Seeds and Fruit 734
Seeds and Seed Predators 735
35-F Germination 737
35-G Breeding Programs 737
35-H Vegetative Reproduction 738
Summary 742
A Journey into Evolution: *Coevolution of Flowers and Their Pollinators* 740
The Educated Citizen: *Altered Vegetable States* 744

PART 6
THE WORLD OF LIFE: ECOLOGY 747

CHAPTER 36 Distribution Of Organisms 748
36-A Climate and Vegetation 750
Biomes 752
36-B Tropical Biomes 752
Tropical Forest 752
Tropical Savanna and Tropical Thornwood 753
36-C Desert 754
36-D Temperate Biomes 755
Temperate Forest 755
Temperate Shrubland 756
Temperate Grassland 756
Temperate Desert 757

36-E Taiga 757
36-F Tundra 757
36-G Aquatic Communities 758
Lakes and Rivers 759
The Edges of the Ocean 760
Coral Reefs 761
The Open Ocean 761
36-H Ecological Succession 762
Primary Succession 762
Secondary Succession 763
Fire-Maintained Communities 764
36-I Why Are Organisms Where They Are? 764
Summary 768
A Journey into the Environment: *The Disappearing Soil* 766

CHAPTER 37 Ecosystems And Communities 770
37-A The Basic Components of Ecosystems 772
37-B Food Webs and Energy Relationships 772
Pyramids of Energy 772
Why So Few Trophic Levels? 776
37-C Productivity 776
Primary Productivity 776
Secondary Productivity 778
37-D Cycling of Mineral Nutrients 778
The Carbon Cycle 778
The Nitrogen Cycle 781
The Phosphorus Cycle 781
An Experimental Ecosystem 782
37-E Species Diversity in Communities 783
Species Turnover in Communities 784
Summary 785
A Journey into the Environment: *Salt Marsh and Seafood* 774

CHAPTER 38 Populations 788
38-A Population Growth 790
Reproductive Strategies and Survivorship 791
Carrying Capacity 793

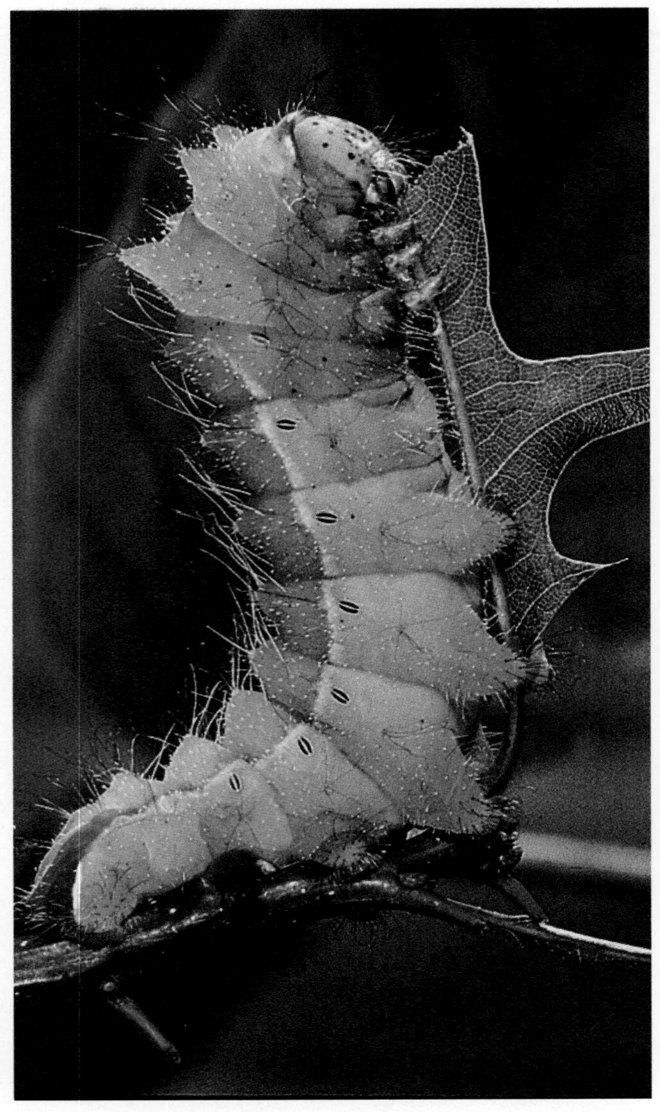

38-B Regulation of Population Size 794
38-C Competition 795
Niche 796
38-D Predation and Pest Control 798
38-E Extinction and Endangered Species 800
38-F Human Population Growth 802
Declining Death Rates 802
The Demographic Transition 802
The Demographically Divided World 803
Summary 807
A Journey into Evolution: *Camouflage* 804

CHAPTER 39 Human Impact On The Environment 810
39-A Hunter-Gatherer Populations 812
39-B The Agricultural Revolution 813
39-C Feeding the Human Population 815
The Efficiency of Agriculture 816
The Green Revolution 817
39-D Deforestation 820
Agriculture in Tropical Forest Biomes 820
39-E The Industrial Revolution 821
39-F Looking Ahead: Two Views of Environmental Problems 823
Overpopulation 823
Global Warming 824
Acid Rain 825
Ozone Hole 826
Deforestation and Loss of Species 826
39-G Evaluating Our Environmental Problems: A Suggestion for Change 827
Education, Imagination, and Caring 830
The Future 831
Summary 831
A Journey into the Environment: *The Environmental Movement and the Tragedy of the Commons* 828
The Educated Citizen: *How Many People Can Earth Hold?* 834

APPENDIX A Periodic Table of the Elements A-1
APPENDIX B The Metric System A-2
APPENDIX C Geologic Time Scale A-4

ANSWERS TO SELF-QUIZZES ANS-1

GLOSSARY G-1

INDEX I-1

Study Techniques for Biology Classes

Students take introductory biology classes for different reasons. You may be genuinely interested in the subject and *want* to take the course to learn more about it. Or you may not be especially interested but *have* to take the course to fulfill graduation requirements. You may be taking introductory biology to find out whether you like it well enough to consider a biology-related career. Whatever your reasons for taking biology, you should know that it is very different from any other college course you will take.

For one thing, biology has a separate, technical language, that is quite precise and thus is slightly different from everyday English. Second, although biological concepts have many practical applications, the concepts themselves are quite abstract. Finally, the information of biology is cumulative and full understanding of the subject requires a view that builds from biological molecules to encompass all living and nonliving systems on Earth. Unlike some other college subjects, opinion has little place in biology. Your task as a student is to learn and correlate the new vocabulary and new concepts, integrating them into personal mental pictures that build as the course progresses.

But, just as biology is a challenging subject, it is rewarding in unique and wonderful ways. Once you have mastered introductory biology, your perceptions of the world change and expand. You begin to understand all the life around you, as well as your own life, too. But introductory biology does not give up its secrets easily. Learning this subject takes work, in and out of class. Here are some suggestions on how to tailor your study skills so that you learn more biology and learn biology more meaningfully. These are techniques that have worked for many of our students, and we hope they work for you.

Step One: Getting Organized

You need to organize both your time and your workspace to study efficiently. Start by making a chart of your schedule. It will show you how much time you actually have to devote to studying biology. Schedule your classes, travel time, and time devoted to job or family, as well as any extra-curricular activities that are important to you. By making this simple schedule, you have already helped yourself to succeed, because you can see how much time you actually have to devote to study each day. Now you're ready to schedule your studying time.

How much studying is necessary to succeed in biology? This depends upon how strong your pre-college training has been, and upon how efficient your work habits are. A good rule of thumb is that college classes require between 2 and 3 hours of private study for each hour spent in class. (Some professors even suggest 4 hours.) For example, if your biology class meets for 3 hours of lecture and 4 hours of laboratory per week, you should put in between 14 and 21 hours of study time per week, or between 2 and 3 hours of study per day. This may sound like an incredible amount of study time, especially as you will be carrying other demanding courses, but we suggest that you try 2 to 3 hours each day. After the first exam you will be better able to judge the appropriate amount of time for private study. Until then, spending 2 to 3 hours is a safe bet. Don't be discouraged if you find yourself "off schedule" from time to time. When other pressures out compete your studying priorities, try to remember that your study schedule is an ideal that may require revision as the semester progresses. Try to get back on track as soon as you can.

Schedule study time for all of your classes. Try to use all of your time wisely. If, for example, you commute to class by public transportation, use your travel time for studying. Students have reported success when they have used their commutes for review of class notes, for review of flash cards (see below), and for review of the day's class lecture using a portable cassette recorder.

Everyone has different biological rhythms that cause them to feel more alert at different times in the day. As much as possible, schedule your hardest tasks when you are most alert. If you are a "morning person," wake up an hour earlier than normal and devote this time to your studies. "Night owls," on the other hand, will benefit more from intensive study sessions scheduled later in the day.

You might want to make several copies of your daily schedule. Put one on the wall that faces your workspace; put another in the front of your biology notebook. Now that your daily schedule is made, it's time to look ahead and schedule the entire semester. By this time, your professors will have distributed their syllabi and course schedules. Enter reading assignments and exam dates on your calendar. You might want to use a different color of ink for each course you are taking. Two weeks before each exam, write a note that will remind you to begin reviewing for it. By the time you have finished filling in all of the information for all of your courses, your calendar will look very full. Hang it on the wall near your desk where you can glance up and see what you need to do each day as well as the deadlines that you must meet. You might want to reduce-photocopy your calendar and tape a miniature copy in your biology notebook.

An organized workspace is another key that will help you to study efficiently. Ideally it should have nothing to distract you from the task at hand. Here are some suggestions that may help you get better organized:

1. Organize your books so that they are close at hand. Ideally, you should be able to put out your hand and grasp the spine of any book that you need.
2. Get the dictionary habit. A standard collegiate dictionary is one of the tools you should have on your desk top. Form the habit of referring to it whenever you come to a new word, rather than trying to guess the meaning from context. Some words will not be in this dictionary and a dictionary of biology (look for an inexpensive, paperback version at your college bookstore) is extremely useful and highly recommended. All of the boldface terms in this textbook are defined in the Glossary at the back of the book. Form the habit of using it, too.
3. Treat yourself to a selection of your favorite pens, pencils, markers, and papers. A supply of "post-its," blank 3" x 5" and 5" x 8" cards, an eraser, a bottle of white-out, some transparent tape, a stapler, a ruler, scissors, and a wastebasket are other minimal supplies.
4. You will be spending hundreds of hours at your workspace. Consider investing in a light that illuminates it properly and find a chair that comfortably supports your back, yet allows you to sit at the right height, with your feet planted firmly on the floor.
5. Give yourself some inspiration. Find a quotation, a photograph, or a cartoon that pleases you and tape it to the wall before you.

Assuming that you began with a quiet room, you have now created a workspace that will enhance, rather than hinder, your studying efforts.

Step Two: Reading This Textbook

A biology book is not a normal book and should be read in a different way from a novel. Although you may be an excellent reader in other subjects, you may be unfamiliar with the technique called "reading to learn content." You will need a separate notebook, a supply of post-its, a pencil, and a highlighter marker to try this technique.

Familiarize Yourself with the Chapter Organization

The Third Edition of *Biology: A Journey Into Life* has been designed to aid "reading to learn content." Each chapter has the following sections:

1. Curiosity Questions. These are real life, not merely academic questions, designed to pique your interest in the topics covered in the chapter. Ideally, the curiosity questions will make you think, "Yeah! I've always wondered about that."
2. Concept Guides. These are our suggested objectives for the chapter. Some students skip over these and hurry into the body of the text, anxious to get the reading assignment finished as quickly as possible. In so doing, they are like motorists who don't read maps or ignore road signs, relying, instead on dead reckoning to get them to their destinations. Sometimes it works; mostly, though, they get lost. Form the habit of reading the "openers" (Curiosity Questions and Concept Guides) and "closers" (Summary and Self-Quiz) *before* leaping into the text. This way you'll organize your thoughts before delving into details and you'll have an idea of what's important in the 20—odd pages of text and illustration that lie ahead.
3. Every chapter is subdivided into sections, each of which treats a slightly different aspect of the topic. Each section ends with a Concept Capsule that restates the most important concepts of the section. *Before* beginning to read a chapter, page through it, reading each Concept Capsule to get an idea of the main ideas that you will encounter.
4. Each chapter contains many figures and some tables. As you are browsing through, reading the Concept Capsules, examine each figure and table. Now you are reinforcing the ideas in the chapter with visual representations. You've already been studying a great deal, and yet, it will seem as though you haven't really started working yet. So far, "reading to learn content" should have been nearly as painless as reading a murder mystery.

There is one simple thing left to do before you read the chapter in detail: find the last page of the chapter and mark it with a book mark. Now you won't have to waste time trying to estimate when the chapter will be done. As you work, you'll see the thickness of pages that remain between your place and the bookmark steadily decrease.

You now have a real overview of the task ahead. All of this should have taken you about an hour of study time, perhaps less. Depending upon how quickly you work, it should take you another 1 to 2 hours to tackle the body of the text. Read actively, not passively. As you read each section, you may want to make marginal notes with your pencil and jot down separate notes into your notebook. The act of physically writing down notes involves more of your senses and more parts of your brain and enhances learning. Make a 3" x 5" flash card for every key (boldfaced) term (with the term on one side and its definition on the other). Alternatively, use a 5" x 8" card and write all key terms and definitions on it. Important concepts (those in Concept Capsules) should be noted down, either on 5" x 8" cards or in your notebook. Use post-its to flag anything that you cannot understand. Work through each section, reading, noting key terms and concepts, and examining each illustration.

Pay special attention to the figure legends. Read them carefully and make sure you understand each illustration. Many students don't know that figure legends in biological texts are different from those in other illustrated books in that they usually give additional information that supplements the text. Most of this information is not duplicated in the body of the text, so it is important to read every figure legend.

Use a highlighting marker sparingly, and try to avoid using it at all until you've read a section and can judge what's important. This will help prevent "yellow page syndrome."

Keep in mind that if this is your first college-level science course, you are encountering a species apart from literature, arts, and philosophy courses. In these, evaluation by opinion and conjecture are important and the task of the professor is to aid you in developing educated, discriminating, thoughtful, clear, even bold and controversial views. While biologists are some of the most opinionated individuals on the planet, opinion and taste are of small importance in biology. Facts, observations, data, correlations, conclusions, and prediction of events are important. Vocabulary is important to describe events with precision.

When you've finished each section, you should have:

- a set of cards of key terms and definitions,
- a set of cards with key concepts,
- a firm idea of the main points made in the text, amplified visually through illustrations,
- notes of important ideas in your notebook,
- any hard-to-understand sections flagged with post-its.

Work your way through each section until you have reached the end of the chapter. Allow yourself a break every hour. Get up and stretch, but don't be away from your desk for too long. This method of studying will take longer at first, but you will gain speed as you become more familiar with it and your ability to concentrate grows. The benefits of this method are that it automatically gets you ready for review and preparation for exams.

When the chapter is finished, turn back to the Concept Guide and see how well you can accomplish these objectives. Test yourself with the Self-Quiz. Once all of this is completed, turn back to the difficult sections that you marked. Some of your problems will have been cleared up. If questions remain, ask your professor for help.

Preparing for Exams

Everyone dreads exams. It is an open secret that many teachers dislike preparing, administering, and grading them almost as much as students loathe taking them. Yet, even though neither teachers nor students look forward to exams, they are an integral part of most courses. Here are suggestions for studying to succeed on exams:

1. *Start early.* One week of intense studying is usually a minimum for a biology exam.
2. *Make and use flash cards.* Start to make and use your flash cards the first week of classes. Keep them in your book bag and review them whenever you have a chance: supermarket lines, commuting time, any waiting time can be used in productive review.
3. *Try to understand, not merely memorize.* Learning becomes real and much easier when you try to think of mental hooks to hang new information from. Get in the habit of making mental pictures and visually associate the terms and concepts of biology. There are many levels of biological knowledge and simple memorization results in only the most short-lived knowledge. Understanding biological principles and being able to apply your understanding involves higher mental processes and results in longer-lasting learning.
4. *Form a study group.* If you can teach the material, you really know it. Use study groups for support and reinforcement of what you've learned. Avoid

study groups that wander away from the work at hand or degenerate into The Dating Game.

5. *Ask for professional help.* Your professor(s) and laboratory instructor(s) weren't born with an innate understanding of biology. Each was once enrolled in introductory biology, just as you are now. Ask for their suggestions about how best to study. Some professors offer copies of old exams as guides. Others don't believe this is an effective teaching tool, fearing that many students will limit their study to the set of questions offered, rather than using them as a means to allay pre-test jitters.
6. *Ask other students.* Upperclassmen may have had your same professor and may give you suggestions of what to expect on examinations and how best to study.
7. *Rest before an exam.* Studies have demonstrated that students who are well-rested perform better on exams than students who've stayed up all night studying.

Has This Helped?

We'd like some feedback.

We're establishing **Postcards!,** a user's newsletter and bulletin board for *Biology: A Journey Into Life.* Send us your favorite studying strategies and the tricks that help you learn. Clever mnemonic devices or mental hooks that helped you learn respiration, meiosis, etc. will be shared with users all over the country and we'll see that you and your school get credit for your ideas.

Send your ideas to:

Journey Postcards
C/O Saunders College Publishing
The Public Ledger Building
620 Chestnut Street
Suite 560
Philadelphia, PA 19106

BIOLOGY

A Journey Into Life

Introduction

You are about to embark upon a journey that will take you to strange and wondrous places—places as exotic as the treetop pools in tropical rain forests and as remote as snow-swept penguin rookeries in Antarctica. We will travel to spaces as small as the nuclear landscape within a single nerve cell and as large as the **biosphere:** the entire fabric of life that covers the planet we call home. The purpose of our journey is to show you the diversity of living creatures and to help you to understand how they go about the business of life. You will learn how life on Earth is organized at many different levels: subcellular, cellular, and all the various layers of complexity that culminate in a living **organism:** an individual living creature. Along the way you will learn much about the workings of your own body and how it is similar to and different from those of other animals. You will study the lives of plants, fungi, bacteria, and single-celled organisms called **protists.** We shall discuss the history of life on Earth, how we think it originated more than 4 billion years ago, and the ways living things have changed in that time. You will learn about **natural selection,** the major evolutionary mechanism that has produced the variety of living organisms, and **ecology,** the study of interactions of living things with one another and with their environment. The overall plan of our journey is to travel from small to large: we will begin at the subcellular level with biological chemistry and work toward the most-encompassing level, ecology and conservation biology.

In the days when steamships were the mode of overseas travel, it was widely recognized that travel teaches you things. While on a trip, chaperones would typically murmur, "Travel is so broadening," as they watched their teenaged charges play shuffleboard. Similarly, the main goal of our journey into life is to display the entire field of biology before you, introducing you to the broad concepts of the science of life, as well as embroidering on much of the detail that makes living things so fascinating. In this introductory chapter, we will discuss why biology has recently moved from merely a broadening educational experience—a discipline full of facts that every literate citizen should understand—to a science that is crucial to the continued existence of all forms of life on Earth, including humans. Biology has become more than just another part of a good education; today's biology class might save your life and, through you, it will have an impact upon future generations of all living things as well.

KEY CONCEPTS

- Recent advances in technology have led to a better understanding of many aspects of biology.
- The theory of evolution is the central principle unifying all biology.
- Using the scientific method, biologists are currently working on many problems of great human importance. These include the AIDS epidemic, loss of biodiversity, and the global environmental impact of the expanding human population.

Curiosity Questions

1? What evidence do we have that evolution occurs in living organisms? (See *A Journey into Evolution:* Natural Selection, page 10)

2? Why do current global environmental problems require the immediate attention of your generation? (See page 4)

3? How can we tell whether something is alive? (See page 12)

1–A BIOLOGY: A MORE COMPLETE PICTURE

Right now is an exciting time to be making this journey into biology. For one thing, knowledge of living processes and their interactions has greatly expanded. Today we know more than ever about the biological world, and, because we know more, we can form a better, more integrated picture of how living things operate and relate to one another. Through the hard work of biologists all over the world, new pieces of the biological puzzle continue to fall into place and, although there is still much to learn, the scope of our knowledge steadily increases. Recently we have learned much about the biology of cells. For example, in your laboratory classes you will probably study the protist, *Amoeba* (Figure 1–1). Fifty years ago, the outer edge of *Amoeba* was merely a dark line in images seen through the most powerful microscopes. Today, improved technologies in the form of electron microscopes and intricate biochemical techniques have shown us that this line is actually a complicated, chemical sandwich, studded with molecules that control the movement of substances into and out of *Amoeba* (Figure 1–1).

The advances in cell biology are only a beginning. We now have a better grasp of how heredity works and how cells make the proteins needed for life and growth. We know more about the biochemistry of processes like respiration and photosynthesis and understand more of the interlocking functions of digestion, respiration, circulation, excretion, movement, nerve activity, and hormonal control. We are beginning to understand how the brain works, how predators and prey interact, what birds hear, and how dinosaur societies might have been organized. We have better ideas of what caused the massive die-offs of many forms of life that appear in the fossil record. We have glimmerings of how complex the relationships are between plants and insects, and between plants and fungi. We are beginning to appreciate the central role that bacteria play in the operation of the biosphere. Finally, we have a better understanding of the complex web of relationships that links all life on our planet.

All of the investigations that have resulted in this explosion of biological knowledge have had one thing in common: they all have used the scientific method to arrive at their conclusions. Most people have only a hazy idea of the scientific method and mentally connect the phrase with images of white-coated, wild-eyed, bespectacled eggheads peering into test tubes in dim and faintly sinister laboratories. To make your journey into life more meaningful, you will need a thorough understanding of this method for attacking problems, answering questions, and making decisions. *A Journey into Science in Process:* The Scientific Method will provide the basic information that you need.

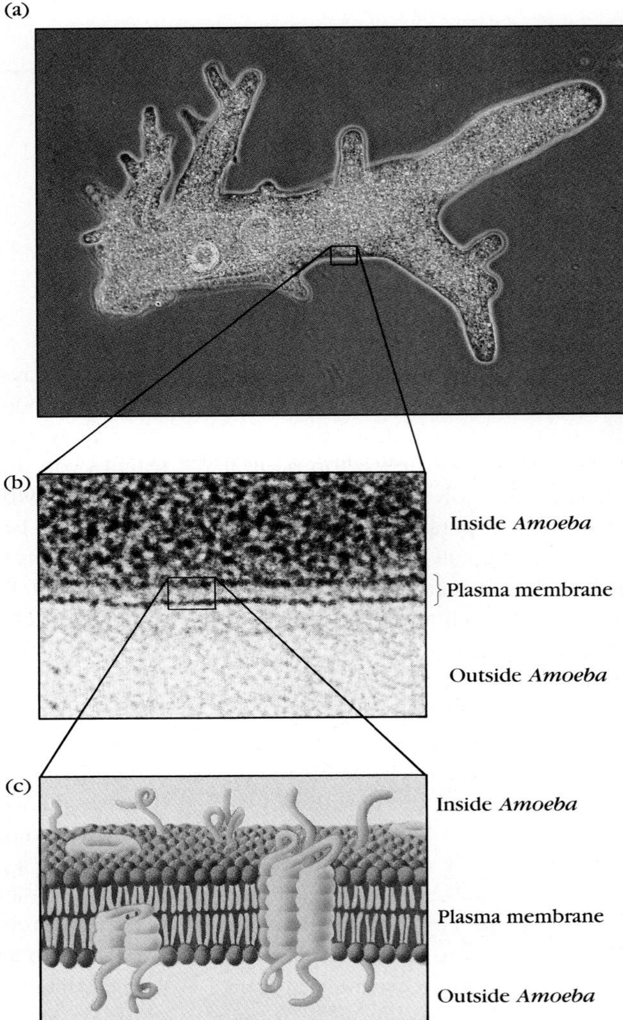

FIGURE 1-1

The plasma membrane of an *Amoeba*. (a) A light microscope photograph of an entire *Amoeba*, similar to those you might see in lab. **(b)** A high magnification transmission electron micrograph of the plasma membrane of an *Amoeba* using the light microscope. Here the membrane appears as a double line. **(c)** As a result of the improved clarity and higher magnification provided by electron microscopes, the plasma membrane appears as a double-layer with proteins as indicated in this drawing. (a, Robert Brons/Biological Photo Service; b, Biophoto Associates)

BIO-BIT

An adult human typically has more than 50 trillion cells (= 50,000 billion)!

Theory of Continental Drift

(a) Pangaea: Late Paleozoic Era, 230 million years ago

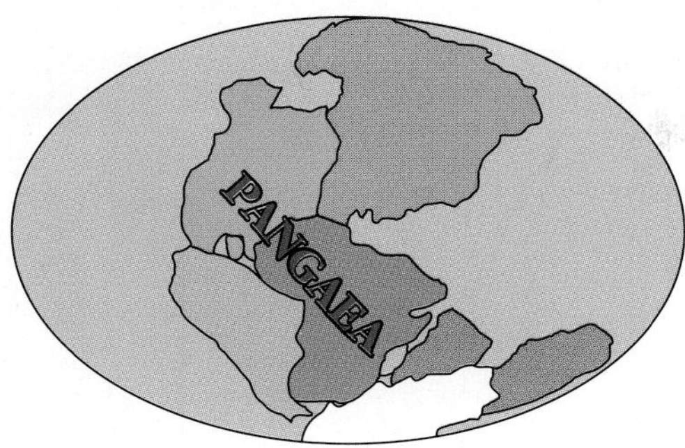

(b) Laurasia and Gondwanaland: Mesozoic Era, 180 million years ago

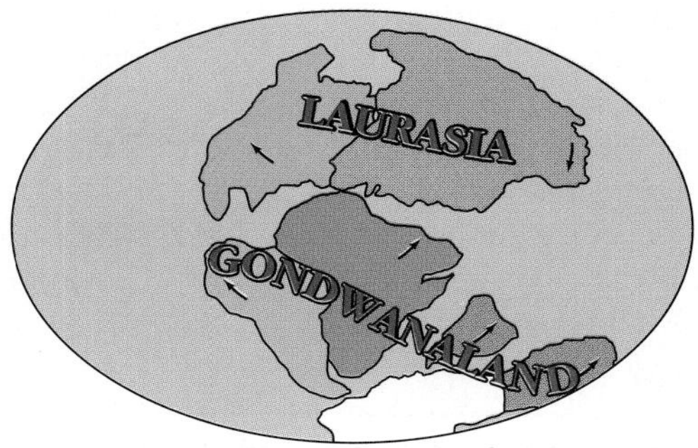

(c) Modern continents (with minor exceptions) had formed by the end of the Mesozoic Era, 110 million years ago

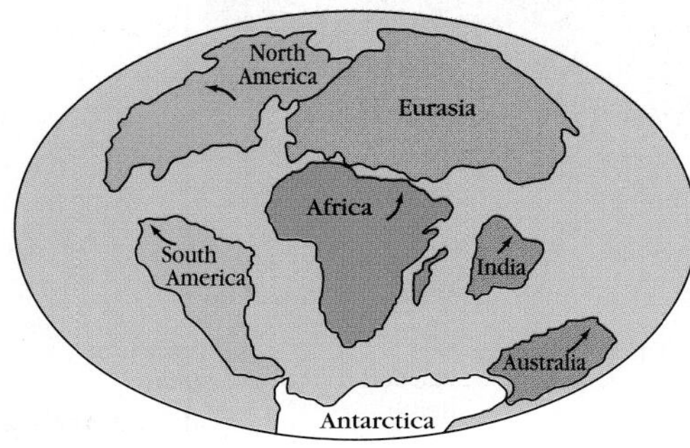

(d) Present day

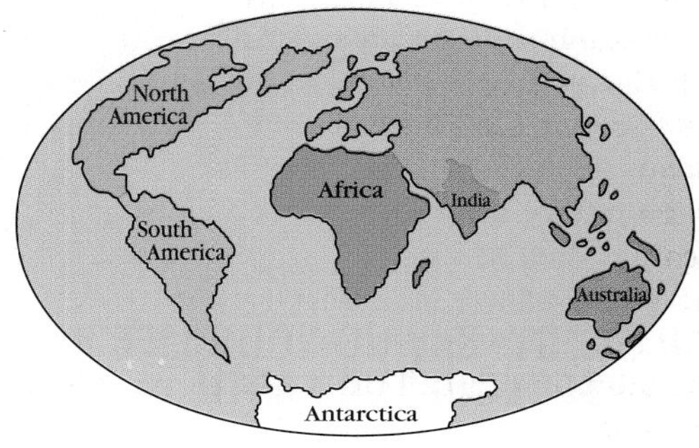

FIGURE 1-2

The continents of today resulted from the gradual break-up of the super-continent Pangaea. Arrows indicate the direction of land mass movement.

New information isn't the only reason that your journey will be more meaningful than ever. Older pieces of the biological puzzle have been incorporated with new information to yield a fuller picture of life on Earth. For example, 50 years ago the theory of **continental drift** (Figure 1-2) was mainly a geographical and geological fairy tale. Then, people viewed continental drift with the bemused scorn that we have toward unicorns, sirens, and other fantastic mythological creatures. Today, explorations of the sediments of the ocean floor have demonstrated that continental drift is a reality that helps to explain the present geographical distribution of many groups of plants and animals (Figure 1-3).

The **theory of evolution of species as directed by the process of natural selection,** jointly proposed by two Englishmen, Charles Darwin and Alfred Russel Wallace in 1858, provides the theoretical foundation for all biological studies. For nearly 150 years this theory has been tested by observation and experimentation, and it remains a central and viable idea in biology today. The relationship that the theory of evolution of species by means of natural selection has to biology is similar to the relationship that gravity has to physics. Like gravity, evolution of species by means of natural selection is a basic, natural phenomenon. But, unlike gravity, which is more or less intuitively understood by anyone who has ever learned to walk, evolution of species by means of natural selection is widely *misunderstood.* Before you begin your journey into life, you should familiarize yourself with the process of natural selection and its relationship to evolution, presented in *A Journey into Evolution:* Natural Selection. Natural selection and evolution are such important topics that we will devote a full chapter to them (Chapter 15).

The scientific method is the basic process by which scientists investigate the world around us. In recent years, great advances have been made in understanding cellular and molecular biology, ecological relationships, geological events, and how organisms change over time. The theory of evolution is central to all of biology and is supported by a wide variety of scientific experiments and observations.

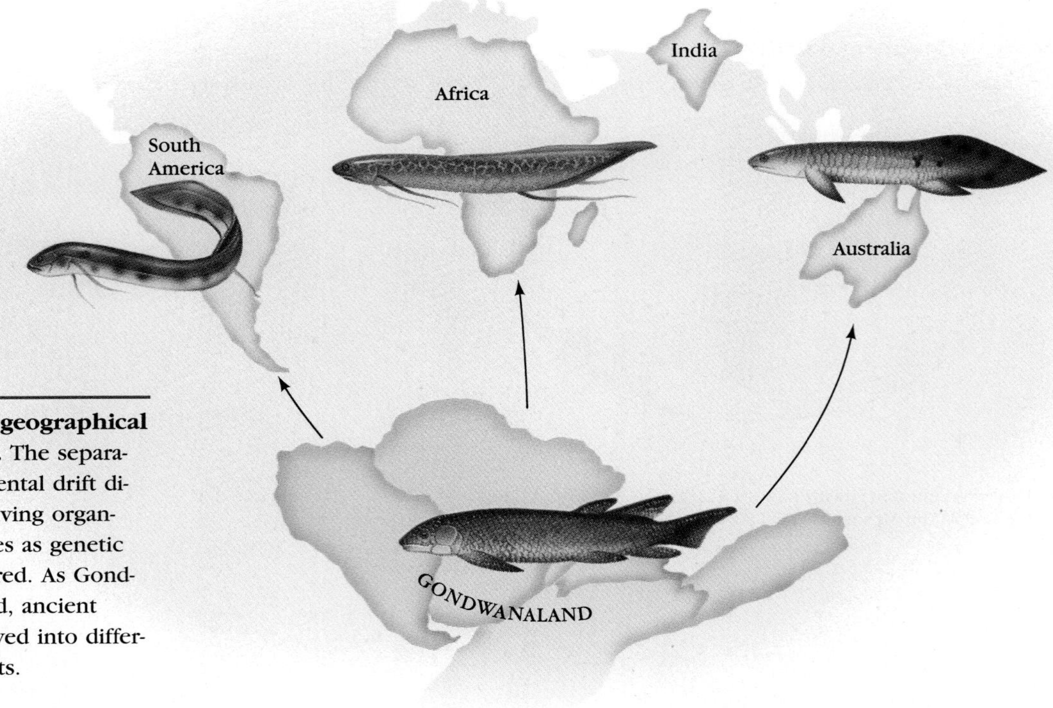

FIGURE 1-3

Continental drift and the geographical distribution of lungfishes. The separation of continents by continental drift divided many populations of living organisms, resulting in new species as genetic changes were no longer shared. As Gondwanaland became subdivided, ancient populations of lungfish evolved into different forms on three continents.

1–B BIOLOGY: SAVING YOUR LIFE AND PRESERVING THE EARTH

No one needs to tell you that we live in precarious times. AIDS (autoimmune deficiency syndrome) threatens the entire human population, and a host of ecological disasters looms larger with each passing year. On our journey into life we will review the known biology of HIV, the virus that causes AIDS, and view the current crisis in the biological framework of host-parasite relationships (Chapter 25). Until a cure is found, your knowledge of the biology of AIDS and the HIV virus that causes it may save your life. When a cure is found for AIDS, it will be biologists who find it, even though they may call themselves by a specialist's title. Immunologists, oncologists, even cytologists are all biologists.

2? AIDS is a grave health problem: a fast-spreading, incurable disease that usually kills within ten years. We are, however, faced with even more serious problems that are the result of the accumulated neglect and exploitation of the biosphere. After 100 years of abuse, these problems are now beginning to have a serious impact. These environmental problems will take perhaps 200 years to become critical, but, by that time, it will be too late to fix many of them: the biosphere will have been irreparably damaged. Life will go on and the mechanism of natural selection will still operate; species will still evolve, but many individuals of our own species will have been selected *against* (Figure 1–4). Many familiar and charismatic species such as elephants, mountain gorillas, and Bengal tigers will have disappeared (Figure 1–5). If present trends con-

(text continues on page 12)

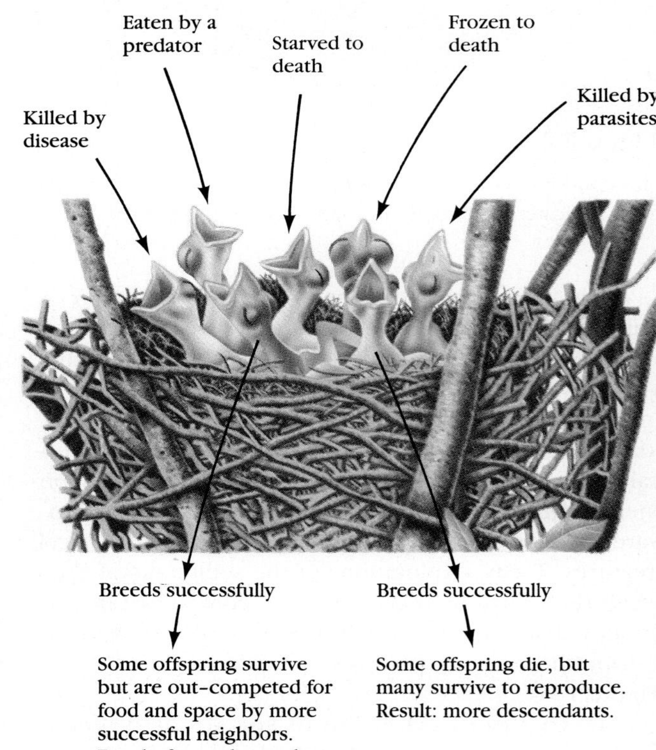
Eaten by a predator
Starved to death
Frozen to death
Killed by disease
Killed by parasites
Breeds successfully
Breeds successfully
Some offspring survive but are out-competed for food and space by more successful neighbors. Result: fewer descendants.
Some offspring die, but many survive to reproduce. Result: more descendants.

FIGURE 1-4

Selection against young. Many factors select against the numerous offspring that most animals produce. Those babies that do survive to reproductive age must still find a mate, prepare and defend a breeding site, and guard and provide for their young.

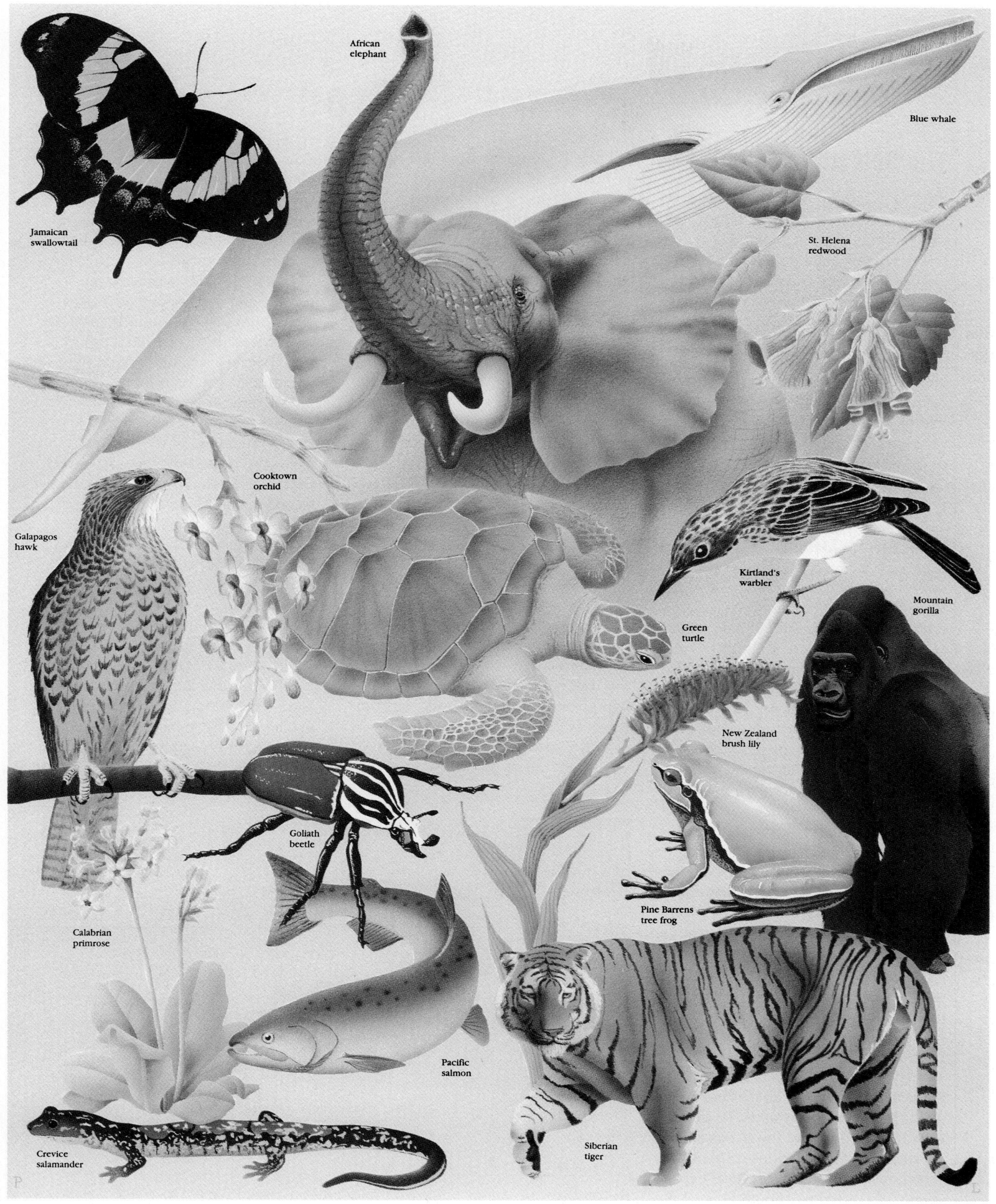

FIGURE 1-5

A few of the world's endangered species.

A Journey into Science in Process

The Scientific Method

The scientific method is a way of answering questions about cause and effect. It is a logical way to try to solve problems and similar reasoning is used by each of us in everyday life. Once we understand the reasoning and procedures scientists use, we are better able to judge for ourselves whether the conclusions of the latest scientific study or advertising claims are justified based on the information presented. We can ask for further tests if a claim does not appear to be well supported by the evidence, and we can agree or disagree with predictions based on such claims.

Experimental Method

When we speak of the scientific method, we usually mean a procedure that involves experiments. This experimental method has three main steps (although in practice, scientists work in many different ways). The first, and key step is to **collect observations,** not only by sight, but perhaps using other senses, too—hearing, smell, taste, and touch. Scientists often use instruments to extend human senses or to detect things our senses cannot. Microscopes, radar, sonar, voltmeters, and Geiger counters are examples of these sorts of mechanisms. Second, the scientist thinks of several **alternative hypotheses** (singular: hypothesis) or proposed answers to questions about what has been observed. The third step is **experimentation,** performing tests designed to show that one or more of the hypotheses is more or less likely to be *incorrect.* (Note that experimentation tests whether an hypothesis is incorrect. Most people don't understand that the scientific method does not—indeed cannot—prove things. Rather, it tests hypotheses, and those that fail are incorrect explanations of the observations.) As a result of these experiments, the scientist should be able to draw some conclusions about why the originally observed events occurred. Let us see how this works in practice.

Scientists usually start with observations that stimulate questions. Some years ago, one of your authors was part of a group of biologists discussing the clusters of butterflies that seemed to be everywhere that June.

"Today," said one, "I saw about 20 yellow sulfur butterflies by a stream and some black swallowtails on a manure heap. What are they doing?"

"It's called 'puddling behavior,'" replied another. "You find puddling butterflies in groups in open places such as the edges of drying puddles, or sandbars (Figure 1-A). I don't think anyone knows what they are doing. Another odd thing is that in many species only the males puddle."

These observations of puddling led us to ask what the butterflies were doing and why. To answer these questions, we had to think of some hypotheses that would account for the observations. That evening, the hypotheses came thick and fast from our armchair scientists.

"An article I read suggested it was a method of population control. Coming together permits the males to count each other. A newcomer can see if there is likely to be enough land for him to set up a territory in the area. Puddling saves them having to fight over territories."

"That sounds wrong to me," replied one of the company. "How can a butterfly figure out the density of males in the area from a group like that? Besides, swallowtails do fight for territories—I've seen them."

"I think it is more likely they're feeding," another contributed. "It was called 'puddling' in the first place because the butterflies often have their proboscises [tongues] out and seem to be sucking something up from the ground" (Figure 1-A).

"I wonder if they are feeding on substances that contain nitrogen. In our lab we've shown that butterfly caterpillars grow faster if you feed them extra nitrogen, and there is lots of nitrogen in a manure pile."

"But not in sand," came the objection. "And if they are after nitrogen, you'd expect females to puddle, not males. The females lay the eggs that hatch into caterpillars, and extra nitrogen in the egg might be very use-

(a) (b)

FIGURE 1-A

Butterflies. (a) Sulfur butterflies puddling on a sand bank. (b) A tiger swallowtail probing the sand with its proboscis. (a, Luiz Claudio Marigo/Peter Arnold, Inc.; b, John Gerlach/Animals Animals)

ful, but it's not the females that puddle."

"It sounds to me," chipped in another, "as if they're after salts—perhaps salts containing sodium. All the puddling places contain quite a lot of salts: manure piles have salts from urine, and puddles have salts at the edges, left behind by evaporation of water. Lots of animals that feed on plants are short of sodium because plants contain so little of it. We put out salt blocks for cows and horses and end up attracting deer and rabbits as well. Perhaps male butterflies need more sodium than females do."

We could test these alternative hypotheses only by doing experiments. Some hypotheses are of no use because they cannot be tested. For instance, the hypothesis "puddling butterflies count each other" is probably untestable because it is hard to imagine an experiment that could show us whether or not an animal has counted its neighbors. Even a testable hypothesis usually cannot be tested directly. We must first develop a testable **prediction** from it. From the hypothesis that butterflies sucked up sodium when they puddled, we predicted that if we put out trays containing sodium, butterflies would be attracted to puddle on them. The hypothesis that puddling butterflies suck up nitrogen generated the prediction that butterflies would puddle on trays of amino acids, substances that contain nitrogen. These predictions can be tested and, in this case, both can be tested at the same time, in the same experiment.

We must design experiments to make their results as clear-cut as possible. For this reason, experiments have to include **control treatments** as well as **experimental treatments.** The two differ only by the factor(s) being investigated. For instance, to test our hypotheses, we had to show that butterflies would puddle on an experimental tray containing amino acids or one containing sodium but would not puddle on control trays that were identical except that they did not contain the amino acids or sodium.

Suppose we put out three trays—one containing sodium, another containing amino acids, and a third containing something butterflies are most unlikely to eat, such as plain sand or sand and water (the control). We would predict that if butterflies are attracted to puddle on sodium, they would come to puddle on the sodium tray but not on the other two. If they are attracted to amino acids, they would puddle only on that tray. If they are attracted to both, they would puddle on both of these but not on the control tray, and if they are attracted neither to amino acids nor to sodium, they would not puddle on the trays at all. Note that there are dozens of other possible reasons for the last result. If no butterflies turned up to puddle on our trays, we would have learned nothing. Butterflies might not puddle on trays because they won't come near trays for some reason, or because they never see the trays, or because they avoid the human watchers nearby, or for any one of a number of other reasons.

So that our experiment would not fail for lack of butterflies, we put our trays on a sandbank by a lake where tiger swallowtail butterflies often puddled in large numbers. We filled the trays with clean sand for the butterflies to stand on, and in each tray we pinned a dead male tiger swallowtail as a decoy, because we thought butterflies might be attracted to puddling places by seeing other butterflies there. We put out ten trays of sand and poured the same volume of solution (substances dissolved in water) into each one. Then we sat nearby, with binoculars, notebooks, and watches, to see what would happen.

Soon dozens of tiger swallowtails were hovering over the trays. Whenever a butterfly landed on a tray, it stuck its proboscis into the sand. At times, as many as 30 butterflies were on a tray together. Most of the butterflies spent a few seconds on every tray, but they puddled (which we defined as staying for more than 15 seconds) on only a few trays: all those containing sodium in any form and those containing amino acids (Figure 1-B).

We were satisfied that these results were accurate because we had taken another precaution: the people recording the butterflies' visits did not know which tray contained which solution. Making an experiment "blind" in this way is important. Psychologists have shown that, even in a carefully controlled experiment, experimenters tend to find the results they want to find. This is also why scientists try to form many hypotheses to explain their observations: it's too easy to bend the truth, without even realizing it, to support the only available hypothesis.

Those of us who favored the hypothesis that butterflies puddle in response to sodium were disappointed that they also puddled on amino acids. But prejudice can sometimes be useful, even in science! Not only were we disappointed by the results, we were inclined to think they were wrong. Back we went to our bottle of amino acids. We now made an observation that should have preceded the experiment: the label said, "Prepared in sodium citrate." According to popular myth, scientists are calm and objective, but we were very excited as a technician analyzed our amino acids: they were chock-full of sodium! There followed frantic phone calls and special deliveries to obtain amino acids free from sodium. At last came a suspenseful experiment, which showed that butterflies did not puddle on our new, sodium-free amino acids.

We had now conducted a well-controlled scientific experiment. What conclusions could we draw? Had we proved the hypothesis that butterflies puddle so as to obtain sodium? No. We had not even shown that the butterflies actually drank the sodium solution. All we had shown was that male tiger swallowtail butterflies would puddle on sand containing sodium salts but not on sand containing various other solutions. Many more hypotheses and experiments were needed if we were to learn more.

One peculiarity of the scientific method is that an hypothesis can never formally be proved but can only be disproved. A correct hypothesis leads

Science in Process (continued)

to predictions that are borne out by experiments, but an incorrect hypothesis may also produce correct predictions (that is, the prediction was right, but for the wrong reason). Therefore, if the results of an experiment agree with the prediction, we are still not sure that the hypothesis is correct. For instance, the hypothesis that butterflies puddle to obtain sodium for food is not proven by the experimental finding that butterflies puddle on sodium. They might puddle because wherever there is sodium in nature there is also nitrogen and they really obtain nitrogen from puddling. We have not even disproved the hypothesis that puddling is a means for the butterflies to "count" each other. They might puddle on sodium merely as a convenient rendezvous (although the fact that the butterflies appear to feed when they puddle makes this hypothesis unlikely). The more alternative hypotheses we disprove or cast doubt on, however, the greater the likelihood that the remaining hypothesis is correct.

Scientists also hesitate to accept the results of an experiment until they have tested its repeatability. Repeating an experiment guards against two kinds of errors. The first is **human error** (a polite term for mistakes); we might have inadvertently switched the solutions, written our results in the wrong column of our data notebook, or alarmed the butterflies. (Even in this simple experiment, the possibilities are endless.) Second, any experiment is subject to **sampling error,** error due to using a relatively small number of subjects. Organisms are notoriously variable. Our experiment sampled only a few dozen butterflies on six days. These butterflies might not have been representative of all tiger swallowtails. We could be more confident of our results if we were to repeat the experiment, using more butterflies (that is, a larger sample) and following precisely the same procedure. How many butterflies do we need? The more the better, but we could not possibly test all the butterflies in the world. In practice, we can use statistical tests to tell how "sure" we are of our results with a given sample size.

An hypothesis supported by many different lines of evidence from repeated experiments is generally regarded as a **theory,** and after even further testing it comes to be generally accepted.

FIGURE 1-B

Arrangement of trays on one day of the puddling experiment. (a) Each tray contained the same volume of sand. Each of eight trays also contained 1.5 liters of water or solution. Different solutions were placed in different trays on subsequent days. (Sugar was tested because swallowtail butterflies eat sugar-filled nectar from flowers, and therefore we wondered if they might be attracted to puddle on sugar.) The red bar on each tray shows the number of "sampling" visits (lasting less than 15 seconds) by butterflies. Blue bars show the number of butterfly-minutes spent puddling on the tray in visits lasting more than 15 seconds. The numbers make it obvious that butterflies puddled on the trays containing sodium and those containing amino acids but not on any of the other trays. (b) Observers watching puddling trays. (c) Butterflies visiting the trays. In the middle of each tray is pinned a dead butterfly to serve as a decoy. (Paul Feeny)

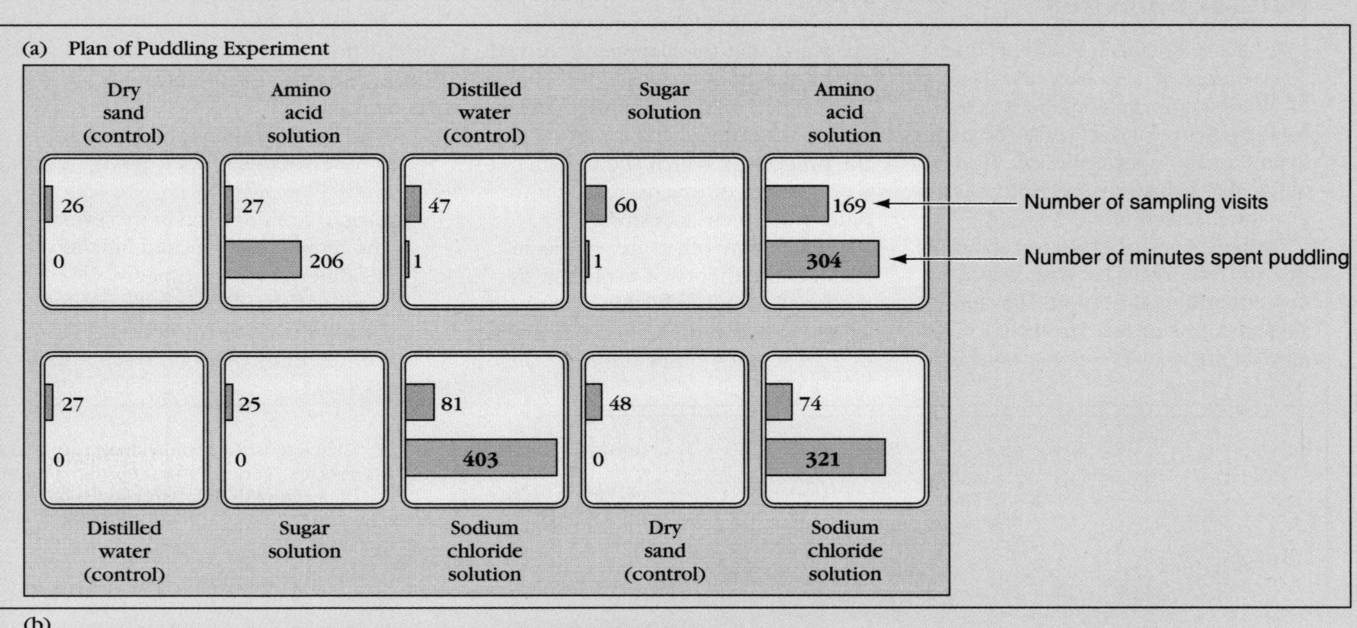
(a) Plan of Puddling Experiment
Dry sand (control)
Amino acid solution
Distilled water (control)
Sugar solution
Amino acid solution
26
0
27
206
47
1
60
1
169
304
Number of sampling visits
Number of minutes spent puddling
27
0
25
0
81
403
48
0
74
321
Distilled water (control)
Sugar solution
Sodium chloride solution
Dry sand (control)
Sodium chloride solution

(b)

(c)

1?

A Journey into Evolution

Natural Selection

A living organism is the product of interactions between its genetic information and its environment. This interaction is the basis for the most important concept in biology, that organisms evolve by means of natural selection.

The theory of evolution states that today's organisms have arisen by descent and modification from more ancient forms of life. For instance, most biologists believe that human beings evolved from now-extinct animals which looked something like apes, and that this happened through accumulation of changes from generation to generation. In more modern terms, we can say that evolution is the process by which the members of a population of organisms come to differ from their ancestors.

Like many other great ideas in science, the theory of evolution by means of natural selection presents a simple explanation that makes sense of a great many observations of the natural world. Soon after Charles Darwin proposed this theory, his champion, Thomas Huxley, remarked, "How extremely stupid not to have thought of that!"

The theory is based on three familiar observations (Figure 1-C):

1. **Organisms are variable.** Even the most closely related individuals differ in some respects.
2. **Some of the differences among organisms are inherited.** Inherited differences between

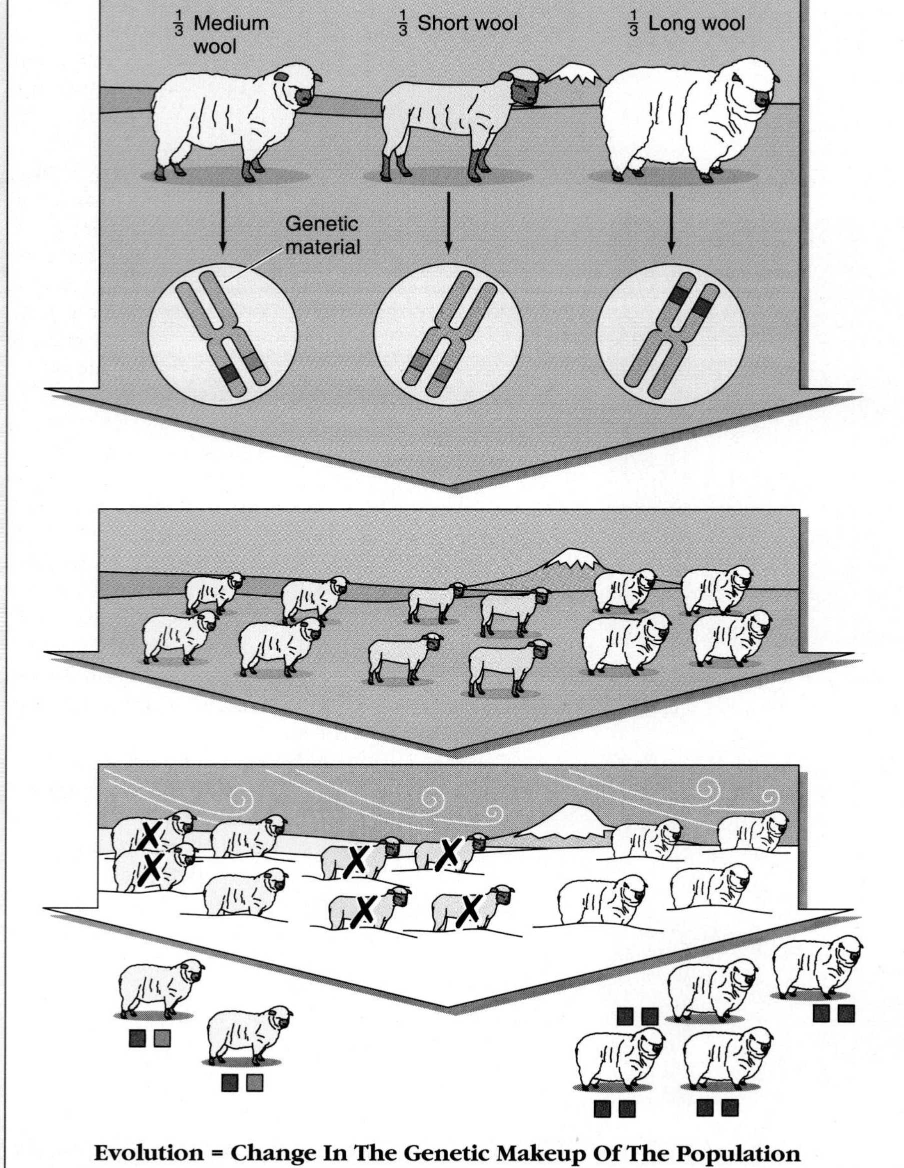

1. Characteristics of individuals vary within species. For example, consider a populat of sheep with different lengths of wool.

2. Wool length is inheritable.

3. More young are produced than survive.

4. Climate is one factor that selects for some individuals and against others.

5. The result of selection is adaptation. Here, an environmental factor has selected for $^1/_3$ medium wool and $^2/_3$ long wool.

FIGURE 1-C

The theory of evolution by natural select

FIGURE 1-D

A black-face sheep with shaggy fur. (Fran Allan/Animals Animals)

individuals result from differences in the genetic material that they inherited from their parents. Because parents and offspring have very similar genetic material, they tend to resemble each other more closely than they resemble organisms to which they are less closely related.

3. **More organisms are produced than live to grow up and reproduce.** Fish and birds may produce hundreds of eggs, oak trees thousands of acorns, but only a few of these survive to reproduce in their turn.

Some of the inherited variations among organisms are bound to affect the chances that an individual will live to reproduce. Individuals with some genetic variations produce more offspring (which inherit this genetic material) than do others. This is called **natural selection,** and it produces **evolution:** a change in the proportions of different genes from one generation of a population to the next.

To take an example of natural selection producing evolution, the length and thickness of an animal's hair is largely determined by its genes. A very cold winter may kill many individuals with short, sparse hair. Individuals with longer, thicker fur are more likely to survive the winter and reproduce in the following spring (Figure 1-D). Because more animals with thicker fur breed and pass on the genetic material that dictates the growth of thick hair, a larger proportion of individuals in the next generation of the population will have genes for thick fur. The genetic makeup of the population has changed somewhat from one generation to the next, and that is evolution. The agent of natural selection in this case is low temperature, which acts as a **selective pressure** against those individuals with short, sparse hair.

The result of natural selection is that populations undergo **adaptation,** a process of accumulating changes appropriate to their environments, over the course of many generations. The selective pressures acting on a population "select" those genetic characteristics that are adapted, or well suited, to the environment. For instance, through selection, populations living in cold areas evolve so as to become better adapted to withstand the cold.

In this discussion of evolution, we use "environment" as a catchall word meaning much more than merely whether an organism lives in a forest rather than a desert and whether or not it can obtain enough food. Environment includes all the external factors that affect the number of offspring the organism produces.

An organism's environment includes its external environment as an embryo, juvenile, and adult. Let us, for example, consider a frog (Figure 1-E). Whether it successfully meets the pressure of its environment depends on the speed and normality of its embryonic development, whether bacteria penetrate the jelly coat of the egg and destroy it during development, whether as a tadpole it can find enough food and avoid being eaten by a predator, whether the pond in which it lives as a tadpole dries up before it becomes a frog, and whether as a small frog it avoids death by disease or predation. To make things more complicated, environmental pressures are often contradictory. For instance, a hot summer benefits our frog in one way because frog embryos develop faster at higher temperatures, but a hot summer also increases the chance that the tadpole's pond will dry up before it is ready to live on land. Even worse, environmental pressures frequently change. The frog must have features that allow it to withstand both the heat of summer and the cold of winter; it should remain still to be safe from some predators and move quickly to escape from others; and so forth. So the frog's genetic makeup is a compromise brought about by selection for a number of opposing characteristics.

Although natural selection is probably the most far-reaching agent of evolution, it is not the only one. We shall discuss other mechanisms that change a population's genetic makeup in Chapter 15. ■

FIGURE 1-E

Survival. This table shows when most deaths occur within the lives of many frogs. Very few eggs develop into tadpoles that live to reproduce.

BIO-BIT

AIDS is the leading cause of death of adults living in some cities in Africa.

tinue, the atmosphere may be radically different, and life for humans may be quite unpleasant. ■

One of the main goals of this journey into life is to alert you to the existence, causes, and far-reaching effects of the ten worst environmental problems, and to suggest ways that you can help rescue yourself, your children, and their grandchildren—as well as the biosphere. We emphasize environmental problems in our journey into life because they have already had, and will continue to have a grave impact upon the biological world. Refer to Table 1-1 for a summary of causes, potential and realized negative effects, suggested solutions, and references to fuller discussion in subsequent Journey Boxes.

We are now nearly ready to embark upon this journey. You know our itinerary and you have packed two essential understandings into your mental suitcase: the scientific method of inquiry and the theory of evolution by means of natural selection. But before we depart, it would be good to more fully define what we mean by "life." What are the major features of living organisms? (Figure 1-6).

3?

1. *Living things are organized into units called cells, the units of structure, function, and reproduction in organisms.* Most cells are so small that we must use a microscope to see them. Many small organisms, such as bacteria and protists, consist of one cell each, while larger organisms, such as grasses and humans, contain up to hundreds of millions of cells. Each cell is a discrete packet of living material, a biochemical factory that shows all the features of life listed below.
2. *Living things are highly ordered.* All organisms contain very similar kinds of chemicals, and the proportions of these chemical elements in living things are very different from those in the nonliving environment. A living organism's chemical composition, structure, and function are all more complex and more highly organized than those of nonliving things.
3. *Living things obtain and use energy from their environments to grow, to reproduce, and to maintain and increase the high degree of orderliness of their bodies.* Most organisms depend, directly or indirectly, on energy from the sun. Green plants use solar energy to make food. This supports the plants themselves and is also used by all organisms that eat plants, and eventually by those that eat the plant-eaters.
4. *Living organisms respond to stimuli from their environments.* Most animals respond rapidly to environmental changes by making some sort of movement—exploring, fleeing, or even rolling into a ball. Plants respond more slowly, but still actively: stems and leaves bend toward light, and roots grow downward. The capacity to respond to environmental stimuli is universal among living things.
5. *Living things develop.* Everything changes with time, but living organisms change in particularly complex ways called development. A nonliving crystal grows by addition of identical or similar units, but during its life cycle a plant or animal develops new structures, such as leaves or teeth, that differ in chemistry and organization from the structures that produced them.
6. *Living things reproduce themselves.* New organisms arise only from the reproduction of other, similar organisms. New cells arise only from the subdivision of other cells.
7. *The information each organism needs to survive, develop, and reproduce is segregated within the organism and passed from each organism to its offspring.* This information is contained in the organism's genetic material—its chromosomes and genes—and it specifies the possible range of the organism's physical, biochemical, and behavioral features. An organism passes genetic information to its offspring, and this is why offspring are similar to their parents. Genetic information does vary somewhat, though, so parents and offspring are usually similar, but not identical.
8. *Living things evolve and are adapted to their environments.* Today's organisms have arisen by evolution, the descent and modification of organisms from more ancient forms of life. Evolution proceeds by way of natural selection, the differential survival and reproduction of those organisms that carry the genes best suited to their environment. This process of adaptation works so well that we can predict roughly how a given organism lives merely by examining its structure. ■

Although we intuitively think that we can tell if something is alive or not, it is often difficult to do so. It is important to emphasize that *all* of these characteristics *taken together* define life.

AIDS and environmental exploitation of the biosphere are just two of the many serious biological problems facing humans today. Addressing these problems will require broad education, serious investigations, and coordinated efforts to change human behavior. There are at least eight characteristics that collectively help us to define life.

Table 1-1
Our Current Environmental Problems

Problem	Cause	Potential Negative Effects	Realized Negative Effects	Solutions	Chapter Section(s) and *A Journey Into* Boxes
Overpopulation	too many people having too many babies	stress, disease, conflict, death, pollution, environmental destruction, starvation, war	stress, disease, conflict, death, pollution, environmental destruction, starvation, war	education, increase in the status of women, population control	**38-F, 39-F, and Ch. 16 Box**
Global warming	greenhouse gases, primarily CO_2	climatic disruptions, rise in sea levels	0.5° rise in temperature	decrease emissions of CO_2 and methane	**39-F and Ch. 6 Box**
Acid rain	SO_2 and NO_2 emissions	forest collapse and acidified lakes and ponds, loss of aquatic species	loss of forest trees, frog and fish populations down	clean up emissions, reduce population	**37-G, 39-F, and Ch. 2 Box**
Ozone hole	chlorofluorocarbons	skin cancers increase, death up food chain in Antarctic waters	ozone holes over Arctic and Antarctic, ozone thinning elsewhere	cease manufacture and use of CFCs, reduce population	**37-G, 39-D, and Ch. 17 Box**
Deforestation	clear-cutting, greed, people trying to feed and support their families	loss of species, climatic disruptions	loss of species, climatic disruption	habitat protection, population control	**37-G, 39-F, and Ch. 7 Box**
Loss of species	habitat loss in forests and wetlands, competition from introduced species, overhunting	impoverished biosphere	impoverished biosphere	habitat protection, population control	**37-F, 39-F, and Ch. 14 Box**
Loss of topsoil	careless farming and irrigation	loss of soil fertility	loss of crops, siltation of estuaries, death of reefs	better farming practices, population control	**39-C and Ch. 36 Box**
Loss of water	overpopulation	increased desertification, war, economic strife	water shortages, desertification, economic strife	population control	**39-C**
Polluted water	toxic metals, toxic landfill leachate, careless farming, inadequate sewage treatment	water shortages, illness, death	illness, death, water shortages	remodeling factory and waste-disposal practices, population control	**37-G**
Overfarming	poor education, greed, desperation	loss of soil fertility, food shortages, hunger	loss of soil fertility, food shortages, hunger	better education, population control	**39-C and Ch. 36 Box**

Living things:

(a) Have cells

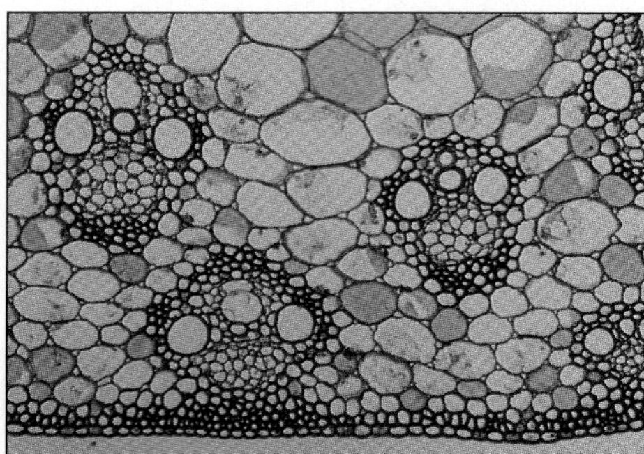

(b) Are highly ordered

(c) Obtain energy to increase the order of their bodies

(d) Respond to stimuli

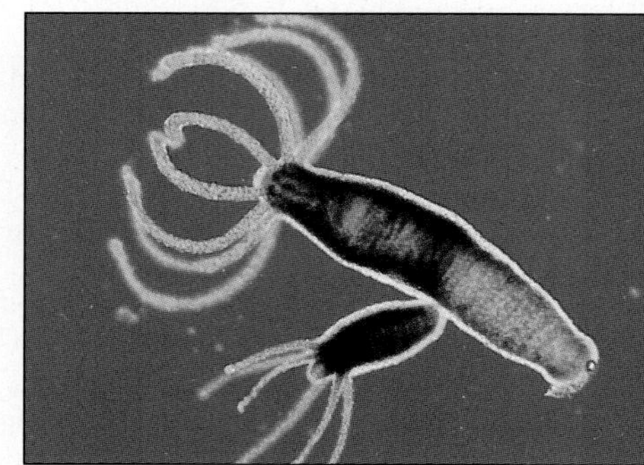

(e) Develop

(f) Reproduce

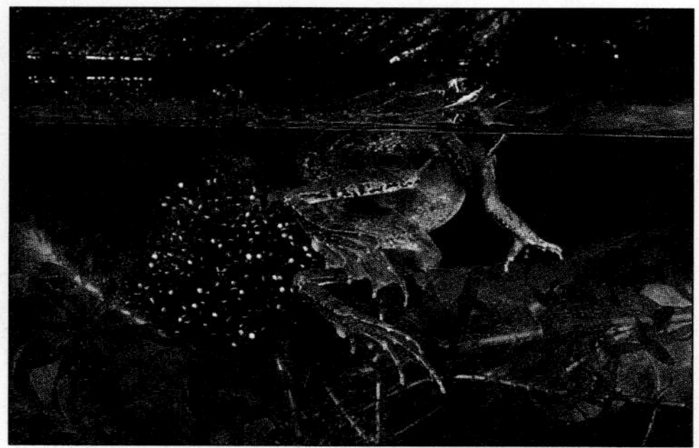

(g) Have genetic material

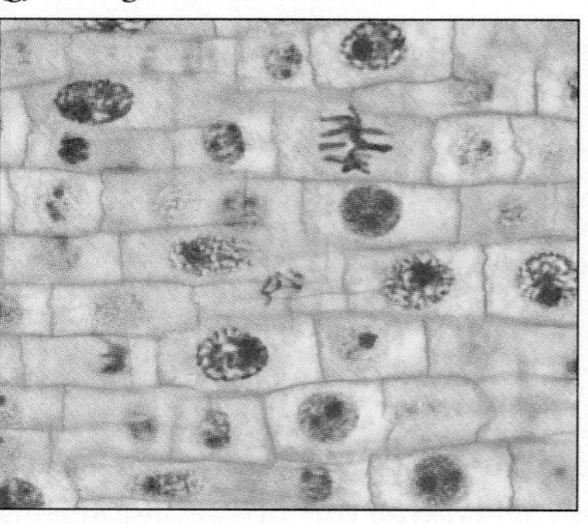

(h) Evolve and are adapted to their environment

FIGURE 1-6

Characteristics of living organisms. (a) Section of a corn stem. **(b)** False color SEM of the skeleton of a single-celled radiolarian. **(c)** Silkworm eating a leaf. **(d)** *Hydra,* a freshwater animal related to sea anemones. **(e)** A cluster of fish embryos. Developing eyes are visible as black dots. **(f)** A male frog grasping a female frog from behind. He releases sperm at the same time that she releases small black and white eggs. **(g)** Cells dividing in the tip of an onion root. The chromosomes are stained red. **(h)** The shape of a hummingbird's bill and its ability to hover in midair are evolutionary adaptations which enable this species to survive by feeding on flower nectar. (a,c, E.R. Degginger; b, Stanley Flegler/Visuals Unlimited; d, Biophoto Associates; e, G.I. Bernard/Animals Animals; f, R. Andrew Odum/Peter Arnold, Inc.; g, Bruce Iverson; h, Stephen Dalton/Photo Researchers, Inc.)

FIGURE 1-7

The colorful scales of a Panther chameleon. (Kevin Schafer)

1–C LIFE: ONLY ON EARTH

Surrounded by life, we seldom stop to consider how remarkable it is. Not only does the beauty in living organisms outshine any human works of art (Figure 1-7), but living things move and act independently. Busy on their own errands, they live their lives in ways that we fail to understand. Think for a moment what life must be like for a dragonfly. Don't just imagine yourself as a human encased inside an armor-plated, winged, dragonfly suit, but instead try to imagine how the big insect perceives the world. What does it see or smell as it flies above the pond? Does it have thoughts? How do the insects it snatches out of the air taste to it? How does it recognize prey? How does it recognize members of its own species? How does it learn to fly? Does it sleep at night? If you're like most of us, you will have no answers for these questions and you will appreciate the limitations of our current knowledge.

Except for the handful of domesticated creatures that rely upon humans in one way or another, the vast majority of living organisms that share our world lead lives as remote from ours as we are from the mountains of the moon. Blind and deaf to the textures of the lives of most other organisms, we coexist in connected, but separate, spheres. Civilization further disconnects us from nature. Our meats, fruits, and vegetables come not from animals and plants, but in plastic-wrapped packages from the supermarket. Experiences with "wild" animals are limited to those on display at zoological parks and captured on nature specials we watch on television. If people think of plants at all, it is as outdoor decoration, and trees are confined to street margins or parklands. In the last 5000 years—equivalent to the blink of an eye in geological time—the human relationship with living creatures has changed from worship and reverence to exploitation and eradication.

This careless attitude is especially strange when you remember that there is no evidence that living things exist anywhere else in the universe. Conditions are too fiery on Mercury, too poisonous on Venus, and too cold on the rest of the planets of our solar system for life to thrive. Only Earth's unique oceans fostered the earliest living organisms. Furthermore, each species is the current product of millions of years of evolution. The chemical products made by each species' genes have been fine-tuned by natural selection so that it can cope with the normal range of fluctuations in its environment. Although there are many species called living fossils because apparently

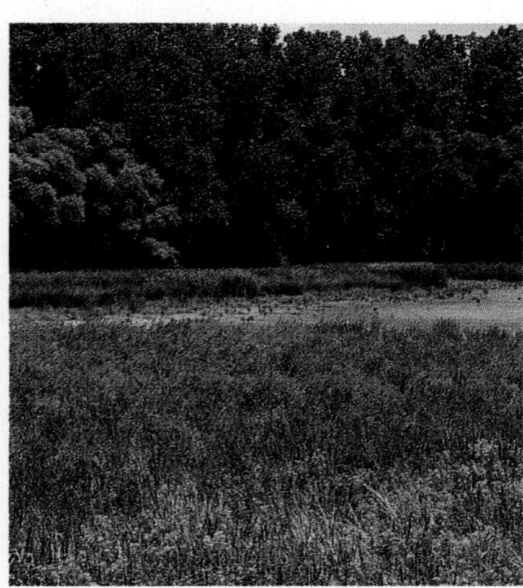

FIGURE 1-8

Tough survivors. (a) American cockroaches clustering behind a kitchen wall. **(b)** A Norway rat, rummaging through garbage. **(c)** Purple loosestrife looks beautiful until you learn that it invades wetlands and crowds out native plants. (a, Ken Lucas/Biological Photo Service; b, Tom McHugh/Photo Researchers, Inc.; c, Biological Photo Service)

identical representatives have been found preserved in rocks that are tens, even hundreds of millions of years old, most species don't last that long. Paleontologists—scientists who study fossils—have estimated that the life span of a typical species ranges between one and ten million years. And they further speculate that between one to ten species naturally go extinct each year. Yet, humans have so little regard for other forms of life that it is estimated that *each day between 10 and 100 more species are forced into extinction.* Loss of habitat to the demands of an ever-expanding human population is the most common, but not the only, reason for this tragic occurrence. (See *A Journey into the Environment:* Loss of Biodiversity, Chapter 14, for a more complete discussion of our current loss of biodiversity.) If we allow it to continue, this daily erosion of life (estimates vary from 4000 to 90,000 species per year) will result in a biological impoverishment of unparalleled proportions. Soon we will be left with only those species that are tough enough to coexist in the margins of human society (Figure 1-8).

A species is an irreplaceable thing, the culmination of millions of years of evolution, an object of fascination and wonder even if it is physically unattractive. Moreover, each species is a piece of the biological puzzle. Each has an ecological role, each is full of history and information, and each has a separate way of life.

FIGURE 1-9

The Carolina parakeet was the only North American small parrot. Once found in huge numbers in the eastern United States, this small, gregarious bird was hunted to extinction. The last captive bird died in 1914. (Will and Deni McIntyre/Photo Researchers, Inc.)

When the European colonists first came to America, the trees were full of strange, colorful birds (Figure 1-9). Recorded by explorers as early as 1588, the Carolina Parakeets were well described by the ornithologist Alexander Wilson in 1828:

> At Big Bone Lick, thirty miles above the mouth of the Kentucky River, I saw them in great numbers. They came screaming through the woods in the morning, about an hour after sunrise, to drink the salt water, of which they, as well as the Pigeons [Passenger Pigeons, another species that is now extinct], are remarkably fond. When they alighted on the ground, it appeared at a distance as if covered by a carpet of richest green, orange, and yellow: they afterward settled, in one body, on a neighboring tree, which stood detached from any other, covering almost every twig of it, and the sun, shining strongly on their gay and glossy plumage, produced a very beautiful and splendid appearance. Here I had an opportunity of observing some very particular traits of their character: Having shot down a number, some of which were only wounded, the whole flock swept repeatedly around their prostrate companions, and again settled on a low tree, within twenty yards of the spot where I stood. At each successive discharge, though showers of them fell, yet the affection of the survivors seemed rather to increase, for after a few circuits around the place, they again alighted near me, looking down upon their slaughtered companions with such manifest symptoms of sympathy and concern, as entirely disarmed me.

FIGURE 1-10

This serving of sauteed frogs' legs (surrounded by a garnish of toasted bread) may be a sign of serious ecological problems for the people and wildlife of India. (Dennis K. Purse/Photo Researchers, Inc.)

Although Wilson's account is sentimental, it is accurate. And, because of their habits, it is little wonder that millions of the small, gregarious, green and yellow parrots were shot for food, for feathers to decorate women's hats, and for "sport." Farmers killed Carolina Parakeets because they attacked grain and fruit crops, and professional bird-catchers captured and sold many thousands as cage birds. By the 1880s, the parakeets were extremely rare and the last wild specimen was shot in 1901. The Carolina Parakeet had been bred successfully in captivity, but unfortunately the last captive bird died in 1914. There are too many similar stories that detail the demise of too many species.

Loss of a single species is not a huge ecological disaster, but the current relentless loss of species pulls strands from the fabric of life that supports us all. Perhaps the zoologist and explorer William Beebe stated the importance of a species best:

> The beauty and genius of a work of art may be reconceived, though its first material expression be destroyed; a vanished harmony may yet again inspire the composer; but when the last individual of a race of living things breathes no more, another heaven and another earth must pass before such a one can be again.

We hope that humans are becoming more concerned over the future of this planet. Although we often act as though only *Homo sapiens* matter, we do not inhabit this planet alone. Every species of animal, plant, fungus, and bacterium has a role to play. In India this lesson is beginning to be better understood. Two species of Indian bullfrogs are currently threatened with extinction because, each year, tons of frozen frogs' legs are exported to the dinner tables of Western Europe (Figure 1-10). But the frogs are worth

BIO-BIT

Over 99% of all of the species of life that ever existed are now extinct.

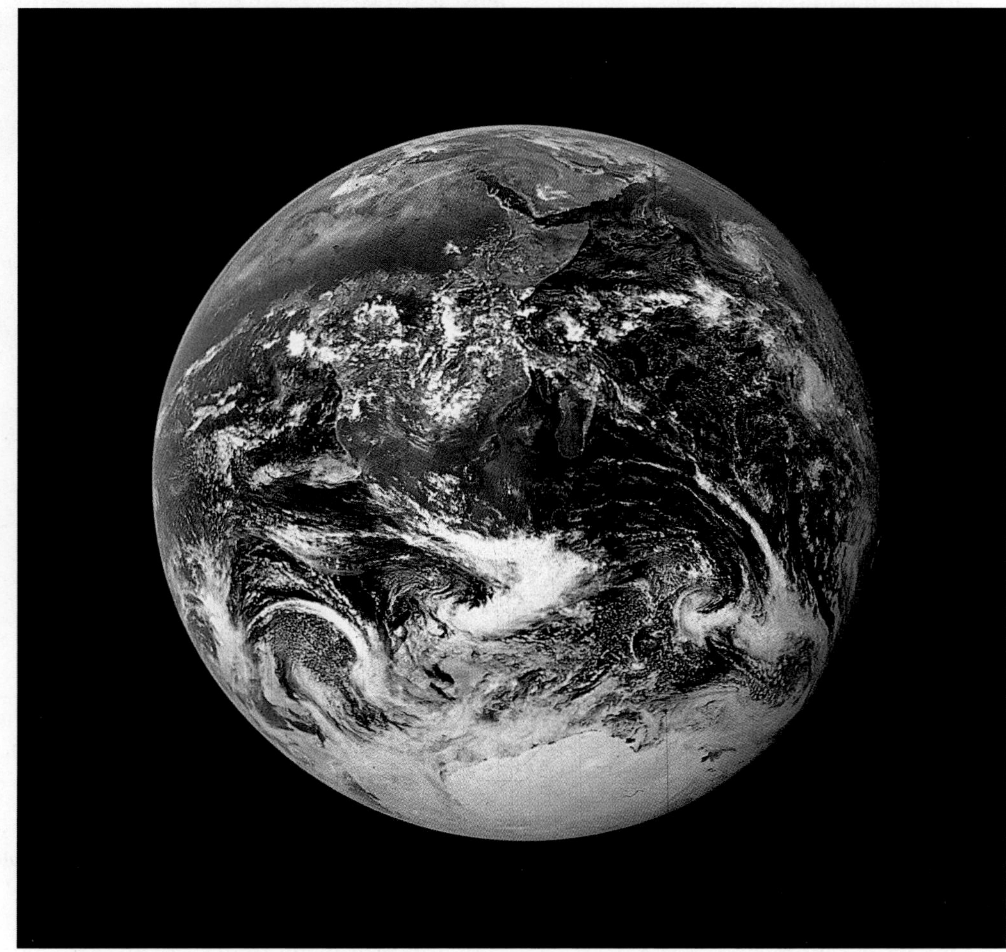

FIGURE 1-11

A view of Earth as seen from the moon. (NASA photo, research by Grant Heilman Photography)

more alive than dead, because an adult bullfrog usually eats its weight in insects *each day.* India's vanishing bullfrogs are linked to higher rates of malaria and greater losses of crops to insects. Malaria is a potentially lethal disease carried by mosquitoes that is the leading cause of death worldwide. Although there are medicines to prevent malaria, few Indians can afford them. And in its struggle to feed its huge population, India cannot afford to lose any food crops to insect pests.

From the 1930s to the 1950s, before space exploration became a reality, there was a common perception, rooted in the American immigrant and frontier heritage, that one day, when this planet became uninhabitable because of pollution, we would all put on silver space suits and climb into needle-nosed ships that would take us off to colonize brave, new planets. Interestingly, our space program showed us a very different scenario—the view of Earth from the moon (Figure 1-11). We learned that we live on a lovely blue and white planet. Veiled with swirling clouds, it floats in isolated grandeur—a tiny sapphire against the velvet blackness of space. From this vantage point we can see no national boundaries, no sign of humanity. Even more important: as far as we know, we are alone in space.

This image has begun to change the way we think of Earth. Many now realize that we all *are* space travelers, and that this unique and fragile planet is our spaceship. Like all our fellow earthlings, we were born on board. Although a handful of us will become astronauts, most will never climb into that needle-nosed ship to escape a polluted Earth. We are only beginning to comprehend how complicated, interconnected, and irreplaceable our home planet is.

Today, our environmental problems have reached the stage at which we're nearly out of time. We have only a few years to change the habits that worsen global warming—to rein in the human population explosion. Our highest task is to educate ourselves and new generations. We must become convinced that all life—not only human life—is unique and valuable. We must learn that our own species has special responsibilities not to scuttle the ship that carries all of us through space, because there is nowhere else to go.

With these ideas in mind, our journey begins.

Human disregard and neglect have resulted in monumental increases in the number of species of life forced into extinction. The loss of a single species can have disastrous, unforeseeable effects. A dramatic change in human attitudes and behaviors will be necessary to reduce the impact that we humans have on other life forms.

SUMMARY

1. Scientists rely upon the scientific method to investigate the world around us.
2. Knowledge of the biological world has expanded greatly in the last 50 years. This is due primarily to advanced technology.
3. The theory of evolution of species as directed by natural selection is the central concept in all of biology.
4. Biological education can help address serious problems such as the AIDS epidemic and the global impact of humans on the environment.
5. All living organisms:
   a. are organized into units called cells,
   b. keep their internal environments fairly constant,
   c. obtain and use energy from their environments to grow, to reproduce, and to maintain and increase the high degree of orderliness of their bodies,
   d. respond to stimuli,
   e. develop,
   f. reproduce themselves,
   g. pass the information to survive, develop, and reproduce to their offspring, and
   h. evolve and adapt to their environments.
6. Humans have seriously affected other life forms by destroying or altering their habitats. As a result, the rate of extinction has increased dramatically.

THINKING CRITICALLY

1. After every hard rain you find dead earthworms lying on the sidewalk. What experiments would you perform to show the cause of death?
2. To what extent should scientists be held responsible for the social and moral consequences of their discoveries?
3. Many professional scientific societies have adopted ethical conduct guidelines for their members and have pledged legal aid to members who "blow the whistle" on employers who make dangerous products or dispose of hazardous materials unsafely. Nevertheless, employees who bring valid protests often find themselves out of a job (management can always find an excuse to eliminate a person's position, or a way to make an employee so uncomfortable that he or she resigns). Why do company managers act this way? What might our society do to ensure its own safety by guaranteeing security to these whistle-blowers?
4. Is some scientific information too dangerous to know?
5. Many characteristics of life can be found in some nonliving things. Can you think of examples of these?
6. What might you expect was the selective pressure that resulted in each of the following adaptations?

   an elephant's trunk — the scent of honeysuckle
   a leopard's spots — the bark of a tree
   human language

SELECTED KEY TERMS

biosphere, *p. 1*
continental drift, *p. 3*
ecology, *p. 1*
evolution, theory of, *p. 3*
natural selection, *p. 1*
organism, *p. 1*
protists, *p. 1*

SUGGESTED READINGS

Arms, K., P. Feeny, and R. C. Lederhouse. "Sodium: Stimulus for puddling behavior by tiger swallowtail butterflies, *Papilio glaucus*." *Science* 185:372, 1974. The story of the puddling experiments described in this chapter.

Mayr, E. *The Growth of Biological Thought.* Boston: Belknap Press of Harvard University Press, 1982. A historical perspective from an eminent evolutionary geneticist.

Roszak, T. *Where the Wasteland Ends.* Garden City, N.Y.: Doubleday, 1973. Critique of modern science by a man who believes science dominates Western society and causes much of its malaise.

PART 1

The Unity of Life: Cells

Chapter 2
Some Basic Chemistry

Chapter 3
Biological Chemistry

Chapter 4
Cells and Their Membranes

Chapter 5
Cell Structure and Function

Chapter 6
Energy and Living Cells

Chapter 7
Food as Fuel: Cellular Respiration and Fermentation

Chapter 8
Photosynthesis

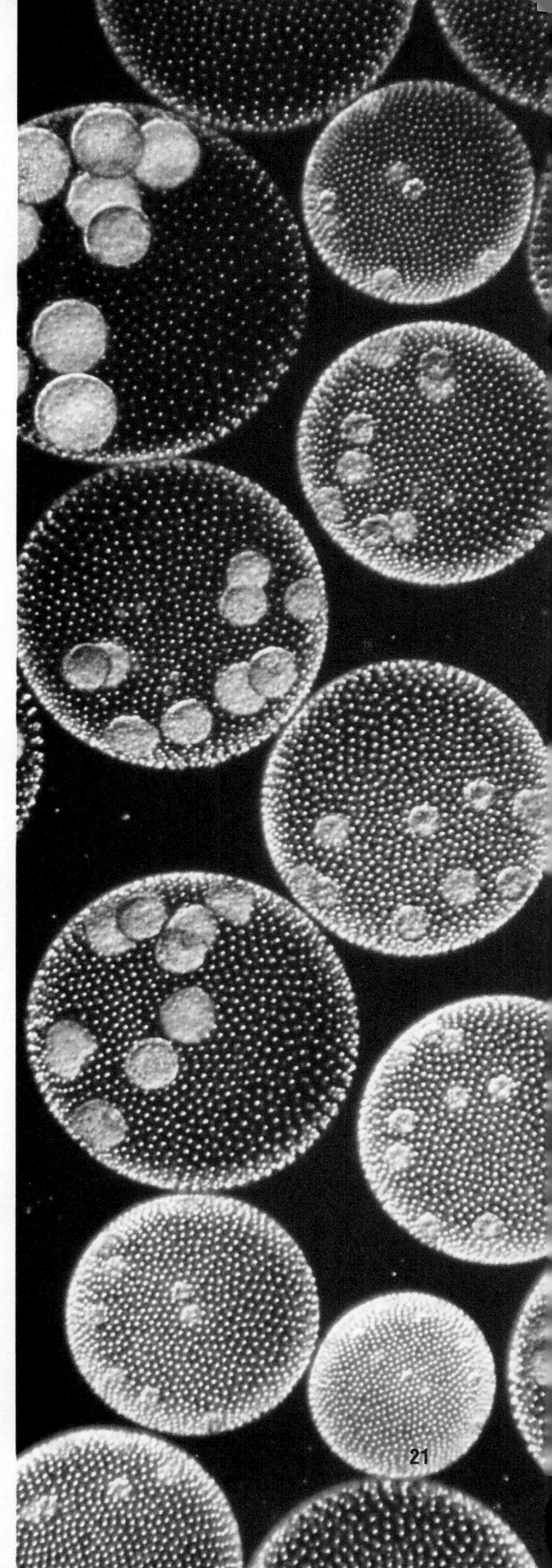

***Volvox* colonies.** Each of these spheres is composed of a few hundred photosynthetic cells joined together to form a hollow ball. Each cell has a pair of flagella, which beat in a coordinated fashion to move the colony through its freshwater environment. *Volvox* colonies reproduce by producing and releasing the small daughter colonies seen within some of these spheres. (M.I. Walker/Science Source/Photo Researchers, Inc.)

CHAPTER

CONCEPT GUIDE

After reading this chapter, you should be able to:

1. Define the term atom. Describe the location and charges of an atom's three main parts. (Section 2–A)
2. Compare the structure and properties of covalent, ionic, and hydrogen bonds. Give an example of a molecule that contains each. (Section 2–B)
3. Explain the difference between a molecule and a compound. (Section 2–C)
4. Compare the activity and arrangement of the molecules in a solid, liquid, and gas for a substance like butter. (Section 2–D)
5. Describe the roles of the reactants, products, and activation energy in a chemical reaction. (Section 2–E)
6. Describe seven of the important properties of water. Explain the significance of each property to living organisms. (Section 2–F)
7. Explain the importance of buffers and how they work in living organisms. (Section 2–G)

All living and nonliving things are composed of molecules that interact in similar ways. Read about how atoms and molecules combine together to form the world around us in Sections A and B. (Stephen J. Krasemann/Photo Researchers, Inc.)

Some Basic Chemistry

It's only the second week of school, but you need to get away from it all. You find a deserted beach and sit on the sand, letting the ocean lap at your toes. You stare at the glints of light on the waves. The only sound is the boom and rush of water. And, for a moment, everything recedes and it's just you, the sand, and the sea.

But you are far from alone in this picture (Figure 2–1). Not only do you carry your personal populations of skin and gut bacteria, but even this seemingly deserted beach is full of life. Vast numbers of tiny animals and some microscopic plants live in the water between sand grains. Larger animals such as worms and mole crabs burrow there, too. The myriad living and nonliving components of this picture have two very basic things in common: they are all composed of chemicals and all these chemicals follow a single set of rules.

But, living and nonliving things also differ in two important ways: while water may or may not be a component of nonliving systems, living organisms are composed mainly of water; and, while carbon is a rare element in nonliving systems, living organisms build their physical structures from chemicals that contain carbon. Figure 2–1 compares the chemical composition of living creatures with that of the atmosphere, the earth's crust, and sea water. The differences in the distribution of chemical elements will be obvious.

In this chapter we shall examine some of the basic properties of matter, paying particular attention to the characteristics of water as background for our study of biology. Building on this chemical theme, Chapter 3 considers the carbon-containing chemicals made by living organisms.

KEY CONCEPTS

- The chemistry of living and nonliving things follows the same rules.
- The chemistry of life has two notable features:
  1. Living things are composed mainly of water.
  2. The large chemical components characteristic of living things have structures based on "skeletons" of carbon.

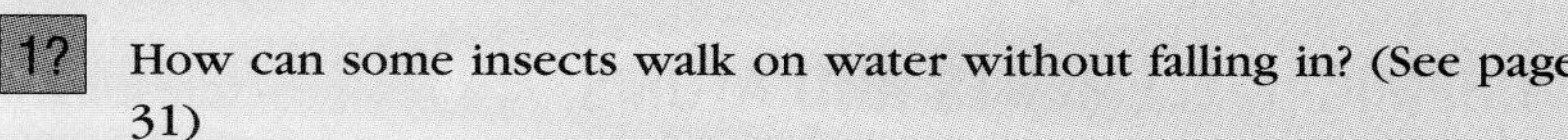

1? How can some insects walk on water without falling in? (See page 31)

2? Why does sweating cool our bodies? (See page 31)

3? Why does ice float? (See page 32)

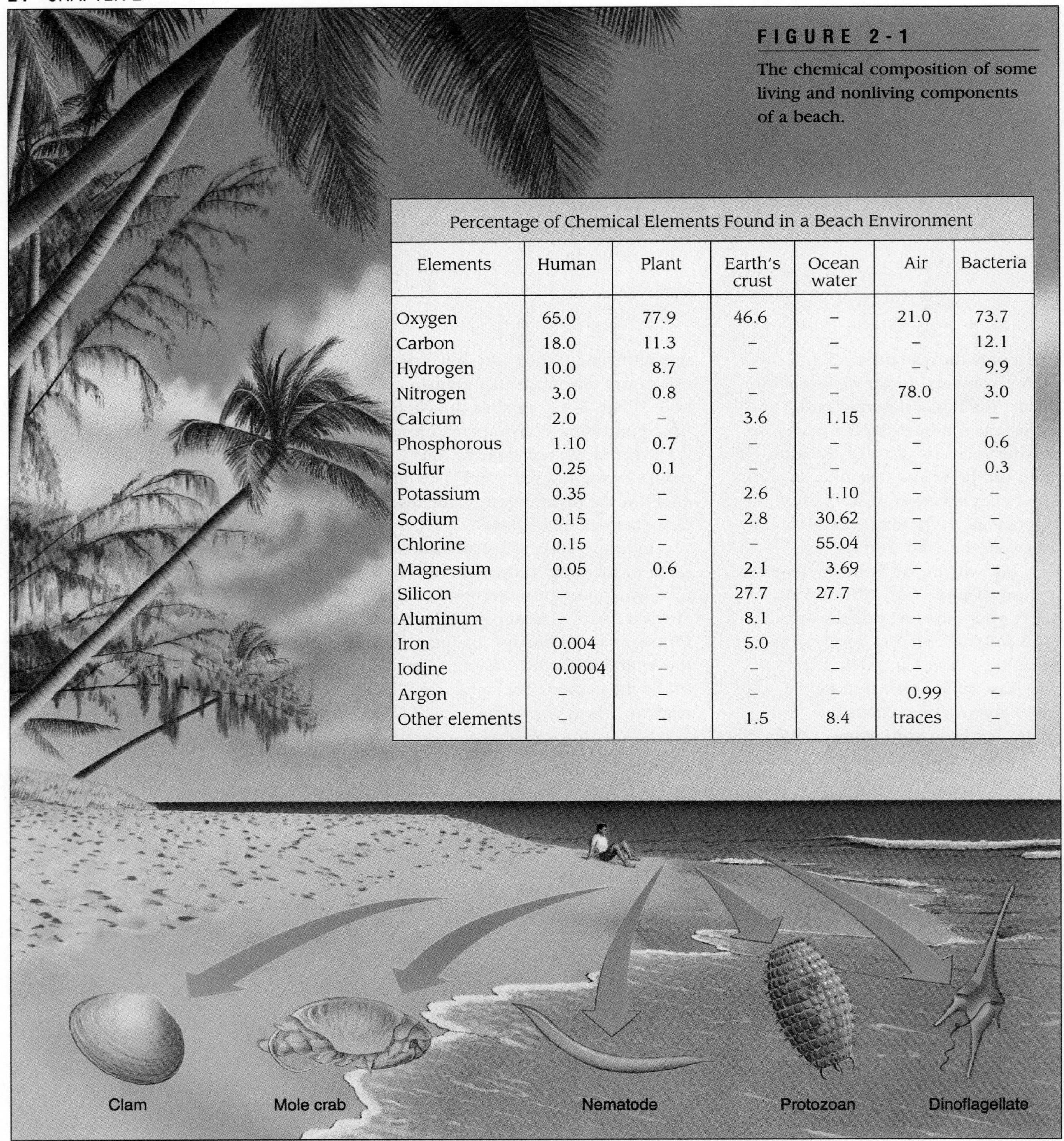

Percentage of Chemical Elements Found in a Beach Environment						
Elements	Human	Plant	Earth's crust	Ocean water	Air	Bacteria
Oxygen	65.0	77.9	46.6	–	21.0	73.7
Carbon	18.0	11.3	–	–	–	12.1
Hydrogen	10.0	8.7	–	–	–	9.9
Nitrogen	3.0	0.8	–	–	78.0	3.0
Calcium	2.0	–	3.6	1.15	–	–
Phosphorous	1.10	0.7	–	–	–	0.6
Sulfur	0.25	0.1	–	–	–	0.3
Potassium	0.35	–	2.6	1.10	–	–
Sodium	0.15	–	2.8	30.62	–	–
Chlorine	0.15	–	–	55.04	–	–
Magnesium	0.05	0.6	2.1	3.69	–	–
Silicon	–	–	27.7	27.7	–	–
Aluminum	–	–	8.1	–	–	–
Iron	0.004	–	5.0	–	–	–
Iodine	0.0004	–	–	–	–	–
Argon	–	–	–	–	0.99	–
Other elements	–	–	1.5	8.4	traces	–

FIGURE 2-1

The chemical composition of some living and nonliving components of a beach.

2–A CHEMICAL ELEMENTS AND ATOMS

Chemical elements are substances that cannot be broken down by ordinary chemical processes into other kinds of substances. Each of the 109 known elements has a unique set of chemical properties, but living organisms use only about 20 of these elements. Interestingly, these are not the most common elements in the atmosphere or in the soil (Figure 2–1), but rather those whose special properties have made them compatible with life on Earth. Carbon is one of these elements.

You are probably familiar with four highly carbonaceous substances: gasoline, oil, soot, and coal, but

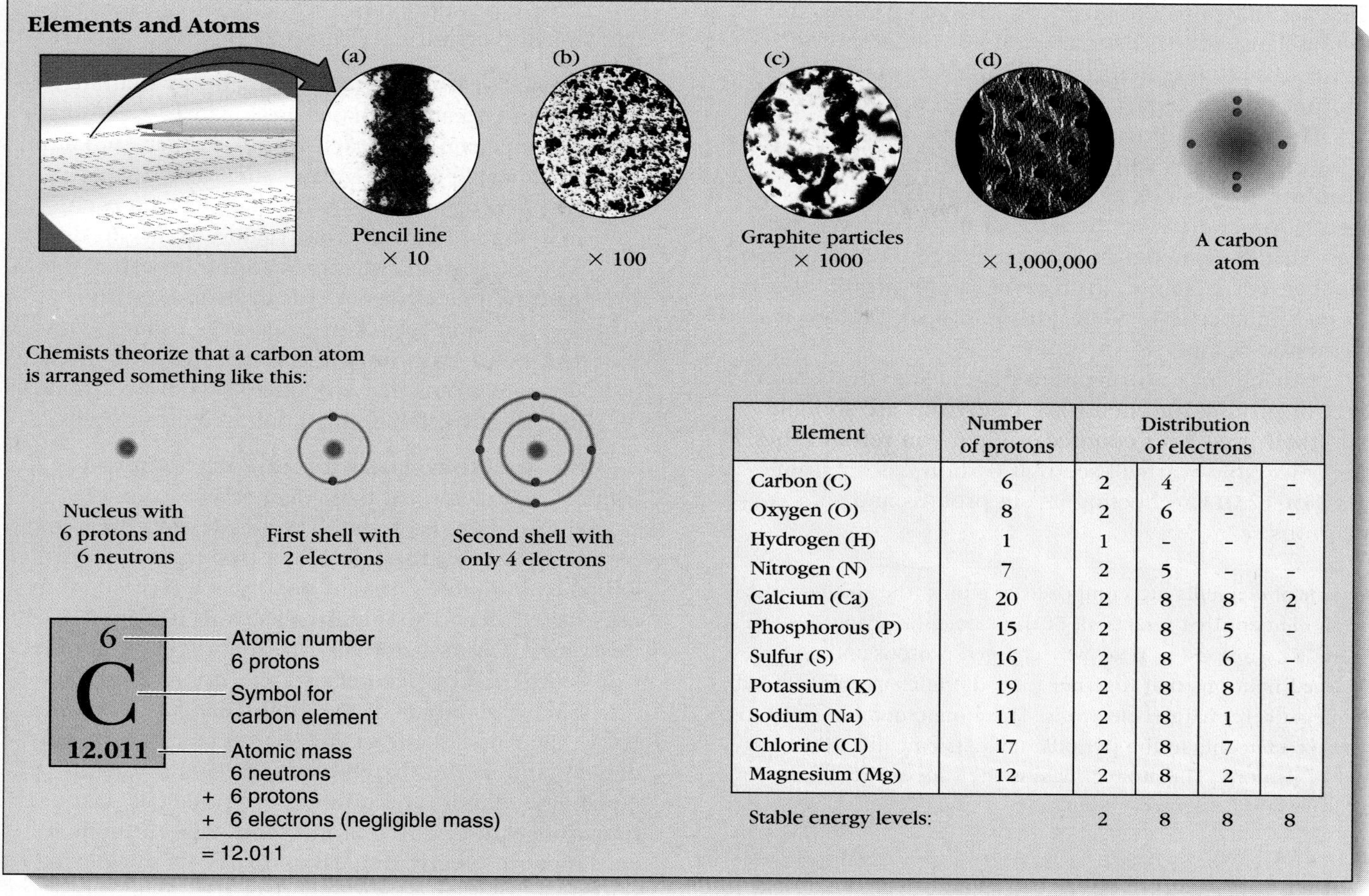

Element	Number of protons	Distribution of electrons			
Carbon (C)	6	2	4	–	–
Oxygen (O)	8	2	6	–	–
Hydrogen (H)	1	1	–	–	–
Nitrogen (N)	7	2	5	–	–
Calcium (Ca)	20	2	8	8	2
Phosphorous (P)	15	2	8	5	–
Sulfur (S)	16	2	8	6	–
Potassium (K)	19	2	8	8	1
Sodium (Na)	11	2	8	1	–
Chlorine (Cl)	17	2	8	7	–
Magnesium (Mg)	12	2	8	2	–
Stable energy levels:		2	8	8	8

FIGURE 2-2

Elements and atoms. Pencil lead is made mostly of carbon, one of the most important elements of life. The basic structure of carbon and all atoms is a nucleus composed of neutrons and protons, and electrons that orbit the nucleus. The atomic mass is the sum of the masses of the protons, neutrons, and the relatively small electrons. The table shows that chemical elements all differ in their number of these three atomic components. (a,b,c Bruce Iverson; d,e IBM Watson Research Lab/Peter Arnold, Inc.)

pure, elemental carbon occurs in only two common forms: diamond and graphite. Graphite is part of the "lead" in pencils. Imagine dividing a piece of graphite into ever-smaller pieces until you have finally separated it into carbon atoms. An **atom** is the smallest unit of an element that retains all of the element's properties (Figure 2-2).

Using the extraordinary physical forces of nuclear accelerators, atoms can be split even further, into three main kinds of particles: **protons, neutrons,** and **electrons.** Imagine an atom the size of a football field with an orange on the center of the fifty-yard line. The orange represents the atom's **nucleus,** where positively charged protons and uncharged neutrons cluster together. The much tinier, negatively charged electrons whiz around the nucleus at nearly the speed of light. They move so fast that, just as the whirring blades of a helicopter appear as a solid blur, the zipping electrons would appear to occupy most of the football field's space if we could see them. Electrons move in orbits or spaces called **electron shells,** and each electron shell can hold only a certain number of electrons (Figure 2-2). Because electrons have negative charges,

BIO-BITS

If you were the nucleus of an atom, your electrons would be about 40 miles away.

they are drawn to the positively charged protons in the nucleus, and this attraction holds the atom together. An atom contains equal numbers of electrons and protons, so its net electric charge is zero.

The number of protons in an atom determines what element it is and forms the element's **atomic number.** If you look at the Periodic Table of the Elements, Appendix A in the back of this book, you will see that the elements are arranged according to atomic number. For example, hydrogen has one proton and its atomic number is 1, while carbon has six protons and its atomic number is 6.

An element's **atomic mass** is the sum of its numbers of protons and neutrons. (Electrons are so light that their mass isn't counted.) When you refer to the Periodic Table you will see that carbon has an atomic mass of 12.011, reflecting its six protons and six neutrons.

Chemical elements are composed of atoms, the smallest unit of an element that retains all of the element's properties. An atom is composed of positively charged protons and uncharged neutrons that together form the nucleus, surrounded by negatively charged electrons. The atomic number, used to arrange elements in the periodic table, is equal to the number of protons. The atomic mass is the sum of the number of protons and neutrons.

2–B BONDS BETWEEN ATOMS

The number of electrons orbiting the nucleus of an atom influences how readily and in what manner it will react with other atoms. Elements are *least* reactive when their outermost electron shell contains a stable number of electrons. The number of electron shells present varies from atom to atom, but in any atom the electron shell closest to the nucleus (called the first electron shell) can hold only two electrons, and the next (second) electron shell can hold a maximum of eight electrons. Subsequent shells can contain more than eight electrons, but are stable when only eight are present. A stable or chemically inert atom does not react with other elements. This is why helium gas is used in blimps and balloons instead of lighter, but unstable (indeed, explosive!) hydrogen gas. Both helium and hydrogen have only one electron shell, but helium has two electrons filling its single shell, while hydrogen has only one electron. Like many other atoms that do not have exactly enough electrons to stabilize their outermost electron shells, hydrogen may take part in chemical reactions or **bond** with another atom to achieve a stable outer electron shell.

Three types of bonds between atoms are important in living organisms (Figure 2–3):

1. **Covalent bonds.** A **covalent bond** is a link between two atoms that share a pair of electrons—one electron from each atom—so that each atom has a stable outer electron shell. For instance, two hydrogen atoms, with one electron each, may share their electrons (Figure 2–3a). In this way, each atom ends up with the first shell filled with two electrons. A pair of hydrogen atoms forms a molecule of hydrogen gas using a covalent bond. A **molecule** is a unit made up of two or more atoms that are joined by covalent bonds, and so have stable, filled outer electron shells.

In a **double covalent bond,** each atom contributes two electrons, for a total of two pairs of shared electrons. Each time you inhale, you draw in oxygen molecules that consist of two oxygen atoms linked by a double covalent bond. Each oxygen atom uses its partner to supply the electrons it is missing. In this way it fills its outer shell with a total of eight electrons, and thereby becomes stable (Figure 2–3a).

When two atoms of the same element bond covalently, their nuclei attract the shared pair of electrons equally, and so the shared electrons spend roughly equal amounts of time orbiting each nucleus. If the two atoms are of different elements, one of them is usually more **electronegative,** that is, its nucleus attracts electrons more strongly than the nucleus of the other element. Hence, the shared electrons spend more time near this atom and give it a partial negative charge. Because of the off-center position of the shared electrons, the other atom bears a partial positive charge. Such an electrically lopsided covalent bond is said to be **polar.**

Oxygen and nitrogen are much more electronegative than hydrogen. So, when oxygen (or nitrogen) bonds with hydrogen, the bond is polar (Figure 2–3a). Within the bonded molecule, oxygen has a partial negative charge, and hydrogen has a partial positive charge. In comparison, carbon and hydrogen are about equally electronegative. Therefore, a carbon-to-hydrogen bond is **nonpolar,** with the average position of

FIGURE 2-3

Types of chemical bonds. (a) Covalent bonding results from the sharing of electrons. **(b)** Ionic bonding involves the transfer of electron(s) from one atom to another. **(c)** Hydrogen bonds form as a result of weak electric attractions between charged molecules.

Types of Chemical Bonding

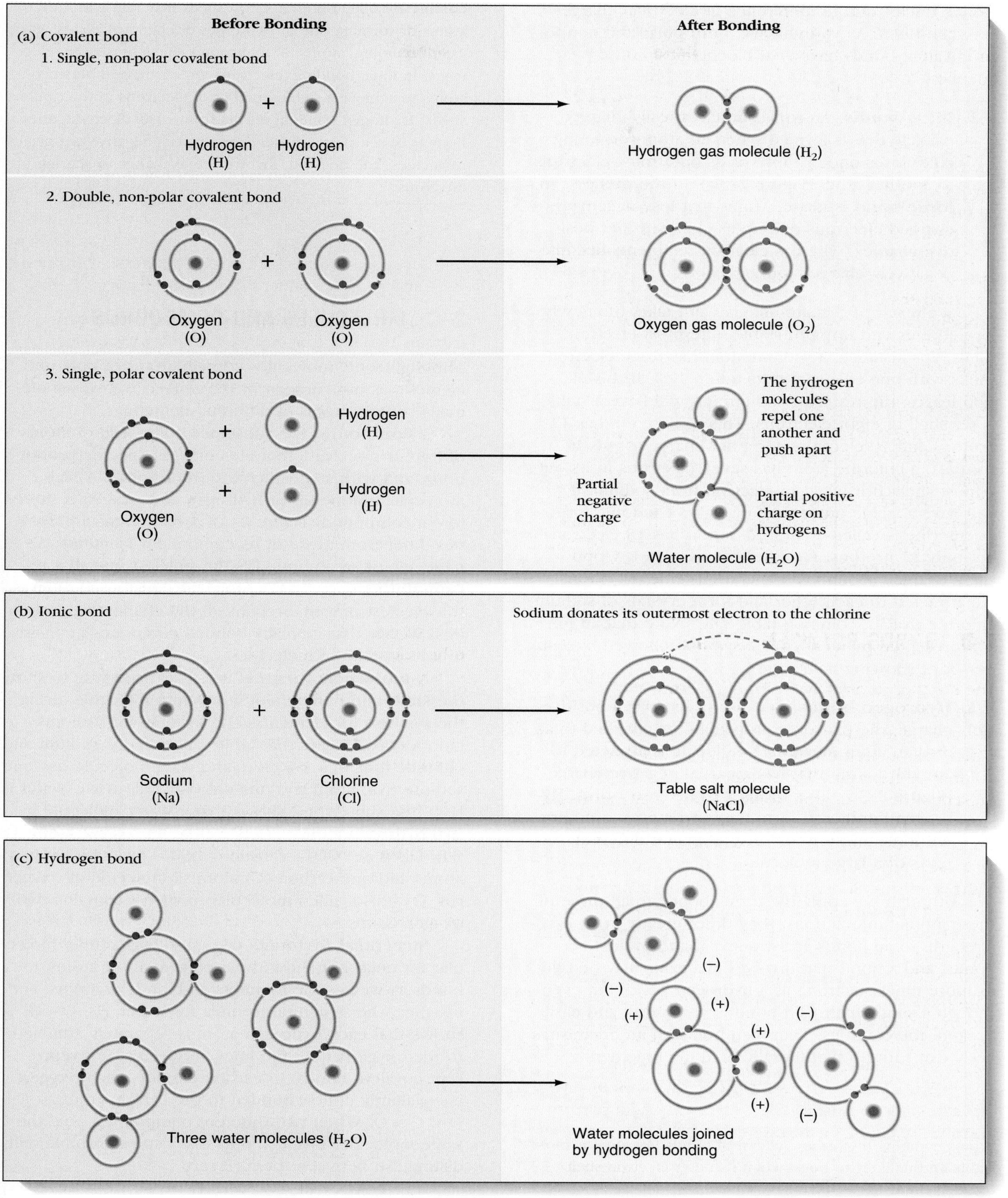

Before Bonding | **After Bonding**

(a) Covalent bond

1. Single, non-polar covalent bond

Hydrogen (H) + Hydrogen (H) → Hydrogen gas molecule (H_2)

2. Double, non-polar covalent bond

Oxygen (O) + Oxygen (O) → Oxygen gas molecule (O_2)

3. Single, polar covalent bond

Oxygen (O) + Hydrogen (H) + Hydrogen (H) →

The hydrogen molecules repel one another and push apart

Partial negative charge

Partial positive charge on hydrogens

Water molecule (H_2O)

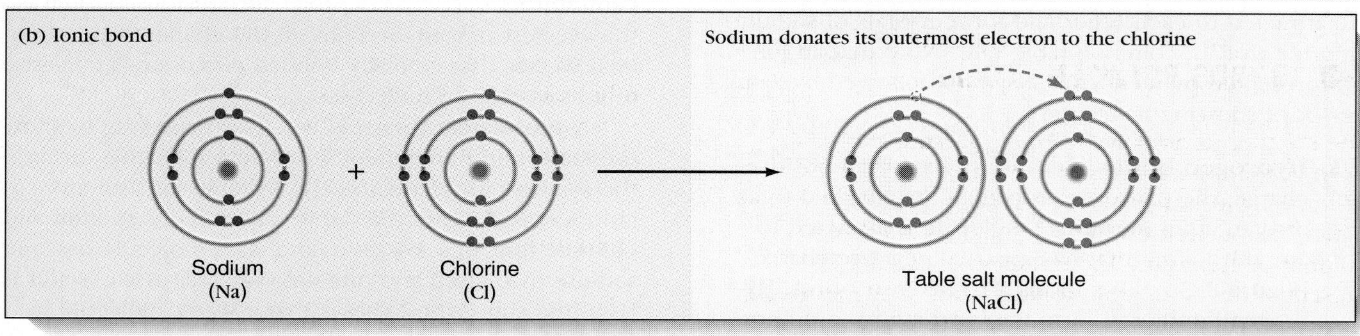

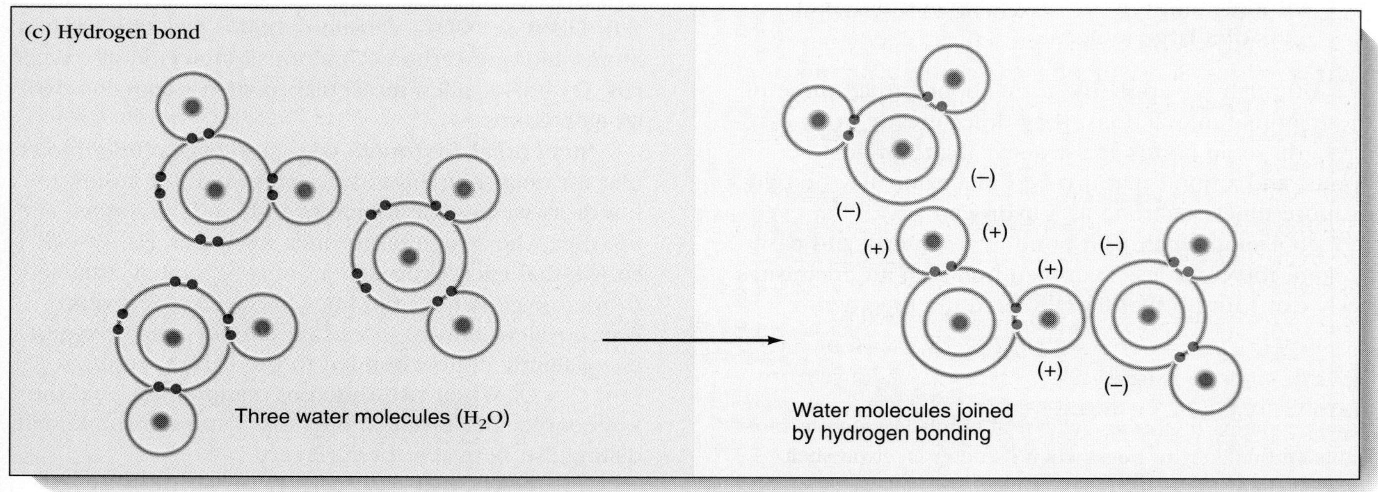

the shared electrons about midway between the two atomic nuclei, and no difference in electrical charge between them. As you will see, both polar and nonpolar covalent bonds have vital biochemical consequences.

2. **Ionic bonds.** An **ion** is an electrically charged particle that is formed when an atom (or a molecule) loses one or more of its outermost electrons to another atom (Figure 2–3b). In the process, an **ionic bond** is made. Atoms that lose negatively charged electrons end up as ions with net positive charges, whereas electron recipients become negatively charged ions.

As a result of this atomic give and take, the newly formed ions end up with stable outermost electron shells. For instance, a sodium atom has three electron shells with one electron in its outer shell. If this electron leaves, the resulting sodium ion will have a stable outer shell of eight electrons. This sodium ion has 11 protons and 10 electrons for a net charge of +1. In contrast, a chlorine atom has seven electrons in its outermost shell. If it takes an electron from sodium, it will have a stable outer shell of eight. The ionic form of chlorine is called a chlor*ide* ion; it has 18 electrons but only 17 protons, for a net charge of −1. Oppositely charged sodium (Na^+) and chloride (Cl^-) ions are attracted to each other and form crystals of sodium chloride (NaCl), common table salt. [Note that an ion is represented by its chemical symbol followed by a superscript showing its charge.]

3. **Hydrogen bonds.** Because of its partial positive charge, the pair of hydrogen atoms attached to an oxygen atom in a water molecule is attracted to any third atom that has a partial negative charge (Figure 2–3c). This forms a **hydrogen bond.** Hydrogen bonds can form between atoms in different molecules or between atoms on different parts of a large molecule.

Although we apply the term "bond" to all three of these atomic interactions, they differ greatly in strength. Ionic bonds are stronger than covalent bonds, and a bond that involves electrons may be two or more times as strong as a hydrogen bond. But even though a single hydrogen bond is very weak and easily broken, the countless hydrogen bonds in an organism's body exert forces that literally hold life together.

Atoms combine by forming chemical bonds. An atom is most stable and unlikely to bond when its outer electron shell contains a specific number of electrons. Atoms that have incompletely filled outer electron shells may join with other atoms by forming one of two types of chemical bonds. In covalent bonding, atoms fill their outer shells by sharing electrons. In ionic bonding, electrons are exchanged between atoms to achieve a stable number of electrons in their outer shells. Hydrogen bonding results from weak electrical attractions between partial positive and partial negative charges of molecules, and does not involve the exchange or sharing of electrons.

2–C MOLECULES AND COMPOUNDS

Although some molecules contain atoms of only one element, as in hydrogen or oxygen gases, many molecules contain atoms of different elements.

A **compound** is a substance composed of atoms of two or more different elements, in specific proportions, and with a specific pattern of bonds. A water molecule, composed of hydrogen and oxygen is therefore a compound (Figure 2–3). A compound's properties differ from those of its component elements. A molecule of a compound is the smallest unit that retains all the compound's properties, just as an atom is the smallest unit that retains all the element's properties. We say that ionically bonded compounds consist of ions instead of molecules.

A **molecular formula** is a shorthand way to show the kinds and numbers of atoms in a molecule, using the symbols for elements. The formula for sodium chloride, NaCl, says that table salt contains sodium and chloride ions in a 1:1 ratio, and each molecule has one sodium (Na) atom and one chloride (Cl) atom. Water is H_2O; the subscript 2 shows that a water molecule has two hydrogen atoms (H) as well as an oxygen atom (O). Likewise, CO_2—carbon dioxide—has two oxygen atoms and one carbon (C) atom. A molecule of oxygen gas, O_2—also called molecular oxygen—contains two oxygen atoms.

Structural formulas take more space than molecular formulas but show the arrangement of atoms and bonds as well as the numbers and kinds of atoms. For instance, the structural formula for water, H—O—H, shows that each hydrogen atom is separately attached to the oxygen atom; the lines between atoms represent covalent bonds. In carbon dioxide, each oxygen is covalently double-bonded to the carbon atom: O=C=O. When two different compounds have the same molecular formula, only a structural formula will distinguish between them (Figure 2–4).

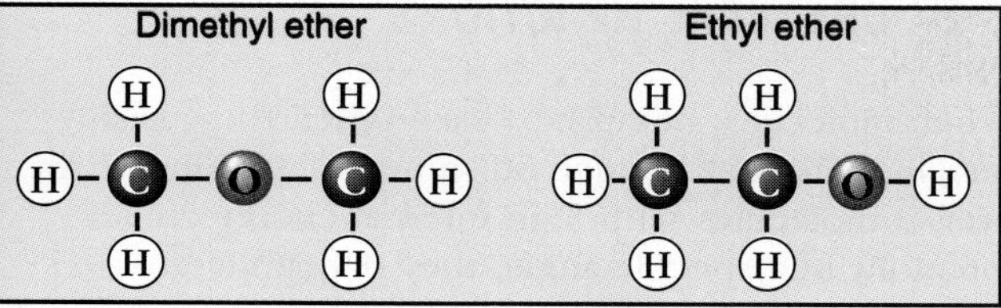

FIGURE 2-4

Structural formulas of dimethyl ether and ethyl ether. Although these two molecules have the same chemical molecular formula, their structural formulas show that they are arranged differently.

A compound is composed of atoms of two or more different elements. A molecule of a compound is the smallest unit that still retains all of the compound's properties. Molecules may be represented by molecular formulas that indicate the types and ratios of the atoms in the molecule or by structural formulas that show the location of bonds and the types of atoms in the molecule.

2–D MOVEMENT OF MOLECULES

All molecules are in constant, random motion. A **gas** consists mostly of space, and the scattered molecules move quickly and freely, occasionally colliding with one another (Figure 2-5). In a **liquid** the molecules slide past each other and change places; they jostle one another constantly. In a **solid,** the molecules occupy fixed positions, and each vibrates in its own space. They knock into each other constantly, much like passengers in a crowded bus.

In any substance, some molecules move faster than others. The faster a particle moves, the greater its **kinetic energy,** or energy of motion. **Temperature** is a measure of the average kinetic energy of molecules; the faster the average speed, the higher the temperature. Heating a substance increases the energy of its molecules (Figure 2-6), hence their average speed and their temperature also increase. If we add enough heat to a solid such as a stick of butter, the molecules will begin to move so fast that the solid melts into a liquid. The fastest molecules will even reach escape velocity and vaporize into the gaseous state, entering the air. It is in this gaseous state that we smell them.

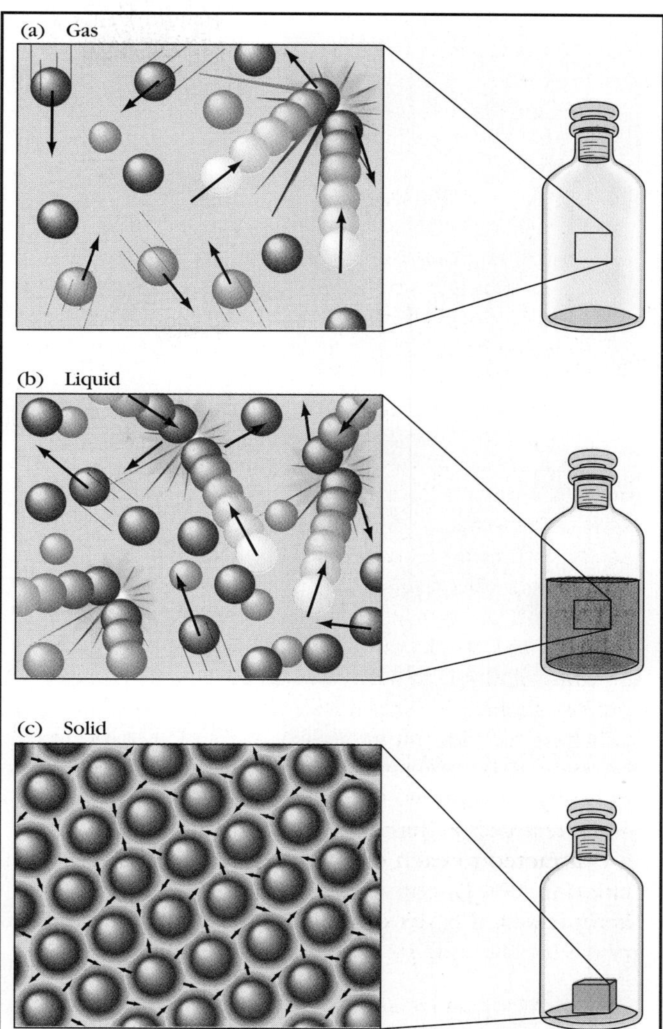

FIGURE 2-5

The three phases of matter. (a) In the gaseous phase molecules are spaced far apart and move at high speeds into and past each other. **(b)** In liquids molecules are closer together and move slower, but still collide and pass each other. Some molecules in a liquid reach even higher speeds and escape the liquid as a gas. **(c)** In solids molecules occupy fixed positions and their motion is limited to vibrations.

The speed at which molecules move and the size of spaces between individual molecules are both related to their temperature. At higher temperatures, molecules are most active and may escape liquids to form gases. As the temperature decreases, molecules move more slowly and become more closely packed together, eventually changing from a liquid to a solid state.

(a)

Solid

(b)

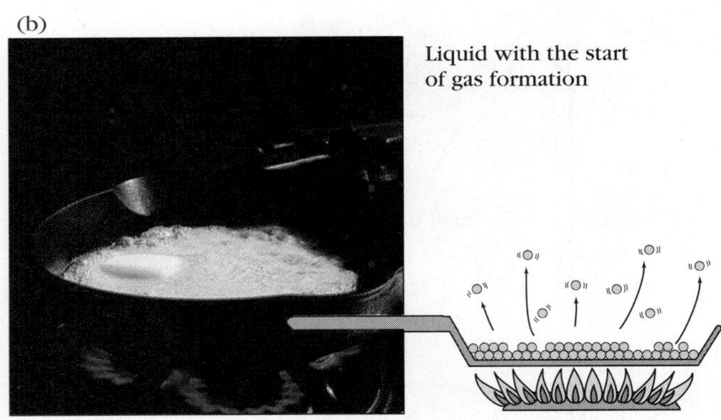

(c)

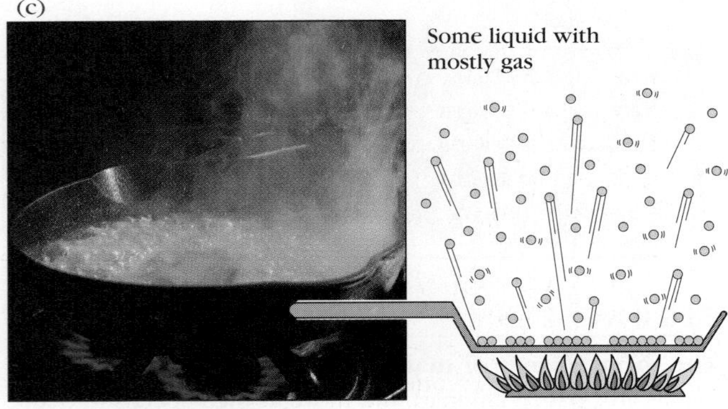

FIGURE 2-6

Phase changes of butter. (a) As a solid, butter maintains a fixed shape and the activity of the molecules is limited. **(b)** As heat is added, the molecular motion increases. The molecules begin to move. They collide and move past each other and the overall shape of the solid butter is lost as a liquid forms. **(c)** As heat energy is added, butter molecules reach even higher speeds and escape the liquid as they form a gas. (Dennis Drenner)

2–E CHEMICAL REACTIONS

When molecules bump into each other, they usually remain intact but bounce off in new directions. However, if molecules with high internal energy collide forcefully at a specific angle, they may undergo a change. The energy of the impact distorts the electron shells, raising the molecules into an unstable, high-energy transition state. One of two things can happen next: either the molecules settle back to their original state, or the electrons rearrange themselves further, forming a new set of bonds and, as a result, making new substances. This is called a **chemical reaction.** The energy needed to raise the molecules to the transition state is the **activation energy.** At normal temperatures on Earth, most molecules don't have enough energy to reach activation, and so relatively few collisions produce spontaneous reactions.

Reactions can be written as equations, like this one for the burning of marsh gas (methane):

$$\underbrace{\underset{\text{methane}}{CH_4} + \underset{\text{oxygen}}{2\,O_2}}_{\text{reactants}} \;\underset{\text{yield}}{\underset{\text{yields}}{\longrightarrow}}\; \underbrace{\underset{\text{carbon dioxide}}{CO_2} + \underset{\text{water}}{2\,H_2O}}_{\text{products}}$$

The **reactants** (starting materials) are shown to the left of the reaction arrow, the **products** after it. This equation says that two molecules of oxygen combine with one of methane, and that for each carbon dioxide molecule produced, two water molecules are also formed. The arrow indicates the direction of the reaction and should be read as "yields." Note that the equation is balanced: the products contain all the atoms of each element from the reactants, rearranged into different molecules. The number of molecules denotes the proportions of reactants and products.

The arrows in a chemical equation may point in both directions:

$$\underset{\text{carbon dioxide}}{CO_2} + \underset{\text{water}}{H_2O} \;\underset{\text{reversibly yields}}{\rightleftharpoons}\; \underset{\text{carbonic acid}}{H_2CO_3}$$

This means that the reaction is **reversible:** it can go either from left to right (forward) or from right to left (backward), depending on the conditions.

Chemical reactions rearrange atoms, forming new molecules from old. They occur when a collision between reactant molecules reaches the activation energy needed to form a transition state, which may or may not go on to form new products.

2–F WATER

Living cells perform a continuous series of chemical reactions, most of which must take place in aqueous (watery) solutions. The peculiar properties of water make it a suitable environment for these reactions. Water also creates a congenial external environment for living cells. In fact, biologists recognize that abundant water is one of the major factors that made the evolution of life on Earth possible.

Water's unique properties result from the structure of the water molecule in which an atom of oxygen is covalently bonded to two atoms of hydrogen (See Figure 2–3c.). The water molecule is polar and the electronegative oxygen attracts the electrons it shares with hydrogen, giving oxygen a partial negative charge and each hydrogen atom a partial positive charge (Figure 2–7). The polarity of the water molecule explains many of water's properties.

A water molecule's partially negative oxygen is attracted to the partially positive hydrogens of other molecules (including other water molecules), so water molecules attach to one another by hydrogen bonds (Figure 2–7). Note that each water molecule can form a maximum of four hydrogen bonds at a time, but this only happens when the motion of the molecules is slowed by decreasing the temperature to the point where ice forms. Otherwise, these weak hydrogen bonds form, break, and re-form rapidly as the molecules tumble past each other in liquid water.

The water molecule's ability to form and break hydrogen bonds gives water several properties important to life:

1. **Water is cohesive and adhesive. Cohesion** occurs when like substances hold together, while **adhesion** is the attachment of different substances. Because of cohesion you can fill a glass of water slightly above its brim; and some aquatic insects, like water striders and whirligig beetles, can skate on the "skin" of a pond's surface (Figure 2–7). Such feats are possible because of water's **surface tension,** the result of the cohesion of water molecules that are more strongly attracted to one another than they are to the molecules in air or the insect's foot. ■

BIO-BITS

About 70% of the Earth's surface is covered by water, with 97% of the water contained in the oceans.

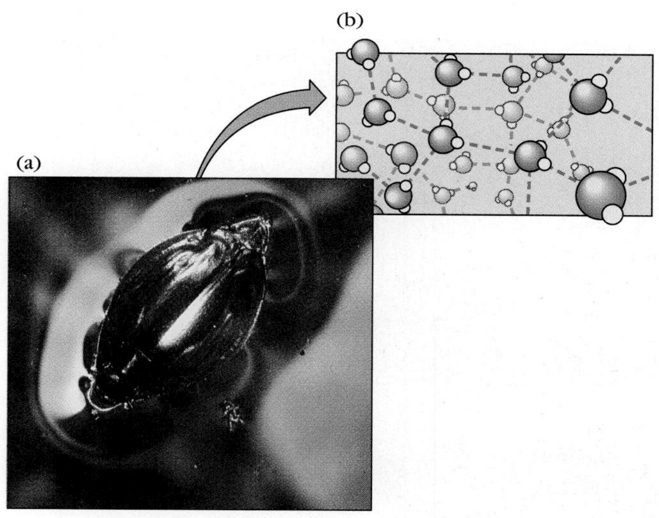

FIGURE 2-7

The adhesive properties of water. (a) A whirligig beetle "walks" on water. **(b)** Because of the adhesive properties of hydrogen bonds between water molecules, water can support the weight of insects. (P. Ceisel/Visuals Unlimited)

2?

2. **Water has a high specific heat.** This means that it takes a lot of heat to raise the temperature of water. This property allows a body of water to heat *and* cool more slowly than the surrounding environment. For aquatic organisms, this means more gradual changes in the temperature of their environment.
3. **Water has high thermal conductivity.** Heat applied to one part of a body of water rapidly spreads to all the rest. Organisms are composed largely of water, and this characteristic of high thermal conductivity carries heat away and prevents the heat produced in an organism's body from creating destructive "hot spots."
4. **Water has a high boiling point.** It takes a lot of heat energy to break all of the hydrogen bonds between water molecules and thereby change water from a liquid to a gas, in which each molecule is separate. Earth's surface temperatures reach the boiling point of water (100°C) only in volcanic vents and thermal springs, so organisms do not ordinarily face the prospect of boiling away.
5. **Water is a good evaporative coolant.** As previously mentioned, it takes a lot of heat to transform liquid water molecules into gaseous water vapor. Those that reach escape velocity and leave

(a)

(b)

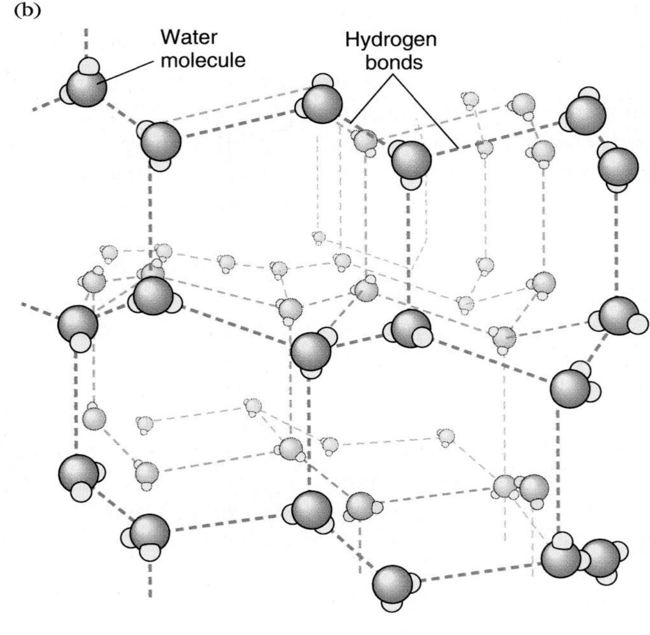

FIGURE 2-8

The crystalline patterns of ice formation. (a) Ice crystals. **(b)** The crystal structure of ice is due to the regular arrangement of water molecules. (William Johnson/Stock Boston)

the body carry away the heat that they have absorbed. Sweating in humans and panting in dogs are mechanisms that employ evaporative cooling. ■

3? 6. **Water has a high freezing point and is less dense as a solid than as a liquid.** As warm water cools, it contracts and becomes more dense. However, unlike most substances, water is peculiar in that it is most dense and thus heaviest at 4°C, when it is still a liquid. As water cools from 4°C to 0°C, it expands again, and becomes less dense as wet water molecules form ice crystals.

Ice is a regular latticework, with each water molecule hydrogen-bonded to four others. As you might guess, ice is less dense than liquid water because its molecules are packed less closely; therefore, an ice crystal is larger than the volume of water it replaces, and ice floats in water. The low density of ice has an advantage for aquatic organisms: in the winter, floating ice forms an insulating blanket between the water below and the colder air above. This blanket slows the formation of more ice from the remaining water and so protects overwintering aquatic organisms from freezing. In the spring, the sun shines directly on the ice and melts it. ■

Because water expands when it freezes, if ice forms within an organism it is likely to destroy delicate internal structures and cause death (Figure 2-8). Some organisms have adaptations that allow them to avoid freezing, such as fishes with natural antifreezes like glycerol in their blood. Other organisms have hardy tissues that are not damaged by ice crystals. Organisms without such adaptations—especially tropical garden plants such as tomatoes and basil—are killed by freezing and must complete a generation of growth and reproduction in the summer months between frosts.

7. **Water is an excellent solvent.** More substances dissolve in water than in any other known liquid. A **solution** consists of a **solvent,** most commonly water, plus the substances dissolved in it, called **solutes.** The **concentration** of a solution is a measure of the proportion of solutes it contains.

When a substance dissolves, its individual molecules or ions separate from one another and mingle with molecules of the solvent (in this case, water). The partial electrical charges of polar water molecules are attracted to charged ions and to partially charged polar molecules, so water readily surrounds and dissolves these solutes.

Nonpolar molecules, such as those made up mainly of carbon and hydrogen, do not dissolve in water because they lack an electrical charge necessary to interact with water molecules. While water and its solutes form one big, "friendly" crowd of molecules all connected by many electrical attractions, these "shy" nonpolar molecules are elbowed aside. Here the nonpolar molecules form groups, not by mutual attraction, but by default. They hang around together because they are all excluded from the mass of water molecules. So, instead of dissolving in water, nonpolar molecules form **interfaces** with it, such as the interface in salad dressing between polar water and nonpolar oil. Similar interfaces are the basis for membranes in cells. Hence, water's inability to dissolve nonpolar substances is also necessary to life.

Water is the most vital molecule for all living organisms. Its structure gives it several properties important to life: (*a*) it absorbs heat and disperses it throughout an organism's body, (*b*) it carries body heat away through evaporative cooling, (*c*) it is denser as a liquid than a solid, (*d*) it dissolves polar and ionic molecules, and (*e*) it forms interfaces with nonpolar molecules.

2–G DISSOCIATION AND THE pH SCALE

Many substances come apart, or **dissociate,** into ions when they dissolve in water. Some compounds dissociate completely, others only partially, so that some molecules of the compound remain intact, while other molecules of the compound dissociate. Water itself dissociates partially, most commonly into hydrogen ions (H^+) and hydroxyl ions (OH^-):

$$H_2O \rightleftharpoons H^+ + OH^-$$

water reversibly yields hydrogen ions + hydroxyl ions

Because water molecules carry both partial negative and partial positive charges, they can assist dissociation by forming "shells" around ions. The watery shells shield the ions from the attraction of oppositely charged ions in the solution and allow them to move independently (Figure 2-9).

Substances are classified by the particles they yield when they dissociate in water. An **acid** releases H^+ when it dissociates in water. For example, when hydrogen chloride gas (HCl) dissolves in water and forms hydrochloric acid, it yields hydrogen ions (H^+) and chloride ions (Cl^-) in solution.

A **base** is a substance that either releases hydroxide ions (OH^-) in water, or accepts H^+. The base sodium hydroxide (NaOH), used in drain cleaners, dissociates into sodium and hydroxide ions (Na^+ and OH^-) in solution. (Note that water is both an acid and a base!)

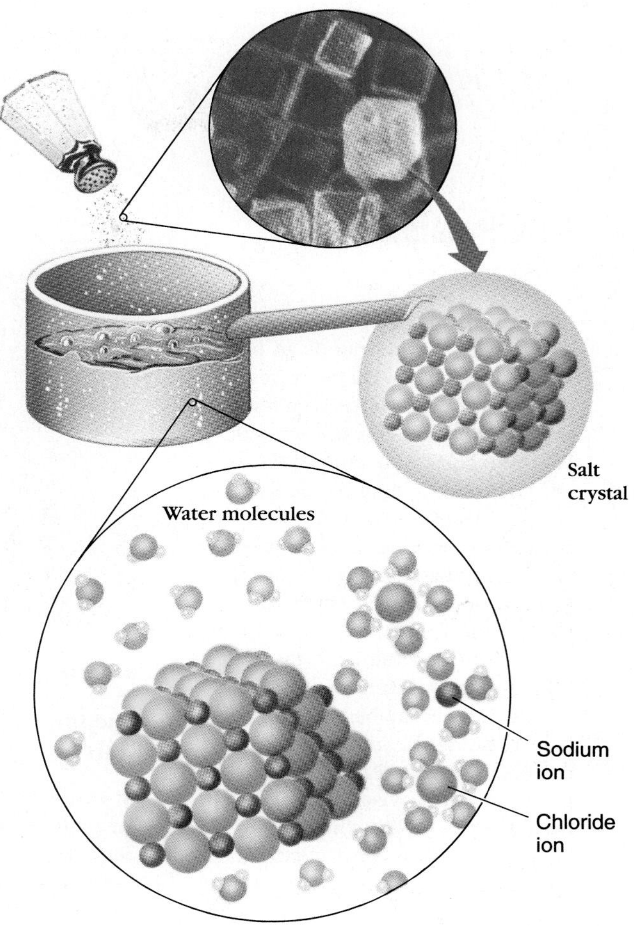
Water molecules
Salt crystal
Sodium ion
Chloride ion
A salt crystal attracts water molecules. The water molecules pull sodium and chloride ions away.

FIGURE 2-9

Table salt dissolving in water. As salt is added to the hot water, the charged water molecules attract sodium and chloride away from each other. Sodium and chloride atoms are then surrounded by a water capsule, preventing the pieces from recombining. (Bruce Iverson, BSc.)

A **salt** is a substance in which the H^+ of an acid has been replaced by another positive ion. A salt dissociates into oppositely charged ions, as when sodium chloride (NaCl) separates into sodium and chloride ions (Na^+ and Cl^-) (Figure 2-9).

A solution's acidity or **alkalinity** is indicated by its **pH,** a measure of the concentration of H^+. (pH stands for *pouvoir hydrogène,* a French phrase meaning "hydrogen power," a reflection of the number of hydrogen and hydroxyl ions in a substance.) The pH scale ranges from 0 to 14, and a pH of 7 is neutral, neither acidic

(text continues on page 36)

A Journey into The Environment

Acid Precipitation

A forest that has been damaged or killed by acid rain is an odd, quiet place. There are few of the usual sounds: birds aren't singing and the wind moves through stiff, bare branches. You stare up at the skeletal trees, feeling the sun on your face, aware that something is wrong—something subtle that you can't immediately identify. And then you realize what it is: there's too much light. Coniferous forests, where evergreen trees dominate the vegetation, are supposed to be dim, shady places. There may be shafts of sunlight piercing the gloom, illuminating the place where a large tree has fallen, but otherwise, high overhead, the needles and branches usually block and filter most of the light. But this forest is different. The screen of needles is thin and transparent; the trees on this mountainside seem to be dying (Figure 2–A).

FIGURE 2-A

The impact of acid rain. (a) A forest in 1963 and **(b)** again in 1983, after 20 years of acid rainfall. (Tim Scherbatskey)

More than a decade ago, foresters in Central and Eastern Europe reported that something was killing their trees. At first the tree death was thought to be confined to evergreen trees on forested slopes on high mountains, but more recently it has begun to include broad-leafed trees at lower elevations. The symptoms of *Waldsterben* (a German word meaning "forest death"), include yellowing of needles and leaves, death of young needles, leaves, and new shoots, increased light piercing the crowns of trees, loss of the beneficial fungi that inhabit the roots of trees and enable them to better absorb nutrients, decrease in growth of tree trunks, clumping of leaves at tips of branches, deformed leaves, excessive growth of shoots along branches, and excessive production of seeds and cones. Years later this phenomenon of tree death was reported in forests in eastern North America, where similar patterns of abnormalities were observed. The most recent reports from both Western Europe and eastern North America indicate that now the symptoms of forest death are seen in broad-leafed forests at lower elevations. As we would expect, in all the affected forests there have been corresponding declines in wildlife populations. The pattern of damage is so widespread that acid precipitation seems to be the most likely culprit.

Although "acid rain" had been recognized since 1872, it wasn't until the 1960s that the work of biologists, soil scientists, foresters, chemists, and meteorologists from all over the world merged into a full picture of the characteristic effects of acid precipitation. Even then, although there were lakes so acidic that they killed fishes, and forests of gnarled, dead and dying trees, it took many more years to convince the public that acid rain was real. Acid precipitation not only kills trees and aquatic ecosystems, it changes the chemistry of soil, removing nutrients that are essential for plant growth and exposing plant roots to damaging heavy metals. Acidified soils are usually poor in calcium and magnesium because these minerals are soluble in acidic water and are washed away. Simultaneously, aluminum and manganese become soluble in acidic soils, and roots absorb toxic amounts of these heavy metals. To make matters worse, as soil becomes more acidic it kills the beneficial fungi that live on and in some cases, within plant roots. These fungi are essential to a plant's efficient uptake of nutrients from the soil and plants that are experimentally deprived of these fungal partners don't grow nearly as well as plants that have fungi on or in their roots

Acidified soils not only damage trees, they also have a negative effect on animals that live within the forest. Salamanders, frogs, worms, as well as many soil invertebrates are sensitive to changes in pH. As these forest creatures become scarce, the larger animals that feed on them must switch to other food sources or migrate. Acid precipitation removes calcium from soils, and this affects the amount of calcium that is available to be concentrated in plant and animal tissues. For example, insects that feed on calcium-poor vegetation store less calcium in their tissues and become, in turn, a less nutritious food source for animals who prey upon them. As animal feeds on animal, these calcium losses show up in the top predators and eventually may culminate in calcium-depleted birds who lay eggs with shells so fragile that they are crushed by the incubating parent. Thus there are many possible reasons that a forest damaged by acid precipitation is so quiet: perhaps the birds have gone elsewhere to search for plant and animal food, or perhaps their numbers are reduced by poor reproductive success, or perhaps birds avoid leafless trees where their predators can easily spot them.

Forests aren't the only ecosystem affected by acid deposition. Many aquatic animals that live in freshwater ponds, lakes, and streams are especially sensitive to shifts in pH. In New York State, for example, approximately 25% of the lakes and ponds in the Adirondack Mountains are so acidic that they kill fishes. Another estimated 20% have absorbed nearly as much acid as they can without becoming acidic, and any future additions of acid will push them into the lethal pH range, too. In Pennsylvania it is estimated that by the year 2000, 50% of the streams will be too acidic for fishes to survive in them. The situation is similar in many states along the U.S. eastern seaboard, and the pattern repeats in Scandinavia, Western Europe, the British Isles, and Brazil.

Source of Acid Deposition

In Europe and North America the first forest die-offs were reported from mountaintops. These "forest graveyards" were especially evident on slopes that faced the prevailing winds. Although the detailed interactions of the process are unclear, acidity in clouds that cover these mountainous forests seems to be a major factor in tree damage. For example, the acid content of the clouds that cover Mount Mitchell (at 6684 feet, the tallest peak in eastern North America) has been measured at a pH of 2.12 to 2.9—more acidic than vinegar. Exactly how this acid kills trees is not completely understood and some scientists believe that acid precipitation doesn't kill trees outright, but instead weakens them so that they are less able to withstand drought, frost, wind damage, and attacks by insects, fungi, and disease.

Acid that is deposited in precipitation comes from burning fossil fuels, especially coal, and from the industrial release of gaseous sulfur dioxide (SO_2). In the atmosphere sulfur dioxide mixes with nitrogen oxides and other compounds, forming acid that then falls to Earth as acid rain, fog, snow, or even as dry particles of acid. Power stations that are fired by fossil fuels, especially coal, and industries that burn fossil fuels are primary sources of sulfur dioxide, but nitrogen oxides are also factors in acid precipitation. Nitrogen oxides are primarily produced by automobile exhausts.

Combating Acid Precipitation

Now that it has been identified, acid precipitation should be an easy problem to solve: reduce emissions of sulfur and nitrogen oxides, and acid deposition will be correspondingly reduced. Unfortunately, because of the huge economic consequences, acid rain isn't an easy problem to solve. Air pollutants do not recognize national or state boundaries and the pollutants travel in the prevailing winds to damage ecosystems that can be several hundreds of miles from the source of pollution. For example, British power plants burn coal and release smokestack exhausts that travel east with the prevailing winds, mix in the atmosphere to form acids, and are deposited as acids that damage forests in Norway and Sweden. Similarly, Canada receives acids formed by emissions that originate in electrical power plants in the United States, and Japan receives acids formed by emissions from China's smokestacks. Which country should pay for the damage to forests? How should those damages be computed? In the value of lost timber? How can you compute the loss of populations of a species of spring wildflower, for example? How do you compute the loss of bird song on a spring morning? Finally, who should pay for the considerable costs involved in installing the smokestack scrubbers needed to reduce emissions of sulfur and nitrogen oxides?

While we have yet to solve these international problems associated with acid deposition, in the United States an amendment to the Clean Air Act was passed in 1990 that will begin to slow acid deposition and force utility companies, the auto industry, and industrial plants to alter their methods. This amendment specifies that sulfur dioxide emissions from the country's most polluting utility companies be halved by 1995 and that total emissions from all power plants be halved by the year 2000. After this time the amendment specifies an upper limit on total emissions. In addition, nitrogen oxides from vehicles and industries must be halved by the year 2000.

Doing Your Part

The most positive thing that most of us can do to reduce acid deposition centers around the automobile and energy consumption. Until cars that do not depend on the combustion of fossil fuels are commercially available, driving less, carpooling, bicycling, and walking to work or to do errands are all alternatives that will reduce per capita emissions of nitrogen oxides.

Another thing that you can do is become more aware of the problem. Convince yourself that acid deposition is real. Get a small supply of pH paper from your biology or chemistry teacher and learn how to use it. Wrap it up carefully so that it isn't damaged in your wallet or purse and keep it handy so that the next time it rains or snows you can test the acidity of the precipitation in your area. Are there nearby forests that are damaged by acid deposition? Visit these to see an acid-damaged forest first hand. Go in the early morning and listen. Then, for contrast, visit a similar forest that is not damaged by acid precipitation. You may be able to hear the difference pH can make.

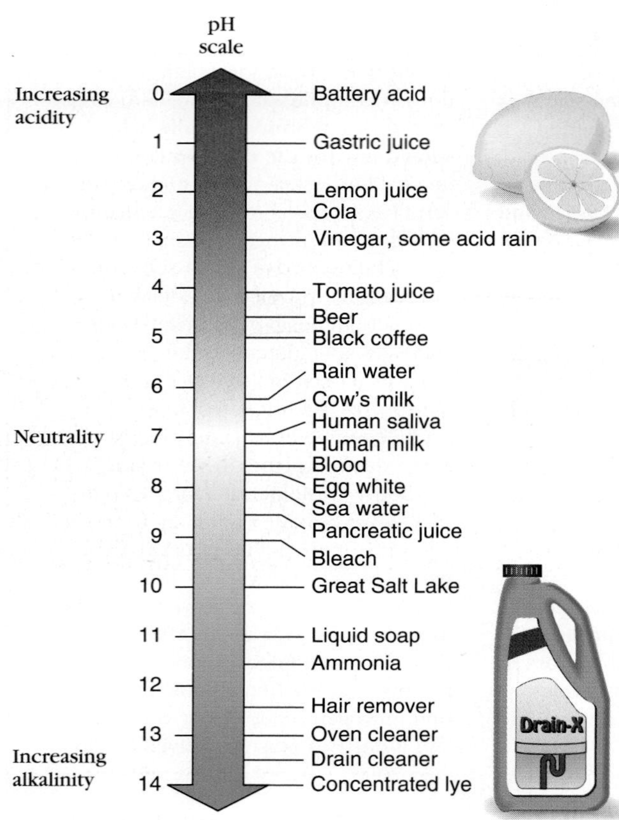

FIGURE 2-10

The pH scale, with the pH readings of some familiar substances.

nor basic (Figure 2-10). Pure water is neutral because it produces equal numbers of H^+ and OH^- ions when it dissociates. A pH value below 7 is acidic, and one above 7 is alkaline.

The numbers on the pH scale come from the exponents of the hydrogen ion concentrations in solutions. For example, a solution with 10^{-5} moles (see Toolbox 2-1) of hydrogen ions per liter has a pH of 5, a solution with 10^{-6} moles of hydrogen ions per liter has a pH of 6, and so on. It is important to understand that this is a logarithmic, *not* an arithmetic scale. Thus, a solution of pH 5 is ten times more acidic than a solution with a pH of 6, and 100 times more acidic than a solution of pH 7! Another "catch" to the pH scale is that *lower* pH values mean *greater* acidity. Try repeating this to yourself a few times until it becomes second nature.

Most chemical reactions in living organisms occur most rapidly at a pH near the neutral point. In human bodies, the pH of blood and most other fluids is about 7.4. A notable exception is the contents of the stomach during digestion of a meal—when the stomach lining secretes hydrochloric acid—with a pH of 1 or less. Acid precipitation is a serious environmental problem that is disrupting life in forests, freshwater lakes, ponds and streams, and may even be killing salamanders, frogs, and birds. (See *A Journey into the Environment*: Acid Precipitation, in this chapter.)

If we add drops of acid or base to a small volume of water, its pH changes rapidly. But if we add the acid or base to blood instead of to water, the pH remains steady until we have added a great excess of acid or base. Blood and other body fluids contain **buffers,** mixtures of salts that tend to keep the pH constant by absorbing or releasing H^+ or OH^- as needed. One of the most important buffers in many body fluids is the bicarbonate ion, HCO_3^- which takes up excesses of either H^+ or OH^- in the reactions:

$$\underset{\text{hydrogen ion}}{H^+} + \underset{\text{bicarbonate ion}}{HCO_3^-} \underset{\text{reversibly yields}}{\rightleftharpoons} \underset{\text{carbonic acid}}{H_2CO_3} \underset{\text{reversibly yields}}{\rightleftharpoons} \underset{\text{water}}{H_2O} + \underset{\text{carbon dioxide}}{CO_2}$$

OR

$$\underset{\text{hydroxyl ion}}{OH^-} + \underset{\text{bicarbonate ion}}{HCO_3^-} \underset{\text{reversibly yields}}{\rightleftharpoons} \underset{\text{carbonate ion}}{CO_3^{2-}} + \underset{\text{water}}{H_2O}$$

In each case, the bicarbonate ion removes the added ion from the solution and the pH does not change. Both equations are reversible, as long as either H^+ or OH^- is present to make up any deficit if some other reaction removes H^+ or OH^-.

The first equation shows that the carbonic acid formed from bicarbonate and hydrogen ions can break down to water and carbon dioxide. In the body, carbon dioxide, usually thought of as something we breathe out, functions as an important buffer in the bloodstream.

Many of the body's chemical reactions produce acids or bases, but buffers such as bicarbonates and carbon dioxide prevent wide swings in pH. Buffers are vital because the chemical reactions of living organisms work best at specific pH values, partly because enzymes, important molecules that facilitate chemical reactions in living organisms, usually work best in a very narrow range of pH (See Section 3-E).

Many substances dissociate into ions when placed into water, and some water molecules are also dissociated into H^+ and OH^-. The pH of a solution is a measure of the concentration of hydrogen ions it contains. Acidic solutions have high concentrations of hydrogen ions and a pH below 7, while bases have low hydrogen ion concentrations and pHs above 7. Buffers tend to keep the pH constant, by absorbing or releasing hydrogen or hydroxyl ions.

TOOLBOX

2–1
Moles

In the same way that bakers count in dozens, chemists like to use standard quantities of molecules called *moles*. One mole of any substance contains $6.023 \times 10^{(23)}$ molecules. The gram molecular mass of a substance is the mass of one mole of molecules measured in grams. Because some individual molecules have more mass than others, moles of molecules also have different masses. For example, a mole of table sugar (342 grams) and a mole of ethanol (46 grams) contain the same number of molecules, but have different masses, because sugar molecules are heavier and larger than ethanol molecules (just as a dozen ostrich eggs have more mass than a dozen chicken eggs).

BIO-BITS

Fish and other aquatic organisms usually die in waters with a pH less than 5.

SUMMARY

1. Living organisms are subject to the same rules that govern nonliving systems. Like nonliving matter, organisms are made up of atoms, which bond in various ways, forming compounds.
2. Covalent bonds form when atoms share electron pairs. Covalent bonds may be polar or nonpolar, depending on the average position of the shared electrons between the ends of the bond.
3. Ionic bonds form when one atom takes one or more electrons from another atom, and the resulting ions are attracted to each other by their opposite electrical charges.
4. Hydrogen bonds are weak electrical attractions between slightly positive and negative charges on polarly bonded atoms of different molecules.
5. Chemical reactions rearrange the bonding of atoms, ions, and molecules and so form different compounds. Living organisms constantly carry out a variety of chemical reactions, forming different compounds as required.
6. Water is the most abundant substance in living things, and is necessary for life as we know it. The water molecule's structure and hydrogen-bonding ability give water a unique set of properties that make it essential to life:
   a. it forms interfaces with nonpolar substances,
   b. it absorbs heat and evenly disperses it throughout the body,
   c. it carries away body heat when it vaporizes from the body surface,
   d. it is denser as a liquid than as a solid, and
   e. it dissolves polar and ionic substances.
7. Many substances dissociate when they dissolve in water.
8. The pH of a solution is a measure of its hydrogen ion concentration. The pH value indicates whether a solution is acidic or basic.
9. Buffers, chiefly bicarbonate ions, keep the body fluids of living organisms at a nearly constant pH.

SELF-QUIZ

1. If the pH of a solution changes from 2 to 5, it has become more (acidic, alkaline); its hydrogen ion concentration has (increased, decreased, remained constant).
2. A positively charged ion has:
   a. more protons than electrons
   b. more electrons than protons
   c. equal numbers of neutrons and electrons
   d. equal numbers of protons and electrons
   e. more neutrons than electrons
3. Write the chemical formulas for:
   a. water ___
   b. table salt ___
   c. carbon dioxide ___
   d. oxygen gas ___

4. What kind of bond involves the sharing of electrons between atoms such that each atom completes its outer electron shell? (covalent, ionic)
5. In a water molecule, the hydrogen atoms are joined to the oxygen atom by (ionic, covalent, hydrogen) bonds.
6. In the dissociation of NaCl in water:
   a. water exerts forces that induce dissociation
   b. water is a passive solvent, accepting particles that dissociate because of their own internal forces
   c. water molecules lose hydrogen ions
   d. twice as many H^+ ions are formed as Na^+ ions
   e. equal numbers of H^+ and Na^+ ions are formed

Matching: for each event listed below, select the property of water responsible from the list (a–h) at the bottom of this column.

___ 7. Heat applied to the bottom of a kettle spreads evenly through the water in the kettle
___ 8. Ionic and polar substances dissolve in water
___ 9. Some insects can stand on the surface of water
___ 10. Freezing kills begonia plants
___ 11. Lake water remains warm in the autumn after the air above it cools

a. adhesion
b. high boiling point
c. cohesion
d. denser as liquid than as solid
e. evaporative coolant
f. polar molecules
g. specific heat
h. thermal conductivity

THINKING CRITICALLY

1. Water makes up the bulk of an organism's body. A water molecule contains two atoms of hydrogen but only one of oxygen. Why, then, does oxygen account for 62% of an organism's weight and hydrogen only 10%?
2. Do oxygen molecules move faster in air or in water?
3. What effect does wind have on the movement of molecules? Using your knowledge of molecular movement and evaporation, explain the phenomenon of the "wind chill factor."
4. Fabrics are sometimes made "water repellent" by coating them with substances that cause water to form beads instead of spreading out on the fabric surface. What do you suppose happens on a molecular level when a surface repels water in this way?
5. Imagine that water, like most other substances, was denser as a solid than as a liquid. How would the freezing of water in winter, and melting of ice in spring, be different? How would these differences affect organisms living in lakes?

SELECTED KEY TERMS

acid, *p. 33*
activation energy, *p. 30*
adhesion, *p. 31*
alkalinity, *p. 33*
atom, *p. 25*
base, *p. 33*
buffer, *p. 36*
cohesion, *p. 31*
compound, *p. 28*
covalent bond, *p. 26*
electron, *p. 25*
electron shells, *p. 25*
hydrogen bond, *p. 28*
ion, *p. 28*
ionic bonds, *p. 28*
kinetic energy, *p. 29*
molecule, *p. 26*
neutron, *p. 25*
nonpolar, *p. 26*
pH, *p. 33*
polar, *p. 26*
proton, *p. 25*
reversible reactions, *p. 30*
solution, *p. 32*
solvent, *p. 32*
solute, *p. 32*
surface tension, *p. 31*
structural formula, *p. 28*
temperature, *p. 29*

SUGGESTED READINGS

Henderson, L. S. *The Fitness of the Environment.* Boston: Beacon Press, 1958. How the earth's physical and chemical conditions support life. Chapter 3, on water and its relationship to life, is especially good.

Hill, J. W. *Chemistry for Changing Times,* 5th ed. New York: Macmillan, 1988. An excellent and entertaining "chemistry for poets," useful also for Chapter 3.

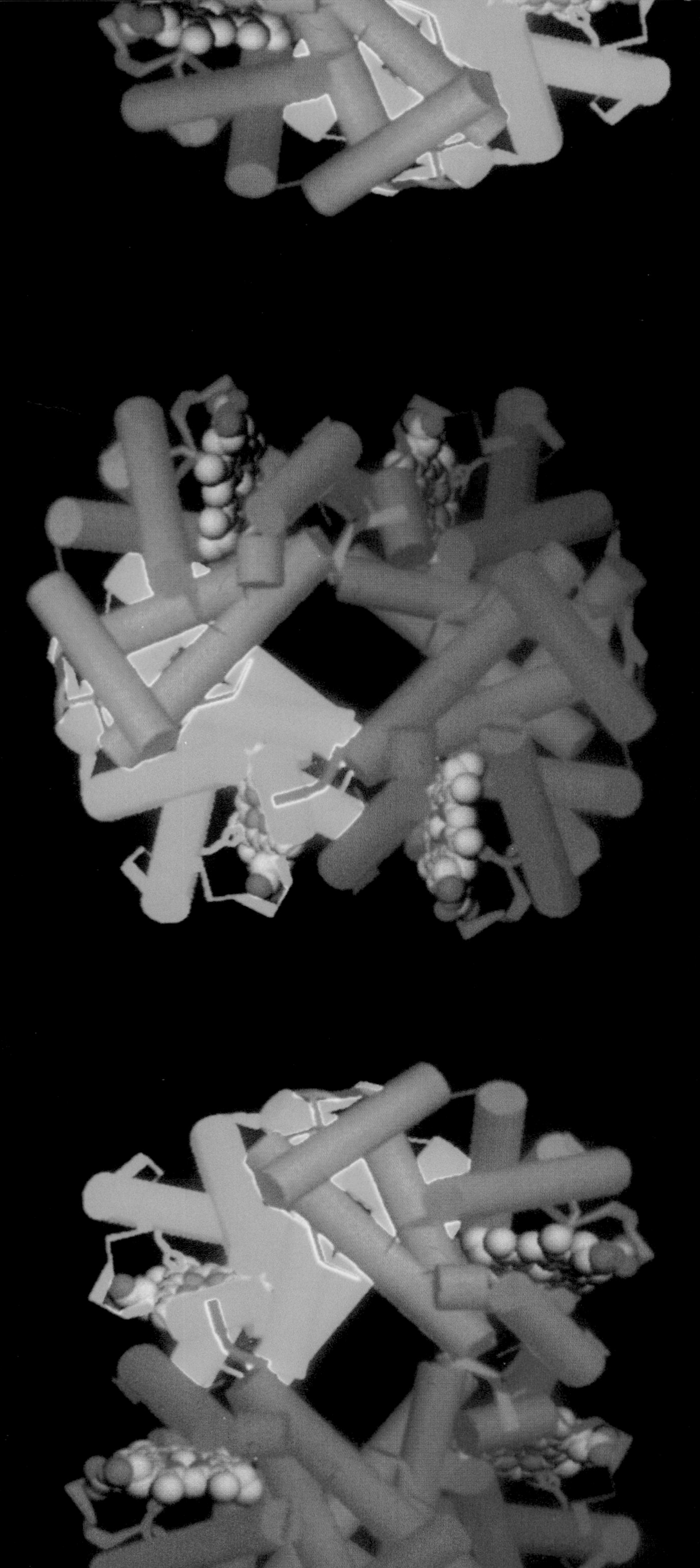

CHAPTER 3

CONCEPT GUIDE

After reading this chapter, you should be able to:

1. Describe the general structure of organic molecules. (Section 3–A)
2. Compare the structures, properties, and functions of lipids, carbohydrates, nucleic acids, and amino acids in living organisms. (Sections 3–B, 3–C, 3–D, and 3–F)
3. Explain how enzymes speed up biological reactions. Describe the effects of pH, temperature, inhibitors, and substrate concentration on enzyme activity. (Section 3–E)
4. Define metabolism and explain how metabolic pathways are controlled. (Section 3–G)

How do red blood cells carry oxygen from your lungs to the rest of your body? The answer is by using hemoglobin molecules, like this one shown here as a computer generated three-dimensional model. Oxygen in the lungs binds to hemoglobin molecules, which act like railroad cars to transport oxygen to cells that need it. Read more about other molecules important to living organisms in Sections 3 B–F. (Laboratory of Molecular Biology, MRC/Science Photo Library/Custom Medical Stock)

Biological Chemistry

A living organism's body is built of, and run by, thousands of different kinds of molecules. Because these molecules are made chiefly by living organisms, they are called **organic molecules.** Each has distinctive properties that have proven advantageous and so each has been selected during the course of evolution. Organic molecules are easy to recognize: all have "skeletons" of carbon atoms bonded into chains or rings, and most of these carbons are also bonded to one or more hydrogen atoms. Most of the "recipes" for organic molecules usually feature oxygen, carbon, and hydrogen, and some also require nitrogen, sulfur, or phosphorus.

Organisms make a variety of small organic molecules that play important roles in the chemistry of life. Many of these small organic molecules also serve as **monomers** (mono = one; meros = part), building units that are joined together like molecular tinker toys to form larger molecules. Some of these larger molecules contain only a few monomers, but many are built from large numbers and so are called **polymers** (poly = many) or **macromolecules** (macro = large).

There are four main classes of biologically important organic compounds:

1. **Lipids:** nonpolar substances such as fats, oils, waxes, and steroids that do not dissolve in water.
2. **Carbohydrates:** sugars, starches, cellulose, and related compounds.
3. **Nucleic acids:** the genetic material (DNA and RNA) and other molecules that help assemble proteins.
4. **Proteins:** molecules that make up muscle, silk, hair, tendons, and similar structures. Proteins are also used in structures that carry out cell movements; they are a major component of muscles and act as hormones, transport substances in the blood, and fight infections. **Enzymes** are proteins that are specialized to trigger biochemical reactions, allowing them to occur at the relatively low temperatures that exist within living cells.

This chapter introduces all four groups and considers enzymes in some detail. Each group has different chemical features, and, as you might expect from our study of water and the chemical consequences of its unique molecular structure, the chemical properties of lipids, carbohydrates, nucleic acids, and proteins make each group distinctive and uniquely useful to living organisms.

KEY CONCEPTS

- Living organisms have a distinctive chemistry based on large molecules with skeletons made up of strings and rings of carbon atoms. These organic molecules also contain hydrogen and oxygen, and often nitrogen, sulfur, or phosphorus.
- Organisms make and use thousands of kinds of organic molecules, which fall into four main classes: lipids, carbohydrates, nucleic acids, and proteins.
- Biological macromolecules are composed of many similar or identical monomer subunits joined together.
- A cell's enzymes convert organic molecules into different forms, step by step, in complex, controlled pathways.

Curiosity Questions

1? Why does a tablespoon of butter have 100 calories while a tablespoon of sugar has only 46? (See page 45)

2? When you are grocery shopping, what is the easiest way to tell whether a fat or oil is saturated or unsaturated? (See page 47)

3? Why does cooking change an egg's texture? (See page 53)

4? How does refrigeration keep food fresh longer? (See page 58)

3–A STRUCTURE OF ORGANIC MOLECULES

In this chapter you will find many diagrams of molecular structure. We do not expect you to learn the name and formula of each molecule, but they are included so that you can see the logical organization of organic molecules. Initially, all these details may be confusing. When you meet a new molecule, first try to get an idea of its size by examining the number of carbon atoms in its carbon skeleton. Then look at its other atoms or groups of atoms—the molecule's **functional groups**: clusters of atoms that behave in particular ways regardless of the molecule to which they are attached. Be alert for patterns, and after a while, these confusing structural diagrams will seem much less threatening. They will even start to make sense.

Carbon Skeletons

Carbon is a versatile element that is able to form a multitude of complex, stable compounds and thus produce the wide variety of molecules found in living organisms. Carbon's versatility stems from its atomic structure as well as its bonding patterns (Figure 3-1). Its structure enables a carbon atom to form four covalent bonds with other atoms. Carbon atoms can join in several spatial patterns, and short or long chains are common configurations. These can have shorter carbon chains as side branches, or the ends of carbon chains can join to form rings. No matter what its shape, the carbon chain is termed the "carbon skeleton" of the organic molecule.

Functional Groups

We will sort our sample of organic molecules by their functional groups, which affect the solubility of a molecule. For example, if the functional group is nonpolar, as are hydrocarbon chains [$(CH_2)nCH_3$)], the molecule will not be soluble in water in the area of the functional group. (Remember: a nonpolar molecule has no net electrical charge, while a polar molecule has a positive or negative electrical charge. See Section 2-B to review polar and nonpolar bonds.) In contrast, if the functional group is polar, as is carboxyl (−COOH), the molecule will be soluble in water in the area of the functional group.

Functional groups are easier to understand if you think of them as similar to the attachments that can be added to an electrical drill to transform its function. Drill bits allow you to bore holes, but if you remove the drill bit and replace it with a sandpaper disc, the same electrical mechanism can transform splintery wood to a smooth, sanded surface. If you replace the sanding disc with a polishing pad, the same electrical mechanism will produce a glossy surface. In each instance, the attachment that is added to the body of the electrical drill changes its function from drilling holes to sanding or polishing. The drill bit, sandpaper, and polishing attachments will have the same function if you attach them to a hand-powered drill, or even if you try to use them manually. Chemical functional groups are much the same. Table 3-1 lists various kinds of polar and nonpolar functional groups and gives common examples of each.

Building Biological Polymers

Organic monomers, such as amino acids and glucose, are joined to form larger molecules. This occurs by a kind of reaction called a **condensation reaction:** two molecules become joined as one loses a hydrogen atom (—H) and the other loses a hydroxyl group (—OH). These, in turn, join to form a water molecule (H + OH ⟶ H_2O) and the term "condensation" refers to this loss of water (Figure 3-2).

Condensation:

Monomer—H + HO—Monomer ⟶ Monomer—Monomer + H_2O

Condensation reactions are readily reversible. Our two-monomer product can be split into its original components by adding a water molecule into the bond that links them. This is a **hydrolysis reaction** (hydro = water; lysis = loosening):

Hydrolysis:

Monomer—Monomer + H_2O ⟶ Monomer—H + HO—Monomer

Hydrolysis and condensation are everyday events in many biological reactions. For example, digestive enzymes in organisms as different as a mushroom and a sequoia tree use hydrolysis to chemically break down polymers, separating them into monomers. These monomers are then sent to all parts of the organism and later are assembled into new macromolecules by condensation reactions. Sooner or later these new macromolecules, too, are hydrolyzed as a part of the continuous turnover of molecules in living tissues. The resultant monomers are either broken down or are used to build still other macromolecules in the ceaseless cycles of biochemical refurbishing that are characteristic of living organisms.

FIGURE 3-1

Chains or rings of carbon atoms serve as the skeletons for most molecules in living organisms.

(a) Carbon atoms can form up to four covalent bonds.

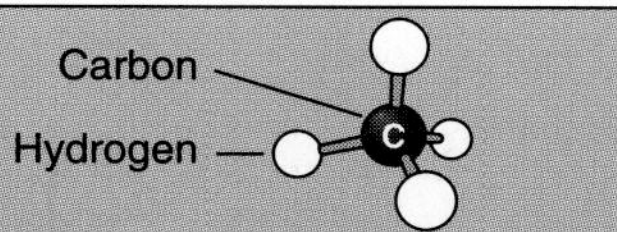

(b) Carbon atoms form carbon skeletons.

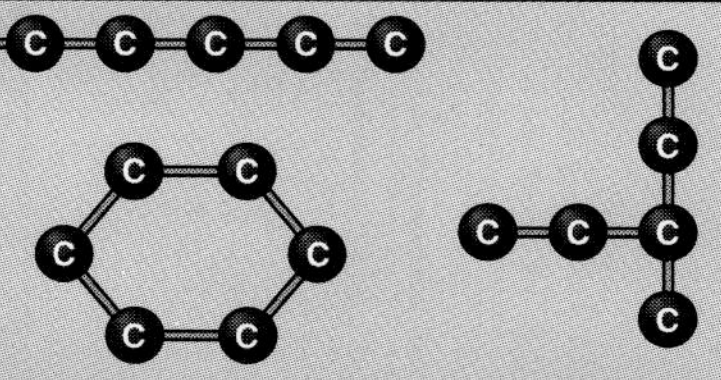

(c) Carbon skeletons have many attached hydrogen atoms.

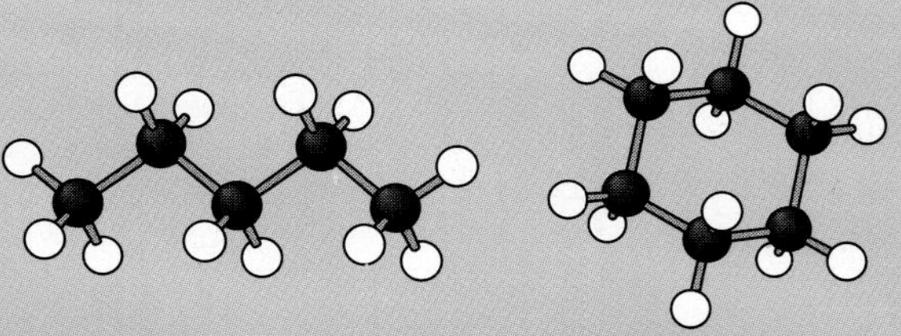

(d) Many organic molecules also have other atoms attached, including oxygen, nitrogen, sulfur, and phophorus. The molecule drawn here is the amino acid isoleucine.

H
H—C—H
H—C—H
H—C—CH_3
H—O C C N—H
O H H

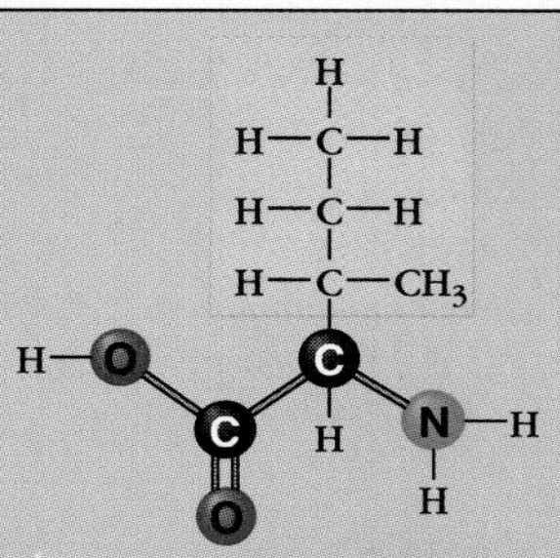

(e) Many monomers join together to form polymers. The amino acid isoleucine above is a monomer, that will join with other amino acid monomers to form polypeptide polymers.

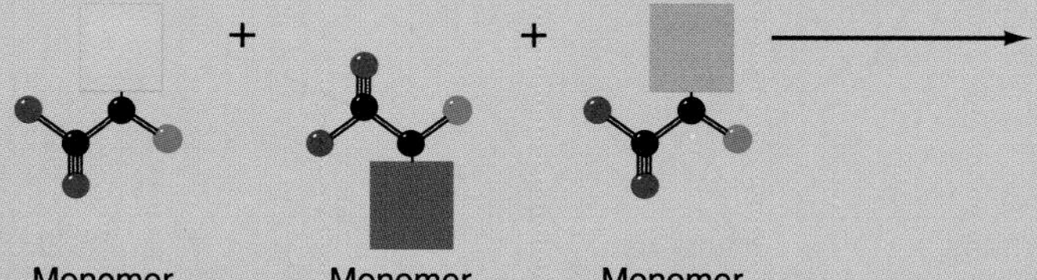

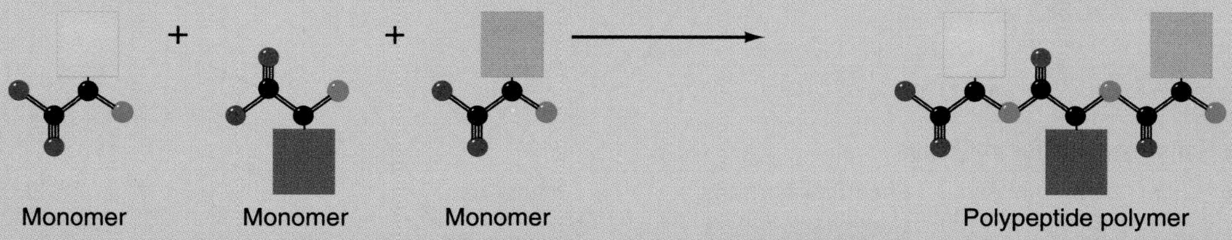

Polypeptide polymer

(f) Monomers join together with polymers to form macromolecules. If these were the polypeptide polymers from above, they would join together to form protein macromolecules.

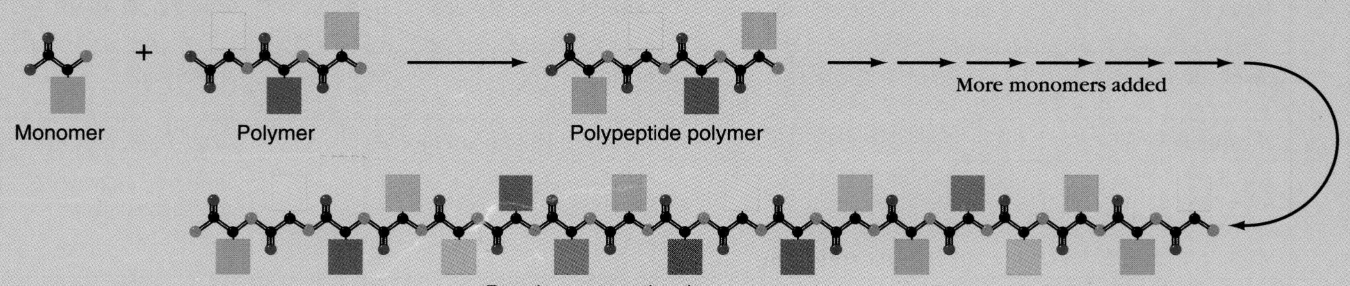

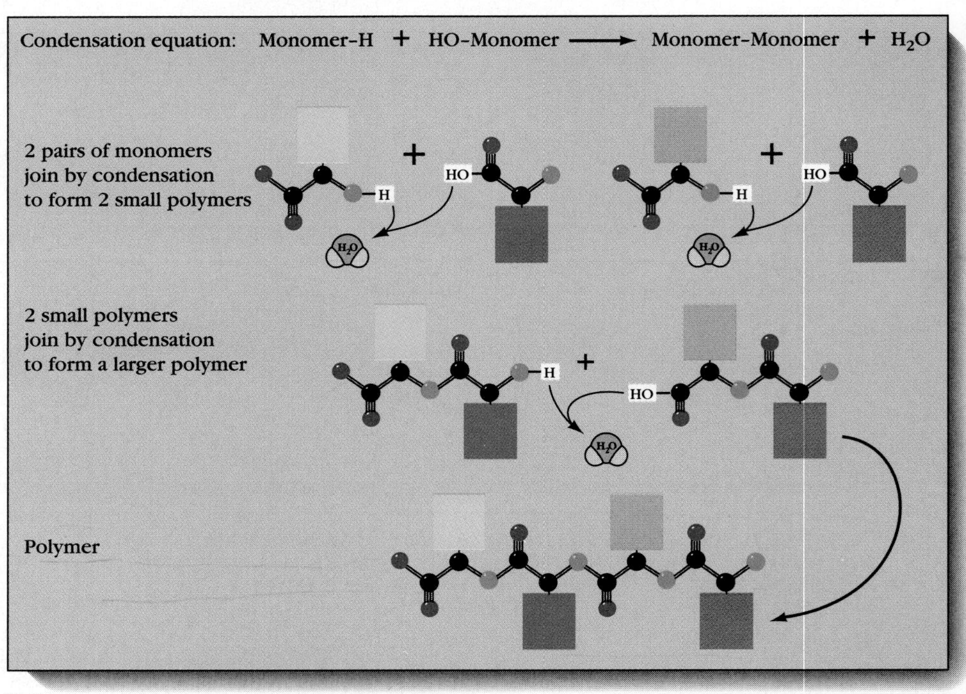

FIGURE 3-2

Condensation and hydrolysis of organic molecules.

Table 3-1
Some Common Functional Groups

Group*	Chemical Formula	Found in	Example
Hydrocarbon chains	$-(CH_2)_nCH_3$	Fatty acids, some amino acids	Cooking oils
Carboxyl	$-C{=}O$, $\backslash OH$	Fatty acids, amino acids	Fats and oils Hair
Alcohol	$-C-OH$	Alcohols, e.g., glycerol	Rubbing alcohol, vodka
Aldehyde	$-C{=}O$, $\backslash H$	Sugars	Table sugar
Phosphate	PO_4^{2-}	Nucleotides, phospholipids	DNA, RNA, and ATP
Amino	$-NH_2$	Amino acids	Tendons, enzymes, and fingernails

* ■ Nonpolar (water-insoluble)
■ Polar or ionic (water-soluble)

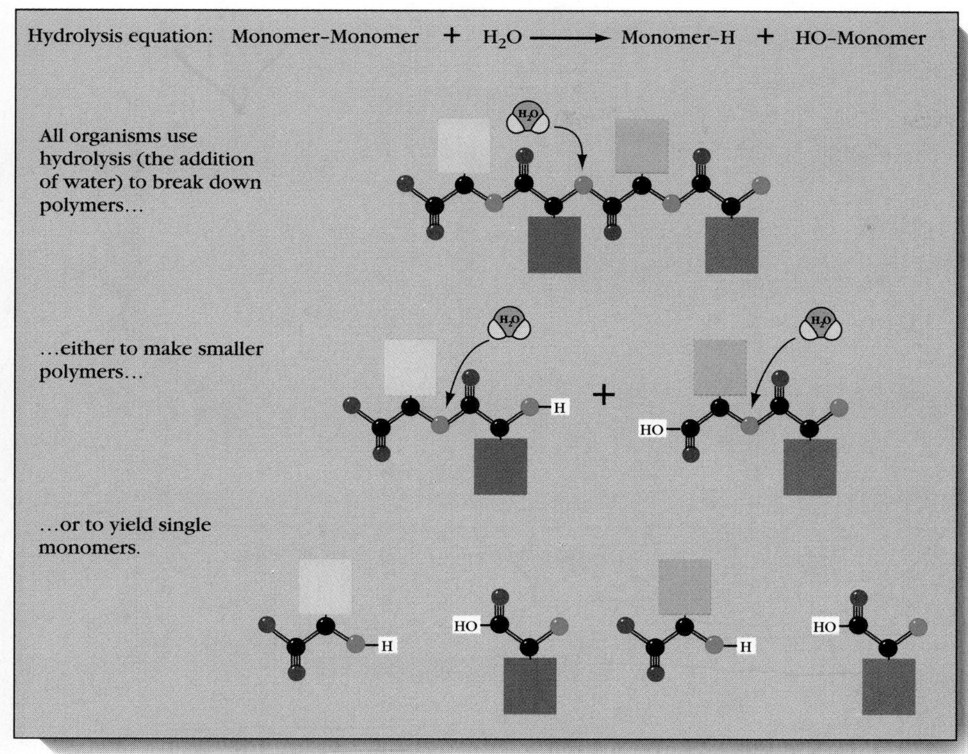

Chains or rings of carbon are the "skeletons" of organic monomers. Attached to the carbon skeleton are atoms, or groups of atoms, of other elements: hydrogen (H), oxygen (O), and sometimes nitrogen (N), sulfur (S), or phosphorus (P). Organic monomers can be linked together by condensation: removal of the components of a water molecule between the two units. They can be separated by hydrolysis: addition of a water molecule into the bond joining the monomers.

3–B LIPIDS

Lipids are organic compounds that share one distinguishing property: They are nonpolar and so do not dissolve appreciably in water. Lipids contain mostly carbon and hydrogen, with a very small portion of oxygen. Some also incorporate phosphorus and nitrogen. Because they repel water, lipids such as oils and waxes are found as waterproof coatings on the outer surfaces of many organisms. For example, leaves, wool, feathers, and many insects are coated with lipids.

1? Because lipids are insoluble in water, they are vital components of the membranes that separate living cells from one another and from their surroundings. Lipids offer a unique way to store energy. They contain a high proportion of energy-rich, carbon-hydrogen bonds. This makes lipid molecules insoluble in water, so they can be stored in a concentrated form within cells, without the addition of water. In contrast, carbohydrates, which are also used for energy, are polar molecules and must be stored surrounded by water. So, equivalent masses of carbohydrate-plus-storage-water and lipid are *not* equivalent in energy yield: the lipid contains six times as much energy! This is undoubtedly why lipids have become increasingly important food reserves for plants and animals during the course of evolution. Before making their fall nonstop migrations to warmer climates, many songbirds gain twice their mass in fat reserves. Some premigratory birds are so fat that they can hardly open their eyelids. If birds had to rely upon carbohydrates plus the necessary storage water for this same amount of stored energy, they would be so heavy that they couldn't fly. ■

Fatty Acids Are Lipids That Form Fats

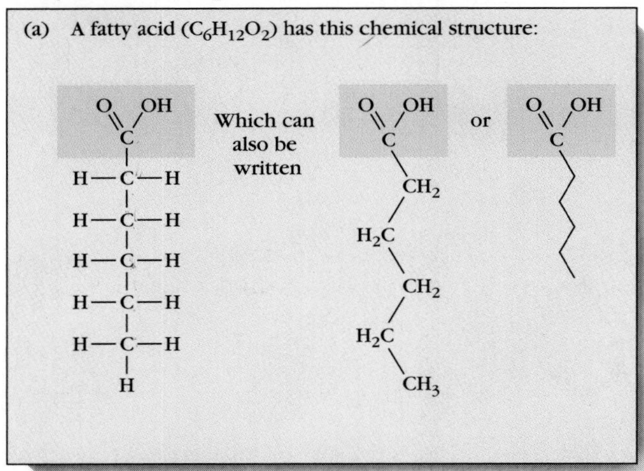

(a) A fatty acid ($C_6H_{12}O_2$) has this chemical structure:
Which can also be written
or

(b) Fatty acids have a hydrophobic end and a hydrophilic end.

Hydrophilic

Hydrophobic

Palmitic acid

Oleic acid

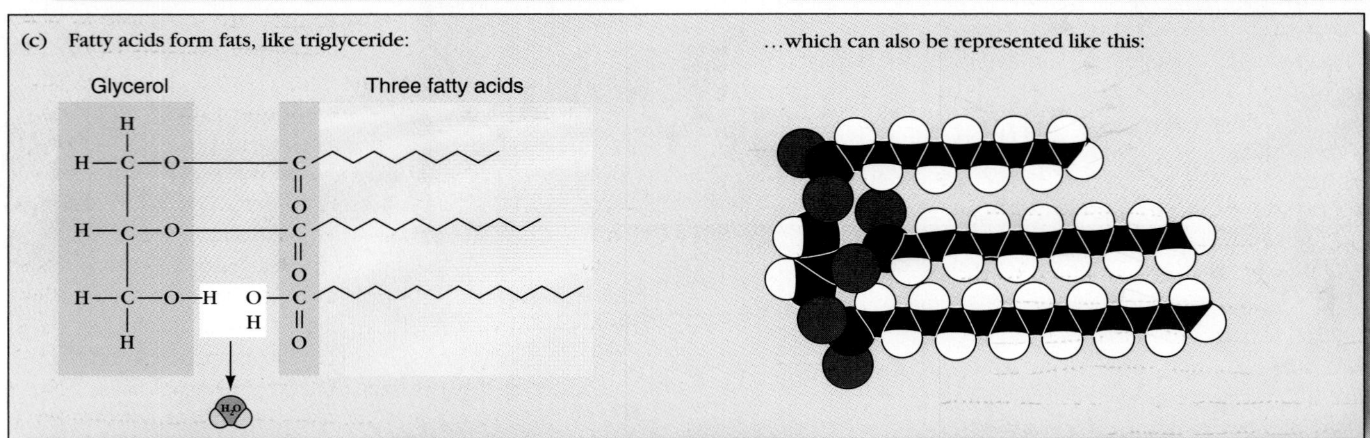

FIGURE 3-3

Properties of fatty acids.

Fatty Acids

Fatty acids are the simplest lipids. We will examine a small fatty acid that contains only six carbons (Figure 3-3). Note these features of this molecule:

1. Each carbon atom (C) forms four bonds to other atoms.
2. The fatty acid's six carbon atoms are linked to one another and form a chain down the center of the molecule.
3. Most of the carbon atoms are attached only to hydrogen (H) atoms, forming a functional group called a **hydrocarbon chain.** Because carbon-hydrogen bonds are nonpolar, this chain of five carbons will not easily dissolve in water.
4. At the bottom of the molecule is another functional group: a carboxyl (—COOH). It contains the sixth carbon atom, double-bonded to an oxygen, and also bonded to a hydroxyl group (—OH). The carboxyl group is the acidic part of the molecule because it can ionize to release $—COO^-$ and H^+. Because the carboxyl group contains the polar C=O and O—H groups, it tends to dissolve in water even though the rest of the fatty acid will not.

In summary, a **fatty acid** is a simple lipid molecule consisting of a long hydrocarbon chain with a carboxyl group at one end. The bonds in a carboxyl group are polar, and so this end of a fatty acid is **hydrophilic** (hydro = water; philic = loving) and attracts water molecules, forms hydrogen bonds with them, and thus dissolves in water. In contrast, the carbon-hydrogen bonds in the carbon chain are nonpolar and the chain is **hydrophobic** ("water-fearing"). The polar head and nonpolar tail of a fatty acid give the molecule a distinctive orientation in respect to other molecules. Because of this characteristic orientation, some lipids that contain fatty acids are important components of cell membranes.

Fats and Oils

Fats and **oils** are lipids used to store energy reserves in plants and animals. They are formed from fatty acids by condensation reactions that join the carboxyl groups of three fatty acids to a molecule of glycerol (Figure 3-3). This structure gives fats and oils another name that you may have heard: **triglycerides.** Note that a molecule of a fat is mostly a large, nonpolar, hydrocarbon chain.

You are probably aware of **saturated** and **unsaturated fats** as dietary factors that may influence the development of fatty deposits in arteries. These terms refer to the chemical structure of fatty acids. In a saturated fatty acid, the hydrocarbon chain is filled with as many hydrogens as it can possibly hold, while an unsaturated fatty acid contains one or more double bonds and could hold more hydrogen atoms if the double bond between carbon atoms were broken and two hydrogen atoms attached to the carbons instead.

Saturated fats such as butter, lard, suet, or coconut oil, are usually solid at room temperature, whereas unsaturated fats, like those from olive, corn, safflower, and peanut, are liquid at room temperature. Unsaturated oils occur most commonly in plants, while animals typically use saturated fats. Animals that live in cold latitudes, such as Arctic and Antarctic fishes, are an exception to this general rule. They produce a relatively high proportion of unsaturated fatty acids that keep their bodies flexible even in the coldest waters. ■

Phospholipids

Phospholipids are similar to fats, except that one or two of the fatty acids are replaced by a phosphate group, which in turn is usually linked to a nitrogen-containing group (Figure 3-4a). Phospholipids are the chief lipid components of biological membranes, and lie with their polar, phosphate, and nitrogenous groups facing watery areas and their nonpolar, fatty acid tails buried in the membrane's nonpolar interior (Section 4-A).

BIO-BIT

When ducks preen themselves, they are actually spreading oils from a gland above their tail out onto their feathers. This makes the feathers less likely to get wet and so increases the duck's buoyancy and insulation.

BIO-BIT

One gram of fat has about 9.5 calories, while a gram of carbohydrate or protein has about 4.3 calories.

Steroids

Steroids differ from other lipids in structure, but are grouped with them because they are insoluble in water. In steroids, the carbon skeleton consists of four contiguous carbon rings (Figure 3-4b). As previously mentioned, **hormones** are chemical messengers carried by the bloodstream to different parts of the body. Some hormones, such as cortisone, secreted by the adrenal glands, as well as sex hormones, such as estrogen, secreted by the ovaries, and testosterone, secreted by the testes, are steroids (Figure 3-4). **Cholesterol** is the most abundant steroid and is an essential component of animal cell membranes. Cholesterol also serves as the raw material for production of Vitamin D and of steroid hormones. The widely publicized role of excess dietary cholesterol associated with cardiovascular disease is controversial, because about 85% of the body's cholesterol is produced by the liver, and does not come from the diet.

Lipids are assorted organic molecules containing many carbon-hydrogen bonds and little oxygen. Hence, they are nonpolar and tend not to dissolve in water. Their insolubility helps form a water-resistant cell membrane that limits what molecules can enter or leave the cell. In addition, lipids make excellent energy reserves because they can be stored in high concentrations without the addition of water. A fatty acid molecule is composed of a nonpolar carbon chain and a polar carboxyl group, the only part of the molecule soluble in water. Fats and oils are energy-storing molecules composed of three fatty acids and glycerol. Phospholipids are the chief components of cell membranes, and are similar to fats in structure. Steroid molecules, though chemically much different from other types of lipids, are also insoluble in water. Many steroids play important roles in living organisms, including cholesterol molecules, which help form part of an animal's cell membranes, and hormones, which carry chemical messages between cells.

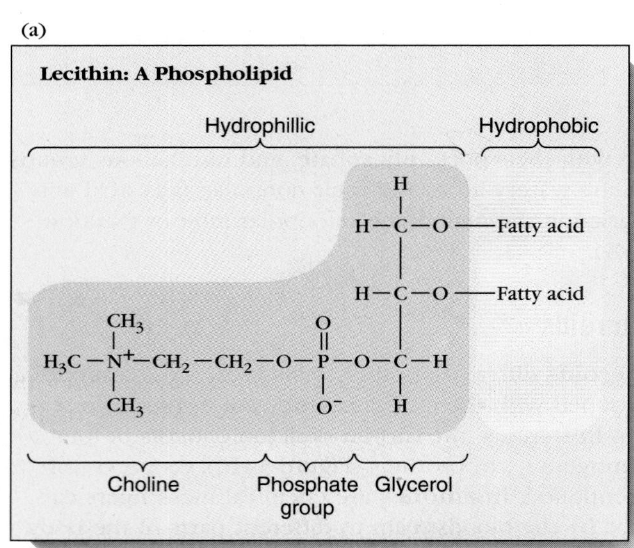

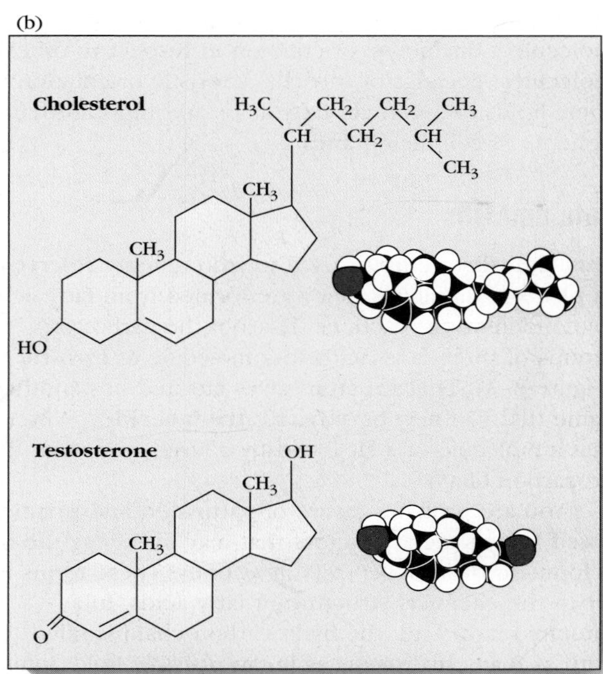

FIGURE 3-4

Phospholipids and steroids. (a) Phospholipids resemble fats with one or two of the fatty acids replaced by a phosphate group that is, in turn, linked to a nitrogen-containing group. **(b)** Steroids have a carbon skeleton consisting of four contiguous carbon rings. Testosterone and cholesterol are both kinds of steroids.

3–C CARBOHYDRATES

Carbohydrates include sugars and starches used as energy-storage molecules, and cellulose, the structural molecule that forms plant cell walls.

Monosaccharides (mono = one; saccharide = sugar), like glucose and fructose, are the simplest carbohydrates. They may occur singly, but they are also the units used to build larger carbohydrates. All sugars have polar aldehyde (H—C=O) or keto (C=O) functional groups, so they dissolve readily in water. When a monosaccharide with five or more carbons is dissolved in water, as it always is in a living system, the bonds become rearranged and the molecule takes the shape of a ring (Figure 3-5).

Monosaccharides play an important role by providing ready energy. Glucose, the most common monosaccharide, is a six-carbon sugar. Cells take it up and extract the energy from its chemical bonds in the process of cellular respiration (a topic discussed in Chapter 7). Other five-carbon sugars are important components of nucleic acids.

Two monosaccharides may be joined together by a condensation reaction, forming a **disaccharide** (di = two) (Figure 3-6). The disaccharide can be hydrolyzed back into its component monosaccharides by adding a molecule of water into the oxygen bridge linking them. Sucrose (table sugar) and lactose (milk sugar) are the most commonly encountered disaccharides.

A series of condensation reactions can join many monosaccharides into a polymer called a **polysaccharide** (poly = many). Four important polysaccharides occur in living things: glycogen, starch, cellulose, and chitin (Figure 3-7). These differ in the arrangement of bonds between the glucose subunits, in the branching patterns of the polymer, and in the total number of glucose subunits per chain.

Glycogen is the molecule mainly used by animals to store readily available energy (Figure 3-7a). The liver and muscles remove glucose from the bloodstream and use it to assemble glycogen. Glycogen is later converted back into glucose as it is needed for energy. A typical adult human may have up to two pounds of glycogen stored in his or her liver, and these two pounds of glycogen can be converted to a supply of glucose that will last for several hours.

Plants use **starch** (Figure 3-7b) as an energy-storage polysaccharide. Starch consists of amylose and amylopectin, two glucose polymers. When you boil

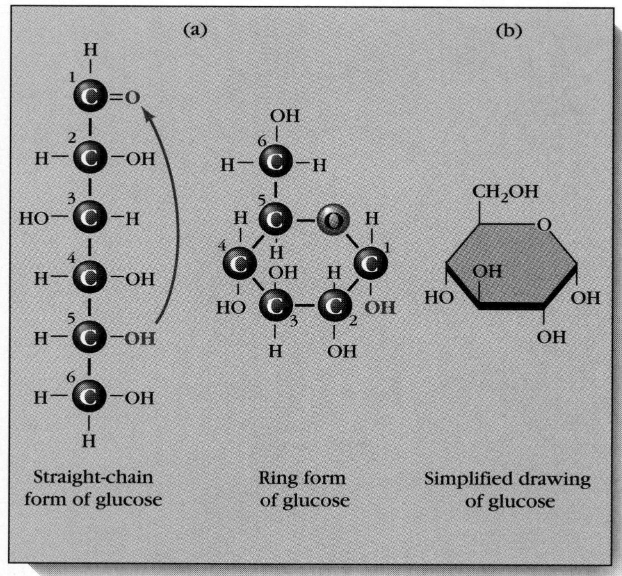

FIGURE 3-5

Monosaccharides. (a) Straight chain monosaccharides with five or more carbons form rings when placed into water. **(b)** A simplified way to represent glucose rings.

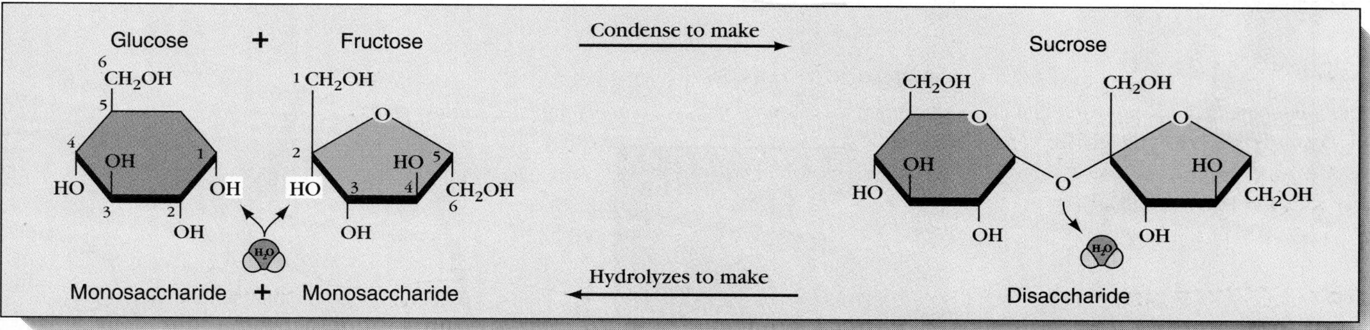

FIGURE 3-6

Two monosaccharides join by condensation to form a disaccharide. A disaccharide can be separated again into two monosaccharides by hydrolysis.

potatoes, the water becomes cloudy as amylose dissolves in it. Amylopectin stays in the potatoes and is later digested to glucose subunits in your intestine. Left to itself, a living potato would eventually break down its starch to glucose and use it for energy, growth, and reproduction.

Plants use glucose to make their structural polysaccharide, **cellulose** (Figure 3-7). Because it is a major component of all plants, cellulose is probably the most abundant organic material on Earth. Cellulose is made up of long, straight chains of glucose monomers linked together end to end by covalent bonds. Several of these polymers are held together by hydrogen bonds to form cellulose fibers. Cotton is almost entirely cellulose (Figure 3-7). Each plant cell surrounds itself with a tough, external cell wall composed of several layers of cellulose fibers that are often reinforced with other substances. Cell walls help to stiffen and support a plant.

Cellulose is indigestible to humans, and to most multicellular organisms, but is nevertheless important in the diet because it provides bulk, also called "fiber" or "roughage," that stimulates the intestines to keep things moving along. Cellulose can be digested by some single-celled organisms, by some bacteria, and by some snails and insects.

Chitin is another important polysaccharide that serves as a major structural component of the external skeletons of arthropods (crabs, insects, spiders) (Figure 3-7). Chitin is also found in the cell walls of fungi.

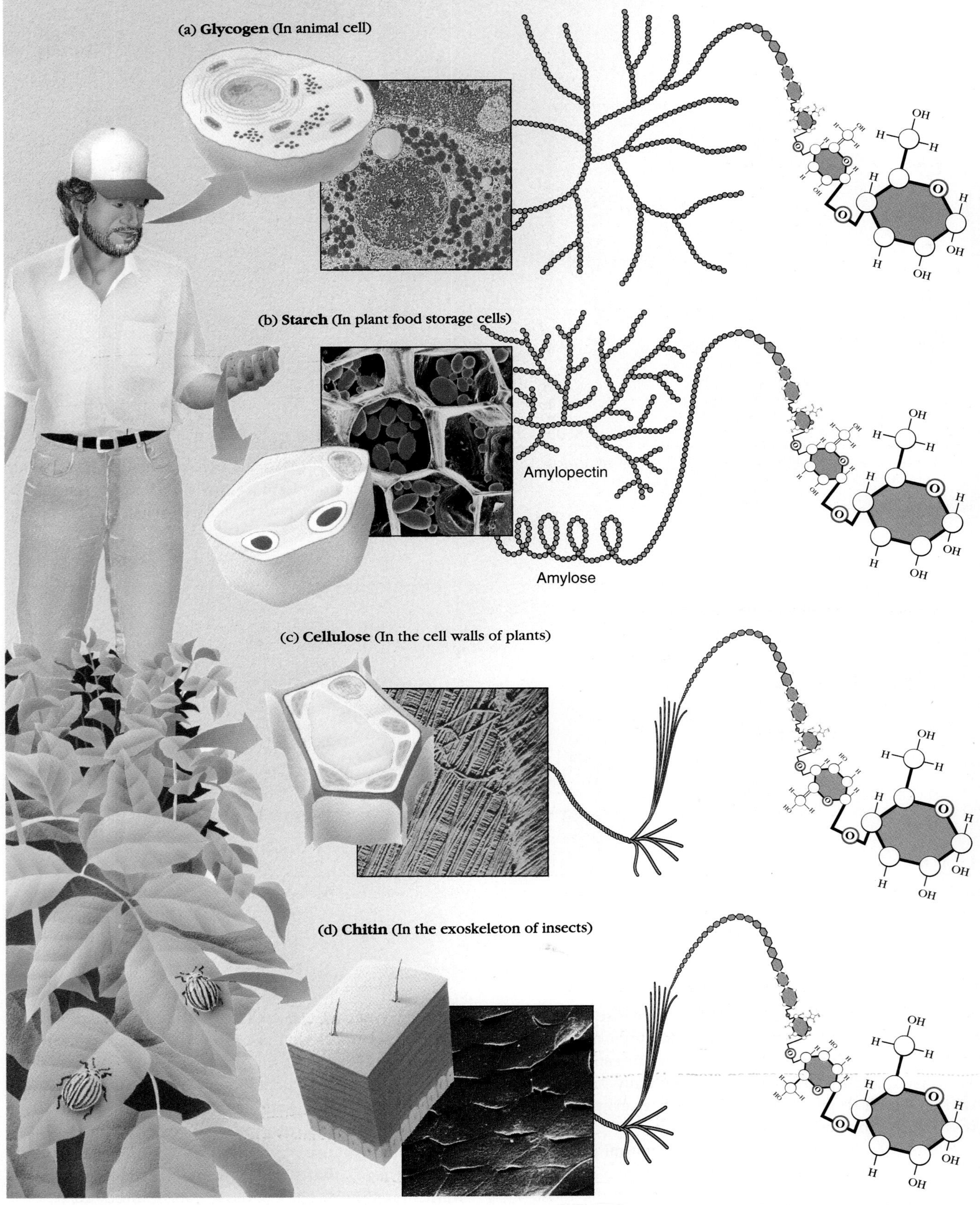

FIGURE 3-7

Important polysaccharides in living organisms. (a) Glycogen provides energy storage that can be used to drive metabolic reactions. **(b)** Starch stores energy for later use by plant cells. **(c)** Cellulose forms strong cell walls to support plant cells. **(d)** Chitin forms tough exoskeletons of many insects. (a, CNRI Science Photo Library/Photo Researchers, Inc.; b, Dr. Jeremy Burgess/Science Photo Library/Photo Researchers, Inc.; c, Biophoto Associates/Photo Researchers, Inc.; d, Len Lessin/Peter Arnold, Inc.)

Table 3-2
Some Functions of Proteins

Type of Protein	Example	Function
Enzymes	amylase	Converts starch to glucose
Structural proteins	keratin, collagen	Hair, wool, nails, horns, hoofs, Tendons, cartilage
Hormones	insulin, glucagon	Regulate glucose metabolism
Contractile proteins	actin, myosin	Contractile filaments in muscle
Storage proteins	ferritin	Stores iron in spleen and egg yolk
Transport proteins	hemoglobin, serum albumin	Carries O_2 in blood Carries fatty acids in blood
Immunological proteins	antibodies	Form complexes with foreign proteins
Toxins	neurotoxin	Cobra venom blocker of nerve function

Table 3-3
Chemical Composition (Excluding Water) of a Common Bacterium

Type of molecule	Percent of Total Dry Weight	Comments
Small molecules	10	Inorganic ions, monomers, coenzymes
Polysaccharides and lipids	16	Protective outer wall and membrane; some glycogen stored inside bacterium
DNA	4	One or two molecules per bacterium; each molecule is about 1 millimeter (mm) long and highly folded; the bacterium itself is only about 0.002 mm long
RNA	20	About 3000 different kinds
Proteins	50	About 2500 different kinds: about ⅓ structural protein, ⅔ enzymes

Carbohydrates include many types of energy-storing sugars, as well as chitin and cellulose—structural molecules of animal exoskeletons and plant cell walls, respectively. Monosaccharides, such as glucose, are the smallest sugars, typically composed of just five or six carbon atoms. Disaccharides, such as sucrose, are composed of two monosaccharides. Polysaccharides, such as glycogen, starch, cellulose, and chitin, are composed of many monosaccharides joined together.

BIO-BIT

In crabs, large amounts of calcium carbonate within the chitin layer make the exoskeleton more brittle than in insects.

3–D PROTEINS

Proteins make up more than 50% of the dry mass of animals and bacteria, and perform many important functions in living organisms (Tables 3-2 and 3-3). Proteins contain the elements carbon, oxygen, hydrogen, nitrogen, and usually some sulfur and are built from monomers called **amino acids** (Figure 3-8). The structure of an amino acid capitalizes on the four bonds that a carbon atom can form. Each amino acid has a carboxyl group (—COOH), an amino group (—NH_2), and a hydrogen atom bonded to a central carbon atom. The fourth component is one of 20 different side chains called R groups, which contribute unique properties to each of the 20 different amino acids. All organisms contain the same 20 common amino acids, and like all of the biological molecules

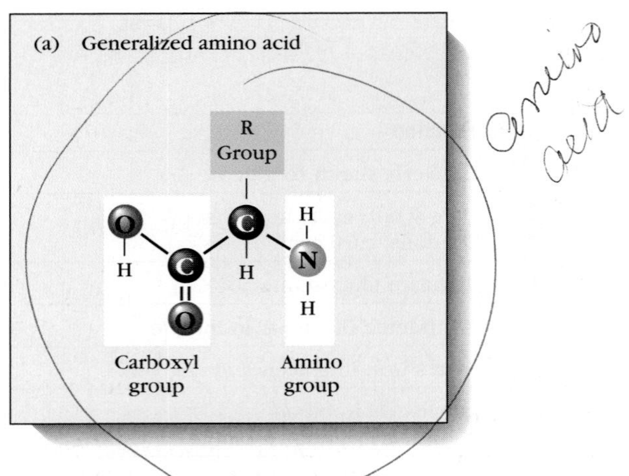

FIGURE 3-8

Amino acid structure and bonding. (a) The general structure of amino acids consists of an amino and a carboxyl group, both attached to the same carbon atom, which is also bonded to a hydrogen atom. The fourth bond is between a carbon and the R group. **(b)** The amino acid serine. **(c)** The amino acid cysteine. **(d)** Amino acids joined by the addition of water during condensation to form a peptide bond.

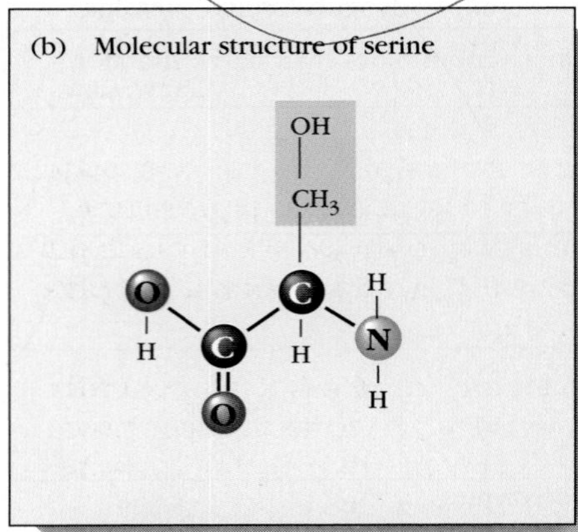

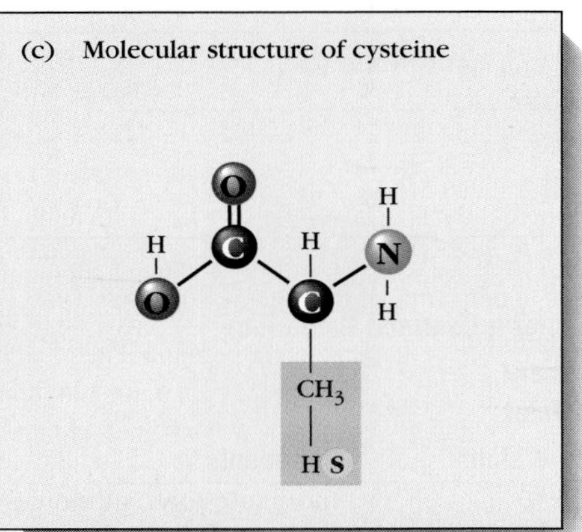

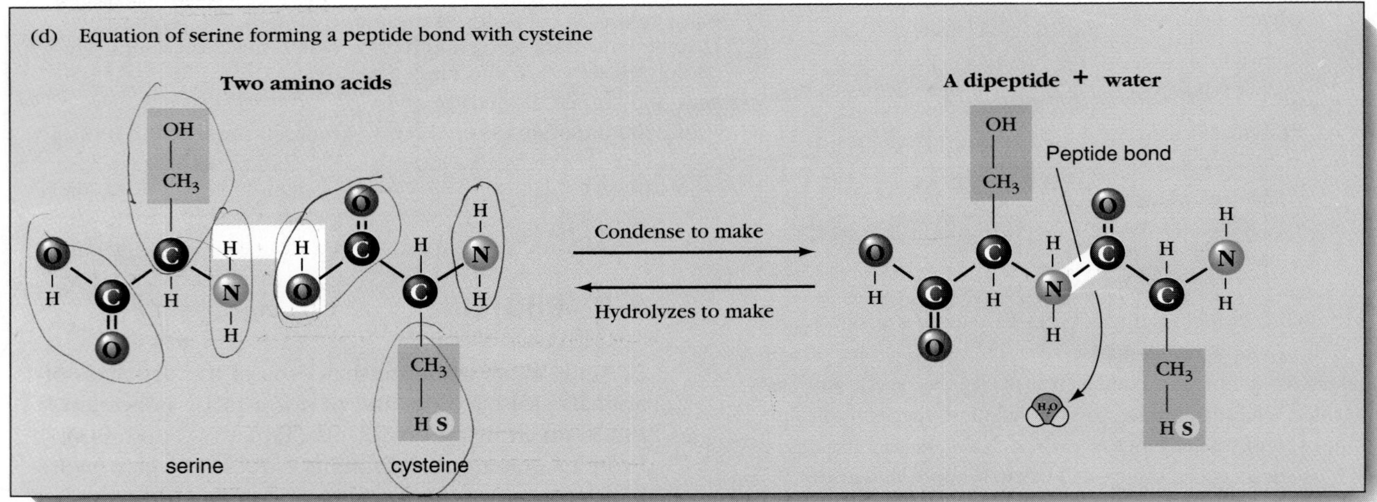

we have encountered, the properties of the amino acids are rooted in their chemical structures.

To form proteins, condensation reactions link amino acid monomers into long chains. A **peptide bond** forms between the carboxyl carbon of one amino acid and the amino nitrogen of another (Figure 3-8). The condensation of two amino acids forms a **dipeptide.** Long strings of amino acids, containing between 100 and 300 amino acids, are called **polypeptides.** A **protein** is a functional unit made of one or

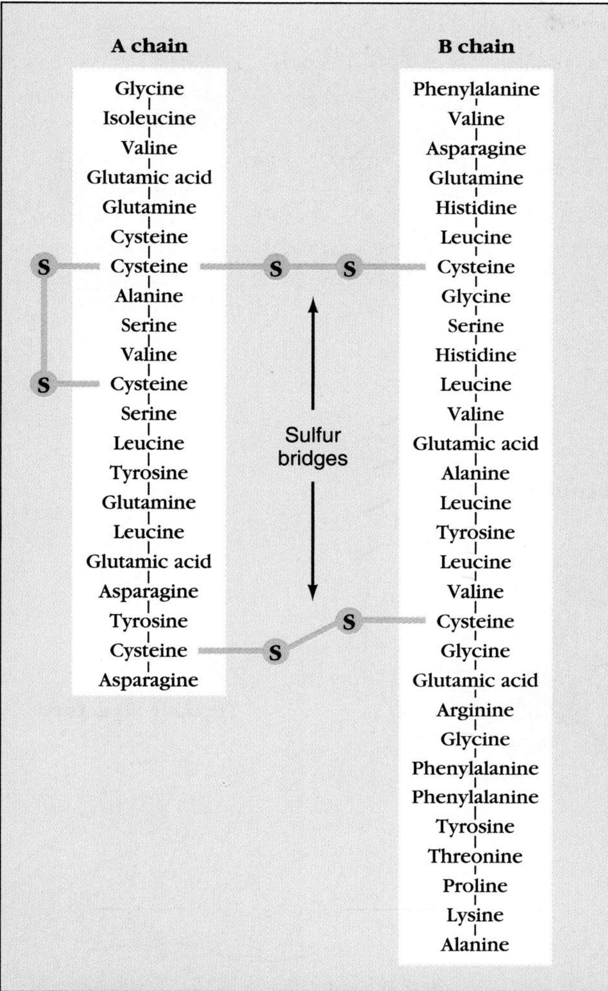

FIGURE 3-9

Insulin. This important molecule helps regulate blood sugar levels. It is formed by two polypeptide chains linked by sulfur bridges.

more polypeptides. For instance, the protein insulin (a hormone that triggers removal of glucose from the bloodstream) is made of two linked polypeptide chains (Figure 3-9). With only 51 amino acids, insulin is among the smallest proteins.

Protein Structure

Proteins are long, unbranched chains of amino acids, but they may fold up into complex shapes. There are four aspects of protein structure that will help you to understand how proteins function:

1. A protein's **primary structure** is the unique sequence of its amino acids (A Journey Through Protein Structure).
2. The **secondary structure** is the shape imparted to local areas of the chain by the amino acids there. The most common type of secondary structure is the alpha helix, a coil assumed by parts of some polypeptide chains.
3. A protein's **tertiary structure** is the three-dimensional shape it assumes because of molecular interactions between the R groups of amino acids in different parts of the chain. These interactions include ionic bonds, hydrogen bonds, disulfide bonds, and hydrophobic interactions—all of which influence the shape of the tertiary structure.
4. Proteins that are made up of two or more polypeptide chains have an additional, **quaternary structure,** the spatial arrangement in which the individual chains fit together to form a complete functional protein.

3? Once amino acids are linked in the proper order to form the primary structure, the subsequent structures follow automatically, governed by the attractive and repulsive forces within different parts of the polypeptide. Gentle heating or certain chemical treatments can loosen these linkages, but many proteins will re-form their original structure once normal conditions are restored. Harsher treatments (stronger chemicals and higher temperatures) make polypeptides permanently lose their shape. These proteins are said to have been **denatured** (*A Journey into Science in Process:* Kitchen Chemistry). This is what happens to the proteins in an egg when it is cooked, resulting in the change in color and appearance. ■

Having certain amino acids in certain positions is crucial to the protein's overall shape, and thus to its function. For example, a change of just one amino acid dramatically alters the shape and makes the difference between normal hemoglobin and the hemoglobin of sickle cell anemia (Section 13-B). Proteins are masterpieces of molecular engineering, and they are tailored to their functions by hundreds of millions of years of natural selection.

Proteins compose nearly 50% of the dry mass of most animals and bacteria. They are composed of strings of amino acids that fold upon themselves to form complex structures. The sequence of the amino acids determines the overall shape and properties of the protein.

A Journey Through Protein Structure: Hemoglobin

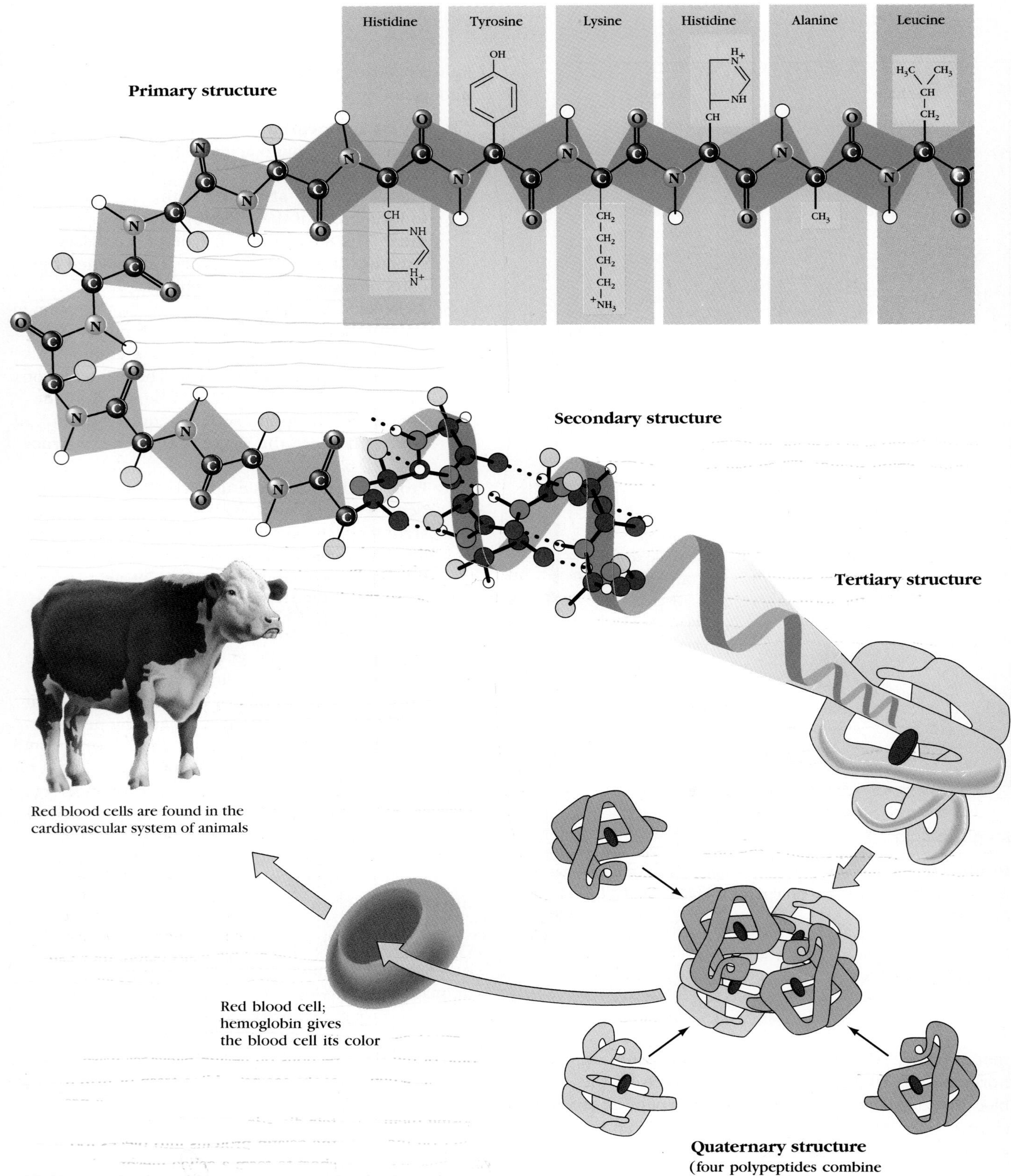

Four levels of protein structure. The amino acid sequence forms the first level of protein structure. The secondary structure results from the angled shapes of amino acids in the protein. Molecular interactions of the R groups of amino acids produce additional folding that results in the tertiary structure. The quarternary structure results from the spatial arrangement of two or more polypeptides.

A Journey into Science in Process

Kitchen Chemistry

A protein's overall three-dimensional shape is determined by interactions between its amino acids. The protein can be denatured, that is, made to lose its characteristic shape, by various treatments that disrupt these interactions. For example, a human hair contains hundreds of strands of the protein keratin. The strength of a hair comes from abundant sulfur bridges linking molecules of the amino acid cysteine in neighboring keratin strands. When hair is given a permanent wave, the first solution used breaks the sulfur bridges. This allows the hair structure to be distorted as the hair is wound around curlers. Then another solution is applied, allowing the formation of new sets of sulfur bridges, which hold the keratin strands in the configuration imparted by the curlers.

FIGURE 3-A

Fresh pineapple and gelatin. The gelatin in the glass on the left containing canned pineapple has set just as firmly as the glass containing only plain gelatin (center). But the gelatin made with fresh pineapple (right) did not solidify, and can be poured out into a bowl. Enzymes in fresh pineapple have digested the proteins in the gelatin, so that the gelatin proteins are too short to form the meshwork that makes it set. Pineapple is heated in the canning process, denaturing the enzyme, so that it cannot affect the gelatin. (Dennis Drenner)

Perhaps the most common way to denature a protein is heat, and the most familiar place this happens is the kitchen. Many proteins in food fold up into roughly globular shapes, with water molecules surrounding the outside of the protein and also lying among some of its internal loops and folds. At high temperatures, the protein's atoms, and the associated water molecules, have so much energy that their motion disrupts the hydrophobic, hydrogen, and ionic bonds that give the protein its normal shape. The protein unfolds and the loose ends form new bonds to other protein molecules, which have also been denatured. As the proteins form a meshwork with one another, there is less room available for water molecules, so water is squeezed out, and some of it is lost by evaporation.

When you roast meat, much of the water is squeezed out into the tissue spaces. If you carve a roast fresh from the oven, this juice runs out as the knife slices through, and the slices of meat are quite dry. However, if you let the roast sit for 15 to 20 minutes, the cooling proteins undergo a partial reversal of their denaturation, allowing water to move back among them. The result: moister meat.

The same thing happens when you cook eggs. The unfolding of globular proteins allows them to interact with each other and eventually form one big, tangled solid network, moist with infiltrating water molecules. But if cooking is not stopped at this crucial moment, the mesh tightens further and squeezes the water out. Overcooking eggs has two possible outcomes: the proteins coagulate in lumps, floating in the squeezed-out liquid, or they form a single, rubbery mass, with the water either separated and floating on top or simply evaporated off altogether.

Another interesting form of denaturation occurs with some of the globular proteins of egg whites. When you whip egg whites, you force air in next to the proteins, exposing them to air on one side and water on the other. The proteins' hydrophilic regions are attracted to the liquid, while the hydrophobic areas are not, but tend to associate with the air pockets instead. So the proteins unfold, and the hydrophilic stretches of different proteins bond to each other. They form a lacy network, reinforcing the liquid wall of the bubble around the trapped air. This structure is not strong enough to withstand baking, however. In the heat of the oven, the air bubbles expand, and they can burst the protein bonds in their walls, collapsing the whole structure. This does not happen in a properly cooked meringue, soufflé, or angel food cake, thanks to still other proteins, which did not participate in forming the foam itself. These proteins are denatured by heat, and they form a stronger meshwork that stabilizes the bubble walls before they can be ruptured by escaping hot air.

Gelatin also consists of a loose protein meshwork, with water held in the spaces between proteins. The directions on gelatin packages warn cooks against adding fresh or frozen pineapple. Pineapple contains a protein-digesting enzyme, which chops the gelatin proteins into pieces too short to form a gelled meshwork. Gelatin made with fresh pineapple never sets, but remains a soupy liquid. Canned pineapple is fine to use because heat applied in the canning process denatures the enzyme, so it cannot attack the gelatin molecules (Figure 3-A).

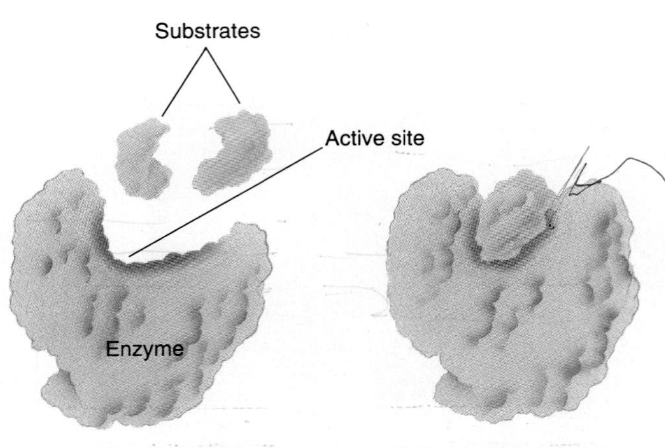

(a) Substrates approach and bind with enzyme's active site.

(b) Enzyme attracts and distorts substrates. The active site also changes shape.

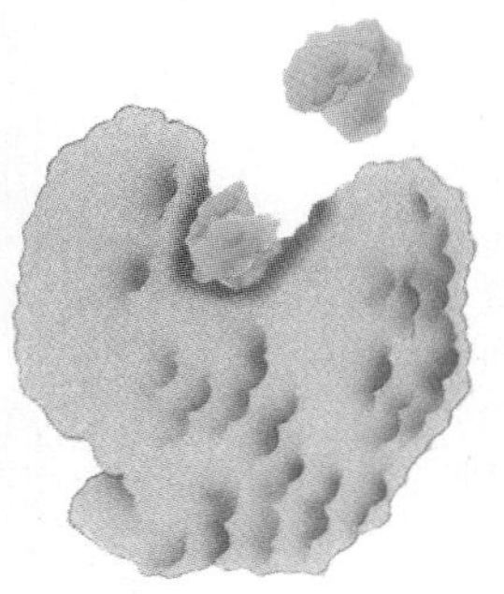

(c) Enzyme pulls atoms off each substrate. The substrates bind to each other and leave.

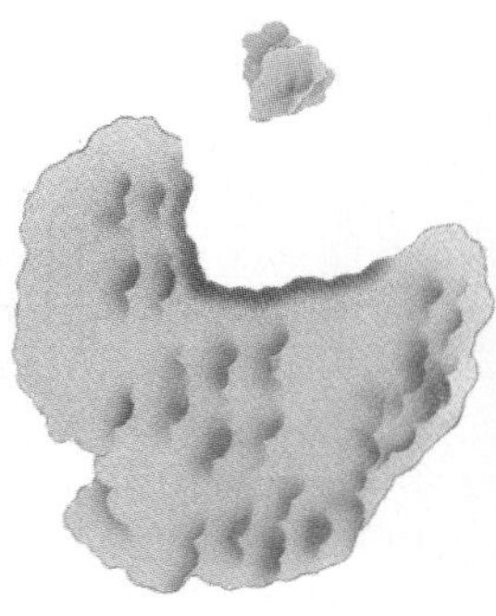

(d) The detached atoms combine and leave. The enzyme is ready to accept new substrate molecules.

FIGURE 3-10

How enzymes work.

3–E ENZYMES

Many of an organism's proteins are **enzymes,** special proteins that **catalyze** or increase the rate of chemical reactions. Each of the approximately 2000 known enzymes catalyzes a particular reaction (Figure 3–10). Some carry out condensation or hydrolysis reactions, linking monomers into larger molecules, or vice versa. Other enzymes transfer whole groups, such as amino groups, from one molecule to another. Enzymes that catalyze oxidation-reduction reactions transfer electrons or hydrogen atoms from one molecule to another.

The reactants in an enzyme-catalyzed reaction are called the enzyme's **substrates.** Like human matchmakers, catalysts are not part of the reactions they facilitate and are not permanently changed by them. Enzymes are named according to their substrates and the reactions they catalyze, and all recently identified enzymes have names that end with the suffix "-ase." For example, RNA polymer*ase* links nucleotides to form RNA polymers, and sucr*ase* hydrolyzes sucrose. However, some enzymes may be called by their old-fashioned names—for example, digestive enzymes such as trypsin and pepsin.

All enzymes speed up reactions by combining with their substrates and holding them at the correct angle for the reaction to occur (Figure 3–10). The enzyme also pulls on the substrates' bonds and "loosens" them. This lowers the amount of energy necessary to allow the reaction to proceed, so the reaction occurs readily, at much lower temperature than would be necessary in the absence of a catalyst. Figure 3–11 demonstrates a familiar enzymatic action: the production of "cat-box odor."

Enzyme-Substrate Complexes

Each enzyme catalyzes only particular reactions of one or a few kinds of molecules. Enzymes are specific because a substrate actually binds to an area called the enzyme's **active site,** a small groove formed as the enzyme molecule folds up (see Figure 3–10). The size, shape, and electrical charge of amino acid R groups at the active site determine which substrates can fit there. The specific fit of enzyme and binding site has been compared to a "lock and key" fit, but this analogy gives a misleading image of unyielding hardware. Actually, both enzyme and substrate change shape slightly when they combine, and the interaction between the two is often compared to the way a hand changes the form of a glove. When the reaction is over, the enzyme releases the products and emerges exactly as it started, ready to catalyze another reaction.

Some enzymes cannot bind their substrates unless **cofactors** are present. Some cofactors are inorganic ions, such as zinc or iron. Others are nonprotein organic molecules called **coenzymes.** Many vitamins are coenzymes or components of coenzymes. We need only scant amounts of coenzymes because, like enzymes, they can be used over and over again.

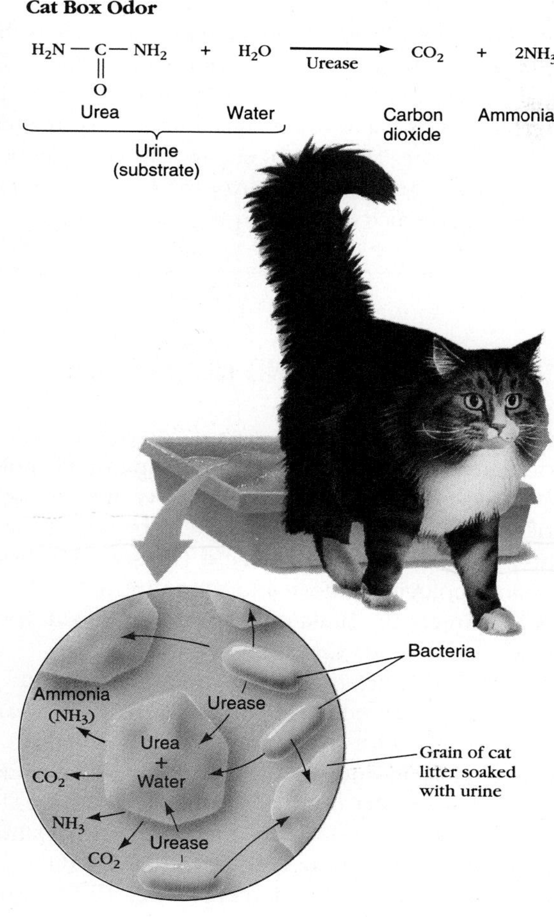

FIGURE 3-11

The enzyme reactions within a cat-litter box. Bacteria on particles of cat litter produce the enzyme urease, which catalyzes the breakdown of urea (from urine) into carbon dioxide and ammonia.

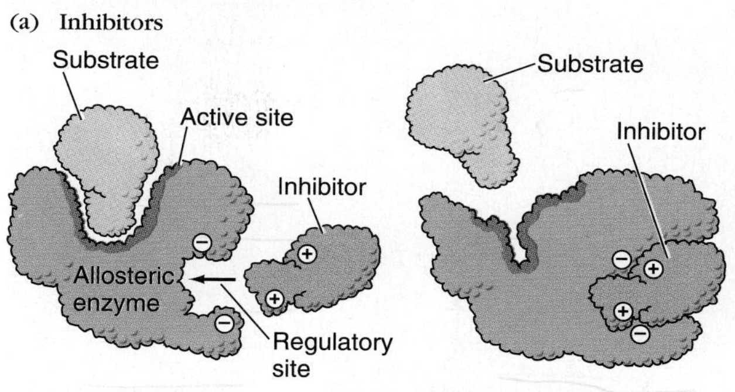

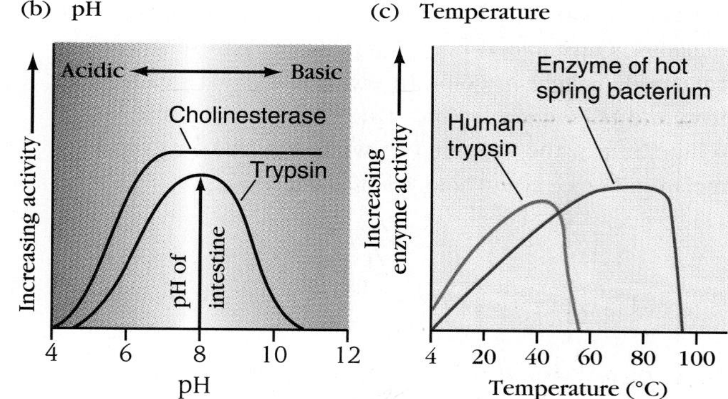

FIGURE 3-12

Factors affecting enzymes. (a) Inhibitors block the ability of enzymes to bind to substrates. **(b)** High basic and low acidic levels slow down or stop the function of enzymes by disrupting the bonds in the enzyme's R groups. **(c)** High temperatures damage and permanently alter the structure of an enzyme, destroying its function.

Factors That Affect Enzyme Activity

1. The *concentration of a substrate* affects the rate of enzyme action, and as substrate concentration increases, the reaction rate does, too. Enzymes differ in their speed of reaction. For example, lysozyme is very slow, catalyzing one reaction every two seconds, whereas urease can catalyze 30,000 reactions per second. And carbonic anhydrase, the enzyme found in red blood cells that speeds the transformation of carbon dioxide and water into bicarbonate, works even faster, catalyzing an astounding 600,000 reactions per second.
2. *Inhibitors* decrease an enzyme's reaction rate, either by blocking the active site and physically preventing the enzyme from binding to its substrate, or by disrupting the enzyme's three-dimensional structure and destroying its function (Figure 3-12).
3. *pH* can negatively affect the activity of enzymes by changing their three-dimensional structure and preventing proper enzyme-substrate fit (Figure 3-12). This can occur when H^+ ions in an acid solution tend to combine with negatively charged R groups on the enzyme. This electrically neutralizes the R groups and disrupts ionic bonds in the enzyme's folding pattern, thus changing its shape.
4. *Temperature* also affects the rate of enzyme reactions (Figure 3-12). Warm temperatures cause molecules to move around faster and collide harder and more often, but high temperatures

FIGURE 3-13

A Siamese cat. The enzyme that makes the dark pigment melanin is unstable at body temperature, and so fur on the cat's body is light in color. In contrast, the cat's ears, nose, feet, and tail are cooler and darker because at these lower temperatures, the enzyme can work and make its product, melanin. (Renee Lynn/Photo Researchers)

(usually above 60°C) denature proteins, permanently changing their three-dimensional shapes and thus destroying their ability to function. Cooking helps to preserve food by destroying the enzymatic activity of bacteria that cause decay.

4? At the other extreme, chemical reactions are slowed at low temperatures, because molecules move so slowly that few collisions occur between enzyme and substrate molecules. Refrigeration preserves food by slowing the activity of enzymes in the food itself, or in the organisms that cause decay. ■

Evolution has produced enzymes adapted to function within a particular range of temperatures. For each organism, this range depends on the normal temperatures in its environment. For example, some cyanobacteria live on the surface of glacial ice and are adapted to temperatures close to the freezing point of water. Other cyanobacteria inhabit the hot springs of Yellowstone Park, which may be at temperatures of 80 to 85°C (176–185°F). In a more familiar example, the characteristic color pattern of Siamese cats is due to a temperature-sensitive enzyme (Figure 3–13).

An enzyme is like a robot on an assembly line, performing the same task over and over on a particular set of parts—the enzyme's substrates—without itself being permanently changed. Technically, most enzymes are protein catalysts that permit organisms to carry out chemical reactions rapidly, at the relatively low temperature in their bodies. An enzyme is specific, because only a few kinds of substrate molecules can fit into its active site. The rates of enzyme reactions can be affected by many factors including (*a*) the concentration of substrates, (*b*) presence of inhibitors, (*c*) pH, and (*d*) temperature.

3–F NUCLEIC ACIDS AND NUCLEOTIDES

Nucleic acids are some of the largest biological molecules and are possibly the most fascinating of all molecules. **Deoxyribonucleic acid (DNA)** is the genetic material and contains the organism's instructions for making proteins. **Ribonucleic acid (RNA)** is synthesized using molecular information found in DNA. In turn, RNA directs the building of proteins, based upon molecular instructions that DNA provides.

DNA and RNA are built from **nucleotides,** and each of these has three basic parts (Figure 3–14): (*a*) a five-carbon sugar with (*b*) a number of phosphate groups attached on one side, and (*c*) a ring-shaped nitrogenous base on the other. The nucleic acids are named after the sugars in their nucleotides: RNA's nucleotides contain the sugar **ribose** and DNA's nucleotides contain the sugar **deoxyribose.** These two sugars differ in the number of oxygen atoms they contain.

Besides serving as the building blocks for nucleic acids, nucleotides play other important roles. **Adenosine triphosphate (ATP)** supplies the energy necessary for many chemical reactions in most living organisms. Other nucleotides act as coenzymes, which, as you recall, are molecules needed for enzymes to work properly.

Nucleic acids, which form DNA and RNA, are some of the largest and most important molecules in the body because they direct the production of proteins. Nucleic acids are made up of nucleotides. ATP and other nucleotides serve as energy carriers.

3–G METABOLISM

You have now met members of all four main classes of biological molecules: lipids, carbohydrates, nucleic acids, and proteins. Now let's put them all together—in a living cell.

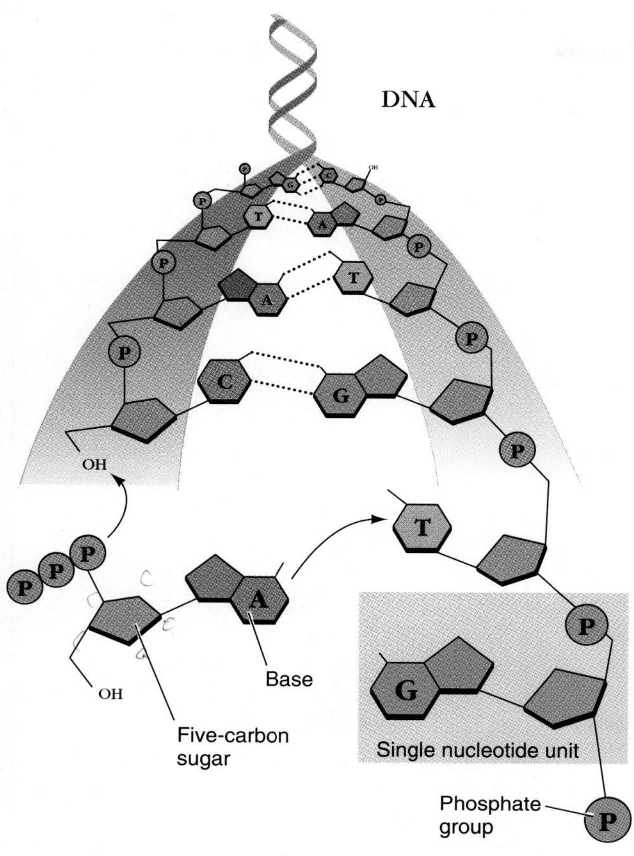

FIGURE 3-14

The structure of DNA. The double helix of DNA consists of two strands of nucleotides. Each nucleotide consists of a sugar, a phosphate group, and a nitrogenous base.

A living cell is a busy biochemical factory that imports, manufactures, uses, recycles, and exports thousands of molecules in a process called **metabolism.** To help you to understand metabolism we will compare it to something that everyone is familiar with: the production of pizzas in a pizza shop (Figure 3-15). In the cell, organic monomers such as sugars, amino acids, and nucleotides are metabolized, or converted from one form to another. Similarly, in the pizza shop, food ingredients (see recipe in Figure 3-15) are processed into pizzas. The energy required to run the cell's molecular factory is extracted from food molecules, while the chef, the delivery boy, and electricity provide the energy needed to make pizzas. In the cell, each reaction is carried out by an enzyme, while in the pizza shop, people are like enzymes—preparing ingredients, assembling them into pies, baking, taking orders, as well as packing and delivering pizzas.

Metabolism encompasses all of a cell's biochemical reactions. Like the work in the pizza shop, these do not occur in chaos, but are organized into sequences similar to assembly lines called **metabolic pathways.** Just as raw tomatoes are physically changed as they are processed into tomato sauce, each pathway consists of several enzyme-facilitated reactions that convert substrates, step-by-step, into a final molecular product. The pathway is organized so that one enzyme's product is the next one's substrate. In animals, for example, one pathway converts the six-carbon sugar glucose into the five-carbon sugar ribose. This pathway contains four different enzymes. Other pathways may have more or fewer steps.

To make pizza dough, the chef keeps the necessary spoons, bowls, measuring cups, knives, and rolling pins within arm's reach and all at one workspace. Pizza preparation is done in a different spot than baking or packaging. Similarly, in the cell, different metabolic pathways may be confined to specific locations. This increases efficiency by isolating and concentrating all of the enzymes and substrates of a metabolic pathway in one place. In cells, metabolic pathways may be isolated within membranous spaces, or all of the enzymes needed for a pathway may be embedded in a particular membrane or attached to its surface.

Pizza production is controlled by demand: the chef stops baking when there are no more orders to fill, an example of **negative feedback.** Metabolic pathways, too, are often controlled by negative feedback: the regulation of the rate of a process by the concentration of its product. Often the enzyme at the beginning of a metabolic pathway changes shape when an inhibitor (often the final product of the pathway) binds to it, "switching off" the pathway and preventing it from working. One way to visualize this enzyme is to compare it to the chef who quits work whenever pizzas accumulate faster than orders are coming in.

The cell is a more efficient organization than the pizza shop because it commonly recycles, dismantling old macromolecules and using their components to make new ones, while the pizza shop cannot recycle uneaten pizzas into fresh pies.

Cellular metabolism encompasses all of the enzyme-catalyzed activities occurring within a cell. Metabolic pathways consist of a stepwise series of specific chemical reactions, catalyzed by enzymes, which build, break down, or rearrange substrates. Metabolic pathways may be isolated within membranous sacs or concentrated near membranes containing the relevant enzymes. Negative feedback inhibits a metabolic pathway when too much product has accumulated.

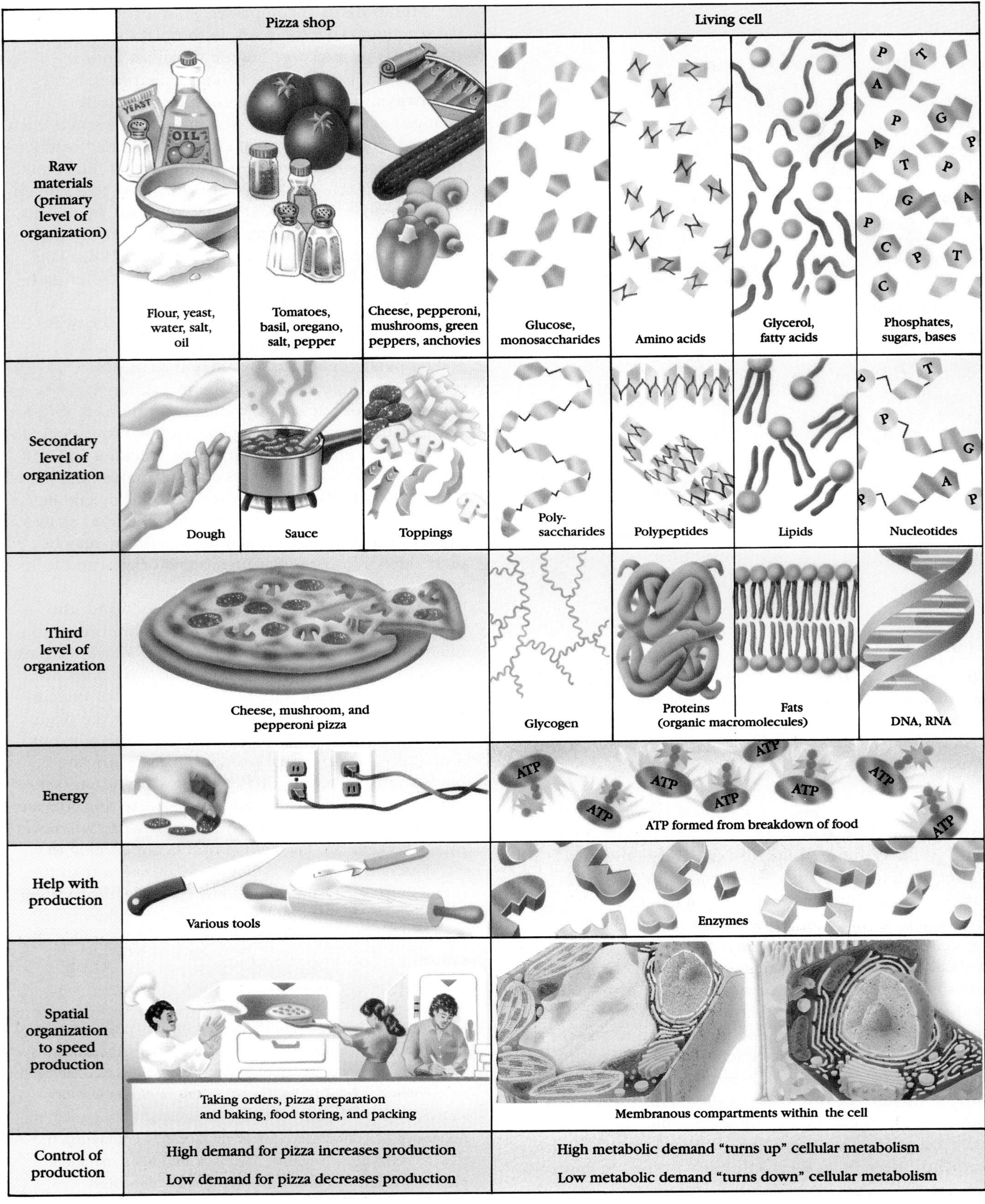

Pizza shop
Living cell
Raw materials (primary level of organization)
Flour, yeast, water, salt, oil
Tomatoes, basil, oregano, salt, pepper
Cheese, pepperoni, mushrooms, green peppers, anchovies
Glucose, monosaccharides
Amino acids
Glycerol, fatty acids
Phosphates, sugars, bases
Secondary level of organization
Dough
Sauce
Toppings
Poly-saccharides
Polypeptides
Lipids
Nucleotides
Third level of organization
Cheese, mushroom, and pepperoni pizza
Glycogen
Proteins
Fats
(organic macromolecules)
DNA, RNA
Energy
ATP formed from breakdown of food
Help with production
Various tools
Enzymes
Spatial organization to speed production
Taking orders, pizza preparation and baking, food storing, and packing
Membranous compartments within the cell
Control of production
High demand for pizza increases production
Low demand for pizza decreases production
High metabolic demand "turns up" cellular metabolism
Low metabolic demand "turns down" cellular metabolism

FIGURE 3-15

Levels of organization in a pizza shop and in the metabolism of a cell. The lowest level of organization includes the ingredients, or molecules of life. In the same way that pizza ingredients are combined to make pies, the molecules of life combine to form basic cell structure. As knives and rolling pins are used to make a pizza, energy and enzymes work together to create the cell structure. Within a pizza shop various tasks are isolated into work spaces and the job of making pizza is compartmentalized, just as it is in the organelles and cytoplasm of cells. Finally, the inner workings of cells and pizza shops are controlled by additional factors, such as inhibiting enzymes, demand for pizzas, or a temperamental chef.

SUMMARY

1. Aside from water, the main chemical components of living organisms are organic molecules. Organic molecules are based on carbon, a versatile element able to form molecular skeletons of myriad sizes and shapes.
2. Most of the organic molecules in a living body are macromolecules. These polymers are made up of many monomeric subunits. A polymer contains identical or similar monomers.
3. Monomers are joined together to form polymers by condensation reactions, in which a bond is formed by removing the components of a water molecule (H^+ and OH^-) from the subunits.
4. Macromolecules are broken down by hydrolysis, the addition of water molecules between the subunits, causing them to become separated.
5. Biological molecules fall into four main groups: lipids, carbohydrates, nucleic acids, and proteins.
6. Lipids do not form molecules large enough to be called polymers, but the other three groups contain polymers formed from monomers as follows:
   a. multiple carbohydrate monomers, called monosaccharides, join together to form polysaccharide polymers.
   b. multiple protein monomers, called amino acids, join together to form polypeptide polymers.
   c. multiple nucleic acid monomers, called nucleotides, join together to form DNA and RNA polymers.
7. Lipids and carbohydrates are composed mainly of carbon, hydrogen, and oxygen. Some lipids and carbohydrates are important energy-storage compounds that may be broken down to release energy.
8. Unlike members of the other three groups, lipids are nonpolar and so do not dissolve in water. Some lipids are vital components of all biological membranes, while others are important hormones.
9. Structural polysaccharides include cellulose in plants and chitin in arthropods and fungi.
10. Nucleic acids and proteins play vital roles in directing an organism's growth, activity, and reproduction.
11. Nucleic acids contain the elements carbon, hydrogen, oxygen, nitrogen, and phosphorus.
12. Some nucleotides, such as ATP, serve as energy carriers.
13. Proteins contain carbon, hydrogen, oxygen, nitrogen, and some sulfur. Important proteins include enzymes, structural and transport proteins, and hormones.
14. Each of the approximately 2000 kinds of enzymes is adapted to catalyze reactions between specific substrates.
15. Enzymes permit vital chemical reactions to occur at the relatively low body temperatures of living organisms.
16. The activity of metabolic enzymes is also affected by substrate concentration, cofactors, inhibitors, pH, and temperature.
17. Enzymatic reactions are organized into various metabolic pathways that convert one kind of molecule to another, build up or break down polymers, and break down food to release energy.

SELF-QUIZ

Make a summary table of this chapter for yourself by filling in the blanks numbered 1–13. (For example, in #1, fill in the class that contains the elements C, H, O, N, P; in #2, name that class's monomer subunits, etc.)

Summary of the Major Classes of Biological Compounds

Class	Chemical elements	Monomer subunits	Main roles
1. __________	C, H, O, N, P	**2.** __________	**3. a.** __________ **b.** __________
4. __________	**5.** __________	Fatty acids, glycerol, etc.	**6. a.** __________ **b.** __________ **c.** __________
Proteins	**7.** __________	**8.** __________	**9. a.** __________ **b.** __________ **c.** __________
10. __________	**11.** __________	**12.** __________	**13. a.** __________ **b.** __________

14. The molecules shown below:
a. both contain amino groups
b. both contain carboxyl groups
c. both belong to the same major class of organic compounds
d. could both serve as monomer subunits of polymers
e. all of the above

$H_3C-\underset{H}{\overset{NH_2}{C}}-C(=O)OH$ $\qquad$ $H-\underset{H}{\overset{H}{C}}-\underset{H}{\overset{H}{C}}-\underset{H}{\overset{H}{C}}-C(=O)OH$

15. The molecule shown below is a:
a. fatty acid
b. dipeptide
c. disaccharide
d. nucleotide
e. steroid

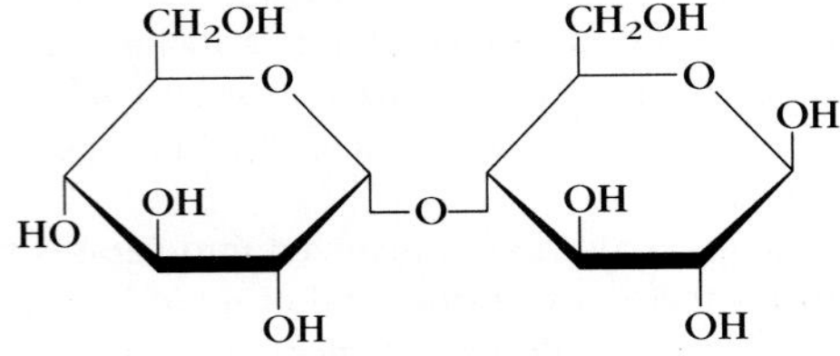

16. Which of the following is *not* made up of monosaccharides?
a. sucrose
b. starch
c. glycogen
d. insulin
e. cellulose

17. You have a solution of an enzyme. You put half of it into each of two beakers containing identical substrate at equal concentrations. After waiting a while, you test both solutions and find that the substrate in beaker A has been changed but the substrate in beaker B has not been acted upon by the enzyme. Suddenly you notice that beaker B has been sitting on a hot plate with the switch turned to "high." The enzyme in beaker B probably did not work because it had been:
a. hydrolyzed
b. denatured
c. condensed
d. catalyzed
e. dehydrated

18. Which of the following statements about enzymes is *false?* Enzymes:
a. catalyze only a particular reaction of specific substrates
b. usually work only in a particular pH range
c. increase the energy of the reactant molecules
d. work better at moderate temperatures than at very high or low ones
e. bind their substrates and hold them in a particular orientation

THINKING CRITICALLY

1. Science fiction tales sometimes feature life forms based on silicon rather than carbon. Silicon is much more abundant on Earth than carbon, and like carbon its atoms can bond to four other atoms. Bonds between two silicon atoms are unstable in the presence of O_2, but bonds between silicon and oxygen atoms are extremely stable and difficult to break. What implications would these properties have for silicon-based life forms?
2. You go on a journey, taking your cat along but leaving its litterbox at home. Draw a graph to show the rate of hydrolysis of the urea in the litterbox to ammonia and carbon dioxide during your absence (see Section 3-E).
3. During the winter, you are too lazy to take your cat's box outside to empty it, until finally the stench is so overpowering you take the box outside and leave it in the snow. What happens to the rate of hydrolysis of urea by the enzyme urease, and why?
4. When you cut an apple or banana, phenol oxidase enzymes in the injured areas quickly begin a "wound reaction," which results in the cut surfaces turning brown. Good cooks sprinkle lemon juice on sliced fruit to prevent it from discoloring in this way. Why does this work?

SELECTED KEY TERMS

adenosine triphosphate (ATP), *p. 58*
amino acid, *p. 51*
carbohydrate, *p. 41*
catalyze, *p. 56*
condensation reaction, *p. 42*
denatured, *p. 53*
deoxyribonucleic acid (DNA), *p. 58*
enzyme, *p. 56*
functional group, *p. 42*
hydrolysis reaction, *p. 42*
hydrophilic, *p. 46*
hydrophobic, *p. 46*
lipid, *p. 41*
metabolism, *p. 59*
monomer, *p. 41*
negative feedback, *p. 59*
nucleic acid, *p. 41*
nucleotide, *p. 58*
organic molecule, *p. 41*
phospholipid, *p. 47*
polymer, *p. 41*
polypeptide, *p. 52*
protein, *p. 41*
ribonucleic acid (RNA), *p. 58*
substrate, *p. 56*

SUGGESTED READINGS

Dickerson, R. E., and I. Geis. *The Structure and Action of Proteins.* Menlo Park, Calif.: Benjamin-Cummings, 1984. Excellent illustrations, highly readable.

Lehninger, A. L. *Principles of Biochemistry.* New York: Worth Publishers, 1982. A readable text, useful background for many chapters in this part of the book.

McGee, H. *On Food and Cooking: The Science and Lore of the Kitchen.* New York: Charles Scribner's Sons, 1984. Great for browsing.

Scientific American, October 1985 issue, *"The molecules of life."*

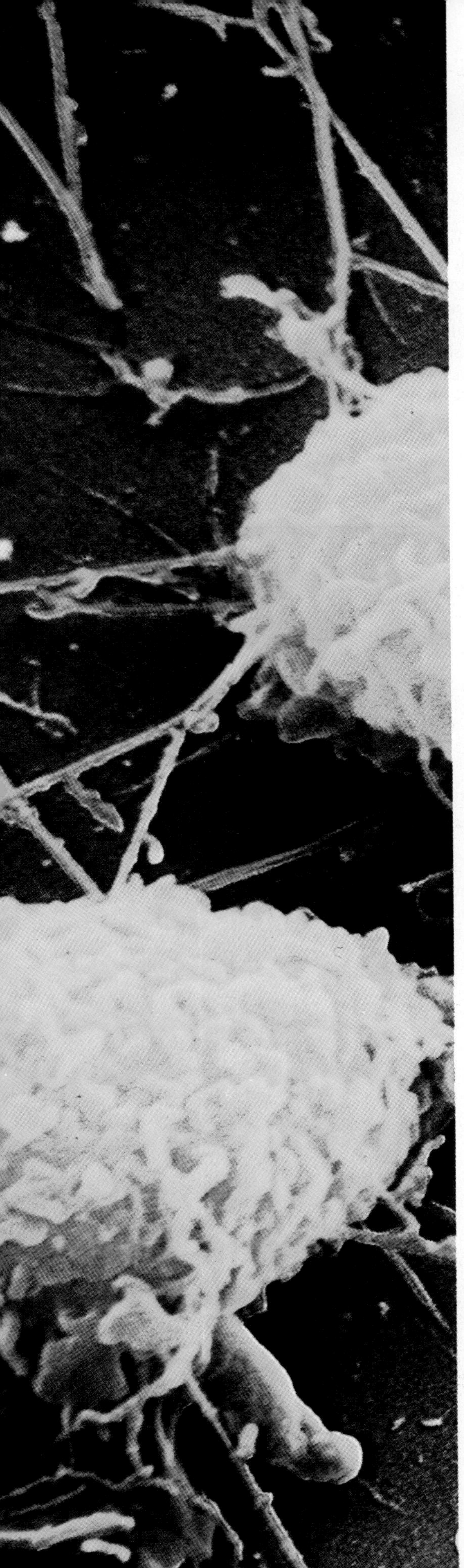

CHAPTER 4

CONCEPT GUIDE

1?

After reading this chapter you should be able to:

1. Describe the arrangement of proteins, carbohydrates, and lipids within a biological membrane and relate this structure to the general functions of biological membranes. (Sections 4–A and 4–B)
2. Compare the processes of diffusion and osmosis. Explain how these processes are affected by the structure of biological membranes. (Section 4–C)
3. Describe the structures and functions of channel and carrier proteins. Compare their mechanisms and their use of energy to move solutes across membranes. (Section 4–D)
4. Compare the processes of endocytosis, exocytosis, and pinocytosis. Describe the importance of each of these processes to cells. (Section 4–E)
5. Relate the structure of tight junctions, desmosomes, gap junctions, and plasmodesmata to their cellular functions. (Sections 4–F and 4–G)

How does the body defend against disease-causing organisms like these green colored bacteria? One way is to use white blood cells (like these colored blue in this photo) that identify and destroy infectious agents. Read more about how cells "eat" in Section 4-E. (Manfred Kage/Peter Arnold, Inc.)

Elephants are the largest living land animals. Male African elephants tower as much as 13 feet (4 meters) at the shoulder and can weigh up to 12,000 pounds (5000 kilograms). Mice are at the opposite end of the size spectrum. The familiar house mouse fits into a teaspoon and usually weighs between 18 and 30 grams. Although it isn't the world's smallest mammal (that distinction goes to the Etruscan shrew, which is about half the size of a house mouse), nevertheless, the house mouse is tiny in comparison to an elephant. Four fully grown mice could comfortably cuddle in a space as big as an elephant's toenail. All of the internal organs of a mouse are correspondingly tiny, and at first thought it might seem logical that the cells of a mouse would be minute while the cells of an elephant would be gigantic: a reflection of the difference in their sizes. But this is not so. The cells of an elephant and the cells of a mouse are about the same size, even though an elephant is thousands of times larger. In this chapter we will discuss cells and their membranes. We will explain this puzzling phenomenon of cellular size, but first we should answer the question, "Why do organisms have cells?" ■

The metabolic reactions of a living organism can take place only in a delicately balanced environment that is different from any found in the nonliving world. Cells are the life-support chambers that contain this special environment. A living cell keeps its chemical composition steady within narrow limits, a condition known as **homeostasis** (homeo = same; stasis = standing). In the controlled environment of a cell, all the activities of life can occur. An organism acquires energy through metabolic processes within a cell, and it then uses this energy to maintain the cell's internal

Cells and Their Membranes

chemical environment, to build organic molecules, to grow, and to reproduce by division into two new cells.

Cells were first scientifically observed and reported in 1655 by the Englishman, Robert Hooke. What Hooke actually saw in the bark of the cork oak tree were empty, dead **cell walls,** lacking the living matter they once contained. Other early microscopists soon observed cells in all kinds of plants. They found similar structures in animals, too, but animal cells were harder to distinguish because they lack the thick cell walls that surround plant cells. Observers also reported the existence of many tiny organisms consisting of only one cell each.

Eventually, biologists recognized the main features of the **cell theory:**

1. All organisms consist of one or more cells.
2. Cells are the fundamental units of life—the smallest entities that can be called "living."
3. Cells arise only by division of existing cells.

Many organisms are **unicellular** (consisting of only one cell). However, a cell's size is limited. The biochemical reactions of its metabolism require raw materials from outside the cell and generate waste products that must be expelled. So, the cell is constantly trading chemicals with its environment. This exchange of raw and waste materials occurs through the cell's outer surface. As a cell grows larger, its surface area, where chemicals are exchanged, grows more slowly than does its volume, the inner portion of the cell where chemicals are used (See Toolbox 4-1). As the cell grows, the surface gets farther from the innermost areas that need chemicals, and it takes longer for things to reach where they are needed. Thus, a cell's size is limited by its metabolic function, and the cells of a mouse and an elephant are roughly the same size because the efficient exchange of substances with the external environment dictates an optimal size for a cell. At some point, the growing cell reaches the upper limit of this optimal size. It then divides into two new cells, each with a lower (and more favorable) ratio of surface area to volume. Large organisms, such as animals and most plants, are **multicellular** (composed of many cells derived by repeated division from one original cell).

All cells must perform certain basic tasks. In addition, each cell of a multicellular organism makes a specialized contribution to the body as a whole. For example, a muscle cell in the heart is specialized to contract and help pump blood. Because it is deep inside the body, it cannot capture its own food or obtain oxygen from the air, but must rely on other specialized cells, such as those of the lungs and blood, to provide the food and oxygen it needs. Thus, there is division of labor among the cells of a multicellular organism.

Most cells have three main parts:

1. The **plasma membrane,** covering the outside of the cell and controlling what enters and leaves. (In plants, this is surrounded by a tough cell wall.)
2. The **cytoplasm** ("cell fluid"), containing water, various salts, and organic molecules, including many metabolic enzymes. The cytoplasm also contains a variety of larger structures, collectively called **organelles,** which perform various tasks. Many of these "little organs" are surrounded by membranes very similar to the plasma membrane.
3. The cell **nucleus** (in bacteria, the **nuclear area**), housing the cell's genetic material (DNA and associated RNA and proteins). The genetic material contains directions for making the cell's proteins.

In this chapter, we examine the structure and function of the plasma membrane. In the next chapter we discuss the cell's other components and look at the differences among the cells of plants, animals, and bacteria.

KEY CONCEPTS

- Cells are the basic structural, functional, and reproductive units of life.
- To remain alive, a cell must maintain homeostasis: keeping its internal chemical composition fairly constant within the narrow limits suitable for life.
- Every cell is surrounded by a membrane, which helps control what enters and leaves the cell.

Curiosity Questions

1? Are the cells of an elephant bigger than the cells of a mouse? (See page 64)

2? How can some cancer cells resist the drugs used to kill them? (See Figure 4-6, page 75)

3? How do the cells in one area of the heart all contract at the same time? (See page 79)

4–A STRUCTURE OF BIOLOGICAL MEMBRANES

The plasma membrane is too thin to be seen with a regular light microscope, such as you might use in the laboratory. In photographs made with more powerful electron microscopes (see *A Journey into Science in Process: Looking at Cells: Microscopy*, this chapter), the membrane appears as a continuous double line around the cell (Figure 4-1). However, long before electron microscopes were invented, biologists had deduced the plasma membrane's existence by observing cell behavior, and they had collected and studied membrane material.

All biological membranes have similar structures and functions, whether they are plasma membranes or membranes of organelles inside the cell. These mem-

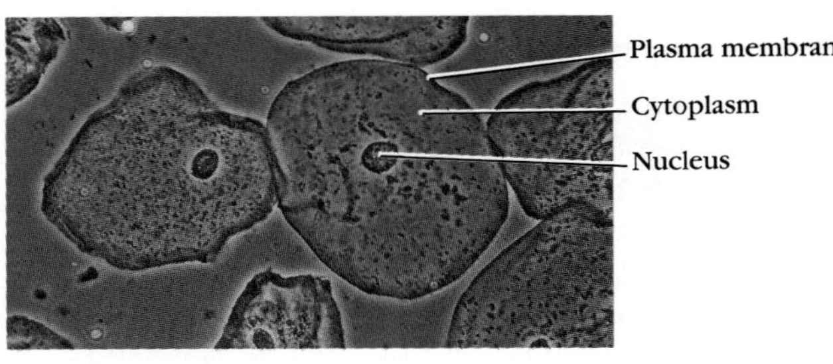

FIGURE 4-1

Human cells obtained by scraping the inside of the cheek with a toothpick. The cells were then placed in a drop of salt solution on a glass slide and viewed through a light microscope. Each cell is surrounded by a plasma membrane and contains a large, oval cell nucleus. The rest of the cell is filled with cytoplasm. (Jim Solliday/Biological Photo Service)

TOOLBOX 4–1

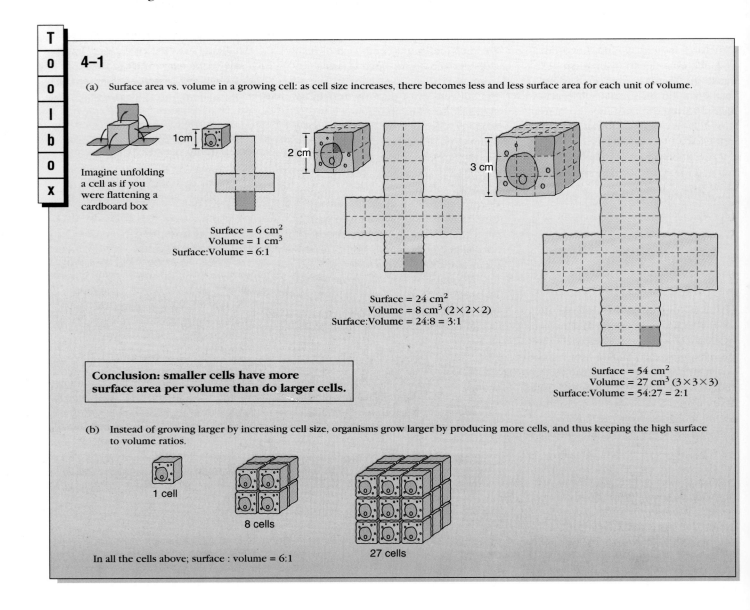

(a) Surface area vs. volume in a growing cell: as cell size increases, there becomes less and less surface area for each unit of volume.

Imagine unfolding a cell as if you were flattening a cardboard box

1cm

Surface = 6 cm^2
Volume = 1 cm^3
Surface:Volume = 6:1

2 cm

Surface = 24 cm^2
Volume = 8 cm^3 (2×2×2)
Surface:Volume = 24:8 = 3:1

3 cm

Surface = 54 cm^2
Volume = 27 cm^3 (3×3×3)
Surface:Volume = 54:27 = 2:1

Conclusion: smaller cells have more surface area per volume than do larger cells.

(b) Instead of growing larger by increasing cell size, organisms grow larger by producing more cells, and thus keeping the high surface to volume ratios.

1 cell

8 cells

27 cells

In all the cells above; surface : volume = 6:1

branes consist mainly of lipids and proteins, which vary from one type of membrane to another.

The main lipids in biological membranes are phospholipids (see Figure 3–4). In addition, plasma membranes usually contain small amounts of glycolipids (lipid + carbohydrate). Membranes of animal cells also contain a lot of cholesterol. All of these are long, asymmetrical molecules with one hydrophilic (polar) end and one hydrophobic (nonpolar) end (Section 3–B). When surrounded by water, these molecules tend to form groups with their hydrophilic heads exposed to the water and their hydrophobic tails huddled together as far from the water as possible. They can do this either by forming spheres or by forming **lipid bilayers** (two layers, each one molecule thick), with the hydrophobic tails sandwiched between the hydrophilic heads (A Journey Through the Plasma Membrane). It is this lipid bilayer arrangement that occurs in biological membranes.

Lipid bilayers are fluid: individual lipid molecules can move about, changing places with their neighbors. This fluidity allows the membrane to stretch under stress and to reseal itself if it is disrupted. The membrane's fluidity also permits some of the proteins to move within it.

A Journey Through the Plasma Membrane summarizes the current model of biological membrane structure. This is called the **fluid mosaic model** because the lipid layers form a three-dimensional fluid that acts as a solvent for various proteins. Some proteins move around in the membrane; others are relatively immobile, some because they are attached to structures in the cell's cytoplasm.

A membrane's inner and outer surfaces contain different molecules and have different functions. Some lipids are more common in one layer or the other, and particular proteins may associate with only one surface. Some proteins that span the entire membrane have distinct inner and outer ends, and must be aligned right to work properly. In addition, many membrane lipids and proteins are combined with carbohydrates to form glycolipids and glycoproteins. The attached carbohydrates are short chains of sugars (oligosaccharides: oligo = few) found only on the outside surface of the plasma membrane (A Journey Through the Plasma Membrane).

BIO-BIT

The longest cells in a human are nerve cells that extend from the base of the spine to the bottom of the foot, over a meter in length!

Glycoproteins on the cell surface act as **receptors,** structures that recognize and bind specific molecules in the cell's surroundings. The molecules to be recognized may be part of another cell's membrane, or hormones or other chemical signals. For example, cell-surface receptors on eggs and sperm bind specifically to each other during sexual reproduction. These receptors are very precise: sperm recognize and fertilize eggs of only their own species (or of closely related species, which produce very similar receptor molecules).

Now that we have seen how molecules are organized into membranes, we can go on and see how the plasma membrane's structure is related to its function.

A biological membrane consists of a fluid lipid bilayer, with various proteins embedded and floating in it. This basic structure has two properties crucial to membrane function. First, lipid bilayers spontaneously form closed compartments. This keeps the solutions inside and outside the membrane separate. Second, the membrane has different lipid and protein molecules on each side of the bilayer. Hence the membrane has distinct "cytoplasmic" and "exterior" sides, each with a different function.

4–B ROLES OF THE PLASMA MEMBRANE

The plasma membrane lies at the frontier between the living cell and its environment. In this strategic location, the membrane performs many vital roles. First, it forms a continuous, closed covering that keeps the cell's contents separate from the external environment. The dual roles of *containment* and *separation* are performed by the lipid bilayer.

Some membrane proteins are enzymes that catalyze chemical reactions as part of the cell's *metabolism.* Others mediate the *exchange* of various substances between the cytoplasm and the external environment and so help to control the cell's chemical homeostasis. The membrane's various receptor molecules carry out the task of *recognition* of other cells

BIO-BIT

By its nature, alcohol has the ability to cross membranes and enter the bloodstream faster than most other molecules that we digest. This is because it is absorbed by the cells of the stomach wall, unlike most nutrients which are absorbed in the small intestine.

A Journey into Science in Process

Looking at Cells: Microscopy

Most cells are too small to be seen without magnification. Much of our knowledge of cells has depended upon the gradual improvement of microscopes since 1590, when Dutch lens grinders Hans and Zacharias Janssen mounted two lenses in a tube to produce the first microscope.

The **compound light microscopes** used today contain two main lenses. The **objective lens,** close to the object being viewed, forms a magnified image of the specimen. The image is further magnified by the **ocular lens,** near the viewer's eye (Figure 4-A).

Surprisingly, the most important factor determining how small an object may be viewed with a microscope is not its magnifying power but its **resolving power,** its ability to distinguish the separateness of two objects that are close together. Without good resolving power, a microscope produces a fuzzy image, and more magnification only produces a larger fuzzy image. The resolving power of a lens system is limited by diffraction (scattering) of light as it passes through the lens opening. Since diffraction at the objective lens enlarges the image of the specimen, small objects close together will have overlapping images that cannot be resolved as separate.

The most important way to increase the resolving power of microscopes is by reducing the wavelength of the light used. Light of short wavelength (such as violet light) is diffracted less than light with a long wavelength (such as red light).

Entire cells and their larger components can be seen with a light microscope, but many smaller cell parts cannot be seen with visible light. Electron microscopes overcome this problem by using beams of electrons, which have shorter wavelengths than visible light. Hence electron microscopes have higher resolving powers than light microscopes.

In **transmission electron microscopes,** invented in the 1930s, electrons pass through the specimen, just as light does in a light microscope. An electron microscope's working parts are much like those of a light microscope assembled upside down (Figure 4-A). The lenses are not glass but electromagnets, which can deflect the negatively charged electrons. A beam of electrons is produced by heating a tungsten filament in the electron gun. The beam is accelerated through the **condenser lens,** which focuses it, and the focused beam then passes through the specimen and the objective lens. The **projector lens** in an electron microscope is the equivalent of the ocular lens in a light microscope. Because our eyes cannot detect electrons, the projector lens focuses the final electron beam onto a photographic film or fluorescent screen, where it produces a visible image.

One disadvantage of transmission electron microscopes is that electrons are easily deflected or absorbed by air molecules or by the specimen itself. For this reason, specimens must be observed in an almost complete vacuum and must also be sliced very thin.

Scanning electron microscopes were invented during the 1930s and 40s and first manufactured in 1963. Here electrons do not pass through the specimen. Instead, the electron beam hitting the object causes atoms at its surface to emit lower-energy **secondary electrons,** which are collected and used to vary the intensity of a spot on a television screen that scans in synchrony with the electron beam. The resolving power is less than with a transmission electron microscope, but the scanning electron microscope has other advantages. First, the specimen needs less preparation. Second, since the microscope has an extraordinary depth of focus, the three-dimensional surface of an intact specimen can be observed in great detail. In addition, some living organisms, such as hardy insects, can withstand the high vacuum of a scanning electron microscope and can be viewed alive.

Electron microscopes certainly do not make light microscopes obsolete. Light microscopes are much better for examining larger biological specimens and can also be used to study live organisms and tissues, which is rarely possible with an electron microscope. Perhaps the most important advantage of a light microscope, though, is that it produces a colored image. Photographs taken through microscopes are called **micrographs.** Light micrographs are usually colored, but electron micrographs are always black and white because electron beams have no color. The colored electron micrographs in this book were produced by coloring a black and white photograph to emphasize certain parts.

A microscope produces an image with light or dark areas because light or electrons pass through some parts of the specimen but are absorbed or deflected by others. Light microscopes also show the specimen's colors. To increase the contrast in the image, most specimens are specially prepared and stained for microscopy. Specimens are first fixed (killed) and then embedded in wax or resin so that they can be sectioned into thin slices with a glass or metal knife.

Stains for microscopy give contrast to the image by absorbing light or electrons. The chemicals used for staining react specifically with certain cell components. For instance, the alkaline proteins attached to DNA in eukaryotic cells react with a blue stain that does not affect the rest of the cell. In this way, genetic material can be stained blue for the light microscope. For transmission electron microscopy, structures are stained with heavy metal ions, which absorb electrons and so produce dark areas in the final image. A specimen might be stained with lead solution, which re-

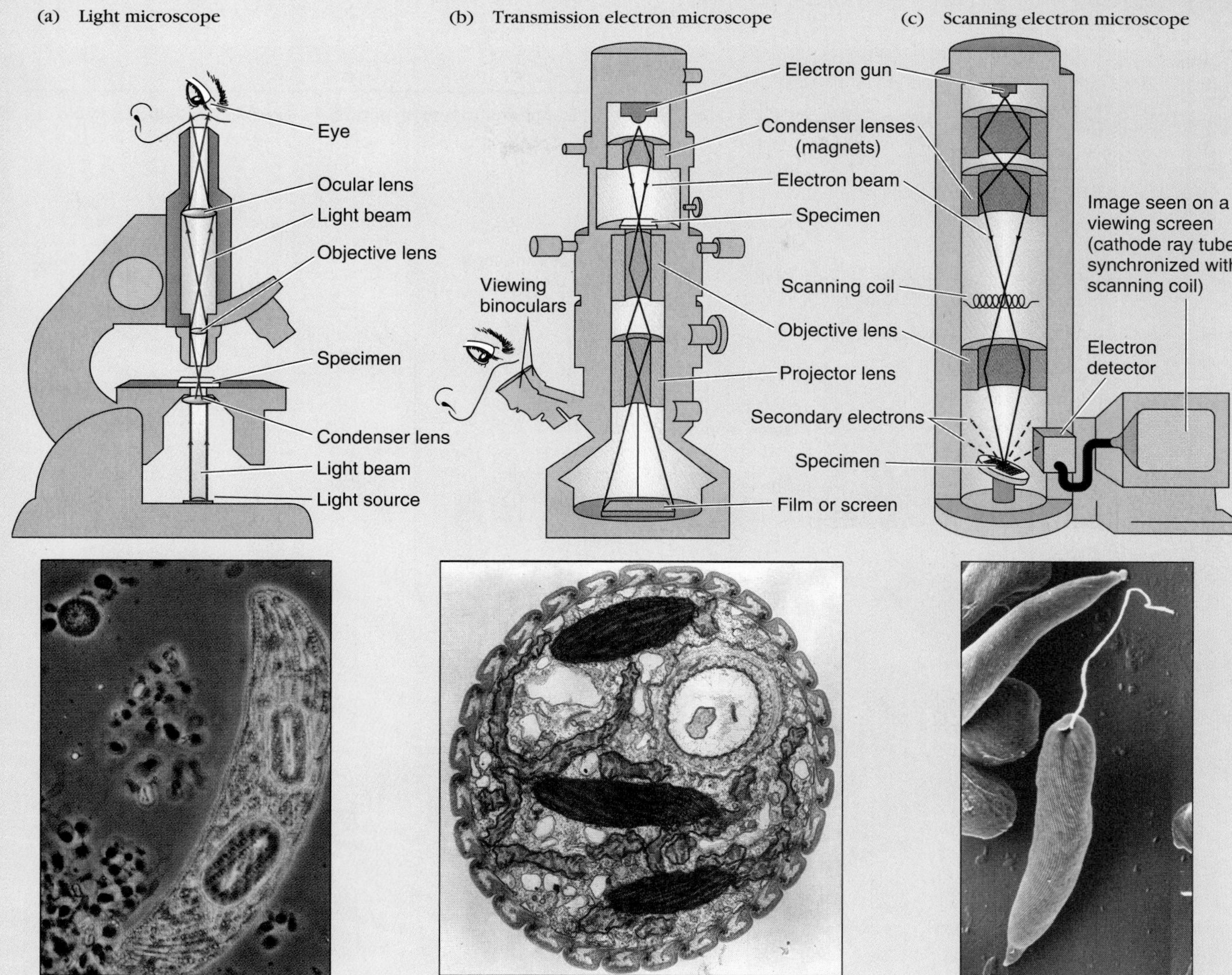

FIGURE 4-A

Microscopes. Comparison of a light microscope **(a)** with a transmission electron microscope, or TEM **(b)**. Electron microscopes use electrons instead of light, and its lenses are electromagnets rather than curved pieces of glass. Because the human eye cannot detect electrons, the image is projected onto a screen or special photographic film. The scanning electron microscope (SEM) in **(c)** also uses electrons and electromagnets. The scanning coils deflect a fine beam of electrons so that it travels rapidly across the specimen, in synchrony with the spot on a cathode ray tube. The user views the specimen as a "television" picture on the screen of the cathode ray tube.

The photographs show views of the protist *Euglena* taken with each kind of microscope. The light micrograph shows a whole cell; the TEM shows a thin cross-section taken by slicing the cell in the light micrograph from left to right; and the SEM shows the cell's outer surface, with its pattern of fine ridges and its flagellum (the white line at the top). (Biophoto Associates)

acts with acid structures to leave a deposit of electron-absorbing lead. For scanning electron microscopy, the whole surface of the dried specimen is coated with a thin layer of gold, platinum, or some other good emitter of secondary electrons. In effect, the viewer "sees" the metal coating, not the specimen itself.

Several types of **scanning probe microscopes** were invented in the 1980s. Instead of aiming light or electrons at a sample, these microscopes feel their way across an object's surface with an extremely fine needle, much as a blind person may explore the surface of the ground with a cane. In many cases, the resolution of these microscopes is good enough to reveal individual atoms. More important, they permit us to look at specimens in new ways, measuring such features as magnetic fields, ion flow, electrical charge, or temperature of the object.

A Journey Through Plasma Membrane Structure

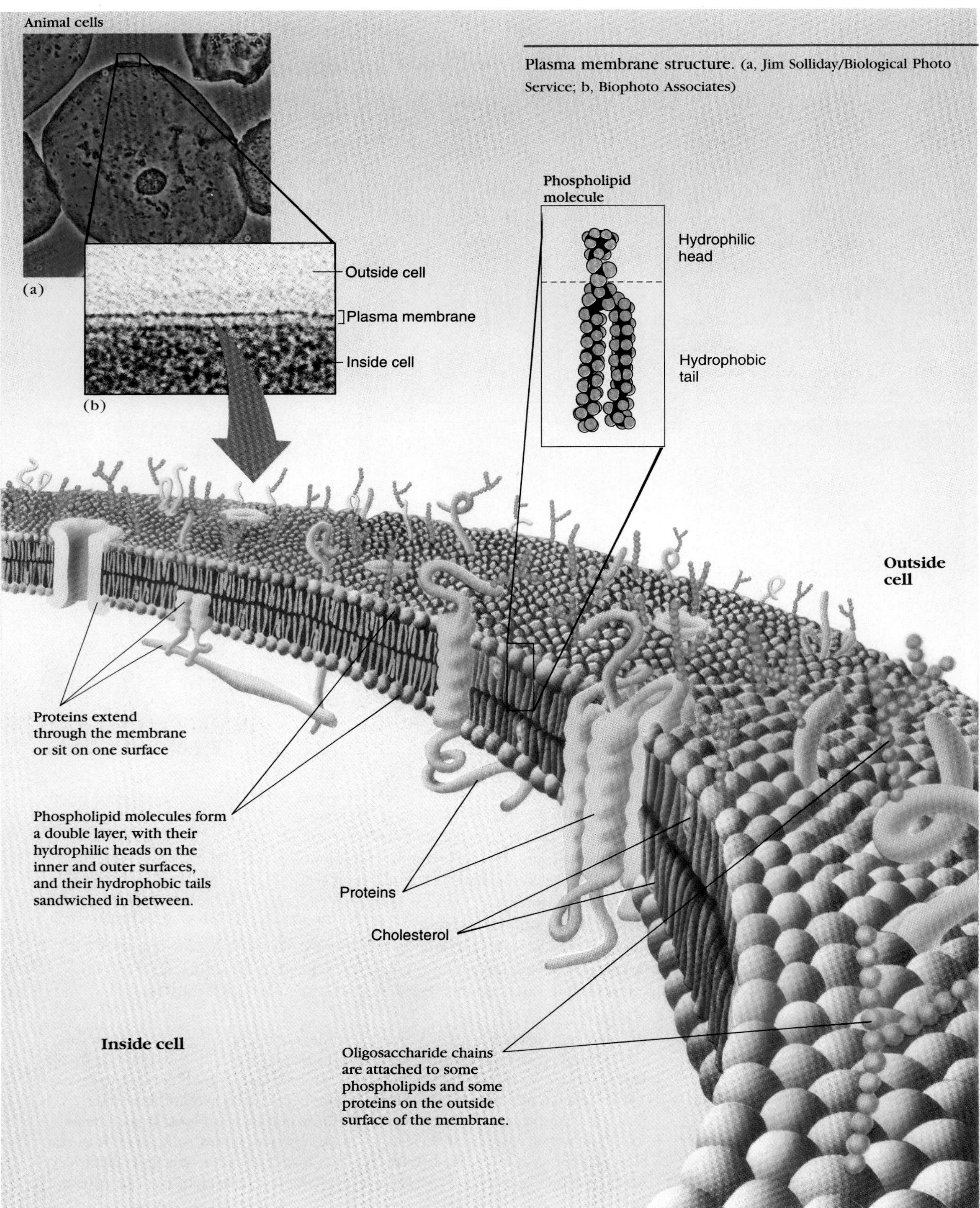
Animal cells
(a)
Outside cell
Plasma membrane
Inside cell
(b)
Phospholipid molecule
Hydrophilic head
Hydrophobic tail
Outside cell
Proteins extend through the membrane or sit on one surface
Phospholipid molecules form a double layer, with their hydrophilic heads on the inner and outer surfaces, and their hydrophobic tails sandwiched in between.
Proteins
Cholesterol
Inside cell
Oligosaccharide chains are attached to some phospholipids and some proteins on the outside surface of the membrane.

Plasma membrane structure. (a, Jim Solliday/Biological Photo Service; b, Biophoto Associates)

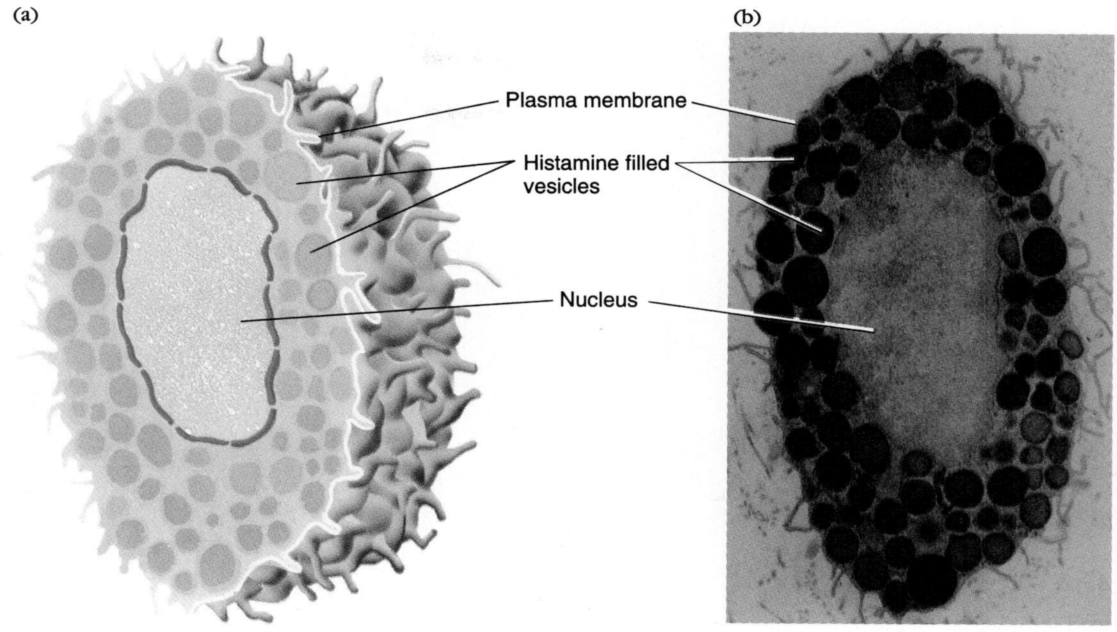

FIGURE 4-2

Mast cell. A mast cell contains many granules of histamine, seen here in numerous vesicles located throughout the cytoplasm. Upon proper chemical stimulation of its plasma membrane, the cell rapidly releases these granules, causing allergy symptoms. (D.W. Fawcett, Visuals Unlimited)

or of substances in the environment. Some receptors also participate in another membrane function, *irritability,* that is, response to stimuli impinging on the cell. Figure 4-2 details membrane function in a mast cell. For example, in nerve and muscle cells, stimulation of certain receptors results in a brief change in the rate at which sodium and potassium ions are permitted to pass through the membrane. This leads to other changes in the cells, causing nerve cells to transmit messages or muscles to contract (Chapters 28 and 29).

We now know that substances cross biological membranes in three distinct ways: some pass straight through the membrane, some are transported by proteins in the membrane, and some move within a sac formed from part of the membrane. We shall consider each of these methods in turn.

The plasma membrane forms a physical barrier around the cell, promotes specific chemical reactions, regulates what enters or leaves the cell, reacts with extracellular materials, and responds to external stimuli.

4–C HOW SMALL UNCHARGED MOLECULES CROSS MEMBRANES

Diffusion

All molecules are in constant, random motion (Section 2-D). Because of this motion each different kind of molecule or ion tends to move from areas where it is more abundant to those where it is scarcer. It is said to move down its **concentration gradient,** which is a gradual decrease in its concentration over a distance.

Consider a sugar cube dissolving in a cup of coffee (Figure 4-3). A sugar molecule moves in one direction until it bumps into another molecule, either another sugar molecule or a water molecule. Both molecules bounce off in new directions. The sugar molecule may bounce back toward the sugar cube, but most of the possible new directions will carry it still farther from the original cube. So, on the whole, sugar molecules tend to move down their concentration gradient, toward areas where they are less concentrated, until they disperse throughout the solution. At the same time, water molecules also move and spread evenly

FIGURE 4-3

Diffusion. Placed into hot coffee, the sugar molecules quickly dissolve and move from an area of high concentration to areas of lower sugar concentration.

throughout the beaker, including the part once occupied by the sugar cube. This process, whereby molecules of two or more substances move about at random and become evenly mixed, is called **diffusion.**

The structure of the plasma membrane does not permit molecules to diffuse through it freely. The membrane is **selectively permeable,** meaning that some substances can pass through it more readily than others. The membrane even prevents the passage of certain kinds of molecules, to which it is **impermeable.** A lipid bilayer's hydrophobic interior makes it relatively impermeable to ions and to many polar molecules. As a result, the plasma membrane prevents most of the water-soluble cell contents from escaping.

Small uncharged molecules cross the plasma membrane by diffusion, each moving down its own concentration gradient. These substances can slip between the hydrophilic heads of the membrane lipids and pass through the bilayer. In essence, these small molecules dissolve in the lipid on one side of the membrane and emerge at the opposite face. For instance, oxygen molecules (O_2) are usually more concentrated outside a cell than in the cytoplasm because the cell constantly uses oxygen. So, on the whole, oxygen molecules dissolve outside of the plasma membrane and eventually reach the cell's interior, where O_2 is less concentrated. Carbon dioxide, produced when the cell uses oxygen, leaves the cell by the reverse route.

How fast a substance can diffuse through the lipid bilayer depends on its solubility in lipids and its molecular size. Small, nonpolar molecules such as oxygen, nitrogen (N_2), and ether (see Figure 2–4) cross membranes rapidly. Uncharged polar molecules also cross the lipid bilayer rapidly if they are small enough. For example, urea (the main waste product in human urine) and ethyl alcohol (from alcoholic beverages) cross rapidly. Glycerol (an antifreeze), which is also uncharged but larger, crosses much more slowly, and the sugar glucose, twice the size of glycerol, hardly crosses a lipid bilayer at all. The bilayer is also virtually impermeable to ions, even such small ones as hydrogen, sodium, and potassium. This is partly because of the ions' electric charge, and partly because ions are surrounded by a layer of water molecules, which in effect makes them much larger (see Figure 2–9).

Osmosis

The most abundant substance in a typical cell is water, the solvent in which most of the cell's other molecules are dissolved. Because water does not dissolve readily in lipid, it is somewhat surprising that water crosses lipid bilayers quite rapidly. This is partly because of the water molecule's small size, but it may also be that the molecule's unique bipolar structure (see Figure 2–7)

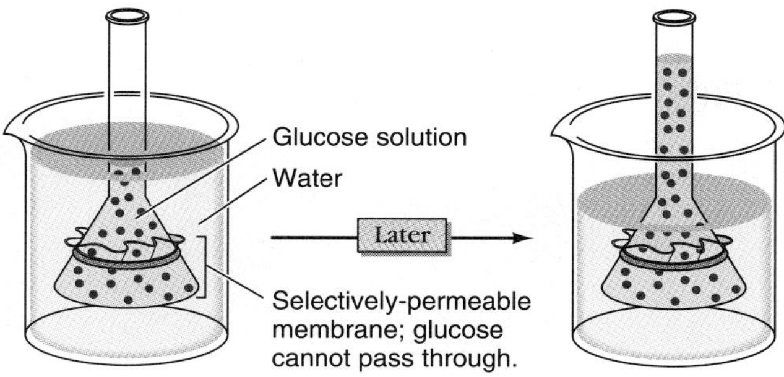

FIGURE 4-4

A funnel filled with glucose solution and covered by a selectively permeable membrane is inverted into a beaker of pure water. Only water can pass through the membrane. What will happen next? Water moves through the membrane into the funnel by osmosis, and the water level within the flask rises as the glucose solution becomes more dilute.

somehow permits it to pass the bilayer's hydrophilic outer layers especially easily.

Osmosis is the process by which water moves through a selectively permeable membrane. Osmosis is a special case of diffusion because it involves the movement of a *solvent* (water) rather than a solute, and because the water is moving *through a membrane.* In osmosis, the net diffusion of water through the membrane is down its concentration gradient: from a weak, or dilute, solution, into a strong, or concentrated solution—that is, from higher to lower concentration of *water.*

A simple way to demonstrate osmosis is to separate distilled (pure) water from an aqueous solution with a membrane permeable to water but not to the solute (Figure 4–4). As time passes, the volume of the solution increases and the volume of the distilled water decreases. Some of the water molecules move by osmosis from the pure water, through the membrane, and into the solution.

Cells as Osmotic Systems

Cells are osmotic systems. A living cell has a selectively permeable plasma membrane, which encloses the cell's internal solution of various particles dissolved in water. To remain alive, the cell must have a covering of water, which also contains solutes. If the internal and external solutions are in osmotic balance, no net exchange of water occurs between them, and the cell is said to be living in an **isotonic** (iso = same; tonus = tension) solution. Solutions that are isotonic with the body fluids are used for such purposes as washing contact lenses and injecting drugs into the bloodstream.

If the solution outside the cell is more concentrated, so that the cell loses water to its environment, this external solution is said to be **hypertonic** (hyper = exceeding) to the cell contents. And if the cell is placed in a solution dilute enough for the cell to gain water from outside, this environment is said to be **hypotonic** (hypo = lower) to the cell. Plant cells in a hypotonic solution can gain only a limited amount of water before their rigid cell walls prevent them from expanding further. After this, the presence of the cell wall squeezes water back out as fast as it enters. In the same hypotonic solution, an animal cell may swell so much that it bursts (Figure 4–5).

Many animals live in fresh water, which is hypotonic to their cells' cytoplasm. Why don't these animals take up so much water by osmosis that they swell up and burst? Most of a freshwater animal's body surface is covered by a layer of rather impermeable material, which retards water uptake. Such layers include the mucus of fish and worms or the chitinous armor of aquatic insects and spiders. In addition, these organisms have well-developed excretory systems that produce large volumes of very dilute urine, thereby ridding their bodies of excess water (Chapter 26).

Because water moves freely across the plasma membrane by osmosis, a cell can control its water content only indirectly. The cell can create a difference in osmotic potential across its membrane by moving solutes from one side of the membrane to the other (as we shall see in the next section). Water then enters or leaves the cell, moving by osmosis toward the side of the membrane with the lower osmotic potential.

Molecules tend to diffuse from an area of higher concentration to an area of lower concentration. However, plasma membranes do not permit molecules to diffuse freely and some molecules pass through faster than others. How quickly a molecule moves through the lipid bilayer of a membrane depends upon its solubility in lipids, its molecular size, and the difference in concentration. In osmosis, water diffuses across semipermeable membranes from areas of high to low water concentration. Net movement of water into or out of a cell depends on the solute concentrations on both sides of the membrane.

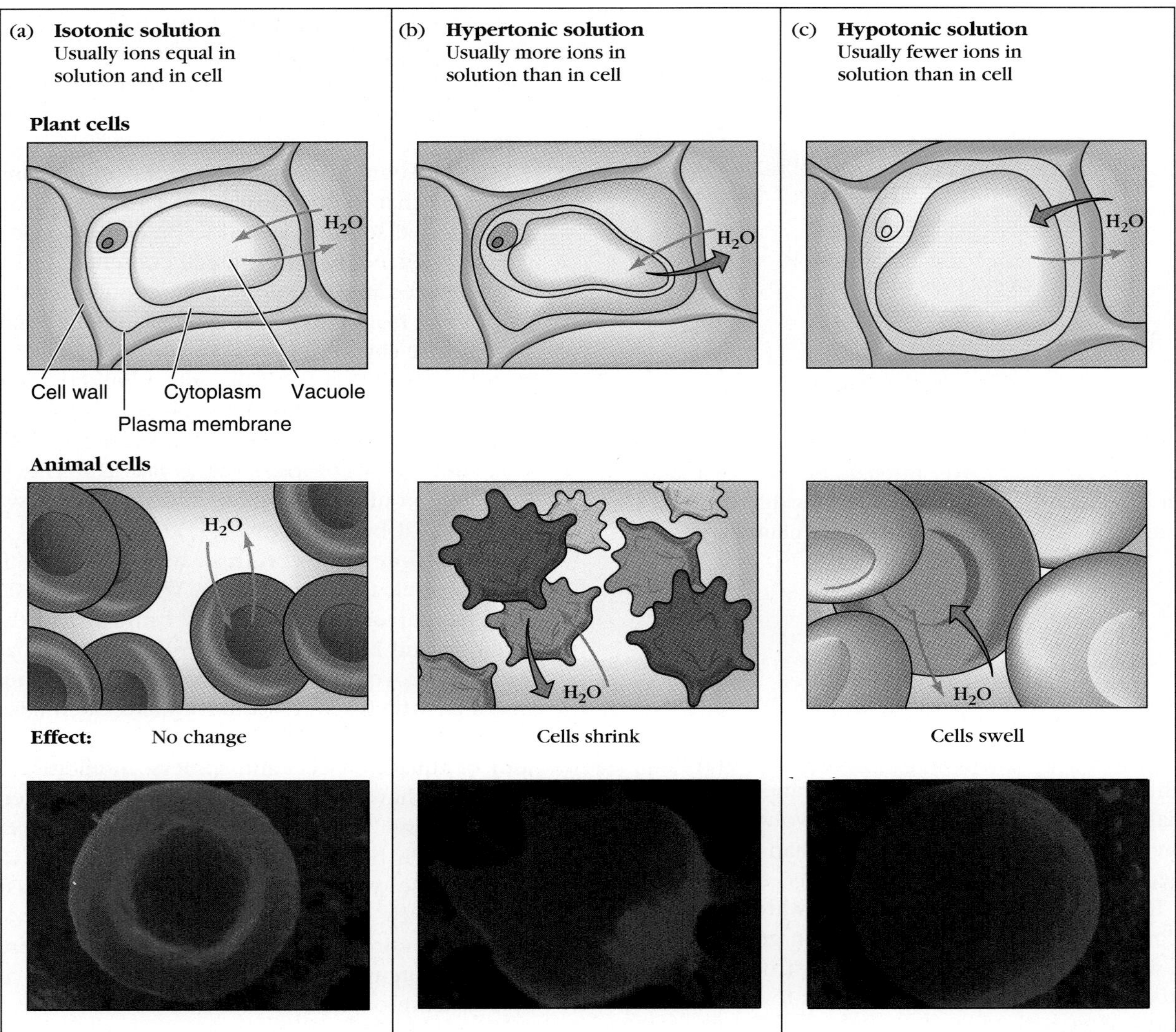

FIGURE 4-5

The effects of isotonic, hypertonic, and hypotonic solutions on plant and animal cells. **(a)** Isotonic solutions result in no net movement of water in or out of a cell, and the cells maintain a stable size. **(b)** Hypertonic solutions cause the cell to shrink. In plant cells, the plasma membrane may even pull away from the cell wall. **(c)** In hypotonic solutions, the net movement of water is inward, as the cells expand. Plant cells become tightly packed within the confines of the cell wall, but animal cells can easily burst in hypotonic solutions. (David M. Phillips/Visuals Unlimited)

4–D TRANSPORT BY MEMBRANE PROTEINS

Many molecules cross biological membranes rapidly even though they are not very soluble in lipid. Examples include various small ions, glucose, and amino acids. These substances are transported by membrane transport proteins either down their concentration gradients using **passive transport** or against their concentration gradients using active transport. All transport proteins extend entirely through the lipid bilayer (or are part of protein complexes that do). Each transport protein is also specific for only one or a few chemically similar substances. Here we consider three

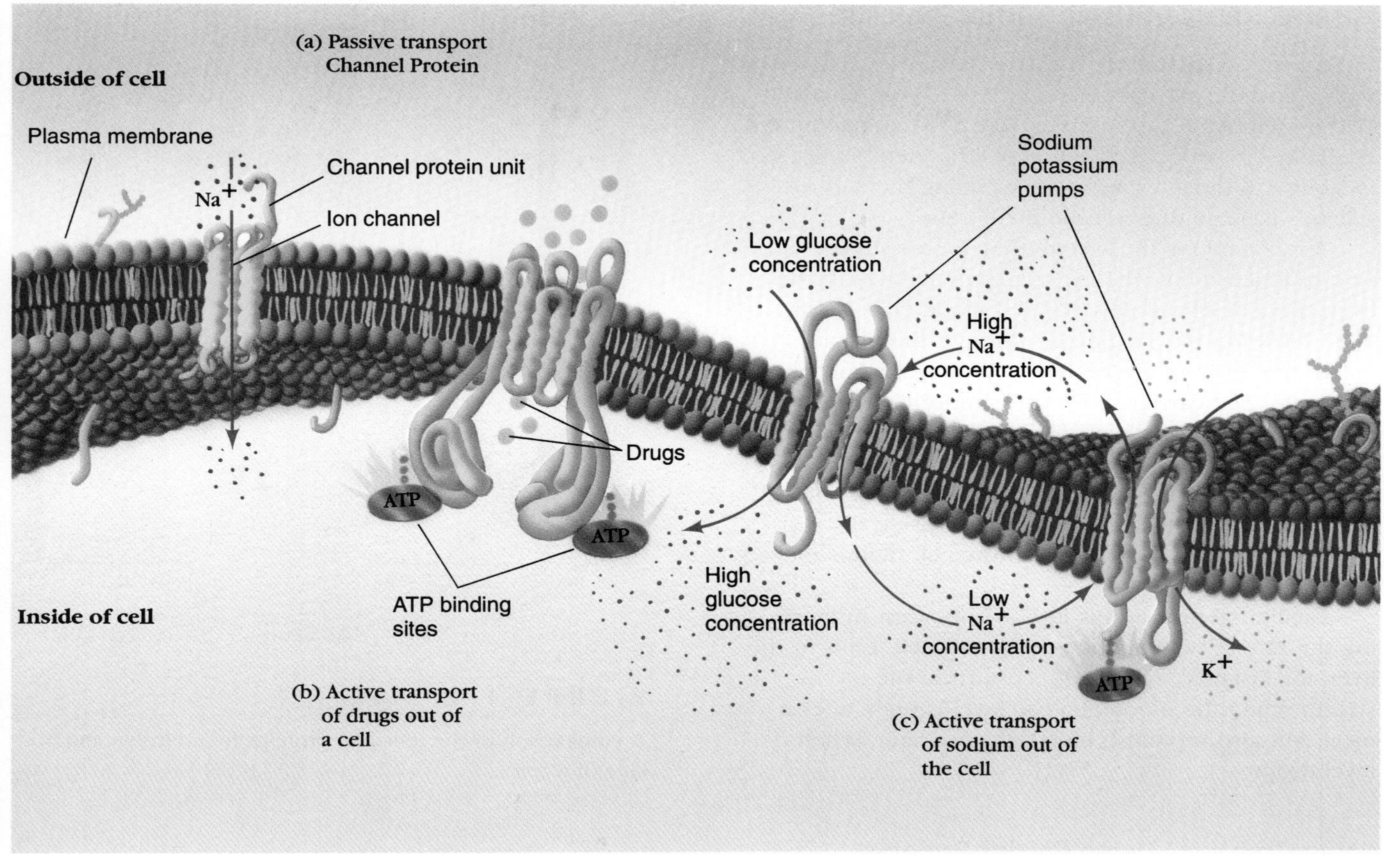

FIGURE 4-6

How molecules get into and out of a cell. (a) Some small molecules simply move down their concentration gradient through special membrane proteins that form channels through the membrane. **(b)** Some cells are resistant to certain drugs because they are able to use carrier proteins and energy to actively move the drugs out of the cell. **(c)** The sodium-potassium pump is a special carrier protein that uses energy to move sodium out of the cell and bring potassium back into the cell. ■

important types of transport proteins, which have some features in common. First, all transport proteins extend entirely through the lipid bilayer (or are part of protein complexes that do). Each transport protein is also specific: it carries only one or a few chemically similar substances.

Passive Transport Using Channel Proteins

Perhaps the simplest case occurs where proteins form channels through the lipid membrane. These tunnel-like structures permit molecules that are soluble in water to pass through the membrane by simple diffusion, avoiding the membrane's hydrophobic interior.

While some channels are open all the time, others behave as if they have "gates" that open and close. These **gated channels** let certain substances through only when the gates are open. Hence the membrane's permeability changes from time to time, depending on whether the gates are open or not. The opening and closing of gates are thought to involve changes in the shape of the channel proteins in response to some signal (Figure 4-6a). This feature is a vital part of the membrane's irritability—its ability to react to stimuli. It plays a key role in, among other things, the working of nerves and muscles (Chapters 28 and 29).

Passive Transport Using Carrier Proteins

In **facilitated diffusion,** a carrier protein combines with a specific substance and moves it from one side of the membrane to the other, down its concentration gradient. This effectively increases the membrane's permeability to the substance, by speeding the substance's passage through the membrane.

An example is the system that speeds up the diffusion of glucose into the cells of some tissues. In the liver, the lens of the eye, and red blood cells, facilitated diffusion moves glucose through the plasma membrane in both directions by means of a carrier molecule. The carrier molecule is more likely to bind a glucose molecule on that side of the membrane where glucose is more plentiful. When the cell is using glucose quickly, the glucose concentration inside the cell falls. Glucose is then more plentiful outside the cell, and more glucose is moved into the cell than is moved out.

Facilitated diffusion is just as important in increasing the rate at which glucose leaves a cell. Cells in the liver, for instance, remove glucose from the bloodstream when the blood glucose level is high, after a meal, but also replenish blood glucose later, when its level drops.

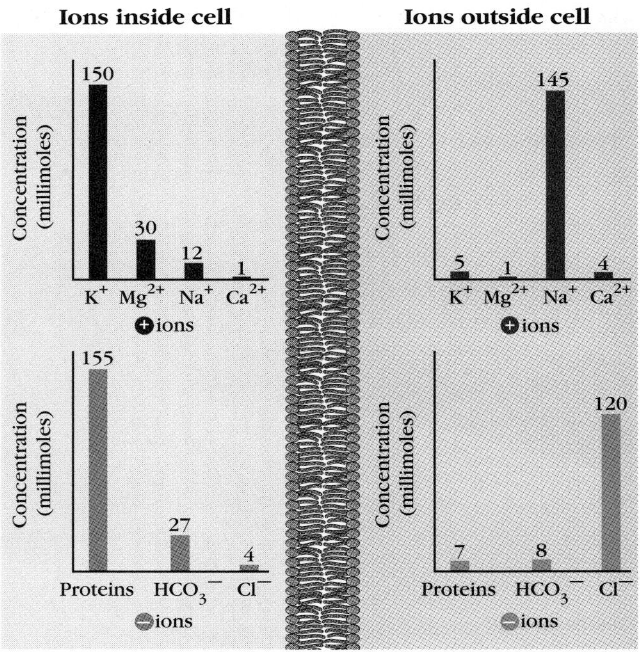

FIGURE 4-7

A comparison of the ions and proteins found inside and outside of a cell.

Active Transport Using Carrier Proteins

Active transport differs in two ways from the two passive transport mechanisms discussed above. First, active transport can move substances against their concentration gradients. Second, active transport requires energy, usually provided by the nucleotide ATP (see Chapter 3) or by a concentration gradient of ions.

The plasma membranes of many cells contain "calcium pumps," which actively transport calcium ions (Ca^{2+}) out of the cell and so keep the Ca^{2+} concentration much lower inside than outside the cell. Some cells in the stomach wall secrete stomach acid via another active transport pump. It uses energy to export hydrogen ions (H^+) from the cell into the stomach fluid against a concentration gradient of about a million to one!

Still another active transport pump occurs in some cancerous cells that are not harmed by chemotherapy (Figure 4-6b). Drug-resistant malaria parasites have a similar pump that rids them of the antimalaria medicine chloroquine.

By far the most important active transport mechanism, though, is the **sodium-potassium pump,** often called the **sodium pump.** It uses energy from ATP to expel sodium (Na^+) from the cell and bring potassium (K^+) in, moving both ions against their concentration gradients (Figure 4-6c). Transport of these ions is linked: the pump moves three Na^+ ions out of the cell for every two K^+ ions that it moves in.

The sodium-potassium pump is enormously important to cells. An estimated one third of all our energy goes to power this pump! The ability of nerves to conduct electrical impulses, of muscles to contract, of the digestive tract to absorb sugars and amino acids from food, and of the kidneys to form urine all depend on the working of this ionic pump. Cells also use the sodium-potassium pump to control their water content. The pump adjusts the concentration of sodium and potassium ions inside and outside the cell by active transport, and water then moves passively through the plasma membrane by osmosis (Section 4-C).

All of the active transport pumps just mentioned use energy from ATP. Other active transport mechanisms use a different source of energy: the movement of ions down a steep gradient across the membrane (Figure 4-7). In our own cells, the sodium-potassium pump builds a steep gradient of Na^+ across the plasma membrane. The Na^+ gradient, in turn, powers the active transport of glucose and amino acids into cells. For example, glucose is transported into cells of the intestine and kidney when sodium and glucose outside the cell bind to sites on a membrane carrier protein. This binding causes the carrier to change shape, pushing the sodium and glucose into the cell. The rate of glucose transport depends on the sodium gradient, which in turn depends on how much sodium the sodium-potassium pump has pumped out of the cell (Figure 4-6c).

The sodium-potassium pump also maintains the cell's **membrane potential,** the difference in electrical charge across the plasma membrane. The inside of the membrane is negatively charged compared to the outside. Most of the negative charge inside the cell comes from proteins and other organic molecules too large to escape through the membrane. Most of the positive charge outside the membrane comes from Na^+ pushed out of the cell by the sodium-potassium pump.

Because of the relatively impermeable lipid bilayer, most ions and polar molecules can cross a membrane only with the aid of the membrane's transport proteins. These substances are transported down their concentration gradients using passive transport or against their concentration gradients using active transport. Channel proteins form gated aqueous channels that permit ions to diffuse through the membrane. Carrier proteins bind to the substances before transporting them across the membrane. Some carrier proteins are used in facilitated diffusion to move a solute down its concentration gradient, while other carrier proteins are used in active transport to move a solute against its concentration gradient. The sodium-potassium pump, powered by ATP, pumps sodium out of a cell and potassium in. The sodium gradient then powers active transport of sugars and amino acids into the cell. The sodium pump also regulates the cell's water content indirectly by actively transporting sodium and potassium. Water follows the solutes passively by osmosis.

4–E MEMBRANE TRANSPORT OF LARGE PARTICLES

We have now seen how small molecules and ions can cross biological membranes. Sometimes a cell must take in or expel very large molecules, or even bigger particles of matter. Items of this size cannot pass through membranes either by penetrating the lipid bilayer or via membrane proteins. However, the transfer of such large particles is possible because of the membrane's fluid nature: the membrane can change shape and fuse with, or pinch off, small membrane-enclosed sacs. When this happens, the membrane automatically seals itself.

In the process of **endocytosis,** cells take in material from their surroundings. The unicellular organism *Amoeba* feeds by extending projections called pseudopods from the cell body and surrounding the food (Figure 4-8a). The pseudopods then fuse so that the food ends up inside the *Amoeba* in a membranous sac called a **vacuole.** There are three main kinds of endocytosis and the kind just described is called **phagocytosis** ("cell eating")—the engulfing of large particles, such as an entire bacterium or a fragment of a disintegrating cell, into a vacuole. Phagocytosis is a major feeding method of many unicellular organisms and simple multicellular animals (Figure 4-8b).

In most animals, phagocytosis rids the body of debris such as dead cells and also plays a part in defense against disease. For example, some human white blood cells are phagocytes, which engulf and digest invading bacteria. Some of the body's cells normally undergo "programmed death," and their remains are also cleaned up by phagocytes. In your body, phagocytes remove 100 billion worn-out red blood cells per day.

Phagocytosis of, say, a bacterium involves specific binding of protein receptors on the phagocyte's plasma membrane to complementary molecules on the bacterium's surface. Some strains of *Streptococcus* ("strep") bacteria surround themselves with a capsule of carbohydrates that inhibits binding and ingestion by phagocytes; these strains are apt to cause illness, whereas phagocytes easily dispose of strains lacking the capsule, before they can cause disease. In contrast, the bacteria that cause leprosy and tuberculosis are easily engulfed but have adaptations making them resistant to being digested by phagocytes.

Many cells take in some of the external fluid, and whatever solutes it contains, by a second kind of endocytosis called **pinocytosis** ("cell drinking").

In contrast, the third kind of endocytosis, **receptor-mediated endocytosis,** takes up small particles selectively. The particles first bind to specific protein receptors on the plasma membrane. Next, the membrane forms a depression around many of these receptors and the molecules they have bound (Figure 4-9). The membrane then pinches off a small sac, containing the loaded receptors, into the cytoplasm.

Materials can be released from cells as well as engulfed. In **exocytosis,** the membrane of an internal sac or vacuole fuses with the plasma membrane, which then opens and allows the sac's contents to escape from the cell. Substances released in this way may be indigestible food particles, or secretions such as hormones.

Figure 4-9 shows how a patch of plasma membrane becomes a vacuolar membrane in the cytoplasm by endocytosis and eventually returns to the plasma membrane during exocytosis. Likewise, membrane material that merges with the plasma membrane during exocytosis is reclaimed by endocytosis and recycled into more vacuoles. This membrane-recycling process is called the endocytic cycle.

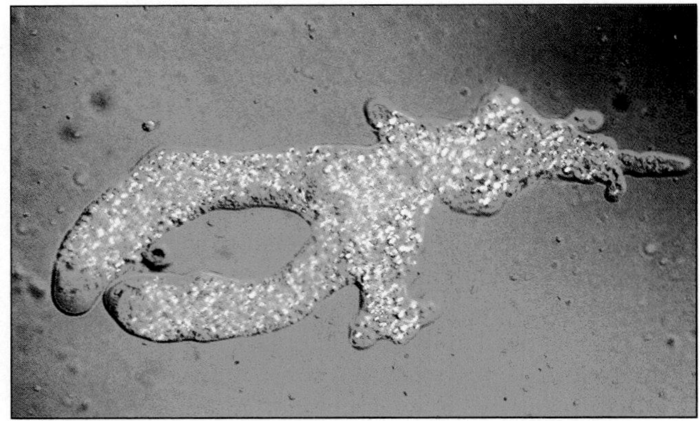

(a)

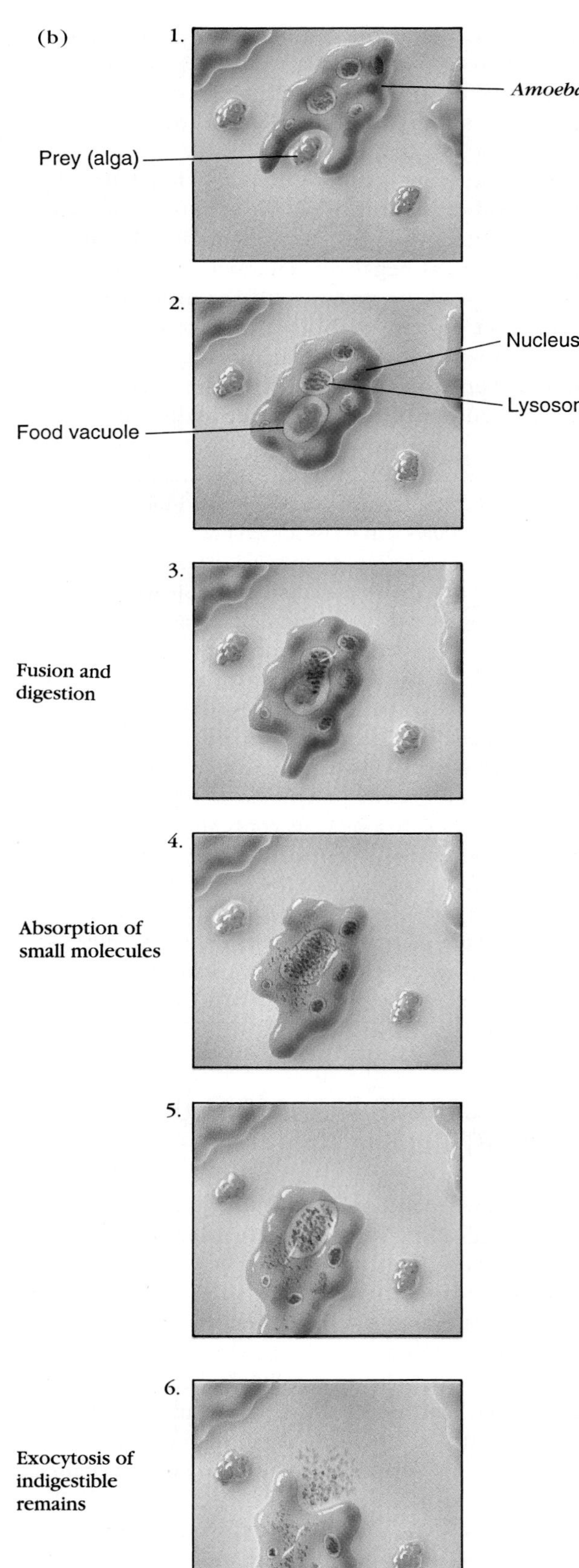

FIGURE 4-8

***Amoeba* feeding. (a)** An *Amoeba* engulfing some algae. **(b)** Phagocytosis is a form of endocytosis. An *Amoeba* engulfs its prey, and part of the plasma membrane pinches off to form a food vacuole inside the cell. The vacuole fuses with a lysosome, a membrane-enclosed sac full of enzymes that digest the prey. The resulting small food molecules are absorbed into the cytoplasm before the indigestible remains are expelled by exocytosis. (M. Abbey/Visuals Unlimited)

Molecules and particles too large for membrane transport proteins move into a cell by endocytosis. The cell's plasma membrane surrounds the particles and pinches off to form a vesicle inside the cell. Materials can be discharged from a cell by the reverse process of exocytosis. A segment of membrane may cycle between various cellular compartments and the plasma membrane by endocytosis and exocytosis.

4–F MEMBRANE ATTACHMENTS BETWEEN CELLS

The many cells of an animal's body are held together in various ways. Cells commonly produce protein fibers and polysaccharide chains, export them by exocytosis, and attach them to the meshwork around themselves using some of their membrane proteins. The plasma membranes of adjacent cells may also attach to each other.

Tight junctions are areas where plasma membranes of animal cells are sealed together, forming such a tight barrier that even small molecules cannot

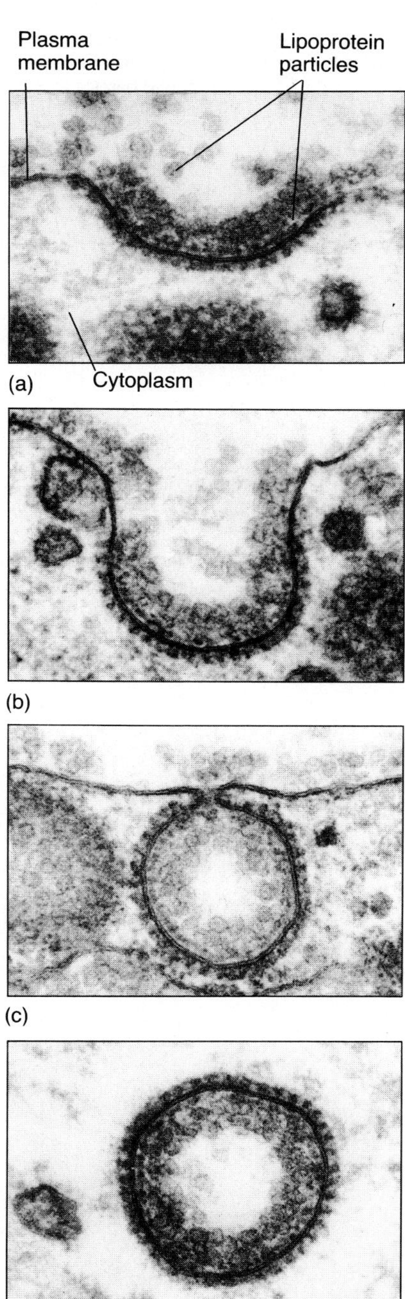

FIGURE 4-9

Receptor-mediated endocytosis. This sequence of photographs is thought to show the stages by which the yolk of a forming chicken egg takes in lipoprotein material. **(a)** Lipoprotein particles bind to receptors in the plasma membrane, which **(b)** folds inward and **(c)** pinches off, **(d)** becoming a separate membrane-enclosed sac (a vesicle) in the cytoplasm. (M.M. Perry and A.B. Gilbert, *J. Cell Sci.* 39:257-272, 1979)

move through the spaces between the cells (A Journey Through Cell Junctions).

Even stronger attachments between cells are provided by **desmosomes,** which occur in regions of high mechanical stress. In these areas, circular patches of the membranes are held together by the interaction of proteins that extend through each membrane into the space between the cells. On the cytoplasmic side of each membrane is a dense plate of proteins that provides mechanical support (A Journey Through Cell Junctions).

One place where tight junctions and desmosomes occur is between the cells lining the small intestine. These attachments keep food from seeping indiscriminately into the body by slipping between cells. Instead, the cells of the lining selectively absorb food and pass it to the bloodstream.

The plasma membranes of adjacent cells may attach to each other. Tight junctions between some animal cells seal the membranes together and prevent the seepage of substances between cells. Desmosomes provide additional mechanical strength by firmly attaching the membranes of adjacent cells.

4–G COMMUNICATION BETWEEN CELLS

Substances often have to pass from one cell to another. This happens when a substance moves out through the membrane of one cell and then in through the membrane of its neighbor. However, it is much faster to have a direct bridge linking the cytoplasm of the two cells, with no membranes to cross. In many cases we do find such connections.

3? Many kinds of animal cells have direct cytoplasm-to-cytoplasm connections at **gap junctions.** These are areas where an array of protein "pipes" permits ions and small molecules to pass directly from cell to cell without leaking into the space between them. Each pipe consists of a ring of six roughly cylindrical proteins that sticks through the plasma membrane and butts against a similar pipe in the adjacent cell's membrane (A Journey Through Cell Junctions). Heart or cardiac muscle cells have many gap junction tunnels that allow the electrical currents of moving ions to pass quickly between them. The gap junctions between cardiac muscle cells help coordinate contractions, allowing all of the muscle cells in certain areas of the heart to contract simultaneously. ■

In plants, neighboring cells are often connected by thousands of strands of cytoplasm called **plasmodes-**

A Journey Through Cell Junctions

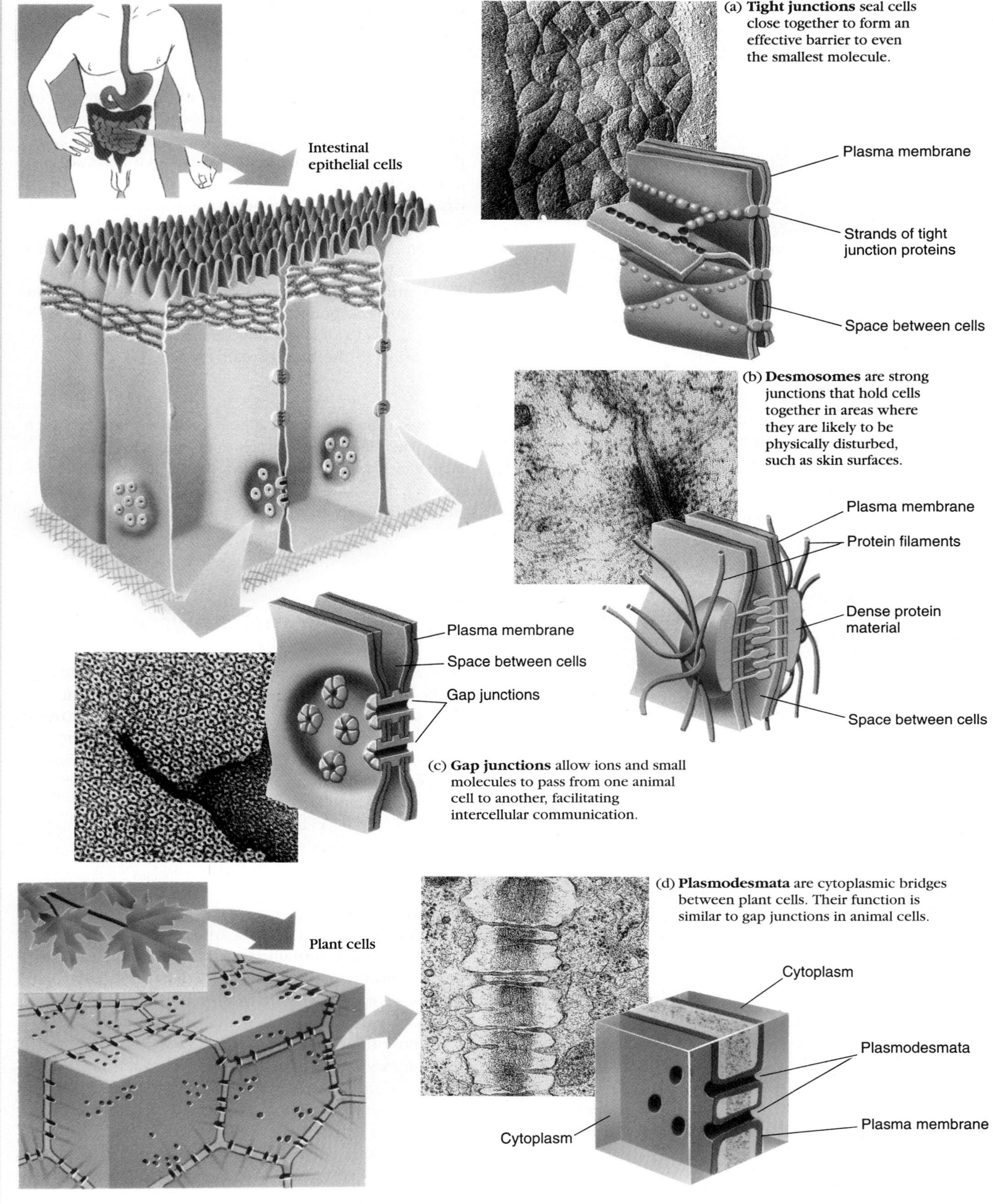

(a) **Tight junctions** seal cells close together to form an effective barrier to even the smallest molecule.

(b) **Desmosomes** are strong junctions that hold cells together in areas where they are likely to be physically disturbed, such as skin surfaces.

(c) **Gap junctions** allow ions and small molecules to pass from one animal cell to another, facilitating intercellular communication.

(d) **Plasmodesmata** are cytoplasmic bridges between plant cells. Their function is similar to gap junctions in animal cells.

◀ Intercellular junctions. (a, Hull, Staehelin, Fawcett/Visuals Unlimited; b, Farquhar, Palade, Fawcett/Visuals Unlimited; c, Gilulap, Fawcett/Visuals Unlimited; d, Visuals Unlimited)

mata (singular: **plasmodesma**). These cytoplasmic bridges pass through openings in the cell walls between the two cells, and both the cytoplasm and the plasma membranes of these cells are essentially continuous with each other (A Journey Through Cell Junctions).

A gap junction consists of many protein "pipes" that directly connect the cytoplasm of adjacent cells, permitting ions to move quickly between the cells. In plants, plasmodesmata provide direct cytoplasm-to-cytoplasm connections between cells.

SUMMARY

1. Organisms are composed of one or more cells, the units of life. Under the general direction of its genetic material, a cell maintains chemical homeostasis, carries on its metabolism, and usually reproduces by dividing into two cells. While a cell is conducting these internal affairs, it must also interact with its environment.
2. The plasma membrane is responsible for interactions between the cell and its environment. It *(a)* serves as a barrier between the cell and its environment, *(b)* regulates what enters or leaves the cell, and *(c)* detects and responds to changes in its surroundings.
3. A biological membrane consists of a fluid lipid bilayer, with various proteins embedded and floating within it. This basic structure has two properties crucial to membrane function:
   a. Lipid bilayers spontaneously form closed compartments, thereby separating the solutions inside and outside the membrane.
   b. The membrane has different lipid and protein components on each side of the bilayer. Hence the membrane has distinct "cytoplasmic" and "exterior" sides, each with different functions.
4. Biological membranes are selectively permeable. Most are freely permeable to water and to small, lipid-soluble molecules, which diffuse through the lipid bilayers down their concentration gradients. Ions and polar molecules cannot cross the lipid bilayer. These solutes are transported through cell membranes by membrane transport proteins either down their concentration gradients by passive transport, or against their concentration gradients by active transport. Three types of membrane transport proteins are recognized based upon whether they bind to the solute and whether they use energy:
   a. Channel proteins form aqueous channels through the membrane; they do not bind to solutes and they allow passive transport.
   b. Passive transport carrier proteins bind specific solutes and passively move them through the membrane, down the solute's concentration gradient.
   c. Active transport carrier proteins also bind specific solutes but move them through the membrane against their concentration gradient using energy (ATP) or a steep concentration gradient of ions.
5. The sodium-potassium pump is a specific example of an active transport carrier protein. Using ATP, it pumps Na^+ out of a cell and K^+ in. The Na^+ gradient then powers active transport of sugars and amino acids into the cell. The sodium pump also regulates the cell's water content by actively transporting these ions through the plasma membrane. By osmosis, water passively follows the solutes, moving toward the area with lower osmotic potential.
6. A cell takes in macromolecules or particles by endocytosis. The membrane surrounds the particle and pinches off to become an intracellular vacuole. Substances can be discharged from many cells by the opposite process of exocytosis. The membrane material cycles between the plasma membrane and some membranous compartments in the cytoplasm by endocytosis and exocytosis.
7. The plasma membranes of adjacent cells may interact to form cell junctions that vary in function.
   a. Tight junctions between some animal cells seal their membranes together and prevent seepage of substances between the cells.
   b. Desmosomes provide mechanical strength by attaching membranes of adjacent cells.
   c. Gap junctions are "molecular pipes" through plasma membranes of adjacent animal cells. They provide for direct transfer of ions from cell to cell.
   d. Plasmodesmata provide direct transfer between the cytoplasm of adjacent cells in plants.

SELF-QUIZ

1. The U-shaped tube in the figure below is divided by a membrane that is impermeable to starch but permeable to water. A 10% starch solution is put into the right-hand half of the tube and an equal amount of 6% starch solution is put into the left-hand half of the tube. In this situation, which of the following occurs?
   a. water will move from the right to the left
   b. water will move from the left to the right
   c. starch will move from the right to the left
   d. water will move in both directions, but more from left to right than right to left
   e. water will move in both directions, but more from right to left than left to right

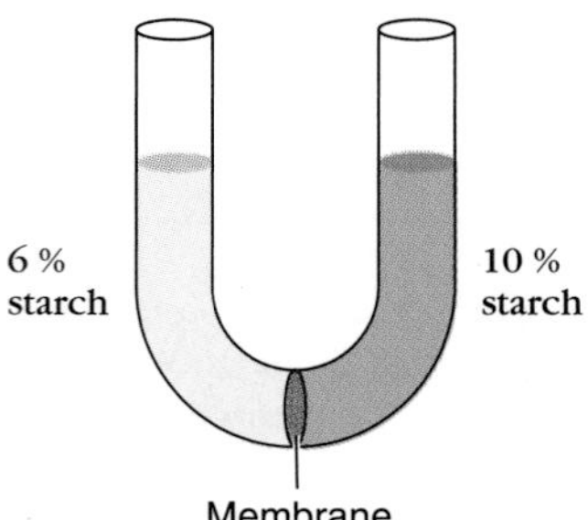

2. *Cambarus* is an animal that excretes a very dilute urine. Therefore you would expect that *Cambarus* lives in an environment that is (hypertonic, isotonic, hypotonic) to its body fluids. Compared to its body fluids, this environment has a (higher, lower) osmotic potential, and the net movement of water is (from the animal into the environment, from the environment into the animal). The habitat of *Cambarus* is most likely which of the following?
   a. a freshwater pond (contains little salt)
   b. the ocean (saltier)
   c. Great Salt Lake (even saltier)
3. A nerve cell sends messages to other cells by means of a special transmitter substance. Membrane-enclosed sacs containing transmitter molecules fuse with the nerve cell's plasma membrane and then open, releasing the transmitter outside the cell. This is an example of:
   a. exocytosis
   b. endocytosis
   c. active transport
   d. facilitated diffusion
   e. phagocytosis
4. Which two of the following cell junctions share a similar function?
   a. tight junction
   b. gap junction
   c. desmosome
   d. plasmodesmata

Match the substances from the list (a–d) to the way(s) they cross membranes.

___	**5.** Active transport	a. ions
___	**6.** Diffusion through bilayer	b. macromolecules
___	**7.** Diffusion through channels	c. small uncharged molecules
___	**8.** Endocytosis	d. water
___	**9.** Osmosis	

Which process(es) in questions 5–9 depend(s) on:

___ **10.** The membrane's fluidity

___ **11.** Membrane proteins

THINKING CRITICALLY

1. Why is it important for cells to maintain chemical homeostasis?
2. Hydrogen cyanide (HCN) and carbon monoxide (CO) are poisons that penetrate cell membranes readily. By what route do these molecules probably cross the plasma membrane into cells? Can you think of an explanation for the fact that no cells have evolved adaptations to keep these molecules out?
3. Most people have noticed the large wrinkles that form when we soak our hands or feet in water. How do we explain these changes in our skin? And why does soapy water increase the rate at which these wrinkles form?
4. When a plasma membrane is bent, cholesterol molecules flip from the inside of the lipid bilayer to the outside. What advantage does this serve?

SELECTED KEY TERMS

active transport, *p. 76*
cell theory, *p. 65*
concentration gradient, *p. 71*
cytoplasm, *p. 65*
diffusion, *p. 72*
endocytosis, *p. 77*
exocytosis, *p. 77*
facilitated diffusion, *p. 76*
fluid mosaic model, *p. 67*
homeostasis, *p. 64*
hypertonic, *p. 73*
hypotonic, *p. 73*
isotonic, *p. 73*
lipid bilayer, *p. 67*
membrane potential, *p. 77*
nucleus, *p. 65*
organelle, *p. 65*
osmosis, *p. 73*
passive transport, *p. 74*
phagocytosis, *p. 77*
pinocytosis, *p. 77*
plasma membrane, *p. 65*
receptor mediated endocytosis, *p. 77*
sodium-potassium pump (sodium pump), *p. 76*

SUGGESTED READINGS

Bretscher, M. S. "The molecules of the cell membrane." *Scientific American,* October 1985.

Bretscher, M. S. "How animal cells move." *Scientific American,* December 1987.

Dautry-Varsat, A., and H. F. Lodish. "How receptors bring proteins and particles into cells." *Scientific American,* May 1984.

Kartner, N., and V. Ling. "Multidrug resistance in cancer." *Scientific American,* March 1989.

Todorov, I. N. "How cells maintain stability." *Scientific American,* December 1990.

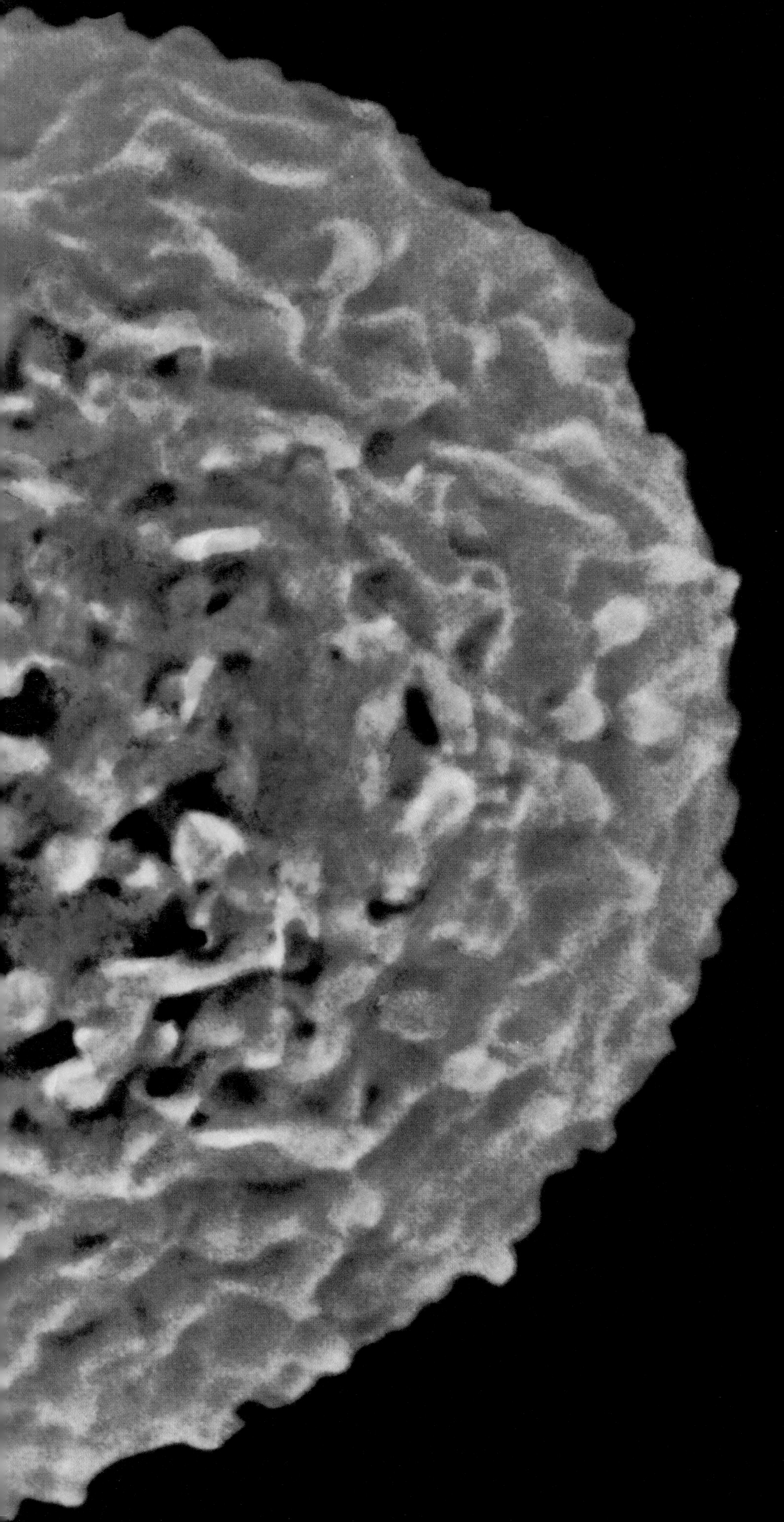

CHAPTER 5

CONCEPT GUIDE

After reading this chapter, you should be able to:

1. Compare the general structure of prokaryotic and eukaryotic cells. (Sections 5-A and 5-B)
2. Describe the general structure and functions of the eukaryotic nucleus. (Section 5-C)
3. Describe the locations, general structures, and functions of ribosomes, endoplasmic reticula, and Golgi complexes. (Sections 5-D, 5-E, and 5-F)
4. Discuss the similarities between community garbage recycling programs and the general functions of lysosomes within cells. (Section 5-G)
5. Compare the structures and functions of mitochondria and plastids. (Sections 5-H and 5-I)
6. Describe the structures and functions of cell walls and vacuoles in plants. (Sections 5-J and 5-K)
7. Describe the structures and functions of microtubules, intermediate filaments, and microfilaments in eukaryotic cells. (Section 5-L)
8. Explain the relationship between the terms cell, tissue, and organ and provide examples of each in plants and animals. (Section 5-M)

Like most of the cells of our bodies, this cell is specialized to perform certain duties. This is a white blood cell that searches out and destroys foreign materials and damaged cells. Tiny organelles within the cell permit these specific functions. Learn more about the structure of organelles and their specific cellular roles in Sections 5-C to 5-K. (CNRI/Science Photo Library/Custom Medical Stock)

Cell Structure and Function

In Chapter 4, we saw that cells are the structural, functional, and reproductive units of life. We also examined the plasma membrane, the thin, flexible covering of the cell, and saw how it controls the vital exchange of substances between the cell and its environment. Now it's time to delve further and see what's going on inside the cell.

The viscous, fluid cytoplasm forms the ground substance in a living cell. It contains food molecules, metabolic enzymes, and so on. In the cytoplasm of virtually all cells we find a nucleus or nuclear area containing the cell's genetic material (DNA). In the cytoplasm we also see many ribosomes, structures that make the cell's proteins according to instructions from the DNA.

What else we find inside the cell depends on which of two basic cell types we have chosen: prokaryotic or eukaryotic. **Prokaryotic cells** (pro = before; karyon = nucleus) have a simpler structure. In particular they lack a discrete nucleus that is separated from the cytoplasm by a membrane and also lack any membrane-bounded organelles. In addition, prokaryotic cells are usually smaller than eukaryotic cells, and they usually have a rigid cell wall outside the plasma membrane. Most **prokaryotes** are also unicellular—in function if not in structure—because when a cell divides, its offspring may stick together and form clumps or strings of attached but independent cells. For these reasons, prokaryotic cells are believed to have evolved earlier than eukaryotic cells. Organisms with prokaryotic cells are the bacteria (including the photosynthetic cyanobacteria), often simply called **prokaryotes**.

A **eukaryotic cell** (eu = true) contains a nucleus bounded by membrane, as well as other membrane-bounded organelles. Eukaryotic cells make up the bodies of all organisms other than bacteria. These **eukaryotes** include the unicellular **protists** and the multicellular fungi, plants, and animals.

The chapter opening photo shows a human white blood cell called a leukocyte, a small eukaryotic cell. It has all the machinery it needs to live. Its plasma membrane takes in food, such as sugars and amino acids, and expels wastes. Some of its cytoplasmic enzymes, and its mitochondria, break down sugars and use their energy to make ATP, the energy source for many biochemical reactions. The leukocyte's ribosomes use amino acids taken in by the plasma membrane, ATP from mitochondria, and message molecules sent out from the genetic information in the nucleus, to make its proteins. Some of these proteins perform the "housekeeping" activities needed in every cell. Others are made especially by the leukocyte as its contribution to the whole body. These special proteins are sent to the surface of the plasma membrane. Here they act as sentinels, ready to detect invasion of the body by "foes" such as disease-causing organisms.

Because our leukocyte's job is patrolling the body, it has no fixed address, although it spends a lot of time hanging out in lymph nodes (such as those in the neck and armpits that swell up during illness). Most other kinds of cells are attached to their neighbors or to a framework of material outside the cell, forming an organized structure of many similar cells, called a tissue. Various kinds of tissues arranged in certain ways form organs, such as the intestine. Blood is also a tissue, but one in which the material outside the cells is a fluid.

KEY CONCEPTS

- Cells can be divided into two groups—prokaryotic and eukaryotic—based on fundamental differences in their organization and size.
- All cells contain components essential for life: a plasma membrane, genetic material (DNA), ribosomes, and cytoplasm. In addition, most prokaryotic cells have a cell wall.
- In addition to the basic cell components, eukaryotic cells have a membrane-bounded nucleus, containing the genetic material, and many other membrane-bounded organelles.
- Eukaryotic organisms may be unicellular or multicellular. Each cell of a multicellular organism must carry on its own life processes and in addition perform some specialized task that contributes to the whole body as a whole.
- In most multicellular organisms, cells are organized into tissues, tissues into organs, and organs into the various organ systems of the body.

Curiosity Questions

How does a tadpole lose its tail? (See page 95)

Why does a plant wilt when it doesn't have enough water? (See page 97)

Why aren't poisonous plants hurt by their own poisons? (See page 99)

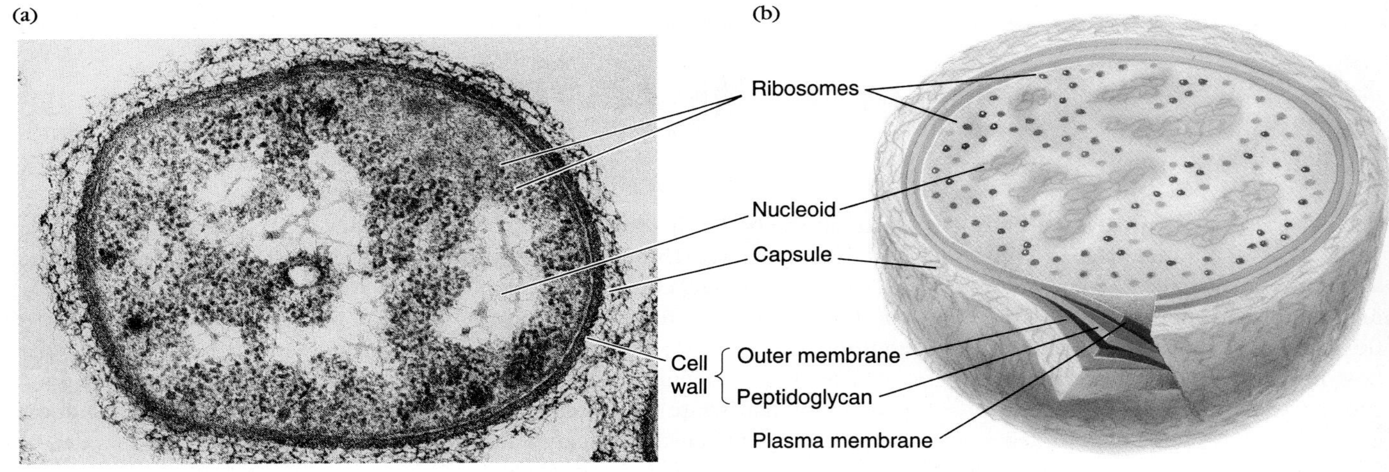

FIGURE 5-1

The bacterium, *Pseudomonas aeruginosa.* (a) A transmission electron micrograph of a *Pseudomonas* bacterium. **(b)** Bacteria are prokaryotes. Prokaryotic cells lack the internal organelles found in eukaryotic cells of plants and animals. Instead, the nucleoid and its DNA float within the cell along with other cellular materials. (Biological Photo Services)

5–A PROKARYOTIC CELLS: CELLS WITHOUT NUCLEI

Prokaryotic cells are so small that a light microscope, such as you might use in the laboratory, shows little of their structure. The fact that bacterial cell structure differs from that of other organisms was revealed after the invention of the more powerful transmission electron microscope (see *A Journey into Science in Process,* Chapter 4). One major difference is prokaryotes' lack of a membrane-bounded nucleus and most other organelles.

The DNA of a prokaryote occurs as a large, circular molecule folded up in a **nuclear area** (Figure 5-1). The cytoplasm contains many ribosomes, clusters of RNA and protein that carry on protein synthesis. A prokaryote's ribosomes are smaller than those of eukaryotic cells.

The plasma membrane, present in all cells, is the only membrane found in many prokaryotes. However, some prokaryotes also contain various internal membranes. A membrane known as the **mesosome** appears as a pocket in the plasma membrane. It has been postulated that mesosomes function in cellular respiration. Most photosynthetic bacteria have a system of internal membranes involved in capturing the light energy needed for this process (see Figure 19-2).

The cytoplasm of a prospering bacterial cell contains granules of storage polymers. These include glycogen, or other food (energy) reserves, and polyphosphate granules, which store phosphorus reserves.

A prokaryotic cell is usually surrounded by a thick cell wall, which performs essentially the same functions as a plant cell wall: it protects the cell, gives it shape, and keeps it from bursting in hypotonic media (Section 4-C). Prokaryote cell walls contain unique polymers of amino sugars (sugars with amino groups) and amino acids. Penicillin and related drugs interfere with the building of these walls and therefore inhibit the growth of bacteria but not of their host organisms, which neither need nor make these kinds of polymers (Figure 5-2). The walls of some bacteria also contain toxic substances. Many diseases caused by bacteria can be duplicated by injecting only these toxic cell-wall chemicals into an animal.

Some bacteria produce a polysaccharide or polypeptide **capsule** outside the cell wall. In nature, a dense felt-like mat of capsule polysaccharides enables bacteria to stick to surfaces—soil particles, rocks in streams, or cells of host animals. For example, *Streptococcus mutans,* the chief agent of tooth decay, glues itself to teeth in this way, using sucrose (but not other sugars) as a raw material. This is why candy and other sugary foods are so bad for our teeth.

Some bacteria produce hundreds of hollow protein strands called **pili,** which serve as means of attachment. In some species, pili are used to attach to

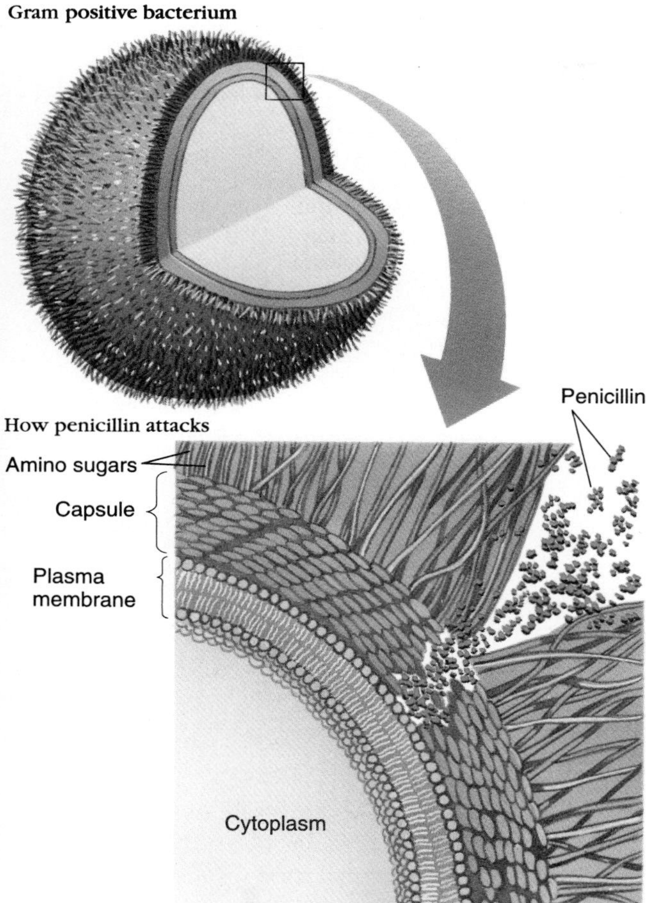

FIGURE 5-2

How antibiotics work. A bacterium produces cell wall molecules that help to maintain its shape and protect it against injury and chemical changes. Antibiotics kill bacteria by interfering with the construction of the cell wall.

another cell and transfer DNA to it during mating. In *Neisseria gonorrhoeae,* only strains that produce pili can attach to host cells and cause gonorrhea. Bacteria of many species move by the rotation of one or more long, thin, helix-shaped **flagella** (singular: **flagellum**) (see Figure 19-1).

BIO-BIT

At the cellular level, humans are more closely related to mushrooms than we are to bacteria.

Because of the small size and simple structure of prokaryotic cells, most biologists believe that they arose earlier in evolution than eukaryotic cells. *A Journey into Evolution:* How are Prokaryotes and Eukaryotes Related? discusses how eukaryotes may have originated.

Prokaryotic cells are usually smaller than eukaryotic cells and lack a membrane-bounded nucleus and most other organelles. The cytoplasm contains ions, food storage deposits, metabolic enzymes, and ribosomes that help assemble proteins by following instructions derived from the DNA in the cell's nuclear area. Most prokaryotes also have a cell wall, which affords shape and protection. Some prokaryotes attach to each other or other cells by pili or move using flagella.

5–B EUKARYOTIC CELLS: THE CELLS OF HIGHER ORGANISMS

Most eukaryotic cells are much larger than prokaryotic cells and contain a greater variety of components (Table 5-1). In addition to the plasma membrane, cytoplasm, and nucleus, eukaryotic cells contain many membrane-bounded organelles and various food storage inclusions such as lipid droplets and glycogen granules (A Journey Through An Animal Cell, A Journey Through A Plant Cell). There is also a framework of protein fibers, the so-called **cytoskeleton,** running through the cytoplasm, giving shape and support to the cell, and serving as tracks for the movement of various organelles to different parts of the cell. All these components can be seen with the light microscope if the cell is prepared suitably.

Cells also contain other components too small to be seen even with the most powerful electron microscope. These make up the **cytosol,** the soluble portion of the cytoplasm. It is a watery suspension that contains ions, molecules, and molecular aggregates of submicroscopic size, and it surrounds the visible structures in the cytoplasm. About half the cell's volume is occupied by the cytosol. Most of the cell's general metabolism and protein synthesis occur here, so it is not surprising that the cytosol is about 20 percent protein—largely metabolic enzymes.

The other half of the cell's volume is split up among the various membrane-bounded organelles.

BIO-BIT

There are over 200 different types of cells in the human body.

Table 5-1
Comparison of Eukaryotic and Prokaryotic Cells

Feature	Function	Eukaryotic Cells: Animal	Eukaryotic Cells: Plant	Prokaryotic Cells
Average size		10–20 μm	30–50 μm	1–10 μm
Cell wall	External support and protection	–	+ (mainly cellulose)	+ (amino sugar, amino acid polymers)
Plasma membrane	Containment; exchange of substances with environment	+	+	+
Nucleus (nuclear area)	Housing of genetic material	+	+	Nuclear area
Nuclear envelope	Enclosure of chromosomes	+	+	No membranes around DNA
Chromosomes	Storage of genetic information	Many, linear	Many, linear	One, circular
Nucleoli	Production of ribosomes	+	+	–
Ribosomes	Protein synthesis	+	+	+ (smaller, different)
Endoplasmic reticulum	Segregation of proteins to be secreted; site of new membrane synthesis	+	+	–
Golgi complex	Modification, sorting, packaging of cell products	+	+	–

Toolbox

5–1
Units of Size Used in the Study of Cells

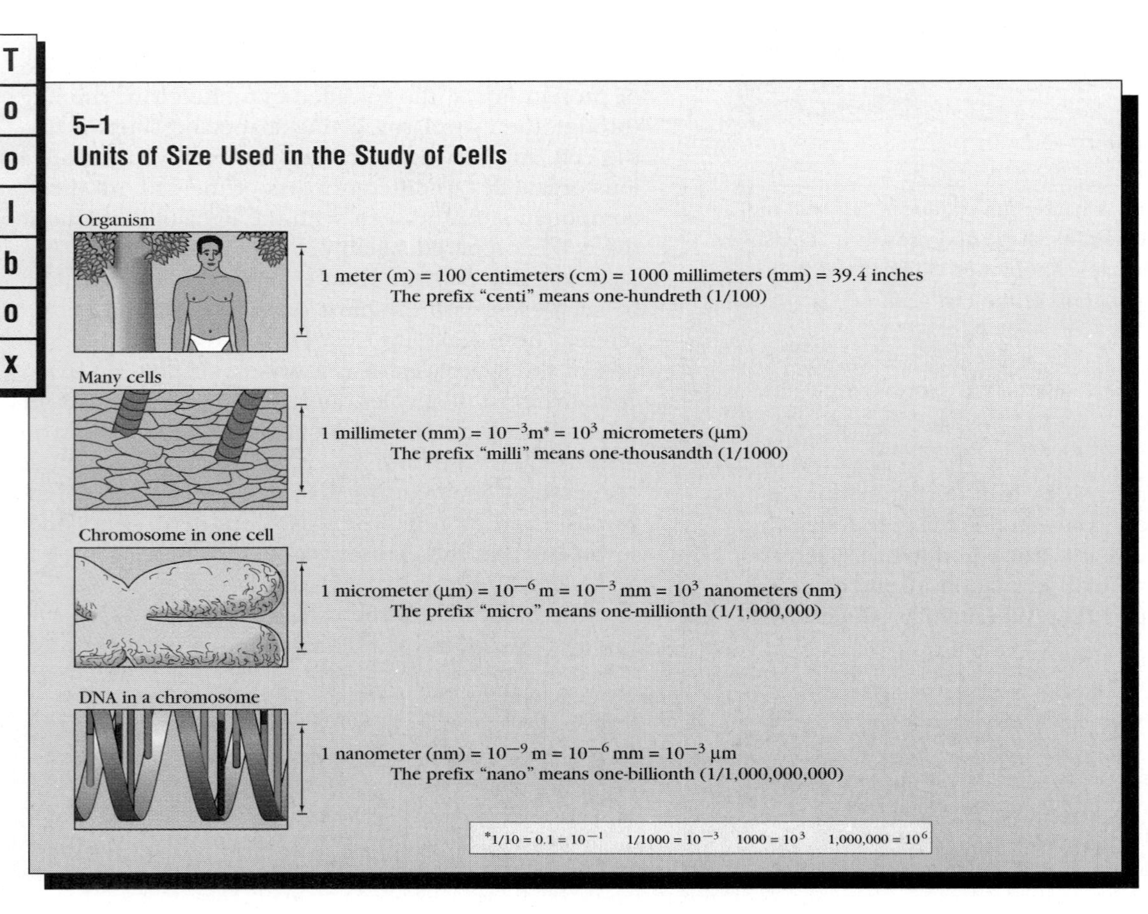
Organism
1 meter (m) = 100 centimeters (cm) = 1000 millimeters (mm) = 39.4 inches
The prefix "centi" means one-hundreth (1/100)
Many cells
1 millimeter (mm) = 10^{-3}m* = 10^3 micrometers (μm)
The prefix "milli" means one-thousandth (1/1000)
Chromosome in one cell
1 micrometer (μm) = 10^{-6} m = 10^{-3} mm = 10^3 nanometers (nm)
The prefix "micro" means one-millionth (1/1,000,000)
DNA in a chromosome
1 nanometer (nm) = 10^{-9} m = 10^{-6} mm = 10^{-3} μm
The prefix "nano" means one-billionth (1/1,000,000,000)
*1/10 = 0.1 = 10^{-1} 1/1000 = 10^{-3} 1000 = 10^3 1,000,000 = 10^6

Table 5-1
Comparison of Eukaryotic and Prokaryotic Cells

Feature	Function	Eukaryotic Cells: Animal	Eukaryotic Cells: Plant	Prokaryotic Cells
Lysosomes	Digestion of food and worn-out cell components	+ (in many cells)	+ (some cells)	−
Mitochondria	Respiration of food to produce ATP for energy	+	+	−
Plastids	Photosynthesis; food storage	−	+	−
Vacuoles	Storage of fluid, food, pigments	+ (some)	+ (most)	−
Cytoskeleton	Provision of cell shape; movement	+	+	−
Microtubules	Cell shape, spindle for chromosome separation during cell division	+	+	−
Centrioles	Organization of microtubules and basal bodies of cilia	+	Only in lower plants	−
Cilia, flagella	Locomotion of cell or movement of fluid past cell	+	Only in lower plants	+ (in some, but different type)
Intermediate filaments	Strengthening of cytoskeleton	+	+	−
Microfilaments	Cell motion and changes in shape	+	+	−

Each of these is a compartment separated from the cytosol by its membrane and contains the proteins and other molecules needed to perform a specific task. When the task is complete, the products move to other parts of the cell where they are used or processed further.

In the rest of this chapter we shall consider the most common components of eukaryotic cells and the elements of the cytoskeleton, and we finally consider the organization of cells into tissues.

Eukaryotic cells differ from prokaryotic cells in that they are larger and possess many membrane-bounded organelles, food storage inclusions, and a cytoskeleton. Nearly half of the eukaryotic cell's volume is made up of the cytosol, where most of the general metabolic processes and protein synthesis occur.

5–C THE NUCLEUS: GENETIC MESSAGE CENTER

When we look at an animal cell with the light microscope, the most obvious structure is usually the **nucleus** (A Journey Through An Animal Cell). A plant cell nucleus is often less obvious but is visible with proper staining (A Journey Through A Plant Cell). The genetic material in the nucleus of a eukaryotic cell is organized into chromosomes. Each **chromosome** consists of a long, linear DNA molecule and associated RNA and proteins. The chromosomes coil up into short, threadlike structures just before and during division of the nucleus, which precedes division of the cytoplasm. Most of the time, though, the nucleus is not dividing, and the chromosomes are uncoiled in a loose, indistinct tangle called **chromatin** (Figure 5–3).

DNA in the chromatin determines what RNA is made in the nucleus. The RNA travels to the cytoplasm, where it directs protein synthesis. Therefore, DNA also determines what proteins the cell makes. This, in turn, affects most of the cell's structures and activities. Hence the nucleus has been called the cell's "control center." However, control is a two-way street. Substances in the cytoplasm enter the nucleus and influence the DNA in such a way that it changes the particular mix of RNA (and hence proteins) produced. So the kinds and amounts of proteins made in the cell change depending on its needs.

With the light microscope, the most obvious features in the nuclei of nondividing cells are the nucleoli. **Nucleoli** (singular: **nucleolus**) are areas where parts of **ribosomes**, sites of protein synthesis (Section 5–D), are made. Nucleoli appear as one or more dense areas, which disappear when the cell divides (Figure 5–3).

(text continues on page 92)

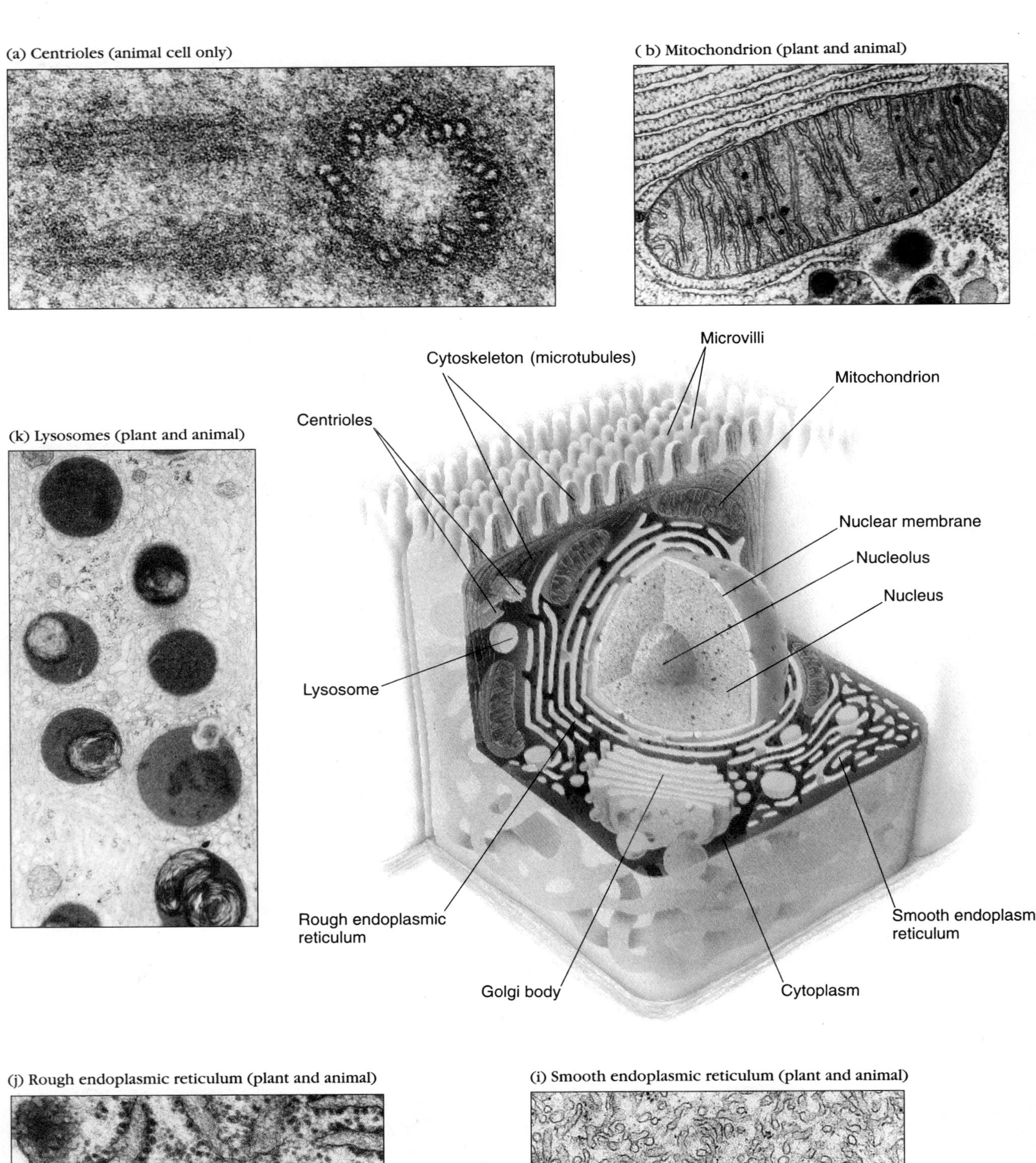

(j) Rough endoplasmic reticulum (plant and animal)

(i) Smooth endoplasmic reticulum (plant and animal)

Cell structure. Animal and plant cells have many types of internal organelles, each with separate functions, that together maintain a cell and allow it to perform specialized tasks. These two drawings illustrate the variety of organelles commonly found in animal and plant cells. (**a, b,** K. R. Porter/Photo Researchers, Inc.; **c,** Dennis Kunkel/PhotoTake NYC; **d,** Biophoto Associates; **e,** Visuals Unlimited; **f,** M. Schliwa, Visuals Unlimited; **g,** David M. Phillips, Visuals Unlimited; **h, i, k,** D.W. Fawcett/Visuals Unlimited; **j,** Dwight Kuhn)

(c) Golgi body (plant and animal)

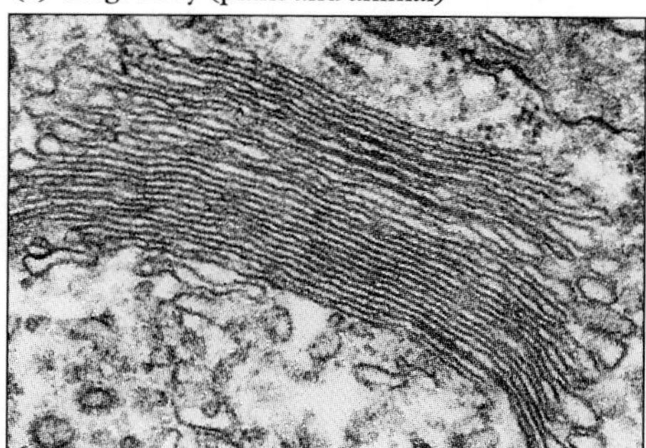

(d) Chloroplast (plant cell only)

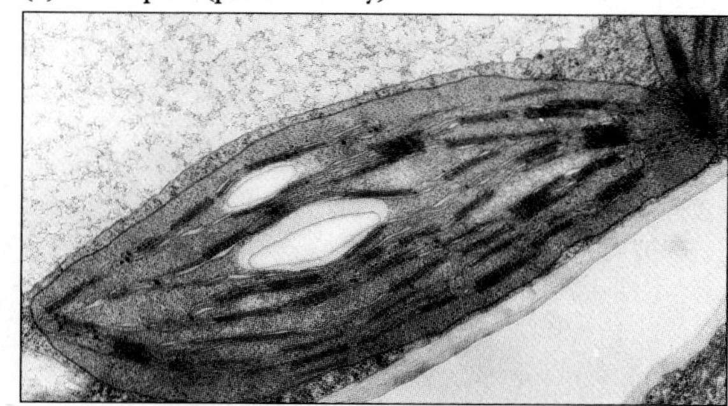

Lysosome
Smooth endoplasmic reticulum
Nuclear membrane
Nucleolus
Nucleus
Mitochondrion
Chloroplast
Rough endoplasmic reticulum
Golgi body
Vacuole
Plasma membrane
Cellulose walls
Pectin
Cell wall

(e) Plant cell wall (plant cell only)

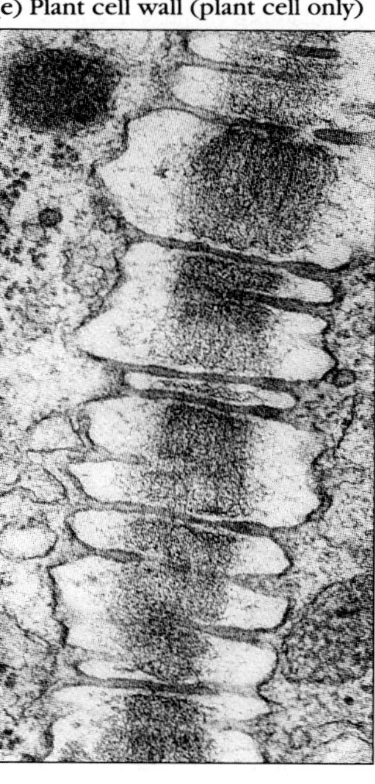

(h) Nucleus (plant and animal)

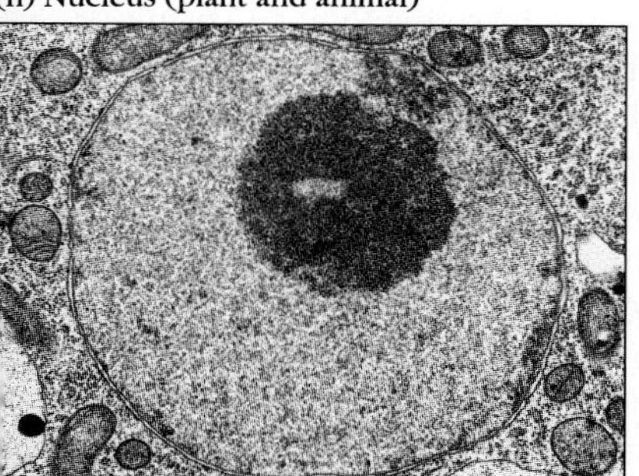

(g) Microtubules (plant and animal)

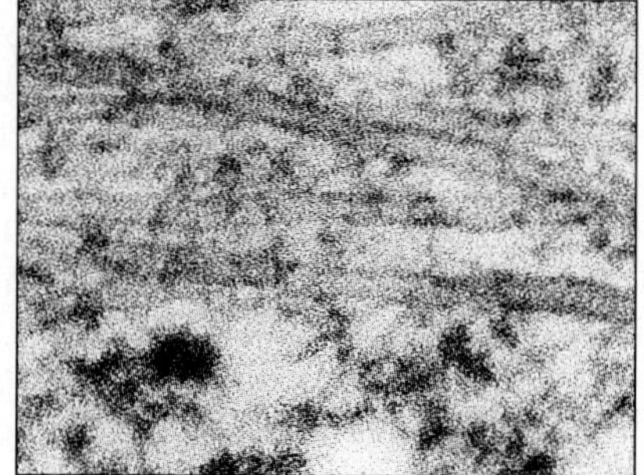

(f) Microtubules (end on view)

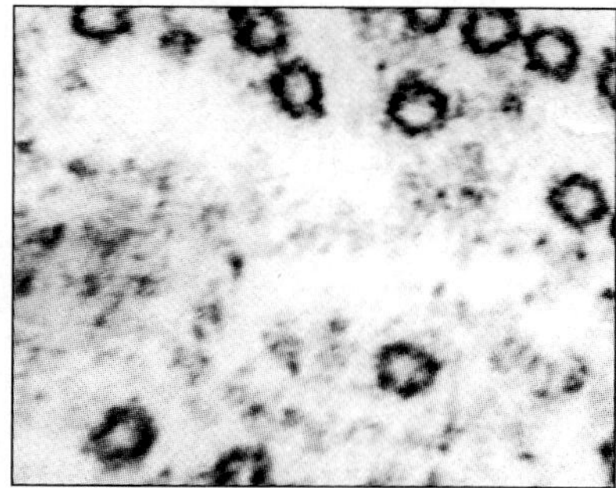

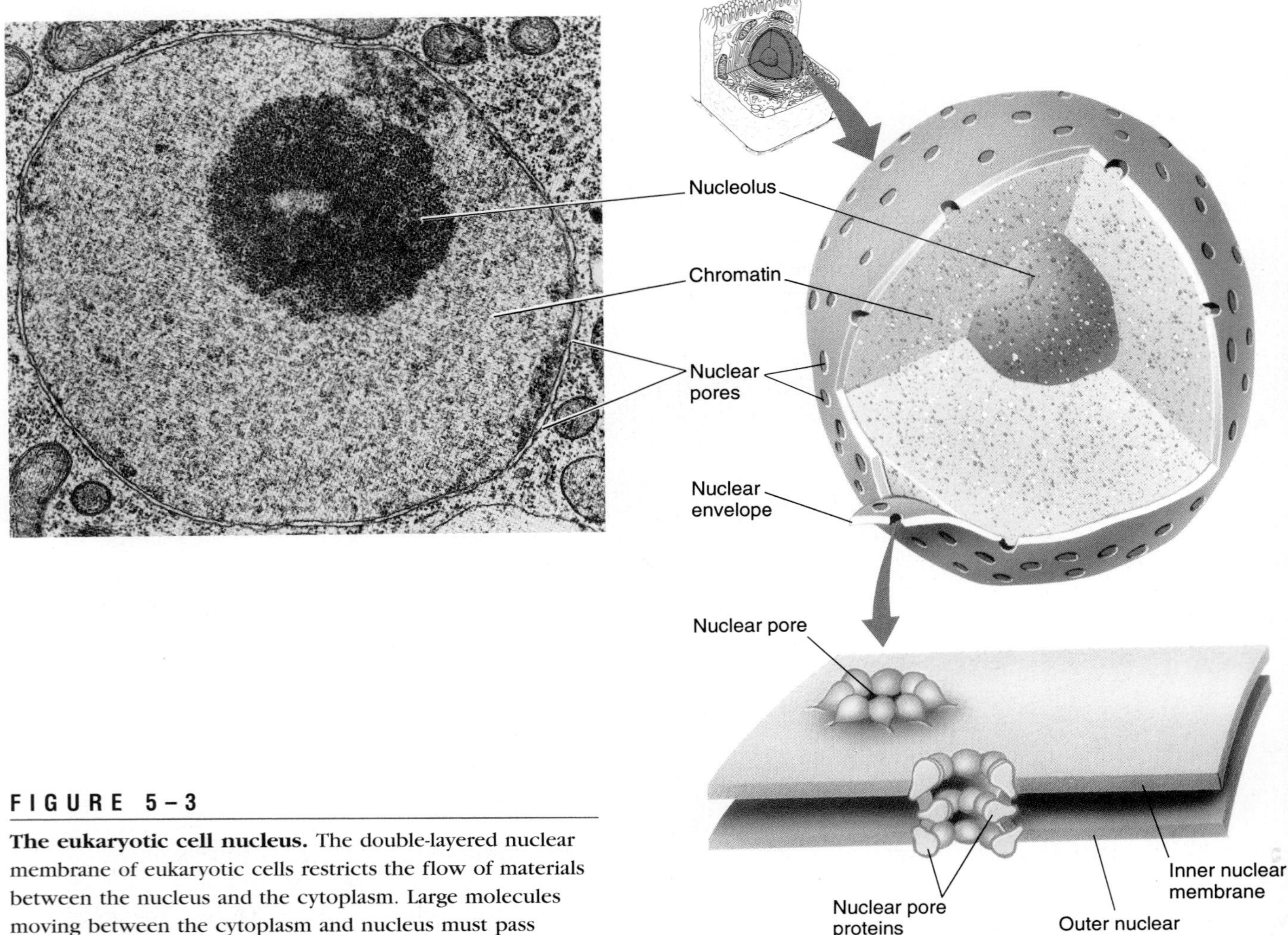

FIGURE 5-3

The eukaryotic cell nucleus. The double-layered nuclear membrane of eukaryotic cells restricts the flow of materials between the nucleus and the cytoplasm. Large molecules moving between the cytoplasm and nucleus must pass through nuclear pores. (D. W. Fawcett/Visuals Unlimited)

The nucleus is surrounded by two membranes, the **nuclear envelope,** which is perforated by numerous pores (Figure 5-3). RNA leaves the nucleus via these pores. RNA molecules are very large as well as hydrophilic. They could not pass through the lipid layers of the nuclear envelope, so the pores are critical for their passage.

The inner and outer membranes of the nuclear envelope connect with each other through the nuclear pores, and the outer membrane, in turn, is continuous with a system of membranes in the cytoplasm called the endoplasmic reticulum (Section 5-E). These interconnections allow the nuclear envelope to shrink or grow rapidly by losing material to, or gaining it from, the endoplasmic reticulum. In most eukaryotes, the nuclear envelope disappears completely during cell division and re-forms afterward. The envelope may expand very rapidly when a dormant cell gears up to produce DNA or RNA.

A eukaryotic cell's nucleus houses its chromosomes within a double-membraned nuclear envelope. Chromosomal DNA directs the synthesis of RNA molecules, which leave the nucleus via pores in the envelope and participate in protein synthesis in the cytoplasm. Portions of ribosomes are produced in the nucleolar region of the nucleus.

BIO-BIT

Groups of heart cells isolated in culture dishes all continue to contract in coordinated rhythms.

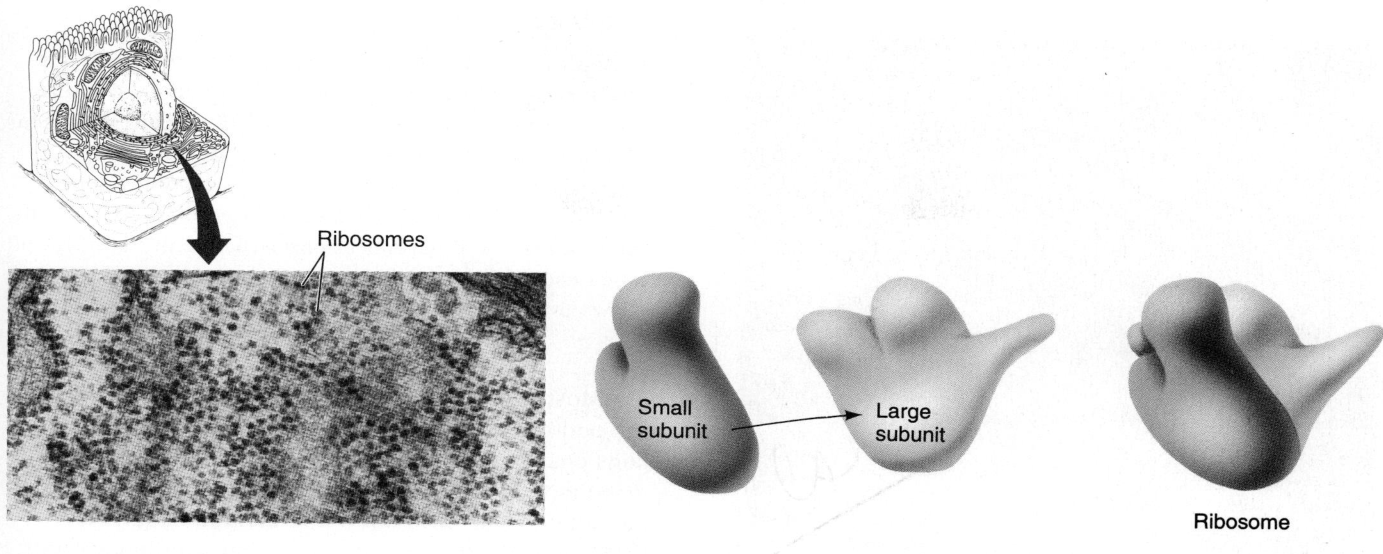

FIGURE 5-4

Ribosomes. Each ribosome is formed by the combination of a large and small subunit. Each subunit is composed of a highly folded ribosomal RNA molecule and many attached proteins. Ribosomes serve as the site of protein synthesis and may be found floating freely in the cytoplasmic fluids or attached to an endoplasmic reticulum. (Dwight Kuhn)

5-D RIBOSOMES: WORKBENCHES FOR PROTEIN ASSEMBLY

Ribosomes are necessary components of cells because they are the sites of protein synthesis. In eukaryotes, portions of ribosomes are made in the nucleolus area(s) of the nucleus and then travel to the cytoplasm through the nuclear pores. Eukaryotic ribosomes are larger than prokaryotic ones; however, they are still so tiny that even electron micrographs show them as little more than small granules (see Figure 5-4) Most of our knowledge of ribosomes comes from laboratory research using ribosomes that have been isolated from cells. The function of ribosomes in protein synthesis is discussed in Chapter 10.

A cell may contain up to half a million ribosomes, the number varying with how much protein the cell makes. Some ribosomes appear to be attached to part of the cytoskeleton (Section 5-L), and some are attached to membranes in the cell, especially those of the endoplasmic reticulum, discussed next.

Ribosomes serve as sites of protein synthesis in the cytoplasm.

5-E ENDOPLASMIC RETICULUM

The **endoplasmic reticulum (ER)** is a system of membranous tunnels and sacs found in most eukaryotic cells. Often it lies just outside the nucleus, since it is continuous with the outer membrane of the nuclear envelope. The endoplasmic reticulum usually accounts for half or more of the cell's total membrane. In electron micrographs, the endoplasmic reticulum appears as piles and sacs of membrane, but it is believed to consist of one continuous sac surrounded by a single, highly convoluted membrane (A Journey Through An Animal Cell, A Journey Through A Plant Cell). The interior of this sac, the **lumen** of the ER, provides the cell with a compartment to contain substances that must be kept separate from the cytosol.

Some of the ER is called **rough endoplasmic reticulum** because ribosomes attached to the outer (cytoplasmic) surface give it a bumpy appearance in electron micrographs (Figure 5-5a). Rough endoplasmic reticulum is especially abundant in cells that make proteins which will be secreted from the cell (for example, cells of the pancreas that make digestive enzymes for export to the small intestine). This is because the endoplasmic reticulum and ribosomes

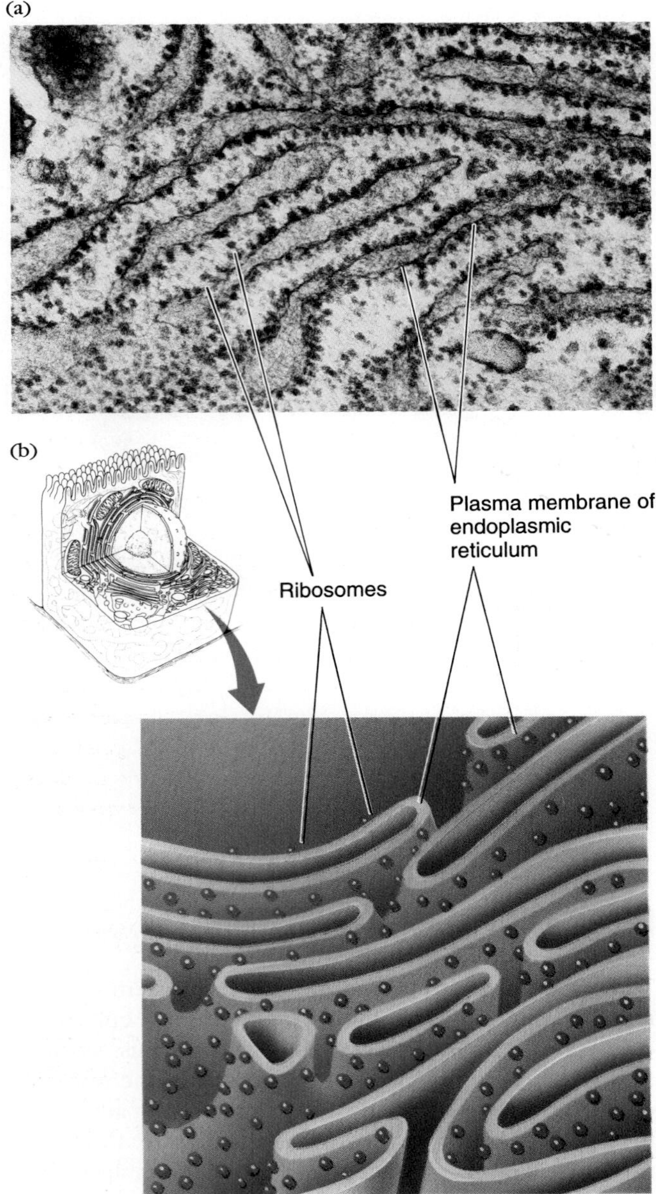

FIGURE 5–5

Endoplasmic reticulum. The endoplasmic reticulum is a series of highly folded and flattened membranous sacs that can be found with or without attached ribosomes. **(a)** When ribosomes are attached, as in this electron micrograph, it is called a rough endoplasmic reticulum. A smooth endoplasmic reticulum does not have attached ribosomes. **(b)** This diagram shows the three-dimensional structure of a segment of rough endoplasmic reticulum. (Dwight Kuhn)

cooperate in solving the problem of getting these large protein molecules out through the plasma membrane. As the proteins are made, they are pushed through the ER membrane into the lumen. Here, enzymes modify the proteins. Next, the proteins move on into a transitional area, where patches of the ER membrane bud off to form membranous sacs called **vesicles,** in this case transport vesicles carrying the proteins (see Figure 5-6). These vesicles move to a Golgi complex (Section 5-F, next), where the proteins are modified further before being packaged into other vesicles bound for exocytosis at the plasma membrane.

Most of the cell's new membrane is produced in the endoplasmic reticulum. The ER membrane contains enzymes that use substrates from the cytosol to make new lipids. These become inserted into the ER membrane, and this is how the membrane grows. We have seen that the ER membranes are continuous with the nuclear envelope, and that parts of the ER can pinch off as vesicles, which may fuse with other membranous structures such as Golgi complexes. New membrane material can thus be added to existing membranes either by direct transfer from the ER or by the integration of vesicles into existing membranes (see Figure 5-6).

Smooth endoplasmic reticulum is ER without attached ribosomes. Most cells contain little smooth ER, but it is abundant in cells responsible for lipid metabolism, such as those making cholesterol and steroid hormones.

The endoplasmic reticulum provides a large surface area where ribosomes can attach and produce proteins to be exported from the cell. New membrane is also made in the ER. Transport vesicles carry proteins and membrane material from the ER to a Golgi complex.

5–F GOLGI COMPLEX: MOLECULAR FINISHING AND SORTING AREA

Most transport vesicles pinched off from the ER soon fuse with larger sacs that are part of a **Golgi complex**—a stack of flattened, membranous sacs like a pile of pita bread. Around the edges of the stack, swarms of small, round transport vesicles carry molecules to or from these large sacs (Figure 5-6). Like the ER, the Golgi often lies near the nucleus. A cell may have one large Golgi complex or up to hundreds of much smaller ones.

The Golgi's overall role appears to be to modify, sort, and package molecules made in other parts of the cell, before they travel to their final destinations in other organelles or outside the cell. Molecules being

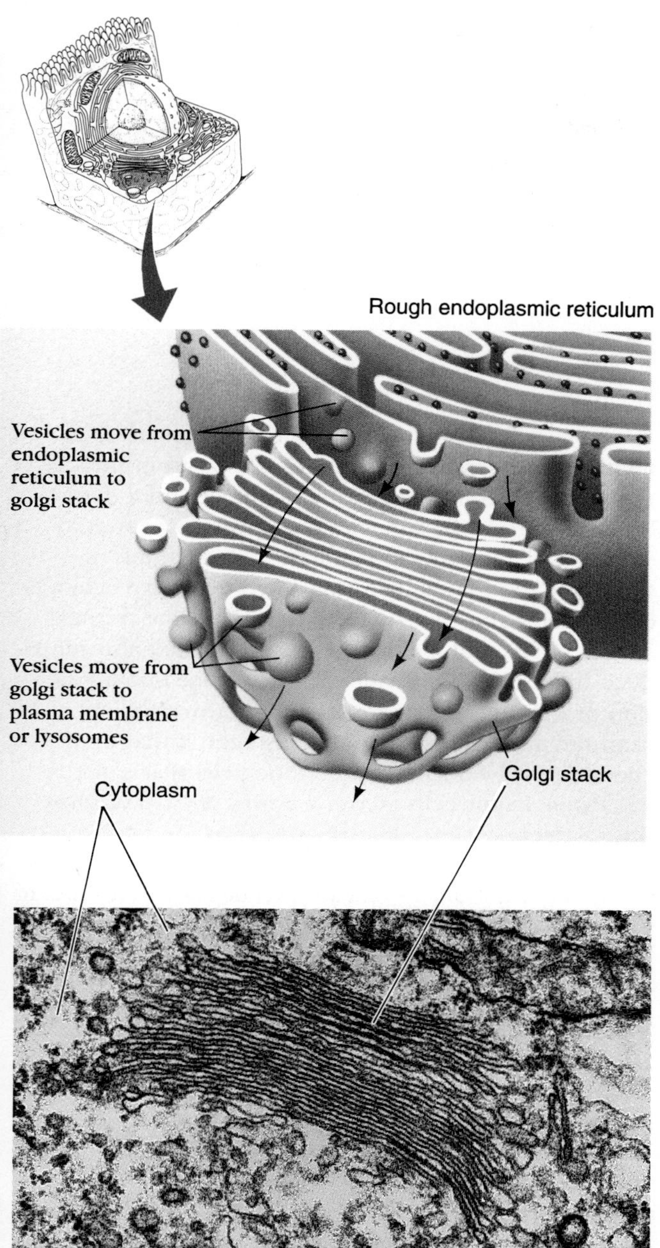
Rough endoplasmic reticulum
Vesicles move from endoplasmic reticulum to golgi stack
Vesicles move from golgi stack to plasma membrane or lysosomes
Golgi stack
Cytoplasm

FIGURE 5-6

Golgi stack. (a) This diagram illustrates the two sides of the Golgi stack where vesicles from the endoplasmic reticulum attach and where Golgi transport vesicles depart. **(b)** In this transmission electron micrograph, we can see the many continuous membranes of one Golgi stack that look like a pile of pita bread. Many interconnected Golgi stacks form the Golgi complex. (Dennis Kunkel/PhotoTake NYC)

processed in the Golgi move from one of the large sacs to another in sequence, carried by transport vesicles. In each sac, the molecules are modified by enzymes and then enclosed in another transport vesicle, which fuses with a sac containing enzymes for the next biochemical steps. The last sac finishes, sorts, and packages the molecules into transport vesicles according to chemical "address labels" that have been attached in the Golgi. For example, proteins en route to lysosomes are labeled with phosphate groups at their first stop in the Golgi. In the last sac these proteins bind to phosphate-specific membrane receptors and become organized into a transport vesicle that will fuse with a lysosome. Molecules to be exported from the cell are enclosed in vesicles that pinch off from the Golgi sac membranes, move to the plasma membrane, and discharge their contents by exocytosis. These molecules include things like mucus and digestive enzymes, secreted by cells lining the digestive tract, or hormones from gland cells. Other vesicles leaving the Golgi carry new protein and lipid material to be added to the plasma membrane itself.

The flattened membranous sacs of the Golgi complex modify, sort, and package molecules made in the endoplasmic reticulum. Chemical labels are added, directing the product into transport vesicles, specific to the proper destination within the cell.

5-G LYSOSOMES: SACS OF HYDROLYTIC ENZYMES

A **lysosome** is a membrane-bounded sac containing hydrolytic (digestive) enzymes (Figure 5-7). These enzymes are made in the endoplasmic reticulum and then modified and packaged in the Golgi. Vesicles filled with lysosomal enzymes bud off from the Golgi and fuse with vesicles containing material to be digested.

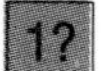

Lysosomes can be thought of as cellular recycling centers. They digest a variety of things: food molecules, disease-causing viruses brought into the cell by endocytosis, damaged organelles, or macromolecules in the cell. The entire cell may even die and then be digested by its lysosomes. For example, as a tadpole develops into a frog, lysosomes in the tadpole's tail digest the tail, and the molecules released are absorbed back into the body and reused by other cells. ■

Lysosomes are membrane-bounded vesicles that contain many enzymes that can break down food, unneeded macromolecules, or damaged cell components to form simpler molecules that the cell can recycle to create new molecules.

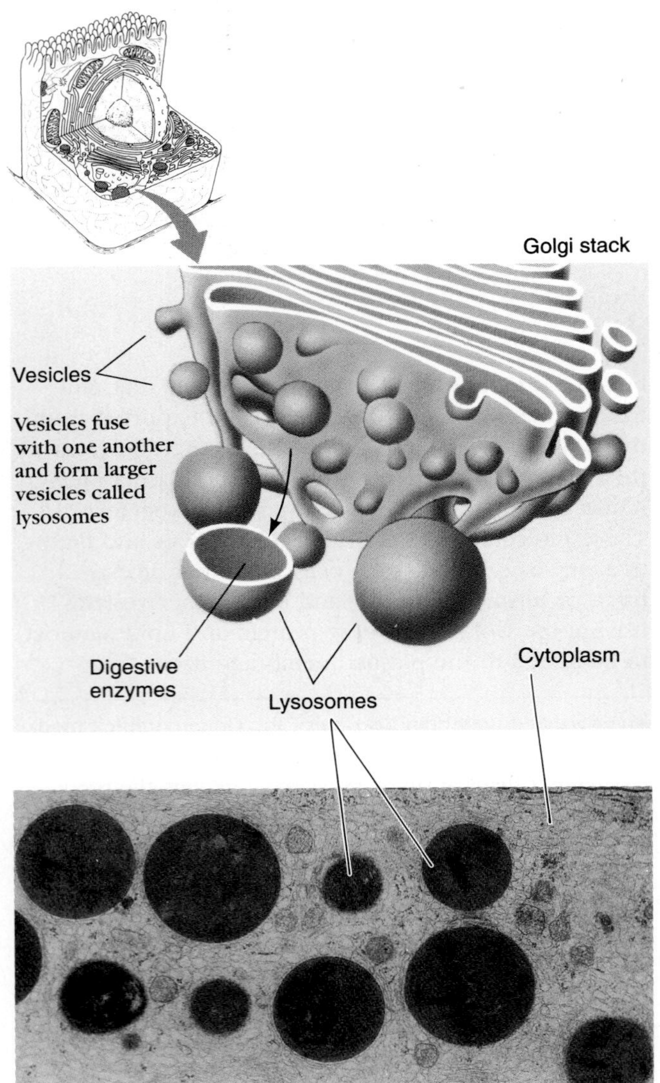

FIGURE 5-7

How lysosomes are formed. Lysosomal enzymes are synthesized in the endoplasmic reticulum, modified in the Golgi complex, and released in vesicles called lysosomes. Lysosomes fuse with other vesicles containing materials to be digested. The digestive enzymes in lysosomes speed up the rate of destruction of materials in these other cytoplasmic vesicles. (D. W. Fawcett/Visuals Unlimited)

5–H MITOCHONDRIA: MILLS TO MAKE ATP

Mitochondria (singular: **mitochondrion**) are large organelles that produce most of a cell's adenosine triphosphate (ATP), the energy supply for many of its metabolic reactions (Section 6–E). Mitochondria get the energy they need to make ATP by **cellular respiration,** a series of reactions that use oxygen in the breakdown of small food molecules to carbon dioxide and water. Cells that use a lot of energy, such as muscle cells in animals, or cells in the growing root tips of plants, have many mitochondria. A liver cell, which is a busy biochemical factory, may contain about 2500 mitochondria.

A mitochondrion has two membranes: an outer membrane separates the mitochondrion from the cytoplasm, and within this is a highly folded inner membrane (Figure 5–8). We shall see how mitochondrial structure contributes to respiration in Chapter 7.

Mitochondria contain their own genetic machinery (DNA, RNA, and ribosomes) and make some of their own proteins and membrane material. They also reproduce themselves: new mitochondria arise only by division of existing ones, and cells cannot make them from raw materials. Many biologists think that mitochondria evolved from prokaryotic cells that came to live inside larger cells (see *A Journey into Evolution*, this chapter).

Mitochondria are large organelles with double membranes, in which the inner membrane is highly folded. These cellular "power plants" produce most of a cell's energy in the form of ATP by using oxygen in a complex series of chemical reactions called cellular respiration.

5–I PLASTIDS: FOOD FACTORIES AND STOREHOUSES

The organelles mentioned so far in this chapter can be found in both plants and animals, because they perform functions essential to nearly all cells. However, if we were to examine a cell under the microscope, we could tell that it came from a plant rather than an animal by the presence of three features: plastids, a cell wall, and a large vacuole—the subjects of the next three sections.

Like mitochondria and bacteria, plastids contain DNA, RNA, and ribosomes, and reproduce themselves. Plastids are bounded by two outer membranes separating them from the cytoplasm, and they also contain a third separate system of internal membranes. There is good evidence that plastids, like mitochondria, may be descendants of free-living bacteria that set up housekeeping inside larger cells (Section 19–E). While virtu-

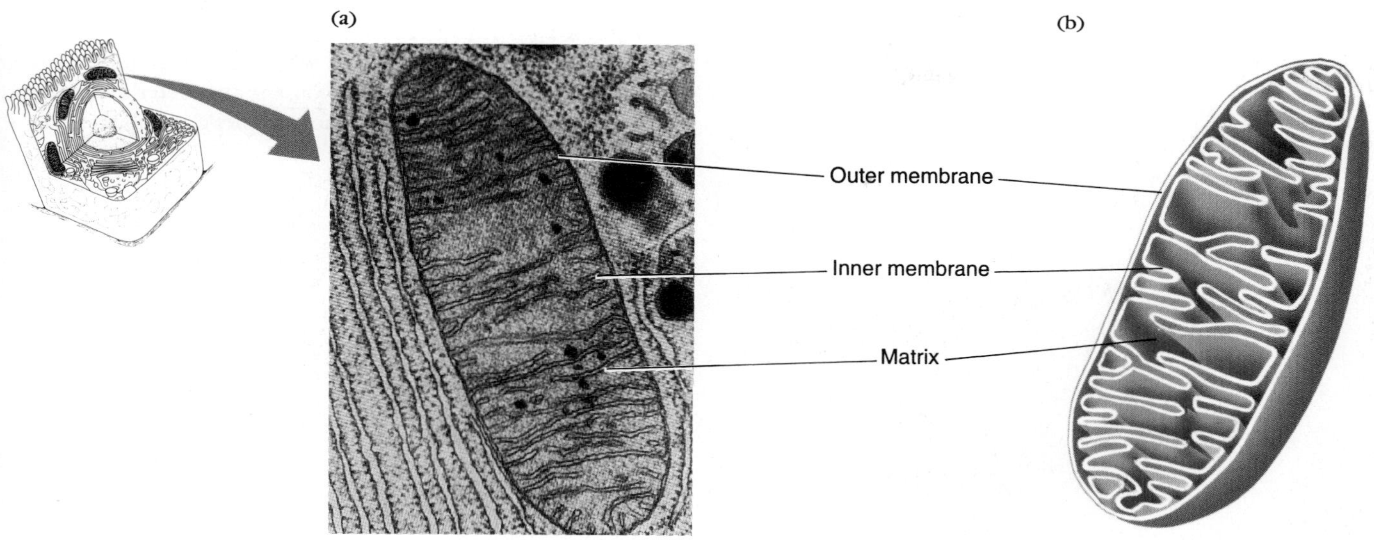

FIGURE 5–8

Mitochondria. (a) In this electron micrograph, we can see how the inner mitochondrial membrane is highly infolded, to increase the surface area for the electron transport system which occurs here. The outer membrane is much smoother, and forms the outer edge of the mitochondrion (singular). **(b)** This diagram illustrates the key features of a mitochondrion. (Keith Porter/Photo Researchers, Inc.)

ally all eukaryotic cells contain mitochondria, plastids are found only in photosynthetic eukaryotes: plants and some protists.

Chloroplasts are the green plastids that carry out photosynthesis—the manufacture of food using carbon dioxide, water, and light energy. The internal membrane of a chloroplast is highly folded (Figure 5-9). Chloroplasts are green because this membrane contains the green pigment chlorophyll, which traps the light used to power photosynthesis. Leaves are green because of the green chloroplasts in some of their cells. The details of chloroplast structure are covered in Chapter 8.

Chromoplasts are plastids that make and store the yellow and orange pigments that give their colors to many flowers, fruits, or roots. Other common plastids are **amyloplasts,** which store starch as a plant's reserve food supply. Amyloplasts are abundant in the cells of potatoes and in roots (Figure 5-10).

Plastids are characteristic of plants. Like mitochondria, plastids are surrounded by two membranes and contain their own DNA, RNA, and ribosomes. Chloroplasts are the plastids containing chlorophyll that carry out photosynthesis. Other plastids store starch or make and store pigments other than chlorophyll.

5–J CELL WALLS: PROTECTION AND SUPPORT

A plant cell is surrounded by a thick but porous **cell wall,** lying just outside the plasma membrane (Figure 5-11).

Composed of cellulose and other fibers, the cell wall is porous enough to allow water and dissolved substances to pass through it freely, tough enough to give the plant body structure and support, and flexible enough to permit the plant to bend in the wind instead of breaking.

Because every plant cell is irrevocably cemented to its neighbors, the cells and organs of plants cannot move much with respect to one another. This, along with the relative rigidity of cell walls, accounts for many of the special characteristics of plants and plant cells. Cell walls contribute largely to the structural support of a plant's cells and of the entire plant body (stems, leaves, and roots). The cell walls of soft-bodied plants are thinner and more flexible than those of woody plants. The cells must also be well filled with water, so that they exert an outward pressure on their walls (see Figure 4-7). If the plant loses too much water, the cells do not fill the space enclosed by their walls completely, and the plant wilts. ■

(a)

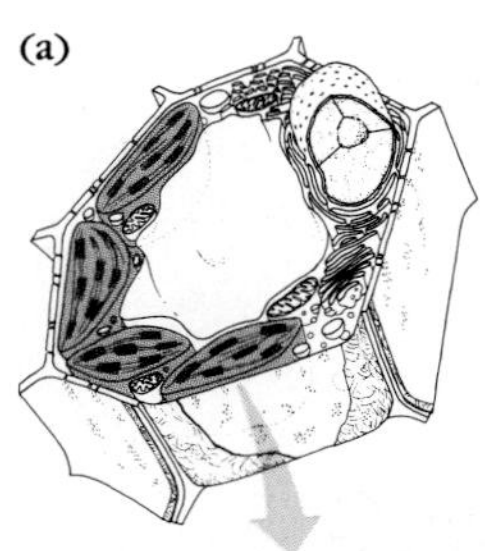

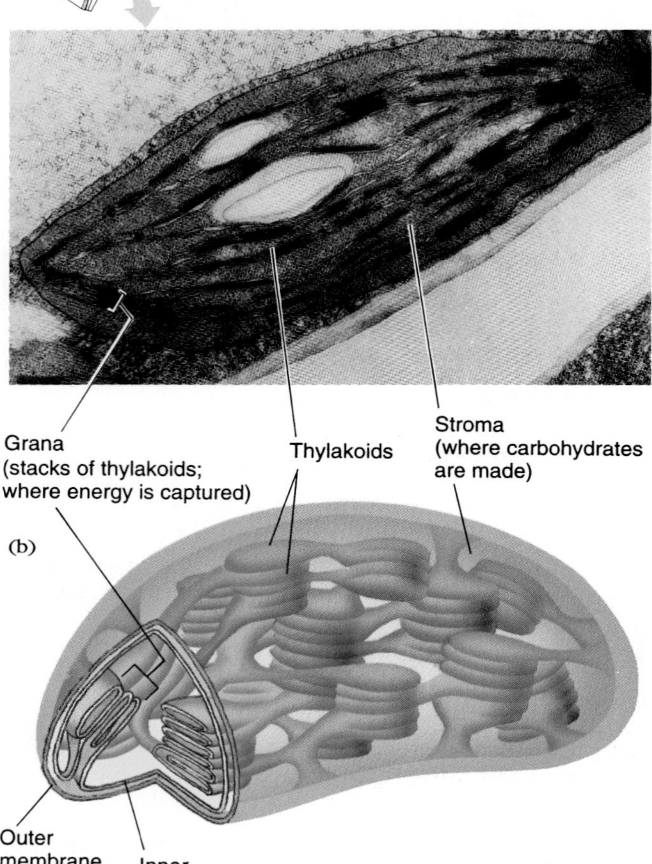

FIGURE 5–9

Chloroplasts. (a) A plant cell with many chloroplasts. **(b)** Drawing of a chloroplast, showing the two outer membranes and the highly folded green photosynthetic membranes inside. (Biophoto Associates)

FIGURE 5–10

Amyloplasts. Amyloplasts are starch containing organelles that are commonly found in plant roots. (J. Burgess/Science Photo Library/Photo Researchers, Inc.)

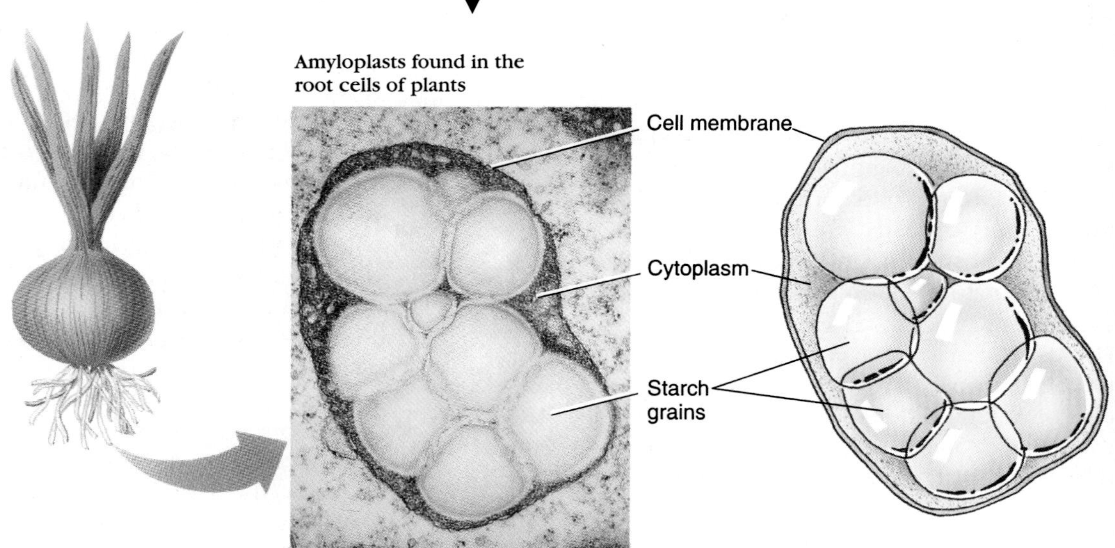

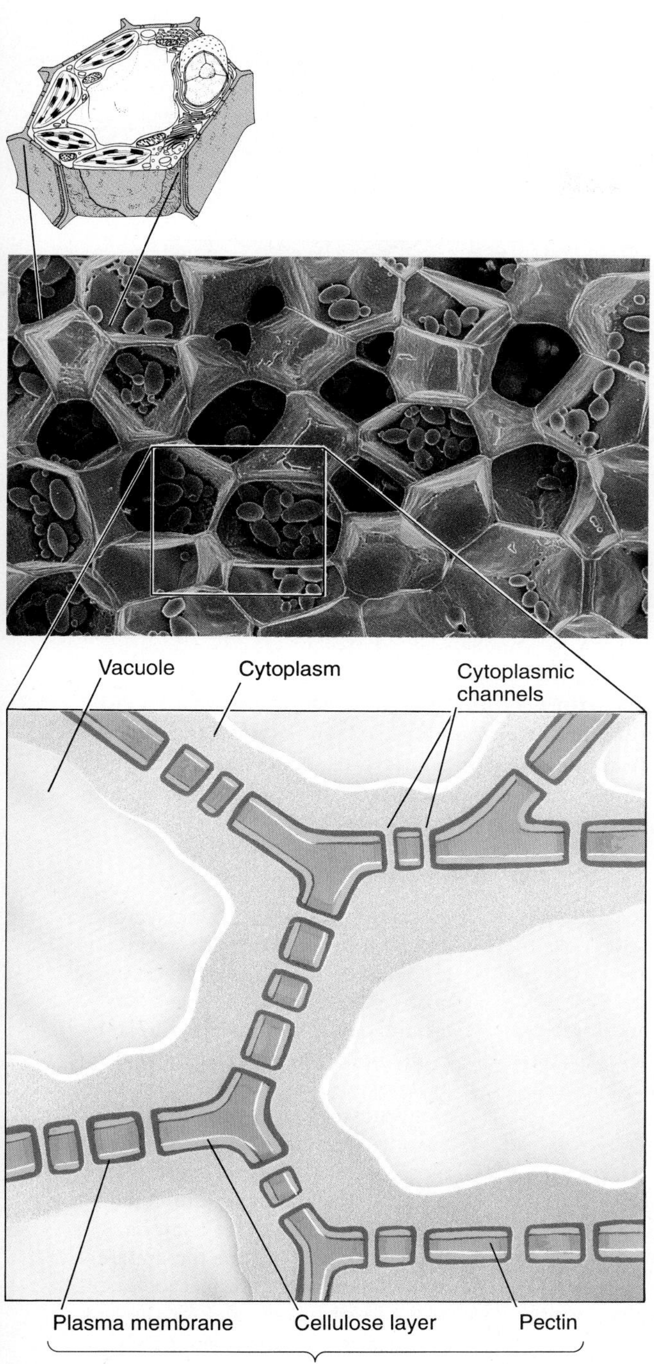

FIGURE 5-11

Plant cell walls. Cell walls surround plant cells to support the cells and help maintain a plant's shape. Pores in cell walls permit plant cells to communicate with each other through cytoplasmic channels called plasmodesmata. (J. Burgess/Science Photo Library/Photo Researchers, Inc.)

Unlike animal cells, plant cells have thick walls composed of cellulose and other fibers. The cell walls hold the enclosed cell in shape, and the cell walls of adjacent plant cells are firmly cemented to each other.

5–K VACUOLES: SACS FULL OF FLUID

A vacuole is a sac of fluid surrounded by a membrane. Vacuoles occur in many cells but are particularly prominent in plant cells. Typically, most of a plant cell's volume is occupied by a single, large, central vacuole, and as a result the nucleus, plastids, mitochondria, and other organelles in the cytoplasm are crowded around the edges (see A Journey Through A Plant Cell).

3? The vacuole holds stored food, pigments, and other substances. It is also convenient as a "storage locker" for toxic substances. For instance, some acacia trees produce and store cyanides (which make them poisonous to animals that eat plants) inside their vacuoles. If the cyanides were in the cytoplasm, they would poison the rest of the cell. ■

Vacuoles are fluid-filled sacs surrounded by a membrane and are especially large in some plant cells. Vacuoles store food, pigments, and other cell products.

5–L THE CYTOSKELETON: CELL SHAPE AND MOVEMENT

The world of living cells is filled with constant motion. Membrane material is exchanged among the plasma membrane, vesicles, Golgi complexes, endoplasmic reticulum, and nuclear envelope. In the cells of a pondweed, the green chloroplasts circulate around the central vacuole, moved by movements of cytoplasm in a process called **cytoplasmic streaming.** Heart cells, whether in an intact heart or isolated in culture dishes, contract rhythmically several times a minute. Cells growing in a culture dish may change shape, putting out long filaments and moving around (Figure 5-12). And nerve cells in an embryo's brain and spinal cord grow long, thin processes that reach to other nerve cells and to the fingers, toes, and other distant parts of the body.

Such movements and changes of shape are still poorly understood because it is difficult to see great detail in living cells. However, we can now use the electron microscope to see some of the delicate structures responsible for this movement. The cytoplasm of

(text continues on page 102)

A Journey into Evolution

How Are Prokaryotes and Eukaryotes Related?

The fossil record shows that, although prokaryotic cells appeared on Earth about 3.6 billion years ago, the earliest known eukaryotic cells do not appear until about 1.45 billion years ago. In the eukaryotic cell, specialized organelles such as the nucleus, mitochondria, and chloroplasts effectively separate the cell into functional areas. The nucleus directs protein synthesis, mitochondria are specialized for respiration, and chloroplasts focus on photosynthesis. The membranous boundaries give each organelle the advantage of separating its chemical reactions from those of the cytoplasm, thus preventing cellular reactions from interfering with one another. Finally, prokaryotes have only a single DNA molecule, while eukaryotic cells have much more genetic information packed into chromosomes. Thus, the evolution of the first eukaryotic cells was a tremendous advance. It not only permitted a dramatic increase in cell size, but also increased the amount of genetic information in a eukaryotic cell. Once meiosis and mitosis had evolved, sexual reproduction, with its varied range of new genetic combinations became possible. This, in turn, permitted the tremendous adaptive radiation of eukaryotes. But where did the first eukaryote come from? How did it evolve?

Many biologists think that plastids and mitochondria evolved from prokaryotes that came to live inside larger host cells in an intimate relationship called a **symbiosis** (together-living). The host and its symbionts are thought to be the ancestors of modern eukaryotes. This **endosymbiont theory** (endon = within) is supported by the following facts:

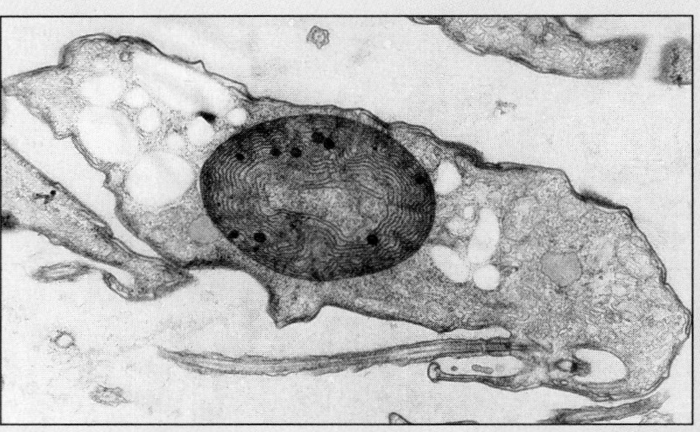

FIGURE 5-A

The protist *Cyanophora* with a symbiotic cyanobacterium in its cytoplasm. The cyanobacterium produces food for itself and its host; *Cyanophora* uses its flagellum to move to areas where light is available for the photosynthesis of its symbiont. Notice the many layers of photosynthetic membranes in the cyanobacterial cell. (Biophoto Associates)

1. Some prokaryotes live as symbionts inside eukaryotic cells today, either as invaders or as undigestible meals (Figure 5-A).
2. Plastids and mitochondria have certain similarities with bacteria. For example, plastids and mitochondria:
   —are about the same size as bacteria;
   —contain circular DNA without the surrounding accessory proteins that are found in eukaryotic chromatin;
   —reproduce by binary fission;
   —make some of their own proteins;
   —have ribosomes similar to prokaryote ribosomes in both size and function;
   —depend on host nuclear genes to supply the information for making some of their proteins.

The endosymbiont theory contends that mitochondria originated from symbiotic bacteria and that much later, photosynthetic symbionts moved into some cells that contained mitochondria. What was the original host cell? The best guess is a bacterium similar to those that live in acidic hot springs because these prokaryotes have proteins like actin and accessory DNA proteins called histones, features that are similar to those of eukaryotes. Because it lacks a cell wall, such a bacterium could easily have engulfed other bacteria. This theory can account for the apparently rapid appearance of eukaryotic cells in the fossil record, and for the lack of organisms intermediate between prokaryotes and eukaryotes.

But all the experts do not agree. The theory's critics argue that first, mitochondrial and prokaryotic protein synthesis are somewhat different, second that the mitochondrial genetic code differs slightly from the usual system, and third that mitochondrial genes sometimes contain segments of DNA called introns (intervening DNA sequences that do not code for parts of proteins), while prokaryotic genes do not.

These critics offer an alternative hypothesis: an **endogenous** (within-generated) origin of mitochondria. The membranes of some bacteria contain molecules needed for respiration as well as attachment sites for DNA. Mitochondria could have originated when fragments of such bacterial membranes broke off and formed separate sacs within the bacterium's cytoplasm, enclosing attached DNA (perhaps in a small loop called a plasmid), ribosomes, and respiratory molecules. Cytoplasmic and mitochondrial DNA and ribosomes then would have evolved at different rates.

Chloroplasts are more similar to bacteria than mitochondria are. This leads some biologists to concede that chloroplasts began as bacterial symbionts, but that mitochondria originated endogenously. Others think, on the contrary, that this reflects a history in which mitochondria became symbionts long before chloroplasts. They point out, first, that all organisms with plastids also have mitochondria, but not vice versa. And that we would expect mitochondria to be less like bacteria if they had been symbionts for a longer time than chloroplasts.

To return to our original question: if the endosymbiotic theory is correct, all eukaryotes are the descendants of prokaryotes who engulfed, but did not digest, other, distinctly specialized prokaryotes (Figure 5-B). The endosymbiotic prokaryotes gave their hosts an advantage over their uninhabited brethren, and so natural selection perpetuated the relationship. Although there is disagreement on the historic sequence in which mitochondria and chloroplasts became endosymbionts, we see relics of their once-independent existence in their DNA, which is independent of nuclear DNA.

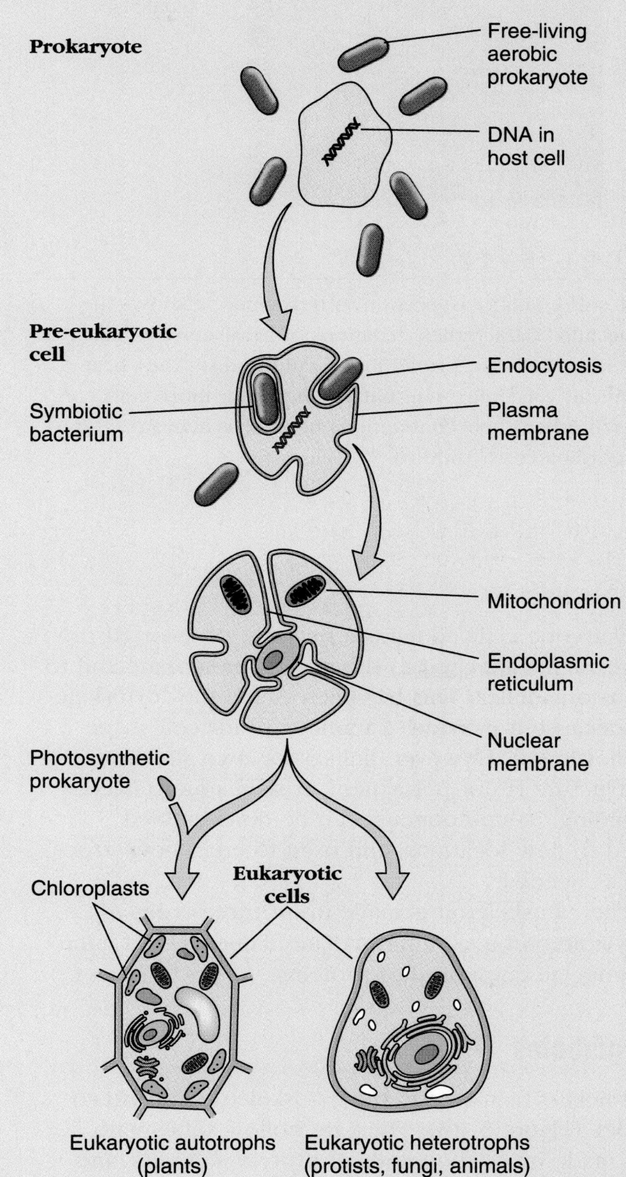

FIGURE 5-B

The endosymbiont theory of eukaryotic cell origins. According to this theory, mitochondria originated from free-living aerobic bacteria that developed a symbiotic relationship inside of early eukaryotic cells. In addition, the endoplasmic reticulum and Golgi complex may have evolved from internal folds of the plasma membrane.

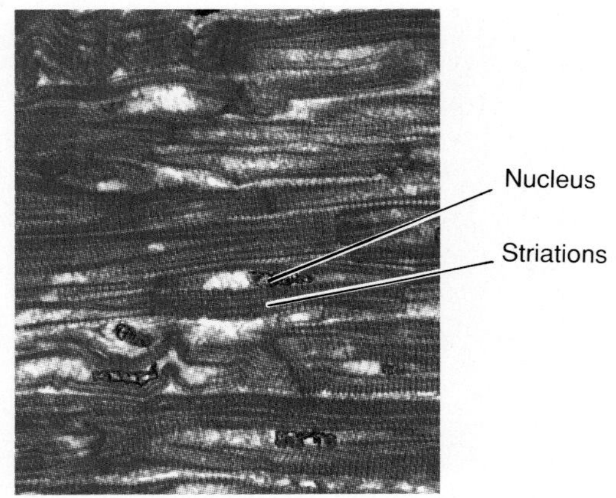

FIGURE 5-12

Heart cells. This cross-section of heart muscle shows the nucleus and characteristic striations of heart muscle cells. When heart cells are placed into a culture dish, they beat rhythmically on their own. But when two or more cells contact each other, their contractions become synchronized into one coordinated rhythm! (G. W. Willis/BPS)

all eukaryotic cells contains a network of assorted protein filaments attached to the plasma membrane and to various organelles. This has been called the "cytoskeleton" because it provides a framework for cell shape and movement. However, unlike our own skeletons, the structure is not permanent. It seems more like a scaffolding: its components can be disassembled, moved to new locations, and used to erect new structures as needed.

The cytoskeleton is made up of three types of fibers: microtubules, intermediate filaments, and microfilaments, all composed of proteins.

Microtubules

The thickest filaments in the cytoskeleton are **microtubules** (Figure 5-13). They are hollow tubes with walls made up of thousands of protein subunits, and they can be assembled and dismantled rapidly from subunits that are always present in the cytoplasm.

Microtubules serve both as a skeletal framework and as tracks for the movement of organelles. Proteins associated with the microtubules use energy from ATP to move the organelle along.

Microtubules play an important part in cell division (Chapter 11). A great many eukaryotic cells also have other structures composed of complex arrangements of microtubules: cilia or flagella, and centrioles.

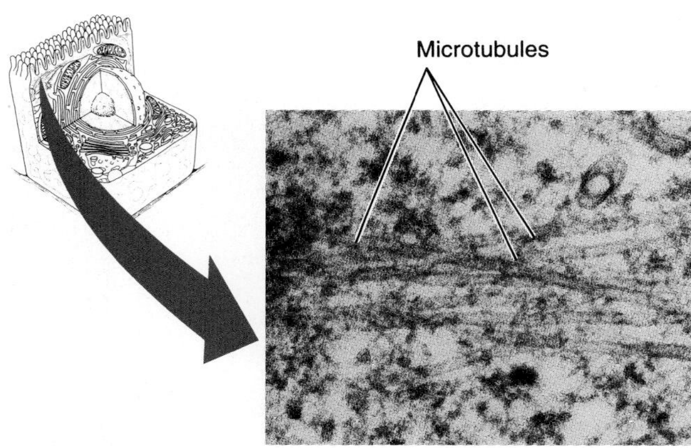

Microtubule structure

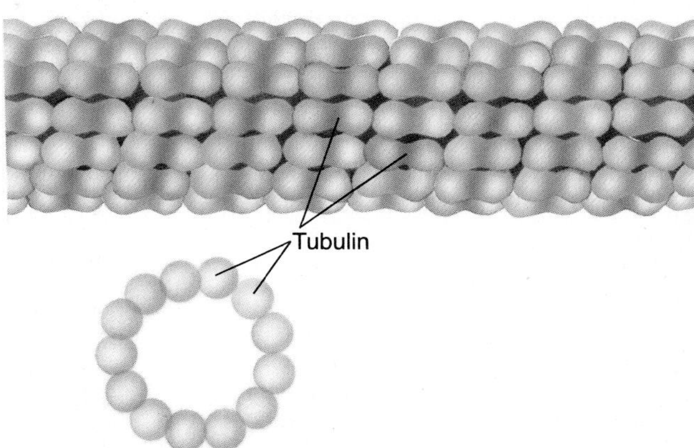

FIGURE 5-13

Microtubules. Microtubules are microscopic tubes composed of many thousands of protein subunits. Microtubules play important roles in the sorting of chromosomes in mitosis and meiosis, the movement of organelles, and in changes in cell shape. (David M. Phillips/Visuals Unlimited)

Cilia and Flagella

Cilia (singular: **cilium**) and **flagella,** are thread-like organelles present on the surfaces of many eukaryotic cells. Cilia are generally shorter and more numerous than flagella, but both have the same basic structure.

Cilia and flagella are organelles of locomotion, serving either to propel the cell through its environment or to move something past the cell. For example, many protists move by the beating of their cilia or flagella, and a human sperm moves by lashing the single flagellum that forms its tail. Cilia also move mucus and debris up and out of the human air passages, thereby helping to keep the lungs clear.

Cilia on cells lining the oviduct

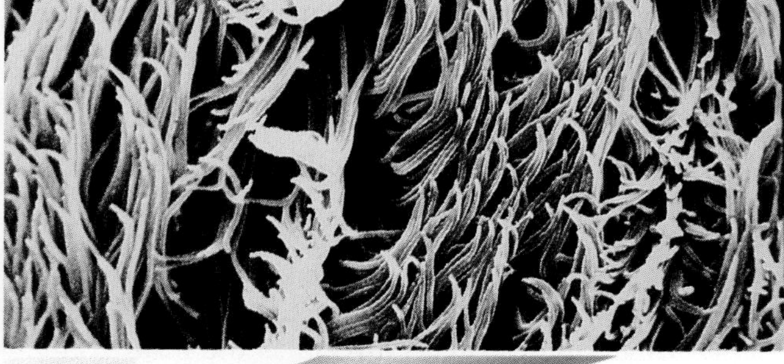

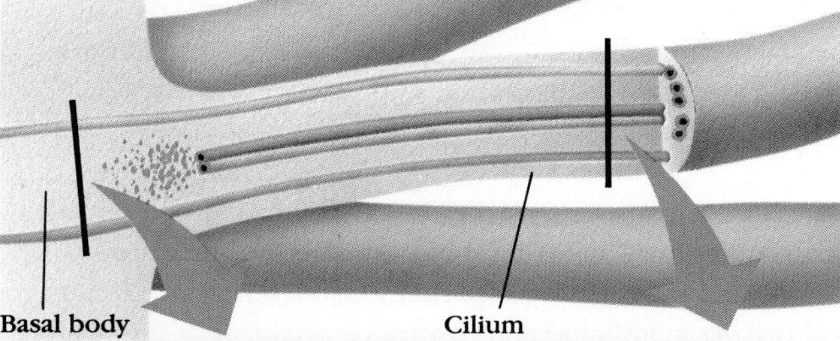

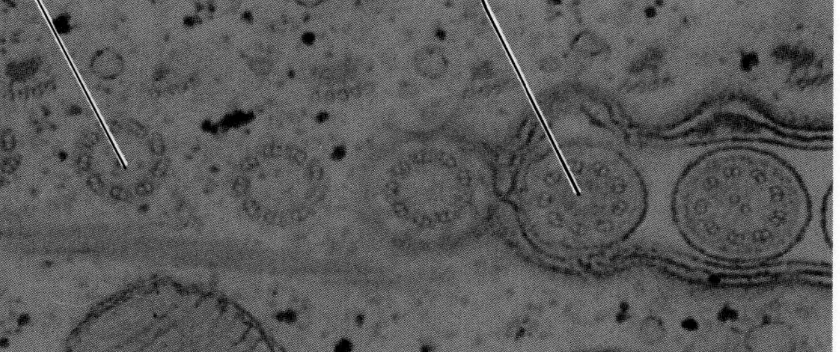

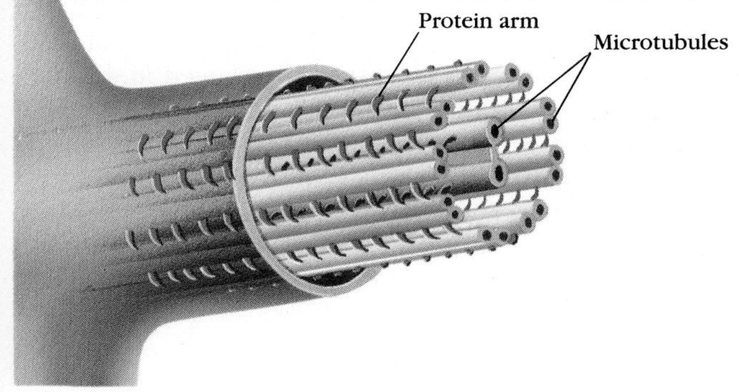

FIGURE 5–14

Cilia. Cilia are composed of long extensions of the cytoplasm that each contain a core of microtubules. Microtubules are organized into nine pairs forming a circle with another pair in the center. Protein arms extending between microtubules help to generate the force that causes cilia to bend, which results in a whip-like action that can move materials past the cell surface. (David M. Phillips/Visuals Unlimited)

Electron micrographs of eukaryotic cilia and flagella show that each contains a circle of nine pairs of microtubules, with two single microtubules in the center. All this is covered by an extension of the plasma membrane. A cilium or flagellum grows from its **basal body,** found where the organelle joins the cell body. The basal body consists of a circle of nine microtubule triplets (instead of pairs), and there are no microtubules in the center of the circle (Figure 5–14).

A cilium (or flagellum) moves by the action of "arms" of a protein that extend from one microtubule of each pair. These protein arms attach briefly to a microtubule in the next pair and use energy from ATP to produce a sliding force. This bends the whole bundle of microtubules in the cilium, and the cilium pushes against the fluid outside the cell (Figure 5–14).

Centrioles

Eukaryotic cells (except cells of higher plants) contain a pair of **centrioles,** oriented at right angles to each other. Each centriole has the same arrangement of microtubules as the basal body of a cilium (Figure 5–15).

Before cell division, the two centrioles move apart and each somehow directs the formation of a new partner at right angles to itself. When the cell divides, the two pairs of centrioles separate and each pair ends up in one of the two new cells (Chapter 11). Centrioles also give rise to the basal bodies of cilia.

Microfilaments

Microfilaments are made up of the globular protein actin. Long before its role in the protein framework of eukaryotic cells was recognized, actin was familiar as one of the contractile proteins in muscle cells.

We now know that it is the most abundant cytoskeletal protein, and also the most abundant protein inside many eukaryotic cells. As with microtubules, free actin molecules in the cytoplasm can be assembled into filaments as needed and later disassembled into actin subunits, which can be reused. Actin filaments can also serve as tracks for the movement of organelles.

Microfilaments are especially plentiful just under the plasma membrane. Here they are attached to the membrane and can strengthen, pull, or push the membrane during changes in cell shape. Associated with microfilaments are molecules of **myosin,** the protein that interacts with actin in muscle cells to produce

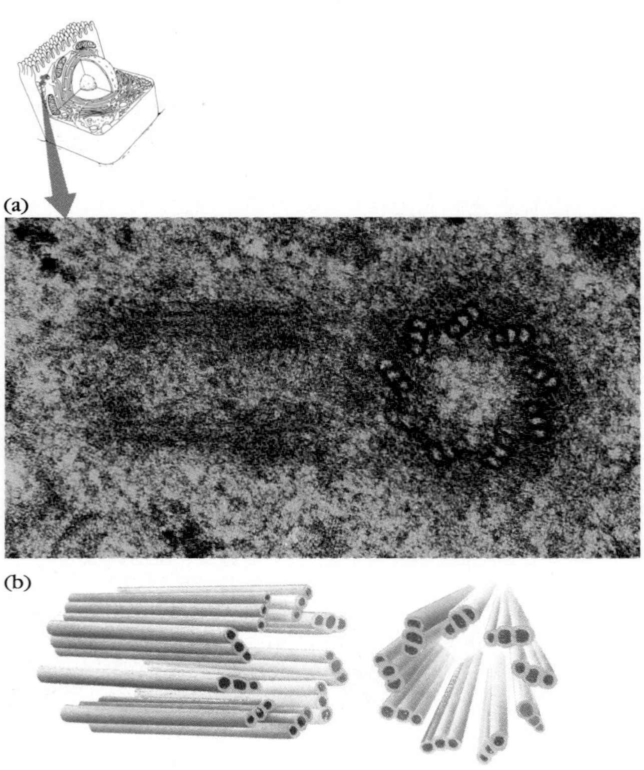

FIGURE 5-15

Centrioles. (a) Centrioles are usually found in pairs that are at right angles to each other. **(b)** Each centriole is composed of a barrel-shaped arrangement of nine groups of three microtubules. (K. R. Porter/Photo Researchers, Inc.)

contraction. Actin filaments and myosin are responsible for many cellular motions, such as cytoplasmic streaming in plant cells, endocytosis, and exocytosis, as well as for muscle contraction. Microfilaments also play an active role in ameboid movement and other changes in cell shape. When an animal cell divides, bundles of microfilaments constrict the cell around the middle and divide it into two. Actin filaments can also give shape to protrusions of the cell surface. For example, in cells of the small intestine, bundles of actin filaments form the cores of microscopic projections called microvilli that increase the efficiency of absorption of digested foods.

Organization of the Cytoskeleton

Most of what we know about the cytoskeleton has been discovered since 1970, and we do not yet have a complete picture of how it is built and how its components interact. With this in mind, we can make a tentative outline to encompass the evidence now available.

Fibers of the cytoskeleton extend throughout the cytoplasm of eukaryotic cells, and some kinds can be broken down and reassembled as required. Microtubules give the cell its general shape, act as tracks on which various organelles move, and make up the framework of cilia and flagella. Next smaller are the intermediate filaments, which are thought to provide mechanical strength. Even thinner are the microfilaments, which interact with myosin as the main contractile elements of the cytoskeleton. These are the fibers responsible for much of the actual movement within cells and changes in cell shape, and they can also serve as tracks for transport of organelles.

The eukaryotic cell's cytoskeleton is composed of a network of protein fibers including microtubules, intermediate filaments, and microfilaments. The cytoskeleton is attached to the plasma membrane and to various organelles. This scaffolding is responsible for the shape, strength, and movement of the cell or of the organelles within it. In addition, microtubules form the functional core of cilia and flagella and microfilaments help stiffen microvilli.

5-M TISSUES AND ORGANS: CELLS ORGANIZED INTO WORK PARTIES

All but the simplest multicellular organisms contain different types of cells, many of them arranged in groups specialized for different functions. A **tissue** is a group of cells of one or a few types and their intercellular substances held together in a characteristic pattern and performing a particular function. **Organs** are functional units of the body made up of more than one type of tissue. Examples of organs are eyes, kidneys, muscles, leaves, and roots.

Animal Tissues

In animals, most cells are surrounded by a space containing **extracellular fluid**—the cell's immediate environment. It is the source of nearly all the substances the cell takes in, as well as being the immediate sink for the cell's wastes.

Animal tissues are divided into four main types:

1. **Epithelial tissues** form coverings and linings (Figure 5-16). Epithelia cover the outside of the body and line the cavities of tubes such as the digestive tract, lungs, vagina, and mouth. In keeping with its function, this type of tissue forms sheets one to several cells thick, with the cells tightly packed to-

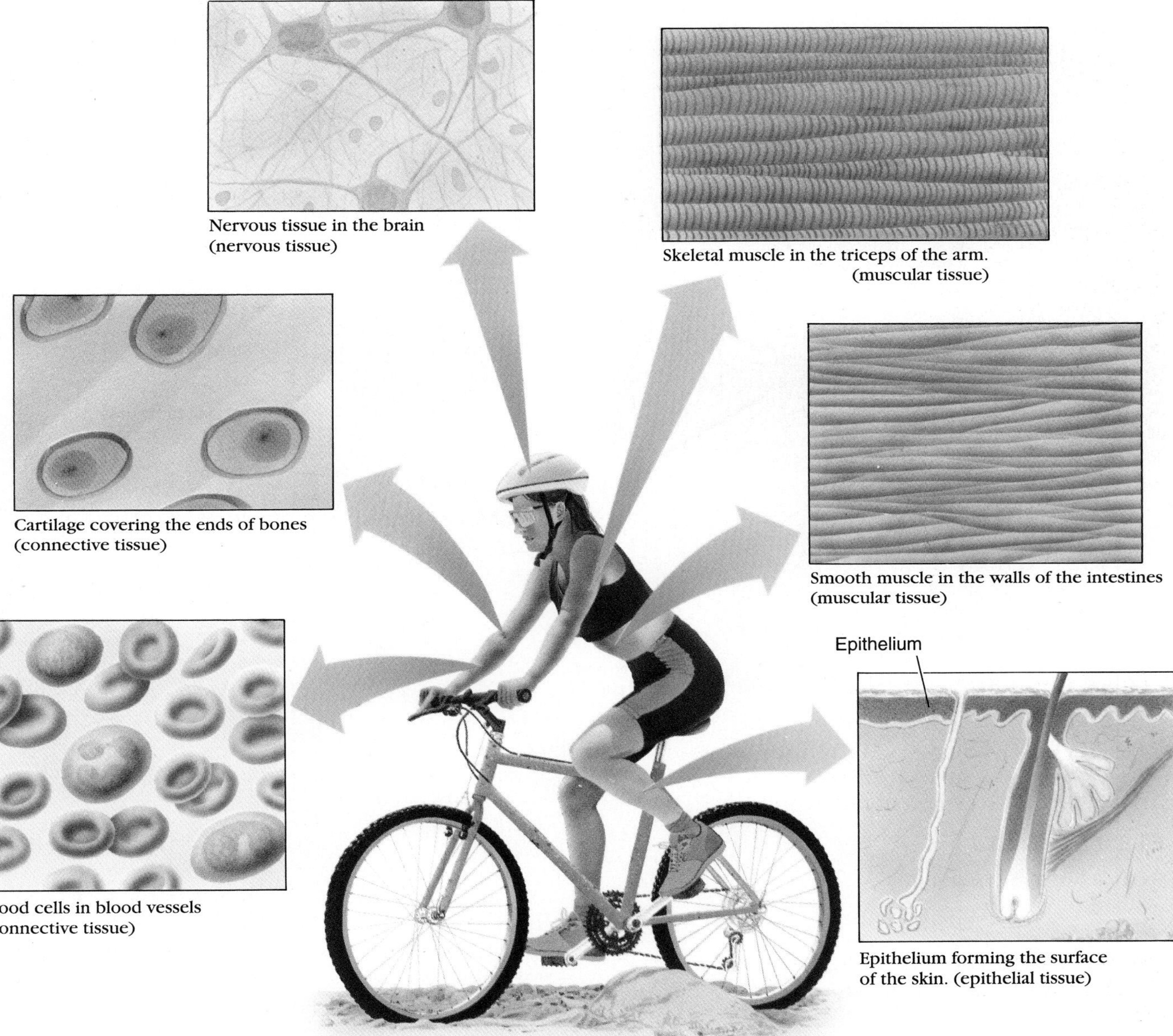

Nervous tissue in the brain (nervous tissue)

Skeletal muscle in the triceps of the arm. (muscular tissue)

Cartilage covering the ends of bones (connective tissue)

Smooth muscle in the walls of the intestines (muscular tissue)

Blood cells in blood vessels (connective tissue)

Epithelium forming the surface of the skin. (epithelial tissue)

FIGURE 5–16

Examples of animal tissues.

gether. Tight junctions and desmosomes often strengthen the attachments between epithelial cells (Section 4-F).

2. **Connective tissue** is probably the most abundant type of animal tissue. Adipose tissue (fat), cartilage, and bone are familiar examples. Connective tissue has a great deal of material between the cells, usually in the form of fibers or gelatinous substances secreted by the cells.
3. **Nervous tissue** contains nerve cells, which conduct electrical impulses.
4. **Muscle tissue** is made up of cells that can both conduct electrical impulses and contract (Chapter 29).

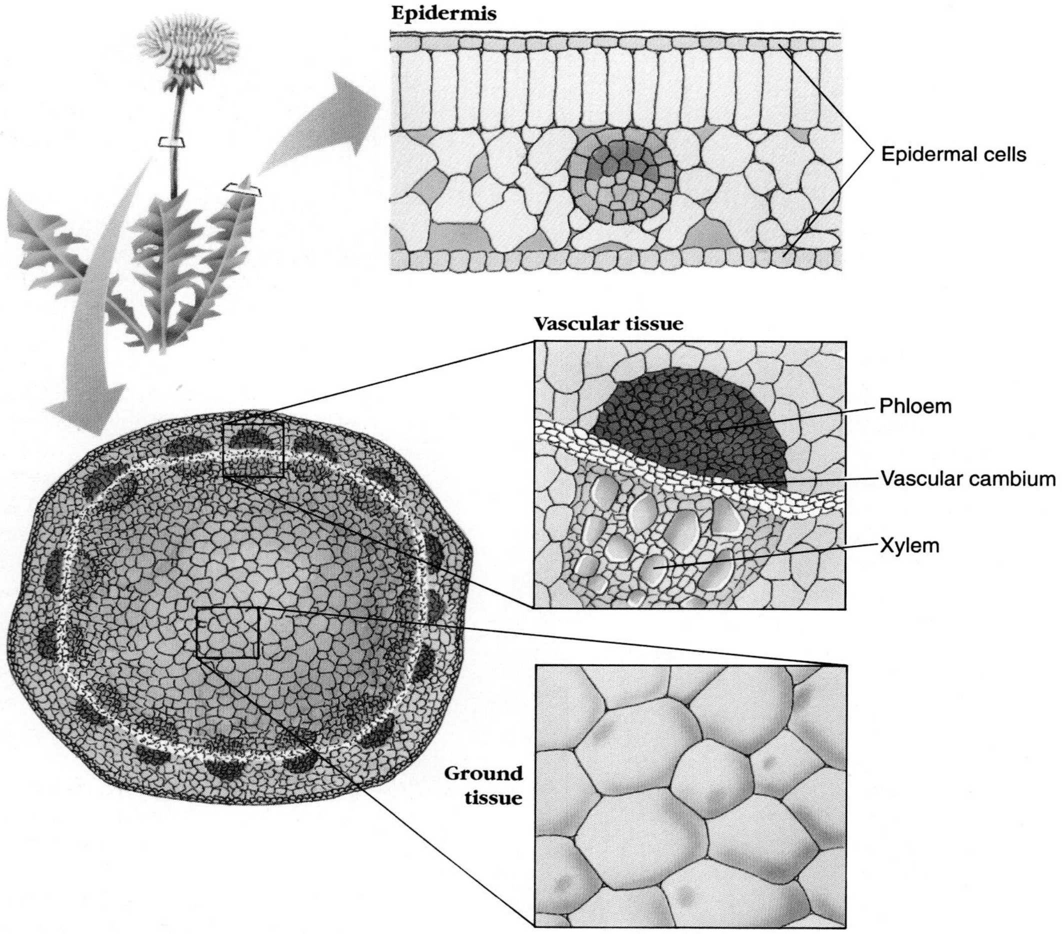

Epidermis
Epidermal cells
Vascular tissue
Phloem
Vascular cambium
Xylem
Ground tissue

FIGURE 5-17

Examples of plant tissues.

Plant Tissues

Plant cells surround themselves with thick but porous cell walls, which are firmly cemented to the walls of neighboring cells. Cells exchange substances with the fluid that fills the spaces between cells and seeps into the pores of the cell walls.

There are three main types of plant tissues:

1. **Epidermis,** the equivalent of epithelium in animals, covers the outsides of leaves and of young stems and roots (Figure 5-17).
2. **Vascular tissue** transports water, food, hormones, and so on between different parts of the plant. It is familiar as the veins in leaves and the wood of trees.
3. **Ground tissue** fills the spaces between the epidermis and vascular tissue inside leaves and in nonwoody stems and roots. It consists mostly of **parenchyma** cells, which are often loosely packed, with many spaces between the cells. Some parenchyma cells contain chloroplasts, organelles specialized for photosynthesis.

Tissues are groups of similar cells that perform a specific function. Several types of tissues function together to form organs, such as the heart or liver in animals or leaves and roots in plants.

SUMMARY

1. The prokaryotic cells of bacteria contain all structures necessary for life. The plasma membrane separates the cell contents from the environment and controls the passage of substances into and out of the cell. The cytoplasm contains small molecules and ions, food storage deposits, metabolic enzymes, and ribosomes, where the cell's proteins are made following instructions from the DNA in the cell's nuclear area. Most prokaryotes have a cell wall, which gives the cell shape and protection.
2. Eukaryotic cells have all of the above features, except that animal cells and some protists lack cell walls. In addition, they have a cytoskeleton and contain a number of separate, interior compartments in the form of membrane-bounded organelles. Table 5-1 summarizes the important structures found in eukaryotic cells.
3. The hallmark of eukaryotic cells is a nucleus surrounded by a nuclear envelope, which consists of two membranes. The nucleus contains genetic material in the form of the DNA. All RNA is made in the nucleus.
4. RNA leaves the nucleus and participates in protein synthesis in the cytoplasm. Some ribosomes attach to the endoplasmic reticulum if they are making proteins for lysosomes, the plasma membrane, or export from the cell. Ribosomes making other kinds of proteins remain free in the cytosol.
5. Many membranous compartments in the cell are connected, either physically or by way of the membranes of transport vesicles. These bud off from one, move to another, and then fuse to become part of a new compartment.
6. The endoplasmic reticulum sends newly assembled proteins in transport vesicles to the flattened, membranous sacs of the Golgi complex, where they are finished, sorted, and packaged into new transport vesicles for their final destinations. Some of these vesicles from Golgi complexes expel material by exocytosis, and the vesicle membrane becomes part of the plasma membrane.
7. Membrane material is returned to structures inside the cell via endocytosis. Some endocytotic vesicles fuse with lysosomes, which digest their contents.
8. In all of these transfers, the contents of the membranous compartments are kept separate from the cytosol.
9. Mitochondria are membrane-bounded organelles that do not take part in the membrane exchanges described above. They produce most of the cell's supply of ATP.
10. Eukaryotic plant cells, including photosynthetic protists, have three features not found in animal cells: (*a*) plastids, such as photosynthetic chloroplasts and starch-storing amyloplasts, (*b*) a cell wall that is a porous but fairly rigid, protective, and supportive structure made largely of cellulose fibers and located outside the plasma membrane, and (*c*) a large, prominent vacuole, which stores fluid and various cell products.
11. The eukaryotic cell's cytoskeleton is responsible for shape, strength, and movement of the cell or of organelles inside the cell. It is composed of a variety of protein fibers: microfilaments, intermediate filaments, and microtubules.
12. Microtubules also form the framework of cilia and flagella, thread-like projections at the cell surface that move the cell itself or move substances past the cell's surface.
13. In multicellular organisms, cells are organized into tissues, such as epithelial and vascular tissue. Several types of tissue may be organized to form organs, like kidneys or roots, each with its own particular function.

SELF-QUIZ

Questions 1 to 11 describe cell components. From the list of structures (a to q) below, choose one *or more* that matches each description.

a. cell walls
b. centrioles
c. chromosomes
d. cilia
e. endoplasmic reticulum
f. flagella
g. Golgi complexes
h. intermediate filaments
i. lysosomes
j. mesosome
k. microfilaments
l. microtubules
m. mitochondria
n. nucleoli
o. plastids
p. ribosomes
q. vacuoles

___ 1. Sites of protein synthesis.
___ 2. Used to propel a cell through a fluid, or to move a fluid past the surface of a cell.
___ 3. Rigid coverings of some cells.
___ 4. Carry hereditary information of cell.
___ 5. Finish and package products for export from cell.
___ 6. Contain digestive enzymes of cell.
___ 7. Impart color to leaves.
___ 8. Storage compartments in plant cells.
___ 9. Sites of ribosome synthesis.
___ 10. Make most of the cells' ATP by cellular respiration.
___ 11. Involved in cell movement.

For the items listed below, place check marks indicating whether each is found in cells of animals, cells of plants, and/or prokaryotic cells.

	Animal	Plant	Prokaryote
12. ribosome	___	___	___
13. flagellum	___	___	___
14. cell wall	___	___	___
15. chromosome	___	___	___
16. mitochondrion	___	___	___

17. Which of the following is *not* found in the cells of higher plants?
    a. plasma membrane
    b. cell wall
    c. chloroplast
    d. ribosome
    e. centriole

THINKING CRITICALLY

1. Would you expect cells that produce hair to contain more ribosomes than cells that store fat? Why?
2. It has been said that animals, as we know them, could not exist if they had cell walls. Why not?
3. What is the advantage to cells of keeping microtubule protein subunits on hand in the cytoplasm, rather than making them anew from amino acids each time they are needed?
4. Tobacco smoke reduces the activity of cilia in the air passages between the throat and lungs. How does this contribute to "smoker's cough" and lung disease?
5. What might be some of the advantages of having a nuclear envelope around the genetic material of a cell?
6. Based upon the information in Section 5-H, what evidence suggests that mitochondria and plastids may have been free-living organisms that became incorporated into eukaryotic cells?

SELECTED KEY TERMS

cellular respiration, *p. 96*
cell wall, *p. 97*
centriole, *p. 103*
chloroplast, *p. 97*
chromosome, *p. 89*
cilia, *p. 102*
connective tissue, *p. 105*
cytoplasmic streaming, *p. 99*
cytoskeleton, *p. 87*
cytosol, *p. 87*
epidermis, *p. 106*
epithelial tissue, *p. 104*
eukaryote, *p. 85*
flagella, *p. 87*
Golgi complex, *p. 94*
ground tissue, *p. 106*
lysosome, *p. 95*
mesosome, *p. 86*
mitochondria, *p. 96*
muscle tissue, *p. 105*
nervous tissue, *p. 105*
nucleoli, *p. 89*
organ, *p. 104*
parenchyma, *p. 106*
pili, *p. 86*
prokaryote, *p. 85*
protist, *p. 85*
rough endoplasmic reticulum, *p. 93*
smooth endoplasmic reticulum, *p. 94*
vascular tissue, *p. 106*
vesicle, *p. 94*

SUGGESTED READINGS

Alberts, B., D. Bray, J. Lewis, M. Raff, K. Roberts, and J. D. Watson. *Molecular Biology of the Cell,* 2d ed. New York: Garland Publishing, 1989. A comprehensive but readable text.

Allen, R. D. "The microtubule as an intracellular engine." *Scientific American,* February 1987.

Feldman, M., and L. Eisenbach. "What makes a tumor cell metastatic?" *Scientific American,* November 1988.

Glover, D. M., C. Gonzalez, and J. W. Raff. "The centrosome." *Scientific American*, June 1993.

Kabnick, K. S., and D. A. Peattie. "Giardia: A missing link between prokaryotes and eukaryotes." *American Scientist,* January 1991. An interesting analysis of the classification of this single-celled intestinal parasite.

Koch, A. L. "Growth and form of the bacterial cell wall." *American Scientist,* July 1990.

Oldstone, M. B. "Viral alteration of cell function." *Scientific American,* August 1989.

Stossel, T. P. "How cells crawl." *American Scientist,* September 1990.

Young, J. D., and Z. A. Cohn. "How killer cells kill." *Scientific American,* January 1988.

CHAPTER 6

CONCEPT GUIDE

After reading this chapter, you should be able to:

1. Describe the laws of thermodynamics and explain how they apply to living organisms. (Section 6–A)
2. Describe three differences between exergonic and endergonic reactions. (Section 6–B)
3. Explain why respiration can be thought of as "the undoing of photosynthesis." (Section 6–C)
4. Describe the roles of redox reactions and ATP in photosynthesis and respiration. (Sections 6–D and 6–E)
5. Describe how most ATP is made by photosynthesis and respiration. (Section 6–E)

Like the up and down ride on a roller coaster, energy-producing reactions provide the power to drive energy-consuming reactions. Read more about how the chemical reactions in a cell are closely integrated in Section 6–B. (Michael Tamborrino, The Stock Market)

Energy and Living Cells

You can have ten U.S. dollars in many forms: one $10 bill, two $5 bills, ten $1 bills, 40 quarters, 100 dimes, 200 nickels or 1000 pennies. Although these forms of currency are physically distinct, they are interchangeable and have equivalent monetary value. **Energy,** defined as the ability to perform work, is much the same in that it also has many interchangeable forms.

In general, energy is either termed **potential energy,** if it is stored and available for use, or **kinetic energy,** if it is associated with motion. Potential energy is associated with position. For example, the dictionary that is about to fall off your desk has mechanical potential energy because of its position. When the dictionary tips over the edge of your desk and falls to the floor, this potential energy is transformed into mechanical kinetic energy, the form in which it can do work. (In this example, "work" could be sending loose papers flying, squashing a bug, or making a sound that frightens your cat.) **Chemical potential energy** is stored in the bonds between atoms in a molecule. It is released and made available to do work when those chemical bonds are broken. Finally, electrical potential energy is present when a barrier separates oppositely charged particles. When this barrier is breached, the electrical charge moves because of the attraction between the opposite electrical charges (see Figure 6–1). All of the forms of potential and kinetic energy are biologically important, but there is a cost involved when one form of energy is transformed into another.

In this chapter we introduce the rules of energy transactions and show how they apply to the chemical reactions carried out by cells.

KEY CONCEPTS

- Energy enters the living world when green plants capture solar energy during photosynthesis. Plants use this energy to build food molecules from carbon dioxide and water.
- Energy is released and made available to do useful things during respiration as food is broken back down to carbon dioxide and water.
- Organisms use energy to combat the universal tendency toward increasing disorder. They preserve their organization by using energy to make molecules needed for growth, repair, and reproduction; to move substances within the body; and often to move the body itself.

Curiosity Questions

1? Why do we get hot when we exercise? (See Figure 6-3, page 115)

2? Why is photosynthesis important to humans? (See page 116)

3? How does the food that we eat produce energy for a cell? (See page 116)

6–A ENERGY TRANSFORMATIONS

Just as when you break your $10 bill into change at the laundromat and receive equivalent monetary value in quarters, changing the form of energy does not increase or decrease the amount of energy. But, like a bank charge on a check, there is a cost involved in an energy transformation. Available energy always goes downhill. Energy is always converted from a more concentrated, more useful form to a less concentrated, less useful form. For example, the motor of a washing machine converts electrical energy (a concentrated form of kinetic energy) into the mechanical kinetic energy that turns the washer's agitator and sloshes your dirty clothes around in the water. Have you ever noticed the warmth given off by a washing machine, even when you're using cold water? That heat is the cost of converting electrical kinetic energy into mechanical kinetic energy. The heat given off by a washing machine is impossible to reconcentrate and use to do subsequent work, and the wasted heat generated from any energy transformation is seldom enough to power another energy transformation. Heat is a diffuse sort of energy that results from random movement of molecules. It is not very usable because it is difficult to gather together to use again.

Energy transformations are governed by two **laws of thermodynamics.** The first law is related to the origin of different forms of energy. It states that energy can be neither created nor destroyed, but only changed from one form to another. This is like the bank giving you exactly one $10 bill—no more, no less—in exchange for 100 dimes. The second law of thermodynamics states that energy always goes from more useful to less useful forms, such as heat. Once it has been degraded into heat, energy cannot be recovered and reconcentrated to power work. Because more useful forms of energy are stored in chemical bonds of molecules that are highly organized and structured, and because heat is a diffuse kind of energy, physicists say that there is an increase of disorder that accompanies each energy conversion. The technical term for this increase of disorder is **entropy.** Thus, whenever energy is transformed from one form to another, heat is produced and entropy increases.

The laws of thermodynamics apply to everything in the universe, including living creatures. Organisms constantly process energy—obtaining it, storing it, or releasing it from food and using the released energy to power activities that require energy. Plants get a regular supply of energy from the sun and store it in the chemical bonds of food molecules. Animals get their energy from the foods they eat, which come directly or indirectly from plants. It is important to understand that the driving force of an energy transaction is the *decrease* in useful energy that occurs during the transformation. Furthermore, cells can use energy only as it is released and "runs downhill" to a lower energy level as an energy transformation occurs.

Living cells use energy in all the forms shown in Figure 6–1. For example, nerve cells perform electrical work as they transmit an electrical nerve impulse, perhaps from the brain to a muscle, telling the muscle to contract. A nerve cell sets up an electrical potential similar to the one shown in Figure 6–1c, by separating positively and negatively charged ions across the barrier of its plasma membrane. This membrane potential (Figure 6–1c) is transformed into the kinetic electrical energy of a nerve impulse when some of the ions are permitted to move through the membrane. The attraction between oppositely charged particles is the main force that moves the ions. However, the useful kinetic energy of the moving ions is dissipated as the ions mingle, leaving only the intrinsic, random kinetic energy (heat) of the ions themselves. Entropy also has increased because the ions are now mixed more completely, that is, they are less orderly. To restore its membrane potential, the cell must expend energy on the active transport of ions against their electrical and chemical gradients.

Muscles perform mechanical work, using potential chemical energy stored in molecules of ATP (adenosine triphosphate) (Figure 6–1b). Breaking a chemical bond in ATP releases energy, and while some of this energy turns into heat, most of it is used to "cock" certain protein molecules in the muscle, causing the muscle to contract. The cell must continually expend energy to keep the muscle contracting. Fresh supplies of ATP are needed constantly, and making new ATP is one of the many kinds of chemical work the cell must perform.

Energy, the ability to do work, cannot be created or destroyed, but only changed from one form to another. In any energy conversion, some or all of the useful energy is converted to less useful energy, usually heat, and entropy increases. Organisms use energy in many forms to build, maintain, and move their tissues and cells.

6–B CHEMICAL REACTIONS AND ENERGY

In our study of biology, we have described a cell as a "busy biochemical factory." By now you have some ideas of how a living cell is organized and of the kinds of chemical changes that take place within it. We now add energy transactions to the picture of cellular metabolism we are building.

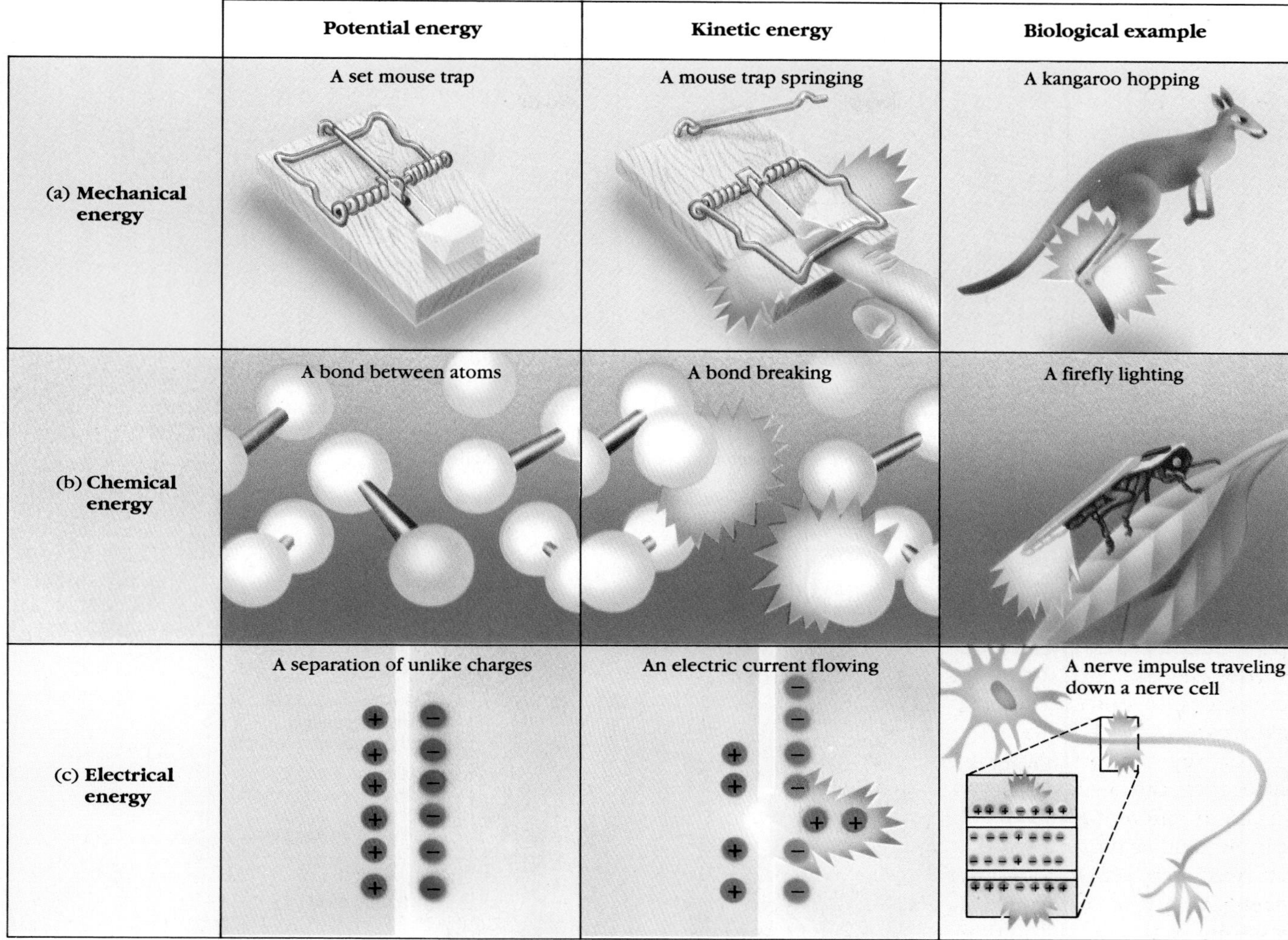

Potential energy
Kinetic energy
Biological example
(a) Mechanical energy
A set mouse trap
A mouse trap springing
A kangaroo hopping
(b) Chemical energy
A bond between atoms
A bond breaking
A firefly lighting
(c) Electrical energy
A separation of unlike charges
An electric current flowing
A nerve impulse traveling down a nerve cell

FIGURE 6–1

Some important forms of energy in biological systems. The stretched tendons in a kangaroo leg and the set-spring of a mousetrap both hold potential energy. Once released, the potential energy is converted to kinetic energy resulting in action. Chemical potential energy is stored in chemical bonds. When these bonds are broken, the energy is released. In nerves, charged particles are kept separate until a signal travels down a nerve, releasing their potential energy and contributing to the nerve impulse.

Each of a cell's metabolic chemical reactions is an energy transformation that follows the laws of thermodynamics. There are two broad categories of reactions, based upon their effect upon the amount of useful energy that remains when they are finished. **Exergonic reactions** (ex = out of; ergon = work) occur spontaneously, and result in products containing less energy than the materials they started with (Figure 6–2). **Endergonic reactions** (endon = within) are not spontaneous; they require energy inputs and result in products containing more energy than the materials they started with (Figure 6–2). Exergonic reactions can provide the energy to power endergonic reactions. This may be easier to understand if you compare these two kinds of reactions to a roller coaster. As the coaster plummets down that first heart-stopping grade (exergonic reaction), it gains enough inertial force to power it up the next steep hill (endergonic reaction). But think about a roller coaster for a minute: no hill that it climbs is *ever* as high as the first one. The potential

Exergonic reactions:

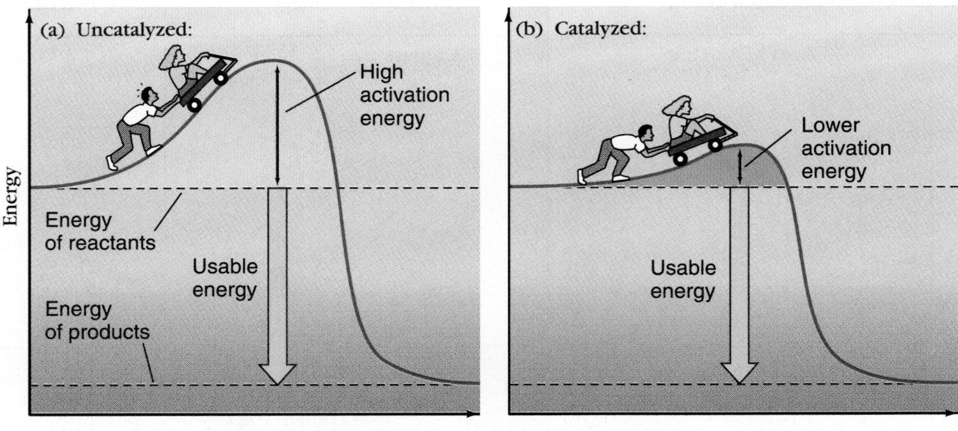

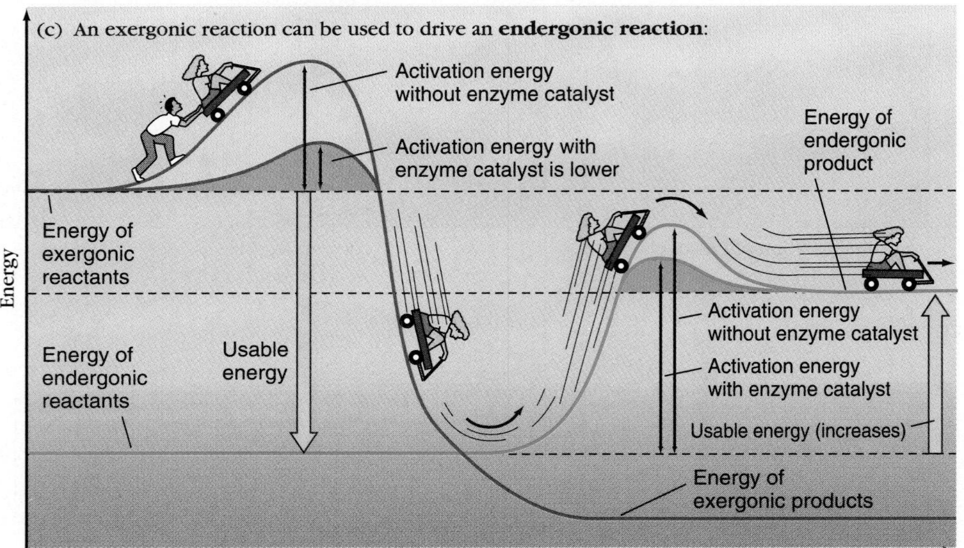

FIGURE 6–2

Exergonic and endergonic reactions. (a) Exergonic reactions release energy and may occur spontaneously. But like this example with a wagon, exergonic reactions need a certain amount of energy added, the activation energy, to get "over the hump" before the process can begin. **(b)** Catalysts can reduce the amount of activation energy needed to begin a reaction (lowering the hump). **(c)** The energy released from an exergonic reaction can be used to raise the energy level of the product of an endergonic reaction. In the wagon analogy, the speed of the wagon at the bottom of the exergonic hill pushed the girl up the endergonic hill. Like the exergonic reaction, an activation energy "hump" must be exceeded as the girl ends her endergonic ride. The height of this hump, or the amount of activation energy, is also lowered by catalysts in endergonic reactions.

energy from this climb powers the first plunge (exergonic reaction) and the pattern repeats, with the peaks and valleys flattening out until the downward pull of gravity overcomes the force of the forward movement of the coaster. Eventually, all the potential energy from the first climb is converted to heat. With this roller coaster image in mind, you can appreciate how pairs of exergonic and endergonic reactions work together within cells. Although both kinds of reactions require an input of **activation energy** to get them started, endergonic reactions usually need the burst of additional energy from an exergonic reaction. The smaller amount of residual energy from an endergonic reaction will often bring an exergonic reaction to activation energy.

At the temperatures found in living organisms, few molecular collisions would have enough force to reach activation energy. So, you would expect chemical reactions to occur very slowly. As we have learned, however, living cells can increase the rate of reactions because their enzymes act as catalysts and lower the activation energy needed to start a reaction.

Strongly exergonic reactions, such as the oxidation of glucose to carbon dioxide and water, produce heat and will continue until almost all of the reactants are used up (Figure 6-3).

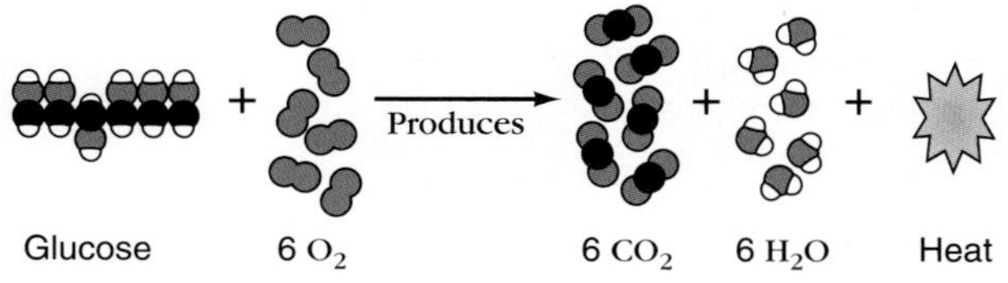

Where the energy change in a reaction is not great, however, the reaction has a tendency to go in both directions, as in the conversion of glucose-phosphate to another sugar phosphate, fructose-phosphate.

$$\underset{\text{(reactants)}}{\text{glucose-phosphate}} \rightleftharpoons \underset{\text{(products)}}{\text{fructose-phosphate}}$$

Under standard laboratory conditions, this reaction is exergonic in the direction of the longer arrow, and endergonic in the direction of the shorter arrow. As a reaction proceeds, the concentration of reactant decreases, and that of product increases. At the same time, the energy released by the reaction declines until it reaches zero. Think of a roller coaster that has a downhill run followed by a climb of equal magnitude. After entering the downhill run and rolling up the other side, the coaster would rock back and forth between the two. The coaster is at an equilibrium. Chemical reactions are much the same. At **equilibrium** the rates of the forward and backward reactions are equal and there is no net change in the proportions of reactants and products.

Like a roller coaster back in the "barn," a reaction at equilibrium is at an end. It releases no energy and cannot be used to do more work. To get the reaction working again, we must either add more reactants, remove products, or both, upsetting equilibrium in the process. This is what happens in a cell's metabolism: reactions are arranged in series, so that one reaction's products are the next one's reactants, and so on. The cell can get energy from some of its reactions because as long as the cell is alive, they *never* reach equilibrium. And, as we have said, enzymes make all of this happen faster. An enzyme lowers the energy barriers of activation to both the forward and reverse reactions, but does not alter a reaction's equilibrium point.

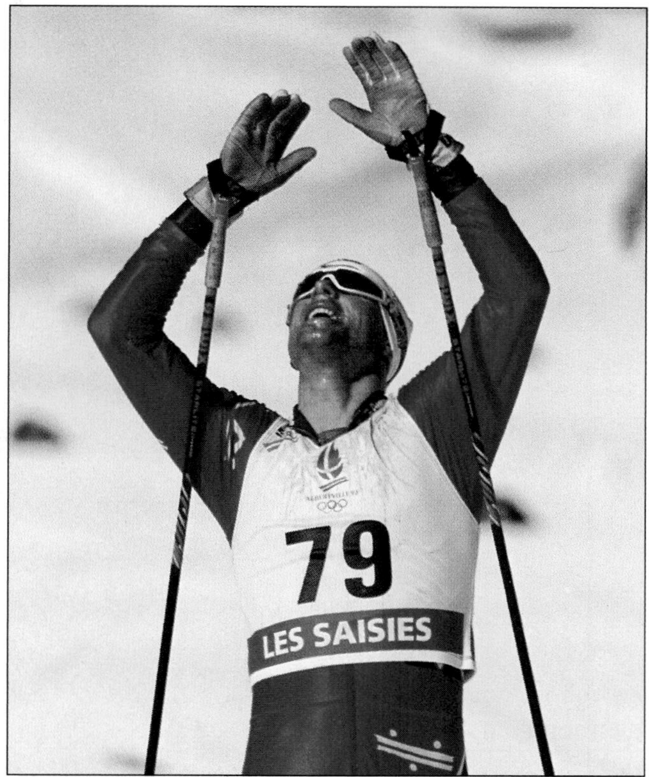

1? **FIGURE 6-3**

The heat of respiration. When we exercise, we need extra energy produced by cellular respiration to contract our muscles. In addition to producing energy-carrying ATP molecules, respiration produces carbon dioxide, which we exhale, water, and a great deal of heat energy. When the environment is cool, this heat helps to maintain our body temperature. In fact, when we are cold, shivering contracts muscles to produce heat. But when we are warm, we must sweat to lose the heat through evaporative cooling. (William R. Sallaz, Duomo)

Exergonic reactions occur spontaneously and release energy, while endergonic reactions require an energy input. Enzymes lower the activation energy levels and thus speed up reactions. To extract energy from reactions, a cell must keep substrates and products in disequilibrium by obtaining new energy sources and expelling end products.

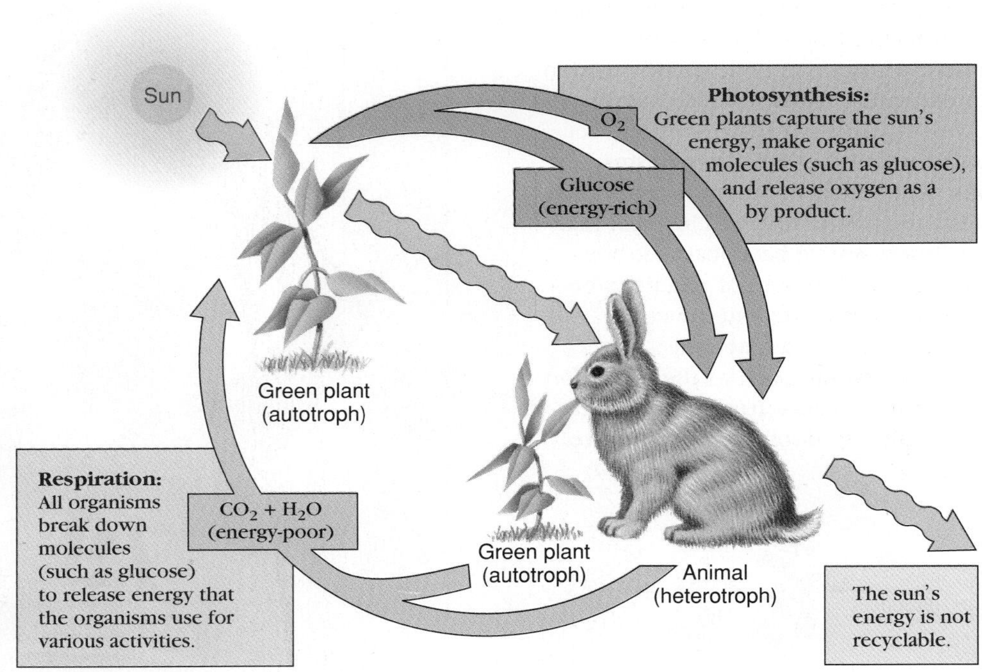

FIGURE 6–4

Photosynthesis and respiration. Although matter cycles indefinitely between photosynthesis and respiration, energy does not. The sun must constantly supply fresh energy to drive these processes.

6–C PHOTOSYNTHESIS AND RESPIRATION

As you have learned, cells get the energy they need from organic food molecules that contain chemical potential energy in the form of the bonds between atoms. This energy can be released by breaking the bonds, and the useful energy they contain can be used by the cell. But, where do these organic food molecules come from?

2? All food is made directly or indirectly by **photosynthesis** (Figure 6–4), a biochemical process that captures solar energy and stores it in the chemical bonds of organic molecules, usually carbohydrates $(CH_2O)_n$. (If you are puzzled by the "n" in this formula, it stands for "number." The formula means that a carbohydrate usually has twice the number of hydrogen atoms as it does carbon and oxygen atoms. For example: glucose $(CH_2O)_6$ has the formula $C_6H_{12}O_6$.) Carbon dioxide and water are the raw materials of photosynthesis. In this endergonic process, these simple, low-energy inorganic molecules are assembled into more complex, high-energy food molecules, and oxygen is given off as a by-product. Virtually all the oxygen in our atmosphere today was produced by photosynthesis. Burning organic matter releases carbon dioxide into the atmosphere and has environmental consequences. (See *A Journey into the Environment: Global Warming*.) ■

Photosynthetic organisms include plants, some bacteria, and some protists. They are all called **autotrophs** (auto = self; trophe = food or "self-feeding"), because they do not require food molecules from other organisms to meet their energy needs. Food produced by autotrophs is eaten by other organisms who use the energy stored in food molecules.

3? Most organisms, including photosynthetic ones, can break down the chemical bonds between atoms of

BIO-BIT

Of the energy released during cellular respiration in humans, only about 40% is used to make ATP. Nearly 60% is released in the form of heat energy, which allows us to maintain body temperatures that are higher than the temperature of our environment.

food molecules by **cellular respiration,** releasing energy that their cells can use. Respiration uses oxygen to dismantle food molecules and releases carbon dioxide and water (plus the stored energy) as its end products (Figure 6–4). Thus, respiration has the overall effect of reversing photosynthesis. Organisms that cannot make their own food molecules are called **heterotrophs** (hetero = other), and they must obtain their food from other organisms, either other heterotrophs or autotrophs. It is important to remember that even though some predatory animals, like lions, wolves, and sharks, may feed exclusively upon other animals, most animals depend ultimately upon plants and the process of photosynthesis for food. ■

The overall equations for photosynthesis and respiration are:

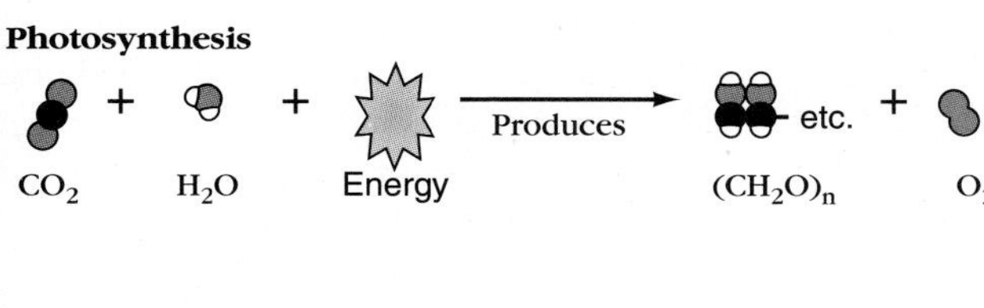

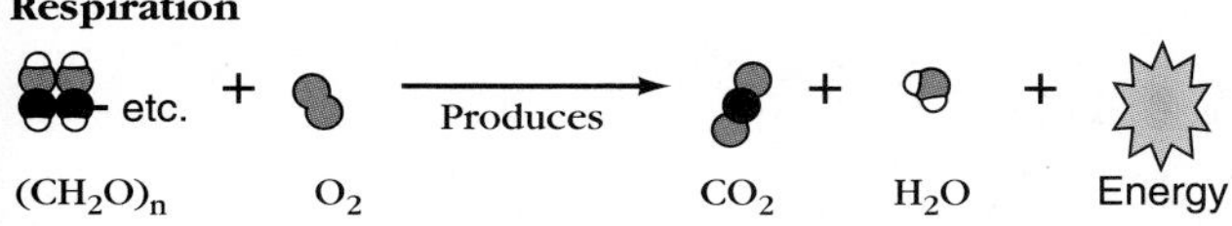

These equations seem to be direct opposites: the raw materials of each are the end products of the other; and photosynthesis uses energy, while respiration releases it. But don't be misled: the two processes are not simple opposites. They are far more complicated, and each will be discussed in detail in the following two chapters.

Life on Earth depends on the energy of sunlight to power photosynthesis, the process by which green plants, protists, and many bacteria store energy in food molecules. These food molecules are then broken down within cells using the exergonic process of respiration, which releases energy that the cell uses to power many endergonic reactions.

6–D OXIDATION-REDUCTION REACTIONS

Photosynthesis and respiration are similar in that they are both metabolic processes that use long strings of oxidation-reduction reactions. Some of these reactions release a lot of energy—enough to drive endergonic reactions that require considerable energy input.

Oxidation-reduction reactions (nicknamed **redox reactions**) involve the transfer of one or more electrons (e^-) from an electron donor molecule to an electron acceptor molecule (Figure 6–5). Sometimes, the transfer involves entire hydrogen atoms, which contain not only an electron but also a proton. The molecule that loses the electron is **oxidized,** and the one that gains the electron is **reduced.** Oxidation and reduction are complementary. Because electrons cannot float around on their own, for every oxidation there is a corresponding reduction. A redox reaction can be summarized:

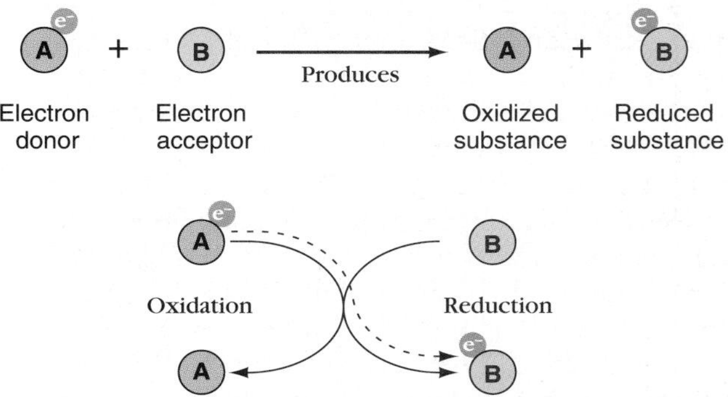

In general, energy-rich molecules are highly reduced (hydrogen-rich), and energy-poor molecules are oxidized (have lost an electron). Respiration and photo-

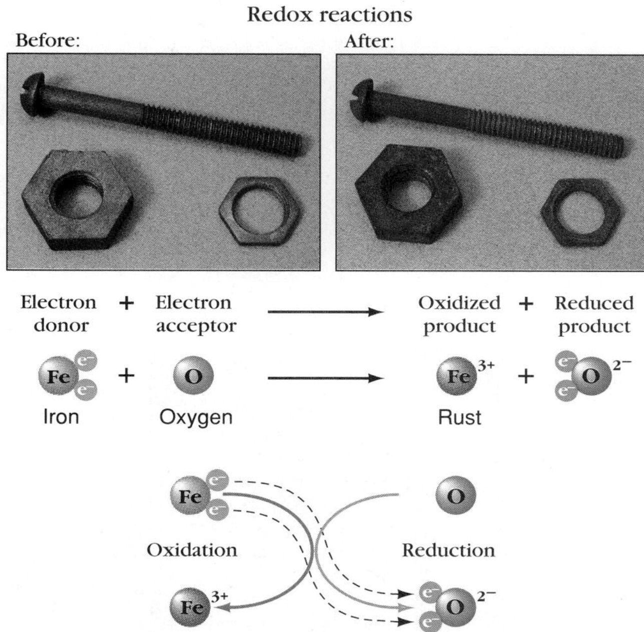

FIGURE 6–5

Oxidation-reduction reactions. Oxygen from the air has accepted electrons from the metal of this hardware, forming a discolored layer of iron oxide (rust). (E. R. Degginger)

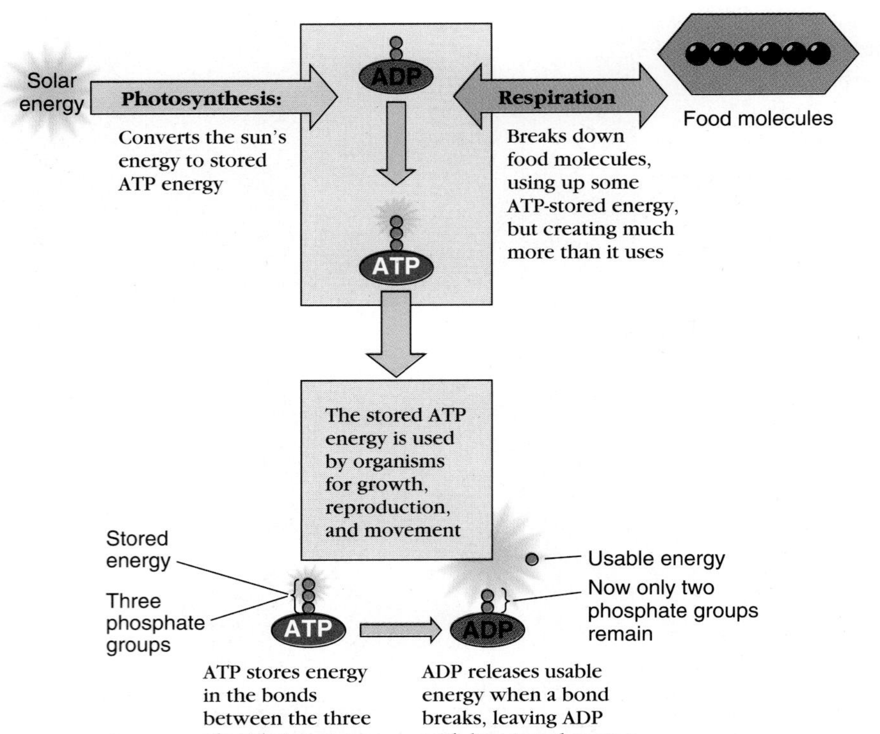

FIGURE 6–6

The production and uses of ATP. Photosynthesis and respiration release energy that is used to add one phosphorus molecule to ADP, temporarily storing this energy in ATP molecules. The stored ATP molecules can then be transferred to other locations within the cell to power other cellular activities.

synthesis are similar in that during both processes, electrons or entire hydrogen atoms are transferred from one molecule to the next. At some steps in these sequential reactions, energy needed to do the cell's work is stored in the form of molecules called energy intermediates.

Oxidation-reduction (redox) reactions involve the transfer of one or more electrons or complete hydrogen atoms from a donor molecule to an acceptor molecule. Both respiration and photosynthesis consist of long series of redox reactions.

6–E ENERGY INTERMEDIATES

Cells must perform many endergonic chemical reactions, which go "uphill" in terms of energy. These reactions are catalyzed by enzymes, but enzymes cannot alter the energy changes that occur during a reaction and they cannot force an endergonic reaction to proceed without the necessary input of energy. As we have seen, cells "jump-start" endergonic reactions by coupling them with exergonic reactions, but in so doing, they must handle these "live wires" with care. Releasing too much extra energy is not only wasteful, but it is potentially dangerous at the cellular level. Remember that unused energy is converted to heat; a burst of heat within a cell could damage the three-dimensional folding pattern of enzymes or disrupt membranes. So, reactions that release a lot of energy, such as the oxidation of organic molecules during respiration, are carried out in a series of small steps that release energy gradually. At various steps in these strings of sequential reactions, some of the released energy is stored as **energy intermediates,** which transfer moderate amounts of energy between highly exergonic reactions and energy-requiring endergonic reactions. In living cells the usual energy intermediates are the electrical energy of a membrane potential and the chemical energy of **ATP** (adenosine triphosphate). We shall discuss ATP first.

The role of ATP in the cell's energy economy has been compared to that of money in the human economy. Removing a phosphate group from an ATP molecule releases enough energy to drive many endergonic biochemical reactions. Various enzymes called **ATPases** break down ATP and provide energy for activities such as muscle contraction, active transport, waste excretion, and synthesis of new macromolecules. During photosynthesis, ATP is also used as an energy intermediate in the building of carbohydrate molecules—the cell's equivalent of a savings account of stored energy that can be broken down as needed to make the ready cash of ATP. Figure 6–6 summarizes the central role of ATP in the economy of life. Both photosynthesis and respiration can be regarded as means to the same end: the synthesis of ATP.

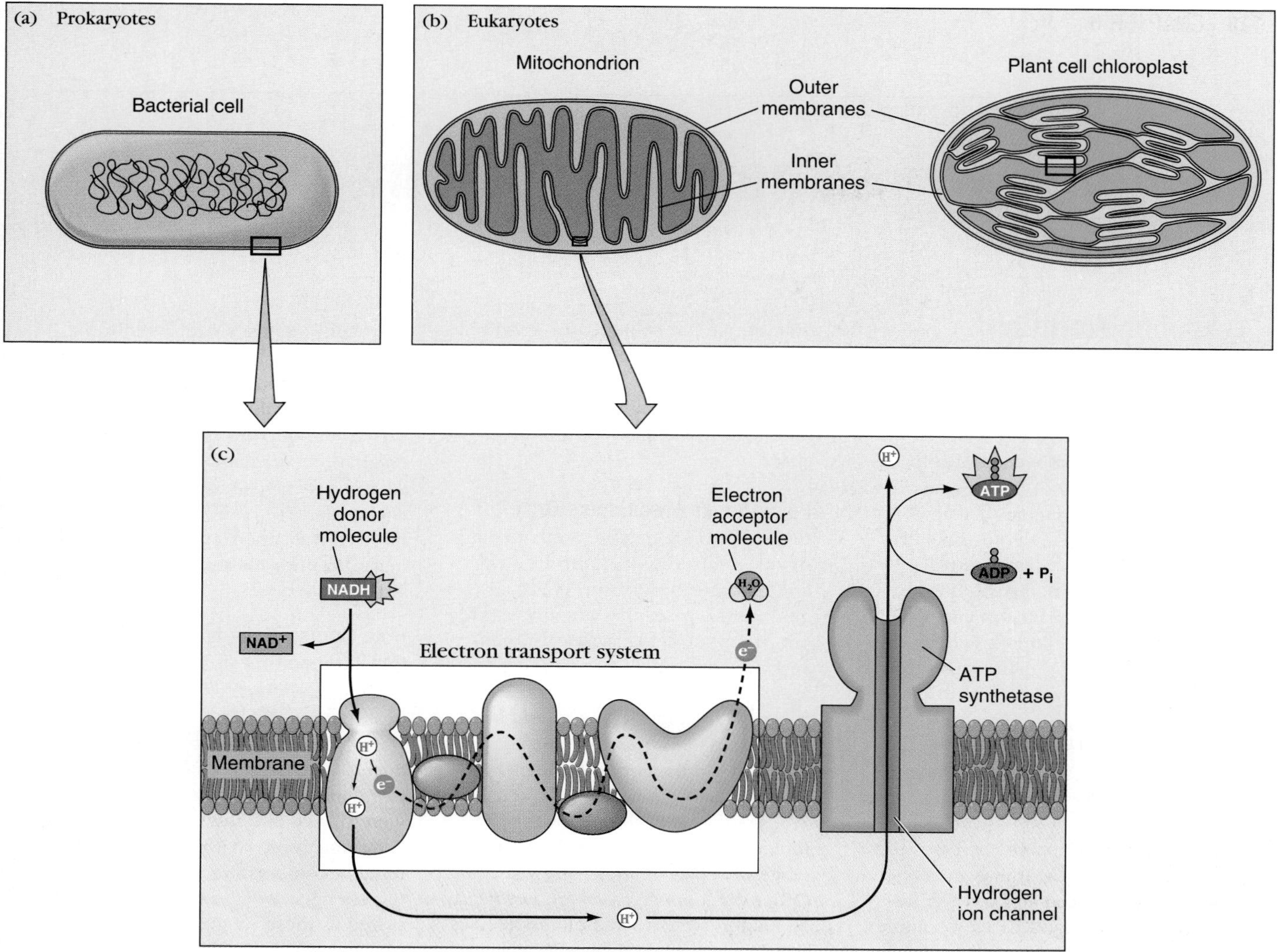

FIGURE 6–7

Locations of specialized membranes and H^+ gradients involved in chemiosmotic ATP synthesis. The membranes of a bacterial cell **(a),** animal (or plant) mitochondrion and plant chloroplast **(b)** all contain electron transport systems. The electron transport system in the bacterial or mitochondrial membrane **(c)** accepts hydrogen atoms donated from NADH, splits them into H^+ and electrons (e^-), and returns the electrons to the cytoplasm (where they bind with an electron acceptor molecule). Meanwhile, H^+ is expelled on the other side of the membrane, creating a high H^+ concentration. This gradient of H^+ ions provides the energy to produce ATP as follows. As the ions pass back through the membrane and down their concentration gradient, ATP synthetase enzymes attached to hydrogen channels use the electrical kinetic energy of the ions to make ATP, in a process called chemiosmotic ATP synthesis. A similar electron transport system, working in the opposite direction, operates in the inner membrane system of chloroplasts during photosynthesis.

There are relatively few ATP molecules in a cell, and all the energy-requiring reactions of metabolism are continually breaking ATP down. Therefore, energy-yielding reactions (mainly those of cellular respiration) must constantly renew the supply of ATP by rejoining a phosphate group to ADP.

Most ATP is made using the other energy intermediate, a membrane potential. This electrical energy always involves concentration gradients of hydrogen ions (H^+) across plasma membranes. Bacterial cell plasma membranes establish hydrogen ion membrane potentials; so do the inner membranes of chloroplasts and mitochondria of eukaryotes (Figure 6–7a and b).

Each of these membranes contains sets of proteins called **electron transport systems** (Figure 6–7c), so named because they pass electrons from one protein

A Journey into The Environment

Global Warming

What if all the world's coastal cities were like Venice, where people move about in boats in canals instead of walking or driving along streets? Imagine that new weather patterns turn forests into deserts, and wheat and corn fields into dry, barren badlands. Imagine ice caps melting and Boston and New York with temperatures that are typical of Miami. Does all of this sound far-fetched? It may happen in the near future, as a result of global warming. To understand how these changes might happen, let's consider some facts about the gases that form our atmosphere.

Once water vapor is removed, the air that we breathe is mostly composed of nitrogen and oxygen (respectively, 78% and 21% by volume). Argon gas makes up most of the remaining one percent of air, but other gases are also present. Tiny quantities of carbon dioxide, methane, nitrous oxide, helium, krypton, xenon, and hydrogen are always found in dry air, while sulfur dioxide, nitrogen dioxide, ozone, ammonia, carbon monoxide, and iodine are often, but not invariably found. Some of these gases that are present in trace amounts have the interesting property of allowing visible light to pass, yet blocking the passage of long-wave, infrared radiation. Thus, they act as a one-way gate, and allow the sun's radiation to warm the Earth, but prevent heat that is radiated from the Earth's surface from escaping out to space. Glass also allows visible light to pass and blocks infrared radiation. Even if you've never visited a greenhouse and experienced how surprisingly warm these glass houses can be, even on a cold day, you are probably familiar with the warmth that is trapped within an automobile left standing for a while in the sun with the windows rolled up. In the atmosphere "greenhouse gases," like glass, absorb infrared radiation and prevent its escape into space.

Facts About Greenhouse Gases

Carbon dioxide, ozone, methane, ammonia, sulfur dioxide, nitrous oxide, and chlorofluorocarbons (CFCs) are greenhouse gases that are always present in dry air, while water vapor, a variable component of air, also acts as a greenhouse gas. Without these gases, heat from the sun would be bounced back into space and the Earth would be a cold, perhaps lifeless planet.

Once it is in the atmosphere, a molecule of carbon dioxide has an average life span of about 500 years. Other greenhouse gases are not nearly so long lived. For example, molecules of CFCs last from 65 to 110 years, while those of nitrous oxide last 140 to 190 years, and those of methane last 7 to 12 years. Ozone and other trace gases last only hours to days in the troposphere (the area from sea level to 6 miles or 10 kilometers above the surface of the Earth).

Excess carbon dioxide normally is absorbed into several kinds of natural "sinks" or storage areas. The oceans are by far the greatest absorber of carbon dioxide and the gas dissolved in the oceans eventually is incorporated into the calcium carbonates of coral reefs and crustacean and mollusc shells. Some of this calcium carbonate goes into long-term storage in ocean sediments and oceanic rock formations. A second major carbon dioxide sink is the photosynthesis of plants, including the tiny, free-floating ocean organisms known as plankton, as well as plants rooted on land. The wood of trees is a longer term carbon dioxide sink and the peat, coal, and petroleum deposits deep within the Earth are even longer term storage sinks for carbon dioxide. Before the Industrial Revolution, the amount of carbon dioxide given off by animal respiration, volcanoes, fires, and other natural processes was roughly balanced by these carbon dioxide sinks. But this was all changed once humans began to burn fossil fuels for large-scale industry.

How Human Activity Has Changed the Composition of the Atmosphere

Organic fuels such as coal, gasoline, peat, natural gas, and oil are called "fossil fuels" because they are produced from preserved deposits of plant and animal tissues that long ago were buried and then subjected to slow, anaerobic (without oxygen) bacterial decomposition. Thus, much of the chemical potential energy remains stored in these tissues, although the plants and animals themselves are chemically and physically transformed. When fossil fuels are burned, they release this concentrated energy, but they also release carbon dioxide and other greenhouse gases into the atmosphere as waste products. Since the beginning of the Industrial Revolution 150 years ago, when an abrupt increase in the use of fossil fuels began, we have theoretically burned enough fossil fuels to increase the amount of carbon dioxide in the atmosphere by about 25%.

When humans began clearing forests on a large scale for agriculture and living space, they removed the living, actively photosynthesizing carbon dioxide sink. Currently in the tropics, massive forest fires are set to clear land quickly and inexpensively. This technique delivers a double whammy. Burning not only removes the sink, it also intensifies levels of greenhouse gases in the atmosphere by releasing the CO_2 stored in the trees. And, if present trends of deforestation continue, matters will grow much worse.

Carbon dioxide is not the only greenhouse gas that is being added to Earth's atmosphere. Chlorofluorocarbons (CFCs) from air conditioning and refrigeration coolants, industrial solvents, and from chemicals released during the manufacture of foamed plastic products, are very stable in the lower atmosphere and are not broken down into ozone eroding compounds until they rise into the stratosphere 6 to 28 miles (10 to 45 kilometers) above the surface of the Earth. In the troposphere CFCs are efficient greenhouse gases. Methane, another greenhouse gas, originates from respiration of anaerobic bacteria that thrive in oxygen-poor mud in bogs, swamps, landfills, and wet soils. Methane-emitting bacteria flourish in rice paddies, and within the guts of termites and cattle. Tropical deforestation has increased the termite population worldwide, because forest clearing generally leaves a lot of waste wood lying about—the raw material that these insects feed upon. As the human population has increased, the amount of land cleared for rice paddies, the number of cattle, and the amount of methane burped into the atmosphere from ruminating cattle have all correspondingly increased. Ozone and nitrous oxide originate from the burning of fossil fuels. Automobile emissions are a major source of both of these greenhouse gases.

How Can We Be Sure the Problem Is Real?

No one was monitoring levels of trace atmospheric gases 150 years ago, but nevertheless, we do have evidence that the atmospheric proportions of these gases have increased. Analysis of bubbles of "fossilized air" (some samples are 160,000 years old) trapped in Antarctic ice and 33 years of atmospheric measurements taken at the Mauna Loa Observatory in Hawaii demonstrate that we have increased atmospheric carbon dioxide levels enough to raise the global temperature an average of about one degree Fahrenheit (0.5°C). In addition, methane levels have doubled, and methane is a much better heat absorber than is carbon dioxide.

It is clear that we have increased the levels of greenhouse gases in the atmosphere and we know that greenhouse gases warm the atmosphere. What is unclear is what will be the result of these increased levels of greenhouse gases. Even our best computer models cannot simulate the complexities and details of all the processes that affect climate. For example, no one knows exactly how much more carbon dioxide and heat the oceans can absorb. In addition, no one completely understands and can predict the temperature lowering effect of the reflectivity of Earth's cloud cover. Gases and particles pumped into the atmosphere by volcanoes also have unpredictable effects. Finally, we are only at the beginning of our knowledge of what controls global climatic patterns. The Earth may have some unknown mechanism that operates on negative feedback (like the thermostat that shuts the furnace off when the room gets too warm) to control global temperature. But based on what we do know, and using computer simulations, most climatologists predict that at some unknown point in the near future the Earth will warm 1 to 3.5°F (0.56 to 1.94°C) beyond the 1°F it has already risen, but no one can predict just how soon this will happen. Some scientists estimate that this will occur in 10 to 50+ years, while computer models predict that at the present rates of carbon dioxide emissions, the Earth will experience a global warming of 2°F by 2025 and 6°F by 2100 (1.12°C and 3.33°C, respectively). One thing to remember is that as the human population continues to explode, all the human activities that generate greenhouse gases will continue to pump them into the atmosphere. And the Earth will continue to get warmer and warmer. The projected 1 to 3.5°F rise will be just the beginning of a general warming trend. Here are some of the possible results of global warming:

As the planet warms:

1. Climatic patterns will shift northward. For example, if global warming trends continue, New York may one day have a climate like Miami has today, and the plants and animals now characteristic of New York will have to migrate further north, adapt, or go locally extinct.
2. Agriculture will be disrupted, especially in the United States. Major droughts are predicted for many agricultural areas. Food shortages, civil unrest, famine, and wars may result.
3. In the Arctic the permanently frozen layer of subsoil, called permafrost, will melt. This will release even more methane and further intensify atmospheric warming.
4. Glaciers, ice sheets, and snow caps will melt (Figure 6-A).
5. Sea levels will rise, inundating low-lying countries and coastlines.
6. Warming of the oceans will generate more and more powerful hurricanes and typhoons and will affect the health of coral reefs, mangrove swamps, and coastal wetlands (Figure 6-B).
7. Warming may cause drastic shifts in the plant and animal populations as natural habitats are altered. As populations of animals shift northward, the nature reserves that have been set aside to conserve natural habitats may lose many of the animals and plants they are supposed to protect.

What You Can Do To Reduce Global Warming

The obvious solution to decreasing the threat of global warming is to cut off the increased supply of greenhouse gases to the atmosphere. Using less fossil fuel and using it more efficiently and wisely are two obvious strategies that are within everyone's

The Environment continued

FIGURE 6–A

In September of 1991, the remains of this "Iceman" were discovered by hikers in the melting glacial ice on a mountain bordering Italy and Austria. The initial forensic team that removed the body thought that it was a recently murdered or lost mountaineer. Upon closer examination, the body was found to be that of an ancient man about 5000 years old. As the Earth continues to warm, causing glaciers and ice fields to melt, we can expect more and more discoveries of this sort to "come in out of the cold." (Sygma)

FIGURE 6–B

In 1992, Hurricane Andrew destroyed homes in large regions of Florida, as seen in this aerial view of a trailer park in Homestead. As the Earth continues to warm, we can expect more monster hurricanes like Andrew. (The Stock Market)

reach. Substituting renewable energy sources for fossil fuels is another solution. So drive less, and when you do drive, do it in a fuel efficient vehicle. Walk more, use less electricity, and look for ways to decrease your dependence upon fossil fuels and the products made from them.

The following organizations will teach you how to help stop global warming. Many of them have free literature that will explain how this problem is connected to other environmental problems, especially overpopulation:

Greenhouse Crisis Foundation. 1130 17th Street NW Suite 630, Washington, DC 20036; (202) 466-2823

National Audubon Society. 950 Third Avenue, New York, NY 10022; (212) 832-3200

Rainforest Action Network. 300 Broadway Suite 28, San Francisco, CA 94133; (415) 398-2732

Rainforest Alliance. 270 Lafayette St. Suite 512, New York, NY 10012; (212) 941-1900

Sierra Club. 730 Polk St., San Francisco, CA 94109; (415) 776-2211

Union of Concerned Scientists. 26 Church St., Cambridge, MA 02238; (617) 547-5552

Worldwatch Institute. 1776 Massachusetts Ave. NW, Washington, DC 20036; (202) 452-1999

to the next. In the process, the electron transport systems also carry out active transport of hydrogen ions to one side of the membrane. The membrane is a barrier that is resistant to hydrogen ions, except where special channels permit hydrogen ions to pass through the membrane, going down their concentration gradient. So a gradient of hydrogen ions across the membrane results from the work of the electron transport system. This gradient of hydrogen ions provides energy to make ATP. As the ions pass through the channels, **ATP synthetase** enzymes attached to the channels use the energy of the hydrogen ion gradient to synthesize ATP. This method of making ATP, using a chemical (H^+) passing through a membrane, is called **chemiosmotic ATP synthesis.** This important concept will be further discussed in Chapters 7 and 8.

Membrane potentials and the chemical energy stored in ATP are energy intermediates in all living things. These energy intermediates store moderate-sized amounts of energy released by highly exergonic reactions and are then used to power endergonic reactions. Most ATP is made by chemiosmotic ATP synthesis, using energy from a gradient of H^+ ions across a membrane.

SUMMARY

1. Living organisms require energy to maintain their chemical composition, as well as to move, repair damage, grow, and reproduce.
2. Energy cannot be created or destroyed. Each time energy is converted from one form to another, some usable energy is converted into a less usable form.
3. To remain alive, organisms must constantly acquire fresh supplies of energy. The central energy-processing pathways of life are photosynthesis and respiration.
4. Photosynthetic organisms capture the sun's energy and store it in the chemical bonds of food molecules, which can later be broken down during respiration to release trapped energy.
5. Both photosynthesis and respiration are essentially oxidation-reduction (redox) reactions. Many steps in each of these processes involve transferring electrons (or hydrogen atoms, which contain electrons). This releases a great deal of useful energy.
6. The ultimate task of both photosynthesis and respiration is to trap this energy and use it to produce energy intermediates, which can then drive endergonic reactions.
7. The most common energy intermediate is ATP.
8. Gradients of ions across membranes also serve as energy intermediates for some energy-requiring processes. In fact, most ATP is formed by using the energy of such a gradient to join phosphate groups to ADP.

SELF-QUIZ

1. Which of the following will re-start a chemical reaction that is currently in a state of equilibrium?
   a. adding more enzymes
   b. increasing the temperature
   c. adding more reactants
   d. adding ATP
2. Does the following equation describe photosynthesis or respiration?

$$CO_2 + H_2O \longrightarrow C_6H_{12}O_6 + O_2$$

3. In the above equation, is carbon oxidized or reduced as a result of the reaction?
4. ATP is made by joining ____ and ____. This reaction is (endergonic, exergonic). Hence it requires ____ in addition to the reactants.
5. The electron transport system:
   a. makes ATP
   b. contains ATP synthetase
   c. is responsible for separating hydrogen atoms into their components
   d. can work against the second law of thermodynamics
   e. cannot perform redox transfers if the membrane is broken

THINKING CRITICALLY

1. Why must photosynthetic plants carry on respiration?
2. Of the light energy reaching the Earth from the sun, the Earth's plants are believed to convert less than 1% into the form of potential energy stored in the chemical bonds of food molecules. What happens to the rest of the energy?
3. Organisms cannot use heat energy to drive their energy-requiring processes. Does this mean that the heat released by metabolism is of no use to them? Why or why not?

SELECTED KEY TERMS

activation energy, *p. 114*
autotroph, *p. 116*
cellular respiration, *p 117*
chemical potential energy, *p. 111*
chemiosmotic ATP synthesis, *p. 122*
electron transport system, *p. 119*
endergonic reaction, *p. 113*
energy intermediate, *p. 118*
entropy, *p. 112*
equilibrium, *p. 115*
exergonic reaction, *p. 113*
heterotroph, *p. 117*
kinetic energy, *p. 111*
laws of thermodynamics, *p. 112*
oxidation-reduction (redox) reaction, *p. 117*
photosynthesis, *p. 116*
potential energy, *p. 111*

SUGGESTED READINGS

Atkins, P. W. *The Second Law.* Scientific American Books, 1984.

Hinkle, P. C., and R. E. McCarty. "How cells make ATP." *Scientific American,* March 1978. Compares the chemiosmotic mechanism for ATP synthesis in bacteria, mitochondria, and chloroplasts. [Note: to understand this article, you must know that biochemists ignore the outer mitochondrial membrane. When they say, "outside the mitochondrion," they mean "outside the inner membrane."]

CHAPTER 7

CONCEPT GUIDE

After reading this chapter, you should be able to:

1. Describe the starting molecules, main molecular end products, and locations within the cell of the following four stages of cellular respiration:

   a. glycolysis
   b. preparation for the citric acid cycle
   c. citric acid cycle
   d. electron transport and chemiosmotic ATP synthesis (oxidative phosphorylation)

   (Sections 7-A, 7-B, 7-C, and 7-D)
2. List the total number of ATP and the molecular end products of the complete oxidation of one glucose molecule. (Sections 7-A and 7-D)
3. Relate the structure of mitochondria to their role in respiration. (Section 7-C)
4. Compare the two types of fermentation and describe how each is important to humans. (Section 7-E)
5. Explain how proteins and fatty acids can be broken down to produce ATP. (Section 7-F)

The core body temperature of humans is about 98.6 °F (37 °C). So why do we feel warm when the air temperature is in the 80-90 °F range . . . shouldn't we feel cold? The reason that we do feel warm relates to cellular respiration. As you will learn in Sections 7-A to 7-D of this chapter, most living cells break down sugars and other molecules to produce energy and run the machinery of life. But this same reaction produces carbon dioxide, water, and heat. On a warm day, we humans have trouble getting rid of the extra heat produced by respiration, and so feel warm. (Joseph Wood)

Food as a Fuel: Cellular Respiration and Fermentation

Think of a long, hushed, hazy summer afternoon.

You have nowhere to go; no one to see; no classes, no exams, no papers. No deadlines. You're lying in a shady hammock, nearly asleep. A paperback novel slides from your fingers and lies forgotten on your chest. You dimly hear the drone of bees nuzzling the honeysuckle, or is someone cutting grass? You're floating. Completely relaxed.

Imagine that you can peer within a single cell of your body to spy on its activities. What would be happening there, even as you rest on a summer afternoon in that shady hammock? If we looked, for example, into a skin cell, the metabolic pathway that powers the cell would probably be busy shuffling molecules in the process of **cellular respiration,** removing electrons in stepwise fashion, oxidizing high-energy food molecules into low-energy molecules of carbon dioxide and water. Each step of cellular respiration releases only a little of the food molecule's energy. Thus, respiration has been called a "slow burn," and its controlled oxidation of food allows the cell to capture and store more energy than if all the energy were released in one big burst.

Respiration is called an **aerobic** process because it requires molecular oxygen (O_2). Some cells live in **anaerobic** conditions, with little or no O_2 available. Many of these carry out **fermentation**—the breakdown of food molecules using organic molecules instead of inorganic oxygen as the electron acceptors. This is less efficient and yields less ATP.

In this chapter we will examine how eukaryotic cells **respire:** how they break down food molecules in the presence of oxygen to obtain chemical energy and then store this energy within the chemical bonds of ATP. We will compare **fermentation,** an alternate, less efficient way to break down food molecules, with aerobic, cellular respiration. Finally, we will see how respiration and fermentation connect with other metabolic pathways that provide each cell with lipids and proteins that it can respire or ferment.

KEY CONCEPTS

- Cellular respiration breaks down food molecules and releases their stored energy in small steps. Some of this is stored as chemical potential energy in ATP.
- Fermentation makes less ATP than respiration, and breaks food molecules down only partway.
- Other metabolic pathways, such as those for proteins and lipids, feed molecules into the respiratory pathway.

Curiosity Questions

1? How are champagne and dry and sweet wines produced? (See page 135)

2? What do your muscles and yogurt have in common? (See page 135)

3? Why do animals store energy in the form of fat? (See page 138)

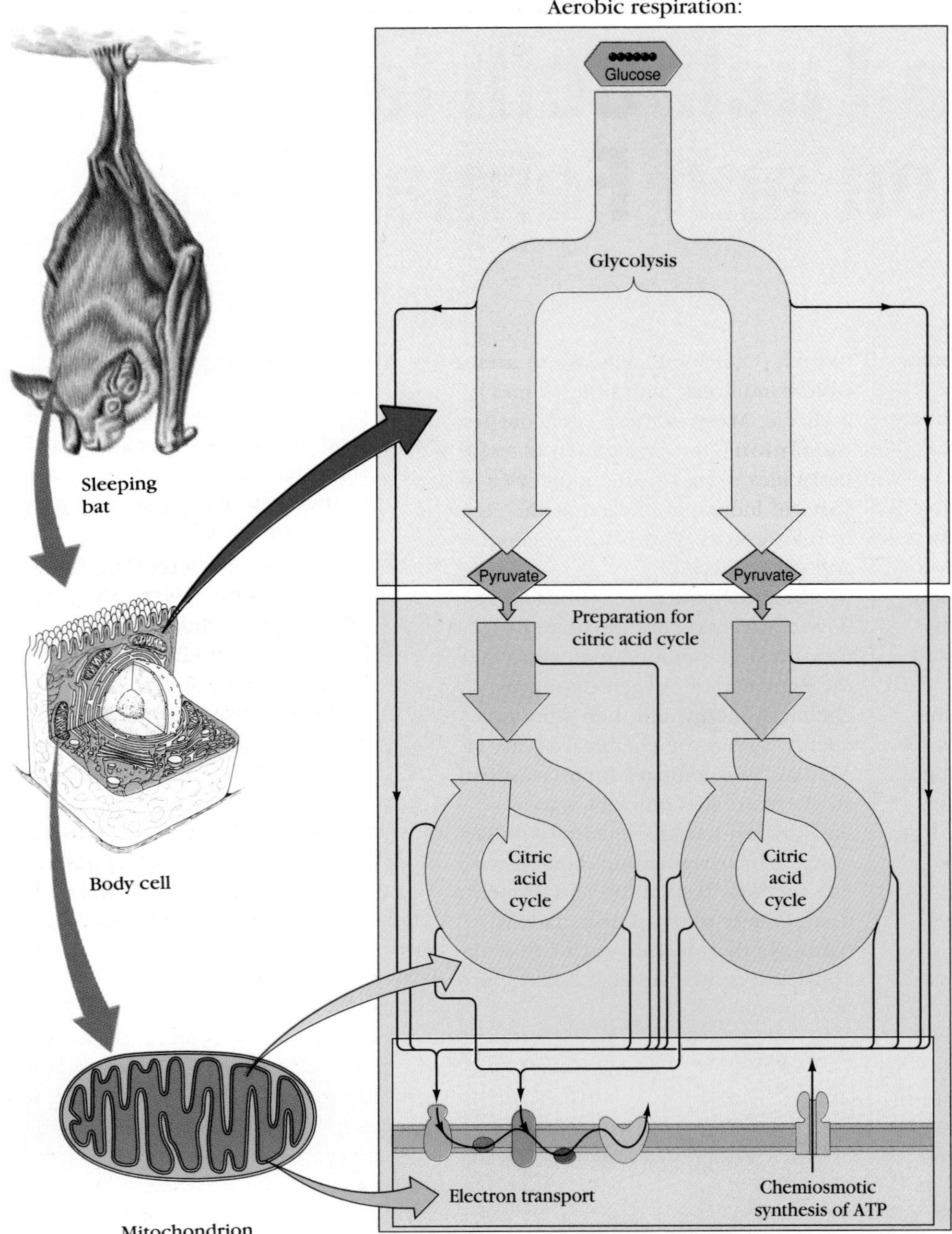

FIGURE 7-1

An overview of aerobic respiration. Glycolysis occurs within the cytoplasm, with the breakdown of a six-carbon sugar into a pair of three-carbon pyruvate molecules. Inside a mitochondrion, preparation for the citric acid cycle and the cycle itself continue the breakdown of the three-carbon pyruvate molecules. Finally, most ATP is generated by the electron transport system and chemiosmotic ATP synthesis across the inner mitochondrial membrane.

7-A RESPIRATION—IN A NUTSHELL: IT'S *WHY* YOU BREATHE, BUT IT'S *NOT* BREATHING

We begin with a molecule of glucose, a six-carbon sugar that is an end product of digestion of most carbohydrates. Glucose is the main fuel molecule used by cells and some of the energy once stored in the glucose molecule's chemical bonds is eventually used to build ATP. But before the energy becomes available to do this, glucose has to be chemically dismantled—in a series of reactions. Most of the transfer of energy occurs in the final phase of the process (see #4) as hydrogen atoms (or their electrons) that have been removed from the breakdown products of glucose are passed through a pathway called the electron transport system in a series of redox reactions (Section 6-D). You will recall that electron transport systems are located in membranes. Refer to Figure 7-1 to help you visualize the process of respiration.

Table 7–1
Coenzymes Used in Respiration

Coenzyme			Function	Made from This Vitamin
Nicotinamide adenine dinucleotide	NAD+	(NAD^+)	Carry hydrogen atoms to electron transport chain	Niacin
Flavin adenine dinucleotide	FAD+	(FAD^+)		Riboflavin
Coenzyme A	CoA	(CoA)	Carries acetyl groups to citric acid cycle	Pantothenic acid

Each reaction in the pathway is catalyzed by an enzyme. Some of these enzymes require **coenzymes,** organic molecules that shuttle substances from one reaction to another. As we have previously noted, many coenzymes are made from dietary vitamins. Refer to Table 7–1 to see where some of the ingredients in a daily multiple vitamin supplement enter the process of respiration.

The overall reaction for the complete oxidation of glucose is:

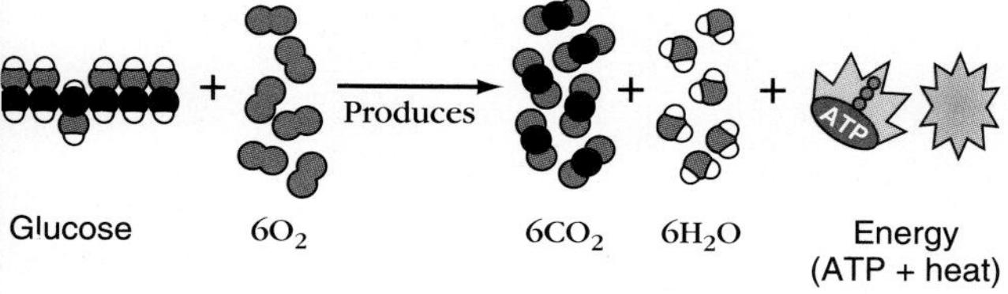

This summarizes a long, complex process that we can divide into four parts (A Journey Through Metabolism):

1. **Glycolysis** (Figure 7–1a) breaks the six-carbon sugar, glucose, into a pair of three-carbon **pyruvate** molecules. Along the way a small amount of ATP and hydrogen atoms are also produced. The hydrogen atoms are picked up by a coenzyme, **nicotinamide adenine dinucleotide (NAD^+),** forming **NADH.** Glycolysis is an anaerobic process that is not strictly part of respiration. The following three aerobic pathways together constitute respiration.
2. **Preparation for the citric acid cycle** (Figure 7–1b) is completed as each molecule of pyruvate gives up one carbon atom, (in the form of a molecule of CO_2) leaving a two-carbon **acetyl group.** The CO_2 is eventually released as a waste product. Hydrogen atoms are also released and are picked up by coenzyme NAD^+, forming more NADH.
3. The **citric acid cycle** (Figure 7–1c) in effect dismantles each two-carbon acetyl group into two CO_2 molecules. This cycle's important products are more ATP and a wealth of hydrogen atoms, which are picked up by coenzymes NAD^+ and **FAD^+ (flavin adenine dinucleotide),** forming NADH + H^+ and $FADH_2$.
4. In the final phase of respiration, **electron transport and chemiosmotic ATP synthesis** (Figure 7–1d), NADH and $FADH_2$ pass the hydrogen atoms they carry to a system of electron transport molecules that are housed in a mitochondrial membrane. Here the hydrogen atoms are separated into hydrogen ions (H^+) and electrons usually represented as: e^-. The electrons' energy is, in effect, used for the active transport of the hydrogen ions, forming a hydrogen ion gradient across the membrane. This gradient supplies energy necessary to make ATP from ADP and an inorganic phosphate group, P_i. Electrons leaving the system join with oxygen atoms and with still other hydrogen ions, forming water.

In the process of aerobic respiration a molecule of glucose, in the presence of oxygen, is chemically broken down in a series of stepwise reactions to yield ATP, carbon dioxide, water, and heat. These reactions begin in the cytoplasm with the anaerobic process of glycolysis. The resulting pyruvate molecules then move into mitochondria, where they undergo the aerobic process of respiration; preparation for the citric acid cycle, electron transport, and chemiosmotic ATP synthesis.

BIO-BIT

Aerobic exercises (aerobics) emphasize activities that permit muscles to obtain their energy through aerobic metabolism. Such exercises improve the oxygen-carrying ability of the lungs and circulatory system.

A Journey Through Metabolism

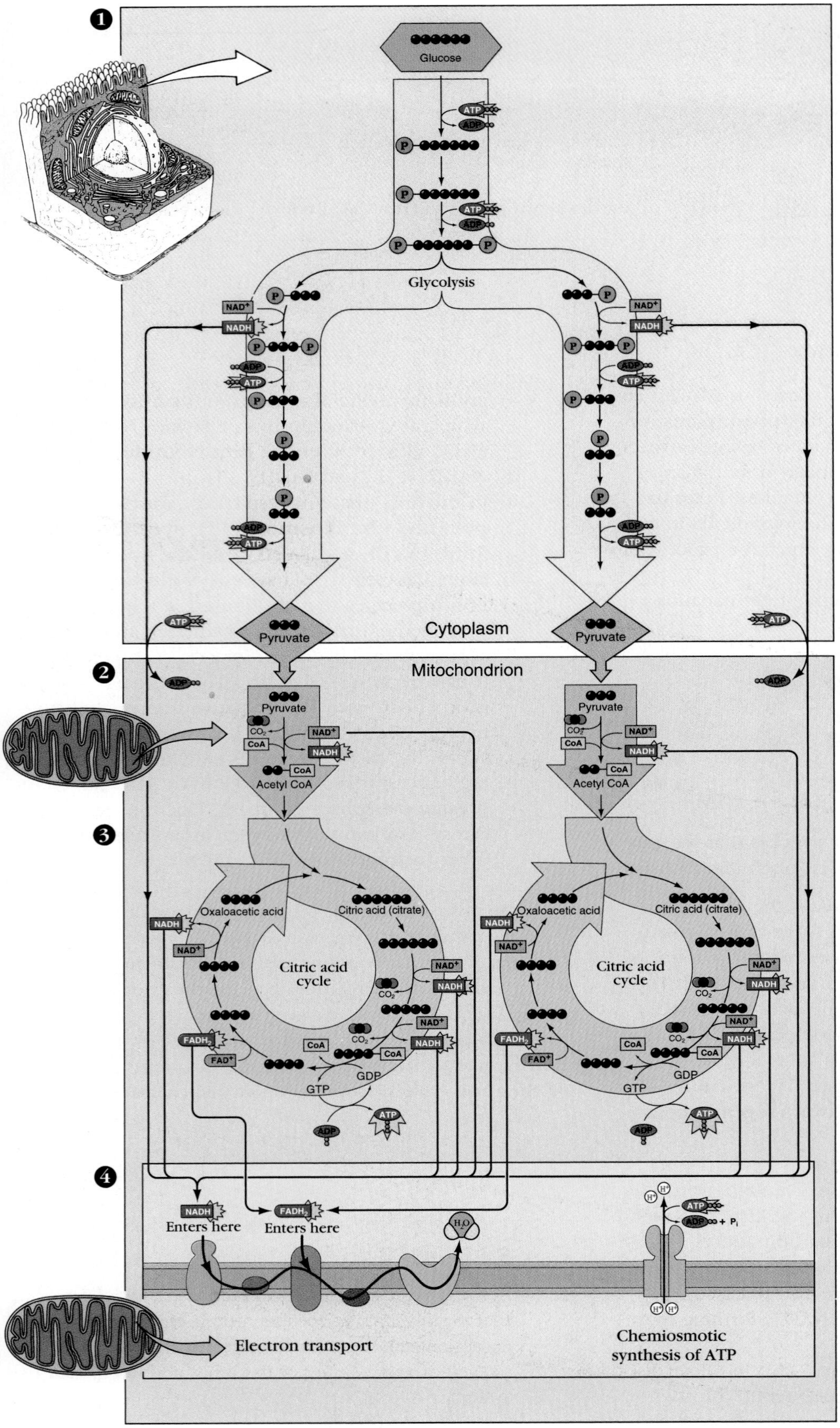

1
Glucose
Glycolysis
ATP
ADP
NAD+
NADH
Pyruvate
Cytoplasm
2
Mitochondrion
Pyruvate
CO2
CoA
Acetyl CoA
3
Oxaloacetic acid
Citric acid (citrate)
Citric acid cycle
FADH2
FAD+
GTP
GDP
4
NADH
Enters here
FADH2
Enters here
H2O
H+
ADP + Pi
Electron transport
Chemiosmotic synthesis of ATP

ENERGY YIELD FROM ONE GLUCOSE MOLECULE

In the cytoplasm:

Yield of ATP from glycolysis.......4 ATP

ATP used in transporting NADH into mitochondrion..................–2 ATP

Inside the mitochondrion:

ATP from citric acid cycle...........2 ATP

Inside the inner mitochondrial membrane:

2 NADH from glycolysis*......4 or 6 ATP

2 NADH from preparation for citric acid cycle......................6 ATP

6 NADH from citric acid cycle..18 ATP

2 $FADH_2$ from citric acid cycle..4 ATP

Total Energy Yield-36 or 38 ATP

* In some cells the 2NADH molecules produced during glycolysis yield only 2 ATP each. This changes the total number of ATP produced from NADH to 28 and changes the net energy yield from one molecule of glucose to 36 ATP.

An overview of cellular metabolism. Glycolysis, the citric acid cycle, and the electron transport system all interrelate in the process of cellular metabolism.

7–B GLYCOLYSIS

As its name implies, **glycolysis** (glyco = sugar; lysis = loosening) splits a single molecule of glucose into two molecules of **pyruvate,** a three-carbon compound. Glycolysis is an anaerobic process—one that takes place without using molecular oxygen—and makes only a small amount of ATP. This makes sense if you think of glycolysis as a very small combustion: if it is starved for oxygen, it cannot produce much (if any) energy. For each glucose molecule broken down, glycolysis produces a *net* energy gain of only two molecules of ATP. (Four ATP molecules are actually formed in glycolysis, but two are used to start the process.) Glycolysis also transfers hydrogen atoms to two molecules of NAD^+, which become NADH. These coenzymes will pass their hydrogen atoms on to the electron transport system, where energy is extracted to produce more ATP.

Some organisms live in anaerobic environments and make all their ATP by glycolysis. Many aerobic bacteria and fungi also use glycolysis to survive temporary anaerobic periods. However, these bacteria and fungi, and all so-called "higher" organisms, also possess the citric acid cycle and electron transport pathways and use them to make ATP when oxygen is available.

The enzymes of glycolysis are found in the cytoplasm of each cell. Refer to Figure 7–2 for the details of glycolysis. Note that a pair of three-carbon molecules, each containing two phosphate groups, donate these phosphate groups to ADP (adenosine diphosphate), forming a total of four ATPs. Note also that a pair of three-carbon molecules called pyruvate emerge from glycolysis.

Glycolysis is an anaerobic process that splits a molecule of glucose into a pair of three-carbon pyruvate molecules and generates a net energy gain of two ATP molecules. Glycolysis also releases hydrogen atoms, which are picked up by coenzymes NAD^+ to form NADH. When oxygen is available, NADH passes these hydrogen ions to the electron transport system, where their energy is used to generate additional ATP.

BIO-BIT

Some single-celled eukaryotic organisms lack mitochondria but contain aerobic bacteria in their cytoplasm that serve the same functions.

7–C MITOCHONDRIA: ATP FACTORIES

Glycolysis occurs in the cytoplasm. In eukaryotes, pyruvate and NADH formed during glycolysis enter mitochondria, the cellular organelles where respiration occurs. The structure of the mitochondrion plays an important role in carrying out the chemical processes of respiration.

A mitochondrion is separated from the cytoplasm by its outer membrane. Its highly folded inner membrane forms a closed compartment that contains a protein-rich solution, the **mitochondrial matrix.** Here lie many of the enzymes of the citric acid cycle. Others are attached to the inner face of the inner membrane. This membrane also contains the electron transport molecules, which create a hydrogen ion gradient by transporting hydrogen ions into the space between the inner and outer membranes. In addition, the inner membrane contains ATP synthetase enzymes, which make ATP (Figure 7–3). Aerobic bacteria lack mitochondria and their plasma membrane carries out these functions.

A eukaryotic cell's mitochondria carry out respiration and make most of the cell's ATP.

7–D RESPIRATION

Respiration, the stepwise oxidation of high-energy food molecules to low-energy molecules of carbon dioxide and water, occurs inside the mitochondria of eukaryotes and in the cytoplasm of bacteria.

Preparation for Respiration

Glycolysis produces two pyruvate molecules. In eukaryotes, these enter a mitochondrion where a large complex of enzymes carries out a series of reactions that prepare each pyruvate for the citric acid cycle (Figure 7–4).

1. The pyruvate loses one carbon attached to two oxygen atoms. This results in a carbon dioxide molecule that is eventually removed from the organism.

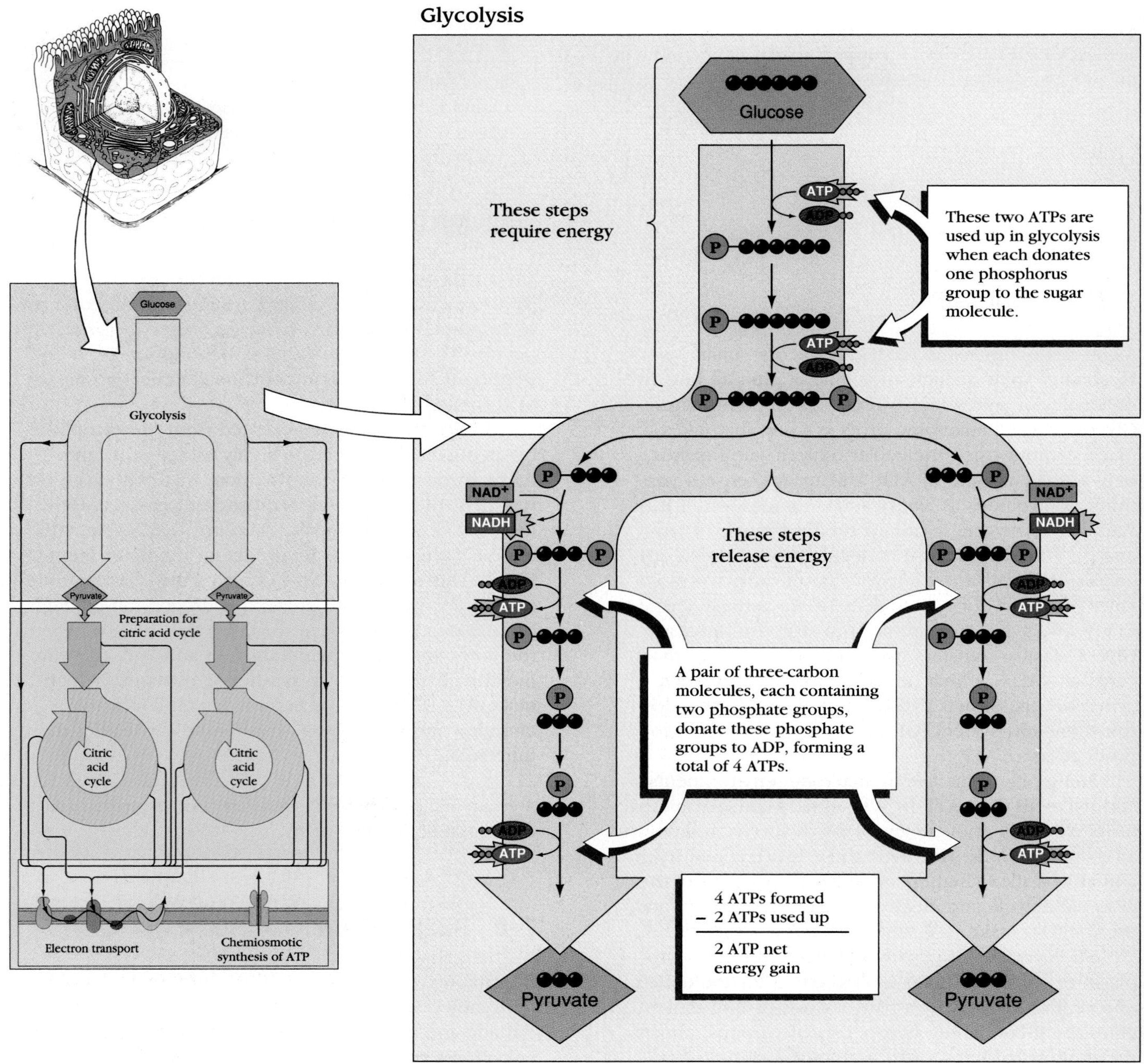
Glycolysis
Glucose
These steps require energy
ATP
ADP
These two ATPs are used up in glycolysis when each donates one phosphorus group to the sugar molecule.
NAD+
NADH
These steps release energy
A pair of three-carbon molecules, each containing two phosphate groups, donate these phosphate groups to ADP, forming a total of 4 ATPs.
4 ATPs formed
– 2 ATPs used up
2 ATP net energy gain
Pyruvate
Glycolysis
Pyruvate
Preparation for citric acid cycle
Citric acid cycle
Electron transport
Chemiosmotic synthesis of ATP

FIGURE 7–2

Within the cytoplasm, a six-carbon glucose molecule is broken down into a pair of three-carbon pyruvate molecules. Although four ATP molecules are generated, two ATP molecules are used up in the process, yielding a net energy gain of two ATP molecules from each glucose molecule entering glycolysis.

2. The remaining two-carbon acetyl group is attached to a molecule of coenzyme A (CoA, for short), forming **acetyl CoA.**
3. Meanwhile, a molecule of NAD^+ is reduced to NADH.

Citric Acid Cycle

The citric acid cycle is also called the **Krebs cycle,** after Sir Hans Krebs, who worked out the cycle in 1937 and received a Nobel Prize for this work in 1953.

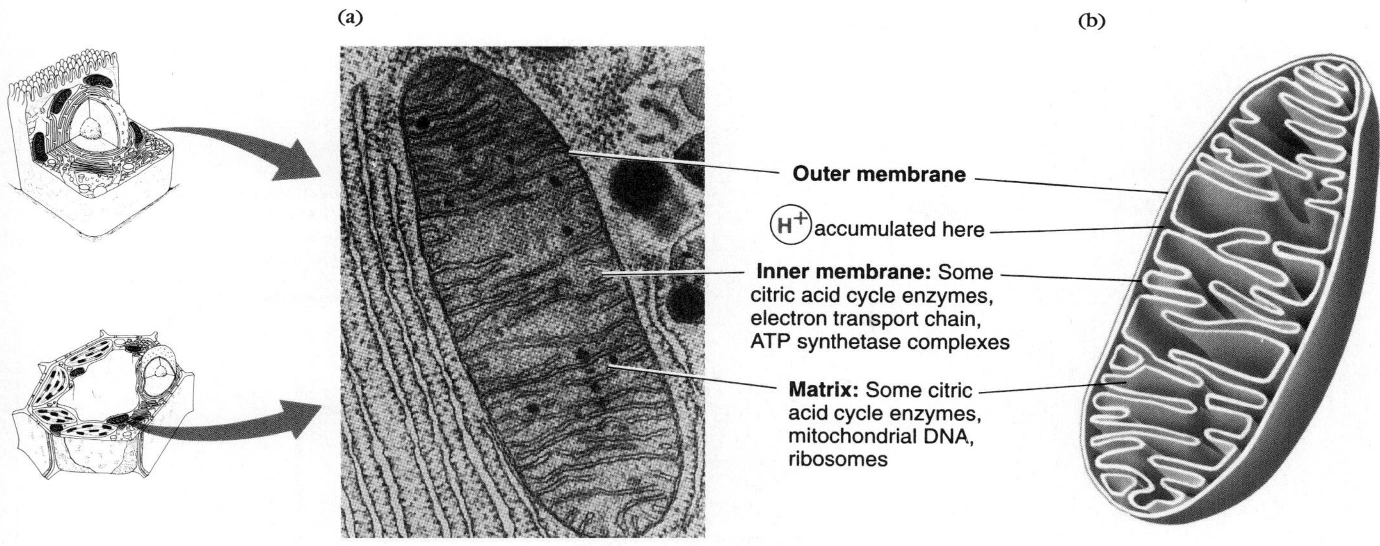

FIGURE 7–3

Mitochondria. (a) Electron micrograph of a mitochondrion from the pancreas of a bat. The inner membrane forms narrow folds extending deep into the matrix inside the organelle; the outer membrane is much smoother. **(b)** Interpretive drawing of mitochondrial structure to show the highly folded inner mitochondrial membrane. (K. R. Porter/Photo Researchers, Inc.)

During the citric acid cycle, the two-carbon acetyl group is combined with a molecule of a four-carbon compound, forming a six-carbon molecule, citric acid. Like a restless housecleaner rearranging furniture and eventually ending up with things nearly the way they started out, the enzymes of the citric acid cycle rearrange and manipulate this molecule in a stepwise series of reactions (Figure 7–4). They sequentially form three kinds of six-carbon compounds, then a five-carbon compound, then five kinds of four-carbon compounds. The end product, a four-carbon compound, oxaloacetic acid, is attached to a new acetyl CoA group and proceeds through the cycle again.

Several of the stepwise reactions that transform citric acid to oxaloacetic acid are redox reactions. They release energy, in the form of hydrogen atoms, to coenzymes NAD^+ or FAD^+. One reaction also releases enough energy to attach a molecule of a phosphate group to ADP, thus creating a molecule of ATP.

Two acetyl groups enter the citric acid cycle from each original glucose molecule. Each goes through the same series of reactions. Return to Figures 7–1 and 7–4, and when you have a better visual picture of the citric acid cycle, continue reading the text below.

Coenzyme A transfers its two-carbon acetyl group to a four-carbon molecule, oxaloacetic acid, forming a six-carbon compound, citric acid. Acting like a shuttle, each available coenzyme A molecule goes back to pick up another acetyl group, derived from new pyruvate that has entered the mitochondrion from continuing glycolysis in the cell's cytoplasm. For each acetyl group entering the cycle:

1. Two molecules of carbon dioxide leave.
2. Also, four pairs of hydrogen atoms are removed at various stages and picked up by the coenzymes NAD^+ and FAD^+, forming NADH and $FADH_2$. In terms of energy, this is the most important outcome of the cycle, because these hydrogens are carried to the electron transport system where their protons are used to power the formation of most of the ATP derived from the original glucose molecule.
3. One molecule of ATP is formed.

Each glucose molecule gives rise to two pyruvate molecules, each of which yields one acetyl group to enter the citric acid cycle. Therefore, it takes the equivalent of two turns of the cycle to break down the remains of one molecule of glucose. At the end of these two cycles, the equivalents of all six carbon atoms from the original glucose molecule have been released as carbon dioxide. In humans, this carbon dioxide leaves the cell and enters the blood. When it reaches the lungs, we breathe it out as a waste product.

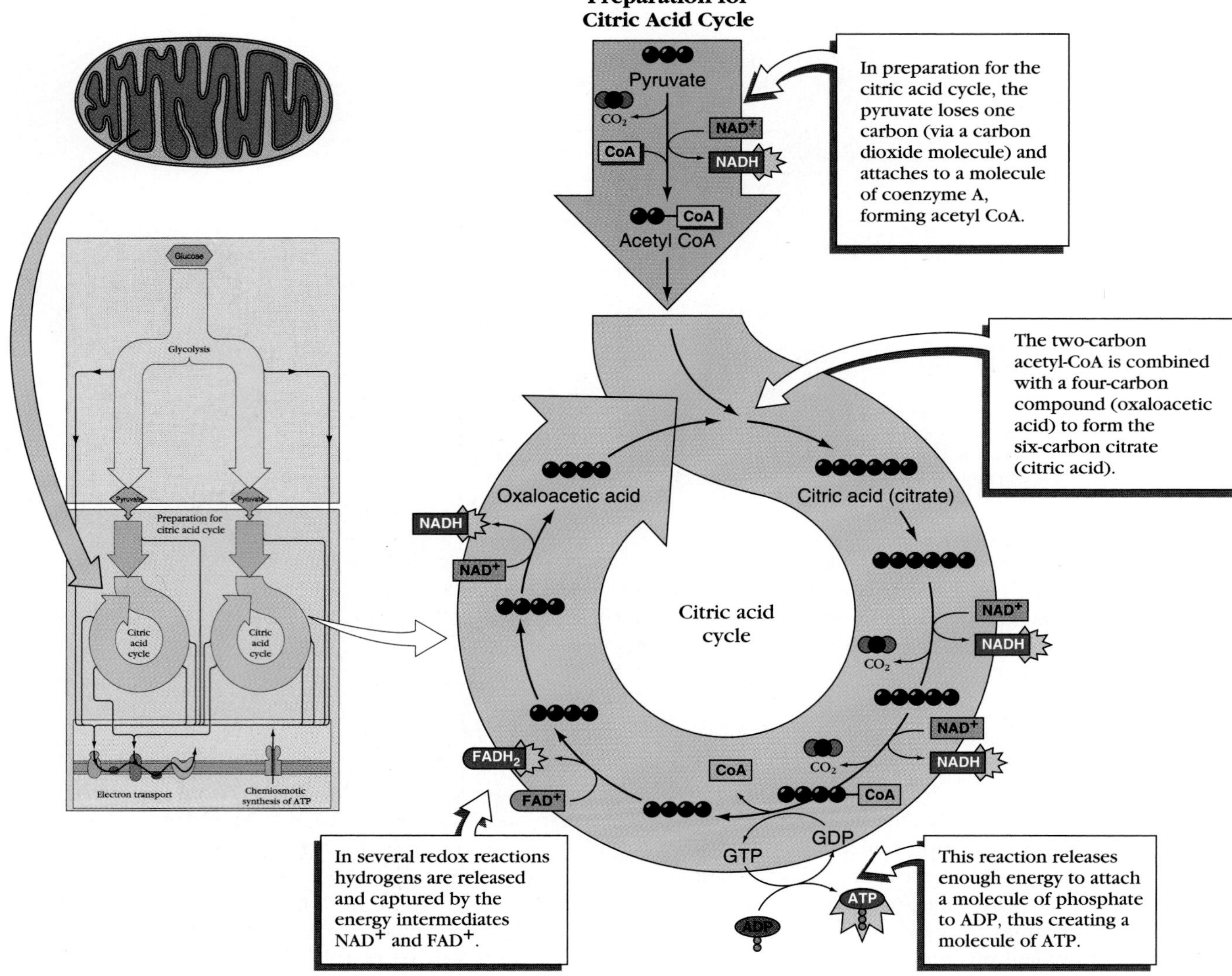

FIGURE 7–4

Now inside the mitochondrial matrix, each pyruvate molecule loses a carbon and attaches to Coenzyme-A forming acetyl-CoA in preparation for the citric acid cycle. In the citric acid cycle within the mitochondrial matrix, acetyl-CoA combines with oxaloacetic acid to form the six-carbon citric acid from which the cycle gets its name. Citric acid is degraded in a series of redox reactions finally producing oxaloacetic acid, two ATP, six NADH, and two FADH molecules. The oxaloacetic acid is now ready to bind to another acetyl-CoA molecule. The NADH and FADH molecules next move to the electron transport chain where additional ATP are produced.

We have examined how acetyl groups derived from glucose enter the citric acid cycle, but other molecules from other metabolic pathways can also enter the cycle. Proteins and fats can enter the cycle, too and, using the citric acid cycle like a traffic circle, or an express stop on the subway, molecules can be transferred to other metabolic pathways, according to the cell's requirements (see Figure 7–7).

Electron Transport and Chemiosmotic ATP Synthesis

With two turns of the citric acid cycle, the molecule of glucose has been completely dismantled. Some of its energy has produced ATP during glycolysis and the citric acid cycle, but most of the energy remains in the electrons carried by coenzymes NADH (from glycoly-

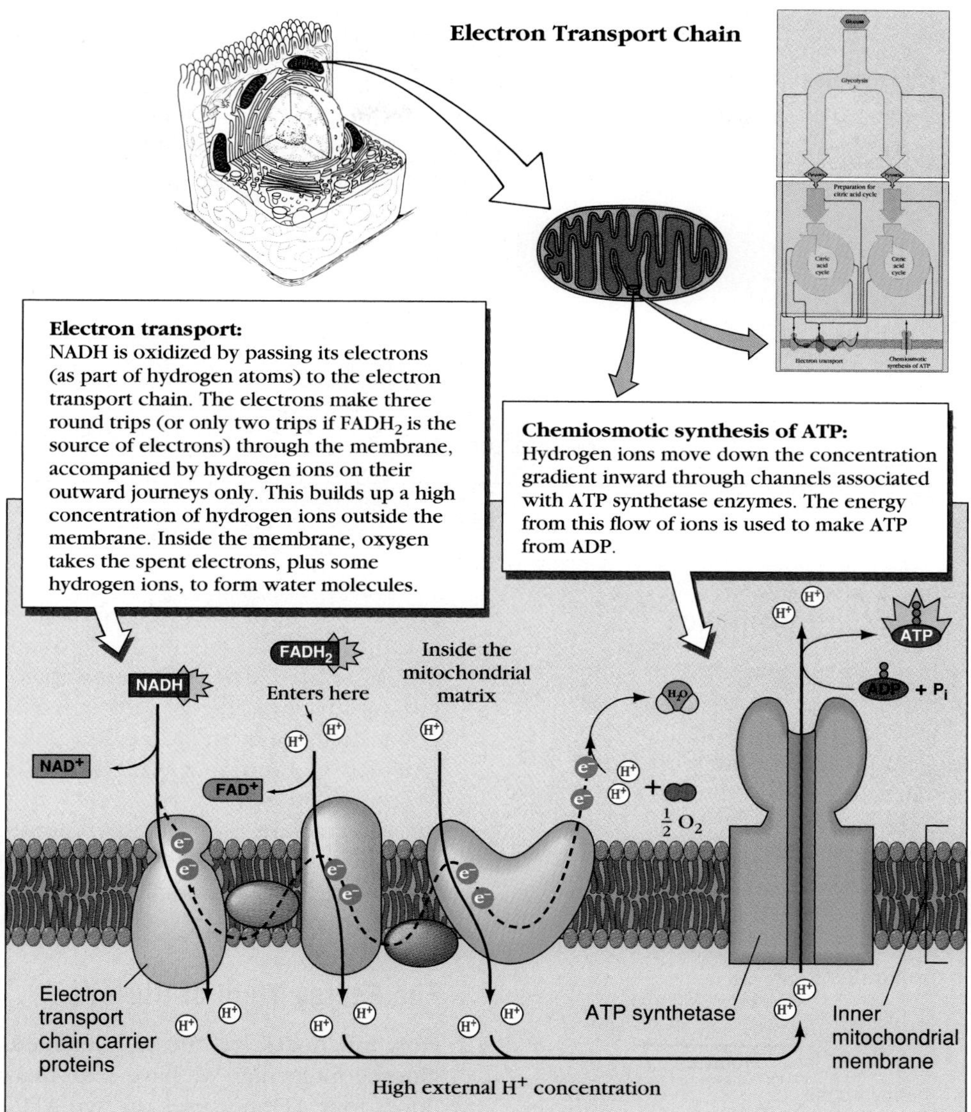

FIGURE 7-5

Hydrogen ions released from NADH and FADH molecules pass through the electron transport chain within the inner mitochondrial membrane generating a high concentration of H^+ on the outside of the membrane. As the H^+ passes back through the ATP synthetase molecules within the membrane, the energy from the flow is used to produce ATP.

sis, formation of acetyl CoA, and the citric acid cycle) and $FADH_2$ (from the citric acid cycle).

Now these coenzymes are oxidized by passing their electrons (as part of hydrogen atoms) to the **electron transport system,** a series of electron carrier proteins in the inner mitochondrial membrane (Figure 7-5). Here, the hydrogen atoms are passed through the system in a series of redox reactions, and their energy is gradually released and used to transport hydrogen ions out through the membrane. This creates a high concentration of hydrogen ions outside the mitochondrial matrix. Meanwhile, coenzymes NAD^+ and FAD^+, having released their hydrogen atoms, can shuttle back and pick up more hydrogen atoms (Figure 7-5).

Table 7–2
Maximum ATP Yield From One Molecule of Glucose

Process	Number of Reduced Hydrogen Carriers		Number of ATP per Carrier		Number of ATP by Chemiosmotic Synthesis	Number (net) of ATP Made Directly
Glycolysis	2 NADH + H^+	×	3	⟶	6 ATP	
						2 ATP
Preparation of acetyl CoA	2 NADH + H^+	×	3	⟶	6 ATP	
Citric acid cycle	6 NADH + H^+	×	3	⟶	18 ATP	
	2 $FADH_2$	×	2	⟶	4 ATP	
						2 ATP
					34 ATP	4 ATP
			GRAND TOTAL:		38 ATP	

Role of Oxygen in Cellular Respiration

Oxygen that is breathed into the lungs has an important role in cellular respiration. From the lungs, oxygen enters the blood, which takes it to the body's cells. Oxygen diffuses into a cell and then on into a mitochondrion. There it takes part in the last step of respiration.

Electrons arriving at the end of the electron transport system have given up most of their energy in the system's redox reactions. An oxygen molecule takes these spent electrons, plus some hydrogen ions, forming two new water molecules.

O_2 + 4 electrons + 4 hydrogen ions —Produces→ 2 H_2O

Without oxygen, the last molecule in the electron transport system would have no way to release its electrons. It would be stuck holding electrons, so the whole electron transport system would soon back up and we could not extract further energy from our food by respiration (Figure 7–5).

Chemiosmotic ATP synthesis uses the H^+ gradient set up by the electron transport system as an energy source in making ATP. Hydrogen ions move inward through the membrane through special ion channels associated with **ATP synthetase** enzymes. These enzymes use the energy of the hydrogen ions moving down their concentration gradient to make ATP from ADP and P_i. Each time that ATP is made this way, the hydrogen ion gradient loses some of its stored energy. But the gradient is constantly renewed by the continuous flow of electrons arriving from NADH and $FADH_2$, and passed along the electron transport system.

BIO-BIT

Cyanide is a metabolic poison that disrupts the flow of electrons to oxygen in the electron transport system and thus interferes with aerobic metabolism.

The Energy Yield of Glucose

How much ATP has the cell obtained from our original glucose molecule? We have seen that glycolysis produces four ATP but uses up two ATP to start the process. So, the net gain from glycolysis is two ATP per glucose molecule that is broken down. The citric acid cycle produces another two ATP, for a total of four ATP made directly per glucose molecule.

The electrons passed from one NADH through the electron transport system to oxygen provide enough energy to form up to three ATP. Electrons donated by $FADH_2$ provide energy to make up to two ATP. So, hydrogen ions coming from glycolysis yield six ATP by chemiosmotic synthesis, those from preparation of acetyl CoA yield six ATP, and those from the citric acid cycle yield 22 ATP. Thus, the maximum possible number of ATP derived from one glucose molecule is 38 (see Table 7–2).

Respiration is a series of redox reactions, using oxygen as a final electron acceptor, that break down organic molecules and release their energy. It occurs in the mitochondria of eukaryotes and in the cytoplasm and plasma membranes of bacteria. Two pyruvate molecules, produced by the breakdown of one glucose molecule in glycolysis, undergo chemical modifications within mitochondria in preparation for the citric acid cycle. The citric acid cycle completes the breakdown of glucose to carbon dioxide, which ultimately diffuses away, and hydrogen atoms, which are sent to the electron transport system within the inner mitochondrial membrane. These electron carrier proteins transport hydrogen ions into the space between the two mitochondrial membranes, creating a hydrogen ion gradient across the inner membrane. In chemiosmotic ATP synthesis, hydrogen ions move back down the gradient through special ion channels, providing energy used in joining ADP and P_i to form ATP. The oxygen that we breathe is crucial to life because it is the final electron acceptor for the electrons leaving the electron transport system. One glucose molecule can yield a maximum number of 38 ATP molecules.

7–E FERMENTATION

Fermentation makes bread rise and produces wine and beer. This fermentation is accomplished by yeasts—unicellular fungi—that use glycolysis to break down sugars in grape juice, for example, to pyruvate molecules. Yeasts then dismantle each pyruvate molecule into a molecule of carbon dioxide and a molecule of the two-carbon compound acetaldehyde. Acetaldehyde is next reduced by accepting two hydrogen atoms from NADH + H^+, forming the two-carbon alcohol, ethanol, also called ethyl alcohol—the active ingredient in alcoholic beverages (Figure 7–6). This transfer of hydrogen atoms has the important effect of freeing NAD^+ to go back to glycolysis and pick up more hydrogen, allowing the yeast cell to keep making ATP.

If fermentation continues until the yeast cells have used up all the sugar, a dry wine results. A sweet wine results when yeast cells produce enough alcohol to inhibit fermentation before they use up all the sugar. Stoppering a wine bottle before fermentation has finished yields a bubbly liquid because carbon dioxide is still being given off. To make a fizzy wine like champagne, one must use a very strong bottle and cork it up before fermentation has finished. The carbon dioxide dissolves in the wine under pressure and is released as bubbles when the bottle is opened. ■

Yeasts produce alcohol only when little or no oxygen is present. (Given plenty of oxygen, they use respiration to break sugar down completely to carbon dioxide and water.) When wine is fermenting rapidly, it produces carbon dioxide fast enough to drive off the air above the wine, and this prevents oxygen from dissolving in the wine. But when fermentation slows down, wine must be sealed up immediately, before oxygen can enter.

A second familiar kind of fermentation occurs in muscles during strenuous exercise. The muscle cells make ATP by respiration, using oxygen as fast as the bloodstream can deliver it, but this does not provide enough ATP. So, when oxygen is not delivered fast enough, the muscles make a small additional supply of ATP by fermentation. Pyruvate from glycolysis is reduced by accepting hydrogens from NADH + H^+ and so it becomes another three-carbon compound, lactate (Figure 7–6). Again, this frees NAD^+ to pick up more hydrogen, allowing glycolysis to proceed. This accumulated lactate (lactic acid) causes the pain of over-exercised muscles.

2? Lactate produced in the muscles is picked up by the blood. The body cannot tolerate a great buildup of lactate, but must eventually oxidize it back to pyruvate. After strenuous exercise we still keep breathing heavily for a time, to repay the body's **oxygen debt.** The liver takes up lactate and converts it back to pyruvate. When additional oxygen is available again, some of the pyruvate is broken down to yield energy by respiration. This provides the energy to convert the remaining pyruvate back to glucose by reverse glycolysis. The glucose re-enters the blood and is carried back to the muscles, where it is converted to glycogen and stored.

Fermentation that produces lactate as an end product is also carried out by various kinds of bacteria. Some of these are cultured and used to produce fermented foods such as cultured buttermilk and sour cream, yogurt, sauerkraut, pickles, green olives, and some sausages. ■

Most cells can carry on glycolysis, but not all organisms can carry on respiration. For this reason, and because the atmosphere of the Earth probably contained little oxygen when life first evolved, it is generally believed that glycolysis is the more primitive pathway of food breakdown and that the more complicated pathways using oxygen evolved later.

Fermentation is an anaerobic metabolic process that breaks down food and yields a small amount of ATP. Alcoholic fermentation is used in the production of alcoholic beverages and in bread-making.

Lactate fermentation permits muscle cells to produce small amounts of additional energy when not enough oxygen is available.

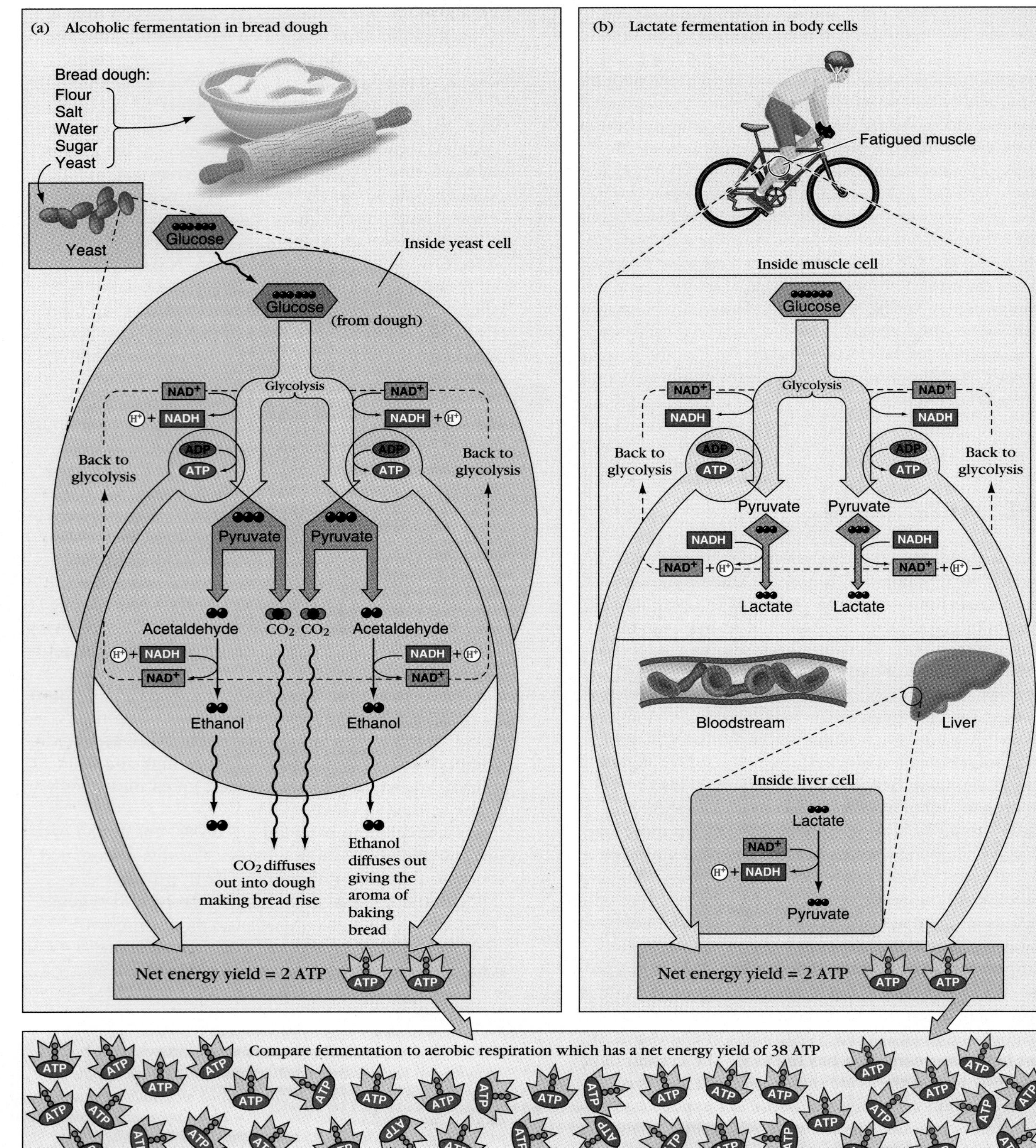
Anaerobic Metabolic Processes
(a) Alcoholic fermentation in bread dough
Bread dough:
Flour
Salt
Water
Sugar
Yeast
Yeast
Glucose
Inside yeast cell
Glucose
(from dough)
Glycolysis
NAD+
H+ + NADH
NADH + H+
ADP
ATP
Back to glycolysis
Pyruvate
Pyruvate
Acetaldehyde
CO2 CO2
Acetaldehyde
H+ + NADH
NAD+
Ethanol
Ethanol
CO2 diffuses out into dough making bread rise
Ethanol diffuses out giving the aroma of baking bread
Net energy yield = 2 ATP
(b) Lactate fermentation in body cells
Fatigued muscle
Inside muscle cell
Glucose
Glycolysis
NAD+
NADH
Back to glycolysis
ADP
ATP
Pyruvate
Pyruvate
NADH
NAD+ + H+
Lactate
Lactate
Bloodstream
Liver
Inside liver cell
Lactate
NAD+
H+ + NADH
Pyruvate
Net energy yield = 2 ATP
Compare fermentation to aerobic respiration which has a net energy yield of 38 ATP

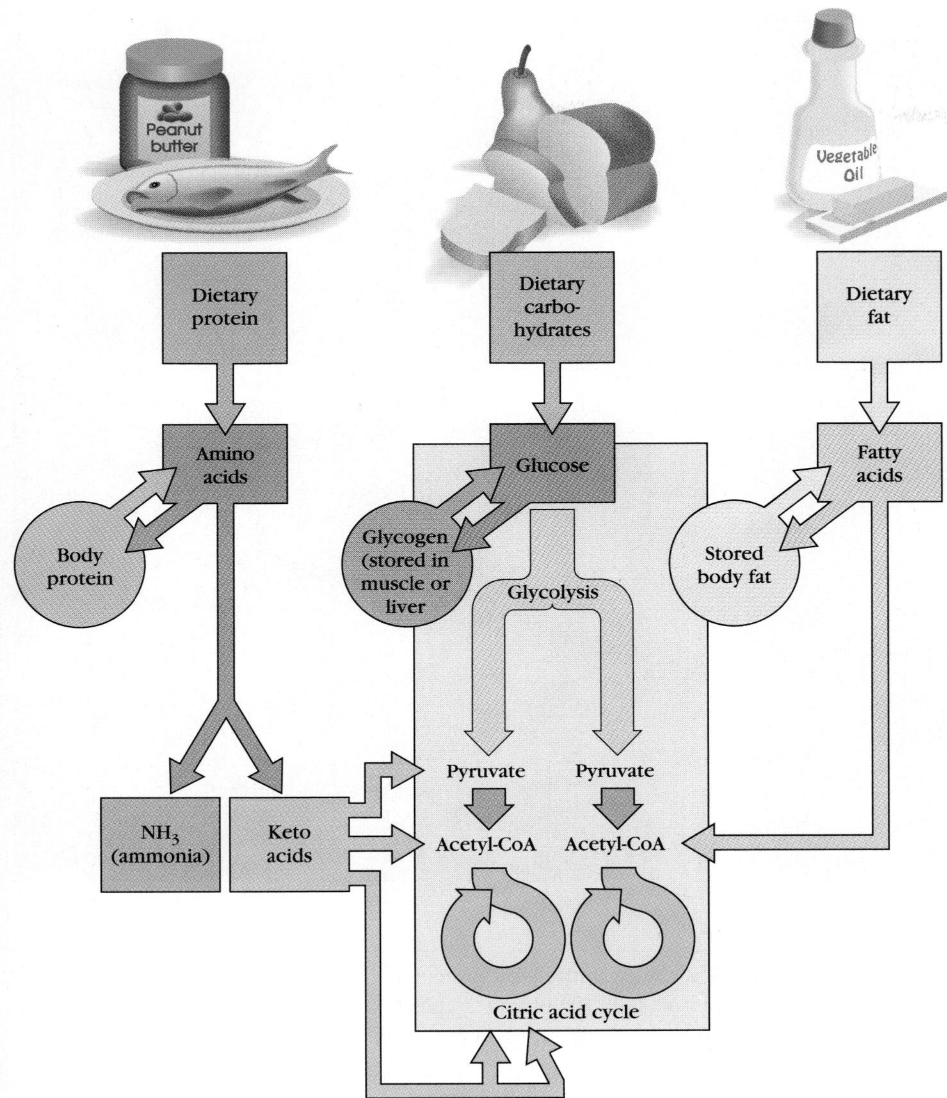

FIGURE 7–7

Major metabolic pathways. Proteins, carbohydrates (starch), and fats in the diet may become part of the body or may be processed to release energy. Cells usually obtain energy via glycolysis, which breaks down carbohydrates to release energy. Fatty acids and some keto acids made from amino acids enter cellular respiration at the level of acetyl CoA. Other keto acids enter as pyruvate or as molecules in the citric acid cycle. Both proteins and carbohydrates can also follow pathways that result in accumulation of body fat.

FIGURE 7–6

Alcoholic and lactate fermentation. In the absence of oxygen, these processes can produce only a fraction of the ATP produced in aerobic metabolism. **(a)** In alcoholic fermentation, pyruvate produced in glycolysis first gives off a molecule of carbon dioxide, leaving a two-carbon compound, acetaldehyde. This is then reduced by accepting two hydrogens from NADH + H^+ The process releases NAD^+, which is needed for glycolysis to continue, and produces ethanol. **(b)** In a tired muscle that lacks O_2, NADH accumulates. Pyruvate produced by glycolysis accepts hydrogens from NADH + H^+ and becomes lactate, freeing NAD^+ to return to glycolysis and accept more hydrogens. Lactate diffuses into the bloodstream where it is carried to the liver, and broken down into pyruvate.

7–F ALTERNATIVE FOOD MOLECULES

We have discussed how cells use glucose to make ATP. But all organic molecules contain stored energy, and many of them may be broken down to release the energy needed to make ATP.

Polysaccharides, either from food or from the body's stores of glycogen, can be broken down to glucose. Sugars other than glucose can be converted to glucose or fructose and fed into glycolysis.

Fatty acids from food or body fat stores do not go through glycolysis, which breaks down only carbohydrates. Instead, fatty acids are broken into two-carbon acetyl groups and enter aerobic respiration at the point where acetyl CoA forms (Figure 7–7).

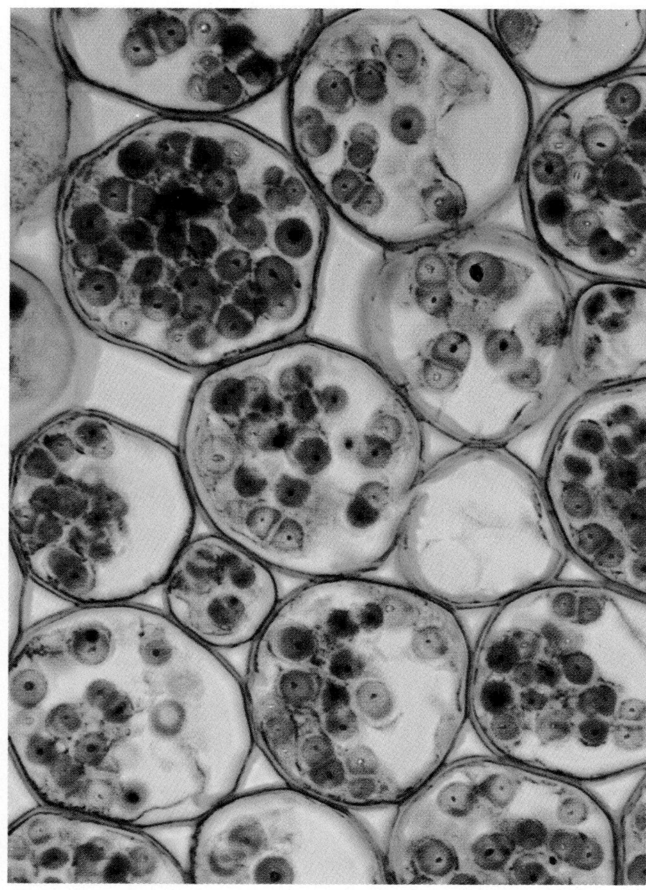

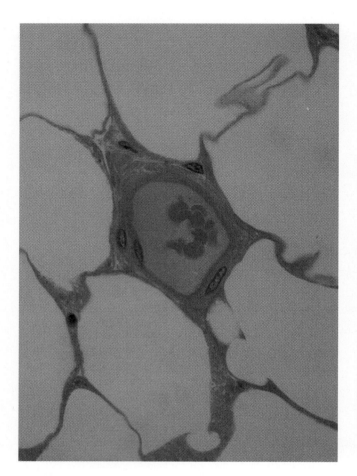

FIGURE 7–8

Food storage. (a) Adipose (fat-storing) cells. Each cell is occupied almost entirely by the colorless fat, with the nucleus pushed against the plasma membrane at one side. **(b)** Starch, stained purple, is stored in numerous amyloplasts in these cells from a buttercup root. (**a,** David Phillips/Visuals Unlimited; **b,** Ed Reschke/Peter Arnold, Inc.)

Proteins that are already incorporated into the body are not used for energy except during advanced starvation, after the body's carbohydrate and fat reserves have been depleted. Proteins in food are broken down into amino acids, and any amino acids that are not needed to build new proteins may be **deaminated** by the removal of their amino groups. The rest of the protein molecule is converted into either pyruvate, acetyl CoA, or one of the molecules of the citric acid cycle, and enters the pathway at the appropriate place (Figure 7–7).

3? Thus, the citric acid cycle and electron transport chain are a final common metabolic pathway for the breakdown of most organic molecules to yield energy. But not all foods are equivalent in energy yield. Table 7–3 shows that mass for mass, fats contain more than twice as much energy as carbohydrates or proteins. This is because fats contain a higher hydrogen:oxygen ratio than do carbohydrates or proteins. As we have seen, most of the energy from food molecules is gained by passing the electrons from hydrogen atoms along the electron transport chain. So, it is logical that molecules with a higher percentage mass of hydrogen atoms are higher in stored chemical potential energy per unit mass. And, as we have seen, because fats are nonpolar and don't require storage water, they are ideal substances for storing energy reserves (Figure 7–8a). ■

Table 7–3
Energy Yield of Major Food Components

Class	Chemical Elements	Energy Yield	Storage
Carbohydrates	CH_2O	4 kcal/gram	Hydrophilic: attracts much water
Fats	CHO	9 kcal/gram	Hydrophobic: concentrated fat droplets
Proteins	CHON(S)	~4 kcal/gram	Not used for energy storage

By contrast, plants store most of their energy reserves in the form of carbohydrates, not fats. Plants store mainly starches (Figure 7–8b), which are polar molecules due to the presence of the hydroxyl groups of carbohydrates. Consequently, when carbohydrates are stored in the aqueous environment of a plant cell, they inevitably attract storage water. This makes carbohydrates heavy and bulky to store, but because plants do not have to move around, the water's mass does not inconvenience them as it would an animal. One positive aspect of using carbohydrates to store energy reserves is that they are much more easily broken down into usable energy than are fats. These properties are probably the reasons that plants store their energy reserves as carbohydrates.

Molecules other than glucose can also be broken down to produce ATP. Polysaccharides can be broken down to glucose, and other sugars can be broken down or converted to glucose or fructose. Fatty acids are broken down into two-carbon acetyl groups which combine with coenzyme A and enter the citric acid cycle. Proteins can be broken down into various molecules that enter the citric acid cycle at several locations. Because of their high proportion of hydrogen atoms and their hydrophobic structure, lipids contain nearly twice as much energy per gram as carbohydrates and proteins and are often used to store energy in animals. Plants typically store their energy as starches, which are polysaccharides.

SUMMARY

1. Cellular respiration is the process that extracts useful energy from the energy stored in the chemical bonds of food molecules. This is done in a series of redox reactions, using oxygen as the final electron acceptor. The released energy is used to make ATP.
2. ATP donates the energy to various energy-requiring processes, such as metabolic reactions, active transport, muscle contraction, or production of new molecules.
3. During glycolysis, glucose is broken down anaerobically, producing two molecules of pyruvate, which are processed further in respiration. Also, in glycolysis, NAD^+ picks up hydrogen ions and becomes NADH + H^+, and some ATP is made directly.
4. Each pyruvate formed during glycolysis loses a carbon dioxide molecule and becomes an acetyl group, which is carried to the citric acid cycle by coenzyme A. More NADH + H^+ is also formed.
5. During one turn of the citric acid cycle, the equivalents of the acetyl group's two carbon atoms are removed as carbon dioxide, one of the end products of respiration. Some ATP is also produced. NAD^+ and FAD^+ accept hydrogen atoms and become reduced to NADH + H^+ and $FADH_2$, which carry hydrogen atoms to the electron transport system.
6. Most of the ATP derived from respiration is produced by electron transport followed by chemiosmotic ATP synthesis. NADH + H^+ and $FADH_2$ (from glycolysis, formation of acetyl CoA, and the citric acid cycle) pass pairs of hydrogen atoms to the electron transport system. Here energy from the hydrogen atoms' electrons is used in building a gradient of H^+ across the inner mitochondrial membrane. This gradient can be used to join ADP and P_i to form ATP. At the end of the electron transport system, the electrons combine with oxygen and H^+ and form water.
7. Neither glycolysis nor the citric acid cycle requires oxygen directly. However, if a cell has too little of the final electron acceptor—oxygen—most of the NAD^+ in the

cell will be tied up as NADH, unable to release its electrons to the electron transport chain. Under such anaerobic conditions, some cells continue to produce ATP during glycolysis by carrying out fermentation. Pyruvate accepts electrons from $NADH^+ + H^+$, forming ethanol or lactate. This releases NAD^+ so that glycolysis can continue. No such mechanism exists for the citric acid cycle, which therefore cannot function under anaerobic conditions.

8. Many other metabolic pathways feed into glycolysis, the citric acid cycle, and the electron transport chain, enabling cells to use food sources other than glucose to generate ATP.

SELF-QUIZ

1. NAD^+ functions in cellular respiration as a(n):
   a. energy intermediate
   b. enzyme
   c. coenzyme
   d. oxidizable substrate
   e. hydrogen donor
2. Which of the following statements is *not* true?
   a. Most of the ATP in an aerobic cell is formed via the electron transport chain and chemiosmotic synthesis.
   b. In eukaryotes, the chemiosmotic formation of ATP requires that the inner mitochondrial membrane remain intact.
   c. NAD^+ is a carrier molecule that travels down the electron transport chain to release ATP.
   d. In eukaryotes, the electron transport chain and the enzymes of the citric acid cycle are located in mitochondria, whereas the enzymes of glycolysis are located in the cytoplasm.
   e. The role of oxygen is to act as an acceptor of electrons.
3. The respiratory electron transport chain is found in the ________________ membrane in eukaryotes.
4. Give the end products of the following reaction sequences:
   a. glycolysis
   b. the citric acid cycle
   c. yeast fermentation
   d. electron transport chain
   e. lactate fermentation in muscle
5. While a muscle is in the process of reducing an oxygen debt:
   a. lactate is converted into pyruvate
   b. all the NAD^+ is in the reduced form
   c. pyruvate is converted into lactate
   d. NADH acts as an oxygen acceptor
6. True or False? Both yeast cells and muscle make two ATP (net) per glucose molecule fermented anaerobically.
7. True or False? Carbohydrate is unnecessary in the human diet since the products of fat and protein breakdown can enter the citric acid cycle to generate energy.

THINKING CRITICALLY

1. The insecticide rotenone inhibits transport in the first part of the electron transport chain. Cyanide inactivates cytochromes, transport molecules near the end of the chain. Explain why these poisons cause death.
2. Why do foods that are rich in fat tend to be expensive (in dollars and cents) compared with carbohydrates?

SELECTED KEY TERMS

acetyl CoA, *p. 130*
acetyl group, *p. 127*
anaerobic, *p. 125*
ATP synthetase, *p. 134*
cellular respiration, *p. 125*
chemiosmotic ATP synthesis, *p. 134*
citric acid cycle, *p. 127*
coenzyme, *p. 127*
deamination, *p. 138*
electron transport system, *p. 133*
FADH, *p. 127*
fermentation, *p. 125*
glycolysis, *p. 129*
Krebs cycle, *p. 130*
mitochondrial matrix, *p. 129*
NADH, *p. 127*
oxygen debt, *p. 135*
pyruvate, *p. 129*

SUGGESTED READING

Krebs, H. A. "The history of the tricarboxylic acid cycle." *Perspectives in Biology and Medicine* 14:154, 1970. An engaging account of how Sir Hans Krebs worked out the citric acid cycle.

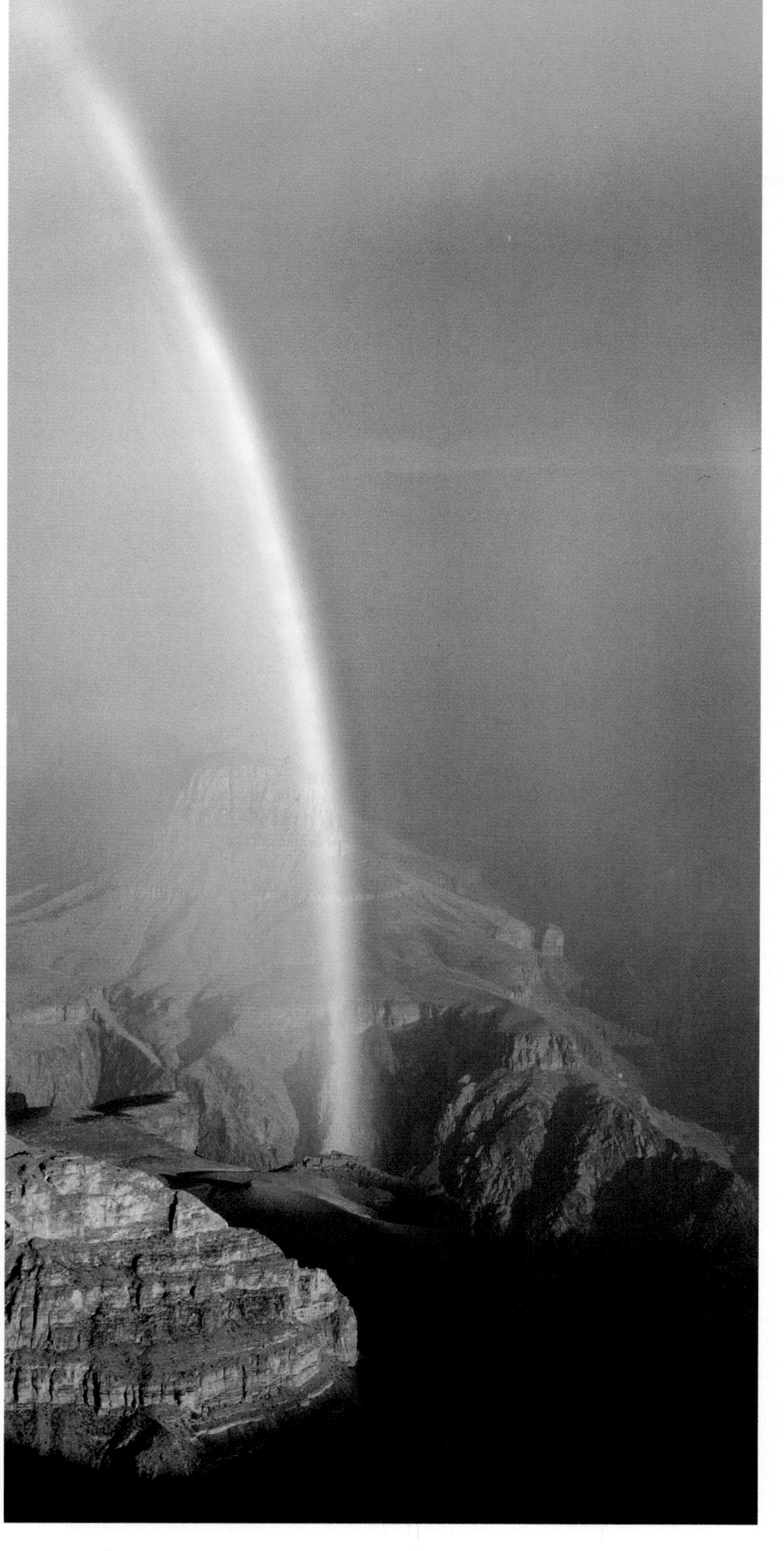

CHAPTER 8

CONCEPT GUIDE

After reading this chapter, you should be able to:

1. Explain the properties of visible light that make it the best-suited radiation for photosynthesis. (Section 8–A)
2. Compare the functions of carotenoids and chlorophylls *a*, *b*, and *c*. (Section 8–B)
3. Relate the structure of a leaf to its role in photosynthesis. Identify where in the leaf photosynthesis occurs, raw materials are brought in, and end products are removed. (Section 8–C)
4. In a diagram of a chloroplast, indicate the location of the electron transport system molecules, photosynthetic pigments, and DNA. (Section 8–C)
5. Describe, in general, the three energy-capturing steps of photosynthesis and their relation to the Calvin cycle. (Section 8–D)
6. Describe the location, main events, and products of the Calvin cycle. (Section 8–D)
7. Describe the effects of light intensity and temperature on the rates of photosynthesis. (Section 8–E)
8. Relate the photosynthetic adaptations of C_4 plants, cacti, and shade plants to their specific environmental conditions. (Section 8–F)

A rainbow shows the range of colors that combine to form white visible light, the visible portion of the spectrum. Read more about how plants use light to produce sugars and stay alive in Section 8–A. (Jeff Gnass/The Stock Market)

Photosynthesis

It is a well-kept secret that plants make animal life on Earth possible. Most of us are only dimly aware that plants make the oxygen necessary for cellular respiration. Furthermore, most people don't realize that, directly or indirectly, plants provide *all* of our food. For their entire lives, from seedlings until they wither and die, most plants work every day that weather permits, shuffling molecules within their cells and making sugar and oxygen from sunlight, carbon dioxide, and water.

Photosynthesis is the process that allows plants to capture the sun's energy and store it as chemical potential energy in the covalent bonds of carbohydrate molecules. Humans have learned to manipulate crop plants so that they will produce more food. We can grow burpless cucumbers and square tomatoes, hybrid corn that outproduces its ancestors, and strains of rice, wheat, and cotton that are resistant to diseases and insects. But with all our technological expertise, humans can't duplicate photosynthesis under laboratory conditions. And because we shall probably never be able to do this on a vast, cost-effective scale, humans and all of the other heterotrophs on the planet will continue to depend upon photosynthesizing plants for food and oxygen.

Photosynthesis consists of a complex series of reactions with the overall equation as shown below. When light is present, the green pigment, chlorophyll, captures light energy to power this process, in which water and carbon dioxide are turned into sugar and oxygen. Eukaryotic land plants, algae, and the prokaryotic cyanobacteria all photosynthesize in this way. Some bacteria that photosynthesize using different pigments and different compounds as a source of hydrogen atoms live in anaerobic environments like the muck of swamps and salt marshes.

KEY CONCEPTS

- Photosynthesis makes carbohydrates from carbon dioxide and water, using light energy trapped by chlorophyll.
- Nearly all of the oxygen in the air comes from photosynthesis. The vast majority of organisms need this oxygen for respiration.
- All our food, and the food of nearly all other organisms, comes directly or indirectly from photosynthesis.

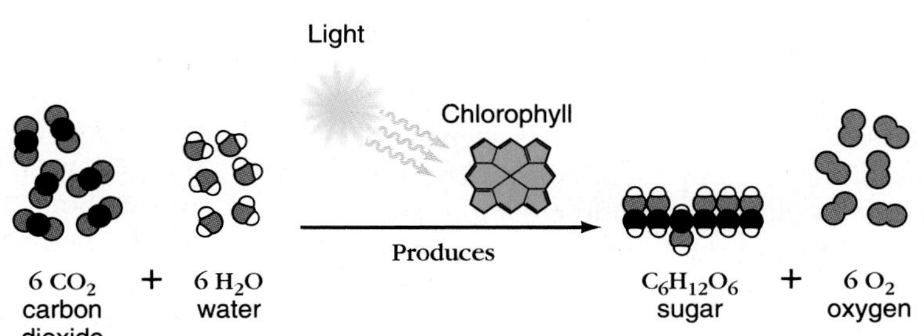

Curiosity Questions

1? Why do leaves change color in the fall? (See page 146)

2? Could exhaling on a plant make it grow faster? (See page 153)

3? Why don't some plants need as much sunlight as others? (See page 154)

(a)

(b)

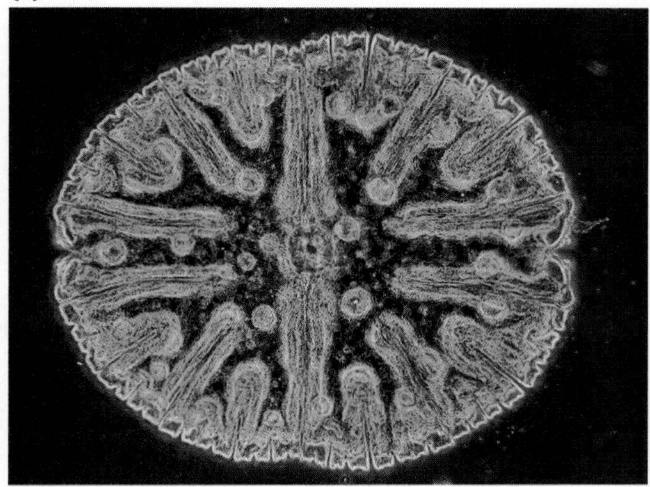

(c)

FIGURE 8–1

Photosynthetic life. (a) The lush Amazon tropical rain forest of Brazil next to the Amazon river. **(b)** A single-celled marine alga. **(c)** Many plants using a fallen tree as a source of nutrients. (**a,** Will and Deni McIntryre/Photo Researchers, Inc.; **b,** M. I. Walker/Photo Researchers, Inc.; **c,** E. R. Degginger)

8–A LIGHT

Anyone who has ever seen a rainbow has seen white light fragmented into its component colors. The sun emits a vast range of electromagnetic radiation, and only part of it is visible to the human eye (see chapter opening photo). Although we can detect radiation such as gamma rays, x-rays, microwaves, ultraviolet rays, radio waves, and infrared rays with various technological gadgetry, these portions of the electromagnetic spectrum are invisible to humans (Figure 8–2a). Instead, our eyes are adapted to see only the radiation in a narrow band of the spectrum: visible light.

Light behaves as if it travels in waves, and different parts of the electromagnetic spectrum have different **wavelengths.** Short wavelengths are measured in nanometers (nm), the distance from the top of one wave of light to the next, and longer wavelengths like radiowaves are measured in meters. Radiation with shorter wavelengths than violet light [about 350 nanometers (nm)] contains so much energy that it breaks hydrogen bonds and thus disrupts the structure of many biological molecules. This is one of the reasons that ultraviolet (UV) radiation causes skin cancer (see *A Journey into the Environment*: Ozone Depletion, Chapter 17). On the other hand, wavelengths above about 750 nanometers contain so little energy that it is rapidly absorbed by the water in the plant cell's cytoplasm before it reaches photosynthetic molecules. Low-energy radiation, like the microwaves that pop corn in your microwave oven, has little effect on the molecules in an organism, except to heat them up. Only visible light, with intermediate wavelengths (from 400 to 750 nanometers) has enough energy to cause chemical changes without destroying biological molecules.

The sun emits many wavelengths of radiation. Radiation with wavelengths shorter than those of visible light disrupts the structure of most biological molecules. Radiation with wavelengths longer than those of visible light contain little energy, which is quickly absorbed by water in living tissues. Only visible light provides enough energy to cause chemical changes without destroying biological molecules.

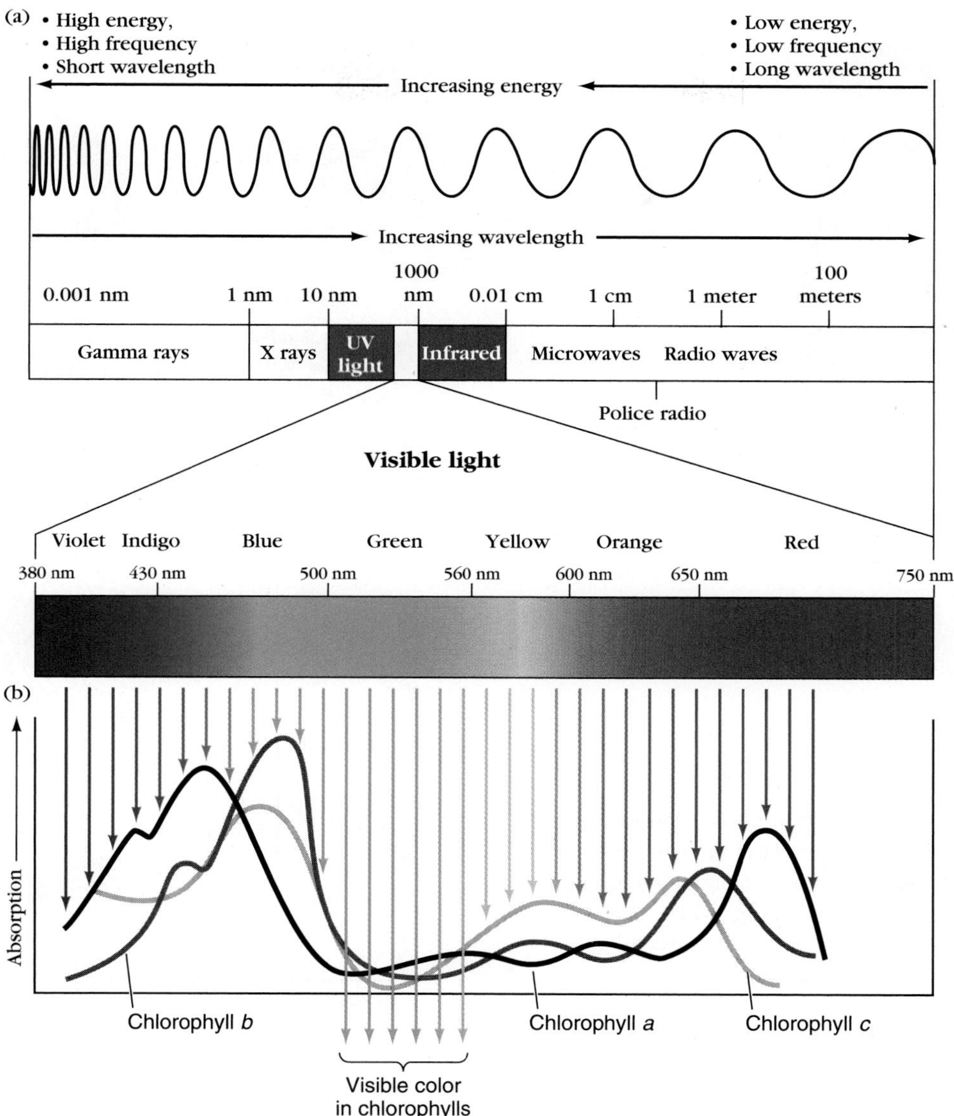

FIGURE 8-2

Electromagnetic radiation. (a) Our eyes can detect light with wavelengths of about 400 to 750 nanometers (nm). This is why light in this range is called visible light. Wavelengths just shorter than visible light are called ultraviolet light. Even shorter are x-rays (which overlap with ultraviolet and gamma rays), then gamma rays. Wavelengths on the longer side of the visible spectrum are known as infrared, then microwaves, and radio waves. **(b)** Different types of chlorophyll are adapted to absorb specific wavelengths of light. Chlorophyll *a* occurs in all photosynthetic plants, protists, and cyanobacteria. Chlorophyll *b* is found in all land plants and some algae. Chlorophyll *c* occurs in some other algae.

8-B TRAPPING LIGHT ENERGY: PHOTOSYNTHETIC PIGMENTS

Plants capture light energy using photosynthetic pigments (Figure 8-2). **Pigments** are molecules that appear to be colored because they absorb some wavelengths of light and reflect others. The photosynthetic pigment, **chlorophyll,** looks green because it absorbs light of colors other than green, and allows green light to be reflected or transmitted to our eyes (Figure 8-2b). In plants, chlorophyll occurs in **chloroplasts,** the organelles of photosynthesis. This becomes obvi-

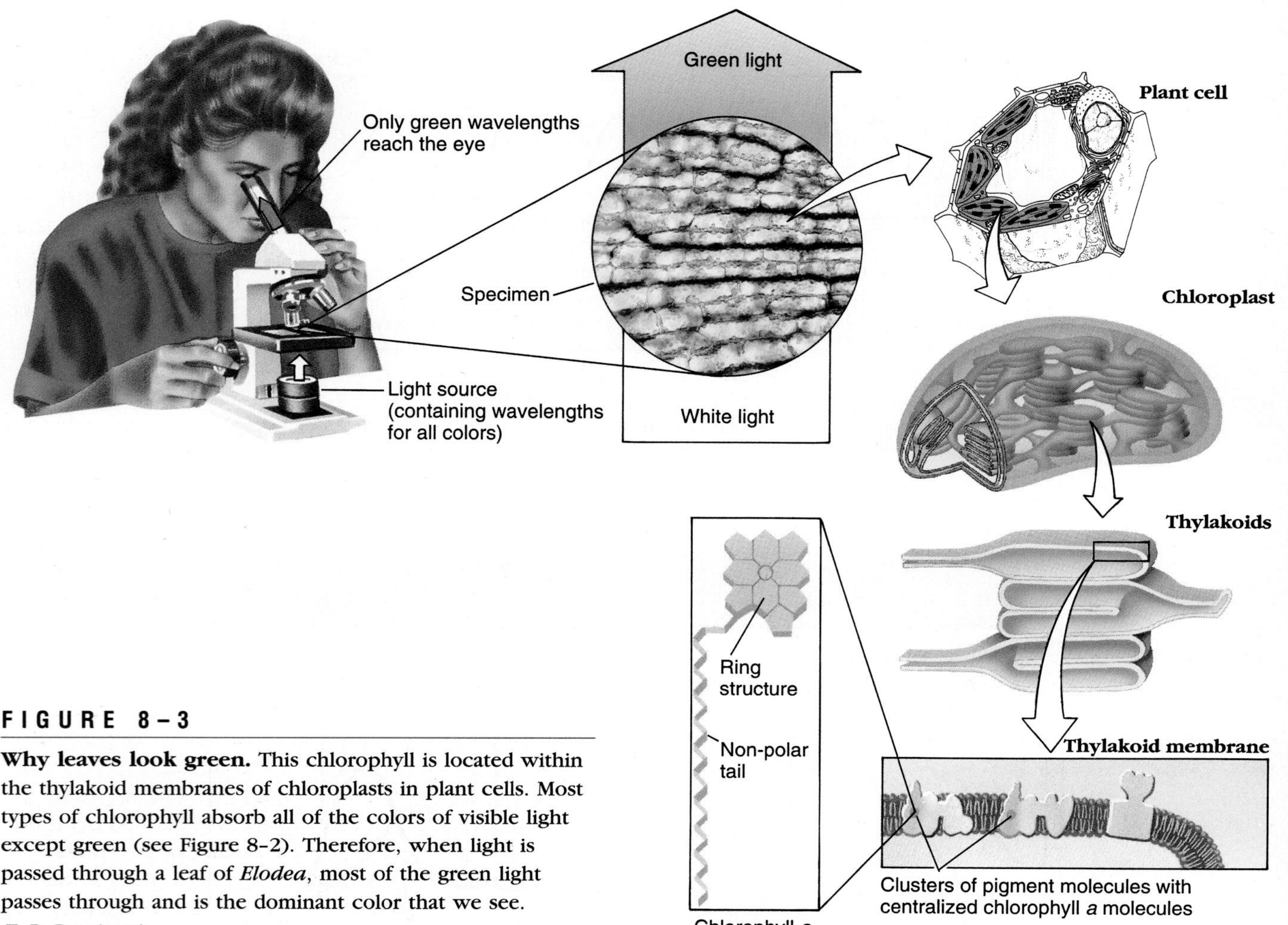

FIGURE 8–3

Why leaves look green. This chlorophyll is located within the thylakoid membranes of chloroplasts in plant cells. Most types of chlorophyll absorb all of the colors of visible light except green (see Figure 8–2). Therefore, when light is passed through a leaf of *Elodea*, most of the green light passes through and is the dominant color that we see. (E. R. Degginger)

ous when you examine a cell of a photosynthetic plant through a microscope (Figure 8–3) and see that only the chloroplasts are green.

There are several types of chlorophyll, with slightly different molecular structures and different properties, but all chlorophyll molecules have two main parts. At one end is a complex ring structure with a magnesium ion (Mg^{2+}) bound in the center. This is where light energy is trapped. The rest of the molecule is a long, nonpolar "tail" anchored in the lipid of the photosynthetic membrane. Chlorophyll *a* is the main photosynthetic pigment in all green plants; it absorbs red and blue light.

But sunlight contains wavelengths other than the blue and red ones absorbed by chlorophyll *a*. These wavelengths do not go to waste because plants also produce various **accessory photosynthetic pigments** that absorb light energy of other wavelengths and transfer it to chlorophyll *a*. Many plants contain chlorophylls *b* or *c*, which absorb different wavelengths in the blue and red parts of the spectrum (Figure 8–3). **Carotenoids** are another important group of accessory pigments found in all green plants (Figure 8–4). They absorb blue and green wavelengths and so impart yellow or orange hues to the plant.

1? In many plants, chlorophyll is broken down in the autumn, and its magnesium and nitrogen are moved to storage areas elsewhere in the plant before the leaves fall. This conserves hard-to-obtain elements, which can be used again the following spring. The breakdown of chlorophyll makes the yellow, brown, and orange colors of carotenoids visible in autumn leaves (Figure 8–4). The red color of some autumn leaves comes from a different class of pigments (anthocyanins).

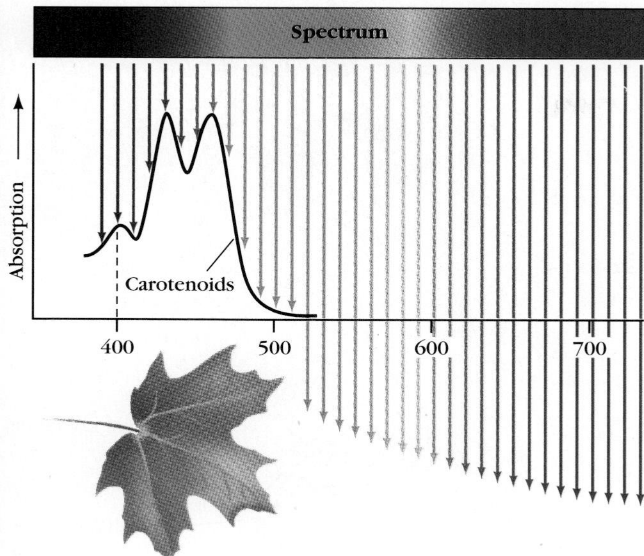

FIGURE 8-4

Why do many leaves have other colors in the autumn? In autumn, many plants break down chlorophyll and store the resulting magnesium and nitrogen elsewhere for leaf production the following spring. With the green chlorophyll gone, light passing through the leaf is mostly absorbed by the remaining carotenoid pigments, resulting in a variety of colors, depending upon the plant's particular combination of these leaf pigments.

These are not photosynthetic pigments. Rather, they are produced when sugar is trapped in the leaf as the tree prepares to shed it. ■

Photosynthetic pigments absorb the energy of different wavelengths of visible light used to drive photosynthesis. Chlorophyll *a* is the main photosynthetic pigment and uses the energy in red and blue light. Accessory photosynthetic pigments absorb wavelengths not absorbed by chlorophyll *a* and transfer the energy to chlorophyll *a*.

8–C TOUR OF A LEAF

Any part of a plant that has chloroplasts can photosynthesize, but in most plants, the main photosynthetic organs are the leaves. A typical leaf's broad, flat shape presents a maximum surface area for light absorption at a minimum weight. A layer of tightly packed cells, the **epidermis,** covers the upper and lower leaf surfaces (A Journey Through the Chloroplast). These cells secrete a waxy, waterproof **cuticle,** which reduces loss of water vapor from the leaf into the air. The epidermis contains pores, called **stomata** ("mouths"; singular: **stoma**), that admit the carbon dioxide needed by the photosynthetic cells in the interior of the leaf. In most plants, the stomata are open during the day, thus permitting the uptake of carbon dioxide and release of oxygen, but also inevitably allowing water vapor to escape. Stomata are usually closed at night, and they also close during the day if the plant has lost too much water.

Cells with chloroplasts form two distinct layers that are sandwiched between the upper and lower epidermis of the leaf. The **palisade layer** consists of closely packed columnar cells, standing upright just beneath the upper epidermis. Below them lie the loosely arranged cells of the **spongy layer,** with air spaces between them (A Journey Through the Chloroplast). Air enters these spaces via the stomata. Since molecules diffuse faster in a gas than in a liquid, the air spaces allow for rapid gas exchange between the air and the photosynthetic cells.

The leaf's **veins** carry water arriving from the roots and remove the newly made sugars to parts of the plant where they are used or stored.

Each of the leaf's photosynthetic cells contains many chloroplasts. A chloroplast is separated from the cytoplasm by its envelope, consisting of two membranes. Inside the chloroplast, a third system of membranes, called **thylakoids,** contains the molecules that trap and use light energy during photosynthesis (A Journey Through the Chloroplast). Some of the thylakoids occur in stacks, called **grana** ("grains"), because a light microscope shows them as little specks. All the thylakoids in one chloroplast appear to be continuous, surrounding a single interior space. This interior space contains a reservoir of hydrogen ions (H^+), built up by electron transport systems housed in the thylakoid membranes. The thylakoids also contain the photosynthetic pigments—chlorophylls and carotenoids. The thylakoid membranes give the photosynthetic machinery a great deal of surface area, which is arranged to intercept as much light as possible.

The thylakoids are surrounded by a protein-rich solution called the **stroma.** The stroma contains the enzymes that make carbohydrates during photosynthesis, as well as the chloroplast's DNA and ribosomes (A Journey Through the Chloroplast).

Most photosynthesis in plants occurs within their leaves. The epidermis covers the upper and lower leaf surfaces and contains stomata, which regulate the movement of water vapor,

A Journey Through the Chloroplast

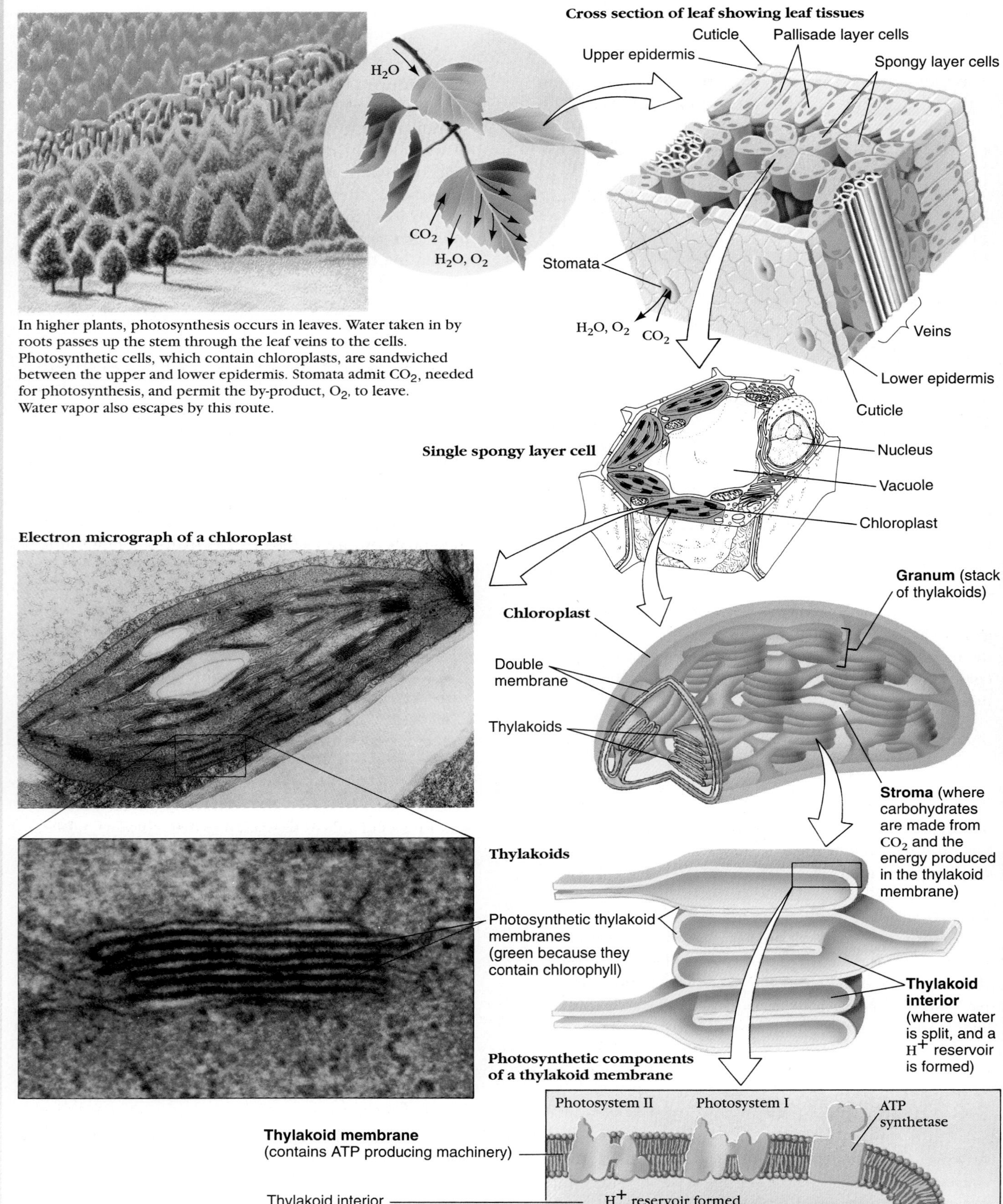

The location of photosynthesis. (c,e, Biophoto Associates)

carbon dioxide, and oxygen between the leaf's interior and the surrounding air. Within the leaf, photosynthetic cells form the dense palisade layer and more loosely arranged spongy layer. Photosynthesis occurs within chloroplasts inside the photosynthetic cells. The thylakoids contain the chloroplasts' photosynthetic pigments and electron transport molecules.

OVERVIEW: WHAT HAPPENS IN PHOTOSYNTHESIS?

The reactions of photosynthesis take place in two main phases. In the first phase, energy is captured; in the second, this energy is used to make sugar. Figure 8-5 is a general overview of the process of photosynthesis. Study it for a moment, examine inputs and outputs of the process, and make sure you have a good general idea of the process of photosynthesis before reading further.

8–D PHOTOSYNTHESIS IN DETAIL

Energy Capture

The energy-capturing stage of photosynthesis involves three steps. All of these occur in the chloroplasts' thylakoid membranes (A Journey Through Photosynthesis).

1. Light absorption. Light is trapped during **photochemical reactions** (chemical reactions that are powered by light energy). Such reactions are familiar in photography, where light striking the film causes chemical reactions that produce an image. The brighter the light, the more reactions occur per unit of time. In the photochemical reactions of photosynthesis, light energy transforms pigment molecules, exciting them and changing their energy levels so that an electron is freed from one pigment molecule and begins moving through an electron transport system.

Light absorption occurs in two kinds of **photosystems.** Both contain chlorophyll *a,* surrounded by different photosystem proteins that give the chlorophyll *a* molecules different absorption maxima. The chlorophyll *a* in Photosystem I is called pigment 700 (or P700 for short), because it absorbs light wavelengths of 700 most efficiently. The chlorophyll in Photosystem II is called pigment 680 (or P680).

Allied with each photosystem are hundreds of chlorophyll molecules called **antenna pigments** (A Journey Through Photosynthesis). Just as a television antenna intercepts television signals and sends them to a TV set, antenna pigments gather light energy from photons and funnel it to the chlorophyll *a* reaction center, either P700 or P680.

When light of the proper wavelength strikes an antenna pigment, it causes a rapid vibration of the pigment molecule. Like "the wave" passing along a row of football fans, this vibration energy migrates from molecule to molecule in the photosystem until it reaches chlorophyll *a* (P680 or P700) in the photosystem's **reaction center,** where electron transport begins. The light absorption phase of photosynthesis allows the energy from sunlight to be transferred to chlorophyll causing one of its electrons to jump into a higher unoccupied shell, from which it can be taken by other molecules. Where the electrons go and what they do depends upon which photosystem we are discussing. The two photosystems work simultaneously, but for clarity we will discuss them as though they worked in sequence.

2. Electron transfer. Photosystem I sends an electron from P700 along the **electron transport system** to $NADP^+$, also located at the thylakoid membrane's outer face (A Journey Through Photosynthesis). When two photons of light have been absorbed, two electrons are passed through these transport molecules and they have enough energy to reduce a molecule of $NADP^+$ to NADPH. NADPH is a coenzyme and energy intermediate that shuttles electrons about in photosynthesis, just as NADH does in the mitochondrion during cellular respiration.

Let us return to P700 in Photosystem I. Because of the loss of one electron, P700 can no longer function as a reaction center chlorophyll, and this is where **Photosystem II** comes into play. The antenna pigments of Photosystem II funnel absorbed light energy to P680. This pigment molecule is excited and transfers this absorbed energy to an electron, causing it to break free and travel through a separate series of electron transport molecules until it reaches P700, replacing its lost electrons. Thus, P700 is regenerated by Photosystem II and is ready to absorb more light energy.

You may be wondering, "What about P680? How is its lost electron replaced?"

Look back at the overall reaction for photosynthesis and note that the reaction specifies an input of water. The lost electrons from P680 are replaced by extracting electrons, one at a time, from two water molecules. This leaves four hydrogen ions, which are released into the thylakoid space, and two oxygen atoms, which combine and form some oxygen, O_2. This by-product diffuses out of the cell and eventually out of the plant.

(text continues on page 152)

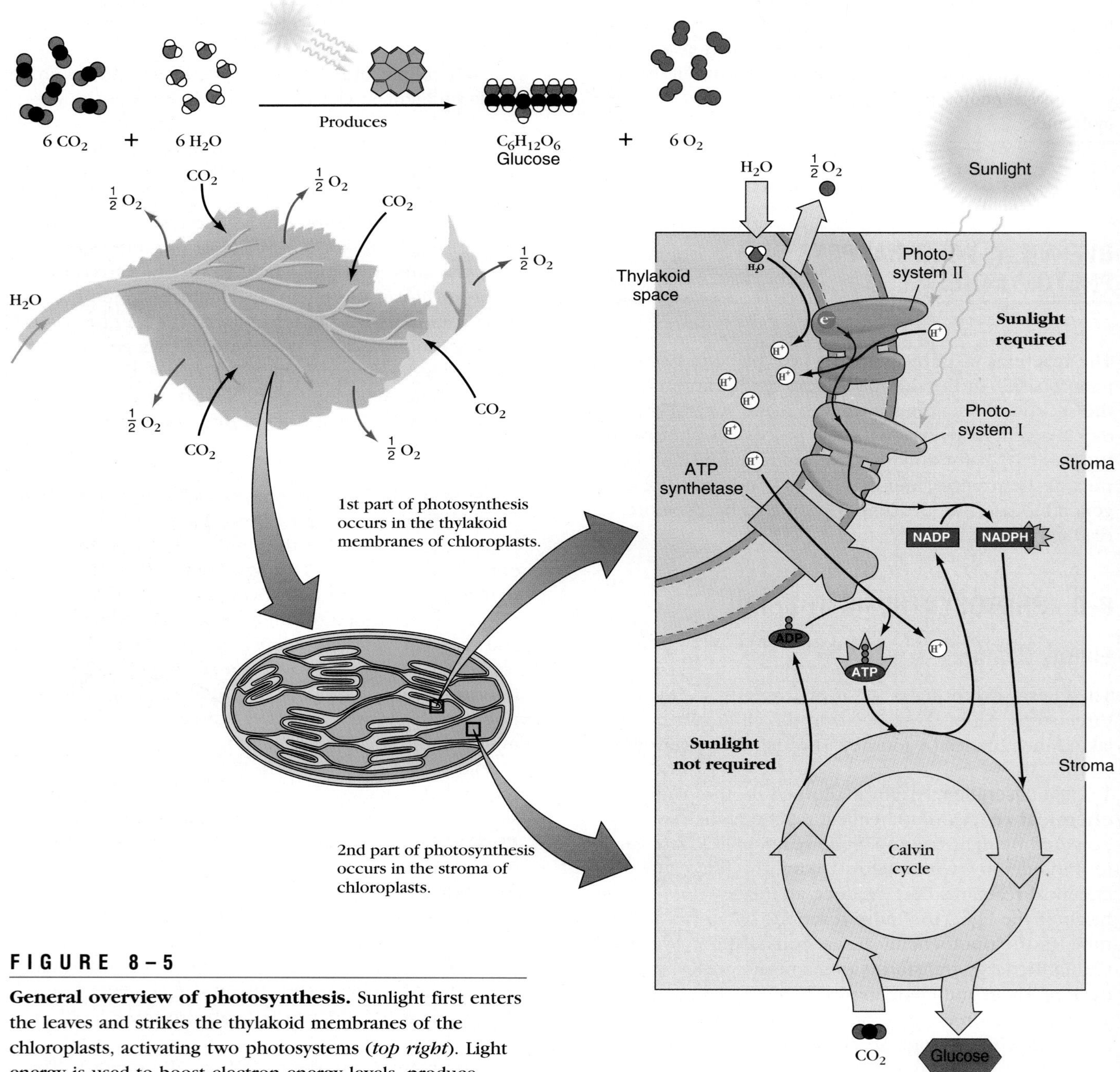

FIGURE 8-5

General overview of photosynthesis. Sunlight first enters the leaves and strikes the thylakoid membranes of the chloroplasts, activating two photosystems (*top right*). Light energy is used to boost electron energy levels, produce NADPH, and build a H^+ reservoir in the thylakoid interior. Finally, H^+ moves down their concentration gradient into the stroma and produce ATP. In the Calvin Cycle (*lower right*), NADPH is used to produce carbohydrates by combining atmospheric carbon dioxide and water.

An overview of photosynthesis. Energy-storing molecules, ATP and NADPH, are produced by photosystems I and II and chemiosmotic synthesis, all within the thylakoid membranes. These energy carriers are then transferred to the stroma of the thylakoids where they power the Calvin Cycle in the production of carbohydrates.

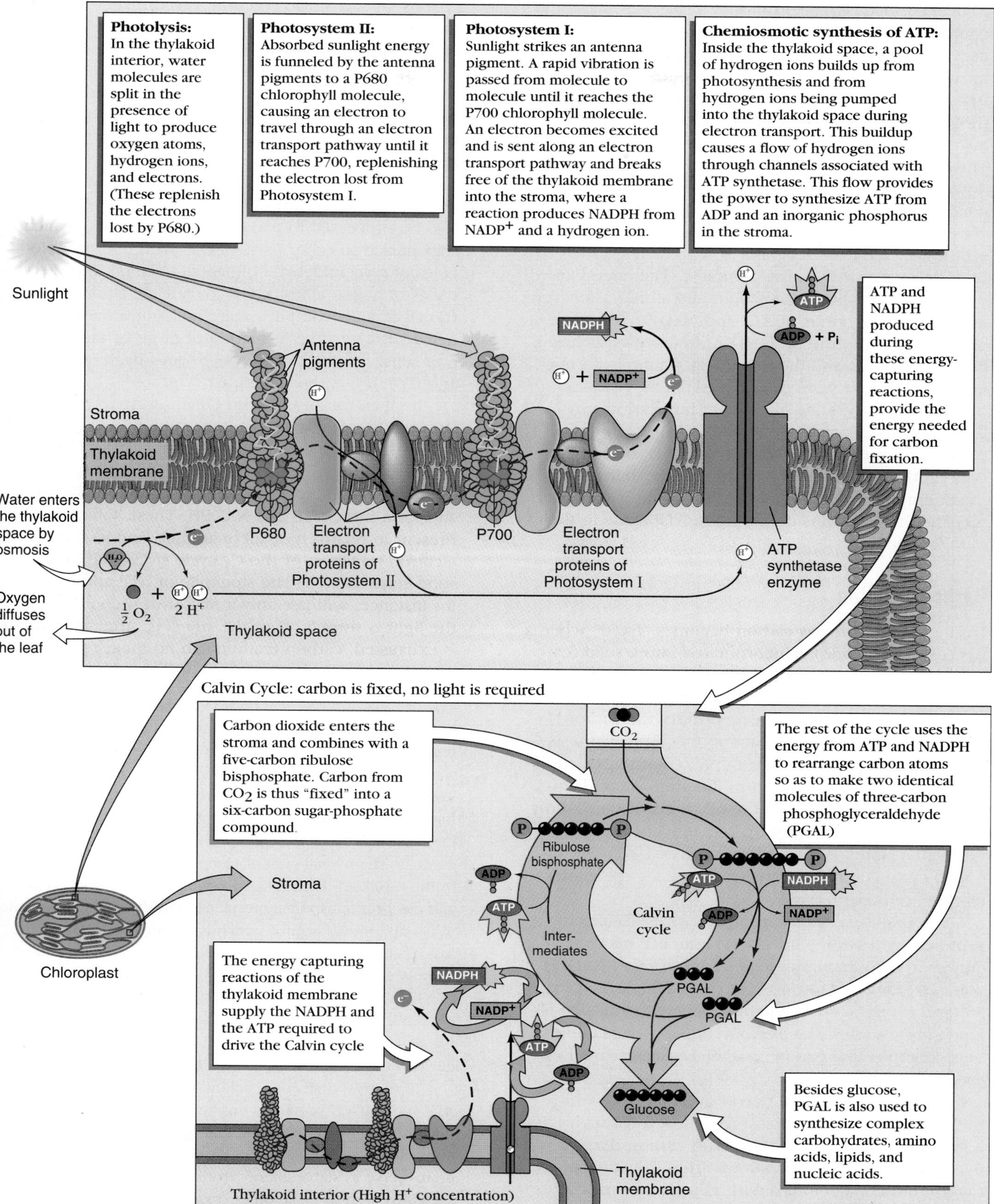

A Journey Through Photosynthesis

3. Chemiosmotic Synthesis of ATP. The final phase of energy capture, **chemiosmotic synthesis of ATP,** uses the energy from the pool of hydrogen ions built up within the thylakoid space as a result of electron transport. These trapped hydrogen ions pass out through the thylakoid membrane in channels associated with ATP synthetase enzymes (A Journey Through Photosynthesis). This flow of hydrogen ions down their concentration gradient, from a high concentration within the thylakoid space to a lower concentration in the stroma, provides the energy to synthesize ATP from ADP and P_i.

Energy capture is now complete. The energy in photons of light has been caught and stored as energy in the chemical bonds of ATP and NADPH. The role played by light in the process of photosynthesis has been completed, and the remaining reactions do not need light. These were once called "dark reactions," but that term gives the misleading impression that these processes occur in the dark. At night, once plants have used up the NADPH and ATP from the energy-capturing phases of photosynthesis, their cells get the energy they need to live until daylight in the same way that heterotrophs do: by using ATP made in their mitochondria.

Carbon Fixation

During photosynthesis carbon becomes "fixed" when a gas (carbon dioxide) is incorporated into a solid (carbohydrate, usually a sugar). ATP and NADPH, produced during the energy-capturing reactions, provide the energy and hydrogen needed to fix the carbon. Look again at the overall reaction for photosynthesis and find the input of carbon dioxide that the reaction specifies. Carbon dioxide from the atmosphere is incorporated into glucose in the reactions of the **Calvin cycle,** named after its discoverer, Melvin Calvin, who received a Nobel Prize for this work in 1961. The Calvin cycle is often referred to as the **C_3 cycle** after its three-carbon products.

You might wonder why a plant that has already captured light energy in NADPH and ATP needs to fix energy in carbon compounds. First, ATP and NADPH don't last long. A food store of carbohydrate that can be used for respiration whenever energy is needed is much more useful. Also, carbon fixation builds up carbon skeletons that can be modified to make other organic molecules.

The enzymes of the Calvin cycle are located in the stroma of the chloroplast (A Journey Through Photosynthesis). The cycle begins with carbon dioxide that has diffused into the stroma from the atmosphere. Carbon dioxide combines with **ribulose bisphosphate** (RuBP), a five-carbon sugar with two phosphate groups. In this single reaction, carbon has been "fixed" in a six-carbon sugar-phosphate compound, but this carbon is still in an oxidized, energy-poor form. In the rest of the cycle it is reduced, using H from NADPH and energy from both NADPH and ATP, yielding two identical molecules of a three-carbon compound, **glyceraldehyde phosphate.** Two molecules of glyceraldehyde phosphate can be processed and joined to make one glucose. The glucose can be polymerized into complex carbohydrates like starch or cellulose; it can also be processed to make the sugar sucrose, which is transported to other parts of the plant. Other molecules of glyceraldehyde phosphate can go into the synthesis of amino acids, lipids, or nucleic acids. Most of the glyceraldehyde phosphate molecules must be channeled back into the formation of more of the five-carbon sugar, ribulose, which then goes back to receive and fix more carbon dioxide.

It takes six turns of the Calvin cycle to produce the equivalent of one (six-carbon) glucose molecule, that is, to fix six carbon atoms from carbon dioxide into a solid, organic form. The ADP, P_i, and $NADP^+$ released by the C_3 cycle are recycled to the thylakoid and form more ATP and NADPH. These substances are present in relatively small quantities, and so if either electron transport or the C_3 cycle stops, the other soon stops as well. The stockpile of ATP and NADPH, for instance, will last only a matter of seconds once the light is turned off or the sun sets. After this supply is exhausted, carbon fixation can no longer proceed.

The energy-capturing reactions take place within the thylakoid membranes and involve three steps. 1. In photosystems I and II, light is absorbed by antenna pigments that transfer its energy to chlorophyll *a* in the reaction center. This light energy boosts an electron in chlorophyll *a* to a higher energy level, so high that the electron can leave the molecule and enter the adjoining electron transport molecules. 2. The overall path of electron transport is from water in the thylakoid interior or space, through the photosystems and electron transport molecules in the thylakoid membrane to $NADP^+$ in the stroma, which is reduced to NADPH. The loss of electrons from water forms a high concentration of hydrogen ions in the thylakoid space and produces oxygen as a by-product. They will eventually diffuse to the atmosphere. 3. In the final step, the pool of hydrogen ions within the thylakoid interior move down their concentration gradient into the stroma and produce ATP from ADP and P_i.

In the carbon-fixing reactions of the Calvin cycle, the ATP and NADPH produced during the energy-capturing reactions provide the energy and hydrogen to build carbohydrates using atmospheric carbon dioxide. These carbohydrates serve as stable energy stores and as the carbon skeletons used to make other organic molecules.

FIGURE 8-6

Oxygen production. In bright light, the energy-capturing reactions of photosynthesis proceed rapidly, as evidenced by the vigorous bubbling of the by-product oxygen from this aquatic plant. (E. R. Degginger)

8–E WHAT CONTROLS THE RATE OF PHOTOSYNTHESIS?

Photosynthesis is the sum of many different chemical reactions. Several factors play key roles in some of these reactions and so can affect the rate of the whole process.

Photosynthesis begins with photochemical reactions, which trap light energy and go at increased rates when the light intensity increases (Figure 8-6). However, the remaining reactions of photosynthesis are like the chemical reactions we saw in previous chapters, called **thermochemical reactions** because their rates are increased by heat. Because photosynthesis is made up of both photochemical and thermochemical reactions, both light and temperature influence its rate.

In dim light, the rate of photosynthesis is limited by lack of light. Photochemical reactions that initiate electron transport, and hence chemiosmotic ATP synthesis, go too slowly to produce NADPH and ATP as fast as carbon fixation can use them. As the light becomes brighter, the rate of photosynthesis increases until ATP and NADPH become so plentiful that carbon fixation cannot keep up, and the rate of photosynthesis levels off. This is what happens on bright, cool days. On warmer days, the rate of photosynthesis increases still further. At these higher temperatures, molecules move faster, and the rate of the thermochemical reactions of carbon fixation speeds up.

The rate of a reaction can also be decreased by scarcity of raw materials or excess of products. On bright, warm days, light and temperature are optimum for photosynthesis, but a low concentration of the raw material carbon dioxide often limits its rate. Adding more carbon dioxide to the air in a greenhouse increases the rate of photosynthesis (and so of plant growth) of some greenhouse crops. Above a certain concentration, however, carbon dioxide inhibits photosynthesis.

2? During a bright warm day, exhaling onto a plant could increase the rate of photosynthesis and thus plant growth by increasing the carbon dioxide levels and temperature. However, significant gains would depend upon the degree to which photosynthesis was limited by temperature and carbon dioxide levels. How would you design an experiment to test this hypothesis? ■

The other raw material of photosynthesis, water, is so abundant in living tissue that the amount channelled into photosynthesis is negligible. However, a water shortage does limit photosynthesis indirectly: if the plant loses too much water, the stomata of the leaves close. This water-conservation measure also cuts down the rate of CO_2 uptake and hence slows the carbon fixation reactions of photosynthesis.

The carbohydrate end products of photosynthesis are quickly changed into other chemical forms or removed from the chloroplast, and so they do not accumulate and inhibit the reaction. The other end product, oxygen, diffuses out of the cell and leaves the plant. However, when oxygen is being produced rapidly, some of it accumulates at levels high enough to slow photosynthesis, by inactivating electron transport molecules and by interfering with the carbon dioxide fixation enzyme.

Photosynthesis involves photochemical reactions that speed up as light intensity increases and thermochemical reactions that speed up as the temperature warms. Low levels of carbon dioxide can limit photosynthesis by restricting carbon fixation.

8–F ECOLOGICAL ASPECTS OF PHOTOSYNTHESIS

In plants in unusual habitats, some interesting variations on photosynthesis have evolved.

As we have seen, the stomata permit carbon dioxide to enter the leaves, but at the same time, some of the plant's water inevitably evaporates through the stomata into the air. If the leaves lose water faster than the roots replace it, the plant starts to wilt and the stomata close somewhat, slowing the loss of water but also reducing the entry of carbon dioxide.

The C_3 cycle enzyme that fixes CO_2 works better at high carbon dioxide levels. However, some plants can carry on photosynthesis rapidly even at low carbon dioxide levels because they can trap and store CO_2. An enzyme attaches available CO_2 to a three-carbon compound, forming a four-carbon molecule that holds the CO_2 temporarily. These plants are called **C_4 plants** after the four-carbon compound. The C_4 molecules are transported to a central location, where their CO_2 is removed and accumulated at high enough levels for the C_3 cycle to re-fix it into carbohydrate.

The C_4 pathway uses five ATP per molecule of carbon dioxide fixed instead of the three ATP used in C_3 photosynthesis alone, but at high temperatures C_4 plants can photosynthesize faster than C_3 plants. This sacrifice of efficiency for speed is advantageous because it allows plants to grow and reproduce faster. However, this profligate use of energy is worthwhile only when energy is abundant compared to other resources. C_4 photosynthesis has evolved in plants of warm, sunny, somewhat dry climates such as grasslands, where there is plenty of light energy to drive ATP synthesis, but water stress is a frequent problem. In such a situation, the C_4 plant's stomata may close partially, reducing loss of water vapor from the leaves. This also reduces uptake of CO_2 from the air, but the C_4 pathway allows the plant to capture enough carbon dioxide for a high rate of photosynthesis anyway. Not surprisingly, C_4 plants include crabgrass, other important weeds, and some major fast-growing crops such as sugarcane and corn.

Because the rapid growth of C_4 plants depends on a high rate of photosynthesis in bright light, C_4 plants are at a relative disadvantage in cool or shady situations, and C_3 and C_4 plants coexist, with neither having a noticeable edge, in many habitats (Figure 8–7).

3? Even ordinary C_3 plants can be separated into two groups, sun plants and shade plants. While **sun plants,** such as soybeans, cotton, and tomatoes, show increased rates of photosynthesis as light intensity increases, **shade plants** do not. Shade plants, including many ferns, African violets, and philodendrons, simply do not photosynthesize rapidly, even in bright light. On the other hand, shade plants are much more efficient at using very dim light (see Figure 8–7).

Some plants use a variation of C_4 photosynthesis called **crassulacean acid metabolism (CAM).** CAM occurs mainly in cacti, euphorbs, and other succulents—plants with fleshy, water-storing stems or leaves, adapted to desert conditions (Figure 8–8). The stomata of CAM plants are closed during the hot day. In the cooler night, when water evaporates more slowly, the stomata are opened and carbon dioxide is taken up and fixed into organic acids. During the day, when the stomata are closed, carbon dioxide is removed from the acids, and light energy is used to re-fix it by way of the C_3 pathway. Although CAM plants receive plenty of sun, they must conserve water very strictly, and so they have very low photosynthetic rates. However, their ability to conserve water enables them to survive in habitats too dry for the "fast food" plants that would otherwise crowd these slow growers out.

Plants with C_4 photosynthesis are able to concentrate carbon dioxide around their Calvin cycle enzymes, an adaptation that enables them to maintain a high photosynthetic rate, at the same time that they reduce their loss of water through the stomata. Other plants, including succulents such as cacti, take up carbon dioxide through their stomata only at night and store it until it is used in photosynthesis during the daytime. Shade plants can use light more efficiently than sun plants, but do not photosynthesize as rapidly.

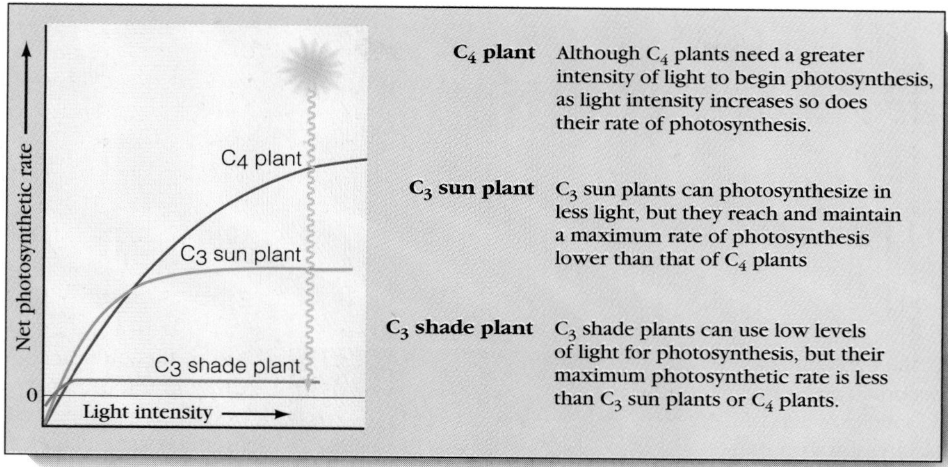

Net photosynthetic rate
Light intensity
0
C4 plant
C3 sun plant
C3 shade plant
C₄ plant Although C₄ plants need a greater intensity of light to begin photosynthesis, as light intensity increases so does their rate of photosynthesis.
C₃ sun plant C₃ sun plants can photosynthesize in less light, but they reach and maintain a maximum rate of photosynthesis lower than that of C₄ plants
C₃ shade plant C₃ shade plants can use low levels of light for photosynthesis, but their maximum photosynthetic rate is less than C₃ sun plants or C₄ plants.

FIGURE 8-7

Photosynthetic rates of C_3 shade plants, C_3 sun plants, and C_4 plants at different light intensities. Shade plants use light more efficiently than sun plants at low light intensities, but reach their maximum photosynthetic rate at relatively low light intensity. The maximum photosynthetic rate of C_3 sun plants occurs at much higher light intensities. Even here, C_4 plants still show increased rates of photosynthesis.

(a)

(b)

FIGURE 8-8

C_4 photosynthesis. (a) Ice plant, a CAM plant indigenous to South Africa, was introduced to California and now covers countless acres along the California coastline. Note the thick, succulent leaves. **(b)** Sugarcane, an important C_4 crop plant. (**a,** Laura Dwight/Peter Arnold, Inc.; **b,** Kunz and Okapia/Photo Researchers, Inc.)

A Journey into Evolution

Coevolution of Plants and Herbivores

Coevolution is the evolutionary change that occurs in two or more different species when they act as selective pressures on one another. This process provides some fascinating examples of evolution in action.

Acacia trees grow in tropical areas throughout the world, and many species have spines that protect them against herbivorous (plant-eating) mammals. Some Central American acacias have a more unusual protection: a mutually beneficial relationship with the ants that live on them. Ant acacias have several structures that benefit the ants. These include hollow thorns in which the ants live, glands on the leaves that produce sugary nectar, and Beltian bodies, swollen, nutrient-rich leaf tips, which the ants cut off and feed to the larvae (Figure 8–A).

It is advantageous for the acacia tree to host an ant colony because the ants reduce the damage done to the tree by other herbivores. The ants react to anything that touches the tree. They remove dust, fungal spores, pollen grains, and spider webs. They destroy the seedlings of other plants that sprout under their tree, and sting other insects or mammals that try to eat the tree (Figure 8–B). When its ants are removed, the tree usually dies after a few months. Fungi invade it, and it grows more slowly and becomes choked with vines.

A few species of insects have evolved defenses that permit them to survive on an acacia tree guarded by ants. Some seem immune to ant stings and ignore the ants. Others can pick up the ants and throw them off the tree. Still others have hard cuticles that an ant's sting cannot penetrate.

FIGURE 8–A

Ant and plant. An acacia ant feeds on the yellowish-brown Beltian bodies on an acacia plant. A nectar gland is also visible on the stem below the ant. (Paul Feeny)

The coevolution of ant acacias and their insect populations probably went something like this: ants invaded an acacia and fed on the leaf parts and nectar of the tree. The ants also removed other insects, allowing the acacia to grow faster, and making it able to shade out other plants and to produce more offspring. Acacias that were more attractive to ants reproduced more rapidly. The availability of food and shelter, in turn, exerted selection on the ants to protect the tree with increasing efficiency. Of the insects that fed on the tree before the ants arrived, most species were expelled, but a few evolved defenses against ant attacks.

The leaves of acacia species that are not defended by ants contain cyanides and other chemicals toxic to herbivores. These seem to have been lost in the ant acacias, presumably because in the presence of ants the tree need no longer expend the energy to produce toxic chemicals. There is selection for the tree to use the energy to produce ant food or more offspring, instead.

We can find examples of coevolution much closer to home. Cooked cabbage, broccoli, or mustard greens give off the distinctive odor of mustard oils—the group of toxic chemicals characteristic of the crucifer plant family. In the intact plant, mustard oils are usually bonded to sugar molecules. This makes them much less toxic and allows them to be stored in the plant without damaging the tissues they are defending. When a cell is damaged—as it is when an insect bites into it—an enzyme cleaves off the sugar molecule. This is analogous to pulling the pin on a hand grenade, and the mustard oil is released.

How toxic are mustard oils? They are plainly not very poisonous to humans or to the cabbage white butterfly caterpillars that sometimes wipe out entire plantings of crucifers in the home garden. The main difficulty in answering this question is that insects usually do not eat anything except their normal food plants. You cannot take a caterpillar from an oak tree and plunk it on a cabbage leaf to see if it will be poisoned, because the caterpillar will not eat the cabbage. In an experiment to get around this problem, black swallowtail butterfly larvae, which normally feed on plants of the carrot family, were raised on some rather special carrot leaves. These leaves were cultured in solutions containing various concentrations of mustard oils. The larvae were therefore

(a)

(b)

FIGURE 8-B

Ants defending the plants they live on. (a) The ground around this young ant-defending acacia in Costa Rica is kept bare by the ants, thus protecting the tree from competing plants and from fires in the dry season. **(b)** This armyworm caterpillar, placed on an ant acacia, was stung to death in minutes by the resident ants. Also seen here are the hollow swollen thorns in which acacia ants make their nests. (Paul Feeny)

feeding on their usual carrot diet, plus compounds from a family of plants that the larvae do not normally eat.

At mustard oil concentrations that occur naturally in crucifer plants, the larvae lost a lot of fluid in their feces and soon died. Clearly, then, mustard oils can be an effective defense against insects that do not normally attack the plants that contain them. These compounds are also toxic to various fungi and bacteria. The selective pressure for mustard oils in crucifers seems to be the protection they give the plants against several natural enemies.

However, insects that are not poisoned by mustard oils can eat crucifers, and they have evolved the ability to detect mustard oils and thereby find the plants. For example, some flea beetles can home in on the odor of their crucifer food plants in a field containing many other crops. Another insect that moves toward the scent of mustard oil is a wasp that does not eat crucifers but is a parasite of an aphid that does. Moving toward mustard oils permits the wasp to find the aphid. Experiments have shown that the aphid escapes the wasp when it is not on a crucifer, for instance when it lives on a beet plant. When it is on a crucifer, or on a beet plant smeared with mustard oil, it is attacked by the parasitic wasp.

These and other experiments have shown that the same compound in a plant can act as both an attractant and a repellent. Thus mustard oils are feeding stimulants to flea beetles, and are egg-laying stimulants to the cabbage white butterfly, who lays her eggs on crucifers. Both are attracted by the mustard oils that crucifers produce. On the other hand, mustard oils repel those herbivores that cannot eat crucifers efficiently.

Other families of plants have their own, different kinds of defensive chemicals and groups of insect species adapted to cope with them. Examples are carrot and parsley (umbellifer family); onion, leek, and garlic (lily family); and the mint family.

These examples show that plants and the animals that eat them influence each other's evolution. Furthermore, the selective pressure exerted by one species may result in a variety of different adaptations in other species.

SUMMARY

1. Photosynthesis is the process in which green plants store the energy of sunlight by converting carbon dioxide and water into organic compounds.

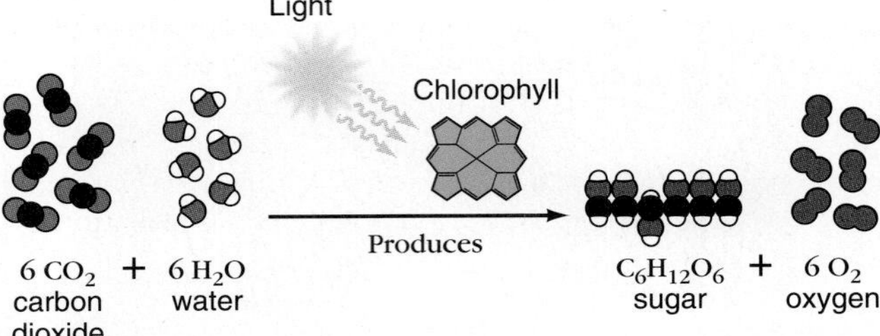

2. Organic compounds produced by photosynthesis are used by plants, and by the animals that eat plants, to build cells and to power other energy-requiring processes.
3. Organic molecules are eventually broken back down to carbon dioxide and water via respiration, and these materials can then be recycled in photosynthesis.
4. The sunlight that drives photosynthesis is the ultimate source of energy for nearly all life on Earth.
5. Photosynthesis may be considered in two parts:
   a. Energy capture.
      i. Solar energy is trapped by chlorophyll and other photosynthetic pigments in the thylakoid membranes, initiating a flow of electrons through the membrane's electron transport chain.
      ii. The overall flow of electrons is from water molecules split inside the thylakoid space, through the electron transport systems and photosystems in the thylakoid membrane, to $NADP^+$ in the stroma. This flow of electrons provides energy used in reducing $NADP^+$ to NADPH and creates the H^+ reservoir used to make ATP. Oxygen from water is released as a by-product.
      iii. The NADPH and ATP are released into the stroma of chloroplasts, where they are used to fix carbon dioxide.
   b. Carbon fixation. During the C_3 cycle, carbon dioxide becomes attached to a five-carbon sugar, and is then reduced using ATP and NADPH. The resulting ADP and $NADP^+$ are recycled.
6. Photosynthesis involves both photochemical and thermochemical reactions, whose rates are increased by light and heat, respectively. Low levels of carbon dioxide can also limit the rate of photosynthesis by slowing down the rate of carbon fixation.
7. C_4 photosynthesis permits plants to conserve water by trapping and storing CO_2 for use during photosynthesis when the openings of the stomata are reduced.

SELF-QUIZ

Match each item with its location in the chloroplast:

___ 1. Chlorophyll
___ 2. Enzymes for carbon fixation
___ 3. ATP synthetase
___ 4. H^+ reservoir for ATP synthesis

a. stroma
b. chloroplast envelope
c. thylakoid membranes
d. thylakoid interior

Match (give all correct answers):

___ 5. Ribulose bisphosphate
___ 6. $NADP^+$
___ 7. glyceraldehyde phosphate
___ 8. O_2
___ 9. CO_2
___ 10. ATP

a. raw material of energy-capturing reactions
b. end product of energy-capturing reactions
c. raw material of carbon fixation
d. end product of carbon fixation

11. The oxygen from H_2O is incorporated into:
    a. oxygen gas
    b. water
    c. carbohydrates
    d. NADPH
    e. ATP
12. Red and blue light support the highest rates of photosynthesis because:
    a. these are the only wavelengths reaching the Earth from the sun
    b. these are the only wavelengths that carotenoids cannot absorb
    c. chlorophyll absorbs these wavelengths more than other wavelengths
    d. these wavelengths have the highest energy in the visible spectrum
    e. these wavelengths activate the ATP synthetase enzyme

THINKING CRITICALLY

1. In the early 1930s, C.B. van Niel found that purple sulfur bacteria use light to make carbohydrates. They require hydrogen sulfide as a raw material and give off sulfur:

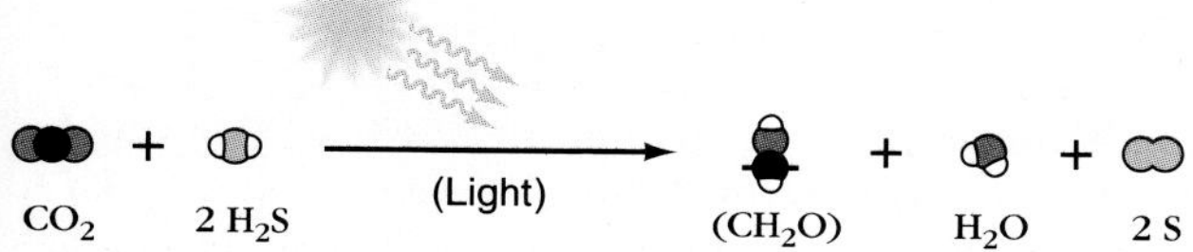

Other kinds of bacteria produce food by processes called chemosynthesis, using energy from inorganic chemical reactions rather than light energy. Why is the kind of photosynthesis outlined in this chapter so much more prevalent among living organisms today than the kind of photosynthesis used by sulfur bacteria or the chemosynthesis used by bacteria?

2. Why is photosynthesis only about 1% efficient in converting the energy in sunlight that strikes a leaf into energy stored in organic molecules? What happens to the rest of the energy?

SELECTED KEY TERMS

accessory photosynthetic pigment, *p. 146*
antenna pigment, *p. 149*
carotenoid, *p. 146*
C_3 cycle, *p. 152*
chemiosmotic synthesis of ATP, *p. 152*
chlorophyll, *p. 145*
crassulacean acid metabolism (CAM), *p. 154*
cuticle, *p. 147*
electron transport system, *p. 149*
epidermis, p. *147*
glyceraldehyde phosphate (PGAL), *p. 152*
grana, *p. 147*
palisade layer, *p. 147*
photochemical reaction, *p. 149*
photosynthesis, *p. 143*
Photosystem I, *p. 149*
Photosystem II, *p. 149*
ribulose bisphosphate, *p. 152*
spongy layer, *p. 147*
stoma, *p. 147*
stroma, *p. 147*
thermochemical reaction, *p. 153*
thylakoid, *p. 147*
vein, *p. 147*

SUGGESTED READINGS

Govindjee and W.J. Coleman. "How plants make oxygen." *Scientific American,* February 1990.

Wald, G. "Life and light." *Scientific American,* October 1959. A classic article covering how organisms use light, not only in photosynthesis but also in phototropism (the bending of plants toward light), vision, and production of light.

Youvan, D. C., and B. L. Marrs. "Molecular mechanisms of photosynthesis." *Scientific American,* June 1984.

The Educated Citizen

No Longer Human

This article by Lori Oliwenstein was excerpted with permission from the December 1992 issue of Discover *magazine.*

Henrietta Lacks achieved a kind of immortality on February 9, 1951. On that day a sample of cancerous cells from her cervix was transferred to a culture dish, doused with nutrients, and left to grow. Lacks, a 30-year-old mother of four from Baltimore, had one of the most aggressive cervical cancers her doctors had ever seen, and the cells culled form her tumor grew avidly, doubling their number each day. Then they escaped. Small spills are always happening in laboratories; what distinguished Lacks's cells was their ability to survive after they were somehow spilled. They were so hardy that if just one of them fell on a petri dish it would outgrow and overwhelm anything else living on that dish within a month.

HeLa cells came from a human being, but for 40 years now they've been evolving on their own, in petri dishes. They may be a new species.

Soon Henrietta Lacks's cells were traveling from lab to lab, either deliberately sent—many cancer researchers had taken to using them in their experiments—or as an unseen contaminant tagging along in another cell line. Some researchers who thought they were looking at something completely different—a line of liver cells, say—ended up studying Henrietta Lacks's cervical cells by accident. The cells even slipped through the iron curtain and into Russia.

Lacks died in October 1951, but her cells lived on. Now some biologists are saying that those cells, called HeLa cells for short, have lost more than their connection to Henrietta Lacks. HeLa cells, these researchers claim, are no longer human at all: they are single-celled microbes—closely related to us, to be sure, but their own distinct species.

How so, you ask? "HeLa cells are not connected in any way to people," explains evolutionary biologist Leigh Van Valen of the University of Chicago. "They have an extremely different ecological niche from us. They don't mate with humans; they probably don't even mate with human cells. They act just like a normal microbial species. They are evolving separately from us, and having a separate evolution is really what a species is all about."

The process of evolution is much the same for HeLas as it is for humans, although the former usually reproduce asexually, by cell division. As the cells divide, genetic mutations inevitably occur, and the ones that make the cells better adapted to their ecological niche—the petri dish—are preserved by natural selection. When Henrietta Lacks's cells first became cancerous, they also acquired the ability to survive indefinitely in a culture medium; that massive genetic transformation made them substantially different from ordinary human cells, and after four decades of evolution they have become moire different still. Different strains of HeLa cells, analogous to different races of human beings, have even developed in some of the geographically separated lines.

"These little unicellular organisms have crossed oceans, spread their range, got into other cultures and outcompeted them," says Richard Strathmann, a marine biologist at the University of Washington's Friday Harbor Laboratories who dabbles in evolutionary theory. "They're only different from other single-celled organisms in that a human being gave rise to them."

Strathmann and Van Valen (the latter with his colleague Virginia Maiorana) put forth these ideas separately, in two papers in the same issue of the journal *Evolutionary Theory,* which Van Valen edits. (Both papers, he points out, were independently reviewed before publication.) Van Valen and Maiorana not only declared that HeLa may not be *Homo sapiens*, they gave the new species a name: *Helacyton gartleri—Hela*, after the HeLa cells themselves; *cyton* from the Greek *cytos*, meaning cavity or cell; and *gartleri* after geneticist Stanley Gartler, who was the first to document the cells' remarkable success.

While Van Valen is willing to name the new species, he is unwilling to suggest which higher taxonomic category it might fall into. "Beyond the family name there are problems," he says. Since a HeLa cell can't survive outside a culture medium, it obviously isn't a primate in the usual sense. At the same time, says Van Valen, you can't call it a protist—a member of the kingdom of all single-celled organisms, which includes protozoans, algae, and fungi—since that would mean that the same group had evolved twice, once sometime before 3.5 billion years ago and again today. It's a fundamental tenet of evolutionary theory that evolution doesn't repeat itself.

But that's exactly what has happened, says Strathmann. And to him, HeLa cells are just a particularly aggressive and successful example of an evolutionary transition that has happened numerous times recently. Many

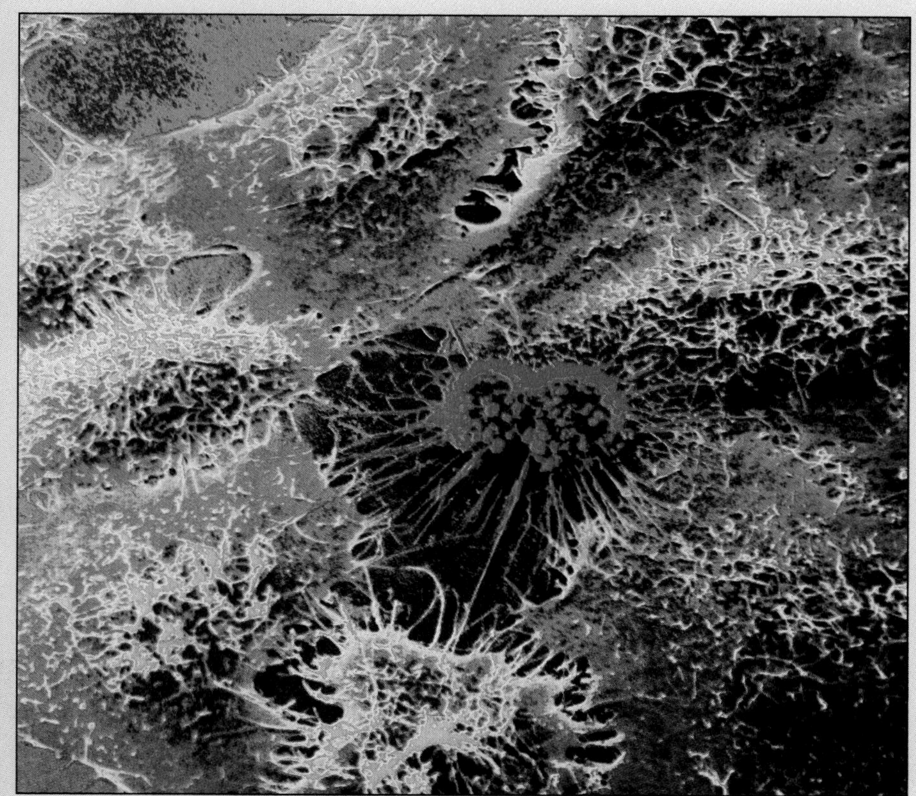

HeLa cells (red) divide and prosper in a petri dish. (Bill Longccore/Science Source/Photo Researchers, Inc.)

cancer cells, in becoming cancerous, undergo the same type of genetic transformation that Henrietta Lacks's cells did and thereby acquire the potential to be immortal; and many different lines of these cells are now surviving in petri dishes all over the world. All of them, according to Strathmann, have made the huge evolutionary leap from being metazoans—multicellular creatures with organs and tissues—to being single-celled protists. What's most amazing, he says, is how fast they did it; it took nearly 3 billion years for the first metazoans to evolve after life originated but just a handful of years for HeLa and other cell lines to take exactly the same step in the other direction.

If this modern-day transition from human being to unicellular blob sounds like far-out fiction to you, you're not alone. Some biologists consider the survival of HeLa cells a purely artificial phenomenon and argue that evolution in a petri dish has little relevance to evolution in nature. Indeed, Strathmann's paper was rejected by other journals for just that reason before Van Valen agreed to publish it along with his own. Van Valen and Strathmann, of course, reject that criticism. "The perception is that if human beings are manipulating the situation, it's not natural," says Strathmann. "But biomedical researchers are part of nature."

"Organisms live in all sorts of odd places, including ones humans have created," adds Van Valen. "Parks and cities are environments that we created, and organisms have become adapted to them." Human beings have even created new species before, albeit not from their own flesh. Modern corn, for instance, is a product of selective breeding by generations of farmers, and like HeLa cells, it can't survive without human help. "If HeLa had not been derived from human tissue," Van Valen says, "there would be no question about its being a new species."

Connecting the Concepts

The following questions will help you to connect the issues discussed in this article with the concepts you have learned in Part 1.

1. Discuss the legal and ethical aspects surrounding the use of human cells in research. For example, should a person who donates cells for medical research (a) be required to give permission, (b) benefit financially if the cells are eventually used to produce a product (perhaps large quantities of a medically important chemical), and (c) be allowed to decide how the cells are used in research?
2. Given your understanding of cell biology from Chapters 4 and 5, discuss the evidence supporting and arguing against the idea that HeLa cells are now a new microbial species.

PART 2

The Unity of Life: Genetic Information and Its Expression

Chapter 9
DNA and Genetic Information

Chapter 10
RNA and Protein Synthesis

Chapter 11
Reproduction of Eukaryotic Cells

Chapter 12
Mendelian Genetics

Chapter 13
Inheritance Patterns and Gene Expression

Computer-generated image of DNA molecule. (Will and Deni McIntyre/Photo Researchers, Inc.)

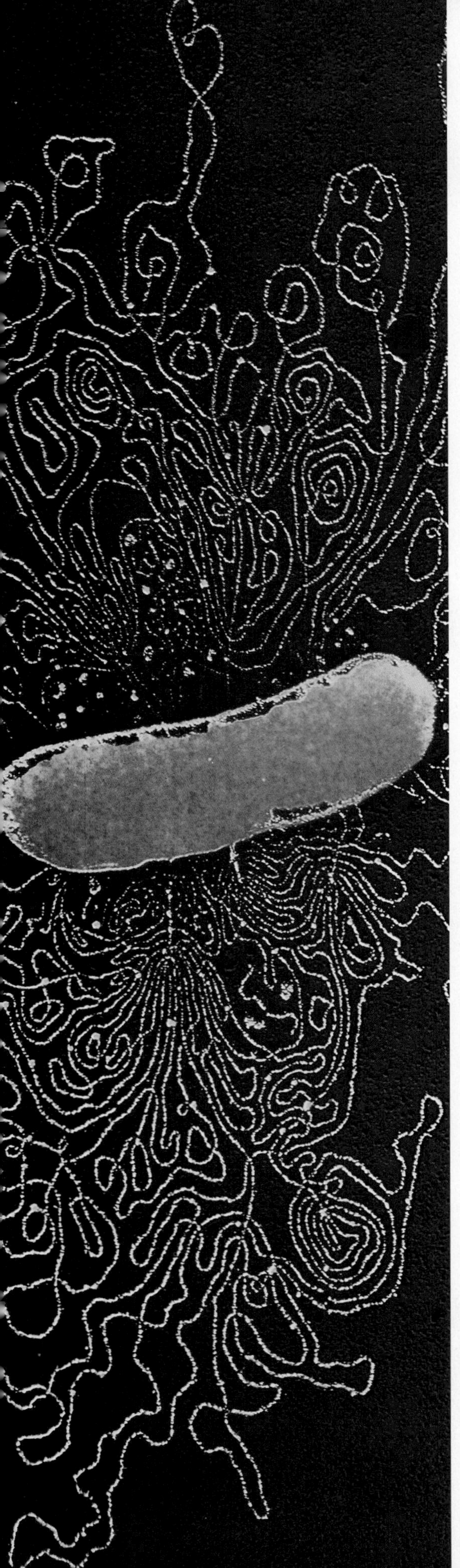

CHAPTER 9

CONCEPT GUIDE

After reading this chapter you should be able to:

1. Describe two experiments that provided evidence that DNA is the genetic material. (Section 9-A)
2. Describe the structure and chemical composition of a DNA molecule. (Section 9-B)
3. Describe and compare the process of DNA replication in prokaryotes and eukaryotes. (Section 9-C)
4. Explain how damaged segments of DNA are identified and corrected. (Section 9-D)
5. Define mutations and explain their importance in evolution. (Section 9-E)
6. Describe the arrangement of DNA and histones in chromosomes. (Section 9-F)
7. Compare the instructions carried on the genomes of prokaryotic cells with those of eukaryotic cells. (Section 9-G)

This electron micrograph shows the circular DNA spilling out of a damaged bacterium. If this DNA were stretched out into a single straight line, it would be about a thousand times as long as the bacterium. Read more about the structure of DNA in Section 9-B. (Gopal Murti, Science Photo Library/Photo Researchers, Inc.)

Each of us originated from a single cell—an egg, fertilized by a sperm. What directed this fertilized egg to divide into more and more new cells, and the resulting mass of cells to move around, grow, absorb nourishment, and take shape as a unique individual? What makes each of us distinct from other people but gives us all a basic similarity as members of the human species? And what allows discerning relatives hovering over a new arrival to proclaim that it has its father's nose and its mother's shy smile? The answer to all these questions is **genetic information.**

The units of genetic information, units governing inherited characteristics such as hair color, blood type, and embryonic development, are called **genes.** A gene is a length of DNA, part of a much longer DNA molecule that is associated with proteins, forming a unit called a **chromosome.** The nucleus of a eukaryotic cell contains several to many chromosomes. Each chromosome's DNA molecule may contain hundreds or thousands of genes.

Many genes contain information that determines what proteins the cell can make. The order of nucleotide monomers in the DNA of a gene constitutes a "code" that dictates the order in which amino acids are assembled to form a protein. So DNA codes for protein structure, and a protein's structure, in turn, determines its function in carrying out some part of the cell's activities (Section 3-D). This, in turn, has an effect on the organism as a whole, such as what color flowers it has, or whether its heart develops normally. The whole process whereby

DNA and Genetic Information

genes direct the production of proteins, which in turn affect the whole organism, is known as **gene expression.**

During the last century, biologists learned that an individual inherits half its genetic information from each parent. When they studied cells dividing to form eggs and sperm, they noted that each egg or sperm ends up with half of the number of chromosomes from the nucleus of the original cell (Section 11-F). When egg and sperm unite at fertilization, only the sperm nucleus, containing chromosomes, enters the egg. The new individual receives a full set of chromosomes: half from its mother's egg and half from its father's sperm.

This behavior of chromosomes was exactly what biologists expected of structures bearing genetic information. They became convinced that chromosomes do in fact contain the genetic information. By 1940, biochemists knew that chromosomes contain two substances: proteins and DNA (short for deoxyribonucleic acid, Section 3-F). But which one carried the genetic information? Or were both involved?

Biologists reasoned that organisms must contain a variety of genetic information. They knew that proteins are a diverse and complex group of polymers, made up of 20 different kinds of amino acid subunits.

On the other hand, DNA polymers contain only four kinds of nucleotide monomers. It seemed as though chromosomal proteins must be more complex than DNA, and most scientists therefore thought that these proteins carried the genetic information. But this turned out to be wrong: DNA contains the genetic information.

A chromosome's DNA contains hundreds or thousands of units of genetic information, or genes. Each gene is a stretch of the chromosomal DNA molecule containing hundreds or thousands of nucleotide monomers. Many genes contain instructions for making proteins. The order of their nucleotides constitutes a **"code"** of directions for the order in which amino acids should be joined to make proteins. Each protein then performs some job, such as making eye pigment or forming part of a hair. So the gene's information is said to be **expressed** in the form of protein, which in turn contributes to the structure or function of the whole organism.

All the cell nuclei in your body contain the same set of genes, but not all genes are expressed in every cell. For example, our eyes, tongues, and internal organs do not normally produce the proteins that would make them sprout hair (although this can happen in some medical disorders). Some genes do not code for proteins but instead regulate the expression of other genes.

In this chapter we examine the evidence that the genetic material is indeed DNA, and how the structure of DNA was worked out. This structure is very simple, yet elegant, and it has the remarkable property that it dictates the production of exact copies of itself—copies that are then passed on to future generations of cells.

KEY CONCEPTS

- The units of genetic information are genes—portions of DNA molecules.
- In sexual reproduction, a new individual receives half of his or her genetic information from each parent in the form of DNA in the chromosomes of the egg and sperm.
- A cell's DNA controls what proteins are made and in this way controls most of the cell's activity.
- A DNA molecule contains the information needed to produce exact copies of itself.

Curiosity Questions

How do we know that DNA is really the genetic material? (See page 166)

What are mutations and what causes them? (See page 171)

What is the difference between DNA and chromosomes? (See page 172)

1?
9–A EVIDENCE THAT DNA IS THE GENETIC MATERIAL

The Riddle of Bacterial Transformation

In 1928, Fred Griffith was studying pneumonia caused by bacteria—a serious disease, especially in the days before antibiotics. He worked with two strains of bacteria, each containing different genetic information. One strain's genetic information directed the production of an external capsule that protected the bacteria from attack by an animal's immune system. This strain was **virulent:** when injected into mice, it caused fatal disease (Figure 9–1a). The other strain, which could not produce capsules, was **nonvirulent** and did not kill the mice.

If virulent, heat-killed bacteria were injected into mice, there was no effect, but if virulent heat-killed bacteria were mixed with living, nonvirulent bacteria, some of the mice died. Even though these mice had received no living, virulent bacteria, their corpses contained living, virulent bacteria. Griffith concluded that some of the genetic material from the dead, virulent bacteria had entered the living, nonvirulent bacteria, changing them to the virulent form. This phenomenon was named **bacterial transformation,** the transfer of genetic information from one bacterium into another.

What factor had transformed the bacteria? Oswald Avery and his colleagues spent a decade on this problem, growing tons of bacteria which they separated into chemical components for testing. In 1944, they reported that the transformation studied by Griffith could be produced only by DNA from virulent bacteria, but not by bacterial proteins or any other substance. Therefore, this meant that DNA was the genetic material of bacteria.

Avery's conclusion that DNA was the genetic material of bacteria was controversial, because most biologists at that time did not regard bacteria as "real" organisms. Their attitude was, "So what if DNA is the genetic material of bacteria?" Many biologists felt this information was irrelevant to understanding the genetic material of higher organisms. We now know that DNA is the genetic material of plants, animals, fungi, and protists as well as of bacteria and some viruses. Indeed, most of the major discoveries in molecular genetics were made using bacteria, which can be grown quickly and easily for experiments. *Escherichia coli,* a common bacterium that inhabits the human intestine, has been particularly well-studied.

Bacteriophages

More evidence that genetic material is DNA was obtained from a study of **bacteriophages.** Nicknamed **phages,** these are viruses that parasitize bacteria. A phage takes over a bacterium's metabolic machinery, causing it to produce new phages and then to burst, releasing the new phages to infect more bacteria (Figure 9–1b).

The phages used in this study consist of a DNA molecule inside a protein coat. In 1952, Alfred Hershey and Martha Chase tested the hypothesis that these phages do not enter bacteria intact, but rather that the phage's protein coat attaches to the bacterial cell wall and the phage then injects its DNA into the bacterial cell. They reasoned that if this is so, and the

BIO-BIT

The human genome is composed of nearly 6 billion nucleotide pairs!

FIGURE 9–1 ►

The identification of DNA as the genetic material. (a) In this experiment, Frederick Griffith demonstrated that bacteria that normally would not kill mice (nonvirulent) were able to become deadly (virulent) in the presence of heat-killed, virulent bacteria. This suggested that something from the virulent bacteria that had been killed must be transferred to the nonvirulent bacteria to make them lethal. But what sort of information molecule could cause this change? Later workers were able to confirm that DNA from the deadly bacteria could be transferred to the nonvirulent strain, making them deadly. This was good evidence that DNA molecules inside the bacteria were responsible for the deadly changes. **(b)** Infection of a bacterium by a bacteriophage. **(c)** Another experiment by Alfred Hershey and Martha Chase used viruses that attack bacteria, called bacteriophages, to determine whether the genetic material was the internal DNA or the protein coat. Bacteriophages attach to bacteria, inject their DNA, and leave their protein coats on the outside of the bacteria. Inside the bacterium the bacteriophage DNA multiplies and each strand forms a new protein coat. In the experiment, one group of bacteriophages had their protein coats labeled with radioactive molecules while another group had their DNA molecules labeled. Only the group with labeled DNA produced new bacteriophages that still had labels. This indicated that during reproduction of the bacteriophage, the labeled DNA was passed to the next generation, but the labeled protein coat was left behind, further supporting the theory that DNA is the genetic material that gets passed from one generation to the next. (Lee D. Simon/Photo Researchers, Inc.)

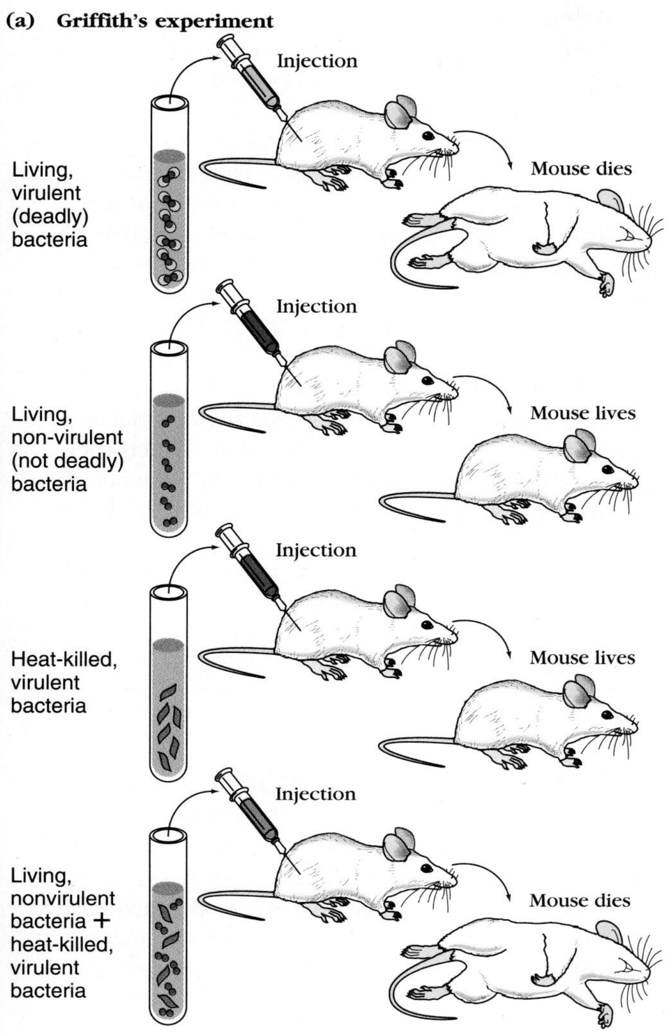
(a) Griffith's experiment
Injection
Living, virulent (deadly) bacteria
Mouse dies
Injection
Living, non-virulent (not deadly) bacteria
Mouse lives
Injection
Heat-killed, virulent bacteria
Mouse lives
Injection
Living, nonvirulent bacteria + heat-killed, virulent bacteria
Mouse dies

(b)

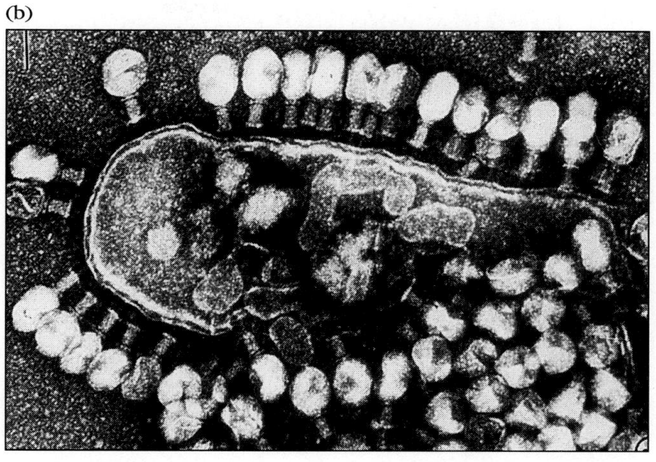

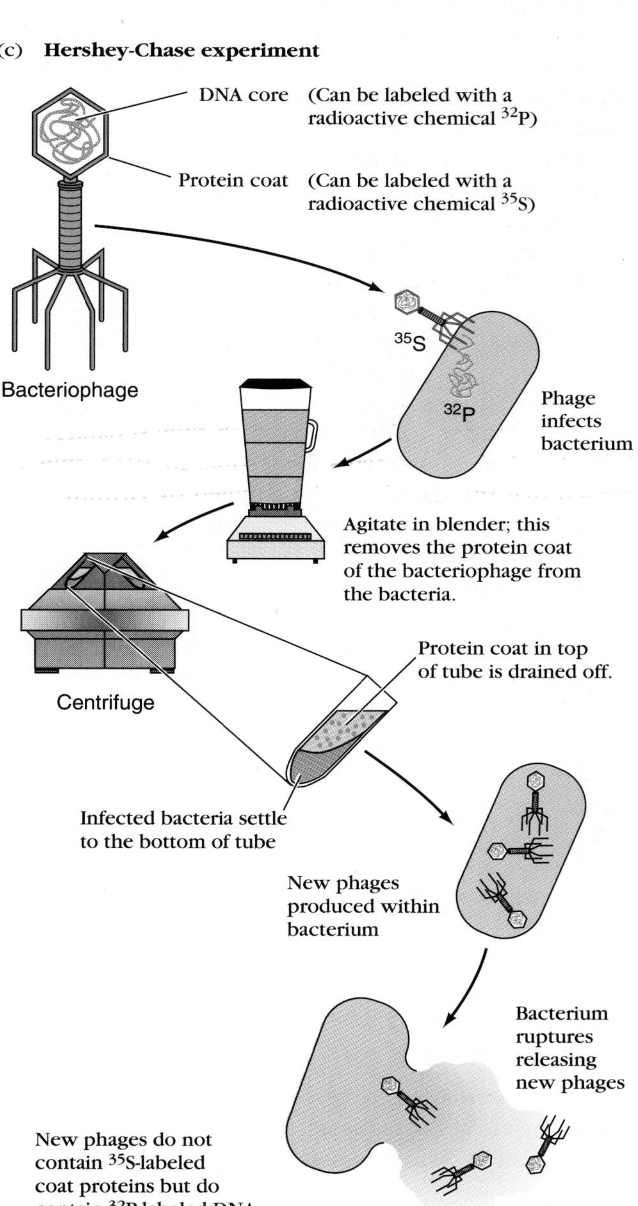
(c) Hershey-Chase experiment
DNA core (Can be labeled with a radioactive chemical ^{32}P)
Protein coat (Can be labeled with a radioactive chemical ^{35}S)
Bacteriophage
^{35}S
^{32}P
Phage infects bacterium
Agitate in blender; this removes the protein coat of the bacteriophage from the bacteria.
Centrifuge
Protein coat in top of tube is drained off.
Infected bacteria settle to the bottom of tube
New phages produced within bacterium
Bacterium ruptures releasing new phages
New phages do not contain ^{35}S-labeled coat proteins but do contain ^{32}P-labeled DNA

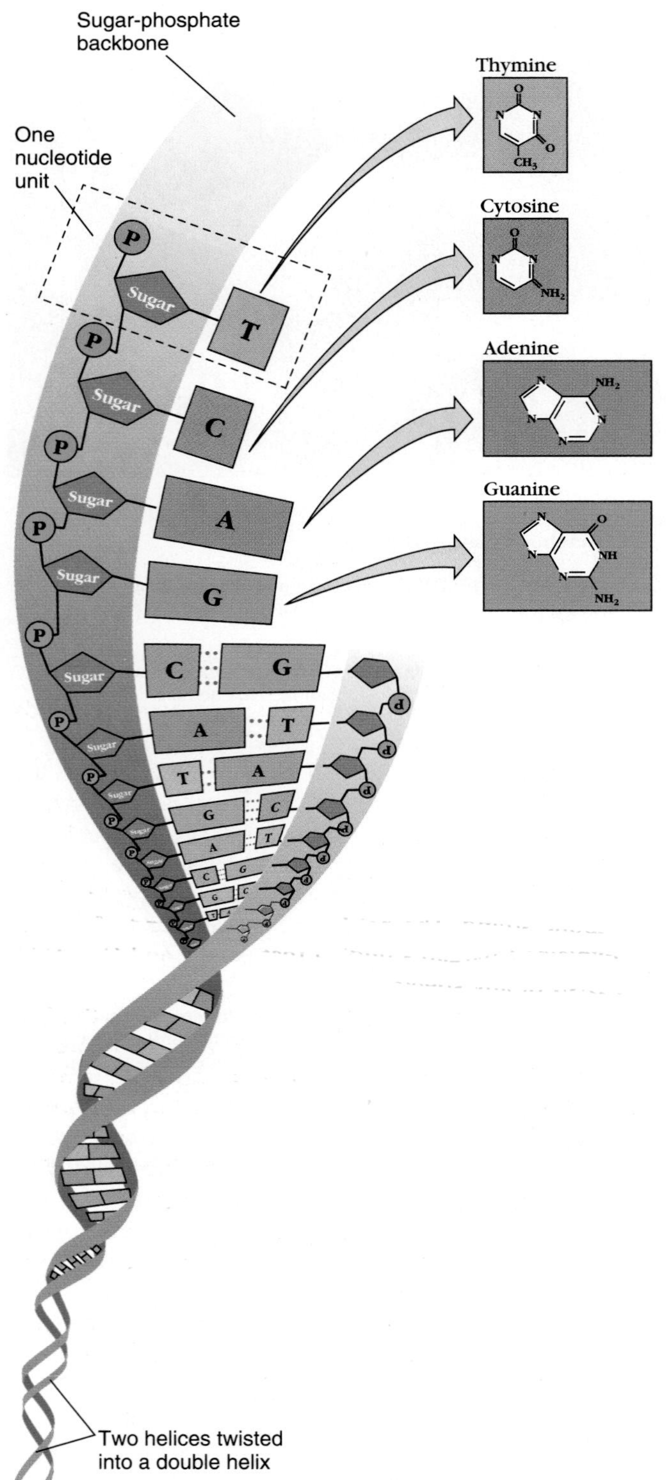

bacterium then produces new phages, DNA must be the phage's genetic material because its protein coat never entered the bacterium (Figure 9–1b).

Hershey and Chase used radioactive isotopes of sulfur and phosphorus to mark and later distinguish between phage protein and DNA. They capitalized upon the fact that proteins contain sulfur but not phosphorus, whereas DNA contains phosphorus but not sulfur. They prepared phages that had proteins marked with radioactive sulfur and DNA marked with radioactive phosphorus (Figure 9–1c). When these specially marked phages infected bacteria, the radioactive phosphorus entered the bacteria, while the radioactive sulfur remained outside. The phages' DNA, but not their protein, had entered the bacteria, which then produced new phages. This was strong evidence that the bacteriophage genetic material consisted of DNA but not protein.

Experiments conducted on bacteria by Griffith and Avery and on bacteriophages by Hershey and Chase provided evidence that DNA is the genetic material of both.

9–B THE STRUCTURE OF DNA

By the early 1950s, biologists were convinced that DNA indeed carries a cell's genetic information, and many people were trying to work out the structure of the DNA molecule. Any model of DNA structure had to take several experimental findings into account:

1. DNA is made up of nucleotide subunits. Each DNA nucleotide has three parts: a five-carbon sugar (deoxyribose), one or more phosphate groups, and one of four possible nitrogen-containing bases: **adenine (A), thymine (T), cytosine (C),** or **guanine (G)** (Figure 9–2).
2. When the nucleotides are linked together in a strand of DNA, sugar and phosphate groups alternate to form a sugar-phosphate "backbone" held together by covalent bonds. The bases stick out to one side (Figure 9–2).

FIGURE 9–2

Basic DNA structure. In the 1950s we knew that DNA was made of four nucleotides: adenine, thymine, cytosine, and guanine. We knew that a strand of DNA was linked in a precise manner: each nucleotide bonded to the phosphate group of the nucleotide below, forming a sugar-phosphate backbone. We knew that the nucleotides followed a pattern: there were equal amounts of adenine and thymine and of guanine and cytosine. Finally, x-ray diffraction studies had shown that the molecule was twisted into a double helix.

Table 9–1
Composition of DNA from Different Organisms

Organism		Percent of Nucleotide Molecules			
		A	T	G	C
Animals:	Human	30.9	29.4	19.9	19.8
	Chicken	28.8	29.2	20.5	21.5
	Locust	29.3	29.3	20.5	20.7
Plant:	Wheat	27.3	27.1	22.7	22.8
Fungus:	Yeast	31.3	32.9	18.7	17.1
Bacterium:	*Escherichia coli*	24.7	23.6	26.0	25.7
Bacteriophage:	T_4	26.0	26.0	24.0	24.0

3. Erwin Chargaff had shown that the number of nucleotides containing adenine (A) equals the number containing thymine (T), and the numbers containing guanine (G) and cytosine (C) are also equal to each other. In the shorthand popular with biologists, A = T, and G = C (Table 9–1) (Figure 9–2).
4. The most direct evidence for the structure of DNA came from x-ray diffraction pictures, made by passing x-rays through crystals of DNA. This produces a pattern of dots that gives information about the molecule's shape. In 1952, Rosalind Franklin produced photographs showing that DNA is twisted into a spiral or **helix** (pl. **helices**), with the bases perpendicular to the length of the molecule. These pictures also gave evidence that the sugar-phosphate backbone is on the outside of the helix, with the bases inside. Furthermore, the diameter of the helix showed that the DNA molecule must be composed of more than one such strand.

Two main questions about DNA's structure remained: how many strands are there in the DNA molecule, and how are they put together?

James Watson and Francis Crick fitted all this evidence together, using a set of scale models of nucleotides to build possible structures until they found one that fitted all the data. The Watson and Crick model of DNA structure consists of two strands of DNA. (To a biologist, the number two is satisfying because both cells and chromosomes reproduce by the formation of two new entities from the original one.) The two strands are arranged like a ladder, with the ladder's sides being the sugar-phosphate backbones of the two strands and the rungs being the bases (Figure 9–2).

A rung consists of either adenine paired to thymine, or guanine paired to cytosine. The atoms in each pair match up in such a way that hydrogen bonds form between the bases, and these hydrogen bonds hold the two bases in the rung together. Therefore, the bases in these pairs are said to be **complementary.** Each rung has one single-ring base (T or C) and one double-ring (A or G) base so all the rungs have equal widths. In each rung, either base may be on either backbone strand. The pairing of A and T, and of G and C, explains Chargaff's finding that A = T and G = C in the DNA of any species.

Finally, the whole ladder is twisted to form the helix detected by Franklin's x-ray photographs. Because the helix is composed of two strands wound around each other, the DNA molecule is referred to as a **double helix.**

Tremendously excited by this simple yet elegant structure, Watson and Crick published their model. Their paper was only two pages long, but it became a cornerstone of modern molecular genetics. As Watson remarked, "It was too pretty not to be true."

A prokaryote's genetic material is one long double helix of DNA with its ends joined to form a circle. The circular bacterial DNA is folded many times and occupies a nuclear area about one tenth of the cell's volume.

Eukaryotic DNA is organized into a number of chromosomes, each containing one long DNA double helix (Section 9–F). Eukaryotic cells contain more DNA than prokaryotic cells do.

BIO-BIT

The animals with the greatest amount of DNA per cell are amphibians.

The structure of the DNA molecule is like a twisted ladder, with the sides composed of alternating sugar and phosphate molecules and the rungs composed of complementary base pairs.

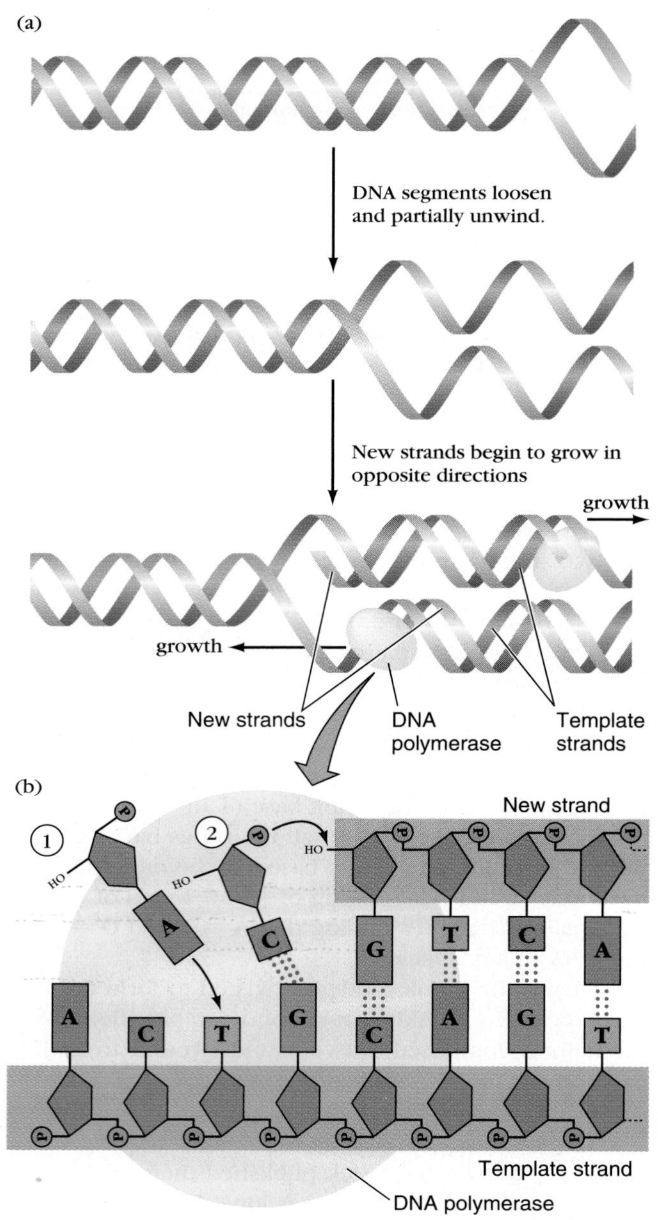

FIGURE 9–3

DNA replication. (a) As DNA replication begins, the two strands of DNA begin to unwind and two new strands form next to the existing ones. **(b)** To form the new DNA molecules, new nucleotides are added according to the sequence of nucleotides in each existing strand. Each base is held to a base on the opposite strand by hydrogen bonds (light dots).

> **BIO-BIT**
>
> The accuracy of DNA replication is equal to a person copying 100 large (1000-page) dictionaries word for word, and symbol for symbol, while making only one error in the entire process!

9–C DNA REPLICATION

Before a cell divides to form two new cells, its DNA is **replicated,** or duplicated. Each new cell will then receive a copy of the original cell's genetic information. Because the two strands of a DNA molecule have complementary base pairs, the nucleotide sequence of each strand automatically supplies the information needed to produce its partner. For example, if one strand runs A-A-T-G-C-C, then its partner *must* run T-T-A-C-G-G. If the two strands of a DNA molecule are separated, each can be used as a mold, or **template,** to produce a complementary strand. The template and its complement together then form a new DNA double helix, identical to the original molecule. Watson and Crick suggested that this was in fact how DNA replicated.

Before replication can occur, the length of the DNA double helix about to be copied must be unwound. In addition, the two strands must be separated, much like the two sides of a zipper, by breaking the weak hydrogen bonds that link the paired bases. (To get a rough idea of the complications involved, think of how difficult it is to straighten out a tangled telephone cord. Imagine the task involving a pair of helical cords.) Once the DNA strands have been unwound, they must be held apart to expose the bases so that new nucleotide partners can hydrogen-bond to them (Figure 9–3).

The enzyme **DNA polymerase** moves along, joining newly arrived nucleotides into a new DNA strand complementary to the template strand.

In a prokaryote, replication begins at one point on the DNA circle, the replication origin. From this point, replication travels in both directions around the molecule until the whole circle is copied. Linear eukaryotic chromosomes, in contrast, have many replication origins. Replication may start at as many as 1000 places at once and continue until all the DNA has been copied.

DNA replication is extraordinarily accurate. DNA polymerase makes very few errors, and most of those that are made are quickly corrected by DNA polymerase and other enzymes that "proofread" the nucleotides added into the new DNA strand. If a newly added nucleotide is not complementary to the one on the template strand, these enzymes remove the nucleotide and replace it with the correct one.

Thanks to the accuracy of replication and proofreading, a cell's DNA is copied with less than one mistake in a billion nucleotides added to the growing chain.

The complementary strands of DNA automatically direct replication. After the two strands are "unzipped," each strand serves as the template for the formation of a new complementary partner strand. "Proofreading" ensures that few errors remain.

9–D DNA REPAIR

Like all other biological polymers, DNA is subject to damage from agents that include the body's own heat and the aqueous environment inside the cell. Any form of damage to DNA can alter its information content and therefore could result in disastrous changes to the cell's proteins (see *A Journey into The Environment:* Ozone Depletion, Chapter 17).

Although thousands of changes occur in a DNA molecule every day, not more than two or three *stable* changes accumulate in a cell's DNA each year. The vast majority are eliminated by the coordinated effort of a squad of 20 or more different kinds of DNA repair enzymes.

DNA repair depends on the existence of two copies of the genetic information, one in each strand of the double helix. As long as one strand remains undamaged, the repair enzymes can use it as a template to replace a damaged segment on its partner. So most damage is remedied unless both strands are altered beyond recognition at the same time.

Enzymes identify and repair damaged DNA by using the undamaged side as a template.

9–E MUTATIONS

Mutations are inheritable changes in DNA molecules. They may result from uncorrected errors in replication, from failure to repair damage properly, or from spontaneous rearrangements of segments of DNA (Table 9–2). In any of these cases, the new, "wrong" DNA sequence is copied just as accurately as "right" ones. ■

The amount of change in the mutated DNA is not necessarily correlated with its effect on the organism.

Table 9–2
Some Types of Mutations

Substitution of one nucleotide for another (most common mutation)
Insertion of one or more nucleotides into DNA sequence
Deletion of one or more nucleotides from DNA sequence
Inversion of part of nucleotide sequence (so that part of the DNA is "backwards")
Breakage of chromosome and loss of fragment
Attachment of part of one chromosome to another
Loss of one or more entire chromosomes
Extra copies of one or more chromosomes
Polyploidy: extra copies of all chromosomes

For example, a change of one nucleotide pair for another in a gene coding for a protein may have effects so slight as to be undetectable, or so severe as to cause death. This depends on how the change affects the protein encoded by that gene. Sickle cell anemia results from a change in a single pair of nucleotides, but has drastic and wide-ranging physiological consequences. Cystic fibrosis results from a mutation that deletes the code for one amino acid. Half of the victims of this mutation die by age 21 (See Section 13–B).

Many mutations are brought about by **mutagenic agents**, often called **mutagens.** Various kinds of radiation cause mutations. X-rays and radioactive particles may cause breaks in the DNA molecule, and have been implicated as causes of some kinds of cancer. Certain chemicals such as nicotine and acridine, found in dyes made from coal tar, also alter DNA.

Mutations are inherited when they are copied during DNA replication and passed on to the cell's descendants. Mutation in a body cell may cause changes in the hereditary characteristics of the cell and of body parts made up of that cell's descendants. Mutations in cells destined to form eggs or sperm can be passed on to an organism's offspring. Typical mutation frequencies for a single gene in a human egg or sperm range from 1 to 250 per million eggs or sperm, depending on the gene involved.

A small amount of mutation is advantageous, because it produces variation in the genetic material, and this is the raw material of evolution. However, the genetic material of modern organisms has resulted from hundreds of millions of years of natural selection. Most of the changes that occur are more apt to harm than to help the delicate balance of living cells. However, there is always the chance that some new mutation will prove beneficial. Genetic engineering, in a variety of ways, makes artificial mutations, manipulates genes, and produces, for example, improved crops that have resistance to disease. We hope that genetic engineering may some day eliminate genetic defects (see *A Journey into Science in Process*: Genetic Engineering, this chapter).

Mutations are inheritable changes in DNA that can be caused by a wide variety of mutagens, including x-rays and many types of chemicals.

9–F STRUCTURE OF EUKARYOTIC CHROMOSOMES

Eukaryotic cells contain many chromosomes. For example, human body cells normally contain 46 chromosomes, with a total of about six billion nucleotide pairs divided among them. Each chromosome contains a single DNA molecule extending from one end of the chromosome to the other. The DNA is associated with various proteins, forming a substance called **chromatin.** Chromatin contains roughly equal amounts of DNA and proteins. ■

A chromosome's long DNA molecule is protected from breaking and tangling by chromosomal proteins called **histones,** which wrap the DNA tidily up into compact chromosomes. Histones are small proteins containing a lot of the alkaline amino acids arginine and lysine. The positively charged R groups of these amino acids bind strongly to the negatively charged phosphate groups of DNA. Histones occur in enormous amounts: there may be 60 million copies of a single type of histone in the chromatin of one cell!

The DNA from the 46 chromosomes of a single human cell nucleus has a total length of about 2 meters, an average of 5 centimeters per chromosome. Chromosomes in a dividing cell are ten thousand times shorter than this. Histones and other proteins are responsible for packing these DNA molecules so tightly. The DNA is wound around clusters of histones, forming a string of bead-like particles called **nucleosomes** (Figure 9–4b). H1 histones lie between nucleosomes on the DNA molecule. They pack DNA more tightly. Most of the nucleosome "beads," in turn, are wound up still more tightly, forming a helical fiber about 30 nanometers (nm) in diameter. Even though it has been wound and re-wound, this fiber is still about 200 times longer than the diameter of a cell nucleus. It is further folded into large loops, called "looped domains," that shorten the chromosome further (Figure 9–4c). Chromatin packed up as tightly as possible is said to be **condensed** (Figure 9–4d).

In the nucleus, DNA molecules are wrapped around histones and other chromosomal proteins in a highly organized way.

9–G ORGANIZATION OF THE GENOME

A cell's **genome** is the total of all its genes. In 1977, researchers found methods of determining the sequences of nucleotides in DNA (and RNA) molecules. This provided the tools to describe precisely how genes are arranged within a cell's DNA molecules. Many of the findings produced by this new technique, however, were completely unexpected and are still unexplained (see *A Journey into Science in Process:* The Human Genome Project, Chapter 11).

Some parts of the genome are genes that carry instructions for making proteins. Other genes dictate the sequence of nucleotides in RNA molecules found in ribosomes. Still others regulate the activity of some of their fellow genes. In prokaryotes, genes directing RNA and protein synthesis make up most of the genome. However, in many plants and animals, including humans, less than 10% of the genome serves these functions. We simply do not know what most of the rest of the DNA of these organisms does. However, we can guess that some of it probably helps to maintain chromosome structure, and some may signal where genes begin and end, or tell which kinds of cells should make the protein encoded by a gene. (For instance, red blood cells make hemoglobin but other cells do not.)

Jumping Genes

Geneticists once envisioned the genome as a fixed number of genes arranged in specific sequences on the chromosomes. In the 1940s, Barbara McClintock cast doubt on this picture when she discovered transposable genes in maize (corn) plants (Figure 9–5). A transposable or "jumping gene" can make a copy of itself that may become incorporated in another part of the genome. The original gene remains in place and does not "jump." A gene's position is important because genes often affect their neighbors' activity.

In the 1970s, it became clear that jumping genes are both widespread and important. For instance, bacteria sometimes become resistant to drugs (such as antibiotics) that once killed them. Worse still, from a medical point of view, this drug resistance may be passed from one bacterium to another. The genes that make a bacterium drug-resistant are passengers on small transposable elements that can move from the

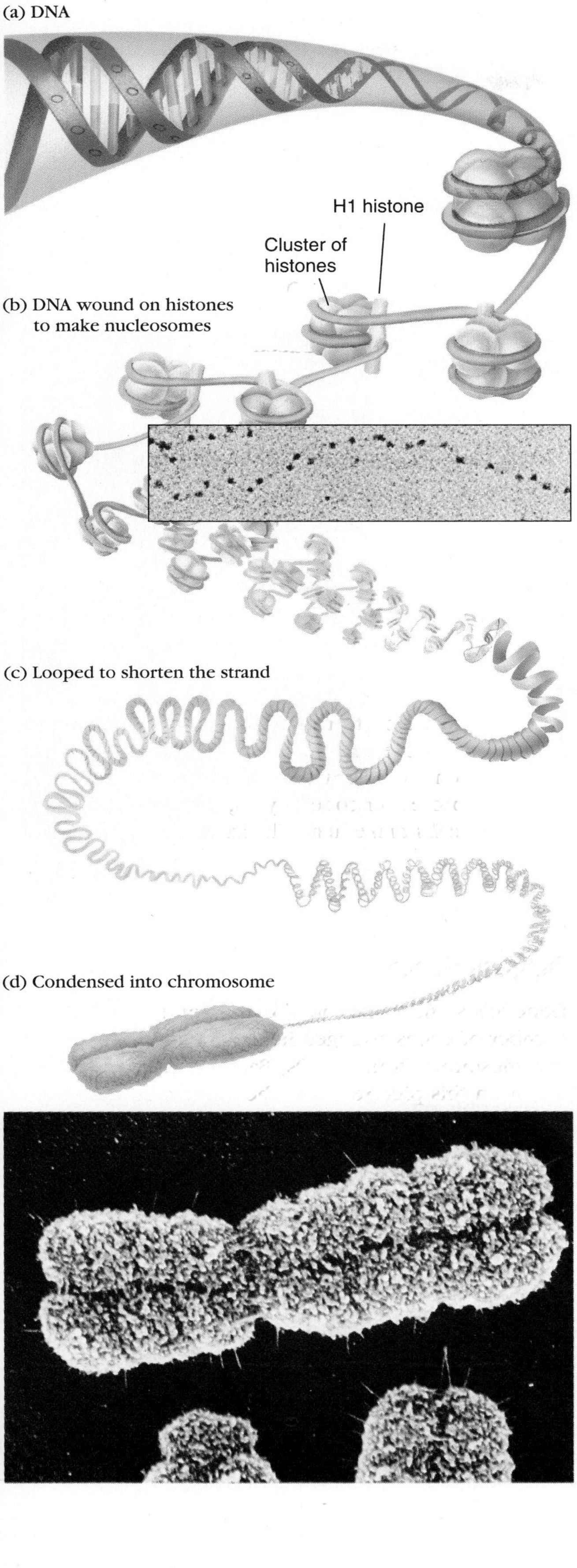

FIGURE 9–4

How DNA is packed into chromosomes. The DNA strand **(a)** is coiled around histone proteins, forming nucleosomes **(b).** An electron micrograph of chromatin shows nucleosomes as dark granules. The thin line between the bend-like nucleosomes is DNA. Large loops **(c)** further shorten the DNA, producing condensed chromatin **(d).** Thus the appearance of this chromosome undergoing mitosis reflects many folded layers of DNA. (b, B. Hamkalo; d, Biophoto Associates)

FIGURE 9–5

Indian corn. The kernels grow in many different colors, which are genetically determined. McClintock discovered that the spotted and streaked patterns seen in many of these kernels result from the interaction of transposable elements with genes that govern pigment production. McClintock received a Nobel Prize for her work on transposable elements in 1983. (Spencer Grant)

A Journey into Science in Process

Genetic Engineering

Although we still have much to learn about genes, recently developed techniques have already given rise to a new technology of molecular genetics. We can isolate a desired gene and grow millions of copies of it. We can analyze these copies to find out the gene's nucleotide sequence. We can also decode this nucleotide sequence to find out the sequence of amino acids in the corresponding protein (Section 10-B). In several cases, we have even transferred functioning genes into cells of bacteria, yeasts, plants, and animals.

We can also make DNA to order, using "gene machines" that can be programmed to produce short strands of DNA in any desired sequence. This tailor-made DNA is a useful tool for studying DNA. It can also be used in protein synthesis experiments. By changing the genetic code so as to eliminate particular amino acids from a protein, we can determine how the amino acids affect the function of the protein as a whole.

These feats are all part of **genetic engineering,** the deliberate manipulation of genetic material. Its applications, today or in the future, include: making safer vaccines by engineering a weaker version of the disease-causing agent; producing chemicals by harnessing the metabolism of microorganisms; producing enzymes for industry; cleaning up wastes from industry, oil spills, and pesticide accidents by engineering bacteria with enzymes that convert the waste to harmless substances; replacing defective genes in human beings; and creating improved strains of crops and farm animals.

All of these applications rely on our ability to transplant genes into a cell's genome. The new gene may come from another organism, of the same or a different species, or it may contain DNA produced in the laboratory.

Before genes can be transplanted into cells of higher organisms, such as plants and animals, they must first be spliced into a **vector,** a carrier such as a transposable element, a virus, or a **plasmid,** a small circular molecule of DNA that occurs naturally in some bacteria and yeasts (Figure 9-A). A vector must be able to enter a cell. It must then become part of the cell's genome so that it obeys the cell's normal controls over gene expression. It is also vital, of course, that any extra genes the vector may bring into the cell be harmless to the cell. Genes are now routinely transplanted into cells in laboratory culture. However, we still have much to learn about the control of gene expression, that is, how genes control protein synthesis. This is holding up efforts to make transplanted genes express themselves normally (Figure 9-B).

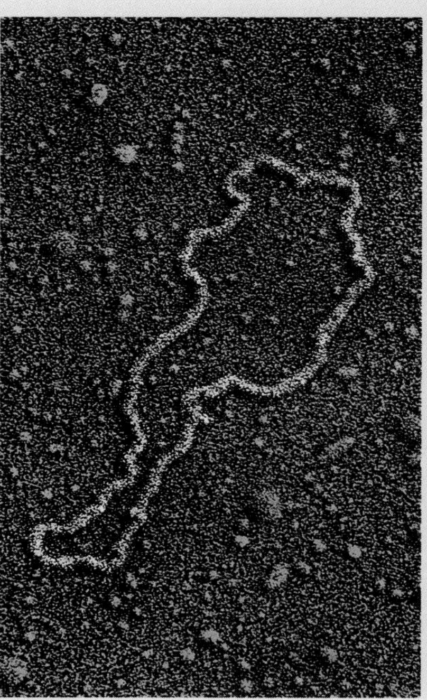

FIGURE 9-A

A plasmid from an *E. coli* bacterium. (Stanley N. Cohen, Science Source/Photo Researchers)

The first practical application of the new genetic technology was the production of useful proteins on a commercial scale using **recombinant DNA,** DNA produced by combining genes from more than one organism. In 1982, human insulin, produced by bacteria containing transplanted human insulin genes, became the first recombinant DNA protein to reach the marketplace. Insulin is a hormone needed daily by millions of people with diabetes. Previously all the insulin available was extracted from the pancreas glands of butchered cattle and pigs, a long and expensive process. Although bacteria-grown insulin is still expensive to produce, it is better for some patients, who cannot tolerate the slight differences between human insulin and that of other eukaryotic species.

Another protein produced in this way is human interferon, a protein that interferes with replication of viruses (Section 25-A). So far, it is used mostly in medical research and to treat a few rare cancers.

If you own a cat, you may already have purchased genetically engineered vaccine against feline leukemia, one of the first commercially available products of recombinant DNA technology.

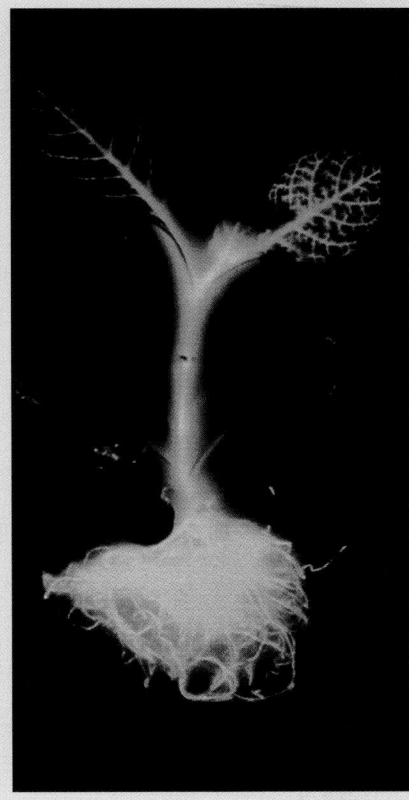

FIGURE 9–B

A luminous tobacco plant grown to demonstrate gene transplantation in plants. Fireflies produce light when the enzyme luciferase acts on luciferin. The gene for luciferase has been transplanted from a firefly into this tobacco plant. When the plant is "watered" with luciferin, it produces light. This shows that the transplanted gene is behaving normally by causing the tobacco plant to produce protein—the enzyme luciferase. (Courtesy of Dr. Marlene DeLuca, University of California, San Diego. From *Science* 234:856-859, 14 November 1986. © 1986 by The American Association for the Advancement of Science.)

How Safe Is Genetic Engineering?

The ability to manipulate genes obviously has many potential rewards for human society. On the negative side, the likelihood that we shall one day be able to control the genetic makeup of human beings and other organisms poses ethical problems that are new to our experience and must be faced.

Many people also worry about the possibility of an accident in a genetic engineering laboratory. Suppose a strain of bacteria with a gene for a dangerous toxin were let loose on the world? Most workers feel that the chance of this happening is slight, because safeguards are already in place. The bacteria used in many recombinant DNA experiments are *Escherichia coli,* a species universally found in the human intestine. However, the genetic strains used in the laboratory have been specially developed to be unable to survive outside their test-tube homes. The danger is further reduced by regulation of laboratories doing recombinant DNA research. Government, scientific organizations, and citizens' groups all participate in drawing up the rules that researchers must follow.

Several experiments are already underway in which genetically engineered bacteria have been applied to outdoor test plots. Researchers then watch to see whether they function as designed under field conditions, whether they die out or become established, and whether they stay put or spread beyond the site of application. Genetically engineered crop plants are also being field-tested. There has been much argument about the safety of such experimental releases of genetically engineered organisms into the environment, and they are being closely monitored.

Another concern is whether transgenic crops (those containing recombinant DNA) will be safe to eat. Since the new genes and the proteins they encode contain the same nucleotides and amino acids in all our food, there seems to be little risk from most new genes. However, new crops must be checked to make sure that any new protein they produce does not interact with the plant's normal chemistry to produce toxic substances. In addition, any plant engineered to produce toxins that fend off insects or diseases must be tested to see if its toxin content endangers human consumers. Finally, some plants are being engineered so that they can survive spraying with herbicides, which are used to kill weeds growing in the same field. The use of herbicides on these crops must be regulated to ensure that crops do not reach the table containing dangerous levels of herbicides.

DNA of one bacterium to that of another. One such transposable element carries a passenger gene that makes any bacterium containing it resistant to the antibiotic ampicillin. No one has yet found transposable elements that move between individual human beings or other vertebrates (animals with backbones). However, we do have transposable elements that move within the genome of a cell.

Some transposable elements apparently lie dormant for many generations, but when they do move they may exert profound effects on the genome. As a result of their activities, genes can be moved around, duplicated, lost, split, or merged.

Repetitive DNA

Most genes are present in only one or a few copies in a genome. However, every eukaryotic cell carries many copies of the genes needed to make ribosomal RNA and histones. For example, the human genome contains 400 copies of ribosomal RNA genes and 30 to 40 copies of histone genes. This is adaptive because the cell needs large quantities of these molecules, and their production is speeded by having multiple copies of these genes.

The genome also contains other repeated DNA sequences of unknown function. About 10% of the human genome consists of a huge number of copies of two kinds of transposable elements. One of these, the *Alu* sequence, is about 300 nucleotide pairs long and is repeated a million times in each body cell. Another 10% of the DNA, called **satellite DNA,** consists of millions of copies of very short sequences of nucleotides. Most satellite DNA occurs near the centromere, the region where two replicated chromosomes are held together.

The DNA of prokaryotes mostly codes for proteins or ribosomal RNA, but such genes are only a small fraction of the genome in plants and animals. Most of the genome of eukaryotes appears to be inactive or related to yet undiscovered functions. In both prokaryotes and eukaryotes, copied segments of DNA called transposable genetic elements can be relocated within or between chromosomes.

SUMMARY

1. The evidence that DNA is the genetic material in all organisms came from several lines of inquiry.
   a. DNA is the substance that transfers genetic information from one cell to another during bacterial transformation.
   b. When a phage takes over the genetic machinery of a bacterium, only its genetic material, DNA, enters the cell.
2. The DNA molecule is a double helix, with two sugar-phosphate backbones forming the sides of a twisted ladder. These strands are connected by crosswise rungs consisting of the base-pairs adenine and thymine or guanine and cytosine, with each base hydrogen-bonded to its complement on the opposite strand.
3. DNA contains the information that dictates its replication. Each strand then serves as the template for the formation of a complementary strand of DNA by the enzyme DNA polymerase. Enzymes "proofread" the newly formed DNA and correct any errors of replication.
4. Damaged DNA is repaired by enzymes that identify and repair errors in the nucleotide sequence of a DNA strand, using information from the undamaged, complementary strand.
5. Mutations are inheritable changes in DNA. Various chemicals and x-rays are among the mutagens that may cause loss or duplication of part of the DNA, or changes in the sequence of nucleotides, which are passed on to new cells in future replications of the DNA.
6. Most of a prokaryote's DNA codes for proteins or for ribosomal RNA, but such genes are only a small fraction of the DNA of eukaryotes. The eukaryotic genome contains much DNA whose function (if any) is unknown. Transposable genetic elements can relocate within or between chromosomes.

SELF-QUIZ

1. DNA is believed to be the genetic material because:
   a. all the body cells of an individual seem to have identical amounts and compositions of DNA, while reproductive cells have half the amount of DNA found in body cells
   b. the proteins are the same from cell to cell in an individual, but the DNA differs; thus the DNA must be the material that makes different tissues different
   c. DNA is the largest type of macromolecule found in living organisms
   d. DNA is found in the cell nucleus
2. A nucleotide consists of:
   a. A, G, T, and C
   b. nitrogenous bases
   c. a sugar, a phosphate group, and a nitrogen-containing base
   d. a sugar-phosphate backbone
3. In a DNA molecule:
   a. nitrogenous bases bond covalently to phosphate groups
   b. sugars bond ionically to nitrogenous bases
   c. sugars bond to nitrogenous bases by hydrogen bonds

d. nitrogenous bases bond to each other by hydrogen bonds

4. Write the sequence of nucleotide bases that would be found in a strand of DNA complementary to this template strand: A-T-C-T-G-T-A-T-G-A ______________

5. The number of adenine bases in a DNA molecule equals the number of thymine bases because:
   a. whenever DNA polymerase places a thymine base into a new DNA strand, it always puts an adenine directly after it
   b. a DNA strand consists of alternating adenine and thymine bases
   c. adenine on one strand hydrogen-bonds to thymine on the other strand
   d. DNA contains equal numbers of each of the four nitrogenous bases

6. A mutation may result from:
   a. addition or loss of one or more nucleotides in a DNA strand
   b. change of one nucleotide to another
   c. loss of an entire chromosome
   d. part of a DNA strand getting turned around "backwards"
   e. all of the above

Tell whether each of the following would be found in the nuclei of your own cells, in the cells of the intestinal bacterium *Escherichia coli,* or both:

___ 7. Circular DNA molecules

___ 8. DNA with adenine base-paired to thymine, and guanine to cytosine

___ 9. DNA in 46 linear molecules

___ 10. DNA wound into nucleosomes

___ 11. Genes carrying instructions for proteins and ribosomal RNA

___ 12. Histones closely bound to DNA

THINKING CRITICALLY

1. Why is the constancy of DNA content from cell to cell in an organism considered to be evidence that DNA is the genetic material? Is it necessary for all cells of an organism to contain identical genetic information? Is it possible for an organism to have different genetic information in different cells of the body?
2. What is the biological importance of the fact that the sugar-phosphate backbones of the DNA double helix are held together by covalent bonds, and that the cross-bridges between the two strands are held together by hydrogen bonds?
3. Why is it necessary to limit the amount of x-rays a person is exposed to over a given period of time? Which organs must be especially well shielded from x-ray exposure?
4. Is transplanting genes into people for medical reasons ethically equivalent to transplanting a replacement kidney into someone whose own kidneys have failed?
5. Manipulation of the human genome raises many horrifying possibilities, including populations with many more of one sex, or clones of identical individuals. What ethical guidelines exist to help us determine what we should and should not permit? What additional guidelines would you propose?
6. Growth hormone produced by genetic engineering techniques is now being used to treat children whose bodies do not produce enough. The added hormone enables them to grow normally. How might such substances be abused? What rules do we need to prevent this from happening?

SELECTED KEY TERMS

bacterial transformation, *p. 166*
bacteriophage, *p. 166*
chromatin, *p. 172*
chromosome, *p. 164*
complementary base pairing, *p. 169*
double helix, *p. 169*
gene, *p. 164*
gene expression, *p. 165*
genetic information, *p. 164*
genome, *p. 172*
histone, *p. 172*
mutagen, *p. 171*
mutation, *p. 171*
template, *p. 170*

SUGGESTED READINGS

Beardsley, T. "Trends in biology: Smart genes." *Scientific American,* August 1991.

DeLisi, C. "The human genome project." *American Scientist,* September 1988.

Fedoroff, N. V. "Transposable genetic elements in maize." *Scientific American,* June 1984.

Gilbert, W., and L. Villa-Komaroff. "Useful proteins from recombinant bacteria." *Scientific American,* April 1980. Gives further details of recombinant DNA techniques.

Radman, M., and R. Wagner. "The high fidelity of DNA duplication." *Scientific American,* August 1988.

Verma, I. M. "Gene therapy." *Scientific American,* November 1990.

Watson, J. D. *The Double Helix.* New York: Atheneum, 1968. A personal story of the discovery of DNA structure. Highly readable and human.

Watson, J. D., and F. H. C. Crick. "Molecular structure of nucleic acids. A structure of deoxyribose nucleic acid." *Nature* 171:737, 1953. A classic Nobel Prize–winning paper.

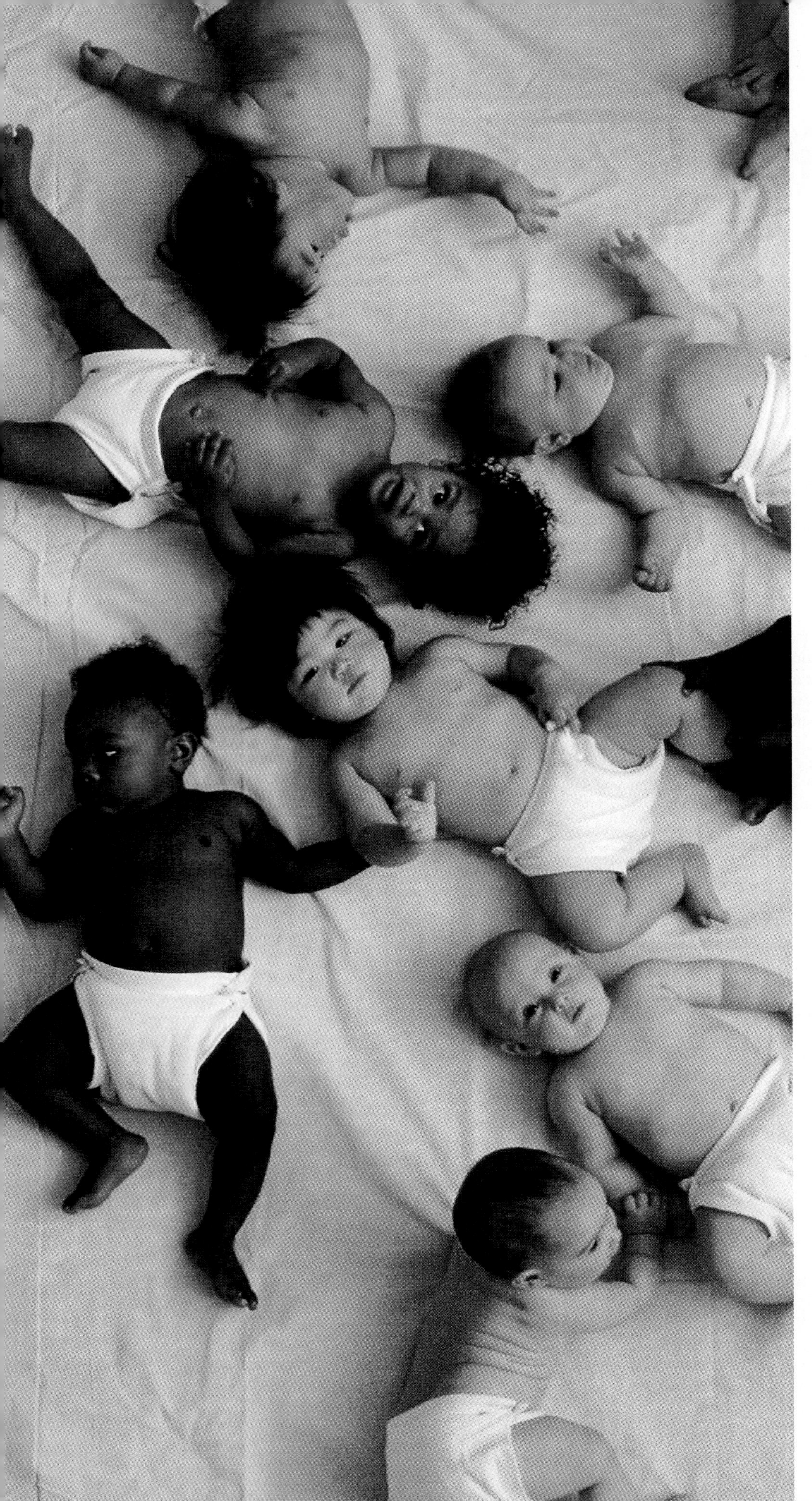

CHAPTER 10

CONCEPT GUIDE

After reading this chapter you should be able to:

1. Describe three structural differences between DNA and RNA molecules. (Section 10–A)
2. Explain why the genetic code must be a triplet code and describe the experiments that tested the triplet code hypothesis. (Section 10–B)
3. Describe the roles of mRNA, rRNA, and tRNA in the process of protein synthesis. (Sections 10–C and 10–D)
4. Compare the mechanisms that control protein synthesis in prokaryotes and eukaryotes. (Section 10–E)
5. Compare the control of metamorphosis in amphibians and in insects. (Section 10–F)
6. Explain what must happen for a cancer to form, and why all mutations do not produce cancerous cells. (Section 10–G)

DNA did it! DNA directed the processes that transformed fertilized human eggs (about the size of the period at the end of this sentence) into these recently born babies. Read how the genetic code is translated to produce the proteins of our bodies in Section 10–B. (B. Campbell/Liaison International)

RNA and Protein Synthesis

You are now familiar with DNA, the molecule that composes the chromosomes in each cell of your body. You know that all the characteristics you have inherited from your parents—and their parents and all your ancestors stretching back for the 3.6 billion years that life has existed on Earth—are somehow bound up in DNA's molecular code. But DNA is a gigantic molecule. Located in chromosomes inside each cell, it is much too big to squeeze through the pores of the nuclear envelope. How does DNA produce the nose you've inherited from your paternal grandfather? How did it give you hair that has the same reddish highlights as your mother's? How does DNA cause the nine months of changes in a fertilized egg that produce a fully formed, squalling newborn who has her father's chin, her mother's long, artistic fingers, and mind of her own? (See Chapter opening photo.)

In the late 1940s it became clear that the nucleic acid RNA (see Section 3–F) plays an important role in protein synthesis. For one thing, cells that make a lot of protein also make RNA rapidly. In a eukaryotic cell DNA is found primarily in the nucleus, but RNA occurs in the nucleus, in the mitochondria, and in the cytoplasm, where proteins are made. It seemed likely that RNA acted as a go-between, carrying genetic information from nuclear DNA into the cytoplasm, where it could be used to make proteins. We now know that this is, in fact, the case. The transfer of information from DNA to RNA to proteins in the process of **protein synthesis** is one way genes are expressed.

Important as genes are, it has always been rather difficult to define a gene, partly because our understanding of the term keeps changing. Like "truth," the word "gene" describes a concept that is fuzzy around the edges. For our purposes, we can define a **gene** as a length of DNA that serves as a functional unit. Most of the genes discussed in this chapter are **structural genes**, those that contain information needed to make proteins. Genes with other functions also exist. Some contain the information needed to make molecules of transfer and ribosomal RNA (Section 10–A). Some are regulatory genes that control the activity of structural genes. Regulatory genes act like switches, and when they are expressed, they turn on structural genes that, in turn, make proteins.

In this chapter we shall examine the structure of RNA and the roles of various types of RNA in protein synthesis. Then we shall see how proteins are assembled and examine how a cell controls the production of the many proteins encoded by its genes.

KEY CONCEPTS

- Proteins are made on ribosomes using genetic information from DNA.
- The genetic information for protein synthesis is carried from the DNA to the ribosomes by RNA molecules.
- Genes are not always expressed: a cell is programmed to make only the proteins it needs.

Curiosity Questions

1? How can DNA direct the production of thousands of kinds of proteins? (See page 182)

2? How does a tadpole change into a frog, or a caterpillar become a butterfly? (See page 195)

3? What causes cancer? (See page 196)

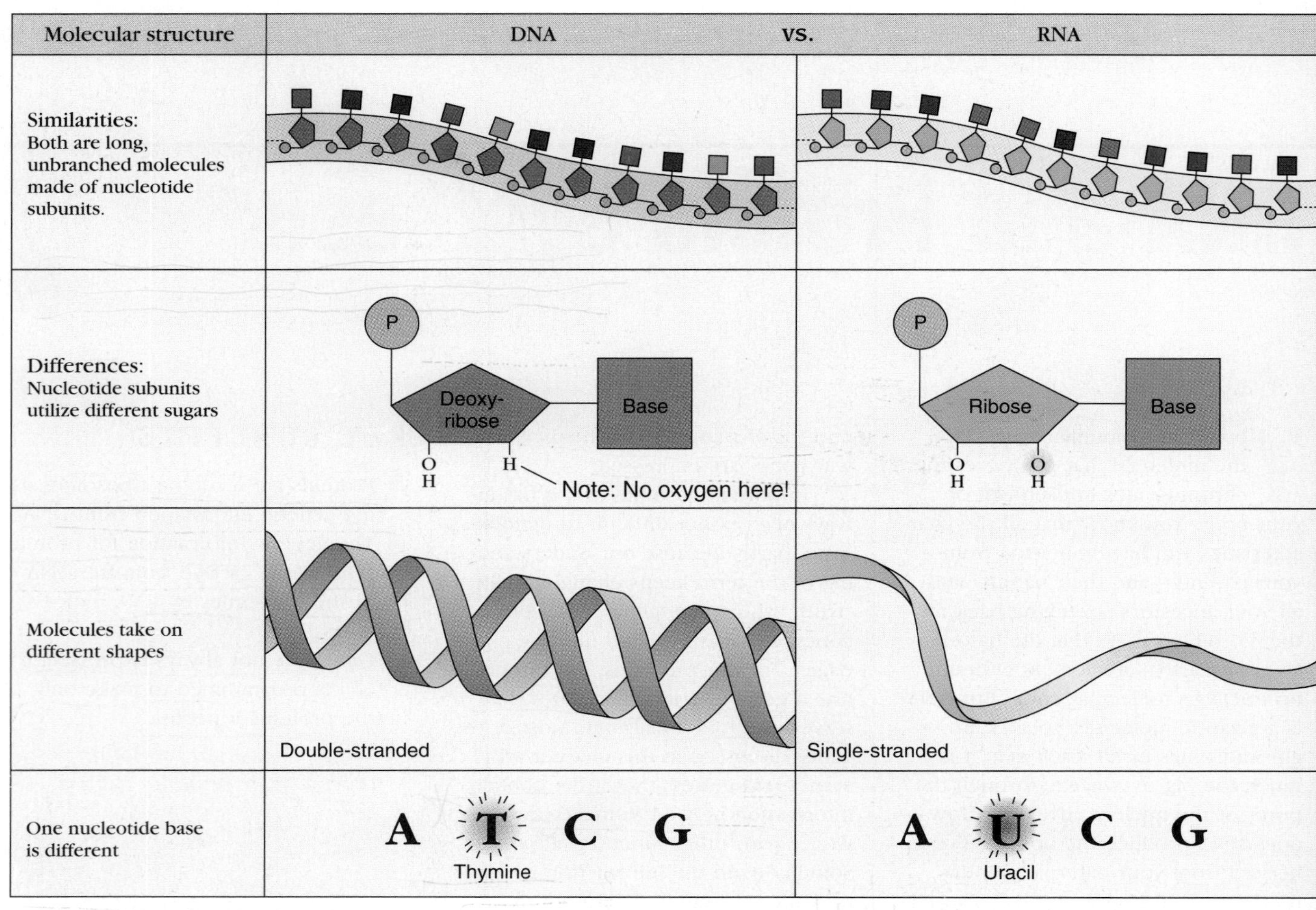

	Messenger RNA (mRNA)	Transfer RNA (tRNA)	Ribosomal RNA (rRNA)
Types of RNA			
Function	Codes for the order of amino acids, a blueprint for protein synthesis	Carries amino acids to mRNA, helping to make proteins	The workbench where amino acids are bonded to form proteins
Location in the cell	In the nucleus near DNA and in the cytoplasm near ribosomes	In cytoplasm near ribosomes	In cytoplasm in ribosomes
REMEMBER:	TRANSCRIPTION PRODUCES RNA FROM DNA!		

FIGURE 10–1

The unique structure and function of RNA molecules.

10–A RNA

Like DNA, RNA comes in long unbranched molecules made up of nucleotide subunits. Each nucleotide is made up of a sugar, a nitrogen-containing base, and a phosphate group (Figure 10–1).

However, RNA differs from DNA in several respects:

1. RNA usually consists of a single strand of nucleotides, while DNA is double-stranded, with two complementary chains of nucleotides.
2. The sugar in RNA is ribose, whereas the sugar in DNA is deoxyribose, with one less oxygen atom than ribose.

3. DNA and RNA differ in the kinds of bases they contain. In DNA we find adenine, guanine, cytosine, and thymine. RNA also contains adenine, guanine, and cytosine; but uracil (which is very similar to thymine) takes the place of thymine and binds to adenine.

Three types of RNA participate in protein synthesis:

1. **Messenger RNA (mRNA)** contains the code for the order of amino acids in a protein. It carries this information from the DNA of a structural gene to ribosomes, where the protein is made.
2. **Transfer RNA (tRNA)** molecules carry amino acids to the mRNA at ribosomes and fit them into the proper place in the growing protein chain.
3. **Ribosomal RNA (rRNA)** is a major component of ribosomes that is capable of making the peptide bond characteristic of protein molecules (see Figure 3-8). One rRNA has a role in recognizing the beginning of the message in mRNA.

Transcription of DNA Into RNA

All RNA is made using information from a DNA template. RNA synthesis is called **transcription** ("written across"), because it rewrites the genetic message coded in DNA into an RNA molecule. In each gene, only one strand of the DNA acts as the template for formation of a complementary strand of RNA. (The template strand may be different in adjacent genes.)

In the process of transcription the enzyme RNA polymerase binds to a specific DNA sequence called a **promoter,** which signals where RNA synthesis should start. The promoter also determines which DNA strand is transcribed. RNA polymerase separates nearby DNA into two strands. It moves along the DNA, using the coding strand as a template and joining the complementary nucleotides (A, C, G, or U) together one by one, to form an RNA strand. The process is similar to the formation of a complementary strand of DNA (see Figure 9-3). When the polymerase reaches a **termination signal** on the DNA, it leaves the DNA, and the newly transcribed RNA strand also detaches. A particular stretch of DNA may be transcribed by several RNA polymerase molecules at once, forming many copies of the complementary RNA (Figure 10-2).

The molecular structure of RNA differs from that of DNA in three important ways: (1) RNA is single-stranded, rather than double-stranded; (2) the sugar in RNA is ribose, rather than deoxyribose; and (3) RNA uses the base uracil instead of thymine. RNA is made on a DNA template in the process of transcription.

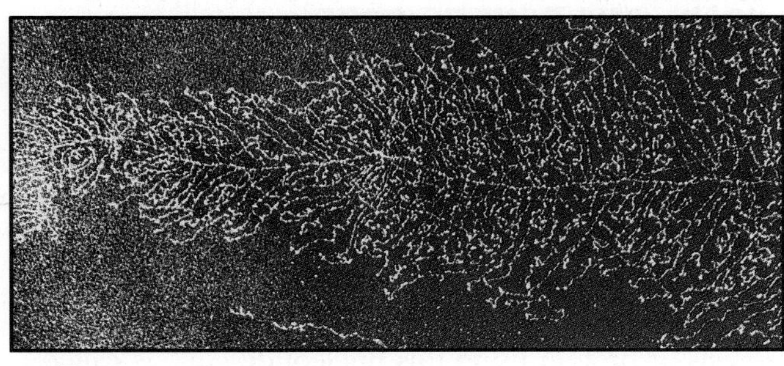

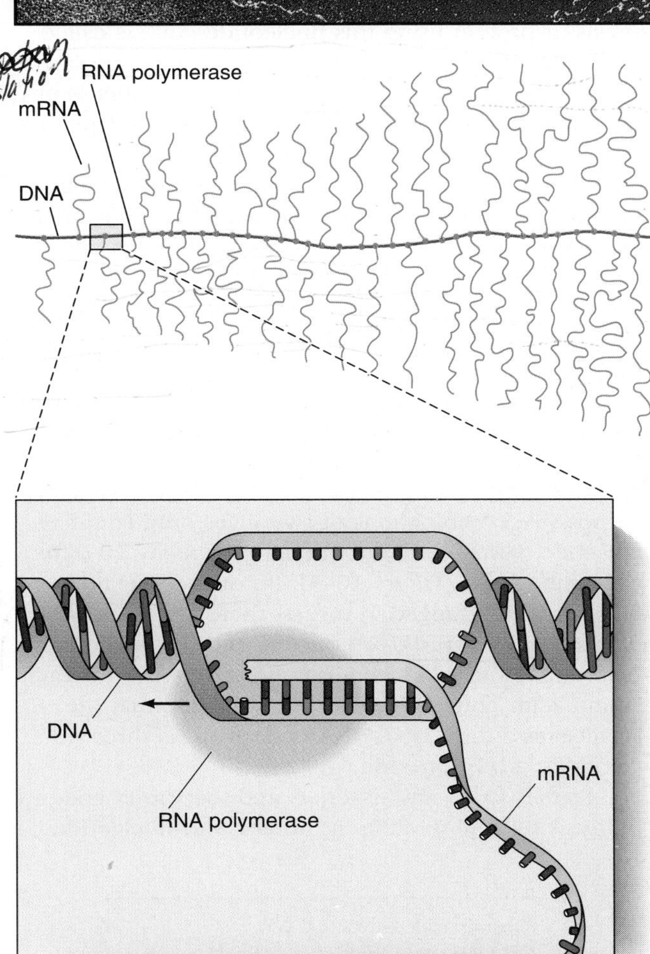

FIGURE 10-2

Gene transcription. In the photo and drawing, the squiggly lines perpendicular to the DNA molecule are many molecules of mRNA. Each is still attached to the RNA polymerase enzyme that is moving along the DNA, and aiding the production of the strand of mRNA. We can tell that each RNA polymerase enzyme is moving from left to right in this photograph because as mRNA molecules are produced the nucleotide chain grows longer, and the longest mRNA is on the right. (O. Miller/Science Photo)

10–B THE GENETIC CODE

1? Each structural gene carries the code or genetic instructions for making one **polypeptide,** a long chain of amino acids. A functional **protein** consists of one or more polypeptides. In transcription a sequence of nucleotides in DNA is transcribed to a sequence of nucleotides in messenger RNA, which, in turn, codes for a sequence of amino acids in a polypeptide. The synthesis of protein using this nucleotide code is called **translation** because genetic information is translated from the DNA/RNA "language," whose alphabet is nucleotides, into the protein "language," whose alphabet is amino acids. But the translation of DNA/RNA nucleotide information into the amino acid sequences that compose proteins is complicated by a fundamental difference between the DNA/RNA alphabet and the protein alphabet: genetic information of DNA/RNA is written in a four-letter nucleotide alphabet, while proteins are written in the 20-letter alphabet of amino acids.

Biologists reasoned that there must be a code of nucleotides that corresponds to each of the 20 amino acids. Each code word of nucleotides cannot be only two nucleotides long, because four letters arranged in all possible combinations of two gives only 16 different code words—still not enough to specify 20 different amino acids (Figure 10–3). But if the four nucleotides are arranged in threes, there are 64 possible different code words, or **codons,** and this is more than enough to produce a unique code word for each amino acid. Thus, biologists hypothesized that the smallest size for a DNA/RNA code word is three nucleotides: a **triplet code.** ■

Francis Crick and others tested the triplet code hypothesis by adding different numbers of nucleotides into the DNA of bacteriophages. There were two steps to their thinking.

1. They reasoned that if the code involves three nucleotides, then inserting one or two nucleotides into the middle of a gene will change the entire message after that point. For example, if we insert one or two nucleotides containing guanine (G) into the DNA message

   CAT—CAT—CAT,

   the message might become

   CA**G**—TCA—TCA—T, or

   CA**G—G**TC—ATC—AT.

2. But if a trio of nucleotides were inserted into a gene, it would only create a short disruption. After it, the message would be identical to the original:

   CAT—**GGG**—CAT—CAT, or

   CA**G—GG**T—CAT—CAT, or

   C**GG—G**AT—CAT—CAT.

Experiments supported these predictions. When a string of three extra nucleotides was added into DNA, a slightly altered polypeptide was produced. In contrast, adding one, two, or four nucleotides changed all subsequent code words, so the polypeptide made according to these new instructions contained a completely different string of amino acids.

The next phase of research focused on cracking the triplet code. Biochemists made artificial messenger RNAs with known sequences of nucleotides. These were put into test tubes along with ribosomes, transfer RNAs, amino acids, and everything else needed for protein synthesis, and the resulting products were analyzed to identify polypeptides. Just as the Rosetta Stone helped crack the code of Egyptian hieroglyphics, a long string of uracil (U) nucleotides that translated into a polypeptide containing only the amino acid phenylalanine helped Marshall Nirenberg and H. Gobind Khorana work out the genetic code for all 64 possible combinations of nucleotide triplets. They received a Nobel Prize in 1968 for this work.

The mRNA code words appear in Table 10–1. Note these features of the genetic code:

1. The codons that are shown are the code words found in messenger RNA. The DNA triplets are the complements of these in RNA codons, except that DNA uses thymine instead of uracil.
2. Three of the 64 codons do not code for amino acids. UAA, UAG, and UGA are *STOP* codes that signal the end of a polypeptide chain.

The four code letters: A C G U
can make 16 possible combinations

	A	C	G	U
A	AA	AC	AG	AU
C	CA	CC	CG	CU
G	GA	GC	GG	GU
U	UA	UC	UG	UU

FIGURE 10–3

Two-letter words. The four different kinds of nucleotides in RNA can be arranged in pairs to form only 16 different possible combinations; not enough to code for 20 different types of amino acids.

Table 10–1
Codons Found in Messenger RNA*

First Base	Second Base U		Second Base C		Second Base A		Second Base G		Third Base
U	UUU	Phenylalanine	UCU	Serine	UAU	Tyrosine	UGU	Cysteine	U
	UUC	Phenylalanine	UCC	Serine	UAC	Tyrosine	UGC	Cysteine	C
	UUA	Leucine	UCA	Serine	UAA	STOP	UGA	STOP	A
	UUG	Leucine	UCG	Serine	UAG	STOP	UGG	Tryptophan	G
C	CUU	Leucine	CCU	Proline	CAU	Histidine	CGU	Arginine	U
	CUC	Leucine	CCC	Proline	CAC	Histidine	CGC	Arginine	C
	CUA	Leucine	CCA	Proline	CAA	Glutamine	CGA	Arginine	A
	CUG	Leucine	CCG	Proline	CAG	Glutamine	CGG	Arginine	G
A	AUU	Isoleucine	ACU	Threonine	AAU	Asparagine	AGU	Serine	U
	AUC	Isoleucine	ACC	Threonine	AAC	Asparagine	AGC	Serine	C
	AUA	Isoleucine	ACA	Threonine	AAA	Lysine	AGA	Arginine	A
	AUG	Methionine (START)	ACG	Threonine	AAG	Lysine	AGG	Arginine	G
G	GUU	Valine	GCU	Alanine	GAU	Aspartic acid	GGU	Glycine	U
	GUC	Valine	GCC	Alanine	GAC	Aspartic acid	GGC	Glycine	C
	GUA	Valine	GCA	Alanine	GAA	Glutamic acid	GGA	Glycine	A
	GUG	Valine	GCG	Alanine	GAG	Glutamic acid	GGG	Glycine	G

*To use the table, find the letter of the first base of the codon in the column at the left, and go across this row until you are in the column headed by the letter of the second base. Then find the third base, marked at the far right of the table. The three *STOP* codons signal positions where the ribosome stops reading and terminates the polypeptide chain. The codon AUG initiates synthesis of a polypeptide.

3. There is more than one codon for most amino acids. Because of this the code is said to be **degenerate.** This degeneracy is biologically useful because, for one thing, it makes mutations less damaging. If the code were not degenerate, 20 codons would specify amino acids and 44 would code for nothing. In effect, they would act as *STOP* codes, and most mutations would lead to *STOP* codes, which would halt protein synthesis.
4. A codon's third base is often less specific than the first two. For instance, all of the codons for the amino acid proline have CC as the first two bases, but any other base can be in the third position.

While there are code words to stop protein synthesis, there is no codon for punctuation that would signal the beginning or end of a code word. As a consequence, the code must be read from a particular starting point or the whole sequence will have a different meaning. For instance, the RNA sequence UCUAGAGCUA will produce the amino acid sequence serine—arginine—alanine if it is read from left to right, beginning with the first U. But if you begin reading with the second nucleotide from the left (C) it will produce leucine—glutamic acid—leucine. It is important for messenger RNA to have an initiation point that says in effect "start here." The initiation codon is the sequence AUG, which codes for the amino acid methionine.

Some mutations exchange one nucleotide for another. By examining Table 10–1, you can see that a mutation that changes the third nucleotide in a codon often will not change the amino acid specified by that

BIO-BIT

The human genome project is an attempt to determine the base sequence of all of the nuclear DNA in one human cell.

codon. However, changing the first or second nucleotide is likely to result in placing a different amino acid into that slot, while leaving the rest of the amino acid sequence unchanged.

Mutations that change the genetic material by adding or deleting nucleotides are called **frameshift mutations,** because they alter the entire reading frame of the message. This usually changes the sequence of amino acids produced beyond the mutated area, as in the CAT—CAT—CAT example above. Muscular dystrophy, one of the most common genetic disorders, is characterized by progressive muscular degeneration that leads to death in early adulthood. It is hypothesized that the most severe form of the disease is caused by a frameshift mutation that causes the premature termination of the translation of the protein dystrophin.

With minor exceptions, the genetic code is universal. The same codons specify the same amino acids in virtually all viruses, bacteria, plants, animals, and fungi. This is compelling evidence that all organisms on Earth today evolved from a common ancestor. The major groups of organisms have had separate evolutionary histories for hundreds of millions of years, and so it seems that the code must have been established shortly after life originated, and that it has continued almost unchanged for the billions of years since. The main exceptions to this rule are mitochondria, organelles that contain their own DNA and ribosomes. Of the 64 codons, up to six may mean something different in mitochondria.

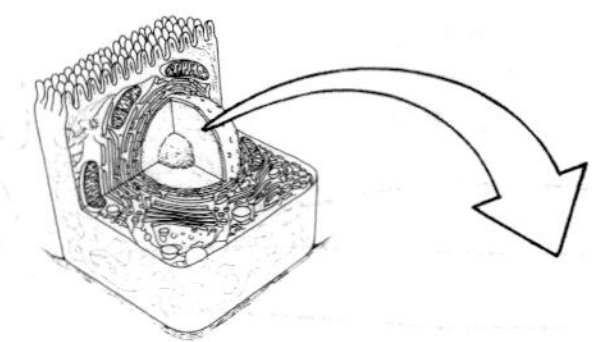

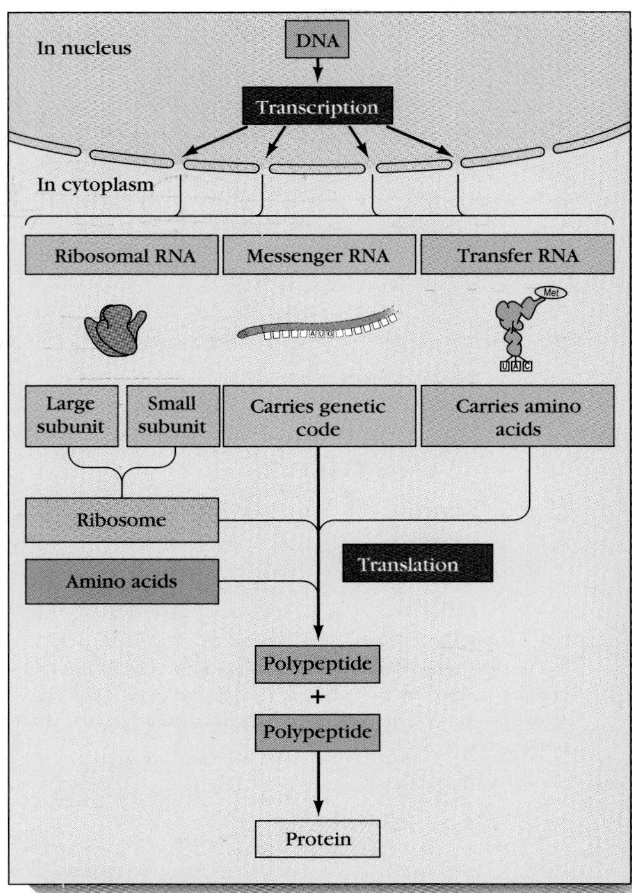

FIGURE 10–4

Transcription and translation. The flow of genetic information from DNA to RNA and from RNA to the amino acid sequence of a polypeptide, and eventually to a protein. RNA molecules are produced by transcription within the nucleus. These RNA molecules then move through nuclear pores into the cytoplasm where they interact with amino acids, enzymes, and ATP to produce proteins in the process of translation.

Messenger RNA (mRNA) molecules, produced during transcription, code for a specific sequence of amino acids that form a polypeptide. Every three bases forming the mRNA molecule either code for one specific amino acid or else signal the start or end of the polypeptide. This process of converting the linear information of mRNA, in the form of a specific sequence of bases, into a linear sequence of amino acids to form polypeptides, is called translation. This system of translating the genetic code is common to all forms of life.

OVERVIEW OF PROTEIN SYNTHESIS

Before we look at the molecules involved in protein synthesis in more detail, let us briefly sum up the whole process. RNA is made on a DNA template. The genetic code for a polypeptide is copied from DNA into messenger RNA (mRNA) molecules. The mRNA carries this code to ribosomes, where protein synthesis takes place (Figure 10–4). The mRNA code specifies the order in which amino acids are joined to form a polypeptide. Messenger RNA cannot recognize an amino acid directly. An adaptor molecule, transfer RNA (tRNA), is needed to bring the two together. Transfer RNA molecules bring the proper amino acids to the ribosome and fit them into their coded positions. As the genetic code in a mRNA molecule is "read" on a ribosome, the amino acids are joined together, one by one.

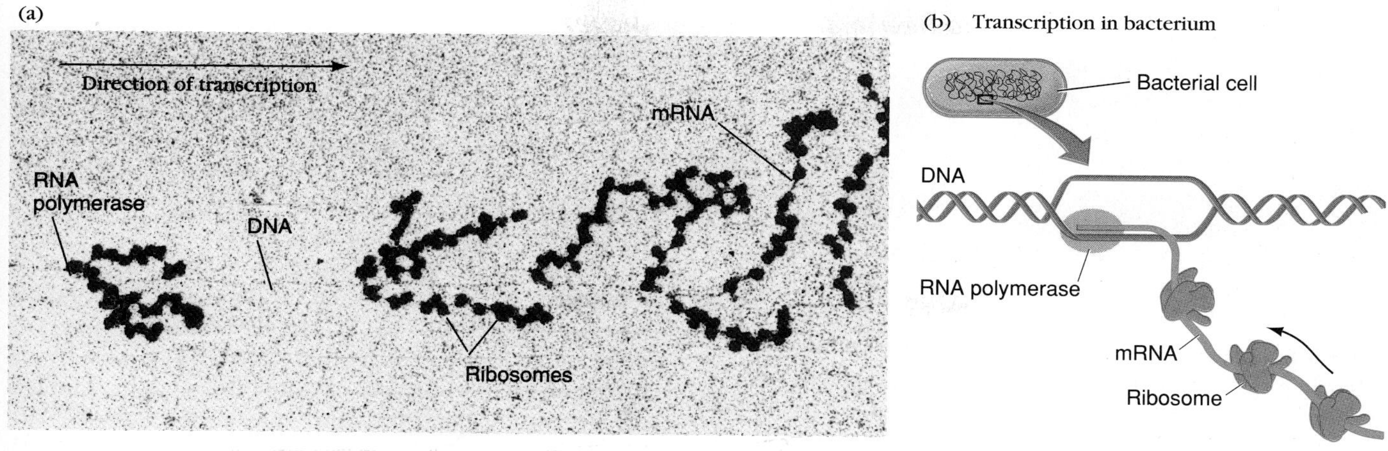

FIGURE 10–5

Simultaneous transcription and translation. (a) In this electron micrograph of *Escherichia coli* genes, RNA polymerase molecules are transcribing DNA into mRNA. **(b)** As the front end of each mRNA is formed, a ribosome binds to it and starts to translate it. Ribosomes move along the mRNA as they translate it. As soon as there is room, another ribosome attaches to the RNA. (Courtesy O. L. Miller. O. L. Miller, Jr., B. A. Hamkalo, and C. A. Thomas, Jr. *Science* 169:392, 1970)

10–C KINDS OF RNA

Messenger RNA

A prokaryote's DNA lies in the cytoplasm. As messenger RNA is transcribed, ribosomes may attach to the front end of the mRNA and start to translate it into protein while the rest of the molecule is still being made on the DNA (Figure 10–5).

In eukaryotes, on the other hand, transcribed RNA must be processed in the nucleus before it becomes mature mRNA, ready to enter the cytoplasm for protein synthesis. First, a special nucleotide "cap" is attached at the front end. This is the signal that will bind the mRNA to a ribosome. A "tail" of adenine-containing nucleotides is attached to the other end. The tail is thought to extend the lifetime of the RNA somewhat, before it is degraded by ever-present RNA-digesting enzymes.

Many eukaryotic structural genes contain sections called **intervening sequences** or **introns,** which do not code for parts of proteins. Introns are transcribed, but they are then "spliced" out of the RNA molecule, leaving a shorter, mature mRNA that can be translated into protein. Some eukaryotic genes contain no introns, but others have as many as 50 introns. The parts of the gene that are represented in the mature mRNA are called **exons** (because they are *ex*pressed as protein) (Figure 10–6). The function of introns is not well understood. Some think they may play a role in the evolution of proteins, but others regard introns as parasites or as useless "junk" DNA.

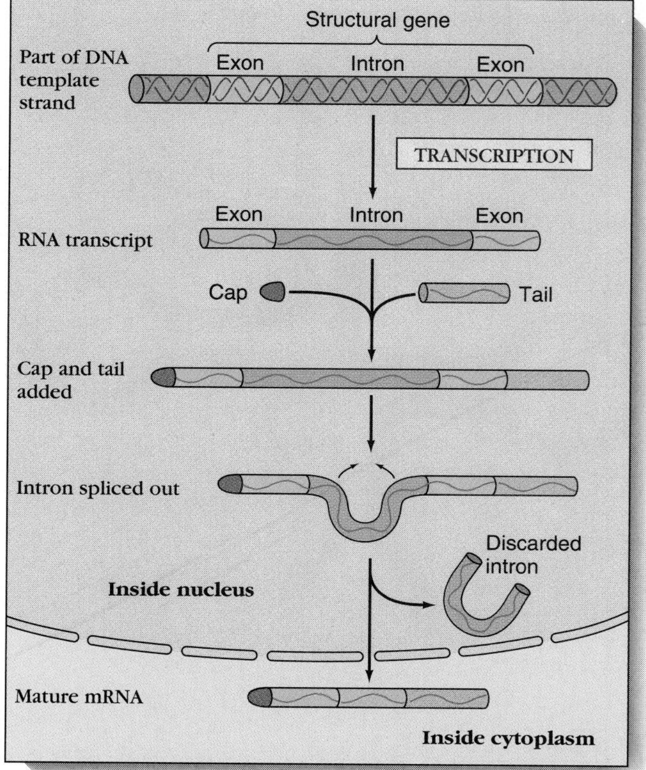

FIGURE 10–6

How a mature messenger RNA molecule is produced in a eukaryotic cell.

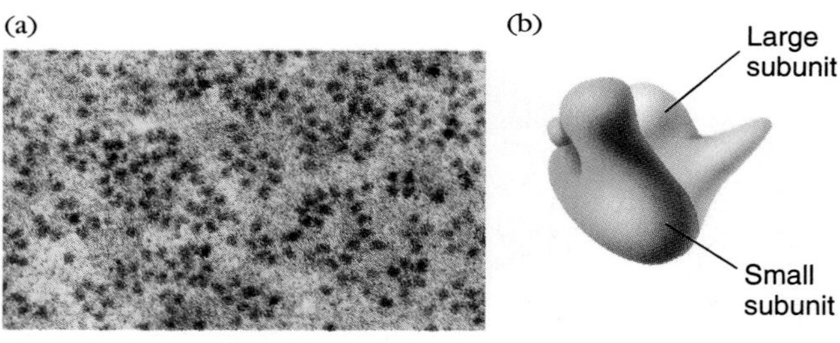

FIGURE 10–7

Working out the structure of ribosomes. (a) Electron micrograph of small ribosomal subunits isolated from human tumor cells grown in cell culture. The irregular shape of the subunits is difficult to distinguish. To determine the shape of a specific type of ribosome, hundreds of photographs are taken from all sides and fed into a computer to generate an overall image. **(b)** Computer generated maps allow us to appreciate the three-dimensional aspects of ribosomes. Comparisons of ribosomal structures indicate that related ribosomes have somewhat different shapes in different organisms. (D. Kunkel/Phototake)

Ribosomal RNA and Ribosomes

Ribosomes, the sites of protein synthesis, consist of several types of ribosomal RNA and about 70 kinds of polypeptides. In eukaryotes, ribosomes are made in an area of the nucleus called the **nucleolus,** a place that bustles with activity, sometimes churning out several hundred thousand ribosomes per hour.

A functional ribosome consists of two subunits, one large and one smaller, each containing RNA and polypeptides (Figure 10–7). The two parts join together only for protein synthesis. The ribosomes of prokaryotes, chloroplasts, and mitochondria are similar to, but smaller than, the ribosomes in eukaryotes.

Transfer RNA

Transfer RNA (tRNA) carries amino acids to the ribosomes during protein synthesis. Each of the 20 amino acids has a specific kind of tRNA molecule to transport it. All tRNA molecules have the same general shape. Parts of the molecule fold into loops, held in shape by base-pairing between different areas of the molecule (Figure 10–8). The most important parts of the tRNA

Three Ways To Represent a tRNA Molecule

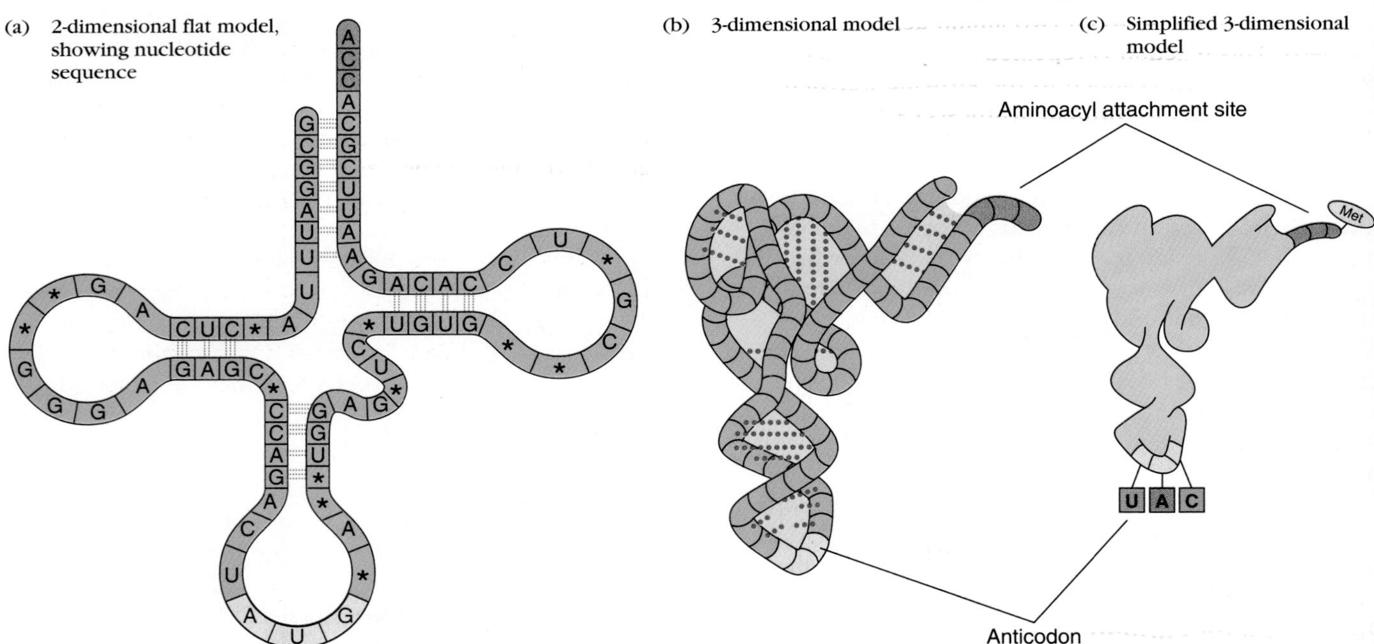

FIGURE 10–8

Transfer RNA. Transfer RNA has two important features: an anticodon that will base-pair with the complementary sequence of nucleotides on mRNA (codon) and an attachment site for the amino acid specified by its triplet of nucleotides. This tRNA carries methionine, the sequence of nucleotides that begins protein synthesis. The asterisks indicate unusual bases unique to tRNA molecules.

molecule are the **aminoacyl attachment site,** where the amino acid is attached, and the **anticodon,** a row of three nucleotides that base-pairs with the complementary codon of a mRNA molecule attached to a ribosome. In this way the tRNA places the amino acid it carries into the correct position in the peptide chain.

Three kinds of RNA molecules, all transcribed from DNA, contribute to the process of translation. Messenger RNA (mRNA) contains the nucleotides that determine the type and sequence of amino acids within the growing polypeptide. Unlike the mRNA of prokaryotes, eukaryotic mRNA is processed between transcription and translation. Nucleotides are added at each end and noncoding intron segments within the molecule are spliced out. Ribosomes serve as the site for protein synthesis. They consist of ribosomal RNA (rRNA), produced in the nucleolus, and many kinds of polypeptides. Transfer RNA (tRNA) molecules carry specific amino acids. Base-pairing between the mRNA codon and tRNA anticodon fits each amino acid into its proper place in the growing polypeptide.

10–D PROTEIN SYNTHESIS

The basic reaction of protein synthesis is the formation of a peptide bond between two amino acids (see Figure 3–8). This reaction is repeated many times, as each amino acid in turn is added to the growing polypeptide chain. But before protein synthesis can begin, several events must occur.

Initiation

First mRNA binds to a small ribosomal subunit (A Journey Through Protein Synthesis). The first codon (AUG = start) then base-pairs with the anticodon of a tRNA carrying the amino acid methionine. This methionine eventually becomes the first amino acid in the polypeptide chain. (After the first AUG, any other AUG codons on the mRNA are translated by inserting the amino acid methionine into the chain.) Now a large ribosomal subunit binds to the complex. Initiation is complete, and the peptide chain can be made.

Peptide Chain Formation

A ribosome has two sites, called the A and P sites, where codons are translated. At the A site, tRNA attaches with an amino acid attached to its aminoacyl attachment site. At the P, or peptidyl site, a tRNA is attached, holding the growing peptide chain. To differentiate these sites, remember that the A site always contains a tRNA with an amino acid attached to it, and the P site contains a tRNA attached to a growing peptide chain.

Initiation has brought the mRNA's initiation codon, AUG, to the ribosome's P site, and the mRNA codon for the second amino acid is lined up at the A site. Three steps then bring in the next amino acid and join the first one to it (A Journey Through Protein Synthesis):

1. A tRNA with a complementary anticodon binds to this second mRNA codon, at the A site. The amino acid carried by this second tRNA will become the second amino acid in the peptide chain.
2. An enzyme joins the first two amino acids in the chain to each other by a peptide bond. The first tRNA is now empty and the second is holding both amino acids.

The empty (methionine) tRNA now moves away from the ribosome. It eventually picks up another molecule of methionine in the cytoplasm.

3. During **translocation,** the second tRNA and its mRNA codon move along the ribosome so that they are now at the vacated P site.

Translocation brings the third codon into the ribosome's A site, where the appropriate tRNA attaches by its anticodon, bringing the third amino acid into position. The growing peptide chain is attached to the newly arrived amino acid on this third tRNA, and the sequence repeats. This sequence of three steps adds amino acids to the peptide one by one. In the bacterium *Escherichia coli,* a peptide grows at the rate of about 20 amino acids per second.

As the leading end of the mRNA emerges from the ribosome, it may bind to another small ribosomal subunit, which initiates protein synthesis again. Each mRNA molecule typically has several to over 100 ribosomes attached to it and transcribing its message as they move along (see Figure 10–5). A mRNA molecule does not last long. Its average life span in a bacterium is 2 minutes.

Termination

The peptide chain grows until the ribosome reaches a *STOP* codon on the mRNA. A special protein, called a releasing factor, then binds to the *STOP* codon and causes the mRNA to leave the ribosome (A Journey Through Protein Synthesis).

During translation, mRNA, rRNA, and tRNA work closely together. Transfer RNA molecules (tRNA) bring amino acids to the mRNA-ribosome complex according to the mRNA codon sequence. Each tRNA in turn adds its amino acid to the growing polypeptide chain. As each successive peptide bond is formed between the lengthening polypeptide and the

(text continues on page 190)

A Journey Through Protein Synthesis

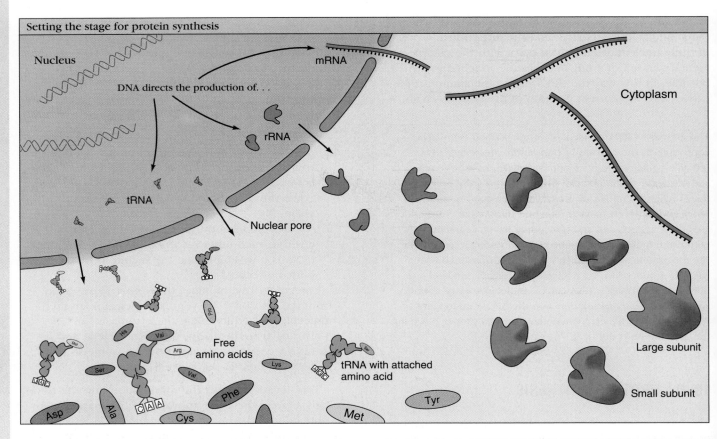

Setting the stage for protein synthesis

Nucleus

DNA directs the production of. . .

mRNA

rRNA

tRNA

Nuclear pore

Cytoplasm

Free amino acids

tRNA with attached amino acid

Large subunit

Small subunit

Met
His
Val
Arg
Pro
Ser
Lys
Val
Phe
Asp
Ala
Cys
Met
Tyr
UAC
CAA
AGC

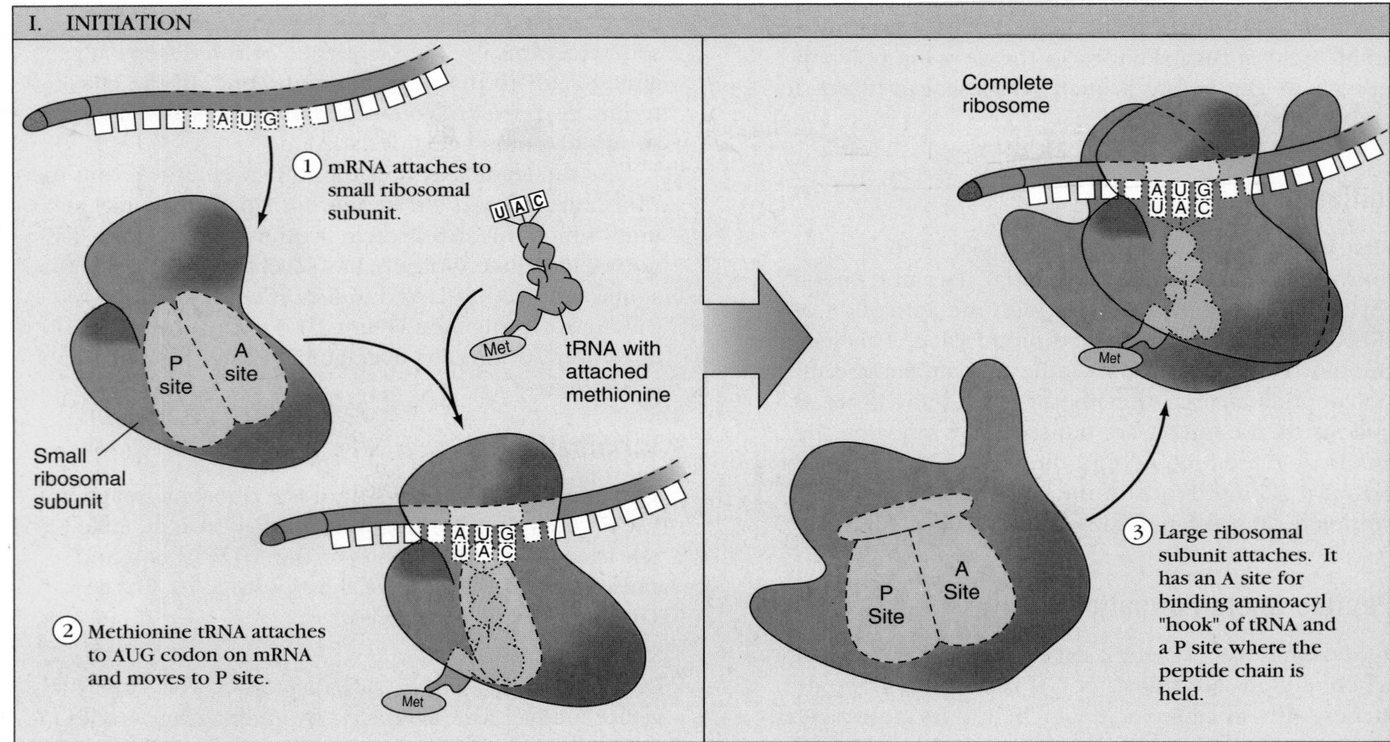

Translation. The three types of RNA involved in protein synthesis are first produced (transcribed) in the nucleus from DNA. In translation, these RNA molecules then interact in specific and coordinated ways with amino acids, enzymes, and ATP to produce polypeptides that join to form proteins.

II. PEPTIDE CHAIN FORMATION

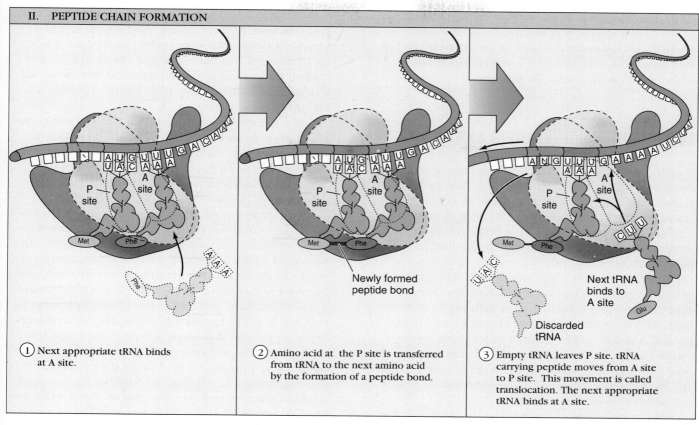

(1) Next appropriate tRNA binds at A site.

(2) Amino acid at the P site is transferred from tRNA to the next amino acid by the formation of a peptide bond.

(3) Empty tRNA leaves P site. tRNA carrying peptide moves from A site to P site. This movement is called translocation. The next appropriate tRNA binds at A site.

III. TERMINATION

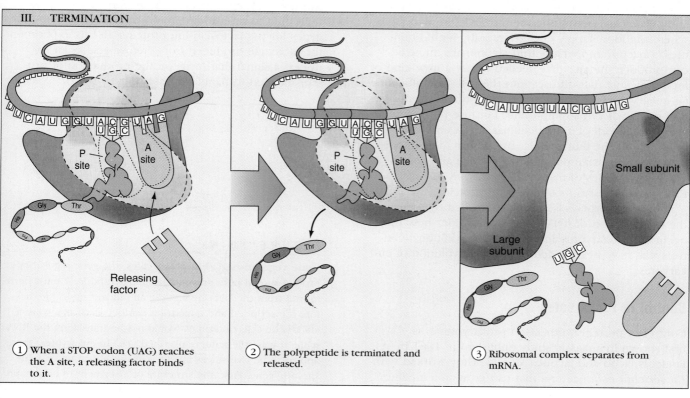

(1) When a STOP codon (UAG) reaches the A site, a releasing factor binds to it.

(2) The polypeptide is terminated and released.

(3) Ribosomal complex separates from mRNA.

newly arrived amino acid, the ribosome moves along the mRNA. This brings the next codon onto the ribosome, where it can bind to the anticodon of the tRNA carrying the next amino acid. The completed polypeptide is released when a *STOP* codon reaches the ribosome.

10–E CONTROL OF PROTEIN SYNTHESIS

A cell's DNA contains instructions for making hundreds of proteins, but it makes only certain proteins at certain times. A gene may be thought of as being "turned on," or expressed, when its DNA is being transcribed into RNA and the RNA is being translated into protein. Genes that are not being transcribed can be thought of as "turned off."

In nearly all cells, large numbers of "housekeeping genes" remain switched on throughout life. These genes control functions that most cells perform all the time, such as protein synthesis, glycolysis, and food uptake. A much smaller number of specialist genes makes one type of eukaryotic cell different from another. For example, some specialist genes give a nerve cell its distinctive functions, while others make a liver cell different from a muscle cell.

How rapidly can an organism switch genes on and off? As you might expect, bacteria can switch genes on and off quickly, permitting them to exist in places where the food supply changes rapidly. Within minutes, a bacterium living within an animal's intestine can produce the enzymes needed to use a new kind of food molecule that diffuses into the cell. In multicellular eukaryotes, the ability to change protein production quickly is not so important, because the surroundings of each cell of a multicellular organism change only slowly. The multicellular organism's most important task is to make proteins to produce and coordinate the activities of many different types of cells. Each cell expresses its own mix of proteins.

Protein synthesis is usually controlled by regulating the transcription of RNA. Here we consider how this occurs in prokaryotes and discuss the additional controls that have been added during the evolution of eukaryotes.

Control in Prokaryotes

Control of protein synthesis in prokaryotes was first shown in a classic study published in 1961 by François Jacob and Jacques Monod. They worked with *Escherichia coli,* bacteria that can grow in a medium containing only salts and a source of carbon—for example, glucose. The bacteria made enzymes that metabolize glucose and used up most of the glucose in the medium. When Jacob and Monod added a different sugar, lactose, to the medium, the bacteria stopped growing briefly while they started to produce a new set of proteins that handle lactose. Jacob and Monod studied the control of this change, and their work led to a general model of how transcription is turned on and off in bacteria.

The agents that switch genes on and off are **gene-regulatory proteins** that bind to DNA and start or stop transcription. Gene-regulatory proteins bind to a specific DNA sequence that is next to a gene's promoter. Some of these proteins physically block the binding of RNA polymerase to the promoter and so keep the gene turned off. Others help RNA polymerase bind and so turn the gene on. Each kind of gene-regulatory protein is specific for the promoter site of one to several genes, so it only turns these genes on or off.

Cells always contain gene-regulatory proteins, and the presence of other chemicals in the cell determines if and when these proteins control transcription. In Jacob and Monod's experiment, while the bacteria were living on glucose, a gene-regulatory protein that was bound near the promoter of the lactose-using genes blocked the promoter from binding RNA polymerase and beginning transcription of the gene. Lactose binds to this regulatory protein, changing its shape and causing it to unbind from the DNA. Because the promoter site was no longer blocked, RNA polymerase could then bind to the promoter and begin to transcribe the genes for the lactose-using proteins. In this experiment the molecule produced from lactose acts as an **inducer,** a signal that turns on the production of proteins needed to handle lactose (Figure 10–9).

FIGURE 10–9

Gene regulation. (a) In prokaryotes, the gene regulatory protein shown here prevents protein synthesis. It must be removed from the DNA by binding an inducer molecule, before transcription and translation can occur. (If this were a eukaryotic cell, ribosomes would not be translating the RNA while it was still being transcribed.) **(b)** In eukaryotes, there appear to be two stages involved in the activation of a eukaryotic gene. First, the structure of a segment of chromatin is altered in preparation for transcription. Second, gene regulatory proteins bind to specific sites on the altered chromatin, which allows RNA polymerase to start transcribing the DNA.

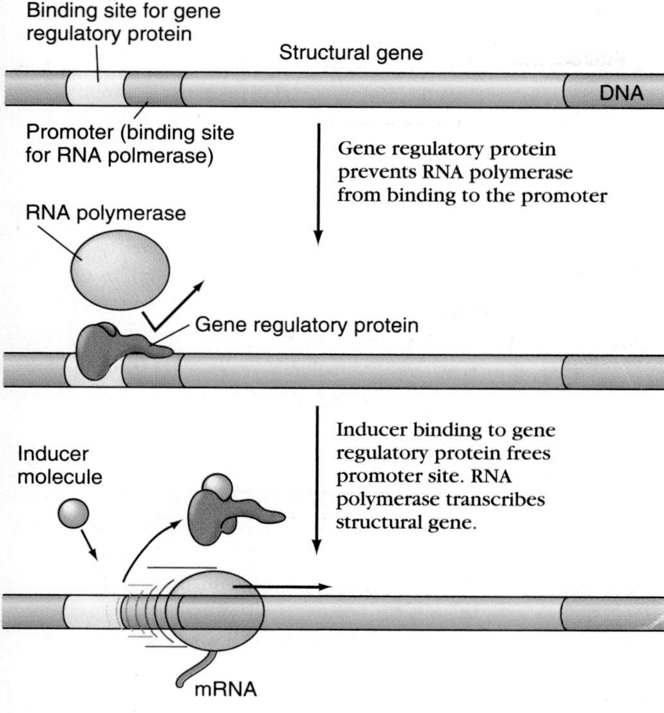
(a) Transcription control in prokaryotes
Binding site for gene regulatory protein
Structural gene
DNA
Promoter (binding site for RNA polmerase)
Gene regulatory protein prevents RNA polymerase from binding to the promoter
RNA polymerase
Gene regulatory protein
Inducer molecule
Inducer binding to gene regulatory protein frees promoter site. RNA polymerase transcribes structural gene.
mRNA

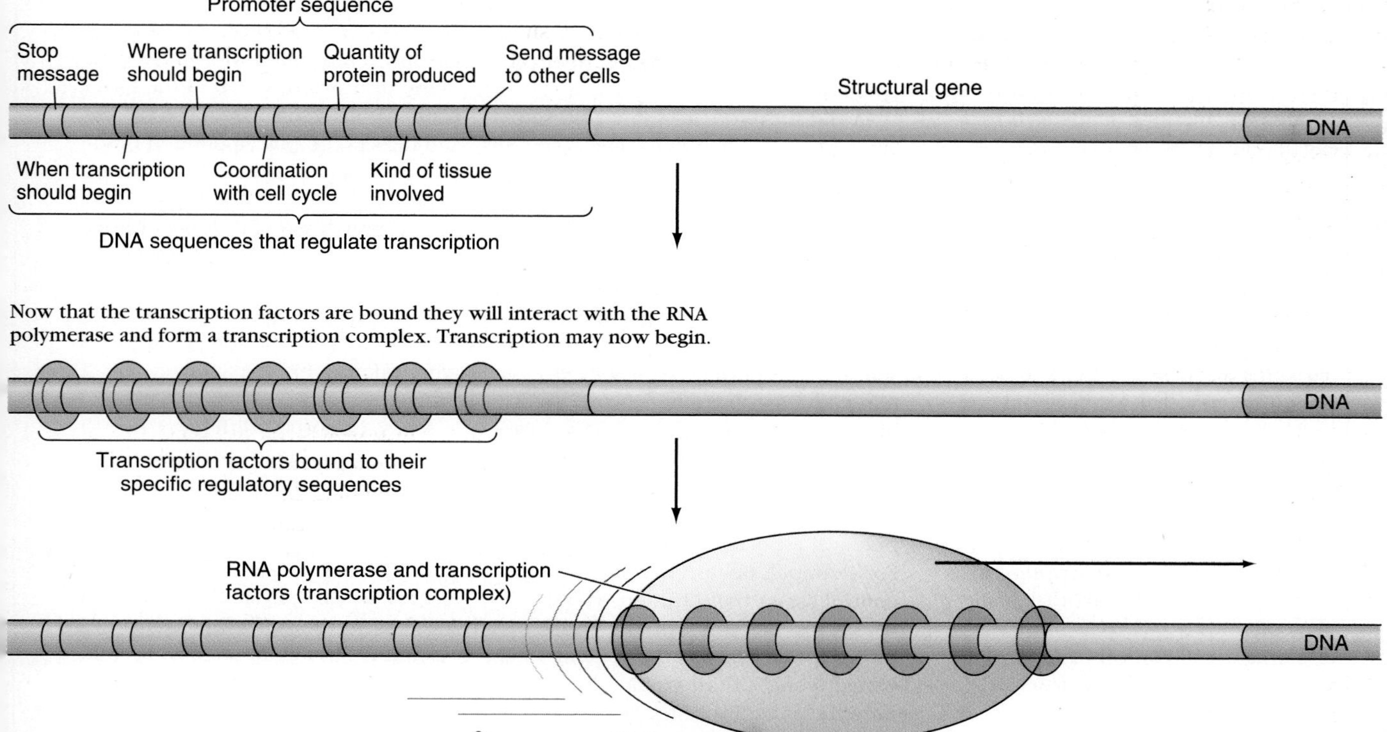
(b) Transcription control in eukaryotes
Many transcription factors (eukaryotic gene regulatory proteins) must first bind to their specific regulatory sequences within the promoter before RNA polymerase can bind to the promoter and begin to transcribe the structural gene.
Promoter sequence
Stop message
Where transcription should begin
Quantity of protein produced
Send message to other cells
Structural gene
DNA
When transcription should begin
Coordination with cell cycle
Kind of tissue involved
DNA sequences that regulate transcription
Now that the transcription factors are bound they will interact with the RNA polymerase and form a transcription complex. Transcription may now begin.
DNA
Transcription factors bound to their specific regulatory sequences
RNA polymerase and transcription factors (transcription complex)
DNA
mRNA

A Journey into Science in Process

Smart Genes

We have a fairly good understanding of how DNA transcription and RNA translation work, but there is much that we do not know, especially about events that take place in the early life of an embryo. Your body, for example, has approximately 250 different types of cells. Each is a variation on a theme, having a slightly different structure than the idealized animal cell (see Chapter 5, A Journey Through An Animal Cell). Most cells are quirky. Many have specialized organelles and weird membrane structures. Most make unique sets of proteins that give them highly specialized functions. One cell is a creeping scavenger, another can contract and relax, another is good at binding oxygen or carbon dioxide, another is an oozing killer, still another can secrete a substance as dense as bone. Amazingly, all these varied cells are descendants of a single fertilized egg and all share identical copies of that cell's original DNA. Events during embryonic development shape the great array of cell types, but the precise control over these events is still unknown.

You know that DNA contains the genetic instructions for making and operating all these different kinds of cells. One area that we have only recently begun to understand is how genes cooperate and coordinate to produce a living organism from a fertilized egg. Recent research has shown that, in eukaryotic organisms, the control of DNA transcription involves groups of proteins that work like committees, whereas in prokaryotic organisms, control of DNA transcription is more like a battery of switches (Figure 10-9).

Everyone understands how committees work. If you want to get something done in your town, neighborhood, church, school, or even work place, the usual first step is to form a committee. Members of a committee usually have slightly different interests—different axes to grind or separate reasons to work toward a common goal. For example, one member of the PTA's Education Committee wants teachers to be paid more because she has three school-age children and is worried about the quality of their schooling; another wants teacher salaries to remain unchanged so that real estate taxes won't increase; a third wants to dictate exactly how much salaries will rise; and a fourth wants to determine the time when this will happen. Some members of the committee do more than their share of the work; others seldom attend meetings.

Researchers have learned that the committees of proteins that control DNA transcription are somewhat similar to human committees. For each gene there may be as many as 20 different proteins on the committee that turns DNA on for transcription of a gene, or keeps it turned off. Like a quorum needed for a committee to vote, five protein messages seem to be the minimum number that must be sent for transcription to begin. Protein committee members "vote" by binding to specific sites on DNA. Like committee members with special interests, these transcription factors seem to send signals concerned with practical aspects of gene expression. For example, one protein factor might control *when* transcription would take place, another would identify *where* in the embryo the protein would work, a third would *coordinate* the activities of the gene with those of the cell's cycle of growth and division of genetic material, a fourth would control the *quantity* of protein produced, a fifth would specify the *kind* of tissue involved and a sixth might *send messages* to adjacent cells.

Once this committee of transcription factors has met and enough members have "voted" by binding to sites on the DNA molecule, they form a transcription complex. This, in turn, activates proteins that influence the enzyme DNA transcriptase to begin copying the gene into a strand of RNA.

The committee-controlled gene is called a "smart gene" because it responds to signals sent from one gene to another in a control network. Hormones may act as committee members, casting their "vote" by binding to specific sites on the DNA molecule. Sound complicated? It is! Figure 10-9 summarizes these events that lead to transcription of a single gene. Think of how complicated the biochemical machinery is that produces and directs the placement and functioning of all the genes that form an organ.

Although we have hints of how transcription is controlled, most of what goes on to produce an embryo from a fertilized egg remains a mystery. No one is sure, for example, what role the histones play in the process.

The important difference between the workings of switches in prokaryotes and committees of proteins in eukaryotes is that the switches are usually much simpler. In prokaryotes they are either on or off and are usually physically close to the mechanism they affect. In contrast, in eukaryotes the committee of protein transcription factors may be quite far from the gene they influence.

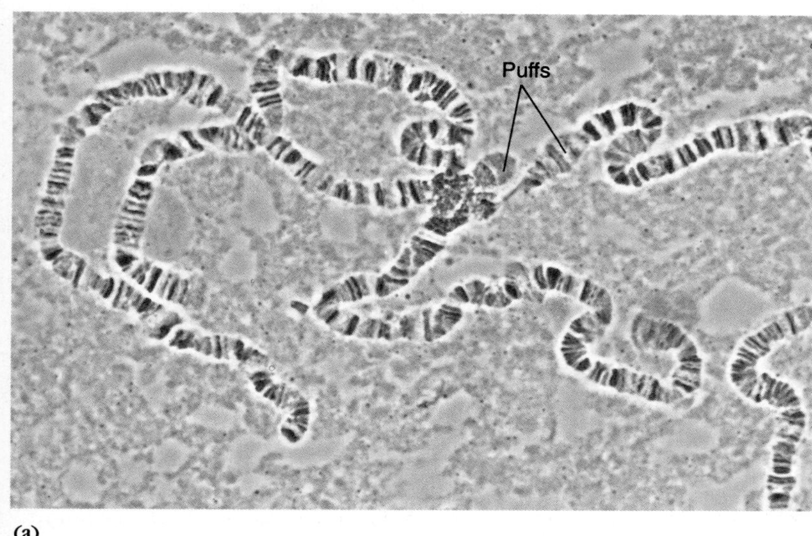

(a)

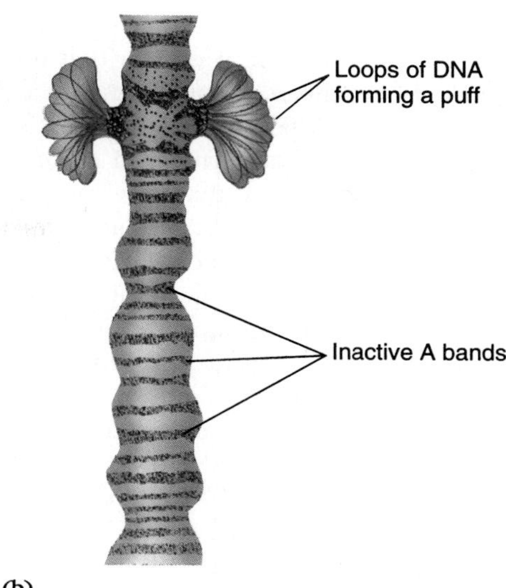

(b)

FIGURE 10-10

Active segments of DNA. (a) Chromosomes from cells in a salamander show enlarged puffs where DNA is being expressed. **(b)** The many loops of DNA undergoing transcription form the chromosomal puffs. (Biological Photo Service)

Control in Eukaryotes

In eukaryotes, once a gene has been switched on, it tends to remain that way for a long time, whether it is a housekeeping gene or a specialist gene. Once a eukaryotic cell has differentiated into, say, a liver cell, its descendants remain liver cells, with the same specialist genes switched on, throughout the many cell divisions that occur during the animal's life.

Eukaryotes produce gene-regulatory proteins that work in the same way as those of prokaryotes, but, as you might predict, regulation in eukaryotes has some additional complications. For one thing, transcription of many genes is also affected by a DNA sequence that is not physically close to the gene and its promoter. This distant sequence, called an **enhancer,** is the binding site for other, different gene-regulatory proteins. Many enhancers can bind more than one gene-regulatory protein at a time, and the rate of transcription depends on which regulatory proteins are bound. This permits precise control over the rate of protein synthesis (*A Journey into Science in Process:* Smart Genes).

An example that may illuminate how eukaryotic genes are regulated comes from studies of humans that were genetically male but developed into females (a process that involves the production of many different proteins). These individuals turned out to lack a single master gene regulatory protein, the receptor for the male hormone testosterone. In a normal male, testosterone binds with this receptor in the cytoplasm. The hormone-plus-receptor combination enters the nucleus, where it binds to the appropriate enhancer and controls transcription of the many genes that make an individual male rather than female.

Gene-regulatory proteins exert precise control over protein synthesis, but other, more general controls seem to determine whether regulatory proteins can reach the DNA in the first place. Control of transcription in eukaryotes seems to be a two-stage process. Before a gene can be expressed, the structure of the chromosome must be altered. Then gene regulatory proteins can bind to the DNA and turn transcription on and off (Figure 10-9b).

Changes in Chromosome Structure

Chromosomes are made up of a combination of DNA and protein called **chromatin** (Section 9-F). Tightly condensed (coiled up) chromatin is inactive, probably because it is so tightly coiled that polymerase enzymes cannot move along the DNA. Looser areas of chromatin contain areas of DNA that are being transcribed (Figure 10-10). The histones that hold nucleosomes together do not prevent transcription, which can go on right through the nucleosome. Above the level of nucleosomes, however, it is clear that the way DNA is packed into chromosomes does affect gene activity, but we know little about how this is controlled.

Many structural changes in chromosomes happen on such a small scale that they cannot be studied with a microscope. Other changes are sufficiently large-scale to be visible. Some insect tissues grow by an increase in cell size rather than in number of cells. As these cells grow, their chromosomes replicate up to 10 times, to produce giant **polytene chromosomes** that contain hundreds of strands of DNA. Each polytene chromosome has a banded structure. Sometimes the bands appear as "puffs" where the DNA is partly unraveled (Figure 10–10b). For instance, in the midge *Chironomus* (a small fly), four salivary gland cells produce granules while the rest of the cells do not. Only the cells that produce granules have a puff at one end of one chromosome. We now know that puffs represent active genes. They are regions where the structure of a chromosome has changed, permitting RNA to be transcribed.

In bacteria, a small amount of RNA is transcribed from genes that are supposedly turned off. These genes are said to be "leaky." Higher eukaryotes have a mechanism that seems to reduce the leakiness of their genes: **methylation,** the addition of a methyl group (CH_3) to some of the cytosine bases in DNA. Later, as DNA is replicated, a methylating enzyme detects methyl groups on the old DNA strand and attaches methyl groups to its new partner. In this way cells pass down the methylation of genes from one generation to the next, and genes that have been turned off during embryonic development stay turned off. This mechanism would probably not be worthwhile for a bacterium that lives for only a few hours. However, in a vertebrate's much longer life, it saves a lot of energy if cells are prevented from making proteins they do not need.

Protein synthesis is usually controlled by turning transcription on or off. This occurs by either binding or preventing the binding of gene regulatory proteins at specific DNA sites. In prokaryotes, these DNA sites are near the gene's promoter. In eukaryotes, there appear to be two steps in transcription: (a) the chromatin structure loosens, as the DNA coiling is partially unwound, and (b) as in prokaryotes, regulatory substances interact with gene regulatory proteins to switch genes on and off. These gene regulatory proteins may control specific genes, by way of their promoters, or groups of genes, by affecting their enhancers. In eukaryotes, genes can be securely turned off through methylation.

10–F CONTROL OF GENE ACTIVITY DURING DEVELOPMENT

Genes are switched on and off throughout life, but the most dramatic examples occur during development of an embryo. Most multicellular organisms start life as single cells, endowed with a particular set of genetic information, from which the walking, talking, or photosynthetic, flowering, adult develops (Chapter opening photo).

As development unfolds, genetically identical cells descended from the original cell **differentiate,** that is, become distinct from one another, as different genes are switched on and off in different cells. We can view development as a precisely coordinated series of events. Each gene is switched on in its turn, and then something resulting from the expression of that gene, or some environmental influence, switches on the next gene in the program.

Embryonic development is complex, with many different things happening at once, and this makes it difficult to study. Clues as to how differentiation occurs in an embryo have come from the study of less complex systems. We will consider only one example here.

Metamorphosis

Metamorphosis is a more-or-less abrupt alteration in an animal's anatomy and physiology as it changes from a larva, such as a caterpillar, maggot, or tadpole, into a very different adult (Figure 10–11).

The biological significance of metamorphosis is that it permits the young stage of an animal to have a way of life very different from the adult's, reducing competition for food and other resources. The change from one form to another must be abrupt when the intermediate between the two stages is not adapted to either way of life. The adult stage usually is the one that eventually becomes sexually mature.

Metamorphosis in most amphibians (frogs and most salamanders) reflects their incomplete adaptation to life on land. Although most adult amphibians are terrestrial, nearly all must return to water to deposit their eggs. Tadpole larvae typically eat aquatic plants, while the transformed adults are carnivores, who usu-

BIO-BIT

Some salamanders never undergo metamorphosis and instead become sexually mature in their larval form.

Table 10–2
Changes That Occur During Frog Metamorphosis

Function	Aquatic Larva	Terrestrial Adult
Locomotion	Tail with fins	Legs, no tail
Obtaining O_2	External gills	Lungs and skin
Transporting O_2	Larval hemoglobin	Hemoglobin has different amino acid sequence and properties
Feeding	Sucking mouth	Big mouth with jaws, sticky tongue
Digestion	Long coiled intestine for digesting algae	Digestive tract short; insect food easily digested
Sensing environment	Small eyes; lateral line organ (row of pressure-sensitive pits along side of body)	Huge eyes, with different visual pigment; ears
Excretion	Ammonia (NH_3)	Urea ($H_2N—CO—NH_2$)

ally eat insects and worms. Salamander larvae and adults are both carnivorous, but larvae eat aquatic invertebrates, while adults eat mostly terrestrial invertebrates. Table 10–2 lists the main changes that occur as a result of metamorphosis.

All of these changes result from changes in gene activity during metamorphosis. They are brought about by the hormone **thyroxin,** secreted by the thyroid gland. This hormone must be present for metamorphosis to occur. Removing the thyroid gland prevents a tadpole from going through metamorphosis. Thyroxin acts directly on tissues that change during metamorphosis. If a tadpole's tail is removed and placed in a bath containing thyroxin, the white blood cells in the tail will digest it so that it gets smaller, just as they would in the intact tadpole. In the control experiment, where the bath contains no thyroxin, the tail remains unchanged. ■

(a)

(b)

FIGURE 10–11

Metamorphosis in a salamander. (a) A larval spotted salamander underwater. The large filamentous gills just behind the head are the primary sites of gas exchange. **(b)** This adult spotted salamander has undergone metamorphosis, lives on land, and now uses lungs for most of its gas exchange. (R. W. VanDevender)

> **BIO-BIT**
>
> Ethanol (the alcohol humans consume) is a strong carcinogen. It is no surprise, then, that alcoholics have much higher rates of cancers of the throat, esophagus, stomach, and pancreas.

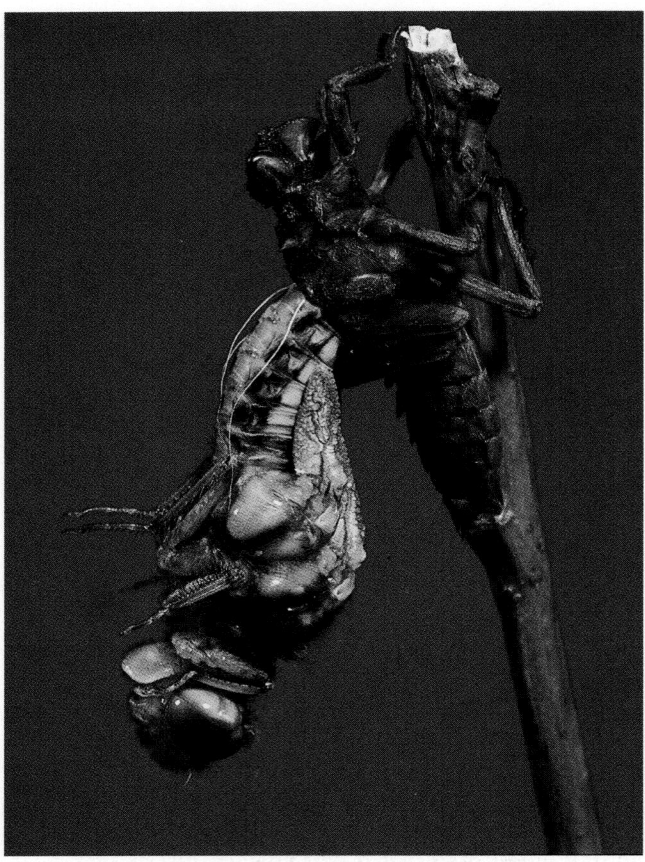

FIGURE 10–12

Metamorphosis in an insect. A dragonfly emerging from its old exoskeleton as it undergoes its final molt. (Hans Pfletschinger/Peter Arnold Inc.)

Metamorphosis in insects is also controlled by hormones. In this case, the presence of **juvenile hormone** prevents metamorphosis into the adult animal (Figure 10–12). So in insects, synthesis of a hormone must be turned off, rather than on, to produce the change from larva to adult.

As an embryo develops, its cells begin to differentiate and assume more specific roles as various genes are selectively turned on or off. The presence or absence of some hormones such as thyroxin or juvenile hormone can dramatically affect development.

10–G CANCER

Most multicellular organisms replace lost or worn-out cells by division of remaining healthy cells. Normally, the division of these cells is under strict control. New cells are produced only when and where they are needed. However, these controls sometimes go awry.

A **tumor** is a clump of cells that grows and divides abnormally. Some tumors are harmless, like the common wart or fibroid cysts of the uterus. Other kinds of tumors may eventually become **malignant,** which means that (1) their cells divide without the normal restraints, and (2) they invade and destroy healthy tissues, often in distant parts of the body. A "cancer" is a malignant tumor.

A cancer starts when a single cell is transformed into a cancer cell by undergoing mutations in some of the genes that control cell division. This cell then divides repeatedly, producing a **clone** of genetically identical cells, which form a tumor. Later, some of these cells undergo further changes and become malignant. Cancer cells may **metastasize,** that is, they detach from their neighbors, travel to other parts of the body in the bloodstream, and start new tumors.

How is a normal cell transformed into a cancerous one? The control of cell division may be compared to a high-security system with multiple safeguards. If one part fails, others will still keep the cell from dividing improperly. Hence, it takes more than one change—three to seven, according to some studies—for a cell to escape from the controls on division and become cancerous. A cancer therefore develops in steps, which are not necessarily the same from one case of cancer to another, even for the same kind of cancer.

The first step in development of a cancer is a mutation, which makes a cell abnormal and which is passed on to the cell's descendants. Later, additional changes may occur in some cells of this lineage and also be passed on. Eventually a cell bearing accumulated changes sustains one more that transforms it to a cancer cell. This may take years.

3? What causes cancers? You can see that many different factors may contribute to a single case of cancer, and some of them may be long gone by the time the cancer appears. The traditional way to probe the causes of cancers is by correlation studies: comparing a group of cancer victims with a matched group of healthy people to see what factors are more common in the background of cancer victims. As long ago as 1775, Percival Potts noted that chimney sweeps suffered from a high incidence of cancer of the scrotum.

(a)

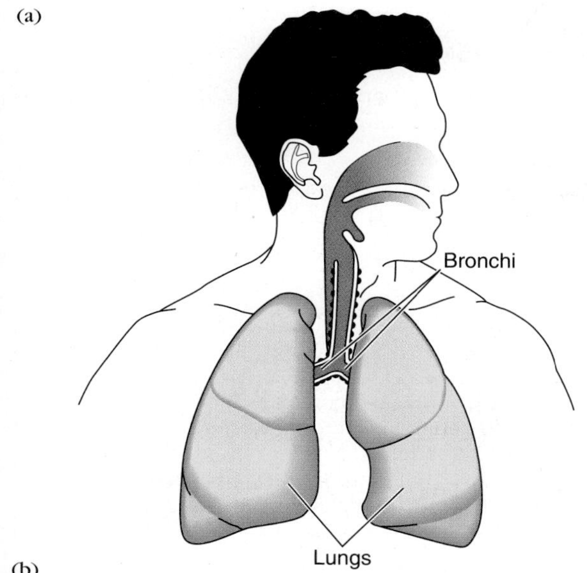

(b)

FIGURE 10–13

Cancer. (a) Diagram of the human respiratory tract. **(b)** False color electron micrograph of cancerous and normal cells that line the bronchi. Normal cells (orange) are covered with cilia, which move mucus that contains trapped dust particles away from the lungs. Cancerous cells (green) have lost their cilia but are covered with numerous shorter finger-like microvilli. (Lennart Nilsson/Boehringer Ingelheim International, GmbH)

In the 1960s, cigarette smoking was linked to most cases of lung cancer and it is now known to increase the risk of several other cancers (Figure 10–13). Some cancers occur only in the presence of particular viruses. Some tend to run in families, suggesting a genetic predisposition: an inherited mutation puts all of the body's cells one step closer to cancer from the time of conception. ■

A **carcinogen** is a factor that increases the risk of cancer (Table 10–3). Most carcinogens are **mutagens,** agents that cause genetic mutations. However, several cancers are linked to substances that are **carcinogenic** (cancer-causing) without being mutagenic. These apparently contribute to cancer by promoting the expression of already existing mutations in genes that control cell division.

Because mutations are largely governed by chance, each case of cancer is genetically unique. However, mutations of certain genes tend to be common in some kinds of cancer.

The mutated genes found in cancerous cells belong to two groups. One group, for lack of a better name, is called **tumor suppressor genes.** Normally these genes code for proteins that inhibit cell division. When these genes mutate, the corresponding protein may no longer be produced correctly, and cell division may occur when it should not.

The other, and better known, group of genes contains the **oncogenes,** genes with the potential to cause cancer (onco = cancer). Oncogenes arise by mutation of normal cellular genes, called **proto-oncogenes,** which control the production of proteins that stimulate growth and cell division. A proto-oncogene is converted to an oncogene by a mutation that causes the gene to be overexpressed. Often such a mutation does not change the gene itself but produces extra copies of the gene, or brings the gene under the control of a regulatory gene that increases its expression. Overexpression means that too much of the gene's protein is made, and so the cell divides more rapidly than usual and forms a tumor. About 60 proto-oncogenes are known, including genes that code for growth hormones and for dozens of other proteins that control cell division.

Viruses are one of the many factors that can contribute to the development of cancer by causing mutations. They may insert their genetic material into the chromosomes of host cells or move host genes to different chromosomal locations. The first human cancer shown to be caused directly by a virus was a rare form of leukemia, caused by HTLV-1, a member of the retrovirus group, which also includes the AIDS virus.

After joining the host genome, some viruses enter a **latent** period and are not expressed until activated

BIO-BIT

Skin cancer is the most common form of cancer. Over 90% of all skin cancers occur on parts of the skin that are regularly exposed to the sun's ultraviolet radiation.

Table 10–3
Some Known Carcinogens

Carcinogen	Comments
Asbestos dust Chromium compounds Some petroleum products	Workers in these industries have high risk of lung cancer
Tobacco	12–15% of cigarette smokers die of lung cancer. Especially risky when combined with exposure to asbestos or radon
Estrogen	Mammalian hormone. In large amounts, can contribute to uterine and breast cancer
X-rays	Many people have unnecessary medical x-rays
Benzene	Until recently, was a common solvent in labs
Nitrates and nitrites	Converted into carcinogenic nitrosamines in digestive tract. Common as food preservatives and in most green vegetables. Also common water pollutants in agricultural areas, where nitrogen fertilizers run off farmland
Aflatoxins	Produced by fungus *Aspergillus flavus* when growing on food. First found in peanuts. Causes liver cancer
Vinyl chloride	Causes one type of liver cancer

days or years later. For instance, the virus that causes hepatitis B may cause liver cancer 20 or 30 years after the attack of hepatitis. Papilloma virus infection, which causes harmless warts, is also necessary for development of cancer of the cervix (the neck of the uterus). But most women infected with papilloma viruses will never develop cervical cancer. For one thing, the virus will not cause cancer without additional (unknown) factors; for another, the disease-fighting immune system probably destroys most cancerous cells before they form sizable tumors (Chapter 25).

A potential cancer may be stopped by surgical removal of the tumor while it is still a locally contained mass of cells. Once the tumor metastasizes, it is very difficult to locate and destroy all the cells that may be able to cause tumors throughout the body. Researchers are studying the possibility of using various molecules, or antibodies or cells of the immune system, to search out and destroy cancerous cells selectively. They are also identifying some environmental factors that increase the chances of developing a tumor in the first place and they are urging changes in workplaces and homes, and in habits, so that people can avoid these factors.

Cancers are the second most common cause of death in the United States, accounting for 20% of all deaths. This figure scares many people but, to put it into perspective, note that an American has almost as great a chance of dying of homicide as of cancer. Before the age of 45, homicide is a much more likely cause of death (although perhaps this is not a particularly comforting thought!). Deaths from cancer have increased in the twentieth century, mainly because cancers tend to develop later in life, and people are living longer instead of dying of infectious bacterial diseases in early life as they used to. (Life expectancy in the United States is now nearly 80 years, compared with 45 years in 1900.) There has also been an increase in cancer because carcinogens are becoming more common in our environment. For example, workers in vinyl chloride factories have an increased chance of dying of one form of liver cancer. And cancer of the prostate is much more common in areas with acute air pollution than in other areas.

A cancer arises from a single cell that has accumulated mutations in the genes governing cell division. Mutations are known to be caused by radiation, viruses, and mutagenic chemicals. A tumor is a clump of cells that does not divide or grow normally. Tumors become cancerous when their cells divide more frequently, spread to other parts of the body, and invade and destroy healthy tissues. We can reduce our risk of death from cancer by reducing our exposure to mutagens and detecting cancers early in their development.

BIO-BIT

Smoking is responsible for nearly 30% of all cancer deaths.

A Journey into Science in Process

Gene Therapy

In the 1970s we met David, the Boy In the Bubble, a seven-year old afflicted with an immune system so debilitated that he could not survive normal contact with other humans and had to live isolated within a plastic bubble that protected him from most, but unfortunately not all, disease causing organisms (Figure 10-A). David suffered from SCID (severe combined immunodeficiency), a rare condition that has no current cure, although we do know much more about the biochemistry of this genetic abnormality than we did 20 years ago. A quarter of the children with SCID have a defect in the gene that makes the enzyme adenosine deaminase. Lack of this enzyme leaves them defenseless against cold viruses, bacteria, fungal spores, and the thousands of other common pathogens that a healthy immune system recognizes and destroys. Even a mother's hug is potentially deadly to a child like David who has SCID.

Although cases of this genetic immune dysfunction are rare, it is the first genetic disorder to be combated with gene therapy: replacement of a non-functional gene with a healthy one. Clinical tests of a treatment that introduces the gene that makes the missing enzyme began in 1990. Not only does this give hope to other children suffering from SCID, but it signals the beginning of a major revolution in medicine. By the time you read this, preliminary results of the treatment for SCID should be available and with knowledge gained from the Human Genome Project (see *A Journey into Science in Process:* The Human Genome Project, Chapter 11) biologists will have learned much more about how to repair non-functioning genes.

Until now disease has been fought on two levels: surgery that physically removes or repairs damaged tissue, and pharmacology that administers drugs to repair, boost, or alter cellular mechanisms, and kill pathogens. Gene therapy adds a completely new level to the medical arsenal: as means to adjust the DNA of a cell and foster the production of normal, healthy proteins in cases where a deficient gene has resulted in a lack of proteins causing suffering, disease, and death.

FIGURE 10-A

The boy in the bubble. Gene therapy may eliminate 25% of the cases of SCID. (Peter Arnold, Inc.)

Gene therapy is a controversial technology. Once biologists learn how to repair and tailor genes, some argue that it is conceivable that this technology may do us more harm than good. For children like David, and the hundreds of thousands suffering from other diseases caused by defective genes, there is hope that gene therapy will enable them to live a normal life, but you should understand the ethical questions involved.

Implications of Gene Therapy

Few would question the worth of saving the lives of millions of children and, perhaps if gene therapy had no further applications, the only questions would revolve around the commitment of so much research money to cure diseases that affect a relatively small number of humans. But perhaps the most controversial aspect of gene therapy is the possibility of adding novel properties to cells that will enhance their natural abilities. For example, it will soon be possible to manufacture lymphocytes that not only can recognize, destroy, and engulf invading cells, but also lymphocytes that can secrete tumor-killing substances.

One of the major rules that ecology and our current environmental crises have taught us is that it is impossible to do just one thing. While it is laudable to attempt to cure the children who will die from genetic disorders, there may be other consequences of the technology to consider. Once we begin to change the human genome, we open a universe of options. At present only nonreproductive, body cells are proposed as candidates for gene therapy, but in the future, it will be possible to produce cultures of therapeutically altered human reproductive cells. Thus, we may be able to culture sperm and eggs that are not only free from genetic defects like hemophilia, thalassemia, sickle-cell anemia, cystic fibrosis, and SCID, but manipulate the genes of these cultured reproductive cells to allow artificial selection of sex and physical characteristics of a human child. It may become possible to eliminate many of the less-than-perfect aspects of humans, and domestic animals and plants may be altered as well. Even as a beginning biologist, you will see the possibilities inherent in this new technology, as well as the ethical problems it poses.

What are the advantages of the *imperfect* human genome? What risks do we run when we begin to tamper with it? Who should get gene therapy? Who *will* get gene therapy? How can we spend billions of dollars on a technology that will help only a small number of children who are destined to die anyway, when hundreds of millions of medically less-interesting people die each year from malnutrition, malaria, schistosomiasis, and sleeping sickness? What are the biological and ethical differences between altering the genome of a somatic cell as compared with altering the genome of a germ cell? These are only a few of the questions that we will grapple with in the decades to come as gene therapy moves out of the laboratory and into use in medical practice.

SUMMARY

1. DNA carries the genetic information that determines the order in which amino acids must be joined to produce proteins.
2. RNA is transcribed from the cell's DNA and thus it receives a complementary nucleotide base sequence.
3. The three main types of RNA in a cell are:
   a. messenger RNA (mRNA), whose base sequence is translated into the sequence of amino acids in a polypeptide,
   b. ribosomal RNA (rRNA), which makes up part of the structure of ribosomes, and
   c. transfer RNA (tRNA), which carries amino acids to the ribosome for protein synthesis and brings them into their proper position to be joined to the polypeptide chain.
4. A sequence of three nucleotides in mRNA codes for each amino acid. This genetic code is degenerate, in that most amino acids are coded for by more than one triplet combination. An initiation codon signals the beginning of an amino acid sequence. Mutations in DNA are also transcribed into RNA and may change the protein produced.
5. During protein synthesis, the code carried by the sequence of nucleotide bases in mRNA is translated into the sequence of amino acids. The mRNA attaches to a ribosome, and tRNAs carrying amino acids attach to the mRNA-ribosomal complex by means of base-pairing between the mRNA codons and the tRNA anticodons. Each tRNA in turn donates its amino acid to the growing polypeptide chain, until a *STOP* codon reaches the ribosome, and the completed polypeptide chain is released.
6. Protein synthesis is usually controlled by turning transcription on or off. In prokaryotes, gene-regulatory proteins bind or unbind DNA near the gene's promoter and start and stop transcription, thus regulating protein synthesis. The presence of food molecules may control the ability of gene-regulatory proteins to bind to DNA.
7. Transcription in eukaryotes is thought to involve two steps:
   a. The structure of part of the chromatin changes, by loosening of the tightly coiled DNA.
   b. As in prokaryotes, regulatory substances interact with gene-regulatory proteins to switch genes on and off. These gene-regulatory proteins may control specific genes, by way of their promoters, or groups of genes, by affecting their enhancers.
8. As eukaryotic cells differentiate during development, gene-regulatory proteins turn off genes they will not need, but like a damaged faucet, this control is leaky. Later, these genes may be turned off more securely by being methylated.
9. Various signals inside and outside an organism cause some of the changes in gene activity that make up differentiation. Amphibian and insect metamorphosis are convenient, nonembryonic systems for studying differentiation.
10. Cancers arise by the progressive accumulation of mutations in genes controlling cell division. These mutations may be brought about by viruses, radiation, or mutagenic chemicals. A tumor arises from a single mutated cell that has escaped the controls on division. Some tumors become malignant and metastasize, detaching from their tumor neighbors, invading healthy tissues elsewhere in the body, and starting new tumors there.

SELF-QUIZ

1. Using the base-pairing rules, fill in the mRNA sequence that would be transcribed from the following strand of DNA. Then, use Table 10-1 to determine the amino acid sequence that would be translated when the mRNA combines with a ribosome:
   T—A—C—A—A—G—T—A—C—T—T—G—T—T—T—C—T—T
   mRNA ______________________________
   amino acids ______________________________
2. Suppose the two guanine (G) nucleotides in Question 1 were changed to cytosine (C) nucleotides. How would this mutation affect the amino acid sequence translated from the mRNA?
3. Suppose the G nucleotides were removed from the DNA in Question 1. How would this mutation affect the amino acid sequence translated from the mRNA?
4. List three differences between the structures of DNA and RNA.
5. According to current ideas concerning protein synthesis:
   a. transfer RNA molecules specific for particular amino acids are synthesized along a messenger RNA template in the cytoplasm
   b. amino acids line up with their mRNA codons on the ribosome and are then linked together by transfer RNA
   c. enzymes that catalyze protein-synthesizing reactions in the cytoplasm are transcribed from regulatory genes
   d. transfer RNA molecules transport mRNA from the nucleus to the ribosomes
   e. messenger RNA, synthesized on a DNA template in the nucleus, provides information that determines the sequence in which amino acids are linked during translation
6. Transfer RNA is synthesized:
   a. on a DNA template

b. from a messenger RNA template on a ribosome
c. on ribosomes without a template
d. in the nucleolus by the interaction of messenger RNA and chromosomal DNA

7. During differentiation, cells with the same DNA:
   a. must develop similarly
   b. divide at equal rates
   c. contain different genes
   d. may transcribe different genes
8. How do food molecules induce prokaryotic cells to make enzymes that metabolize the food? Food molecules:
   a. cause an inducer to bind to the DNA and attract RNA polymerase, which transcribes mRNA coding for the needed enzymes
   b. cause repressor proteins to leave the DNA, which can then be transcribed to mRNA
   c. bind to tRNA, which carries them to the ribosome for protein synthesis
   d. bind to RNA polymerase enzymes and activate them
   e. change the structure of the DNA so that regulatory substances can bind to it

THINKING CRITICALLY

1. Why is it important for each type of tRNA to have its own type of enzyme to bind it to an amino acid?
2. Suppose a cell's DNA contained a mutation that changed one of the nucleotides in an anticodon of tRNA. How might this mutation affect protein synthesis?
3. We have seen why the genetic code could not consist of codons with fewer than three nucleotides each. What factors might have selected against codons of more than three nucleotides?
4. Even though some 20% of Americans die of cancer, life expectancy in developed countries is about 80 years. Would the enormous sums spent on cancer research, therefore, be better spent on diseases such as AIDS that kill most of their victims when they are much younger, or diseases such as Alzheimer's that damage the quality of the patients' (and their relatives') lives for much longer?
5. Why are treatments that prevent viruses from reproducing not likely to cure many types of cancer caused by viruses?

SELECTED KEY TERMS

anticodon, *p. 187*
codon, *p. 182*
differentiate, *p. 194*
enhancer, *p. 193*
exon, *p. 185*
frameshift mutation, *p. 184*
gene regulatory protein, *p. 190*
inducer, *p. 190*
intron, *p. 185*
malignant, *p. 196*
messenger RNA, *p. 181*
metamorphosis, *p. 195*
metastasize, *p. 196*
methylation, *p. 194*
mutagen, *p. 197*
oncogene, *p. 197*
promoter, *p. 181*
ribosomal RNA, *p. 181*
transcription, *p. 181*
transfer RNA, *p. 181*
translation, *p. 182*

SUGGESTED READINGS

Alberts, B., et al. *Molecular Biology of the Cell.* New York: Garland Publishing, Inc., 1989. Chapters 5, 10, and 21 provide clear, modern treatment of the topics of protein synthesis, control of gene expression, and cancer.

Browder, L. W. *Developmental Biology,* 3rd ed. Philadelphia: Saunders College Publishing, 1991. An embryology textbook with a good section on differentiation.

DeRobertis, E. M., et al. "Homeobox genes and the vertebrate body plan." *Scientific American,* July 1990.

Halliday, R. "A different kind of inheritance." *Scientific American,* June 1989. How the addition of methyl groups to DNA may affect gene expression during development.

Hunter, T. "The proteins of oncogenes." *Scientific American,* August 1984. How proteins encoded by oncogenes make cells containing them cancerous.

Liotta, L. A. "Cancer cell invasion and metastasis." *Scientific American,* February 1992.

Ptashne, M. "How gene activators work." *Scientific American,* January 1989. The action of gene-regulatory proteins.

Ross, J. "The turnover of messenger RNA." *Scientific American,* April 1989.

Weinberg, R. A. "Finding the anti-oncogene." *Scientific American,* September 1988. Mutation of a gene that normally suppresses the development of cancer can increase susceptibility to cancer of the eye.

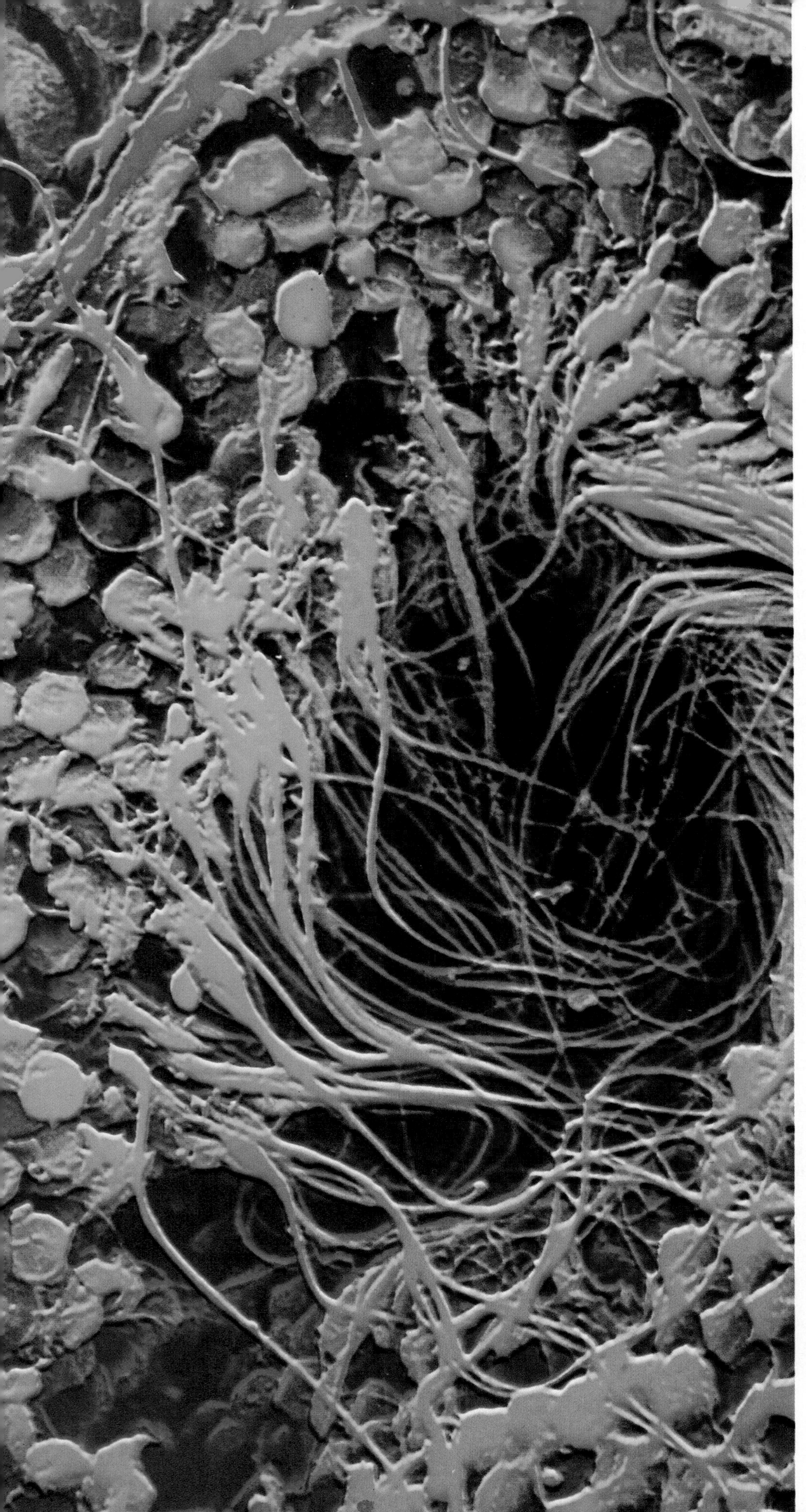

CHAPTER 11

CONCEPT GUIDE

After reading this chapter, you should be able to:

1. Distinguish between each of the following pairs of terms: (a) sex chromosomes and autosomes, (b) diploid and haploid, (c) germ cells and somatic cells, and (d) haploid and tetraploid. (Section 11–B)
2. Describe the major events that take place in G_1, S, and G_2 of the cell cycle. (Section 11–C)
3. Diagram the division of a single pair of homologous chromosomes by mitosis and by meiosis. Explain how mitosis and meiosis are similar and how they are different. (Sections 11–A, 11–D, and 11–F)
4. Compare cytokinesis in plant and animal cells. Explain the significance of these differences. (Section 11–E)
5. Describe two genetic mechanisms that occur during meiosis and result in increased genetic diversity. (Section 11–G)
6. Compare the size and number of gametes produced in spermatogenesis and oogenesis. Explain why these differences exist. (Section 11–H)

Sexually mature human males produce millions of sperm cells every day using meiosis. In this cross-section of a seminiferous tubule within a human testis, we see the many round cells that divide to produce the sperm in the center of the tube. Learn more about how cells reproduce in Sections 11–D, 11–E, and 11–F. (Secchi, Lecaque, Roussel, Uclaf, CNRI/Science Source/Photo Researchers, Inc.)

Reproduction of Eukaryotic Cells

Life is handed down from one generation of organisms to the next in the form of new cells. This legacy of new cells is made by division of existing cells and there are two ways that this is accomplished. In the first process, an existing cell replicates all of its structures and divides into two new cells. Unicellular organisms, like *Amoeba,* or yeast make new cells by dividing in two in this way. A multicellular organism, like you, your dog, or its fleas, begins life as a single cell—the fertilized egg. Subsequent bouts of cell division produce the many cells of the body. In multicellular organisms we see the second method of transmitting a legacy of new cells when eventually, some cells in the reproductive organs divide using a different process that forms reproductive cells. These generally have half of the original cell's genetic legacy and carry that legacy to the next generation of humans, dogs, or fleas.

When a cell divides, it must pass on the genetic information needed to specify the kinds of proteins it can produce. In a eukaryotic cell, the DNA carrying this genetic information is divided among the many chromosomes in the nucleus. Before the cell divides, its chromosomes must be duplicated and then distributed precisely into two new nuclei, so that each receives a complete set of chromosomes. Each new cell must inherit not only a nucleus containing all the genetic information it will need, but also the cytoplasmic components required to express the genetic information, such as ribosomes to make proteins and mitochondria to supply the necessary energy.

The most complicated part of cell division is nuclear division. Various cell components must interact in such a way that they divide the replicated chromosomes accurately into two complete sets.

KEY CONCEPTS

- A cell reproduces by dividing into two new cells.
- Before a eukaryotic cell divides, its chromosomes are distributed precisely into two new nuclei.

Curiosity Questions

1? What types of cells in our bodies do not divide? (See page 210)

2? How can parents produce children who do not look very much like each other or like their parents? (See page 218)

3? Why are eggs so much larger than sperm? (See page 221)

11–A MITOSIS AND MEIOSIS: AN OVERVIEW

The two main types of nuclear division are mitosis and meiosis (Figure 11–1). **Mitosis** produces two new nuclei with the same number of chromosomes as in the original nucleus. This ensures that each new cell inherits a complete set of the parent cell's genetic information. In unicellular organisms, mitosis produces two genetically identical new individuals.

Embryos grow using mitosis, and mitosis occurs all the time in our bodies, as new cells replace old ones—such as worn-out blood cells or skin cells injured by cuts or burns.

Meiosis is the type of nuclear division that produces new combinations of chromosomes and genes, packaged in nuclei containing only half the number of chromosomes found in the original nucleus. Meiosis is associated with reproduction. In animals, it gives rise to the **gametes**—the sexual reproductive cells—sperm and eggs (Figure 11–1). In plants, meiosis produces **spores,** reproductive cells that are asexual (a = not). Spores divide and produce structures that give rise to the sexual gametes (Section 20–D).

Meiosis is vital to any organism that reproduces sexually. Without meiosis, the gametes would contain as many chromosomes as the other cells of the parent, and the fertilized egg would contain twice as many. Hence the number of chromosomes would double in each generation.

Before considering the reproduction of eukaryotic cells, we must learn more about how their chromosomes are organized.

Mitosis produces two new nuclei that both contain the same genetic information as could be found in the original nucleus. Meiosis produces new nuclei with only half the number of chromosomes found in the original nucleus, and with new genetic combinations.

FIGURE 11–1

Mitosis versus meiosis. The diploid number for the cell shown here is four. Mitosis produces two new cells that are genetically similar to the original cell. Meiosis produces four cells, each with half as much genetic information as the original cell. As a result of meiosis, sperm and egg cells contain only half as many chromosomes as other cells in the plant or animal's body. During fertilization a sperm fuses with an egg, producing a cell that contains the same number of chromosomes as in the normal body cells of the species. This new cell divides by mitosis to give rise to generations of cells that form the new organism.

11–B EUKARYOTIC CHROMOSOMES

Viewed through a microscope, a cell's chromosomes usually appear as a single, diffuse mass. Individual chromosomes can be distinguished only right before and during cell division, when they **condense**—that is, coil up tightly into short, thread-like structures (Section 9–F). By this time, the chromosomes have already been replicated as described in Section 9–C. Each consists of two copies, attached at a region called the **centromere.** As long as the two copies remain attached, they are called **sister chromatids** (Figure 11–2). When they ultimately separate, they are called chromosomes again.

Haploid and Diploid Chromosome Numbers

In most higher plants and animals, chromosomes from the body cells can be matched up in pairs. The two chromosomes of a pair are called **homologous chromosomes,** or simply **homologues.** Most homologous chromosomes look alike: they are the same length, their centromeres are in the same position, they show the same pattern of light and dark bands when stained, and they carry genes for the same inherited characteristics, lined up on the chromosome in the same order (Figure 11–2). For example, human chromosome #1 contains the genes for the Rh blood protein and for a starch-digesting enzyme in the saliva. However, the corresponding genes on the two homologues need not be identical. For instance, some chromosomes have a gene for the protein that makes a person Rh-positive, and some have a gene coding for a different version of this protein (Rh-negative), at the Rh location (see *A Journey into Science in Process:* The Human Genome Project). The number of chromosome pairs varies from one species to another.

Humans have 46 chromosomes in most of their body cells. These can be arranged in homologous pairs according to their length and the position of the centromere. In human males, 22 of these pairs contain chromosomes that look similar, but the twenty-third pair is odd, with two unlike chromosomes, called X and Y (Figure 11–2). In the cells of a human female, both chromosomes in the twenty-third pair are X chromosomes. The X and Y chromosomes are called **sex chromosomes,** because this pair of chromosomes differs between members of each sex and because they play a role in determining their owner's sex (Section 13–F). The other 22 pairs are called **autosomal chromosomes,** or **autosomes.**

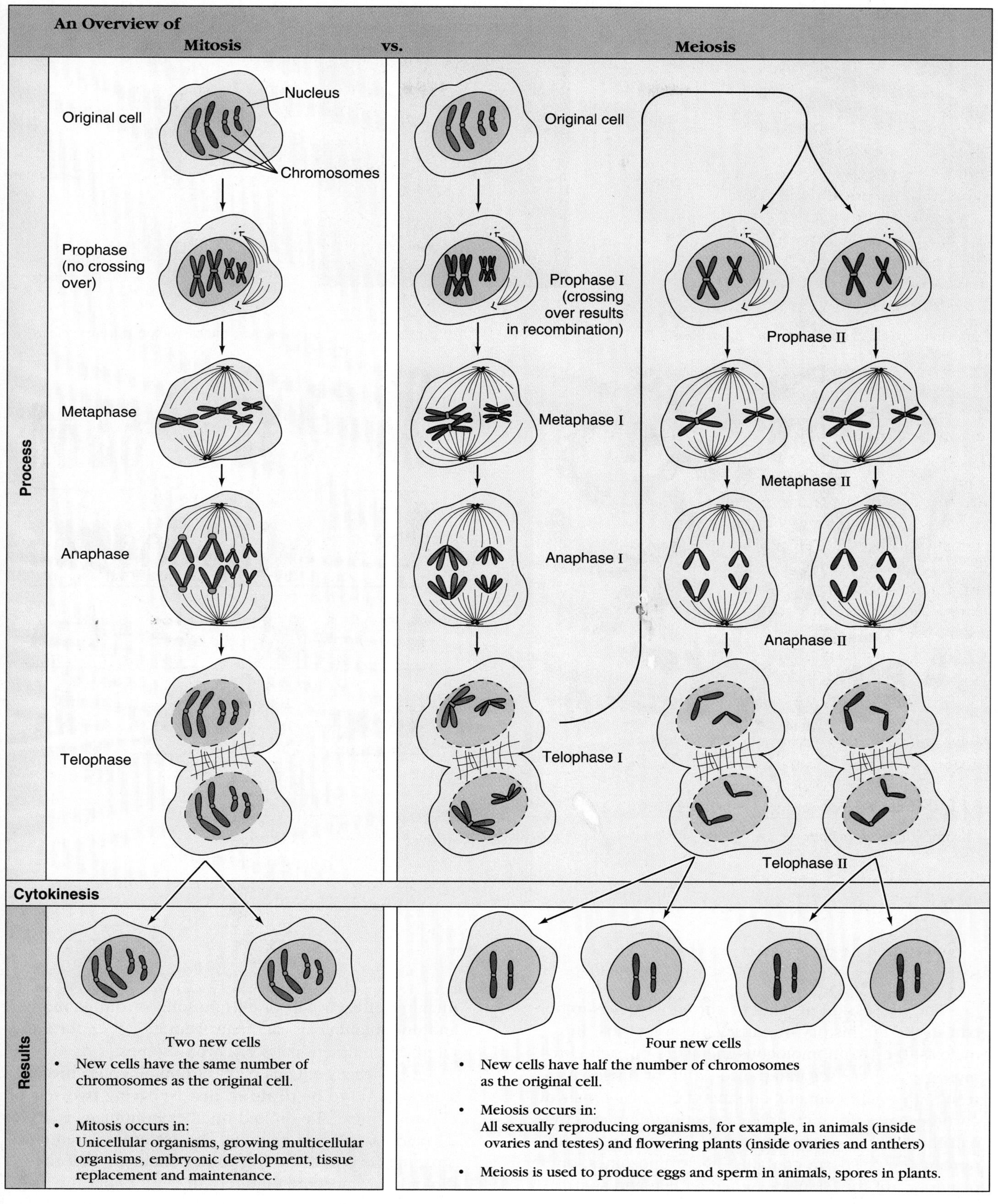

- New cells have the same number of chromosomes as the original cell.
- Mitosis occurs in:
  Unicellular organisms, growing multicellular organisms, embryonic development, tissue replacement and maintenance.

- New cells have half the number of chromosomes as the original cell.
- Meiosis occurs in:
  All sexually reproducing organisms, for example, in animals (inside ovaries and testes) and flowering plants (inside ovaries and anthers)
- Meiosis is used to produce eggs and sperm in animals, spores in plants.

(a) Diffuse chromosomes

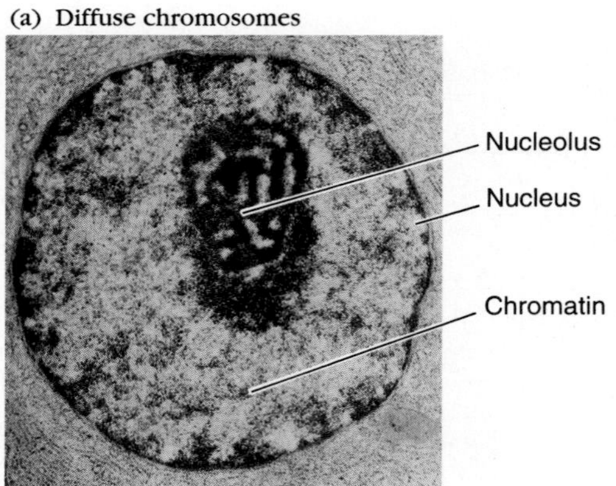

(b) Condensed chromosomes

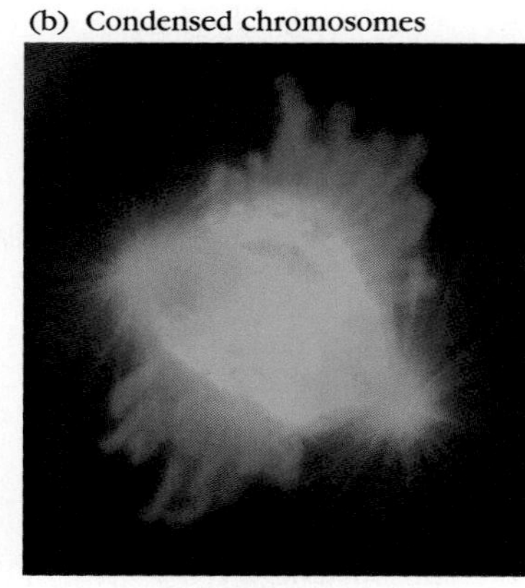

(c) Condensed and duplicated chromosome structure after DNA replication in preparation for mitosis or meiosis.

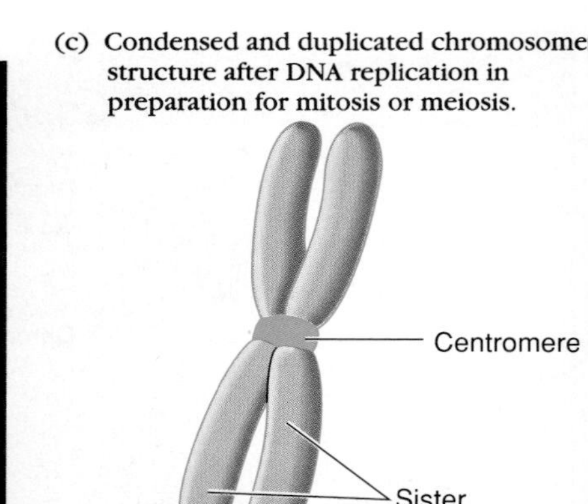

(d) Human male chromosomes during meiosis

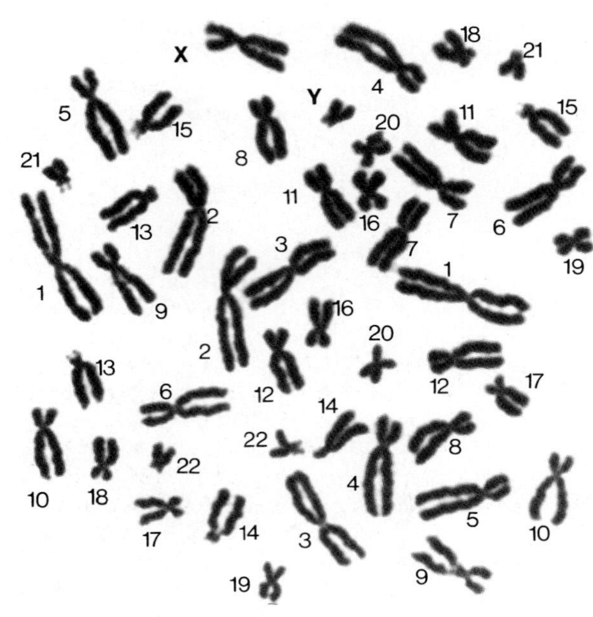

(e) Human male chromosomes, the karyotype, during meiosis

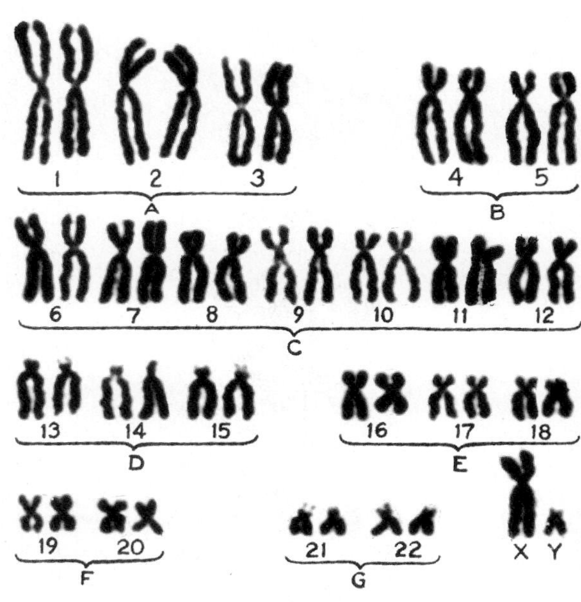

FIGURE 11–2

Chromosomes: a primer. (a, D. W. Fawcett/Visuals Unlimited; b, Rieder/BPS; d, e, Biophoto Associates)

The possession of pairs of chromosomes is important in the life history of eukaryotic organisms. In meiosis, the two homologous members of each chromosome pair are separated into different nuclei. As a result, a gamete contains one member of each pair of chromosomes, for a complete set containing exactly half the number of chromosomes. For example, each human egg or sperm contains 23 chromosomes, one from each of the 23 pairs. When an egg and sperm join at fertilization, the new individual receives one member of each pair of chromosomes from its mother, and one member of each pair from its father, for a total of 23 complete pairs of chromosomes.

A cell that contains pairs of homologous chromosomes is said to be **diploid,** that is, having two sets of chromosomes. The diploid number in humans is 46, or 23 pairs. A cell that contains one set of unpaired chromosomes is said to be **haploid,** containing half the diploid chromosome number. An egg or sperm is haploid, and in humans the haploid number of chromo-

A Journey into Science in Process

The Human Genome Project

In 1988 biology entered the arena of "big science," as the Human Genome Project got underway. Previous genetic research had been done on projects of individual interest to biologists in universities, medical schools, private and public foundations, and biomedical industries. In contrast, the Human Genome Project coordinates the efforts of many genetics labs in the U.S., England, France, Australia, South Africa, Japan, Canada, and Wales as they focus on separate facets of a single, mammoth research project: locating the position of genes in human DNA. The Human Genome Project aims to map human genes and determine the sequences of the nucleotide bases that compose them. We currently know of about 3500 genetic disorders that are caused by defective DNA. These include familiar diseases like muscular dystrophy, Huntington disease, Down syndrome, and less well-known disorders such as mental retardation caused by fragile-X syndrome. It is speculated that at least some cancers, susceptibility to cardiovascular disease, Alzheimer's disease, alcoholism, and schizophrenia may have their roots in gene defects. The hope is that the gene maps and eventually the genes sequenced by the Human Genome Project will allow defective human DNA to be repaired, eliminating these and many other diseases, turning science fiction into science fact. But the dream of routinely preventing disease by correcting errors in DNA is far in the future. Right now biologists are focusing on the first step toward this goal: making maps of all human genes.

Why Is This Research So Difficult?

In the nucleus of every cell, we humans have about 100,000 genes buried within about three feet (one meter) of tightly coiled DNA divided among 46 chromosomes. The average gene is approximately 1500 pairs of nucleotides long. As far as we know, genes seem to fall into one of two categories: structural genes that code for proteins, or regulatory genes that turn the structural genes on and off. A structural gene is characterized by two things: each produces a protein and each begins and ends with the universal start and stop codes of nucleotides. So, at first glance, it would seem to be easy to recognize structural genes—and in organisms with straightforward gene sequences or "lean" DNA, such as single-celled baker's yeast (*Saccharomyces cerevisiae*) the task is fairly simple. Each gene on yeast DNA is bracketed by start and stop codes, each gene translates into a portion of a protein code, or into one of the regulatory genes that controls it, and like local stops on a train line, each yeast gene is positioned one after the other on the DNA molecule. In contrast, human DNA is more like an express train on a very long run in that it has enormously long stretches of (currently) meaningless nucleotide bases which intervene between genes. And, to make the problem of mapping the human genome even worse, there are very few stops on this express line. *Ninety-five percent* of human DNA is made up of these mysterious, repetitive stretches of nucleotide bases. These used to be thought of as "junk DNA," but this term has fallen out of favor as we have learned that many accessory proteins and molecules associated with DNA (such as histones) have essential functions. The current view is that as our knowledge of DNA increases, we will eventually unravel the meaning of these repetitive lengths of DNA.

How Are Genes Mapped?

The initial task of investigators is to locate the five percent of the human genome that translates into genes. To do this, workers have developed DNA sleuthing techniques that give them clues as to where genes are located. Genome research is one of the hottest fields of biological research and new techniques are being devised all the time. By the time you read this, today's methods may be out-of-date, but they will still show how new science is built. Laboratory wizards survey the array of biotechnology and research techniques to create inventive, hybrid methods that fuse biochemical and computer analysis with robotics to speed the drudgery of repetitive work. Many of the present developments in genome research have capitalized on some little-known life processes of viruses, fungi, and bacteria.

For example, **restriction enzymes** are one of the main tools used to find genes. Bacteria make these highly specific enzymes to destroy DNA that has been injected by a parasitic virus. Think of restriction enzymes as molecular scalpels that cut DNA only in certain places. Just as there are scalpels with curved blades, broad blades, pointy tips, blunt ends, etc., each specialized for a single surgical technique, there are restriction enzymes that cut DNA into fragments only at specific sequences of bases (between A and G for example, or between A and A). Restriction enzymes are not fussy about whose DNA they fragment. Their advantage is that they will cut human DNA into fragments that are of smaller size, but do not alter the structure of the DNA molecules.

Gel electrophoresis is used as a molecular strainer, separating DNA fragments by size and making them available for subsequent study. This method introduces a sample containing DNA fragments into a film of starch gel and then sends an electric current through the gel. Like an ink

(continued)

Science in Process (continued)

stain bleeding onto wet fabric, the fragments of DNA move through the gel and will be sorted by size and molecular weight and electrical charge. Small molecules will move farthest from the source (being small they can move fairly rapidly through the starch gel); and larger molecules will remain closer to the source (larger molecules will move more slowly through the starch gel). Once the substances have been sorted, the gel is stained for further analysis (Figure 11-A).

Even if you've never played golf, you may know that flags mark the hole on each putting green. Most often these flags are visible from the tee, and the golfer proceeds through the course, scanning the horizon for these flags that help locate each target. Genetic **markers** are similar. They are physical or biochemical landmarks that are usually located near genes. The easiest markers for geneticists to locate are physical abnormalities of chromosomes that are visible with the microscope. For example, cri-du-chat syndrome in humans results from the loss of one arm of chromosome 5. Children with this syndrome have malformed intestines, hearts, and are often mentally retarded. The syndrome, which is estimated to occur in 1 of 50,000 live births worldwide, is named because a child who has this chromosomal abnormality has a cry that sounds like a meowing cat.

FIGURE 11-A

Gel electrophoresis showing RFLPs for DNA fingerprinting. (Philippe Plailly/Science Photo Library)

Before the Human Genome Project began, many genetic disorders had been traced to individual chromosomes, but their precise locations were unknown. One of the primary aims of each research group is to locate molecular markers—short sequences of DNA in which the same two or three letters are repeated a number of times and RFLPs (pronounced "rif-lips"—short for restriction fragment length polymorphisms), those minute variations in DNA sequences that may cause the same chromosome from two people to be cut into different-sized fragments by a particular restriction enzyme. These individual differences give each person a distinctive "DNA fingerprint."

Yacs and Maps

One of the problems with human genome research is that human DNA comes in very small amounts and bioassay techniques typically analyze DNA in quantities. One of the new hybrid techniques amplifies DNA by capitalizing upon the ability of yeast cells to incorporate foreign DNA into their own genome. In this technique human chromosomes are cut with restriction enzymes and incubated with actively growing yeast cells. Some of these incorporate human DNA into their own DNA, making a **yeast artificial chromosome (yac** for short). As each unicellular yeast cell grows and divides in two, the yacs are multiplied, and after several generations, the amount of human DNA is greatly amplified. Once there is enough DNA to analyze, yacs can be cut with restriction enzymes and the fragments studied by electrophoresis. These techniques show geneticists which

somes is 23. The haploid number of chromosomes in a species is generally designated as **N** or **1N.** In human beings N = 23, and **2N** (diploid) = 46. Table 11-1 shows chromosome numbers for several organisms.

In most animals, the fertilized egg and the cells that arise from it by mitotic division contain the diploid number of chromosomes. Some cells in the ovaries or testes eventually undergo meiosis, which produces haploid nuclei. Only the egg and sperm cells have haploid nuclei. Cells that can undergo meiosis are known as **germ cells;** the rest of the body's cells are called body or **somatic** (soma = body) **cells.**

Not all organisms are diploid. Many lower organisms are haploid (1N). Familiar examples include moss

genes are linked together and produces a **linkage map.**

Think of what you might draw if someone handed you a paper napkin and asked you to quickly show the geographical relationship between Boston, New York City, and Philadelphia. This sort of crude map is similar to a linkage map. And, just as it would be difficult to actually get from Boston to New York City relying on it alone, linkage maps are rough-and-ready approximations of where genes are located on chromosomes in relationship to one another. They are useless for actually locating a gene.

A **physical map** of a chromosome is more detailed. One technique used to make physical maps is to cut a chromosome with restriction enzymes, amplify the DNA with yacs, treat the yacs with restriction enzymes, apply electrophoresis analysis, and then analyze the electrophoretic results for overlaps. Piecing the overlaps together will give the linear order of the genes on the chromosomes. As of 1993, the Human Genome Project has produced physical maps of two chromosomes: 21, where the gene for Down syndrome is located, and the Y chromosome found only in males and the site of many genetic disorders that are usually observed only in males (Figure 11-B).

FIGURE 11-B

Some of the computers and equipment used in the human genome project. (Philippe Plailly/Science Photo Library)

While physical maps are more detailed than linkage maps, the ultimate in gene mapping is a **sequence map** that gives the exact location and composition of each specific gene. In this sort of map each letter is equivalent to a single nucleotide base. But, don't hold your breath waiting for the Human Genome Project to complete sequence mapping of all 23 human chromosomes. Because the human genome is so immense, and because the vast majority of it consists of mysterious, repetitious sequences of nucleotides, the production of an entire sequence map is almost unimaginable. There are some 3 million base pairs of DNA in the human genome and if you printed a sequence of 3 million bases, with only one letter (A, T, C, or G) representing each base, the entire string of 3 million letters would fill 10 sets of the *Encyclopedia Britannica*! It is estimated that, using present methods, this task will take 40 years to finish, so investigators are concentrating on sequence maps of only the 5% of the human genome that has been identified as actual structural or regulatory genes thus far. The long stretches of repetitive, mysterious DNA are being ignored for the present.

Watch the news for continuing developments of the Human Genome Project.

plants, many algae and fungi, and male honeybees (drones). Some organisms, especially many plants, are **tetraploid** (4N), with four homologous chromosomes of each type. Other ploidy numbers are also found, but more rarely.

The two chromatids of a replicated chromosome remain attached at the centromere until they separate as new chromosomes during mitosis or meiosis. Chromosomes typically occur in homologous pairs, and the members of each pair bear comparable genes. Humans and many other organisms have several pairs of autosomes and one pair of sex chromosomes. Most body cells are diploid, but reproductive cells are typically haploid.

Table 11–1
Haploid (1N) and Diploid (2N) Numbers of Chromosomes in Some Organisms

Organism	N	2N
Adder's tongue fern	over 600	over 1200
Pea plant	7	14
Corn	10	20
Potato	24	48
Fruit fly	4	8
Chicken	39	78
Cat	19	38
Dog	39	78
Chimpanzee	24	48
Human	23	46

BIO-BIT

The early embryonic cells of some frogs and insects have an extremely rapid cell cycle that skips the G_1 and G_2 stages.

11–C THE CELL CYCLE

1? Each kind of cell has a typical lifespan, which begins when the cell is formed by division of the parent cell and ends when the cell itself divides or dies. Under good growing conditions, the lifespan of lower eukaryotes varies from about 2 hours for yeast (a unicellular fungus) to a few days for *Amoeba* (a protist). After fertilization, animal embryos also divide rapidly, as often as every 15 or 20 minutes. Most dividing cells in an adult have lifespans of about 8 hours to over 100 days, depending on the cell type. Some types of cells cannot divide once they have reached their final differentiated state, and they must eventually die. Examples in our bodies are nerve, skeletal muscle, and red blood cells. ■

A newly formed cell will usually not divide until it has approximately doubled in size. To do this, it must absorb nutrients and use them to produce more ribosomes, enzymes, cytoskeleton molecules, and membrane material. Mitochondria and chloroplasts reproduce themselves by dividing in two.

Many unicellular organisms grow and divide as fast as they can obtain enough nutrients to do so. This often makes for evolutionary success because the sooner a cell divides, the more descendants it will leave over a period of time. During periods of growth, most cells of multicellular organisms divide rapidly. Once adult size has been achieved, cells stop dividing so rapidly and growth slows or ceases.

Control of cell division is a complex process that we are only beginning to understand. In animals, cell division seems to require at least two things in addition to nutrients. First, a cell's membrane receptors must bind a critical number of one or more kinds of **growth factors**—highly specific proteins found in the body fluids in minute amounts. Cells compete for molecules of the right kinds of growth factors for their cell type. So, at any one time, only a limited number of cells in any tissue have enough growth factors to begin dividing. The second requirement is attachment to something outside the cell, such as neighboring cells or the membranes or fibers between cells. This requirement may help to ensure that, if a cell becomes detached from its proper place in the body, it cannot divide and form a tumor wherever it comes to rest. Once an attached cell does begin to divide, however, it loses much of its attachment and assumes a rounded shape until division is complete. The two new cells then settle into the parent cell's space, reattach to their surroundings, and resume their typical shape.

Cells that can and do divide have a typical life history, called the **cell cycle,** lasting from the time the cell is formed by division until it divides in its turn (Figure 11–3). The cell cycle has four distinct periods. During the period of mitosis **(M)** and cytokinesis, the nucleus and cytoplasm divide and form two new cells. The rest of the cycle, known as **interphase,** is divided into the remaining three periods. The period from a new cell's "birth" until it begins to replicate its DNA is called the first gap period, or $\mathbf{G_1}$. During this time, the cell is usually growing and carrying on the business of life. The middle period, the **S** (synthesis) period, is the time of DNA synthesis, when the chromosomes are replicated in preparation for the next cell division. The second gap period, $\mathbf{G_2}$, lasts from the end of DNA synthesis until the beginning of the next period of mitosis.

The length of G_1 is the most variable part of the cell cycle. The key point in the cycle occurs late in G_1: some unknown signal molecule(s) is believed to switch the cell into the S period. After this point of no return, the cell is committed to proceed completely through the next mitosis.

During the S period, DNA replication begins at specific points in the DNA and follows a definite pattern until all the DNA is copied. Histone proteins for the new chromosomes are also made during the S period, and the centrioles double (except in higher plants, which lack centrioles) (Section 5–L).

When replication is finished, the cell enters G_2, the time of final preparation for division. Organelles are replicated and, during G_2, the cell is believed to make some of the proteins used in mitosis itself. One

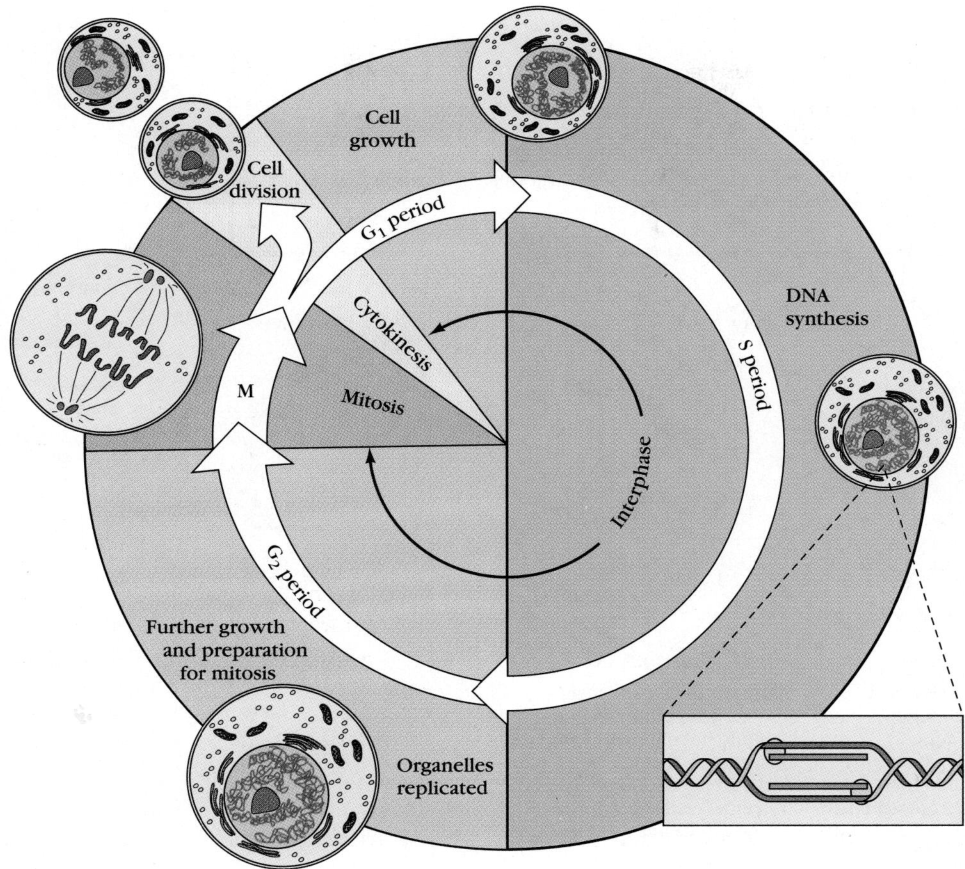

FIGURE 11-3

A typical cell cycle. The two gap periods (G_1 and G_2) are separated by the synthesis (S) period, during which the chromosomes are replicated. Mitosis, or nuclear division, is followed by cell division (cytokinesis).

> **BIO-BIT**
>
> Most of the cells lining the stomach last only about 3 days before they succumb to the high acidity of the stomach and are replaced by rapid division of underlying cells.

of these activates another protein, which prepares the cell for mitosis. In particular, it causes the chromosomes, which have been spread out in a loose mass during interphase, to condense very tightly. The cell is now ready for mitosis.

The length of the cell cycle can vary greatly, depending upon the availability of nutrients and growth factors. The four periods of the cell cycle include mitosis and cytokinesis followed by a period of normal cell activity (G_1), a period of DNA replication (S), and the period of time between DNA replication and mitosis (G_2).

11-D MITOSIS

In mitosis, the replicated, condensed chromosomes are separated into two equal groups. Two new nuclei are formed, each containing a complete set of the genetic information present in the original nucleus. Mitosis is a continuous process, but for convenience it is divided into four phases according to the appearance of the chromosomes as viewed through a light microscope: prophase, metaphase, anaphase, and telophase (A Journey Through Mitosis).

MITOSIS
Mitosis results in two genetically identical nuclei which are also identical to the nucleus from which they are formed. This process is diagrammed below using a simplified animal cell model with a diploid number of 4. Most organisms have more chromosome pairs. Humans, for example have 23 pairs or a total of 46 chromosomes.

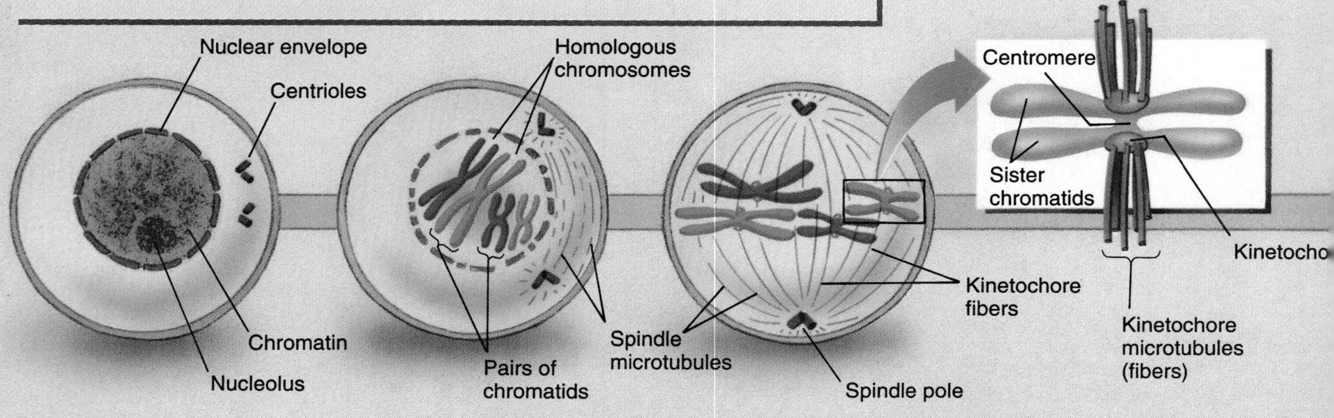

INTERPHASE
Chromatin spread out in indistinct mass. Nucleus and nucleolus distinct. DNA replication occurs so that the cell has twice the number of chromosomes.

PROPHASE
Chromosomes condense, homologous chromosomes pair and become visible as sets of sister chromatids. Nucleolus and nuclear envelope disappear. In animal cells, centrioles migrate to opposite poles of the cell. Spindle microtubules appear.

METAPHASE
Mitotic spindle complete. Chromatid sets move to spindle equator.

Mitosis in an animal cell (whitefish)

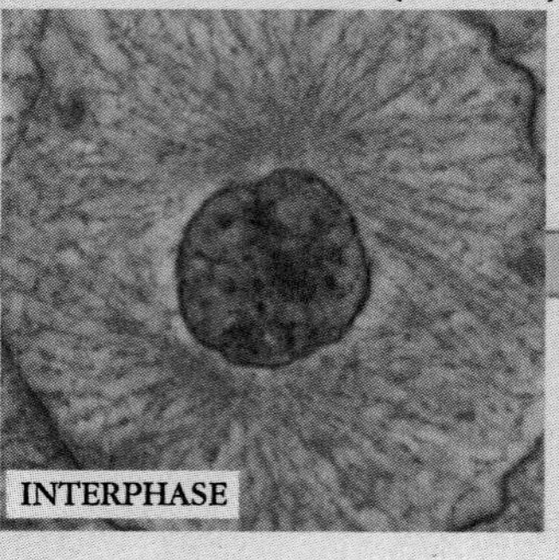

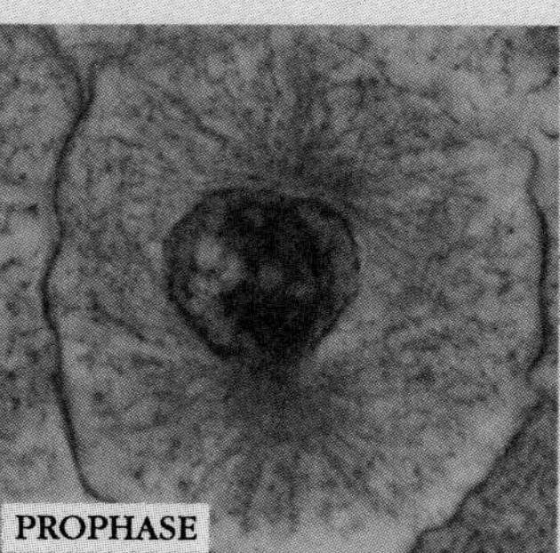

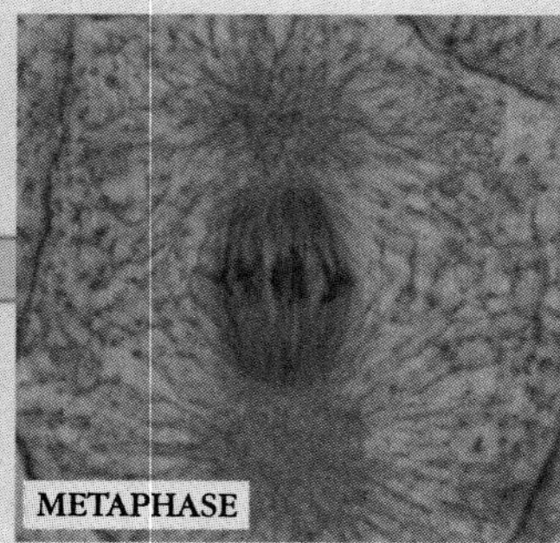

Mitosis in a plant cell (onion)

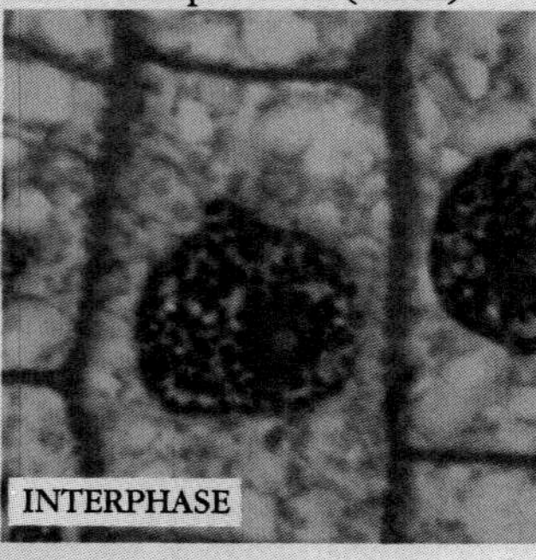

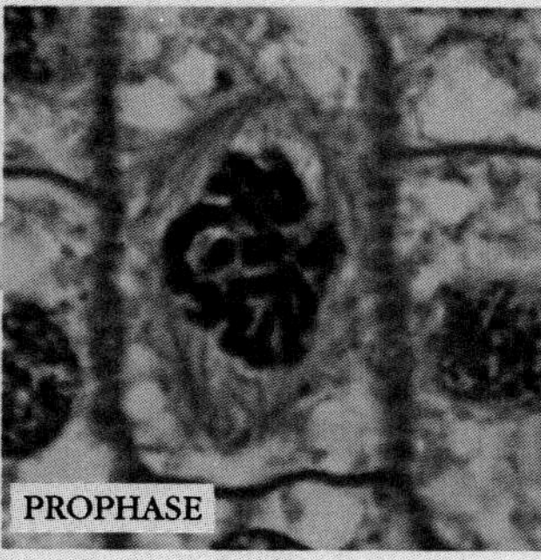

A Journey Through Mitosis. (Whitefish mitosis, Carolina Biological Supply Company/Phototake NYC; Onion mitosis, Visuals Unlimited/R. Calentine)

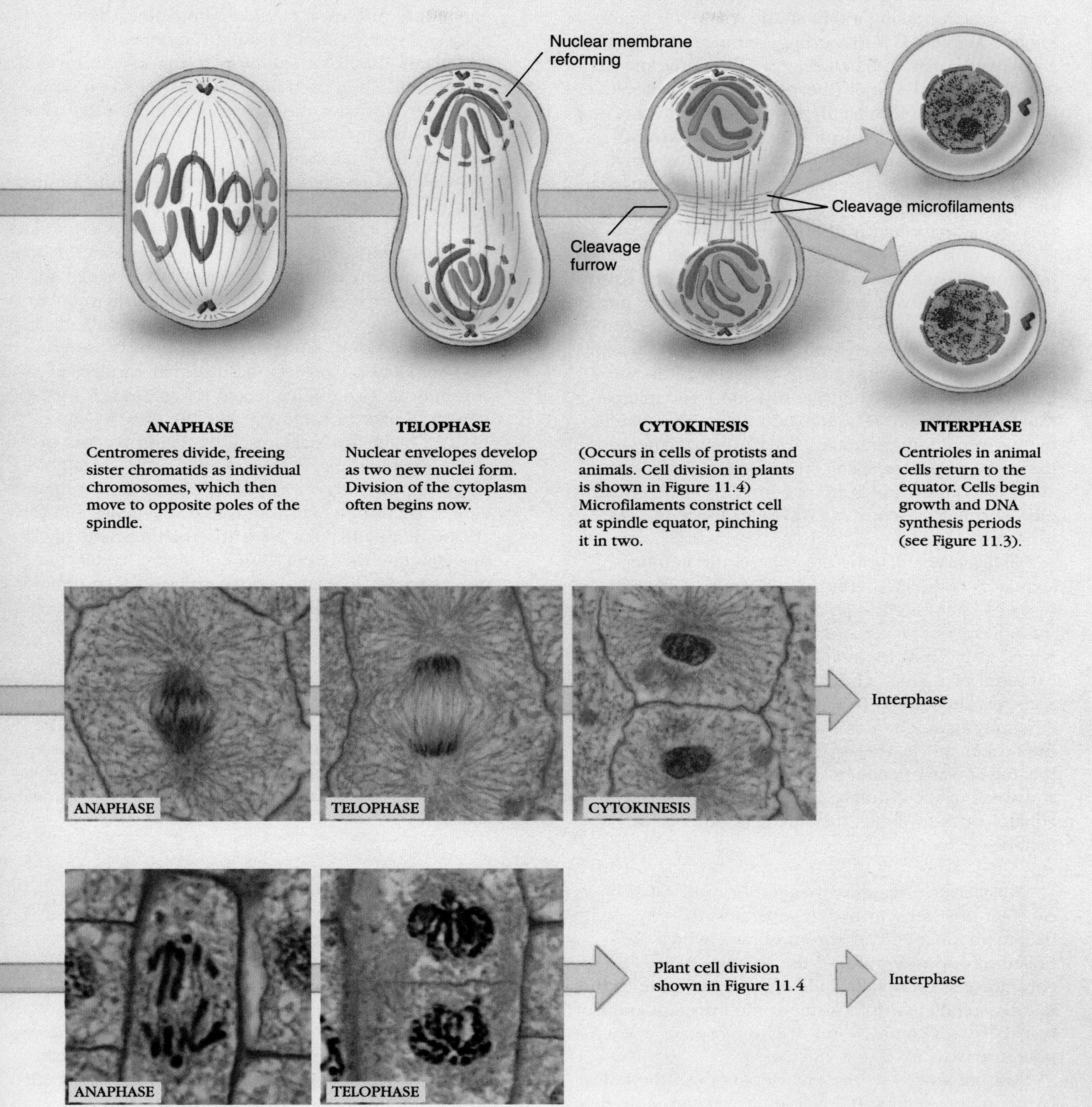

Nuclear membrane reforming
Cleavage microfilaments
Cleavage furrow
ANAPHASE
Centromeres divide, freeing sister chromatids as individual chromosomes, which then move to opposite poles of the spindle.
TELOPHASE
Nuclear envelopes develop as two new nuclei form. Division of the cytoplasm often begins now.
CYTOKINESIS
(Occurs in cells of protists and animals. Cell division in plants is shown in Figure 11.4) Microfilaments constrict cell at spindle equator, pinching it in two.
INTERPHASE
Centrioles in animal cells return to the equator. Cells begin growth and DNA synthesis periods (see Figure 11.3).
ANAPHASE
TELOPHASE
CYTOKINESIS
Interphase
ANAPHASE
TELOPHASE
Plant cell division shown in Figure 11.4
Interphase

Prophase. Looking through a microscope, you can first tell that a cell is about to divide during prophase, when the loose mass of interphase chromatin condenses into distinct chromosomes, visible as sets of sister chromatids.

This condensation is an impressive process. It is comparable to taking a thin strand some 200 meters long and coiling it into a cylinder about 1 millimeter across and 8 mm long. During prophase, the nucleolus, which is the site of ribosome synthesis, usually disappears because the material in the nucleolus becomes scattered. A complex of proteins, called a **kinetochore,** assembles on each chromatid, in the centromere region. The nuclear membrane disappears at the end of prophase. It breaks down into small vesicles and is re-formed from them after mitosis.

The other notable change during prophase is the beginning of a framework of microtubules, the **mitotic spindle,** which will eventually take part in the movement of the chromosomes. As mitosis begins, microtubules of the cytoskeleton break down into their protein subunits. (This loss of the cytoskeleton is why cells become rounded during division.) The microtubule subunits are reassembled to form the spindle. As the spindle microtubules are assembled, they push the ends, or **poles,** of the spindle apart. In animal cells, the spindle poles are occupied by pairs of centrioles, also composed of microtubules (Section 5-L).

Metaphase. During metaphase, the mitotic spindle is completed. The spindle consists of two types of microtubules: polar fibers, which extend from the poles past the equator, and kinetochore fibers, which grow from the poles until their free ends are captured by a kinetochore. The two kinetochores of sister chromatids capture fibers from opposite poles.

Each kinetochore is pulled toward its fibers' pole: the farther this is, the stronger the pull. The forces in this tug-of-war balance out midway between the poles, and so all of the chromatids become lined up at the equator of the spindle, the sure sign of a cell in metaphase.

Anaphase. Anaphase begins abruptly. All at once, each set of sister chromatids separates, thereby becoming independent chromosomes, which are pulled to opposite poles of the spindle. Each chromosome eventually ends up near one pole of the mitotic spindle, with its sister at the opposite end, so that there is a complete set of chromosomes at each pole, the basis for a new nucleus.

During anaphase, the polar fibers push the poles farther apart, making the cell longer. At the same time, the chromosomes are pulled toward the poles as the kinetochore fibers shorten by losing the subunits nearest the kinetochore. (Exactly how this works without the chromosome's falling off the fiber is not understood.)

Telophase. In the last stage of mitosis, telophase, two new nuclei are organized. The chromosomes, now in two groups at the poles of the mitotic spindle, uncoil into masses of tangled chromatin. A new nuclear envelope forms around each mass of chromatin. Ribosome synthesis resumes, and nucleoli become visible in each new nucleus. The two new nuclei are ready for the normal activities of interphase.

Cells of any ploidy (haploid, diploid, tetraploid, etc.) can undergo mitosis.

Colchicine, a chemical derived from the autumn crocus plant, prevents formation of a mitotic spindle and so blocks mitosis and cell division. It is sometimes used in attempts to prevent cancer cells from dividing. In colchicine-treated cells, sister chromatids can still separate from each other, thus doubling the number of chromosomes in the cell. In this way, diploid cells can become tetraploid. Colchicine treatment is used to stop cell division at metaphase so that condensed chromosomes can be collected for analysis (for instance, to make a karyotype like the one in Figure 11-2). Colchicine is also used by plant breeders to make tetraploid plants from diploid ones. Tetraploid plants are often larger and more vigorous than their diploid ancestors. Many cultivated vegetables and flowers are tetraploids that have arisen either naturally or by deliberate treatment with colchicine. Water-processed decaffeinated coffee comes from a diploid species with strongly flavored beans. The milder species used for regular coffee is an artificially produced tetraploid.

Mitosis is a type of nuclear division that results in twin nuclei with chromosomes identical to their parent cell. The process of mitosis can be divided into four distinct phases according to the arrangement of the chromosomes. The key events of each phase are summarized in A Journey Through Mitosis.

11–E CYTOKINESIS

Mitosis is now complete, but the two nuclei still lie in the same cytoplasm. Division of the cytoplasm is called **cytokinesis**: the original cell forms two new cells, each housing one of the newly formed nuclei (Figure 11-4). In animal cells, cytokinesis begins during early anaphase. A ring of microfilaments, made up of the contractile proteins actin and myosin, forms around the cell's equator, just beneath the plasma membrane. These filaments constrict the cell to form a

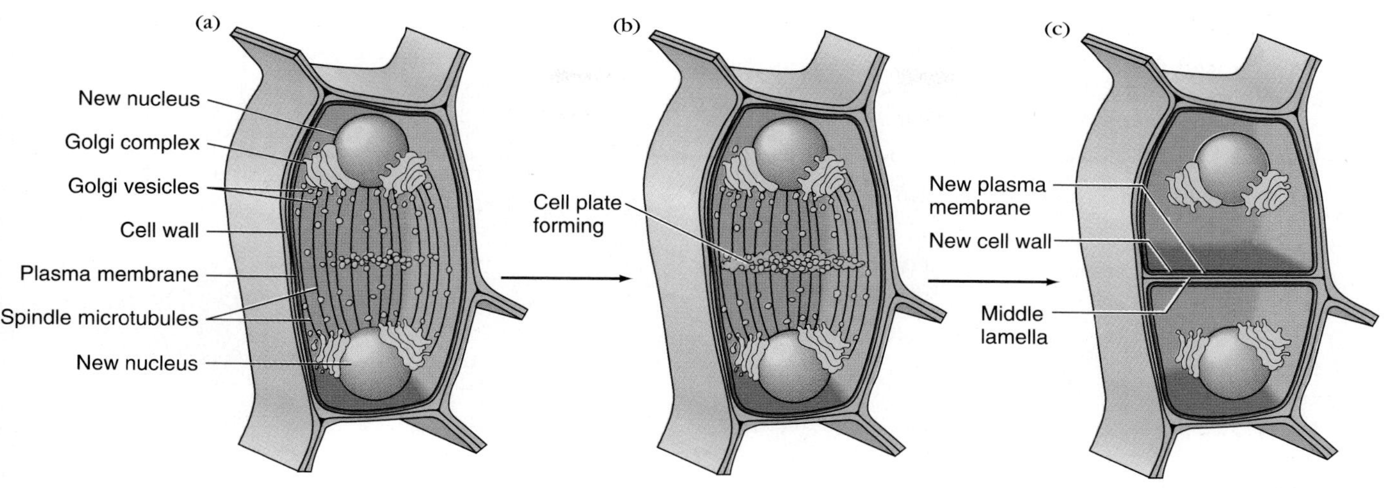

FIGURE 11–4

Cytokinesis in a plant cell. (a) Vesicles containing material for the cell plate bud off from Golgi complexes. The lingering spindle microtubules serve as tracks for the vesicles to move to the equator. **(b)** The vesicles fuse, and the material they contain is built into a cell plate, which forms the middle lamella. The fused vesicle membranes on either side of the plate will become the plasma membranes at the ends of the new cells. **(c)** Each new cell lays down a wall between the middle lamella and the new portion of its plasma membrane.

cleavage furrow and eventually pinch the cytoplasm in two.

Plant cells are surrounded by a rigid cell wall, and cytokinesis occurs by a completely different method. The Golgi complexes release vesicles containing material for a new partition between the cells-to-be. The vesicles move along the spindle microtubules to the middle of the cell. Here they fuse together and the material they contain forms a flat disc, the **cell plate,** enclosed by a membrane made up of the fused vesicle membranes. The cell plate and its surrounding membrane grow around the edges by addition of more vesicles (Figure 11–4). Soon the cell plate extends completely across the cell, cutting it in two. The material in the cell plate forms the **middle lamella,** the common partition between the two new cells, and the membrane on either side of the cell plate becomes part of the new cells' plasma membranes. Each new cell builds a new cell wall on its side of the middle lamella (Figure 11–4).

Cytokinesis divides up not only the cytoplasm, but also the various structures within it, such as ribosomes, Golgi complexes, mitochondria, plastids, and cytoskeleton molecules. Most cells contain many of each kind of structure, distributed throughout the cytoplasm, so that each new cell is bound to receive at least some of every component it needs.

During cytokinesis in animals, a ring of microfilaments just beneath the plasma membrane constricts around the equator of the cell, pinching the cytoplasm in two. However, the rigid cell wall of plants prevents the cells from dividing in the same manner. Instead, a partition is formed between the two new cells, which then build new end cell walls on either side.

BIO-BIT

In many insects, the early nuclei undergo many cycles of mitosis without complete cytokinesis, resulting in an embryo that is a large, multinucleated cell. Later, cytokinesis separates the nuclei into individual cells.

11–F MEIOSIS

Meiosis is the process of nuclear division in which haploid nuclei are formed from diploid nuclei. DNA synthesis occurs before meiosis as well as before mitosis. Therefore, a nucleus enters meiosis with enough DNA to make four haploid nuclei. Because a nucleus cannot divide into more than two new nuclei at any one division, it takes two divisions during meiosis to reduce the DNA content of each nucleus to haploid. We summarize the movements of two pairs of homologous chromosomes through meiosis in A Journey Through Meiosis.

MEIOSIS I
Meiosis is a two stage process. In the first stage, homologous chromosomes separate from their partners.

(a) (DNA replication occurs prior to meiosis)

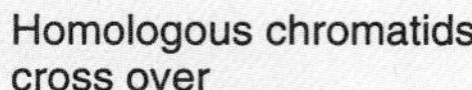

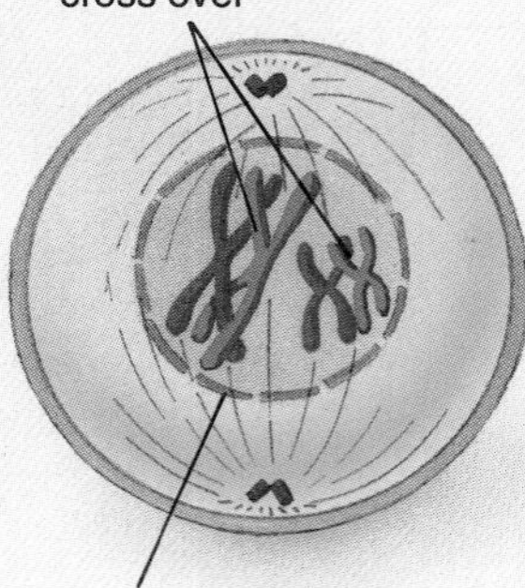

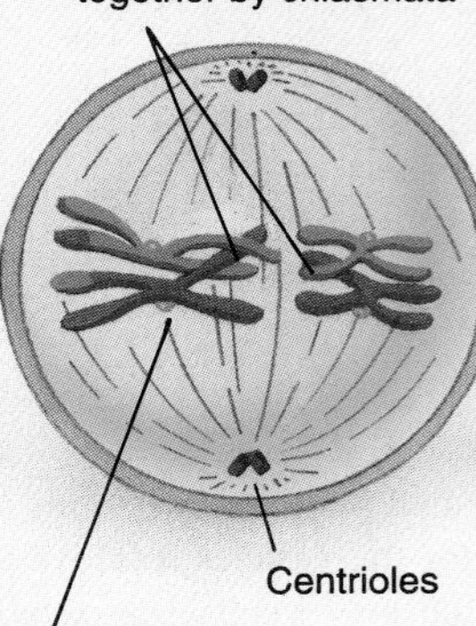

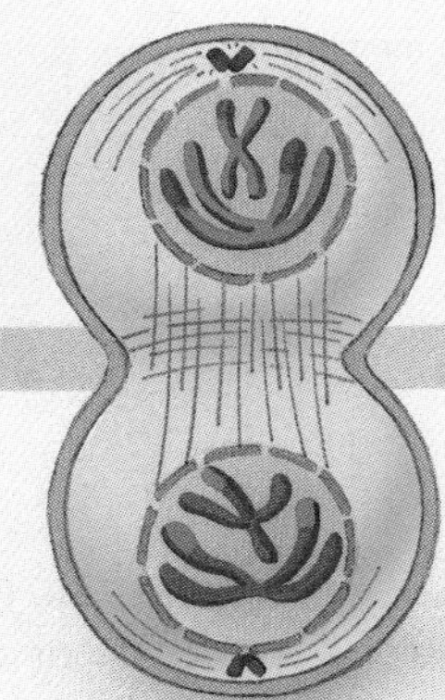

PROPHASE I

Replicated chromosomes condense and pair with their homologues to form tetrads. Pairing is necessary for separation of members of each homologous pair in the first meiotic division, so that each resulting nucleus receives one member of each pair. Crossing over occurs during Prophase I.

METAPHASE I

Nuclear envelope has dispersed. Tetrads are held together by chiasmata. Kinetochores of homologous sets of sister chromatids attach to spindle fibers from opposite poles, and tetrads move to the equator.

ANAPHASE I

Chiasmata separate but centromeres do not. Each set of sister chromatids moves toward a pole of the spindle, as its homologue travels toward the opposite pole. Sister chromatids travel as a pair and do not separate until anaphase II.

TELOPHASE I

The chromosomes have formed two groups. In some species, nuclear envelopes reappear and the cytoplasm divides. In others, nuclear envelopes remain absent and metaphase II starts immediately.

(b)

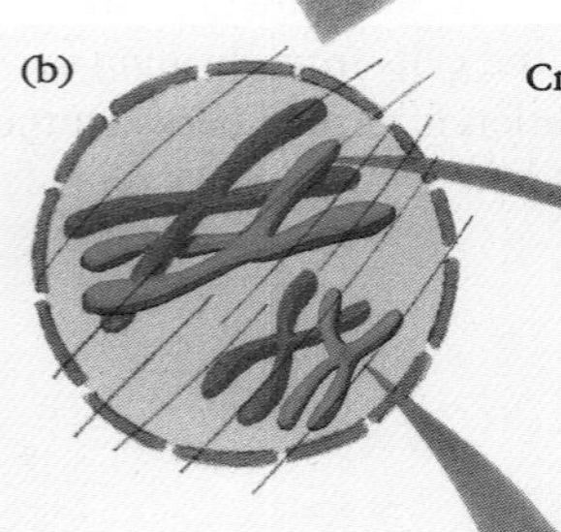

Crossing-over shuffles genes. Each pair of homologous chromosomes may have one or more cross-overs resulting in recombination.

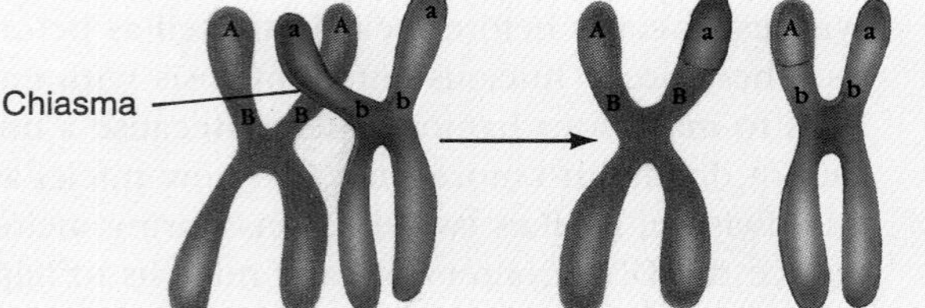

A single crossover Recombinant chromosomes

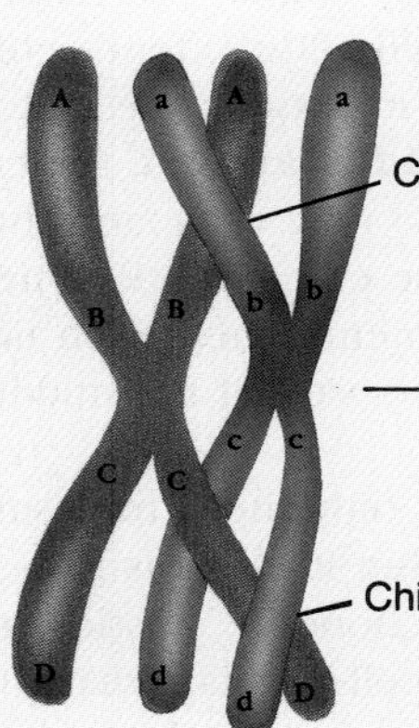

Two crossovers, one each on two different chromatids

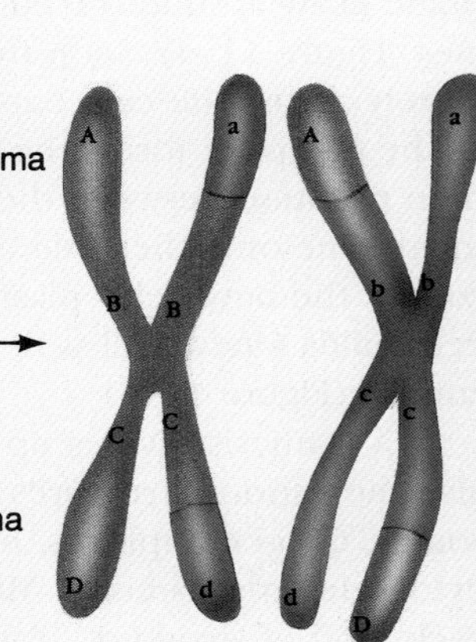

Recombinant chromosomes

MEIOSIS II
Each nucleus divides again.
Centromeres divide, and sister chromatids become separate chromosomes.
The net result is four new cells with haploid nuclei.

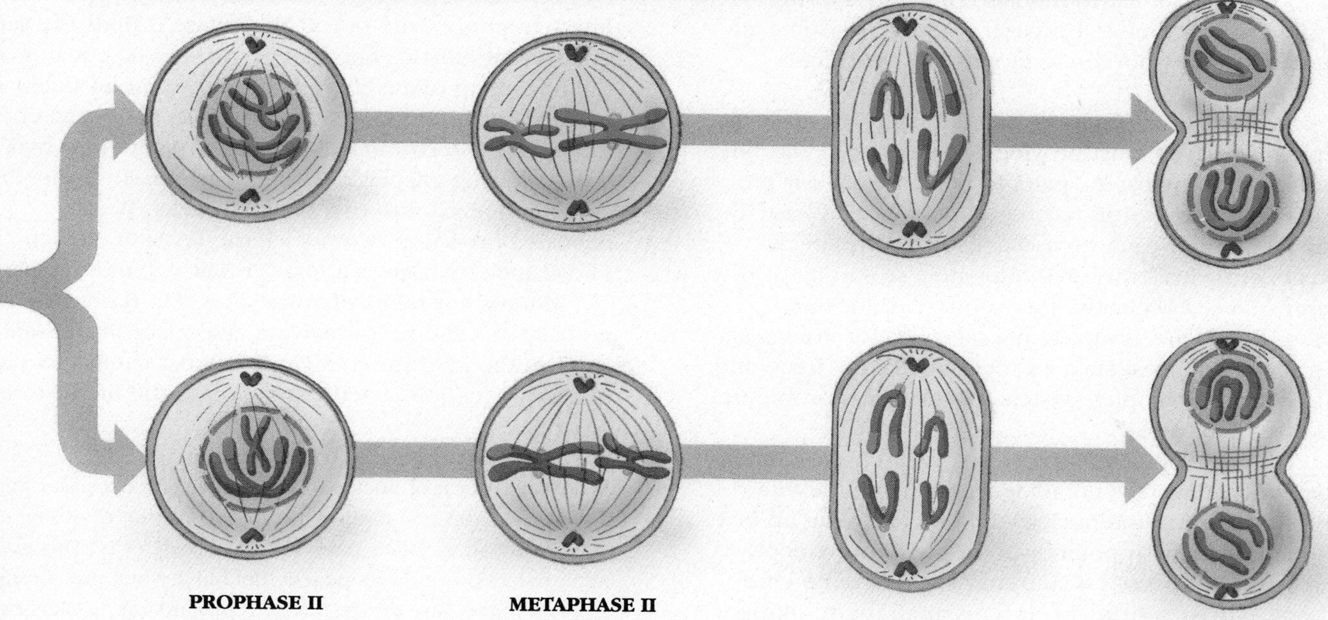

PROPHASE II

In those organisms (particularly plants) with a period of interkinesis between the two meiotic divisions, the chromosomes must condense again before the second division.

METAPHASE II

Spindles form again. The kinetochores of each set of chromatids attach to spindle fibers from opposite poles. The centromeres then divide, just as they do in metaphase of mitosis, and each set of sister chromatids becomes two separate chromosomes.

ANAPHASE II

The newly separated chromosomes move to opposite poles of the spindle.

TELOPHASE II

Four haploid nuclei are formed, each with one member of each pair of chromosomes from the original nucleus that entered meiosis. Nuclear envelopes form and cytokinesis occurs.

The two divisions in meiosis are unimaginatively called meiosis I and meiosis II. Like mitosis, both meiotic divisions involve formation of a spindle and movement of chromosomes to the spindle's poles, so meiosis looks very similar to mitosis. The names of the stages are also similar. However, meiosis has some additional features not found in mitosis (see Figure 11-1).

Since meiosis produces haploid nuclei from diploid nuclei, it must provide a way for the cell's homologous chromosome pairs to be parcelled out precisely into two groups, each group containing exactly one member of each homologous pair. The special events of meiosis that allow this precise sorting of the chromosomes occur during prophase of meiosis I. These events are complex and take a lot of time, making this the longest stage of meiosis. Meiosis frequently takes days to complete instead of the hours or minutes required for mitosis.

During prophase I of meiosis, each chromosome somehow "finds" its homologue among all the other chromosomes in the nucleus, and the two line up next to each other with point-by-point precision, a poorly understood process called **synapsis.** Since the chromosomes have already been replicated, the resulting group consists of four chromatids altogether and is called a **tetrad.** During this tetrad stage, portions of the chromatids are exchanged between the homologous chromosomes, a phenomenon called **crossing over** (see Section 11-G). This is one source of genetic variation that occurs as a result of sexual reproduction. For a while, the chromatids remain joined at the crossover exchange point, called a chiasma ("cross"; plural: **chiasmata**). This holds the homologous pair of chromosomes together, while the centromeres hold the two sister chromatids of each chromosome together; hence the entire tetrad moves as one (A Journey Through Meiosis). Research during the 1980s showed that crossing over is a common event. Normally, each tetrad contains at least one chiasma.

In metaphase I, all the tetrads line up at the spindle equator. Each set of sister chromatids has a kinetochore attached to spindle fibers from one pole, and the homologous set, just across the equator, is attached to the opposite pole. This arrangement, much like couples lined up opposite their partners for a barn dance, allows the partners to be separated from one another at anaphase I. As anaphase I begins, the chiasmata come apart, whereas the centromeres remain intact. Hence each set of sister chromatids moves as a unit toward one spindle pole, while the homologous set of chromatids moves to the opposite pole. Each group of chromosomes is then organized into a new nucleus during telophase I.

There is no DNA replication between meiosis I and II.

During prophase II, a new spindle forms in each of the two new cells. Each set of chromatids now has two separate kinetochores, which attach to spindle fibers from opposite poles. Metaphase II finds the sets of sister chromatids lined up at the spindle equator. At the beginning of anaphase II the centromeres finally divide, releasing the sister chromatids as individual chromosomes. The chromosomes then separate into two groups during anaphase II, and at telophase II they become organized into two haploid nuclei. Because meiosis I produces two nuclei, the division of each one at meiosis II gives a total of four haploid nuclei.

Meiosis is vital in all eukaryotes that reproduce sexually, but it does not always take place at the same stage in the life history. Meiosis is most familiar as part of the process of gamete formation in the life histories of animals.

Meiosis is a form of nuclear division that produces four haploid nuclei from one diploid nucleus. Thus, the resulting cells have half as much DNA as typical body cells. This genetic reduction prevents genetic doubling when the sperm and egg nuclei fuse during sexual reproduction. It takes two cellular divisions for meiosis to produce four daughter cells. The key events are summarized in A Journey Through Meiosis. Meiosis and mitosis are compared in Table 11-2.

11-G GENETIC REASSORTMENT

22? Besides reducing the chromosome number from diploid to haploid, meiosis also shuffles the genetic material, forming new combinations of genes and chromosomes that become the genetic information of the next generation. Meiosis produces this **genetic reassortment** in two ways: by producing new combinations of genes on chromosomes (crossing over) and by producing new assortments of chromosomes. Because of these mechanisms of genetic reassortment, children with the same parents can have quite different genetic combination. ■

The production of chromosomes with new combinations of genes occurs during crossing over (Section 11-F). This results from the exchange of segments of DNA between homologous chromosomes, a process called **genetic recombination.** Crossing over occurs while the chromosomes are in tetrads during prophase I: two chromatids, one from each homologue, cross each other and are broken off and joined to the opposite strand. This rearranges genes that were on the same chromosome so that they are on two different

Table 11–2
Differences Between Mitosis and Meiosis

Mitosis	Meiosis
Occurs in haploid (N) and diploid (2N) cells	Occurs in diploid (2N) cells
Nucleus divides once (sister chromatids separate)	Nucleus divides twice: I reduction division (sister chromatids remain together) II mitotic division of haploid nucleus (sister chromatids separate)
No synapsis of chromosomes	Synapsis, tetrad formation, and crossing over during prophase I
Produces two daughter nuclei of same ploidy as original nucleus	Produces four haploid daughter nuclei
Daughter nuclei genetically identical to each other and to original cell, with same chromosome content	Daughter nuclei genetically different from each other and from original cell, with half its chromosome content
Occurs in somatic cells	Occurs in reproductive cells
Produces new individuals in unicellular organisms; new cells in growth and repair of multicellular organisms	Produces gametes in animals, spores in plants and fungi

chromosomes, and vice versa (see A Journey Through Meiosis). Because one of the original chromosomes was inherited from each parent, crossing over combines genes from both parents in each of the recombinant chromatids. In humans, crossing over occurs an average of two or three times in each pair of homologous chromosomes during gamete formation. When more than one crossover occurs, these may involve the same two chromatids or different combinations of chromatids.

Another source of genetic reassortment occurs because chromatid tetrads can line up at metaphase I with either set of chromatids nearer either pole. Then, in anaphase I, each set of chromatids is separated from the homologous set and goes into a new nucleus with the members of other homologous pairs that were lined up on the same side of the equator as itself (Figure 11-5). The lining up at metaphase I is random, so there is an equal chance that any one chromosome will end up in a new cell with either member of any other pair of homologous chromosomes.

Additional genetic reassortment occurs during the random fusion of gametes at fertilization. In this way, a steady supply of new genetic combinations arises in sexually reproducing species, furnishing the raw material—variation—for evolution by means of natural selection.

Prior to meiosis, one member of each pair of homologous chromosomes can be traced back to the parent who contributed it during fertilization. But genetic reassortment during meiosis produces new combinations of chromosomes and genes. Crossing over during meiosis results in the exchange of corresponding segments of DNA between chromatids of homologous chromosomes, blurring the distinction between chromosomes donated from either parent. In addition, chance determines which sister chromatids from homologous chromosomes line up at metaphase and pair up at fertilization. This creates even greater genetic diversity.

11–H GAMETE FORMATION IN ANIMALS

The formation of gametes (sperm and eggs) is similar in most animals, although details vary among species. However, the two processes—formation of sperm and of eggs—differ somewhat from each other. Let us begin with the production of sperm, which is in some ways simpler.

Sperm, or spermatozoa, are the male gametes. Because a sperm contains little cytoplasm, it is very small, and in nearly all species the sperm can swim, using the flagellum that forms its tail.

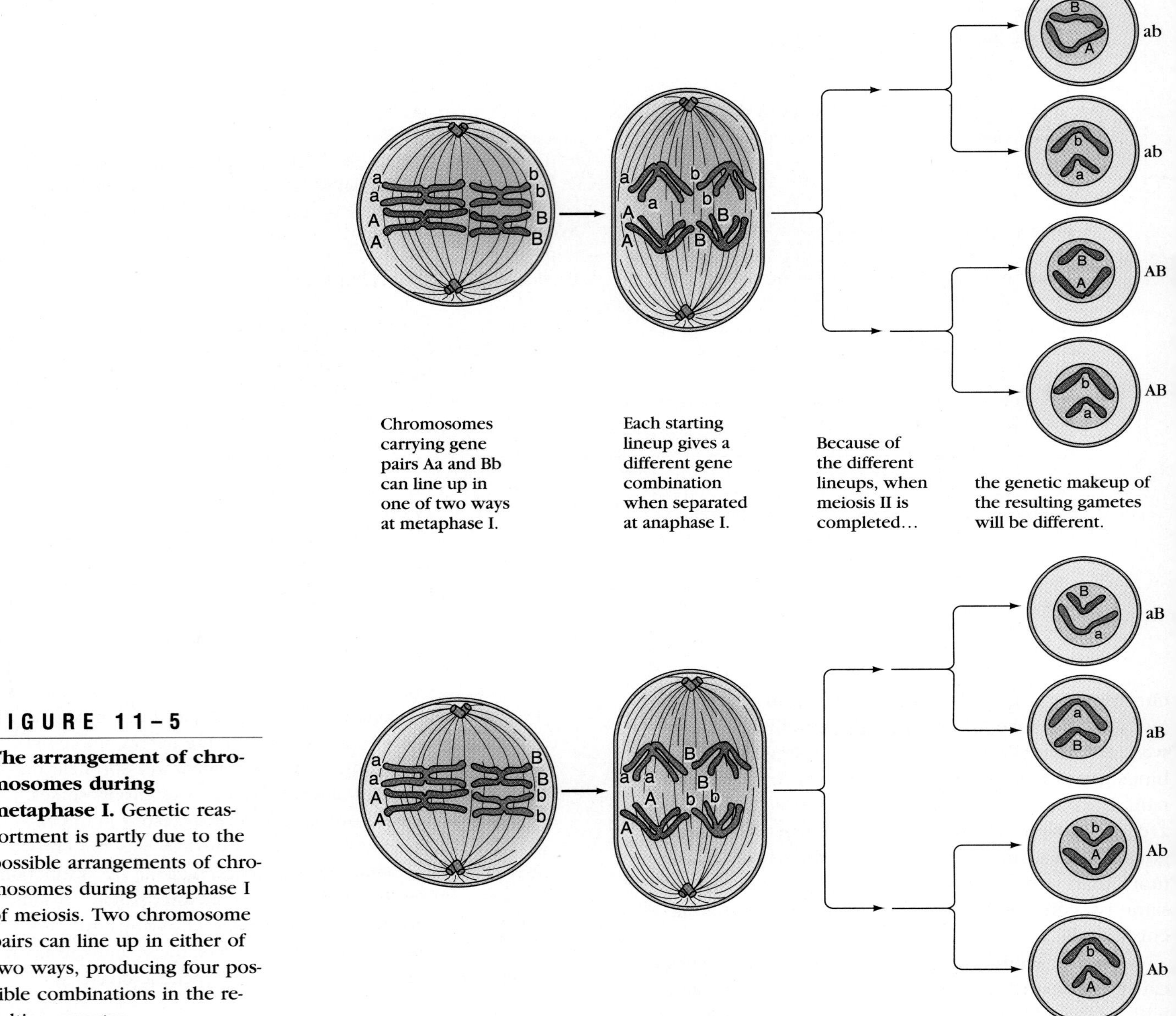
a
a
A
A
b
b
B
B
Chromosomes carrying gene pairs Aa and Bb can line up in one of two ways at metaphase I.
Each starting lineup gives a different gene combination when separated at anaphase I.
Because of the different lineups, when meiosis II is completed...
the genetic makeup of the resulting gametes will be different.
ab
ab
AB
AB
aB
aB
Ab
Ab

FIGURE 11-5

The arrangement of chromosomes during metaphase I. Genetic reassortment is partly due to the possible arrangements of chromosomes during metaphase I of meiosis. Two chromosome pairs can line up in either of two ways, producing four possible combinations in the resulting gametes.

Sperm are produced in the testes by the process of **spermatogenesis** (Figure 11-6). The male's germ cells, called **spermatogonia,** divide continuously by mitosis. Some of the new cells become **spermatocytes,** the cells that undergo meiosis. Primary spermatocytes go through meiosis I, which produces two secondary spermatocytes. During meiosis II, the two secondary spermatocytes divide again and produce a total of four haploid **spermatids.** Although meiosis is now complete, the spermatids must undergo further differentiation into spermatozoa. A mature sperm has a head, which contains the nucleus with its haploid set of chromosomes; a long tail, or flagellum, which propels the sperm through its fluid surroundings; and between these a midpiece containing many mitochondria, which supply the ATP necessary for flagellar motion.

BIO-BIT

An average human male's ejaculate contains between 200 and 300 million sperm, while a human female will release only about 400 to 500 eggs in her entire lifetime.

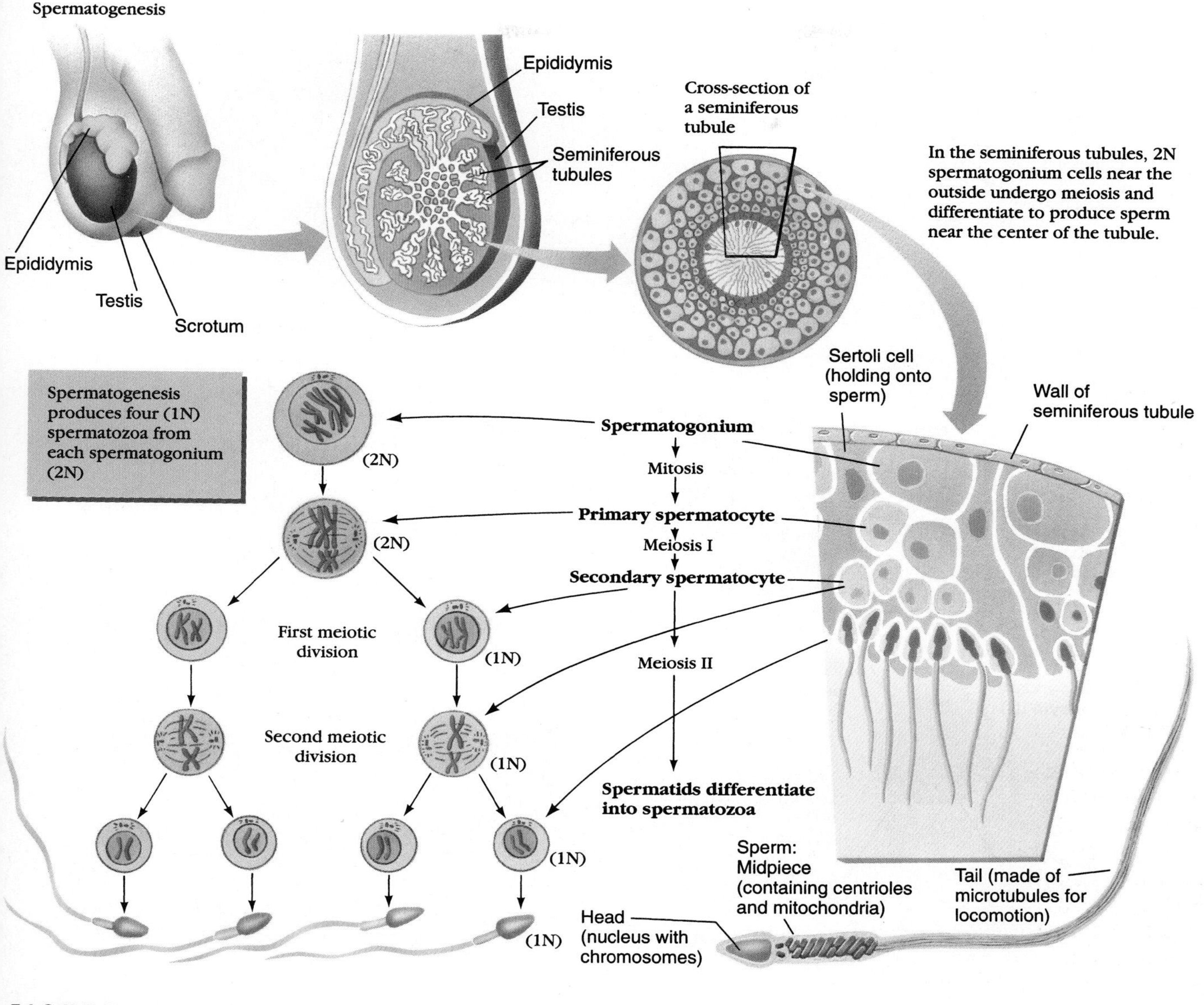

FIGURE 11-6

Spermatogenesis. As a result of spermatogenesis, four sperm are produced from one 2N spermatogonium.

Oogenesis is the formation of female gametes, the eggs or **ova** (singular: **ovum**). Whereas sperm are often the smallest cells in a male animal's body, eggs are the largest cells in a female. This is because the egg cell is the main source of stored food, ribosomes, messenger RNA, and other cytoplasmic components that support the embryo's early development. Oogenesis ensures that the mature egg contains as much as possible of these components. Meiotic nuclear division is accompanied by unequal cytokinesis, so that the original diploid cell produces only one large ovum and two or three tiny cells called **polar bodies** (Figure 11-7). ■

In the ovary, cells called **oogonia** divide by mitosis for a time. Eventually, they stop dividing and differentiate into primary oocytes. The DNA is replicated, and the primary oocytes enter prophase I, proceeding through the formation of tetrads and crossing over. In humans and other mammals, all this occurs while the female is still an embryo. At this point in prophase I, meiosis is arrested for days or years, depending on the

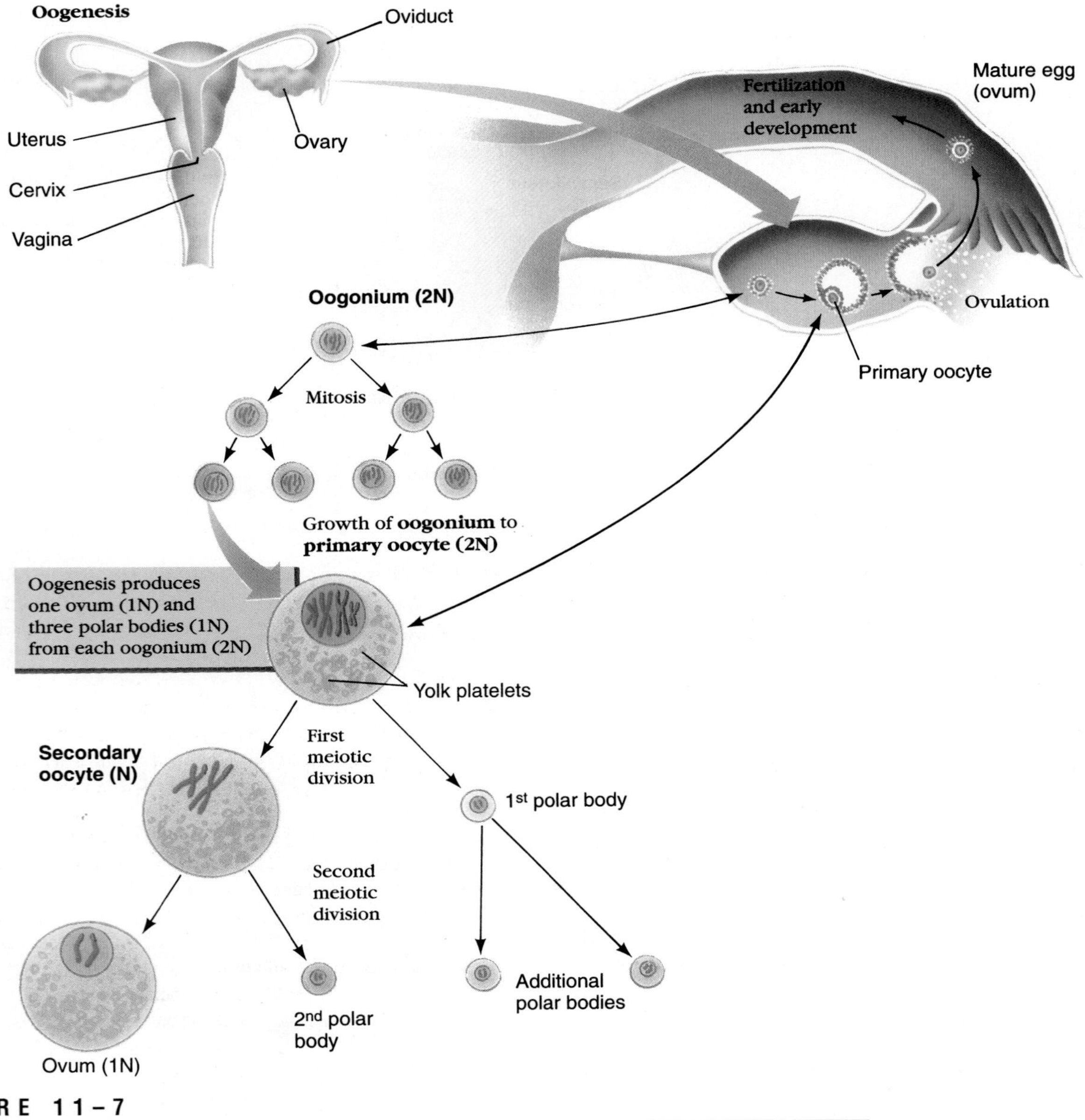

FIGURE 11-7

Oogenesis. In oogenesis, only one functional egg cell and three nonreproductive polar bodies result from one 2N oogonium. The two uneven cytoplasmic divisions that produce only one functional egg are believed to be adaptations to concentrate as much yolk as possible in one egg instead of distributing it evenly among four cells.

species, and it does not resume until the female reaches sexual maturity. During this time the cell absorbs nutrients from neighboring somatic cells and stockpiles the materials needed for early embryonic development.

In meiosis I of oogenesis, the chromosomes separate in the usual manner, but during cytokinesis the cytoplasm divides unequally. One nucleus is pinched off with a minimum of cytoplasm, forming the first polar body which may or may not undergo its own division, while the other nucleus is left with most of the cytoplasm, in a cell called the secondary oocyte. This secondary oocyte goes through the second meiotic division. Again two haploid nuclei are formed according to the normal events of meiosis, but cytokinesis is extremely unequal, forming a tiny second polar body and an enormous ovum. The polar bodies are really just a means of shedding excess chromosomes from the developing egg, and they soon disintegrate.

The process of meiosis is similar in both sperm and egg formation. The process of cytokinesis, however, is different. In egg formation, the cytoplasm of the original parent cell is divided unequally, producing only one viable egg. In sperm production, cytokinesis is equal, forming four spermatozoa, each with an equal chance of fertilizing an egg.

SUMMARY

1. Cells are the basic units of life. New cells are produced when existing cells divide in two. These divisions are of two kinds:
   a. mitotic division, in which a cell gives rise to two new cells with sets of chromosomes identical to those of the parent cell, and
   b. meiotic division, in which a diploid cell divides twice, forming four haploid cells with new genetic combinations.
2. The time from one mitotic division to the next is known as the cell cycle. It can be divided into interphase (G_1, S, and G_2) and mitosis and cytokinesis. The initiation of DNA synthesis in the S period of interphase is the key event that commits the cell to undergo mitotic division.
3. Mitosis is a nuclear division in which precise events ensure that the two new nuclei inherit chromosomes identical to those in the nucleus of the parent cell. Although mitosis is a continuous process, it can be divided into four stages:
   a. Prophase: the replicated chromosomes, each consisting of two sister chromatids, condense and become visible under the light microscope. The nucleolus and nuclear membrane disperse, and microtubules are assembled into the mitotic spindle.
   b. Metaphase: all of the sets of sister chromatids are lined up at the equator of the spindle. In each set, the two sisters' kinetochores are attached to spindle fibers from opposite poles.
   c. Anaphase: each centromere splits into two, releasing the sister chromatids from one another and allowing them to travel to the opposite poles of the spindle.
   d. Telophase: at each pole, the nuclear membrane reforms around the chromosomes, the nucleolus reforms, and the chromosomes unravel from their condensed form.
4. Mitosis is usually followed by cytokinesis, in which the cytoplasm and its components are divided to form two separate cells. In animal cells, a band of microfilaments pinches the cell in two. Cytokinesis in plants involves the assembly of a partition between the two newly formed nuclei, which then build new cell walls on either side.
5. Meiosis is the series of two nuclear divisions that produces four haploid nuclei from a diploid nucleus. Meiosis reduces the number of chromosomes by half. Each new nucleus receives only one member of each pair of homologous chromosomes.
6. Meiosis also results in genetic reassortment. This occurs from both crossing over, in which homologous chromosomes exchange genes, and from the formation of new chromosome combinations resulting from the way that the chromosomes line up and separate at metaphase I. Additional genetic diversity results from the random combination of gametes at fertilization.
7. Synapsis and crossing over occur during prophase I of meiosis. Then, in metaphase I, the tetrads of sister chromatids line up at the spindle equator in such a way that the homologous chromosomes are separated from each other during the first meiotic division. Each of the two resulting nuclei contains one member of each pair of homologous chromosomes. Not until the second division do the centromeres divide, permitting sister chromatids to move into different nuclei.
8. Gamete formation in animals involves both meiosis and differentiation to form specialized reproductive cells. Each spermatocyte gives rise to four sperm. These male gametes are stripped down to the bare necessities: a haploid set of genetic material and the locomotory apparatus to deliver it to the egg.
9. Oogenesis involves unequal cytokinesis. It produces one relatively large egg, with sufficient nutrients to support the early embryo, and two or three tiny polar bodies, which cannot be fertilized and contain little more than the excess chromosomes being shed from the forming egg.

SELF-QUIZ

1. A cell cycle is:
   a. the time from the formation of a cell until its death
   b. the series of events that takes place from the formation of a cell until it divides again
   c. the sequence of events that assures each new cell of a set of chromosomes identical with that of its parent cell (mitosis)
   d. the growth of a cell until it is large enough to divide again
2. A diploid somatic cell:
   a. cannot undergo division again
   b. can undergo mitosis but not meiosis
   c. can undergo mitosis or meiosis
   d. can undergo meiosis but not mitosis
3. A cell in prophase of mitosis can be distinguished from a cell in prophase I of meiosis by:
   a. the presence of only half as many chromosomes in the meiotic cell
   b. the formation of tetrads in the meiotic cell
   c. the presence of twice as many chromosomes in the meiotic cell
4. The function of mitotic cell division in the life history of an organism is:
   a. reproduction of identical individuals if the organism is unicellular
   b. growth of an individual if the organism is multicellular
   c. repair of injured tissue
   d. all of the above
5. Substances that interfere with microtubule function interfere with cell division because:
   a. microtubules must be distributed equally to the new cells
   b. microtubules are involved in the precise separation of the chromosomes, which ensures that a complete set of chromosomes gets into each daughter cell
   c. without microtubules, crossing over cannot take place, and a cell with two identical nuclei is formed
   d. microtubules are essential to the disappearance of the nuclear membrane, and without them the chromosomes have to stay too close together within the nuclear membrane to be able to separate into two new nuclei
6. The importance of crossing over during meiosis is:
   a. it assures that one member of each homologous pair ends up in each new nucleus
   b. it results in chromosomes containing new combinations of genes
   c. it results in nuclei with too much or too little genetic material
   d. it ensures that the developing egg receives most of the cytoplasm from the oocyte
7. Both oogenesis and spermatogenesis involve equal division of the ______. However, unequal division of the ______ occurs during production of ______, whereas in production of ______ this division is equal.

THINKING CRITICALLY

1. Tetraploid plants are frequently larger and have larger fruits and flowers than their diploid relatives. What might account for the fact that octaploid plants (8N) tend to be tiny and scrawny, and produce few offspring?
2. Why is it necessary for cytokinesis to occur in such a way that each new cell receives some ribosomes, mitochondria, and, in plants, plastids?
3. Because the genetic information is carried equally by egg and sperm, what do you suppose to be the selective advantage of the inequality of size that has evolved between the tiny mobile sperm and the large immobile egg?
4. How would leakage from nuclear waste repositories affect the organisms that come into contact with it? Why is it difficult to design safe nuclear waste disposal facilities?

SELECTED KEY TERMS

anaphase, *p. 214*
autosome, *p. 204*
cell cycle, *p. 210*
cell plate, *p. 215*
centromere, *p. 204*
chiasma, *p. 218*
cleavage furrow, *p. 215*
crossing over, *p. 218*
cytokinesis, *p. 214*
diploid, *p. 206*
gamete, *p. 219*
genetic reassortment, *p. 218*
genetic recombination, *p. 218*
germ cell, *p. 208*
haploid, *p. 206*
homologue, *p. 204*
interphase, *p. 210*
kinetochore, *p. 214*
meiosis, *p. 204*
metaphase, *p. 214*
mitosis, *p. 204*
mitotic spindle, *p. 214*
oogenesis, *p. 221*
polar bodies, *p. 221*
prophase, *p. 214*
sex chromosome, *p. 204*
sister chromatid, *p. 204*
somatic cell, *p. 208*
spermatogenesis, *p. 220*
synapsis, *p. 218*
telophase, *p. 214*
tetrad, *p. 218*
tetraploid, *p. 209*

SUGGESTED READINGS

McIntosh, J. R., and K. L. McDonald. "The mitotic spindle." *Scientific American,* October 1989.

Murray, A. W., and M. W. Kirschner. "What controls the cell cycle." *Scientific American,* March 1991.

CHAPTER 12

CONCEPT GUIDE

After reading this chapter, you should be able to:

1. Compare the following pairs of terms: (a) homozygous and heterozygous, (b) dominant and recessive, and (c) genotype and phenotype. (Section 12-A)
2. Use a Punnett square to determine the genotypic ratio of offspring from parents who are heterozygous and homozygous dominant. Calculate the phenotypic ratio of the dominant and recessive traits. (Section 12-B)
3. Explain how to use a test cross to determine the genotype of an organism with a dominant phenotype. (Section 12-C)
4. Explain why a dihybrid cross with an independent assortment of genes produces a different phenotypic ratio than a monohybrid cross. (Section 12-D)
5. Compare the phenotypes of heterozygous genotypes that show complete dominance, incomplete dominance, or codominance. (Section 12-E)
6. Explain how gene pairs showing linkage differ from gene pairs that show independent assortment. Explain how we can determine from a phenotypic ratio if a pair of genes is linked or independently assorting. (Section 12-F)
7. Explain how the relative distance between genes on the same chromosome can be determined by studying recombinant chromosomes. (Section 12-G)

Gregor Mendel, studying the inheritance of pea characteristics, established many of the basic principles of genetics. Read more about his experiments and what he learned in Section 12-A. (Peter Arnold, Inc.)

Mendelian Genetics

It is common knowledge that plants and animals inherit most characteristics from their parents. Prehistoric people doubtless recognized a child's resemblance to parents, bred calves from the cows that gave the most milk, and saved some of the most productive grain for seed. However, most of this was practical, anecdotal lore, and the breeding of plants and animals was not analyzed scientifically until the twentieth century. Genetics is the study of patterns of inheritance, the manner in which characteristics (also called **characters** or **traits**) are passed from parents to offspring. This information is applied to the practical goal of breeding economically useful varieties of plants and animals.

Genetics is based on the work of Gregor Mendel, a monk (and later the abbot) at the monastery of Brünn, in what was then Austria. In 1866, Mendel published a completely new and thoroughly documented model of inheritance. However, the influential scientists then studying inheritance were absorbed in a maze of complex hypotheses, and the few who read Mendel's paper dismissed his model as trivial because it was so simple. Hence Mendel's work received little attention until after his death. It was rediscovered in 1900—almost simultaneously by three different people.

In the meantime, chromosomes had been named and their movements during mitosis and meiosis observed and described (see Sections 11-D and F). In 1902, several scientists realized that the chromosomes moved precisely as would be expected of the structures responsible for the patterns of inheritance Mendel reported. Once this connection between chromosomes and heredity was established, the science of genetics entered a period of productive research.

Mendel was the first person to recognize that genetic traits are inherited as separate particles. He did not actually see these hereditary particles, but he reasoned that they must exist because that would explain the patterns of inheritance shown by genetic traits. He proposed that organisms have a pair of particles for each inherited trait, one from each parent. We now know that the particles of inheritance are segments of chromosomal DNA molecules, and we call them **genes.** Many genes code for proteins, and in this way the submicroscopic gene manifests itself as some trait of the organism, such as curly hair. Genes are replicated and passed on to new cells as parts of DNA molecules, and this accounts for the observation that offspring inherit genetic traits from their parents. In sexual reproduction, a new individual receives half of its genes from the mother's egg and half from the father's sperm.

KEY CONCEPTS

- In most familiar organisms, each individual has a pair of genes for each trait, one inherited from each parent.
- Each of the individual's gametes (eggs or sperm), and hence each offspring, receives one of the individual's two genes for each trait.
- The pattern of inheritance of genes from generation to generation reflects the behavior of chromosomes in meiosis and fertilization.

Curiosity Questions

1? Why are peas and other plants often used in genetic studies? (See page 228)

2? How can people who are not albinos have children who are albinos? (See page 230)

3? Why do closely related people (such as brothers, sisters, and cousins) who mate with each other have increased chances of producing children with genetic diseases? (See page 230)

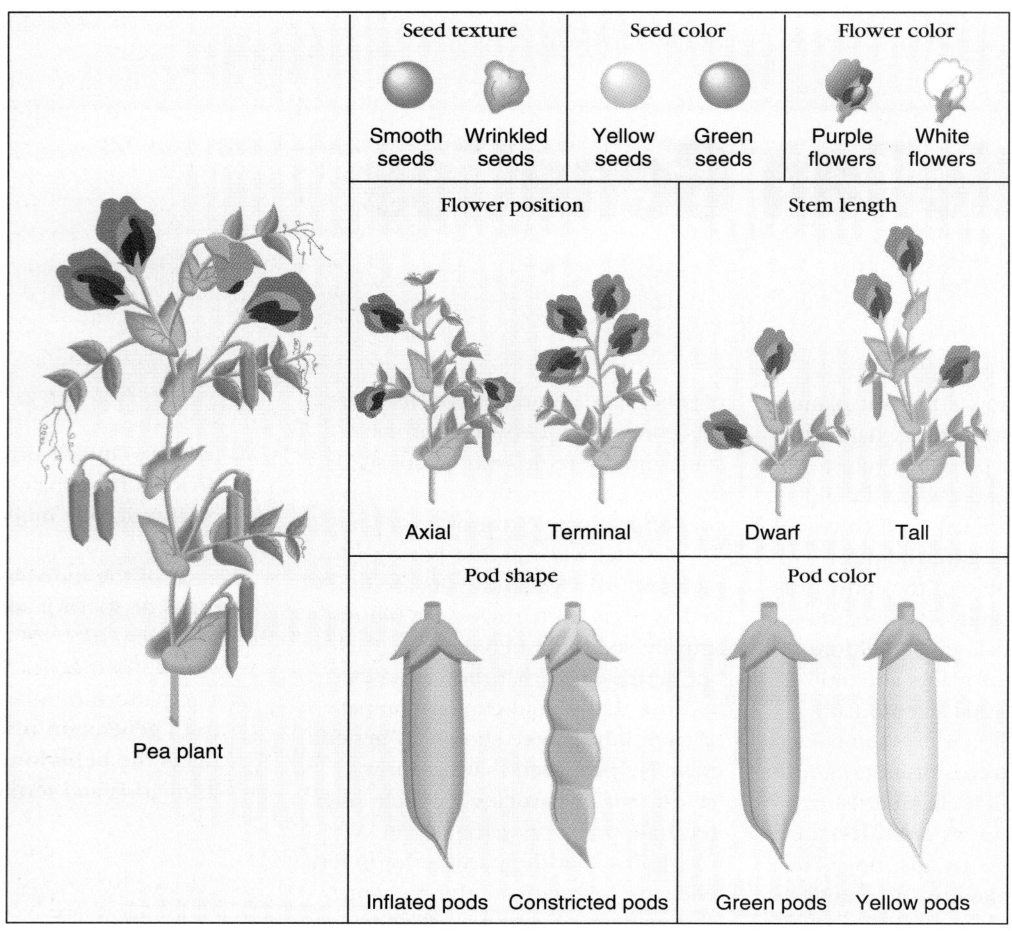

FIGURE 12–1

Genetic traits. Mendel studied seven different traits of pea plants, each of which appeared in two different forms. In each pair, the dominant form is the one on the left.

12–A A SIMPLE BREEDING EXPERIMENT

Mendel worked with garden peas, which were available in many different varieties. Each variety bred true to type. For example, tall pea plants of one variety always produced tall offspring, and plants of a dwarf variety always produced dwarf offspring. Pea flowers contain both male and female parts, and normally the flower pollinates itself. Hence, each plant is both male and female parent to its own seeds. Over the generations, this leads to considerable genetic uniformity. But it is possible to cross-pollinate peas artificially by transferring pollen from the male flower parts of one plant to the female flower parts of another. By crossing (that is, breeding together) plants of two varieties with contrasting traits, such as tall and dwarf varieties, Mendel could trace the inheritance of these traits. In all, Mendel worked with seven traits, each of which occurs in two distinct forms (Figure 12–1). ■

Mendel began by studying crosses involving only one trait at a time. Let us follow one such experiment, on the inheritance of flower color. In this experiment, Mendel crossed a pure-breeding strain of purple-flowered pea plants with a pure-breeding strain that produced white flowers. These plants are referred to as the **parental,** or **P_1, generation.** A cross between different parental strains, such as these, produces genetically mixed offspring known as **hybrids.** Mendel collected the hybrid seeds and planted them to see what traits this **first filial (F_1) generation** had inherited from the P_1 parents. When the F_1 hybrid plants matured, they all produced purple flowers. Mendel allowed these purple flowers to self-pollinate, and from them he collected over 900 seeds of the **second filial**

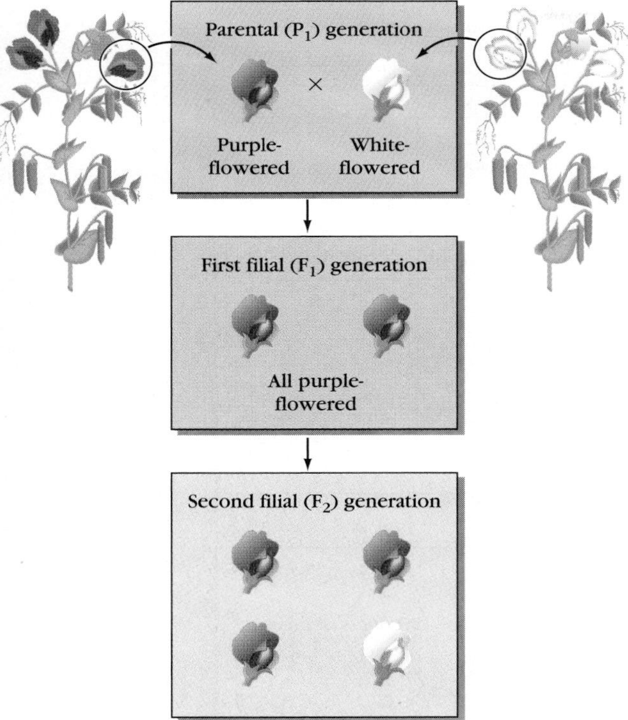

FIGURE 12–2

A cross between pure-breeding, purple-flowered and pure-breeding, white-flowered pea plants (P_1). All the offspring of the first filial (F_1) generation had purple flowers. Self-pollination of these F_1 offspring produced an F_2 generation of about ¾ purple-flowered and ¼ white-flowered plants.

(F_2) generation. Most of these F_2 seeds grew into purple-flowered plants, but about a quarter of them produced white-flowered plants (Figure 12–2).

Gene Pairs

Mendel saw that these results could be explained if an inherited trait, such as flower color, was governed by two "factors," which we now call genes. A plant received two genes for each of its traits, one from each parent. In turn, each plant passed on one of its two genes at random to each offspring.

If each purple-flowered parent had two genes for purple flowers, and each white-flowered parent had two genes for white flowers, then each offspring of the cross between the two received one purple-flower gene and one white-flower gene. When these plants reproduced, each egg or pollen grain would receive one of the two genes, so that half the eggs and half the pollen would contain each kind of gene. When the genes from egg and pollen combined, at random, a quarter of the offspring would have two purple-flower genes, a quarter would have two white-flower genes, and half would have one purple and one white.

This conclusion fits in with what we know about chromosomes and genes (Section 11–B). Diploid eukaryotic cells contain pairs of homologous chromosomes. **Homologous chromosomes** are usually of the same length, their centromeres are in the same position, and they bear genes for the same traits in the same locations. Since chromosomes come in pairs, so do the genes carried by the chromosomes.

A genetic trait occurs in two or more different forms. For example, the trait of color for a pea flower may be either purple or white. Therefore, the genes that govern the trait must come in more than one form. Alternative forms of a single gene are called **alleles.** In Mendel's pea plants, one chromosome may have the purple flower allele at the flower-color location, and its homologue can have either the purple-flower or the white-flower allele at the same location.

Any one pea plant may have two alleles for purple flowers, or two alleles for white, or one of each. An individual with two of the same allele is said to be **homozygous** for that allele. Plants with two alleles for purple or two for white are homozygous for flower color. An individual with two different alleles for a trait is said to be **heterozygous;** for example, plants with one purple and one white allele are heterozygous for flower color.

In Mendel's crosses, the original P_1 generation came from pure-breeding stock. This means that all the purple-flowered plants were homozygous for purple flowers, and the white-flowered plants were homozygous for white flowers. Each member of the F_1 generation must have received one allele for purple flower color from the purple-flowered parent and one allele for white flower color from the white-flowered parent. The F_1 generation was therefore heterozygous with respect to flower color.

Dominant and Recessive Alleles

Mendel found that all of the F_1 plants bore purple flowers. What had happened to the white alleles? Since self-crossing of the F_1 plants produced both purple- and white-flowered plants, the alleles for white must have been present in the F_1 plants, but masked. Mendel concluded that one allele of a gene may express itself (that is, appear as an observable trait in the organism) and mask the presence of the other allele, when the two occur together in a heterozygote. The allele that expresses itself is called the **dominant** allele, and the masked allele is said to be **recessive.**

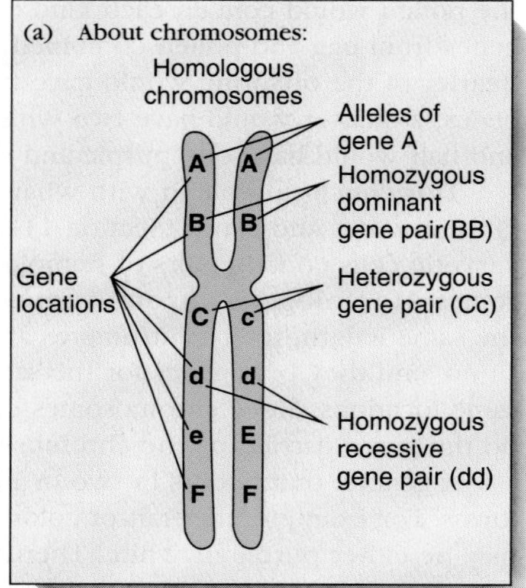

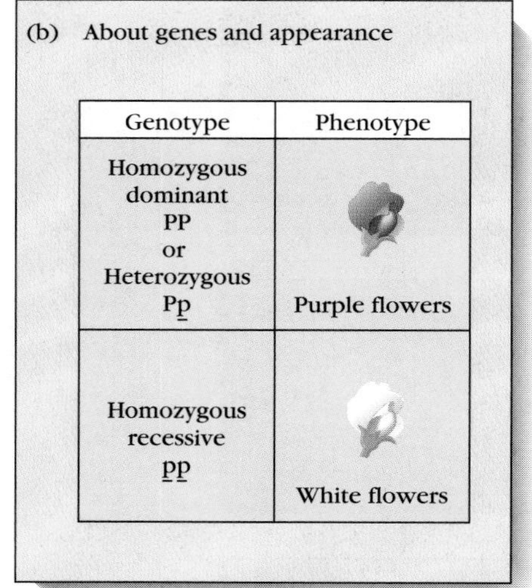

(b) About genes and appearance
Genotype
Phenotype
Homozygous dominant PP or Heterozygous Pp
Purple flowers
Homozygous recessive pp
White flowers

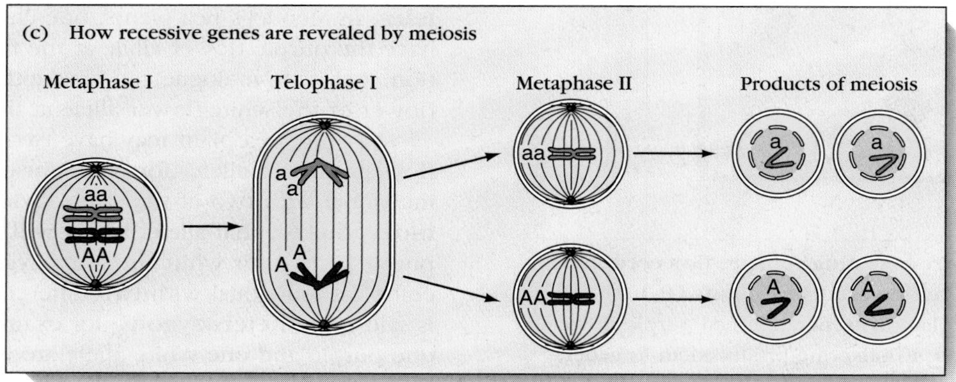

FIGURE 12–3

A genetics vocabulary. (a) Chromosome structure. Homologous chromosomes look similar, but may carry different alleles in their genes. **(b)** How phenotypes and genotypes are represented. Recessive phenotypes are produced when both alleles are recessive. **(c)** Revealing recessive genes. Organisms that are heterozygous for a trait can produce gametes that carry just the recessive allele. If two gametes carrying the same recessive allele combine, the offspring will express the recessive phenotype.

Homozygous and heterozygous purple-flowered plants cannot be told apart just by looking at them. The recessive allele can be detected only in the homozygous condition, when the dominant allele is not present.

Geneticists often use a shorthand, in which genes are designated by letters of the alphabet—capital letters for dominant alleles, and the lower case of the same letter for recessive alleles (Figure 12–3a). In the flower-color example, we can use *PP* for the purple-flowered parent, *pp* for the white-flowered parent, and *Pp* for the heterozygous F_1 plants.

2?3 Human albinism (Figure 12–4) is an example of simple dominance in which the pigmented skin allele (*P*) is dominant over the recessive albino skin allele (*p*). If two people who are both heterozygous (*Pp*) have a child resulting from gametes that both carry the recessive allele, there is a 25% chance that the child will be homozygous recessive (*pp*) and show albinism even though both parents were normally pigmented. Because *P* is much more common than *p* in the human population as a whole, the incidence of albinism in large, diverse populations is relatively low. Among people of European ancestry, 1 in 40,000 children is born with albinism. In contrast, among Arizona's Hopi, where it was traditional for tribal members to intermarry, albinism has occurred at a rate as high as 1 in

FIGURE 12–4

Human albinism. Due to marriages within the tribe and a high occurrence of the allele for albinism, Hopi Indians in Arizona have produced albino children at a rate as high as 1 in 200. (Field Museum of Natural History, Chicago)

200. In most populations, the chances of two carriers meeting and having a child is low, depending upon the overall frequency of the recessive allele. But, if a recessive allele is more common in a particular family or tribe, as is the albinism allele for the Hopi, then inbreeding among this group greatly increases the chances that this genetic defect will appear in children. In many cases, families are unaware of the presence of a recessive allele. In the past, marriages to close relatives often revealed these defects by producing children with two recessive alleles. Therefore, most human societies have long discouraged marriages of close relatives as a general preventive policy even before the genetic basis of inheritance was understood. ■

Genotype and Phenotype

Because of dominance, we cannot tell the **genotype,** or genetic makeup, of an individual that shows the dominant trait merely by inspection: pea plants with genotypes *PP* and *Pp* look alike in that both have purple flowers. In this case, both kinds of plants have a purple-flowered **phenotype:** the expression of their genes (Figure 12-3b). The phenotype can be observed in some way, perhaps visually, as in flower color, or chemically, as in the tests used to find out the blood types of people whose blood looks identical, and so on. An individual with a dominant phenotype may have a genotype that is either **homozygous dominant** (homozygous for the dominant allele) or heterozygous. An individual with a recessive phenotype, however, must have a genotype that is **homozygous recessive** (homozygous for the recessive allele).

Both an organism's genes and its environment can affect its phenotype. For example, a plant may be short because it has "dwarf" genes or because it is so poorly nourished that it cannot grow to the height dictated by its "tall" genes.

Law of Segregation and Meiosis

Mendel recognized that a pea plant's paired genes must separate from each other when the plant reproduces. This is now called Mendel's **law of segregation.** He also saw that the gametes containing the single genes must combine at random to form new gene pairs at fertilization.

Scientists in Mendel's time had not yet discovered the steps of meiosis (Section 11-F). However, we now know that the events of meiosis account for the law of segregation: the members of each pair of homologous chromosomes separate into different nuclei during meiosis, and so the genes carried by these chromosomes also become separated. For example, consider a diploid cell containing a pair of homologous chromosomes, one chromosome carrying the *A* allele and its homologue carrying the *a* allele (Figure 12-3c). As the chromosomes proceed through meiosis, *A* and *a* are separated into different nuclei, and so they end up in different gametes. Each chromosome is replicated before meiosis begins, but the two copies remain attached. During the first division of meiosis, each chromosome lines up with its homologue to form a **tetrad** (a group containing four chromosome copies) (Figure 12-3c). Then the homologous chromosomes are separated, one going to each pole of the spindle. Two nuclei are formed, each containing one of the two alleles (*A* and *a*). At the second meiotic division, the two copies of each chromosome separate, forming a total of four nuclei (Figure 12-3c).

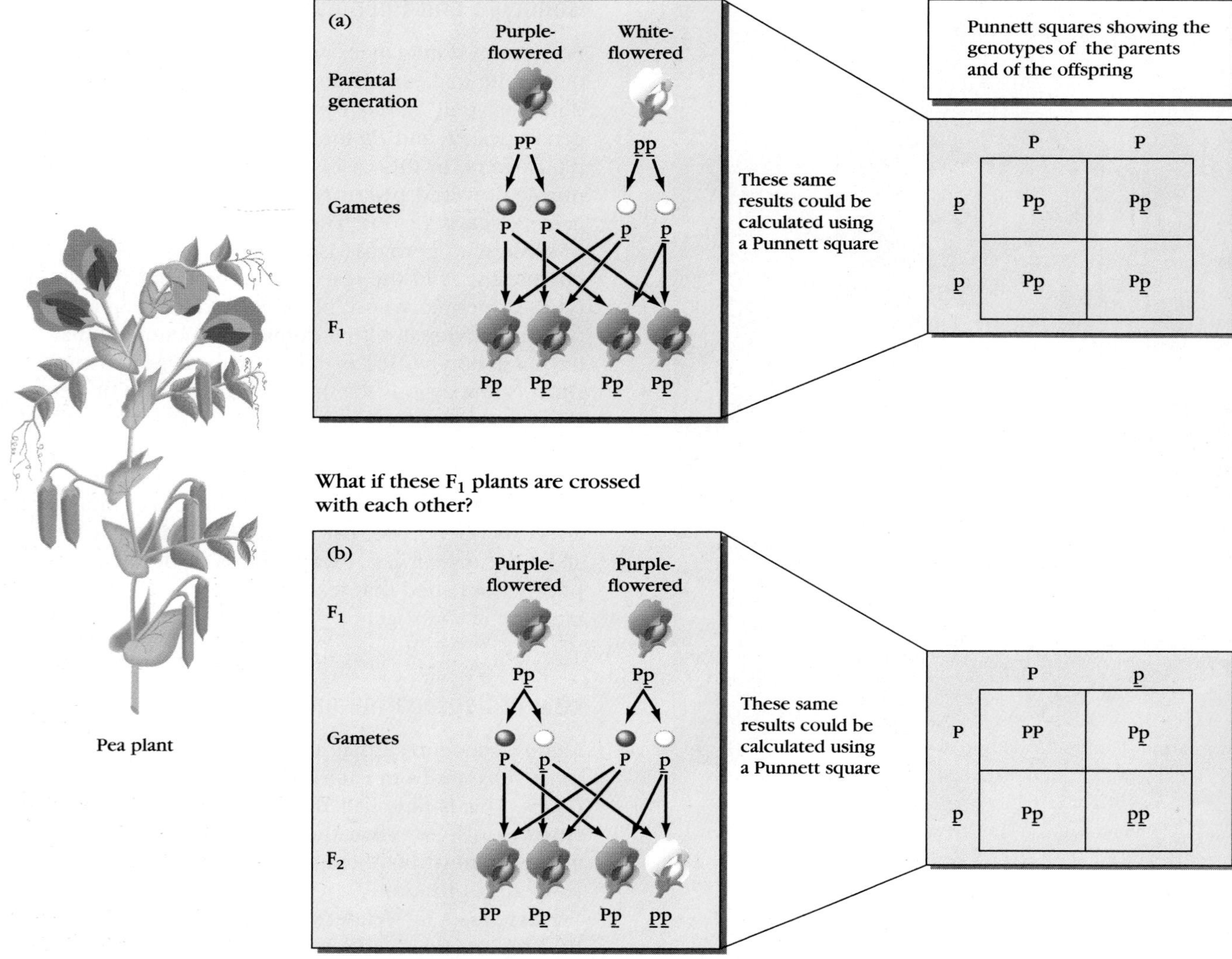

FIGURE 12–5

A monohybrid cross. (a) This diagram shows genotypes of gametes and offspring when a plant homozygous for purple-flower color is crossed with one homozygous for white-flower color and **(b)** the results if the hybrid offspring are then allowed to self-pollinate. Punnett squares of these two monohybrid crosses show the genotypes of the parents and the genotypes of the resulting offspring.

Monohybrid Cross

Now let us return to Mendel's flower-color cross. Such a genetic cross, in which only one trait of the parents (flower color, in this case) is of interest, is called a **monohybrid cross.** We can now diagram the pattern of inheritance as the genes are passed from one generation to the next (Figure 12–5a). As we just saw in our review of meiosis, the purple-flowered parent, *PP,* forms gametes containing a single *P* allele, and the white-flowered parent, *pp,* forms only *p* gametes. At fertilization, the F_1 offspring inherit a pair of flower-color genes, *P* from the purple flowered parent and *p* from the white-flowered one, so that they have the genotype *Pp.* These F_1 plants are purple-flowered because *P,* the dominant allele, expresses itself as purple flower color and masks the recessive allele.

When the F_1 generation reproduces, *P* and *p* segregate into separate cells at meiosis: half the gametes carry the allele for purple flowers (*P*) and half carry the allele for white (p). An egg has an equal chance of receiving either a P or a p allele, and it is equally likely to be fertilized by a sperm from a pollen grain containing either P or p. Therefore, four combinations are possible to produce the F_2 generation (Figure 12–5b).

(1) a P egg and P sperm (PP) (3) a p egg and P sperm (pP)
(2) a P egg and p sperm (Pp) (4) a p egg and p sperm (pp)

Once fertilization is complete, (2) and (3) are indistinguishable. Hence the possible genotypes in the F_2 generation are PP, Pp, and pp. We expect to find these in a ratio of $1PP : 2Pp : 1pp$ (or, equivalently, $\frac{1}{4}PP : \frac{1}{2}Pp : \frac{1}{4}pp$), since there are two ways to obtain the Pp combination and only one way to obtain each of the others. The ratio of phenotypes is three purple-flowered plants to one white-flowered plant, since the PP and the Pp plants all have purple flowers. (For convenience, we can write the genotype of a purple-flowered plant as P__. This notation indicates that at least one dominant allele is present. We use the underline in place of the second gene either to show that we don't know which allele is present, or to show that we are lumping together both PP and Pp individuals, which are phenotypically indistinguishable.)

Each organism receives two genes for each trait, one from each parent. During reproduction, each parent contributes, at random, one of these two genes to each offspring. Alleles are alternate forms of a gene, which are responsible for variations in a trait. Many pairs of alleles show a dominant-recessive relationship, with the dominant allele expressing itself and masking the presence of the recessive allele in the heterozygous condition. An individual's genotype is fixed at the time of fertilization, but its phenotype results from the interaction of all of the individual's genes with one another and with factors in the environment.

BIO-BIT

If you have brown eyes, you might have children with blue eyes.

The combinations produced by filling in all of the boxes show the possible genotypes of the F_2 individuals and the ratio in which they are expected to occur. In the flower-color cross, the F_2 generation includes PP, Pp, and pp individuals in the ratio of 1:2:1. If we know that one allele is dominant to the other, we can also predict the phenotypes of the F_2 generation. In this case, since P is dominant to p, the F_2 generation is expected to have three times as many purple-flowered (P__) as white-flowered (pp) individuals.

We could also calculate this outcome directly from the probabilities of each type of gamete. The gametes produced by each F_1 heterozygous parent can be written as ($\frac{1}{2}P + \frac{1}{2}p$). To find the distribution of genotypes in the next generation, we must multiply the proportions of gametes of each sex together, in the same way we multiply two binomials in an algebra problem:

$$\underbrace{(\tfrac{1}{2}P + \tfrac{1}{2}p)}_{\text{male gametes}}\ \underbrace{(\tfrac{1}{2}P + \tfrac{1}{2}p)}_{\text{female gametes}} = \underbrace{\tfrac{1}{4}PP + \tfrac{1}{4}Pp + \tfrac{1}{4}pP + \tfrac{1}{4}pp}_{\text{offspring genotypes}}$$

Combining the middle two terms, we come out with $\frac{1}{4}PP + \frac{1}{2}Pp + \frac{1}{4}pp$. This is the same 1:2:1 ratio as the genotypes found using the Punnett square.

Punnett squares allow us to determine the genotypic ratio of offspring quickly when we know the genotypes of the parents. If we know which allele is dominant, we can also predict the phenotypic ratio.

12–B PREDICTING THE OUTCOME OF A GENETIC CROSS

When we know the genotypes of parents used in a genetic cross, we can predict the genotypes of the offspring and their expected ratios. One way to do this is by drawing a **Punnett square** (named after geneticist Reginald Crundall Punnett). Each box of a Punnett square represents an individual offspring, and to predict the outcome of a genetic cross you first determine the alleles in the gametes made by each parent. The alleles are entered above and to the left of the boxes of the Punnett square and the boxes are filled in by combining alleles (Figure 12–5b).

12–C TEST CROSS

If both PP and Pp plants have the same dominant purple-flowered phenotype, how can we find out the genotype of a particular purple-flowered plant? The usual method is to cross such a plant with a plant of known genotype and observe the phenotypes of the offspring.

In a **test cross,** an organism of dominant phenotype but unknown genotype is crossed with one that is homozygous recessive for the trait in question. A white-flowered pea plant, in our example, has the homozygous recessive genotype, pp, and must pass on a p allele to each of its offspring. If the purple-flowered

Is this pea plant's genotype PP or Pp?

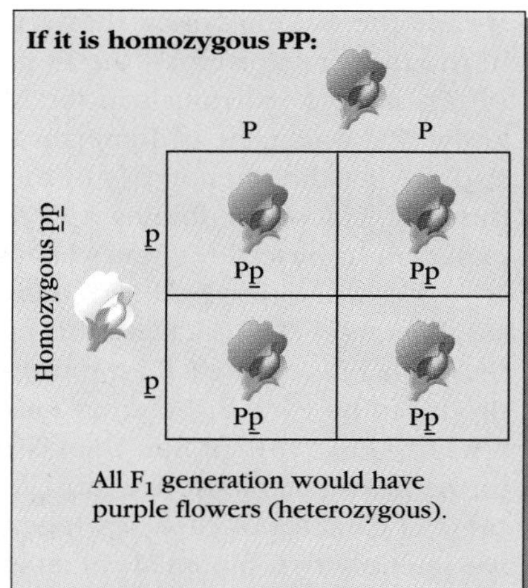

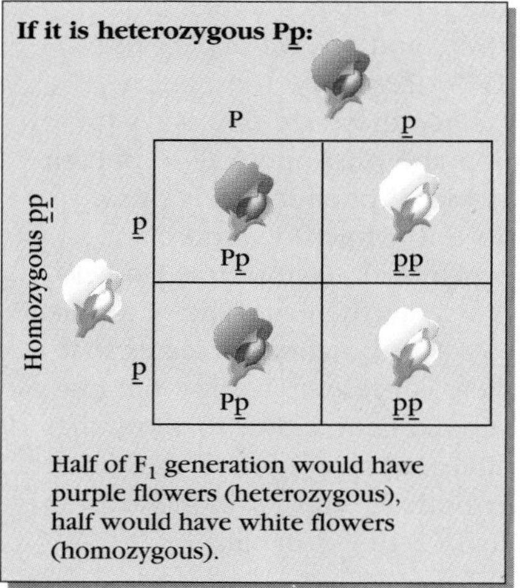

FIGURE 12-6

Test cross. A purple-flowered plant of dominant phenotype but unknown genotype is crossed with a white-flowered (homozygous/recessive) plant. By examining the offspring, it should be possible to determine the genotype of the purple-flowered parent plant.

plant of unknown genotype were actually heterozygous (*Pp*), we should expect half of the offspring of the test cross to be white-flowered and half purple-flowered. On the other hand, if the purple-flowered parent were homozygous (*PP*), it could pass only the *P* allele to its offspring and they would all be heterozygous, with purple flowers (Figure 12-6). So, if a purple-flowered × white-flowered test cross produces any white-flowered offspring, the purple-flowered parent must be heterozygous, because white-flowered offspring must have obtained a *p* allele from each parent. However, if all the progeny have purple flowers, it is not absolutely certain that the purple-flowered parent was homozygous *PP.* It is possible, though unlikely, for a plant to have all purple-flowered offspring even if its genotype is *Pp.*

The ratios shown in diagrams of genetic crosses represent *expected* proportions of offspring of different genotypes. The *actual* proportions of genotypes resulting from a particular cross depend on chance events of meiosis and fertilization. For example, if more gametes carrying *P* happen to fertilize more eggs than normally would be expected, the proportion of purple-flowered plants will be greater than is usually expected. It is also true that the more offspring—all purple-flowered—we get from such a cross, the more nearly certain we can be that the parent with the dominant phenotype is not heterozygous.

The genotype of an individual with a dominant phenotype can be determined from the phenotypic ratio of its offspring in a test cross with a homozygous recessive individual.

12-D THE DIHYBRID CROSS: INDEPENDENT ASSORTMENT OF GENES

In addition to his monohybrid crosses, Mendel performed **dihybrid** crosses. These involved plants with two different pairs of contrasting alleles. In one experiment, Mendel crossed plants homozygous for seeds that were both smooth and yellow with plants ho-

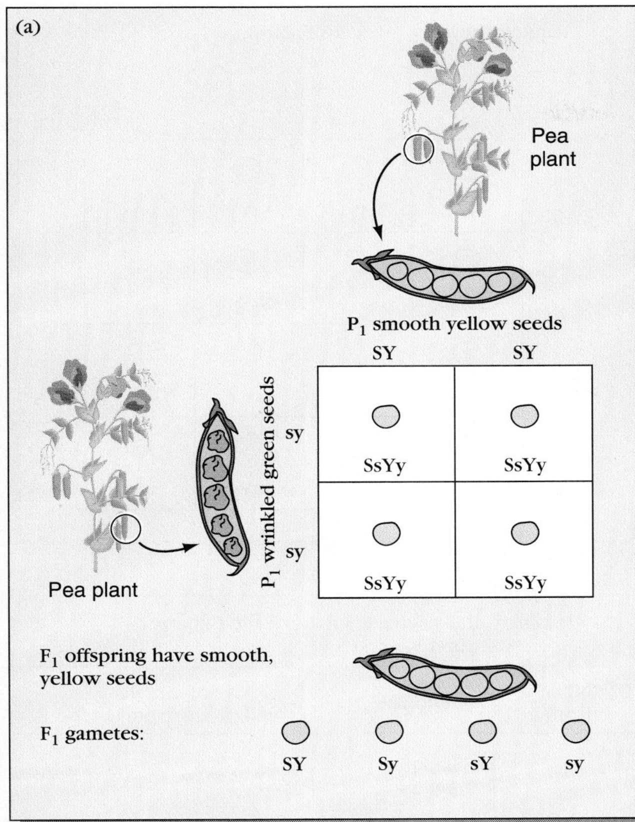

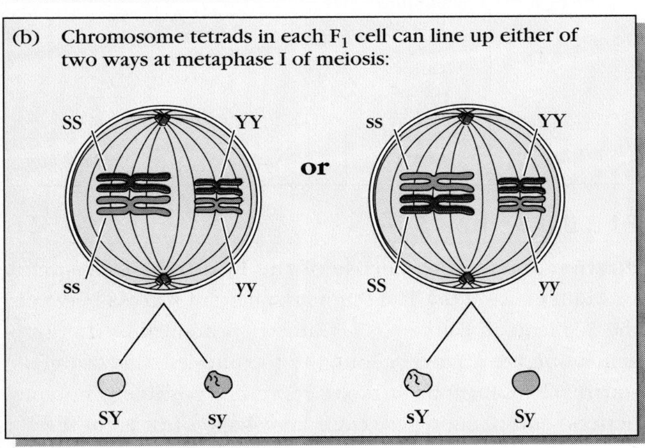

FIGURE 12-7

Dihybrid cross. (a) A pea plant with the two dominant traits, smooth and yellow seeds, is crossed with another plant that is recessive for both of these traits, wrinkled and green seeds. The resulting F_1 offspring are all heterozygous for both traits. **(b)** The genotypes of the gametes produced by the F_1 offspring vary according to how the chromosomes line up during metaphase I of meiosis.

mozygous for wrinkled, green seeds (Figure 12-7). All the F_1 offspring were smooth and yellow, showing that smooth was dominant to wrinkled and yellow was dominant to green.

Self-fertilization of the F_1 plants produced an F_2 generation of seeds with the following phenotypes:

315 smooth yellow	101 wrinkled yellow
108 smooth green	32 wrinkled green.

To find the ratio among these F_2 phenotypes, we take the number of offspring in the smallest category—32—and divide it into the number of offspring in each category. Then we round the quotient to the nearest whole number. We find that the phenotypic ratio in the F_2 generation is about 9:3:3:1. This ratio is now known to be typical of a dihybrid cross in which both pairs of alleles show a dominant-recessive relationship.

Mendel explained these data by assuming that genes governing seed color and seed texture move independently during reproduction. In this process of **independent assortment,** each pair of alleles behaved as it would in a monohybrid cross, without reference to the other pair. For example, when we consider only smooth versus wrinkled seeds, we find:

$$315 + 108 = 423 \text{ smooth, and } 101 + 32 = 133 \text{ wrinkled.}$$
$$423/133 = 3.18{:}1.$$

A ratio of 3.18 to 1 is quite close to the 3:1 ratio of a monohybrid cross. Similarly, the inheritance of yellow versus green seeds behaves like a monohybrid cross. In other words, this dihybrid cross is the product of two separate monohybrid crosses:

(3 smooth + 1 wrinkled) (3 yellow + 1 green) = 9 smooth yellow + 3 smooth green + 3 wrinkled yellow + 1 wrinkled green.

In this example, let S = smooth, s = wrinkled, Y = yellow, and y = green. The P_1 plants must have had genotypes *SSYY* (smooth, yellow) and *ssyy* (wrinkled, green). Since each gamete receives just one member of each gene pair, these parents must have produced gametes *SY* and *sy* respectively. All members of the F_1 generation have the genotype *SsYy,* giving a smooth yellow phenotype.

Self-fertilization of the F_1 plants produces a more complex situation. According to Mendel's "law of independent assortment," the members of each gene pair are sorted into gametes independently of the members of the other gene pair. We can see that this must be so from our study of meiosis: when the chromosome

BIO-BIT

In humans, there are 2^{23} different genetic combinations that can be produced in one gamete by independent assortment alone.

tetrads line up for the first division of meiosis (at metaphase I), the four alleles can be arranged in either of two ways (Figure 12–7). If *S* and *Y* line up opposite *s* and *y*, the gametes formed are *SY* and *sy*. If *S* and *y* line up opposite *s* and *Y*, the gametes *Sy* and *sY* form. Either arrangement is equally likely, and so each F_1 plant produces four kinds of gametes—*SY*, *sy*, *Sy*, and *sY*—in equal proportions. Each gamete always receives *one member of each pair of genes.*

To find all possible genetic combinations in the F_2 offspring, the F_1 gametes formed by independent assortment can be written along the sides of a Punnett square. Since any female gamete can be fertilized by any male gamete, there is a total of nine possible genotypes, falling into four phenotypes, in the F_2 generation (Figure 12–8).

(Independent assortment is not the invariable rule. It applies to genes carried on different chromosomes. In Section 12–F, we shall see what happens if the genes are on the same chromosome.)

Mendel's law of independent assortment states that the members of a gene pair are sorted into gametes independently of members of other gene pairs. Thus, the 9:3:3:1 dihybrid phenotypic ratio is the product of two separate 3:1 monohybrid ratios, one for each gene pair.

(a) Punnett Square of Mendel's F_1 Dihybrid Cross

F_1 Gametes	SY	Sy	sY	sy
SY	SSYY	SSYy	SsYY	SsYy
Sy	SSYy	SSyy	SsYy	Ssyy
sY	SsYY	SsYy	ssYY	ssYy
sy	SsYy	Ssyy	ssYy	ssyy

(b) Summary of Expected Phenotypes and Genotypes in This Dihybrid F_1 Cross

Number of offspring	Phenotype	Genotypes that could produce phenotype
9	Smooth, yellow	1 SSYY, 2 SSYy, 2 SsYY, 4 SsYy
3	Smooth, green	1 SSyy, 2 Ssyy
3	Wrinkled, yellow	2 ssYy, 1 ssYY
1	Wrinkled, green	1 ssyy

FIGURE 12–8

Punnett square for a cross of the F_1 dihybrid offspring in Figure 12–7. (a) The Punnett square of a cross between the F_1 plants in Figure 12–7. Gametes produced by the F_1 generation are arranged along the top and left side of the square. Combinations in boxes represent possible genotypes (letters) and phenotypes (color and shape of peas) in the F_2 generation. **(b)** A summary of the expected phenotypes and genotypes from this dihybrid F_1 cross.

12–E INCOMPLETE DOMINANCE AND CODOMINANCE

The pairs of alleles studied by Mendel all showed a dominant-recessive relationship. Indeed, Mendel chose his pairs because they appeared to behave as distinct alternatives. Since Mendel's time, geneticists have found many allelic pairs that do not behave this way. Instead, they show **incomplete dominance:** neither allele masks the presence of the other, and so the heterozygote has a different phenotype (as well as a different genotype) from homozygotes for either allele.

For example, in snapdragons (Figure 12–9a) flower color is controlled by alleles that show incomplete dominance. Plants with red or white flowers are homozygous. When a red-flowered plant is crossed with a white-flowered plant, the F_1 plants all have pink flowers, and they are all heterozygous for red and white flower color. Half their gametes will contain the allele for red flowers, and half will contain the allele for white flowers. The F_2 phenotype and genotype ratios will both be 1:2:1, just the same as the F_2 *genotype* ratios for any other monohybrid cross (Figure 12–9b). Because the heterozygote produces pink flowers, the expected phenotype ratio for the F_2 plants is:

1 red-flowered:2 pink-flowered:1 white-flowered

In incomplete dominance the heterozygote has a phenotype that is intermediate. In **codominance,** the situation is somewhat different. Here both genes are expressed independently in the heterozygote, rather than producing an intermediate phenotype. A good example of codominance is sickle cell anemia, a genetic defect that may cause death in homozygous recessive individuals. The sickle cell allele directs the production of a different "recipe" for hemoglobin that causes af-

(a)

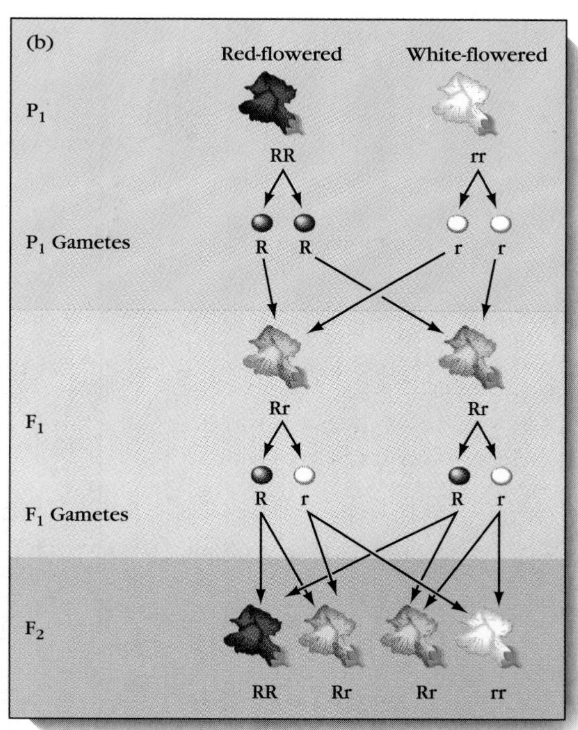

FIGURE 12–9

Incomplete dominance. (a) In snapdragon plants, the red and white flower alleles show incomplete dominance: thus heterozygotes have pink flowers. **(b)** The genotypes and phenotypes of snapdragon plants that result from monohybrid crosses. (J. D. Cunningham/Visuals Unlimited)

fected red blood cells to elongate, often forming crescent-shaped cells. People who are heterozygous produce some normal and some sickling hemoglobin. Simultaneous expression of these codominant alleles gives them resistance to malaria (because the protistan parasite that causes malaria has difficulty continuing its life cycle within sickled red blood cells—see Figure 19–17) *and* gives them normal oxygen transport. (Sickle cell anemia is more fully discussed in Section 13–B.)

In both incomplete dominance and codominance, heterozygotes have a different phenotype from homozygotes for either allele. In incomplete dominance, the phenotype is a sort of average of the homozygous phenotypes, while in codominance, the phenotypes of both alleles are expressed.

12–F LINKAGE GROUPS

In individuals heterozygous for two pairs of genes (for example, *SsYy*), Mendel had found that four types of gametes—*SY, Sy, sY,* and *sy*—occurred with equal frequency. This led to the law of independent assortment. However, later researchers found many pairs of genes that did not assort independently: offspring with two of the combinations showed up in higher proportions than expected, and those with the other two combinations were much rarer than expected.

How can we explain this? Looking back at Figure 12–7, we can see that the alleles *S* and *s* assort independently from *Y* and *y* because the S and Y gene pairs are on different pairs of homologous chromosomes. But there are many genes on each chromosome (see Figure 12–3). What happens if a cross involves two gene pairs carried on the same pair of chromosomes?

The chromosome moves as a unit during meiosis. Hence, we would expect genes on the same chromosome to stay together throughout the process and end up in the same haploid nucleus, rather than assorting independently, as genes on different chromosome pairs do. In other words, they will act as though they are linked together. Cases of **linkage**, in which genes are inherited as a pair or group, are common because every chromosome carries many genes.

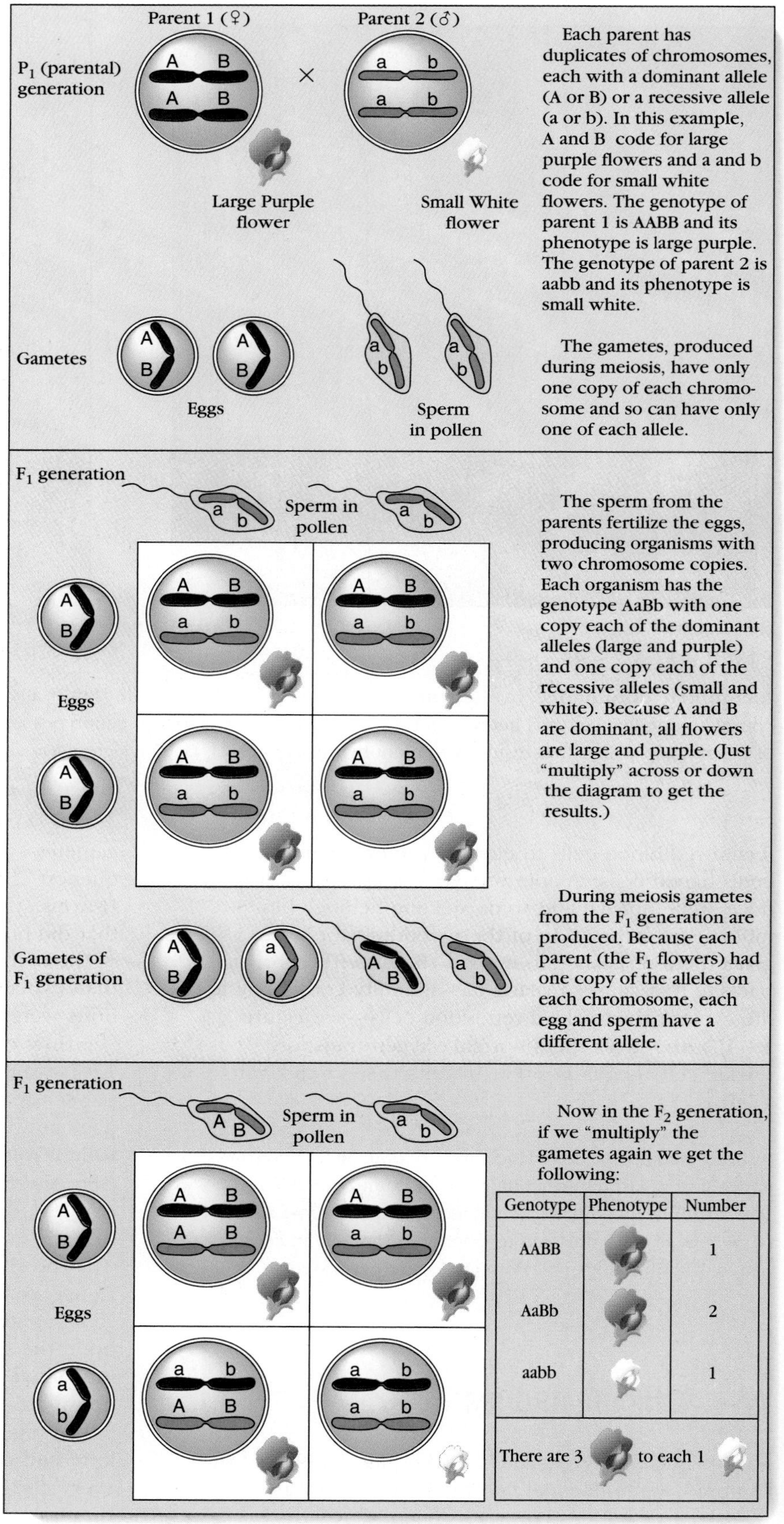

Parent 1 (♀) Parent 2 (♂)

P_1 (parental) generation

Large Purple flower

Small White flower

Each parent has duplicates of chromosomes, each with a dominant allele (A or B) or a recessive allele (a or b). In this example, A and B code for large purple flowers and a and b code for small white flowers. The genotype of parent 1 is AABB and its phenotype is large purple. The genotype of parent 2 is aabb and its phenotype is small white.

Gametes

Eggs

Sperm in pollen

The gametes, produced during meiosis, have only one copy of each chromosome and so can have only one of each allele.

F_1 generation

Sperm in pollen

Eggs

The sperm from the parents fertilize the eggs, producing organisms with two chromosome copies. Each organism has the genotype AaBb with one copy each of the dominant alleles (large and purple) and one copy each of the recessive alleles (small and white). Because A and B are dominant, all flowers are large and purple. (Just "multiply" across or down the diagram to get the results.)

Gametes of F_1 generation

During meiosis gametes from the F_1 generation are produced. Because each parent (the F_1 flowers) had one copy of the alleles on each chromosome, each egg and sperm have a different allele.

F_1 generation

Sperm in pollen

Eggs

Now in the F_2 generation, if we "multiply" the gametes again we get the following:

Genotype	Phenotype	Number
AABB		1
AaBb		2
aabb		1

There are 3 to each 1

FIGURE 12–10

Dihybrid cross involving linkage. Since the *A* and *B* gene locations are on the same chromosome, the alleles on each chromosome move as a unit rather than assorting independently (compare with *S* and *Y* pairs in Figure 12-7). Hence the genotypic and phenotypic ratios in the F_2 generation are the same as for a monohybrid cross rather than for an unlinked dihybrid cross.

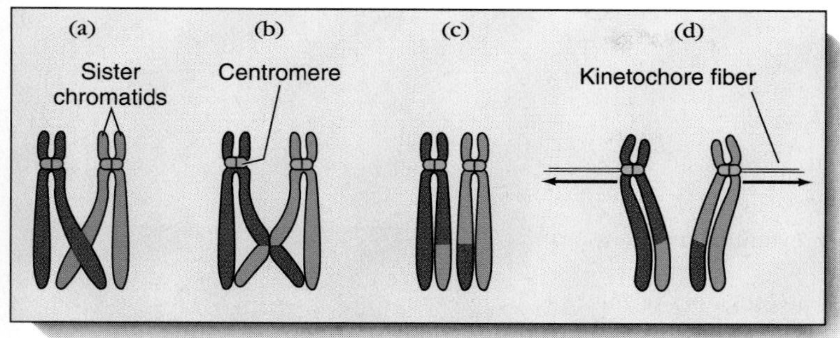

FIGURE 12-11

Crossing over. (a, b) During prophase I of meiosis, pairs of homologous chromatids may cross each other. **(c)** The crossed regions may break off and rejoin onto the opposite chromatid, creating new gene combinations that **(d)** separate during metaphase I of meiosis.

Let us consider an example of genetic linkage (Figure 12-10). In a cross between parents *AABB* and *aabb* in which *A* and *B* (and *a* and *b*) are linked, the only possible gamete from the homozygous dominant parent is *AB* because *A* and *B* are linked. Similarly, the other parent can form only gametes containing *ab*. The F_1 individuals are heterozygous *AaBb,* but the gametes they form must be like those of the P_1 individuals because *A* is still linked to *B* and *a* to *b;* unless crossing over occurs, these pairs will not assort independently. Instead, *A* and *B* are carried into one gamete by the chromosome that bears them both, while *a* and *b* are carried into the opposite gamete by the homologous chromosome (Figure 12-10).

The genotypic and phenotypic ratios from a cross involving linked genes differ from those expected if the genes were not linked. This deviation from the expected results is the clue showing that two gene pairs are linked. Hence we can identify linked genes by studying the ratios of offspring obtained in the F_2 generation. If the ratio is the Mendelian ratio 9:3:3:1 for a two-character cross, then the genes are assorting independently, and they are probably located on different chromosomes. However, if we find a different ratio, we are looking at linked genes.

The term **linkage group** refers to all the genes with inheritance patterns that show they are linked to each other, so a linkage group is really all the genes on one chromosome. This may be hundreds of genes.

In genetic linkage, genes on the same chromosome are inherited together. In fact, most of the genes on one chromosome end up in the same gamete, depending on the extent of crossing over that occurs during meiosis.

12-G CROSSING OVER

During meiosis, chromosomes exchange segments of DNA by **crossing over**. Crossing over rearranges genes that were previously linked so that they are now on opposite chromosomes, and vice versa (Figure 12-11). The resulting chromosomes are **recombinants,** bearing new combinations of alleles.

As far as we know, with the exception of the area around the centromere, a crossover is equally likely to occur at any point along the chromosomes. Hence, the closer together two genes are on a chromosome, the fewer possible points of crossover there are between them, and the less frequently such a crossover will occur. Two genes that are close together are likely to stay on the same chromosome and be inherited together. It follows that genes that are farther and farther apart on the same chromosome are more and more likely to be swapped between homologous chromosomes by crossing over.

We can estimate the relative distances between genes on the same pair of chromosomes by performing a large number of crosses, counting the offspring with each phenotype, and calculating the percentage of recombinant offspring, those showing new, "crossed-over" gene combinations. For instance, we could perform a test cross of *AaBb* dihybrid offspring from our previous example with homozygous recessive *(aabb)* individuals. We would then determine the percentage of offspring with phenotypes showing that they had received recombinant *Ab* or *aB* chromosomes in gametes from the heterozygous parent, rather than the original *AB* or *ab* chromosomes. If 10% of the offspring have crossed-over chromosomes (*Ab* + *aB*), then the A and B gene locations are said to be 10 map units apart on the same chromosome.

By compiling data from a large number of such crosses, using different combinations of gene pairs,

BIO-BIT

The scientific name of the fruit fly, *Drosophila melanogaster,* means "black-bellied sweet-lover" in Latin.

A Journey into Science in Process

Secrets of Mendel's Success

It is one of the ironies of science history that between 1860 and 1900 many biologists (including Charles Darwin) struggled in vain with the question of how characters are inherited: Mendel had the answer by 1866. Mendel's 1866 paper put forth a simple, yet clear, mathematical model containing the basic rules of heredity. His work has withstood the test of time, although others added to it after it was rediscovered in the first decade of the twentieth century.

Why did it take biologists 34 years to realize that Mendel had solved the problem?

First, Mendel himself may not have realized that he had discovered the general rules of heredity, because this was not exactly what he had set out to do. In Mendel's day, plant breeders had begun to study the inheritance of variation in plants such as melons and peas. Artificial hybridization to obtain novel or useful varieties of organisms was practiced by many agriculturists. However, most professional biologists were more interested in evolution and the differences between species. Franz Unger, Mendel's botany professor at the University of Vienna, believed that new species evolved from variants within existing species. Others were investigating the idea that new plant species evolve by hybridization, but their experiments produced confusing results. Mendel set out to learn how many different types of descendants pairs of hybrids could produce, and their proportions, in each generation. In the process he discovered the essential principles of heredity. However, his paper did not consider his results from this angle. Rather, he focused on the mathematical relationships he deduced and on evolutionary questions of speciation.

Previous investigators of this problem had no grasp of scientific method. In contrast, Mendel applied the methods he had learned as a student and teacher of experimental physics, especially the need to obtain numerical data from a large sample size. He was fascinated with numbers and kept records of weather, sunspots, and other phenomena throughout his life.

Mendel considered the methods he used to be his chief contribution:

1. He kept track of which generation each plant belonged to.
2. He determined how many different forms (genotypes) of offspring were produced by hybrids and their descendants in each generation.
3. He kept track of the ratios among these different forms and used them as clues to genotypes.

Because Mendel was interested in species and evolution, he studied populations of plants, as well as individuals. This was important because it revealed patterns that would not have shown up if he had studied only a few plants. This aspect of his work was partly a result of his having studied with Unger, but the remainder of his success was due to his own genius backed by willingness to put in long hours of tedious labor.

Mendel knew from the start that it was very important to choose the right subject for his experiments. He decided that he needed a plant in which (a) reproduction could be readily controlled; (b) there were pure-breeding varieties with contrasting traits; and (c) the offspring of crosses between different varieties were just as fertile as their parents.

The reproduction of peas is easy to control because of their flower structure. Most familiar flowers have male parts **(stamens)** and female parts **(pistils)** exposed to the air, and the pistils can receive pollen blown or rubbed off neighboring plants or carried by insects. However, in peas and their relatives, a modified petal, the **keel,** completely surrounds the reproductive parts, separating them from the outside world. Pollen from other flowers cannot enter, and each flower normally pollinates itself (Figure 12–B).

FIGURE 12–A

Gregor Mendel. Mendel discovered many of the essential principles of heredity through careful and meticulous breeding experiments. (Leslie Holzer/Science Source/Photo Researchers)

Mendel could permit self-pollination or he could cross-pollinate by hand, taking pollen from flowers of one variety of peas and placing it on the pistils of flowers of another variety. In order to do this, Mendel had to open one flower and pluck off a stamen. Then he had to open another flower and dust some of the first flower's pollen onto the second flower's pistil. (To be certain that the pistil was not fertilized by pollen from its own flower, Mendel had opened these flowers earlier and amputated the stamens before they produced pollen.) In order to obtain large numbers of offspring, Mendel hand-pollinated hundreds of flowers. This painstaking labor ensured that Mendel knew the parentage of every seed he collected. Equally time-consuming but important, he kept meticulous records of these crosses and their outcomes.

Mendel had studied mathematics and probability theory. Hence he realized the importance of obtaining a large number of offspring in his experiments in order to minimize the effects of "sampling error," which may result from looking at too few cases. He also began by studying just one genetic trait at a time, and he followed each trait through many generations. In this way he was able to discern the patterns of dominance and recessiveness, segregation, and the inheritance of one factor for each character from each parent. When he came to study two traits at a time, Mendel's mathematical ability quickly helped him to grasp the essentials of independent assortment.

Perhaps Mendel's most important contribution, however, was his recognition of the discrete nature of inherited characters. This is far from obvious. Many hereditary traits, such as human height, intelligence, and skin color, are continuous over a broad range. Today we realize that this is because each of these traits is controlled by many pairs of alleles, which interact with each other and make each offspring's phenotype somewhat different from those of its parents. However, virtually all biologists before 1900 had the mistaken notion that the inheritable characters of parents blended in their offspring. They also thought that each character was determined by an indefinite number of particles (genes) in each cell. If this were so, no consistent ratios would ever be found in crosses, and it is hard to see how any theory of genetics could ever be developed.

This blending theory presented a problem for Charles Darwin and other evolutionists. The theory of evolution by natural selection required that inherited variations be maintained from generation to generation, giving natural selection different traits to "select" among. If these variations blended with each other in each generation, they would eventually merge into some great average, and differences among individuals would disappear.

Why did these mistakes persist even after Mendel's work was published? Mendel's modesty did not help. He made little effort to publicize his work and once referred to his seven years' labor, involving more than 30,000 plants, as "one isolated experiment"!

It also seems that even nineteenth century biologists did not keep up with their reading. Although Mendel published very little, his most important paper went to 115 libraries. Ironically, a copy of Mendel's treatise was found among Darwin's papers, but the pages were still uncut: Darwin had never read it, and he never knew that Mendel's work would have removed one of the chief objections to the theory of evolution by natural selection.

A major reason Mendel was ignored was the arrogance of a professional biologist toward an amateur. Mendel sent his paper to an influential botanist, Carl Nägeli, with a cheerful and enthusiastic letter. Nägeli either did not understand Mendel's theory or, more likely, rejected it because it conflicted with his own theory of blending inheritance. Instead of supporting Mendel, Nägeli suggested that Mendel should repeat his pea plant experiments using hawkweeds. We now know that hawkweeds do not always follow the rules of sexual reproduction: they sometimes produce seeds without benefit of pollen, in which case the offspring have no male parent. It would obviously be extremely confusing to try to sort out patterns of heredity if you think you know which plants are parents but really don't! Unfortunately, Mendel took Nägeli's advice, to his great confusion. Nägeli's influential book on evolution and inheritance, published in 1884, did not mention Mendel or his work.

In addition to the hawkweed debacle, Mendel attempted to study inheritance in honeybees, which also have an aberrant sexual system (Section 16–E). After two such disasters, his earlier success with peas may well have seemed interesting but a minor fluke, irrelevant to the general problem of inheritance. It would be small wonder if Mendel became discouraged about the true significance of his earlier work on peas.

Mendel became abbot of the monastery where he spent most of his life, a position that left little time for his research. For years much of his energy went to resisting a tax on monasteries, which he regarded as unjust. It was not until after his death that his work was recognized and became the foundation of modern genetics.

Although we usually think of the basis of Mendelian genetics as being all Mendel's own work, later authorities deserve credit for some of the basic concepts and for almost all the terminology we use today. The terms "dominant" and "recessive" were coined by Mendel, but there is evidence that he didn't have the concept of gene pairs or alleles (although he worked with both and could predict and explain the results of crosses properly). He also phrased his results in terms of combinations of abstract mathematical symbols: he did not indicate any notion of a physical entity responsible for carrying inherited characters from parent to offspring.

geneticists have been able to construct **chromosome maps** for several organisms. That is, they have determined which genes are together on which chromosomes, in what linear order, and approximately how far apart. We now know a great deal about the chromosome maps of such organisms as the fruit fly *Drosophila* (Figure 12-12), laboratory mice, corn, and the pink bread mold *Neurospora,* all popular subjects for genetic experiments. Somewhat different techniques have also allowed geneticists to prepare genetic maps for some viruses and for the circular DNA of some bacteria.

Ethics forbids setting up controlled crosses of humans. Therefore few features of the human chromosome map were known until recently. However, it is now being filled in rapidly, using genetic engineering techniques to produce millions of copies of DNA segments, and then analyzing these copies to determine their nucleotide sequences.

In 1986, the United States committed itself to the formidable undertaking of mapping the entire human genome. This is expected to take another seven years and cost $3 billion. To print the nucleotide sequence for the entire human genome will require the equivalent of 200 Manhattan telephone books. A large number of researchers are working on this project, using high-technology equipment to sequence the DNA, as well as powerful computers to handle the resulting data. (See *A Journey into Science in Process:* The Human Genome Project, Chapter 11.)

By 1992, about 2400 genes had been mapped according to the particular chromosome they occupy in the human genome, and many had been localized to specific parts of their chromosome. New genes were being discovered at the rate of one every other day. One finding that surprised some researchers was that genes with related functions are often located near each other on a chromosome. With the broad outlines of a human chromosome map established, research will fill in more detail between major landmarks already known.

During meiosis, crossing over exchanges segments of DNA between chromosomes. This can rearrange and "unlink" genes that previously were inherited together. Because it appears that crossing over can occur almost anywhere along the length of a chromosome, genes that are closer together are less likely to be separated. This relationship of the distance between genes and the likelihood of separation is used to construct chromosome maps showing the general positioning of genes along the length of a chromosome.

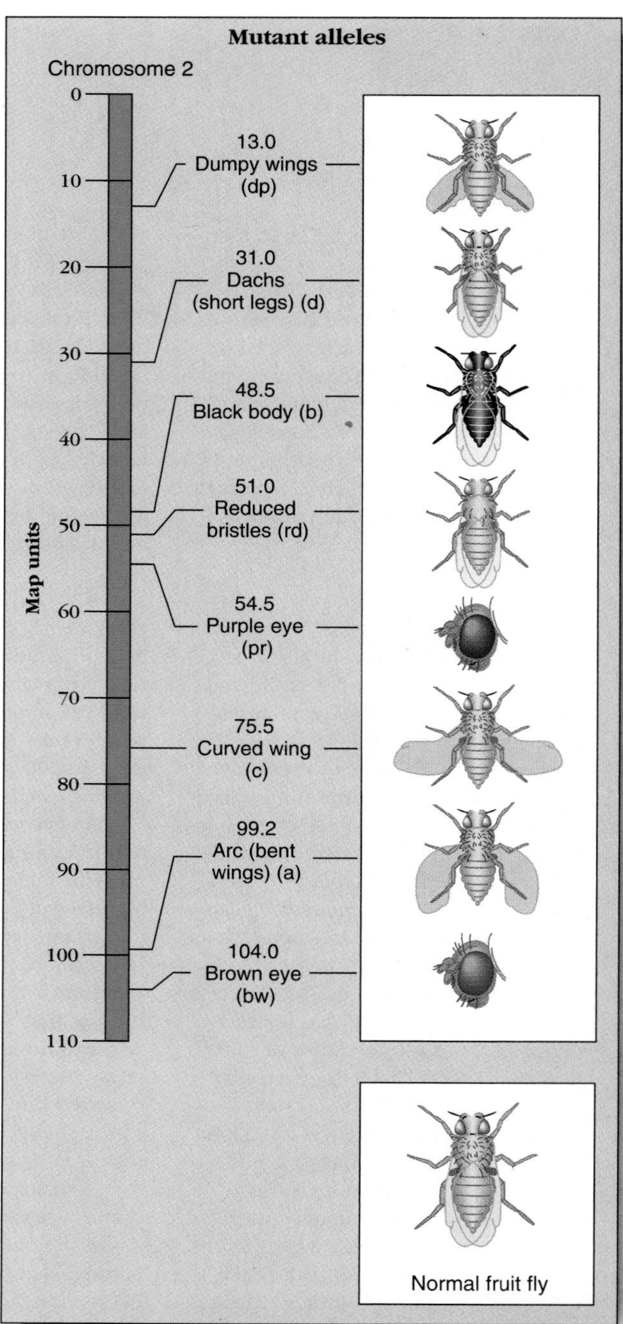

FIGURE 12-12

A map of some of the genes on chromosome 2 of the fruit fly *Drosophila.* The map shows mutant alleles that have been identified at the gene sites shown, along with their distance in map units from one end of the chromosome. Note that the mutations are all recessive (although a few fly mutations are dominant).

SUMMARY

1. Gregor Mendel's experiments formed the foundation of modern genetics, the study of patterns of inheritance of genetic traits. Mendel succeeded in discovering the principles of inheritance largely because of (a) his choice of an experimental organism, the garden pea plant, (b) his painstaking breeding of hundreds of plants, and (c) his use of mathematics to analyze his results. The inheritance of genetic traits can be summarized as follows:
   a. Genetically based traits are determined by discrete units, called genes, which are passed from parent to offspring during reproduction.
   b. A plant or animal contains pairs of genes that determine its genetic characteristics.
   c. During meiosis, the two members of each gene pair are separated from one another into different cells (law of segregation).
   d. At fertilization, each offspring receives a pair of genes for each characteristic. One member of each pair comes from the gamete of each parent.
   e. Genes for a trait may occur in different allelic forms, and one allele of a gene (dominant) may mask the presence of another allele (recessive) with which it is paired in a heterozygous individual.
   f. The genes from each parent remain distinct in the offspring. Even if they are masked in intervening generations by the phenomenon of dominance in some individuals, they may reappear in the phenotype of later generations.
   g. During meiosis, the genes of one pair assort independently of genes of other pairs, so long as they are located on different chromosomes (law of independent assortment).
2. The behavior of genetically determined traits in breeding experiments is paralleled by the behavior of the chromosomes during meiosis. This parallelism provides part of the evidence that genes are carried on chromosomes.
3. Genes located on the same chromosome are linked and are inherited together, except when they are separated by crossing over during meiosis.

SELF-QUIZ

The following problems will test your understanding of the ideas in this chapter.

1. In humans, the ability to taste phenylthiourea (PTU) is dominant. "Tasters" (*TT*) or (*Tt*) perceive an extremely bitter taste from very dilute solutions of PTU, while "non-tasters" (*tt*) experience no sensation even at much higher concentrations.
   a. What are the genotypes of Mr. and Mrs. Gagglebud, who can taste PTU, and who have three children, one of whom is a non-taster?

   What offspring phenotypes would be expected from the following crosses, and in what ratios?
   b. heterozygote × heterozygote
   c. homozygous taster × heterozygote
   d. heterozygote × non-taster
2. Two *Drosophila* (fruit flies) with normal wings are crossed. Among 123 progeny, 88 have normal wings and 35 have "dumpy" wings.
   a. What inheritance pattern is shown by the normal and dumpy alleles?
   b. What were the genotypes of the two parents?
3. If a dumpy-winged female (from Question 2) is crossed with her father, how many normal-winged flies will be expected among 80 offspring?
4. A number of plant species have a recessive allele for albinism. Homozygous albino (white) individuals are unable to make chlorophyll. If a tobacco plant heterozygous for albinism is allowed to self-pollinate and 500 of its seeds germinate:
   a. how many of these offspring will be expected to have the same genotype as the parent plant?
   b. how many seedlings will be expected to be white?
5. Sniffles, a male mouse with a colored coat, was mated with Esmeralda, an alluring albino. The resulting litter of six young all had colored fur. The next time around, Esmeralda was mated with Whiskers, who was the same color as Sniffles. Some of Esmeralda's next litter were white.
   a. What are the probable genotypes of Sniffles, Whiskers, and Esmeralda?
   b. If a male of the first litter were mated with a colored female of the second litter, what phenotypic ratio might be expected among the offspring?
   c. What would the expected results be if a male from the first litter were mated with an albino female from the second litter?
6. A kennel owner has a magnificent Irish setter, which he wants to hire out for stud. He knows that one of its ancestors was Erin-go-bragh, who carried a recessive allele for atrophy of the retina. In its homozygous state, this gene produces blindness. Before he can charge a stud fee, he must check to make sure his dog does not carry this allele. How can he go about this?
7. For a long time, human eye color was thought to be controlled by a single gene: brown (*B*) dominant over blue (*b*). Using this assumption,
   a. can brown-eyed parents have a blue-eyed child?
   b. can blue-eyed parents have a brown-eyed child?

   It is now known that at least two other pairs of genes can affect eye color, making it possible for blue-eyed parents to have a brown-eyed child, although this is rare.

8. In cats, the allele for black fur (*B*) is dominant to the allele for brown (*b*) and the allele for short hair (*S*) is dominant to the allele for long hair (*s*). Make a Punnett square for each of the following crosses:
   a. *BbSs* × *Bbss*
   b. *BBSs* × *Bbss*
   c. *BbSs* × *bbss*
   d. What proportion of the offspring from the cross shown in part b would be expected to be black with short hair?
9. In tomato plants, the gene for purple stems (*A*) is dominant to its allele for green stems (*a*) and the gene for red fruit (*R*) is dominant to its allele for yellow fruit (*r*). If two tomato plants heterozygous for both traits are crossed, state what proportion of the offspring are expected to have:
   a. purple stems and yellow fruit
   b. green stems and red fruit
   c. purple stems and red fruit
10. If 640 seeds resulting from the cross in Question 9 are collected and planted, determine how many are expected to grow into plants with:
   a. red fruit
   b. green stems
   c. both green stems and yellow fruit
11. If one of the parents from Question 9 is crossed with a green-stemmed plant heterozygous for red fruit, what proportion of the offspring would you expect to have:
   a. purple stems and yellow fruit?
   b. green stems and yellow fruit?
   c. green stems and red fruit?
12. A peony plant with straight stamens and red petals was crossed with another plant having straight stamens and streaky petals. The seeds were collected and germinated, and the following offspring were obtained:
   62 straight stamens, red petals
   59 straight stamens, streaky petals
   18 incurved stamens, red petals
   22 incurved stamens, streaky petals
   a. Which allele in each pair (straight vs. incurved stamens, red vs. streaky petals) is dominant?
   b. What were the genotypes of the parental plants?
   c. What further crosses would you make in order to get a definite answer for part a?
13. In a plant heterozygous for two pairs of genes (*AaBb*), state the chance that a pollen grain it produces will carry:
   a. an *A* allele
   b. an *a* allele and a *b* allele
   c. an *a* allele and a *B* allele
14. In cattle, the gene for straight coat (*S*) is dominant to its allele for curly coat (*s*). The gene pairs for red (*RR*) or white (*R′R′*) coat color show codominance; heterozygotes have a roan coat (*RR′*) (red lightened by intermixed white hairs).
   a. If a curly red cow is mated to a homozygous straight white bull, what will the genotype and phenotype of the calf be?
   b. If the calf is mated to a roan animal with curly hair, what are the possible offspring phenotypes?
15. A farmer has three groups of cows: white ones in the clover patch, red ones in the alfalfa field, and roan in the cornfield. He has a roan bull, Ferdinand, who services the cows in all three fields. (Refer to Question 14 for more information.)
   a. What color calves should he expect in each field, and in what proportions?
   b. Ferdinand dies from a bee sting and the farmer decides to make his herd of cows exclusively roan coat in memory of his beloved bull. He sells all the red and white cows, and vows to sell any red or white calves born later. What color bull should he buy to replace Ferdinand, if he wants to sell as many calves as possible?
16. The allele for pea comb (*P*) in chickens is dominant to the allele for single comb (*p*), but the alleles for black (*B*) and white (*B′*) feather color show codominance, *BB′* individuals having "blue" feathers. If birds heterozygous for both pairs of genes are mated, determine what proportion of the offspring are expected to be:
   a. single-combed
   b. blue-feathered
   c. white-feathered
   d. white-feathered and pea-combed
   e. blue-feathered and single-combed
17. A female *Drosophila* heterozygous for the recessive genes sable body and miniature wing was mated with a sable-bodied, miniature-winged male, and the following progeny were obtained:
   249 sable body, normal wings
   20 normal body, normal wings
   15 sable body, miniature wings
   216 normal body, miniature wings

   From these results, would you conclude that the two gene pairs involved are linked or unlinked?

   If you decided they are linked, which statement below describes the linkages found in the female parent?
   a. The genes for sable body and miniature wings were on one chromosome, and the genes for the normal forms of these traits were on its homologue. Some crossovers occurred during meiosis.
   b. The genes for sable body and normal wings were on one chromosome, and the genes for normal body and miniature wings were on its homologue. Some crossovers occurred.
18. In *Drosophila,* the gene for red eyes is dominant to the gene for purple eyes and the gene for long wings is dominant to the gene for dumpy wings. A female fly heterozygous for both traits is crossed with a male that has purple eyes and dumpy wings. The F_1 are:
   109 red eyes, long wings
   114 red eyes, dumpy wings
   122 purple eyes, long wings
   116 purple eyes, dumpy wings

   From these results, would you conclude that the two gene pairs involved are linked or unlinked?

   If you decided they are linked, which of the following statements describes the linkages found in the female parent?

a. The genes for red eyes and long wings were on one chromosome, and the genes for purple eyes and dumpy wings were on its homologue. Some crossovers occurred during meiosis.

b. The genes for red eyes and dumpy wings were on one chromosome, and the genes for purple eyes and long wings were on its homologue. Some crossovers occurred.

THINKING CRITICALLY

1. In performing a cross to determine the genotype of an organism having a dominant phenotype (*A*__), why is it preferable to mate it with a homozygous recessive individual rather than a known heterozygote?
2. Figure 12-7 shows that two pairs of chromosomes can line up two different ways during meiosis. Therefore, an individual heterozygous for two gene pairs on different pairs of chromosomes (*AaBb*) can form four different kinds of gametes. Consider an individual heterozygous for three gene pairs (*AaBbCc*) on three different chromosome pairs. How many ways can the chromosomes line up at metaphase I, and how many different kinds of gametes will be formed? How many different kinds of gametes are possible from a human being, with 23 chromosome pairs?
3. Mendel worked with two pairs of genes that were on the same chromosome but gave results not much different from the 9:3:3:1 ratio expected of unlinked genes. How frequently must such genes cross over in order for their linkage to go undetected in experiments like Mendel's? How could you tell that these genes were really linked after all?
4. Evaluate the saying "alike as two peas in a pod" in light of your study of this chapter.
5. Many scientists question the value of committing so much money and talent to the human genome mapping project: are we doing it simply because it is now technically feasible? They feel we would do better working as in the past: focusing on specific genetic defects, pinpointing the responsible genes, and studying these genes and their neighbors to discover ways to alleviate the suffering of people born with the defect. How do you feel our resources should be allocated?

SELECTED KEY TERMS

allele, *p. 229*
chromosome map, *p. 242*
codominance, *p. 236*
crossing over, *p. 239*
dihybrid, *p. 234*
dominant allele, *p. 229*
gene, *p. 227*
genotype, *p. 231*
heterozygous, *p. 229*
homologous chromosome, *p. 229*
homozygous, *p. 229*
hybrid, *p. 228*
incomplete dominance, *p. 236*
independent assortment, *p. 235*
law of segregation, *p. 231*
linkage, *p. 237*
monohybrid cross, *p. 232*
phenotype, *p. 231*
Punnett square, *p. 233*
recessive allele, *p. 229*
recombinant, *p. 239*
test cross, *p. 233*
trait, *p. 227*

SUGGESTED READINGS

Mendel, G. J. *Experiments in Plant Hybridisation.* Edinburgh, Scotland: Oliver and Boyd, 1965. An English translation of Mendel's original paper, with comments and a biography of Mendel by others.

Miller, J. A. "Mendel's peas: a matter of genius or of guile?" *Science News* 125:108, February 18, 1984. Discusses the accusation by modern statisticians that Mendel fudged his data.

Strickberger, M. W. *Genetics,* 3d ed. New York: Macmillan, 1984.

White, R., and J. M. Lalouel. "Chromosome mapping with DNA markers." *Scientific American,* February 1988.

CHAPTER 13

CONCEPT GUIDE

After reading this chapter you should be able to:

1. Explain why most mutations produce recessive alleles. (Section 13-A)
2. Explain why most dominant lethal alleles are rapidly eliminated from a population, while recessive lethal alleles can become quite common. (Section 13-B)
3. Explain what phenylketonuria and albinism have in common. (Section 13-C)
4. List all of the possible parental genotype combinations that could produce a child with AB blood type. (Section 13-D)
5. Explain how a trait that is controlled by multiple alleles is different from a polygenic trait. (Section 13-E)
6. Explain how the sex of an offspring is determined in most animals. (Section 13-F)
7. Explain why recessive alleles of sex-linked traits are more likely to be expressed in the sex that has heterozygous sex chromosomes. (Section 13-G)
8. Explain the difference between sex-linked and sex-influenced genes. (Section 13-H)
9. List five factors that may affect the expression of a gene. (Section 13-I)
10. Describe how and when nondisjunction and translocation occur and the results of these abnormal events. (Section 13-J)

This black rat snake is anything but black! Its genetic code does not permit the formation of pigment and results in its white scales and red-colored eyes. Read about the causes of albinism in **Section 13-C.** (Jim Merlt/Visuals Unlimited)

Inheritance Patterns and Gene Expression

Lack of pigmentation (see chapter opening photo)—like many abnormalities of humans and other organisms—is a phenotypic expression of mutant alleles of genes. Much research has been devoted to finding out what these mutations are, how the mutant alleles are inherited and expressed, and how their expression is affected by factors in the organism's environment.

Genes are lengths of DNA that act as units of hereditary information. Many genes code for the sequences of amino acids in polypeptides and proteins (Chapter 10). Amino acid sequences determine the protein's functions, how it folds, and how it interacts with other molecules (Section 3-D). Many mutations change the DNA of a gene in such a way that it ends up coding for a different sequence of amino acids. As a result, a different protein (or perhaps no protein) is made. This, in turn, may alter the organism's metabolism or structure, and this is how a mutated DNA sequence results in a difference in the observed phenotype.

In this chapter we consider how some genetic differences between individuals produce different phenotypes. We shall look at some of the factors that control phenotypic expression of genes and introduce some new genotypic and phenotypic ratios that provide clues to the nature of particular genes.

KEY CONCEPTS

- The effect of a mutation of a gene depends on how much it changes the protein encoded by the gene and on how important the protein is to life.
- A gene's expression depends on the other genes present in the genome and on factors in the organism's external environment.

Curiosity Questions

1? How can two short parents have a very tall child? (See page 255)

2? Why are almost all hemophiliacs male? (See page 260)

3? Why aren't there as many bald women as men? (See page 262)

13–A PHENOTYPIC EXPRESSION OF MUTATIONS

A **mutation** is a rare, random, and inheritable change in a cell's genetic material (Section 9-E). The mutation becomes part of the genotype of the cell and of all its descendants, and this may result in an abnormal phenotype. Mutations in somatic (body) cells may cause damage, including cancer, to the parts of the body that arise from the mutated cells. Mutations in germ (reproductive) cells may cause no noticeable abnormality in the individual in which they occur, but they will be passed on to offspring and may be expressed in the offspring's phenotype.

Whatever the change wrought by a mutation, it is much more likely to harm than to help the delicate evolutionary engineering of the protein encoded by the mutated gene. Why is this? Even though mutations are rare, evolution has been going on so long that a particular mutation is likely to have occurred many times before. If the protein produced by the mutation was superior to the protein it replaced, the superior protein would be preserved through the process of natural selection and passed on to future generations. Therefore it would now be a common allele in the population.

Most mutations produce recessive alleles (see Figure 12-13). Although opinions differ about why this is so, we do understand some mutations on a molecular level. For example, one study analyzed the various mutations found in seven men with **hemophilia,** a genetic disorder in which blood fails to clot normally. Three of these men had mutations that resulted in premature *STOP* codons, causing production of an incomplete version of a protein needed for clotting; three others had mutations that deleted thousands of nucleotides from the gene. The seventh man had a mutation that substituted one amino acid for another, resulting in a milder form of hemophilia.

An allele that codes for no protein or for an inactive protein will be at least partly recessive to an allele that codes for a functional protein molecule. In a heterozygote, the normal allele directs the production of the normal protein, while the recessive allele contributes no functional protein. If enough normal protein is produced, the individual's phenotype appears normal, and the normal allele is dominant. In an individual homozygous for the inactive allele, no protein is made. Therefore, the trait is not expressed, and the recessive phenotype is the absence of the normal trait (Figure 13-1).

If one copy of the normal allele does not ensure synthesis of enough protein to produce the normal phenotype, a heterozygote might be noticeably different from either homozygote. Hence the normal and mutant alleles would show incomplete dominance.

In codominance, both alleles code for functional proteins, but these may have different properties. As a result, the phenotype of the heterozygote may be distinct from that of either homozygote.

BIO-BIT

In a condition called Congenital Insensitivity to Androgen Syndrome (CIAS), a gene mutation on the X chromosome of a genetic male makes most body tissues unresponsive to testosterone. The person develops breasts, female genitalia, a shortened vagina, but no uterus or oviducts. The testes remain in the body cavity.

Most mutations produce alleles that are both unfavorable and recessive. Many mutations result in no proteins or defective proteins being produced. If the individual is heterozygous for the mutant allele, the normal allele may produce enough of the normal protein to result in a normal phenotype. If the individual is homozygous for the mutant allele, the organism will lack that protein and, therefore, will lack its important function.

13–B LETHAL ALLELES

Suppose a mutation destroys the genetic code for a protein that is essential to life. An organism that fails to produce an active form of that protein will die prematurely. The allele that fails to encode a vital protein is called a **lethal allele.** Dominant lethal alleles are possible, but most are rapidly eliminated because they cause death before the individual carrying them reproduces. (Exceptions are those not expressed until later in life, as in the case of Huntington's disease in humans.) However, recessive lethal alleles are eliminated by selection only when they occur in homozygotes. More often they occur heterozygously, masked by functional dominant alleles that permit the individual to survive and reproduce. So, nothing prevents recessive lethal alleles from spreading to future generations and becoming quite common. In fact, it has been calculated that the average human is heterozygous for perhaps three to five lethal recessive alleles. This is higher than the figure for many other organisms, and is part of the reason that marriages between close relatives produce a higher proportion of offspring with lethal inherited traits in humans than in most other species.

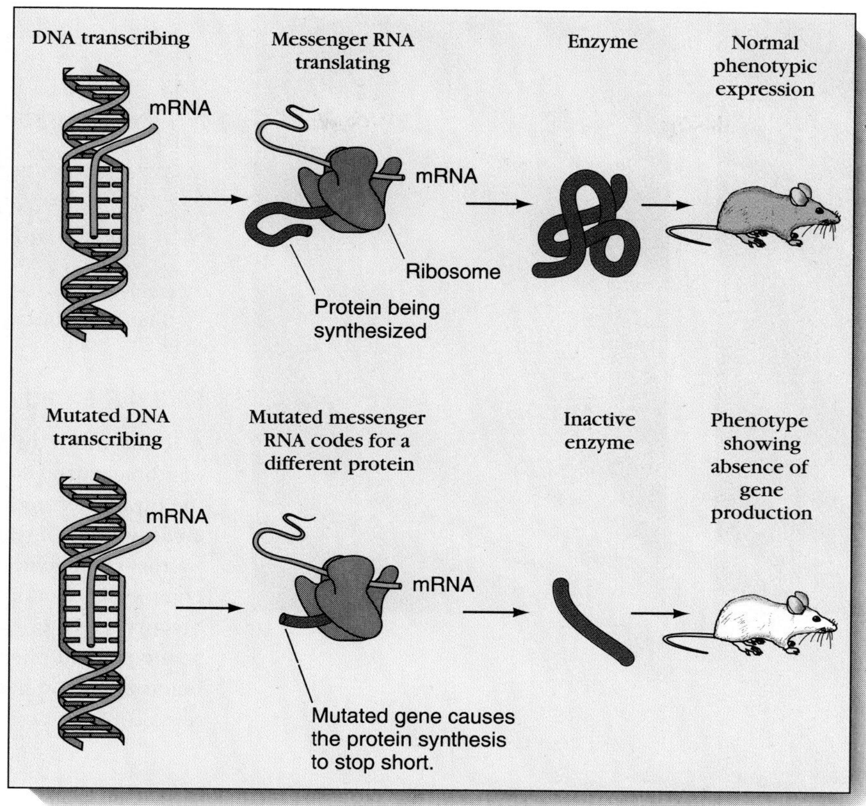

FIGURE 13–1

How some mutations may result in recessive phenotypes. The DNA in this example codes for an enzyme needed to make a pigment in mouse fur. The mutant makes an incomplete enzyme that does not permit the mouse to produce pigment. Thus, an albino animal results.

If just one copy of a normal allele does not produce enough of its protein for normal body functioning, the heterozygote has a different phenotype from either homozygote. An example in humans is the lethal allele that causes shortening of the middle bone in the fingers (brachydactyly) in heterozygotes. This makes the fingers appear to have only two bones instead of three. In homozygotes, this allele results in abnormal development of the skeleton. Homozygous babies lack fingers and show other skeletal defects that cause death in infancy.

In a marriage between two brachydactylic people, one out of every four children would be expected to be homozygous for the lethal allele and die during infancy; half would be expected to be heterozygous and show brachydactyly; and one fourth would be expected to be normal (Figure 13-2). This 1:2:1 ratio among offspring is typical of lethal alleles when the normal allele does not mask the mutant one completely.

Some lethal alleles are mutations of genes that code for proteins so essential that without them the embryo does not develop normally. In pregnancies with more than one offspring, (dogs, mice, cats—and humans, too), embryos that die early may be resorbed back into the uterus, and a 2:1 ratio may be observed when the remaining offspring are born: two thirds heterozygotes to one third homozygous normal offspring (Figure 13-3). In mice, for example, the short-tail allele) *(t)* causes early embryonic death in the homozygote. The embryo is then resorbed. If such embryos are taken from the uterus early in pregnancy, before they can be resorbed, they are seen to have no backbone and none of the tissue that later forms the muscles, kidneys, and many other important organs. Heterozygotes *(Tt)* have shorter tails than normal mice *(TT)*.

A similar lethal allele exists in cats. A Manx cat is heterozygous for this allele, and its backbone is so short that the cat has no tail. The last vertebrae of the

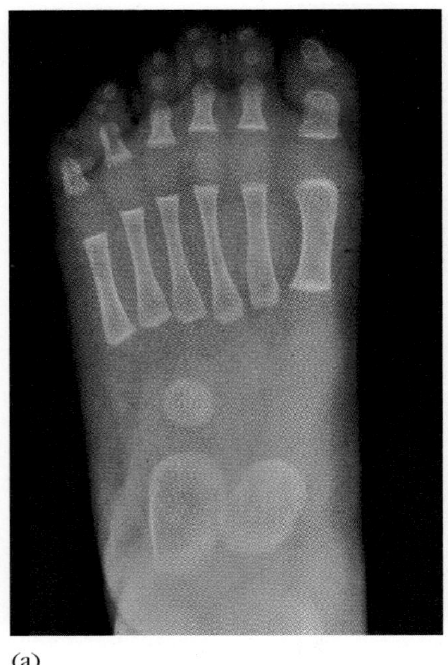
(a)

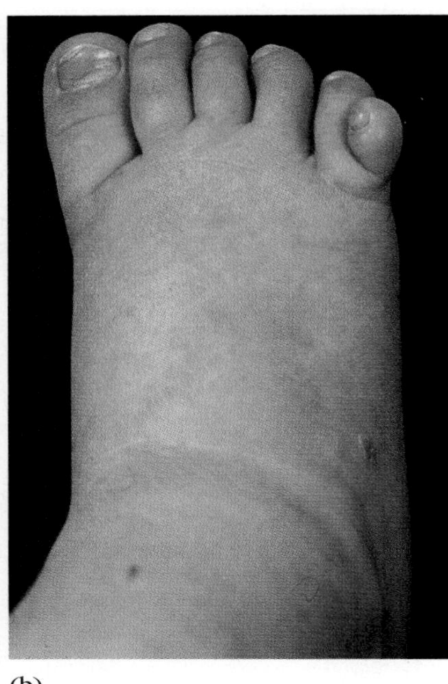
(b)

	B	b
B	BB Normal skeleton	Bb Short fingers
b	Bb Short fingers	bb Abnormal skeleton -Dies in infancy

1 normal, 2 brachydactylic, 1 dies
1 : 2 : 1 ratio

FIGURE 13-2

A lethal allele in humans. (a) An x-ray of a brachydactylic foot. Note the extra toe on this left foot. **(b)** A right foot of a child with brachydactyly, again with an extra toe. **(c)** The normal allele, *B*, is incompletely dominant to the allele for brachydactyly, *b*. Note the characteristic genotypic and phenotypic ratios for this heterozygous × heterozygous cross. (Patricia Barber/Custom Medical Stock)

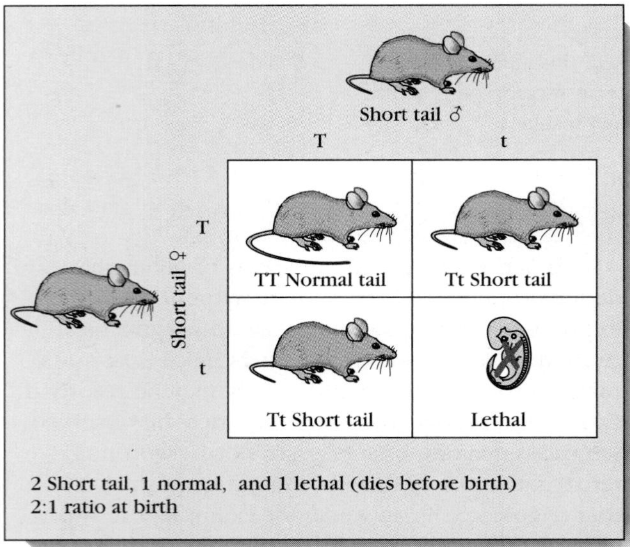

FIGURE 13-3

A cross involving a lethal allele that shows a 2:1 ratio among the offspring. The short-tail allele (*t*) in mice causes death and resorption of homozygous recessive embryos early in development. Thus, homozygous recessive embryos never appear among offspring born to short-tailed parents. One-third of the progeny are normal long-tailed (*TT*), and two-thirds are short-tailed (*Tt*).

back and the last part of the digestive tract may be abnormal, and in this case the cat may have problems that prevent it from living out a normal life span (Figure 13-4).

Sickle Cell Anemia. A famous human allele that is frequently lethal in the homozygous condition is the one responsible for **sickle cell anemia.** The gene involved codes for the beta polypeptide chain of hemoglobin—the oxygen-carrying protein found in red blood cells and responsible for their red color (see A Journey Through Protein Structure, Chapter 3). The sickle allele results from a change in just one nucleotide pair. This results in the substitution of the amino acid valine for glutamic acid as the sixth amino acid in the hemoglobin beta chain (also see Table 10-1).

This seemingly small change has drastic consequences. When red blood cells containing the abnormal hemoglobin are exposed to low oxygen levels, the hemoglobin molecules aggregate and form fibers. These fibers distort the cells into odd shapes, such as sickles (Figure 13-5). Sickled cells become stuck in the smaller blood vessels and restrict circulation to the areas supplied by these vessels. The sickled cells also break down easily, leaving the victim with fewer red blood cells than normal, a condition known as anemia.

(a)

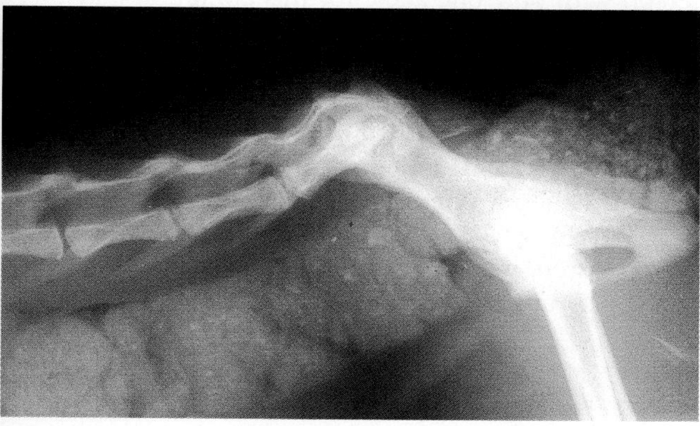

(b)

FIGURE 13-4

A Manx cat. (a) Manx cats are heterozygous for the short-tail gene, and thus have no tails. **(b)** This x-ray shows missing tail vertebrae in a Manx cat. (a, Bob Riedlinger/Custom Medical Stock; b, courtesy of Dr. Nathan Dykes)

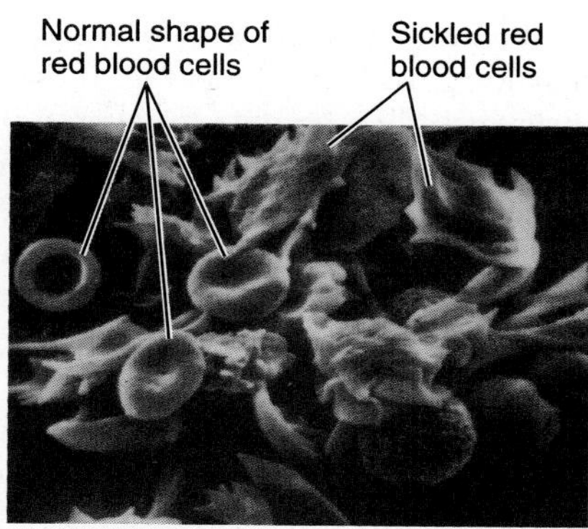

FIGURE 13-5

Blood cells from a person with sickle cell anemia. When the blood is low in oxygen, the red blood cells assume long, thin, jagged "sickle" shapes. Here, some red blood cells still have their normal shape: round, flattened discs indented in the middle. (David M. Phillips/Visuals Unlimited)

Poor circulation and anemia deprive the tissues of needed oxygen, producing symptoms such as tiredness, headaches, muscle cramps, poor growth, and possibly failure of organs such as the heart and kidneys.

The sickle allele shows codominance with the normal allele: heterozygous people produce both normal and abnormal beta chains. Their red blood cells sickle only when the oxygen level is extremely low—for instance, at very high altitudes. Without special blood tests, heterozygotes may not know that they carry the sickle allele. People homozygous for the sickle allele are much more severely affected because all of their beta chains are abnormal.

An individual heterozygous for a genetic condition is referred to as a **carrier,** whereas an individual homozygous for the condition is called an **affected individual.** Sickle cell carriers are sometimes referred to as "having sickle cell trait." This terminology is unfortunate, since it suggests that the carrier is less fit than the normal homozygote, which is not usually the case.

About half of the people homozygous for the sickle allele die by the age of 20. Furthermore, women in this group have fewer babies than do heterozygous or homozygous normal women. We might expect natural selection to keep such a lethal allele at a very low frequency in the population, as many people homozygous for the sickle allele die without producing offspring. Yet in large areas of tropical Africa, 20 to 40% of the people are heterozygous for the allele. This strongly suggests that heterozygotes have some selective advantage so great that it offsets the disadvantage of recessive homozygotes. In 1953 it was noted that these same African regions are precisely those with the highest rates of death from a virulent form of malaria, a disease caused by a parasite of red blood cells.

In the United States, where cases of malaria are few, African Americans still carry the trait in very high numbers, averaging one case of sickle cell anemia for every 500 people. Cases of sickle cell anemia in Americans of non-African descent are very rare.

Having at least one copy of the sickle allele lowers a person's chances of developing malaria. Red blood cells containing abnormal hemoglobin sickle more readily when they are infected with malaria parasites. When a cell sickles, the parasites inside it die. The body's defenses may then be able to destroy the remaining parasites before a full-blown case of malaria develops. In malaria-infested regions, therefore, it is advantageous to be heterozygous for the sickle allele, which protects against a common deadly disease, even though the sickle allele is usually lethal in the homozygous state.

The same explanation may account for the high frequency of thalassemias, a group of genetic disorders in which too little hemoglobin is produced, in districts of Italy, Greece, and other areas with high incidences of malaria.

Tay-Sachs Disease. **Tay-Sachs disease,** a metabolic disorder resulting in deterioration of the brain and death by about the age of four, is also the result of a lethal recessive allele. A homozygous recessive child lacks an enzyme that metabolizes a certain lipid in the brain's nerve cells. Without this enzyme, the lipid accumulates and destroys the cells' ability to function. So far, this condition is incurable. One in 30 people of East European Jewish extraction is a carrier (heterozygous) for this disorder. However, about one-third of the Tay-Sachs cases in the United States occur in non-Jewish people.

Cystic Fibrosis. The most common lethal allele in the Caucasian population of the United States is the one responsible for **cystic fibrosis.** About one in 20 Caucasians is a carrier, and one in 2000 babies is affected by this disorder. This disease is apparently caused by a defect in control of the cell-membrane channel proteins that permit chloride ions to pass through the membrane. Normally, chloride ions leave cells in the lung surface via these channels, and water follows by osmosis. In persons with cystic fibrosis, the movement of chloride (and hence of water) is impeded. As a result, the mucus on the lung surface is unusually thick, because it contains little water. Victims suffer from poor lung function, and 50% die of respiratory infections by age 21, with few surviving past age 30. In addition, thick mucus in the digestive tract may interfere with digestion and absorption of food, and victims may become poorly nourished despite an adequate diet. The defective channel activity also produces abnormally salty sweat, which is often the first clue in diagnosing cystic fibrosis.

In 1989, cystic fibrosis genes and their normal counterparts were isolated and analyzed. About 70% of alleles from cystic fibrosis patients have a mutation in which the code for one amino acid is deleted. This amino acid is thought to play an important role in control of the channel's activity. Various other defects occur in the remaining cystic fibrosis alleles. As researchers learn more about the defects involved, they are devising some promising drug therapies.

BIO-BIT

About 1400 dominant genetic diseases are known for humans, but most are rare. In general, they cause severe medical problems, including mental retardation, grotesque disfigurement, or psychosis.

Huntington's Disease. **Huntington's disease** is unusual because it is a dominant lethal allele. Hence, a person with only one allele for this condition may develop the disease. Its symptoms include involuntary twitching, degeneration of part of the brain, depression, and irritability. The disease progresses slowly for 10 to 20 years, finally causing death. The gene that controls Huntington's disease was identified in 1983. The first symptoms usually do not appear until 35 to 45 years of age. By that time, most victims have children, who in turn have a 50:50 chance of having also inherited the allele.

Most dominant lethal alleles do not last long in a population, because affected individuals typically die before reproducing. However, carriers (heterozygotes) of recessive lethal alleles may have some intermediate traits but still live normal lives, and may pass on their deadly genes. In some cases, as in sickle cell anemia carriers, the intermediate phenotype may even be adaptive.

13–C INBORN ERRORS OF METABOLISM

Many genes code for proteins that are enzymes in the body's metabolism. Mutation of such a gene may result in an inherited genetic abnormality known as an **inborn error of metabolism.** We have seen that the metabolic disorders of Tay-Sachs and Huntington's diseases are lethal, but others are less severe, and some do little or no apparent harm to affected individuals.

Phenylketonuria (PKU) and **albinism** are two human hereditary disorders that happen to be on the same metabolic pathway (Figure 13-6). Since this pathway is not vital, the responsible alleles are not lethal.

PKU-affected individuals are homozygous recessives who lack the enzyme that normally converts the amino acid phenylalanine to another amino acid, tyrosine. Without this enzyme, phenylalanine builds up, perhaps to 50 times its normal level. Minor metabolic pathways convert some of this phenylalanine to vari-

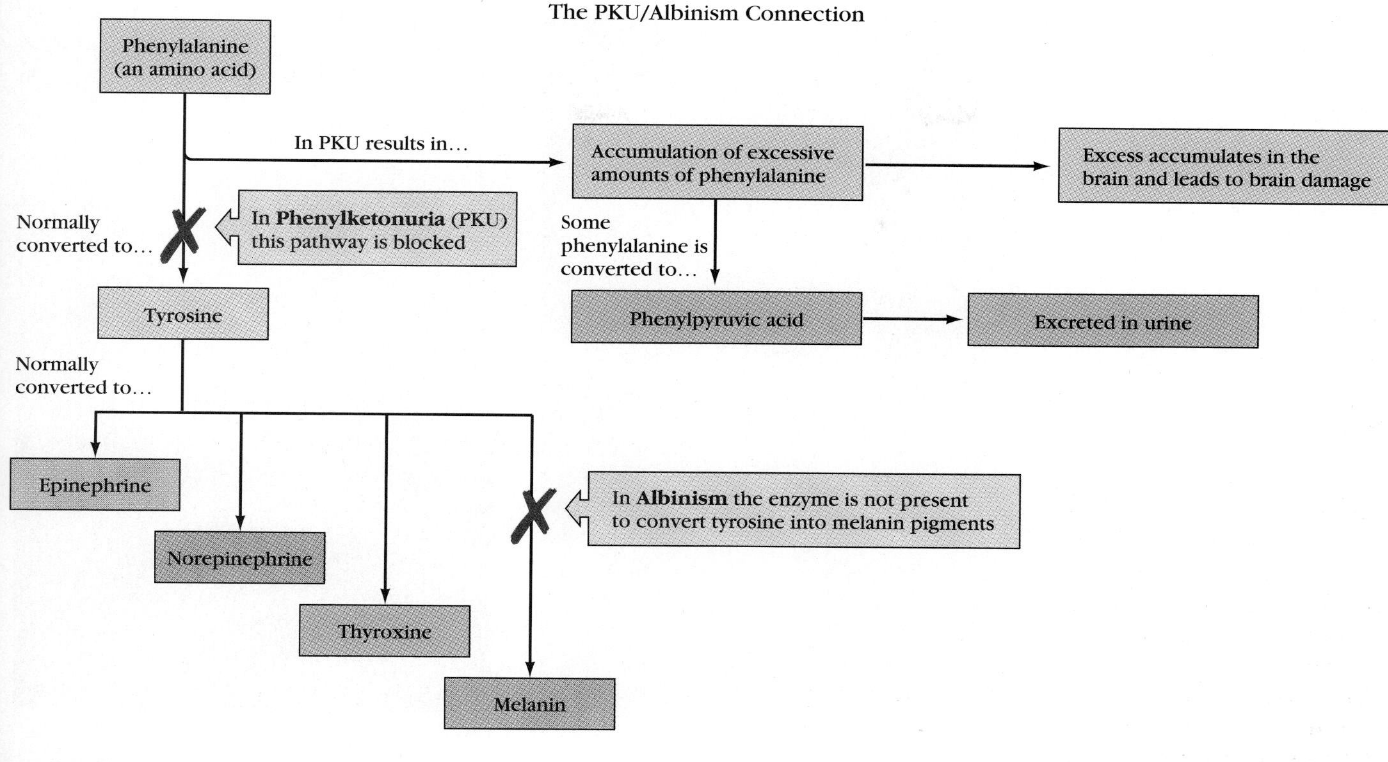

FIGURE 13-6

Inborn errors of metabolism. This diagram shows the metabolic pathway that converts the amino acid phenylalanine to tyrosine, which, in turn, can be converted to several other substances. The arrows indicate enzyme-catalyzed steps in the pathway. "Metabolic blocks" (red X's) result from the absence of the corresponding enzymes. Two metabolic blocks are shown here—those responsible for phenylketonuria (PKU) and albinism.

ous other products, such as phenylpyruvic acid, which is excreted in the urine, giving it a characteristic odor. High concentrations of phenylalanine and its products inhibit the activity of many metabolic enzymes. This damages various organs, especially the brain, and, without treatment, victims of PKU become mentally retarded. PKU can now be controlled by a special diet low in phenylalanine during childhood. This prevents most brain damage, but victims may still have learning disabilities. Since this treatment must begin within a few weeks of birth, many states now require that newborns receive a blood test for PKU (and for several other metabolic disorders). When brain development is complete, the PKU victim can adopt a normal diet.

If a woman homozygous for PKU becomes pregnant, the high phenylalanine level in her blood is transferred to the fetus through the placenta. This puts the fetus at risk of mental retardation or microcephaly (small head). Some such women have returned to a low-phenylalanine diet during pregnancy, but it is not yet clear whether this eliminates the risks to the fetus. Since the mother is homozygous for PKU, her children must inherit one copy of the recessive allele from her. Hence they will all be PKU carriers (or homozygotes if they also receive a PKU allele from their father).

Albinism is a condition characterized by absence of melanin, the dark pigment that makes eyes, hair, and skin brown or black. True albinos have white hair (or feathers, as in Figure 13-7) and very light skin and red pupils. They lack functional enzymes to convert the amino acid tyrosine to melanin.

BIO-BIT

Among European Americans, albinism occurs only once in about every 40,000 births.

(a)

(b)

FIGURE 13-7

Albinism, a homozygous recessive condition. (a) A normally colored male peacock displaying his magnificent feathers. **(b)** An albino peacock has all white feathers. (a, Chris Gulker/Picture Group; b, Joe McDonald/Visuals Unlimited)

You may wonder whether victims of PKU are also albino, since they cannot make the tyrosine that is eventually converted to melanin. The answer is no, because tyrosine can be obtained in the diet as well as from conversion of phenylalanine. However, people homozygous for PKU usually have light coloring. A person could, of course, be both an albino and PKU-affected if he or she were homozygous for both pairs of alleles.

Phenylketonuria (PKU) and albinism are nonlethal recessive genetic disorders that result from defects of metabolic enzymes. Because PKU can result in mental retardation unless treated soon after birth, most states now test all newborns for this disease.

13-D MULTIPLE ALLELES

Up to this point, we have considered only genes with two distinct alleles. However, a gene contains hundreds of nucleotides. If mutations occur in different parts of the gene in different individuals, the population as a whole will contain a number of different alleles, known as **multiple alleles.** Each allele may produce a different phenotype, and various combinations of multiple alleles produce an array of genotypes and phenotypes in the population as a whole. Any one individual can contain no more than two of the different alleles of a gene, one on each chromosome of the homologous pair carrying that gene.

A familiar case of multiple alleles is that of the human ABO blood groups, with three main alleles, I^A, I^B, and i. (I and i stand for dominant and recessive forms of the immunogen gene.) I^A and I^B code for two different enzymes, each of which attaches a different sugar to a protein on the surface of red blood cells. Both enzymes are produced in a person having both the I^A and I^B alleles. That is, I^A and I^B are codominant. The i allele does not code for an enzyme; it is recessive to both I^A and I^B. Table 13-1 shows the possible genotypes and phenotypes in this blood group system.

Blood types must be matched when a person receives a blood transfusion. The ABO and rhesus (Rh positive or negative) blood groups are the best known and the most medically important, but more than 20 different human chromosome locations are known to carry genes coding for various blood group proteins.

In the past, blood groups were sometimes used to decide questions of parentage, such as in paternity lawsuits or in cases of suspected mix-ups of babies in a hospital nursery. Only a few drops of blood are needed to determine the blood types of the child and

Table 13–1
Human ABO Blood Groups

Blood Group (Phenotype)	Genotype
A	I^AI^A or I^Ai
B	I^BI^B or I^Bi
AB	I^AI^B
O	ii

its supposed parents. This genetic evidence reveals whether a particular person or couple could have had a child of a particular blood type. Such evidence can never be used to establish that a particular person definitely is the father or mother of a particular child, but it can rule out a particular person as the child's parent. For instance, a man with blood type AB has the genotype I^AI^B, and so he could not have been the father of a baby with blood type O. A baby with blood type O must have the genotype *ii,* and its father must therefore have had at least one *i* allele to pass on (see Table 13–1). If the baby's blood type is A or B, the same man could have been the father. However, there are many other men in the world with blood types such that they could have fathered the baby, so the baby's parentage can never be established conclusively on ABO blood group evidence.

Modern paternity testing uses proteins coded by another series of multiple alleles: the MHC (short for major histocompatibility complex) antigens, proteins found on the surfaces of most human cells (Section 25–D). There are six different gene pairs, each with several to many different possible alleles, coding for these proteins. This results in such a large number of possible genotype combinations that everyone (except identical twins) has an essentially unique "chemical fingerprint" on his or her cells' plasma membranes. Because of their great variety, these proteins can usually be used to settle questions of paternity when ABO blood tests are inconclusive. For convenience, white blood cells are used for these tests. This technique was first used in 1980 to determine the paternity of identical twins born to a rape victim. The twins' MHC antigens matched those of the woman and her husband and were distinctly different from those of the accused rapist. The MHC technique is very accurate, and, as a result, many more fathers are now being forced to accept financial responsibility for their offspring.

A population may contain multiple alleles of a gene, but any one individual has at most only two alleles (one pair) of each gene. ABO blood types are based upon multiple alleles of one gene. Because MHC antigens have so many different alleles, they provide a unique chemical fingerprint that can be used to determine parentage.

13–E POLYGENIC CHARACTERS

In the case of multiple alleles, a particular location on a chromosome can be occupied by any one of several different alleles of a single gene. In contrast, the term **polygenic character** describes the case of a single phenotypic trait governed by more than one pair of genes. These genes may occupy two or more different locations on the same homologous chromosome pair or on nonhomologous chromosomes. Familiar examples in humans include height, intelligence, body build, and hair and skin color, all determined by the interactions of many genes.

Human skin color is a polygenic trait. There is some debate whether three or four different gene pairs are involved. Very dark-skinned people have alleles coding for production of melanin at all of their skin-color-gene locations, whereas in light-skinned people these locations are occupied by alleles that don't code for melanin production. For example, if there are three gene pairs involved, a very dark-skinned person would have six alleles for melanin *(AABBCC),* whereas a very light-skinned person would have none *(aabbcc).* The alleles are thought to have additive effects: the more alleles for melanin production a person has, the more melanin is produced, and the darker the skin.

A similar example is known in wheat. Three different gene pairs govern the color of kernels, from white to deep red. Figure 13–8 shows a cross of two strains of wheat that differ in only two of these gene pairs: *AABB* × *aabb* (in both strains, the third pair is assumed to be *cc,* coding for no pigment).

1? Polygenic characters are difficult to study because it is hard to disentangle the effects of the many interacting alleles that influence a single phenotypic character. Environment may further muddy the waters; for instance, in our example, whether a person has light or dark skin, his or her color can become lighter or darker depending on exposure to the sun. Traits such as human height (Figure 13–9), human intelligence, and the size of an ear of corn are strongly influenced by environmental factors, such as nutrition, as well as by the many genes that determine the possible range of variation in the phenotype. ■

Polygenic characters are so common that early observers were misled into believing in blending inheri-

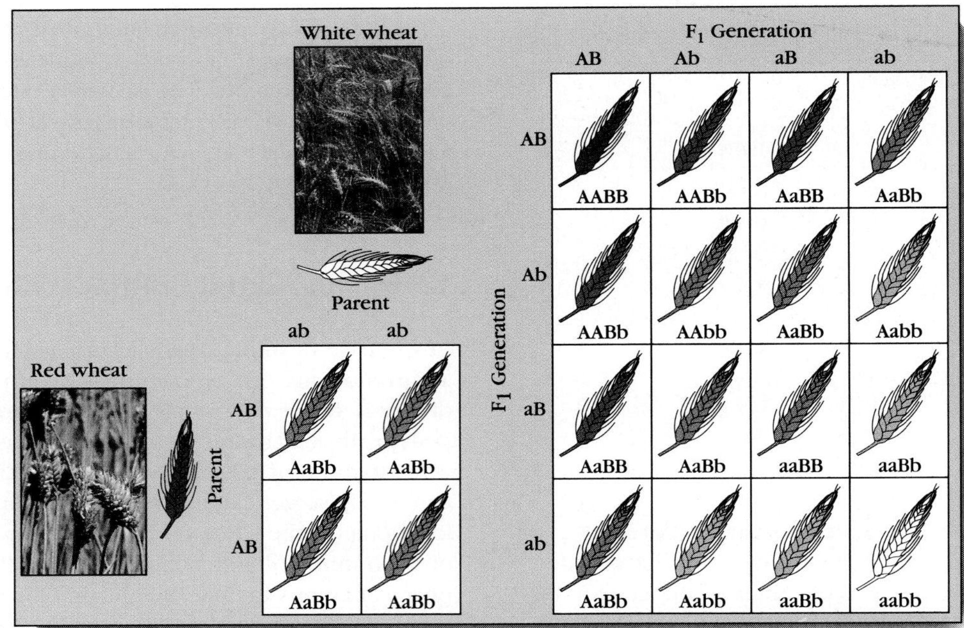

FIGURE 13–8

A cross involving a polygenic trait. The two parental (P_1) strains of wheat differ in two gene pairs for kernel color. One strain has four alleles for uncolored kernels (*aabb*), and the other strain has four alleles for red color in the kernels (*AABB*). F_1 plants have two red-color alleles (*AaBb*), and so their color is midway between those of the two parental strains. F_2 plants have kernels of five different colors (white and four shades of red), depending on whether they have zero, one, two, three, or four red color alleles. (Red wheat, Link/Visuals Unlimited; White wheat, Pat Armstrong/Visuals Unlimited)

tance. Polygenic traits frequently appear to blend because the phenotypes of various individuals show a wide range, and the offspring of parents at the extreme ends of the range generally come out about midway on the scale.

Polygenic characters are single phenotypic traits that are controlled by more than one pair of genes. Because of the many possible genotypes, polygenic characters show a wide range of phenotypes.

13–F SEX DETERMINATION

In many species, the most obvious difference in phenotype between individuals is their sex. It is also one of the most far-reaching differences, for the sex hormones influence the phenotypic expression of many other genes, hence sex affects many organs that are not directly involved in sexual reproduction. In most familiar animals, sex is genetically determined, and about half the individuals are male and half female.

FIGURE 13–9

How can a child grow taller than either parent? The genotype for height is a polygenic trait, in which many genes act together. Parents can have a child taller than either one of them if both give that child mostly their alleles for tall height. In addition, better nutrition may allow the child to reach more of the genetic potential and grow even taller. (Jeff Isaac Greenberg/Photo Researchers, Inc.)

The simplest cross to produce half males and half females (a 1:1 ratio in the offspring) is one between a homozygote and a heterozygote. In humans and most other mammals, and in birds, this is what happens: one sex is heterozygous and one homozygous for a pair of chromosomes, called the **sex chromosomes** (Figure 13–10). In most mammals, males are heterozygous for the sex chromosomes, having one X and one Y chromosome (see Figure 13–10). Females have two X chromosomes. Most birds and some reptiles are the other way around: females are heterozygous, with one Z and one W chromosome, while males are homozygous ZZ. Many insects are like mammals in having XX females and XY males, but some have a reversed system much like that of birds. Still others have haploid males and diploid females. Higher plants often have reproductive organs of both sexes in the same individual, but those with separate sexes tend to have XX females and XY males.

During the formation of gametes in a man (or in XY males of other species), the X and Y chromosomes segregate during meiosis and end up in different sperm. All eggs produced by an XX female contain one of her two X chromosomes. At fertilization, an X-bearing egg can combine with an X-bearing sperm (forming a female) or with a Y-bearing sperm (forming a male).

Apparently all individuals have the genes needed to develop into a member of either sex. For example, embryonic birds and mammals have been induced to grow into members of the genetically "wrong" sex by hormone treatment. However, under normal circumstances, hormones from the ovary or testis maintain the correct sex of each individual.

How do the sex chromosomes influence sex? The answer is not the same for all organisms. In humans and most other mammals, rudimentary "indifferent gonads," which can become either testes or ovaries, form

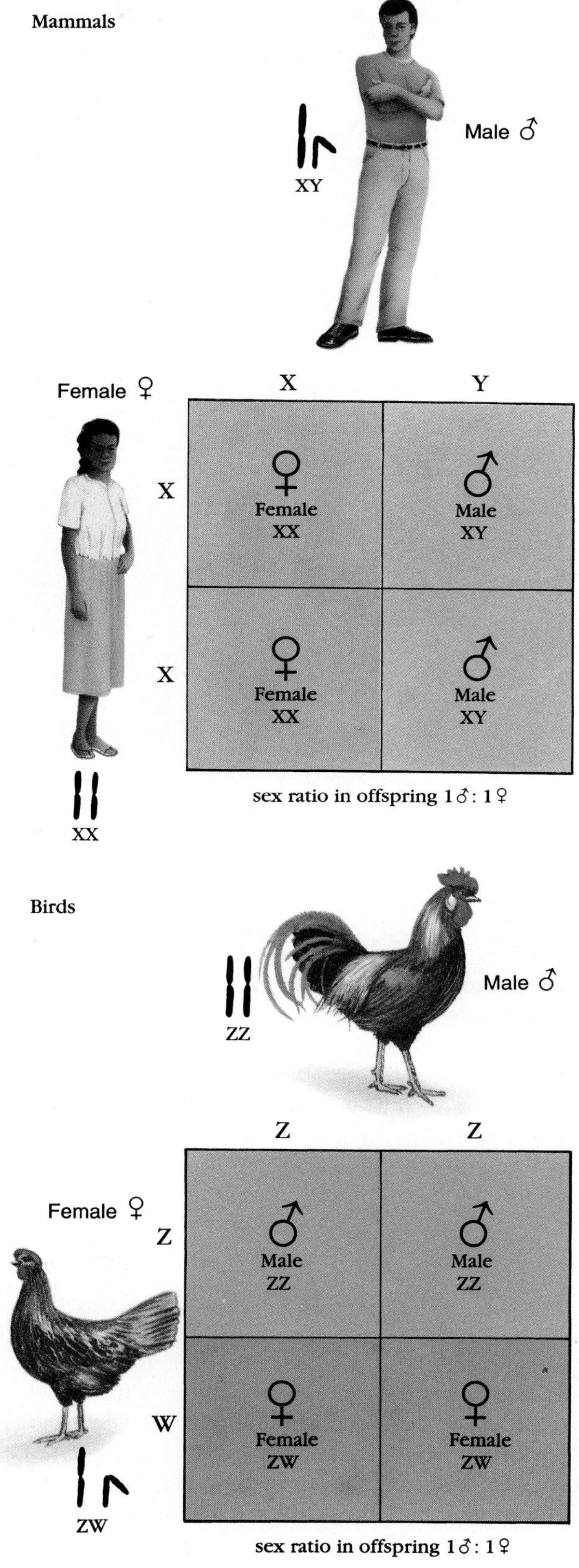

FIGURE 13–10

Systems of sex determination in mammals and birds. In mammals, females have a pair of X chromosomes; males have one X and one Y chromosome. In birds, males have a pair of like chromosomes, called Z, to emphasize that the male/female homozygote/heterozygote system is reversed from the situation in mammals. While male birds have a pair of Z chromosomes (ZZ), female birds have one Z chromosome and one W chromosome.

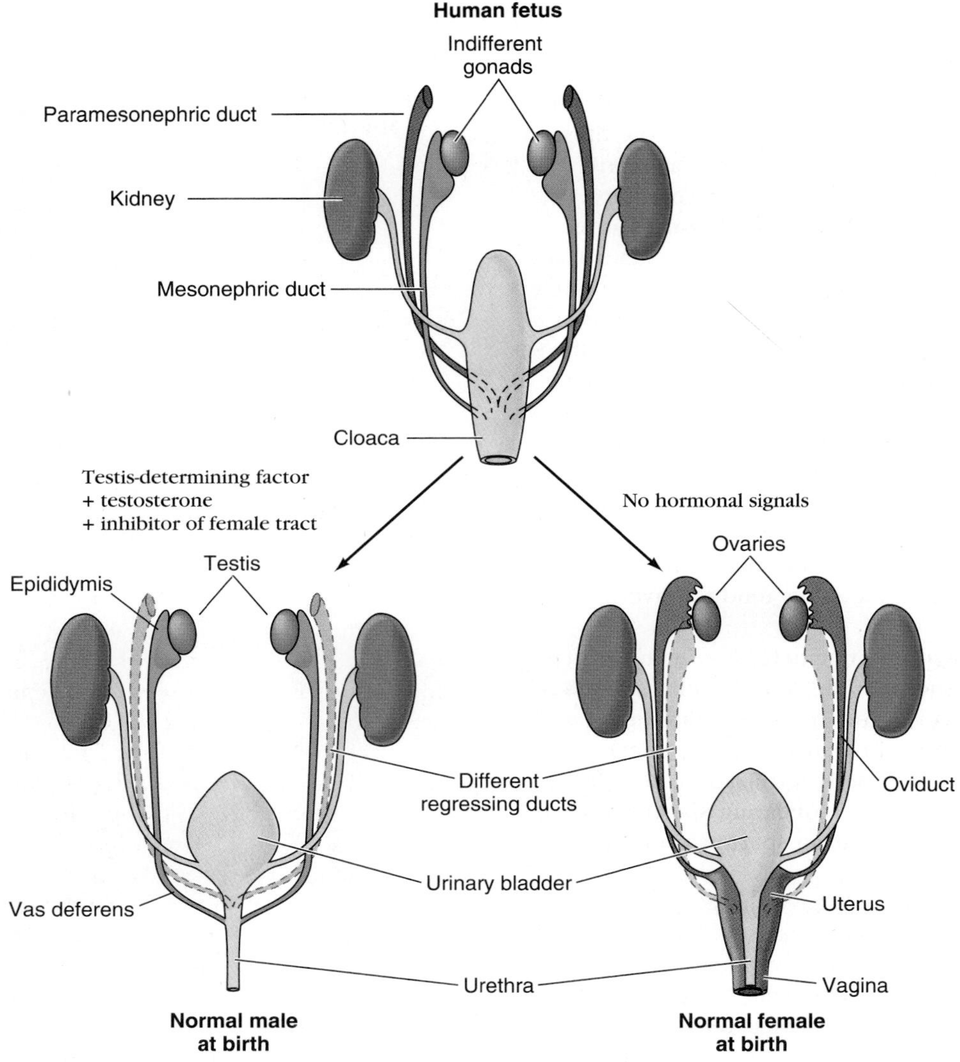

FIGURE 13–11

Sex differentiation in human embryos. Early embryos have indifferent gonads. Near them lie two sets of undifferentiated ducts (mesonephric and paramesonephric ducts). The testes produce the male hormone testosterone, which stimulates the mesonephric ducts to differentiate into ducts of the male reproductive tract. The end of the duct near the testis becomes the epididymis, and the remainder becomes the vas deferens. Females lack this chemical stimulation and the gonads develop into ovaries and the paramesonephric ducts become the oviducts of the female tract. In both sexes, the unused pair of ducts regresses.

in the early embryo (Figure 13–11). Which way they develop depends on the sex chromosomes present. The mammalian Y chromosome carries at least one gene that makes the embryo develop as a male. This gene codes for testis-determining factor, a gene regulatory protein, which causes the gonads to differentiate into testes. The testes, in turn, produce the hormone testosterone, which must be present to induce differentiation of the male reproductive tract. They also produce an inhibitory substance that causes the rudimentary female tract to regress.

In human embryos, the testes begin differentiating in the sixth week of development. If this does not occur, the gonads differentiate into ovaries in the following week. In this case, the rest of the female reproductive tract develops automatically, without hormonal

Table 13–2
Phenotypes for Various Sex Chromosome Complements in Humans

Sex chromosomes	Phenotype*
XX	Normal female
XY	Normal male
XXX	Female; fertile or sterile; usually normal
X (Turner's syndrome)	Female; short, with webbed neck; sterile, ovaries rudimentary or absent
XXY (Klinefelter's syndrome)	Male; sterile; underdevelopment of testes, enlarged breasts, and/or mild mental retardation
XXXY	Male
XYY	Male; unusually tall, acne-prone; impaired fertility; possible mild mental retardation

*Defects in various genes involved in hormone production can alter the phenotype normally exhibited by a particular sex chromosome combination.

signals from the ovaries. If a mammalian embryo's ovaries or testes are removed before the reproductive tract differentiates, the embryo develops a female tract. So, female is the default sex. Having a Y chromosome leads to maleness, while embryos with only X chromosomes become phenotypically female. This is true even in the rare cases of people born with more or fewer than two sex chromosomes (Table 13–2). The genes on the Y chromosome that trigger differentiation of the testes seem to be the only genes involved that are on the sex chromosomes. Many other genes come into play later, and all of these seem to be located on the **autosomes,** the nonsex chromosomes.

Various unusual methods of sex determination occur in some organisms. In certain animals, sex is determined by environmental factors. In the American alligator, snapping turtle, and some other reptiles, sex is determined by the environmental temperature during a particular stage of development. (This discovery led to speculation about the mysterious extinction of the dinosaurs and other ancient reptile groups. If these reptiles had a similar mode of sex determination, a prolonged change in climate could have resulted in production of offspring of only one sex. Eventually this would have doomed any species whose members could reproduce only sexually.)

Environmental determination of sex may prove useful to animals that cannot move far to find a mate. For instance, in the marine worm *Bonellia,* the developing larva swimming in the sea belongs to neither sex. Eventually it drifts to the bottom and becomes an adult. If it settles down alone, it develops into a relatively large female, but if it lands near an existing female, it is attracted to her, and she produces a chemical that causes the larva to develop into a microscopic male. The male migrates into the female's reproductive tract and lives there as a parasite.

The snail-like slipper shell lives in stacks of individuals. Young individuals are males, which turn into females as they grow larger and older. Chemicals appear to influence sex determination in this situation as well. If a stack consists entirely of males, some of them turn into females. In contrast, some fish, such as the saddleback wrasse and bluestreak cleaner, may start out as females, but when there are no larger males around, the largest females become males. All of these systems guarantee that when two or three are gathered together, some will be male and some female.

In most species, one sex is heterozygous and the other sex is homozygous for one pair of chromosomes, the sex chromosomes. The sex of offspring is determined by which sex chromosome it receives from the heterozygous parent. In some unusual situations, the sex of an animal is determined by environmental factors such as temperature or the sex of members of the same species in the immediate area.

13–G SEX LINKAGE

Like other chromosomes, the sex chromosomes carry genes. We know of several genes that are located on the Y chromosome: one determines whether testes are present and another affects the size of the teeth.

In mammals, part of the X chromosome is homologous with part of the Y chromosome. This permits the X and Y chromosomes to pair at the beginning of

Red-green Colorblindness

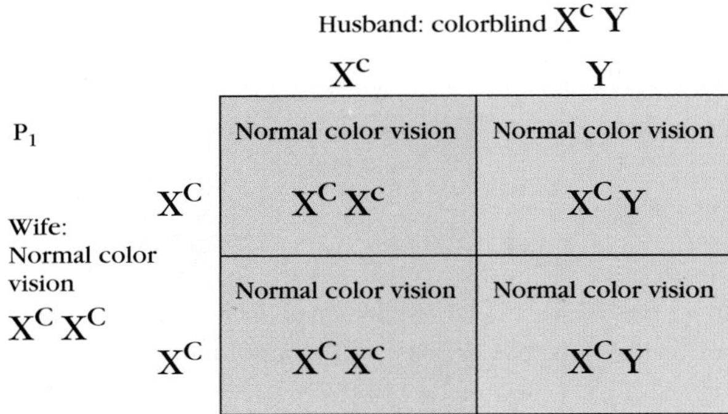

2 girls with normal color vision
2 boys with normal color vision

F_1 Cross

	X^C	Y
X^C	Normal color vision $X^C X^C$	Normal color vision $X^C Y$
X^c	Normal color vision $X^C X^c$	Colorblind $X^c Y$

2 girls with normal color vision
1 boy with normal color vision
1 colorblind boy

FIGURE 13-12

Inheritance of red-green colorblindness, controlled by genes on the X chromosome. In a marriage of a woman homozygous for normal color vision and a colorblind man, all the children have normal color vision. If the children marry people with the same genotypes as their siblings, half of the male offspring are expected to be colorblind, inheriting the maternal X chromosomes bearing the color blindness allele. Girls that inherit this maternal X chromosome receive an X chromosome with a normal allele from their fathers, and so they are not colorblind.

meiosis and segregate into different sperm cells (Section 11-F). Because genes located in these homologous areas of the sex chromosomes are paired, they behave like the autosomal genes we have studied before, and only detailed chromosome mapping will reveal that they are on the sex chromosomes.

Mammalian X chromosomes also have large **nonhomologous portions,** which contain genes that have no mates on the Y chromosome (Figure 13-12). Genes located in these nonhomologous areas of the X chromosome are said to be **sex-linked.**

In male mammals, any recessive allele on a nonhomologous part of the X chromosome will be expressed in the phenotype, since there is no gene on the Y chromosome that could mask it. Therefore, it is possible for a single recessive allele to express itself in the male. A female must have two copies of such a recessive allele before it shows in her phenotype. Recessive sex-linked phenotypes, therefore, are more common in male mammals than in females. Since many recessive alleles are harmful, this is one of the reasons that slightly more male than female mammals of any age die.

Red-green colorblindness, hemophilia, and the Duchenne type of muscular dystrophy are well-known recessive sex-linked traits in humans. Let us consider a cross between a woman homozygous for normal color vision and a colorblind man. All the children of this marriage will have normal color vision, since all receive an X chromosome bearing a normal allele from their mother (Figure 13-12). Imagine that the daughters in this family marry men with normal color vision (like their brothers), while the sons marry women who are carriers of the gene for colorblindness (like their sisters). In the next (F_2) generation, we would find the ratio of three offspring with normal vision to one colorblind, as expected from a monohybrid cross. Our results have this added twist, however: all the colorblind children are male! Girls can also be colorblind, but only if they inherit an X chromosome with an allele for colorblindness from their fathers as well as from their mothers. Therefore, colorblind girls are much rarer than colorblind boys.

2? Certain forms of hemophilia are also caused by a recessive allele on the X chromosome. A person with hemophilia lacks a protein needed to make blood clot, and so may bleed to death after even a slight cut (Section 13-A). If a woman has one recessive hemophilia allele, she usually also has a dominant, normal allele on her other X chromosome, and so she does not have hemophilia. A man with the allele, on the other hand, has no second X chromosome bearing a normal allele to code for the vital clotting protein. Hence his hemophilia allele will be expressed. In the past, men with hemophilia seldom lived to become fathers, and therefore few hemophiliac females are known. However, hemophilia can now be controlled (but not cured) by regular injections of clotting factor extracted from normal blood. Genetically engineered clotting factor will also be available soon. As a result, more men with hemophilia now live to grow up and reproduce. More hemophiliac females can be expected, as some men with hemophilia may marry women heterozygous for the allele. ■

Queen Victoria of England was the world's most famous hemophilia carrier. Her hemophiliac son,

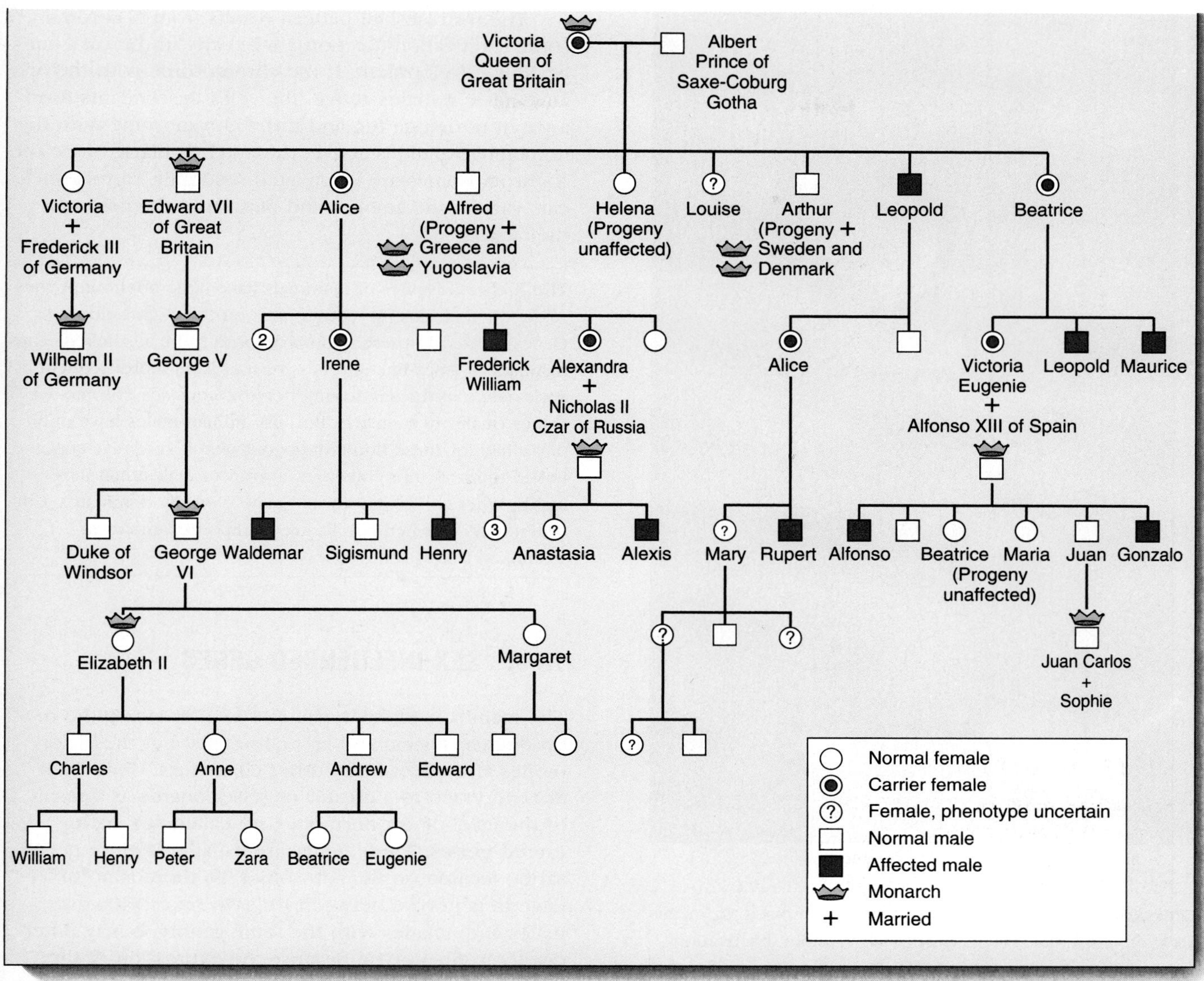

FIGURE 13-13

Pedigree of hemophilia among the descendants of Queen Victoria. Queen Victoria was the first identified carrier of this allele in her family. She passed it on to at least three of her children. Only some of her descendants are shown, including the royal family of Great Britain, which has not inherited the allele, and those lineages that did contain it. Notice the standard symbols used to denote affected and normal males and females in pedigree diagrams (key at bottom right). Numbers inside symbols indicate more than one offspring showing the phenotype.

Leopold, Duke of Albany, and her two carrier daughters, Princesses Alice and Beatrice, spread the gene to the royal houses of Russia, Prussia, and Spain. For a time hemophilia was called the "royal disease," but fortunately none of Queen Victoria's modern descendants appears to have inherited the allele (Figure 13-13).

Another familiar example of sex linkage occurs in one of the many gene pairs that affect coat color in cats. The sex-linked orange allele, located on the X chromosome, diverts molecules into a metabolic pathway that makes them into orange pigment instead of black. A male cat with the orange allele on his X chro-

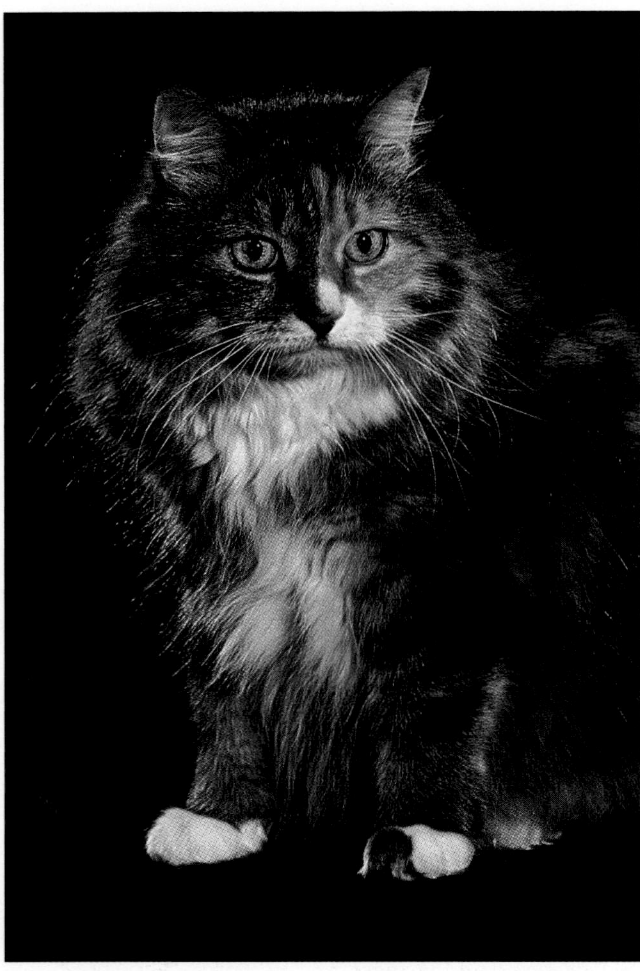

FIGURE 13–14

A Persian calico cat. Patches of orange and black fur in this heterozygous female resulted from the random inactivation of one X chromosome in each cell during embryonic development. (E. R. Degginger)

mosome, or a female with orange on both X chromosomes, is some shade of orange. The non-orange allele allows black pigment to form, and the non-orange male or homozygous non-orange female is some shade of black (including brown or gray, depending on its other genes). A cat with an orange allele on one X chromosome and a non-orange allele on the other has a mottled pattern of random orange and black spots called tortoiseshell (Figure 13–14). Such a cat is almost always female, since only females normally have two X chromosomes (see Table 13–2 for human males with extra X chromosomes).

The tortoiseshell pattern results from X chromosome inactivation (Section 13–F) early in the cat's embryonic development. If the chromosome with the orange allele remains active, the cell's descendants form a patch of orange fur, and if the chromosome with the non-orange allele is active, the patch is black. Since the X chromosomes are inactivated randomly, tortoiseshell cats vary in the amount and pattern of the colors in their coats.

The X chromosomes of mammals have large nonhomologous segments that have no counterpart on the Y chromosomes. Genes located in these regions are said to be sex-linked. The hallmark of a sex-linked recessive trait is its more frequent appearance in the sex having heterozygous sex chromosomes (males in humans). Because human males have only one allele for these nonhomologous genes, recessive sex-linked traits are more likely to appear in males than in females, where a second normal gene typically dominates. Colorblindness and hemophilia are common examples of sex-linked traits in humans.

13–H SEX-INFLUENCED GENES

The main role of sex hormones is influencing the reproductive system and related organs, but these hormones also affect many other characters. Genes that are expressed to a greater or lesser degree as a result of the level of sex hormones are called **sex-influenced genes.** These genes are usually (but not necessarily) located on the autosomes, so there is no difference in genotype between the two sexes. However, males and females with the same genotype may differ greatly in phenotype because the expression of the genes depends on the levels of sex hormones, and hence on the individual's sex. For example, a bull may have genes for high milk production, but he will not produce milk because he has only low levels of female hormones. However, these genes would make him a useful sire for a dairy herd. Similarly, males and females both have the genetic potential to produce the organs characteristic of the opposite sex, but they develop organs typical of their own sex because they have more of the appropriate hormones. (Both males and females do have the hormones characteristic of the opposite sex, but at lower levels.)

3? In humans, the allele for male pattern baldness is autosomal, and its expression is influenced by the presence of male hormones (Figure 13–15). A man will become bald if he has only one allele for baldness. In other words, the allele acts as a dominant in men because the male sex hormones somehow stimulate expression of the baldness allele. However, in women,

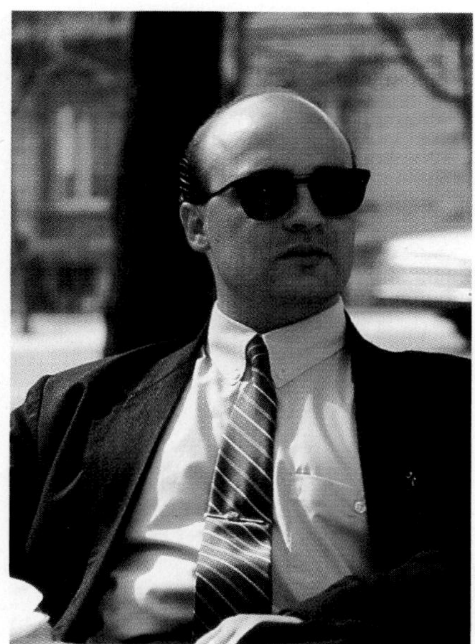

FIGURE 13-15

Baldness. Baldness occurs more often in men than in women because it is a sex-influenced trait. (Bob Newman/Visuals Unlimited)

the allele acts as a recessive, so that a female must have two alleles for baldness before she loses her hair. ■

Gout is another trait whose expression is influenced by sex. In gout, painful deposits of uric acid salts build up in the tissues, especially in the joints of the big toes. The allele for gout is expressed much more in the presence of male than of female sex hormones. In Victorian literature, gout figured largely as a reason for the temper tantrums of irascible old men. Avoiding red wine and rich and spicy foods was supposed to alleviate the condition, but this treatment tried the tempers of its victims still further. Gout can now be treated, but it is still a painful nuisance to many sufferers.

In sex-influenced genes, males and females of the same genotype may differ greatly in phenotype as a result of the effects of sex hormones on gene expression.

13–I SOME FACTORS THAT AFFECT GENE EXPRESSION

All of the genes possessed by an individual determine its genetic potential: what might be. What actually happens is another matter. The expression of a gene is influenced by the other genes present, either by way of proteins encoded by these other genes, or more indirectly. The influence of enzyme-produced sex hormones on gene expression is one example of such an indirect effect.

Hormone production varies with age, and so age may play a part in gene expression. Consider the many changes accompanying puberty, such as voice change and growth of the testes in males; breast enlargement and the characteristic pattern of body fat deposition in females; and growth of hair in the armpits and pubic area in both sexes.

Modifier genes influence traits that are controlled by other pairs of genes. It was long believed that eye color in humans was controlled by a single pair of genes, with brown eyes dominant to blue. It is now known that there are also at least two pairs of modifier genes involved, and it is possible, though extremely uncommon, for blue-eyed parents to have brown-eyed children.

The external environment also plays an important role, both in embryonic development and in later life. In the last few decades there have been several documented cases in which drugs taken by pregnant women have caused improper development of the fetus, or cancers later in the baby's life.

Many other external factors also affect gene expression. For instance, a good diet is necessary if a person is to reach the height made possible by his or her genes. In many countries, young adults tower above their parents or grandparents as a result of improved nutrition. Farmers and gardeners know that proper nourishment is just as important for plants. One species of caterpillar even develops a phenotype that mimics the appearance of the food it eats, thereby blending in with it (Figure 13-16).

Light also influences gene expression. The development of plants is especially sensitive to light. A human being becomes darker (or redder!) when exposed to bright sunlight for a time.

Temperature affects the expression of some genes. Himalayan rabbits and Siamese cats are normally light-colored, with black feet, ears, nose, and tail. What makes the fur different colors on different parts of the body? We now know that the darker color is due to the activities of an enzyme that is unstable at higher temperatures. The extremities of these animals are cool enough for the enzyme to function and produce dark fur, but the body itself is warm enough to inactivate the enzyme (Figure 13-17).

Gene expression is influenced by products of other genes and by environmental factors such as diet, temperature, and light.

(a) (b)

FIGURE 13–16

Phenotypes dependent on diet. (a) This oak-feeding caterpillar has been fed a diet of oak flowers and has developed a resemblance to one of the male flower clusters, complete with dots that look like the flowers' pollen sacs. **(b)** A sibling of the first caterpillar, fed oak leaves, resembles the oak tree's twigs. In nature, caterpillars that hatch in the spring look like the flowers they eat; those hatched after the flowers wither eat leaves and become twig mimics. (Erick Greene)

FIGURE 13–17

Effect of skin temperature on expression of coat color genes in the Himalayan rabbit. Black fur grows on parts of the body with skin temperature below 33° C. If fur is shaved from a warmer part of the body and an ice pack applied while the fur grows back, the new fur is also black.

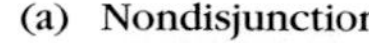

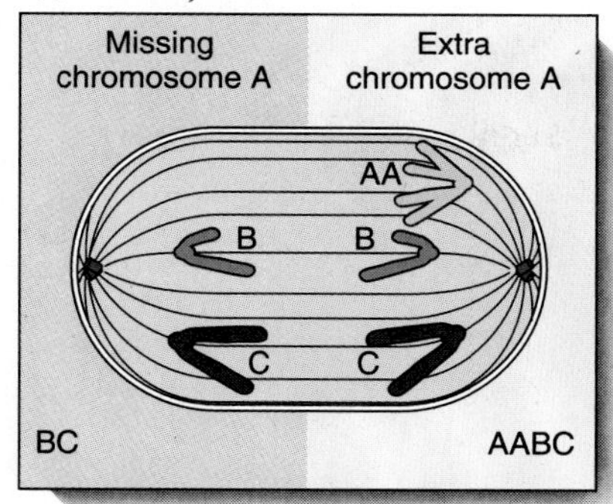

(b) Translocation

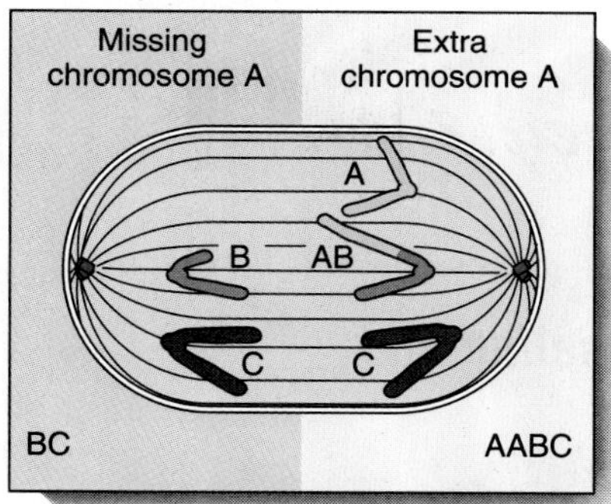

FIGURE 13–18

Abnormal chromosome movements. Both cells are in the second division of meiosis. **(a)** Nondisjunction. In chromosome A, the sister chromatids fail to separate, but remain attached and move to the same pole, eventually ending up in the same gamete. **(b)** Translocation. One copy of chromosome A becomes attached to chromosome B, which moves into a gamete along with a separate copy of chromosome A. In both cases, the gametes formed on the left will lack chromosome A, whereas those on the right will contain two copies of chromosome A.

13–J NONDISJUNCTION AND TRANSLOCATION

Occasionally, chromosomes behave abnormally during meiosis. **Nondisjunction** is the failure of chromosomes to separate properly. **Translocation** is the attachment of all or part of a chromosome to a nonhomologous chromosome. Either of these abnormal events produces gametes with one chromosome too many or too few (Figure 13–18). Most of the resulting fetuses die, but sometimes an individual with an extra chromosome in each cell survives. Such a person is usually mentally retarded if the extra chromosome is an autosome.

A common cause of mental retardation in the United States is Down syndrome, which results from having an extra copy of all or part of chromosome 21 (see Figure 11–2). The symptoms of Down syndrome include a small brain and mental retardation, a heavy fold of flesh over the eye, impaired immunity to disease, short stature, and flaccid muscles. Defects in the heart and the lenses of the eyes are also common. It has long been known that older women have an increased risk of bearing children with Down syndrome. Recent studies show that older men are also more likely to father such infants, regardless of the mother's age. Translocations of chromosome 21, or variants of chromosome 21 with a tendency to nondisjunction, run in some families and produce Down syndrome children regardless of the parents' ages. About one baby in 700 live births has Down syndrome. The incidence of Down births increases to approximately 1 in 50 when the mother is 40 years old. It is now routine for older mothers to have amniocentesis to determine whether the fetus they are carrying has Down syndrome or other chromosomal abnormalities (see *A Journey into Healthy Living:* Genetic Counseling and Fetal Testing).

Failure of the sex chromosomes to segregate properly during meiosis results in such abnormal sets of sex chromosomes as XXY (Klinefelter syndrome), XYY, XXX, or a single X chromosome (Turner syndrome). Some of these conditions produce sterility or mental retardation (see Table 13–2).

During meiosis, the improper separation of chromosomes (nondisjunction) or the attachment of all or part of a chromosome to a nonhomologous chromosome (translocation) typically results in a spontaneous abortion or a newborn with some degree of mental retardation, such as Down syndrome.

A Journey into Healthy Living

Genetic Counseling and Fetal Testing

Genetic abnormalities can bring much pain and suffering to the victims and to their families. Parents of victims may have feelings of guilt that lead to alcoholism, drug addiction, or divorce. The time, energy, and money needed to care for afflicted children may also deprive the family's other children of a normal home life.

Genetic counseling can help couples to determine their chances of having children afflicted by a particular defect, an event that is more likely if the couple or their relatives have already had such a child. Blood tests can now determine whether or not prospective parents are carriers for such traits as Tay-Sachs disease, cystic fibrosis, or sickle cell anemia, or are destined to develop Huntington's disease.

Some couples, faced with the knowledge that each child has a 25% chance of being homozygous recessive for a condition that will bring years of suffering or incapacity followed by an early death, choose not to have families. Others begin pregnancy and have the fetus tested to determine whether or not it is affected; if it is, they may choose abortion and hope that a later pregnancy will have happier results. In fact, the availability of such testing has increased the birth rate among at-risk couples.

How is a fetus tested for genetic defects? In the technique of **amniocentesis,** the tip of a syringe is inserted through the mother's abdominal wall and uterus into the sac of fluid (amniotic sac) surrounding the fetus (Figure 13-A). Cells that have sloughed from the fetus's skin or respiratory passages into the fluid are sucked into the syringe. These cells can be examined for chromosomal abnormalities, such as those resulting from translocation or nondisjunction of chromosomes during meiosis (Section 13-J). The cells can also be cultured in the laboratory, and in about two weeks will produce enough cells for geneticists to examine for disorders such as PKU, sickle cell anemia, cystic fibrosis, and Huntington's disease. A drawback of this technique is that it cannot be performed until the sixteenth week of pregnancy, when the amniotic sac is large enough for a doctor to extract the required fluid and its cells without accidentally damaging the fetus with the needle. By the time the cells have been grown in laboratory culture and analyzed, it may be too late for a legal abortion to be performed.

A newer technique can be performed during the eighth to tenth weeks of pregnancy, when abortion is safer for the woman. In **chorionic villus sampling (CVS),** cells are removed from chorionic villi, part of the fetus's life support system attached to the wall of the uterus. These cells are sucked into a catheter inserted through the cervix (the neck of the uterus) (Figure 13-A). Unlike amniocentesis, this supplies a mass of many rapidly dividing fetal cells, and so a diagnosis of chromosomal or genetic defects can be made quickly, within a few days. However, chorionic villus sampling cannot detect a common class of birth defects resulting from failure of the neural tube to close (Section 27-G). These defects can be found by amniocentesis.

Recently, some human fetuses have been treated before birth for deficiency of the B vitamin biotin, an inborn error of metabolism. Other fetuses have had surgery to correct blockage of the urethra, the tube that empties the urinary bladder, or diaphragmatic hernia, a hole in the diaphragm that permits abdominal organs to protrude into the space where the lungs should be developing. We can expect that more genetic disorders and other conditions will be detected and treated before birth in the future.

FIGURE 13-A

Diagnosing genetic defects in a fetus. (a) Amniocentesis. A needle is inserted through the mother's abdomen into the fluid-filled amniotic sac surrounding the fetus. Fluid which includes some fetal cells is withdrawn into the syringe. The fetal cells are cultured and the DNA is examined for signs of chromosomal abnormalities. The fluid and cells can also be tested for metabolic defects. When early delivery of the baby is necessary, amniocentesis may be done late in a pregnancy to see whether the lungs are producing the chemicals that will enable the baby to breathe air. If so, the fetus is ready to be delivered by Caesarean section. **(b)** Chorionic villus sampling. By about three weeks after fertilization, the fetus's chorionic membrane has developed many branched villi, forming part of the placenta. At 8 to 10 weeks there are enough fetal cells here to allow for some to be sucked up to provide a sample for genetic analysis. Although both methods are possible, each carry slight risks to the developing fetus.

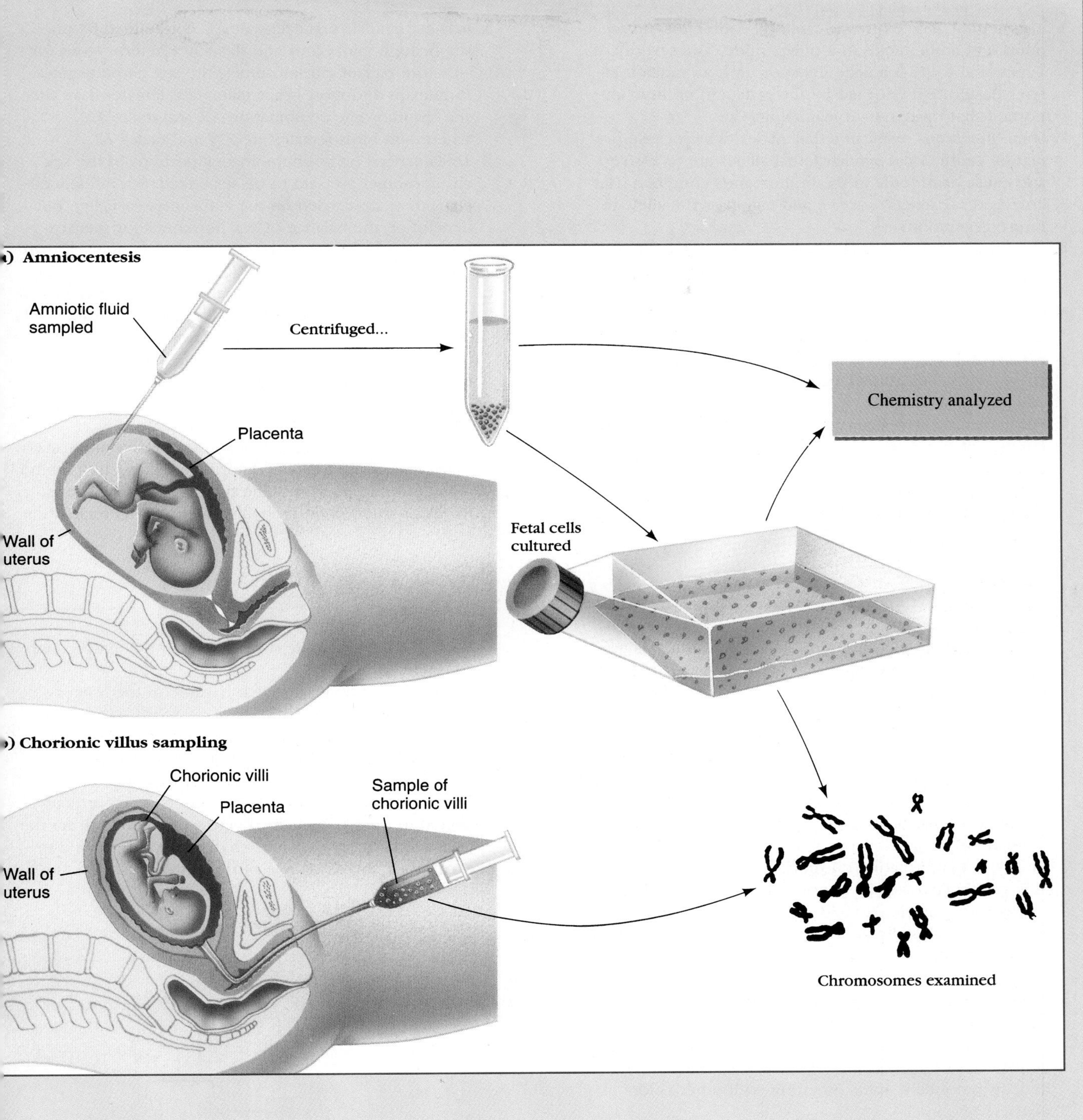
) Amniocentesis
Amniotic fluid sampled
Centrifuged...
Chemistry analyzed
Placenta
Wall of uterus
Fetal cells cultured
) Chorionic villus sampling
Chorionic villi
Placenta
Sample of chorionic villi
Wall of uterus
Chromosomes examined

SUMMARY

1. A structural gene expresses itself by coding for the sequence of amino acids in a polypeptide. The severity of a mutation to such a gene depends on how much it affects the protein encoded by the gene and on how important that protein is in maintaining life.
2. Some mutations result in lethal alleles, which cause premature death. Most familiar lethal alleles are recessive and cause death only in the homozygous condition. Heterozygotes (carriers) survive and may pass the allele to future generations.
3. Changes in less vital enzymes may cause metabolic disorders (inborn errors in metabolism), such as albinism and phenylketonuria.
4. Several different alleles of a gene may exist in a population as a result of different mutations in individuals. Such multiple alleles are found in the human ABO blood group and in cell surface proteins of the major histocompatibility complex.
5. Polygenic characters are phenotypes determined by the interaction of several different gene pairs. These polygenic characters show a wide range of phenotypes because different individuals have varying mixtures of these genes governing the trait.
6. In most familiar organisms, sex is determined by one sex being homozygous and the other heterozygous for an entire pair of chromosomes, the sex chromosomes. In humans and most other mammals, females have the sex chromosome combination XX and males XY, whereas in birds females are ZW and males ZZ.
7. Traits carried on nonhomologous portions of the sex chromosomes are said to be sex-linked. Sex-influenced characters are carried on the autosomes (usually), but depend on the balance of sex hormones for their expression, and hence are more common in one sex than the other.
8. An individual's phenotype depends on his or her mix of genes, how the expression of these genes is influenced by the products of other genes (enzymes, enzyme products such as hormones or pigments, or nonenzyme proteins), and the factors encountered in the external environment. External factors influencing gene expression include nutrition, light, and temperature.
9. In meiosis, chromosomes sometimes behave abnormally, either separating improperly or by attaching to a nonhomologous chromosome. The typical result is a spontaneous abortion or an individual with mental retardation.

SELF-QUIZ

1. In the homozygous condition, a recessive lethal allele in cattle produces "amputated" calves with malformations of the limbs, skull, and internal organs. These calves die soon after birth.
   a. What proportion of the normal offspring from a cross of two heterozygotes would be expected to be carriers for this trait?
   b. How could a farmer eliminate this trait from his herd if some "amputated" calves have been born to his cows?
2. Review the information on brachydactyly, Section 13-B.
   a. If two brachydactylic people marry, what are their chances of having a child with normal fingers?
   b. If a brachydactylic person marries a normal person, what phenotypic ratios are expected in their offspring?
3. A geneticist studying the various gene pairs that govern coat color in mice is trying to develop true-breeding strains of each possible coat color. He carries out several generations of matings among mice with yellow coats and always obtains some offspring with other colors of coats.
   a. What does this indicate about the genotype of yellow mice?
   b. The geneticist tallies up his results over several generations and finds that he has obtained a total of 184 yellow mice and 95 of other colors. What does this suggest about the nature of the yellow allele?
   c. Why did the geneticist never obtain a homozygous yellow mouse?
   d. How could he prove what became of the homozygous yellow offspring?
4. Below is a pedigree of ABO blood groups for several generations of humans. Circles represent females, squares males. Marriages are shown by horizontal lines directly connecting two people, and children are connected to their parents by a vertical line down from the marriage line. For example, (b) and (c) are married to each other, and (d) is one of their two sons. Give the

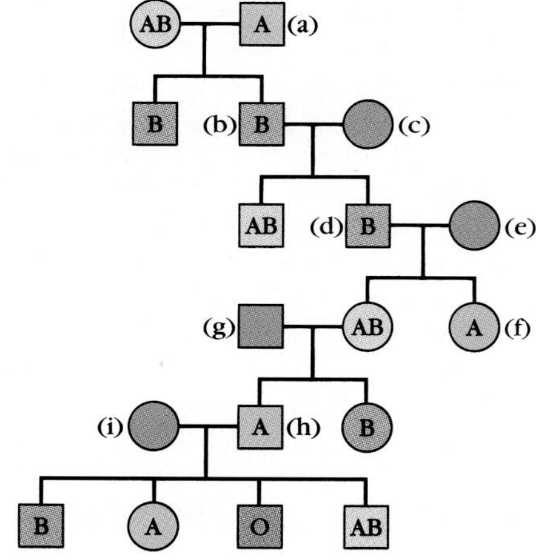

possible genotype(s) for each individual marked with a letter. (Hint: start at the bottom of the diagram and use what you know about children's genotypes to determine those of their parents.)

5. Ms. Smith and Ms. Jones gave birth to baby boys (named John and Tom, respectively) on the same day in a large city hospital. After Ms. Smith took her baby home, she began to suspect that it was Ms. Jones's baby, and that the hospital had somehow mixed the infants up. Blood tests revealed that Mr. Smith had blood type O, MN, and Rh^+; Ms. Smith had blood type B, N, Rh^+; and John Smith had blood type B, M, Rh^-. Mr. Jones had blood type A, M, Rh^+; Ms. Jones had blood type AB, MN, Rh^+; and Tom Jones had blood type O, MN, Rh^+. The Rh^+ allele is dominant to the Rh^- allele; the M and N alleles are codominant. Had a mixup occurred?
6. In rabbits, normal coat color (C) is dominant to chinchilla (c^{ch}), which is dominant to Himalayan (c^h), which is dominant to albino (c). What offspring are expected from the following crosses, and in what ratios?
   a. $Cc^h \times c^{ch}c^h$
   b. $c^{ch}c \times c^hc$
   c. $c^{ch}c \times c^{ch}c$
7. In chickens a sex-linked dominant allele causes a feather pattern known as "barred." If a barred hen is mated with a nonbarred rooster, what will be the feather pattern and sex of the offspring?
8. What are the expected genotypic ratios among the children of a normal woman whose father was a hemophiliac, and whose husband is normal?
9. Under what circumstances is it possible for both a father and his son to be hemophiliacs?
10. Red-green colorblindness in humans is a sex-linked recessive trait.
    a. In a large family in which all the daughters have normal vision and all the sons are colorblind, what are the probable genotypes of the parents?
    b. If a normal-sighted woman whose father was colorblind marries a colorblind man, what is the probability that their son will be colorblind?
    c. What is the probability that the couple in part b will have a colorblind daughter?
11. If a species of mammal has some members which carry a sex-linked lethal trait that causes early death and resorption of the embryo, what sex ratio would be expected among the offspring of a female carrier and a normal male?
12. It is often said that men inherit baldness from their maternal grandfathers via their mothers. In light of what you have learned about this trait, is this a valid statement? Explain.

THINKING CRITICALLY

1. One problem with genetic counseling is that people who learn they are carriers for genetic diseases such as hemophilia, Tay-Sachs disease, sickle cell anemia, or phenylketonuria may consider this a terrible stigma. Men have been known to deny paternity of their children and divorce their wives for infidelity when told that a child had inherited a harmful recessive gene from each parent. What kinds of arguments and counseling would you use, if you were a genetics counselor, in an attempt to induce a healthier, more productive response to such a discovery?
2. People who are carriers for sickle cell anemia face the risk that their red blood cells may sickle when they are in environments with low oxygen levels. Hence these people are probably exposed to greater than usual risks if they become divers, jet pilots, or mountaineers. Otherwise they have no physical handicaps; nevertheless, they have frequently been denied access to various professions as a result of ignorance and prejudice against genetic disorders. Since this is the case, a proposed nationwide screening for sickle cell carriers might well do more social harm than good. Is it better for carriers to remain in ignorance of genetic conditions for which there is no cure at the moment? If not, why not?
3. Until the advent of modern technology, hemophiliac men usually died before they reached reproductive age. Nowadays they can be provided with "clotting factor," a blood extract that permits them to lead normal lives and live to have children. The treatment costs about $6,000 to $10,000 a year per person, and there are about 20,000 hemophiliacs living in the United States. Can society or should society insist that such men be sterilized, so that they cannot perpetuate their disease, if taxpayers have to pay the bill for their medication?
4. The genes for both normal clotting factor (missing in hemophiliacs) and normal chloride channels (defective in cystic fibrosis patients) have been isolated and cloned. Hence both proteins can now be produced in the laboratory. Injections of the normal clotting protein will be an effective treatment for hemophilia. Why will it not be possible to treat cystic fibrosis with laboratory-produced chloride channel proteins?
5. Table 13-2 shows that individuals with a single X chromosome are known to occur, but not individuals with only a Y chromosome. Why do you think this is?
6. Every so often the newspaper advice column has a letter from a mother whose husband or in-laws have been chiding her for having daughters instead of sons. Is this censure justified? Why?
7. Describe some factors besides those mentioned in the chapter that may influence gene expression.

8. Propose an explanation of the genetic basis for the phenotypes shown in each of these photographs:

(a)

(b)

Individual genetic variations. (a) The lion's mane and the black tips of its hairs. **(b)** Variations of color and pattern in the leaves of a *Caladium* plant. (a, Kjell B. Sandved, Visuals Unlimited; b, John D. Cunningham, Visuals Unlimited)

SELECTED KEY TERMS

albinism, *p. 252*
autosome, *p. 259*
carrier, *p. 251*
cystic fibrosis, *p. 252*
hemophilia, *p. 248*
Huntington's disease, *p. 252*
lethal allele, *p. 248*
modifier gene, *p. 263*
multiple allele, *p. 253*
phenylketonuria (PKU), *p. 252*
polygenic character, *p. 255*
sex chromosome, *p. 257*
sex-influenced gene, *p. 262*
sex-linked trait, *p. 260*
sickle cell anemia, *p. 250*
Tay-Sachs disease, *p. 252*
translocation, *p. 265*

SUGGESTED READINGS

Carlson, E. A. *Human Genetics*. D.C. Heath and Co., 1984. An excellent book that includes a number of ethical dilemmas and social issues related to human genetics.

Friedmann, T. "Prenatal diagnosis of genetic disease." *Scientific American,* November 1971. A thoughtful article explaining the technology of amniocentesis and pointing out the ethical problems it presents.

Greene, E. "A diet-induced developmental polymorphism in a caterpillar." *Science* 243:643, 1989. Easily understood experiments show that what this caterpillar looks like depends on what it eats.

Hartl, D. L. *Human Genetics.* New York: Harper and Row, 1983.

Lawn, R. M., and G. A. Vehar. "The molecular genetics of hemophilia." *Scientific American,* March 1986.

Murray, J. D. "How the leopard gets its spots." *Scientific American,* March 1988. A speculative article presenting a model of how coat patterns might develop in spotted and striped animals.

Nathans, J. "The genes for color vision." *Scientific American,* February 1989. How we see color, how this ability evolved, and how mistakes in crossing over during meiosis result in colorblindness.

Patterson, D. "The causes of Down syndrome." *Scientific American,* August 1987.

Sayers, D. L. *Have His Carcass.* London: Harcourt, Brace Jovanovich, 1932. A mystery novel about a human genetic trait.

The Educated Citizen

Gay Genes

This article by Denise Grady was excerpted with permission from the January 1993 issue of Discover *magazine.*

Homosexuality may be more than a state of mind. Recent studies have offered tantalizing clues that the brains of gay men are physically different from the brains of heterosexual men. The studies are controversial, but if the differences are real, researchers would love to know when they came about: during puberty, in the womb, or perhaps even earlier, in the genes.

Now there's evidence that, in both gay men and lesbians, genes play an important role. Michael Bailey, a psychologist at Northwestern University, and Richard Pillard, a psychiatrist at the Boston University School of Medicine, have conducted two highly provocative studies of gay and straight adults—the first on men, and the second, presented during the summer of 1992, on women. All these people had either twin or adopted siblings of the same sex. Bailey and Pillard found that if one of a pair of identical twins (who share the same genes) was gay, then the odds of the other twin being gay jumped dramatically. Among the 335 men in the first study, 52 percent of the identical twins were both gay, as compared with only 22 percent of the fraternal twins and just 11 percent of the adoptive brothers. (Estimates of homosexuality in the general population range from 4 to 10 percent.) The figures were similar for the approximately 300 women in the second study.

In both studies, because the percentage was so much higher in identical twins than in fraternal ones, it's safe to assume that conditions shared in the womb were not solely responsible for sexual orientation. What's left is genetics. But the researchers find it hard to explain why there should be genes for homosexuality at all, let alone why they should occur so often. In evolutionary terms, such genes don't seem to make sense. Natural selection favors traits that enhance reproduction or at least don't interfere with it—certainly not traits that preempt it. "That paradox is the thing that most makes me doubt my own findings," says Bailey.

Yet the paradox may have a logical explanation. When deleterious genes flourish in any population, it's possible they also carry some hidden advantage. The best known example of this is the gene for sickle cell anemia, a severe blood disorder that strikes African Americans. People who inherit two copies of the gene—one from each parent—develop life-threatening symptoms of the disease. But their parents, who each carry a single copy, remain symptom-free—and immune to malaria. Presumably the sickle cell gene has some malaria-defeating function, and single copies of it served carriers well in regions of Africa where malaria was common.

In evolutionary terms, genes for homosexuality don't seem to make sense, but there may be some hidden advantage.

"If there are genes for homosexuality, they are so prevalent that there must be some beneficial effect to somebody," says Bailey. "But I have no idea what the particular benefit is."

Sociobiologists such as James Weinrich of the University of California at San Diego and Edward O. Wilson of Harvard offer another explanation for the persistence of genes for homosexuality. "They have suggested," says Bailey, "that maybe some people don't have very good reproductive futures"—and that somehow those people sense that—"so instead of electing to reproduce, they adopt an alternative life strategy, and perhaps they help their siblings reproduce more children by refraining from reproducing themselves." But Bailey rejects what he calls "the kind gay uncle hypothesis" because no evidence supports it.

If one accepts the idea that sexual orientation has a strong genetic component—and some researchers doubt it—the next logical step is to try to track down the responsible genes. Among the scientists taking that step is geneticist Cassandra Smith, who is collaborating with Pillard. One way to approach the problem, says Smith, "is to take some educated guesses about what might be involved in homosexuality and sexuality. Men and women are different; maybe it's something on the X or Y chromosome. There's not much on the Y. But several candidates have been isolated recently on the X." She hopes to study codes for a substance known as the androgen receptor protein, which enables the body, including the brain, to respond to male hormones. "It's a long shot," she says, "and there's a high probability it's wrong. But this gene is interesting. Everyone is excited about it." Recent studies have shown the gene to have three unstable spots that might make it vulnerable to mutation. "Because there are

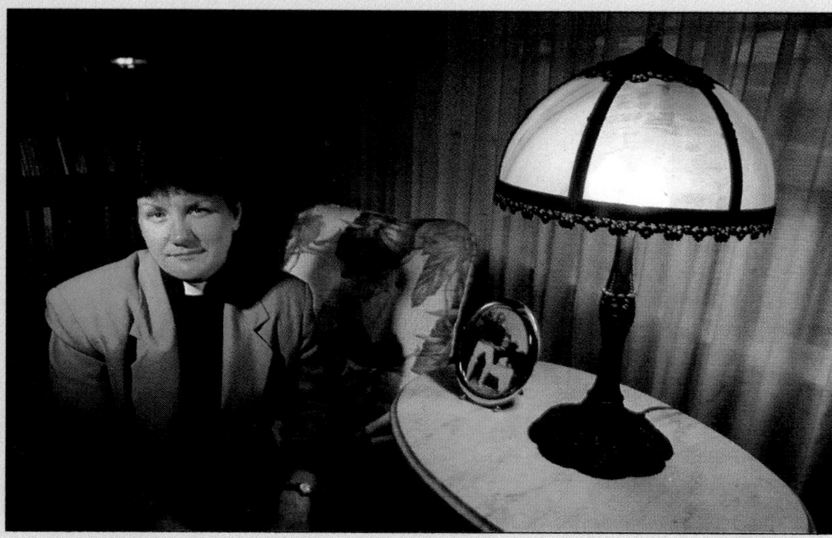

DNA may help explain why identical twins Jan (left) and Judy Dahl, participants in Bailey and Pillard's study, are lesbians. (© Eduardo Citrinblum)

three independent regions," Smith says, "there's potential for enormous variation between individuals."

Like Bailey, she suspects that any gene found to cause homosexuality will turn out to have a hidden advantage. "Nature," she says, "is keeping it around for some reason."

Connecting the Concepts

The following questions will help you to connect the issues discussed in this article with the concepts you have learned in Part 2.

1. If gay genes exist, by their very nature they reduce the likelihood of sexual reproduction. How can a population afford to have genes that decrease the rate of reproduction? Will the frequency of gay genes continually decrease in the human population until they no longer exist?
2. Discuss the various ways in which the discovery of gay genes might change or influence social, medical, legal, and political ideas and attitudes.
3. If the existence of genes that determine homosexuality is substantiated, how then might bisexuality be genetically determined?

PART 3

Evolution and the Diversity of Life

Chapter 14
Evolution and Natural Selection

Chapter 15
Population Genetics and Speciation

Chapter 16
Evolution and Reproduction

Chapter 17
Origin of Life

Chapter 18
Classification of Organisms and the Problem of Viruses

Chapter 19
Bacteria, Protists, and Fungi

Chapter 20
The Plant Kingdom

Chapter 21
The Animal Kingdom

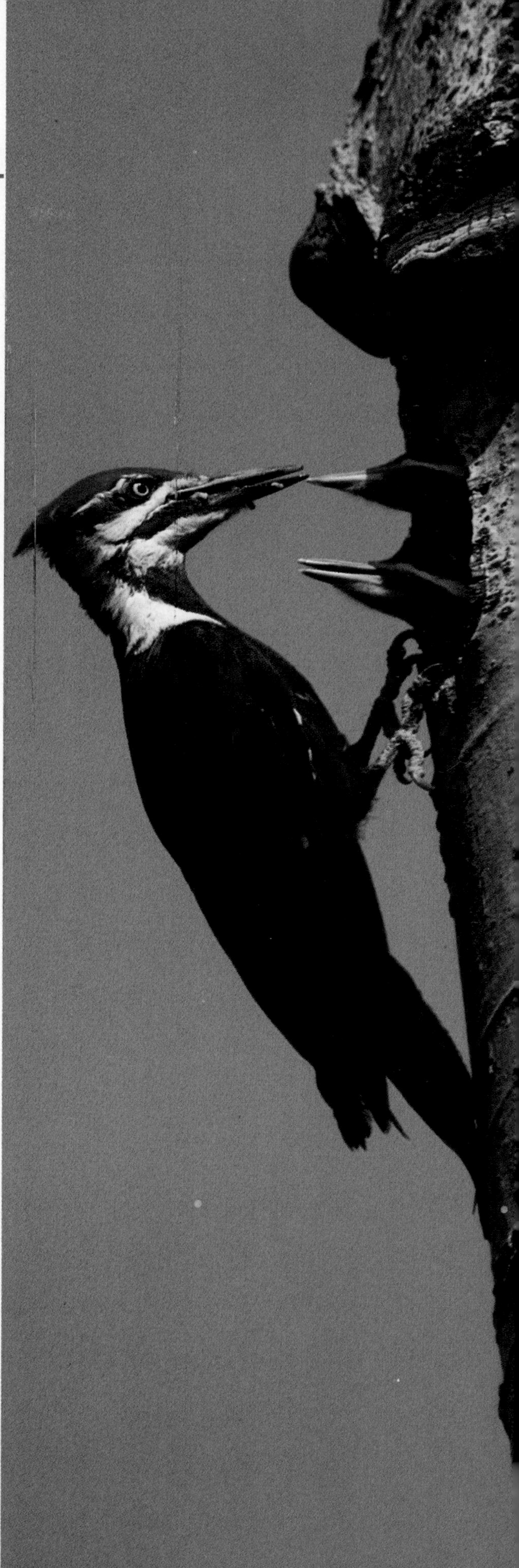

Pileated Woodpecker and young at nest cavity. Nearly everything about a Pileated Woodpecker is an evolutionary adaptation to its lifestyle of chiseling into tree trunks for food and shelter. Its' tail feathers form a spiky prop, giving increased stability as the bird's large feet with pairs of forward- and backward-pointing toes grip a tree trunk. The woodpecker uses its strong beak to chisel into wood and its exceptionally long, barbed tongue harpoons beetle grubs that tunnel deep into tree trunks. Here, a parent brings a meal to its three hungry-looking young. (Alan G. Nelson/Animals Animals)

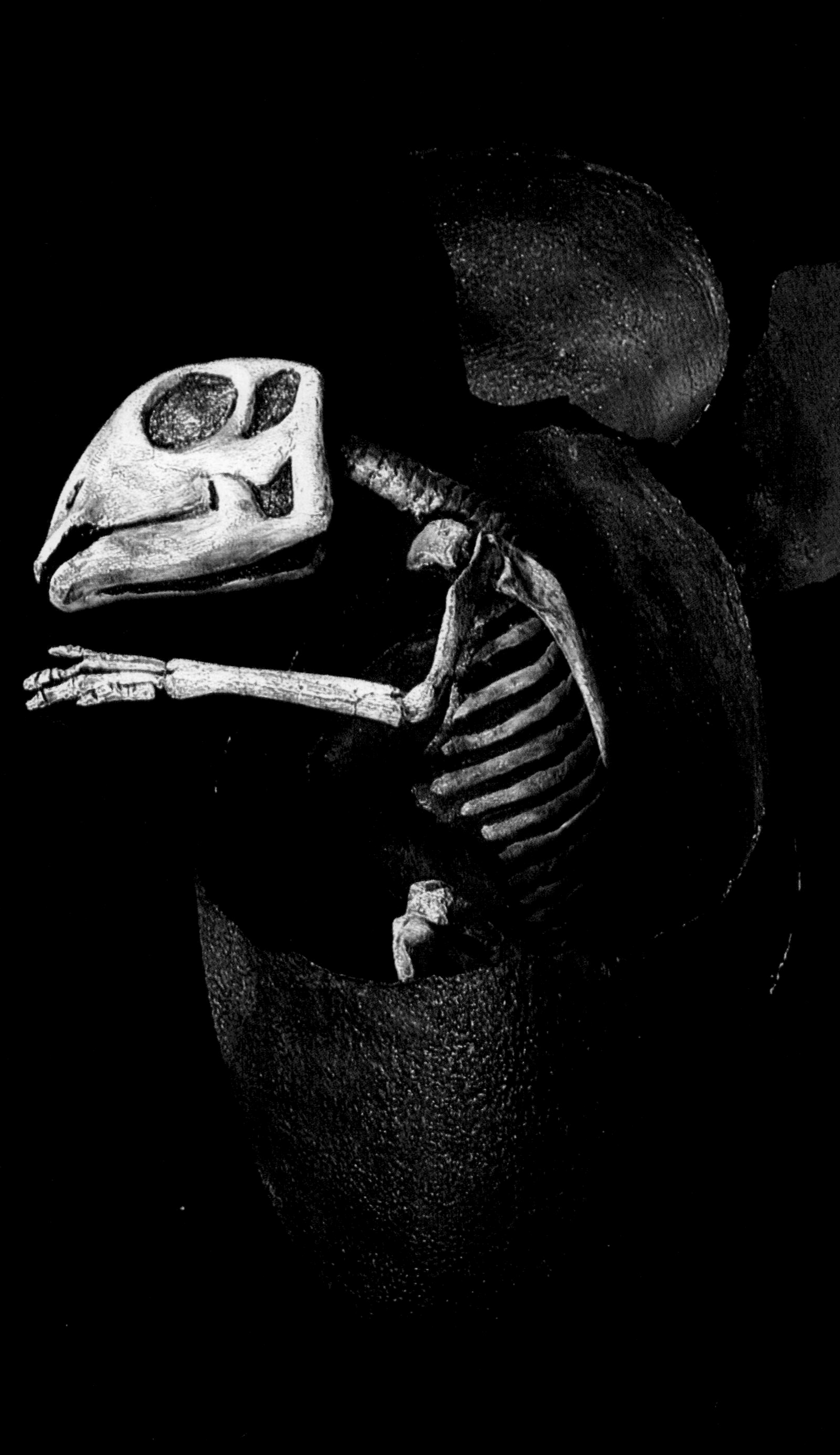

CHAPTER 14

CONCEPT GUIDE

After reading this chapter you should be able to:

1. Compare the theory of natural selection to Lamarck's ideas about how evolution occurs. (Sections 14-A,B)
2. Explain how the following provide evidence for evolution, and give or recognize examples in each category: artificial selection, the fossil record, comparative anatomy, comparative biochemistry, developmental biology, and biogeography. (Section 14-C)
3. Explain the four main assumptions upon which natural selection is based. Describe an example of natural selection seen in a wild population. (Section 14-D)
4. Explain how the phrase "survival of the fittest" can be a misleading description of evolution. Use the example in Table 14-2 in your explanation. (Section 14-E)
5. Explain how humans have contributed to the widespread production of antibiotic-resistant bacteria. (Section 14-F)

Some fossils reveal much more than details about an organism's skeleton. This baby Hadrosaur was found still within its egg! Read more about the importance of fossils in Section 14-C. (E. R. Degginger)

Evolution and Natural Selection

We have seen how genes are passed from grandparents to parents to offspring. In the next few chapters we consider these same events, but on a much larger scale. We shall no longer focus on pairs of genes passed through a few generations of a single family, but instead we shall consider all the genes passed on over thousands of generations and within a *whole population* of a particular species.

We shall see that the proportions of various genes in a population may change over this broad sweep of time. Sometimes, genetic changes produce changes in the population so slight that they are scarcely noticeable (for example, an increase in the proportion of individuals in a population who carry the gene for a particular enzyme). In some cases, however, genetic changes accumulate and result in organisms that are quite different from their ancestors. This is evolution on a broader scale: the origin of new species by descent and modification from previously existing forms of life. New species result when some members of a population have evolved sufficient differences from others in the population so that the two groups can no longer interbreed. Such changes in the overall genetic makeup of a population over time are called **evolution**.

KEY CONCEPTS

- The theory of evolution states that organisms arise by descent and modification from previously existing organisms.
- Evolution is a change in proportions of one or more alleles in a population from one generation to the next.
- Natural selection, the differential reproduction of genotypes, is the most important cause of evolution.
- Populations of organisms evolve adaptations to their environments.

Curiosity Questions

1? Why do animals have structures that don't seem to be necessary, such as the appendix in humans? (See page 288)

2? How do insects become resistant to pesticides? (See page 295)

3? What are the dangers of regularly feeding antibiotics to cows? (See page 296)

14–A OVERVIEW: THE THEORY OF EVOLUTION

Notice three things about evolution:

1. *An individual cannot evolve.* Rather it is the population that evolves—that changes over time—not its individual members. Each individual's genetic makeup is fixed at fertilization, when egg and sperm unite. The individual's significance in the evolutionary story is to pass on to its offspring some of its own particular genetic composition, slightly different from that of any other individual—and so to contribute genes to the next generation. Evolution occurs as a result of the accumulation of these new genetic variations in populations over long periods of time (Table 14–1).
2. *Natural selection acting on genetic variability is the mechanism of evolution.* As we saw in Chapter 1, every natural environment presents a variety of hazards for the organisms that are born there. Individuals survive in their environments with varying degrees of success—or do not survive at all. In evolutionary terms, an individual is only truly successful if it can survive the pressures of its environment long enough to reproduce. In Chapter 1 we saw that the process of **natural selection** is the way that the environment interacts with genetic variability, resulting in differences in reproductive success of individuals. When we say that a trait has been "selected for" or "selected against," we are *not* referring to an intentional act of selecting. We mean only that a particular trait, among all others that may arise, is one that has been passed on through generations (Table 14–1). This leads to our third, and most important, point.
3. *Evolution by natural selection occurs by chance.* There is no purposeful action that changes organisms, allowing them to evolve new structures or abilities. No matter how much a trait may *appear* to be a deliberate response to the environment, it is not. Birds did not develop wings because they wanted to fly, or even because flying was an asset to their survival. Rather, by chance, at some point in time, genetic changes occurred that produced feathers in certain individuals. It is possible that once feathers had appeared, they conferred survival advantages to the animals who possessed them. The environment favored feathered individuals and allowed them to reproduce successfully.

Similarly, the octopus did not develop arms because they would be good for manipulating things. Rather, over time, natural selection among random genetic changes resulted in the modification of a gliding foot into flexible tentacles, and under the existing environmental conditions, tentacles turned out to be an advantage. Perhaps many other traits appeared in other individuals, but they did not survive. Possibly tentacled animals were more successful at finding food, evading predators, or tickling mates—we cannot know for sure. Whatever the case, the ancestors of octopi did survive, reproduced successfully, and passed on the trait of tentacles.

One final example of evolution by means of natural selection should help integrate these ideas. Take, for example, the color of an octopus. An ability to change color to match the surroundings is largely determined by random reassortment of its genes. An individual that blends into its natural background is less likely to be noticed by predators, and therefore is more likely to survive and produce offspring than is a conspicuously colored individual. Because more octopi with good camouflage survive to breed and pass on the genes that dictate this coloring, a larger proportion of individuals in the next generation of the population will be camouflaged. The genetic makeup of the population has changed somewhat from one generation to the next, and that is evolution.

Imagine that at some time in the future, the predator's eyes or other senses become sharper, or perhaps a new octopus predator appears and hunts in a different manner. As a result, a camouflaged octopus may not have the same survival advantage and some other genetically derived characteristic (escaping behind a dense cloud of ink, for example) may provide a survival advantage. Octopi that can squirt ink will be more likely to survive to reproduce. This will result in a shift in the octopus population toward ink-squirting.

Finally, if random genetic changes in the predator population produce an individual with hyperacute sense of smell or sonar that would allow it to track its prey within a cloud of dark ink, some other trait that has randomly appeared in the octopus population may be selected for—highly toxic venom, for example. Indeed, the blue-ringed octopus that is found in Australian tidal pools (Figure 14–1) possesses all these adaptations: camouflage, ink-squirting, and venom—the result of evolution by means of natural selection working on random genetic changes within a population.

It is important to note, however, that random genetic changes may *not* result in survival, and that it is equally possible that a population may never develop an effective defense against predators or other environmental pressures. An entire species may die out completely and become **extinct** (See Section 38–E). In

Table 14–1
Evolution: What, Who, How

	Random Genetic Changes	Natural Selection	Evolution
What Is It?	The raw material acted upon by natural selection	Difference in reproductive success between genetically different individuals of a population	Genetic change in a population from generation to generation
Who Does It Affect?	Only individuals	Individuals and populations	Only populations
How Does It Work?	Produces an organism with an individual genetic makeup that: 1. is unique 2. may be inherited by the organism's offspring	Produces an organism with genetic differences that: 1. are advantageous under the existing environmental conditions 2. make it more likely to survive and reproduce, passing these advantages on to offspring	Natural selection affects generations and produces a population that is better adapted to existing environmental conditions

Wrong: "Individuals can evolve."
Right: "Individuals do not evolve."

Look at this white moth on a dark tree trunk. The genes that determine color cannot be modified in an individual. Therefore, this moth has no way to change its color to hide from predators in this dark-colored environment, and it will likely be selected against—eaten!

Wrong: "Evolution produces perfection."
Right: "Organisms are not perfectly adapted."

Almost all adaptations represent a compromise, in which the adaptation has advantages and disadvantages. For example, a turtle's shell provides good protection against the attack of foxes, raccoons, badgers, etc..., but this same shell makes for relatively slow locomotion, decreased maneuverability over obstacles like logs, and makes reproduction quite challenging.

Wrong: "Evolution has a purpose."
Right: "Evolution involves chance."

Selective factors are contradictory and unpredictable, and so is evolution. For example, when the weather is bad in the north, geese that migrate south for the winter produce more offspring than those that don't. But in a mild winter, those that stay in the north do better, passing stay-at-home genes on to more members of the next generation.

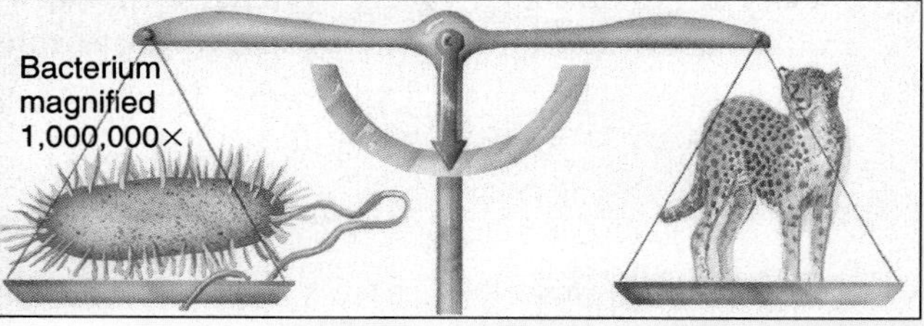

Wrong: "More evolved equals better."
Right: "New species are not better than older species."

Any species alive today is successful (so far) whether it originated 200 years ago or 200 million years ago. Is a bacterium that evolved a billion years ago better than a cheetah? Less beautiful to human eyes perhaps, but the bacterium is just as successful. Indeed, bacteria may very well outsurvive the cheetah species and all of its relatives.

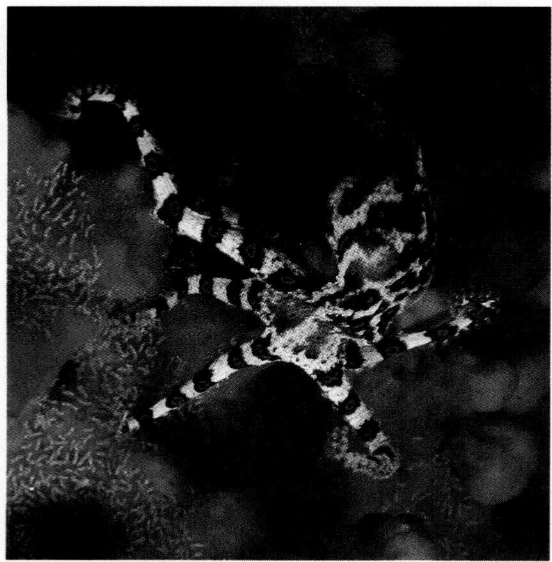

FIGURE 14–1

A blue-ringed octopus from Australia. As a result of evolution by means of natural selection, this octopus species uses camouflage, ink-squirting and a highly toxic venom in defense against predators. (F. Bavendam)

other situations, predators and environmental conditions may change so rapidly that the population cannot adapt quickly enough. Consider the rapid and drastic changes that human tech-nology can produce in a natural habitat, by clearing large areas of forest, for example. The pace of human activity can easily outstrip the many generations required by the process of evolution, causing organisms that are not highly adaptable to go extinct (see *A Journey into Evolution:* Going . . .Going . . . in this chapter).

Natural selection among random genetic changes is the mechanism by which populations evolve.

14–B HISTORY OF THE THEORY OF EVOLUTION

For thousands of years, most people believed that each separate species of organism had been specially created. From time to time philosophers proposed that the living world changed over time, but until the mid-seventeenth century this idea gained little ground in the Western world. From about 1750 on, however, many people became convinced that species changed over the ages.

Lamarckism

In 1809 the French biologist Jean Baptiste de Lamarck blended some of the common beliefs of the day into his own version of how evolution occurs. Today he is remembered more for his *incorrect* explanation than for his belief in evolution.

Lamarck suggested that organisms could acquire traits that made them better adapted to their environments, and could pass such traits on to their offspring. Hence a population changed from one generation to the next. This is the theory of evolution by the inheritance of acquired characters. Lamarck's most famous example was the long neck of the giraffe (Figure 14–2). He suggested that giraffes had evolved their long necks because they strained to reach leaves growing above their heads as they ate, thereby stretching their necks, and that this added length was passed on to their offspring. This idea dovetailed nicely with pre-1900 beliefs, which held that different parts of the body contributed to eggs and sperm by sending minute particles through the bloodstream to a collection point in the reproductive organs.

Darwin did not refute this idea when he proposed natural selection as the mechanism of evolution. He did not understand that individuals inherit discrete genes, and he thought that the inheritance of acquired characters might have a minor role in evolution. This view was discarded when Mendel's work on genetics was rediscovered and expanded (Chapter 12).

This is not to say that nothing an organism does in its lifetime can affect its offspring's genotype. Taking drugs that destroy chromosomes, or being exposed to high levels of radioactivity, may alter the genes passed on to the offspring. However, it is clear that, with a few possible exceptions, nothing an organism does will make its offspring inherit the same characteristic that it has acquired during its lifetime.

Darwin and Wallace

The theory of evolution by natural selection was put forward in a joint presentation of the views of Charles Darwin and Alfred Russel Wallace before the Linnaean Society of London in 1858. As we have seen, Darwin and Wallace were not the first to suggest that evolu-

BIO-BIT

The basic idea of evolution—that organisms change over time—was first suggested by Greek philosophers nearly 2500 years ago.

(a) Lamarck:

Giraffes strained to reach leaves growing above their heads, thereby stretching their necks—an acquired trait.

This acquired physical trait (increased neck length) was passed on to offspring, which changed the population after many generations.

Long-necked giraffes resulted.

(b) Darwin:

Individuals of a species vary. (In giraffes, neck length varied.)

Individuals with some genetic variations (such as longer necks) were more likely to survive and reproduce because they could better reach their food.

Thus, long-necked giraffes became more common in a population from one generation to the next.

FIGURE 14–2

Why do giraffes have long necks? Lamarck suggested that giraffe ancestors stretched their necks to browse on leaves high up in trees, and that this behavior affected the anatomy of succeeding generations. This appears to be wrong, however, in part, because there is no current evidence that adaptive behavior such as this can change the genetics of an individual and be passed on to the next generation.

tion occurred. Their names are linked with evolution because they proposed natural selection as the mechanism that brought it about. We are more likely to believe in a process when people give a convincing explanation of *how* it happens than if they merely assert that it *does* happen. Darwin's explanation of natural selection eventually convinced the world that evolution occurred.

Darwin and Wallace came to the same conclusion about evolution as a result of very similar experiences.

(a)

(b)

(c)

(d)

(e)

FIGURE 14–3

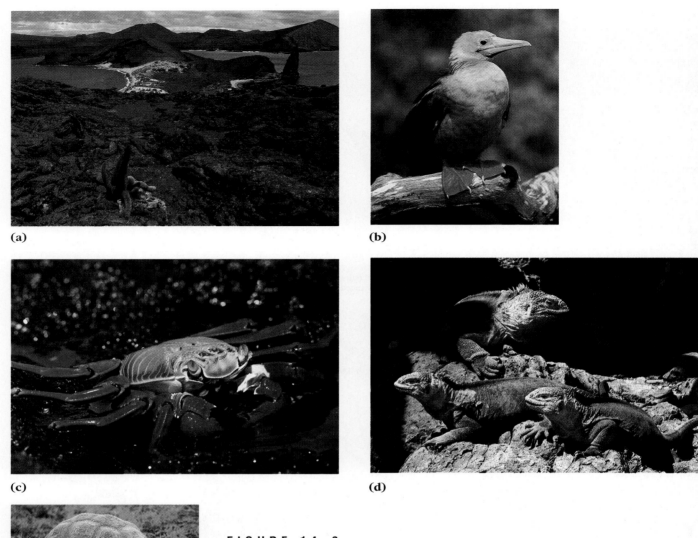

The Galapagos Islands. (a) Many of the islands look barren, dominated by the volcanic rock from which they first formed. A Red-footed Booby **(b)**, crab **(c)**, land iguana **(d),** and giant tortoise **(e)** represent a few of the unusual species found in the Galapagos Islands. (a, G. Holton/Photo Researchers, Inc.; b, F. Lanting/Minden Pictures; c, J. McDonald, Animals Animals; d, F. Bavendam/Peter Arnold; e, K. Schafer & M. Hill)

First, both Wallace and Darwin were influenced by reading the works of geologist Charles Lyell and economist Thomas Malthus. Lyell wrote that the world was an ancient arena in which rock formations slowly appeared, changed, and disappeared. He recognized that competition between species leads to a "struggle for existence" and even discussed the extinction of species caused by human activities. All the information needed to formulate the theory of evolution was present in Lyell's work. Malthus similarly argued that there is competition between organisms. He wrote that every human population must eventually outgrow its food supply and then be reduced by disease, starvation, or war.

Second, both Wallace and Darwin observed plant and animal life in several parts of the world. Wallace traveled in South America, and later in the islands of Indonesia. It was here, in 1854, that the idea of natural selection came to him as he lay in bed with a fever. In the 1830s, Darwin obtained a position on *H.M.S. Beagle,* a British naval ship embarking on a five-year mapping and collecting expedition. This trip took Darwin to South America and the nearby Galapagos Islands (Figure 14–3), where he collected much of the evidence he later used to support the theory of evolution by natural selection.

Upon his return to England in 1837, Darwin settled down to a lifetime of writing and thought. By the next year, he had formulated the theory of evolution by means of natural selection, but he pondered it and accumulated supporting evidence before presenting it publicly. In 1845, Darwin published *The Voyage of the Beagle,* an account of his travels. He showed that he already held the clue to how evolutionary change was brought about, a mechanism that he was not to publish until more than a decade later. He wrote, "some check is constantly preventing the too rapid increase of every organized being left in a state of nature. The supply of food, on average, remains constant; yet the tendency in every animal to increase by propagation is geometrical."

In 1858 Darwin received a manuscript from Wallace, describing natural selection. Wallace had written his paper in three days. Darwin passed Wallace's paper to Lyell and to the botanist Joseph Dalton Hooker, who persuaded Darwin to let them present a version of his theory and Wallace's paper at a scientific meeting in 1858. Darwin then worked feverishly to finish his book, *The Origin of Species by Means of Natural Selection,* which was published in 1859. In it, he marshalled an impressive array of evidence to support his theory, the result of a quarter of a century of observation and inquiry. The book sparked immense controversy, a fitting tribute to the most original and important biology book ever written. Although evolution was accepted in Darwin's day, most biologists did not fully accept the idea that evolution occurs by means of natural selection until the twentieth century.

Evolution results in the origin of new forms of organisms by modification from previously existing forms of life. Although the idea that living organisms change over time had been around long before Darwin's work, the mechanism of evolution had yet to be defined. Lamarck suggested the mechanism of the inheritance of acquired characteristics, which cannot be true with our present understanding of genetics. Darwin and Wallace arrived at the theory of natural selection independently, and presented their ideas in public together in 1858.

14–C THE EVIDENCE FOR EVOLUTION

Several different lines of evidence convinced Darwin and Wallace, and many of their contemporaries, that modern organisms have arisen by evolution from more ancient forms of life.

The Evidence from Artificial Selection

Darwin illustrated selection with examples drawn from the selective breeding of domestic plants and animals. These organisms do not usually breed randomly. Breeders and gardeners save seed only from the largest, most attractive flowers and the tastiest melons. Dairy farmers mate the cows that produce the most milk with bulls whose mothers were good milk producers. Modern breeds of dogs look distinctly different from one another and from the hypothetical ancestral dog {Figure 14–4). Breeders and farmers exert **artificial selection** on domesticated animals and plants by determining which members of the population shall reproduce and which shall not. The striking changes produced over relatively few generations are powerful proof that organisms can evolve.

However, this evolution results from deliberate manipulation by breeders with definite ends in view. It is more difficult to show that a similar process accounts for changes in natural populations. A weakness of Darwin's evidence for evolution was that he never provided a convincing demonstration that selection actually occurs in nature. His detractors pointed out that nature has no mind, no goal or purpose. How could a haphazard series of accidents result in organisms that appeared as though they were designed specifically for the place they hold in nature? The examples of selection in wild populations described in Section 14–D were not worked out until a century later.

The Evidence from the Fossil Record

Usually, when an organism dies, scavengers and decay organisms rapidly destroy it. Occasionally, however, a body may come to rest in an acid bog, or be buried under a layer of mud that cuts off oxygen—conditions that prevent decay and may permit the body to be preserved. A **fossil** is any preserved evidence of life long past: a body or body part, an impression of the surface of the body such as a footprint; organic molecules such as oil, which are chemical remains of organisms; or even a coprolite ("fossilized feces").

> **BIO-BIT**
>
> About 250 million years ago, more than 90% of all of the existing marine animal species became extinct. This mass extinction is used by geologists to mark the boundary between the Mesozoic and Paleozoic eras.

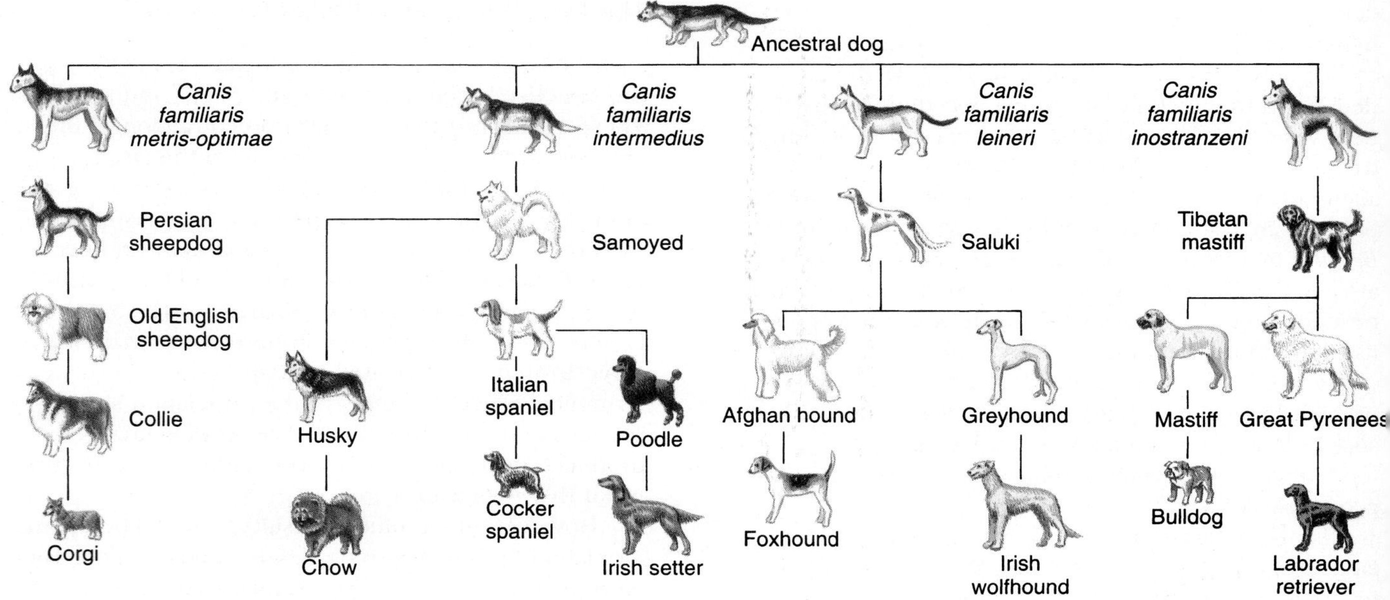

FIGURE 14-4

Artificial selection in dogs. All of the dog breeds known today have resulted from tremendous inbreeding and artificial selection of a single ancestral dog species. This diagram illustrates the ancestry of a few of the many dog breeds that currently exist.

FIGURE 14-5

Stratification. These canyon walls reveal the many multicolored layers of soil that were deposited over millions of years. Each layer, like a chapter in a book, contains its own fossil history of life at the time the layer was deposited. (A. Sanfedele/Earth Scenes)

Fossil-hunting became a popular pastime in nineteenth-century England. Drawings in Victorian magazines portray ladies in long skirts, and gentlemen in jackets and ties, scrambling over rocks with geological hammers in their hands and fanatical gleams in their eyes. Most of the important fossils of the time were found by amateurs such as these.

These finds captured the popular imagination, and newspapers printed articles and letters arguing about the religious and scientific implications of fossils. Some people suggested that God had fashioned the fossils and scattered them in the rocks to delight fossil-hunters. However, geologists were beginning to produce a very different explanation. The growing collections of fossils in North America and Europe provided strong evidence that organisms had changed over the centuries.

First, nearly all of the fossils ever found are of species that are now extinct. Second, some species are older than others. Many fossils occur in formations made up of several layers of rocks, and geologists realized that the bottom layer had usually been formed first and contained older fossils than the overlying layers (Figure 14-5). Assigning relative dates to fossils in this way made it clear that some groups of species were older than others. (In this century, it became

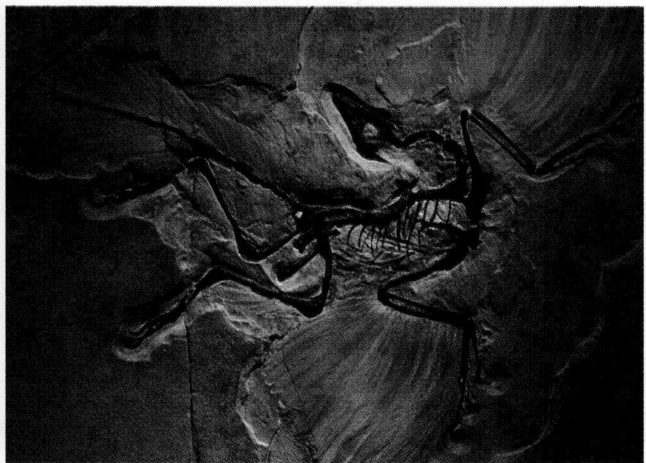

FIGURE 14-6

A fossil of *Archaeopteryx*. This extinct animal from the Jurassic period of the Mesozoic Era is one of the first known birds. Although its skeleton is very lizard-like, the impressions of feathers (characteristic of birds) are clearly seen in the area of the wings and tail. The head of this animal has been bent backwards during the fossilization process. (J. L. Amos/Photo Researchers, Inc.)

possible to calculate the actual age of rock formations by using radioactive isotopes.)

The abundance of members of different groups also changed with the age of the rock. It appears that some groups of organisms originated much earlier than others, and that some of the more recent groups have largely replaced more ancient ones over the course of time (see Appendix A).

In a few cases, fossils allow us to reconstruct the family tree of a particular group. For instance, we can trace the origin of mammals from reptiles in great detail in the fossil record. We can also follow the changes from dinosaur-like reptiles to ancient birds with feathers, but also with teeth (Figure 14-6), and eventually to modern birds with feathers and no teeth.

The classic case of fossil genealogy is the story of horse evolution, published by Othniel C. Marsh in 1879. Marsh examined a series of fossils linking modern horses with the tiny dog-sized *Hyracotherium,* the "dawn horse" found in Eocene rocks. Through the fossil record he traced the major changes in the teeth, legs, and feet of ancestral horses (Figure 14-7). Many similar examples have now been described in which the ancestry of modern species can be traced through successive rock layers, with the youngest rocks containing those fossils most like the modern forms.

The Evidence from Comparative Anatomy

Even without fossil evidence that different organisms have lived at different times in the past, we might suspect that organisms had evolved by comparing the structures of species alive today. Not surprisingly, similar kinds of organisms have very similar structures. For example, the skeletons, teeth, and muscles of different members of the cat family are very similar, and the same is true of different species of bats or of whales. However, a comparison of a cat's bones with those of a bat or a whale reveals that these three groups of animals all have skeletons composed of quite similar groups of bones, despite their adaptations to very different ways of life. The forelimb bones of cats, bats, and whales are arranged in the same pattern: a bat's wing, a cat's front leg, and a whale's flipper all contain bones identifiable as humerus, radius, ulna, and so on (Figure 14-8). Indeed all mammals, birds, reptiles, and adult amphibians have forelimbs with this same basic framework, although the limbs may perform very different functions in animals as different as a pigeon, a penguin, a turtle, or a human. Furthermore, all of these forelimb bones originate from the same part of the embryo. Such structures, with the same origin, but occurring in different species, are said to be **homologous** to each other. Homologous organs may perform the same or different functions.

An alternative to homologous organs are **analogous** organs, which have similar functions but are constructed differently and appear unrelated. The wings of birds and of insects, for instance, may both be used for flying, but they are completely different structurally (Figure 14-9).

Darwin saw that homologous and analogous organs posed problems to a creationist viewpoint. It made no sense that several different types of wings should be invented. Even the homologous wings of birds and bats differ somewhat in structure. Surely one design must be superior to the other. Why create several different sorts of wings? Similarly, why did so many animals contain apparently inefficient homologous structures? Why do whales have heavy bones, like those of terrestrial mammals, in their flippers, instead of the lighter, folding fins of a fish, apparently so much better designed for propulsion in water?

Eventually Darwin came to realize that evolution made sense of all these paradoxes. Organisms were not created from a clean slate. They arose from ancestors with characteristics already determined by their own evolutionary histories. Whales have bony flippers because they evolved from land mammals with bony forelimbs. Insects have no bones to support their wings because they evolved from animals with

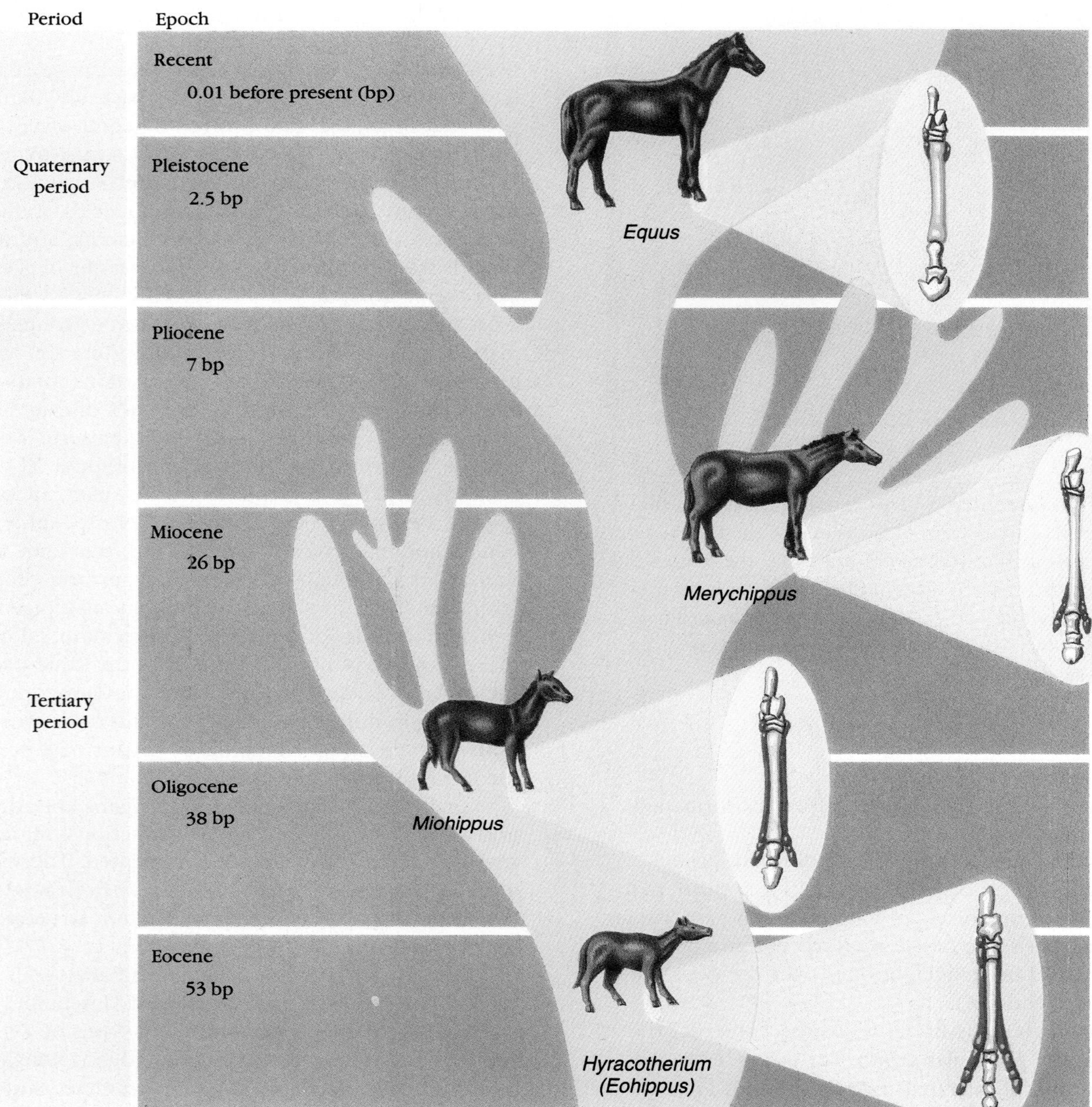

FIGURE 14-7

Evolution of the modern horse. The ancestral *Hyracotherium* gave rise to many species, of which a few are still alive today. Most of these species of horse-relatives became extinct, represented by green branches that do not continue to the top of the diagram. The legs of the four species drawn here show progressive enlargement of the central toe and loss of the side toes. The overall size of the animal is also progressively larger, although some evolutionary branches that are not shown produced species smaller than their ancestors. The number given under each epoch is a date (million of years ago).

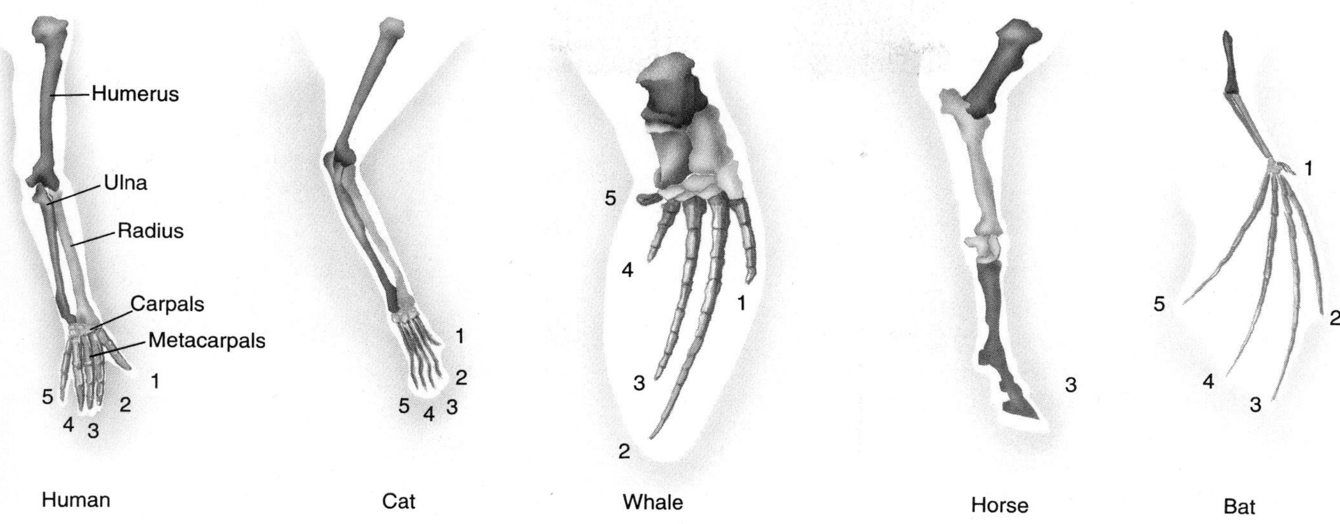

FIGURE 14–8

Homologous structures in the forelimbs of a human, cat, whale, horse, and bat. All of these forelimbs show the same basic skeletal pattern. In different vertebrate groups, the various bones have been modified as the limbs became specialized for different functions. Note that the horse has lost all of its phalanges (finger bones in humans) except those associated with its "middle finger."

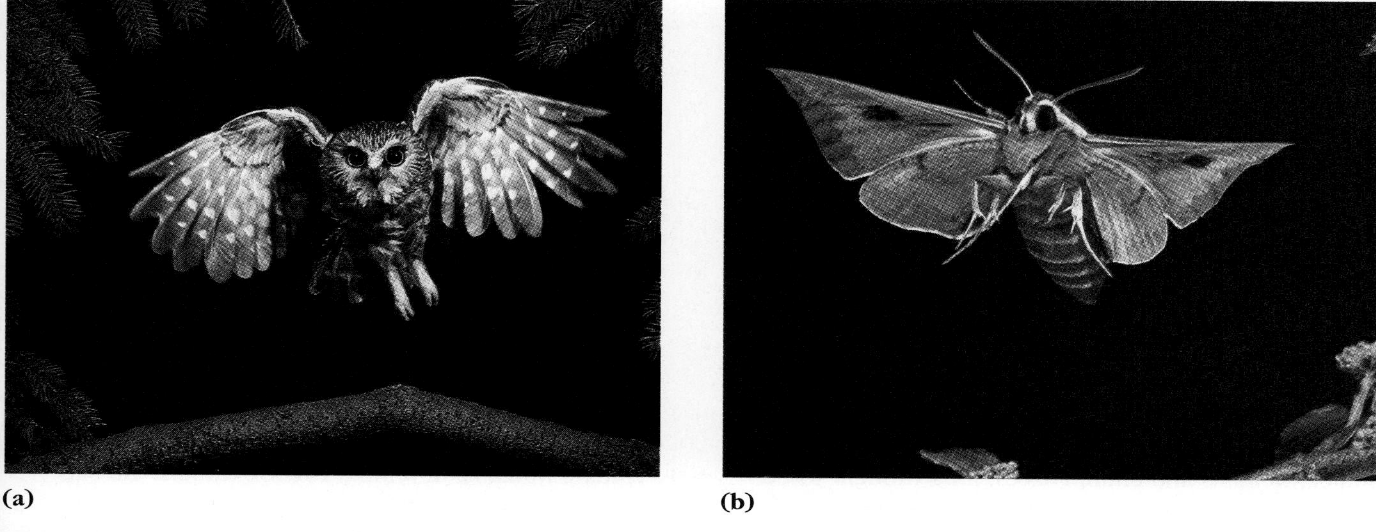

(a) **(b)**

FIGURE 14–9

Analogous structures. The wings of a bird **(a)** and of a moth **(b)** share similar functions but develop in completely different ways. The wings of a bird result from the modification of the forelimbs while the wings of a moth develop independently from the legs. (a, D. Kuhn; b, S. Dalton/ Photo Researchers, Inc.)

BIO-BIT

The incus, malleus, and stapes bones within the ears of mammals are homologous to bones within the jaws of bony fish, amphibians, reptiles, and birds.

external skeletons of chitin and without bones. The very imperfection of adaptations, the feeling that so many of them could have been designed better, became, to Darwin, the most convincing evidence that evolution has occurred.

1? Anatomy provides a further argument for evolution in the form of **vestigial structures**: organs useless to their present owners but homologous with structures that serve important functions in other species. The most familiar example is the human appendix, a worm-like blind sac near the junction of the small and large intestines (see Chapter 22, A Journey Through Human Digestion). The appendix is homologous to the caecum, a large, blind chamber where leaves and grasses are digested in many other mammals. Other examples are the minute pelvic and hindlimb bones in the skeletons of whales and of boa constrictors, even though these animals have no true hindlimbs. All these vestigial organs are the evolutionary remnants of organs that were larger, and useful, in their owners' ancestors. ■

The Evidence from Comparative Biochemistry

In more modern times, the arguments for evolution based on comparative anatomy have been paralleled by evidence from other fields. Studies of the similarities and differences in the structures of homologous proteins and genetic material show that organisms known to be related are very similar in these respects. For instance, humans and chimpanzees, long thought to be closely related, have proteins that are 99% alike, and the number, shape, and banding patterns of chromosomes also show relatedness. DNA and proteins are much less similar when they come from organisms that are only distantly related.

The Evidence from Developmental Biology

A similar line of evidence for evolution comes from studying embryonic development, especially of animals. In many instances, the embryo contains structures that will not be found in the adult. For example, the early embryos of reptiles, birds, and mammals, including humans, develop a row of vestigial gill slits just behind the head (Figure 14–10). This suggests that these groups of animals descended from the fishes, in which gill slits persist and function throughout life. Similarly, the embryos of baleen (whalebone) whales and of birds develop tooth buds, even though the adult animals are toothless. Sometimes human babies are born with short tails, or with several nipples in two rows down the front of the body, characteristics that are common in other mammals but not in human adults.

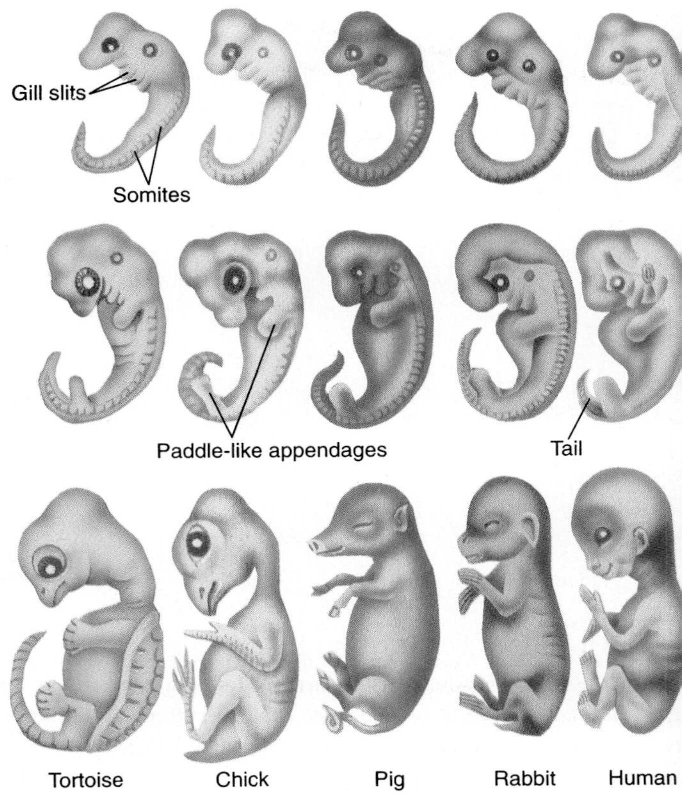

FIGURE 14–10

Homologous components of vertebrate embryos. Despite the differences in the appearance of the adult animals, these vertebrate embryos appear very similar because of their homologous structures. For example, each at some stage in its development has a tail that projects behind the anus, paddle-like appendages, segmented muscle masses called somites on either side of the spinal column, and gill slits behind the eye.

Developmental biology also reveals that certain apparently "new" features of higher vertebrates developed, not from scratch, but from the remodeling of ancestral structures. For instance, some of the embryo's gill arches, which in fish develop into structures supporting the gills, become in mammals parts of the lower jaw, the ear bones, and parts of the air passages of the respiratory system.

The Evidence from Biogeography

Biogeography is the study of the distribution of organisms across the face of the globe. In their travels, both Darwin and Wallace noticed that the present-day distribution of organisms made no sense seen from a creationist point of view, but could be explained by evolution. Why did the Galapagos, a group of small islands off the west coast of South America, contain more different species of finches than the entire South American continent? Another puzzle was the distribution of mammals. Why were marsupial (pouched) mammals found only in Australia and South America? (Opossums are marsupials, but have colonized North America from South America.) Why did Australia contain none of the placental mammals found throughout the rest of the world?

Both men became convinced that these puzzles, and dozens of others could be explained as the result of the evolutionary histories of these modern organisms—including where their ancestors lived. From an ancestral group living in a particular place, descendant populations could spread, or radiate, into other areas. In doing so, they would encounter new environmental conditions that would bring about the evolution of new adaptations. Such an evolutionary process, giving rise to new species adapted to new habitats and ways of life, is called **adaptive radiation.** In Australia the adaptive radiation of marsupials gave rise to a variety of species that closely resemble equivalent placental mammals elsewhere. Australian marsupials include the rabbit-like bandicoot, the woodchuck-like wombat, the Tasmanian "wolf," and the flying-squirrel-like flying phalanger, as well as unique forms such as koalas and kangaroos. However, the spread of organisms may be limited by geographical boundaries. For instance, marsupials in Australia could not move to other continents because oceans barred the way.

Darwin was profoundly struck by the flora and fauna of islands, particularly the Galapagos Islands, about 1000 kilometers west of South America. What caught his attention were the remarkable numbers of **endemic species,** species found nowhere else, even on other apparently similar islands nearby (Figure 14-11).

That the tiny, relatively barren Galapagos Islands (and other isolated islands visited by the *Beagle*) housed such large numbers of endemic species seemed a profligate waste of effort by the Creator. Darwin saw that the existence of so many endemic organisms could be explained by assuming that members of certain mainland species had colonized the islands, and there evolved into new species. Some of the original settlers had even given rise to several new species through adaptive radiation, as different groups of their descendants became adapted to different habitats on the same or different islands. Such was the case with the 13 species of Darwin's Galapagos finches: the ancestral finches must have come from the South American mainland, and their descendants underwent adaptive radiation in the Galapagos, becoming adapted to living in different habitat zones and to eating different sorts of food (Figure 14-12). A fourteenth species evolved on Cocos Island, about 800 kilometers away. Local conditions on islands differ from those on the mainland, so natural selection on an island inevitably produces different adaptations. Islands are isolated gene pools, and the random nature of genetic variation in each ensures that mutations arising in one island population will be different from those arising on a neighboring, but isolated, island.

FIGURE 14-11

Two unique species of the Galapagos. The pads of the endemic *Opuntia* cactus are a favorite food of the land iguana. A pair of iguanas may live under their own cactus, which may grow to be the size of a small tree. During his visit to the Galapagos, Darwin wrote the following about these iguanas: "we could not for some time find a spot free from their burrows on which to pitch our single tent . . . they are ugly animals, of a yellowish orange beneath, and of a brownish red colour above. These lizards, when cooked, yield a white meat, which is liked by those whose stomachs soar above all prejudices." (F. Lanting/Minden Pictures)

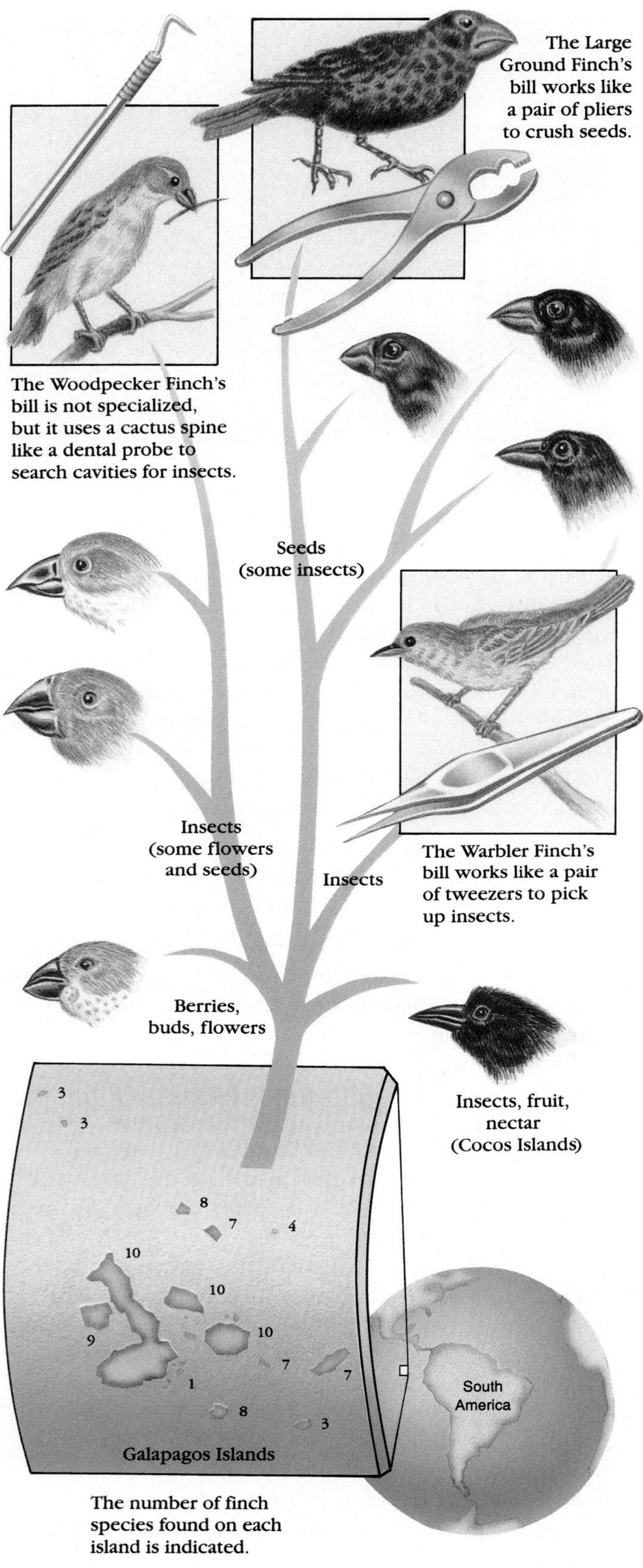

The Large Ground Finch's bill works like a pair of pliers to crush seeds.
The Woodpecker Finch's bill is not specialized, but it uses a cactus spine like a dental probe to search cavities for insects.
Seeds (some insects)
Insects (some flowers and seeds)
Insects
The Warbler Finch's bill works like a pair of tweezers to pick up insects.
Berries, buds, flowers
Insects, fruit, nectar (Cocos Islands)
3
3
8
7
4
10
10
10
9
7
7
1
8
3
Galapagos Islands
South America
The number of finch species found on each island is indicated.

FIGURE 14-12

Darwin's finches. This map of the Galapagos Islands shows the number of finch species found on each island, and a probable family tree of some of the finches that have evolved in the Galapagos Islands and on the Cocos Islands, 800 kilometers away. The birds probably all evolved from a common ancestor. They look very similar and differ mainly in the adaptations of their bills for eating different types of food.

Evidence supporting the idea that organisms undergo evolutionary change comes from many sources. Using artificial selection, farms and breeders have produced rapid evolution in domesticated animals and plants. The fossil record shows that different organisms have lived at different times during the Earth's history, and that organisms have changed over geological time. Comparative anatomy and developmental biology provide evidence that, as new lifestyles evolved, many ancestral structures were either modified, reduced, or lost. Comparative biochemistry demonstrates that the more closely related two organisms are, the more similar their DNA and protein structure will be. Finally, the uneven distribution of organisms around the world suggests that various groups have arisen at different times and different places, from which they radiated out as they became adapted to new habitats and ways of life.

14–D EVOLUTION BY MEANS OF NATURAL SELECTION

All of this evidence for the occurrence of evolution is quite convincing, but the feature of Darwin and Wallace's theory that convinces most people is the idea that natural selection produces evolution. We shall see in Chapter 15 that factors other than natural selection can also produce evolution, but natural selection is by far the most important reason that evolution occurs.

Natural selection is a simple idea. What it amounts to is that some genotypes are reproduced more frequently than others. What is not so simple is grasping how natural selection affects populations and brings about evolution. This is one area of biology in which thinking about a subject will teach you more than reading about it. (The Thinking Critically questions at the end of this chapter list some examples to consider.)

We start with the logical argument that natural selection occurs and brings about evolution. This may be summarized as follows:

1. Individuals of a species vary.

2. Some variations are genetically determined.
3. More individuals are produced than live to grow up and reproduce.
4. Individuals with certain genetic variations are more likely to survive and reproduce than those with others.

Conclusion: From the preceding four premises, it follows that those hereditary traits that make their owners more likely to grow up and reproduce will become more common in a population from one generation to the next.

To take an example, if part of a child's genetic variation is an inherited mutation that causes a severe liver disease, the child has much less chance of living to grow up and reproduce than someone without this mutation. And only by reproducing does an individual pass on its inherited characteristics. If an organism does not reproduce, it plays no direct role in the evolution of future generations.

Inherited characteristics that improve an organism's chance of living and reproducing will be more common in the next generation than those that decrease its chance of reproducing. Various combinations of genes will be naturally selected for or against, from one generation to the next, depending on how they affect survival and reproduction. For natural selection to cause a change in a population from one generation to the next (that is, to cause evolution), it is not necessary that all genes affect survival and reproduction. The same result occurs if there are just some genes that make an individual more likely to grow up and reproduce.

The Peppered Moth

A classic example of natural selection in the wild, documented by observation and experiment, is the case of the peppered moth, which lives in all parts of England. In nineteenth-century England, many people collected moths and butterflies, and collectors avidly sought rare specimens of the peppered moth that were a dark, almost black color rather than the usual pale, mottled gray. We now know that each moth's genes determine whether it is the normal gray form or the black form—called the **melanic** form after the black pigment melanin. By looking at collections made from about 1850 to 1950, biologists found that melanic moths became more and more common during that time, and gray ones scarcer, particularly near industrial cities. This change in a population of organisms over time is, in itself, evolution.

Moths fly, feed, and mate at night. During the day they rest on tree trunks or other surfaces, protected from predators by camouflage. Biologists proposed that before industrial pollution, the typical gray form of the peppered moth had been well camouflaged against tree trunks covered with pale, plant-like lichens. In polluted areas, however, where industrial smoke had killed the lichens and blackened the tree trunks, the gray form stood out in contrast to its background (Figure 14–13). Here, many more gray than melanic moths would be found and eaten by predators. The most likely predators were birds, which hunt by sight and against whom camouflage, or lack of it, would be important. The evolution of darker populations of an animal in the presence of industrial pollution is known as **industrial melanism.**

FIGURE 14–13

Different forms of the peppered moth. The light moth is much more visible than the dark one against this background of a tree trunk blackened by soot. (M. Tweedie/Photo Researchers)

In the 1950s, Bernard Kettlewell decided to use these moths to study natural selection experimentally. He raised large numbers of both black and gray forms of the moth in the laboratory, marked them, and released them in two places: one an unpolluted rural area where the black form was more visible, the other a polluted industrial area where the gray form was easier to see against the blackened tree trunks. Kettlewell then recaptured as many of the marked moths as he could. The percentage of melanic moths recovered was twice that of gray moths in the industrial area, but only half that of gray moths in the unpolluted countryside. This agreed with the prediction that the gray moths were more likely to survive (and so to be recaptured) in the country, and melanic moths were more likely to survive near the town.

This experiment was done with a human "predator" (the person catching the moths), but humans are not normally much of a threat to survival of the pep-

(Text continues on page 294)

A Journey into The Environment

Going...Going...

Ornithologists and bird-watchers agree that the news about northeastern songbirds is unsettling: there seem to be fewer of them than ever before. Every spring "our" songbirds migrate north from their winter homes in Mexico, the Caribbean islands, and Central and South America to the forests and fields of North America where they raise their young. Months later, the birds make the return trip south to their winter ranges.

The annual spring migration brings more than 120 species north and historically has been a spectacular event, especially along the migration corridors of the northeastern states and the Midwest where birds are funneled into flyways by geographic features such as mountain ranges and ocean shores. Most birders can recount exciting "wave days" when warblers, those tiny, active, colorful, insect-eating migrants (some of the most challenging birds to spot and identify) seemed so numerous that it was as though they were falling out of the trees. But in the last few years there seem to be fewer birds.

Since the 1970s, bird-watchers have reported that numbers of migrant songbirds seemed to decrease and now breeding songbirds are being counted in widespread research efforts. Results are still preliminary, but in recent years Wood Thrush populations seem to have dropped by 40%. The conspicuous absence of numbers of warblers, orioles, tanagers, vireos, flycatchers, and other songbirds is an early indication of the loss of species diversity that is becoming an increasingly common phenomenon.

Squeezed North and South

From a plane you can see the reason for songbird decline. In the eastern United States, real estate development has gobbled up temperate and coniferous forests, replacing them with suburbs, shopping malls, and highways. Less suitable breeding habitat is available for songbirds and, to make matters worse, the clearing of the forests that used to blanket the eastern states, from the foothills of the Appalachians and Blue Ridge Mountains to the Mississippi River has brought songbirds into contact with the Eastern Cowbird, a species that parasitizes their nests. Cowbirds were originally restricted to the plains and prairies of the midwestern states, but as land development has cleared forests, they have expanded their range eastward. The result of cowbird nest parasitism is fewer songbirds.

When songbirds migrate south, the tropical forests where they spend the winter have often been cleared for pasture and cropland. Although data are still preliminary, it is estimated that many songbird populations are currently dropping at the rate of 1 to 2% per year. This sounds like an insignificant figure, but if the trend continues, the reduction of songbird populations will inevitably have a ripple effect as populations of the insects they eat begin to expand. This, in turn, will affect many other plant and animal species, most in ways that we have yet to understand.

Amphibians Are Also Disappearing

Because they migrate between two shrinking habitats, songbirds are squeezed especially hard, but the trend leading to the loss of species is not restricted to songbirds. All over the world populations of amphibians seem to be declining, too, but here the causes are more difficult to pinpoint. Habitat loss due to development, introduction of competitive or predatory alien species, pesticide contamination, acid rain, and water pollution are some of the reasons postulated for the decline of amphibians. Some experts suggest that frog populations may be declining because their immune systems are unable to cope with the stresses imposed by pollutants in air, water, and soil. Amphibians have delicate, mucous-covered skin that is extremely sensitive to environmental toxins and their disappearance may be a warning sign of low levels of pollutants that otherwise might go unnoticed (Figure 14–A). Whatever the cause, the trend is clear: human impact is eliminating populations of many of the world's frogs, toads, and salamanders. As populations are eliminated, the chances for a species' successful survival diminish and, again, no one can predict the ripple effect of declining populations of amphibians with certainty.

FIGURE 14–A

Golden Toad. This frog species from Costa Rica has undergone a dramatic decline in numbers in recent years. (E. R. Degginger)

Widespread Loss of Biodiversity

The statistics are different but the trend is similar with animals such as freshwater fishes and marine fishes, reptiles, and mammals of all kinds, especially whales, elephants, rhinoceros, carnivores of all kinds, and primates. Large animals are especially at risk because they need lots of space and because they typically produce fewer offspring during a lifetime than do smaller animals. Fewer offspring means fewer chances for adaptations to changing habitats. As wilderness is cleared for human enterprise, all of the plants and small organisms that compromise those habitats disappear. When habitat is lost to chain saws and bulldozers, local populations of bacteria, fungi, protists, microscopic arthropods that live in the soil, insects, molluscs, and worms are lost, too.

FIGURE 14–B

Dung beetle. (G. G. Dimijian)

Why Should We Care?

People will miss the small birds that enliven parks and gardens. Many will mourn the loss of beautiful and compelling large mammals such as tigers, elephants, and wolves. Some will notice the lack of frog choruses on wet spring nights, and a few will remember that once hundreds of salamanders used to emerge from their subterranean burrows and flood across woodland roads on their annual journey to mating ponds. Hardly anyone may notice the loss of soil invertebrates, insects, worms, and plants, yet these may be much more important to the preservation of the environment for humans than those species whose absence we *will* notice. Why should we be concerned about losing them? A look at the life story of a dung beetle, an animal that is not listed as endangered, but whose ecological role is fairly well understood, should provide an answer.

A kind of scarab, the dung beetle is especially noticeable in the tropics, and feeds upon animal droppings (Figure 14–B). In many species of dung beetle, the males fly to masses of fresh dung and begin to scoop out a walnut-sized portion. Pushing and patting with their legs and snouts, each male separates a bit of dung from the mass of droppings. Working with his hind legs, the male rolls the ball along the ground, perhaps attracting a female who may land on top of the rolling ball and ride it to a burying site. The male excavates the earth from beneath the ball of dung and it slowly descends into the soil. When the proper depth has been reached, the male joins the female on top of the ball of dung. They mate, she deposits a single, fertilized egg upon the dung, the pair bury the ball, and both fly away. The larva that hatches from the fertilized egg feeds on the ball of dung, changes into a pupa, and is eventually transformed into an adult dung beetle that burrows out of the soil and flies away to begin the cycle again.

It is important to notice that the dung beetle *removes* animal droppings almost as quickly as they appear in the environment. If you watch these insects you will be impressed with the speed with which they land at a pile of droppings, cut out a ball, and roll it away. These are insects in a hurry. As they eliminate dung from the habitat, and use this concentrated source of nutrients to complete their life cycle, they affect the lives of other organisms as well. By eating and burying dung they kill the roundworms and other intestinal parasites found in animal droppings, and check population growth of these organisms. There are also carnivorous mites that hitch rides on the backs of dung beetles, using them as transportation from dung pile to dung pile. These mites are predators of the larvae of flies that feed on dung. When dung beetles are absent, populations of flies increase, because the predatory mites have no means of transport.

From this example, it should be clear that the lives of many diverse species are inextricably intertwined and alterations in any of the populations would have a widespread ecological effect. Consider how the activities of dung beetles affect the lives of the animals whose dung they bury, the parasites that live in the digestive tracts of these animals, the other animals (particularly snails, grasshoppers, and earthworms) involved in the life cycles of these parasites, the predatory mites who use dung beetles for transportation, and the fly species preyed upon by the predatory mites.

We are only beginning to appreciate interrelationships between organisms, even in the comparatively well studied temperate zone habitats. Much more remains to be learned about interrelationships among tropical species and, unfortunately, given the current rate at which species and habitat are being lost, we may never know the true extent of our loss. Why should we care if we lose species? On a self-interested level, because the species might be good for something that humans need or want, now or in the future. Most of the cancer-fighting drugs now in use had their origins in plants, and the forests of the world are vast unexplored pharmacological storehouses waiting to be developed. Plants may have a practical use in crop improvement. On an ecosystem level, it is the insects, bacteria, worms, protozoans, and plants that make life possible on Earth as they play roles in chemical, nutrient, and energy cycles. Finally, we should be dismayed by the loss of a species because even if it is unknown to science (and the vast majority of species are unknown), each species is irreplaceable and each contributes to the stability of its ecosystem.

Table 14–2
Survival in Swiss Starlings in Relation to Number of Eggs Laid*

Brood Size (Number of Eggs in Nest)	Number of Young Marked	Recoveries per 100 Birds Marked†
1	65	0
2	328	1.8
3	1278	2.0
4	3956	2.1
5	6175	2.1
6	3156	1.7
7	651	1.5
8	120	0.8
9, 10	28	0

*The number of eggs laid during one nesting period is genetically regulated and, like other genetic variations, is acted upon by natural selection. David Lack marked all the nestlings in all the nests he could find, and then recaptured them months later when they had left the nest.

†The only recoveries scored are those for birds over three months old when they were recaptured.

Source: Lack, D. *Ecology* 2, 1948

pered moth. Does the differential camouflage work against the moths' real predators? To find out, Kettlewell hid in a blind and watched moths he had placed on tree trunks. On one occasion, he watched equal numbers of gray and black moths in an unpolluted area. Birds caught 164 of the melanic and only 26 gray moths.

In a polluted area, a larger proportion of melanic than of gray moths will live long enough to reproduce. Since the color of the moths is inherited, the next generation will contain proportionally more melanic moths. In other words, the frequency of the gene for black color increases in the population with time—and that is evolution.

The selective pressure that brings about this evolution is clear: in polluted areas birds kill a higher percentage of moths with the gene for gray color than of moths with the gene for black color. Natural selection over many generations has produced populations of the peppered moth that are well adapted to survive in their environments, either black or gray depending on the environmental selection factor.

On the basis of this evidence, we would predict that if pollution were reduced, melanic moths would become rarer and gray forms more common in industrial areas. In fact, the Clean Air Act of 1952 reduced air pollution in England, and collections of peppered moths from industrial Manchester in the next 20 years revealed a dramatic decrease in the relative number of melanic individuals in the moth population. The ability to use our understanding of evolution to predict events in this way is impressive evidence for this scientific theory.

Natural selection as the mechanism of evolution is based upon the principles that genetic variation plays an important role in determining which of many offspring will survive and reproduce in a particular environment. The first well documented case of natural selection leading to evolution in a wild population was observed in the selective predation by birds upon peppered moths.

14–E GENETIC CONTRIBUTION TO FUTURE GENERATIONS

The phrase "survival of the fittest," often used in discussion of evolution, suggests that natural selection selects mainly for survival. It does not. It selects for the contribution of genes to future generations—that is, reproduction resulting in viable offspring. Survival is important, in that an individual that dies young will not reproduce, but even reproduction is no guarantee of evolutionary success.

Consider Table 14–2, which shows how many young starlings survived for three months after hatching. The female starlings that seemed to be reproducing most efficiently—those laying nine or ten eggs in one brood—could actually be doomed to evolutionary failure and strongly selected against because hardly any of their young survived. Females laying four or five eggs per brood had a higher number of offspring surviving for at least three months after they hatched.

Young birds from the larger broods weigh less than those from the smaller broods, presumably because the parents could provide adequate food for no more than five or six nestlings. A shortage of food was probably a major cause of death of young from larger broods. Table 14–2 also shows that the most common brood sizes produce the nestlings with the lowest mortality rates, as we would predict from the action of natural selection. It seems reasonable to suppose that in years when there is more (or less) food available to the birds than in the year studied in this example, selection would favor birds with broods larger (or smaller) than the average. This accounts for the fact that the population contains birds producing broods larger and smaller than the average brood: these genes persist because in some years they are favored.

Plainly, the reproductive success of a starling is not fully told by the story of one brood. Selection optimizes reproductive success over a lifetime, and the adaptations that produce this success are many.

Natural selection selects for individuals that are able to produce the greatest number of young that, in turn, survive and reproduce.

14–F ADAPTATIONS

Adaptations are the result of natural selection acting upon a population. They are physical, physiological, and/or behavioral traits that enhance an organism's chances of surviving in its environment.

For example, let us consider an acorn. Whether it successfully resists the selective pressure of its environment depends on the speed and normality of its germination and development, whether bacteria or fungi infect it as a seed or seedling and destroy it at this stage, whether as a seedling it has enough stored food for rapid growth, whether it escapes being eaten, whether the soil in which it grows can support a large plant, and whether the young tree avoids death by disease, trampling, or browsing. The genome of a successful oak will contain genes that adapt it to withstand all these selective pressures (Figure 14–14).

Resistance to Pesticides and Antibiotics

Several dramatic examples of natural selection in action today, and the adaptations it produces, are pro-

Will this seedling survive:

1. Bacterial and fungal infections
2. Insufficient nutrients
3. Drought
4. Predators
5. Parasites
6. Trampling or browsing

FIGURE 14–14

A future oak tree? To grow and reproduce, this sprouting acorn will need to survive many challenges of its environment. (P. Murray/Earth Scenes)

vided by the evolution of resistance to pesticides and antibiotics.

2? A scale insect feeds on citrus trees in California. In the early 1900s, growers sprayed the trees with cyanide gas, and this killed the scale. But in 1914 some of the insects survived the spraying. The cyanide did not kill them because they possessed a single gene, newly apparent in the population, that permitted them to break cyanide down into harmless compounds. As spraying continued, more insects with the new gene than without it survived to reproduce, and they passed on the gene to their offspring. The frequency of the new gene in the population increased until the whole population was resistant to the spray. Because scale insects, like many other insects, have more than one generation a year, they evolve quickly. Resistance to pesticides is a very expensive problem for agriculture. To combat the evolution of resistance, growers are encouraged to spray pesticides on their crops only when necessary and to use different chemicals in different months and years. ■

Precisely the same thing happens with antibiotics used to kill bacteria that cause human disease. When a bacterial population meets a particular drug, bacteria susceptible to that drug are killed (Figure 14–15). Sometimes a population happens to contain one or a few individuals with mutations that confer resistance to the drug; they will survive, and they multiply rapidly once competing bacteria have died. In addition, many genes conferring resistance to antibiotics are now known to be carried in plasmids, small loops of extra DNA found in some bacteria and fungi, which can be duplicated and passed to other members of the population that previously lacked genetic resistance. Soon these genes become widespread. Since antibiotics and disease-causing bacteria frequently meet in hospitals, it is not surprising that some hospitals har-

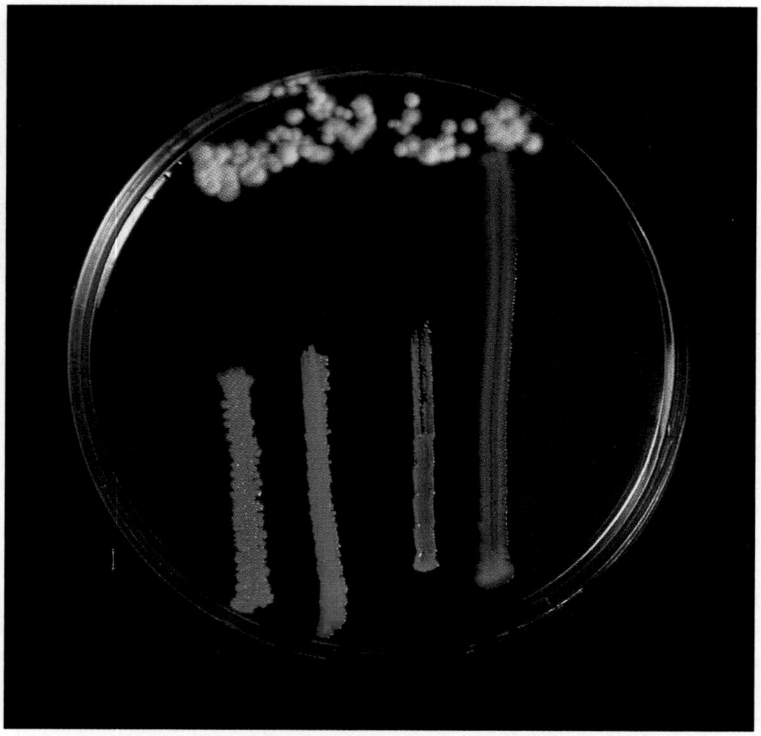

FIGURE 14–15

Resistance to antibiotics in bacteria. The fuzzy dots across the top of this dish are clumps of the fungus *Penicillium*, which produces the antibiotic penicillin. The penicillin is spreading across the dish. The four lines are rows of different varieties of bacteria. Three of the bacterial varieties were killed as the penicillin reached them; the fourth (far right) is penicillin-resistant, and it continues to grow. (Biophoto Associates)

bor drug-resistant bacteria. In many countries, women are now encouraged to give birth at home whenever possible, because mother and infant are safer from bacterial infection at home than in the hospital.

3? Most countries have outlawed the use of antibiotics in cattle feed. Cattle fatten faster if fed antibiotics, but they also become breeding grounds for antibiotic-resistant bacteria. Antibiotics are still added to cattle feed in the United States, and drug-resistant bacteria in cattle are becoming increasingly common. ■

Disease-causing organisms don't have things all their own way. In 1915, nearly all the oysters in the Malpeque Bay of Prince Edward Island, Canada, were killed by a disease. However, a few oysters survived, and began to re-establish the population. Fifteen years later, the disease-causing organism was still present, but most of the oysters now had genetic resistance to it; only one oyster in 1000 was susceptible. By 1938, the oyster harvest was higher than it had been before the disease struck, and when the disease appeared elsewhere, oysters from Malpeque Bay were sent to contribute their genetic resistance to the newly afflicted populations.

These examples of the evolution of adaptations illustrate merely a few of the less obvious selective pressures that are always acting on all organisms and the adaptations that have evolved in response to them. Adaptations, appear ingenious, but are really the result of natural selection among randomly produced variations of genes.

Adaptations are an organism's many anatomical, physiological, and behavioral traits that have survived the natural selection process and which increase an organism's chance of survival in its environment. In many organisms, such as bacteria and insects, the unique genetic composition of a few individuals allows them to survive when others are killed by a new environmental threat. The survivors then reproduce and pass along these adaptations, which permit widespread survival of the population with the new characteristic.

SUMMARY

1. According to the theory of evolution, new life forms arise by modification of previously existing forms of life.
2. Members of any species differ from one another, and some of their differences are inherited. Natural selection results from the different rates of successful reproduction in genetically distinct individuals. It leads to evolution, a change in the proportion of genes in a population from one generation to the next.
3. Lamarck suggested that evolution may occur through the inheritance of acquired characteristics. However his mechanism is incorrect and cannot occur, given our current understanding of genetics.
4. The theory of evolution by natural selection was advanced in 1858 by Charles Darwin and Alfred Russel Wallace. Their thinking was stimulated by the writings of Charles Lyell and Thomas Robert Malthus as well as by observations that they made during their own travels.
5. Biologists recognize many lines of evidence that support the idea of evolution:
   a. artificial selection by farmers and breeders has produced rapid evolution in domesticated plants and animals,
   b. the fossil record demonstrates that throughout geological time, animals and plants have changed,
   c. comparative anatomy and developmental biology provide evidence that various structures in ancestral organisms have been modified in their descendants and have become adapted to different functions or have been lost when a new way of life rendered them unnecessary,
   d. comparative biochemistry has indicated that the more closely related two organisms are, the more similar their DNA and protein structure will be, and finally,
   e. biogeography shows that different animal groups have originated in isolated places and spread out from these original locations.
6. A classic example of evolution by means of natural selection was reported in the 1950s. In this study, bird predation was shown to be the likely selective pressure that led to the evolution of dark color in populations of the peppered moth in polluted areas of England.
7. The anatomical, behavioral, and physiological traits that survive the process of natural selection may be thought of as adaptations that fit an organism to its particular environment. Many types of adaptations exist. The only consistent result of selection is that it maximizes the genetic contributions of a "successful" individual to future generations.

SELECTED KEY TERMS

adaptation, *p. 295*
adaptive radiation, *p. 289*
analogous structures, *p. 285*
artificial selection, *p. 283*
endemic species, *p. 289*
evolution, *p. 277*
fossil, *p. 283*
homologous structures, *p. 285*
industrial melanism, *p. 291*
Lamarckism, *p. 280*
natural selection, *p. 278*
vestigial structures, *p. 288*

SELF-QUIZ

1. In light of the definition of evolution, which of the following is *not* capable of evolving?
   a. a population of deer
   b. the color of a population of moths
   c. your biology teacher
   d. a population of chickadees
   e. the millions of bacteria in your large intestine
2. Which of the following did Kettlewell conclude from his studies on industrial melanism in moths?
   a. a black moth lays more eggs than a gray moth in industrial areas
   b. black moths are more resistant to pollution than are gray moths
   c. pollution caused some moths to become darker than others
   d. black moths are more likely to survive in polluted areas than are gray moths
   e. birds prefer the taste of black moths over gray moths
3. Which bird is most evolutionarily successful?
   a. lays 9 eggs, 8 hatch and 2 reproduce
   b. lays 2 eggs, 2 hatch and 2 reproduce
   c. lays 5 eggs, 5 hatch and 3 reproduce
   d. lays 9 eggs, 9 hatch and 2 reproduce
   e. lays 7 eggs, 5 hatch and 4 reproduce
4. Suppose that you have a pack of 50 assorted dogs. You select the largest male and the largest female, mate them, and sterilize the other dogs. Assuming that food supplies remain adequate, you should expect that, in the next generation, the young dogs will grow to be, on average:
   a. smaller than their two parents
   b. larger than most adult members of the pack
   c. the same size as the adult dogs of the pack
   d. smaller than the adult dogs of the pack

5. Explain how Darwin would have accounted for the evolution of the long necks of giraffes.
6. Penicillin and other antibiotics were introduced in the 1940s and were effective in combatting infections caused by *Staphylococcus* bacteria. In 1958, however, there were several outbreaks of *Staphylococcus* infection. People with the infections did not respond to treatment with any antibiotic, and many people died. The most likely explanation for this situation is:
   a. the bacteria reproduced in hosts that were not contaminated by antibiotics
   b. bacteria from other animals (such as deer, birds, and cats) migrated into human hosts
   c. the bacteria exposed to nonlethal doses of antibiotics quickly learned to avoid them
   d. each generation of bacteria acquired the ability to use the antibiotics as nutrients
   e. bacteria containing a gene for antibiotic resistance survived and multiplied and these were the forms causing the lethal infections

THINKING CRITICALLY

For Questions 1 to 5, consider Table 14-2.

1. From what brood size do the greatest number of young survive?
2. Is this also the most frequent brood size? (Assume that the experimenter marked every bird that could be found.)
3. What do you suppose is the disadvantage to a starling of laying a smaller than average clutch of eggs?
4. Suppose the environment changed so that only half as much food was available to the starlings. Would you expect a gradual change in the most frequent brood size? How would this change be brought about?
5. Which female starlings will leave more young in the population and hence make the greatest contribution to the genes of the next generation?
6. Are all causes of death natural selection? For example, when organisms die in an earthquake, have they been selected against?
7. The embryologist Charles H. Waddington treated fly larvae with heat shock. As a result of this treatment, some of the adult flies showed the abnormal condition "crossveinless" (some of their wing veins were missing). After many generations of this treatment, he let a generation of flies develop without heat treatment and many of them were also crossveinless. Does this experiment provide convincing proof of Lamarckism? If not, what other explanation can you suggest, and what experiments would you perform to test your suggestion?
8. Is human evolution subject to the same pressures as the evolution of other species? Why or why not?
9. Is there any time in its life history when an organism is not subject to selective pressure? Are gametes subject to selective pressure? Are eggs? Embryos? Is there selective pressure on young animals that are fed and protected by their parents?
10. Some insects lay eggs on more than one species of larval food plant. There is some evidence that a female is more likely to lay her eggs on the plant species on which she grew as a larva than on any other kind of plant. Is this an example of Lamarckian inheritance? Why?
11. Scientists are beginning to breed crop plants to have built-in chemical defenses against insect pests. In your opinion, how well will this work?
12. What is the adaptive advantage to a plant of a contact irritant (such as the oil on poison ivy leaves that makes a rash on the skin of passing animals)?

SUGGESTED READINGS

Bishop, J. A., and L. M. Cook. "Moths, melanism, and clean air." *Scientific American,* January 1975. The peppered moth story and how the moth can be used to monitor air pollution.

Cook, L. M., G. S. Mani, and M. E. Varley. "Postindustrial melanism in the peppered moth." *Science* 231:611, 1986. A follow-up to the peppered moth story.

Darwin, C. *The Origin of Species by Means of Natural Selection.* New York: The Modern Library, Random House, Inc., 1982. A reprint of the 1859 first edition in one volume together with the sequel, *The Descent of Man.*

Grant, P. R. "Natural selection and Darwin's finches." *Scientific American,* October 1991.

Mayr, E. "Darwin and natural selection." *American Scientist* 65:321, 1977. An eminent geneticist's discussion of the logical argument for evolution by natural selection.

Nelkin, D. *The Creation Controversy: Science or Scripture in the Schools.* New York: W. W. Norton, 1982. The controversy between creationism and evolution. Nelkin examines the tactics by which creationists attempt to impose their views on educational systems. She contends that the battle is political (rather than scientific or religious).

Judge Overton's decision in the Arkansas case is discussed and reprinted.

Scientific American, September 1978 issue, *Evolution.*

Stebbins, G. L., and F. J. Ayala. "The evolution of Darwinism. *Scientific American,* July 1985. Two eminent geneticists discuss how views of evolution have changed since Darwin's day.

Stone, I. *Origin.* New York: Plume/New American Library, 1981. A very readable historical novel based on the life of Charles Darwin.

CHAPTER 15

CONCEPT GUIDE

After reading this chapter, you should be able to:

1. Describe the Hardy-Weinberg Principle and the conditions under which it is valid, and explain the significance of this principle. (Section 15-A)
2. Compare the effects of stabilizing selection, directional selection, disruptive selection, gene flow, genetic drift, and the founder effect on the gene pool. (Section 15-B)
3. Describe four mechanisms that help to maintain genetic diversity in populations. (Section 15-C)
4. Define a species. (Section 15-D)
5. Distinguish between allopatric and sympatric modes of speciation and give examples of each. (Section 15-E)
6. Distinguish between microevolution and macroevolution. (Section 15-F)

All of these shells belong to one species of snail, and the variations are due to individual differences. Learn why individual variation is an important part of natural selection in Section 15-B. (C. Clark)

Population Genetics and Speciation

In Chapter 14 we saw that populations (not individual organisms) are the units that evolve. This is because evolution involves changes in the mix of genes present in entire populations of organisms. The difference between a population of ancient dinosaurs and a population of their descendants, modern birds, lies in the different genes of members of the two populations.

A **population** consists of all the members of a species that occupy a particular area at the same time—for example, the perch population of a lake, the dandelion population of a hillside, or the penguin population of an island. The members of a population are much more likely to breed with one another than with members of other populations of the same species. Therefore, populations form breeding groups, and genes tend to stay within the same population for generation after generation.

All the genes in all the members of a population are collectively called the population's **gene pool.** According to one definition, evolution is the change in the frequency of genes in a population's gene pool from one generation to the next. So, if we can discover how a population's gene pool changes with time, we shall understand how evolution occurs.

Different populations of the same species sometimes become isolated from one another. When this happens, one of the most important processes in evolution may occur: the formation of a new species, whose members, by definition, do not exchange genes with members of other species. They are reproductively isolated from all other groups.

KEY CONCEPTS

- Populations (not individuals) are the units that evolve.
- When a population evolves in isolation, it may become a new species.

Curiosity Questions

1? How did an animal as flashy as a peacock ever evolve? (See Figure 15-3)

2? Why do farmers buy hybrid corn seed every year instead of saving seeds from the last year's crop? (See Figure 15-9)

3? Why don't bananas have seeds? (See Figure 15-13)

15–A THE HARDY-WEINBERG PRINCIPLE

One way to see how a population evolves is to construct a model of a population that does not change genetically from one generation to the next, and then see how a real population differs from this model. The **Hardy-Weinberg Principle** provides such a model (Figure 15-1). Devised by British mathematician Godfrey H. Hardy and the German physician Wilhelm Weinberg in 1908, it shows that mathematically and theoretically there are situations where evolution *does not* occur. This condition of no evolution is called Hardy-Weinberg equilibrium. As we will see, Hardy-Weinberg equilibrium is seldom achieved in nature, and this mathematical model has great implications for our assumptions about when evolution can occur.

Consider a population whose gene pool contains two alternate alleles, *A* and *a*, either of which can occupy one particular location on a chromosome. Every member of the population has one of three possible genotypes: *AA, Aa,* or *aa.* The Hardy-Weinberg Principle shows that the next generation will contain the two alleles in the same frequencies (their proportions relative to each other in the population's gene pool). This will remain true through all successive generations, if the population meets *all of the following conditions:*

1. **No mutations.** The alleles in question must not mutate. (If they do, they become other alleles, and so their frequencies automatically change.)
2. **No selection pressure.** There must be no natural selection with respect to the alleles in question (no genotype has a reproductive advantage over the others).
3. **No mating preferences.** The population must reproduce sexually, and mating must be random with respect to genotype (so that, for instance, an *AA* female does not prefer *aa* to *AA* or *Aa* males when she mates).
4. **Isolation.** There must be no exchange of genes (gene flow) between the population and any other population.
5. **Large size.** The population must be very large, because the law is based on statistical probabilities. Random sampling errors are more likely to occur in small populations.

If all these conditions are fulfilled, *A* and *a* will remain in the population indefinitely at the same frequencies and there will be *no evolution.*

The Hardy-Weinberg Principle shows that sexual reproduction, with its reshuffling of genes, is not by itself enough to cause evolution. Evolution is a change in allele frequencies from one generation to the next, and under the conditions of the Hardy-Weinberg Principle there is no change.

The model has a useful application. Since it gives the conditions under which evolution will not occur, it implies that if any one of these conditions is not met, evolution is likely to occur. Because the conditions for Hardy-Weinberg equilibrium are seldom met, evolution can occur whenever the environment favors a change in allelic frequencies.

According to the Hardy-Weinberg Principle, evolution will not occur in a large, genetically isolated population with no mutations, no selective pressures, and no mating preferences.

15–B CAUSES OF EVOLUTION

Mutation violates the conditions for Hardy-Weinberg equilibrium because it changes one gene into another and therefore alters the frequency of that gene in the next generation. Let us see how the other factors produce evolution.

Natural Selection

Some combinations of traits allow the organisms that possess them to survive and successfully reproduce, while other combinations are less successful, or not successful at all. **Natural selection** is this phenomenon of nonrandom survival and reproduction of phenotypes. In natural selection, the environment—or some agent of the environment such as a predator, or severe weather—allows some organisms to survive and reproduce, while restricting (or eliminating) reproduction in others.

Lethal recessive alleles (for example, those for sickle cell anemia) provide an example of natural selection. Whenever a lethal recessive allele occurs homozygously, the affected individual dies before it can reproduce, and therefore those particular copies of the allele do not have a chance to be passed on to the next generation. This violates condition 2 of the Hardy-Weinberg Principle. Homozygous dominant individuals and those heterozygous for a lethal allele have a reproductive advantage over homozygous recessives, which are said to be **selected against.** Hence the frequency of the lethal recessive allele decreases over the generations.

Individuals in a population show a range of phenotypes. When an agent of natural selection, such as predation or competition, is at work, some of these phenotypes are more likely to survive and reproduce than

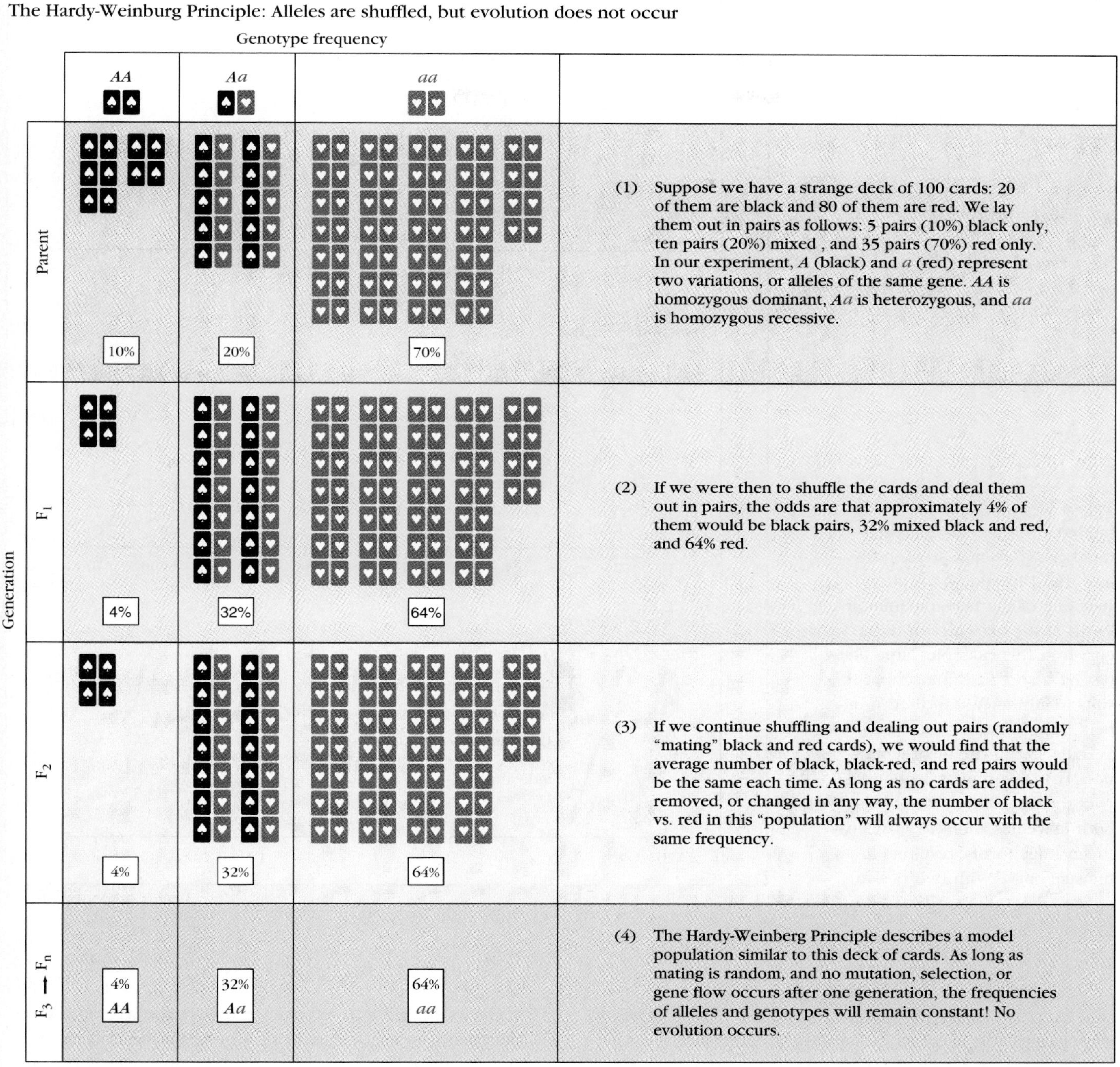

The Hardy-Weinburg Principle: Alleles are shuffled, but evolution does not occur

Genotype frequency

Generation	*AA*	*Aa*	*aa*	
Parent	10%	20%	70%	(1) Suppose we have a strange deck of 100 cards: 20 of them are black and 80 of them are red. We lay them out in pairs as follows: 5 pairs (10%) black only, ten pairs (20%) mixed , and 35 pairs (70%) red only. In our experiment, *A* (black) and *a* (red) represent two variations, or alleles of the same gene. *AA* is homozygous dominant, *Aa* is heterozygous, and *aa* is homozygous recessive.
F_1	4%	32%	64%	(2) If we were then to shuffle the cards and deal them out in pairs, the odds are that approximately 4% of them would be black pairs, 32% mixed black and red, and 64% red.
F_2	4%	32%	64%	(3) If we continue shuffling and dealing out pairs (randomly "mating" black and red cards), we would find that the average number of black, black-red, and red pairs would be the same each time. As long as no cards are added, removed, or changed in any way, the number of black vs. red in this "population" will always occur with the same frequency.
$F_3 \rightarrow F_n$	4% *AA*	32% *Aa*	64% *aa*	(4) The Hardy-Weinberg Principle describes a model population similar to this deck of cards. As long as mating is random, and no mutation, selection, or gene flow occurs after one generation, the frequencies of alleles and genotypes will remain constant! No evolution occurs.

FIGURE 15–1

An experiment using black and red cards to demonstrate the Hardy-Weinberg Principle. If no mutation, selection, mating preferences, or gene flow occurs after the first generation of a population, in all future generations the frequencies of alleles remain constant. Thus, there will be no evolution.

others. Usually the surviving traits are at least partly under genetic control. Hence the genes responsible for these phenotypes will be more abundant in the next generation. These genes are said to have been **selected for,** which means the same as "not selected against." In the simplest case of a single pair of alleles, if possession of one allele confers even a slight reproductive advantage, its occurrence in the population

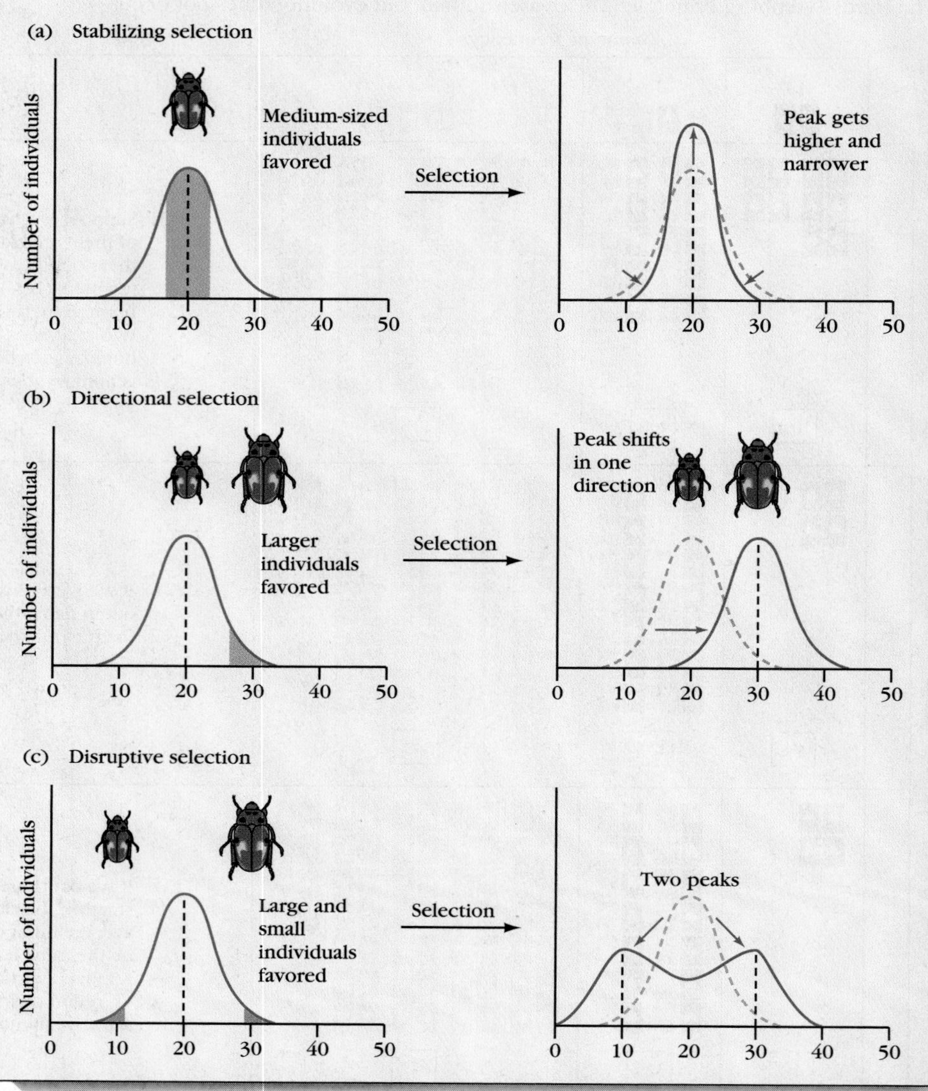

FIGURE 15–2

Types of selection. (a) Stabilizing selection produces greater numbers of average-sized individuals. **(b)** Directional selection favors one of the two uncommon forms at the extremes of the curve. In this example, large beetles have some advantage that results in an increase in their numbers, resulting in an increase in average size in the next generation. **(c)** Disruptive selection favors the uncommon forms at both extremes and selects against the average forms, resulting in an unusual curve with two peaks.

will increase from one generation to the next, at the expense of the less favorable alternate allele. The more favorable allele is said to have greater **fitness** than the less favorable allele. In this way, selection can change the gene frequencies in a population from one generation to the next, resulting in evolution. Natural selection is by far the most important and potent cause of evolution.

Two common forms of natural selection are stabilizing selection and directional selection. In **stabilizing selection,** average phenotypes have a selective advantage over extremes in either direction. It is exceedingly common in nature. For instance, several studies have shown that human babies of average birth weight (between 7 and 8 pounds) have a much higher chance of surviving to the age of 5 than babies with weights significantly above or below the average. **Directional selection** occurs when the phenotypes at one extreme have a selective advantage over those at the other extreme. Figure 15-2 shows the actions of these kinds of selection, using a population with a normal (bell-shaped) distribution of phenotypes from a polygenic trait—a trait controlled by many genes, such as human height, size of beetles, or the weight of seeds.

As an example, let us consider a population of seeds. Those of average size have a better chance of germinating and of growing than those that are unusually large or small, and if seed size is inherited, stabilizing selection is acting. The next generation will contain a lower proportion of unusually large or small seeds. On the other hand, if birds tend to eat large

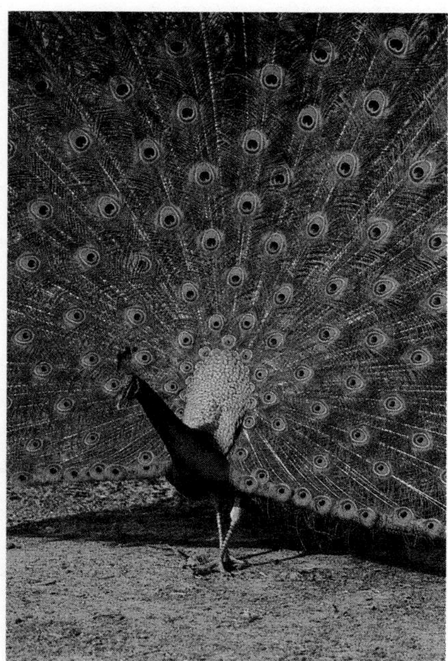

FIGURE 15–3

Courtship display of a peacock. The peacock's tail is the product of mate selection. Although the tail reduces the male's chance of survival because it is heavy and takes much energy to produce, females prefer to mate with males who display fabulous tails during courtship. So the genes for producing this remarkable decoration survive. (D. Cavagnaro/Visuals Unlimited)

FIGURE 15–4

White clouds of pollen blow away from a pine tree. Gene flow between populations of plants may occur when some pollen grains are blown away and carry their genes from one population to the next. However, most plants will be fertilized by pollen grains from plants within their own population. (M. Cooper/Peter Arnold, Inc.)

seeds and ignore small ones, they will exert directional selection in favor of small seeds.

Disruptive selection is less common. It takes place when the extremes of a range of phenotypes are favored relative to intermediate phenotypes (Figure 15–2c). It might happen to our seeds, for example, if a particular kind of beetle specialized in feeding only on seeds of intermediate size, ignoring the very small and very large seeds.

Mating Preferences

Mating that is not random with respect to genotype can also bring about evolutionary change. If females consistently choose to mate with males with certain genetic traits, they exert selection in favor of the alleles for those traits. Such nonrandom mating can have bizarre results. For instance, over the centuries peahens have preferred to mate with peacocks who can produce brilliant displays with their tails (Figure 15–3). This has selected for ever larger and more colorful tails. Such mating preference is really only a form of natural selection because it gives one genotype a reproductive advantage over another. However, the agent of selection is different from the kind of selective pressure we usually think of. This points up what is probably a universal situation: a population's gene pool represents a balance between opposing selective forces. In this case, female choice favors males with large gaudy tails, whereas predation tends to eliminate such males. The colorful tail makes the male more conspicuous to predators, and the tail's size doubtless hampers his attempts to escape. The outcome is stabilizing selection held in balance by two opposing selective agents: female preference and predation.

Gene Flow

The gene pools of most populations of the same species exchange genes, resulting in **gene flow** between the populations. Animals may leave one area and contribute their genes to the gene pool of a neighboring population, or a high wind may disperse plant seeds or pollen far beyond the bounds of the local population (Figure 15–4). Gene flow between populations is generally second only to selection as a cause of evolution in local populations. Gene flow between populations tends to increase their similarity. Natural selection has the opposite effect: it tends to make every population uniquely specialized for its particular

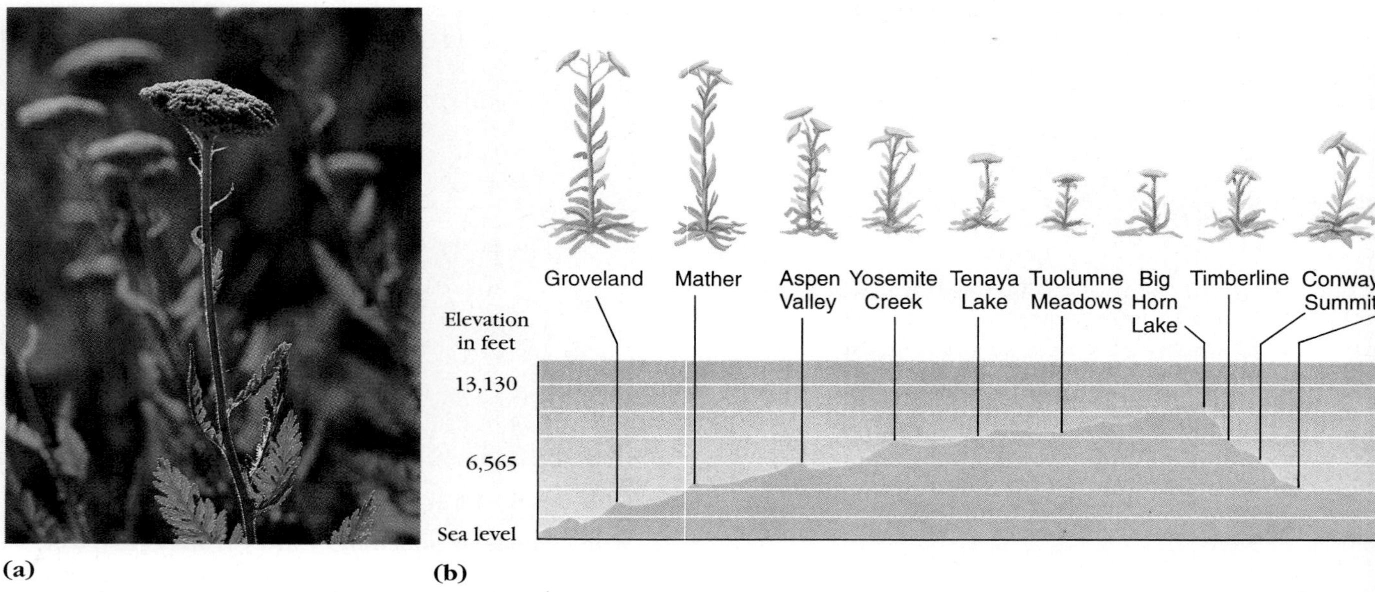

FIGURE 15-5

Genetic variation between populations. (a) A Fernleaf yarrow plant *Achillea filipendulina*. **(b)** These yarrow plants were collected from different populations in California and Nevada and then grown under uniform conditions to reveal genetic differences between them. The plants show gradients of differences from their neighbors. (Lefever/Grushow/Grant Heilman)

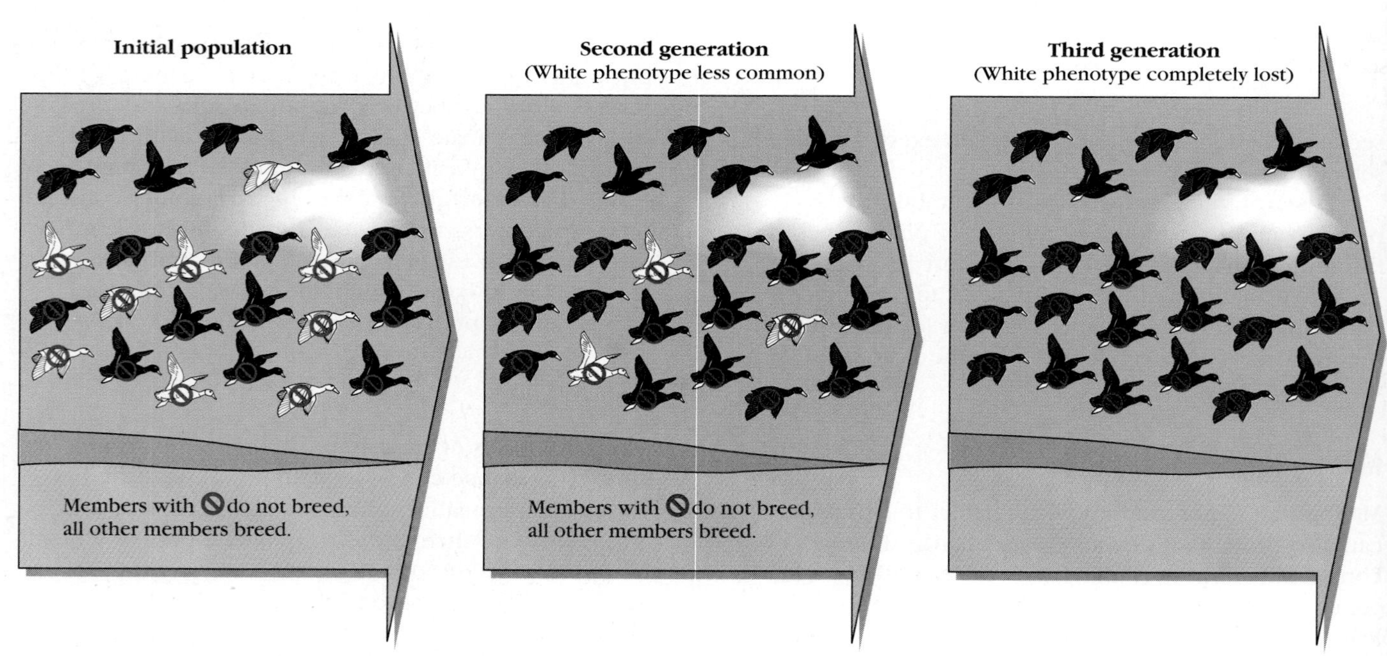

FIGURE 15-6

Genetic drift. The white phenotype was lost from this population as a result of factors of natural selection that were unrelated to feather color.

habitat. One possible outcome of these two conflicting forces is a gradient of variation from one population to the next (Figure 15-5). The closer together two populations are, the more genetically similar they are likely to be.

Genetic Drift

Evolution can occur simply by chance. Some random event may bring death to some individual, while another random event may result in parenthood. For example an earthquake, tornado, or asteroid impacting the Earth may eliminate a sizeable proportion of a population, allowing some organisms to reproduce more than they normally might have. The resulting random change in the gene pool is called **genetic drift** (Figure 15-6).

Genetic drift is much more important in a small population than in a large one. Consider a population of five individuals, in which only two breed. The chances are good that any particular allele is represented in only one member of the population. If this individual does not breed, the allele will not be present in the next generation. If the individual does breed, the frequency of the allele may increase in the next generation. In either case, because a change in gene frequency from one generation to the next has taken place, evolution has occurred. By contrast, if the population is large—say we multiply each of these numbers by 1000—then the 2000 who breed are likely to include about 400 who contain the allele. This is the same proportion as in the general population, and so evolution by genetic drift is much less likely in a large population than in a small one.

An interesting case of genetic drift is the **founder effect,** in which the individuals establishing a new population do not contain all the genes present in the old one. For instance, when a few individuals leave a large population and colonize a new area, the chances are good that the founders of the new population do not carry a representative sample of all the genes in the old population. The gene pools of the old and new populations will be different. The founder effect is thought to play an important role in the evolution of island species (Figure 15-7). It may also come into

The Founder Effect

(a) A few individuals become separated from a population.

(b) These individuals, with their own unique genetic makeup, arrive at a new environment where they establish a new population.

(c) After many generations, the new population looks quite different from the old one. Some of these differences (large crests and stripes) were inherited from the founding individuals. Others (webbed feet and long tails for swimming) are adaptations to the new environment.

FIGURE 15-7

The founder effect and rafting. As a result of the founder effect, a new population is established which is genetically distinct from the parent population because of two factors. First, the colonizing individuals possess only a fraction of the genetic diversity of the parent population. And second, the new environment selects for characteristics that are different from those of the parent population.

A Journey into Evolution

Double Trouble for Cheetahs

The rapid growth of the human population today is causing many problems in our environment, including drastic declines in the wild populations of most other large mammals. In attempts to save these species from extinction, zoos and game parks have established programs to breed many of them in captivity. But efforts to breed cheetahs have fared poorly (Figure 15–A). While seeking the reason for this, researchers discovered that cheetahs face an additional threat from within: a high degree of genetic uniformity. In fact, cheetahs have less genetic variability than most laboratory mice or other *deliberately* inbred livestock.

This suggested to researchers that modern cheetahs are descended from a very small population: their lack of genetic variation is a result of the founder effect. The fossil record shows that there were once several species of cheetahs, and our present-day species was distributed worldwide. About 10,000 to 12,000 years ago (late Pleistocene Epoch), many species of mammals became extinct. It is thought that cheetahs narrowly escaped the same fate. Very few animals, and a correspondingly limited gene pool, survived. These survivors repopulated Africa, and the Near East from Arabia to India. Then, heavy hunting about a century ago exterminated the cheetah from much of this range, leaving populations only in parts of Africa south of the Sahara Desert.

The lack of genetic variation in cheetahs has had three serious consequences:

FIGURE 15–A

Cheetahs. (M. Barlow/Dembinsky Photo Associates)

1. **Low fertility.** The males produce sperm of poor quantity and quality. The semen of cheetahs has a sperm concentration only 10% of that found in other members of the cat family, such as housecats, and 71% of cheetah sperm cells are abnormal (compared to 29% in housecats). Sperm quantity and quality are believed to be under genetic control because they are adversely affected by inbreeding in many species.
2. **High death rate among cubs.** In the wild, cheetah cubs suffer a 70% death rate. Even in captive breeding programs, where predators (such as lions and leopards) and starvation are no threat, 30% or more die by six months of age. This suggests that deleterious recessive genes are taking an unusually high toll among cheetahs.
3. **Susceptibility to disease.** Most populations of animals have a high degree of variability among the genes involved in fighting diseases. When a disease strikes a population, it kills those individuals whose genes do not enable them to fight that particular disease effectively. Other individuals, with a different mix of disease-fighting genes, survive. The genetic uniformity of cheetahs means that a very high proportion of the population is susceptible to any disease that their genes do not equip them to fight. An outbreak of feline infectious peritonitis in an Oregon wildlife park killed more than half the cheetahs. Yet ten lions in the same compound did not even fall ill, and this disease is only fatal to housecats.

With less than 15% of cheetahs in U.S. zoos actively reproducing, and with the danger of disease epidemics among relatively dense captive populations, it is clear that human efforts to sustain captive cheetah populations face many challenges.

Wild cheetah populations face the same problems from the lack of genetic variability. They also face the dangers of predation on cubs, starvation, drought, and parasite infestations in nature. A further threat is loss of natural habitat. When humans take over wild land for ranching, cheetahs and other predators are killed to prevent their killing livestock. In 1988, researchers convinced farmers in Namibia to trap cheetahs alive and turn them over to captive breeding programs. By studying the cheetah's diet and behavior, providing veterinary care, keeping careful pedigree (family tree) records, and promoting outbreeding, researchers hope to be able to increase the genetic diversity of the cheetah population and hence its chances of survival.

Table 15–1
Factors that Increase and Decrease Genetic Variation in a Natural Population

	Factor	Effect
Increasing variation	Mutation	Introduces variation
	Sexual reproduction	Genetic reassortment occurs at gamete formation and at fertilization
	Polymorphism, disruptive natural selection, and heterozygote superiority	Retain more than one genetic form of a character in the population
	Immigration and outbreeding	May introduce new genes or gene combinations
	Increased population size	Occurs when selective pressures are relaxed; hence more variants survive in the breeding population
	Geographic variation	Adaptation to several different habitats increases variation
Decreasing variation	Natural selection (both stabilizing and directional)	Limits number of genotypes passed on to the gene pool of the next generation
	Inbreeding	Reduces number of heterozygotes
	Emigration	May remove genotypes from gene pool
	Decreased population size	Usually due to increased selection so there is less variation in breeding population. (Also, loss of variation by genetic drift is more likely in small populations.)

play when a population is reduced to very few individuals, who then become the ancestors of a later, larger population (see *A Journey Into Evolution:* Double Trouble for Cheetahs, this chapter).

A new population's environment will inevitably be somewhat different from the one its founders left, so the new population will experience different selective pressures, and therefore evolve in a new direction. In practice, it is usually impossible to tell how much of the genetic difference between the old and new populations results from the founder effect and how much results from different selective pressures in the two environments. The founder effect will have a great influence on a population of plants that populate an island from a single seed, or animals such as domestic hamsters, most of which have descended from one original pregnant wild female.

The most important mechanism of evolution is natural selection. Stabilizing selection and directional selection are the two most common forms of natural selection. Disruptive selection is a third, less common pattern. The death of individuals by random events can quickly change gene frequencies in the phenomenon of genetic drift. The founder effect is a special case of genetic drift, in which new populations are established by a few individuals resulting in new gene pools that are dramatically different from the ancestral population.

15–C WHAT PROMOTES AND MAINTAINS VARIABILITY IN POPULATIONS?

The theory of evolution predicts that a population will become increasingly well adapted to its environment. Therefore we might expect all the members of a population to end up with the same genotype, as selection eliminates the less fit alleles. However, most populations contain variants of many genetic traits, at levels so high that they cannot be due simply to recurring mutation. How can we explain this **genetic polymorphism** (poly = many; morph = form)? The answer must be that factors promoting genetic variation are constantly at work in natural populations (Table 15–1). Polymorphism can result from different genotypes be-

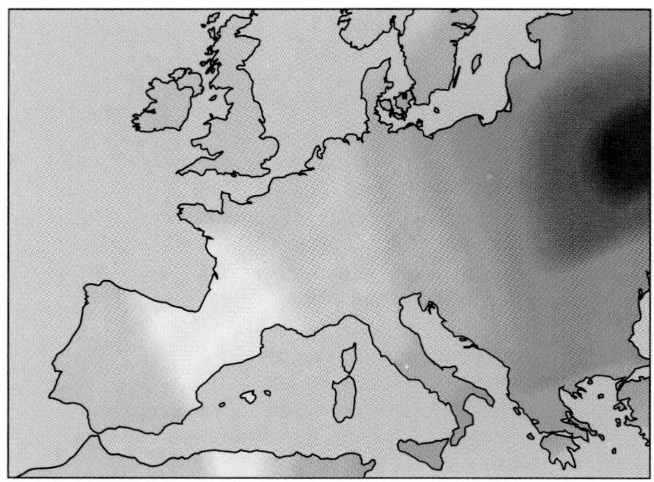

Frequency of B allelle:

4% Lower frequency | Higher frequency 16%

FIGURE 15-8

Frequency of the B allele of the human ABO blood group system in various European populations. In general, the frequency of the allele decreases progressively from east to west. Populations in central Asia and India, to the east of the area shown, have especially high frequencies of this allele.

ing favored at different times because of environmental changes. Or, it can result from the advantage of having more than one form of a genetic trait in the population.

Sex is probably the most widespread polymorphic trait. Most individuals are either male or female, and both forms are almost always present in the population at the same time. The human ABO blood group is a polymorphism involving three main alleles (see Table 13-1). The proportions of each blood type vary widely in different human populations. For instance, the frequency of the B allele is highest in central Eurasia—India, Mongolia, and western Siberia—and generally decreases with increasing distance from these areas (Figure 15-8).

To explain the distribution of ABO blood groups, we might suggest gene flow as populations mix, a gradient of selective pressures, or perhaps a combination of these factors. There is some evidence that people of blood type A are more susceptible to smallpox. This fits in with the finding that in India, where smallpox has been a persistent threat, only about 27% of the people have the A allele, compared with 46% in England and 48% in Germany. Smallpox was officially declared extinct in the late 1970s. It will be interesting to observe any changes in the frequency of blood groups over the next few generations in India, as might be expected if smallpox was one of the selective pressures determining the frequency of blood groups.

The phenomenon of heterozygote advantage is another cause of genetic polymorphism in a population.

Heterozygote Advantage and Hybrid Vigor

Sometimes individuals heterozygous for a particular gene are more common in a population than the Hardy-Weinberg Principle would predict. This suggests that these heterozygotes have a selective advantage, called **heterozygote advantage,** over homozygotes. The best documented example in humans is the case of sickle cell anemia (Section 13-B), where the heterozygote is at a selective advantage over either homozygote in areas with a high incidence of malaria.

Heterozygote advantage can arise in several possible ways. For instance, each allele may contribute its beneficial effects to the phenotype of the heterozygote. This is the case with heterozygotes for sickle cell anemia in areas where malaria is common. Neither homozygote has the advantage of both normal hemoglobin and resistance to malaria.

Heterozygotes are rare in populations with a high degree of inbreeding (mating between close relatives or self-fertilization). This is because close relatives usually inherit many of the same alleles from their common ancestors. The production of domestic plants and animals involves extensive inbreeding, beginning with a small number of individuals that show the desired traits. Such breeding programs often produce populations with genetic disadvantages as well as the advantages for which they were bred. Strawberries and tomatoes bred to resist bruising during the journey to market may have miserably little taste, and race horses often have especially delicate legs. Individuals from inbred populations are frequently homozygous for nearly all their alleles. These often include disadvantageous homozygous recessive alleles which are expressed in the phenotype, reducing health and vigor.

Matings between members of two different inbred strains produce **hybrid** offspring, which may be superior to their parents in many ways. Mongrel dogs tend to be healthier than "pure-bred" dogs, which often suffer from genetic ailments. Hybrid corn is valued for its reliable uniformity as well as for the specific qualities of the parental lines (Figure 15-9).

What is the genetic basis of this **hybrid vigor?** Since hybrids are heterozygous for many gene pairs where their inbred parents were homozygous, their superiority might result from heterozygote advantage in many gene pairs. Perhaps more often, hybrid superiority results simply from the fact that hybrids have a

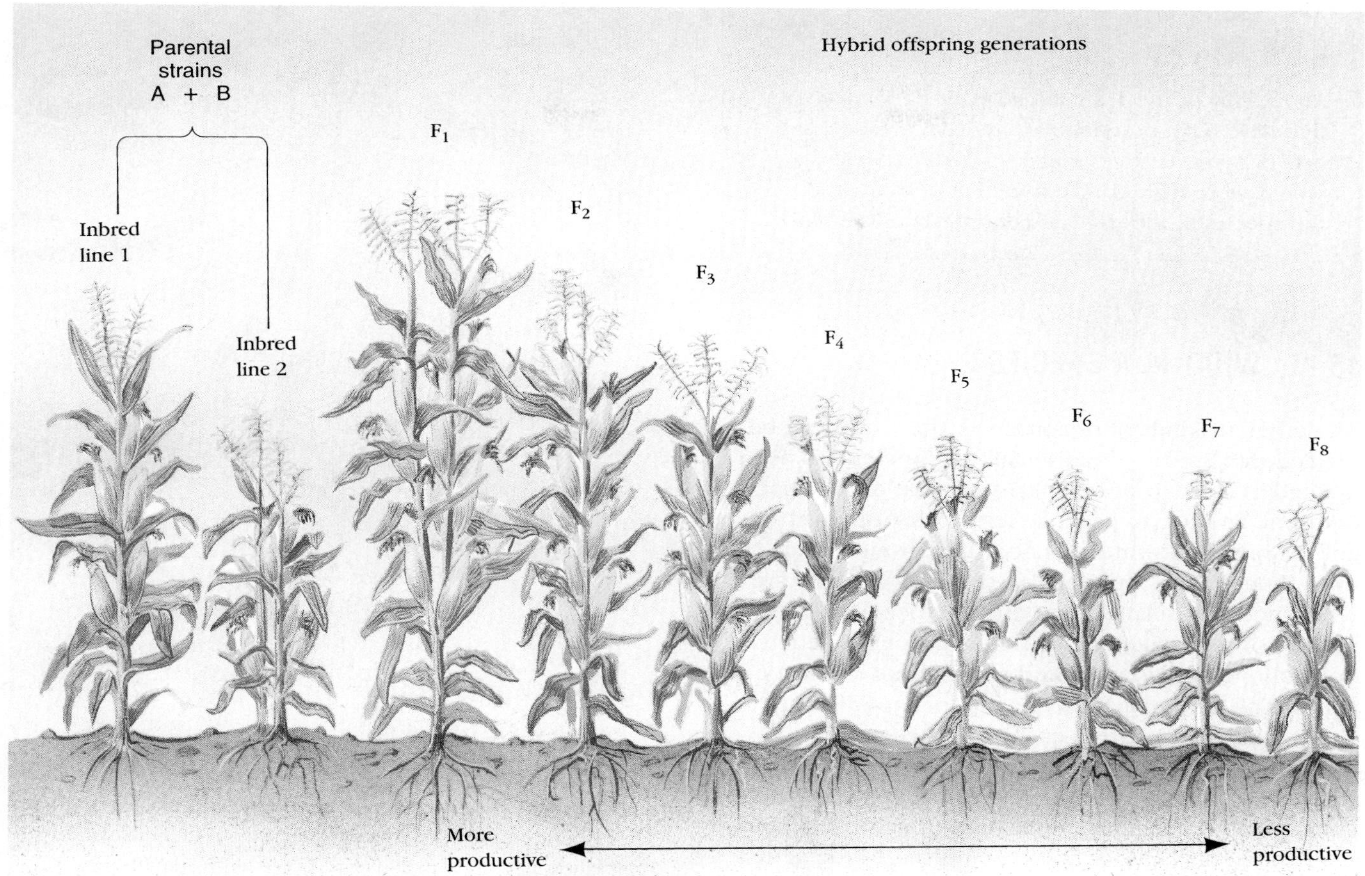

FIGURE 15-9

Hybrid corn. Two parental strains (P_1) are crossed to produce hybrid corn, the F_1 generation. The parental strains are homozygous for different alleles which results in heterozygous F_1 hybrids. For example, one P_1 may be homozygous recessive, "*aa*", while the other P_1 may be homozygous dominant, "*AA*", producing a heterozygous F_1 generation that is all "*aA*." Heterozygous hybrids that are superior to the parental strains are then used to produce seed for farmers. Inbreeding of the F_1 and succeeding generations produces many individuals homozygous for various traits and results in decreasing genetic qualities (like fewer kernels of corn) of the average plant.

high percentage of gene pairs with at least one dominant, advantageous allele. Each dominant allele will mask disadvantageous effects of its recessive partner.

The pitfalls of inbreeding make it clear that inbreeding is often selected against, and explain why adaptations that promote outbreeding are common. For example, many plants have adaptations promoting cross-pollination rather than self-pollination. Most human societies have taboos against incest. In many animals, such as monkeys and lions, young males leave the social group in which they were born and, with luck, eventually join another group, where they breed with the resident females. In chimpanzees and some rodents, it is the females who emigrate and join other groups.

A population evolving adaptations to a specific environment does not become genetically homogeneous over time. Instead, genetic variation is maintained by mutations, sexual reproduction, polymorphisms, heterozygote advantage, hybrid vigor, and changing selective pressures.

BIO-BIT

About 70% of the 1.2 million species of animals that have been described are insects! Most scientists believe that the number of insect species would more than double if we could identify all of the species in the tropical rain forests of the world.

15–D WHAT IS A SPECIES?

The different kinds of organisms in the world can be divided into species. Traditionally, a **species** has been defined as a group of organisms capable of breeding with one another to produce fertile offspring, yet unable to breed with members of other groups. Thus, even though many species live side by side, they are reproductively isolated from one another. The consequence of reproductive isolation is genetic isolation: populations that do not interbreed cannot exchange genes. This means that the gene pools of different species are isolated from each other, whereas members of the same species share a common gene pool.

A more modern definition states that a species consists of one or more populations that share a common gene pool. This definition emphasizes the genetic continuity of related individuals and populations. The members of a species that reproduces sexually usually share a common gene pool because there is gene exchange among them due to interbreeding. Members of other species (including species that reproduce only asexually) share a common gene pool because they are descended from a common ancestor.

As long as gene flow occurs between two populations, they belong to the same species even if their members cannot breed together (Figure 15–10). For instance, stretching across Europe is a string of populations of the European cherry fruit fly, with interbreeding and gene flow between the populations. However, when flies from eastern and western Europe are brought together to breed, the resulting eggs fail to develop. Nevertheless, these populations are members of the same species because there is, at least in theory, some exchange of genes between them, and so they share a common gene pool.

In practice, it is seldom possible to test directly whether two populations share the same gene pool. However, the range of phenotypes a species' gene pool can produce is limited, and therefore members of the same species usually look alike. The eighteenth-century naturalist Carolus Linnaeus developed the first working definition of a species, based on **morphological** (structural, anatomical) differences. If two organisms were sufficiently different, they were considered to belong to different species. In practice, this is how most organisms are classified today.

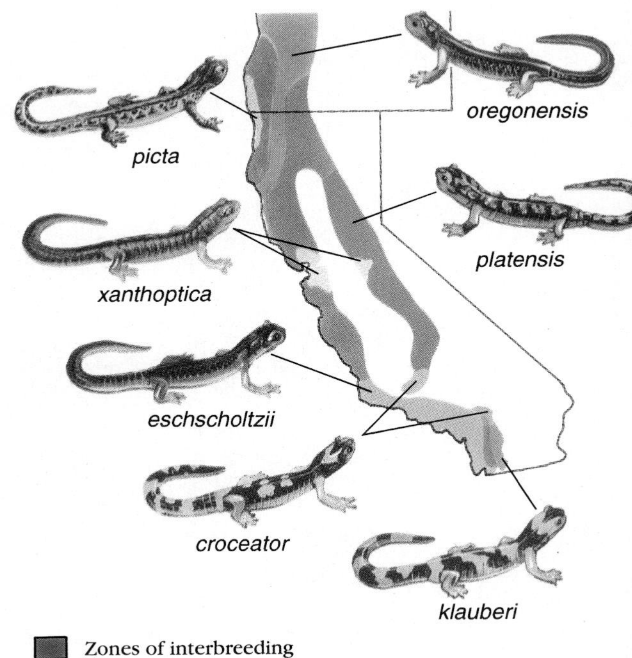

FIGURE 15–10

The *Ensatina* salamander cline. These seven subspecies of *Ensatina eschscholtzi*, located in California, Oregon, and Washington form a series or cline of populations. Each subspecies may occasionally interbreed with neighboring subspecies, but it does not naturally interbreed with more distant populations.

Morphological characters are convenient to work with and they allow for clear communication. People who discover a new species can write a precise description of it. They can also select **type specimens** of the species, individuals that are preserved in a museum for future reference. The morphological characters that distinguish species are also the natural criteria to use in constructing a dichotomous key, a list of paired, mutually exclusive statements that eventually lead to the identification of a species (see Toolbox 15–1). Dichotomous keys are one of the most important tools of a field biologist.

Today, other characteristics, such as DNA sequences, can also be used to describe species. We can store genomes, in the form of eggs, seeds, or tissues, at low temperatures. In theory, we could deposit gene samples from all the millions of species of organisms alive today in immense gene banks. The specimens could be used as type specimens for the concept of a species based on a shared gene pool.

Tool box

15–1
Key to the common classes of Arthropoda (adult specimens)

I

II

To use the key, begin at the first pair of statements and decide which one pertains to the specimen you are trying to identify. Each member of the pair indicates either a class or the number of the next pair of statements to consult. Continue until you arrive at a statement that gives the class of the specimen.

1a.	Three pairs of jointed legs on the thorax (part of the body between the head and the abdomen)............	**Class Insecta**
b.	Other............	2
2a.	Four pairs of jointed legs............	**Class Arachnida**
b.	Other............	3
3a.	Two pairs of antennae............	**Class Crustacea**
b.	One pair of antennae............	4
4a.	Two pairs of jointed legs per body segment............	**Class Diplopoda**
b.	One pair of jointed legs per body segment............	**Class Chilopoda**

III

IV

V

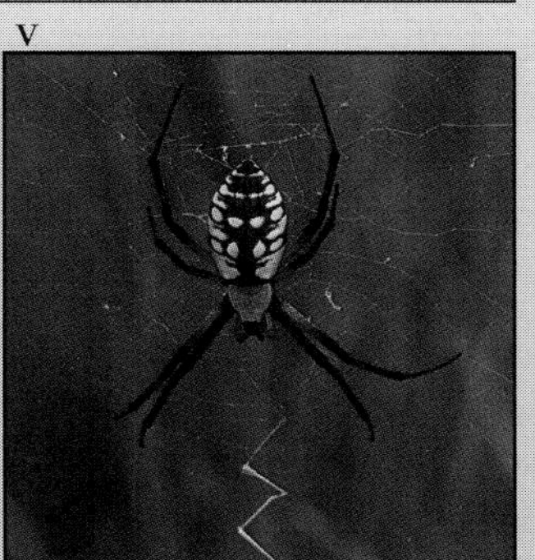

A dichotomous key, so called because each step gives two alternative choices. See how it works by following the instructions given at the top. The pictures are all animals from the large group known as arthropods, which have external skeletons and jointed appendages. Use this key to identify the classes of the specimens in the photographs around the key. For the answers, see the Chapter 15 Self-Quiz answers at the back of the book. (I, G. I. Bernard/Oxford Scientific Films/Animals Animals; II, T. McHugh/Photo Researchers, Inc.; III, E. R. Degginger; IV, D. Kuhn; V, J. Lepore/Photo Researchers, Inc.)

We tend to talk as if all organisms can be neatly divided into species. In reality, some populations are only partially isolated from their neighbors, genetically or sexually.

When two different populations live in the same place, they are said to be **sympatric** (sym = same; patria = homeland). If they live in different places, they are **allopatric** (allo = other). In nature, sympatric populations of different species usually do not interbreed because each species has its own, unique reproductive mechanism, and the mechanisms of different species are not compatible. Reproduction involves many adaptations of anatomy, physiology, and behavior. Members of the two sexes must come into breeding condition at the same time, usually as a result of hormonal changes within their bodies. Then they must often produce the appropriate steps in the courtship and copulation behavior of the species before the male and female gametes come together. Fertilization involves biochemical recognition between sperm and egg or between pollen and female flower parts. Finally, on a molecular level, the genetic information and cyto-

Table 15–2
Some Barriers that Prevent Interbreeding Between Members of Different Species

Barrier	Effect
Prezygotic isolation	Prevents mating (and therefore fertilization to form a zygote) between species
Habitat differences	Individuals of two species never meet
Different breeding times	Members of two species not in breeding condition at the same time (see Section 16–C)
Mechanical barriers	Shape of genitalia (copulatory structures) prevents fertilization by members of other species (common in insects)
Behavioral specificity	Mating cannot occur without species-specific behavior (see Section 16–C)
Postzygotic isolation	Prevents successful reproduction after fertilization (and therefore zygote formation)
Hybrid inviability	Hybrid offspring dies before reaching sexual maturity
Hybrid sterility	Hybrid offspring survives but is sterile
Hybrid breakdown	Hybrid offspring fertile but many of its offspring are not

plasmic messengers of sperm and egg must be compatible if embryonic development is to proceed normally and the resulting offspring survive.

The types of reproductive barriers that usually prevent members of different species from breeding with each other are listed in Table 15–2.

The members of a species share a common gene pool that is different from the gene pools of other species. As a result of this genetic similarity, members of a species are usually more similar to each other in structure, physiology, and behavior than they are to members of other species.

15–E SPECIATION

Speciation is the formation of one or more new species from an existing species.

Allopatric Speciation

Allopatric speciation occurs when a population becomes geographically separated from the rest of its species and then changes so much that it becomes a new species. This can occur when a small number of individuals colonize a new area. For instance, in the uplands of Hawaii, there is a species of goose found nowhere else. This goose closely resembles the Canada Goose of North America and almost certainly evolved from Canada Geese that migrated to Hawaii (Figure 15–11).

Suppose a few seeds or a few birds are blown onto an island where they grow and breed. The most likely fate of this new population is that it will become extinct because its members are poorly adapted to their new island home. But sometimes a few individuals will survive and start a new, isolated population.

The new population will tend to become genetically distinct from the rest of the species for two main reasons. First, the founder effect (Section 15–B) will often apply, and the new population will have a unique gene pool from the start. Second, the new population will be subject to a new set of selection pressures (Section 15–B) and so it will evolve adaptations to its new home. Its reproduction is one of the many features that may be changed as a result of this selection.

Eventually, the differences between the original population of the species and the new population on the island may become so great that the two can no longer interbreed. At this stage, it may be impossible to tell, just by looking at them, whether the two populations are different species or not. The test of speciation comes if the two populations ever come together again. If they do not then interbreed, the two populations are considered distinct species.

It may occur to you that new species might also form in a short period of time as when one large, widespread species becomes split into two, for in-

(a)

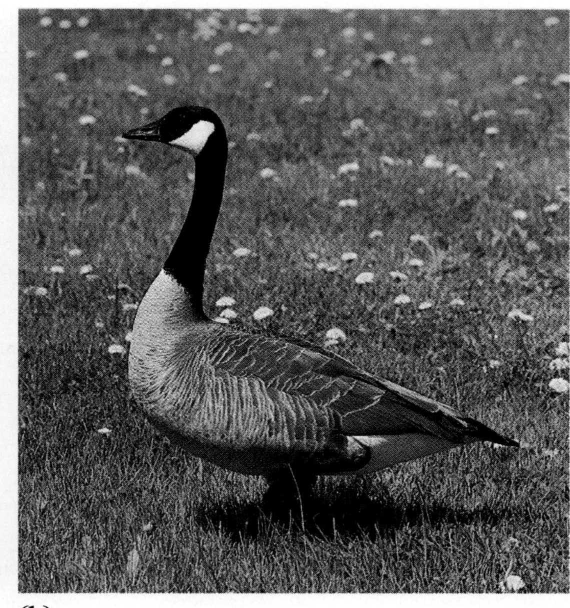

(b)

FIGURE 15-11

Parent and derived species. (a) The Hawaiian Goose (nene, pronounced "nay-nay"), Hawaii's state bird, is adapted to life on rugged upland lava flows far from water. This species is thought to be derived from a small population of **(b)** the Canada Goose of North America. (S. J. Krasemann/Peter Arnold, Inc.)

stance by a landslide, by construction of a highway, or over a long period of time as when continents drift apart (see Figure 1-2). Apparently, however, this does not usually occur. For instance, North American and European sycamore trees have been geographically separated for at least 30 million years, but they have not formed separate species. The two are very similar, and they can interbreed to produce normal, fertile offspring. It seems that species containing large numbers of individuals and covering wide areas of a continent may remain essentially unaltered for millions of years. This is because they are subject to gene flow and stabilizing selection. The genetic changes that lead to the formation of new species are much more likely to occur under the influence of directional selection acting on a small population of a species adapting to a new habitat. (Directional selection acting on *all* the members of a species changes that species, but it does not produce a new species.)

Pleistocene Glaciations

A dramatic example of directional selection producing speciation in small isolated populations comes from the effects of the great glaciations of the Pleistocene Epoch. Over the past million years, and ending only a few thousand years ago, four major glaciations (Ice Ages) covered most of Canada, the northernmost United States, and the western mountains (as well as Northern Europe) with ice, often thousands of feet deep.

Each time the ice sheets spread south from the Arctic, they pushed many species of animals and plants into new habitats and cut them off into isolated populations. During the glaciations, separated populations of a single species evolved adaptations to their new habitats. The glaciers then retreated, permitting some of these populations to come back into contact. In some cases, the populations had evolved into two or more distinct species from one species present before the glaciation.

The wood warblers of North America underwent impressive allopatric speciation in this way. Two dozen or more species can be found hunting insects in the foliage of even a single forest in parts of North America. The ancestral Black-throated Green Warbler was probably distributed across the continent before the Pleistocene. At each glacial advance, its range was pushed down to the southeast, pinching off a popula-

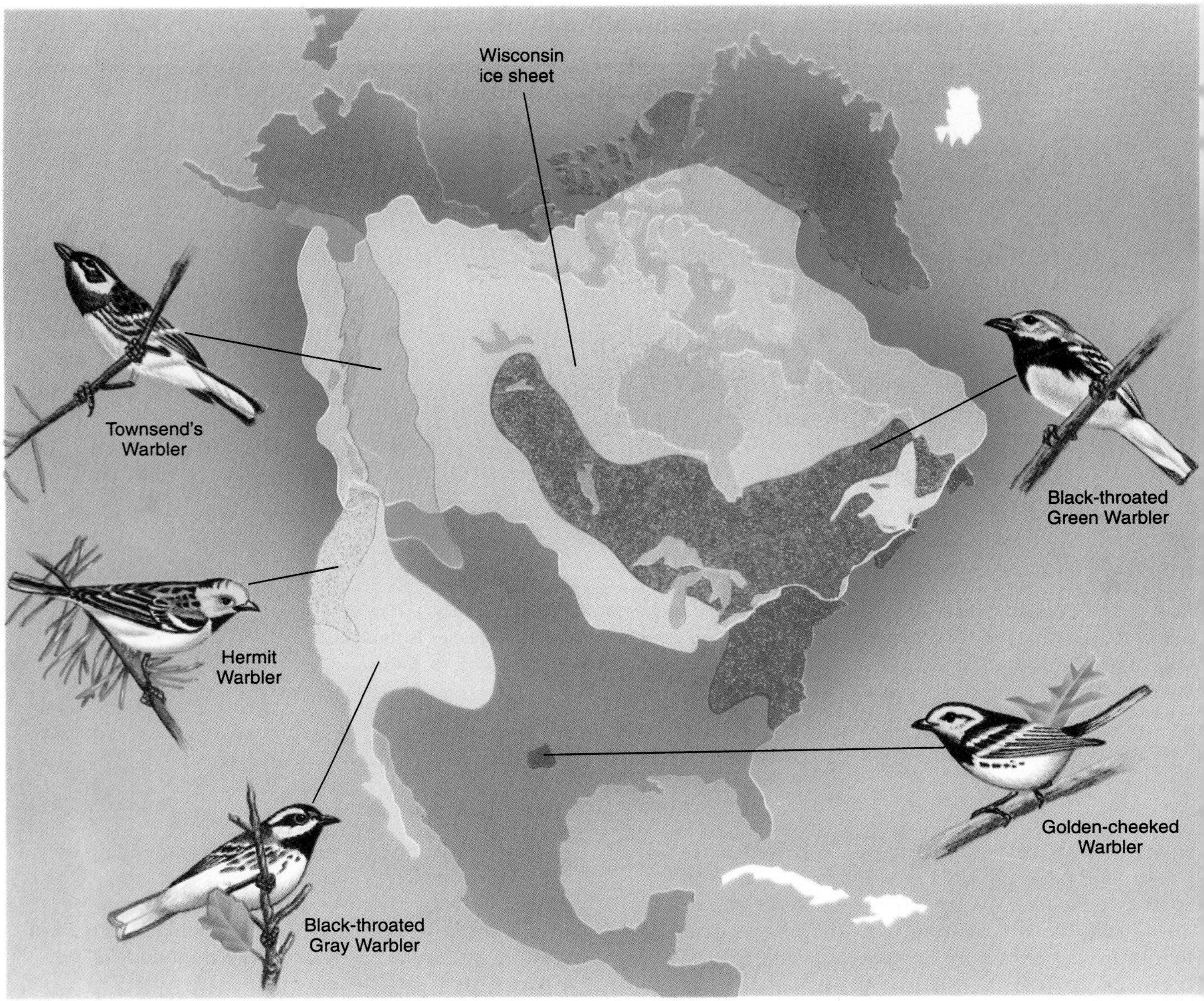

FIGURE 15–12

Speciation of warblers during the Pleistocene glaciations. The last glaciation of North America subdivided what had once been a single ancestral population of warblers into five separate species. The breeding ranges of five species of the Black-throated Green Warbler group are shown on the map. Note that there is very little overlap between breeding ranges.

tion in the west that evolved into a new species (Figure 15–12). When the glacier retreated, the Black-throated Green Warblers again spread across most of the continent, and small populations were again isolated by the next glacial advance, and so on.

Sympatric Speciation

Sympatric speciation is the formation of new species within a single population without geographical isolation. The example of sympatric speciation usually cited is that of polyploidy. **Polyploidy** is multiplication of the normal chromosome number. This can happen when chromosomes fail to segregate at meiosis, producing diploid instead of haploid gametes, or when a cell in a plant stem replicates its chromosomes without undergoing mitosis. The resulting polyploid cell may later give rise to a stem bearing polyploid flowers, which in turn form diploid gametes. (Some geneticists do not think polyploids should be consid-

FIGURE 15-13

A red banana flower and many young green bananas. Most cultivated bananas are triploid (3N). With this uneven number of chromosomes within a set, they cannot undergo the delicate process of meiosis (Section 11-F). The bananas develop without fertilization, and hence without seeds. Banana plants multiply vegetatively (asexually). (G. Grant/Photo Researchers, Inc.)

ered separate species from their diploid relatives. They contain *more* genes than their diploid parents, but not *different* genes, at least at first.)

Polyploid plants are often able to tolerate some adverse environmental condition, such as cold or drought, better than their normal parents. Indeed, the number of polyploid species increases with increasing latitude (that is, with colder climate). Probably more than a third of all plant species have arisen by polyploidy. Nearly all domesticated varieties of plants with larger fruits and flowers than those of their wild ancestors are polyploid. Polyploidy is rare among organisms that must be fertilized by another individual in order to reproduce. In such organisms, a polyploid individual would usually be an evolutionary dead end for lack of a genetically compatible mate. Therefore, polyploidy is most common in self-fertilizing plants and in plants that reproduce only vegetatively, such as bananas (Figure 15-13). It also occurs in **parthenogenetic** animals, those that reproduce by means of unfertilized eggs. Some parthenogenetic animals are polyploid: various beetles, moths, shrimp, goldfish, and lizards fall into this category.

In theory, sympatric speciation without polyploidy can occur in animals. If a population is polymorphic such that two varieties are advantageous but hybrids between them are selected against, the two varieties may evolve into separate species. This has been demonstrated in an experiment using artificial selection in a laboratory population of fruit flies. However, there is no known example of this situation in nature.

Unhappily for those with tidy minds, it will probably never be possible to produce convincing evidence that any animal species actually has arisen sympatrically. This is because sympatry is so difficult to demonstrate in natural conditions. Many pairs of populations appear to live together but close examination shows that they are effectively isolated. For instance, cichlid fish inhabit the Great Lakes of the African Rift Valley. There are about 126 species of cichlid in Lake Tanganyika and more than 200 species in Lake Malawi. The species are distinguished mainly by their feeding adaptations. Some feed in deep and some in shallow water, some eat algae, some molluscs, some plankton, and some other cichlids. Despite the fact that all are swimming in the same lake, these fish species are isolated from each other by specific habitat within the lake and by reproduction. To demonstrate sympatric speciation, one would have to show that two species belonged to one population before speciation occurred, and this would involve showing gene flow or reproduction between them—difficult or impossible to do.

Selection Against Hybrids

When two previously separated populations of related organisms come together, they may interbreed. This situation has various possible outcomes. First, if the two are already distinct, stable species, they will not interbreed at all (or may interbreed in some areas to produce populations of hybrids). Second, they may interbreed freely so that they merge into one big population and all genetic distinction between them disappears. A third possibility might theoretically exist but probably occurs very seldom: members of the two populations may interbreed at first but later form separate species. This might occur if the hybrid offspring between the two populations are at a selective disadvantage compared with members of either parent population. In this case, selection would favor those individuals that mate with members of their own population over those that produce hybrids.

Most new species form by allopatric speciation, when a small population is cut off from the rest of the species and evolves adaptations to its new habitat in isolation. This may occur when a small number of individuals colonizes a new area that is physically isolated from the ancestral populations. Or it may result from a subdivision of an existing range, as a result of a physical barrier. New species may also form by sympatric speciation, in which a new species forms within a single population without isolation. Members of sympatric species remain reproductively isolated because of differences in their reproductive anatomy, physiology, and/or behavior.

FIGURE 15–14

Look-alikes. This fossil of a ginkgo leaf about 50 million years old (right) is strikingly similar to the leaf from a living ginkgo (left). (B. P. Kent/Earth Scenes)

15–F HOW QUICKLY DO NEW SPECIES FORM?

Biologists sometimes divide evolution into microevolution and macroevolution. **Microevolution** is evolution of small genetic changes within populations and the formation of new species—the kinds of evolution we have discussed so far in this chapter. **Macroevolution** is the evolution of new groups of species, such as flowering plants or dinosaurs. The main way of investigating macroevolution is to study the fossil record of life on Earth. Examining fossils reveals patterns in macroevolution which we might not predict from what we know of microevolution. For instance, when we measure the rate at which new species appear in the fossil record, we find that evolutionary changes may be fast or slow, or anywhere in between.

Many fossil species existed for millions of years with very little change and without giving rise to new species. Then, in a "short" period of time (which in geological terms may mean thousands of years), related but different species appeared. Sometimes these new species replaced the older ones, and sometimes the old and new species coexisted at least for a while. The evolution of the horse is sometimes shown as a gradual process, with body size steadily increasing and the number of toes becoming reduced from three to one (see Figure 14–7). Closer study shows that speciation occurred many times during the evolution of horses, leading both to species with smaller bodies and to species with larger bodies. Since larger-bodied species survived longer than smaller-bodied species and gave rise to more new species, what we see in the fossil record is the *appearance* of a gradual increase in body size. If smaller horse species had survived, our perception would be different.

In many cases, fossils millions of years old are strikingly similar to species alive today (Figure 15–14). For example, half of the fossil seashells from seven million years ago apparently belonged to species still alive today. The sycamores discussed in Section 15–E are another example. At other times, many new species have evolved in a relatively short time (one to a few thousand years), as in the cases of the warblers (see Figure 15–12).

This "quick and slow" pattern of speciation is called **punctuated equilibrium.** We can see how it probably comes about. A widespread species exists unchanged for a long time under the influence of stabilizing selection. Eventually, one of its small, isolated populations evolves into a particularly successful new species. This new species spreads rapidly, either driving the parent species to extinction or existing with it.

This model helps to explain a riddle that puzzled Darwin and many other people: the sudden appearance of various major groups of organisms in the fossil record. Any new group of organisms evolves from a previously existing one, but often we do not find fossils of gradually changing intermediate forms spanning the time until the new group is clearly distinct from its ancestors. Instead, some groups, such as the flowering plants, seemed to appear almost full-fledged in the fossil record, with little evidence to show how they originated or which organisms were their ancestors.

Major groups such as this usually have evolutionary novelties that distinguish them from their ancestors. It is noteworthy that these often include reproductive adaptations, such as flowers in plants, and waterproof eggs in reptiles. These new adaptations proved so successful that the species in which they evolved underwent wide adaptive radiation, spreading across the globe and spawning hundreds of new populations, many of which also formed new species. This swift burst of speciation appears as rapid change in the group's fossil record. When organisms with the new adaptations had spread into most of the habitats they were equipped to exploit, speciation became less frequent. This is reflected in a slower rate of change in the group's fossil record. Since such evolutionary novelties usually arise in small, local populations, the chances of the intermediate forms being preserved as fossils, and of our finding them if they were, are very small.

Evolution occurs at many different rates. During some periods of time, groups change rapidly, while in other periods, there appears to be little change. These periods of quick change usually occur in small, local populations that spread rapidly.

SUMMARY

1. Natural selection acts on individuals, but only populations can evolve. This is because the population is the smallest unit with a gene pool in which the frequency of alleles can change.
2. The Hardy-Weinberg Principle shows that the proportions of different alleles and genotypes in a population will remain the same as long as all of the following conditions are met:
   a. there is no net mutation,
   b. there is no selection for or against the traits being considered,
   c. mating is random with respect to the genotype, and
   d. there is no gene flow to or from other populations, and the population is large.

   However, if any one of these conditions is not met, evolution is likely to occur.
3. Populations of the same species in different geographical areas tend to differ somewhat in their genetic makeup. Adaptations to local areas increase this difference; gene flow between adjacent populations decreases it.
4. Random changes in the gene pool as a result of genetic drift may be an important cause of evolution in small populations that are not subject to strong selective pressures.
5. Genetic variation in a population can be increased by mutation, by polymorphism maintained by heterozygote advantage, by gene flow between populations, and by sexual reproduction.
6. The most important factor that decreases variation in a gene pool is natural selection, which adapts a population to local conditions. Genetic drift may also decrease genetic variation by randomly eliminating some alleles from a small population.
7. A species is an interbreeding (actually or potentially) group of organisms with a common gene pool and similar morphology, physiology, and behavior.
8. A population of an existing species may evolve into a new species. Allopatric speciation occurs when a population is separated from the rest of a species, cutting off gene flow between the two groups. This isolated population evolves under the influence of local directional selection, and may become so different from the parent population that it is considered to be a new species.
9. New species may also evolve by sympatric speciation, in which a new species forms within a population without isolation, for example, by polyploidy.
10. In a process called punctuated equilibrium, many species appear to remain similar for long periods of time and then give rise to a new species as a subpopulation undergoes rapid and dramatic change.

SELF-QUIZ

1. The Hardy-Weinberg Principle allows us to predict that:
   a. sexual reproduction is necessary for evolution
   b. sexual reproduction may be a cause of evolution
   c. sexual reproduction plays no role in evolution
   d. sexual reproduction will cause evolution if individuals prefer mates with one genotype over those with other genotypes
2. In certain parts of Africa, people with one normal hemoglobin and one lethal sickle hemoglobin allele are more likely to survive than homozygotes for either allele. These populations are *not* experiencing:
   a. natural selection
   b. heterozygote advantage
   c. polymorphism
   d. genetic drift
   e. violations of the Hardy-Weinberg Principle

3. Suppose that mosquito control measures completely eliminate the threat of malaria from an area where it was once prevalent. If you followed human allele frequencies for the next 30 generations, what changes would you expect to see in the frequencies of the normal and sickle alleles, once selection against the homozygous normal individuals is removed?
4. Selection will not eliminate a lethal recessive allele from a large population of diploid organisms because:
   a. there will always be some heterozygote carriers for the allele
   b. heterozygotes are at a selective advantage
   c. the allele will have some good effects
   d. the rate of mutation producing new copies of the lethal allele is higher in a larger population
5. Genetic drift is more likely to cause evolution in a small population because:
   a. mating is nonrandom in small populations
   b. random events are more apt to happen to small populations
   c. there is no natural selection in small populations
   d. deviations from averages are more likely to be seen in small populations than in large ones
6. In each of the following situations indicate whether genetic variability in a population would increase, decrease, or remain the same.
   a. increased mutation rate
   b. decreased natural selection
   c. increased variability in the environment
   d. sexual reproduction
7. Which of the following situations will *not* produce a new species?
   a. Several individuals colonize a new area and can no longer reproduce with the ancestral population
   b. Several individuals join an existing population and introduce new alleles to the population
   c. An existing species becomes subdivided by the effects of a glacier and the resulting groups remain physically isolated from each other for thousands of years
   d. A new group of polyploid organisms arises from abnormal meiosis and can only reproduce with each other
8. All of the following conditions would result in a change in the frequency of a specific allele in a population *except:*
   a. selection against the homozygous recessive phenotype
   b. selection against the dominant genotype
   c. genetic drift
   d. random mating
   e. mutation of the dominant allele to the recessive allele

THINKING CRITICALLY

1. What, in biological terms, is a "race" of people or any other animal?
2. Evolutionist Theodosius Dobzhansky wrote that a totally uniform physical environment could support only one species of organism. Do you agree? Why or why not?
3. What is the effect on the human gene pool of the discovery of expensive medical treatments that permit individuals with previously lethal phenotypes to live? Should society permit research to find ways to treat more such conditions?
4. Huntington's disease is a genetic condition caused by a dominant gene. The brain deteriorates, and the victim loses control over both mind and muscles. A period of insanity accompanied by jerky movements of the face and limbs is finally followed by death. (Interestingly, at least seven women accused of witchcraft in New England during the 1600s were related to families now known to have had Huntington's disease.) This condition is unusual because typically, it does not set in until the victim is in his or her 30s or 40s, and by then he or she may have produced children, half of whom will also receive the allele and are therefore doomed to develop the disease. Can selection operate against such a late-acting genetic trait?
5. What characteristics would you expect to see in the genetic makeup of a species that has the ability to colonize a variety of habitats that may become available?
6. Suppose an impassable barrier cuts off gene flow between two populations that inhabit areas with somewhat different climates. What change in the two populations' gene pools would you expect to observe as time passes?

SELECTED KEY TERMS

allopatric speciation, *p. 314*
directional selection, *p. 304*
disruptive selection, *p. 305*
fitness, *p. 304*
founder effect, *p. 307*
gene flow, *p. 305*
gene pool, *p. 301*
genetic drift, *p. 307*
genetic polymorphism, *p. 309*
Hardy-Weinberg Principle, *p. 302*
heterozygote advantage, *p. 310*
hybrid vigor, *p. 310*
macroevolution, *p. 318*
microevolution, *p. 318*
natural selection, *p. 302*
polyploidy, *p. 316*
population, *p. 301*
punctuated equilibrium, *p. 318*
species, *p. 311*
stabilizing selection, *p. 304*
sympatric speciation, *p. 316*

SUGGESTED READINGS

Ayala, F. J., and J. A. Kiger. *Modern Genetics,* 2d ed. Menlo Park, CA: Benjamin/Cummings, 1984. Several good chapters on population genetics.

Gould, S. J. *The Panda's Thumb.* New York: W.W. Norton & Company, 1980. A wonderful series of short essays exploring many facets of evolution.

O'Brien, S. J., D. E. Wildt, and M. Bush. "The cheetah in genetic peril." *Scientific American,* May 1986.

Stanley, S. M. *The New Evolutionary Timetable.* New York: Basic Books, 1981. Layperson's guide to evolutionary thought from Darwin to the present by a leading punctuational evolutionist.

Stone, C. P., and D. B. Stone, Eds. *Conservation Biology in Hawaii.* Honolulu: University of Hawaii Press, 1988. A resource on the biology of Hawaii, highlighting the ignorance of visitors and residents about these remarkable islands, which are rapidly being destroyed, and stressing the need for research and education.

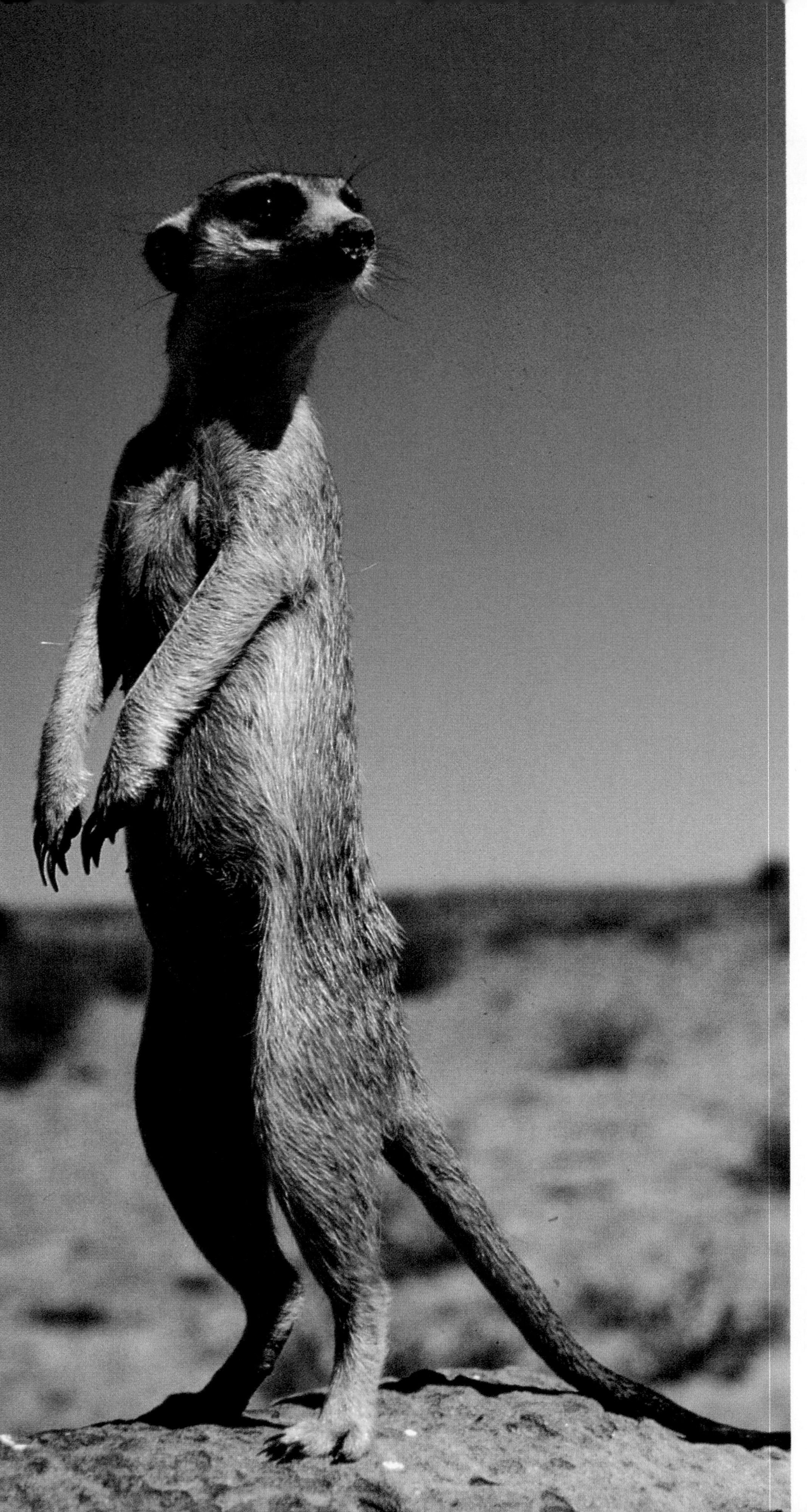

CHAPTER 16

This meerkat is placing itself at risk to look for danger. Why would an animal risk its life for the safety of its relatives? Read about altruistic behavior in **Section 16-E.** (D. MacDonald/Oxford Scientific Films)

Evolution and Reproduction

CONCEPT GUIDE

After reading this chapter, you should be able to:

1. Compare the advantages and disadvantages of sexual reproduction and asexual reproduction. (Section 16-A)
2. Explain how sexual reproduction might have first occurred and how it has changed in most living plant and animal species. (Section 16-B)
3. Explain the adaptive advantage of hermaphroditism. (Section 16-C)
4. Explain the adaptive advantages of polygyny, polyandry, and monogamy to males and to females. Describe the circumstances under which each of these mating systems would be expected to evolve. (Section 16-D)
5. Describe the circumstances under which altruistic behavior would be expected to evolve. (Section 16-E)

An organism's genes may enable it to obtain food, avoid being eaten, and cope with the climate, but ultimately there is only one measure of its evolutionary success: the proportion of its genes present in future generations. Without genes that make it reproduce, all the adaptations resulting from its other genes are useless, in evolutionary terms. The reproductive adaptations of organisms are many and varied. We shall consider some of them in this chapter, as examples of the remarkable effects of natural selection.

Successful reproduction automatically selects for perpetuation of the genes that brought it about—and the various modes of reproduction in the living world are indeed wonderful.

KEY CONCEPTS

- An organism can be viewed as a vehicle by which genes are passed from one generation to the next.
- The reason sexual reproduction is so widespread is probably that it produces more genetic variation in organisms than does asexual reproduction.
- The ecology of a species plays a major role in determining the type of sexual system it develops.
- A species' sexual system determines how members of the species are related to each other and thus the types of behavior toward other individuals that can evolve.

Curiosity Questions

1? Do hermaphrodites really exist? (See page 328)

2? What are the advantages to having just one mate instead of many mates? (See page 332)

3? Do animals ever risk their lives to save the lives of other animals? (See page 333)

16–A IS SEX NECESSARY?

At first glance it seems that sex *is* necessary. Most animals, and many higher plants, rely exclusively on sexual reproduction to perpetuate their genes. On the other hand, sexual reproduction is much less common among protists, fungi, and lower plants, which usually reproduce in other ways. Perhaps, in these organisms, sex is not necessary but is valuable under some circumstances.

Asexual reproduction is any means of multiplying that does not involve the combination of eggs and sperm. Many unicellular organisms reproduce asexually simply by dividing into two identical, smaller cells. Many plants can reproduce vegetatively, using runners or tubers (Section 35–H). Asexual reproduction in some lower animals occurs by the budding off of smaller individuals, which eventually detach from the parent (Figure 16–1). **Parthenogenetic** organisms develop from unfertilized eggs (see Figure 16–2) and are therefore genetically identical to the mother.

When an organism reproduces asexually, each offspring represents a considerable investment of the parent's energy. However, the energy is efficiently used in that nearly all of it goes into the growth of a new individual. Furthermore, the relatively large size of the offspring produced by many asexual methods gives them a good chance of surviving to reproductive age.

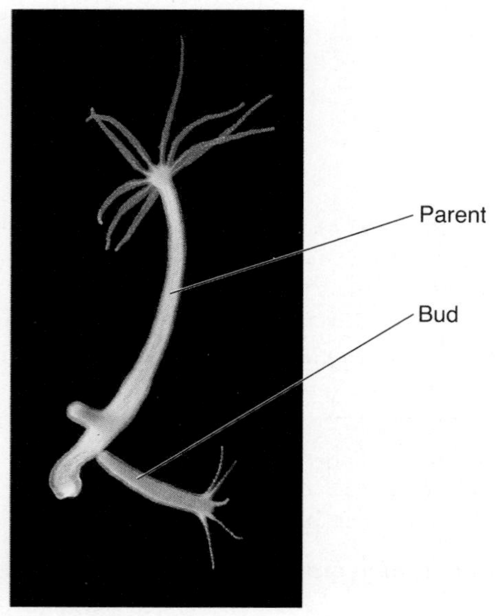

FIGURE 16–1

Asexual budding. This adult *Hydra* is reproducing by budding. When the bud is large enough, it breaks off and starts an independent life. (Biological Photo Services)

By contrast, **sexual reproduction,** any means of multiplying that involves gametes, requires much more energy. With **external fertilization** in which sperm, pollen, or eggs are released into the water or air to find each other by chance, millions of gametes fail to find mates or fall prey to other organisms, and the energy spent to make them is wasted. With **internal fertilization,** sperm are introduced directly into the female's body near the eggs. Fewer gametes are lost, but the organism must invest energy in other ways. For example, plants may produce flowers and nectar, thereby attracting animals that carry pollen to the female parts of other flowers. Animals must spend a lot of time and energy finding and courting mates.

Overall, attempts to reproduce sexually are more likely to fail than are attempts at asexual reproduction. Thus, many asexually reproducing organisms can produce more surviving offspring per season than can similar organisms that reproduce sexually. Why, then, do so many living things use so much energy and make so many wasted cells in the process of sexual reproduction? Sexual reproduction must have tremendous adaptive value, or it could not have become so common.

The main biological difference between sexual and asexual reproduction is that sexual reproduction involves genetic recombination and reassortment during meiosis and fertilization, and so it produces more genetic variation in a population (Section 11–G). Genetic variation allows a population more genetic options, and some of these may become favorable options if the environmental conditions change. In most organisms that reproduce asexually, the offspring result from mitosis. Hence the parent passes on all of its own genes to all its offspring, and therefore they are all alike genetically—that is, the original organism plus its descendants are said to form a **clone** (Figure 16–2). The only genetic variation that can arise in such a population comes from mutations. (An exception is the asexual production of male hymenopterans [ants, bees, and wasps] [Section 16–E], who receive half of their mothers' genes as a result of meiosis.)

Some organisms can reproduce both sexually and asexually. Many protists, algae, and small invertebrates (such as the aphids in Figure 16–2) reproduce asexually during the summer and then reproduce sexually when the temperature drops and days become shorter in the fall. These organisms switch from asexual to sexual reproduction when environmental conditions deteriorate.

Under this system, an organism that is well adapted to its environment can produce numerous, equally well-adapted copies of itself while conditions remain constant during the summer. When conditions then change in the fall, the organism reproduces sexu-

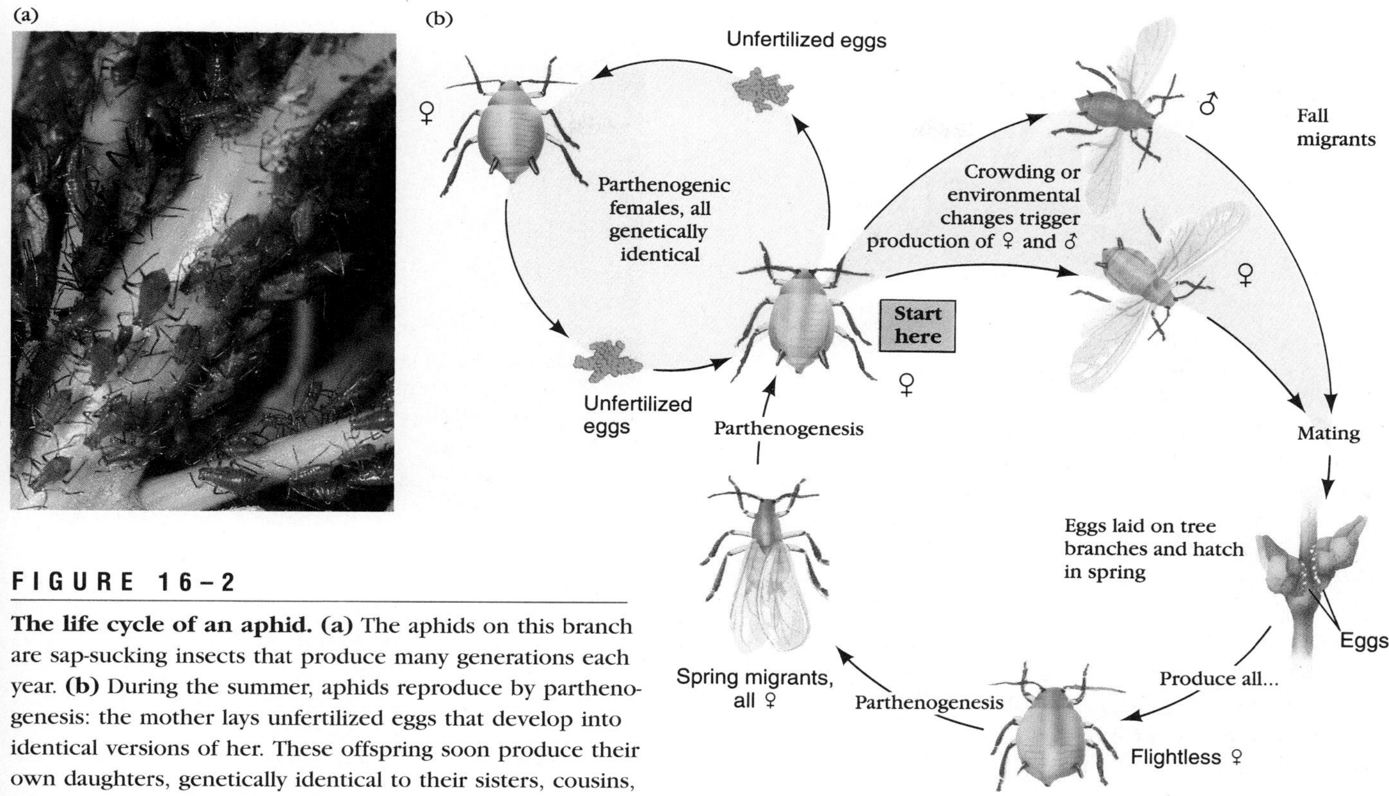

FIGURE 16–2

The life cycle of an aphid. (a) The aphids on this branch are sap-sucking insects that produce many generations each year. **(b)** During the summer, aphids reproduce by parthenogenesis: the mother lays unfertilized eggs that develop into identical versions of her. These offspring soon produce their own daughters, genetically identical to their sisters, cousins, and aunts, who are all descended from the same ancestor. Eventually, overcrowding or other environmental cues may cause a female to lay eggs which develop into winged males and females that disperse to new locations where sexual reproduction and egg-laying occur. In the spring, these eggs hatch to produce a new generation of females who reproduce parthenogenetically, forming winged clones who disperse and continue parthenogenesis until the fall, when males usually appear again. (A. C. Twomey/Photo Researchers, Inc.)

ally, creating many genetically different offspring, some of which will probably survive. Often, sexual reproduction gives rise to an egg or cyst in a weatherproof covering, which protects the enclosed individual until good growing conditions return.

Many kinds of bacteria and other "lower" organisms have existed virtually unchanged for more than 500 million years. Such organisms can tolerate a wide range of conditions. It is hard to imagine an environmental change so catastrophic and widespread that it would threaten them with extinction. Most higher plants and animals, on the other hand, can live only in the few places that supply their particular needs. A specialized species with only asexual reproduction (the common dandelion is a good example) may do very well for a while, but is ultimately doomed to extinction much more surely than is a species that reproduces sexually. A change in the environment that kills dandelions will probably kill all dandelions in the area because they are all very similar genetically (Figure 16–3).

There may be occasions when asexual reproduction would be selectively more favorable even in higher organisms. But living things are trapped by their evolutionary history. Sexual reproduction may be so entrenched that it is hard to get rid of.

Asexual reproduction is found among many common, widespread species that can tolerate a broad range of environmental conditions and reproduce rapidly. More specialized species are in danger of dying out, leaving no descendants, if their members do not have the genetic diversity that sexual reproduction provides.

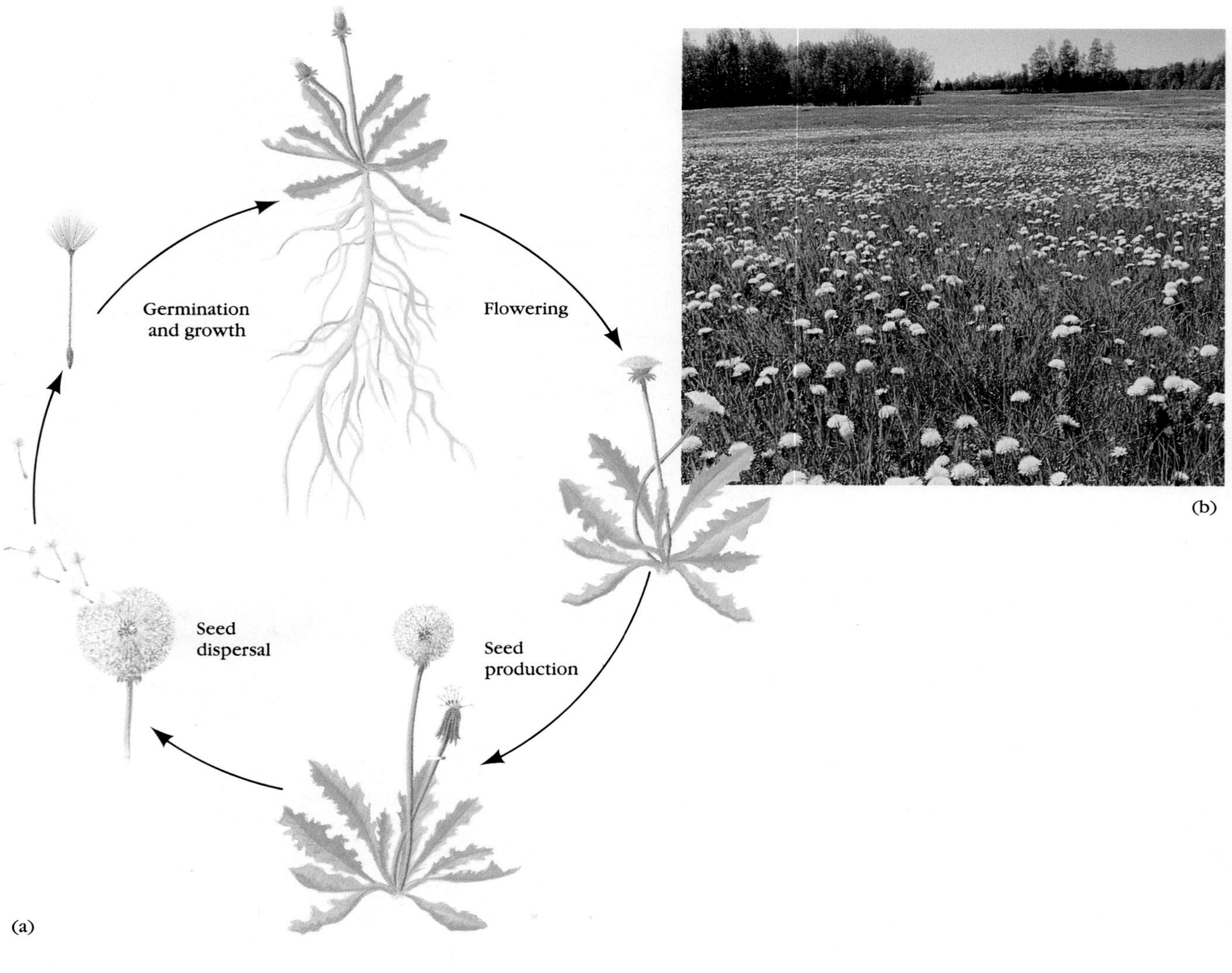

FIGURE 16–3

Dandelions. (a) These plants only reproduce asexually and are therefore nearly identical genetically. Dandelion pollen is sterile and the eggs develop without fertilization. Therefore, each seed carries the same genes as the parent plant. **(b)** A field of dandelions. (D. Kuhn)

16–B EVOLUTION OF SEXUAL REPRODUCTION

Sexual reproduction, involving meiosis and fertilization at some stage in the life history, occurs only in eukaryotes. The first type of sexual reproduction probably resembled that seen even today in organisms such as the filamentous alga *Ulothrix,* which can also reproduce asexually. This alga consists of a string of haploid cells (having unpaired chromosomes). During the growing season, *Ulothrix* reproduces asexually by means of small flagellated cells called **zoospores.** The zoospores are released through a pore in the parent cell. They swim to suitable locations, where they settle and divide, starting new filaments (Figure 16–4).

Sexual reproduction in *Ulothrix* probably evolved as a modification of this process. Some cells of the filament divide to produce flagellated gametes, cells smaller than zoospores. All the *Ulothrix* filaments in an area release their gametes at the same time, and they swim around in the water. When two gametes collide, they stick together and fuse to become one. This is a sexual event, and it forms a diploid **zygote**

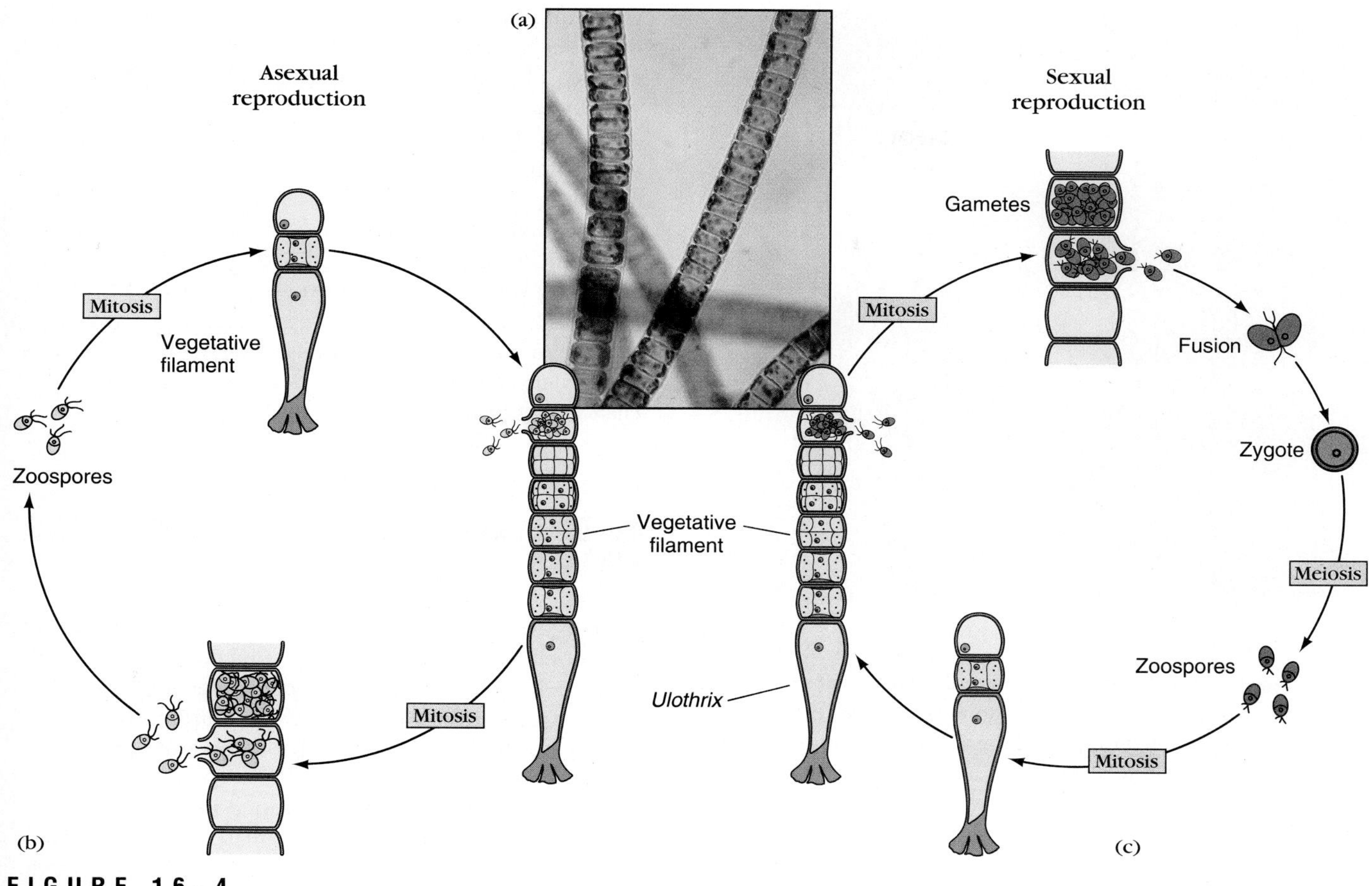

FIGURE 16–4

Ulothrix. (a) *Ulothrix* is a filamentous alga that reproduces both asexually and sexually. **(b)** In asexual reproduction, each zoospore germinates into a vegetative filament with the same genetic makeup as the parent filament. **(c)** In sexual reproduction, gametes are produced instead of spores. These gametes fuse to form a diploid zygote, which undergoes meiosis to form four zoospores of new genetic types. Each zoospore then germinates into a new vegetative filament. (E. Reschke/Peter Arnold, Inc.)

(having pairs of homologous chromosomes). The *Ulothrix* zygote eventually undergoes meiosis to produce haploid zoospores, which can develop into new filaments. This is a very simple form of sexual reproduction in that all the gametes look alike. Fertilization may have originated as an accidental fusion of two undersized spores which, as a result, gained the selective advantage of larger size by pooling their resources.

In the sexual reproduction of most plants and all animals, the egg produced by the female is a very large cell that is incapable of independent movement. The male sperm is a small cell that travels to the egg. The egg contains not only stored food but also messenger RNA carrying the information necessary to direct the process of protein synthesis that is important in the early stages of embryonic development. The major advantage of this system is that the embryo can be larger and better developed before it has to provide its own food. Higher animals could probably never have evolved without an egg containing stored food and genetic information for embryonic development. It is hard to imagine even a worm developing if the tiny embryo had to form a mouth and feed itself when it contained only two or four cells.

Thus, an egg that contains food and information, and is therefore too large to be independently motile, is of enormous selective advantage. But a nonmotile egg is no use without a sperm that can find and fertilize it (Figure 16–5).

The first type of sexual reproduction probably resembled that seen in some filamentous algae today, in which gametes

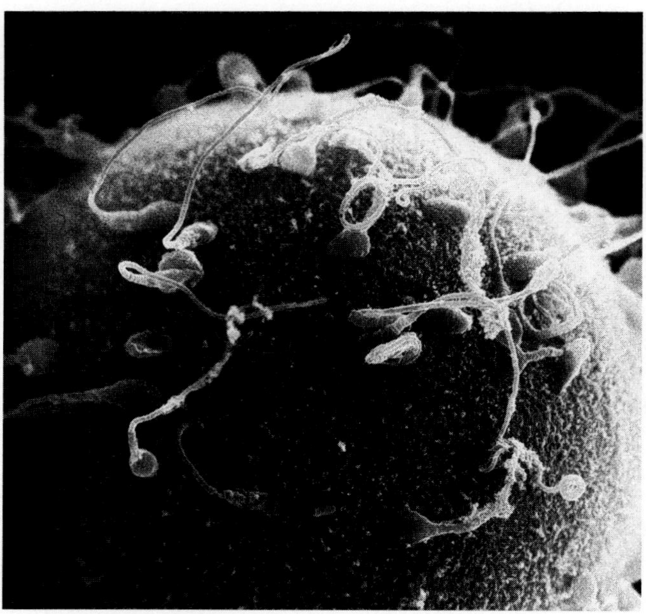

FIGURE 16–5

The relative sizes of eggs and sperm. This hamster egg (about the size of a human egg) is much larger than the human sperm swarming around it. Hamster eggs are often used to test for human sperm viability. (D. M. Phillips/The Population Council/Science Source)

of equal size fuse to form a zygote. However, in most currently living plant and animal species, small sperm travel to the larger egg, which contains stored nutrients and mRNA to assist the growth and development of the early embryo.

16–C EVOLUTION OF MECHANISMS THAT ENSURE FERTILIZATION

When an organism reproduces asexually, it is independent of other individuals. In sexual reproduction, however, males and females must cooperate. One mechanism that coordinates male and female activity is the existence of breeding seasons, whereby members of a population all come into breeding condition in response to some environmental cue, such as temperature or length of daylight. Another is mating behavior, including chemical secretions by some members of the population that stimulate others to come into breeding condition or indicate readiness to breed.

1? An adaptation found among organisms that are fixed in place, or at least cannot move very far to find mates, is the existence of reproductive organs of both sexes in the same individual. This **hermaphroditism** is common in plants, and it also occurs in some animals, such as leeches and earthworms (Figure 16–6). It ensures that any other member of the species in the vicinity is a potential mate, and gives each organism a chance to be both a mother and a father! Hermaphroditic animals and plants may be able to self-fertilize, but there is often selective pressure to mate with other individuals. This reduces the chance of inbreeding, which would eliminate the genetic variation produced by sexual reproduction (Section 15–C). ■

Sexually reproducing organisms use a variety of mechanisms to coordinate reproductive behavior and increase the chances of finding a mate.

16–D EVOLUTIONARY ROLES OF MALE AND FEMALE

A male's evolutionary success is usually limited, not by the number of sperm he can produce, but by his ability to deliver his sperm to as many eggs as possible. The female's evolutionary success is limited by the number of her eggs that survive to become part of the breeding population.

The fact that a female's parental investment in each future offspring is usually greater than the male's has a fascinating consequence: the selective pressures acting on a female may conflict with those acting on a male of the same species. While it may be advantageous for a male animal to copulate with as many females as possible in order to raise his chances of fathering surviving offspring, it is apt to be advantageous for a female to be much more choosy. She produces fewer eggs and so has fewer second chances to reproduce successfully if her first mate is genetically unfit.

Under this selective pressure, it is not surprising that females of all species of animals studied show discrimination in their choice of mates. The female who discriminates, and copulates only with genetically fit males, will be at a selective advantage. On the other hand, it may be to a male's advantage to appear genetically fit even when he is not, because females may then be deceived into mating with him.

Sexual Differences

A female's reproductive success is not usually limited by her inability to find mates. It is more likely to be limited by her inability to rear her young. A male who demonstrates that he can contribute to raising offspring will be attractive to females. Among birds, the male of choice is often the holder and defender of a territory that also provides food and shelter needed by

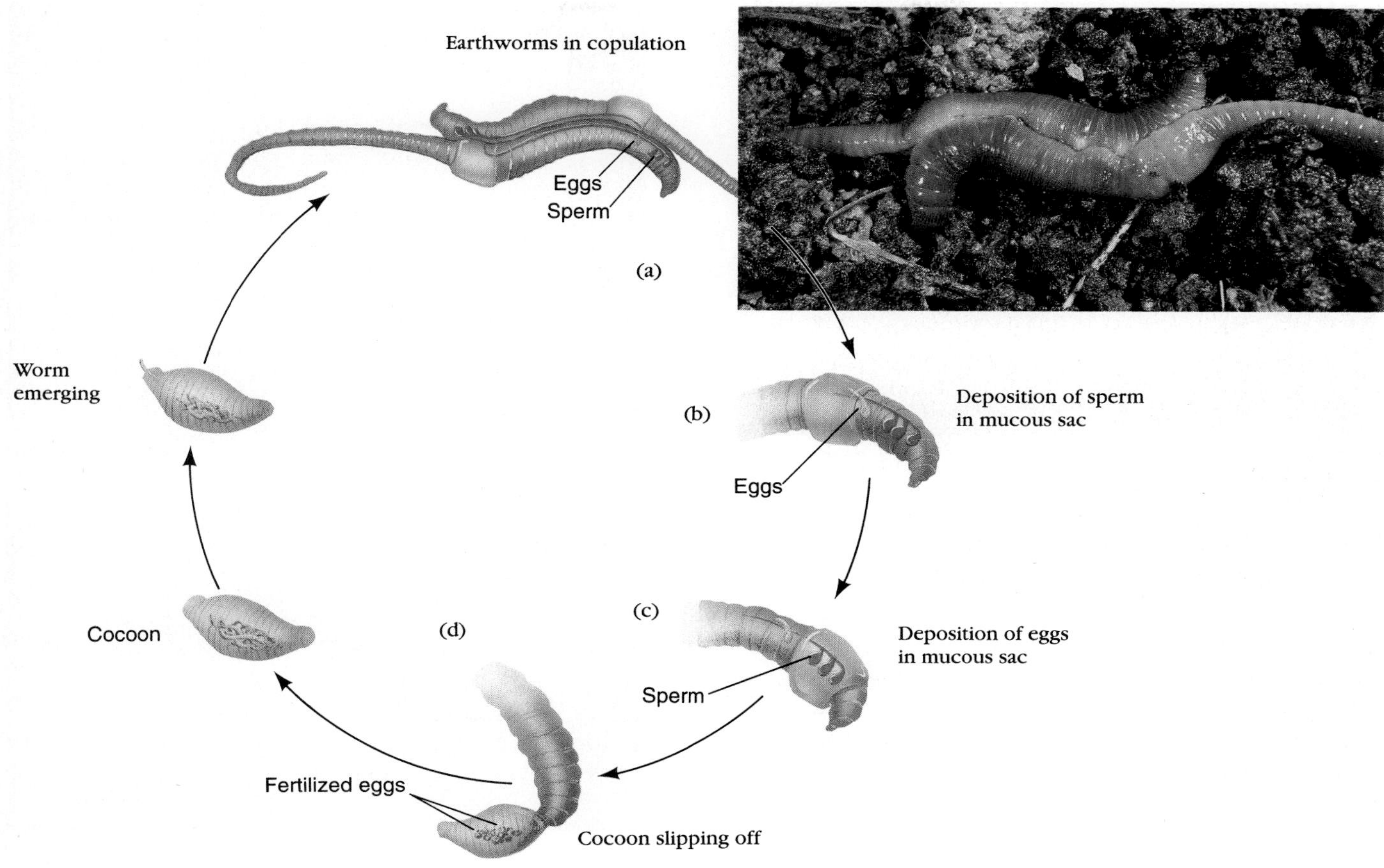

FIGURE 16–6

Earthworm reproduction. Earthworms are hermaphroditic, and each individual has both testes and ovaries. **(a)** During copulation, they exchange sperm and each stores the sperm of its partner. **(b)** Later, a mucous sleeve is produced, which begins to pass down the body of the worm. The mucous sleeve picks up eggs as it is pushed past the ovaries. **(c)** As it continues to move, stored sperm are added and the eggs are fertilized. **(d)** Finally, the mucous sac containing fertilized eggs rolls off the end of the worm and forms a cocoon where the eggs develop into worms. (Runk & Schoenberger/Grant Heilman)

the female and her young. In addition, males may compete for control over valued resources other than territories, because possession makes them irresistibly appealing to females.

The different sexual roles of male and female may be reflected in different appearances, a phenomenon called **sexual dimorphism.** For example, female birds are more likely than males to have drab colors (Figure 16–7). Because females are vulnerable as they sit on their eggs, it is advantageous for them to be inconspicuous. This camouflage works. When the male is the more conspicuous sex, mortality is invariably higher among males than among females. When defending a territory, a male may flaunt vivid coloration or may make unusual, exaggerated postures that render him more visible not only to other males who might think of invading, but also to females, who may notice what a nice territory he has.

An interesting variation on males' use of color to advertise their valuable property to females is found among some bowerbirds and weaverbirds. The male African Village Weaverbird, for instance, is dull colored, but he builds a colorful nest and jumps up and down beside it saying, in effect, not "look at me" but "look at the gorgeous nest I have built for you." If no females are attracted, or if the color of the nest starts to fade, he will tear it to pieces and build another one.

Another type of sexual dimorphism is the possession of weapons by the male but not the female. Features such as large antlers or horns in many hoofed

FIGURE 16-7

Sexual dimorphism. Male and female Siamese fighting fish are dramatically different in appearance. This red male uses his color to attract a mate, like this dark female below him. (D. Kuhn)

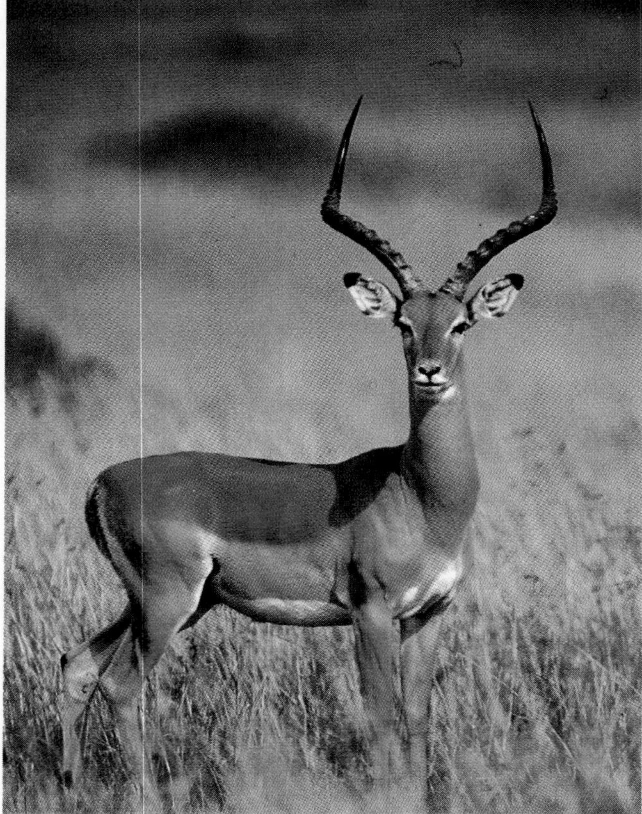

FIGURE 16-8

A male destined for death or glory. The horns of this male impala are used in fights with other males for the right to mate, as well as to identify him to females. Horns take a lot of energy to produce and are heavy to carry. (F. Lanting/ Minden Pictures)

animals, long tusks in boars, and the enormous size of male seals give a male an edge in combat against other males for mates or breeding territories (Figure 16-8). Hence there is selective pressure for males, but not for females, to possess these traits.

Mating Systems

Many species of animals have a characteristic courtship behavior. One of the functions of this behavior is to ensure that prospective mates recognize each other as members of the same species, so that a female does not attempt to copulate with a member of the wrong species. In polygamous species, those in which each animal may mate with more than one partner, males attempt to copulate with little discrimination. They will court almost anything vaguely appropriate, and females must recognize and pick out the right male. The male's appearance, physique, and courtship behavior assist in this. As a corollary, in monogamous animals, which mate with a single partner, courtship is mutual and equally selective, and the sexes are often indistinguishable in behavior and appearance (Figure 16-9).

Polygamy may be divided into polyandry, in which one female mates with more than one male, and the much more common polygyny, in which one male mates with many females.

Polygyny

Polygyny is the mating system in which one male mates with many females. It may evolve in situations where a female gets a better share of some limited resource for her offspring by joining a mated pair, or a male and his harem, than by mating with an unmated male. The resource she gains may be nothing but better genes for her offspring. For instance, male swallowtail butterflies fight for and defend territories at the tops of hills. This is purely a mating area, and any male that can defend it will mate many times. Females fly up the hill, fighting off males who attempt to mate with them on the way. If a female makes it to the top, she will mate with the dominant male there. As a result, her male offspring will inherit good genes for fighting and will have a good chance of fathering the next generation of butterflies.

In a polygynous social system, one or a few males live in a group with a number of females. The dominant male is usually the only one that mates with the

FIGURE 16-9

Fischer's lovebirds. Monogamous animals, such as these lovebirds, are often indistinguishable in behavior and appearance. (B. Everett Webb)

FIGURE 16-10

Rivals. Male bighorn sheep fight for leadership of a herd of females, a prerequisite for reproduction. The energy a male puts into fighting is part of the parental investment in his offspring. (Pat & Tom Leeson/Photo Researchers, Inc.)

females when they are at their most fertile, and he must defend his harem not only against predators but also against other males who try to depose him as head of the harem (Figure 16-10). Defending his dominant position requires a lot of energy and constant vigilance. In many species the dominant male is displaced by a rival several times in a year.

Dominance is worth fighting for because dominant males have an enormous reproductive advantage. For instance, male elephant seals guard harems of females as they haul out of the ocean onto rocky coasts to bear their pups and then to mate. In one study, 4% of the males were responsible for 88% of copulations observed in such a polygynous group.

All other factors being equal, polygyny is a more favorable system than monogamy for the male of a species (or at least for the few males who succeed in reproducing). However, polygyny is possible only where the female does not need the male's full-time help in bringing up the young. Its presence or absence, therefore, usually depends on the advantages or disadvantages to the female.

Polyandry

Polyandry, the mating of one female with more than one male, is much less common than polygyny, but there are at least five examples of polyandry in different groups of birds. Consider the case of the jacana, a long-legged bird that runs around on the waterlily pads covering some lakes in Central America. These birds defend small territories in which females lay eggs and then abandon them to their mates to incubate and raise. A female may mate with, and lay eggs for, several males in one breeding season.

There has been much debate as to how polyandry evolves. The most widely accepted theory is that it evolves from a monogamous situation, common in birds, in which both parents share equally in nest-building, incubation, and parental care. If it became advantageous for only one parent to be at the nest at any one time, pure chance might determine which parent leaves and which stays.

Polyandry also evolves in environments where eggs and young are frequently destroyed. This is the case with some birds that breed at the edges of rivers and streams where flooding often destroys the nest. The female's energy is more usefully devoted to laying replacement eggs than to incubating existing clutches. The male incubates the eggs and the female can devote most of her energy to producing a new clutch of eggs. These will be fertilized and incubated by the

BIO-BIT

Human males release approximately 200 to 300 million sperm in each ejaculate to assure the fertilization of a single egg by a single sperm.

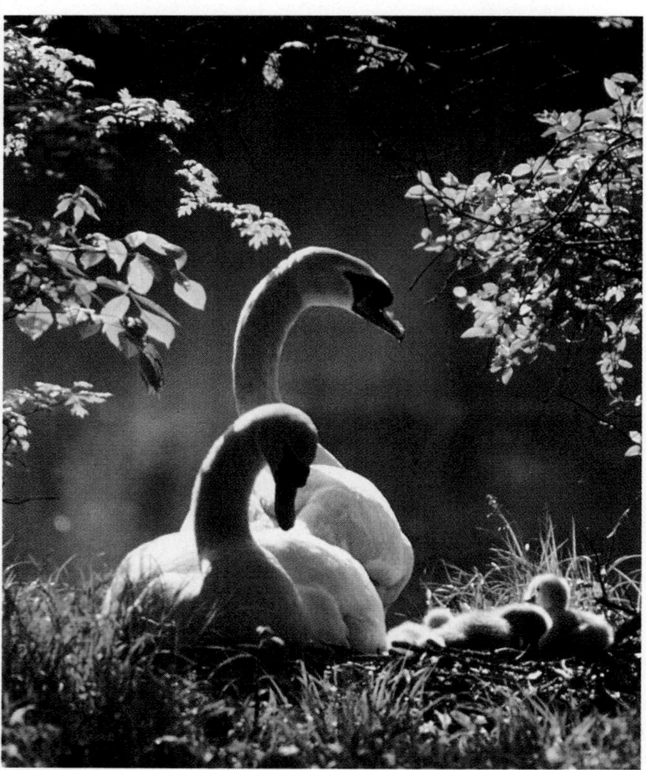

FIGURE 16–11

Monogamy. Like most other birds, swans mate for life, and both parents care for their young. (E. Robinson/Oxford Scientific Films)

original male if the original clutch has been destroyed, or by another male if it has not. In this case, the female also benefits from having her eggs in several different nests, because this increases the chance that at least one clutch will survive. Females compete for males, which are generally scarce because many of them are busy incubating eggs at any one time.

Monogamy

2? When the combined energy of both parents is needed to raise the young, there will be selection for males with **monogamous** behavior (Figure 16–11). For example, monogamy is the most common form of human sexual system (although polygamy arises quite frequently). We can infer that monogamy predominates because the human infant is so demanding to raise. Humans mature more slowly than most other animals, and it seems likely that, throughout much of human history, both parents have had to do their share if they were to raise offspring to sexual maturity with consistent success. It is in the man's interest as much as the woman's that his offspring reach maturity, and so the man, too, will be better off monogamous unless he can provide for the children of more than one wife. ■

The slow maturation of humans is related to the large size of the human brain and the long time needed to educate it. The large body of culture in any human society is a major factor in the survival and reproduction of its members. However, this culture is not transmitted genetically but must be passed down by example and by language. A particular aspect of culture, unique to humans, that probably reinforced the tendency toward monogamy in early humans is the possession of material goods, which are of vast importance to human evolutionary success. It would be strongly advantageous to hand down things like clothes, a cave or house, or land to your genetic children, and the majority of men throughout history have not amassed enough to split up among the children of more than one wife. Whether humans are monogamous or polygamous, the worldwide human population is increasing at an unprecedented rate (see *A Journey into the Environment:* Overpopulation and the Human Animal, this chapter).

Ecology and Mating Systems

An important factor in determining the mating system of a species is its **ecology,** the relationship of the organism to its environment. Factors such as the distribution of food, water, nesting sites, and shelter in the environment affect the distribution and social behavior of individuals.

Where food is scarce and found in small, isolated pockets, individuals tend to be solitary and come together only for a short time, in the breeding season. (Bears, badgers, and moose behave like this.) Couples may come together only to mate, or they may form short-lived pairs or colonies while they raise the young.

Alternatively, some animals may live as monogamous pairs defending a territory with widely scattered resources. In this way they can help each other hunt or watch for predators, and they need not spend time and energy searching for a mate when the time comes. Many wild members of the dog family fall into this group, as do various other mammals.

On the other hand, most monkeys and apes can live in troops because their diet is mainly plentiful plant food. The all-important (in evolutionary terms) females and young can eat well and still enjoy the protection of living in groups. As we would expect from this, in situations where food is occasionally inadequate, groups containing only one male are the rule and there is considerable competition among males to

FIGURE 16–12

Members of a baboon troop. The troop is a social group of related individuals that feed and travel together. (E. R. Degginger)

enter a troop, because this is the only way they can breed.

An interesting illustration of this point came from a study of baboon troops in Africa (Figure 16–12). Most troops live on the open plain and travel from one feeding ground to another. They are not limited by food and water, but they are exposed to many predators, and females must stay close to their young to protect them. Males protect the troop as a whole, a role that makes extra males welcome. The troop can afford to let them eat some of the food, and the extra males also mate occasionally.

In contrast, one troop lived in a woodland area that contained food and water all year. This troop treated their woodlot as a territory, and *both* females and males defended its boundaries. Individual members came and went from year to year, but the size of the troop remained the same, and there were fewer males than in any other troop studied. This is almost certainly because the territory provided food for a limited number of individuals, so that nonreproductive males were regularly driven out of the troop.

Because males have more but smaller gametes than females and are usually less involved in rearing young, it is to their advantage to mate with as many females as possible. But because females invest more time and energy per offspring, they must be more selective, typically seeking mates who can provide resources that will help to raise the young. Sexual dimorphism evolves in response to different selective pressures on the sexes because of their different roles: males attract a mate or defend a reproductive territory, whereas females produce and nourish the young.

The sexual system of a species reflects its ecology, because the location of resources also determines where the organisms spend most of their time and how they behave. Polygyny gives dominant males greater reproductive potential than does monogamy, but it can evolve only when the female can raise the young alone. Polyandry is much less common, and requires that the male raise the young. Monogamy exists when the combined energy of both parents is necessary to raise the young. Human monogamy reflects the energy it takes to raise and educate children, as well as the selective advantages of passing on cultural and material legacies.

16–E SELFISHNESS AND ALTRUISM

We have seen that a species' ecology and evolutionary history are often reflected in its mating system. Mating systems have evolutionary effects of their own. For one thing, the mating system has a profound effect upon how closely members of a species are related to one another. In a polygynous herd of horses where only one male mates, all of one year's offspring will be more closely related than they would be in a monogamous group where the year's offspring have different fathers. The degree to which members of a species are related can have some unusual evolutionary results.

Altruism is self-sacrifice, and an altruistic individual helps others but decreases his or her own chances of survival or success. Altruistic behavior favors the reproductive success not of the altruistic individual, but of another member of the species. (As far as we know, other animals do not consciously think of their behavior as altruistic, but they may act in ways that we would describe by this term.)

3? We can understand altruism by thinking of selection as acting, not on individuals, but on genes. An allele for altruism may spread through a population at the expense of a particular individual that carries it. For instance, an allele that favors altruistic behavior toward close relatives enhances its own survival in the population because there is a good chance that closely related individuals also contain the allele. To take an extreme example, if an individual dies to save ten close relatives, one copy of the "kin-altruism" allele is lost, but ten or more other copies are saved. ■

This kind of indirect selection, for genes present in related individuals, is known as **kin selection.** Kin selection is the basis of parental care (which is a dramatic example of altruistic behavior). Suckling her young costs a female mammal energy and benefits her nothing directly, but she is enhancing the survival of her own genes, which are carried by her offspring.

A Journey into The Environment

Overpopulation and the Human Animal

Ecologists call the ability of the environment to support a population of any species its **carrying capacity.** They recognize, for example, that the resources of a forest are not infinite; it can provide food for only so many white-tailed deer, only so many cottontail rabbits, only so many bobcats, only so many great horned owls.

When the population of any of these species produces so many young that it overshoots the carrying capacity of the forest, several things can happen: (1) the surplus animals can emigrate to adjoining forests where resources may be more plentiful; (2) the surplus animals can stay in the forest and continue consuming until they are eliminated by shortages of resources (food, water, shelter or nesting sites), by an influx of predators, by social disruptions, or by disease and parasites. All of these are agents of natural selection, and when one of them causes the death of an individual we say that the environment has *selected against* that excess animal in the population.

If we were to graph the fluctuations in population numbers of any of the species in our forest example, we might find that the population fluctuates around a certain level and that from year to year there are relatively small changes in population numbers. Dramatic population booms and crashes are normal in some species, but aren't the general rule. In general, population numbers tend to be fairly constant, and populations are kept from growing at a maximum rate by **environmental resistance:** any factor in the environment that limits population growth.

But what about humans? How is the growth of the human population different from that of other species? Let's consider how human activities have affected the selective factors that control the lives and numbers of other animals.

Social insects construct paper and wax nests, polar bears dig ice caves for hibernation, penguins huddle together to socially thermoregulate, but no other animal insulates itself quite as much as humans do from fluctuations in weather and climate. Clothing, housing, air conditioning, heating, dams, and levees all protect humans from the accidents of weather that are some of the most common selective factors. Leaf-cutter ants feed exclusively on fungi cultivated in underground "ant farms," specialist browsing by various species of African antelopes encourages new growth of grass shoots, but no animal gardener can compare with the technological expertise in agriculture that has enabled humans to extend the "normal" carrying capacity of the Earth's farms and gardens. There are folk tales of foxes encasing a broken leg within a thick layer of mud that forms a crude sort of cast, and every cat owner knows that felines eat grass to purge themselves of parasites or a bad meal, but advances in medicine, hygiene, water treatment, and sanitation have further allowed the human population to overcome many of the accidental injuries, infectious diseases, and parasites that decimate populations of other species. In addition, for all practical purposes, we humans have eliminated most of the large predators that evolved along with us. In summary, humans are unusual animals in that we have some control of selective factors in our external environment (Figure 16–A). This is not to imply that humans are immune to forces of natural selection. Selection still acts on the human population, but it is mainly through diseases caused by microscopic parasites: viruses, bacteria, and protozoans.

As you would expect of any animal population that is freed from the selective factors of weather, food shortages, many diseases, and predation, year by year the human population is rising, experiencing the exponential growth of a **population explosion.** How will we know when there are too many humans? How many people can the resources of the Earth sustain? This is a highly controversial topic, with no sure answers, but ecologists use three criteria to define overpopulation for any animal species: (1) overcrowding; (2) a high rate of premature death; and (3) an environment that is so degraded that its ability to support the population has deteriorated. All these criteria are observed in animals such as lemmings and white-tail deer that experience periodic population explosions. In relationship to humans, we do know that some areas of the Earth are overcrowded, and that social disruptions and tensions flourish in these places. We do know that in some areas of the world, particularly sub-Saharan Africa, and much of India, life expectancy is shorter than in places where food, medicines, and adequate sanitation are available. Perhaps most important, though, it is possible that the wide range of environmental crises that confront us (ozone depletion, global warming, air and water pollution, loss

(a)

(b)

FIGURE 16-A

With technology, humans may have removed most of the factors that limit population size in other species, but have we created a whole host of new selective factors for ourselves in the process? Both this herd of wild horses **(a)** and this "herd" of Chicago commuters **(b)** are subject to environmental resistance, though of vastly different kinds. (a, Takeshi Kawamoto; b, Robert Frerck)

of biodiversity, loss of topsoil) are warning signals that the human population has reached the upper limit of the Earth's carrying capacity for the demands of our species. Our current environmental crises may be equivalent to environmental resistance in the growth of populations of other animal species.

Unfortunately we will not know the answers to our questions concerning the Earth's ability to support more than 6 billion people until it is too late. Unlike other animals, the rate at which humans reproduce is governed by more than selective factors. Social mores, religious beliefs, emotions, traditional values, and economics all influence whether an individual becomes a parent. All of our questions about the ability of the Earth to sustain an exploding human population probably will be answered by the end of the next century. We can only hope that we will have time to adjust the social, economic, and belief systems that influence human reproductive rate before the "excess" human population (and many of us along with it) is cruelly eliminated by natural selection.

The more closely individuals are related, the more likely altruism is to evolve between them because the more likely it is that they share some of the same genes. The relatedness between a parent and its offspring, for instance, is ½, because half the parent's genetic material is inherited by its offspring. This figure, describing how closely two individuals are related, is called an **index of relatedness.**

Indices of relatedness can be calculated for other relatives. In a monogamous species, the chance that a brother or sister also has a particular gene of yours is ½ (50%). The chance that a grandparent, uncle, aunt, nephew, or niece also has a particular allele of yours is ¼. The index of relatedness with a first cousin or great grandchild is ⅛, with a second cousin 1/32, and with an identical twin 1. Third cousins are much less special: the index of relatedness is only 1/128, probably not much greater than the chances of sharing the allele with any other individual in the whole population.

We would therefore expect the degree of altruistic behavior to tail off toward relatives who are less close. An allele that prompted an individual to lay down his or her life for a relative would have to save more than two brothers, sisters, or children, more than four uncles and aunts, or more than eight first cousins, and so on, to be selectively advantageous. Otherwise it would not survive, on the average, in enough other bodies to compensate for its loss in the altruistic individual.

The altruism of parental behavior was a prerequisite to the evolution of animals such as birds and mammals, who produce young that do not survive unless their parents invest a lot of energy in caring for, feeding, and protecting them. There is also strong selection for such behavior, because individuals who do not help raise young do not contribute genes to future generations.

However, some biologists question whether other altruistic behaviors truly exist. Many cases of apparent altruism, when examined more closely, turn out to benefit the "altruistic" individual at a later time.

The Social Insects

Biologists have long been intrigued by the specialized division of labor in an insect colony and by the degree of cooperation between individuals. Most members of a colony are sterile workers, who will never leave any offspring. Instead, the workers devote their lives to raising the offspring of other individuals. This is altruism on a grand scale.

Colonies of wasps, ants, termites, and social bees, especially honeybees, are impressive societies. Food is shared communally, and information is exchanged in elaborate ways, including a great range of chemical signals and the "dances" by which worker honeybees indicate the direction and distance of a food source to their nestmates. Workers are fearless in the defense of a colony; many sacrifice their lives on its behalf.

A colony of social insects is a huge family, sometimes numbering several million, all descended from the same mother—the reproductive female, or **queen.** Among the ants, bees, and wasps (all in the insect order Hymenoptera), **workers** are infertile females (Figure 16–13). Termite workers also include infertile males. How do the workers gain an advantage great enough for such altruistic, even suicidal, individuals to have evolved?

In ants, bees, and wasps, the answer lies in a curious aspect of sex determination. The single mature queen in the colony receives enough sperm during her mating flight to last for the rest of her life. She uses these sperm to fertilize eggs as she lays them, producing female offspring, most of which develop as workers (a few become new queens). But not all of the queen's eggs are fertilized. The unfertilized ones develop as males, which thus have no father. All they have is a single (haploid) set of chromosomes derived from their mother. Thus a male's sperm cells all contain the same haploid set of chromosomes, and his degree of relatedness to his mother is 100%. His mother, on the other hand, is diploid and her son receives only 50% of her genes so, strange as it may seem, her relatedness to him is only 50%.

A colony contains workers who are full sisters, sharing the same (haploid) father (whose sperm are genetically identical), so any allele a worker obtained from her father is also present in each of her sisters. However, there is only a 50% chance that any allele a worker obtained from her mother will also be present in any one sister. In other words, the relatedness between sisters averages 75% (not 50% as in normal sexual animals).

This remarkable situation means that a worker bee is related more closely to her sisters than to her mother! She can therefore increase her evolutionary success more by farming her mother as a sister-making machine than she can by reproducing. More accurately, an allele that promotes making sisters will replicate more rapidly than will an allele for making offspring. This selects for sterility of workers, which has

> **BIO-BIT**
>
> Some army ant queens will produce about 6 million eggs in a typical 6-year life span.

(a)

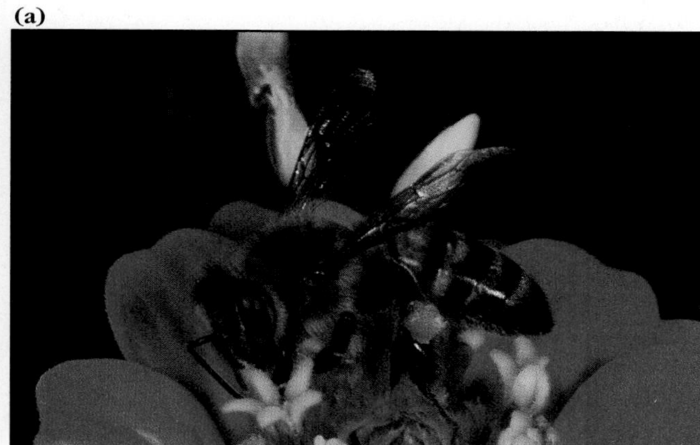

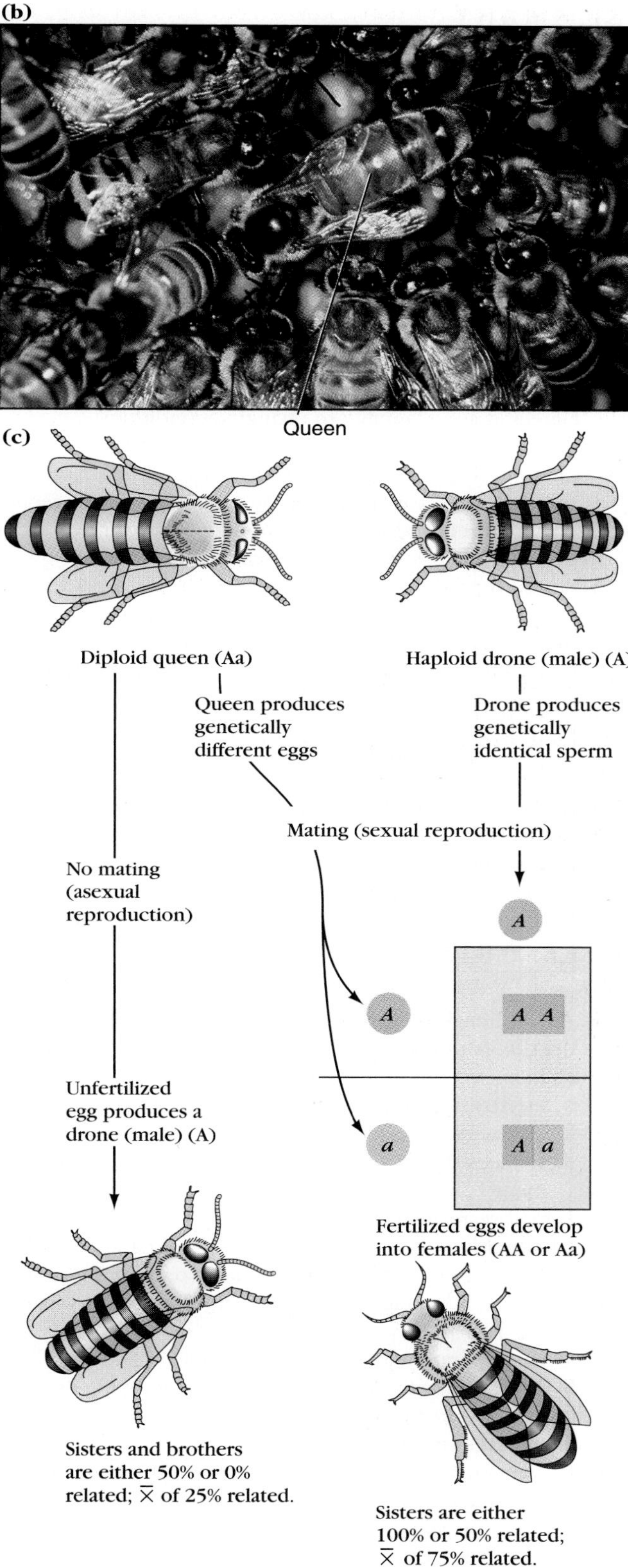

FIGURE 16–13

The genetic relationships of bees in a hive. (a) A worker foraging for nectar to contribute to the hive's communal honeycombs. **(b)** A queen (largest bee in center) surrounded by workers, the daughters who raise her offspring, their sisters, and their brothers. **(c)** Each sister receives half of her genes from the male and half from the queen. Because the male contributes identical sperm and the queen produces two genetically different eggs, sisters are either 100% or 50% related, depending upon whether or not they received the same type of egg from the queen. Thus, there is an average 75% genetic similarity between sister worker bees, while drones are usually only 50% related to their sisters. (a, Ed Kanze; b, John Cancalosi/Peter Arnold, Inc.)

evolved independently at least 11 times in the Hymenoptera and once elsewhere (the termites), an extraordinary example of the complex effects that can be produced by a particular sexual system.

A species' mating system determines the degree of relatedness among individuals. This, in turn, determines the extent of selection for altruistic behavior. In many of the social insects, the unusually close genetic relationship of the colony selects for sterile female workers who support the production of their closest relatives—sisters.

SUMMARY

1. All species have a certain amount of genetic variation among their members as a result of mutation. The amount of such variation is slight, and members of a species that reproduces asexually are genetically very similar to one another.
2. Members of sexual species are much more variable than those of asexual species because genes are reshuffled during meiosis and form new combinations at fertilization.
3. Asexual reproduction uses energy more efficiently than sexual reproduction. It is common in species throughout the world that produce many offspring in a short time and are adaptable to changing conditions.
4. For more localized and specialized species, the great amount of energy expended in sexual reproduction is worthwhile because, usually, at least some of the genetically different individuals so produced can survive and evolve in deteriorating environmental conditions.
5. Sexual reproduction probably originated as the accidental fusion of two asexually reproducing cells. As small, motile sperm developed that could swim to an egg, larger immobile eggs evolved with larger food supplies for the embryo.
6. Adaptations such as specific mating seasons and hermaphroditism help to overcome the difficulties encountered when two individuals must act in concert to achieve reproduction.
7. The sexual system of a species is determined by the amount of energy each sex puts into producing and rearing offspring, and by ecological factors such as the distribution of food and the prevalence of predators. Because it takes more energy to produce eggs than sperm, females are usually more selective than males in their choice of mate.
8. Sexual dimorphism may arise when the selective pressures on the two sexes are different, or when males and females have different roles in reproduction. Sexual dimorphism tends to be more pronounced in polygamous species.
9. In monogamous species, the members of both sexes must choose their mates with care. Overall, the role and behavior of monogamous males and females are more alike than they are in polygamous males and females.
10. A species' sexual system determines the degree of relatedness among members of a population. An allele that causes an individual to perform altruistic behaviors detrimental to itself can be selected for when the behavior enhances the survival of other copies of the allele in the individual's relatives, allowing that allele to spread through the population.
11. Altruistic behavior is most likely to arise in species in which closely related individuals spend much time together. Parental behavior is the best example.
12. Social insects typically consist of closely related populations of many infertile workers who support the production of their close relatives.

SELF-QUIZ

1. An advantage of sexual reproduction over asexual reproduction is that it:
   a. increases the mutation rate
   b. increases genetic variability in a population
   c. produces larger offspring
   d. reduces the offsprings' risk of death during development
2. Some organisms reproduce asexually when environmental conditions are (favorable, unfavorable) and sexually when conditions are (favorable, unfavorable) for growth.
3. The main advantage of having a large egg and a small sperm is that it:
   a. has separate male and female sexes
   b. provides for the nourishment of the growing embryo
   c. assures cross-fertilization
   d. involves two immobile gametes
4. It is generally true that males increase their chances of evolutionary success by trying to mate with as many females as possible. One possible exception to this generalization might occur when:
   a. there are many more females than males
   b. there are about equal numbers of males and females
   c. the father's care is required to raise the offspring
   d. there are many predators
   e. the male holds a territory against other males
5. When food is distributed in such a way that an animal must spend a large part of its day traveling from one place to another to find enough to eat, what type of mating system would you expect it to have?
   a. monogamy
   b. polyandry
   c. polygamy
   d. polygyny
6. A childless adult human male is most likely to enhance his evolutionary success by altruistic behavior toward:
   a. the children of his sister(s)
   b. the children of his brother(s)
   c. his sister(s) and his brother(s)
   d. the children of his niece(s)
7. Monogamy is most likely to be found among birds and mammals whose young are:
   a. born or hatched helpless and in need of much parental care
   b. born or hatched precocial (able to care for themselves, like a horse)
   c. nourished on milk
   d. part of a litter or clutch of more than 50
   e. carried by the female from conception to birth

THINKING CRITICALLY

1. List as many possible selective advantages as you can think of for courtship rituals.
2. Do humans have courtship rituals? Is courtship in humans mutual or is it carried out predominantly by one sex? Why?
3. Female walruses must bear their young on land, but suitable stretches of beach are scarce. Similarly, seaside nest sites for gulls are scarce. Both have crowded breeding grounds. Walruses eat mussels, clams, and so forth, whereas gulls will eat almost anything—including the egg or chick next door. Explain what mating system you would expect each of these animals to show.
4. What selective pressure might have led to evolution of species with separate sexes? Why are there never more than two different sexes in a species?
5. Some people do not reciprocate a favor when it is their turn. What might be some possible outcomes if such a nonaltruistic allele were present in a population together with an altruistic allele?
6. Explain sibling rivalry for parental favors.
7. Do you think that altruistic behavior in humans is genetically controlled?
8. Researchers Trivers and Hare weighed the fertile males and females produced in colonies of 20 species of ants and found that the investment in females was three times the investment in males (by weight). Can you explain how this situation may have been selected for?
9. How might the invention of birth-control methods and labor-saving household appliances affect the monogamous mating system of humans?
10. Martin Daly and Margo Wilson analyzed statistics on homicides within human families in the United States, Canada, and elsewhere. Can you explain the genetic or evolutionary theory that might account for the following examples of their findings?
    a. In nonindustrial societies, infants are most often killed by parents for the following reasons: (1) doubt that the child is the parent's own; (2) conviction that the child is weak and unlikely to produce offspring as an adult; (3) external pressures such as food scarcity and burdensome demands from older siblings that reduce the baby's chance of surviving.
    b. In industrial societies, parents are more likely to kill infants than to kill older children.
    c. Mothers are more likely than fathers to kill infants.
    d. Disproportionate numbers of child killings are by stepparents.

SELECTED KEY TERMS

altruism, *p. 333*
asexual reproduction, *p. 324*
clone, *p. 324*
ecology, *p. 332*
hermaphroditism, *p. 328*
index of relatedness, *p. 336*
kin selection, *p. 333*
monogamy, *p. 332*
parthenogenetic reproduction, *p. 324*
polyandry, *p. 331*
polygyny, *p. 330*
queen, *p. 336*
sexual dimorphism, *p. 329*
workers, *p. 336*
zoospores, *p. 326*
zygote, *p. 326*

SUGGESTED READINGS

Borgia, G. "Sexual selection in bowerbirds." *Scientific American,* June 1986. A study of female choice of mates in a species that attracts females by building elaborate nests.

Cole, C. J. "Unisexual lizards." *Scientific American,* January 1984. An interesting study of parthenogenetic vertebrates.

Dawkins, R. *The Selfish Gene.* New York: Oxford University Press, 1976. Natural selection from the gene's point of view; good discussion of altruism.

Eberhard, W. G. "Animal genitalia and female choice." *American Scientist* 78, 1990.

Grandjean, E. C. "Love among the icebergs." *The Living Bird Quarterly,* Summer, 1986. A discussion of the ecological and evolutionary basis of polyandry.

Holldobler, B., and E. O. Wilson. *The Ants.* Cambridge, MA: Harvard University Press, 1990. A landmark account of all aspects of ants from around the world.

Ryan, M. M. "Signals, species, and sexual selection." *American Scientist* 78, 1990.

Seeley, T. D. "The honey bee colony as a superorganism." *American Scientist* 77, 1989.

Sherman, P. W., J. U. M. Jarvis, and S. H. Braude. "Naked mole rats." *Scientific American,* August 1992. A mammal with a social system similar to that of many insect societies.

Wilson, E. O. *Sociobiology: The New Synthesis.* Cambridge, MA: Harvard University Press, 1975. After years of work on social insects, Wilson produced this large book on social behavior and its genetic basis. The book caused a furor because it treats human behavior as just another example of the social behavior of animals. Fascinating, but not very easy reading.

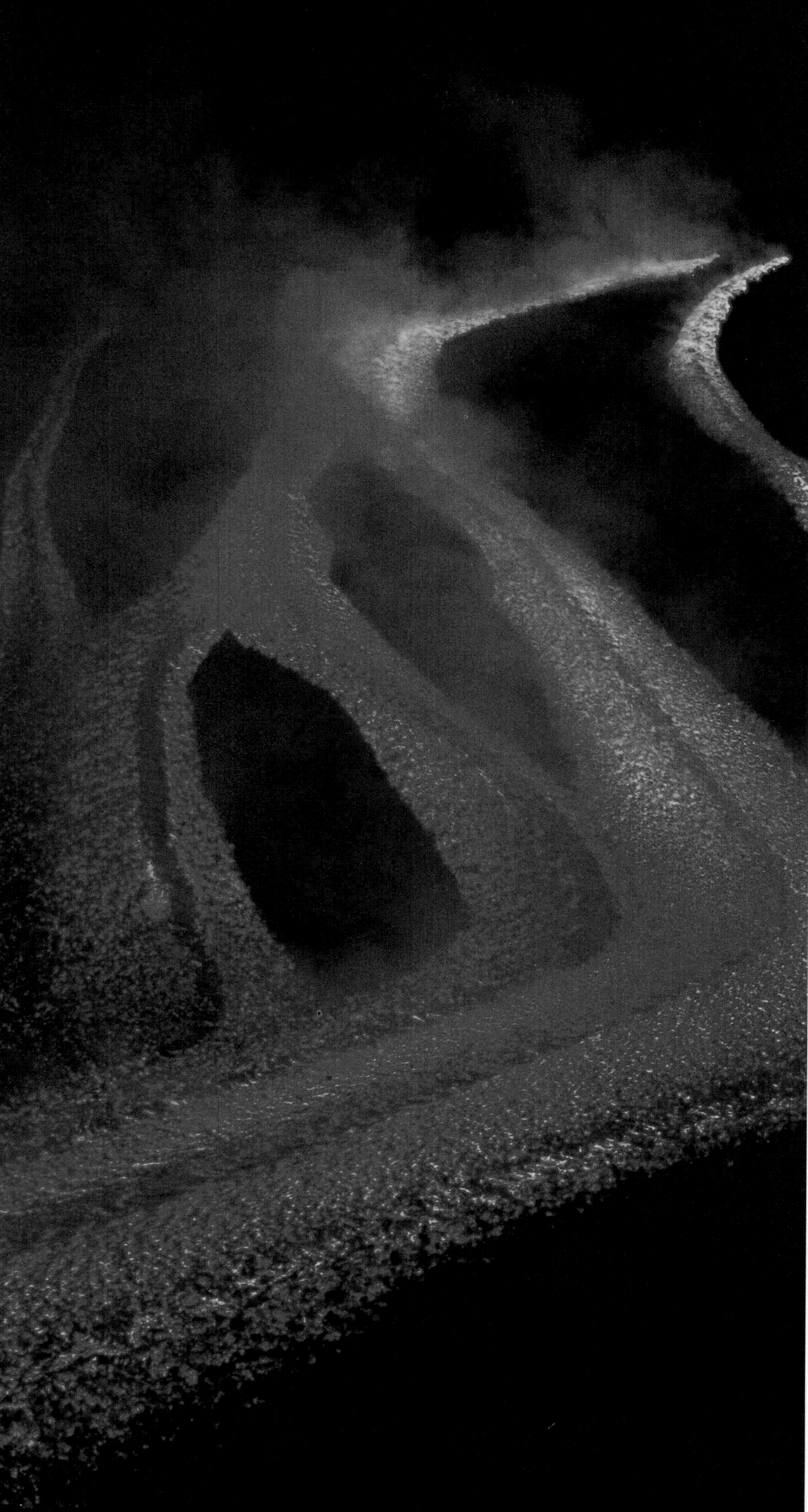

CHAPTER 17

CONCEPT GUIDE

After reading this chapter you should be able to:

1. Explain why it was crucial for the first life forms to evolve in the absence of free oxygen. (Section 17-A)
2. Describe the atmospheric conditions of the prebiotic Earth. (Section 17-B)
3. Explain what astronomy, geology, and experiments like Stanley Miller's indicate about the formation of life on the prebiotic Earth (Section 17-C)
4. Describe the molecules that could have catalyzed early organic reactions. (Section 17-D)
5. Explain how cell membranes might have first formed and functioned. (Section 17-E)
6. Explain how metabolic pathways evolved in early aggregate systems. (Section 17-F)
7. Explain how early transcription and translation were catalyzed. (Section 17-G)
8. Explain why the evolution of autotrophy was important in the evolution of life. (Section 17-H)
9. Explain how the accumulation of atmospheric oxygen affected the evolution of early organisms. (Section 17-I)
10. Explain why eukaryotic cells were able to evolve faster than prokaryotic cells. (Section 17-J)
11. Describe what the fossil evidence tells us about the early history of life on Earth. (Section 17-K)

The primitive Earth was a violent and spectacular place with extensive volcanic activity, lightning, high levels of ultraviolet light, and little free oxygen in the atmosphere. How could life have originated in such a hostile place? Read more about how life could have originated on primitive Earth in Section 17-B. (Ric Ergenbright Photography)

Origin of Life

If we could trace the ancestry of living organisms, we would find a long line of cells stretching back billions of years. Each cell came from division of a previously existing cell . . . but where did the first cell come from?

Some people say the first organisms came to Earth in spaceships or meteorites, but this only moves the question of how life began to a more distant arena, beyond our reach to study. Most scientists think that life on Earth started here. If we assume this is true, we can search our Earth for evidence of how and when life arose (Figure 17–1).

Geologists who study ancient rocks tell us that the early Earth was very different from what we see now—even if we were to subtract all the living things. In fact, conditions on the early Earth would have killed the vast majority of modern organisms almost instantly. The atmosphere was toxic, the heat intense. Yet this poisonous inferno gave rise to life: chance chemical events, over a period of hundreds of millions of years, built the simple compounds of the primordial Earth into organic monomers, then polymers, then aggregates of many polymers. By a few billion years ago, some of these aggregates had become cells. This scenario was first proposed in 1924 by a Russian, Alexander Oparin. In 1929 Oparin's ideas, not yet translated from the Russian, were echoed by J. B. S. Haldane in England.

Research supports these predictions. Scientists have simulated the **prebiotic** ("before life") world, using their best guesses about the temperature and the available chemicals and energy sources. Amazingly, the nonliving systems formed in the laboratory show many features typical of life.

What properties would we seek in a chemical aggregate on its way to becoming a cell? In earlier chapters, we saw the basic features of present-day cells. The forerunners of true cells should have shown at least the rudiments of these:

1. A large number and variety of organic molecules, many of them joined into long, polymer chains.
2. A lipid-protein membrane that separates the cell from its environment, selectively controls the exchange of materials between the two, and maintains a difference in the concentration of molecules and ions between them.
3. Proteins that help to exchange substances with the environment, form cell structures, and most importantly, catalyze the cell's metabolic reactions.
4. Nucleic acids that contain precise instructions for making proteins.

In this chapter, we consider what the prebiotic Earth was like and outline how nonliving chemicals may have become organized into living cells. Many interesting experiments shed light on aspects of the problem and show how various features of life could have evolved. This chapter omits a great deal of evidence that requires advanced knowledge of chemistry and of the metabolism of obscure types of bacteria. Hence the evidence given here looks much scantier than it is. But it is true that the subject matter of this chapter, perhaps more than any other in the book, is highly speculative. Future work will fill more gaps in the picture, but we shall never really know whether or not it is a good likeness of the events that actually took place.

KEY CONCEPTS

- Chemical and physical conditions on the early Earth differed from those here today.
- Random chemical reactions built up supplies of organic chemicals, which assembled into large aggregates. Some of these eventually came to have the organization, function, and reproduction characteristic of life.
- Living organisms themselves are responsible for many of the differences between early and present-day environmental conditions on Earth.

Curiosity Questions

1? Weren't the chances of life first forming on primitive Earth too slight for it to have ever happened? (See page 342)

2? What did Earth's first life forms use for food? (See page 349)

3? How long ago do scientists think life first evolved on Earth? (See page 354)

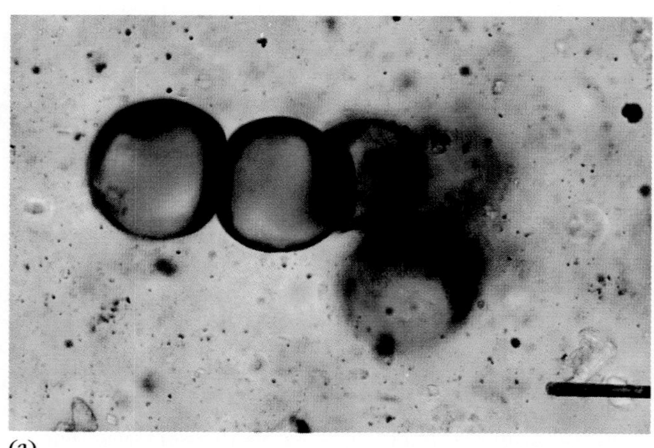
(a)

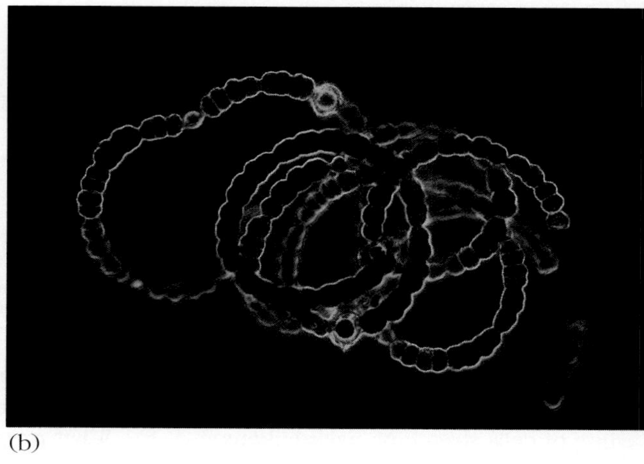
(b)

FIGURE 17-1

Fossil bacteria. (a) These 700 million-year-old cyanobacteria were found by microscopic examination of thin sections of rock. **(b)** Modern blue-green algae have many similar characteristics. (a, Biological Photo Services; b, D. Kuhn)

17-A CONDITIONS FOR THE ORIGIN OF LIFE

Under what conditions can life arise? Scientists believe there are four basic requirements: (1) the right chemicals, including water, various inorganic ions, and organic molecules; (2) an energy source; (3) little or no oxygen gas (O_2); and (4) eons of time. Of the necessary chemicals, water is abundant on Earth, and the inorganic ions occur in rocks, volcanic gases, and the atmosphere. How were organic molecules produced from these simple chemicals without the enzymes of living organisms, which catalyze most of these reactions today? Before we answer this question, let us look at the last two conditions.

Absence of Molecular Oxygen. The origin of life required an atmosphere with little or no oxygen because O_2 is a powerful oxidizing agent. Oxidation would have broken down organic molecules, or at least made them useless to a prebiotic system, at a relatively rapid rate. Organic molecules exposed to O_2 on the early Earth would not have lasted long enough to form more complex structures. This is one reason why we do not find new organisms arising from organic matter on Earth today. (Another is that free organic molecules are usually absorbed and used as food by bacteria and fungi even before oxygen can damage them.) The Earth's atmosphere originally contained much less oxygen than it does now. However, some oxygen may have been present.

1? **Time.** It may take millions of years for a given quantity of some chemical to undergo a reaction that an enzyme could catalyze in a second or two. In the prebiotic era, inorganic chemicals reacted to form organic molecules without the help of enzymes—that is, extremely slowly. Once made, the simple organic chemicals had to come together in larger, more complex structures. The chances of this happening are minuscule. ■

Given enough time, however, even highly improbable events are almost bound to occur. For example, if the probability that an event will not occur in one year is 99.9% or (0.999), the probability that it will not occur in two years is $(0.999)^2$; in three years, $(0.999)^3$, and so on. Table 17-1 shows that as time goes on, the chance of this event not happening becomes less. There is a 1% chance of its happening in a million years and after a hundred million years its probability has increased to 63%. After 1 billion years it is virtually certain that the event will have happened at least once, and once may have been enough for the origin of life on Earth.

The origin of life required a supply of the proper chemicals, an energy source that promoted their reaction, the near-absence of molecular oxygen, and eons of time. All of these conditions existed on the early Earth.

Table 17–1
What's the Probability?

If the probability that an event will occur in 1 year is 1 in 100 million, then the probability that it will not occur is 99.999999%. In 10 years, the percentage has decreased to 99.99999.

Years	Probability (%)	High probability that the event will NOT occur
100	99.9999	↑
1000	99.999	
10,000	99.99	
100,000	99.9	
1,000,000	99.0	
10 million	90.5	
50 million	60.65	
100 million	36.8	↓
500 million	0.7	Low probability that
1 billion	0.005	the event will NOT occur

Notice that 1 million years must pass before there is even a 1% chance of the event occurring, after 10 million years however, there is more than a 10% chance of the event occurring. In other words, as time passes, the probability that the event **will** happen increases:

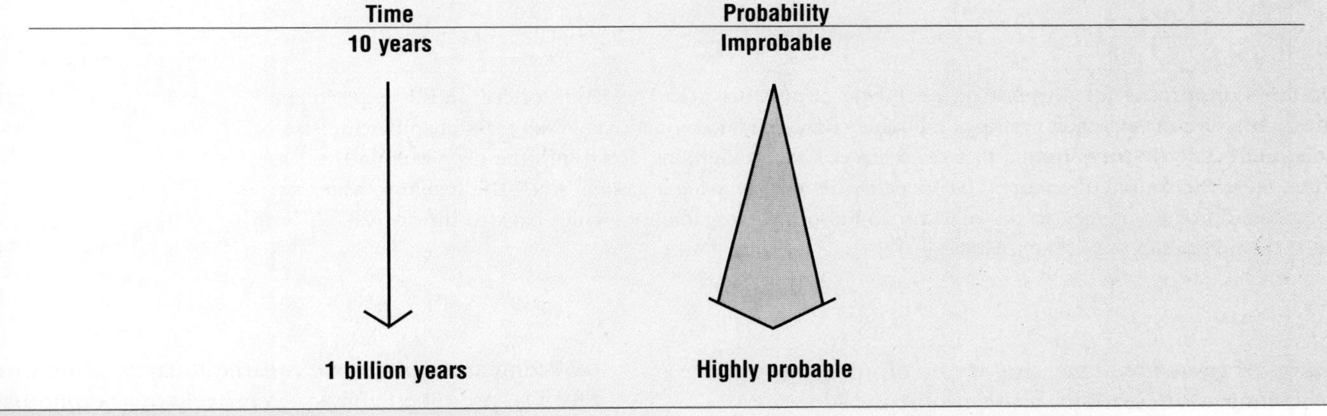

17–B THE PREBIOTIC EARTH

The Earth formed by solidification and accumulation of matter from space about 4.6 billion years ago. Chaos prevailed for the next 0.2 to 0.3 billion years, as bombardment by meteorites added material to the forming planet. About 4.3 billion years ago, conditions began to stabilize.

At this time the early Earth had a **reducing atmosphere,** so called because hydrogen, an element that tends to donate electrons to other substances, reducing them in the process (see Section 6–D), was the early atmosphere's most common element. Scientists believe that this reducing atmosphere consisted of H_2 (molecular hydrogen) and hydrogen compounds of other elements, such as H_2O (water vapor), NH_3 (ammonia), and CH_4 (methane).

This view is based on several lines of evidence. Hydrogen is enormously abundant in the solar system, especially in the gases that formed the sun and planets. Even today the atmospheres of Jupiter and Saturn are mostly H_2, water vapor, and ammonia. The compo-

(a)

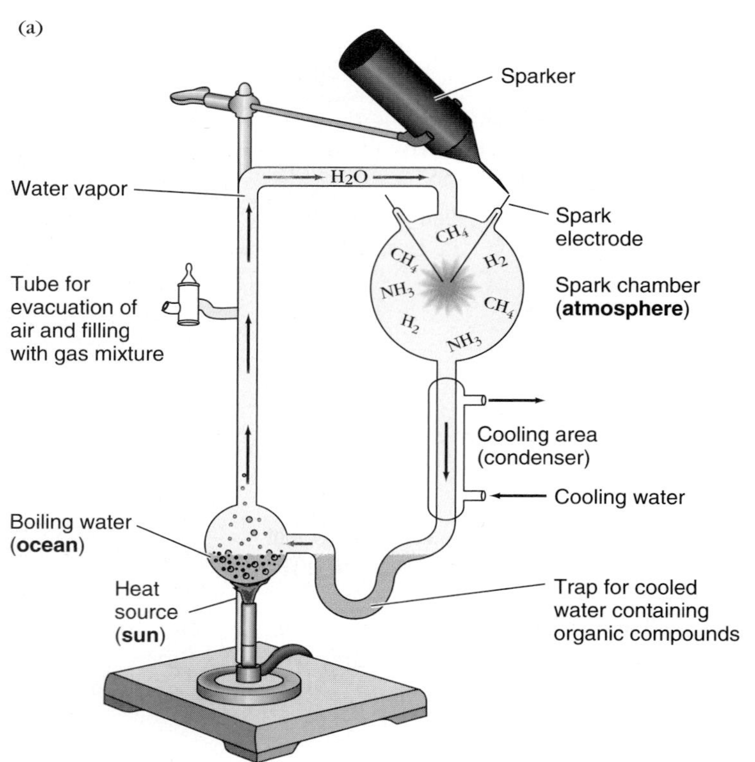

(b)

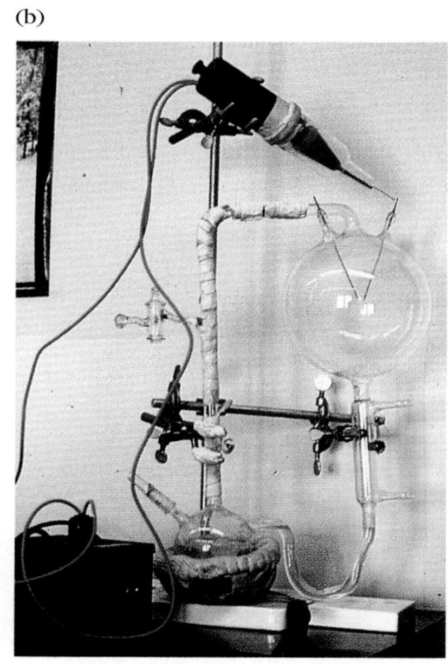

FIGURE 17–2

Miller's apparatus for simulating prebiotic conditions. (a) The "atmosphere" in this experiment was a mixture of hydrogen gas (H_2), methane (CH_4), and ammonia (NH_3) in a glass chamber the size of a soccer ball. **(b)** Sparks from a pair of electrodes seen extending down into the glass bulb in this picture, were the source of energy. The "ocean" was heated using a gas-jet "sun." The resulting water vapor could provide oxygen to the reaction and also carry organic molecules back to the "ocean" in "rainfall." (Photo courtesy of Dr. Stanley Miller)

sition of gases now emerging from volcanoes, as well as chemical calculations, also support the idea of a primitive reducing atmosphere.

However, H_2 is so light that it soon escaped the Earth's gravity and went off into space. Sunlight, much brighter on Earth than on the outer planets, decomposed the ammonia into H_2 (which also escaped) and nitrogen gas (N_2). Similarly, methane was replaced by CO_2.

By the time life was evolving, the atmosphere was only mildly reducing. The most abundant gas was probably water vapor, which later condensed to form the oceans. Next were carbon dioxide and carbon monoxide, which later became locked in rocks in the form of carbonates, where they are today. Also present were nitrogen (which makes up about 80% of the air today), smaller amounts of other gases, and negligible O_2. Sunlight striking water or carbon dioxide released small amounts of O_2, but this oxygen did not last long: it soon reacted with iron or sulfur compounds. The oxidizing atmosphere we breathe today is about one fifth O_2, produced almost exclusively by the photosynthesis of green plants.

When life first evolved, the Earth's atmosphere consisted mostly of hydrogen, water, ammonia, and methane. Atmospheric oxygen did not become abundant until an advanced form of photosynthesis evolved much later.

17–C PRODUCTION OF ORGANIC MONOMERS

In 1953 Stanley Miller, then a graduate student, built a small-scale model of conditions on the early Earth, including an "ocean" and a primitive reducing "atmosphere" (Figure 17–2). Electrodes in the "atmosphere" chamber gave off electric sparks, representing lightning, a possible source of energy to drive chemical re-

Table 17–2
Important Precursors of Organic Compounds Found Beyond the Earth

Molecule	Name
NH_3	Ammonia
$HC{\equiv}N$	Hydrogen cyanide
$N{\equiv}C-C{\equiv}N$	Cyanogen
CH_4	Methane
H_3C-CH_3	Ethane
$H_2C{=}CH_2$	Ethylene
$HC{\equiv}CH$	Acetylene
CO	Carbon monoxide
H_2CO	Formaldehyde
CH_3CH_2OH	Ethanol

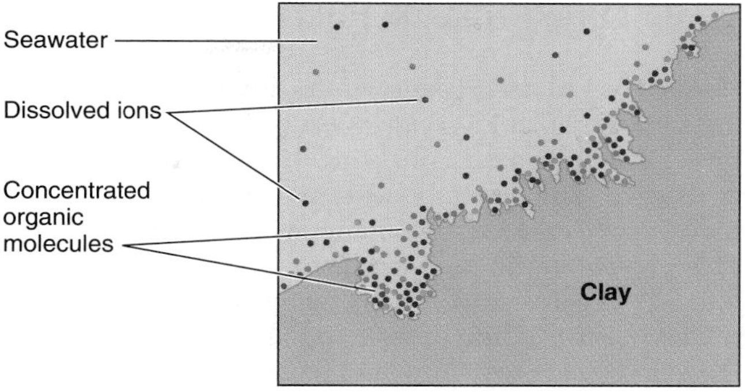

FIGURE 17–3

How clay concentrates organic molecules. Clay and other minerals have irregular, electrically charged surfaces. In this computer model, dissolved ions or polar molecules (colored dots) are concentrated at the interface between the clay (brown) and seawater (blue). These minerals not only attract and hold other molecules, but also act as catalysts for chemical reactions between them.

actions on the early Earth. After a week, the "ocean" contained many small organic compounds, including amino acids, carried from the "atmosphere" by the condensation of "rain."

Miller's amino acids generated much excitement. Amino acids are the building blocks of proteins, and biochemists recognized the enormous variety of proteins and their overwhelming importance in the activities of living cells. Indeed, proteins were regarded as *the* class of substances necessary for life.

Current evidence indicates that the composition of Miller's "atmosphere" wasn't accurate, and more recently many people have done variations of Miller's experiment, using different proportions of starting gases, and sometimes including hydrogen sulfide (H_2S), carbon dioxide, or inorganic ions. Besides electric sparks, other energy sources have been used including heat, bright sunlight, ultraviolet light, and radioactivity, all possible sources of energy on the prebiotic Earth. As long as no O_2 is present, these simulations yield a variety of organic products, including a host of compounds not formed in Miller's mixture, such as nucleic acids.

Astronomy and geology also provide evidence that organic monomers can form without the presence of living organisms. Several important small molecules that are raw materials or intermediates in the synthesis of organic monomers are found in stars, dust clouds, space, and the atmospheres of other planets (Table 17–2). The European Space Agency's Giotto spacecraft, which flew past Comet Halley, even detected polymers of one of these small molecules (formaldehyde). Meteorites, chunks of material that fall from space, contain a wide variety of more complex organic monomers, including amino acids, alcohols, sugars, the nitrogenous bases found in nucleic acids, and lipid-like molecules capable of forming films similar to membranes. Small amounts of six common amino acids were also found in material brought back from the moon. Even now, organic compounds are formed on Earth **abiotically** (without life: neither within living cells nor under their influence). Hot metallic carbides in volcanic gases and lava form hydrocarbons when they react with water, but this occurs at a very low rate.

All these lines of evidence support the contention that organic compounds could have formed on the prebiotic Earth by the action of available forms of energy. Without oxygen to destroy them or organisms to absorb them, these compounds would have accumulated. Haldane suggested that eventually the sea had the composition of a "hot, dilute soup."

Nowadays, many scientists favor a different view: organic molecules formed not a soup, but a sludge—attracted to the complex, charged surfaces of clay or other minerals in the pores of underwater rocks (Figure 17–3). These minerals could serve both as templates holding organic molecules in place, and as catalysts of chemical reactions. Indeed, the chemical industry today uses mineral catalysts to produce many organic compounds, and many enzymes also use mineral ions as cofactors.

The evidence that organic monomers could have formed on the prebiotic Earth comes from laboratory simulations and from observations of these kinds of molecules in extraterrestrial bodies.

17–D FORMATION OF POLYMERS

Sidney Fox, an American scientist, found that heating a dry mixture of amino acids produced **proteinoids,** protein-like polymers consisting of about 100 amino acids each. The heat (about 60°C) quickly evaporated the water released as the amino acids linked together. This prevented the proteinoids from hydrolyzing back into amino acids. Such events might have happened on the early Earth if tide pools trapped seawater containing amino acids, which would have polymerized as the water evaporated on hot, sunny days.

Clay and other minerals also enhance the formation of polymers. Many kinds of monomers adhere to the surfaces of mineral particles. This increases their local concentration. Polymers form readily when the minerals are dried and then warmed.

Some proteinoids exhibit enzyme-like properties. For example, they can catalyze some chemical reactions, and their catalytic properties can be destroyed by overheating and by chemicals that inhibit enzymes. The conclusion must be that molecules similar to enzymes, which are so vital to life, could have been produced on Earth before living organisms existed.

Clay and other minerals could have catalyzed early organic reactions to form proteinoids. Some proteinoids have enzyme-like properties.

(a)

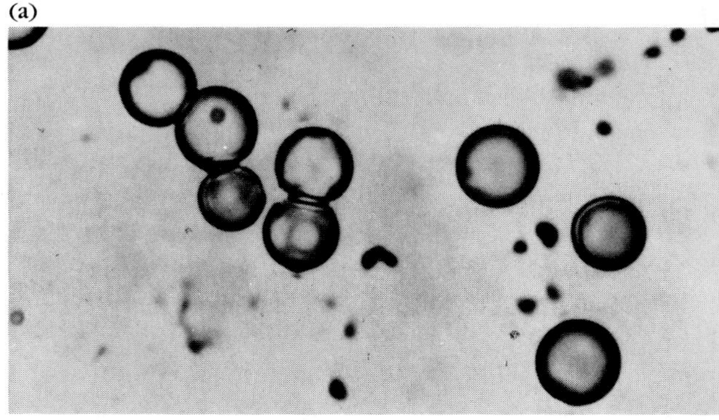

(b)

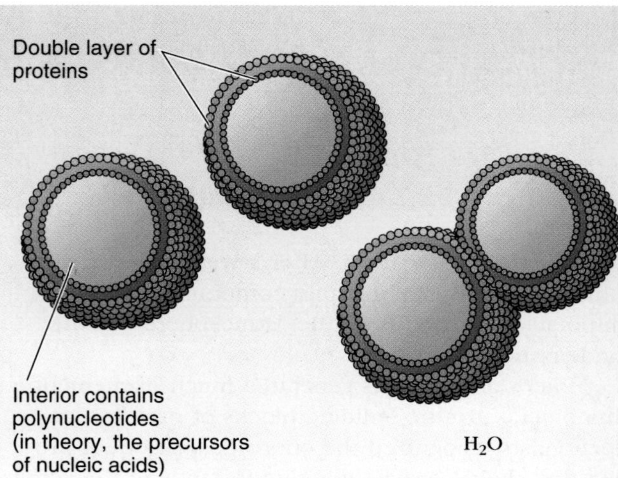

FIGURE 17–4

Proteinoid microspheres. (a) These microspheres formed when water was added to proteinoids, protein-like polymers of about 100 amino acids each. Each is 1 to 2 micrometers in diameter. **(b)** The microspheres have a double layer of proteins that form a semipermeable boundary between the interior and the environment. Like living cells, microspheres have internal structure: watery areas, lipid-like areas, and boundary-layer areas which serve as sites for specific chemical activities. (Sidney Fox)

17–E FORMATION OF AGGREGATES

Natural or artificial polymers placed in water may aggregate and form larger structures (Figure 17–4). These aggregates show some features of living cells. Any lipids present form membrane-like coatings around the aggregate's exterior. Aggregates are selectively permeable and accumulate some kinds of monomers, but not others, from the surrounding medium. They also catalyze certain reactions in their interiors, and they grow, eventually breaking up into smaller aggregates.

These aggregates illustrate the saying, "the whole is greater than the sum of its parts." An aggregate has both structures (such as the "membrane") and functions (such as the ability to "choose" molecules, collect them, and grow) not found in its component molecules.

Aggregates are extremely limited in the function of their "membranes" and "metabolism," and they lack a reliable means of reproducing any advantageous molecules they might contain. Nevertheless, they do show that organic polymers can associate and interact in discrete units, set apart from their surroundings by distinct boundaries. Such units are prerequisites for the evolution of life, because selection must have entities to "choose" among, and it seems likely that these entities must have been combinations of interacting molecules with potentially life-like properties, rather than individual polymers.

When placed in water, polymers may cluster into aggregates with features of living cells. Lipids may form membrane-like coatings that surround the aggregates and function as selectively permeable barriers.

17–F BEGINNINGS OF METABOLISM

Aggregates studied in the laboratory have a simple metabolism of one or a few reactions—a far cry from a living cell with its thousands of reactions. Oparin wrote: "The path followed by nature from the original systems of protobionts [pre-living aggregates] to the most primitive bacteria . . . was not in the least shorter or simpler than the path from the amoeba to man."[1] Along the way, the protobiont must have added hundreds of chemical reactions to its metabolism; feedback mechanisms evolved to regulate the various metabolic pathways; each metabolic enzyme became more efficient; and mechanisms for protein synthesis and replication of genetic material evolved, allowing the protobiont to make many copies of its metabolic enzymes and other proteins.

Protobionts arrived at the threshold of life by **chemical selection,** a process similar to natural selection but acting on nonliving systems. At first, selection probably favored mere longevity: aggregates with the most stable combinations of chemicals lasted longer before disintegrating. As more and more of these aggregates accumulated, it was a real asset for them to have catalysts for chemical reactions that made them more stable.

At first, the raw materials for these reactions were probably abundant compared to the few aggregates that needed them. However, as these successful, stable aggregates grew and fragmented, and as new ones formed spontaneously, more and more similar systems competed for fewer and fewer raw materials. Selection favored the aggregates that were most efficient at competing for the now-scarce material, and for those that could convert a second, abundant material to the first, now-scarce one. For example, suppose that the crucial molecule A was in short supply because of intense competition among aggregates taking it up faster than abiotic forces were producing it. Any aggregate containing a catalyst that converted abundant molecule B to the now-scarce A would have an advantage—until there were so many systems using the reaction B $\longrightarrow$ A that B also became scarce. Now a system that evolved a catalyst for the reaction C $\longrightarrow$ B as well as B $\longrightarrow$ A would survive, while B-users would disappear, and so on. Hence metabolic pathways became ever longer. In addition, they evolved "backwards," from the useful end product to less directly useful raw materials (Figure 17–5).

[1]This statement displays poetic license and should not be taken to imply that an amoeba was a direct ancestor of humans.

(a) Molecule "A" is absorbed by a chemical aggregate.

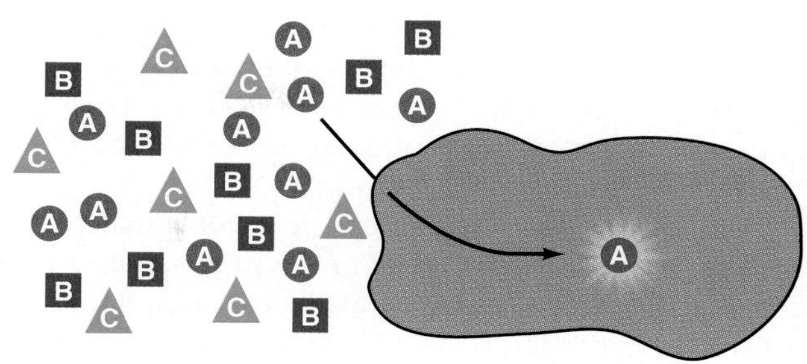

(b) Molecule "A" becomes scarce. Aggregates containing enzyme "L," which converts "B" to "A," have a survival advantage over those that don't have the enzyme.

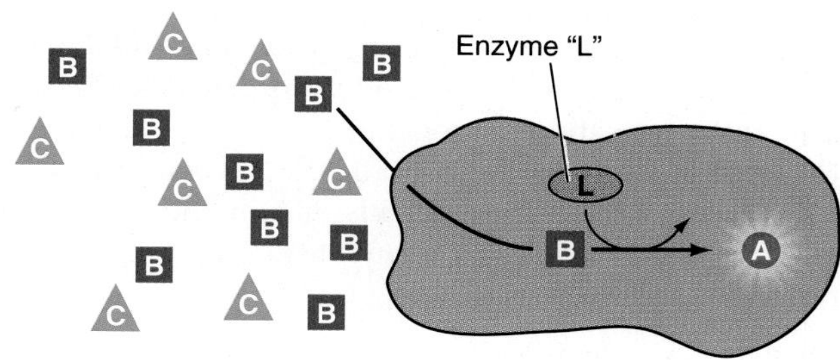

(c) Eventually both B and A become scarce. Aggregates containing both enzyme "Z," which converts "C" to "B," *and* enzyme "L" are selected for.

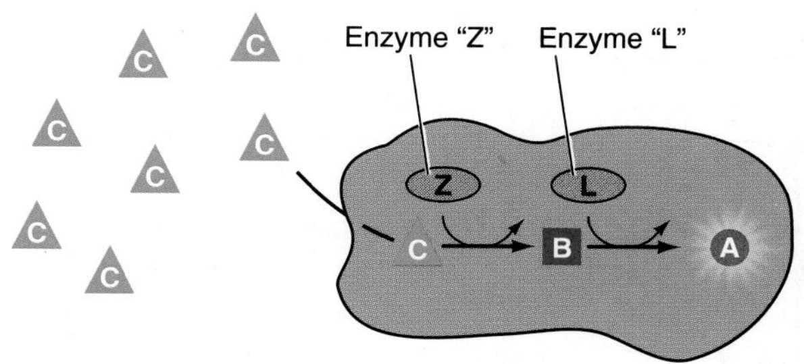

FIGURE 17–5

How metabolic pathways evolved. This diagram illustrates how new enzymes and extensive metabolic pathways evolved to produce molecules that became scarce.

Origin of Energy Metabolism

One of the first requirements of a metabolic system is to trap and use energy to drive various reactions. All organisms living today have hydrogen ion (H^+) pumps housed in membranes. Perhaps similar pumps provided energy that primitive aggregates or cells used to transport materials through their membranes. Such pumps might have been driven by light, an abundant energy source used by simple H^+ pumps in some kinds of bacteria today. The modern electron transport systems of respiration and photosynthesis are more elaborate H^+ pumps.

An older, more conventional view holds that energy originally came from ATP in the primordial soup or sludge. At first, ATP was a relatively common and available energy source. As ATP became scarce, systems that could use their H^+ membrane potentials, or energy released by chemical reactions, to make ATP from ADP and P_i (inorganic phosphate) would have had a selective advantage.

The need for an ATP-regenerating system probably selected for anaerobic pathways of fermentation, such as glycolysis, the fermentation of sugars (Section 7–B). Glycolysis is undoubtedly extremely ancient, judging by its presence in nearly all modern organisms. The citric acid cycle and electron transport chain of aerobic organisms provide highly efficient ways to extract further energy from the products of fermentation. This aerobic respiration arrived relatively late in the evolution of metabolism.

Chemical selection favored aggregate systems that could use catalysts, compete efficiently for raw materials, and generate their own supply of essential but scarce materials. Once the initial supplies of ATP began to run low, selection favored systems that could generate ATP using their H^+ membrane potentials.

17–G THE BEGINNINGS OF BIOLOGICAL INFORMATION

Self-organized aggregates of polymers are similar to modern cells in some ways, but they cannot be called "living" because they cannot reproduce. In modern organisms, genetic information allows cells to make proteins and to pass to their offspring copies of the information needed to make proteins. In contrast, when self-organized polymer aggregates split into smaller aggregates, these "offspring" may end up without some molecules crucial to the success of the "parent."

When modern cells reproduce, each offspring receives a set of the parent's DNA, containing the genetic information for making all of the parent's proteins. DNA directs the formation of RNA, which in turn directs the linking of amino acids into proteins (Chapter 10). Biologists diagram this information flow in modern organisms as:

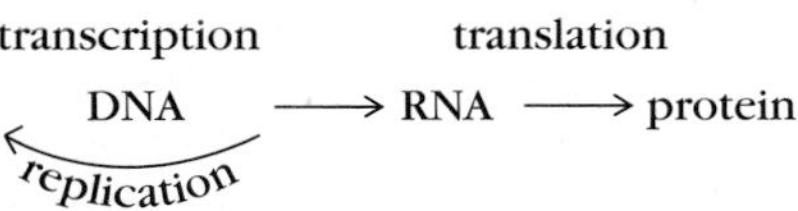

In this system, the three kinds of substances show division of labor. DNA is very stable, and so it can serve as a file copy of the information needed to make many RNA copies. DNA also spends some time being replicated before cell division. This frees RNA to spend all of its time in protein synthesis. Proteins, in turn, perform the actual work of the cell's metabolism.

When we ask how such a complex system could have evolved, we find a problem. In the synthesis of nucleic acids and proteins, the nucleotide and amino acid monomers must be joined together by enzymes. However, these enzymes must first be made according to instructions provided by already-existing nucleic acids! Which came first, the enzymes or the nucleic acids?

The most likely answer seems to be that the two evolved together. Amino acids form and polymerize much more easily than nucleotides do under prebiotic conditions, but they show no sign of self-replication. RNA polymers self-assemble less readily but, surprisingly, they catalyze reactions involving RNA and its monomers, including replication and splicing. If some early aggregates contained both, the proteinoid catalysts might help the aggregate survive and help the RNA reproduce. In turn, the RNA might interact with proteinoids and amino acids in a way that promoted production of rough copies of the same, useful proteinoids.

As time went on, chemical selection would preserve not only certain types of proteinoids and RNA alone, but rather those aggregates that lasted longer because of interactions between the two. Indeed, experiments have shown possible steps in this association, and have indicated that the genetic code may not be random but based on slight mutual chemical attractions between certain amino acids and RNA nucleotides.

From this and other evidence, it seems likely that the first genetic information was RNA, not the DNA used by all modern cells. DNA was added onto the leading end of an existing (or evolving) RNA-protein synthesis system. DNA's double-stranded structure gave it two advantages over RNA: DNA is more stable than RNA, and it is easier to copy without error. RNA molecules remain as the "go-betweens" that carry instructions from DNA to proteins today.

Although we have discussed them separately, reproduction and metabolism must have evolved together. Reproduction relies on metabolism to provide the energy and raw materials needed to replicate the genetic information and to produce many copies of proteins. Metabolism, in turn, relies on genetic information to direct the production of all the enzyme catalysts needed to make these raw materials.

Once a reliable means of protein synthesis and gene replication evolved, the story of the origin of life was complete. However, the story we have outlined here is mostly ideas, not evidence, and, so far, experiments have not produced impressive support. Is this because researchers haven't yet found the right set of conditions or because the ideas are too complex? Some feel research should begin with simpler, self-replicating catalytic substances, perhaps small organic molecules or even inorganic crystals. These could have attracted RNA and protein helpers, which eventually became so proficient that the original replicators are now lost to us.

Early aggregates of RNA polymers and proteinoids may have catalyzed each other's replication. Selection would have favored those aggregates that lasted longer because of the interaction between these two components. In this theory, the first genetic information would have been RNA, and only later DNA.

17–H HETEROTROPHS AND AUTOTROPHS

The advent of the first true organisms marked the end of the era of chemical selection and the beginning of the era of natural selection. Competition grew more intense, and primitive cells evolved faster ways to get energy and use it to reproduce. For a long time all organisms were anaerobic, and they were all **heterotrophs,** feeding on organic molecules made outside their own bodies. At first, they absorbed food from the surrounding soup (or sludge). But as this food became scarce, some cells began to devour their neighbors to obtain nutrients. Others evolved ways to acquire energy from other sources. They became **autotrophs,** organisms that make their own food from inorganic molecules. ■

BIO-BIT

Some chemosynthetic sulfur bacteria use the inorganic compounds released from deep-sea hydrothermal vents as a source of energy.

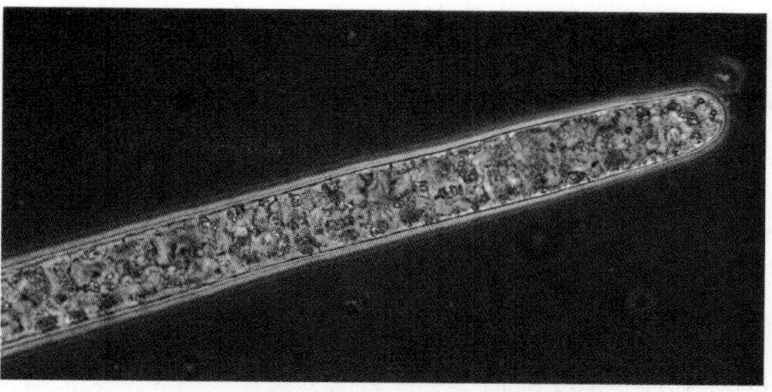

FIGURE 17–6

A chemosynthetic bacterium. These relatively large cells grow in long filaments. They get energy, not from light, but by oxidizing H_2S to sulfur, which builds up as yellow particles visible in the cell. Some species of these bacteria are important in the culture of rice because they remove toxic H_2S from the water where rice grows. (Biological Photo Service)

The evolution of autotrophy was tremendously important to the evolution of life. It freed the living world from dependence on the slow production of food by abiotic means. An organized metabolic assembly line made food much faster, and in great abundance. Autotrophic organisms could therefore grow and reproduce more rapidly. Inevitably, some of them fell victim to their heterotrophic neighbors, whose population also grew as they exploited this new food supply.

The photosynthesis of green plants is a very advanced form of autotrophy (Chapter 8). Other, less successful evolutionary "experiments" in do-it-yourself food production are still found among some modern bacteria. **Chemosynthetic** bacteria use energy released by various inorganic chemical reactions to make CO_2 into organic compounds, even in the absence of light (Figure 17–6).

Various kinds of bacteria also show a range of ability to trap light energy. Studying these bacteria allows us to reconstruct the steps in the evolution of photosynthesis. At first, light energy was probably used to create an H^+ membrane potential, which supplied energy to make ATP and transport substances through the membrane. This met some energy needs, but the cell still had to get food from its surroundings. The next step was using light energy to obtain hydrogen atoms to reduce CO_2 as it was fixed in storage molecules such as glucose—the hallmark of true photosyn-

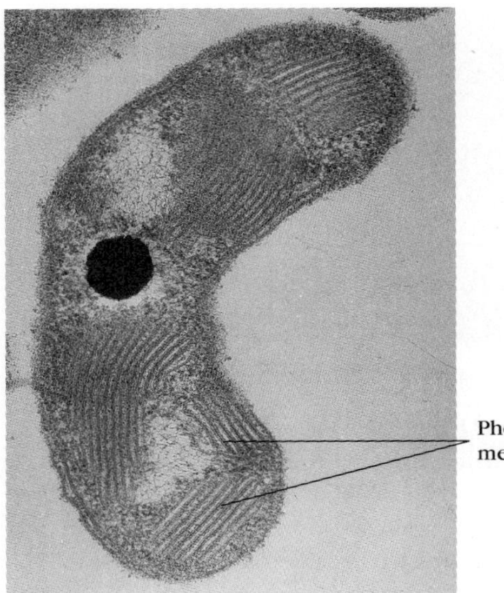

FIGURE 17–7

Photosynthetic purple sulfur bacterium in length-wise section. The many sets of parallel lines are internal photosynthetic membranes, which contain the pigment molecules used to trap light energy. (Compare these with the photosynthetic membranes of the chloroplast in A Journey Through the Chloroplast, p. 148.) (S. C. Holt, Univ. of Mass./BPS)

thesis. Early photosynthetic bacteria used easy hydrogen donors: their own organic waste products, or inorganic hydrogen gas or hydrogen sulfide (H_2S) (Figure 17–7). The use of water as a hydrogen donor came quite late because it is hard to extract hydrogen atoms from water. However, water is much more abundant than other hydrogen donors, and so organisms that can use water as a hydrogen source have an enormous selective advantage.

The first life forms were probably heterotrophs that absorbed organic molecules from their surroundings. The ability of autotrophs to produce their own food allowed them to grow and reproduce faster, provided additional food for heterotrophs, and thus vastly increased the amount of life the Earth could support. Eventually, some photosynthetic autotrophs developed the ability to use light energy to make their own food.

17–I RESPIRATION

Photosynthesis that uses water as a hydrogen donor also releases O_2. Indeed, virtually all of the O_2 in the air today has come from water-splitting photosynthesis. When early organisms first evolved this type of photosynthesis, they produced small amounts of O_2. At first, the oxygen reacted with iron-rich minerals dissolved in the surrounding shallow seas, and the resulting oxidized minerals settled to the bottom in layers that later formed rock. Once these minerals were exhausted, oxygen began to accumulate in the atmosphere and in the oceans (Table 17–3). This produced a grave environmental crisis: O_2 destroys flavins, an important group of coenzymes, and without their flavins organisms could not survive. The threat posed by oxygen was probably the selective pressure that brought about the next important evolutionary advance, respiration.

You will recall that respiration is a complex metabolic pathway that involves many molecules functioning as enzymes and cofactors. Each cell's DNA produces this array of necessary chemicals, and at first glance it may seem impossible for an organism to evolve all these different and highly specific molecules. But, perhaps all these molecules necessary for respiration did not evolve at once.

Studies of metabolism in living bacteria show that many molecules of the citric acid cycle and electron transport probably already existed in ancient anaerobic autotrophic bacteria, but played different biochemical roles. In respiration, these pathways run backward from their original direction, and some molecules have been modified or added. Emerging aerobic bacteria had half a billion years to perfect these changes as O_2 slowly accumulated to modern levels.

Respiration permitted organisms to release much more energy from each food molecule. Whereas anaerobic glycolysis yields two ATP molecules per glucose molecule, respiration produces more than ten times that much. This tremendous energy bonus made it possible for organisms to grow and reproduce much faster. It also allowed them to "experiment" with new enzymes and new structures that used a lot of energy but made them superior competitors, able to outstrip their anaerobic neighbors. Today anaerobic organisms are restricted to habitats without enough oxygen to support aerobic forms of life.

The accumulation of atmospheric oxygen released by water-splitting photosynthetic autotrophs selected for cells that used aerobic respiration.

17–J ORIGIN OF EUKARYOTES

The next big jump in evolution was the rise of eukaryotic cells, containing membrane-bounded organelles (Figure 17–8). Mitochondria and chloroplasts probably originated as bacteria that came to live within other

Table 17-3
Some Important Milestones in the Origin of Life

Time (billions of years ago)	Event
4.6	Earth originates
4.3	Conditions on Earth stabilize
3.8	Ocean mineral content similar to today's; atmosphere like today's but without O_2; carbon as CO_2 (Isua formation rocks, Iceland)
3.5	Earliest known stromatolites (mats of prokaryotic cells, Australia)
3.5–3.3	Probable cyanobacteria with oxygen-generating photosynthesis
≈3.0	Oxygen-generating photosynthesis (for sure)
2.7–2.0	Deposits of banded iron, oxidized by O_2 from photosynthesis
2.0	Great diversity of bacteria (Gunflint chert, Ontario)
	Atmospheric oxygen reaches 1%
	Respiration appears
1.45	Eukaryotic cells; sexual reproduction allows more rapid evolution
1.4	Multicellular algae (seaweeds)
0.7	Soft-bodied animals (jellyfish, worms)
0.6	Hard animal skeletons (Cambrian Period begins)

Subsequent events are outlined in Appendix C, *"Geologic Time Scale."*

(a)

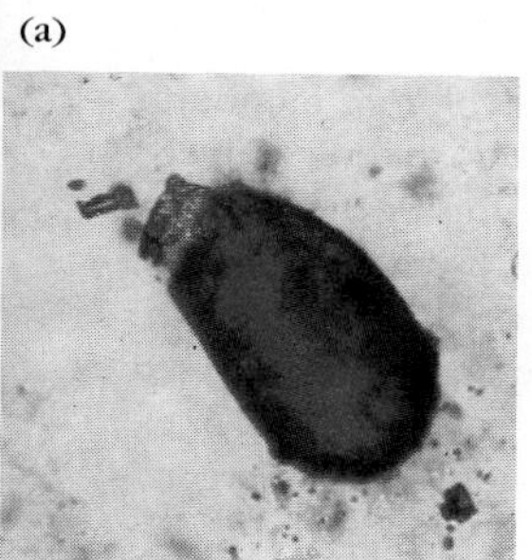

(b)

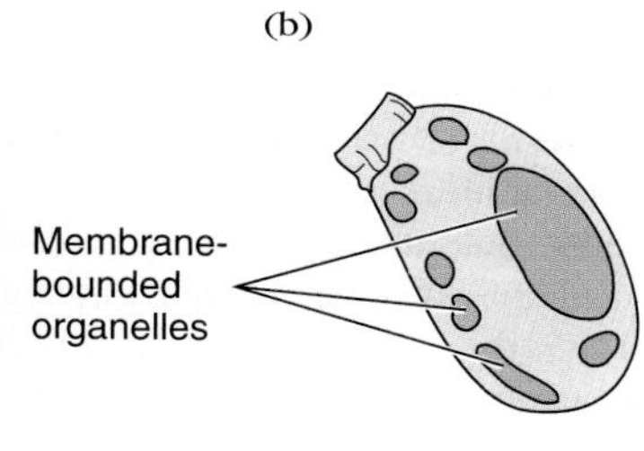

FIGURE 17-8

The earliest known eukaryote. (a) This single-celled organism lived 800 million years ago. It was found in rock from the Backlundtoppen Formation, Spitsbergen (an island several hundred miles north of Norway). **(b)** Membrane-bounded organelles are seen within this organism. (A. H. Knoll. Knoll and Calder, Palaeontology 26:467, 1983)

BIO-BIT

Eukaryotic cells evolved nearly 2 billion years after the first prokaryotic forms of life appeared on Earth.

cells (see Chapter 5: *A Journey into Evolution:* How Are Prokaryotes and Eukaryotes Related?).

A major advance shown by eukaryotes is sexual reproduction, which involves meiosis and fertilization. This gives rise to genetic variation by reshuffling existing genes, in addition to the variation produced by mutation to form new genes. Genetic variability is the raw material for natural selection, and therefore for evolution. The fossil record indicates that eukaryotes evolved much faster than prokaryotes, diversifying into organisms with a variety of sizes, shapes, and lifestyles. Although eukaryotes have dominated the Earth for at least 700 million years, prokaryotes have also survived and flourished. Many of them appear to have changed very little in more than a billion years.

Eukaryotic cells with membrane-bounded organelles appeared about 700 million years ago. Their ability to reproduce sexually allowed them to diversify more rapidly than prokaryotes.

A Journey into The Environment

Ozone Depletion

The history of the environmental problem of ozone depletion reads like the plot of a Grade B sci-fi horror flick. Imagine this scenario:

Scene One: While no one is paying attention, a dire problem mushrooms, seemingly overnight, to menace all living things on planet Earth. The culprits are widely used synthetic compounds, chlorofluorocarbon (CFCs for short). These compounds are inert and harmless on the Earth's surface, and at first no one but a few "crackpot" chemists and "way-out" environmentalists are alarmed when CFCs, first synthesized in the 1930s, find wide application in all sorts of industries. CFCs are used as coolants in air conditioners and refrigerators. They are used in styrofoam products, in foamed insulation, as propellants in aerosol products, and as cleaners of computer circuit boards. The exhaust from each space shuttle launch emits CFCs. These seemingly harmless, quite useful molecules eventually diffuse into the atmosphere and six miles up they reach the stratosphere where they undergo a Dr. Jekyl-and-Mr. Hyde transformation, degrading into highly reactive chlorine monoxide (ClO) molecules. These act as catalysts that speed the breakdown of ozone (O^3) into molecular oxygen (O^2).

The ozone layer (O^3) in the stratosphere (located 6 to 28 miles or 10 to 45 kilometers above the Earth's surface) has formed from the reaction of ultraviolet solar energy with molecular oxygen. Ozone gas normally deflects the sun's powerful and harmful ultraviolet radiation, but as more and more CFCs form more and more chlorine monoxide molecules, the ozone layer gets thinner and thinner. Eventually holes appear in the ozone layer and deadly ultraviolet radiation streams through the stratosphere to strike the surface of planet Earth.

Scene Two: In the 1970s the chemists and environmentalists continue to sound the alarm, but they are not taken seriously. No one can conceive of human activities having a negative effect on something as remote and majestic as the ozone layer. In the United States the earliest satellite reports of damage to the ozone layer are mistaken as computer errors and, in true Grade B movie fashion, they are ignored. Data accumulate, and one day in 1985, British researchers startle the world with the news that over Antarctica there is, indeed, a hole in the planet's protective ozone shield.

Scene Three: This takes place 100 years in the future and the script has yet to be written. The events are uncertain and what will happen is up to us. To understand the problem of ozone depletion, we begin with the chemistry that creates the ozone hole.

How CFCs Create the Ozone Hole

It was not chance, but the result of chemical, climatic, and geographical factors that caused the ozone hole to first appear over the South Pole. Since its initial appearance in 1985, the Antarctic ozone hole has become a seasonal phenomenon that begins in April or May, (Fall in Antarctica) when a cyclone of high-speed stratospheric winds, called the polar vortex, begins to whirl around in the south, polar stratosphere. As the Antarctic winter progresses the extreme low temperatures and furious cyclonic winds of the polar vortex act like a high altitude blender in which CFCs, chlorine monoxide, and ozone are whirled. The net effect is to speed the degradation of ozone and eventually, depending upon the severity of the winter and the relative amounts of CFCs and ozone present, the ozone hole appears. By September and October (Spring in Antarctica) the Antarctic ozone hole is at its maximum. As the warmth of the Antarctic summer dissipates the extreme cold temperatures and fierce, cyclonic stratospheric storms of the polar vortex, the ozone hole begins to close. Without the icy, high-speed blending of chlorine monoxide and ozone, the annual accelerated erosion of ozone slows. Notice that the ozone hole's peak season coincides with Spring in Antarctica: exactly the worst time for plants and animals to be exposed to high levels of deadly ultraviolet radiation.

Extent of the Ozone Hole and Its Dangers

The extent of ozone loss is alarming. In ten years the annually appearing hole has widened until in 1992 it stretched to a record 9 million square miles. Covering most of the south polar region, it extends nearly to the tip of South America (Figure 17–A). Because chlorine monoxide is a catalyst, it has an extremely long life. Like all catalysts, it speeds chemical reactions, but does not enter into them. Thus, each molecule of chlorine monoxide remains in the stratosphere, actively degrading ozone, for a century or more. As a result, even though international agreements have legislated that CFCs will be completely phased out by 2030, the south polar ozone hole is probably going to get larger for many years to come because of the long life of CFCs. In addition, a second, smaller ozone hole has appeared over the North Pole. Unlike the Antarctic ozone hole, this Arctic ozone hole did not appear in 1992, because the winter was unexpectedly warm and the Arctic polar vortex was dissipated before ozone erosion could occur. Ozone measurements over nonpolar areas of the globe have revealed that ozone is thinning over Europe and over North America. Some researchers estimate that over Europe the ozone layer is being eroded at a rate of 1% per day and, like the polar holes, the ozone layer over Europe and North America has recovered each Spring.

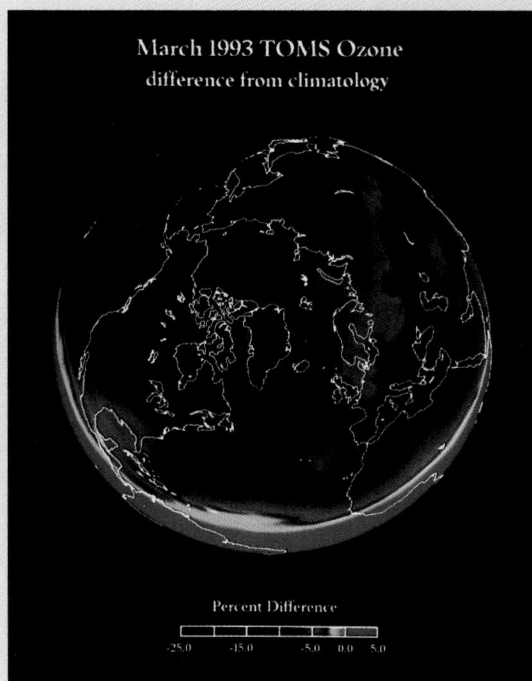

FIGURE 17–A

Ozone depletion. This false color image illustrates the total ozone difference between March 1993 and the average March ozone level from 1979–1990. Blue/purple colors indicate decreased values in the ozone levels while increased values are shown in red. No measurements were made in the small black circle at the North Pole. (NASA)

Effect of Increased Ultraviolet Radiation

In humans ultraviolet radiation causes skin cancer, retinal cancer, and cataracts. Skin cancers are expected to increase by 26% for every 10% drop in the density of the ozone layer. Ultraviolet radiation is also known to depress the immune system. Especially in the early Spring, when the ozone layer is most degraded, humans are advised to protect themselves from ultraviolet radiation with protective clothing, sunscreens, broad-brimmed hats, and sunglasses that block UV rays. Children should be especially protected, and everyone should be aware that exposure to UV radiation is worst between 10:00 AM and 3:00 PM, when the sun's rays are most direct. Prolonged sunbathing is not advisable.

Humans have the option of protecting themselves from UV radiation, but other organisms aren't so fortunate. No one knows exactly what the effects of increased exposure to UV radiation will be on a global scale, but many are predicting a drop in photosynthesis that may lead to a collapse of the Antarctic food chain as the photosynthetic bacteria, protists, and algae that are the base of the food chain are damaged or die off from exposure to UV radiation. Without this photosynthesizing base, the krill population (shrimp-like crustaceans that are the primary food source for many larger marine animals) will decrease. As krill becomes scarce, populations of Antarctic animals such as penguins, killer whales, great whales, seals, and fish of all kinds will decrease.

Just as UV radiation will damage populations of marine phytoplankton, land plants will also be damaged. Although some alpine plants have accessory pigments that shield them from UV radiation, low altitude plants have no such protective pigments. No one knows exactly what will happen as increased UV radiation damages natural vegetation and agricultural crops.

Everything that we have learned about ozone depletion underscores two truths about the mechanism of planet Earth:

1. We have seen that, although it is difficult to comprehend, everything on our home planet is interrelated.
2. Because of this interrelationship, it is impossible to do just one thing on Earth. Everything has widespread environmental consequences.

What You Can Do To Protect Your Ozone and Yourself

1. Stay out of the sun. Realize that your "healthy, all-over, nut-brown tan" may be the first step toward skin cancer. If you're going to be out in the sun for a long time, protect yourself. Wear sunscreen and UV blocking sunglasses.
2. When it's time to change the coolant in your car's air conditioner, insist that these CFCs be "vampired" or drawn into a leak-proof container instead of being expelled into the atmosphere.
3. Switch to pumped products instead of aerosol products. Even the "ozone-friendly" aerosol propellants may be damaging. Some may harm the ozone layer, while others may contribute to global warming.
4. Use home insulation that does not contain CFCs.
5. Avoid foamed products (foam mattresses, foam furniture, styrofoam, etc.). They are made with and contain CFCs.
6. Read product ingredient labels and avoid other products that are not ozone-friendly: propellants such as halon in some fire extinguishers and methyl chloroform in spot removers and fabric protectors. (Avoid any product containing 1,1,1-trichlorethane.)

(a)

(b)

FIGURE 17–9

Stromatolites. (a) These mounds in shallow seawater are living stromatolites nearly 1 meter wide. The rock forms from mats of autotrophic prokaryotes that grow in successive layers. Each layer contains a vast population of microscopic prokaryotes. **(b)** Fossil stromatolites are seen as they appear to the naked eye in this section of rock that is 3.5 billion years old. The layers where autotrophic prokaryotes lived are clearly seen. (a, F. Bavendam/Peter Arnold, Inc.; b, K. Schafer & M. Hill)

17–K EARLY FOSSILS

Until 1954, the oldest known fossils came from the Cambrian Period of geological time, which began about 600 million years ago. Most major groups of organisms had evolved by this time. Their origins were veiled in mystery because most Precambrian rocks have been either deeply buried, extensively eroded, altered by heat and pressure great enough to destroy fossils, or unidentifiable until recent methods of identification were developed.

3? By using microscopes to examine thin sections of rock, scientists have found many microfossils of unicellular organisms even in Precambrian deposits (see Figures 17–1 and 17–8). Thus far, the oldest definite fossils are stromatolites (mats of cyanobacteria), about 3.5 billion years old, similar to those shown in Figure 17–9. Fossils believed to be photosynthetic cyanobacteria, possibly able to carry out oxygen-generating photosynthesis, also date from up to 3.5 billion years ago. A 1-billion-year-old rock formation contains prokaryotes (and possibly eukaryotes) of about 30 different species. Some of the presumed cyanobacteria in this assemblage look exactly like forms alive today! ■

Further microscopic and chemical analysis of ancient rocks will give us a better picture of the early history of life on Earth.

The earliest evidence of life on Earth is in the form of fossils at least 3.5 billion years old. Some 1-billion-year-old fossil cyanobacteria look exactly like forms alive today.

SUMMARY

1. The conditions when life began were very different from those on Earth today. Evidence from geology and astronomy suggest that the Earth then probably had a mildly reducing atmosphere, composed of most of the gases in today's atmosphere except oxygen. Such an atmosphere could have been a place where organic compounds could form.
2. Gradually, the small organic molecules could have polymerized and evolved systems of metabolism, information transfer, and reproduction—eventually becoming living organisms.
3. Some important events during the early history of life were the evolution of photosynthesis and respiration, the evolution of eukaryotes, the acquisition of intracellular organelles, and the beginning of sexual reproduction

SELF-QUIZ

1. For each gas listed below, tell whether it was *more* or *less* abundant in the mildly reducing atmosphere of the early Earth than it is today:
   ___ oxygen
   ___ carbon dioxide
   ___ water vapor
2. In Stanley Miller's classic experiment:
   a. nucleic acids were formed
   b. ultraviolet radiation was used
   c. oxygen was one of the starting ingredients
   d. water was strictly excluded from the system
   e. amino acids were formed
3. Number the following structures and processes in the order in which they are believed to have evolved:
   ___ respiration
   ___ polymers
   ___ water-splitting photosynthesis
   ___ organic monomers
   ___ acquisition of intracellular organelles
   ___ fermentation
4. The earliest autotrophs were important to the evolution of life on Earth because:
   a. they blocked harmful ultraviolet rays from the sun
   b. they provided a self-renewing food supply
   c. they rid the environment of toxic substances
   d. they provided oxygen for respiration
   e. all of the above
5. List two changes in the environment that resulted from evolution of water-splitting photosynthesis. What effects did these have on the later evolution of living organisms?
6. Respiration was important to early life on Earth because:
   a. it rid the environment of toxic ozone
   b. it provided much more energy than photosynthesis
   c. it provided much more energy than fermentation
   d. it permitted experimentation with the genetic code
   e. all of the above

THINKING CRITICALLY

1. Suppose that a scientist claimed to have produced life from nonliving materials under laboratory conditions. What criteria must such an "organism" meet before *you* agree that it was truly living?
2. Some organic compounds formed in simulations such as Miller's are not found in living organisms. Can you explain this?
3. What is the evidence that respiration evolved before the eukaryotic condition?
4. How might the destruction of the ozone layer today affect life on Earth?

SELECTED KEY TERMS

abiotic, *p. 345*
autotrophs, *p. 349*
chemical selection, *p. 347*
chemosynthetic bacteria, *p. 349*
heterotrophs, *p. 349*
prebiotic, *p. 341*
proteinoids, *p. 346*
reducing atmosphere, *p. 343*

SUGGESTED READINGS

Cairns-Smith, A. G. "The first organisms." *Scientific American,* June 1985. A view of the origin of life very different from the one in this chapter.

Cech, T. R. "RNA as an enzyme." *Scientific American,* November 1986.

Cloud, P. "The biosphere." *Scientific American,* September 1983. How geological and fossil evidence provides information about early life on Earth.

Dickerson, R. E. "Chemical evolution and the origin of life." *Scientific American,* September 1978.

Gurin, J. "In the beginning." *Science 80,* July–August, 1980. A lively account of recent finds of ancient fossil microorganisms.

Horgan, J. "Trends in evolution: in the beginning . . ." *Scientific American,* February 1991. A thorough account of how life might have first evolved.

Schopf, J. W. "The evolution of the earliest cells." *Scientific American,* September 1978. An authority on early microfossils explains the evidence for current beliefs about the early history of life.

CHAPTER 18

CONCEPT GUIDE

After reading this chapter, you should be able to:

1. Explain how the scientific name of a species is assigned. (Section 18-A)
2. Compare the classification system of Linnaeus and the methods of phylogenetic taxonomists. (Section 18-B)
3. Explain what is meant by saying that characteristics are ancestral, derived, conservative, homologous, or convergent. (Section 18-C)
4. Describe the five kingdoms of organisms used in this textbook and what criteria are used to define each. (Section 18-D)
5. Describe how viruses are similar to and different from cellular organisms. (Section 18-E)
6. Explain how viruses play an important role in the evolution of cellular organisms. (Sections 18-F and G)

How should we classify whales within the animal kingdom? The classification of animals is not always apparent. Whales, like fish, live entirely in water and have fin-like appendages. Yet unlike fish, they regulate their body temperatures very closely, lack scales, breathe air, and provide milk for their young. Read more about the difficult task of classifying animals in Section 18-C. (Francois Gohier/Photo Researchers, Inc.)

Classification of Organisms and the Problem of Viruses

No one is sure of how many species of organisms are alive today. Estimates vary from 5 to 30 million, depending upon the methods used to estimate the number of unknown species. So far, scientists have officially described only about 1.5 million species. We are presently destroying natural habitats so rapidly—by clearing, draining, filling, damming, and polluting—that many of the existing species will probably be extinct before they can be discovered and catalogued. We are now losing an estimated four species of plants and animals per day!

The first comprehensive system for classifying organisms was invented by the Swedish botanist Carolus Linnaeus (see *A Journey into Science in Process:* The Name Game in this chapter). Linnaeus based his classification on morphology (structure, anatomy). For example, two species of trees with similar leaves and flowers fell close together in his scheme. After Darwin's theory of evolution gained acceptance, biologists moved toward classifying organisms according to their evolutionary relationships. Fortunately, the morphological features used by Linnaeus usually reflect evolutionary relationships. Hence biologists have been able to retain many of the names and groupings of Linnaeus's thorough and meticulous system.

Our system of naming and classifying organisms is inadequate to the task. The sheer number of species strains its capacity. More importantly, classification attempts to force the natural world of life into artificial "bins," and the boundaries of these bins can never be defined to the satisfaction of all. However, until better systems of classification come along, we must make do with what we have. Scientists need official names and definitions for organisms in order to know when they are talking about the same species. We also need a classification scheme to make sense of the vast diversity of organisms all around us.

In this chapter we consider how living things are classified, and how biologists try to decide where in the classification scheme particular organisms belong. We then introduce the five-kingdom system of classification that we use in this book. Finally, we look at a peculiar group, the viruses, which show some features of life and yet are not living.

KEY CONCEPTS

- Classification of organisms is based on evolutionary relationships.
- Every species of organism known to biologists has been given an official two-word Latin name designating its genus and species.
- Viruses are tiny particles made of nucleic acid enclosed in a protein coat. They have orderly structures, and they can evolve and become adapted to their environment. They show none of the other features of life, but rely on the cells they invade and parasitize to provide the energy and materials needed for their reproduction.

Curiosity Questions

1? Why aren't mushrooms considered to be plants? (See page 364)

2? Are viruses living organisms? (See page 364)

3? Why don't people become permanently immune to flu and cold viruses? (See page 370)

FIGURE 18–1

***Ursus arctos* (Northern Bear).** These grizzly bears show the dish-faced profile, white-tipped ("grizzled") fur, and hump above the shoulders characteristic of members of this species. (M. Hoshino/Minden Pictures)

18–A BINOMIAL NOMENCLATURE

The basic unit for classifying organisms is the species. In the case of organisms that reproduce sexually (and many do not), a **species** is a group of organisms that share a common gene pool and do not interbreed with members of other species under natural conditions (Chapter 15).

Linnaeus began the practice of giving every species its own, unique Latin **binomial,** a two-word name. The first word in this binomial is the **genus** (plural, **genera;** adjective, **generic**). A genus contains one species or a group of very similar species. The second word in the binomial denotes the species itself. For example, the binomial for the grizzly bear is *Ursus arctos* (Figure 18–1). Another species in the genus *Ursus* is *Ursus maritimus,* the polar bear.

Note that the genus is always capitalized while the specific name is not. Both are italicized (or underlined), as is any foreign-language word found in an English text, and it takes *both* names to indicate the name of a species. In the grizzly bear, for example, the name of the species is *Ursus arctos,* not just *arctos.* Often the generic name is abbreviated by using only its first initial, as in *U. arctos.*

BIO-BIT

The translation of the scientific name for humans means "wise man."

The biological classification system assigns each species a Latin binomial name that denotes genus and species.

18–B TAXONOMY

Taxonomy is the branch of biology concerned with the classification of organisms. Linnaeus arranged organisms into a hierarchy of ever larger and more inclusive categories, a system borrowed from the highly disciplined Swedish military of his day. The most inclusive categories are the **kingdoms.** The other main categories, in descending order, are **phylum, class, order, family, genus, and species.** (You can remember this sequence by memorizing the sentence "**K**eep **P**ots **C**lean **O**r **F**amily **G**ets **S**ick.") Botanists use **divisions,** instead of phyla, as categories in the plant kingdom.

A **taxon** (plural, **taxa**) is a group of organisms defined by the classification scheme, such as a particular species or class. For example, Ursidae (a family) is a taxon including the species in the genus *Ursus* as well as those in other genera of bears, such as the South American spectacled bear, *Tremarctos ornatus,* and the southeast asian sun bear, *Helarctos malayanus.* Some taxa contain only one group at the next lower level; for example, many families contain only one genus and one species.

Table 18–1 gives the classification for human beings, and here you can see the seven important levels of the hierarchical classification.

Linnaeus named and classified all of the plants and animals known to him in his massive books *Systema Naturae* and *Species Plantarum.* He believed that a species could be described by listing the morphological characteristics of a "perfect" member of the species. This led to the practice of selecting and preserving a typical individual of each newly described species, which became the official **type specimen** of the species. Today, authors of new species preserve several specimens showing a typical range of the species's characteristics. Type specimens then become the definition of the species. If subsequent investigators/workers need to know whether they are really working on the same species of daisies that the original author described, they compare their daisies with the type specimen(s).

Currently, some microorganisms are preserved alive, as clones of individuals with identical genomes,

Table 18–1
Classification of *Homo sapiens*

Category	Taxon	Characteristics
Kingdom	Animalia	Heterotrophic, multicellular organisms lacking cell walls, and possessing a motile stage in the life history
Phylum	Chordata	Animals with a dorsal, hollow nerve tube, a notochord, and pharyngeal gill slits at some stage in life
Class	Mammalia	Chordates with only one bone in each side of the lower jaw, hair, or fur, young nourished by milk from the mother's mammary glands
Order	Primates	Originally arboreal (tree-living) mammals with flattened fingers and nails, vision the most important sense, poor sense of smell
Family	Hominidae	Primates with bipedal locomotion, flat faces, binocular color vision
Genus	*Homo**	Hominid with large brain, speech, long childhood
Species	*Homo sapiens*†	High forehead, body hair reduced, prominent chin

**Homo* Latin: man.
†*sapiens* Latin: knowing, wise.
Note that a species is a two-word name.

by freezing them in liquid nitrogen (−196°C). Researchers can then order a culture of a particular species, and even of particular genetic strains within a species, for comparison.

Thousands of Linnaeus's species and taxon names are still in use. However, an understanding of evolution has led modern biologists to classify organisms by evolutionary relationships rather than by structure. An organism's evolutionary history is its **phylogeny.** Phylogenetic taxonomists try to classify organisms by their phylogenetic relationships.

The two methods can be distinguished by their goals. Linnaeus's classification was an artificial system, designed to be helpful in organizing and retrieving information about organisms. The goal of the phylogenetic system is to produce a classification system that is both easy to use and informative about evolutionary history. The phylogenetic method amounts to drawing an evolutionary family tree for an organism. In many cases, this leads to the same result as an artificial classification, since organisms that have evolved from a common ancestor are likely to be more similar than those that have not.

Phylogenetic taxonomy has some inherent problems. Even if we can work out the true evolutionary history of a group of organisms, no taxonomist can classify them in a way that everyone will accept. Phylogeny depicts the natural and continuous process of evolution, but the taxonomist must carve this continuum into artificial and separate taxa, for human convenience. There are bound to be disagreements about how, where, and why boxes should be drawn around parts of the evolutionary tree. Are all species of the dog family close enough relatives to be in the same genus? If not, how many genera should we use? How alike must different members of a family be? There are no simple answers to such questions. As a result, there are many different systems of classification in use today.

The Linnaean system of classification ranks species within a hierarchy of categories. When a new species is described, one specimen is set aside as the type specimen, which becomes a physical definition of the species. Current classification schemes try to organize species according to their phylogenetic relationships.

18–C INTERPRETING THE CHARACTERISTICS OF ORGANISMS

Every organism has a variety of different features, often giving conflicting information about its phylogenetic position. Biologists need ways to interpret the significance of different features, giving some more weight than others, as they reach a decision. Suppose your biology teacher asked you to classify a flea, a frog, and a kangaroo in a phylogenetic scheme. We hope you would ignore the fact that all three have hind legs strongly developed for jumping, and give more emphasis to features like the general plan of the skeleton, the

(a)

(b)

FIGURE 18–2

Ancestral and derived characters. (a) This alpine newt has a tail, an ancestral characteristic, but only four fingers on the forelimbs, a derived trait. **(b)** In contrast, the orangutan and other apes have five fingers, an ancestral trait, but have lost their tails, a derived feature. (a, R. Maier/Animals Animals; b, L. Migdale /Photo Researchers)

type of body covering, and the mode of reproduction. By this reasoning you might arrive at the conclusions made by biologists: fleas belong among the insects, close to the flies; frogs among the amphibians, with salamanders and toads; and kangaroos among the marsupial mammals, with the koala and opossum.

As a species evolves, some characters change faster than others. Every organism shows some **ancestral characters,** which have persisted essentially unchanged from its remote ancestors, and some **derived characters,** which have evolved more recently. For instance, let us compare apes and newts with their common ancestors, the amphibians of the Carboniferous Period. We find that newts have a tail, an ancestral feature, but they have a derived number of fingers (four) on the forelimb, whereas apes lack a tail, a derived trait, but have the ancestral number (five) of fingers (Figure 18–2).

Traits that have changed little during evolutionary history are called **conservative characters.** These are usually features that cannot change very much if the organism is to survive. They are therefore useful for defining the various taxa. The shape, size, and number of teeth is the conservative character used to define many mammalian orders. For instance, beavers, kangaroo rats, and some fossil species all have similar incisors (front teeth) and are therefore all classified as rodents (Order Rodentia). Similarly, embryonic development is conservative, and similarities between developmental stages reveal relationships that might not be obvious in adult individuals.

To identify ancestral and derived characters, we must decide which ones share a common evolutionary origin. Characters found in two species and derived from the same part of a common ancestor are said to be **homologous.** Homologous characters have the same genetic basis in the two species, but not necessarily the same appearance or function. A classic example is the forelimbs of vertebrates. A human arm, a horse's foreleg, a whale's flipper, and a bird's wing all look different and do different jobs. However, they share a similar underlying bone structure and have remarkable physical resemblances in embryonic stages. Further, the fossil record shows that they all have a common evolutionary origin from the forelimbs of ancient amphibians. The limbs are homologous, their basic bone structure is ancestral, and their detailed structures are derived. For phylogenetic purposes we conclude that all animals with this type of forelimb are related. However, it is not always easy to disentangle the derived, ancestral, and homologous characters of organisms.

One difficulty is sorting out cases of **convergent evolution,** the development of similar adaptations by organisms of different ancestries, in response to similar environmental pressures. Striking examples occur among plants. Many desert-dwelling members of the New World family Cactaceae (cactuses) resemble

BIO-BIT

The similarity in appearance between bottle-nosed dolphins and fish is a result of convergent evolution.

FIGURE 18–3

Convergent evolution. Old World vultures have strong, hooked bills, and small naked heads adapted to poking into animal carcasses for food. These vultures are closely related to eagles. Also present at the feast are taller storks, members of the groups most closely related to New World vultures. New World vultures have features so similar to those of Old World species that they have traditionally been classified with them, near eagles and hawks, instead of with their stork relatives. The similarities between these distantly related vultures is due to their common lifestyles, and not the shared traits of recent ancestors. (F. Lanting/Minden Pictures)

desert-adapted members of the Old World family Euphorbiaceae.

In both groups, the advantage of being able to conserve water in desert habitats has led to the evolution of thick, water-storing stems and of spiny leaves that deter animals from using the stems as the source of their own water. Among animals, hares and rabbits (mammalian Order Lagomorpha) were once placed with the similar-looking rodents (Order Rodentia), but the fossil record shows that the two groups had separate origins and have converged. Recent evidence also confirms that the Old World and New World vultures evolved their distinctive features independently (Figure 18–3).

To classify a newly discovered organism, many kinds of evidence in addition to appearance must be considered, such as:

1. Life history. The genetic program for development tends to be conservative. Hence, similarities between embryonic or larval stages often give better clues to relationships than do resemblances (or nonresemblances) between adults.
2. Biochemical studies. Comparing the pigments, proteins, or nucleic acids of different species may provide useful information.
3. Behavior.
4. Geographic distribution.
5. The fossil record.

Even with all this information, taxonomists still face the old problem of how much weight to give characters that point to different conclusions about how the organism should be classified.

Monophyletic or Polyphyletic?

Many taxonomists consider that the most useful taxon is a **clade,** containing a common ancestral species and all the species descended from it. Such a taxon is **monophyletic,** meaning that it represents one evolutionary line. In practice, many taxa in use today are **polyphyletic,** made up of several evolutionary lines but not including their common ancestor (if there is one). One example is the mammals. The class Mammalia contains all vertebrates with only one bone in each side of the lower jaw. By this definition, mammals probably arose at least three, and possibly five, times from different groups of reptiles in ancient times. Mammals also differ from reptiles in the way the jaw is hinged to the skull, and in the way the limbs attach to the rest of the skeleton. The differences arose during the evolution of fast-moving, dog-like carnivorous mammals from reptiles that had less efficient locomotion and weak jaws liable to be broken by struggling prey. Several different groups of reptiles evolved the same method of strengthening the jaw and of moving faster, all of them giving rise to descendants classified as mammals. Mammals, then, are not members of a monophyletic clade but of a polyphyletic **grade** of organization attained more than once by related, but separate, evolutionary lines (Figure 18–4).

In order to determine phylogenetic relationships among organisms, scientists must carefully assess whether features are ancestral or derived. In theory, taxonomy reflects phylogeny. In practice, many taxa are actually based mainly on morphology, for convenience, and are in fact polyphyletic.

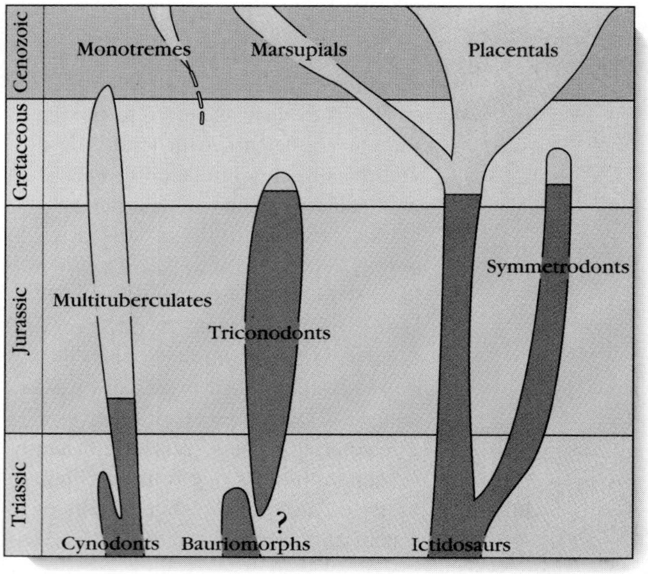

FIGURE 18–4

The evolutionary lines leading to three groups of modern mammals. Although the relationships of the monotremes continue to be controversial, marsupials and placentals (along with extinct symmetrodonts and ictidosaurs) belong to a single clade, or evolutionary lineage. Multituberculates and cynodonts form a separate distinct clade while the extinct triconodonts and bauriomorphs may form a third mammalian clade.

18–D THE FIVE KINGDOMS

The most inclusive taxa are the kingdoms. Linnaeus's system of classification had two kingdoms, the plants (Plantae) and animals (Animalia). This seemed reasonable in his day, since the familiar land plants and animals were clearly very different. Plants did not move around; they did not eat, but seemed to need only water in order to grow. Animals were **motile;** that is, they could move from place to place. Animals had to eat plants, or each other, in order to stay alive. On a microscopic level, plants could be seen to have cell walls, which were lacking in animal cells. Fungi seemed to be aberrant plants, because they had cell walls and root-like structures but lacked the green pigments of the other "plants."

Today, however, it is apparent that many forms of life do not fit neatly into either the plant or the animal camp. Some organisms, such as *Euglena,* seem to fit both descriptions. *Euglena* (see Figure 19–13) has a rather stiff covering, not as thick as a plant cell wall but certainly giving more protection than a plasma membrane. *Euglena* also has chloroplasts and carries on photosynthesis. However, it also has animal-like features: it has a flagellum that it uses to swim, and it can engulf other organisms and digest them as food. Bacteria present another taxonomic problem, because they have cell walls but may also have flagella, used to move around. Most cannot make their own food, but some can carry on photosynthesis. These and other organisms give evidence that the division into plant and animal kingdoms is artificial, with a confusing zone between the two.

Modern attempts to revise biological classification at the kingdom level have been many and varied. For consistency, this book uses a scheme based on the five-kingdom system popularized by ecologist Robert Whittaker. In this system the kingdoms are separated according to two main criteria: degree of cellular complexity and mode of nutrition. The kingdom Prokaryotae includes organisms with prokaryotic cells, while eukaryotes are divided into the other four kingdoms. The kingdom Protista is composed of unicellular (one-celled) eukaryotic organisms (Figure 18–5). Eukaryotes with multicellular (many-celled) bodies are placed into one of the other three kingdoms—kingdom Fungi, kingdom Plantae, and kingdom Animalia—largely on the basis of how they obtain food. Most of the members of the kingdom Plantae are photosynthetic and make their own food. Most members of the kingdom Animalia are **ingestive,** engulfing or swallowing food and digesting it internally. Members of the kingdom Fungi are **absorptive.** They absorb organic molecules from outside their bodies directly through their exterior plasma membranes. We shall consider each kingdom separately.

Kingdom Prokaryotae

All prokaryotic organisms—the bacteria and cyanobacteria—are placed in the kingdom Prokaryotae. Prokaryotes lack nuclear membranes and mitochondria, chloroplasts, and other membrane-bounded organelles found in eukaryotic cells. Their DNA consists primarily of one circular double helix, and they divide and reproduce without the nuclear divisions of meiosis or mitosis found in eukaryotes. Prokaryotes also differ from eukaryotes biochemically in such things as the materials and structure of their cell walls, the size and composition of their ribosomes, and some of their metabolic pathways.

Kingdom Protista

The Protista are microscopic organisms that are common in fresh and salt water. They live almost everywhere there is moisture—in the film of water that coats each particle of soil as well as in your mouth and gums. Most are harmless to humans, but a few cause

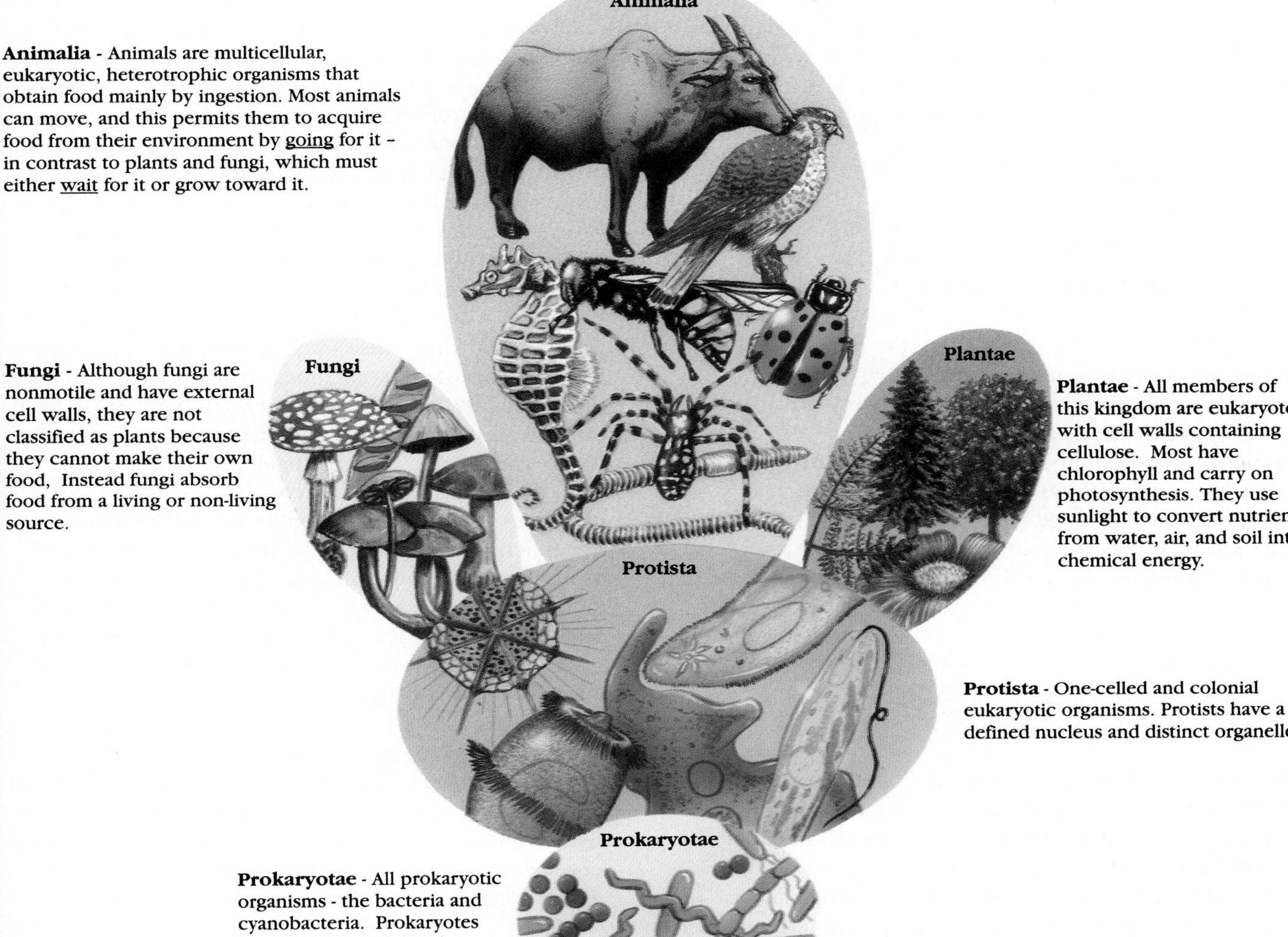
Animalia
Animalia - Animals are multicellular, eukaryotic, heterotrophic organisms that obtain food mainly by ingestion. Most animals can move, and this permits them to acquire food from their environment by going for it - in contrast to plants and fungi, which must either wait for it or grow toward it.
Fungi
Fungi - Although fungi are nonmotile and have external cell walls, they are not classified as plants because they cannot make their own food, Instead fungi absorb food from a living or non-living source.
Plantae
Plantae - All members of this kingdom are eukaryotes with cell walls containing cellulose. Most have chlorophyll and carry on photosynthesis. They use sunlight to convert nutrients from water, air, and soil into chemical energy.
Protista
Protista - One-celled and colonial eukaryotic organisms. Protists have a defined nucleus and distinct organelles.
Prokaryotae
Prokaryotae - All prokaryotic organisms - the bacteria and cyanobacteria. Prokaryotes have no true nuclei and contain no organelles.

FIGURE 18-5

The five-kingdom classification system.

diseases such as malaria, dysentery, and sleeping sickness. Some of the macroscopic, colonial protists are slime molds. As eukaryotes, they have nuclear membranes and linear chromosomes (composed of DNA and proteins) that can go through mitosis and, in most forms, meiosis. The members of the three higher kingdoms undoubtedly evolved from protist ancestors. Modern protists show ways of life that closely parallel those of higher eukaryotes, some being photosynthetic, some ingestive, and some absorptive, some motile and some nonmotile, some with cell walls and some without. All combinations of these features occur in various members of this group.

Kingdom Fungi

Fungi include mushrooms, toadstools, puffballs, bracket fungi, rusts, molds, and yeasts. They were once classified as plants because they are nonmotile and because they have external cell walls. However, fungi cannot make their own food; they must absorb food from a living or nonliving organic source. In many cases, they secrete digestive enzymes that digest food outside their bodies before they can absorb it. As nutrients around their bodies become depleted, they must grow outward into fresh food sources. Ecologist Whittaker felt that fungi should be separated from plants because they are completely absorptive, whereas most plants absorb only the simple raw materials they need to make their own food. In addition, fungi and green plants differ in cell wall composition, body plan, and reproduction.

Kingdom Plantae

The plant kingdom includes the multicellular algae as well as all the familiar multicellular land plants—the mosses, ferns, grasses, shrubs, and trees. All members of the plant kingdom are eukaryotes with cell walls containing cellulose. Most contain chlorophyll and carry on photosynthesis inside chloroplasts, although a few species have lost their chlorophyll and obtain all of their nutrients by absorption. Most plants are immobile and remain in one place throughout life. They depend on water and air to bring nutrients to them, and they also grow out and intercept more sunlight and nutrients.

Kingdom Animalia

Animals include organisms as diverse as humans and sponges, jellyfish and sea anemones, slugs, beetles, and slime hags. Animals are multicellular, eukaryotic, heterotrophic organisms that obtain food mainly by ingestion. Most animals can move, and this permits them to acquire food from their environment by *going* for it—in contrast to plants and fungi, which must either *wait* for it or *grow* for it! All but the simplest animals produce gametes (eggs and sperm) in multicellular organs, and the fertilized eggs develop into multicellular embryos.

Difficulties With the Five-Kingdom System

We have seen that all classification schemes pose problems, and the five-kingdom system is no exception. There is clearly a distinction between prokaryotes and eukaryotes. However, the four eukaryotic kingdoms are unsatisfactory for two main reasons. First, modern taxonomy should aim to classify organisms so that the members of a group are related more closely to each other than to members of other groups. The kingdom Protista violates this rule by containing organisms that are related more closely to some plants than to any other protists. Second, many different protistan lines gave rise to multicellular groups, and so none of the three higher kingdoms contains the single ancestor (if there is one) of all its members. For instance, some biologists think that different groups of fungi arose from different protistan groups rather than from a common fungal ancestor. In fact, multicellular lines originated from protists at least 17 different times, and it seems likely that the eukaryotic condition itself originated independently more than once. Breaking up the world of life into truly monophyletic groups would give us more kingdoms than we wish to bother with for most purposes. Despite its inadequacies, the five-kingdom system is widely used today, not because it is natural but because it is convenient.

In the five-kingdom system of classification, organisms are grouped according to their degree of cellular complexity and mode of nutrition. Although this five-kingdom plan is convenient, it probably includes several evolutionary lines in each kingdom, often grouping together organisms that are not closely related.

18–E THE PROBLEM OF VIRUSES

Viruses are tiny particles composed of nucleic acid inside a protein coat, which is sometimes surrounded by a membrane. However, they lack many of the features of living cells and occupy a strange limbo somewhere between the living and nonliving worlds. Viruses are like living organisms in that they possess genetic material, are composed of nucleic acids, and are capable of mutation and recombination. Viruses can therefore evolve and adapt to their changing environments. On the other hand, viruses are not made up of cells, and

Table 18–2
Differences Between Viruses and Cells

Character	Viruses	Cells
Structure	Virus particle: nucleic acid core inside protein coat	Cell containing nucleic acids, lipid-protein membrane, ribosomes, cytoplasm, etc.
Nucleic acids	DNA or RNA, but not both	Both DNA and RNA
Enzymes	One or a few; e.g., lysozyme (digests bacterial cell wall), polymerase (replicates viral genome)	Many enzymes; diverse functions
Metabolism	None; relies on host cell metabolism for monomers, protein synthesis machinery, and some enzymes of nucleic acid synthesis	Makes own ribosomes and enzymes needed for synthesis of proteins, nucleic acids, etc.
Reproduction	Nucleic acid genome and coat proteins produced separately, then assembled into virus particle	Division into two similar cells following growth

BIO-BIT

Some viruses have been shown to cause mutations in host cells that can lead to cancer.

they have no ribosomes, nor the metabolic machinery for protein synthesis and energy generation. ■

Lacking these components, viruses are clearly parasites. They can reproduce only inside living cells, and even here their reproduction is unique. Cells reproduce by growing and eventually dividing into two new cells. By contrast, viruses are disassembled into their separate components: nucleic acid genomes and protein coats. The virus takes over the host cell and causes its metabolic machinery to produce dozens to hundreds of new viral genomes, and thousands of protein subunits to make new viral coats. Then these components are assembled into new viruses, the same size as the original ones (Figure 18–6): unlike cells, viruses do not grow (Table 18–2).

Another bizarre feature of viruses is that many of them can be crystallized, a common enough property of minerals and even of fairly complex organic molecules, but certainly not of living cells. Furthermore, when dry, crystallized viruses are dissolved in solutions and exposed to living host cells, they soon establish infections and get back to the business of causing the cell to make more viruses.

Because of all these odd characteristics, viruses are not considered real living organisms and do not belong in any of the five kingdoms. Nevertheless, because viruses are active only inside living cells, and indeed may have devastating effects on their hosts (see *A Journey into Human Health*: AIDS, Chapter 25), the study of viruses is clearly within the province of biology.

Viruses are tiny, parasitic particles that can be produced only inside living cells. Viruses do not eat, grow, metabolize, make proteins, or reproduce by themselves, but they do evolve.

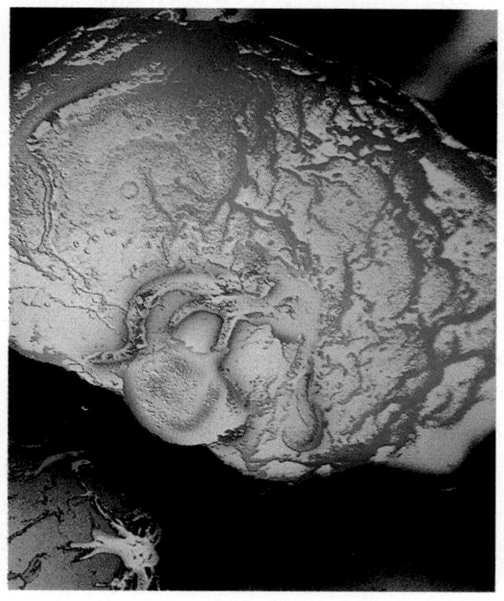

FIGURE 18–6

A false-colored image of a white blood cell infected with the AIDS virus (shown in yellow and orange). (CDC: Science Source/Photo Researchers, Inc.)

Table 18–3
Major Groups of Viruses That Infect Animals

Nucleic acid	Group	Some diseases caused
DNA	Poxviruses	Smallpox, cowpox, myxomatosis in rabbits, diseases in fowl
	Herpesviruses	Human oral and genital infections, Epstein-Barr infections, tumors
	Adenoviruses	Human respiratory and intestinal infections, conjunctivitis, sore throat, tumors
	Papovaviruses	Human warts; cancers in other animals
RNA	Paramyxoviruses	Human rubeola, mumps; canine distemper; Newcastle disease of chickens
	Myxoviruses	Influenza of humans, other animals
	Retroviruses	Rous sarcoma of chickens; mouse mammary tumor; feline leukemia; AIDS
	Rhabdoviruses	Rabies, various infections
	Reoviruses	Vomiting and diarrhea in children; Colorado tick fever
	Togaviruses	Human rubella, yellow fever, dengue, equine encephalitis, etc.
	Picornaviruses	Intestinal infections (enteroviruses), poliomyelitis, common cold (rhinoviruses)

18–F VIRAL REPRODUCTION

Viruses cause many diseases (Table 18–3), and it is interesting to understand how they go about the business of pirating and parasitizing host cells, mainly because that host might be you. Three main types of reproductive cycles have been found among viruses.

1. A **lytic cycle** occurs when a virus invades a cell, destroys the cell's DNA, takes over its metabolic machinery, and causes the cell to make as many as several thousand new virus particles. Viral enzymes then cause the cell to break, or **lyse** (Figure 18–7). The released viruses disperse to infect new host cells. A cell invaded by a lytic virus is almost invariably killed by it within a very short time. This is typical of many phages and also of viruses that cause colds and poliomyelitis ("polio").

2. A **lysogenic cycle** is typical of some bacteriophages, parasites of bacteria, called temperate phages. These may either go through a lytic cycle and destroy the host cell, or may instead enter a dormant phase in which their DNA is joined to the host's (Figure 18–7). Here the viral DNA is replicated with the host's in each cell generation. Certain external stimuli can cause a lysogenic cell's phage DNA to enter the lytic cycle, releasing intact phages. The herpes viruses that parasitize animals, such as those that cause cold sores in humans, show comparable cycles.

Some lysogenic bacterial cells are of importance to human health. The bacteria that cause diphtheria, botulism, and scarlet fever produce the toxins responsible for these diseases only when they contain particular phages, which carry the genes encoding the toxins.

Phages that are released when a lysogenic cell finally lyses may carry along part of the bacterial DNA, which is later inserted into the DNA of a new host bacterium. This process, called **transduction** ("leading across"), produces genetic recombination in the new host. As we have previously seen, viruses that can carry DNA from one cell to another are used to study bacterial genetics. In the lytic and lysogenic cycles, viruses are assembled inside the host cell and released when the cell lyses.

FIGURE 18–7

Reproductive cycles in a virus. (a) Bacteriophage reproduction. In a lysogenic cycle, the host may survive for many generations with the phage genetic material incorporated into the host genome, until some condition triggers the phage to become lytic. In a lytic cycle, the phage takes over the host cell and destroys it. **(b)** Continuous production of viruses. An animal cell becomes infected by a virus and produces many new viral genomes and protein coats, which combine in the cytoplasm to form nucleocapsids. These move to the plasma membrane, where they associate with new viral envelope proteins, also produced by the host under the direction of viral RNA. The finished virus particle buds off from the plasma membrane.

(a)

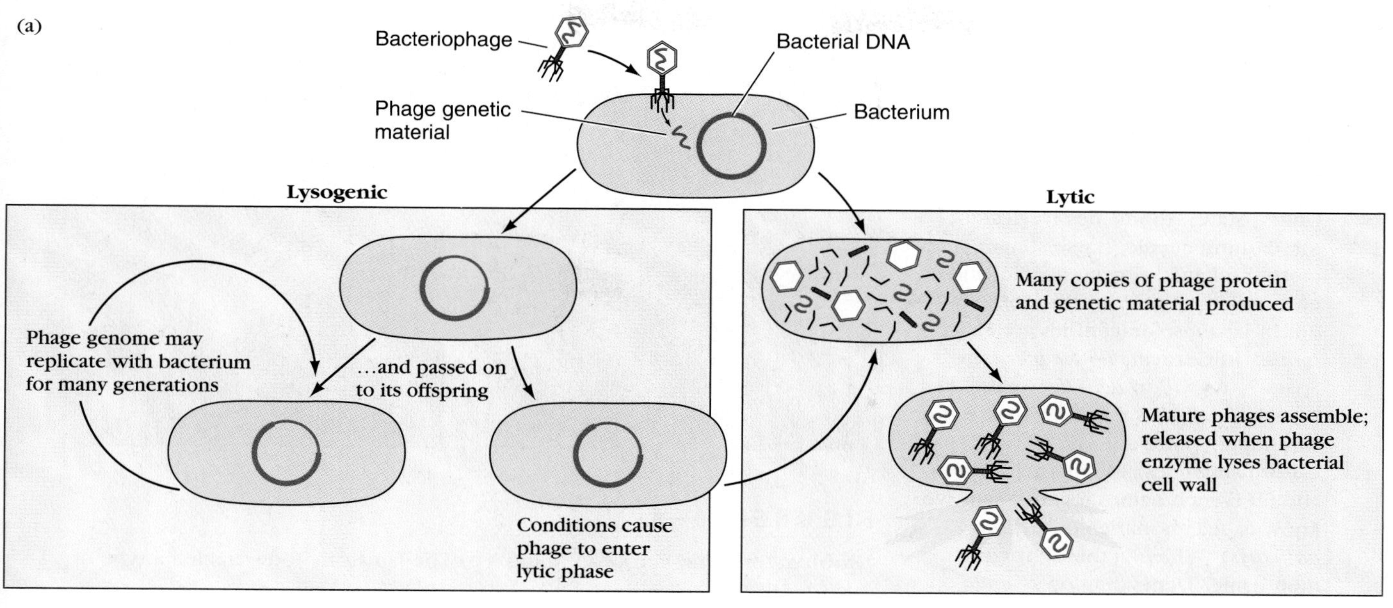

(b)

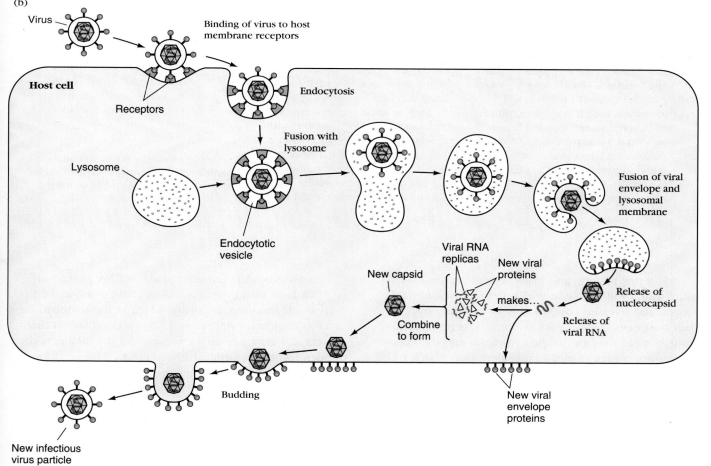

Virus
Binding of virus to host membrane receptors
Host cell
Receptors
Endocytosis
Fusion with lysosome
Lysosome
Endocytotic vesicle
Fusion of viral envelope and lysosomal membrane
Viral RNA replicas
New viral proteins
New capsid
makes...
Release of nucleocapsid
Combine to form
Release of viral RNA
Budding
New viral envelope proteins
New infectious virus particle

A Journey into Science in Process

The Name Game

Common names can be colorful, but they are also ambiguous because there can be many names for one organism. For example, in the United States, the names Devil's doctor, darning needle, horse stinger, mosquito hawk, sewing needle, and blue darner all refer to dragonflies. But in Japanese dragonflies are called *tombo,* in German, *Teufelspferd,* in Spanish, *Caballito del Diablo,* and in French, *la libellule.* And this is a mere beginning. There are said to be 90 local British names, 140 German names, and 60 French names for the plant we know as marsh marigold (*Caltha palustris*). Which is the right common name? Depending on where you live, any of these names might be correct, but in a strict scientific sense, none of them is right.

(a)

(b)

FIGURE 18–A

"Robins." (a) The European Robin. **(b)** The Jamaican Tody, labeled a type of robin by British colonists. (a, F. Lanting; b, B. Brower Hall)

This name game becomes even crazier when we consider another problem with common names—the same name is often applied to different organisms. Take robins, for example. The European Robin is a chickadee-sized bird. It has a greenish gray back, a red-orange face and breast, and white underparts. It has a roly-poly, no-neck look and usually is found in gardens, hedges, and woods (Figure 18–A). But once British colonists reached North America, they seemed to forget everything about the European Robin except for its red breast. Accordingly, they named nearly every new bird they encountered that had any orange or red on its body as some sort of robin. So, in addition to the American Robin, the medium-sized, lanky, worm eater that has a slaty gray back and brick red breast (which, by the way, looks nothing like the European Robin), they also named these other birds:

the ground robin (officially recognized* as the Rufous-sided Towhee, a bird the size of an American Robin, which has a black head and back, white breast, and brick red sides),

*Don't be puzzled by the capitalizations. Birds are the only animals that have official common names designated by an international group of ornithologists.

3. A few phages, and many animal viruses have a third kind of life cycle in which new viruses are produced and released continuously by budding from intact host cells. New copies of the viral genome and protein coat combine in the cytoplasm and then move to the host's plasma membrane. Here they attach to viral envelope proteins that the host has made and inserted into its own membrane. The host's plasma membrane then bulges out around the forming virus particle, until at last the virus, surrounded by its new membranous envelope, buds off from the host cell (Figure 18–7). Enveloped animal viruses that bud from their host cells in this way include influenza, measles, mumps, and rabies viruses.

Viruses with genomes made of RNA instead of DNA have novel mechanisms of replication. If the RNA is single-stranded, it usually serves as the template for a complementary RNA strand. This second strand then acts as a template for new copies of the viral genome identical to the original RNA strand.

Other viruses with single-stranded RNA genomes code for an unusual enzyme called **reverse transcriptase.** Contrary to the usual flow of genetic information, this enzyme uses the viral RNA as a template to make a complementary DNA strand! Next, the DNA acts as the template to make double-stranded DNA, which is inserted into the host cell genome. Here it is transcribed into many new viral RNA molecules.

the Canadian robin (now officially recognized as the Cedar Waxwing, a sleek, brown, crested bird with a black mask; the only red on its body are the tiny, waxy, red tips on some of its wing feathers),

the golden robin (officially known as the Northern Oriole, a medium-sized, black and orange bird),

the robin snipe (the Red Knot, a short-billed, long-legged, chunky-bodied wading bird; in the spring its breast and head are brick red),

. . . and strangest of all, the sea robin (the Red-breasted Merganser, a green-headed, sea-going *duck,* whose breast is rust-colored).

In Jamaica this plague of "robinism" reached the ultimate lunacy, as the British colonists to this West Indian island labeled the Jamaican Tody, a tiny, leaf-green bird with a bright red bill and red throat, "Robin Redbreast" (Figure 18–A).

At least all of these are birds. Things get even more confusing when one common name refers to two completely different organisms. For example, in Jamaica, "bats" are those colorful insects that we call butterflies and moths. To distinguish the furry, insect-eating mammals that we know as bats, Jamaicans call them "rat-bats." Such different applications of names led to worldwide confusion as to what plant or what animal a person might be talking about.

Scientific Names

Carolus Linnaeus must have been fascinated with naming and identifying things because he gave a two-word, standard, "scientific" name to everything in the world—rocks and minerals, as well as plants and animals. These names for rocks and minerals soon disappeared, but we still use binomials for plants and animals. To the uninitiated, scientific names may be baffling, hard-to-pronounce, and complicated. But once you begin to try to say them, you will find that they aren't nearly as difficult as they first appear. Because these names are often long and in Latin, many people are afraid of saying them wrong and so avoid saying them at all. There are all sorts of rules about proper pronunciation of scientific names, but until someone corrects you, be fearless. Pronounce every consonant and vowel of a scientific name and, with a little practice, you will surprise yourself as the names roll grandly off your tongue. Try these:

Liriodendron tulipifera (pronounce it "leerio-dend-ron tuli-**pif**-erah"), the name of the tulip tree, translates into "lily-tree that has tulips."

Liquidambar styraciflua (pronounced "liquid-am-bar sty-ras-ih-**floo**-ah"), the name of the sweet gum tree.

Rana catesbeiana (pronounced "ray-nah ka-tez-bee-**a**-nah"), the name of the American Bullfrog, named for Marc Catesby, a Colonial naturalist who first brought many American species to the attention of the European world. *Rana catesbeiana* means "Catesby's frog."

Mephitis mephitis (pronounced "mef-ee-tus mef-**ee**-tus") is the striped skunk. This binomial translates into something like "bad-smelling bad-smelling," or "skunky skunk."

The major advantage of Latin binomials is that they are unambiguous. No two species of animals and no two species of plants can have exactly the same binomial. When correctly used, they ensure that scientists, zookeepers, horticulturalists, etc., whether they are Brazilian, French, or Chinese, are referring to exactly the same organism. The taxonomic rules that govern the scientific naming of animals are different from those that govern the scientific naming of plants, though. Because the two systems are completely separate, a plant and an animal may occasionally share the same Latin name. At first glance this might seem to make the system of Latin binomials as ambiguous as common names, but in practice it does not, mainly because it is usually difficult to mistake a plant for an animal.

Viruses that produce reverse transcriptase are called **retroviruses.** Since their genomes spend some time joined to the host DNA, new retrovirus particles may also incorporate host genes and carry them to new host cells. More than 20 different vertebrate **oncogenes,** mutated genes from a former host cell that can cause cancer in new host cells, have been identified in various retroviruses (Section 10–G).

The retrovirus that causes AIDS (acquired immune deficiency syndrome) can go through reproductive cycles resembling all three types we saw earlier. The AIDS virus enters the bloodstream and binds specifically to receptors on a T helper cell (part of the immune system; Chapter 25). Inside the cell, the virus loses its protein coat, releasing its RNA genome and reverse transcriptase enzyme into the cytoplasm. The enzyme makes a DNA molecule complementary to the RNA genome. This DNA enters the nucleus and inserts itself into a cell chromosome. After this, one of three things may happen. First, the virus may quickly direct the cell to produce a flood of new AIDS viruses, which are released so rapidly that the cell is destroyed. Or, the virus may instead enter a latent period lasting up to ten years before it is activated and causes the cell to churn out a destructive horde of viruses. Third, the virus may cause a persistent infection, with production of new viruses slow enough so that few host cells are killed. AIDS occurs when so many T helper cells have been destroyed that the body's immune system can no longer fight off diseases. (See *A Journey into Healthy Living:* AIDS in Chapter 25.)

Viruses pirate the cellular components of their host cells and use them to reproduce new viruses. The host cell may or may not be killed by the invading virus.

18–G VIRUSES AND EVOLUTION

The peculiarities of viruses raise the question: what is their evolutionary origin? Several answers have been proposed. The first is that viruses are evolutionary relics, descended from ancestors that never evolved into true cells. When biologists realized that viruses depend totally on the very complex protein-synthesis and energy-generating machinery of living cells, most discarded this idea. Second, viruses may be reduced cells, which became parasites inside other cells and eventually jettisoned most of their own cell components and genes. These things were, after all, readily available in their host cells. Some fairly large viruses, containing dozens of genes and surrounded by a lipid-protein membrane, might be viewed as stripped-down cells. A third view is that viruses are neither retarded pre-cells nor regressed cells, but renegade genes that must return "home" to be replicated. This view is supported by the fact that the genetic similarity seems to be much closer between virus and host than between one virus and another. Or, this similarity could be explained by assuming that viruses have captured host cell genes during their evolution. In this view, viruses may have started as **plasmids,** small independently replicating nucleic acid molecules found today in some bacteria, yeast, and mammalian cells (see Figure 9–A).

Whatever the case, viruses clearly play an important role in the evolution of cellular organisms. First, they exert selective pressure. Second, many viral genomes are inserted into the host DNA and later released from it, often carrying part of the host's genome along into new host cells of the same or different species. Third, studies show that viral genes have become permanent parts of most species' genomes.

3? Viruses themselves often evolve very rapidly. Those that cause colds or influenza mutate often, producing offspring with novel genes. Hence some of the viruses can always find some hosts that have not yet built up immunity to the proteins produced by their particular genes. ■

The evolutionary origin of viruses is unclear. However, viruses do play an important role in the evolution of cellular organisms by exerting pressure and by altering the genome of the host.

SUMMARY

1. Taxonomy is the branch of biology concerned with relationships among organisms and with their classification.
2. The basic unit of classification is the species; each species is given a unique Latin binomial, denoting its genus and species.
3. Species are grouped into progressively more inclusive taxa. The main levels in the taxonomic hierarchy, from most to least inclusive, are: kingdom, division or phylum, class, order, family, genus, and species. A taxon in each higher level contains one or more taxa of the next lower level.
4. Taxa are not units of intrinsic nature, waiting to be discovered by humans, but instead are artificial inventions that reflect evolutionary relationships, to help us think about living organisms in an orderly manner.
5. Biologists often disagree about how the rules of taxonomy should be applied and where the lines should be drawn to define taxa.
6. In theory, living things are classified by phylogenetic relationships, but these are often difficult to disentangle, and the sheer number of existing species precludes drawing up a phylogenetic tree that encompasses all known organisms. In practice, therefore, living things are usually classified by morphology. Other features, such as physiology, biochemistry, behavior, geographic distribution, and fossil evidence, are also used.
7. This book uses the taxonomic system that divides organisms into five kingdoms: Prokaryotae, the prokaryotes; Protista, the unicellular eukaryotes; Fungi, the fungi; Plantae, the plants; and Animalia, the animals.
8. The definitions of the kingdoms are based largely on modes of nutrition and cellular organization. Several evolutionary lines or organisms are almost certainly grouped into each kingdom.
9. Viruses do not share all the features of cellular organisms and do not belong in any of the five kingdoms of living organisms. They resemble cells in that they contain nucleic acid and protein molecules, and some are surrounded by membranous envelopes of lipid and protein. Unlike cells, viruses lack the metabolic machinery to make proteins and to generate energy, they do not grow, and many viruses can be crystallized. Cells reproduce by dividing in two, but hundreds of viruses may be produced in a host cell after infection by a single virus.
10. The evolutionary origin of viruses is uncertain. However, it is clear that they play an important evolutionary role by acting as selection pressures and by transferring genes between hosts.

SELF-QUIZ

1. Modern classification is based on:
   a. taxonomy
   b. phylogeny
   c. morphology
   d. fossils
   e. autotrophy
2. All of the following make it difficult to construct acceptable classification schemes *except:*
   a. deciding where to impose artificial cutoffs in the midst of a naturally continuous series of organisms
   b. convergent evolution
   c. differences in rates of evolution for different characters
   d. persistence of conservative characters
   e. the large numbers of species that must be accommodated
3. The Latin binomial for the common dog is properly written:
   a. canis familiaris
   b. Canis Familiaris
   c. Canis familiaris
   d. *Canis familiaris*
   e. *canis familiaris*
4. In which of the following lists are the levels of the taxonomic hierarchy *not* arranged in correct descending order?
   a. phylum, order, family
   b. class, family, genus
   c. class, order, family
   d. family, class, order
   e. order, family, genus
5. Characters of two different organisms that have evolved from the same structure in an ancestral form but now have very different appearance and function can properly be termed (indicate all correct answers):
   a. derived
   b. homologous
   c. polyphyletic
   d. conservative
   e. convergent
6. You are given a microscope slide on which is mounted some biological material. On examining it, you observe that there are numerous individual cells containing chloroplasts and swimming around rapidly. This material belongs in the kingdom ___.
7. One reason that viruses are considered nonliving is that:
   a. they lack replicable nucleic acids
   b. their nucleic acids do not code for proteins
   c. they cannot make their own food molecules
   d. they cannot carry out their own reproduction
   e. they do not undergo mutation and so do not become adapted to changes in their environment

THINKING CRITICALLY

1. Although modern taxonomy tries to classify organisms on the basis of their phylogenetic relationships, in practice most organisms are classified according to their morphology. Why?
2. Review the characteristics of life listed in Section 1-B. What features do viruses share with cellular organisms? Which do they lack?
3. Why is it easier to classify organisms that reproduce sexually than it is to classify organisms that only reproduce by asexual division?

SELECTED KEY TERMS

absorptive, *p. 362*
ancestral characters, *p. 360*
binomial, *p. 358*
clade, *p. 361*
class, *p. 358*
conservative characters, *p. 360*
convergent evolution, *p. 360*
derived characters, *p. 360*
division, *p. 358*
family, *p. 358*
genus, *p. 358*
grade, *p. 361*
homologous characters, *p. 360*
ingestive, *p. 362*
kingdom, *p. 358*
monophyletic, *p. 361*
motile, *p. 362*
order, *p. 358*
phylogeny, *p. 359*
phylum, *p. 358*
plasmid, *p. 370*
polyphyletic, *p. 361*
species, *p. 358*
taxon, *p. 358*
taxonomy, *p. 358*
type specimen, *p. 358*

SUGGESTED READINGS

Margulis, L., and K. V. Schwartz. *Five Kingdoms: An Illustrated Guide to the Phyla of Life on Earth,* 2d ed. San Francisco: W. H. Freeman, 1988.

May, R. "How many species inhabit the Earth?" *Scientific American,* October 1992.

Sibley, C. G., and J. E. Ahlquist. "Reconstructing bird phylogeny by comparing DNA's." *Scientific American,* February 1986. Direct studies of genes have revealed that some birds long thought to be close relatives have instead developed close resemblances by convergent evolution.

Simons, K., H. Garoff, and A. Helenius. "How an animal virus gets into and out of its host cell." *Scientific American,* February 1982.

Weiss, R. "The viral advantage." *Science News* 136:200, September 23, 1989. A chilling account of recent outbreaks of viral diseases.

CHAPTER 19

CONCEPT GUIDE

After reading this chapter you should be able to:

1. Explain how prokaryotes achieve genetic diversity. (Section 19–A)
2. Compare the lifestyles of autotrophic and heterotrophic bacteria. (Section 19–B)
3. Explain why some scientists place the archaeobacteria in their own kingdom. (Section 19–C)
4. Describe three types of symbiotic relationships. (Section 19–D)
5. Describe the different modes of locomotion found among protozoans. (Section 19–E)
6. Compare the structure and habitats of members of the three phyla of photosynthetic protists. (Section 19–F)
7. Compare the structure and habitats of members of the seven phyla of heterotrophic protists. (Section 19–G)
8. Describe the adaptive advantages of multicellularity. (Section 19–H)
9. Explain how fungi are different from plants and animals. (Section 19–I)
10. Describe the major differences between the four divisions of fungi. (Section 19–J)
11. Explain why mycorrhizae are valuable to many plants. (Section 19–K)
12. Explain how humans use bacteria and fungi in food production. (Section 19–L)

The pool of a sulfur spring in Yellowstone National Park. The unusual color patterns seen within the waters of these pools are due to bacteria (*Sulfolobus*) that tolerate these high temperatures and use sulfur compounds for energy. Read more about the unusual lifestyles of bacteria and their interactions with humans in Sections 19 B–D. (J. Brandenburg/Minden Pictures)

Bacteria, Protists, and Fungi

The kingdoms Prokaryotae, Protista, and Fungi contain mostly microscopic organisms (**microorganisms**). Despite their small size, many of them have specialized cell structures, metabolism, or ways of life not found among larger organisms. Studying them, we also find clues to two important evolutionary advances, the origin of eukaryotic cells and of multicellularity.

These points are of biological interest, but there is an even more compelling reason to study microorganisms: their importance in the economy of nature and of human society. Autotrophic bacteria and protists are the main source of food for many other organisms, particularly in aquatic habitats, where they are eaten by many small animals and heterotrophic protists. Other heterotrophic microorganisms live by recycling: they are **saprobes**, organisms that break down dead organic matter and release nutrients that are reused by other organisms. Without the activities of these decomposing bacteria and fungi, plants (and animals that eat plants) could not exist. Nor could they survive without nitrogen-fixing bacteria, which provide nitrogen in a form plants can use to make amino acids for their proteins.

Although many microorganisms are independent, others cannot live without forming close associations with members of other species. These symbiotic relationships may be mutually beneficial, or one member may be a parasite that damages its larger host. Some of these parasites cause devastating diseases of humans and of domestic animals and crops.

KEY CONCEPTS

- Microorganisms include free-living autotrophs (bacteria, cyanobacteria, and unicellular algae), saprobes (bacteria and fungi), and predators (protozoa). Many members of all three kingdoms live in symbiotic relationships with other species, as parasites or mutualistic symbionts.
- All three kingdoms contain many disease-causing members, some of which have changed the course of human history and caused untold loss of life, labor, and property.
- Bacteria and fungi play important ecological roles as decomposers. Some bacteria can remove nitrogen gas from air, incorporate or "fix" it in proteins and so ultimately support all plant life on Earth.

Curiosity Questions

1? Why does swamp mud smell so bad? (See page 379)

2? What causes leprosy? (See page 379)

3? What causes food poisoning? (See page 400)

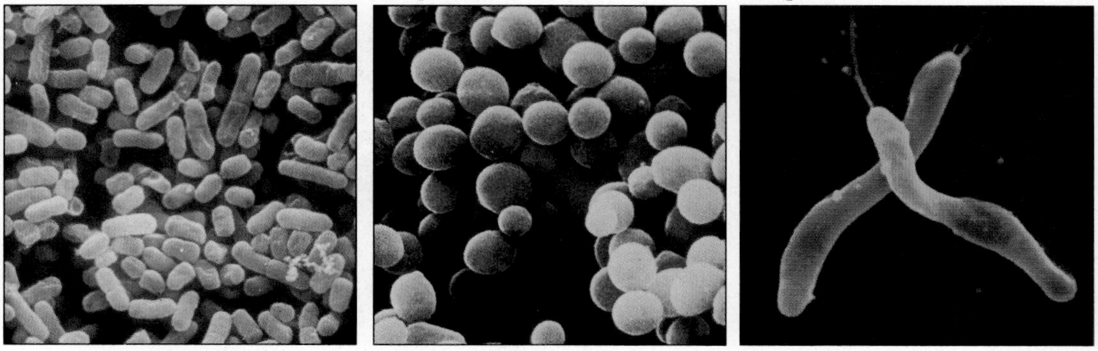

(a) Rods (b) Spheres (c) Spirals

FIGURE 19–1

Cell shapes of bacteria. (a) Rods: *Klebsiella pneumoniae.* **(b)** Spheres: cells of *Staphylococcus aureus* live on human skin, where they stick together in clumps. **(c)** Spirals: *Aquaspirillum sinosum* has a flagellum at each end that it uses in locomotion. (a, R. Kessel & G. Shih/Visuals Unlimited; b, D. Phillips/Visuals Unlimited; c, S. Flesler/Visuals Unlimited)

KINGDOM PROKARYOTAE: THE BACTERIA

The kingdom Prokaryotae[1] contains the bacteria. Fossil bacteria dating from 3.5 billion years ago are the earliest signs of life on Earth. Today, bacteria still outnumber other organisms, thriving in almost every conceivable habitat. But because of their tiny size, they were the last major group of organisms to be discovered—some large ones were first seen in 1676. In the 1860s and 1870s, Louis Pasteur and Robert Koch discovered the role of bacteria in causing food spoilage and many diseases. To most people, bacteria mean disease and decay—enemies of human health and wealth. However, we are now coming to appreciate the broader roles of bacteria in the environment. Bacteria change and recycle mineral nutrients, help clean up pollution, and combat many organisms harmful to humans.

The prokaryotic cells of bacteria are generally much smaller, and much simpler in structure, than eukaryotic cells. Their definitive feature is the absence of a nuclear envelope separating the genetic material from the cytoplasm (Section 5-A). Bacterial cells come in three general shapes: rods, spheres, and spirals (Figure 19-1). In some species, new cells stick together after the parent cell divides, forming characteristic filaments (strings) or clusters. These cells are usually independent, but in some species of prokaryotes the cells cooperate, with a limited degree of division of labor.

[1]Kingdom Prokaryotae is the current name for Kingdom Monera, used in previous editions of this book. The new name conforms with the latest edition of *Bergey's Manual of Systematic Bacteriology,* the authoritative source on nomenclature of prokaryotes.

19–A REPRODUCTION AND EVOLUTION IN PROKARYOTES

Prokaryotes usually reproduce by **binary fission,** division into two genetically identical new cells (Figure 19-2). In the ideal environment of a laboratory culture, some bacteria divide every 20 to 30 minutes. However, such good growing conditions rarely occur in nature.

Bacteria do not reproduce sexually, as eukaryotes do, by fusion of two gametes with roughly equal amounts of genetic material. In prokaryotes, genetic information from two individuals can be combined into one by three means (Figure 19-3):

1. **Transduction.** In the most common means of genetic recombination, a bacteriophage (virus) carries bacterial DNA from one bacterium to another, attached to the viral genome (Section 9-A).
2. **Transformation.** DNA from a broken cell is taken up by a living one.
3. **Conjugation.** DNA passes from one cell to another by way of a protein strand called a **pilus.** The amount of DNA transferred varies from part of a gene to an entire genome.

(a)

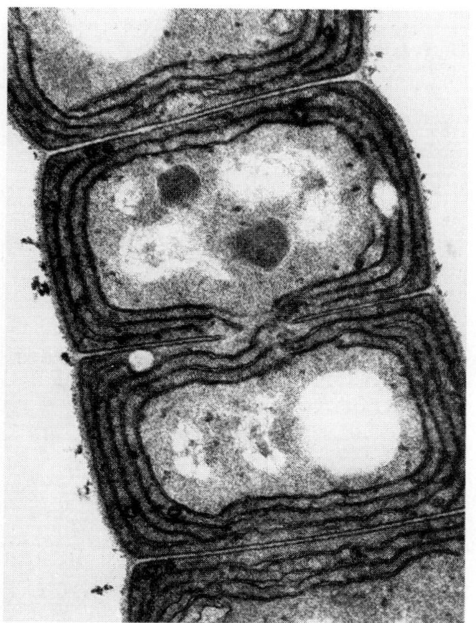
(b)

FIGURE 19-2

Binary fission. (a) *Oscillatoria* is a filamentous cyanobacterium. **(b)** This electron micrograph shows a segment of this one filament. The cell in the middle is dividing into two new cells of equal size as the new cell wall forms at the edges and moves toward the center. Notice the several layers of photosynthetic membranes inside each cell. (a, Walker/Photo Researchers, Inc.; b, M. Ledbetter/Biophoto Associates)

New gene combinations are the raw material of natural selection. The transfer of genes from cell to cell must play a role in prokaryote evolution, but mutations are probably more important. Mutations occur rarely—a given gene mutates once in every 10 thousand to 10 billion individuals. How can this provide enough variability to explain, say, the widespread evolution of resistance to antibiotics in the last few decades?

The secret of prokaryote evolution lies in rapid reproduction: under ideal conditions, a single bacterium can produce millions of offspring in a week. Given the incredible reproductive rate of a single bacterium, the chances are high that even rare mutations will appear within an entire population of bacteria. This rapid appearance of mutations creates genetic variability.

Prokaryotes typically divide by binary fission, but genetic information from two individuals can be combined by transduction, transformation, or conjugation. Genetic diversity in prokaryotes is achieved by these three mechanisms and by mutations, which accumulate quickly in rapidly reproducing prokaryotes.

19-B BACTERIAL METABOLISM AND WAYS OF LIFE

Bacteria are the most numerous organisms on Earth; a handful of soil may contain billions of them. They occur in habitats as remote as the sea floor, icebergs, and hot springs, or as close as your mouth (Figure 19-4).

This far-flung distribution is possible because many bacteria have metabolic abilities not found in any eukaryote. Some carry on fermentation, others respiration (Chapter 7). **Aerobes** require O_2 for respiration; **obligate anaerobes** are killed by O_2; **facultative anaerobes** can live in the presence or absence of oxygen. Some anaerobic bacteria use nitrate (NO_3^-), sulfate (SO_4^{2-}), or iron (Fe^{3+}) instead of O_2 in their respiration. Most of these forms of **anaerobic respiration** yield more energy than fermentation but less than aerobic respiration.

Because of their rigid cell walls, bacteria cannot take in large food particles. Rather, they absorb small molecules that filter through their walls. Many bacteria can make any organic molecule they need, starting with only one kind of organic molecule, such as glucose or fatty acids. Others require a more complex diet.

When conditions are unfavorable for growth, some bacteria form thick-walled, resting **spores** that are resistant to heat and drying. Most bacterial spores are not a means of reproduction, because no increase in cell number occurs. Rather, spore formation permits cells to survive adverse conditions and to disperse to new locations, where a new cell may develop from the spore.

Bacteria undoubtedly owe their biological success to their varied metabolic abilities, coupled with small size, rapid reproductive rate, and the ability to form re-

Three Forms of Genetic Recombination in Bacteria

(a) Transduction
1 Infecting phage with DNA from a previous host
Bacterial DNA
2 Phage attaches to bacterium and injects DNA
3 Host chromosome breaks down and chromosomes replicate
4 Phage components assemble, some contain recombinations of DNA from host bacteria
5 New phages are released and may transfer recombined DNA to a new bacterium
(b) Transformation
1 DNA from broken bacteria enters a recipient cell
Recipient cell
Bacterial DNA
2 Recombination of donor DNA and recipient DNA
3 Genetically transformed cell
(c) Conjugation
Donor
Recipient
Bacterial DNA
Plasmid
Replication and transfer of plasmid
Pilus
Transfer complete

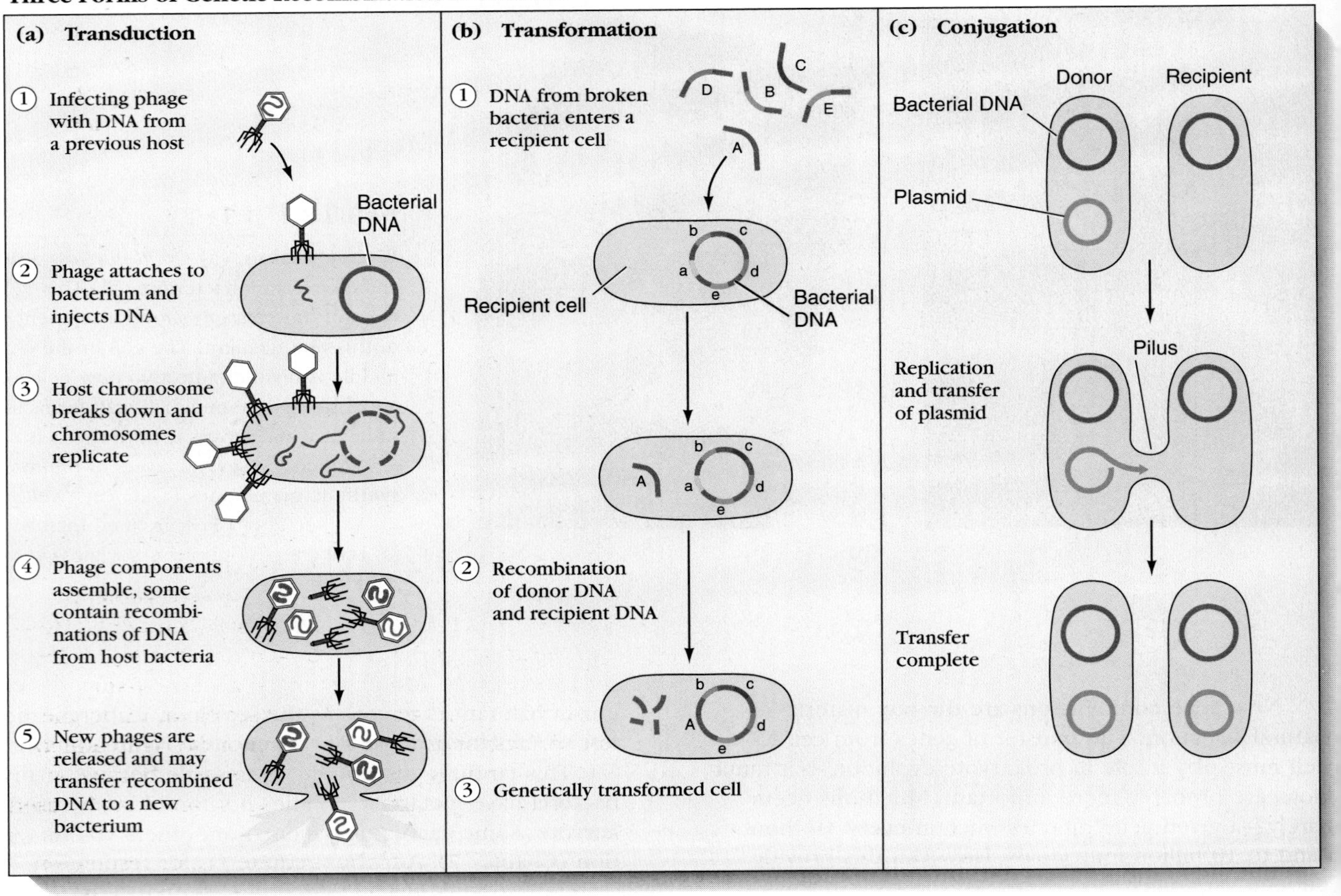

FIGURE 19–3

Genetic recombination in bacteria. (a) In transduction, the DNA from one bacterium is transferred to another bacterium via an infecting phage. The new phage DNA directs the production of new phage particles which may eventually infect other bacteria. **(b)** During transformation, free DNA from broken bacterial cells becomes integrated into the DNA of the recipient bacterium. **(c)** During conjugation, a replicated plasmid is transferred to a recipient cell through a pilus.

sistant spores. These features permit bacteria to live in many habitats that are here today and gone tomorrow. A raindrop on a leaf evaporates in a few hours, but by then a bacterium has divided several times and its descendants have formed spores that can blow away to new habitats.

As we discuss the metabolism of bacteria, notice how the various species interact: one microorganism's wastes are another's food and drink.

Autotrophic Bacteria

Autotrophs can make their own food from simple inorganic compounds, such as carbon dioxide, mineral ions, and water. Purple and green bacteria use light energy to drive primitive forms of photosynthesis, which differ from that of green plants. First, bacterial photosynthesis does not obtain hydrogen by splitting water molecules. Hence it does not produce O_2. Various photosynthetic bacteria obtain hydrogen from hydrogen gas, hydrogen sulfide, fatty acids, or alcohols. Second, the chlorophylls of purple and green bacteria absorb, not visible light, but light in the near-infrared range (see Figure 8–2). Hence they can photosynthesize in what we would consider darkness. Third, many photosynthetic bacteria are anaerobic.

Cyanobacteria use the kind of photosynthesis found in eukaryotes. They get hydrogen ions by split-

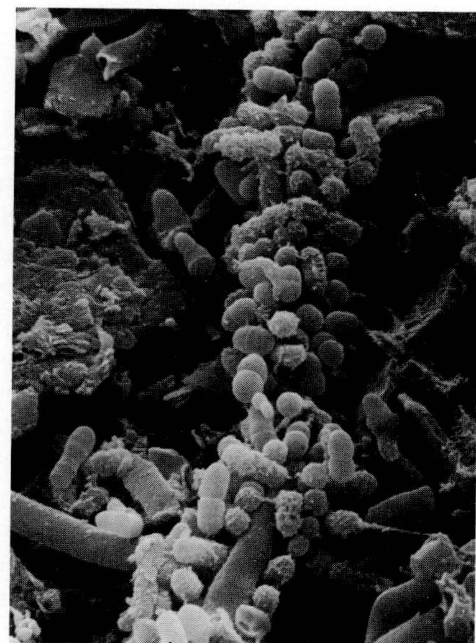

FIGURE 19-4

Oral bacteria. Plaque that have accumulated around human teeth. The round bacteria are *Streptococci*. (Courtesy of Drs. Kessel and Shih)

ting water molecules and release O_2 as a by-product (Chapter 8).

Chemosynthesis is a form of autotrophy found only in certain bacteria. They make food using energy, not from light, but from inorganic chemical reactions involving the oxidation of ammonia, nitrites, sulfides, or iron. Chemosynthetic bacteria are the sole food producers of some areas near vents in the deep sea floor that spew forth warm sulfide-laden water (Figure 19-5).

Nitrifying bacteria are chemosynthetic bacteria that oxidize ammonia (NH_3) or ammonium (NH_4^+) to nitrites (NO_2^-), and nitrites to nitrates (NO_3^-). These bacteria play a crucial role in the nitrogen cycle (Section 37-D).

Nitrogen-Fixing Bacteria

Nitrogen fixation is a complex set of reactions that reduce nitrogen gas (N_2) from the air to ammonia (NH_3). The only organisms that can fix nitrogen are some bacteria and cyanobacteria. All other life on Earth is directly or indirectly dependent on their activities. Green plants need nitrogen, from ammonium or nitrates, to produce amino acids for their proteins (some of which later become animals' proteins). However, most of the Earth's nitrogen exists as N_2 in the air, which plants cannot use. The nitrogen-fixing prokaryotes link the vast N_2 supply of the air to the rest of the living world. They transform N_2 to ammonia, which nitrifying bacteria (discussed above) convert to nitrites and then to nitrates, which can be used by plants.

Nitrogen-fixing bacteria have great importance to agriculture. Some live in nodules on the roots of **legumes** (plants such as peas, beans, clover, alfalfa,

(a)

(b)

FIGURE 19-5

Animals dependent on autotrophic bacteria. (a) Near a thermal vent in the sea floor, entire communities are based upon chemosynthetic bacteria which use hydrogen sulfide as an energy source to make their own food. The tube worms with red tips live in a mutually beneficial symbiotic relationship with these bacteria. **(b)** Flamingos feast on filaments of the cyanobacterium *Spirulina*. (a, J. Frederick Grassle/Woods Hole Oceanographic Institution; b, G. Ziesler/Peter Arnold, Inc.)

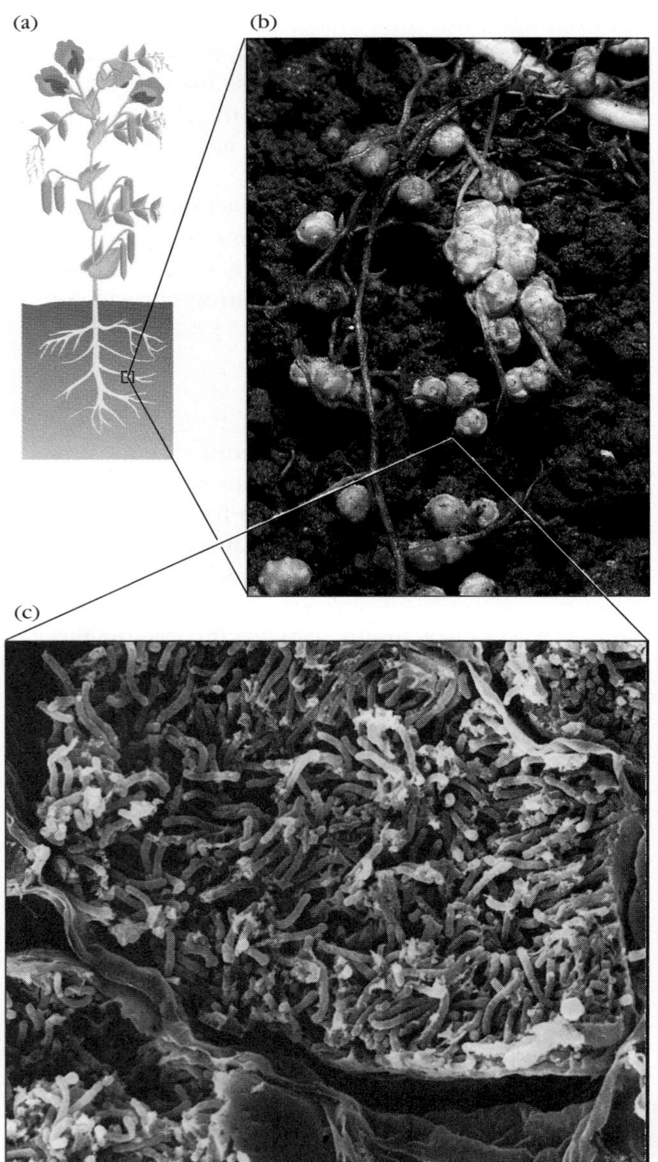

FIGURE 19-6

Root nodules. **(a)** A bean plant. **(b)** Root nodules on bean roots. **(c)** A root nodule that has been cut open to reveal the nitrogen-fixing *Rhizobium* bacteria. (b, D. Kuhn; c. Courtesy of Drs. Kessel and Shih)

vetch) (Figure 19-6). These bacteria use sugars produced by the legume's photosynthesis and supply the plant with ammonium. Growing legumes as part of crop rotation adds nitrogen compounds to the soil and so improves its fertility. Genetic engineers are working to make other crops able to house such nitrogen-fixing bacteria. In addition, many blue-green algae (also called cyanobacteria) are nitrogen-fixers. These are important in the cycling of nutrients in aquatic environments and salt marshes.

Heterotrophic Bacteria

Most bacteria are **heterotrophs,** using food made by other organisms. **Saprobes** live on dead organic material, such as dead animals or plants, feces, leaves, or bark. They secrete hydrolytic (digestive) enzymes into the food around them and absorb the resulting small organic molecules and inorganic ions. Saprobic bacteria and fungi are the most important **decomposer** organisms. When a dead leaf drifts to the forest floor or an animal dies, fungal and bacterial spores floating in the air have already settled on it. The spores quickly begin to grow and break down the dead organism, recycling it in the form of carbon dioxide, water, and minerals that autotrophs can use to produce more food. Many heterotrophic bacteria are **parasites,** obtaining food directly from the bodies of living organisms.

Bacteria are the most numerous and widespread organisms on Earth, inhabiting areas with and without oxygen. When conditions are unfavorable, some bacteria form spores by producing a thick-walled coat that allows them to survive until conditions improve. Unlike plants, photosynthetic purple and green bacteria are anaerobic, do not produce oxygen, and use different wavelengths of light. In contrast, cyanobacteria have a type of photosynthesis similar to that used by green plants. Other autotrophic bacteria obtain energy from nitrification or other inorganic chemical oxidation reactions. Most bacteria are heterotrophs and consume food made by other living things. Decomposers recycle minerals and carbon dioxide from dead organisms while many other heterotrophic bacteria are parasites of living organisms.

19–C CLASSIFICATION OF PROKARYOTES

Biologists classify most organisms by their structure. However, many kinds of bacteria look so much alike that they cannot be told apart by structure alone. Because bacteria reproduce asexually by dividing in two, the usual criterion for identification of a species does not apply (bacteria cannot cross-breed and produce fertile young; see Section 18-A).

Bacteria have long been classified by both structural and metabolic features: cell size, shape (rods, spheres, or spirals), and arrangement in filaments or clumps; number and position of flagella; ability to use oxygen or to ferment various organic compounds; type of photosynthesis; reaction to Gram staining (Figure 19-7), and so on. This gives us a useful way to identify

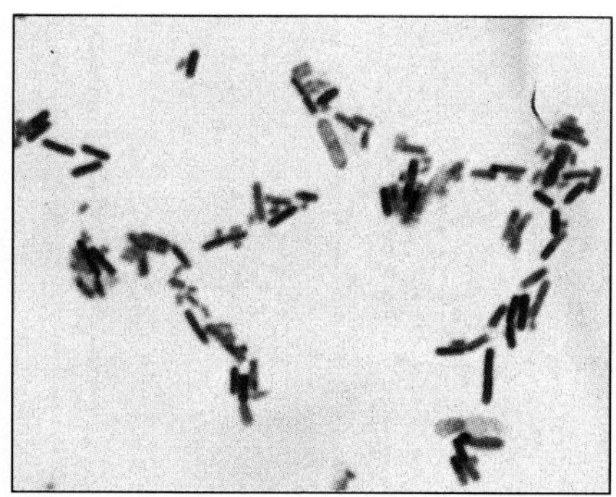

FIGURE 19–7

Gram-stained bacteria. The Gram stain, invented by H. C. Gram, is used to identify bacteria. Because of differences in their cell walls, the Gram staining procedure turns Gram-negative cells pink and Gram-positive cells purple. The difference between Gram-positive and Gram-negative bacteria is medically important: Gram-positive forms are more susceptible to most antibiotics and to lysozyme, an enzyme in human body fluids that destroys bacterial cell walls. (Biophoto Associates, Photo Researchers, Inc.)

unknown bacteria, a necessary first step in practical situations such as diagnosing a disease or finding the cause of food spoilage. However, such characters do not always reflect evolutionary relationships.

Recently, researchers have drawn up an evolutionary tree for prokaryotes, based on nucleotide sequences in their nucleic acids.

Archaeobacteria

Archaeobacteria are so different from other prokaryotes that some biologists think they belong in a sixth kingdom by themselves. In structure they resemble other bacteria, but in some details of protein synthesis they are more like eukaryotes. Still other features are like nothing else on Earth today, including their membrane lipids, cell walls, coenzymes, and transfer RNA molecules. Such basic biochemical features tend to be highly conservative (Section 18–C). Therefore, many biologists think that archaeobacteria diverged from other organisms very early in evolution, after the origin of the genetic code but while the machinery of metabolism and protein synthesis was still evolving.

There are three main types of archaeobacteria. One group makes methane (CH_4) in anaerobic habitats such as the bottoms of bogs, lakes, and sewage treatment ponds, and the digestive tracts of animals, especially cows and other ruminants (Section 22–D). The other two groups are aerobic. Salt-loving bacteria inhabit extra-salty environments such as salt evaporation ponds, the Great Salt Lake, and the Dead Sea. The third group lives in hot, acidic springs, at temperatures of 80 to over 90°C and pH of less than 2! ■

Eubacteria

The vast majority of prokaryotes falls into a second main evolutionary group, the **eubacteria** ("true bacteria"). Although these include the photosynthetic and chemosynthetic autotrophs, most eubacteria are harmless saprobes. Some are normal residents of humans or other eukaryotes, and still others (considered as parasites by some) cause diseases such as plague, tetanus, botulism, gangrene, diphtheria, tuberculosis, and leprosy. Leprosy is the most widespread human bacterial disease in the world today, with over 20 million cases. The leprosy bacterium divides only once every 12 days, and so the disease takes a long time to develop. A few groups of eubacteria deserve special mention. ■

Actinomycetes produce branching, multicellular filaments that resemble fungi. They are the source of many valuable antibiotics such as streptomycin.

Mycoplasmas have two notable features: they have lost their cell walls, and they are easily the smallest living cells, only 0.1 to 0.25 micrometer in diameter. These features permit them to pass through filters that trap other bacteria. The lack of cell walls also makes them resistant to penicillin (Section 5–A). Mycoplasmas live as parasites in plant or animal cells. They cause one kind of human pneumonia, and many diseases of other animals.

Spirochetes are spiral bacteria with distinctive flagella, called axial filaments, between layers of the cell wall. Parasitic spirochetes include those that cause syphilis, yaws (a tropical skin disease), and Lyme disease.

Rickettsiae are tiny, parasitic bacteria that usually live inside other cells. They are carried by ticks or insects, which transmit them to mammals by bites. Rickettsial diseases include typhus, one of the all-time great killers of humans, transmitted by lice, and Rocky Mountain spotted fever, carried by ticks.

Most **cyanobacteria (blue-green bacteria)** are photosynthetic. Like eukaryotic plants, they obtain hydrogen by splitting water and give off O_2. Most are blue-green because they contain green chlorophyll *a*

BIO-BIT

Lyme disease is caused by a bacterium transmitted mainly by deer ticks, which are about half the size of the more common wood tick.

plus blue accessory photosynthetic pigments called phycobilins. Cyanobacteria may form filamentous or clustered colonies. Some colonies show division of labor: besides photosynthetic cells, there may be spore-producing cells, or cells specialized for attachment to a rock or other substrate, or cells specialized for nitrogen fixation.

Prochlorophytes are a group of photosynthetic bacteria first discovered in the late 1970s. Like cyanobacteria, they carry on water-splitting, oxygen-producing photosynthesis. Unlike cyanobacteria, however, prochlorophytes contain two chlorophylls—*a* and *b*—which also occur in most chloroplasts. For this reason, it seems likely that some member(s) of the group gave rise to the chloroplasts in most members of the plant kingdom. They are also important ecologically: one tiny species of prochlorophyte, discovered in 1988, is one of the two most numerous photosynthetic organisms in the world's oceans (the other is a cyanobacterium). Ten gallons of water contain about 5 billion of these cells—the same as the Earth's current human population!

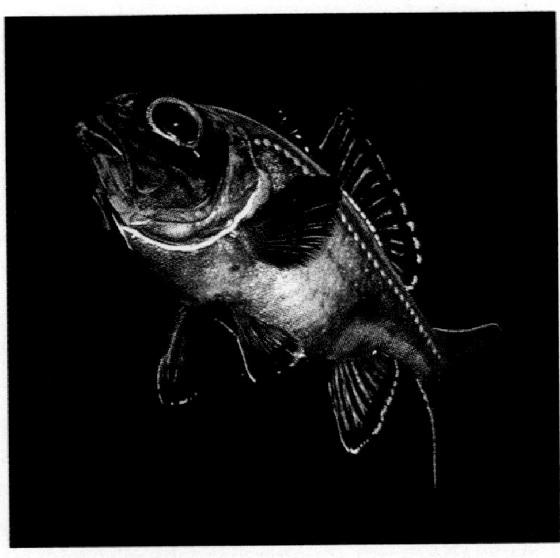

FIGURE 19-8

A symbiotic relationship. This deep sea flashlight fish emits light thanks to bioluminescent bacterial symbionts. (K. McCarthy/Offshoot Stock)

Bacteria are classified into two main groups: the archaeobacteria and the eubacteria. Unlike other prokaryotes, archaeobacteria have unique membrane lipids, cell walls, coenzymes, and RNA molecules, which lead some scientists to place them in their own kingdoms. Eubacteria are by far the most common bacteria and include photosynthetic, chemosynthetic, and heterotrophic forms.

19–D SYMBIOTIC BACTERIA

We have seen that many bacteria exploit the resources of the nonliving environment. The living environment—other organisms—also contains a wealth of resources. Many members of all five kingdoms have evolved ways to tap their living neighbors' resources by forming close associations with them. Such an intimate relationship between members of different species is called **symbiosis** ("together living"). It is often divided into three categories:

1. **Parasitism.** In this symbiotic relationship, one species (the parasite) benefits at the expense of the other (the host). Bacterial diseases are examples of parasitism, and all viruses are parasites.
2. **Commensalism.** (Com = together; mensa = table) Here, one species benefits, and the other is not harmed. The countless billions of bacteria that live on human skin are commensals. They receive a habitat and food (scales of skin and dead tissue) and do no harm to human skin.
3. **Mutualism.** Both species benefit from this association. When one species is much smaller than the other—such as a bacterium or protist associated with a plant or animal—it is called a **symbiont** of the larger host species.

Because of their small size, bacteria can form especially intimate symbioses with a variety of other organisms. Hence the symbiotic relationships of bacteria make a good introduction to a topic that will be a recurrent theme in our study of all five kingdoms.

Many symbiotic bacteria are part of the **microbiota** of an animal: organisms that normally live on or within it. They may be saprobes, living on dead skin cells or digested food in the intestines, or parasites or symbionts absorbing food from living tissue. An animal's normal microbiota may benefit the host by producing enzymes that can digest cellulose, by making vitamins, by producing antibiotics, or by competing with disease-causing bacteria (all organisms grow better without competition for food and space).

On the other hand, members of an animal's microbiota may cause disease if they settle down in the wrong places. *Escherichia coli* (from the human intestine) causes cystitis if it gets into the urinary bladder. *Staphylococcus aureus* (from human skin) can cause serious infections if it gets into a wound; this is one reason why surgeons wear gowns, masks, and gloves in the operating room.

Other symbiotic bacteria include those that live in the root nodules of legumes and in bioluminescent fish (Figure 19-8). Perhaps the most fascinating symbiosis is that between the anaerobic protist *Mixotricha paradoxa* and four different kinds of bacteria—one that

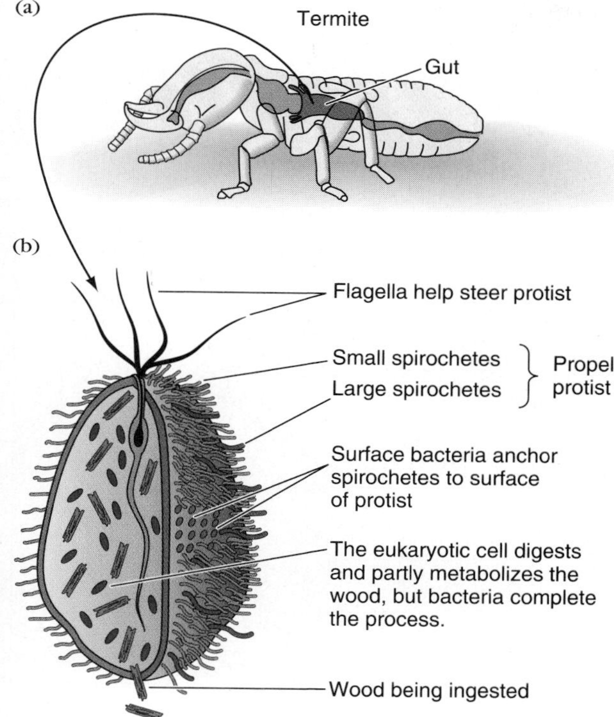
(a)
Termite
Gut
(b)
Flagella help steer protist
Small spirochetes
Large spirochetes
Propel protist
Surface bacteria anchor spirochetes to surface of protist
The eukaryotic cell digests and partly metabolizes the wood, but bacteria complete the process.
Wood being ingested

FIGURE 19-9

A symbiont's symbiont. (a) Termites can digest cellulose as a result of symbiotic protists living within their guts. **(b)** The cell's four flagella do not move it, but serve as rudders. Instead, the cells move by the activities of up to half a million spirochetes on its outer surface. Still other bacteria, inside the cell, are thought to play a metabolic role, breaking down the end products of the cell's fermentation.

acts much like mitochondria, breaking down the protist's fermentation products, and others that serve in locomotion (Figure 19-9). The whole assemblage lives in the gut of a termite: *M. paradoxa* itself is a symbiont, digesting the termite's meals of wood! The termite cannot produce wood-digesting enzymes, and without its symbionts it would starve.

Three different types of symbiotic relationships are recognized, based upon how the other organism is affected. These include parasitism, commensalism, and mutualism. Many prokaryotes live with other prokaryotes or with eukaryotes in all three types of symbiotic associations.

ORIGIN OF EUKARYOTES AND THE KINGDOM PROTISTA

The evolution of the first eukaryotic cells, about 1.45 billion years ago, was a tremendous advance. Specialized organelles, surrounded by membranes, kept different cell activities separate in distinct areas, where they did not interfere with each other. They also permitted a dramatic increase in cell size. As we have seen, mitochondria and chloroplasts may have evolved as symbionts inside eukaryotic cells (see *A Journey into Evolution:* How Are Prokaryotes and Eukaryotes Related?, Chapter 5). The membranes of the nuclear envelope, endoplasmic reticulum, and Golgi complexes probably evolved from invaginations of the plasma membrane.

The hallmark of eukaryotic cells is the nuclear envelope around the genetic material, separating it from the cytoplasm. Eukaryotic cells have a number of chromosomes, containing much more genetic information than a prokaryote's single DNA molecule. The evolution of mitosis and meiosis guaranteed an orderly sorting of chromosomes into new nuclei during cell division. Most eukaryotes can engage in true sexual reproduction, with its varied range of new genetic combinations. This in turn permitted the tremendous adaptive radiation of eukaryotes. (Refer to Table 5-1 for a more complete comparison of prokaryotes and eukaryotes.)

19-E KINGDOM PROTISTA: UNICELLULAR ALGAE AND PROTOZOA

Most unicellular eukaryotes are classified in the kingdom Protista (Figure 19-10). However, some are so closely related to particular animals, plants, or fungi that they are placed in those kingdoms, with their multicellular relatives. Conversely, although most protists live as solitary cells, many form **colonies** of similar, but largely independent, cells.

Protists have radiated into many different habitats and ways of life. They live wherever there is water: in the sea, in fresh water, in moist soil, or in the bodies of animals.

Photosynthetic protists, and simple multicellular plants, are often called **algae.** Other protists, called **protozoa,** ingest their food and lack chloroplasts and cell walls. Before protists were separated into a kingdom by themselves, the autotrophs were considered plants (unicellular algae), and the heterotrophs were considered animals (protozoa).

Physiology of Protists

The Protista are so heterogeneous that few generalizations can be made about their physiology (how they

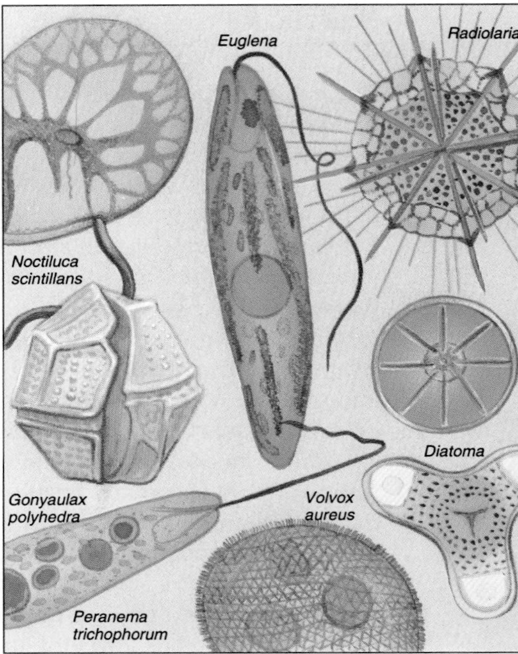

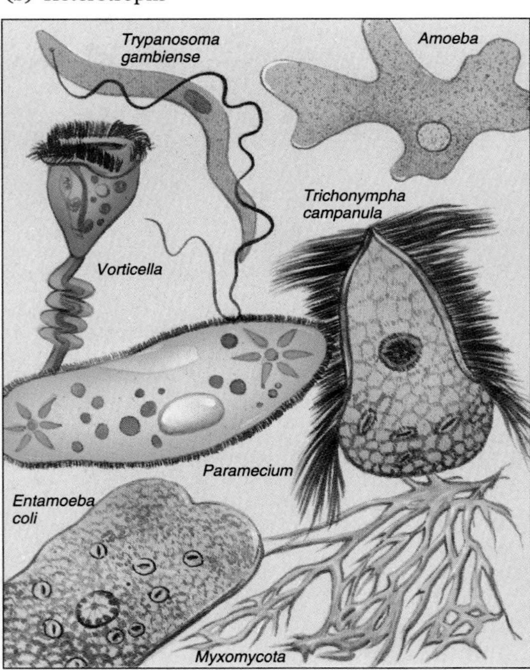

FIGURE 19–10

Protozoan overview.

work). Some have a cellulose cell wall, a shell, or other covering that provides protection and support and may also keep the cell from taking in too much water by osmosis (Section 4–C). Many protists without cell walls have **contractile vacuoles** that collect excess water and expel it from the cell (see Figure 19–19). Nearly all protists can carry on respiration, but many can also live indefinitely using fermentation when oxygen is not available.

Under adverse conditions, many protists can enclose themselves in thick walls, forming **cysts**, that resist desiccation (drying out) and temperature changes. Here the cell can survive until favorable conditions return, or until it is carried to a new home.

Some protists are **sessile** ("sitting"), living attached to objects such as rocks or plants. Others float passively in their watery homes. However, many can move by means of cilia, flagella, or **pseudopods** (flowing extensions of the cytoplasm).

In flagellar locomotion, wave-like bending motions pass from one end of the flagellum to the other. This pulls or pushes the cell through the water, depending on whether the flagellum is at the anterior or posterior (front or rear) of the cell (Figure 19–11).

Cilia are shorter and generally more numerous than flagella. Ciliary locomotion can be compared to rowing a boat. Each cilium bends on its return stroke and offers the least possible resistance to the water.

In pseudopodial or amoeboid locomotion, an amoeba extends pseudopods ("false feet") from its body, and the rest of the cytoplasm flows into these forward extrusions. This type of locomotion involves actin microfilaments (Section 5–L), but is not well understood.

Protists feed in various ways. Heterotrophs ingest food by endocytosis (see Figure 4–10) or absorb small organic molecules from their environment. Many protists have more than one type of nutrition: some species can switch from photosynthesis to endocytosis to absorption, as conditions dictate.

Most protists can sense stimuli such as light, temperature, touch, gravity, and chemicals. **Eyespots** consist of light-sensitive pigments in small organelles. Many protists can detect objects that touch their cilia or flagella. They also detect chemicals in the environment, probably by way of changes these substances produce in proteins in their plasma membranes. Protists respond to stimuli by moving toward or away from them as appropriate for their particular way of life. For example, a protist with flagella and chloroplasts swims toward light, which it needs for photosynthesis. A heterotroph that eats decaying organic matter might avoid light and so move toward a food supply at the bottom of the pond.

Protists usually reproduce asexually by dividing in two after a mitotic division of the nucleus. Most can also undergo some kind of sexual reproduction.

Protists are unicellular organisms that live singly or in colonics in a variety of habitats including water, soil, or the bodies of other organisms. Protists are an incredibly diverse group. They differ according to their mode of nutrition (pho-

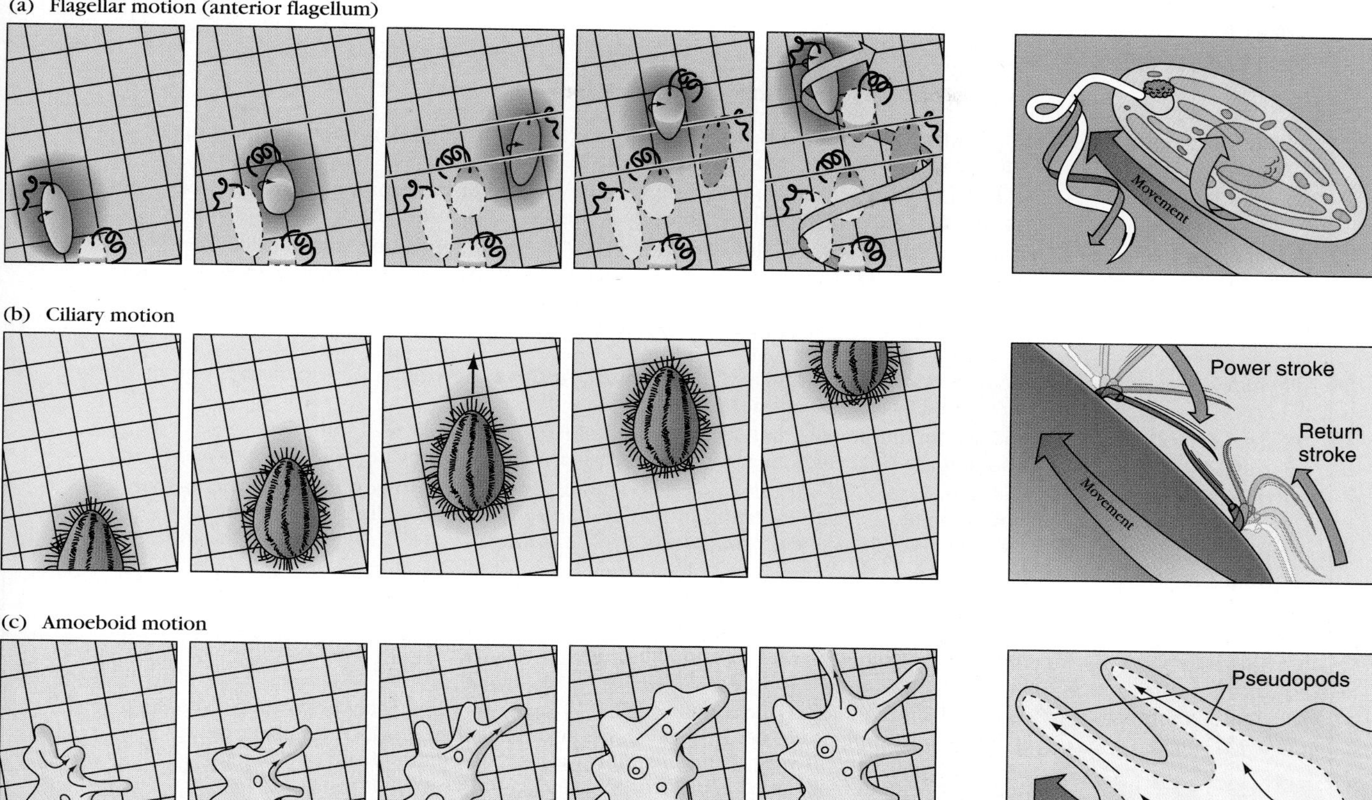

FIGURE 19–11

How protozoans move. (a) Flagellar motion. The anterior flagellum of a protozoan whips around and pulls the cell forward through the water. **(b)** Ciliary motion. Just like the oars of a boat, coordinated strokes of cilia propel the protozoan through the water. By bending on the return stroke, the cilia avoid pushing the cell backward. This recovery stroke is similar to lifting the oar out of the water when preparing for another stroke. **(c)** Amoeboid motion. Arrows show the movement of the cytoplasm. In the same way that a baby crawls across the floor on its belly, pushing the arms forward while drawing up the hindlegs, streams of the cytoplasm push out temporary pseudopods while the trailing end is pulled back up.

tosynthesis, ingestion, or absorption), locomotion (cilia, flagella, pseudopods, or more), and covering (cell wall, shell, etc.). Most protists can detect light and chemicals in their environments and reproduce both asexually and sexually.

19–F PHOTOSYNTHETIC PROTISTS

Many cyanobacteria (Prokaryotae) and algae (Protista) are members of the **phytoplankton** (phyton = plant; plankton = wanderer), photosynthetic organisms floating near the surfaces of oceans, lakes, and ponds. Phytoplankton are ecologically important as food for many aquatic animals. They produce much of the oxygen dissolved in aquatic habitats and also produce an estimated 30 to 50% or more of the oxygen in the atmosphere.

Phytoplankton need light, but because water absorbs much light, they can survive only if they remain near the surface. Some use flagella to swim upward. Others have projections that keep them afloat by expanding their surface area and hence providing resistance to sinking. Still others store their extra food as oil, which buoys them up near the surface.

Table 19–1
The Kingdom Protista

	Unicellular Algae (Mainly Autotrophs)
Phylum Pyrrophyta* (~2000 species)	Dinoflagellates. Pectin and cellulose cell walls; mostly marine; two flagella; chlorophylls *a* and *c* and carotenoids; food stored as starch
Phylum Euglenida* (~800 species)	*Euglena* and its relatives. No cell wall; flexible pellicle of protein; mostly freshwater; usually two anterior flagella; chlorophylls *a* and *b* and carotenoids
Phylum Chrysophyta (~10,000 species)	Diatoms and their allies. Cellulose and pectin cell walls, containing silica in diatoms; marine or freshwater; diatoms lack flagella, allied forms may have one or more flagella; chlorophylls *a* and *c* and carotenoids; food stored as oils
	Protozoans (Mainly Heterotrophs)
Phylum Zoomastigina*	Flagellates without chloroplasts. No cell wall; freshwater, symbiotic, or parasitic; one or more flagella; e.g., *Trypanosoma*
Phylum Sarcodina* (~40,000 species)	Amoebas, foraminiferans, heliozoans, and radiolarians. No cell wall but some secrete shells containing silica, calcium carbonate, etc.; marine and freshwater; pseudopods
Phylum Apicomplexa (~3,900 species)	Parasitic protists. Cell with distinctive, complex arrangement of several components (apical complex) at one end; locomotion by bending or gliding, flagella in male gametes; complex life history; e.g., *Toxoplasma, Plasmodium*
Phylum Ciliophora (~8,000 species)	Ciliates. Two types of nuclei; cilia for locomotion and food collection; no cell wall; freshwater and marine, some parasites; e.g., *Paramecium, Didinium*
Phylum Oomycota (~475 species)	Water molds. Hyphae produce flagellated zoospores that grow into a mycelium that is coenocytic. Sexual reproduction occurs when environmental conditions degenerate; e.g., *Saprolegnia*
Phylum Myxomycota (~475 species)	Plasmodial slime molds. Have a multinucleate plasmodium that may be brightly colored; when food supply dwindles, plasmodium forms stalked reproductive structures that produce haploid spores; flagellated myxamoebae emerge from spores, fuse, and zygote undergoes mitosis to form multinucleate plasmodium; e.g. *Physarum*
Phylum Acrasiomycota (~26 species)	Cellular slime molds. Haploid amoeba-like forms aggregate into "slugs" for reproduction. Spores are produced and a new generation of amoebas is released from them. No flagellated life cycle stages; e.g., *Dictyostelium*

*All groups marked * are sometimes combined into the Phylum Sarcomastigophora: protists moving by means of flagella or pseudopodia.

When oil-storing phytoplankton die and sink to the bottom, they may be covered by sediment and water that exert enormous pressure. Such phytoplankton, of hundreds of millions of years ago, made the oil we use as fossil fuel today.

Photosynthetic protists are placed in three phyla on the basis of conservative characters such as the type of cell wall, flagella, photosynthetic pigments, and the form in which food is stored (Table 19–1).

Phylum Pyrrophyta (Dinoflagellates)

Phylum Pyrrophyta ("fire plants") takes its name from some species that are bioluminescent, able to give off light (like fireflies). Members of this phylum have two flagella and are commonly called dinoflagellates (dinos = whirling) (Figure 19–12). Most are **marine** (ocean-dwelling). Some forms lose their flagella and live as symbionts of corals, sea anemones, and clams, supplying food made by photosynthesis in exchange for protection. However, most are free-living and make up one of the two main groups of eukaryotes in the marine phytoplankton. Ancient planktonic dinoflagellates formed large oil deposits.

Some members of the genus *Gonyaulax* produce a nerve poison deadly to vertebrates. During population explosions of these forms, shellfish that have consumed large quantities of these cells by filter feeding become unfit for human consumption. Because *Gonyaulax* contains a red pigment, such population explosions may tinge the water pink, a fitting warning signal called a "red tide." Only filter-feeding molluscs or crustaceans are affected.

Phylum Euglenida

The Euglenida are named after the common genus *Euglena* (Figure 19–13). Most members of this group live

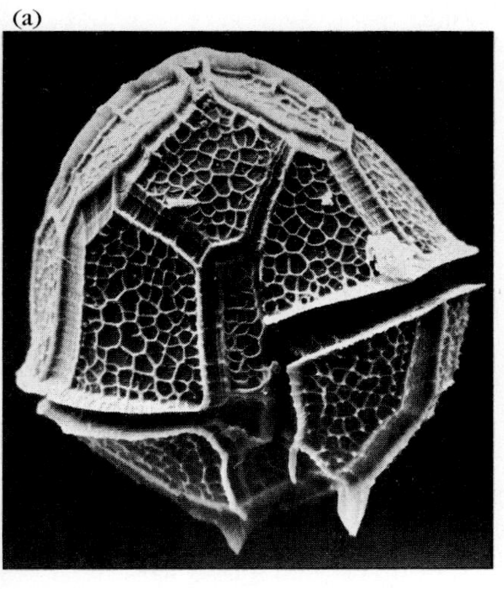

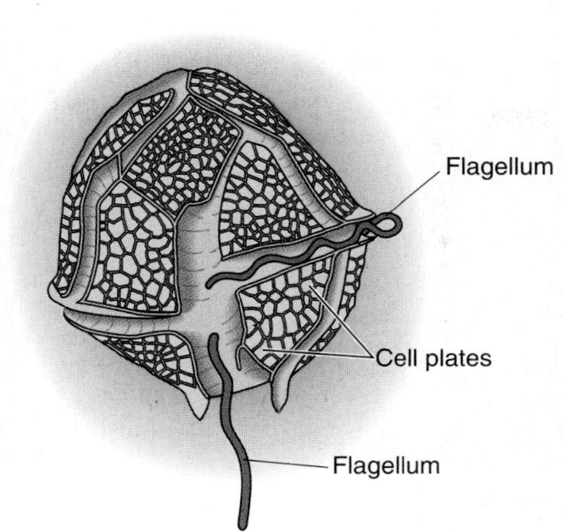

FIGURE 19–12

A dinoflagellate. (a) This scanning electron micrograph, at more than 2000x magnification, shows the many cellulose "plates" that lie under the plasma membrane. **(b)** One of the two flagella lies in the groove around the cell's middle, the other beats at the rear. (Biophoto Associates, Photo Researchers Inc.)

in fresh water, especially water polluted by excess nutrients. There are usually two flagella. These protists lack cell walls but may have an elastic, transparent **pellicle,** made of protein, just beneath the plasma membrane.

About one-third of euglenids are photosynthetic and have a light sensor at the base of the flagella. However, if *Euglena* is raised in the dark, it loses its green color and becomes a heterotroph, ingesting food.

Phylum Chrysophyta (Diatoms and Golden Algae)

Chrysophytes include the diatoms and the "golden algae" (chrysos = golden). They live as single cells or simple colonies in the sea, in fresh water, and in wet spots on land. Diatoms and pyrrophytes are the two main groups of eukaryotic marine phytoplankton. In unpolluted fresh water, diatoms (along with green algae, Chapter 20) are the most important group of phy-

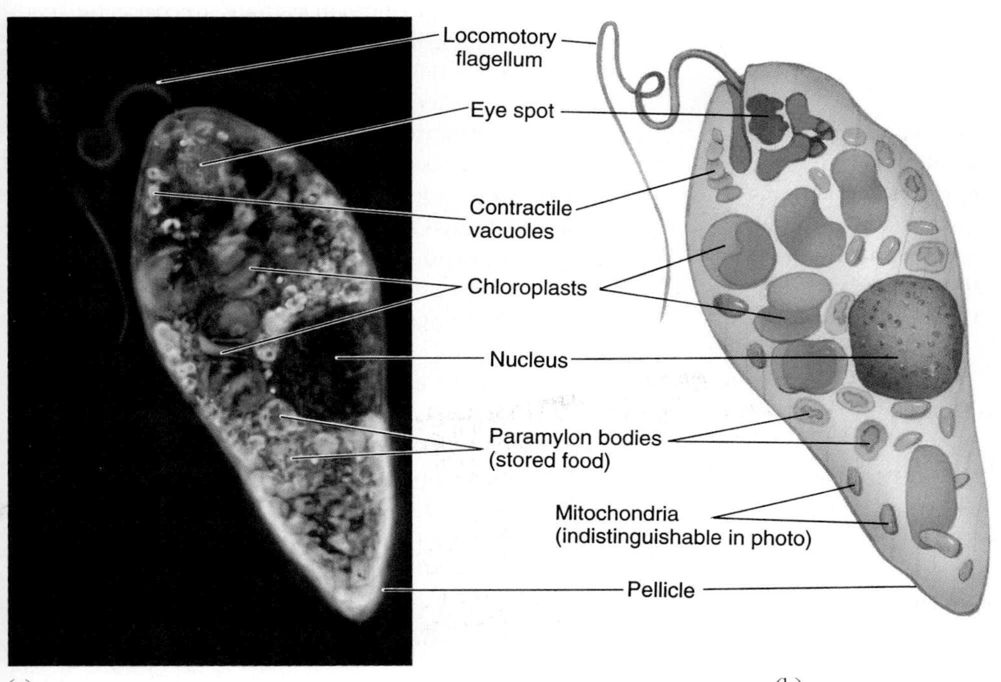

FIGURE 19–13

***Euglena* (Euglenida).** **(a)** A photograph of a living *Euglena.* **(b)** The eyespot (a cluster of pigment) plays a role in sensing and responding to light. The cell moves toward light, using its locomotory flagellum. This positions the chloroplasts favorably for photosynthesis to make the cell's food. The contractile vacuoles expel excess water taken in by osmosis. (Biophoto Associates, Photo Researchers, Inc.)

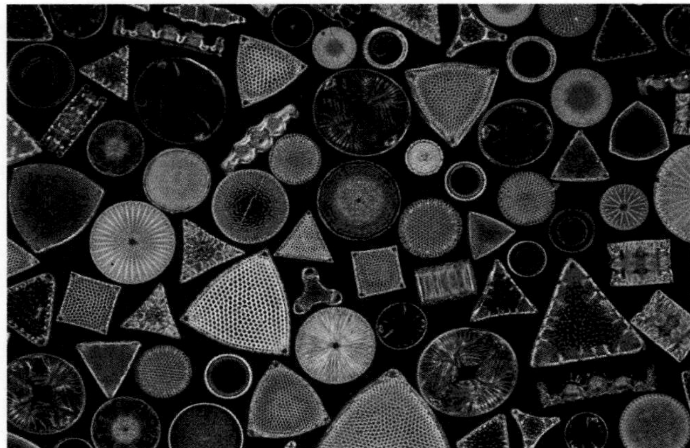

FIGURE 19–14

Diatoms (Chrysophyta). The intricately patterned walls of diatoms were once used to test the resolving power of microscopes. Victorian microscopists painstakingly arranged diatoms to form designs on microscope slides. This photo shows part of one such arrangement. (M. I. Walker/Science Source)

toplankton. Chrysophytes store much of their surplus food as oil and were important in the formation of petroleum deposits.

Diatoms have a rigid cell wall impregnated with silica (SiO_2), a compound also found in glass and sand. Often the cell wall has an intricate pattern of pits or ridges (Figure 19–14). These walls are very resistant to decay and accumulate on the ocean floor in enormous numbers. Such deposits may later be raised above sea level by geological activity. The resulting "diatomaceous earth" is used as a fine abrasive in silver polish and toothpaste and as the packing in air and water filters.

Many photosynthetic protists (algae) and cyanobacteria float near the surface of large bodies of water. There they form an important source of food for many other organisms and produce much of the Earth's atmospheric oxygen and dissolved oxygen in aquatic ecosystems. Photosynthetic protists are classified into three phyla based upon the type of cell wall, flagella, photosynthetic pigments, and the form in which food is stored.

19–G PROTOZOA (HETEROTROPHIC PROTISTS)

Phylum Zoomastigina (Zooflagellates)

The phylum Zoomastigina (zoon = animal; mastix = whip) contains heterotrophic flagellates. These include free-living organisms, symbionts, and parasites. The parasites often have complex life histories with two host species. Termites and wood roaches are totally dependent on their flagellate symbionts to digest their meals of wood.

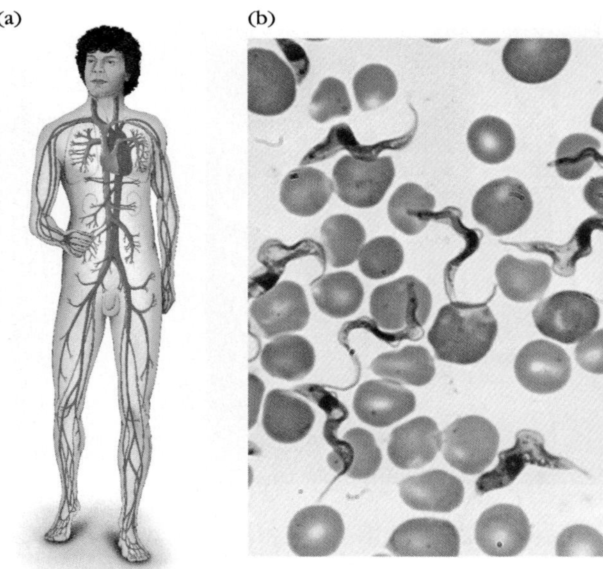

FIGURE 19–15

Phylum Zoomastigina: trypanosomes, parasites of vertebrate blood. Here, members of the species that causes African sleeping sickness swarm among red blood cells. (Biophoto Associates, Photo Researchers, Inc.)

A major threat to human welfare are parasitic trypanosomes, which live in the blood of vertebrates, mainly mammals (Figure 19–15). Human trypanosome diseases include sleeping sickness (Africa) and Chagas' disease (Latin America). Sleeping sickness is transmitted by bites of tsetse flies, hosts for part of the trypanosome life history. In Latin America, Chagas' disease affects about 12 million people. It too is transmitted by bloodsucking insects, which hide in houses or other buildings, crawling out at night to feed on people, dogs, cats, or guinea pigs (which are raised for food).

Trypanosomes are especially difficult for the body to conquer because they keep changing the cell-surface proteins (antigens) that the immune system must recognize in order to kill the parasites. As quickly as the host develops a defense against one antigen, the parasite switches to another. The immune system must start all over to combat what is essentially a new disease.

Leishmaniasis is caused by parasitic flagellates transmitted by sand flies. This disease afflicts millions of people in Africa and Asia. Symptoms range from

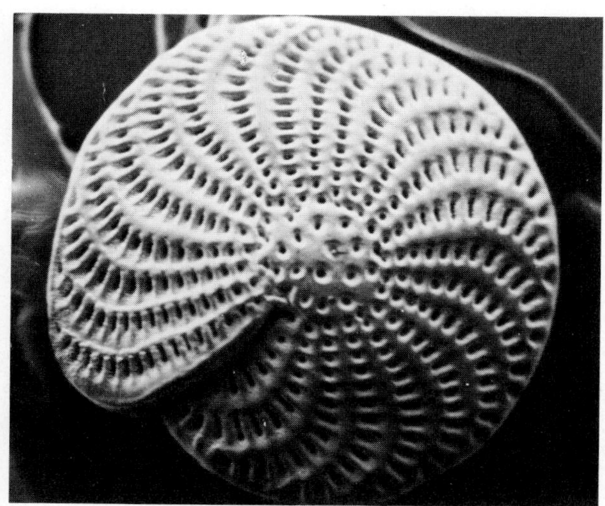

FIGURE 19–16

A foraminiferan. This scanning electron micrograph shows the details of this member of the phylum Sarcodina. (Biophoto Associates)

skin sores to loss of soft tissues of the face, and the disease may result in death.

Vaccines against these flagellate diseases will take years to develop. The best available countermeasure is to control the insect carriers necessary for the parasites to complete their complex life histories.

Closer to home for most of us is *Giardia,* carried to streams, rivers, lakes, and reservoirs by expanding populations of beavers and muskrats. To avoid the diarrhea, cramps, fatigue, and weight loss of giardiasis, campers should boil all water before drinking it.

Phylum Sarcodina

The Sarcodina move and engulf their prey with pseudopods. *Amoeba proteus,* a freshwater species, has been thoroughly studied. It is easy to raise on a diet of bacteria and is large enough to be a good subject for experiments on amoeboid movement, endocytosis, and nucleus-cytoplasm interactions. However, this amoeba is not typical of the phylum. Most sarcodines have shells, made of calcium carbonate (lime), silica, or bits of sand, through which they extrude long, thin pseudopods to catch prey (Figure 19–16). Some forms are colonial.

The limy shells of one group, the **foraminiferans,** sink to the bottom of the sea when the organism dies. Millions of years' worth of this debris has formed chalk rocks or limestone, such as the famous white cliffs of Dover in England. Foraminiferan fossils are also common in deposits where oil has accumulated from ancient phytoplankton. As an oil well is drilled, the bit passes through layers of foraminiferans that lived at different times. By identifying the species in a particular layer, a geologist can estimate the age of the rock and decide where oil deposits are likely to be found.

Phylum Apicomplexa

All apicomplexans are parasites, usually with the complicated life histories typical of this way of life. Some need two different host species. Most move by body flexion or gliding; some have flagellated gametes. The phylum is named for the "apical complex," a distinctive arrangement of organelles and cytoskeletal fibers used to penetrate into new host cells.

Toxoplasma may well be the most common human parasite. Half of all adults in the United States have probably been infected at some time. This protozoan invades body cells and usually causes mild symptoms (enlarged lymph nodes) that pass unnoticed. However, it can cause serious disease in a fetus, newborn, or HIV-infected person, leading to blindness, mental retardation, or death. *Toxoplasma* is usually spread in the feces of cats or in undercooked meat from infected pigs.

Human malaria, caused by four species of the genus *Plasmodium,* illustrates the complex life histories typical of parasites. Malaria parasites require two different hosts: humans, and female mosquitoes of the genus *Anopheles,* which transmit the parasite from one person to another (Figure 19–17).

Malaria was attacked vigorously earlier in this century by draining swamps where mosquitoes breed, by spraying houses with DDT to kill mosquitoes, and by treating victims with drugs to destroy the parasites. The disease was eradicated from much of its former range, including much of the United States, and drastically lowered in most other areas. But just when victory seemed near, some parasites evolved resistance to the drugs, and mosquitoes became resistant to DDT, bringing a resurgence of malaria to many areas of Asia, Latin America, and Africa. About 3.5 million people are infected each year and 2 million die each year, mostly children. Most victims live in developing nations, and neither they nor their governments can afford new insecticides nor drainage programs to combat mosquitoes. Current work on a vaccine against the sporozoite stage may take years to finish. Furthermore, sporozoites are vulnerable only on the short trip from the

BIO-BIT

Pregnant women should not clean out cat boxes because of the risk of contracting *Toxoplasma* from the cat's fecal material.

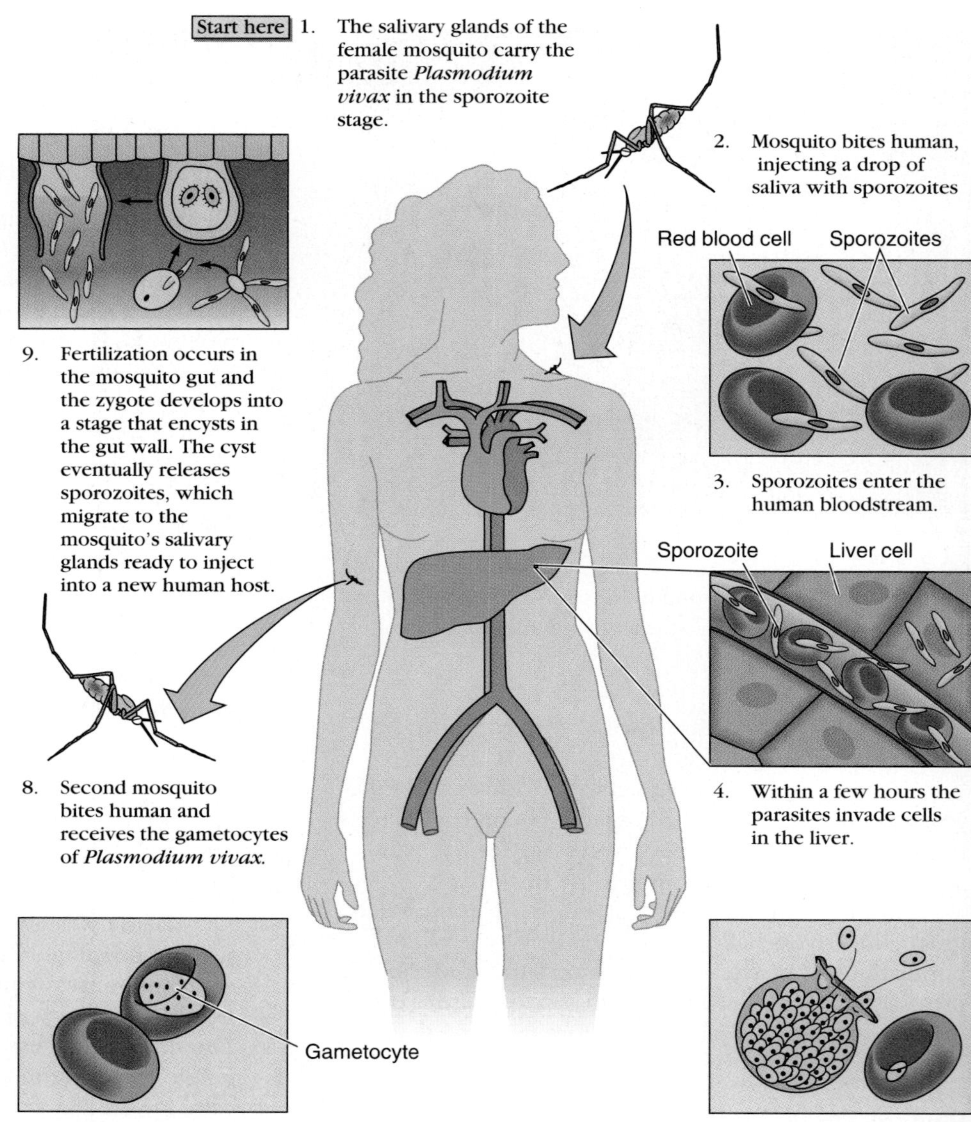

Start here
1. The salivary glands of the female mosquito carry the parasite *Plasmodium vivax* in the sporozoite stage.
2. Mosquito bites human, injecting a drop of saliva with sporozoites
Red blood cell
Sporozoites
3. Sporozoites enter the human bloodstream.
Sporozoite
Liver cell
4. Within a few hours the parasites invade cells in the liver.
5. For a week or two the sporozoites grow, divide and eventually release about 20,000 merozoites. Each merozoite enters a red blood cell and repeats the sequence of development.
6. Repeated cycles of rupture and invasion occur every two or three days, producing the symptoms of malaria: periodic bouts of fever, chills, shaking, and sweating.
Gametocyte
7. Some of the parasites in the red blood cells develop into sexual gametocytes, which may be taken up in the blood meal of another female mosquito.
8. Second mosquito bites human and receives the gametocytes of *Plasmodium vivax*.
9. Fertilization occurs in the mosquito gut and the zygote develops into a stage that encysts in the gut wall. The cyst eventually releases sporozoites, which migrate to the mosquito's salivary glands ready to inject into a new human host.

FIGURE 19–17

Life history of *Plasmodium vivax*, which causes a potentially lethal form of malaria in humans.

bite site to the liver, and they continuously shed their surface proteins and make replacements. The immune system ends up attacking the empty coats rather than live parasites.

Phylum Ciliophora (Ciliates)

Ciliates are complex, heterotrophic protists with many cilia (Figure 19–18). The body wall contains a pellicle and often numerous **trichocysts,** barbed or poisoned thread-like organelles that can be discharged to the outside. Trichocysts serve for anchorage, defense, or capture of prey.

Most ciliates prey on bacteria, small animals, or fellow protists; some eat organic particles from the water, some are symbionts, and a few are parasites. Cilia around the mouth sweep food into a gullet. Here food enters a vacuole, which fuses with a lysosome full of

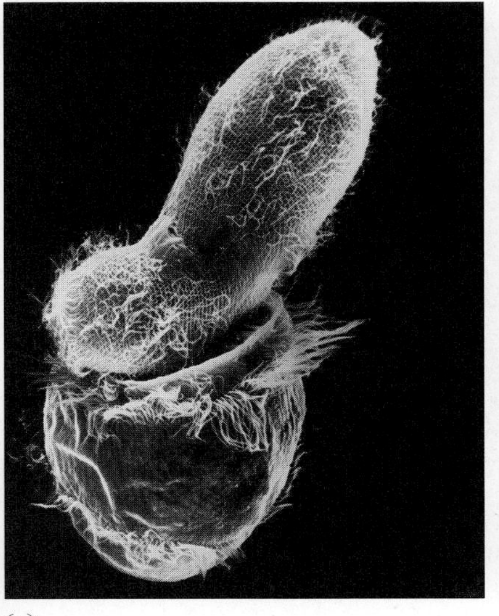

(a)

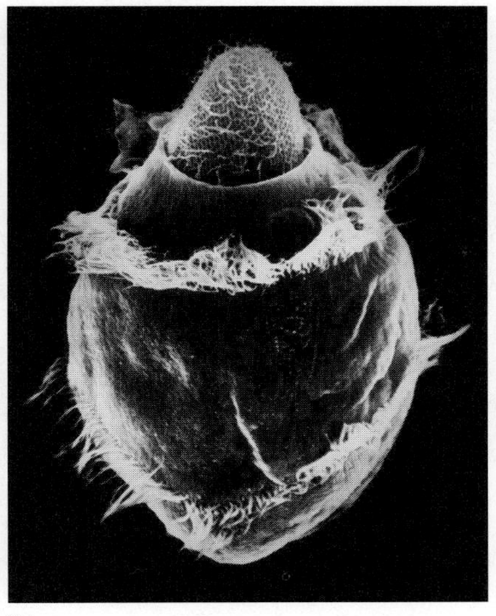

(b)

FIGURE 19–18

Ciliate eats ciliate. (a) A *Didinium* (at bottom) attacks a *Paramecium* that looks much too big for it. **(b)** The *Didinium* has stretched enormously and engulfed all but the tip of the *Paramecium*. (Biophoto Associates, Photo Researchers, Inc.)

digestive enzymes. The products of digestion are absorbed into the cytoplasm. Contractile vacuoles discharge excess water at specific sites on the cell surface.

Ciliates have more than one nucleus: each cell has one or more small **micronuclei,** each containing one copy of the genome, and a large **macronucleus,** containing up to 500 times more DNA. Apparently the micronucleus controls sexual reproduction and heredity, while the macronucleus controls growth, metabolism, and asexual reproduction. *Paramecium* is the most familiar genus of freshwater ciliates (Figure 19–19).

Slime Molds: Protists or Fungi?

Slime molds are a classic problem in taxonomy. They spend part of their lives in a mobile, amoeba-like state, engulfing organic matter and bacteria, much as protozoa do. Hence many people would assign slime molds to the kingdom Protista. On the other hand, slime molds also produce fungus-like reproductive structures: some cells form a stalk, and others become spores with cellulose walls. This cooperation and division of labor show a primitive degree of multicellularity. At present, slime molds have been placed in kingdom Protista.

Phylum Myxomycota

The feeding stage of an **acellular slime mold** is a mass of cytoplasm containing many nuclei but not separated into individual cells (Figure 19–20).

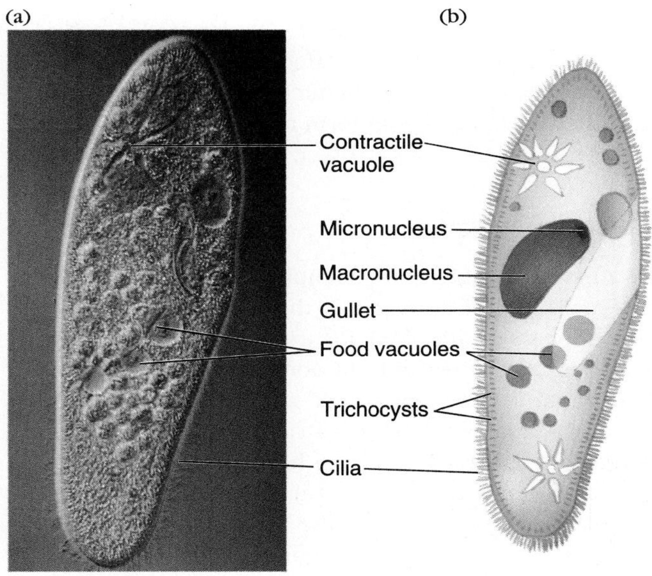

FIGURE 19–19

***Paramecium*, a ciliate. (a)** A photograph of a living *Paramecium.* **(b)** The body is completely covered with cilia, used in locomotion and also to sweep food particles into the gullet. The cytoplasm at the end of the gullet engulfs food into vacuoles by phagocytosis. Food is digested by enzymes in the food vacuoles as they move around in the cytoplasm. The two star-shaped contractile vacuoles collect water taken up by osmosis and expel it from the cell. Trichocysts are defensive, barbed threads that the cell discharges at its surface when it is disturbed. (M. Abbey/Visuals Unlimited)

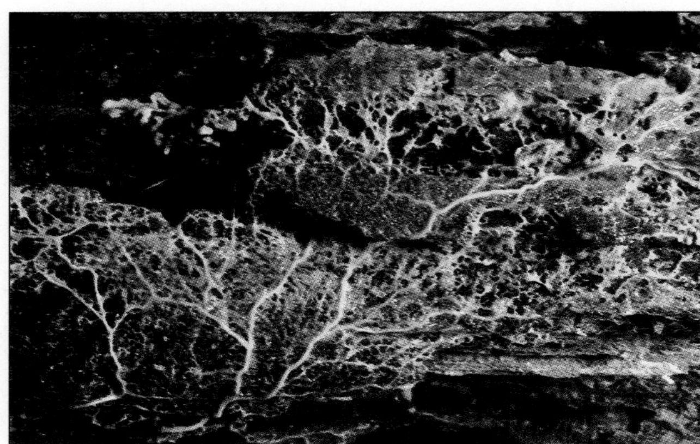

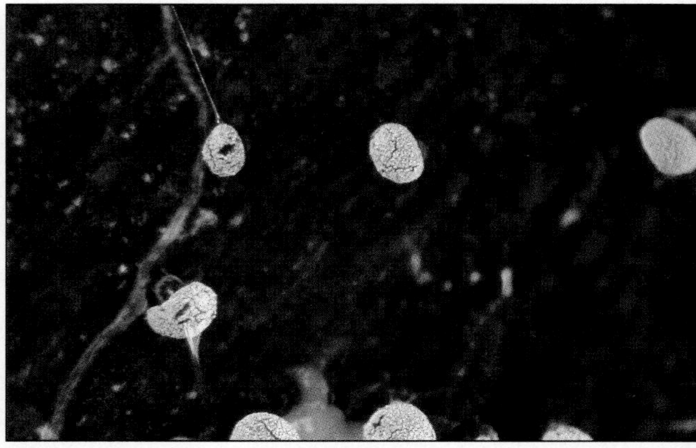

FIGURE 19–20

Acellular slime molds. (a) The slime mold *Physarum polycephalum*. **(b)** The bulbous fruiting bodies of the same species of slime mold. (a, E. R. Degginger; b, Runk/Schoenberger from Grant Heilman)

Phylum Acrasiomycota

In **cellular slime molds** the feeding stage is a uninucleate amoeba. When these amoebas run out of food, some of them secrete a chemical that attracts others, and they congregate and form a reproductive structure (Figure 19–21).

Phylum Oomycota

You have probably seen examples of this phylum without realizing it. For example, the familiar white fuzz that appears on diseased aquarium fish or on organic matter sitting in water is an oomycete. The name, Oomycota, means "egg fungi," and refers to the appearance of sexual reproductive structures that produce flagellated spores. This group has only recently been added to kingdom Protista. It includes organisms that are plant parasites, such as the late blight of potatoes (responsible for the Potato Famine in Ireland) and downy mildew.

The heterotrophic protists are a diverse group divided into seven phyla based largely upon locomotor organelles. The zoomastigophorans have flagella and include free-living forms, symbionts, and parasites. Many of these parasitic forms cause important human diseases, including trypanosome infections and leishmaniasis. Sarcodines typically move and engulf their food with pseudopods. Many have limy shells, whose fossilized remains have formed huge limestone deposits. Apicomplexans move by gliding or body flexion. All are parasites causing diseases such as malaria and toxoplasmosis. Ciliates are mostly freshwater protists that have numerous cilia, used for locomotion and feeding. Oomycota organisms have flagellated spores. Slime molds (phyla Myxomycota and Acrasiomycota) have feeding stages like amoeboid protists but reproductive strategies resembling those of fungi.

19–H ON TO MULTICELLULARITY!

The first organisms were unicellular. Wherever such life was abundant, there must have been strong selective pressure for increased size. A large organism could eat more of its neighbors and be eaten by fewer. However, a single cell cannot just become larger and larger. Eventually the center of the cell is too far from the outside to obtain the substances it needs from its environment fast enough. As cell size increases, the ratio of cell surface area to cytoplasmic volume decreases (See Toolbox in Chapter 4). Because food, oxygen, and wastes are exchanged through the surface, the decrease in the surface-to-volume ratio also decreases the rate at which supplies reach each unit volume of cytoplasm. Thus the size of single cells is limited. Body size must be increased by an increase in cell number.

Some protists are colonial, with many independent cells joined into a larger unit. Each cell is small and exchanges substances with its environment efficiently, but the cells are still more or less identical (see Figure 20–4c). True multicellularity implies division of labor among the cells of an organism.

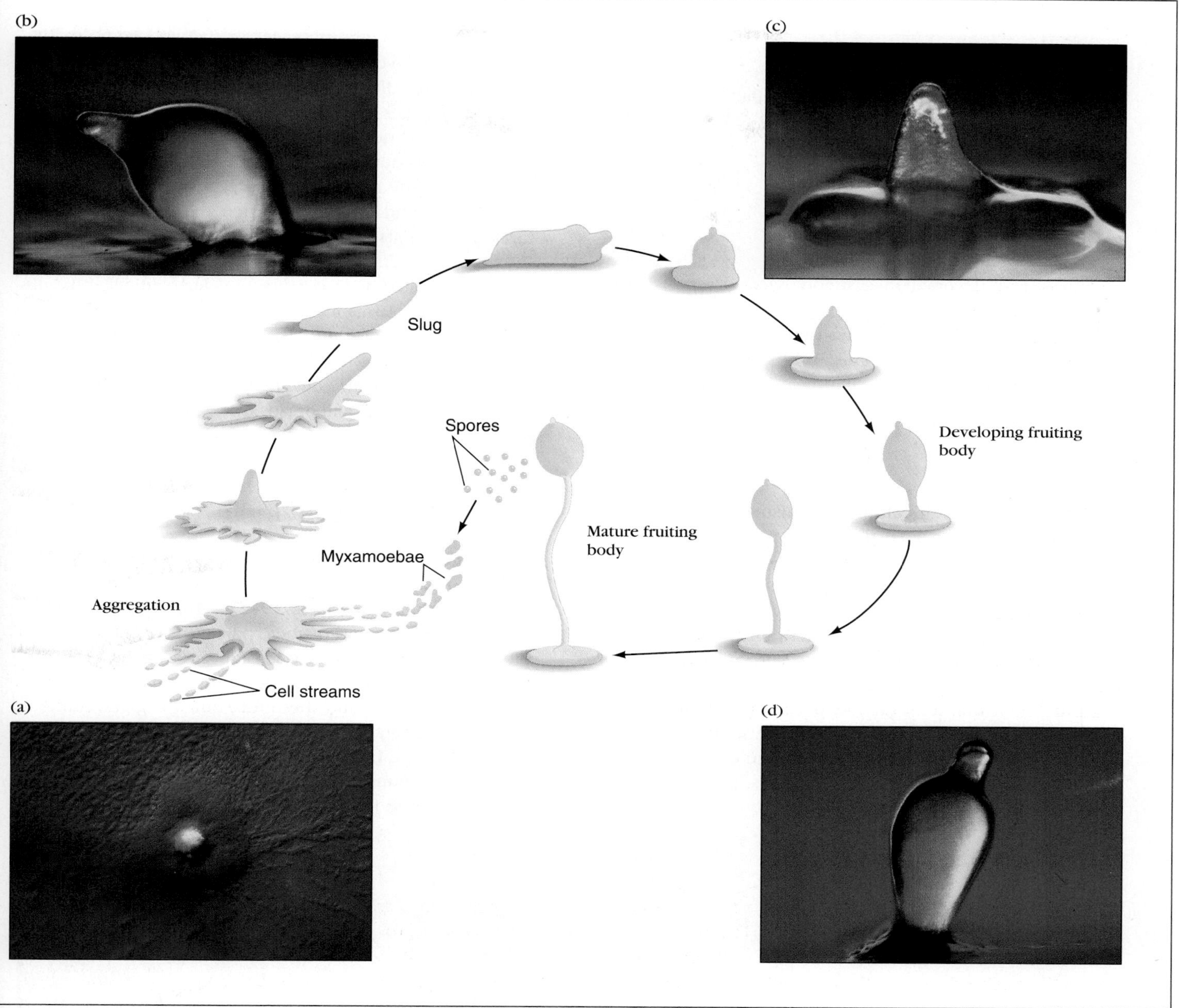

FIGURE 19–21

Reproduction in a cellular slime mold. (a) Individual amoeba-like cells congregate into a single mass, visible as a spot in the center of the frame. **(b)** The slug-like mass of cells crawls about until it **(c)** settles in one spot and **(d)** forms a reproductive structure, a stalk topped by a rounded sporangium (spore-producing structure). (Carolina Biological Supply Company/Phototake)

Every cell and every organism must carry out certain basic life functions (Table 19–2). Usually each cell can carry out its own basic functions, but in multicellular organisms each cell is also specialized to help carry out one of these functions for the entire body. The story of the kingdoms Fungi, Plantae, and Animalia can be viewed as the evolution of increasingly specialized groups of cells to form tissues and organs. The first sign of division of labor is usually the specialization of some cells for reproduction, protection, or anchorage, while others acquire energy by feeding or photosynthesis.

Table 19–2
Basic Functions That Must Be Carried Out by Every Species

Feeding or making food
Gas exchange
Waste removal
Internal transport of food, gas, etc.
Sensing environmental stimuli
Dispersal (locomotion, scattering seeds or larvae)
Support and protection
Coordination of all functions (nerves, hormones, etc.)
Reproduction

The simplest multicellular organisms are small, with no cell very far from the watery environment that provides food and removes wastes. As organisms became larger, the inner cells were farther from the environment. Organisms that evolved ways to transport substances between the environment and these cells were able to grow larger, but those without such systems had to remain small. Finally, a large-bodied organism must coordinate its various parts into a working system. Coordination exists in even the smallest of single cells, but it is most impressive in the complex systems of sense organs, nerves, hormones, and muscles found among the higher animals.

We tend to view increases in size and complexity as "progress" over yesteryear's smallness and simplicity. However, unicellular organisms do have some advantages over larger creatures. A single cell can live in a tiny space and needs only a little food before it is ready to reproduce. This allows unicellular organisms to exploit many habitats that larger forms could not. The current widespread presence and diversity of bacteria and protists attest to their evolutionary success.

Cell size is limited by the surface to volume ratio and the distance from the plasma membrane to the cell interior. Multicellularity permitted size to increase while maintaining favorable cell dimensions. It also permitted division of labor among cells. Eventually, some organisms evolved entire systems of cells, tissues, and organs that carry out specialized functions.

19–I KINGDOM FUNGI

The kingdom Fungi contains eukaryotic, mostly multicellular organisms that absorb food molecules from their surroundings (Figure 19–22). All fungi are heterotrophs, living as saprobes, parasites, or partners in a mutualistic symbiosis. They obtain food as bacteria do: they secrete digestive enzymes, which hydrolyze the organic matter around them into small organic molecules and minerals that the fungus can absorb. As decomposers, the fungi are vitally important to plants and animals. Fungi are also important in the human economy. Some fungi cause tremendous losses of food

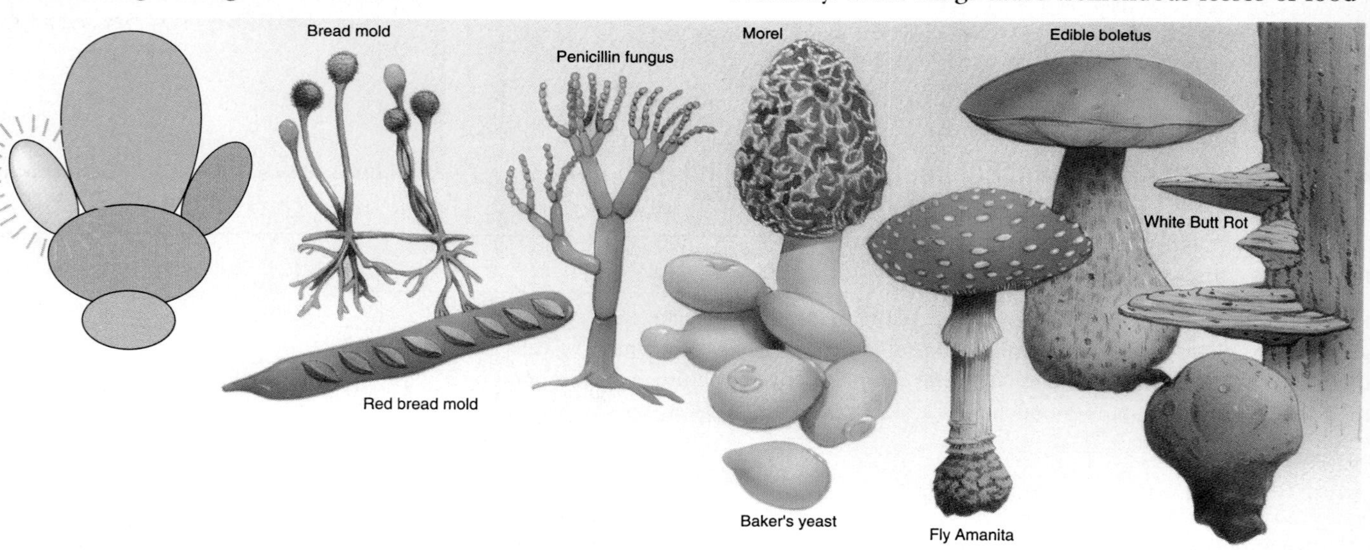

FIGURE 19–22

Overview of the kingdom Fungi.

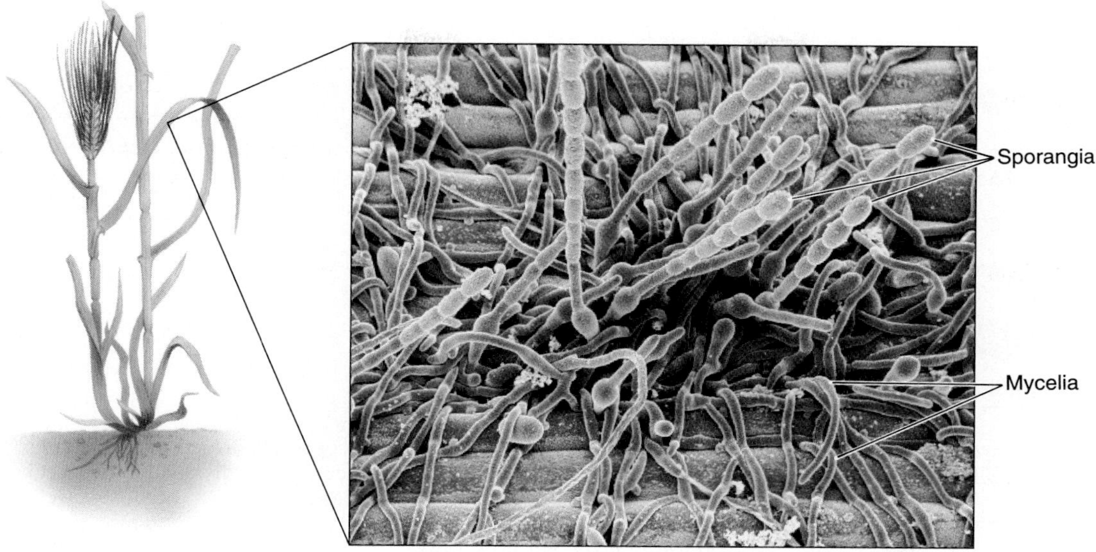

FIGURE 19–23

Mycelia of powdery mildew on the surface of a barley plant. The tips of the vertical, aerial hyphae are pinching in and forming spores (powdery mildew) that can spread to other cereal plants. (Biophoto Associates)

and crops every year, but fungi are also used to produce foods and medicines.

Fungi almost always have cell walls, and they reproduce and disperse to new habitats by means of various types of spores.

Body Plan

The absorptive lifestyle of fungi is intimately linked with two important characteristics: production of spores and mycelial growth. A **spore** is a tiny, usually haploid, cell that disperses the fungus to new habitats, usually by floating through the air. The production of many tiny spores increases the chance that at least a few will fall onto a suitable food source. When this happens, the spore germinates, starts absorbing food, and grows into a thread-like **hypha.** The hypha grows rapidly and branches to form a tangled mass called a **mycelium.**

The mycelium, with its high surface-to-volume ratio, is well suited to absorbing food. A hypha grows from the tips and releases chemicals that cause other hyphae to grow away from it. As a result, the fungus spreads out through its food source. Parasitic fungi absorb nutrients from the host's body fluids, and parasites of plants may produce specialized hyphae called **haustoria** that penetrate the plant's cell walls and lie against the plasma membranes, where they can absorb food.

Some fungi are **coenocytic,** with many nuclei lying in the same cytoplasm, not divided into separate cells. Others are divided by **septa** into compartments containing one or more nuclei.

Fungi have rigid cell walls. Most groups have walls containing the polysaccharide chitin, also found in the external skeletons of insects and their kin. Some fungi have cellulose cell walls, like those of plant cells.

Reproduction

Fungi may spread vegetatively, by growth or fragmentation of the mycelium. Spores may be formed asexually or as a result of sexual processes. Spores are often produced on aerial structures that hold them away from the food source, up where they can catch an air current for a ride to a new home.

The parts of a fungus we normally see are reproductive structures. If you use a microscope to examine the green, white, pink, or black fuzz on moldy food or diseased plants, you will see that it consists of masses of hyphal threads tipped with **sporangia:** structures bearing strings or spherical clusters of rounded, thick-walled spores (Figure 19–23). Similarly, the above-ground parts of mushrooms and cup fungi are **fruiting bodies**—large, complex reproductive structures composed of many hyphae. Fruiting bodies disperse spores produced by sexual processes; the vegetative mycelia grow hidden in the food source.

Table 19–3
The Kingdom Fungi and Its Divisions

Division Zygomycota (~600 species) 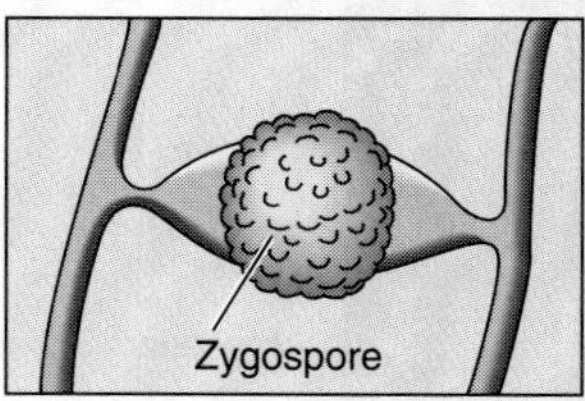	Sexual reproduction by zygospores; coenocytic hyphae; chitinous cell walls; e.g., *Rhizopus* (black bread mold), *Entomophthora muscae* (parasite on the housefly)
Division Ascomycota (~30,000 species) 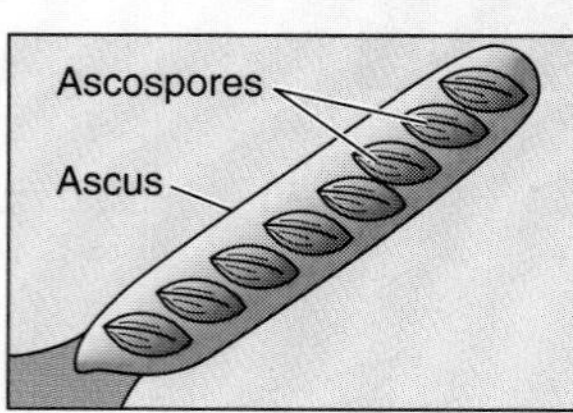	"Sac fungi." Sexual reproduction by ascospores formed in asci; some unicellular (yeasts); hyphae of multicellular forms divided by perforated septa; e.g., pink bread mold, bread and wine yeast, ergot disease of grasses, cup fungus, chestnut blight, *Aspergillus* (a common mold on foods; one species used to ferment beans for soy sauce); *Penicillium* (various species used for production of penicillin and of Roquefort and Camembert cheeses)
Division Basidiomycota (~25,000 species) 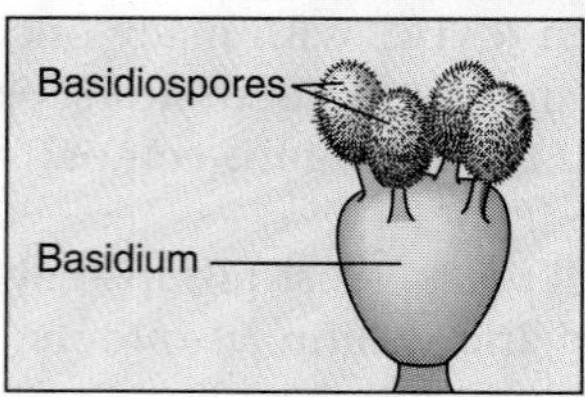	"Club fungi." Sexual reproduction by basidiospores borne by basidia. Septate hyphae. Mushrooms and toadstools, puffballs, bracket fungi, rusts, and smuts
Division Deuteromycota (~10,000 species)	"Imperfect fungi," not known to reproduce sexually; e.g., ringworm, athlete's foot, spear rot of asparagus, root rot of wheat

Most fungi are multicellular eukaryotes with cell walls. They are heterotrophs and can be either saprobic, parasitic, or mutualistic. In fungi, reproduction and dispersal to new habitats are accomplished vegetatively by spore production.

19–J CLASSIFICATION OF FUNGI

Fungal taxonomy follows the rules for plants, which use the term **division** instead of phylum in the taxonomic hierarchy. The names of fungal divisions end in -mycota (myco = fungus), but common names of these groups end in -mycetes, harking back to an older classification scheme. Fungi are assigned to divisions on the basis of the characteristic form of microscopic sexual reproductive structures—one of the few consistently conservative features in each division (Table 19-3).

Fungi are mostly terrestrial, with chitinous walls, haploid nuclei, haploid spores, and no flagellated cells. Most fungi commonly reproduce by forming asexual spores atop aerial hyphae (hyphae that stick out in the air, such as those of bread mold) (Figure 19-24).

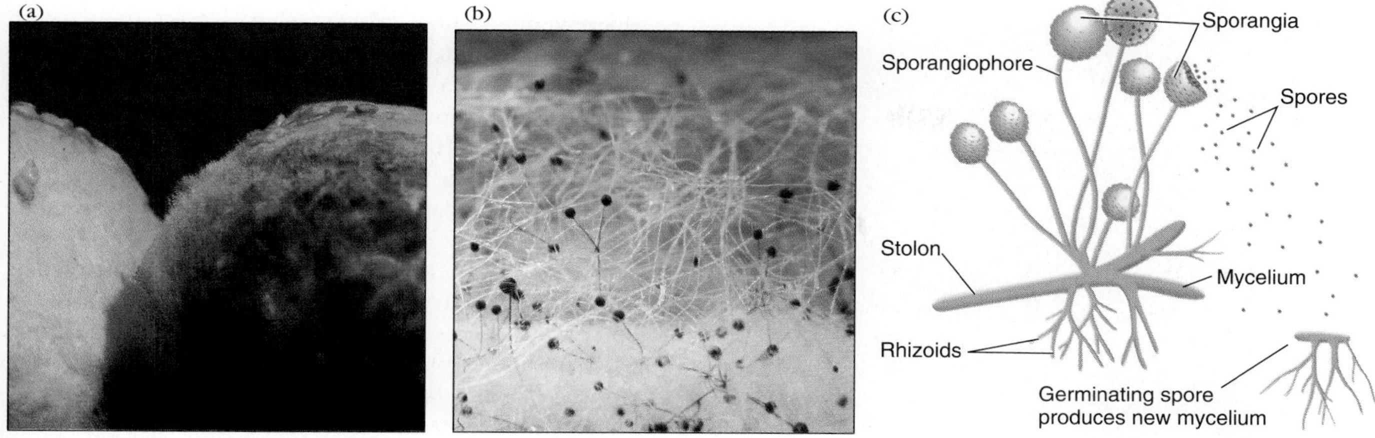

FIGURE 19–24

Asexual reproduction in *Rhizopus* (black bread mold). (a) Moldy bread. **(b)** Fungal hyphae and sporangia on the surface of the food source. **(c)** Sporangiophores are aerial hyphae bearing sporangia. Spores are shed into the air, which wafts them to new habitats. The mycelium is colorless; the common name of this organism comes from the masses of black spores it produces. The organism usually reproduces by this asexual process. (a, V. Tyler/OffShoot Stock; b, Oxford Scientific Films)

Division Zygomycota

The sexual reproduction of the Zygomycota is characterized by formation of diploid zygote nuclei enclosed in thick-walled zygospores. This group includes many saprobes, some parasites, and some mutualistic mycorrhizal fungi that grow in plant roots (Section 19–K).

Division Ascomycota

Among the more than 30,000 species of Ascomycota are unicellular forms, called yeasts, and a great variety of multicellular forms, including morels, cup fungi, ergots and other plant parasites, and the common molds *Aspergillus* and *Penicillium* (Figure 19–25). The char-

(a)

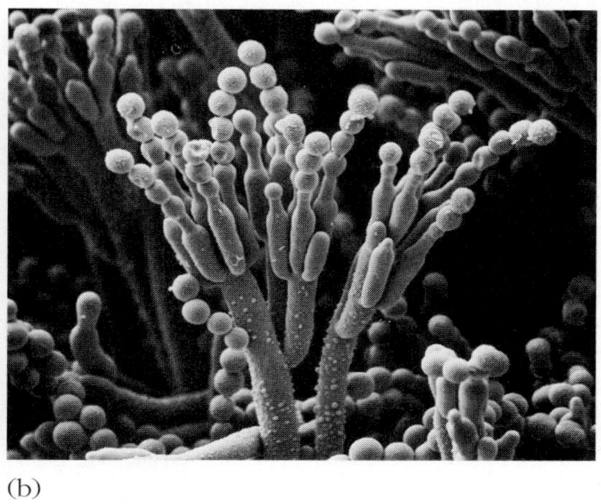
(b)

FIGURE 19–25

The ascomycete *Penicillium*. (a) Asexual spores called conidia impart the green color seen here. While some hyphae were producing these spores, others grew out into new areas of the food source. The white fuzz consists of immature reproductive hyphae that have not yet formed conidia. Vegetative parts of the mycelium are buried in the fruit, turning it to mush. **(b)** Hyphae and conidia of *Penicillium* as seen through the electron microscope. (a, Sidney Molds/Custom Medical Stock; b, Biophoto Associates)

FIGURE 19-26

Ascomycetes with chemical defenses. An ergot in an ear of wheat. (H. Reinhard/Oxford Scientific Films)

acteristic sexual structure is an **ascus** ("sac"), which usually contains a stack of eight haploid spores, called ascospores. Asexual reproduction is also common.

In 1927, Alexander Fleming noticed that his bacterial cultures had been killed by the ascomycete *Penicillium notatum.* This led to the discovery of the world's most widely used and effective antibiotic, **penicillin.**

Various other ascomycote fungi are also used to produce vitamins, amino acids, enzymes, and other organic compounds on a commercial scale.

An ergot in a head of rye or other cereals is an ascomycete containing many extremely toxic substances (Figure 19-26). Humans may be poisoned by eating bread made from infected rye. Ergotism, also called St. Anthony's Fire, caused the death of thousands in medieval Europe. Ergot supplied the chemicals from which lysergic acid diethylamide (LSD) was first synthesized. Ergot toxins, and the notorious poisons of some mushrooms, protect these fungi from predators. The substances produced by ergots are also valuable drugs. Although they are deadly in large amounts, small doses are used to induce labor, control bleeding, and treat migraine headaches, high blood pressure, and varicose veins.

Division Basidiomycota

The Basidiomycota (basidi = a small pedestal) include rusts, smuts, mushrooms, puffballs, bracket fungi, and coral fungi (Figure 19-27). The sexual reproductive structure is a **basidium** ("club") bearing four haploid basidiospores.

Mushrooms are the most familiar basidiomycetes. In these fungi, a well-fed mycelium forms an underground mass of hyphae with a bulbous base, a stalk, and a knob-like cap. Some morning after a heavy rain we awake to find that the hyphae of the stalk have swelled with moisture and elongated, lifting the cap above ground. The cap opens like an umbrella, and numerous basidia along the edges of the gills or pores beneath the cap shed their spores (Figure 19-28).

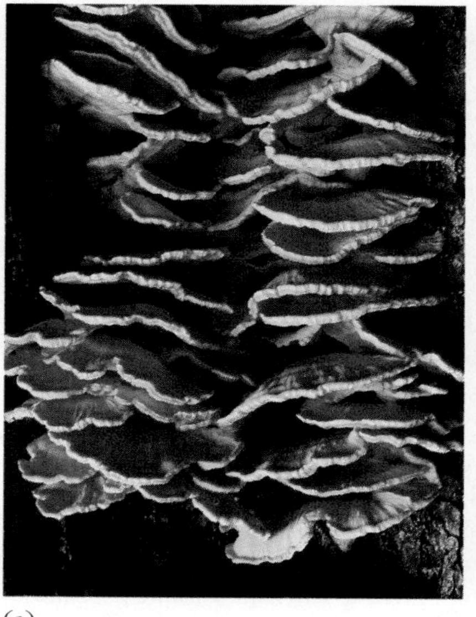

(a)

(b)

FIGURE 19-27

Basidiomycete fruiting bodies. **(a)** Bracket fungi on the surface of wood which is being decomposed by the fungal vegetative mycelium growing inside it. **(b)** Coral fungus, the reproductive structure of a mycelium growing on dead plant matter in the soil. (a, F. Lanting/Minden Pictures; b, R. Lubeck/-Animals, Animals)

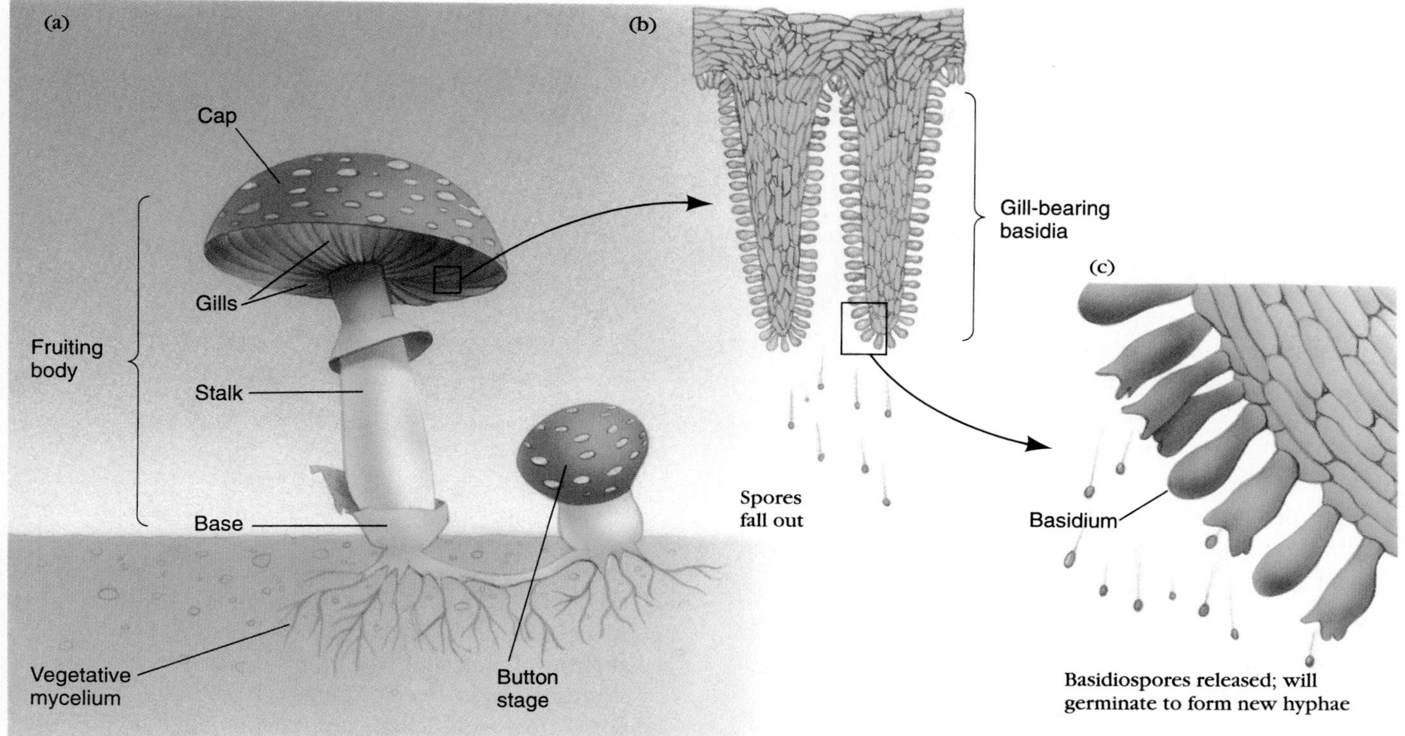

FIGURE 19–28

Reproduction of a mushroom. (a) A mushroom is the fruiting body of a basidiomycete. Both the fruiting body and the vegetative, absorptive mycelium are made up of masses of hyphae. **(b)** Cross-section of two gills, showing basidia with basidiospores all over the outer surfaces. **(c)** Close-up of basidia and basidiospores.

Division Deuteromycota

The division Deuteromycota (deutero = second) contains a rummage-sale collection of fungi that cannot be assigned to any other group because their sexual reproduction (if they have any) has never been observed. Fungi that cause ringworm, athlete's foot, and other skin infections belong to this group. Other deuteromycetes cause important diseases of crops, including strawberry leaf blight and bitter rot of grapes.

Fungi are classified into three main divisions on the basis of sexual reproductive structures. A fourth division, Deuteromycota, contains fungi not known to reproduce sexually.

19–K SYMBIOTIC RELATIONSHIPS OF FUNGI

Many fungi are harmful parasites, but some form mutualistic associations, in which both partners benefit.

Mycorrhizae

Many fungi grow associated with plant roots in a symbiosis called a **mycorrhiza** ("fungus root") (Figure 19–29). The fungal hyphae branch out into the soil and use their superior absorptive ability to take up mineral nutrients. By using radioactive elements as tracers, researchers found that the fungi pass on some of these minerals to the plant roots and receive food from the plant. Most land plants form mycorrhizal associations, usually with zygomycete or basidiomycete partners. When pine trees were introduced into new areas, such as Puerto Rico and Australia, they grew very poorly until supplied with soil from pine forests, containing the appropriate mycorrhizal fungi; after this, they grew rapidly.

Some fungi help protect their plant partners from acid rain. Acid promotes two changes in the soil unfavorable to plants: leaching (washing away) of required nutrients, and increased solubility of toxic minerals such as zinc, copper, aluminum, and manganese. The

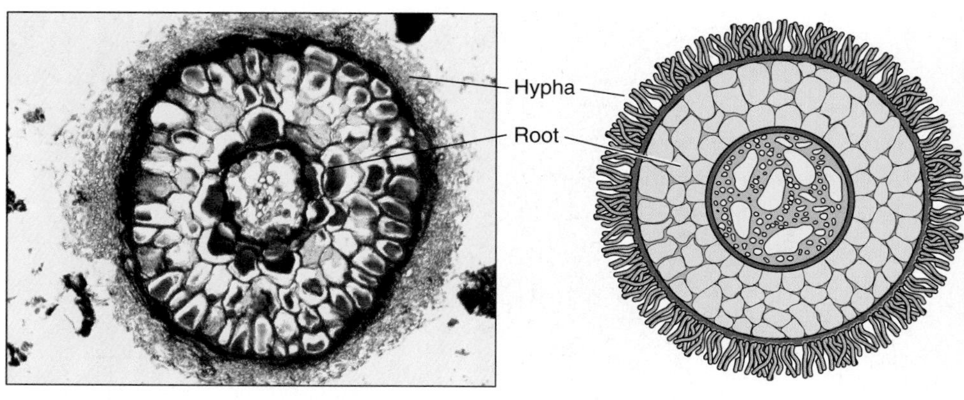

FIGURE 19-29

A mycorrhiza. Fungal hyphae, stained aqua, surround a cross-section of a plant root. The fungus absorbs soil minerals and passes them to the plant, receiving organic substances made by the plant in return. (Carolina Biological Supply Company, Phototake)

right mycorrhizal fungus can absorb nutrients from depleted soil water and pass them to the plant, and can also protect the plant from toxic substances in the soil.

Lichens

Lichens look like plants (Figure 19-30). However, a lichen is really an intimate symbiosis between two kinds of organisms: ascomycete or basidiomycete fungi, and photosynthetic cells—either cyanobacteria or green algae (Sections 19-C and 20-C). The fungus obtains organic compounds from the photosynthetic partner, but it is unclear what it provides in return. Possibly it absorbs water and minerals.

Lichens have slow metabolism and growth, but they are extremely resistant to drought and cold. They are the most important autotrophs in the low-growing vegetation of the **tundra,** found in the Arctic and also at high altitudes on mountains. Lichens absorb minerals from the air and so can grow without soil, on stones or rocky ocean islands where nothing else survives. However, lichens are quickly killed by air pollution, and the state of the lichens in an area can be used as an indication of its air quality.

People have found other odd uses for lichens. Archaeologists have estimated the age of the mysterious stone heads of Easter Island by measuring patches of lichens growing on them. Lichens also produce pigments traditionally used to dye Harris tweed. They are also used by Kurds to make bread.

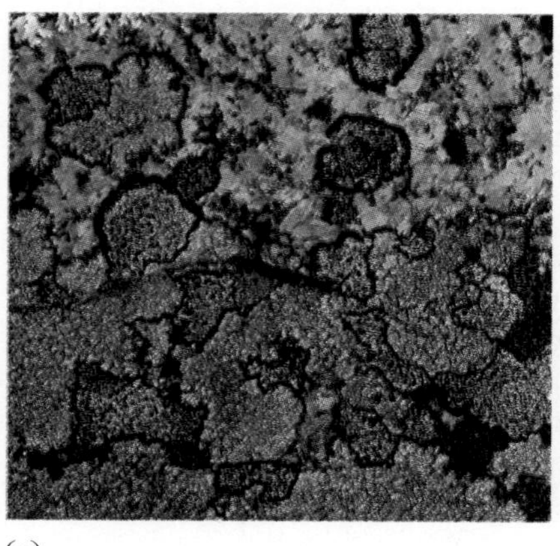
(a)

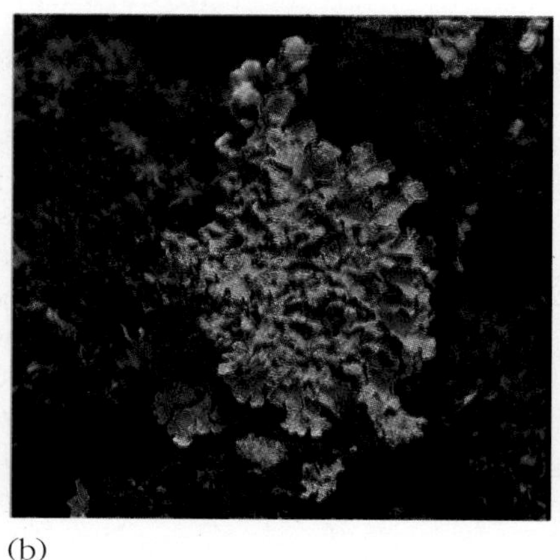
(b)

(c)

FIGURE 19-30

Lichens come in three main body forms. (a) Crustose forms resemble thick blotches of paint. **(b)** Foliose forms resemble leaves, flat with curling edges. **(c)** "British soldiers" is a colorful example of the fruticose (shrubby, branched) form. (a, D. Kuhn; b, E. Reschke/Peter Arnold, Inc.; c, D. Specker/Animals, Animals)

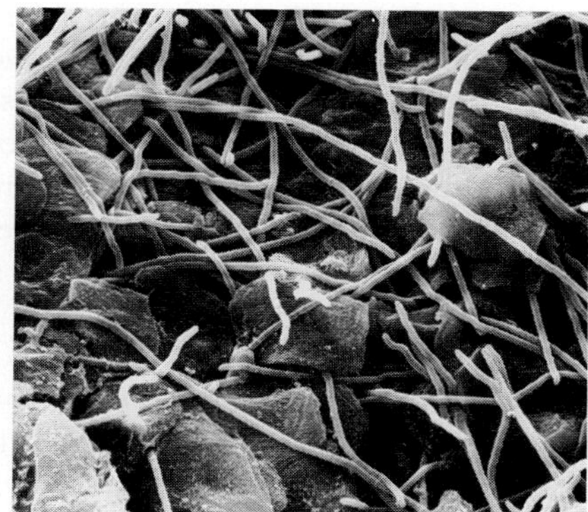

FIGURE 19-31

Fungal infection of human skin. Slender, tubular hyphae grow over and under flat human epidermal cells. (Biophoto Associates, Photo Researchers, Inc.)

Mycorrhizal fungi form a mutualistic symbiosis in which their superior absorptive abilities supply minerals from the soil to the roots of a plant. Lichens are formed by an intimate symbiosis between fungus and photosynthetic cells. Lichens absorb minerals from the air and therefore can grow on bare rocks where nothing else can survive.

19–L BACTERIA AND FUNGI: FRIENDS AND FOES

Diseases

A **pathogen** is an organism able to produce disease. Pathogens are specific for particular hosts because infections begin with the specific binding of certain pathogen and host surface molecules, much like the binding between an enzyme and its substrate (Section 3-E). Some pathogenic bacteria destroy the host's cells, but most cause disease by producing **toxins** (poisonous substances). Often the toxin alone can cause symptoms typical of a bacterial disease.

Most fungi require more oxygen than bacteria do. Hence, human fungal infections, such as ringworm, athlete's foot, and vaginal infections, are nearly all restricted to moist body surfaces that are exposed to air (Figure 19-31). Fungal diseases of plants are much more numerous. Fungi can easily enter plants through injured areas or natural pores in leaf surfaces. Air spaces within leaves permit the fungus to obtain all the oxygen it needs and still send specialized hyphae into the plant's cells for food.

The Irish potato famine was a major disaster caused by a fungal disease, late blight of potatoes. The potato, native to South America, was brought to Europe in the sixteenth century. Potatoes require little labor and produce high yields of one of the most nutritious crops. By the nineteenth century, they were almost the only crop grown in Ireland. However, **monocultures** (plantings of only one crop) are especially susceptible to the rapid spread of disease. In 1845 and 1846, late blight destroyed virtually the whole Irish potato crop, leading to a devastating famine. From 1845 to 1851, a million people died in Ireland and a million and a half emigrated, mainly to the United States and Canada. The European potato shortage stimulated agriculture in North America, where grain for export has been grown in large quantities ever since.

Control of Disease

Preventing the spread of pathogenic bacteria depends on knowing how they disperse. Diseases like diphtheria, scarlet fever, whooping cough, and tuberculosis are caused by airborne bacteria, usually released in cough or sneeze droplets. This is why victims are often quarantined. The most important preventive measures against bacterial disease are hygiene and sanitation. In the last century, maternal deaths following childbirth were reduced about tenfold when doctors and midwives learned to wash their hands and their instruments between patients. Also, Joseph Lister developed aseptic surgical techniques. Even today, keeping food and water reasonably free of bacteria saves many more lives than antibiotics do.

Heating to 60°C for 30 minutes destroys protein toxins and kills most bacteria. For this reason pasteurization (heating) has proved both easy and effective in protecting against botulism in canned food, brucellosis and tuberculosis in milk, and dysentery in drinking water contaminated with human feces. In addition, antibiotics have saved thousands of lives since about 1940. During all wars before World War II, more lives were lost to disease (from unsanitary camps and infected wounds) than to enemy action.

Once a fungus establishes itself in a plant or an animal, or in wood, paper, or leather, it is virtually impossible to eradicate. Measures that prevent spores from germinating, such as keeping things dry and using fungicides, are the most effective defenses against fungi.

Bacteria, Fungi, and Food

Bacterial and fungal spores are everywhere, and some of them inevitably land on our food. Milk spoils when bacteria ferment the milk sugar lactose and produce

FIGURE 19-32

A common morel, an edible fruiting body of an ascomycete. It releases spores produced by sexual reproductive processes. (J. Lepore)

lactic acid, which coagulates the milk proteins. Pasteurization retards spoilage by killing many of these bacteria. On the other hand, bacteria are necessary in making dairy products such as cottage cheese and yogurt, and other foods such as sauerkraut, pickles, and vinegar (see Section 7-E). Fermented foods like these keep longer, and are often more nutritious, flavorful, and digestible than their raw materials.

Brewers' yeast, a unicellular ascomycete fungus, ferments sugars to make alcohol in the production of wine and beer. A different strain of yeast is used in bread-making. Here the carbon dioxide it gives off forms bubbles in the dough and makes the bread rise.

English Stilton and French Roquefort, Brie, and Camembert cheeses all get their flavors from specific ascomycete fungi. In the Orient, soy sauce is traditionally made by fermenting boiled soybeans and wheat with another ascomycete for about a year. This produces a flavorful sauce rich in vitamins and amino acids, a valuable addition to a low-protein diet of rice.

Growing edible mushrooms (basidiomycetes) is a million-dollar industry in many parts of the world. Morels (Figure 19-32) and truffles are ascomycetes. In France, truffle hounds and pigs are trained to hunt the
3? underground truffles by smell. In 1990, the most prized truffles sold for $720 per pound.

"Food poisoning" comes from toxins produced by bacteria growing in food. The most serious (usually fatal) form of food poisoning is botulism. Since botulism bacteria are obligate anaerobes, they grow in canned goods but not in fresh and frozen foods. Most cases of botulism are due to inadequate heating of home-canned goods. The most effective safeguard is to can only acid foods like fruit and pickles, since acid kills the bacteria.

Salmonella bacteria in pork, poultry, or eggs cause diarrhea shortly after the contaminated food is eaten. The culprits are living bacteria, not toxins, and so this is called a food infection, not food poisoning. ■

Many fungi and bacteria can cause serious economic and health problems by destroying crops, foods, and materials, and by causing diseases in animals and plants. However, many fungi and bacteria are used in the production of food and antibiotics and some fungi produce edible fruiting bodies.

SUMMARY

1. Kingdom Prokaryotae contains the bacteria. Mutation and rapid reproduction, coupled with small size and metabolic diversity, account for the evolutionary success of bacteria.
2. Three billion years ago, bacteria were the Earth's only inhabitants, living in communities of many interdependent species. Some were autotrophs, making their own food. Photosynthetic forms used solar energy, while chemosynthetic forms obtained energy from chemical reactions. Other bacteria exploited their neighbors for food. Dead cells or cell wastes provided raw materials for other bacteria, and so nutrients were recycled within the bacterial community.
3. Today bacteria are still the most numerous organisms on Earth and play all of these roles. However, eukaryotes have become the dominant photosynthetic and heterotrophic organisms.
4. Nitrogen-fixing bacteria and cyanobacteria support eukaryotic life by converting nitrogen in the atmosphere into a form that plants can use to make amino acids.

Saprobic bacteria play the vital ecological role of decom posers.

5. Many bacteria are free-living, while others form parasitic or mutualistic symbiotic relationships with other bacteria or with eukaryotes.
6. Some bacteria are pathogens, producing toxins that cause disease, while other bacteria are part of the normal microbiota of animals.
7. Eukaryotic cells appeared about 1.5 billion years ago and evolved into a wide array of unicellular heterotrophs and autotrophs currently grouped in the kingdom Protista.
8. Autotrophic protists live either free as phytoplankton, attached to moist surfaces, or as symbionts of larger organisms. The clues to their phylogeny include such conservative features as photosynthetic pigments, the form of food storage, cell wall structure, and type of flagella (if any).
9. Heterotrophic protists may be free-living predators on other protists, bacteria, and small multicellular organisms, or may be symbionts or parasites of animals.
10. Seven major phyla of heterotrophic protists are characterized by their type of locomotion:
    a. Zoomastigophora move by means of flagella.
    b. Sarcodina use pseudopods.
    c. Apicomplexa may move by body flexion, or by gliding, or they may rely upon motile gametes.
    d. Ciliophora move with cilia.
    e. Oomycota usually use flagella.
    f., g. Myxomycota and Acrasiomycota feed and move like sarcodines but have reproductive stages resembling those of fungi.
11. Some parasitic protists have an enormous impact on human health and economies.
12. Small organisms, such as protists and bacteria, are well adapted to exploiting habitats and ways of life that provide only limited food and space. Tiny organisms can reproduce rapidly, and many of them can form spores or cysts that are resistant to adverse conditions such as drying or cold.
13. Selective pressure for large size probably favored the evolution of some ancient protists into multicellular fungi, plants, and animals. Here, division of labor among cells is added to division of labor among each cell's organelles.
14. Fungi are eukaryotic saprobes and parasites that obtain food by absorption and usually have cell walls. They reproduce by sexual and asexual spores. Three divisions of fungi have been established according to sexual reproductive structures: Zygomycota, Ascomycota, and Basidiomycota. A fourth division, Deuteromycota, contains fungi not known to reproduce sexually. Some fungi form mycorrhizal associations with the roots of higher plants, to which they supply minerals.
15. Lichens—symbioses between fungi and photosynthetic cells—are important producers of food and soil in cold or barren areas.
16. Fungi and bacteria cause great economic harm by destroying crops, food, and material possessions, and by infecting humans and livestock. On the positive side of the ledger, some fungi produce tasty fruiting bodies, and other fungi and bacteria are used to make fermented foods, drugs, antibiotics, and various organic chemicals.

SELF-QUIZ

1. An organism should be placed in the kingdom Prokaryotae if it:
    a. consists of a single cell
    b. has a cell wall
    c. forms spores
    d. lacks a nuclear membrane separating its genetic material from the cytoplasm
    e. causes diseases
2. Newly started rice paddies produce poor crops until they have established a flourishing population of cyanobacteria. This is probably because:
    a. the rice needs nitrogen fixed by the cyanobacteria
    b. the rice cannot compete with weeds, which are poisoned by toxins produced by the cyanobacteria
    c. the cyanobacteria use up surplus nutrients from sewage in the rice paddies
    d. the cyanobacteria provide plasmids carrying genes that increase fertility
    e. cyanobacteria form a protective coating on the rice plants
3. True or False. Many bacteria protect their hosts from pathogens by producing antibiotics.
4. True or False. All food containing bacteria is unsafe for consumption and should be thrown out.
5. Bacteria are used in the production of:
    a. wine
    b. vinegar
    c. gelatin
    d. pasteurized milk
    e. marshmallows
6. An organism should be placed in the kingdom Protista if it is ___, ___, and is not clearly related to members of the kingdom Fungi, Plantae, or Animalia.
7. State whether each description below applies to locomotion by means of cilia, flagella, or pseudopodia:
    ___a. cytoplasm flows out into temporary extrusions from the cell body
    ___b. oar-like beating propels the cell through the water
    ___c. wave-like undulations pull or push the cell through the water
8. Commercial mushrooms are grown in soil enriched with old straw (stems of dead plants). These mushrooms are:
    a. autotrophic
    b. parasitic
    c. saprobic
    d. chemosynthetic

9. The organism below that would have hyphae is:
   a. a slime mold
   b. black bread mold *(Rhizopus)*
   c. yeast
   d. a trypanosome
10. The taxonomy of the fungi is based on:
   a. life history
   b. sexual reproductive structures
   c. mode of nutrition
   d. complexity of vegetative structures
11. The bracket fungi found on trees are:
   a. fruiting bodies of mycelia growing hidden in the tree trunk
   b. mycelia absorbing nutrients from the exposed surface of the wood
   c. sporangia
   d. lichens
12. In the mycorrhizal association between a pine tree and a fungus, the fungus:
   a. eventually depletes the tree's mineral supply
   b. secretes toxic materials that inhibit the growth of nearby trees
   c. absorbs nutrients from the soil
   d. converts nitrogen into a form the tree can use
13. Fungi are probably most widely spread by:
   a. airborne spores
   b. being eaten by animals and later deposited in their feces
   c. spores or bits of hyphae stuck to insects
   d. fragmentation of vegetative mycelia
   e. water currents

THINKING CRITICALLY

1. Why are fossil prokaryotes more difficult to find and study than fossils of other organisms?
2. What is the adaptive value to an organism of producing a toxin? What is the disadvantage?
3. Many prokaryotes can make all the molecules they need when provided with a nutrient medium containing a supply of inorganic salts and one type of small organic molecule (for example, a monosaccharide or fatty acid) as an organic carbon source. How can prokaryotes be so self-sufficient if their DNA is so small compared to even one of the many chromosomes found in eukaryotic cells?
4. Explain how a bloom of photosynthetic cyanobacteria can lead to shortage of oxygen in a lake or pond.
5. Can you think of selective pressures other than the one discussed in this chapter that might have selected for evolution of multicellular organisms? What other advantages do multicellular organisms have?
6. How do the protists and fungi you studied in this chapter carry out each of the functions listed in Table 19–2?
7. How is the structure of fungi related to their way of life?
8. When plants are moved from their original habitat into a new one, a disease that was a minor nuisance in the home country may become a major disaster. This was the case with the late blight of potatoes, downy mildew on grapes, and white pine blister rust. What possible reasons can you think of for this?
9. Why do you think monocultures of an agricultural crop are more susceptible to disease than mixed plantings?
10. Fresh water heavily polluted with industrial wastes contains few fungi, compared to unpolluted waters. How might this affect the life in a lake or stream?
11. Is it valid to conclude that *Penicillium* secretions aid the fungus by reducing competition from bacteria in nature because they do so in the laboratory? What experiments could you do to investigate this question?

SELECTED KEY TERMS

aerobe, *p. 375*
archaeobacteria, *p. 379*
ascus, *p. 396*
autotroph, *p. 376*
basidium, *p. 396*
binary fission, *p. 374*
chemosynthesis, *p. 377*
commensalism, *p. 380*
conjugation, *p. 374*
cyanobacteria, *p. 379*
decomposer, *p. 378*
eubacteria, *p. 379*
facultative anaerobe, *p. 375*
fruiting body, *p. 393*
heterotroph, *p. 378*
hypha, *p. 393*
lichen, *p. 398*
mutualism, *p. 380*
mycelium, *p. 393*
mycorrhiza, *p. 397*
nitrifying bacteria, *p. 377*
nitrogen fixation, *p. 377*
obligate anaerobe, *p. 375*
parasitism, *p. 380*
pathogen, *p. 399*
penicillin, *p. 396*
phytoplankton, *p. 383*
protozoa, *p. 381*
pseudopod, *p. 382*
saprobe, *p. 378*
slime mold, *p. 389*
sporangium, *p. 393*
spore, *p. 375*
symbiotic relationships, *p. 380*
transduction, *p. 374*
transformation, *p. 374*

SUGGESTED READINGS

Baker, D. *"Giardia!" National Wildlife,* August-September 1985.

Carson, R. L. *The Sea Around Us.* New York: Oxford University Press, 1951. Contains a delightful discussion of the importance of algae in the sea.

Curtis, H. *The Marvellous Animals.* Garden City, NY: Natural History Press, 1968. A delightfully written introduction to the kingdom Protista.

Habicht, G. S., G. Beck, and J. L. Benach. "Lyme disease." *Scientific American,* July 1987.

Hardy, A. *The Open Sea.* Part I, The World of Plankton. Boston: Houghton Mifflin, 1965. Anecdotal and wonderfully illustrated.

Margulis, L. *Symbiosis in Cell Evolution.* San Francisco: W. H. Freeman, 1981. The case for the theory of symbiotic origin of organelles.

Sagan D. and L. Margulis. *Garden of Microbial Delights.* Boston: Harcourt Brace Jovanovich, Publishers, 1988.

Shapiro, J. A. "Bacteria as multicellular organisms." *Scientific American,* June 1988.

Woese, C. "Archaebacteria." *Scientific American,* June 1981.

CHAPTER 20

CONCEPT GUIDE

After reading this chapter you should be able to:

1. Compare red, brown, and green algae on the basis of: (a) habitat, (b) body form, (c) type of chlorophyll, and (d) type of photosynthetic pigment. (Sections 20–A, B, and C)
2. Describe the typical life history in a plant that reproduces by alternation of generations. (Section 20–D)
3. Describe four challenges faced by plants as they first moved onto land. (Land Plants Section)
4. Explain why bryophytes might have generally remained small and confined to moist habitats. (Section 20–E)
5. Describe the adaptive advantage of vascular tissue and explain how it affected the evolution of vascular plants. (Section 20–F)
6. Explain why most lower vascular plants still live in moist areas. (Section 20–G)
7. Describe three adaptations of gymnosperm plants that were not seen in earlier terrestrial plants. (Section 20–H)
8. Describe the parts of a flowering plant's life history that represent the gametophyte and sporophyte stages. (Section 20–I)

The lush, dense vegetation of the tropical rain forests are some of the most diverse habitats on Earth. Relatively unexplored, the plant life there may hold cures for cancer and raw materials for products never imagined. Read more about flowering plants in Section 20–I. (R. Frerck/Odyssey)

The Plant Kingdom

The plant kingdom contains the multicellular, photosynthetic eukaryotes (and their very close unicellular relatives). In all plants, the cells are surrounded by cell walls containing cellulose. Plant cells also contain plastids. Most plants have chloroplasts in at least some of their cells, but a few kinds have lost the ability to carry on photosynthesis. This way of defining the plant kingdom was used in Whittaker's proposal for a five-kingdom system. Some people now feel that only land plants (embryophytes) belong in the plant kingdom. They classify the multicellular algae as protists (using a different definition of that kingdom from the one used in this book).

What kinds of selective pressures do plants meet? First and foremost, plants must obtain light for photosynthesis. The multicellular bodies of many plants have specialized cells that hold the photosynthetic parts in positions where they receive a reliable supply of sunlight. Second, most plants cannot move about. They are too large to swim with flagella, and their stiff cell walls preclude the evolution of muscle tissue. This makes it doubly important for a plant to grow where it can obtain enough sunlight, and many plants have adaptations that increase their offsprings' chances of starting life in favorable habitats. Because they cannot move, plants also cannot come together to mate, and so they must have other ways to bring their gametes together for sexual reproduction. However, many plants do not face this problem because they engage in only asexual, or vegetative, reproduction.

During the course of evolution, plants have become able to exploit just about every habitat with enough light, moisture, and minerals to support the growth of photosynthetic organisms (Figure 20–1).

KEY CONCEPTS

- Most members of the plant kingdom are multicellular, photosynthetic, and immobile, and all plants have cellulose cell walls.
- Because most plants cannot move around, obtaining light for photosynthesis and bringing gametes together for sexual reproduction have been dominant themes in their evolution.
- Plants originated in the sea, and some forms later colonized land. The most successful land plants have vascular tissues that transport food and water through the plant and support the plant body in the air.
- The most successful vascular plants have adaptations freeing their reproduction from dependence on water.

Curiosity Questions

1? What are the largest living plants? (See Figure 20–15)

2? Why do some plants have colorful flowers? (See Figure 20–20)

3? What do brown algae and ice cream have in common? (See page 408)

FIGURE 20–1

Plant overview.

Table 20–1
Divisions of Multicellular Algae

Division	Common Name	Type(s) of Chlorophyll	Accessory Pigments	Flagellated Cells?	Size	Habitat	Food Storage	Example
Rhodophyta (~4,000 species)	Red algae	*a*	Phycobilins and carotenoids	None	Unicellular or multicellular	Mostly marine, some in freshwater, a few are terrestrial	Starch-like polymer	Irish moss, nori
Phaeophyta (~1,500 species)	Brown algae	*a* and *c*	Fucoxanthin and other carotenoids	Yes	All multicellular some very large	Mostly marine	Carbohydrates and lipids	*Fucus,* giant kelps
Chlorophyta (~7,000 species)	Green algae	*a* and *b*	Carotenoids	Yes	Unicellular to multicellular	Mostly freshwater, many marine, some terrestrial	Starch	*Spirogyra, Ulothrix, Ulva*

THE MULTICELLULAR ALGAE

The groups of algae with at least some multicellular members are placed in three divisions: Rhodophyta (red algae), Phaeophyta (brown algae), and Chlorophyta (green algae) (Table 20–1). (Divisions in the plant kingdom are equivalent to phyla in the animal kingdom.) The members of these three divisions are believed to have evolved from different groups of protistan ancestors.

Two of the three divisions, the red and brown algae, are mostly marine, with only a few freshwater forms. Both groups include species that have attained some size and complexity. The third division, the green algae, is well represented in both marine and freshwater environments, and also in moist areas on land. Multicellular algae provide food for animals such as snails and sea urchins.

Multicellular algae vary a lot in structure, and so they are classified by conservative biochemical characteristics such as kinds of photosynthetic pigments and food storage molecules. Like cyanobacteria and photosynthetic protists, all multicellular algae contain chlorophyll *a,* but the other forms of chlorophyll (*b, c*) and the accessory photosynthetic pigments (which gather other wavelengths of light energy) are used to divide the algae into their main groups.

20–A DIVISION RHODOPHYTA: RED ALGAE

The division Rhodophyta contains single-celled forms as well as plants that grow as filaments, branching structures, and broad flat plates or ruffles (Figure 20–2a). Some of the more complex red algae grow up to a meter long, but most are small and delicate. All live attached to some surface: a rock, coral reef, animal shell, or even a larger alga. No flagellated cells occur among red algae.

The chloroplasts of red algae show strong evidence of descent from cyanobacteria (Section 19–C). Both have chlorophyll *a* as the only chlorophyll, and both contain similar phycobilin accessory pigments. The arrangement of photosynthetic membranes is also similar.

Red algae are prominent residents of tropical coral reefs, where they play an important part in reef-building. The construction of reefs used to be attributed solely to coral animals, which secrete calcium carbonate (lime) tubes around themselves. However, many algae, especially red algae, become encrusted with calcium carbonate which they extract from the sea. This adds material to the reef.

Some red algae are eaten by people in the Orient and in coastal areas of Europe (Figure 20–2b). Carrageenan, a substance used in puddings, candies, and ice cream, comes from the red alga called Irish moss. Agar, also from a red alga, is used as a base for nutrient media in the laboratory culture of microorganisms.

BIO-BIT

There are approximately 266,000 currently living species of plants.

(a)

(b)

FIGURE 20–2

Red marine algae. (E. R. Degginger)

Red algae (Rhodophyta) are primarily marine and grow in a variety of forms attached to rocks on other surfaces. Their chloroplasts are thought to be direct descendants of the cyanobacteria.

20–B DIVISION PHAEOPHYTA: BROWN ALGAE

Members of the division Phaeophyta are all multicellular, ranging from microscopic filaments to Pacific kelps that may be 70 meters long—the biggest and most complex algae. The carotenoid pigment fucoxanthin imparts the typical brown to olive drab hue of most phaeophytes. They also contain chlorophylls *a* and *c*. These pigments also occur in members of the protistan phylum Chrysophyta (see Table 19–1). Both groups have similar food storage molecules as well, and so brown algae are thought to have evolved from chrysophyte ancestors.

Brown algae are especially noticeable in cool, shallow waters along the sea coast in temperate and subpolar areas. The larger members of the group are interesting because of their tissue differentiation. Members of the genus *Fucus* are good examples. A foot or more long, they grow on rocks in the intertidal zone, attached by a specialized **holdfast** (Figure 20–3). A stalk-like **stipe** connects the holdfast to the flat, photosynthetic **blades.** When the tide is in, the blades float near the surface of the water, close to the light, buoyed up by gas-filled **air bladders.** Numerous swellings at the ends of the blades contain chambers where reproductive cells develop. The plant has a gelatinous covering, which reduces evaporation and keeps it from losing too much water when it is exposed to the air at low tide.

Some kelps grow much larger than *Fucus,* and can live anchored in deeper water and still have parts that float at the surface, near the sunlight. Their holdfasts keep them from being swept out to sea.

The brown algae include many forms of economic importance, used as fertilizer and as food for livestock or humans. Some kelps are processed to extract a component of their cell walls called alginate. This substance is an ingredient in about half the ice cream produced in the United States; it imparts smoothness and helps prevent formation of ice crystals. Alginate is also used in various drugs and cosmetics as a suspension or emulsifying agent. ■

The mostly marine brown algae (Phaeophyta) are multicellular and range in size from microscopic to kelp forms over 70 meters long. Phaeophyta algae are thought to have evolved from the protistan phylum Chrysophyta.

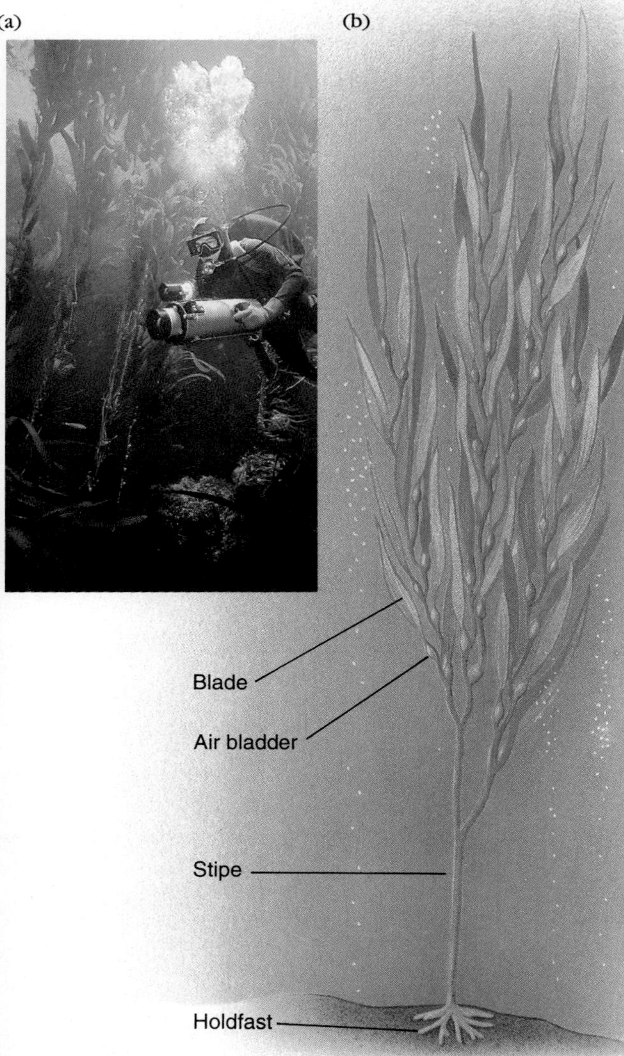

FIGURE 20-3

The brown algae, giant kelp. (a) A diver in a kelp forest. **(b)** The plant is buoyed upright in the water by its air bladders. (M. Snyderman)

20–C DIVISION CHLOROPHYTA: GREEN ALGAE

Members of the division Chlorophyta (green algae) show great diversity of form and live in a variety of habitats. Most are quite small and relatively simple compared with the larger brown algae. Many chlorophytes are single-celled; others are simple or branched filaments of cells, with or without holdfasts (Figure 20–4). Most Chlorophyta inhabit fresh water, both still and running, and some live on moist rocks, soil, and tree trunks on land.

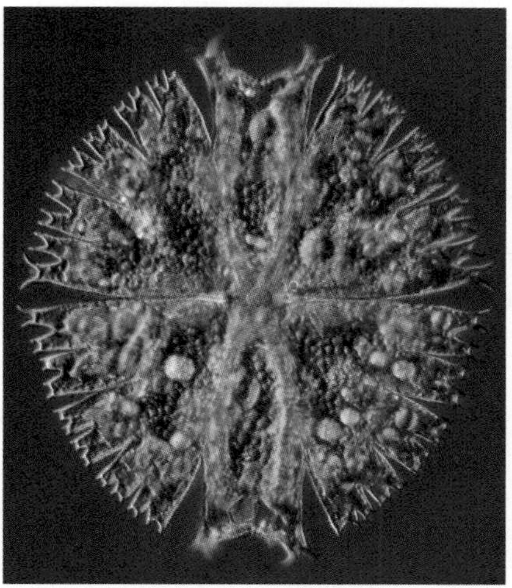

(a)

(b)

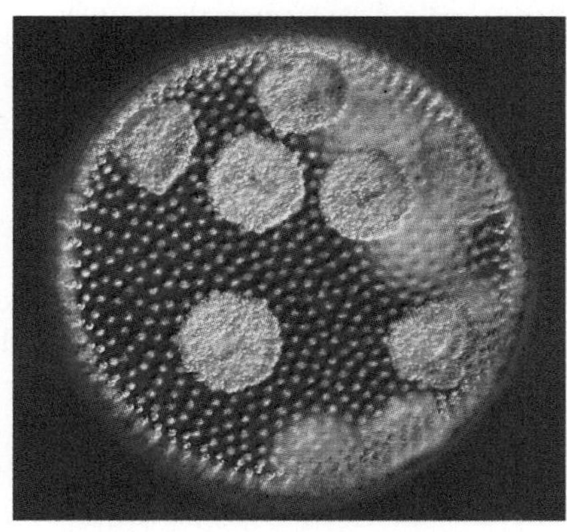

(c)

FIGURE 20-4

Members of the division Chlorophyta. (a) A desmid, a single-celled form with two mirror-image halves. **(b)** A calcium-secreting marine form. **(c)** *Volvox*, a colonial form. The colony rotates in the sunlight by the beating of flagella on the individual cells. Daughter colonies form inside the colony and are released by rupture of the parent colony's single layer of cells. (a, M. I. Walker/Photo Researchers, Inc.; b, A. Kerstitch; c, Carolina Biological Supply Co./Phototake)

Table 20–2
Major Groups of Plants

Common Name	Division	Number of Species	Important Characteristics
Algae			
Red algae	Rhodophyta	4,000	Unicellular or multicellular algae, found in marine, fresh water, or terrestrial habitats; phycobillin is an accessory pigment to chlorophyll *a*
Brown algae	Phaeophyta	1,500	Multicellular algae, some very large; mostly all marine; chlorophylls *a* and *c* and fucoxanthin as an accessory pigment
Green algae	Chlorophyta	7,000	Unicellular or multicellular algae, most are fresh water, many marine, some terrestrial; chlorophylls *a* and *b* and carotenoids as accessory pigments
Embryophytes			
Nonvascular plants			
Mosses, Liverworts, Horn Worts	Bryophyta	23,000	Small plants, anchored by rhizoids, most found in moist habitats; the gametophyte generation is dominant; flagellated sperm swim to egg
Vascular plants with spores			
Club Mosses Ground Pine	Lycophyta	1,200	Mostly small plants with scale-like leaves, with a single vein; flagellated sperm swim to egg; found in wet or shady habitats
Horsetails	Sphenophyta	25	Jointed, hollow, photosynthetic stems with vertical ribs; tiny, single-veined leaves around joints of stems; wet or moist habitat
Ferns	Pterophyta	12,000	Small to tree-like; large, many-veined leaves, often with divided shapes (fronds); anchored with rhizomes in moist habitats
Vascular plants with naked seeds (gymnosperms)			
Cycads	Cycadophyta	100	Palm-like shrubs and small trees with woody stems; tropical and subtropical; seeds in cones
Gnetum, Ephedra, Welwitschia	Gnetophyta	70	Desert plants and woody shrubs; pollen cones and seed cones on separate individuals
Conifers	Coniferophyta	550	Woody trees and shrubs; most are evergreen, with needle-like or scale-like leaves; pollen cones and seed cones on a single individual
Ginkgo	Ginkgophyta	1	Tree with broad, fan-shaped leaves; smooth, naked seeds; separate sexes
Vascular plants with flowers, seeds enclosed in fruits (angiosperms)			
Flowering plants	Magnoliophyta	235,000	Tiny to large; efficient vascular tissue; most have broad leaves; reproductive parts are in flowers; pollen carried by wind or animals; dispersal by seeds

The Chlorophyta appear to belong to the same evolutionary line as the land plants. The chloroplasts of both groups contain chlorophylls *a* and *b*. Both also have many carotenoids and store their food as starch.

The 7,000 species of green algae include several lines that clearly show evolutionary trends from unicellular to multicellular forms. Another trend is in sexual reproduction, from forms in which two similar gametes unite at fertilization, to those with large, immobile eggs and small, motile sperm.

Some unicellular, flagellated green algae look much like the flagellated sperm or spores of larger species. Not all green algae have flagellated stages, however. Those without flagella include the single-celled desmids (Figure 20–4) and filamentous forms such as *Spirogyra.*

The green algae (Chlorophyta) are a diverse group including unicellular and multicellular forms with simple or filamentous structures. They primarily live in fresh water or moist habitats. Green algae are most closely related to the land plants.

20–D LIFE HISTORIES

Life histories show some important trends in plant evolution. In the familiar human life history, the body's cells are diploid, and the only haploid cells are the gametes (egg and sperm) produced by meiosis. The same situation occurs in the brown alga *Fucus* and in many other algae. A completely opposite type of life history is also found in some algae, for example, in the green alga *Ulothrix* (see Figure 16–4). The cells of a *Ulothrix* filament are haploid, and the zygote is the only diploid cell in the life history. Most plants have life histories intermediate between these two types: a haploid stage alternates with a diploid stage, a situation known as **alternation of generations.**

In these plants, as in *Ulothrix,* meiosis does not give rise to gametes. Rather, it produces **spores,** haploid cells that divide by mitosis to produce multicellular haploid plants. Eventually, the haploid plants produce gametes. Since the plants are already haploid, however, the gametes are produced by mitosis rather than by meiosis. The haploid gametes then undergo fertilization and produce diploid zygotes. The zygotes divide by mitosis and grow into multicellular diploid plants. Some of the resulting diploid cells undergo meiosis to produce haploid spores.

Figure 20–5 shows a life history with both a multicellular diploid form and a multicellular haploid form, in the green alga *Ulva* (sea lettuce), which grows near the tide line. The multicellular diploid body is called a **sporophyte** because it produces spores. Each haploid spore grows into a multicellular haploid plant, called a **gametophyte** because it produces gametes. Fusion of two haploid gametes produces a diploid zygote, which grows into a multicellular diploid sporophyte:

sporophyte (2N)
(meiosis)
↘ **spore (1N)**
(mitosis)
↘ **gametophyte (1N)**
(mitosis)
↘ **gamete (1N)**
(fertilization)
↘ **zygote (2N)**
(mitosis)
↘ **sporophyte (2N)**

This is basically the life history found in all higher plants.

In plant life histories with alternation of generations, a haploid stage alternates with a diploid stage.

LAND PLANTS

Both land plants and green algae have the same photosynthetic pigments, chlorophylls *a* and *b* and beta-carotene, and store most of their food reserves as starch. In addition, most groups of green algae have flagellated sperm, as do the so-called "lower" land plants (those that do not produce seeds).

Land plants probably evolved from green algae growing at the edges of oceans or lakes. Such plants face constant danger from extreme conditions. When the water recedes, they may lose too much water to the air and die. On the other hand, a high tide or a heavy rain may sweep them into an unsuitable habitat, or bury them under mud and silt. Such conditions select for several adaptations: structures that anchor the plants in one place, a large size that keeps them from being buried completely, and a surface coating that reduces water loss. Plants that evolved such adaptations could survive better not only at the water's edge but also higher up on the shore, on dry land.

Land offers many advantages to plants that can function out of water (Table 20–3). More light is available on land because water itself absorbs much of the sun's energy. Carbon dioxide for photosynthesis is also

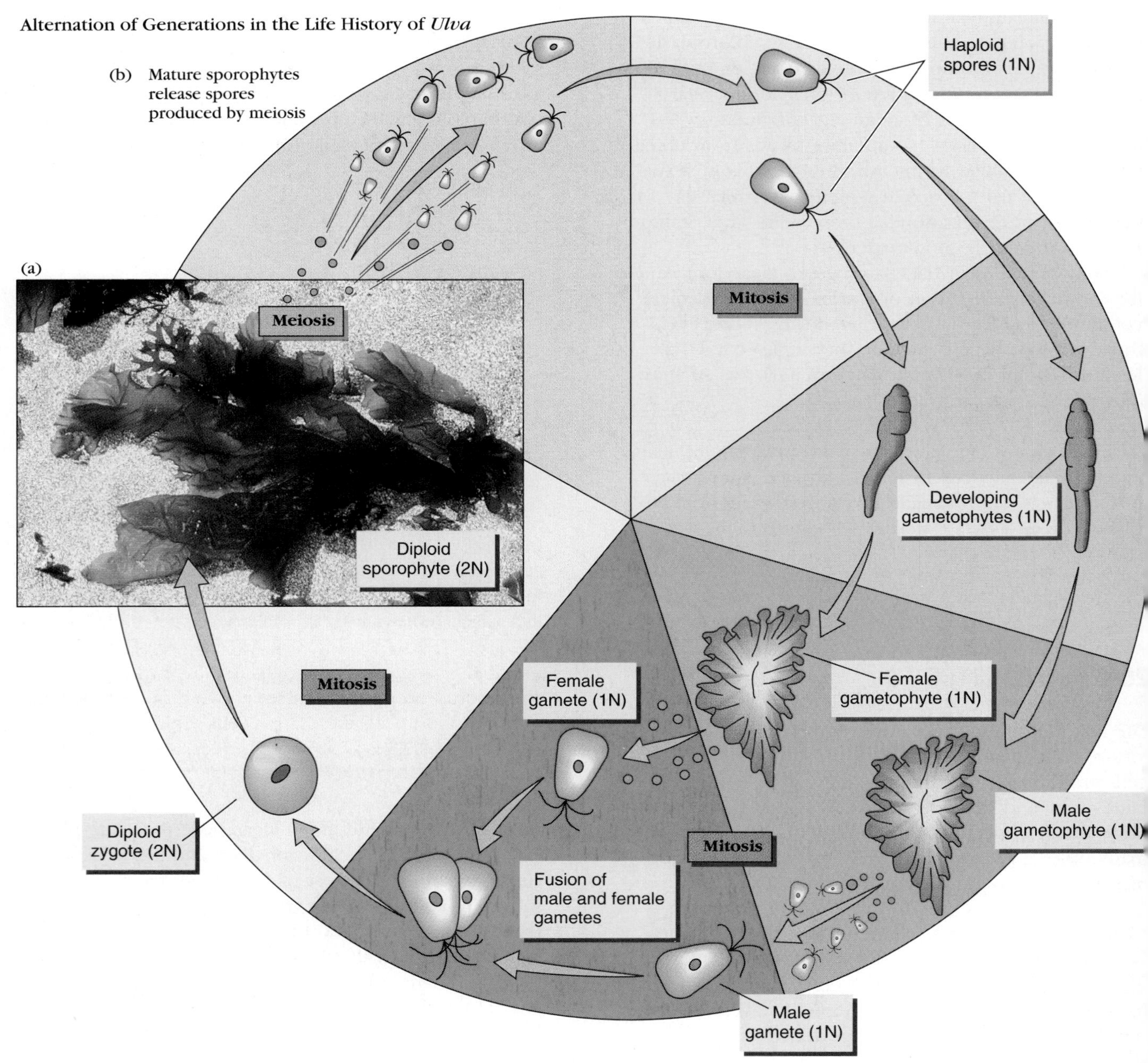

FIGURE 20–5

Life history of *Ulva*. (a) *Ulva* or sea lettuce is a common marine alga. **(b)** The mature diploid (2N) sporophyte of *Ulva* produces haploid (1N) spores using meiosis. These haploid spores develop into either male or female gametophytes. Mature gametophytes produce haploid (1N) gametes that fuse to form a single celled diploid (2N) zygote that develops into a mature diploid sporophyte.

present in much higher concentrations in air than in water. In addition, the first land plants had to compete for these resources only against their fellow colonists, and predation was probably limited to a few animals venturing out of their watery homes to feed at night.

The main disadvantage of moving out of water is that water becomes hard to obtain and is easily lost by evaporation from the plant's surface. Also, air is much less dense than water, and so it gives virtually no support to the plant body. Finally, water is no longer avail-

Table 20–3
Comparison of Water and Land as Habitats for Plants

	Water	Land
Water	Close to each cell	Under land surface; evaporates quickly above surface
Minerals	Close to each cell	On or under land surface
Gases	Dissolved at low concentrations	Plentiful in the air
Support	Provides buoyancy, support	Much less support for parts in air
Light	Cuts out some wavelengths, and lowers intensity	More light available
Temperature	Changes slowly, narrow range	Changes more rapidly, wider extremes
Reproduction	Motile gametes swim	Water seldom available for swimming gametes
Dispersal	Water carries offspring to new locations	Water seldom available to carry offspring to new locations

ble for reproduction. Many algae require water for heir flagellated sperm to swim to eggs, and for the young plants to disperse to new locations. On land, plants must carry out these functions without a constant supply of water, either by taking advantage of rain or dew to carry the reproductive cells or by providing these cells with a waterproof coating before hey travel through the air.

On land, the resources a plant needs are segregated: water and minerals lie below the surface of the soil, while light and air are above it. The division of labor in the bodies of land plants reflects this division of resources. Underground structures serve as anchors and absorb water and minerals for the entire plant. The photosynthetic structures above ground produce enough food for all the plant's cells (Table 20–4).

With this separation of resources, and of the plant parts specialized for obtaining them, comes a new problem: moving substances between the two areas. Most land plants have **vascular tissue,** which trans-

Table 20–4
Adaptations of Land Plants to Terrestrial Environments

Problem	Adaptation
1. Obtaining water and mineral nutrients when they no longer surround the entire plant	Roots
2. Transporting water within the plant	Xylem
3. Transporting food from sites of manufacture to sites of use	Phloem
4. Preventing evaporation from surfaces exposed to air	Cuticle
5. Obtaining gases for photosynthesis and respiration	Stomata
6. Obtaining sunlight for photosynthesis	Leaves
7. Supporting body in medium lacking buoyancy	Xylem
8. Coordinating growth and response to environment	Hormones
9. Getting gametes together without reliable supply of water for sperm	Pollen
10. Dispersing new individuals to suitable locations	Airborne spores; seeds

ports substances within the plant, especially between the food-making parts above ground and the water-absorbing parts below. There are two types of vascular tissue. **Phloem** conducts organic materials, mainly food, from sites of manufacture to sites of use or storage. **Xylem** transports mainly water and minerals from the roots to the stems and leaves. Xylem also provides support, so that the plant body stands up in the air, where the leaves can be deployed effectively for photosynthesis. A plant with xylem may be compared to a building whose plumbing pipes double as supporting columns.

Above ground level, the plant's surface is covered by a waxy **cuticle,** which is nearly impermeable to water and reduces evaporation of precious water from the plant into the air. However, the cuticle is also impermeable to carbon dioxide and oxygen, which a plant must exchange with the air. The leaves and stems of vascular plants can exchange gases through tiny pores called **stomata** ("mouths;" singular: **stoma**). Stomata are surrounded by pairs of guard cells, which can regulate the size of the opening (Figure 20-6). Some water is inevitably lost through the stomata, but much less than if evaporation proceeded freely from the entire above-ground surface of the plant.

Land plants also produce a variety of **hormones,** chemicals that coordinate the activities of the plant and its response to environmental cues. Hormones make roots grow down into the soil, and stems turn up toward the light, and they initiate reproduction and dormancy (Chapter 34).

One division of land plants, the Bryophyta contains nonvascular plants (plants without vascular tissue), and the other divisions contain vascular plants. Botanists speculate that bryophytes and vascular plants evolved in different directions from green algal ancestors.

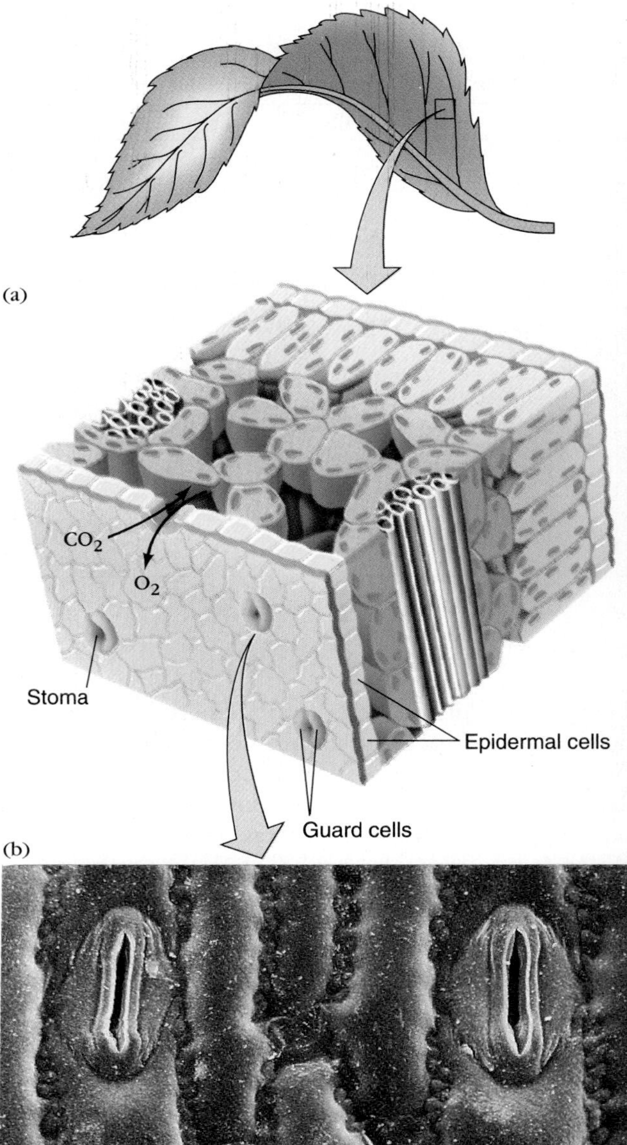

FIGURE 20-6

Cells from the epidermis on the underside of a leaf. **(a)** Each stoma is a pore between two guard cells, which resemble lips. **(b)** Guard cells and stomata are seen in this photograph taken using an electron microscope. (Courtesy of Drs. Kessel and Shih)

Reproduction on Land

All land plants have multicellular reproductive structures. Here developing reproductive cells—sperm, eggs, and spores—are protected, at least for a time, by a surrounding jacket of sterile cells. Land plants are sometimes called **embryophytes** because the zygote remains within the parent plant as it develops into an embryo.

We shall encounter a few important trends in the evolution of the life histories of land plants:

1. The earliest land plants probably had life histories with gametophytes and sporophytes similar in size and appearance and living independently of one another, as in *Ulva.* In the bryophytes—the mosses and their relatives—the haploid gametophytes became the dominant generation: that is, they are larger and longer-lived. The diploid sporophytes became reduced to smaller forms growing on the gametophytes. The vascular plants show a trend in the opposite direction: from forms in which the small gametophyte and much larger sporophyte are independent (as in ferns), to the condition in "higher" (seed-bearing)

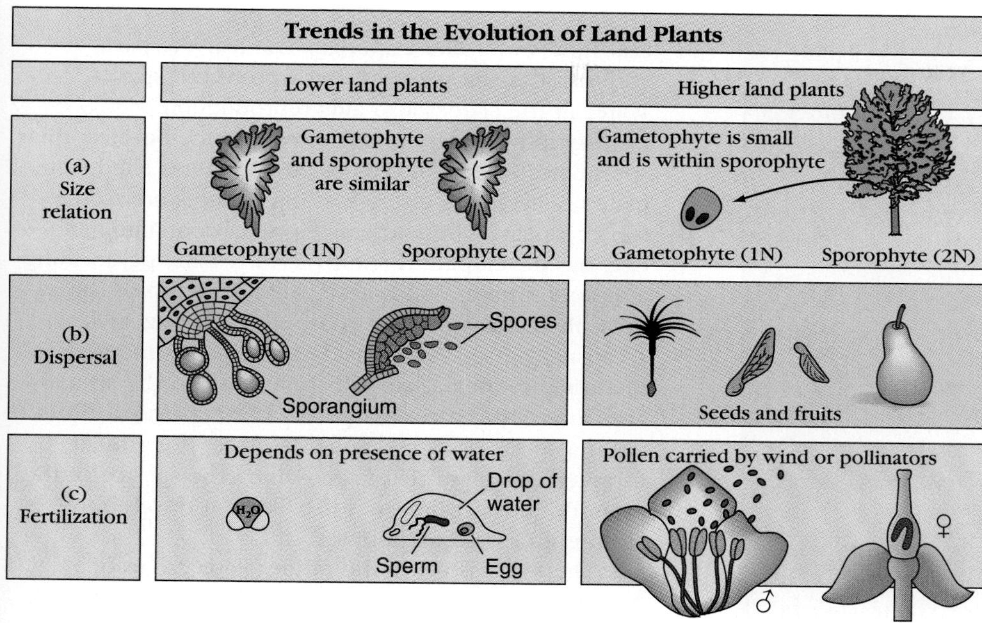

FIGURE 20–7

Trends in the evolution of land plants include **(a)** a reduction in the size of the gametophyte and an increase in the size of the sporophyte, **(b)** a shift to multicellular seeds, instead of single-celled spores, as dispersal mechanisms, and **(c)** a shift from motile sperm that swim to reach the eggs, to the use of pollen that wind or animals transfer from male to female flower parts.

plants where the sporophyte is dominant and the greatly reduced gametophyte grows within the sporophyte and is dependent on it.

2. In the lower land plants, spores are the main means of dispersal; in higher plants, there is a new structure, the seed, which is well adapted for dispersing a new individual to a new location and establishing it there.
3. Lower land plants produce flagellated sperm that must swim to the eggs in a liquid medium; in seed-bearing plants, pollen grains are carried to female reproductive structures by the wind or by animals (Figure 20-7).

Some of the adaptations that permitted plants to live on intermittently dry shores may have evolved to the point at which the plants could live on land permanently. Advantages to life on land include more light and carbon dioxide and fewer predators for the first land plants. However, once plants were on land, water was no longer available for external support or reproduction and they lost internal water quickly through evaporation. The overall trend in the evolution of land plants is toward greater independence of water, with increasing size and division of labor in the vegetative body as well as increasing protection and nourishment of small reproductive stages. Table 20-4 lists the specific adaptations involved.

BIO-BIT

Sphagnum or peat moss forms deep, dense masses in boggy environments. The ability of peat moss to hold water makes it useful in gardening.

20–E DIVISION BRYOPHYTA: MOSSES AND LIVERWORTS

The most familiar members of the division Bryophyta are the mosses. The haploid gametophyte (the familiar green moss plant) is the dominant generation (Figure 20-8). Many bryophytes have conducting cells that carry out slow transport of water and food in both gametophyte and sporophyte. However, they do not have true vascular tissue. Because of this, and because their swimming sperm require water to reach the eggs, bryophytes have remained small and generally confined to moist habitats. Some bryophytes, though, survive surprisingly hot or dry conditions.

Bryophytes have root-hair-like organs called **rhizoids,** which anchor the plant. The green, leaf-like structures where photosynthesis occurs are only one or two cells thick, and very susceptible to drying out. Moss gametophytes grow close together, holding each

(a)

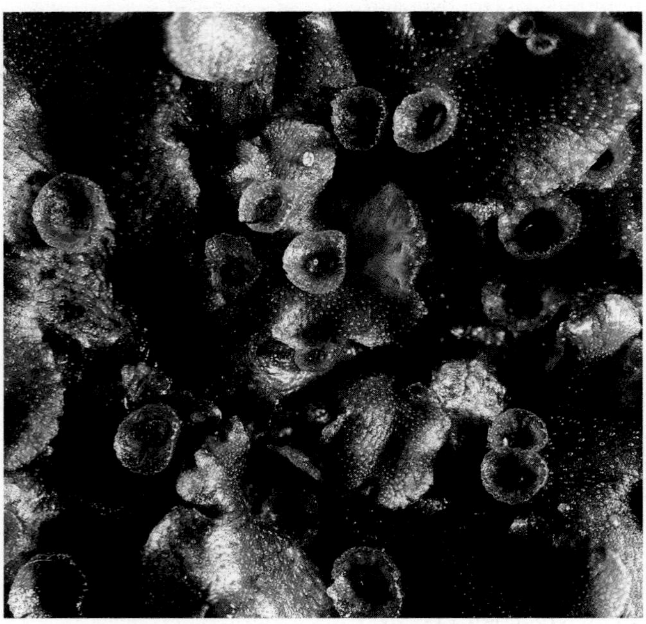

(b)

FIGURE 20-8

Bryophytes. (a) Green gametophytes of a moss, with sporophytes (red stalks with pale capsules) growing on them. Spores shed from the sporophyte capsules are carried away in the breeze. **(b)** The liverwort *Marchantia* growing in a bed of moss. Note the flat, branching body and the little gemmae cups, asexual reproductive structures containing balls of cells that may be splashed out and grow into new plants. (a, M. Gadamski/Photo Researchers, Inc.; b, G. I. Bernard/Animals, Animals)

other up and also creating a network of tiny spaces that hold water like a sponge. The leaf-like parts absorb most of the plant's water from these spaces. Water in the spaces also aids reproduction.

Bryophytes need water to reproduce because their sperm are flagellated (Figure 20-9). Sperm shed from male organs swim through a film of moisture to the top of a plant where a female organ, containing an egg, has developed. A sperm fertilizes an egg, forming a diploid zygote. The zygote, still at the top of the gametophyte, grows into a sporophyte, a long stalk with a capsule on top. Cells inside the capsule undergo meiosis, producing haploid spores. When the spores are mature, the capsule opens and flings them into the air. If a spore lands in a suitable place, it germinates, forming a filament that grows along the surface of the ground. Buds arising from the filament develop into a clump of new gametophytes.

The liverworts are the other major group of bryophytes. Most liverworts look much like flattened mosses with round-lobed leaves. However, some have flat, ribbon-like gametophytes with rhizoids on the underside, including the species that are commonly studied in introductory biology classes. Liverworts have sexual reproduction similar to that of mosses.

Division Bryophyta contains mosses and liverworts, in which the dominant generation is a gametophyte. Bryophytes do not have vascular tissues, and they require the presence of water for their sperm to swim to the egg. Hence they generally live in moist environments.

20–F VASCULAR PLANTS

The vascular plants include the most advanced and complex forms of plant life. For convenience, we can arrange the divisions of vascular plants into three groups: lower vascular plants, gymnosperms, and angiosperms (flowering plants) (see Table 20-2). Although the divisions in each group evolved independently, their members are similar in life history, reproductive structures, and vascular tissue. The adaptations shown by each group can be seen as major evolutionary jumps, conferring tremendous advantages over previously existing groups. The vascular tissue of the sporophyte generation was the first such evolutionary leap. By providing rapid transport and firm support, it permitted plants to grow taller, a big advantage in competition for sunlight. Vascular tissue runs throughout a plant's roots, stems, and leaves—in fact these three familiar terms cannot properly be applied to plant parts lacking vascular tissue.

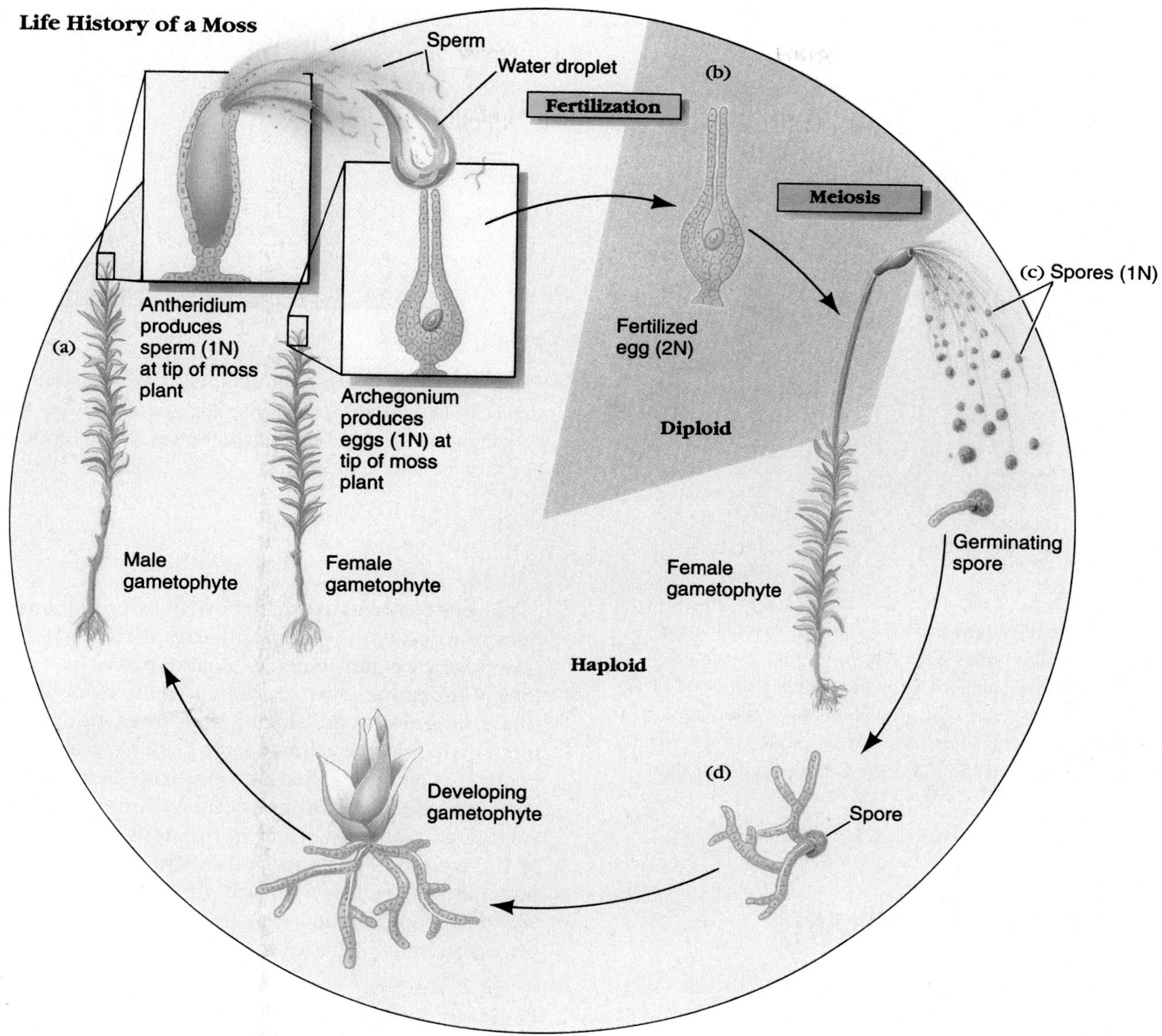

FIGURE 20-9

Life history of a moss. The large gray area shows haploid (1N) life stages, while the blue area includes the diploid (2N) structures. **(a)** Sperm formed in the antheridia of the haploid (1N) male gametophyte use their flagella to swim to haploid (1N) eggs that develop in organs called archegonia at the top of (1N) female gametophytes. **(b)** Fertilization forms a diploid (2N) zygote which develops into a sporophyte, consisting of a stalk bearing a capsule. **(c)** Cells inside the capsule undergo meiosis and form haploid (1N) spores, which are shed into the air when the tip of the capsule opens. **(d)** A spore germinates to form either adult male or female gametophytes.

The fossil record shows that early vascular plants had no roots or leaves. The sporophyte consisted of an underground stem, called a **rhizome,** anchored by nonvascular rhiz*oids* similar to those of bryophytes. The rhizome produced aerial photosynthetic stems (Figure 20-10).

The division of vascular plants can be arranged into three groups: lower vascular plants, gymnosperms, and angiosperms. The sporophytes of vascular plants all have vascular tissue that runs throughout the roots, stems, and leaves and transports water, minerals, and organic substances.

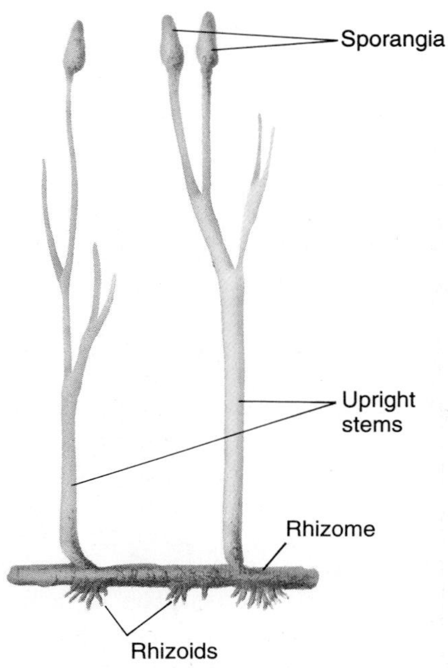

FIGURE 20–10

The sporophyte of *Rhynia*. This primitive vascular plant lived over 350 million years ago. The plant had an underground stem, the rhizome, with rhizoids similar to those of bryophytes. Upright photosynthetic stems bore sporangia, which produced spores. There were two species of *Rhynia*, one standing about knee-high and the other about half that height.

FIGURE 20–11

Club moss. The green photosynthetic stems, covered with tiny scale-like leaves, sprawl over the ground. Leaves bearing sporangia are arranged in golden clusters held aloft on stems like a branched candelabra. (D. Kuhn)

20–G LOWER VASCULAR PLANTS

Lower vascular plants include the members of the divisions Lycophyta (club mosses), Sphenophyta (horsetails), and Pterophyta (ferns). Their sporophytes have small, slender roots, which grow from an underground rhizome or from a trailing above-ground stem. Aerial stems, with leaves, grow from the rhizome.

Lycopods have small leaves, each containing a single **vein,** a strand of vascular tissue that brings water to the leaf from the roots and carries away excess food made by the leaf. Ferns and seed-bearing plants have large leaves with many veins, often in highly branched, intricate patterns. The sturdy veins support the more delicate photosynthetic tissue.

Division Lycophyta: Club Mosses and Ground Pines

The sporophytes of club mosses and ground pines have aerial stems covered with many small, scale-like leaves (Figure 20–11). **Sporangia** (singular: **sporangium**), the spore-producing structures, appear on the upper surfaces of leaves.

Lycopods reached their heyday as prominent members of forests in the Carboniferous Period which are now fossilized and mined as coal deposits (see the geological timetable in Appendix C). Many lycopods of that time grew to the size of large trees, up to 30 meters tall. When the climate later became drier, these enormous lycopods died out, probably because their vascular systems could not cope with the demands of such large plants for water in the new climate. Many of the lycopods that survived this period are still with us today. All are less than half a meter high, but they are quite common and easy to find in fields and forests once you know what to look for.

Division Sphenophyta: Horsetails or Scouring Rushes

The members of the division Sphenophyta are nicknamed "horsetails," because some of them resemble horses' tails, or "scouring rushes," because pioneers of the American West used them to scrub their dirty pots and pans. Horsetails were ideal for this because some of their cells are impregnated with silica, a very abrasive material.

Like lycopods, horsetails have an underground rhizome that produces tiny roots and aerial stems. The stems are mostly hollow and are easily recognized by their jointed appearance and the fine vertical ribs between joints. A ring of leaves, so small that they are often overlooked, grows around the stem at each joint. Some species have slender branches that look like the needles of young pine trees. Depending on the species, sporangia grow in clusters atop green vegetative stems or nongreen reproductive stems (Figure 20–12).

(a)

(b)

FIGURE 20–12

Horsetails. (a) ***Equisetum arvense***, with its thin green branches, is often mistaken for a clump of pine seedlings. **(b)** Fertile shoots of *E. arvense* are non-photosynthetic. They grow in early spring, before the vegetative shoots; this gives the spores more free air space for their travels. Note the jointed stems ringed by tiny dark leaves. (a, B. Kent/Animals, Animals; b, W. Hodge/Peter Arnold, Inc.)

Like lycopods, horsetails flourished in the Carboniferous forests, with some members attaining heights of about 15 meters. However, only one genus, *Equisetum,* with about 25 species, has survived to represent the group today.

Division Pterophyta: Ferns

The ferns are the first plants we have met that have large, many-veined leaves. Fern leaves are called **fronds,** and they usually arise directly from a rhizome, which also has many small roots. Ferns in temperate areas seldom grow more than a meter high, but tropical tree ferns may be quite tall (Figure 20–13). Although they were abundant in the Carboniferous coal forests, ferns have probably never been a dominant part of the vegetation.

The life history of ferns is similar to that of lycopods and horsetails, and can be taken to represent all three groups. In ferns, sporangia usually form on the undersides of the green vegetative fronds (where worried owners sometimes mistake them for a dis-

(a)

(b)

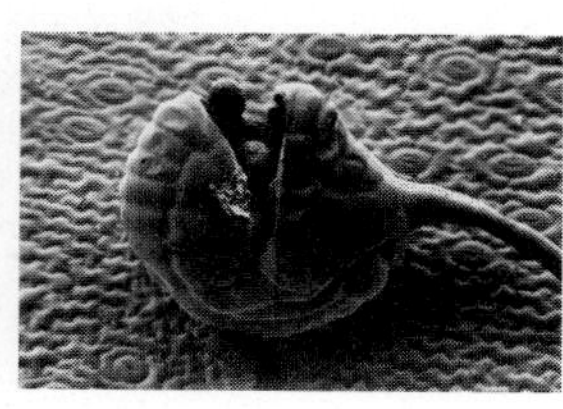

(c)

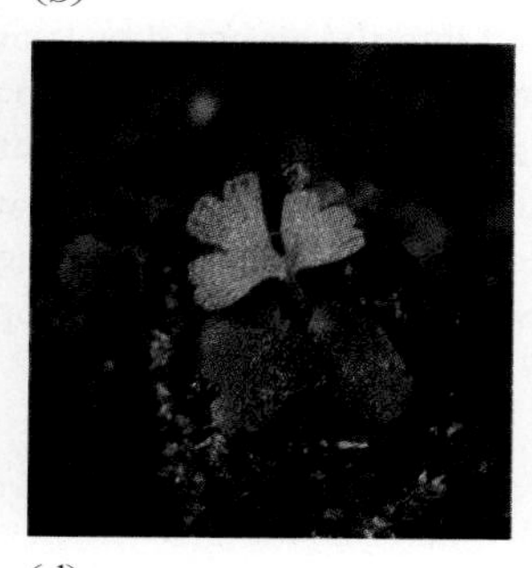

(d)

FIGURE 20–13

Ferns. (a) A tropical tree fern. **(b)** Sporangia clustered on the back of a fern frond. **(c)** Close-up of one sporangium, which has split open to release its spores. **(d)** A fern gametophyte. Note the photosynthetic cells with their green chloroplasts and the colorless rhizoids. (a, E. Kanze; b, A. Kerstitch; c, Biophoto Associates; d, B. Kent/Animals, Animals)

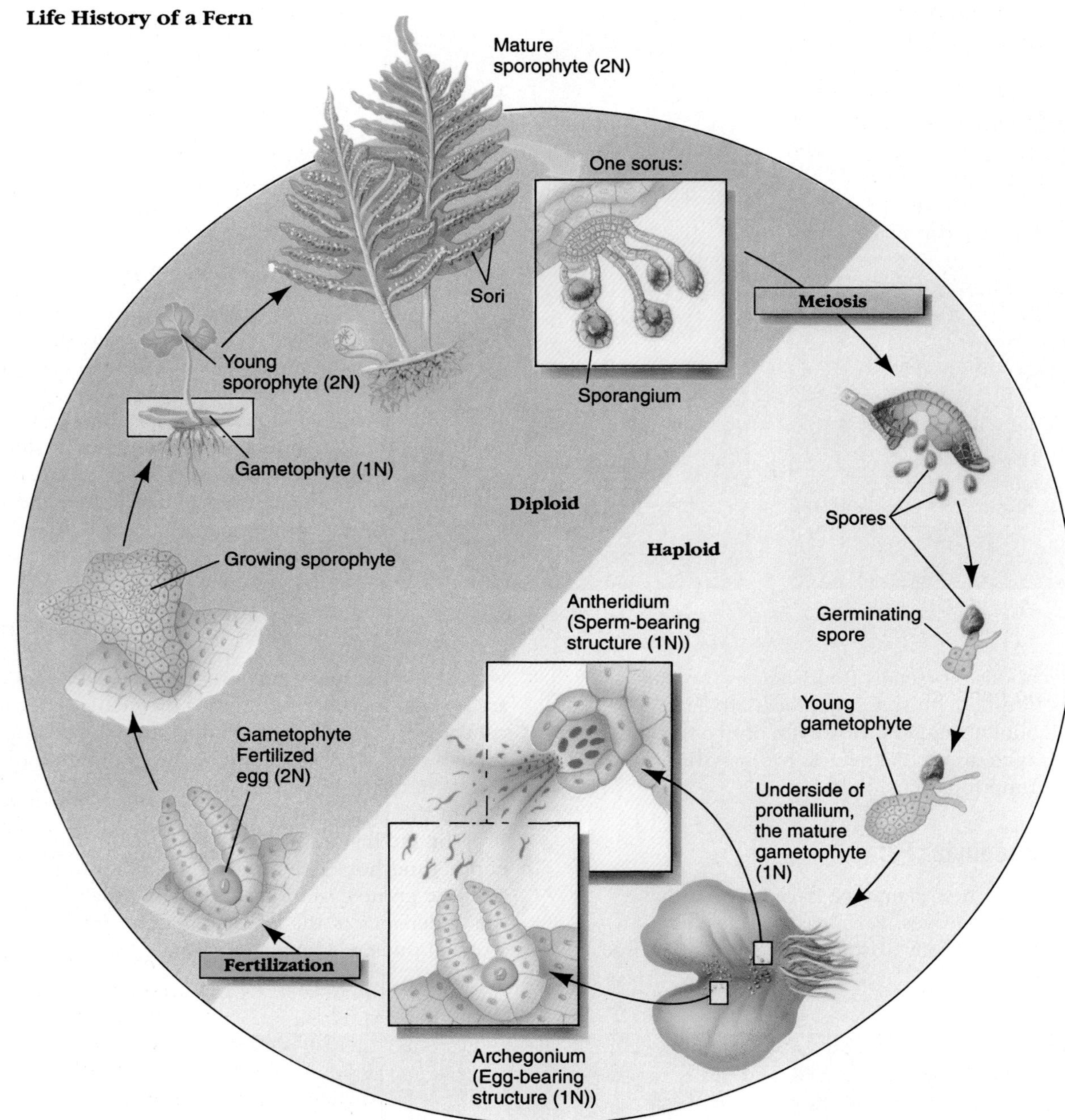

FIGURE 20–14

Life history of a fern. Diploid sporophytes bear sporangia in clusters called sori. Meiosis of cells inside the sporangia forms haploid spores, which are shed into the air and then grow into tiny gametophytes. Male and female organs (antheridia and archegonia) typically develop on the underside of the same gametophyte. (However, the antheridia and archegonia usually mature at different times to prevent self-fertilization.) The antheridia produce sperm with many flagella. These sperm swim through their moist environment to an archegonium to fertilize the eggs. The diploid zygotes then develop into new sporophytes that grow out of the haploid gametophytes.

ease), or on separate nongreen fertile fronds. Cells within the sporangia undergo meiosis, forming haploid spores, which are shed into the air. A spore grows into a small, green photosynthetic gametophyte, anchored to the soil by rhizoids (Figure 20–14). The gametophytes produce sperm and eggs. Fern sperm have flagella, and must have moisture on the surface of the soil to swim through. The zygote remains within the (female) gametophyte while it develops into a sporophyte embryo. The young sporophyte pushes a leaf up into the air and a root down into the soil, and it soon establishes itself as an independent plant.

The lower vascular plants include club mosses, horsetails, and ferns. Their sporophytes have roots that take up water from the soil, and vascular tissue that transports water to leaves and stems. However, these plants must still live in somewhat moist areas because their roots are not deep and sexual reproduction depends on the presence of water.

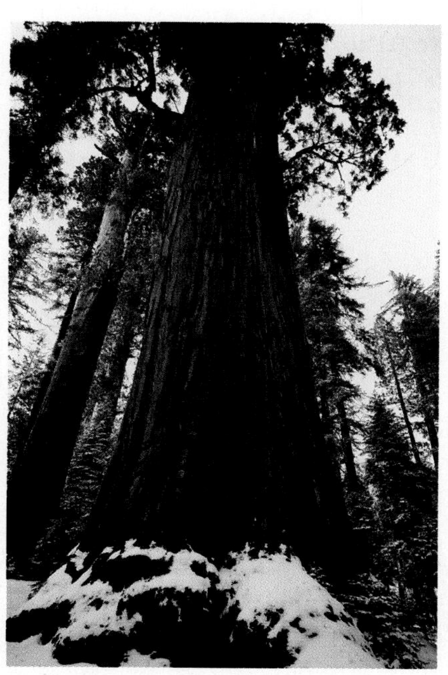

1? FIGURE 20–15

Gymnosperm giants. Mature giant sequoias are the largest of living plants. (M. Muench/D. Muench Photography, Inc.)

20–H GYMNOSPERMS

The redwoods of California's coastal mountains, Canadian pines and hemlocks, cypress standing knee-deep in Southern swamps, *Ginkgo* trees on city streets, wiry *Ephedra* in the western deserts, and palm-like cycads of the tropics are all members of the divisions lumped together under the term gymnosperm ("naked seed"). What is probably the oldest tree in the world, a 4900-year-old bristlecone pine in the mountains of eastern Nevada, is a gymnosperm. So are the tallest tree, a coast redwood over 100 meters high, and the tree with the greatest bulk, a giant sequoia nicknamed "General Sherman," which is over 80 meters tall, 20 meters around at its base, and 3500 to 4000 years old (Figure 20–15).

In contrast to lycopods, horsetails, and ferns, gymnosperms are fully adapted to life on land and can live in dry places. One of their evolutionary achievements is strong, woody tissue, made up of xylem. New wood is added each year so that the plant grows in diameter as well as in height. Wood strengthens the stem and allows the plant to grow tall and compete for sunlight. And, since xylem is the water-transporting tissue, it also delivers water to the leaves efficiently. As a further adaptation to dry habitats, the leaves produce a thick cuticle, a layer of thickened and/or waxy cells that prevents desiccation.

Gymnosperm reproduction also has two new evolutionary advances that free it from requiring liquid water: wind-borne pollen, and seeds. The reproduction of a pine tree may serve as an example (Figure 20–16).

The tree forms two kinds of cones, in which meiosis produces two kinds of spores (small **microspores** in the pollen cones and larger **megaspores** in the seed cones). Megaspores remain in the seed cones, and develop into female gametophytes, each containing two to several eggs. Microspores develop into immature male gametophytes, otherwise known as **pollen grains.**

Pollination occurs when wind carries the pollen to a seed cone on another tree. A seed cone's shape creates eddies that cause pollen to settle out of wind currents. Each species produces cones and pollen of a distinct size and shape, such that cones tend to filter more pollen of their own than of other species out of the air. As a result of these adaptations, pollen of the same species is deposited at the bases of the cone scales, near the entrance to the female gametophyte.

The pollen grain now completes its development, growing a slender pollen tube toward the female gametophyte. As the tube approaches its destination, one of its haploid nuclei divides by mitosis and forms two sperm nuclei. **Fertilization** occurs when a sperm nucleus leaves the pollen tube and fuses with an egg nucleus. The resulting zygote and the surrounding female structures now develop into a **seed,** with three main parts. Innermost is the multicellular diploid **embryo** of the new sporophyte generation, which develops by

Life History of a Pine

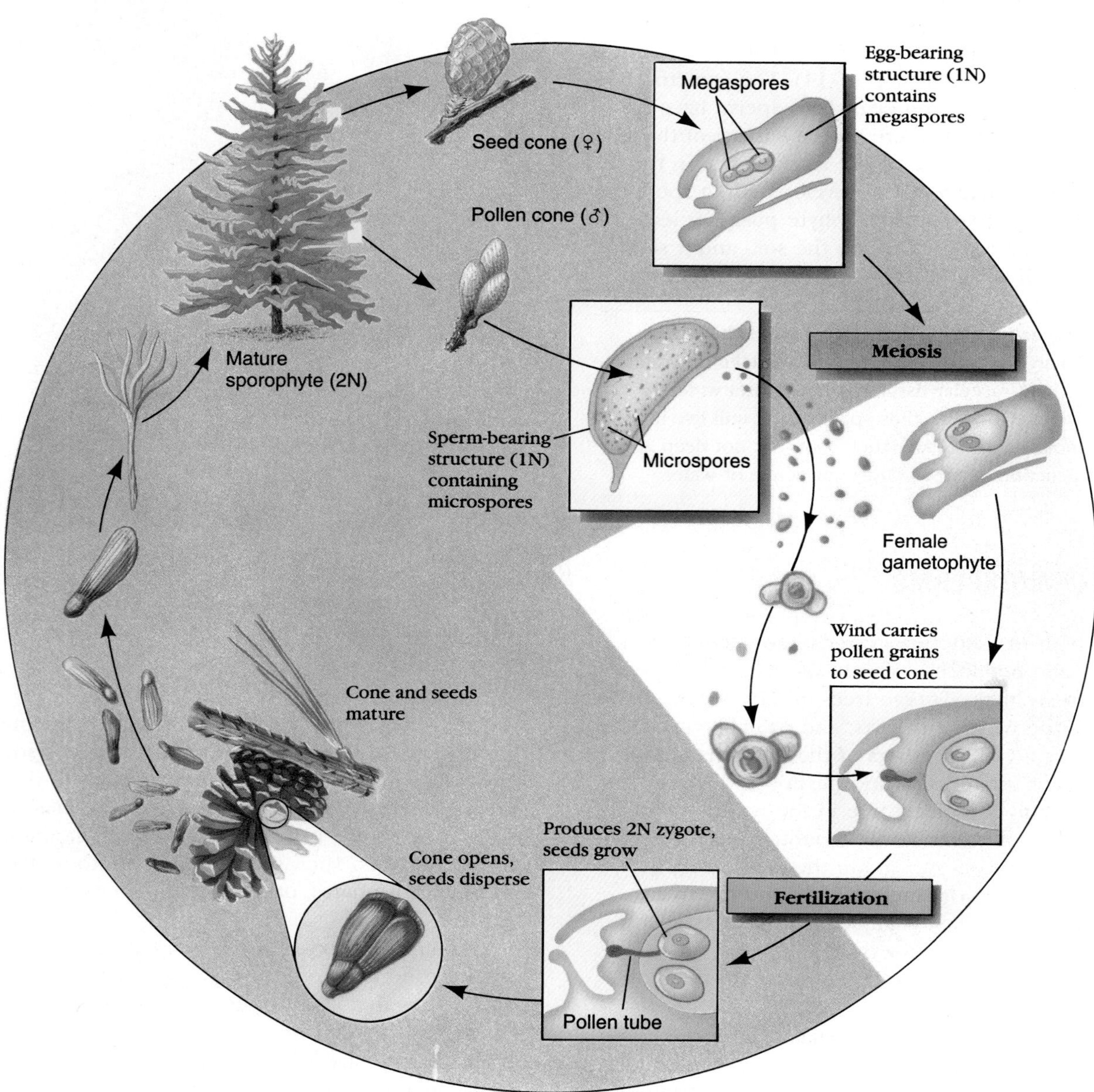

FIGURE 20-16

Life history of pine. The mature sporophyte (pine tree) produces megaspores in the seed cones and microspores in the pollen cones. Megaspores develop into female gametophytes within the seed cone. Microspores develop into pollen grains (male gametophytes), which are carried by the wind to seed cones. Here, each pollen grain grows a pollen tube, which releases sperm. After fertilization, the zygote grows into the embryonic sporophyte, surrounded by a food supply and a protective seed coat. The winged, wind-blown seed is shed when the seed cone scales open and serves as the dispersal unit in the life history.

mitosis from the zygote. Around the embryo is a **food supply,** which will provide energy for the embryo to grow until it is large enough to make its own food. Around the outside is the **seed coat,** which develops from the parent sporophyte. The seed is an extraordinary evolutionary invention that permits terrestrial plants to distribute their offspring widely without relying on water.

(a)

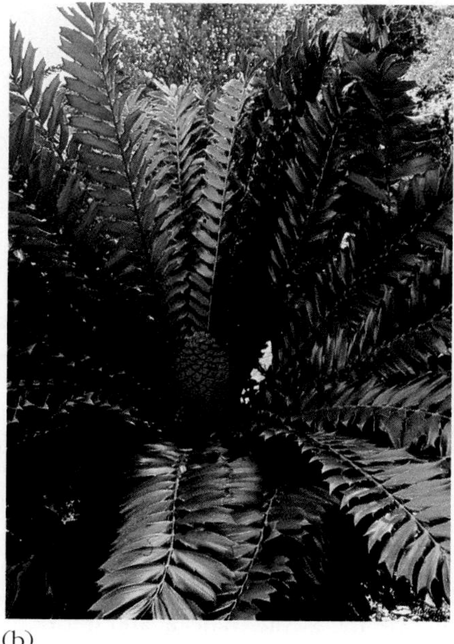

(b)

FIGURE 20-17

Cycads. Cycads **(a)** look like palm trees, but they have cones **(b)**. (W. H. Hodge/Peter Arnold, Inc.)

(a)

(b)

FIGURE 20-18

Ginkgo. (a) A *Ginkgo* tree. **(b)** A branch, showing a seed and the distinctive fan-shaped leaves. (a, W. E. Ferguson; b, Biophoto Associates)

Division Cycadophyta

Cycads are gymnosperms of tropical and semitropical regions. With their large, palm-like leaves they might easily be mistaken for palm trees, except that they have cones (Figure 20-17). Some human populations eat cycad stems and seeds, but some of these must be prepared specially to eliminate toxins. The living cycads are mere remnants of a group that was once much more abundant.

Division Coniferophyta

The most familiar gymnosperms are conifers—pines, firs, cedars, yews, hemlocks, junipers, larches, spruces, redwoods, and cypresses. Most have cone-like reproductive structures and needle-like or scale-like leaves with little surface area and thick cuticle. These drought-resistant leaves appeared during the Permian Period and were probably a major reason for the conifers' success during those times of global aridity. Many conifers do well in poor or shallow soil, where nutrients are scarce, and in areas subject to regular cold or dry spells. This is probably one reason conifers are so common today.

Division Ginkgophyta

Ginkgo biloba (Figure 20-18) is the sole surviving species of a group that flourished during the Mesozoic Era. With its broad, fan-shaped leaves and round, smooth seeds, it bears little resemblance to the

A Journey into Healthy Living

A Tree That Changed History

The discovery of a plant that controlled malaria had dramatic effects on human history. Malaria is one of the most devastating human diseases (Section 19-G). Carried from one person to another by blood-sucking female mosquitoes, it was endemic in southern Europe, Asia, and northern Africa during the Middle Ages. European travellers took malaria all over the world. Thousands died of it during the settlement of the United States. In nineteenth-century India, it killed about 2% of the population each year and made another 2% too ill to work. Even in this century, 10 million cases were reported during an epidemic in the Soviet Union in the 1920s, and each year 2 million people worldwide die from it.

The treatment for malaria came from Peru, where malaria was introduced by European settlers. In 1638, the Spanish Countess of Cinchon recovered from a bout of malaria after taking an extract of the bark of the tree now called *Cinchona* (Figure 20-A). Native Peruvians had recommended this to Jesuit missionaries as a cure for fever. The bark was soon imported to Europe, but many Protestants, such as English leader Oliver Cromwell, distrusted the Catholic Jesuits, refused their drug, and died of malaria. In 1852, the indefatigable Louis Pasteur discovered that quinine, the active principle in the bark, was a mixture of substances similar to strychnine and morphine. Not surprisingly, quinine is somewhat toxic; it can cause temporary ringing in the ears when taken, and large quantities can cause permanent hearing loss.

We have seen (Section 13-B) that some west Africans (and some southern European populations) had already evolved sickle cell hemoglobin, which affords some protection from malaria. Genetic resistance to malaria was a major reason west Africans were prized as slaves in the southern United States and the Caribbean, because quinine was expensive. During the nineteenth century, the cost of quinine for the British Army occupying India was over $320 million each year (in 1990 dollars).

FIGURE 20-A

Cinchona. (M. Balick/Peter Arnold, Inc.)

Tonic water is quinine dissolved in carbonated, sweetened water. Few who enjoy "gin and tonic" today realize that this drink was invented to make daily doses of bitter quinine palatable to people living in malarial areas. Quinine permitted northern Europeans, who had no natural defenses against the disease, to establish vast empires in tropical areas better defended by malaria than by any army.

Quinine also permitted Europeans to move about 20 million Indian and Chinese laborers to tropical areas where they would have died without quinine. The great European empires of the nineteenth century in Africa, Madagascar, Malaya, and Ceylon were based on plantations worked by the cheap labor of these immigrants, the ancestors of large modern populations in these areas.

Huge new industries were based on these population movements: sugar in the Indian Ocean and the Caribbean, tin and rubber in Malaysia, and tea in India and Ceylon were all made possible by quinine. The drug also probably permitted the United States to win the Second World War in the Pacific against Japan. During that war, 25 million Allied troops travelled to areas where malaria was epidemic, including most of the Pacific, the Mediterranean, and northern Africa, areas where they could not have survived without quinine.

A German scientist, Paul Ehrlich, first treated malaria with an artificial substitute in 1881. His efforts to make quinine led to a wealth of artificial fertilizers, pharmaceuticals, and plastics which, in the 1920s, became the basis for the modern chemical industry.

However, most quinine is still extracted from the bark of *Cinchona* grown in plantations. The natural product is cheaper and pleasanter tasting than the synthetic variety and remains one of the world's most important drugs. If one South American tree affected world history in all these ways, no wonder scientists are sure that in destroying unexplored tropical forest, we are depriving ourselves of products whose value we cannot even imagine.

(a)

(b)

(c)

(d)

(e)

(f)

FIGURE 20–19

The variety of flowering plants. (a) This cactus receives little water in its arid home. **(b)** At the opposite extreme, water lilies are surrounded by water. **(c)** Male red maple flowers. **(d)** Bull thistle. **(e)** Heliconia from Hawaii. **(f)** An orchid. (a, R. Winslow; b, P. Slocum/Animals, Animals; c, E. Kanze; d, R. Shiell/Animals, Animals; e. J. Brown/OffShoot Stock; f, C. Clark)

conifers but has similar vascular tissue. Wild *Ginkgo* trees may still exist in remote mountainous parts of China but for centuries this species was known only from cultivated specimens in oriental gardens and temples. Today it is very popular for street plantings in many American and European cities, because it is remarkably resistant to urban smog and to insect pests. Those in the know prefer to plant only male trees, because the females produce foul-smelling, fleshy seeds.

Gymnosperms include pine, cypress, *Ginkgo,* and cycads. Gymnosperm adaptations include stronger and more efficient vascular tissue, windborne pollen in place of swimming sperm, and dispersal by seeds rather than by spores.

20–I DIVISION ANTHOPHYTA: ANGIOSPERMS OR FLOWERING PLANTS

Flowering plants include an estimated 275,000 species, six times the number of species of all other plants! They are impressively varied, ranging from duckweeds to dogwoods, onions to oak trees. They range in size from tiny *Wolffia,* about 1 millimeter long, to towering *Eucalyptus* trees that vie with redwoods for botanical height records. Representatives of the Anthophyta grow in deserts, on mountain tops, and in polar regions, salt marshes, lakes, and streams. Their flowers borrow every hue of the rainbow (Figure 20-19).

Flowering plants are crucial to the existence and economy of human beings (Table 20-5). We may occasionally snack on pine seeds or sauté a batch of fern fiddleheads (the unrolling fronds) as a novel spring vegetable. However, almost all of our plant food comes from flowering plants, as does most of the food for domesticated animals. Flowering plants also provide housing, clothing, dyes, medicines, and spices (See *A Journey into Healthy Living*: A Tree That Changed History, this chapter).

The fossil record provides little evidence about the origin of the Anthophyta. The oldest identified fossil flower dates from the early Cretaceous Period, 110 million years ago, although fossil flower pollen several million years older is known. Flowering plants radiated rapidly into a variety of different forms during the late

Table 20–5
Economic Importance of Land Plants

Bryophytes	Sphagnum moss: used as fuel (peat) in Ireland, Scotland, etc.; used as mulch and planting medium in gardening and nursery industries
Lycopods	Formerly used as Christmas greens, the ground pine is now rare and protected in most areas
	Waterproof spores once used to dust pills so that they would not stick together in humid weather, highly inflammable, they were also used in fireworks
Ferns	Foliage used in florist industry; plants sold for house and garden
Gymnosperms	Lumber: Douglas fir, hemlock, spruce, various pine species, cedar, redwood
	Turpentine: distilled from pine trees
	Pulp for paper: various conifers
	Christmas trees: spruces, pines, firs, eastern red cedar
	Landscape plants: spruces, junipers, yews, cedars, cypress, hemlock, pines, cycads, *Ginkgo*
	Gin: sometimes flavored by redistilling spirits with juniper "berries"
Flowering plants	Food: fruits, berries, seeds, nuts, grains, stalks, leaves, roots, tubers; extracted juices, syrups, fats
	Clothing: cotton, linen
	Lumber: oak, maple, ash, birch, poplar, walnut, cherry, pecan
	Fuel: wood, charcoal
	Landscaping: grass, oak, maple, magnolia, birch, and many other trees; flowering shrubs, annual and perennial herbaceous flowers
	Beverages: coffee; tea; fermentation of many angiosperm species to make beer, wine, and liquors
	Drugs and medicines: tobacco; aspirin (originally derived from willow bark); morphine, opium; marijuana; atropine (from *Belladonna* plant); digitalis (from foxglove); various tonics, from sassafras, dandelion, coltsfoot, etc.: quinine (from *Cinchona* bark) used to treat malaria

Cretaceous, around 70 million years ago. The dinosaurs, most of the cycads, and some conifers became extinct at this time, while mammals and flowering plants started to dominate the living world.

Why were flowering plants so successful? At least part of the reason comes from their coevolution with animals. When earlier land plants were evolving, terrestrial animals were few, but when flowering plants appeared, land animals were already well established and diversified. Many flowering plants depend on animals, especially insects, for pollination (Figure 20–20). They may also rely on animals to disperse their seeds. However, many flowering plants rely on the wind for one or both of these functions.

The most distinctive feature of angiosperms is the **flower**. Its parts, including the male and female

(a)

(b)

2? **FIGURE 20–20**

Angiosperm flowers. (a) Colorful flowers attract animal pollinators such as this Black Swallowtail. **(b)** The flowers of a grass are wind-pollinated. Pollen is released from the dangling yellow anthers into the breeze.
(a, S. Cummings/Dembinsky Photo; b, OXS, Animals, Animals)

reproductive structures, are modified leaves of the sporophyte generation. The pollen grains, as in gymnosperms, are immature male gametophytes, transported to female flower parts by animals or wind. The female gametophytes, enclosed inside the female flower parts, are even smaller and simpler than those of gymnosperms.

The term angiosperm means "hidden seed," a reference to the fact that the seeds develop inside **ovaries,** which in turn grow from female flower parts. Fruits are not necessarily juicy and delicious. The pods of peas and milkweed are fruits, and so are pumpkins and peanut shells.

A third reproductive feature found among flowering plants is **double fertilization.** The pollen tube releases two sperm nuclei. One of them fertilizes the egg, forming a zygote which develops into an embryo in the seed. The other sperm nucleus fuses with two nuclei near the egg, in the female gametophyte. The resulting nucleus then divides and develops into **endosperm** tissue, which provides the seed's food supply. We shall study reproduction of flowering plants in detail in Chapter 35.

BIO-BIT

With more than 20,000 described species, orchids account for nearly 10% of all flowering plants.

The flowering plants (Anthophyta) are by far the most diverse group, with more than six times as many species as all other plants combined. Compared to earlier land plants, flowering plants have more efficient vascular tissue, faster growth, and more protection for their seeds. Their unique features include sexual organs contained in flowers, seed production inside ovaries, and nutritive endosperm in the seeds as a result of double fertilization.

SUMMARY

1. The three groups of multicellular algae have not evolved the adaptations that enable other plants to live on land, and so they are restricted to living in water (or very moist places on land).
2. Because they require light for photosynthesis, algae live only in shallow water where enough light is available. Holdfasts or adhesive secretions attach most algae to rocks or other surfaces, and air bladders in some allow the photosynthetic parts of the plant to float near the surface where there is adequate light.
3. Alternation of diploid and haploid generations, with similar or different body structure, is characteristic of many algae and all land plants.
4. Land offers plants advantages that are not found in water: more sunlight and more carbon dioxide. However, land plants must have adaptations that allow them to cope with the problems of a terrestrial existence: scarcity of water, evaporation of the water that is available, lack of support, and the need to get gametes together without constantly available water to transport sperm. Vascular plants have several adaptations to a life on land:
   a. Vascular tissue transports water taken in by the roots, supports the stems and leaves in the air, and transports food from the photosynthetic parts to the roots.
   b. A waxy cuticle retards evaporation from the leaves and stems, and in some plants stomata allow gases to enter the leaves with minimal water loss.
   c. Leaves expose large surface areas for photosynthesis.
   d. Hormones coordinate the activities of different parts of the plant with one another and with cues from the environment.
   e. In the more advanced vascular plants—the gymnosperms and angiosperms—pollen grains and pollen tubes permit sexual reproduction without the need for water for sperm transport, and seeds supply food and protection to young individuals, increasing their chances of survival.
5. The adaptations of plants to life on land show several evolutionary trends:
   a. In bryophytes, the gametophyte became dominant, with the small sporophyte growing on it. In vascular plants, the (vascular) sporophyte became the dominant stage in the life history. During the evolution of vascular plants, the size of the sporophyte progressively increased, while that of the gametophyte decreased.
   b. The increase in size of the vascular plant sporophyte was accompanied by an increase in strength and efficiency of its vascular tissue. In addition to the stems found in the earliest forms, specialized water-absorbing and photosynthetic organs (roots and leaves) evolved, and larger forms also evolved strong, woody xylem.
   c. The male gametophyte evolved into a waterproof pollen grain, which traveled to the female gametophyte before releasing the sperm. Sperm no longer faced a long, dangerous swim to the egg.
   d. The female gametophyte of all land plants retains the egg and protects the zygote as it develops into an embryo. In seed-bearing plants, the female gametophyte itself is retained on or in the sporophyte parent, which protects and nourishes it. The sporophyte also contributes food and protective coatings to the

seed—a new dispersal structure containing the embryo of the next sporophyte generation.

e. A spore is a single haploid representative cell that is the dispersal stage of bryophytes, lycopods, horsetails, and ferns. In gymnosperms and angiosperms, megaspores give rise to female gametophytes, and microspores give rise to pollen grains (male gametophytes). Instead of being shed, the megaspores remain in the sporophyte, which protects the female gametophyte and plays a major role in the production of the seed. Seeds replaced spores as the dispersal stage in the plant's life history.

f. As plant structure and reproduction became more and more independent of moisture in the environment, plants underwent adaptive radiation and spread into an increasing variety of habitats.

SELF-QUIZ

For each adaptation of algae on the left, pick its function from the list on the right. The functions may be used more than once or not at all.

Adaptation	*Function*
___1. Accessory photosynthetic pigment	a. maintains a favorable location
___2. Gelatinous secretion	b. obtains more nutrients
___3. Holdfast	c. protects from drying out
___4. Air bladder	d. obtains more energy

5. A haploid reproductive cell that may divide and give rise to a new plant is called a(n):
   a. spore
   b. gamete
   c. zygote
   d. embryo
6. Which of the following stages in the life history of a plant are diploid?
   a. gamete
   b. zygote
   c. gametophyte
   d. sporophyte
   e. spore
7. Beginning with the zygote stage, arrange the terms in Question 6 in the order found in a plant's life history.
8. In the evolution of land plants, sporophytes became dominant over gametophytes due to what adaptation?
   a. airborne pollen
   b. seeds
   c. vascular tissue
   d. diploidy
   e. stomata
9. A seed encloses an individual of the next sporophyte generation in the form of a(n):
   a. spore
   b. gamete
   c. zygote
   d. embryo
   e. pollen grain
10. An adaptation of land plants that reduces evaporation of water from the body surface into the air is:
    a. roots
    b. rhizoids
    c. stomata
    d. cuticle
    e. wood
11. One way that mosses have adapted to life on land is by:
    a. reproducing vegetatively
    b. using alternation of generations
    c. having dependent sporophytes
    d. holding water near their bodies
    e. producing seeds
12. An advance shown by gymnosperms that was not found in previous groups of plants was:
    a. growth exceeding a few meters high
    b. dominance of the sporophyte over the gametophyte generation
    c. protection of the gametophyte within the sporophyte body
    d. presence of large leaves
    e. production of reproductive structures at tips of the plant
13. Where would you expect to find the gametophyte of: (a) a moss, (b) a fern, (c) a pine tree, or (d) a lily?

THINKING CRITICALLY

1. Nonmotile plants must be able to get gametes together for fertilization. How does *Ulva* (Figure 20-5) accomplish this?
2. Red algae have no flagellated cells; how do you think their gametes might get together for sexual reproduction?
3. What modifications in the structure and reproduction of aquatic algae would be necessary for them to live permanently on land?
4. Algae lack roots; how do they obtain nutrients?
5. Parental care of the young was an adaptation that played a major part in the evolutionary success of birds and mammals. What parallels can you find in the plant kingdom?
6. Most conifers in temperate climates keep their leaves through the winter. Most angiosperm trees in the same environment drop their leaves each fall and produce a new set each spring. What are the advantages and disadvantages of being evergreen? Of dropping and regrowing the leaves?
7. What energy expenditures must a plant make if it is pollinated by animals? What energy savings does the plant gain by having animal pollination? What is the adaptive advantage to a plant of being pollinated by animals rather than by wind?

SELECTED KEY TERMS

air bladder, *p. 408*
blade, *p. 408*
cuticle, *p. 414*
double fertilization, *p. 427*
embryophyte, *p. 414*
endosperm, *p. 427*
flower, *p. 426*
frond, *p. 419*
gametophyte, *p. 411*
holdfast, *p. 408*
megaspore, *p. 421*
microspore, *p. 421*
ovary, *p. 427*
phloem, *p. 414*
pollen grain, *p. 421*
rhizoid, *p. 415*
rhizome, *p. 417*
seed coat, *p. 422*
sporangium, *p. 418*
spore, *p. 411*
sporophyte, *p. 411*
stipe, *p. 408*
stoma, *p. 414*
vascular tissue, *p. 413*
xylem, *p. 414*

SUGGESTED READINGS

Hobhouse, H. *Seeds of Change: Five Plants That Transformed Mankind.* New York: Harper and Row, 1986. The story of quinine—and potatoes, tea, sugar cane, and cotton.

Jensen, W. A., and F. B. Salisbury. *Botany: An Ecological Approach,* 2d ed. Belmont, CA: Wadsworth, 1984. A fresh approach, enjoyable reading.

Niklas, K. J. "Aerodynamics of wind pollination." *Scientific American,* July 1987. Despite its daunting title, this article is more biology than physics.

Wilkins, M. B. *Plantwatching: How Plants Remember, Tell Time, Form Relationships, and More.* New York: Facts on File, 1988. A survey of various plant groups and how plants live and work.

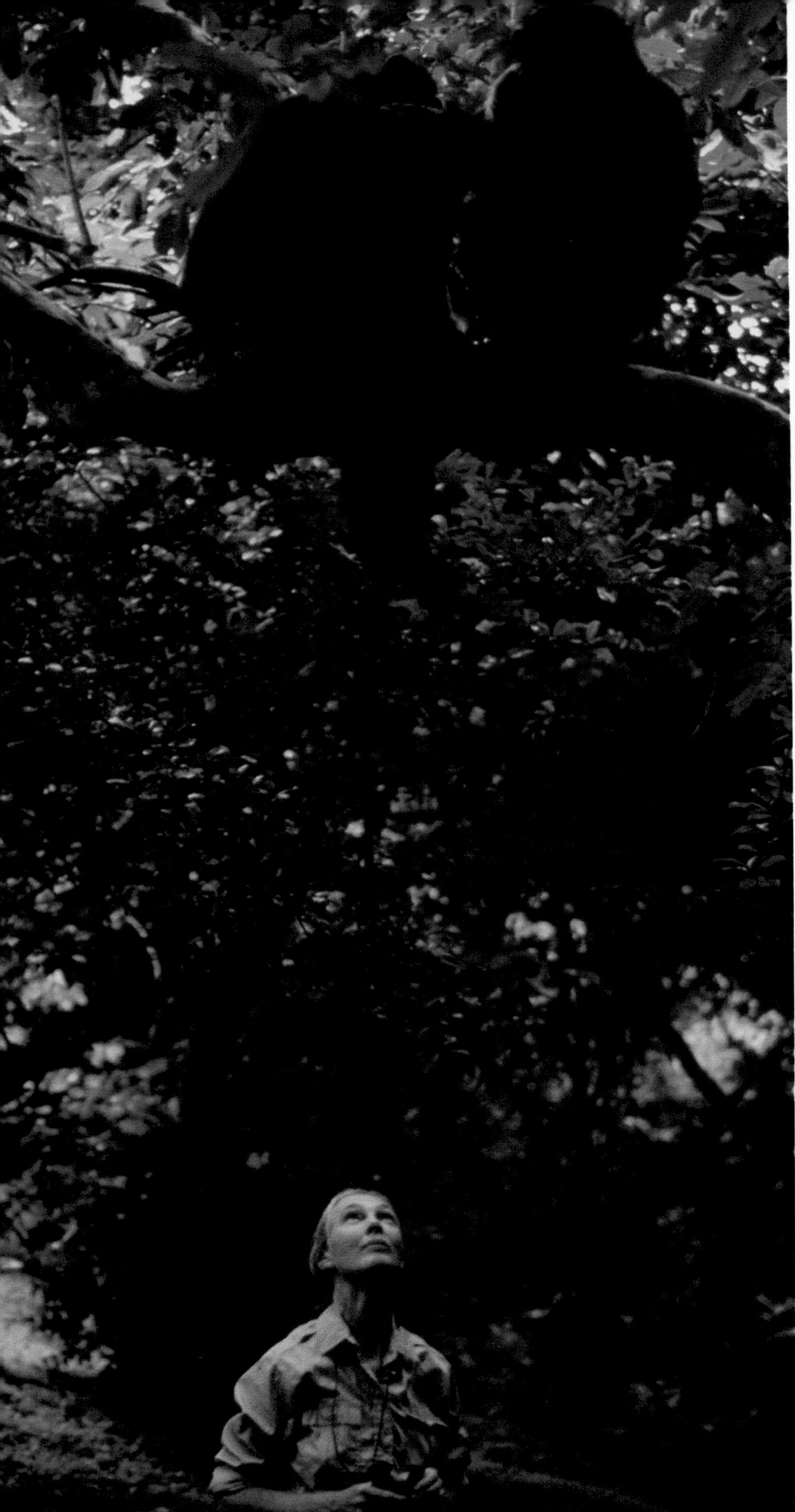

CHAPTER 21

CONCEPT GUIDE

After reading this chapter you should be able to:

1. Describe the advantages and disadvantages of living in the sea. (Section 21–A)
2. Distinguish between radial and bilateral symmetry and give two examples of organisms with each type of symmetry. Describe the advantages and disadvantages of possessing each of these body patterns. (Section 21–B)
3. Explain how the arrangement of the endoderm, mesoderm, and ectoderm establish the basic body pattern. (Section 21–B)
4. Describe how the cells of sponges are different from those of all other animals. (Section 21–C)
5. Explain how cnidarians and nematodes are each different from flatworms. (Sections 21–D, E, and F)
6. Compare the anatomy and habitats of the nematode worms and annelid worms. (Sections 21–F and G)
7. Compare the body plans of gastropods, bivalves, and cephalopods. (Section 21–H)
8. Describe how insects are specially adapted to life on land. (Section 21–I)
9. Describe four characteristics common to all echinoderms. (Section 21–J)
10. Describe three features that are common to all chordates. (Section 21–K)
11. Compare sharks and bony fish on the basis of gill structure, source of buoyancy, and intestinal structure. (Section 21–L)
12. Describe the problems that the first terrestrial vertebrates faced as they moved onto land. (Section 21–M)

The life-long studies of chimpanzees by zoologist Dr. Jane Goodall have provided enlightening insights into our nearest relatives. Read about the different types of mammals in Section 21–M. (C. S. Perkins/Magnum)

The Animal Kingdom

Visit any zoo anywhere in the world and you will see diverse and unusual animals: polar bears, pythons, parrots, even porcupines. Monkeys scamper about their enclosures, gazelles graze, tigers and lions sleep a lot, elephants flap their huge ears, birds dart about their aviaries, while most reptiles and amphibians lie motionless within their cages. A trip to a zoo can give an idea of the diversity of the animal kingdom, but zoos usually concentrate on large and charismatic land **vertebrates** (animals with backbones). Thus, they display only a small fraction of the animal kingdom. If a zoo did display a true sample of the animal kingdom, it would be mostly devoted to insects, because more than 70% of all animal species are insects (Figure 21-1). An additional 20% of our zoo would be marine tanks housing clams, sponges, jellyfish, sea anemones, crabs, and other marine **invertebrates** (animals without backbones). Only the last 10% of our zoo would look like most zoos today, devoted primarily to vertebrate groups: fishes, amphibians, reptiles, birds, and mammals.

All the organisms in the animal kingdom have some things in common. They are all multicellular heterotrophs, and all produce **gametes** (eggs and sperm), most in multicellular structures called **gonads** (ovaries and testes). After fertilization, the zygote develops into an embryo inside the egg or inside its mother's body, until it is mature enough to be hatched or born. Most animals ingest their food—eat now, digest later—rather than digesting food externally and then absorbing it, as fungi do. Animals are also typically **motile**, able to move around in at least one stage of the life history. Even those that spend most of their lives in one place can move parts of the body and so obtain food.

The first animals arose in the sea during the Precambrian Era. Many new animal phyla appeared at the start of the Cambrian Period, about 570 million years ago. These included most phyla alive today, and many long since extinct.

Depending on who is counting, the animal kingdom contains up to 33 living phyla, each with a distinct body plan. Here we cover only the nine major ones. One subphylum of the phylum Chordata contains all the vertebrates, all other animals are invertebrates. We will look at some common selective pressures and evolutionary trends in the animal kingdom, and examine the general body plans and ways of life in members of the nine major animal phyla. Later chapters cover the evolution of some animal structures and functions more specifically.

KEY CONCEPTS

- Life evolved in the sea. The few animal groups that successfully colonized fresh water and land evolved many adaptations of structure, function, and reproduction to these difficult environments.
- Most animals are made up of three layers: an outer protective and sensory tube, an inner digestive tube, and between them the organs of motion, reproduction, excretion, and transport.
- Active carnivores made many of the major evolutionary advances among animals, but most groups then radiated to produce members with other lifestyles such as herbivory, sessile filter feeding, or parasitism.

Curiosity Questions

1? How does a jellyfish sting a human? (See page 436)

2? Why do some physicians use leeches even today? (See page 442)

3? Why are most insects so small? (See page 449)

4? Why did the dinosaurs become extinct? (See page 459)

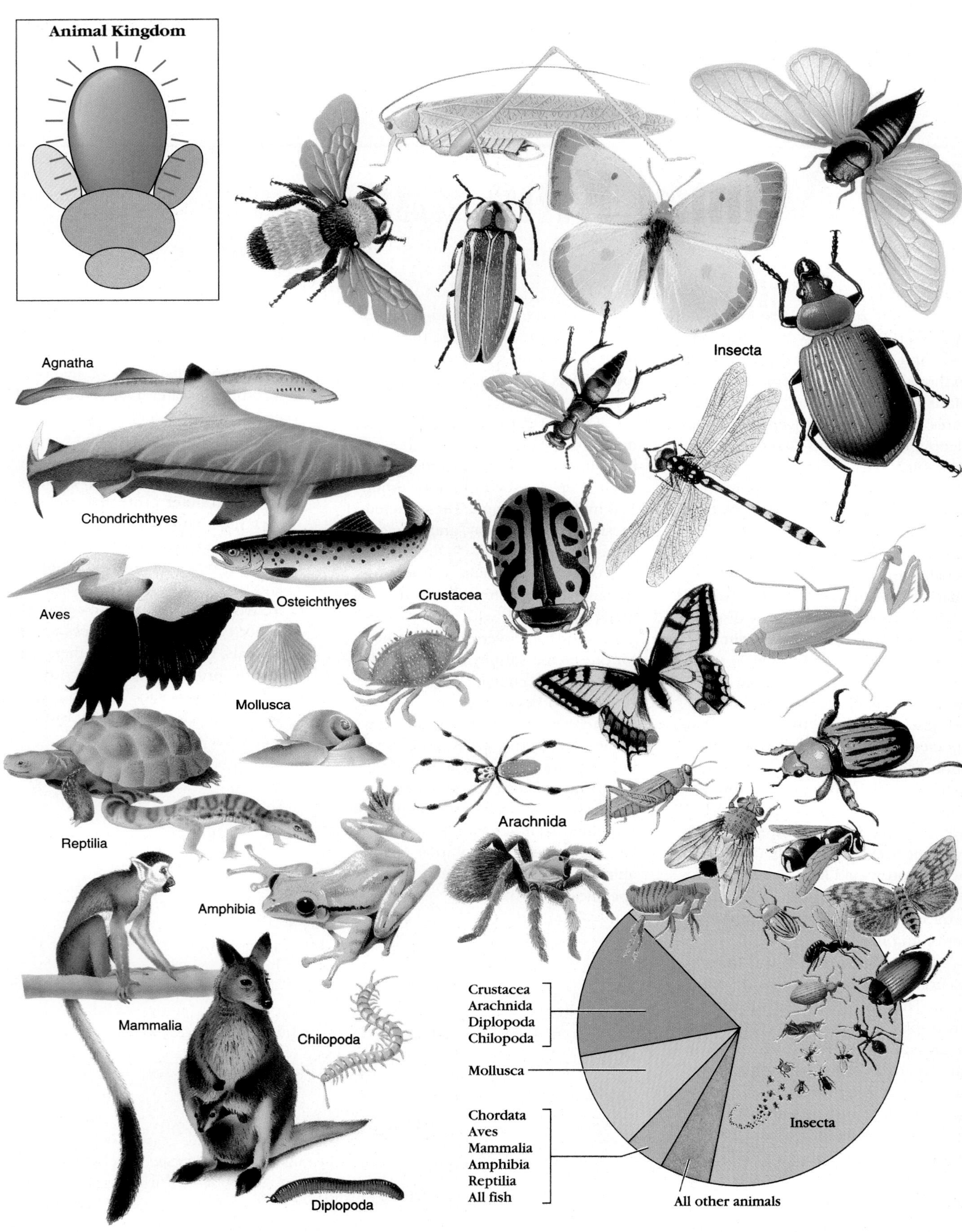

FIGURE 21–1

The relative species diversity of major animal groups.

Table 21–1
The Major Invertebrate Phyla

Phylum	Common names	Characteristics
Porifera (~10,000 species)	Sponges	Simple, sessile animals with specialized cells but no tissues or organs; no mouth or digestive cavity; collar cells used in filter feeding; most marine
Cnidaria (~9,000 species)	Jellyfish, sea anemones, corals, *Hydra*	Free-swimming medusa or sessile polyp; two tissue layers; mouth surrounded by tentacles bearing nematocysts; gastrovascular cavity; no anus; predatory; radially symmetrical; solitary or colonial; most marine
Platyhelminthes (~12,700 species)	Flatworms: turbellarians, flukes, tapeworms	Free-living or parasitic; bilaterally symmetrical; three body layers, with organs; gut with mouth but no anus; most marine or freshwater
Nematoda (~10,000 species)	Roundworms	Free-living or parasitic; gut with mouth and anus; fluid-filled body cavity; marine, freshwater, and damp soil habitats
Annelida (~6,200 species)	Segmented worms: polychaetes, earthworms, leeches	Segmented body with coelom and circulatory system (except leeches); setae (bristles) in some; gut with mouth and anus; gas exchange through skin or gills; marine, freshwater, or damp soil
Mollusca (~87,000 species)	Snails, slugs, nudibranchs, clams, oysters, scallops, mussels, squids, octopuses	Segmentation reduced; body covered by a mantle, which may secrete a shell; head and muscular foot usually present; gas exchange by gills or lining of mantle cavity; circulatory system with heart; marine, freshwater, or terrestrial
Arthropoda (~1,000,000 species)	Spiders, mites, crustaceans, insects, millipedes, centipedes	Segmented, with jointed exoskeleton containing chitin; jointed appendages; gas exchange by body surface, gills, or tracheae; marine, freshwater, or terrestrial
Echinodermata (~6,000 species)	Sea stars, brittle stars, sea urchins, sea cucumbers	Spiny calcareous skeletons; tube feet; pentaradial symmetry; all marine
Chordata (~1,300 invertebrate species)	Tunicates, lancelet	Notochord; hollow dorsal nerve tube; pharyngeal gill slits; post-anal tail at some stage; ventral heart; marine

21–A ANIMALS AND THEIR ENVIRONMENTS

The salt concentration of the sea is very similar to that of a cell; for this reason the sea is a more stable and hospitable environment for life than either fresh water or land. Most marine invertebrates, therefore, have no osmotic problems of water gain or loss. Although light and temperature vary, largely with depth, the temperature in any one area of the sea changes slowly, within fairly narrow limits. Furthermore, the small algae, protozoa, and animals in the **plankton** (floating organisms) provide a constantly renewed source of food.

In contrast, fresh water usually contains lower concentrations of nutrients and so supports less life. It is also hypotonic to living cells: a freshwater animal must constantly expend energy to retain its salts and expel the excess water that enters its body by osmosis.

Land is an even more difficult environment because water is often in short supply, making death by dehydration a constant danger. Relatively few animals have invaded fresh water successfully, and only two groups, the terrestrial arthropods (spiders, insects, and their kin) and the higher vertebrates (reptiles, birds, and mammals), have really solved the problems of life on land (Table 21-1). Many members of these groups move about freely even in the dehydrating conditions of warm, sunny days. The terrestrial representatives of other animal groups must retreat to cool, moist places on such days, emerging only at night.

Despite its advantages, life in the sea poses certain difficulties. Photosynthesis requires light, but water absorbs light; thus most plant life—the main food for marine animals—lives only near the surface, and is constantly tossed about by water currents. Fish, which are

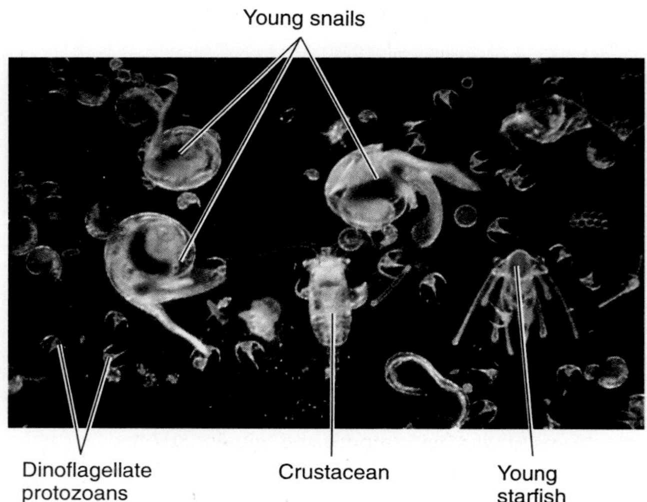

FIGURE 21-2

Plankton in a drop of seawater. Plankton is composed of an array of microscopic plants, animals, and single-celled organisms (protists). (D. P. Wilson/Science Source/Photo Researchers, Inc.)

vertebrates, swim well enough to maintain their positions despite the waves, but few invertebrates are powerful enough swimmers to do this. Invertebrates have evolved two sorts of adaptations that permit them to cope with the problem of being swept away by the sea. One is to remain small enough to float near their food supply (Figure 21-2). The alternative is for an animal to burrow in the sea floor, to cling to rocks or other stable objects using structures such as claws or suckers, or to be **sessile** ("sitting"), anchored in one place.

Most sessile animals are **filter feeders**. They use cilia or muscles to set up water currents past (or through) their bodies and filter out or seize any food carried by the current. Other sessile animals hang out a net of tentacles or mucus and trap passing food. A sessile organism needs protection from mobile predators. It may have active protection such as stingers, or passive protection such as a coating of toxic mucus or a thick shell.

Sessile animals cannot move around to find a mate. Instead, most of them have behavioral adaptations in which all members of a species release their sperm and eggs at the same time in response to environmental cues, such as a particular water temperature and a full moon.

In most sessile or slow-moving marine invertebrates, the embryo develops into a tiny, motile animal that can disperse the organism to new habitats. This **larva** is a stage in the life history that differs from the adult in its structure, habitat, and diet. The disadvantage of a larval stage is the high mortality rate of larvae. Predators take many, and many others do not end up in a suitable habitat. These selective pressures have favored the production of vast numbers of offspring in many invertebrates. The danger of being carried into unfavorable habitats by currents is especially great in freshwater. Most freshwater invertebrates no longer have a larval stage, but hatch from the egg as miniature adults.

Many sessile animals such as sea anemones can also reproduce asexually by dividing, usually into two parts, or by budding. This permits an individual that has found a good spot to populate the area with a clone of individuals containing its own genes.

Sea water is isotonic with the body fluids of most marine organisms, and therefore poses no significant water regulation problem. Fresh water, however, is hypotonic to animal fluids and freshwater animals have osmotic adaptations to this habitat. Animals living on land have adaptations that prevent loss of water by evaporation. Many animals have adapted to the turbulence of the ocean by adopting a sessile way of life: clinging to rocks or other stable objects and filtering food out of the water. A motile larva helps to distribute most sessile species.

21-B ANIMAL STRUCTURE

As we study the major animal phyla, we shall see some evolutionary trends in body structure. Let us preview some of these.

Body Symmetry

Some lower animals have body plans with **radial symmetry,** that is, the animal has a central axis and can be divided into mirror-image halves by several different planes passing through this axis (Figure 21-3). Radial symmetry permits an animal to detect food or danger approaching from any side. However, animals with well-developed muscular systems can move and obtain food more efficiently if they are **bilaterally symmetrical,** with only one plane that divides the body into two mirror-image halves. Bilateral symmetry allows an animal to have a more streamlined shape and to concentrate the power of its muscles and appendages into producing motion in one direction.

Along with the trend toward bilateral symmetry in animals goes the evolution of **cephalization**—the development of a head, with a concentration of sensory and nervous tissue that monitors the area the animal is entering.

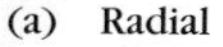

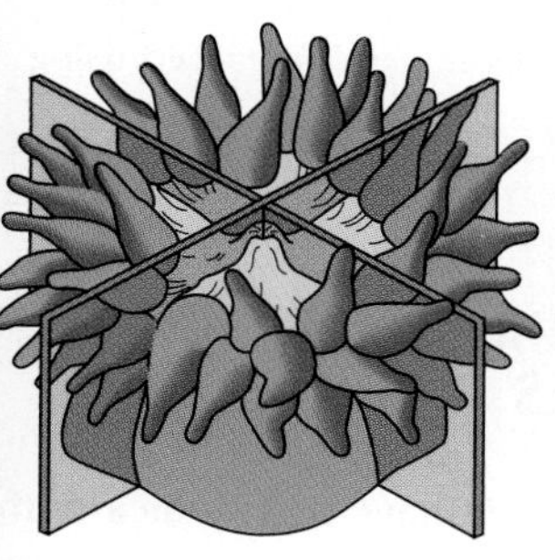

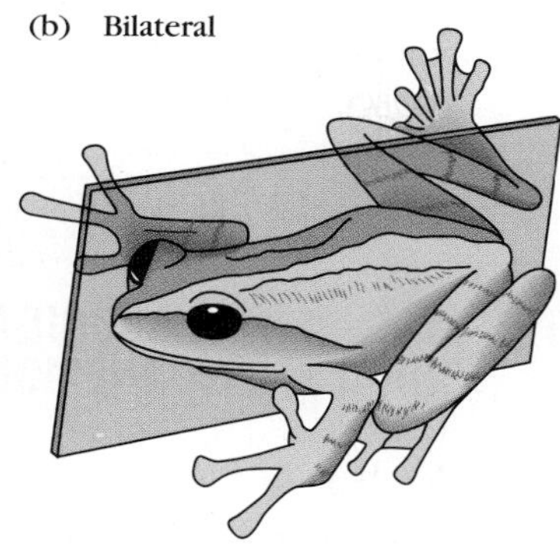

Body Layers and Coelom

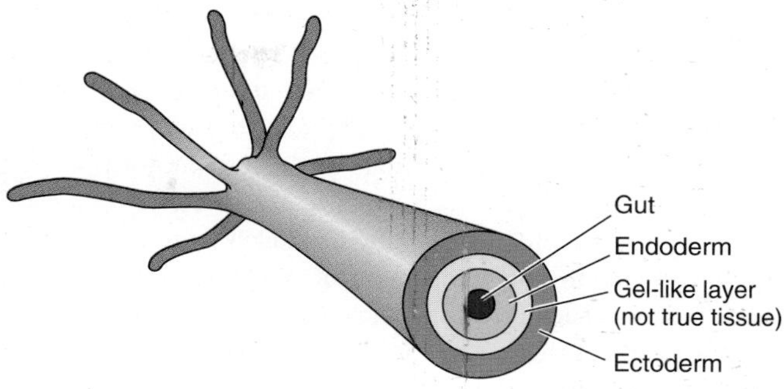

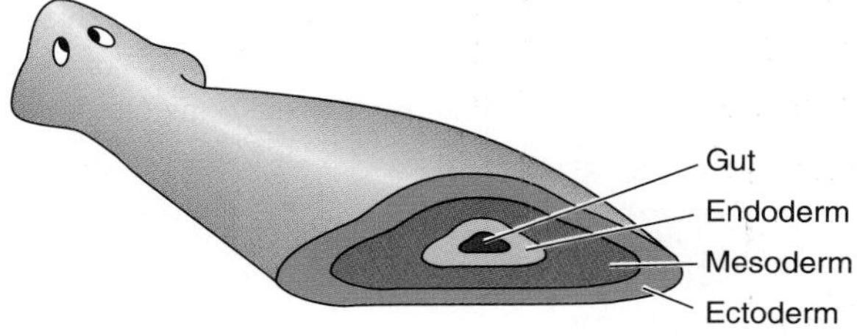

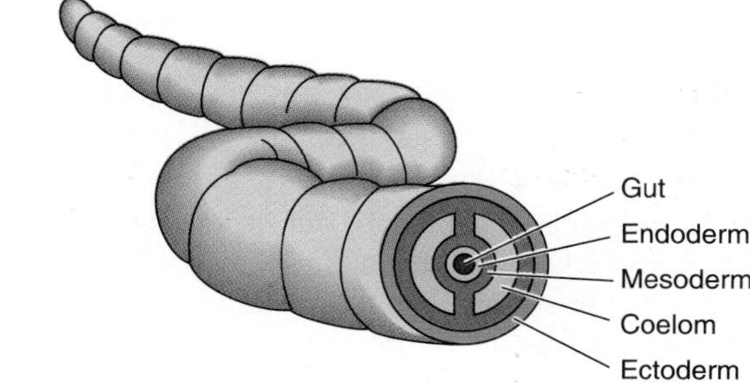

FIGURE 21-3

Evolutionary trends in animals. (a) Radially symmetric animals are able to respond to food or danger from any side. **(b)** Bilaterally symmetric animals move more efficiently and have most sensory organs located near the head end of the body. **(c)** As animals evolved, they developed three tissue layers and eventually a body cavity (coelom) surrounding the internal organs. *Hydra* represent the primitive condition. *Hydra* have no coelom and just a two-layered body with a gel-like layer between the endoderm and ectoderm. **(d)** Flatworms have three fully formed tissue layers but no coelom. **(e)** Earthworms and all higher groups have three layers, with a coelom that forms as a space in the mesodermal layer.

Body Layers

The structure of most animals is basically a tube within a tube. The inner tube is the digestive tract. The cells lining this tube are specialized to digest food, absorb it, and push it through the body. The outer tube consists of tissues specialized for dealing with the outside world. Here we find protective structures, such as skin, shells, and horns, as well as sense organs and the nervous system, which tell an animal what is going on around it. Between these inner and outer body layers are packed the other organs of the body: the muscles, blood vessels, reproductive organs, and so on. These deal with internal functions and have no direct contact with the outside world or the gut.

In all except a few primitive invertebrate phyla, this basic body structure develops in the early embryo from three layers of cells: the **endoderm,** which will form the lining of the gut; the **ectoderm,** which will form the body's outer layer, and the **mesoderm,** which forms between the other two layers (Figure 21-3) (endon = within; ektos = outside; mesos = middle; derma = skin).

The Coelom

The **coelom** is a fluid-filled body cavity that develops as a space in the mesoderm of higher animals (Figure 21-3). The origin of the coelom was one of the most important steps in animal evolution. First, the coelom separates the muscles of the gut from the muscles of the body wall. As a result, the gut can move independently of the body wall, and movement of food down the gut does not depend on locomotory movements. Second, the fluid that fills the coelom may act as a simple circulatory system, transporting waste, food, and gases around the body. More importantly, the coelom provides space where a true circulatory system with blood vessels can develop. A constant blood flow would be impossible if the heart were squashed by other organs every time the animal moved a muscle.

With these principles of animal structure and ways of life in mind, let us look at the major phyla and see what they show about animals and their evolution.

Most animals are either bilaterally or radially symmetrical. In most animal embryos, three layers of cells give rise to basic body systems, such as skin, muscle, and gut. The arrangement of each of these layers in the embryo determines where these adult systems will form, and so the basic body plan is established in the early life of the embryo. The coelom separates the muscles of the gut from the muscles of the body wall. It allows the gut to move independently and permits body fluids to circulate with each muscle contraction without being blocked by the internal body organs.

INVERTEBRATES

21–C PHYLUM PORIFERA: SPONGES

Sponges are simple, sessile animals living singly or in colonies (Figure 21–4). Their bodies do not have the kinds of layers described in the previous section. They have some cell specialization, but no tissues, and their cells are remarkably independent of each other. For instance, a living sponge can be pushed through fine silk, which breaks it up into individual cells and cell debris. If these cells are left to stand in sea water, they will re-form into a functional sponge!

Sponge sizes range from a few millimeters to the size of a barrel. The body is held in shape by a skeleton of fibrous protein (as in natural sponges sometimes used for bathing) or of little spikes (spicules) of calcium carbonate or silica. The sharp, brittle spicules also serve as a defense against predators; in addition, the sponge may produce toxic chemicals. Such defenses are vital for sessile animals. The simplest sponge body form is like a vase with porous sides. Water enters through these pores and exits through the large opening at the top (Figure 21–4).

The distinctive feature of sponges is a type of cell called a **collar cell** or choanocyte, with a flagellum that beats and propels the water current through the body (Figure 21–4). Food particles swept into the body with this current are strained out by the collar cells, which pass food on to amoeba-like cells that digest it and distribute it to other cells. The water current provides the cells with oxygen and removes wastes. Most sponges are marine.

Sponges are some of the simplest animals. Their cells do not form the layers or tissues seen in all other animal groups. Sponges vary greatly in size, are sessile, and filter-feed using specialized collar cells.

21–D PHYLUM CNIDARIA: JELLYFISH, CORALS, SEA ANEMONES

Members of the phylum Cnidaria have soft bodies with a basic radial symmetry. The inner and outer layers are well developed, but the middle layer is not. The two body layers contain several cell types, but there are no full-fledged organs. The gut is not a complete tube, with mouth and anus, but rather a blind sac, with one opening serving both to ingest food and to expel indigestible remains. It is called a **gastrovascular** cavity (gastro = stomach; vascular = vessel) because it also serves as a circulatory system, distributing food around the body.

Cnidarians come in two basic, and essentially similar, body forms: **polyp** and **medusa** (Figure 21–5). In both forms, a ring of tentacles surrounds the mouth, which leads into the gastrovascular cavity. Polyps are sessile, and most medusas are free-swimming, moving by weak contractions of their bell-shaped bodies. Jellyfish are medusas. *Hydra,* a species often studied in biology class, is a polyp, as are sea anemones and corals (Figure 21–5). Some cnidarians have both polyp and medusa stages in their life history (Figure 21–5), but others have lost one stage or the other. Many are colonial, consisting of numerous polyps and/or medusas attached to one another.

1? Most cnidarians are **carnivores,** animals that eat other animals. They do not chase their prey but sit (or float) waiting to trap victims. The tentacles bear special offensive and defensive structures called **nematocysts,** unique to cnidarians (Figure 21–5). The nematocyst may entangle or lasso prey animals, or may secrete a sticky or paralytic substance. Nematocyst toxins can cause a nasty sting and are occasionally fatal to swimmers. ■

In most cnidarians, the tentacles pull prey trapped by the nematocysts and stuff it into the mouth. Having nematocysts and a mouth and gastrovascular cavity permits cnidarians to eat much larger prey than a protozoan or sponge can manage.

Cnidarians are composed of two cell layers, derived from the endoderm and ectoderm of the embryo. Their bodies are radially symmetrical and have a pocket-shaped gut with only one opening. Two body forms, polyp and medusa, occur within the phylum. Nematocysts can inject poison for feeding or defense and assistance in subduing prey.

Sponge Diversity

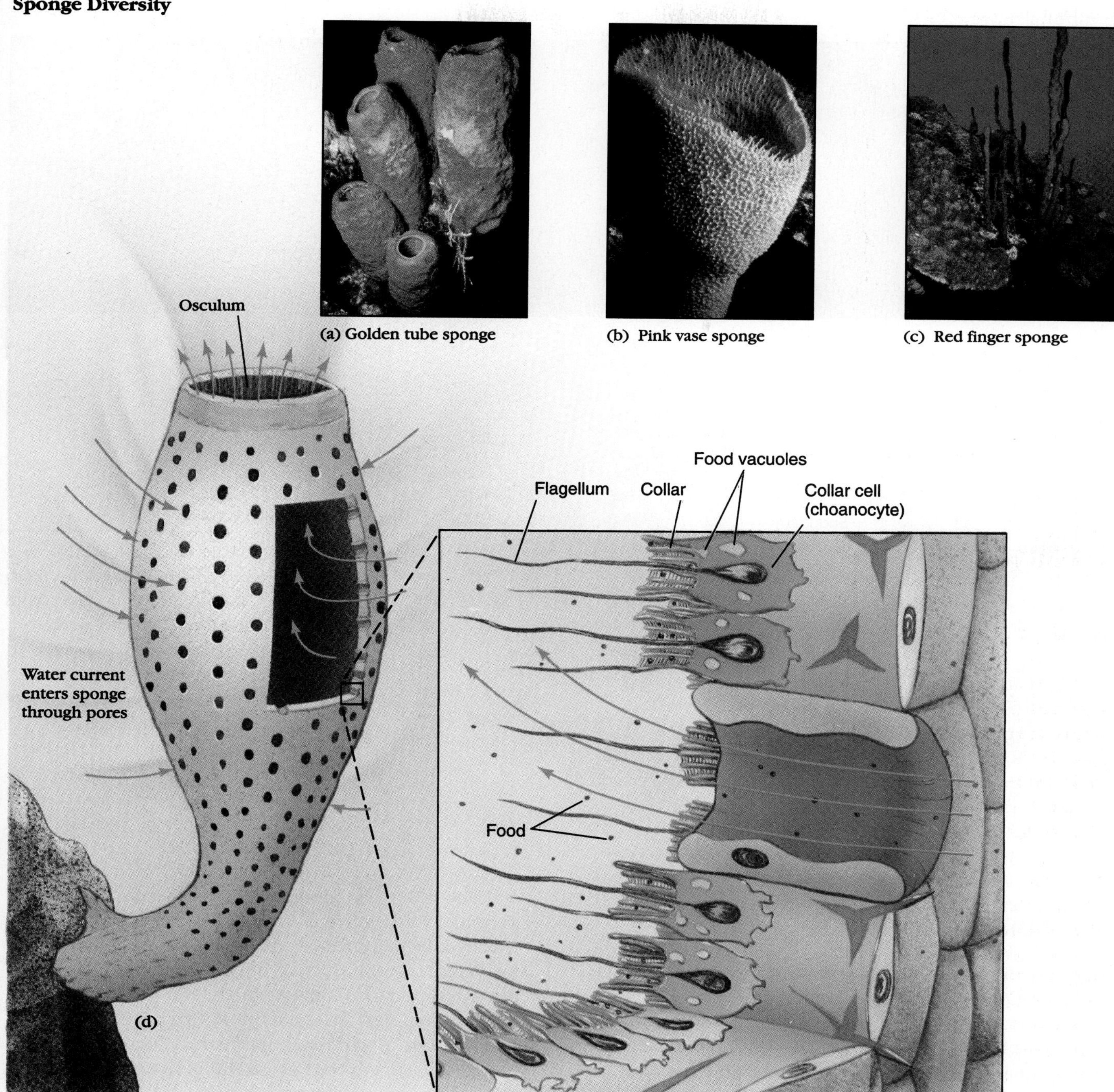

FIGURE 21-4

Phylum Porifera. (a) These golden tubes are sponges, growing attached to coral on the sea floor. **(b)** A pink vase sponge. **(c)** Red finger sponges. **(d)** The sponge body plan. Water is swept from outside the sponge, through its body wall, and up out the top opening (the osculum) by special choanocyte cells using their beating flagella. In this type of sponge, the choanocytes are located along the inside walls. The collective beating of their flagella creates water currents that sweep water and suspended food past the sieve-like collar of each choanocyte, where food is strained out. (a, C. Seaborn/Odyssey; b, M. Kazmer/Dembinsky Associates; c, R. Frerck/Odyssey)

Cnidarian Diversity

(a) Jellyfish

(b) Northern red anemone

(c) Reef coral

(d)

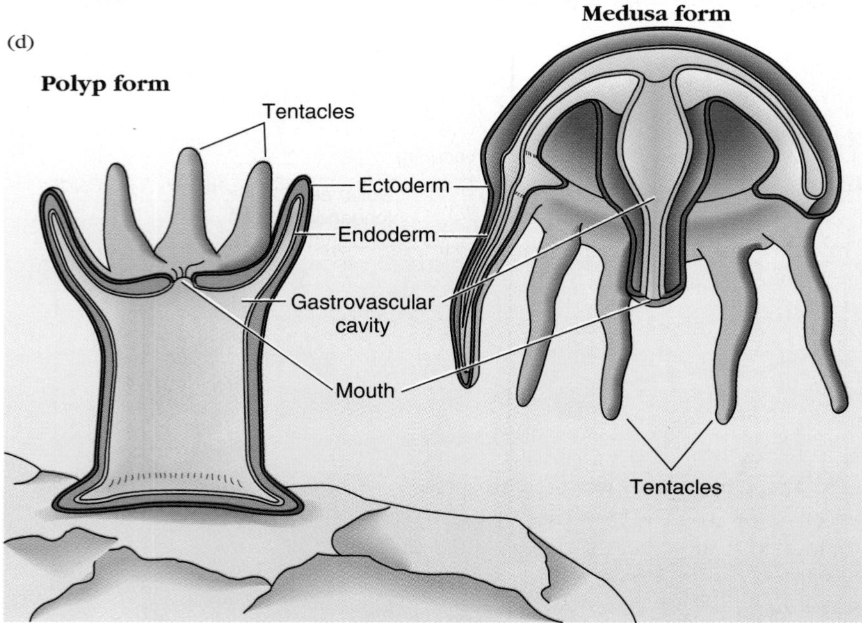

FIGURE 21-5

Phylum Cnidaria. (a) A jellyfish, showing the medusa body form. **(b)** A northern red anemone from the Gulf of Maine, with the polyp body form. **(c)** A reef coral, with many polyps. **(d)** The two body forms found in cnidarians. The free-floating medusa is an inverted version of the sessile polyp. Both have tentacles surrounding the mouth, which leads into a gastrovascular cavity (gut). (a, R. Frerck/Odyssey; b, A. Martinez/Photo Researchers, Inc.; c, C. Seaborn/Odyssey)

21–E PHYLUM PLATYHELMINTHES: FLATWORMS

The flatworms are the most primitive group of animals with bilateral symmetry, the beginnings of cephalization, the development of true organs, and three fully-developed body layers (Section 21-B). The middle layer contains the reproductive organs, an excretory system, and distinct layers of muscles. However, the digestive tract still has only one opening (Figure 21-6). There is no coelom, circulatory system, or blood to transport food around the body. Hence food is distributed by the digestive system, which branches throughout the body. Gas exchange and waste excretion occur over the entire surface of the extremely flat body.

Turbellarians are free-living flatworms with cilia over the entire body surface (Figure 21-6). They feed on smaller organisms or dead organic matter.

The other two groups of flatworms are entirely parasitic. **Flukes** live as external or internal parasites, usually attached to the host by suckers. As is typical for parasites, reproductive organs occupy most of the body, reproductive rates are high, and the life history is often complex, involving more than one host (see *A Journey into Evolution:* Parasitism). Blood flukes of the genus *Schistosoma* cause the devastating disease schistosomiasis, which affects some 200 million people in over 70 tropical nations. These flukes live in freshwater snails for part of the life history. They cycle between their human and snail hosts wherever people drink and wash in water that also contains wastes from infected people.

Tapeworms are highly specialized parasites found in the intestines of probably every species of vertebrate. A tapeworm has no head, mouth, or digestive system. It absorbs digested food from the host's gut

Platyhelminth Diversity

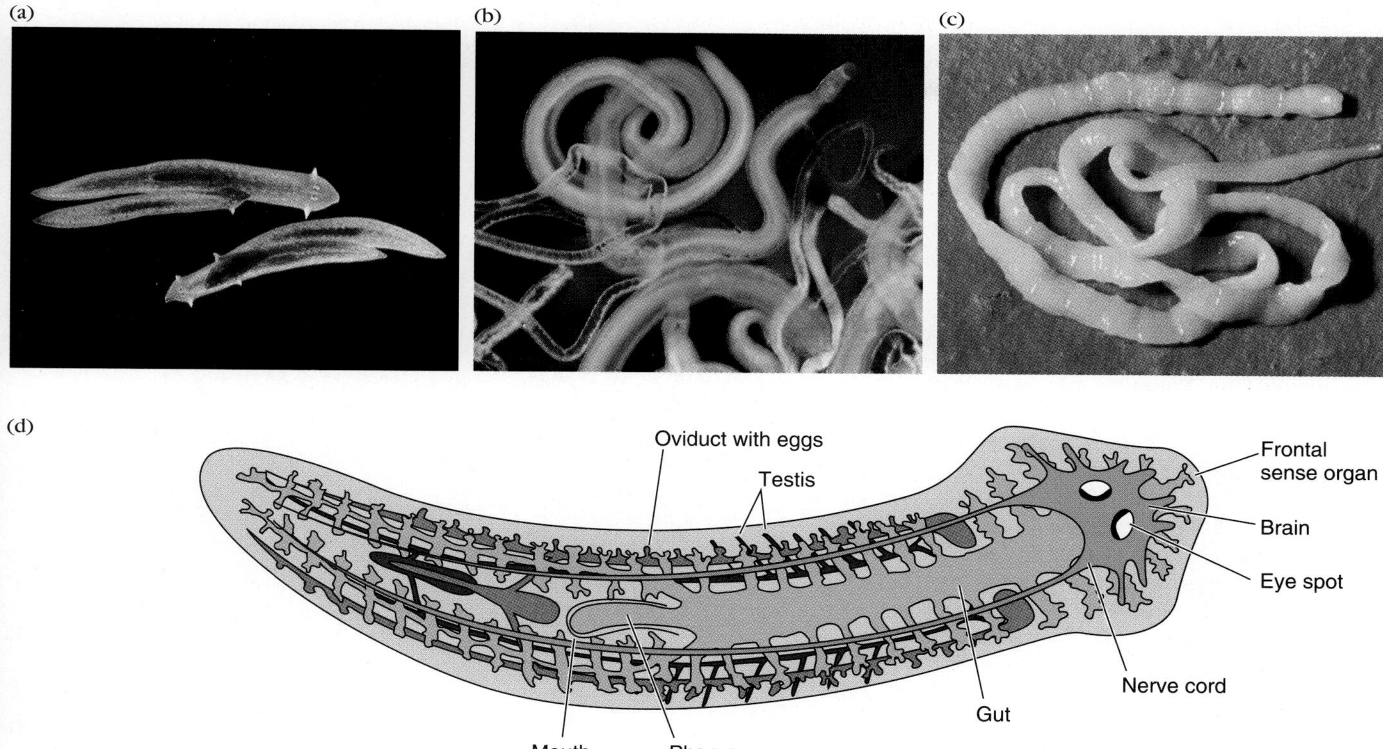

FIGURE 21-6

Phylum Platyhelminthes. (a) The planarian *Dugesia*, a free-living form found under rocks along freshwater streams and at the edges of lakes. **(b)** The nematode *Schistosoma mansoni*, a parasite of humans. **(c)** A dog tapeworm *Dipylidium caninum*. **(d)** The body plan of a planarian. The digestive tract has only one opening, the mouth, located at the tip of the pharynx, and all undigested food must be expelled out of this same opening. Planarians use their senses of smell, touch, and light detection to carefully explore their environment to find and feed upon small invertebrates and decaying animal matter. Many classical studies have shown that planarians are capable of learning and memory. (a, T. E. Adams/Visuals Unlimited; b, Science Photo Library; c, J. H. Robinson/Photo Researchers, Inc.)

through its body surface, and its body contains little except reproductive organs. The secondary hosts for the most common tapeworms that infect humans are cattle, pigs, and fishes. Tapeworm eggs or larvae enter these hosts with the food. The larvae burrow through the host's gut wall, enter the blood, and travel to the muscles, where they lodge as immature worms. Tapeworms infect humans who eat this meat. These parasites can be killed by thorough cooking.

Flatworms are bilaterally symmetrical, show the beginnings of cephalization, and have true organs and three fully developed body layers. However, they still have a primitive digestive tract with only one opening, and there is no coelom. Flatworms can be either parasitic or free-living.

21-F PHYLUM NEMATODA: ROUNDWORMS

Most nematodes, or roundworms, are so small that they attract little notice. However, they can be found by the millions in every environment. Any handful of good soil contains thousands of tiny white or transparent roundworms (Figure 21-7).

Nematodes have advanced over the flatworms in possessing a complete digestive tract, with an anus as well as a mouth, but they lack a circulatory system. A tough, nonliving cuticle covers the body.

About 50 species of roundworms parasitize humans. One causes trichinosis, often contracted by eating undercooked, infected pork, although an outbreak in Canada was traced to eating undercooked bear

(Text continues on page 442)

A Journey into Evolution

Parasitism

Many phyla of animals include parasites, from microscopic body lice to tapeworms several meters long. **Parasites** are organisms that extract their food from living **hosts.** Some are external parasites **(ectoparasites)** that attach to the outside of the host's body. These include leeches (annelids) and ticks, lice, and fleas (arthropods) (Figure 21–A). Others are **endoparasites** that live inside the host. Flukes and tapeworms (flatworms) and many roundworms belong in this group (Table 21–A).

Finding food is often difficult for parasites because appropriate hosts may be few and far between. Many tapeworms and flukes compensate by producing hundreds of thousands to millions of eggs, the large numbers ensuring that at least a few find a host; most starve to death. Pinworms have a hand-to-mouth method of transmission: these small, short-lived worms infect mainly young children, who scratch the worms' eggs from the anal area and transfer them to the mouth on the fingers, allowing the next generation of worms to reach the digestive system of the same host. Female fleas have an interesting adaptation: they are most attracted to pregnant females of host mammal species. The flea feeds on the pregnant female's blood and lays eggs, which hatch into larvae that live on scraps of skin or dung in the burrow of the mother-to-be. When the young mammals leave their mother, the next generation of fleas hops aboard, provided with new hosts and a means of dispersal.

Parasite life histories are often complex and involve more than one host. Some immature parasites change the behavior of their host, making it more apt to be eaten by a second host in which the parasites can become adults. For example, worms that form cysts in muscles may make the host slower and less likely to escape a predator. Larvae of the canine tapeworm invade the nervous system of hosts such as sheep, causing them to totter around in circles and become separated from the herd. They are easily picked off by wolves, which then provide the tapeworm with its adult home.

FIGURE 21–A

An ectoparasite. This scanning electron micrograph shows a cat flea, an insect that lives as an ectoparasite during adulthood. The body is very flat, so that the flea looks very thin when viewed from above. This is an adaptation to running quickly between hairs on the host's skin. (Photo Researchers, Inc.)

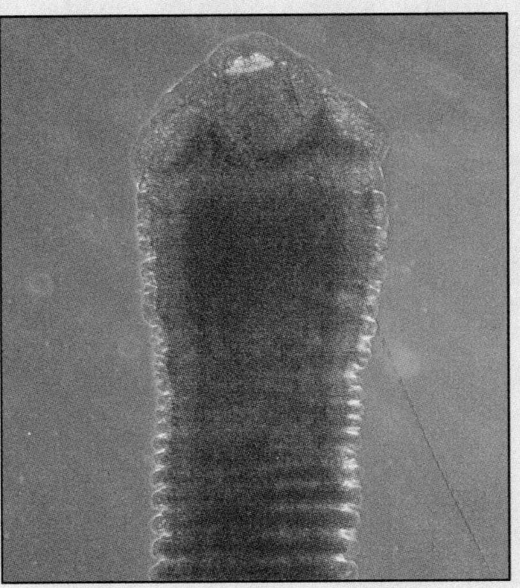

FIGURE 21–B

An endoparasite. This photograph of one end of a tapeworm shows that there is no mouth. Instead, tapeworms use a crown of hooks and suckers to attach to the wall of the host's digestive tract and absorb nutrients directly through their own body walls. The long string of short, wide sections contains mostly reproductive organs (not visible here). Ripe sections detach from the other end of the worm and are shed in the feces. (London Scientific Films/Animals Animals)

Intestinal parasites are surrounded by digested food, and other parasites feed on nutritious blood. There is not much for the parasite's digestion to do, and most parasites have reduced digestive systems. The energy freed by this savings is devoted to expansion of the reproductive system. In fact, tapeworms have no digestive system whatever: they absorb all their food through the body

Table 21–A
Some Parasitic Worms Common in Humans

Name	Symptoms	Means of infection
Platyhelminthes		
Chinese liver fluke	None in mild cases; destruction of liver, bile stones, and clogging of liver ducts in severe cases	Eating raw fish
Blood fluke (schistosomes)	Enlargement of liver and spleen Urinary disorders Bloated abdomen, wasted arms and legs	Drinking or wading barefoot in water containing infected person's urine. Infects about 200 million people in 70 nations. Not found where there are modern sewage disposal systems
Swimmer's itch	Itching after exposure of skin to infested water	Burrowing of fluke larvae of species that cannot successfully infect humans
Bladder worm (immature stage of a worm that lives in dogs when adult)	Cysts up to the size of an orange; symptoms depend on part of body invaded	Infected dogs licking people's hands or faces or contaminating drinking water
Pork, beef, and fish tapeworms	Immature worms: cysts Adult worms may cause diarrhea, loss of weight, perforation of intestine	Eating undercooked meat containing worm cysts
Nematoda		
Pinworm	Anal itching	Females lay eggs around anal opening; hands may transfer eggs to mouth, maintaining infection in same person. Physical contact may also transfer to other people
Hookworm	Anemia, lethargy	Young worms burrow through skin (bare feet) from moist soil and grass contaminated by feces of infected humans

wall, and their bodies are little more than egg factories (Figure 21–B).

From the parasite's point of view, the ideal host-parasite relationship is one in which the host remains alive at least long enough to permit the parasite to complete its development and reproduce. Thus there is often strong selection for parasites not to kill their hosts. The worst outbreaks of diseases occur when parasites—be they viruses, bacteria, protists, fungi, or animals—first come into contact with a particular population of hosts. History is full of examples of parasites, such as the plague or syphilis bacteria, that killed huge proportions of new-found host populations, but this also, of course, killed most of the parasites. Such a first encounter selects for those hosts with defense mechanisms against the parasite, and for those parasites that are less virulent, until, after a while, the parasite will do less damage to its new-found host population than it did initially.

A slightly different way of life is found among some insects known as **parasitoids** ("parasite-like"). The female lays an egg in a host (usually the larva, pupa, adult, or egg of another insect species), and the egg hatches and uses the host as a food source in its own development. Just as the young parasitoid reaches adulthood, it kills its host and cuts its way out of the host's body. So, although the developing parasitoid does feed on a living host, killing the host is a programmed part of its life rather than an accident. Females of some species of parasitoids can tell whether a host is already parasitized, and lay eggs only on uninhabited hosts. This ensures that their own offspring will have enough food to develop. Many kinds of parasitoids attack only one or a few closely related species of hosts. Parasitoids of pest insects are sometimes raised and released as part of pest management programs.

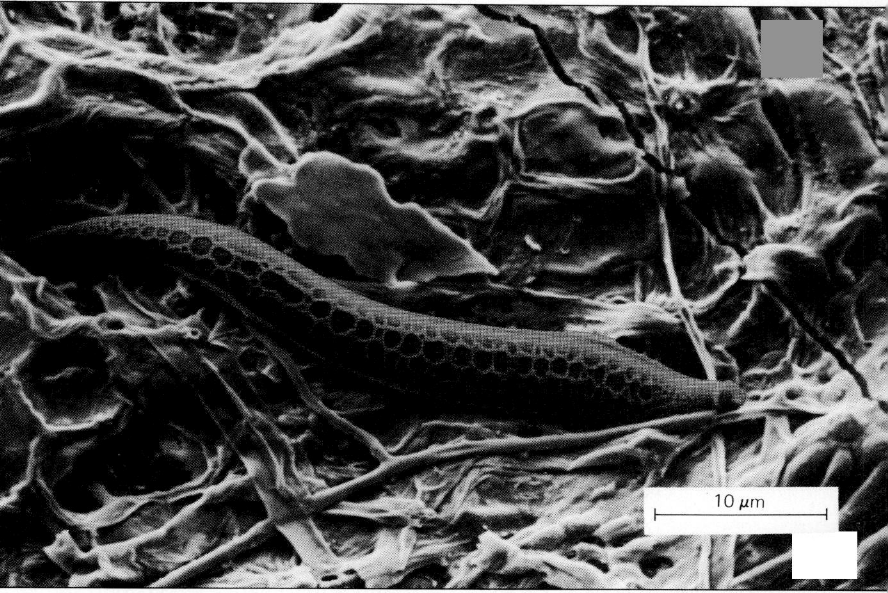

FIGURE 21–7

Phylum Nematoda. This scanning electron micrograph shows a soil nematode (roundworm), false-colored for clarity. (Biophoto Associates)

meat. Among the more harmful species are the intestinal roundworm *(Ascaris lumbricoides),* the guinea worm, and the filaria worm, which lodges in the lymph nodes, and in extreme cases causes elephantiasis (enormous swelling of the leg). Hookworms enter the skin of people walking barefoot on soil contaminated with feces from infected humans. These worms caused the anemia, weakness, and fatigue—formerly attributed to laziness—of thousands of poor laborers in the southern United States.

Nematodes are similar to flatworms, except that nematodes have evolved a complete, one-way digestive tract with a mouth and anus. Some species of nematodes are parasitic. About 50 species infect humans.

21–G PHYLUM ANNELIDA: SEGMENTED WORMS

The advances shown by annelid worms include the development of a coelom and a circulatory system with closed blood vessels, some of them enlarged as pumping "hearts" (Figure 21–8). Their bodies are **segmented**, divided into repeated sections by partitions called **septa.** Annelids exchange gases with their environment through the body surface. The area available for this function may be increased by **gills,** extensions of the body surface.

The largest group of annelids, the **polychaetes** ("many bristles"), are mostly marine. Polychaetes burrow in the sea floor or paddle lightly over reefs. Some are tube-dwelling carnivores or filter feeders (Figure 21–8).

Earthworms and their kin make up a second group of annelids. Most earthworms burrow in mud or soil. They ingest soil as they burrow, digest any organic matter, and void the rest as worm casts, which can often be found in a garden. Earthworms play a crucial role in loosening, aerating, and mixing soil.

2? **Leeches**, the third annelid group, live mainly in freshwater although some are terrestrial in moist habitats like cloud forests. Some prey on other invertebrates or scavenge dead matter, but the most famous are external parasites of aquatic vertebrates. Only a few species have the rasping "teeth" needed to break through the tough skin of a mammal. In the last century, doctors often used medicinal leeches to bleed patients as a treatment for all manner of diseases. Indeed, "leech" was used as a nickname for doctors, and doctors kept aquaria with leeches that could be popped into containers in their medical bags when they went out on housecalls. Leeches are still occasionally used to prevent bruising and blood pooling after surgery (Figure 21–8). ■

Annelids are the only worms with a coelom, a circulatory system, and body segmentation. The annelids include marine polychaete worms, earthworms, and leeches.

21–H PHYLUM MOLLUSCA

Molluscs include chitons, snails, octopuses, and "shellfish"—clams, oysters, limpets, scallops, and their kin. The molluscan body consists basically of a muscular **head-foot,** named for the large muscular mass that moves the animal and contains the mouth, sensory or-

Annelid Diversity

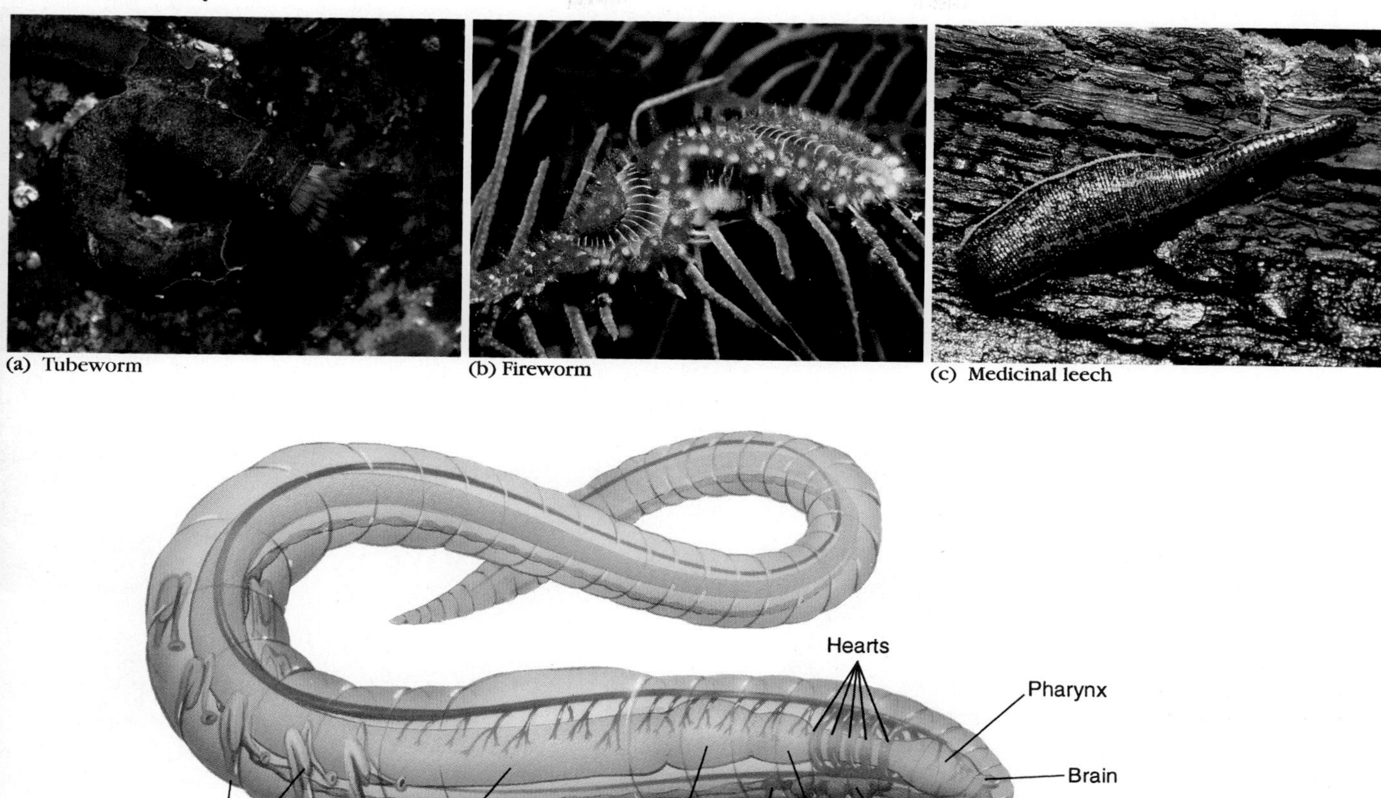

FIGURE 21-8

Phylum Annelida. (a) A polychaete worm that lives in a tube. Its feathery, spiral tentacles filter food out of the water. **(b)** This polychaete fireworm is crawling across some coral. The separate tufts of white bristles are hollow and filled with venom that stings humans if they break off in the skin. **(c)** Medicinal leeches inject an anesthetic and anticoagulant as they suck blood away from an incision, reducing the sensation of pain and preventing blood clots from accumulating at the wound. **(d)** Anatomy of an earthworm. Soil is brought in the mouth and temporarily stored in the crop. After being ground up by the gizzard and mixed with digestive enzymes, decayed organic matter is absorbed out of the soil by the walls of the intestine. Blood is circulated by a series of hearts that join the lower and upper blood vessels. The nervous system is located on the belly side of the animal and expands into a small brain near the mouth. This nervous system allows coordinated muscular movements and the sensation of pain. Paired tubes called nephridia lie along the sides of the worm and expel extra water and waste products out of the body. (a, M. Snyderman/Visuals Unlimited; b, G. Retherford/Photo Researchers, Inc.; c, R. Cramm/Photo Researchers, Inc.)

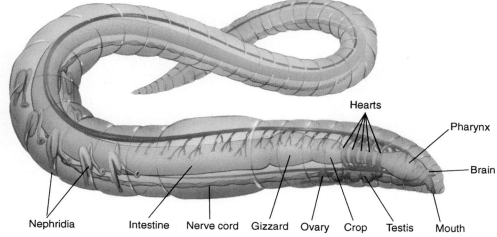

gans, and nerves. A flattened piece of tissue, called the **mantle,** covers the head-foot and in many species secretes a shell containing calcium and/or acts as a site of gas exchange (Figure 21-9).

The **gastropods** include marine, freshwater, and terrestrial snails and slugs. Most gastropods have a well-defined head, usually with eyes and tentacles, and an elongated, flattened foot on which they creep around. Most have gills inside the mantle cavity. However, in some forms, such as terrestrial snails, the mantle cavity acts as a lung, permitting the animal to obtain oxygen from air instead of from water. Some of the world's loveliest animals are **nudibranchs**—commonly known by the unlovely name of sea slugs. They

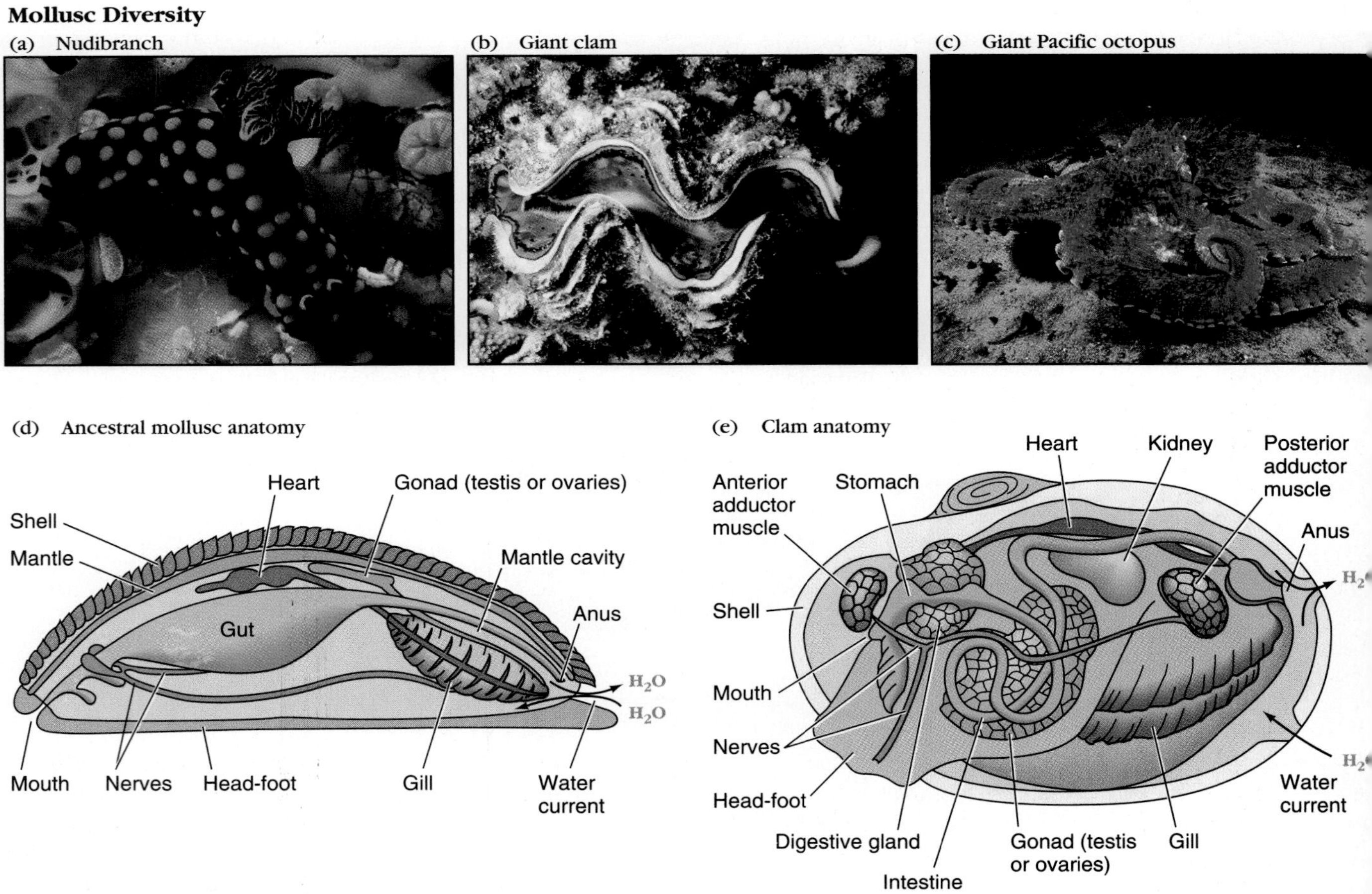

Mollusc Diversity

(a) Nudibranch (b) Giant clam (c) Giant Pacific octopus

(d) Ancestral mollusc anatomy

(e) Clam anatomy

FIGURE 21-9

Phylum Mollusca. (a) A nudibranch, a marine gastropod without a shell. The tufts are its gills, giving rise to the name (nudi = naked; branch = gill). **(b)** A marine bivalve, *Tridacbna*. Note the blue edges of the mantle visible within the shell. **(c)** A giant Pacific octopus off the coast of British Columbia, Canada. **(d)** A hypothetical ancestral mollusc. This diagram shows the general anatomy of members of the phylum Mollusca. The muscles of the lower body form a continuous mass from which both the head and foot are formed in animals that evolved later. **(e)** Anatomy of a clam. The left side of this clam has been removed to reveal the internal anatomy. Water is swept into the body by cilia. As the water passes over the gills, food particles stick to the mucous covered surface. This food and mucous mass is transferred to the mouth. Undigested food is expelled near the excurrent siphon as the water passes out of the clam. Adductor muscles help keep the shell closed to avoid predation. (a, A. Wood/Photo Researchers, Inc.; b, P. Degginger; c, F. Bavenda/Peter Arnold, Inc.)

have lost the mantle, shell, and gills, leaving a naked body that is often brilliantly colored (Figure 21-9).

The **bivalves** are a large group of marine and freshwater molluscs with the body flattened between the two valves (halves) of a hinged shell (Figure 21-9). The edge of the mantle is drawn out to form two **siphons,** one for water entering the mantle cavity and one for water leaving. Most bivalves are filter feeders. Cilia draw a water current across the gills, where food particles are strained out and trapped in strands of mucus. The gill cilia then move these strands to the mouth.

The marine, carnivorous **cephalopods** (octopuses, squids, and nautiluses) are among the most advanced of all invertebrates. Behavioral studies have shown that octopuses are quite intelligent and can solve many problems. Cephalopods appear to be quite closely related to gastropods, but the body has been rearranged. The mouth is now in the middle of the foot, whose edges are drawn out to form tentacles lined with rows of suckers (Figure 21-9). Most cephalopods swim by jerky jet propulsion. They take water into the mantle cavity and force it out through a siphon, which the animal can point in various direc-

Table 21–2
The Phylum Arthropoda and Its Major Classes

Phylum Arthropoda	Segmented animals with jointed exoskeletons containing chitin; jointed appendages; respiration through body surfaces or by gills or tracheae; marine, freshwater, and terrestrial
Class Arachnida (~57,000 species)	Body with 1 or 2 main parts; 6 pairs of appendages (chelicerae, pedipalps, 4 pairs of walking legs); most terrestrial; e.g., spiders, scorpions, ticks, mites
Class Crustacea (~25,000 species)	Body of 2 or 3 parts; antennae, chewing mouthparts, 3 or more pairs of legs; most marine; e.g., shrimp, krill, lobsters, crabs, barnacles, ostracods, copepods
Class Insecta (~700,000 species)	Body divided into head, thorax, and abdomen; antennae; mouthparts modified for chewing, sucking, or lapping; adults with 3 pairs of legs and usually 2 pairs of wings; breathing by tracheae; most terrestrial; e.g., beetles, flies, butterflies, ants, termites, dragonflies, aphids
Class Diplopoda (~7,000 species)	Body with distinct head bearing antennae and chewing mouthparts; most segments of body grouped in pairs covered by a single skeletal plate, each apparent segment bearing 2 pairs of walking legs; breathing by tracheal system; terrestrial, eating dead or living plants. The millipedes
Class Chilopoda (~2,000 species)	Body with distinct head bearing large antennae and chewing mouthparts; appendages of first body segment modified as poison claws; remaining segments bearing a pair of walking legs each; terrestrial in damp areas including houses; predaceous on insects. The centipedes

ions to determine which way it will move. More often, however, octopuses use their tentacles to crawl along the ocean floor.

The phylum Mollusca contains a diverse group of species that have a muscular head-foot, a broad, flattened mantle, and they often have a shell. The molluscs include marine, freshwater, and terrestrial species.

21–I PHYLUM ARTHROPODA

The phylum **Arthropoda** ("jointed foot") contains more than three times as many species as all other animal phyla combined (Table 21–2). Arthropods include familiar animals such as butterflies, beetles, spiders, crabs, and shrimp. Some groups of arthropods have solved the problems of life on land more completely than any other group except higher vertebrates. Their major step forward was the improvement of the cuticle, which also covers the body in many other invertebrates. In arthropods, the cuticle is composed of layers of protein complexed with the strong, flexible polysaccharide, **chitin.** It forms an **exoskeleton** (external skeleton) consisting of a set of plates, which cover not only the body but also the jointed appendages. Hinges between adjacent plates permit movement, while the tough plates protect against attack and injury (Figure 21–10). In many species, wax in the outer layer also prevents water loss.

Arthropod survival often depends upon the efficiency of their waterproofing. Since Roman times, people in Africa and Europe have mixed fine dust with stored grain to keep it free of insects. The dust abrades the soft cuticle between segments, destroying the insect's waterproofing so that it dries up and dies.

One result of possessing an exoskeleton that cannot expand is that arthropods must **molt,** or shed their cuticles, in order to grow. In many arthropods, successive molts take the animal through a series of larval stages.

The primitive arthropod body plan has one pair of jointed appendages per body segment. The appendages became variously modified during the course of evolution into specialized antennae, mouthparts, walking legs, claws, or swimming paddles. Segmentation of the body also became modified. In some arthropods, the segments became grouped to form distinct parts of the body, such as the **head, thorax,** and **abdomen** in insects.

Class Arachnida

The **arachnids** include the spiders, ticks, mites, and scorpions (Figure 21–10). Arachnids have six pairs of jointed appendages. The first pair are adapted for feeding, and often have associated poison glands that inject a substance that anesthetizes or kills the prey. The second pair act as sense organs to detect touch and chemicals and also help hold food. The other four pairs are walking legs.

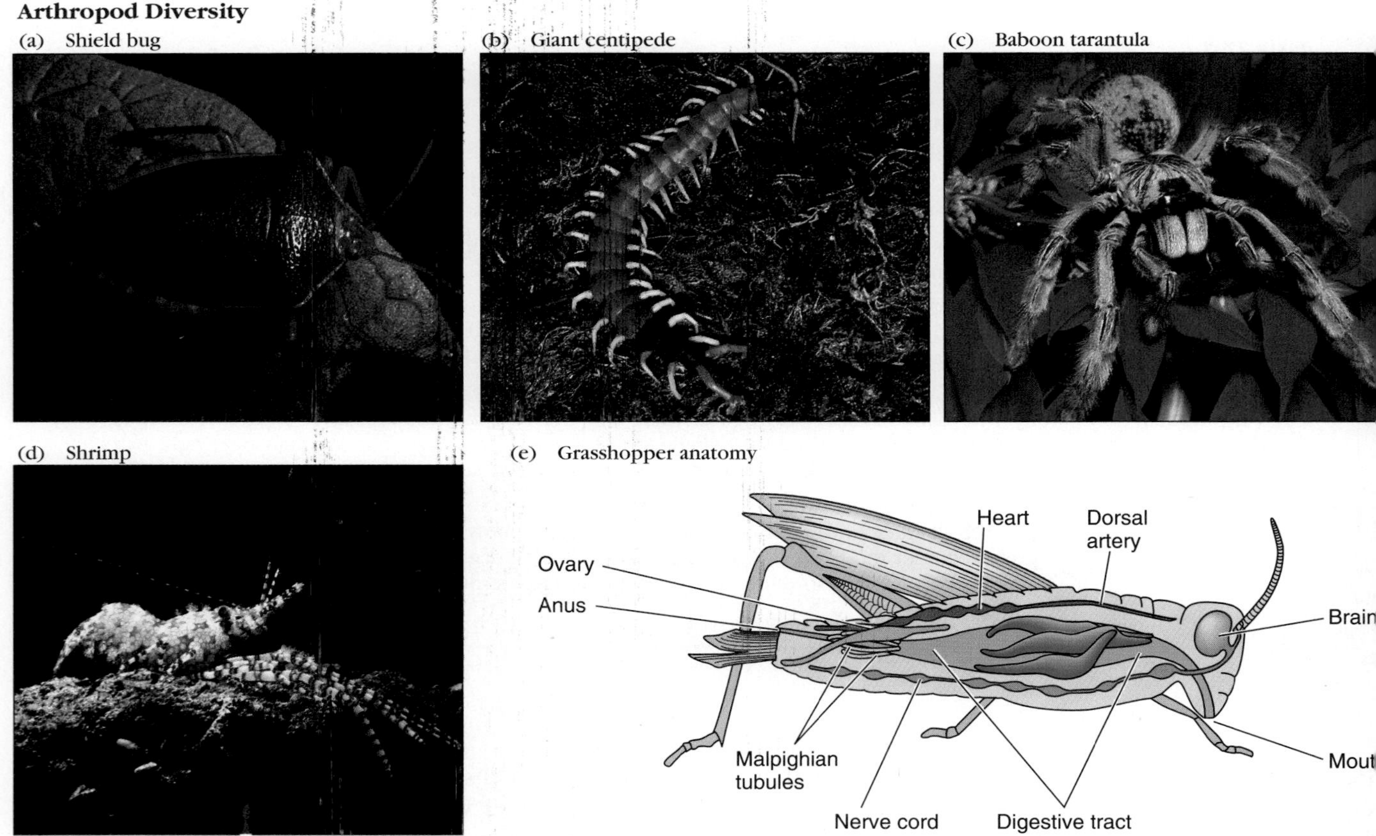

FIGURE 21–10

Phylum Arthropoda. (a) A shield bug, a typical insect with six legs and a single pair of antennae. **(b)** Centipedes have fewer legs than millipedes but can move faster. They are typically carnivores that use venoms for defense and to disable their prey. **(c)** This baboon tarantula shows 6 pairs of appendages typical of the class Arachnida. **(d)** A shrimp, with a cuticle that is reinforced with calcium carbonate. **(e)** Grasshopper anatomy. Grasshoppers feed by chewing up leaves or other plant parts and then passing this food into the intestine, where it is mixed with digestive enzymes and absorbed by the walls of the intestine. Malpighian tubules add chemical waste products to the feces as the undigested material is passed out of the body. A heart circulates blood in an open circulatory system in which blood outside of blood vessels freely passes amongst the tissues. The front end of the nervous system is expanded into a brain to interpret a broad range of sensory information and coordinate muscle movements. (a, K. Schafer; b, A. Kerstitch; c, T. McHugh/Photo Researchers, Inc.; d, A. Kerstitch)

All spiders have spinnerets with which they spin silk, and which may be used to make webs that trap prey and the cocoons that protect their eggs. All spiders are carnivorous, usually preying on insects.

Some ticks and mites transmit diseases, including Rocky Mountain spotted fever, Lyme disease, and Asian scrub typhus. Mites themselves cause the itching of mange and scabies. Some mites are also serious plant pests. The vast majority of mites live around us unnoticed because of their tiny size—a species that reaches a millimeter or two is enormous as mites go.

BIO-BIT

Soft-shelled crabs are individuals that have just molted, and have not yet produced another hardened exoskeleton.

Class Crustacea: Lobsters, Crabs, Wood Lice, and Their Relatives

The class **Crustacea** includes lobsters, shrimp, crabs, crayfish, wood lice, pill bugs, water fleas, and barna-

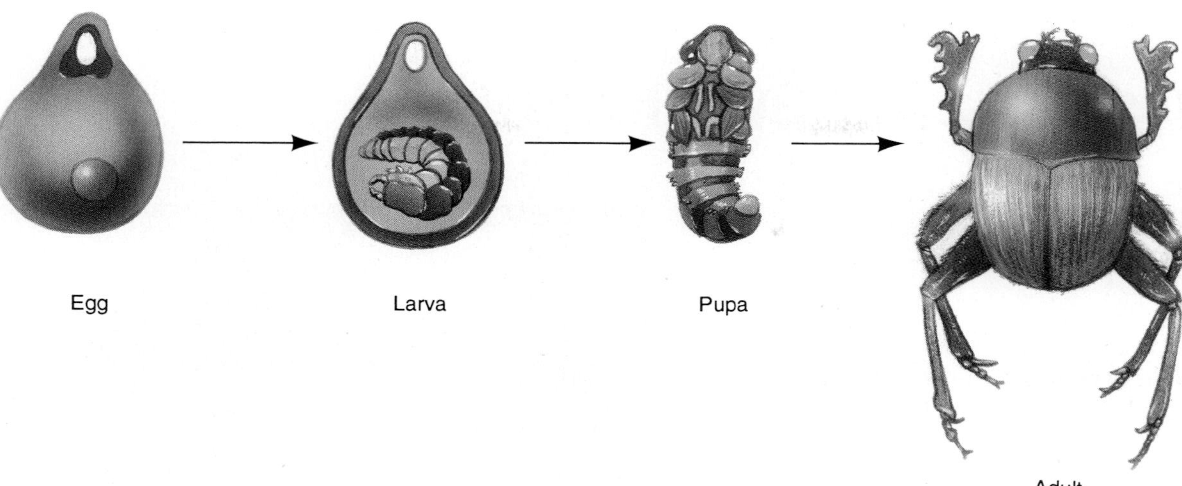

FIGURE 21-11

ife stages of a dung beetle (Phylum Arthropoda, Class Insecta, Order Coleoptera). Adult lung beetles (*Canthon pilularis*) chew off a piece of dung and roll it into a ball. The eggs are de- osited inside and the larva feed upon the surrounding dung. The larvae develop into the pupa tage before assuming the adult form.

les (Figure 21-10). In general, crustaceans are aquatic rthropods with forked appendages: two pairs of an- ennae, three pairs of feeding appendages formed for ood handling and chewing, and several pairs of legs. ypically, anterior appendages are modified to grasp ood and convey it to the mouth. Planktonic crus- aceans are the main food of many vertebrates and in- lude the shrimp-like krill upon which the biggest vhales feed.

lass Insecta

nsects are, without doubt, the most successful terres- rial invertebrates and the only major competitors with umans for dominance on land (See *A Journey into volution:* Insects in Our Environment, this chapter.) t has been estimated that the insects on Earth weigh 2 times as much as the humans, and that there are 00 million insects for every person alive. Nearly a mil- ion insect species have been described, more than the umber of all other animal species put together. In- ects, bats, and birds are the only living animal groups vith members that can fly, although members of many ther groups can glide.

Insects range in size from tiny beetles only 0.1 mil- meter long to tropical moths with a wingspan of 30 entimeters. The insect body is usually divided into he head, thorax, and abdomen. The thorax of most dult insects bears three pairs of walking legs and two pairs of wings (flies and mosquitoes have only one pair, and fleas, lice, and silverfish have none). The head bears one pair of antennae, specialized mouth- parts, and, usually, compound eyes (Figure 21-10).

Insects show several adaptations to life on land, the first being the waxy, waterproofed cuticle, which also supports the body in the air. Insects have an inter- nal system of **tracheae,** air-filled tubes that branch throughout the body and carry oxygen to the cells. Specialized excretory tubules produce nearly solid wastes, minimizing the loss of body water. Since in- sects are basically terrestrial, they have internal fertil- ization. The eggs are laid with a waterproof covering, which protects them from dehydration. Females lay their eggs where the young will find food when they hatch. The young undergo a series of molts, and the higher insects pass through larval stages (such as cater- pillars or maggots) that are completely different from the adult in appearance and way of life (Figure 21-11).

BIO-BIT

Many insects died from the fine ash produced by the 1980 blast of Mt. St. Helens volcano in Washing- ton state. The dust scratched holes in the insects' exoskeletons when they rubbed themselves, which caused the insects to die of dehydration.

A Journey into Evolution

Insects in Our Environment

Insects perform many roles vital to human life. Without bees and other insects, for instance, many flowering plants would never be pollinated—a prerequisite to producing crops such as apples, citrus fruits, berries, and cucumbers. Many beetles, ants, and flies are important decomposers, breaking down the dead bodies of plants and animals.

Nevertheless, people have devoted more time to killing insects than to praising them. It is perhaps unduly gloomy to conclude that we are losing the battle against insects, who will one day inherit the Earth, but those who believe this have good reason for their opinion.

Insects attack human beings directly with bites and stings. Much more important, blood-sucking insects transmit many diseases. Malaria, river blindness, and sleeping sickness carried by insects blind and kill millions of people a year. Insects probably do more damage indirectly, however, by transmitting plant diseases, such as Dutch elm disease and many viral diseases of crop plants and by eating crops and killing trees (Figure 21–C). In the United States alone, during 1975 the gypsy moth, tussock moth, southern pine beetle, and spruce budworm destroyed enough forest trees to build nearly a million houses.

Insects destroy more than 10% of all crops grown in the United States, but the damage is even worse in the tropics, where hot weather throughout the year permits insects to grow and reproduce faster. In Kenya, officials estimate that insects destroy 75% of the nation's crops. A locust swarm in Africa may be 30 meters deep along a front 1500 meters long, and will consume every fragment of plant material in its path, leaving hundreds of square kilometers of country devastated.

Pesticides have not solved the insect problem. This is partly because pesticides act as selective pressures for the evolution of resistant strains of insects, which evolve too fast for expensive pesticide research to keep up. The list of pesticide-resistant insects nearly doubled between 1970 and 1980. Workers now direct much of their effort to using a combination of chemical and biological methods to control damaging outbreaks of insects. Biological controls include raising, sterilizing, and releasing large numbers of males (used for species in which females will mate only once), using sex attractant chemicals (pheromones) to attract males to traps instead of to females, breeding pest-resistant plants, and introducing specific predators and parasites of pest insects. Yet despite the fact that human beings have waged war on insects since the two have existed together, human efforts have apparently not succeeded in exterminating even a single species of unwanted insects.

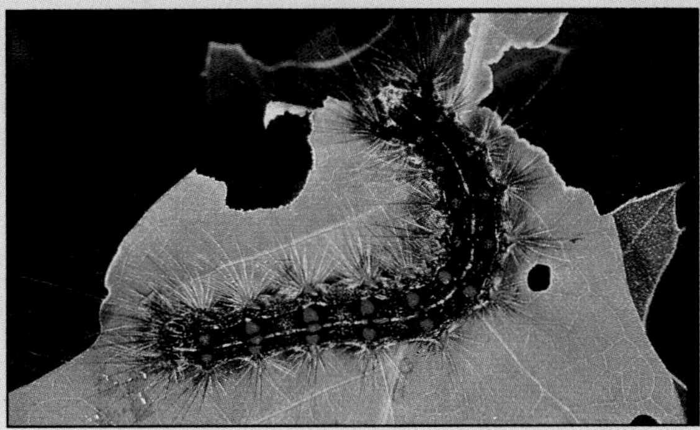

FIGURE 21–C

Insect devastation. (a) The voracious caterpillar of the gypsy moth, a species imported into Massachusetts in the nineteenth century in hopes of starting a silk industry. Escapees were the ancestors of populations now found throughout most of the United States and Canada. During periodic population explosions, hordes of caterpillars strip leaves from millions of acres of forest. **(b)** A gypsy moth outbreak in the late 1980s defoliated these trees on Maryland's eastern shore. (a, John Burnley/Photo Researchers, Inc.; b, James L. Amos/Photo Researchers, Inc.)

Why are most insects so small? The mechanics of an exoskeleton, and the fact that flight requires less energy with a lighter body, must impose some theoretical upper limit on the size of a flying insect. Nevertheless, modern insects are considerably smaller than some of their extinct ancestors, who were quite capable fliers. The answer appears to be that small size permits insects to occupy habitats where vertebrates, the other main group of land animals, cannot compete with them. ■

Several species of insects have been domesticated, including honeybees, which produce honey and wax, and lac insects, source of the main ingredient in shellac. Silk, from the cocoon of the silkworm pupa, has been a major product of China for centuries. Cochineal insects are the source of a bright red dye that the Aztecs and Incas, and their Spanish conquerors, prized nearly as much as gold and silver. The Star-Spangled Banner and the red coats of the invading British Army were both colored with this dye.

Crime investigators have also found uses for insects. A thirteenth-century Chinese law enforcement manual recounted how a murderer was caught when his sickle (the murder weapon) was the only one in the village to attract flies sensitive to the odor of decaying flesh. In New Zealand, the bodies of Asian insects in a load of confiscated marijuana provided the crucial evidence to convict drug dealers on charges of importation (a more serious offense than mere possession). Police have also found that the age of maggots, and the other kinds of insects present, can sometimes provide a remarkably accurate estimate of how long a corpse has been dead.

The chitinous exoskeleton of arthropods, with its many jointed appendages modified for a variety of jobs, proved so versatile that the arthropods have undergone an impressive adaptive radiation, with more species and individuals than any other animal phylum. The main groups within the phylum Arthropoda are the spiders, crustaceans, and insects.

21–J PHYLUM ECHINODERMATA

Sea stars, brittle stars, sea cucumbers, sea lilies, sea urchins, and sand dollars belong to the phylum **Echinodermata.** From looking at the adult animals, we would hardly guess that vertebrates are more closely related to echinoderms than to annelids, molluscs, and arthropods. Yet studies of embryonic development suggest that such is indeed the case.

> **BIO-BIT**
>
> When disturbed, sea cucumbers will expel their guts to discourage predators.

The name Echinodermata, meaning "spiny-skinned," refers to the spines and plates of calcium carbonate that form a skeleton just under the skin in all members of the phylum. Another characteristic feature is suction-cup tube feet, used for locomotion, for gas exchange, and, in predatory forms, for feeding. Most adult echinoderms also have an unusual kind of radial symmetry: the body is divided into five parts around a central area where the mouth lies (Figure 21–12). All echinoderms are marine. Most are bottom-dwelling and able to move about slowly.

Most sea stars are carnivorous, using their tube feet to grip their prey (Figure 21–12). Those that are predators on bivalves can exert enough suction to pry open a very narrow slit between the valves. The animal then everts its stomach, which squeezes into the shell and digests the prey. The "crown of thorns" sea star feeds on cnidarian polyps and is notorious for the damage it does to such coral reefs as the Great Barrier Reef of Australia.

Sea urchins and sand dollars live mouth-down on the bottom of the sea, protected from intrusion by brittle calcareous (calcium carbonate) spines, or plates. The calcareous skeletal plates are fused into an envelope around the animal, pierced by holes for the mouth, anus, and tube feet. The sausage-shaped sea cucumbers also lack arms (Figure 21–12). The mouth, at one end of the long body, is sometimes surrounded by modified tube feet called tentacles, used for filter feeding.

Animals within the phylum Echinodermata all possess a skeleton of spines or plates of calcium carbonate that develop under the skin. In addition, echinoderms have tube feet for locomotion, gas exchange, and feeding. All echinoderm species live in the ocean.

21–K PHYLUM CHORDATA

The phylum **Chordata** contains some invertebrates and all the vertebrates. Although we have a remarkable fossil record of the evolution of vertebrates into a wide range of forms, details of the origin of the earliest chordates are lost, probably forever, owing to the poor fossil record from the Precambrian and Cambrian geological periods. The only way to reconstruct early vertebrate history is to study the invertebrate groups—especially the echinoderms and invertebrate chordates—most closely related to the vertebrates.

Echinoderm Diversity

(a) Sea cucumber

(b) Sea urchin

(c) Giant sun star

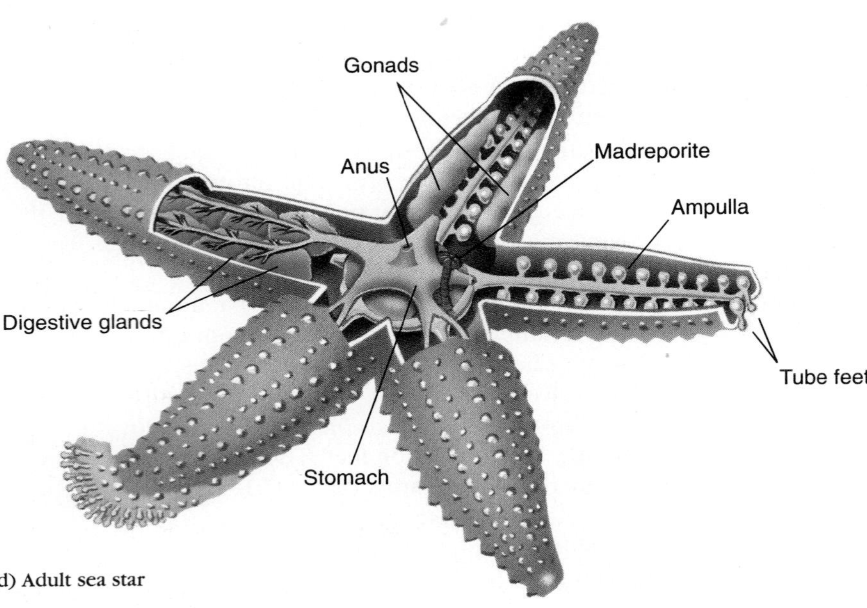

(d) Adult sea star

FIGURE 21-12

Phylum Echinodermata. (a) A sea cucumber. The sharp projections are stiff and help prevent predation. **(b)** A sea urchin creeping over a coral. **(c)** A giant sun star. Many starfish have more than five arms. **(d)** Sea star. Sea stars evert the stomach and digest their prey outside the body. Later they draw the partially digested prey back into the digestive tract where absorption of nutrients occurs. Undigested food is expelled out the anus on the side opposite the mouth. The madreporite is an opening through which water is taken into the body and strained before entering the water-vascular system (yellow). This system uses fluid pressure to manipulate the tube feet extending along each arm. (a, c, C. Seaborn/Odyssey; b, A. Kerstitch)

All chordates share several important features (Figure 21-13):

1. At some stage in the life history, all chordates have a stiff, rod-like **notochord,** which serves as an internal skeleton. In the embryonic development of vertebrates, the notochord is surrounded or replaced by a column of vertebrae that form the backbone.
2. At some time in their lives, all chordates have **pharyngeal gill slits** leading from the **pharynx,** the throat cavity behind the mouth, to the exterior.
3. The nerve cord forms a hollow tube running from head to tail on the dorsal side of the body, in contrast to most invertebrates, whose main nerve cord is solid and ventral.

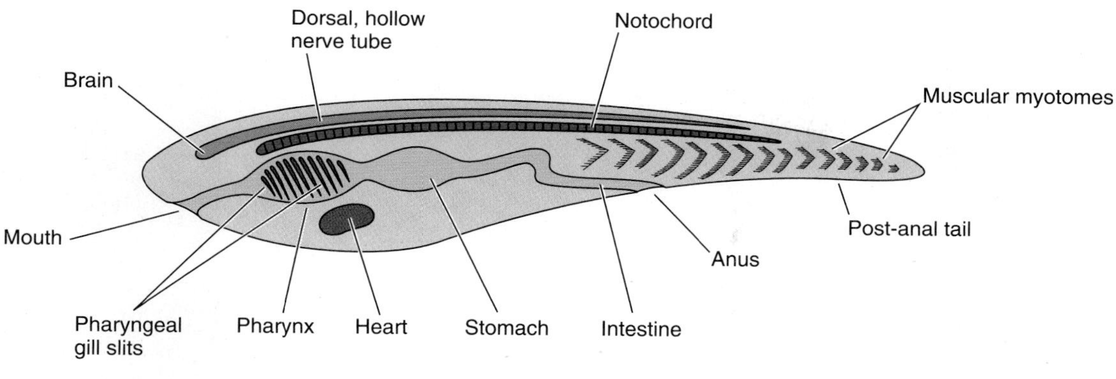

FIGURE 21-13

Chordate characteristics. This diagram of a generalized, primitive chordate shows the notochord, with the dorsal, hollow nerve tube above it; pharyngeal gill slits; segmentally arranged blocks of muscle (myotomes) for tail movements; and a tail extending behind the anus.

These three features define the chordates. But, in addition, most chordates have more or less segmented bodies, an endoskeleton (internal skeleton), and, at some stage of life, a tail that extends beyond the anus.

These features comprise a particularly successful set of adaptations. The internal notochord, working with segmented blocks of muscle allowed early chordates to swim quickly and efficiently by side-to-side wiggles of the body. As they swam forward, they took in food and water through the mouth and let the extra water escape through the gill slits. On its way out, the water gave up oxygen to the blood passing through the gills. Sense organs in the head detected where the animal was going and found food, and the brain and nervous system became well developed.

Chordate Subphylum Urochordata: Tunicates

Urochordates, the sea squirts and their relatives, are a group of entirely marine invertebrates. Most sea squirts are sessile filter feeders. Many live in colonies, which may share a common mouth. The tadpole-like larval stage could almost have posed for our drawing of a generalized chordate (Figure 21-14). Upon hatching, it swims to the surface with efficient fish-like wriggles, using the action of its muscles against its notochord. It drifts a short distance, turns, and swims down to search for a suitable rock or dock piling. Here it attaches by adhesive projections on the tip of its nose and metamorphoses into an adult, losing its notochord and tail, while the gill slits expand tremendously.

Since this larva is the most primitive known chordate, it looks as if the characters typical of chordates evolved, not as adaptations of an adult to its way of life, but in a larva. Most biologists think that vertebrates evolved from a tadpole-like creature that failed to metamorphose but became sexually mature while still a larva. This idea is supported by the facts that a number of living animals develop some degree of sexual maturity while still larvae, and that groups other than vertebrates also seem to have originated in this way.

Chordate Subphylum Cephalochordata

The lancelet is the only cephalochordate. Adults look like tunicate tadpoles with the gill system vastly expanded to form an enormous pharyngeal gill basket (Figure 21-15). This leaves little room for swimming muscles, and these animals swim poorly. A lancelet lives buried in sand with only its head end protruding. Here it feeds, like a urochordate, by using cilia to pull a current of water into its mouth. Any food in the current is filtered and trapped in a mucous net which passes down the pharynx into the gut, where the food is digested.

VERTEBRATES

Members of the subphylum Vertebrata differ from other chordates in having a backbone, the vertebral column, which replaces the notochord to a greater or lesser extent (Table 21-3). In addition to the basic chordate features listed before, all vertebrates have some sort of a liver, endocrine organs that secrete hormones, kidneys, a ventral heart, closed blood vessels, and some degree of segmentation. Cephalization is pronounced, with sense organs and nerves concentrated at the front end of the body so that vertebrates

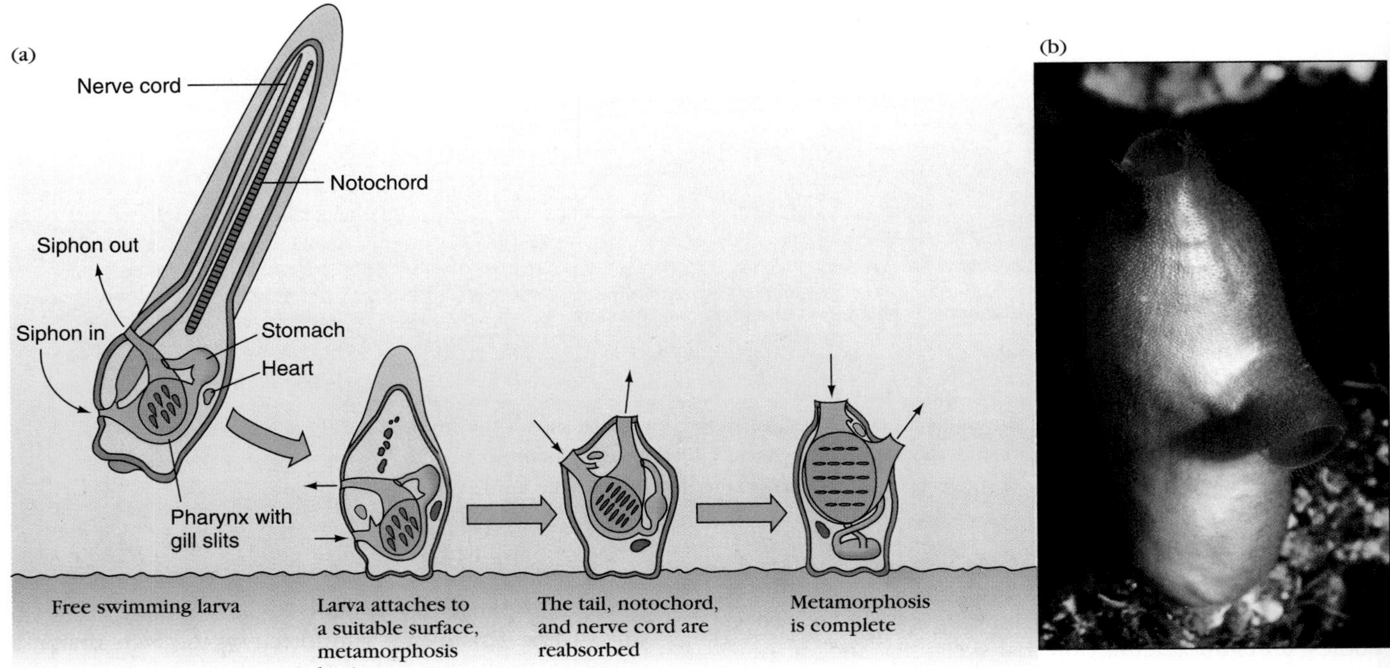

FIGURE 21-14

Subphylum Urochordata: Tunicates. (a) The metamorphosis from a motile larva to a sessile adult. The larva shows all four chordate characteristics (dorsal hollow nerve cord, notochord, gill slits, and postanal tail) while the adult has only one (gill slits). **(b)** Adult urochordate or sea squirt. (b, T. McHugh/Photo Researchers, Inc.)

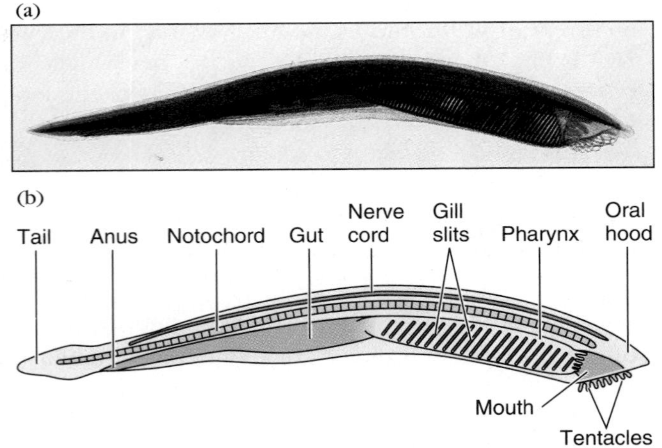

FIGURE 21-15

Subphylum Cephalochordata. (a) Lancelet. **(b)** Lancelet anatomy. Note the four major chordate characteristics. (Cabisco/Visuals Unlimited)

have very obvious heads. Early vertebrates, known only from fossils, lacked movable jaws and probably fed like lancelets.

The evolution of a backbone in vertebrates permitted rapid, efficient locomotion. Primitive vertebrates were fishes that swam as sharks do today, by throwing the body into S-shaped curves, with the segmented muscles pulling against the vertebral column and pushing the paddle-like tail against the water. An animal that can move rapidly can be carnivorous, feeding on other animals, which weight for weight are more nutritious than plants. Many vertebrate evolutionary advances were associated with a carnivorous way of life, starting with the adaptive radiation of the fishes (Figure 21-16).

At some point in their lifetimes, all chordates possess a notochord, pharyngeal gill slits, and a dorsal hollow nerve cord. In addition, most chordates have segmented bodies, an endoskeleton, and a post-anal tail. The phylum Chordata in-

Table 21–3
The Classes of the Subphylum Vertebrata

Class Agnatha (~45 species)	Jawless fishes; gill openings separate; skeleton cartilaginous; notochord persists throughout life; marine and freshwater. Lampreys and hagfishes
Class Chondrichthyes (~275 species)	Cartilaginous fishes; cartilaginous skeletons; jaws; notochord replaced by vertebrae in the adult; gill openings separate; paired pectoral and pelvic fins; tail fin usually asymmetrical; most marine. Sharks, skates, and rays
Class Osteichthyes (~25,000 species)	Bony fishes; bony skeletons and jaws; gill openings all covered by a single operculum; paired pectoral and pelvic fins; tail fin usually symmetrical; many have a swimbladder; marine and freshwater; e.g., herring, salmon, sturgeon, eels, sea horse, electric eel
Class Amphibia (~2,500 species)	Tetrapods that lay eggs without an amnion or shell; respiration via lungs and skin; scales absent; most freshwater or terrestrial. Salamanders, newts, frogs, and toads
Class Reptilia (~6,000 species)	Tetrapods with amniotic eggs and scaly skin. Snakes and lizards, turtles, and crocodilians
Class Aves (~8,600 species)	Birds. Tetrapods with feathers; oviparous, laying amniotic eggs; high body temperature; bipedal, most species have more than one mode of locomotion; forelimbs usually modified to form wings; e.g., sparrows, penguins, ostriches
Class Mammalia (~4,400 species)	Tetrapods with young nourished by milk from mammary glands of females; most viviparous; high body temperature; body usually covered with hair; only one bone in each side of lower jaw, teeth differentiated and specialized. The monotremes (echidnas and platypus), marsupials (e.g., opossum, kangaroo), and placental mammals (e.g., humans, bats, whales, rodents, dogs, cattle, elephants)

cludes the subphyla Urochordata, Cephalochordata, and Vertebrata. Unlike the other two subphyla, vertebrates have a vertebral column which replaces the notochord. Vertebrates may have evolved from a tadpole-like invertebrate larva, similar to the present-day urochordate larva.

21–L THREE CLASSES OF FISHES

Class Agnatha: Jawless Fishes

The most primitive vertebrates were **ostracoderms.** Known only as fossils, these agnathan fishes were covered with heavy bony plates. Their modern relatives are the jawless **cyclostomes** ("round mouths"), the lampreys and hagfishes. Adult lampreys and hagfishes are long, cylindrical creatures without paired fins. The adult lamprey is a semiparasite, and hagfishes are scavengers.

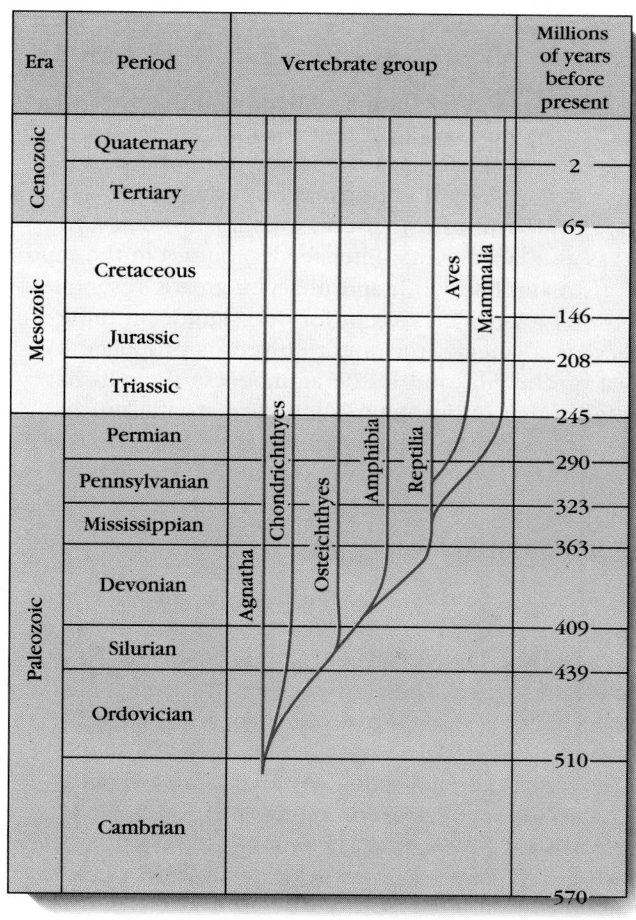

FIGURE 21–16

Fossil record and probable family tree of the classes of vertebrates. Note that birds are the class that originated most recently and that birds and mammals arose independently from reptiles.

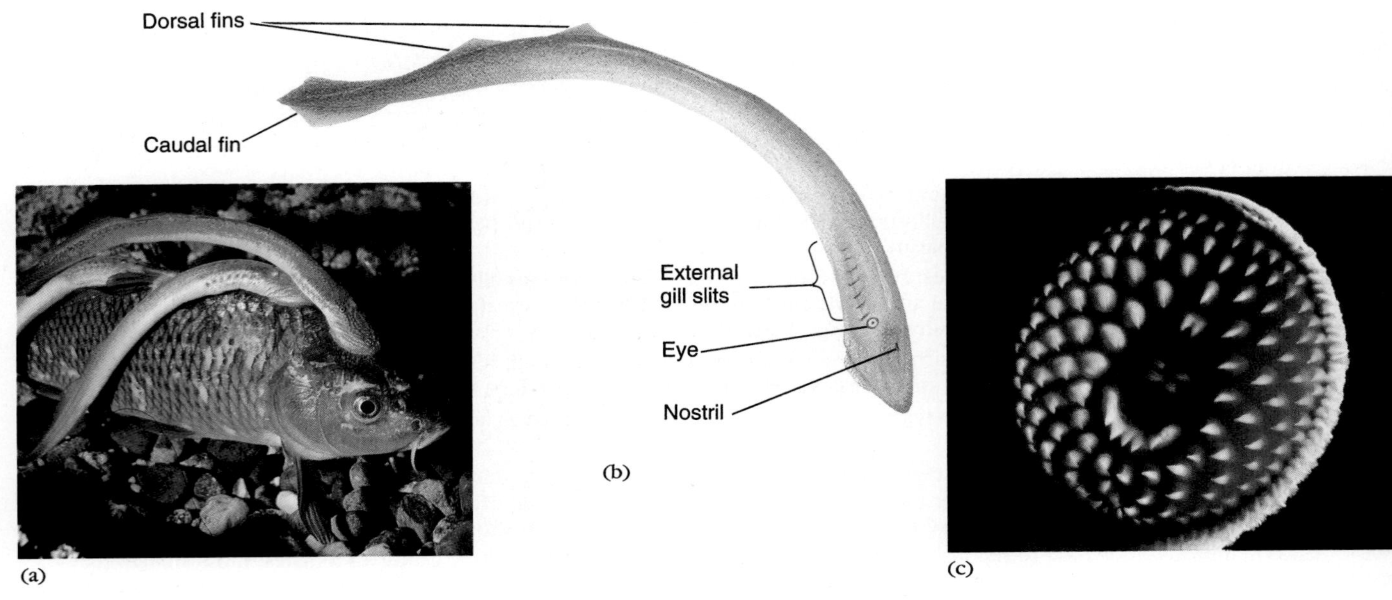

FIGURE 21-17

Lampreys. (a) Three lampreys attached to a carp by their round, sucking mouths. Their attack will weaken, but usually not kill the fish. **(b)** This diagram shows the external gill slits, eye, nostril, and unpaired fins. **(c)** The round, suction-cup mouth contains many rasping teeth. (a, Tom Stack/Tom Stack & Associates; c, Patrice Ceisel/Visuals Unlimited)

Adult lampreys have a sucking mouth and a rasping tongue covered with teeth, which are used to break the skin and suck the blood of bony fishes (Figure 21-17). They do not usually kill their prey. The larva, called an **ammocoete,** lives for up to seven years as a lancelet-like filter feeder, buried in the mud of a stream. Its mouth and pharynx greatly resemble those of a lancelet, except for one significant difference: in a lancelet the water current is propelled by cilia on the gills, but in the ammocoete the gills have muscles that pump the feeding current, at a much faster rate. Hence an ammocoete takes in food faster, and it can grow to a greater size on filtered food.

Evolution of Fishes

Agnathans gave rise to two other large and successful groups of fishes: the **Chondrichthyes,** the sharks and rays, and the **Osteichthyes,** or bony fishes, such as salmon and tuna. These two groups made two major evolutionary advances over their agnathan ancestors, increasing their efficiency as fast-moving carnivores. First, part of the gill skeleton moved forward and evolved into **jaws,** permitting the fish to bite and chew its food instead of sucking or filtering it. Second, both groups have two pairs of **lateral fins: pectoral fins** at the front and **pelvic fins** further back, as well as other, unpaired fins (Figure 21-18). Paired fins allowed these fishes to balance and maneuver in new ways. Paired fins also played an important role in evolution: they eventually gave rise to the paired forelimbs and hindlimbs of terrestrial vertebrates.

Class Chondrichthyes: Cartilaginous Fishes (Sharks and Rays)

Sharks, dogfish, rays, and skates have skeletons composed entirely of cartilage, rather than bone, and almost all are marine (Figure 20-18). Sharks swim in primitive vertebrate fashion, by sinuous waves of the body, using their segmental muscles and jointed backbones.

The sharks are notorious carnivores and scavengers. One shark killed in the Adriatic Sea had in its stomach two raincoats, part of a horse, an automobile license plate, and a piece of rope. The stomach of a small dogfish shark, such as you might dissect in the laboratory, is more likely to contain crustaceans and bony fishes. The two largest sharks (like the largest whales) are not predaceous but gentle filter feeders. The whale shark lives mainly on plankton, filtering more than a million liters of water an hour.

The flattened skates and rays, which live on the sea floor, feed mostly on invertebrates. Their pectoral fins are greatly enlarged and are used for locomotion (Figure 21-18). Some rays have poison spines on the back or tail, which they use to defend themselves. The electric ray repels intruders with an organ that can produce quite a powerful electric shock.

Class Osteichthyes: Bony Fishes

The number of species of Chondrichthyes has declined since the Permian Period, but the bony fishes are still expanding and diversifying, thanks to their versatile anatomy and physiology. Most modern bony fishes are members of one very successful group, the **teleosts.**

The diversity among teleosts is almost as amazing as that of the insects. Among them are filter feeders like herrings, parrot-fish that crunch up coral, insect-eaters such as trout, and predaceous carnivores such as barracuda and blennies. Teleosts come in all shapes, sizes, and colors. Boxfish are nearly spherical, moray eels are snake-like, and a stonefish looks like a rock.

The impressive adaptive radiation of teleosts depends largely on a few evolutionary innovations. One of the most important was the **swimbladder,** a gas-filled sac formed as an outgrowth of the pharynx. By altering the gas pressure in the bladder, a fish can alter its buoyancy so that it floats at any depth in the water without exerting its muscles. In some bony fishes, the swimbladder is used, like a lung, to breathe.

In most teleosts, the tail provides much of the push during swimming, while the paired fins provide fine control. The pelvic fins are usually farther forward and higher on the body than those of a shark (Figure 21-18). You can always tell a bony fish from a shark because the gills of a bony fish do not open separately to the exterior, as do those of a cartilaginous fish. Instead, the bony fish's gills are all covered by a common operculum. Water for respiration moves in through the mouth and out through the gills, pumped by muscles in the head and at the base of the operculum. The gills of bony fishes also have osmotic adaptations that help to control the body's water content (Section 26-B). Bony fishes occur in both fresh and salt water, and forms such as trout, salmon, and eels can travel from one to the other.

BIO-BIT

There are more species of bony fishes than all land vertebrate species combined.

Many teleosts have sharp, protective spines on their dorsal fins. In some species, these spines are connected to poison glands and can inject poisons powerful enough to kill human beings or large fishes.

Many deep-sea fishes are luminescent. Light flashes are probably used to signal the opposite sex and to startle attackers. In the deep-sea angler fishes, the luminous tip of a fin is used as a lure to attract prey. In addition, many fishes can change color; tiny muscles in the skin alter the sizes of different chromatophores (color cells).

When the surrounding water is low in oxygen, some fishes come to the surface and gulp air into their swimbladders. Other fishes have lungs or other arrangements for getting oxygen from air. It is not difficult to imagine that some air-breathing fishes, using their pectoral and pelvic fins to move about on land, were the ancestors of the first terrestrial vertebrates. In fact, some modern fishes, such as mudskippers, do leave the water and crawl about on land.

The agnathan fishes are the most primitive group of vertebrates. The only present-day examples are lampreys and hagfish. The agnathans gave rise to two other large groups of fishes: the Chondrichthyes (sharks and rays) and the Osteichthyes (bony fishes). Both of these classes differ from agnathans in possessing true jaws and paired fins. Sharks and rays have skeletons made of cartilage, a series of gill slits, a spiral valve in the intestine, and fins that are fixed in position. Bony fishes have bony skeletons and a covering over all of the gills. They lack a spiral valve, and have movable fins that assist in delicate maneuvering.

21-M THE MOVE TO LAND: TETRAPODS

Many selective pressures probably contributed to the evolution of vertebrates that could live, at least part-time, on land. In the Devonian Period, the seas teemed with carnivorous fishes, and any fish that could move itself or its eggs onto land would lower its mortality rate impressively. Plant life on land was well established, and terrestrial insects were evolving and multiplying rapidly. A vertebrate that ate plants or insects, and that could survive on land, would have had little competition for food. In addition, air contains more oxygen than water does. However, any fish that survives on land for even a short time must have adaptations to existing surrounded by air.

To support the body on land, a fish needs sturdy bones, especially in its pectoral and pelvic fins and in its backbone, along with strong muscles to move these bones. Furthermore, gills cannot be used to breathe on

Fish Diversity
Cartilaginous fish

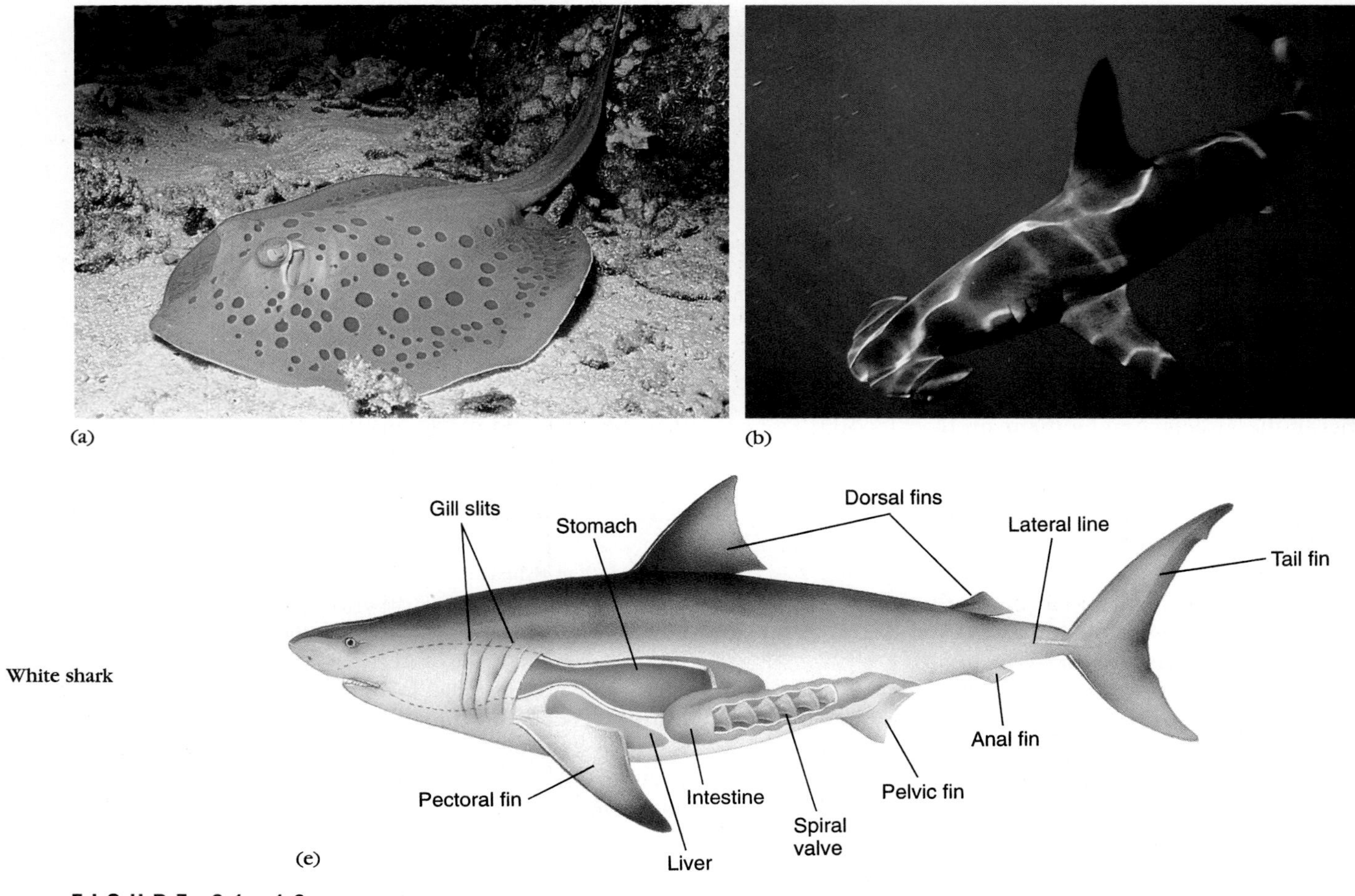

FIGURE 21–18

Cartilaginous and bony fish. (a) A reef stingray with broad front (pectoral) fins for maneuvering near the ocean floor. **(b)** A hammerhead shark. **(c)** A flame angelfish. **(d)** A school of cardinal fish. **(e)** Externally, the shark has a row of separate gill slits, while the gills of bony fish **(f)** are covered by a bony plate (an operculum) with one large opening at the rear edge. Both fish have the same types of fins, but the pectoral fins of the shark are rigid, while the pectoral fins of the bony fish can rotate to steer and propel the fish. A lateral line system helps both fish detect vibrations in the water. Internally, the bodies of both fish consists mostly of swimming muscles, with a relatively small body cavity. The skeleton of a shark consists entirely of calcified cartilage while the bony fish has both cartilage and bone in its skeleton. The shark's large, fat-filled liver and the bony fish's swimbladder both provide buoyancy. A bony fish has a long, thin intestine, whose walls provide a large surface area for absorption of digested food. The shark's intestine is shorter but its internal surface area is increased by a winding spiral valve. (a, M. Kazmers/Dembinsky Associates; b, F. Nicklin/Minden Pictures; c, A. Kerstitch; d, C. Seaborn/Odyssey)

land. The surface tension of water makes the feathery gill filaments stick together when a fish comes out of water into the air. A respiratory surface that keeps its shape in air is necessary.

The body surface must be waterproofed to reduce dehydration, but this is not enough. The respiratory surface must be kept moist because gases can cross plasma membranes only in solution. Internal respiratory surfaces—lungs and swimbladders—lose less water by evaporation than do gills.

The fossil record shows that there evolved in the Carboniferous Period (about 300 million years ago) a

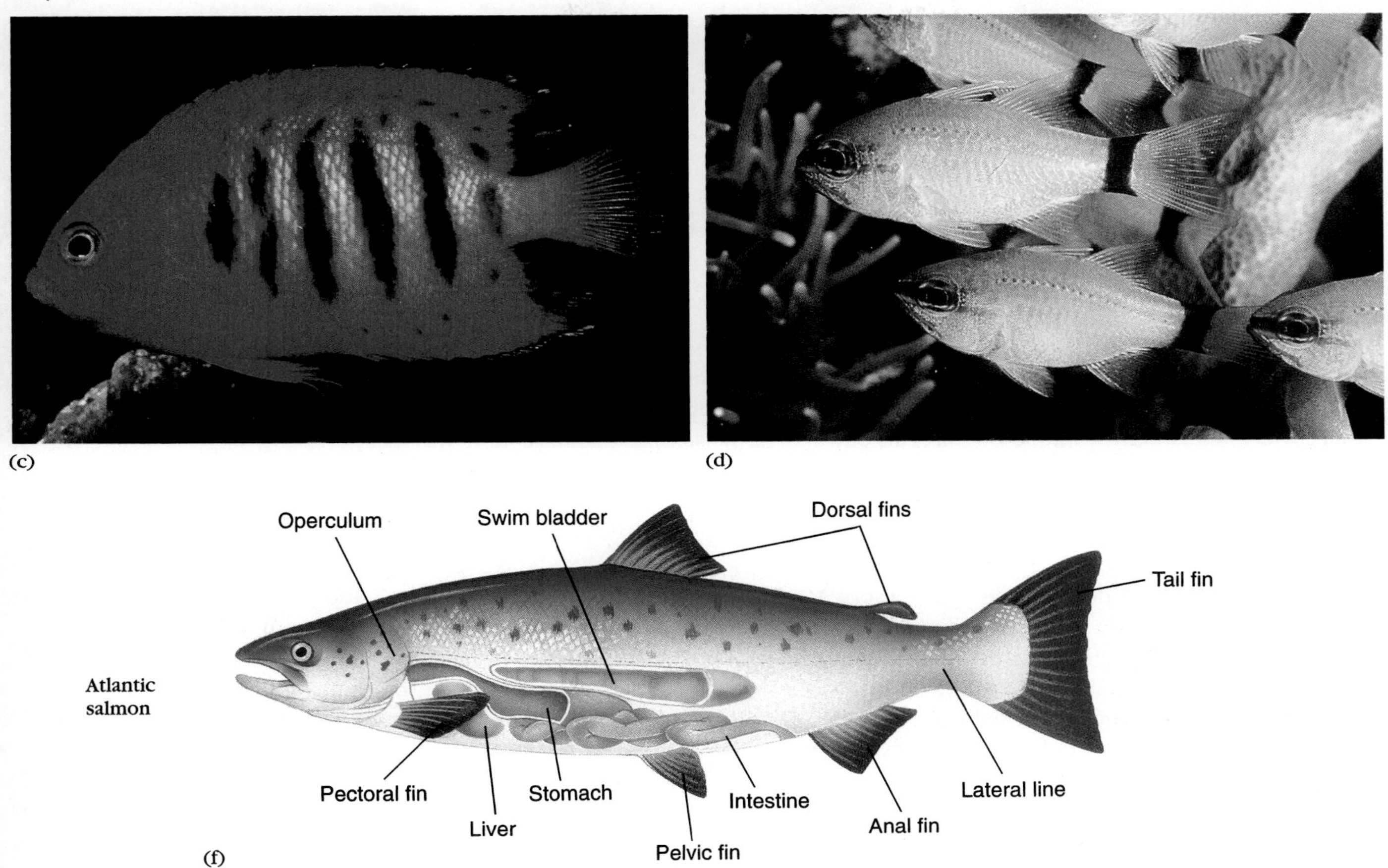

number of animals that looked much like modern lungfish. These creatures had fish-shaped bodies, short stubby legs, and no gills. They were the first land vertebrates, ancestors of modern amphibians and reptiles.

Modern land vertebrates—amphibians, reptiles, birds, and mammals—are called tetrapods ("four feet"). Only reptiles, birds, and mammals are fully adapted to life on land. Most amphibians are still largely dependent on water.

Class Amphibia: Frogs, Salamanders, Toads, Newts

The amphibians are still tied to water because their eggs dry out easily and because most still have a fully aquatic larval stage. Most amphibians must return to water to reproduce (Figure 21-19).

The two largest groups of living amphibians are the **urodeles** (newts, salamanders, hell-bender, mud puppy) and the **anurans** (frogs and toads). The urodeles are more generalized and show the transition from fish to tetrapod more clearly. Their limbs contain small bones and muscles, like those found at the base of the pectoral and pelvic fins of lungfish. In addition, they have pectoral and pelvic limb girdles (shoulder and hip girdles), which eventually evolved until they formed a strut between the backbone and limbs in all higher vertebrates. Adult anurans have very specialized skeletons, with shortened backbones, loss of the tail, limb girdles firmly attached to the backbone, and leg bones and muscles developed for jumping.

Amphibians have a soft glandular skin, which is used for gas exchange in most species, despite the fact that most amphibians also have small lungs. (The aquatic larvae have gills.) Amphibians were also the first vertebrates with true tongues. In most frogs and toads, the tongue is long and sticky and can be shot out rapidly to catch insects.

Biologists are concerned because populations of most amphibians in North America (and many else-

(a)

(b)

(c)

FIGURE 21–19

Amphibian diversity. (a) This marbled salamander looks very similar to a lizard, which is a reptile. However the salamander's skin is moist and glandular while reptiles have dry scaly skin. **(b)** A red-eyed treefrog. **(c)** A caecilian. Some of these legless and worm-like amphibians live in water while others burrow into loose soil. This one lives in the soils of Columbia. (a, b, E. Degginger; c, J. M. Renjifo/Animals, Animals)

where) have declined dramatically since about 1975, and no one knows why. Since so many species are affected, some environmental change is probably to blame. The increase in acid rain and in periods of drought, human destruction of habitat, stocking of lakes with fishes that eat tadpoles, and accumulation of toxic chemicals are all known or suspected threats to one species or another of vanishing amphibians.

Class Reptilia: Lizards, Snakes, Turtles, Crocodiles

The Mesozoic Era is sometimes known as the "Age of Reptiles" because reptiles of all shapes and sizes are the main animals found in marine, freshwater, and terrestrial fossil beds laid down during this time. With the insects, reptiles dominated animal life on land for about 200 million years and are still very much with us now. Today there are about 6,000 living species.

Reptiles are better adapted to life on land than are amphibians. Their main advantage is an egg that can be laid on land because it is protected from dehydration. Reptiles, birds, and a few mammals lay **amniotic eggs** (Figure 21–20). The developing embryo is surrounded by a membrane, the **amnion,** enclosing amniotic fluid, which protects the embryo from dehydration and from being jolted around. Two membranous sacs are attached to the embryo. The **yolk sac** contains yolk, the embryo's food. The **allantois** stores the embryo's nitrogenous waste until hatching and is found only in reptiles, birds, and mammals. Blood vessels grow out from the embryo through the membranes of the yolk sac and allantois until they come close to the surface of the egg, where they take in oxygen from the environment and release carbon dioxide. The embryo, amnion, yolk sac, and allantois are all surrounded by a membrane called the **chorion,** which controls the overall permeability of the egg. The egg is permeable to gases, but relatively impermeable to water. Around the chorion is the outer egg shell. Because the egg is laid in a leathery shell or the young develop within the mother's body, reptiles have internal fertilization, and the male has a penis (or even two).

BIO-BIT

A turtle shell is formed by the fusion of a turtle's ribs.

The first reptiles were carnivores that looked rather like small dogs. Their limbs were stronger and were tucked further under their bodies than those of amphibians, enabling them to move faster. Their jaws were more firmly attached to the skull, permitting them to subdue and eat larger prey, and their skins were waterproofed and scaly, minimizing water loss. Without a moist skin, reptiles had to breathe entirely with their lungs.

The adaptive radiation of the early reptiles is a fascinating story. There were reptiles that swam, walked, and flew—some the size of small airplanes. The dog-like ancestors of the mammals became fast-running carnivorous quadrupeds (animals walking on four legs). Many members of the group that gave rise to the birds had a tendency to bipedalism (walking on two legs) and had reduced forelimbs. Some reptiles were insect-eaters, some carnivores, and some placid herbivores of enormous size.

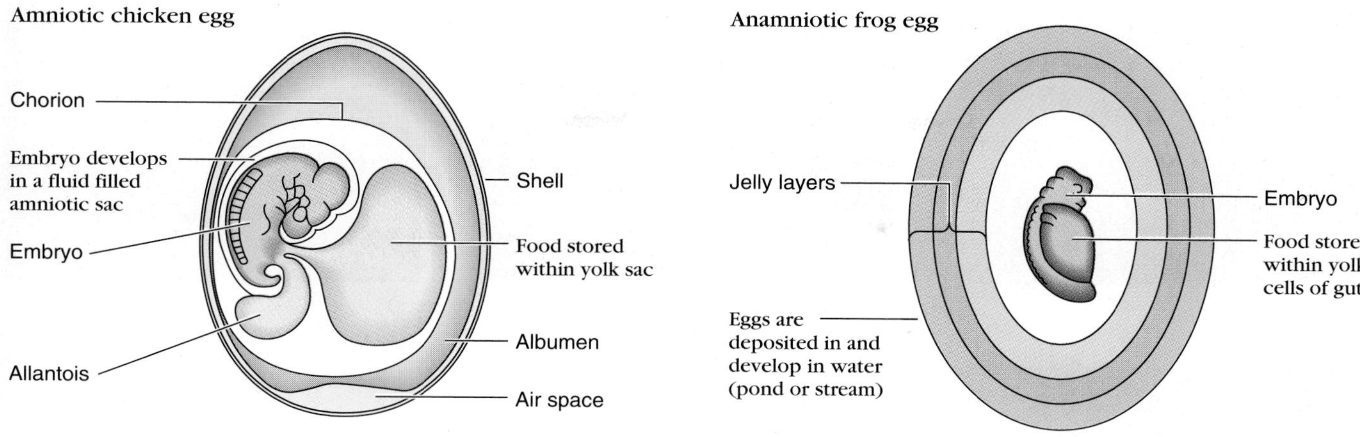
Amniotic chicken egg
Chorion
Embryo develops in a fluid filled amniotic sac
Embryo
Allantois
Shell
Food stored within yolk sac
Albumen
Air space
Anamniotic frog egg
Jelly layers
Eggs are deposited in and develop in water (pond or stream)
Embryo
Food stored within yolk cells of gut

FIGURE 21–20

Amniotic and anamniotic eggs. Embryos within amniotic eggs develop within a fluid-filled pouch, the amniotic sac. This adaptation to dry land allows reptiles, birds, and mammals to live in areas with little water. The allantois collects metabolic wastes, the yolk sac stores nutritious yolk, and the chorion assists in gas exchange. Albumen surrounding these membranes provides additional moisture and protection from physical damage. Animals with anamniotic eggs (eggs that lack an anniotic sac, for example, those of amphibians and fish, must be deposited into water, even if the adults live on land. Energy reserves are stored within yolk cells. A series of jelly layers deposited around the egg protects the embryo from predation and quick chemical changes.

There is still some mystery about the extinction of many of the reptiles. Large numbers of them, including all the dinosaurs and all the flying reptiles, disappeared from the fossil record during a short time at the end of the Cretaceous Period. Some people think these extinctions resulted from competition with early mammals. Others speculate that they were caused by climate changes, perhaps resulting from the collision of a large meteorite with the Earth. This would have sent huge dust clouds into the atmosphere, cutting off sunlight from the Earth's surface, which, in turn, would have both killed many plants and cooled the global climate. It is quite likely that more than one factor contributed to these extinctions. ■

It is incorrect to call reptiles "cold-blooded." Most maintain a body temperature considerably higher than their surroundings, but they are **ectothermic,** taking most of their heat from the environment. Many reptiles must lie in the sun before they warm up enough to be active. Birds and mammals, on the other hand, are **endothermic,** generating most of their heat by their metabolism. Thus they can be active at any time, which gives them an enormous advantage over reptiles.

Of the vast array of prehistoric reptiles, only three major groups have members alive today: lizards and snakes; turtles and tortoises; and the crocodile clan (Figure 21–21).

The lizards and crocodiles are the modern reptiles most like their prehistoric ancestors. Many people confuse lizards with salamanders. However, lizards are reptiles and have dry, scaly skin, while salamanders are amphibians with moist, glandular skin and no scales. Most snake species move by using their muscles to throw the body into curves. Scales on the ventral surface, or the curves of the body itself, provide traction. The group includes expert swimmers, burrowers, and tree climbers. The backbone is greatly elongated, and most of the vertebrae bear long, flexible ribs that hold the body in shape.

A snake's tongue flicks in and out, carrying chemicals from the air or ground to sense organs located in the roof of the mouth. Although snakes' ears have no external openings, they are well developed, responding mainly to vibrations of the ground detected through the lower jaw. Pit vipers and some boas also have heat-detecting organs on the head, which allow them to strike warm-blooded prey accurately on dark nights or in deep burrows.

Crocodiles and their kin are the closest living relatives of the extinct ruling reptiles and of their descen-

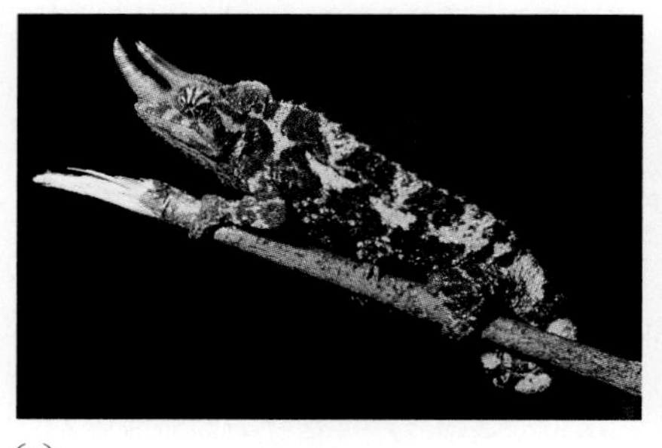
(a)

(b)

(c)

(d)

FIGURE 21-21

Reptile diversity. (a) This Jackson's Chameleon shows the dry scales and clawed toes characteristic of reptiles. **(b)** This spectacled caiman represents an ancient form of reptiles that have changed little over millions of years. **(c)** This green sea turtle and **(d)** scarlet kingsnake represent the broad range in body plans found within the reptiles. (a, A. Kerstitch; b, K. Schafer & M. Hill; c, C. Seaborn/Odyssey; d, E. Degginger)

dants, the birds. Although crocodiles are quadrupeds, they plainly belong to this line of evolution toward bipedalism because their hind limbs are longer than their forelimbs. Crocodiles, alligators, and gavials all spend much of their time in water, and have a special arrangement of their nostrils that permits them to breathe while the rest of the body is submerged (Figure 21-21). All are carnivorous.

The turtles, terrapins, and tortoises are one of the most ancient reptilian groups, specialized by the development of a protective bony or leathery shell. Most are herbivorous. Various species are adapted to life on land, in fresh water, and in the sea. Sea turtles are famous for their annual migrations to the beaches where they lay their eggs, and all species are endangered, many of them on the verge of extinction (Figure 21-21).

Birds and mammals both originated from different, early groups of reptiles.

Class Aves: The Birds

Birds can be simply defined as the only organisms with feathers. In addition, all birds are oviparous, meaning the females lay eggs. (In contrast, many female fishes and reptiles retain their eggs in their bodies until the embryo is developed enough to survive on its own.)

The best-known fossil bird *Archaeopteryx*, lived 150 million years ago, in the Jurassic Period. *Archaeopteryx* was a bird because it had feathers, but in other ways it looked much more like one of the small, bipedal dinosaurs from which the birds presumably originated. *Archaeopteryx* had a long tail, with separate vertebrae (modern birds have a greatly reduced tail skeleton), teeth (birds have a beak), and small wings with claws on the ends of the toes (the forelimb skeleton had not completed the change from legs into wings) (Figure 21-22).

How did bird flight begin? One theory holds that birds used their forelimbs at first to stabilize themselves when jumping from branch to branch, and later as parachutes (as in flying lizards and flying squirrels). Another theory holds that birds were bipedal insectivores, running along the ground catching insects and waving their arms to jump higher after escaping prey. A third theory merges the first two.

Birds owe much of their success to feathers, a remarkable evolutionary innovation. For their weight, feathers are among the strongest known materials, and they also provide flexibility, excellent insulation, and an admirable covering for a flying surface (Figure 21-23). In the other flying vertebrates, the bats, the web of skin that forms the wing stretches from forelimb to hindlimb. Birds use only the forelimbs for flight. The legs are free to be used for running or swimming, and nearly all birds have two different types of locomotion (such as flying and swimming, running and flying). The colorful feathers of many birds are used in courtship displays and for other social signals.

The strongly social nature of many birds and their complex behavior patterns result from two main factors. First, a bird's brain is quite large and complex, as it must be to control the intricate muscular movements of flying. Second, birds are endothermic, and they use a lot of energy to generate heat for themselves and for their eggs, which must also be kept warm. Therefore birds must feed more often than most reptiles, and they must take time off from babysitting to do so. As a result, parent birds usually collaborate in building nests, incubating eggs, and feeding the young.

Bird anatomy is conservative (Figure 21-23), without the great range of structural modifications found in other vertebrate classes. This is undoubtedly because most birds fly, and flight is structurally demanding. Bird flight ranges from flapping flight, such as that of

FIGURE 21-22

***Archaeopteryx*, the best known fossil bird.** This drawing is based on several fossil skeletons that show feather imprints. *Archaeopteryx* was the size of a pigeon. The skeleton shows many characteristics including a heavy tail, teeth, and claws resembling a reptile. Yet the feather imprints and development of wings shows that it was a bird. (After a painting by Rudolf Freund, Carnegie Museum of Natural History)

(a)

(c)

(d)

(e)

(b)

FIGURE 21-23

Bird diversity. (a) Lilac-breasted Roller. **(b)** A group of wild turkeys. **(c)** A Great Grey Owl. **(d)** A Keel-billed Toucan. **(e)** King Penguins. (a, F. Polking/Dembinsky Associates; b, J. McDonald/Visuals Unlimited; c, R. Planck/Dembinsky Associates; d, e, F. Lanting/Minden Pictures)

sparrows, robins, chickadees, and so forth, to the soaring flight of hawks, vultures, and albatrosses. Soaring birds ride the thermal currents in the air, like a glider (though more efficiently), and flap their wings infrequently.

Birds that live on land usually eat seeds and fruit (parrots, fowl, grosbeaks), insects and their larvae (thrushes, swifts, woodpeckers), smaller vertebrates (owls, eagles, hawks), or carrion (vultures and crows). Many birds find most of their food in water, either by wading (sandpipers and herons) or by swimming and diving, with feet modified as paddles (gulls, pelicans, ducks, geese, and cormorants). Penguins, the most highly specialized water birds, cannot fly because their wings as well as their feet are modified as paddles. The other flightless birds are typified by ostriches, which rely on their running speed to escape predators.

Class Mammalia

Mammals originated from early reptiles some 200 million years ago, before the first birds. Early mammals were about the size of small mice, with teeth adapted to eat mainly insects. Their large eye sockets suggest that they were nocturnal (active at night). Well on the way to becoming endothermic, they could be active during the cooler night, avoiding competition with reptiles.

A burst of mammalian evolution followed the extinction of many reptile species over a span of several million years at the end of the Cretaceous Period (see Figure 18–4). Many forms became larger, and mammals came to exploit many of the resources formerly monopolized by reptiles. Two features—fast quadrupedal locomotion and new adaptations to carnivory—are the secrets of mammalian success.

Most mammals are **viviparous** ("alive-bearing")—the young develop in the mother's uterus, nourished and supplied with oxygen by her blood, which flows through vessels close to those of the embryo. Viviparity permits the mammalian mother to remain mobile while incubating embryos that must be warm to survive. All female mammals nourish their young with milk produced in mammary glands.

Mammals, like birds, are endothermic. The body is insulated by hair or fur and by a layer of fat beneath the skin. Endothermy gives a carnivore an enormous advantage, making it ready for action at all times. It also permits birds and mammals to live in extreme temperatures that other land vertebrates cannot survive: penguins and polar bears inhabit polar areas, and camels and vultures are among the few animals active at noon in the desert.

An additional reason for mammalian success is the **integument,** consisting of the skin and associated structures. The integument helps control body temperature. The hair on the skin, and the fat under it, provide insulation, and the sweat glands permit cooling. In many species, individuals communicate via chemical signals produced by glands in the skin. Other integumentary structures include claws, nails, hoofs, horns, and antlers.

Another mammalian advance was the evolution of specialized teeth (Figure 21–24). Whereas reptiles and fish have teeth all roughly the same size and more or less conical, early mammals evolved different kinds of teeth: chisel-like incisors for cutting, pointed canines for gripping and tearing, and grindstone-like molars for crushing and breaking. During vertebrate evolution, the number of separate bones that make up the lower jaw has been steadily reduced until mammals have just one bone on each side (the **mandible**). These alterations in the jaw and teeth permit a carnivore to grip a

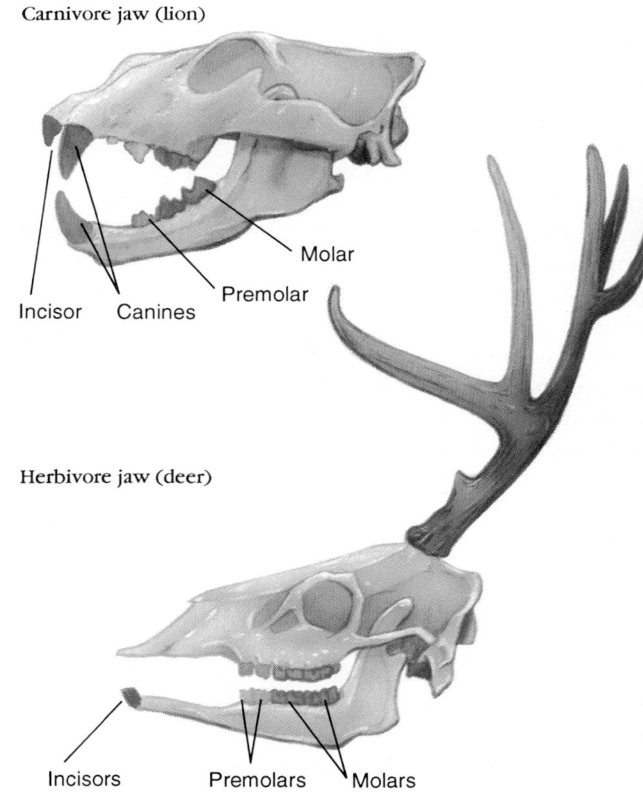

FIGURE 21–24

Teeth. A lion's teeth show specializations for its carnivorous diet: sharp incisors for grabbing and biting, large dagger-like canines for stabbing and holding prey, and sharp scissor-like premolars and molars for slicing flesh into bite-sized pieces. Deer teeth are adapted for a plant diet. The incisors have sharp tips for biting off grasses. The premolars and molars have blunt tips for grinding and shredding the tough cell walls of the plants.

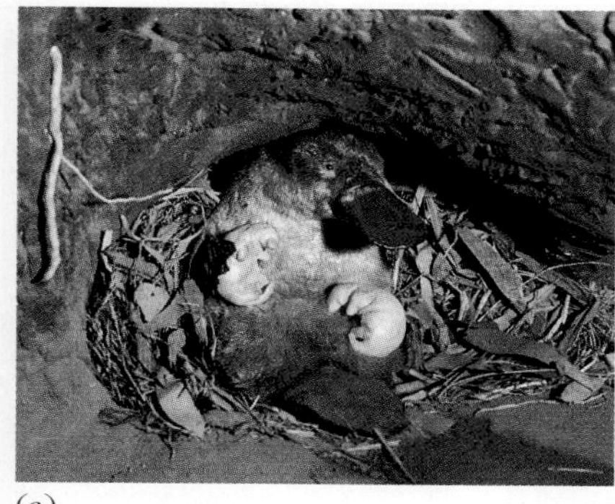

(a)

(b)

(c)

FIGURE 21-25

The three mammalian groups. The three groups of mammals all differ in their modes of reproduction. **(a)** Monotremes, like this duck-billed platypus, lay eggs. Their young hatch from them. **(b)** Marsupials, such as this kangaroo, give birth to immature young that are then raised in a pouch. **(c)** Placental mammals, including elephants and humans, give birth to young that are able to live away from the mother's body. (a, J. P. Vàrin/Photo Researchers, Inc.; b, M. James/Photo Researchers, Inc.; c, D. W. Fawcett, Visuals Unlimited.)

struggling victim more firmly, with less chance of damaging the jaw (Figure 21-24).

Although the earliest mammalian design is that of a fast, quadrupedal carnivore, many modern mammals are neither fast, quadrupedal, nor carnivorous. The success of this group led to extensive adaptive radiation, during which herbivory, bipedalism, flight, and many other adaptations evolved.

Monotremes The monotremes are one of the three groups of modern mammals. The only three species now living are the duck-billed platypus and two kinds of spiny anteaters (echidnas). Unlike other mammals, monotremes lay eggs, which are very similar to reptilian eggs. Females suckle the hatched young with milk from mammary glands in typical mammalian fashion (Figure 21-25).

Marsupials Marsupials are mammals whose young are born at an early stage of development and finish developing in a pouch, or marsupium. They live only in Australia and America, although fossil evidence shows that they once inhabited Europe. The most familiar American marsupial is the opossum, which has colonized North America from Central America. Other marsupials live in Central and South America, but the greatest variety, including kangaroos and the koala, inhabit Australia (Figure 21-25).

Placental Mammals Most modern mammals are placental. The egg has very little yolk, and the first organ to form during embryonic development is a **placenta,** formed from tissues of both mother and embryo. Two membranes found in reptile eggs, the chorion and allantois, evolved in mammals into the embryonic part of the placenta. This grows into the wall of the mother's uterus, and its blood vessels carry food and gases between the embryo and the mother's blood (Figure 21-25).

Hedgehogs, shrews, and moles are members of the order **Insectivora,** small active mammals that feed on insects and other invertebrates (Figure 21-26). Because

(a) (b) (c) (d) (e) (f)

FIGURE 21-26

Mammalian diversity. (a) Burchell's zebras, Kenya. **(b)** An elephant. **(c)** Timber wolf. **(d)** Pig-tailed macaque. **(e)** Spotted dolphins. **(f)** Mountain lion, native of western North America. (a, G. Dimijian, Photo Researchers, Inc.; b, F. Lanting/Minden Pictures; c, J. Brandenburg/Minden Pictures; d, F. Lanting/Minden Pictures; e, M. Thomas/Dembinsky Associates; f, M. Thomas, Dembinsky Associates)

their small bodies lose heat rapidly, the insectivores must eat almost constantly to provide metabolic fuel. Insectivores evolved early in mammalian history and have changed very little since the Cretaceous Period. Some early insectivores are believed to have given rise to the bats and primates.

Bats are the only mammals with true flapping flight (Figure 21-26). Their wings consist of a web of skin stretched over very long thin bones of the forelimb digits and attached also to the hind limbs. Bat flight is slower but more maneuverable than that of most birds, enabling most species of bats to live by catching insects. Most bats are nocturnal. They navigate and catch food in the dark using an echolocation system, in which they detect echoes of their own voices that have bounced off other objects. These habits reduce competition between bats and birds, most of which fly by day, using vision as the primary sense. Surprisingly, the 850 species of bats account for 22% of the species of mammals. Adaptive radiation among bats has produced species that catch fishes or frogs, blood-feeding vampires, pollen and nectar feeders, and large diurnal fruit eaters with good vision but lacking echolocation.

Primates include lemurs, monkeys, apes, and humans (Figure 21-26). They retain many features of early mammals, with additional adaptations to living in trees.

Rodents—mice, squirrels, and their relatives, with gnawing teeth—are the most widespread and generally successful modern mammals apart from humans (Figure 21-26). Most rodents have remained small and reproduce very rapidly.

BIO-BIT

The largest animal in the world, the blue whale, eats only very small and microscopic organisms (plankton).

Rabbits, hares, and **pikas** belong to another group with rodent-like features, but they lack the ever-growing teeth characteristic of rodents.

Cetaceans are mammals highly adapted to a permanent life in the sea. Toothed whales, including porpoises, sperm whales, and killer whales, feed mainly on fish and large invertebrates (Figure 21-26). Baleen whales are filter feeders. They engulf huge mouthfuls of plankton and water and use the sieve of whalebone or baleen plates lining the jaws to trap planktonic crustaceans, while their huge tongues push the water out. Surprisingly, these are the largest whales, and include the blue whale, the largest animal that has ever lived. Whales are intelligent, sociable animals, able to communicate with one another by sound. Echolocation helps them identify food and other objects in the water.

The most specialized mammalian hunters belong to the Order **Carnivora:** the cats, dogs, skunks, bears, and so on (Figure 21-26). Their behavior is complicated and involves the ability to learn much of their hunting skill.

BIO-BIT

A giraffe has exactly as many neck vertebrae as you do, seven.

The **ungulates,** mammals that walk on the tips of their toes, are the ultimate vertebrate herbivores. One group includes the elephants, and two others contain the hoofed mammals, divided according to whether they walk on an even number of toes (deer, cattle, and so forth) or an odd number (horses, rhinoceroses, and tapirs) (Figure 21–26). Because their teeth are flattened and used to crush and grind tough plant material, many ungulates are not well equipped to fight a potential predator. Most rely on running fast to escape their enemies, and they also tend to feed in herds, where every animal watches out for danger.

The even-toed ungulates owe much of their evolutionary success to a digestive system in which bacteria break down plant cellulose (Section 22–D). In addition, most are keen of sense and fleet of foot. Their adaptive radiation is impressive: they have given rise to such diverse forms as pigs, hippopotamuses, and camels. Cattle, sheep, and deer often have horns or antlers, which probably evolved as defensive weapons, although escape is usually the preferred course of action.

The first land vertebrates evolved adaptations that permitted some fishes to breathe air and support their bodies on land. Because amphibian eggs are not waterproof, most modern forms must still reproduce in water, whereas waterproof reptile eggs can be laid on land. This change in reproduction was accompanied by many other adaptations to terrestrial life, including a waterproof skin, stronger limbs, and jaws. The fossil record shows that birds and mammals had separate origins from reptilian ancestral stocks. Unlike reptiles, birds have feathers—adaptations to flight. Feathers also provide insulation that helps to regulate the body temperature of birds, which has permitted them to colonize most parts of the world. The first mammals were small, mouse-sized insect eaters, but later, a burst of mammalian diversification followed the decline of the reptiles. One early successful mammalian plan was a fast-moving, endothermic carnivore that protected its developing embryo within the female's body.

SUMMARY

1. The lower invertebrates evolved in the ocean, where most of them still live. These animals are either planktonic, or bottom-dwelling, with planktonic larvae. Some have adapted to life in fresh water, in damp terrestrial habitats, or in the moist interior of a host's body.
2. Porifera (sponges) and Cnidaria (jellyfish, sea anemones, and corals) display the least specialization and cooperation among cells. Sponges are sessile filter feeders, and cnidarians are sessile or slow-moving predators with stinging, radially arranged tentacles.
3. In most Platyhelminthes and Nematoda the main organ systems found in most animals are present, except for a circulatory system and a skeleton. These flatworms and roundworms can be free-living or parasitic, with bilateral symmetry and some cephalization.
4. The coelom, a fluid-filled body cavity, permitted the evolution of more efficient modes of digestion and circulation. The most abundant invertebrate animal phyla possessing a coelom are:
   a. Annelida, segmented worms including polychaetes, earthworms, and leeches.
   b. Mollusca, largely unsegmented animals with a muscular foot, and a mantle covering the body and usually secreting a calcareous shell: gastropods (snails and slugs), bivalves (shellfish), and cephalopods (nautiluses, squids, and octopuses). Most molluscs are marine, but others are freshwater or terrestrial.
   c. Arthropoda, the most successful animal phylum, includes animals with segmented bodies, chitinous exoskeletons, and a varied array of jointed appendages. The crustaceans are mainly marine, and the arachnids and insects are mainly adapted to terrestrial life.
   d. Echinodermata includes slow-moving marine animals with a spiny skeleton, tube feet, and (usually) radial symmetry. Echinoderm embryology suggests that they are close relatives of the chordates.
5. The phylum Chordata includes animals with a notochord, a dorsal hollow nerve cord, and pharyngeal gill slits at one point in their life cycle. Living invertebrate chordates include tunicates and the lancelet. Most chordates living today are vertebrates, with the notochord surrounded or replaced by a vertebral column of cartilage or bone.
6. Vertebrates probably evolved from animals resembling the tadpole-like larva of tunicates, using their gills for filter feeding and their segmental muscles, attached to the notochord, to throw the body into curves as they swim.
7. The agnathan fishes, the first vertebrates, gave rise to the cartilaginous and bony fishes. These two groups of fishes showed two major advances over the agnathans:
   a. jaws that could snap and bite their food, and
   b. paired fins that provided balance while the body or tail muscles still gave them the main thrust for swimming.
8. Amphibians, the first terrestrial vertebrates, evolved from bony fishes. Terrestrial adaptations included air-breathing lungs and sturdy paired fins able to support the body on land. With their thin, moist skin used as a

major gas exchange surface, and eggs that must be laid in water, amphibians have remained in moist habitats.

9. With the evolution of well-developed air-breathing lungs, waterproof integuments, and the amniotic egg, the reptiles began the adaptive radiation of land vertebrates.
10. Reptiles gave rise to both mammals and birds. With the extinction of many reptiles toward the end of the Cretaceous Period, mammals and birds became dominant and diverse.
11. Both birds and mammals are endothermic, using rapid metabolism to generate body heat, which is retained by a layer of fat under the skin and an outer layer of feathers or hair. Parental care is well developed in both groups.

SELF-QUIZ

1. An animal that must move a great deal will experience selective pressures favoring (bilateral, radial) symmetry.
2. The chief function of the larval stages of marine invertebrates is ___ .
3. The evolutionary importance of a coelom is that it:
   a. permitted animals to have a circulatory system and other internal organs that move
   b. permitted animals to move onto land with an internal storage place for extra body fluid
   c. provided the possibility of evolving a hard, protective exoskeleton
   d. allowed organisms to have excretory systems
   e. paved the way for evolution of locomotory appendages
4. Which of the following is *not* an advantage of living in the sea, compared to freshwater or on land?
   a. more stable temperature in any one area
   b. salt concentration similar to that in cells
   c. less risk of dehydration
   d. greater access to sunlight

Matching: Match the one group on the left to its set of characteristics on the right. Each set of characteristics matches only one group on the left.

___ 5. Urochordates	a. jawless fishes
___ 6. Chondrichthyes	b. fishes with a swimbladder and scales
___ 7. Agnathans	c. no vertebrae, adult lacks notochord
___ 8. Amphibia	d. fishes with cartilaginous skeleton and jaws
___ 9. Osteichthyes	e. no vertebrae, adult has all chordate features
___ 10. Mammalia	f. no scales, glandular skin, jaws
___ 11. Reptilia	g. has one pair of jaw bones
___ 12. Aves	h. feathers
___ 13. Cephalochordates	i. ectothermic, has scales, females produce amniotic eggs

14. Which one of the following animals is radially symmetric?
    a. bird
    b. caterpillar
    c. jellyfish
    d. earthworm
    e. snail
15. Which of the following embryonic layers gives rise to the gut lining?
    a. mesoderm
    b. endoderm
    c. ectoderm
16. Filter feeding is *not* found in:
    a. adult sea squirts (tunicates)
    b. adult lancelets
    c. adult agnathans
    d. larval agnathans
    e. sponges
17. You would be most likely to find an adult tunicate:
    a. in a mountain stream
    b. in a large river such as the Mississippi
    c. preying on clams
    d. in a seacoast town, attached to the piling of a dock
18. Which of the following chordate characteristics contributes *least* to its efficiency of locomotion?
    a. myotomes
    b. pharyngeal gill slits
    c. notochord
    d. post-anal tail
    e. streamlined body shape
19. Which of the following is *not* a vertebrate?
    a. lancelet
    b. lamprey
    c. shark
    d. kangaroo
    e. duck-billed platypus
20. What is the most characteristic feature of Aves, found in no other class of living vertebrates?
21. List three problems of terrestrial life that the previously aquatic vertebrates had to overcome before they could invade the land.
22. The first major vertebrate class to be totally independent of bodies of water during reproduction was the class ___.
23. List at least two differences in body structure and at least two differences in reproduction between reptiles and amphibians.

Matching: Match the invertebrate phylum on the right to its characteristic on the left. Items on the right will only be used once.

___ **24.**	Has the greatest number of species	a. Nematoda
___ **25.**	Uses tube feet for locomotion	b. Cnidaria
___ **26.**	Many parasitic forms, one causes trichinosis	c. Porifera
___ **27.**	Segmented worms with a complete digestive tract	d. Arthropoda
___ **28.**	Body is mostly a head-foot	e. Annelida
___ **29.**	Have no true tissues	f. Mollusca
___ **30.**	Use nematocysts in predation	g. Platyhelminthes
___ **31.**	Unsegmented worms with an incomplete digestive tract	h. Echinodermata

THINKING CRITICALLY

1. Why is filter feeding such a common way of life among invertebrates?
2. Many invertebrates reproduce by parthenogenesis, budding, or other asexual means. Why is this so common?
3. How does the formation of the endoderm, mesoderm, and ectoderm layers in an embryo establish the basic body plan?
4. Groups that evolved later are often referred to as advanced. Does this mean that they are better animals?
5. Cephalization is pronounced in the vertebrates but not in the tunicates, amphioxus, or echinoderms. What differences in selective pressures may have caused this difference in degree of cephalization?
6. Many mammals have adapted to a life permanently at sea. Why haven't they gone back to using gills for respiration?
7. What are the advantages of social and parental behavior that have made it profitable for organisms to spend some of their energy in these activities? What are the drawbacks?
8. Why do so few birds live as grazers on leaves and grass?

SELECTED KEY TERMS

amniotic egg, *p. 458*
Annelida, *p. 442*
Arachnida, *p. 445*
Arthropoda, *p. 445*
bilateral symmetry, *p. 434*
cephalization, *p. 434*
Chordata, *p. 449*
coelom, *p. 435*
Crustacea, *p. 446*
Echinodermata, *p. 449*
ectoderm, *p. 435*
ectothermic, *p. 459*
endoderm, *p. 435*
endothermic, *p. 459*
exoskeleton, *p. 445*
gamete, *p. 431*
gastrovascular cavity, *p. 436*
gonads, *p. 431*
Insecta, *p. 447*
invertebrate, *p. 431*
mantle, *p. 443*
marsupial, *p. 463*
medusa, *p. 436*
mesoderm, *p. 435*
Mollusca, *p. 442*
monotreme, *p. 463*
motile, *p. 431*
nematocyst, *p. 436*
Nematoda, *p. 439*
notochord, *p. 450*
pharyngeal gill slit, *p. 450*
placenta, *p. 463*
plankton, *p. 433*
Platyhelminthes, *p. 438*
polyp, *p. 436*
Porifera, *p. 436*
radial symmetry, *p. 434*
sessile, *p. 434*
tracheae, *p. 447*
vertebrate, *p. 431*
viviparous, *p. 462*

SUGGESTED READINGS

Brusca, R. C., and G. J. Brusca. *Invertebrates.* Sunderland, Mass.: Sinauer Associates, Inc., 1990. One of the best technical accounts of invertebrates currently available.

Buchsbaum, R. *Animals Without Backbones.* 2d ed. Chicago: University of Chicago Press, 1976. A classic elementary textbook on invertebrates.

Ehrlich, P. R., D. S. Dobkin, and D. Wheye. *The Birder's Handbook: A Field Guide to the Natural History of North American Birds.* New York: Simon and Schuster, 1988. A field guide embellished with essays on all aspects of bird biology and conservation.

Evans, H. E. *Life on a Little-Known Planet.* New York: E. P. Dutton, 1978. An entertaining and enlightening account of our insect neighbors.

Glenn, W. "What killed the dinosaurs." *American Scientist,* July 1990. Examines the various theories that are an attempt to explain the sudden extinction of the dinosaurs.

Goreau, T. F., N. I. Goreau, and T. J. Goreau. "Corals and coral reefs." *Scientific American,* August 1979.

McFarland, W. N., F. H. Pough, T. J. Cade, and J. B. Heiser. *Vertebrate Life,* 3d ed. New York: Macmillan Publishing, 1989. A readable general text.

Moore, J. "Parasites that change the behavior of their host." *Scientific American,* May 1984.

O'Toole, C., ed. *The Encyclopedia of Insects.* New York: Facts on File Publications, 1986. Includes all terrestrial arthropods.

Weiss, R. "Incrimination by insect." *Science News* 134:90–91 (August 6, 1988). How insects help police analyze evidence of crimes.

Wellnhofer, P. "*Archaeopteryx.*" *Scientific American,* May 1990. What the few known fossils of this organism tell us about the evolution of bird flight.

The Educated Citizen

Can We Wipe Out Disease?

This article by Jerold M. Lowenstein was excerpted with permission from the November 1992 issue of Discover *magazine.*

Are health and sickness, like life and death, inevitable antitheses that will always circumscribe our fate as humans? Or will a diseaseless society one day be within our reach on Earth?

Certainly the prospects of finding cures for most of our ills have never looked better. The history of medicine until fairly recently was a deplorable tale of ignorance, hocus-pocus, and guesswork. In the 1800s physicians were still starving, purging, and bleeding their patients to cure nearly any disease, much as physicians had done in ancient Greece. (The Greeks believed that such treatments evacuated the excessive "humors," or body fluids, that were upsetting the healthy equilibrium of the body.) Patients were still operated on without anesthesia. Many, if they survived that ordeal, died of sepsis because surgeons, operating in ignorance of germs, plunged into their patients with unsterilized instruments and filthy hands. With the exception of digitalis for heart failure and quinine for malaria, there were almost no effective drugs.

Looked at from this perspective, the high-tech medicine with which we're rocketing into the twenty-first century seems astounding. For the first time we are beginning to understand the diseases that come from the genes inside our very own cells—inherited disorders and genetic anomalies that make up a major share of the still unconquered scourges of humankind. In fact, we've come so far so fast that some medical researchers predict that the next couple of decades will be a mopping-up operation for those diseases not yet overcome—cancer, AIDS, malaria, Alzheimer's, and schizophrenia. But will it be that simple? Let's consider for a moment two of those problems: AIDS, a lethal viral infection, and malaria, caused by a protozoan.

AIDS is one of the newest infectious diseases of humankind and malaria one of the oldest. AIDS has killed close to a million people since it was first observed about ten years ago. Malaria is still responsible for at least one to two million deaths every year. Why haven't we developed a vaccine against AIDS or a chemical that can eradicate the persistent and deadly parasite that causes malaria?

There's no doubt that we are smart enough to conquer many diseases. But like it or not, we are also part of the process that produces them.

This question introduces a new variable into our task of controlling disease—the evolutionary factor. Retroviruses like the human immunodeficiency virus (HIV) of AIDS evolve at about a million times the rate of most other viruses. As a result, they change their protein coat fast enough to present a moving target to the human immune system that is charged with destroying them. Meanwhile the virus attacks and destroys the immune system, leaving the body defenseless, and the victim usually dies of opportunistic infections like *Pneumocystis* pneumonia.

As for malaria-causing parasites, like many bacteria they initially seemed to succumb to our chemical warfare: for a few years drugs like chloroquine were effective in preventing and curing the disease. Then in the 1950s drug-resistant strains of parasites evolved. Whatever new drugs or drug combinations we have employed since, some varieties of malaria have found a way to sidestep their effects.

So when we think about eliminating disease, we must consider that the evolutionary process practically guarantees an endless conflict between the defenders and the would-be invaders of the human body—with the tide of battle surging first one way and then the other. Having said that, however, it's heartening to realize that we humans have never been better equipped to gain the upper hand. The more we know of the life cycles, genetics, and biochemistry of infectious agents like HIV and malaria parasites, the more likely it seems that our science will triumph over their evolutionary evasiveness and keep us one step ahead in the evolutionary rat race.

Cancer and genetic diseases, the enemies that bore from within, will be harder nuts to crack, though we are getting a grip on them too. In 1989 Michael Bishop and Harold Varmus of the University of California at San Francisco received a Nobel Prize for their discovery of cancer-causing oncogenes in humans. Oncogenes are normal genes gone wild because of mutations or dislocations in the DNA. If we can find the mistakes in their molecular structure—the keys to their ability to stimulate abnormal cell growth—then we should be able to correct them. Similarly several hundreds of gene defects are known to cause inherited illnesses, ranging from sickle-cell anemia to Huntington's disease.

But once more we encounter an evolutionary paradox, for sickle cells are actually a rather successful defense against killer malaria. These red blood cells, with their odd scimitar shape, are resistant to the malaria parasite. Ten percent of African Americans have some protective sickle cells in their blood, a result of inheriting a sickling gene from one parent. It's only when children inherit the gene from both parents that they develop sickle-cell disease, which may involve anemia, impaired growth, and painful crises when the spiky cells form clumps in their blood vessels.

In 1949 Linus Pauling traced sickle-cell disease to a tiny defect in the molecular structure of hemoglobin, the oxygen-carrying pigment in red blood cells. A single mutation in the long DNA chain that codes for hemoglobin results in a single amino acid change that causes sickling. We now know of more than 300 other abnormal hemoglobins that occur sporadically throughout the general population. Many don't transport oxygen as well as normal hemoglobin and shorten the life of red cells. But they confer no known advantage or resistance to malaria or other diseases.

What these many variations on the hemoglobin theme show once again is how busy evolution is at the molecular level. And all that change provides the raw material for potential defenses against new diseases. If, say, hemoglobin D, one of the variant blood pigments, were to prove resistant to a lethal disease not yet inflicted on our species, future populations would doubtless show a huge increase in the hemoglobin D gene.

From the perspective of molecular evolution, then, it is not so easy to distinguish between disease, normal variation, and adaptation to changing circumstances. Ridding the gene pool of sickle hemoglobin would be beneficial to African Americans because in this country the trait has become a liability. But it would be decidedly harmful to Africans, who still have to cope with malaria.

Questions like these are still hypothetical, but they soon may not be if genetic engineering becomes a reality. If, say, sickle-cell disease is diagnosed prenatally, it might be possible to replace part of a baby's bone marrow, the organ that makes red cells, with borrowed bone marrow cells that have the normal hemoglobin gene. In this way children destined to have genetic abnormalities might be able to avoid their fate. But suppose by genetic engineering we could successfully purge the human population of all its pathological hemoglobin? Might we not then be in the same situation as food crops during the green revolution of the 1960s, when the widespread planting of "superior" monocultures of rice and other grains made them particularly vulnerable to fungi and insects? No longer faced with a diversity of resistant variants, plant diseases swept through crops worldwide.

The very concept of a static healthy human population runs contrary to the incessant mobility of evolutionary history. Based on the common genetic language of DNA and the many genes shared by all living organisms, it is virtually certain that the 10 million or more species on Earth today descended from a common one-celled ancestor that emerged from the hot broths of the early planet some 3.5 billion years ago. Obviously tremendous changes had to take place just to get from a single-celled organism to a multicellular one, let alone to the complicated creatures we've become.

The cells of humans and other animals bear witness to one of the earliest and most extraordinary of these changes. They are permanently "infected" by mitochondria, small bodies that originated as bacterial invaders of those ancient unicellular organisms. The aerobic invaders enabled their previously anaerobic hosts to capitalize on oxygen as a new energy source. Now we're completely dependent on these prehistoric trespassers. Mitochondria are the "batteries" that power our cells.

At any point in this evolutionary history, a perfectly healthy, perfectly adapted species would have been unlikely to evolve into something different. After all, why change if you're doing well? It takes changes in environment and climate, overcrowding, and disease to drive the evolutionary process, to encourage new adaptations such as the movement from life in the water to life on land, or back from land to water. Genetic abnormalities, maladaptive to the old environment but adaptive to the new, made these movements possible. This process of flux is going on all the time, though mostly at the unseen molecular level. Organisms are constantly changing their relationship to one another and to the environment, and what we call diseases are part of that process. Whether we like it or not, we humans are part of it, too.

Connecting the Concepts

The following questions will help you to connect the issues discussed in this article with the concepts you have learned in Part 3.

1. What factors make it unlikely that any one pathogen will quickly kill all humans on Earth? Conversely, why is it unlikely that humans will be able to rid the Earth of all bacteria and viruses that cause human disease?
2. What are the adaptive advantages and disadvantages of genetic diversity in any species of organism? Under what circumstances is there selection *for* genetic diversity?

PART 4

Animal Biology

Chapter 22
Animal Nutrition and Digestion

Chapter 23
Gas Exchange in Animals

Chapter 24
Internal Transport

Chapter 25
Defenses Against Disease

Chapter 26
Excretion

Chapter 27
Sexual Reproduction and Embryonic Development

Chapter 28
Nervous Systems and Sense Organs

Chapter 29
Muscles and Skeletons

Chapter 30
Animal Hormones and Chemical Regulation

Chapter 31
Behavior

Beautiful beetles. There are more arthropods than any other animal phylum and, of the arthropods, the Coleoptera, the Order of beetles, has the most species. This harlequin beetle has ectoparasitic mites. (Raymond A. Mendez/Animals Animals)

CHAPTER 22

CONCEPT GUIDE

After reading this chapter, you should be able to:

1. Define macronutrients and micronutrients and explain why each is important in the diet of animals. (Section 22–A)
2. Explain the advantages of having a digestive tract instead of just absorbing nutrients through body surfaces. (Section 22–B)
3. Describe the major events of digestion within the human stomach, small intestine, and large intestine. (Section 22–C)
4. Explain how symbiotic microorganisms help herbivores digest land plants. (Section 22–D)
5. Compare the teeth of herbivores and carnivores, and describe how the teeth of each are adapted for their particular diets. (Section 22–E)
6. Explain why birds have a gizzard and a crop. (Section 22–F)
7. Describe the major functions of the liver. (Section 22–G)
8. Describe the types of molecules that are used for immediate and long-term energy supplies in a starving animal. (Section 22–H)

Extended polyps of a sea fan. These polyps filter small animals and particles out of the water using their tentacles to trap the food. Unlike most other animals, these animals do not have an anus, and must therefore ingest food and expel waste out of the same opening. Read more about variations in animal digestive systems in Section 22–B. (D. B. Fleetham/Visuals Unlimited)

Animal Nutrition and Digestion

Animals are heterotrophs—organisms that cannot make their own food from inorganic substances, but must ingest organic molecules from the environment. Animals can be broadly divided into **herbivores,** which eat plants; **carnivores,** which eat animals; and **omnivores,** which eat both. Each mode of nutrition involves digestive systems suited to handling and digesting the type of food eaten.

In many respects, feeding is a necessary evil. Animals must feed to supply energy for all of their life processes, yet it takes energy to obtain food, and feeding may expose an animal to predators. The longer an animal spends eating, the less time it has for more advantageous activities, such as reproduction and raising its offspring. Thus, there is strong selective pressure for an animal to feed as rapidly as possible.

Digestion is the mechanical and chemical breakdown of food into small organic molecules, which can be absorbed from the gut. In this chapter we shall consider what food animals need, how animals obtain it, and the role of the digestive system and liver in processing food so that it can be used by the body.

KEY CONCEPTS

- Animals need macronutrients (proteins, carbohydrates, and fats) and micronutrients (vitamins and minerals) in their diets.
- Animal digestive systems have evolved into disassembly lines where food is broken down first mechanically, and then chemically by digestive enzymes, after which it is absorbed and distributed throughout the body.

Curiosity Questions

1? Do we need to take vitamins? (See page 475)

2? What keeps food out of our lungs when we swallow? (See page 480)

3? Why doesn't the stomach digest itself? (See page 482)

22–A NUTRIENTS

The nutrients that any animal must ingest as food may be divided, for convenience, into **macronutrients**—nutrients needed in large quantities—and **micronutrients**—nutrients required only in small amounts.

Macronutrients

The macronutrients are proteins, fats, and carbohydrates. All three can serve as energy sources because they can be broken down into molecules that are respired to produce ATP (Chapter 7). The amount of energy available from a given amount of a macronutrient is commonly measured as the number of Calories of heat it yields when fully oxidized (Table 22-1). All three classes of macronutrients also provide carbon atoms used by the body to form organic polymers. In addition, proteins supply amino acids for building the body's own proteins.

Macronutrients can be stored until the body needs them for energy. Carbohydrates are stored as the polysaccharide glycogen, in muscle and liver, and fats are stored as fat. Proteins cannot be stored. Excess protein is broken down to amino acids, which are deaminated—that is, their amino ($—NH_2$) groups are removed. The rest of the molecule is then processed like fat or carbohydrate.

Table 22–1
The Caloric Values of Macronutrients

Macronutrient	Calories per gram*
Protein	~4.0
Fat	9.0
Carbohydrates	4.0

*A calorie is the amount of heat needed to raise the temperature of 1 gram of water by 1°C. The "calories" in food are actually kilocalories, often indicated by a capitol C.

1 kilocalorie (Kcal) = 1 Calorie = 1000 calories.

Problems With Macronutrients

The most common dietary problem for people in industrialized countries is obesity: if people eat more calories than their bodies use, the excess is stored as fat. Overeating and lack of physical activity are the usual causes of obesity. Once obesity sets in, it may alter the body's metabolism in such a way as to perpetuate itself in a vicious circle of overeating and inactivity. Women have a greater tendency than men to store fat under the skin. Although this is currently unfashionable, it is biologically useful because it provides food reserves that can carry not just the woman but also her unborn child or nursing infant through times of shortage.

Probably the most common dietary problem for most animals is protein deficiency. This usually occurs, not because the diet contains too little total protein, but because it does not contain enough of the eight essential amino acids. All animals can convert some amino acids into others, but the essential **amino acids** must be supplied in the diet because an animal cannot make them from other amino acids.

One of the best-known protein deficiency diseases is kwashiorkor, often found in African populations where the diet consists primarily of cornmeal. Such a diet contains very little of the essential amino acids tryptophan and lysine. Victims of kwashiorkor (particularly growing children, who need much protein) show symptoms such as failure to grow normally, lethargy, and edema (swelling due to excessive retention of fluid) in parts of the body.

Fat deficiencies can also cause nutritional diseases. Certain fatty acids are essential in the diet. They are constituents of cell membranes and of some hormones. People who live on diets of fish, rice, or fruit, which are all very low in fat, develop an intense desire for fats and treat them as a delicacy.

We are deluged by advertising for foods such as high-fiber cereals, polyunsaturated fats, and cholesterol-free foods. Claims that eating these foods improves health are unsubstantiated, and indeed illegal in the United States (since they have not been approved by the Food and Drug Administration). The National Academy of Sciences was asked to review all the studies on diet and health and to produce recommendations for improving the American diet. The Academy reported that there was not enough evidence to draw conclusions about precise relationships between diet and health. However, there was considerable evidence that the typical Western diet contained too much fat for optimal adult health. (Children require more fat than adults.) The Academy conclusion was simple: Americans might improve their health if they ate more fresh fruits and vegetables. (See *A Journey into Human Health:* Diet and Cardiovascular Disease.)

Micronutrients

Micronutrients are the substances an organism must have in its diet in small quantities because it cannot make them for itself or because it cannot make them as fast as it needs them. Micronutrients can be divided

Table 22–2
Functions and Deficiency Symptoms of Water-Soluble and Fat-Soluble Vitamins

Water-Soluble Vitamins

Vitamin	Metabolic role	Deficiency symptoms
Thiamine (B_1)	Coenzyme in carbohydrate metabolism	Beriberi, loss of appetite, fatigue
Riboflavin (B_2)	Part of FAD, coenzyme in respiration and protein metabolism	Inflammation and breakdown of skin
Niacin (B_3)	Part of NAD and NADP, coenzymes in energy metabolism	Pellagra, fatigue
Pantothenic acid (B_5)	Part of coenzyme A, in carbohydrate and fat metabolism	Similar to other B vitamins
Pyridoxine (B_6)	Coenzyme in amino acid metabolism	Anemia, nerve problems
Cobalamin (B_{12})	Coenzyme in formation of proteins and nucleic acids	Anemia
Ascorbic acid (C)	Helps build intercellular cement for bones, cartilage, skin	Scurvy, anemia, slow wound healing
Biotin	Coenzyme in addition of carboxyl groups	Fatigue, dermatitis, muscle pain
Folic acid	Coenzyme in formation of nucleotides and hemoglobin	Anemia

Fat-Soluble Vitamins*

Vitamin	Metabolic role	Deficiency symptoms
A (retinol)	Part of visual pigments in eye	Night blindness, drying of mucous membranes
D (calciferol)	Absorption of calcium and phosphorus, building bones	Rickets in children
E (tocopherol)	Protects blood cells, vitamin A, etc., from oxidation	Lysis of red blood cells, anemia
K (menadione)	Needed in synthesis of prothrombin for blood clotting	Hemorrhage in newborns who lack gut bacteria that produce vitamin K

*Vitamins A, D, and K are toxic in large amounts.

into **vitamins,** which are organic compounds, and **minerals,** which are inorganic. Various deficiency diseases result from shortage of certain vitamins and minerals in the diet.

Vitamins

The vitamin needs of various animals differ. For instance, many other animals can make ascorbic acid from other molecules, and so they do not need to eat vitamin C as humans do. The vitamins humans need are generally divided into two categories: water-soluble and fat-soluble.

Most water-soluble vitamins (Table 22–2) are coenzymes needed in metabolism. They are easily excreted by the kidneys. The fat-soluble vitamins (Table 22–2) have various poorly understood functions. Fat-soluble vitamins that are not needed immediately may be stored in fatty tissue. Because these vitamins are not soluble in water, the body's enzymes must process them before they can be excreted by the kidneys. As a result, some of them can accumulate to toxic levels if consumed in amounts larger than the body can handle. Some diet plans recommend dangerous doses of certain fat-soluble vitamins. (In 1978, for instance, two people on high-vitamin diets died of vitamin A poisoning.) The key to avoiding such dangers is common sense and moderation. Almost anything can be poisonous if consumed in excessive amounts.

Vitamin-rich foods include fresh fruits and vegetables and whole or enriched grain products, as well as fortified milk, liver, and fish oil. ■

Minerals

We need some minerals in relatively large amounts. Sodium and potassium, for instance, are vital to the working of every nerve and muscle in the body (Table

A Journey into Human Health

Diet and Cardiovascular Disease

Cardiovascular diseases, those that affect the heart and blood vessels, cause about half the deaths in North America. These diseases include **hypertension** (high blood pressure) and atherosclerosis. In **atherosclerosis,** deposits of lipids develop on artery walls, reducing the diameter of the blood vessel and making its walls less elastic (Figure 22–A). Cardiovascular diseases cause death in many ways. They increase the risk of strokes and heart attacks (including "coronaries," blockage of the arteries that supply blood to the heart muscles).

Hypertension can be controlled by losing weight or, if that fails, by drugs. Doctors often tell hypertension patients to eat less sodium. However, studies published in 1990 show that the amount of sodium chloride (table salt) in the diet does not affect blood pressure. No matter how much salt the diet contains, the sodium content of the blood is controlled within narrow limits by hormones that regulate how much sodium the body excretes (Section 26–E).

Susceptibility to heart attacks is correlated with blood levels of cholesterol-carriers called HDLs and LDLs. Cholesterol is a small lipid that the body needs to produce new membranes. Most cholesterol is carried by low-density lipoproteins (LDLs). Some is carried by high density lipoproteins (HDLs). Studies of men of many nationalities and races have shown that the risk of heart attack is greater the higher the LDL concentration and the lower the HDL concentration in the blood. People with high levels of HDL or low levels of LDL because of their genetic makeup apparently never die of atherosclerosis.

The factors correlated with high HDL levels are those long known to be associated with a low risk of heart attack—being slim, exercising, not smoking, consuming little alcohol, and being female. (Women's HDL levels average about 10% higher than men's, probably because of the female hormone estrogen, which raises HDL levels).

You will notice that these studies talk about correlations, not about cause and effect. It is difficult to show cause and effect for these diseases. This is because cardiovascular diseases take a long time to develop, and few experiments can be performed on human subjects. Even an obvious correlation may not prove much. For instance, runners have a low incidence of cardiovascular disease. Is this because people with high HDL levels are predisposed to take up running, or does running raise HDL levels? Runners are also more likely to consume alcohol and less likely to smoke than other members of the population. Is this cause, or effect, or neither?

Since 1965, deaths from heart attacks in North America have decreased by 34%. Researchers do not know why this has happened. It would seem that some change in lifestyle must be responsible. Some have suggested that people now eat less fat, exercise more, and smoke less than in the 1960s. Most of these changes are known to reduce the risk of heart attack, but their effects are too minor to account for a 34% decrease in deaths from heart attacks.

The most effective way to reduce the risk of a heart attack is for people with a couch potato lifestyle to get more exercise. Even mild exercise, such as gardening or walking regularly, has been shown to reduce the death rate from cardiovascular disease significantly. However, it is hard to believe that people are much more active now than they were in 1965, because the average American is now more overweight.

Among doctors, lowering blood cholesterol levels is the fashionable approach to preventing heart attacks, but it is not very effective. Lowering blood cholesterol does reduce the risk of heart attack slightly, but it does not lead to longer life. In 1990, researchers reviewed studies on more than 27,000 people who had lowered their blood cholesterol by diet or drugs. These people suffered slightly fewer heart attacks than members of a control group with uncontrolled blood cholesterol levels, but they did not live any longer. (They died, on average, slightly sooner.)

It is generally true that reducing deaths from heart attacks does not increase life expectancy (the average number of years that a newborn baby can be expected to live). Life expectancy in developed countries is now about 80 years. If infant deaths, homicides, and accidents were eliminated, it would be even higher. People would die of old age—the general failure of their organ systems—at between 50 and 120 years of age. Interestingly, even if cardiovascular disease and cancer were cured, life expectancy would not increase by more than a few years.

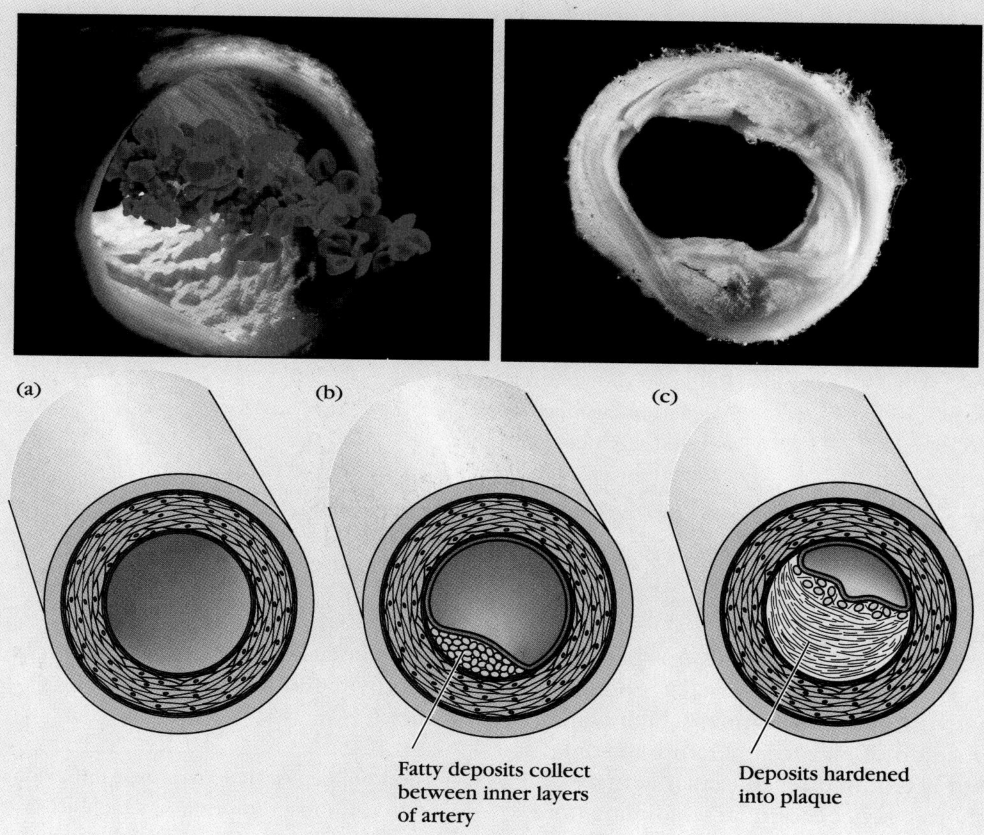

FIGURE 22-A

Atherosclerosis. (a) A photograph and drawing of a healthy blood vessel. **(b)** Fat accumulating in the walls of a blood vessel. **(c)** A photograph and drawing of an artery with hardened deposits of fat which reduce the flow of blood past this region. (a, Goivaux/Phototake NYC; b, R. Kirby/Photo Researchers)

Table 22–3
Physiological Roles of the Important Minerals and Trace Minerals

Important Minerals

Mineral	Major physiological roles
Sodium (Na)	Main extracellular positive ion; active transport, osmotic and acid-base balance, nerve and muscle activity
Potassium (K)	Major intracellular positive ion; acid-base balance, nerve and muscle activity
Calcium (Ca)	Component of bones and teeth; membrane permeability, blood clotting, nerve and muscle activity
Phosphorus (P)	Bone formation; part of DNA, RNA, ATP, etc; energy metabolism
Magnesium (Mg)	Bones and teeth; carbohydrate and protein metabolism
Chlorine (Cl)	Major extracellular negative ion; osmotic and acid-base balance; stomach acid
Sulfur (S)	Protein structure; detoxification reactions

Important Trace Minerals

Mineral	Major physiological roles
Iron (Fe)	Component of heme group in hemoglobin, cytochromes
Copper (Cu)	Needed to make hemoglobin and bone; part of cytochromes
Iodine (I)	Component of thyroid hormone
Manganese (Mn)	Needed in urea formation, protein metabolism, glycolysis, citric acid cycle
Cobalt (Co)	Component of vitamin B_{12}; required for red blood cell formation
Zinc (Zn)	Component of many enzymes; senses of smell and taste
Molybdenum (Mo)	Component of some enzymes
Fluorine (F)	Helps prevent tooth decay
Selenium (Se)	Needed in fat metabolism
Chromium (Cr)	Needed in glucose metabolism
Boron (B)	Needed for calcium metabolism and use of copper

22–3). Large quantities of these minerals (particularly sodium) are excreted in the urine every day. Sodium excretion is also a vital part of sweating, which is necessary to the regulation of body temperature in some mammals. Calcium is required for muscular activity and, with phosphorus, is needed in large amounts for bone formation.

Other minerals are known as **trace minerals** (Table 22–3). Some of them are needed in tiny amounts for the activities of enzymes in various metabolic pathways. The functions of other trace minerals are poorly understood.

Foods rich in minerals include meats, milk and cheese, grains, nuts, legumes (members of the pea, bean, and peanut family), and spinach.

Should you take megadoses of vitamin and mineral supplements? Despite the current fad for megadoses of vitamins, there is no clear evidence that these have any beneficial effect on a normally healthy individual. The best vitamin and mineral insurance is to eat a well-balanced diet, rich in natural sources of vitamins and minerals. Taking a daily multiple vitamin tablet is good vitamin insurance if you don't eat a well-balanced diet.

Food molecules that an animal must consume may be divided into macronutrients, which are needed in large quantities, and micronutrients, which are required in only small amounts. Macronutrients supply energy, carbon atoms, essential amino acids, and essential fatty acids. Micronutrients include vitamins and minerals. Vitamins often function as coenzymes. Some minerals are vital to muscle and nerve activity while others work with enzymes. A well-balanced diet appears to be the best way to obtain necessary vitamins.

22–B DIGESTIVE SYSTEMS

Digestion is the hydrolysis of food macromolecules into monomers, particularly amino acids, monosaccharides, glycerol, and fatty acids. Food can be digested either **intracellularly,** that is, in vacuoles within cells, or **extracellularly,** in a digestive cavity in the animal's body.

Sponges have mainly intracellular digestion. They are limited to food items small enough to be taken into a cell, and this often means that they must feed continuously.

Extracellular digestion becomes more specialized and efficient in more complex animals, although intracellular digestion remains an important final stage in digestion. Using their tentacles, cnidarians, such as *Hydra* and sea anemones, pull surprisingly large prey into the **gastrovascular cavity** (see Figure 21-5). Cells lining the cavity secrete digestive enzymes into it and hydrolysis begins. As small particles separate from the mass of food, they are engulfed by cells lining the cavity, and digestion is completed within food vacuoles.

Free-living flatworms have a muscular **pharynx** that allows them to suck up food, and a highly branched intestine, with a large surface area for secretion of digestive enzymes and absorption of food (Figure 22-1). Again digestion begins extracellularly and is completed intracellularly. Both cnidarians and flatworms have only one opening to the digestive cavity.

In all higher animals, the **digestive tract** is basically a tube with two openings: a mouth for ingestion and an anus for **egestion.** This tubular digestive system has a one-way flow of nutrients and lends itself to

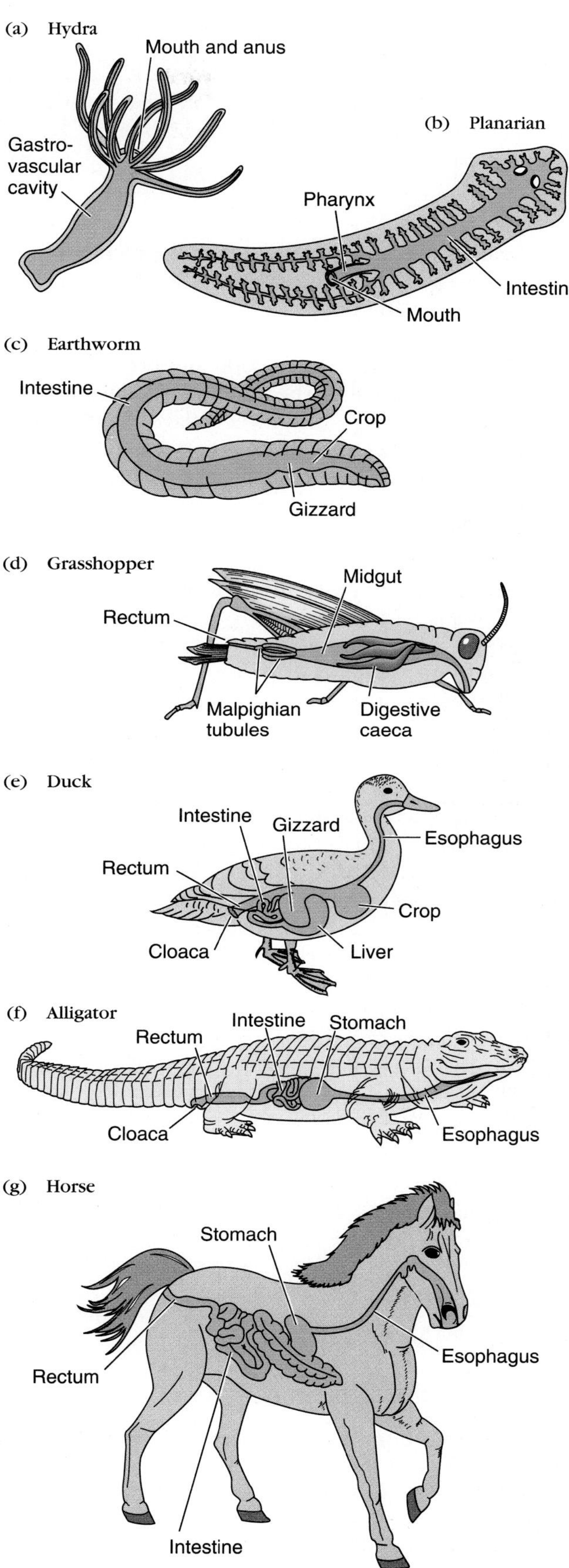

FIGURE 22-1

Animal Digestion. (a) The gastrovascular cavity in a *Hydra* has only one opening through which food is taken in and remaining wastes are expelled. **(b)** A freshwater flatworm (platyhelminth) has a mouth but no anus, and similarly must expel wastes out the pharynx, where food is brought in. The branched gut of a planarian also acts like a circulatory system, distributing food to all parts of the body. **(c)** The muscular pharynx of an earthworm pulls soil into the mouth. This soil passes down the esophagus to the crop, a storage chamber that passes small amounts of soil to the grinding surfaces in the gizzard. Organic matter and small organisms within the soil are broken down by enzymes and absorbed by the walls of the long digestive tract. **(d)** In a grasshopper, chewed plant parts are mixed with digestive enzymes. The resulting nutrients are absorbed by the walls of the gut. Malpighian tubules release metabolic wastes into the gut, where they pass out of the body along with the feces. **(e)** Like the gut of an earthworm, a duck uses a crop to store food and a gizzard for grinding. **(f)** Alligators eat large chunks of food which are stored in a large stomach until released into the intestine. **(g)** In a horse, food is temporarily stored in the stomach and slowly released down the intestine.

BIO-BIT

In the housefly, salivary secretions are released onto the food and then sucked back up with special spongy mouthparts.

specialization for different functions. The food is broken down step by step into nutrients small enough to be absorbed into the body.

Somewhere near the front end, many animals have the tools necessary to capture food and break it into smaller parts. Mammals, fishes, and sharks have teeth; turtles and birds have beaks; insects have mouthparts and molluscs, whales, and many others have filter-feeding mechanisms. Animals without teeth usually have a muscular **gizzard** containing stones or grinding edges. This compartment is found in earthworms, insects, most birds, crocodiles (Figure 22–1), and even in some dinosaurs. A thin-walled **crop** often stores food and releases it in small amounts to the gizzard, where it is ground up.

The intestine is the next part of the digestive tract. The anterior part of it may be specialized as a **stomach** to store food and secrete digestive enzymes, while subsequent portions are increasingly specialized to secrete other enzymes, absorb food molecules when digestion is completed, and absorb excess water. Intestines of many animals have outpocketings called **caeca** (singular **caecum**), blind sacs that hold food destined for a longer stay in the intestine.

Digestion occurs both intracellularly and extracellularly. The least complex animals, such as sponges, have no digestive tract and rely upon intracellular digestion. More complex organisms, including the cnidarians and free-living flatworms, have a tubular gut that allows digestive enzymes to be mixed with food. Small molecules are absorbed by the cells lining the gut, and undigested particles are expelled through its single opening. Still more complex animals, including annelids, arthropods, echinoderms, and chordates have a second opening to the gut which allows a one-way flow of foods, as well as specialization of regions along the length of the gut.

22–C HUMAN DIGESTION

The Human Digestive Tract

The **gut, alimentary canal,** and the **gastrointestinal tract** are all aliases for the digestive tract. We will consider its physical features and functions before discussing the chemical disassembly of food.

Food enters the human gut by manipulations of the mouth. Besides taking in food, the mouth begins to dismantle it, using lips, tongue, teeth, and jaw muscles. **Incisors,** chisel-like teeth in the front of the mouth, cut bite-sized pieces of food from a larger portion (Figure 22–2). The **tongue,** a slippery, mobile platform, manipulates food during chewing, pushing it back to the **molars,** the millstones that mash food into small particles. All the while, **saliva** is released into the mouth to moisten the food, stick it together into a **bolus** (ball), and lubricate the food so that it does not scratch the delicate mucous membranes of the digestive tract. Chewing and swallowing are conscious, voluntary acts, but every other process connected with digestion is automatic until unabsorbed materials reach the rectum and are defecated.

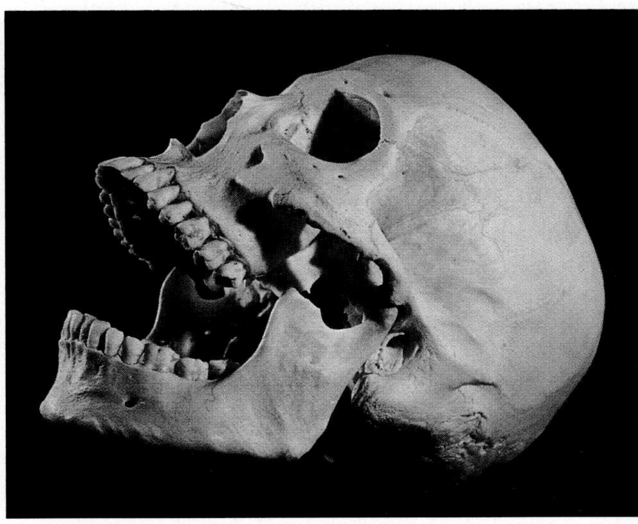

FIGURE 22–2

Human teeth are specialized for various roles in the mechanical breakdown of food. Starting from the top or bottom front and working back, there are: two pairs of chisel-shaped incisors which help bite off chunks of food, one pair of sharply pointed canines which puncture tough surfaces; and two pairs of premolars and two or three pairs of molars which collectively form broad grinding surfaces. (CNRI/Phototake, NYC)

2? Try this: swallow, and analyze the familiar movements in your mouth and throat. First, the tongue lifts the bolus to the roof of the mouth and pushes it back so it can be swallowed. The bolus passes through the **pharynx** (rear of the throat where nasal and oral passages meet) (See A Journey Through Human Digestion) and glides over the **epiglottis,** a sort of trap door that prevents food from entering the trachea (windpipe) where it could block off the air supply to the lungs. Food then drops into the **esophagus,** a long tube with thin, muscular walls. ■

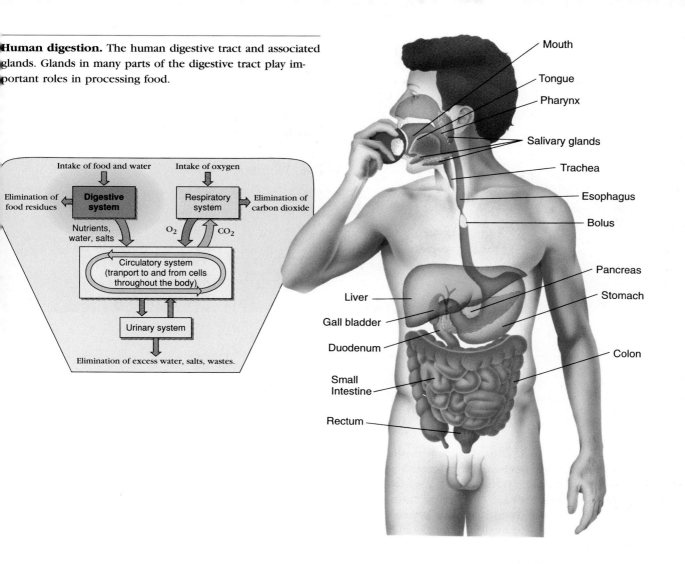

Human digestion. The human digestive tract and associated glands. Glands in many parts of the digestive tract play important roles in processing food.

Intake of food and water
Intake of oxygen
Elimination of food residues
Digestive system
Respiratory system
Elimination of carbon dioxide
Nutrients, water, salts
O_2
CO_2
Circulatory system (tranport to and from cells throughout the body)
Urinary system
Elimination of excess water, salts, wastes.

Mouth
Tongue
Pharynx
Salivary glands
Trachea
Esophagus
Bolus
Pancreas
Stomach
Liver
Gall bladder
Duodenum
Colon
Small Intestine
Rectum

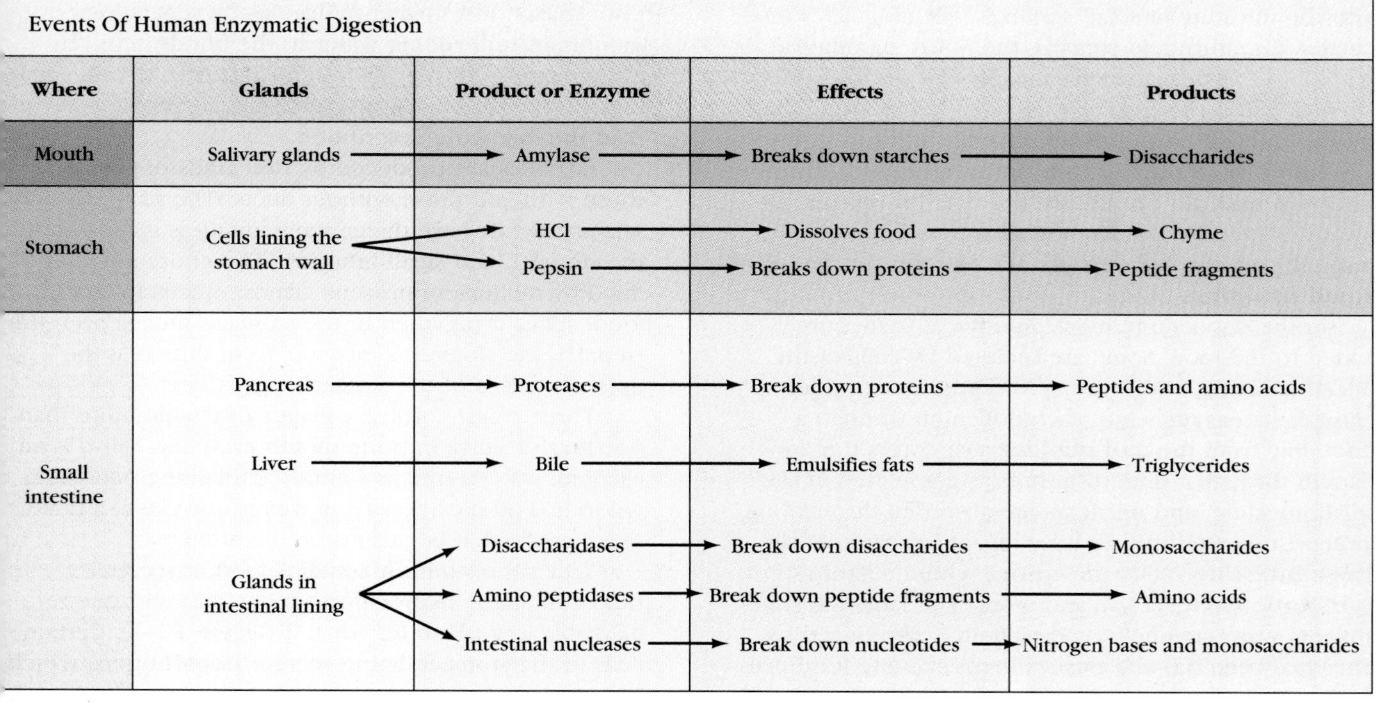

Events Of Human Enzymatic Digestion

Where	Glands	Product or Enzyme	Effects	Products
Mouth	Salivary glands →	Amylase →	Breaks down starches →	Disaccharides
Stomach	Cells lining the stomach wall →	HCl →	Dissolves food →	Chyme
	→	Pepsin →	Breaks down proteins →	Peptide fragments
Small intestine	Pancreas →	Proteases →	Break down proteins →	Peptides and amino acids
	Liver →	Bile →	Emulsifies fats →	Triglycerides
	Glands in intestinal lining →	Disaccharidases →	Break down disaccharides →	Monosaccharides
	→	Amino peptidases →	Break down peptide fragments →	Amino acids
	→	Intestinal nucleases →	Break down nucleotides →	Nitrogen bases and monosaccharides

(a)

(b)

(c)

FIGURE 22-3

Signs of peristalsis in an ostrich. The ostrich takes a mouthful of grass **(a)**, which is formed into a bolus in the mouth. The bolus is swallowed into the esophagus, where it is pushed down to the stomach by peristalsis **(b, c).**

The muscular walls of all parts of the gut contract in wave-like motion, called **peristalsis,** that starts at the end closest to the mouth (Figure 22-3). Assisted by gravity, peristalsis is the primary force that moves the bolus down the esophagus and into the stomach. Thus theoretically, you can eat or drink while standing on your head (but we don't suggest that you try it).

If the gut is a canal, the **stomach** is a lake on this canal. It serves as a storage chamber that releases food into the intestine in small servings. The stomach's muscular walls churn and squeeze the bolus, mixing it with acid and digestive enzymes that begin to digest any protein in the food. This action changes the bolus into a liquefied paste called **chyme.**

A ring of muscle, or **sphincter,** leads to the intestine. It is normally closed, except when it relaxes briefly, allowing a small amount of chyme to squirt through into the **duodenum,** the first part of the **small intestine.**

In the duodenum, more digestive enzymes are added to the food. Some are secreted by cells of the intestinal lining, and some by the **pancreas,** which empties its enzymes into the duodenum through a duct. Bile from the **gall bladder** also enters the duodenum through a duct. Digestion is completed in the small intestine, and nutrients are absorbed through its lining. Undigested food passes into the portion of the **large intestine** called the **colon,** where millions of **symbiotic bacteria** live and work. The large intestine absorbs water, minerals, and vitamin K (produced by intestinal bacteria) and pushes the remaining fecal matter into its last portion, the **rectum,** where it is held until it is voided. Defecation, or expulsion of feces, depends on the contraction of the walls of the rectum and abdomen and the relaxation of the anal sphincter, a circular muscle at the very end of the digestive tract.

Human Digestive Enzymes

Digestive enzymes break down polymers in the food into smaller and smaller molecules by separating monomers and adding water to the bonds between monomers. A Journey Through Human Digestion summarizes the major digestive enzymes. Refer to it as you read the following description.

3? Enzymes are produced by two kinds of glands: those with and those without ducts. The pancreas and salivary glands have ducts, while ductless glands line the stomach and small intestine. The entire gut is also lined by millions of mucous glands; mucus lubricates food, sticks it together. In the stomach mucus prevents **pepsin,** the stomach's enzyme, from digesting the muscular walls of the stomach. ■

Three pairs of salivary glands discharge more than one liter of saliva into the mouth each day. Saliva is an alkaline, watery, mucus solution containing potassium, chloride, and bicarbonate as well as **amylase,** a digestive enzyme that begins starch digestion.

When the stomach contains food, it secretes a fluid containing hydrochloric acid (HCl) with the remarkably low pH of less than 2 (Figure 22-4). Certain cells in the stomach lining secrete **pepsinogen,** which

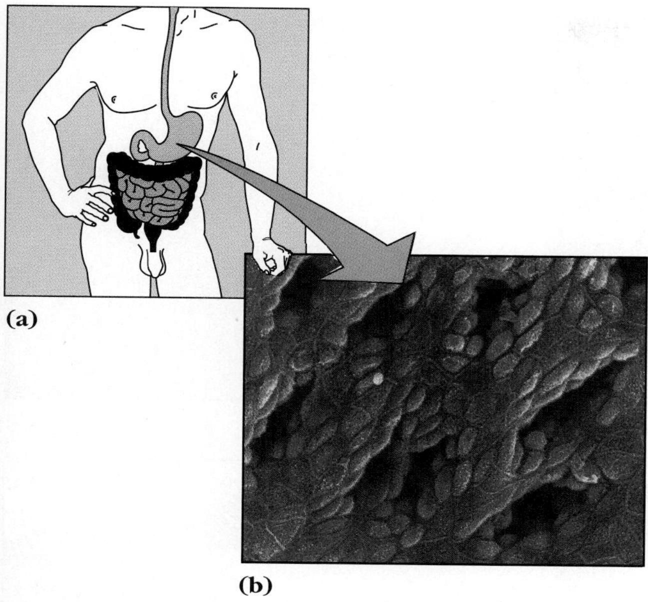
(a)
(b)

FIGURE 22–4

The interior of the human stomach. The small openings are gastric pits, leading into narrow, dead-end tunnels lined by gland cells that secrete mucus, acid, or digestive enzymes. (F. Hossler/Visuals Unlimited)

quickly becomes activated as pepsin upon contact with HCl. Pepsin will work only at a very low pH. One of the most important actions of pepsin is to digest collagen, a major constituent of fibrous tissue in meat.

Most of the digestive enzymes are produced in the pancreas and empty into the small intestine close to the stomach. Because pancreatic enzymes work best at a pH of 7 to 8, the first role of the pancreas is to change the acidic pH of the chyme, that squirts out of the stomach into the duodenum. It does this by secreting pancreatic juice, a concentrated solution of sodium bicarbonate, which neutralizes the acidic chyme entering the intestine from the stomach. Pancreatic juice also contains enzymes that digest all three major food types. Protein-digesting enzymes or proteases are made and stored in inactive forms, so they do not digest the pancreas itself. The inactive enzymes become active only after they have reached the intestine. For instance, **trypsinogen** is released and becomes activated in the intestine as **trypsin** when another enzyme cleaves off part of the inactive molecule. Trypsin itself activates other pancreatic enzymes.

BIO-BIT

Heartburn occurs when stomach acids bubble up into the esophagus and burn the unprotected cells lining this tube.

Food entering the small intestine is mixed, not only with pancreatic fluid and digestive enzymes, but also with bile. **Bile** is made in the liver, is stored in the gallbladder, and enters the small intestine through a duct. Bile has two functions in the intestine. It acts as a detergent, emulsifying fats into small globules that can be attacked by digestive enzymes. Bile salts also aid in the absorption of lipids from the intestine. Gallstones sometimes develop when large amounts of water are absorbed from bile, leaving behind solids that block the bile duct and cause great pain. Removal of the gallbladder sometimes causes difficulty with fat absorption.

Cells lining the intestine secrete other enzymes such as sucrase, maltase, and lactase, that digest small food molecules into even simpler sugars.

Absorption of Nutrients

As we have seen, digestion is a process in which food is continually acted upon mechanically and chemically. Digestion is completed once proteins, carbohydrates, and fats have been broken down into amino acids, simple sugars, fatty acids, and glycerol. It is interesting to note that the act of eating gets food into your mouth and gut, but not into your body. Hours after you swallow a mouthful of food, it has been reduced to a nutrient soup, but in order to get into the body's cells, these nutrients must be **absorbed** through a cell membrane. Technically, the contents of the gut are not within the body. Figure 22–5 summarizes the events of food absorption.

Although small, lipid-soluble molecules are absorbed by cell membranes all along the digestive tract, everything we digest is absorbed into the body from the small intestine, a 4.6- to 9-meter-long tube that is highly adapted for this purpose. The surface of its lining has finger-like extensions called **villi** (Figure 22–5). In turn, the surface of each villus is covered with a carpet of tiny **microvilli** (Figure 22–5), formed by the extensions of the plasma membranes of cells lining the lumen. The net effect greatly increases the surface area available to absorb food molecules.

In this large, membranous surface is an array of transport molecules that selectively absorb food molecules from the **lumen** (hollow space) of the intestine. Within the villus is a network of blood vessels that picks up absorbed molecules and transports them to cells throughout the body.

Lipid-soluble molecules enter cells of the intestinal lining by diffusing through the plasma membrane. Here they are converted into fats and then released from the far end of the cells as tiny fat droplets, called

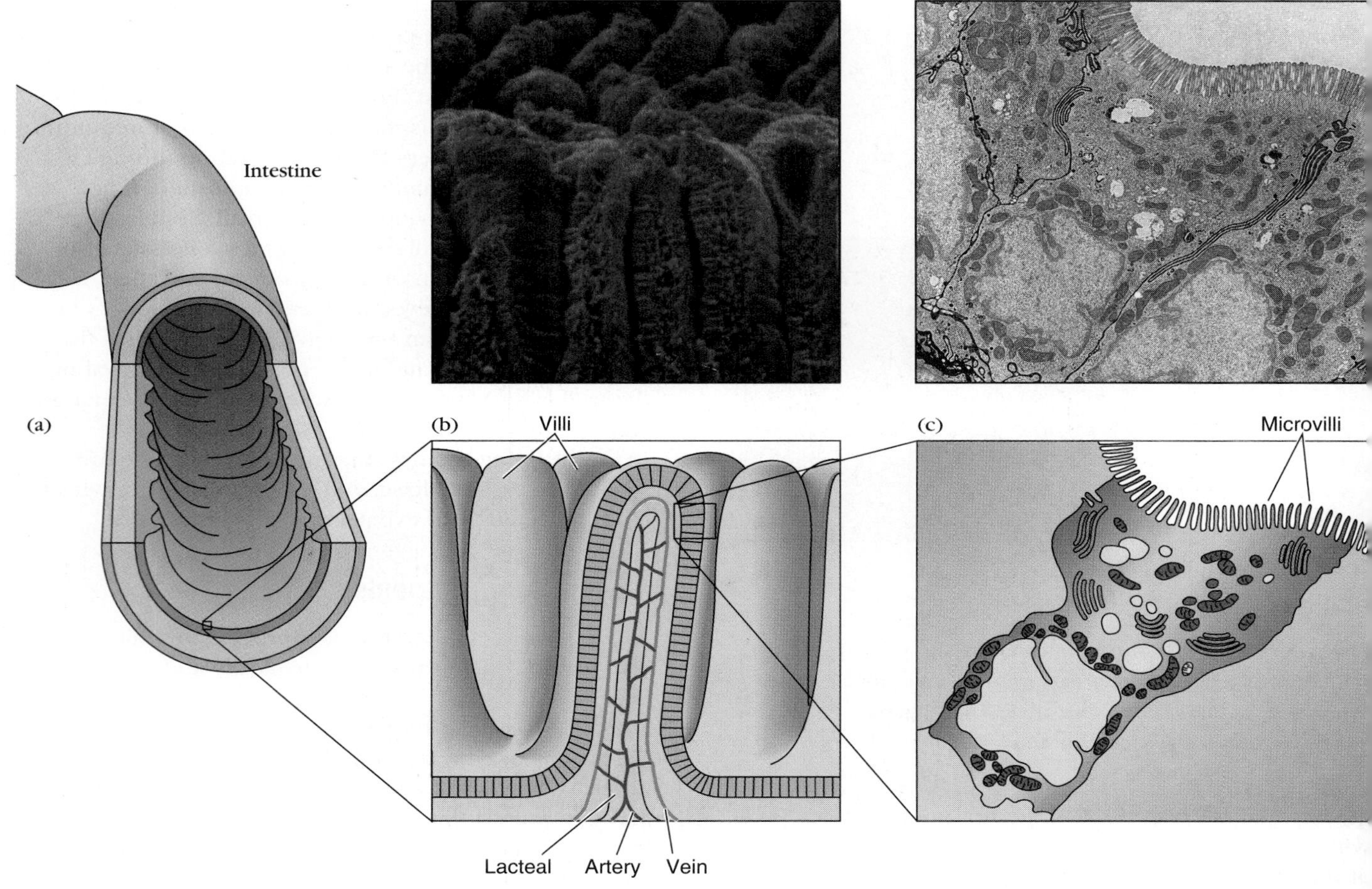

FIGURE 22-5

The large surface area of the small intestine. (a) The lining of the intestine is highly folded. **(b)** Finger-like villi line the intestine. The villi contain blood capillaries and lymphatic vessels (lacteals), both of which transport food absorbed from the intestinal lumen. **(c)** The plasma membranes of cells covering the villi are folded into microvilli, which further increase the absorptive surface areas facing the lumen. (b, Courtesy of Drs. Shih and Kessel; c, R. Rodewald, Univ. of Va./BPS)

chylomicrons. These are coated by a layer of protein which makes them water-soluble and easily transported in the blood. Glucose and amino acids are absorbed from the lumen by facilitated diffusion and active transport. Intestinal cells also absorb short peptides and digest them intracellularly before passing them on to the bloodstream.

In addition to all these food molecules, every day about 10 liters of fluid must be absorbed from the digestive tract. Of this, about 1.5 liters consists of fluid we have drunk, and about 8.5 liters consist of the fluid secreted into the lumen as digestive enzymes and mucus. Most of these 10 liters are absorbed in the small intestine.

The main substances absorbed in the large intestine are sodium, small amounts of other ions, vitamins K, B_1, B_2, and water. This absorption ensures that only about 100 milliliters of water and small amounts of inorganic ions are lost in the feces every day. The feces are about three-fourths water and one-fourth solid matter. Of the solid matter, about 30% is bacteria (normal residents of the intestine), about 30% is undigested roughage, about 20% is fat, about 15% is inorganic matter, and about 3% is protein.

The entire length of the human digestive tract is highly specialized. The mouth begins the process of digestion by breaking down food and shaping it into a bolus before sending it through the esophagus to the stomach. Peristalsis moves the food through the digestive tract. The stomach stores a meal and mixes it with an enzyme and a strong acid released from cells lining the stomach wall. The food is then expelled into

(a) (b) (c)

FIGURE 22-6

Feeding methods of herbivores. (a) The sabellid, a marine annelid worm, feeds by filtering out food from the sea water with its feathery tentacles. **(b)** An aphid uses its tubular mouthparts to suck sap from plants. **(c)** The caterpillar of a hornworm chews parts of a tomato plant. (a, Biophoto Associates; b, R. Walters/Visuals Unlimited; c, M. Tierney/Visuals Unlimited)

the small intestine, where added bicarbonate neutralizes the pH. Here the food is also mixed with enzymes produced by the pancreas and cells lining the small intestine and with salts secreted by the liver. Villi and microvilli increase the surface area of the small intestine available for absorption of food into the body. Finally, in the large intestine, water, minerals, and vitamins B_1, B_2, and K are absorbed as the fecal matter becomes concentrated. The rectum holds the fecal material until expulsion.

22–D FEEDING AND DIGESTION IN HERBIVORES

Most of the aquatic plant life on Earth consists of tiny floating plants. It is therefore not surprising that most aquatic herbivores are filter feeders, straining these minute plants out of large volumes of water (Figure 22-6). The filtering system usually traps the plants in mucus, which is then moved to the gut by cilia.

Animals that eat land plants must cope with the tough cell walls that help support plants in air. These cell walls are largely cellulose, but animals do not produce enzymes that digest cellulose. Hence, breaking cell walls and digesting cellulose are major problems for terrestrial herbivores.

Most herbivorous insects have mouthparts adapted to breaking or piercing cell walls so that they can feed on the cytoplasm. One of the reasons herbivorous insects such as locusts and grasshoppers are so destructive of crops is that they are voracious feeders, but digest only a small fraction of the food they eat. Termites use their food more efficiently, thanks to symbionts living in their guts (see Figure 19-9). These symbionts secrete cellulose-digesting enzymes and enable termites to live on wood, which is mostly tough cell walls—a food source most insects cannot use.

Herbivorous mammals have greatly enlarged molars for grinding plant cell walls (see Figure 22-7). But their most effective adaptation is a collection of symbiotic microorganisms in their guts. These include cellulose-fermenting bacteria, which can use polysaccharides as food under anaerobic conditions. Because the food must be exposed to the bacteria for fairly long periods of time, it must pass through the gut slowly. Consequently, herbivores tend to have longer intestines than do omnivores or carnivores of similar

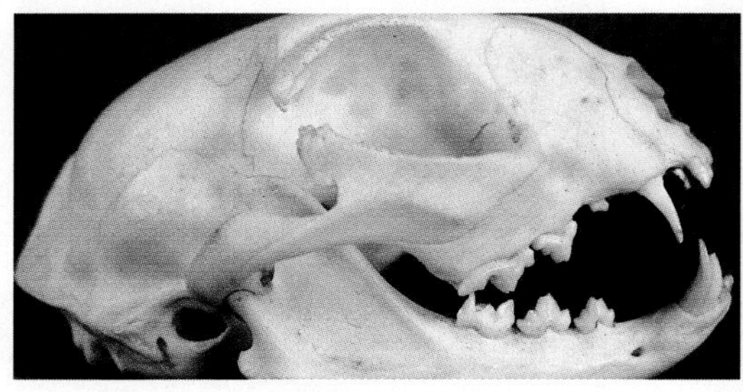

(a)

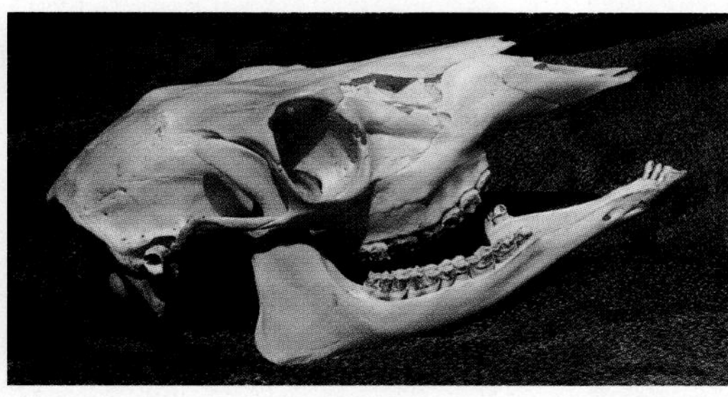

(b)

FIGURE 22–7

Tooth form and function. Skulls of a carnivorous cat and herbivorous deer, showing how their teeth are adapted to specific diets. **(a)** The carnivore's incisors and canines are pointed and blade-like, specialized for cutting and puncturing. The saw-like molars and premolars use a scissor-like action to shred animal tissues into bite-sized pieces. **(b)** In the deer, the incisors are specialized to clip vegetation. The molars and premolars are broad and flattened, and the jaws move sideways to grind leaves and grasses. (a, D. Fawcett/Visuals Unlimited; b, J. D. Cunningham/ Visuals Unlimited)

sizes. (Some bacterial fermentation of food probably occurs in the intestines of all terrestrial vertebrates, including omnivores such as pigs, rats, and humans.)

Many vertebrate herbivores house their gut microorganisms in a caecum, a sac set off to one side at the junction of small and large intestines, and used as a fermentation chamber. (In humans, the caecum is a blind sac just behind the junction of the small and large intestines. The narrow appendix was probably once a functioning caecum, which has become reduced in size as we evolved toward a less herbivorous diet.)

Digestion aided by microorganisms has reached its greatest complexity in the **ruminants.** In these mammals, the end of the esophagus and beginning of the stomach form a large sac, the **rumen,** containing microorganisms in an alkaline fluid. Most ruminants are even-toed ungulates, including sheep, cattle, and deer. Ruminant digestion has also evolved in unrelated animals, including some marsupials, colobus monkeys, sloths, and even in the hoatzin, an unusual bird.

There are many advantages to using symbionts for digestion. Some microorganisms can make amino acids from urea and ammonia. (Animals do not have enzymes that can do this.) Hence microorganisms are valuable when the diet is low in protein. In addition, symbionts produce many vitamins, especially B vitamins, which the host can use. Herbivores such as baboons, which do not have ruminant digestion, have to eat meat occasionally (grasshoppers, snakes, baby monkeys) to replenish their B vitamins. A ruminant needs few dietary vitamins except vitamin A (which can be made from a molecule common in plants) and vitamin D (which is less common).

Animals that eat land plants must cope with tough cell walls containing cellulose, which animals cannot digest. Instead, many animals have developed symbiotic relationships with microorganisms that live within their gut. In addition to digesting cellulose, many of these symbionts can make amino acids from urea and ammonia, and can produce many vitamins important to the host.

22–E ADAPTATIONS OF MAMMALIAN CARNIVORES

Herbivores eat food that is hard to digest. Carnivores, on the other hand, encounter their main nutritional problems in catching their food in the first place. Since the food of carnivores consists of other animals, its chemical composition is very similar to that of their own bodies, with little of the waste that results from eating plants with thick cell walls and a high water content.

Carnivorous mammals have teeth adapted to killing their prey and shredding it into bite-size pieces (Figure 22-7). The canine teeth are often elongated into fangs that can inflict extensive damage. The muscles and bones of the skull are powerful, enabling these animals to subdue their meals without damaging their jaws. The molars are modified so that they resemble the blades of short saws, adapted to shredding

FIGURE 22-8

Bird bills. **(a)** A flat bill and specialized tongue help ducks to filter small organisms out of the water. **(b)** The chisel-like woodpecker beak helps it chip away wood to locate hidden insects. **(c)** A pelican's bill is huge, with a soft, fleshy bag on its undersurface. The bill is used to scoop up water and fish when the bird dives. When it surfaces, the pelican points its bill down to drain out the water and then tilts it up to swallow the fish. **(d)** The thick, stout bill of a parrot is suited to crushing seeds. **(e)** Eagles use their sharp beaks to rip apart their prey. **(f)** The great blue heron uses its long pointed bill to snap up small animals in the water. **(g)** The beak of a hummingbird matches the shape of the flowers where it gathers nectar.

meat into chunks that can be swallowed. Extensive chewing is not necessary, because there are no thick cell walls to break. The strong stomach acid and powerful protein-digesting enzymes make quick work of the food, and the intestine is short compared with that of herbivores and omnivores of the same size.

Herbivores eat plants, which are abundant but often difficult to digest. In contrast, carnivores eat animals that are more difficult to capture, but more nutritious and easier to digest. Herbivore teeth are broad and flat and are used to grind plants, whereas carnivore teeth are pointed and bladelike, and are used to puncture and tear flesh.

22–F FEEDING IN BIRDS

A bird's jaws are composed of a beak (or bill) made up of bone and keratin, a protein also found in hair and fingernails. The beak is modified according to the feeding habits of the species (Figure 22-8). Modern birds have no teeth, and so birds that eat hard food grind it in a muscular gizzard, the hindmost part of the stomach. The gizzard usually contains stones that the bird picks up and swallows. This arrangement moves weight from the head (where teeth would have been) closer to the center of gravity, an adaptation that makes flying more efficient.

A special organ found particularly in seed-eating birds is the crop, a storage sac between the esophagus and the stomach. Birds use a lot of energy in flying and maintaining a high body temperature. Storing food in the crop ensures an almost continuous supply of food to the stomach, and so the bird does not have to spend all its time eating. Birds generally use their food more efficiently than do mammals. A three-week-old stork converts about 33% of the weight of its diet of fish into stork. This compares with a food-use efficiency of about 10% in a young mammal.

Birds' beaks are tools specialized according to the species' particular feeding habits. All modern birds lack teeth and must rely upon a muscular gizzard containing stones for the physical breakdown of their food. Some food is stored in the crop, an arrangement that ensures a continuous supply of food for the stomach and thus a ready supply of energy for flight and metabolism.

22–G FUNCTIONS OF THE MAMMALIAN LIVER

The liver's main function is to control the level of many substances in the blood. Digested food molecules absorbed into the bloodstream from the intestine pass directly to the liver by way of the **hepatic portal vein.** Before these molecules pass to the rest of the body, the liver may change their concentrations and even their chemical structures. The liver detoxifies some otherwise poisonous substances. For instance, it produces bile as well as enzymes that metabolize various drugs. It also stores food molecules, converts them biochemically, and releases them back into the bloodstream as they are needed. For instance, the liver removes glucose from the blood under the influence of the hormone insulin and stores it as glycogen. When the level of glucose in the blood falls, the hormone glucagon causes the liver to break down glycogen and release glucose into the blood.

Liver cells make many of the blood proteins. In addition, they convert nitrogenous wastes into urea, which can be excreted by the kidneys. Together with the kidneys, the liver is vital in regulating what the blood contains when it reaches all the other organs of the body. Because the liver is the body's major organ for making all these biochemical adjustments, severe liver damage or loss of the liver is rapidly fatal. This is one reason that liver transplants are an active medical frontier.

The liver is a major chemical regulatory center, controlling the organic components in the blood and other fluids.

22–H STORED FOOD AND ITS USES

The body's carbohydrate stores (glycogen in the liver and muscles) would supply its energy needs for only about 12 hours if they were used alone. However, a human being of normal weight can usually survive without food for about six weeks, using fat reserves for energy (Figure 22–9). A hibernating animal, with a lowered body temperature, can survive for months on the fat reserves that it built up by eating extra food during the autumn.

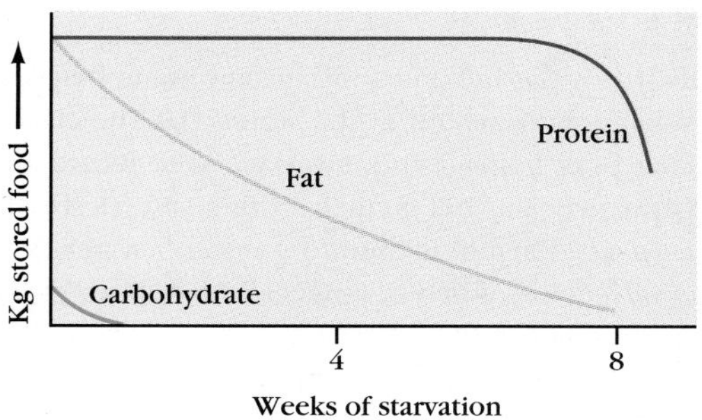

FIGURE 22–9

The fate of stored food in a starving adult human being with an initial body weight of 15% fat. A human being of average weight takes weeks to die from starvation, although death from thirst can occur within a few days.

Because a given weight of fat provides about twice as many calories as the same weight of carbohydrate or protein, energy is stored most compactly in the form of fat. Fat is stored in the fat cells of **adipose tissue,** a storage tissue found under the skin and elsewhere in the body. Fat is constantly exchanged between the bloodstream and adipose tissue: every molecule of fat in adipose tissue is replaced about every three weeks.

Animals store energy in the form of carbohydrates and fats. Carbohydrates serve as short-term energy supplies and are quickly depleted. Fats, with more calories per gram, last longer and thus are used as energy reserves for longer periods of time.

SUMMARY

1. Because animals are heterotrophs, their diets must contain all of the organic and inorganic substances they need for metabolism, growth, and energy production.
2. Animals obtain fats, carbohydrates, proteins, vitamins, and minerals in their food.
3. The function of digestion is to break down food into molecules that can be absorbed from the digestive tract into the body.
4. In the vertebrates, digestive enzymes are made in the salivary glands, pancreas, and glands in the lining of the stomach and small intestine.
5. Many animals, particularly herbivores, harbor symbiotic microorganisms that digest food in the gut.
6. Digested food is absorbed into the body by diffusion, facilitated diffusion, and active transport across the enormous surface area of the small intestine.
7. The liver plays a major role in controlling the fate of newly absorbed food molecules. It stores excess glucose as glycogen, produces many blood proteins, and converts nitrogenous and other wastes into a form that can be excreted by the kidneys.
8. Energy is stored most efficiently as fat in adipose tissue.

SELF-QUIZ

Matching: For each numbered phrase below, choose the letter of the correct class of nutrient on the right. More than one letter may be correct; choose all that apply.

___ 1.	Inorganic nutrients	a. protein
___ 2.	Macronutrient that cannot be stored in the body	b. carbohydrate
___ 3.	May be the source of energy for the body's metabolism	c. fat
___ 4.	Source of material for cell membranes	d. water-soluble vitamins
___ 5.	Coenzymes for metabolic enzymes	e. fat-soluble vitamins
___ 6.	Digested by enzyme in saliva	f. minerals
___ 7.	Absorbed in large intestine	

8. The main advantage of having a digestive tract with a mouth and anus is:
   a. it permits different parts of the gut to become specialized to perform different parts of the digestive process in turn
   b. it permits an animal without teeth to have a means of grinding its food
   c. it permits animals to eat a great deal at once and digest it while doing something else
   d. it permits animals to eat larger organisms as food
   e. it permits animals to eat food in larger chunks
9. In humans, digestion of food is completed in the:
   a. small intestine
   b. mouth
   c. large intestine
   d. stomach
   e. rectum
10. In humans, protein digestion is carried out by enzymes secreted by the:
    a. stomach, pancreas, and salivary glands
    b. liver, salivary glands, pancreas, and small intestine
    c. salivary glands, stomach, pancreas, and small intestine
    d. liver, stomach, pancreas, and small intestine
    e. stomach, small intestine, and pancreas
11. A portion of the stomach that has evolved extremely thickened muscular walls and is quite efficient at grinding hard food is called a(n):
    a. rumen
    b. gizzard
    c. crop
    d. pancreas
    e. caecum
12. Which of the following is probably *not* an action of symbiotic microorganisms of the gut?
    a. use of the host's food for its own nutrition
    b. extracellular digestion
    c. photosynthesis
    d. breakdown of substrates that the host cannot digest
    e. manufacture of vitamins needed by the host animal
13. Which of the following is *not* a function of the mammalian liver?
    a. secretion of digestive enzymes for export to the gut
    b. regulation of blood glucose and amino acid content
    c. production of the nitrogenous waste urea
    d. production of plasma proteins for the blood
    e. detoxification of poisonous substances

THINKING CRITICALLY

1. Herbivores can seldom survive by eating only one species of plant (for example, corn is low in the amino acids tryptophan and lysine; many plants contain too little sodium). This is probably no evolutionary accident. What's in it for the plant?
2. Why does it take longer to become hungry after a protein-rich meal than after a meal that is mostly carbohydrate?
3. Some kinds of stress can upset an animal's normal nutrient balance. For example, infection increases the rate at which vitamin C is used. How might the organism compensate for this disturbance?

SELECTED KEY TERMS

absorption, *p. 483*
amino acid, *p. 474*
bile, *p. 483*
bolus, *p. 480*
caecum, *p. 480*
carnivore, *p. 473*
chylomicron, *p. 484*
crop, *p. 480*
extracellular digestion, *p. 479*
gastrovascular cavity, *p. 479*
gizzard, *p. 480*
herbivore, *p. 473*
intracellular digestion, *p. 479*
macronutrient, *p. 474*
micronutrient, *p. 474*
microvilli, *p. 483*
omnivore, *p. 473*
peristalsis, *p. 482*
villi, *p. 483*

SUGGESTED READINGS

Brown, M. S., and J. L. Goldstein. "How LDL receptors influence cholesterol and atherosclerosis." *Scientific American,* November 1984.

Cohen, L. A. "Diet and cancer." *Scientific American,* November 1987. Discussion of the National Research Council recommendation that a U.S. diet containing less fat and refined sugar and more fiber would lower the risk of cancer. Discusses what we can learn about our dietary needs from the diets of our Stone Age ancestors.

Livingston, E. H., and P. H. Guth. "Peptic ulcer disease." *American Scientist,* November–December 1992. How the stomach wall is normally protected and what goes wrong when these defenses are breached.

Sanderson, S. L., and R. Wassersug. "Suspension-feeding vertebrates." *Scientific American,* October 1990.

The Surgeon General's Report on Nutrition and Health, 1988. Washington, DC: Science News Books, 1988. A summary of recent research on the relationship between diet and health.

Uvnas-Moberg, K. "The gastrointestinal tract in growth and reproduction." *Scientific American,* July 1989. A treatment of the digestive tract as the largest hormone-secreting organ in the body. Discusses the role it plays in adjusting digestion and diet to pregnancy and fetal and infant growth.

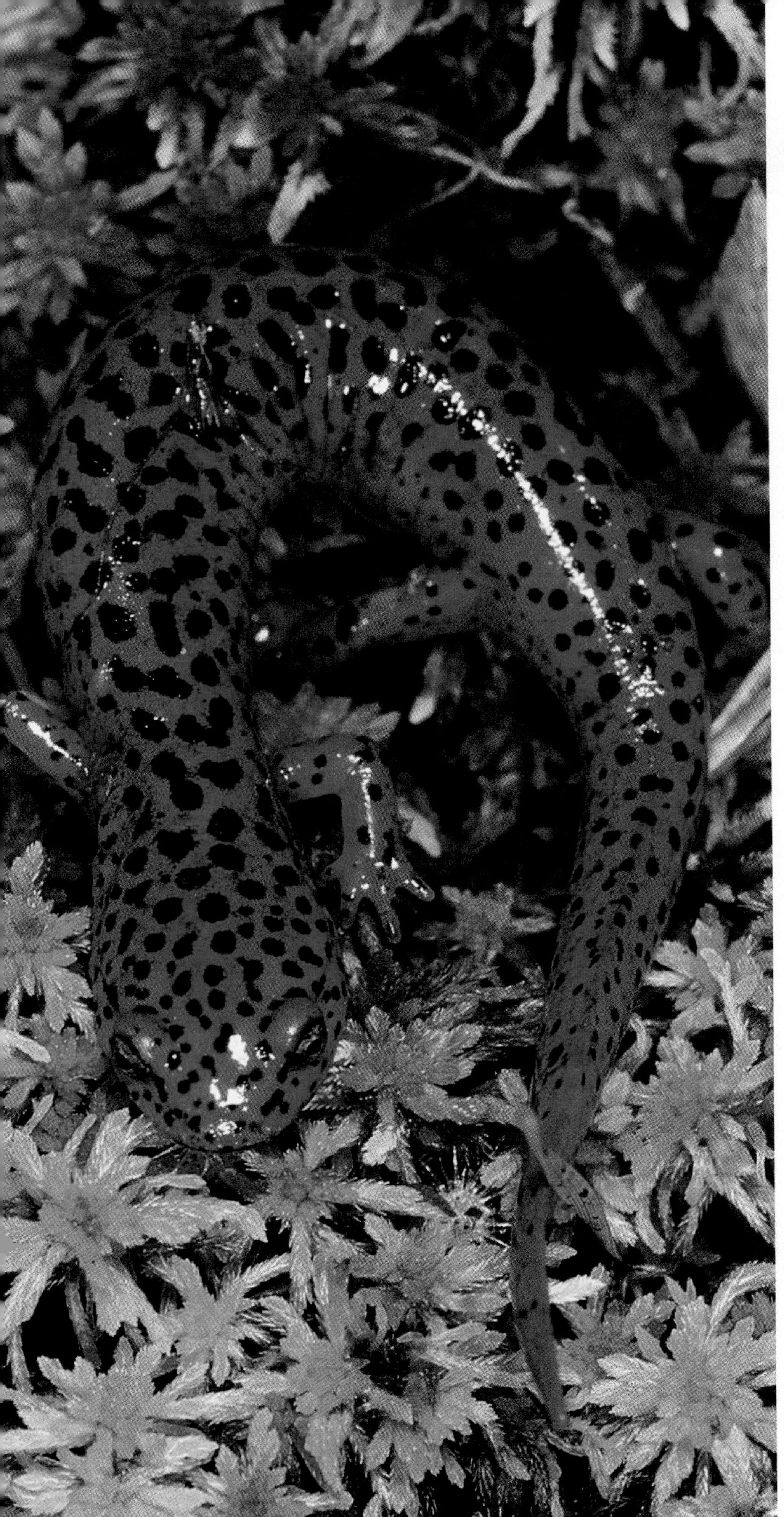

CHAPTER 23

CONCEPT GUIDE

After reading this chapter, you should be able to:

1. Describe the steps involved in moving oxygen from either air or water into gills or lungs. (Section 23-A)
2. Discuss the advantages and disadvantages of using air instead of water as a respiratory medium. (Section 23-B)
3. Describe the four main types of respiratory surfaces used by animals. (Section 23-C)
4. Explain how oxygen is carried within blood. Describe the factors that influence how quickly oxygen is released from blood. (Section 23-D)
5. Explain what happens to the carbon dioxide produced during respiration. (Section 23-E)
6. Explain how the rate of breathing is controlled in humans. (Section 23-F)

This adult salamander lives under logs in moist soil yet does not have lungs or gills. Instead, it carries out all of its gas exchange only through its skin. Read more about the different ways that animals exchange gases in Section 23-C. (J. Burnley/Photo Researchers, Inc.)

Gas Exchange In Animals

From a climber struggling to the top of Mount Everest with an oxygen cylinder, to a shark gliding through the depths of the ocean, every animal is continually exchanging gases with its environment. Why is this constant interchange necessary? As we saw in Chapter 7, living cells obtain energy to drive their activities from the oxidation of food molecules, usually by cellular respiration. Respiration requires oxygen, which must be obtained from the environment, and produces carbon dioxide, which the body must expel. Small animals can obtain oxygen and give off carbon dioxide directly through plasma membranes on the outer surfaces of their bodies. These gases do not have far to go to reach any cell in the body. The evolution of larger animals has been possible, however, only because they have evolved specialized respiratory and circulatory systems. The respiratory system provides a special, large surface area to take up oxygen from the environment and eliminate carbon dioxide from the body. The circulatory system distributes oxygen to the body's cells and collects their carbon dioxide for the return trip to the respiratory surface. This chapter looks at the behavior of gases and describes some of the arrangements for gas exchange found in animals.

KEY CONCEPTS

- Gas exchange between cells and their environment occurs by diffusion. In large or active animals, gas exchange is speeded up by ventilation of a respiratory surface, facilitated diffusion, and transport by respiratory pigments in a circulatory system.
- The main respiratory organs in animals are the body surface, gills, tracheae, and lungs.

Curiosity Questions

1? Why do we need to breathe? (See page 492)

2? Does the color of blood change when it picks up oxygen? (See page 501)

3? How is our rate of breathing controlled? (See page 502)

(a) Two different ways to exchange respiratory gases:

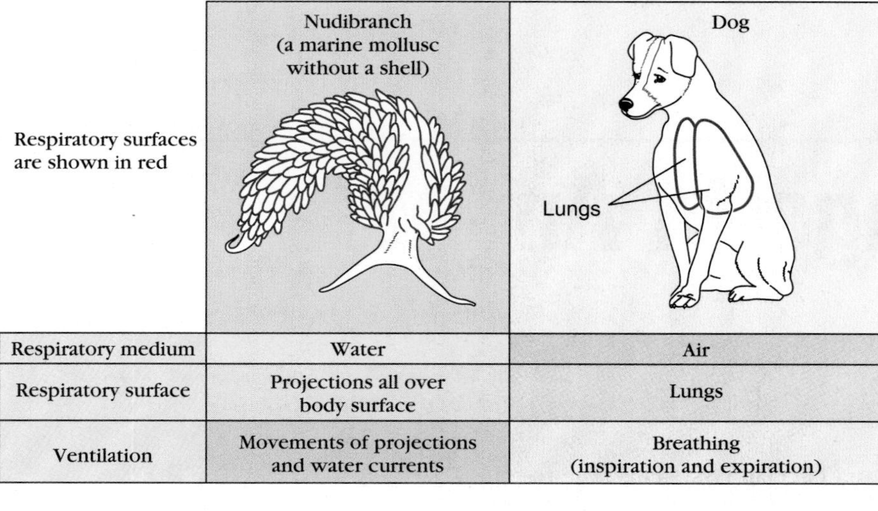

	Nudibranch (a marine mollusc without a shell)	Dog
Respiratory medium	Water	Air
Respiratory surface	Projections all over body surface	Lungs
Ventilation	Movements of projections and water currents	Breathing (inspiration and expiration)

(b) How respiration happens:

FIGURE 23–1

Gas exchange. (a) The finger-like projections of a nudibranch increase its body surface area, which is used as a respiratory surface, while most terrestrial vertebrates, including this dog, breathe air and use lungs which are sites of respiration. **(b)** This general diagram demonstrates how most respiratory surfaces function, whether they are gills or lungs. In both cases, oxygen diffuses into the blood at the respiratory surface. The animal's blood then transports the oxygen to the cells of the body and picks up carbon dioxide on the return trip, to be released at the respiratory surface.

23–A SUPPLYING OXYGEN

Most of the Earth's oxygen is in the air, but some is dissolved in water. Either water or air may serve as an animal's **respiratory medium,** the immediate source of oxygen.

The respiratory medium gives up oxygen to the body at the body's **respiratory surface,** which may be the general body surface or a specialized area such as gills or lungs. Most of an animal's cells lie some distance from the respiratory surface and obtain their oxygen from the **extracellular fluid,** the fluid that bathes every cell. They rely on the blood to bring oxygen from the respiratory surface to the extracellular fluid and to remove carbon dioxide by the reverse route (Figure 23–1).

1? In many animals, the process of **ventilation** moves the respiratory medium over the respiratory surface so that a fresh supply of oxygen is made available. Ventilation using air as a respiratory medium is generally called breathing. Ventilation is necessary because gas exchange depends on diffusion, and the greater the concentration gradient of gases across the respiratory surface, the faster gases will diffuse between the respiratory medium and the blood. Near the respiratory surface, oxygen is quickly removed from the respiratory medium, and carbon dioxide rapidly builds up. This reduces the concentration gradient of both gases and slows their diffusion across the respiratory surface. Ventilation creates a current that brings a fresh supply of air or water, which renews the oxygen and removes the carbon dioxide at the respiratory surface, permitting gas exchange to occur as rapidly as possible. ■

Oxygen moves from the environment into the blood partly by diffusion and partly by **facilitated diffusion,** using a carrier protein (Section 4–D). The carrier is a cytochrome, P450, which speeds up diffusion, allowing the blood to pick up oxygen faster.

P450 also oxidizes toxic substances, rendering them less harmful to the body. (Although the liver is the main organ of detoxification, the lungs also contribute to this activity.) The ready supply of oxygen at the lung surface makes it an ideal site for oxidizing foreign substances.

All animals have respiratory surfaces where oxygen is obtained from a respiratory medium and waste carbon dioxide is released. Ventilation moves the respiratory medium, bring-

BIO-BIT

The air that we breathe contains about 21% oxygen and only about 0.04% carbon dioxide, with nitrogen composing almost all of the rest.

ing in more oxygen and removing carbon dioxide. Oxygen moves into the blood by diffusion and by facilitated diffusion using a carrier protein.

23–B FACTORS THAT INFLUENCE GAS EXCHANGE

What factors are involved in moving oxygen from the external medium to the extracellular fluid surrounding a cell? First, the amount of oxygen available from the environment varies. Compared to air, water contains little oxygen, and the warmer or saltier the water, the less oxygen it can hold (Table 23–1). Oxygen also diffuses half a million times faster in air than in water; hence, oxygen is less readily available in water. On the other hand, there are difficulties with breathing air. Gas molecules must cross plasma membranes in solution, and so respiratory surfaces must always be moist. Air-breathing animals can lose a lot of their precious water by evaporation from the respiratory surface into the air.

A second factor revolves around the rate of gas exchange. How fast an animal can absorb oxygen depends on the amount of oxygen available and on the size of the respiratory surface exposed to the respiratory medium. All but the smallest animals have adaptations that increase the area of the respiratory surface. Because carbon dioxide is much more soluble in water than is oxygen, any respiratory surface large enough to supply an animal with oxygen is well able to dispose of carbon dioxide quickly enough.

A third consideration in gas exchange is how much oxygen an animal needs. This depends on its size (that is, how many cells must be supplied with oxygen) and on its activity. An active animal uses energy at a higher rate than a sluggish one, and so must obtain oxygen at a higher rate. A human being uses oxygen 15 to 20 times faster during exercise than at rest, so the respiratory and circulatory systems must be able to respond to the increased oxygen demand when an animal is active.

Table 23–1
Oxygen Content of Some Respiratory Media

Medium	Oxygen content (milliliters per liter)
Sea water at 5°C	6.4
Freshwater at 5°C	9.0
Freshwater at 25°C	5.8
Air	209.5

An animal's **metabolic rate** is the rate at which it releases energy from its food to drive its metabolism. Because most energy is released by cellular respiration using oxygen, metabolic rate is usually measured as the volume of oxygen used per unit of body weight per unit of time. **Homeothermic** mammals and birds maintain relatively stable body temperatures that are usually higher than those of the surroundings. They have high metabolic rates because they continually oxidize food. As a byproduct, homeothermic animals produce heat that replaces what is lost across the body surface. Smaller animals have a greater proportion of body surface area to their body volume than do larger animals because the ratio of body surface area-to-volume increases with decreasing size (Figure 23–2 and

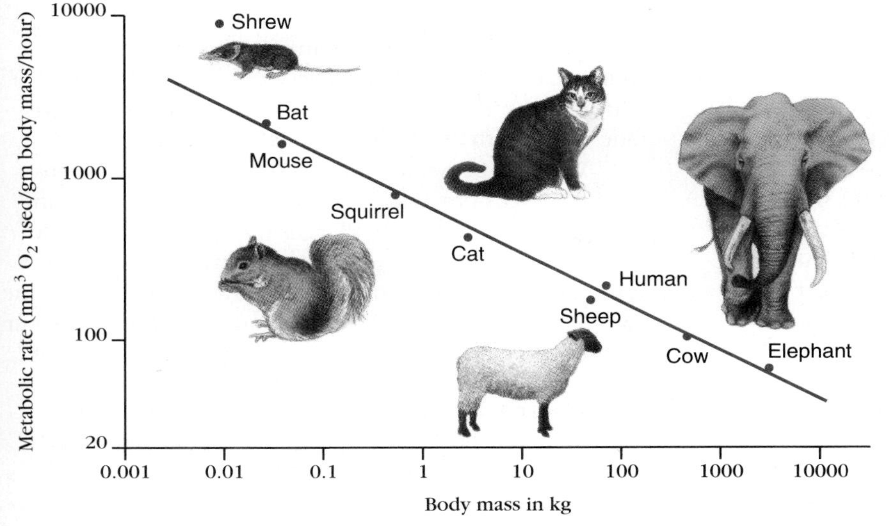

FIGURE 23–2

Size and metabolic rate. The metabolic rates of homeothermic animals are inversely proportional to their size. Notice that both axes in this graph have logarithmic scales, in which increments increase by a factor of 10 times. Both body mass and metabolic rate range over several orders of magnitude for the mammals indicated.

Metabolic rate increases as environmental temperature increases...

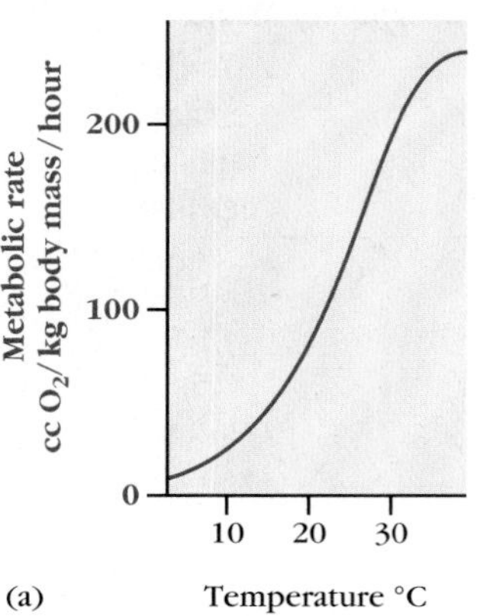

...but, in aquatic environments less O_2 is available as temperature increases.

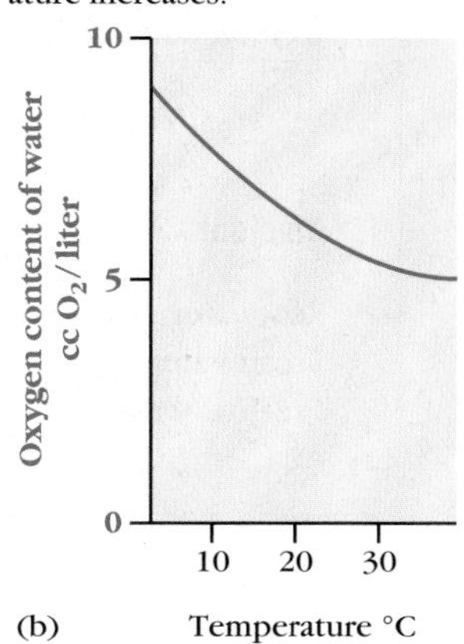

FIGURE 23-3

How temperature affects the metabolic rates of poikilotherms. (a) In these animals, metabolic rates increase as temperature increases. **(b)** But this creates a problem for animals that live in aquatic environments because as water temperatures increase, the amount of oxygen in the water decreases. Thus, as an animal's metabolic rate increases, it requires additional oxygen, but less oxygen is available in the water around it.

see Toolbox, Chapter 4). Small, homeothermic animals lose relatively more heat than do large ones, which is why shrews have the highest metabolic rates of all mammals. (However, a large animal still requires more total oxygen per unit of time than does a small one.)

Poikilotherms, including most fishes, amphibians, and reptiles, allow their body temperatures to fluctuate. They may show a drastic change in oxygen consumption with changing environmental temperature (Figure 23-3). The higher the temperature, the faster the chemical reactions of metabolism occur, the higher the animal's metabolic rate, and the more oxygen it needs. An animal that obtains its oxygen supply from water has a dual problem as the temperature rises. It needs more oxygen because of the rise in its metabolic rate, but less oxygen is dissolved in water at higher temperatures (see Table 23-1). A few active fish, such as trout, can obtain enough oxygen only in very cold water. Thus the discharge of waste heat (for instance, from a power plant) into a lake or stream may ruin the trout fishing.

Oxygen can diffuse more quickly and can occur in higher concentrations in air than in water. Cold water can hold more oxygen than warm water. To increase the rate of gas exchange, most animals have adaptations to increase the area of the respiratory surface. As animals grow larger and become more active, they must increase the size of the respiratory surface and the rate of ventilation. Homeothermic animals have respiratory systems with high rates of gas exchange to support their high metabolic rates. The amount of oxygen required by poikilothermic animals depends upon the environmental temperature and is usually much less than the oxygen levels required by homeotherms.

23-C RESPIRATORY SURFACES AND VENTILATION

Four main types of respiratory surfaces are used by animals: the body surface, gills, lungs, and tracheae (Figure 23-4).

The Body Surface

Some animals, such as earthworms, obtain all the oxygen they need across the general body surface. These animals must be fairly small, so that they have high ratios of body surface area-to-volume. They must have a fairly low metabolic rate, the body surface must be kept moist at all times, and must be protected from injury. This thin, moist surface is often covered with a slimy mucus that makes it too slippery to be damaged by sharp objects.

The body surface is also used to supplement gas exchange by some vertebrates who have lungs or gills, especially amphibians (frogs, salamanders, and caecilians). Many amphibians obtain 25% or more of their oxygen across their thin, moist skin. Indeed, the largest group of salamanders have no lungs or gills as adults and obtain all of their oxygen through the skin.

Gills

Animals such as fish and many aquatic arthropods, molluscs, and amphibians carry out gas exchange through the thin membranes of **gills,** the feathery tissue outgrowths that are exposed to water.

Water is heavy for the amount of oxygen it contains. It takes considerable energy to push a constant current of water across the gill membranes. It would take even more energy to stop the water, reverse its direction, and pass it back out of the gill area the same way it came in. Instead, in most aquatic animals, water enters the gills through one opening and exits by another in a continuous one-way flow.

The gills of a bony fish are located behind the head (Figure 23-5). Water enters through the mouth, passes across the gills, and leaves via the opening behind the **operculum,** which covers the gills.

Types of Respiratory Surfaces

(a) Worm

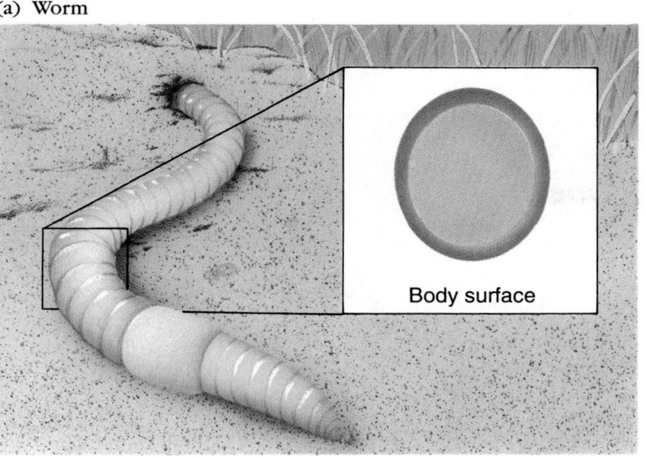

(b) Fish

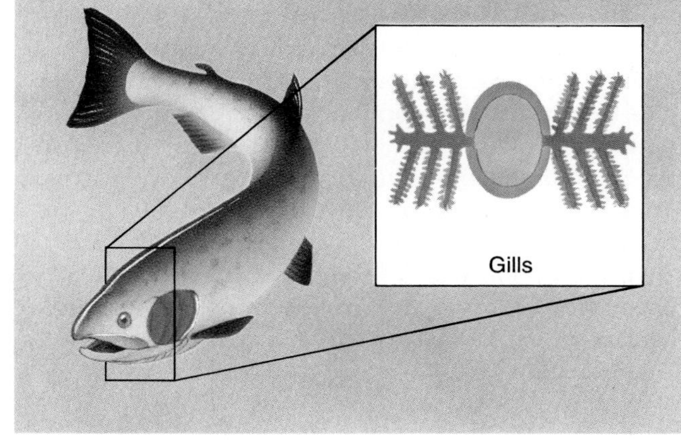

(c) Insect

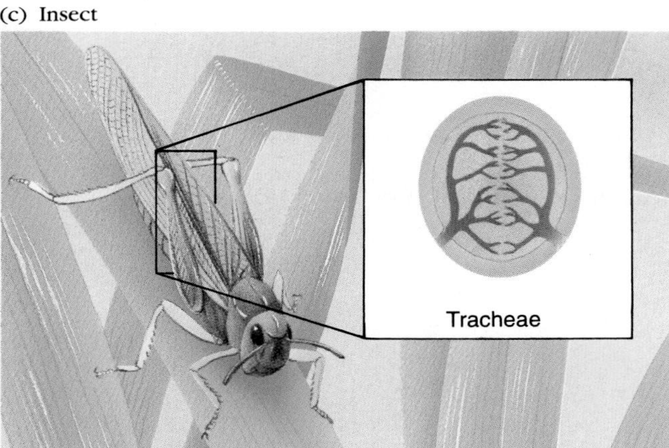

(d) Mammal

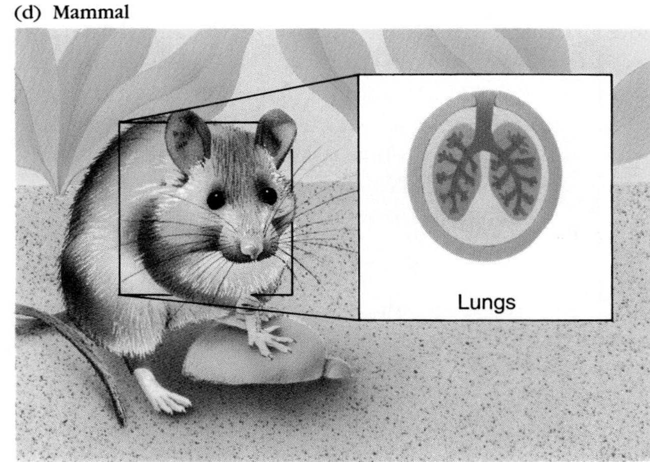

FIGURE 23–4

The four main types of respiratory surfaces. (a) Animals such as this earthworm, sponges, cnidarians, and amphibians use the general body surface as the respiratory surface. **(b)** Gills typically have expanded surface areas that are rich in blood vessels. They may be exposed directly to water (as in the nudibranch in Figure 23-1) or covered by other body parts such as the operculum in this fish. **(c)** Lungs and tracheal systems are internal respiratory surfaces. Tracheae branch throughout the body, bringing air close to all cells. **(d)** Lungs provide localized gas exchange.

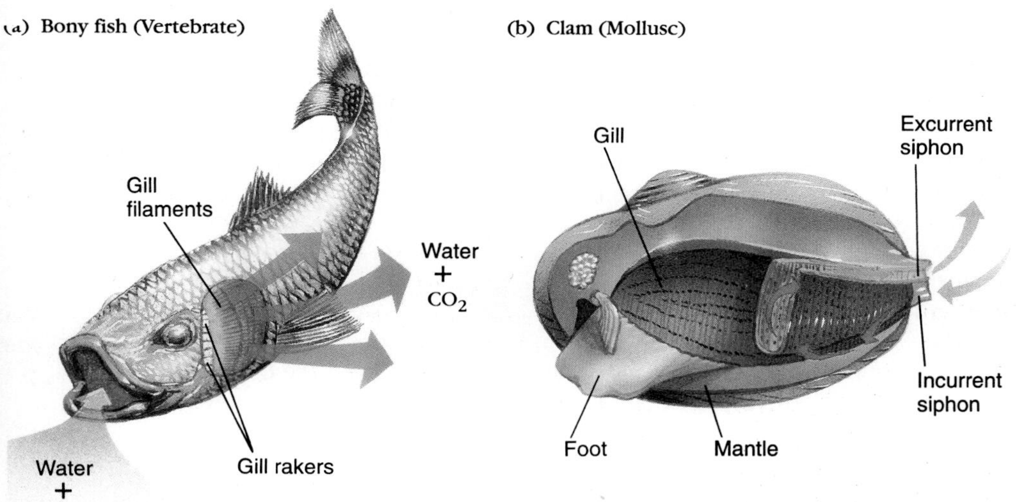

FIGURE 23–5

Gill-breathing animals. A one-way current of water crosses the respiratory surface. **(a)** In a fish, water enters through the mouth and passes out across the gills. **(b)** In a clam, water enters through the incurrent siphon, passes through pores in the gills, and leaves through the excurrent siphon.

In many bivalve molluscs and lower chordates (amphioxus, tunicates [Section 21-K]), ventilation of the gills is linked to feeding; food is filtered out of water drawn into the gill area, and gas exchange occurs at the same time. Some animals use respiratory water currents for locomotion by squeezing water forcibly out of the gill area. Squid, for instance, can eject water from the siphon with considerable force, creating a jet-propulsion stream that moves them rapidly forward or backward, depending on which way the siphon is aimed (Figure 23-6).

Lungs

Because air is less dense than water, it takes less energy to move it to and from the respiratory surface. In addition, air contains about 30 times as much oxygen as the same volume of cold water, and oxygen molecules diffuse 500,000 times faster in air than in water. Thus, air-breathing animals can ventilate their lungs by a **tidal** (in-and-out) **flow** rather than by the one-way stream usual with gills. This makes it unnecessary to have two respiratory openings on the surface of the body. On the debit side, breathing air presents the problem of losing water by evaporation from the respiratory surface.

As an example of a respiratory system with lungs, let us examine our own (see A Journey Through Human Respiration). Air enters the body through the nose or mouth. It passes through the **pharynx,** a common passageway for both air and food, and enters the **trachea** (windpipe) by way of the **larynx,** also known as the voice box or Adam's apple. The walls of the trachea contain rings of cartilage which hold the tube open. On the inner surface of the trachea, cilia keep the air passages clear. They move foreign particles that have been trapped in mucus up into the pharynx, where they can be coughed up or swallowed. The lower end of the trachea divides into two **bronchi,** which divide into finer and finer tubes, the **bronchioles.** The smallest bronchioles end in a myriad of tiny sacs, the **alveoli,** whose thin walls are the actual respiratory surfaces. A vast network of capillaries surrounds the alveoli. Blood in these capillaries picks up oxygen for transport to the rest of the body and gives up carbon dioxide.

Much air stays in the alveoli even when we breathe out as much as possible. This air keeps the walls of the alveoli from sticking together and collapsing.

How does an air-breathing vertebrate ventilate its lungs? There are two main ways, positive and negative pressure breathing, each named for the force that moves air into the lungs.

(a) Water filling mantle cavity of squid

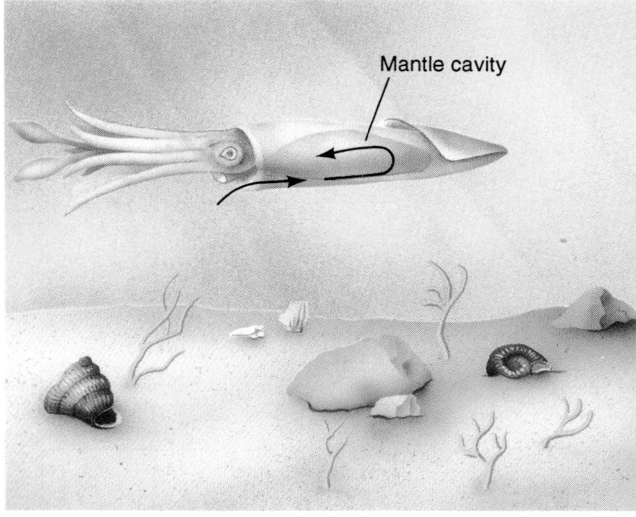

(b) Water is pumped out, squid moves forward

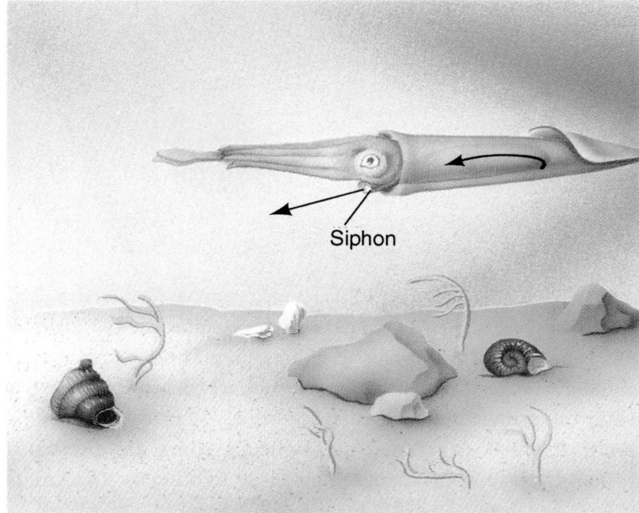

FIGURE 23-6

Jet propulsion in a squid. Squid can move quickly by forcefully ejecting jets of water out through the excurrent siphon. **(a)** Water is first brought into the mantle cavity and then **(b)** forcefully expelled out the siphon to generate a quick forward motion.

BIO-BIT

When we clear our throats, we bring up and then swallow mucus that lined the trachea and trapped particles of dust and dirt before they reached our lungs.

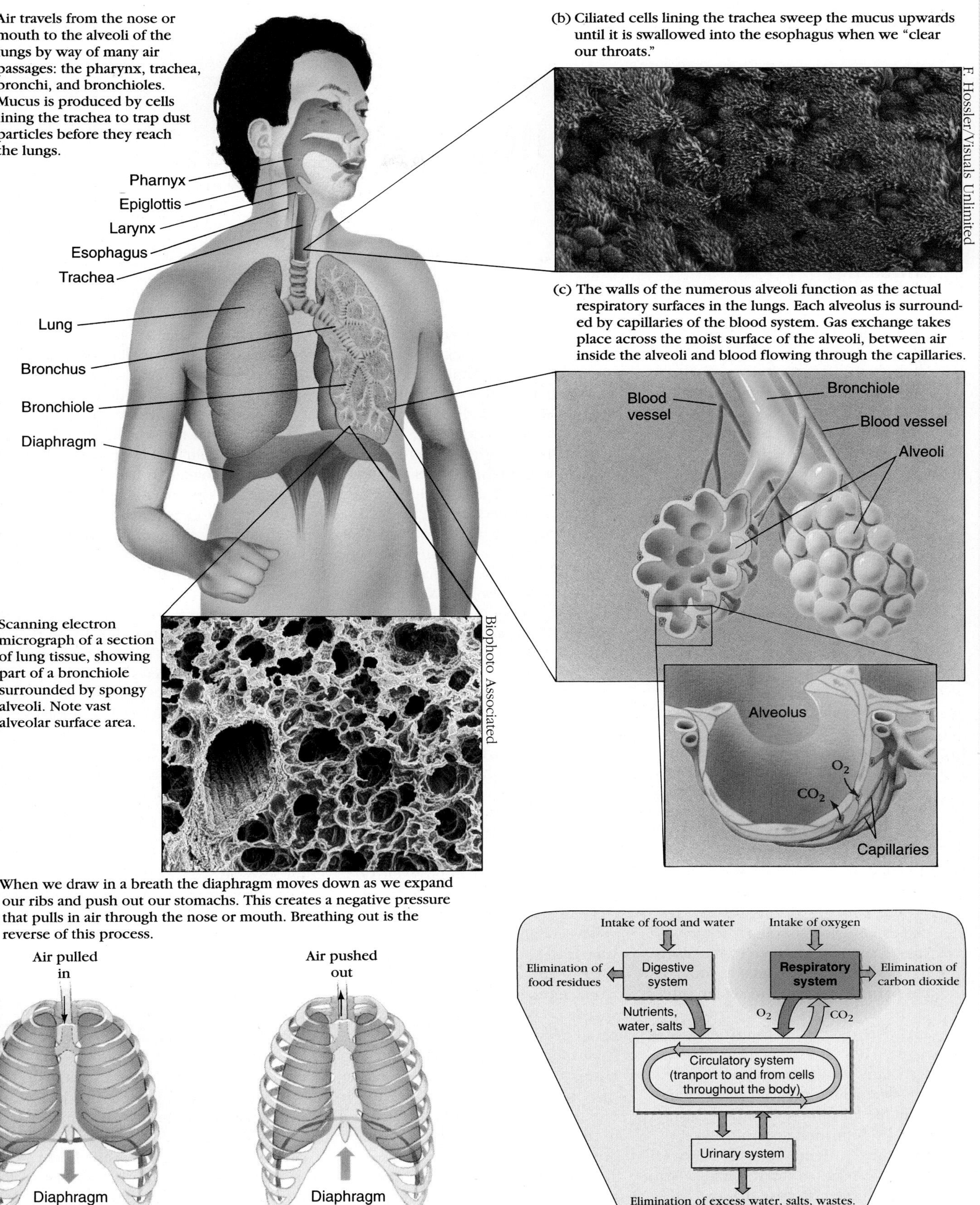

(a) Air travels from the nose or mouth to the alveoli of the lungs by way of many air passages: the pharynx, trachea, bronchi, and bronchioles. Mucus is produced by cells lining the trachea to trap dust particles before they reach the lungs.

(b) Ciliated cells lining the trachea sweep the mucus upwards until it is swallowed into the esophagus when we "clear our throats."

(c) The walls of the numerous alveoli function as the actual respiratory surfaces in the lungs. Each alveolus is surrounded by capillaries of the blood system. Gas exchange takes place across the moist surface of the alveoli, between air inside the alveoli and blood flowing through the capillaries.

(d) Scanning electron micrograph of a section of lung tissue, showing part of a bronchiole surrounded by spongy alveoli. Note vast alveolar surface area.

(e) When we draw in a breath the diaphragm moves down as we expand our ribs and push out our stomachs. This creates a negative pressure that pulls in air through the nose or mouth. Breathing out is the reverse of this process.

A Journey into Healthy Living

Air Pollution and Smoking

In Athens, Greece, six times as many people as usual die on days when the air is heavily polluted. The Hungarian government attributes at least one in 17 deaths to outdoor air pollution and the American Lung Association blames at least 120,000 U.S. deaths each year on this cause. Most urban air pollution comes from motor vehicles and industry. But the air inside a building is nearly always more polluted than the air outside it. Sources of indoor air pollution include fungi and dust in air conditioning and heating ducts, glue in chipboard and other building materials, plastic in furniture and carpets, tar, particles, and smoke from open fires, stoves, cooking ranges, and furnaces. All these pollutants can damage our bodies.

Most of the research on health effects of air pollution has focused on smoking tobacco and marijuana, because smoking introduces pollutants directly into the lungs at high concentrations.

Because smoke is drawn into the lungs by way of the respiratory air passages, the respiratory system comes into closest contact with smoke and is most directly affected by the chemicals it contains. Most of the particulate matter in smoke settles in the lungs. **Tars** form a brown, sticky substance that can damage lung tissue. They also contain carcinogenic (cancer-causing) chemicals.

Nicotine in tobacco smoke paralyzes the cilia in the bronchi. This permits debris to accumulate in the air passages. Hence smokers suffer excessive damage to the respiratory system from the effects of tobacco smoke and other air pollutants. One such effect is **emphysema,** destruction of the gas exchange surface of the lungs' alveoli. Another is **chronic bronchitis,** a low-level infection of the bronchioles. Both reduce the rate of gas exchange in the lungs and can lead to disabling shortage of breath. Bronchitis and other respiratory infections result partly from failure of paralyzed cilia to sweep away disease-causing bacteria, and partly from smoke's inhibition of bacteria-fighting immune cells that reside in the lungs.

All smoke also contains carbon monoxide, produced by incomplete burning. This gas occurs in smokers' blood at 4 to 15 times the level found in nonsmokers (depending on how much the person smokes). Carbon monoxide combines with hemoglobin and reduces the amount of oxygen the blood can carry. Hemoglobin binds carbon monoxide very tightly, and so the gas lingers in the blood, robbing the body of oxygen, for as long as 6 hours after the cigarette is finished.

The effects of carbon monoxide can be seen from smoking just one cigarette. Smoking one cigarette provides enough nicotine to cause blood vessels to constrict (become narrower). This cuts down the flow of blood and oxygen to the body. It also speeds up the heart rate and increases blood pressure. A steady smoking habit renews these effects with each cigarette. This can eventually damage the circulatory system and lead to cardiovascular disease. (Smoke damage to the circulatory system may have other consequences. For instance, smokers are usually not eligible for procedures such as kidney or liver transplants. Transplant organs are in short supply and must be given to those with healthy circulatory systems and, therefore, a greater chance of recovering from the operation.)

Another effect of breathing polluted air is that the body is forced to detoxify inordinately large amounts of foreign substances. Tobacco smoke, anesthetic gases, or other pollutants in the lungs force cytochrome P450 (Section 23-A) to use up a lot of oxygen detoxifying these substances, instead of transferring this oxygen to the blood.

Pregnant women are especially affected by pollutants in the air. Not only do pollutants reduce the amount of oxygen transported from the lungs into the mother's bloodstream, but they also restrict the oxygen supply to the fetus. If the P450 in the placenta is involved in detoxifying substances from the mother's blood, the fetus loses a large percentage of its oxygen supply. This may account for the fact that the rate of miscarriage and birth defects is 30% higher for nurses who work with anesthetics than for women in other occupations, and that women who smoke tend to give birth to small babies.

Fetuses are not the only nonsmokers affected by smoking. The smoke from an "idling" cigarette contains more tar, nicotine, and cadmium (a toxic metal) than does the smoke inhaled by a smoker. People inhaling cigarette smoke in the air around them are affected by the same kinds of changes as the smoker, but to a lesser degree. Children exposed to such "passive smoking" in the home have twice as many respiratory infections as children of nonsmokers. And adults living or working in smoky

FIGURE 23–A

Effects of smoking on human lungs. Normal human lungs are on the left and the lungs of a cigarette smoker are on the right. (A. Glauberman/Photo Researchers, Inc.)

places have an increased risk of developing lung cancer.

In 1919, doctors in a Boston hospital hastily called medical students to an autopsy to see a type of cancer so rare that they might never witness it again: lung cancer. Today, lung cancer is one of the most common types of cancer. There appear to be many carcinogens in tobacco smoke. They cause several types of lung cancer, most of them highly malignant and rapidly fatal. Cells in the lungs of smokers show a number of chromosomal abnormalities, and cells in the nearby larynx and esophagus, and even more distant organs such as the kidneys and liver, also show DNA damage. In addition, the breakdown products of the radioactive gas radon become attached to particles in smoke and lodge with them in the lung. Because smoking exacerbates the effect of radon in this way, about 85% of the cancers attributed to radon occur in smokers. Smokers are similarly much more likely to die of lung cancer caused by inhaling asbestos fibers than are nonsmokers. Studies have implicated smoking in various cancers other than lung cancer. Benzene in cigarette smoke is blamed for about 550 cases of adult leukemia every year.

In summary, smoking is a major cause of emphysema, chronic bronchitis, cancer, and heart disease. Smokers can expect to live about ten years less than nonsmokers. About one in every 15 smokers can expect to die of lung cancer (some 40,000 people each year), and several times that number will die prematurely of cardiovascular disease. In fact, smoking tobacco causes greater health damage than any other drug, legal or illegal.

With all these disadvantages, why do so many people smoke? For reasons that are not fully understood, smoking is highly addictive. However, after learning of the adverse affects of smoking, over 50 million Americans have managed to stop. In people who quit, the physiological damage done by smoking is gradually reversed. Ten years later, a person's chance of suffering adverse health effects from a previous smoking habit are hardly any higher than those of a lifelong nonsmoker.

Mammals have **negative pressure breathing** (see A Journey Through Human Respiration). They have a respiratory muscle, the **diaphragm,** not found in other vertebrates. The diaphragm extends across the bottom of the chest cavity, beneath the lungs, closing off the chest cavity from the abdominal cavity below. During inhalation, the muscles between the ribs contract and lift the ribs outward. At the same time, the diaphragm contracts and so moves lower. These movements increase the volume of the chest cavity and so decrease the pressure within it. Air then rushes into the nose or mouth, down the trachea, and into the lungs, *pulled* by the negative pressure created in the chest cavity. During exhalation, relaxation of the rib muscles and diaphragm decreases the volume of the chest cavity, increasing the pressure inside it and forcing air back out of the lungs. (You can get an idea of how negative pressure breathing works by closing your mouth and holding your nose while you expand your rib cage and lower your diaphragm. You will feel the partial vacuum created. Then, remove your fingers from your nose and you can hear and feel the air rushing in as the pressure equalizes.)

Negative pressure breathing allows an animal to eat and breathe at the same time. If it were necessary to push air from the mouth into the lungs, any food in the mouth might be pushed into the trachea and cause an obstruction. Negative pressure breathing creates a more gentle stream of air, which is less apt to pull food along into the air passages.

In contrast, a frog breathes by a **positive pressure** mechanism (Figure 23–7). At the beginning of a ventilation cycle, the frog opens its nostrils and lowers the floor of its mouth. Enlargement of the mouth cavity creates a partial vacuum there, and air enters through the nostrils. (This part of the frog's breathing cycle actually operates on the negative pressure principle, just described.) The frog then closes its nostrils and raises the floor of its mouth, *pushing* the air into the lungs. After holding the air in its lungs, the frog pushes the air back out by *opening* its nostrils and contracting its abdominal muscles.

(a) Nostrils open

(b) Floor of mouth lowered, pulling air in through the nostrils

(c) Nostrils closed, floor of mouth raised, forcing air into the lungs

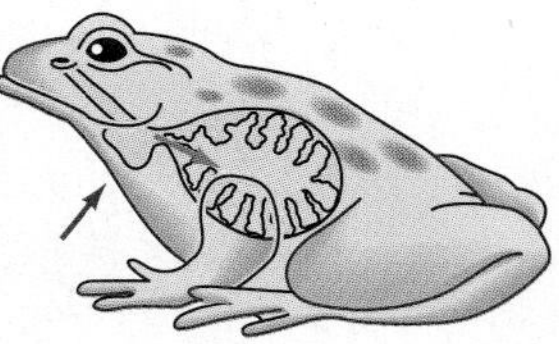

(d) Nostrils opened, abdominal muscles contracted, forcing air out of the lungs and out through the nostrils.

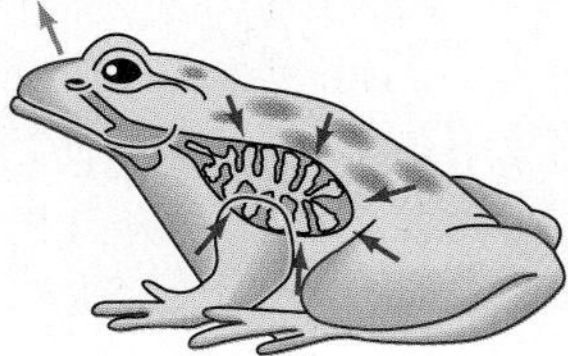

FIGURE 23–7

Positive pressure breathing in a frog.

Tracheal Systems

Air-breathing vertebrates have lungs. The other major group of land animals, the terrestrial arthropods (insects, centipedes, millipedes, and some spiders), breathe air by means of **tracheae,** air tubes that extend throughout the body. The tracheal system does not rely on the circulatory system to transport gases. Instead, the tracheae start at **spiracles,** tiny openings at the body surface, and branch into all parts of the body, ending close to every cell (Figure 23–8). Chemicals given off by cells that are short of oxygen induce tracheae to grow branches into that area. Thus, every cell is close to an oxygen-containing branch of the tracheal system.

Large or active insects ventilate the tracheal system by pumping air in and out with their abdominal muscles. The smallest insects need only open their spiracles and let diffusion bring oxygen to their cells.

Animals use four main types of respiratory surfaces. Small animals with high body surface area-to-volume ratios may use moist body surfaces. Most aquatic animals ventilate water over thin gill membranes. Air-breathing terrestrial vertebrates use lungs. Terrestrial arthropods use an extensive system of tubes that extend from the body surface to cells throughout the body.

FIGURE 23-8

Tracheal system. As this cockroach splits and molts its dark-colored, outgrown exoskeleton, a new, light-colored exoskeleton is revealed beneath it. Exposure to air will harden and darken the new exoskeleton, but while it is still semi-transparent, the components of the tracheal system are visible as a network of fine, white lines that branch into each part of the insect's body. Spiracles, the openings at the body surface on the sides of the body, are not visible from this angle. (R. and L. Mitchell)

23-D RESPIRATORY PIGMENTS

Blood is largely salt water, which cannot carry much dissolved oxygen. Many animals have adaptations that increase the amount of oxygen their blood can carry. The most common adaptation is to possess **respiratory pigments,** large protein molecules that bind and release oxygen. The respiratory pigment hemoglobin in the red blood cells of a mammal carries 98% of the oxygen in the blood.

Hemoglobin (Hb) is the general name for a group of oxygen-carrying molecules that all contain a heme group (Figure 23-9). The iron atom in the center of the heme group is the actual binding site for oxygen. When oxygen is bound to the iron atom, the pigment is said to be **oxygenated** (not oxidized) and is called **oxyhemoglobin (HbO).** Oxyhemoglobin appears bright red, and deoxygenated hemoglobin is a darker, purplish red. The hemoglobin of vertebrates consists of four polypeptide chains, each with a heme group attached (see Chapter 3: A Journey Through Protein Structure: Hemoglobin). ■

Hemoglobins are found in almost all vertebrates, and in various invertebrates, including earthworms and some insects. Other invertebrates, such as many arthropods and molluscs, have other kinds of respiratory pigments—copper-containing hemocyanins, for example.

How Does Blood Carry Oxygen?

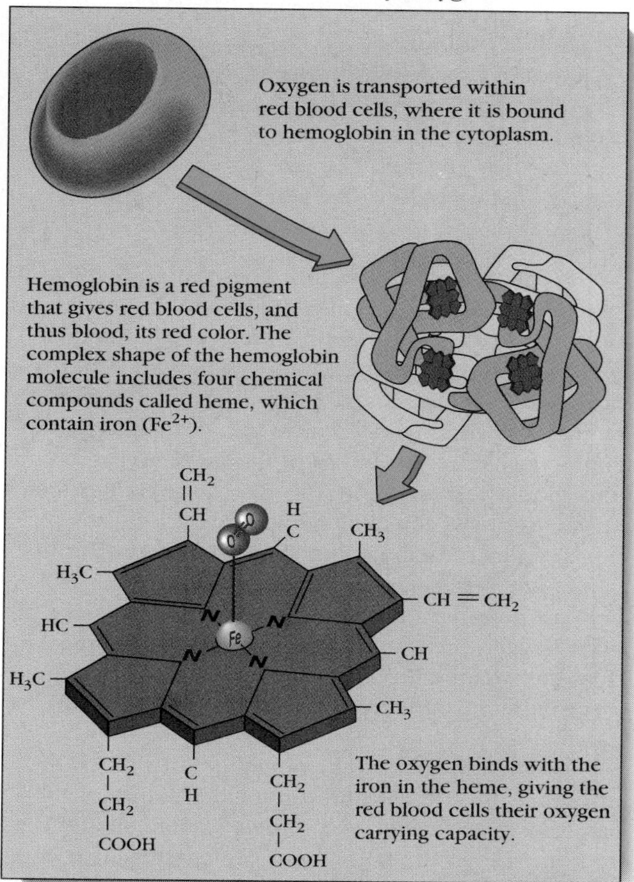

Oxygen is transported within red blood cells, where it is bound to hemoglobin in the cytoplasm.
Hemoglobin is a red pigment that gives red blood cells, and thus blood, its red color. The complex shape of the hemoglobin molecule includes four chemical compounds called heme, which contain iron (Fe^{2+}).
CH_2
CH
H
C
CH_3
H_3C
$CH=CH_2$
HC
Fe
N
CH
H_3C
CH_3
CH_2
CH_2
COOH
C
H
CH_2
CH_2
COOH
The oxygen binds with the iron in the heme, giving the red blood cells their oxygen carrying capacity.

FIGURE 23-9

How blood carries oxygen.

As blood moves through the lungs, its hemoglobin picks up oxygen until it is nearly saturated. That is, almost all the iron atoms in all the heme groups have bound oxygen molecules. The degree of saturation depends on the **partial pressure of oxygen:** the portion of the total air pressure attributable to oxygen. The higher that oxygen pressure, the more oxygen binds to hemoglobin in the blood, up to the saturation point.

The oxygen pressure is lower in the rest of the body than in the lungs because the body's cells continually use up oxygen in respiration. So blood moving through the tissues encounters lower oxygen pressures, which cause hemoglobin molecules to give up some of their oxygen—normally about 37%. During exercise, when the cells use more oxygen than usual, the oxygen pressure drops lower and the hemoglobin releases almost 20% more of its oxygen.

> **BIO-BIT**
>
> Carbon monoxide is an odorless, colorless, and potentially fatal gas produced by cars and cigarettes. It binds to hemoglobin molecules at levels 200 times greater than oxygen does, and prevents blood from carrying the necessary oxygen levels.

Oxyhemoglobin gives up oxygen more readily in an acid environment, such as that found in active tissues. An exercising muscle releases carbon dioxide from aerobic respiration, which reacts with water to form carbonic acid. It also releases lactic acid from fermentation (Section 7-E). These acids cause oxyhemoglobin to give up 10% more of its oxygen. Hence, oxyhemoglobin releases more oxygen to the tissues that are the most oxygen-depleted and need it most.

An animal's hemoglobin shows adaptations to its way of life. For instance, the hemoglobin of a small mammal gives up more oxygen to the body tissues at a given oxygen pressure than that of a large mammal. This is an adaptation to the higher metabolic rate and correspondingly greater oxygen demands of small mammals.

An animal may have hemoglobin with different properties at different times in its life. Before birth, human fetuses produce several kinds of hemoglobin that are not made in the body of an adult. After birth, fetal hemoglobin is gradually replaced by adult hemoglobin. All of a fetus's oxygen comes from its mother's bloodstream. If fetal and adult hemoglobins had the same affinity for oxygen, the fetus could not pick up very much of the oxygen released by the mother's blood. But a fetus's hemoglobin has a higher affinity for oxygen than the mother's, permitting it to pick up oxygen at oxygen pressures low enough to cause the mother's hemoglobin to release oxygen.

Oxygen is carried within red blood cells where it binds to the iron atom of a heme group in hemoglobin. Hemoglobin's oxygen-saturation level depends mainly on the oxygen pressure in the surrounding fluid and, to a lesser extent, on the pH.

23–E CARBON DIOXIDE TRANSPORT

Carbon dioxide produced during respiration must be carried by the blood to the lungs, where it is excreted during exhalation. Some of this carbon dioxide travels combined with hemoglobin and other proteins in the blood.

Most of the carbon dioxide in the blood is carried as dissolved bicarbonate ions. The enzyme **carbonic anhydrase,** in red blood cells, forms carbonic acid from water and carbon dioxide. Carbonic acid readily dissociates into hydrogen and bicarbonate ions:

$$\underset{\text{water}}{H_2O} + \underset{\substack{\text{carbon}\\\text{dioxide}}}{CO_2} \xrightarrow{\substack{\text{carbonic}\\\text{anhydrase}}} \underset{\substack{\text{carbonic}\\\text{acid}}}{H_2CO_3} \longrightarrow \underset{\substack{\text{hydrogen}\\\text{ion}}}{H^+} + \underset{\substack{\text{bicarbonate}\\\text{ion}}}{HCO_3^-}$$

If all of these hydrogen ions remained in the bloodstream, the blood would be very acidic indeed. But most of the hydrogen ions are neutralized by combining with hemoglobin, which is negatively charged, to form **acid hemoglobin.** This reduces the acidity of the blood:

$$\underset{\substack{\text{hydrogen}\\\text{ion}}}{H^+} + \underset{\substack{\text{bicarbonate}\\\text{ion}}}{HCO_3^-} + \underset{\substack{\text{hemoglobin}\\\text{ion}}}{Hb^-} \longrightarrow HCO_3^- + \underset{\substack{\text{acid}\\\text{hemoglobin}}}{HHb}$$

If the blood becomes too basic, acid hemoglobin dissociates, releasing hydrogen ions.

The blood gives up only about 10% of its carbon dioxide as it passes through the lungs. The other 90% is retained, mostly in the form of bicarbonate ions, which act as important blood buffers, substances that keep the pH from fluctuating (Section 2-G). Therefore, although carbon dioxide is a waste product, its presence is essential in regulating the pH of the blood.

Most carbon dioxide is carried in blood plasma. The carbon dioxide dissolves to form bicarbonate ions, the most important buffer in the blood.

23–F REGULATION OF VENTILATION

The partial pressures of oxygen and carbon dioxide in our blood must be continually adjusted so that they stay within so-called physiological limits. This adjustment is made by altering the depth and rate of breathing. As blood passes through the capillaries of the lungs, the gases in the blood come into equilibrium with those in the alveoli. The faster and deeper we breathe, the greater the percentage of oxygen in the alveolar air, and the lower the percentage of carbon dioxide, because the air in the lungs is replaced more completely and more frequently. With slow or shallow breathing the reverse is true. The breathing rate is controlled by the carbon dioxide and oxygen content of the blood and cerebrospinal fluid (the fluid around the brain and spinal cord). Receptor organs monitor the pressures of these two gases in the body fluids. The brain responds to this information by sending nerve signals to the diaphragm and rib muscles, speeding or slowing the rate of breathing (Figure 23–10).

Surprisingly, the level of carbon dioxide has more effect on breathing than does the level of oxygen. If the carbon dioxide content of the blood drops below a certain critical level, breathing is inhibited. By purposely hyperventilating, that is, taking several deep breaths in swift succession, thus decreasing your blood's CO_2 content, you can hold your breath longer. Swimmers often do this so that they can swim underwater for a longer time. At each breath, however, the carbon dioxide content of the blood goes down, and if it goes too far, you lose consciousness. This important mechanism prevents you from lowering the carbon dioxide pressure of the blood to a dangerous level.

Similarly, when you hold your breath, the carbon dioxide level in the blood rises, and if it goes above a certain level, you will lose consciousness. Once this happens, the breathing reflexes take control again, and cause you to inhale.

Breathing during sleep is governed by somewhat different mechanisms, which sometimes permit breathing to lapse. Periods of **apnea** a (cessation of breathing) occur several times each night, causing the person to wake up briefly. The mechanisms that govern breathing while the person is awake then take over and cause breathing to resume. The frequency of apnea is increased by consumption of alcohol, antihistamines, and tranquilizers. ■

The partial pressures of carbon dioxide and oxygen in the blood and cerebrospinal fluid are measured by receptor centers that help regulate the rate of breathing. Carbon dioxide levels are more closely monitored than oxygen levels. Low carbon dioxide levels cause slower rates of breathing, while high carbon dioxide levels increase breathing rates.

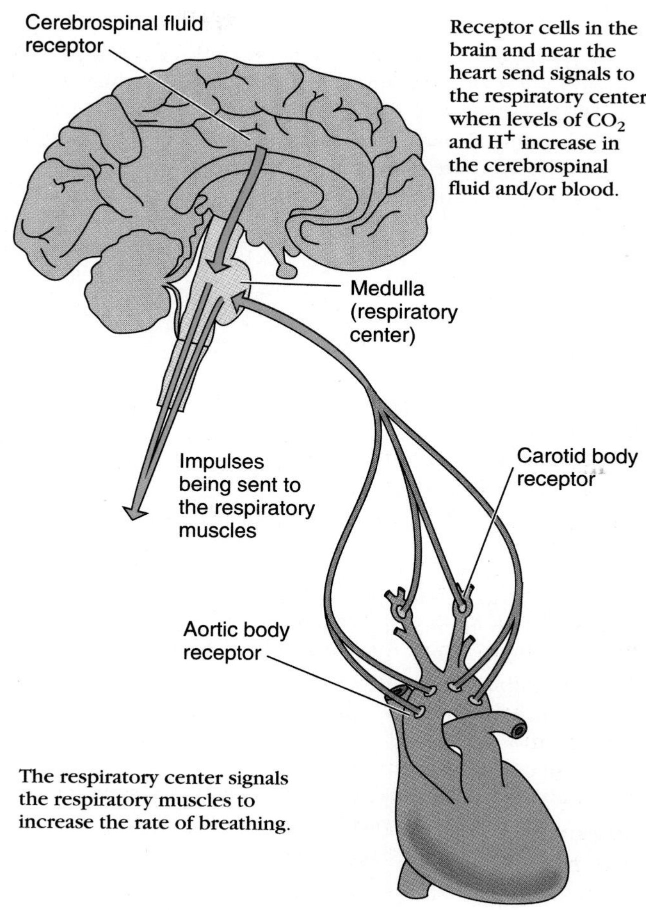
Cerebrospinal fluid receptor
Receptor cells in the brain and near the heart send signals to the respiratory center when levels of CO_2 and H^+ increase in the cerebrospinal fluid and/or blood.
Medulla (respiratory center)
Impulses being sent to the respiratory muscles
Carotid body receptor
Aortic body receptor
The respiratory center signals the respiratory muscles to increase the rate of breathing.

FIGURE 23–10

How breathing rate is regulated.

SUMMARY

1. All animals must obtain oxygen from their environments and expel carbon dioxide. For aquatic organisms, obtaining enough oxygen is a more difficult task because oxygen is much less water-soluble than is carbon dioxide.
2. The total amount of oxygen an animal requires depends largely on its total number of cells and on its metabolic rate. The metabolic rate increases with activity. In warm-blooded animals, the oxygen demand increases as the ratio of body surface area-to-volume increases. Therefore, the smallest animals have the greatest oxygen requirements, but large animals still require more total oxygen at any given time.
3. Gas exchange with the environment can take place only across a moist surface.
4. Only small or relatively inactive animals can obtain all of their oxygen by diffusion across the general body surface. (A high body surface area-to-volume ratio is found in animals whose bodies are small, flattened, or covered with projections.)
5. Larger or more active animals usually have part of the body specialized as an expanded respiratory surface and they transport gases to and from this surface using a circulatory system. An exception is the tracheal system of insects, in which a series of branching tubes transport gases from the surface directly to the cells throughout the body.
6. The four main types of respiratory organs are the body surface, gills, lungs, and tracheal systems.

7. In animals with lungs, the gas content of the blood is regulated by controlling the depth and rate of breathing and, therefore, the amount of oxygen and carbon dioxide in the alveoli.
8. Many animals have respiratory pigments in their blood that vastly increase its oxygen-carrying capacity. The oxygen saturation of the respiratory pigment hemoglobin depends on the oxygen pressures of the nearby fluid and air.
9. Carbon dioxide is transported mainly in the form of bicarbonate ions, important buffers that help keep the pH of the body fluids at a constant level. Normally, carbon dioxide levels in the body determine the breathing rate.

SELF-QUIZ

1. Under *normal circumstances,* why does unconsciousness occur?
   a. change in carbon dioxide levels in the blood
   b. change in oxygen levels in the blood
   c. loss of hemoglobin from red blood cells
   d. distress of the lungs
   e. excess release of oxygen from hemoglobin
2. Most oxygen in the blood is carried in the form of ___; most carbon dioxide in the form of ___.
   a. carbonic acid
   b. bicarbonate ion
   c. oxyhemoglobin
   d. oxygen pressure
   e. hemoglobin ion
3. A disadvantage of using air as a respiratory medium is:
   a. it carries less oxygen than water does
   b. it increases the risk of drying out
   c. oxygen diffuses faster in air than in water
   d. air contains nitrogen as well as oxygen
   e. air pressure changes more than water pressure with changes in temperature
4. The metabolic rate of a poikilothermic animal increases:
   a. with increasing environmental temperature
   b. with decreasing environmental temperature
   c. with increase in size
   d. with decrease in muscular activity
   e. with increase in age
5. The main difference between the insect tracheal system and most other types of respiratory systems is:
   a. tracheal systems do not rely on the blood to transport oxygen to the tissues
   b. insects do not ventilate their tracheal systems
   c. insects do not dispose of carbon dioxide via their tracheal systems
   d. insects exchange both carbon dioxide and oxygen via their tracheal systems
   e. oxygen need not be in solution to cross the membranes in the tracheal systems of insects
6. The main factor that determines the saturation of hemoglobin with oxygen is:
   a. oxygen concentration in the blood
   b. carbon dioxide concentration in the blood
   c. pH of the blood
   d. hemoglobin concentration in the blood
   e. breathing rate
7. All of the following affect the amount of oxygen the blood delivers to the tissues *except:*
   a. the concentration of red blood cells in the blood
   b. the pH of the blood
   c. the concentration of CO_2 in the blood
   d. the concentration of white blood cells in the blood
   e. the oxygen pressure in the air

THINKING CRITICALLY

1. Ice fish are a family of bony fishes that inhabit Antarctic waters. These fish have no hemoglobin in their blood. How do you think they are able to survive without this respiratory pigment that all other adult vertebrates possess? What characteristics would you expect a member of this family to show?
2. Some adult amphibians use gills for breathing. There are many reptiles, birds, and mammals that spend almost all their time in water. Why do you think it is that none of them has evolved so that it retains the embryonic gills and uses them for gas exchange in the adult stage?
3. The evolution of coverings for the gills necessitated arrangements for openings to and from the gill area as well as muscles to draw a current of water across the gills. Yet very few animals have gills without some sort of covering. What is the advantage of a gill covering that has selected for evolution of the covering plus all these accessory arrangements?
4. Do you consider that the respiratory surfaces of your lungs are exposed to the environment? Why or why not?
5. A long-term smoking habit can destroy the cilia in the lining of the air passages. How might this affect health?

SELECTED KEY TERMS

alveoli, *p. 496*
bronchi, *p. 496*
bronchiole, *p. 496*
diaphragm, *p. 500*
extracellular fluid, *p. 492*
facilitated diffusion, *p. 492*
gill, *p. 494*
hemoglobin (Hb), *p. 501*
homeotherm, *p. 493*
larynx, *p. 496*
metabolic rate, *p. 493*
negative pressure breathing, *p. 500*
operculum, *p. 494*
oxyhemoglobin (HbO), *p. 501*
pharynx, *p. 496*
poikilotherm, *p. 494*
positive pressure breathing, *p. 500*
respiratory medium, *p. 492*
respiratory pigment, *p. 501*
respiratory surface, *p. 492*
spiracle, *p. 500*
tidal flow, *p. 496*
trachea (windpipe), *p. 496*
tracheae (air tubes in arthropods), *p. 500*
ventilation, *p. 492*

SUGGESTED READINGS

Avery, M. E., N-S. Wang, and H. W. Taeusch, Jr. "The lung of the newborn infant." *Scientific American,* April 1973. Describes the physiological changes in the lung just before birth and efforts to speed these changes in premature infants.

Comroe, J. H., Jr. "The lung." *Scientific American,* February 1966. Anatomy and physiology of the human lung and respiratory tract, describing physiological measurement techniques and applications.

Houston, C. S. "Mountain sickness." *Scientific American,* October 1992. Describes the physiological problems associated with life at high atmospheric levels.

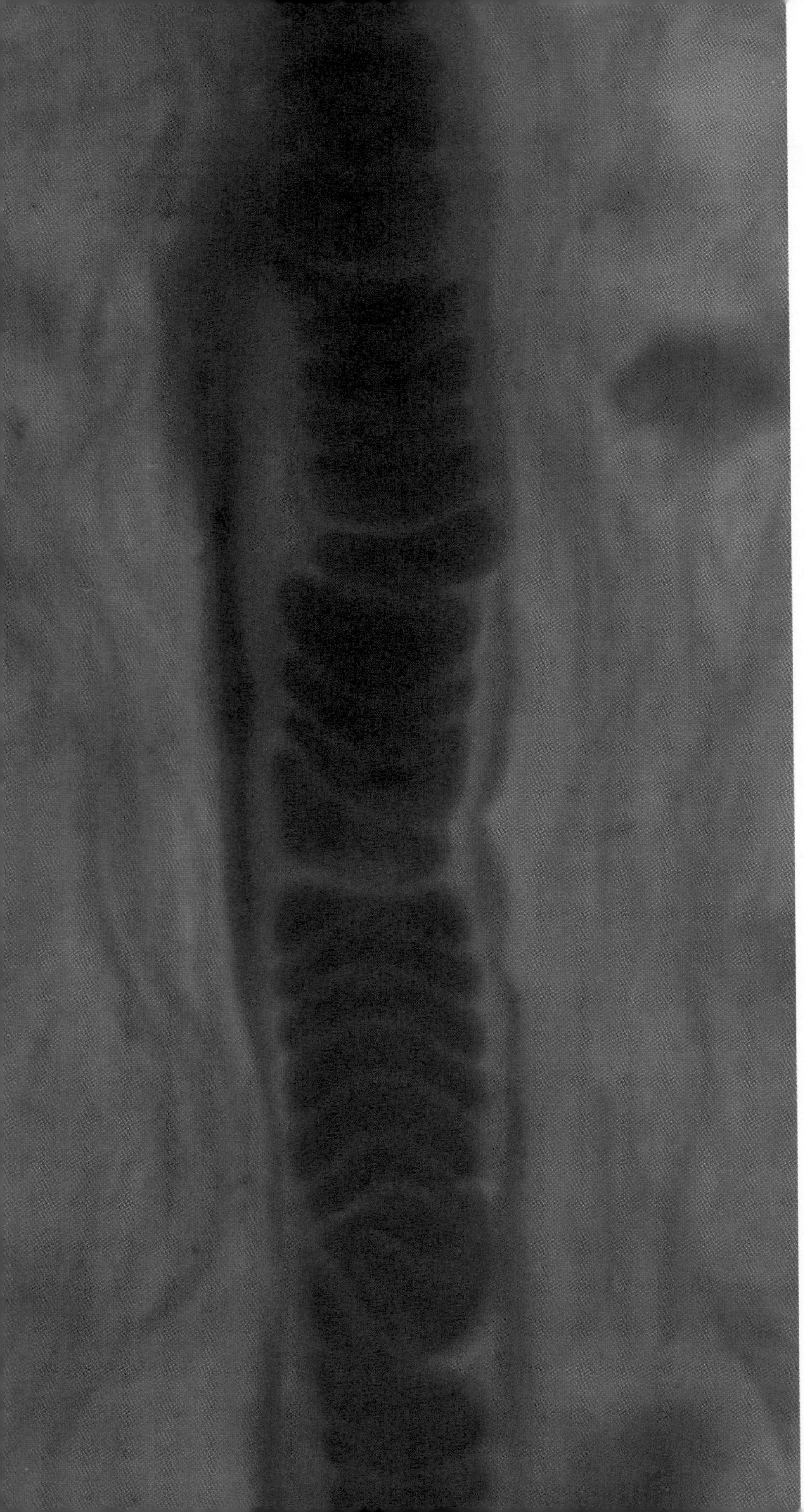

CHAPTER 24

CONCEPT GUIDE

After reading this chapter, you should be able to:

1. Compare and contrast circulation in cnidarians, planarians, annelids, and arthropods. (Section 24-A)
2. Describe the differences between single and double circulatory systems and give the advantages and disadvantages of each system. (Section 24-B)
3. Describe the structural and functional differences between arteries, capillaries, and veins. (Section 24-C)
4. Describe the primary functions of red blood cells, white blood cells, and platelets. (Section 24-D)
5. Describe the major functions of the lymphatic system. (Section 24-E)

Capillaries are the smallest of all blood vessels, carrying blood cells in single file through respiratory surfaces. Red blood cells in capillaries give skin its pink tones. Read more about how blood is transported in vertebrate bodies in Section 24-B. (A. Owczarzak/Biological Photo Service)

Internal Transport

The earliest organisms lived in the sea, which provided them with oxygen, carried away carbon dioxide and other wastes, and surrounded them with a relatively constant environment. The **extracellular fluid** that surrounds every cell of a higher animal is like a tiny captive sea. Cells obtain food and oxygen from this fluid and discharge their wastes into it. The body's transport system supplies the extracellular fluid with fresh food and oxygen and removes wastes from it, ensuring that the fluid remains an environment where cells can flourish.

Even the smallest animal must transport substances within its body. Oxygen must be moved into the body, and then to the fluid around each cell. Food molecules must move from the site of digestion to the extracellular fluid, and waste products must be removed from it and expelled. Various fluid systems, called **vascular systems,** transport substances in most members of the animal kingdom. A **circulatory system** is a vascular system in which the transport fluid moves continuously in one direction, usually because it is propelled by a muscular, pumping heart.

Large, active animals have circulatory systems that transport not only gases, food, and wastes, but also hormones, as well as molecules and cells that help protect the body from disease (Chapter 25). A circulatory system also distributes heat generated by metabolism or absorbed from the environment. The circulatory systems of birds and mammals transport high levels of oxygen to meet these animals' high metabolic requirements.

KEY CONCEPTS

- Vascular systems transport substances and heat from one part of the body to another and between the animal's external environment and the extracellular fluid, which surrounds every cell.
- Animals with faster, more direct transport systems can metabolize faster and lead more active lives than those with less efficient transport.

Curiosity Questions

1? What causes varicose veins and hemorrhoids? (See page 512)

2? What causes heart attacks and strokes? (See page 516)

3? How does blood clot when we get a cut? (See page 518)

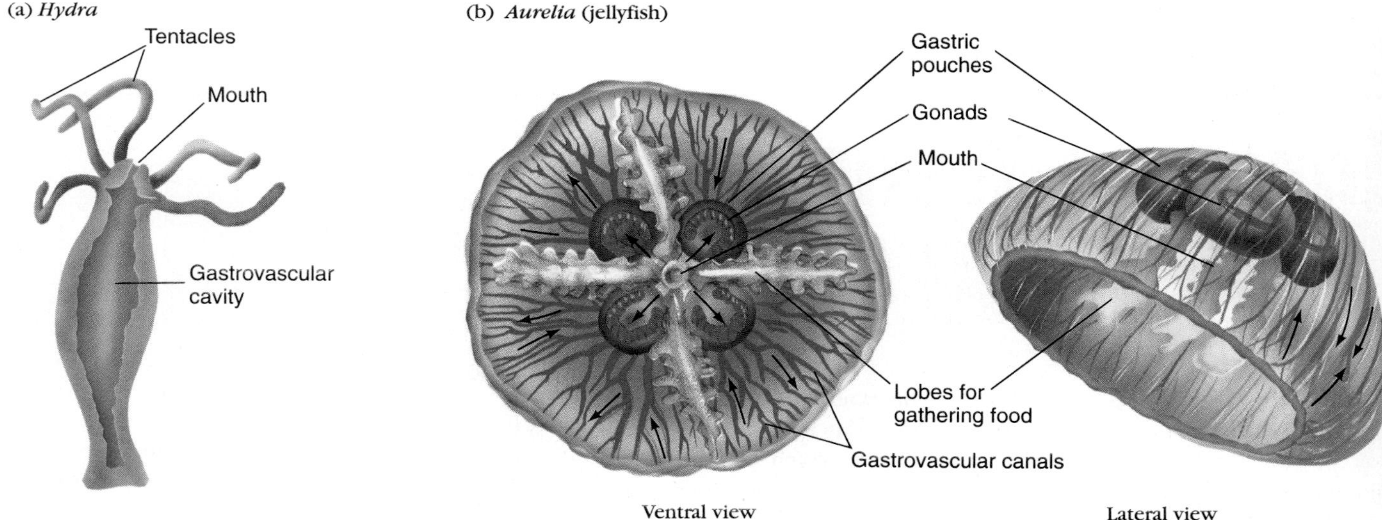

FIGURE 24–1

Gastrovascular cavities of two cnidarians. (a) In a *Hydra*, food is distributed throughout the body within the broad gastrovascular cavity and narrow canals within the tentacles. **(b)** Arrows show the movement of digested food in a jellyfish, (*Aurelia*), where a branching system of small canals carries digested food from the gastrovascular cavity to the tentacles and other outlying parts of the body.

24–A TRANSPORT IN INVERTEBRATES

Cnidaria

Cnidarians are slow-moving creatures with low metabolic rates. Most of their oxygen enters through the body's thin outer layer of cells, and soluble wastes diffuse out across the general body surface. Food is transported by the **gastrovascular cavity,** which does double duty as a digestive (gastro) and transport (vascular) system.

In *Hydra* the gastrovascular cavity extends into each tentacle from the center of the body (Figure 24–1). In larger jellyfish, the cavity extends into a system of canals that branch throughout the body. The fluid in the system is pushed in defined pathways, propelled by cilia. A cnidarian's metabolic rate increases when it moves or feeds, but the contraction of its muscle fibers during these activities also speeds the flow of fluid through the canals, automatically speeding delivery of food as the demand increases.

Planaria

Free-living flatworms have higher metabolic rates than cnidarians. However, because they have flattened bodies, the general body surface still provides enough surface area for gas exchange. Flatworms have no separate circulatory system or blood. Food is distributed by the digestive cavity, which branches throughout the body, providing a large surface area from which nutrients can be absorbed. The excretory system of a flatworm also acts as its own transport system. It also branches throughout the body and collects waste substances that must be expelled (Figure 24–2).

Annelida

An annelid has a **coelom,** a fluid-filled body cavity in which organs can move independently of one another. The coelom provides room for blood vessels and space for a heart to expand and contract. The beating of an earthworm's dorsal blood vessel and five pairs of muscular "hearts" (enlarged blood vessels) moves the blood even while the animal is at rest. Muscles are more powerful than cilia, and so the circulation is faster in an earthworm's blood vessels than in the ciliary canals of cnidarians.

An earthworm has a simple **closed circulatory system,** in which the blood never leaves the vessels (Figure 24–3). The walls of the blood vessels are thin, and substances can diffuse easily between the blood and the extracellular fluid bathing the cells. The blood of earthworms contains a type of hemoglobin, which allows the blood to transport more oxygen.

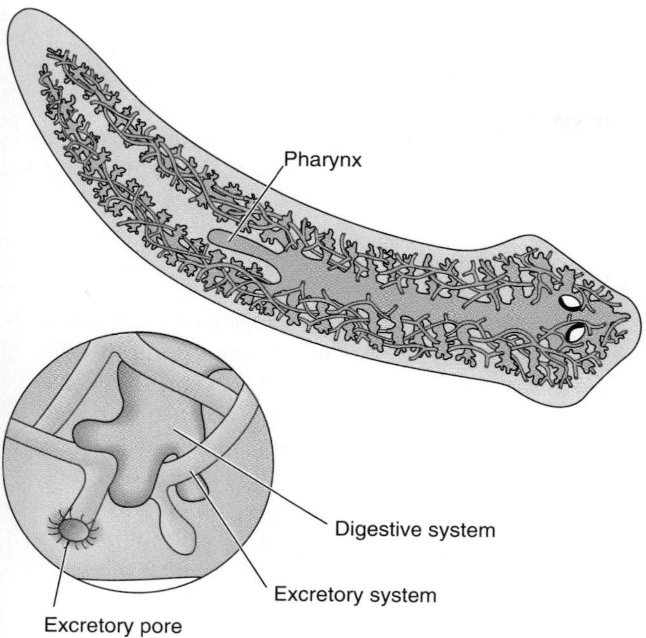

FIGURE 24-2

Transport in a planarian. The interwoven digestive and excretory systems branch throughout the body and perform their separate functions. The excretory system collects excess water and waste products and excretes these through excretory pores. The digestive system processes food that has been brought in through the pharynx and transports it throughout the body.

FIGURE 24-3

Open and closed circulatory systems. (a) Blood always remains within tubes as it travels through the circulatory system of a worm. **(b)** But in an insect, the blood is propelled around the body by a heart which opens into a large sinus or space that surrounds the body tissues.

(a)

Blood is pumped through a series of hearts

Dorsal vessel

Body wall

Veins

Subneural vein

Gut

Ventral vessel

Hearts

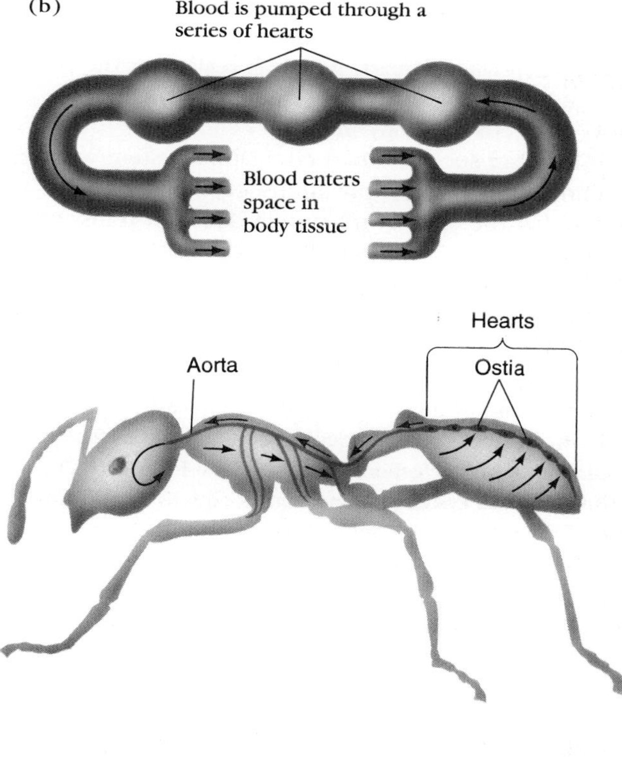

A circulatory system that moves the blood steadily at all times permits division of labor among organs and tissues. In flatworms, which have no circulatory system, the gut and excretory organs branch throughout the body, serving as their own circulatory systems. In an annelid, by comparison, the gut is a simple unbranched tube, specialized only for digestion. The separate circulatory system carries the digested food to all of the body cells.

The localization of function in specific organs is one of the characteristics of higher animals. An outstanding example is the clumping of nerve cells to form a brain. Information is processed more rapidly and the animal responds faster and more precisely than it could if messages had to travel around a nerve network scattered throughout the body. However, since nerve cells are easily damaged by temporary shortages of food and oxygen, a brain with many nerve cells is possible only with a circulatory system that supplies these needs rapidly and reliably.

Insects

The insect circulatory system is of special interest. Many insects are very active, and so we would expect to find an efficient circulatory system. Surprisingly, most insects have **open circulatory systems,** with only one open-ended blood vessel. This vessel releases blood into the body cavity, where it bathes the tissues directly. An insect's blood is moved by contractions of a long, thin-walled heart, which is really just the posterior end of the blood vessel (Figure 24–3), and by movements of the body muscles.

How can such an open circulatory system work efficiently enough to supply an insect's needs? The secret is that the circulatory system does not transport oxygen. Oxygen reaches cells by way of the **tracheal system,** a series of air-filled tubes that branch throughout the body. Other substances can travel more slowly via the blood.

Cnidarians and planarians have sufficiently large surface areas and low metabolic rates to permit gas and fluid exchange to occur through their general body surfaces. In insects, an extensive tracheal system channels gases directly to and from the vicinity of the cells, and an open circulatory system transports dissolved substances. Annelids have a coelom and a closed circulatory system that transports gases and fluids quickly and efficiently.

24–B CIRCULATION IN THE VERTEBRATES

All vertebrates have closed circulatory systems. Exchange of substances between the blood and the extracellular fluid occurs only across the thin walls of the **capillaries,** the narrowest blood vessels. Contractions of a strong, muscular heart exert the pressure needed to force the blood through the capillaries. Blood travels from the heart to the capillaries through large vessels called **arteries** and returns from the capillaries to the heart through **veins.**

Fishes

In fishes, the heart consists of a series of three chambers, which collect the blood and then pump it out (Figure 24–4). Blood leaves the heart via a short, muscular artery, the **ventral aorta** (pl. **aortae**), and travels to the gills. Here the aorta branches into smaller vessels, and eventually capillaries, where the blood picks up oxygen. From the gill capillaries, blood flows into the **dorsal aorta,** whose branches distribute blood to the capillaries of all the body organs (Figure 24–4). Blood returns to the heart through the veins.

This type of circulation, in which blood passes through the heart only once in a complete circuit around the body, is called a **single circulation.** Such a system has the advantage that all of the blood going to the body has already been oxygenated in the gills. A disadvantage is that the narrow gill capillaries slow down the blood flow, so that blood leaves the gills at a much lower pressure than when it entered. No matter how hard the heart pumps, the blood in a fish's dorsal aorta is travelling at a relatively low pressure, because it has had to pass through the gill capillaries. This slows the rate of oxygen delivery to the cells and limits the metabolic rate that fish can attain.

Amphibians and Reptiles

In higher vertebrates, the problem of low blood pressure in the body capillaries is overcome by a **double circulation,** which passes blood through the heart twice in each complete circuit around the body. Blood is first pumped from the heart to the lungs, where it is oxygenated. Blood then returns to the heart and is pumped through it a second time. This raises the blood pressure again before the blood goes out to the rest of the body. The hearts of birds and mammals are divided into two sides, right and left. Each side has an **atrium** (pl: **atria**) or receiving chamber and a **ventricle** or pumping chamber. The right side of the heart receives deoxygenated blood from the body and sends it to the lungs. Oxygenated blood returns to the left side of the heart, which then pumps it to the body (Figure 24–4).

The hearts of amphibians and most reptiles are not completely divided into left and right halves (although crocodilians *do* have completely divided atria and ventricles). Blood returning from the body enters the right

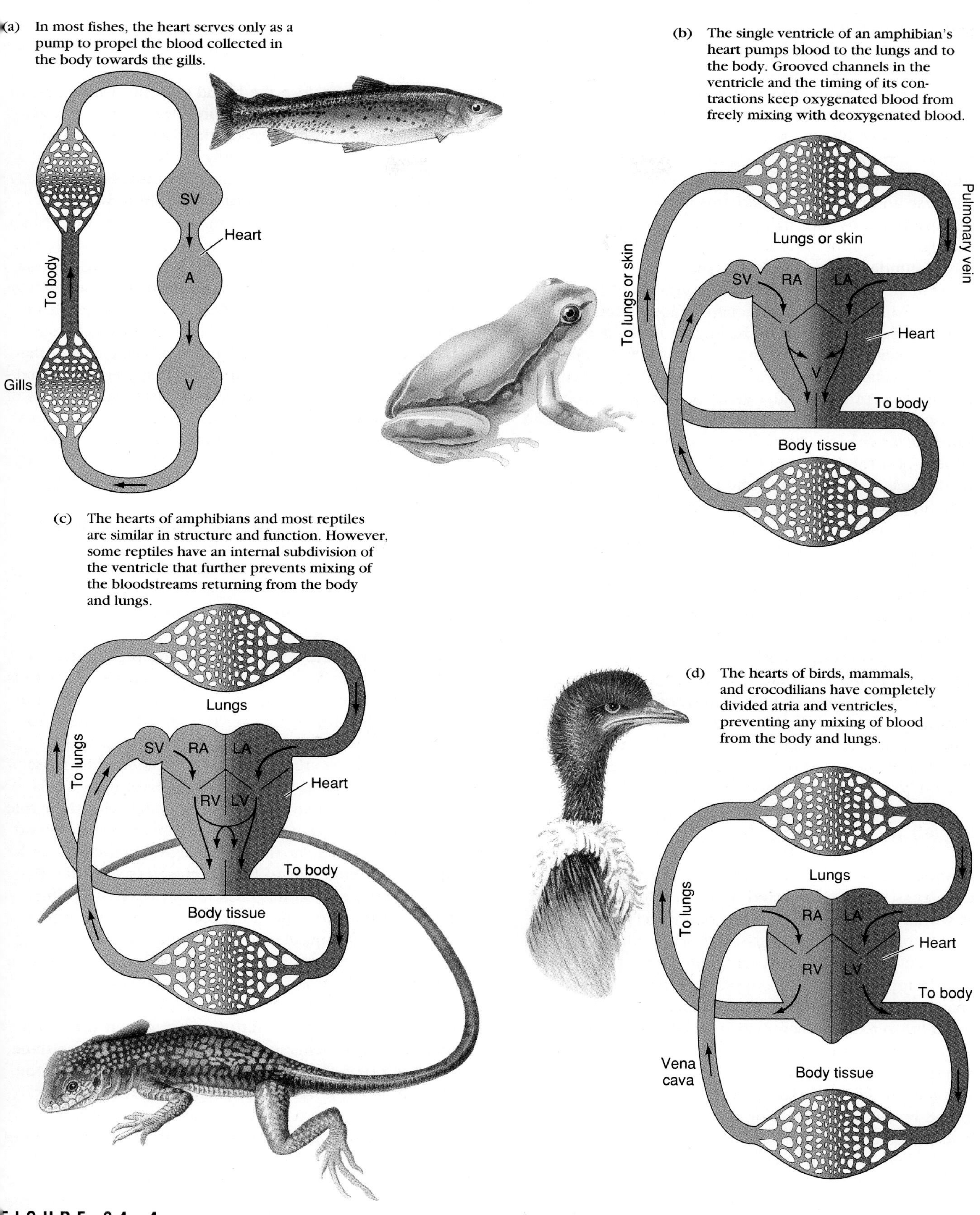
(a) In most fishes, the heart serves only as a pump to propel the blood collected in the body towards the gills.
SV
Heart
A
V
To body
Gills
(b) The single ventricle of an amphibian's heart pumps blood to the lungs and to the body. Grooved channels in the ventricle and the timing of its contractions keep oxygenated blood from freely mixing with deoxygenated blood.
Lungs or skin
Pulmonary vein
To lungs or skin
SV
RA
LA
Heart
V
To body
Body tissue
(c) The hearts of amphibians and most reptiles are similar in structure and function. However, some reptiles have an internal subdivision of the ventricle that further prevents mixing of the bloodstreams returning from the body and lungs.
Lungs
To lungs
SV
RA
LA
Heart
RV
LV
To body
Body tissue
(d) The hearts of birds, mammals, and crocodilians have completely divided atria and ventricles, preventing any mixing of blood from the body and lungs.
Lungs
To lungs
RA
LA
Heart
RV
LV
To body
Vena cava
Body tissue

FIGURE 24-4

Blood flow through the hearts of vertebrates. Blood with low oxygen content returning from the body is colored blue and blood with high oxygen content returning from respiratory sites is colored red.

atrium and is pumped to the incompletely divided ventricle, where it is channelled to the gills, lungs, or skin (Figure 24-4). Blood returning from the lungs enters the left atrium and then the undivided ventricle, where it is pumped out the ventral aorta to the body. In these animals, the two streams in the ventricle are kept separate by internal, grooved channels and the timing of the contractions.

An undivided ventricle is somewhat inefficient because it permits some mixing of oxygenated blood from the lungs with deoxygenated blood from the body. So what is gained by having an undivided ventricle? The answer requires that we consider the environments in which these animals live. Most amphibians and many reptiles spend a great deal of time in or under water, where they cannot breathe air. By having an undivided ventricle, blood from the body can be routed past the lungs, and sent out to the tissues for further gas exchange with the watery environment. And, in most amphibians, a considerable amount of gas exchange occurs in blood vessels under the skin, an added advantage when the animal is submerged.

In the majority of reptiles the movement of blood through the heart is somewhat different. The blood obtains oxygen only in the lungs, and is immediately returned to the heart. There are two atria, but in most reptiles the ventricle is only partially divided. However, the valves in the ventricle work in such a way that there is little mixing of blood from the two sides. Thus, the heart is functionally, if not structurally, divided. In reptiles, blood travels from the heart to the body through paired dorsal aortae, one on each side of the body.

Mammals and Birds

Blood leaving the heart for the body by way of two aortae, as it does in reptiles, loses pressure faster than it would if it were travelling in one large vessel. It is not surprising that mammals and birds, with their very high metabolic rates, have only one aorta: the other has disappeared. Birds have retained the right, and mammals the left, of the reptilian paired aortae.

Both birds and mammals have double circulations with the ventricles completely separated. This has two important effects. First, keeping oxygenated and deoxygenated blood separate in the heart ensures that blood reaching the body organs from the aorta contains as much oxygen as possible. Second, to animals with a high metabolic rate, it is important for the blood in the aorta to be under considerable pressure. The blood loses pressure as it passes through the capillaries of the lungs. Returning it to the heart after it passes through the lungs permits the heart to raise the pressure again before the blood goes out to the rest of the body. Higher blood pressure means faster circulation. Oxygen and food reach the tissues faster, and waste is removed more rapidly.

Vertebrates have closed circulatory systems in which the heart pumps and directs blood through arteries to capillaries, where gases and fluids are exchanged. Blood is collected by veins and returned to the heart. Fishes have a single circulation, in which blood returning from the body passes through the heart to the gills, and on to the body. In the double circulation of amphibians, reptiles, birds, and mammals, the blood passes through the heart twice in each circuit of the body. As a result, blood pressure is high in every artery and substances are transported rapidly. The right and left sides of bird and mammal hearts are divided, resulting in the complete separation of oxygenated and deoxygenated blood.

24–C THE MAMMALIAN CIRCULATORY SYSTEM

Blood Vessels

Arteries, capillaries, and veins are the pipes through which blood travels to the tissues. **Arteries** are vessels that carry blood away from the heart. Their walls are muscular and highly elastic. The arteries branch and rebranch into capillaries.

Capillaries are so narrow that blood cells must pass through them in single file. The capillary walls are only one cell layer thick, in keeping with their role as the sites where substances pass between the blood and the extracellular fluid. The highly branched capillaries come together to form larger vessels which finally combine to form **veins,** blood vessels leading back to the heart. The walls of veins contain connective tissue and muscle, as do those of arteries, but veins are much less elastic. They have thinner walls and tend to have larger internal diameters (Figure 24-5).

Another important difference between veins and arteries is that veins in the lower body contain **valves,** flaps of tissue that help to keep blood flowing in one direction. These valves open under the pressure of blood going toward the heart, and close when the blood begins to go backward in response to the pull of gravity (Figure 24-5).

When the walls of a vein are weakened, blood may collect in the vein and distend it so much that the valve flaps cannot meet. Because the valve cannot now prevent blood from flowing backwards, pools of blood

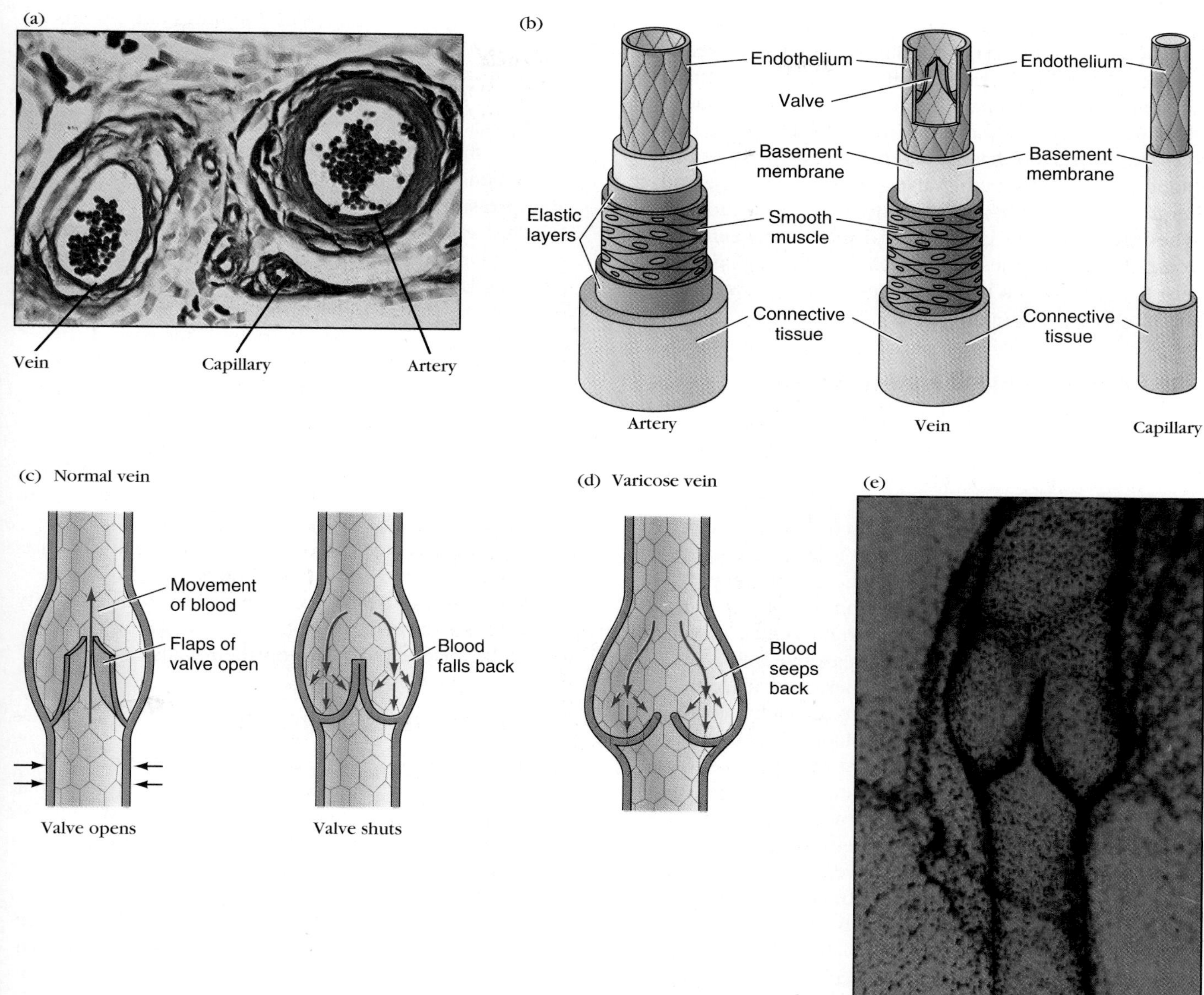

FIGURE 24-5

Blood vessels. (a) Cross-section of a vein, capillary, and artery . **(b)** A comparison of the layers that form the walls of blood vessels. A single layer of epithelium, called the endothelium (endo = within), forms the inner lining of all blood vessels. In veins and arteries the endothelium is surrounded by muscle and connective tissue. The walls of veins are thinner and more flexible than those of similarly sized arteries. The walls of capillaries are the thinnest, and are mostly composed of the endothelium. **(c)** When blood is squeezed through a vein toward the heart, the valve swings open and allows it to pass. If blood moves in the reverse direction, it fills the cup-like flaps of the valve and presses the edges together, preventing backflow. **(d)** The walls of a varicose vein are weak and allow blood to spread the edges of the valve flaps so that they cannot meet. Blood may then flow backward through the valve instead of moving towards the heart. **(e)** A valve in a vein. (a, E. Reschke/Peter Arnold, Inc; e, John D. Cunningham/Visuals Unlimited)

collect in the weakened vein. Such **varicose veins** can be very painful if the weakness is in a large vein. **Hemorrhoids** are varicose veins in the walls of the rectum. These veins have been damaged by pressure from conditions such as constipation or pregnancy. ■

The heart, too, has valves that direct the flow of blood in a one-way path. Valves between the atria and ventricles prevent backflow of the blood into the atria when the ventricles contract, and valves between the ventricles and arteries prevent blood from falling back into the heart when the ventricles relax after pumping the blood out (see Figure 24-5).

The Pathway of Blood Flow in the Body

Closed blood vessels, with strategically placed valves preventing backflow, ensure that the blood of a vertebrate flows in only one direction and in definite channels. Blood returns to the heart from the body via two large veins, the **venae cavae;** it then flows through the right atrium, and continues on into the right ventricle (see A Journey Through Human Circulation). Contraction of the right atrium sends more blood in to "top off" the ventricle. When the right ventricle contracts, it pushes the blood through a valve into the **pulmonary artery,** whose branches carry it to the lungs. In the lungs, the blood flows through capillaries surrounding the air-filled alveoli. Blood picks up oxygen and loses carbon dioxide across the thin walls of the alveoli and lung capillaries. The freshly oxygenated blood then flows through the **pulmonary veins,** back to the heart. This time, the blood enters the left atrium and passes through the valve into the left ventricle. When the left ventricle contracts, the oxygenated blood passes through a valve into the **aorta,** the main artery to the body. The wall of the left ventricle is much thicker and more muscular than the wall of the right ventricle; it must push the blood throughout the body, not just on the short journey to the lungs.

The aorta gives rise to many branch arteries, which take blood throughout the body. Blood leaving capillaries in the body enters veins, all of which empty into the venae cavae before they join the right atrium of the heart.

The Heart Cycle

As we have seen, the heart of a bird or mammal is really two pumps joined side by side. Each pump consists of a thin-walled atrium, which receives blood from veins and pumps it into the adjoining ventricle, and a thick-walled ventricle, which pumps blood into arteries. The right side of the heart receives blood from the body and pumps it to the lungs; the left side receives blood returning from the lungs and pumps it to the rest of the body.

The heart beats continuously throughout an animal's life. Each heartbeat is initiated by a "pacemaker," a small mass of tissue called the **sinoatrial node,** located at the entrance to the right atrium.

When the ventricles contract, they exert considerable pressure on the blood. This point in the heartbeat is called **systole,** and the pressure of the blood during ventricular contraction is known as the **systolic pressure.** Following contraction, the heart relaxes and blood rushes in from the venae cavae and pulmonary veins, partially filling the ventricles. This part of the heartbeat cycle is known as **diastole,** and the blood pressure at this time is called the **diastolic pressure.**

The brachial artery in the arm, just above the elbow, is usually used for measuring blood pressure. Blood pressure is expressed as the ratio of the systolic pressure over the diastolic pressure (both measured in millimeters of mercury). For instance, 120/80 is considered the "average" blood pressure. Systolic pressure indicates the force with which the left ventricle pushes blood. Diastolic pressure indicates the elasticity of the blood vessels; it is useful in diagnosing hardening of the arteries or strain on their walls.

Blood Pressure and Circulation

Blood pressure is determined mainly by the rate and force of the heartbeat, the volume of blood pumped at each stroke, and the resistance of the blood vessels to the flow of blood. An increase in any of these factors can cause an increase in blood pressure.

As the ventricles push blood into the arteries, the artery muscles relax and the elastic walls expand to accommodate the blood. As the blood passes, the muscular artery walls contract and exert pressure on it, helping the heart to push the blood along. After the blood leaves the main arteries, the walls of the other blood vessels offer only frictional resistance to its flow. Muscles in the walls of a blood vessel may contract or relax, changing the vessel's diameter, and this changes the blood pressure by making it harder or easier for the blood to pass through the vessel. The narrower the vessel, the greater the resistance, and so the speed of blood flow drops very low in the capillaries. The veins are flabby and do not help to push the blood back to the heart. In addition, blood in the veins below the heart must be returned against the pull of gravity. Thus, blood tends to collect in the veins.

Blood eventually does return to the heart, propelled mainly by the muscles of the body. When muscles contract, they squeeze against the outsides of veins, forcing blood to move along inside. The valves in the veins permit blood to flow only toward the

We are looking at the side of the heart that is closest to the front side of the body. Therefore, the right side of the heart appears to our left. The atria receive blood from veins and propel it into the ventricles. The ventricles contract to force blood into arteries.

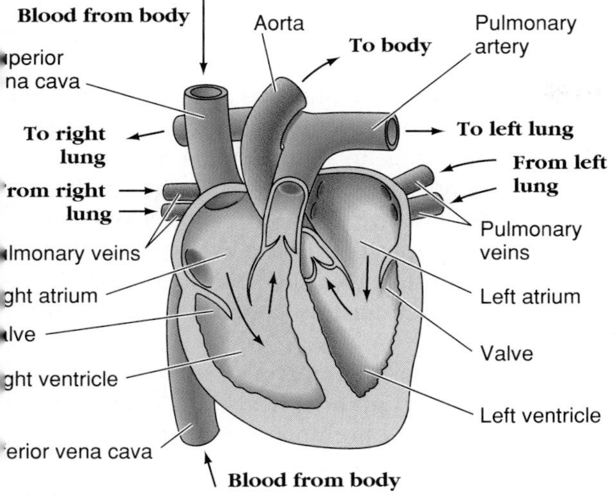

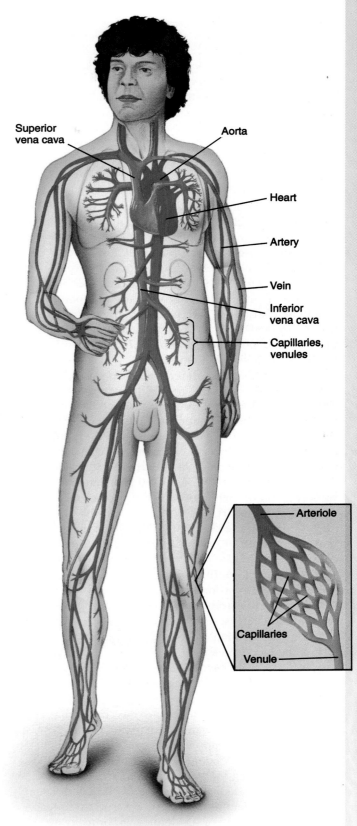

The path of blood in the mammalian circulation can be summarized:
right atrium ⟶ right ventricle ⟶ pulmonary artery ⟶
lung capillaries ⟶ pulmonary vein ⟶ left atrium ⟶
left ventricle ⟶ aorta ⟶ body arteries ⟶ body capillaries ⟶
veins ⟶ vena cavae ⟶ right atrium

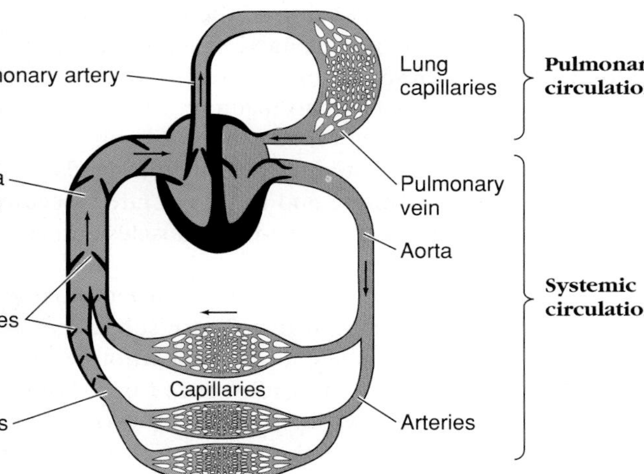

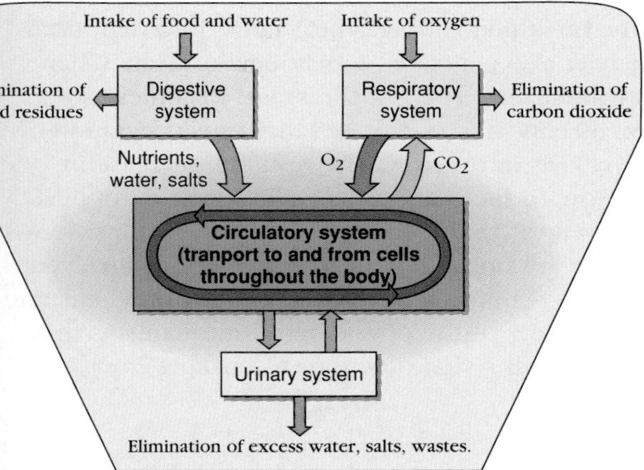

(c) A more realistic view of blood flow through a human, showing the major blood vessels associated with the heart.

A Journey Through Human Circulation. The heart propels blood throughout the thousands of large and small blood vessels of the body to distribute oxygen and nutrients and collect carbon dioxide and metabolic wastes.

BIO-BIT

High blood pressure is often called a "silent killer" because it has few if any noticeable symptoms and can be diagnosed only through a medical exam in which the pressure is measured.

heart. When the muscles again relax, the valves keep the blood from falling back. Because muscular contraction is needed to push blood through the veins, it is more tiring to stand still than to walk for an equal period. Standing allows blood to collect in the veins of the feet and legs. The feet swell with stranded blood, and the body temporarily loses the use of blood that should be distributing oxygen and nutrients to other tissues. Studies have shown that students who jiggle their feet are more alert, and perform better on long exams, than their peers who sit still.

Several different short-term mechanisms raise and lower the pressure in the arteries to maintain homeostasis. When the blood pressure rises above 60 millimeters of mercury, it stimulates **baroreceptors,** pressure receptors in the walls of arteries in the head and chest. Acting by way of the nervous system, the baroreceptors trigger (a) **vasodilatation,** or widening, of the small arteries, (b) decreased heart rate, and (c) vasodilatation of the veins (which decreases the return of blood to the heart). All of these lower the blood pressure. When the blood pressure falls rapidly, for instance as a result of heavy bleeding, several hormones constrict the blood vessels and usually restore the blood pressure to normal within a few minutes.

One of these hormones, vasopressin, is also involved in the long-term regulation of blood pressure. Vasopressin decreases the volume of water excreted by the kidneys, retaining more water in the blood and raising the blood pressure (see Section 26-E). Many people suffer from **essential hypertension,** long-term high blood pressure of unknown origin. Hypertension increases the likelihood of a stroke or heart attack. The usual treatment is **diuretic drugs,** drugs that increase the amount of water excreted by the kidneys. A diet very low in sodium may have the same effect. Drugs or diet both counteract the effect of vasopressin, increasing the volume of water excreted by the kidneys. The kidneys of people with hypertension require higher than normal blood pressure to function properly. The causes of this kidney abnormality are not known, but the condition appears to be hereditary and to involve damage to the blood vessels that supply the kidneys.

The Circulatory System's Adjustment to Exercise

The circulatory system adjusts in various ways to changes in physiological conditions. These adjustments are usually controlled by **negative feedback,** a mechanism whereby a change in some condition, such as blood pH, stimulates activity that brings the condition back to its normal range. Negative feedback systems ensure that the composition of the extracellular fluid remains almost constant. We shall consider, as an example, some of the circulatory system's responses to vigorous exercise.

As exercise begins, the nervous system sends impulses to the **adrenal glands,** near the kidneys, causing them to release the hormone **epinephrine** (also called **adrenalin**) into the bloodstream. Epinephrine causes blood vessels in the skin and abdominal organs to constrict, decreasing the blood supply to these organs and sending blood that is normally "stored" in these areas into more active circulation. This in effect increases the volume of blood available. Epinephrine also causes local vasodilatation of the small arteries and capillaries in the muscles and heart, increasing the blood supply to these organs (Figure 24-6). This tradeoff of blood supplies helps to maintain the blood pressure. There is not enough blood to fill the whole circulatory system in the dilated state. Epinephrine also stimulates faster breathing and heartbeat rate, speeding both delivery of extra oxygen to the muscles and removal of wastes.

Exercising muscles produce more carbon dioxide and lactic acid than those at rest. These substances make the blood more acidic as it passes through the muscles, and an increase in acidity does three things: 1. it makes the blood give up more of its oxygen in the muscles, 2. it increases the dilatation of the blood vessels in the muscles, and 3. it also stimulates the nervous system to increase the secretion of epinephrine, and the breathing and heartbeat rates. Intense muscular activity also generates a great deal of heat. When the hypothalamus (part of the brain) becomes too warm, it sends nerve impulses that cause dilatation of blood vessels in the skin. The resulting increase in blood flow to the body surface allows the extra heat to be given off to the environment.

These are only a few of the interactions involved in the body's adjustment to exercise, but they illustrate the complexity of the physiological mechanisms that adjust the body's vital functions to changes in its activity.

Diseases of the Circulatory System

2? Cardiovascular diseases are diseases of the heart and blood vessels (cardio = heart). More than half of all

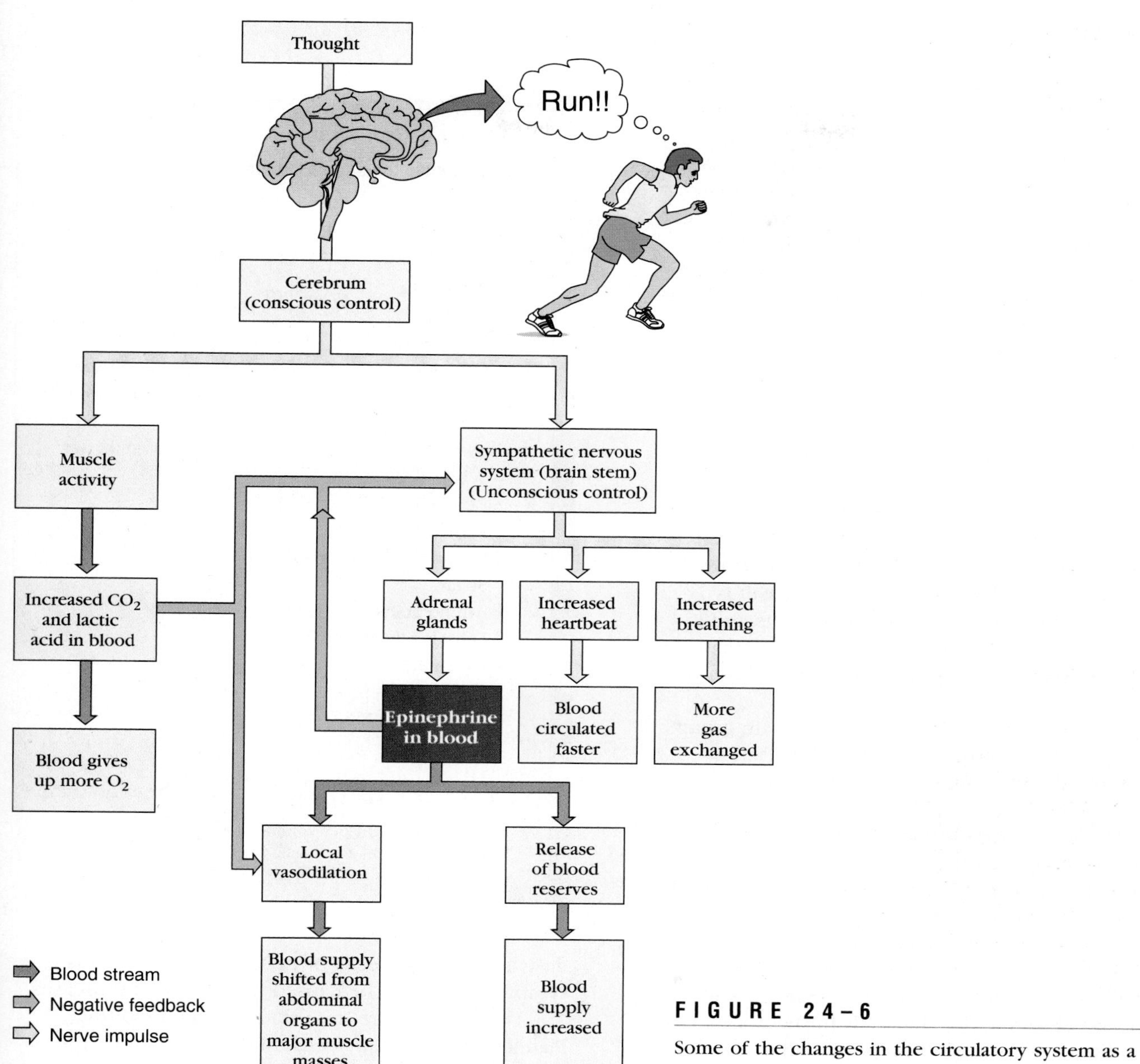

FIGURE 24-6

Some of the changes in the circulatory system as a response to exercise.

cardiovascular deaths are caused by heart attacks. A heart attack occurs when the blood supply to part of the muscle that makes up the heart fails. With their blood supply cut, the cardiac muscle cells stop contracting and may die. A heart attack may occur when one of the heart's arteries is obstructed by a blood clot. It can also be caused by atherosclerosis, a condition in which the artery walls are thickened, and the passageway for blood narrowed by the growth of cells and deposits of lipids and other materials (see *A Journey into Healthy Living:* Diet and Cardiovascular Disease, Chapter 22). Even if the patient recovers from the heart attack, part of the heart muscle may have been killed and the heart permanently weakened.

A stroke occurs when the blood supply to some part of the brain is damaged. As with a heart attack, this is apt to kill the cells deprived of the blood's life-giving oxygen. A stroke may result from a blood clot in one of the brain's blood vessels, or from the rupture of a weak blood vessel. Its severity depends on what part, and how much, of the brain is damaged. ■

BIO-BIT

People do not always know when they are having a heart attack. Heart attacks are confirmed by the presence of certain enzymes in the blood that have been released from damaged heart muscle.

Arteries carry blood away from the heart and have muscular and elastic walls. Veins return blood to the heart and have thinner walls that are less muscular and elastic. Capillaries carry blood from arteries to veins, and their thin walls serve as the site of gas and fluid exchange. Valves within veins and between heart chambers limit blood flow to a single direction. Blood pressure is a measure of the greatest force exerted when the ventricles contract and the lowest pressure when the ventricles are relaxed. It is affected by the rate and force of the heartbeat, the volume of blood pumped, the flexibility of the walls of the blood vessels, and the volume of water retained within the body. Negative feedback systems ensure that the blood pressure as well as the temperature and composition of extracellular fluid remain almost constant.

24–D BLOOD

The familiar red fluid called **blood** is really a tissue made up of a liquid containing several types of cells (Table 24–1). About half the volume of blood is made up of a fluid called **plasma,** and the other half is blood cells. The plasma contains various salts and a great variety of plasma proteins. **Serum** is plasma from which the proteins involved in clotting have been removed.

Blood cells can be divided into three main groups: the **white cells, red cells,** and **platelets** (Figure 24–7). Each microliter (millionth of a liter) of blood contains 4.7 to 9.7 thousand white blood cells. Most of the many types of white blood cells help protect the body from disease (Chapter 25).

Red blood cells are by far the most numerous cells in the blood (3.6 to 5.5 million per microliter). Their main function is oxygen transport. Mature mammalian red cells have no nuclei and contain mostly hemoglobin, a protein that binds oxygen. Red cells are produced from nucleated, dividing cells in the bone marrow. Red blood cells usually survive for about four months in the bloodstream. Then they break up, and certain white blood cells destroy their remains by phagocytosis. If the number of red blood cells falls, the resulting oxygen shortage causes kidney cells to secrete the hormone **erythropoietin** into the blood. This hormone stimulates the bone marrow to increase red blood cell production. These new cells boost the blood's oxygen-transporting capacity. As the blood's oxygen level returns to normal, erythropoietin production stops and red blood cell production returns to normal.

Anemia is a condition in which the blood contains fewer red blood cells or less hemoglobin than usual as a result of unusually slow production or fast destruction of red cells or hemoglobin. Anemia is a symptom that may be caused by a variety of diseases.

BIO-BIT

In an average adult human, 2.5 million red blood cells are produced every second.

Table 24–1
Main Components of the Blood

Water	45–54% vv*
Salts	
Sodium	2400 mg/l
Potassium	80 mg/l
Calcium	80 mg/l
Magnesium	26 mg/l
Chloride	2600 mg/l
Bicarbonate	1500 mg/l
Plasma Proteins	7–9% wv†
Blood Cells	40–50% wv
White cells	
Red cells	
Platelets	
Substances Transported by Blood	
Sugars	
Amino acids	
Fatty acids, glycerol	
Hormones	
Nitrogenous wastes	
Carbon dioxide	
Oxygen	

*vv means volume per volume; e.g., 12 ml per 100 ml is 12% vv.

†wv means weight per volume; e.g., 13 g per 100 ml is 13% wv.

The platelets are important in blood clotting. Platelets are not really cells, but are formed by the pinching off of parts of the cytoplasm of large cells in the bone marrow.

Blood Clotting

Clotting begins when the wall of a blood vessel is broken or damaged. The injured cells release substances that attract blood platelets. When the platelets come into contact with fibers of the structural protein collagen exposed by the injury, they disintegrate and form a temporary plug for the injured vessel. The platelets also release two substances. The first is **serotonin,** which causes the muscles in the blood vessel wall to contract and constrict the vessel, reducing blood loss. Platelets also release the enzyme **thromboplastin,** which changes one of the plasma proteins, **prothrombin,** into **thrombin.** Thrombin is also an enzyme; it changes another plasma protein, **fibrinogen,** into **fibrin.** Strands of fibrin form a meshwork around the disintegrated platelets. Still another plasma protein

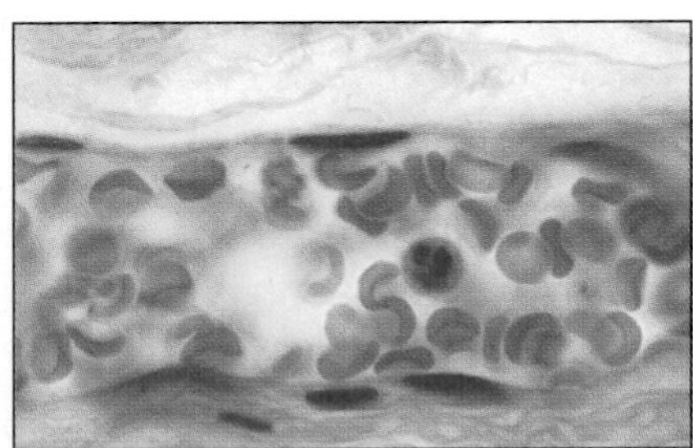

FIGURE 24-7

Red and white blood cells within a blood vessel. One white blood cell can be seen near the right center of the photograph. Its nucleus appears as dark regions in its center. Red blood cells from an adult mammal, such as those pictured here, have lost their nuclei during their maturation process prior to entering the circulatory system. (A. Owczarzak/BPS)

converts the loose fibrin meshwork into a tough, hard, permanent plug or clot, which seals off the injured part of the blood vessel from the exterior (Figure 24-8). ■

Blood is composed of a fluid called plasma, in which red blood cells, white blood cells, and platelets are suspended. Red blood cells transport oxygen and are the most numerous blood cells. White blood cells help protect the body from disease. Platelets play a vital role in blood clotting.

24-E THE LYMPHATIC SYSTEM

In many ways, the body's capillary networks are the most important parts of the circulatory system, for it is here that the exchange of substances between blood, extracellular fluid, and cells takes place. Most substances, such as glucose and oxygen, leave the blood and enter the extracellular fluid by diffusing down the concentration gradient between the two. Wastes and carbon dioxide enter the blood in the same manner. In addition, water and larger molecules, such as hormones and small proteins, enter and leave the blood either by moving through spaces between the cells of the capillary walls or by passing through the cell itself by way of endocytotic vesicles.

Water leaves the capillaries under the pressure generated as blood is forced through a tube of small diameter. Toward the end of a capillary, so much water has been lost that the proteins left behind are quite concentrated and most (but not all) of the water returns to the capillary by osmosis.

The remaining fluid is collected and drained away through the **lymphatics,** thin-walled vessels with valves that ensure one-way flow. The lymphatics eventually join to form the **thoracic duct** and the **right lymph duct,** which empty into veins near the heart (see Figure 25-5). Often these are the only lymph vessels large enough to be visible. The lymphatics perform several vital functions:

1. They drain excess tissue fluid from the extracellular fluid back into the circulatory system.
2. They temporarily store fluids taken into the body. Some of the fluid absorbed from the digestive tract finds its way into the lymphatic system, which re-

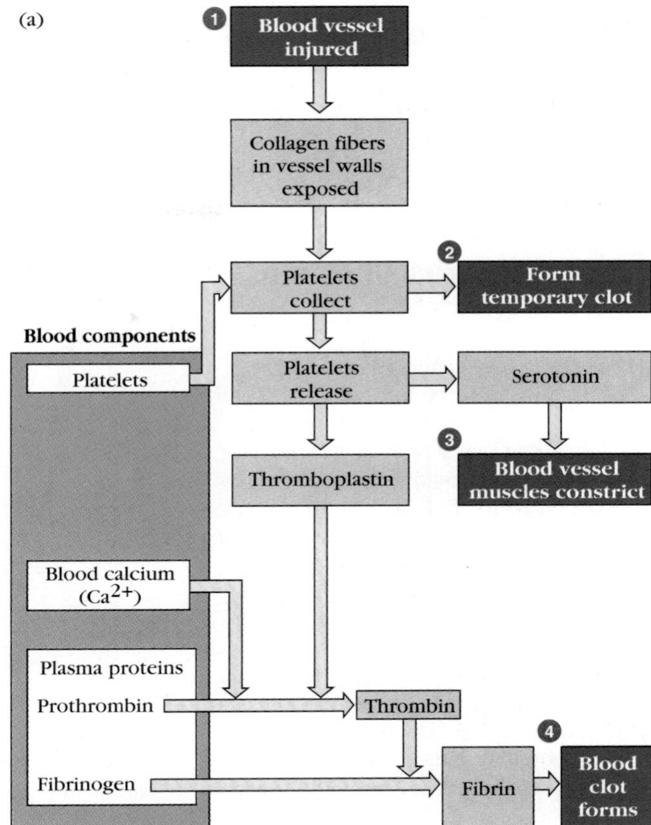

(b)

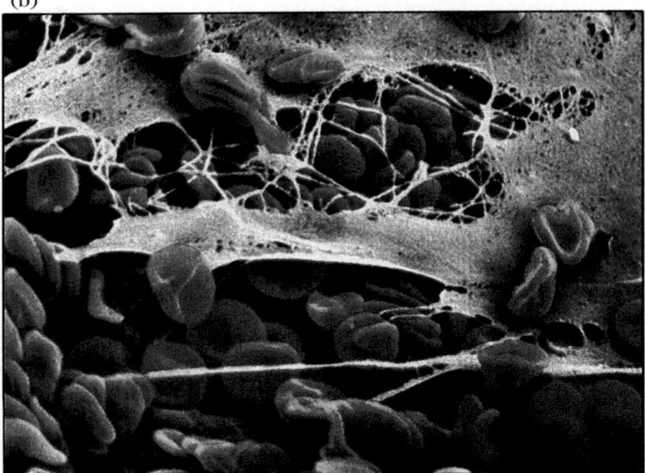

FIGURE 24-8

Blood clotting. (a) A simplified diagram of some of the reactions involved in the clotting of blood. **(b)** False color scanning electron micrograph of part of a blood clot, showing red blood cells caught in a network of fibrin. (Lennart Nilsson, Boehringer Ingelheim International Gmbh)

A Journey into Evolution

Emperor Penguins: Unanswered Questions

At first glance, penguins don't seem to be birds at all. They are flightless, and from a distance they don't appear to have feathers, the unifying characteristic shared by all birds. As they float at the water's surface, these immaculate white and black birds look as smooth as seals, and their flippers look scaly. An even closer examination reveals that these tiny scales are actually small, flattened feathers that cover almost every inch of a penguin's skin. Penguin behavior is also superficially un-bird like. When traveling long distances they swim like small whales, repeatedly "porpoising" up out of the water, grabbing a breath, and in a single movement, arcing down below the surface again. Their aerial leaps cover 3 to 4 meters; average porpoising speeds have been measured at 24.1 kph (15 mph), and it is thought that penguins can double this "cruising" speed for short bursts. On land penguins have an erect, waddling gait, and to move more quickly over ice, they flop down on their bellies and "toboggan," propelling themselves with their feet and flippers.

All penguins are oceanic carnivores, superb high speed underwater swimmers and divers who "fly" through the water in pursuit of shrimp-like crustaceans called krill, as well as squids, and fishes. Penguin wings are highly adapted flippers that are moved up and down through the water with great power, nearly clapping together on the apex of the upstroke. While ostriches, kiwis, and other flightless birds have lost the deep keel of the breastbone (sternum) and thus the sites for the attachment of heavy flight muscles, penguins retain the sternal keel and have massive flight muscles. They also add a third adaptation to underwater flight: fused wing bones. Unlike all other birds, penguins cannot fold their wings. Penguin wing bones are flattened and fused to form stiff flippers, while its feet act as rudders. Penguins dive to pursue their prey and typically hunt in tight groups whose movements are synchronized. Although no one is exactly sure how penguins locate their prey, we surmise that their spiny tongues help them hold onto it once they have caught it.

FIGURE 24-A

Emperor Penguin and chick. This two- to three-week-old chick is insulated from the Antarctic ice by the warm feet of its parent, while the parent's body blocks the wind and keeps the small chick warm. As the chick grows it becomes too big to snuggle beneath the loose fold of feathered skin just above the parent's feet, and it huddles with other chicks to keep warm. (Doug Allen/Animals Animals)

Diving Adaptations of Emperors

Although all penguins have these specializations for diving, Emperor Penguins are perhaps the most extremely specialized of all. Standing about three feet fall and weighing between 40 and 90 pounds, depending upon the time of year, Emperor Penguins are the largest living penguins. They are the only birds who never come to solid land, raising their young on ice fields (technically called fast ice) that break up in the Antarctic spring, taking the young birds out to sea. While typical birds have skeletal adaptations to decrease weight (absence of teeth, light bones full of air spaces, fusion of many skeletal parts), Emperor Penguins have adaptations to increase weight and, presumably, thereby increase the efficiency of diving. Ten pounds of stones were discovered in the crop of one Emperor, presumably used as ballast as well as to grind the shells of shrimp-like krill. As the largest penguins, Emperors dive the deepest and stay down the longest. Although Emperor Penguins usually only dive for three to eight minutes, one investigator reported a record-breaking dive of 18 minutes. An Emperor Penguin carrying a depth recorder and watched from an underwater chamber was observed to dive to a record 256 meters (78 feet). Emperor Penguins feed and dive in the Antarctic winter darkness, hunting squid that are up to a meter long. Unlike many other penguins who are found in ice-free waters, Emperors dive beneath thick pack ice as well as under stationary, fast ice.

Problems Associated with Diving

Like all diving animals, Emperor Penguins face a variety of problems. First, diving animals must have a supply of air. Second, as the animal dives and then resurfaces, it encounters changes in pressure of about one atmosphere (760 millimeters of mercury) for every 10 meters of depth. At high external pressures the gas in the lungs is compressed and forced into solution in the blood. Then, when the animal resurfaces, pressure is reduced and the dissolved gases, mainly nitrogen, come out of solution and form bubbles that may block blood vessels (caisson disease or "the bends"). Finally, water, especially in polar regions, absorbs heat from an animal's body very rapidly, threatening a warm-blooded animal with death from hypothermia: rapid loss of body heat.

Penguins and seals show different adaptations to oxygen carrying capacity, perhaps reflecting their separate evolutionary lineages. In mammals the trend seems to be to decrease lung volume anatomically or behaviorally. These are adaptations that help prevent caisson disease: the less air in the lungs, the less nitrogen will dissolve in the blood during a dive (air is 79% nitrogen). Furthermore, just before a seal dives, it exhales and this compression of the lungs forces much of the air into the tracheae and bronchi, air passages whose walls are impermeable to gases. This air cannot dissolve in the blood. (Some nitrogen inevitably does enter the blood, but not enough to do damage.) In contrast, penguins *inhale* before making a dive. Like all birds, penguins have a system of air sacs that may have an important function as oxygen stores during dives.

Most diving animals show **bradycardia** (slowing of the heart rate) when they dive. For example, a seal's heart rate drops from 150 to 10 beats per minute. In penguins the heart rate drops to one-fifth of the normal, resting rate. Bradycardia is seen in other animals, including humans when they submerge, and, interestingly, fish undergo bradycardia when they are removed from water. The advantage of bradycardia is probably that it saves the body energy and oxygen. In addition, blood vessels constrict and reduce the circulation to the kidneys, gut, and so forth. This conserves oxygen for use by the brain, which must not be deprived of oxygen. Furthermore, although circulation to the head is maintained, the respiratory center in the nervous system of a diving animal tolerates relatively high levels of carbon dioxide, and thus it does not stimulate the animal to breathe while submerged. Other body organs carry out fermentation rather than respiration of their food, and most of the carbon dioxide and lactic acid they produce is retained in the tissues; when seals and other diving animals resurface, a surge of metabolic wastes enters the blood stream. In seals, this anaerobic metabolism occurs only during occasional long dives. During more common, short dives the muscles continue to use oxygen.

Many diving mammals can store extra oxygen for use during a dive. They have more red blood cells than do non-diving animals which permits their blood to carry more oxygen. Their muscles also contain extra quantities of **myoglobin,** an oxygen-carrying pigment related to hemoglobin. Seals carry most of their oxygen store in the blood, whereas whales store more oxygen in myoglobin. It is estimated that a penguin stores 60% more oxygen per kilogram than a human does. In addition, the temperature of peripheral body surfaces drops during a dive, an adaptation that many mammals also have.

All penguins and diving mammals reduce heat loss by reducing the flow of blood to the skin during a dive. A penguin's unusual, waterproof coat of feathers traps air and provides 80% of the bird's insulation, while the thick layer of blubber beneath its skin provides the balance. In addition the low surface-area-to-volume ratio characteristic of the torpedo-shaped bodies of penguins, seals, and whales, and the heat conserving arrangement of blood vessels within their appendages both diminish heat loss.

Unanswered Questions

Although it seems that we know quite a lot about Emperor Penguins, there are many mysteries that surround them. We think that Emperors return to the vicinity where they hatched to raise their young, but no one is sure exactly how they accomplish this feat of navigation. While salmon home in on the smell of the stream where they hatched, and other penguins may also use the scent of "home," this cue will not work for Emperor Penguins because their rookeries are on ice that melts in the summer of each year. We know that Emperors hunt for squid in the black waters that surround Antarctica, often diving exceptionally deep for their prey, and we know that they have special eye muscles that pull on their eyeballs, allowing them to focus underwater—a built-in face mask—but we do not know how Emperors locate these schools of squid. Nor do we understand how the movements of an underwater "squadron" of Emperors are coordinated and synchronized.

In Emperors and many penguins, both parents take turns incubating and brooding the egg(s) and chick(s). We know that each Emperor parent spends long periods at sea, feeding, while its mate incubates their egg and later the chick that hatches from it. Each is insulated from the ice and sub-zero temperatures by the parent's warm, naked feet, and covered by a fold of warm, feathered skin. We do not know how the feeding parent (who is typically gone for more than a month, feeding at sea) relocates the rookery, which can be a considerable distance from the edge of the ocean, or how it recognizes the call of its mate among all the other identical-sounding (to human ears) calls of neighbors in the rookery. We believe that the black and white markings on the chick's head (that make it look as though it is wearing the helmet and goggles of a World War I flying ace) help the feeding parent locate its mouth and feed it in the Antarctic darkness, but we still don't know how each parent finds its own large chick once it has joined a group of identical-looking companions in a mob of young birds.

So many questions, so few answers—and perhaps so little time. Emperor Penguins and all other Antarctic wildlife are protected from hunting by international agreements. Unfortunately nothing protects them from the increased levels of UV radiation that penetrates Earth's atmosphere through the newly formed hole in the ozone layer. It is predicted that increased UV radiation will cause corneal damage to penguins as well as weaken the base of the Antarctic food chain where they act as top predators. Increased UV radiation kills the photosynthetic algae and cyanobacteria, the food eaten by krill, and thus ultimately threatens all animals in the Antarctic food web.

leases it gradually so that the kidneys do not have to perform sudden surges of urine excretion.

3. They carry large molecules, such as proteins and hormones, from the cells where they are produced to the bloodstream. Such molecules are too large to cross capillary walls and so cannot reach the bloodstream directly.
4. Some food molecules, especially fats, move into the lymph rather than into the blood when they are absorbed from the small intestine. The lymphatics form the main route by which such molecules reach the blood.
5. Lymph nodes occur in several areas of the body. These nodes are an important part of the body's defense against disease (Chapter 25).

The lymphatic system consists of a series of ducts that collect body fluids, proteins, and digested fats from the extracellular fluid and empty them into the venous system.

SUMMARY

1. Most animals have transport systems that move substances within the body.
2. In some primitive animals, transport of food is carried out by the gastrovascular cavity.
3. Most animals with a coelom have true circulatory systems. An open circulatory system has few blood vessels, and blood directly bathes the cells. The closed circulatory systems of many invertebrates and of all vertebrates have blood vessels through which blood is pumped by the heart.
4. The circulatory systems of vertebrates show an evolutionary trend from single to double circulation. The double circulation of birds and mammals provides complete separation of oxygenated and deoxygenated blood and raises the blood pressure in the body's capillaries.
5. The quick and orderly flow of blood is accomplished by a muscular pump, the heart, and a set of tissue pipes, the blood vessels. Blood flows through the circuit from the region of high pressure, the contracting ventricles of the heart, through the vessels at progressively lower pressure, until it returns to the heart. Valves in veins and in the heart prevent backflow.
6. Blood pressure and blood supply in various parts of the body may be regulated by dilation and contraction of small arteries and capillaries, and by changes in heartbeat rate and volume of blood pumped by the heart at each beat.
7. The circulatory system responds to changes in the body's activities so that the body's new needs are met. During exercise, the amount of blood flowing and the rate of flow are increased, and more blood is diverted to the active muscles.
8. Blood is a liquid tissue consisting of a watery fluid that carries salts, proteins, and blood cells.
9. White blood cells defend the body against disease.
10. Red blood cells contain hemoglobin, a protein that binds oxygen and transports it from the lungs to the capillaries of the body tissues.
11. The blood platelets are important components in the clotting mechanism of the blood. Clotting helps to plug the vessel walls after injury, preventing loss of vital fluids or entry of disease organisms.
12. The lymphatic system consists of vessels that collect extracellular fluid, proteins, and digested fats, and empty them into the venous half of the circulatory system.

SELF-QUIZ

1. Place an "X" in the boxes to indicate which transport systems exhibit the feature mentioned:

	cnidarian	earthworm	insect	fish	mammal
food transport					
oxygen transport					
high-pressure fluid picks up food					
muscular circulatory pump(s)					

2. Select *two* advantages of a double circulation over a single circulation:
   a. In the double circulation, all the blood going to the tissues is oxygenated, whereas in the single circulation it is not
   b. In the double circulation, the blood can transport twice as many types of substances
   c. In the double circulation, the blood is at higher pressure when it enters the body tissues
   d. In a double circulation, the blood travels around the body faster
   e. In a double circulation, there are twice as many blood vessels servicing the body tissues
3. The greatest amount of oxygen will be lost from the blood while it is travelling through:
   a. the capillaries around the alveoli
   b. the left atrium of the heart
   c. the arteries
   d. the capillaries in the body
   e. the veins

4. If you were asked to dissect an animal so as to reveal a valve, all of the following places would be good to try *except:*
   a. the opening between the right atrium and the right ventricle
   b. the fork where the pulmonary artery splits and one branch goes to each lung
   c. the base of the aorta where it leaves the left ventricle
   d. a vein in the leg
   e. a lymph vessel that empties into the thoracic duct
5. Which of the following organs will receive a decreased flow of blood during strenuous exercise?
   a. brain
   b. skin
   c. liver
   d. heart
   e. lungs
6. When a person exercises hard, all of the following occur *except:*
   a. blood glucose decreases
   b. ADP increases
   c. glycogen increases
   d. lactic acid increases
   e. CO_2 increases

THINKING CRITICALLY

1. What forces, besides contraction of the heart, may move fluids in the bodies of animals?
2. What restrictions in size and activity are imposed on animals that possess an open circulation combined with a tracheal system?
3. Can you think of any reasons why cephalopods (squid, octopus, and so forth) are the only molluscs with closed circulatory systems, and why other molluscs (snails, clams, and so forth) manage with open systems?
4. Birds and mammals have four-chambered hearts and maintain high body temperatures. In what way might these two characteristics be linked?
5. Arteries usually lie deep in the body, whereas veins lie near the surface. What is the advantage of this arrangement?
6. Why does blood flow to the skin increase during strenuous exercise?
7. Misinformed people often define arteries as blood vessels that contain oxygenated blood, and veins as vessels that contain deoxygenated blood. What is wrong with these definitions?
8. Why do you suppose the body core of an alligator basking in the sun warms up faster when the animal is moving than when it lies still?

SELECTED KEY TERMS

aorta, *p. 510*
artery, *p. 510*
atrium, *p. 510*
baroreceptor, *p. 516*
capillary, *p. 510*
closed circulatory system, *p. 508*
coelom, *p. 508*
diastolic pressure, *p. 514*
erythropoietin, *p. 518*
extracellular fluid, *p. 507*
lymph duct, *p. 519*
open circulatory system, *p. 510*
plasma, *p. 518*
platelet, *p. 518*
red blood cell, *p. 518*
serum, *p. 518*
systolic pressure, *p. 514*
tracheal system, *p. 510*
valve, *p. 512*
vascular system, *p. 507*
vein, *p. 510*
ventricle, *p. 510*
white blood cell, *p. 518*

SUGGESTED READINGS

Kanwisher, J. W., and S. H. Ridgway. "The physiological ecology of whales and porpoises." *Scientific American,* June 1983.

Lawn, R. M. "Lipoprotein (a) in heart disease." *Scientific American,* June 1992.

Ramsay, J. A. Chapter 2: Circulation, *Physiological Approach to the Lower Animals,* 2d ed. New York: Cambridge University Press, 1968. A brief, but charming, comparative invertebrate physiology text.

Schmidt-Nielsen, K. "Countercurrent systems in animals." *Scientific American,* May 1981.

Wood, J. E. "The venous system." *Scientific American,* January 1968.

Zapol, W. M. "Diving adaptations of the Weddell seal." *Scientific American,* June 1987.

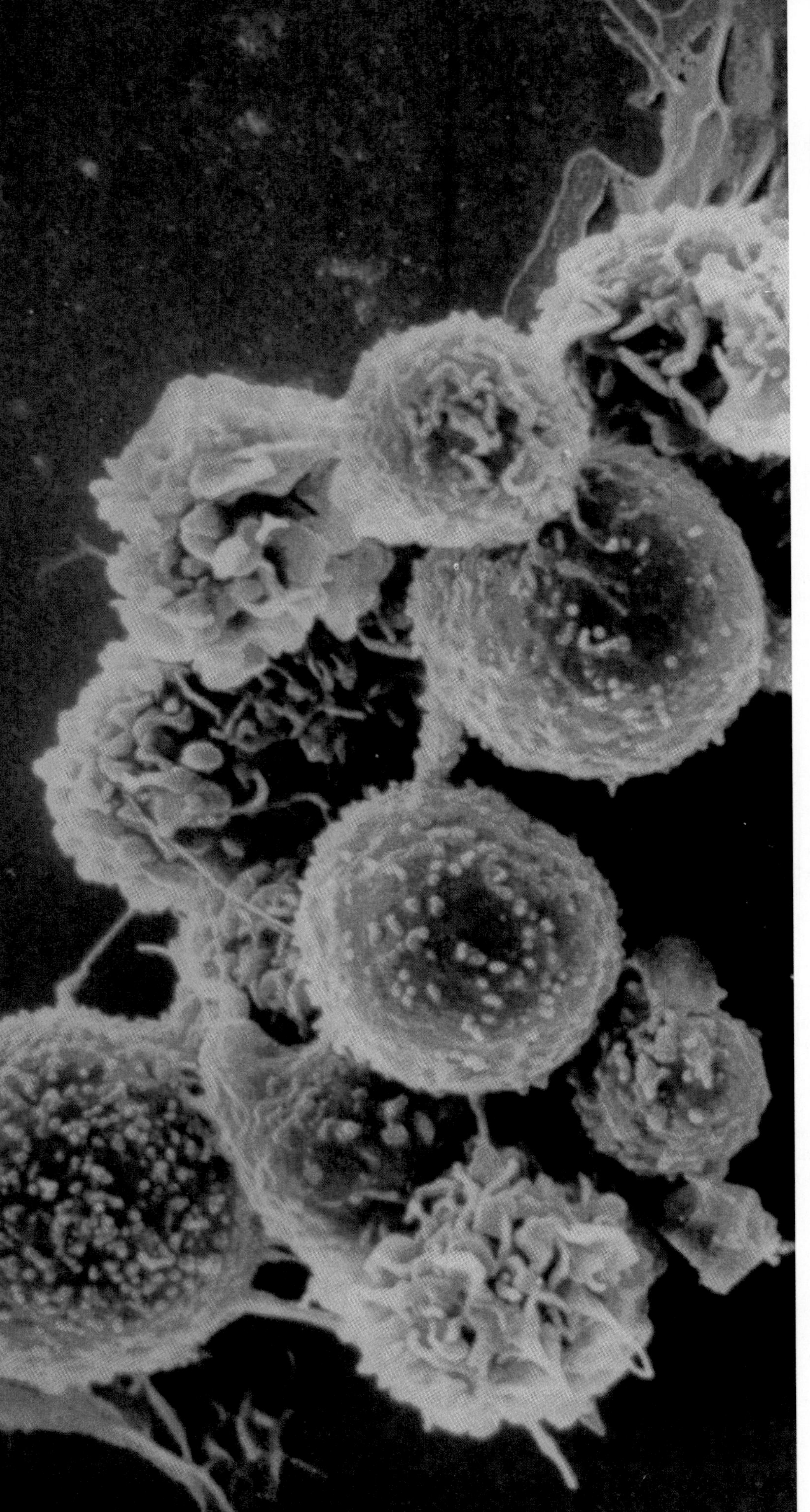

CHAPTER 25

CONCEPT GUIDE

After reading this chapter, you should be able to:

1. Describe how an organism's external and internal defenses resist invasion by other organisms. (Section 25-A)
2. Describe three basic properties of immune responses. (Section 25-B)
3. Compare the functions of phagocytes and lymphocytes. (Section 25-C)
4. Explain the general function of the constant and variable regions of an antibody. (Section 25-D)
5. Compare the processes of cellular and humoral immunity. (Section 25-E)
6. Explain how a secondary immune response is different from a primary immune response. (Section 25-F)
7. Explain how vaccination and passive immunity provide protection against disease. (Section 25-G)
8. Explain the causes of autoimmune diseases and allergies. (Section 25-H)

Scanning electron micrograph of macrophages swarming around cancer cells. The immune system protects us against foreign invaders as well as cancerous and other diseased cells. Read more about how the immune system helps fight disease in Sections 25-A and B. (W. Johnson/Visuals Unlimited)

Defenses Against Disease

A living body is an ideal incubation chamber, providing food, shelter, and just the right combination of water, minerals, and temperature for cells. It is no wonder that many small organisms have adapted to life inside the bodies of their larger neighbors. Some, like symbiotic bacteria in digestive tracts, cause their hosts no trouble or are beneficial. Other organisms, called **pathogens**, cause disease.

Probably all organisms have defenses against pathogens, but here we consider only the defenses of animals. Some defenses are **nonspecific**, that is, they protect the animal from many different diseases. For instance, nearly all animals contain **phagocytes,** wandering cells that engulf and destroy any pathogens and debris they encounter. Other defenses are **specific,** working against particular pathogens. These include the reactions of the **immune systems** of vertebrates, whose cells can mount an attack tailored specifically to each different disease-causing invader. When the disease has been overcome, cells of the immune system "remember" how to fight that disease more effectively in the future.

Occasionally, the immune system is damaged or fails to perform its normal functions. Two main types of disease may result: either the immune system does not protect the body against pathogens, or its cells damage parts of its own body.

KEY CONCEPTS

- Animals defend themselves against disease by nonspecific mechanisms and by immune reactions, which recognize specific foreign substances in the body, destroy them, and remember them in order to fight them more efficiently in the future.
- Failures of the immune system often lead to devastating diseases, either because the immune system fails to defend the body against the attack of foreign organisms or because it attacks its own body.

Curiosity Questions

1? Is there anything good about running a low fever when we are sick? (See page 527)

2? How does a vaccination protect against developing a particular disease? (See page 536)

3? Why do some people have allergies, while others do not? (See page 541)

25–A NONSPECIFIC DEFENSES

All animals have protective measures that resist invasion by other organisms, or that kill those foreign organisms that do manage to enter the body.

External Barriers

The human body surfaces in contact with the outside world include the skin, and the mucous membranes lining the eyelids, nose, mouth, digestive tract, vagina, and urethra. These external and internal surfaces are barriers that prevent most pathogens from entering the body. In addition, millions of bacteria live on these surfaces (Figure 25–1). These resident bacteria produce substances that protect their homes—our skin and mucous membranes—from foreign organisms, many of which might cause disease.

Skin is the human body's largest organ and its main protection against infection (Figure 25–2). The **epidermal cells** of the skin's surface are constantly dying and sloughing off. We notice this when a sunburned nose peels, but it also goes on all over the body in undamaged skin. The lost cells are replaced by division of cells that lie under them. In this way the body soon repairs minor scrapes and cuts and maintains a constantly renewed barrier against infection.

The skin also produces chemical defenses: oil and wax from its sebaceous glands, and sweat from its sweat glands. These secretions contain lactic acid and fatty acids, which make the pH acidic enough to kill or slow the growth of many fungi and bacteria.

Mucous membranes cover body surfaces that must be kept moist, such as the linings of the nose, mouth, and vagina. Mucous membranes secrete **mucus,** a fluid that traps microorganisms and that contains bacteria-killing enzymes such as **lysozyme,** found in tears, nasal mucus, and saliva. (Lysozyme was discovered by Alexander Fleming when he noticed that bacterial cultures died after he sneezed on them.) Frequent douching of the vagina is unhealthy because it removes the bacteria-killing mucus.

Pathogens in the respiratory tract are likely to be trapped in mucus and swept into the pharynx by cilia. They are then usually swallowed and enter the stomach, where they encounter protein-digesting enzymes and an extremely acid environment. If they survive this and reach the large intestine, they are attacked by gut microorganisms, which secrete antibiotics that kill many pathogenic bacteria.

Despite these defenses, some pathogens do enter the body, usually by crossing the mucous membranes, which are thin, moist, and therefore more vulnerable than dry, oily skin.

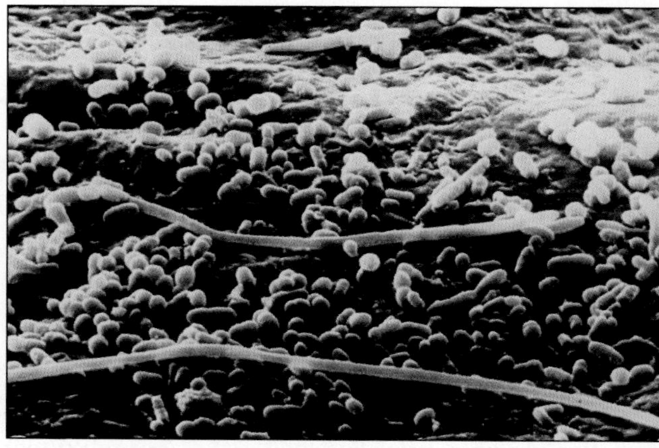

FIGURE 25–1

Our resident bacteria. The many small and round structures are bacteria on the surface of human skin.
(D. M. Phillips/Visuals Unlimited)

Internal Defenses

Once past the barrier of the skin or mucous membranes, pathogens face several nonspecific physiological reactions, capable of destroying a variety of organisms. In many situations, these nonspecific reactions work together with specific immune responses (Section 25–B).

1. **Inflammation.** A swollen, red area occurs where the skin or a mucous membrane has been wounded, and pathogens have entered the body (Figure 25–3). A clogged pore or a boil is the result of inflammation from such a local infection. Inflammation helps to fight infection and heal the wound. Chemicals released by the wound attract **mast cells,** which are found in most organs and which contain a variety of hormone-like substances (Section 30–D). The mast cells release **histamine,** an amino acid derivative that causes increased capillary permeability and thus increased blood flow, and results in redness and swelling at the site of the wound. Histamine attracts large numbers of phagocytes, a kind of white blood cell, that engulf bacteria and dying body cells. Eventually the phagocytes die and may accumulate as pus.

   The **complement reactions** complement (that is, round out) the pathogen-destroying effects of inflammation. **Complement** consists of about 20 kinds of proteins, produced mainly by **macrophages,** a kind of white blood cell, which

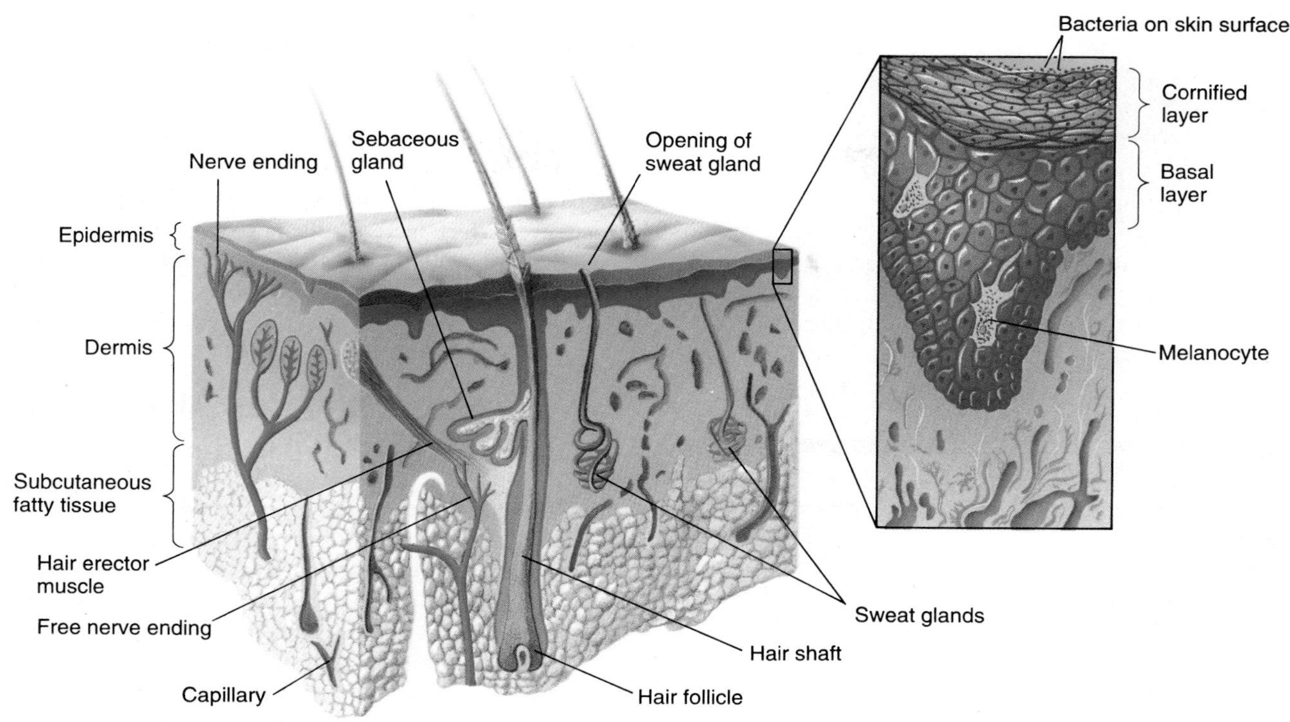

FIGURE 25-2

A section through human skin. The cells in the outer cornified layer of the epidermis constantly die and slough off. They are replaced by cell division in the basal layer. Chemical defenses of the skin include acids found in the oily secretion of sebaceous glands and in sweat from the sweat glands. These acids combat bacteria (shown on the skin surface in the upper right) and fungi. Evaporation of sweat cools the body. The hair erector muscles can raise the hairs to form "goose pimples," creating a thicker cushion of warm air next to the skin which helps to control the body temperature.

circulate in the body fluids. When they encounter a microorganism, these proteins undergo a series of reactions that may break down the microorganism's plasma membrane or attract phagocytes, which engulf the pathogen.

2. **Fever.** An infected area often feels warm to the touch. Heat is one of the body's ways to fight pathogens. Normally, the brain keeps the human body between 97.2 and 100.4°F (36.2–38°C). However, when the body is infected by pathogens, some white blood cells respond by releasing hormones that act as **pyrogens** ("fire-producers"). If enough pyrogens reach the brain, the body's thermostat is reset to a higher temperature, allowing the temperature of the entire body to rise, which we call a **fever.**

   Very high fevers are dangerous and must be reduced. But a study of children with influenza ("flu") showed that a few *degrees* of fever helps the body fight infection. In this study, children treated with aspirin to reduce their fever were compared with those whose temperature was permitted to rise naturally. The children whose fever was kept down were ill longer, and had other symptoms that were more serious than those with the natural fever.

   Cells metabolize faster at higher temperatures, so fever increases the rate at which the immune cells fight infection. In addition, many bacteria require more iron in order to reproduce at these temperatures. Pyrogens not only raise the body temperature but also reduce the concentration of iron in the blood, slowing bacterial reproduction even further. ■

3. **Interferons.** Fever increases the production of virus-fighting proteins called **interferons.** When some cells in the body are invaded by viruses, they produce interferons, which help to protect

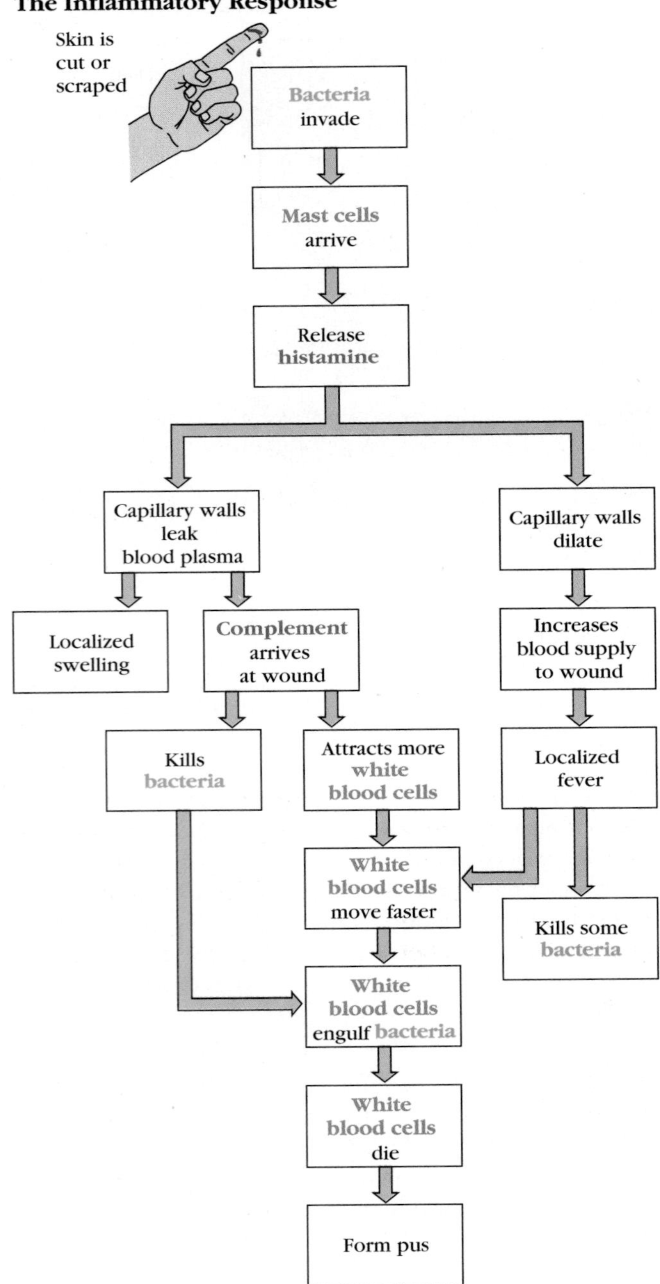

FIGURE 25–3

An overview of the inflammatory response. When the skin is damaged, the wound may permit the invasion of bacteria and other microorganisms into the surrounding tissues. The initial response to this infection is the release of histamine by mast cells that migrate to the wound. Additional responses include localized swelling, increased blood flow to the wound, and localized fever that together help to overcome the infectious agents.

healthy neighboring cells from viruses. Interferons stimulate cells to produce substances that interfere with viral replication.

Skin and mucous membranes provide the first lines of defense against infectious organisms. Internal defenses include inflammation, which increases the number of phagocytes, and fever, which increases the rate of production of disease-fighting cells, reduces the ability of bacteria to reproduce, and produces virus-fighting interferon.

25–B SPECIFIC DEFENSES: AN OVERVIEW OF IMMUNE RESPONSES

Few people suffer twice from diseases such as measles, chickenpox, and mumps. The body's first encounter with the pathogens that cause these diseases equips it to get rid of the same kinds of pathogens when they next invade. A response that is bigger and faster-acting the second time around tells us that this reaction is produced by the immune system.

If you have had measles, you are immune to further attacks of measles but not to rubella (German measles) or mumps: the immune response to measles is specific for the measles virus. The body's attack on the measles virus, but not on its own cells, shows that the body can distinguish the measles virus as **foreign,** that is, not a normal part of the body. The immune response to a bout of measles confers **immunity,** protection against measles viruses encountered later in life. This response shows that immunity involves some sort of **memory:** the body "remembers" that it has previously encountered this type of virus.

The role of the immune system is to recognize and destroy foreign antigens that invade the body. An **antigen** is any substance that can stimulate an immune response against it. The most common antigens are substances from another organism, such as toxins produced by bacteria or the protein coats of viruses.

An immune reaction is specific because it involves the binding of antigen by receptors produced by white blood cells called **lymphocytes.** Each lymphocyte produces only one kind of antigen-binding receptor, which can bind only one kind of antigen. Different lymphocytes produce different receptors. Among all the lymphocytes in the body, there are receptors that can bind almost any antigen (Figure 25–4).

The binding of antigens to receptors on its surface may stimulate a lymphocyte to reproduce, forming a clone of cells all making receptors that bind that antigen. These cells can fight the pathogen that formed the antigen, which is also multiplying in the body and

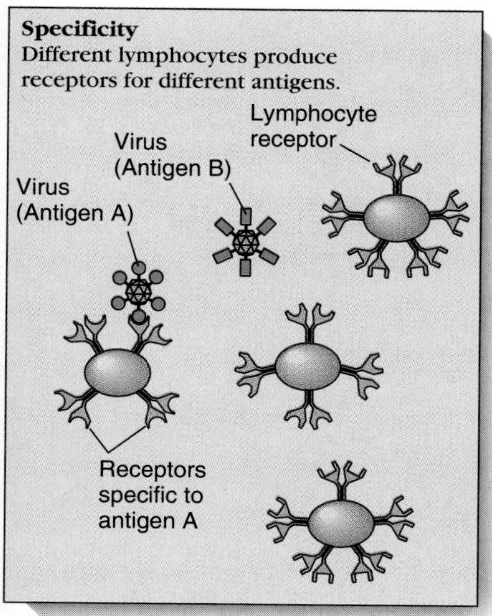

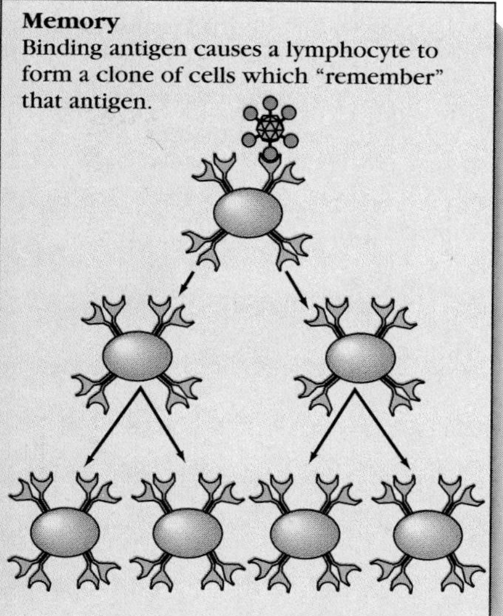

Memory
Binding antigen causes a lymphocyte to form a clone of cells which "remember" that antigen.

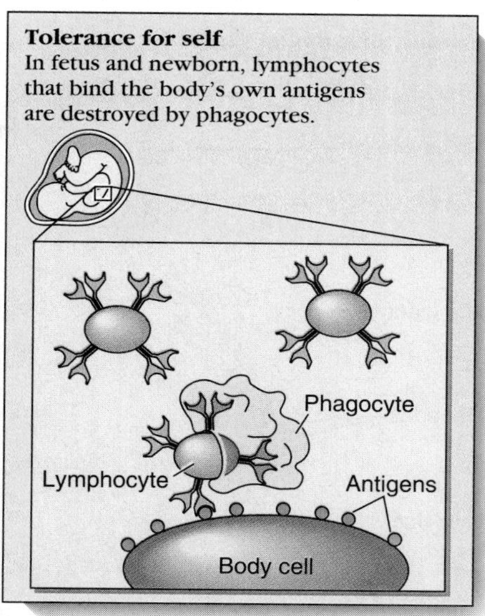

FIGURE 25-4

The three main features of an immune response. The response is specific to a particular antigen, permits the body to remember an antigen, and distinguishes foreign antigen from normal parts of the body.

causing disease. The clone of lymphocytes also accounts for the body's memory of a pathogen. If the pathogen infects the body again, the many cells of the memory clone can mop up the pathogen before it has a chance to reproduce and damage the body.

It is essential that the immune system not attack the body's own cells. The body is said to have **tolerance** for its own antigens. Tolerance develops during embryonic life. As a fetus or newborn, an animal destroys developing lymphocytes if they produce receptors that bind any of the body's own antigens.

Immune responses are specific for certain antigens, recognize foreign materials or cells, and use a sort of cellular memory that allows a quicker response upon reinfection.

25–C THE IMMUNE SYSTEM

The immune system is not a set of organs like the digestive or respiratory systems. Instead, it consists of many sites in various parts of the body, where the cells involved in immune reactions are produced and function. **Bone marrow,** the soft tissue within many bones, produces nearly all of the blood cells (Section 24–D). Some of the white blood cells produced here are part of the immune system. Once formed, these cells move out to live and be active throughout the body, travelling by way of the blood and lymph vessels.

Many immune responses take place in lymph nodes (sometimes called lymph glands), in the spleen, and in the primary locations where pathogens invade the body—the linings of the respiratory, digestive, genital, and urinary tracts.

In Chapter 24 we saw that fluid filters out of the blood in the body's capillary beds and joins the extracellular fluid, or lymph, that surrounds all cells. The lymph drains slowly into thin-walled lymphatic vessels, which drain back into the blood via the thoracic duct near the heart (Figure 25–5). At various places, lymph travelling toward the heart in lymphatic vessels passes through **lymph nodes,** which contain a meshy network lined with white blood cells. The lymph nodes

BIO-BIT

Bubonic plague is a bacterial infection, transferred to humans from the bite of a rat, that localizes within lymph nodes. Unless treated with antibiotics, the bacteria enter the bloodstream, multiply, and spread throughout the body. Plague kills between 70 to 90% of those infected.

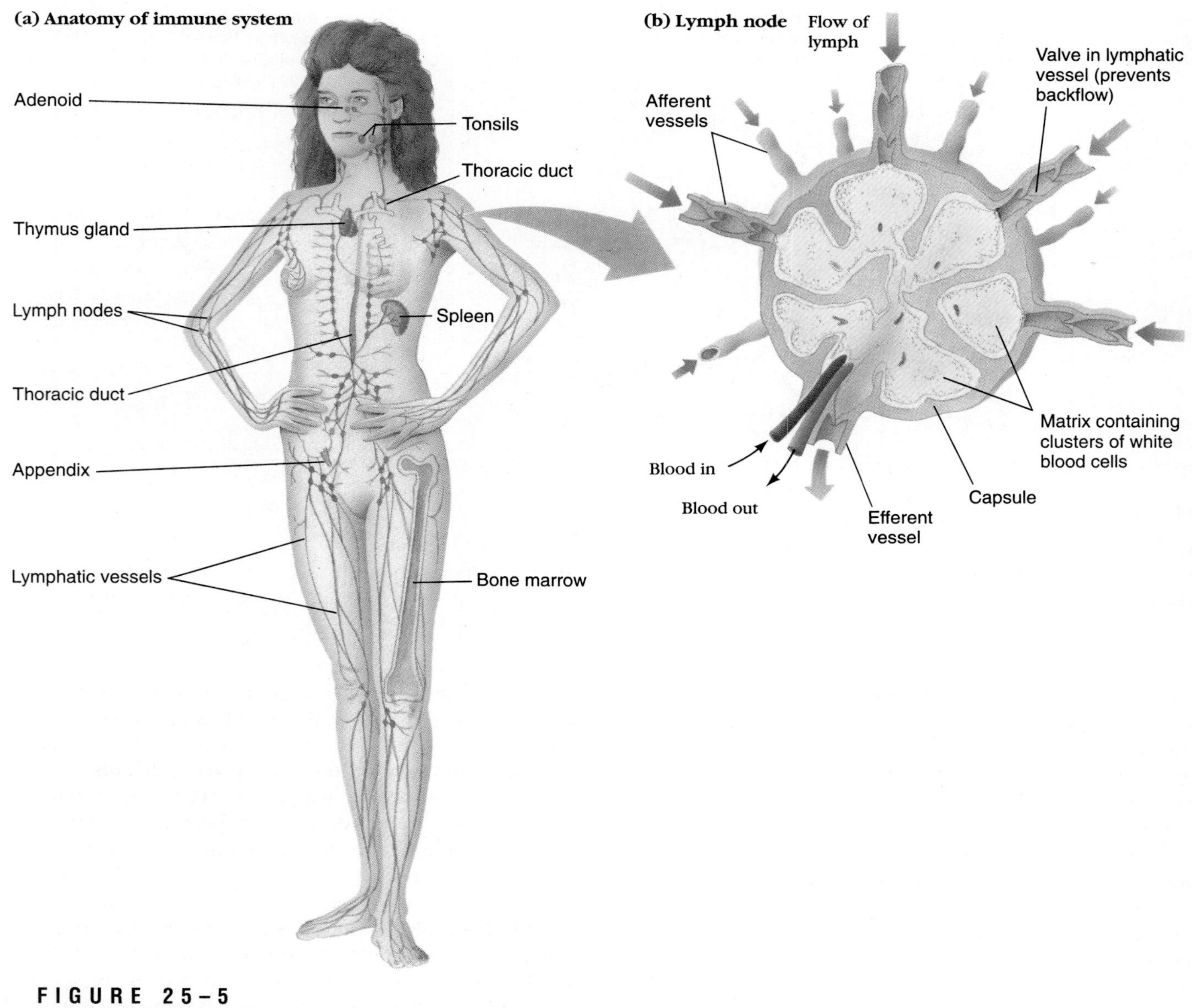

FIGURE 25-5

The human immune system. (a) Lymphatic vessels extend throughout the body and drain into and out of the lymph nodes. Lymph nodes in the groin, armpits, and neck often feel tender and swollen during illness. **(b)** A lymph node. Lymph enters the node through several afferent lymphatic vessels. In the node, the lymph filters through a network of spaces containing large clusters of white blood cells until it reaches the one or two efferent lymphatic vessels that carry it away from the node.

filter pathogens out of the lymph for attack by the white blood cells of the immune system. Tonsils and adenoids are lymph nodes in the throat and nose, respectively. Lymph nodes also occur in the armpits, neck, and groin. The spleen acts as a similar filter for the blood.

Cells of the Immune System

Two major groups of white blood cells take part in immune responses.

1. **Phagocytes** are cells that engulf and destroy pathogens. These are of two main types: **neu-**

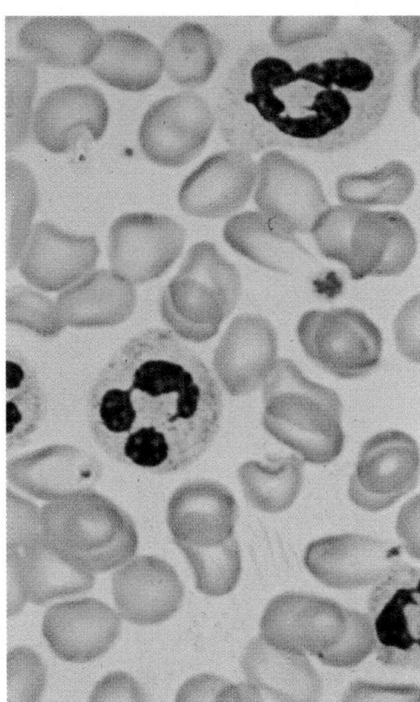

FIGURE 25-6

Human white blood cells. In this sample of human blood, only the two neutrophils pictured have nuclei. Their nuclei are stained deep purple and have many joined lobes, giving rise to another name for these white blood cells, polymorphonuclear (poly = many, morpho = form, nuclear = nucleus) leukocytes. Red blood cells such as those in this photograph lose their nuclei during development that occurs before they enter the circulatory system. (Biophoto Associates)

trophils, found in the blood (Figure 25-6), and larger **macrophages,** which can leave the blood and enter the tissues and body cavities. Macrophages are big enough to engulf large microorganisms, including protists.

2. **Lymphocytes** are smaller white blood cells responsible for the immune system's recognition and memory of foreign antigens. Lymphocytes circulate throughout the body, from the bloodstream through the lymph and back into the blood, with extended stays in the lymph nodes and spleen.

There are two main types of lymphocytes, both of which originate in the bone marrow. As they develop, **T lymphocytes** (or simply **T cells**) move to the thymus gland, a lymph gland at the base of the neck under the breastbone (see Figure 25-5). Here they mature, becoming either helper T cells, suppressor T cells, or killer T cells. In contrast, **B lymphocytes** (or **B cells**) undergo similar developmental processes in the bone marrow or other immune tissues.

The immune system is located throughout the body and includes the bone marrow, lymphatic system (spleen, lymph nodes, lymph ducts), and white blood cells. The two main categories of white blood cells are (a) phagocytes, which engulf and destroy pathogens, and (b) lymphocytes, which recognize and remember antigens. Immune responses occur in the spleen, in lymph nodes, and in the main places where pathogens occur.

There are three main steps in the immune system's response to an invader: recognizing the invader, attacking it, and remembering it. These steps are the subjects of the next three sections.

25-D RECOGNIZING FOREIGN ANTIGENS: RECEPTORS AND ANTIBODIES

Foreign antigens are recognized by the body when they bind to specific receptor proteins produced by lymphocytes. The first of these receptor proteins to be discovered were **antibodies** produced by B cells. B cells produce some antibodies that remain attached to the B cell surface and some that are released into the blood or lymph. T cells, in contrast, produce only **T cell receptors,** which remain attached to the T cell surface.

The body produces millions of different antibody molecules, but all have the same basic Y-shaped structure (Figure 25-7). Each is made up of four peptide chains: two identical heavy chains and two identical light chains, all joined by disulfide bonds. Each of the four chains consists of a so-called constant region at one end, and a variable region at the other. While constant regions are chemically similar, variable regions differ from antibody to antibody. The variable regions form binding sites so specific that each kind of antibody can bind only one (or a few closely related) antigens. This specificity of variable sites determines the specificity of antibodies.

Antibodies are classified into five main groups—A, M, G, E, and D—according to which constant regions occur in their heavy chains. An antibody's group determines its general biological function (Table 25-1). For instance, group G antibodies combine with bacteria and viruses in the blood. The group G constant region

Table 25–1
The Antibody Groups

Group	Major characteristics
A	Found in mucous secretions and gut; defends external body surfaces
M	First antibody produced during immune response; sticks bacteria together and immobilizes them; stimulates complement reactions and macrophages
G	Main antibody in blood; combats microorganisms and their toxins; stimulates complement reactions and phagocytes; can cross the placenta
E	Responsible for symptoms of allergy; effective against parasitic worm infections
D	Rare; found on the surfaces of lymphocytes; function unknown

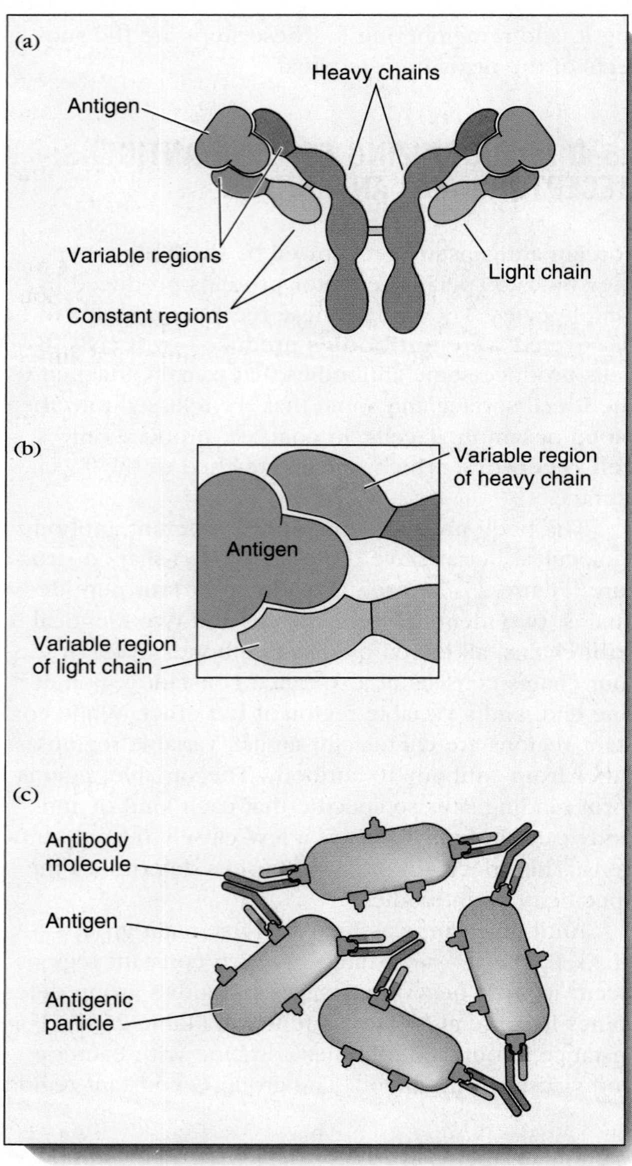

FIGURE 25–7

Antibodies. (a) An antibody molecule bound to two molecules of antigen (blue). Each antibody molecule consists of two identical light peptide chains and two identical heavy chains joined by disulfide bonds (yellow). Part of each chain has a variable amino acid structure (orange) and part is constant (purple). **(b)** Enlarged view of the antibody's antigen-binding site. **(c)** The forked nature of an antibody permits clumping. Each antibody can bind to two antigen molecules (blue), which may be either loose in the body fluids or exposed on the surface of a particle such as a bacterium as indicated in this figure.

attracts macrophages that engulf and destroy an invading bacterium. It also enables a G antibody to cross the placenta, so that antibodies produced by the mother's body can reach the fetus and protect it from disease.

The Genetics of Antibodies and T Cell Receptors

Each lymphocyte produces only one type of antibody or receptor. However, the mammalian body can recognize millions of different antigens because it produces millions of different antibodies. This poses an interest-

BIO-BIT

Each circulating B lymphocyte that is inactive may have as many as 100,000 copies of its specific antibody incorporated into its plasma membranes, ready to bind with an antigen.

ing genetic problem. Antibodies are proteins, and a traditional rule of thumb in genetics is "one gene, one polypeptide"—that is, there should be one gene for each different kind of polypeptide produced. But a mammal produces more different antibodies than it has genes. How can this be? The answer lies in the fact that antibody genes come in "kits" of several different interchangeable pieces. These lengths of DNA can be spliced together in various combinations to generate the vast number of different antibodies found in the body. Another set of genes provides a similar "kit" of DNA segments that can be spliced in different ways to produce a great variety of T cell receptors.

During its development, each lymphocyte splices some of these lengths of DNA together in a sequence that becomes the code for the T cell receptor or antibody that the lymphocyte and the clone of its descendants will produce.

The immune response begins with the recognition of an antigen that binds to specific antibodies or T cell receptors. Antibodies are Y-shaped molecules with constant and variable regions. An antibody's variable region determines which antigen(s) it will bind, and the constant region determines its general role in the body. The production of millions of different antibodies allows the immune system to respond to most types of antigens.

25–E ATTACKING PATHOGENS

The immune system produces several different kinds of immune responses. For instance, the body destroys eukaryotic cells bearing foreign cell-surface antigens. These include cancer cells, whose cell-surface antigens are often altered when a normal cell is genetically transformed into a cancer cell. The antigen-bearing cell may be eaten by macrophages or killed by **killer T cells** (lymphocyte-like immune cells), which are specialized to destroy abnormal body cells (see this chapter's opening photo). Killer T cells are also called natural killer (NK) cells.

The best-understood immune responses are those produced by T and B cells—cellular and humoral immunity. Both are involved to some extent when the body is fighting most types of infection. We can illustrate the difference between the two by considering what may happen when a virus invades the body (see A Journey Through Human Immunity).

1. **Cellular immunity.** If the virus invades a body cell, that cell can be recognized, attacked, and destroyed by T cells. Destroying the cell prevents the virus from replicating.
2. **Humoral immunity.** If the virus has not yet invaded a body cell, it may be bound by antibody molecules that have been secreted into the body fluids by B cells. The virus-antibody complex will then be engulfed and destroyed by a phagocyte.

> **BIO-BIT**
>
> It is estimated that during an immune response, each plasma cell can produce nearly 2000 antibodies per second.

Cellular Immunity

Killer T cells destroy body cells that have been invaded by viruses. This includes cells invaded by cancer-causing viruses, believed to be responsible for about 20% of human cancers.

How does a T cell distinguish between a healthy body cell and a viral invader? The surface of each normal body cell bears glycoproteins that identify the cell as belonging to a particular tissue in a particular individual. In mammals, many of these cell-surface molecules are specified by a group of genes called the **major histocompatibility complex (MHC)** (histo = tissue). When a cell has been invaded by a virus, viral proteins are often attached to the MHC molecules on the cell's plasma membrane. This gives the cell a double identity: "self" MHC markers show that it is a member of the body, but at the same time viral antigen indicates that it is foreign.

In the **cellular immune response,** a killer T cell recognizes this combination as a body cell infected by virus. This T cell bears one of the body's many different T cell receptors, which binds the combined virus-MHC marker. The killer T cell may then destroy the cell by punching holes in its plasma membrane.

Humoral Immunity

Humoral immunity is the body's main defense against pathogenic bacteria, viruses, and fungi in the blood or other body fluids (humor = fluid). These pathogens are bound by antibodies secreted into the body fluids by B cells, and are then engulfed and destroyed by phagocytes.

Throughout life, the bone marrow produces large numbers of B cells. Each B cell produces one type of antibody and displays copies of this antibody on its surface. Many B cells circulate in the blood, and most of them die within a month or two. However, if a B cell encounters an antigen that binds to its surface antibody, it is activated in a series of steps. As an example, consider what happens when a bacterium gets into the bloodstream after the initial contact with the antigen.

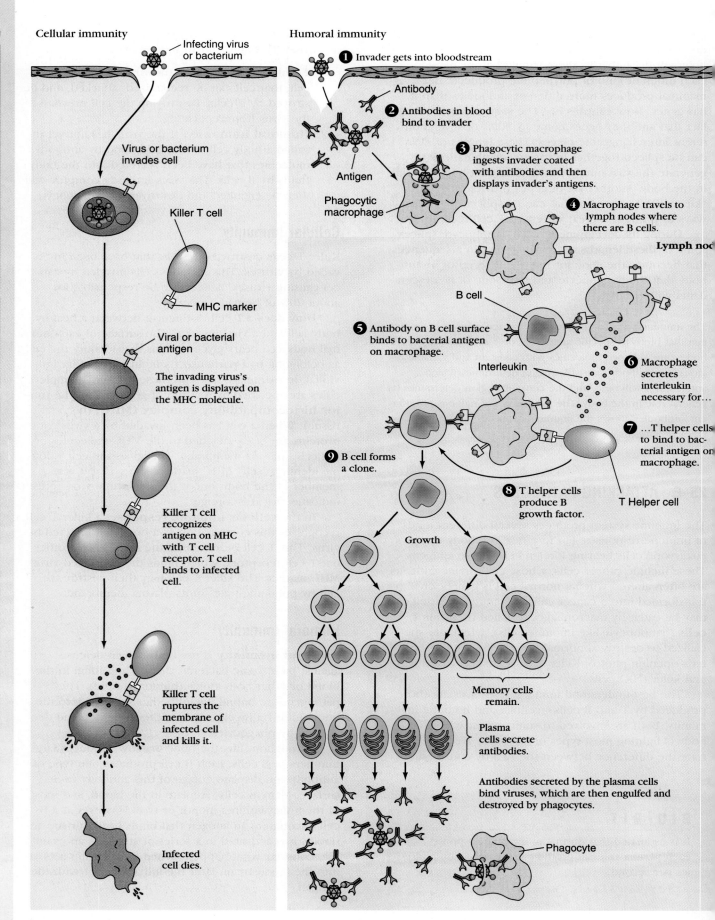

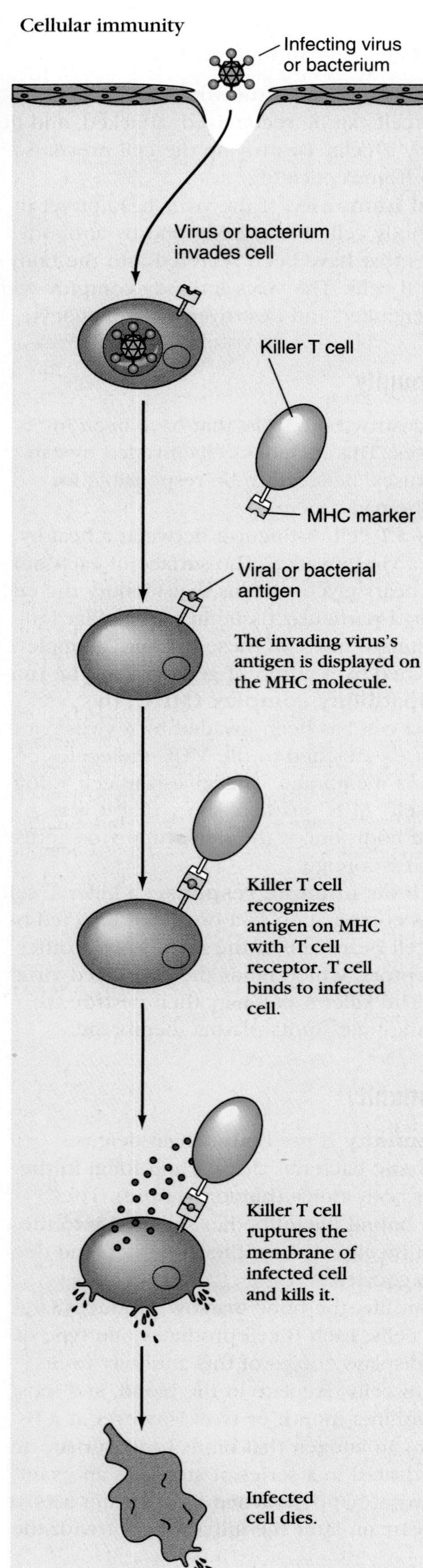

Humoral immunity

1 Invader gets into bloodstream

Antibody

2 Antibodies in blood bind to invader

Antigen

3 Phagocytic macrophage ingests invader coated with antibodies and then displays invader's antigens.

Phagocytic macrophage

4 Macrophage travels to lymph nodes where there are B cells.

Lymph nod

B cell

5 Antibody on B cell surface binds to bacterial antigen on macrophage.

Interleukin

6 Macrophage secretes interleukin necessary for…

7 …T helper cells to bind to bacterial antigen on macrophage.

8 T helper cells produce B growth factor.

T Helper cell

9 B cell forms a clone.

Growth

Memory cells remain.

Plasma cells secrete antibodies.

Antibodies secreted by the plasma cells bind viruses, which are then engulfed and destroyed by phagocytes.

Phagocyte

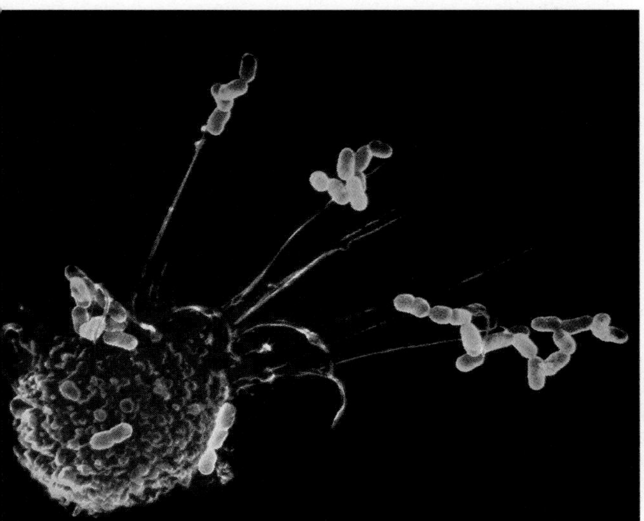

FIGURE 25–8

Phagocytosis. A macrophage (colored red) engulfing bacteria (colored green). (Photo courtesy of Lennart Nilsson, © Boehringer Ingelheim International GmbH)

1. One or more of the millions of different antibodies in the blood binds to antigen on the surface of the bacterium.
2. A phagocytic macrophage ingests the bacterium coated with antibody (Figure 25-8). The macrophage then displays the bacterium's antigen by pushing it out onto its own plasma membrane. In its travels through the body, the macrophage reaches a lymph node, where many B cells spend a lot of their time.
3. Antibody on the surface of one of the B cells binds the bacterial antigen exposed on the macrophage surface. This holds the macrophage and the B cell together so that the B cell can be activated to divide.
4. The macrophage now secretes **interleukin,** a protein essential to B cell activation.
5. T lymphocytes known as "helpers" are also needed for activation. Helper T cells bearing the appropriate T cell receptors bind to the bacterial antigen on the macrophage. Once bound, these helper T cells produce **B-cell growth factor,** another protein necessary for B cell activation.
6. Each activated B cell divides to form a clone (see A Journey Through Human Immunity).
7. Some members of the B cell clone set to work churning out antibody specific to the bacterial antigen and secreting it into the body fluids. These cells, called **plasma cells,** can be recognized by their large areas of rough endoplasmic reticulum, which produces antibody for secretion from the cell. Within a few days of the bacterial infection, a great deal of antibody to the bacterium enters the blood. The bacteria will be bound by the antibody and engulfed and destroyed by macrophages.

Other members of the dividing population of B cells will produce **memory cells** that are highly specific for the invading bacterium, as we will see in the next section.

Immune responses. The body fights viruses (or other pathogens such as bacteria, fungi, protists, or foreign cells) in two main ways. Invaders that manage to penetrate a body cell will usually provoke cellular immunity: attack by killer T cells. Invaders traveling in the body fluids provoke humoral immunity: attack by soluble antibodies produced by B cells (an event that usually occurs after a phagocyte such as a macrophage has engulfed some of the viruses).

The second step in the immune response is destroying the foreign antigen. Cellular and humoral immunity are the two best understood immune responses. In cellular immunity, killer T cells recognize and destroy infected body cells bearing combined self- and foreign-surface markers. In humoral immunity macrophages and helper T cells activate B lymphocytes to secrete antibody specific for an invading antigen.

25–F IMMUNOLOGICAL MEMORY: PRIMARY AND SECONDARY RESPONSES

A cellular or humoral response to the body's first encounter with a foreign antigen is called a **primary immune response.** During a primary response, the antigen will eventually disappear from the blood, bound by antibody and destroyed by killer T cells or macrophages. The plasma cells that secreted antibody will also die. However, the **memory cells,** descendants of the clone of B cells produced as a result of the initial immune response, remain in the body for life. If the same antigen enters the body again, the memory cells permit the immune system to mount a **secondary immune response,** faster and more extensive than the primary response (Figure 25-9). The secondary immune response quickly eliminates the antigen again.

Memory cells from the humoral response display a sample of their antibody on their surfaces. If the antigen invades the body again, the memory cells with

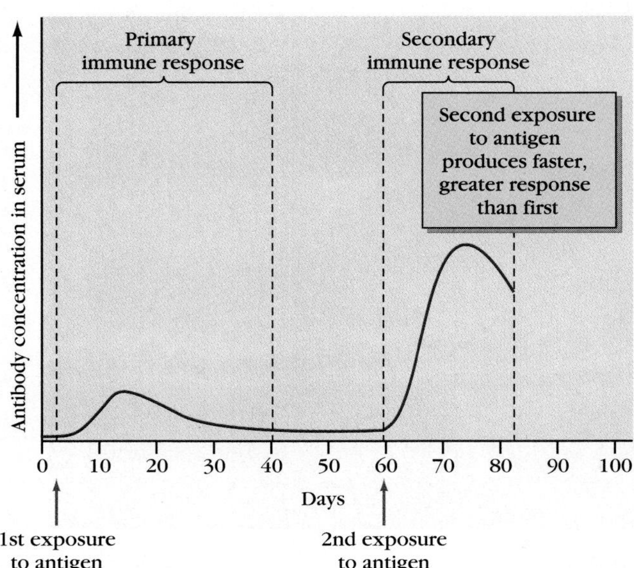

FIGURE 25-9

Primary and secondary immune responses. The purple curve shows the amount of antibody to a specific antigen found in the blood of a rabbit at different times. Red arrows indicate the times of a first and then a second exposure to the antigen. After the second exposure, the rabbit produces the specific antibody more rapidly and in greater amounts.

matching surface antibodies bind the antigen. This stimulates them to divide rapidly and produce a clone of many antibody-secreting plasma B cells.

The body must build up a memory clone for each antigen it encounters before it can produce secondary responses to most microorganisms. This is why babies and children starting school have so many colds and infections: they must encounter many antigens, and build up many clones of memory cells, before they are immune to as many diseases as the average adult.

The final phase of an immune response is the production of memory cells, formed from B cell clones during the primary immune response. If the same antigen later reinfects the body, memory cells remaining from the primary immune response will again produce antibody in an overwhelming secondary response.

25–G MEDICAL ASPECTS OF IMMUNE RESPONSES

Vaccination

2? Vaccination against a specific disease works by inducing the immune system to mount a primary immune response and to produce memory cells, ready to trigger a secondary response at the body's first real battle against the disease antigen. The practice of vaccination, however, began long before people understood how it worked. Arabic and Chinese manuscripts more than a thousand years old refer to vaccination against smallpox. Lady Mary Wortley Montagu, wife of the British ambassador to Turkey, introduced this ancient custom into England in 1718. She had her children vaccinated by rubbing part of the scab from a healed smallpox sore into a small wound in the skin. This introduced a few live smallpox viruses into the body, stimulating a primary immune response and thereby conferring immunity to smallpox in later life. The snag was that vaccination with even a small amount of live virus sometimes caused a case of smallpox, which could be fatal.

Edward Jenner, an English physician, found a way around this problem in 1796. Jenner noticed that dairy workers who had caught the relatively mild disease cowpox from cows seemed to be immune to smallpox. He found that rubbing pus from cowpox sores into scratches in the skin prevented people from coming down with smallpox later. In this case, the antigens of smallpox and cowpox are so similar that the same antibodies work against both of them. Almost a century later, Louis Pasteur found a safer way to prepare vaccines. He discovered how to disable microorganisms by heat and other treatments. This damaged the microorganisms so that they could no longer cause disease but left their antigens intact and able to stimulate a primary immune response.

Nowadays, we have vaccines for a number of bacterial and viral diseases. The first successful immunizations against cancer (in cats and chickens) are based on the fact that cells infected with cancer-causing viruses usually display two new antigens on their surfaces, one of them specific to the virus. Unfortunately, attempts to produce antibodies to cancers not caused by viruses have had very little success. Several important infectious diseases also remain without effective vaccines, including malaria, trypanosome infections, and Acquired Immune Deficiency Syndrome (AIDS; see

A Journey into Healthy Living: Understanding AIDS, this chapter). All these diseases are caused by pathogens whose cell surface antigens change frequently. As a result, no single antibody is effective against very many cases.

Smallpox was not only the first disease to be prevented by vaccination, but also the first disease to be officially declared wiped out by human efforts. The last known outbreaks of smallpox occurred in India and Africa in the late 1970s. International vaccination programs had greatly reduced the number of smallpox cases. The final conquest came after health officials adopted a different strategy: searching out pockets of infection (people were rewarded for each case they reported), quarantining the victims, and vaccinating their friends and relations. ■

Passive Immunity

An animal is said to be passively immune when it contains antibodies that were not produced in its own body. A newborn baby is passively immune, temporarily protected from disease by antibodies that reached it from the mother's blood before birth. Bolstered by antibodies found in breast milk, these antibodies are used up during the first few months of life, during which the baby's immune system becomes sufficiently mature to take over.

Passive immunity can also be used medically. Some antigens are so virulent that the body's own primary immune response has little chance of preventing serious damage or death. If by some mischance such an antigen enters the body, the victim can sometimes be protected temporarily by injections of antibodies. Antibodies against potent antigens such as tetanus toxin or snake venom may be prepared by injecting the antigen into a horse and later collecting samples of the horse's blood, which now contains antibodies to that antigen. A more recently-developed system is to harvest the antibodies from laboratory-grown clones of antibody-producing cells.

However they are acquired, the antibodies involved in passive immunity eventually disappear from the recipient's body, and the immunity is lost because the antibodies were not produced by the body's own plasma cells.

Tissue and Organ Transplantation

The rejection of transplanted organs is caused by cellular immune responses. Immunologists study this process using skin grafts, which are easy to work with and do not harm the recipient. If skin is transplanted from one mouse to another, it looks healthy for several days. However, the graft is eventually invaded by T cells, and within a day or two it sloughs off. A second graft from the same donor to any part of the same host is rejected faster than the first one.

Organs can be transplanted from one animal to another without being rejected only if the two animals have compatible MHC antigens (Section 25-E). Identical twins have identical MHC antigens, but even close relatives such as nonidentical siblings, or parent and child, often do not have antigens similar enough to permit a successful transplant and prevent the recipient's T cells from making a cellular immune response. So when someone needs a skin or organ transplant, the first step is to find a donor with antigens that match those of the recipient as closely as possible.

Rejection of transplanted tissue is normally prevented by **immunosuppressant drugs,** drugs that suppress the body's immune responses. Such drugs are always used after heart, liver, or kidney transplant operations. Most immunosuppressant drugs work by preventing lymphocyte cell division.

Vaccination causes a primary immune response against a harmless form of a disease-causing antigen, so that the body will produce a powerful secondary response if the live pathogen that makes the antigen again invades the body. Passive immunity protects against disease or toxins by using antibodies produced by another animal. Organ transplants between individuals that do not have identical MHC antigens will likely be rejected unless immunosuppressant drugs are used.

25-H MALFUNCTIONS OF THE IMMUNE SYSTEM

The immune system is vital in protecting the body from disease. As we know in the case of AIDS, when something goes wrong with the immune system, the consequences are often fatal. For instance, if the thymus gland is abnormal, T lymphocytes fail to develop. Without helper T cells, B lymphocytes cannot form clones. A baby born without T lymphocytes fails to produce B cell clones to combat invading organisms, and so it usually is killed by the first pathogens it encounters. A few such babies have been saved by keeping them in sterile environments and by transplanting bone marrow cells and thymus tissue into them, which may permit them to make antibodies.

Autoimmune Diseases

Autoimmunity is a dangerous condition in which the self-recognition system breaks down and the body develops antibodies to some of its own antigens. In some cases this happens because the body is stimulated to produce antibody in response to foreign antigen very

A Journey into Healthy Living

Understanding AIDS

Acquired immune deficiency syndrome (AIDS) is a deadly disease that was first identified in 1981. AIDS is caused by human immunodeficiency virus (HIV or AIDS virus), which may have passed from a monkey host into human populations in Africa during the 1960s.

Most people infected with the AIDS virus have no immediate symptoms. They go about their lives, passing the virus to others, sometimes before the virus can be detected by blood tests. AIDS develops years later. How quickly this happens seems to depend upon the general state of the immune system. The young and healthy may take 10 years to develop the disease. Older people, those with other infections, and newborn babies may develop the disease within a year. As far as we know, AIDS is always fatal, usually several years after symptoms of the disease first appear.

The AIDS virus is transmitted almost exclusively via semen and blood. The virus can pass from one person to another during anal or vaginal intercourse. Blood-to-blood transmission may occur when the same hypodermic needle is used to inject drugs or vaccines into more than one person. It can also occur when someone receives a transfusion of infected blood or blood products, or even when someone accidentally passes blood to another person through a lesion, such as a cut in the skin. Similarly, blood vessels break during childbirth, allowing blood from an infected mother to reach her baby. Blood donors are in no danger in developed countries, because a new needle is used for each donor. Since mid-1985, blood for transfusions has been screened for AIDS virus antibodies in most countries and is largely safe (although an individual may not develop viral antibodies for as much as 3 years after infection, which means that some virus particles escape detection and may be present in transfused blood).

The AIDS virus is not spread by casual contact, hugging, changing diapers, using the same toilet seats, or even sharing toothbrushes. Doctors, nurses, friends, and family members who live with AIDS patients almost never catch the disease. The virus has been found in urine, tears, saliva, breast milk, and vaginal secretions, but it seems not to be transmitted by these fluids unless it gets into a cut. The virus apparently will not cross intact mucous membranes (in the mouth, vagina, or rectum) or skin. However, there are often gaps in mucous membranes which have been stretched, or which have been injured by ulcers or by fungus, herpes, or gonorrhea infections. These gaps in the membrane form a route by which the virus can pass from one person to another.

In Western nations, infection from a man was the first method of transmission recognized, particularly between male homosexuals. AIDS has now spread to heterosexual men and women, and to babies, mainly by way of intravenous drug users who share needles. In one study, 56% of female prostitutes in New Jersey were found to carry the AIDS virus. All of them had been infected by intravenous drug use, sex with bisexual men, or sex with male drug users. Infected women can pass the virus to their sexual partners, especially during menstruation.

In Africa, the virus is often transmitted by sexual promiscuity, the vaccination of many people with the same needle, and transfusion of unscreened blood into malaria patients. Africa also has a high incidence of sexually transmitted diseases, which produce lesions in genital areas. For all these reasons, as many African women as men are already infected with AIDS.

Today, between 13 and 15 million people worldwide are infected with the AIDS virus and, worldwide over 2 million people have died. In the United States, a million people are infected and more than 180,000 people have died. The most common means of infection is now heterosexual intercourse.

The AIDS virus is a retrovirus (a virus with an RNA genome). It attacks cells in the brain, which is why many AIDS patients experience brain damage and insanity before they die. However, the usual cause of death from AIDS in adults is that the virus kills lymphocytes (Figure 25–A). AIDS patients have normal numbers of macrophages and B lymphocytes, but the number of helper T lymphocytes is drastically reduced. With a disabled immune system, the AIDS patient dies from pathogens that a normal immune system can control. The patient may die of diarrhea, rare forms of cancer, pneumonia, or tuberculosis. (The speed with which AIDS patients die of these infections shows what a good job the immune system usually does of defending us from the pathogens that always surround us.)

Drugs to combat AIDS are slowly being developed. Antibiotics, which help the immune system combat bacterial infections, have no effect on viruses. The drug zidovudine (also called azidothymidine [AZT]) resembles the nucleotide thymidine and prolongs the life of many AIDS patients. It also postpones the disease in people infected with the virus but not yet showing symptoms of AIDS. The drug kills replicating viruses, probably by being incorporated into their genetic material. This increases the number of lymphocytes that survive the virus and restores some immune functions. However, some patients' AIDS viruses have already mutated to

FIGURE 25-A

AIDS infection. A false-color electron micrograph of an AIDS virus as it leaves an infected lymphocyte. (C. Daguet/Pasteur Institute/Petit format/Science Source)

AZT-resistant forms. Other drugs with effects similar to AZT are now being tested. None of them is a cure for the disease.

We are left with three defenses against the AIDS virus, or any other virus: vaccination, immune responses, and prevention. It is not easy to produce a vaccine to the AIDS virus because the virus evolves rapidly and changes its antigens frequently, showing great variation from patient to patient and even in the same patient at different times. A vaccine is a sample of the antigen that makes up part of the virus's surface coat. It is no use vaccinating someone with a particular antigen if they will later be exposed to a different one.

The AIDS virus does not completely disable the immune system. People infected with the virus produce antibodies to the virus, but somehow these antibodies do not protect them from the virus. Some people infected with the virus apparently never develop AIDS. Some of these people, presumably, already have immune systems that defend them against the virus. This is natural selection in action, and we can see that even if a cure for AIDS is never found, a human population resistant to the virus will eventually evolve.

However, this is no consolation to those now infected by the virus or to health officials overwhelmed by the scope of the problem. In parts of West Africa, one quarter of the population is infected, and millions of children have been orphaned by AIDS. The U.S. Public Health Service estimates that more than 2 million people in the United States are already infected with the virus. At current rates of infection, health officials predict that, worldwide, up to 120 million people will be infected with the virus by the year 2000.

AIDS is a staggering public burden. AIDS patients require frequent hospitalization and treatment for their many symptoms, and their average hospital stay is twice as long as that of other patients. The psychological strain on those caring for so many patients with no hope of recovery is another inestimable burden.

Most cases of AIDS result from sex with an infected partner or sharing hypodermic needles. No one is completely safe from AIDS if, during the last ten years, they have had sex with anyone who has had sex with a third person. Chastity, lifelong monogamy, and a clean needle for each injection are the only ways to prevent the spread of AIDS. Even these practices would not stop AIDS for many years (because of the risk from blood transfusions and the number of individuals already infected who do not know it).

The most effective way to combat AIDS is education. This was started by homosexual organizations in the United States and is now recommended for the entire population by scientists and the U.S. Surgeon General. Education aims to (1) explain how the virus is transmitted, (2) promote chastity and monogamy, (3) encourage the use of condoms, and (4) point out the dangers of sharing needles. People who use condoms properly, throughout sexual intercourse, are almost completely safe from transmitting the virus and from being infected. It also seems that people infected by the AIDS virus but who do not have AIDS are less likely to develop AIDS if they are not exposed to the virus again. Condoms reduce the chance of a second exposure. Another measure, common in Europe, that drastically slows the spread of the disease, is to make disposable hypodermic needles readily available so that drug users are less likely to share needles.

Barring a miracle, because researchers do not predict development of a vaccine until after the year 2000, the AIDS epidemic is bound to get worse, mainly because AIDS education has not had much effect on the behavior of people at risk of contracting AIDS. A 1989 study of sexually active U.S. college students showed that fewer than half were monogamous and that condom use (by about 40% of the study group) had increased only slightly in the previous five years. Matters do not seem to have significantly changed since this study was done.

Although we do not know precisely how fast AIDS is spreading through the population, it is difficult to exaggerate the danger of this disease. Unless clean needles are made available to drug addicts, and unless millions of people change their sexual habits almost immediately, we may well face the modern equivalent of the medieval plague—which killed a quarter of the population of Europe.

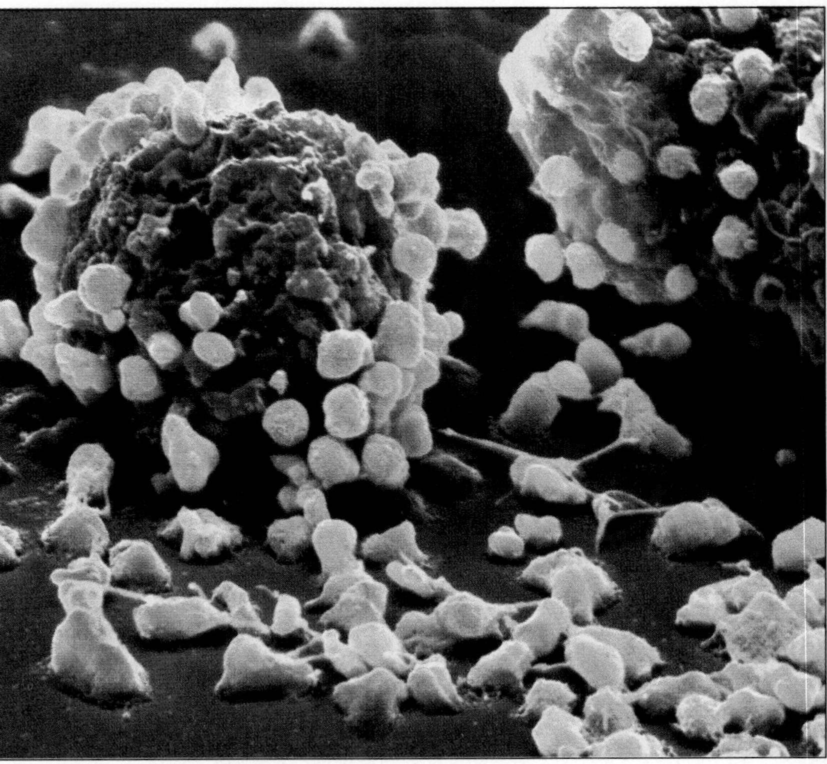

FIGURE 25-10

An allergic reaction. A mast cell contains many granules of histamine. During an allergic reaction, mast cells release these granules rapidly, causing allergy symptoms. Antihistamine drugs provide some relief to allergy sufferers by interfering with histamine molecules. (Lennart Nilsson/Boehringer Ingelheim International GmbH/National Geographic Society)

similar to one of the body's own antigens. In this case, the antibody may destroy the body's similar protein as well as the foreign antigen.

For example, when the body fights infections by streptococcal bacteria, it forms antibodies that may break down the body's own proteins, frequently damaging or destroying the heart valves. This is why "strep throat" is a serious disease that should be treated quickly with antibiotics. A number of other devastating (although fairly rare) diseases are also thought to be caused by autoimmunity, including insulin-dependent diabetes, rheumatoid arthritis, multiple sclerosis, pernicious anemia, Addison's disease, myasthenia gravis, and ulcerative colitis. Some of these diseases occur almost exclusively in people with specific MHC antigens—presumably ones that are easily mistaken for foreign antigens by the body's immune cells. Thus, people with the genes for these MHC antigens have an inherited risk of developing the corresponding autoimmune disease.

BIO-BIT

Antihistamine drugs reduce allergy symptoms by neutralizing histamines released from mast cells.

Allergies and Anaphylaxis

About 10% of the human population suffers from allergies: inappropriate immune responses to harmless substances encountered in the environment, or in food or medicine—for example, milk, chocolate, pollen, penicillin, cat saliva, or mites in house dust. Generally the first exposure to the allergy-producing antigen (an **allergen**) produces no symptoms, but it **sensitizes** the body by evoking a primary immune response. The next encounter with the allergen evokes a secondary immune response known as an allergic, or hypersensitive, response.

Allergic reactions are due to group E antibodies. In most people, the first time an allergen enters the body, T cells recognize it as harmless and prevent B cells from responding to it. In a person who will produce an allergic reaction, however, something goes wrong. The B cells produce group E antibodies, which bind to surface receptors on mast cells (Section 25-A). When the same allergen next enters the body and reaches such a bound mast cell, the antigen binds to the E antibody and the mast cell self-destructs, releasing histamine, which causes a hypersensitivity reaction (Figure 25-10). Histamine makes blood vessels dilate and increases the permeability of capillary walls, so that fluid escapes and swells the tissues.

The normal role of E antibodies is not to cause allergy but to protect the body from infection by platyhelminth parasites (tapeworms and flukes). Flatworm infestations are not a major threat to human health in developed countries today, largely due to improved hygiene in the last two hundred years. Such parasites are still important however, in many developing countries.

Anaphylaxis is a severe secondary allergic reaction. The first time an antigen such as egg albumin is injected into a guinea pig, it has no obvious effect. If the injection is repeated three weeks later, the sensitized animal produces the symptoms of general anaphylaxis: the muscles of the bronchiole walls contract, constricting the air passages to the lungs, and the capillaries dilate. The animal will probably die unless injected with the hormone epinephrine, which will counteract all of these symptoms. Similar **anaphylactic shock** sometimes occurs in human beings who are allergic to such things as penicillin or insect stings.

We do not know why some people have allergies while others do not, but allergies tend to run in families, suggesting that they have a genetic basis. Studies also suggest that breast-fed infants are less prone to develop some kinds of allergies later in life, compared to infants fed on baby formula. Those who suffer from allergies may derive some consolation from the thought that they are apparently less likely to develop tumors than are other members of the population. ■

Allergies are sometimes treated by attempting to induce tolerance to allergens (Section 25-B). Small amounts of the allergen are repeatedly injected, in the hope that the body will cease to react to it. This treatment does not always work. Part of the explanation probably lies in the fact that the immune response and the genes that control the immune system vary enormously among individuals.

Autoimmune diseases result when the immune system accidentally produces antibodies that react with some of the body's own cells. Allergic reactions are secondary immune responses to substances that have provoked inappropriate primary responses.

SUMMARY

1. All animals have nonspecific defense mechanisms that protect them from a variety of diseases. These nonspecific defenses include the skin and mucous membranes, the inflammatory reaction, fever, and interferons.
2. Vertebrates are also protected by immune responses, which are defenses against specific diseases. Immune responses are characterized by the ability to recognize antigens either as part of the body or as foreign, and by formation of a "memory" that the body has encountered a particular foreign antigen before.
3. The bone marrow produces the white blood cells that interact to produce immune responses. These cells travel in the blood and lymph to all parts of the body, but are particularly concentrated in the lymph nodes.
4. The body does not normally produce antibodies to its own molecules. The immune system develops a tolerance to the body's own molecules during embryonic development.
5. The vast diversity of antibodies and T cell receptors in the body is produced by recombination of relatively few genes, each of which specifies part of an antibody or receptor molecule. The variable regions of these molecules bind foreign antigens with great specificity.
6. In cellular immunity, killer T cells recognize and kill abnormal body cells such as cancer cells.
7. Humoral immunity is the main defense against invading pathogens in body fluids. In humoral immunity:
   a. B cells specific to an invading antigen are activated by macrophages and helper T cells to divide to form a clone containing plasma cells and memory cells.
   b. Plasma cells from the B cell clone make and release antibody molecules.
   c. The foreign antigen is destroyed by phagocytes, which engulf the antigen when it is bound by this antibody.
8. The immune system's first exposure to an antigen causes a primary immune response, in which lymphocytes specific to that antigen are stimulated to form a clone of cells capable of combating that antigen.
9. After the invasion has been defeated, memory cells remain, so that a secondary immune response to the same antigen is greater and more rapid than the first, often destroying the antigen before it can cause illness.
10. Vaccination stimulates a primary immune response to a pathogenic antigen so that the body responds with an effective secondary response if it later encounters the pathogen itself.
11. Tissue and organ transplants are usually rejected because the host's T cells recognize the donor's antigens as foreign and destroy the transplanted organ. Matching MHC antigens of donor and host and drugs that suppress immune responses are needed to prevent transplanted organs from being rejected.
12. Autoimmune diseases, which destroy body tissues, occur when the body develops antibodies to some of its own antigens.
13. Allergic reactions or anaphylactic shock occur when an allergen inappropriately induces mast cells bound to group E antibodies to release histamine.

SELF-QUIZ

Match the following structures with the function each performs:

___ 1.	A foreign macromolecule that may endanger the body	a. antibody
___ 2.	Site of a filter that removes invaders from the body	b. antigen
___ 3.	Long-lived cell that helps the body respond quickly to previously encountered antigens	c. B lymphocyte
___ 4.	Macromolecule that binds foreign molecules in the bloodstream	d. lymph node
		e. memory cell
		f. thymus gland

5. The skin protects the body against disease by:
   a. repairing breaks in its surface
   b. secreting acid
   c. forming a barrier between the body and the external environment
   d. all of the above
6. The introduction of a bacterial antigen into the body triggers a response specifically against that antigen by:
   a. causing antibody molecules to assume a shape that permits them to bind the antibody
   b. causing mutations in cells so that they produce antibodies to the antigen that caused the mutation
   c. causing cells with the proper antibody to disintegrate and release the antibody
   d. stimulating reproduction of cells that make the antibody to that antigen
7. Skin can be grafted from one identical twin to another, time after time, without being rejected because:
   a. antibodies in the blood do not react to antigens on the other twin's cells
   b. the twins have the same MHC genes and antigens, and so the twins' cells do not stimulate cellular immune responses in each other
   c. the twins have been exposed to each others' MHC antigens as fetuses and, as a result, do not mount immune responses against each other
   d. macrophages cannot destroy the twins' cells because the cells are not bound to antibody
   e. B cells do not react to the cell surface antigens of an individual with the same MHC genes
8. Which of the following is *not* a role of lymphocytes?
   a. producing antigens
   b. producing B-cell growth factor
   c. forming immunological memory
   d. destroying body cells that are infected by viruses
   e. producing plasma cells
9. Vaccination protects the body against catching a disease because:
   a. it provides antibodies made by another animal
   b. it makes the disease organism histocompatible with the body
   c. it produces an enlarged clone of memory cells against that disease
   d. it builds up an immunological tolerance for the disease antigen
   e. it releases large amounts of nonspecific defensive secretions

THINKING CRITICALLY

1. What might be the selective advantage of a baby's being born before its immune system has matured?
2. Studies have shown that poliomyelitis, mononucleosis, and Hodgkin's disease are more common among children and young adults who have few or no siblings, few playmates, uncrowded homes, and well-educated, well-to-do parents. How might these factors affect their immune systems' ability to fend off the viruses that are probably responsible for these diseases?
3. Breast cancer nearly always develops in women past childbearing age. Does this mean that natural selection cannot increase the immune response to tumors of the breast?
4. The blood of people who use large amounts of opiates (heroin, opium, codeine) contains fewer T cells than the blood of nonusers. What characteristics would you expect users to show as a result?
5. If a mouse (#1) that has rejected a skin graft from mouse #2 is then given a new graft from mouse #3, would you expect the rejection of this new graft to follow the same pattern as would be seen with a second graft from mouse #2? Explain.

SUMMARY

allergen, *p. 540*
antibody, *p. 531*
antigen, *p. 528*
autoimmunity, *p. 537*
B lymphocyte (B cell), *p. 531*
bone marrow, *p. 529*
cellular immunity, *p. 533*
complement reaction, *p. 526*
histamine, *p. 526*
humoral immunity, *p. 533*
inflammation, *p. 526*
interferon, *p. 527*
interleukin, *p. 535*
killer T cell, *p. 533*
lymph node, *p. 529*
macrophage, *p. 526*
major histocompatability complex (MHC), *p. 533*
mast cell, *p. 526*
memory cell, *p. 535*
neutrophil, *p. 530*
nonspecific defenses, *p. 525*
pathogen, *p. 525*
phagocyte, *p. 525*
primary immune response, *p. 535*
pyrogen, *p. 527*
secondary immune response, *p. 535*
specific defenses, *p. 525*
T cell receptor, *p. 531*
T lymphocyte, *p. 531*
tolerance, *p. 529*

SUGGESTED READINGS

Ada, G. L., and G. Nossal. "The clonal-selection theory." *Scientific American,* August 1987. The history of experiments showing how the body responds to antigen by producing specific antibodies.

Buisseret, P. "Allergy." *Scientific American,* August 1982. Why allergies appear to be malfunctions of the immune system.

Cohen, I. R. "The self, the world and autoimmunity." *Scientific American,* August 1988. A discussion of autoimmune diseases and the possibility that the immune system can be induced to control them.

Marrack, P., and J. Kappler. "The T cell and its receptor." *Scientific American,* February 1986.

Scientific American, October 1988 issue, "What Science Knows About AIDS." An entire issue devoted to AIDS, including articles on the prospects for a vaccine, origin of the virus, epidemiology of the disease, international issues, treating AIDS, social and public health aspects of the disease.

Scientific American, September 1993 issue, "Life, Death, and the Immune System." An entire issue addressing all major aspects of the immune system.

Von Boehmer, H., and P. Kisielow. "How the immune system learns about self." *Scientific American,* October 1991.

Young, J. D.-E., and Z. A Cohn. "How killer cells kill." *Scientific American,* January 1988.

CHAPTER 26

CONCEPT GUIDE

After reading this chapter, you should be able to:

1. List the three forms of nitrogenous waste and specify an example of an animal group that uses each form. (Section 26–A)
2. Explain how osmoregulation, habitat, and energy expenditure all relate to the form of nitrogenous waste used by animals. (Section 26–B)
3. Compare the general structures and functions of flame cells, nephridia, Malpighian tubules, and nephrons. List examples of animal groups that use each of these types of excretory organs. (Sections 26–C and D)
4. Describe the formation of urine in the vertebrate kidney, including the processes of filtration, resorption, and secretion. (Section 26–D)
5. Describe the effects of vasopressin, aldosterone, renin, and angiotensin on blood sodium levels, blood pressure, and urine composition. (Section 26–E)

Marine iguanas feed on underwater beds of marine algae and ingest large quantities of salt. To rid themselves of this extra salt, they have a special salt-secreting gland that empties into the front part of each nostril. Marine iguanas clear away this fluid by periodically blowing out showers of extremely salty vapor. The salt dries and accumulates, resulting in the "frosted" appearance of their heads and backs. Learn how other marine mammals regulate their salt levels by reading Section 26–B. (J. Rotman/Peter Arnold, Inc.)

Excretion

An animal's body fluids must be maintained as a medium in which its cells can live. This requires regulation of temperature, pH, and the amounts and proportions of salts and water in the body fluids. It is relatively easy for animals that live in the sea to maintain **homeostasis** ("same-standing") of their body fluids because the sea has a relatively stable salt composition and pH, and its temperature changes little and slowly.

You will recall (Chapter 4) that homeostasis is the maintenance of the chemical composition of a cell within narrow limits and that living cells provide the environment that is necessary for all the activities of life. A great deal of evidence suggests that life began in the sea. First, most invertebrates and primitive plants are marine. Second, the body fluids of most invertebrates resemble sea water in their salt concentration. When organisms invaded fresh water and land, however, body fluid homeostasis demanded new adaptations. For instance, early vertebrates invaded fresh water, which contains few salts. They evolved the ability to live with body fluids containing a low salt concentration. Today, the body fluids of all vertebrates have a salt composition similar to that of sea water, but only about one-third as concentrated. The early freshwater fish evolved organs, the kidneys, that permitted them to excrete excess water that entered the body by osmosis. Thus, kidneys are found only in vertebrates.

All the body's fluid compartments are connected. Blood and extracellular fluid (which surrounds all cells) exchange substances as the blood passes through the capillaries of the circulatory system. Blood is also the source of fluid and solutes in the lymph and in the **cerebrospinal fluid,** which bathes the brain and spinal cord. Both of these fluids drain back into the veins and return to the heart. Because all the body's fluid compartments connect with one another, an animal can regulate the composition of all its body fluids by controlling the content of any one of them. In the vertebrate body, the liver and kidneys are the most important organs that monitor and adjust the composition of the blood and thus keep the composition of all the body fluids constant. The liver regulates the blood's content of food molecules (Section 22-G). The kidneys dispose of nitrogenous and other wastes and regulate salts and water.

Homeostasis might be easier if an organism were a self-contained system, but every organism must constantly take in substances from its environment, use them in the chemical reactions of metabolism, and discharge the resulting wastes back into the environment. With this constant flow of substances, the task of maintaining the constant composition of the fluid surrounding the cells is formidable. It is further complicated by the fact that wastes and substances needed by the body are mingled in the body fluids. Thus, the excretory system has the complex task of removing waste, especially toxic nitrogenous wastes from the breakdown of proteins, while maintaining the body fluids' pH, water, and salt balance.

KEY CONCEPTS

- Living cells require an almost constant chemical composition in their immediate environment, the extracellular fluid.
- Because an animal's body fluids are in contact with one another, regulating the content of any one body fluid (usually the blood) therefore maintains homeostasis in all of them.
- An animal's excretory organs must rid its body of nitrogenous wastes while keeping salts and water in proper balance.

Curiosity Questions

1? How can the use of illegal drugs be detected by urine tests? (See page 546)

2? Why should a person who is lost at sea not drink sea water nor eat fish? (See page 550)

3? Why does drinking an alcoholic beverage increase the need to urinate? (See page 555)

26–A SUBSTANCES EXCRETED

1? The body's chief wastes are carbon dioxide and water from oxidation of organic molecules, and nitrogenous wastes from the breakdown of proteins (Figure 26–1). Carbon dioxide is excreted across the body's respiratory surfaces. Excretory organs such as kidneys have the two major functions of removing nitrogenous wastes and regulating the body's salt and water content. In addition, excretory organs control the excretion of substances like spices, drugs, and hormones. Onions, garlic, and some other spices have volatile components that leave the body through the lungs, whereas the rest leave through the kidneys. Penicillin and other drugs are removed primarily via the kidneys, and the drugs or the products of their breakdown can be detected by chemical tests of the urine. The kidneys, liver, and lungs also carry out **detoxification,** converting toxic substances into forms that are not poisonous to the body. ■

Although the kidneys control the salt concentration and composition of the blood, some salts are also excreted through the skin in sweat, and some leave with the feces. When a lot of water and salt exits through the skin as sweat, the kidneys form less urine. A person working in the desert may lose more than 1 liter of sweat per hour, with a loss of 10 to 30 grams of sodium chloride per day. This heavy loss of salt causes no immediate difficulty because water is also lost, and so the salt concentration of the body fluids is maintained. However, drinking water after such heavy sweating dilutes the extracellular fluid, a condition sometimes called "electrolyte imbalance." This may lead to muscle cramps—an example of the importance of maintaining the composition of the body fluids.

Undigested food from the gut does not appear on our list of substances excreted by the body. Food that passes down the digestive tract and out the anus is not excreted but **egested**—that is, it travels through and is expelled from the body without ever passing through a plasma membrane to become part of the body. The term "excretion" applies only to substances that must cross plasma membranes to leave the body.

BIO-BIT

While doing hard work in high temperatures, a human may excrete from 3 to 4 liters of sweat an hour!

Amount of Vital Substances Exchanged Daily

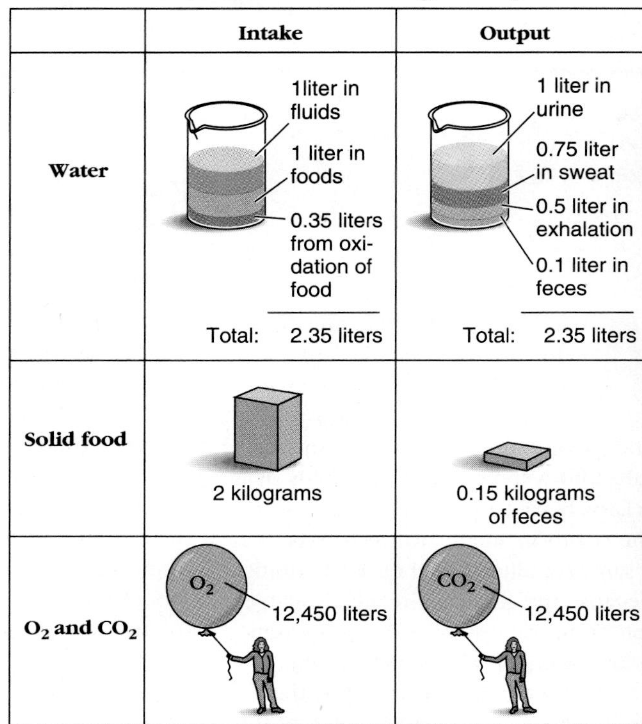

Substances excreted	Excretory organs
Nitrogenous wastes	Kidneys, Skin (small amount in sweat)
Water	Kidneys, Skin, Lungs
Salts	Kidneys, Skin (in sweat)
CO_2	Lungs

FIGURE 26–1

Daily exchange and excretion. This example of daily human intake illustrates the conversion of ingested water and solid food to urine, sweat, and feces. In addition, oxygen and carbon dioxide are exchanged in the lungs. The major sites of human excretion include the kidneys, skin, and lungs.

Nitrogenous Wastes

Nitrogenous wastes are a by-product of the breakdown of proteins. First, proteins are hydrolyzed to amino acids. This occurs in the digestive tract as food proteins are digested and in the body cells, where proteins are constantly made and destroyed. Fats and carbohydrates are stored for future use, but the body cannot store proteins or amino acids. Amino acids that the body cannot use immediately are **deaminated;** that is, their amino ($—NH_2$) groups are removed. The remaining organic acid can be used for energy or converted into carbohydrate or fat and stored. Each $—NH_2$

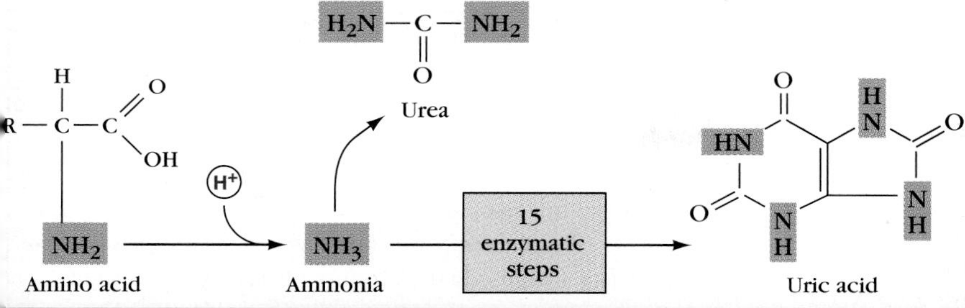

Nitrogenous wastes	Advantages	Disadvantages	Distribution
NH_3 Ammonia	Costs little energy to produce	Toxic if not diluted	Many aquatic animals
$H_2N-C(=O)-NH_2$ Urea	Less toxic than ammonia, can accumulate in body without harming tissues, less water lost in excretion	Costs more metabolic energy to produce	Mammals and adult amphibians
Uric acid	Conserves nearly all body water	Requires a lot of metabolic energy to produce	Insects, birds, many reptiles

FIGURE 26–2

Nitrogenous wastes. An amino acid and the three main forms of nitrogenous waste that animals produce from amino groups removed from excess amino acids. Urea and uric acid molecules contain nitrogen and hydrogen atoms derived from more than one amino group.

group removed picks up another hydrogen atom and becomes NH_3, ammonia. Because ammonia is the first metabolic breakdown product of amino acids, it can be produced with very little energy.

Ammonia is toxic except in a very dilute solution. Many aquatic animals dissolve their ammonia in large quantities of water from their environment and excrete this dilute solution.

Land-dwelling animals are often short of water. Mammals and adult amphibians conserve water by converting ammonia to urea and excreting the urea (Figure 26–2). The additional metabolic steps cost energy but are still worthwhile because they save water. Because urea is much less toxic than ammonia, it can accumulate at higher levels without damaging the tissues and can be excreted in a more concentrated form, using less water.

In mammals, most urea is formed by the liver, which takes excess amino acids out of the blood, deaminates them, and incorporates them into urea molecules. The brain and kidneys also form urea, but in lesser amounts.

Other land animals, notably reptiles, birds, and insects, incorporate their ammonia into **uric acid.** The process consists of about 15 enzymatic steps and requires a lot of energy. However, this energy investment pays off in terms of water conservation because uric acid can be excreted in almost solid form.

Because the kidneys can handle nitrogenous waste only in solution, birds and reptiles pass a dilute solution of uric acid from the kidneys into the **cloaca** ("sewer"), a common reservoir at the end of the urinary, digestive, and reproductive tracts. Here most of the water is resorbed, and uric acid crystals precipitate and mix with the feces; the two are then voided together. Insects have a similar arrangement.

Most mammals do not excrete uric acid, but small amounts do appear in human urine. Our bodies produce uric acid by the breakdown of adenine and guanine (nucleotide bases), not of proteins. Persons with certain metabolic disorders produce more uric acid than usual, and this causes the disease known as gout. In gout, uric acid crystals accumulate in some of the joints of the body, especially in the toes, causing great pain.

Less complex animals with access to a lot of water excrete most of their nitrogenous wastes as ammonia. Some groups of more complex animals evolved the

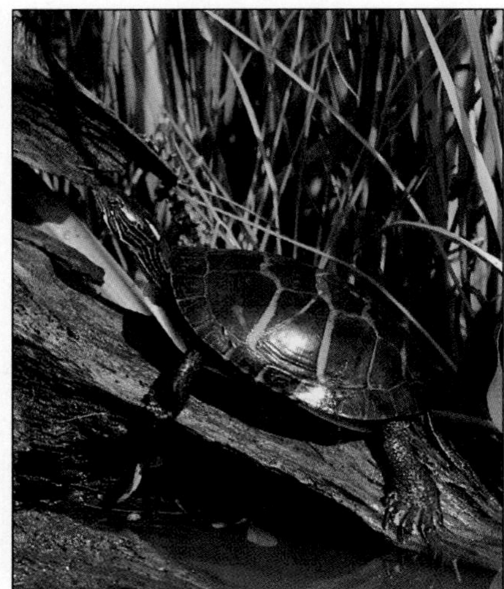

FIGURE 26–3

An exception to the rule. Although the aquatic painted turtle is a reptile, it excretes up to half its nitrogenous wastes as ammonia rather than the uric acid typical of land-dwelling reptiles. (D. Kuhn)

metabolism and excretory organs to produce and excrete urea or uric acid, and many members of these groups were able to move into drier habitats. However, animals do not always excrete the form of nitrogenous waste predicted by their taxonomy; habitat plays an important part, too. For example, most reptiles excrete uric acid, but aquatic turtles also excrete a good deal of ammonia and urea (Figure 26–3).

Why do reptiles and birds excrete uric acid, whereas mammals turn their nitrogenous wastes into urea? The production of urea or uric acid is linked to the mode of reproduction. Birds and reptiles lay amniotic eggs, with membranes and shells that enclose all the water the embryo will have until it hatches (see Figure 21–20). The embryo produces uric acid, which sits as a solid mass, leaving the water in the egg free for other uses. A mammalian embryo, by contrast, has access to its mother's fluids via the placenta, and so it can rely on her system to dispose of its wastes. Any water available to the mother can also be used by the embryo. Mammals appear to have lost the metabolic pathway by which birds and reptiles produce uric acid from amino acids.

BIO-BIT

The white paste-like material in bird droppings consists of acid precipitate and feces.

Body waters are excreted by the kidneys, lungs, and skin. The kidneys regulate the body's salt and water content and remove nitrogenous wastes. The production and excretion of nitrogenous wastes involves a trade-off between energy expenditure and water conservation. Excreting ammonia takes little energy but uses a lot of water, whereas excreting uric acid costs lots of energy but uses very little water. Urea excretion uses intermediate amounts of energy and water.

26–B OSMOREGULATION IN DIFFERENT ENVIRONMENTS

The removal of nitrogenous wastes from the body fluids is inextricably tied to **osmoregulation**—regulation of the fluids' osmotic properties, that is, the balance between salts and water. The form of nitrogenous waste excreted depends on the availability of water, which in turn depends on the water balance between the animal's body fluids and its environment.

The body fluids of most marine invertebrates are **isotonic** (same chemical tension) with the sea water in which they live (that is, no net gain or loss of water occurs between the body fluids and sea water across the membranes separating them). Freshwater invertebrates and most vertebrates have body fluids about one-third as concentrated as those of marine invertebrates. Hence their fluids are **hypotonic** (less chemical tension) to sea water, but **hypertonic** (greater chemical tension) to fresh water. These animals tend to lose water from their body fluids by osmosis if they are placed in sea water, and to gain water from fresh water (Section 4–C).

Marine fish and other marine vertebrates must prevent osmotic water loss to their hypertonic environment and uptake of too many salts by diffusion. Freshwater organisms have just the opposite problems: they must prevent loss of salts by diffusion and uptake of water by osmosis. They do this in part by excreting large volumes of dilute urine, but they must also conserve salts while ridding their bodies of nitrogenous wastes.

How do vertebrates live with these osmotic problems? Freshwater fish are covered with mucus, which retards passage of water and salts through the body surface. Freshwater bony fish do not drink water, but they must pass water over the gills to get oxygen, and water inevitably enters through the permeable gill membranes. These fish eliminate water by producing copious, dilute urine, but they lose salts both via the urine and via diffusion from the gills. This is counteracted by active transport of salts into the body by special cells in the gills (Figure 26–4a). Freshwater fish also take in salts as part of their food.

FIGURE 26-4

Osmotic adaptations of vertebrates. (a) Freshwater bony fish gain water from their environment by osmosis. **(b)** Marine bony fish tend to lose water by osmosis, because the salt concentration in the surrounding water is higher than that in their bodies. **(c)** Sharks face the same problem of water loss but have adapted differently. Retention of urea makes their body fluids hypertonic to sea water, and so they gain water through their gills. **(d)** Marine birds often have a salt gland on their heads that secretes a highly concentrated solution of salt taken in with sea water. The kidneys produce a concentrated urine. These features permit the bird to conserve water for use in the body. **(e)** Marine mammals drink sea water. They conserve body water by excreting salts in a very concentrated urine.

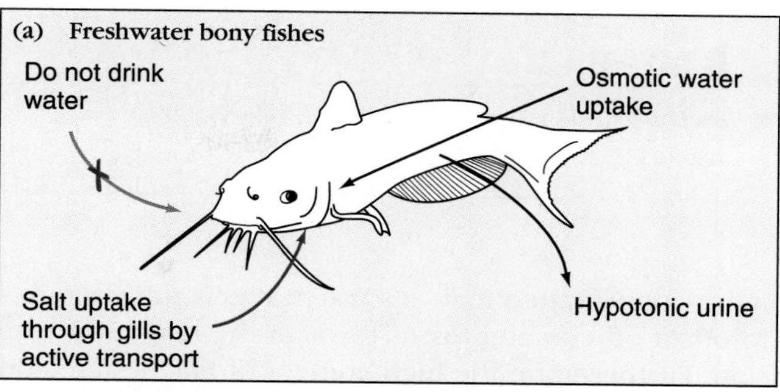

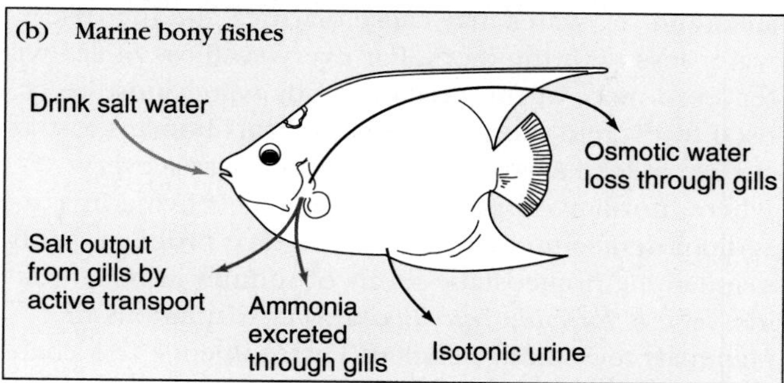

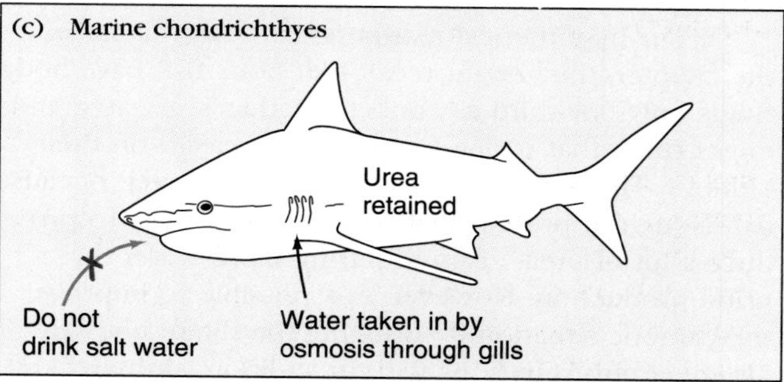

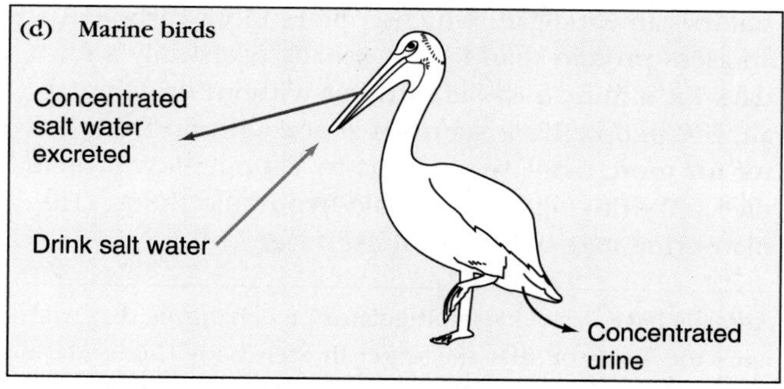

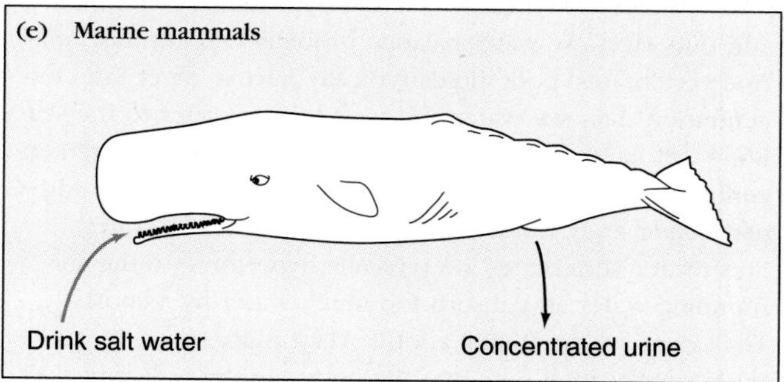

In a sense, although a marine bony fish is surrounded by water, it actually lives in a physiological desert, because it tends to lose water to its hypertonic environment. How does the fish conserve water and survive in this desert? The marine fish loses water both through the gills and in its urine. It takes in salt and water as part of its food. It achieves osmotic balance by excreting much of the salt by active transport through its gills (Figure 26-4b).

The Chondrichthyes (sharks, skates, and rays) have evolved an interesting and unusual method of coping with the marine environment. Like most other vertebrates, they have body fluids with a salt concentration about one-third that of sea water, but they also produce and retain large quantities of urea (Figure 26-4c). Their tissues have become adapted to levels of urea that would kill most other organisms. The combination of salts plus urea makes their body fluids slightly hypertonic to sea water: these fish actually absorb some water from the sea through their gills by osmosis and use it for excretion.

Seafaring birds drink sea water, and get rid of their excess salt by way of a **salt gland** in the head. This excretes a very concentrated salt solution, which drips out of the nostrils. Birds also excrete uric acid, conserving as much water as possible (Figure 26-4d).

Marine mammals, such as whales and porpoises, take in sea water along with their food. Their kidneys can produce a urine several times as concentrated as sea water (Figure 26-4e). This is especially important for carnivorous marine mammals, because their high-protein diet yields a lot of urea to excrete.

Some land vertebrates can also produce highly concentrated urine. For example, laboratory rats can live indefinitely when all they are given to drink is sea water. Sea water is too concentrated to support human life. Although the human kidney can produce urine slightly

BIO-BIT

Sweat glands excrete 5 to 10% of all metabolic wastes.

more concentrated than sea water, this is not enough to offset other water losses through the lungs and skin. Furthermore, the high content of magnesium and sulfate in sea water may cause diarrhea and increase water loss with the feces. For every swallow of sea water, even more of the precious body water must be used to excrete the salts taken in. Thus humans lost at sea are indeed surrounded by "water, water everywhere, nor any drop to drink." Mammals living in a sodium deficient environment also face problems with maintaining homeostatic levels of sodium in body fluids. (See *A Journey into Evolution:* Adaptations of Mammals to Sodium-Deficient Environments, this chapter.)

2? If our hypothetical mariners had read this far in the chapter, they might recall that bony fish have body fluids only one-third as concentrated as sea water, and might think that eating fish would be easier on their kidneys. This would be of little help, however, because fish is high in protein that will force the body to produce a lot of urea, again requiring more water for urine production. However, it is possible to improve the osmotic situation by drinking the dilute body fluids squeezed from bony fish. In addition, shipwrecked sailors can eat algae, which contain more carbohydrate and less protein than fish. It usually takes only a few days for a human to die of thirst without drinking at all, but people have survived at sea with no fresh water for more than two months by eating a low-protein diet and drinking any available hypotonic fluids. (Human urine may be one of these.) ■

Animals have various osmoregulatory mechanisms that balance the levels of salts and water in their body fluids. Marine invertebrates are typically isotonic with their environment and thus face few water balance problems. In contrast, marine vertebrates' body fluids typically have a lower salt concentration than sea water and tend to lose water to the sea. Bony fish have adapted by secreting salt and producing concentrated urine, whereas cartilaginous fish retain urea and make their body fluids slightly hypertonic to sea water. Freshwater vertebrates are typically hypertonic to the surrounding water and absorb too much water by osmosis. They cope by producing a lot of very dilute urine, by actively transporting salts into the body, and by reducing skin permeability.

OVERVIEW: HOW EXCRETORY SYSTEMS WORK

Any excretory system does three things:

1. It collects fluids from somewhere inside the body, usually from the blood or from spaces between organs.
2. It modifies the fluid by resorbing substances the body needs to keep, and by transporting waste substances into the excretory product.
3. It provides a way to expel the excretory product from the body.

During excretion, an organism expends metabolic energy. First, it uses energy to break down proteins and in many cases to form urea or uric acid. Second, energy is used in active transport to modify fluids collected from the body into final excretory products. Although human kidneys make up less than 0.5% of the body weight, they use 7.2% of the oxygen consumed by the body. Pumping blood from the heart to the kidneys takes another 2.7%, so that about 10% of the human body's energy is spent just moving blood to the kidneys and cleansing it.

26–C EXCRETORY ORGANS OF INVERTEBRATES

Corals and other members of the marine phylum Cnidaria live in an isotonic environment and lose most of their wastes by diffusion. Freshwater flatworms (phylum Platyhelminthes), on the other hand, have organized, multicellular excretory systems, which expel excess water (Figure 26–5a). Body fluids are collected into **flame cells** by the beating of cilia. The fluid then passes through a series of tubules to an excretory pore at the body surface.

The functional unit of excretion in an earthworm is the **nephridium** (Figure 26–5b). The nephrostome, an open, ciliated funnel, draws fluid from the coelom into the long, thin tubule. As fluid flows through the nephridium, substances that the body needs are reclaimed and passed into surrounding capillaries of the circulatory system. Fluid expelled at the body surface contains water, nitrogenous wastes, and any salts that have not been resorbed.

Like the earthworm, most other invertebrates have nephridia. Insects, however, have evolved a completely different system: long, slender, **Malpighian tubules,** attached at one or both ends to the gut (Figure 26–5c). Nitrogenous wastes from the body fluid are converted into uric acid, which is then moved down the Malpighian tubule into the gut. Cells in the rectum

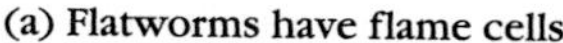

FIGURE 26-5

Invertebrate excretory organs. (a) Flame cells form the excretory system of flatworms. Beating cilia in flame cells propel fluids down a system of excretory ducts to pores on the surface of the body. **(b)** The excretory organ of an earthworm is called a nephridium. The funnel-shaped opening of the nephrostome collects excess body fluid. The rest of the nephridium lies in the next body segment back. Capillaries of the circulatory system intertwined with the tubule resorb needed substances before the fluid reaches the enlarged bladder, from which the urine is discharged through the nephridiopore. Each segment, except a few in the anterior end, contains a pair of nephridia. **(c)** In insects, Malpighian tubules discharge uric acid into the gut, where it is voided with the feces.

(a) Flatworms have flame cells

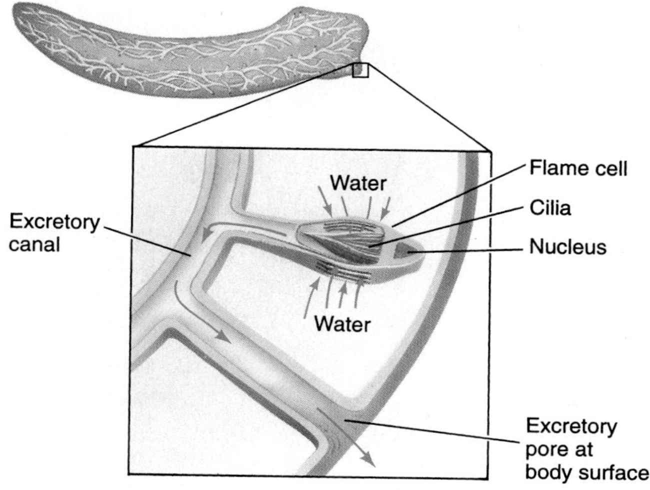

(b) Earthworms have nephridia

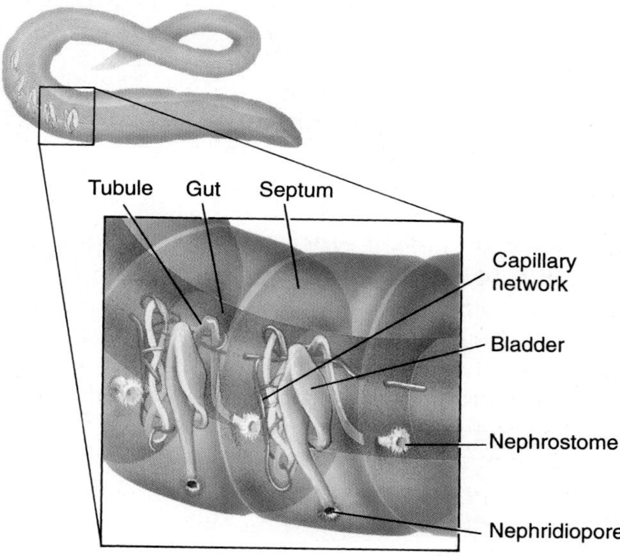

(c) Insects have Malpighian tubules

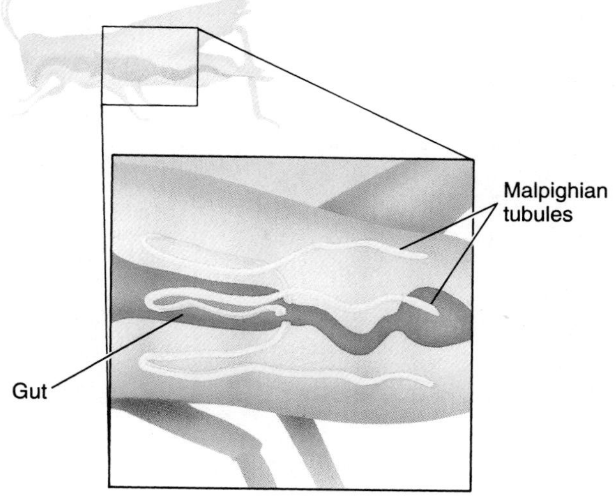

resorb water from the combined nitrogenous wastes and undigested foods before they are eliminated as fairly dry fecal pellets. Water conservation is particularly critical for insects because, unlike nearly all other invertebrates, they are predominately terrestrial organisms.

The excretory organs of invertebrates range from simple to complex. Flatworms and earthworms use open-ended tubes that collect, modify, and expel the fluids. Insects convert the nitrogenous wastes to uric acid and release it into the gut.

26–D THE VERTEBRATE KIDNEY

Every vertebrate has a pair of kidneys. A kidney's functional units are **nephrons,** which closely resemble the nephridia of earthworms, except that nephrons collect fluid filtered from the blood. A human kidney contains more than a million nephrons, each intertwined with capillaries of the circulatory system.

In the human excretory system, the **renal artery** carries blood from the aorta into the kidney. Here fluid, salts, and wastes are removed from the blood to form urine. Purified blood leaves the kidney via the **renal vein,** whereas urine leaves via a **ureter** and is stored in the **urinary bladder.** Eventually the urine is expelled from the body via the **urethra** (A Journey Through Kidney Structure and Function).

The human body contains about 5.6 liters of blood; 1.2 to 1.3 liters pass through the kidneys each minute, for a daily total of about 1800 liters. This is nearly one quarter of all the blood pumped by the heart, and so a very high proportion of all the blood is

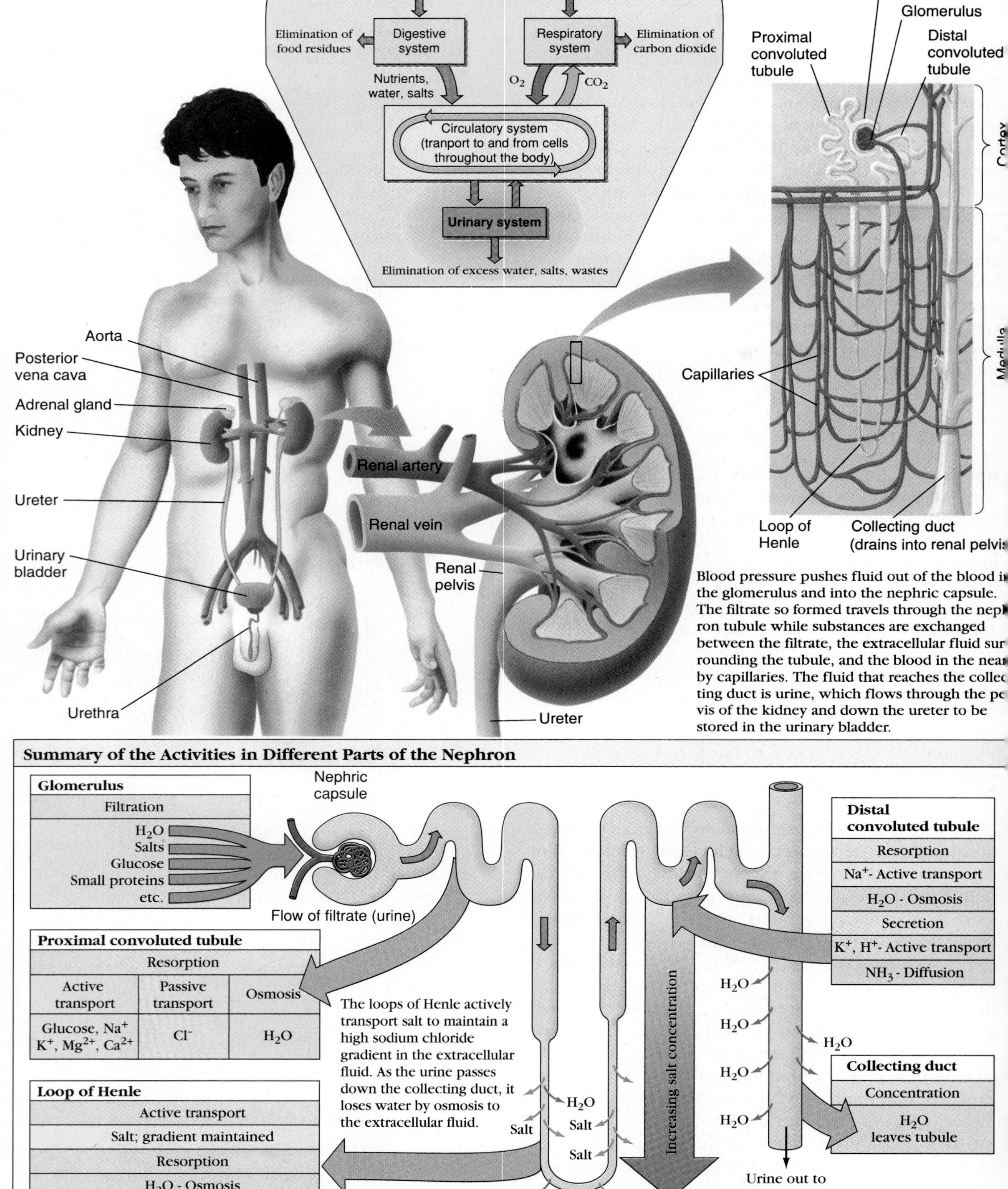

Blood pressure pushes fluid out of the blood in the glomerulus and into the nephric capsule. The filtrate so formed travels through the nephron tubule while substances are exchanged between the filtrate, the extracellular fluid surrounding the tubule, and the blood in the nearby capillaries. The fluid that reaches the collecting duct is urine, which flows through the pelvis of the kidney and down the ureter to be stored in the urinary bladder.

Summary of the Activities in Different Parts of the Nephron

Glomerulus
Filtration
H_2O
Salts
Glucose
Small proteins
etc.

Proximal convoluted tubule		
Resorption		
Active transport	Passive transport	Osmosis
Glucose, Na^+ K^+, Mg^{2+}, Ca^{2+}	Cl^-	H_2O

Loop of Henle
Active transport
Salt; gradient maintained
Resorption
H_2O - Osmosis

Distal convoluted tubule
Resorption
Na^+- Active transport
H_2O - Osmosis
Secretion
K^+, H^+- Active transport
NH_3 - Diffusion

Collecting duct
Concentration
H_2O leaves tubule

The human excretory system and associated blood vessels. Located in the small of the back, the kidneys play a vital role in the regulation of blood levels of water, salts, sugars, proteins, and nitrogenous wastes.

Table 26-1
Concentration of Various Substances in Human Urine

Substance	Concentration
Sodium (Na^+)	128 mmol/L*
Potassium (K^+)	60mmol/L*
Calcium (Ca^{2+})	5 mmol/L*
Magnesium (Mg^{2+})	15 mmol/L*
Chloride (Cl^-)	134 mmol/L*
Bicarbonate (HCO_3^-)	14 mmol/L*
Phosphate	50 mmol/L*
Sulfate (SO_4^{2-})	33 mmol/L*
Glucose	none (normally)
Urea	1820 mg/100 ml†
Uric acid	42 mg/100ml†
Creatinine	196 mg/100 ml†

* mmol/L = millimoles per liter (see Toolbox, Chapter 2)

† mg/100 ml = milligrams per 100 milliliters

passing through the kidneys at any time. Each day about 180 liters of fluid filter out of the blood and enter the nephrons. Most of this fluid is resorbed, leaving a daily urine output of about 1 liter. Table 26-1 shows the approximate composition of urine. This varies a lot depending upon what has been taken into the body and what needs to be excreted to restore the body fluids to within normal range.

Functions of the Nephron

Let us follow the process of urine formation in a nephron (A Journey Through Kidney Structure and Function). The nephron's cup-shaped **nephric capsule** surrounds a cluster of capillaries called a **glomerulus.** Because the blood is under pressure and the walls of the capillaries are permeable, much of the fluid from the blood filters into the capsule. Left behind are large proteins and whole blood cells, which are too large to pass through the filter, along with the rest of the blood fluid (plasma).

From the capsule, the **filtrate** (filtered fluid) passes into the nephron tubule. Meanwhile, the remainder of the blood follows along in capillaries outside the nephron. The cells of the nephron tubule modify both fluids, **resorbing** some substances from the filtrate and returning them to the blood, and **secreting** others from the blood into the urine-to-be.

The nephron tubule has four main parts: the **proximal convoluted tubule,** the U-shaped **loop of Henle,** the **distal convoluted tubule,** and finally the **collecting duct.**

In the proximal convoluted tubule, a considerable amount of resorption takes place. Small proteins, glucose, and ions such as sodium, potassium, magnesium, and calcium are returned to the blood by active transport. Negatively charged chloride ions follow passively after the positively charged ions. Water follows these solutes passively by osmosis. About 75% of the salt and water in the filtrate returns to the blood from the proximal convoluted tubule. Normally, all of the glucose also returns here. However, if glucose in the filtrate exceeds the **kidney threshold level,** some glucose will remain and appear in the urine. This happens when the active transport carriers for glucose, working at top speed, still cannot pick up all the glucose and move it back to the blood as quickly as the filtrate passes through the proximal tubule.

In many mammals and birds the loops of Henle are quite long, and their activities (discussed later) permit the production of relatively concentrated urine. As the fluid passes through the distal convoluted tubule, more sodium is resorbed. This is the principal area of secretion into the tubule, chiefly of potassium and hydrogen ions by active transport, and of ammonia by diffusion. Secretion of hydrogen ions adjusts the pH of the blood. Various drugs, such as penicillin, are also secreted into the urine here. The collecting duct, along with the loop of Henle, plays a vital role in water balance.

Concentration of the Urine

The actual concentration of urine takes place in the collecting ducts, but the process depends on the activities of the loops of Henle. To understand how this works, it is necessary to realize that the loops of Henle and collecting ducts lie in the **medulla** of the kidney. The other parts of the nephron lie outside the medulla in the kidney's outer region, the **cortex.** The extracellular fluid in the medulla contains an osmotic gradient, with solutes steadily more concentrated in the direction away from the cortex. The gradient contains two kinds of solutes: salt (sodium and chloride ions) and urea.

The salt gradient is created by the loops of Henle, which actively transport salt ions out of the filtrate. Active transport starts when the filtrate reaches the thick section of the ascending part of the loop. As the filtrate moves up, less salt is available to be transported out, and so the gradient has a higher salt concentration near the bottom of the loop (A Journey Through Kidney Structure and Function). This part of the loop of

A Journey into Evolution

Adaptations of Mammals to Sodium-Deficient Environments

Many arid and mountainous areas contain little sodium. Although no animal can survive any but the slightest drop in the body's normal sodium level, most mammals have adaptations that permit them to survive in such sodium-deficient areas. Some of these adaptations increase the dietary intake of sodium, whereas others reduce loss of sodium from the body.

Herbivores, including butterflies, reindeer, elephants, gorillas, elk, moose, sheep, and kangaroos, have been seen eating soil, drinking sea water, or consuming other improbable things whose only redeeming value appears to be their high sodium content (Figure 26-A). Plainly an appetite for salt when the body is deficient in sodium is a useful adaptation, and it is common in mammals.

A number of physiological differences are found between kangaroos, sheep, foxes, cattle, and rabbits in sodium-deficient mountainous areas of Australia and members of the same species in coastal areas, where there is plenty of sodium (from the sea). Whenever possible, animals in mountainous areas select food plants containing sodium, and in addition hold sodium loss from the body to a minimum. For instance, the urine of mountain animals contains virtually no sodium, whereas kangaroos and wombats on the coast excrete a lot of sodium in their urine. This difference is undoubtedly due to the higher levels of renin, angiotensin, and aldosterone found in the mountain-dwelling animals. During the day rabbits produce soft feces, which they eat to extract further sodium (and vitamins). The fecal pellets egested by highland rabbits after the second digestion contain practically no sodium.

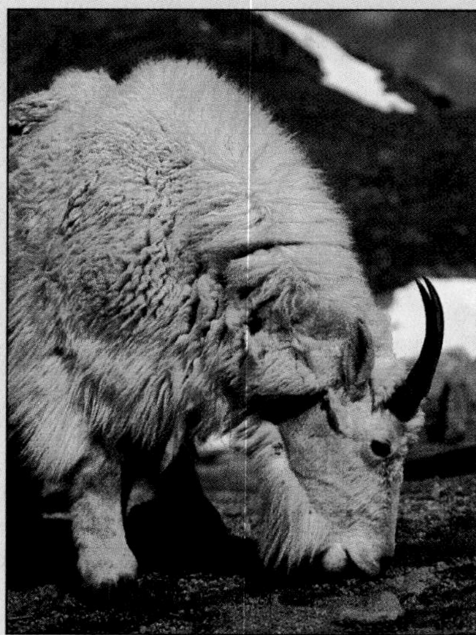

FIGURE 26-A

A mountain goat licking minerals from the soil. (R. Thom/Visuals Unlimited)

Sodium in the feces may be reduced by resorption from the gut. Sodium-deficient mammals with ruminant digestion (Section 22-D) also reduce the sodium loss in their feces by changing the composition of their saliva. Ruminants such as cattle and sheep produce many liters of saliva a day. This usually contains a high concentration of sodium bicarbonate, which produces an alkaline environment in the stomach for the microbial symbionts that ferment the animal's food. Much of this sodium bicarbonate is absorbed back into the body as the feces pass down the intestine, but some is lost with the feces. Aldosterone decreases the amount of sodium in the saliva and causes potassium to be secreted in its place. Although the need to secrete large quantities of sodium in the saliva every day seems to impose a sodium-supply problem on ruminants, the effect may, in fact, be the other way around. Sheep survive temporary sodium shortages better than do nonruminants such as humans or foxes. This is probably because the ruminant stomach contains a large store of sodium, which can be steadily replaced by potassium and used for more vital functions in the body during times of sodium shortage.

Henle is relatively impermeable to water, and so water cannot follow the ions out here.

From the loop of Henle, urine passes through the distal convoluted tubule to the collecting duct, where the actual concentration of the urine takes place. The walls of the collecting duct are permeable to water. As the urine passes down the collecting duct, it re-enters the medulla of the kidney. Here water moves from the urine into the extracellular fluid by osmosis in response to the high concentration of solutes in the medulla. From here it returns to the blood. The last part of the collecting duct is also permeable to urea, which by now is quite concentrated in the urine. Some urea diffuses out into the medulla, down its own gradient, adding to the solute concentration there. Urine leaving the collecting duct passes through the pelvis of the kidney and down the ureter to the urinary bladder, where it is stored.

Vertebrate kidneys are composed of many nephrons that collect and modify fluid filtered from the blood. The close association of capillaries and nephrons permits the exchange of substances between the blood and the filtrate as it passes through the nephrons. In this way, the kidneys can move substances that are needed by the body back into the blood and secrete substances that are not needed by the body into the filtrate. The loop of Henle establishes a high concentration of sodium chloride in the medulla of the kidney. This high sodium concentration draws water from the tubule and the collecting ducts, thereby concentrating the urine and retaining water in the body.

26–E REGULATION OF KIDNEY FUNCTION

Kidney function is under the minute-by-minute control of many systems. For instance, the filtration rate remains nearly constant because certain cells in the kidneys can detect changes in blood pressure and adjust the contraction of the muscles in the blood vessel walls, maintaining the proper blood pressure in the glomeruli. On the other hand, the rate of urine formation and the urine composition change dramatically, depending on circumstances. For instance, drinking a lot of water dilutes the blood but raises the filtration rate only slightly. The major change is a drop in the rate of water resorption, so that the excess water is excreted from the body. Urine composition and the rate of urine formation are largely regulated by the hormones vasopressin, aldosterone, and angiotensin, and the enzyme renin.

Vasopressin (ADH)

Vasopressin (also known as **antidiuretic hormone, ADH**) is a hormone released from the posterior pituitary gland in the brain (Section 30–C). Its presence increases resorption of water from the urine. Loss of water from the body stimulates vasopressin secretion and so slows the loss of water via the urine. The body may detect a decrease in its water content in one of two ways: either as a reduction in blood volume (for example, caused by severe bleeding) or as an increase in the concentration of blood plasma due to loss of water (for example, from sweating).

Vasopressin increases the permeability of the collecting ducts to water. In the absence of vasopressin, the walls of the ducts are nearly impermeable to water, and very little water is resorbed from the urine.

Insufficient vasopressin production results in diabetes insipidus, a disease characterized by thirst and production of large amounts of dilute urine. It is much less common than diabetes mellitus, characterized by glucose in the urine.

3? Alcohol decreases the levels of vasopressin. This results in less water being resorbed and too much water being released in urine. Beverages high in alcohol therefore cause excessive urination and dehydration, making them very poor drinks to consume after physical activity involving much sweating. However, beer contains such a high water-to-alcohol ratio that the extra urine formation caused by drinking beer results from the need to rid the body of excess water. ■

Aldosterone, Renin, and Angiotensin

Sodium accounts for about 90% of all the positively charged ions in the body fluids. Because water moves passively (by osmosis) following the movement of salt ions and other solutes, the volume of water in the extracellular fluid depends to a large extent on the amount of salt. Thus, the amount of sodium in the body is the chief factor determining the volume and concentration of the blood and extracellular fluid.

The body's sodium content (and hence the volume of body water) depends upon the balance between intake in the diet and loss in the urine. (Sodium is also lost in feces and sweat, but the body has less control over these.) Vertebrates have nervous and hormonal means of regulating how much sodium they eat and how much leaves the body.

Angiotensin and aldosterone are hormones that together control how much sodium is resorbed from the filtrate in the nephron. **Aldosterone** is one of several steroid hormones secreted by the cortex of the **adrenal gland,** which is attached to the kidney (Figure 26–6). Aldosterone promotes resorption of sodium by the distal tubule. The rate of aldosterone secretion is determined by the blood's salt content. A slight decrease in the sodium content of blood causes the adrenal cortex to increase its aldosterone secretion. This leads to resorption of more sodium from the filtrate, and hence to a decrease in the urine's sodium content.

While aldosterone secretion can be controlled directly by the blood's sodium concentration, it is also controlled indirectly by renin and angiotensin. The kidneys secrete the enzyme **renin** in response to a decrease in blood pressure or in sodium or potassium in

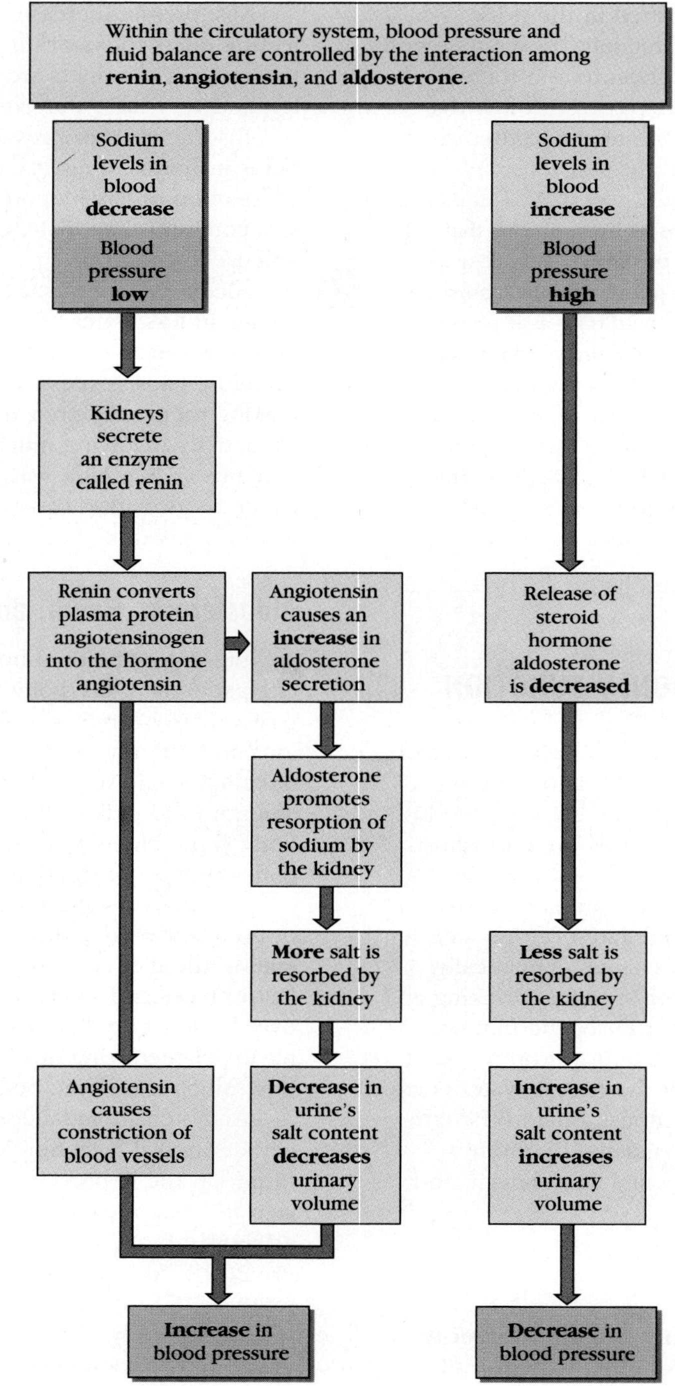

FIGURE 26–6

How the sodium content of blood is regulated by renin, angiotensin, and aldosterone. Changes in the sodium concentration in blood and blood pressure both trigger a series of hormonal events that bring the levels back to normal.

the blood. Renin converts the protein angiotensinogen in the plasma into the hormone angiotensin. **Angiotensin** causes constriction of blood vessels and an increase in aldosterone secretion. Both of these actions increase the blood pressure, either directly or by raising its sodium concentration. An increase in blood sodium concentration decreases loss of water in the urine and therefore maintains the volume, and hence the pressure, of the blood (Figure 26–6).

Vasopressin, aldosterone, and angiotensin work together to conserve body water and maintain blood pressure. Vasopressin increases the permeability of the collecting ducts to water and increases the amount of water resorbed from the urine. Aldosterone also promotes water conservation by increasing the resorption of sodium by the distal tubule, maintaining sodium (and thus water) levels in the blood. Angiotensin causes blood vessel constriction and increases aldosterone secretion, thereby raising the blood pressure.

SUMMARY

1. Living cells require a relatively constant chemical environment.
2. Cells continuously alter their environment by taking things from it and converting them into new substances. Some of these new substances are useful, while others are potentially toxic wastes that must be removed from the cell and from its immediate environment.
3. Animals have evolved various specialized mechanisms that maintain a fairly constant internal chemical environment in the face of this continuous flow of materials between the external environment, the extracellular fluid, and the interiors of cells.
4. All of an animal's body fluids are in contact with one another. By removing wastes from any one of these fluids, the excretory system can keep the fluid environment of the body's cells fairly constant.
5. All excretory systems work by collecting fluid (from the extracellular fluid or blood plasma), retrieving the substances needed by the body, and expelling the remaining fluid with the waste it contains.
6. An animal's excretory system must sacrifice some water to rid the body fluids of toxic nitrogenous wastes. However, many animals live in habitats where it is difficult to obtain or keep water or salts. These animals have adaptations for disposing of nitrogenous wastes, while at the same time maintaining the proper osmotic balance of salts and water in their body fluids.
7. Animals that must conserve water spend most of their excretion energy retaining the substances they need. Gills, salt glands, or kidneys use energy for active transport of salts into or out of the body as necessary. Most of these animals also use energy to convert the nitrogenous waste, ammonia, into urea or uric acid, which can be excreted using less water.
8. The vertebrate kidney maintains the composition of all of the body's fluids within the narrow limits necessary for the body's cells to function.
9. The basic unit of the kidney is the nephron, a long tube closely associated with capillaries of the circulatory system. Blood plasma is filtered, under pressure, into one end of this tube.
10. As the filtrate passes through the nephron, substances needed by the body are resorbed by the cells of the nephron tubule and passed into the extracellular fluid. From there they diffuse or are actively transported into the capillaries. Substances that the body does not need are secreted from the blood into the filtrate and are excreted.
11. After being changed during its passage through the tubule, the filtrate is collected as urine and held in the bladder until it is released from the body.
12. Homeostatic mechanisms under hormonal control regulate the amount and composition of the urine produced. The hormones vasopressin and aldosterone help the body conserve water and sodium, respectively, by reclaiming more of these substances from the urine.

SELF-QUIZ

1. The pack rat, a rodent, often goes for long periods without drinking, eats leaves of juicy plants, and moves about in the open only in the evening and at night. From these habits, you can guess that it lives in:
   a. desert areas
   b. the Arctic tundra
   c. the woodlands of the eastern United States
2. The main nitrogenous waste substance excreted by the pack rat will probably be:
   a. ammonia b. urea c. uric acid
3. An advantage of excreting nitrogenous wastes in the form of uric acid is that:
   a. uric acid can be excreted in almost solid form
   b. the formation of uric acid requires a great deal of energy
   c. uric acid is the first metabolic breakdown product of amino acids
   d. uric acid may be excreted through the lungs
   e. uric acid is highly toxic, so it is important for the animal to get rid of it

4. The main excretory structure in houseflies is the:
   a. Malpighian tubule
   b. flame cell
   c. nephron
   d. loop of Henle
   e. nephridium
5. Salmon have gills that are more permeable to water than to salts. Salmon hatch in freshwater streams, and then migrate to the ocean. Once they reach the ocean, you would expect the rate of uptake of water into their bodies through the gills to:
   a. increase
   b. decrease
   c. remain the same
6. Urine leaves the kidney via:
   a. the renal vein
   b. the urethra
   c. the bladder
   d. the ureter
   e. the collecting duct

In questions 7 through 11, match each structure on the left with its function from the list at the right.

___ 7.	Loop of Henle	a.	carries blood into the kidney
___ 8.	Renal artery	b.	area where a considerable amount of resorption takes place
___ 9.	Proximal convoluted tubule	c.	main area of secretion
___ 10.	Glomerulus	d.	filtration of blood
___ 11.	Distal convoluted tubule	e.	plays a role in concentration of urine

12. Filtration into the kidney tubule is accomplished by means of:
    a. active transport
    b. blood pressure
    c. an osmotic gradient
    d. secretion
    e. diffusion
13. Severe dehydration causes an increased concentration of solutes in the blood. This causes a(n) (increase/decrease) in the amount of urine produced. This change in urine production is caused primarily by:
    a. an increase in the amount of water filtered out of the blood
    b. a decrease in the amount of water filtered out of the blood
    c. an increase in the amount of water resorbed
    d. a decrease in the amount of water resorbed

THINKING CRITICALLY

1. Why does an increase in aldosterone secretion increase the volume of extracellular fluid and increase blood pressure?
2. If we divide animals into marine and freshwater invertebrates, marine and freshwater bony fish, cartilaginous fish (class Chondrichthyes), amphibians, and terrestrial vertebrates, which animals are in the following osmotic situations?
   a. approximately in osmotic equilibrium with their environment
   b. must have adaptations to guard against dehydration (water loss)
   c. must have adaptations to prevent gain of excess water from the environment
   d. in danger of taking up too many salts from the environment
   e. must have adaptations to prevent loss of salts to the environment
3. Many fish lack a urinary bladder, but urinary bladders are found in amphibians and all higher vertebrates. What is the advantage of having a urinary bladder?
4. Certain reptiles have kidneys with no glomeruli in the nephric capsules. How would this affect urine formation? What type of habitat are such animals adapted to?

SELECTED KEY TERMS

adrenal gland, *p. 555*
aldosterone, *p. 555*
cloaca, *p. 547*
collecting duct, *p. 553*
detoxification, *p. 546*
egest, *p. 546*
filtrate, *p. 553*
flame cell, *p. 550*
glomerulus, *p. 553*
homeostasis, *p. 545*
hypertonic, *p. 548*
hypotonic, *p. 548*
isotonic, *p. 548*
Malpighian tubule, *p. 550*
nephric capsule, *p. 553*
nephridium, *p. 550*
nephron, *p. 551*
osmoregulation, *p. 548*
renal artery, *p. 551*
renal vein, *p. 551*
resorption, *p. 553*
salt gland, *p. 549*
secretion, *p. 553*
ureter, *p. 551*
urethra, *p. 551*
uric acid, *p. 547*
urinary bladder, *p. 551*
vasopressin, *p. 555*

SUGGESTED READINGS

Baldwin, E. *An Introduction to Comparative Biochemistry,* 4th ed. New York: Cambridge University Press, 1964. A short and provocative book with emphasis on osmoregulatory problems and their solutions by animals.

Guyton, A. C. *Physiology of the Human Body,* 6th ed. Philadelphia: W. B. Saunders, 1984. A brief textbook covering all aspects of human physiology.

Schmidt-Nielsen, K. "Salt glands." *Scientific American,* January 1959. An account of the structure and function of salt glands and their contribution to osmoregulation, particularly in sea birds.

Smith, H. W. *From Fish to Philosopher.* Garden City, NY: Doubleday, 1961. Evolutionary history of vertebrate internal homeostasis delightfully written by an eminent renal physiologist.

Solomon, E. P., R. R. Schmidt, and P. J. Adragna. *Human Anatomy and Physiology.* 2nd ed. Philadelphia: W. B. Saunders, 1990. A well-illustrated introductory text covering most aspects of human anatomy and physiology.

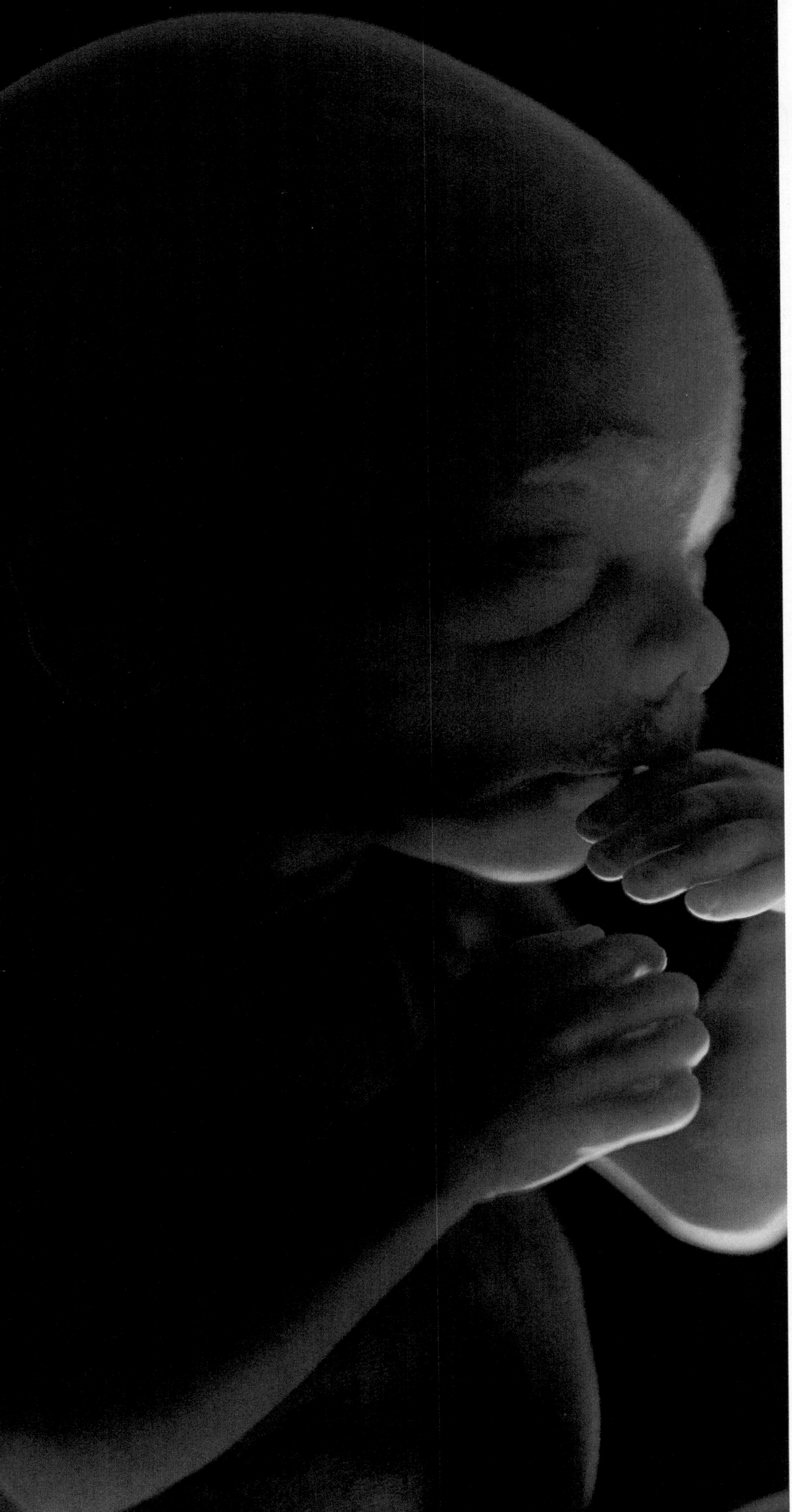

CHAPTER 27

CONCEPT GUIDE

After reading this chapter, you should be able to:

1. Compare the three general sexual reproductive patterns on the basis of how sperm is transferred and where development occurs. (Section 27–A)
2. Describe the major structures and functions of the female and male human reproductive systems. (Section 27–B)
3. Describe the physiological changes that occur within males and females during sexual arousal and orgasm. (Section 27–C)
4. Compare the types and functions of male and female hormones. Explain how the menstrual cycle is controlled. (Section 27–D)
5. Compare the different types of birth control methods, noting which techniques also provide protection against sexually transmitted diseases. (Section 27–E)
6. Describe the main functions of the three extra-embryonic membranes. (Section 27–F)
7. Describe the four main stages of embryonic development and key events of each stage. (Section 27–G)
8. Describe the three stages of labor. (Section 27–H)
9. Compare the processes of maturation and aging. (Section 27–I)

The human body is formed of trillions of cells of many hundreds of different types, all resulting from the union of an egg and a sperm. This five-month-old fetus has already developed into a recognizable human form. Read Sections 27–G and 27–H to learn how this everyday "miracle" of life can occur. (J. Stevenson/Science Photo Library/Custom Medical Stock Photo)

Sexual Reproduction and Embryonic Development

Animals have evolved many adaptations for obtaining food, exchanging gases, regulating body temperature, and retaining body water and salts. These adaptations allow some animals to survive in a variety of challenging environments. But the only true measure of evolutionary success is reproduction. Many animals can reproduce asexually (Chapter 16), but sexual reproduction is much more common.

The crucial event in sexual reproduction is **fertilization,** the union of two haploid reproductive cells, the **gametes,** to form a diploid **zygote.** In animals, the gametes are always different from each other. The female gamete is a large, nonmotile **egg** while the male gamete is a **sperm,** which actively moves toward the egg.

After fertilization, the zygote undergoes a period of embryonic development before it is born or hatched. Embryonic development involves cell division, cell movement, and differentiation into various cell types, as the genetic information in the zygote expresses itself and forms the young animal.

In most animal species, the two gametes that unite at fertilization come from different parents. The parents' anatomy, physiology, and behavior are coordinated so that their gametes are produced at the same time and brought together in the same place. Hormones often play an important role in this coordination. Many animals also engage in **courtship behavior** before they release gametes, increasing the chance of fertilization.

KEY CONCEPTS

- Sexual reproduction involves (1) getting sperm and egg together for fertilization, and (2) ensuring that the fertilized egg has a suitable environment where it can develop until the young animal is ready to survive on its own. In animals, fertilization and development may be external or internal, or fertilization may be internal, followed by external development.
- Human fertilization and embryonic development are both internal. The male reproductive tract produces sperm and the penis inserts them into the female tract. The female tract produces eggs and provides the sites of fertilization and embryonic development.
- Hormones coordinate production of gametes in both sexes. In females, hormones maintain pregnancy and, after childbirth, stimulate milk production.
- Understanding how pregnancy occurs can help people avoid unwanted pregnancy or enhance the chances of beginning a desired pregnancy.
- Embryonic development involves division of the fertilized egg into many new cells, movement of these cells into new positions, and differentiation of cells to form the tissues and organs of the body.

Curiosity Questions

1? How does a penis become erect? (See page 566)

2? How do twin babies form? (See page 572)

3? How does a single fertilized egg develop into a human? (See page 576)

(a)

(b)

FIGURE 27-1

External and internal fertilization. **(a)** External fertilization. A white cloud of sperm is released into the water from this basket sponge. Fertilization occurs in the water when some of these sperm contact eggs released by another sponge. **(b)** Internal fertilization. As these Chinese mantids mate, sperm is passed from the male to the female to fertilize eggs within her body. (a, M. Snyderman/Visuals Unlimited; b, D. Kuhn)

27–A REPRODUCTIVE PATTERNS

Animal reproduction falls into three main patterns, depending on whether fertilization and embryonic development occur within or outside the body of the (female) parent:

1. External fertilization and development.
2. Internal fertilization followed by external development.
3. Internal fertilization and development.

In the first pattern, found in many aquatic animals, eggs and sperm are released from the parents' bodies into the surrounding water, where they must find each other (Figure 27-1a). For this to succeed, male and female must release their gametes at the same time and place.

In some animals with internal fertilization, sperm are passed from the male to the female. Male squids and octopi have a modified arm that transfers packets of sperm into the female's body. The paper nautilus takes this pattern one step further: the reproductive arm of the male paper nautilus *breaks off* within the female's body ensuring sperm delivery.

In other invertebrates, sharks, reptiles, birds, and mammals, sperm is transferred by **copulation,** direct passage of sperm from inside the male's body to inside the female's body (Figure 27-1b). In most animals, this transfer involves an intromittent ("into-sending") organ, such as a penis.

Internal fertilization has several advantages. The female reproductive tract provides a confined, protected space where sperm and egg can get together without danger of being eaten or washed away. If development is internal, fertilization must also be internal. But even if the egg develops externally, internal fertilization gives the sperm easy access to the egg before the final preparations for laying. As the fertilized egg then passes down the female reproductive tract to the exterior, it can be surrounded with secretions, membranes, or shells that will protect the developing embryo.

Internal development gives an animal embryo even more advantages. The mother's body provides exactly the right chemical conditions and, in mammals, warmth as well. Because the mother carries the embryo everywhere she goes, it is not vulnerable to the predators that plunder the egg masses or nests of animals that develop externally. In most species other than mammals, internal development simply means keeping the egg inside the body instead of laying it. As with animals that develop externally, the food stored in the egg must support the entire embryonic development. In mammals, however, the embryo shares whatever food its mother takes in during her pregnancy. Such a continuously renewed food supply may permit the embryo to reach a larger size before birth.

Animal reproduction falls into three main patterns. In many aquatic invertebrates and aquatic vertebrates, males and females release gametes into the water, and fertilization and development occur outside the body. In other invertebrates and many vertebrates, males transfer sperm through copulation, and fertilization is internal, but the embryos develop outside the mother's body. In other vertebrates, including all mammals, sperm is transferred through copulation, and fertilization and development occur internally.

HUMAN REPRODUCTION

27–B HUMAN REPRODUCTIVE ORGANS

We will illustrate reproduction in animals with internal fertilization by considering human reproduction. The human reproductive organs can be roughly divided into the internal organs and the external, or accessory, sex organs. In addition, each sex has **secondary sexual characteristics,** such as breasts in the human female or a higher metabolic rate in the human male, which are not part of the actual reproductive apparatus.

Female Reproductive Organs

Figure 27–2a shows the external sex organs of a woman, collectively known as the **vulva.** The **labia majora** and **labia minora** (labia = lips) cover and protect the urinary and genital openings. Note that the openings of the urethra, from the urinary tract, and of the vagina, from the reproductive tract, are separate. Human females are born with a membrane, the **hymen,** covering the vaginal orifice. A slit in the hymen allows the menstrual flow to pass. The hymen presumably performs a protective function, but it is essentially absent in many women. During embryonic development, the **clitoris** arises from the same structure that develops into part of the penis in a male embryo. Like its male counterpart, the clitoris produces a pleasurable sensation when it is stimulated by touch and, indeed, this seems to be its only function.

The internal female reproductive organs consist of the ovaries and oviducts (fallopian tubes), uterus, and vagina (Figure 27–2b).

The two **ovaries** are the female **gonads,** or gamete-producing organs. The ovaries produce eggs. An ovary contains many follicles, each consisting of an immature egg surrounded by nutritive follicle cells (Figure 27–2c). At **ovulation,** a mature follicle ruptures and the egg pops out into the coelom. This release of a ripe egg stimulates the oviduct's finger-like endings to surround the egg. The beating of cilia lining the oviduct draws the egg into the tube and on toward the uterus. Fertilization usually occurs about halfway down the oviduct. An unfertilized egg disintegrates as it passes through the uterus about 72 hours after ovulation.

The **uterus** is a highly elastic organ whose main function is to hold a developing embryo and expel it during childbirth. The external opening of the uterus is the **cervix,** made up largely of the biggest, most powerful circular muscle in the body. The strength of the cervical muscle is necessary to hold about 15 pounds of fetus and fluid in the uterus against the pull of gravity during pregnancy. The cervix protrudes into the upper end of the vagina.

The **vagina** is the receptacle for the penis during copulation, or sexual intercourse, and the pathway to the exterior for the baby during childbirth. In keeping with these functions, it has extremely elastic walls.

Male Reproductive Organs

The gonads of a human male are the **testes** (singular: **testis**). They produce sperm from the time of sexual maturity, at puberty, until death. The testes of most mammals lie in the **scrotum,** a sac outside the body cavity, which provides the cool temperature required for sperm production.

Sperm develop from cells lining the **seminiferous tubules** of the testes (Section 11–H). Muscular action in the walls of these tubules carries sperm to the **epididymis.** Here they are stored and complete their maturation. During sexual stimulation, the sperm move through the **vas deferens** (plural: **vasa deferentia**) by contraction of its walls. The sperm then move into the urethra, where they are joined by secretions from the **seminal vesicles** and the **prostate** and **bulbourethral glands** (Cowper's glands). These secretions activate the sperm and provide a fluid medium for them to swim in. The sperm and their attendant secretions, collectively called **semen,** leave the penis via the urethra (Figure 27–3). The **penis** is an external sex organ that introduces semen into the vagina during sexual intercourse. Urine from the bladder also leaves the male's body via the urethra, but the two fluids cannot pass through the urethra at the same time.

In females the urinary and reproductive systems have separate openings to the outside of the body. Pregnancy results when a mature egg is released from an ovary, travels through the oviduct where it is typically fertilized, and embeds in the wall of the uterus. Males produce sperm within seminiferous tubules of the testes. During ejaculation, sperm are released from the epididymis. They mix with secretions from the prostate, seminal vesicles, and bulbourethral glands, and are expelled out the end of the penis through the urethra.

27–C PHYSIOLOGY OF SEXUAL INTERCOURSE

During copulation (often called sexual intercourse in humans), the male's penis introduces sperm into the female's vagina. Before the penis can enter, it must become at least partly erect under the influence of sexual stimulation. Sexual stimulation can be brought about by any of the senses, usually most effectively by

(a) Female external genitalia

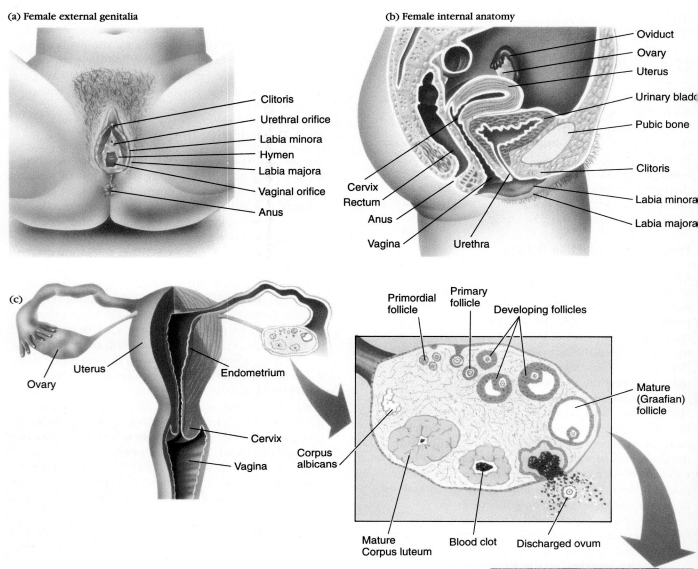

Clitoris
Urethral orifice
Labia minora
Hymen
Labia majora
Vaginal orifice
Anus

(b) Female internal anatomy

Oviduct
Ovary
Uterus
Urinary blad
Pubic bone
Clitoris
Labia minora
Labia majora
Cervix
Rectum
Anus
Vagina
Urethra

(c)

Uterus
Ovary
Endometrium
Cervix
Vagina
Primordial follicle
Primary follicle
Developing follicles
Mature (Graafian) follicle
Corpus albicans
Mature Corpus luteum
Blood clot
Discharged ovum

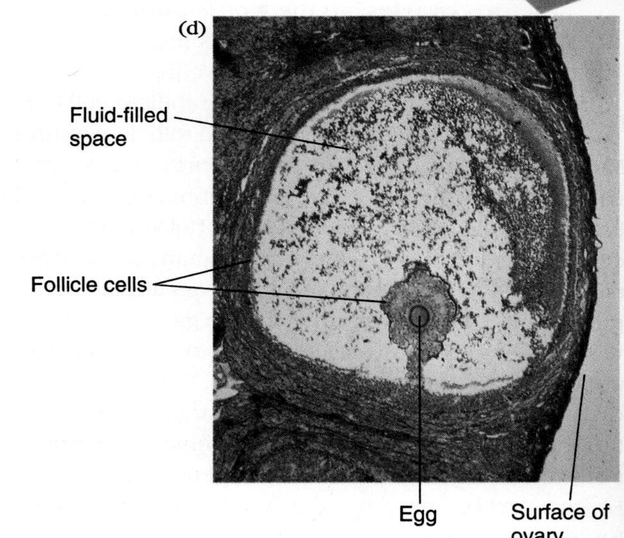

FIGURE 27-2

Female reproductive system. (a) External female genitalia. Note the three separate openings characteristic of females of higher mammals: the urethral orifice from the bladder, vaginal orifice from the uterus, and anus from the digestive tract. The clitoris is anterior to the urethral orifice. **(b)** The internal reproductive organs of a woman in a cutaway view from one side. **(c)** Ovulation. The ovary, suspended at the end of the oviducts, contains hundreds of eggs in various stages of development. As primordial follicles mature into larger stages (successive stages are indicated by the blue arrow), the egg undergoes some meiotic stages and the follicle accumulates fluid. At ovulation, the follicle bursts, releasing the egg near the opening of the oviduct. The residual follicle, called the corpus luteum, includes hormone-secreting cells. **(d)** A mature ovarian follicle. (Biophoto Associates)

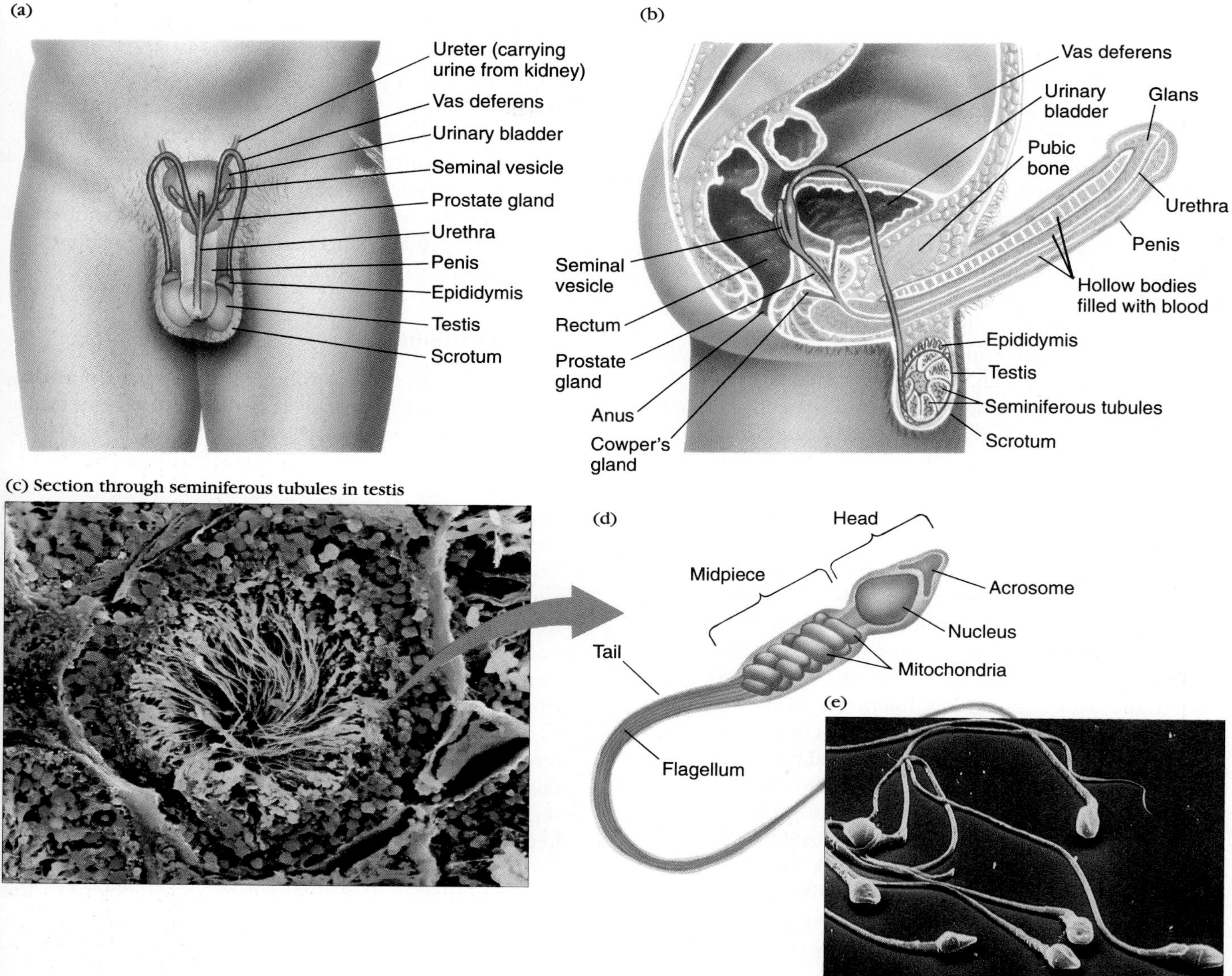

FIGURE 27-3

Male reproductive tract. (a) Position of the reproductive organs in a human male. **(b)** The internal reproductive organs of a man in a cutaway view from one side. During erection, the hollow bodies fill with blood, partly blocking the veins so that little blood can leave the penis. The engorged penis thus becomes enlarged and stiff due to fluid pressure. Sperm are formed in the testes and stored within the epididymis. During ejaculation, sperm and secretions from the prostate gland, seminal vesicles, and Cowpers gland travel through the vasa deferentia (singular, vas deferens) to the urethra, leaving the body through the opening at the tip of the penis. Notice that the urethra and vasa deferentia pass through the prostate gland. Enlargement of the prostate can constrict the urethra and thus interfere with urination. **(c)** Cross section of a human testis. Sperm (yellow) are produced in the walls of the seminiferous tubules and released into the central cavity of the tubule. **(d)** Human sperm. The sperm nucleus contains the male's genetic material that will combine with the female's genetic material in the egg to form the nucleus of the embryo during fertilization. Mitochondria at the base of the tail produce ATP to power the beating of the flagellum. **(e)** A photograph of human sperm. The enlarged ends are the sperm heads. (c, P. Motta/Science Photo Library/Photo Researchers, Inc.; e, D. M. Phillips/Visuals Unlimited)

touch. The external genitals, and especially the glans of the penis, are the most sexually sensitive areas.

1? The penis consists of three cylinders of spongy tissue that extend into the male's body. Within these cylinders are many hollow bodies, which are usually empty and collapsed. During sexual stimulation, about ten times the usual volume of blood is carried from the arteries into the hollow bodies. The blood filling the spaces presses against the outsides of the veins, blocking much of the flow of blood leaving the penis. The penis thus enlarges and becomes rigid. ■

Sexual arousal of a human female can also result from many different stimuli. The clitoris is the organ most sensitive to touch. The vulva contains erectile tissue like that in the male's penis. In a sexually aroused woman, the vulva becomes swollen because of an increased blood supply, and the walls of the vagina exude fluid. Glands just inside the vaginal opening secrete mucus, which acts as a lubricant for entrance of the penis.

When the erect penis has been inserted into the vagina, stimulation by the movement of the genitals against each other may result in **orgasm.** Orgasm is characterized by an increase in heart rate and blood pressure, engorging of various tissues with blood, and faster and deeper breathing, finally resulting in an explosive burst of involuntary muscular contractions. In men, orgasm is accompanied by **ejaculation,** the forceful ejection of semen, propelled by peristaltic waves of contraction of the muscles in the sperm ducts. In women, orgasm is characterized by rhythmic spasms of the muscles surrounding the vagina.

Fertilization can occur without either sex experiencing orgasm. Often a small amount of semen is released before ejaculation, and this may contain enough viable sperm to fertilize an egg.

An erection results from the simultaneous increase of blood flowing into the penis and a decrease in blood leaving the penis. Sexual arousal in a female also results in an increase in the blood flow to the external genitals as well as additional lubrication near the opening of the vagina. Orgasm in both sexes is a highly pleasurable, involuntary reaction accompanied by a series of muscular contractions and deeper, faster breathing. In males, orgasm typically accompanies ejaculation.

BIO-BIT

The firmness of an erect penis is entirely due to the hydraulic pressures of blood filling the internal arteries.

27–D HORMONES AND REPRODUCTION

Hormones play a major role in the reproduction of nearly all animals. Hormones control sexual maturation and gamete production. They also control the superficial characteristics that differentiate the sexes, and they are necessary for sexual behavior. In addition to producing the hormones of their own sex, members of both sexes produce small amounts of the hormones characteristic of the opposite sex (Table 27–1).

Male Hormones

Hormones with masculinizing effects are called **androgens.** The most important is **testosterone,** a steroid hormone secreted primarily by the testes.

A boy's testes remain dormant until the onset of puberty, at the age of 10 to 14. From then on, the pituitary gland, located beneath the brain, releases a gonadotropic ("gonad-changing") hormone into the circulation. This is **luteinizing hormone (LH).** Another gonadotropic pituitary hormone, **follicle-stimulating hormone (FSH),** also helps to regulate sperm production. (We shall see that these two hormones play important roles in females as well, and their names come from their effects on the ovaries.)

Luteinizing hormone promotes secretion of testosterone by the testes. Testosterone, in turn, causes a reduction in the pituitary's secretion of LH. As a result, the level of testosterone in the blood is constantly kept within narrow limits.

Testosterone, with the help of FSH, promotes sperm formation. It also stimulates development of the male secondary sexual characteristics, features that are not directly related to reproduction but are characteristic of a sexually mature male. The secondary sexual characteristics of human males include deepening of the voice, development of male musculature, and growth of hair on the face and other parts of the body.

Female Hormones

Hormones control women's secondary sexual characteristics and menstrual cycles. As in the male, the pituitary hormones FSH and LH are secreted at puberty, triggering the development of the secondary sexual characteristics and the onset of menstrual cycles. The hormonal progress of the menstrual cycle determines a woman's fertility.

The secondary sexual characteristics of human females include breasts and the rounded contours imparted by a thick, widespread layer of subcutaneous ("under skin") fat. This fat provides food during periods of starvation, and insulation against cold, giving women (and the unborn babies or nursing infants that

Table 27–1
Hormones Involved in Human Reproduction

Source	Hormone	Role
Pituitary	Follicle-stimulating hormone (FSH)	Production of gametes in both sexes
	Luteinizing hormone (LH)	Secretion of sex hormones by gonads in both sexes Ovulation in females
	Oxytocin	Contractions of uterus during childbirth Letdown of milk from mammary glands to nipple
	Prolactin	Growth of mammary glands and milk production (lactation) Inhibition of FSH and LH
Testes	Testosterone	Sperm formation, with FSH Development of male secondary sexual characteristics
Ovaries	Estrogen	Development of female secondary sexual characteristics Preparation of uterine lining for implantation of embryo
	Progesterone	Preparation of uterine lining for implantation of embryo Enlargement of breasts for lactation
Placenta	Human chorionic gonadotropin (HCG)	Maintenance of pregnancy

depend on the mother's body for food) a survival edge over men. Subcutaneous fat may be unfashionable, but it is nevertheless vital for fertility. Women with insufficient fat do not ovulate and cannot become pregnant. Female athletes in training commonly stop menstruating when their bodies lose excessive amounts of fat.

The Menstrual Cycle

At puberty (usually 10 to 14 years of age), the pituitary gland starts a series of hormonal cycles that periodically render a woman fertile (capable of becoming pregnant) until the cycles cease at the menopause, some 30 to 40 years later. These hormonal changes and the effects they produce are called menstrual cycles.

Human menstrual cycles are notoriously variable, but the "model" cycle lasts 28 days. The days are numbered from the first day of blood flow in the menstrual period (Figure 27–4). At the beginning of the cycle, the pituitary secretes increasing amounts of follicle-stimulating hormone (FSH). FSH causes an ovarian follicle to mature and produce the steroid hormone **estrogen.** A surge of estrogen from the follicle stimulates a dramatic increase in secretion of luteinizing hormone (LH) by the pituitary. This LH, together with FSH, brings about the final maturation of the follicle, culminating in ovulation, when the follicle ruptures and releases a ripened egg. Ovulation occurs on about the fourteenth day of the cycle.

Still under the influence of LH, the cells of the ruptured follicle grow and form a **corpus luteum,** which secretes more estrogen and yet another steroid hormone, progesterone. Progesterone and estrogen prepare the endometrium, the lining of the uterus, to receive a fertilized egg. The levels of progesterone and estrogen at this time inhibit secretion of LH and FSH from the pituitary. If the egg is not fertilized and implanted in the uterus, the corpus luteum degenerates and the production of estrogen and progesterone drops. This causes the surface of the endometrium to die and slough off in a menstrual period. Without the corpus luteum hormones to inhibit it, the pituitary increases secretion of FSH once more, and the cycle repeats. This sequence is interrupted if fertilization and pregnancy occur.

Hormones of Pregnancy

Early in pregnancy, the developing embryo and the lining of the uterus jointly form a special organ, the **placenta,** which provides for nourishment of the embryo. It also secretes **human chorionic gonadotropin (HCG).** Increased levels of HCG, as of LH, maintain the corpus luteum and therefore production of progesterone, which inhibits the pituitary's secretion of FSH

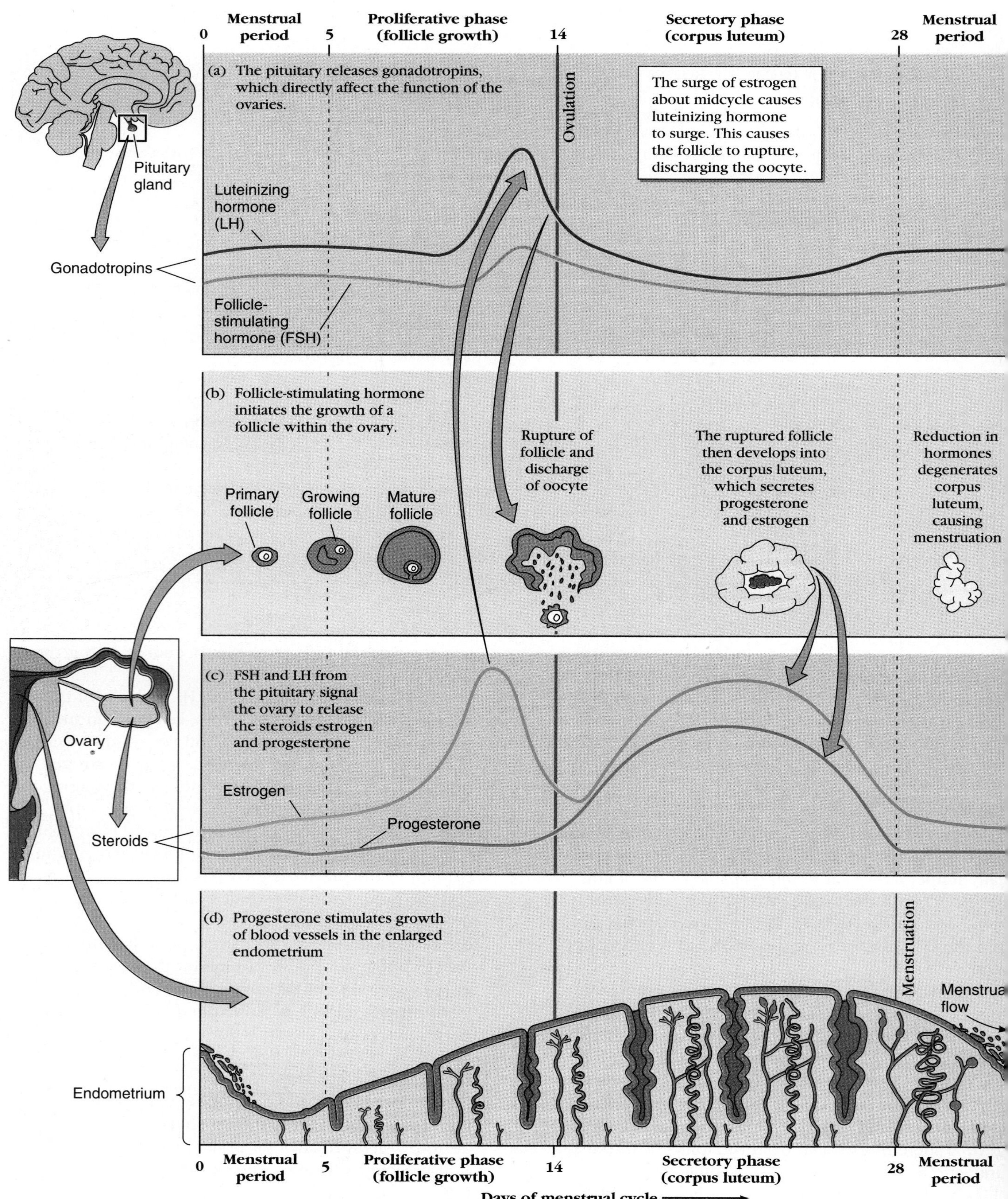
Menstrual period
Proliferative phase (follicle growth)
Secretory phase (corpus luteum)
Menstrual period
0
5
14
28
(a) The pituitary releases gonadotropins, which directly affect the function of the ovaries.
Ovulation
The surge of estrogen about midcycle causes luteinizing hormone to surge. This causes the follicle to rupture, discharging the oocyte.
Pituitary gland
Luteinizing hormone (LH)
Gonadotropins
Follicle-stimulating hormone (FSH)
(b) Follicle-stimulating hormone initiates the growth of a follicle within the ovary.
Rupture of follicle and discharge of oocyte
The ruptured follicle then develops into the corpus luteum, which secretes progesterone and estrogen
Reduction in hormones degenerates corpus luteum, causing menstruation
Primary follicle
Growing follicle
Mature follicle
(c) FSH and LH from the pituitary signal the ovary to release the steroids estrogen and progesterone
Ovary
Estrogen
Progesterone
Steroids
(d) Progesterone stimulates growth of blood vessels in the enlarged endometrium
Menstruation
Menstrual flow
Endometrium
0
Menstrual period
5
Proliferative phase (follicle growth)
14
Secretory phase (corpus luteum)
28
Menstrual period
Days of menstrual cycle

FIGURE 27–4

Events of the menstrual cycle. (a) Sex hormones released by the pituitary during the phases of one menstrual cycle in which pregnancy does not occur. **(b)** Development of an ovarian follicle during the cycle. A fluid-filled space develops as the follicle grows. At ovulation the follicle ruptures, releasing the egg (oocyte). The remaining follicle walls then develop into a corpus luteum ("yellow body"), which secretes progesterone and estrogen for a time and then degenerates. **(c)** Levels of hormones released by the ovary. **(d)** Phases of development of the endometrium, the lining of the uterus. The endometrium thickens during the proliferative phase, developing long narrow glands and a rich blood supply. In the secretory phase, after ovulation, the endometrium continues to thicken and the glands secrete nutritive material in preparation to receive an embryo. When no embryo implants, the new outer layers of the endometrium disintegrate and the blood vessels rupture, producing the menstrual flow.

and LH. This prevents the next ovulation, and thus prevents formation of any new embryo while the first one is developing. If the embryo is abnormal, or dies, secretion of HCG by the placenta stops, permitting the body to **abort** or **miscarry** the pregnancy as the uterine lining degenerates and sloughs, along with the embryo. It is estimated that as many as three out of five human embryos that implant are abnormal and abort naturally in this manner. If a hormone from the mother's body maintained pregnancy, there would be no way to discard dead or abnormal embryos.

Likewise, a hormone produced by the developing fetus determines when birth occurs. The average period of **gestation,** or pregnancy, is 270 days in humans, but this varies a lot. It is important that the baby be born when it is mature, and not when the mother feels like it. The fetus signals that it is mature by secreting hormones from its adrenal glands. These fetal hormones diffuse across the placenta and build up in the mother's bloodstream until they cause secretion of the hormone **oxytocin** by the mother's pituitary. Oxytocin stimulates the muscles of the uterus to contract and cause birth. Later, it also causes milk to be let down from the mammary glands to the nipples in response to the baby's suckling.

The placenta is expelled within an hour after the baby is born. Many mammalian mothers eat their placentas, thereby re-using the nutrients and avoiding the risk of attracting predators. In addition, hormones eaten in the placenta trigger maternal behavior.

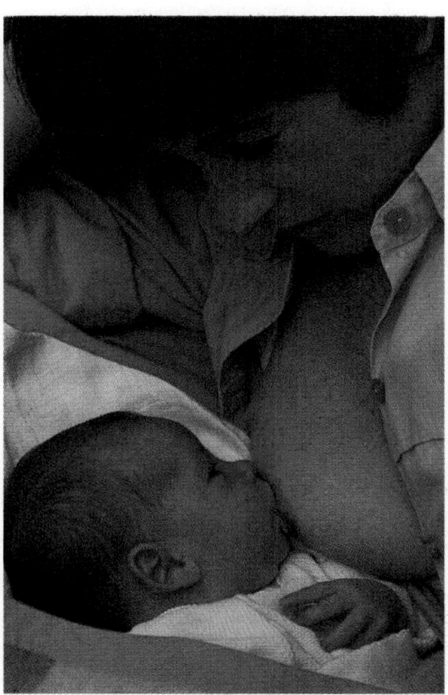

FIGURE 27–5

A lactating woman feeding her baby. (H. Gritscher/Peter Arnold, Inc.)

Whether or not women would make better mothers if they ate their placentas has not, as far as we know, been tested!

Expulsion of the placenta during childbirth causes the mother's pituitary to secrete **prolactin,** a hormone that initiates **lactation** (milk production) by the mammary glands (Figure 27–5). Prolactin also inhibits the release of LH from the pituitary and counters the effects of FSH and LH on the ovarian follicles. Hence in nursing mothers the menstrual cycles are suppressed. In some cultures breast feeding is an important means of birth control, although it is not as reliable as most methods of artificial birth control. Breast feeding is also the best way to ensure that the baby receives adequate nutrition and avoids intestinal and other bacterial infections.

In males and females FSH and LH trigger puberty. In males this begins the process of sperm production and the development of secondary sexual characteristics such as a deeper voice, greater muscle mass, and increased growth of hair on the body. In females these same hormones begin the menstrual cycle and result in secondary sexual characteristics such as enlarged breasts, more subcutaneous fat, and extra growth of body hair.

Table 27–2
Contraceptive Use in the United States*

Method	Estimated % Use	% Accidental Pregnancy in One Year of Use†
Male sterilization	15	0.15
Female sterilization	19	0.4
Oral contraceptive pill	32	3
Condom	17	12
Diaphragm + spermicide	5	18
"Rhythm" (periodic abstinence)	4	20
IUD	3	6
Contraceptive sponge	3	18
Vaginal foams, jellies	2	21

* Data from *Developing New Contraceptives: Obstacles and Opportunities.* Washington, DC: National Academy Press, 1990.
† About 89% of women using no contraceptive become pregnant within one year.

The menstrual cycle results from the interaction of pituitary and ovarian hormones. Pituitary FSH and LH act on the ovary, promoting maturation of an egg follicle, ovulation, and formation of the corpus luteum. The ovarian hormones estrogen and progesterone prepare the uterus for pregnancy. They also inhibit secretion of the pituitary hormones. Hence, if pregnancy does not occur, the cycle begins anew.

27–E BIRTH CONTROL

Human beings and other (particularly social) animals regulate their reproduction, so that the number of offspring bears some relation to the parents' ability to raise them. A bird may abandon her eggs if her neighbors of the same species have many young already hatched. Severely malnourished women are infertile. Rabbit and mouse embryos die when their mother is stressed, as by overcrowding or the presence of a strange male.

Strictly speaking, **contraception** refers to birth control methods that prevent fertilization. The only methods of contraception that are 100% reliable are abstention from sexual intercourse and sterilization.

The condom is the contraceptive device most commonly used by men. A condom is rolled onto the erect penis shortly before intercourse. It catches the semen so that sperm do not enter the female reproductive tract. The condom is also the only form of birth control that can reduce or prevent the spread of AIDS and other sexually transmitted (venereal) diseases from one partner to the other. Female condoms have more recently been developed. (See *A Journey into Healthy Living:* Sexually Transmitted Diseases, this chapter.)

Birth control pills (oral contraceptives) are very reliable contraceptives if they are properly used (Table 27–2). "The pill" contains synthetic estrogen and progesterone at a dose high enough to prevent ovulation. Pregnancy is avoided because there is no egg to be fertilized. The pill does not reliably prevent ovulation and pregnancy until it has been taken regularly for at least two weeks.

The medical risks of using the pill are largely the same as those of pregnancy. The pill increases the risk of such things as high blood pressure and excessive blood clotting, mainly among women over 35 who smoke. Conversely, oral contraceptives reduce their users' chances of getting certain kinds of cancer.

Another method of contraception is the use of a rubber diaphragm, which is smeared with a spermicidal (sperm-killing) jelly or cream each time it is used. The woman then inserts it into the vagina so that it covers the cervix. The diaphragm blocks the entrance to the uterus so that sperm cannot reach the egg. The cervical cap is similar.

The much less reliable rhythm method consists of avoiding intercourse during the woman's "fertile period," the part of the menstrual cycle when there is an egg present to be fertilized. The difficulty is in deciding when ovulation will occur, because the menstrual cycle is so variable.

A Journey into Healthy Living

Sexually Transmitted Diseases

Sexually transmitted diseases (STDs) are transmitted from one person to another primarily by contact between the genital organs during sexual activity. A dozen or more such diseases occur in humans.

Historically, the most famous sexually transmitted disease was syphilis, caused by a spirochete bacterium. In some people the disease causes damage to the nervous and circulatory systems. Nervous degeneration caused by syphilis is believed to have been responsible for the strange behavior of England's King Henry VIII, the insanity of the composer Smetana, and the blindness of the composer Frederick Delius.

Syphilis is not very common because, even if it is not treated, it is not very contagious. Like herpes infections, syphilis may become **latent,** producing no symptoms. Syphilis cannot be transmitted to another person when it is in its latent stages, which is most of the time. Furthermore, the chance of catching syphilis from a single sexual encounter with someone who has an active case is only about 1 in 40. However, new cases of syphilis increased 23% in early 1987, reversing a previous downward trend. More recently the incidence of syphilis has continued to rise. In urban areas, there have been local outbreaks of syphilis associated with the practice of trading sex for crack cocaine.

Gonorrhea, one of the most widespread STDs, is somewhat more contagious than syphilis. It too is caused by a bacterium. In men the disease is readily detected because it produces a discharge of pus from the penis and a burning sensation during urination. Most infected men therefore seek treatment. The disease is widespread because it produces few or no symptoms in women, where the cervix is the area usually infected. In both sexes gonorrhea can cause sterility. Gonorrhea has usually been treated with penicillin. Predictably, penicillin-resistant gonorrhea bacteria have evolved, as only those that happened to have mutations conferring resistance to penicillin survived and reproduced. There are now some strains of the disease that are incurable because they are resistant to all known antibiotics.

Infections caused by bacteria of the genus *Chlamydia* are also widespread, although many cases involve no symptoms, and others are mistaken for gonorrhea. However, *Chlamydia* is not susceptible to the penicillin used to treat gonorrhea, and so *Chlamydia* infections cannot be cured until properly diagnosed. As with gonorrhea, infections can cause scarring of the reproductive tissues and hence sterility. In the United States, *Chlamydia* is more common than either syphilis or gonorrhea.

Many people today are concerned about infectious genital blisters caused by Type 2 herpes simplex virus, very similar to the virus that causes cold sores and fever blisters on the lips. The blisters rupture, leaving painful ulcers. These heal, and the disease then becomes dormant. The virus is still there, however, and the disease may recur at any time. Herpes is infectious most when it is active. People with herpes can therefore easily avoid transmitting the disease by avoiding sexual contact during the relatively short periods when they have herpes blisters. The greatest danger posed by herpes lies in the high mortality rate (about one third) among those newborn babies who catch the disease from their mothers during childbirth. (Most women with genital herpes, however, do not communicate the disease to their infants.)

Acquired immune deficiency syndrome (AIDS) was added to the list of STDs in the 1970s. This particularly terrifying STD is discussed in *A Journey into Healthy Living*: AIDS Awareness, Chapter 25).

Various other sexually transmitted diseases are caused by viruses, bacteria, yeasts, and protists. It is extremely difficult to stop the spread of these diseases because victims often pass a disease on to others before they learn that they are themselves infected. The growing resistance to antibiotics among pathogens also makes treatment increasingly difficult.

Intrauterine devices (IUDs) are small plastic or metal objects inserted into the uterus by a doctor. IUDs are not true contraceptives. They somehow act on the uterine lining so that the already developing embryo cannot implant. IUDs are as effective as contraceptive pills in preventing pregnancy. On the other hand, a disquieting number of IUDs become deeply imbedded in the wall of the uterus, causing dangerous abdominal infections and requiring surgery to remove the device. Such infections may result in infertility.

Contraceptives implanted under the skin of the upper arm will prevent pregnancy for three to five years, and West Germany, Mexico, and China make injectable one-month contraceptives that are marketed in 40 countries. The French have an abortion pill, RU 486. In contrast, birth control methods in the United States have changed little since 1960, largely because of our legal liability system, which makes developing and marketing new birth control devices financially risky.

Vasectomy and Tubal Ligation

Sterilization is any more or less permanent change that prevents an animal from reproducing sexually. Sterilization of either sex is the fastest-growing method of contraception in the world. The most common sterilization operation for men is **vasectomy,** that is, severing and tying off the vasa deferentia. This is a simple operation, usually performed under local anesthesia. Afterwards, sperm are still produced but are resorbed into the body, and the fluid ejaculated contains only the secretions of various glands.

Sterilization of a woman usually involves **tubal ligation,** the cutting and tying off of the oviducts. This operation can also now be performed under local anesthesia. After tubal ligation the ovary continues to function as it did before, but sperm cannot reach the eggs, and thus fertilization cannot occur.

Abortion

Induced abortion is one of the oldest human birth control methods. Abortion was largely accepted, and fairly common, in the United States and Europe until the early nineteenth century. Before this time, a woman had up to a one-third chance of dying in childbirth, whereas abortions were less risky for her, so that abortions often saved women's lives. After midwives and doctors found that washing their hands and clothes improved their patients' survival rates, childbirth became less dangerous to a woman than a nineteenth-century abortion, and doctors started to oppose abortion. Most religious and ethical opposition did not develop until some time later. The situation has changed since World War II, now that the danger of abortion to a woman is again less than the risk of a completed pregnancy, and abortion is, once again, legal in most countries.

Sexual abstinence and sterilization are the surest methods of contraception. A condom combined with a contraceptive sponge or vaginal foam provide good protection against pregnancy, and the condom is the only device that protects against sexually transmitted diseases (STDs), including AIDS. The pill uses synthetic hormones to prevent pregnancy and is a very reliable contraceptive, but it provides no protection against STDs. Other, less reliable birth-control methods that also do not prevent the transmission of STDs include the diaphragm, contraceptive sponge, and the rhythm method. The IUD does not prevent conception, but it does prevent the developing embryo from implanting in the uterus. Induced abortion is one of the oldest birth-control methods and is presently considered to be safer for women than completing pregnancy and giving birth.

27–F FERTILIZATION AND IMPLANTATION

Sperm released into the vagina during ejaculation swim through the cervical opening and up the uterus into the oviduct, where fertilization usually occurs. (The description of human pregnancy in this section applies generally to other mammals as well. See A Journey Through Fertilization and Implantation.)

As a sperm penetrates the egg cell membrane, a rapid electrical reaction, followed by a slower chemical change, runs through the membrane. After this, the egg cannot be penetrated by another sperm (see *A Journey into Science in Process*: Test Tube Babies and Surrogate Mothers, this chapter).

2? During evolution, selection has strongly favored the birth of only one human baby at a time. Babies borne singly tend to be larger and healthier, with a better chance of survival, than those with wombmates. However, multiple births result from a small percentage of pregnancies. Identical twins are produced when the mass of cells that will develop into an embryo separates into two groups sometime during the first week of development. Each group of cells develops into a separate embryo. Because these cells contain identical genes, the resulting embryos are genetically identical, and so must be of the same sex. Identical twins are also called monozygotic twins because they originate from a single zygote. Multiple births may also result when more than one egg is produced and fertilized. Each develops into a separate embryo, and they are no more genetically alike than any other siblings. ■

The fertilized egg moves toward the uterus, dividing rapidly into many cells, at which stage it is called a **morula.** Within a few days it has become a hollow ball of cells with inner and outer layers that is called a **blastula** in other organisms and a **blastocyst** in humans.

The inner cells of the blastocyst will become the **embryo,** the group of cells that develop into the adult. The outer cells give rise to the placenta and to three membranes found only in reptiles, birds, and mammals, called the **extra-embryonic membranes.** An aquatic embryo is surrounded by water, which protects the embryo, keeps it moist, removes wastes, and permits gas exchange. In terrestrial vertebrates these functions are taken over by the extra-embryonic mem-

Fertilization occurs after a sperm penetrates the egg and releases its nucleus into the egg's cytoplasm. This triggers meiosis II in the egg.

Cleavage is the process by which the fertilized egg (zygote) divides. The first cleavage produces a two-cell stage embryo. The next cleavage produces a four-cell stage embryo and so on.

Morula is the term used to describe an embryo that has undergone many stages of cleavage and is now a ball of cells.

The embryo is termed a **blastocyst** when a cavity forms within the center of the morula. This results in a ball of cells with a surface layer and an inner cell mass.

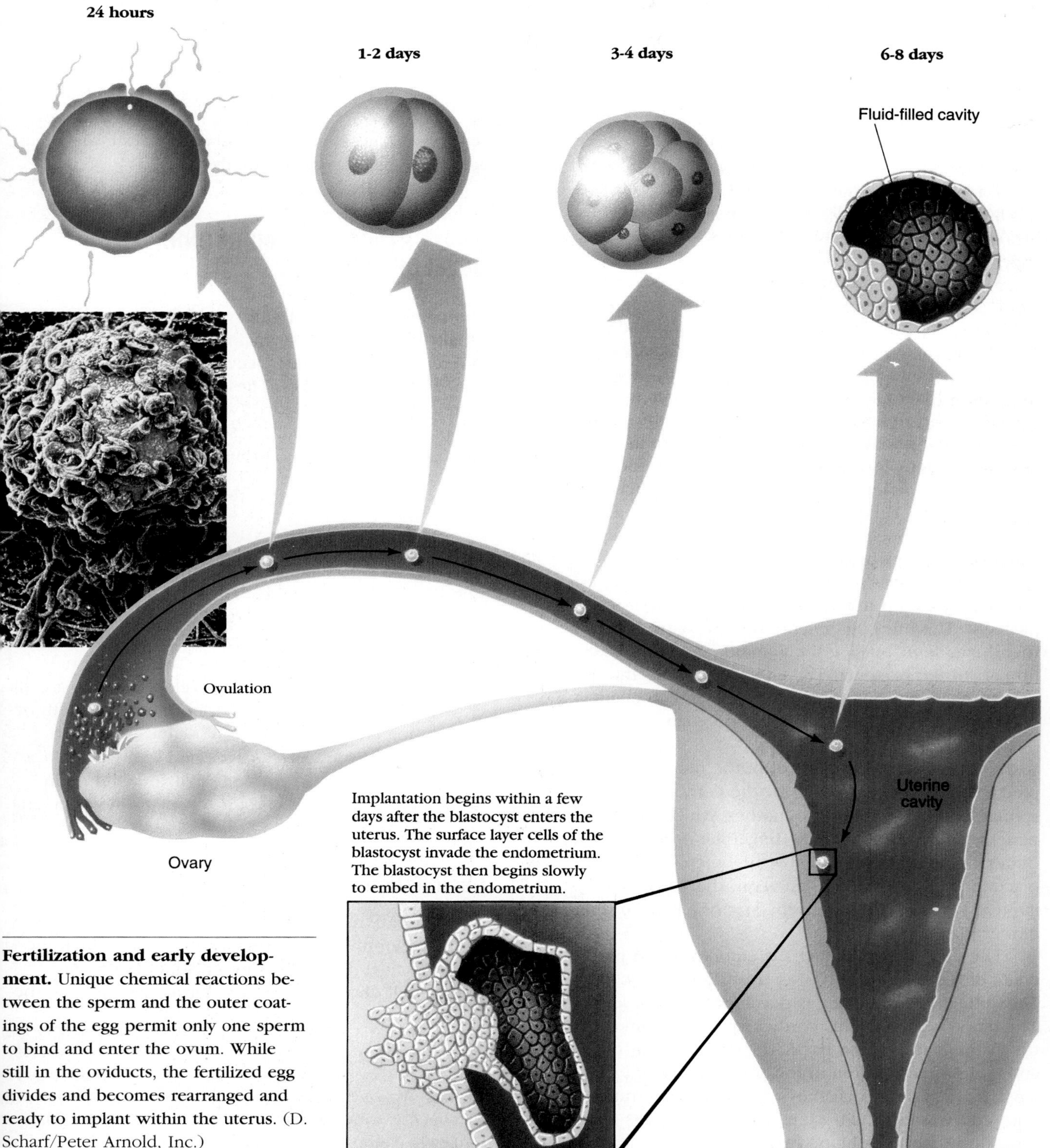
24 hours
1-2 days
3-4 days
6-8 days
Fluid-filled cavity
Ovulation
Ovary
Uterine cavity
Implantation begins within a few days after the blastocyst enters the uterus. The surface layer cells of the blastocyst invade the endometrium. The blastocyst then begins slowly to embed in the endometrium.

Fertilization and early development. Unique chemical reactions between the sperm and the outer coatings of the egg permit only one sperm to bind and enter the ovum. While still in the oviducts, the fertilized egg divides and becomes rearranged and ready to implant within the uterus. (D. Scharf/Peter Arnold, Inc.)

branes, parts of which form the placenta in mammals. Because a mammalian embryo is so dependent upon the mother's body, the placenta develops before the embryo.

About a week after fertilization, outer cells of the blastocyst start to grow into the endometrium of the uterus. These cells make up the **trophoblast,** a tissue with remarkable properties. Because the trophoblast develops from the fertilized egg, which is genetically different from the mother, we would expect the mother's immune system to recognize the trophoblast as foreign and reject it as if it were a transplanted organ. Obviously, this does not happen. The trophoblast invades the uterine wall almost as if it were a cancerous growth. Occasionally something goes wrong and the trophoblast forms one of the most malignant and invasive of cancers.

As the trophoblast invades the uterus, the extra-embryonic membranes, the amnion, chorion, and allantois, start to form. The **amnion** becomes a fluid-filled sac around the embryo that cushions it against bumps and bacterial infections. This is the sac that bursts at the onset of labor. Since amniotic fluid contains some cells from the embryo, drawing fluid from the sac into a hypodermic needle **(amniocentesis)** permits a geneticist to check for chromosomal defects in the embryo (see *A Journey into Healthy Living*: Genetic Counseling and Fetal Testing, Chapter 13). The **allantois** is a sac connected to the embryo's gut. In reptiles and birds it stores waste products until the egg hatches. In humans it becomes invaded by blood vessels and makes up most of the embryonic part of the placenta. The third membrane, the **chorion,** surrounds the fetus outside the amnion, and also makes up part of the placenta. The chorion secretes the hormone human chorionic gonadotropin, which maintains pregnancy and can be detected in pregnancy tests (Section 27–D).

Soon, a quarter of the wall of the uterus is occupied by the placenta formed by the combination of the spongy endometrium of the mother, which has been penetrated by chorionic villi, fingerlike projections of the embryonic trophoblast. By the time of birth, the placenta will develop into an organ that looks like a large raw hamburger, about the size of a stack of four dinner plates.

The membranes joining the embryo to the placenta develop into a cord, the **umbilicus,** which grows thicker and longer as development proceeds. The blood vessels in this cord carry blood from the fetus to the placenta and back. The capillaries of mother and embryo do not join in the placenta, but they lie so close to each other that they exchange gases, nutrients, and wastes by diffusion. Other substances present in the mother's blood may also enter the fetal circulation. Many drugs can retard development or, if they interfere with differentiation, cause birth defects, and doctors urge pregnant women to avoid smoking, drinking alcohol, and taking unnecessary medication.

A fertilized human egg begins dividing into a multicellular blastocyst that implants in the wall of the uterus. Some cells of the blastocyst develop into three extra-embryonic membranes. The amnion is a fluid-filled sac providing moisture and mechanical protection to the embryo. The allantois forms most of the placenta, which anchors the embryo to the wall of the uterus and provides for nutritional support and gas exchange. The chorion produces human chorionic gonadotropin and helps to form the placenta.

27–G STAGES OF EMBRYONIC DEVELOPMENT

The complex events of embryonic development are usually divided into four main stages, involving rather different cellular events: cleavage, gastrulation, neurulation, and organogenesis. These stages are found in all vertebrate embryos, but we will focus on human development. As you read this section, refer to the accompanying A Journey Through Fetal Development.

Cleavage

After fertilization, the zygote divides to form 2, 4, 8, and then 16 cells, and so on. This period of cell division is known as **cleavage.** The main functions of cleavage are to produce a large number of cells by rapid division and to segregate different substances in the cytoplasm into different cells. These substances determine how the various cells develop later. Most cells grow in size before they divide, but no cell growth occurs during cleavage: the zygote divides into ever smaller cells. Eventually, a fluid-filled cavity called the **blastocoel** appears in the center of the embryo in most animals. After the blastocoel has formed, the embryo is known as a blastocyst.

Gastrulation

After formation of a blastocyst the next stage in embryonic development is **gastrulation,** during which the cells rearrange themselves into distinct layers. Although cell division continues during gastrulation, the

BIO-BIT

"It is not birth, marriage, or death, but gastrulation, which is truly the most important time in your life." Lewis Wolpert

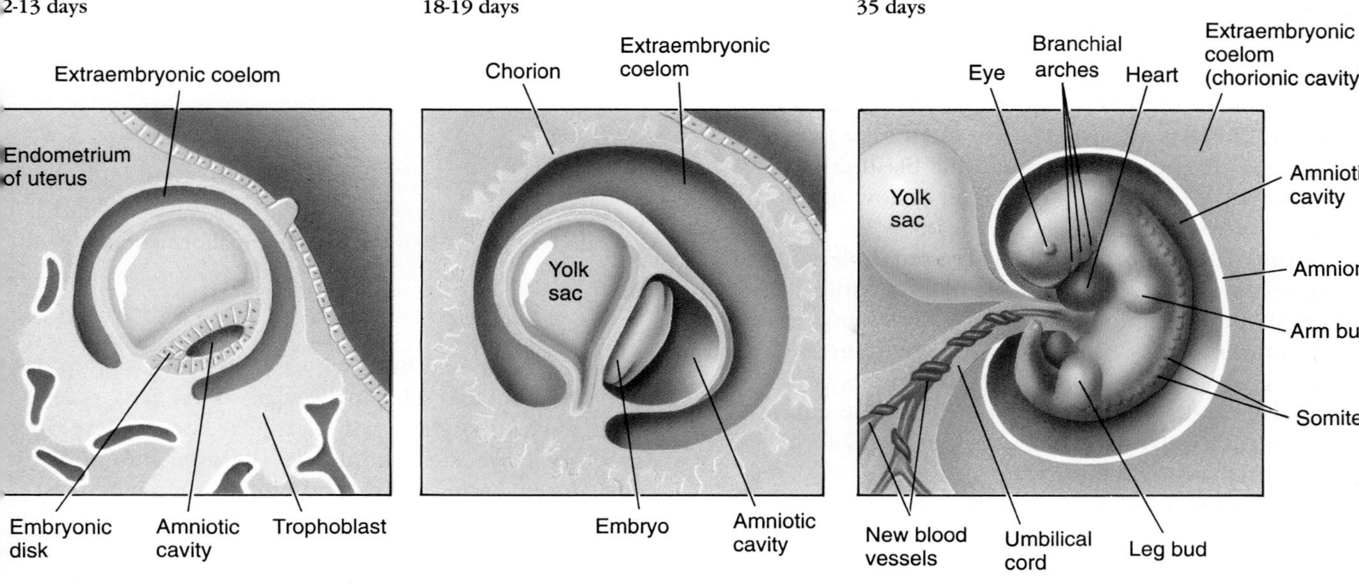

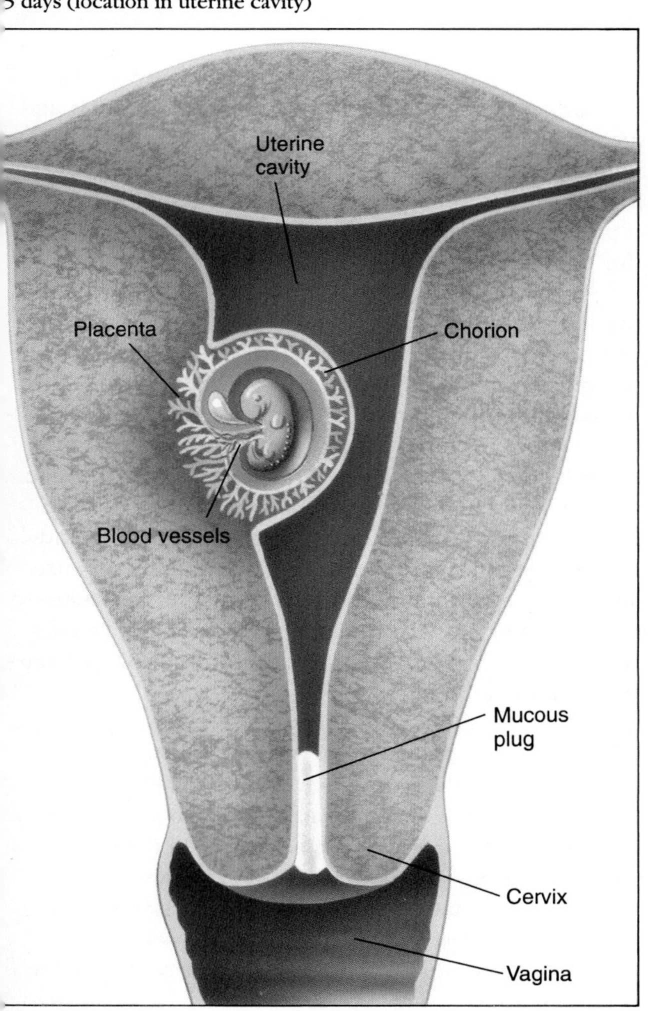

6 weeks ($\frac{1}{2}$ inch long)

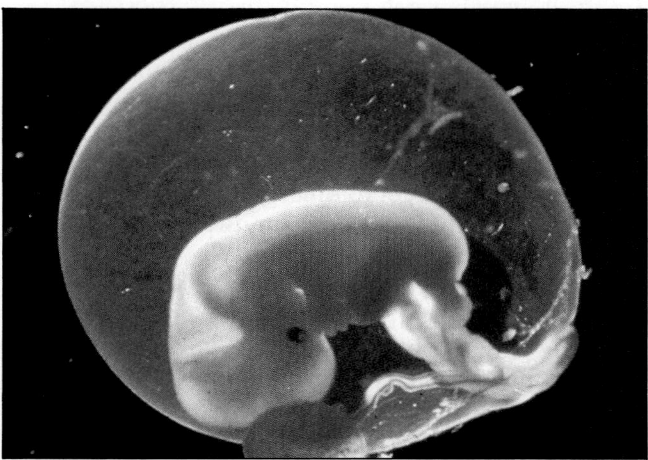

4 months (7 inches long)

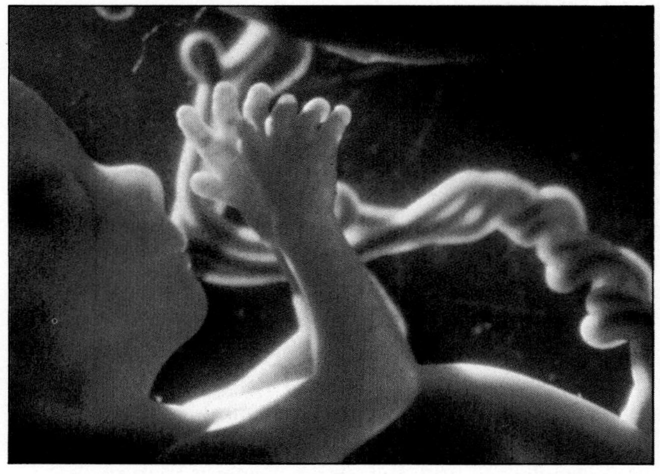

arly human development. (Petit Format/Nestle/Science Source)

A Journey Through Fetal Development

most obvious event is cell movement. An opening called the **blastopore** forms at the side of the blastocyst. Cells from the surface around the blastopore move through the blastopore into the hollow interior of the blastocyst.

During gastrulation, the single-layered, hollow blastocyst becomes a **gastrula** and the cells eventually arrange themselves into three layers called the **germ layers.** (Germ in this context means something small that will give rise to more elaborate structures later.) The gastrula contains a new central cavity, the **gastrocoel** (or **archenteron**), which will eventually form the lumen of the digestive tract.

The outermost germ layer is the **ectoderm.** This will form both the **neurectoderm,** which develops into the nervous system, and the **epidermis,** which gives rise to skin, hair, nails, sweat glands, and other structures on the outside of the body. The innermost layer of cells is the **endoderm,** next to the gastrocoel. From the endoderm will come the lining of the gut, the digestive glands, and other internal structures. Between the endoderm and the ectoderm lie the cells of the third germ layer, the **mesoderm,** which will form the skeleton, muscles, gonads, kidneys, and allied structures in the adult. At a later stage, yet another cavity forms in the embryo, this time in the middle of the mesoderm. This is the **coelom,** the body cavity in which internal organs are suspended.

Neurulation

After the three germ layers have formed, the neural tube and head begin to develop in the process of **neurulation.** The part of the ectoderm destined to form the nervous tissue first forms two parallel folds that rise up and finally join to form the **neural tube,** from which all of the nervous system develops.

Organogenesis

The final stage of development is **organogenesis,** organ formation. The nervous, digestive, and circulatory systems begin to form. The **notochord** forms from mesoderm. The notochord is a cartilage rod that runs from the embryo's head to its tail. Later in development it becomes encased in cartilage and bone and is replaced by the backbone. Muscle blocks (somites) appear in a regular row along the backbone, and then in other parts of the body. The heart, shaped like a lumpy tube, starts to pulsate. After the beginning of organogenesis, the embryo grows in size and the nervous, circulatory, and respiratory systems mature.

Organogenesis begins immediately after neurulation, and by 4 weeks after fertilization the embryo has a simple heart, and buds where limbs will develop. It also has a tail and gill slits, vestiges of its early vertebrate ancestors which will disappear later in development. By 6 weeks, a simple eye is visible and the tail is still present. After the second month, the embryo is recognizable as a primate because it has features, such as a neck and large head, that distinguish it from other vertebrate embryos. From this stage on, the embryo is sometimes referred to as a fetus. By the end of the first trimester (first three months) the embryo is about 9 centimeters long and all the major organ systems have formed.

The second trimester is occupied mainly by further organ development. During this period, limb and face muscles move, and the heart becomes large enough so that it can be heard through a stethoscope placed on the mother's abdomen. Bone continues to replace the cartilage that forms the early skeleton, and the body is covered with hair (the lanugo).

3? At the beginning of the third trimester (6 months), the fetus is still tiny. It will normally more than triple in size before birth. By 7 months, the fetus looks from the outside like a tiny normal baby, but it is not. The lungs and temperature regulation systems are not functional. More important, the central nervous system is not fully developed. Nearly all babies born at this age will be severely handicapped if they survive. By the end of the third trimester (9 months), the lungs and nervous system will develop. Then the fetus will usually be born, with almost a 99% chance of surviving to live a normal life. At birth, the full-term baby weighs about 3000 grams (7 pounds) and is about 52 centimeters (20 inches) long. ■

BIO-BIT

Between the second and ninth weeks of development, the human embryo is most sensitive to outside influences that could cause malformations.

Embryonic development begins with the cleavage stage, a period of rapid cellular growth. The resulting cells then become rearranged in the gastrula stage, forming three layers that establish the animal's basic body plan. In the neurula stage, the outermost (ectodermal) layer pinches up and forms the neural tube, which will eventually develop into the brain and spinal cord. In the final embryonic stage, organogenesis forms the body's various organ systems. The remainder of the gestation period is a time of maturation of these organ systems, accompanied by a tremendous increase in weight and size.

Table 27–3
Outline of Human Development

Medical Name	Days After Fertilization (approximate; varies a lot)	What's Happening?
First trimester (Embryo)	0–8	Cleavage
	6	Implantation begins
	14	Gastrulation
	21	Neurulation
	24	Nervous system, gut, and blood vessels start to develop
	28–35	Embryo most susceptible to damage by rubella, thalidomide, most drugs, etc.
(Fetus)	45	Testes differentiating in male
	70	Arms and legs move
	75	Primary oocytes enter first meiotic division in female
Second trimester	90	All major organ systems formed but tiny and growing in size
	140	Heart can be heard with a stethoscope
Third trimester	180	Temperature regulation, central nervous system, and lungs complete development; fetus most susceptible to damage by alcohol
	270	Birth (parturition)

27–H BIRTH

The process in which uterine contractions expel the baby and the placenta is called **labor** and can be divided into three main stages (Figure 27–6). The first stage, **dilation,** usually lasts from 2 to 20 hours and ends when the cervix of the uterus is fully open or dilated. The second stage, **expulsion,** lasts from about 2 to 100 minutes. It begins with **full crowning,** the appearance of the baby's head in the cervix, and continues while the baby is pushed, head first, down through the vagina into the outside world, where it draws its first breath.

The third, or **placental,** stage begins when the baby has been born. The uterus continues to contract while the umbilical cord is clamped, and some 5 to 45 minutes after the baby is born the uterus expels the placenta. The umbilical cord can now be severed, and the baby's independent existence begins.

The process of labor can be divided into three stages: (1) dilation, as the cervix expands, (2) expulsion, as the baby moves from the uterus out of the vagina, and (3) the placental stage, when the placenta becomes dislodged and is expelled from the uterus.

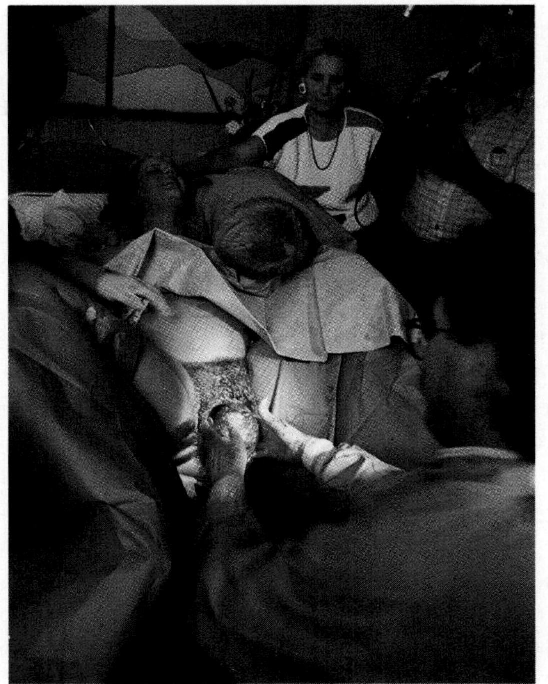

FIGURE 27–6

Human birth. The baby's head is emerging. (D. Plummer)

A Journey into Science in Process

Test Tube Babies and Surrogate Mothers

After fertilization but before implantation, the embryo lives free from attachment to the mother. Medicine and animal husbandry take advantage of this brief period to intervene. Eggs or developing morulas can be removed from the ovaries or oviducts, treated in the laboratory, and then returned to the uterus without damage, if they are provided with a suitable fluid environment during their stay in the outside world. The biggest problem with these techniques has been finding a suitable chemical solution. Mammalian eggs and early embryos, in particular, are easily damaged. (Sperm survive better because they are protected by the semen—their special fluid environment.)

"Test tube" babies do not develop in test tubes, but merely undergo fertilization and a few rounds of cell division outside a woman's body. The scientific term for this is *in vitro* ("in glass") fertilization. If a woman has blocked oviducts or if her vagina produces spermicidal secretions, or if her husband's sperm count is low, fertilization may not occur naturally in her body. However, a doctor may be able to remove eggs from her body, add the husband's sperm in a laboratory dish, check that development has begun, and then return the morulas to the uterus. This technique succeeds less than 20% of the time. To increase the chances of success, doctors usually treat the prospective mother with fertility drugs. These usually cause several ovarian follicles to mature at the same time, so that more than one egg is released. Hence this technique results in a high proportion of multiple births.

FIGURE 27-A

Surrogate parenthood. A female bongo calf born to a surrogate mother eland. As a young embryo, the bongo was transplanted into the eland's uterus, where it implanted and developed. Bongos are a rare and elusive species inhabiting dense forests in Africa. The larger and more common elands, members of the same genus, inhabit open areas. (Cincinnati Zoo)

Once the embryo is outside the female's body, it can just as easily be inserted in the uterus of a different female, provided her hormones are in the proper phase of the reproductive cycle for implantation to occur. To date, human babies and young of many other species have been born to such surrogate mothers. This permits a woman who has a normal uterus but damaged ovaries to bear a child (which of course will have the genes of the woman who donated the egg).

However, the greatest use for embryo transplants is not for humans but for other mammals. With it, a breeder can obtain many embryos from a genetically superior cow or mare, for instance, and use inferior stock as surrogate mothers, increasing the rate of production of desirable young. It also turns out that female mammals can successfully gestate embryos of closely related species. This allows zoo keepers to produce more young of endangered species, using surrogate mothers of species that are common and easily bred in captivity (Figure 27-A).

27–1 MATURATION, AGING, AND DEATH

Animals continue to change in various ways after leaving the protection of the egg or of the mother's body at hatching or birth. Some animals undergo metamorphosis, a change in body form, and most animals take some time to become sexually mature. Maturation involves the same processes of cell differentiation, growth, and change of form as those that occurred in the embryo. Aging appears to be different.

Aging is the sum of changes that accumulate with time and make an organism more likely to die. It begins even before an individual officially completes development and sexual maturity. Slower healing, for instance, is one of the signs of aging, and even a human teenager's broken bones heal less rapidly than those of a young child.

Aging and death are genetically programmed (Figure 27-7). A human dies of old age at about 80 to 100 years of age, and some insects at a few days or weeks.

FIGURE 27-7

Four generations. (D. Plummer/Animals Animals)

Aging assures death at this characteristic age if disease or predators do not kill the individual first.

Although we can describe many processes that are characteristic of aging, it is still unclear which of these are cause and which effect. For instance, with age, the immune system becomes less efficient at defending the body against disease. Is this why disease becomes more common with age, or vice versa?

As people grow older, their bodies cope less effectively with stress and disease. Because the ability to survive changing conditions depends largely on the immune, nervous, and hormonal systems, researchers have concentrated on these systems, using the hypothesis that degeneration here results in the loss of adaptability seen in the rest of the body: slower healing, hardening connective tissue, brittle bones, and so forth.

One example of this comes from studies of people with diabetes, who age faster than normal. Diabetics have more glucose in their blood than other people, and one symptom of aging is the formation of glucose links between structural proteins such as collagen in the skin, tendons, ligaments, and the walls of blood vessels. The glucose linkages make these tissues rigid and inelastic. The immune system's macrophages destroy cells with glucose-linked proteins on their surfaces. However, it appears that macrophages in older individuals have fewer receptors for these proteins and are less efficient at removing the damaged cells from the body.

Despite a long list of hypotheses, searches for a single cause of aging have all ended in failure. It seems likely that the "aging genes" do not control one single system, but instead control many minor degenerations. Because the body's systems all interact with one another, minor deficiencies anywhere in the body can accumulate to produce aging of the body as a whole and of its individual systems.

Animals continue to develop long after birth or hatching. This process includes attaining reproductive maturity and may involve the metamorphosis or other dramatic changes in shape, size, and proportions. By contrast, aging is a genetically related process that decreases the overall health of the individual over time.

SUMMARY

1. Most animals reproduce sexually by means of large nonmotile eggs fertilized by small, motile sperm. Sexual reproduction usually involves the coordinated anatomy, physiology, and behavior of two parents.
2. In a human female, an egg is released from a follicle in the ovary and travels through the oviduct to the uterus. If it is fertilized during this journey, it starts to divide, and will implant in the endometrial lining of the uterus.
3. The male's testes, which produce sperm, lie in the scrotum. After sperm leave the testes, they pass sequentially through the epididymis, vas deferens, and urethra as they move out of the body. During this passage, glandular secretions are added to sperm, forming the semen.
4. The penis is composed of spongy tissue that becomes engorged with blood during sexual stimulation. It can then be inserted into the vagina.
5. Sexual stimulation may eventually result in orgasm by both sexes and in ejaculation of semen by the male.
6. Hormones control puberty, maturation of gametes, and the female menstrual cycle. Hormones also regulate pregnancy, birth, and lactation.
7. Contraceptive techniques prevent fertilization. Other birth control methods prevent a developing embryo from completing its growth in the uterus.
8. Abstinence and sterilization are the only fully effective forms of contraception. Contraceptive pills work nearly as well, as do condoms and diaphragms when they are used properly, along with a contraceptive sponge or vaginal foam. Condoms are the best protection against STDs.
9. Embryonic development involves several kinds of processes: cell division; differentiation into various types

of cells due to activation of different sets of genes; growth of the embryo; and changes in the shape of the body as a whole and of the organs that form within it.

10. The stages of development are:
    a. Cleavage, when the embryo divides rapidly into ever-smaller cells and forms a hollow blastula.
    b. Gastrulation, when the cells rearrange themselves into ectoderm, mesoderm, and endoderm.
    c. Neurulation, when the neurectoderm rolls up and forms the neural tube.
    d. Organogenesis, which produces all the organs of the body by a complicated series of cell interactions.
11. Part of the blastula forms the amnion, chorion, and allantois, the membranes surrounding the embryo. In humans, the chorion and allantois become the fetal part of the placenta.
12. The placenta anchors the fetus and secretes the hormones that permit pregnancy to continue. Nutrients, wastes, gases, and hormones pass between the fetal and maternal blood vessels in the placenta.
13. Aging is a genetically related process that results in a decrease in the overall health of the individual.

SELF-QUIZ

Associate the reproductive organs on the right with the descriptions on the left:

___ 1. Tube for conducting sperm	a. cervix
___ 2. Receptacle for penis	b. bulbourethral gland
___ 3. Produces seminal secretions	c. ovary
	d. oviduct
___ 4. Conducts eggs	e. prostate gland
___ 5. Holds baby in uterus	f. testis
___ 6. Produces sperm	g. urethra
___ 7. Prepares nutritive lining for embryo	h. uterus
	i. vagina
	j. vas deferens

8. From the list of reproductive organs and passages above, construct a (correct) route for the passage of sperm from the site of production to the site of fertilization.

____

For each of the following birth-control methods, choose the correct means of interference with reproduction:

___ 9. Diaphragm and jelly	a. prevents fertilization of egg
___ 10. Vasectomy	b. prevents embryo implantation
___ 11. "The pill"	c. prevents completion of embryonic development of implanted embryo
___ 12. IUD	
___ 13. Tubal ligation	d. prevents ovulation
___ 14. Induced abortion	e. prevents sperm formation
	f. prevents release of sperm into seminal fluid
___ 15. Condom	

From the list of hormones on the right, choose the ones with the functions listed on the left:

___ 16. Maintains pregnancy	a. estrogen
___ 17. Ovulation occurs in response to a surge of ___, which in turn is stimulated by a surge of ___.	b. follicle-stimulating hormone
	c. human chorionic gonadotropin
___ 18. Primary hormone in sperm production	d. luteinizing hormone
___ 19. Induces labor	e. oxytocin
___ 20. Produced by the placenta	f. progesterone
	g. testosterone

Match the correct stage(s) of development with each of the following characteristics:

C = Cleavage N = Neurulation
G = Gastrulation O = Organogenesis

___ 21. Pattern depends on amount and distribution of yolk
___ 22. Embryonic nerve tube first forms
___ 23. Results in formation of skeleton and muscles from mesoderm
___ 24 Rapid cell division with no increase in size
___ 25. Produces gastrocoel

THINKING CRITICALLY

1. Does vasectomy affect male potency (ability to engage in sexual intercourse)?
2. Can abortion affect a woman's subsequent fertility?
3. Can a woman become pregnant the first time she has sexual intercourse?
4. Why do you think the venereal disease gonorrhea is now epidemic in the United States and Western Europe?
5. A bird's egg has a hard, dry shell. How do sperm manage to fertilize these eggs?
6. Most countries have rules that extraordinary attempts should not be made to save babies born before 8 months of gestation because nearly all those born before 7½ months are severely handicapped, both mentally and physically. They become a severe burden to the state and to their families. The American medical profession devotes staggering amounts of time and money to saving babies born before 8 months of gestation. Why do you think this is? Do you think this is a wise use of our medical resources?

SELECTED KEY TERMS

allantois, *p. 574*
amniocentesis, *p. 574*
amnion, *p. 574*
androgen, *p. 566*
blastocoel, *p. 574*
blastula, *p. 572*
bulbourethral gland, *p. 563*
cervix, *p. 563*
chorion, *p. 574*
cleavage, *p. 574*
coelom, *p. 576*
corpus luteum, *p. 567*
courtship behavior, *p. 561*
ectoderm, *p. 576*
endoderm, *p. 576*
epididymis, *p. 563*
estrogen, *p. 567*
fertilization, *p. 561*
gamete, *p. 561*
gastrocoel, *p. 576*
gastrula, *p. 576*
gastrulation, *p. 574*
germ layer, *p. 576*
gestation, *p. 569*
labor, *p. 577*
lactation, *p. 569*
mesoderm, *p. 576*
morula, *p. 572*
neurulation, *p. 576*
notochord, *p. 576*
organogenesis, *p. 576*
ovary, *p. 563*
placenta, *p. 567*
seminal vesicles, *p. 563*
seminiferous tubules, *p. 563*
testes, *p. 563*
umbilicus, *p. 574*
uterus, *p. 563*
vas deferens, *p. 563*
zygote, *p. 561*

SUGGESTED READINGS

Aral, S. O., and K. K. Holmes. "Sexually transmitted diseases in the AIDS era." *Scientific American,* February 1991. An analysis of many of the sexually transmitted diseases facing humans today.

Avery, M. E., N-S. Wang, and H. W. Taeusch, Jr. "The lung of the newborn infant." *Scientific American,* April 1973. Describes the physiological changes in the lung just before birth and efforts to speed these changes in premature infants.

Browder, L. *Developmental Biology,* 3rd ed. Philadelphia: Saunders College Publishing, 1991. A straightforward, well-illustrated text.

Cerami, A., H. Vlassara, and M. Brownler. "Glucose and aging." *Scientific American,* May 1987. Explains how the nonenzymatic linkage of glucose to structural proteins, and possibly DNA, might account for many symptoms of aging.

Djerassi, C. *The Politics of Contraception.* New York: W. W. Norton, 1980. Why are modern contraceptives relatively primitive compared with our other technology? The author argues that this results from the control of contraceptive development by drug companies and governments.

Frisch, R. E. "Fatness and fertility." *Scientific American,* March 1988.

Lein, A. *The Cycling Female.* San Diego: University of California Press, 1979. A readable book on menstrual cycles, their variations, and how they affect women.

Riddle, J. M., and J. W. Estes. "Oral contraceptives in ancient and medieval times." *American Scientist,* May 1992.

Rusting, R. "Trends in biology: why do we age?" *Scientific American,* December 1992.

Short, R. V. "Breast feeding." *Scientific American,* April 1984.

Ulmann, A., G. Teutsch, and D. Philibert. "RU 486." *Scientific American,* June 1990. An excellent review of this controversial method of birth control.

Wassarman, P. M. "Fertilization in mammals." *Scientific American,* December 1988.

CHAPTER 28

CONCEPT GUIDE

After reading this chapter, you should be able to:

1. Explain how an action potential is generated. Describe the effects that myelin sheaths and nodes of Ranvier have on action potentials. (Section 28-A)
2. Compare nerve impulse transmissions across chemical and electrical synapses. (Section 28-B)
3. Describe and provide examples of the general functions of neurotransmitters. (Section 28-C)
4. Describe the advantages of having a nerve cord, specialized nerve cells, and a brain. (Section 28-D)
5. Compare the general functions of the three main brain regions. (Section 28-E)
6. Explain how reflexes work and why they are important. (Section 28-F)
7. Describe the general structure and functions of cranial and spinal nerves. (Section 28-G)
8. Compare the functions of the sympathetic and parasympathetic divisions of the autonomic nervous system. (Section 28-H)
9. Describe the three steps of an animal's response to a stimulus. (Section 28-I)

The eyes of this great horned owl collect light and send action potential signals to the brain where the information is interpreted to form an image. Read more about the important functions of the nervous system of many different animals throughout this chapter. (J. Lepore/Photo Researchers, Inc.)

The Nervous System and Sense Organs

Digestion, respiration, circulation, excretion, and reproduction may be studied separately, but in a living animal they must all work together. An animal's activities are coordinated mainly by its nervous and hormonal systems. We have already seen several examples of this type of coordination, such as control of breathing and of body fluid volume.

The remaining chapters on animal biology deal with the nervous system and with the sense organs, muscles, and glands that work with it in coordinating the animal's activities. The body's **receptors,** in the sense organs, detect **stimuli**—changes in the body's internal or external environment, such as blood chemistry, sound, or light. This information is converted into electrical impulses, the form in which information is handled by the nervous system. The cells that carry these electrical impulses are called **neurons.**

In an animal's nervous system, neurons are arranged to carry messages in pathways. **Sensory neurons** receive information from receptors in the sense organs and transmit it to **interneurons,** neurons that relay messages from one neuron to another. Often a nervous pathway involves many interneurons, perhaps receiving and processing information from various parts of the body. Eventually, the message passes to **motor neurons,** which send out signals to the body's **effectors,** the organs that carry out a response. The most common effectors are glands, which secrete hormones, digestive enzymes, and so on, and muscles, which contract and move parts of the body. The responses made by effectors do two main things: they maintain internal homeostasis, or they make suitable behavioral responses to the external environment.

KEY CONCEPTS

- The nervous system works with the hormonal system to coordinate the physiology and behavior of the various organs and systems in the body.
- The progress of information through any nervous system occurs in three main steps:
  1. Sensory activities, the collection of information from outside and inside the body,
  2. Central processing and/or relaying this information in the nervous system, and
  3. Responding appropriately to the results of the first two steps.
- Sense organs gather information about changes in the animal's body and in its external environment and pass it to the nervous system for evaluation.
- The particular collection of sense organs in each animal species gives its members a unique perception of their bodies and of their environment.

Curiosity Questions

1? What is a "runner's high"? (See page 592)

2? How can a jellyfish function without a brain? (See page 593)

3? What is a spinal tap and why is it performed? (See page 595)

NEURONS

Neurons are the cells that transmit messages in the nervous system. Every neuron has a **cell body** containing the nucleus and most of the cell's organelles. A neuron also has long, thin extensions of two types (Figures 28-1). The neuron's many, branching **dendrites** are extensions that receive information from other cells or from the external environment. A typical neuron also has one **axon,** the extension that carries information to other cells. The end of the axon divides up into many terminals.

A neuron's axon terminals make connections to other neurons (or, if it is a motor neuron, to muscle or gland cells). Each neuron may make connections with many others, and so a single neuron may receive information from, or send information to, many parts of the body. In the vertebrate brain, the dendrites of a single cell receive thousands of connections from the axon terminals of other neurons.

A neuron may carry messages over a considerable distance. For example, the axons of some neurons extend from the spinal cord to the toe of an elephant or the flipper of a whale. The longest cells in any animal are some of its neurons.

In addition to neurons, the nervous system contains **glial cells,** collectively referred to as **neuroglia** ("nerve-glue"). These support the neurons, convey food to them, and perform other service functions. In fact, the human nervous system contains about ten times as many neuroglial cells as neurons. Neuroglial cells account for up to half the brain's volume.

OVERVIEW: HOW NEURONS WORK

A neuron's distinctive feature is its **electrical excitability,** that is, its ability to generate and transmit a rapidly moving electrical impulse. It can do this because oppositely charged ions are distributed unequally on the two sides of its membrane. These ions are poised to rush through the membrane, down their concentration gradients, when the membrane's permeability changes and allows them to pass. The result is an electrical **nerve impulse,** the form in which information travels in the nervous system. The transmission of a nerve impulse is summarized in Figure 28-1. Nerve impulses can also be artificially generated by applying electrical impulses to neurons, a common experimental technique.

28–A ELECTRICAL PROPERTIES OF NEURONS

The Resting Potential

Like most other cells, a neuron has an asymmetric distribution of ions across its plasma membrane. This is mainly due to active transport by **sodium-potassium pumps** in the cell membrane which use energy to pump sodium ions (Na^+) out of the cell and potassium ions (K^+) in (Section 4-D). By making these two stockpiles of ions—sodium outside and potassium inside—the pump has stored electrical potential energy. This potential energy can be converted into useful energy—an electric current carrying information—if the ions are permitted to flow back through the membrane, down their concentration gradients (Section 6-A).

In a resting neuron the membrane is largely impermeable to sodium ions; however, it does leak slightly, and potassium ions escape from the cell faster than sodium ions enter (Figure 28-2). In this way a net positive charge builds up outside the membrane, until it repels any further exodus of positively charged potassium ions.

At this balance point, called the membrane's **resting potential,** the inside of the neuron is more negative than the outside. Because of the difference in electrical charges on the two sides of the membrane, the membrane or cell is said to be **polarized.**

When a neuron is at its resting potential, the area immediately inside the neuron's membrane is slightly more negative than the immediate area outside the membrane. The negative interior results from an imbalance in the distribution of ions from the sodium pump and various large and negatively charged molecules inside the cell.

When a stimulus is received by a neuron, the resting potential is temporarily upset. The membrane's permeability changes at the point where the stimulus is received when some of the sodium channels in the membrane open briefly. (The gates are thought to be channel proteins that work by changing shape.) The opening of sodium gates permits sodium to rush into the cell, down both a concentration gradient and an electrical gradient. The result of this influx of positively charged ions is that the area just inside the membrane becomes neutral and thus **depolarized.** But within a fraction of a second, the sodium gates close and the potassium channels open. Positively charged potassium now rushes out of the cell, making the inside of the cell more negative. The actions of the sodium potassium pump then complete the reestablishment of the resting potential.

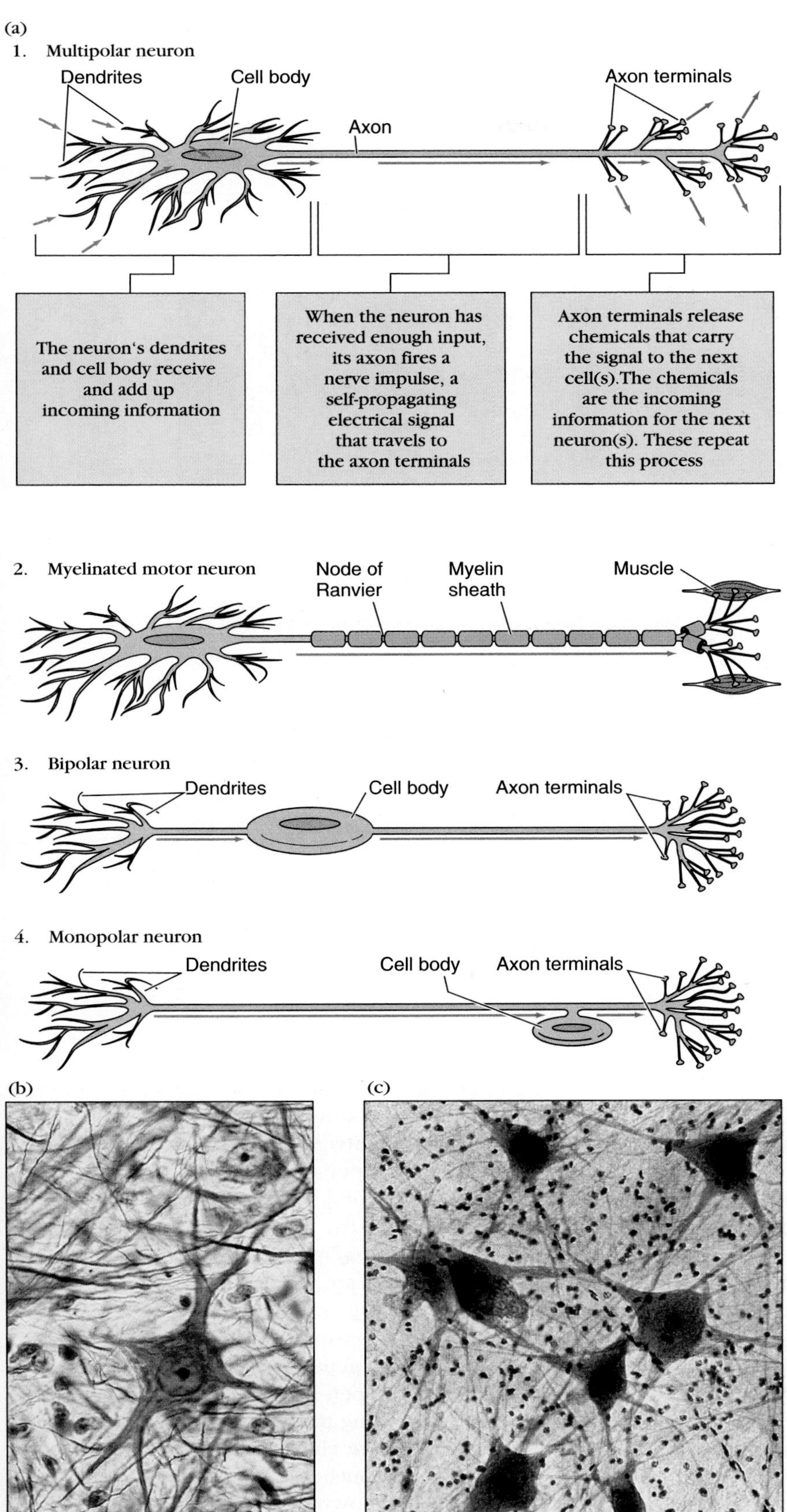
(a)
1. Multipolar neuron
Dendrites
Cell body
Axon
Axon terminals
The neuron's dendrites and cell body receive and add up incoming information
When the neuron has received enough input, its axon fires a nerve impulse, a self-propagating electrical signal that travels to the axon terminals
Axon terminals release chemicals that carry the signal to the next cell(s). The chemicals are the incoming information for the next neuron(s). These repeat this process
2. Myelinated motor neuron
Node of Ranvier
Myelin sheath
Muscle
3. Bipolar neuron
Dendrites
Cell body
Axon terminals
4. Monopolar neuron
Dendrites
Cell body
Axon terminals
(b)
(c)

FIGURE 28-1

Structure of the different types of neurons. The arrow shows the direction in which nerve impulses are transmitted in all four of the neurons. **(a)** Neuron structure. 1. A multipolar neuron (i.e., a motor neuron) has a large central cell body, many dendrites, and a branched axon. These are the most common type of neuron, found extensively throughout the brain and spinal cord. 2. A motor neuron on its way to a vertebrate muscle. This neuron has an axon surrounded by myelin. The nodes of Ranvier are gaps in the myelin sheath that help increase the speed of the signal. 3. A bipolar neuron, found in sensory organs, has two main branches on opposite sides of the cell body. 4. Most sensory neurons in sense organs (i.e., taste buds on the tongue) are monopolar, with only a single attachment to the nerve cell body. **(b)** This section of gray matter shows several cell bodies of multipolar neurons and some of their extensions—dendrites and axons. The cells have been stained and photographed using a light microscope. **(c)** A false color scanning electron micrograph of a motor neuron axon (light green) as it contacts several muscle fibers (pink). (b, Biophoto Associates; c, D. Fawcett, Science Source/Photo Researchers, Inc.)

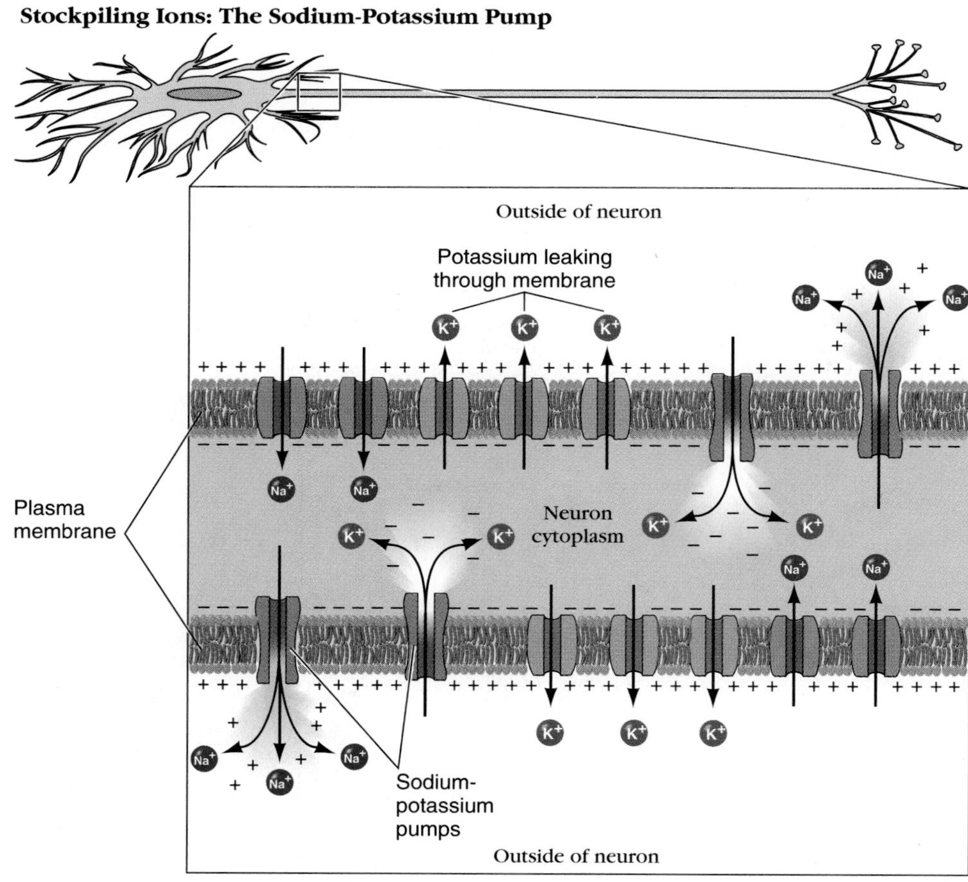

FIGURE 28–2

Source of the resting potential. Active transport by the sodium-potassium pump keeps sodium more concentrated outside the cell and potassium more concentrated inside. Large, negatively charged ions remain inside the cell because they are too large to move across the plasma membrane.

Local Potential and Action Potentials

When a potential is initiated, it triggers neighboring sodium channels to open, causing the potential to spread in a wave. When the stimulus is weak, a **local potential** is generated and the signal does not spread over the entire surface of the cell. But if the stimulus is sufficiently strong, perhaps due to the addition of many local potentials, an **action potential** is generated. Action potentials travel like a wave over the entire cell membrane surface without losing strength. This is how a nerve conducts an action potential over a distance (sometimes more than a meter). During an action potential, the membrane depolarization is initiated at one end of the nerve. The depolarization then travels along the length of the nerve cell's membrane to the opposite end, where it may contact yet another cell and initiate an action potential.

Some local potentials depolarize the membrane (decrease the difference in electrical charge across it). They do this by opening sodium gates. Positive sodium ions enter and make the membrane potential temporarily less negative (Figure 28-3). Stimuli that depolarize the membrane are **excitatory,** making the neuron more likely to fire a nerve impulse.

Some local responses are **inhibitory,** making the neuron less likely to generate a nerve impulse. Inhibitory responses open more potassium and chloride (Cl^-) gates, permitting the escape of more potassium and the entry of more chloride, which is normally more concentrated outside than inside the cell. Movement of these ions down their gradients causes the

Local Potentials Can Initiate or Inhibit an Action Potential

Excitatory responses make an impulse more likely

0
Membrane potential (millivolts)
Na+
Axon
K+
Resting potential
Depolarization
Recovery from depolarization
-70
Time

Inhibitory responses make an impulse less likely

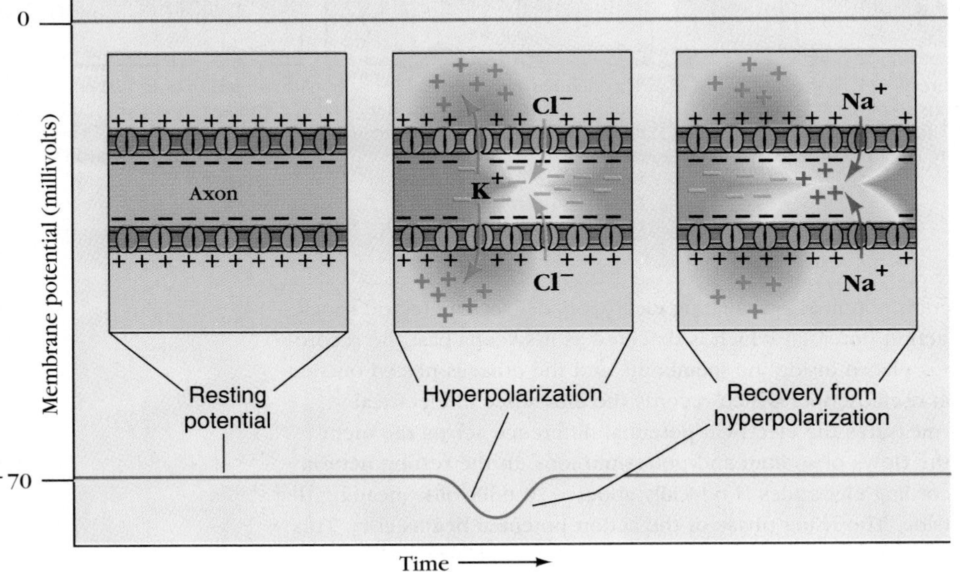

FIGURE 28-3

Changes in membrane potential caused by excitatory and inhibitory stimuli. The red line shows membrane potential changes during excitatory (depolarizing) changes in local potential and inhibitory (hyperpolarizing) changes in potential. The diagrams show the flow of sodium, potassium, and chloride ions through the membrane as the resting potential is disturbed, and then the flow of ions in the opposite direction as the resting potential is restored. In addition, chloride moves into the cell during hyperpolarization.

membrane potential either to become more negative **(hyperpolarization)** or to remain near the resting potential despite excitatory input.

Local changes in membrane potential spread along the membrane from the point of stimulation, much as ripples spread when you throw a stone into a pond. However, the cytoplasm does offer some resistance to ion flow. Hence a local potential dissipates within a short distance—unless it is reinforced by other local potentials.

Nerve Impulses

An action potential is a brief reversal in membrane charges in a local area and within a fraction of a second, the resting potential is restored. However, the change in membrane charges causes neighboring sodium channels to open which, in turn, results in an action potential as sodium ions flow into the cell. This effect continues along the length of the neuron, so that the action potential is **self-propagated.** This self-

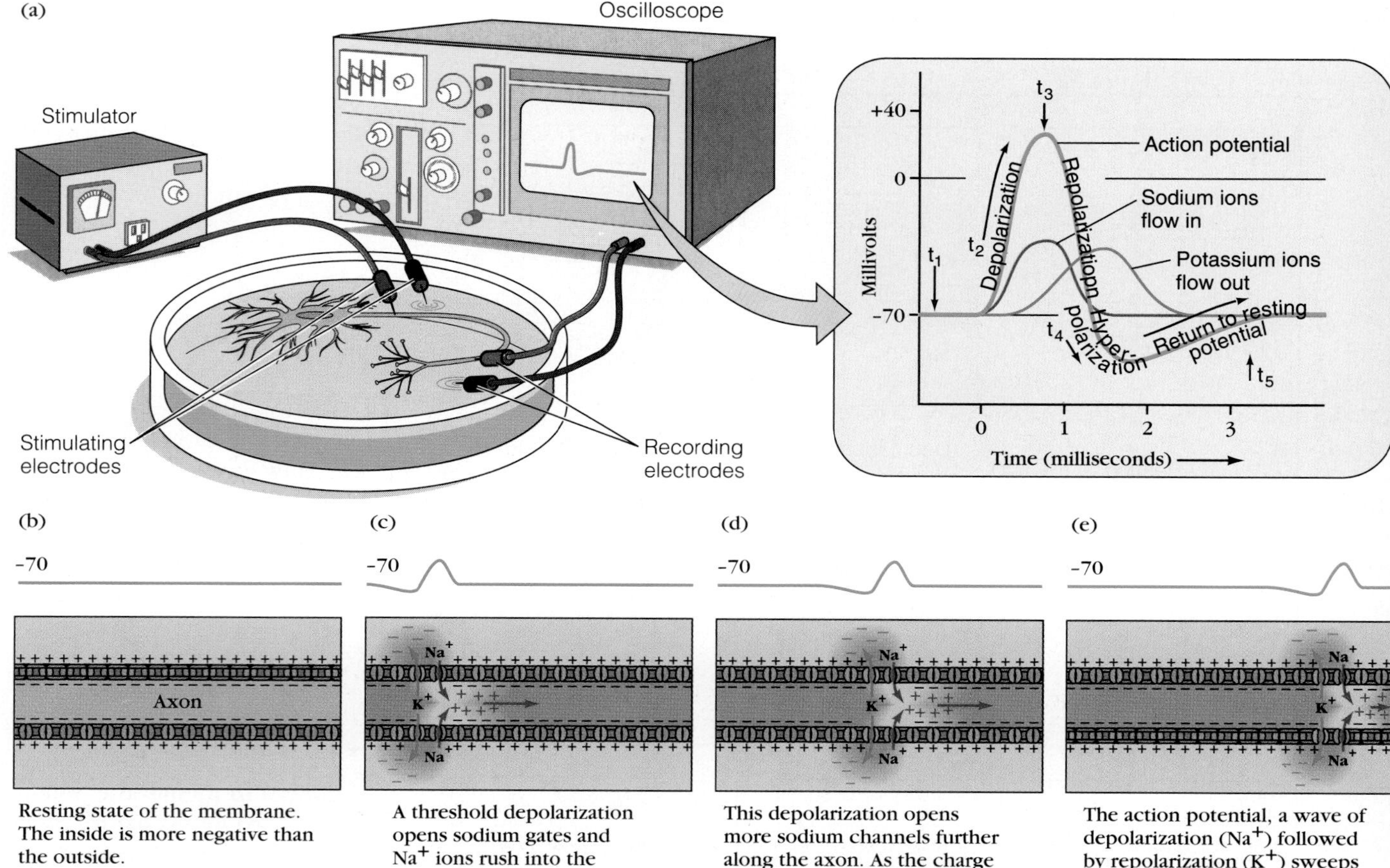

(b) Resting state of the membrane. The inside is more negative than the outside.

(c) A threshold depolarization opens sodium gates and Na^+ ions rush into the more negative axon.

(d) This depolarization opens more sodium channels further along the axon. As the charge moves forward, potassium ions move out of the axon behind it to restore the resting potential of the axon.

(e) The action potential, a wave of depolarization (Na^+) followed by repolarization (K^+) sweeps down the axon toward the terminals.

FIGURE 28-4

Action potentials. (a) Recording an action potential. Stimulating electrodes deliver an electric shock to the axon of a neuron, generating an action potential which is detected as it sweeps past the recording electrodes. One recording electrode is placed inside the membrane and the other is placed outside of it. These electrodes are attached to an oscilloscope, which records the difference in electrical charges at the two electrodes, and thus measures the electrical potential difference across the membrane. Superimposed on the graph are the flows of sodium and potassium ions. In the resting neuron (t_1), the difference between the two recording electrodes is typically about −70 millivolts, meaning that the inside is more negative than the outside. The rising phase of the action potential begins at t_2. This signals depolarization of the membrane as sodium rushes in. The potential difference across the membrane declines from −70 to 0, and reverses briefly as the inside becomes positive with respect to the outside (at t_3). Now potassium rushes out, and the inside of the cell once more becomes negative with respect to the outside. It reaches a point of hyperpolarization at t_4 where the sodium-potassium pump restores the membrane potential to −70 millivolts (t_5). **(b)** Stages in the spread of an action potential.

BIO-BIT

When your arm "goes to sleep," you have pinched a nerve and have lost the sensations in those regions affected by that nerve.

propagating wave of depolarization down an axon is called a **nerve impulse** (Figure 28-4).

Excitatory local potentials in a neuron's dendrites and cell body may add together and cause the axon to fire a nerve impulse. If inhibitory local potentials are occurring, it takes more excitatory input to overcome them. Each axon has a **threshold,** a level of depolarization that must be exceeded before it will fire. Once this threshold is reached, the resulting depolarization is **self-propagated.**

All axons transmit nerve impulses rapidly, but this speed can be increased in two ways. First, axons of larger diameter have less internal electrical resistance and so transmit impulses faster. Animals have axons of various diameters. The thickest are the "giant" axons involved in escape reactions in invertebrates. When you swat at a cockroach and miss, you are foiled by such a giant fiber, which runs from the cerci (the roach's twin "tailpipes") to the brain, and informs the brain of the air currents that disturb the cerci on your downstroke.

The second adaptation, found mainly in vertebrates, is a **myelin sheath** around the axon. Each segment of the sheath consists of several layers of lipid-rich membrane, produced by a glial cell and wrapped around the axon (Figure 28–5). The naked axon is exposed between these cells in very short gaps called **nodes of Ranvier** (see Figure 28–1).

In axons without myelin sheaths, the speed of impulse conduction is limited by how fast the membrane's ion gates can open and close, and by how quickly the ions flow through them and then spread sideways to open nearby gates. The fatty layers of myelin act like insulation around an electric wire. The tight sheath prevents ions from flowing through the membrane. The electrical impulse is forced to move down the interior of the axon from one node of Ranvier to the next, passing through the ion-rich fluid as a much faster electric current. (In this situation, the ion-rich axon is analogous to a pipe tightly packed with marbles: if you force a marble [ion] in one end, a marble at the far end pops out almost instantly.) When the current reaches a node of Ranvier, it opens sodium gates and generates another action potential (Figure 28–5).

The result is an extremely fast form of conduction termed **saltatory** ("jumping") **conduction.** The action potential at each node regenerates the original signal so that it reaches the next node strong enough to start an action potential there. Although some unmyelinated giant axons in squid conduct action potentials at speeds of nearly 20 meters per second, myelinated mammalian axons of much smaller diameter may transmit impulses at up to 100 meters per second.

Bundles, or tracts, of myelinated, fast-conducting axons occur in various parts of the nervous system. These areas, called **white matter,** owe their color to the abundant lipid in their myelin sheaths. **Gray matter** consists of unmyelinated nervous tissue, including cell bodies and axons. When axons of a number of neurons are tightly bound into bundles, these are called **nerves.**

A neuron processes and transmits information in the form of spurts of ions moving through the membrane, briefly disturbing its resting potential. Local potentials may add together and initiate an action potential, which travels down an axon. Action potentials travel faster along axons that have a large diameter or a myelin sheath.

28–B SYNAPTIC TRANSMISSION

When an action potential reaches the axon terminals, its message passes to other cells by way of **synapses,** junction areas between neurons. Most synapses are chemical synapses, where the signal is carried between cells in the form of chemical messengers called **neurotransmitters.** These chemicals occur in tiny membranous sacs, or **vesicles,** in the axon terminals (Figure 28–6). The plasma membrane of the axon terminal is called the **presynaptic membrane.** It lies close to the **postsynaptic membrane,** part of another neuron, or part of a muscle or gland cell. Between the two membranes lies a very narrow space, the **synaptic cleft.**

As an action potential reaches the end of an axon, it causes some of the neurotransmitter vesicles to discharge their contents into the synaptic cleft. Transmitter molecules cross the cleft and bind to receptor molecules in the postsynaptic membrane. Many postsynaptic receptors have gated ion channels, which open when neurotransmitters bind to the receptor. This allows ions to flow through the channel, causing a local potential in the postsynaptic membrane.

Synapses may be excitatory or inhibitory, depending on the postsynaptic receptors. Receptors at excitatory synapses have sodium channels, which let sodium ions enter and depolarize the postsynaptic cell. Often the cell must receive excitatory signals from more than one presynaptic cell before it will fire an action potential.

When inhibitory synapses are active, it takes more depolarization than usual to reach the threshold for an action potential. Examples of inhibitory synapses occur in sense organs, where they suppress some incoming information and so sharpen the ability to pick out important features of stimuli.

A neuron can transmit an action potential in both directions from its starting point. However, information passes through the nervous system in only one direction because of the layout of chemical synapses.

(text continues on page 592)

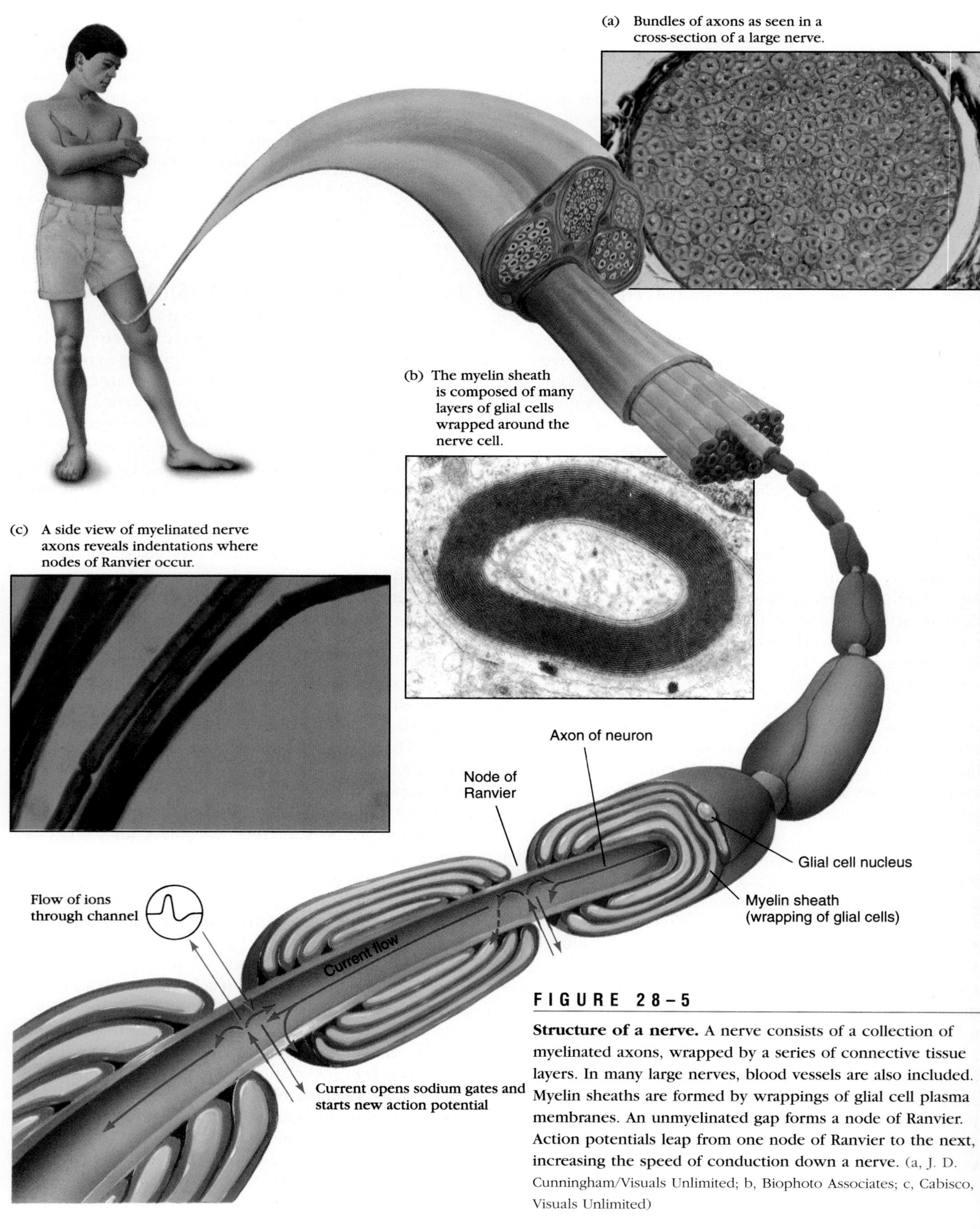
(a) Bundles of axons as seen in a cross-section of a large nerve.
(b) The myelin sheath is composed of many layers of glial cells wrapped around the nerve cell.
(c) A side view of myelinated nerve axons reveals indentations where nodes of Ranvier occur.
Axon of neuron
Node of Ranvier
Glial cell nucleus
Myelin sheath (wrapping of glial cells)
Flow of ions through channel
Current flow
Current opens sodium gates and starts new action potential

FIGURE 28–5

Structure of a nerve. A nerve consists of a collection of myelinated axons, wrapped by a series of connective tissue layers. In many large nerves, blood vessels are also included. Myelin sheaths are formed by wrappings of glial cell plasma membranes. An unmyelinated gap forms a node of Ranvier. Action potentials leap from one node of Ranvier to the next, increasing the speed of conduction down a nerve. (a, J. D. Cunningham/Visuals Unlimited; b, Biophoto Associates; c, Cabisco, Visuals Unlimited)

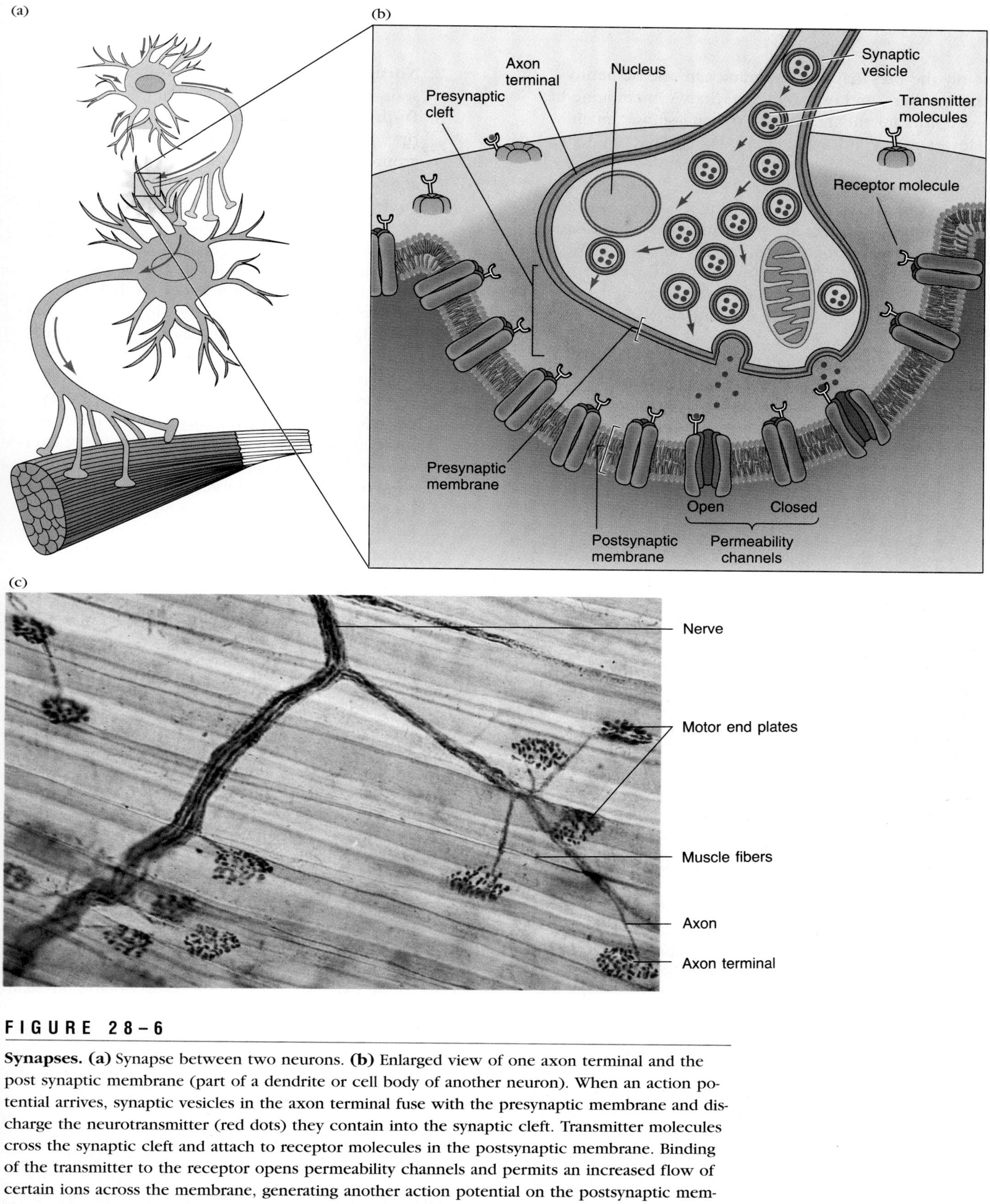

(a)

(b)

Axon terminal
Nucleus
Synaptic vesicle
Presynaptic cleft
Transmitter molecules
Receptor molecule
Presynaptic membrane
Postsynaptic membrane
Open
Closed
Permeability channels

(c)

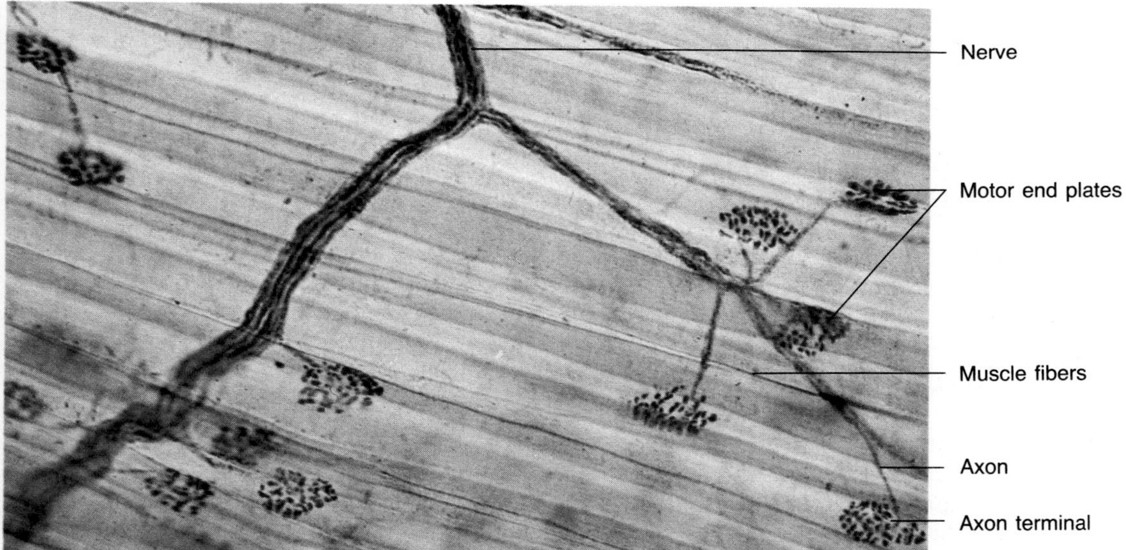

FIGURE 28-6

Synapses. (a) Synapse between two neurons. **(b)** Enlarged view of one axon terminal and the post synaptic membrane (part of a dendrite or cell body of another neuron). When an action potential arrives, synaptic vesicles in the axon terminal fuse with the presynaptic membrane and discharge the neurotransmitter (red dots) they contain into the synaptic cleft. Transmitter molecules cross the synaptic cleft and attach to receptor molecules in the postsynaptic membrane. Binding of the transmitter to the receptor opens permeability channels and permits an increased flow of certain ions across the membrane, generating another action potential on the postsynaptic membrane. **(c)** Neuromuscular junctions. A nerve is composed of the axons of many neurons. Near the muscle, the nerve divides into individual axons. Each axon may divide into several branches, each of which ends in a cluster of axon terminals called a motor end plate. These sites are used to study the action of the transmitter acetylcholine.

Only the presynaptic membrane can release neurotransmitters, and only the postsynaptic membrane has receptors for them. Synaptic transmission is much slower than conduction along the axon. So, in general, the more chemical synapses in a neural pathway, the more slowly information passes along that pathway.

In an **electrical synapse,** the membranes of the two cells are connected by gap junctions (see Chapter 4, A Journey Through Plasma Membranes). Electric current flows directly from one cell to the other in the form of ions moving in the cytoplasm. In contrast to chemical synapses, electrical synapses work rapidly and in both directions.

If electrical synapses are so efficient, why do animals have chemical synapses at all? The reason seems to be that chemical synapses permit more complexity and variety in information processing. Electrical synapses can only be excitatory, but chemical synapses can be either excitatory or inhibitory. By activating different combinations of the excitatory and inhibitory input coming to a particular neuron, it is possible to block, enhance, shorten, or prolong the neuron's response to a signal. This makes it possible for the nervous system as a whole to adjust the intensity of the animal's response, or to make a different response, depending on circumstances.

Messages usually travel between neurons in the form of chemical transmitters that cross a synaptic cleft from one neuron to the next. Electrical synapses permit signals to be transmitted faster through direct flow of ions between the cytoplasm of adjacent neurons.

28–C NEUROTRANSMITTERS

About 30 different neurotransmitters have been identified. Each occurs in specific sets of neurons, and each neuron typically makes and releases one kind of transmitter. Many transmitters can be either excitatory or inhibitory, depending on which receptors occur at the synapse where they are released. Other transmitters appear to occur only at excitatory or at inhibitory synapses. (Presumably only receptors with excitatory or inhibitory effects exist for these transmitters.)

Some important neurotransmitters are:

1. **Acetylcholine,** the best-known transmitter. It occurs at many synapses in the brain and in other parts of the nervous system, and also at **neuromuscular junctions,** synapses between neurons and skeletal muscles (muscles that move parts of the skeleton and are under our conscious control).
2. **Norepinephrine,** important in the body's response to stress (Section 28–H).
3. **Dopamine,** a molecule similar to norepinephrine. Dopamine is the transmitter for a small group of neurons concerned only with muscular activity. Parkinson's disease, with its bursts of uncontrollable muscular movement, may sometimes be caused by lack of this transmitter; it is often treated with L-dopa, the substrate of the enzyme that makes dopamine. Dopamine also occurs in neurons in some other areas of the brain.
4. **Serotonin,** produced by a group of cells in the medulla, part of the brain just above the spinal cord (see Figure 28–9). Serotonin seems to be concerned with functions such as sensory perception, regulation of body temperature, sleep, consciousness, and emotions. Deficits of serotonin are linked to depression and anxiety; excess can cause nausea.
5. **Neuropeptides,** the largest group of neurotransmitters, each containing 2 to 39 amino acids. Some neuropeptides are identical to local chemical messengers in other body areas (Section 30–D) or to hormones released from the pituitary gland, hypothalamus (both shown in Figure 28–9), or gut. Their roles in the nervous system are poorly understood.

   1? The larger **endorphins** and smaller **enkephalins** have been called the brain's "natural morphine" because the brain's receptors for them also bind morphine, presumably with the same effect on the neurons involved. This explains why the brain is so sensitive to morphine, opium, and related drugs, such as heroin, codeine, and caffeine (see *A Journey into Healthy Living*: This Is Your Brain on Drugs, this chapter.) Enkephalins occur in neurons that process information relating to emotion, mood, and pain. They apparently act by suppressing the response to other neurotransmitters. Endorphins are responsible for the "runner's high" experienced by people who run or perform other intense exercise to the point of producing physical pain. ■
6. **GABA** (short for gamma-aminobutyric acid) is a transmitter at inhibitory synapses throughout the brain. Studies of GABA suggest that as much as 90% of the brain may be devoted to inhibition.

After transmitter molecules have acted on the postsynaptic membrane, they are removed or destroyed. If they weren't, their action would continue indefinitely, and all useful information would be lost. Norepinephrine is resorbed by the presynaptic membrane and reused. Acetylcholine is broken down by the enzyme **acetylcholinesterase.** Many insecticides inhibit this enzyme, permitting acetylcholine in the synapses to keep on stimulating the postsynaptic membranes. The nervous system soon goes wild, firing one nerve impulse after another. These impulses in turn cause contraction of the muscles in uncontrollable spasms and, eventually, death. (Such substances must be used with care. They are also toxic to other organisms, such as people and pets, that use acetylcholine as a transmitter.)

Various neurons produce different synaptic transmitters. A neurotransmitter released from one neuron may either inhibit or excite the receiving cell, depending on the properties of the cell's receptor molecules. The transmitter must then be removed from the synapse or destroyed, thereby terminating the effect.

28–D ORGANIZATION OF NEURONS INTO NERVOUS SYSTEMS

In an animal's nervous system, neurons are "wired" to each other at synapses, forming nervous pathways. Some synapses are excitatory, others inhibitory. Any one neuron may synapse with many others, and may make excitatory synapses with some and inhibitory synapses with others. The number, arrangement, and synaptic connections of neurons determine how the animal responds to stimuli and the kinds of behavior it can perform.

During the course of animal evolution, nervous systems became increasingly complex as animals became larger and more mobile. Both trends require more neurons. A tiny, sluggish, parasitic roundworm may have as few as 160 neurons, and a correspondingly small range of behaviors. An octopus, with precise control over its eight tentacles and considerable ability to learn new behavior patterns, has more than a billion. A human has about 100 billion neurons.

Large numbers of neurons are not enough; their organization is also crucial. The simple nerve nets of cnidarians have neurons scattered throughout the body (Figure 28-7). Though primitive, this arrangement serves the needs of a radially symmetrical animal, whose food or enemies may approach from any direction. A cnidarian's reaction to most stimuli is generalized because each neuron transmits impulses to all of its neighbors. The animal's reaction also depends on the strength of the stimulus; only part of the body responds to a weak stimulus, the entire animal to a strong one. Surprisingly, even, an animal with this primitive nerve net is capable of simple forms of learning. ■

The platyhelminths (flatworms) show the beginnings of some important trends. The first is consolidation: instead of the diffuse nerve net of cnidarians, some neurons became arranged into nerve cords running the length of the body, with cross-connections between them. This permits faster, more direct processing of information collected from outlying areas, and quick signalling of actions to be taken by effectors. A second advance is specialization: cells took on distinct roles as sensory or motor neurons or as interneurons. The patterns of synapses became more precise, so that incoming information is passed to specific neurons. A third important trend seen in flatworms is **cephalization,** formation of a head. Having a particular front end correlates with the evolution of bilateral symmetry. A head bears the major sense organs, such as the eyes and ears, which detect what is happening in the outside world; the animal's leading end can sample the new environment for food or safety as it moves. The closer the "decision-making" neurons are to these sense organs, the faster the animal can react. Consequently, nervous tissue became more concentrated in the head, with neurons forming clusters called **ganglia** (singular: **ganglion**). The ganglia in the head are generally called a **brain,** the body's main nervous control center.

In annelid worms, arthropods, and other higher invertebrates, ganglia also occur along the nerve cord and govern particular parts of the body. The nerve cord itself is now single, running down the body's ventral midline.

Some invertebrate nervous systems have relatively few neurons, with a correspondingly limited repertoire of possible behaviors. In a few species of leeches, slugs, and crayfish, we now have maps showing exactly which neurons do what. In a leech, for instance, one particular pair of motor neurons everts the penis, and four neurons in each ganglion sense "pain." Researchers use these animals to examine basic principles of nervous system organization and its effects on behavior. But even these "simple" nervous systems are challenging to study, and we are nowhere near ready to tackle such a cell-by-cell analysis of the vastly more complex nervous systems of more complicated animals. Instead, we approach the nervous systems of vertebrates mostly by determining the functions of groups of cells in various areas.

The nervous systems of animals have become larger, more centralized, and more specialized to interpret additional sensory information and control new behaviors.

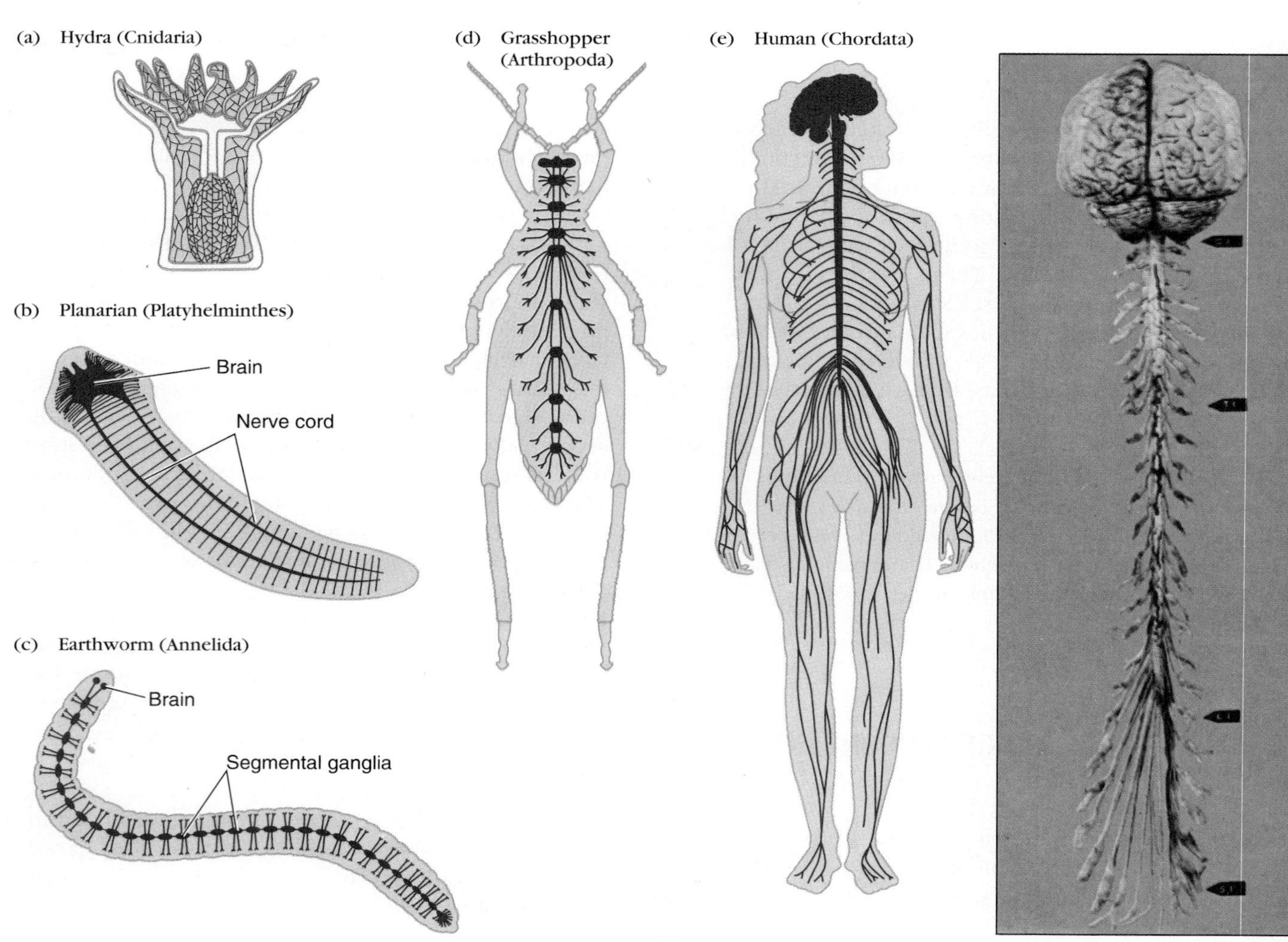

FIGURE 28–7

Nervous systems. (a) Radially symmetrical cnidarians have a diffuse nerve net in which neurons make many interconnections. However, the nervous systems of other invertebrates **(b, c, d)** and humans **(e)** are arranged in well-defined systems along the length of the body. (e, Dissection by Dr. M. C. E. Hutchinson, Department of Anatomy, Guy's Hospital Medical School, London, England. From Williams and Warwick (eds.): *Grays Anatomy*)

THE VERTEBRATE NERVOUS SYSTEM

The vertebrate nervous system can be divided into the central and peripheral nervous systems. The **central nervous system** consists of the brain and spinal cord. The vast majority of all neuron cell bodies, and in most cases the dendrites and axons too, lie in the central nervous system. Most of these neurons are interneurons, which relay information from one neuron to another. In the process, sensory information is **integrated**—compared, changed, added up, or suppressed—assuring that the response made will be appropriate for existing conditions. (For example, your response to a plate of pastries depends on how full your stomach is, as well as how delicious the pastries look and smell.) The central nervous system is also responsible for **association**—channelling sensory input into appropriate motor pathways.

In the human nervous system, some of these activities go on automatically, without our being aware of them. Others are carried out by the "higher centers," parts of the nervous system at the level of consciousness, where we are aware of the existence of sensory stimuli and of our own thoughts and emotions, and may deliberately choose one course of action over another. We have no way of knowing whether other animals have distinctly conscious nervous functions.

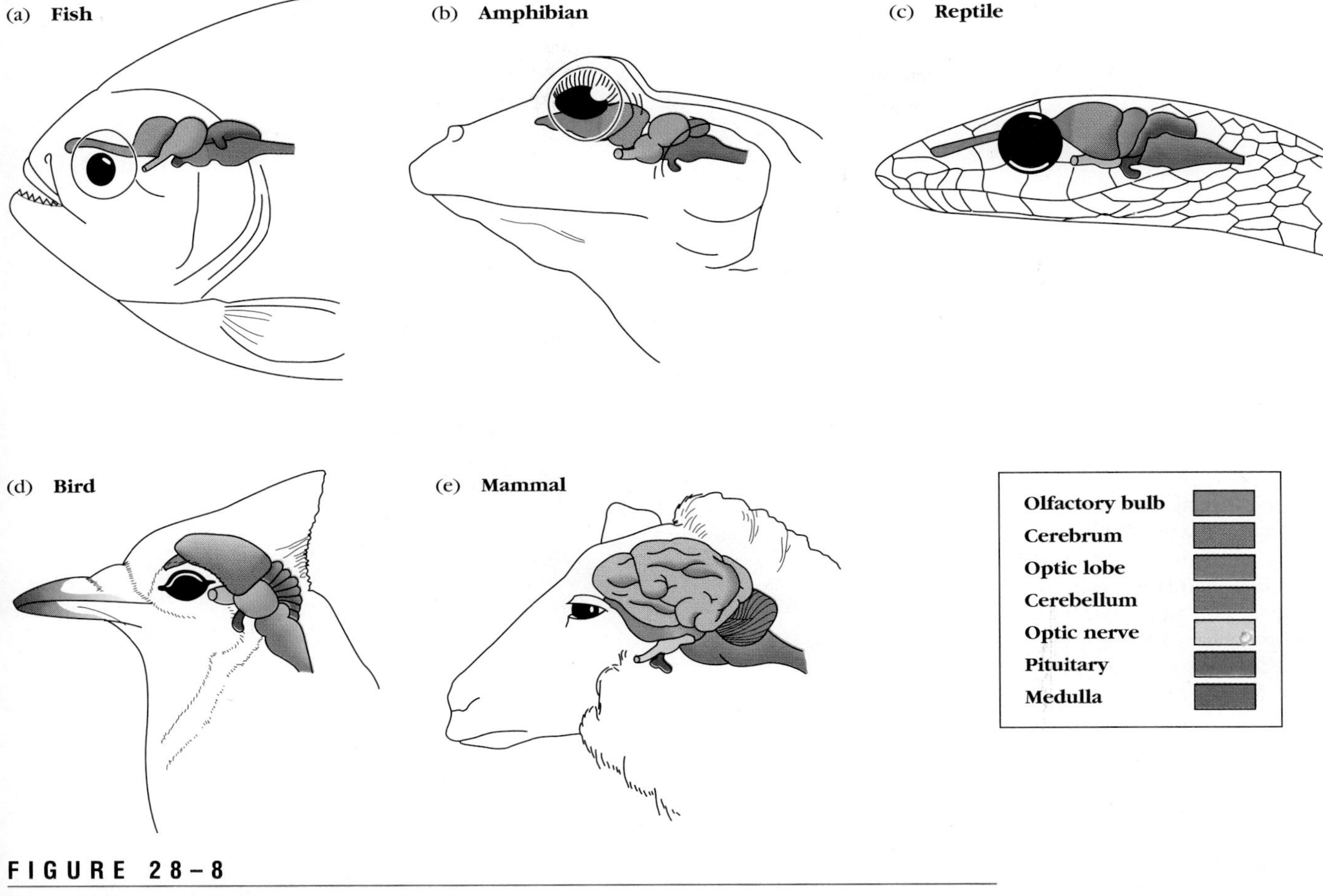

FIGURE 28-8

Comparison of several vertebrate brains. Note the increase in the size of the cerebrum and cerebellum from fish to birds and mammals. Human brains differ from those of other vertebrates only in the relative size of each brain region.

The brain is protected by the skull and the spinal cord is protected by the vertebral column. Inside these bony coverings, the brain and spinal cord are covered by three layers of membranes, the **meninges.** Within the meninges, the **cerebrospinal fluid** bathes the central nervous system and cushions it from jarring. Disorders of the central nervous system can sometimes be diagnosed by doing a "spinal tap" to withdraw a sample of the fluid around the spinal cord. Inflammation of the meninges (meningitis) is often a serious condition because the meninges are so close to vital nervous tissues. ■

The **peripheral nervous system** connects the central nervous system with the rest of the body, including sense organs that detect stimuli in the outside world. It consists of **nerves,** bundles of axons covered with protective sheaths of connective tissue, and some **peripheral ganglia,** clusters of neuron cell bodies lying outside the central nervous system.

The peripheral nervous system has two parts, which play different roles. The **somatic nervous system** contains both motor and sensory neurons. It controls the activity of the skeletal muscles, those under "conscious" control, such as the muscles we use to smile, run, sing, or draw. The **autonomic nervous system** contains only motor neurons. It controls muscles and glands that usually operate without our being aware of them. These effectors control things like blood pressure and the movement of food in the gut.

A Journey into Healthy Living

This Is Your Brain On Drugs

You may have seen the television commercial. It opens with a shot of a raw egg in a frying pan. "This is your brain," says an aggressive, young, male voice. The egg begins to sizzle. The white and yolk congeal and we watch as a sputtering, blackened lace forms around the egg's edges. "This is your brain on drugs," the voice continues. The camera pulls back to reveal the gritty spokesman. He looks directly into the camera's eye and with undisguised scorn, asks, "Any questions?" This anti-drug commercial clearly communicated the idea that drugs chemically alter the human brain. But exactly how do drugs produce alterations in mood, perceptions, and behavior? As we will see, it depends upon what drug we are talking about and how it interacts with neurons.

Many drugs affect the nervous system, in most cases by regulating the activity at synapses. However, some have more general effects. For example, caffeine (in coffee, tea, cola, and chocolate) increases the metabolic rate of neurons, thereby increasing alertness and thought. Alcohol is a depressant, acting on all neurons and affecting many functions such as orientation, alertness, coordination, judgment, memory, and mood. Nicotine in all forms of tobacco (cigarettes, pipes, cigars, snuff, and chewing tobacco) volatilizes when the leaves are burned or make contact with mucous membranes of the mouth, nose, and respiratory tract. Low concentrations of nicotine act like acetylcholine and stimulate nerve cells, while high concentrations paralyze nerve cells.

Opium, from the seedpod of a poppy, has been used as a drug since ancient Greek times, not only because it is the most effective painkiller known, but also because it induces a state of euphoria, an exaggerated feeling of well-being, also known as a

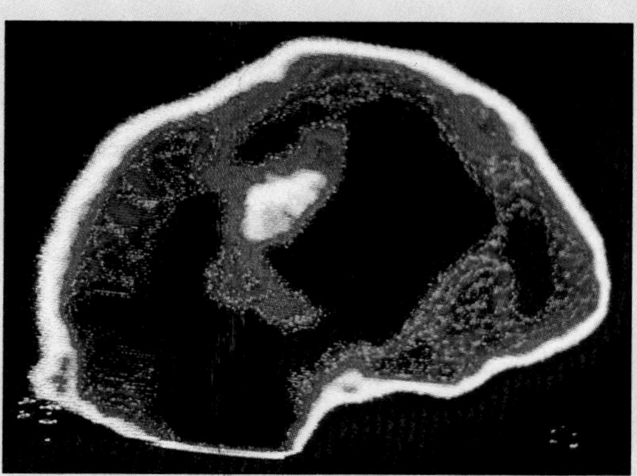

FIGURE 28–A

Neurological damage caused by cocaine use. In this color-enhanced SPECT (single photon emission computer tomography) scan of the brain of a cocaine abuser, an area of decreased metabolism near the frontal region appears as a white patch. (Howard Sochurek, courtesy of Brigham and Womens Hospital, Boston)

28–E THE VERTEBRATE BRAIN

The central nervous system grows from a hollow tube formed during early embryonic development. The walls of the front part of the tube enlarge in a series of uneven bulges, with three main parts, unimaginatively called the forebrain, midbrain, and hindbrain, followed by the long, straight spinal cord.

All vertebrates, from fish to mammals, have brains with the same basic structures (Figure 28–8). In the course of evolution, some parts of the brain changed very little, while others grew enormously. Some areas of the brain retained their primitive functions, whereas others took on new functions as vertebrates evolved (Table 28–1).

Hindbrain

The most obvious part of the hindbrain of a fish is the **medulla,** the enlargement where the spinal cord enters the brain (Figure 28–8). Through it pass many neurons carrying messages to or from parts of the brain that integrate sensory information. The medulla also contains many neurons that receive sensory input and send out motor signals for automatic, or **reflex,** functions such as breathing, swallowing, vomiting, constriction of blood vessels, and regulation of heart-

"high." Addiction to opiates (opium and related compounds) has been a social problem in the United States ever since the Civil War, when they were used as painkillers. The search for a nonaddictive opiate has been intense, but all known opium derivatives—including morphine, Demerol, Percodan, Darvon, methadone, codeine, and heroin—may produce addiction in many people who take them. They are therefore most useful as pain-killers for the terminally ill, in whom relief from pain outweighs possible addiction.

Opiates bind to postsynaptic receptors in the brain and block the binding of neurotransmitters. This prevents the transmission of nerve impulses along a tract of nerves by which the body normally "tells" the brain that it is in pain. (Pain is a useful biological reaction, for when the brain is informed that some part of the body is in pain, perhaps from a cut or burn, it causes the body to move away from the source of the damage.) Opiates also depress the immune system, especially cells that fight cancer and virus infections.

Marijuana's active hallucinogen, THC, interferes with short-term memory by binding to receptors in the part of the brain involved in the formation of memories. It also binds to receptors in the cerebral cortex where it impairs thought and reasoning, heightens sensory perception, changes the perception of time, and produces mild euphoria, relaxation, and relief from anxiety. Marijuana also lowers the levels of sex hormones, interferes with the menstrual cycle, and suppresses the immune system.

Many other drugs also act on synapses in the brain. For example, barbiturates such as phenobarbital, nembutal, and quaalude, depress the activity of inhibitory synapses, permitting more excitatory activity, which produces mild euphoria. LSD and mescaline are thought to produce hallucinations by binding with receptors for the neurotransmitter serotonin, a molecule they resemble. Cocaine, crack, and amphetamines (Benzadrine, Dexedrine, Methedrine, "speed," and "ice") mimic or enhance the effects of dopamine and norepinephrine. Cocaine in all its forms prevents the normal removal of the transmitter dopamine from synapses in the pleasure center of the brain, so that the synapses continue to fire long after they should have stopped. At low doses it stimulates behavior and elevates mood, but use of cocaine causes the brain to produce less dopamine of its own, and addiction results. Overdoses are fatal.

Addiction to drugs has two components, one physiological, the other psychological. Addicts who stop using drugs, or gradually decrease their dosage, go through a period of withdrawal with unpleasant physical reactions such as shaking, vomiting, hallucinations, pounding heart, and pain. But after their bodies have adjusted to working without the drug and returned to normal, many people still have a psychological craving for the drug that never goes away.

Drug-taking by humans has parallels among other animals. Animals as different as birds, rabbits, deer, and elephants have been observed to seek out and eat such things as fermented fruit (containing alcohol) or intoxicating mushrooms.

beat rate. The medulla has changed little during evolution.

The **cerebellum** is an outgrowth of the medulla. During evolution it has grown noticeably and taken on much of the central control of balance and movement.

Midbrain

During evolution, the midbrain has changed more in function than in size or structure. In fish and amphibians it is the principal area for association of sensory input with suitable motor output. In these lower vertebrates, a major part of the midbrain is the **optic tectum,** which receives signals from the **optic nerves,** carrying visual information from the eyes.

In mammals, the analysis of vision has moved out of the midbrain and into part of the forebrain. The midbrain of mammals controls reflexes of the iris of the eye and the eyelids, and analyzes and relays information coming in from the ear via the auditory nerve.

Forebrain

The forebrain has changed a great deal during vertebrate evolution. It has two major parts, the diencephalon and the telencephalon. Lying just in front of the midbrain, the **diencephalon** contains the thalamus, the hypothalamus, and the posterior lobe of the

Table 28–1
Functions of Major Parts of the Vertebrate Brain

Derivation	Name	Function
Hindbrain	Medulla	Passage of messages between brain and spinal cord; control of visceral reflexes
	Cerebellum	Coordination of equilibrium and movement
Midbrain	Tectum (in lower vertebrates)	Association of sensory and motor pathways
	Anterior colliculi (mammals)	Reflexes of iris and eyelid
	Posterior colliculi (mammals)	Receive sensory information from ear
Forebrain		
Diencephalon	Thalamus	Relays olfactory messages to midbrain (in fishes) Area of sensory integration (in higher vertebrates)
	Hypothalamus	Controls emotional states and drives (pleasure, pain, thirst, sex, rage); secretes hormones from neurosecretory cells
	Posterior pituitary	Stores and releases hormones (Section 30–C)
Telencephalon	Olfactory bulb	Receives olfactory information (most important telencephalon area in fishes)
	Corpus striatum	Complex behavior patterns (in birds)
	Cerebral hemispheres (cerebral cortex)	Well developed only in mammals. Sensory and motor association, visual and auditory processing, seat of conscious thought, "intelligence," and, in humans, of ability to use language, both written and spoken

pituitary gland (Figure 28–8). In fishes, the **thalamus** relays information to the midbrain from the **olfactory** (sense of smell) organs. However, in other vertebrates it is one of the centers that integrates all sensory information. Immediately below the thalamus lies the **hypothalamus,** a vitally important area where the nervous and hormonal systems interact (Section 30–D). This area controls homeostatic functions: regulation of body temperature, growth, sexual drive and maturity, hunger, thirst, and salt and water balance.

BIO-BIT

Epileptic seizures result from random electrical activity in the brain that interferes with its normal electrical functioning. Epilepsy may be inherited or caused by severe physical damage to the brain, and affects about 1% of humans. It does not affect intelligence.

While the diencephalon slowly expanded to handle increased sensory input, the **telencephalon,** the front part of the forebrain, grew astoundingly in both size and complexity. In fish and amphibians it handles mostly olfactory information, which plays a major role in the lives of these aquatic animals. In birds, however, the most important part of the brain is the **corpus striatum,** which is responsible for their complex behavior patterns. The corpus striatum of birds occupies much of the inside of the **cerebrum,** which is divided by a fissure into the right and left **cerebral hemispheres.**

In mammals, there is a progressive increase in the size and importance of the **cerebral cortex,** the outer layer of the cerebral hemispheres. The original, deeper layers of the hemispheres, the **hippocampus** and other **limbic structures,** regulate emotional state and probably short-term memory. Above these areas, the cerebral cortex lies like the cap of a wrinkled mushroom over the rest of the brain. The gray matter of the cerebrum is composed of thick layers of unmyelinated cells.

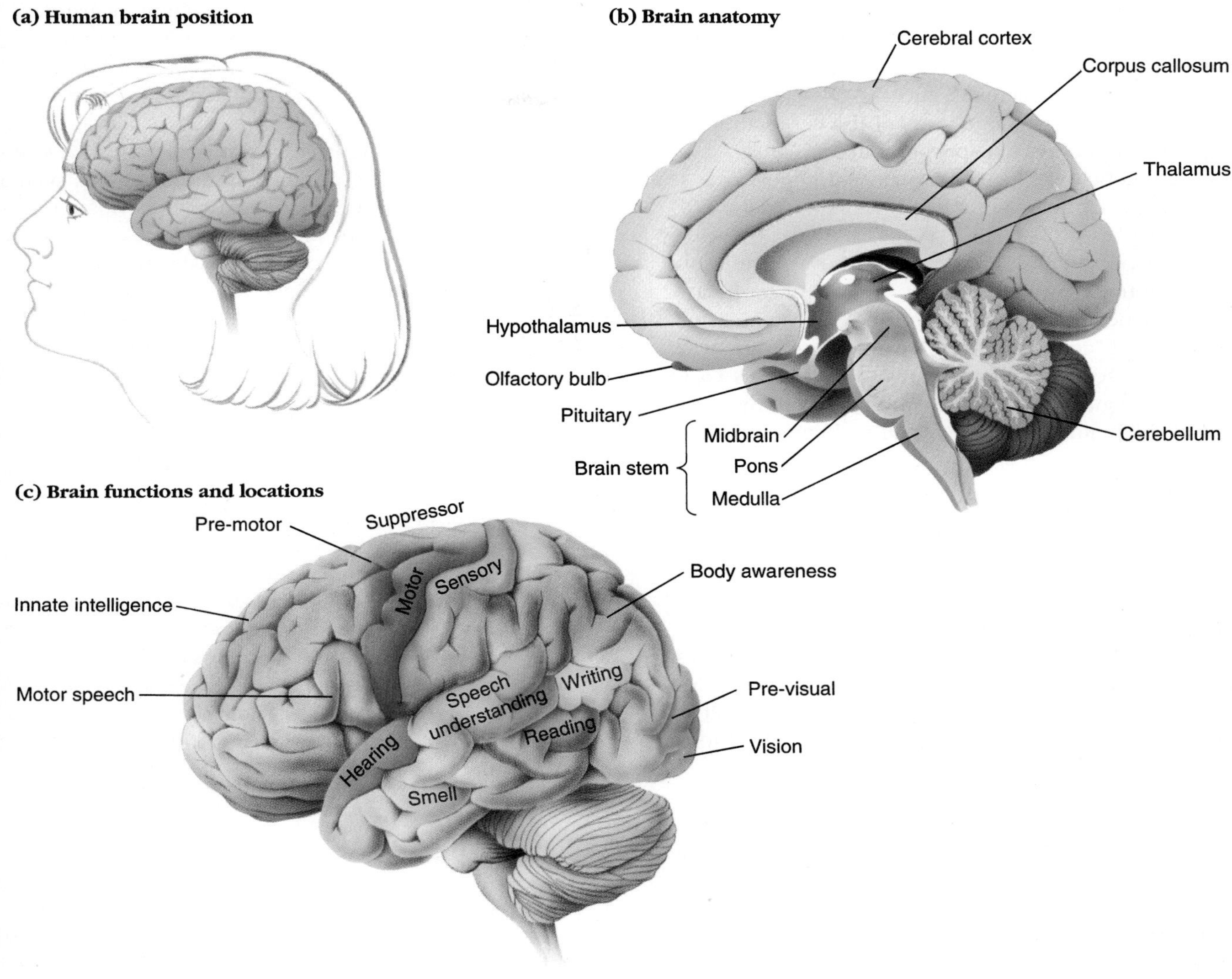

FIGURE 28–9

The human brain. (a) The human brain as it sits within the skull. Human brain size is correlated with the size of the person, and not intelligence. **(b)** A sagittal section of the brain showing the major structures mentioned in the text. **(c)** Brain regions and their general functions. The functions assigned to several areas are written on this view of the brain's surface.

How do we know which parts of the brain do what? Using laboratory animals or human patients, researchers can stimulate different areas of the brain with electricity, to make the neurons fire, and observe what happens. Also, they can observe what functions are lost when certain parts of the brain are deliberately or accidentally destroyed. A newer method is to determine which parts of the brain accumulate an experimental chemical during a particular mental activity.

It turns out that many functions involve cells in several areas of the brain. However, certain areas tend to be "in charge" of certain functions. For example, the cerebral cortex has primary sensory areas and primary motor areas (Figure 28-9). Other areas of the cerebral cortex are involved in perception of visual or auditory stimuli and, in humans, in use of symbols and language.

In primitive mammals, each area of the cerebral cortex has specific sensory or motor functions, and this is true to some extent in more advanced mammals. However, in mammals such as primates (monkeys, apes, humans), large areas of the cerebral hemispheres have no known specific function. Most researchers believe that these areas are important to versatility of behavior, abstract thinking, and personality.

Interspersed with the gray matter of the cerebrum are many tracts of white matter, bundles of myelinated axons connecting with other areas of the brain. The right and left cerebral hemispheres are linked by way of a large tract of myelinated axons, the **corpus callosum** (see Figure 28–9). Its function is to tell the right half of the brain what the left half is doing (and vice versa).

Roger W. Sperry and his co-workers showed that the two cerebral hemispheres can operate as two different brains; for this work, Sperry shared in a 1981 Nobel Prize. To understand it, you must know that each half of the brain controls structures on the opposite side of the body (the nerve tracts cross over at lower nervous centers). Sperry severed the corpus callosum in experimental animals, and also worked with people who had undergone similar surgery to treat severe cases of epilepsy. When tasks are presented to the left eye and hand, which are controlled by the right side of the cerebral cortex, the person may show no sign of recognizing them even though the right eye and hand, and left cerebral cortex, had previously mastered the task.

Furthermore, the two sides of the cerebral cortex do not have exactly the same functions. For instance, a person may be able to perform a task set for the right half of the brain correctly, but not be able to speak or write about it, whereas the left half of the brain can both learn to perform the task and generate a verbal explanation of what it is doing. In general, the left side has a strong tendency to assume control of language, whether written, spoken, or the sign language of the deaf. However, attempts to divide functions neatly between the two halves of the brain are too simplistic. To produce normal functioning in most activities, both sides must work together, each contributing expertise in certain aspects of the activity.

Motor Pathways

In the cerebral cortex, information about sensory stimuli is passed to association neurons, which connect with the motor pathways that will eventually cause appropriate action by effectors.

The motor area of the cerebral cortex sends out commands for voluntary movements. These signals do not directly stimulate muscle contraction or gland secretion. Rather, the command from the cortex may cross anywhere from three to thousands of synapses, allowing the message to be adjusted on its outward path.

The cerebellum plays an important role in coordinating movements, especially rapid movements such as running or typing. It receives sensory information about where each part of the body is, makes moment-by-moment predictions of where each part will be next, and sends out instructions for fine adjustments so that actions are carried out smoothly. People with damaged cerebellums move jerkily, and their motions invariably overshoot the intended final position. For instance, a hand may reach too far to pick up a desired object, and then the next movement, intended to compensate, will bring the hand too far back.

The brains of all vertebrates consist of the same three regions with common basic functions. The hindbrain conducts sensory input between the body and the brain, controls basic physiological reflexes like breathing and swallowing, and coordinates motion. The midbrain is a major association area in lower vertebrates, but this function has moved to the forebrain in mammals. The forebrain interprets sensory information, stores most memories, links the nervous and endocrine systems, and initiates most voluntary actions.

28–F THE SPINAL CORD

The spinal cord extends from the base of the hindbrain to the end of the vertebral column (Figure 28–9). It is a relay system carrying information between the brain and the peripheral nervous system. It is also the seat of the many **spinal reflexes** that allow the body to make quick responses.

Reflex Arcs

Although we have all learned to perform many complex activities, our responses to certain stimuli are simple, unvarying, and quick. For instance, it is hard to hold your lower leg still when the doctor hits you below the knee with a little rubber mallet. The knee jerk, and the rapid withdrawal of a hand from a hot stove, are controlled by **reflex arcs,** pathways of a few neurons each, under little control from higher centers. Reflex arcs contain sensory and motor neurons, and usually one or more interneurons between them. A reflex pathway results in rapid responses to stimuli because it has a small number of synapses. The fewer synapses between receptor and effector, the faster the effector responds to a stimulus. In addition, the mes-

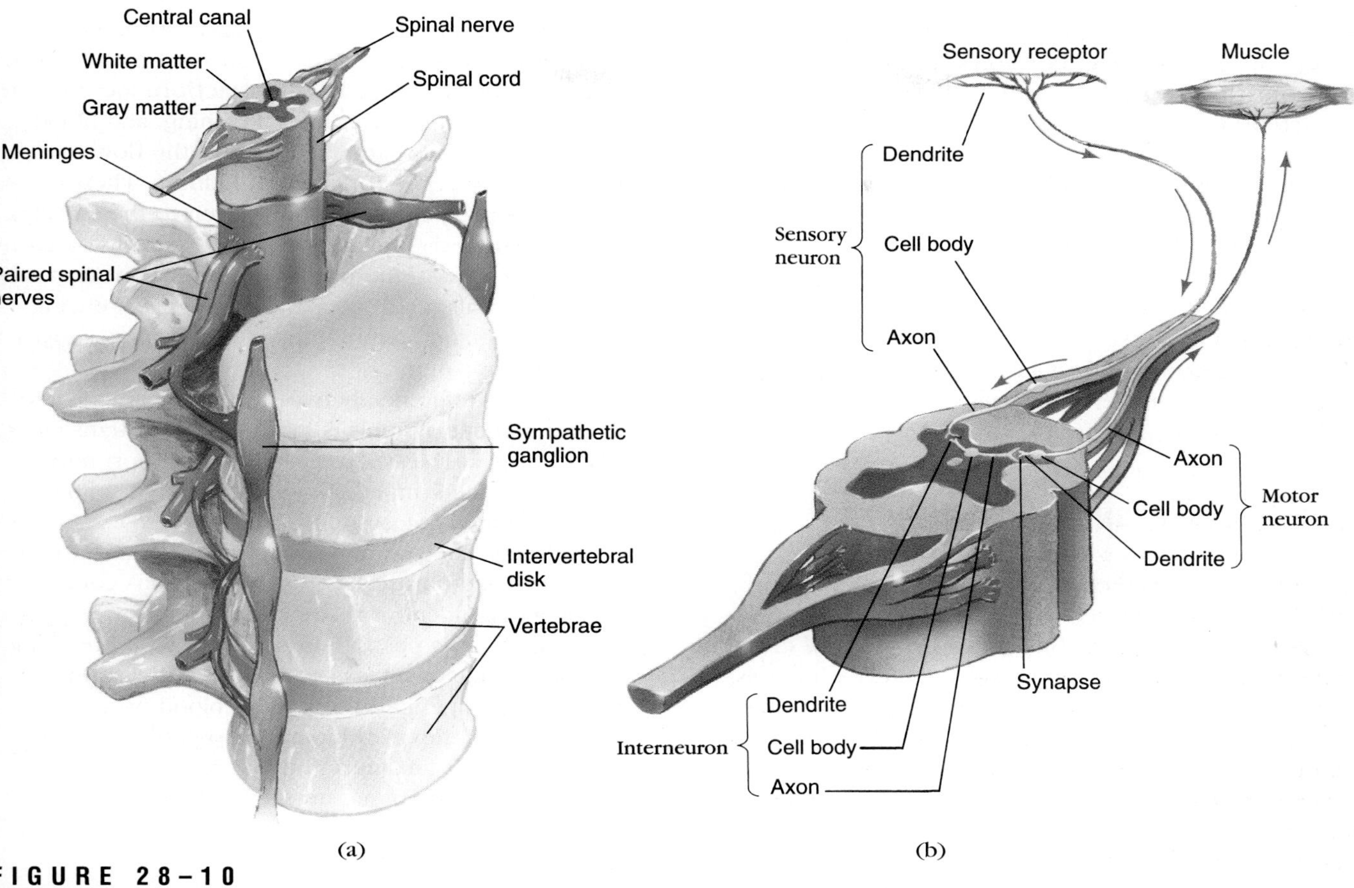

FIGURE 28–10

The human spinal cord. (a) The spinal cord is surrounded and protected by the meninges and by the vertebral column. The paired spinal nerves protrude through spaces between the vertebrae. The butterfly-shaped area of gray matter is composed of cell bodies. White matter is formed by bundles of myelinated axons. **(b)** A simple reflex arc in the spinal cord (enlarged view). The arc shown here contains two synapses and three neurons: sensory neuron, interneuron, and motor neuron. Sensory neurons such as the one shown here are some of the few neurons whose cell bodies lie outside the central nervous system. Ganglia just outside the spinal cord house the cell bodies of its sensory neurons.

sage need not make the lengthy trip to areas of consciousness in the brain and back before an appropriate motor response begins. By the time you actually feel the mallet below the knee, your foot has already begun to swing out.

The organs that regulate physiological homeostasis are controlled by more or less simple reflexes of the autonomic nervous system. Heart rate, breathing, dilation of the pupils, and digestion are all controlled in this way. More complicated activities, such as posture control, locomotion, sexual behavior, and defensive responses, which are governed mainly by the somatic nervous system, also involve many reflex arcs.

A reflex may be quite complex, using many neurons and moving many muscles. For example, if you raise your arm to protect your face, muscles in your back must also adjust their contraction or you would lose your balance.

The spinal cord plays an important role in the integration of reflex behavior. For instance, a "spinal" animal, one whose brain has been destroyed or removed, still shows reflexes. If a piece of acid-soaked paper is touched to the back of a spinal frog, one leg will come up and kick it away; the behavior is repeated no matter how many times the paper is placed on the skin. This response, involving the coordinated action of many muscles, clearly demonstrates one of the chief characteristics of a reflex: unvarying repetition. A frog with an intact brain might make the response two or three times, but eventually the higher centers would intervene and the frog would do something else—perhaps hop away.

The spinal cord carries messages between the body and brain. Some of its neurons mediate reflexes that use less time and energy by performing routine actions without involving higher centers of the nervous system.

28–G CRANIAL AND SPINAL NERVES

The vertebrate peripheral nervous system consists of paired nerves that branch from the central nervous system. In reptiles, birds, and mammals, 12 pairs of **cranial nerves** connect the brain with various structures, mostly in the head and neck; fish and amphibians have only the first 10 pairs. The thickest cranial nerves are the olfactory, optic, and auditory, which carry only sensory information coming to the brain from the major sense organs—the nose, the retinas of the eyes, and the ear. The other cranial nerves carry both motor and sensory information to and from the tongue, muscles of the eye and face, and so on (see Figure 28-8).

Human beings have 31 pairs of **spinal nerves,** which branch out from the spinal cord between adjacent pairs of vertebrae (Figure 28-10). Each spinal nerve serves the skin, muscles, and internal organs in its particular segment of the body and also overlaps adjoining segments. So, if a spinal nerve is cut, the corresponding part of the body does not completely lose sensation and the ability to move.

The cranial and spinal nerves of the peripheral nervous system relay sensory and motor information to and from the brain.

28–H THE AUTONOMIC NERVOUS SYSTEM

The autonomic nervous system governs most of the body's homeostasis: it regulates the heartbeat and controls contraction of the muscles in the walls of the blood vessels and the digestive, urinary, and reproductive tracts. Autonomic nerves also stimulate glands to secrete mucus, tears, and digestive enzymes.

The autonomic system has two divisions, sympathetic and parasympathetic, with different functions. The **sympathetic system** dominates in time of stress. It initiates the **"fight or flight" reaction:** increases in blood pressure, heartbeat rate, breathing, and blood flow to the muscles, and decreases in the flow of blood to the digestive organs and kidneys. These changes increase the supply of oxygen to the muscles at a time when they may be called upon to use a lot of energy. In contrast, the **parasympathetic system** acts as a counterbalance. It conserves energy by slowing the heartbeat and breathing rates and promotes digestion and elimination.

Another difference between the sympathetic and parasympathetic systems is in the chemical transmitters released at the synapse with the effector: norepinephrine in the sympathetic system, acetylcholine in the parasympathetic system.

Although the autonomic nervous system can carry out its tasks automatically, it is by no means completely independent of the animal's voluntary control. For example, it is possible to decide to stop breathing for a short time. Humans and animals can also be trained to change their heart rates, blood pressures, and digestive reflexes voluntarily. However, any voluntary control that endangers life quickly disturbs homeostasis of the brain tissue, resulting in unconsciousness. Then the autonomic system takes over again and restores normal functions.

The sympathetic and parasympathetic divisions of the autonomic nervous system regulate most vital functions of the body and thus maintain the body's homeostasis.

28–I SENSE ORGANS AND THEIR FUNCTIONS

Sense organs permit an animal to detect changes in its own body and objects and events in the world around it. Information collected by sense organs is passed to the nervous system, which determines and initiates an appropriate response.

A popular expression holds that we have five senses. In fact, we have more than a dozen different

BIO-BIT

Chili peppers and other "hot" foods actually stimulate pain receptors in the mouth.

'IGURE 28-11

\n optical illusion. Each stripe in this series is of uniform larkness. However, the human visual system is set up in uch a way that it accentuates edges. Hence it misinterprets hese stripes, perceiving each one as having a brighter edge vhere it meets its darker neighbor.

ypes of sense organs, which monitor conditions both outside and inside our bodies. Internal sense organs letect and report changes in conditions such as body emperature, osmotic relationships, and pH. This provides information used to maintain homeostasis. Sense organs that report sights, sounds, and chemicals in the outside world are used in feeding, finding mates, avoiding enemies, and making other adaptive responses to the environment.

The key elements of sense organs are **receptor cells.** In some sense organs, the receptor consists merely of the dendrites of the sensory neuron. In others, the receptor is a separate, non-nervous cell close o the sensory neuron's dendrites (see A Journey Through Animal Senses). Either way, receptor cells respond to stimuli by producing electrical activity in the nervous system. A **stimulus** is some form of energy; various receptors respond to energy in the form of pressure, light, electric current, chemical changes, osmotic potential, and heat.

Each sense organ is particularly sensitive to one ype of stimulus. The eye, for example, responds to light but not to sound waves or chemicals in the air, which also strike the eye. Sense organs also function well over a very wide range of stimulus levels. Many sense organs are also incredibly sensitive to stimuli. The receptor cells called rods in our eyes can detect a single photon (a unit of light), and a male gypsy moth can detect a single molecule of the female's sexual attractant chemical that comes into contact with his antenna!

Most sense organs are geared to alert the nervous system to *changes* in stimuli. Our eyes send particularly strong signals to the brain when they detect movement, changes in light intensity (such as a sudden shadow or flash of light), and edges between light and dark areas (Figure 28-11). This is adaptive. Detecting movements of predators or prey, and outlines of objects in the surroundings, is more important to an animal than detecting a constant stimulus.

An animal responds to a stimulus in a three-step process:

1. **Receptor cells** are changed by the energy of a stimulus in such a way that the stimulus energy is converted into electrical activity, which can travel in the nervous system.
2. The **nervous system** processes the electrical impulses it receives from all of the sense organs. Eventually the resulting signals pass to appropriate effectors.
3. **Effectors** produce suitable responses. Effectors are usually muscles or glands. Muscles contract, or glands secrete chemicals, as a result of the information passed to the nervous system by the sense organs (Chapters 29 and 30).

"What about when I just sit and listen to music?" you may ask. "The music goes in (step 1) and the nervous system processes it (step 2), but no response occurs—I just enjoy." Not so; in fact, tiny muscles in your ears react to noises and adjust the ears' sensitivity, depending on the loudness of the music. In addition, depending on the type of music, your heart may beat faster or slower, and your other muscles may become more tense or relaxed. Even stopping the natural urge to tap your feet to the beat is a response!

(text continues on page 606)

BIO-BIT

Seventy percent of all of the sensory receptors in the human body are in the eye.

Mechanoreceptors

Lateral line of fish

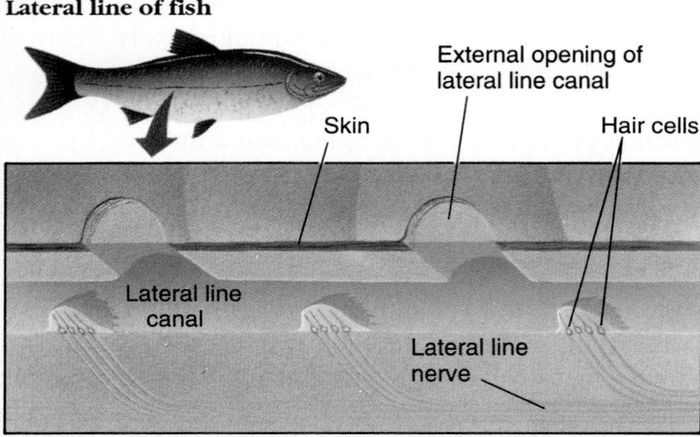

Both lateral line organs of fish and the inner ears of all vertebrates have mechanoreceptors called hair cells. Each of these remarkable cells has a tuft of large microvilli of graduated lengths. Bending this tuft appears to open gated ion channels, producing the electrical activity of a receptor potential in the hair cell. This electrical potential is transmitted to the dendrites of a sensory neuron near the base of the hair cell and on to the brain for further processing.

Lateral line organs detect such things as water currents, the movements of other animals in the water, and pressure waves bouncing off stationary objects nearby. Within the human inner ear the semicircular canals and utriculus, the organs of equilibrium, contain hair cells that detect both gravity and head movement, while a separate set of hair cells in the cochlea detect pressure waves in the fluid of this coiled tube. In humans, each pitch of sound produces a maximum vibration in hair cells located at one point along the cochlea. The brain "knows" the pitch of the sound because it knows the location of the sensory neurons that are firing in response to signals from hair cells.

Human ear

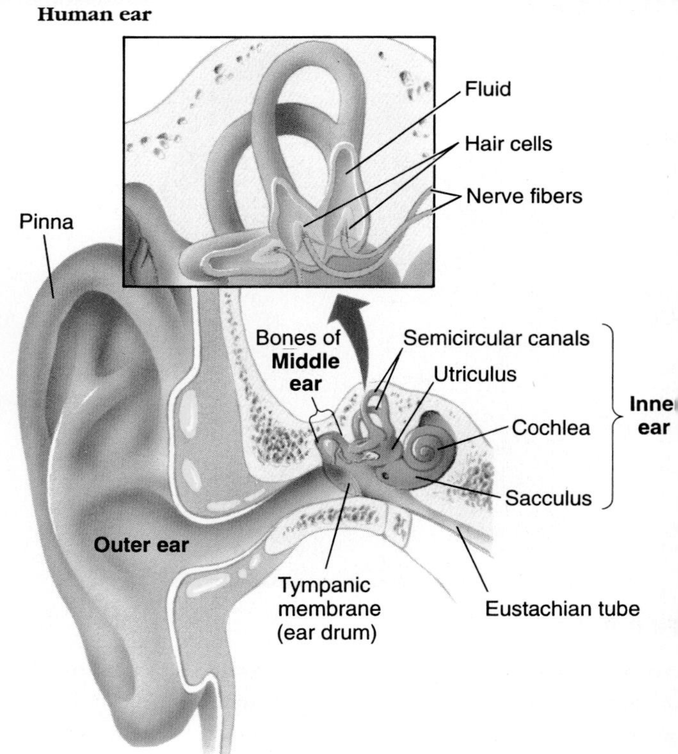

Chemoreceptors

Taste

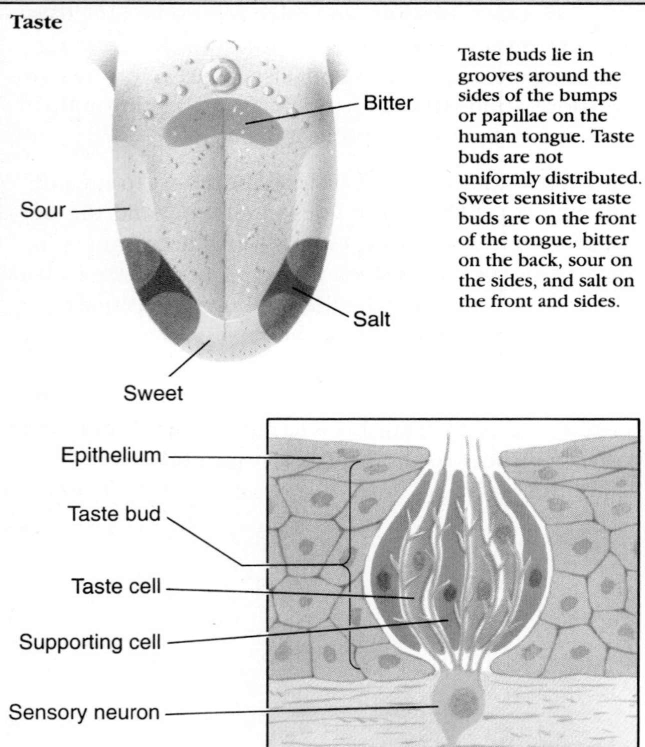

Taste buds lie in grooves around the sides of the bumps or papillae on the human tongue. Taste buds are not uniformly distributed. Sweet sensitive taste buds are on the front of the tongue, bitter on the back, sour on the sides, and salt on the front and sides.

Smell

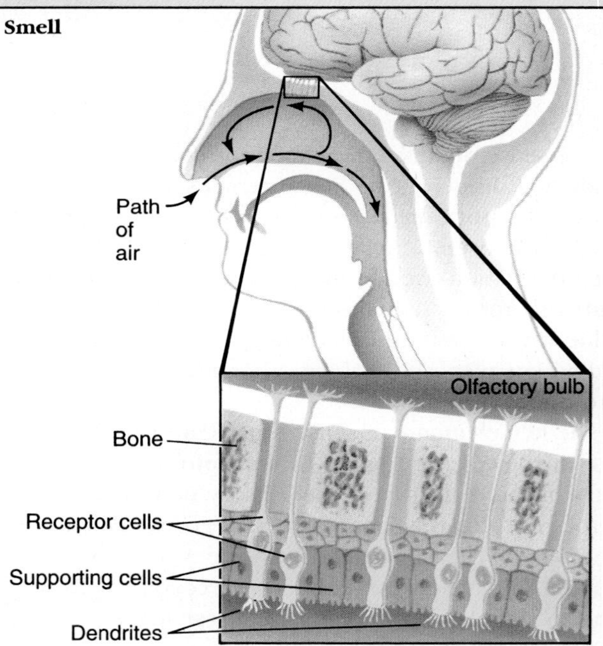

Olfactory receptors lie in the top of the nasal cavity. Sniffing moves air more rapidly into contact with this small patch of tissue that is packed with dendrites of neurons that connect with the nearby olfactory lobe of the brain.

A Journey Through Animal Senses. (N. A. S./M. W. F. Tweedie)

Photoreceptors

Human vision

The cornea and lens focus light (black lines) on the retina that contains the rods and cones that detect light. These send messages to the brain via neurons that lead to the optic nerve. A layer of black pigment surrounds the eye and absorbs light that has passed through the retina, preventing it from reflecting back to the rods and cones. The choroid layer contains blood vessels and the ciliary muscles adjust the curvature of the lens.

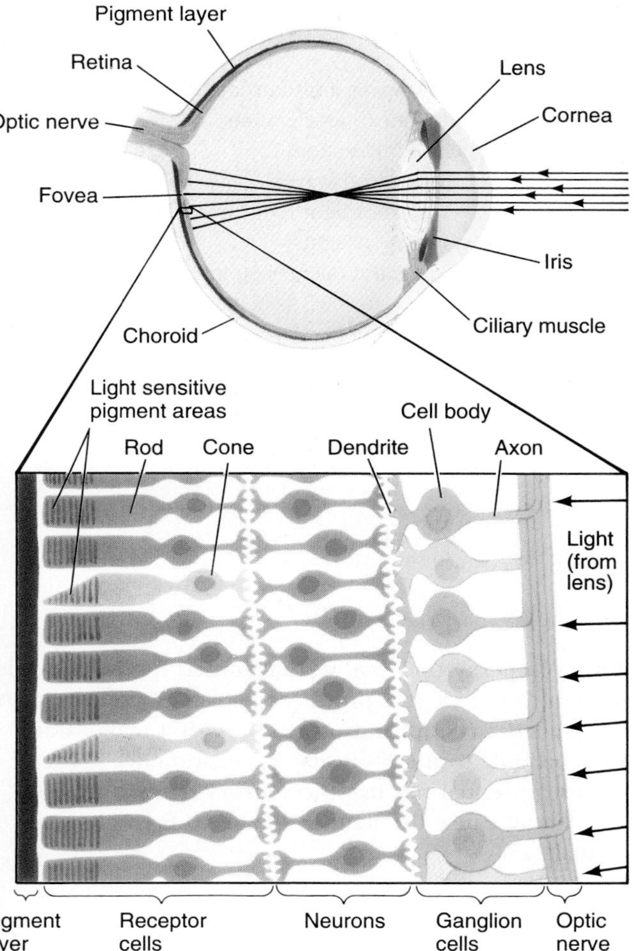

The human eye sees a marigold with no markings. (D. Kuhn)

Insect vision

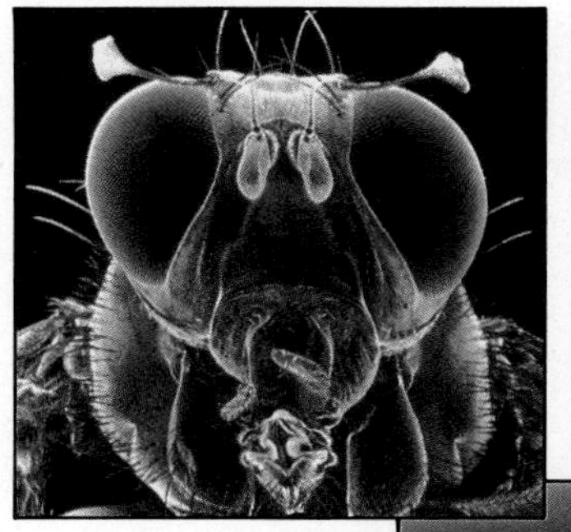

Insects have compound eyes made up of many ommatidia. Each ommatidium has a cornea, lens, and a light sensitive receptor surrounded by retinal cells that transmit the sensory stimulus. Pigment cells around each ommatidium prevent light from passing into surrounding ommatidia. Compound eyes are especially sensitive to movement because each ommatidium responds independently. (D. Scharf)

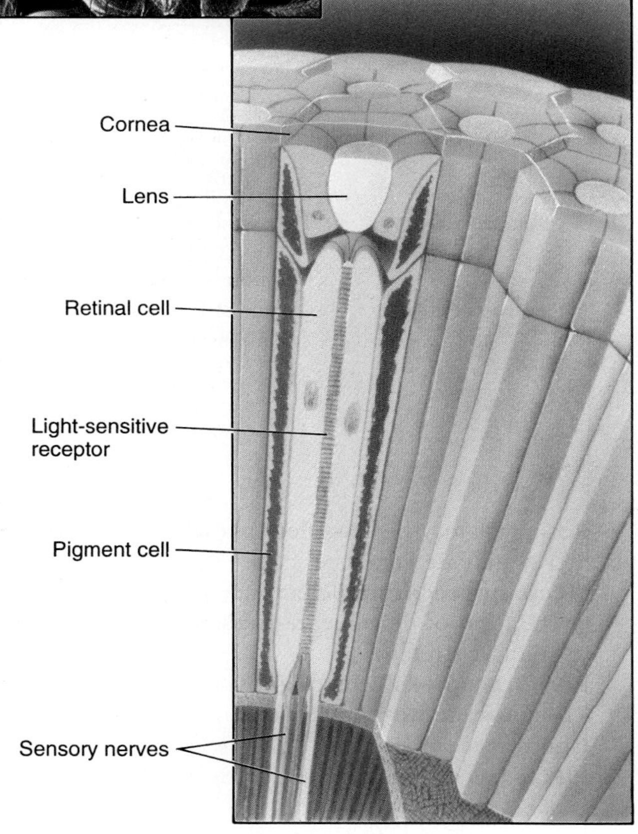

An insect sees big patches on the same flowers (here taken with ultraviolet sensitive film) because its eyes react to ultraviolet light energy, whereas ours do not. (D. Kuhn)

Table 28–2
Classification of Some Receptors by Stimuli

General name	Examples	Effective stimulus
Mechanoreceptors	Meissner's corpuscles	Touch on skin
	Proprioceptors	Position of parts of body
	Hair cells of vertebrates	
	Lateral line organs in fish	Pressure waves and currents in water
	Inner ear	Gravity; linear acceleration
		Angular acceleration
		Airborne sound waves
Photoreceptors	Rods and cones of vertebrate retina	Wavelength of light
	Ommatidia of arthropods	Wavelength of light
Thermoreceptors	Pit organs of pit vipers; nerve endings in the tongue and skin of mammals	Increasing and decreasing infrared radiation
	Krause's end bulbs	Cold on skin
Chemoreceptors	Taste buds and olfactory organs of vertebrates	Unknown features of the chemistry of specific molecules in air or water
	Chemoreceptors of invertebrates	Same as above
Electroreceptors	Organs in the skin of some fish; bill of platypus	Electric currents in surrounding water

Receptors may be classified by the form of stimulus energy they detect:

Mechanoreceptors detect mechanical energy in the form of movement, pressure, or tension (pull or stretch).
Photoreceptors detect light energy.
Thermoreceptors detect heat.
Chemoreceptors detect chemicals.
Electroreceptors detect electrical energy (Table 28–2).

A Journey Through the Senses details the anatomy of the major kinds of receptor cells. Many animals, such as pigeons, dolphins, and honeybees, can also respond to the Earth's magnetic field.

A receptor's most important role is to convert the energy of a stimulus into electrical energy, the only form of energy that can be transmitted by the nervous system. The receptor produces a local electrical potential, called a **receptor potential,** with a magnitude proportional to the intensity of the stimulus. If the receptor potential reaches the sensory neuron's threshold, it will cause the neuron's axon to fire action potentials.

Sensory receptors consist of either neurons or non-nervous cells that are particularly sensitive to one type of stimulus. Receptors absorb stimulus energy and convert it to electrical energy, the form "understood" by the nervous system. The nervous system processes the sensory information and responds through effectors that produce the suitable response.

SUMMARY

1. Information passes along a neuron in the form of fast-moving electrical changes across the plasma membrane.
2. To transmit its messages, a neuron uses the stored energy of its resting membrane potential, the result of a relatively high external concentration of sodium ions (Na^+), a relatively high internal concentration of potassium ions (K^+), and a high internal concentration of various negatively charged molecules.
3. A stimulus applied to the neuron changes its membrane's resting potential briefly by opening sodium ion channels through the membrane. This allows sodium ions to cross the membrane and depolarize that part of the membrane. Such small local changes in the membrane potential in the dendrites or cell body may add up until they exceed the threshold needed for the axon to produce an action potential.
4. An axon transmits action potentials faster if it has a large diameter or if it is electrically insulated by a myelin sheath.
5. When an action potential reaches a synapse, its message is usually transmitted as a chemical that crosses the synaptic cleft and changes the membrane potential of the next cell(s) in line.
6. During the evolution of the nervous system, progressive cephalization resulted in the formation of a brain and major sense organs at the front end of the body. A vast increase in the number of neurons permitted better control of the many muscles in a large body as well as more flexibility of response to stimuli in a complex and changing environment.
7. The nervous system of a vertebrate consists of the brain and spinal cord, which together compose its central nervous system, and the peripheral nervous system in the rest of the body.
8. The vertebrate brain has three major parts: the forebrain, midbrain, and hindbrain, whose functions are summarized in Table 28–1.
9. During evolution, the vertebrate brain increased in size and complexity. Some parts of the brain retained their primitive functions, while others took on new functions as body structure and behavior became more complex, and as intelligence increased.
10. The brain's main role is to "make decisions." Using information coded as patterns of action potentials coming from the external or internal environment via the sense organs, the brain produces a set of directions coded as another set of action potentials that cause the effector organs to respond.
11. Information passes through various levels of organization in both sensory and motor areas as the brain analyzes and integrates sensory input and determines and executes appropriate responses.
12. The spinal cord is primarily a relay station connecting the brain with the peripheral nervous system, although there are spinal reflexes in which sensory and motor components interact through the spinal cord without input from the brain.
13. Reflexes make quick, local responses to potentially harmful stimuli without waiting for analysis by higher levels in the brain.
14. The peripheral nervous system is divided into the somatic nervous system, which largely serves the muscles under conscious control, and the autonomic nervous system, which carries motor impulses to the muscles and glands of the internal organs, under little conscious control.
15. An animal must be able to detect changes inside its body and in the world around it in order to maintain homeostasis and to produce suitable behavior.
16. Receptors are collections of cells specialized to react to particular forms of energy by producing electrical impulses in the nervous system.

SELF-QUIZ

1. Impulses leave a neuron via the:
   a. dendrites
   b. nucleus
   c. myelin sheath
   d. axon
   e. cell body
2. The myelin sheath around the axons of some vertebrate neurons:
   a. is rich in lipids because it is formed by many layers of membranes
   b. is a secretory product of glial cells
   c. is produced inside the axon and extruded out through the membrane
   d. is continuous all along the length of the axon
   e. secretes neurotransmitter substances for release at the axon terminals
3. Write a short sentence describing the function and importance of each of the following components of a synapse:
   a. neurotransmitter substance
   b. neurotransmitter vesicle
   c. receptor molecules
   d. enzymes that destroy neurotransmitters
4. In an inhibitory synapse:
   a. information travels from the postsynaptic to the presynaptic cell
   b. the neurotransmitter used is different from the neurotransmitter used in excitatory synapses
   c. there are no receptor molecules on the postsynaptic membrane

d. the postsynaptic receptor molecules may cause hyperpolarization of the membrane rather than depolarization when they bind neurotransmitters
e. the stimulus is transmitted electrically, not chemically

5. Regulatory control of deep body temperature, osmoregulation, thirst, and hunger occurs in the:
a. anterior colliculi
b. hypothalamus
c. thalamus
d. cerebellum
e. tectum

6. In looking at the evolution of the brain from fishes to humans, the greatest increases in size are seen in the:
a. medulla and cerebellum
b. cerebellum and tectum
c. tectum and cerebral hemispheres
d. cerebral hemispheres and cerebellum
e. medulla and thalamus

7. Label the parts of the reflex arc indicated by letters A through E.

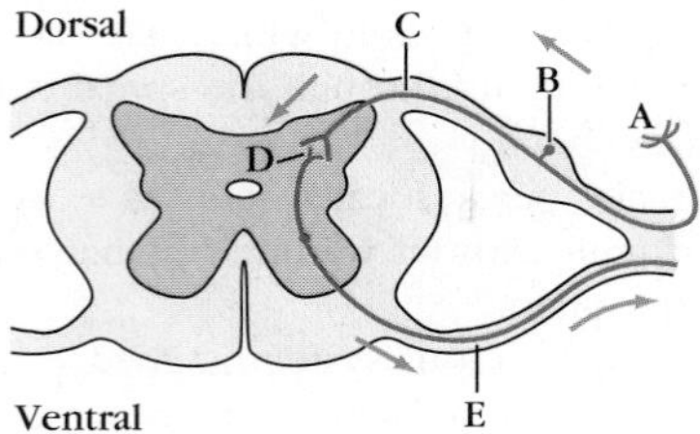

8. The parasympathetic nervous system:
a. uses norepinephrine as the neurotransmitter at the synapse with the effector
b. is part of the central nervous system, located in the spinal cord
c. tends to lower blood pressure, slow the heartbeat rate and breathing, and promote digestion
d. predominates over the sympathetic system in times of stress
e. is composed of sensory neurons running parallel to the motor neurons of the sympathetic system

9. The major transmitter between neuron and effector in the sympathetic system is ___, while in the parasympathetic system it is ___.

Match the type of receptor on the right to its description on the left.

___ 10. detect movement, pressure, & tension	a. electroreceptors
___ 11. detect light	b. photoreceptors
___ 12. detect chemicals	c. mechanoreceptors
___ 13. detect electrical energy	d. thermoreceptors
___ 14. detect heat	e. chemoreceptors

THINKING CRITICALLY

1. After an action potential arrives at the presynaptic membrane, what factors determine whether the postsynaptic cell will fire an action potential?
2. Why don't all the neurons in an animal's nervous system have either giant or myelinated axons? What is the selective advantage of having some neurons with axons that conduct impulses more slowly?
3. Tetrodotoxin is a chemical produced by puffer fish that interferes with the opening of sodium gates involved in transmission of an action potential. It has no effect on chemical synapses. Why is tetrodotoxin a powerful poison?
4. Multiple sclerosis (MS) is characterized by patchy destruction of myelin. What symptoms would you expect this to produce?
5. What possible disadvantages are there in the evolutionary trend toward cephalization of the nervous system? What advantages?
6. During embryonic development of mammals, the nervous system forms all the neurons an animal will ever have. After birth, neurons never divide to form new neurons. However, synapses do form and decay. What is the adaptive advantage of this state of affairs? What are some drawbacks?
7. Thermoreceptors of mammals are particularly concentrated on the tongue. These receptors keep human beings from burning their mouths with hot food, but cooking is a recent (in evolutionary terms) invention, practiced only by humans. What is the advantage to a wild mammal of having so many thermoreceptors on the tongue?

SELECTED KEY TERMS

action potential, *p. 586*
autonomic nervous system, *p. 595*
axon, *p. 584*
cerebellum, *p. 597*
cerebral cortex, *p. 598*
central nervous system, *p. 594*
cephalization, *p. 593*
cerebrum, *p. 598*
chemoreceptor, *p. 606*
cranial nerve, *p. 602*
dendrite, *p. 584*
depolarized, *p. 584*
diencephalon, *p. 597*
electroreceptor, *p. 606*
ganglia, *p. 593*
gray matter, *p. 589*
hippocampus, *p. 598*
hypothalamus, *p. 598*
local potential, *p. 586*
mechanoreceptor, *p. 606*
medulla, *p. 596*
meninges, *p. 595*
motor neuron, *p. 583*
myelin sheath, *p. 589*
neuroglial cell, *p. 584*
neuromuscular junction, *p. 592*
neurons, *p. 583*
neurotransmitters, *p. 589*
parasympathetic nervous system, *p. 602*
peripheral nervous system, *p. 595*
photoreceptor, *p. 606*
receptor cells, *p. 603*
reflex, *p. 596*
resting potential, *p. 584*
saltatory conduction, *p. 589*
sensory neuron, *p. 583*
sodium-potassium pump, *p. 584*
somatic nervous system, *p. 595*
spinal nerve, *p. 602*
sympathetic nervous system, *p. 602*
synapse, *p. 589*
telencephalon, *p. 598*
thalamus, *p. 598*
thermoreceptor, *p. 606*
white matter, *p. 589*

SUGGESTED READINGS

Adrian, R. H. "The nerve impulse." *Carolina Biology Reader.* Burlington, NC: Carolina Biological Supply Company, 1980.

Axelrod, J. "Neurotransmitters." *Scientific American,* June 1974. Covers neurotransmitters and aspects of drugs affecting the nervous system.

Franklin, J. *Molecules of the Mind.* New York: Atheneum, 1987. A thought-provoking look at the neurochemical revolution, mental illness, and society's attitudes towards both.

Julien, R. M. *A Primer of Drug Action,* 5th ed. New York, W H. Freeman, 1988.

Konishi, M. "Listening with two ears." *Scientific American,* April 1993.

Lent, C. M., and M. H. Dickinson. "The neurobiology of feeding in leeches." *Scientific American,* June 1988. An interesting and readable case study of the nervous system's role in a vital activity. Outlines experiments that detected the nervous pathway responsible for feeding.

Mishkin, M., and T Appenzeller. "The anatomy of memory." *Scientific American,* June 1987.

Scientific American. "The Brain." September 1992 issue. A collection of articles about neurons and brain structure and function.

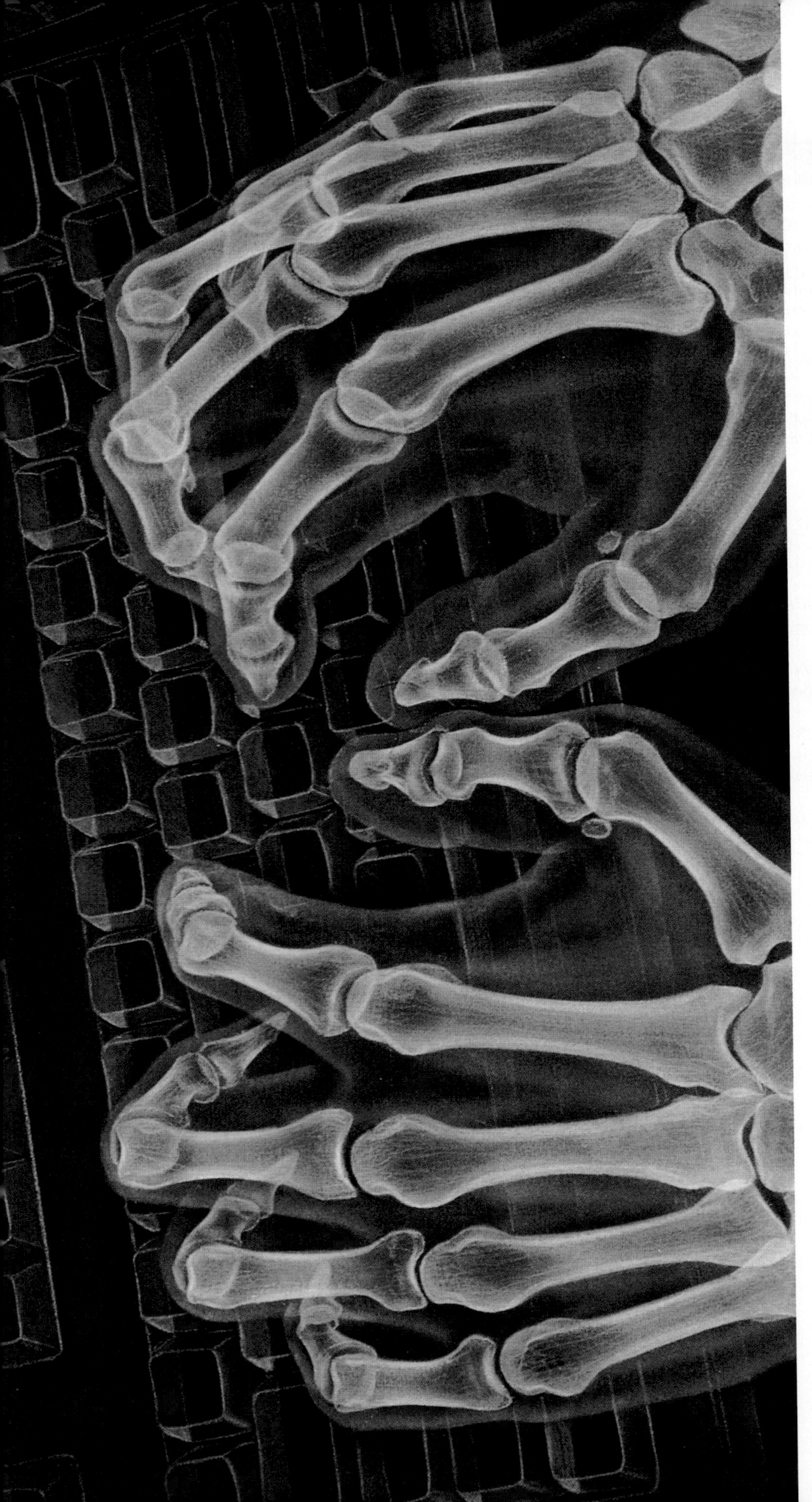

CHAPTER 29

CONCEPT GUIDE

After reading this chapter, you should be able to:

1. Compare the locations, structures, and functions of the three types of muscle found in humans. (Section 29-A)
2. Explain the sliding filament mechanism of muscle contraction. (Section 29-B)
3. Describe the advantages of having muscles arranged in antagonistic pairs. (Section 29-C)
4. Describe the major functions of the vertebrate skeleton. (Sections 29-D and 29-E)
5. Compare the structure and properties of cartilage and bone. (Section 29-E)

Nerves, muscles, and bones work together to allow delicate, highly coordinated actions like typing. Read more about the function and control of the muscular system in Section 29-B. (T. Buck/Custom Medical Stock Photo)

Muscles and Skeletons

An animal detects things with its sense organs, processes the information in its nervous system, and reacts with its **effectors:** organs that do things in response to signals from the nervous system. Muscles and glands are the most widespread effectors, but electric organs that allow electric eels to stun their prey and light-emitting organs of a variety of animals from lightning bugs to angler fishes also respond to information gathered by sense organs. Glands are considered in Chapter 30. In this chapter, we concentrate on muscles.

Muscles react to signals from motor neurons by contracting. Muscle contractions may move part of the body, or even the whole body during locomotion. Muscles also play important roles in maintaining the body's homeostasis, by circulating blood, pushing food down the digestive tube, and so on. However, muscle contraction does not always result in movement. Sometimes it maintains the status quo, as when a scallop contracts the muscle that keeps its shell closed and a would-be predator out. In all these cases, muscle contraction makes adaptive responses to conditions in the animal's internal or external environment.

The force of muscle contraction cannot do useful work without something to pull or push against: the scallop's shell, the bones of the human body, blood in the heart, or food in the stomach, for instance. An animal's skeleton provides places for its muscles to attach, and a system of levers and pivots for them to pull against, permitting them to perform useful work in moving body parts or in locomotion. Even the muscles of the internal organs, which are not attached to the skeleton, have more or less firm attachments to other structures, which serve the same sort of "skeletal" functions.

The rigidity of a skeleton also provides support for the body and protects delicate parts such as the brain and other soft organs.

KEY CONCEPTS

- Actions of animal's effectors—most commonly muscles and glands—help the animal to maintain homeostasis or make suitable responses to events in the environment.
- Whether executing a motion or holding something steady, muscles and skeletons work together: the muscle provides a force, and the skeleton provides an object for the force to work against.
- In addition to their role in movement, skeletons provide support and protection to other organs.

Curiosity Questions

1? What causes our hearts to beat? (See page 612)

2? Why do the muscles of a dead animal become locked in *rigor mortis?* (See page 615)

3? What is osteoporosis? What causes this disease? (See *A Journey into Healthy Living:* Preventing Osteoporosis, page 625)

29–A MUSCLE TISSUE

Muscle tissue has two distinguishing properties: contractility and electrical excitability. Excitability is also a characteristic of neurons (Chapter 28). In both muscle and nervous tissue, electrical excitability is due to the energy stored in an electrical potential difference across the membrane. The excitable muscle membrane depolarizes in response to a chemical transmitter released at a **neuromuscular junction,** the synapse between a motor neuron and a muscle. This excitation in the membrane initiates contraction in the muscle. Muscle cells are packed with contractile proteins, which interact in such a way as to cause the cell to grow shorter.

Vertebrates have three types of muscle (Table 29–1):

Smooth muscle lines the walls of blood vessels and many of the internal organs.

Cardiac muscle makes up the heart.

Skeletal, or **striated muscle** is responsible for locomotion and change of position. It is generally the only type of muscle under the animal's voluntary control.

Smooth Muscle

Smooth muscle is made up of sheets of individual muscle cells. It forms the muscular layer in the walls of many internal organs, including the walls of the arteries, veins, digestive tract, urinary bladder, and reproductive organs. Smooth muscle usually exerts pressure on the contents of these organs: it constricts blood vessels, moves food along the gut, and expels urine from the bladder, semen from the seminal vesicles, or a baby from the uterus.

Some smooth muscles contract spontaneously, but others must first be stimulated by nerves or hormones. Most smooth muscle receives signals from motor neurons in the sympathetic part of the autonomic nervous system, which use the neurotransmitter norepinephrine (Section 28–H). These muscles also contract when a very similar chemical, the hormone epinephrine, arrives from the adrenal glands via the bloodstream. The parasympathetic system also sends signals to some smooth muscles, using acetylcholine as a transmitter.

Smooth muscle produces a gradual contraction of variable force. A smooth muscle cell's electrical activity is not an all-or-none event but a graded response, and how much the cell contracts depends on how much its membrane is depolarized. Smooth muscle contraction often lasts for a long time without fatigue.

Cardiac Muscle

Cardiac muscle occurs only in the vertebrate heart. Here, the individual muscle cells are arranged in long columns of fibers.

Cardiac muscle illustrates the similarity between nerve and muscle tissue. Some cells in the heart are so specialized to conduct electrical impulses that they have lost their contractile proteins and behave much more like neurons than like muscle cells. These cells are part of the mechanism that controls the heartbeat.

1? Each heartbeat is initiated by spontaneous electrical activity of the heart's pacemaker, the **sinoatrial node** in the wall of the right atrium (Figure 29–1). The pacemaker cells fire in their rhythmic pattern because of a special feature: they are unusually leaky to sodium ions and so they rapidly become depolarized and reach the threshold to fire an action potential. This resets them, and the process repeats itself. ■

An impulse generated at the sinoatrial node spreads to all parts of the atria, and then to the **atrioventricular node,** on the base of the right atrium near the partition between the two ventricles. From here the impu1se spreads rapidly through the ventricular walls, triggering simultaneous contraction throughout the ventricle.

The electrical impulse travels by way of the membranes of the heart muscle cells themselves, and it passes between these cells electrically, not by way of chemical transmitters. The cells are specialized so that their electrical transmission is particularly fast. They are joined together by structures in their plasma membranes called **gap junctions** (Section 4–G). These tiny tunnels permit ions, and therefore electric current, to pass from one cell to its neighbor rapidly, with very little electrical resistance.

The main nerve to the heart is the tenth cranial nerve, the **vagus.** It contains nerves from the parasympathetic system that slow the heart rate, while nerves from the sympathetic system speed up the heart. However, if all of the nerves from the nervous system to the heart are removed, a vertebrate can survive in an apparently normal condition, and its heartbeat alters with the changing demands of exercise just as it does in an intact animal. The rate and force of heartbeats are governed partly by hormones that reach the heart through the bloodstream, and partly by reflexes in a system of nerves that lie completely within the heart and work even though they are not in contact with the rest of the nervous system.

Cardiac muscle must clearly be highly resistant to fatigue if it is to contract throughout an animal's life-

Table 29–1
Distribution and Functions of Muscular Tissues and Organs in the Human Body

Function	Structures	Muscle Type
Circulation	Heart	Cardiac
	Walls of arteries, veins	Smooth
Excretion	Walls of renal pelvis, ureter, bladder	Smooth
	Internal sphincter between bladder and urethra	Smooth
	External sphincter near exit from body	Striated
Digestion	Tongue, muscles of jaw and pharynx	Striated
	Walls of esophagus, stomach, intestines	Smooth
	Internal sphincter of anus	Smooth
	External sphincter of anus	Striated
Ventilation (Breathing)	Diaphragm, intercostal muscles	Striated
Ejaculation	Walls of genital ducts	Smooth
	Skeletal muscles at base of penis	Striated
Parturition (childbirth)	Uterine wall, cervix	Smooth
	Abdominal muscles, diaphragm	Striated
Heat production	Skeletal muscles (exercise, shivering)	Striated
Maintenance of posture Change of position Locomotion	Skeletal muscles	Striated

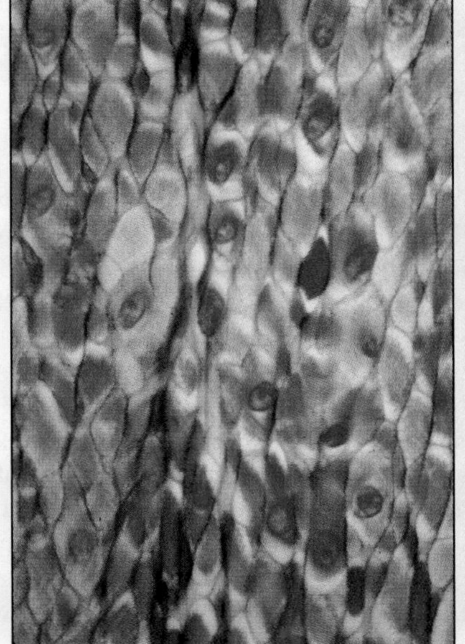
(a) Smooth muscle

(b) Cardiac muscle

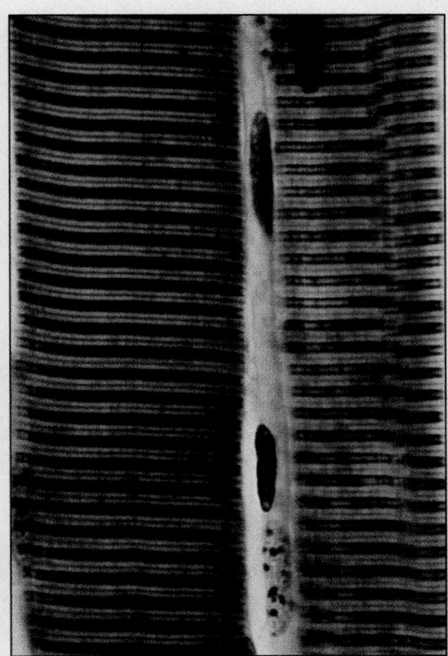
(c) Striated or skeletal muscle

(a, Biophoto Associates)

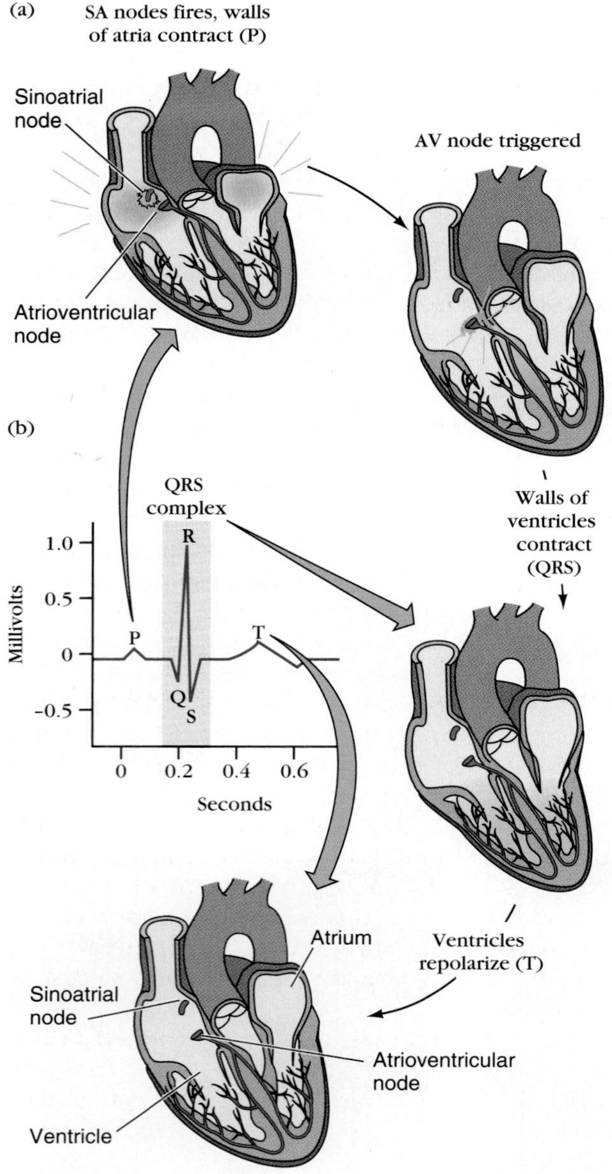

FIGURE 29-1

The electrical conduction system of the human heart. (a) The sinoatrial node, located on the right atrium of the heart, initiates each heartbeat (contraction). The impulse from the sinoatrial node spreads throughout the walls of the atria and to the atrioventricular node, located near the atria on the partition between the two ventricles. From here, the impulse spreads rapidly and triggers a simultaneous contraction throughout the walls of both ventricles. **(b)** An electrocardiogram (EKG). The electrical currents that cause the heartbeat are also conducted throughout the body in the body fluids and so they can be monitored by electrodes attached to the skin. An EKG is used to determine whether the heart's electrical activity is normal. The P wave is produced by depolarization of the atria; the Q, R, and S waves result from depolarization of the ventricles. The T wave is a result of ventricular repolarization.

time. Not surprisingly, cardiac muscle has abundant mitochondria (producing ATP which supplies energy for muscle contraction), and its comparatively large blood supply guarantees adequate oxygen.

Skeletal Muscle

The muscles that attach to, and move, the skeleton of a vertebrate are called skeletal, or striated (striped), or voluntary muscles. Most of the research on vertebrate muscle has been done on this readily available tissue.

A skeletal muscle is made up of many **muscle fibers,** each running the entire length of the muscle. Each skeletal muscle fiber is a single cell that has arisen from the fusion of many embryonic muscle cells and contains all of their nuclei. Their plasma membranes coalesce to form a continuous membrane, the **sarcolemma,** around the whole muscle fiber. The many nuclei lie along the outside of the fiber, just under the sarcolemma (see *A Journey Through Human Skeletal Muscle*). The axon terminals of a single motor neuron may form neuromuscular junctions with several to over a hundred muscle fibers.

A muscle fiber consists of a bundle of **myofibrils.** Each myofibril is made up of units called **sarcomeres** arranged end to end along its length. A sarcomere is part of the myofibril between two adjacent **Z lines,** protein-containing structures extending across the myofibril. Each sarcomere contains a well-developed cytoskeleton, consisting of a precise arrangement of two kinds of protein filaments. Attached to each side of a Z line are **thin filaments,** which extend less than halfway to the center of the sarcomere. Their free ends overlap **thick filaments,** which are centered in the sarcomere. Viewed with a microscope, the Z lines and the areas where thick and thin filaments overlap appear dark. The sarcomeres of neighboring myofibrils are lined up side by side, so that their light and dark areas produce a visible, striped pattern extending across the muscle fiber.

Electrical excitation of muscle cell membranes initiates muscular contractions. Vertebrates have three muscle types, distinguished by differences in their structure, function, and location. Smooth muscle is found in the walls of many body organs and blood vessels, is organized into broad sheets, and provides slow but prolonged contractile forces. Cardiac muscle is found in the heart, occurs in long columns of fibers, and uses quick but strong contractions to propel blood throughout the body. Skeletal muscle provides most of the strength needed to support and move the skeleton and is composed of long, thin, multinuclear cells that extend the full length of the muscle. The rate and strength of smooth and cardiac muscle contraction are controlled by hormones and autonomic nerves, while skeletal muscle contraction is controlled by voluntary motor nerves.

29–B MUSCLE CONTRACTION

The contraction of a skeletal muscle fiber results from the movement of the protein filaments of the cytoskeleton in the sarcomeres. To better understand how muscles contract, refer to A Journey Through Human Skeletal Muscle as you read this section.

When a sketetal muscle fiber contracts, each sarcomere's protein filaments slide along each other and mesh together more closely. In this **sliding filament mechanism** of muscle contraction, the filaments stay the same length, but the free ends of the thin filaments move closer to the center of the sarcomere. Since the other ends of the thin filaments are attached to the Z lines, this sliding moves the Z lines of each sarcomere closer together, shortening the sarcomeres, and therefore the whole muscle. Contraction does not make a muscle smaller. Rather, it gets shorter and thicker, but its volume remains the same.

Each thin filament is made up of many molecules of the globular protein **actin,** with smaller amounts of the proteins **troponin** and **tropomyosin**. A thick filament contains hundreds of molecules of the protein **myosin,** each shaped like a golf club with a double head.

How do these filaments slide past each other? The heads of the myosin molecules in the thick filaments form cross-bridges to the actin in the thin filaments. The cross-bridges swivel, pushing the thin filaments toward the center of the sarcomere. The cross-bridges then detach, reattach farther along the thin filaments, and push them still farther toward the center of the sarcomere. As the myosin heads complete one swiveling cycle, the sarcomere is shortened by about 1% of its original length. Because there are many cross-bridges and they do not all move at the same instant, the thin filament cannot slip back while individual myosin heads are detached.

Muscle contraction is powered by ATP, produced mainly in the muscle's many mitochondria. ATP binds to the heads of myosin molecules and is broken down to provide energy for the swiveling of the cross-bridges. This process is very efficient, wasting only 30 to 50% of the energy from ATP as heat. In contrast, car engines waste 80 to 90% of the energy available from gasoline.

Curiously, muscles need ATP to relax as well as to contract. Myosin heads cannot detach from the thin filaments until they have bound new ATP molecules. When an animal dies, its muscles soon run out of ATP and lose the ability to contract or relax. They become rigidly locked in whatever position they occupied when the ATP was used up, a phenomenon called **rigor mortis.** ■

Control of Contraction

A muscle usually contains ATP, and yet it contracts only some of the time. Between contractions, the protein tropomyosin blocks the actin molecules' binding sites for myosin heads. As a result, cross-bridges cannot form and the myosin heads cannot swivel and cause contraction.

A muscle fiber contracts only when calcium ions are present to unblock the actin binding sites. Calcium ions (Ca^{2+}) are stored inside the membranes of the endoplasmic reticulum, which in skeletal muscle fibers is called the **sarcoplasmic reticulum.** Calcium leaves the sarcoplasmic reticulum when a muscle fiber is activated and is pumped back in when it relaxes. A muscle fiber contracts if its cytoplasm contains enough Ca^{2+} and ATP.

A skeletal muscle fiber is normally activated by the arrival of an action potential in the axon of a motor neuron. The axon terminals release the neurotransmitter acetylcholine, which crosses the neuromuscular junction, binds to receptors on the sarcolemma, and causes the sarcolemma to depolarize. An electrical impulse spreads through the sarcolemma, which has extensions that form an internal membrane system (the T tubules) that is in intimate contact with the sarcoplasmic reticulum. The electrical signal causes opening of calcium channels in the sarcoplasmic reticulum, and this releases calcium ions into the cytoplasm. Here they bind to troponin. This causes tropomyosin to move, exposing the actin binding sites for myosin. Myosin heads attach to the actin and swivel, so that the muscle contracts.

Calcium is continually pumped back into the reservoirs of the sarcoplasmic reticulum by active transport. When most of the calcium ions have left the troponin, tropomyosin moves out and blocks the binding sites on the actin again, and the fiber relaxes.

Tetanization and Fatigue

A single nerve impulse arriving at the neuromuscular junction of a mammalian skeletal muscle fiber causes an all-or-nothing muscle twitch, a swift contraction followed immediately by relaxation. The electrical events of depolarization and restoration of the resting potential in the sarcolemma take a short time compared to the mechanical events of contraction and relaxation (see Figure 29–1).

The smooth, sustained muscle contractions that allow you to carry a cup of coffee across the room without mishap result when the nervous system sends continuous trains of nerve impulses through the motor neurons. Now muscle fibers receive new electrical stimulations before they have relaxed from the previous impulse, and the individual muscle twitches fuse

Skeletal Muscle Structure

The muscle is made up of bundles of muscle fibers. In each fiber, many myofibrils and nuclei are enlcosed within a common sarcolemma. Each myofibril contains numerous sarcomeres (the area between one Z line and the next, see below), arranged end to end in single file.

The sarcoplasmic reticulum is a reservoir containing calcium ions (Ca^{2+}). Calcium leaves the sarcoplasmic reticulum when a muscle fiber is activated and is pumped back in when it relaxes. A muscle fiber contracts if its cytoplasm contains enough Ca^{2+} and ATP.

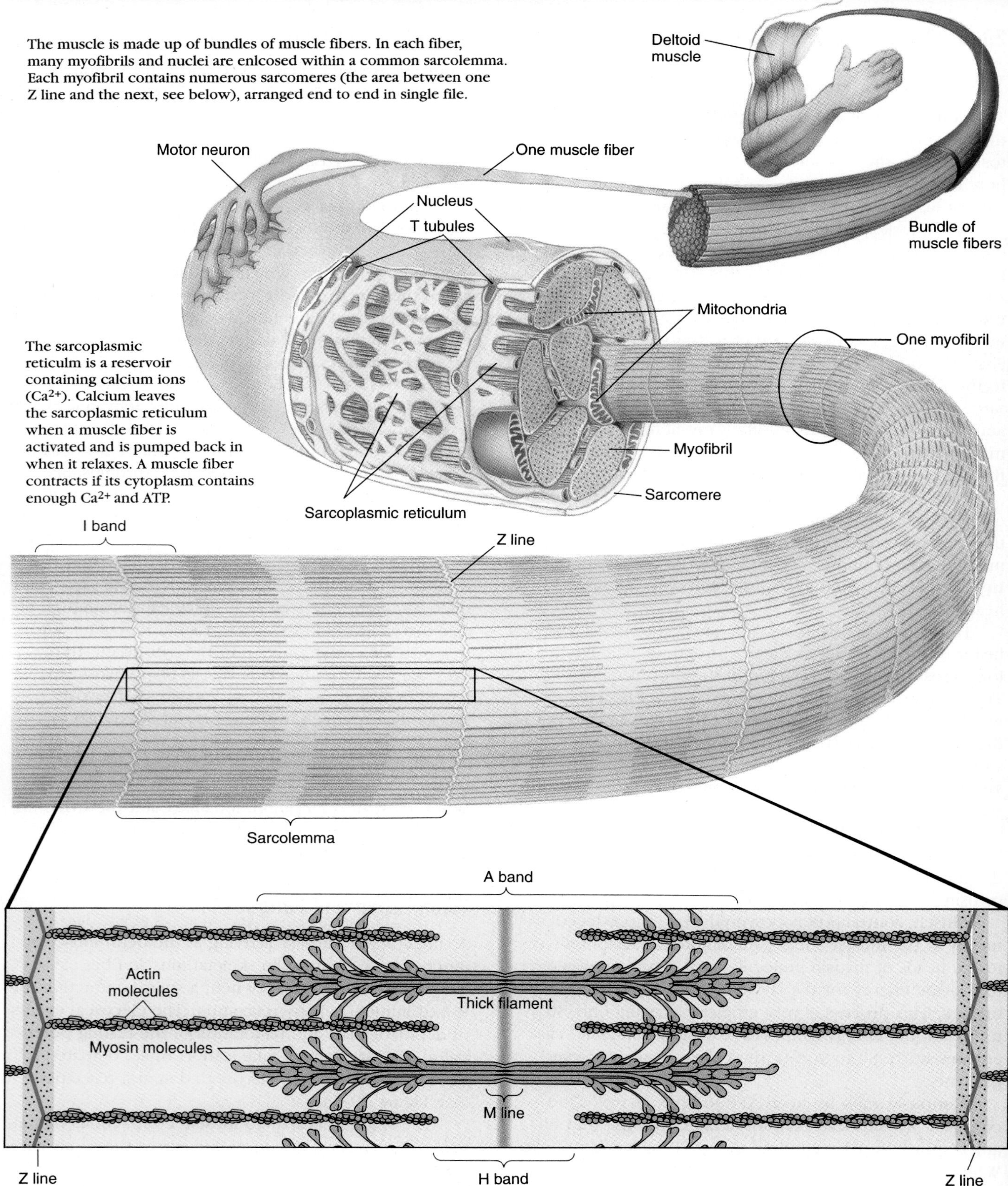

Muscle Contraction

An electron micrograph of part of one skeletal myofibril, showing one complete sarcomere. Sarcomeres are the contractile units of the fibril. Compare this photograph to the diagram at the bottom of the previous page.

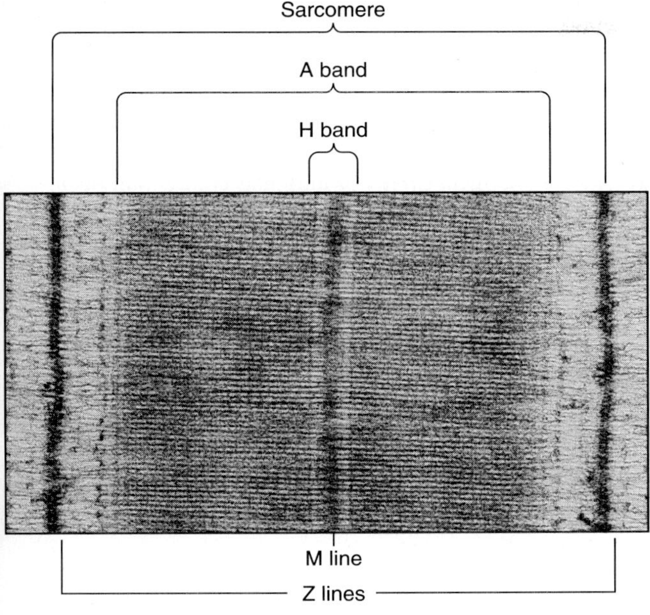

The sliding filament model of muscle contraction

Diagram of the filaments of contractile proteins responsible for the striped appearance of the myofibril. The thin filaments are attached to the Z lines. The thick filaments lie between the thin filaments.

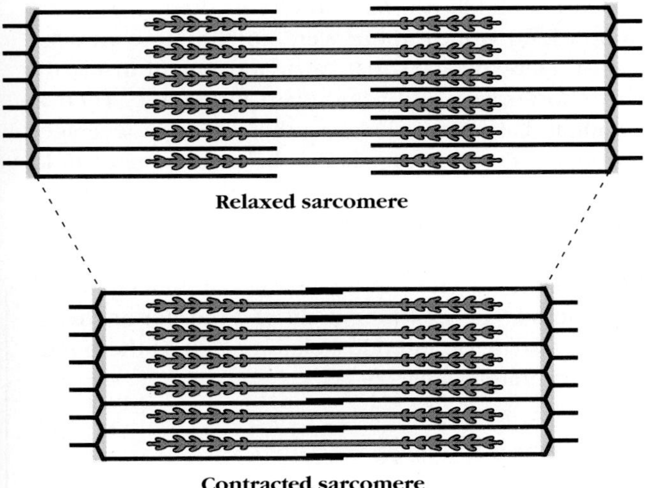

During contraction, the thick and thin filaments slide past each other, reducing the distance between adjacent Z lines. The filaments themselves do not shorten. The coordinated contraction of all of the sarcomeres in a myofibril results in the shortening of the myofibril and thus the muscle.

How filaments slide during muscle contraction attachment

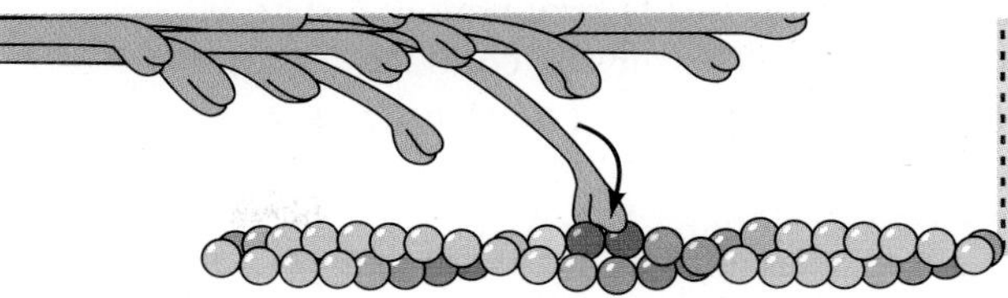

The head of the myosin molecule in a thick filament attaches to an actin molecule in a thin filament, forming a cross-bridge.

Power stroke

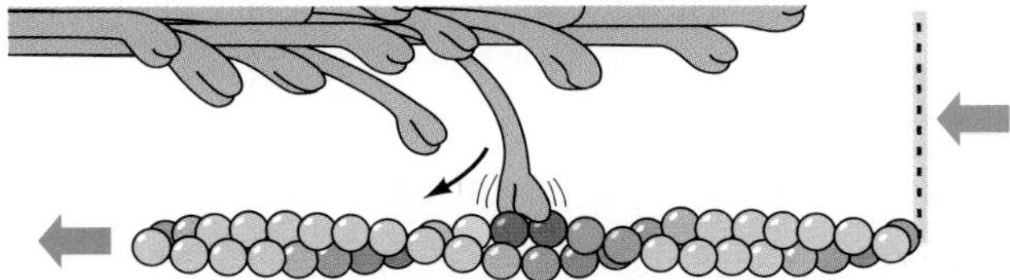

The cross-bridge swivels and pushes the thin filaments toward the center of the sarcomere.

Release and reattachment

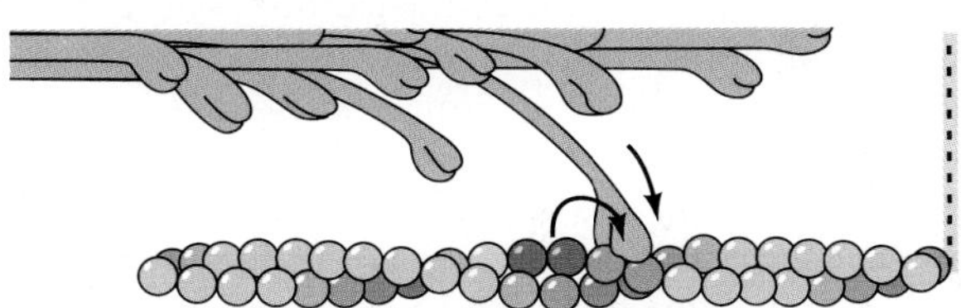

The cross-bridge then detaches and reattaches farther along the thin filament and the above steps are repeated.

How muscles sustain a contraction

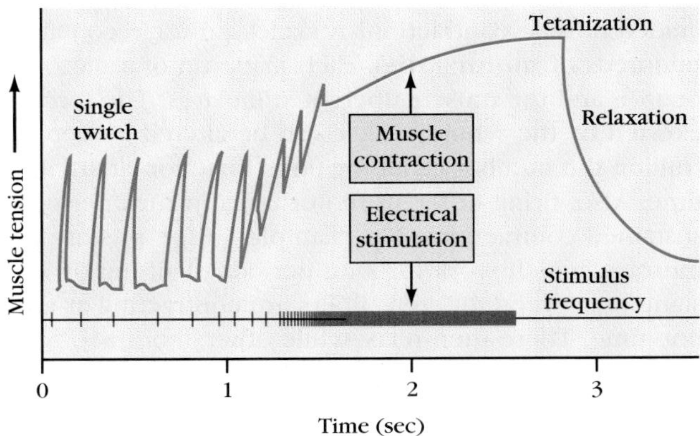

Tetanic contraction. The muscle's motor neuron is stimulated electrically at an increasing frequency (red line). When the rate of stimulation is low, the muscle contracts in individual twitches (blue curve). As the stimulus frequency increases, the muscle does not have time to relax fully between stimuli. Summation occurs, finally resulting in a smooth, sustained tetanized contraction.

Skeletal muscle structure and function. This diagram of the deltoid muscle of the shoulder shows serial enlargements of the muscle components, from the intact muscle to the pattern of contractile protein filaments that gives the muscle fiber its striped appearance. Thick and thin filaments slide past each other as muscle fibers contract. (J. A. Dennis/Phototake)

into a smooth, steady contraction, a state known as **tetanization.** The muscle contractions also undergo **summation,** producing more force than a single twitch would generate by itself; how this happens is not understood.

A skeletal muscle that is stimulated to contract repeatedly for a long period will eventually become unable to respond to further stimulations, and it gradually returns to its resting length. This phenomenon, known as **fatigue,** occurs only in striated muscle. Complete fatigue seldom occurs in an intact organism, but it is easily induced in a muscle that has been removed from the body.

Graded Response of an Intact Muscle

It takes more force to carry a whole tray of dishes across the room than to carry just one cup of coffee. How is the strength of muscle contraction controlled? Different animals have different mechanisms for producing this graded response.

Most invertebrate skeletal muscle fibers produce graded contractions, which increase as more impulses arrive via the motor neuron. Similar muscle fibers, called "slow" fibers, are also common in lower vertebrates—fishes, birds, and reptiles. However, most mammalian skeletal muscle fibers are of the "fast," or "twitch," variety, which respond to stimulation in an all-or-nothing manner, as described above. These fibers contract faster than "slow" fibers and so permit rapid movement. The flight muscles of some insects contract even more rapidly than mammalian "fast" muscles.

How can "fast," all-or-nothing fibers produce a graded muscle contraction? A skeletal muscle contains hundreds of **motor units,** each made up of a motor neuron and the muscle fibers it stimulates. The force exerted by the whole muscle can be altered by controlling the number of motor units in action at any one time. Also, firing different motor units in turn permits sustained contractions. For example, in the posture muscles, which work for long periods of sitting or standing, several different fibers are contracting at any one time. These then relax while others contract. As a result, the entire muscle remains partially contracted, but no one fiber stays contracted until it has exhausted its supply of ATP. This system has the disadvantage of requiring many more motor neurons than does a "slow" muscle system, because many different motor units are necessary.

BIO-BIT

In some small fly species, the wing muscles contract faster than 1000 times per second.

In skeletal muscle contraction, the thick and thin filaments in each sarcomere slide past each other, meshing together more closely and shortening the entire muscle fiber. Sliding occurs when the heads of myosin molecules in thick filaments attach to actin molecules in thin filaments and push the free ends of the thin filaments closer to the center of the sarcomere. The main events of muscular contraction occur in response to electrical excitation of the muscle membrane by the neurotransmitter acetylcholine released from motor neurons. A continuous train of nerve impulses causes a mammalian skeletal muscle fiber to produce a sustained contraction. The strength of a muscle contraction is controlled by the number of motor units called into action.

29–C HOW MUSCLES AND SKELETONS INTERACT

Muscles can move parts of the body only because they work against skeletons. A **skeleton** may be defined as anything on which a muscle exerts force. Some structures that act as skeletons according to this definition are not part of an animal's "official" skeleton. For instance, the muscles in the wall of a blood vessel work against the surrounding tissue, and against the blood itself, when they contract and reduce the vessel's diameter. During childbirth, the muscles of the uterus could do no useful work if there were no baby to contract against. Many muscles that cause movement inside the body shorten very little, and exert little force. Their attachment to nearby cells gives them enough leverage to operate. On the other hand, the muscles used in locomotion may contract forcefully and dramatically, and they usually exert their force on what we generally think of as a skeleton: a hydrostatic skeleton, such as the fluid-filled cavities of a cnidarian or annelid; an exoskeleton, such as the cuticle of an insect or the shell of a snail; or an endoskeleton, such as the bones of a vertebrate (Figure 29–2).

Antagonistic Muscles

All movements are controlled by the action of two sets of **antagonistic muscles**—muscles with opposite effects. In systems lacking hard skeletons, such as soft-bodied invertebrates and the internal organs of vertebrates, these antagonistic muscles are arranged in circular and longitudinal sheets. Longitudinal muscles run lengthwise along the body (or organ) and make it shorter and wider when they contract. Contraction of circular muscles, which run around the body (or or-

(a)

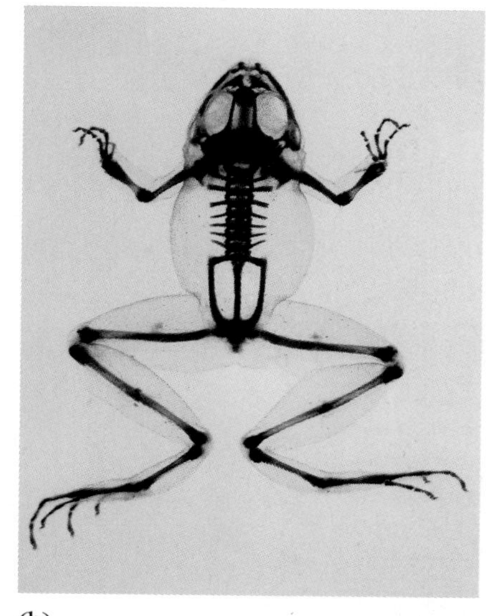

(b)

FIGURE 29-2

Skeletons. (a) The relatively thin but tough exoskeleton (outside skeleton) of an arthropod, such as this 17-year cicada, provides a lot of strength for its weight while it supports and protects soft body parts. Because the arthropod exoskeleton cannot stretch and grow, the animal (right) sheds, or molts, its exoskeleton (left) several times during its life. **(b)** In contrast, the endoskeleton (internal skeleton) of a frog is made of bone and cartilage. The skeleton grows as the animal grows, and bone replaces most of the cartilage as the animal matures. (a, D. Jackson/Visuals Unlimited; b, Cabisco/Visuals Unlimited)

gan), makes it longer and narrower. The antagonistic action of longitudinal and circular muscles against the contents of the alimentary canal moves food down the gut in most animals. Similar forces exerted against the fluid-filled coelom of an earthworm permit the worm to move (Figure 29-3).

Muscles attached to hard skeletons, such as the exoskeletons of arthropods or molluscs or the endoskeletons of vertebrates, may be arranged in bundles rather than sheets. A skeletal muscle is attached to the skeleton either directly or by way of a **tendon.** In some cases—for instance, with the muscles that move our fingers—the tendons may be almost as long as the muscle.

Antagonistic muscles run parallel to each other across a joint in the skeleton. The muscle that causes the joint to stretch out is called an **extensor,** and its antagonist, which causes the joint to close up, is called a **flexor.** The biceps is the main flexor across the human elbow joint, and the triceps is its antagonistic extensor (Figure 29-4).

Locomotor muscles are so powerful that if a flexor and its antagonistic extensor were both to contract strongly at the same time, they could easily break a bone. This is prevented by **reciprocal inhibition,** a process involving a reflex arc from each muscle to its antagonist. Sense organs in the muscle detect how far it is contracted. Their sensory neurons send this information to interneurons in the spinal cord, which in turn inhibit the firing of motor neurons to the antagonistic muscle. Thus any stimulus that causes a muscle to contract also inhibits the contraction of its antagonist, and the joint moves smoothly.

This is the simplest reflex involved in locomotion. Much more complex reflexes come into play when we walk. For instance, reflexes from the opposite limb ensure that one leg moves after another. Such reflexes are even more complicated in animals that have many legs or that use their tails for balance.

The muscular and skeletal systems work together to cause movement or perform other work. In most cases, muscles occur in antagonistic pairs, with opposing functions. Reflexes help to coordinate the actions of antagonistic muscles and ensure smooth contractions.

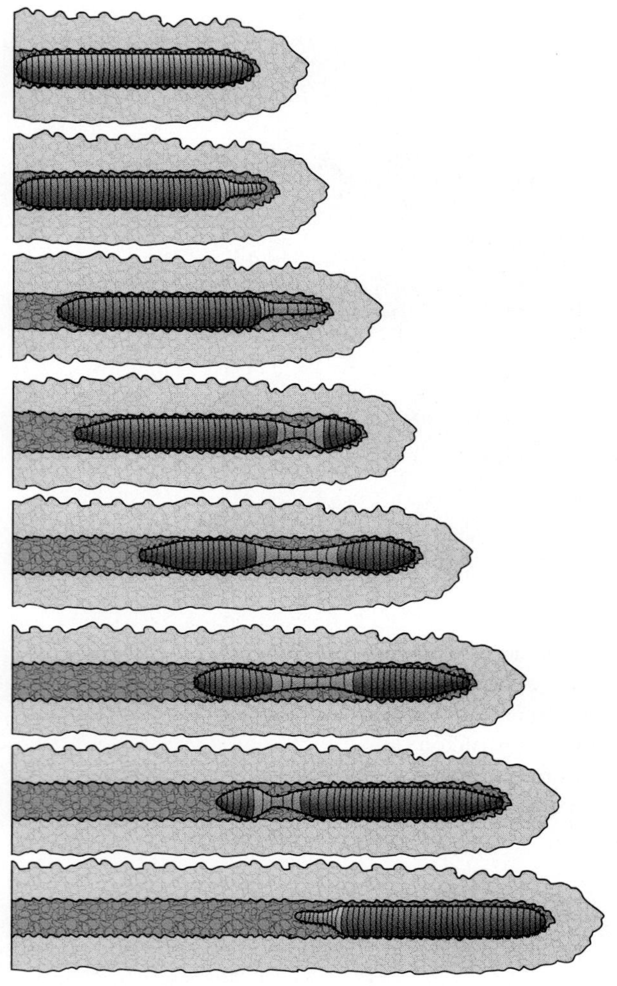

FIGURE 29–3

Earthworm movement. Each segment acts as a closed sac of fluid surrounded by two sets of muscles. These can contract alternately shortening or lengthening the segment. When the segment is short and wide, the bristles anchor the worm as it pushes against the soil to move forward. Then the thin segments can be pulled up past the soil with little friction as the worm moves forward.

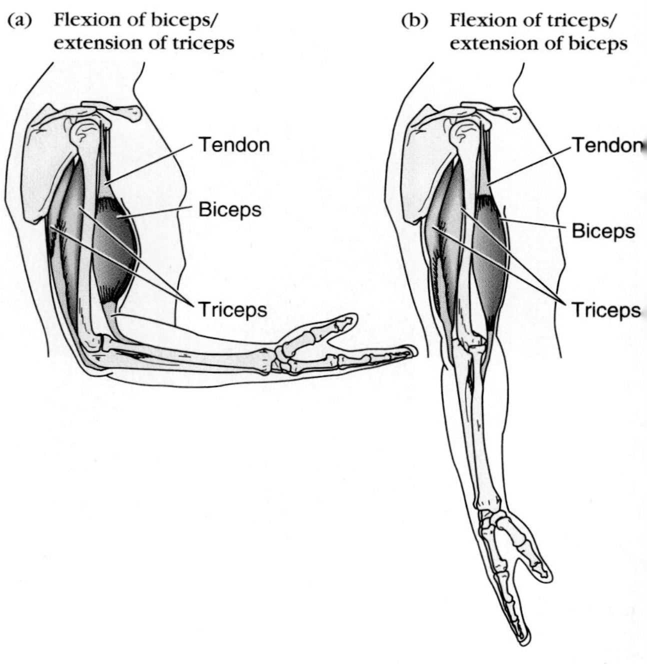

FIGURE 29–4

Skeletal muscles in the human upper arm. Each muscle is attached to the skeleton by tough tendons.

29–D THE VERTEBRATE SKELETON

The vertebrate skeleton has two major portions. The **axial** (along the body's main axis) skeleton consists of the skull, backbone, ribs, and tail. The **appendicular** (limb) skeleton includes the bones in the limbs and in the girdles that attach the limbs to the backbone. In humans the **pectoral girdle** includes the collarbones (clavicles) and shoulder blades (scapulas), and the **pelvic girdle** includes the large, fused hip bones (ilium, ischium, and pubis) (Figure 29-5).

The bones of the skeleton anchor various muscles and permit them to move the body, but they serve many other functions as well. First, they provide a framework that supports the body against the pull of gravity, and land vertebrates must have skeletons more substantial than those of vertebrates that live supported by the buoyancy of water. Parts of the skeleton also protect delicate internal organs. The skull protects the brain and major sense organs, the backbone surrounds and protects the spinal cord, and the pectoral girdle and ribs protect the heart and lungs.

Joints in the Vertebrate Skeleton

Joints between bones in the skeleton can be of many different types and degrees of rigidity. At one extreme are **sutures,** wiggly lines in the skull where bones meet at interlocking projections that hold them very

BIO-BIT

Because of the absence of gravity in space, the skeletons of astronauts become measurably reduced in mass after trips as short as a week.

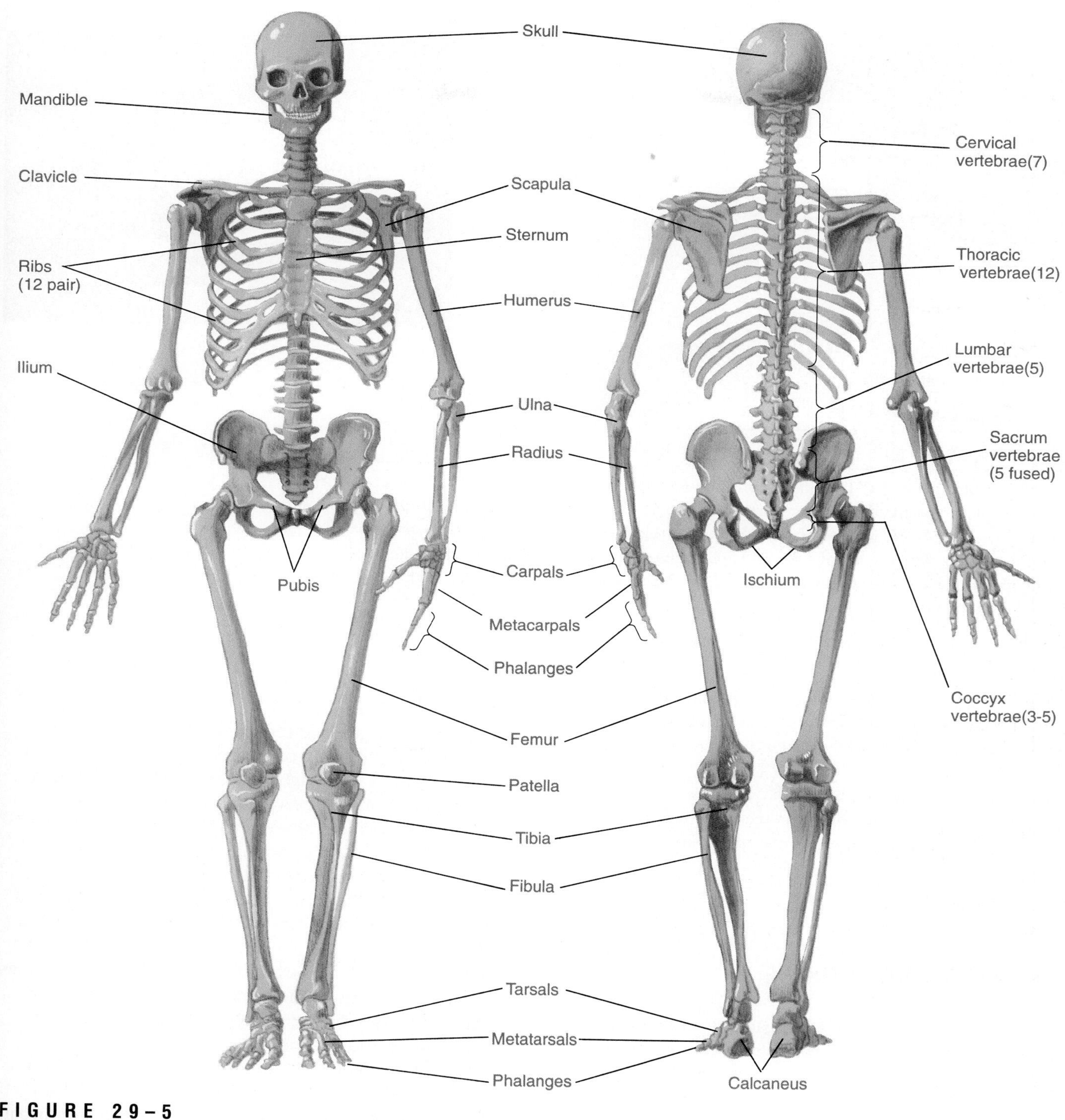

FIGURE 29-5

Major bones of the human skeleton. Names of the bones of the axial skeleton are printed in red; bones of the appendicular skeleton are printed in blue.

tightly together. However, most bones are joined to each other by **ligaments.** An extremely flexible joint in females is the one between the two bones in the front of the hip girdle, which are held loosely together by a tough disc of cartilage that stretches considerably and allows the bones to separate during childbirth.

The joints of the limbs must permit smooth movement in various directions. The ends of the bones in

(a)

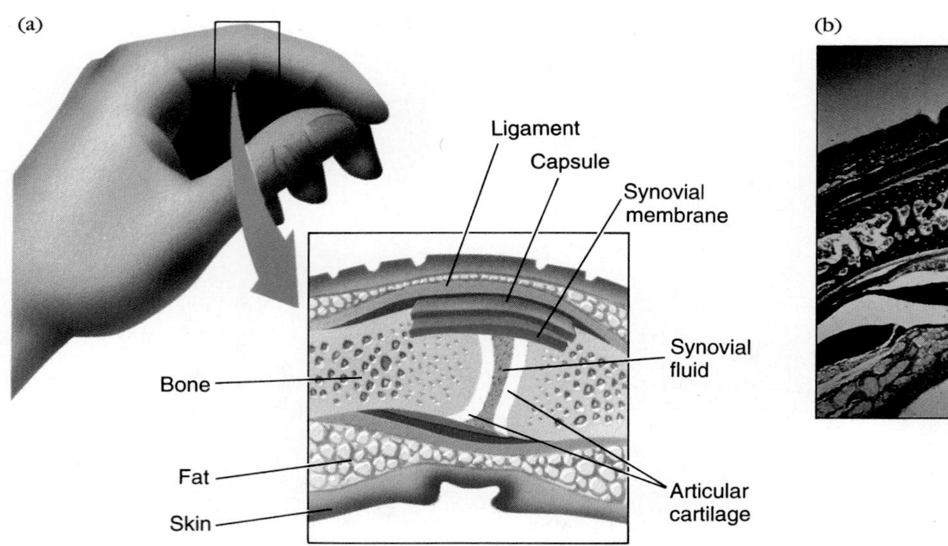

(b)

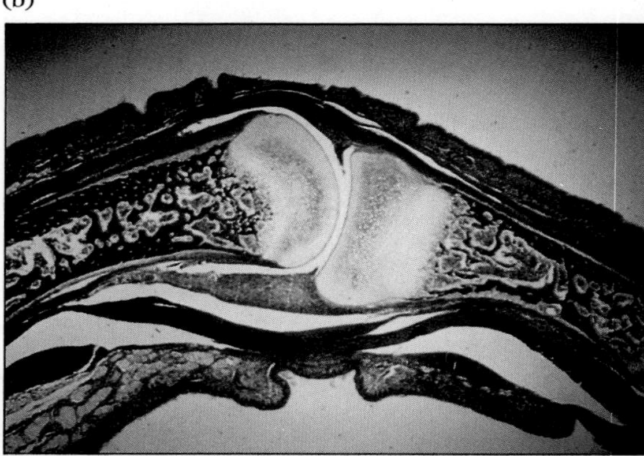

FIGURE 29-6

Structure of a joint. (a) The finger joint of an adult. The space between the bones is filled with synovial fluid, kept in place by the surrounding synovial membrane. A tough, fibrous capsule surrounds the membrane. Ligaments bind the bones together and limit the movement between them. Cartilage, which cushions the ends of bones, is present in the adult, although in relatively small amounts. Compare with **(b)**, which shows large, pale areas of cartilage at the ends of the bones in this finger joint of a young child. (Biophoto Associates)

these joints are covered by smooth, hyaline cartilage, and the whole joint is surrounded by a sac filled with **synovial fluid,** which lubricates the joint (Figure 29-6).

The skeleton supports the body, protects internal organs, and provides attachment sites, pivots, and levers for the muscles to move body parts.

29–E CONNECTIVE TISSUE

Bones, cartilage, tendons, and ligaments are all made up of various types of **connective tissue,** composed mainly of intercellular ("between cells") substances secreted by scattered cells. The intercellular substance contains fibers of a tough, elastic protein, **collagen.** Collagen is the most common protein in mammals, accounting for about one fourth of all the body's protein (it is also the main ingredient of gelatin). Bone is hard and brittle because it has mineral deposits in addition to these fibers. Cartilage has a great deal of a firm intercellular jelly surrounding the protein fibers and cells, while tendons and ligaments are mostly composed of fibers, with very few cells.

Cartilage

Unlike bone, cartilage receives no blood supply; its cells rely on diffusion of nutrients from capillaries in nearby tissues. Cartilage makes up the entire skeleton in members of the class Chondrichthyes (sharks, skates, and rays) and in all early vertebrate embryos. During embryonic development of vertebrates with bony skeletons, minerals are deposited in most of this embryonic cartilage, a blood supply and other typical features of bone appear, and the cartilage is replaced by bone (Figure 29-7). Cartilage persists in the adult only in areas where flexibility is necessary. These areas include: the ends of the bones where they form synovial joints (see Figure 29-6); the discs between the vertebrae in the backbone (see Figure 28-10); the ends of the ribs where they join the breastbone; the rings that keep the walls of the trachea (windpipe) from collapsing; the larynx ("voice box") in the throat, at the front end of the trachea; the external ear; the eustachian tube, which connects the throat to the middle ear; and the tip of the nose.

Bone

Although the mineral matter of the dried human skeleton weighs only about 5 kilograms, the skeleton of a

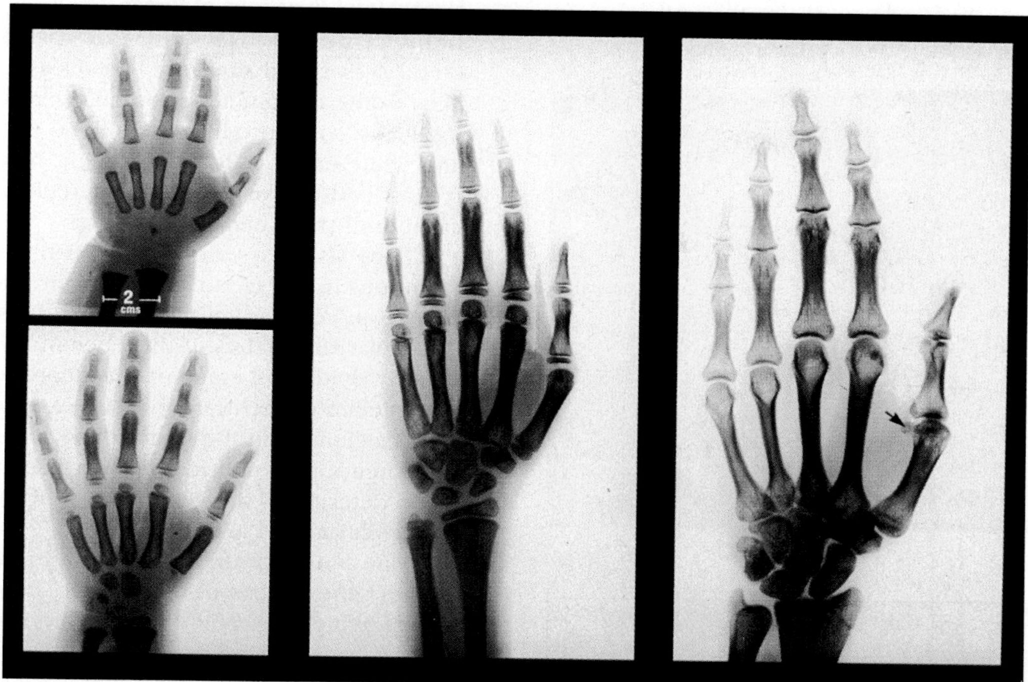

FIGURE 29-7

Growth of a human hand. This series of x-ray photographs shows the hand at 2 years (top left), 2 years 9 months (bottom left), 12 years (center), and 16 years (right). Note that at 2 years of age, there appear to be large gaps between the bones of a finger. These gaps actually contain the cartilaginous ends of the bones and the finger joints, but cartilage shows up only faintly in an x-ray. In the hand of a 12-year-old (center), the cartilage on either side of the joints has been almost completely replaced by bone, leaving only a thin cartilage cap over the end of the bone at the joint. In the fully developed hand (right) the ends of the bones are enlarged into knobs. The arrow points to a small spur of bone on the thumb. Many clinical conditions, from injury to arthritis, can cause such abnormal deposition of minerals in the body. When such deposits occur in joints, they may be painful and interfere with movement. (Biophoto Associates)

living human being is much heavier because bone is a living tissue that contains cells, blood vessels, nerves, fluid, and fat deposits.

Bone varies in structure depending upon its position and function in the body. At one extreme is **spongy bone,** composed of an irregular network of mineralized bars, and at the other is **compact bone,** composed of tubular units called Haversian systems (Figure 29-8).

The hard part of bone is made up of organic matter and inorganic salts. The organic matter is mainly collagen fibers, which give bone some flexibility and most of its ability to withstand pull (tension). The inorganic salts are mainly calcium phosphate and smaller amounts of other ions. These give bone its considerable ability to withstand compression and side-slippage.

Distributed throughout a bone are the cells that lay down the collagen and mineral deposits as the bone grows. These cells last into adult life and are responsible for repair and replacement of broken bone and for the formation of **calluses,** which develop at points of pressure on a bone. People who perform one action many times, such as squatting instead of sitting, throwing a baseball, or carrying a baby on the hip, develop particular alterations to their bones. Anthropologists studying ancient humans can look for such bone-structure markers as clues to how these people lived.

Bones grow throughout life as the body increases in size and strength, and as they grow they are also remodeled. New material is added to the outer surfaces and the ends, while old material is destroyed to enlarge the internal cavities for the bone marrow. The differentiation of bone is controlled by a number of lo-

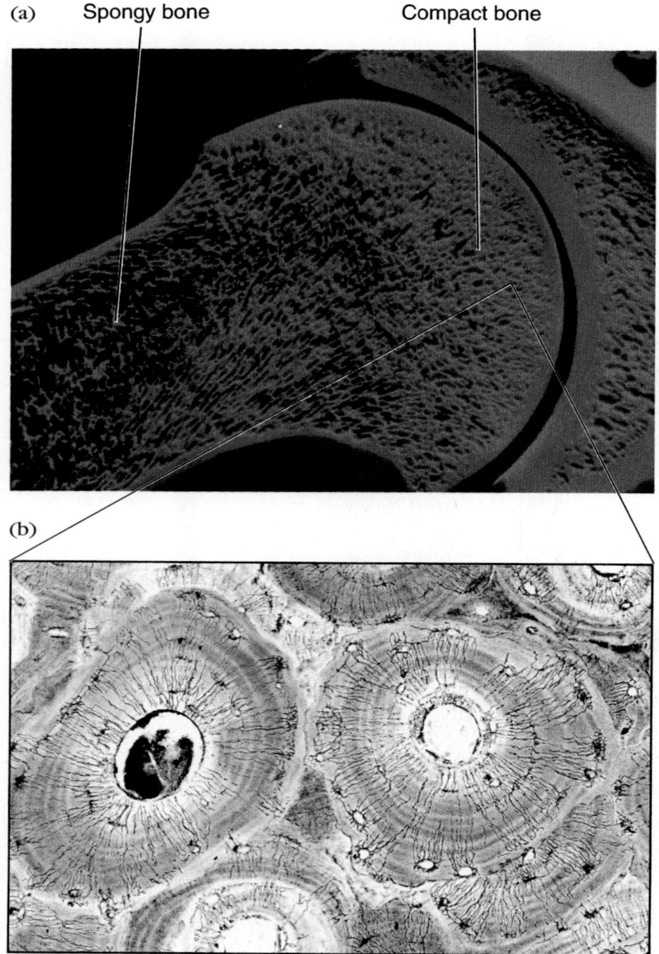

FIGURE 29–8

Two types of bone. (a) This photo of a joint shows both spongy and compact bone. **(b)** A cross-section of compact bone. Each set of concentric rings is a Haversian system, surrounding a central channel, the Haversian canal. In living bone, the Haversian canals contain blood vessels and nerves. The many tan rings around each Haversian canal are mineral deposits, laid down by living cells that inhabit the small, purple, spider-shaped spaces. (a, Don W. Fawcett/Visuals Unlimited; b, Biophoto Associates)

cal hormones, produced in growing animals and also in response to physical stress, fractures, and other stimuli that indicate the need for more bone growth. **Bone growth protein,** for instance, is a hormone that occurs in the intercellular material of bone. When released, by a fracture or similar trauma, it induces undifferentiated cells to differentiate into bone-forming cells. The amount of bone growth protein declines with age, contributing to osteoporosis (see *A Journey into Healthy Living,* Preventing Osteoporosis, this chapter).

Nonskeletal Functions of Bones

In the centers of many bones are cavities filled with blood vessels and **marrow,** a soft tissue with a number of different functions. Some marrow is primarily a fat depot; some produces red blood cells, platelets, and white blood cells. Interestingly, blood is another type of connective tissue. Its intercellular matrix is composed predominantly of water.

Bone also plays a role in regulating the level of calcium ions in the blood. The calcium phosphate in bone is in equilibrium with that in the surrounding extracellular fluid. Thus, if the calcium level in the extracellular fluid rises, some of it is deposited in bone; if the calcium concentration in the extracellular fluid falls, calcium from the bones dissolves into the fluid. The chemical equilibrium between solution and deposition determines the general level of calcium in the extracellular fluid and in the blood.

Fine tuning of the calcium level in the body fluids is under the control of hormones. The parathyroid glands, which lie behind the thyroid gland in the neck (see Figure 30–1), secrete **parathyroid hormone.** This stimulates the cells in bone to take additional calcium out of the bone and release it into the blood. Parathyroid hormone also causes the kidneys to reclaim more calcium from the forming urine. Third, it stimulates cells in the kidneys to convert vitamin D to an active form, which promotes absorption of calcium from food in the intestine. A drop in the blood calcium level stimulates the parathyroid glands to produce this hormone; calcium is then released from bone, resorbed from the urine, and absorbed from food to make up the deficit.

A rise in the blood calcium level is counteracted by the hormone **calcitonin,** secreted by the thyroid gland. Calcitonin inhibits the removal of calcium from the bone deposits. This permits the opposite process, deposition of new mineral material in the bone, to bring the blood's calcium level back down to normal.

The connective tissue that makes up ligaments, tendons, cartilage, and bone consists of a great deal of collagen and other intercellular materials secreted by scattered cells. Cartilage has no blood vessels and is much more flexible than bone. In most vertebrates, cartilage forms the embryo's skeletal elements, which are later replaced by bone. Bone contains bone cells, nerves, and blood vessels surrounded by a matrix of collagen stiffened by inorganic salts. The specific shape of a bone is continually being modified in response to the forces acting upon it. In addition to skeletal support and protection of inner organs, bones serve as the body's calcium repositories and contain marrow where blood cells are produced.

A Journey into Healthy Living

Preventing Osteoporosis

In the United States, more than 600,000 bone fractures a year are attributed to brittle bones. Researchers hope to reduce this expensive toll by studying the formation of bone in the embryo and the healing of bones in later life.

Part of the reason osteoporosis is so common is that people live much longer nowadays, and degeneration of bone is a serious problem in many older people. **Osteoporosis** is a condition in which bones have lost so much inorganic mineral matter that they become light, brittle, and easily broken (Figure 29-A). The disorder has at least two distinct forms. Postmenopausal osteoporosis, involving loss of bone in the spine and forearms, afflicts women 50 to 65 years old. An important contributing factor is the great reduction in the hormone estrogen at the time of menopause. This may be exacerbated by smoking, which reduces the body's estrogen level. Senile osteoporosis occurs in people of both sexes aged 75 and older, and involves loss of bone in the hips and legs. The most effective ways to prevent or treat osteoporosis appear to be exercise, which stimulates addition of material to bone, and increasing calcium in the diet, especially from organic sources such as dairy products. It is more effective to drink milk than to take calcium pills because very little of the calcium from pills is taken up by the body. ■

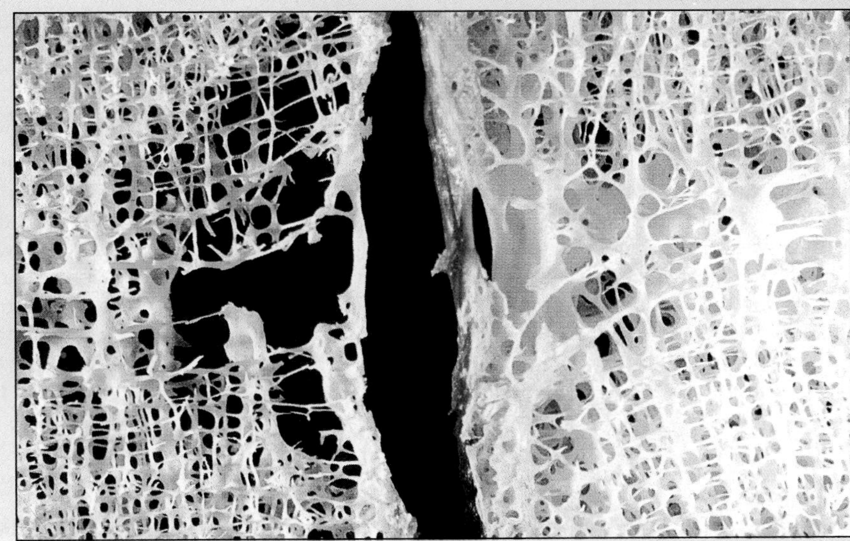

FIGURE 29-A

Osteoporosis. A portion of a human vertebra weakened by osteoporosis (left) and normal bone (right). (M. Klein/Peter Arnold, Inc.)

SUMMARY

1. Most movement in an animal's body is due to the action of its muscles. A muscle works by shortening and pulling against the skeleton or against adjacent tissues.
2. Vertebrates have three kinds of muscle tissue: smooth muscle in the walls of blood vessels and other internal organs, cardiac muscle in the heart, and skeletal muscle in the voluntary muscles such as those attached to the bones and used in locomotion.
3. In the presence of ATP and calcium ions, filaments of the protein myosin form cross-bridges to filaments of the protein actin. The swiveling of the cross-bridges moves the filaments past each other, reducing the distance between Z lines of the sarcomeres.
4. ATP supplies the energy needed for contraction.
5. Contraction of a muscle is stimulated by depolarization of the membranes of muscle cells (fibers). This happens when a neurotransmitter is released from the axon of a motor neuron or, in smooth or cardiac muscle, when the hormone norepinephrine is present.
6. In mammalian skeletal muscle, the arrangement of motor neurons and muscle fibers into motor units allows for graded muscle contractions. Smooth, sustained con-

tractions are brought about by a series of closely spaced impulses from motor neurons.

7. The vertebrate skeleton supports the body, protects internal organs, and permits the muscles to move parts of the animal's body.
8. Pairs of antagonistic muscles move bones and joints. Reciprocal inhibition prevents the two members of a pair of antagonistic muscles from contracting strongly at the same time.
9. The vertebrate skeleton is made up of relatively flexible cartilage and bone.

SELF-QUIZ

For each phrase in questions 1 through 5, give the type(s) of muscle that show the characteristic.

___ 1. Multinucleate fibers	a. Cardiac
___ 2. Receives nerves from autonomic nervous system	b. Skeletal
___ 3. Can contract without nervous stimulation	c. Smooth
___ 4. Typically found in sheets rather than in bundles	
___ 5. Unicellular organization	

6. Indicate which of the following items would be found at each of the places indicated on the following diagram:

actin	troponin
calcium	Z line
mitochondria	
myosin	
tropomyosin	

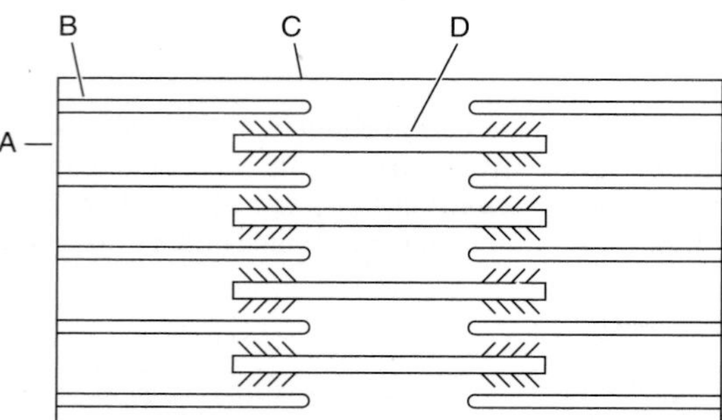

7. During the time that a muscle is in tetanic contraction, what are the motor neurons to the muscle doing?
8. Why doesn't a muscle relax between the arrival of nerve impulses when it is in tetanic contraction?
9. The extent of contraction of a mammalian skeletal muscle is controlled by:
   a. interaction of excitatory and inhibitory nervous input to individual fibers
   b. contraction of some sarcomeres in a fiber but not others
   c. contraction of some entire muscle fibers while others remain relaxed
   d. reciprocal inhibition
   e. all of the above
10. Which of the following is *not* a role of bones?
    a. maintenance of simple chemical equilibrium between dissolved and deposited calcium without intervention of living cells
    b. removal of calcium deposits carried out by living bone cells in response to a hormone
    c. production of red blood cells
    d. production of white blood cells
    e. production of hormones to regulate calcium levels in body fluids
    f. storage of fat
    g. providing attachment sites for muscles
    h. protection of internal organs

THINKING CRITICALLY

1. If you read detective stories, you know that it takes a variable length of time for *rigor mortis* to set in after death. What are some reasons for this variation?
2. *Rigor mortis* lasts for several hours and then disappears. Why does it eventually go away?
3. What is the advantage of having cartilaginous tissue in each area of the human body where it is characteristically found? (See Section 29-E.)
4. What are some advantages of having muscles attached to bones via tendons rather than directly?
5. Comment on the biological validity of the sayings "I can feel it in my bones" and "dry as a bone."
6. The disease rickets is characterized by bending of the bones due to lack of calcium deposits to keep them stiffened into the proper shape. Why is it advantageous for the bones to give up these calcium deposits even though doing so results in permanent skeletal deformity?
7. Would you expect levels of parathyroid hormone or of calcitonin to be elevated in a victim of rickets? In a pregnant woman? In her fetus?
8. Pumiliotoxin B, produced by dart-poison frogs, promotes the release of Ca^{2+} from muscle storage areas and inhibits its return. What symptoms would you expect to see in an animal dying after being shot with a dart tipped with this poison?

SELECTED KEY TERMS

actin, *p. 615*
antagonistic muscle, *p. 618*
appendicular skeleton, *p. 620*
axial skeleton, *p. 620*
cardiac muscle, *p. 612*
collagen, *p. 622*
compact bone, *p. 623*
connective tissue, *p. 622*
extensor, *p. 619*
flexor, *p. 619*
ligament, *p. 621*
marrow, *p. 624*
motor unit, *p. 618*
muscle fiber, *p. 614*
myofibril, *p. 614*
myosin, *p. 615*
neuromuscular junction, *p. 612*
pectoral girdle, *p. 620*
pelvic girdle, *p. 620*
rigor mortis, *p. 615*
sarcomere, *p. 614*
sarcoplasmic reticulum, *p. 615*
sinoatrial node, *p. 612*
skeletal muscle, *p. 612*
smooth muscle, *p. 612*
spongy bone, *p. 623*
tendon, *p. 619*

SUGGESTED READINGS

Caplan, A. I. "Cartilage." *Scientific American,* October 1984.

McLean, F. C. "Bone." *Scientific American,* February 1955. A brief, clear description of bone structure, growth, and function.

McMahon, T. A. *Muscles, Reflexes, and Locomotion.* Princeton, NJ: Princeton University Press, 1984. Chapters on the mechanics and neural control of locomotion.

Schiefelbein, S., et al. *The Incredible Machine.* Washington, DC: National Geographic Society, 1986. Fascinating account of human biology and life.

Vogel, S. *Life's Devices: The Physical World of Animals and Plants.* Princeton, NJ: Princeton University Press, 1989. An entertaining book on how organisms adapt to the physical forces of the world they live in.

CHAPTER 30

CONCEPT GUIDE

After reading this chapter, you should be able to:

1. Compare the functions of neurotransmitters, neurohormones, local chemical messengers, and pheromones. (Section 30: Overview)
2. Explain how negative feedback control promotes homeostasis. (Section 30–A)
3. Compare the mechanisms of action of lipid-soluble and water-soluble hormones. (Section 30–B)
4. Compare the mechanisms by which the endocrine and nervous systems control other body systems. Describe how these two systems work together in the "fight or flight" response. (Section 30–C)
5. Describe how local chemical messengers are different from other hormones. (Section 30–D)
6. Describe how an animal's physiology and behavior can be influenced by environmental changes. (Section 30–E)
7. Compare circadian and annual rhythms and give examples of each. (Section 30–F)
8. Define pheromone and describe three different ways that animals use pheromones. (Section 30–G)

When a cat rubs its head and body up against you, it is greeting and trading scents with you. The cat has scent glands on its temples, at the corners of its mouth, and at the base of its tail. As it twines around your ankles, the cat marks you with its scent and simultaneously rubs your scent onto its fur. Although humans cannot appreciate the cat's individual scent, when the cat later licks and grooms its fur, it tastes your individual scent. (Animals Animals)

Animal Hormones and Chemical Regulation

Wherever there is division of labor, there must also be an exchange of information to coordinate the work of different units. On a construction site, dozens of people may perform different jobs. They must know what their co-workers are doing and interact appropriately with one another to build the structure properly. In any multicellular organism, the jobs to be done are divided among many different types of cells. Like the different workers on a construction site, these cells must exchange information and work together if the organism is to survive. The nervous system and the body's chemical messengers work together in coordinating the activities of all the different cells in the body.

We can define a **chemical messenger** as a substance produced by one cell that affects other cells. **Hormones** are chemical messengers (such as estrogen and insulin) that travel in the blood and affect cells some distance from those that produced them. Hormones are produced by cells that are usually organized into **endocrine glands** (such as the ovary and parts of the pancreas), glands whose secretions move to their sites of action via the blood rather than through ducts.

Chemical messengers carry out several kinds of regulation. First, many contribute to the body's physiological homeostasis by acting as part of a negative feedback control system. In this way they raise or lower the body fluids' level of some chemical—including other hormones. Second, some chemical messengers are involved in repair, growth, and differentiation of the body, particularly during embryonic development. Third, hormones play a role in adaptive responses to events outside the body. For instance, hormones ensure that an animal will be in reproductive condition when environmental conditions favor the survival of the young. The most obvious interactions between the nervous and hormonal systems occur in this type of control. An animal can detect environmental events only through its sense organs, and the nervous system must convey this information to the glands that produce hormonal changes. In the vertebrates, the hypothalamus in the brain (part of the nervous system) communicates with the pituitary gland (part of the endocrine system), which in turn regulates most of the other endocrine glands.

A wide variety of chemical messengers occurs throughout the body, but each cell responds only to some of the messages. This is because different kinds of cells have receptor molecules that recognize and bind particular kinds of messenger molecules, initiating changes in the cells' activities.

KEY CONCEPTS

- Neurons and chemical messengers coordinate activities within an animal's body by affecting cells in various ways.
- Chemical messengers are involved in maintaining homeostasis within the body, in growth and differentiation, and in the body's response to outside stimuli.
- The body maintains homeostasis of various chemical components by negative feedback control, using hormones as messengers to raise or lower the levels of other chemicals, including other hormones.
- The need to secrete a hormone is determined, in many cases, by the nervous system, which stimulates secretion by some glands.
- Only those cells with receptors for a hormone can respond to that hormone.

Curiosity Questions

1? Why do students often get sweaty hands and faster heartbeats just before an exam? (See page 634)

2? How is the timing of sleep controlled? (See page 637)

3? Why do dogs urinate on trees and fire hydrants? (See page 642)

CHEMICAL MESSENGERS: AN OVERVIEW

We now know that animals have dozens of chemical messengers, including:

1. **Hormones** (produced by neurons and by endocrine cells in endocrine organs or in other tissues such as the stomach and kidneys [Figure 30–1]). In Chapter 28 we saw that neurotransmitters, such as acetylcholine and norepinephrine, carry nerve impulses across synapses. Some neurons, however, respond to stimulation by releasing neurohormones (such as oxytocin, which induces labor during childbirth) from their axons into the bloodstream.
2. **Local chemical messengers.** Some chemical messengers have only local effects, within a few millimeters of the cell that produced them, because they are rapidly removed from the extracellular fluid. These local messengers include histamine, which participates in inflammatory and immune reactions (Section 25–A); growth factors, which stimulate growth by particular tissues; prostaglandins, lipids with a variety of effects; and short-range neuroregulators released into a local area by neurons.
3. **Pheromones.** These are chemical signals produced by one animal that affect the behavior of other individuals of the same species.

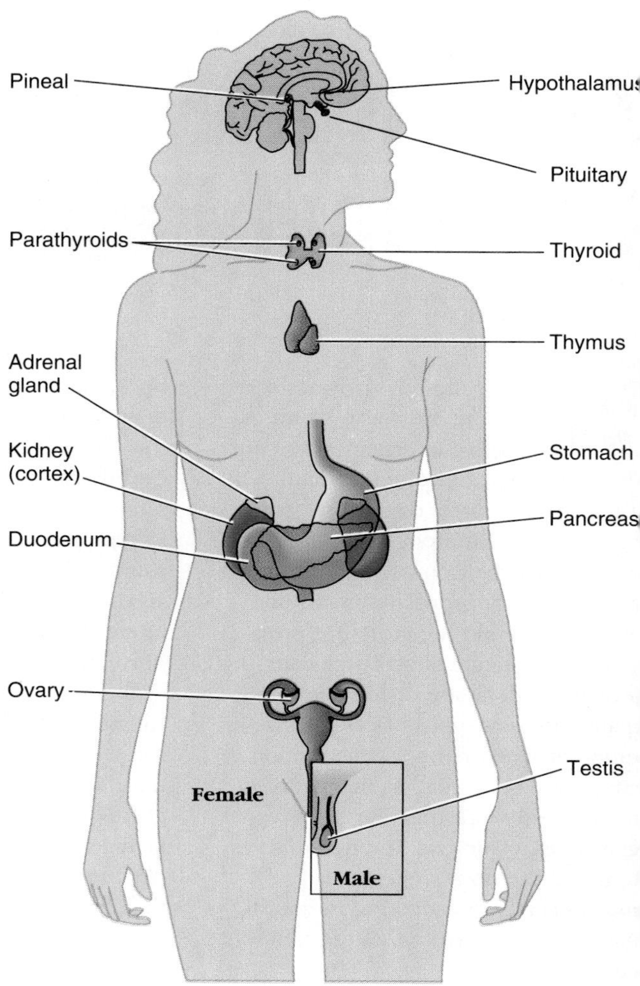

FIGURE 30–1

Endocrine glands. The locations of the major endocrine glands and other organs with endocrine functions in humans.

30–A HORMONES

How do we know that a hormone exists and what it does? Most hormones have been discovered by removing the relevant endocrine gland and observing how this affects the animal. For instance, an amphibian tadpole without a thyroid gland never metamorphoses into a frog; it is reasonable to hypothesize that a hormone produced by the thyroid causes metamorphosis. This can be tested by transplanting a thyroid gland into a tadpole that lacks one. Since this operation is followed by metamorphosis, the theory is strengthened.

More convincing evidence for the hormone can then be produced by breaking down thyroid tissue into various chemical fractions and showing that some of these, but not others, produce metamorphosis after they are injected into a tadpole. Finally, it may be possible to isolate the pure hormone from the thyroid gland.

Such experiments are much harder when the endocrine gland involved, such as the vertebrate pituitary gland, produces many hormones (see Table 30–1). In such situations, the most common approach is to take blood from an animal that is thought to contain the hormone and transfuse it into an animal that does not. For instance, the vertebrate pituitary was long suspected of producing vasopressin, a water-conserving hormone, if the animal became dehydrated. To test this, blood from a dehydrated dog was transfused into a dog that had an adequate water supply. Urine production in the second animal decreased; it returned to normal only when the transfusion was stopped. This suggested that the blood of the first dog did in fact contain a water-conserving hormone. Showing that the hormone comes from the pituitary is more difficult. However, such an assumption would be strengthened if chemical analysis showed that the hormone is found only in the pituitary and in the bloodstream.

Some of the most important hormones secreted by vertebrates are listed in Tables 30–1 and 30–2.

Table 30–1
Some Vertebrate Pituitary Hormones, Their Sources, and Main Actions*

Hormone	Source	Effect	See Section
Releasing factors (various)	Hypothalamus	Release of hormones from anterior pituitary	30–C
Oxytocin	Hypothalamus via posterior pituitary	Uterine contractions in mammals Letdown of milk to nipple in mammals	27–D
Vasopressin (=ADH, antidiuretic hormone)	Hypothalamus via posterior pituitary	Water resorption by nephron tubules Increase in permeability of skin to water in amphibians	26–E
Adrenocorticotropic hormone (ACTH)	Anterior pituitary	Secretion of corticosteroids by cortex of adrenal gland	30–C
Thyrotropin (TSH, thyroid-stimulating hormone)	Anterior pituitary	Secretion of hormones by thyroid	30–C
Follicle-stimulating hormone (FSH)	Anterior pituitary	Production of gametes in both sexes	27–D
Luteinizing hormone (LH)	Anterior pituitary	Secretion of sex hormones by gonads in both sexes Ovulation in females	27–D
Prolactin (=LTH, luteotropic hormone)	Anterior pituitary	Mammary gland growth and lactation in mammals Maintenance of corpus luteum in mammals Migration to water in amphibians Reproductive functions in birds	27–D
Somatotropin (growth hormone)	Anterior pituitary	Body growth in reptiles and mammals Increased blood sugar in mammals	30–C

*All pituitary hormones are proteins or parts of proteins.

Table 30–2
Some Vertebrate Hormones, Their Sources, and Main Actions*

Hormone	Source	Effect	See Section
Thyroxin	Thyroid	Stimulation of growth and metabolism Metamorphosis in amphibians	10–F
Calcitonin	Thyroid	Decrease in blood calcium by suppressing its resorption from the bones	29–E
Parathyroid hormone	Parathyroids	Increase in blood calcium by release of calcium stored in bones	30–A, 29–E
Insulin	Pancreas	Decrease in blood sugar	22–G
Glucagon	Pancreas	Increase in blood sugar	22–G
Gastrin	Stomach	Secretion of HCl by stomach	
Epinephrine (adrenalin)	Adrenal medulla	Dilation of blood vessels Increase in blood pressure Increase in blood sugar	24–C, 30–C
Norepinephrine (noradrenalin)	Adrenal medulla	Same as epinephrine Also serves as a neurotransmitter	28–C, 28–H
Cortisol, etc.	Adrenal cortex	Metabolism of carbohydrate, protein, fat	
Aldosterone	Adrenal cortex	Na^+ and K^+ retention by kidney	26–E
Chorionic gonadotropin	Placenta	Maintenance of all body functions necessary for pregnancy	27–D
Progesterone	Corpus luteum of ovary	Maintenance of uterine endometrium in mammals Enlargement of breasts during pregnancy in mammals	27–D 27–D
Estrogen	Ovary	Initiation and maintenance of sexual maturity and behavior in female mammals	27–D
Testosterone	Testis	Initiation and maintenance of sexual maturity and behavior in male mammals; sex drive in males and females	27–D

*Unless a particular class of vertebrates is specified, the action applies to members of all classes as far as we know.

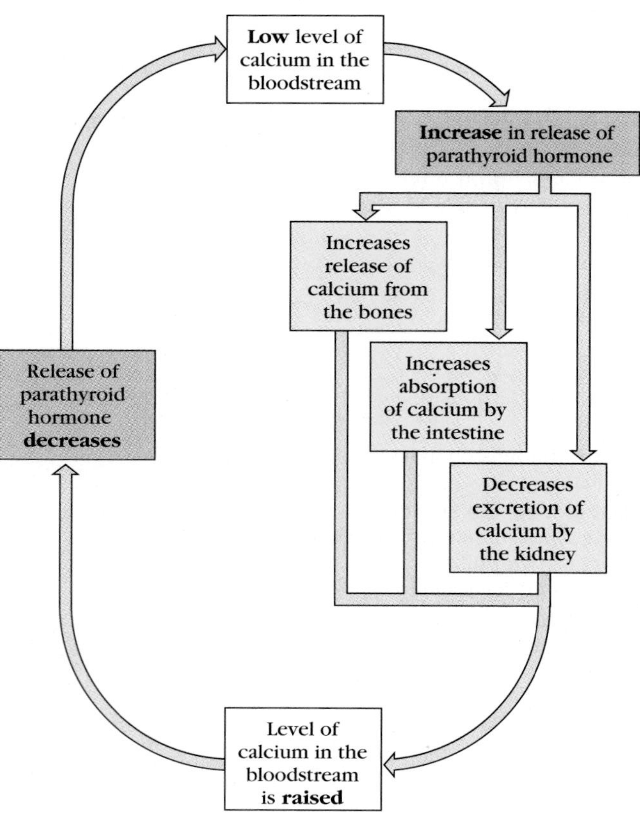

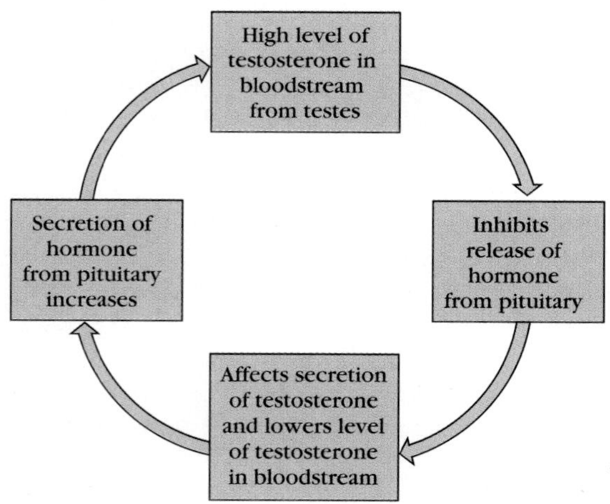

FIGURE 30-2

Two examples of negative feedback control: (a) between parathyroid hormone and calcium levels in the blood; **(b)** between two hormones. In both cases, each of the two substances serves to regulate the level of the other in the body fluids.

Feedback Control of Secretion

Many of the hormones listed in Tables 30-1 and 30-2 control the composition of the body fluids or the rate of metabolism. For these homeostatic mechanisms to function properly, the hormones must be secreted into the bloodstream only when they are needed. Their secretion is under **negative feedback control.** Like the thermostat that turns a furnace off when the proper temperature is reached, secretion of hormones is automatically limited.

A thermostat controls a familiar negative feedback system (Figure 30-2). The thermostat responds to a drop in temperature by turning the furnace on. When the thermostat detects that the temperature has risen to the set level, it turns the furnace off. As a biological example, a drop in the level of calcium in the blood causes secretion of **parathyroid hormone** by the parathyroid glands. Parathyroid hormone stimulates release of calcium from the bones, decreases excretion of calcium by the kidneys, and increases absorption of calcium from the intestine. These actions usually increase the calcium concentration of the blood so that it is back to normal within a few hours. The rise in calcium, in turn, tends to decrease the secretion of parathyroid hormone (Figure 30-2a). This negative feedback loop is one of the control systems ensuring that the level of calcium in the blood remains constant.

A hormonal feedback loop may involve two or more hormones, instead of a hormone and some other substance (for example, calcium). This is how hormones secreted by the pituitary gland control the release of other hormones. For instance, a male vertebrate must have the right levels of both **luteinizing hormone (LH)** and **testosterone** in the blood for the testes to produce sperm and function properly. LH, secreted by the pituitary, stimulates testosterone secretion by the testes, but testosterone inhibits the secretion of LH. So, if the level of testosterone in the bloodstream rises, it inhibits the pituitary's secretion of LH, and less testosterone is produced until the testosterone level falls low enough that the inhibition of LH

BIO-BIT

The use of anabolic steroids (androgens) by male athletes could lead to liver disease, prostate enlargement, heart disease, strokes, and infertility. Women who use these steroids may develop masculine characteristics such as a deeper voice and reduced breast size, and they may at least temporarily become sterile.

; turned off. LH secretion then rises again, stimulating ie secretion of more testosterone. Thus, a rise or fall 1 the level of either hormone is automatically cor-ected by way of the other (Figure 30–2b).

'he secretion of hormones involved in homeostasis is con-rolled by negative feedback in a mechanism similar to a iermostat regulating a furnace.

0–B HOW CHEMICAL MESSENGERS AFFECT ELLS

ndocrine cells release hormones into the extracellular luid. From there, the hormones diffuse into the blood-tream, which carries them throughout the body. How-ver, hormones are specific, influencing only certain ells—called their **target cells.** Only the target cells f each chemical messenger have receptor molecules o bind that messenger. For instance, although the hor-ione estrogen travels throughout the body, diffusing ito many cells, only certain types of cells respond to s chemical message. For example, cells in the ovaries, terus, and breasts have estrogen receptors in their cy-oplasm and respond by producing the changes charac-eristic of the menstrual cycle (see Figure 27–4).

A hormone delivers its "message" to the target cell y changing the shape of the receptor that binds it.)ther molecules in the cell are already set up in such way that the receptor's new shape initiates certain hanges in the cell, such as changes in permeability, nzyme activity, or gene transcription. All of these hanges involve large numbers of ions or molecules, ompared to the number of messenger molecules the ell received. In effect, the altered receptor **amplifies** ie signal of a tiny amount of hormone into a much irger response by the cell.

The reaction of the target tissue to a hormone dif-rs at different times. For example, injection of the ormone prolactin into a sexually immature female vill not cause lactation. Presumably, her mammary tis-ue has not yet produced receptors for the hormone.

We can divide hormones and local messengers ito two groups, which differ in the way they affect irget cells: (1) lipid-soluble steroid and thyroid hor-iones, and (2) water-soluble hormones.

teroid and Thyroid Hormones

teroid hormones are small lipids made from choles-erol (see Figure 3–4). The even smaller thyroid hor-iones are made from iodine and amino acids. (We eed iodine in our diets, from seafood or iodized salt, in order to make thyroid hormones.) Both kinds of hormones control many developmental and physiological processes in animals, and can enter any cell of the body.

Both steroid and thyroid hormones are lipid-soluble. They diffuse through the plasma membrane and bind to receptors inside the target cell (Figure 30–3a). Only target cells produce receptors and this binding activates the receptor so that it can bind with DNA. The hormone-receptor complex enters the nucleus, where it binds with genes known as enhancers (Section 10–E). The effect of a small number of hormone molecules is amplified by turning on or off the cell's production of many RNA and protein molecules. Amplification of the hormonal message may be increased still more if the product of transcription is a gene-regulatory protein that goes on to activate other genes (Section 10–E).

Water-Soluble Hormones

Most water-soluble hormones and local messengers are proteins (including amino acid derivatives and glycoproteins). They cannot cross the target cell's plasma membrane but bind to receptors on the outside of the cell. This may alter the cell's permeability or cause changes in enzyme activity within the cell. For instance, when the hormone epinephrine binds to its receptors on muscle and liver cells, it activates the enzymes that break down the cell's store of glycogen into glucose.

Most cell-surface receptors that bind these hormones act by way of small signalling molecules within the cell known as **second messengers.** Two of the most important second messengers are calcium ions (Ca^{2+}) and the nucleotide cyclic adenosine monophosphate (cyclic AMP) (Figure 30–3b).

Cyclic AMP is made by a roundabout process. First, the hormone binds to its receptor. This activates another membrane protein, which in turn activates a membrane enzyme to convert ATP into cyclic AMP. Cyclic AMP may cause many changes within the cell: it activates enzymes, alters membrane permeability, and even initiates protein synthesis, depending on the type of cell.

Calcium is another widespread second messenger, often acting in concert with cyclic AMP. For instance, binding of the hormone angiotensin to cells in the adrenal cortex triggers aldosterone secretion. The first effect of angiotensin is to increase the cells' permeability to calcium ions, which flow into the cells from the extracellular fluid. Calcium then alters the activities of various enzymes, including the one that produces cyclic AMP! The triggering of muscle contraction is an-

(a) **How lipid-soluble hormones work**

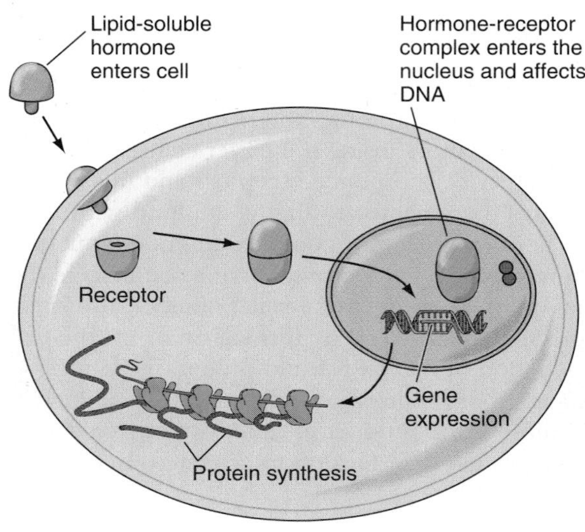

(b) **How water-soluble hormones work**

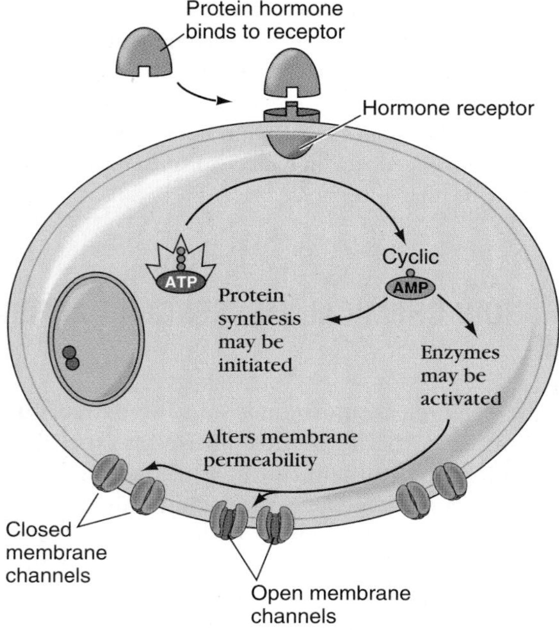

FIGURE 30–3

Two ways hormones affect cells. (a) Lipid-soluble hormones can cross the plasma membrane. Bound to their receptors, they usually cause the cell's DNA to transcribe messenger RNA for production of new proteins. **(b)** Some water-soluble hormones bind to receptors on the cell surface. This binding causes ATP inside the cell to be converted into cyclic AMP. The cyclic AMP may have many effects, such as altering the cell's metabolism and its membrane permeability.

other example of calcium as a second messenger (Section 29–B).

A hormone circulates through the body but affects only its target cells because only they have receptors for the hormone. Lipid-soluble and thyroid hormones enter their target cells and change gene transcription in the nucleus. Water-soluble hormones and local messenger proteins bind to surface receptors of target cells and activate second messengers such as cyclic AMP and calcium ions.

30–C HORMONAL AND NERVOUS CONTROL

It usually takes longer for a hormone to act on its target cells than for the nervous system to activate an effector, a muscle, or a gland. This is because it takes longer for a hormone to reach its target tissue, and because the hormone causes slower responses in the target tissue.

By having both nervous and endocrine control systems, the body is equipped to cope with a variety of situations. The nervous system enables an animal to escape from an enemy in a fraction of a second, and some hormones also act relatively swiftly. For example suckling by young mammals causes milk to let down (Figure 30–4). Suckling of the nipple sends sensory signals to the hypothalamus in the mother's brain. Cells in the hypothalamus extend into the pituitary, where they release the hormone oxytocin into the blood. Oxytocin reaches the mammary glands by way of the bloodstream and stimulates smooth muscles there to contract, pushing milk toward the nipple. This whole sequence of events takes less than a minute.

At the other extreme, other hormones keep the body pregnant for many months. Hormones are more suitable than nerves for controlling long-term changes involving many different organs, whereas nerves are more suitable for rapid reactions involving relatively few organs. Often the two systems acting together can control a situation more efficiently than either of them acting alone, as in the next example.

Fight or Flight

1? A hot, red face, cold perspiring hands, and a rapidly beating heart commonly precede a stage appearance

FIGURE 30-4

The letdown response. Letdown of milk from the mammary gland is a comparatively rapid hormone response to suckling by the young. (B. S. Rauch)

or an important exam. These symptoms are part of the "fight or flight" reaction, which prepares the body to meet stress or danger. We shall mention only a few aspects of this reaction here (Figure 30-5).

When a vertebrate senses danger or stress, the central nervous system stimulates the adrenal medulla (inner part of the adrenal glands) to release the hormones epinephrine and norepinephrine into the bloodstream. In addition, much of the sympathetic nervous system is activated, releasing more norepinephrine. Epinephrine and norepinephrine cause the heart to beat faster and to increase the volume of blood pumped per stroke. This raises the blood pressure and circulates the blood more rapidly. In addition, extra glucose is released into the bloodstream by the liver, raising the blood sugar level. Blood vessels in the muscles dilate, increasing the muscles' blood supply and preparing them for action by supplying them with extra oxygen and glucose. At the same time, constriction of the vessels supplying the kidneys and digestive tract reduces their supply of blood.

Nervous control evokes these reactions very rapidly in time of danger. Hormones provide a backup that can maintain the response for a long period. This explains why the state of "nervous energy" persists even after the performance is over or the exam paper handed in. ■

The Hypothalamus–Pituitary Connection

Most of the endocrine glands in the vertebrate body are controlled, directly or indirectly, by the brain. The nervous and endocrine systems interact by way of the connections between the hypothalamus in the brain and the pituitary gland. The **hypothalamus** is a small area of the forebrain lying in front of the midbrain. It receives input from all parts of the brain. Stimulation of various neurons in the hypothalamus elicits sensations and behaviors such as sex drive, pleasure, rage, fear, satiation, hunger, and thirst.

The hypothalamus produces some hormones, but it does not release them directly into the bloodstream.

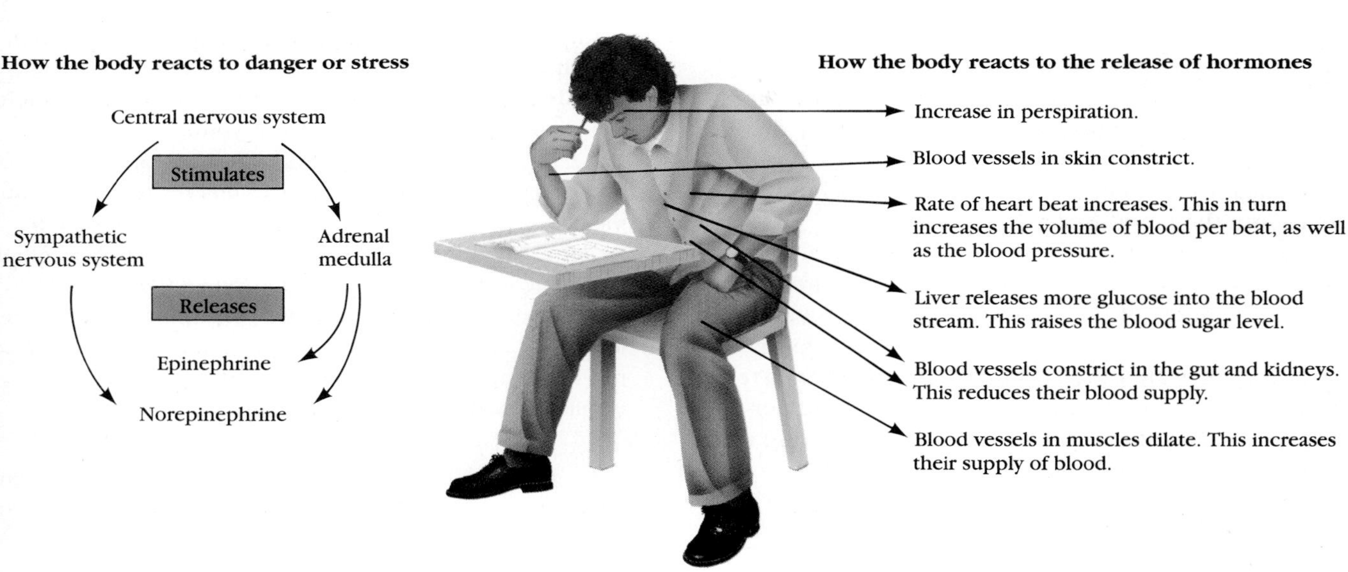
How the body reacts to danger or stress
Central nervous system
Stimulates
Sympathetic nervous system
Adrenal medulla
Releases
Epinephrine
Norepinephrine
How the body reacts to the release of hormones
Increase in perspiration.
Blood vessels in skin constrict.
Rate of heart beat increases. This in turn increases the volume of blood per beat, as well as the blood pressure.
Liver releases more glucose into the blood stream. This raises the blood sugar level.
Blood vessels constrict in the gut and kidneys. This reduces their blood supply.
Blood vessels in muscles dilate. This increases their supply of blood.

FIGURE 30-5

Fight or flight reaction. The stress of an examination may trigger the "fight or flight" reaction, mediated by both the nervous system and hormones.

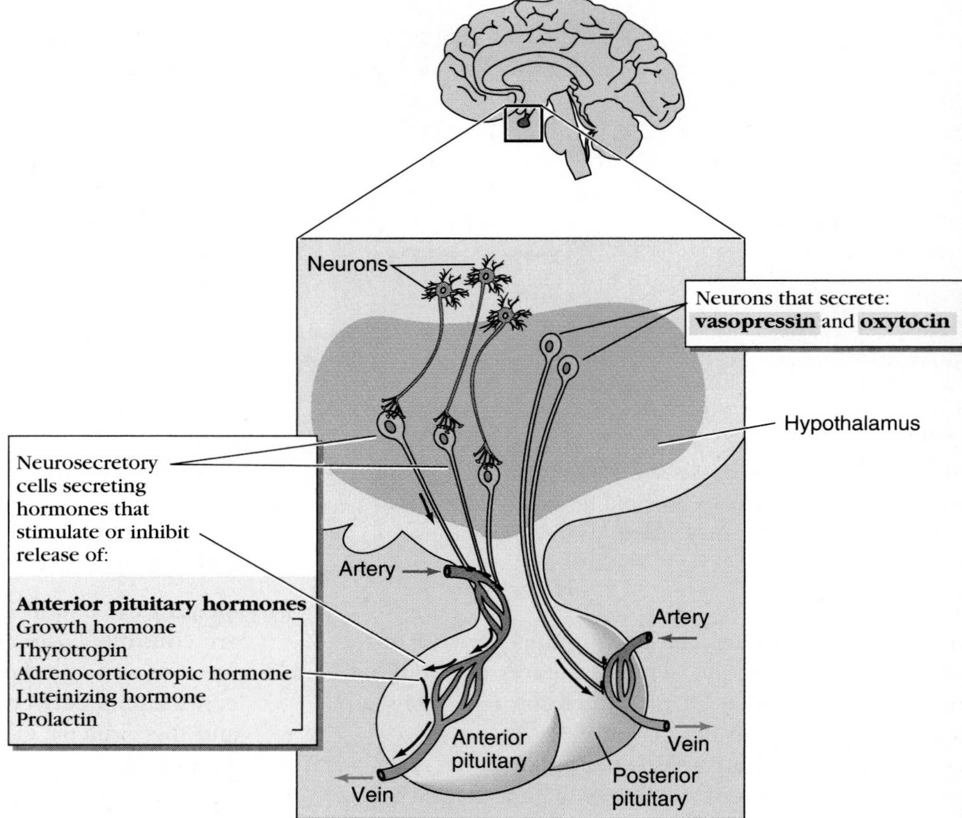

FIGURE 30–6

The interrelations of hypothalamus and pituitary in the vertebrate brain. The left part of the diagram shows neurons that release hormones into the blood as it passes through capillaries in the stalk of the pituitary gland. This blood then travels to the anterior pituitary, where it passes through a second set of capillaries. Here the hypothalamic hormones enter the extracellular fluid and stimulate release of corresponding pituitary hormones into the bloodstream. At the right are other neurons whose cell bodies lie in the hypothalamus, while their axons extend into the posterior pituitary, where they release hormones into the bloodstream. Most of the hormones shown here are described in Table 30–1.

Rather, these hormones travel down neurons from the hypothalamus to the posterior lobe of the pituitary gland, where they are released into the blood. The hypothalamus also produces hormones called **releasing factors,** which control the release of other hormones from the anterior lobe of the pituitary. When stimulated, secretory neurons in the hypothalamus discharge the releasing factors into capillaries, which carry them to the anterior pituitary (Figure 30–6). Here each releasing factor causes its target cells to release hormones into the bloodstream.

The hypothalamus is the body's single most important control center. Messages from sensory neurons, and the chemistry of the surrounding fluid, provide the hypothalamus with a continuous flow of information about the state of the body. The hypothalamus reacts to these stimuli by producing activity in the nervous system, by initiating behaviors such as feeding or mating, and by controlling the pituitary's secretion of hormones.

The **posterior pituitary gland** releases hormones such as **vasopressin** and **oxytocin,** which are produced in the hypothalamus. Many of the hormones secreted by the **anterior pituitary** induce other endocrine glands in various parts of the body to secrete their particular hormones. For instance, **thyrotropin** and **adrenocorticotropic hormone** cause hormone secretion by the thyroid and adrenal glands, respectively. The anterior pituitary also produces follicle-stimulating hormone (FSH) and luteinizing hormone (LH), which are necessary for hormone production by the gonads in all vertebrates, and growth hormone, which is essential for normal growth of young vertebrates.

The nervous system usually provides quicker responses than the endocrine system, but circulating hormones invoke responses that generally last longer than those from the nervous system. The nervous and endocrine systems are linked by the connections between the hypothalamus and the pituitary gland. The hypothalamus receives information from the rest of the nervous system and stimulates the secretion of circulating hormones by the pituitary.

30–D LOCAL CHEMICAL MESSENGERS

Hormones travel in the bloodstream and so they reach, even though they do not affect, most of the body's cells. By contrast, many other chemical messengers are secreted into the extracellular fluid and absorbed by neighboring cells or destroyed so rapidly that they never reach the bloodstream. Many examples of these local messengers have been discovered since the 1960s, and more probably remain to be described.

Some local messengers are secreted only by cells specialized for the purpose. In wounded tissue, for instance, histamine is secreted almost entirely by mast cells. Other local messengers are probably produced by all cells under certain conditions.

Prostaglandins

Prostaglandins are a group of about 20 molecules made from fatty acids. They occur in most vertebrate tissues and have many different actions as local chemical messengers.

Some prostaglandins stimulate smooth muscle to contract, and others cause it to relax. Some cause constriction of capillaries, and others cause dilation. Prostaglandins are involved in many different aspects of reproduction, and in menstrual cramps, allergic reactions to food, and the inflammatory response to infection. Aspirin inhibits prostaglandin synthesis, and this is thought to be related to aspirin's ability to inhibit inflammation, lower fever, and reduce pain.

Some prostaglandins have medical uses: to induce labor in childbirth, to promote healing of stomach and duodenal ulcers, to relieve asthma, and to synchronize the reproductive cycles of livestock for breeding.

Unlike most other chemical messengers, prostaglandins are not stored. They are produced in membranes, and are continuously released to the exterior of the cell. They are also continuously destroyed in the extracellular fluid. When a cell is activated by changes in its environment, however, it may increase the rate of prostaglandin synthesis and release enough prostaglandin so that the messenger influences both the cell that produced it and its neighbors.

Neurotransmitters As Local Messengers

Many neurotransmitters are deactivated as soon as they have crossed the synapse to the postsynaptic cell, and so they never escape from the synapse (Section 28-C). If all signals in the nervous system were directed at such nearby targets, like this, very few neurotransmitters would be needed because there would be no chance of confusion. But in fact more than 30 different neurotransmitters have already been identified in the vertebrate brain alone. Biologists are beginning to realize that many neurotransmitters diffuse out of synapses and so come into contact with many different cells, some of which have receptors for them. Neurotransmitters that behave in this way really act as local hormones and may be distinguished by the special name of **neuroregulators.**

2? Sleep is at least partly controlled by a neuroregulator. If cerebrospinal fluid is extracted from a sleeping animal and injected into one that is wide awake, the second animal promptly goes to sleep; the fluid appears to contain a "sleep neuroregulator." Sleep appears to require the suppression of a large group of specific nerve cells in a particular part of the brain. A neuroregulator can cause this suppression by affecting all the neurons that bear receptors for the sleep neuroregulator. The neurons need not be completely "turned off" by the neuroregulator: they may remain responsive to other neurotransmitters. ■

Growth Factors

In many tissues, cell division is controlled by conventional circulating hormones, such as the anterior pituitary hormone **somatotropin** (also called **growth hormone**). Infants who produce too little somatotropin become dwarfs.

Other growth factors do not circulate in the blood but act as local messengers. For example, **bone growth protein** (Section 29-E) and **nerve growth factor** are produced locally during development and are necessary for bones and nerves to develop normally.

Local chemical messengers act near the cells that secrete them. They include prostaglandins, neuroregulators secreted by some neurons, and some growth factors.

30–E HORMONES AND SEASONAL CHANGES

Animals respond to information about the outside world that is sent to the nervous system by way of the sense organs. Although most reactions, such as eating or running away, are carried out by muscle contractions, some are mediated by hormones. Most of the responses to the environment involving hormones are slow changes, such as occur when an animal comes into breeding condition in the spring.

Environmental Control of Reproduction

There is selective pressure for animals to reproduce under conditions that favor survival of their offspring. For most animals this means birth or hatching in the spring, when warm weather and plentiful food offer the best possible conditions. Breeding often involves dramatic changes in an animal's anatomy, physiology, and behavior.

Animals come into breeding condition in response to environmental conditions such as daylength, temperature, rainfall, food, and so on. Animals detect the external stimuli that induce breeding by way of their sensory receptors. For instance, light may be detected by the eyes or by the **pineal gland,** the "third eye" that lies in the middle of the top of the head in many lower vertebrates. The appropriate stimulus causes hormone production by way of the hypothalamus. The hypothalamus does two things: first, it stimulates the pituitary to release hormones that cause the gonads to grow and produce sex hormones. Second, the sex hormones feed back to the hypothalamus, which then initiates reproductive behavior by sending out the appropriate signals in the nervous system.

The sensory systems of animals detect changes in the environment. Under the proper conditions, the endocrine system is signaled to release hormones that will prepare the animal for the reproductive season.

30–F BIOLOGICAL RHYTHMS

Reproductive cycles are examples of rhythmic or cyclical events in an animal's life. A number of other physiological and behavioral cycles also exist.

Circadian Rhythms

All eukaryotes, even unicellular protists, show daily cycles in many physiological processes. For instance, in most vertebrates the metabolic rate, body temperature, blood sugar level, urine composition, general level of nervous activity, and many other functions alter regularly in a 24-hour cycle (Figure 30–7).

Because 24 hours is the period of the Earth's rotation and of one light-dark cycle, it might seem obvious that daily cycles would be controlled by the onset of light or of dark. Is this the case, or is the rhythm **endogenous** ("built into" the animal)? To answer this question, investigators isolate an animal from cues that might tell it the time of day, and see whether the daily rhythm persists. When the animal is kept in darkness with constant temperature and humidity, it still maintains these rhythms, but the cycles are no longer exactly 24 hours. Because the rhythms repeat approximately, but not exactly, every 24 hours in the absence of external cues, they are called **circadian rhythms** (circa = about; dies = day). These endogenous rhythms must be related to an "internal clock."

In a normal environment, light from the sun keeps our body clocks set on a 24-hour cycle. The eyes detect light, which stimulates events in the nervous system that adjust the metabolism of certain cells in the hypothalamus of the brain. When the eyes are exposed to light at night, changes in DNA transcription can be detected in these hypothalamic cells, suggesting that protein synthesis may be involved in resetting the clock.

It is possible to reset the circadian clock artificially. For instance, if a vertebrate is kept in a controlled environment where the light is switched on only during the night, night becomes day and vice versa for the animal, and its daily rhythm soon becomes reset exactly 12 hours later. Many zoos do this to nocturnal animals (those that are usually active only at night) so that they become active during the day when visitors want to see them. These animals are exposed to bright light at night, and to a dim red light during the day. Their internal clocks are set so that the animals act as though it is night during what is really daytime. (See *A Journey into Healthy Living:* Our Daily Spread, this chapter.)

Why did circadian rhythms evolve in the first place? Night and day always follow each other on a 24-hour cycle, and so there is a perfectly good stimulus available to trigger daily cycles. Why should an organism also have an internal cycle of its own? The answer is probably that an animal's internal rhythm permits it to anticipate regular daily events before environmental cues appear. This ability is valuable in many ways. It permits a bat to start hunting at the time of day when its insect prey will also be active, without wasting the energy it would take to check on the light level or insect activity in the area every few hours.

Annual Rhythms

In addition to circadian rhythms, some animals have yearly cycles. Many mammals continue to hibernate at roughly the right time even if they are deprived of environmental cues that could tell them the time of year. Deer kept under constant conditions grow, and later shed, their antlers at the same time every year. Several species of birds start migratory behavior at the same time of year even if they are deprived of environmental cues. In this last case, the adaptive advantage of the annual rhythm is probably that it permits a bird to return north for the breeding season even though it receives few cues of seasonal change in the relatively constant tropical environment where it winters.

Biological Clocks

The existence of circadian and annual rhythms raises the question, "How does it work?" Organisms plainly have internal "clocks" of some sort that keep track of

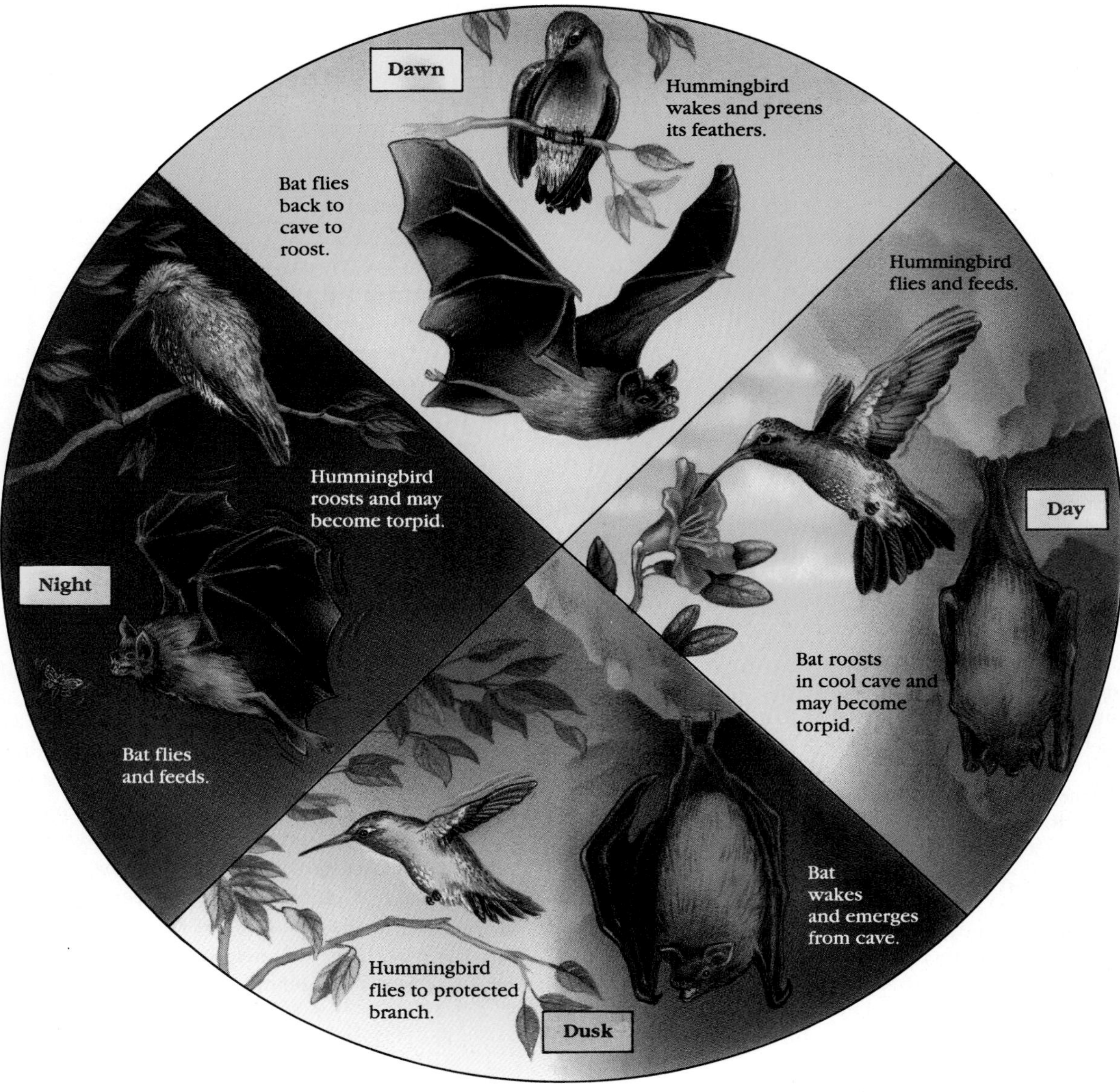
Dawn
Hummingbird wakes and preens its feathers.
Bat flies back to cave to roost.
Hummingbird flies and feeds.
Day
Hummingbird roosts and may become torpid.
Night
Bat roosts in cool cave and may become torpid.
Bat flies and feeds.
Bat wakes and emerges from cave.
Hummingbird flies to protected branch.
Dusk

FIGURE 30-7

Circadian rhythms. Many animals, such as hummingbirds and bats, restrict certain activities to particular parts of the day.

> **BIO-BIT**
>
> During the dark winter days in extreme northern countries, some people suffer from periods of lethargy, depression, and a craving for carbohydrates resulting from seasonal affective disorder (SAD). The symptoms seem to be related to an overproduction of the hormone melatonin, as a result of insufficient exposure to sunlight.

time and can act as "pacemakers" to run these rhythms independently of environmental stimuli.

The vertebrate body appears to contain many different clocks (including at least two in the hypothalamus), each with its own endogenous rhythm. Normally, environmental cues keep all the body's clocks set to the 24-hour daily cycle but we are still far from understanding how all this works.

Circadian rhythms may be remarkably persistent. For example, a reptile continues to show daily fluctua-

A Journey into Healthy Living

Our Daily Spread

An American will tell you that the normal adult human body temperature is 98.6°F, or 37°C. European fever thermometers, on the other hand, give the normal temperature as 98.4°F. In fact, normal body temperature is anywhere from 36 to 37.6°C (96.8 to 99.7°F). If you take your temperature first thing in the morning, it is likely to be at the low end of this range, and if you keep taking it every hour or two, it will probably reach a high point late in the evening.

Many other physiological factors vary during the day, including blood pressure, heartbeat rate, excretion of various ions in the urine, secretion of different hormones, alertness and reaction time, and the ability to detect faint sounds. Each factor has its own predictable daily pattern, which may differ from the patterns of other factors (Figure 30–A). For example, one hormone may peak during the middle of the night, and another first thing in the morning, whereas body temperature peaks shortly before bedtime.

Human Circadian Rhythms

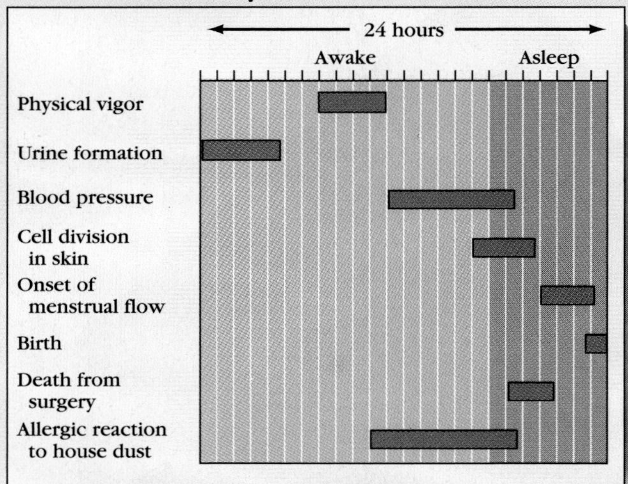

FIGURE 30–A

Human circadian rhythms. Activities of the human body are not uniform throughout the day: different factors or events tend to be higher or more frequent at different times in most people. The 24-hour time span shown begins, at the left, at the time when people get up in the morning (this varies from person to person, so no clock times are given). The red bars show the periods during the day when the high value of each function occurs. For example, urine formation is highest soon after rising, and cell division in the skin is highest just after going to sleep. Likewise, babies are not born uniformly throughout the day; there is an excess of births in the hours just before the mother would normally wake up.

Are these regular daily patterns in human body functions triggered by external cues, or are they endogenous? To distinguish between these two possibilities, experimenters have put human volunteers into isolation in caves, underground apartments, or windowless rooms for periods of several weeks. The subjects have no clocks or watches, no sunlight, no radio or television—nothing to tell them what time of day it is. They eat, sleep, read, or exercise when they want to. Under these conditions, people drift from their almost exactly 24-

tions in body temperature even when it has been kept in the dark at a constant temperature for months. Another intriguing feature of circadian rhythms is their ability to adjust to changes in temperature. Since the rates of biochemical processes usually vary with temperature, we would expect that at lower temperatures the rhythm would be slower. When the temperature of the environment is lowered, the animal's rhythm does slow down, but within a few days the rhythm adjusts to the new temperature and resumes its natural frequency. How this temperature compensation is made is an interesting puzzle.

hour pattern of normal daily life into a new rhythm that depends on the person, frequently about 25 hours.

Ignorance of circadian rhythms can be hazardous to our health. For instance, blood pressure tends to be low in the morning and to rise during the day. People who always visit the doctor in the morning may have blood pressure readings that fall in the normal range, but their blood pressure may rise dangerously in the afternoons and evenings. These people will not receive the treatment needed to control their blood pressure.

Our circadian rhythms probably make themselves felt most acutely when we try to reset our internal clocks. People who have travelled by airplane across several time zones experience "jet lag," and people who work rotating shifts, such as airline flight crews, air traffic controllers, police, and military and hospital personnel, have similar problems. The trouble is that some rhythms reset to the "new time" more quickly than others, and so the body goes through a period when its rhythms are out of synchronization with one another. This can have serious consequences. The famous accident at the Three Mile Island nuclear power station would probably have been quite minor if the people on duty had been more alert. Their errors in handling the emergency may well have arisen from the shift rotation schedule followed at the plant. In one study, more than half of rotating-shift workers reported falling asleep on the job, including truck drivers and nuclear power plant operators. Some people have also developed ulcers because of the continuous stress of having desynchronized body rhythms.

The study of circadian rhythms has produced information that can help in the scheduling of shift rotations for the 20 to 30 million people in the United States who work at such jobs. For example, it is easier to reset the internal clock to a later time than to an earlier time. This explains why we adjust more easily to travel from east to west than in the opposite direction. It also suggests that rotating shifts should be scheduled so that people work the day, evening, and night shifts in that order, rather than switching to nights and then evenings after the day shift. Furthermore, since it takes up to a week for a person to adjust to the new schedule, workers should be left on each shift for as long as possible. Workers at a mineral mine in Utah were switched from spending a week on each shift and then rotating to an earlier one, to a new schedule, on which they worked three weeks at each shift and then rotated to a later one. They reported increased job satisfaction and better health on the new schedule, and no one fell asleep during work hours anymore. The employer gained, too, in less employee turnover and in increased output.

Related research shows that people tend to wake up when their body temperature begins to rise, regardless of how long they have slept or how long they were awake before going to sleep. This explains why people whose travel or work schedules force them to go to sleep at unusual times may wake up exhausted: their internal alarm, which is linked to deep body temperature, wakes them up when the temperature begins to rise though they have not had enough rest to recover from previous waking periods.

Researchers are experimenting with ways to reset the human biological clock using meals, bright lights, or exercise at specific times of the day. This may eventually help people adjust to work shifts and jet lag more quickly and with less discomfort and danger.

Animals display daily and annual rhythms of physiological and behavioral activities that persist even when changes in daylight, temperature, and other cues are removed, suggesting that an internal clock regulates these cycles. Current theories suggest that more than one physiological clock is involved in determining biological rhythms and that these clocks are reset in response to environmental cues such as length of daylight and changes in temperature.

30–G PHEROMONES

A **pheromone** is a chemical that travels outside the body, carrying information to other members of the same species (Figure 30–8). The first pheromones described were sex attractants from insects. In many species of moths, beetles, cockroaches, and flies, the female releases a chemical that attracts the male. He

FIGURE 30–8

Pheromones. Pheromones help organize the activity of individuals in insect societies, such as this termite colony. (J. P. Jackson)

finds his mate by flying or crawling up the odor gradient toward her.

3? Many vertebrates, and particularly mammals, use pheromones in urine or feces, or from special scent glands, to mark trails and territories. When a dog urinates on a fire hydrant, he is depositing a pheromone that tells other dogs that the hydrant is part of his territory. Pheromones also permit members of one sex to distinguish which members of the opposite sex are in breeding condition. ■

The best evidence for human reproductive pheromones comes from anecdotal evidence that when numbers of women live together, in dormitories or similar situations, their menstrual cycles eventually synchronize. Other than this, there is no convincing evidence for human pheromones, despite the large number of reports on the subject that appear periodically.

In the insect societies of bees, ants, and termites, pheromones organize not just the reproduction but also the behavior and social structure of a colony (Section 31–J).

Pheromones function like hormones except that they act on other individuals of the same species, instead of acting on different cells in the body of the animal that secretes them.

BIO-BIT

In some moth species, the male can detect the female's pheromones from as much as 2.7 miles (4.3 km) away!

SUMMARY

1. The nervous and hormonal systems carry messages that travel between an animal's cells and coordinate their activities.
2. Neurons affect specific cells briefly. Hormones may act on a wide range of cells over a long time.
3. Animals have three types of chemical messengers:
   a. hormones, produced by endocrine cells or by secretory neurons, and carried all over the body by the blood,
   b. local hormones, such as histamine, prostaglandins, growth factors, and neuroregulators, which usually act near the cells that produce them, and
   c. pheromones, which carry information between different individuals of the same species.
4. Hormones are involved in homeostatic mechanisms within the body, in growth and differentiation of tissues, and in many of an animal's responses to its environment, such as reproductive cycles.
5. The effects of each hormone or other chemical messenger are limited to those specific target cells with receptors for the hormone.
6. Lipid-soluble and thyroid hormones attach to receptors in the target cell's cytoplasm and enter the nucleus, where they affect gene expression.
7. Water-soluble hormones do not enter the cell but instead cause their cell-surface receptors to make changes inside the cell, often by way of second messengers such as calcium ions or cyclic AMP.
8. Hormones usually change the cell's permeability or enzyme activity. Thus, tiny amounts of hormones can drastically affect the target cell's activity.
9. Hormone secretion is usually controlled by negative feedback in response to some disturbance in the body or to the level of another hormone in the blood.
10. Hormones and the nervous system often interact with one another to control both long- and short-term aspects of an animal's response to a stimulus. The interaction between the two systems occurs mainly in the hypothalamus.
11. The hypothalamus receives nervous signals from the sense organs by way of other areas in the brain, and also detects changes in blood chemistry. It initiates appropriate responses by way of the nervous system and pituitary gland, which release hormones that carry messages to many of the body's other endocrine glands.
12. Animals have internal biological clocks that tell them the time of day. The 24-hour clock is reset every day by stimuli that include light.

SELF-QUIZ

1. Which of the following techniques would be *least* likely to help in determining the function of a hormone produced by an endocrine gland?
   a. removal of the gland and subsequent analysis of what functions are lost
   b. transplantation of the gland into an animal that lacked the gland
   c. transfusions of blood from an animal lacking the gland into an animal that has the gland and observation of its effects
   d. observing effects of gland extract on various tissues grown in culture
   e. observing the condition of animals with a tumor that causes the gland to be overactive
2. All of the following commonly serve as signals that stimulate hormone secretion *except:*
   a. conditions outside the body
   b. rising levels of another hormone
   c. rising levels of the hormone in question
   d. falling levels of the hormone in question
   e. falling levels of another hormone
3. Hormones are known to cause all the following changes in target cells *except:*
   a. changes in genetic makeup
   b. changes in permeability
   c. changes in metabolic rate
   d. increase in cyclic AMP concentration
   e. synthesis of different messenger RNA and proteins
4. An advantage to having the endocrine system as well as the nervous system involved in the "fight or flight" response is:
   a. the endocrine system responds faster
   b. the endocrine response usually lasts longer
   c. the endocrine system is tuned more precisely to the degree of need
   d. the endocrine system affects only the target organs whose response is needed to meet the emergency
   e. response by the endocrine system frees the nervous system to think of a way out of the situation instead of simply maintaining the body in an alert state
5. a. Thyroid-stimulating hormone (TSH) is released from the anterior pituitary in response to low levels of the hormone thyroxin in the blood. Higher levels of thyroxin reduce the release of TSH. This regulation of levels of TSH and thyroxin is called ___.
   b. Receptor proteins for TSH exist only on cells in the thyroid gland. These thyroid cells are thus designated the ___ of TSH.
6. Using the secretion of thyroxin as an example, diagram the feedback control system for hormone production (see Question 5).

THINKING CRITICALLY

1. Thyroxin levels are generally at about 100 units. A patient has only 80 units in the bloodstream. Normal therapy for this situation is to inject the extra 20 units. Why might this therapy not be effective?
2. Why does the blood of a male mammal contain a much higher concentration of luteinizing (and follicle-stimulating) hormone after he has been neutered than it did before the operation? (Hint: see Figure 30-2)

SELECTED KEY TERMS

adrenocorticotropic hormone, *p. 636*
chemical messenger, *p. 629*
circadian rhythm, *p. 638*
endocrine gland, *p. 629*
endogenous rhythm, *p. 638*
growth hormone, *p. 637*
hypothalamus, *p. 635*
local chemical messenger, *p. 630*
luteinizing hormone, *p. 632*
negative feedback control, *p. 632*
nerve growth factor, *p. 637*
neuroregulator, *p. 637*
oxytocin, *p. 636*
parathyroid hormone, *p. 632*
pheromone, *p. 630*
pineal gland, *p. 638*
pituitary gland, *p. 636*
second messenger, *p. 633*
target cell, *p. 633*
testosterone, *p. 632*
vasopressin, *p. 636*

SUGGESTED READINGS

Berridge, M. J. "The molecular basis of communication within the cell." *Scientific American,* October 1985. Second-messenger pathways in cells.

Coleman, R. M. *Wide Awake at 3:00 A.M. By Choice or by Chance?* New York: W. H. Freeman, 1986. An entertaining account of research on human biological clocks and recommendations for coping with time shifts.

Crapo, L. *Hormones: Messengers of Life.* San Francisco: W. H. Freeman, 1985.

Stricker, E. M., and J. G. Verbalis. "Hormones and behavior: the biology of thirst and sodium appetite." *Scientific American,* May 1988.

Uvnas-Moberg, K. "The gastrointestinal tract in growth and reproduction." *Scientific American,* July 1989. The GI tract treated as the largest endocrine organ in the body and the role it plays in adjusting metabolism to pregnancy and infant growth.

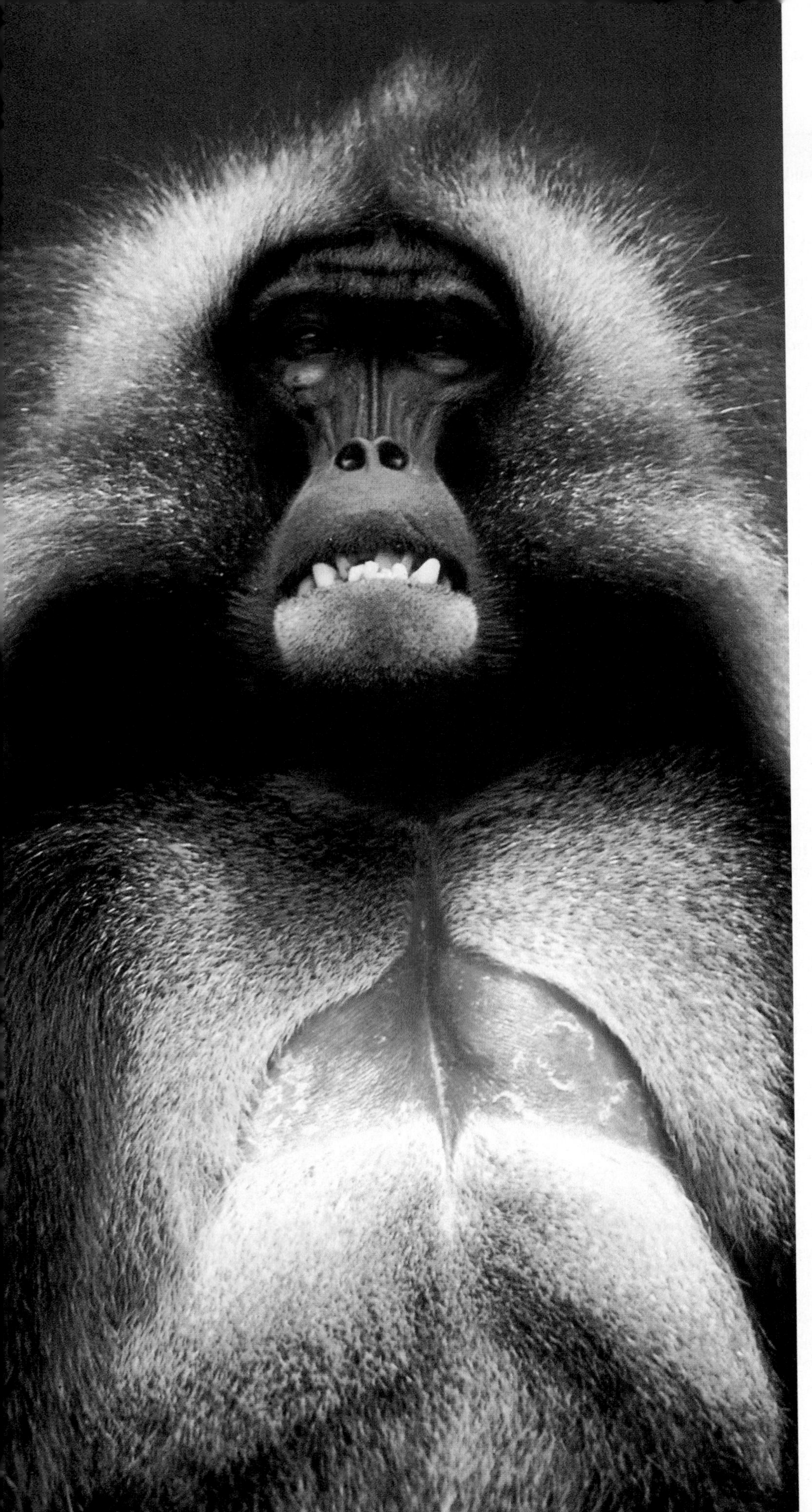

CHAPTER 31

CONCEPT GUIDE

After reading this chapter, you should be able to:

1. Explain how genes and the environment can influence behavior. (Sections 31-A, 31-B, and 31-C)
2. Discuss the circumstances that favor instinctive behavior and those that favor learned behaviors. (Section 31-D)
3. Describe two unique characteristics of stereotyped behaviors. Explain how sign stimuli and motivation affect behavior patterns. (Section 31-E)
4. Describe how latent learning and insight learning are different from types of associative learning. (Section 31-F)
5. Explain the relationship between conflict and courtship behaviors. (Sections 31-G and 31-H)
6. Describe four adaptations that permit many long-distance migratory animals to find their way to precise locations. (Section 31-I)
7. Describe the adaptive advantages of animal societies. (Section 31-J)

A threatening male gelada baboon. Read about territorial behavior and courtship in Sections 31-G and 31-H. (R. Van Nostrand/Photo Researchers, Inc.)

Behavior

We are prone to conclude that a dog is "ashamed" when it puts its tail between its legs and sneaks into a corner after a spanking, and "happy" when it wags its tail. This type of thinking is called **anthropomorphism,** ascribing human emotions to animals. At the opposite extreme, we may say that a bird sings from instinct, because it is incapable of behaving intelligently. The view of human behavior as intelligent and that of other animals as instinctive is reflected in the tendency to ascribe actions of which we are ashamed to "animal instincts." Neither of these approaches to animal behavior gives much useful insight into why animals behave as they do. Recent research has attempted to study animal behavior with as little bias as possible from our human perspective.

In many ways, natural selection acts more directly on behavior than on anything else. Dozens of different behavior patterns may distinguish the individual that reproduces from the one that does not, and all the adaptations of an animal's anatomy and physiology are useless if the animal does not feed itself, escape predators, and find a mate.

An animal's genes determine the range of characteristics it can develop, but just which traits develop depends upon interactions between the genes and the environment. In this chapter we shall consider how behaviorists may try to disentangle the genetic and environmental influences on an animal's behavior, and the kinds of selective pressures that have produced the varied behavioral repertoires of different animals.

KEY CONCEPTS

- An animal's genes determine the range of behavior patterns it can develop in response to environmental stimuli.
- Behaviors that must be produced perfectly the first time they are performed are usually innate. Learning produces behavior that must be flexible to meet local or changeable conditions.
- Many behavior patterns become programmed into the nervous system.
- An animal responds to a particular stimulus only when it is in an appropriate physiological state.
- Some insect and some vertebrate species have evolved societies made up of related individuals, which communicate and cooperate with each other.

Curiosity Questions

1? Why do some animals defend territories? (See page 653)

2? Why do some animals have elaborate courtships? (See page 654)

3? How do birds migrate over great distances and arrive at the proper destination? (See page 655)

31–A SHORT- AND LONG-TERM CAUSES OF BEHAVIOR

A frog is sitting in the grass when a fly buzzes past. Zip! The frog's tongue flicks out and pulls the fly into the frog's mouth. How and why does the frog do this? The question "how" can be answered by describing the frog's sensory, nervous, and muscular systems. The stimulus of a moving fly before the eyes sends impulses along sensory neurons to the central nervous system, which in turn activates and directs the tongue muscles used to catch the fly.

The question of "why" the frog catches the fly is different because it can be answered on two levels. The immediate reason is that the behavior pattern results from a nervous reflex. Seeing a fly activates a reflex that results in the frog's striking at the fly. However, there is also a more fundamental answer to the question "why?" That particular behavior pattern exists because it has been selected for during the course of evolution.

Three main selective pressures have brought about the behavior patterns we see today:

1. Ultimately, an animal's behavior patterns will be selected for as they contribute to its reproductive success. It is occasionally possible to see how (why) a behavior pattern has evolved by showing that not performing the behavior is selected against. For instance, many birds remove the empty egg shell from the nest after the young have hatched (Figure 31–1). Niko Tinbergen, a well-known animal behaviorist, showed that adult Black-headed Gulls that did not remove the shells lost more chicks to predators; the white inside of the egg shell allowed certain predators to discover the otherwise camouflaged nest and chicks.
2. Behavior patterns must allow an animal to solve immediate problems. Hungry animals must feed, and hunted animals must escape predators, if they are to survive and reproduce.
3. Sights, sounds, and other environmental stimuli continually bombard all animals. Behavioral adaptations permit an animal to detect stimuli that are important to survival or reproduction, and then to carry out behavior patterns appropriate to those stimuli (Figure 31–2). Means for discriminating between stimuli and for ensuring that an animal completes a behavior pattern are crucial parts of any animal's behavioral makeup.

FIGURE 31–1

Adaptive behavior. Black-headed Gulls remove empty egg shells from their nests.

FIGURE 31–2

Reacting to a stimulus. This time-lapse photograph shows a lacewing taking off when frightened by movement. Many behavior patterns are immediate responses to environmental stimuli. (S. Dalton/Oxford Scientific Films)

An animal's behavioral repertoire consists of activities that were selected for during evolution because they promoted the survival and reproduction of its ancestors in similar circumstances.

31–B GENES AND ENVIRONMENT

Most genes influence behavior because they control an animal's anatomy and physiology, and therefore determine how it can and does respond to its environment. In laboratory populations of the fruit fly *Drosophila,* single genes have been identified that cause a fly to do such things as court members of the same sex, copulate for much longer than normal, or follow a 19-hour rhythm of activity instead of the usual 24-hour cycle.

These behaviors result from mutations that cause some abnormality in the sense organs or in the nervous or muscular systems.

Environment affects behavior in two main ways. In the short run, animals perform many behavior patterns only when they are induced to do so by environmental stimuli. In the long run, the environment influences gene expression in the development of many behavior patterns.

One reason behavior is difficult to study is that, in animals with complex nervous and muscular systems, genes and environment continue to interact and alter an animal's behavior throughout its life.

Genes influence behavior by affecting the structure or function of sense organs, nerves, muscles, hormones, and other physiological systems.

FIGURE 31-3

Imprinting. Sexual partners are often determined by imprinting. This Peregrine Falcon is in a mating stance, displaying to this human researcher. (F. Lanting/Univ. of California at Santa Cruz)

31-C DEVELOPMENT OF BEHAVIOR

An animal's behavior, like its anatomy and physiology, forms during its development, through the interaction between its genes and its environment.

There are often critical periods when a particular environmental influence must be present if a particular behavior pattern is to appear. An example occurs during **imprinting** of young animals. Goslings (young geese) and ducklings learn to follow their parents, and to respond to their parents' signals, during a critical period after they hatch. Konrad Lorenz, another pioneer of animal behavior, found that young birds would follow him as if he were their mother if they saw him, rather than their mother, during the critical period. Many animals learn what their future mates will look like by a similar process of sexual imprinting during a critical period (Figure 31-3). (Lorenz had a tame jackdaw [Figure 31-4] that unfortunately became sexually imprinted on him before he understood how the process worked. The jackdaw caused Lorenz great inconvenience by stuffing regurgitated worms into his ear during its "courtship feeding.")

FIGURE 31-4

A European jackdaw. Male jackdaws offer morsels of predigested worms to potential sexual partners as part of their courtship behavior. (Leonard Lee Rue, III)

The interactions between genetics and environment in the development of behavior have been studied intensively in the case of bird song. Peter Marler and Masakazu Konishi showed that male White-crowned Sparrows reared in isolation sing only the inherited song of their species, whereas in the wild, the sparrows learn the dialect of their own local population by listening to adult birds sing. To complicate matters, a bird must hear the dialect during a critical period when it is about three months old if it is to produce the dialect when it first begins to sing, at the age of one year. And even if it has heard the dialect during the critical period, it will never sing this dialect correctly if it is deafened before it has also sung the dialect. Once a bird has sung the full dialect song, however, deafening has no effect on its further performance (Figure 31-5). This example shows that many factors may be involved in the development of a normal adult behavior pattern.

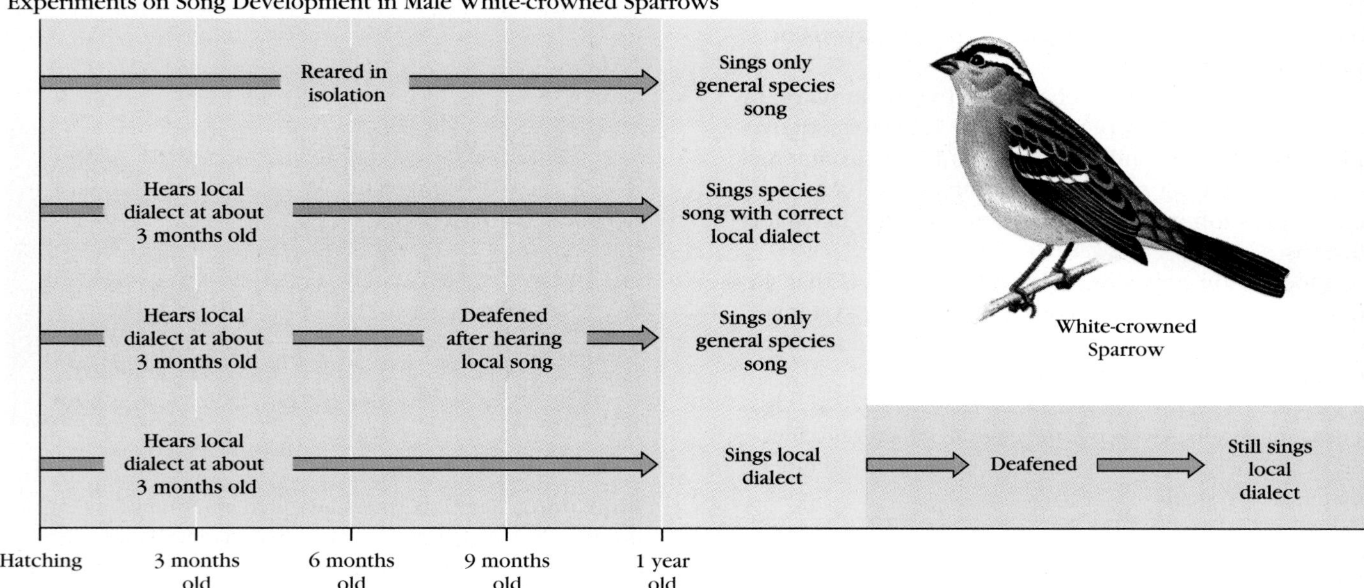

FIGURE 31-5

Development of behavior. Summary of Marler and Konishi's findings on song development in White-crowned Sparrows.

One further conclusion that can be made from this work is that an animal inherits a tendency to learn some behavior patterns but not others. White-crowned Sparrows learn the dialects of their own species, but exposing them to the songs of other (even closely related) species during the critical period does not make them learn the songs of these other species. Birds treated in this way end up with a song like that of the completely isolated male.

The environmental stimulus necessary to the normal development of a behavior pattern may be very precise, or it may be rather general. Rats or mice that have been picked up and returned to the nest once or twice a week in their youth mature more quickly, in behavioral terms, than do those never handled. This is a rather unspecific stimulus to the maturation of behavior.

Environment influences behavior by providing stimuli that may evoke a behavioral response immediately or may have longer-term effects on the development of the behavior. Many behavior patterns originate as a result of interactions between the developing nervous system and environmental factors normally encountered during this period.

31–D INSTINCT VERSUS LEARNING

The idea that a behavior pattern is either instinctive or learned (or, more often, a combination of the two) is common among behaviorists and in everyday life. **Instinctive** or **innate** (inborn) behavior is genetically programmed into the nervous system and is difficult to alter. Learned behavior is acquired or lost as a result of experience.

On the one hand, experimental psychologists have demonstrated that most animals are capable of learning many things. On the other hand, field behaviorists have shown that members of the same species tend to show identical behavior patterns in the wild, suggesting that much of behavior is instinctive. "Instinct" is a difficult concept, however, because it is hard to define except by negatives: instinctive behavior develops without the animal's having to learn it. Such negative definitions are notoriously difficult to use. Furthermore, the only possible experiment to determine whether or not a behavior pattern is instinctive is to deprive the developing animal of as many environmental stimuli as possible and see if the behavior pattern still appears. Even if the pattern does appear under such circumstances, it may still not be instinctive. The experimenter might merely have failed to remove the stimuli that permit the animal to learn the behavior.

BIO-BIT

Spiders instinctively know how to build their elaborate and complex webs.

FIGURE 31–6

An example of invertebrate learning. This ghost crab learns how to find its home and how to locate feeding sites on the beach. For example, it will dig down into a loggerhead turtle nest several feet below the surface of the sand, and return, night after night, to eat the eggs. (S. J. Krasemann)

Adaptive Value of Learned and Innate Behaviors

One way to make sense of the diversity of behavior is to consider the selective advantages of the two extremes: innate and learned behavior.

As an example, Kittiwakes are sea birds that nest on narrow ledges. The chicks keep still from the moment they hatch, whereas related Herring Gull chicks move around. This innate behavior (or nonbehavior) of Kittiwakes is clearly adaptive because a false step means death to a Kittiwake chick. There is no room for learning. Innate behavior also saves energy. There is clearly a selective advantage to the animal that does not have to waste energy learning responses that are sure to be required frequently and without variation.

With all of these advantages of innate behavior, why are so many behavior patterns learned? In particular, why are vital behavior patterns, such as recognizing a mate or learning to fly, so often partly learned? Learning gives an animal the flexibility to adapt to a changing environment by acquiring new behavior patterns as they become appropriate, or by responding in new ways to old stimuli. For animals such as vertebrates, which have relatively long lifespans and so experience changing environmental conditions, this flexibility often means the difference between life and death.

Learning is also necessary whenever a stimulus differs for individual members of a species. For instance, every mobile animal with a home base must learn to find that home. No amount of genetic programming will permit a crab to find its own burrow among all the holes in a sandy beach (Figure 31–6). In addition, many social animals live in environments where the relationships between individuals change constantly, and such relationships must almost always be learned.

A species' way of life also determines whether its members evolve learned or innate behaviors. Consider a solitary wasp, which hatches alone, develops as a larva and matures without interacting with other members of her species. The behavior by which she finds a male, mates, builds a nest, and lays her eggs must be largely innate in order for her to perform each action perfectly the first, and perhaps the only, time in her life. On the other hand, a social animal such as a wolf can learn much of its behavior from observing other members of its group. It would, however, be an enormous oversimplification to say that the behavior of an insect is innate, and the behavior of a mammal is learned. Even the solitary wasp learns to search for food, to find her way back to her nest, and many other behavior patterns during her short life. Figure 31–7 demonstrates one type of learned behavior in a digger-wasp. Similarly, mammals have many innate behavior patterns.

Innate behavior typically develops where the stimulus is always the same, speed of reaction is important, and the cost of an initial mistake is high. Learning plays a large part in developing appropriate responses to local or changeable conditions, particularly in animals with long life spans.

31–E THE NEURAL BASIS OF BEHAVIOR

At a physiological level, a behavior pattern is the action of an animal's effectors (muscles, glands, and so on) in response to a stimulus detected by its receptors. Between receptor and effector lies the nervous system, which determines what information travels from one to the other. In many ways, the nervous system is still the "black box" of behavior. The stimulus that goes in and the behavior that comes out can often be defined, but precisely what goes on inside the nervous system is, in most cases, a mystery. However, the characteristics of a behavior pattern must reflect the organization of the nerve cells that control it.

Many workers have looked for simple behavior patterns that they hope will reveal the essential features of more complex activities. We are now finding out how certain individual neurons function in locomotion and escape reactions of invertebrates such as cockroaches, crayfish, the sea slug *Aplysia,* and leeches, but the picture is still far from complete.

Reflexes and more complex behavior patterns share a number of properties that result from the way neurons operate. For instance, both show **latency,** a time delay between stimulus and response. Latency is due to the time necessary for reception by the sense organ, conduction through the nervous system, and excitation of the effector.

(a)

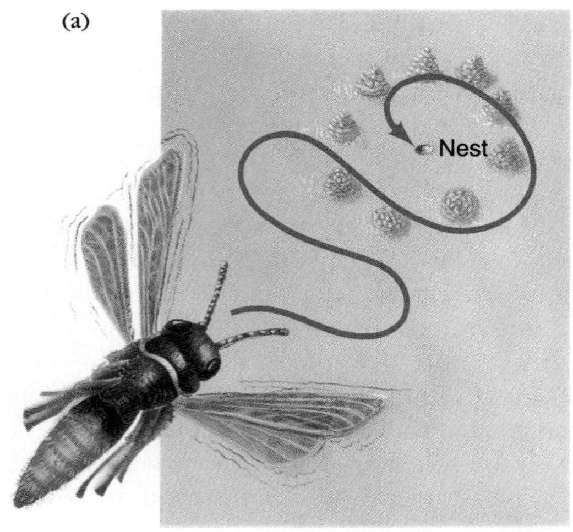

(b)

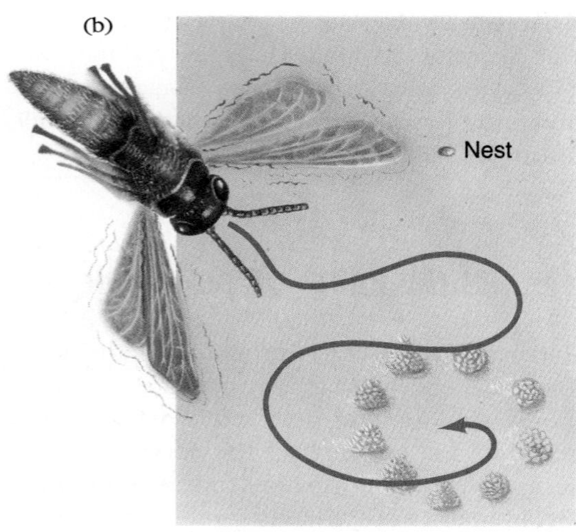

FIGURE 31–7

Learning by a solitary insect. A digger wasp finds her nest by visual landmarks. **(a)** The wasp makes an orientation flight over a nest entrance that the investigator has surrounded by pine cones. **(b)** When the investigator moves the ring of cones to a new location, the wasp returns to the center of the new ring and looks for the opening to her nest.

Stereotyped Behavior

A striking feature of the behavior of any animal is its repertoire of **stereotyped behaviors**—acts, involving the use of many muscles in a precisely timed sequence, that are always performed in an essentially identical pattern. Reflexes are the simplest examples, but more complicated activities such as locomotion, sound production, breathing, and feeding also fall into this category. For instance, a cockroach or cricket will leap forward in a standard escape reaction when receptors on its abdomen are stimulated by a puff of air. This is an example of an innate stereotyped behavior, sometimes known as a **fixed action pattern.** Other stereotyped behaviors are learned.

Studies of invertebrate neurophysiology suggest that stereotyped behaviors differ from other behavior in two ways: they are controlled by very few neurons in the central nervous system, and they can occur without feedback from the sense organs, although such feedback is available and is often used. For example, the fixed action patterns by which a crayfish flexes its abdomen when it swims are induced by the activity of single, identifiable neurons in the central nervous system. At least some fixed action patterns are "hard-wired" into the nervous system. They are always produced in identical fashion because only one or a few control cells trigger all of the motor neurons for the entire behavior pattern.

By contrast, consider what happens when you pick up a glass of water. This behavior, which is not stereotyped, is a complicated series of interactions between the sense organs and muscles in the arm. The sense organs signal how much the water is sloshing about and how far and how fast the glass is rising. These sensory messages reach thousands of neurons in the central nervous system, and hundreds of motor neurons respond, controlling the muscles in the arm so that the glass rises steadily and the water does not spill. Hundreds of central neurons and continual feedback from the sense organs are necessary for this behavior pattern.

An example of a learned stereotyped behavior is a rat's pressing a lever for food. Each rat presses the lever with a characteristic gesture. One uses a fist, another one finger, and each uses its own gesture time after time. Similarly, how you hold your pen, walk, play the piano, or ride a bicycle is a learned sterotyped behavior that is unique to you and performed in a nearly identical way each time you do it (Figure 31–8).

The selective advantage of behavior patterns programmed into the nervous system is probably that they reduce the number of neurons used in a relatively complex task that must be performed perfectly and often. Some stereotyped behaviors, such as escape movements, must be performed perfectly to work at all. Others save energy because they ensure that the muscular movements of, say, walking or feeding need not be worked out with sensory feedback every time they are performed. All animals seem to be able to "write" programs for behavior patterns into the nervous sys-

(a)

(b)

FIGURE 31–8

Two types of behavior. (a) Walking is a stereotyped behavior. These giraffes need little feedback from their sense organs to stroll across the savanna. **(b)** Landing is not a stereotyped behavior. This Galapagos Hawk needs all the information it gets from its superb binocular vision, from sense organs in the wing muscles that detect air pressure against the wings, and from other sensory input to make a safe landing. (a, K. B. Sandved/Visuals Unlimited; b, F. Polking/Dembinsky Photo Assoc.)

tem during embryonic development and, in many cases, in later life.

Sign Stimuli

What sorts of stimuli trigger behavior patterns? In the 1940s, Niko Tinbergen was studying male three-spined sticklebacks that sported the red belly characteristic of the males of this species in breeding condition. Every time a red truck drove past a nearby window, all of the fish made frantic attempts to swim through the glass of their tanks toward the truck, as if they would attack it. A male stickleback in reproductive condition will also attack other breeding males.

What stimulus provoked this attack behavior? Tinbergen presented various models to the sticklebacks to find out. When he showed wooden models of sticklebacks to males in reproductive condition, they attacked crude models having an eye and a red belly in preference to life-like models without the red belly (Figure 31-9). The red belly of the male stickleback in reproductive condition is thus the sign stimulus that triggers the fixed action pattern of attack by another breeding male. A **sign stimulus** (also known as a **releaser**) is that portion of the total stimulus which releases a particular behavior pattern.

An interesting extrapolation of the theory of how sign stimuli work is that it is possible to produce a model called a **supernormal stimulus,** which provokes a behavior pattern more effectively than does the normal stimulus. For instance, Herring Gull chicks peck at the stimulus provided by a red spot on the parent's bill. This induces the parent to regurgitate fish to feed the chick. When models of the stimulus were tested, it was found that a bar with big red and white stripes provoked more pecks from young chicks than did a realistic model of the bill (Figure 31-10). The bar was a supernormal stimulus for the fixed action pattern.

What triggers a male stickleback's attack?

The following models were presented to a male stickleback:

Model	Behavior
Fish-like, but with no red belly	No attack
	Attack
	Attack

FIGURE 31–9

The sign stimulus for attack by a male stickleback. Niko Tinbergen found that male sticklebacks in breeding condition do not attack the life-like model that lacks a red belly, but they will attack either of the crude models with a red undersurface and an eye. The presence of an eye is often necessary for an animal to identify an object as another animal.

(a)

(b) (c)

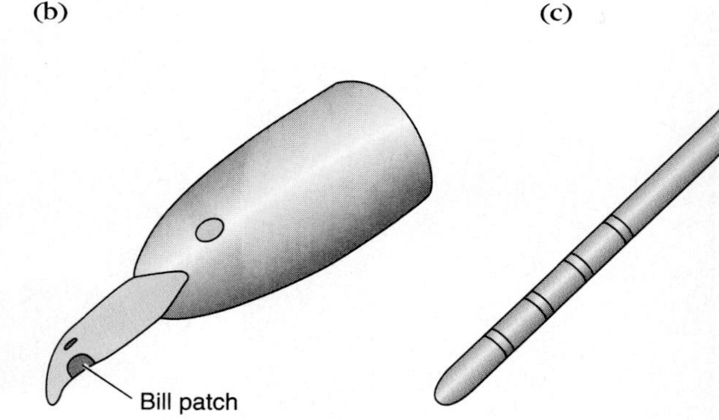

FIGURE 31–10

Sign stimulus vs. supernormal stimulus. (a) Young Herring Gull chicks respond to a sign stimulus, a colored spot on the parent's bill, by pecking at it. This induces the parent to feed the chick. **(b)** A life-like (though flat) model of the parent's head results in fewer pecks by newly hatched chicks than **(c)** a supernormal stimulus, a model that is long and thin and emphasizes the contrast between bill color and bill spot. (Frans Lanting/Minden Pictures)

Drive and Motivation

A particular stimulus may evoke different responses in the same animal at different times. For example, an animal that sees food will eat if it is hungry, but may ignore the food if it has just eaten. Something inside the animal, which we may call **motivation** or **drive,** is different at these two times. Since different behaviors are appropriate at different times even when the stimulus is the same, variations in motivation help to ensure that an animal's behavior changes to fulfill its short-term needs.

In the vertebrate brain, the hypothalamus appears to control motivation. Attack, escape, and sexual behavior can be evoked by electrical stimulation of parts of the hypothalamus. The hypothalamus seldom acts alone to determine motivation. Its activities can be modified by input from other parts of the brain and by hormone levels. For instance, only when they are in breeding condition, with high levels of the hormone testosterone, do male sticklebacks attack other fish with red bellies.

Stereotyped behaviors permit animals to repeat important behaviors accurately using minimal neural control. Many behavior patterns are triggered by specific stimuli, provided the animal is also in a state of nervous or hormonal receptivity for the response.

31–F KINDS OF LEARNING

Learning produces adaptive changes in an individual's behavior as a result of its experiences. It occurs in so many different ways that we have to classify them somehow, although there is no evidence that the categories used here bear any relationship to the largely unknown physiological basis of learning. Evidence suggests that learning involves the cerebral cortex. The proportionately larger cerebral cortexes of primates, especially humans, are directly related to their ability to carry out higher thought processes.

Habituation involves the loss of old responses. Animals may learn not to respond to stimuli that occur often and are unimportant to them. Young animals often show alarm behavior at a variety of stimuli, most of which they rapidly learn to ignore. Similarly, human city dwellers become so accustomed to hearing sirens and shouts that they cease to be alarmed by them.

Conditioned reflexes are behavior patterns evoked by a previously neutral stimulus which an animal has learned to associate with the stimulus that normally elicits the reflex. The Russian physiologist Ivan Pavlov showed that there is a reflex that causes hungry dogs to secrete saliva when they see food. Pavlov rang a bell when he showed food to the dogs, and after several trials the dogs would salivate when the bell was rung, even though he stopped showing them food. The dogs had learned to respond to the new stimulus, the bell, to which they had not previously responded. Pavlov called the bell the **conditioned stimulus.**

Trial-and-error learning is what its name implies. An animal's spontaneous movements may by chance produce a reward, and the animal learns by trial and error to repeat that behavior pattern. The reward may often be the "pleasure" of performing an action more accurately than before. Trial and error is probably the most appropriate category for the learn-

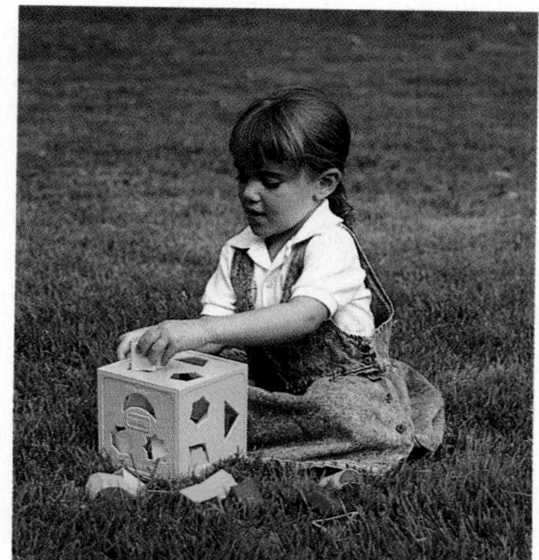

FIGURE 31–11

Trial and error learning. Animals, especially juveniles, learn many behaviors by trial and error. (E. Zalisko)

ing of new motor skills (Figure 31–11). Young mammals and birds perfect their prey-catching movements, and humans learn to play the piano, by a trial-and-error form of practice.

All of these types of learning are varieties of **associative learning. Reinforcement** (reward or punishment) is a central feature of associative learning. Another characteristic of associative learning is that it improves with repetition.

Latent learning occurs without any obvious reward or punishment. It is learning that produces no obvious behavior at the time it occurs. This often happens during exploratory behavior. A recently fed animal may give no sign that it has noticed a new food source until it later returns to feed there.

Insight learning is a form of reasoning that draws on the results of past experiences to arrive at the solution of a novel problem. The classic example of insight in animals came from the work of Wolfgang Kohler on chimpanzees. Presented with a bunch of bananas too high to reach, they would pile up boxes to make a stand from which they could reach the bananas (Figure 31–12). Reasoning of this sort has been shown in many mammals and in some birds, although it is often difficult to distinguish from other forms of learning.

Learning makes adaptive changes in an animal's behavior. Habituation, conditioned reflexes, and trial and error are types of associative learning acquired as a result of reward and punishment in recurring situations. Latent learning and insight learning occur without such repetition and involve the integration of observations and memories of past experiences.

FIGURE 31–12

Insight learning. In Kohler's experiment, a chimpanzee was left in a room with a number of boxes and a bunch of bananas hanging from the ceiling. After a period of time (perhaps for thought), the chimpanzee piled the boxes on top of one another, climbed on them, and reached the bananas.

31–G TERRITORIAL BEHAVIOR

Now that we have seen something of the evolutionary origins and physical basis of behavior, we go on to consider particular behavior patterns that illustrate these ideas.

1? Many animals defend **territories,** areas where they have a monopoly on resources such as food or nesting sites. Holding a territory has a clear evolutionary advantage, because the resources contribute to the successful production of young. A territory holder attacks and drives away other members of the same species (Figure 31–13). It is to an animal's advantage to defend the territory with a minimum of attack behavior, because every attack carries the risk that the attacker will be injured or spotted by predators. Animals have evolved several features that minimize damage during territorial confrontations. For instance, fighting is infrequent because there are "rules" about who wins encounters between two individuals. ■

Consider a male thrush defending a territory before the female arrives in the spring. The male is most

FIGURE 31-13

Territorial behavior. Eastern Meadowlarks typically sit on elevated perches to sing, a song that some birders think sounds like, "Spring-of-the-year." Whenever you hear a meadowlark song, you are actually hearing a proclamation of territorial ownership. (A. Morris/VIREO)

FIGURE 31-14

Conflict behavior. A gull involved in a territorial clash violently pulls up a clump of grass. The bird acts as if it were caught in a conflict between tendencies to attack and to flee. Instead of doing either, it engages in apparently irrelevant "displacement activity"—pulling up grass. A more placid form of grass-pulling is part of its nest-building behavior.

aggressive near the center of his territory. As he moves toward the boundary, his attacks on a trespassing neighbor become less violent, until he reaches a point at which he is as likely to retreat as to attack when he sees another male thrush. This point marks the boundary of his territory. When two neighbors meet at the boundary between their territories, they both act as if they have conflicting retreat and attack motivations. These tendencies are manifested as conflict behaviors.

Conflict behavior usually contains elements of the two conflicting tendencies (in this case movements toward retreat and toward attack), as well as containing movements apparently unrelated to the issue at hand, such as pecking at the ground, pulling up grass (Figure 31-14), or preening (maintenance of the feathers by oiling and smoothing them with the bill). In many species, patterns of conflict behavior appear to have evolved into ritualized **threat displays** that are directed toward intruders (Figure 31-15). Threat is obviously more advantageous than actual fighting in that it does not injure the animal. In the case of a mutual threat display between two animals (which is effectively a ritualized fight), an experienced observer can predict which animal will win by deciding which animal incorporates more attack movements in its display. The loser will eventually move away from the winner.

Many animals defend territories that include limited resources such as food or nesting sites. Most have evolved threat displays, which reduce the risk of injury by discouraging intruders without engaging in actual fights.

31-H CONFLICT AND COURTSHIP

2? Most animals, even those that live in social groups, maintain a minimum **individual distance** from one another (Figure 31-16). For example, swallows sitting on a telephone wire are always a certain minimum distance apart—determined by the reach of the neighbor's bill. The invasion of individual distance is perceived as a threat, and the invading animal is usually attacked. The conflicting tendencies to attack and to permit another animal to come close enough to mate are often evident in **courtship behavior,** the behavior patterns that precede mating in most animals. The courtship displays of many species seem to have evolved from such conflict behavior. ■

In a well-studied example of courtship behavior, the male Black-headed Gull attracts a female to his territory. She alights near him, and both gulls adopt a series of postures that resemble, but are slightly different from, the characteristic threat display of the species. If neither bird attacks the other, both display appeasement gestures, which imply that the hostility between them has lessened. Eventually the female flies off, but

BIO-BIT

Individual distance in humans is reflected in our occasional discomfort about sitting next to a stranger in a lecture hall or on a bus.

FIGURE 31-15

An Emperor Goose at his nest threatens an intruder. (Biophoto Associates, N.H.P.A.)

FIGURE 31-16

Individual distance. Humans and many other animals maintain a minimal distance between each other. (Jacques Jangoux/Peter Arnold, Inc.)

she may return many times, and each bird will display fewer threatening and more appeasement gestures with each visit. Eventually the greeting ceremony ceases entirely, and the male feeds the female. After this, copulation can occur and a permanent pair bond forms (Figure 31-17).

In their initial encounters, the birds are displaying behavior that reveals three conflicting tendencies—to attack, to flee, and to stay together. The behavior patterns that result from the conflict have evolved into an elaborate courtship ritual.

Courtship behavior appears to have evolved from conflicting behavioral tendencies. It permits the invasion of the individual distance without signaling an attack.

31–I MIGRATION AND HOMING

3? Many animals have remarkable navigational ability, sometimes travelling over hundreds of miles of land and sea. Migrating animals can do many things that we cannot do ourselves without the aid of maps and mechanical devices. A Manx Shearwater (a sea bird), which had never been more than 10 miles from home, was removed from her nest on an island off the coast of Wales, flown to Boston, and released. She was back on her nest before the letter announcing her release reached the observers in Wales. To perform an equiva-

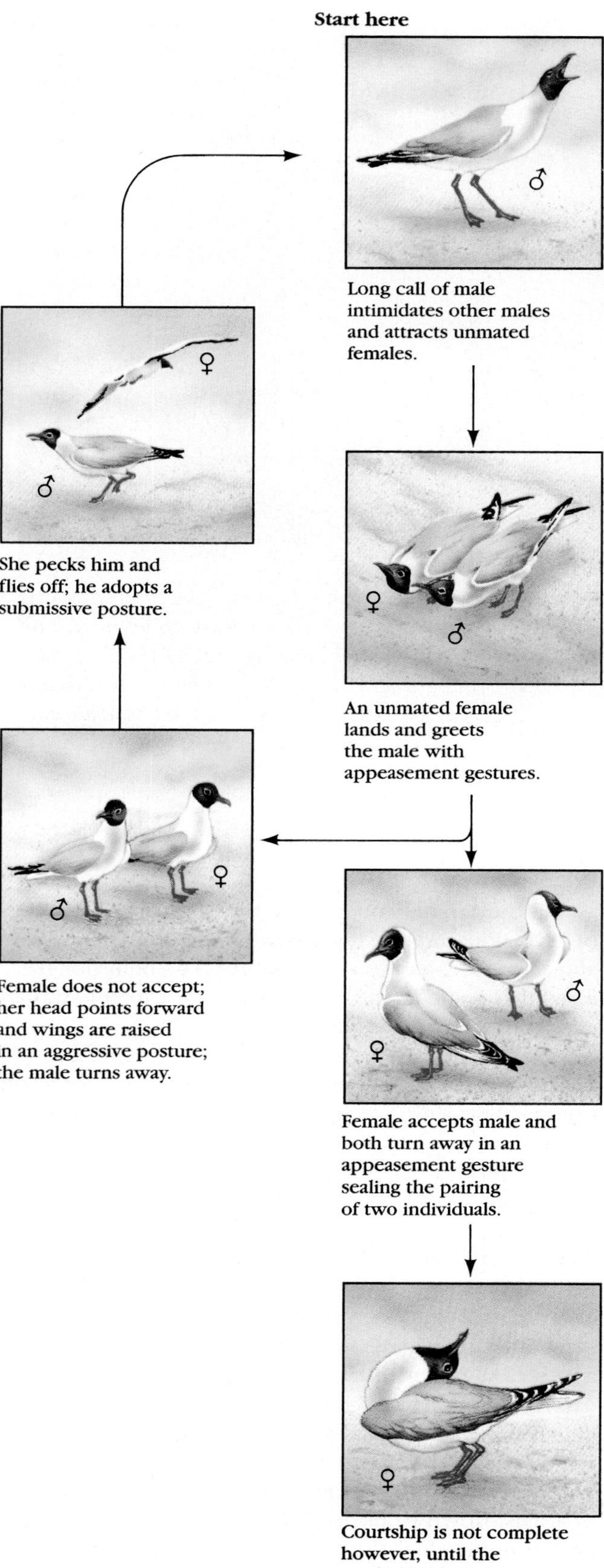
Start here
♂
Long call of male intimidates other males and attracts unmated females.
♀
♂
An unmated female lands and greets the male with appeasement gestures.
♂
♀
Female accepts male and both turn away in an appeasement gesture sealing the pairing of two individuals.
♀
Courtship is not complete however, until the symbolic food-begging ritual takes place.
♂
♀
Female does not accept; her head points forward and wings are raised in an aggressive posture; the male turns away.
♀
♂
She pecks him and flies off; he adopts a submissive posture.
(a)

(b)

FIGURE 31–17

Black-headed Gull courtship. (a) The sequence of courtship behavior in the Black-headed Gull. **(b)** A Black-headed Gull (V. Hasselblad/Academy of Natural Sciences, Philadelphia/VIREO)

lent feat, such as sailing from Boston to Wales, a human being would have to spend hours learning to use navigation instruments to cross the ocean, and would still need a map to find the nest on the other side. Birds, monarch butterflies, fish, and salamanders are some of the animals that perform equivalent journeys without mechanical aids and with little or no learning (Figure 31–18).

Many animals, like horses, dogs, cats, and humans, orient themselves by landmarks that they learn and recognize visually. Many animals can also find their way around by using their chemical receptors. Dogs can follow long and complicated scent trails in unfamiliar territory. Moths find mates, and ants find their nests by following odor gradients. A dramatic case is that of the salmon, which hatches in a freshwater stream and matures hundreds of miles away in the ocean. Years later, when the time comes to spawn, each salmon apparently "smells" its way back to the very stream in which it hatched.

Many animals can move in a specific compass direction. Bees can tell direction from the sun, and birds can orient themselves in a specific compass direction in the same way. Because the sun appears to move from east to west during the day, an animal using the sun as a reference to maintain a constant compass di-

BIO-BIT

The Arctic Tern migrates from the Antarctic to the Arctic and back each year, a 22,000 mile round-trip journey.

(a)

(b)

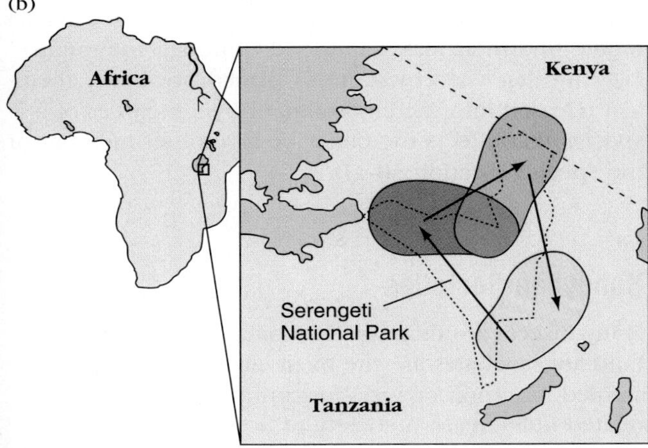

FIGURE 31-18

Migration. (a) Migrating wildebeest crossing the Mara River in Kenya, East Africa. **(b)** Migration patterns of wildebeest in and around Serengeti National Park in Tanzania. (S. R. Maglione/Photo Researchers, Inc.)

rection must also know the time of day. It must also be able to compensate for the sun's apparent movement throughout the daylight hours.

Animals have internal clocks that control their circadian rhythms (Section 30-F). By exposing migrating animals to artificial daylight that began and ended six hours (one fourth of a day) later than natural daylight, it proved possible to "clock-shift," or reset, their internal clocks. The animals then interpreted the sun's position incorrectly and oriented themselves in a compass direction that was 90°(one fourth of a circle) clockwise away from the correct direction for migration. These experiments confirmed that by combining information from the direction of the sun and their internal clocks, birds can point themselves in a particular compass direction. Birds that migrate at night can also use star patterns to find a compass direction.

The ability to fly on a constant compass setting using the sun or stars, however, does not explain the ability of the Manx Shearwater to cross the Atlantic or of a homing pigeon to return to its loft after it has been released in unknown country. In order to get from A to B using a compass you have to know whether A is north, south, east, or west of B. This is called a map sense because it means that you must know the relative positions of A and B on a (hypothetical) map. Many animals obviously have a map sense, which tells them where on Earth they are.

Studies with homing pigeons have shown two other cues that some individuals use to navigate, and that could provide them with a map sense: odors and the Earth's magnetic field. Bits of magnetite, a magnetic iron ore, have been found in the heads of pigeons, as well as in dolphins, whales, monarch butterflies, tuna, mackerel, and some bacteria. These particles are thought to be involved in the ability to sense a magnetic field.

Studies of how pigeons navigate are complicated because pigeons have an innate ability to learn to find their way home using any one of several cues. Apparently local conditions near a pigeon's home loft determine which cues it tends to pay attention to. Because learning plays a part in homing, experienced birds reach home faster than naive ones. Inexperienced birds use a simple hierarchy of navigational cues. If the sun is shining, they use a sun compass, so that if they have been clock-shifted, they fly off in the wrong direction. Experienced birds are less likely to be fooled by a clock shift. They appear to cross-check the information from their sun compass with other cues, such as odor or the magnetic field, which give conflicting information if the birds have been clock-shifted. An experienced bird solves this dilemma by going to sleep in the nearest tree until the effect of the clock-shift wears off. It then flies straight home.

Much has been learned about animal navigation and orientation in the last two decades, but we are still far from understanding this remarkable collection of behavior patterns. ■

Many animals can migrate over long distances to specific locations using the sun, magnetic fields, star patterns, chemical concentration gradients, and other cues.

31-J SOCIAL BEHAVIOR (SOCIOBIOLOGY)

Some animals have little contact with members of their own kind, but in many species, individuals do interact to some degree, either for a short time or throughout life. **Social behavior** is the cooperative interaction of members of the same species.

FIGURE 31–19

Social animals. A lion cub greets its mother. Young lions learn much of their behavior from their elders, and hone their motor skills during trial and error "play." The nucleus of a pride is several related females. A new pride is formed when a group of related males joins a group of females to which the males are unrelated. This pride will last until another group of males comes along and replaces the existing males. Members of the pride cooperate in most activities, notably in hunting. (S. R. Maglione)

It is often difficult to tell whether particular groups of animals are truly social. For instance, Gray Herons spend much of their lives alone or with their mates. Occasionally, however, these birds spend days, and even weeks, together feeding. Is this feeding group social? Do the birds communicate with each other or help each other with the fishing? It turns out that the herons are all in the same place only because each of them has found food there. They do not interact any more than human diners at separate tables in a restaurant do. Thus it is not safe to assume that a group of animals constitutes a society. A group of animals is truly social only when there is considerable cooperation and communication among the individuals (Figure 31–19). Examples of extremely social animals are honeybees, humans, and wolves, which form cooperative, long-lived societies upon which the individual's very life depends.

BIO-BIT

In chicken societies, a dominance hierarchy is established through confrontations that involve chickens pecking each other, leading to the common phrase "pecking-order."

Communication

All animals that live in societies communicate with each other. Human beings, for instance, use sound and hearing (when we speak, clap, or laugh), and visual stimuli and vision (advertising posters, "body language," dressing up, shaking a fist) among our means of communication. Birds, like humans, have highly developed vision. They communicate largely by movement, color, and sound. Communication by sound can be highly elaborate even if it does not involve language, which is the most important aspect of our own sound communication.

Because our sense of smell is relatively poor, we pay little attention to the chemical communication so common in other animals. Many mammals, such as dogs, mark their territories, determine another animal's mood, find their mates and food, and may communicate in other ways by scent. A **pheromone** is a chemical released into the environment by a member of a species that affects the behavior of another member of the species (Section 30–G).

Honeybee Societies

Many insects are more or less social, but honeybee (and ant) societies are the most elaborate and widely studied. The unit of social organization is a family of related individuals. A society of honeybees typically consists of a reproductive female (the queen) and her daughters (and sometimes her sons).

A honeybee hive may contain 80,000 individuals, each with its own job. The queen lays eggs, drones (males) produce sperm needed to fertilize the eggs of new queens, and workers (sterile females) tend larvae, clean the hive, and forage for food (Figure 31–20). The tasks a bee performs, the number of queens produced, and the founding of a new hive are all organized by pheromones released by the queen and other members of the family.

A honeybee queen takes one mating flight, during which she mates with several males. She stores the sperm and uses them to fertilize the thousands of eggs she lays during her life of seven years or more. Her eggs hatch into larvae, cared for by the workers. The diet fed to a larva determines whether it develops into a queen or a worker. A new worker usually first serves as a nurse, preparing cells for eggs and feeding the larvae after they hatch. After about two weeks the worker becomes a house-bee, cleaning, secreting wax for the honeycomb, and guarding the hive. After this she forages outside the hive for the remaining five or six weeks of her life.

As we would expect from such a complex society, honeybees communicate extensively. Karl von Frisch

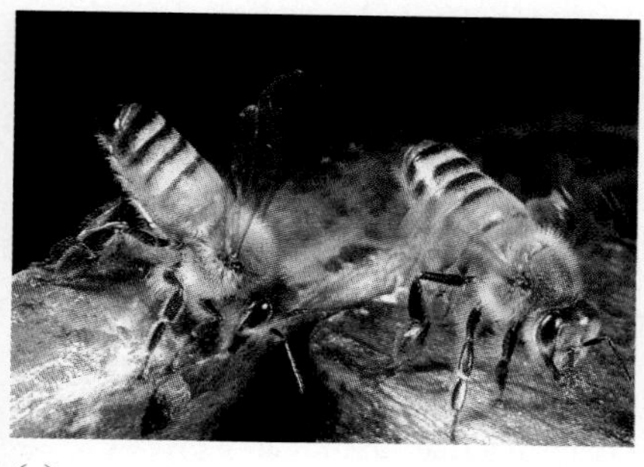
(a)

(b)

(c)

FIGURE 31-20

The life and times of honeybees. (a) A worker "fanning." She stands on her forelegs and uses her wings to disperse pheromones secreted by the scent gland on her abdomen. **(b)** Workers tending larvae (white objects). **(c)** A swarm of honeybees on a tree. (a, S. Camazine/Photo Researchers, Inc.; b, S. Dalton; c, S. Camazine/Photo Researchers, Inc.)

found that foraging bees returning from a successful trip "dance" on the honeycomb, recruiting other bees to harvest a new good food source, and indicating where to find it. Pheromones both permit bees to identify their own hive and serve as alarm signals. A pheromone produced by the queen prevents the workers from producing any more queens and ensures that all the female larvae are fed so that they develop as workers. This continues until the hive is overcrowded, when the queen stops producing that particular pheromone, and the workers start to raise new queens.

Eventually the old queen may leave, for only one queen can survive in a hive. As she leaves, the queen secretes a swarming pheromone, which attracts many of the workers and keeps them with her. The swarm lands somewhere and may remain several days while scouts search for a new site. The scouts return to "dance" a description of the location of a possible new nest. The intensity of her dance conveys the scout's impression of the site's merits. Other workers go to inspect the sites, and finally a consensus emerges when all the scouts are dancing for one site. The swarm then flies to the new site and settles in. This method of making a decision impresses us by its resemblance to the way we like to think humans act.

Vertebrate Societies

Like insect societies, most vertebrate societies consist of genetically related individuals. Unlike insect societies, however, all members of the vertebrate society are fertile, and competition to reproduce is important in determining the social system. A typical vertebrate society consists of genetically related females and their offspring, plus unrelated males who join the society for shorter or longer periods. An interesting exception is chimpanzees: here a territory belongs to a group of related males, and females born in other families move into the area (see *A Journey into Evolution:* Chimpanzee Societies, this chapter).

Most vertebrates learn a much greater proportion of their behavioral repertoires than do insects. How does this difference affect the social behavior of the two groups? Members of vertebrate societies can identify each member of the group individually, whereas insects probably cannot. There is usually a dominance hierarchy, or "pecking order," which ensures that dominant individuals have first choice of desirable but limited things such as food, shelter, or mates. An individual's role in the society is largely determined by its position in the hierarchy, and so ability has more effect on an individual's role in a vertebrate society than in an insect society (Figure 31-21). Individuals may fall in rank as the result of age or disability. In one baboon troop, the top male changed five times in two years. Position in the hierarchy is not determined solely by an individual's size or fighting ability, however. In many species (probably most primates), having a mother of high status gives one an initial boost up the social ladder.

Threat displays (Section 31-G) and related behavior patterns are important in maintaining dominance hierarchies. A dominant individual displaces a subordinate by threat behavior. A subordinate responds with appeasement gestures, which inhibit other animals from attacking. Appeasement gestures show submission, often by turning away weapons (teeth or beak) or by presenting the vulnerable throat or belly to the dominant animal.

A Journey into Evolution

Chimpanzee Societies

The social structure of chimpanzees differs from that of most other social animals. This is of interest not only in itself, but also because chimpanzees are the nearest living relatives of humans: we share about 99% of the same genes. Chimpanzee behavior has some interesting parallels with, and differences from, human behavior.

Chimpanzees live in groups with 50 or more members who occupy a territory of 10 to 30 square kilometers of tropical rain forest in Africa. A chimpanzee spends more than three fourths of its feeding time eating fruits, preferring figs, which are rich in protein. Chimpanzees also eat various other plant parts and insects, and hunt mammals, including monkeys (which also reduces competition for fruit!).

In one study, the territory of a group contained 100 different species of trees, but only ten of these species bore edible fruit. Tropical trees produce fruit at unpredictable intervals, and the crop is ripe only for a short time. Hence chimpanzees spend much of their time travelling in search of trees with fruit. They must also fend off fruit-eating birds and monkeys, which are smaller and more agile in the treetops. Chimpanzees move differently through the forest: they walk on the ground and then climb the fruiting trees, rather than moving through the branches of the forest canopy as monkeys do.

Like other social animals, chimpanzees tend to spend time in large groups. However, few trees bear enough fruit to feed many chimpanzees at once. Hence, when chimpanzees travel in search of food, they go alone or in groups of 3 or 4 animals. By scattering in different directions, each little group manages to find enough food while minimizing the energy spent travelling. If a group finds a tree with more than enough fruit to feed its members, the males give loud cries, which soon attract other chimpanzees to feed on the tree too.

Chimpanzees spend their "resting" time in larger groups. Much of this time is spent grooming: searching another animal's fur and picking lice off the skin. This closeness is thought to promote social bonds, but it also keeps the animals healthy by removing parasites that might carry disease. A grooming animal mostly picks lice off parts of the body that the animal being groomed cannot see or reach easily itself—the head, neck, and back.

In most social animals, males are expelled from the group as they approach maturity. The group's nucleus is the females and their daughters, while males come and go, fighting one another for the chance to join the female group and hence to reproduce. In chimpanzees, however, this situation is reversed.

Male chimpanzees remain in the community of their birth. They become its defenders and the fathers of the group's new offspring. Over the generations, this means that the males in a chimpanzee community are all closely related to each other, and that the territory is in effect held and passed down through the male line.

Females, on the other hand, join new groups on reaching maturity. Hence the females are unrelated to their mates in the new community.

FIGURE 31–21

Primate society. Nearly all primates, like these Barbary apes, live in social groups with dominance hierarchies. Mutual grooming behavior, possible only between social animals, removes parasites and contributes to the health of all members of the group. (T. McHugh/Photo Researchers, Inc.)

The females may or may not be related to each other.

As a result of this social structure, it is to a male chimpanzee's evolutionary advantage to promote the welfare not only of himself, but also of other individuals. The group's other adult males are his close relatives and share many of his genes, and the group's young are his own offspring or the offspring of his near relatives. By helping any of these individuals, who share many of the same genes as his own, the male contributes to the welfare of many copies of his own genes. Hence, a male's calling when he finds an abundant food source, which attracts other troop members, is selected for.

An adult female, on the other hand, can best ensure her own evolutionary success by looking out for herself and her own offspring. Indeed, motherhood takes up a lot of her time and energy: nine months for each pregnancy (starting at about 15 years of age), and four or five years of nursing, protecting, teaching, and carrying the infant on her food-finding expeditions, until it becomes independent and she can begin a new pregnancy (Figure 31-A).

The social structure of chimpanzees probably explains why males tend to spend more time with other males, while females spend more time with other females and their offspring. It also accounts for the fact that male chimpanzees do not compete intensely for opportunities to copulate with females in estrus and so become fathers of the next generation. Males often ignore copulations occurring only a few meters away, and an estrous female may copulate with all the males in a party within a short time. However, a dominant male sometimes takes an estrous female "on safari" apart from the group, thereby excluding other males from mating during her receptive period.

FIGURE 31-A

A chimpanzee and her young. (C. Palek)

Although chimpanzees have ten times more friendly than aggressive interactions with members of their own community, they are hostile to members of other groups. Groups of females have been observed to drive away "foreign" females. However, male aggression is much more frequent and more brutal. A community's males patrol the boundaries of their territory and keep members of other groups out. Males encountering a foreign female and her young have been observed to kill the infant. This behavior is selectively advantageous to the male on two counts: first, it eliminates competition from unrelated genes. Second, having lost her infant, a female quickly comes into estrus, and so the murderer or his relatives may be able to mate with her and father her next offspring if she remains in their territory. Hence natural selection favors perpetuation of genes that make a male more likely to kill young born into other groups, and of genes that make a female more likely to stay well within the boundaries of the territory protected by the males of her community. Males of one community have even been known to kill the males of a smaller group and take over their territory—a form of warfare.

These observations of chimpanzee behavior give much food for thought about our own behavior and its genetic consequences.

The evolutionary advantage of a social hierarchy is probably that it reduces the harmful effects (such as injury from fights) of the inevitable competition between related individuals living in the same area. The society also ensures that at times when resources (such as food) are in short supply, some individuals will get all the food they need to survive instead of the whole group becoming half-starved and likely to die, as happens with honeybees. When members of a group are related, such behavior will be selected for because individuals carry many of the same genes, and an individual that starves while a relative lives to reproduce is actually contributing to the survival of many of his or her own genes in future generations.

Animal societies have evolved in cases where interaction among individuals enhances the reproductive success of the genes they share. These social organizations have specific rules of conduct and means of communication that govern individual behavior.

SUMMARY

1. The genes that an animal inherits determine the range of behavior patterns it can develop.
2. Most behavior patterns, innate or learned, develop normally only if the animal is exposed to the appropriate environmental conditions.
3. Animals are always exposed to a variety of stimuli, which may or may not evoke a response, depending on factors such as the animal's physiological state.
4. Behavior patterns that must be produced perfectly at the first exposure to the stimulus are usually innate.
5. Learning requires time and energy and is reserved for behavior that must be flexible in meeting local or changing conditions.
6. Many behavior patterns, both innate and learned, become programmed into the nervous system. Such stereotyped behaviors may be triggered by sign stimuli and controlled by a small number of neurons with minimal sensory feedback.
7. Conflict behavior is frequently seen in courtship and territorial displays, and includes conflicting behavioral tendencies to fight or flee.
8. Most animals seldom cooperate with other members of their own species, but true societies have evolved in some species of insects and vertebrates. A society usually consists of genetically related individuals whose cooperation enhances their own survival and the survival of the genes that they share. Communication between individuals is most highly developed in social animals.
9. Vertebrate societies are characterized by hierarchies that determine an individual's access to limited resources.

SELF-QUIZ

From the list of types of behavior patterns below, choose the one exemplified by each of the following situations.

___ 1. A male cardinal attacks any other male cardinal that tries to come into your backyard.

___ 2. A puppy rolls on its back when a strange adult dog growls at it.

___ 3. A cat meeting a strange (and not overly large or fierce) dog arches its back, fluffs its fur, and hisses.

___ 4. A student in a typing class makes fewer errors on the tenth homework assignment than on the first.

___ 5. Your signature is the same every time you write it.

___ 6. Newly hatched ducklings follow a windup toy as if it were their mother.

___ 7. One chicken pecks any other member of the flock that gets too close, but no other chicken pecks it back.

___ 8. A skilled musician can play a tune she or he has never heard before, after someone hums a few bars.

a. appeasement
b. dominance
c. imprinting
d. insight
e. conditioned reflex
f. stereotyped behavior
g. territoriality
h. threat
i. latent learning
j. trial and error

9. Courtship behavior is said to show conflict because:
   a. the two mates fight frequently
   b. the mates cannot immediately agree on a nest site
   c. the mates are sexually attracted to each other but do not normally permit another animal to get as close as copulation requires
   d. the mates must choose each other from a large number of members of the opposite sex
   e. the mates are in competition for food in a territory of limited size
10. The part of the nervous system important in the control of drives for such behavior patterns as feeding, drinking and sexual behavior is the:
    a. pituitary
    b. sympathetic system
    c. parasympathetic system
    d. reticular activating system
    e. hypothalamus
11. Which of the following is *not* true of stereotyped behaviors?
    a. They may be triggered by sign stimuli.
    b. They are initiated by one or a few neurons.
    c. They exhibit latency.
    d. They can be learned or innate.
    e. None of the above.

THINKING CRITICALLY

1. Despite everything you have read in this chapter, you probably still think that most of an insect's behavior is innate, and that most of a human's or chimpanzee's is learned. Can you justify this position?
2. What are some of the possible selective advantages to defending a territory, an activity that consumes time and energy and increases the risk of injury?
3. William Dilger studied the nest-building behavior of parakeets of the genus *Agapornis.* He crossed members of a species that carries nest-building material in its beak with members of a species that carries its nesting material tucked under its tail feathers and observed the behavior of the hybrid offspring. These offspring showed hybrid behavior and usually dropped the material

whether they carried it in their beaks or their feathers. One particular bird tried to build a nest 48 times and failed. On the 49th try it was successful. What does this tell you about whether nest-building behavior in this genus is innate or learned?

SELECTED KEY TERMS

anthropomorphism, *p. 645*
associative learning, *p. 653*
conditioned stimulus, *p. 652*
conflict behavior, *p. 654*
courtship behavior, *p. 654*
fixed action pattern, *p. 650*
habituation, *p. 652*
imprinting, *p. 647*
individual distance, *p. 654*
innate, *p. 648*
insight learning, *p. 653*
instinctive, *p. 648*
latent learning, *p. 653*
reinforcement, *p. 653*
releaser, *p. 651*
sign stimulus, *p. 651*
stereotyped behavior, *p. 650*
supernormal stimulus, *p. 651*
territory, *p. 653*
threat display, *p. 654*
trial-and-error learning, *p. 652*

SUGGESTED READINGS

Dilger, W. C. "The behavior of lovebirds." *Scientific American,* December 1962. The mixed up lovebirds referred to in Thinking Critically 3.

Ghiglieri, M. P. "The social ecology of chimpanzees." *Scientific American,* June 1985. A fascinating study of our nearest relatives.

Gould, J. L., and P. Marler. "Learning by instinct." *Scientific American,* January 1987. An excellent account of how the innate tendencies of different animals to learn different kinds of information is adaptive for the species' way of life.

Gwinner, E. "Internal rhythms in bird migration." *Scientific American,* April 1986. A biological clock is involved in migration and navigation.

Honeycutt, R. L. "Naked mole rats." *American Scientist,* January 1992.

Kalin, N. H. "The neurobiology of fear." *Scientific American,* May 1993.

Lorenz, K. Z., *King Solomon's Ring.* London: Methuen, 1942. Delightfully written autobiographical account of life with animals.

Tinbergen, N. *The Animal in its World.* Cambridge, MA: Harvard University Press, 1972. Selections from Tinbergen's work, including experiments and general papers. A very readable description of the types of experiments you might perform if you became a behaviorist.

Tinbergen, N. *Curious Naturalists.* Garden City, NY: Doubleday & Company, Inc., 1969. Autobiographical description of Tinbergen's life as an animal behaviorist.

The Educated Citizen

The Transplant Gap

This article by Jerold M. Lowenstein was excerpted with permission from the June 1993 issue of Discover *magazine.*

We humans don't supply enough organs for transplant. Can animals solve our problem?

Twice during 1992 a surgical team led by Thomas Starzl at the University of Pittsburgh transplanted a liver from a baboon into a human, and by doing so raised anew the question of whether the species barrier to organ grafting can be overcome. As both a researcher and a practicing physician, I'm interested in this experimental surgery from two points of view. In my lab I study molecular evolution, examining proteins from different species to see how closely related they are. One way to compare the proteins of humans and baboons is to inject them into a rabbit, then measure the rabbit's antibody reaction to them—measure, that is, the degree to which its immune system recognizes the proteins as foreign. As it turns out, the rabbit immune system perceives a big difference between human and baboon proteins, so I'm sure the human immune system will perceive that difference, too. That doesn't augur well for the success of baboon-organ transplants.

As a specialist in nuclear medicine, I also do diagnostic tests on patients who have undergone liver, kidney, and heart transplants. Despite our best efforts, sooner or later many of these human-to-human transplants fail; they are rejected by patients' immune systems. When it comes to transplants from baboons—animals that, on the molecular level, are so much more different from us than we are from each other—it's hard for me to see how we can prevent a rejection level approaching 100 percent.

There's no doubt that during the past 30 years organ transplantations have become an increasingly important form of medical therapy. You can classify transplants, in order of ascending difficulty, as autografts, in which an individual's own tissue (such as skin in a burn patient) is moved from one place to another; isografts, between genetically identical twins; allografts, between members of the same species; and xenografts, between members of different species.

Autografts and isografts are almost always successful. That's because the transplanted tissues have the right ID badges—the right particular combination of proteins on their cells—to pass muster by the recipient body's immune surveillance. All cells carry these critical marker proteins, which tell the immune system whether they're compatible with the body—whether they resemble "self" and should therefore be accepted or look like "nonself" and should be cast out.

Allografts have made steady progress since the 1960s. Initially most of them failed, but now, thanks to better tissue typing and other advances, the average one-year patient survival rate for all types of allografts is about 80 percent. We talk about one-year survival rates because many of these patients would have died fairly soon without a transplant, and a year's extension of life often seems like a triumph. But the longer patients live, the greater the chances that their transplants will give out—most seem to falter within ten years. With new patients requiring first-time transplants, and old transplants requiring replacement, there's always a long waiting list of dying people desperately hoping to get the next kidney or heart or liver. That's why transplant specialists, and surgeons in particular, are thinking about xenografts—again.

Chimp and baboon xenografts were first tried in earnest in the 1960s and 1970s by pioneers like Christiaan Barnard, and most were rejected with dismal rapidity. Nowadays, of course, no one would even think of using chimpanzees, our closest living relatives; they are too endangered and (thanks to the consciousness-raising work of Jane Goodall) viewed too sympathetically to be considered as donors. But other, less closely related and less threatened primates, such as baboons, could, in theory, be bred in captivity for their organs—assuming, of course, that baboon transplants could be made to work.

In 1984 Leonard Bailey, a cardiac surgeon at Loma Linda Medical Center in California, took the plunge and put a baboon heart into a child whose own heart had a hopelessly underdeveloped left side. The baboon heart and Baby Fae, as the child was known, survived only 20 days. At the time, I was amazed that anyone would try an experiment so unlikely to succeed. Even with recent advances made in immune-system-taming drugs, the species barrier looked insurmountable.

Compared with other primates, we humans are an unusually homogenous species. Genetic studies suggest that modern humans originated in Africa about 200,000 years ago, an evolutionary eye blink, so there's less variability among us than among our primate predecessors. On average, we differ genetically from each other by only 1 percent. And even so, the genes that code for the ID badges on our cells are sufficiently variable to cause nearly all organ transplants to be rejected—unless they are carefully tissue typed and immunosuppressive drugs are used for the rest of the recipient's life.

Chimpanzees, our nearest cousins, are about 1.5 percent different from us genetically, which means that their genetic material, their DNA, is 15 times more different from that of humans than individual human DNAs are from each other. So it's not too surprising that past attempts to transplant chimp kidneys and hearts into humans met with rejection.

By the same measure, Old World monkeys, which include baboons, are at least 5 percent different from us—that is, 50 times more distant genetically than we are from each other. Unless the human immune system were to be almost totally wiped out, as it is in AIDS, the cells and antibodies of our complex defense system would perceive a baboon transplant as a foreign invader, an enemy, and proceed to attack and destroy it.

At the time Bailey transplanted the baboon heart into Baby Fae, he was asked about the problem of evolutionary distance. According to Norman Swan, an Australian doctor who conducted the interview for the Australian Broadcasting Company, Bailey replied that he was a Seventh-day Adventist; Adventists are creationists—they simply do not believe in Darwinian evolution.

Thomas Starzl presumably does believe in evolution. But he also has faith in the human ability to learn from trial and error. He thinks that by using better drugs and other techniques to manipulate the immune system, surgeons can overcome the species barrier, just as he and his associates found ways of breaking the barrier to transplants between two unrelated people. But the bottom line here is that there are far more people with diseased livers and hearts than there are human donors to provide new organs for them. And so the xenograft dream will not die, and every new technical or immunologic advance in the field is likely to resuscitate the hope that baboon transplants can be made to overcome the mile-high species barrier.

Starzl's hopes are buoyed by an interesting phenomenon called the chimera effect, which he and others have observed when performing autopsies of liver recipients. When a liver is transplanted, some of its cells migrate to other parts of the recipient's body and manage to survive there; meanwhile, cells from the recipient infiltrate the graft. This cell exchange even crosses the species barrier, judging from experimental bone-marrow transplants between rats and mice. Thus the recipient becomes a chimera—a creature made of incongruous parts, like the fire-breathing monster of Greek mythology that had the head of a lion, the body of a goat, and the tail of a serpent. In this case the chimera is made up of the cells of two different organisms. Starzl believes that this cell swap makes the host's immune system more tolerant of the foreign cells and so protects the transplant recipient from rejection.

The patient selected by the Pittsburgh team in 1992 for its first baboon-liver transplant was a man who had nothing to lose. His own liver was being destroyed by the hepatitis B virus. He was not a candidate for another human liver, as the donor liver would also have subsequently become infected by hepatitis B. The xenograft had a definite advantage here: human hepatitis B cannot infect baboon livers. Ergo, the transplant surgeons saw a chance of giving the dying patient a liver that was hepatitis-proof *and* protected from immune destruction by the molecular magic of FK506.

But this same logic can be turned around to reveal the paradox of the xenograft gambit. Hepatitis B does not infect baboon livers, because the molecules in baboon livers are so different from those in human livers that the virus cannot accommodate itself to them. By the same token, the human immune system, conditioned to distinguish self from nonself, will attack these alien liver molecules when it encounters them.

The chimera effect may induce some tolerance to the cell-surface proteins that trigger the rejection process. But I doubt that it can protect against a deluge of alien proteins. From my own work with animals, I'd guess that the only sure way to prevent a patient's immune system from responding to this massive stimulus would be to wipe out the immune system altogether—by whole-body radiation treatment or by powerful immunodestructive drugs. Research by Suzanne Ildstad at Pittsburgh has had some early success in frying mouse immune systems and reseeding them with rat-mouse cells to induce tolerance to rat grafts.

Unfortunately, severe immune suppression or bone-marrow eradication also sets up the patient for overwhelming opportunistic infections. The balance between enough suppression to prevent transplant rejection and not so much as to encourage lethal infection is already hard to achieve with organs from other humans. It could well be impossible to achieve with baboon organs, even with the help of FK506.

As it turned out, Starzl's first patient was perhaps not an ideal test case for FK506 and the three other potent drugs—prednisone, prostaglandin, and cyclophosphamide—that constituted his immune-suppressing cocktail. Seventy days after surgery, when he died of a stroke provoked by a fungus infection of the brain, it was revealed that he'd been infected with the AIDS virus. His immune system was already sick, so the boast of the transplant team that the liver had undergone very little rejection because of the drugs sounded a little hollow.

As for the second baboon-liver recipient, he received not only the immune-suppressing drug cocktail but also an infusion of baboon cells in an attempt to promote chimeric tolerance. He did poorly from the start and died in February, less than a month after surgery, of sepsis resulting from an abdominal infection.

From its original reference to a Greek mythological beast, the word *chimera* has acquired a double meaning. The label now refers to any actual or imagined creature made up of parts from more than one species. It has also come to imply an impossible or idle fancy.

A human with a baboon liver is thus a chimera in the literal sense. Whether creating such a person is an impossible fancy, as evolutionary considerations unfortunately lead me to suspect, remains to be seen.

Connecting the Concepts

The following questions will help you to connect the issues discussed in this article with the concepts you have learned in Part 4.

1. Humans tend to feel greater emotional attachment toward organisms that have closer evolutionary relationships with us. For example, we are more likely to want a dog or cat as a pet instead of a frog. What problems does this present when we consider the best animals to choose for possible xenografic transplants? Do you think that animals have certain rights, and if so, what ethical guidelines should we use to determine what animals can be used as organ donors?
2. Consider the parents of an only child that will die within a year without a kidney transplant, because a donor is unavailable. Is it ethical for them to conceive and give birth to another child just to produce a kidney for possible transplantation? (Most people who have two healthy kidneys can donate one of them without any future complications, except for the usual risks of surgery.)

PART 5

Plant Biology

Chapter 32
Plant Structure and Growth

Chapter 33
Nutrition and Transport in Vascular Plants

Chapter 34
Regulation and Response in Plants

Chapter 35
Reproduction in Flowering Plants

Fiddleheads play fern music. Fiddleheads are young fern fronds. Their compact, often wool-covered heads push up through the soil, appearing in moist woods early in the spring. As the fern fronds grow, the fiddleheads unroll, grow larger, and develop into mature fern fronds. (Linda H. Hopson/Visuals Unlimited)

CHAPTER 32

CONCEPT GUIDE

After reading this chapter, you should be able to:

1. Describe the structure and function of each of the following parts of a bean seed: seed coat, cotyledons, first foliage leaves, and apical meristems. (Section 32-A)
2. Describe the formation and growth of a seedling's roots, stems, and leaves. Indicate the structure and function of each tissue layer. (Sections 32-B, 32-C, and 32-D)
3. Describe secondary growth of tissues in woody dicotyledon stems. (Section 32-E)
4. Compare the structures of monocotyledons and dicotyledons. (Section 32-F)

How do sugars produced in the leaves of these coconut palm trees and water absorbed by its roots make their way to the palm tree's other cells that need them? Read about the food and water transport systems within plants in Section 32-C. (R. Fried)

Plant Structure and Growth

Most plants are autotrophic: they make their own food, using the energy of sunlight and the materials of carbon dioxide from the air, and water and minerals from the soil (Chapter 8). The structure of plants is adapted to their autotrophic way of life. Although simpler kinds of plants lack them, vascular plants (those with internal systems of fluid-transporting tubules) typically have a **root system** that absorbs water and minerals from the soil and a **shoot system** made up of one or more stems with leaves. Stems hold the leaves up where they can intercept sunlight, and the leaves' expanded surface area allows them to capture a great deal of light energy, which they convert into food molecules and store.

Unlike most animals, a vascular plant grows larger and adds new parts throughout its life. Most of the plant body consists of mature, differentiated cells that have specialized functions. Normally, these cells will not divide again. But the plant can grow indefinitely because it has tissues called **meristems,** which contain cells that retain the ability to divide and produce new cells. Some of these cells differentiate and become new parts of the plant. Others remain meristematic and continue to grow and divide, producing new cells.

Plants vary a lot in size and shape. A plant's structure depends on which parts grow, and how much; on where branches arise, and how much they grow. Ultimately, all of this depends on the activities of the meristems, which produce the cells for each part.

The growth of plants can be divided into two aspects, resulting from the activities of different meristems. **Primary growth** is growth in length of the shoots and roots, production of new root and shoot branches, and production of leaves. The meristems responsible for primary growth occur at the ends of the growing stems and roots and the cells laid down during primary growth make up the primary plant body. **Secondary growth** is growth in girth, or thickness, of the stems and roots that have been produced by primary growth. The meristems that produce secondary growth lie a short distance under the surface of stems and roots. The tissues contributed by secondary growth strengthen the plant and provide the support necessary for new primary growth, both in height and in the spread of the branches. Although all vascular plants exhibit primary growth, secondary growth varies a lot: some species have none, while others like trees produce extensive secondary growth.

A plant's meristematic tissues produce cells that grow and differentiate to form three main kinds of mature tissues:

1. **Dermal** ("skin") **tissue** covers and protects the outside of the plant. This layer contains close-fitting cells that often form structures specialized for defense or for obtaining materials from the environment.
2. **Vascular tissue** transports substances over long distances within the plant. There are two types of vascular tissue: **xylem** conducts water and minerals absorbed by the roots up to the stems and leaves, and **phloem** conducts food from sites of production to sites of use or storage. Conducting cells in the xylem and phloem are long and narrow, like sections of a pipe.
3. **Ground tissue** makes up the bulk of the primary plant body, filling the space between the dermal and vascular tissues. Ground tissue consists mainly of a type of cell called **parenchyma,** rather rounded in cross section, with relatively thin cell walls and with many spaces between cells. Parenchyma cells often contain many photosynthetic or storage plastids.

KEY CONCEPTS

- A plant grows as its meristems lay down new cells, which then grow, differentiate, and mature as specialized cells in new plant tissues.
- The vascular plant body consists of roots, stems, and leaves. Each of these organs contains three kinds of tissue:
  1. Dermal tissue, which forms an outer, protective layer.
  2. Vascular tissue, which transports water and food in the plant body.
  3. Ground tissue, which fills in between.

Curiosity Questions

1? How does a young plant get its energy before it develops leaves? (See page 670)

2? How does a plant cutting grow roots when placed into water? (See page 673)

3? What determines whether a leaf will be round or tapered, with jagged or smooth edges? (See page 676)

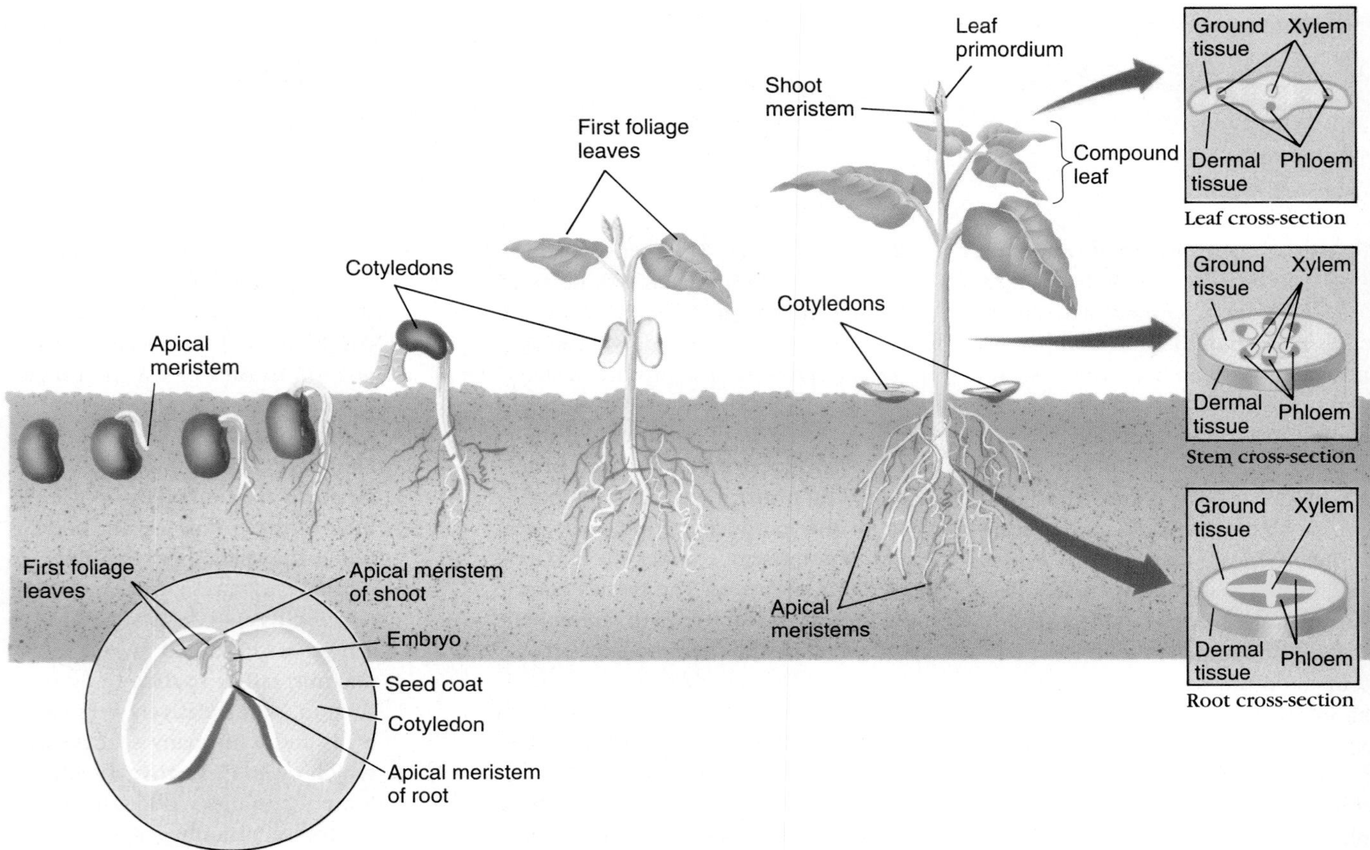

FIGURE 32–1

Stages in the growth of a bean plant. The bean plant's root system starts to develop first. Soon the growing stem forms a loop and pulls the cotyledons free of the soil. Finally, the shoot system grows, foliage leaves are formed, and the green stem and leaves begin photosynthesizing. The location of different kinds of tissue are shown in cross sections at the right.

32–A THE BEAN SEED

1? We begin our study of plant structure by following the growth of a bean seed into a young plant, to see how its meristems or growing tips produce its primary structure (Figure 32–1). A bean seed is a reproductive package containing a plant embryo in an arrested stage of development, along with its food supply. The "skin" of the bean is the protective **seed coat.** Within this coat is the embryonic bean plant. If we peel the seed coat off, we can see the two "halves" of the bean, which are really the embryo's two large "seed leaves," or **cotyledons.** The presence of two seed leaves shows that beans belong to the group of flowering plants called **dicotyledons (dicots** for short). A bean's cotyledons are large because they have absorbed the food supply that will nourish the growing bean seedling. In the seeds of many other species, the food is stored separately, outside the embryo itself (see Figure 32–1). ■

If we separate the cotyledons, we can see that they are part of the embryo, attached to its tiny main axis. Also attached to this axis are the two **first foliage leaves,** which will develop into the first "true" leaves.

The bean embryo has two meristems, one at each end of its axis. Because each of these meristems is located at a tip, or apex, of the plant, they are called **apical** meristems. Cell division in these meristems, followed by growth and differentiation of some of the cells formed, results in growth of the plant's root and shoot systems.

A bean seed is an embryo packaged with food needed to begin the development of a new adult bean plant.

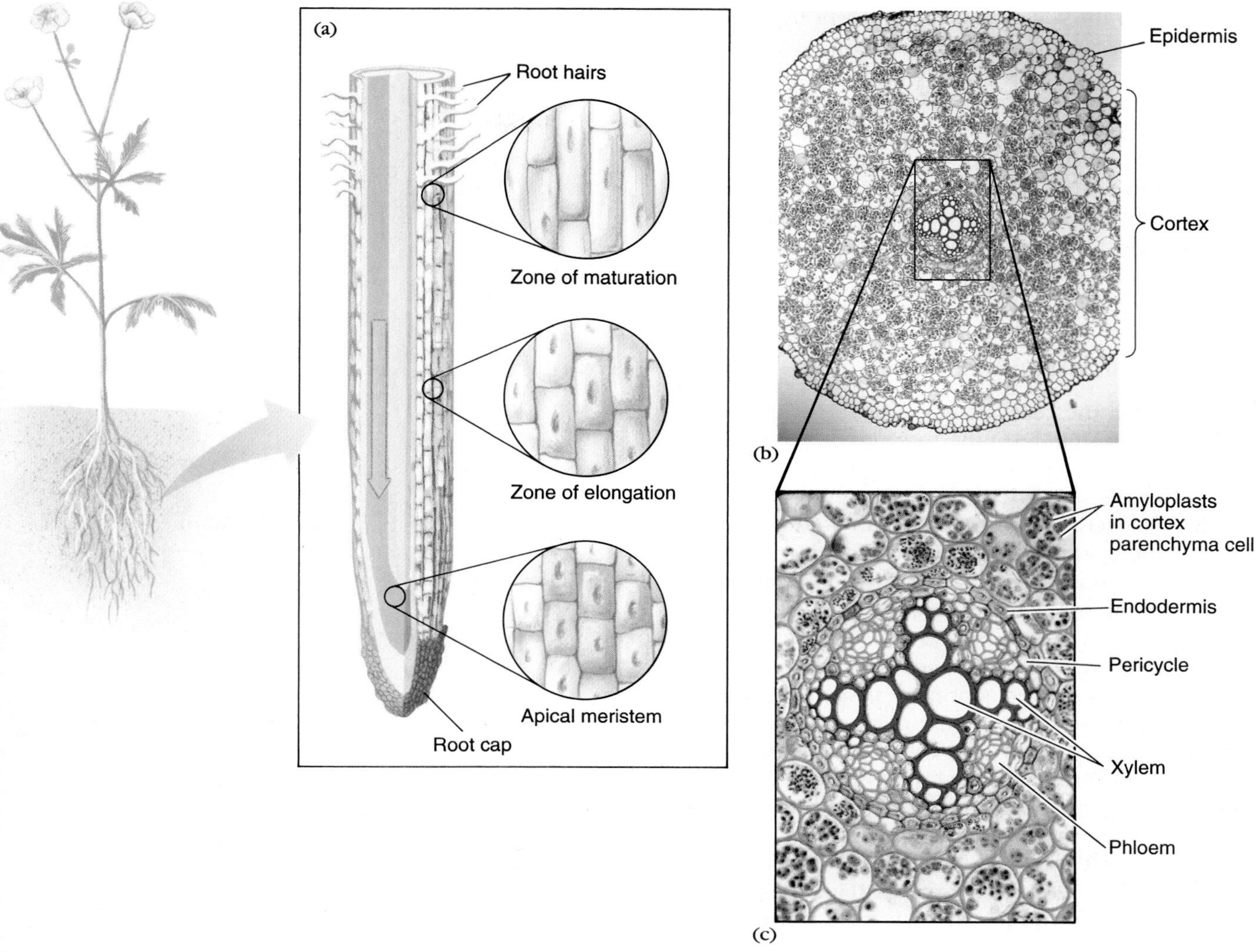

FIGURE 32-2

A growing root tip of a buttercup. Three distinct regions can be identified based upon the cell structure and function. **(a)** Cells in the apical meristem divide and produce new cells for the root and root cap. The root cap protects the underlying apical meristem as the root pushes through the soil. The zone of elongation contains newly produced cells that are growing and pushing the root tip through the soil (arrow). Older cells are differentiating in the zone of maturation to form the tissues shown in b and c. **(b)** Cross-section of cells within the zone of maturation. **(c)** Closer view of cells within the pericycle within the zone of maturation. (b and c, Ed Reschke)

32–B THE ROOT SYSTEM

Primary Growth of Roots: Growth in Length

As the bean seed germinates, the first part of the embryo to start growing is the seedling's first root (see Figure 32-1). The end of the root is covered by the **root cap,** a thimble of cells protecting the growing root tip. Cells in the root cap detect the direction of gravity and ensure that the root grows downward (Section 34-C). The root cap's outer cells are scraped off as the root pushes its way down among the rough soil particles.

The root cap cradles the apical meristem, where cells divide and produce new cells for both the root cap and the root itself (Figure 32-2). Cells destined to become part of the actual root become arranged into definite strands. Above this area, in the **zone of elongation,** the cells are dividing more slowly and are growing longer, pushing the root cap and apical meristem through the soil. The lengthening of these cells is the actual force of growth in the root: the root tip is literally pushed through the soil. Above the zone of elongation, cells in the **zone of maturation** have reached full size and are developing into specialized cells that will perform particular functions. Meanwhile, the apical meristem continues to divide and produce new cells, which follow the same sequence of division, growth, and specialization.

Primary Structure of Roots

When mature, the cells laid down by the activity of the apical meristem form the root's **primary tissues.** The outermost layer consists of dermal tissue called the **epidermis.** Its cells fit together tightly and form a protective covering, one cell thick, around the root. Some of the epidermal cells in the zone of maturation become **root hairs** by forming extensions that grow out among the soil particles (Figure 32-2). Root hairs anchor the plant in the soil and increase the surface area for absorption of water and minerals.

The bulk of the root's primary tissue consists of a thick layer of ground tissue, the **cortex,** just inside the epidermis (Figure 32-2). The cortex is made up of many parenchyma cells, which may contain **amyloplasts,** plastids that store starch. There are typically many spaces between the cells of the cortex.

The **endodermis** ("inner skin") is a single layer of cells at the inner edge of the cortex (Figure 32-3). The endodermis helps control the movement of substances between the root cortex and the root's interior (Section 33-C). Just within the endodermis is another single layer of cells, the **pericycle.** These cells do not differentiate, but remain meristematic, able to divide as needed.

Within the pericycle are the vascular tissues, the root's xylem and phloem, which transport substances to and from the shoot system. Conducting cells of the phloem are among the first cells of the young root to differentiate. They then bring in food for the growth and development of cells near the root tip. Transport cells in the xylem do not start to function until they have grown to their final size and died, leaving a hollow cell wall to conduct water upward. We shall examine the structure and function of vascular tissues in the next chapter. For now, note that they usually occupy the core of the root. Cross sections of dicot roots commonly show a central "star" of xylem, with the phloem in pockets between the arms of the xylem star, while cross sections of monocot roots show a ring of xylem and phloem tubes.

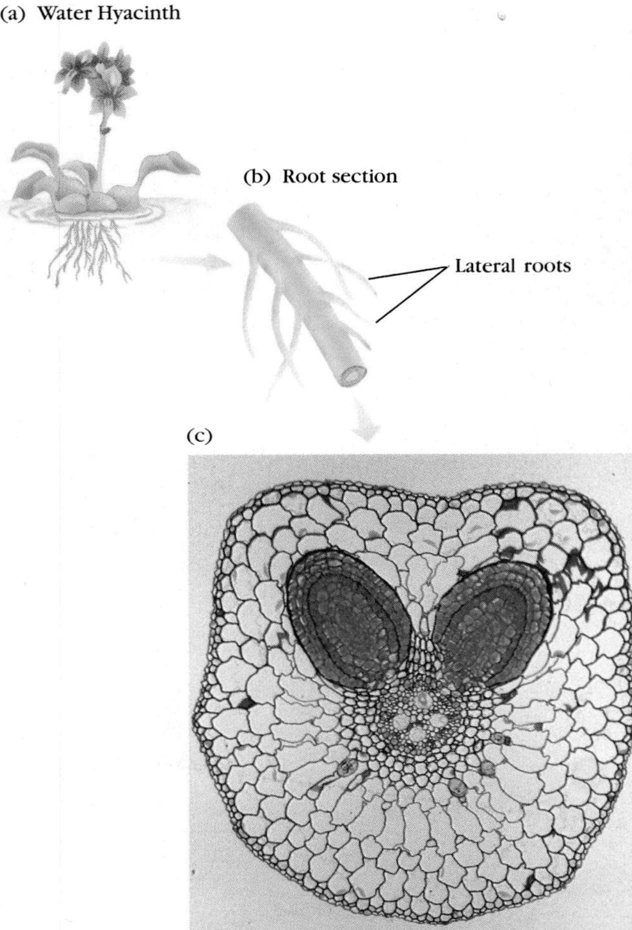

FIGURE 32-3

Root structure. Water hyacinth plants **(a)** have numerous root hairs **(b)**, tiny, branching projections that increase their absorptive surface area. The relationship between these lateral root projections and the vascular tissue within the pericycle are shown in **(c)**. (Carolina Biological Supply Company/Phototake)

Primary Growth of Roots: Production of Laterals

The first root of a bean seedling grows quickly and soon begins to produce side branches, or **laterals.** Cells in the pericycle, which remained meristematic when their neighbors differentiated, divide and form a new apical meristem. As the innermost cells elongate, they push the new meristem out through the endodermis, cortex, and epidermis, and on into the water or soil (Figure 32-3). This new branch root follows the same pattern of growth outlined before, and becomes a lateral root with the same kinds of primary tissues.

(a)

(b)

FIGURE 32–4

Two types of root systems. (a) A taproot system of a dandelion that can penetrate deep into the soil to reach water. (b) A fibrous root system with many hair-like roots that spread evenly throughout the soil. (a, Runk/Schoenberger from Grant Heilman, b, E. R. Degginger, FPSA)

There are two basic types of root systems (Figure 32–4). A **taproot system** has one main root, the **taproot,** by far the plant's longest and thickest root. Its side branches are much shorter and thinner. By concentrating growth into one axis, a taproot may penetrate deep enough into the soil to reach a reliable supply of water. The taproot of an old grape vine may reach 15 meters down and obtain water far beneath the soil surface. Many kinds of trees have taproots as seedlings or saplings and develop a more branching root system later.

Fibrous root systems branch throughout a large volume of soil, from which they absorb water and minerals. Instead of one main root, they have several roots of about equal size, each with smaller lateral roots. Plants with fibrous root systems, such as grasses, are very good at holding soil and therefore controlling soil erosion.

Some roots do not arise from existing roots. **Adventitious roots** are formed by meristems growing from other parts of the plant. For example, corn plants have "prop roots" that arise adventitiously from the base of the stem (Figure 32–5). Many climbing plants form tiny roots on the undersides of their stems as they grow up tree trunks or brick walls. African violets can form roots from the undersides of leaves placed on moist soil. Cuttings of many plants form adventitious roots if the stems are placed in water or moist soil. ■

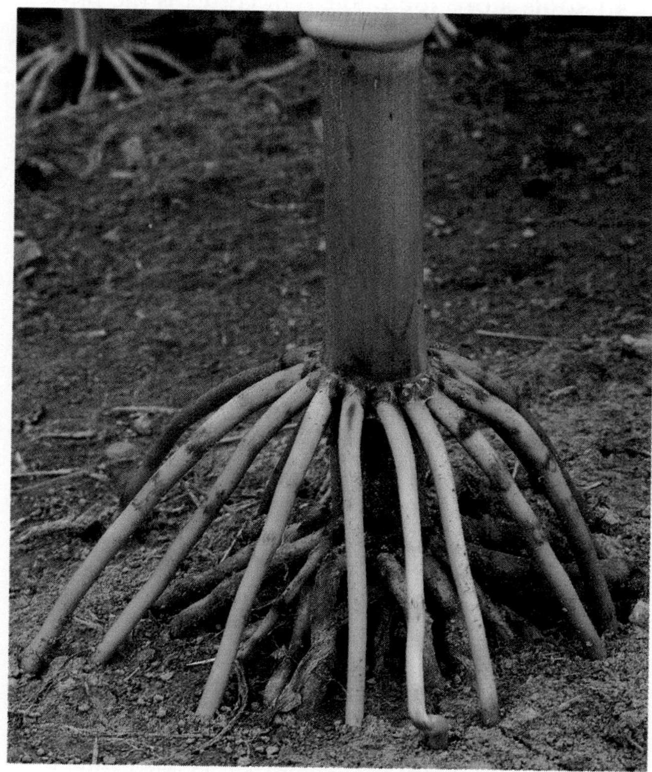

FIGURE 32–5

Prop roots of a corn plant. (J. Grushow from Grant Heilman)

BIO-BIT

Some orchids that live attached high up on a tree have photosynthetic roots.

Functions of Roots

We can now list the four main functions of roots:

1. **Roots anchor the plant in the soil.** Taproots are especially good anchors, since they are long, and so thick and tough that it is hard to break them. Fibrous roots are good at holding soil in place.
2. **Roots absorb water and minerals from the soil.** Fibrous root systems often expose more absorptive surface area to the soil than do taproot systems, but taproots may grow deep enough to reach a more reliable supply of water.
3. **Roots transport water and minerals up to the shoot system.** The xylem serves this function. The endodermal layer exerts some control over the passage of substances into or out of this central core of xylem and phloem.
4. **Roots store food.** The root cortex is the primary food storage area in some plants. In many other kinds of plants, the root stores only a small supply of food for itself.

New root cells are produced in the root apical meristem. These cells elongate and push the tip of the root deeper into the soil. Primary tissue cells laid down in the growing root form (a) a protective skin with root hairs that anchor the plant and increase surface area for absorption, (b) ground tissues for food storage, and (c) vascular tissues that transport substances to and from the shoot system. Additional roots may develop as branches off the first root or they may come off other sites like the stem or leaf. Roots help to anchor a plant within soil, absorb and transport water and minerals, and store food.

32–C STEMS

Primary Growth of Stems: Growth in Length

Once the bean seedling has established roots, the shoot begins to grow (see Figure 32–1). First, the lower part of the stem forms a loop that pushes up through the soil and then straightens, pulling the cotyledons free of the soil. Meanwhile, the apical meristem remains enclosed and protected between the cotyledons. Once the cotyledons are above ground, they spread apart. The first foliage leaves, which were fully formed but tiny in the bean embryo, have already begun to expand. When they are exposed to sunlight they become green and expand more rapidly. Until these leaves can feed the plant by their photosynthesis, the bean seedling continues to use the food stored in the cotyledons. Eventually, the cotyledons shrivel as their food supply is used up, and they fall off.

The shoot's apical meristem, at the stem tip, consists of a little mound of rapidly dividing cells. The cells laid down by the apical meristem continue to divide, and this (rather than division in the apical meristem itself) produces most of the new cells during the primary growth of the stem. As in the root tip, the new cells in the stem eventually elongate and then mature. Growth in both length and diameter by cells in the shoot may last longer than in the root. Hence growth in the shoot can be detected quite a distance behind the growing tip of the apical meristem.

Primary Structure of Stems

Cells laid down by the shoot's apical meristem divide, enlarge, and differentiate to become the primary tissues of the stem. The epidermis of the stem is usually just one cell thick. Its cells fit tightly together, reducing both loss of moisture and damage by invading fungi or insects. These cells also conserve water by secreting a layer of waxy substances, the **cuticle,** which forms a waterproof covering over the stem surface. The cuticle of the stems is continuous with a layer of cuticle on the leaves.

Most of the primary tissue in the stem is ground tissue in the cortex and pith. The cortex is the next layer inside the epidermis (Figure 32–6). Most of the cortex is made up of parenchyma cells. Some of these cells contain chloroplasts and carry out photosynthesis, and some may store starch in amyloplasts. The **pith** tissue, in the center of the stem, is also composed of parenchyma cells. In this respect, stems differ from roots, in which the center is usually occupied by the cylinder of vascular tissue. As the stem continues to grow, the cells in the outer parts may enlarge and expand so much that the pith is pulled apart, and the stem becomes hollow.

The dicot stem's primary vascular tissue occurs as a ring of discrete **vascular bundles**, strands of tissue containing phloem and xylem, rather than as the central core found in roots. Generally the xylem lies in the part of the bundle toward the inside of the stem, with the phloem toward the epidermis. Each vascular bundle is partly surrounded by thick-walled **fiber** cells, which provide extra support. The walls of the conducting cells in the xylem may also be especially thick.

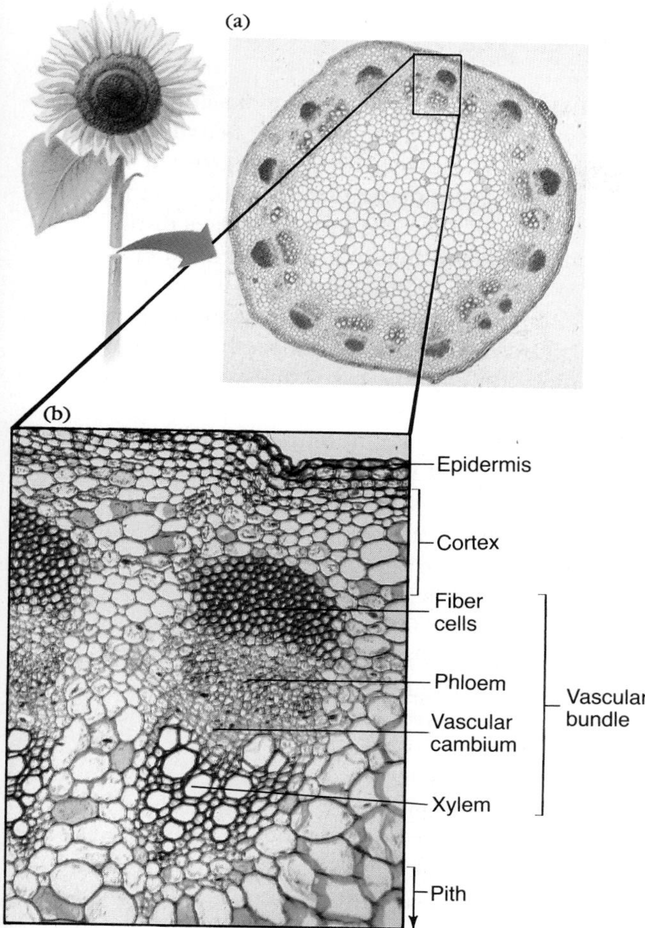

FIGURE 32-6

Primary tissues of a dicot stem. A cross-section of a sunflower stem shows **(a)** that most of the stem consists of pith, which lies inside the ring of vascular bundles. **(b)** Close-up of part of the ring of vascular bundles, with various tissues pointed out. (a, Carolina Biological Supply/Phototake; b, Ed Reschke)

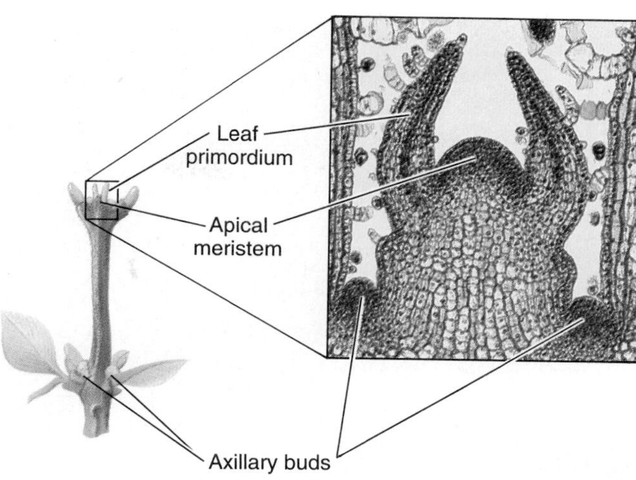

FIGURE 32-7

Growth of a shoot tip. Longitudinal section of a *Coleus* shoot tip, showing the apical meristem flanked by new leaf primordia. Note the axillary buds in the axils of the next older pair of leaves, which are still differentiating. (Ken Wagner/Phototake, NYC)

This extra strengthening material in the vascular bundles helps to support the stem in the air, which gives much less support than the soil that surrounds the roots.

Stems usually lack the endodermis and pericycle found in roots.

Primary Growth of Stems: Production of Laterals

As the shoot's apical meristem grows, it lays down not only the cells that divide and form the primary tissues of the stem but also **leaf primordia,** which develop into leaves (Figure 32-7). The angle between the base of the leaf and the stem is called the leaf's **axil** ("armpit"). Cell division in the axil produces a small group of meristematic cells called an axillary or **lateral bud,** which remains dormant for a time.

A new shoot branch arises when an axillary bud becomes active and begins to grow as the apical meristem of a new stem (Figure 32-7). These branches produce leaves with axillary buds, just as the main stem does.

While some axillary buds produce new branches, others remain dormant. The growth of most of a plant's axillary buds is repressed by hormones from the apical meristems, a phenomenon called **apical dominance** (Section 34-B).

Functions of Stems

We can identify four main functions of stems:

1. **Stems support structures of the shoot system.** The thick cell walls in the vascular tissue of stems form a tough, strong framework. In addition, the parenchyma cells in the pith and cortex contribute **turgor,** the internal pressure of the fluid inside them, which keeps the stem firm. With this tough, resilient structure the stem can hold the leaves up to the sunlight they need for photosynthesis and can bend in the wind without breaking.
2. **Stems transport substances between the roots and leaves.** The vascular tissue of stems is

FIGURE 32–8

Specialized stems. The stems of a prickly pear cactus grow in the form of jointed, green pads, which carry out photosynthesis. They are also succulent (water-storing), swelling as the plant absorbs water after a rain. The leaves are modified as sharp spines that fend off animals who might attempt to consume the stem's food and water. (E. R. Degginger, FPSA)

continuous with that of the roots and leaves. The xylem transports water and minerals taken in by the roots up to the leaves and to the living cells of the stem, and the phloem carries food from the leaves or from storage areas to the living and growing parts of the plant, some of which cannot carry on photosynthesis.

3. **Stems may produce food.** Some stems are green and photosynthetic. In most plants they supplement photosynthesis carried out in the leaves, but in plants such as cacti, stems are the main photosynthetic organs (Figure 32–8).
4. **Stems store substances.** Some stems contain many amyloplasts. A potato is an underground shoot with a large stem specialized for storage. Other stems store lesser amounts of food. Some stems, such as those of some cacti, also store a lot of water.

As the stem develops, cells laid down by the apical meristem divide, enlarge, and differentiate—forming primary tissues of the stem: epidermis, ground tissue in the cortex and pith, and vascular tissue. Other cells laid down by the apical meristem form leaf primordia. Stems help support the shoot system, produce and store food, and transport substances between the roots and leaves.

32–D LEAVES

3? A leaf begins as a leaf primordium, a slender rod of meristematic cells laid down by the activity of an apical meristem. These cells divide and grow in various patterns, producing mature leaves with a shape typical of the species. If the cells near the edges of the developing leaf keep dividing in a uniform manner, the leaf grows in a smooth, rounded shape. However, if cells stop dividing in some areas but not in others, an irregular shape results, with lobes or teeth alternating with indentations. In extreme cases, the cells in some areas near the midrib do not divide at all, and the mature leaf is composed of several **leaflets,** connected only by the midrib. Such leaves are called **compound leaves,** with each leaflet looking like a separate leaf (A Journey Through Leaf Structure). ■

Our bean plant has both simple and compound leaves. The first foliage leaves are simple, somewhat heart-shaped leaves. The leaves formed by the plant later, however, are compound leaves, with three leaflets each (see Figure 32–1).

Most leaves have a broad, flat shape, which maximizes the amount of surface area per unit of volume. A large surface area is important both for capturing the energy of sunlight and for exchanging gases with the atmosphere. The comparatively small volume makes the leaves lightweight; heavy leaves would require more supporting tissue in the stems and roots.

Structure of Leaves

As in the primary structure of the root and stem, a leaf's outer surface, both top and bottom, is covered by the epidermis. The epidermal cells of leaves, like those of stems, secrete a waxy, waterproof cuticle, which is especially thick in plants that must conserve water most strictly. Although the cuticle retards the loss of water to the atmosphere, it also impedes the passage of gases from the air into the plant. Because the leaves need carbon dioxide from the air for photosynthesis, and must release the oxygen they produce, the secretion of a continuous layer of cuticle would solve one problem but create another.

In fact, the cuticle is not continuous but is interrupted at intervals by **stomata** ("mouths"; singular: **stoma**). Each stoma is a pore between a pair of lip-like

BIO-BIT

Venus flytrap plants eat insects that become trapped between the modified leaves.

af Diversity

aves are large, flat solar panels. Their shapes are
termined by different rates of growth and division of
lls along the leaf's axis.

Pin oak

Yellow buckeye

Quaking aspen

Black locust

Ginkgo

af structure

Gnetum

Cuticle

Upper epidermis

Palisade mesophyll

Vascular bundle

Spongy mesophyll

Air space

Chloroplast

Lower epidermis

Cuticle

Stomata

CO_2 in

O_2 out

Wandering Jew

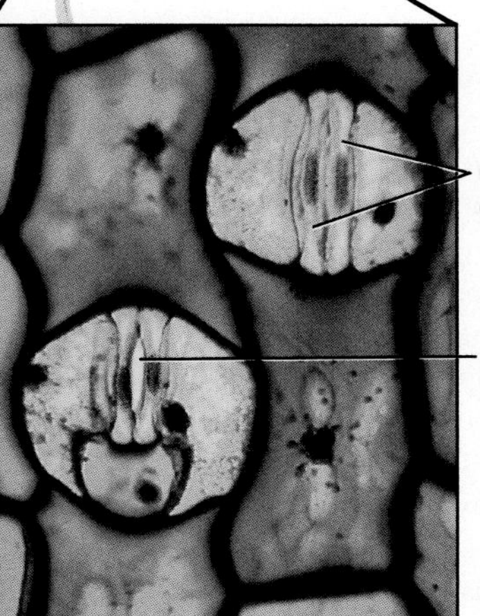

Micrograph of leaf surface with visible stomata.

Veining in a dicot leaf

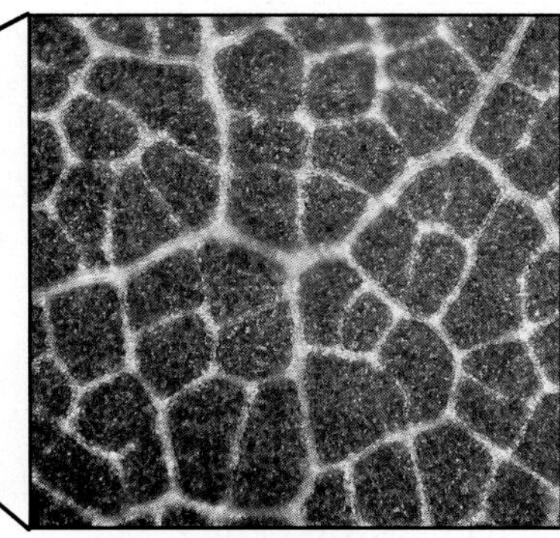

Leaf surface with netted veins.

A Journey Through Leaf Structure. **The leaves of plants appear in many different forms.** (a, b, Runk Schoenberger from Grant Heilman)

guard cells. The size of the pore can be controlled by the guard cells. When the stoma is open, more carbon dioxide can enter the plant, but also more water vapor can escape into the atmosphere. Closing the stoma reduces the exchange of these gases between the leaf's interior and the atmosphere. Most dicot leaves have stomata primarily in the lower epidermis. However, plants such as water lilies have them primarily in the upper epidermis, the only surface exposed to the air. Stomata occur not only in leaves, but also in the epidermal layers of stems and flower parts. All parts of a plant need oxygen for respiration, and some also need carbon dioxide for photosynthesis.

The ground tissue, which makes up the bulk of the leaf, consists of two layers of photosynthetic tissue. The cells in these layers are modified parenchyma, with many chloroplasts. Just beneath the upper epidermis is the **palisade mesophyll:** long, thin cells arranged vertically, perpendicular to the upper epidermis. Below the palisade mesophyll is the **spongy mesophyll,** its name suggested by the network of air spaces among the cells. The air spaces in both the palisade and spongy layers open to the atmosphere via the stomata, allowing gases to circulate to the photosynthetic cells inside the leaf.

The leaf also has vascular tissue running through the mesophyll in strands called **veins.** In dicots the main vein, in the midrib, is a continuation of one of the stem's vascular bundles, and it in turn is continuous with the system of smaller veins in the leaf. The vascular tissue brings water to the leaves and carries away the products of photosynthesis. It also provides a supporting framework that stretches the delicate photosynthetic tissues out where they can intercept the rays of the sun and exchange gases with the atmosphere.

In addition to producing food, leaves function in a plant's defenses against herbivores (See *A Journey into Evolution:* Leaves to the Defense, this chapter).

Leaves are the chief food-producing organs in most plants. The veins carry away food made in the leaf. They also provide the leaf with water, most of which escapes into the air through the stomata.

BIO-BIT

Approximately 99% of all water that is absorbed by the root systems of most land plants is lost by evaporation through the leaves or stems. Nearly all of this water is lost through the stomata.

32–E SECONDARY GROWTH

Secondary growth is growth in the diameter of stems and roots by the addition of new cells. It is most familiar in woody, perennial (living for many years) angiosperms and gymnosperms, plants that enclose their seeds in the ovary of a fruit, and those whose seeds are "naked"—without an enclosing ovary (see Sections 20–H and 20–I). Secondary growth is also found to some extent in nonwoody plants such as alfalfa, sunflowers, and some annuals (plants that live just one season).

The cells that form secondary tissues are produced by **lateral meristems.** The first of these is the **vascular cambium,** which lays down cells that become the **secondary vascular tissues.** In the stem, cells between the primary xylem and primary phloem in the vascular bundles become meristematic and form part of the vascular cambium. Additional cells between the vascular bundles also become meristematic. Hence, the vascular cambium can be seen in a cross section of the stem as a continuous ring of tissue, with the xylem and pith on the inside and the phloem, cortex, and epidermis on the outside (Figure 32–9).

Now cells in the vascular cambium divide, producing new cells. Those produced inside the ring of vascular cambium differentiate into **secondary xylem** tissue, also called **wood.** Most of these cells produce very thick cell walls and then die. As the vascular cambium produces new wood on the inside of the ring, the stem grows in diameter, and the phloem outside the vascular cambium becomes stretched and crushed. Meanwhile, though, cells produced just outside the vascular cambium have differentiated as **secondary phloem,** which transports food in place of the destroyed primary phloem. As more secondary xylem

FIGURE 32–9

Secondary growth of a woody stem. (a) This series of cross-sections shows the changes in stem structure due to growth. Note that as the tree grows, the vascular cambium produces additional layers of xylem. As new xylem layers are added, the (outer) secondary phloem becomes compressed. **(b)** This section of a basswood twig shows secondary growth after three years. The secondary xylem formed during each growing season appears as a distinct band, the annual rings, which can be counted to determine the stem's age. Each ring has large cells on the inside, where the spring growth occurs. Cells laid down later in the season are progressively smaller and thus spring and summer wood can be distinguished. The phloem is part of the bark and can be removed by peeling the bark from the twig. (M. Kage/Peter Arnold, Inc.)

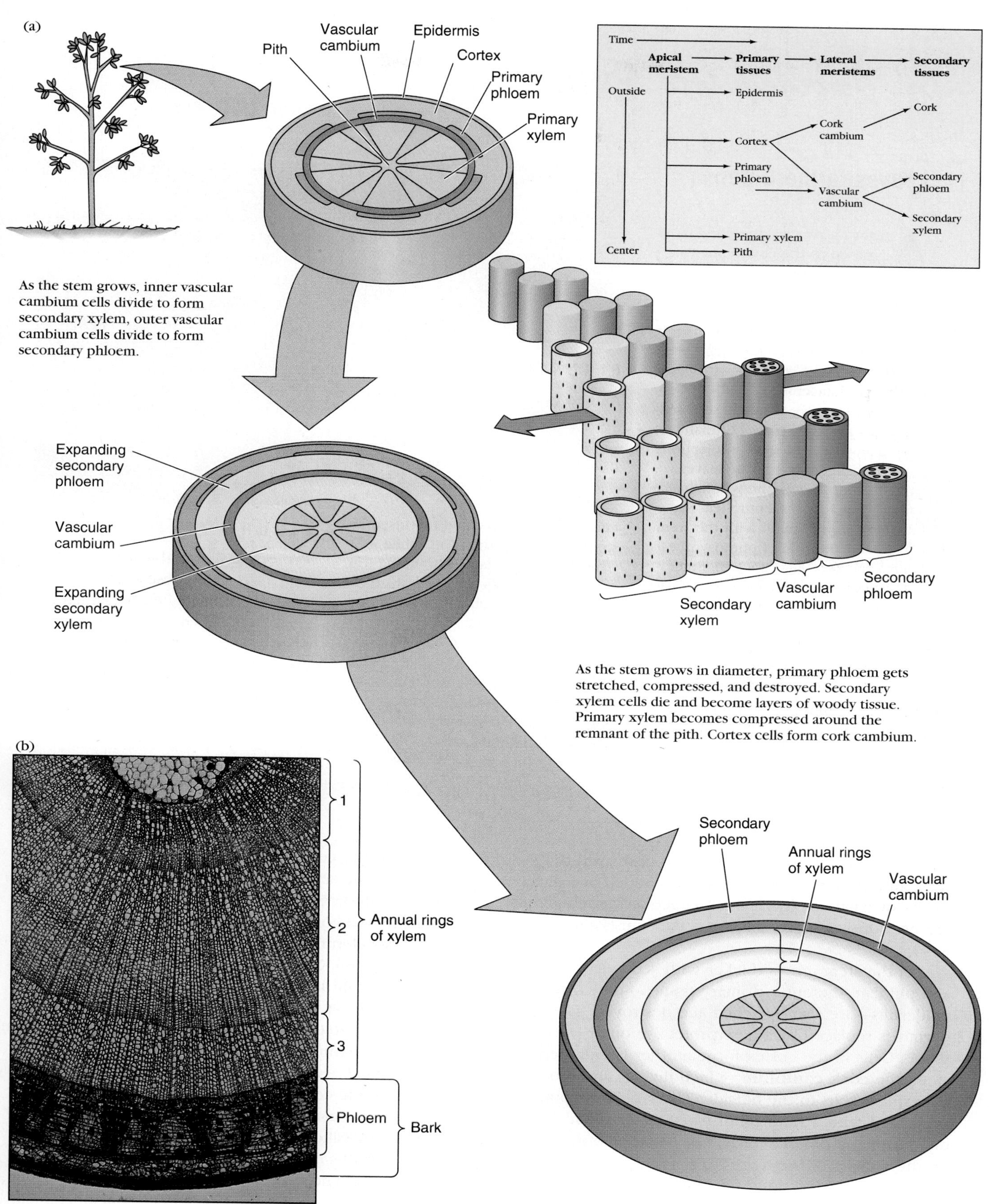
(a)
Pith
Vascular cambium
Epidermis
Cortex
Primary phloem
Primary xylem
Time
Apical meristem
Primary tissues
Lateral meristems
Secondary tissues
Outside
Epidermis
Cork
Cork cambium
Cortex
Primary phloem
Secondary phloem
Vascular cambium
Secondary xylem
Primary xylem
Center
Pith
As the stem grows, inner vascular cambium cells divide to form secondary xylem, outer vascular cambium cells divide to form secondary phloem.
Expanding secondary phloem
Vascular cambium
Expanding secondary xylem
Secondary xylem
Vascular cambium
Secondary phloem
As the stem grows in diameter, primary phloem gets stretched, compressed, and destroyed. Secondary xylem cells die and become layers of woody tissue. Primary xylem becomes compressed around the remnant of the pith. Cortex cells form cork cambium.
(b)
1
2
Annual rings of xylem
3
Phloem
Bark
Secondary phloem
Annual rings of xylem
Vascular cambium

A Journey into Evolution

Leaves to the Defense

A plant's leaves are the parts most exposed to the environment. Here they spread out and absorb sunlight and carbon dioxide for photosynthesis. And here they also catch the attention of hungry herbivores. It is not surprising, therefore, that many plants have evolved leaves with protective mechanisms.

The most obvious defense is having sharp prickles. Leaves of many hollies have several sharp projections along the edges, and thistle leaves carry this motif over the whole leaf surface. In cacti, the entire leaf has become a sharp spine, with the stem taking over the role of photosynthesis completely (see Figure 32-8). Some grasses grow their weapons in miniature, producing saws rather than swords (Figure 32-A).

Many leaves have more subtle defenses, such as specialized epidermal hairs. The itchy juice secreted by the hairs of tomato plants, for instance, deters some animals from brushing against them, much less eating them. The hook-shaped hairs of beans entangle small insects, and the insect stops feeding on the plant while it struggles to free itself. The nettle leaf in Figure 32-B has hairs of both these types. Epidermal hairs of other kinds of plants release glue, which traps small insects and immobilizes the feet and mouthparts of larger ones.

FIGURE 32-A

Armed grass. This scanning electron micrograph reveals the saw-like teeth on leaves of grass that can inflict a painful cut. (J. Burgess/Science Photo Lab.)

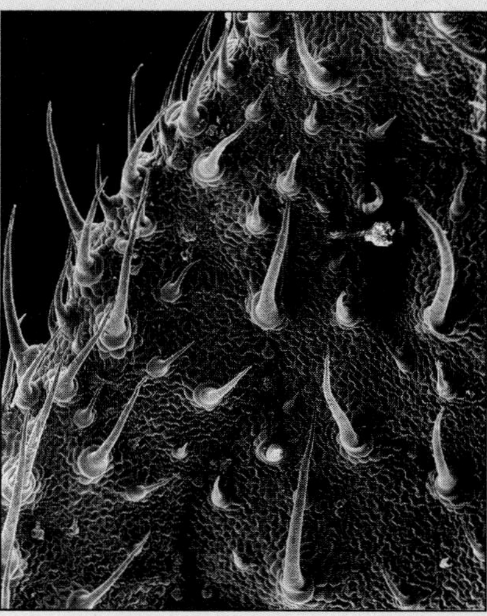

FIGURE 32-B

Protective hairs. This scanning electron micrograph shows that the surface of this stinging nettle leaf has spiky hairs tipped with droplets of irritating fluid. (G. F. Leedale/Biophoto Associates/Photo Researchers, Inc.)

Some leaves' defenses are not external but internal. For example, leaf cells may contain sharp, needle-like crystals of calcium oxalate, called raphides, which ensure that animals will not take a second mouthful of leaves from the plant.

Other plants produce toxic chemicals. For example, milkweed

forms, the first secondary phloem is destroyed in its turn, and more secondary phloem is added just outside of the vascular cambium—that is, immediately *inside* the previously formed secondary phloem (Figure 32-9).

As the stem grows thicker with secondary xylem and phloem, the outer primary tissues—the cortex and epidermis—are also stretched and destroyed. The epidermis is replaced by a new protective outer layer of tissue. Cells in the cortex form another lateral meristem, the **cork cambium,** which divides and produces new cells to the outside. These cells become impregnated with a waterproof waxy material and then die, forming a protective layer of **cork,** or **outer bark,** on the outside of the tree.

In the cross section of a stem with well-developed secondary growth, we would find these tissues, starting from the center and working outward: pith, primary xylem, secondary xylem, vascular cambium, secondary phloem, the crushed remains of primary phloem and cortex (these would be present but perhaps not identifiable), cork cambium, and cork (bark) (Figure 32-9). The primary layer of epidermis has by this time ruptured and sloughed away. When the bark is peeled off a tree, it breaks at the layer of delicate undifferentiated cells in the region of the vascular cambium. The trunk of a tree is almost entirely secondary xylem, with a very slender column of pith and primary xylem in the center (Figure 32-9). The tissues outside the vascular cambium are all in the part of the trunk commonly called the **bark.**

From this discussion, we can see that most of the tree's wood and bark consist of dead cells. In essence, the living cells of the vascular and cork cambia keep

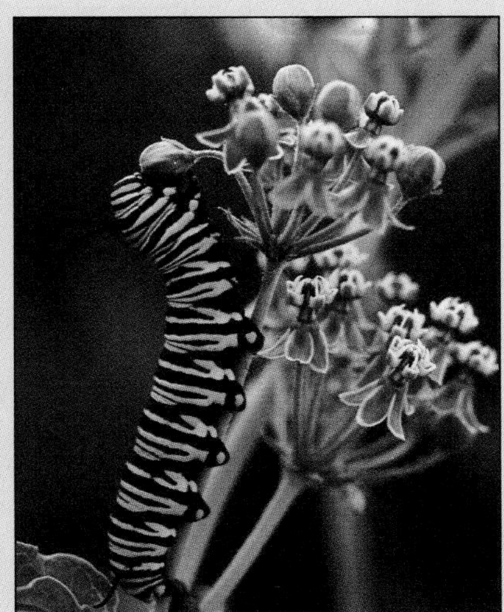

FIGURE 32-C

Chemical defense. Unlike most other animals, the caterpillars of monarch butterflies can feed on milkweed, despite the toxic chemicals in the tissues of the milkweed plant. (Frans Lanting/Minden Pictures)

leaves contain cardiac glycosides, chemicals that cause the heart to beat faster, sometimes leading to death of the animal. However, this defense is not absolute: the caterpillars of monarch butterflies have evolved the ability to eat milkweed leaves without ill effects (Figure 32-C). In fact, these caterpillars store the milkweeds' cardiac glycosides in their own bodies, making themselves unpalatable to predators. These defensive chemicals are retained when the caterpillar metamorphoses into an adult butterfly. Hence, the adult is protected from predators by chemicals from the food plant it ate as a larva.

In some plants, leaves produce their defenses only when needed. For example, tannins are defensive chemicals that precipitate the digestive enzymes of many herbivores; unable to digest its food, the animal starves. Sugar maples produce leaves with little tannin in the spring. In years when this first set of leaves is eaten by a heavy infestation of insects, the tree grows a set of replacement leaves containing more tannins.

A plant must fend off not only hungry herbivores but also other plants that compete with it for sunlight and water. Many plants produce chemicals that inhibit the growth of near neighbors, and what better way to deploy them into a "no trespassing" zone than by washing off the leaves? This is how shrubs of the chaparral in California inhibit the germination of seeds in the surrounding soil. Similarly, it is notoriously difficult to grow anything under a walnut tree because of a substance called juglone, which washes onto the ground and inhibits the growth of other plants.

growing a new tree outside of the dead xylem core of its former self, and inside of its old, dead "skin" (Figure 32-9).

Secondary growth in roots is similar to that in stems. The main roots of a tree are large and woody. They provide support and serve as transport conduits. Encased in their tough bark, these roots cannot absorb water and minerals from the soil, a task performed by young roots at the far ends of the root system.

Wood produced during secondary growth thickens and strengthens stems and roots. New phloem and cork replace those tissues destroyed by this growth of secondary xylem.

32-F A COMPARISON OF MONOCOTYLEDONS AND DICOTYLEDONS

Monocotyledons include grasses, palms, and flowering bulb plants. The large seeds of the grasses known as cereals (rice, wheat, corn, oats, barley, rye, and so forth) are the staples in the diets of most human societies. Monocotyledons differ from dicotyledons, such as the beans described earlier, in the number of cotyledons (seed leaves) possessed by the embryo: monocotyledon embryos have only one cotyledon. The embryos of the two groups also have other structural differences.

The corn kernel is convenient for studying the structure of a monocot seed. A kernel of corn is actually a one-seeded fruit, and its "skin" contains both the

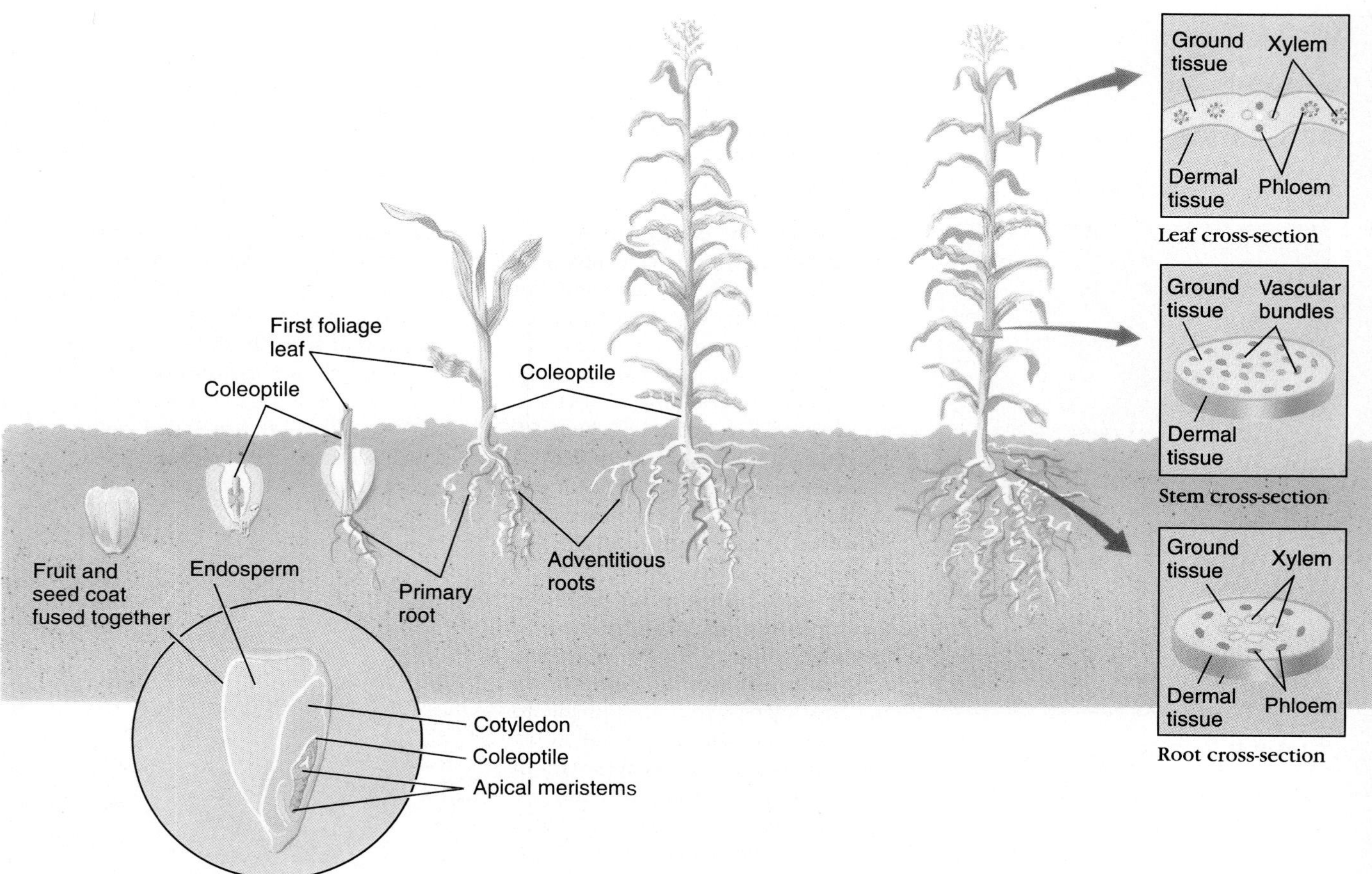

FIGURE 32-10

Structure of a corn seed, a representative monocot. A monocot embryo has only one cotyledon. Notice that the food supply of the endosperm occupies a great deal of the seed. The coleoptile is a sheath covering the leaves, which are rolled above the apical meristem of the shoot. Growth of a corn seedling. Root hairs on the roots anchor the plant in the soil and absorb water. The pointed, golden tip of the coleoptile grows upward, into a position where the enclosed leaves will be able to reach light. Then the leaves will expand, rupture the coleoptile, and begin photosynthesis. In a monocot, the vascular bundles are scattered throughout the leaves, stems, and roots rather than being arranged in a single ring as in dicots (compare with Figure 32-9).

seed coat and fruit parts, closely attached to each other. In the corn seed, the **endosperm,** a nutritive tissue, is separate from the embryo, whereas in beans the food-rich endosperm has been absorbed into the cotyledons. (The placement of endosperm is not consistently different between monocots and dicots, however.) The corn embryo digests the food stored in the endosperm and uses it for growth until it becomes established as a self-sufficient, photosynthetic plant. The corn embryo has several developing leaves, wrapped above the apical meristem. A tough sheath, the **coleoptile,** protects the leaves and the apical meristem from injury as the seedling pushes up through the soil (Figure 32-10). Once the coleoptile has penetrated into the light above the ground, the leaves expand greatly, rupture the tip of the coleoptile, and grow out.

Monocots also differ from dicots in the structure of the primary plant body and the arrangement of its tissues. The xylem of a monocot root usually has many more separate tubes (corn has 20 to 40), and most monocot roots have pith in the center. Monocot stems have numerous small scattered vascular bundles (Figure 32-10) rather than the single ring of bundles found in dicots. Therefore, the ground tissue has no distinct pith and cortex. In monocot leaves, the principal veins of vascular tissue run parallel to each other, rather than in the fan-like or net-like pattern common in dicots (Table 32-1). Monocot leaves often stand more or less upright, like grass, and stomata are plenti-

Table 32–1
Comparison of Monocotyledons and Dicotyledons

Monocots	Dicots
Cotyledons:	
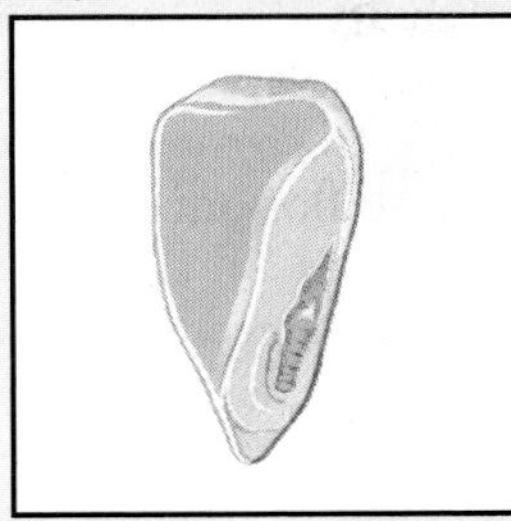 One	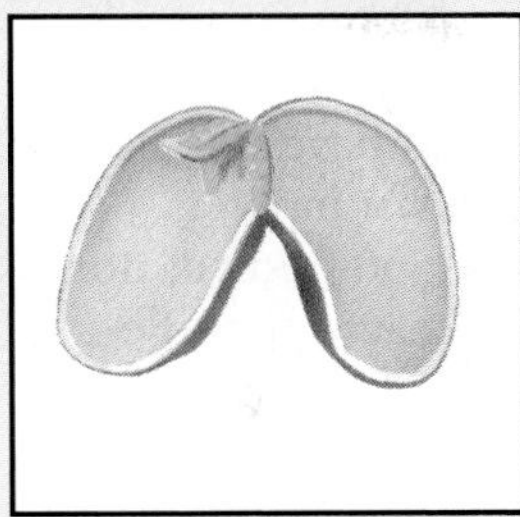 Two
Flower parts:	
 In threes or multiples of three, or can be irregular	 In fours or fives or multiples of these, or can be irregular
Veins of leaves:	
 Usually parallel	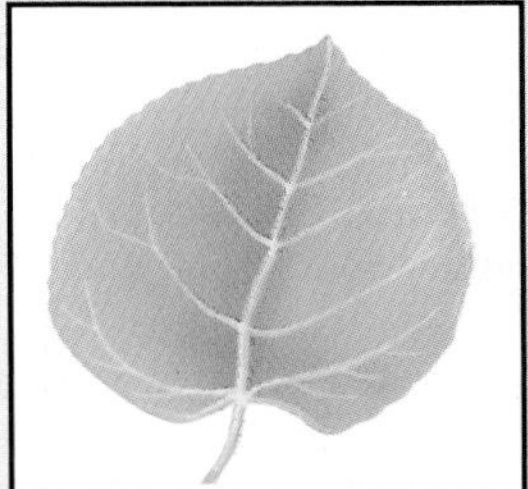 Usually netted or fan shaped
Vascular bundles:	
 Scattered through the stem; usually no cambium	 Arranged in a ring
Examples:	
 Lilies, grasses, daffodils, orchids, irises, crocuses, palms	 Maples, legumes, roses, mints, squashes, daisies, walnuts, cacti, poppies

ful in both upper and lower epidermal layers of the leaf.

Grasses and some other monocotyledons have growth areas besides the apical meristems. **Intercalary meristems** at the bases of leaves permit the plants to keep growing after their tips have been eaten by grazing animals or clipped by a lawn mower.

Most monocots are herbaceous (soft-bodied, non-woody) annuals, or perennials with leaves that die back to the ground after each growing season—especially in areas with freezing winter temperatures. Except for Joshua trees and other yuccas, and dragon trees (which have an anomalous type of secondary growth), most monocots lack secondary growth and secondary tissue. Even the tallest monocots, palm trees, grow in diameter without secondary growth: they have "primary thickening meristems," close behind the apical meristem of the trunk. This is why palm tree trunks are nearly the same diameter from top to bottom.

Monocotyledons differ from dicotyledons in the number of cotyledons in the embryo, the arrangement of vascular tissue in the roots, stems, and leaves, and the arrangement of leaf stomata. In addition, most monocots have no secondary growth, whereas many dicots do.

SUMMARY

1. A vascular plant grows in one place, making its own food and competing with other plants for the basic resources of sunlight, water, and minerals.
2. The plant body consists of roots, stems, and leaves, all containing vascular tissue—which conducts water, minerals, and food rapidly from one part of the body to another. This body plan is adapted to obtaining water and sunlight efficiently. The plant grows and produces new tissues throughout its life.
3. A seed consists of a protective covering, food supply, and plant embryo with a rudimentary root, stem, and leaves.
4. Early in life, all cells of the plant can divide. Later, most cells mature, specialize, and lose their capacity to divide. Cells that retain the ability to divide are found in meristems at specific locations in the plant.
5. Primary growth is principally growth in length and production of new root and shoot branches.
6. The tips of stems and roots grow longer through the activity of apical meristems.
7. Stem branches arise from axillary buds, located at the bases of leaves. An axillary meristem thus becomes the apical meristem of a new branch.
8. Lateral roots arise from the pericycle tissue inside the root. This produces a new apical meristem of a branch root.
9. Secondary growth is growth in girth or diameter by production of secondary tissues.
10. Vascular cambium, a lateral meristem, produces secondary vascular tissues.
11. Secondary xylem adds structural support and increases the capacity for conducting water to the leaves; secondary phloem replaces primary phloem, or older secondary phloem, destroyed as interior tissues grow.
12. The cork cambium, another lateral meristem, produces cork, which replaces the epidermis that has been destroyed by expansion of the interior tissues.
13. Monocotyledons and dicotyledons differ in the number of cotyledons found in the embryo (one versus two, respectively); in the arrangement of vascular tissue, especially in the stems and leaves; and often in the arrangement of leaf stomata. In addition, very few monocots have secondary growth, whereas secondary growth is found even in some annual dicots.

SELF-QUIZ

1. A fibrous root system is apt to perform which function better than a taproot system?
   a. absorption
   b. anchorage
   c. food storage
   d. transport
2. One difference between a bean seed and a kernel of corn is that:
   a. a bean seed has a seed coat but a kernel of corn does not
   b. the bean has two cotyledons; the corn has one
   c. the bean contains stored food; the corn does not
   d. the bean embryo has leaves, while the corn embryo lacks leaves
   e. the bean embryo has two apical meristems; the corn embryo has only one
3. Primary growth of a tree occurs:
   a. through the activities of apical meristems
   b. through the activity of a vascular cambium
   c. through the activity of the root cap
   d. only in the first year of the tree's life
   e. in stems, but generally not in roots

4. Arrange the following events during the growth of a shoot in proper order:
   a. cell division
   b. cell maturation
   c. cell elongation
5. Lateral roots arise from:
   a. root hairs
   b. pericycle
   c. cork cambium
   d. endodermis
   e. axillary buds
6. Secondary xylem and phloem are laid down by:
   a. apical meristems
   b. axillary meristems
   c. vascular cambium
   d. cork cambium
   e. intercalary meristems
7. Which of the following would *not* secrete a cuticle?
   a. leaf epidermis
   b. stem epidermis
   c. root epidermis
8. For each characteristic listed below, tell whether it is characteristic of monocotyledons, dicotyledons, or both:
   ___ a. stomata on both leaf surfaces
   ___ b. roots with root caps
   ___ c. parallel venation in leaves
   ___ d. many vascular bundles scattered throughout a stem cross section
   ___ e. vascular cambium
9. John and Marie visited a forest on their honeymoon. John selected a tree 10 meters tall and 30 centimeters in diameter. He carved their initials into its bark 1.5 m above ground level. On their tenth anniversary, John and Marie return to the forest; the tree is now 12 m tall and 33 cm in diameter. Their initials are now:
   a. 1.5 m above ground level
   b. 2 m above ground level
   c. 3.5 m above ground level

THINKING CRITICALLY

1. List specialized structures in a plant's dermal tissue that (a) exchange substances with the environment; (b) defend the plant.
2. What is the advantage of the arrangement of xylem in a star-shaped pattern in the root, rather than its being internal to the phloem as it is in the shoot?
3. Since the cotyledons of a bean seedling become photosynthetic, one might expect them to be valuable organs of the plant even after their stored food is used up. Why might it be adaptively advantageous to the plant to shed them soon after it becomes well established?
4. Leaf structure varies from one species to another, and it is often closely correlated with the habitat of the plant. In what type of habitat would you expect to find each of the following modifications of leaf structure?
   a. more than one cell layer in the epidermis
   b. extra thick layers of cuticle
   c. little or no cuticle
   d. little or no cuticle; little or no xylem; no stomata
   e. large air spaces in the mesophyll; stomata in the upper epidermis instead of in the lower epidermis

SELECTED KEY TERMS

amyloplast, *p. 672*
apical dominance, *p. 675*
axil, *p. 675*
coleoptile, *p. 682*
cotyledon, *p. 670*
cuticle, *p. 674*
dermal tissue, *p. 669*
dicotyledon, *p. 670*
endodermis, *p. 672*
endosperm, *p. 682*
epidermis, *p. 672*
ground tissue, *p. 669*
meristem, *p. 669*
pericycle, *p. 672*
phloem, *p. 669*
pith, *p. 674*
primary growth, *p. 669*
root hair, *p. 672*
root system, *p. 669*
secondary growth, *p. 669*
shoot system, *p. 669*
stomata, *p. 676*
taproot, *p. 673*
turgor, *p. 675*
vascular tissue, *p. 669*
vein, *p. 678*
wood, *p. 678*
xylem, *p. 669*

CHAPTER 33

CONCEPT GUIDE

After reading this chapter, you should be able to:

1. Distinguish between macronutrients and micronutrients and list several examples of each. (Section 33-A)
2. Describe how particle size, organic matter, living organisms, and pH can affect soil fertility. (Section 33-B)
3. Describe the processes by which water and minerals are absorbed by roots and transported to the rest of the plant. (Section 33-C)
4. Compare the general functions of xylem and phloem. (Section 33-D)
5. Compare the structure of xylem in gymnosperms and angiosperms. Explain how differences in structure alter the way that fluids are conducted. (Section 33-E)
6. Describe the forces that move fluids through the conductive tissues of plants. (Sections 33-F and 33-G)
7. Compare the conducting cells of xylem and phloem. (Sections 33-F and 33-G)
8. Explain how sucrose is transported within phloem from areas of sucrose production to areas of its use. (Section 33-H)
9. Explain the adaptive advantages of underground food storage in plants. (Section 33-I)

Sugars produced in leaves are transported throughout a tree in the form of sap. The system of phloem tissues that transports sap in sugar maple trees can be "tapped" to collect diluted maple syrup, which this man is pouring into a bucket. Read more about the movements of sugars and minerals throughout plants in Sections 33 D–H. (S. Kaufman/Peter Arnold, Inc.)

Nutrition and Transport in Vascular Plants

Plants take inorganic substances from the environment and use them to make their organic food molecules. All living organisms, including plants, need the same major nutrients. Land plants obtain carbon, from carbon dioxide in the air, through their stomata (Section 32-D). However, it is the roots that absorb the wide array of mineral nutrients and the huge amounts of water the plant needs. Taken all together, a plant's roots contain a vast total surface area of plasma membranes that absorb water and minerals from the soil.

Roots absorb water and minerals but cannot make food by photosynthesis in the darkness of the soil. The leaves rely on water and minerals from the roots, and in turn supply the roots with food. This complementary division of labor is possible because of the vascular tissues: xylem, which conducts water and minerals from the roots to the leaves; and phloem, which transports food from the leaves to the roots, as well as to growing buds, flowers, and fruits. Having an efficient transport system permits a plant to have parts specialized for different functions, such as water-gathering, energy-gathering, and reproduction. Vascular tissue also provides strength and support, as seen most clearly in the wood of roots and stems, and the veins of leaves. This support permits the plant to exploit a much larger volume of soil and air, and so increases its ability to compete with other plants.

How does a plant obtain nutrients from the soil? How does it move them up to the leaves, and move the food produced in the leaves to the roots and to growing buds and fruits? To answer these questions, we must study the plant's roots and the soil in which they grow, and the features of vascular tissue. We must also apply our knowledge of active transport, osmosis, and the other properties of water (Sections 4-D, 4-C, and 2-F). This is because transport in plants, as in animals, moves substances down a pressure gradient, from a higher to a lower pressure area. Phloem contents are pushed by a positive pressure generated by osmosis. A similar mechanism may move substances in xylem. However, most often xylem transport results from a negative pressure—a pull—generated by evaporation of water from the leaves.

KEY CONCEPTS

- Roots absorb water and minerals from the soil solution through a large surface area of selectively permeable cell membranes.
- Transport in xylem and phloem follows a fluid pressure gradient.
- Water and minerals move upward in plants through xylem. The main force is the pull exerted when water evaporates from the leaves via the stomata.
- Food in the form of sugar moves through phloem from leaves to sites of use or storage. Phloem transport follows a gradient of pressure built up by active transport of solutes and osmotic movement of water.

Curiosity Questions

1? Can too much fertilizer hurt a plant? (See page 690)

2? What is the function of the thick, sticky fluid sometimes found on the bark of pine trees? (See page 695)

3? Why do trees develop rings within their trunks and branches? (See page 695)

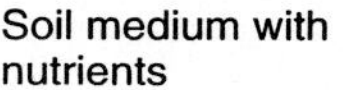

FIGURE 33–1

Sunflower seedlings grown in various nutrient solutions. Left to right: nutritionally complete medium, followed by media lacking only sulfur, nitrogen, phosphorus, and potassium, respectively.

33–A NUTRITIONAL REQUIREMENTS OF PLANTS

Plants obtain carbon, hydrogen, and oxygen from carbon dioxide and water during photosynthesis. Plants also need the elements nitrogen (N), phosphorus (P), potassium (K), calcium (Ca), magnesium (Mg), sulfur (S), and iron (Fe). This list of plant nutrients is easy to remember using a phrase made up of the elements' chemical symbols: C HOPK'NS CaFe, *M*ighty *g*ood.

All of these nutrients except iron are needed in relatively large amounts, and so are called **macronutrients.** Calcium and magnesium are plentiful in many soils. Farmers have long fertilized their soil by adding the other elements in either organic or inorganic form. Nitrogen, phosphorus, and potassium are required in the greatest quantities, and commercial fertilizers are rated by the percentage of each of these elements they contain. For example, a "5-10-5" fertilizer contains 5% nitrogen, 10% phosphorus, and 5% potassium, by weight. (The other 80% is inert materials.)

Plants also need various **micronutrients,** minerals used in small amounts: iron, boron, zinc, manganese, chlorine, molybdenum, and copper. Whereas nitrogen is applied to the soil at the rate of several hundred pounds per acre per year, the treatment for molybdenum-deficient soils in Australia is 2 ounces of MoO_3 per acre, applied once every 10 years.

BIO-BIT

Hydroponics is the growing of plants in aerated water containing dissolved minerals and salts.

Many minerals are vital components of important biological molecules (Table 33–1). An inadequate supply of any one of several minerals may result in rather general deficiency symptoms, such as **chlorosis** (paleness) and poor growth (Figure 33–1). On the other hand, deficiency symptoms of some nutrients may be localized in a specific part of the plant.

Macronutrients and micronutrients are absorbed by roots and used by plants to make organic food molecules.

33–B SOIL

The minerals used by plants come ultimately from the soil around their roots. Soil contains particles of varying sizes formed as the underlying rock breaks down. The type of rock is largely responsible for the minerals present in the soil (except for nitrogen, discussed shortly).

The size of soil particles influences the soil's capacity to hold **soil water.** Soil can hold water by adhesion to the surface of each particle (Section 2–F) and by capillarity. **Capillarity** allows water to fill narrow spaces, such as those between soil particles. This occurs because water adheres to hydrophilic ("water-loving") surfaces on the sides of the space and pulls other water molecules along by cohesion, so that water fills the space. The smaller the soil particles, the more water the soil can hold, because smaller particles have a greater collective surface area and there are more small-sized spaces between them than in an equal volume of larger particles (Figure 33–2). In fact, clay par-

Table 33–1
The Roles of Mineral Nutrients in Plants

Nutrient	Role in Plant	Deficiency Symptoms
Macronutrients		
Nitrogen	Component of proteins, nucleic acids, chlorophyll, some hormones	Mild: older leaves yellow; purplish leaf veins and stems Severe: stunting
Potassium	Not well understood; important in membrane potentials and opening of stomata; activates enzymes in photosynthesis, respiration, and starch and protein synthesis	Various, general; mottled chlorosis and dead spots, starting in older leaves; weak stems
Phosphorus	Component of nucleic acids, ATP, phospholipids	Various, general; dark color; loss of older leaves; stunting; slow maturation
Sulfur	Component of some amino acids, coenzyme A (Section 7–D)	Chlorosis, poor roots
Calcium	Component of middle lamella of cell walls (Section 11–E); ties up waste products as insoluble salts; involved in membrane function	Meristem death; abnormal cell division; deformed tissues; breakdown of membrane structure
Magnesium	Component of chlorophyll; cofactor of many metabolic enzymes	Chlorosis, appearing first in older leaves
Micronutrients		
Iron	Needed for chlorophyll production; part of electron transport molecules and of some enzymes	Chlorosis, appearing first in youngest leaves
Boron	Mostly unknown; nucleic acid synthesis; pollen germination; carbohydrate transport	Various; thick dark leaves; malformations; cell division and elongation and flowering inhibited
Zinc	Synthesis of tryptophan (precursor of auxin); component of some enzymes	Small puckered leaves; reduced stem elongation
Manganese	Activates citric acid cycle enzymes; involved in production of O_2 during photosynthesis	Mottled chlorosis
Chlorine	Production of O_2 during photosynthesis	Small leaves, slow growth, thick stunted roots
Molybdenum	Part of enzymes for nitrate reduction and nitrogen fixation	Same as nitrogen deficiency
Copper	Component of some enzymes in respiration and photosynthesis	Dark misshapen leaves with dead spots

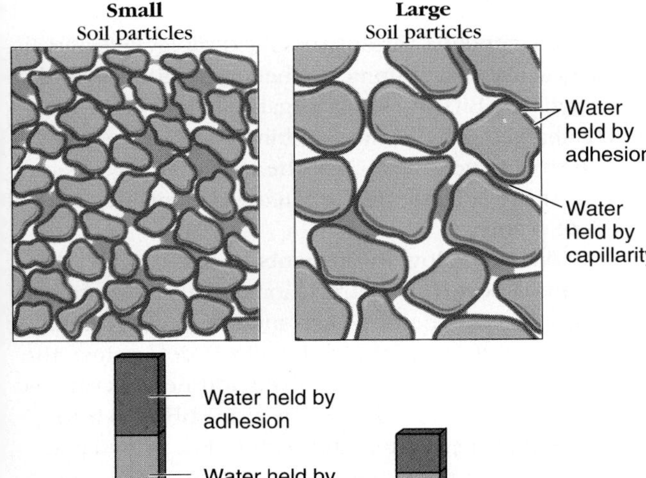

FIGURE 33–2

Soil particle size and water capacity. More water can be held by clay soils (small particles) than by sandy soils (larger particles). Light blue shows water held as a film on the surface of each particle; dark blue is water held by capillarity in the tiny crevices between particles.

ticles, with the smallest particle size, tend to hold water so tightly that plants cannot withdraw much of it. The large particles of a sandy soil, on the other hand, often let so much water drain out that the soil is too dry for most plants. Most plants grow best in soil with a mixture of particle sizes.

Rainfall exceeding the soil's water-holding capacity drains away, carrying dissolved nutrients with it. This is called **leaching.**

Fertile soil contains a great deal of organic matter. Manure, dead leaves, and bits of wood act like sponges, soaking up water and swelling when it rains, and releasing the water slowly later, as the soil dries out. This alternate swelling and shrinking helps loosen the soil, allowing roots to grow through it easily. Organic matter on top of the soil (mulch) acts as a sunshield, reducing soil temperature and retarding evaporation from the soil. More importantly, the organic matter in soil is a reservoir of nutrients, which are released slowly by decomposition.

1? Does organic matter make better fertilizer than inorganic products? Plants can take up nutrients only in certain chemical forms. Inorganic fertilizers usually provide these forms, bypassing the steps of microbial digestion needed to make organic fertilizers available to plants. Organic fertilizers do not provide better nutrition. They are superior because they help hold water in the soil and because they release their minerals gradually, giving plants a sustained source of nutrients. By contrast, nutrients in inorganic fertilizers may leach rapidly out of the soil after each treatment. This wastes money and may pollute water supplies. Too much inorganic fertilizer may also make the soil solution so concentrated that plants lose water by osmosis. Such "fertilizer burn" may dehydrate and kill the plant. ■

An ongoing study is comparing the effects of inorganic and organic fertilizers on two adjacent farms in Washington state. One farm, first cultivated in 1908, has been conventionally managed since 1948, with inorganic fertilizer and pesticides applied as recommended by the state. Crops of winter wheat and spring peas are grown in succession in each field. The "organic" farm next door has received no inorganic fertilizer since it was first plowed in 1909. It too grows wheat and peas, but every third year, each field grows **green manure,** a crop that is not harvested but plowed back into the soil to add organic fertilizer to the soil. The organic farm now has 16 centimeter (6 inches) more topsoil, more soil organic matter, more water storage capacity, and less rain-shedding top crust.

Living organisms are a vital component of soil (Figure 33-3). Bacteria and fungi digest organic matter, slowly releasing minerals that plants can absorb. Nitrogen-fixing bacteria convert nitrogen from the air (N_2), a form plants cannot use, to a form plants can use to make amino acids—ammonium ion (NH_4^+). If plants do not pick up ammonium quickly, other nitrifying bacteria oxidize it to nitrate (NO_3^-), which is easily leached out of the soil by rain. Plants of the legume family (peas, beans, clover, alfalfa) avert this problem by housing nitrogen-fixing bacteria in their root nodules (see Figure 19-6). Because legumes increase the soil's nitrogen content, they are planted as part of crop rotation programs.

The roots of a large majority of plants form **mycorrhizae** (pronounced "my-co-rise-ee"), symbiotic associations with soil fungi (Section 19-K). Fungi absorb soil minerals very efficiently. The hyphae of mycorrhizal fungi extend out into the soil around the root and also grow into the root itself. In this arrangement, they transfer minerals they have absorbed from the soil to the plant, especially phosphorus, which is not very mobile in the soil and is therefore hard to obtain. In return, these fungi receive organic molecules made by the plant. Most mycorrhizal fungi cannot grow without plant partners and are totally dependent on the plant for food. The plants also do much better with fungal partners (Figure 33-4).

Oxygen in the soil is important because most living things, including plant roots, require oxygen for respiration. The burrowing of organisms such as worms and insects mixes and breaks up the soil, allowing air to penetrate.

BIO-BIT

Plants that live in anaerobic environments such as swamps and marshes typically have large air spaces within their tissues and special roots for aeration.

The Soil Solution

Soil water contains various dissolved molecules and ions. However, some minerals become tightly bound to soil particles. Bits of clay or organic matter have negatively charged surfaces and so bind most of the soil's positively charged ions. Only the ions that stay dissolved in the soil solution are actually available to the roots of plants.

Acidic soil is low in nutrients because its H^+ displace other positively charged ions from the surfaces of soil particles. The displaced mineral ions are then readily leached out of the soil. Soil pH determines the mix of nutrients that are dissolved and hence available to plants. Most plants do best at a soil pH of 6.0 to 7.5, although there are plants adapted to growing at higher or lower pH ranges.

Soil pH can be adjusted to suit the plants we wish to grow. Lime is applied to acidic soils to raise the pH,

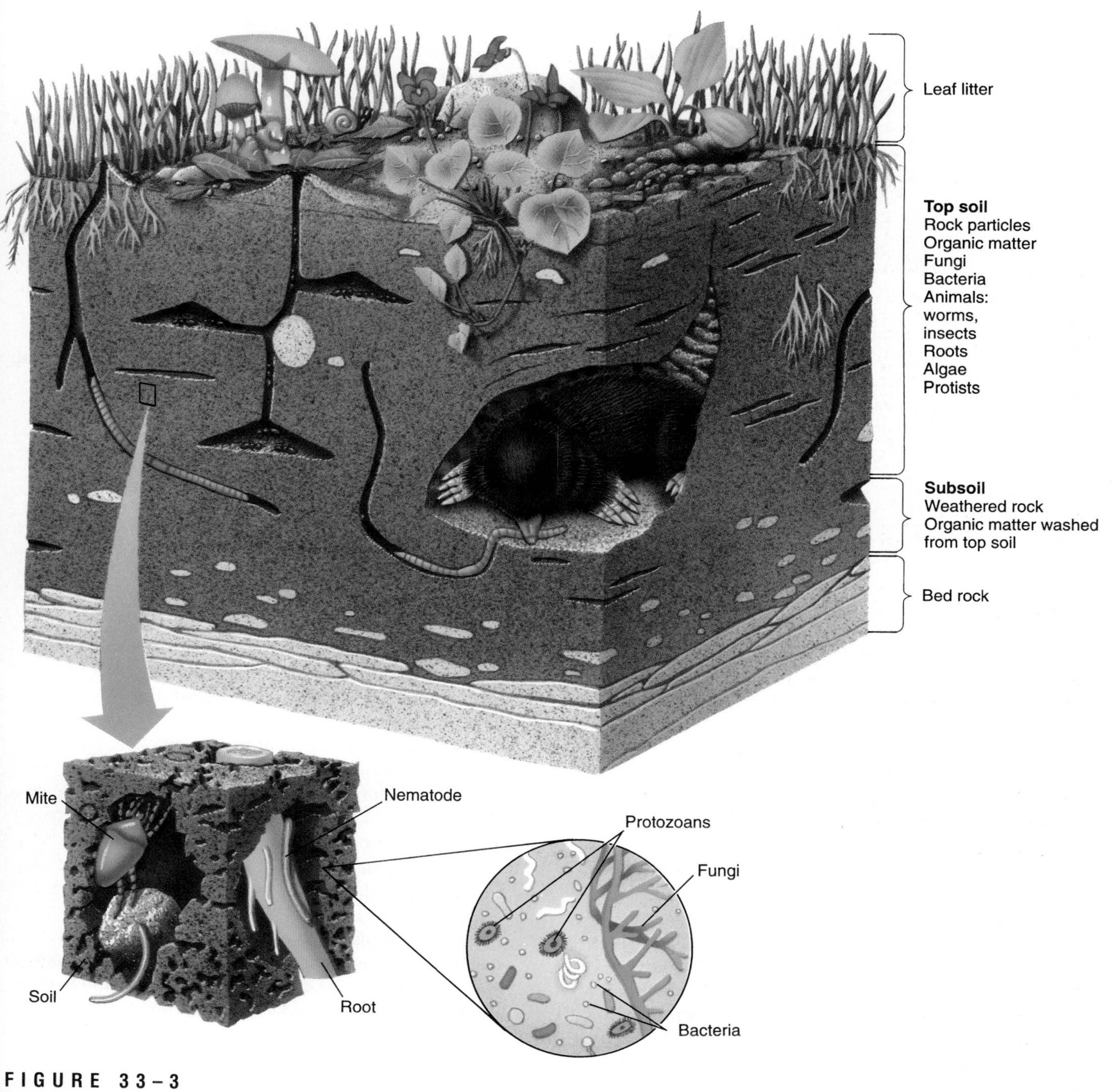

FIGURE 33-3

Soil. Fertile topsoil is created by soil organisms from the rock particles and decaying vegetation that they feed on. The subsoil consists largely of rock particles and contains fewer large organisms.

whereas organic matter, sulfur, or ammonium sulfate is applied to alkaline soils. These treatments have other effects besides changing the pH: lime provides calcium, ammonium sulfate contributes nitrogen and sulfur, and organic matter provides many nutrients. If an acidic soil's pH is raised too much, iron may precipitate out of the soil solution. One way around this is to apply **chelated iron**—iron bound to organic molecules so that it cannot be precipitated—along with lime.

FIGURE 33-4

Which plants have fungi? The two orange seedlings on the left were grown for six months with a mycorrhizal fungus; the two on the right without the fungus. All the plants were given nitrogen-containing fertilizer (calcium nitrate to the left-hand plant and ammonium nitrate to the right-hand plant in each pair). However, only the plants with mycorrhizae absorbed the nitrogen and grew well. (R. Ronacordi/Visuals Unlimited, Inc.)

Soil fertility is affected by particle size, pH, and water content, as well as mineral, nitrogen, and oxygen content. Bacteria and fungi contribute to soil fertility by digesting organic matter and releasing vital minerals. In addition, some bacteria add nitrogen to soil and some fungi help plants obtain minerals.

33–C ABSORPTION BY THE ROOTS

A plant's small, young roots, near the ends of the root system, absorb water and minerals from the soil solution. The external surface area of these roots is increased by root hairs, epidermal cells with extensions that grow out between soil particles (see A Journey Through Root Structure and Function). But the epidermis is only part of a root's absorptive surface. The soil solution also moves freely into the root cortex via spaces between the cellulose fibers in cell walls. Hence a root can take up water and minerals through the entire surface area of the plasma membranes of cells in its epidermis and cortex.

At the inner edge of the cortex, the soil solution reaches a barricade: the root's endodermis. The cells of the endodermis form a layer like a brick wall, separating the epidermis and cortex from the root's interior. A **Casparian strip,** made of an impermeable waxy material, encloses each cell in the endodermis, like a plastic bag around a sandwich. This material blocks the cell wall spaces and butts against the Casparian strips or the adjacent endodermal cells. Since the Casparian strips are impermeable to water and dissolved solutes, no substances can pass through the pores of the cell walls from the outside of the endodermis to the inside, or vice versa.

All substances that travel through the endodermis, therefore, must pass through the cytoplasm of the endodermal cells themselves. Thus the endodermis lives up to the translation of its name, "inner skin," by serving as a selective barrier. Substances in the cortex must pass through the living cells of the endodermis before they can enter the vascular tissue, in the center of the root, and be transported to the rest of the plant.

Some substances from the soil solution enter the plant through the plasma membrane of the endodermis, but most are taken up by the membranes of cells in the epidermis and cortex. All of these living cells form a continuous system of cytoplasm extending throughout the root, connected by strands of cytoplasm called **plasmodesmata.** Hence minerals absorbed by outer cells can be transported through the cytoplasm, which continues right through the endodermis, and into the living cells around the dead xylem conducting cells.

Some minerals, such as calcium and magnesium, enter root cells by diffusion. However, most minerals are taken in by active transport, including potassium, nitrate, phosphate, and sulfate. Active transport uses the energy of ATP, and it stops in the absence of oxygen, which is needed to make ATP during cellular respiration.

Once minerals have passed through the endodermis, they must leave the cytoplasm of living cells, either by diffusion or by active transport, and move into the dead conducting cells of the xylem. From here they are transported upward to the rest of the plant.

In all of this movement of minerals into the root and then into the xylem, water follows the solutes by osmosis.

Although plants take up substances selectively, they do not totally exclude substances that are unnecessary, or even toxic. This can pose problems. For example, nutrient-rich sludge from sewage treatment plants is often recycled by using it as fertilizer. This sludge must be free of toxic substances, because plants will take them up along with nutrients. Wastes from industrial towns often contain high levels of toxic elements, such as antimony, cadmium, and tin, as well as toxic levels of micronutrients such as selenium. Plants fertilized with these wastes may take up so much of these substances that they become unsafe for human consumption.

Some plants inhabit soils with high concentrations of heavy metals, such as the slag heaps of mines. In Wales, a species of the bent grass *Agrostis* grows well in soil containing tailings from copper mines. This

Absorption by the Roots

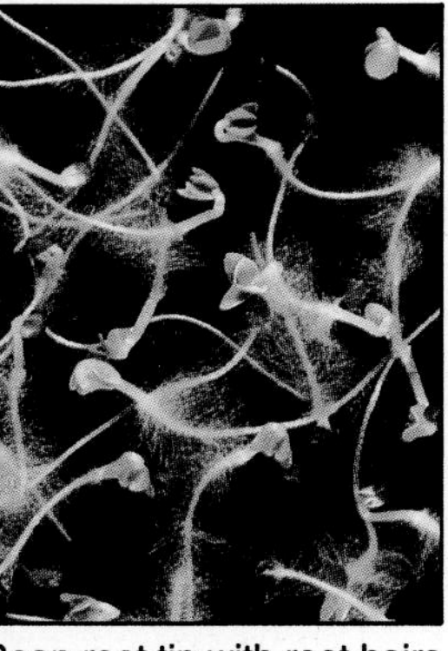

Bean root tip with root hairs

Root hairs

Plasmodesmata

After the nutrients pass through the Casparian strip, they can move freely about using plasmodesmata: the openings between the cell walls.

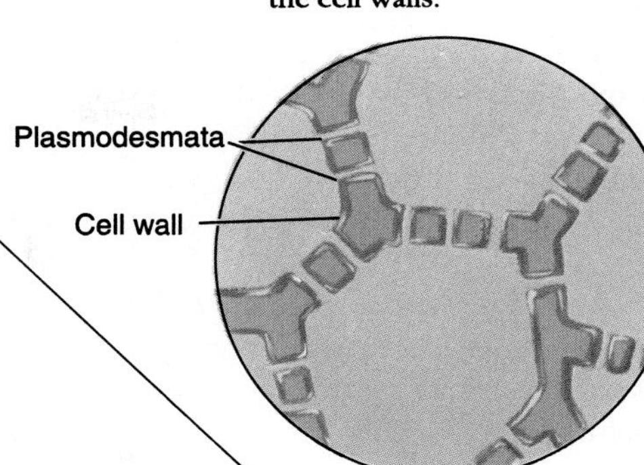

Endodermis

Root hair

Vascular tissue

Cortex

Casparian strip

Casparian strip

Minerals and water are stopped from entering the vascular tissue by the endodermis, which contains the Casparian strip. The Casparian strip is an impermeable barrier that seals off the endodermis and forces liquids to be absorbed through cell membranes.

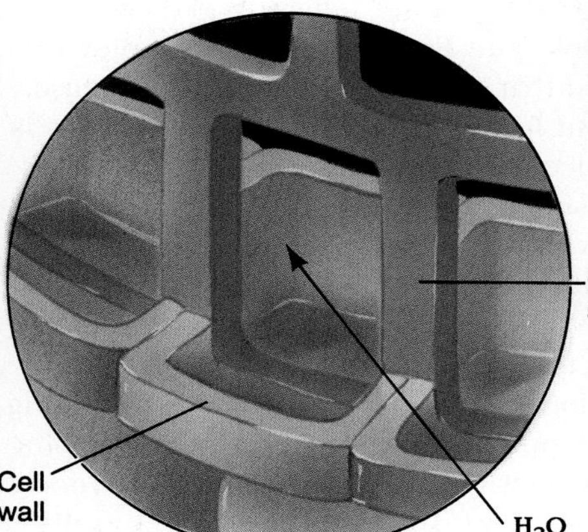

H_2O movement

Absorption

Minerals and water enter the cortex of the root by way of the root hairs. They then make their way to the vascular tissue by passing through the spaces between the cell walls.

A Girdling Experiment

(a)

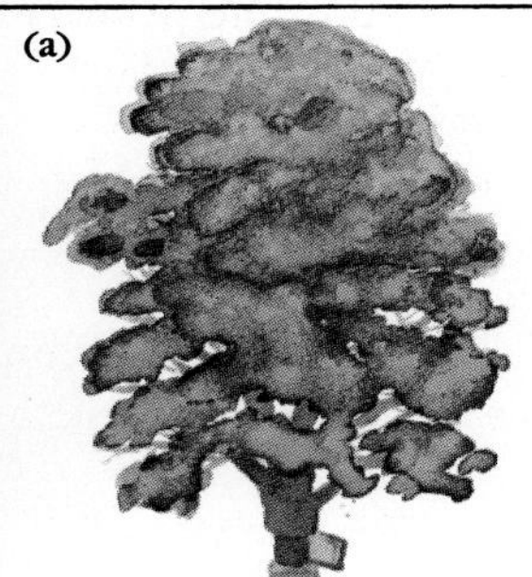

(b)

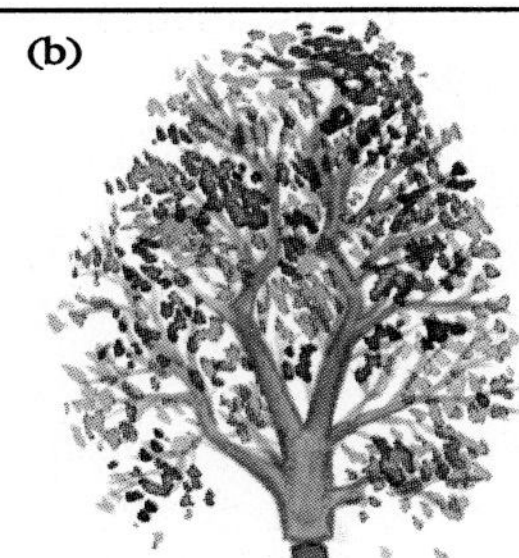

(c)

When the nutrients enter the vascular tissues of the root, the xylem carries the nutrients up to the leaves. At the same time the phloem moves food from the leaves to the roots. When the phloem is removed with the bark as in (a), the flow of food to the roots is interrupted, but because the xylem is still intact the leaves will continue to receive nutrients (b) and will remain alive for some time. Eventually the roots will die from lack of food and then the leaves will die (c).

A plant's root system absorbs water and minerals for distribution throughout the entire organism.
(D. Guravich/Photo Researchers, Inc.)

grass tolerates levels of copper in its body that would kill other plants. Not only does the plant thrive where other species would perish, so that it has little competition from other plants, but also the plant's copper content makes it toxic to herbivorous animals. The ability of some plants to absorb toxins can be used to clean up polluted soil. In California workers are planting a wild mustard from Pakistan on soil polluted with excessive selenium. The plant removes selenium from the soil and can then be fed to cattle, replacing the selenium supplement often added to their fodder.

Spaces between cellulose fibers in cell walls in the root's epidermis and cortex are virtually continuous with the external soil solution. Minerals move into the root in the soil solution that bathes the cells of the epidermis and cortex. Here a large area of selectively permeable plasma membrane absorbs minerals into the cytoplasm. All minerals must pass through the cytoplasm of living cells if they are to reach the xylem and, through it, the rest of the plant.

33–D FUNCTIONS OF XYLEM AND PHLOEM

The young roots of a tall tree that absorb water and minerals may be many meters from the leaves. The roots and leaves are connected by the vascular tissues, xylem and phloem, which form a transport system linking the various parts of the plant and moving substances between them.

In 1679, the Italian scientist Marcello Malpighi performed an experiment that showed the functions of xylem and phloem. He peeled off the bark in a complete ring around the trunk of a tree, a procedure known as **girdling.** This removes the phloem, which makes up the inner bark, but leaves the secondary xylem, or **wood,** intact. After this treatment, Malpighi found that a swelling appeared in the bark just above the stripped area. Fluid exuded from this swelling was sweet (we now know that it contained sucrose). The leaves showed no effects for days or months, but eventually they wilted and then died, and the entire tree was soon dead (see A Journey Through Root Structure and Function).

From these observations, Malpighi concluded that phloem transports food, such as the sugar in the liquid exuded from the bark, to the roots. Without this supply of food, the roots died after they had used the food already stored below the girdle. Leaves deprived of water wilt and die within hours. Since the leaves of the girdled tree remained healthy for much longer, Malpighi concluded that xylem was transporting water to the leaves. Other girdling experiments have since shown that phloem also conducts food to growing buds, flowers, and fruits.

The vascular tissues divide the task of transport: xylem moves water from the roots to the shoot system, and phloem moves food from leaves to roots or other organs that need food.

33–E STRUCTURE OF XYLEM

Xylem tissue contains many different types of cells. Only some of these conduct the mixture of water and solutes called **sap.** Both conducting and nonconducting cells may provide structural support for the plant body.

Xylem structure shows many adaptations that enhance its dual functions of transport and support. Right after being formed by cell division, most plant cells lay down **primary cell walls** of cellulose and other polysaccharides. Later, the cells that will become xylem conducting cells add more material, which strengthens the cell walls. These secondary thickenings vary from disconnected rings to extensive **secondary cell walls,** which cover the cell almost completely (Figure 33-5). The secondary thickenings consist of cellulose and **lignin,** a tough, complex organic compound that makes wood strong and dense.

An important feature of xylem conducting cells is that they die once their cell walls are complete. The cell contents disintegrate, leaving a strong, hollow cylinder filled with water. Because the conducting cells are stacked on top of one another, water can travel in a more or less straight line up the plant.

The conducting cells of primitive vascular plants are **tracheids,** long, extremely thin cells with slanting end walls. The wood of a gymnosperm such as a pine is almost all tracheids (see A Journey Through Wood Structure and Function). Heavy secondary cell walls slow the passage of water from one tracheid to the next, except in thin areas called **pits,** where little or no secondary material has been added to the primary cell wall.

BIO-BIT

In the timber industry, hardwoods come from dicot trees while softwoods come from conifers. However, some hardwoods (such as willow) are actually softer than some softwoods (such as cedar).

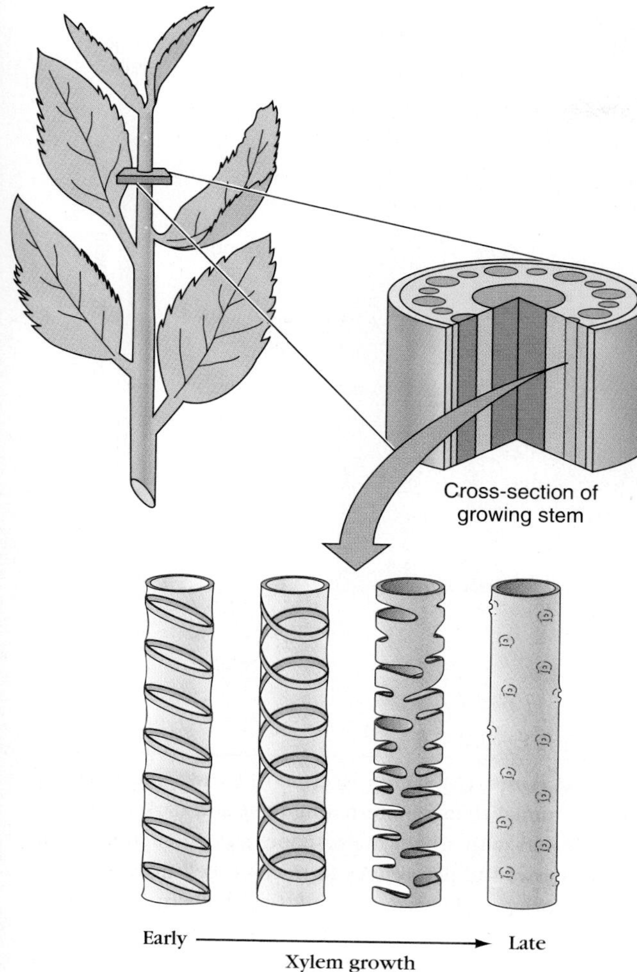

FIGURE 33-5

Secondary thickenings in xylem conducting cells. In a new length of stem, the first xylem cells lay down thickenings in disconnected rings or spirals (see the two early xylem stages at lower left). As this section of the stem develops, the cells elongate by stretching of the primary cell wall between them. Xylem cells formed after the stem reaches its final length lay down more extensive thickenings, which cannot be stretched, and the cell is fixed at this size (see the late stage at lower right).

Wood also contains living parenchyma cells, in **rays** running out from the center of the tree. The ray cells conduct materials laterally, out to the living tissues in the vascular cambium and phloem. They may also serve as food storage depots. In pine wood, other living parenchyma cells secrete **resin,** a sticky, pungent fluid that repels wood-boring insects and inhibits the growth of certain pathogenic (disease-causing) organisms. Hence resin helps protect an injured tree from disease. Resin may be collected and distilled to make turpentine. It is also the raw material for pitch and tars, used to protect wood from rotting. ■

Angiosperms (flowering plants) have even more efficient xylem. Some conducting cells have literally made an evolutionary breakthrough: their last act in life is to digest parts of their end walls, forming real holes, not just thin pit areas. This allows water to flow more rapidly from one cell to the next. In some angiosperms, the end walls are entirely absent, and the cells are like sections of pipe.

Angiosperm conducting cells are also shorter and wider than tracheids. Their greater diameter reduces the cell walls' frictional drag and permits water to travel faster. The hollow xylem tubes of angiosperms are called **vessels,** and the individual cells in the vessels are **vessel elements.**

The wood of an angiosperm such as an oak contains many vessels, but it also consists of other types of cells. Some are **fibers,** long, thin cells with thick cell walls that help support the tree long after the cell contents have died. Angiosperm xylem also contains rays made up of living parenchyma cells, and many have tracheids as well as vessel elements.

Changes in the Xylem of Woody Plants

3? The xylem of a woody plant changes throughout its life. As we saw in Chapter 32, the vascular cambium adds new xylem outside the existing xylem. In climates where good growing conditions alternate with cold winters or dry seasons, the cambium's activity follows the weather, and the new wood forms **growth rings.** In temperate climates, where a new ring is added each year, these rings are called **annual rings** (Figure 33-6). Trees in uniform climates, such as tropical rain forests, grow continuously and do not form distinct rings. ■

A tree "writes its diary" in its wood. Variations in cell size, thickness of growth rings, and so on, record rainfall, fire, insect plagues, or the death of a neighbor. Annual growth rings are sometimes used to date archaeological finds, or to determine the climate at some time in history. A wide growth ring reflects a good growing season, while a narrow band is formed in a poor year. By comparing the pattern of tree rings in building timbers with the rings in large, old trees in the same area, archaeologists can determine when the structures were built. Chemical analysis of tree rings from different years can also provide information on an area's history of air pollution and acid rain.

The latest several years' xylem is called **sapwood** because it actually conducts the sap in trees. Interior

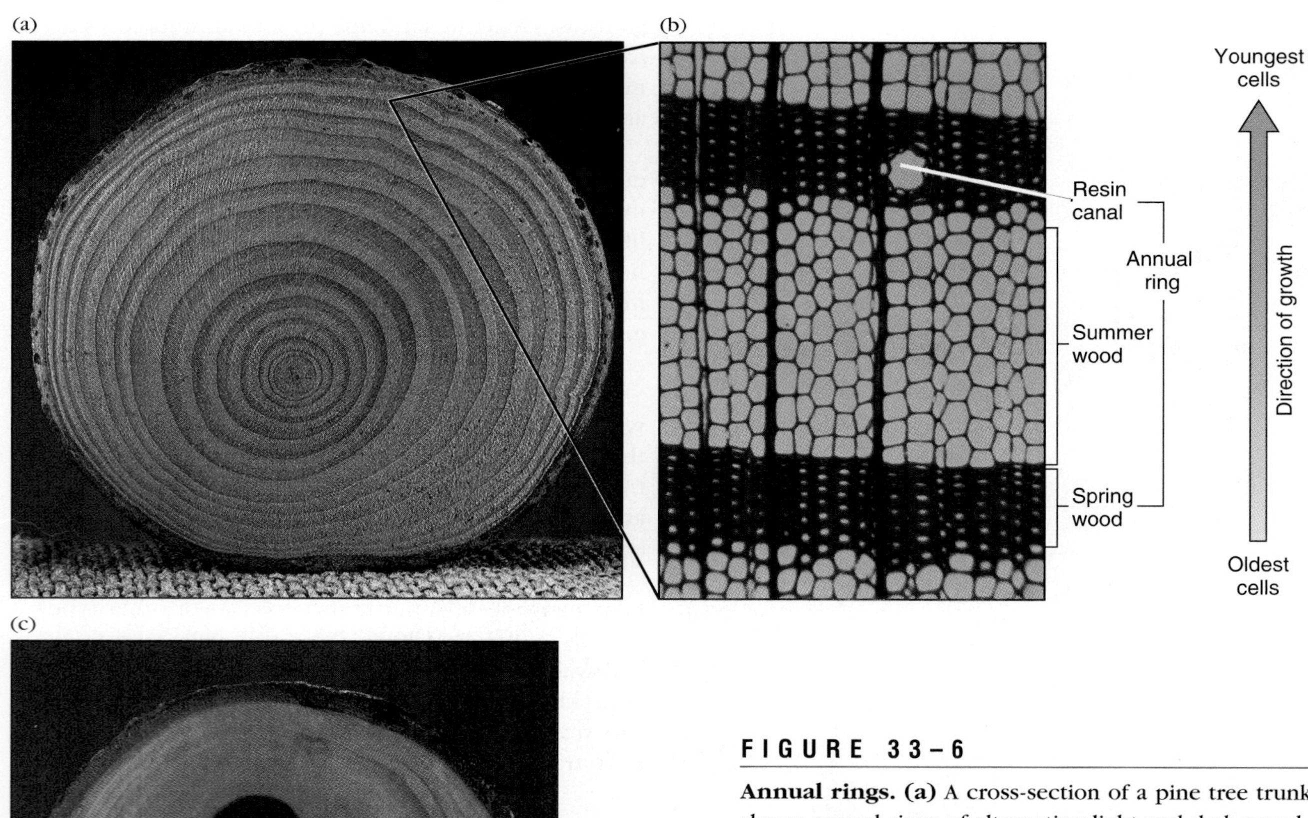

(c)

FIGURE 33-6

Annual rings. (a) A cross-section of a pine tree trunk shows annual rings of alternating light and dark wood. **(b)** Viewed with a light microscope, a cross-section of pine wood shows the differences in tracheid cell diameter between spring and summer wood. **(c)** This elm log shows a marked difference in color between the dark inner heartwood and the pale outer sapwood. (a and c, Runk/Schoenberger from Grant Heilman, b, J. M. Bostrack/Visuals Unlimited, Inc.)

to the sapwood is the **heartwood,** older xylem that no longer conducts sap. The heartwood is often plugged by deposits of waste material. Sometimes it rots away, leaving a hollow tree that is still alive because its sapwood and phloem still function.

Xylem tissue includes an integrated system of vertical and horizontal tubes. Long-distance vertical transport occurs in tubes formed by cell walls that remain from cells that have died. The cell walls of both conducting and nonconducting xylem cells also help to strengthen and support the plant body. The short, wide vessels of angiosperms permit faster sap conduction than do the long, thin tracheids of gymnosperms. Tree rings form as a result of seasonal variations in the growth of xylem.

33–F TRANSPORT IN THE XYLEM

Tracheids and vessels, the conducting cells of xylem, are stacked in columns, an advantage in the transport of water. How does water move through these pipelines of dead cell walls?

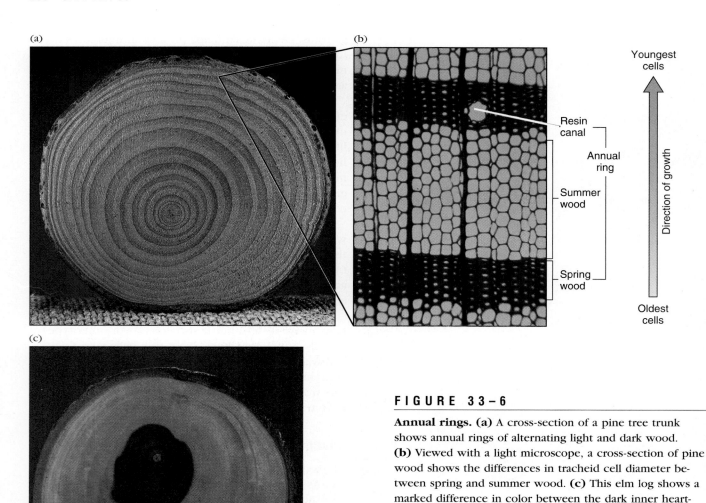

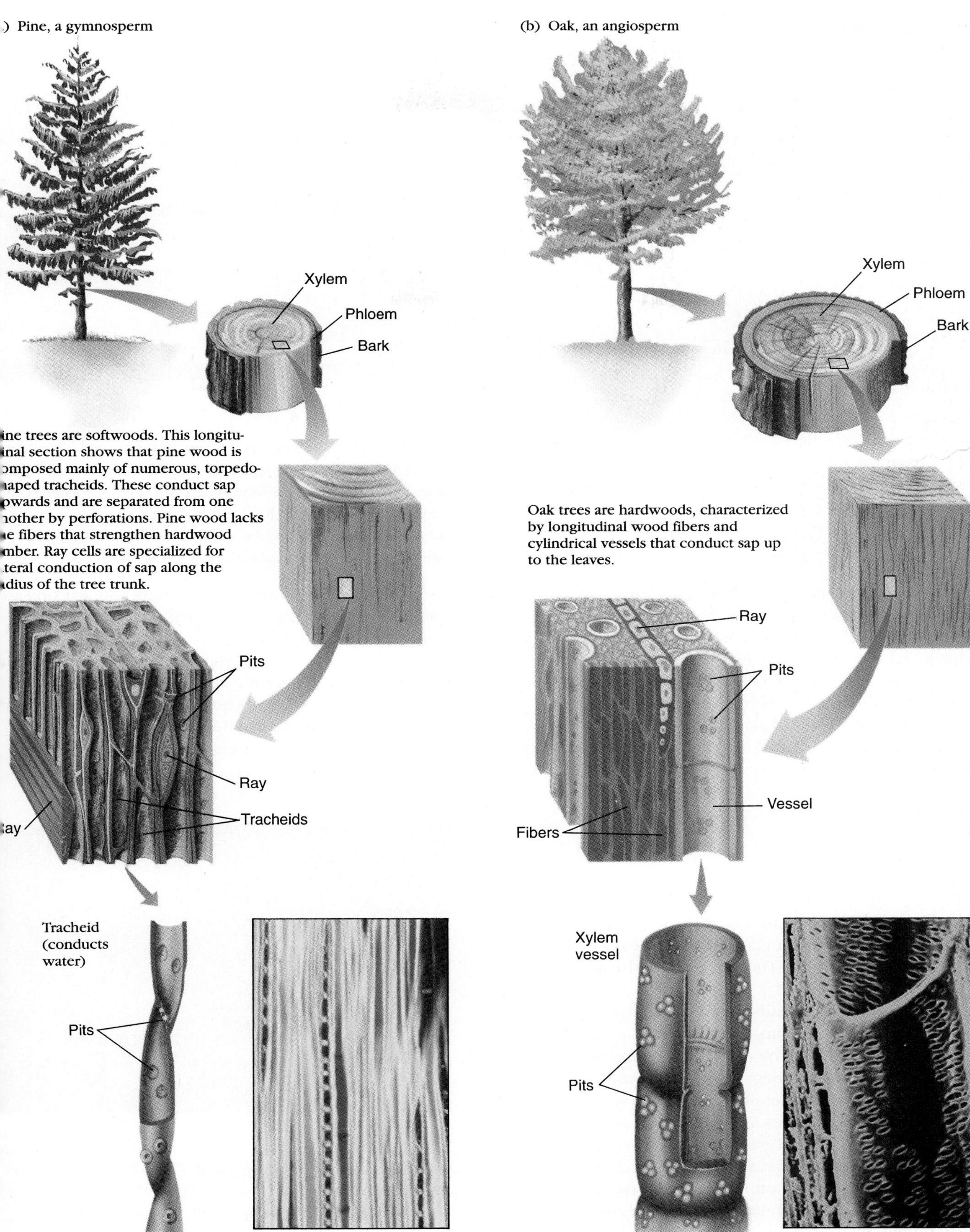

) Pine, a gymnosperm
Xylem
Phloem
Bark
ine trees are softwoods. This longitu-
inal section shows that pine wood is
omposed mainly of numerous, torpedo-
haped tracheids. These conduct sap
pwards and are separated from one
nother by perforations. Pine wood lacks
e fibers that strengthen hardwood
mber. Ray cells are specialized for
teral conduction of sap along the
dius of the tree trunk.
Pits
Ray
Tracheids
ay
Tracheid (conducts water)
Pits
(b) Oak, an angiosperm
Xylem
Phloem
Bark
Oak trees are hardwoods, characterized by longitudinal wood fibers and cylindrical vessels that conduct sap up to the leaves.
Ray
Pits
Vessel
Fibers
Xylem vessel
Pits

he secondary xylem (wood) of gymnosperms and angiosperms permits the transport of sap and provides physical support (a, G. hih, R. Kessel/Visuals Unlimited, Inc.; b, Biological Photo Service)

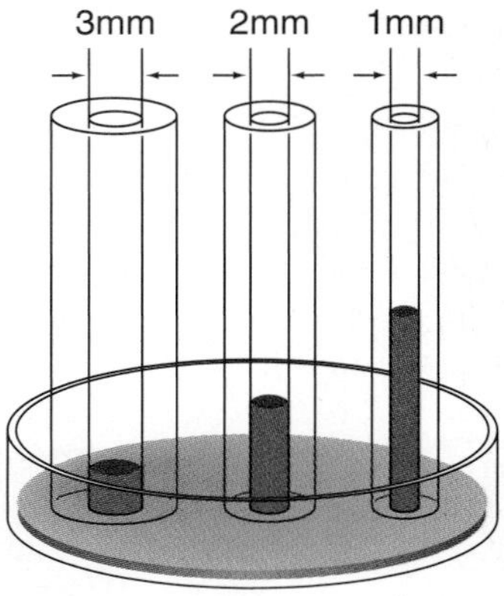

FIGURE 33-7

Capillarity. When the ends of these tubes are touched to red colored water, water molecules are attracted to the walls of the tubes and these molecules pull others up the center of the tubes. The smaller the inside diameter of a tube, the higher the water rises within it.

If you touch the tip of a thin glass tube to the surface of water, water rises into the tube quickly by capillarity (Section 33-B). The narrower the tube, the higher water rises (Figure 33-7). Water should also move in this way up the cell walls in the xylem tubes. However, calculations indicate that water moving by capillarity in the slenderest tracheids can reach a maximum height of less than 2 meters, not high enough to account for the rise of water in many plants.

Root Pressure

If water cannot climb on its own, it must be either pushed up from the bottom of the plant, or pulled from the top. A push from below can be seen in some plants by cutting the plant off near the ground and sealing the stump into a glass tube: water rises into the tube (Figure 33-8). The push is called **root pressure.**

Root pressure depends on active transport of ions from the soil solution into the xylem in the center of the root. Water follows by osmosis, building up hydrostatic pressure in the xylem vessels. The endodermis keeps the water from flowing out of the root back to the soil. So, the water has nowhere to go but up.

Because roots use a great deal of energy for active transport, they must have an adequate oxygen supply. When the soil contains ample water and oxygen, root pressure may be high. If, at the same time, the air is very humid, water does not evaporate readily from the leaves, and water forced up from the roots may be exuded from the tips of the leaves. This **guttation** occurs only in rather short plants, where the leaf tips are relatively close to the root endings, the source of the root pressure (Figure 33-9).

However, the greatest root pressures measured can push water less than a meter, not far enough to reach the tops of many plants. Thus, root pressure alone can account for xylem conduction only in certain short plants growing in certain environmental conditions.

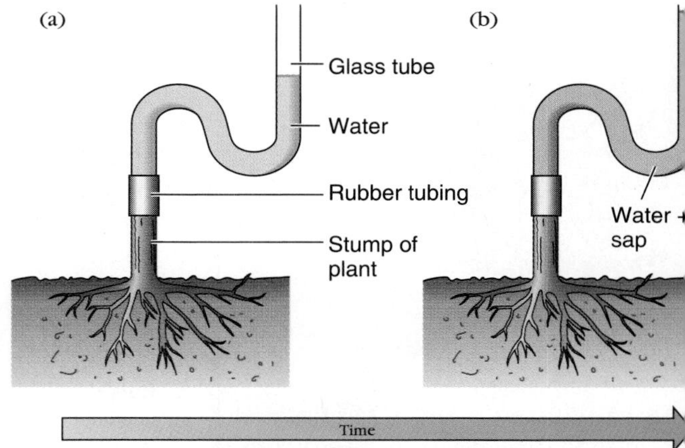

FIGURE 33-8

Demonstration of root pressure. (a) A curved glass tube is sealed to a freshly decapitated plant and is then filled with water. **(b)** Sap rising into the tube from the roots raises the water level in the tube until the hydrostatic pressure in the tube equals the root pressure. The pressure can be calculated using the final height of the fluid in the tube.

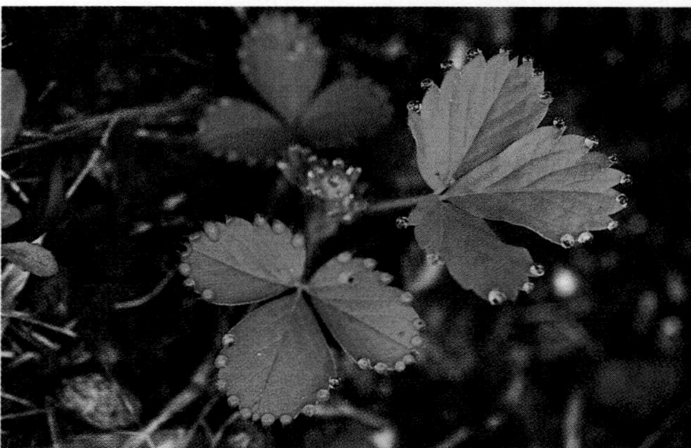

FIGURE 33-9

Guttation in a strawberry plant. Guttation occurs when ample water and oxygen are available in the soil and the air is too humid to permit evaporation of water as fast as it is pushed up from the roots. Sap is exuded from the tips of xylem columns in the leaves. (S. Camazine/Photo Researchers, Inc.)

Transpiration Pull

This evidence forces us to consider another mechanism for xylem transport, the pull from above. In 1727 Stephen Hales, an English clergyman, showed that **transpiration,** the evaporation of water from leaves, can pull water up through the xylem of a plant. Hales cut sets of similar leafy branches and removed different amounts of leaves from some of them. He then set each branch in a container of water and observed that the amount of water removed from the container was roughly proportional to the area of leaf surface on the branch (Figure 33-10). Hales decided that some activity of the leaves caused water to rise in the stem of a plant.

In another experiment, Hales found that the leaves had to be dry and exposed to the air in order for water to move through the branch. A branch with its leaves immersed in water could not "perspire" (or transpire, as we would say today), and almost no water moved through the branch.

In yet another experiment, Hales dug a hole to expose a root of a pear tree and measured how strongly the tree could pull on water (Figure 33-10). The pull was stronger on bright, sunny, dry days than on cloudy or damp days, and it slackened at night. These were exactly the results expected if evaporation of water from the leaves were indeed the force pulling water up through the tree.

The chain of events that moves water upward depends on the structure of plants and the properties of water:

1. Transpiration occurs: water evaporates from the walls of leaf cells into the air spaces inside the leaves and then diffuses out of the leaf into the atmosphere by way of the stomata (Figure 33-10).
2. The loss of water by evaporation creates a water deficit in the cell walls next to the air spaces. Water is quickly replaced by movement of water in from the walls of neighboring cells. These cells get replacement water from their neighbors' cell walls, and so on. Water moves between the cellulose fibers of the porous cell walls by capillarity. Eventually, the water lost from the cell walls of the leaf cells is replaced by water from a conducting cell in the xylem at the tip of a veinlet in the leaf.
3. Since water molecules attract one another strongly, they stick together, or cohere. Pulling water out of the top of a xylem column is like pulling on a rope of water that extends all the way down the thin xylem pipeline to its ends in the roots. All the water in the xylem moves up a bit, and eventually the pull reaches root cells which take in soil water to replace that lost from the leaves.
4. Xylem vessels and tracheids are very narrow tubes. Capillarity of water also plays a role in moving water upward in a plant.

Transpiration sets this process in motion, and the cohesion of water molecules allows it to continue. Hence this is called the **transpiration pull–water cohesion** mechanism of xylem transport.

Transpiration seems to be a necessary evil, the price plants pay for having leaves, which are very efficient structures for obtaining sunlight and carbon dioxide. The cuticle, guard cells, and internal air spaces seem to have evolved as a means of limiting water loss through the leaves.

Evaporation of water creates a negative water pressure in the leaves. So, when a plant is transpiring rapidly, the water in the xylem conducting cells is under tension. Cutting the xylem lets the water column snap apart, just as the two ends of a stretched elastic snap when it is cut. Air rushes into the xylem column, breaking the cohesion of the water and preventing further movement of sap by transpiration pull through the affected tracheids or vessels.

Rate of Transpiration

Hales estimated that, weight for weight, a plant's daily water intake is 17 times that of a human being because the plant transpires so much water to the atmosphere. On a warm, dry day Hales found that a 1-meter-tall sunflower plant lost 0.9 liter of water by transpiration in 12 hours. On dry nights, the plant lost less than 0.1 liter in 12 hours, and on nights with dew, the plant gained water by absorbing the dew through its leaves. A large tree may transpire more than 5 tons of water on a hot, sunny day.

A plant's transpiration rate depends on many factors. Structural features such as the density of stomata and the thickness of the cuticle have an effect. So do environmental factors. The availability of soil water determines how much water the plant can "afford" to lose by transpiration. Sunlight stimulates photosynthesis and causes the opening of the stomata and the faster uptake of carbon dioxide, and so the plant loses water more rapidly in bright sunlight. Low humidity increases transpiration because the concentration gradient between the leaf interior and the air is steeper: water diffuses from the leaf to the air faster. High temperatures increase evaporation. Furthermore, the higher the temperature, the more water vapor the air can gain before it becomes saturated. In still air, the water vapor transpired by a leaf remains nearby, forming a layer of saturated air that retards further evaporation from the leaf. However, if there is a breeze, this layer of still, saturated air blows away, and the leaf loses more water.

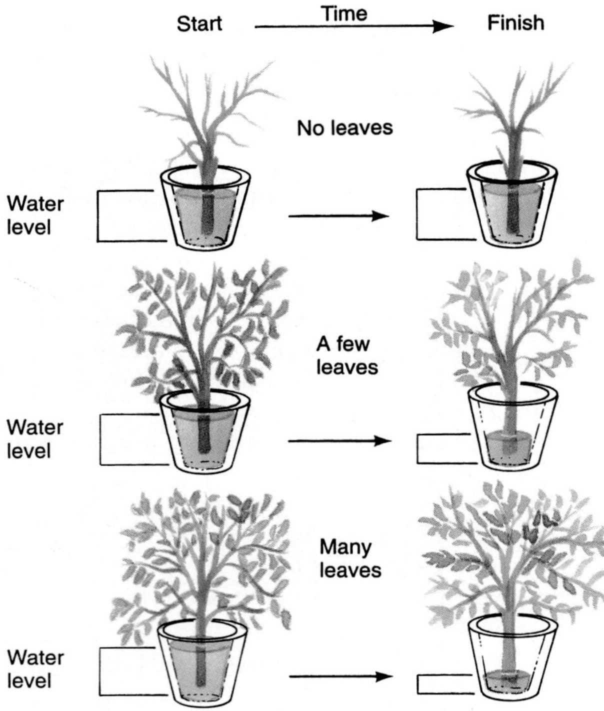

Leaves and water movement. Stephen Hales measured the rates of water absorption by branches with differing amounts of leaf surfaces. He determined that water movement depends on the leaves.

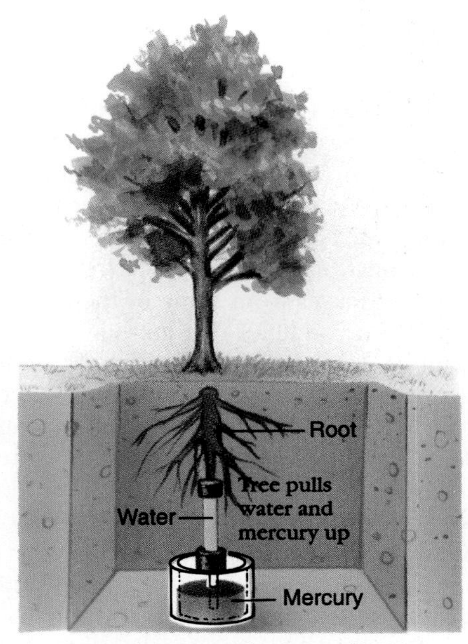

Measuring transpiration pull. Hales attached a tube filled with water to the cut root of a pear tree. He then placed a vessel of mercury in contact with the water and measured how far the mercury was pulled up into the tube as the tree withdrew water.

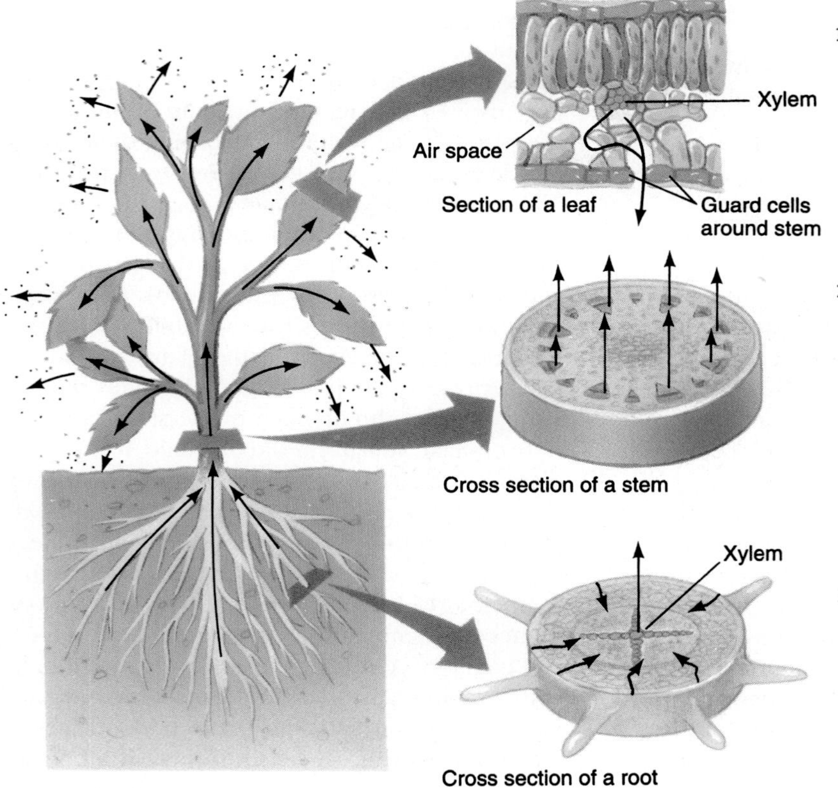

1. Water evaporates from leaf tissue.

2. Pulled by transpiration, water moves upward in xylem vessels by cohesion, adhesion, and capillarity.

3. Soil water is drawn into spaces in root tissue.

FIGURE 33-10

Overview of transpiration.

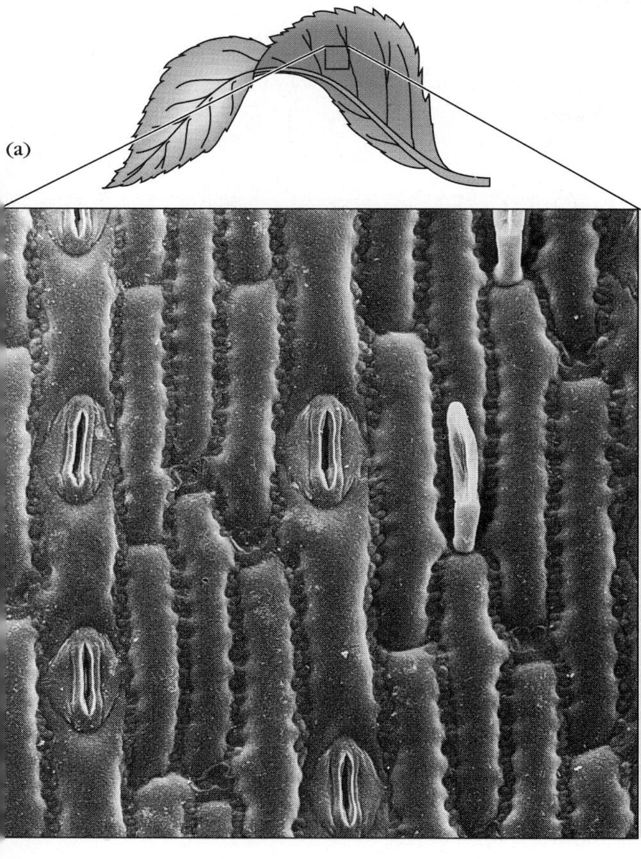

(b) Stoma closed Stoma open

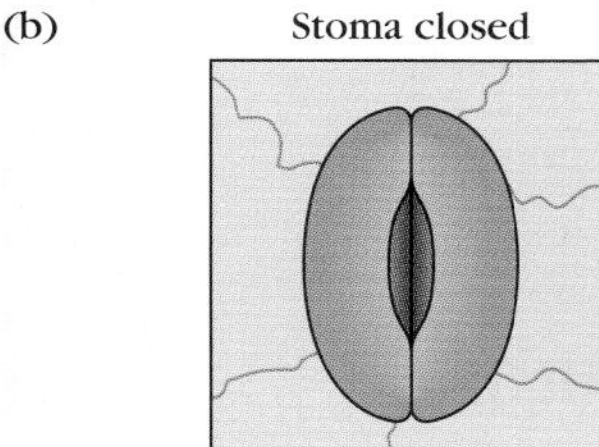

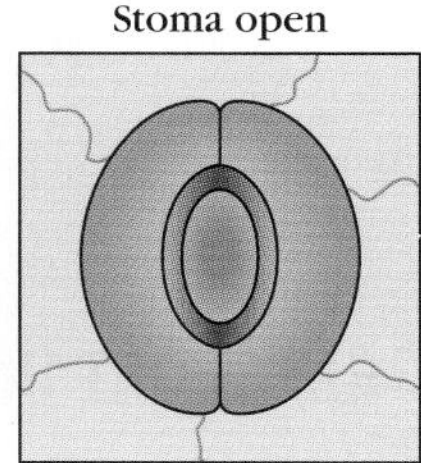

FIGURE 33-11

Guard cells and stomata. (a) The waxy lower surface of a rose leaf, showing stomata varying from closed to well opened. Each stoma is surrounded by a pair of lip-like guard cells. **(b)** Opening and closing of a stoma depends on turgor pressure in the guard cells. The guard cells have thicker walls on the side facing the stoma. Hence, when the cells swell, these inner parts of their cell walls expand less than the ends and outer portions, and the stoma opens. (a, Courtesy of Drs. Kessel and Shih)

Transpiration can be regulated according to environmental conditions. This regulation is performed largely by the guard cells, which open and close the stomata.

Operation of Guard Cells

Each stoma is surrounded by two guard cells (Figure 33-11). Guard cells can control the size of the stomatal opening because of their peculiar structure: their cell walls are thicker next to the stomatal opening than elsewhere. Guard cells open the stoma by accumulating solutes. This causes water to enter the cells by osmosis, but because of the uneven thickness of their cell walls, the guard cells do not swell evenly all around. The thicker parts of the cell wall expand less than thinner parts, and so the guard cells assume a curved shape, with an opening between them.

Guard cells are the only epidermal cells with chloroplasts. The chloroplasts produce ATP that is used for active transport of solutes, mostly potassium ions, into the cell, and this transport leads to the opening of the stoma. The stomata open when the carbon dioxide level in the guard cells is low. This normally happens in the daytime, when photosynthesis is using up carbon dioxide, but it also occurs in the dark if carbon dioxide levels in the air are lowered. We do not know how low carbon dioxide levels stimulate transport of potassium.

Sometimes water is lost by transpiration faster than the xylem can replace it. When the leaves' water content becomes too low, the hormone abscisic acid is quickly released from nearby cells. This hormone causes the guard cells to release potassium ions and other solutes rapidly. As water follows by osmosis, the guard cells shrink, and the stomata close until the xylem's delivery of water catches up with the needs of the leaves.

Capillarity and root pressure may help to move water and minerals through short distances in xylem tissues. However, most transport in xylem occurs by transpiration pull. A continuous "transpiration stream" moves from the plant's roots, up through the xylem, and out through the stomata. Water lost through the leaves is continuously replaced as soil water is taken up through the surfaces of root cells. Transpiration is affected by the plant's anatomy and by environmental factors such as availability of soil water, intensity of sunlight, humidity, temperature, and wind.

33-G PHLOEM STRUCTURE

The other transport tissue, phloem, carries food and other organic compounds throughout the plant from sites of production to sites of use or storage. Most of the food moves in the form of sucrose (table sugar). Phloem may also contain minerals taken back into the plant from dying leaves. Phloem usually lies near the xylem (see Figures 32-1 and 32-10).

In angiosperms, the conducting cells of the phloem are called **sieve tube elements,** and they are stacked in long columns called **sieve tubes.** The walls of these cells have special **sieve areas,** where many

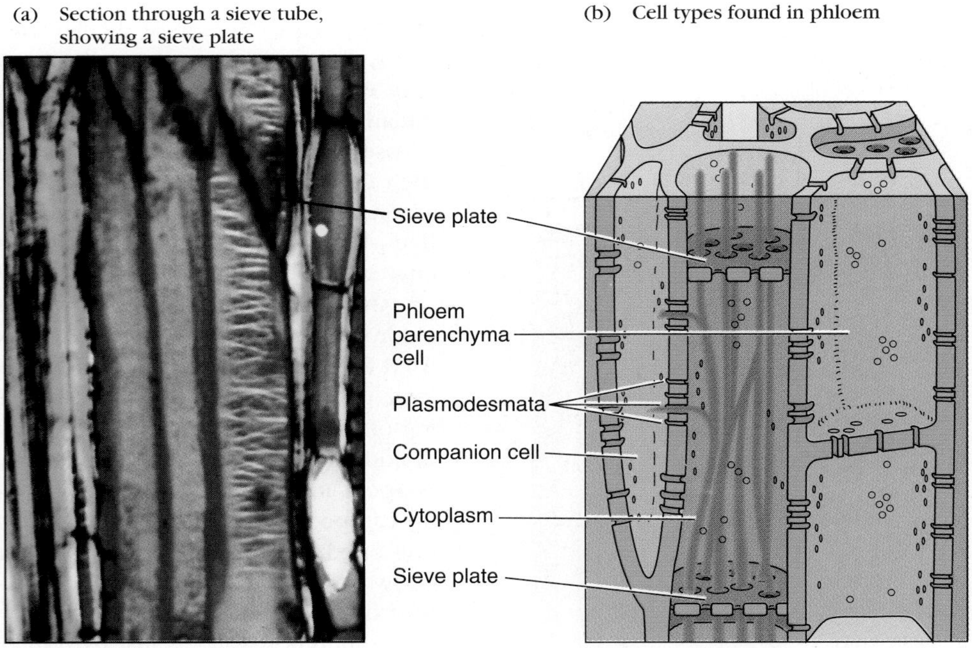

(a) Section through a sieve tube, showing a sieve plate

(b) Cell types found in phloem

FIGURE 33–12

Cell types found in phloem of angiosperms. (a) Micrograph of a longitudinal section through a sieve tube, showing a sieve plate. **(b)** Drawing of the kinds of cells in angiosperm phloem. Materials pass up and down from one sieve tube element to the next through sieve plates, and sideways to neighboring sieve tube elements through sieve areas. These conducting cells contain living cytoplasm, in which substances are transported, but they lack nuclei. "Life support" is provided by the nuclei of neighboring companion cells. Phloem parenchyma cells also contain nuclei and cytoplasm. Note the many plasmodesmata connecting the cells. (G. Wilder/Visuals Unlimited)

pores permit exchange of substances with neighboring cells. Sieve areas in the end walls have evolved as **sieve plates** with rather large pores (Figure 33–12).

The conducting cells of phloem, unlike those of xylem, contain living cytoplasm. Before a cell becomes able to conduct, it loses its nucleus and most other organelles. In addition, there is a great enlargement of its plasmodesmata, which pass through the sieve areas and sieve plates and connect neighboring cells.

After a conducting cell has lost most of its own metabolic machinery, it is maintained by the metabolism of its next door neighbor, a **companion cell,** which remains intact. The sieve tube element and its companion cell are the offspring of a single cell that divided unequally (Figure 33–13).

As in xylem, phloem contains living parenchyma cells. Phloem also contains nonliving **phloem fibers,** similar to the strengthening fibers in xylem.

Phloem conducting cells transport food, other organic compounds, and some minerals throughout a plant. These cells are highly specialized and require the metabolic support of a related companion cell.

33–H TRANSPORT IN PHLOEM

When a phloem sieve tube is injured, plugs of protein and polysaccharide rapidly seal the sieve plates and stop transport. This reaction, like the clotting of blood in an animal, keeps the plant from losing valuable food molecules through the wound. It also makes transport in phloem impossible to study directly, and so investigators have devised several indirect ways to observe phloem transport.

One much-used method is tracing the movement of radioactive materials applied to a plant's roots or leaves. The movement of a substance in a large tree may be measured by following the radioactivity with a Geiger counter. The usual method for small plants is autoradiography: the plant is given time to take up and transport the radioactive substance and is then pressed flat and placed on a sheet of photographic film. Particles emitted by radioactive decay expose (blacken) the film. The plant thus takes its own radioactivity portrait, which shows where the substance has been transported, by what route, and how fast (Figure 33–14).

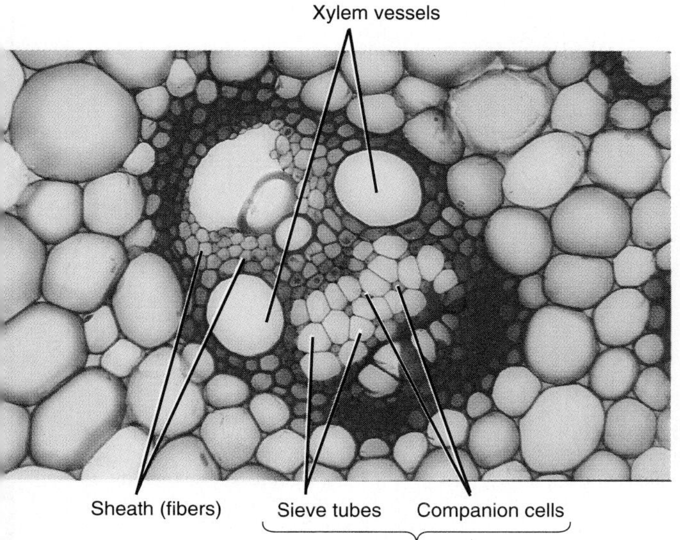

FIGURE 33–13

Cross-section through a vascular bundle of a corn stem. The bundle is surrounded by a sheath of fiber cells. The xylem vessels conduct sap, and the sieve tube elements of the phloem conduct food and other substances. (C. Kingery/PhotoTake)

Observations of phloem transport show that the contents of the conducting cells are under pressure. A huge volume of fluid passes through each cell in a short time, moving at speeds of 30 to 200 centimeters per hour. The direction of flow in a particular cell may reverse from time to time, and neighboring cells may conduct in opposite directions at the same time. Killing the cells stops transport in that part of the plant.

The most widely accepted model of phloem transport is the **mass flow** or **pressure flow** hypothesis, suggested by Ernst Munch in 1926 (Figure 33–15). Munch built a physical model that behaved much like the phloem system in a living plant. Two chambers were connected by a tube, and each chamber had a membrane permeable to water but not to sucrose. One chamber was filled with plain water, the other with a sucrose solution. When the two chambers were immersed in plain water, water entered the chamber

FIGURE 33–14

Autoradiographs showing movement of phosphorus from older to younger plant parts. The roots of living bean plants were placed in a solution containing radioactive phosphorus (^{32}P). After plant **(a)** was allowed time to take up and transport the phosphorus, it was pressed and autoradiographed. The darkest areas represent the parts of the plant with the most concentrated radioactive substance: the roots and youngest leaves. Plant **(b)** was removed from the solution containing ^{32}P at the same time as plant **(a)**, but was allowed to continue growing for a time in nonradioactive solution before it was autoradiographed. Note that the radioactive phosphorus has been moved to the younger leaves, which had not yet formed when the plant was in the radioactive solution. (Susann and Orlin Biddulph)

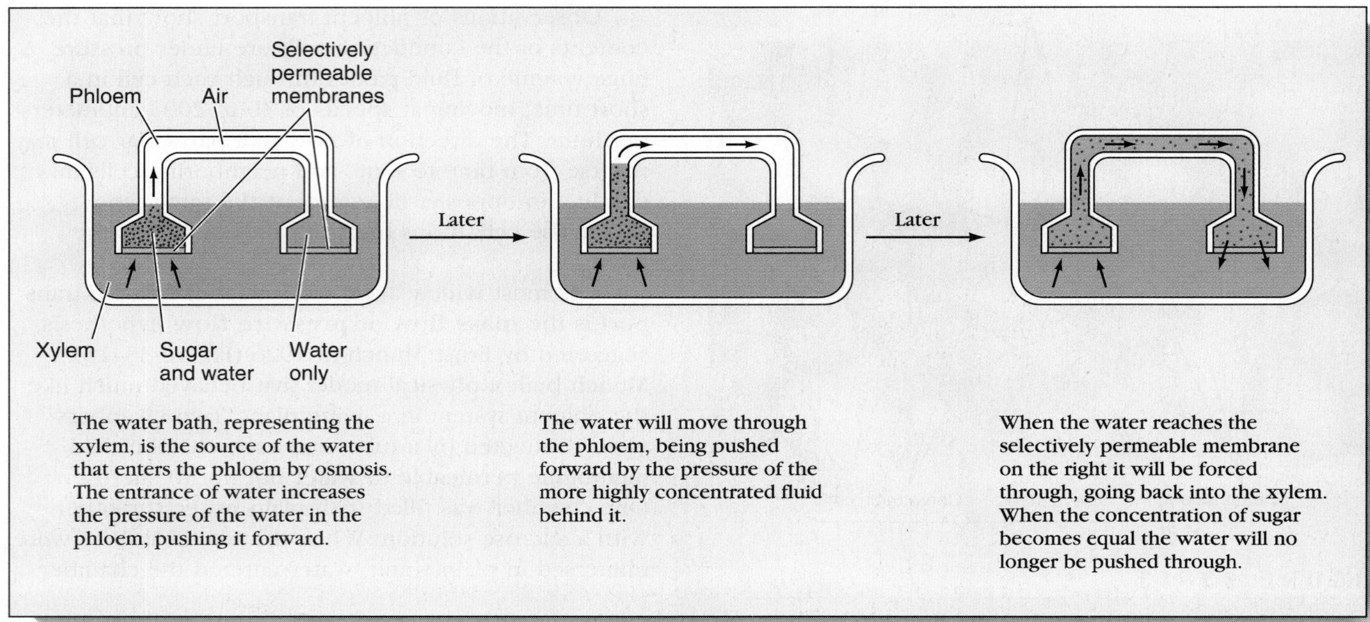

Selectively permeable membranes
Phloem
Air
Later
Later
Xylem
Sugar and water
Water only
The water bath, representing the xylem, is the source of the water that enters the phloem by osmosis. The entrance of water increases the pressure of the water in the phloem, pushing it forward.
The water will move through the phloem, being pushed forward by the pressure of the more highly concentrated fluid behind it.
When the water reaches the selectively permeable membrane on the right it will be forced through, going back into the xylem. When the concentration of sugar becomes equal the water will no longer be pushed through.

FIGURE 33–15

Munch's model of mass flow in the phloem.

with the sucrose solution by osmosis. This forced the solution to rise into the connecting arm and flow across to the other chamber, taking some of the sucrose along. As sucrose solution arrived at the second chamber, it forced some of the water there to flow out through the membrane into the water bath. The flow continued until the sucrose was equally concentrated in the two chambers, and then stopped.

If sucrose could be continuously added to the first chamber, and removed from the second chamber as it arrived, the osmotic gradient would be maintained, and the flow would continue indefinitely. Such conditions are indeed found in the phloem system. In a living plant, some areas are **sucrose sources,** continually making sugar (in leaves or green stems) or releasing it by breaking down stored starch (in roots or stems). Other areas are **sucrose sinks,** where sucrose is consumed as it arrives. This could be any cell that needs energy. Food storage tissues also act as sucrose sinks when sugar molecules are being converted into starch granules, which have less osmotic activity. Sucrose is the main solute in the phloem, and the phloem solution always moves down a sucrose gradient, regardless of the concentration gradients of other substances present.

In Munch's model, the water bath represents the xylem, the source of the water that enters the phloem solution by osmosis. The entrance of water would increase the pressure in the phloem, forcing the solution through the phloem until it arrived at a tissue that removed the dissolved nutrients. The water would then be forced out of the phloem by the pressure of the more concentrated fluid moving behind it and taken back into the xylem.

Sugar enters the phloem sieve tube elements in the smallest leaf veins. This is thought to occur by active transport, carried out by surrounding companion cells, which can build up sugar concentrations of 10 to 25%. There are no further membranes to cross, because strands of cytoplasm connect the cells of the sieve tube to each other through the sieve plates. Once sucrose arrives near a hungry cell, it is removed from the phloem and stored or metabolized.

This theory agrees well with the observations on the speed and pressure at which substances move through phloem. It can also account for observed changes in direction of flow. Flow proceeds from an area of high pressure to one of lower pressure. The pressure is lowest in tissues that withdraw nutrients most actively, which may be different parts of the plant at different times. At one time, the roots may be most active as they store food, but later growing buds or fruits may begin to withdraw nutrients faster, reversing the flow. In this case, the food storage tissues may even switch roles, from sucrose sinks to sucrose sources, as they break down starch reserves and re-

A Journey into Science in Process

How To Care For Cut Flowers

After thanking the donor for a bouquet of flowers and sniffing them appreciatively, the next step is to find a vase and put them in water. The florist usually sends a message to cut off an inch or two of stem first; purists even instruct that this should be done while the stem is immersed in water.

Figure 33-A shows the importance of this minor surgery. Three stems were cut from the same plant at the same time. Stem 1 was placed in water immediately. Stem 2 was left for half an hour, and then a 2-inch length was cut off the bottom end. It was then placed in water. Stem 3 was also left for half an hour and then placed in water without further treatment. The photograph was taken one hour after the stems were cut from the plant. Stem 3 is distinctly wilted, whereas the other two look normal.

FIGURE 33-A

The importance of a continuous column of fluid in the xylem. Results of a demonstration described in the text.

Stems 1 and 2 have fared so much better because in both, the column of water in the xylem is continuous with the water in the container. As the leaves transpire, the water lost from the plant is replaced by pulling more water into the xylem from the container. While Stems 2 and 3 were left out of water, transpiration from the leaves pulled the water column in the xylem up into the stem, leaving room for air to enter the base of the xylem. Cutting Stem 2 the second time removed this air-filled xylem, permitting the water still in the xylem to link up with the water in the container. However, Stem 3 was left with an air bubble at the base of its xylem, which blocks the entry of water from the container. As the leaves transpire, they lose water faster than the dwindling xylem contents can replace it. Water is pulled out of other cells in the leaves until they no longer fill their walls. Without this internal support the walls buckle and the plant wilts (Section 5-J).

ease sucrose to support growth or reproduction in other parts of the plant.

Transport in the phloem occurs down a sucrose gradient from "source" to "sink," moving from a high-sucrose, high-pressure area to a lower-sucrose, lower-pressure area.

33-I DISTRIBUTION OF SUBSTANCES

The two transport tissues, xylem and phloem, between them carry the various substances that must be moved from one part of the plant to another. Water and minerals taken up by the roots are transported by xylem and may be removed by living cells in any part of the plant. Normally, the vast bulk of water moved to the leaves is lost to the atmosphere. However, some water becomes part of the cytoplasm, some serves as a raw material for photosynthesis, and some moves into the phloem (this water eventually returns to the xylem).

Some minerals also eventually make their way into the phloem. As leaves grow old, their minerals may be moved back into the plant for use in younger leaves or for storage over the winter. Phosphorus, potassium, and nitrogen especially are reclaimed, and sometimes iron as well. Calcium is a mineral that does not move once the xylem has delivered it to the leaves.

Most plants do not store much food in the leaves, where it is made, but move it through the phloem to other parts of the plant. When sucrose or other sugars arrive in the roots, some of them are used by the root

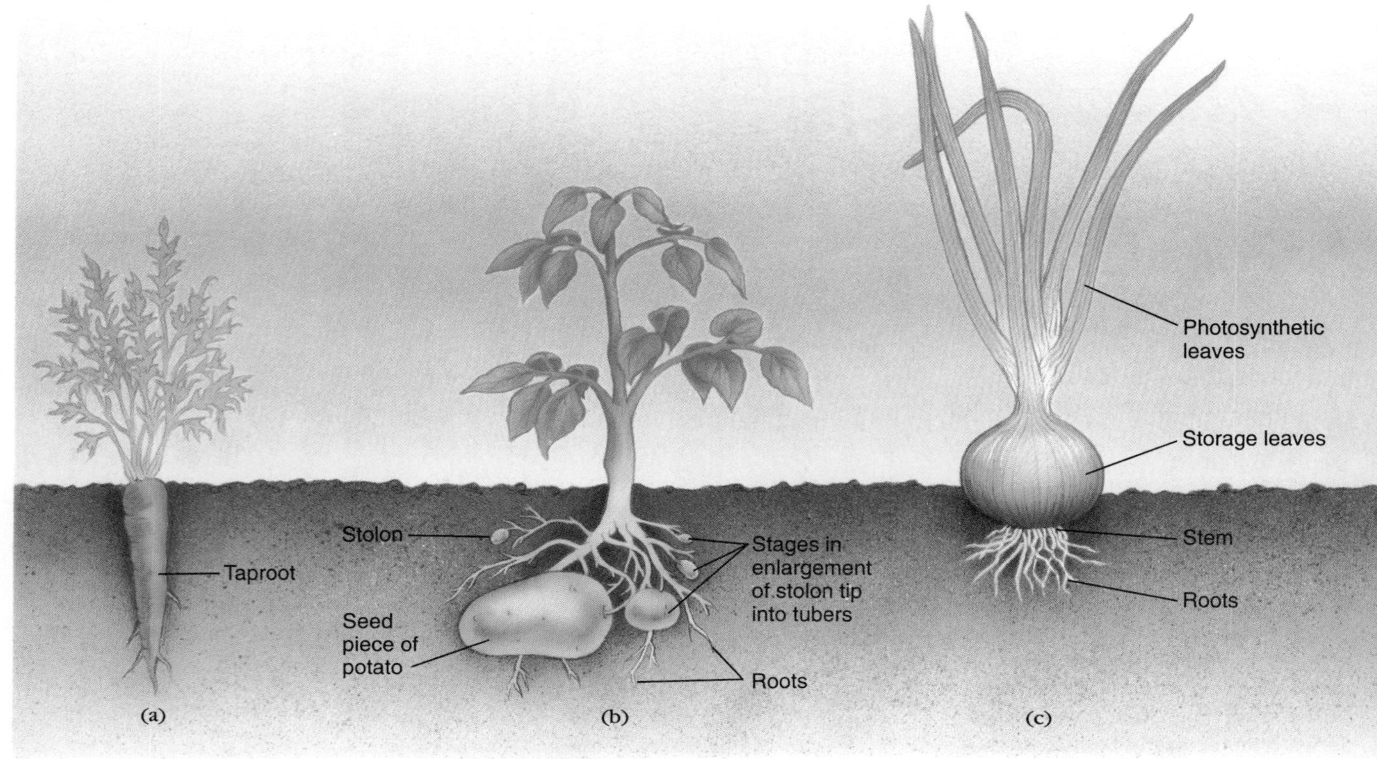

FIGURE 33–16

Underground food storage. (a) Many plants, such as carrots, store food in enlarged taproots. **(b)** A potato is an underground storage stem (tuber). Potato plants are grown from pieces of potatoes, each containing an "eye" (bud) that sprouts to form the stem and leaves. As food is produced it is stored in modified underground stems, the stolons, whose tips swell and form tubers. **(c)** An onion is composed of modified, fleshy, food-storing leaves formed underground at the bases of the green photosynthetic leaves.

cells in respiration, some are stored as starch, and some are combined with ammonium ion (NH_4^+) or nitrate ion (NO_3^-), taken in through the roots, to form amino acids. These amino acids are then moved to other parts of the plant by the xylem or the phloem. Phloem also carries various other organic compounds.

The food storage areas of plants include roots, such as those of radishes, carrots, and rutabagas; underground stems, such as those of potatoes; and sometimes even underground leaves, such as those of onions and other bulbs (Figure 33–16). One adaptive advantage of underground storage may be that it hides the food from hungry herbivores. In addition, underground storage organs are less vulnerable to freezing or drying out.

During reproduction, plants move their food reserves from storage areas to the developing seeds and fruits. The food in a seed nourishes the young embryo as it germinates, before it can make its own food. In some cases, storage of food in fruits is an adaptation to seed dispersal by animals: the animal is attracted by the nutritious fruit, and eats it, depositing the seeds elsewhere.

Practical Applications

Transport in the vascular tissues of plants is of more than academic interest. Knowing how plants handle various substances can be helpful in the management of plants and plant pests. For example, leaves can absorb some substances placed on their surfaces in solution, and move them into the phloem, which transports them throughout the plant. This is the basis of **foliar feeding,** or fertilizing with a nutrient solution sprayed onto the leaves. Other substances must be worked into the soil because only the roots will absorb them.

Systemic pesticides, watered into the soil or sprayed onto the leaves, are absorbed and distributed throughout the plant. Systemic pesticides combat both sucking insects (such as aphids) and internal pests (such as leaf miners, insect larvae that live entirely inside leaves and never come into contact with sprays applied externally). **Topical pesticides** are not absorbed into the plant, but remain on the leaf surfaces and protect the plant by killing leaf-chewing insects or invading fungi.

Vascular tissues distribute nutrients or other substances throughout the entire plant. Plants typically store surplus food in underground roots, stems, or modified leaves.

SUMMARY

1. The minerals required by plants are divided into two groups, macronutrients and micronutrients, depending on the quantities needed.
2. Plants take up most of the water and minerals they need from the soil.
3. The nature of the soil is determined by the type of rock from which it is derived, rock particle size, rainfall, resident organisms, and the content of organic matter and oxygen. All of these components of soil interact and affect the availability of water and minerals in the soil solution.
4. The soil solution moves freely into the root cortex.
5. The endodermis forms a living, selective barrier between the soil solution and the rest of the plant.
6. The cells of the root epidermis and cortex absorb minerals and move them through the endodermis to the xylem, which transports them to the rest of the plant. Water enters these cells by osmosis, following the minerals that were accumulated mostly by active transport.
7. The conducting cells of xylem and phloem show structural adaptations for their rapid transport of substances. Tracheids, found in the xylem of both gymnosperms and angiosperms, have pits in their walls, while vessel cells, found only in angiosperms, may have perforated end walls or no end walls.
8. The sieve tube elements of angiosperm phloem have sieve areas in their cell walls and well-developed sieve plates in their end walls. Companion cells carry on life support for the living but organelle-depleted sieve tube elements.
9. Both xylem and phloem also contain dead, supporting fiber cells and living, parenchyma cells.
10. In xylem, a series of dead, tubular cells conducts water and dissolved minerals upward from the roots to the leaves, flowers, and fruits.
11. Root pressure pushes sap up a short distance in the stems of some plants. In most cases, however, much more upward movement results from the transpiration of water vapor from the leaves.
12. Transpiration creates a pull from the top of the water column down the xylem to the roots and out into the soil. The cohesion of water in the xylem conducting cells is crucial to the transport of water to the top of tall trees.
13. Transpiration is controlled by the opening and closing of the stomata.
14. Phloem transport is still poorly understood. The conducting cells of phloem form continuous tubes filled with living cytoplasm.
15. The mass flow theory of phloem transport suggests that a high concentration of sugar in the phloem cells of leaves causes water to enter these cells by osmosis. This uptake of water creates hydrostatic pressure, which pushes the phloem contents along from one cell to the next.
16. Phloem distributes sugars, amino acids, hormones, and some minerals to the roots, fruits, and growing buds.
17. An understanding of transport mechanisms in plants, as well as a knowledge of the pathways followed by various substances within the plant body, is important in the planning of fertilization, pest control, and hormone treatment programs in modern agriculture.
18. Plants can store both organic and inorganic nutrients in structures such as roots, underground stems, and some leaves, for future use in growth or reproduction.

SELF-QUIZ

Match the following components of the soil with their role in plant nutrition:

___ **1.** Living organism	a. ultimate source of most soil minerals
___ **2.** Organic matter	b. used in breakdown of organic molecules to release energy and minerals
___ **3.** Oxygen	c. dissolves minerals and carries them into roots
___ **4.** Rock particles	d. provides food for fungi and bacteria in soil
___ **5.** Water	e. releases minerals bound in organic molecules

6. Which of the following would *not* make minerals more available to plants?
   a. increasing the rainfall in a wet, forested area
   b. raising the pH of a very acid soil
   c. spreading manure or other fertilizer
   d. introducing mycorrhizal fungi into sterilized soil

7. Soil solution fills the cell wall spaces of the root's ___ and ___, and substances are absorbed by the living cells in these layers. The soil solution is prevented from penetrating the entire plant by the presence of the ___ in the ___ layer of root cells.

8. From the list below, pick the three nutrients needed by plants from the soil in highest amounts.

boron	magnesium	phosphorus
calcium	manganese	potassium
copper	molybdenum	sulfur
iron	nitrogen	zinc

9. The main solute transported by phloem is:
   a. glucose
   b. potassium
   c. sucrose
   d. starch
   e. amino acids

10. The girdling experiments performed by Malpighi supported the theory that:
   a. water moves in a tree by the root pressure mechanism
   b. water moves in a tree by a transpiration-cohesion mechanism
   c. xylem is primarily responsible for conducting water from the roots to the leaves
   d. phloem is primarily responsible for conducting water from the roots to the leaves

11. List two functions of xylem tissue, and describe at least one adaptation of cells found in the xylem that contributes to the performance of each function.

12. Movement of water up through a tree trunk depends on:
   a. the high boiling point of water
   b. capillary movement of water
   c. the vapor pressure of water
   d. attraction between water molecules
   e. osmotic movement of the sap

13. Would root pressure increase, decrease, or remain the same under each of the following conditions?
   ___ a. high humidity
   ___ b. watering dried-out soil
   ___ c. darkness

14. Would the rate of transpiration increase, decrease, or remain the same under the following conditions?
   ___ a. high humidity
   ___ b. increased turgor pressure in the guard cells
   ___ c. increased light
   ___ d. increased wind

15. Xylem that is not conducting water is called:
   a. heartwood
   b. sapwood
   c. rays
   d. vessels
   e. tracheids

16. Which of the following is *not* necessary to the operation of the mass flow mechanism, according to the theory presented in this book?
   a. ATP
   b. root pressure
   c. intact membranes in conducting cells
   d. different osmotic concentrations in different parts of the plant
   e. constant production or release of sugar molecules

THINKING CRITICALLY

1. Explain why chemical analysis of the elements present in a plant is not a good indication of its nutritional needs.

2. Why is vegetation often sparse in soil with many pebbles and boulders?

3. Boron-deficient soils are improved by applying sodium tetraborate, which contains about 2 to 5 pounds of boron, at a recommended rate of 20 to 50 pounds per acre. One study showed that wheat plants took up only 0.3 pound of boron per acre in one growing season. If these values are typical, what happens to the rest of the boron applied to the soil?

4. Design an experiment to determine how carnivorous plants obtain mineral nutrients other than nitrogen.

5. In 1936, Bruno Huber performed an experiment in which he inserted thin heating wires into the xylem of a tree. He placed a thermocouple (a sensitive heat-detecting device) farther up the stem and timed how long it took before the heated water passed the thermocouple. Huber found that the water moved slowly at night. In the morning, the water movement speeded up first in the twigs; later, the water began to rise more quickly in the trunk farther down the tree. Do these results support the root pressure or the transpiration pull theory of xylem transport? Justify your answer.

6. What is the best time of day to cut flowers? Justify your answer.
7. We have seen in this chapter that cohesion of water keeps a column of water travelling up through the xylem of a tall tree. How did the water reach the top in the first place?
8. Why should the ground around evergreen plants be watered thoroughly before the ground freezes for the winter?
9. Contrast the conditions affecting transpiration experienced by a houseplant in winter with those in summer.
10. What is the advantage of bean leaves' having more stomata in the lower epidermis than in the upper epidermis? Why are the stomata of a corn leaf located in both epidermal layers?
11. What does wilting indicate about the movement of water in a plant? What does it indicate about the water content of the soil?
12. The virus disease known as "beet yellow" is transmitted from plant to plant by aphids. Why does the disease spread through the plant rapidly and kill it quickly?
13. Large marine brown algae known as kelps have a system of "sieve filaments" whose cells look remarkably like the phloem conducting cells of vascular plants. These sieve filaments transport carbohydrates from the blades (photosynthetic parts) to the stipe (stalk) and holdfast of the plant. Why do these kelps have phloem-like tissue but no xylem-like tissue?

SELECTED KEY TERMS

capillarity, *p. 688*
foliar feeding, *p. 706*
girdling, *p. 694*
growth ring, *p. 695*
guttation, *p. 698*
heartwood, *p. 696*
leaching, *p. 690*
lignin, *p. 694*
macronutrient, *p. 688*
mass flow, *p. 703*
micronutrient, *p. 688*
mycorrhizae, *p. 690*
plasmodesmata, *p. 692*
pressure flow, *p. 703*
ray, *p. 695*
resin, *p. 695*
root pressure, *p. 698*
sap, *p. 694*
sapwood, *p. 695*
sieve tube, *p. 701*
systemic pesticide, *p. 707*
topical pesticide, *p. 707*
tracheid, *p. 694*
transpiration, *p. 699*
transpiration pull–water cohesion, *p. 699*
vessel, *p. 695*
wood, *p. 694*

SUGGESTED READINGS

Cohen, I. B. "Stephen Hales." *Scientific American,* May 1976. A biographical sketch of the life and experimental work of an energetic pioneer in plant physiology.

Epstein, E. "Roots." *Scientific American,* May 1973. Compares the form of root systems in different types of plants as well as how roots absorb minerals from the soil.

Fritts, H. C. "Tree rings and climate." *Scientific American,* May 1972. How tree ring patterns are used to analyze the climate of bygone times.

Heslop-Harrison, Y. "Carnivorous plants." *Scientific American,* February 1978. Briefly describes the types of carnivorous plants and presents experimental and photographic evidence on how prey is captured and digested.

Reganold, J. P., L. F. Elliott, and Y. L. Unger. "Long-term effects of organic and conventional farming on soil erosion." *Nature* 330:370, 1987.

Wooding, F. B. P. "Phloem." *Carolina Biology Reader.* Burlington, NC: Carolina Biological Supply Company, 1978.

Zimmermann, M. H. "How sap moves in trees." *Scientific American,* March 1963. Outlines various methods used to discover how transport occurs in xylem and phloem.

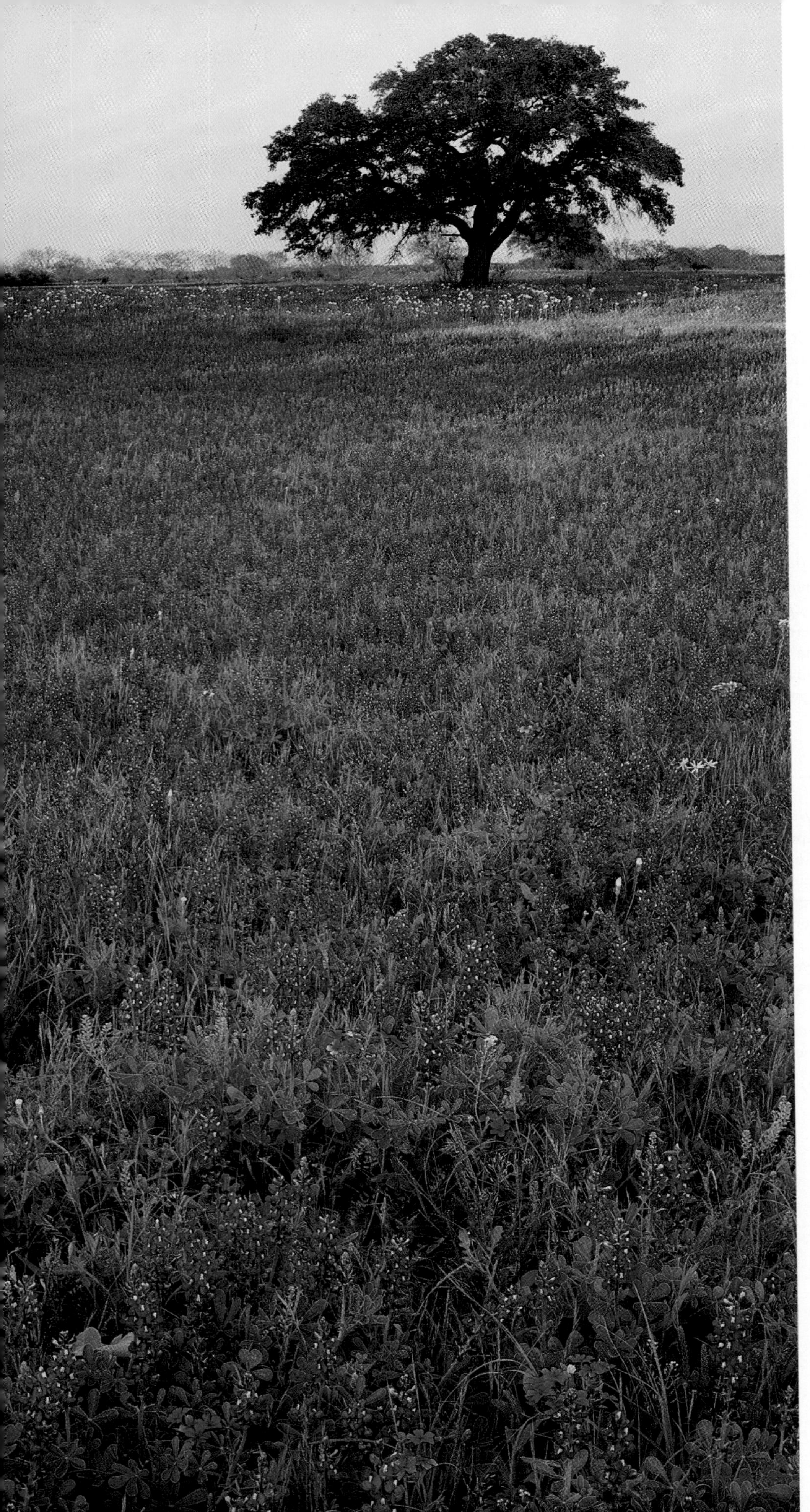

CHAPTER 34

CONCEPT GUIDE

After reading this chapter, you should be able to:

1. Describe the general effects of plant hormones. List the five major types of plant hormones and relate them to these functions. (Section 34–A)
2. Explain the process of apical dominance and how it can be controlled. (Section 34–B)
3. Explain the advantages of tropisms and describe how they are controlled. (Section 34–C)
4. Describe how plant flowering is influenced by changes in periods of daylight and darkness. (Section 34–D)
5. Compare the processes of senescence and abscission. (Section 34–E)

How do these Texas bluebonnet plants all manage to produce flowers at the same time? The general answer is that they are all responding to the same environmental cues. But what specific factors affect the timing of flower production? Read about these specific factors in Section 34–D. (W. Clay/Dembinsky Photo Assoc.)

Regulation and Response in Plants

A plant grows through the activity of its meristems. Meristematic cells divide and produce cells that grow, differentiate, and mature (Chapter 32). How is this growth controlled? Why does one cell differentiate into an epidermal cell while another becomes part of a xylem vessel? Why does the root grow down into the soil and the shoot grow up into the air? What determines whether a lateral bud forms a new branch this year, next year, or never? Why do all the cherry trees in an orchard bloom at the same time in the early spring, and why do their fruits ripen over a short season in early summer? And why do all the thistles in a field bloom at once in the late summer?

All of these events, and more, are under the control of plant hormones, chemicals produced in various parts of the plant and distributed throughout its body, affecting virtually every area in one way or another.

KEY CONCEPTS

- Every aspect of the production, differentiation, growth, and maturation of a plant is regulated by chemicals, the plant hormones.
- Environmental stimuli affect the production and distribution of hormones. The hormones in turn govern responses by which the plant adapts to these external factors.

Curiosity Questions

1? Why does one rotten apple spoil the whole bunch? (See page 715)

2? What happens when you "pinch back" a house plant? (See page 717)

3? What determines when a plant will have flowers? (See page 720)

34–A PLANT HORMONES

Hormones are chemical messengers produced in one part of an organism and transported to other parts, where they exert effects out of proportion to their very small concentrations.

Unlike animals, plants do not have distinct endocrine organs. Plant hormones are produced in various organs that have other functions as well.

Another difference between plant and animal hormones is that plant hormones can affect virtually any tissue in the plant, rather than having specific target tissues. (In both plants and animals, however, different tissues may respond differently to a particular hormone, depending on the tissue and its stage of development, and on the amount of hormone.) Plant hormones do not usually maintain homeostasis, as do many animal hormones. Instead they affect differentiation and growth by causing permanent changes such as cell division, elongation, and death.

How do plant hormones influence differentiation and growth? Applying a hormone to part of a plant often results in changes in the activity of certain enzymes or in membrane permeability. In addition, the plant hormones all have long-term effects on protein synthesis. However, it is not yet clear whether the hormones themselves bring about these changes, or whether they act indirectly, via second messengers.

A cell's differentiation is determined by the hormones and other chemicals it contains, which in turn are affected by the rest of the plant. Hormonal messages passing between different parts of the plant regulate growth of new parts. Thus the root system remains in physiological balance with the shoot system, and the position of a branch or leaf or root determines where others are added.

A plant must also respond to stimuli in its physical environment, such as light, temperature, wind, water, the pull of gravity, and the change of seasons. These responses, too, are controlled by hormones.

The traditional list of plant hormones has five major entries: auxin, gibberellin, and cytokinin, which induce cell division or cell growth; and abscisic acid and ethylene. Each hormone plays a leading role in certain activities, but many plant responses are governed by the interaction of two or more hormones. Recently, investigators have identified a sixth group of regulatory molecules, the **oligosaccharins,** which act in their own unique roles and also may serve as intermediaries for some of the five major hormones.

Auxin

Auxin was discovered in 1926 by the Dutch plant physiologist Frits W. Went (Figure 34–1). It affects many plant activities, such as the promotion of meristematic growth, cell elongation in shoots, and cell division, but it inhibits the growth of lateral buds (Section 34–B). It also induces production of xylem, development of fruits, and differentiation of sexual parts in flowers. Although auxin stimulates formation of new roots, it inhibits elongation of root cells except at very low concentrations. High concentrations of auxin promote production of ethylene.

Auxin also mediates many responses to the environment. It causes roots to grow downward and shoots to grow upward, and it is responsible for the bending of stems and leaves toward light. Auxin also induces trees to produce more secondary xylem in response to mechanical disturbance, such as wind. This additional xylem thickens and strengthens the tree trunk.

We have at least a partial picture of how auxin stimulates growth. Auxin causes the plasma membrane to transport hydrogen ions out of the cell. Here the increased acidity activates enzymes that loosen the crosslinks between the cellulose fibers of the cell wall. As water enters the cell from the plant's dilute sap, the cell expands and pushes its wall outward. This expansion starts in the first half hour after auxin is supplied. Afterward, auxin continues to promote growth by speeding the synthesis of proteins needed for growth.

Synthetic auxins, such as 2,4–D, are used as weed killers in lawns. Dicots, which include broadleaved weeds, are much more sensitive to low levels of auxins than are monocots, which include the grasses. Hence, applying a low dosage of synthetic auxins to a lawn can cause the dicots literally to grow themselves to death. To be most effective, these weed killers must be applied when the plants are actively growing and therefore sensitive to extra doses of hormones.

During the Vietnam conflict, American airplanes sprayed Agent Orange, a herbicide containing synthetic auxins, on forests and crops. This had a drastic effect on the ecology of the countryside. Experiments with laboratory animals showed that a contaminant in the spray (a dioxin) causes malformations of developing animal fetuses. After dioxin was detected in drinking water and in fish, one of the few sources of protein in the Vietnamese diet, the spraying was abandoned. Manufacturers later found ways to reduce the concentration of dioxin in the herbicide, and it was used in the United States to kill unwanted forest trees and to

BIO-BIT

Auxins increase the thickness of tree trunks as a response to mechanical disturbance. Therefore, trees grown indoors where there is little wind movement will grow stronger if they are periodically shaken!

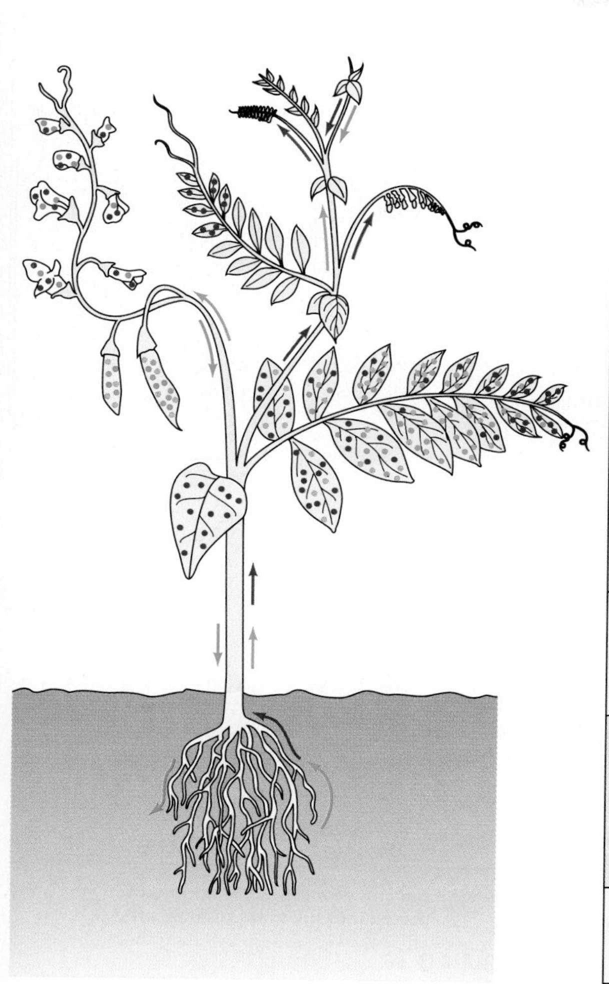

Hormone	Produced In	Action of Hormone	
AUXIN	Young tissues, shoot apex, young leaves, and seeds	In fruit:	develops fruit, ethylene production
		In flowers:	sexual differentiation of flower parts
		In stem:	stimulates stem elongation, responsible for positive phototropism, responsible for negative gravitropism
		In roots:	forms new roots, inhibits elongation of roots, inhibits gravitropism
		In meristems:	growth of meristems, cell division, enlargement of cells
GIBBERELLINS	Young tissues, embryo	In flowers:	sexual differentiation of flower parts
		In stem:	elongation of stem cells, differentiation of phloem
		In leaves:	promotes growth
		In roots:	inhibits formation of roots
		In meristems:	inhibits formation of roots
CYTOKININS	Fruit and entire plant	In fruit:	production of fruits and seeds
		In entire plant:	cell division, growth, inhibits dormancy, inhibits senescence
ABSCISIC ACID	Seeds, shoots, and leaves	In seeds:	promotes dormancy
		In shoots:	promotes dormancy, promotes formation of bud scales
		In leaves:	closure of stomata during dry times
ETHYLENE	Entire plant	In fruits:	ripens fruit
		In entire plant:	counteracts effects of auxins, promotes senescence

FIGURE 34–1

Major plant hormones. The sites of production and effects of the five major categories of plant hormones for a pea plant. Arrows indicate the direction in which each hormone is transported. See Section 34–A for additional details on each hormone's structure, production, transport, and action.

control weeds on grazing ranges. However, further research showed that dioxin is a powerful carcinogen, and probably a mutagen, even in minute concentrations. After dioxin was found in beef fat and in human milk, the herbicide was banned.

Other synthetic auxins are sprayed on pear and apple trees to keep them from dropping fruits before they ripen, or applied to cuttings of plant shoots to promote root formation.

Gibberellins

Gibberellins are an important group of plant hormones. The first gibberellin, discovered in Japan, was actually produced by a fungus, *Gibberella fujikuroi,* which causes "foolish seedling" disease of rice. Rice seedlings with the disease grow abnormally quickly, but are spindly and unhealthy, and seldom yield fruit. In vascular plants, gibberellins are produced in young leaves.

One important role of gibberellins is to stimulate elongation in stem cells. Dwarf or miniature strains of plants are often genetic mutants that do not produce gibberellins, but they will grow as tall as normal varieties if treated with gibberellins. Giving extra gibberellins to normal plants has little effect (Figure 34-2).

Gibberellins also stimulate cell division at the stem apex, leaf growth, development of sexual flower parts, and (with sugars) differentiation of phloem tissue. Ad-

FIGURE 34–2

Effect of gibberellin. The plant on the left side was treated with gibberellin and the plant on the right side was not, and served as the control (E. Webber/Visuals Unlimited, Inc.)

ditionally, they stimulate the production of auxin, as well as making plants more responsive to auxin treatment. On the other hand, they inhibit root formation. Gibberellins are also involved in the responses of plants to their environment. Some plants that require a period of cold, or exposure to light, before flowering or before seed germination will respond after gibberellin treatment even if they have not experienced the proper exposure to cold or light.

Cytokinins

The first known **cytokinin** was a component of coconut milk. This compound proved tricky to isolate, so whole coconut milk became a standard ingredient in plant tissue cultures in the laboratory because it stimulates cell division.

In intact plants, cytokinins are involved with growth and with production of fruits and seeds. Cytokinins also prevent the onset of dormancy and slow the aging of cut leaves or fruits. Holly that is cut early for the holiday season can be kept fresh and green by spraying it with cytokinins.

Cytokinins may interact in various ways with other hormones. This is often studied using plant cells grown in laboratory tissue cultures. For example, the relative concentrations of cytokinins and auxin in tissue cultures determine whether or not cells divide and differentiate, and whether they develop into roots or into shoots with leaves (Figure 34–3). To grow an entire plant from cultured cells, hormones must be applied at certain concentrations, in particular ratios to each other, and in the correct sequence. Presumably, similar hormonal interactions in intact plants govern the switching on and off of genes, resulting in the differentiation of the various tissues and organs during embryonic development and throughout the plant's life.

BIO-BIT

Most commercial tomatoes are picked green and treated with ethylene to cause them to turn red. Unfortunately, this process does not allow the tomatoes to accumulate the acids, sugars, and other flavor molecules that give vine-ripened tomatoes their taste.

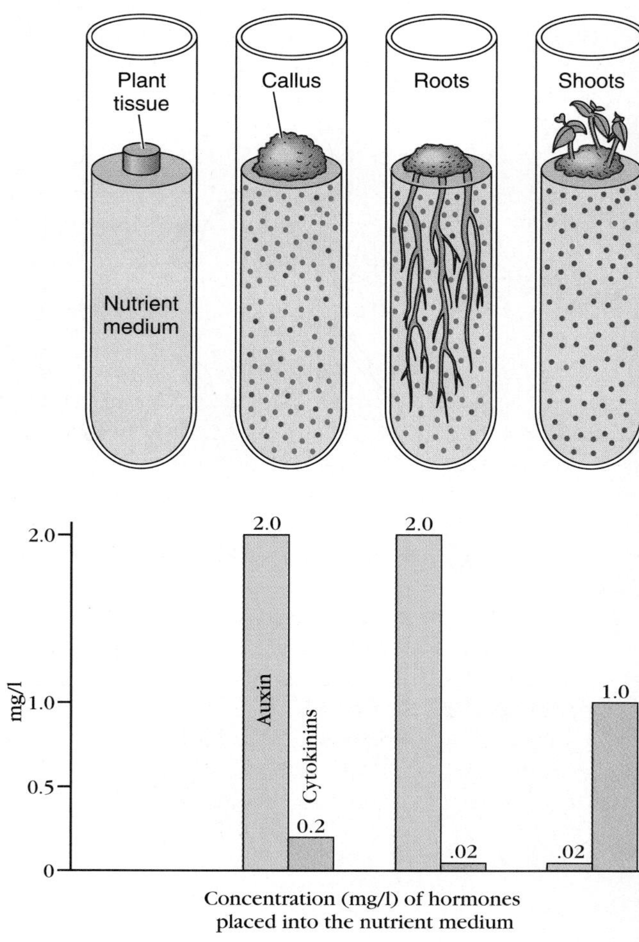

FIGURE 34–3

Responses of cultured plant tissues to auxin and cytokinin. The initial explant is a plug of pith cut from the center of a tobacco stem and placed on nutrient medium. Depending on the concentrations of auxin and cytokinins in the medium, the tissue may develop as an undifferentiated lump of cells (callus) or develop roots or shoots.

Abscisic Acid

Auxin, gibberellins, and cytokinins have many growth-stimulating roles. In contrast, **abscisic acid** is often called the "stress hormone" because of its roles in helping the plant cope with adverse environmental conditions. For example, during periods of water shortage abscisic acid causes stomatal closure, reducing the plant's loss of water by transpiration (Section 33-F).

Abscisic acid is produced in mature leaves. At the end of the growing season, its concentration in twigs is high compared to the levels of gibberellins and cytokinins. This high level of abscisic acid induces the apical meristem to stop dividing, and the newest leaf primordia form into bud scales around the tip instead of becoming normal photosynthetic leaves (Figure 34-4). The bud scales protect the delicate apical meristem from freezing or drying out in the cold of winter. In spring, overwintering twigs are released from dormancy by destruction of abscisic acid, or production of substances that counteract it. Abscisic acid also keeps seeds dormant until it is leached away by water or overcome by a stimulatory hormone, usually gibberellin. Again, this keeps the tender seedling from being blighted by winter's cold. Abscisic acid does, however, promote the embryo's growth and synthesis of storage proteins, making it better prepared for life after it does germinate.

Ethylene

Unlike the other known plant hormones, **ethylene** is a gas, and so it travels in the air as well as within the plant. Hence it affects not only the plant that produced it, but others as well. For example, ethylene given off by apples inhibits the sprouting of potatoes stored in the same bin.

The existence of a gas that affects plant physiology was known long before it was shown to be ethylene. Nineteenth-century growers of greenhouse plants knew that something in the gas used in gas lighting caused blossoms to wither prematurely. Mango and pineapple growers commonly lit fires in their groves because something in the smoke synchronized flowering and fruit ripening.

The production and effects of ethylene are closely tied to auxin. After auxin exceeds a certain level, it stimulates production of ethylene, which in turn counteracts the effect of auxin. Thus production of ethylene by lateral buds as they receive auxin seems to play some part in repressing their growth.

Ethylene also has interesting roles in plant senescence (Section 34-E) and in the ripening of fruits. The

(a) Bud scales form around the shoot apical meristem in fall.

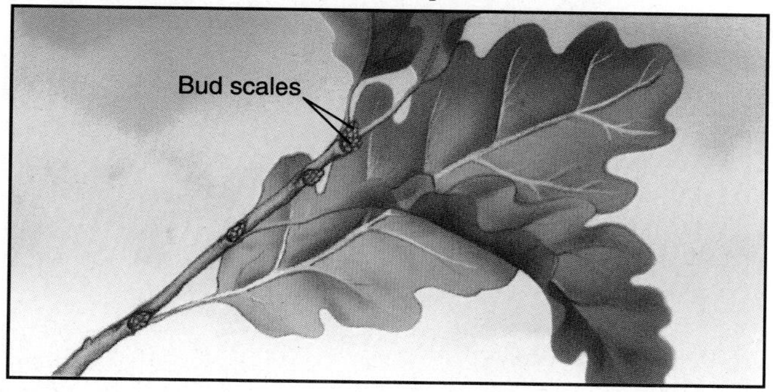

(b) Bud scales protect shoot apical meristem through winter.

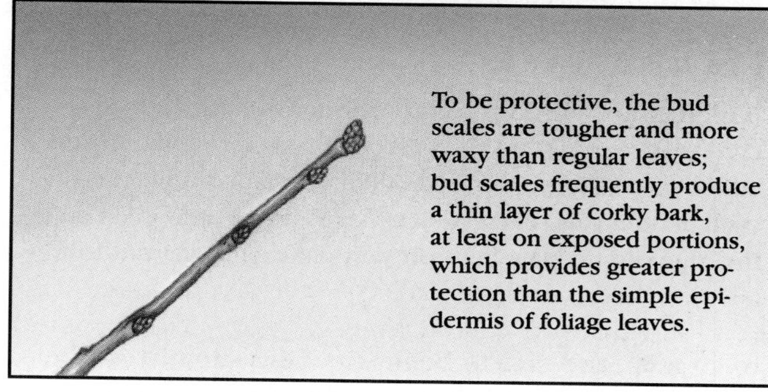

To be protective, the bud scales are tougher and more waxy than regular leaves; bud scales frequently produce a thin layer of corky bark, at least on exposed portions, which provides greater protection than the simple epidermis of foliage leaves.

(c) New leaves burst through bud scales in the spring.

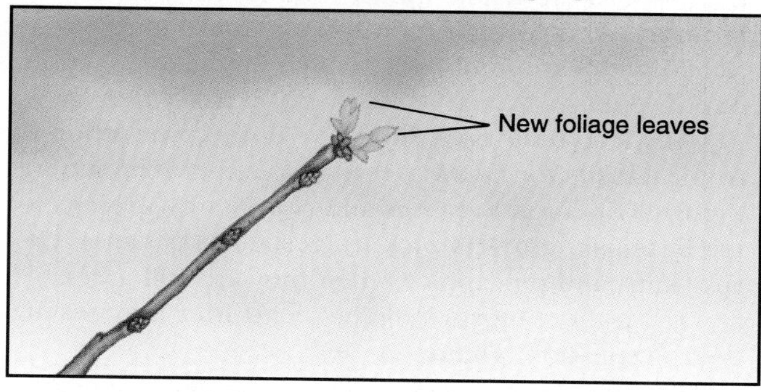

FIGURE 34-4

The structure and function of bud scales.

development of a fruit is stimulated by auxin, gibberellins, and cytokinins. The level of auxin builds up and then drops; this is followed by production of ethylene. Ethylene stimulates the activity of various enzymes, which (1) convert starch and acids of the unripe fruit to sugars, and (2) soften the fruit by breaking down pectins in the cell walls.

1? The release of ethylene has a positive feedback effect: the more ethylene produced, the more the fruit is stimulated to produce additional ethylene. Hence the entire fruit ripens at once, and if there are many fruits

FIGURE 34-5

Bananas growing in a plantation. Fruits ripen in synchrony in response to the gaseous plant hormone ethylene, either produced by the fruits themselves or supplied artificially by humans. Bananas are picked while still green and the blue plastic bags help prevent damage to mature fruit. (M. Balick/Peter Arnold, Inc.)

in an area, the first to begin ripening stimulates ripening of its neighbors. By the same token, overripening is also contagious, and it is quite true that "one bad apple spoils the whole bunch." Apples are now stored in refrigerated, airtight rooms in an atmosphere enriched with carbon dioxide, which inhibits the action of ethylene.

Ethylene finds uses, however, in the production of tropical fruits for far-away markets. Rather than letting the fruit ripen on the trees and risk loss by overripening in transit, growers pick fruits such as bananas, citrus fruits, and pineapples when they are still green and ripen them by applying ethylene after they reach their destination (Figure 34-5). ■

Recent research also shows that ethylene induces plants to make certain enzymes, including chitinase, which defends the plant against fungi by breaking down chitin in the fungal walls.

In general, plant hormones affect the growth, differentiation, cell division, and death of plant tissues. These hormones allow plants to respond to environmental changes. Five major types of plant hormones interact to govern plant cells.

34-B APICAL DOMINANCE

Many plants exhibit **apical dominance,** in which the growing apical meristem represses the growth of lateral buds on the same stem, causing them to remain dormant. This is due primarily to auxin, which is produced in the apical meristem and transported toward the base of the plant. If the apical meristem is cut off, the lateral buds develop into new branches. This can be prevented by placing auxin on the cut stump immediately after the apical meristem is removed (Figure 34-6).

Other plant hormones may interact with auxin in apical dominance. Auxin stimulates lateral buds to produce ethylene, which in turn seems to play some part in inhibiting the buds' growth. In contrast, cytokinins may help to overcome apical dominance. Virtually all of the cytokinins produced in a plant's roots are transported to growing areas such as apical and lateral

Control

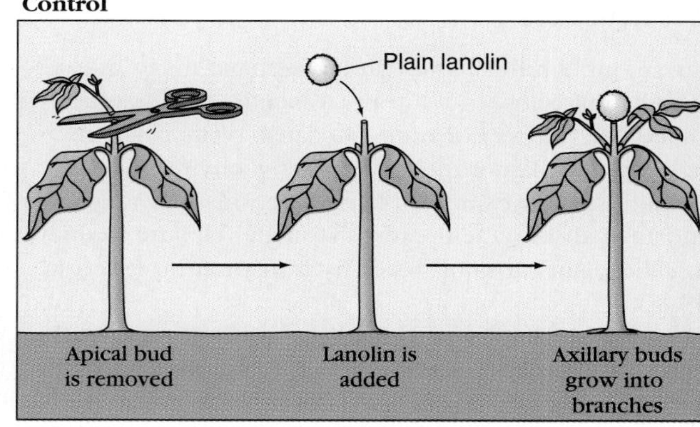

Auxin treatment

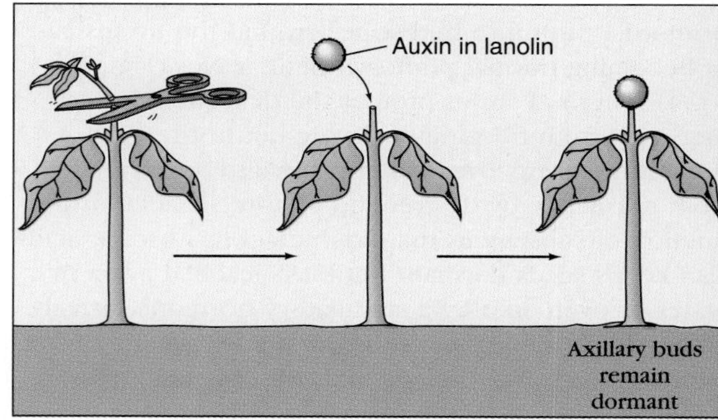

FIGURE 34-6

An apical dominance experiment. The apical bud is removed from all of the plants in this experiment. Then a drop of either plain lanolin or lanolin with auxin is added over the cut end of the plant. In the plain lanolin group, the absence of the apical meristem permits the growth of the axillary buds into branches. But when lanolin with auxin was added, the axillary buds did not grow. This experiment supports the hypothesis that auxin produced by the apical meristem inhibits the growth of axillary buds into branches. Explain why the control was lanolin without auxin instead of nothing at all.

(a)

(b)

FIGURE 34-7

Tree shapes. (a) Conifers with strong apical dominance. The trunk grows much faster than the branches, and the terminal buds repress the nearest branches most strongly. **(b)** Very weak apical dominance results in a rounded, spreading shape, with many large branches as seen in this apple tree. (a, B. P. Kent/Earth Scenes; b, Runk/Schoenberger from Grant Heilman)

meristems, and fruits. When the apical meristem is removed, the lateral buds receive more cytokinins. As a shoot grows, its lower buds break dormancy, and this may be due to the accumulation of more cytokinins from the roots and the reduction in the amount of auxin reaching them from the apical meristem. So it may be more cytokinin, perhaps combined with less auxin, that releases the lateral buds from inhibition.

Along similar lines, it is known that food and water are transported preferentially into areas of the plant where auxin concentration is high. Thus it is possible that the lateral buds fail to grow because of starvation rather than repression. Removal of the auxin-producing apical meristem would reduce auxin levels at the tip of the plant and permit the transport of more food to the lateral buds. These buds could then begin to grow and produce their own auxin, causing them, in turn, to receive more nutrients from the transport system.

It is hard to design experiments to show whether the inhibition of lateral buds comes from too much auxin, from too little cytokinin, from the wrong combination of auxin and cytokinin, from starvation, or from some combination of these.

A plant with strong apical dominance has a distinct main axis, with few side branches, whereas a plant with weak apical dominance has many well-developed branches and a bushy appearance. Gardeners often pinch out apical meristems to make their plants grow into compact, bushy shapes. ■

By looking at the plants around us, we can easily see that the degree of apical dominance varies a great deal (Figure 34-7). Some of this variation is due to genetic differences, but environment can also affect apical dominance. A beech tree in a forest is tall and slender, with short, thin branches, whereas another, growing in the open, is wide and spreading; a third, growing at the edge of the wood, has large branches on the open side, and small, thin branches on the shady side. These variations in shape suggest that light may somehow counteract apical dominance, allowing the tree to grow new branches into any openings in the forest canopy.

Auxin, produced in the apical meristems of shoots, tends to repress growth of lateral buds into new branches. Strong apical dominance produces tall plants with few side branches, while weak apical dominance produces bushy plants.

34-C RESPONSES TO THE ENVIRONMENT

A plant's growth is governed by the interactions of its hormones, produced in various parts of the plant and transported throughout its body in varying concentrations. Factors in the environment influence the production and distribution of hormones. This results in adaptive adjustments in growth, as when light shining on a lateral bud helps to release it from apical dominance: the bud grows into a new branch, with leaves that can use the light for photosynthesis to make more food for the plant.

Many plants require a period of cold before they break dormancy and resume growth. Examples include germination of some seeds, blooming of spring bulbs, and breaking of dormancy in overwintering twigs of some woody plants. In all of these cases, the requirement for a cold period ensures that the plant delays growth until spring, when growing conditions are likely to be favorable. Many of these plant species do not grow well in climates where the winters are not cold enough to provide the necessary stimulus.

Tropisms: Growth in Response to Gradients

Growth responses along the direction of environmental gradients are called **tropisms.** The response of a plant to gravity, for example, is called **gravitropism.** In most plants, the roots are **positively gravitropic,** tending to grow downward, while the shoot is **negatively gravitropic,** growing away from the pull of gravity, or upward. **Thigmotropism** is a response to contact, as when climbing plants wrap themselves around a supporting object (see A Journey Through Plant Tropisms).

Tropisms orient the parts of a plant favorably with respect to the resources it needs. The first root of a germinating seed always grows downward, no matter what the position of the seed in the soil. This positive gravitropism ensures that the root establishes a firm anchorage for the plant. It also gives the best possible chance of finding water and dissolved nutrients. Roots may also respond to water itself. For example, the roots of a willow may grow horizontally a dozen meters or more toward water sources. This strongly positive **hydrotropism** is responsible for clogging many a sewer line or septic system, an expensive botany lesson for unwary homeowners.

Unlike roots, most shoots show negative gravitropism, growing away from the pull of gravity. This ensures that a seedling's shoot grows upward and reaches the soil surface by the shortest route. Once the shoot reaches light, it also shows **positive phototropism,** growing toward light. If the light shines mainly from one side, the plant bends in that direction.

Auxin was first detected because of its role in phototropism in oat seedlings. A sheath, the **coleoptile,** covers the first leaves of the oat seedling and other grasses and cereals. If the coleoptile of a seedling grown in the dark is exposed to light from one side, it bends toward the light. Charles Darwin studied this phototropic response. Although the growth curvature takes place in the region of elongation below the coleoptile tip, Darwin found that covering this area did not affect the phototropic response. However, when he removed the coleoptile tip, or covered it with an opaque cap, the phototropic response did not occur. He concluded that phototropic bending in the zone of elongation was controlled by the tip.

Later experiments showed that phototropism is caused by the chemical we now call auxin. A coleoptile tip was cut off and placed on a block of gelatin-like agar. When this block was later placed on one side of a freshly decapitated coleoptile, the coleoptile grew and bent away from that side.

These results suggest that phototropism in the intact seedling is due to an auxin concentration that is higher on the dark than on the lighted side of the coleoptile. By applying radioactively labelled auxin to plants, it was found that auxin is transported from the lit to the shaded side. Here it causes increased cell elongation—hence the increased bending seen in the phototropic response. So plants actually grow toward light by growing away from the dark. Plants grown in bright light show a decreased sensitivity to light, and are often shorter (and sturdier) than plants grown in dimmer conditions. In this way the plant does not waste energy growing toward light when it already receives enough.

Gravitropism is also mediated mainly by auxin. In roots, the pull of gravity is detected in the root cap, and the root bends farther back, in the zone of elongation (see Figure 32–2). Auxin is transported down through the center of the root tip, and when it reaches the tip of the root cap it makes a U-turn and goes back up the root through the outer cell layers. If the root is vertical, auxin is distributed evenly to all sides, and the root continues to grow straight down. However, if the root is horizontal, more auxin is distributed to the lower side of the root. Unlike stem cells, root cells are inhibited by all but the lowest levels of auxin. Hence cells on the lower side of the root grow less than the cells on the upper side, and the root bends in the zone of elongation until it is growing straight down.

A root's response to gravity depends on the starch-storing **amyloplasts** of root cap cells. The heavy amyloplasts fall to the bottoms of the cells. Here they apparently press on the endoplasmic reticulum and cause it to release stored calcium ions. This leads to activation of auxin pumps in the nearby areas of plasma membrane (on the lower sides of cells). Auxin is pumped out of the lower sides of the cells and soon builds up more on the lower side of the root cap than on the upper side. This difference is maintained while the auxin is transported back up the outside of the root to the zone of elongation.

Environmental factors such as light, gravity, water, and temperature can influence the effect of a hormone, permitting plants to grow along the direction of environmental gradients. Differences in auxin distribution result in positive phototropism of shoots and gravitropism of shoots and roots.

Positive phototropism causes plants to bend toward the light.

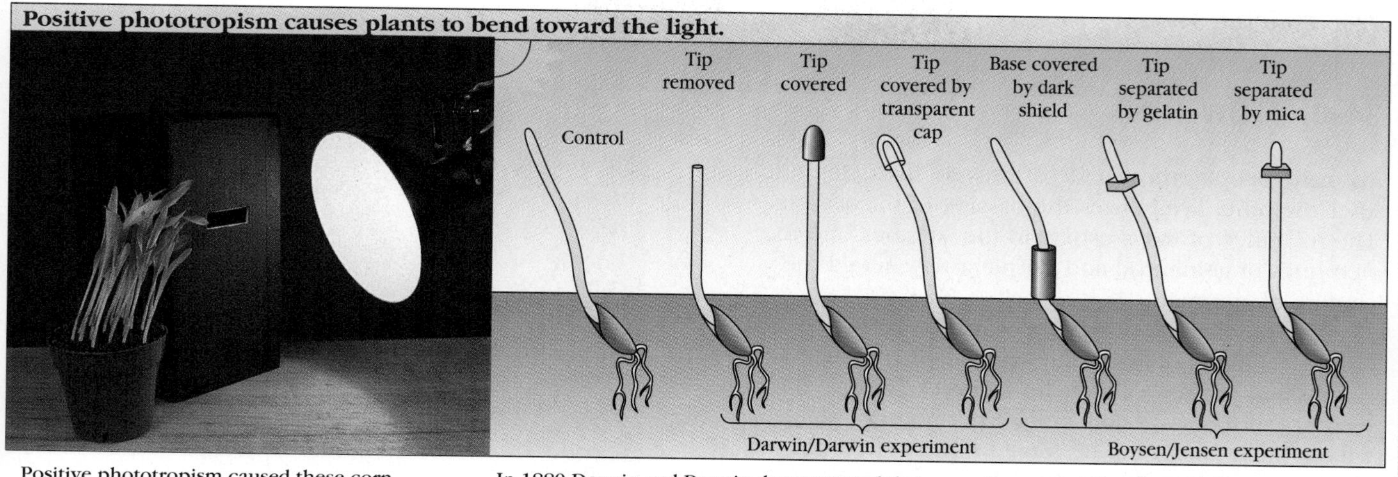

Positive phototropism caused these corn seedlings to grow up and towards the light that entered at the opening in the hood that covered them.

In 1880 Darwin and Darwin demonstrated that removing or covering the coleoptile tip of an oat seedling destroys the phototropic response, whereas covering the zone of elongation (where bending actually occurs) has no effect. In 1913 Boysen and Jensen demonstrated that an agar block can transmit the agent responsible for phototropism from the coleoptile tip to the zone of elongation, whereas a piece of mica rock blocks the transmission of the phototropic agent.

Positive gravitropism causes roots to grow down.

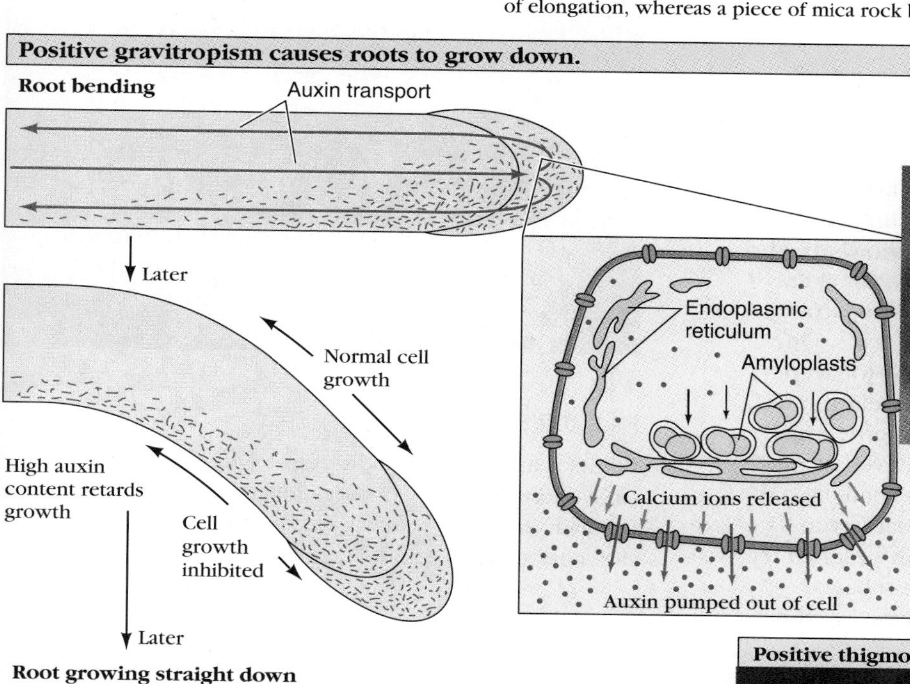

Positive gravitropism causes the roots of these corn seeds to grow down even if they have been planted upside down

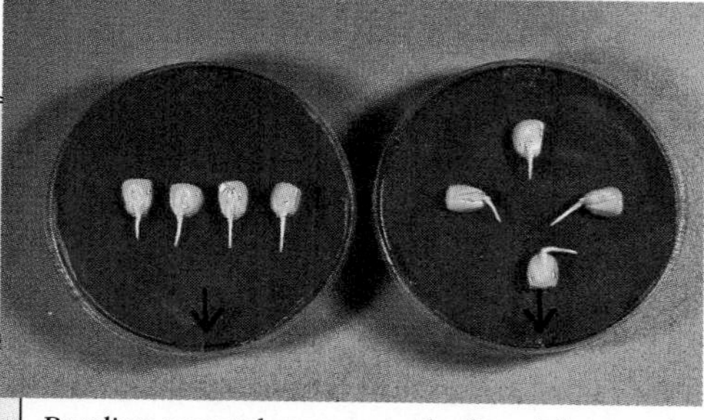

Bending occurs because amyloplasts fall toward the bottom of the cell. This is thought to cause the underlying endoplasmic reticulum to release calcium ions, which activate nearby auxin pumps. Auxin is therefore transported to the bottom of the root where it inhibits elongation of root cells.

Root growing straight down

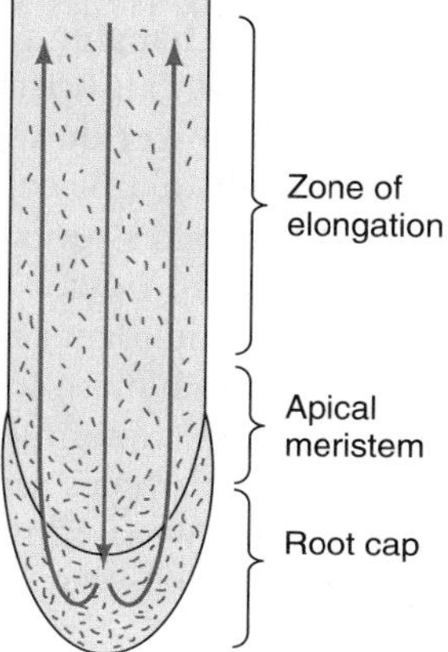

Auxin is transported down the center of the root to the root cap,where it is distributed sideways and sent back toward the stem.

Positive thigmotropism

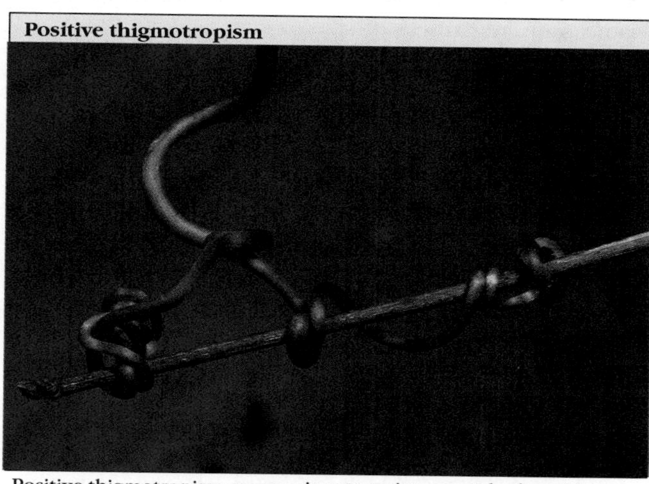

Positive thigmotropism causes vines to twine around other objects, like this twig.

Plant responses to sunlight, gravity, and contact. (phototropism and gravitropism, Runk/Schoenberger from Grant Heilman; thigmotropism, B. Kent/Earth Scenes)

34–D FLOWERING

To many people, the parade of flowers blooming one after the other symbolizes the passing of the seasons. The fragrance of roses ushers in the summer, and the rich hues of goldenrod and purple asters herald the coming of autumn.

The predictable flowering of many plants depends on environmental cues. Some familiar plants, such as carrots and parsley, flower the second year after the seed began to grow—following a period of cold in the winter. Such plants can be forced to flower by placing them in the cold for several weeks, or by applying gibberellins. Evidently a cold period somehow triggers a hormonal change in these plants, switching them from the vegetative to the reproductive phase.

(a)

(b)

FIGURE 34–8

Plants with photoperiodic requirements for flowering. (a) A short-day variety of chrysanthemums, blooming in late October. **(b)** These petunias are long-day plants. (a, R. Shell/Earth Scenes; b, R. F. Head/Earth Scenes)

Photoperiodism: Responses to the Length of Night and Day

In higher latitudes, the period of daylight becomes longer in spring and shorter again in the fall. Many plants flower only when exposed to certain light and dark periods, a response called **photoperiodism.** Surprisingly, this was not recognized until 1920, when W. W. Garner and H. A. Allard of the U.S. Department of Agriculture studied two flowering problems. One was the behavior of "Maryland mammoth" tobacco, which reached prodigious heights in the field but did not flower. However, cuttings rooted and grown in greenhouses during the winter flowered even at much smaller sizes. The second problem concerned "Biloxi" soybeans, which flowered on the same date even when farmers staggered their plantings in order to try to stagger the harvest. Because all of the tobacco plants and all of the soybean plants flowered at the same time, regardless of size or age, it looked as though flowering must be determined by some cue in the environment.

Experiments with different temperatures and light intensities failed to link these factors to flowering. Garner and Allard then turned to another hypothesis, that plants could respond to varying *lengths* of daylight—that is, to different photoperiods. This seemed silly because it meant that plants would have to be able to measure time. But, sure enough, when the researchers artificially changed the photoperiod, they found that the tobacco and soybean plants would flower only if exposed to light for less than a certain maximum amount of time each day! Accordingly, these were dubbed **short-day plants.**

3? We now know that these plants should really be called "long-night" plants because what they need to flower is a certain minimum period of uninterrupted darkness, the **critical night length** (which occurs naturally if the period of light—the **critical daylength**—is short enough). Interrupting the dark period with a flash of light prevents the plants from changing to the flowering mode. That is why, if you own a Christmas cactus or poinsettia that you wish to flower during the holiday season, you must put the plant in a place kept dark from sunset to sunrise. It usually takes several days of such treatment to make the plant flower. ■

Later experiments showed that some plants are **long-day plants,** requiring a certain minimum length of light period (or less than a certain maximum length of dark) for flowering (Figure 34-8). Examples are spinach, radish, barley, and black-eyed Susan. (In some long-day plants, treatment with gibberellins can substitute for the proper light/dark regimen.)

There are also many **day-neutral** plants, such as tomatoes and cucumbers, which begin to flower at a certain stage of growth regardless of daylength.

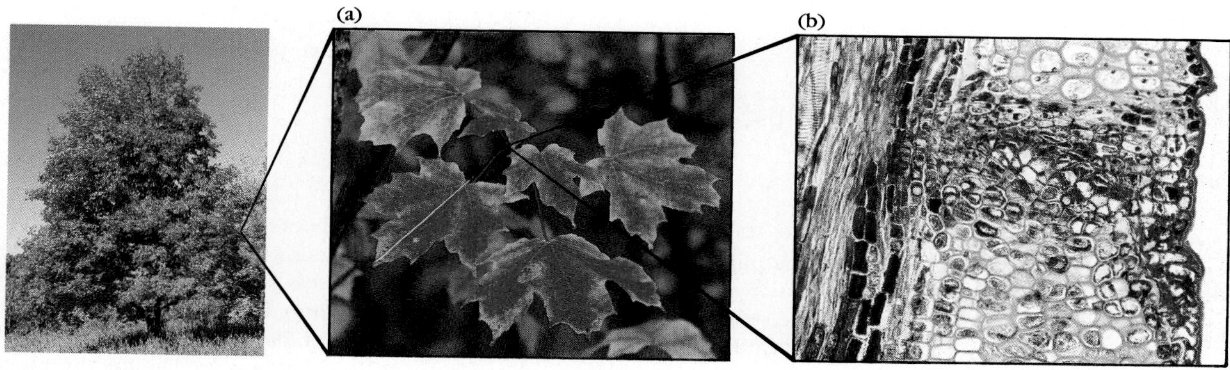

FIGURE 34–9

Leaf senescence and abscission. Senescence occurs regularly each autumn in sugar maple trees. **(a)** Chlorophyll is destroyed, revealing pigments of other colors, and the nutrients of chlorophyll are drawn back into the plant. **(b)** A leaf abscission zone (the horizontal purple band of cells). The attachments of the leaves weaken as the abscission zones form. The leaves eventually separate from the stems at these areas of weakness and the leaves fall to the ground. (Photo at far left, S. Osolinski/Dembinsky Photo Assoc.; a, S. Krasemann/Photo Researchers, Inc.; b, K. Wagner/Phototake, NYC)

Many plants require other factors in addition to the proper daylength: stage of maturity, temperature, day/night temperature changes, moisture, or a sequence of short days after long days or vice versa. The florist industry profits from the study of these factors. We can now obtain chrysanthemum plants in full bloom for Mother's Day and send carnations to our Valentines.

Various plants flower in response to one or more different cues, including stage of maturity or environmental factors such as the length of daylight or darkness, temperature, and moisture.

34–E SENESCENCE

Flowering, setting of seeds and fruit, and germination of seeds are events in a plant's life that occur predictably in response to environmental or genetic factors. **Senescence,** the process of aging that makes all or part of the plant more susceptible to death, is also an integral part of its life history. A wheat plant turns yellow, dries up, and dies after it has set seed; a plum tree drops its fruits during a short period in early summer and loses all its leaves during the fall.

Senescence is under hormonal control, probably by ethylene. In annual plants, the hormonal changes that initiate senescence of the whole plant often seem to be triggered by the setting of seed. In perennial plants, the leaves, fruits, and withered flower parts senesce and drop off each year, a process called **abscission.** Although its name suggests a link with abscisic acid, abscission is promoted mainly by ethylene. (Auxin and cytokinins inhibit senescence and abscission.)

Before abscission, several changes take place. Enzymes that degrade the leaf tissue become active, and nutrients may be withdrawn from the leaf. Cell divisions occur in the **abscission zone,** where the leaf joins the stem. In this region, ethylene stimulates the production and release of the enzyme cellulase, which breaks down the cell walls, thereby creating a zone of weakness (Figure 34–9). Finally, this zone breaks apart, and the leaf drifts away. The leaf scar is quickly healed by the deposition of a corky, waterproof seal that prevents the loss of water or entry of pathogens (disease-causing organisms).

Plant senescence and abscission are under hormonal control and appear to be a normal part of a plant's life cycle.

BIO-BIT

The seeds of many plants in the bean and pea (legume) family have tough, waterproof coats that allow them to germinate after many decades of dormancy.

A Journey into The Environment

Tropical Rain Forest Canopy—Where the Action Is

It is almost impossible for a visitor not to think of the canopy of a tropical rain forest as the high ceiling of a green cathedral. The trunks of huge trees seem like flying buttresses that tower high overhead, supporting the green, leafy cover that intercepts the tropical sun and creates the twilight gloom of the forest floor. The trees dwarf humans, and unless you can climb or fly, the canopy, often as high as a 12-story building (30 to 50 meters), is unreachable. It wasn't until biologists began climbing up into the canopy of tropical rain forests that they began to appreciate the mind-boggling diversity of tropical rain forest species and the complexity of the interactions between them. The image of the tropical rain forest canopy that is now emerging from these research efforts is far different from that of a cathedral ceiling—we now know that the canopy is more like Las Vegas—because in a tropical rain forest, the canopy is where the action is.

Like casino owners, the trees are the biggest players in the tropical rain forest. Using the abundant rainfall of the tropics as capital (Earth's tropical belt, where rain forests grow, receives at least 200 centimeters [more than 78 inches] of rain per year), rain forest trees create the conditions that all other rain forest species have become adapted to through the process of natural selection. Most rain forest trees are very tall, with slender trunks and small crowns. A few trees, called **emergents,** spread their branches *above* the other canopy species. Emergent trees can have gigantic, spreading crowns—more than 20 meters (65 feet) is not uncommon.

Plants must reach the light so that photosynthesis can occur. Canopy trees spread their crowns in the sunlight, allowing little light to reach the forest floor, where it is usually so dim that you need a flashlight to see things clearly. Few plants can photosynthesize in this gloom; thus, beneath the intact canopy, the floor of a tropical rain forest is usually clear of vegetation. **Lianas,** thick, woody vines, take a shortcut to the sun. Instead of investing energy in building strong, supportive woody trunks, lianas use the support of canopy trees to climb toward the light, often spreading luxuriantly over portions of canopy, monopolizing the sunlight. **Epiphytes** are ferns, orchids, mosses, lichens, or other herbaceous (soft-tissued) plants that take an even more direct route to the sun and sprout in the canopy, growing on tree limbs and in crevices of bark. Epiphytes have aerial roots that absorb moisture from the surroundings. Like desert plants, their tissues are often specialized for water storage and drought resistance. Many epiphytes, especially bromeliads (relatives of pineapples), have a hollow, internal cavity that collects water. It used to be thought that epiphytes were parasitic, deriving nutrients from their host trees. However, once rain forests were studied from within the canopy, using equipment and techniques adapted from mountain climbing, we learned that the reverse is often true: some host trees send rootlets *into* epiphytes, which act as arboreal nutrient traps. The water tanks of epiphytes are the equivalent of springs and pools on the forest floor. They provide shelter for a bewildering variety of aquatic life, including tadpoles, frogs, worms, snails, protozoans, insect larvae, and even crabs!

Although climate and plants create the rain forest, animals play major roles as well. Some species of ants and bees act like casino bouncers or grounds keepers, repelling unwanted invaders or competitors (see *A Journey into The Environment:* Coevolution of Plants and Herbivores, Chapter 8). Other species, like the three-toed sloth that spends much of its life in a few trees, feeding on leaves, act more like caretakers (Figure 34-A). Instead of releasing its dung as it feeds, the sloth climbs down from the branches (the only times that it voluntarily leaves its arboreal perch) to dig a hole and bury its dung at the base of its tree. This behavior is risky, because on the ground the sloth is exposed to predators such as jaguars, but it may have beneficial effects for both the sloth and its home tree. Dung-burying prevents the sloth from advertising its presence with droppings, and the tree receives nutrients, which are in short supply in the barren soils beneath a tropical rain forest.

Birds and insects are the costumed entertainers who rely on flash-and-dazzle to advertise their presence, attract mates, and, in some cases, protect them from predators. Heliconian butterflies are some of the most conspicuous species. They are usually black with yellow, red, or orange markings, and their leisurely, fluttering flight demonstrates the protection offered them by the toxins in their tissues derived from the plants eaten by their larvae, or by their resemblance to toxic species. Morpho butterflies are the masters of flash and dazzle (Figure 34-B). At rest, a morpho holds its wings closed, showing the mottled, chocolate and gray surface that mimics a dead leaf. When it flies, though, a very different-looking creature suddenly materializes with iridescent, neon blue wings that glint indigo and purple in the sunlight. Morphos also use the tactic of "now you see me—now you don't." They have the habit of darting into the shadows, closing their glittery wings,

FIGURE 34–A

A three-toed sloth from Brazil. (K. Schafer)

FIGURE 34–B

A morpho butterfly from Peru. (K. B. Sandved/Visuals Unlimited)

and hanging head-downward when pursued. Another variation of flash-and-dazzle creatures are the many butterflies, moths, grasshoppers, mantids, and cockroaches that have eyespots on their hind wings. Suddenly unfurling these eyespots can startle a predator, allowing the eyespotted prey to escape. Clear-winged butterflies are another canopy species that use flash-and-dazzle, but of a kind that human eyes cannot appreciate. Instead, the cellophane wings of these butterflies reflect ultraviolet radiation. Blind to the colors seen by insects, we can only surmise what clear wings may be advertising to those that can see them. Perhaps they are signalling that they, like the heliconians, are bad-tasting or poisonous, while being invisible to other predators.

Appearances are often deceiving in Las Vegas as well as in the tropical rain forest canopy. A patch of lichen-covered bark is actually a camouflaged eyelash viper; a disturbed caterpillar inflates its body to mimic a venomous snake; an oddly moving twig metamorphoses into a caterpillar; the skinny coils of an arboreal, rear-fanged snake masquerade as a loosely curled vine; lizards blush green and blue to disappear against leaves.

The canopy of tropical rain forests is biologically where the action is because it is home to more species than any other habitat. Tropical rain forests occupy just 6% of our planet's surface, but experts agree that they are probably home to *more than half of the species* on Earth. No one knows exactly how many species live there, because exploration and taxonomic inventory of this fascinating habitat have only begun. Unfortunately, humans are gambling with this largely unknown biological legacy, carving it into smaller and smaller portions, as they cut down the trees, mine for gold, drill for oil. Every day, populations of species are diminished and some species are forced into extinction. If current trends continue, it is a safe bet that the world will be a much poorer place.

But the outlook is not entirely bleak. Some tropical countries are beginning to realize the value of their intact and unspoiled rain forests. Reserves have been set aside, and in many countries the "ecotourism" industry is beginning to grow. Once developed, ecotourism promises to be a lucrative source of much-needed revenue for countries that are home to tropical rain forests. One can only hope that attitudes fostering conservation will replace exploitation of these fascinating and unique natural casinos.

SUMMARY

1. Five groups of major plant hormones are known. Auxins, gibberellins, and cytokinins are generally growth promoting; abscisic acid enables the plant to cope with stress; and ethylene promotes ripening of fruits and senescence of leaves.
2. Probably no single effect can be attributed to just one hormone; rather, the interactions of these hormones govern a plant's growth. Hence, different parts, such as the leaves and roots, remain in anatomical and physiological balance.
3. Hormones are also involved in a plant's response to its environment. They enable the plant to respond appropriately to the direction of light, gravity, or prevailing winds, and to changes in daylength and temperature that signal changes in seasons.
4. The timing of flowering, senescence, and abscission is also controlled by plant hormones.
5. How a plant responds to a particular hormone depends on the tissue that receives the hormone, the concentration of the hormone and of other hormones also present, the age and physiological state of the tissue, and environmental factors such as temperature, light, and photoperiod.

SELF-QUIZ

Match the hormones listed below to their effects.

___ 1.	Promotes ripening of fruits	a. abscisic acid
___ 2.	Initiates cell division in tissue culture	b. auxin
		c. cytokinin
___ 3.	High concentrations stimulate ethylene production	d. ethylene
		e. gibberellin
___ 4.	Substitutes for cold period in the flowering of some plants	
___ 5.	Responsible for gravitropic response in roots	
___ 6.	Counteracts the effects of auxin	
___ 7.	Promotes onset of dormancy	

8. In phototropism, auxin:
   a. promotes growth of cells
   b. stimulates differential growth of cells in different sides of the plant
   c. inhibits growth of cells
   d. inhibits cell division
   e. absorbs stimuli and signals the direction of light or gravity to the plant
9. The flowering of certain plants only under "short-day" conditions is an example of:
   a. apical dominance
   b. positive phototropism
   c. negative phototropism
   d. photoperiodism
10. A long-day plant is one that:
    a. requires more than 12 hours of light in order to flower
    b. increases in height when it flowers
    c. needs a certain minimum length of photoperiod in order to flower
    d. is not affected by temperature in its flowering response
    e. will not flower if its dark period is interrupted by a flash of light

THINKING CRITICALLY

1. Propose a mechanism by which hydrotropism might work (see Section 34-C).
2. The fungus that causes the "foolish seedling" disease in rice appears not to need the gibberellin it produces for its own growth. Why might it be selectively advantageous for the fungus to secrete this substance?
3. Why are apples, oranges, and grapefruit sold in plastic bags with holes in them rather than in unperforated bags? Why does it often turn out that produce packaged in market trays with clear wrap is too soft on the underside, which you could not see when you picked it out in the store?
4. Many plants of the forest floor produce seeds that germinate only in the early spring, before the canopy leafs out. Propose an explanation for the timing of this germination, and explain its adaptive advantage.
5. In some species of trees, individuals growing near streetlights become dormant later in the fall than do other individuals. How could you account for this?
6. Exposure to a flash of light during the dark period can change the plant's subsequent flowering response (or lack of it). However, interrupting the light period with an interval of dark has no effect. How can you explain this?
7. Some plants use the end of a cold period as their cue that spring has come and it is time to flower, whereas others use increasing daylength. Which is a more reliable predictor of favorable conditions for flowering and seed production? Why do different plants use different cues?
8. Oligosaccharins originate from the cell wall rather than from structures inside the cell. What is the adaptive value to the plant of having regulatory molecules with such a source?

SELECTED KEY TERMS

abscisic acid, *p. 715*
abscission, *p. 721*
apical dominance, *p. 716*
auxin, *p. 712*
coleoptile, *p. 718*
cytokinin, *p. 714*
ethylene, *p. 715*
gibberellins, *p. 713*
gravitropism, *p. 718*
hydrotropism, *p. 718*
long-day plants, *p. 720*
photoperiodism, *p. 720*
phototropism, *p. 718*
senescence, *p. 721*
short-day plants, *p. 720*
thigmotropism, *p. 718*

SUGGESTED READINGS

Evans, M. L., R. Moore, and K. H. Hasenstein. "How roots respond to gravity." *Scientific American,* December 1986.

Galston, A. W., P. J. Davies, and R. L. Satter. *The Life of the Green Plant,* 3d ed. Englewood Cliffs, NJ: Prentice-Hall, 1980. A clear, well-illustrated plant physiology text.

CHAPTER 35

CONCEPT GUIDE

After reading this chapter, you should be able to:

1. Describe the four kinds of flower parts and indicate the general function of each. (Section 35-A)
2. Compare the processes of self-pollination and cross-pollination. (Section 35-B)
3. Describe the processes of male and female gametophyte maturation, gamete production, and the process of fertilization. (Sections 35-B and 35-C)
4. List the three main parts of a seed, the general functions of each, and how each part was formed. (Section 35-D)
5. Compare the mechanisms of seed dispersal by wind and by animals. Describe adaptations of seeds to each dispersal mechanism. (Section 35-E)
6. List four conditions that may affect seed germination. (Section 35-F)
7. Using the examples of corn and apples, explain why new corn hybrids are easier to develop and test than new apple hybrids. (Section 35-G)
8. Describe the various methods by which plants reproduce asexually and the circumstances under which asexual reproduction would be most advantageous to a plant. (Section 35-H)

The flower of the bee orchid is pollinated by male bees that are attracted to its scent. The male's legs brush the flower's odor-producing surfaces and store the scent. Meanwhile, the orchid releases a pollen packet, which the bee then carries to the next orchid he visits. This unique plant/pollinator relationship ensures that the plant will have its pollen dispersed only to other plants of the same species, while the male bees acquire a scent that makes them more attractive to females. Read about other mutually beneficial plant/animal relationships in Sections 35-B and 35-E. (J. A. L. Cooke/Oxford Scientific Films)

Reproduction in Flowering Plants

Plants acquire resources from their environment as they grow (Chapters 32 and 33). Ultimately, many of these resources are channelled into the plant's reproduction.

Two kinds of reproduction are found among flowering plants. **Vegetative reproduction** is an extension of the kinds of growth we saw in Chapter 32. It gives rise to new individuals with genetic makeup identical to the parent's, and thus perpetuates gene combinations that are well adapted to the local environment. Individuals with the favorable genetic combination quickly spread through the area where the parent plant is growing.

Sexual reproduction involves more complex events: production and growth of a group of structures making up the flower, production and fertilization of gametes, and development of the embryo, seed, and fruit. Sexual reproduction has two main advantages. First, it forms new genetic combinations in each generation. It also produces seeds, which can disperse over a wide area, and which are protected against adverse environmental conditions that might kill the parent plant.

In Chapter 20, we studied the basic sexual plant life history, in which the diploid (2N) (where 1N is the number of chromosomes in a haploid nucleus) sporophyte generation alternates with the haploid (1N) gametophyte generation. We can diagram the life history for flowering plants as shown below.

In flowering plants, the sporophyte generation is dominant and the gametophyte very much reduced. The familiar flowering plants of garden or forest are sporophytes. The male gametophyte consists of a pollen grain and the tube that grows from it, and the female gametophyte is hidden within the female flower parts.

Modern human society depends on our ability to grow plants for food and for many other uses. In this we employ both vegetative and sexual reproduction. Plant breeders manipulate the sexual reproduction of economically important plants in order to produce individuals with more desirable combinations of genetic features. Once such a set of features is achieved, vegetative propagation can be used to increase the number of plants available to farmers and gardeners.

KEY CONCEPTS

- Vegetative reproduction perpetuates combinations of genes suited to the local environment and allows plants with these combinations to spread widely within this local area.
- Sexual reproduction in flowering plants has two important outcomes.
  1. It produces new genetic combinations, the raw material for evolution by natural selection.
  2. It produces seeds, units that disperse offspring to new areas some distance from the parent plant.
- Both kinds of reproduction are important in human attempts to improve the strains of plants that we grow for our own use.

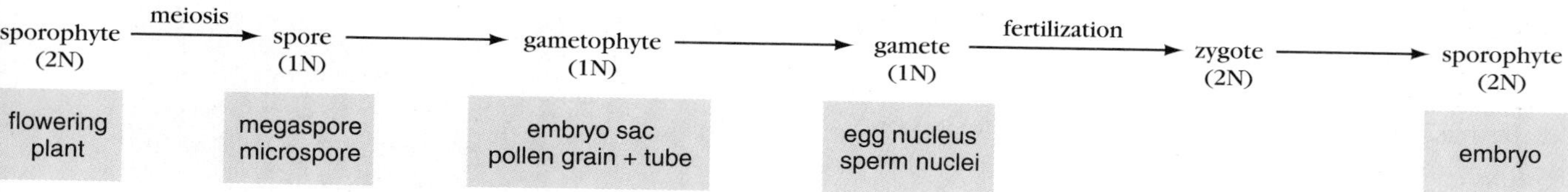

(You may wish to refer back to this diagram as you read.)

Curiosity Questions

1? Do squirrels find all the acorns that they bury? (See page 736)

2? What causes a seed to sprout or germinate? (See page 737)

3? How is it possible to have a tree that produces two different varieties of apples or pears? (See page 739)

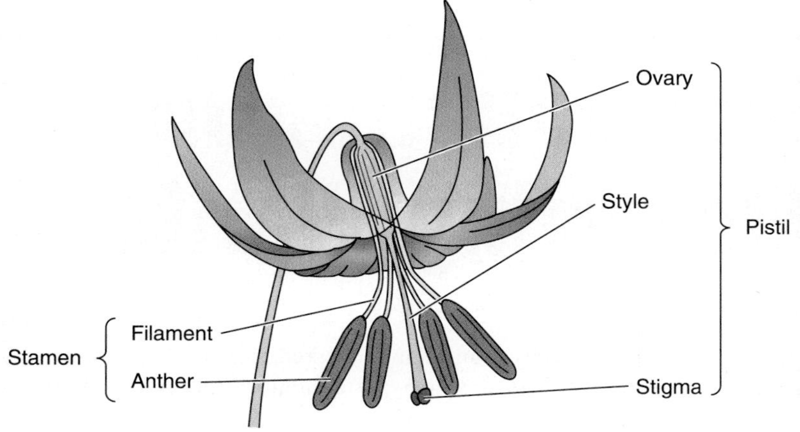

FIGURE 35-1

Anatomy of a flower. (S. Moody/Dembinsky Photo Assoc.)

35-A FLOWERS

Flowers are sexual reproductive structures, produced in response to hormonal changes in the plant. Some plants flower at a certain age, whereas others are induced to flower by certain environmental stimuli (Section 34-D). The new hormone balance makes a meristem develop into an abbreviated shoot with a cluster of many highly modified leaves—the flower parts.

A typical flower has four types of modified leaves. Beginning at the outside, the first flower parts are the **sepals,** which are often green. The sepals develop first and protect the other parts maturing inside the flower bud. Just inside the sepals are the **petals,** which are often large and showy, with bright colors and patterns that attract animal pollinators. Next come the **stamens,** the male, or pollen-producing, parts. These bear **anthers,** chambers where pollen grains develop. In the center of the flower are one or more **pistils,** structures containing the flower's female parts (Figure 35-1). At the tip of the pistil is the sticky **stigma,** which traps pollen grains. Next is the long **style,** and at the base of the pistil lies the **ovary,** enclosing one or more **ovules** (Figure 35-1).

Many variations on these typical flower parts occur among the 275,000 known species of flowering plants. For example, in lilies the three sepals look almost exactly like the three petals (Figure 35-1). In wind-pollinated plants such as grasses, the sepals and petals are often tiny or absent, allowing greater exposure of the stamens as they shed pollen, and of the stigmas as they receive it (Figure 35-2). Some plants produce separate male and female flowers on the same

FIGURE 35-2

Wind pollination. The hazel bush bears separate clusters of female and male flowers. The pendulous male catkins swing in the breeze and shed pollen grains which are intercepted by the tiny pink stigmas of female flowers. Nonreproductive parts of the flowers are small, allowing the stigmas and anthers free access to the air. (R. Maier/Earth Scenes)

a)

(b)

FIGURE 35-3

Non-flower "flowers." (a) Showy, red, leaf-like structures surround the inconspicuous whitish-yellow flowers of bougainvillea. **(b)** The white "petals" of a dogwood are also leaf-like structures surrounding a cluster of inconspicuous flowers. (a, A. Gloor/Earth Scenes; b, E. R. Degginger/Earth Scenes)

lant (for example, corn and members of the squash amily, including cucumbers and pumpkins). Still oth- rs have separate "male" and "female" plants, as in pinach, willows, and some hollies. There are also nany plants in which structures near the flowers act s parts of the "flower." The "petals" of poinsettias, logwood, and bougainvillea, for example, are really nodified leaves (technically termed **bracts**) around lusters of small, inconspicuous flowers (Figure 35-3).

lowers typically have four kinds of flower parts: sepals, etals, male stamens, and female pistils.

5–B POLLEN, POLLINATION, AND OVULE REPARATION

n plants meiosis gives rise to haploid cells called **pores,** rather than to gametes, which are the prod- cts of meiosis in animals. The haploid spores grow nto haploid **gametophytes,** which in turn produce he gametes that take part in fertilization. Flowering lants produce spores of two sizes, microspores and negaspores, which give rise to male and female game- ophytes, respectively.

Microspores are produced in the chambers of the nthers and begin to develop into male gametophytes see A Journey Through Reproduction of Flowering lants). A **pollen grain** is an immature male gameto- hyte, enclosed in a protective wall. Before it can com- lete its development, the male gametophyte must be eposited on the stigma of a flower.

Just as leaf and flower structures vary among lants, so too do the shape and pattern of the pollen grain wall. In fact, experts can easily place a particular pollen grain into the proper genus (and sometimes species) by its distinctive wall pattern (Figure 35-4). Pollen grains may last for millions of years, preserved

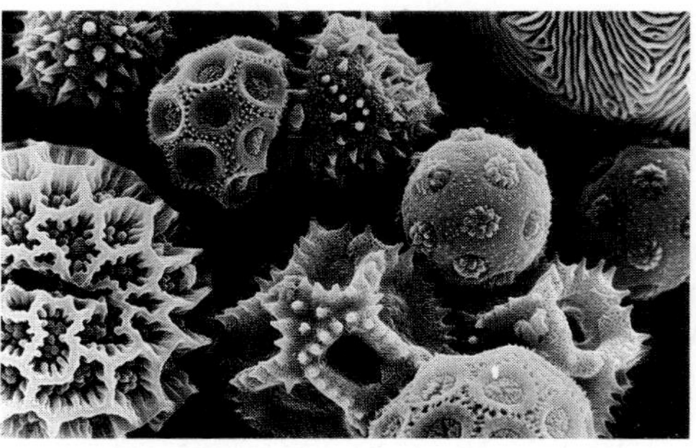

(a)

(b)

FIGURE 35-4

Pollen grains. (a) A variety of plant pollens from phlox, chrysanthemum, smartweed, rooster comb, and geranium plants. **(b)** Ragweed pollen. (a, M. Cooper/Peter Arnold, Inc.; D. Scharf/Peter Arnold, Inc.)

in rock formations or peat deposits. The history of the vegetation in an area can be traced by examining this fossil pollen.

Pollination

Pollination is the transfer of pollen from the (male) anther, where it forms, to the stigma of the (female) pistil. Pollen may simply fall from the anther onto the stigma of the same flower, resulting in **self-pollination.** Some flowers, such as peas and their relatives (see Figure 12–1), are so constructed that their stamens and pistils are completely enclosed within the petals, resulting in a high percentage of self-pollination.

Cross-pollination, the transfer of pollen to another individual of the same species, results in more genetic variety. This is often an evolutionary advantage, and many plants have adaptations that ensure cross-pollination. For example, a flower's pistils may mature only after its anthers have shed their pollen. The existence of separate male and female plants or flowers is probably due to selective pressure for cross-pollination.

Pollen cannot move far on its own power; plants usually rely on wind or animals as agents of pollination. From a plant's point of view, pollination by animals may have advantages over pollination by wind. First, wind pollination wastes a lot of the energy invested in pollen production because much of the pollen never reaches another flower. Second, wind pollination is very inefficient for a plant that does not live in dense populations. If the nearest neighbor of the same species is far away, there is a good chance that no pollen will reach its stigmas. By contrast, an animal that visits only one kind of plant carries pollen directly from one individual to another of the same species. Many flowers have evolved structures such that only one species of animal can pollinate them, and these flowers enjoy highly specific transfer of pollen from one individual to another.

Animal pollinators are attracted by some type of reward, usually a sweet nectar. The reward is made easier for the animal to find by an attention-catching "advertisement," such as the odor, shape, or color of a flower—preferably all three. The reward or apparent reward (plants often "trick" animals into transporting pollen without giving them a reward) is so located that the animal cannot reach it without picking up pollen at the same time. All of this has a cost: the animal-pollinated flower must invest energy in making its nectar and its large, showy petals, even though it need not make the prodigious amounts of pollen required for wind pollination (see *A Journey into Evolution:* Coevolution of Flowers and Their Pollinators, this chapter).

Animals that serve as pollinators include insects—bees, butterflies, moths, wasps, flies, and beetles—and vertebrates such as birds, bats, and even a South African mouse!

Pollen Maturation

A pollen grain completes development into a mature male gametophyte after it has landed on the stigma. The protective coat of the pollen grain ruptures and a **pollen tube** grows out. The pollen grain wall contains glycoproteins that must be compatible with proteins in the stigma if the pollen is to grow. Hence pollen will usually not germinate on the stigma of flowers of a different species. Some species of plants also prevent self-fertilization, by means of a system of compatibility genes: pollen cannot complete its development when it contains the same genetic alleles as the stigma. This assures that the flower is cross-fertilized and so maintains genetic diversity in the population.

If the pollen and stigma are compatible, the pollen tube grows down the style toward the ovule(s) in the ovary at the base of the pistil. Many pollen grains may land and produce pollen tubes in the same style. The pollen tube grows into the ovule through a tiny pore, the **micropyle,** and releases two **sperm nuclei.** The micropyle then closes, preventing the entry of any more pollen tubes. If there are more ovules, other pollen tubes enter through their micropyles. The traffic problem that must exist in the extremely slender style of a cantaloupe flower as hundreds of pollen tubes grow toward the ovules is fearful to contemplate!

Preparation of the Ovule

Before the pollen tube arrives, a megaspore has formed inside the ovule and developed into an **em-**

BIO-BIT

Paleontologists often study fossilized pollen grains found near animal fossils to learn about the types of plants as well as the climate in which the animals lived.

BIO-BIT

Many plants are pollinated by only a single animal species. Thus, the extinction of that animal species would likely result in the extinction of the plant.

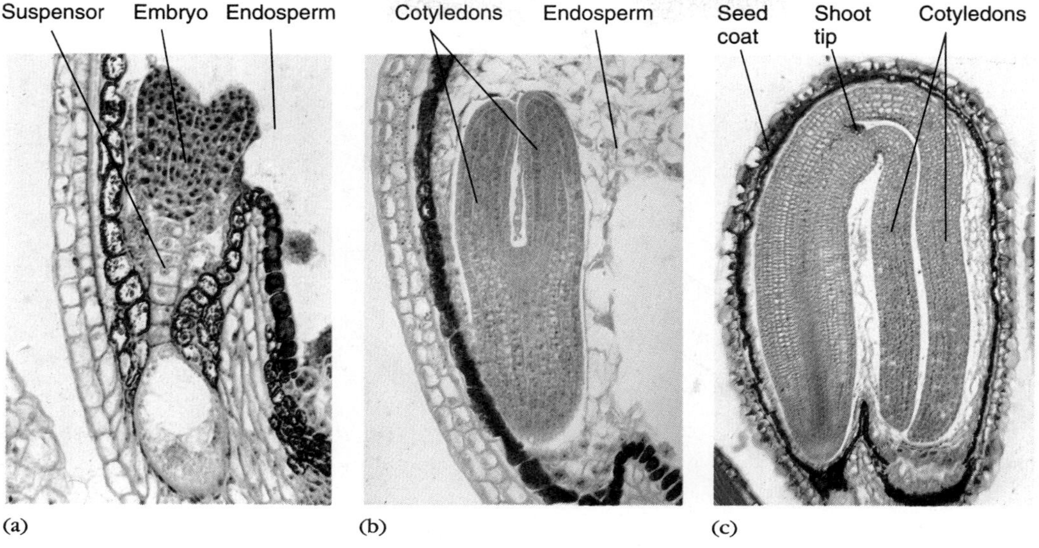

FIGURE 35–5

Early development of the embryo. (a) A series of cell divisions forms a string of cells (the suspensor) that pushes the embryo into the nutritive endosperm. The end cell divides to form an embryo, which grows by cell division and absorption of food from the endosperm. As the embryo grows, its shape changes from "heart" to "torpedo." **(b)** In the torpedo stage the cotyledons begin to form and the end of the embryo near the suspensor develops into the root. **(c)** As the embryo grows larger, the cotyledons fold over so that their ends lie beside the root. The shoot tip is between the cotyledons and remaining patches of endosperm lie around the embryo. The outside wall of the ovule develops into the seed coat. All of these micrographs are of shepherd's purse (*Capsella bursa-pastoris*). (a, Calisco/Visuals Unlimited; b, J. R. Waaland/Biological Photo Service; c. J. M. Bostrack/Visuals Unlimited)

ryo sac, a mature female gametophyte. The embryo sac usually contains eight haploid nuclei: three at the end near the micropyle, one of which is the egg nucleus; three at the opposite end; and two **polar nuclei** in the center. This female gametophyte is now ready to be fertilized.

Microspores produced in the chambers of anthers develop into pollen grains with a cell wall structure characteristic of the genus. Plants that do not self-pollinate typically rely upon external agents such as animal pollinators or wind to transport pollen to other plants of the same species. Once pollen is on the stigma, it grows a pollen tube. The pollen tube eventually releases two sperm nuclei to the female gametophyte that has developed within the ovule.

35–C FERTILIZATION

Fertilization may take place as little as an hour after pollination, as in barley, or as much as several months later, as in witch hazel. The flowers of this shrub appear and are pollinated in late fall but fertilization does not occur until spring, when the sperm nuclei arrive at the micropyle.

During the sexual reproduction of flowering plants, both sperm nuclei take part in fertilization. In this **double fertilization,** one haploid sperm nucleus fertilizes the haploid egg nucleus, forming the diploid zygote. The other sperm nucleus fuses with the two polar nuclei, forming an **endosperm** nucleus that is triploid (3N, where 1N is the number of chromosomes in a haploid nucleus). The adaptive value of this second fertilization is unclear, although the endosperm tissue that arises from this nucleus has a very important role, as we shall see shortly.

In flowering plants, double fertilization produces a diploid zygote nucleus and a triploid endosperm nucleus.

35–D DEVELOPMENT OF THE SEED AND FRUIT

In the next stage of development, the zygote develops into an embryonic plant, and the parent plant supplies it with nutrients that will help it to establish itself as an independent individual. In addition, the wall of the ovule develops into a protective **seed coat,** and the wall of the ovary develops into a **fruit.**

Right after fertilization, the zygote enters a period of dormancy. Meanwhile, the endosperm nucleus divides many times and forms endosperm tissue, which enlarges and absorbs food from the parent plant. When the zygote breaks dormancy, the endosperm tissue supplies it with food.

The first structure formed by division of the zygote is a line of cells called the **suspensor.** The suspensor cells near the micropyle elongate and push the cells at the far end into the nutrient-rich endosperm. Soon the embryo starts to develop at this end of the suspensor (Figure 35–5). The mature plant embryo has

(text continues on page 734)

A Journey Through Reproduction of Flowering Plants

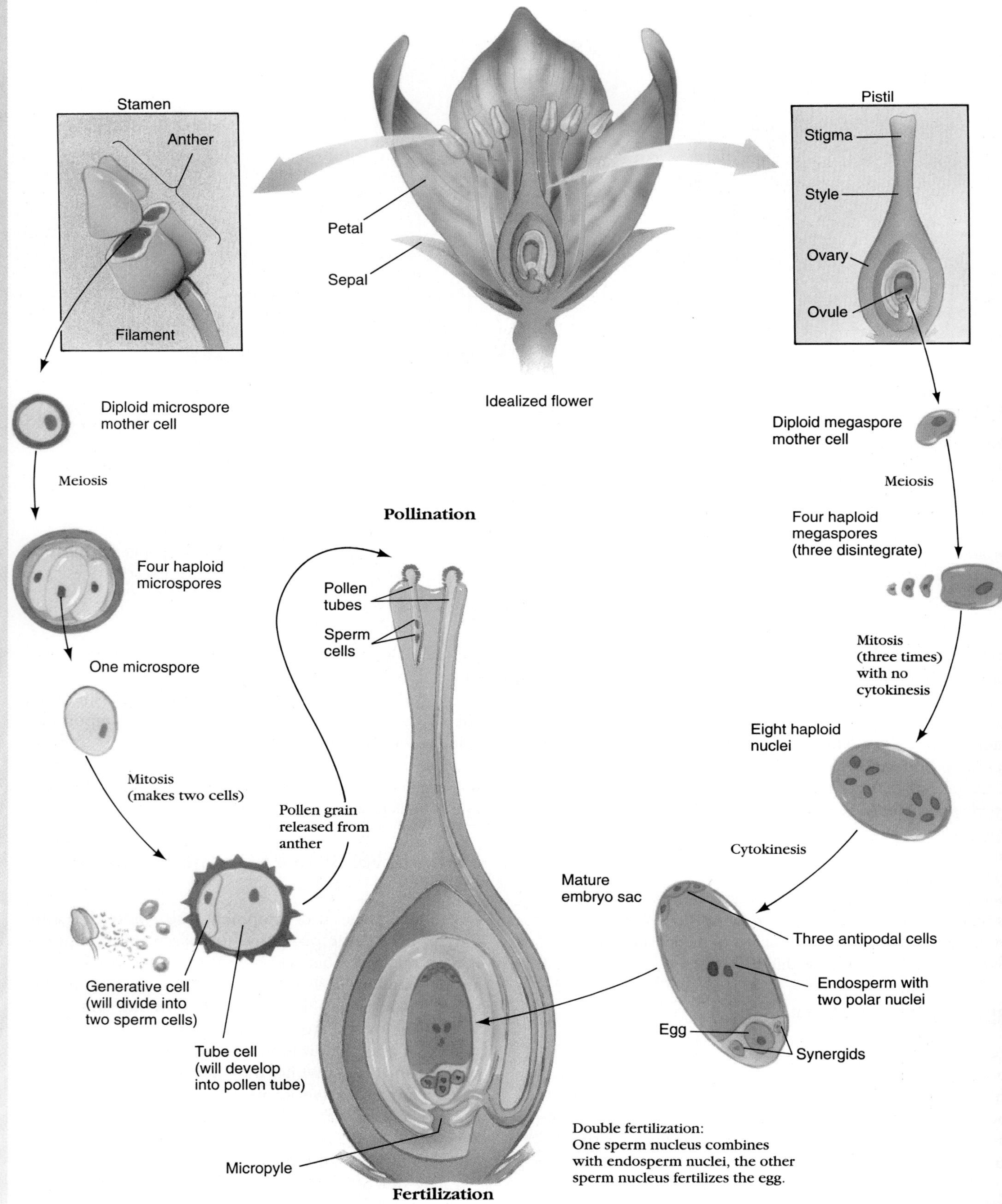

An ideal flower, such as an apple blossom, has both male and female sexual organs. After pollination deposits a grain of pollen onto the stigma, pollen tubes grow down the style toward the ovary, forming a tunnel that allows for the passage of two sperms' nuclei. These travel to the embryo sac within the ovary where one fertilizes the egg while the other combines with the endosperm nuclei. Endosperm will nourish the zygote as it develops into a seed within a mature fruit.

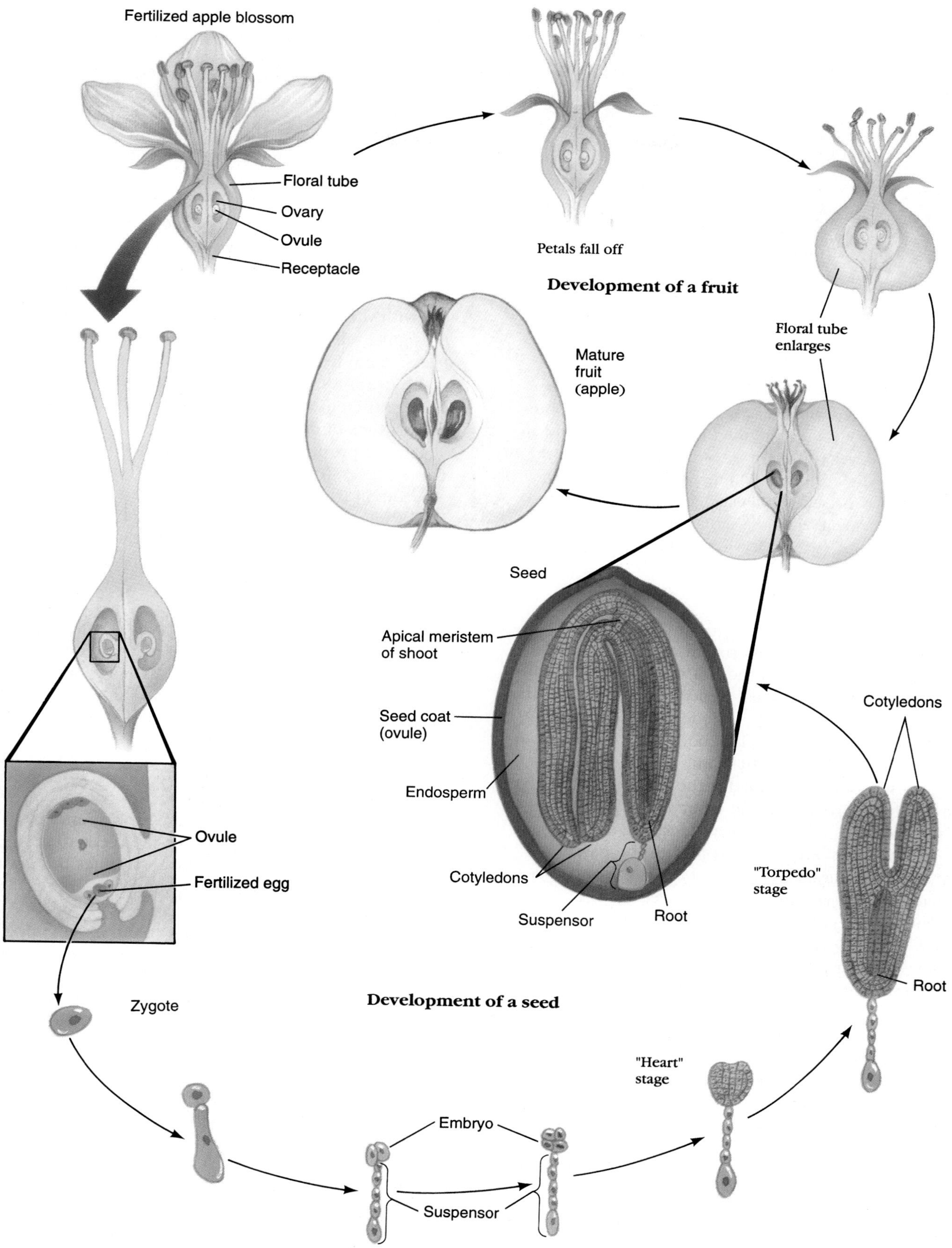
Fertilized apple blossom
Floral tube
Ovary
Ovule
Receptacle
Petals fall off
Development of a fruit
Floral tube enlarges
Mature fruit (apple)
Seed
Apical meristem of shoot
Seed coat (ovule)
Endosperm
Cotyledons
Suspensor
Root
Cotyledons
"Torpedo" stage
Root
Ovule
Fertilized egg
Zygote
Development of a seed
"Heart" stage
Embryo
Suspensor

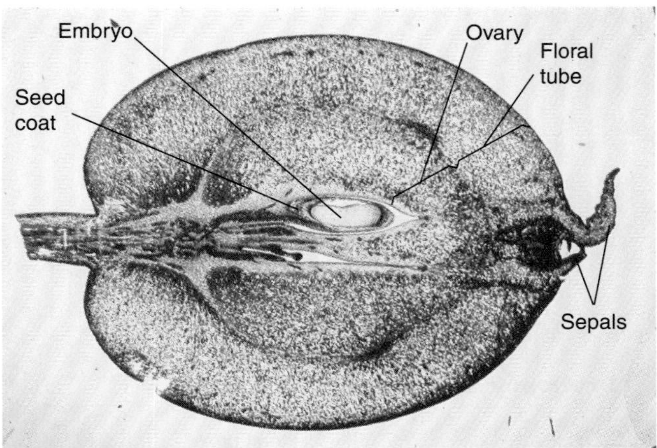

FIGURE 35-6

Developing fruit. The embryo develops inside the ovule, which itself develops into a seed coat. Meanwhile, the ovary wall enlarges greatly as part of the fruit. The outer part of an apple actually develops from parts of the floral tube which surround the base of the ovary in the apple blossom. Notice the remnants of the sepals still clinging at the "blossom end" of the apple. Depending on the variety, it takes the photosynthetic output of 15 to 35 leaves to support the growth of one apple. (Biophoto Associates)

a tiny axis with an apical meristem at each end (one for the primary root, the other for the shoot system) and one or two **cotyledons,** depending on whether it is a monocotyledon or a dicotyledon (see Table 32-1).

As the embryo grows, the endosperm continues to absorb food from the parent plant. The endosperm may persist as a food supply for the embryo or may be completely absorbed into the embryo's cotyledons as the seed matures (as in the bean seed, Figure 32-1). The wall of the ovule, which is part of the parent plant, becomes larger as the embryo grows, and usually hardens to form the protective seed coat.

Outside the seed coat, the wall of the ovary also enlarges and absorbs more nutrients to form a fruit (Figure 35-6). Fruit growth begins when the pollen tube releases tiny amounts of the hormones auxin and gibberellin (Chapter 34). Soon the developing seed begins to produce its own hormones, which continue to stimulate growth of the fruit. In addition, there is enhanced transport of cytokinins, hormones that stimulate cell division, from the parent plant. These three hormones promote fruit growth by both cell division and cell enlargement.

Other parts of the plant near the flower may also develop into fruit-like structures. For example, the outer part of an apple develops from the floral tube, which surrounds the base of the ovary in the apple blossom (Figure 35-6). A strawberry is really an enlarged **receptacle,** the area of the flower stalk that holds all of the flower parts. Its "seeds" are technically the *fruits* of the strawberry plant, each of them having arisen from a separate pistil of the flower.

BIO-BIT

Aggregate fruits, such as raspberries, develop from a flower containing multiple ovaries.

Most types of plants do not set seed and develop fruits unless their flowers have been pollinated and fertilized. In some species, however, spraying the flowers with the proper concentration of auxin or auxin plus gibberellin induces the production of seedless fruits. Some fruits, such as cultivated strains of bananas and pineapples, develop naturally without fertilization and are therefore seedless. Humans have increased the populations of these economically desirable plants by vegetative propagation, and the plants have thus become successful through artificial selection.

Not all fruits are fleshy and delicious. Many are rather minimal protective layers. When the seeds are ripe, the fruit may rupture along its lines of weakness and release the seeds. Pods of peas, beans, and milkweeds are examples of such fruits.

A seed consists of (1) an embryonic plant, developed from the zygote, (2) a food supply, and (3) a protective seed coat, developed from the ovule wall. As the embryo develops within the parent plant, the wall of the ovary outside of the seed coat enlarges to form a fruit.

35–E DISPERSAL OF SEEDS AND FRUIT

Once mature, the seeds are ready for dispersal. In most cases it is advantageous for the seeds to grow far away from the parent. First, this distributes the population of the plant's descendants over a wider area. Second, it avoids competition for light, water, and soil minerals between the parent plant and its offspring.

The usual dispersal agents are wind and animals. Small, lightweight milkweed seeds and dandelion fruits have parachute-like tufts of fiber that enable them to disperse by floating through the air (Figure 35-7). The thin, flat wings of maple fruits whirl in the breeze like the blades of a helicopter. Larger seeds will have a distinct advantage when they start growing because they contain more food for the embryo, but wind cannot waft a large seed far from the parent plant. An animal (usually a bird or mammal), on the other hand, is strong enough to carry even a large seed quite a distance.

Many plants invest a lot of energy in adaptations that promote dispersal by animals. Most commonly,

a)

(b)

FIGURE 35-7

Vind dispersal. (a) Milkweed seeds float off on parachute-like tufts of fibers after the wall of he fruit dries and splits open. **(b)** The "wings" of maple fruits, reminiscent of aircraft propellers, re enormous outgrowths of the ovary wall. The seeds are enclosed in the green swellings at heir bases. (a, Runk/Schoenberger from Grant Heilman; b, Z. Leszczynski/Earth Scenes)

he seeds are protected in indigestible seed coats and urrounded with a tasty, nutritious fruit that an animal vill eat. The seeds then pass unharmed through the nimal and are deposited with a small pile of organic ertilizer. In fact, the seeds of some plants usually will ot germinate unless their seed coats have been roded somewhat by an animal's digestive enzymes.

Fruits are usually protected from being eaten be-ore the seeds have matured. Unripe fruit is often dis-asteful and may even contain toxic chemicals. When he fruit is ripe, its chemical composition changes so hat it tastes good. The fruit may also change color, a isual signal that it is now ready to be eaten (Figure 5-8). Thus, enlarged fruit is often a plant's "ticket" or seed dispersal.

Some fruits or seeds have hook-like extensions that ttach to the feathers, fur, or clothing of passing ani-nals or people, giving the seed a free ride to a new ome (Figure 35-9). These hitchhikers gain the use of he animal's mobility for a small energy investment in he production of hooks.

FIGURE 35-8

Ripening fruits. These raspberries change from green to red to deep blue by way of several intermediate shades, a signal that they are ready to be eaten. (J. Colwell from Grant Heilman)

eeds and Seed Predators

ecause seeds contain the food supply for an embry-nic plant, they also make good food for animals. Hence there is strong selection for adaptations that rotect the seeds from predation (Figure 35-10). The ypes of adaptations that are effective depend on the nain types of predators eating the seeds. Some seed redators gobble up every seed they find. Plants with his type of predator usually produce many small seeds. This gives a good chance for some of the seeds to escape notice, so not all are eaten.

Other seed predators maximize their food intake for the energy they spend. They may attack plants that have the most seeds in a fruit, or the seeds that are largest or easiest to chew. Plants attacked by such predators usually enclose their seeds in a hard cover-ing, such as a nutshell, that discourages the predator

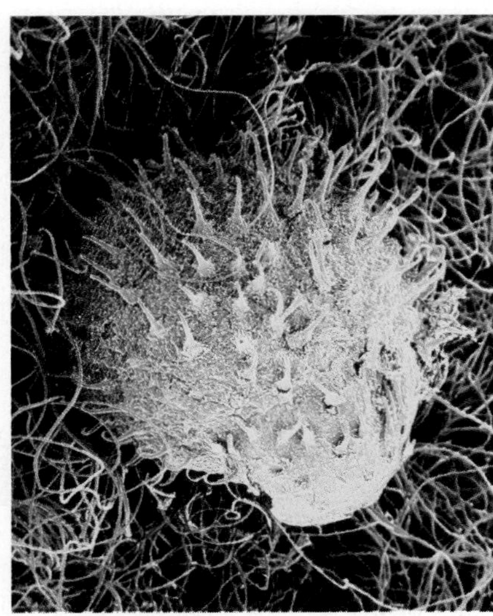

FIGURE 35-9

Hitching a ride. This false color scanning electron micrograph shows a goosegrass fruit, or burr, entangled in Shetland wool. These fruits disperse by hooking onto an animal or onto human clothing when we walk through a meadow in summer. The design for synthetic "Velcro", fastener material, is based on the hooked structure of these fruits. (Dr. Jeremy Burgess/Science Photo Library/Photo Researchers, Inc.)

FIGURE 35-10

An unusual adaptation. Peanut plants bury their own seeds. After the flower petals die, the female part of the flower elongates, forming a "peg" which pushes into the ground. Underground, the end of the peg enlarges into a peanut shell, enclosing the seed. This "self-planting" keeps the peanut hidden from hungry animals. Peanut plants have nitrogen-fixing bacteria in their roots. Keeping the seeds near the parent permits the young plants to benefit from the nitrogen that the parent plant contributes to the soil.

(Figure 35-11). Producing smaller seeds works only if the seeds become smaller than those of another species to which the predator might switch.

Another adaptation is **seed masting,** the simultaneous release of seeds by all the plants of the same species in an area, at intervals of two years or more. This makes seeds available to predators for a minimum time period. Beech trees and some oaks do this, but the most impressive examples are bamboos. Part of a bamboo stand in India was collected and sent to botanical gardens in the United States and Britain at the beginning of the nineteenth century. The plants grew vegetatively for the next 130 years, and then the bamboo stands on all three continents produced seeds in the same year! The advantage of seed masting to the plant is that most of the time there are no seeds to support the growth of large populations of seed-eaters. When seeds are finally shed, there are so many that a small population of seed predators cannot eat them all, and some seeds escape to produce the next generation of plants.

1? Some large seeds, such as nuts and acorns, are actually dispersed by would-be predators. Squirrels and some birds collect these large, nutritious seeds and hide them for later use. However, they do not return to all of their caches. The forgotten seeds sprout the next spring and grow into new trees. In the long run, this predation behavior benefits the tree, which has gained some surviving offspring at the expense of others. Researchers attribute the wide distribution of oak trees to the industry of jays, which may fly several kilometers to bury acorns in soft earth or under moist leaves (squirrels have much smaller home ranges). In one study, 50 jays spent September hiding 150,000 acorns! The diligence of jays is thought to account for the rapid spread of oak trees north, following the retreating glaciers, after the last ice age—a feat the heavy acorns could never have accomplished themselves. ■

BIO-BIT

Seed dispersal in tumbleweeds occurs when the entire plant detaches, releasing seeds as it rolls along the ground.

Wind-distributed seeds are typically smaller and associated with wind-catching structures that aid in their dispersal. Seeds dispersed by animals are larger and have many adapta-

FIGURE 35-11

A defensive arsenal. A large, nutritious horse chestnut (buckeye) seed lies within a tough shell, which is covered by a prickly husk, a deterrent to prospective seed predators. (R. Maier/Earth Scenes)

tions to increase the chances of dispersal and survival, including structures that attach to animal skins, and edible fruits that have tough, digestion-resistant seed coats. Seed masting and the production of many tiny seeds increase the chances that some seeds will not be eaten.

35–F GERMINATION

Many seeds become dormant after they have formed. As a seed enters dormancy, it dries out, and its final water content may be less than 5% of its total weight. In many kinds of seeds, dryness seem to be the main factor assuring the seed's **viability,** or its ability to survive an extended period of dormancy. Many commercial seed suppliers now dry their seeds thoroughly and wrap them in moisture-proof foil packets.

The viability of seeds varies among species, and among individuals within a species. The viability record is held by water lily seeds found in a peat bed in Manchuria and dated at over 1000 years old by radioactive isotope methods. Seeds from even older deposits have been grown, but no dating was done to prove that the seeds were as old as the layer in which they were found. In contrast, sugar maple seeds live less than a week. These seeds do not have a dormant period, and their viability is better if they do not dry out much.

2? In order to **germinate,** or begin growing into a new plant, seeds must be supplied with water. Most seeds also require oxygen, and many require particular temperature and light conditions before they will germinate. Furthermore, germination requirements may vary not only from species to species, but also from individual to individual. Thus a plant does not have all of its seeds "in one basket": germination may spread over months or years, and at least some seedlings are likely to find conditions that favor their survival. ■

Germination of seeds is associated with an increase in the growth-stimulating gibberellin hormones. Applying gibberellins can often overcome a seed's special light or temperature requirements for germination. In some species, other plant hormones are needed for germination.

Germination has been best studied in barley. The barley embryo secretes gibberellin, which induces the secretion of various enzymes. These enzymes break down starch and other stored food in the endosperm, making it available for absorption by the developing embryo. As the embryo uses this food, the dry weight of the seed decreases until the embryo begins to make its own food through photosynthesis. The seedling then grows as described in Chapter 32.

Seeds that pass through a period of dormancy typically dry out. Thereafter, germination can be triggered by a combination of favorable environmental factors, which usually include water, oxygen levels, light, and temperature.

35–G BREEDING PROGRAMS

Probably since the beginning of agriculture, humans have been breeding plants selectively to improve the features that make them useful to us. This starts by choosing plants that produce well in the field and collecting their seeds to plant for the next generation. For thousands of years, this was the only available way to improve plants. Nevertheless, it produced strains of cultivated plants strikingly different from their wild ancestors, especially in improved food yield. More control of the process can be gained by carefully cross-pollinating plants selected for desirable traits, in an attempt to get offspring with all of the desired features in the same plants. Our modern knowledge of genetics guides the selection of parental strains for such breeding programs.

The genetics of some crop plants are fairly well understood, especially for major crops like tomatoes, wheat, and corn. The popular strains of hybrid corn are produced by carefully planned crosses between a number of strains, each bred for particular traits (Fig-

(a)

(b)

FIGURE 35–12

Plant breeding. Generations of selective breeding by commercial vegetable growers have produced many varieties of tomatoes from an original stock. These tomatoes can be large, like beefsteaks **(a)**, or small, like cherry tomatoes **(b)**. (a, D. Cavagnaro/Peter Arnold, Inc.; b, H. Reinhard/Okapia)

ure 35–12). It may take several generations of crosses to produce the seed used to grow hybrid corn, and these crosses must be made anew each year in a continuous breeding program, because the hybrids do not breed true. So seed for hybrid corn must be purchased from breeders each year.

Corn is a fairly easy crop to use in genetics programs. Because male and female flowers grow in large, separate clusters, it is easy to carry out a desired pollination. And, as corn is an annual plant, the success of a cross can be judged within a year.

Other plants pose difficulties. For example, a breeder of apples must wait four to ten years until seeds grow into mature trees and bear fruit. Furthermore, about 99% of such trees carry new genetic combinations that are inferior to old varieties. Any new tree that looks promising must be screened for another ten years before it is ready for marketing—or the woodpile.

Artificial selection in plant breeding programs has permitted the development of new plant hybrids with characteristics that are more desirable to humans.

35–H VEGETATIVE REPRODUCTION

Most flowering plants can reproduce sexually. Sexual reproduction increases the genetic variation in a population, and so provides many "trial" assortments of genes that natural selection can act upon. This has the potential to produce genetic combinations better suited to the environment than those of the parent plants. However, a large percentage of these offspring end up with less favorable genetic combinations. Many of the rest fall prey to animals or fungal disease, while still others end up in habitats unfavorable to their growth.

Asexual reproduction is also common among plants. Dandelions, hawkweeds, and many grasses reproduce asexually by seeds that develop from unfertilized ovules. In some of these plants pollination occurs, but it is only a stimulus to the ovule to develop into a seed, and the pollen contributes no genetic material to the offspring.

Many plants reproduce asexually by **vegetative reproduction,** in which new individuals arise from the parent's roots, stems, or leaves (Figure 35–13). Asexual reproduction allows these plants to perpetuate combinations of genes that are well adapted to their environment. They spread and cover a wide area with a **clone** of plants, a population of individuals with identical genotypes.

Vegetative propagation is often desirable from the human as well as from the plant point of view. Home gardeners root cuttings of *Coleus,* geranium, or ivy shoots by placing them in water, or set leaves of African violets or jade plants on moist soil until they grow roots. Some plants can be rooted more successfully with the use of commercial plant hormone preparations. People also help spread plants such as daffodils and onions by digging up the bulbs when the tops die back and separating those that have multiplied, giving each more room to grow.

Potatoes are an important crop produced by vegetative means. A potato is an underground stem, or **tuber.** Farmers get many offspring from a good potato

(a)

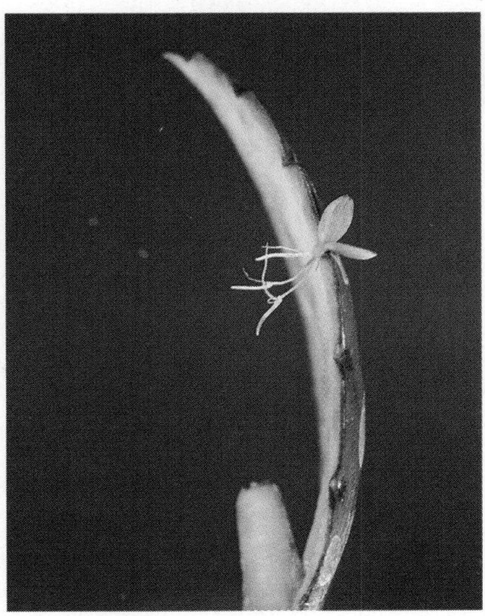

(b)

FIGURE 35–13

Vegetative reproduction. (a) Stolons of strawberry plants begin as slender stems that arch away from the parent plant and form roots and leaves of a new plant where they touch the ground. A new plant is forming at the bottom right of this photograph. **(b)** On the edge of a leaf, this mature air plant is producing a new plantlet that looks somewhat like a flower. (a, b, Runk/Schoenberger from Grant Heilman)

(a)

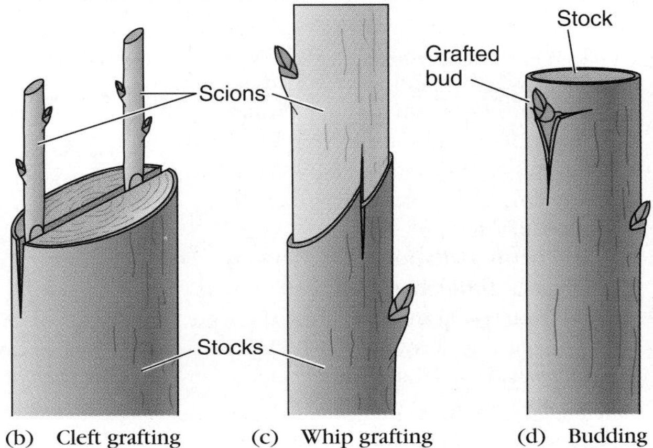

FIGURE 35–14

Plant grafting. (a) The dark branches grafted onto these light-colored apple tree stocks will bear apples of different varieties than those of the root stock they have been grafted onto. **(b-d)** Three grafting techniques. (G. Holscher)

plant by digging up "seed potatoes" (tubers of good quality), cutting them into pieces, each with an "eye" (bud), and replanting them. The seeds produced by potato flowers usually have inferior genetic combinations. However, breeders do grow plants from these seeds in an attempt to produce new strains with desirable traits, such as resistance to certain diseases and pests and the ability to form tubers when grown in tropical climates.

Grafting is another means of artificial vegetative propagation. A **scion,** a twig or bud of a desirable plant, is attached to a **stock,** the root system or stem of another plant, from which a twig or bud similar to the scion has just been removed. Scion and stock are then wrapped closely together, and their cut areas soon produce new cells that merge the two parts into a functioning unit (Figure 35–14). Each part of the graft retains its genetic identity, however.

3? Grafts work only between plants of the same or closely related species. The method is used to produce fruit trees, grape vines, or rose bushes that combine the desirable fruits or flowers of the scion with a sturdy, pest-resistant rootstock. For example, all the Red Delicious, Golden Delicious, and McIntosh apple trees now in existence are derived, by grafting, from trees that happened to have desirable fruits. Dwarf fruit trees are produced by grafting scions onto rootstocks of related species. Thus, dwarf pear trees consist of pear scions grafted onto quince roots. It is even possible to purchase small apple and pear trees with five or six grafted branches, each bearing fruits of a different variety. ■

A Journey into Evolution

Coevolution of Flowers and Their Pollinators

Much of the evolutionary success of flowering plants comes from the fact that they evolved after animal life was well established on land. Flowering plants and animals have exerted strong selective pressures on one another, and each has shaped the evolution of the other in many ways. Pollination systems offer many fascinating examples of such **coevolution.**

The most important pollinating animals are the bees (Figure 35-A). A flower enjoys several advantages in being pollinated by bees, which are widely distributed and numerous. Bees also work very hard at visiting flowers because many bees depend entirely on the food they obtain from flowers, both to nourish themselves and to feed the larvae. The behavior of bees also makes them highly desirable pollinators. Bees quickly learn to tell the different types of flowers apart, and they are faithful to one kind of flower for long periods. Bees are also available throughout the growing season, and they can remain active even at very low air temperatures, which immobilize most other insects.

Various butterflies and moths are important flower pollinators in all parts of the world. Because these insects use nectar only as a supplementary food for their short-lived adult stage, however, they are not as effective pollinators as bees. Most moths fly at night, and the flowers that depend on them for pollination tend to have pale colors visible in dim light. Some flowers, such as *Nicotiana* (a member of the tobacco family), produce scent only at night, when the moths that pollinate them are active. Flowers pollinated by butterflies, on the other hand, are more likely to have bright colors that stand out by day (Figure 35-B).

FIGURE 35-A

Honeybee. Bees are important plant pollinators. (J. Vucci, Peter Arnold, Inc.)

FIGURE 35-B

A giant swallowtail butterfly sucking nectar through its tubular proboscis. (D. Cavagnaro/Visuals Unlimited, Inc.)

Although many butterflies and moths feed on nectar from more than one species of flower, they concentrate on one species at a time. Thus, a hawk moth feeds only on, say, toadflax for as many as five days, and then switches to feeding on nothing but bedstraw. This faithfulness is plainly advantageous to both flowers and insects. The flower benefits because the insect is likely to convey pollen to another flower of the same species. The pollinator benefits by "keying in" on certain cues provided by the flower. The insect can then find more flowers of that species efficiently and ignore the cues from competing "restaurants" (just as some people key in on the "Golden Arches"!).

Many species of birds feed on nectar and supplement their diet with insects. However, birds often pierce the sides of tubular flowers and so obtain the nectar without picking up pollen, a situation that is disadvantageous to the flower. This may be one of the selective pressures that led to the evolution of flowers shaped in such a way that birds can reach the nectar more easily from a position where they also brush against the pollen.

The 300 species of hummingbirds are the largest group of bird pollinators. They nearly always feed while in flight, hovering in front of a flower and using their long bills and tube-like tongues to suck up the nectar deep within the flower (Figure 35-C). Flowers pollinated by hummingbirds usually have long stigmas

FIGURE 35-C

A Rufous Hummingbird visiting a flower. (C. & E. Schwartz, Animals Animals)

FIGURE 35-D

A bat pollinator. This greater short-nosed fruit bat feeds upon the nectar of banana plants and transfers pollen between flowers. (M. Tuttle/Photo Researchers, Inc.)

FIGURE 35-E

A purple carrion flower. These flowers smell like rotting meat. Flies that are attracted to the odor help to pollinate the plants. (E. R. Degginger)

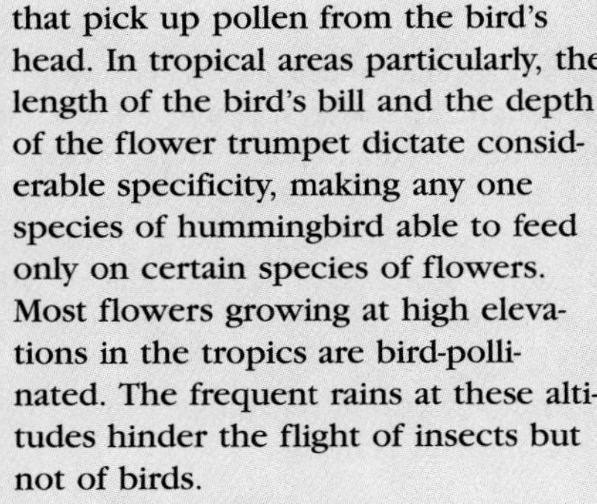

that pick up pollen from the bird's head. In tropical areas particularly, the length of the bird's bill and the depth of the flower trumpet dictate considerable specificity, making any one species of hummingbird able to feed only on certain species of flowers. Most flowers growing at high elevations in the tropics are bird-pollinated. The frequent rains at these altitudes hinder the flight of insects but not of birds.

Plants need to ensure not only that their pollinators are faithful to their species but also that the pollinators visit the flowers frequently. Flowers produce small amounts of nectar, making it necessary for the pollinator to visit many flowers, and to revisit each flower again and again. Another adaptation that ensures frequent visits is the flower's distinctiveness: if the flower looks and smells different from other flowers nearby, the pollinator can find the flower easily and discover more flowers of the same

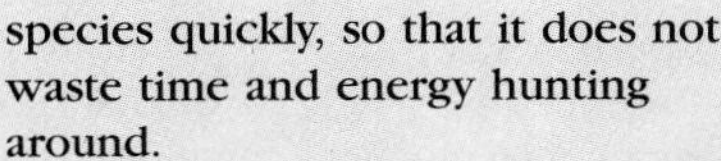

species quickly, so that it does not waste time and energy hunting around.

There is also strong selection for different plant species to bloom at different times, so that each species receives the attentions of pollinators in its turn, rather than having all flowers competing for attention during a brief period. Such staggered blooming also provides pollinators with a steady food supply throughout the growing season.

Although most flower-pollinator relationships are mutually beneficial, some plants have evolved ways to secure animals' services without paying a reward. In many orchids, part of the flower resembles the rear end of a female bee. The flower may even emit the same chemicals used by the bee as her sex-attractant pheromone. Deluded male bees attempt to copulate with the flower and pick up pollen, which they carry to other orchids of the same species. Male bees that have visited orchids have increased success of mating with female bees. These adaptations ensure that the flowers are visited frequently and faithfully.

The hairy red petals and rotting-meat stench of carrion flowers mimic dead mammals well enough to fool a female blowfly looking for a place to lay her eggs (Figure 35-E). The fly travels from flower to flower, transferring pollen as she leaves clusters of eggs. Since her larvae cannot survive without animal protein, this arrangement boosts the plant's evolutionary success but lowers the fly's.

An exciting development in vegetative propagation is the production of entire plants from meristematic cells grown in laboratory culture. The culture medium contains high levels of plant hormones, which produce high rates of mutation. Researchers start with cells from plants with many desirable qualities, allow the cells to multiply and mutate in the laboratory, and then grow plants from them. The resulting plants are very similar genetically—more so than sexually produced offspring—but they may differ in important traits such as resistance to drought or to particular diseases. This method allows plant breeders to "fine-tune" the genetic makeup of crop plants. It also gives a way to produce more replicas of a desirable plant quickly. Only a small lump of meristematic cells is needed to start each new plant, rather than a large, leafy cutting. Cells kept in a flask or two in the laboratory can substitute for acres of plants formerly kept as sources for cuttings. Many plants, such as chrysanthemums, are now grown commercially in laboratory culture from meristematic cells. Laboratory culture is also used to grow genetically engineered cells into complete plants.

Many plants reproduce asexually by means of seeds from unfertilized ovules or by vegetative reproduction from roots, stems, or leaves. Vegetative propagation permits humans to raise many genetically identical plants with desirable genetic combinations.

SUMMARY

1. Plants flower in response to specific cues that differ greatly among the various species, and each species of angiosperm has its own distinctive flower structure. In all this diversity, however, we can find a basic unity in the structure and function of flowers.
2. A flower is an abbreviated shoot, with modified leaves as the flower parts. The male parts, the stamens, produce the haploid male gametophytes in the form of pollen grains. The female parts, the pistils, produce the female gametophytes: the embryo sacs.
3. Double fertilization forms a zygote and an endosperm mother cell in the embryo sac. The endosperm mother cell divides and develops into the endosperm, which absorbs and stores food from the parent plant. The zygote soon develops into the embryo of a new plant.
4. Besides contributing food to the new embryo, the parent plant also protects the embryo and its food supply in a seed coat derived from the wall of the ovule. This, in turn, is surrounded by the fruit, derived from the wall of the ovary, also part of the parent plant.
5. Much of the evolutionary success of flowering plants is undoubtedly due to the fact that they have coevolved with animals. Many species rely on animals, rather than on wind or water, to pollinate their flowers and disperse their seeds. Plants devote considerable energy to attracting animals that will perform these services, and they are rewarded by pollination that is efficient and specific, and by seed dispersal that distributes even large seeds over a relatively wide area.
6. Many seeds enter a period of dormancy following their release from the parent plant. Eventually a seed germinates in response to environmental cues and grows into a mature plant.
7. Sexual reproduction results in individuals with new combinations of genetic characters. Some of these combinations are more desirable and some are less desirable than those of the parents, from either the human or the plant point of view. Many plants have some means of asexual reproduction—which produces genetically identical plants—in addition to, or instead of, sexual reproduction.
8. Humans propagate many plants vegetatively by artificial means such as rooting and grafting.

SELF-QUIZ

Label the structures numbered in the diagram below:

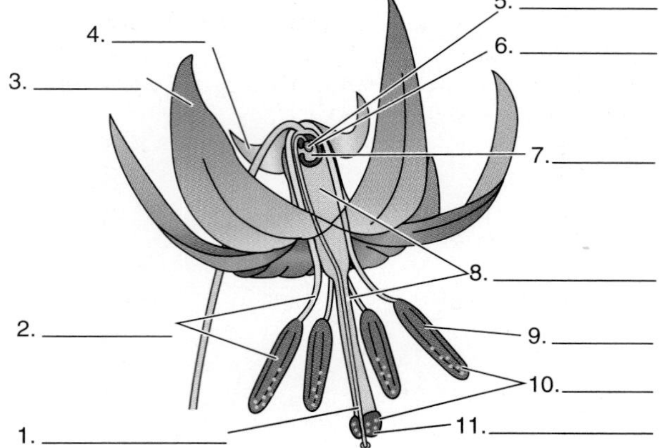

For each of the following descriptions, give the number and name of the structure in the diagram:

____ 12. Site of pollen production
____ 13. Female gametophyte
____ 14. Protective flower part
____ 15. Develops into the seed coat
____ 16. Contains structure that gives rise to endosperm
____ 17. Develops into the fruit

18. True or False: The terms pollination and fertilization can be used interchangeably.
19. Which of the following is *never* required for seed germination?
   a. certain temperature conditions
   b. oxygen
   c. water
   d. light
   e. none of the above
20. Grafting is used to propagate plants because:
   a. it is faster than growing seeds
   b. it maintains a desired set of genetic characteristics
   c. it combines the genetic characteristics of two desirable strains of plants
   d. healthy plants will graft by themselves, so that they reproduce profusely
   e. a plant can produce many more scions than seeds

THINKING CRITICALLY

1. Why do banana plants put so much energy into producing fruits that contain no seeds?
2. Plants given large amounts of fertilizer, especially fertilizer containing much nitrogen, often flower poorly or not at all, and do not accumulate food reserves; instead they engage in vigorous vegetative growth. Is there an adaptive advantage to this?
3. Some plants, such as dandelions and hawkweeds, have lost the ability to reproduce sexually but still produce flowers and set seed by development of the ovule without meiosis or fertilization. What is the advantage of this system over a more orthodox means of vegetative reproduction?
4. Pollen is produced at the tips of the stamens, whereas ovaries lie at the bases of the pistils. What is the adaptive advantage of these differences in position?

SELECTED KEY TERMS

anther, *p. 728*
clone, *p. 738*
cotyledon, *p. 734*
cross-pollination, *p. 730*
endosperm, *p. 731*
fruit, *p. 731*
gametophyte, *p. 729*
germinate, *p. 737*
grafting, *p. 739*
micropyle, *p. 730*
pistil, *p. 728*
pollen grain, *p. 729*
pollination, *p. 730*
scion, *p. 739*
seed coat, *p. 731*
seed masting, *p. 736*
self-pollination, *p. 730*
sepal, *p. 728*
spore, *p. 729*
stamen, *p. 728*
stigma, *p. 728*
style, *p. 728*
tuber, *p. 738*
vegetative reproduction, *p. 738*

SUGGESTED READINGS

Barrett, S. C. H. "Mimicry in plants." *Scientific American,* September 1987. Some plants cheat pollinators or farmers by resembling other species so closely that the animals are duped into helping the plant without reaping the expected reward.

Echlin, P. "Pollen." *Scientific American,* April 1968. Many interesting facts and illustrations.

Faegri, K., and L. van der Pijl. *The Principles of Pollination Ecology,* 3d ed. New York: Pergamon Press, 1979.

Handel, S. N., and A. J. Beattie. "Seed dispersal by ants." *Scientific American,* August 1990.

Heinrich, B. "The energetics of the bumblebee." *Scientific American,* April 1973. The relationships between bumblebees and the flowers they pollinate, viewed in terms of the influence of energy expenditure on evolution of adaptations.

Koller, D. "Germination." *Scientific American,* April 1959.

Meeuse, B., and S. Morris. *The Sex Life of Flowers.* New York: Facts on File, 1984.

Miller, J. A. "Somaclonal variation." *Science News* 128:120, 1985. How new varieties of plants are grown from cells that mutate in tissue culture.

Proctor, M., and P. Yeo. *The Pollination of Flowers.* Glasgow: William Collins Sons, 1973.

Wickelgren, I. "Please pass the genes." *Science News* 136:120, August 19, 1989. A discussion of whether or not genetically engineered food crops pose novel hazards to human diners.

The Educated Citizen

Altered Vegetable States

This article by Rosie Mestel was excerpted with permission from the January 1993 issue of Discover *magazine.*

Our world is full of symbols, from the skull and crossbones on the poison bottle to the looping arrows on the recycling bin. Now developments in agriculture are giving us two more graphics for modern life. A DNA helix with a slash running through it may soon appear in your local restaurant as a protest against gene-engineered food. And a leafy flower in a broken circle is adorning boxes of Florida strawberries. The symbol, which has been mandatory since last January, means "irradiated" (though it is widely taken to mean "fresh").

With those strawberries, irradiated foods finally made their grocery-store debut in 1992, 30 years after the Food and Drug Administration deemed irradiation safe. The process not only slows ripening in fruits and sprouting in potatoes, it disinfects herbs and teas, destroys parasites in pork, and kills *Salmonella,* a food-poisoning bug carried by over half the nation's chickens and turkeys. Still, critics worry about bombarding food with gamma rays. Not that a zapped potato gets "hot"—you'd need rays three times more energetic to make food radioactive. But when rays pass through tissues they create free radicals, chemicals that can damage cells and sometimes form carcinogens. As George Pauli of the FDA notes, however, radicals and carcinogens *normally* occur in food. "Compared with changes from canning, broiling, and cooking in general, those from irradiation are much, much smaller," he says.

Gene-engineered plants aren't with us yet, but in 1992 the FDA paved the way for their arrival. (Playing with genes is nothing new, of course—they have been manipulated for millennia by breeding. But molecular biology permits new tricks. For example, you can transfer genes from one species to another—say, put a flounder antifreeze gene into strawberries.) The FDA decided that gene engineering per se wasn't hazardous. Crops will need review only if engineering increases their natural toxins, depletes important nutrients, introduces an allergy-producing substance, or poses environmental concerns. Otherwise, the FDA said, the products can go—unlabeled—straight onto the market.

Already more than 50 gene-engineered products are in the pipeline, including virus-resistant cantaloupes, herbicide-resistant soybeans, and high-starch potatoes that absorb less fat. First onto our shelves will be Calgene's Flavr Savr tomato. Flavr Savr promises to rid us of a symbol as confusing as any irradiation logo: the nice red color of store-bought tomatoes that's totally divorced from flavor. Vine-ripened tomatoes turn mushy after reaching their prime because an enzyme starts to destroy their solids. Thus most tomatoes are picked and shipped green, then given a whiff of ripening hormone at journey's end. They turn red, but they never load up with the medley of sugars, acids, and small, volatile molecules that gives the fruit full flavor. But in Flavr Savr the gene that turns fruit to mush is essentially shut off. Since fruit stays firm during shipping, growers can let it ripen on the vine. Ergo, red may once again mean "tasty."

You'd think chefs would rejoice, but 1,500 of them (including Wolfgang Puck of the elite Los Angeles restaurant, Spago's) have thrown their toques into the ring with biotech foe Jeremy Rifkin—vowing to boycott the bionic berry and display anti-DNA symbols in their restaurants. As for Rifkin, he wants every engineered gene tested as a new additive. "Any genetically engineered product could be toxic or allergenic," he says. "Without pretesting and labeling, no one will know until they get sick or die." And what, he adds, of the unsuspecting Jew or Muslim who chomps into a pig-gene-tainted vegetable? (So far no nonkosher vegetables are planned, however.)

"Rifkin's got the public looking for killer tomatoes," says Norman Ellstrand, a geneticist at the University of California at Riverside. "I think the effects will be more subtle"—more on the order of invasive weeds. If an herbicide-resistant carrot mated with its wild relative, a weed called Queen Anne's lace, could the offspring spread out of control?

To many plant scientists, though, the fuss in this country seems overblown. The biggest role they foresee for plant biotechnology is elsewhere in the world. Plant diseases and pests here have nothing on those found in developing countries, especially in the tropics, points out Roger Beachy, the Scripps Institute molecular biologist who first engineered virus resistance into plants. "Fat-repellent potatoes are all very well," he says. "But this technology could have a vital impact on feeding the developing world."

Connecting the Concepts

The following questions will help you to connect the issues discussed in this article with the concepts you have learned in Part 5.

1. How is genetic engineering different from artificial selection and the creation of new types of plants by hybridization?
2. How could genetically engineered plants increase world-wide food production?
3. What could the FDA do to determine if a new genetically engineered crop poses health and environmental concerns?
4. Discuss the advantages and disadvantages, to both producers and to the general public, of labeling irradiated and genetically engineered food products.

The greatest thing since sliced bread? Campbell Soup Company thinks so. It owns exclusive rights to use the genetically engineered FLAVR SAVR tomatoes for processed food products. (Courtesy of Calgene Fresh, Inc.)

PART 6

The World of Life: Ecology

Chapter 36
Distribution of Organisms

Chapter 37
Ecosystems and Communities

Chapter 38
Populations

Chapter 39
Human Ecology

Mutualism. A clown fish settles into the tentacles of "its" anemone. Clown fish and anemones have mutualistic relationships. The stinging tentacles of the anemone provide a safe haven for the clown fish. Because the clown fish is a messy eater, the anemone catches scraps of food that the fish drops. (M. J. Thomas/Dembinsky Photo Associates)

CHAPTER 36

CONCEPT GUIDE

After reading this chapter, you should be able to:

1. Describe the two major factors that help determine the climate of a particular area. (Section 36-A)
2. Compare the climate, plant life, and animal life of a rain forest, tropical savanna, tropical thornwood, desert, temperate forest, temperate shrubland, temperate grassland, temperate desert, taiga, and tundra. (Sections 36-B through 36-F)
3. Explain why freshwater littoral zones and marine subtidal zones are some of the most densely populated aquatic communities. (Section 36-G)
4. Compare the circumstances that initiate primary and secondary succession and the main events of each process. (Section 36-H)
5. Describe the main factors that limit the distribution of most animal and plant species. (Sections 36-A and 36-I)

The spectacular fires that swept much of Yellowstone National Park in 1988 were a natural event that opened up large areas for new plant growth. In the years since the fire, a series of plant communities have become established where once there was forest, in a process called succession. Read more about the regular changes that occur in many communities in Section 36-G. (J. Henry/Peter Arnold, Inc.)

Distribution of Organisms

If you were interested in animals as a child, you probably played some version of the game of animal geography, matching animals with their correct environments. For example, in the simplest version of this game, you match polar bears, Arctic foxes, walruses, and seals with the cold, northern, Arctic regions, while penguins belong in the even colder, southern, Antarctic regions. You might have learned that larger, fan-eared African elephants live on grassy African savannas, while smaller, Indian elephants live in dense forests in India. If you were a reptile enthusiast, you might have learned that boas are found only in the New World tropics, while pythons live only in Old World tropics and that giant tortoises tend to be restricted to islands. You might know that alligators lurk in the waters of Chinese and American swamps, while crocodiles are found in Africa, Australia, Asia, and India. But the big lesson that this children's game of animal geography might teach you is that organisms are not scattered randomly about the planet, but rather there are patterns to the distribution of animals and plants. These distributional patterns are one of the concerns of the branch of biology called **ecology,** the study of the interactions of organisms with the world around them.

The word "ecology," coined in 1869, is based on the Greek word *oikos,* meaning "house" or, more loosely, "habitat." The term **ecosystem,** from the same root, is the habitat or environment where organisms live and interact. The ecosystem includes **abiotic** (nonliving) factors such as sunlight, temperature, water, and soil and **biotic** factors (all the other organisms in the ecosystem). Each ecosystem contains several **communities** of organisms, collections of different species living together. Ecologists study the patterns of distribution and abundance of organisms in nature, how these patterns are maintained in the short run, and how they change during the course of evolution.

In this chapter we shall discuss some of the larger ecological units of distribution. **Biomes** contain many ecosystems, but are defined by climate. For example tropical forest, found in several parts of the world, is a biome. So is desert, grassland, or tundra. Because they are defined by climate, biomes look superficially similar, no matter where they occur. For example, if we were walking through a South American tropical forest, we would find tall trees with large leaves and fruits. They would be draped with climbing vines and colorful birds, and butterflies would be flitting through the gloomy shade of the forest. A tropical forest in Asia would look superficially similar, but the biological details—the exact species of trees, vines, butterflies, and birds—would be different. Forests in South America and Asia are similar because they occur in areas with similar climates, high temperatures, and heavy rainfall. Wherever a particular pattern of temperature and precipitation occurs, plants adapted to that climate will be found. The climate and kinds of plants, in turn, influence the animal life.

Aquatic communities are not classified into biomes, but their distribution is also determined by the physical environment. For instance, photosynthetic organisms live only near the water surface because there is not enough light for photosynthesis in deep water.

In this chapter, we consider the question, "Why are organisms where they are?"

KEY CONCEPTS

- The organisms on Earth are found in only a small number of different types of biomes, each of which is found in several parts of the world.
- On land, the biome in an area is determined largely by the pattern of temperature and precipitation.
- The community found in a particular aquatic ecosystem depends largely on light, temperature, salinity, water currents, and the type of bottom sediments in the area.

Curiosity Questions

1? Why do so many animals live around coral reefs? (See page 761)

2? How can fires be helpful to a plant community? (See page 764)

3? Why do most animals live in only certain parts of the world? (See page 764)

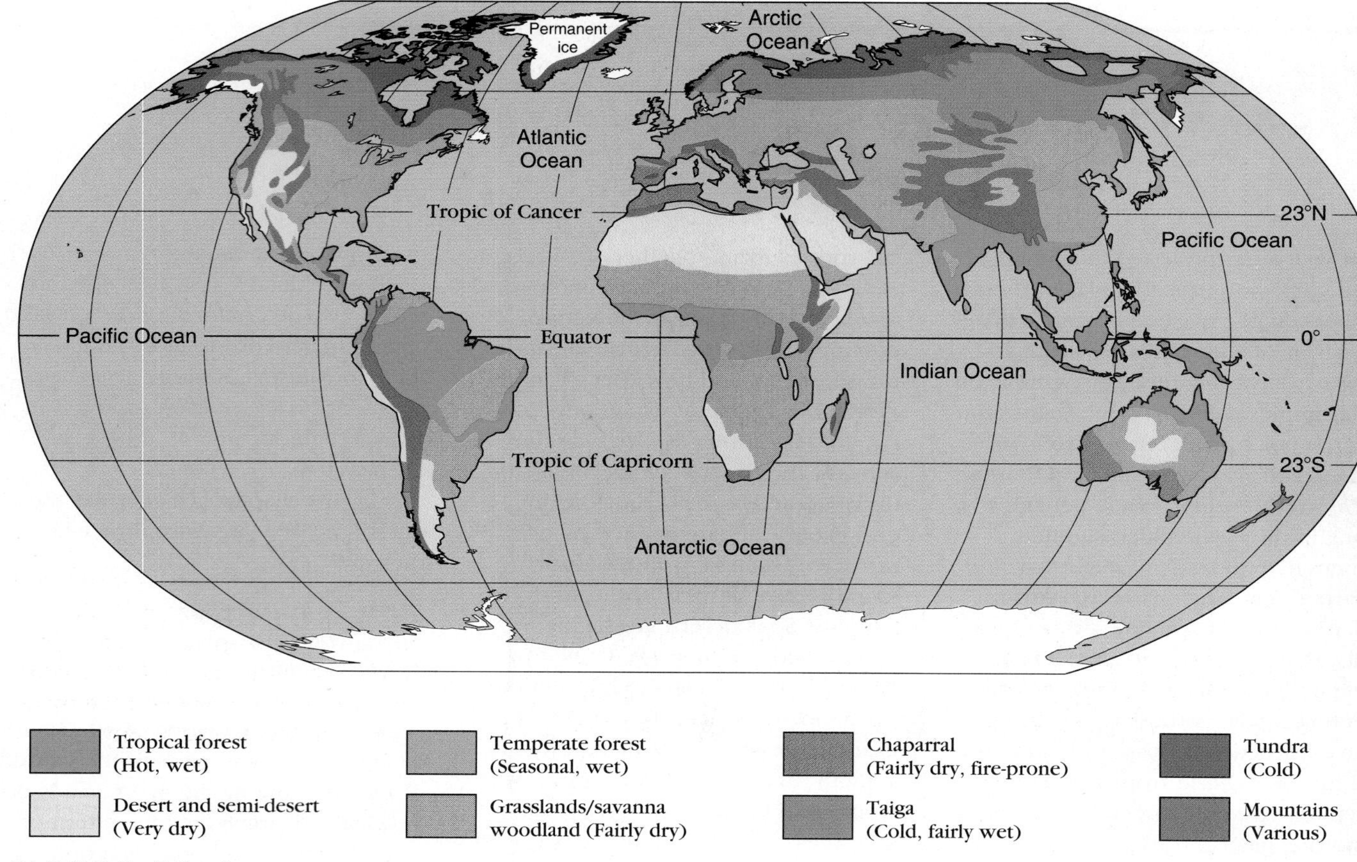

FIGURE 36–1

Biomes of the Earth. This map shows the major biomes, simplified to emphasize the overall pattern. Notice that the order of biomes is governed by latitude; the sequence of biomes north of the equator mirrors the sequence of biomes south of the equator. Oceans, land masses, and mountain ranges also affect climate, and thus vegetation, making the map more complicated than it would be otherwise.

36–A CLIMATE AND VEGETATION

If we look at a map of the world showing the kinds of communities in different places, we find that areas with similar climates have communities of the same type (Figure 36–1). Climate is the main factor determining the type of soil and the types of plants in an area. The plant life and climate, in turn, determine the types of animals and microorganisms present.

Climate depends on the sun. Solar energy provides the heat that determines the average temperature, causes the winds, and powers precipitation. Tropical climates, receiving near-vertical sunlight throughout the year, have fairly steady, high temperatures (Figure 36–2). In other areas, the temperature varies roughly with the amount and intensity of sunlight at different seasons. Temperature varies with altitude (height above sea level) as well as with latitude (distance from the equator). As a result, the plant life on mountains shows changes from base to peak similar to those seen when travelling farther and farther north or south from the equator (Figure 36–3).

In addition to temperature, moisture is the other major factor determining where organisms can live, and this also depends on the sun. Warm air holds more moisture than cool air, and as air cools, some of its moisture condenses as rain, snow, or dew. Air heated by the sun at the equator rises, expanding and cooling as it mounts higher into the atmosphere. This makes it release much of its moisture, producing the teeming rains of tropical jungles. The dry air that remains moves both north and south from the equator, and eventually sinks to Earth again, becoming warmer as it does so. The descent of this dry air creates the world's great deserts. Still further north and south, in the temperate latitudes that include most of the United States and Europe, swirling winds pull masses of air, sometimes from warm tropical areas, sometimes from frigid polar regions, producing the varied weather found between the two.

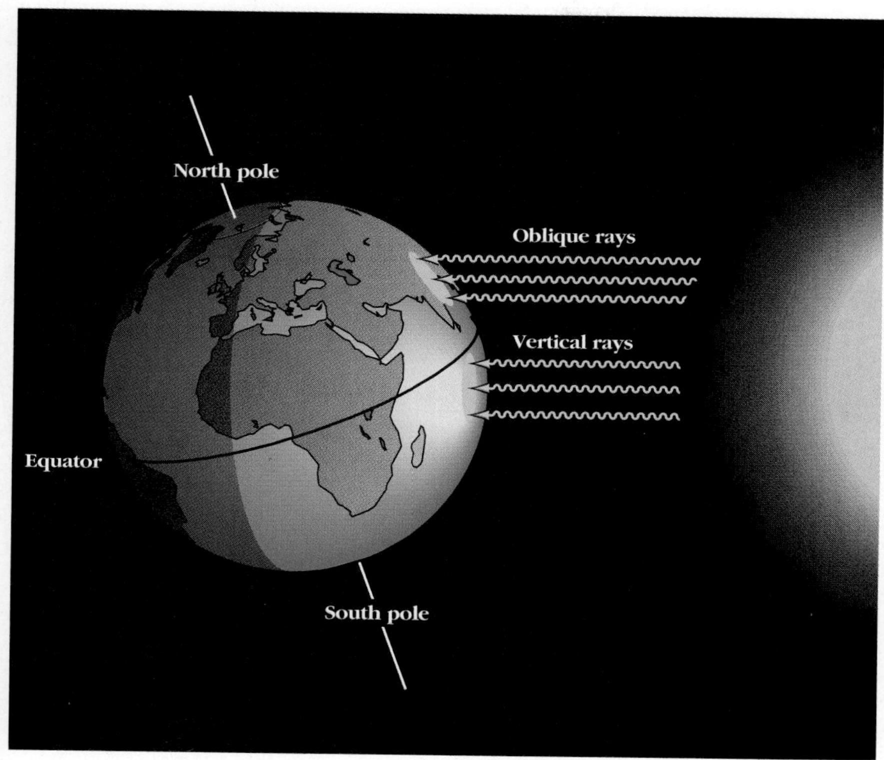

FIGURE 36-2

Vertical vs. oblique rays of sunlight. A beam of sunshine striking the Earth away from the equator is spread over a wider area. It is therefore less intense at any one point than a similar beam near the equator, which strikes the Earth vertically.

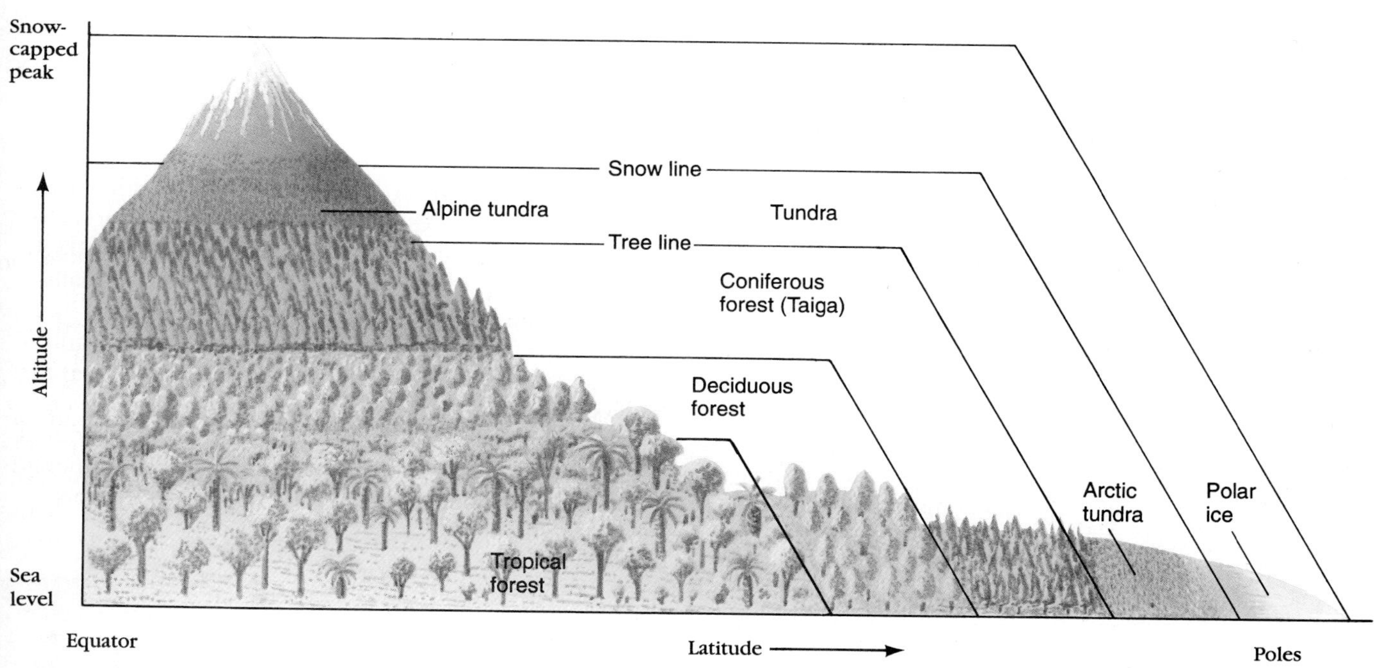

FIGURE 36-3

Similar effects of altitude and latitude on vegetation. As you go up a mountain, the vegetation changes much as it does when you travel north or south from the equator. This is because similar temperatures favor similar types of plants. Vegetation type is also influenced by moisture. This example shows a cross-section through communities with abundant precipitation. (The horizontal axis is greatly compressed.)

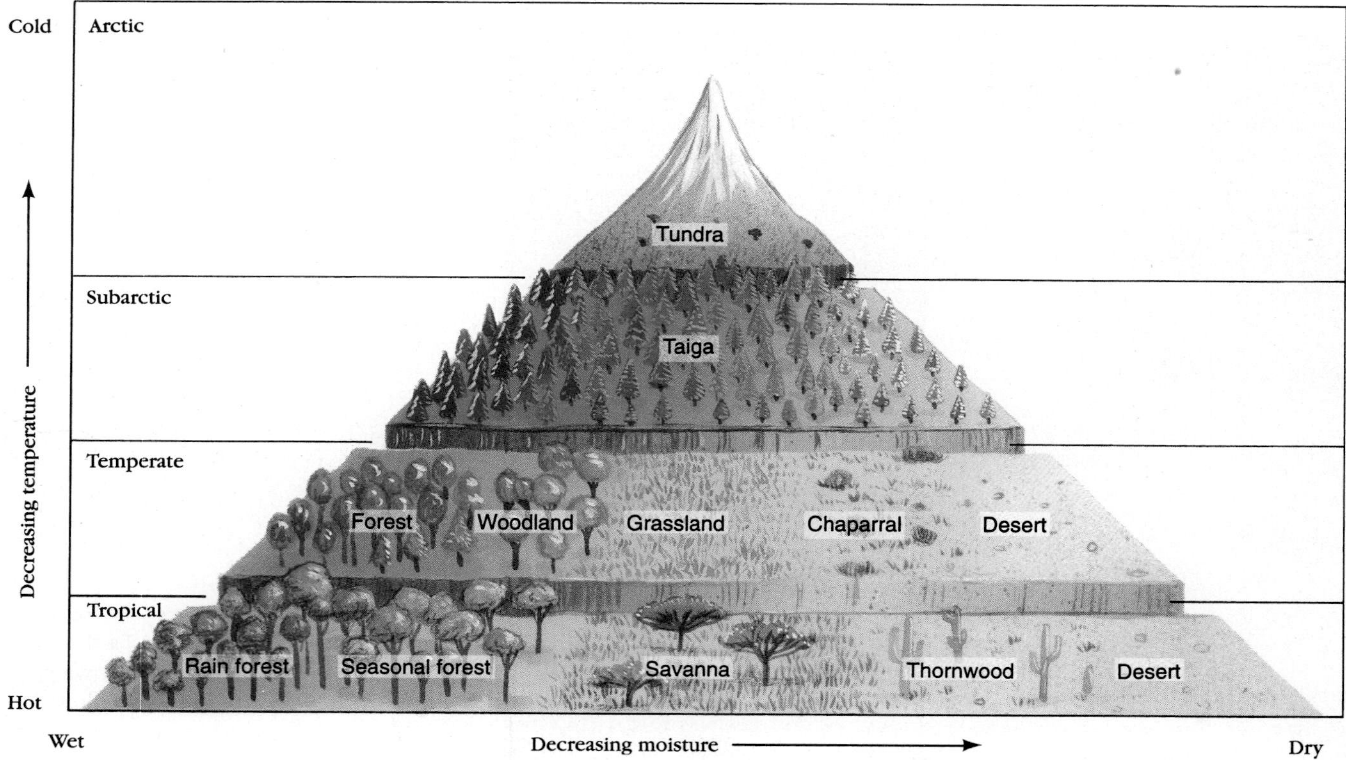

FIGURE 36–4

The relationship between temperature and biome. This diagram shows that temperature and precipitation are the main factors determining what where biomes occur.

Biomes

Each biome has a characteristic type of vegetation, but biomes seldom have sharp boundaries. Instead, they gradually merge into one another, forming gradients of changing community type along gradients of changing climate.

Here we group the world's terrestrial communities into a small number of major biomes (see Figure 36–1). We shall start at the equator and move north and south, discussing terrestrial biomes on a latitudinal basis. In tropical and temperate areas, the biomes can be arranged along gradients of decreasing moisture (Figure 36–4). For instance, trees use a lot of water and can survive only where there is heavy rainfall. Progressively lighter rainfall supports communities dominated by small trees, shrubs, grasses, and finally scattered cacti or other desert plants. In extreme deserts, there is so little rainfall that plants cannot grow at all.

Temperate biomes have lower temperatures than the tropics, as well as distinct hot and cold seasons. These biomes support fewer species and less total mass of living organisms than tropical regions with comparable precipitation. Temperate grassland has produced the world's deepest soils and best agricultural land.

36–B TROPICAL BIOMES

Tropical Forest

Tropical rain forest occurs near the equator where the climate provides excellent growing conditions for plants throughout the year: intense sunlight and high, fairly constant temperature and high rainfall, usually 200 cm (80 inches) per year.

The soil in most rain forests is waterlogged and thin. The high temperature and moisture are ideal for decomposer organisms that break down organic matter. Here, a fallen leaf may decompose in two months, a process that takes one to seven years in a temperate forest. The minerals released by decomposition are

BIO-BIT

It is likely that, within your lifetime, nearly one quarter of all of the plant, animal, and microorganism species will become extinct due to destruction of tropical rain forests. The vast majority of the species lost will never even have been studied or identified.

(a)

(b)

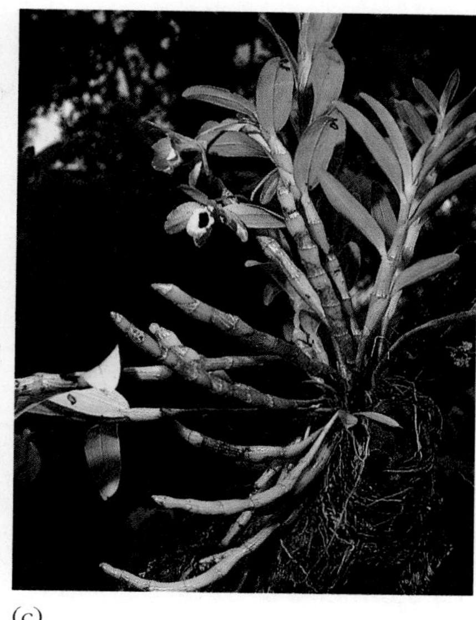
(c)

FIGURE 36–5

Life in a tropical rain forest. (a) Epiphytes, biomeliads, and climbing vines characterize the perennial damp twilight of this patch of rain forest. Some of the larger trees have been cut down, allowing understory plants to grow. **(b)** A Blue-crowned Motmot, found in rain forests in Costa Rica. Motmots are tropical relatives of Kingfishes. **(c)** An orchid growing as an epiphyte in a Hawaiian rainforest. (a, J. D. Cunningham/Visuals Unlimited; b, K. Schafer/Peter Arnold, Inc.; c, E. R. Degginger)

rapidly taken up again by the fast-growing plants, and so almost all the forest's nutrients are inside the bodies of living organisms instead of in the soil.

Tropical rain forest is the richest of all biomes, in that it has the greatest diversity of species in a given area. The dominant plants are tall trees with slender trunks that branch only near the top, covering the forest with a dense canopy of leathery evergreen leaves that shed water rapidly. The tall trees provide surfaces for the anchorage of many other plants called **epiphytes** (epi = upon; phyte = plant; also called air plants) that have aerial roots to absorb water, and trap falling leaves and other organic debris as a source of nutrients. Epiphytes include a great variety of orchids, bromeliads, and ferns. Because the canopy blocks most of the light, few plants grow on the floor of an undisturbed tropical rain forest.

The animal life of rain forests is also exceedingly rich. Birds, butterflies, beetles, and frogs exhibit an almost bewildering diversity of striking color patterns. Because most of the plant food is high in the canopy, most of the animals also live in the trees (Figure 36–5). The difficulty of working in the high canopy is one reason we know so little about the species that inhabit tropical rain forests.

Farther from the equator, climates have distinct seasons, with rainfall concentrated during part of the year and a definite dry season. These areas support **tropical seasonal forest.** As the length of the dry season increases, we find more and more **deciduous trees** (trees that lose their leaves for part of the year). Tropical seasonal forests include the monsoon forests of India and Southeast Asia.

All tropical forests are being cut down rapidly as a result of the enormous growth of human populations in these biomes. The loss of tropical forest and the inevitable loss of species of plants and animals is a pressing world problem (see Chapter 34: *A Journey into The Environment*: Tropical Rainforest Canopy—Going...Going).

Tropical Savanna and Tropical Thornwood

Tropical savanna extends over large areas where annual rainfall (between 90 and 150 cm—30 to 60 inches) is too low to support many trees, or where growth of forests is prevented by recurrent fires. Typical savanna consists of grassland dotted with scattered small trees or shrubs, such as acacias (Figure 36–6). Some savannas are entirely grassland, while others contain many trees.

Savannas are most well known in Africa, where they support a rich variety of grazing mammals, such as zebras, wildebeest, and gazelles, as well as the carnivores that hunt them: lions, leopards, cheetahs, and hyaenas. The spectacular migrations of some of these

(a)

(b)

FIGURE 36-6

Savanna. (a) The African savanna of the Serengeti National Park in Tanzania during the dry season. **(b)** Wildebeests in the African savanna in Kenya during the wet season. (a, S. Osolinski/Dembinsky Photo Assoc.; b, M. Barlow/Dembinsky Photo Assoc.)

species are related to shifting patterns of local rainfall that permit the growth of the young, nutritious foliage of grasses.

Tropical thornwood occurs in many regions too dry to support forest, but with at least a short rainy season each year. Spiny acacias and other trees of the pea family often dominate thornwoods of the Americas and Africa. Many of the plants in a tropical thornwood lose their small leaves during the long dry season, and their growth and reproduction take place entirely during the wet season.

Tropical rain forest, containing an enormous variety of plants and animals, is found where rainfall and temperature are high throughout the year. Tropical savanna and thornwood consist primarily of grassland and occur in many regions that are too dry to support trees.

36–C DESERT

Deserts occur in regions having less than about 20 to 25 centimeters (8 to 10 inches) of rain each year. Typical hot deserts are found around latitudes 20 to 30° north and south, where dry air from the equator falls from the upper atmosphere, warming as it is compressed near the Earth. Because it contains little water vapor, the air over a desert is a poor insulator, and although days can be very hot, nights are often cold because the ground radiates heat rapidly.

The Sahara desert, stretching across north Africa, is the largest hot desert in the world. Hot deserts also occur in southwestern North America, the west coast of South America, and central Australia. Desert areas with less than 2 centimeters of rain per year support little life of any kind, and the terrain is mainly rocks and sand. Less extreme areas have highly specialized plants, many of them annuals that grow, bloom, and set seed in the few days when water is available. Most desert perennials are small woody shrubs that shed their leaves during the dry season, or else **succulents,** plants that store water in their tissues, such as the American cacti (Figure 36-7). Desert animals have adaptations that restrict the loss of water through their skin and lungs and in their urine and feces. Many are nocturnal, avoiding the heat of the day, when they would lose water rapidly, by burrowing into the cooler soil.

Deserts are defined by limited rainfall, and desert organisms have many adaptations for water conservation.

BIO-BIT

Many desert plants have evolved spines and toxins which defend them against grazing in this food-scarce environment.

(a)

(b)

FIGURE 36–7

Desert in Arizona. (a) Organ Pipe Cactus in Organ Pipe National Monument in Arizona. **(b)** Coatimundis are highly social, raccoon-relatives that are natives of American deserts. (a, W. Clay/Dembinsky Photo Assoc.; b, S. J. Krasemann/Peter Arnold, Inc.)

36–D TEMPERATE BIOMES

Temperate Forest

North and south of the tropics and their adjacent deserts lie the world's temperate regions, so called because their climate typically has moderate temperatures (although you may not think so as you struggle to start a car on a February morning in Minnesota).

The **temperate forest** biome occurs in temperate regions with plentiful annual rainfall (75 to 150 cm or 30 to 60 inches). The canopy trees absorb about 40% of the sunlight reaching them. Below the canopy grows an **understory** of smaller trees. Less than 10% of the initial sunlight may reach the next level down, the shrubs. Beneath the shrubs there is usually a layer of low-growing, nonwoody **herbs** that receives less than 5% of the original sunlight striking the forest. Mosses and creeping herbs may provide yet another layer of vegetation close to the ground. Vertical structure continues down into the soil, where the roots of different plants extend to different depths.

Temperate forest falls into three major categories: deciduous, evergreen, and rain forests. **Temperate deciduous forests** occur in moderately humid inland climates where precipitation occurs throughout the year, but where winters are cold, restricting plant growth to the warm summers. Broad-leaved deciduous trees, such as beeches, oaks, hickories, and maples, dominate this kind of forest. There is also a well-developed understory of shrubs and herbaceous plants on the forest floor (Figure 36–8). The soil is rich in minerals and organic matter.

Mammals of North American deciduous forests include white-tailed deer, chipmunks, squirrels, and foxes. Wolves, black bears, bobcats, and mountain lions roamed widely until they were largely eliminated by human activities. As winter draws near, many of the birds migrate south, and many of the mammals hibernate. In the spring, plants such as trilliums, violets, and Solomon's seal produce their leaves and flower before the tree canopy leafs out and prevents most of the light from reaching the forest floor.

Temperate evergreen forests occur where poor soils, droughts, and forest fires favor gymnosperms or broad-leaved evergreens over deciduous trees (Section 36–H). In the United States, temperate evergreen forests include impressive stands of ponderosa and other pines in the west, as well as the pine forests of the southern states. These are now prime areas for commercial timber operations. Elsewhere in the world, temperate evergreen forests occur in eastern Asia, in southern Chile, in New Zealand, and in Australia, where forests are dominated by various species of *Eucalyptus.*

Temperate rain forests occur in cool climates near the sea with abundant winter rainfall and summer cloudiness or fog. They include the forests of giant trees along the Pacific coast of North America, from the mixed coniferous forest of Washington's Olympic Peninsula to the coastal redwood forests of Oregon

(a)

(b)

(c)

FIGURE 36-8

Temperate deciduous forest. (a) Springtime in a Michigan forest. **(b)** Fall in Shawnee National Forest, Illinois. **(c)** Raccoons, common residents of temperate deciduous forests. (a, M. L. Dembinsky, Jr./ Dembinsky Photo Assoc.; b, W. Clay/Dembinsky Photo Associates; c, B. P. Kent/Animals Animals)

and northern California. Although there is little rainfall in California in summer, the foliage of redwoods can absorb water from the frequent fogs.

Temperate woodland occurs in climates too dry to support forests, yet with enough moisture to support more than grassland. Pygmy conifer woodlands of piñon pine and juniper cover extensive areas of the American West. Oak woodlands are common in central California, and extensive evergreen oak and oak-pine woodlands occur in the southwestern states and in Mexico.

Temperate Shrubland

The **temperate shrubland** biome is best represented by the **chaparral** communities in all five areas of the world with a Mediterranean climate: the Mediterranean region, southern Australia, the southern tip of Africa, coastal Chile, and California. These areas have moderately dry climates with little or no summer rain. The shrubs are mainly angiosperms (flowering plants). They are often distinctly aromatic, and their leaves contain volatile and flammable organic compounds. Fires are frequent and pose a constant threat to residents of Santa Barbara and other cities in this biome.

Temperate Grassland

Early visitors to the American West were most impressed not by the forest but by the prairie with its burrowing prairie dogs and large grazing mammals, such as bison and pronghorn antelope. Prairie is **temperate grassland,** which covers extensive areas in the interiors of continents where there is not enough moisture to support forest or woodland. Scattered shrubs may occur, often in depressions or watercourses where extra water is available.

Although grassland vegetation forms only a single layer, many plant species may be present. Rich deep soil underlies much temperate grassland because dead vegetation is added to the soil faster than it decomposes. These regions of deep soil, including the midwestern United States, the Asian steppe, the Argentine pampas, and Ukrainia, have become prime areas for farming. As a result of its agricultural value, prairie has

BIO-BIT

The development of the steel plow and tractors used to break up and turn dense sod has all but eliminated the natural prairie biome as these areas were converted to farmland.

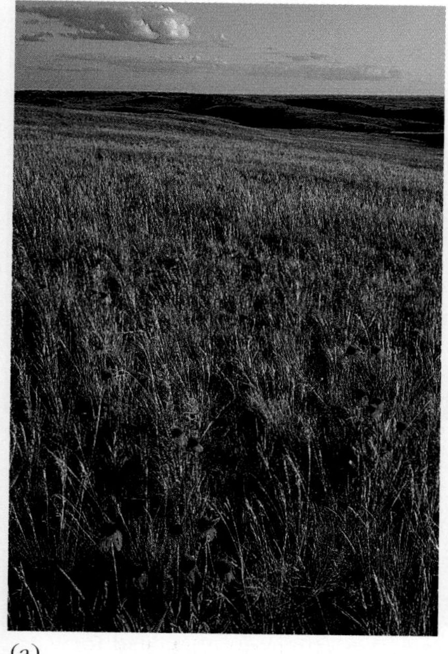
(a)

(b)

FIGURE 36-9

Prairie. (a) Purple coneflower in bloom in a prairie in North Dakota. **(b)** A prairie dog town near Devils Tower (in background), Wyoming. Prairie dogs aren't dogs at all, but instead are rodents related to chipmunks, squirrels, mice and rats. (a, W. Clay/ Dembinsky Photo Assoc.; b, J. Gerlach/ Dembinsky Photo Assoc.)

suffered more complete destruction than any other biome in North America. Some types of prairie have been so completely eliminated that ecologists are not even sure what plants and animals they originally contained. Conservation groups are undertaking the restoration of partly destroyed prairie in several parts of the United States (Figure 36-9). Overfarming is a serious environmental problem that has destroyed the temperate grassland biome. Accompanying overfarming is loss of topsoil, a worldwide ecological problem that threatens food production (see *A Journey into The Environment:* The Disappearing Soil, this chapter.)

Temperate Desert

Temperate desert or semidesert, sometimes called scrubland, occurs in regions too dry to support fertile grassland. Cool semidesert occupies much of the Great Basin east of the Cascade and northern Sierra Nevada mountain ranges in the western United States. Large areas are dominated by sagebrush, interspersed with perennial grasses (Figure 36-10). Typical animals include jackrabbits, sage grouse, pocket mice, and kangaroo rats. Cool, temperate semideserts also occur in central Asia, South America, and Australia.

Temperate forests and woodlands are less productive and diverse than tropical forests because for part of the year the temperature (and sometimes rainfall) is too low for plant growth. Temperate grassland has produced the world's deepest soils and, therefore, its best agricultural land.

36–E TAIGA

The **taiga** biome is dominated by conifers—spruces, pines, and firs—that can survive extreme cold in winter. Trees in the taiga tend to be farther apart than those in a deciduous forest, and light penetrating to the forest floor supports a ground cover of shrubs. The taiga, or **boreal forest,** as it is sometimes called, stretches in a giant circle through Canada and Siberia (Figure 36-11). The forest, containing only a few tree species, is occasionally interrupted by extensive areas of bog, or "muskeg," in poorly drained areas.

Precipitation is abundant (40 to 100 cm or 100 to 250 inches), and heavy snows characterize taiga. Because much of the precipitation in the taiga falls as snow, in the winter many of the animals grow white fur or plumage that blends with the background. Animals of the North American taiga include moose, wolverines, wolves, lynx, bears, spruce grouse, gray jays, crossbills, and (in summer) many species of warblers (birds). There are few species of reptiles and amphibians.

Only a few tree species form taiga forests, in which very low winter temperatures limit animal diversity.

36–F TUNDRA

The **tundra,** a treeless biome, occurs far north in the arctic regions, where winters are too cold and dry to permit the growth of trees. There is little annual pre-

(a)

(b)

FIGURE 36–10

Temperate semidesert. **(a)** Semidesert with sagebrush in central Washington state. **(b)** A whitetail jackrabbit makes its home in this California semidesert. (a, E. Zalisko; b, R. Planck, Dembinsky Photo Assoc.)

cipitation in the tundra (less than 25 cm or 10 inches), and the growing season is very short. In many areas the deeper layers of soil remain frozen, as **permafrost,** throughout the year, and only the surface thaws during the summer. Because the ground is so cold, decomposition is slow, so the soil is shallow, and plant growth slow. As a result, tundra takes a long time to recover when it is destroyed. This is why conservationists are so concerned about the effects of oil spills and oil industry traffic on the wildlife of the tundra.

Tundra vegetation is dominated by sedges, grasses, mosses, lichens, and dwarf woody shrubs. Bogs are common because the permafrost about a meter below the surface prevents water from draining away. The largest animals of the tundra are caribou in North America and reindeer in Europe and Asia. Hordes of mosquitoes, deerflies, and blackflies breed in the wet spots during the brief arctic summer. These insects contribute to the food available for a variety of birds, including various plovers, sandpipers, and horned larks, which migrate north to nest in the tundra and return to warmer climates for the winter.

Neither taiga nor tundra occurs at sea level in the Southern Hemisphere because the continents do not extend far enough south. Antarctica harbors only a few forms of life around its edges.

BIO-BIT

In tundra permafrost, entire mammoths, dead for perhaps 10,000 to 20,000 years, have been found intact, with "edible" meat still clinging to the skeleton!

Tundra is a biome of slow-growing plants found where there is enough light and water for plant growth for only a few months a year.

FIGURE 36–11

Taiga. Male (left) and female moose browsing in a boreal forest. (C. R. Sams, II/J. F. Stoick/Dembinsky Photo Assoc.)

36–G AQUATIC COMMUNITIES

Strictly speaking, the term biome refers only to communities on land. However, there are also many different kinds of aquatic communities, both marine and freshwater, which, like biomes, exhibit similarities wherever in the world they occur. Here we shall consider temperate freshwater lakes and life in the sea.

As on land, environmental conditions influence the distribution of organisms in water. Temperature, nutrient supply, the intensity of sunlight, and salinity (salt concentration) determine what can grow where. Lack of water is not a problem, but in some areas a

FIGURE 36–12

Tundra. Fall foliage on the tundra at the base of Mt. McKinley in Denali National Park, Alaska. (J. Johnson/Earth Scenes)

shortage of minerals dissolved in the water limits plant life. The types of organisms found in an aquatic ecosystem also depend upon the type of bottom (mud, sand, or rock), and upon wave action and water currents.

Lakes and Rivers

The world's large, temperate-region lakes, such as the Great Lakes of North America, and Lake Baykal in Russia, are vitally important sources of fresh water for drinking, agriculture, and industry.

Sunlight is a lake's source of energy. As it passes down through the water, some of it is used in photosynthesis by phytoplankton (free-floating algae) and some is absorbed by the water itself. So, as light passes deeper into the water, it becomes dimmer. In deep lakes there is a **compensation depth** where the available light is just bright enough for green plants to eke out a living: their photosynthesis (production of food and oxygen) exactly offsets their respiration (use of food and oxygen). Above the compensation depth plants produce more oxygen than they use, and so extra oxygen is available for the respiration of other organisms; below it there is not enough photosynthesis to offset respiration, and any available oxygen must come down from the water above.

Rooted aquatic plants such as water lilies and rushes grow in shallow water around the edge of the lake. Often, most of the lake's productivity comes from photosynthesis in this **littoral zone.** Fishes, amphibians, insects and other arthropods, snails, and worms live and feed among the plants. In the lake's open surface waters live floating plants, which need light, and animals that need abundant oxygen, such as fishes and small arthropods.

The amount of oxygen dissolved in the lake's waters is one of its most important features. Oxygen affects nearly every aspect of the lake, including what animals and plants can live where and the solubility of many inorganic nutrients. Oxygen enters the water from the photosynthesis of aquatic plants and by dissolving into the water from the air. Oxygen leaves the water when it is used in respiration. The cooler the water, the more dissolved oxygen it can hold. A lot of oxygen is used by the bacteria that decompose **detritus** (dead organic matter). Many commercial, desirable fish, such as trout, can survive only in waters containing large amounts of oxygen.

Deep in a lake, where there is not enough light for photosynthesis, the main biological process is decomposition. Here the chief input of energy is detritus falling from above, which feeds decomposers, fishes, and invertebrates.

Lakes can be divided into categories based on how much plant life they support. **Oligotrophic** ("few food") lakes are low in nutrients such as phosphorus, calcium, and nitrogen, so they support little plant growth and contain few organisms. Oligotrophic lakes are usually deep, with steep sides and narrow littoral zones. Their water is usually very clear, and the deep waters always contain oxygen because there are few organisms to use it (Figure 36–13). **Eutrophic** ("good food") lakes are rich in nutrients and organisms and are usually shallow. Such lakes contain little oxygen because decomposer organisms rapidly use it up metabolizing the organic matter produced by the lake's many other residents. In the normal course of events, a lake

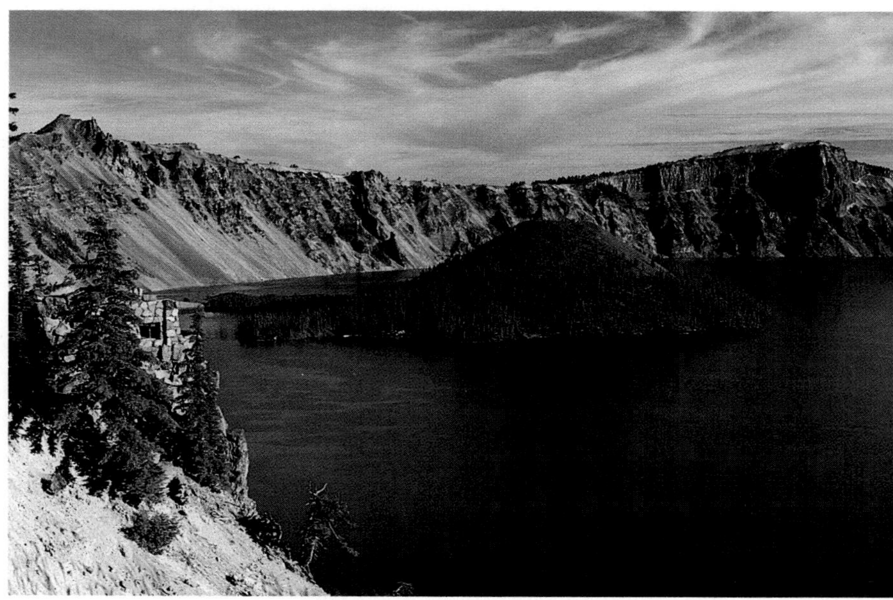

FIGURE 36–13

An oligotrophic lake in the Rocky Mountains. The lake has dark, clear water characteristic of oligotrophic lakes. (B. Berg/Visuals Unlimited)

ages as it is steadily filled in with minerals and organic matter, becoming more eutrophic as it ages. Natural eutrophication takes thousands of years, but the process may be speeded up to take only a few years if the lake becomes polluted. When nutrients in sewage or minerals such as chemical fertilizers wash into a lake, they speed plant growth and hence eutrophication.

The main difference between rivers and lakes in the same climate is that rivers usually have stronger currents. Where the current is strong, only organisms that can anchor themselves or swim against it can survive. A shallow, rapidly flowing river is usually well oxygenated because it has a large surface area to absorb oxygen from the air.

The Edges of the Ocean

The edges of the sea are the hatcheries and nurseries of many important species of marine life. Coastal wetlands and estuaries (areas where rivers enter the ocean) also serve as nesting, feeding, and resting spots for migratory waterfowl, and they reduce erosion and flooding inland. Recognizing these important but indirect contributions to human food, fun, and safety, ecologists are alarmed by the draining and filling of these areas to build towns, marinas, and resorts (see *A Journey into the Environment:* "Salt Marsh and Seafood," Chapter 37).

Along the seacoasts, many kinds of plants and animals thrive in the **intertidal zone,** the area between the high and low tide marks, where they are submerged for part of the day. There are three main types of intertidal zone: muddy, sandy, and rocky shores, which support very different communities.

Mudflats occur where the water moves slowly enough to deposit a sediment of small particles. Algae cover the particles and provide food for a multitude of burrowing molluscs, worms, and crustaceans.

Sandy beaches are less stable than mudflats, for sand shifts constantly and dries out faster than mud when the tide is out. Most of the tiny protists, worms, and crustaceans that live between the sand grains eat marine plankton stranded when the tide goes out, or algae attached to the sand grains. A wide variety of shore birds feeds on the invertebrate inhabitants.

Neither muddy nor sandy shores provide much foothold for sessile animals or anchored seaweed. These are much more common on rocky shores, which support a wider variety of organisms. As the water crashes onto the rocks, motile animals such as crustaceans anchor themselves firmly to rocks or seaweeds by their legs, or hide in crevices. Very few vertebrates live in the intertidal zone, although a number of birds come in at low tide to scavenge or to prey on invertebrates (Figure 36–14).

The **subtidal zone** occupies the **continental shelves**—the edges of continents—extending from the low tide mark to a depth of about 200 meters. Here, temperature fluctuates less, and wave movement is less violent than in the intertidal zone. Mineral nutrients are also readily available, washed from the land by

BIO-BIT

Seventy percent of the Earth's surface is covered with water, with 98% of it in the oceans.

(a)

(b)

FIGURE 36–14

Sea shores. (a) Sanderlings race the waves to feed at the water's edge. **(b)** Anemones, sea stars, barnacles, and kelp in a tidal pool in Olympic National Park, Washington. (a, A. Lowry/Photo Researchers, Inc.; b, B. Kent/Animals Animals)

rivers. Continental shelves are among the most densely populated areas on Earth.

Coral Reefs

Coral reefs are restricted to warm oceans, where the water temperature seldom falls below 21°C. Corals are cnidarians that live in symbiotic association with photosynthetic protists. The reef itself is made up of calcareous material, secreted by the coral animals themselves and by red and green algae. Since photosynthetic organisms are so important to their formation, coral reefs are found only in clear, shallow water where there is enough light for photosynthesis.

Most of a coral reef is submerged, although its top may be exposed at low tide. A reef acts physically like a rocky shore in providing anchorage for algae and sessile animals. A great variety of fish and swimming invertebrates find shelter within the reefs crevices (Figure 36–15). Today, many coral reefs are threatened by the dumping of wastes from nearby tourist areas and by drilling for underlying oil deposits. States such as the American and British Virgin Islands in the Caribbean enforce protective laws because once it has been destroyed, a coral reef takes many years to regrow. ■

FIGURE 36–15

A coral reef in the Red Sea. (M. Kazmers/Sharksong)

The Open Ocean

The open ocean can be divided into the top 100 meters or so, where photosynthesis can occur, the deep ocean, and the ocean floor. In the surface waters live the **plankton,** drifting protists, plants, and animals not powerful enough to swim against currents. They can, however, control their vertical position in the water, thereby moving from one current to another, for currents flow in different directions at different depths in many parts of the ocean. Fish and similar large animals in the ocean make up the **nekton,** creatures that can swim independent of currents. These animals feed mainly on plankton or on each other (Figure 36–16). Because mineral nutrients are scarce in many areas of

(a)

(b)

FIGURE 36-16

Animals of the nekton. (a) A green sea turtle heading out to sea. **(b)** A large school of glass minnows in the Caribbean Sea. (a, F. Lanting/Minden Pictures; b, A. Jones/Dembinsky Photo Assoc.)

the open ocean, the world's major fisheries lie over continental shelves, which receive minerals washed down rivers, and in parts of the open ocean where currents carry minerals up from the bottom.

Seventy-five percent of the ocean's water lies more than 1000 meters deep. Diving techniques permit sampling at depths of more than 6000 meters and reveal fascinating communities on the ocean floor—communities that contain no plants. The **benthos** (depths of the sea), the community of the ocean floor, includes decomposer bacteria that live on dead organisms falling from the surface layers above them or on dead members of the deep-sea community. Larger members of the benthos also feed on falling carcasses or on the decomposers, and filter-feeders strain food out of the water.

In lakes and rivers, the penetration of light into water determines where plants can live, and the concentration of oxygen dissolved in the water determines where animals can survive. Intertidal and subtidal zones are among the most densely populated marine communities because they receive abundant light, and mineral nutrients are readily available to plants. The distribution of life in the ocean depends on water temperature and the availability of sunlight, nutrients, and surfaces to which organisms can attach.

36–H ECOLOGICAL SUCCESSION

Coastal California is predominantly covered, not by chaparral, but by farms, roads, and buildings. Human civilization has disturbed the natural communities, clearing the vegetation to make room for human affairs and their adjunct parking lots. So, when we say that climate determines the type of community in an area, we mean the community that would exist if the area were left alone long enough, rather than what actually exists there. The community that forms if the land is left undisturbed, and that remains as long as no disturbances occur, is called the **climax community.**

After a climax community is disturbed, either by human activities or by natural means such as floods or fires, the area slowly returns to its original state. **Ecological succession** is the progressive series of changes that ultimately produces a climax community.

Primary Succession

Primary succession occurs where there is initially no soil—when a new island rises out of the sea, or when a glacier retreats or a mountainside caves in, leaving a pile of rocks. Consider an area of rock created by a landslide. Lichens adapted to exposed conditions may spread over the rock surface. They produce organic acids, which dissolve some of the rock. Dead lichens contribute organic matter to the forming soil, and mosses may gain a hold in even a thin layer of lichen remains and rock dust. As the mosses break up the rock further and add their own dead bodies to the pile, the seeds of small, rooted plants can germinate and grow. The process continues along similar lines until the climax community becomes established. It may take thousands of years for the soil and the climax vegetation to develop fully (Figure 36-17).

FIGURE 36–17

Primary succession. Orange lichens growing on rocks in the Falkland Islands. (F. Lanting/Minden Pictures)

Primary succession also occurs as an oligotrophic lake or pond ages. It fills up with silt and fallen leaves, becoming more eutrophic as the shoreline creeps toward the center of the lake. Gradually the lake turns into a marsh and then into dry land, eventually colonized by plants of climax species from surrounding communities.

Secondary Succession

Secondary succession is the series of changes that occur in a community that has been disturbed but not totally stripped of its soil and vegetation. Although it may take a hundred years or more for the climax vegetation to return during secondary succession, the process is much faster than primary succession because soil already exists.

A familiar example of secondary succession in the New England area is "old field succession," by which abandoned farms return to the climax temperate forest. When a farmer stops cultivating the land, grasses and annual weeds quickly move in and clothe the earth with a carpet of wild carrot, black mustard, and dandelions. The **pioneers** of newly available habitats, these plants grow rapidly and produce seeds adapted to disperse over a wide area. Soon taller plants, such as goldenrod and perennial grasses, move in. Because these newcomers shade the ground and their long root systems monopolize the soil water, it is difficult for seedlings of the pioneer species to grow. But even as these tall weeds choke out the sun-loving pioneer species, they are in turn shaded and deprived of water by the seedlings of pioneer trees, such as pin cherries, dogwoods, sumac, and aspens, which take longer to become established but command the lion's share of the resources once they reach a respectable size.

Succession is still not complete, for the pioneer trees are not members of the climax forest. After 5 to 30 years, slower-growing oak, maple, and hickory trees will take over, shading out the saplings of the pioneer tree species. After perhaps a century or two, the land is covered with mature climax forest.

In any tract of land, we can always find at least small patches that are undergoing succession following disturbance—a spot where a large tree has fallen, leaving a light gap where pioneer weeds can move in, or a burned forest (Figure 36–18). The existence of various patches undergoing succession ensures that there is a steady supply of **fugitive** plants, the fast-growing, here-today-and-gone-tomorrow pioneers, which include many kinds of weeds.

Animals, as well as plants, may be fugitive species. Insects that specialize in eating a particular plant species may travel far and use their keen senses to smell out new patches of their food plant some distance away. Some of our agricultural pest problems stem from the fact that most crop plants originated as fugitive species. By planting fields exclusively in one crop year after year, farmers create a paradise for fugitive animals such as cabbage worms and cucumber beetles, which no longer have to spend energy to find food and have nothing to do but eat and multiply.

Succession occurs because of progressive changes that make the environment less favorable for the species that are present and more favorable for colonization by others. Some of the changes are purely

(a)

(b)

(c)

FIGURE 36–18

Secondary succession. **(a)** Regrowth of glacial lilies soon after the 1988 fire at Yellowstone National Park. **(b)** Fireweed, one of the plants to first appear after a fire. **(c)** Regrowth 10 years after Mt. St. Helens' volcanic blast devastated this area in Washington. Note the fireweed in the foreground. (a, S. Osolinski/Dembinsky Photo Assoc.; b, c, E. Zalisko)

physical, like the silting in of a lake or the weathering of rock, but many are caused by the organisms themselves. As succession proceeds, the supply of available nutrients in the soil declines as minerals become increasingly locked up in living organisms. Both the community's production of new organic matter (through photosynthesis) and the total weight of all the organisms in the community increase during succession, levelling off as the climax is approached.

Fire-Maintained Communities

Fire is one cause of disturbance and succession that has been particularly well studied. Fires, set by lightning or by human activities, occasionally sweep through large areas of taiga and temperate forest, burning trees and destroying entire communities of animals and plants. Burned areas undergo secondary succession.

In some communities, fire occurs often enough to determine the nature of the dominant vegetation. Such communities include some pine forests, chaparral, and temperate grassland. Grasses readily regenerate after fires that would kill trees; thus recurrent fires may prevent grassland or savanna from turning into woodland.

Seedlings and saplings of deciduous trees are especially susceptible to fire, whereas many pines are adapted to survive, and even to exploit, fires. In some pines, for example, the cones open and their seeds germinate only when exposed to temperatures of several hundred degrees. This ensures that the seeds germinate in areas that have just been burned. If fires are prevented in a fire-adapted pine forest, deciduous trees may become established. In addition, dead wood and litter build up on the ground, adding extra fuel. When a fire eventually does occur, it is more severe than usual, destroying not only any deciduous colonizers but also the pines and other species.

2? Odd though it may seem at first, frequent burning is essential for the preservation of many natural communities (Figure 36–19). This is the reason that the U.S. Park Service adopted the policy of letting fires in National Forests burn if they do not endanger human life or property. This policy caused an outcry when fires burned millions of hectares in Yellowstone National Park in 1988 because most people did not understand the ecology of fire-adapted communities. These fires became an opportunity for visitors to learn about succession following a fire. In 1989, the park had more visitors than ever before—come to survey the damage and to watch pioneer species colonizing the blackened landscape. ■

Primary succession occurs slowly because it takes time for soil to form and for plant life to move into an area. In secondary succession, an area where vegetation has been disturbed or destroyed undergoes a series of species replacements until the soil and organisms of the area's climax community are restored.

36–I WHY ARE ORGANISMS WHERE THEY ARE?

3? Fugitive species get around quickly, with efficient dispersal mechanisms, such as seeds that float on the wind or are carried by animals. However, there is a limit to how far an organism can travel over territory that is unsuitable for its survival. If we return to the question that began this chapter and ask again, "Why are organisms where they are?" we find that at least two factors determine the answer: climate and the organisms' ability to disperse to areas where they can survive.

(a)

(b)

FIGURE 36-19

The role of fire in ecosystems. (a) Controlled burning of hillside prairies in Redwood National Park, California. **(b)** A fire-resistant species of pitch cone. (a, R. Archibald/Earth Scenes; b, J. McDonald/Earth Scenes)

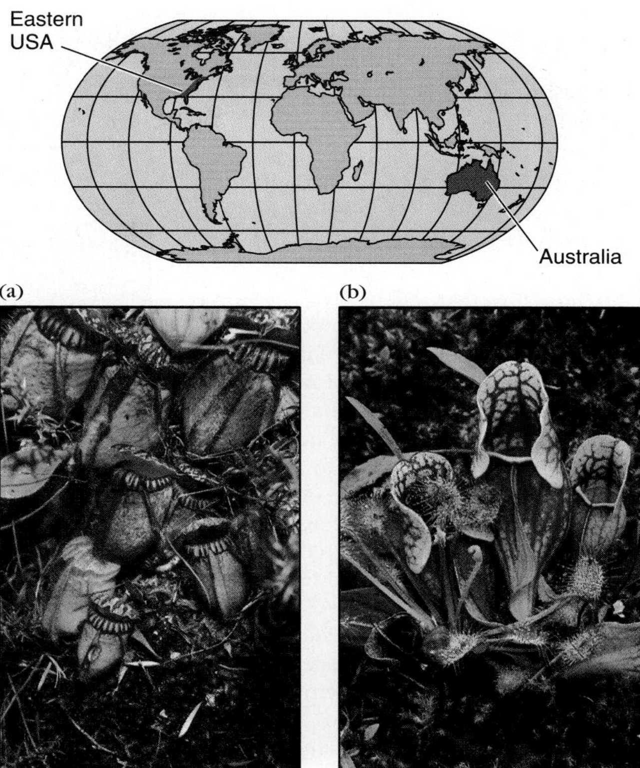

FIGURE 36-20

Convergent evolution in insectivorous plants. (a) The Australian Albany pitcher plant closely resembles American pitcher plants **(b)**, but is unrelated. Both plants have leaves modified into liquid-filled jugs that trap and drown insects to supplement nutrient-poor soil —a remarkable example of convergent evolution. (a, J. Alcock/Visuals Unlimited; b, S. Moody/Dembinsky Photo Assoc.)

When we look again at African and South American rain forests and ask, "Why are there different species in these two areas?" our answer is that, although each area has a climate suitable for growth of the plants and animals of the other area, they are separated by wide stretches of ocean, an impassable barrier to dispersal. There are exceptions: for example, small animals called rotifers form cysts that can be blown almost anywhere in the world, but the animals can live only in very restricted types of environments. As a result, the rotifer species in a puddle of water in a marble cemetery urn in Pennsylvania may be the same as that in a marble urn in a South African cemetery, but different from that in the granite urn on the next grave!

Because most species cannot travel between continents (unless they are transported by humans), there are different species in the tropical rain forests of the different continents. The similarities of form and color of species in each place result from convergent evolution (Section 18-C). For instance, the advantage of being able to conserve water in desert habitats has led to the evolution of plants with thick, water-storing stems, as well as spiny leaves that deter animals from using the stems as the source of their own water, in deserts all over the world (Figure 36-20). ■

A species' distribution depends on the type of climate it requires and on its ability to disperse (or be dispersed) to suitable new areas.

BIO-BIT

Factors that limit the location of populations are often surprising. Ring-necked pheasants imported from the north do not do well in the southern United States because their eggs do not develop properly at the higher temperatures there.

A Journey into The Environment

The Disappearing Soil

Every year, an average of 25 to 30 tons of soil erodes from each hectare (2.5 acres) of American agricultural land. **Soil erosion** is the movement of soil from one location, usually agricultural land, to somewhere else, usually a lake or river. Wind and water are the forces that move the soil. In the United States, about 70% of all farmland has had its productivity reduced by soil loss, and many areas have lost essentially all their soil. The soil of an area is considered destroyed when crop plants will no longer grow there.

Soil destruction threatens our ability to feed the world's growing population, and it also causes other problems. For instance, soil washing into streams and rivers carries fertilizer and pesticides with it and so contributes to water pollution. The soil particles may also block out light needed by photosynthetic organisms in streams and lakes (Figure 36-A).

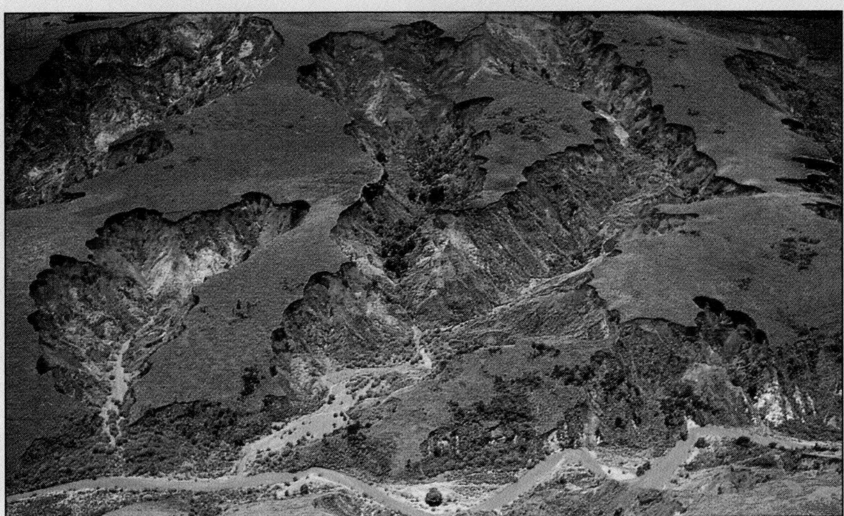

FIGURE 36-A

Soil erosion. Massive erosion and siltation due to deforestation in Madagascar. Notice how eroded soil has colored the river orange in the foreground as well as two streams. (F. Lanting/Minden Pictures)

Erosion is not the only cause of soil destruction. Soil in many parts of the world has become so salty that plants can no longer grow in it, especially in arid regions, including the semidesert and shrubland of the western United States. **Arid regions** are areas where more water leaves the soil by evaporation than reaches it through rainfall. (Water moves up from the water table underground, drawn by plant transpiration and by capillarity.) When water evaporates, the salts dissolved in it are left behind, so arid land soil contains high concentrations of salts. Because arid land receives less than 25 centimeters of rain a year, most agriculture in arid areas is irrigated. When irrigation water evaporates, even more salts are added to the soil. Eventually, the soil may become so salty that plants cannot grow in it. The accumulation of salts in the soil is called **salinization.**

The usual way to grow crops on very salty soil is to add much more water than the plants can absorb. Then the water will run off the land, carrying dissolved salts with it. This salt-laden water is often toxic. Kesterton Wildlife Refuge in California receives such high concentrations of the element selenium from the irrigated land of the San Joaquin Valley that birds and plants in the refuge die. Selenium occurs naturally in the soil in low, harmless concentrations, but evaporation of irrigation water from arid land has concentrated it to a dangerous level. Scientists are attempting to clean up the toxic soil in Kesterton by growing plants and fungi that absorb selenium.

Saline soil can be restored to usefulness without intensive care by planting salt-tolerant trees. These ab-

FIGURE 36-B

Erosion by wind. A Texas cotton farmer in a windstorm. (J. Brandenburg/ Minden Pictures)

sorb salts and hold the soil in place. The leaves and twigs that fall from the trees add organic matter to the soil. Australia has organized its Girl Scouts and other groups to plant more than a billion trees on land turned to saline desert by intensive agriculture. Tree seedlings survive best if they are watered and protected from animals for the first few years, so soil reclamation projects are most successful if residents of the area undertake to care for the trees.

The problem of soil destruction is largely economic: an individual farmer profits for a short period by farming practices that destroy the soil. These practices include planting crops instead of windbreak trees, irrigating arid land, or leaving land without plant cover between crops (Figure 36-B). Because the problem is economic, the solution usually involves altering government regulations and subsidies. For instance, in 1985, the United States passed a soil conservation act that rewards farmers for planting trees and grass on previously plowed land that erodes easily (such as that on steep hillsides). This program greatly reduced the rate of soil erosion.

Soil destruction is sometimes called **desertification,** the formation of desert from previously productive land as a result of human activities. The name arose because many of the world's deserts have been produced or enlarged in this way, and each year desertification adds to them an estimated million hectares (nearly 4000 square miles). In prehistoric times, the Sahara Desert was less than a fifth its present size.

The desert created by destroying soil can be made to bloom again, but this is a long, slow process. The basic solution is to keep the soil covered with plants at all times and let some of this plant material decay into the soil. If only rock or gravel remains when the soil is gone, soil will take thousands of years to form again by primary succession. If sand and clay remain, the outlook is brighter. Both can be made into fertile soil by the addition of organic matter. This can be done intensively, by planting **green manure,** plants that will grow in nutrient-poor soil and are then plowed into the soil to decompose. After only a few years of this treatment, soil will regain its fertility.

Loss of soil is probably the most devastating environmental problem that we face. Governments around the world are beginning to realize this and to design economic incentives that will encourage farmers to conserve and restore the soil.

SUMMARY

1. A worldwide survey of the distribution of organisms reveals two main patterns:
   a. Different areas of the world are inhabited by different species of plants and animals.
   b. Terrestrial communities in different parts of the world can be divided into a fairly small number of categories, or biomes, each with a characteristic array of plant and animal life. These biomes are worldwide, occurring wherever a suitable climate exists.
2. While the actual species found in an area depend on the area's evolutionary history, the biome depends mainly upon the annual pattern of rainfall and temperature. Similar changes in biomes occur with increasing altitude and latitude.
3. The richest biome is tropical rain forest, where high temperatures and rainfall permit plants to grow throughout the year. Most of the plant and animal life is found in the canopy among the broad evergreen leaves of trees. The soil is poor because leaching and decomposition are extremely rapid. Most of the nutrients in a tropical rain forest are locked in the bodies of living organisms.
4. In temperate deciduous forest, the soil is much richer in nutrients because the trees lose their leaves in the fall, creating a litter layer that decomposes and releases the leaves' nutrients only slowly.
5. Deciduous forest is an important biome of North America, Europe, and Asia in areas characterized by warm, moist summers and cold winters.
6. Where the soil is poor or fire is frequent, temperate evergreen forest replaces temperate deciduous forest. Farther north, both are replaced by taiga, a biome dominated by conifers adapted to growing in sparse soil and to resisting extreme cold and water loss in winter.
7. North of the taiga lies the tundra, dominated by cold-resistant grasses, lichens, and slow-growing woody shrubs.
8. Grasslands receive less rainfall than forests but more than deserts, and occur in the drier interiors of continents. Shrubs and trees may be scattered among the tall grasses. Grasses are replaced by small woody shrubs in areas where there is extra water.
9. Deserts have hot days, cold nights, and very little rainfall. Desert plant life is mainly composed of annuals with very short growing seasons, succulent perennials adapted to the low rainfall, and small-leafed, woody shrubs.
10. The distribution of aquatic organisms is determined by water temperature, depth, salinity, and motion, and by the availability of light, oxygen, and minerals.
11. Shallow areas near the coast are well supplied with both light and minerals, and thus support dense communities of life.
12. Coral reefs are specialized communities found only in tropical ocean waters.
13. In the open ocean, the availability of light for photosynthesis restricts plankton to the upper layers of the water, but scarcity of nutrients in these layers may limit the numbers of organisms. Larger nektonic organisms are found primarily where planktonic food is abundant.
14. Dead organisms from the surface layers of the ocean supply food for a benthic community of bacteria and other organisms that live on the sea floor.
15. Although the climate of an area determines the composition of its climax community, patches are always in various stages of ecological succession following disturbances.
16. Organisms adapted to living in the unstable communities of early successional stages have effective dispersal mechanisms and perpetuate themselves by continuously colonizing new habitats as they arise.
17. Most organisms cannot travel far and thus do not colonize all possible habitats. Therefore, in different parts of the world we find similar communities inhabited by similar species, reflecting convergent evolution.

SELF-QUIZ

1. Which of the following has a vegetation structure with only one main level?
   a. tropical rain forest
   b. taiga
   c. grassland
   d. shrubland
   e. desert
2. Which of the following communities would have trees?
   a. taiga
   b. intertidal zone
   c. shrubland
   d. tundra
   e. plankton
3. A biome with high temperature, high rainfall, and poor soil is:
   a. shrubland
   b. coral reef
   c. semidesert scrub
   d. tropical rain forest
   e. temperate evergreen forest
4. Which of the following communities has no living green plants?
   a. a rocky shore
   b. the plankton
   c. a mud flat
   d. the deep ocean floor
   e. a coral reef
5. Compared with a eutrophic lake, an oligotrophic lake contains a greater concentration of:
   a. organic matter
   b. plants
   c. oxygen
   d. bacteria
   e. mineral nutrients

In questions 6 and 7, choose the correct term from each pair in parentheses.

6. Colonization of an abandoned stone quarry would be an example of (primary/secondary) succession.

7. A(n) (early successional/climax) community would have a high proportion of fugitive species.
8. A pond in a deciduous forest becomes filled in with rock particles and dead leaves, creating soil. List, in order, the types of vegetation that would be seen as this area undergoes ecological succession, and name the climax community that would eventually result.
9. The American prairies and the Asian steppes do not have the same species of grasses because ____. However, both are inhabited primarily by grasses because ____.

THINKING CRITICALLY

1. The 30°N latitude line runs through southern Louisiana and northern Florida as well as through desert country in Mexico and Texas. Why is the area in Louisiana and Florida not desert like the area in Mexico and Texas?
2. Why is it proving difficult to carry out large-scale "agribusiness" farming in vast tracts of land cleared of their tropical rainforest vegetation?
3. Why is there less variation in size of vegetation in the tundra than in tropical regions?
4. Why does secondary succession slow down as it proceeds?
5. How can frequent fires increase the species diversity of a region?
6. Why does a light gap contain different species of animals and plants from those in surrounding climax forest?

SELECTED KEY TERMS

biome, *p. 749*
boreal forest, *p. 757*
chaparral, *p. 756*
climax community, *p. 762*
deciduous trees, *p. 753*
desert, *p. 754*
ecological succession, *p. 762*
ecosystem, *p. 749*
intertidal zone, *p. 760*
littoral zone, *p. 759*
permafrost, *p. 758*
plankton, *p. 761*
primary succession, *p. 762*
secondary succession, *p. 763*
subtidal zone, *p. 760*
taiga, *p. 757*
temperate deciduous forest, *p. 755*
temperate desert, *p. 757*
temperate evergreen forest, *p. 755*
temperate forest, *p. 755*
temperate grassland, *p. 756*
temperate rain forest, *p. 755*
temperate shrubland, *p. 756*
temperate woodland, *p. 756*
tropical rain forest, *p. 752*
tropical savanna, *p. 753*
tropical seasonal forest, *p. 753*
tropical thornwood, *p. 754*
tundra, *p. 757*

SUGGESTED READINGS

BioScience 39 (10), 1989. "Yellowstone Fires." An entire issue devoted to the effect of the 1988 fires on the ecology of Yellowstone National Park and the controversy over the extent to which fires should be extinguished or left to burn naturally.

Childress, J. J., H. Felbeck, and G. N. Somero. "Symbiosis in the deep sea." *Scientific American,* May 1987. Tube worms and clams live on the energy from hydrothermal vents in the deep sea floor. They are supplied with nutrients by symbiotic bacteria that live on them.

Colinvaux, P. "The past and future Amazon." *Scientific American,* May 1989. Colinvaux argues that the enormous diversity of species in the Amazon Basin is partly a result of frequent disturbances (of climate and geology). We can, therefore, be reasonably optimistic that the Amazon will survive recent human disturbances—as long as these are not too destructive.

Perry, D. R. "The canopy of the tropical rain forest." *Scientific American,* November 1984. Its inventor tells the story of a climbing method that has made it possible to do research in the once-inaccessible rainforest canopy.

Repetto, R. "Deforestation in the tropics." *Scientific American,* April 1990.

Wilson, E. O. "Threats to biodiversity." *Scientific American,* September 1989.

CHAPTER 37

CONCEPT GUIDE

After reading this chapter, you should be able to:

1. Compare the roles of producers, decomposers, and consumers in an ecosystem. (Section 37-A)
2. Explain why the flow of energy through an ecosystem can be illustrated by a pyramid, with less energy available to higher trophic levels than to lower ones. (Sections 37-B and 37-C)
3. Explain how deforestation and fossil fuel consumption affect the carbon cycle. Also explain how human activity can affect the nitrogen and phosphorus cycles. (Section 37-D)
4. Describe the experiment in which ecologists tested the hypothesis of species turnover on a mangrove island in Florida. Explain the significance of their results. (Section 37-E)

The Florida Everglades are an example of a major U.S. wetland, home to this Roseate Spoonbill and a wide variety of other organisms. Wetlands cover nearly 6% of the world's land surfaces, help control flooding, and produce many commercially valuable products. Read more about the importance of wetlands in *A Journey into The Environment:* Salt Marsh and Seafood (page 774).
(J. Roetzel/Dembinsky Photo Associates)

Ecosystems and Communities

When you gaze up at the sky on a clear night, it is difficult to imagine that you are looking at history, but any astronomer will tell you that this is true. The stars are so far away that their distances are not measured in miles, but in the distance it takes light to travel in a year. Thus, for example, Alpha Centauri (the star that is closest to Earth) is 4.3 light years away and the light you see when you look at this star started on its journey through space more than four years ago. Most stars are much farther away than Alpha Centauri and the Andromeda Nebula, a neighboring galaxy in the constellation Andromeda, sheds light that is about 2.7 million years old—prehistoric in Earth terms. In much the same way, when an ecologist studies an environment, he or she is also looking at history. But here the history reflects events on Earth. Past events have shaped not only animal and plant distributions, and morphology, but they have also created the web of relationships between organisms that we call ecology.

A tropical rain forest or a desert may cover a huge area. For convenience, ecologists usually study smaller units—for example, a hillside, a lake, or a field. Each of these is an **ecosystem,** consisting of the **community** of all the organisms living there, along with their physical environment. We usually treat an ecosystem as an isolated unit, but in fact things invariably move from one ecosystem to another, as when leaves blow from a forest into a lake, or birds migrate between their summer and winter homes.

Not all ecosystems are natural: a space station, an aquarium, and a pot of houseplants are artificial ecosystems. A farm is often considered as an ecosystem because farmers must recognize the interactions between crop plants, fertilizers, pesticides, soil, climate, and the natural plant and animal life in order to manage the farm effectively.

The organisms in an ecosystem interact with each other and with their physical environment. These interactions can be viewed on two different time scales. Ecologists study what is happening here and now: plant growth, animals eating plants, and so on. But over the long term, every environmental event affecting organisms may also be a selective force that shapes their evolution. Each time an owl catches a mouse, it not only feeds itself and reduces the number of mice, but also selects against a set of mouse genes that are not effective at avoiding capture by owls.

KEY CONCEPTS

- The organisms in an ecosystem interact with one another and with their physical environment, influencing each others' lives and evolution.
- Nutrients may cycle indefinitely in an ecosystem, but energy is continuously lost.
- A self-sustaining ecosystem must contain nutrients, producers, and decomposers, and receive a continuous input of energy.

Curiosity Questions

1? Why aren't there organisms that prey upon wolves or tigers? (See page 776)

2? What factors contribute to global warming, and what are the likely results of global warming? (See page 780)

3? How does cutting down the trees in a forest affect that ecosystem? (See page 782)

37–A THE BASIC COMPONENTS OF ECOSYSTEMS

To be sustainable, capable of lasting indefinitely, an ecosystem must contain the resources necessary to support its resident organisms and to dispose of their wastes. The necessary components of an ecosystem are water, various minerals, carbon dioxide, oxygen (in most cases), and various kinds of organisms. An ecosystem must also receive a continuous supply of energy.

The sun, as it drives photosynthesis by green plants, is the ultimate source of energy for almost all ecosystems. (Chemosynthetic bacteria are the exception; they make food using energy from inorganic chemical reactions [see Section 19-B]). Because green plants are **autotrophs,** making their own food from inorganic substances, they are called the **producers** of all the food in the ecosystem (see A Journey Through Energy Flow in Ecosystems).

Plants eventually die. Their remains are usually broken down by **decomposers,** organisms like bacteria and fungi that acquire their food molecules from dead organic material. In the process of extracting energy and nutrients from this material, decomposers release some of the nutrients back into the ecosystem, where they are again available to producers. Nutrients are thus cycled through the ecosystem and may be used again and again in the same small area. Energy, by contrast, is not cycled but is continuously lost from an ecosystem. Most organisms would soon die if the sun's energy were cut off for any length of time.

The basic requirements for a self-sustaining ecosystem are inorganic nutrients, producers, decomposers, and a continuously renewed source of energy. However, most ecosystems also contain **consumers,** animals (and other organisms) that eat plants or each other. Plant-eating animals are collectively known as **herbivores** or **primary consumers.** These may die and pass directly to the decomposers, or they may be eaten by **carnivores,** also called **secondary consumers.** There may be tertiary or even quaternary consumers, carnivores like Great Horned Owls, Snowy Egrets, Ospreys, and polar bears that feed on the secondary and tertiary consumers, respectively. Consumers and decomposers are **heterotrophs,** feeding on organic matter produced by other organisms. (See *A Journey into The Environment:* Salt Marsh and Seafood, this chapter.)

Ecosystems typically involve a complex exchange of energy and inorganic nutrients between producers, consumers, decomposers, and the nonliving environment. The ultimate energy source driving the system is usually the sun.

BIO-BIT

Although most ecosystems are based upon autotrophic plants that use the sun's energy, some exceptions have been discovered. Deep in the ocean, where hot volcanic rock meets the seawater, scientists have discovered entire communities based upon chemosynthetic autotrophic bacteria that derive their energy from hydrogen sulfide in the water. The water these bacteria live in is near 200°C, and would vaporize if it were not under such high pressure deep beneath the ocean's surface.

37–B FOOD WEBS AND ENERGY RELATIONSHIPS

A **food chain** is a series of organisms, each of which is eaten by the next. Examples are the passage of nutrients and energy from leaves to caterpillars that eat them, to chickadees, to hawks, or from dead animals to fly maggots to parasitic wasps. Food chains are interconnected in a complex pattern called a **food web,** which includes all the feeding relationships in an ecosystem or community (see A Journey Through Energy Flow in Ecosystems).

The **trophic level** to which an organism belongs is an indication of how far it is removed from plants, the first link in the food chain. Green plants make up the first (producer) trophic level. The second trophic level contains the plant-eating animals (primary consumers), and higher trophic levels are made up of carnivores (secondary consumers, and so forth). An organism cannot always be assigned to just one trophic level. Thus some plants, such as Venus flytraps, are carnivores as well as producers. A frog tadpole eats diatoms or other plant life, but the adult frog is carnivorous. Many mammals, such as foxes, bears, and humans, are **omnivores,** organisms that belong to several trophic levels because they eat both plants and other animals.

Pyramids of Energy

The flow of energy through an ecosystem can be represented in the form of a **pyramid of energy,** which

BIO-BIT

The complicated nature of food webs often leads to unexpected health risks. Many pesticides sprayed near sites where cows graze wind up in the milk we drink, and even in the milk of human mothers.

(text continues on page 776)

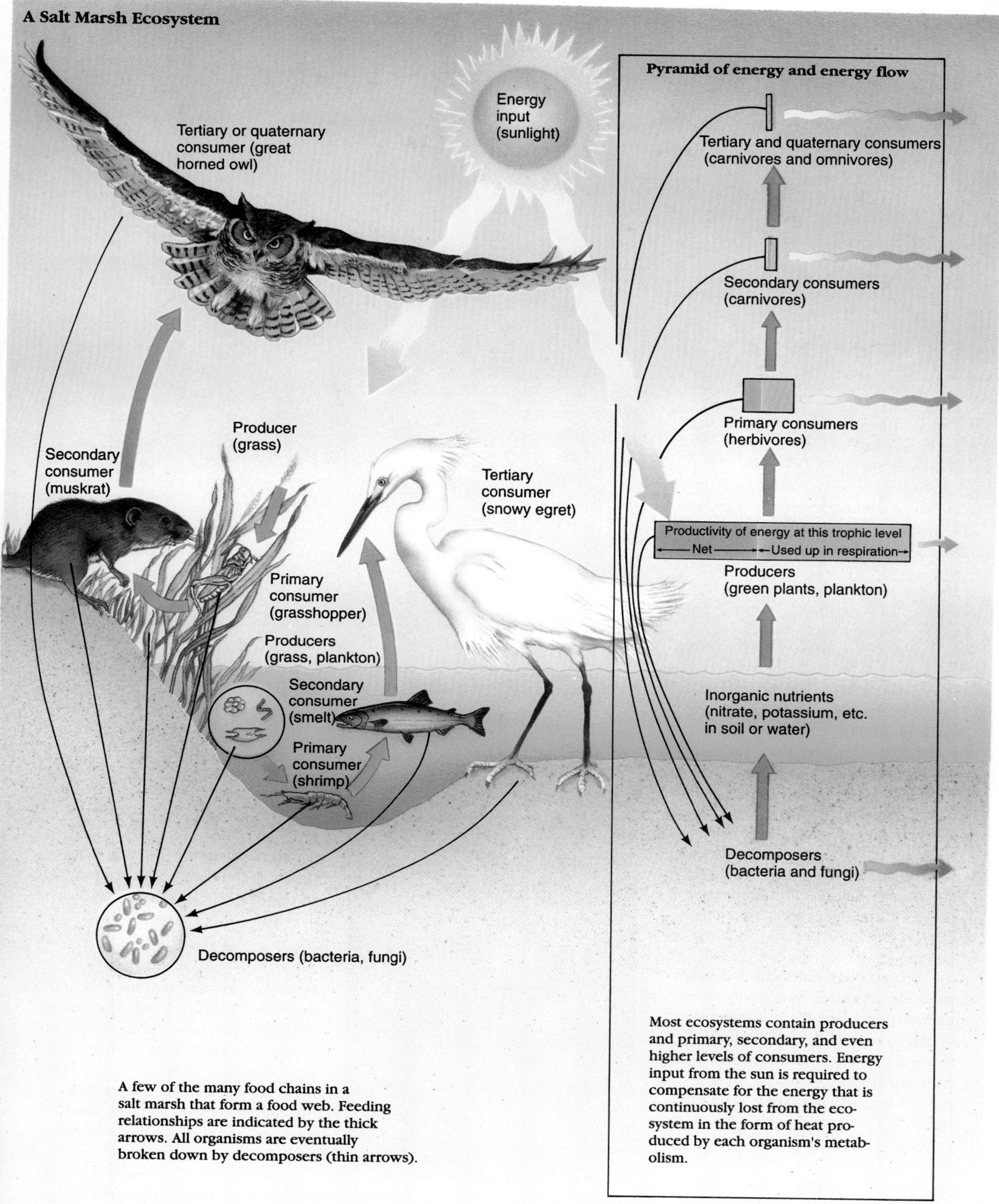

A few of the many food chains in a salt marsh that form a food web. Feeding relationships are indicated by the thick arrows. All organisms are eventually broken down by decomposers (thin arrows).

Most ecosystems contain producers and primary, secondary, and even higher levels of consumers. Energy input from the sun is required to compensate for the energy that is continuously lost from the ecosystem in the form of heat produced by each organism's metabolism.

Energy flow in ecosystems. Energy moves between the many organisms of an ecosystem to maintain the activities of life. Notice that energy is lost in the form of heat as it moves up the food chain indicated on the right.

A Journey into The Environment

Salt Marsh and Seafood

Revellers at clambakes, oyster roasts, or shrimp boils seldom pause to consider that their favorite seafood is on the menu only because the local salt marsh has not yet been converted to highways or condominiums.

Salt marsh dominates much of the flat shoreline of the Gulf of Mexico and Atlantic coasts of the United States. Here rivers, such as the Mississippi, divide into numerous channels and deposit their load of mud and sand as they wend slowly across huge marshy deltas to the sea. Smooth cordgrass *(Spartina alterniflora)* covers hundreds of thousands of hectares of the marsh, with widely scattered stands of other plants on higher, dryer ground (Figure 37-A).

FIGURE 37-A

A Florida salt marsh. (M. and B. Hunn/Visuals Unlimited)

Cordgrass is the mainstay of an ecosystem that is unusual in two ways. First, it contains few species of plants. Salt marshes are flushed twice a day by the tide, and so the soil is made up of wet, salty mud and sand. Few other flowering plants can tolerate these conditions, but cordgrass thrives in them. Second, few animals can eat cordgrass because its leaves contain quantities of glass-like silica.

So, although cordgrass is a major producer in the ecosystem, its energy and nutrients pass to consumers indirectly, by way of decomposers. Because the soil is constantly flushed by water full of nutrients and oxygen, the cordgrass grows rapidly, and salt marsh is a very productive ecosystem. Every year, the dead stalks of last year's cordgrass break off and wash up high into the marsh, where they are broken down by decomposers into **detritus,** molecules and small particles of organic matter. The tide then washes the detritus slowly down toward the many creeks that meander through the marsh. The organic matter provides nutrients for a large biomass of microscopic algae, and food for crabs, mussels, worms, snails, and clams that live in the soil and among the cordgrass stems. At the edges of creeks, oysters feed on detritus and algae that they filter out of the water. All these invertebrates, in turn, feed human collectors, raccoons, marsh rats, and a host of wading birds (Figure 37-B).

FIGURE 37-B

A marsh resident. A Great Blue Heron, a common predator upon fish and crustaceans in salt marshes. (F. Lanting/Minden Pictures)

The marsh is the nursery where a number of transient animals find food and protection while they are small. For instance, southern commercial shrimp live on the ocean floor as

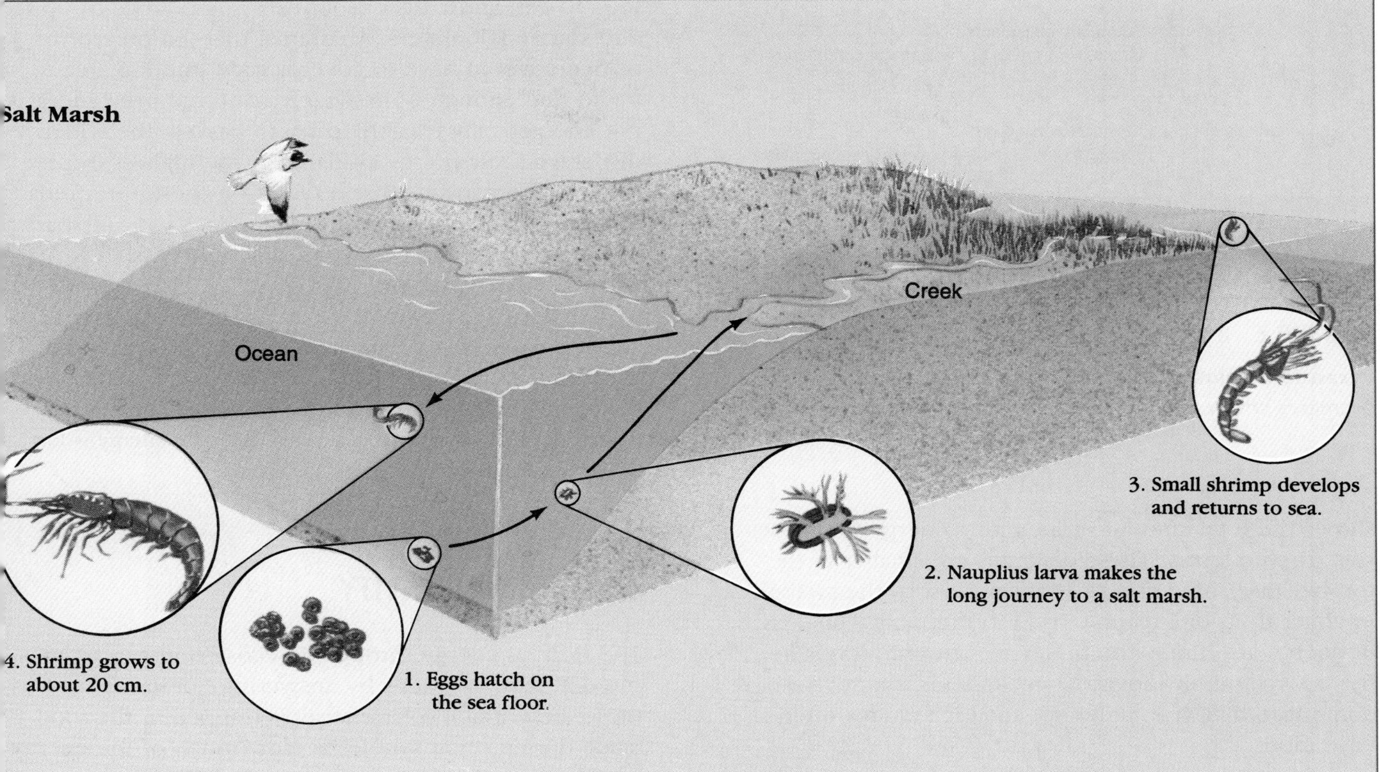
Salt Marsh
Creek
Ocean
3. Small shrimp develops and returns to sea.
2. Nauplius larva makes the long journey to a salt marsh.
4. Shrimp grows to about 20 cm.
1. Eggs hatch on the sea floor.

FIGURE 37–C

The life-cycle of the southern pink shrimp (*Penaeus duorarum*). The nauplius larva hatches from an egg on the sea floor and makes the long journey up a creek into the marsh. Here it develops into a small shrimp, which returns to the sea. These shrimp grow to about 20 centimeters long. Every year, millions of dollars' worth of them are caught for the dinner table by trawlers.

adults. They lay their eggs on the bottom, and the young move in with the tidal currents until they reach the marsh, where they grow to adulthood (Figure 37–C). Other commercially important species that spend part of their lives in the marsh include the young stages of several species of crabs, as well as fishes such as flounders, menhaden, and sea trout. As they grow to maturity, these animals migrate down the creeks and out to sea, where they may end up as seafood on the table or as food for ocean fishes such as swordfish, snapper, grouper, and tuna.

The economic importance of salt marsh was not widely recognized until recently, after large areas of this ecosystem had been destroyed. In Louisiana, which contains about 40% of the United States's remaining coastal wetland, one hectare of marsh is lost every *hour.* (It is built on, washed away, or sinks beneath the rising sea.) Led by ecologist Eugene Odum, Georgia was among the first states to try to save its salt marshes. A 1970 act protects the marsh against all incursions except road-building and the federal government (which has destroyed vast areas of marsh with "flood control" projects and military bases).

Legislators are particularly concerned about protecting the marsh's roles in seafood production and pollution control. Salt marsh has a considerable ability to absorb and detoxify pollutants. A 1987 oil spill in the Savannah River caused considerable damage in the river itself, but closed the oyster beds for only a few days when it leaked into the nearby salt marsh. Large quantities of inadequately treated sewage, farm fertilizer runoff, and even heavy metal industrial waste, which kill fishes and eutrophy lakes in other places, are absorbed by bottom mud, or broken down by organisms, and rendered harmless by the marsh.

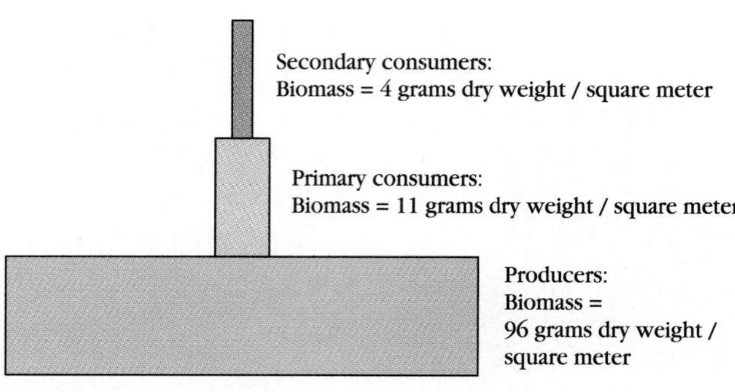

FIGURE 37–1

Pyramid of biomass for a lake in Wisconsin. Biomass decreases with every step up a food chain.

shows the total amount of incoming energy for successive trophic levels. These diagrams are smaller at the top because some energy is always lost to the system in going from one trophic level to the next. This loss of energy also limits the number of trophic levels in the ecosystem: in terrestrial ecosystems, there are seldom more than five, although aquatic systems often have more.

Why So Few Trophic Levels?

There are several reasons ecosystems on land have so few trophic levels. First, not all of the food available at one trophic level is actually eaten by animals in the next level. So, some energy is lost in this way.

Second, not all of the food eaten is useful. For instance, the rate at which herbivores grow is often limited by the content of nutrients such as essential amino acids in their food plants. In the process of eating enough food to extract the amino acids they need, they may excrete and "waste" a lot of energy-rich plant material.

A third source of energy loss is respiration to drive the organism's metabolism. In Chapter 6 we saw that energy is lost in every energy transformation. Cellular respiration of glucose, the major body fuel, is only about 50% efficient. More energy is lost as the ATP formed during respiration is used in the maintenance and repair of body tissues, in functions such as feeding, excretion, and circulation, and in behavior. The energy that remains is stored in new **biomass,** material that makes up the bodies of organisms and is available for other organisms to eat (Figure 37–1).

1? Because of all these energy losses from one trophic level to the next, there is not enough energy left to support higher trophic levels on land. A wolf may have to travel 30 kilometers a day to find enough food to eat, and a tiger requires a home range of up to 240 square kilometers. An animal that fed on wolves or tigers would have to cover a wide hunting area to try to find enough of its widely scattered prey and it is not energetically feasible to try to harvest the widely dispersed food energy available in the highest trophic level. The organisms that do feast on these top predators are parasitic worms and fleas. They eat only part of the predator and get only a tiny crumb of the ecosystem's energy pie. ■

All the organisms in a community are interconnected in a complex food web. Energy from the sun enters a community during photosynthesis. Then it passes through one to five trophic levels in a terrestrial food web. Some energy is lost to the system at each transfer.

37–C PRODUCTIVITY

The flow of energy through an ecosystem can be measured at various points by answering questions such as these: how much solar energy is trapped in the food made during photosynthesis? How much of the energy in plant material can a herbivore use? How much energy does a herbivore use before it is eaten by a carnivore? Similar questions may be asked, proceeding from one trophic level to the next. Let us consider some of these questions.

Primary Productivity

Primary productivity is the rate at which energy is stored in organic matter by photosynthesis. It is usually measured as increases in energy stored or in biomass. Not all of the organic matter made during primary productivity accumulates as plant biomass: about half is used up in the plant's own respiration. What remains is the **net primary productivity,** which appears as plant growth (new biomass) and is available to heterotrophs.

Productivity varies considerably from one type of community to another (Table 37–1). Why should this be? We have seen that plants require a number of resources for growth. If one of these is in short supply, it becomes the **limiting factor** for plant growth. As might be expected, energy is one of the most crucial factors: productivity generally increases from polar regions toward the tropics, because of the increasing sunlight and temperature. Where water is scarce, or nutrients are lacking, productivity is low no matter how much sunlight reaches the area. In dry areas, such as savannas and deserts, water is usually the limiting factor (Figure 37–2). Intensive agriculture can achieve net productivities as high as those of any natural vegetation on land (see *A Journey into The Environment:* The Disappearing Soil, Chapter 36).

Table 37–1
Net Primary Productivity of Some Major Community Types* (In Grams of Dry Plant Material per Square Meter Per Year)†

Community	Average Net Primary Productivity
Coral reef	2500
Tropical rain forest	2200
Temperate forest	1250
Savanna	900
Cultivated land	650
Open ocean	125
Semidesert	90

* See Chapter 36 for characteristics of the various communities mentioned in this table.

† After R. H. Whittaker, *Communities and Ecosystems,* 2d ed. New York: Macmillan, 1975.

FIGURE 37–2

Productivity and precipitation. (a) A graph showing how net primary productivity increases with average annual precipitation. **(b)** Artificial irrigation supports the much higher productivity of this Death Valley golf course. **(c)** The relationship between rainfall and productivity is striking in arid ecosystems, where productivity is limited by scarcity of water. Irrigation enables the floor of this valley (near Fresno, California) to support lush crops, a striking contrast with the surrounding arid lands. (c, M. L. Dembinsky/Dembinsky Photo Assoc.)
▼

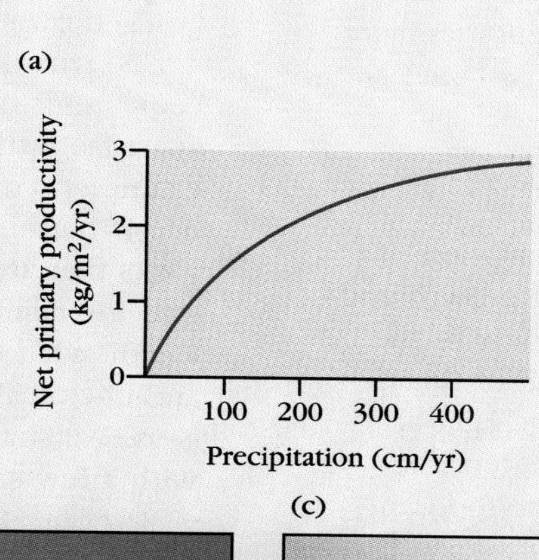

(b)

(c)

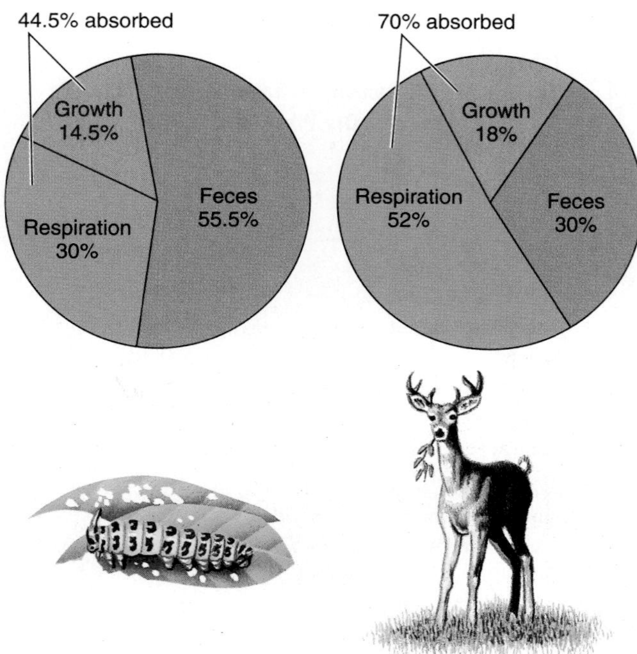

FIGURE 37-3

Food and energy budgets. The fate of energy in black swallowtail butterfly caterpillars and deer. Energy is expressed as percent of calories taken in as food. The deer is more efficient at absorbing food from the gut, but uses more energy in respiration than the caterpillar. (From data courtesy of J. M. Scriber, P. Feeny, and R. L. Cowan)

Because energy is lost at each transfer, productivity decreases at each higher trophic level. On average, animals convert about 10% of energy they eat into growth or reproduction.

Secondary Productivity

Secondary productivity is the rate of formation of new organic matter by heterotrophs (their growth and production of offspring). Herbivores eat some of the plant biomass in any ecosystem, but not all of this becomes secondary productivity. A caterpillar's gut absorbs only about half of the leaf material eaten; the rest is egested as feces. Furthermore, of the food absorbed, about two thirds is used in respiration, so that only about 15% of the food eaten appears as secondary productivity (Figure 37-3). Grasshoppers were found to convert only about 4% of the food they ate into secondary productivity. An average figure for animals is about 10%. The average is lower for herbivores and higher for carnivores, because a given weight of meat contains more of the nutrients an animal needs than the same weight of plant food.

Even if the organisms at each trophic level could find, capture, and eat all of the net productivity from the previous level, the tertiary consumers would receive only about $\frac{1}{10} \times \frac{1}{10} \times \frac{1}{10} = \frac{1}{1000}$th of the energy present in the original producers in their food web. It is clear from this that in times of food scarcity, eating meat is a luxury for omnivores, including humans. By adopting a vegetarian diet, eating "low on the food chain," it is possible to skip the energy losses at one trophic level so that more people can be fed from a given area of land.

37-D CYCLING OF MINERAL NUTRIENTS

Although an ecosystem's productivity may be limited by the supply of sunlight or water, in many cases it is limited instead by the availability of nutrients.

Living organisms require as nutrients eight elements in relatively large amounts: carbon, hydrogen, oxygen, nitrogen, potassium, calcium, phosphorus, and sulfur. These elements are present in the environment. Some occur in rocks and are released by erosion and weathering into soil, rivers, lakes, and the oceans. Some are also present in the atmosphere. The movements of nutrient elements through the biosphere, or through any particular ecosystem, by physical and biological processes, are called **biogeochemical cycles.** They are called cycles because nutrient elements may be used over and over again by living systems.

Nutrients are sometimes recycled rapidly through ecosystems. In other cases, nutrients spend many years apart from the activities of the biological world. For example, remains of marine organisms may sink to the ocean bottom and be incorporated into sedimentary rocks that are lifted and exposed at the Earth's surface only after millions of years. Every nutrient element has a somewhat different fate, depending on its physical and chemical properties and on its role in living organisms. We shall illustrate the concept of nutrient cycling with a few simplified examples.

The Carbon Cycle

Carbon is said to move in an **atmospheric cycle** because carbon occurs in the air, as well as in rocks and dissolved in water. Much of the carbon in terrestrial ecosystems travels rapidly between living organisms and the atmosphere as a result of photosynthesis and respiration (Figure 37-4). Most of the carbon fixed as organic matter by photosynthesis is rapidly broken down and released back into the air as carbon dioxide (CO_2) produced in respiration. In aquatic ecosystems, the exchange occurs between living organisms and CO_2 or bicarbonate (HCO_3^-) dissolved in the surrounding water. Another carbon compound, methane (CH_4), is released into the atmosphere by anaerobic organisms, such as those in sewage treatment plants, the

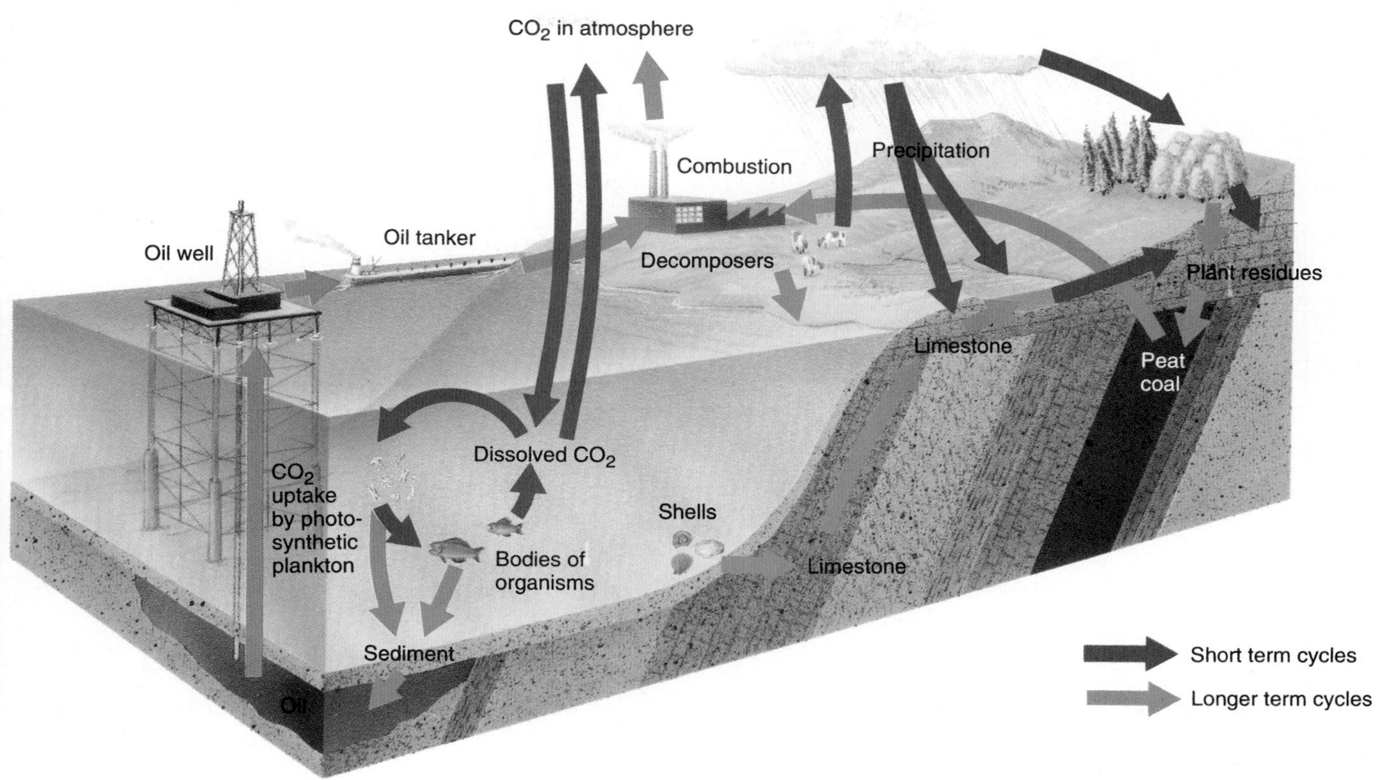

FIGURE 37-4

A simplified carbon cycle. This diagram shows how carbon moves through several ecosystems. Carbon passes through the processes indicated by purple arrows more rapidly than through those indicated by orange arrows.

bottom mud of wetlands, and the digestive systems of ruminants (such as cattle and sheep).

Some carbon enters a long-term cycle. For instance, deposits of coal, oil, and natural gas were part of the net productivity of ancient ecosystems. These fossil fuels are carbon compounds that were buried before they decomposed, and were then transformed by time and geological processes.

The quantity of methane in the air is increasing, partly because the numbers of ruminants (mainly cattle and sheep) and of sewage plants are increasing rapidly. The carbon dioxide content of the atmosphere is also increasing.

Our burning of fossil fuels and other organic matter releases CO_2 into the air. In addition, widespread **deforestation,** the cutting down of trees, has slowed the rate at which CO_2 is removed from the air. Most of the terrestrial vegetation that absorbs CO_2 during photosynthesis is found in wet parts of the tropics. Most of the world's tropical forests have already been destroyed. When tropical forest is cut down to make way for agriculture, the trees are usually burned to get them out of the way, releasing still more CO_2 into the air. However, this produces only about one fifth as much CO_2 each year as does burning fossil fuel in cars, power stations, and factories.

The concentration of CO_2 in the atmosphere is increasing much more slowly than would be predicted from all these human activities. The rest of the CO_2 is almost certainly being absorbed by the oceans, which act as a global "sink" for CO_2. It is not clear how much more CO_2 the oceans can hold. CO_2, methane, and other "greenhouse gases," are important because they absorb infrared radiation (heat) from the Earth. They trap heat within the atmosphere, preventing it from escaping into space (Figure 37-5). Without the greenhouse effect, the Earth would be too cold for life. The more of these gases the atmosphere contains, the more heat is retained. (See *A Journey into The Environment:* Global Warming, Chapter 6.)

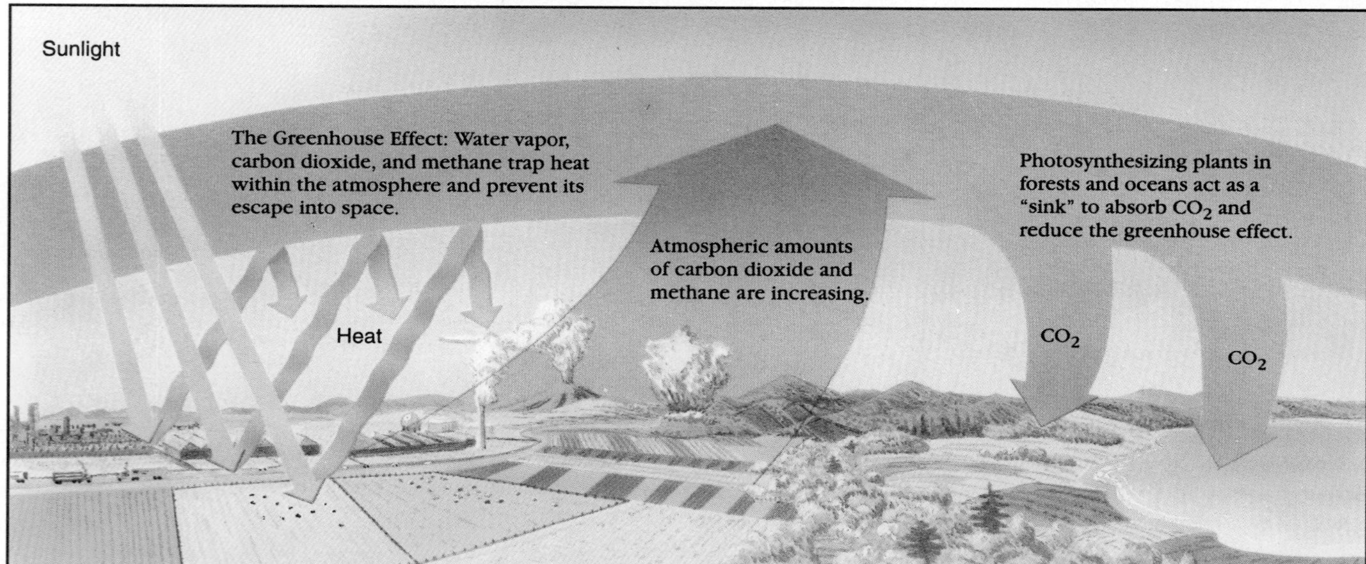

FIGURE 37–5

Global warming. The glass in our cars or in a greenhouse act in the same way as water vapor, carbon dioxide, and methane gases in the Earth's atmosphere. Each permits solar energy to pass through while each prevents passage of most of the heat energy that is produced. The net effect is a warmer car, greenhouse, or Earth. Increases in atmospheric carbon dioxide levels caused by additions to the carbon cycle (Figure 37–4) and additions of methane and other greenhouse gases promote the greenhouse effect.

2? It is difficult to say precisely how much our addition of carbon dioxide and methane to the atmosphere will heat up the Earth. Both heat and CO_2 are stored in various ways, and many factors besides the greenhouse gases affect the average temperature of the atmosphere. Even with these uncertainties, however, most estimates predict a global warming of 1 to 5°C during the next century. This temperature rise would produce many effects that we are not prepared for. Climate zones and ocean currents would shift, agriculture would be displaced, and the world's major vegetation zones would alter. Another dramatic effect might be thermal expansion of the ocean and further melting of the world's great ice caps, which would raise the sea level, flooding many coastal areas. ■

Humankind is not sitting idle while our forests burn. Reforestation is taking place on a massive scale. About 11 million hectares of tropical forest are felled each year, but by the year 2000 about 17 million hectares of land are expected to be reforested each year. Among the examples: China has doubled its forested area in less than 20 years, New England's

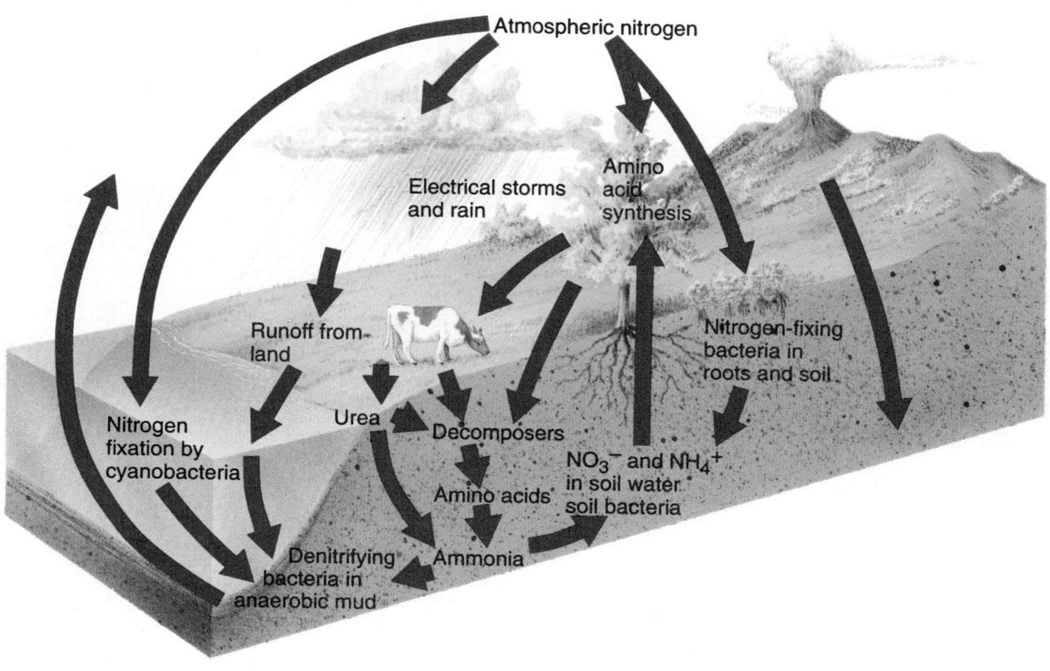

FIGURE 37-6

The nitrogen cycle. Plants must get nitrogen from either nitrogen-fixing microorganisms or indirectly from the limited amounts of ammonium (NH_4^+) or nitrate (NO_3^-) ions found dissolved in water in the soil, rivers, lakes, or oceans.

forested area is 50% greater than it was a century ago, Chile has doubled its area of pine plantations in 15 years, and in Rwanda, trees planted by rural people now cover 200,000 hectares, more than the area of the country's remaining natural forest and artificial plantations put together.

The Nitrogen Cycle

Nitrogen is a vital part of many essential organic compounds, especially proteins and nucleic acids. About 80% of the atmosphere is molecular nitrogen gas (N_2), which is much more abundant than CO_2. However, whereas all producers use CO_2 directly, only certain species of prokaryotes can use N_2. These bacteria capture nitrogen by reducing it to ammonia (NH_3) (Section 19-B). Plants must get this vital nutrient either directly from nitrogen-fixing microorganisms or indirectly from the limited pool of nitrogen found as ammonium (NH_4^+) or nitrate (NO_3^-) ions in soil water, rivers, lakes, and oceans. Nitrogen fixation is slow compared with the rate at which plants can take up fixed nitrogen, and so nitrogen is often the limiting nutrient in plant growth. Nitrogen is one of the elements most commonly applied in crop fertilizers. This additional input of usable nitrogen into the biosphere is based on the industrial manufacture of ammonia-based fertilizers from atmospheric nitrogen, using catalysts and a lot of fossil fuel energy (Figure 37-6).

There is a second difference between the movements of nitrogen and carbon through an ecosystem. Most organisms release carbon dioxide during respiration, but the conversion of nitrogen from organic molecules into inorganic forms such as nitrates involves steps carried out by a series of different microorganisms. Certain bacteria and fungi use the proteins and amino acids in dead organic matter for their own dietary nitrogen. They release the excess as nitrogen compounds that can be absorbed by plants and incorporated into plant tissue.

Nitrogen is returned to the atmosphere by denitrifying bacteria living in the anaerobic mud of fertile lakes, bogs, estuaries, and parts of the ocean floor. Unlike the carbon cycle, the nitrogen cycle is primarily accomplished by bacteria.

The Phosphorus Cycle

Phosphorus is a major constituent of nucleic acids and of membrane phospholipids, and many animals also need a lot of phosphorus for shells, bones, and teeth.

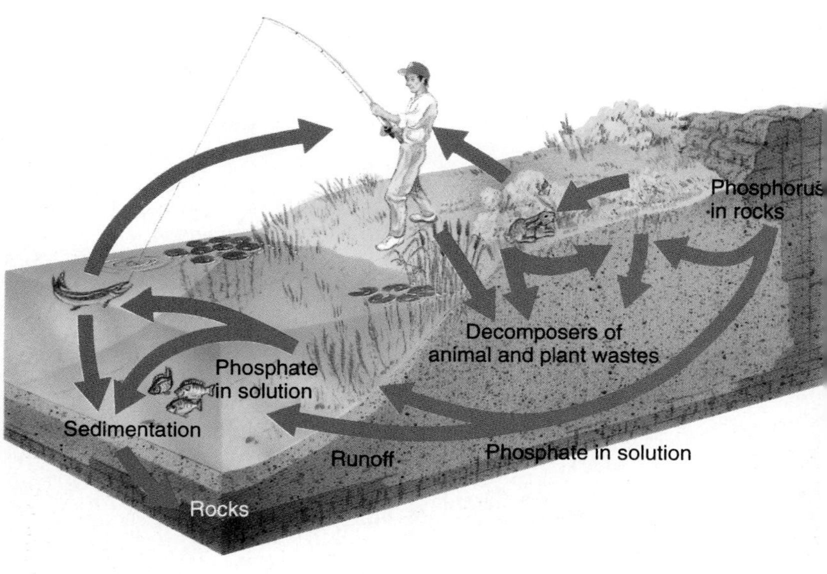

FIGURE 37-7

The phosphorus cycle. Plants take in most phosphorus in the form of phosphate. Fishing and the droppings of sea birds bring trivial amounts of phosphorus from water to land ecosystems. However, most flow is one way, from terrestrial rocks to the sea floor.

Because phosphorus almost never occurs as a gas, its cycle, unlike the atmospheric cycles of nitrogen or carbon, is a **sedimentary cycle** (Figure 37-7). When rocks are eroded by weather, small amounts of phosphorus dissolve, usually as phosphate, and so become available to plants. Much of the phosphorus excreted by animals is in the form of phosphate, which plants can use immediately. Thus the cycling of phosphorus in terrestrial ecosystems is usually very efficient, although small amounts are continually lost downstream and to the oceans.

The phosphorus cycle is, in the short run, a one-way flow—from rocks to land ecosystems to the ocean, and finally to ocean sediments. The only natural way for phosphorus to return to land is by slow geological processes in which sea floor sediments may again become terrestrial rocks. However, terrestrial ecosystems are able to retain much of their phosphorus because soil particles absorb phosphate ions, helping to provide a steady supply of it for plant growth. Soil erosion robs an ecosystem of its phosphorus, and it may take thousands of years to recoup this loss through the weathering of rocks.

An Experimental Ecosystem

Hubbard Brook Experimental Forest in the White Mountains of central New Hampshire is one of several areas where researchers study nutrient cycles and how they are affected by human activities. The forest consists of a group of valleys, each with its own creek running down the middle.

Data from Hubbard Brook reveal that the forest is extremely efficient at retaining nutrients. Nutrients that reach the forest dissolved in rain and snow approximately balance those leaving the forest, washed away in its streams. Both quantities are small relative to the total amounts of nutrients present.

3? Investigators also examined what happens when a forest is cut down. One winter they cut down all of the trees and shrubs in one of the six valleys. Dramatic effects became obvious almost immediately. The rate at which nutrients left the forest dissolved in the stream water increased six- to eightfold. Other experiments revealed that nutrient losses are reduced if the forest is cut in horizontal strips, leaving strips of standing trees, rather than being clear-cut (Figure 37-8). The remaining trees absorb many of the nutrients that enter the soil water after their neighbors are felled. ■

Biogeochemical cycles move the various nutrient elements through the biosphere. Carbon is fixed in organic molecules during photosynthesis, and most of it is soon released by respiration. The carbon balance of the biosphere is moderated by the exchange of CO_2 between the atmosphere and the oceans. Nitrogen is abundant in the air, but in a form few organisms can use. The ability of plants to use much more fixed nitrogen than is usually available makes this vital resource a limiting nutrient in many ecosystems. Phosphorus moves through ecosystems in a one-way, sedimentary cycle. It is usually in short supply but tends to be retained efficiently. Nutrients recycle indefinitely in a natural ecosystem, with small losses into the air and in runoff and sedimentation. Natural or man-made disturbances can deplete an ecosystem of enormous amounts of nutrients.

(a)

(b)

FIGURE 37–8

Hubbard Brook Experimental Forest. (a) This region has been clearcut to study how methods of harvesting trees affect the flow of nutrients. **(b)** A V-notch weir. All the water that runs off one of the valleys is funneled through this weir so that its volume can be measured and its nutrient content analyzed. (J. D. Cunningham/Visuals Unlimited)

37–E SPECIES DIVERSITY IN COMMUNITIES

We turn from mineral nutrients to the types of organisms they sustain. Communities of organisms differ in their **species diversity,** the number of species they contain. The simplest measure of such diversity is the number of species in a given area. Often there are a few dominant species, which are especially abundant or obvious in the ecosystem, and a host of species that are less common.

We might expect that one kind of decomposer would become so efficient at digesting and absorbing dead matter that it would drive competing species to extinction. Competition between species for limited resources sometimes does lead to the extinction of one of them, but this is not always the case. What maintains species diversity? What prevents one or more species in a trophic level from eliminating the others through competition? One answer is that when species are in potential competition for a food supply, they may become specialized to take advantage of a portion of the resource in some way. Different insect-eating birds, for instance, search for food at different heights in a forest. An important reason for species diversity within a particular community, therefore, is specialization in the use of limited resources.

A second contribution to the overall species diversity of a region is the variety of habitats it contains, each characterized by its own diversity. For example, species diversity may vary with height above sea level, soil type, and so forth. Another cause of diversity is the creation of different habitats within a region by periodic disturbance. Light gaps in a forest may be inhabited by different species of birds and insects from those in the nearby climax forest.

Predation also helps to maintain species diversity. This was shown in a study of an intertidal community on the rocky coast of Washington, where mussels, barnacles, limpets, and so forth, were all fed on by the food web's top carnivore, a sea star. When the sea stars were removed from the rocks, barnacles settled and took up about 80% of the available space within three months. Later, the barnacles were gradually crowded out by two species of mussels. After a year or so, one mussel species dominated the experimental area, and the number of species present had dropped from 15 to 8. Thus, in the natural community, the sea star predator maintains diversity by preventing some of its prey species from excluding the others by competition for space. Grazing by herbivores sometimes has similar effects. In several experiments, herbivorous mammals were fenced out of areas of grassland. A few

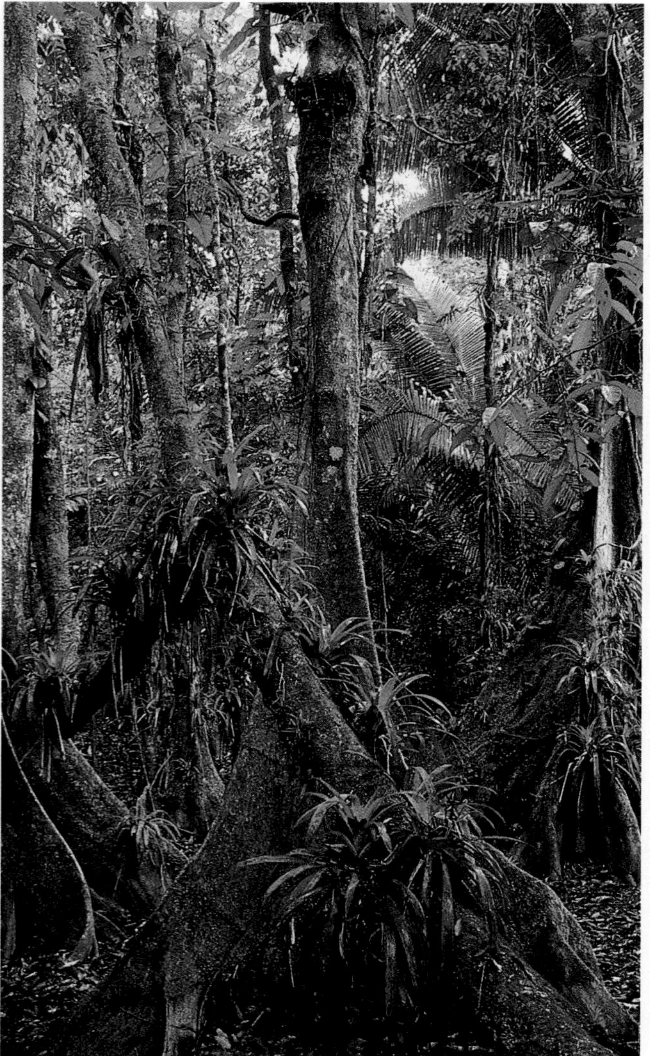

FIGURE 37-9

Species diversity. Plants in a lowland tropical rain forest in Belize show some of the incredible diversity found in this species-rich community. (F. Lanting/Minden Pictures)

species of grasses and herbs took over more and more of the area and the species diversity was reduced.

Specialized herbivorous insects can contribute to diversity. In tropical forests, almost all of the seeds and seedlings of some plant species die if they remain near their parents, because they are attacked by insects. Only if seeds are dispersed some distance do they have a reasonable chance of escaping the seed and seedling predatory insects near the parent plant. This sort of predation probably contributes to the enormous diversity of species in tropical forests (Figure 37-9) because species that become common are most easily discovered by enemies, which quickly eat them so that they become rare again, leaving space for other species.

Species Turnover in Communities

A 1968 survey of the number of species of birds breeding on each of the nine Channel Islands off the coast of southern California showed that the total number of species on each island had changed little since a similar survey 50 years earlier. However, the species composition had changed markedly. On San Nicolas Island, for example, there were 11 species of birds in 1917 and the same number in 1968, but only five of these species were the same. Six species had disappeared from the island, but six other species had colonized it. It seems that each undisturbed habitat can always support the same number of species, although the particular species present may change from time to time.

Ecologists tested this hypothesis experimentally. First, they counted the number of species, mainly insects, on six small mangrove islands in Florida. Next, they completely exterminated all of the arthropods on these islands by enclosing the islands in enormous plastic tents and fumigating them (Figure 37-10). They then removed the tents and sampled the animals regularly. Within six months, the number of species on each island had returned to approximately the number present before fumigation, and remained at about the same level for as long as the islands were observed. As predicted, however, the species compositions were not identical to those before the experiment, and species turnover continued on each island, with losses approximately balanced by immigration.

Species turnover is not restricted to island communities. Turnover is slow in many communities on continents because populations are large and any losses are easily replaced by immigration from nearby areas. Small, isolated communities in places such as mountaintops, bogs, and lakes, however, may have appreciable species turnover. An understanding of island communities promises to be a valuable tool in the design of parks and nature reserves, artificial "islands" where decisions must be made as to what area of habitat should be preserved and whether more species will be saved if the park contains one large or several smaller pieces of the habitat.

Species diversity in an ecosystem results from specialization in the use of limited resources, habitat variation, and predation. In most communities, the total number of species changes little although the particular species present may vary.

(a)

(b)

FIGURE 37–10

Measuring species diversity. (a) Scaffolding completely encloses a small mangrove island in the Florida Keys, in preparation for enclosing the island with plastic sheeting. **(b)** After elimination of the island's insects by fumigation, tent and scaffolding were removed and the process of recolonization of the island was monitored. (E. O. Wilson)

SUMMARY

1. An ecosystem consists of an interdependent community of organisms and their physical environment.
2. Two of the most important factors determining the character of an ecosystem are productivity and nutrient cycles.
3. The productivity of an ecosystem is determined by the temperature and by the availability of light, water, and minerals for photosynthesis.
4. In an ecosystem, energy flows one way, from one trophic level to the next, with only about 10% passing to each successive level. Consequently the biomass that an ecosystem can support at each successive trophic level declines rapidly.
5. Nutrients cycle through an ecosystem. They are taken in by organisms as inorganic substances, many of which are incorporated into organic molecules. Nutrients may pass through the food web for a time, but eventually they are once again released into the environment as inorganic substances.
6. The availability of nutrients often limits the productivity of an ecosystem, even though an ecosystem may be very efficient at conserving and recycling its nutrients.
7. Species diversity in a community is maintained by the adaptations of potentially competing species to use a subset of limited resources (such as food) within habitats, and to become specialized to slightly different habitats.
8. Species diversity may also be maintained by predation, which can prevent one prey species from eliminating others by competition.
9. The study of islands shows that there is species turnover in communities, with new species moving in and resident species becoming locally extinct.
10. The number of species varies from one community to another depending on the size of the area the community occupies, diversity of habitats in this area, and availability of colonists from nearby areas.

SELF-QUIZ

1. The role of decomposers in an ecosystem is ___ .
2. Using the items listed below, diagram a food web and indicate the trophic level of each organism.

deer	herbivorous insect
soil bacteria	spider
shrub	sparrow
wolf	hawk

3. The annual primary productivity of any ecosystem is greater than the annual increase in biomass of the herbivores in that ecosystem because:
   a. plants are more efficient than animals in converting energy input to biomass
   b. energy is lost during each energy transformation
   c. there are always more plants than plant eaters
   d. woody plants live much longer than most herbivores
4. Of the total amount of energy that passes from one trophic level to another in a food chain, about 10% is:
   a. transpired
   b. "burned" in respiration
   c. stored as body tissue
   d. reradiated in the form of heat
   e. passed out in the feces
5. Nutrient cycles may involve:
   a. movement of the nutrient from the organism to the atmosphere
   b. movement of nutrients into the soil
   c. limitations on the number of organisms in the ecosystem due to shortage of some nutrients
   d. loss of the nutrient from the ecosystem
   e. all of the above
6. Wolves and lions may be said to occupy the same trophic level because:
   a. they both eat primary consumers
   b. they both use their food with about 10% efficiency
   c. they both live on land
   d. they are both large mammals
   e. they both eat a wide range of dietary items
7. State whether the species diversity of a community would be likely to increase or decrease under each of the following conditions:
   ___ a. increased frequency of disturbances
   ___ b. extermination of a predator that preys on many other species
   ___ c. partitioning of a shared resource among competing species
8. Which of the following is a likely result of deforestation?
   a. The amount of CO_2 removed from the atmosphere is reduced.
   b. Wind blows the soil away because its plant cover has been removed.
   c. Water washes off the land more rapidly, causing floods.
   d. Water washes soil into rivers and streams, causing them to silt up.
   e. All of the above.

THINKING CRITICALLY

1. Is more energy lost from an ecosystem when an herbivore eats a plant or when a carnivore eats an animal? Why?
2. How is the flow of energy in an ecosystem linked to the flow of nutrients? How do energy and nutrient flows differ?
3. Robert MacArthur found that the number of bird species in forests is correlated not with plant species diversity but with the amount of layering of foliage at different heights. Can you account for these findings?
4. How can frequent fires increase the species diversity of a region?
5. Suppose you were a member of a congressional committee developing legislation to set aside 100,000 acres of land within a particular region for a biological preserve or national park. The committee must decide whether to recommend government purchase of one large area or several smaller areas. What would your advice be if the prime objective was (1) to preserve an endangered large mammal population (such as the Texas red wolf), (2) to preserve as many species as possible, (3) to preserve as many local habitats as possible, and (4) the best possible compromise of these?

SELECTED KEY TERMS

atmospheric cycle, *p. 778*
autotroph, *p. 772*
biogeochemical cycle, *p. 778*
biomass, *p. 776*
community, *p. 771*
consumer, *p. 772*
decomposer, *p. 772*
deforestation, *p. 779*
ecosystem, *p. 771*
food chain, *p. 772*
food web, *p. 772*
heterotroph, *p. 772*
limiting factor, *p. 776*
primary productivity, *p. 776*
producer, *p. 772*
pyramid of energy, *p. 772*
secondary productivity, *p. 778*
sedimentary cycle, *p. 782*
species diversity, *p. 783*
trophic level, *p. 772*

SUGGESTED READINGS

Anderson, J. M. *Ecology for Environmental Sciences: Biosphere, Ecosystems and Man.* London: Edward Arnold (Publishers) Ltd., 1983. The effects of pollutants on the workings of ecosystems, populations, and communities.

Bazzaz, F. A., and E. D. Fajer. "Plant life in a CO_2-rich world." *Scientific American,* January 1992.

Cohn, J. P. "Gauging the biological impacts of the greenhouse effect." *BioScience* 39(3):142, 1989. The high points of a World Wildlife Fund conference on the probable effects of global warming on individual plants and animals and upon the probable distribution of organisms.

Elkington, J., J. Hailes, and J. Makower. *The Green Consumer.* New York: Viking Penguin, 1990. A guide to doing your bit for the environment.

MacEachern, D. *Save Our Planet: 750 Everyday Ways You Can Help Clean Up the Earth.* Washington, DC: Dell, 1990.

Myers, N., ed. *The Gaia Atlas of Planet Management.* London and Sydney: Pan Books, 1985. Our environmental problems portrayed in the form of maps. Some excellent illustrations and loads of interesting information.

Tunnicliffe, V. "Hydrothermal-vent communities of the deep sea." *American Scientist,* July 1992.

CHAPTER 38

CONCEPT GUIDE

After reading this chapter, you should be able to:

1. Describe three factors contributing to biotic potential and explain why each is important. (Section 38–A)
2. Compare the effects of density-dependent and density-independent mortality factors on large and small populations. (Section 38–B)
3. Explain the theory of competitive exclusion. Given this theory, explain how two or more species with overlapping niches can coexist. (Section 38–C)
4. Compare the effects of specialized and generalized predators on their prey. (Section 38–D)
5. Explain how humans have affected the rate of extinction of plant and animal species over the last 200 years. (Section 38–E)
6. Describe the factors that affect infant mortality and birth rates in human populations. (Section 38–F)

Why are these birds so tightly packed together? For many bird species, including flamingos, nesting sites are in short supply, resulting in crowded nest areas. Individuals must defend their nesting territory from invaders, while sitting or standing on their nests. Read more about competition and the defense of territories in Section 38–C. (M. P. Kahl/Photo Researchers.)

Populations

Biologists study the nonliving components of an ecosystem, such as climate, energy input, and nutrient flow, but they are most interested in the populations that make up the ecosystem's community of living organisms. A **population** consists of all the members of a species occupying an area at the same time. Examples are the bass population of a lake and the human population of the United States. In our study of evolution, we saw that each population evolves adaptations suited to its own locale. Hence the population is the unit that evolves. It is also an ecological unit playing a role in the community of organisms in an ecosystem.

A population has characteristics not found in its individual members. For example, each population has a gene pool and a certain pattern of distribution, density, and age structure. **Population density** is the number of individuals per unit area or volume—for example, the number of bison per hectare.

The **age structure** of a population is the percentage of individuals of each age. These and other features can be used to describe populations and to predict their fates.

KEY CONCEPTS

- The community of living things in an ecosystem is made up of populations of various organisms.
- Although a population invading a new area may grow exponentially, a population in nature seldom grows as fast as its biotic potential would allow.
- In nature, most populations remain at some fairly constant size, determined by factors including the supply of resources, competition for those resources, and predation.
- The human population is now growing rapidly despite a recent decline in birth rates. Many people consider this to be the biggest current human problem, as well as the biggest ecological problem.

Curiosity Questions

1? Why do horses have only a few offspring while many kinds of fish have thousands? (See page 791)

2? What animal holds the record for causing the extinction of the greatest numbers of plants and animals? (See page 800)

3? Why has the human population increased dramatically in the last 200 years? (See page 802)

FIGURE 38–1

A swarm of tadpoles. This black cloud consists of thousands of recently hatched tadpoles. But only a few of these tadpoles will survive. Predation, disease, and other causes of death will reduce the number to just a few that will live to reproduce. Thus the population does not grow at its maximum possible rate. (E. Zalisko)

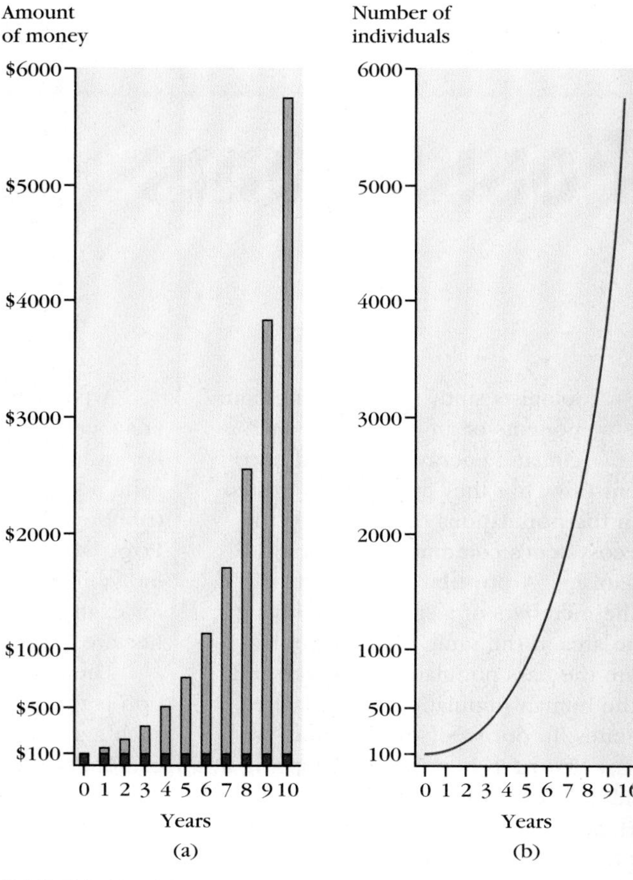

FIGURE 38–2

Like money in the bank. (a) This graph illustrates the exponential growth over ten years of an investment of $100 earning 50% interest compounded annually. **(b)** The growth for a population experiencing exponential growth also has a J-shaped curve.

38–A POPULATION GROWTH

A population gains individuals by birth and immigration, and loses them by death and emigration. Whether a population grows, declines, or remains the same size, depends upon the balance between these factors:

change in population size = (Births + Immigration) − (Deaths + Emigration)

The size of a population of mice in a field or of violets in a woodlot seems, at first sight, to vary little from year to year. Is this really the case? Surely organisms produce so many offspring that their populations could increase greatly from one year to the next (Figure 38–1). What limits the size of natural populations?

To answer these questions, it is convenient first to find out how rapidly populations could increase if nothing stopped their growth. A population's **biotic potential** is its fastest possible rate of growth, given ideal conditions. In nature, conditions are seldom perfect. Predators, disease, or food shortages nearly always prevent populations from growing as fast as their biotic potentials would permit. However, there are times when populations grow very fast.

The Russian ecologist, G. F. Gause, studied the growth of populations of the protist *Paramecium caudatum*. Every few hours a well-nourished *Paramecium* divides into two new individuals. Gause set up tubes containing plenty of bacteria for food and introduced one *Paramecium* into each. He observed that the population of *Paramecium* in each tube showed **exponential growth,** that is, growth in which an ever-increasing number of new individuals is added in each unit of time. Exponential growth, plotted as a graph, produces a J-shaped curve (Figure 38–2).

Populations sometimes grow exponentially when they have a superabundant supply of resources. History records many cases of exponential growth by species imported, by accident or on purpose, into areas where resources were available and natural enemies or competitors were lacking. For example, dandelions, starlings, and water hyacinths introduced into the United States underwent dramatic population explosions. Similar population explosions occur when bacteria invade the intestinal tract of a newborn ani-

BIO-BIT

In 1967, R. MacArthur and E. O. Wilson published *The Theory of Island Biogeography,* in which they demonstrated that the number of species living on an island increases as the size of the island increases. In other words, larger islands permit not only more individuals, but more diversity.

mal or when decomposers invade a newly dead animal or plant.

Exponential growth does not necessarily mean that the population is growing at its biotic potential. The human population started growing exponentially in the mid-eighteenth century, although women were not bearing infants as frequently as is biologically possible. In addition, many people who could have reproduced did not.

Biotic potential differs from one species to another. An individual's contribution to population growth can be increased in any or all of three ways:

1. By producing more offspring at a time (that is, a larger litter size).
2. By having a longer reproductive life, so that the individual reproduces more times.
3. By reproducing earlier in life.

Of these three factors, the last is by far the most important. A bacterium does not have a long life and it produces only two offspring at a time. Nevertheless, a population of bacteria has a higher biotic potential than does a population of dogs. This is because most bacteria can reproduce within an hour after being formed, whereas a dog cannot reproduce until it is about six months old. The shorter the generation time of a species (that is, the younger its members when they first reproduce), the higher its biotic potential (Figure 38-3).

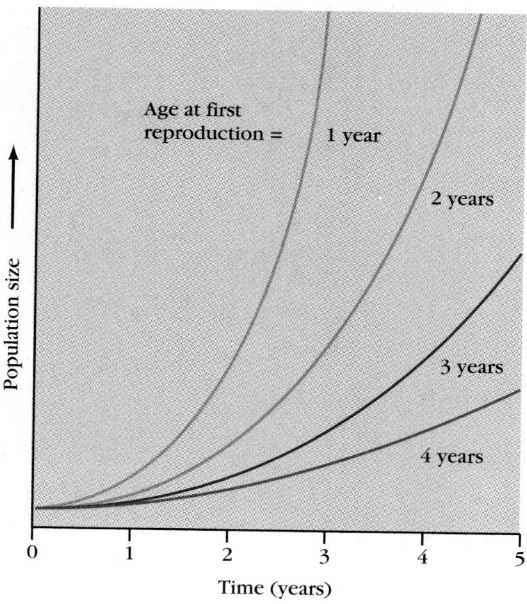

FIGURE 38-3

How (female) age at first reproduction affects population growth. In all of these curves, females produce two offspring per year, but the age at which females first reproduce differs in each curve. After five years, note that changing the age at first reproduction from four to three years of age has the same effect as *doubling* the number of young produced each year. Each curve includes all females in each population.

Reproductive Strategies and Survivorship

In each species, the number of offspring produced and the age at first reproduction result from natural selection. These and other factors are part of the species **reproductive strategy.** There are two extremes in reproductive strategies. At one extreme, parents may produce many small individuals, and provide each with little food or parental care. At the other extreme, an organism may produce very few offspring but invest a lot of energy in each one (Figure 38-4). As we might expect, the rate of survival among the young tends to be higher in the second case. Where many small offspring are produced and receive no care, most of them die at a very young age. Because the parents have produced so many offspring, however, there is a good chance that some will survive to become parents of the next generation. ■

Species with different reproductive strategies tend to have different patterns of **survivorship,** the length of time an individual of a particular age can expect to survive. This can be shown by a graph (Figure 38-5a). In Type I survivorship, most individuals live for a long time, and die as a result of the conditions associated with old age. This type of survivorship curve usually occurs in species where the parents devote considerable energy and care to their offspring. Most human populations in developed nations approach a Type I survivorship curve after the first year of life (when there is a high death rate from genetic or developmental defects or birth accidents). A baby who survives the first year of life is likely to live for another 60-plus years (Figure 38-5b).

In Type III survivorship, most individuals die young, as spores, eggs, or larvae. This type of survivorship is characteristic of many species of invertebrates, bony fishes, plants, and fungi, where large numbers of

(a)

(b)

FIGURE 38-4

Reproductive strategies. (a) A fern produces thousands of tiny spores in the many sporangia on the undersides of its leaves. Some of the spores have caught in a spider's web as they were shed into the air. The spores are at the mercy of air currents to deposit them where they can grow. Needless to say, many do not survive. **(b)** By contrast, mother rhinos produce only a single offspring every couple of years, thus investing much more energy per offspring than the fern. (a, C. Ott/Photo Researchers, Inc.; b, F. Lanting/Minden Pictures)

bony fishes, plants, and fungi, where large numbers of offspring are produced but receive very little care.

The Type II curve falls between Types I and III. There is again an initial period of high mortality (death) due to defective genes or accidents during development, birth, or hatching. However, once past this critical period, an individual is just as likely to die at any age. The chances of dying or being killed are equal throughout life. This type of curve is typical of several birds and of human beings exposed to poor nutrition and hygiene.

Survivorship within a population can also be represented in the form of a life table, a summary of the likelihood of death in groups of individuals of each age

(a)

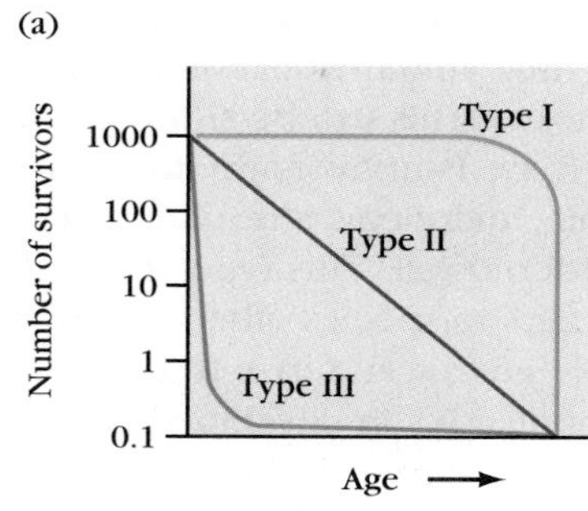

(b)

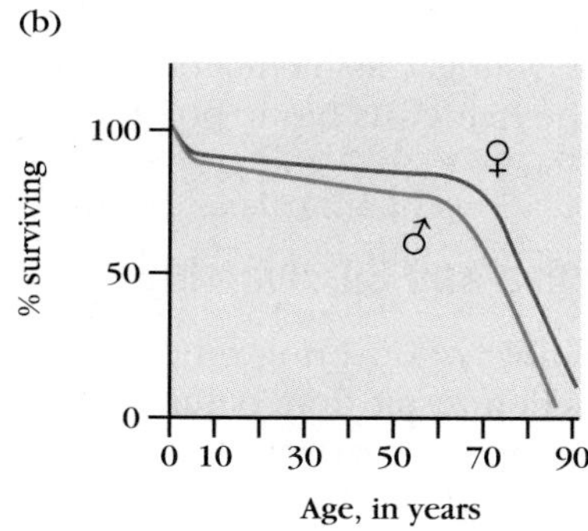

FIGURE 38-5

Survivorship curves. (a) The three main types of survivorship curves. Curves for real populations may fall between those shown here. These hypothetical curves are rarely, if ever, exactly as shown, because there is always an especially high rate of death among very young individuals. **(b)** What type of survivorship curve do we see in human populations? Based upon the 1980 census data, we can see that human survivorship is most like a Type I curve.

Table 38–1
Life Table for the Population of the United States, 1990*

Age Interval	Of 100,000 Born Alive		Average Remaining Lifetime
Period of life between two exact ages (in years)	Number living at beginning of age interval	Number dying during age interval	Average number of years of life remaining at beginning of age interval
0-1	100,000	981	75.3
1-5	99,019	196	75.0
5-10	98,823	88	71.1
10-15	98,735	132	66.2
15-20	98,603	491	61.3
20-25	98,112	550	56.6
25-30	97,564	605	51.9
30-35	96,959	736	47.2
35-40	96,223	935	42.5
40-45	95,288	1219	37.9
45-50	94,069	1764	33.4
50-55	92,305	2724	28.9
55-60	89,581	4229	24.7
60-65	85,352	6210	20.8
65-70	79,142	8706	17.2
70-75	70,436	11,622	13.9
75-80	58,814	14,704	10.9
80-85	41,110	14,879	8.3
85 and over	26,231	26,231	9.2

*Data courtesy of the National Center for Health Statistics

Table 38–1). Life tables for human populations are ised by life insurance companies to predict how much onger people of a given age are likely to live. This deermines the price of insurance for people of various ges. Life tables for populations of animals and plants re useful aids for summarizing and analyzing the efects of different causes of death acting on populaions.

Carrying Capacity

No population can grow exponentially for long. Gause ound that his *Paramecium* populations eventually stopped growing (Figure 38–6a). They had reached heir environment's **carrying capacity,** the number of ndividuals that this particular environment could support indefinitely. When a population reaches this size t may stabilize, with fluctuations above and below the carrying capacity (Figure 38–6b).

Carrying capacity is determined by many factors, including predation, competition, and climate. The factors that limit a population's growth may change, and so the carrying capacity of any area for a population of a given species also changes with time. (See *A Journey into The Environment:* Overpopulation and the Human Animal, Chapter 16.)

A population growing at its biotic potential increases exponentially. However, any growth will eventually stop as the population reaches its carrying capacity. Age at first reproduction is the most important factor determining a species' biotic potential. Reproductive strategies are usually correlated with patterns of survivorship.

(a)

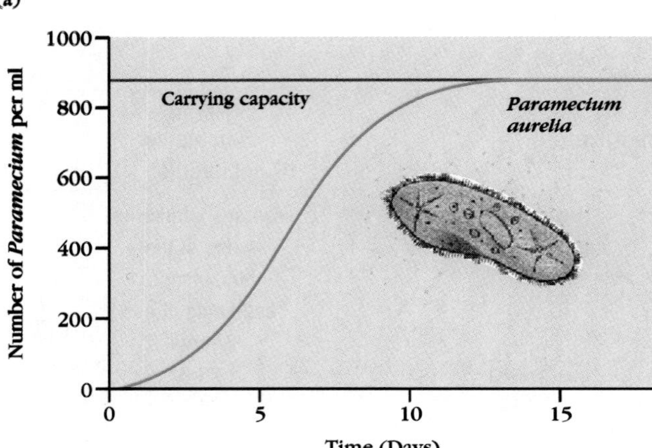

(b)

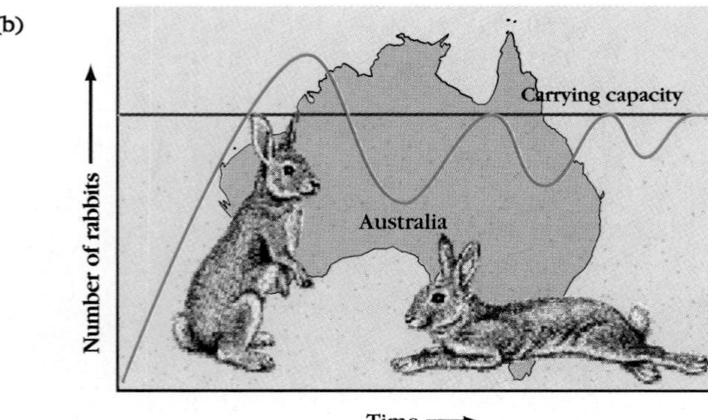

FIGURE 38–6

Carrying capacity. The purple line shows the approximate size of the organism's carrying capacity. **(a)** In a population of *Paramecium aurelia* raised in a culture medium to which Gause added a constant supply of bacteria each day as food, the population first grew exponentially. Then the growth rate decreased and eventually reached zero, so that the population size leveled out as its numbers approached the carrying capacity. **(b)** When rabbits were first introduced into Australia from Europe, their number increased rapidly. Then the population crashed. Now, the population size oscillates around a value that represents the carrying capacity of the environment for rabbits.

38–B REGULATION OF POPULATION SIZE

The number of individuals in a natural population varies, sometimes dramatically. In spite of fluctuations, however, the average size of most large populations changes relatively little over the years (Figure 38–7). This suggests that population sizes are usually regulated in such a way that small populations grow quickly, larger populations grow more slowly, and still larger populations decline.

One reason for this is that at least some of the **mortality factors** (causes of death) affecting populations are **density-dependent.** That is, as the population density increases, these factors kill a larger *proportion* (not just a larger number) of individuals. Several kinds of mortality can act in a density-dependent manner. For instance, predation and disease are density-dependent factors partly because a disease-causing organism is more likely to encounter a host, or a predator its prey, when there are more hosts or prey in the area.

In addition, studies on some wild animal populations have shown that individuals in dense populations have decreased health and vigor. This may decrease their immunity to disease and make them more susceptible to predation. In some extremely crowded populations, stress itself seems to lead to changes in body function that eventually prove fatal. It is not clear whether poor health is caused directly by overcrowding or by some underlying factor, such as too little of some vital nutrient to go around in a dense population. Although air pollution, crime, and urban stresses may be harmful, overcrowding does not appear to harm human health directly. In cities such as Hong Kong, New York, and Mexico City, people live at incredibly high densities with no apparent ill-effects.

Other mortality factors are **density-independent,** that is, they influence a proportion of the population regardless of its density. For example, in several insect species, harsh winter weather kills about 90% of a population regardless of its density. Hurricanes, drought, or earthquakes may also kill a large proportion of individuals, no matter what the population density. However, bad weather may sometimes cause death in a density-dependent manner: if it is possible to survive by finding shelter, and if the number of shelters is limited, then all of the members of a sparse population may survive, whereas only a fraction of a denser population will be protected.

BIO-BIT

The limiting factors of animal populations are not always obvious. Most pinniped populations (sea lions, walruses, and seals) are limited not by food or predators such as sharks and killer whales, but rather by the availability of breeding sites on beaches or ice that are safe from polar bears and wolves.

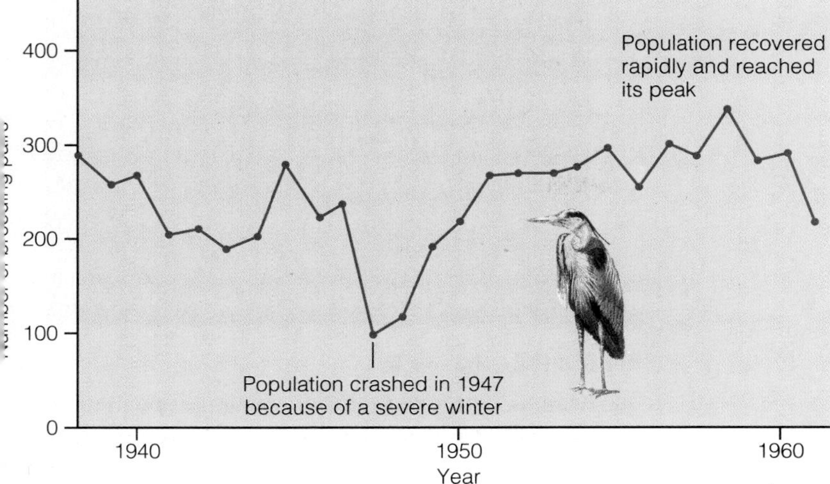

FIGURE 38-7

A stable population. The number of breeding pairs of Gray Herons in part of northwest England fluctuated little over 25 years. (The population recovered rapidly from the severe winter of 1947.) Fluctuations near the carrying capacity are small compared with the heron's reproductive potential (three new birds per breeding pair per year). (After D. Lack, *Population Studies of Birds*. New York: Clarendon Press, 1966)

It would be convenient if we could pinpoint one or two factors and say that they determine the size of a particular population. However, the sizes of natural populations are often affected by many different factors, whose interactions can be complex.

Density-dependent mortality factors are important in limiting large populations. Density-independent mortality factors have similar effects on populations of all sizes and densities.

38–C COMPETITION

Competition is a density-dependent factor that helps to regulate the size of many populations. **Competition** occurs when two or more organisms attempt to exploit a limited resource, such as food or space (Figure 38–8). Competition between individuals of the same species is very common. Generally, members of the same species need the same resources and so they usually compete for them.

In one experiment, seeds of white clover were planted at three different densities. Half of the plants at each density were watered throughout the experiment, but the other half were watered only for the first 18 days. After seven weeks, the densities of the surviving seedlings were measured. Among the seedlings that were watered regularly, mortality was low regardless of density. Among the seedlings deprived of water, however, the proportion of seedlings killed was three times greater in the high-density than in the intermediate-density plots, dramatic evidence of

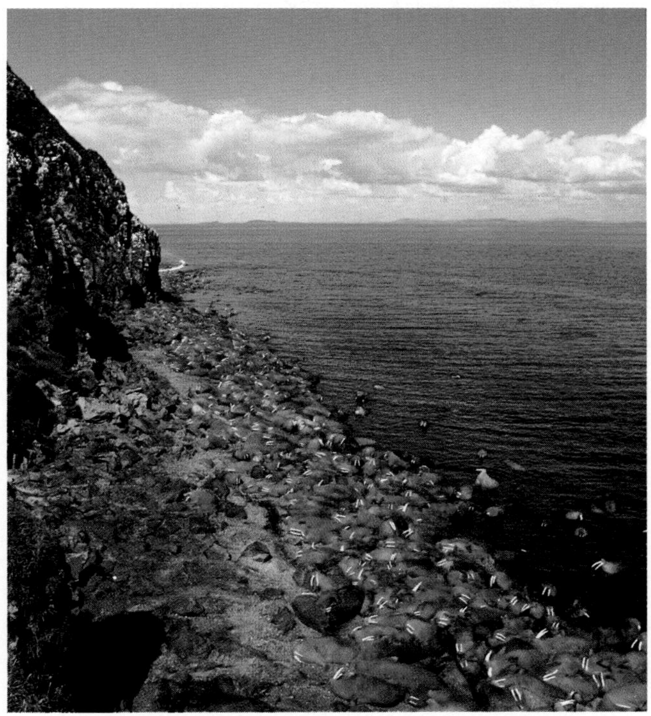

FIGURE 38-8

Competition in a dense population. These walruses compete for space on a limited number of breeding beaches. Lack of space on breeding beaches limits the rate at which the population can grow and the ultimate size of the seal population. We can also see that disease or a predator (such as a human hunter) would kill a higher proportion of this dense population than of a less dense one. (M. Hoshino/Minden Pictures)

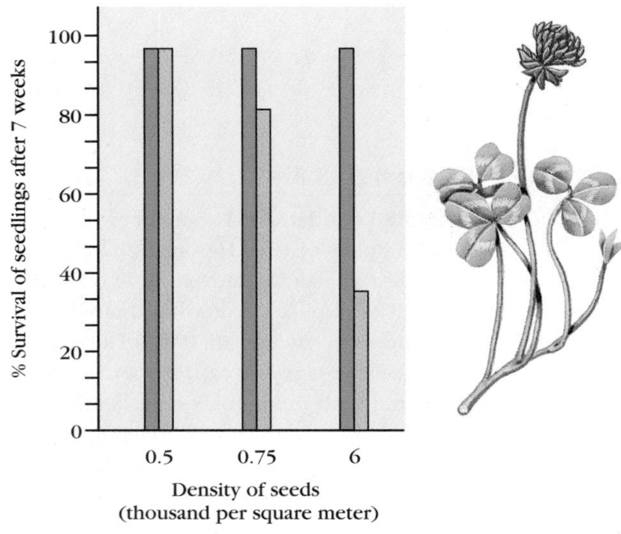

FIGURE 38-9

The effect of seed density on survival of white clover seedlings subjected to the stress of water shortage. When water was available, survival was about the same at all population densities (green bar). Orange bars represent the survival of seedlings that were not watered after the 18th day. Survival for these seedlings was much lower in the dense population. (After J. L. Harper, *Society for Experimental Biology Symposium* 15:1, 1961, Cambridge University Press.)

density-dependent mortality when a resource (water) was scarce (Figure 38-9).

Instead of competing directly for a limiting resource, individuals of many animal species compete indirectly for social dominance or for a territory. A **territory** is an area occupied by one or more individuals and defended by its occupants against other members of the same species (and sometimes of other species). Possessing a territory may guarantee the territory-holder an adequate supply of some limiting resource such as food, space, or a nesting site. In one study, a particular area of marsh in Iowa always contained about 400 adult muskrats. Males compete for territories, and the marsh provided about 180 territories, each with food and a refuge from predators for a pair of muskrats and their young. Animals in their territories are relatively safe from predators because they know their territories well, and the territories have ample cover for escape. Muskrats unsuccessful in the annual competition for territories were forced to live in unfavorable areas at the edge of the marsh, where they and their offspring suffered a high rate of death from overcrowding, predation, inadequate food, and interference by other animals.

FIGURE 38-10

Earth moves. This badger is uprooting plants and turning over soil as it digs a new burrow for itself. Its interactions with the biotic and abiotic factors of its environment form the badger's ecological niche. (J. McDonald/Visuals Unlimited)

Niche

When members of different species compete, they are said to have overlapping niches. The **niche** of a population or species is *not* a place where the species lives rather it is its functional *role* in an ecosystem. To illustrate niche, consider a badger, an inhabitant of grassland or woodland. The badger is a carnivore; to live, it eats mainly rodents, and it eats some species more than others. It alters the soil by digging a burrow to live in, and by digging its prey out of their burrows. It drinks water; it produces feces, little ecosystems in themselves; it attracts flies that feed on its feces; it may be eaten by wolves or other large predators, and so forth (Figure 38-10). Every interaction of the badger with its environment is part of its niche and determines where it can live and what other organisms can live in the same habitat with it.

The theory of **competitive exclusion** states that two species with very similar niches cannot survive together because they compete so intensely that one species eliminates the other. Gause set out to test this theory. He chose two species of *Paramecium* that have similar food requirements and hence similar niches. When he raised the two in the same culture, he found that one species always eliminated the other. The particular conditions in the culture determined which species survived (Figure 38-11).

In nature, it often looks as though several species coexist while competing strongly for the same resources. Whenever such cases have been examined in detail, however, it has always been found that the species divide the resources in some way. For instance Robert MacArthur found that different species of woo warblers in northeastern forests forage for insects in different parts of the trees, reducing the competition between them (Figure 38-12).

(a)

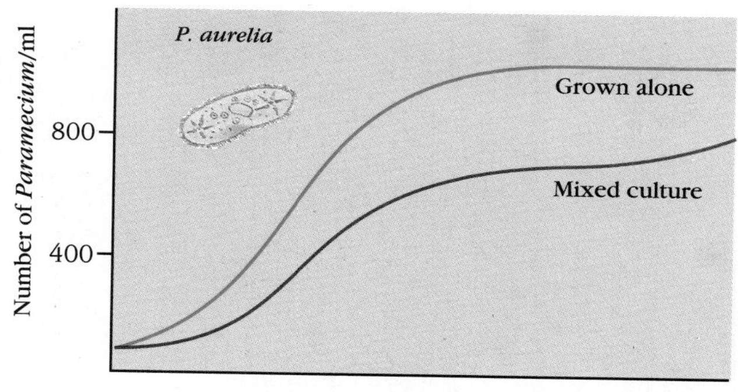

(b)

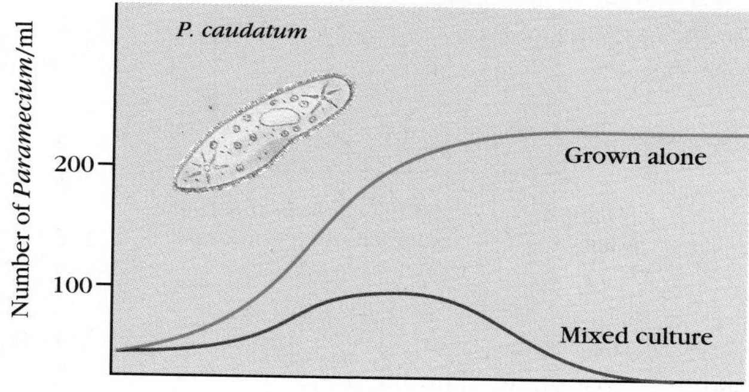

FIGURE 38-11

Gause's experiment with two species of *Paramecium*. Both species grew well by themselves in culture tubes with daily changes of water and inputs of bacterial food. When placed together under these conditions, *P. aurelia* **(a)** survived, while *P. caudatum* **(b)**, the larger and slower-growing of the two species, always declined to extinction. However, if the water was not changed, waste built up and the competitive outcome was invariably reversed.

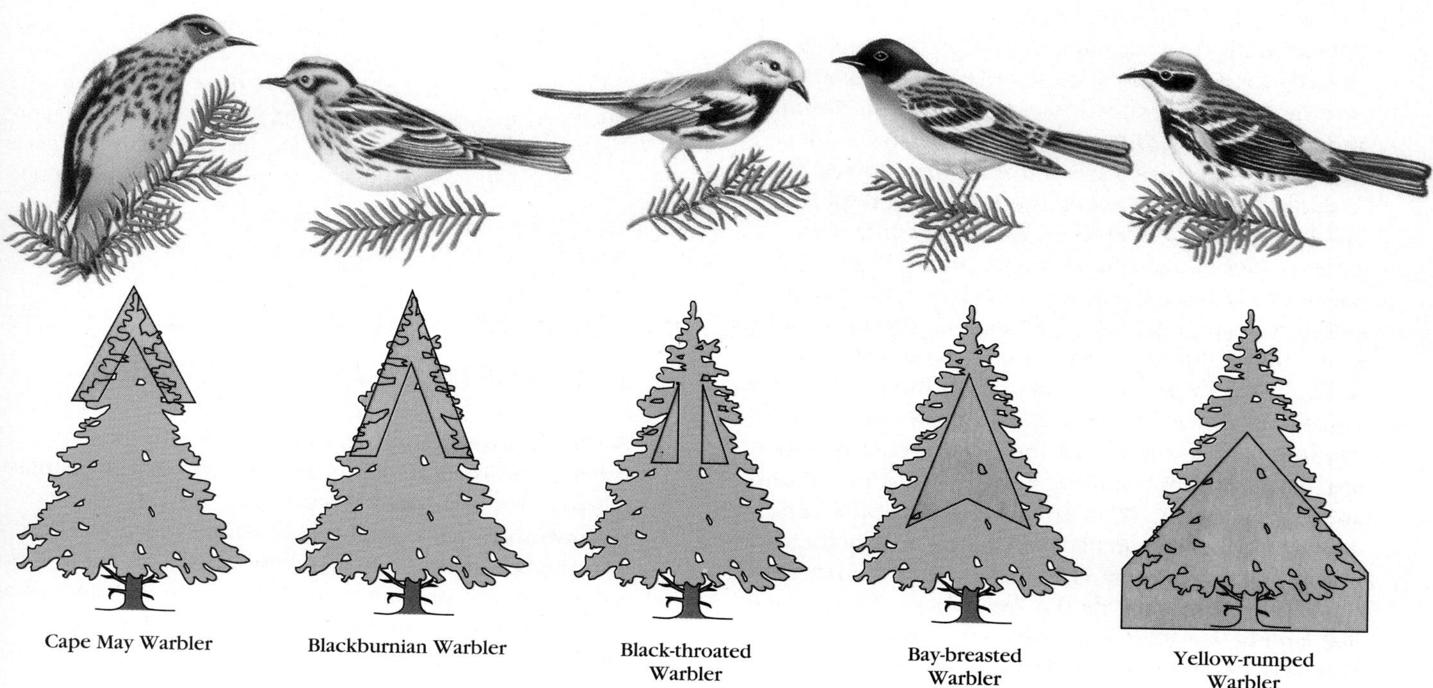

FIGURE 38-12

Coexistence by avoiding competition. Several species of warbler of the genus *Dendroica* hunt for insects in coniferous trees in the same New England forests. Each usually forages in a different part of the tree (shaded region), thus reducing the competition for food.

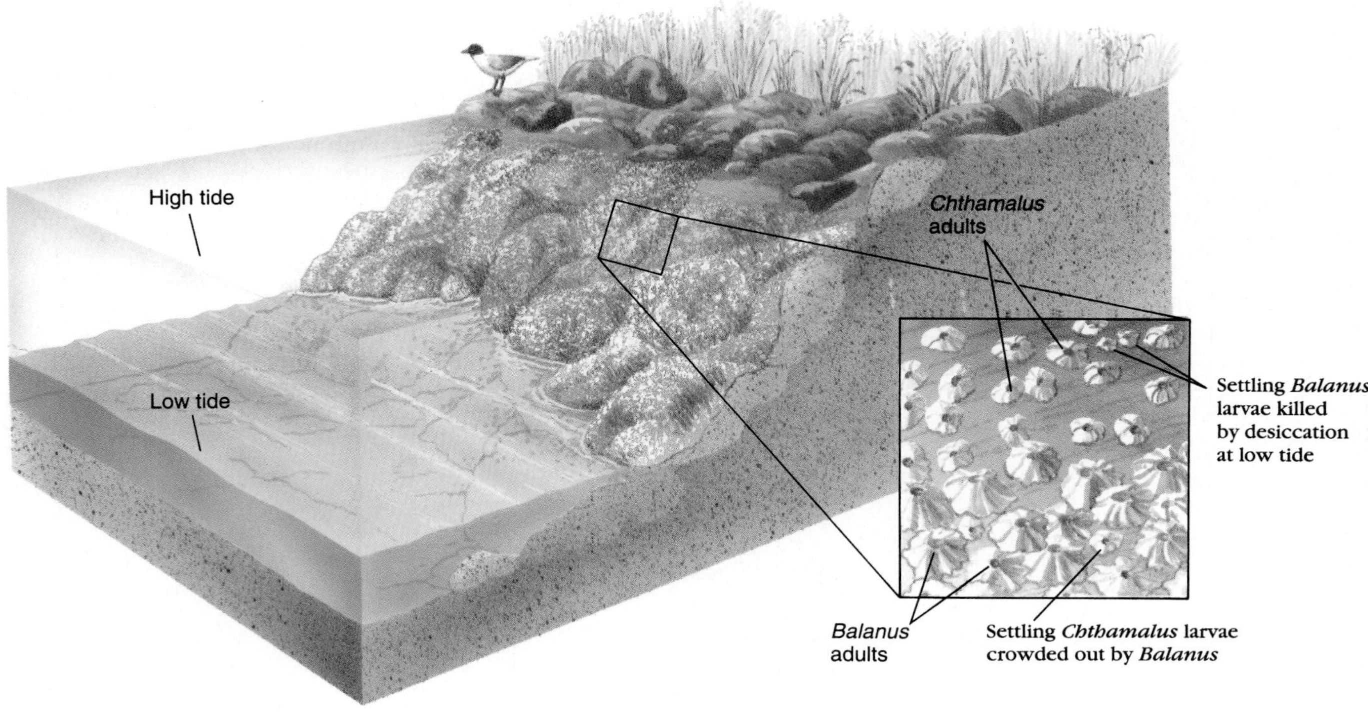

FIGURE 38–13

Division of space between two species of barnacles. On the rocky coast of Scotland, the vertical distribution of *Chthamalus* barnacles is limited by competitive exclusion. *Chthamalus* larvae settling from the plankton are crowded out by growing *Balanus* barnacles. The upper limit of *Balanus* is determined by its lesser tolerance of exposure at low tide.

An example of overlapping niches and division of resources comes from a study of two species of barnacles on the rocky coast of western Scotland, *Balanus* and *Chthamalus. Chthamalus* occupies the upper part of the intertidal zone and *Balanus* a lower zone, with little overlap between them. The planktonic larvae of both species settle on rocks in both zones. However, *Balanus* cannot survive in the upper zone because it is less tolerant than *Chthamalus* of exposure to the air at low tide. By removing *Balanus* larvae as they settled in the lower zone, Joseph Connell showed that *Chthamalus* survives in the lower zone when *Balanus* is absent. *Balanus* grows faster than *Chthamalus,* so that when the species compete for space on the rocks, *Balanus* grows over *Chthamalus* and crowds it out. Thus the lower limit of the zone occupied by *Chthamalus* is determined by whether or not *Balanus* is present (Figure 38–13). Competition of this sort, which restricts the area where a species can live, and hence limits the size of a population, is probably very common in nature.

Organisms compete for limited resources such as food and water, and for territories that include these and other resources such as shelter and nesting sites. When two or more species occupy similar niches, only one species usually survives, unless the species divide up the resources and reduce competition.

38–D PREDATION AND PEST CONTROL

Predation causes considerable mortality in most species and is therefore an important factor in regulating population size. In this context, we can define **predation** broadly, as any case where individuals of one species exploit a living prey species for food. By this definition, **predators** include herbivores and parasites, and **prey** includes plants and the parasites' hosts.

(a)

(b)

(c)

FIGURE 38–14

Biological control proved very effective in reclaiming range land overgrown with Klamath weed. (a) A field with a large population of yellow-flowered Klamath weed in California, June 1948. **(b)** The same field in June 1949, a year after the introduction of beetles adapted to feeding on Klamath weed. **(c)** The Klamath weed beetle. (a, b, F. E. Skinner; c, W. E. Ferguson)

Predators that specialize in eating only one species of prey may control the population size of their prey. For instance, in California, a small beetle controls the numbers of an attractive yellow-flowered plant called Klamath weed. This plant, introduced from Europe, spread throughout the United States. Klamath weed is toxic to cattle, and is also an aggressive competitor, displacing desirable plants from grazing land. By the late 1940s it covered 80% of the ground in many areas of California and Oregon, making it useless for cattle ranching (Figure 38–14).

In its European home, Klamath weed is attacked by several insects, including two species of beetles. Supplies of these beetles were let loose in California in 1945 and 1946. They flourished, and by 1959 Klamath weed had been reduced to less than 1% of its former abundance, chiefly due to the voracious appetites of the beetle larvae. Because the beetles are specialists, they cannot switch to other food plants when the Klamath weed population declines. (If they could, they might become pests themselves.) Both Klamath weed and its beetle predators persist in the United States, but at low densities. This is an example of **biological control** of a pest species by a natural enemy.

BIO-BIT

One method of biological control primarily used on insects begins with the sterilization by x-ray of large numbers of males. These sterile males are then released into the wild to mate with healthy females. If this is done to a species in which the females mate only once, it rapidly decreases the size of the next generation.

FIGURE 38–15

Water hyacinths choke a waterway. (M. Tierney, Jr./Visuals Unlimited)

Today, the search is on for predators of more pest species as the most effective way to control them. For example, imported water hyacinth has clogged southern waterways and resisted all efforts at control. Finally, in 1989, scientists sent to Pakistan for a species of beetle that eats this pest. The beetle offers the first sign of hope that this weed may finally be controlled (Figure 38–15).

Both these examples show that specialist predators, which attack only one or a few species, may limit the populations of their prey. It appears that generalized predators, those that can feed on many different species, seldom have this effect. Such predators can exploit any prey species that becomes common in their habitat. By concentrating on such a prey species for a while, they may slow a population explosion, but as the prey becomes scarce they switch to other foods.

Specialized predators may hold the population size of their prey species at very low levels. Generalized predators seldom do so because they switch from one prey species to another as the abundance of prey species changes.

38–E EXTINCTION AND ENDANGERED SPECIES

A species becomes **extinct** when its last member dies. Local populations of many species do become extinct quite often, but the population is reestablished later by immigration from neighboring populations of the same species. This happens, for example, when butterfly populations in some of the high mountain valleys of Colorado are occasionally exterminated by freak midsummer snowstorms.

When an entire species consists only of one small, restricted population, it is especially prone to extinction. Such species are most common on islands or in small isolated areas of restricted habitat. Here, a disaster may wipe out the species in one blow. On the Caribbean island of Martinique, the Martinique rice rat was exterminated by a volcanic explosion, which also killed every person in a town of 30,000 except a prisoner in an underground jail cell.

Extinction is not new. About 65 million years ago, more than half of all species on Earth, including the dinosaurs, became extinct in a relatively short period. Biologists are still arguing about the reasons. Some 50,000 years ago, the rate of extinctions again started to increase. A large number of animals, including many African game species, the Irish elk, mammoth, steppe bison, woolly rhinoceros, giant sloth, and mastodon, became extinct between 50,000 and 10,000 years ago. The dates of the extinctions coincide with the spread of human populations into new areas. A site at Solutré in France contains the remains of over 100,000 horses! Native Americans slaughtered bison by the thousands by driving them into pits. However, there were still about 60 million bison in North America when European settlers arrived. It took less than 200 years of hunting with firearms to reduce the bison to near-extinction in 1850, when only 250 remained in all of North America.

2? The extinctions caused by humans since 1900 dwarf anything in history. Hunting, for food, fur, livestock protection, or pleasure, has exterminated many species of large mammals and left nearly all large carnivores endangered. But most species become extinct because humans destroy their habitat, converting it into farmland, highways, or suburbs. The fate of a species in the twentieth century depends largely on how compatible it is with human population growth.

As humans occupy more and more of the Earth, animals that need a lot of space are particularly vulnerable. Many of these are large carnivores occupying the top trophic level of a food chain. Mountain lions, Florida panthers, and California condors are examples of animals that need large areas in order to find enough food. Their habitat has steadily disappeared as the human population has increased.

Many species cannot survive competition from the organisms that accompany human invasions: weeds such as dandelions and goldenrod, and animals such as goats, pigs, dogs, cats, rats, and mice. These animals eat eggs, young, and adults of many native species, and compete with them for space. For instance, when the Polynesian discoverers of Hawaii arrived there 1500 years ago, they brought with them dogs, pigs, rats, chickens, and some 30 species of plants. Centuries later, European settlers brought dozens more new species to the islands. Competition with these imports has caused the extinction of hundreds of species. An estimated half of Hawaiian native species are now extinct. ■

Experts believe that humans have exterminated about a million species in the twentieth century, mainly in tropical forests. The tropical rain forest biome contains millions of species that have never been described, most of which will be extinct before they are even named. The area of tropical forest has decreased by more than 50% in the last 200 years. Each year, an area of forest about the size of West Virginia is cut down, and soon little will be left (See Section 1-C).

Probably 99% of all the species that have ever lived are now extinct. Species form and then they become extinct, anywhere from a few weeks to millions of years later. So why worry about what is, after all, a natural process? The reason for concern is that now human activities are causing species to go extinct much faster than new ones can form. Agricultural and medical researchers tell us that extinctions deprive us of irreplaceable drugs and genetic resources for improving crop plants and animals. They feel we should stop our heedless extermination of other species, at least until we have assessed their possible usefulness. Threatened forests have already yielded a wealth of medicines, foods, and new seed stocks for crops. In the 1970s, a strain of disease-resistant wild corn (maize) was discovered in the Mexican forest. The hardy hybrids produced from breeding this corn are already worth billions of dollars to farmers. Many of the prescription drugs sold in the United States are derived from substances discovered in tropical forests. For example, the drug vincristine, used to cure childhood leukemia, comes from the Madagascar periwinkle (Figure 38-16).

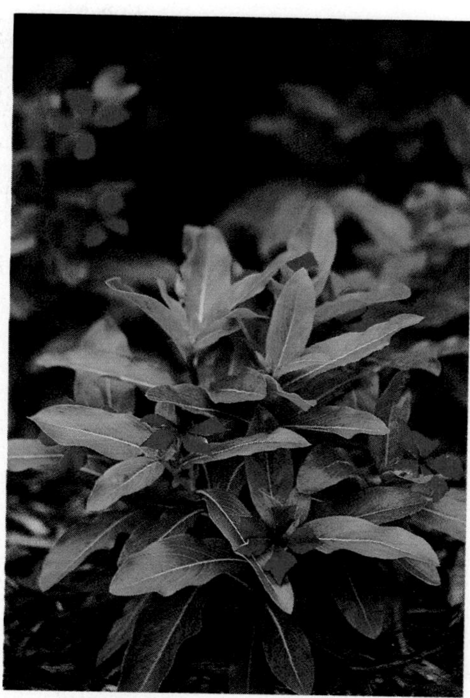

FIGURE 38-16

Madagascar periwinkle, from which the anti-leukemia drug vincristine is extracted. This common ornamental plant is also known as vinca, its old generic name. The international commission on nomenclature for plants has changed its name to *Catharanthus roseus*. (S. Camazine/Photo Researchers)

Ecologists point out that any species may be a vital link in a food web. Its extinction may disrupt an ecosystem or cause the extinction of other species. Others feel that there are more important considerations than the economics of extinction—for example, the esthetic loss we suffer when a species disappears (Figure 38-17).

Most countries have enacted laws and regulations designed to prevent more species from becoming extinct. The United States' Endangered Species Act is considered by many to be the strongest of these laws, but this does not mean that it does the job it was designed to do. Since European settlers arrived, more than 500 species of plants and animals have disappeared from North America, and some 3900 species of American plants and animals are currently endangered, 700 of them in Florida alone.

Human activities have greatly increased the rate of species extinction, mainly by destruction of habitat and competition, or predation by new species introduced by humans.

FIGURE 38-17

Endangered species. A Bald Eagle. (F. Lanting/Minden Pictures)

38-F HUMAN POPULATION GROWTH

Human population growth is responsible for the extinction of other species and for most of our own environmental problems, from traffic jams to pollution.

Declining Death Rates

3? The number of people on Earth has increased steadily for at least 10,000 years. But the greatest population growth has taken place in the last 200 years, and the rate of growth is still increasing (Figure 38-18). During the 1990s nearly one billion people will be added to the human population, more than in any previous decade. They will increase the population by 20%, from about 5.5 billion in 1993 to about 8.4 billion by 2025 and 11.2 billion by 2100.*

The human population explosion is largely a result of reduced death rates, particularly among infants. Infant mortality is the most important factor determining **life expectancy,** the average number of years that a newborn baby can be expected to survive. In 1900, the infant mortality rate in North America, the U.S.S.R., and Western Europe was about 40 per 1000 babies born, and the life expectancy was 46 years. In 1993, the infant mortality rate for these countries was about 10 per 1000 and life expectancy was more than 73 years.*

In developing countries, most deaths are due to respiratory and digestive tract infections in infants, and such deaths are easily reduced without expensive medical care. The simple lack of clean water, nutritious food, and basic education in hygiene is the reason that average life expectancy for most of Africa is 54 years, while the average for Latin America and East Asia (including Japan and China) has risen to nearly 70 years since 1950* (Figure 38-19). ■

The Demographic Transition

Demography is the study of populations, particularly human populations.

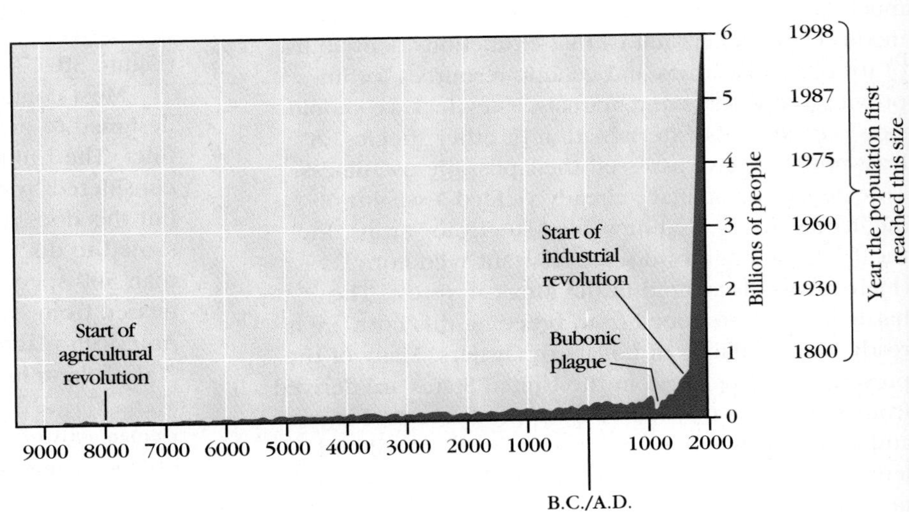

FIGURE 38-18

Growth of the world human population in the last 11,000 years.

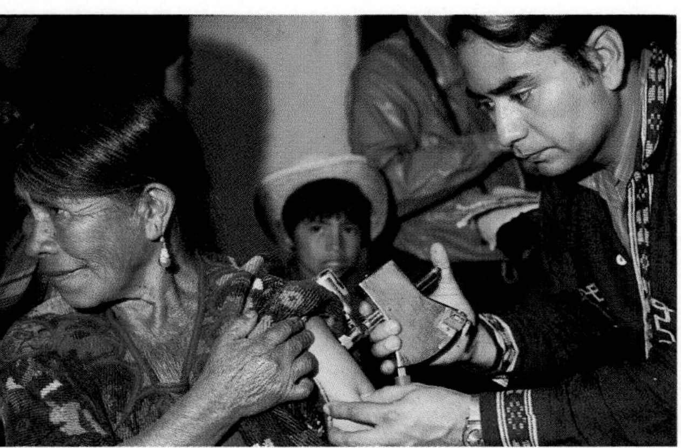

FIGURE 38-19

A doctor immunizes a woman against typhoid at a clinic in Guatemala. (A. C. Twomey)

The theory of the **demographic transition** states that economic and social progress affects population growth in three stages. In the first stage, in preindustrial societies, birth and death rates are both high and the population grows slowly, if at all. It is in the second stage that a population explosion occurs. Death rates fall as public health improves, but birth rates remain high. The population grows very fast, by about 3% per year, and so it doubles every 25 years. In the third stage, birth rates fall until they roughly equal death rates, and population growth slows down and ultimately stops (Figure 38-20).

The theory of the demographic transition was based largely on what had happened to European populations. The developed countries of Europe, North America, and parts of Asia have indeed passed through these stages. Populations in these areas have grown as much as five-fold since 1850, but today they are growing slowly or even declining.

The single factor most clearly correlated with a decline in birth rate during the demographic transition is increasing education and economic independence for women. The spread of knowledge that lowers the death rate is usually part of a general program to improve the educational level. Educated women find that they need not bear many children to ensure that a few survive, and they also learn contraceptive techniques. In addition, women find that they can contribute to the family's increasing prosperity by holding a job and by spending less time and energy on raising children. This is usually attractive to women, even in countries where religious doctrine and tradition dictate large families.

The demographic transition has taken from one to three generations to spread through the populations in most developed countries. During its progress, death rates are low, but birth rates remain high, and so the population grows enormously. The population of the United States will almost have quadrupled during the twentieth century.

The Demographically Divided World

Some countries are not proceeding through the demographic transition as predicted by the theory. They appear to have stopped in the second stage, unable to make the educational and economic gains that would reduce the birth rate. If the vast rate of population growth in the second stage is sustained, it begins to overwhelm food production, medical care, and educa-

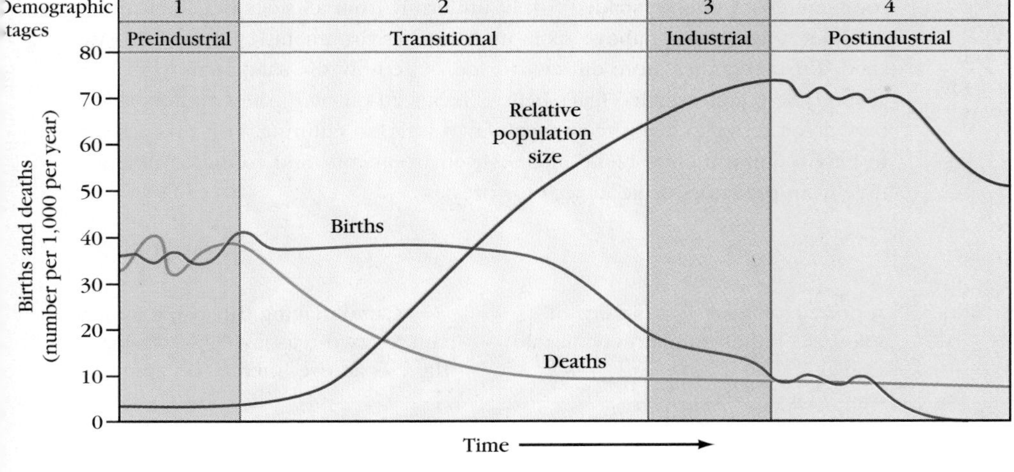

FIGURE 38-20

Changes in a population's characteristics as it progresses through the four stages of economic development.

A Journey into Evolution

Camouflage

Most organisms are vulnerable to predation, and so there is strong selective pressure for adaptations that serve as defenses against predators. Many animals are **camouflaged**—disguised in such a way that they are difficult to perceive even when they are in plain sight. Experiments have shown that camouflage is, in fact, of selective advantage to its owner. We have studied one such case in this book, the evolution of melanism in moths: moths the same color as their background are less likely to be eaten by birds than moths that contrast with their background (see Figure 14-13).

Since vision is the dominant sense in humans, visual camouflage is very obvious to us. We recognize an animal by three main visible features: its silhouette, its eye, and its bulk, or appearance of being rounded. An animal's camouflage must disguise these three features.

Bulk is nearly always disguised by countershading (Figure 38-A). If an object is the same color all over, its underside appears darker when light falls on it from above and makes it appear rounded. The vast majority of animals have light-colored bellies and dark backs. Light falling from above makes them look uniformly colored and therefore flat. A camouflaged gun or plane is also painted with a pattern that countershades it.

Camouflage often involves coloration that disrupts the silhouette (Figure 38-B). Some parts of the body appear the same color and intensity as the normal background, while others contrast with it. Under these conditions, some parts of the object stand out whereas others seem to disappear. The result is a pattern of splotches rather than a recognizable animal.

Camouflaging the eye is important for two reasons. First, where there is an eye, there is an animal.

FIGURE 38-A

Countershading. In aquatic environments, the light source always comes from above. **(a)** A uniformly colored fish lit from above is clearly visible in its environment. **(b)** A countershaded fish lit uniformly from all sides is also clearly visible. But when lit from above, as in its natural environment **(c),** a countershaded fish becomes more difficult to see. Note how the shade that forms around the countershaded fish's belly is balanced out with the countershaded dark color above to give the appearance of uniform coloring. Fish that appear to have a uniform color blend into their environments, and are better able to hide from predators or lie in wait for prey.

FIGURE 38-B

An orchid mimic. This spider hides among these flowers, waiting for prey. (Nuridsany et Perennou)

FIGURE 38-C

Nine cryptic moths. Can you find the nine different moths hidden on this tree trunk in Costa Rica? (S. Krasener/Photo Researchers)

FIGURE 38-D

A female nightjar bird sitting on her nest. Cryptic coloration is common in animals that must remain in view of potential predators for any length of time. (F. Lanting/Minden Pictures)

Second, an eye is always near the brain, one of an animal's most vital organs. The eyes may be disguised in various ways (Figure 38-C).

Camouflage is useless without appropriate behavior patterns. For instance, a "leaf" wandering up and down a twig is apt to be noticed by a predator. Most butterflies are not camouflaged; they fly by day and their motion makes them visible whatever their coloration. Camouflage is much more common in moths, most of which are active at night; during the day they rest motionless and camouflaged on some appropriate surface.

What is the advantage of visual camouflage? First, coloration that hides an animal against its background might have been selected for in animals with predators that hunt by sight. If this is true, we would expect that animals with few predators would be less likely to be cryptically colored. This is in fact the case. Large birds such as swans and gulls, which have few predators, are conspicuous, but their small, vulnerable young are cryptically colored (Figure 38-D).

Plants, too, can be camouflaged. Mistletoes in Australian forests, for example, mimic the leaf shapes of their plant hosts in a most remarkable fashion. This probably makes the mistletoes less apparent to the butterflies (and perhaps to some mammals) that feed on them.

Table 38–2
Population Growth and Wealth in Asian Countries, With Estimates for the Year 2000 (World Bank Data, 1992)

Country	Population (millions)			Total Fertility rates*			GNP per head, $†	Literacy rate, %
	1965	1990	2000	1965	1990	2000		
China	729	1,134	1299	6.4	2.5	2.1	547	73
Hong Kong	4	6	6	4.5	1.5	1.5	11,490	90
Japan	99	123	128	2.0	1.6	1.6	23,558	99
Singapore	2	3	3	4.7	1.9	1.9	11,160	88
South Korea	29	43	46	4.9	1.8	1.8	5400	96
Thailand	31	56	64	6.3	2.5	2.1	1420	93
Afghanistan	12	18	27	7.0	7.2	6.8	150	29
India	495	850	1042	6.2	4.0	3.0	350	48
Indonesia	107	178	219	5.5	3.1	2.4	570	77
Malaysia	10	18	22	6.3	3.8	3.0	2320	78
Nepal	10	19	24	6.0	5.7	4.6	170	26
Pakistan	57	112	162	7.0	5.8	4.6	380	35
Philippines	32	61	77	6.8	3.7	2.7	380	35

*Births per woman during her lifetime.

†In 1990. For comparison, Switzerland had the highest GNP per person in the world in 1990 at $35,081. GNP per person for the United States was $21,863.

tion. When the demands of the population exceed the sustainable production of local farming and grazing land, forests, and water supplies, people begin to consume the resource base itself. Vegetation disappears, soil erodes, wells run dry, and productivity declines. This, in turn, reduces food production and incomes in a downward spiral.

Population trends appear to be driving about half the world's people toward this kind of economic decline and half toward a better future (Table 38–2). The world is divided into countries that have completed the demographic transition and countries stuck in the second stage, with few countries still in the process of passing through the demographic transition.

In more developed countries, mainly Europe, North America, Australia, and Japan, populations are growing on average 0.5% per year, and living conditions are fairly stable. Some ecological resources may actually be increasing, as these countries invest in pollution control, reforestation, and restocking wild populations of plants and animals. In complete contrast, populations in high-growth areas (chiefly Southeast Asia, Latin America, India, the Middle East, and Africa) are growing more than four times as fast—on average by 2.1% per year* (Figure 38–21).

Modern increases in human life expectancy are due mostly to better nutrition and hygiene, which have boosted infant survival. Low birth rates are found where women have at least some education and economic control over their own lives. The greatest increases in human populations currently exist in countries where education and economic gains have not reduced the birth rates.

*Data courtesy of 1993 World Population Data Sheet of the Population Reference Bureau, Inc.

FIGURE 38–21

A street scene in Madagascar, where high birth rates continue to cause overcrowding. (F. Lanting/Minden Pictures)

SUMMARY

1. Populations consist of all members of one species occupying a given area at a given time.
2. Under ideal environmental conditions, the number of individuals in a population increases at its biotic potential. In sexually reproducing organisms, this biotic potential is determined mainly by the age of the female parent at first reproduction, but it is also influenced by the number of offspring produced at each reproductive event and by the parent's reproductive lifespan. A population seldom, if ever, reproduces at its biotic potential even when it is growing exponentially.
3. Exponential growth often occurs when populations are introduced into new environments where resources are abundant and natural enemies and competitors are scarce.
4. Survivorship curves for members of a population reflect the population's reproductive strategy.
5. At the two extremes, the members of a population may produce many small offspring and leave them to fend for themselves, or they may produce a few large offspring that are nourished and trained by the parents.
6. Survivorship data are the basis for constructing life tables—useful in predicting future changes of population size.
7. Most populations remain about the same size from year to year. Populations living in areas with extreme climates may be kept in check by density-independent natural events (such as droughts or freezes). However, most populations are generally limited by density-dependent factors, such as predation, disease, or competition for resources.
8. When growth of a population ceases, under the influence of one or more of these factors, the size of the population stabilizes at approximately the carrying capacity of the environment for that species.
9. Competition for resources may play an important part in regulating the size of a population.
10. Competition between members of different species with similar niches leads either to the extinction of one species, the weaker competitor, within the area of overlap, or to selection for adaptations that reduce the competition. For example, each species may become specialized in such a way that it uses only part of a limited resource.
11. Predation is another factor that may reduce the size of a population. Many specialized predators and parasites are known to keep their prey species at low density.

12. Extinction is the inevitable fate of populations and whole species. Species consisting of only one or a few small populations, living in restricted habitats (such as islands), are particularly vulnerable to extinction.
13. Humans have greatly increased the rate at which species become extinct by introducing predators or competitors into new areas, by hunting, and especially by destroying habitats. We are currently losing many species before we can assess their usefulness as resources or the exact roles they play in their ecosystems.
14. The number of people in the world has grown exponentially in the last few hundred years as a result of improved nutrition and hygiene, which have reduced the death rate, especially among infants. Today the populations of many countries are growing faster than the supply of resources their people need to survive.
15. The nations of the world can be divided into those that have passed through the demographic transition and show little population growth, and those that appear stuck in the middle of the demographic transition, whose populations are doubling about every 25 years.

SELF-QUIZ

1. Draw a graph showing the long-term growth of a population of bacteria on a nutrient medium in a laboratory culture.
2. A population can grow exponentially:
   a. when food is the only limiting resource
   b. when first invading a suitable and previously unoccupied habitat
   c. only if there is no predation
   d. only in the laboratory
3. Which of the following does *not* directly affect biotic potential?
   a. a female's age at first reproduction
   b. carrying capacity of the environment
   c. length of time a female is fertile
   d. average number of offspring per brood or litter
4. If a population exceeds the carrying capacity of the environment:
   a. it will evolve adaptations to avoid a population crash
   b. its numbers will probably decrease rapidly
   c. its food supply will increase in the next generation
   d. the average number of young per individual will increase
5. Which of the following would be *least* likely to act as a density-dependent factor limiting the size of a population of mice?
   a. parasitism
   b. buildup of waste products
   c. predation
   d. unfavorable climate
6. A female elephant bears one offspring every two to four years. Which type of survivorship curve would you expect elephant populations to show: I, II, or III?
7. Studies suggest that:
   a. two species may share the same resource only if both are strong competitors for it
   b. two species that appear to be sharing the same resource are probably specializing so that each uses only a particular part of the resource
   c. if two species are sharing the same resource, neither is capable of exploiting the part of the resource used by the other
   d. no resource can be used indefinitely by more than one species
8. The demographic transition in a country is correlated with:
   a. education for women
   b. improvements in agricultural production
   c. the industrial revolution
   d. improved hygiene and nutrition
   e. improved treatment for illnesses

THINKING CRITICALLY

1. Paul Ehrlich has said, "It is quite possible that the penalty for frantic attempts to feed burgeoning populations may be a lowering of the carrying capacity of the entire planet." Ecologist Lee Talbot has said, "We haven't inherited the Earth from our parents. We've borrowed it from our children." What does each of these statements mean? Do you agree? Why?
2. What are some of the lines of evidence that indicate that the human population has already exceeded the Earth's carrying capacity?
3. How does a project such as filling in a marsh for a housing development or building a four-lane highway affect the populations of organisms in these areas?
4. The 1970s saw a decline in the birth rate in the United States. Why is the population of the United States still growing?
5. Consider two women born in the same year, each of whom will give birth to twin girls as her only children. However, one woman (A) will have her twins at age 18 the other (B) at age 36. Each daughter will have twin

daughters at the same age her mother gave birth, and so on. All mothers will die at age 72.

a. How many descendants does A have when she dies?

b. How many descendants does B have when she dies?

c. Construct a graph to show the growth of populations A and B.

d. How do the rates of increase compare in the two populations? (Find a numerical answer if you can!)

SELECTED KEY TERMS

biological control, *p. 799*
biotic potential, *p. 790*
carrying capacity, *p. 793*
competitive exclusion, *p. 796*
demographic transition, *p. 803*
demography, *p. 802*
density-dependent, *p. 794*
density-independent, *p. 794*
exponential growth, *p. 790*
extinction, *p. 800*
life expectancy, *p. 802*
mortality factors, *p. 794*
niche, *p. 796*
population, *p. 789*
population density, *p. 789*
predation, *p. 798*
reproductive strategy, *p. 791*
survivorship, *p. 791*
territory, *p. 796*

SUGGESTED READINGS

Barrett, S. C. H. "Waterweed invasions." *Scientific American,* October 1989. The story of the damage done by introduced water hyacinth and kariba weed, the failure of mechanical and chemical control, and the hope that natural predators will control the invasions.

Ehrlich, P. R., and A. E. Ehrlich, *Healing the Planet,* Reading, MA: Addison Wesley Publishing Company, 1991. Strategies for resolving the environmental crisis.

Hoage, R. J., ed. *Animal Extinctions. What Everyone Should Know.* Washington, DC: Smithsonian Institution, 1985. The record of a symposium to acquaint the public with the major issues of species extinction and related habitat destruction; dozens of well-documented examples.

Keyfitz, N. "The growing human population." *Scientific American,* September 1989.

Wilson, E. O. *The Diversity of Life.* Cambridge, MA: Belknap Press, 1992.

CHAPTER 39

CONCEPT GUIDE

After reading this chapter, you should be able to:

1. Describe the impact of early hunter-gatherer people on their environment. (Section 39–A)
2. Describe the positive and negative results of the development of agriculture. (Section 39–B)
3. Explain how low-input farming is better than green revolution techniques. (Section 39–C)
4. Describe the effects of global deforestation. Explain why slash-and-burn deforestation techniques are particularly damaging in tropical rain forests. (Section 39–D)
5. Describe the impact of the use of fossil fuels on the environment during the industrial revolution. (Section 39–E)
6. Compare the causes and impacts of overpopulation, global warming, acid rain, ozone depletion, deforestation, and loss of species diversity. (Section 39–F)
7. Explain some of the things you can do to address the many problems discussed in this chapter. (Section 39–G)

What is this woman doing with this wood? She, like billions of other people, relies on wood as a cooking fuel. In fact, over half of all the wood that humans use in the world is used for cooking. Read more about the human impact on the world's forests in Section 39–D. (Dr. N. Smith/Earth Scenes)

Human Impact On The Environment

All organisms interact with their environments. For example, plants withdraw minerals from soil and when they wither and die, their remains enrich the soil with nutrients. Herbivores eat plants and in turn are eaten by carnivores in the constant cycling of energy through the **biosphere**—Earth's layer of air, water, and soil where life can exist. Natural selection favors plants that can evade the attacks of herbivores and successfully reproduce. Plants with spines, thorns, and chemical defenses can inhibit a wide range of herbivores from eating their tissues, but over time, some herbivores adapt to these plant defenses. All of these interactions between populations subtly change the environment.

A few animals, called **keystone species,** are able to make long-lasting changes that alter or maintain their environments. African elephants, for example, are the keystone species that helps keep forests from encroaching on the savannas, the open, park-like fields favored by herds of grazing antelope, giraffes, and wildebeest. An elephant not only browses lower branches of trees, but it can kill trees by stripping their bark or uprooting them. In the southeastern United States, alligators are the keystone species whose swimming movements dig and maintain "gator holes"—free of rooted swamp vegetation. Fishes and other aquatic species survive dry times by congregating in these deeper areas of the swamp.

Like other animals, human beings also interact with the environment, but because of human numbers and technology we affect our environment intentionally as well as more dramatically and permanently than do populations of other living organisms. During the course of history, two changes in the way humans live have been particularly important in altering our relationship with the environment. The first of these was the development of agriculture, as humans shifted from finding food to growing it. The second was the industrial revolution, which has drastically changed the way we live and has even further amplified our effects on the Earth.

In this chapter we consider how human interactions with the environment have changed, how these changes have led to the environmental problems that confront us today, and how human attitudes must change if we are to overcome these problems.

KEY CONCEPTS

- Ecologically, the two most important events in human history were the development of agriculture and the industrial revolution.
- Modern problems that result from agriculture include population growth, habitat destruction, narrower diet, and soil loss.
- Problems that result from the industrial revolution include population growth and many forms of pollution.

Curiosity Questions

1? Why is there a shortage of food in the world today? (See page 815)

2? Why are people cutting down rain forests? (See page 820)

3? How quickly is the world's human population increasing? (See page 823)

(a)

(b)

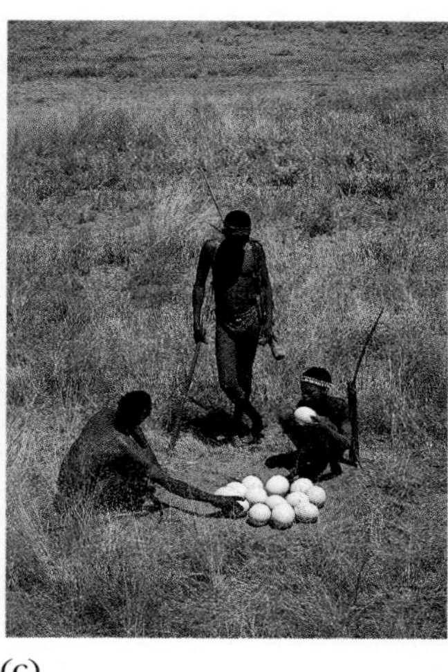
(c)

FIGURE 39-1

Bushmen of the Kalahari desert in southern Africa. (a) Lightweight bows and arrows are used to shoot small game animals. **(b)** Important skills and cultural traditions are passed from one generation to the next. **(c)** Bushmen have an encyclopedic knowledge of their environment and gather many different types of food (like these ostrich eggs). Children do not hunt or gather food for the group because both are skilled activities that take many years to learn. (In agricultural societies, in contrast, very young children usually participate in food production.) (a, F. Lanting/Minden Pictures; b, A. Huber/Okapia/Photo Researchers, Inc.; c, M. P. Kahl)

39–A HUNTER-GATHERER POPULATIONS

For most of our evolutionary history, human beings were **hunter-gatherers,** people who obtain their food by collecting plants and killing wild animals. Our knowledge of hunter-gatherers comes from archeological finds and from studies of modern hunter-gatherer groups such as Eskimos, Australian aboriginals, and the !Kung bushmen of the Kalahari desert in southern Africa (Figure 39-1).

Cooperative efforts permit hunters to spear or trap large game animals. But the main source of food in hunter-gatherer groups is plant materials—seeds, fruits, leaves, nuts, roots, and berries—gathered as they are used, or preserved and stored for later. The use of fire broadened the range of plants that could be used for food. Cooking can remove plant toxins and can also soften plants, making them more digestible.

Seafood is another source of food for hunter-gatherers. Native Americans in the southeastern United States left piles of shells up to 20 feet high from the oysters that formed a major part of their diet. They also caught fish by spearing and trapping them. Although some fish are farmed today, most of our fish still come from "hunting" these animals in their wild habitat.

Many early hunter-gatherer groups were nomadic, moving from place to place according to the availability of plant and animal food. Human populations spread from Europe and Asia to the Americas and Australia some 25,000 years ago. Later, more or less permanent settlememts developed and people began to build more elaborate dwellings and to accumulate possessions such as the decorated tools and pots that began to appear in Europe and Asia about 20,000 years ago, and in America at least 12,000 years ago (Figure 39-2).

FIGURE 39-2

A clay bowl from ancient Egypt. Decorated, breakable pots like this are not characteristic of hunter-gatherer groups, but rather are typical of settled, agricultural people. (British Museum, London/Bridgeman Art Library)

Even the small populations of ancient hunter-gatherers affected their environment in many ways. As they travelled to new areas, early humans carried plant seeds with them, altering the distribution of plants (and possibly animals). We can only guess at the size of the Old Stone Age population (before the invention of agriculture), but estimates put it at around 5 million people. This would seem to be too few to have had very dramatic effects on the environment. Nevertheless, human activities such as starting or spreading fires do not require many people to produce major effects.

A large number of animal species, including many African game species, the Irish elk, the mammoth, steppe bison, woolly rhinoceros, and mastodon became extinct during the Old Stone Age. Many more species became extinct than in earlier or later periods. The traditional explanation has been that these extinctions were caused by rapid changes in climate during the Ice Ages, which occurred between about 2 million and 10,000 years ago. Some scientists, however, believe that many of these extinctions were caused by human hunting and destruction of habitat (by starting fires to drive game animals into traps).

Settlements, permanent or temporary, usually lead to environmental problems such as pollution. Deformities in the skeletons of ancient hunter-gatherers suggest that some of these people suffered from diseases caused by polluted water. A village well or a nearby stream is easily polluted by human waste or the runoff from a garbage pile.

The environmental impact of early hunter-gatherers included changing the distribution of plants, destroying some habitats by fire, and possibly causing the extinction of some game animal species.

39–B THE AGRICULTURAL REVOLUTION

Agriculture is the practice of breeding and caring for animals and plants that are used for food, clothing, housing, and other purposes. Agriculture originated, probably independently, in many different places at about the same time some 10,000 years ago (Figure 39–3). Fossils of domesticated dogs dating from 11,000 years ago have been found in Iraq, and cultivated plants date back to 10,000 years ago in America.

At first, agriculture merely provided part of the food for a hunter-gatherer society. For instance, when Spanish missionaries landed in the southeastern United States in the sixteenth century, they found hunter-gatherer societies living largely on deer, seafood, and local plants. In addition, however, many tribes cultivated a few plants around their huts. They also cleared temporary plots in the forest, where they grew maize (corn) and beans, using seeds that were originally imported from Mexico.

Later, human societies became completely dependent on agriculture for food. Although it occurred slowly, the changeover from hunting and gathering to agriculture had such a dramatic impact on human societies that it is often called the **agricultural revolution.**

We are so used to thinking of agriculture as a superior way of life that the relative advantages of hunter-gatherer culture may come as a surprise. Hunter-gatherers do not face the constant battle with pests, droughts, and famine that beset all agricultural communities. Studies in southern Africa during a drought showed that farmers starved while the population of hunter-gatherer bushmen in the Kalahari desert remained stable in size and the people well-fed. This is probably because hunter-gatherers keep their populations well below the size that their territory can support. This is not a result of a high death rate. The people make a conscious effort to keep their population size down by such practices as abstention from sexual intercourse, abortion, infanticide, late marriage, and late weaning. Furthermore, these bushmen have a more balanced diet than most farmers, and their life expectancies are comparable with those of agricultural peoples in most of the world.

A striking consequence of agriculture is that a new type of division of labor grows up within the group. In the more prosperous agricultural societies, a few people can produce food for everyone. The rest are freed to become builders, bakers, and merchants. Finally, the population may even be able to afford the luxury of poets, scholars, and students, who contribute little to the group's physical well-being, but are the basis of its cultural life.

Once farming had begun anywhere on Earth, it inevitably spread over the face of the globe. Because farming supports larger populations, an agricultural community tends to expand into the land of any nearby hunter-gatherer group, fighting for the territory if necessary, and often driving the hunter-gatherers to extinction or to become integrated into the farming community. Similarly, settled agriculture usually overwhelms nomadic livestock herding as a way of life. For instance, the battle for land between farmers and ranchers in the West is part of American history. In Africa today, expanding agriculture is destroying the traditional life-styles of Masai and Bedouin herders.

The roots of most environmental problems can be traced to the evolutionary success of agricultural societies. No one believes we can return to a preagricultural way of life. The question is how our destructive, agricultural way of life can be changed into a sustain-

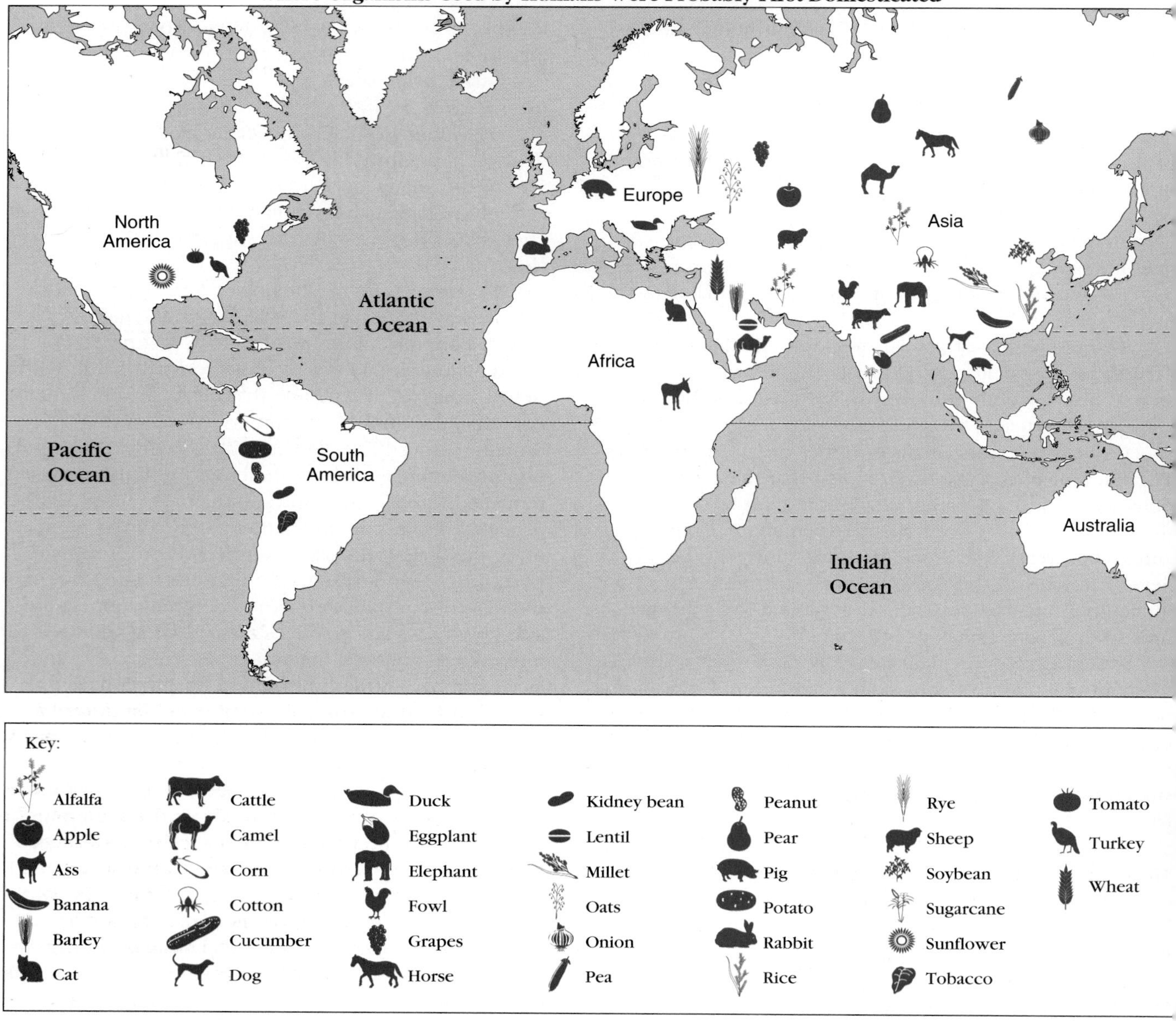

Where Organisms Used by Humans Were Probably First Domesticated

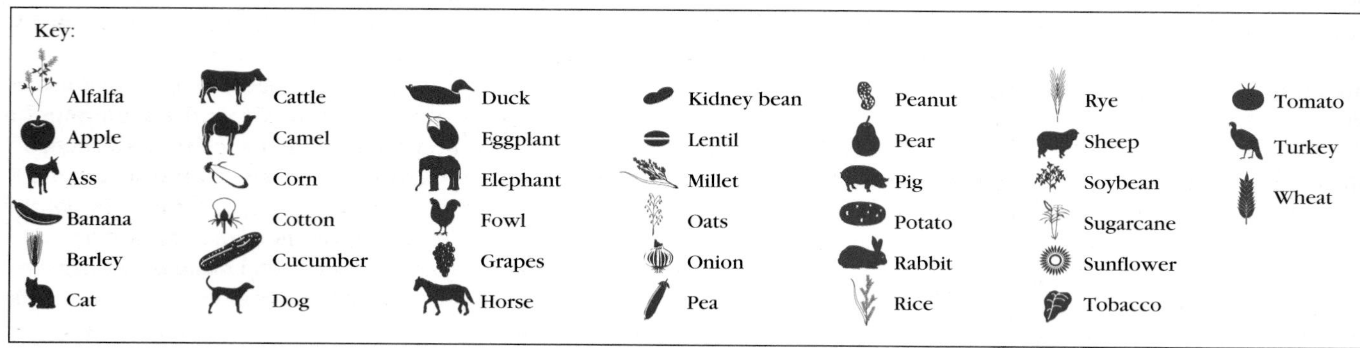

FIGURE 39–3

Areas where plants and animals we use today were probably first domesticated. Some species are believed to have been domesticated independently in two or more areas, and these are shown in both places.

able way of feeding our populations. The most important effects of agriculture can be summarized:

1. **Habitat destruction.** The area of land devoted to farming has increased steadily, with the conversion of natural habitats to farmland causing the extinction of thousands of species of plants and animals. Clearing natural vegetation, particularly forests, to produce farmland has had many effects, including altering the climate, producing a widespread shortage of cooking and heating fuel, and causing floods and water shortages.
2. **Soil erosion.** Agriculture need not destroy the soil, but it usually does. The amount of soil on the surface of the Earth has steadily decreased since the first digging tool was invented thousands of years ago.
3. **Population increase.** Not only can agriculture support more people on a given land area than hunting and gathering, but the population control practiced by hunter-gatherer societies is usually abandoned by agricultural communities. This is partly because children, who are not important food collectors in a hunter-gatherer society, are

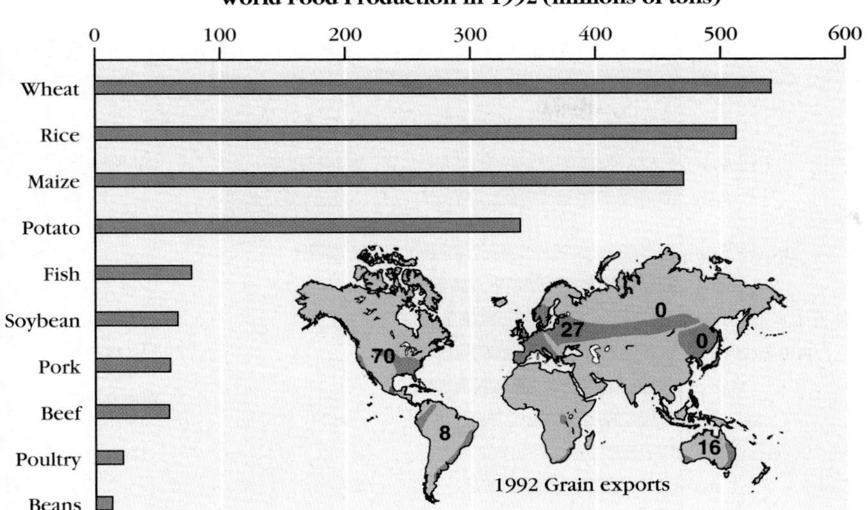

FIGURE 39–4

Annual world production of major foods. The four most important foods are shown at the top. The numbers on the map show 1992 grain exports in millions of tons. Several grain-growing areas have no numbers, meaning either that they produce just enough grain for their own use (India and China) or that they import grain (Eastern Europe, Latin America, Africa). Note that North America produces most of the grain that goes to food-importing nations. As a result, the food security of much of the world is threatened by falling grain yields in North America in recent years. Note the overwhelming importance of the major cereals (wheat, rice, and maize).

valued as a labor force on farms. Only in very recent times has effective population control been practiced by agricultural people.

4. **Diet simplification.** Studies of hunter-gatherers make it clear that humans once ate a much greater variety of plant and animal species than we do today. Several thousand species of plants and several hundred species of animals once formed part of the human diet. The Cherokee Indians of North America alone ate more than 400 different species of plants. Nowadays, few people eat more than 50 species of organisms, and the vast majority of human food comes from just four species of plants: wheat, rice, maize, and potatoes (Figure 39–4). This dependence on a few species lays societies open to famine on a scale unimaginable in earlier times: a new disease or pest affecting any of these crops can rapidly eliminate a large part of our food supply.
5. **Attitude change.** Hunter-gatherers typically view themselves as belonging *to* a particular place, while farmers typically view a parcel of land as *a possession.* With agriculture, the inheritance of land and goods becomes important, and the desire to have children who will inherit the property and care for their aged parents is a recurrent theme in mythology and literature of agricultural societies.

Agriculture supports more people on a given area of land than does a hunter-gatherer lifestyle. However, agricultural practices have led to habitat destruction, soil erosion, overpopulation, and a diet consisting of fewer types of food.

39–C FEEDING THE HUMAN POPULATION

The enormous growth of the human population (discussed in Chapter 16, *A Journey into The Environment:* Overpopulation and the Human Animal), particularly in the last hundred years, has resulted in widespread starvation. Over 41,000 people die of starvation every day, and of the world's 5.5 billion people, between 350 million and 1 billion are so undernourished that their lives are in danger. **Starvation** means death from lack of food. However, most people who are inadequately fed actually die not because they eat too few calories to sustain life, but because their malnourished bodies have little resistance to infectious diseases that would not be fatal if they were properly fed. Most malnourished people get about as many calories as they need, but their food is deficient in essential amino acids and vitamins. The problem of diet, then, is not merely the problem of obtaining enough calories, but of obtaining a balanced diet, one that contains the mixture of macronutrients and micronutrients needed to keep the body healthy.

1? Human starvation is not biologically necessary—yet. The world's farmers produce enough food calories, proteins, and vitamins to keep more than 6 billion people in good health. The trouble is that food, and the income to buy food, are unevenly distributed between the rich and the poor. The problem of hunger today is more economic and political than biological. But we are pushing the limits of the Earth's productivity. Declines in grain production, which began in the 1980s, may be the start of an expected per capita fall in world food production. ■

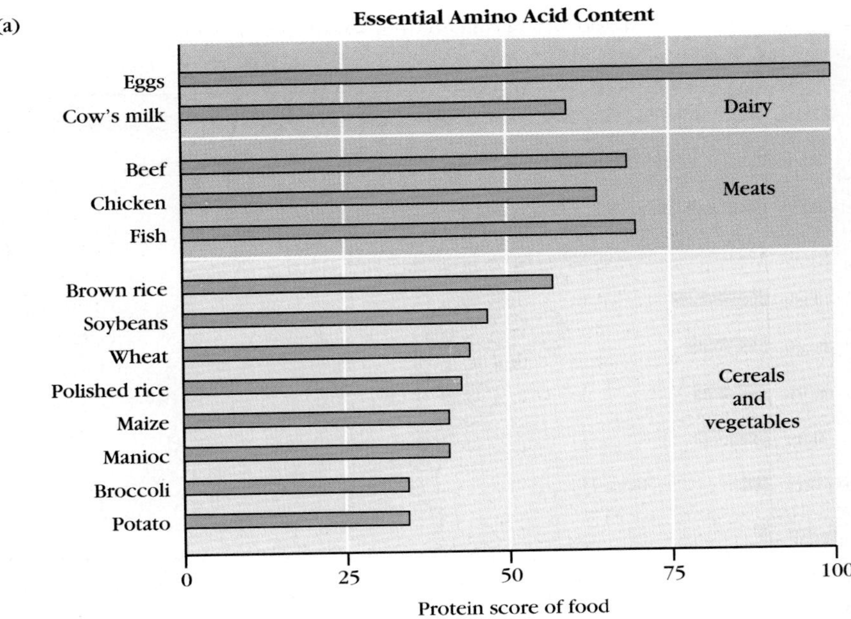

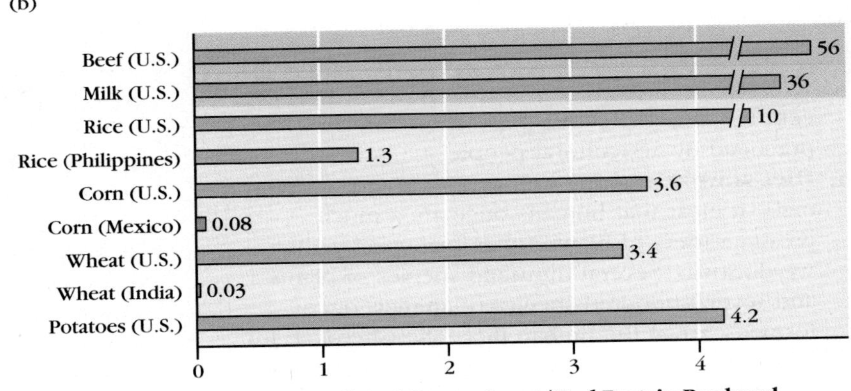

FIGURE 39-5

The nutritional value and cost of production of major foods. (a) Eggs are considered to have an almost ideal amino acid content for human nutrition, and the other foods are rated in comparison with eggs to give a protein score (thus a score of 75 means that the food is 75% as good as an egg, and not that it contains 75% protein). **(b)** Compare the nutritional value of many of the foods above with the energy cost of production. Note that it takes 5.6 times more energy to produce a pound of beef than it does to produce a pound of rice in the United States. The differences in energy costs reflect the degree to which agriculture is mechanized. Rice, for example, is cultivated largely by manual labor in the Philippines, but by machine in the United States. (Data courtesy of David Pimentel)

The Efficiency of Agriculture

Food production depends ultimately upon photosynthesis. However, lack of moisture, nutrients, and warmth usually limits the growth of crops to a fraction of their photosynthetic potential.

Most of the calories in plants are in the form of carbohydrates. The protein content of plants varies from about 6 to 20%. Because a plant must expend more energy to make proteins than to make carbohydrates, and because protein synthesis also reduces the amount of energy available for a plant to store, it is more efficient to grow and eat plants with a low protein content, as long as enough of the amino acids essential to human health are present. Plants such as wheat and rice produce around 10% protein. This is nutritionally adequate for an adult human, although children need a higher percentage of protein if they are to grow normally.

More than 70% of the world's farmland is devoted to growing cereals (grains), such as wheat, rice, maize (corn), barley, rye, and oats, all members of the grass family. Cereals are the fruits of these grass plants and they contain a lot of energy-rich stored starch and some fats, protein, minerals, and vitamins (provided by the parent plant for the nourishment of the young seedling). Cereals produce high yields, are easy to collect, and may be stored for long periods without spoiling. The cereals are excellent sources of human food except that they are deficient in some essential amino acids (Figure 39-5a). Besides feeding humans directly, the grains and leaves of grasses are the main food of most domesticated food animals.

Only about 50 animal species have ever been **domesticated,** meaning that their reproduction is controlled by humans, and this figure includes honeybees, silkworms, and a few aquatic animals, such as oysters, carp, and trout. Aside from cats and dogs, chickens are the most numerous domesticated animals. Then come sheep, and then cattle. Goats, pigs, and water buffalo are also important in many parts of the world. Fifteen of the 22 most important domesticated animals are artiodactyls, the even-toed ungulates. Many of these are ruminants (cud-chewing animals), including three of the most important as well as the first to be domesticated: cattle, sheep, and goats. Symbiotic microorganisms in ruminants' stomachs permit them to digest food, such as the cellulose cell walls of plants, that humans cannot. We can therefore use ruminants to convert plants that we cannot digest—such as grass stems and woody shrubs—into foods that we can digest: the meat or milk of cattle, sheep, or goats.

The value of cattle for food has led to their introduction into parts of the world to which they are not well adapted, such as the semiarid scrub region of the western United States and the tropical forest biome of South America. Advances in breeding and feeding have made cattle more efficient meat producers than they used to be, but beef is still the most expensive meat to produce. More than half the maize (corn) grown in the United States is used to feed livestock.

Food takes a lot of energy to produce (Figure 39–5b). However, in both traditional and industrial societies, the energy used to process, distribute, and cook food is greater than the energy used to produce the food in the first place. For instance, about twice as much energy goes into cooking a kilogram of rice in rural India as was invested in producing it. Human energy demands, especially for cooking fuels, have caused problems, including deforestation, in many parts of the world. In the United States, it has been estimated that each calorie on our dinner tables has cost nine calories to put it here. Half a calorie represents investment on the farm; the rest represents the cost of processing, packaging, distributing, and cooking.

The Green Revolution

Between 1950 and 1970 there was great optimism that a newly developed agricultural technology would eradicate world hunger. This **green revolution** involved a high-input form of agriculture in which specially developed, high-yielding strains of crop plants (especially wheat and rice) are supplied with large amounts of water, fertilizers, and pesticides. Successful green-revolution agriculture depends upon the availability of farm machinery, expert technical advice, and government subsidies or loans to purchase supplies, as well as ready markets for the crops produced.

For a while, the green revolution seemed to work: in Mexico, wheat production increased eightfold from 1950 to 1970, India doubled its production of grain during the same period, and China also experienced enormous gains in grain production. But the biological problems inherent in high-input agriculture soon began to appear.

Green-revolution technology is environmentally expensive, and if any of its components is missing, crop yields are low. For example, the specially developed strains of wheat and rice require liberal applications of fertilizer. Not only are chemical fertilizers an expensive product derived from fossil fuels, but also their use has a hidden environmental cost. An estimated half of all chemical fertilizer is not absorbed by plants, but instead washes off the land, where it pollutes rivers, lakes, and underground water supplies.

Access to an abundant water supply is the second problem associated with green-revolution grains. We have access to only a tiny fraction of the water on Earth, and nearly all of it is water that is purified by the water cycle (Figure 39–6). The water cycle is driven by the sun's heat, which causes water to evaporate. Some of the water vapor in the air eventually descends on the land as rain or snow. This volume makes up our "water income," the recycled supply of purified water upon which life depends.

The water cycle replaces every body of water more or less rapidly: water moves downstream and "new" water from rain or runoff takes its place. For instance, the water in any part of a river is replaced every 10 to 20 days. The water in a deep lake may be completely replaced only once every 100 or more years. The rate of replacement is important to us because it influences how long natural processes will take to replace polluted water once the pollution stops.

More than 80% of all the fresh water used in the United States is used for agricultural irrigation. Many areas face water shortages as a result. The media tend to blame southern California's water shortage on lawns, golf courses, and domestic use, but this is not the main cause. More than twice as much water is used on the farm valleys east of Los Angeles as in the urban area itself. This water is used to grow crops such as alfalfa, which requires tremendous amounts of water and is used to feed cattle.

A third negative aspect of the green revolution is the need for repeated applications of pesticides. Intensive use of pesticides has exerted strong selective pres-

BIO-BIT

Because of selective breeding, individual members of most domesticated plant and animal species are genetically similar to each other. This loss of genetic diversity increases the impact of new diseases and pests far beyond what would be experienced by naturally occurring ancestral populations.

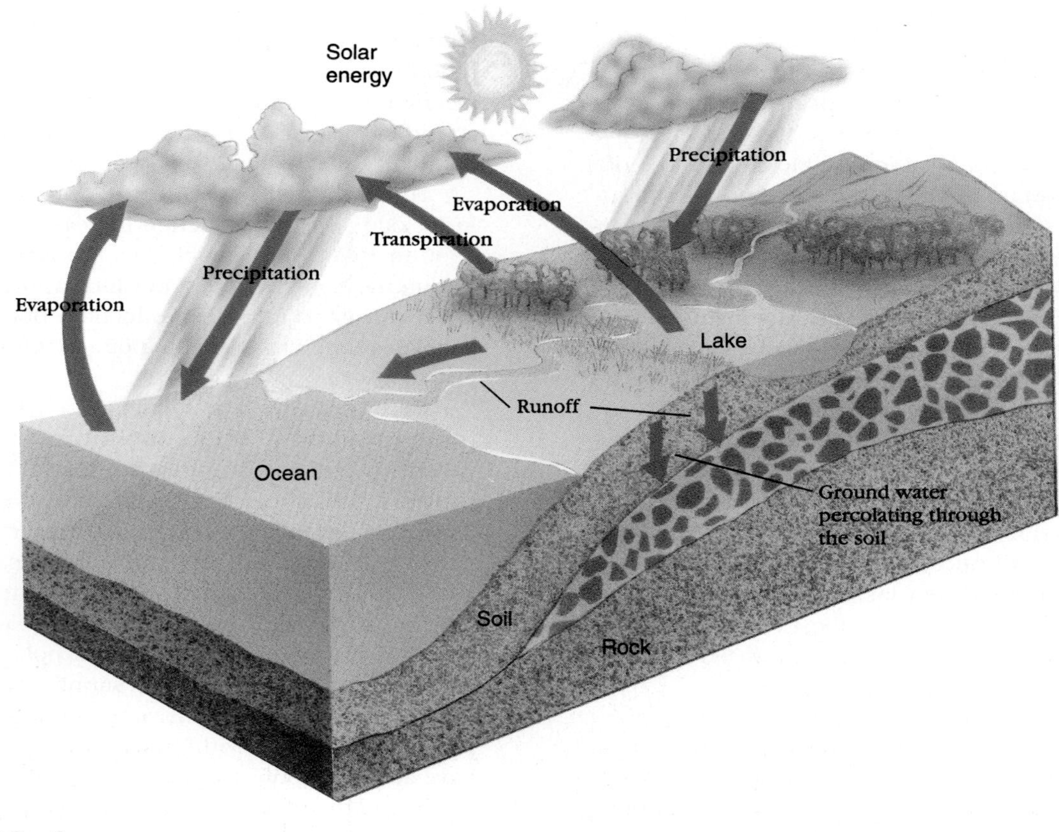

FIGURE 39-6

The water cycle. Water that reaches the atmosphere by evaporation or transpiration is pure, fresh water. But the source of water for many cities, the water that reaches our rivers and streams, may have washed down our streets, over chemical-laden farm fields, or through polluted industrial regions.

sure for insects and fungi to evolve resistance to these poisons. There is also the hidden environmental cost of pesticide-related health problems to farm workers as well as to consumers, and the unforeseen global spread of pesticides from the areas where they are applied.

A traditional practice among farmers in developing countries is to save a portion of each crop as seed that will be planted in the following growing season. In contrast, seeds of newly developed, green-revolution grain crops cannot be saved as seeds because they are hybrids whose progeny will not have the characteristics of the parent generation (see Chapter 12). As a result, farmers must spend additional money to buy seeds as well as fertilizers and pesticides.

Green-revolution crops tend to grow best in situations where farm machines can be used, yet many small farmers in developing countries have no access to farm machines. Finally, green-revolution technology has failed in some areas because it has inappropriately attempted to impose the high-input agricultural crops and practices of wealthy, developed countries onto cultures of developing countries.

Worldwide grain production fell in 1988, the first year that the United States produced less grain than it consumed. The shortfall was only partly made up in 1989. One reason for the decline is that the United States has been forced to take land out of production because its soil is so badly eroded. With the world population (5.5 billion in 1993) growing at an explosive rate, adding nearly 100 million people each year, it is easy to see that our present rates of food-production and food-distribution systems cannot hope to keep pace with demand. Indeed, we are not currently keeping pace with demand for food.

Food production is declining partly because of shortages of water for irrigation and partly because intensive farming is destroying the soil on existing farms. There is very little more land that can be converted into farmland. Modern farming has experienced other

(a)

(b)

FIGURE 39-7

Low-input farming. (a) Harvesting corn by hand in Mexico. **(b)** Harvesting wheat by hand in India. (R. Frerck/Odyssey)

adverse effects. For instance, because pesticide resistance has become increasingly common, pesticide use on American farms is now falling because the pesticides simply do not work well anymore.

In contrast to green-revolution techniques, **low-input farming** uses less pesticide, fertilizer, and irrigation than green-revolution methods (Figure 39–7). It relies upon keeping the soil as fertile as possible by adding manure, compost, and other organic matter, keeping the soil planted at all times to prevent erosion, interspersing different crops to reduce pest infestations, crop-rotation, and similar practices. Low-input farming is often more profitable than conventional farming because the farmer spends less on irrigation, pesticides, and fertilizer. The sums of money that North American farmers spend on chemicals are now falling as the trend toward low-input farming gathers momentum.

Plant breeders are working to produce the crop varieties needed by low-input farmers—varieties that produce high yields without irrigation, pesticides, or fertilizers. Agriculture specialists have also realized that, unlike green-revolution technology that tried to transplant western agricultural methods to nonwestern cultures, the most efficient sustainable agriculture is that tailored to suit a particular geographic area. We have learned that the agriculture of developed nations often cannot be exported efficiently to developing countries where the soil and climate are different. This realization has led to the establishment of agricultural research stations in many developing nations, and particularly in tropical areas, in attempts to find sustainable agricultural methods for their skyrocketing populations.

Although the green revolution failed to fulfill the dream of banishing hunger worldwide, it has taught us useful ecological lessons. It has demonstrated that if an exported technology is to work it must be appropriate, sustainable, affordable, and suited to local growing conditions. It is widely recognized that the emerging field of genetic engineering may hold the key to solving world hunger. It is not far-fetched to visualize strains of yams, cassavas, or maniocs (three kinds of starchy tubers that are basic to the diet in many tropical cultures) that have genetic resistance to fungi and insects, and grow well and produce abundant crops with no applications of fertilizer or supplemental watering. It is also conceivable that these traits will be seen in nonhybrid crops, that breed true generation after generation.

Most human starvation is due to unbalanced nutrient content in food. Death from starvation is usually the result of decreased resistance to disease. Wheat, rice, corn, and potatoes supply more than half of all human food. Cereals are the most important human food because they contain the minimum nutritionally necessary protein content (about 10%). The processing and preparation of food consumes more energy than is used to produce the food initially. The green revolution promoted a type of high-input agriculture to produce large crop yields. However, this form of agriculture increases soil erosion, causes pollution of nearby waters, and requires the extensive use of costly fossil fuels and farm machinery. Newer agricultural techniques emphasize low-input farming, which minimizes costs to farmers, local pollution, soil erosion, and water and fossil fuel consumption.

(a)

(b)

FIGURE 39-8

Deforestation and fuel shortage. Logging companies, operating with heavy vehicles and crews of chain-sawers, cut the best timber from a rain forest, leaving smaller, less valuable trees **(a)**. This practice damages the entire forest and removes an important source of fuel. **(b)** This Ethiopian woman must dry cakes of cow dung in the sun to provide fuel needed to cook food for her family (a, G. Merillon/Gamma Liaison; b, N. Smith/Earth Scenes)

39–D DEFORESTATION

Hand in hand with the spread of agriculture goes the loss of forest as it is converted into farmland. Today, the world's forests and woodlands are being destroyed at an alarming rate. This destruction is changing patterns of rainfall, causing the extinction of forest-dwelling species and the accompanying loss of irreplaceable genetic resources, as well as producing a shortage of wood for building and fuel.

In Europe and North America, the main cause of tree death is acid rain resulting from pollutants emitted by industry and automobiles (see *A Journey into The Environment:* Acid Rain, Chapter 2). In other parts of the world, the usual problem is that trees are being cut down faster than they can regrow.

For most people in the world, wood is the main source of energy, with a consumption of one to two tons per person per year. Half of all wood used in the world is burned as fuel. The extent to which demand exceeds supply can be gauged from some examples. Around Khartoum, the capital of Sudan, nearly every tree has been felled within a radius of 100 kilometers (Figure 39-8a). In Upper Volta, women may have to walk for 6 hours three times a week to find enough wood to cook the evening meal, and urban families spend as much as a quarter of their incomes on fuel wood and charcoal.

The shortage of wood causes a number of other damaging consequences. For lack of wood, many people burn animal dung, which would otherwise have been used to fertilize the land (Figure 39-8b). One expert estimates that burning dung reduces grain production by enough to feed 100 million people each year.

Because deforestation removes plants that hold soil in place, it permits floods and soil erosion. Sometimes the topsoil washes away slowly; often deforestation results in more spectacular landslides and avalanches. In India, 75% of the trees on the Himalaya Mountains have been felled. The few remaining trees cannot absorb the periodic torrential monsoon rains as the forests used to do, thus floods now kill people and animals and wash away soil from farmland every year. The flood water deposits this soil as the water slows down behind irrigation and flood control dams. In 1979, a silted-up dam in India burst, unleashing a 5-meter wall of water and mud that killed 15,000 people. India spends about $250 million each year merely to repair the immediate damage done by floods. The damage is not confined to India. North America also loses millions of tons of soil to erosion caused by cutting down trees (Table 39-1).

Agriculture in Tropical Forest Biomes

2? The traditional method of growing crops in tropical forest biomes is called "slash-and-burn" agriculture. A patch of forest is cut down and burned and crops are grown in the area for a few years. Because most tropical soil is thin, rain quickly washes the minerals away

Table 39–1
Soil Erosion Caused by Cutting Forest and Building Dirt Roads Through Forested Areas*

Site	Volume of Soil Lost (cu m/sq km/yr)
Alder Creek, Oregon	
Uncut forest	45
Clear-cut area†	117
Dirt road	15,565
Coast Mountains, SW British Columbia	
Uncut forest	11
Clear-cut area	25
Dirt road	283

*Data from Swanston, D. N., and F. J. Swanson, 1976. "Timber harvesting, mass erosion and steepland forest geomorphology in the Pacific northwest." In D. R. Coates, ed. *Geomorphology and Engineering.* Stroudburg: Dowden, Hutchinson and Ross.

†A forest is clear-cut when all the trees in an area are felled. The ground still has some vegetation cover, however, because herbs and shrubs in the undergrowth survive.

and the soil becomes less fertile. Except in young, nutrient-rich, volcanic soils, the cleared land rapidly loses its fertility, and agriculture becomes increasingly difficult. After farming the area for a few seasons, the tribe moves on, leaving the surrounding forest to fill in from the edges and return the area to forest once again.

Recently, however, developing nations have been clearing their rain forests on a massive scale, in attempts to provide farmland for their rapidly growing populations. (Ninety percent of the human population growth in the next 20 years will occur in moist tropical forest areas.) If this clearing continues at the present rate, there will soon be very little tropical rain forest left, and many species of organisms, including a large number that scientists have not yet discovered or described, will have become extinct. Loss of tropical forests will also decimate populations of North American birds that spend the winter feeding in the tropics. ■

Almost too late we have begun to appreciate the wisdom of traditional farming methods of tropical rain forest people. The Lacandoan Mayans live in the largest remaining tropical rain forest in North America, in the state of Chiapas in Mexico. Their agricultural methods are distinctly different from typical American farming in that they avoid monoculture (planting a whole field with a single crop) and plant a variety of

BIO-BIT

Of all tropical deforestation in the world, 60% is due to slash-and-burn agriculture, about 20% is related to commercial logging, and about 10% is done to provide open rangeland for cattle.

crops in a small plot that is cleared from the surroundings. Rather than slash-and-burn agriculture, which is only productive for 3 to 8 years before the fragile, tropical soils become exhausted, Lacandoan Mayan methods extend the useful life of a cleared plot for 25 to 30 years. They use subtle environmental cues extensively in timing plantings, and they plant crops in layers both above and below the ground. Once a field has become overgrown by weeds, the tree crops such as mangoes and rubber trees that were planted among the grains and other vegetables and fruits continue to bear, and over time the garden is transformed into a grove of planted trees. Gradually animals are attracted to the grove and the Lacandoan continue to use it as a hunting preserve long after they have moved on to clear another field. In his lifetime, a Lacandoan farmer will clear only three or four such plots.

Results of deforestation include a reduction of forest-dwelling species, altered rainfall patterns, fuel shortages, and increased soil erosion. Soil erosion problems have resulted in life-threatening floods and costly property damage. Large-scale slash-and-burn techniques that clear forested land in order to plant crops result in massive deforestation and produce land that is only fertile for a few years.

39–E THE INDUSTRIAL REVOLUTION

For some 8000 years, most of the people on Earth lived in agricultural societies. Most people lived on farms, and towns were mainly centers of trade and culture. The sources of energy were such things as wood, sunlight, and moving water. These sources are **renewable,** replaced by natural processes (Figure 39–9). Then, some 200 years ago, people began to make extensive use of **nonrenewable** energy sources, particularly fossil fuels. The enormous amounts of energy provided by fossil fuels changed society in ways that are collectively known as the **industrial revolution.**

Like the agricultural revolution, the industrial revolution has reduced the land space required to support each individual and has increased the depletion of natural resources. This accelerating use of resources has led to enormous increases in our standard of living,

(a)

(b)

FIGURE 39-9

Sources of renewable energy. Unlike fossil fuels, which are limited, nonrenewable resources, solar power **(a)** and wind power **(b)** can provide continuous supplies of energy. But they only work as long as the sun is shining, as it usually does in New Mexico where these solar collectors are located, or as long as the wind is blowing, as it often does at this wind farm in Aktamonte, California. (a, Peter Arnold, Inc.; b, S. W. Elems/Visuals Unlimited)

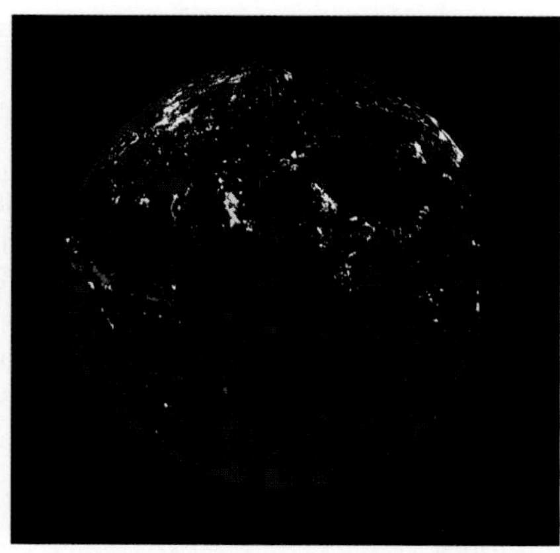

FIGURE 39-10

Satellite image of Africa, Europe, and Asia at night shows the extent of urbanization. The brightly lit British Isles and the Japanese Islands are just visible at the extreme left and right of the globe, respectively. Europe and India are brightly lit, especially when compared with the blackness of much of Eurasia and Siberia. The Sahara desert in Northern Africa is dark, while settlements clustered along the Nile show clearly. Yellow represents natural gas flares at oil wells in Mid-eastern oil fields, while red areas indicate burning vegetation. This is especially visible in the tropical forests of equatorial Africa. (NASA GSFC/Science Photo Library)

but it has also caused many environmental problems. These include the modern population explosion, pollution, and diminishing reserves of the fuel and minerals that are the basis of the industrial revolution itself.

The modern pollution crisis dates from the industrial revolution with its use of fossil fuels. Until about 1950, most air pollution was caused by burning coal. Modern "smogs" are caused largely by the combustion products of gasoline from cars. In cities such as Los Angeles, Mexico City, Denver, London, and Tokyo, smog frequently renders the air unfit for human consumption. Because it is impractical to filter air as people breathe it, air pollution must be controlled at its source.

Since about 1950, pollution from solid waste has also become an acute problem. Part of this is due to the population explosion, which has resulted in a greater total amount of solid waste. In addition, our higher standard of living in a "throw-away" society has increased the amount of waste produced per person. The use of fossil fuel, this time as the raw material for the manufacture of many artificial polymers collectively known as "plastics," has also contributed to solid waste pollution.

We have used modern technology to clean up as well as to pollute our environment. Cities in places such as North America and Western Europe are more pleasant places to live than they have ever been. This is of prime importance because a major effect of the industrial revolution has been to hasten **urbanization**—the movement of people from the countryside, where relatively few people now find jobs on farms, into cities, where industries are established and most jobs are to be found (Figure 39-10).

Industrialization has accelerated every kind of human impact on the environment, including population growth, resource depletion, and pollution.

39–F LOOKING AHEAD: TWO VIEWS OF ENVIRONMENTAL PROBLEMS

People tend to look at our current environmental problems in one of two ways. Economists tend to be "Pollyannas," named after a fictional character who always saw the bright side of even the most desperate situation. Many economists suggest that there are no limits to growth, that each problem is an opportunity, and that human technology will be able to "fix" everything and save us from our excesses.

In contrast, biologists tend to be "Cassandras," named for the mythological character who was first gifted with prophecy, and then later cursed, so that no one would believe her prophecies. Like Cassandra, many biologists warn that we are approaching the limits of the capacity of the Earth to sustain the human population and that steps must be taken to ensure that our own excesses don't lead to a dismal future.

Overpopulation

In thinking about overpopulation, three trends need to be kept in mind:

1. *The size of a population affects how rapidly it uses resources. As the human population grows, it will continually need more food, water, space, and especially energy resources.*

Nearly all biologists view overpopulation as our most pressing environmental problem because it intensifies all of the other environmental problems we face. They think that global warming, acid rain, loss of species, deforestation, and ozone thinning are warnings that we are approaching the limits of the Earth to adequately sustain the expanding human population.

Wildlife experts have ways to estimate how many deer, elephants, quail, or cattle an area can support and sustain, but people are unusual animals whose actions are difficult to predict. Because human technology can modify the environment and because changing attitudes can alter human behavior, no one can really predict the future of the human population. In short, no one really knows just how many people the Earth can support. Opinions vary, and some economists think that each person added to the work force enriches the range of economic opportunities. Following this line of logic, these Pollyannas reason that perhaps the next Einstein may be reared in the slums of Bangkok. They argue that necessity is the mother of invention and that perhaps future adverse conditions will produce a technological solution to our population problems.

2. *The human population has recently grown at an unimaginably rapid rate. World population—now more than 5.5 billion people (in 1993)—has more than doubled in the last 40 years.*

Although we have fairly good information on past trends in human population growth (see Figure 38–18) and have a reasonable estimate of the current size of the human population, estimates of future growth depend upon who is making the projections and upon what factors are considered. Demographers at both the United Nations and the World Bank have projected similar scenarios for world population growth. They agree that if there is no decline in the number of children born to each woman, and if the world average birth rate continues to be around 3.8 children per woman, then by 2089, there will be 32 billion people on Earth. Alternatively, if the birth rate falls and if the average woman has only 2 children, in 2089 there will be 10.4 billion people. Even if a replacement level of population growth (in which there is one child to replace each parent) is achieved, the world population will continue to grow for many decades because a great proportion of the world's population is now of child-bearing age. ■

3. *Most of the increase in numbers will be in unindustrialized, developing countries, but this does not mean that most of the impact of increased population will be in developing countries.*

Each person in a developed country makes much greater environmental demands than does each person in a developing country. Think about it for a minute: how does the lifestyle of a teenager living on his family's farm in rural India compare with that of a teenager living in his family's split-level house in suburban Chicago? (Figure 39–11) The suburbanite will consider himself somewhat deprived if he does not have a car, stereo, portable cassette and/or CD player, television equipped with cable or satellite link, computer, telephone, fast food, movies, and vacation travel by jet (the list could go on and on), while to the rural teenager a bicycle is the ultimate dream. The rural teen's family will not own a car, television, or radio. Fast food is unknown in rural India and jet travel is out of reach, unless one has a very rich uncle. The two teenagers have lifestyles that are vastly different in terms of consumption of both energy and resources.

Although Cassandras warn that we are breeding ourselves out of existence, and Pollyannas contend that overpopulation may be just an illusion, we have seen that analyzing our overpopulation problem is complex because so many factors affect it (see *A Journey into The Environment:* Overpopulation and the Human Animal, Chapter 16). It is certain, though, that the rate at which world population is growing can be changed. Many countries (nearly all of western Europe and Mexico, for example) have lowered their birth rates, as compared to those of recent years. There seems to be a correlation between the status of women and the birth rate. Where women are most val-

(a)

(b)

FIGURE 39-11

Lifestyles reflect different levels of consumption. **(a)** A home-made wooden bike delights these two boys in the Philippines, **(b)** while this American boy concentrates on beating a computer game. (J. Brown; H. Rose/Science Photo Library)

ued for their roles as mothers, with little independence or rights, where there is little education of women, where there is little pre- and neonatal care, birth rates tend to be high (Figure 39-12). Therefore, it appears to follow that if we promote health care, e ucation (especially of women), and attitudes that discourage sexism, the birth rate will thereby be reduce Finally, the birth rate will diminish when small familie become socially desirable and economically feasible.

Global Warming

We have seen that the amounts of greenhouse gases i the Earth's atmosphere have increased as use of fossil fuels has increased (*A Journey into The Environmen* Global Warming, Chapter 6). These gases (mostly carbon dioxide, but also methane, nitrous oxides, ozone, and chlorofluorocarbons) have the effect of preventin heat from the Earth's surface from escaping to space, thus increasing the overall global temperature. Pollyan nas claim that we will find a technological remedy to fix global warming without having to change our habits or energy sources, while Cassandras call for sul stitute energy sources to replace fossil fuels (coal, natural gas, and oil), the major producer of carbon dioxide.

One of the major difficulties that Cassandras have is in communicating the scope and implications of global warming. Climate modelers are predicting that sometime in the twenty-first century, temperatures wi rise between 2° and 5°C or 5°–9°F. A rise of 9°F sounds insignificant until you learn that a 5°F change

FIGURE 39-12

Poverty, social status, and birthrate. These women are harvesting wheat by hand in India. In many developing countries, women receive little education and are most valued for their role as mothers and laborers. (R. Frerck/Odyssey/Chicago)

BIO-BIT

Sulfur emissions from the burning of fossil fuels and volcanic eruptions produce a haze that reflects some of the sun's heat away from the Earth. Although this has an effect that counters global warming, most scientists do not think that it is enough to reverse the general warming trend of the greenhouse effect.

probably enough to melt the polar ice caps. Worse et, this environmental change will not occur over an xtended period of time, but rather will be more like n environmental convulsion—the fastest rate of cliate warming since the last glaciers retreated 10,000 ears ago. As global temperature warms, sea levels will se, coastal cities will be flooded, but most importntly, the usual patterns of rainfall and temperature ill be altered. There will be major effects upon agriulture and natural ecosystems (Figure 39-13). And his is the most important part), unless we change the vels of greenhouse gases that we put into the atmophere, this initial increase will only be the beginning. s more and more carbon dioxide and methane are dded to the atmosphere, the global temperature will keep on getting warmer. Most astonishing though, we will not be able to predict these events with any certainty for a very long time. Computer models that accurately simulate the global atmosphere with all of its variables, are currently beyond our capabilities. It is certain that levels of greenhouse gases are rising. What is uncertain is how rapidly we will experience global warming. As one environmentalist has put it, "The one thing that is certain is that we are in for some nasty surprises."

To address the problem of global warming, we must recognize the reality and magnitude of the problem. Although it is now too late to stop the initial 5° to 9°F (2°-5°C) rise in temperature that is already underway, it is not too late to form energy-saving habits that will help keep global warming from getting worse.

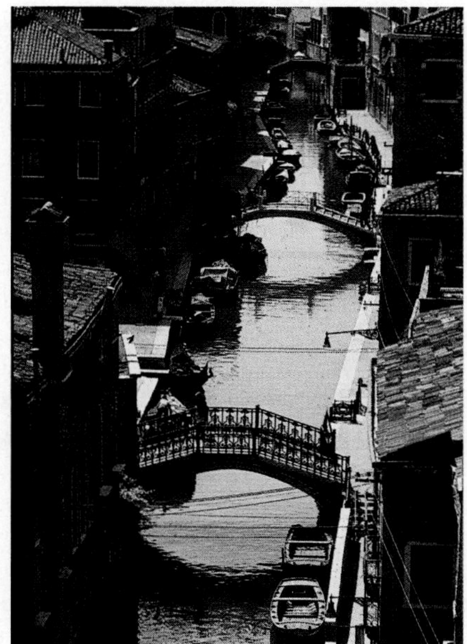

IGURE 39-13

Venice, Italy, a city of the future? Global warming ould result in extensive melting of the Earth's polar icecaps, ising sea levels and flooding coastal cities. (J. Carr/ hoto Researchers, Inc.)

Acid Rain

Acid rain is a term applied to precipitation more acidic than pH 5.6. It is one of the environmental problems that we can control. We know the source of this pollutant: sulfur and nitrogen oxides from burning fossil fuels. Vehicles release most of the nitrogen oxides as they burn gasoline, and most of the sulfur dioxide originates from factories, industries, and electrical power plants. Sulfur dioxides and nitrogen oxides are converted to acids in the upper atmosphere, carried long distances by winds, dissolved in rain, fog, ice, and snow, and deposited downwind.

For two decades, Pollyannas have denied that acid rain is real (Figure 39-14). They have called for more

FIGURE 39-14

Damage caused by acid rain. Acid rain has dissolved the facial features and sculptural detail of this statue. (J. Cunningham/Visuals Unlimited)

data and delayed action until now the problem of acid damage is acute in some areas. Cassandras have observed that acidified lakes, streams, creeks, and ponds are quite different from normal aquatic ecosystems because they have little of the normal diversity of aquatic species. Acid-sensitive species of invertebrates and vertebrates die out or disappear, leaving only acid-tolerant species. We know that acid rain has an effect upon amphibians and larval fishes who are extremely sensitive to the pH of the water in which they breed and grow (see *A Journey into The Environment:* Acid Rain, Chapter 2). We know that acid waters also have several damaging effects upon aquatic birds. We are unsure of the effects of acid rain upon aquatic reptiles and mammals, but it is logical to predict that such a fundamental change would have a negative effect upon the entire food web of a pond, lake, stream, or creek. When acid rain falls on forests, it kills trees by damaging leaves and bark. It also kills trees in more subtle ways, by interfering with the symbiotic fungi that help them absorb water and nutrients. As a result, trees decline and die and, as they do, the forest and all its species of plants and animals similarly decline, die, and disappear.

Although we know how to control acid rain, it continues to be an environmental problem because we have yet to force international pollution controls on the industries that produce emissions and because the sources of pollution are distant from the environments that are damaged. For example, the forests in Canada are damaged by emissions that originate in the United States, and the forests of Scandinavia are damaged by emissions that originate in England and Western Europe. Installation of scrubbing mechanisms that clean sulfur and nitrogen oxides from smokestack emissions, gradually phasing out the use of fossil fuels, using automobiles less, and replacing the internal combustion engine with nonpolluting electrical motors are all ways to reduce acid rain (as long as the electrical power plant doesn't produce these acids in the process of generating electricity).

Ozone Hole

In 1985 the world was shocked to learn that human activities had created a hole in the stratospheric ozone layer that screens Earth from most of the sun's damaging ultraviolet radiation. Although it took several years to respond to this challenge, the knowledge of the damage caused by **chlorofluorocarbons** eventually resulted in two international agreements to significantly reduce CFC production. Recently we have learned that the holes in the ozone layer aren't confined to polar latitudes (see *A Journey into The Environment:* Ozone Depletion, Chapter 17). Ozone depletion at the South Pole will endanger the phytoplankton that form the base of the Antarctic food chain, threatening the world's fisheries, as well as causing skin and retinal cancers (Figure 39–15). Stricter controls have been developed that dictate that CFCs be completely phased out in the United States by 1995. The international cooperation that produced global controls on emissions of CFCs is an environmental success story. It demonstrates that although response is far from instantaneous, Cassandras and Pollyannas can cooperate to reduce a pressing environmental problem.

FIGURE 39–15

Soaking in the rays. The gradual destruction of the ozone layer has led to a corresponding increase in the rate of skin cancers. In addition, ultraviolet ray exposure accelerates the aging process, resulting in dry and tough skin at an earlier age. (J. Halaska/Photo Researchers, Inc.)

Although fewer CFCs are now entering the stratosphere, those that are present will continue to damage the ozone layer for at least another hundred years. You can help solve the problem of ozone thinning by insisting that the service station recycle the CFCs in your car's air conditioner, you can reduce or eliminate use of foamed plastics, and you can avoid purchasing any aerosol product that contains CFCs or halon, a related propellant.

Deforestation and Loss of Species

Deforestation and the resulting loss of many forest species have accelerated to a fever pitch over the last few decades. More than 90% of the original forest of the United States is already gone, and those of the Pacific Coast are all that remains. Half of the original area once occupied by the world's tropical, lowland forest has been cut, burned, or degraded, and what remains is equivalent to an area that is two-thirds the size of

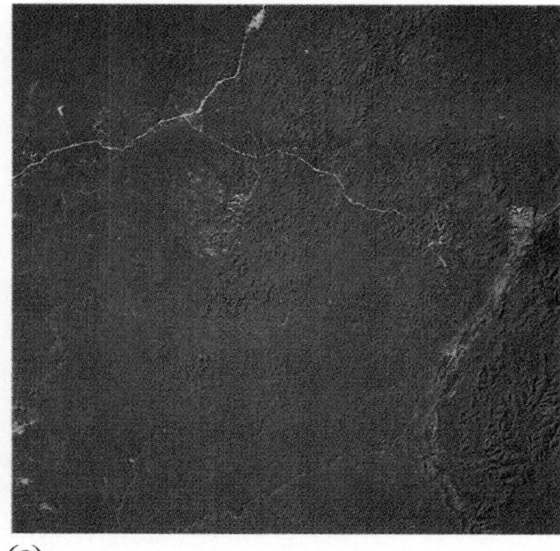

(a)

(b)

IGURE 39–16

leven years of settlement in a tropical rain forest. oth of these pictures were taken by the NASA satellite andsat as it orbited the Earth. The photo at the top shows a egion in the state of Rondonia in Brazil in 1975 and the hoto at the bottom is of the same region in 1986. In these nages, red indicates vegetation and blue indicates cleared nd, agricultural fields, or roadways. Notice how settlement this rain-forested region of Brazil has increased in eleven ears, largely because of a policy of rain forest settlement hat has been fostered by the Brazilian government. (World erspectives)

he United States. These tropical lowland forests,loated in Latin America, Africa, and Asia, are being detroyed so quickly, that if the present rate of destrucion continues, most will be gone in only 30 years. ropical forests are being destroyed for several reasons, nost associated with the demands of the ever-growing uman population for cash, food, fuel, and land (Figre 39–16). As we have seen, most tropical forest soils re not suited for agriculture, and farms cut from tropical lowland forest are doomed to failure (see *A Journey into The Environment:* Loss of Species Diversity. . . Going, . . . Going, . . . , Chapter 14).

As habitat is destroyed, animals and plants are destroyed, too, and because of the intricate links between plants, animals, and microbes in tropical forest communities, most organisms cannot survive when forced out of their habitats. While Cassandras know that each species that is lost is the irreplaceable result of natural selection and evolution, Pollyannas do not understand that there is a web of relationships within forests. They emphasize human needs over the integrity of forest ecosystems. "What good is it?" Pollyannas ask, viewing other species as entities that humans can use for food, fuel, shelter, medicine, or other commodities.

No one really knows how many species are going extinct each day, and estimates vary from six to dozens per day. It's the trend of reduction of species that is important, and the trend of a massive loss of species is clear. While some Pollyannas will argue that we're better off without the millions of nameless crawling creatures and "weeds" that infest the tropical rain forests, some Cassandras see this as out-moded thinking. Genetic engineering techniques are about to transform the way we view even "useless bugs and weeds." As this new technology gathers momentum, every species of plant, animal, protist, fungus, or prokaryote will be valued as a source of genetic material that can be used in very practical ways to develop new drugs, to improve crops and livestock, and to perhaps even enhance the human genome.

Human populations continue to grow at alarming rates, intensifying current environmental problems. New approaches to address some of these problems, including global warming, acid rain, ozone depletion, deforestation, and a reduction in global species diversity must include education, human population control, and better pollution control technologies.

39–G EVALUATING OUR ENVIRONMENTAL PROBLEMS: A SUGGESTION FOR CHANGE

A revolution of human attitudes has begun in western cultures, and especially in the United States. We have been awed by the view of Earth as seen from the Moon, startled by the holes we have made in the ozone layer, disturbed by the fierce storms and irregularities of weather systems that seem increasingly frequent, sick from air and water pollution, frightened that we may be poisoning ourselves with pesticides

A Journey into The Environment

The Environmental Movement and the Tragedy of the Commons

In 1968, ecologist Garrett Hardin published an influential parable entitled *The Tragedy of the Commons.* In it, Hardin argued that the main difficulty in our attempts to solve problems such as overpopulation and pollution is the conflict between the short-term welfare of individuals and the long-term welfare of society.

Hardin illustrated this conflict with the case of commons, such as those of medieval Europe or colonial America. A common was grazing land that belonged to the whole village. Any member of the community could graze cows and sheep there. The tragedy of the commons is that it was in the interest of each individual to put as many animals on the common as possible, to take advantage of the free animal feed. However, if too many animals grazed on the common, they destroyed the grass. Then everyone suffered because no one could raise animals on it. For this reason, common land was eventually replaced by individually owned, enclosed fields. Before the era of subsidized agriculture, owners were careful not to put too many cows on one patch of grass, because overgrazing one year would mean that fewer cows could be supported the next year (Figure 39–A).

The tragedy of the commons is a result of natural selection. During the course of evolution, some people gave priority to increasing the health and wealth of their own families. These people raised more children than those who put the long-term welfare of their village or nation first.

The parable of the commons throws fresh light on many public problems. Congress as a whole deplores the budget deficit, but each member goes on increasing it with expenditures that benefit his or her own district. Each fur trapper or whaler contributes to the extinction of the species from which he or she makes a living. Hardin asserted that the commons (which we can think of as all natural resources) are limited, and that people will pursue their own self-interest, destroying the commons to the point of society's collapse.

(a)

(b)

FIGURE 39–A

Use and abuse of the land in two cultures. (a) Overgrazing in Tanzania. Once fertile land has been grazed clean and compacted by the hooves of cattle. Soil destruction has reduced most of the area to grinding poverty. **(b)** Aftermath of an outdoor concert in New York State, where a day of entertainment has been at the expense of the environment. (a, F. Lanting/Minden Pictures; b, D. McCoy/Black Star)

Hardin was certainly right in thinking that individuals acting separately cannot be expected to solve all environmental problems. If we, as individuals or as corporations, knew we should pay directly for the overpopulation and the pollution each of us causes, we would each have fewer children and contribute less to pollution. However, it is not usually possible to assign responsibility for ecological problems directly to the people who cause them. Future generations will pay most of the price for the fact that we have too many children and cause soil erosion now, and our individual contributions to water pollution cannot be easily distinguished. Thus, although incentives for individual action are effective ways of getting most things done, many environmental problems can be solved only by effective action by groups such as governments that, at least in theory, can look beyond the immediate interests of individuals and plan for the long-term welfare of society.

Hardin pursued his argument in a 1974 essay, *The Ethics of a Lifeboat,* which explores the way the wealthy countries of the world treat their impoverished neighbors. An allegorical lifeboat filled with the world's rich is surrounded by the struggling poor who attempt to clamber aboard. How should the rich behave? They cannot let everyone into the boat because it would sink. If they allow some into the boat, they remove the lifeboat's safety margin and are also presented with the further ethical dilemma of whom to save and whom to condemn. To Hardin, the only sensible course of action is to ignore the pleas for help and maintain the boat's safety margin.

The lifeboat argument proposes that it is counterproductive to supply economic aid to poor countries. The argument is particularly strong in the case of countries that do not attempt to curb their population growth. If a poor country can call on aid in times of need so that its people do not starve to death, its population, and its need, continue to grow indefinitely, reducing still further the resources that are left for future generations everywhere. Lifeboat philosophers argue that the most humane course, in the long run, is to withhold aid so that starving people die. The smaller population that remains will have more chance of developing sustainable agriculture that does not degrade the environment and can feed the population.

If Hardin's analysis of human nature were the whole story, it would make the task of solving environmental problems almost impossible. Acting each with rational self-interest, we should continue to destroy the Earth's resources until we became extinct. However, Hardin ignored important additional facts about the nature of human beings: we are intelligent, deeply social animals, and we are capable of altruistic behavior (Section 16-E).

Altruistic behavior is part of our makeup because we are social animals. We depend upon cooperation with other people for our very survival. This is, ultimately, why we can prevent ourselves and each other from destroying the environment. The chieftain who burns down the local forest to enlarge his own wheat field runs the risk that he will be thrown out of the village, and if he is ejected from human society he will die. The many heads of nations who have abused their citizens and natural resources and then been deposed (and often killed) are obvious examples. So if our urge to work first for ourselves and our own families endangers the environment, our need to be accepted by society can save it. When enough people, in a village or a nation, want a particular reform, the rest must go along with it. You can make almost any change in society if you can convince enough people that it is the right thing to do.

As we would also predict from this, people are more inclined to work for the general welfare if their actions are effective and do not cost them very much. Public opinion polls show that, even during economic recessions, the vast majority of people will accept tax increases to pay for pollution control and the conservation of resources. Politicians are frequently surprised at the results of such polls, but they should not be. People are behaving logically in voting for effective environmental action that will not cost each of them very much.

and other chemicals, and saddened by the predictions that some of our favorite animals may be facing extinction just when we are beginning to understand their lives. Many people are beginning to realize that we must change our attitudes if we are to survive and prosper. Some understand that although our big brains, transmission of culture, and technology set us apart from other animal species—like everything else on this planet, we belong to the Earth and our future is inextricably intertwined with it and all of its inhabitants.

But how can we bring about the changes that must occur if overpopulation, global warming, ozone depletion, acid rain, deforestation, loss of biodiversity, and the pollution of air and water are to be reversed? How can short-term, self-interest ever give precedence to long-term global interest? To find out what and how you think about this, take a moment to ask yourself: "How far into the future do I care?"

FIGURE 39-17

Isle Royale National Park in Lake Superior, Michigan. (Peter Arnold/Peter Arnold, Inc.)

Education, Imagination, and Caring

> "Some men see things as they are and ask why, I dream things that never were and say why not?"
>
> Robert F. Kennedy

There is no easy fix for our global environmental dilemma, but the first step has already begun. You are learning about our global environmental problems, and it is certain that without a widespread awareness of their reality, there will be no solutions to them. You must now take the next step and give serious thought to imagining what the world will be like once we have overcome our common environmental problems.

Imagination sounds very unscientific, but it is a way to initiate the patterns of thought that will lead to the solution of our environmental problems: a way to allow solutions to occur. It is essential when you consider that if one cannot imagine how things will ever be better, one is stuck with things as they are. Vision can counteract adaptability, one of the human traits that has led both to our great success as a species and to the complacent attitude that has allowed our environmental problems to become so serious. Adaptability allows us to make the best of situations, to forget that things were ever any different as we adjust to new conditions, and finally to alter these conditions to suit ourselves. It is human adaptability that allows desperately poor children in Mexico City, Rio de Janeiro, and Calcutta to live on the fringes of garbage heaps, running to leap on top of a newly arriving truck to be the first to glean a broken wristwatch, scraps of spoiled food, and discarded plastic and paper. These children have never known a different life. They play amidst the garbage. Their vision of the possibilities of life is limited to the shelters that their families have built from scavenged materials. Similarly, many city dwellers from New York to Hong Kong have no vision of what the countryside is like. Some are afraid of wild places and imagine danger lurking behind every tree. People who have grown up in the countryside are often familiar with only degraded habitats, those that have been altered by centuries of agriculture. We must educate ourselves, allow our education to feed our imaginations, and envision a future for humanity that is compatible with the needs of the Earth and its other inhabitants.

The final step in the mental preparation that will allow us to overcome our environmental problems is caring: for ourselves, for one another, for other forms of life, and for our home planet. More than anything else, we must learn to be thoughtful and to apply our knowledge and adaptability to sustain, rather than to exploit. We must begin to care for all members of our own species, especially women, children, and minorities. Everyone needs nurturing, and as selfish interests are balanced by generous, thoughtful acts, we will begin to be able to learn, to think, to dream, to care, and to fight our way out of the mess of environmental problems that confront us.

The Future

The success of this process of change starts with each individual. There are books and magazines to read, videos to watch, ideas to consider, organizations to join, and changes to make in our attitudes and habits. This journey into biology has provided a solid foundation. We hope it has enriched your imagination, and given you an understanding of the organization and complexity of living organisms, and we trust that it is just the beginning of your journey into life (Figure 39–17).

SUMMARY

1. Early humans were hunter-gatherers, hunting animals and gathering plants for food. Some were nomadic, and others lived in settled villages or migrated between seasonal homes. These people permanently altered their environment in a number of ways, but agricultural societies have had a much greater impact.
2. Agriculture supports more people on a given area of land than does a hunter-gatherer way of life.
3. Agricultural societies have steadily squeezed hunter-gatherer populations into less productive areas as agriculture has expanded.
4. The most important environmental effects of agriculture are habitat destruction, soil erosion, pesticide and fertilizer pollution, and diet simplification.
5. Humans feed themselves most cheaply when they eat plants that are nutritionally adequate but contain little protein. Most of the world's food supply comes from grains, which fulfill this requirement and are also easy to store.
6. The new plant varieties of the green revolution permitted huge increases in grain production by means of high-input agriculture, dependent on the use of fertilizers, pesticides, irrigation, fossil fuels, and heavy machinery.
7. The environmental damage caused by high-input agriculture includes disproportionate use of water, water pollution, and soil erosion. The current trend is toward low-input agriculture, which is cheaper and less environmentally destructive.
8. We are still producing enough food to feed everyone in the world, but many people are too poor to buy their share.
9. Unless we succeed in curbing our population growth, the loss of farmland to soil erosion, and development, we shall soon exceed the capacity of the world's farmland to support us.
10. The industrial revolution came about when societies started to power technology, including agriculture, with the enormous quantities of energy available from fossil fuels. Industrial development accelerated population growth and raised standards of living, but contributed to modern problems, particularly air pollution, toxic wastes, and the huge volume of solid waste we produce today.
11. The rapid increase in human populations experienced in the last century is predicted to continue well into the next century.
12. Increasing human populations magnify current environmental problems including global warming and acid rain associated with the use of fossil fuels, ozone depletion from industrial pollutants, and deforestation and the loss of species diversity as forests are logged and cleared for short-term economic gains.

SELF-QUIZ

1. As a result of the agricultural revolution, all of the following increased *except:*
   a. human population size
   b. quality of human nutrition
   c. pollution
   d. rate of deforestation
   e. soil erosion
2. The agricultural revolution is believed to have been responsible for a dramatic increase in the human population. Which of the following was *not* a factor in this increase?
   a. Food became more concentrated and thus easier to obtain.
   b. Many methods of birth control were abandoned.

   c. Improved medical knowledge increased life expectancy.
   d. Larger amounts of food could be produced by fewer people.
   e. People could accumulate possessions and wanted more children to pass them to.
3. Humans can eat animals, or plants, or both. Which of the following statements about the feeding of human populations is *not* true?
   a. Because animals eat plants, less land is required to produce plant than animal food.
   b. A diet consisting only of meat contains a higher percentage of protein than a human adult needs.
   c. About 10% of the calories in a plant are lost when an animal eats the plant.
   d. An adult human eating nothing but wheat, rice, or potatoes would receive enough essential amino acids.
4. Which of the following is a likely result of deforestation
   a. The amount of carbon dioxide removed from the atmosphere is reduced.
   b. Wind blows soil away because its plant cover has been removed.
   c. Water washes off the land more rapidly, causing floods.
   d. Water washes soil into rivers and streams, causing them to silt up.
   e. All of the above.
5. The industrial revolution increased all of the following *except:*
   a. use of fossil fuel
   b. water pollution
   c. urbanization
   d. human birth rate
   e. air pollution

THINKING CRITICALLY

1. Do you think that *Homo sapiens* will be extinct within the next thousand years? Why, or why not?
2. The world contains a few remaining hunter-gatherer societies in places such as Australia, New Guinea, Brazil, and Peru. The governments that control the countries where they live generally believe these people should be left alone and perhaps given title to the land where they have lived for centuries. In some places this may actually happen. In others, hunter-gatherers are rapidly disappearing. What pressures determine one fate rather than the other? Is it worth trying to save modern hunter-gatherers?
3. It has been argued that advances in human civilization can be traced to exploitation of new sources of energy. Major steps included the addition of meat to the largely herbivorous diet of our ancestors, the taming of fire, and the use of fossil fuels. Other advances along the way have been domestication of beasts of burden and harnessing of wind and water power. How has each of these affected the ecology of human beings?
4. What evidence suggests that humans are a keystone species?

SELECTED KEY TERMS

agricultural revolution, *p. 813*
chlorofluorocarbons (CFCs), *p. 826*
diet simplification, *p. 815*
domestication, *p. 816*
green revolution, *p. 817*
habitat destruction, *p. 814*
hunter-gatherers, *p. 812*
industrial revolution, *p. 821*
keystone species, *p. 811*
low-input farming, *p. 819*
nonrenewable resources, *p. 821*
renewable resources, *p. 821*
urbanization, *p. 822*

SUGGESTED READINGS

Brown, L. R., and others. *State of the World 1993, A Worldwatch Institute Report on Progess Toward a Sustainable Society.* New York: W. W. Norton, The Worldwatch Institute, 1993. A volume of this book is produced each year by the Worldwatch Institute. It contains about a dozen chapters on environmental problems and progress.

Costanza, R. "Social traps and environmental policy." *BioScience* 37(6):407, 1987. Argues that "social traps" tempt people into situations from which environmental solutions are extremely difficult. Hazardous waste sites, the eroding coast of Louisiana, and nuclear war are the examples used. The tragedy of the commons is seen as an example of a social trap.

Ellis, D. *Environments at Risk: Case Histories of Impact Assessment.* New York: Springer-Verlag, 1989. A series of case studies, including the stories of Minamata, the Amoco Cadiz, Bhopal, Chernobyl, the Thames Estuary, and Hell's Gate. Includes chapters on reducing environmental risk.

Hardin, G. "The tragedy of the commons." *Science* 162:1243, 1968. The thought-provoking argument that ecologically ethical individuals are evolutionarily doomed.

Harlan, J. R. "The plants and animals that nourish man." *Scientific American,* September 1976. An interesting discussion of human food plants and animals: where they originated and how they have become tamed or cultivated.

McKay, B. J., and J. M. Acheson, eds. *The Culture and Ecology of Communal Resources.* Tucson: University of Arizona Press, 1987. A collection of papers evaluating Hardin's parable "The Tragedy of the Commons" in the light of studies of fisheries, grazing lands, and similar commons. The authors conclude that sustainable management of commons is possible if the users of the commons make up a single group of accountable people, and if the limits of the commons are under the group's control.

Quinn, D. *Ishmael.* New York: Bantam, 1993. A novel about the role of humans on Earth.

Reganold, J. P., R. I. Papendick, and J. F. Parr. "Sustainable agriculture." *Scientific American,* June 1990. How American farmers are turning to more environmentally sound practices.

Scientific American, "Managing Planet Earth." September 1989. An entire issue devoted to environmental affairs including management, pollution, biodiversity, population growth, sustainable agriculture, and energy.

The Educated Citizen

How Many People Can Earth Hold?

This article by Joel E. Cohen was excerpted with permission from the November 1992 issue of Discover *magazine.*

According to the United Nations, some 5.3 billion people enlivened our planet in 1990. By the time you read this, that number will have increased to more than 5.6 billion, an addition nearly equal to the population of the United States. Of course no one, including the UN, has a reliable crystal ball that reveals precisely how human numbers will change. Still, people have to plan for the future, and so the UN's analysts and computers have been busy figuring what might happen.

One possibility they consider is that future world fertility rates will remain what they were in 1990. The consequences of this, with accompanying small declines in death rates, are startling. By 2025, when my 16-year-old daughter will have finished having whatever children she will have, the world would have 11 billion people, double its number today. Another doubling would take only a bit more than 25 years, as the faster-growing segments of the population become a larger proportion of the total. At my daughter's centennial, in 2076, the human population would have more than doubled again, passing 46 billion. By 2150 there would be 694,213,000,000 of us, a little over 125 times our present population.

There, in 2150, the projections of the United Nations Population Division stop. Perhaps they stop because the computers grew weary of the thought of so many births to celebrate, so many marriages to consummate, so many dead to bury. At any rate, there, in 2150, the computers—and an unchanging urge to go forth and multiply—leave us, with a hypothetical 12,100 people for every square mile of land, or 3,500 people for every square mile of Earth's surface, oceans included. At this rate of growth the population would, before 2250, surpass 30 *trillion,* more than 200 people for every *acre* of the planet's surface, wet or dry.

The human population must ultimately approach a long-term average growth rate of zero. That is a law from which no country or region is exempt. According to every plausible calculation that's ever been done, Earth could not feed even the 694 billion people that the UN projected for 2150. Though there is tremendous uncertainty about the details of when, where, and how, the long-term constraint of an average population growth of zero is likely to come into play within the next century and a half.

Our urge to go forth and multiply could, a century and a half from now, leave Earth with more than 694 billion people—some 125 times our current population.

Theories regarding the limitations on population growth have come and gone over the years. In an essay published in 1798, the English clergyman Thomas Robert Malthus argued that human numbers always increase more rapidly than food supplies and that humans are condemned always to breed to the point of misery and the edge of starvation. The two centuries since his famous essay have not been kind to Malthus's theory. In that time human numbers have increased from fewer than one billion to today's 5.6 billion. In many parts of the world, food production has grown faster than the population, thanks to the opening of new lands, mechanization, fertilizers, pesticides, better water control, improved breeds of plants and animals, and better farmer know-how. Though many of today's bottom billion people live in misery on the edge of starvation, Malthus would be astonished at the relative well-being of most of a vastly enlarged population.

That Malthus's theory failed widely during the past two centuries does not prove that it will remain wrong for the next two. Some observers see a coming vindication of Malthus in the recent faltering of growth rates of per capita food production in some regions. Many scientists have adopted Malthus's general strategy of supposing that limiting factors constrain populations, and in fact the theory has gained some scientific support from agricultural experiments.

One of the assumptions that may pop up in discussions of population involves the idea of "carrying capacity," which refers to the number of individuals of a species that an environment can support for some period. The human population that could be supported by Earth's capacity to produce food has been estimated many times. In outline, if food is the limiting factor, the potentially supportable population equals the potentially arable land area times the yield per unit of area divided by the consumption per person. Easy enough. But of course, there is much uncertainty about the numerical values of arable area, yield, and consumption per capita. Estimates of agricultural carrying capacity have ranged from a low of 902 million in 1945 to a high of 147 billion in 1967.

Globally, food supply is limited physically by the plant energy available for consumption by animals and decomposers. Ecologists call this quantity the net primary productivity (NPP). It is the total amount of solar energy annually converted into living matter, minus the amount of energy the plants themselves use for respiration. NPP is equivalent to about 225 billion metric tons of organic matter a year, an amount that contains enough calories to feed about 1,000 billion people. But that's only if every other consumer of green plants on Earth (including bacteria) were eliminated and at the same time people learned how to enjoy eating wood.

In 1986 Stanford biologists Peter Vitousek, Paul Ehrlich, and Anne Ehrlich and NASA ecologist Pamela Matson estimated that the 5 billion people then on Earth and their domestic animals directly consumed about 3% of NPP in the form of vegetables and other plants. But they also estimated that humans actually "co-opted" about 19% of NPP, a figure arrived at by adding to what was directly consumed the material indirectly consumed in such activities as clearing land or converting it for human use.

This aggregate figure of 19%, or roughly one-fifth, of NPP does not mean the planet can support about five times as many people as the 5 billion it had in 1986. That's because the 19% itself is an average of 31% of *terrestrial* NPP and 2% of *aquatic* NPP. Since people already consume nearly one-third of terrestrial NPP, Earth could support five times as many people only if we either exploited the oceans much more than at present or greatly increased the NPP of the land. The present terrestrial NPP and present human consumption patterns would permit little more than a tripling of the human population, perhaps to 16 billion people, to the practical exclusion of most other terrestrial species.

Not everyone agrees with these conclusions, especially in political circles. When the U.S. Congress declared the week of October 20, 1991, World Population Week, then President Bush issued a proclamation that stated: "Population growth in itself is a neutral phenomenon . . . . Every human being represents hands to work, and not just another mouth to feed." This statement voices an alluring partial truth.

True, carrying capacity can be extended—sometimes immensely—through technological, social, and economic change. If you can't grow enough food, make computer chips and trade for food. True, every additional human being is one more producer (at least potentially, given good health and enough education) as well as one more consumer (inevitably). But the productivity of each additional pair of hands depends on other factors of production that are not currently in infinite supply and that are, sooner or later, subject to diminishing returns: land to work or live on, air, fresh water, geologic deposits, and others. When all other factors in human productivity are available in excess, an additional human being may be not merely neutral but a great asset. Frontier communities rightly celebrate the birth of children. When other factors in human productivity are already taxed to the point of severely diminishing returns, one more human being may represent one more perennially empty stomach, one more soul stunted before it can realize its share of the glory of being human.

Estimating the human population that Earth can sustain is difficult. To see the major dimensions of this problem, imagine a tetrahedron, a pyramid with a triangular base and three triangular sides. At the top is population, which includes size and growth rate, age structure (How many young people need schooling? How many old people need pensions?), health (Are people free of parasites and malnutrition? Are they in good mental health?), distribution (Are people in cities or rural areas?), and migration (Are people moving from poor countries?). At the three corners of the base, place environment, economy, and culture. The environment includes soil, fresh and salt water, air, all nonhuman living creatures, and Earth's stage of mountains, rivers, plains, oceans, volcanoes, earthquakes, meteorites, and solar flares. The economy includes all the human and material arrangements for the production and exchange of goods and services to satisfy people's wants and needs. Culture includes values (What do people want?), technology (What knowledge and artifacts—machines—do people inherit?), and social institutions (How do people interact in satisfying their wants?).

Obviously the boundaries between these four points are fuzzy. And each corner has its academic devotees. Demographers worship at the temple of population; ecologists are the priests of the environment; economists preside over the economy; and anthropologists and other social scientists claim to interpret culture. But as the pyramidal arrangement graphically emphasizes, each element interacts with the others.

To make credible estimates of Earth's human carrying capacity, then, and of the path by which the human population growth rate will approach zero, scientists must learn more of the interactions of population, the environment, economy, and culture. The problem has at least those dimensions—and that's if we consider only one such abstract pyramid. In fact, the Earth is covered with thousands or millions of such pyramids. Populations, environments, economies, and cultures vary from place to place.

Nearly two decades ago the ethicist Daniel Callahan saw the problem: "Excessive population growth raises ethical questions because it threatens existing or desired human values and ideas of what is good. In addition, all or some of the possible solutions to the problem have the potential for creating difficult ethical dilemmas. The decision to act or not to act in the face of the threats is an ethical decision." Better knowledge of the pyramid of population, environment, economy, and culture will at least help us understand our options, their consequences, and the choices being made by our fellow human beings.

Connecting the Concepts

The following questions will help you to connect the issues discussed in this article with the concepts you have learned in Part 6.

1. Why is it unlikely that the human population will continue to grow at its present rate for the next 200 years?
2. How might global warming, acid rain, and dwindling supplies of fossil fuels affect the Earth's human carrying capacity in the future?
3. What could we humans do to increase the Earth's carrying capacity for humans? What are the likely environmental impacts of these actions and their interrelations?

APPENDIX A Periodic Table of Elements

Atomic number / Symbol for the element / Atomic mass

Metals | Metalloids | Nonmetals

I	II											III	IV	V	VI	VII	VIII
1 **H** 1.0079																	2 **He** 4.0026
3 **Li** 6.941	4 **Be** 9.0122											5 **B** 10.811	6 **C** 12.011	7 **N** 14.007	8 **O** 15.999	9 **F** 18.998	10 **Ne** 20.180
11 **Na** 22.990	12 **Mg** 24.305	Transition Elements										13 **Al** 26.982	14 **Si** 28.086	15 **P** 30.974	16 **S** 32.066	17 **Cl** 35.453	18 **Ar** 39.948
19 **K** 39.098	20 **Ca** 40.078	21 **Sc** 44.956	22 **Ti** 47.88	23 **V** 50.942	24 **Cr** 51.996	25 **Mn** 54.938	26 **Fe** 55.847	27 **Co** 58.933	28 **Ni** 58.69	29 **Cu** 63.546	30 **Zn** 65.39	31 **Ga** 69.723	32 **Ge** 72.61	33 **As** 74.922	34 **Se** 78.96	35 **Br** 79.904	36 **Kr** 83.80
37 **Rb** 85.468	38 **Sr** 87.62	39 **Y** 88.906	40 **Zr** 91.224	41 **Nb** 92.906	42 **Mo** 95.94	43 **Tc** (98)	44 **Ru** 101.07	45 **Rh** 102.91	46 **Pd** 106.42	47 **Ag** 107.87	48 **Cd** 112.41	49 **In** 114.82	50 **Sn** 118.71	51 **Sb** 121.75	52 **Te** 127.60	53 **I** 126.90	54 **Xe** 131.29
55 **Cs** 132.91	56 **Ba** 137.33	71 **Lu** 174.97	72 **Hf** 178.49	73 **Ta** 180.95	74 **W** 183.85	75 **Re** 186.21	76 **Os** 190.2	77 **Ir** 192.22	78 **Pt** 195.08	79 **Au** 196.97	80 **Hg** 200.59	81 **Tl** 204.38	82 **Pb** 207.2	83 **Bi** 208.98	84 **Po** (209)	85 **At** (210)	86 **Rn** (222)
87 **Fr** (223)	88 **Ra** (226)	103 **Lr** (260)	104 **Unq** (261)	105 **Unp** (262)	106 **Unh** (263)	107 **Uns** (262)	108 **Uno** (265)	109 **Une** (266)									

Lanthanide series

57 **La** 138.9	58 **Ce** 140.1	59 **Pr** 140.9	60 **Nd** 144.2	61 **Pm** (145)	62 **Sm** 150.4	63 **Eu** 152.0	64 **Gd** 157.3	65 **Tb** 158.9	66 **Dy** 162.5	67 **Ho** 164.9	68 **Er** 167.3	69 **Tm** 168.9	70 **Yb** 173.0

Actinide series

89 **Ac** (227)	90 **Th** 232.04	91 **Pa** 231.04	92 **U** 238.03	93 **Np** (237)	94 **Pu** (244)	95 **Am** (243)	96 **Cm** (247)	97 **Bk** (247)	98 **Cf** (251)	99 **Es** (252)	100 **Fm** (257)	101 **Md** (258)	102 **No** (259)

APPENDIX B The Metric System

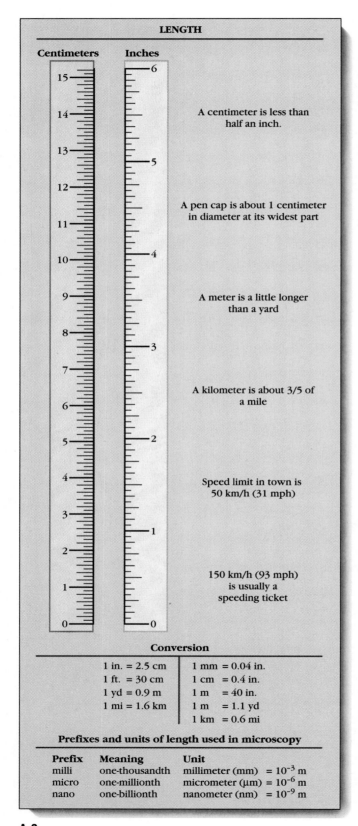

LENGTH

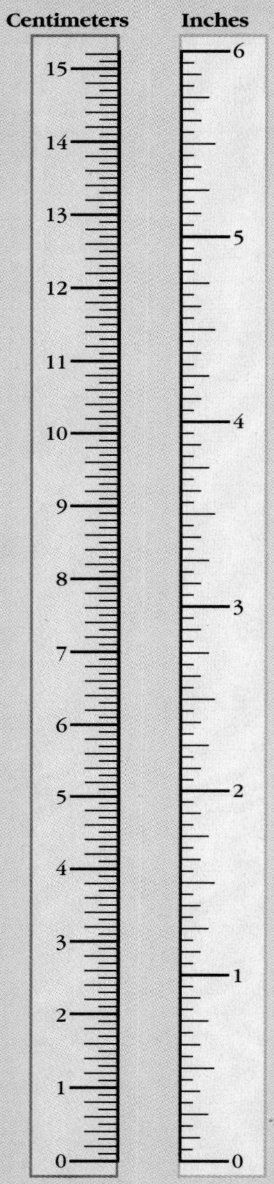

A centimeter is less than half an inch.

A pen cap is about 1 centimeter in diameter at its widest part

A meter is a little longer than a yard

A kilometer is about 3/5 of a mile

Speed limit in town is 50 km/h (31 mph)

150 km/h (93 mph) is usually a speeding ticket

Conversion

1 in. = 2.5 cm	1 mm = 0.04 in.
1 ft. = 30 cm	1 cm = 0.4 in.
1 yd = 0.9 m	1 m = 40 in.
1 mi = 1.6 km	1 m = 1.1 yd
	1 km = 0.6 mi

Prefixes and units of length used in microscopy

Prefix	Meaning	Unit
milli	one-thousandth	millimeter (mm) = 10^{-3} m
micro	one-millionth	micrometer (μm) = 10^{-6} m
nano	one-billionth	nanometer (nm) = 10^{-9} m

MASS

A plastic pen cap has a mass of about 1 gram

A recipe for lasagna calls for 700 grams of ground beef and 113 grams of shredded cheese

A hardback college dictionary has a mass of 1.8-2.25 kilograms

A football player can have a mass of as much as 126 kilograms

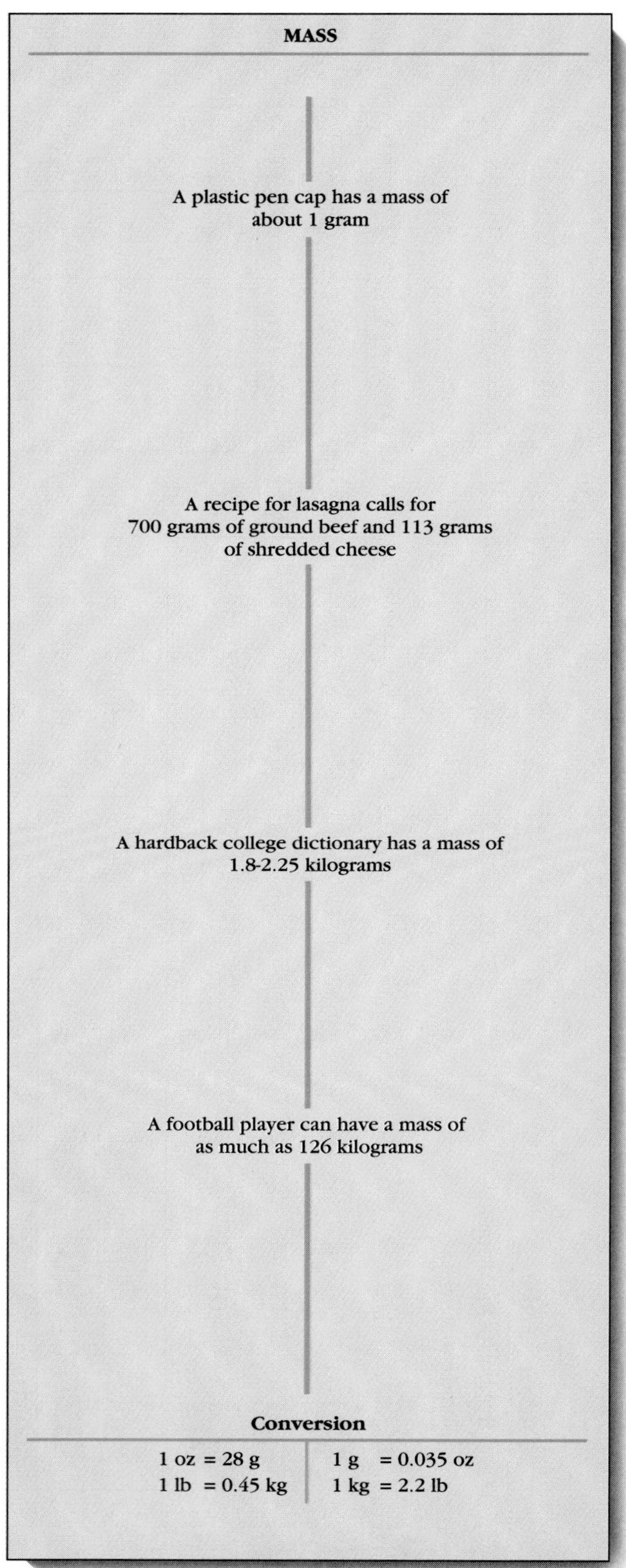

Conversion

1 oz = 28 g	1 g = 0.035 oz
1 lb = 0.45 kg	1 kg = 2.2 lb

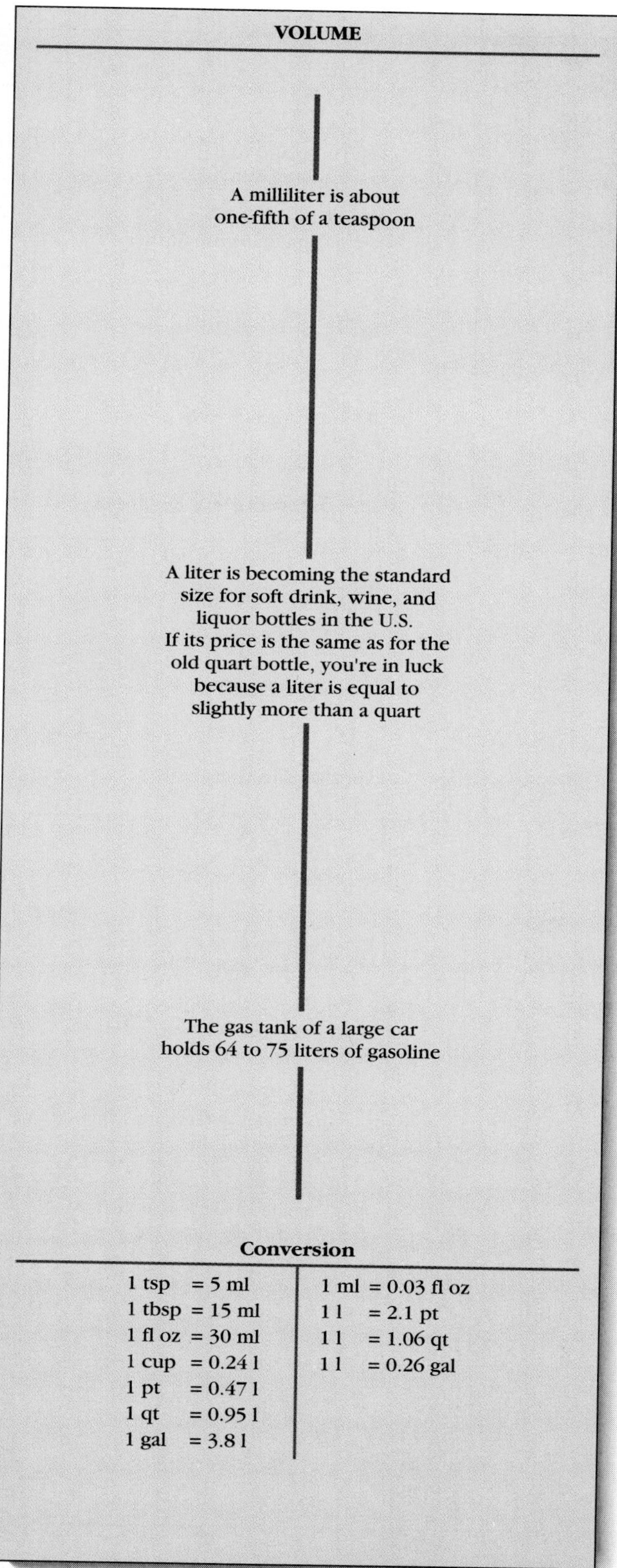

1 tsp = 5 ml	1 ml = 0.03 fl oz
1 tbsp = 15 ml	1 l = 2.1 pt
1 fl oz = 30 ml	1 l = 1.06 qt
1 cup = 0.24 l	1 l = 0.26 gal
1 pt = 0.47 l	
1 qt = 0.95 l	
1 gal = 3.8 l	

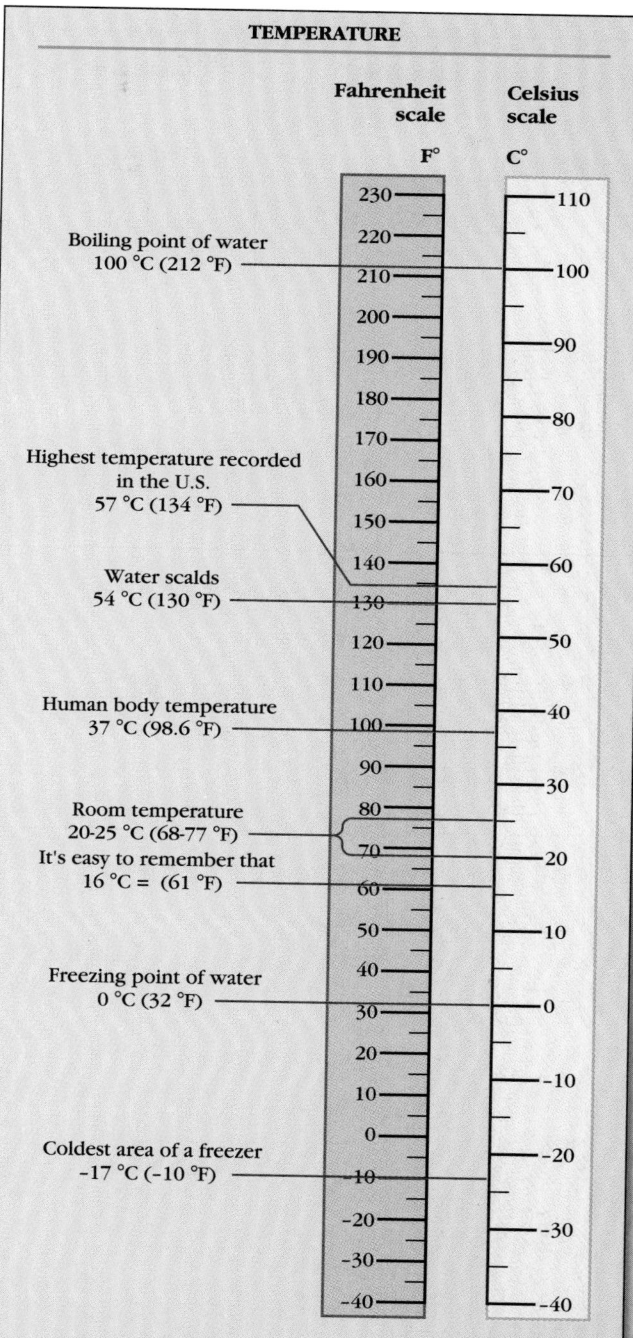

Temperature Conversion

$$°C = \frac{(°F - 32) \times 5}{9}$$

$$°F = \frac{°C \times 9}{5} + 32$$

Interval Equivalents

°F		°C
1.8 °F	=	1 °C
9 °F	=	5 °C
18 °F	=	10 °C

APPENDIX C Geologic Time Scale

Years Ago* (millions)	Era	Period†	Epoch	Climate and Physical Events
	CENOZOIC	Quaternary	Recent	4 ice ages; rise of Sierra Nevada.
			Pleistocene	
2		Tertiary	Pliocene	Rise of Panama; cool; extinction of many species.
5			Miocene	Rocky Mountains rise further.
23			Oligocene	Rise of Alps and Himalayas.
35			Eocene	Mild to tropical weather.
56			Paleocene	Most continental seas disappear.
65	MESOZOIC	Cretaceous		Rise of Rockies reduces rainfall to their east.
146		Jurassic		Much of Europe covered by sea; Breakup of Pangaea.
208		Triassic		Large areas arid and mountainous; Appalachians rising.
245	PALEOZOIC	Permian		Appalachians rising; glaciation in Southern Hemisphere.
290		Carboniferous	Pennsylvanian Mississippian	Land low, covered by shallow seas and swamps; subtropical climate.
363		Devonian		U.S. largely low and sea-covered; Europe mountainous.
409		Silurian		Continents flat; mountains beginning to rise in Europe.
439		Ordovician		Mild climate; shallow seas cover continents, which are flat.
510		Cambrian		
570	PRECAMBRIAN	Precambrian		Planet cooling, shallow seas, many mountains rise.
(The Earth is about $4\frac{1}{2}$ billion years old)				

*Data from *A geologic time scale*, 1989. Harland, W.B., et al., Cambridge University Press.

†Remember the periods with this mnemonic: **P**lease **C**ome **O**ver **S**ome **D**ay, **C**ould **P**lay poker-**T**hree **J**acks **C**onquer **T**hree **Q**ueens.

	Climate and physical events
	*H. sapiens*; extinction of many large mammals; many desert forms evolve. Large carnivores; hominoid apes. Forests dwindle; grassland spreads. Anthropoid apes; ungulates, whales. Earliest tiny horses. First primates and carnivores.
	Angiosperms originate and spread; extinction of many dinosaurs; marsupials present. Origin of birds; reptiles dominant; cycads, ferns. First dinosaurs and mammals, forests of gymnosperms and ferns.
	First conifers, cycads, ginkgos. Origin of reptiles; amphibians dominant. Fungi, first insects, sphenopsids, lycopsids. Fishes dominant, origin of modern vascular plants, sharks. First vascular plants, modern groups of algae and fungi. First (agnathan) fish; plants invade land; invertebrates and marine plants.
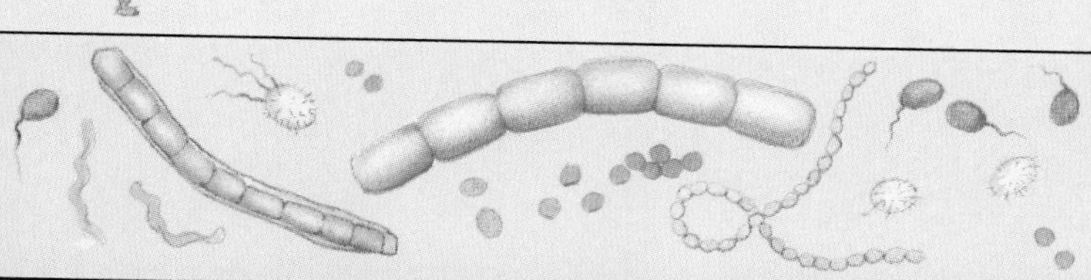	Origin of eukaryotes; cyanophytes and bacteria.

Answers to Self-Quizzes

CHAPTER 2 *Some Basic Chemistry*

1. alkaline; decreased
2. a.
3. a. H_2O
   b. NaCL
   c. CO_2
   d. O_2
4. covalent
5. covalent
6. a.
7. h.
8. f.
9. c.
10. d.
11. g.

CHAPTER 3 *Biological Chemistry*

1. nucleic acids
2. nucleotides
3. *any two of the following:*
   carry genetic information
   direct the synthesis of proteins
   supply energy to enzyme reactions
   act as coenzymes
4. lipids
5. C, H, O (P, N)
6. energy storage
   formation of biological membranes
   hormones
7. C, H, O, N (S)
8. amino acids
9. *any of the following:*
   catalyze reactions (enzymes)
   form structural elements of body
   hormones
   muscle contraction
   defense against disease
   transport of substances
10. carbohydrates
11. C, H, O (N)
12. monosaccharides
13. energy storage
    structural support and protection
14. b.
15. c.
16. d.
17. b.
18. c.

CHAPTER 4 *Cells and Their Membranes*

1. d.
2. hypotonic; higher; from the environment into the animal; a.
3. a.
4. b., d.
5. a.
6. c., d.
7. a.
8. b.
9. d.
10. endocytosis
11. active transport, diffusion through channels, endocytosis

CHAPTER 5 *Cell Structure and Function*

1. p.
2. d., f.
3. a.
4. c.
5. g.
6. i.
7. o.
8. q.
9. n.
10. m.
11. d., f., k.
12. all
13. animals, some plants, some prokaryotes
14. plants, prokaryotes
15. animals, plants (prokaryote DNA is not organized into chromosomes)
16. animals, plants
17. e.

CHAPTER 6 *Energy and Living Cells*

1. c.
2. photosynthesis
3. reduced
4. ADP; P_i; endergonic; energy
5. c.

CHAPTER 7 *Food as Fuel: Cellular Respiration and Fermentation*

1. c.
2. c.
3. inner mitochondrial
4. a. pyruvate, NADH, ATP
   b. CO_2, ATP, NADH, $FADH_2$, oxaloacetic acid, CoA
   c. ATP, NAD^+, C_2O, ethanol
   d. NAD^+, FAD, H_2O
   e. ATP, NAD^+, lactate
5. a.
6. true
7. true

CHAPTER 8 *Photosynthesis*

1. c.
2. a.
3. c.
4. d.
5. c.
6. a., d.
7. d.
8. b.
9. c.
10. b., c.
11. a.
12. c.

CHAPTER 9 *DNA and Genetic Information*

1. a.
2. c.
3. d.
4. T-A-G-A-C-A-T-A-C-T
5. c.
6. e.
7. *Escherichia coli*
8. both
9. your own cells
10. your own cells
11. both
12. your own cells

CHAPTER 10 *RNA and Protein Synthesis*

1. mRNA: A-U-G-U-U-C-A-U-G-A-A-C-A-A-A-G-A-A
   amino acids: methionine—phenylalanine—methionine—asparagine—lysine—glutamic acid
2. The second amino acid would be leucine instead of phenylalanine, and the fourth amino acid would be lysine instead of asparagine.
3. The second amino acid would be leucine instead of phenylalanine, and the chain would terminate after this amino acid, because the next codon is now a *STOP* codon.
4. a. DNA double-stranded; RNA single-stranded
   b. DNA nucleotides contain deoxyribose; RNA nucleotides contain ribose
   c. DNA contains thymine nucleotides; RNA contains uracil nucleotides
5. e.
6. a.
7. d.
8. b.

CHAPTER 11 *Reproduction of Eukaryotic Cells*

1. b.
2. b.
3. b.
4. d.
5. b.
6. b.
7. nucleus; cytoplasm; eggs; sperm

CHAPTER 12 *Mendelian Genetics*

1. a. both are *Tt*
   b. ¾ tasters: ¼ nontasters
   c. all tasters
   d. ½ tasters: ½ nontasters
2. a. dumpy recessive to normal, which is dominant
   b. both heterozygous for dumpy wings
3. 40
4. a. 250 b. 125
5. a. Sniffles: homozygous dominant (colored)
   Whiskers: heterozygous for albino
   Esmeralda: homozygous recessive (albino)
   b. ¾ colored: ¼ albino
   c. ½ colored: ½ albino
6. Mate his dog to bitches known to carry the trait. If any pups show it, the dog is heterozygous for the allele. If none of a large number of pups shows it, the dog is probably homozygous normal.
7. a. yes, if both are heterozygous
   b. no
8. a.

	BS	Bs	bS	bs
Bs	BBSs	BBss	BbSs	Bbss
bs	BbSs	Bbss	bbSs	bbss

b.

	BS	Bs
Bs	BBSs	BBss
bs	BbSs	Bbss

c.

	BS	Bs	bS	bs
bs	BbSs	Bbss	bbSs	bbss

d. ½

9. a. $\frac{3}{16}$ b. $\frac{3}{16}$ c. $\frac{9}{16}$
10. a. 480 b. 160 c. 40
11. a. ⅛ b. ⅛ c. ⅜
12. a. stamens: straight dominant to incurved; petals (red vs. streaky): can't tell from information given
    b. stamens: both parents heterozygous; petals: one heterozygous, one homozygous recessive, but no indication which is which from information given
    c. red × red and streaky × streaky: if red is dominant, red × red will produce some streaky progeny, and vice versa
13. a. ½ b. ¼ c. ¼
14. a. genotype: SsRR'
    phenotype: straight roan
    b. straight red (⅛)
    straight roan (¼)
    straight white (⅛)
    curly red (⅛)
    curly roan (¼)
    curly white (⅛)
15. a. clover patch: ½ roan, ½ white
    alfalfa field: ½ red, ½ roan
    cornfield: ¼ red, ½ roan, ¼ white
    b. it doesn't matter; ½ the calves will be roan in any case
16. a. ¼ b. ½ c. ¼ d. $\frac{3}{16}$ e. ⅛
17. a. The genes are linked.
    b. In the female parent, the genes for sable body and normal wing are on one chromosome, and the genes for normal body and miniature wing are on its homologue. This arrangement is indicated by the large numbers of offspring with the combinations sable body + normal wing and normal body + miniature wings.
18. The genes are probably unlinked; this is shown by the

almost-equal numbers of offspring in each phenotype category.

CHAPTER 13 *Inheritance Patterns and Gene Expression*

1. a. ⅔
   b. sell cows bearing these calves and try a new bull
2. a. ¼
   b. ½ normal: ½ brachydactylic
3. a. Yellow mice are heterozygous.
   b. "Yellow" allele is lethal in the homozygous condition, early in embryonic development (2:1 ratio).
   c. Homozygous yellow individuals die as early embryos and are resorbed by the uterus.
   d. He could carry out yellow × yellow matings and examine contents of females' uteri early in pregnancy to detect defective embryos.
4. a. I^Ai d. I^Bi g. I^Ai or I^Bi or ii
   b. I^Bi e. I^AI^A or I^Ai or I^AI^B h. I^Ai
   c. I^Ai or I^AI^B f. I^Ai i. I^Bi
5. Yes, John Smith is really Tom Jones! A baby with blood type M cannot have a parent with blood type N (Ms. Smith). Also, a parent with blood type AB (Ms. Jones) cannot have a child with blood type O.
6. a. ½ normal: ¼ chinchilla: ¼ Himalayan
   b. ½ chinchilla: ¼ Himalayan: ¼ albino
   c. ¾ chinchilla: ¼ albino
7. ½ barred males: ½ non-barred females (remember, male birds are ZZ and females ZW [see Figure 13–8])

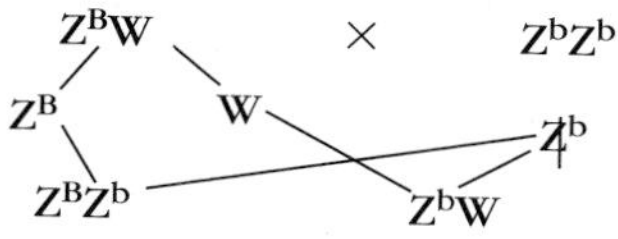

8. ½ the sons hemophiliacs: ½ the sons normal
   ½ the daughters heterozygous carriers: ½ the daughters homozygous normal
9. when the mother is either a hemophiliac or a carrier
10. a. mother homozygous recessive colorblind; father normal allele × Y
    b. ½ of the sons are expected to be colorblind
    c. ¼ (½ of the offspring are expected to be daughters; ½ of these are expected to be colorblind; ½ × ½ = ¼)
11. 2 female offspring: 1 male
12. No; since the baldness trait is carried on the autosomes, a man could inherit the trait from either parent who carried it, and from any of his four grandparents who passed it on to the appropriate parent.

CHAPTER 14 *Evolution and Natural Selection*

1. c.
2. d.
3. e.
4. b.
5. Individuals with longer necks could reach food higher on trees and therefore had a better chance to survive and reproduce when food was scarce nearer the ground. These individuals passed on the genes for longer necks to their offspring. Continued selection pressure of this sort over many generations led to evolution of longer and longer neck length.
6. e.

CHAPTER 15 *Population Genetics and Speciation*

Identification of dichotomous key specimens (Toolbox):
- a. Diplopoda
- b. Chilopoda
- c. Crustacea
- d. Insecta
- e. Arachnida

1. d.
2. d.
3. The normal allele would gradually increase, and the sickle allele would decrease, but not to zero; the sickle allele would remain at low levels in the heterozygous condition.
4. a.
5. d.
6. a. increase
   b. increase
   c. increase
   d. increase
7. b.
8. d.

CHAPTER 16 *Evolution and Reproduction*

1. b.
2. favorable; unfavorable
3. b.
4. c.
5. a.
6. a., b., c.
7. a.

CHAPTER 17 *Origin of Life*

1. oxygen: less
   carbon dioxide: more
   water vapor: more
2. e.
3. 1. organic monomers
   2. polymers
   3. fermentation
   4. water-splitting photosynthesis
   5. respiration
   6. intracellular organelles
4. b.
5. 1. Addition of O_2 to the atmosphere made it oxidizing rather than mildly reducing.
   2. An ozone layer formed.
   Effects:
   1. Addition of O_2 acted as selection pressure for the evolution of respiration.
   2. Ozone layer permitted organisms to move onto land.
6. c.

CHAPTER 18 *Classification of Organisms and the Problem of Viruses*

1. b.
2. d.
3. d.
4. d.
5. a., b.
6. Protista
7. d.

CHAPTER 19 *Bacteria, Protists, and Fungi*

1. d.
2. a.
3. true
4. false
5. b.
6. eukaryotic, unicellular
7. a. pseudopodia
   b. cilia
   c. flagella
8. c.
9. b.
10. b.
11. a.
12. c.
13. a.

CHAPTER 20 *The Plant Kingdom*

1. d.
2. c.
3. a.
4. a.
5. a.
6. b., d.
7. b. zygote, d. sporophyte, e. spore, c. gametophyte, a. gamete
8. c.
9. d.
10. d.
11. d.
12. c.
13. a. growing on moist soil or rocks in shady areas
    b. growing on surface of moist soil in vicinity of fern plants
    c. male: immature is pollen grain, in pollen cone; mature is pollen tube that has grown to female gametophyte in seed cone.
    female: at base of seed cone scales
    d. in male and female flower parts, respectively

CHAPTER 21 *The Animal Kingdom*

1. bilateral
2. dispersal
3. a.
4. d.
5. c.
6. d.
7. a.
8. f.
9. b.
10. g.
11. i.
12. h.
13. e.
14. c.
15. b.
16. c.
17. d.
18. b.
19. a.
20. feathers
21. a. support,
    b. dehydration of body and of eggs laid outside the body,
    c. and extraction of oxygen from (dry) air rather than water.
22. Reptilia
23.

	Amphibians	Reptiles
Body structure:	no claws	claws on toes
	no scales	scales on skin
	legs out to sides	legs under body
	slender bones	robust bones
Reproduction:	eggs lack shells	eggs have waterproof shells
	eggs laid in water	eggs laid on land

24. d.
25. b.
26. a.
27. e.
28. f.
29. c.
30. b.
31. g.

CHAPTER 22 *Animal Nutrition and Digestion*

1. f.
2. a.
3. a., b., c.
4. a., b., c.
5. d.
6. b.
7. f.
8. a.
9. a.
10. e.
11. b.
12. c.
13. a.

CHAPTER 23 *Gas Exchange in Animals*

1. a.
2. c.; b.
3. b.
4. a.
5. a.
6. a.
7. d.

CHAPTER 24 *Internal Transport*

1.

cnidarian	earthworm	insect	fish	mammal
x	x	x	x	x
	x		x	x
				x
	x	x	x	x

2. c., d.
3. d.
4. b.
5. c. (flow to the skin and lungs decreases as exercise begins but later increases)
6. c.

CHAPTER 25 *Defenses Against Disease*

1. b.
2. d.
3. e.
4. a.
5. d.
6. d.
7. b.
8. a.
9. c.

CHAPTER 26 *Excretion*

1. a.
2. b. (it is a mammal)
3. a.
4. a.
5. b.
6. d.
7. e.
8. a.
9. b.
10. d.
11. c.
12. b.
13. decrease; c.

CHAPTER 27 *Sexual Reproduction and Embryonic Development*

1. j.
2. i.
3. b., e.
4. d.
5. a.
6. f.
7. h.
8. f. (testis) ⟶ j. (vas deferens) ⟶ g. (urethra) ⟶ i. (vagina) ⟶ a. (cervix) ⟶ h. (uterus) ⟶ d. (oviduct)
9. a.
10. f.
11. d.
12. b.
13. a.
14. c.
15. a.
16. c.
17. d., a.
18. g.
19. e.
20. c.
21. C
22. N
23. O
24. C
25. G

CHAPTER 28 *The Nervous System and Sense Organs*

1. d.
2. a.
3. a. A neurotransmitter substance carries information between two neurons in that it is released as a result of the arrival of an action potential at the presynaptic membrane of one neuron, and its arrival at the postsynaptic membrane stimulates activity in the postsynaptic cell.
   b. Vesicles store neurotransmitter molecules in the presynaptic terminal and release them when an action potential arrives at the terminal.
   c. On combining with neurotransmitter molecules, receptor molecules change the permeability of the postsynaptic membrane to ions, resulting in a local potential.
   d. Enzymes that destroy neurotransmitter molecules in effect "turn off" the signal brought across the synapse by the neurotransmitter.
4. d.
5. b.
6. d.
7. A. cell body of sensory neuron
   B. dendrite of sensory neuron
   C. axon of motor neuron
   D. axon of sensory neuron
   E. synapse between sensory and motor neuron
8. c.
9. norepinephrine; acetylcholine
10. c.
11. b.
12. e.
13. a.
14. d.

CHAPTER 29 *Muscles and Skeletons*

1. b.
2. a., c.
3. a., c.
4. c.
5. a., c.
6. A. Z line
   B. actin, calcium (during contraction), tropomyosin, troponin
   C. mitochondria, calcium (during relaxation)
   D. myosin
7. During tetanic contraction, motor neurons are firing a continuous train of closely spaced nerve impulses.
8. A muscle does not relax between nerve impulses during tetanic contraction because each new impulse arrives

before the muscle completes the sequence of chemical and mechanical events of contraction initiated by the previous impulse.

9. c.
10. e.

CHAPTER 30 *Animal Hormones and Chemical Regulation*

1. c.
2. c.
3. a.
4. b.
5. a. negative feedback
   b. target cells
6.

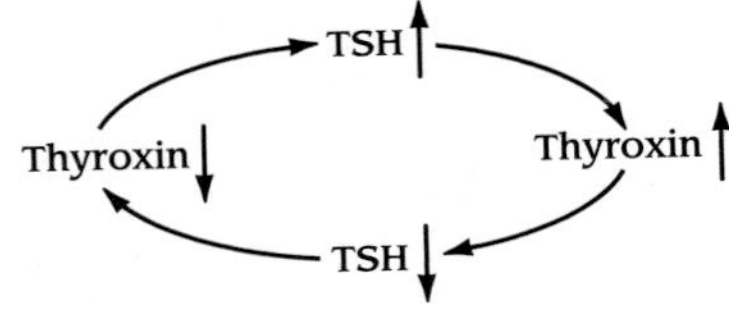

CHAPTER 31 *Behavior*

1. g.
2. a.
3. h.
4. j.
5. f.
6. c.
7. b.
8. d.
9. c.
10. e.
11. e.

CHAPTER 32 *Plant Structure and Growth*

1. a.
2. b.
3. a.
4. a., c., b.
5. b.
6. c.
7. c.
8. a. monocotyledons
   b. both
   c. monocotyledons
   d. monocotyledons
   e. dicotyledons
9. a.

CHAPTER 33 *Nutrition and Transport in Vascular Plants*

1. e.
2. d.
3. b.
4. a.
5. c.
6. a.
7. epidermis, cortex, Casparian strip, endodermal
8. nitrogen, phosphorus, potassium
9. c.
10. c.
11. function: transport of sap
    adaptation: tubular, dead and hollow, pits or open ends, stacked end to end
    function: support of plant body
    adaptation: secondary wall thickenings
12. d.
13. a. remain the same
    b. increase
    c. remain the same
14. a. decrease
    b. increase
    c. increase
    d. increase
15. a.
16. b.

CHAPTER 34 *Regulation and Response in Plants*

1. d.
2. c.
3. b.
4. e.
5. b.
6. d.
7. a.
8. b.
9. d.
10. c.

CHAPTER 35 *Reproduction in Flowering Plants*

1. pollen tube
2. stamen
3. petal
4. sepal
5. embryo sac
6. ovule
7. ovary
8. pistil
9. anther
10. pollen
11. stigma
12. #2: anther
13. #6: embryo sac
14. #5: sepal
15. #11: ovule
16. #6: embryo sac
17. #4: ovary
18. false
19. e. (but not all seeds require all factors listed)
20. b.

CHAPTER 36 *Distribution of Organisms*

1. c.
2. a.
3. d.
4. d.
5. c.
6. primary
7. early successional
8. (1) marsh plants
   (2) short fugitive plants
   (3) taller, perennial plants
   (4) shrubs and pioneer trees
   (5) climax trees
9. they are separated by great distances; they have similar climates, which exert similar selective pressures (in this case, nothing much taller than grasses can grow with so little rainfall)

CHAPTER 37 *Ecosystems and Communities*

1. recycling of nutrients
2.

hawk
spider
sparrow
wolf
herbivorous insect
deer
shrub
soil bacteria

decomposers: soil bacteria
producer: shrub
primary consumers: deer, herbivorous insect
secondary consumers: wolf, spider, (sparrow)
tertiary consumers: hawk, (sparrow)
(sparrow feeds at several possible trophic levels)

3. b.
4. c.
5. e.
6. a.
7. a. increase
b. decrease
c. increase
8. e.

CHAPTER 38 *Populations*

1.

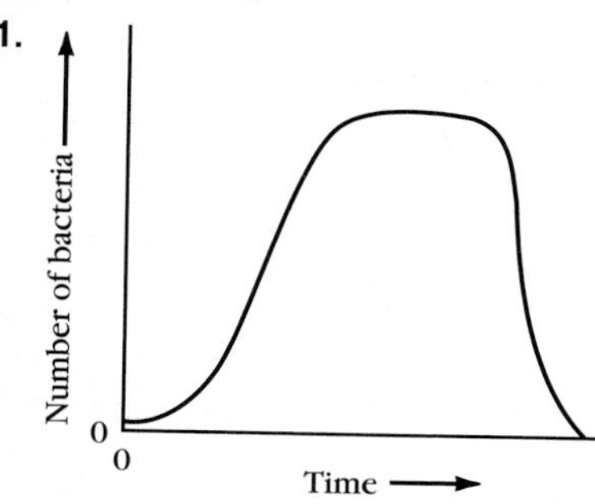

The population grows exponentially, levels off, and then declines rapidly to extinction in this "closed" environment. Mark yourself ½ off if you forgot to label the axes.

2. b.
3. b.
4. b.
5. d.
6. I.
7. b.
8. a.

CHAPTER 39 *Human Impact on the Environment*

1. b.
2. c.
3. c.
4. e.
5. d.

-a-

abdomen In vertebrates and some arthropods, the rearmost part of the body, containing the end of the digestive tract, external genital organs, etc.

abiotic Nonbiological; occurring neither within living cells nor under their influence.

abortion Process whereby a mammalian embryo or fetus becomes detached from the uterine wall and expelled from the female's body either naturally ("miscarriage," spontaneous abortion) or by medical means (induced abortion).

Acacia Large genus of trees of the legume family. Most acacias are tropical. Common name: thornwood.

acetyl CoA Acetyl coenzyme A: a molecule that can donate its (two-carbon) acetyl group to production of fatty acids or to the citric acid cycle, where it is broken down to yield energy.

acetylcholine An important neurotransmitter in vertebrates and invertebrates; the transmitter at some synapses in the brain, many nerve-muscle junctions, etc.

acid 1. A substance that releases hydrogen ions (H^+) in aqueous solution. 2. A substance that accepts electrons.

acid rain Rain containing pollutants that give it an acid pH, usually of less than 5.0. The most common pollutants with this effect are oxides of nitrogen and sulfur which form acids when they dissolve in water.

Acrasiomycota Phylum containing the cellular slime molds.

actin A protein that forms the thin filaments involved in muscle contraction and the microfilaments involved in changes of shape in other kinds of cells.

actinomycetes Filamentous prokaryotic organisms.

action potential A fast-moving, all-or-nothing electrical disturbance in the membrane of a nerve cell; a nerve impulse. Also occurs in the membrane of a muscle when it is stimulated to contract.

activation energy Energy input required before a chemical reaction can proceed.

active transport Process in which energy is expended in moving a substance through a membrane against its concentration gradient.

acylation The attachment of an amino acid to a transfer RNA molecule.

adaptation 1. Process by which populations evolve to become suited to their environments over the course of generations. 2. Characteristic that increases an organism's evolutionary success. 3. Process by which a sense organ ceases to respond to a constant stimulus.

adaptive radiation Formation of two or more new species, macromolecules, or physiological pathways, adapted to different ways of life, from one ancestral species, molecule, or pathway.

adhesion Sticking together (of cells, molecules of different substances, etc.).

adipose tissue Tissue made up mainly of fat-storing cells.

adrenalin See epinephrine.

aerial In the air.

aerobic 1. Requiring molecular oxygen (O_2) to live. 2. In the presence of molecular oxygen.

affected individual In genetics, an individual homozygous for a deleterious recessive allele, with a phenotype expressing the allele.

air bladder Air-filled sac providing buoyancy in fish (= swimbladder) or algae.

albinism (adj. albino) White coloration due to lack of genes for normal pigmentation.

alcohol Organic compound containing an alcohol group (COH).

alcoholic fermentation Biochemical pathway in some yeasts (and other organisms) in which pyruvate is converted into ethyl alcohol (ethanol), releasing NAD; used to make alcoholic beverages.

alga (pl. algae) Photosynthetic organism with a one-celled or simple multicellular body plan and lacking a multicellular embryo protected by the female reproductive structure.

alkali (= base) A substance that releases hydroxide ions (OH^-) in water or that accepts hydrogen ions (H^+) or that gives up electrons.

alkaline Having a (basic) pH of more than 7.0.

allantois An embryonic membrane: sac that grows out of the embryonic gut in higher vertebrates; stores the embryo's nitrogenous waste until hatching in reptiles and birds; forms the main embryonic respiratory surface and part of the placenta in mammals.

allele One of two or more genes carrying the information for contrasting forms of the same genetic trait (e.g., blue eyes and brown eyes).

allopatric Living in different areas.

allosteric protein A protein capable of changing shape in a way that alters its binding sites and its activity.

alpine Of high mountain regions.

altitude Height above sea level.

altruistic behavior Behavior that favors the reproductive success of another member of the same species.

ambient temperature Temperature of the surroundings (equivalent to "room temperature").

amino acid Small organic molecule containing both a carboxyl group and an amino group bonded to the same carbon atom. Amino acids are the monomers from which polypeptides and proteins are made.

aminoacyl attachment site That part of a transfer RNA molecule to which an amino acid attaches.

ammonia NH_3.

ammonium NH_4^+.

amniocentesis Technique for obtaining a sample of fetal cells by inserting a syringe through the mother's abdominal wall and uterus, into the amniotic sac surrounding the fetus, which contains cells sloughed off the fetus.

amnion A fluid-filled sac that surrounds the embryo in reptiles, birds, and mammals.

amniotic egg Type of egg found in reptiles and their descendants (birds and mammals) in which the embryo is surrounded by an amnion and other membranes (allantois and chorion).

amphibian Member of the vertebrate class Amphibia, e.g., frogs, toads, salamanders, newts, and their relatives.

amylase An enzyme that breaks glucose molecules from starch molecules.

amyloplast Plastid that stores starch.

amylose, amylopectin Polymers of glucose that combine to make up starch.

anaerobic (n. anaerobe) 1. Without molecular oxygen (O_2) 2. Not requiring molecular oxygen for extraction of energy from food (respiration).

analogous In biology, of two or more organs with the same function but with different evolutionary origins.

anaphase The stage of nuclear division during which chromosomes move to the poles of the spindle apparatus.

anatomy The physical structure of an organism.

anemia deficiency of hemoglobin or of red blood cells, resulting in a deficiency of oxygen supply to the tissues.

angiosperms Flowering plants.

anion A negatively charged ion.

annelid A segmented worm, member of the phylum Annelida; e.g., earthworm, polychaete, leech.

anterior At or toward the front end of an animal.

anther Organ in flowers that produces pollen.

antheridium Organ that produces male gametes in fungi, algae, mosses, and ferns.

anthocyanins Pigments of red, blue, or purple hue, commonly found in vacuoles of plant cells.

Anthophyta Plant division containing the flowering plants (= angiosperms).

anthropoids Monkeys, apes, and humans, which make up the suborder Anthropoidea of the order Primates.

anthropomorphism Attribution of human characteristics to other animals.

antibiotics Substances that kill microorganisms.

antibody (= immunoglobulin) One of a huge group of proteins, produced by B lymphocytes, which will react with a specific antigen as part of the body's defensive immune response to disease.

anticodon The row of three nucleotides on a tRNA molecule that base-pairs with the complementary codon on a mRNA molecule attached to a ribosome. This allows the amino acid carried by the tRNA to be placed correctly into the peptide chain that is being synthesized on the ribosome.

antigen A substance capable of stimulating an immune response by binding to an antibody or lymphocyte receptor; usually a protein or other substance that is not naturally part of the body.

Anura The order of amphibians containing frogs and toads.

aorta (pl. aortae) Large artery that carries blood from the heart toward the rest of the body.

apical Of the tip of something.

apical meristem Area of dividing cells at the root and stem tips of a plant.

Apicomplexa Phylum of parasitic protists with a distinctive, complex arrangement of components at one end of the cell.

aposematic coloration Conspicuous, "warning" coloration of an animal that is dangerous or distasteful to eat.

aquatic Of water (fresh or salt).

aqueous Containing, or composed largely of, water.

arboreal Living in trees.

Archaeobacteria Ancient group containing the methane-producing, salt-loving, and hot acid-loving bacteria.

arid Dry; of areas where more water leaves the ecosystem (by evaporation and transpiration) than enters it (as precipitation).

arteriole A small artery carrying blood from an artery to a capillary bed.

artery Vessel that carries blood away from the heart.

arthropods Members of the phylum Arthropoda: segmented animals with jointed appendages, and stiff chitin-containing external skeletons; e.g., crabs, lobsters, barnacles, insects, spiders.

ascidian A marine invertebrate chordate (sea squirt).

Ascomycota Division of fungi which reproduce sexually by means of ascospores produced inside asci; sac fungi.

ATP Adenosine triphosphate; most common energy-donating molecule in biochemical reactions.

atrium (pl. atria) A chamber; e.g., part of the heart that receives blood as it enters.

autonomic nervous system The division of the vertebrate nervous system over which the animal usually has no control; composed of the sympathetic and parasympathetic systems; regulates blood pressure, breathing, gut movements, etc.

autoradiography Technique for determining the position of a radioactive element introduced into a system; based on the fact that radiation emitted by radioactive substances exposes (blackens) photographic film that is placed near them.

autosome Any chromosome that is not a sex chromosome.

autotroph (adj. autotrophic) Organism that can make its own (organic) food molecules from inorganic constituents, using energy either from sunlight (green plants) or inorganic chemical reactions (chemosynthetic bacteria). Compare heterotroph.

avian Pertaining to birds, members of the class Aves.

axial Pertaining to an axis.

axil (adj. axillary) In plants, the angle between the stalk (petiole) of a leaf and the stem.

axon (adj. axonal) Extension of a neuron that carries nerve impulses to the next cell(s) in line.

-b-

bacterial transformation Uptake and incorporation of genetic material by a bacterium.

bacteriophage (abbr. phage) A virus that infects, and reproduces within, a bacterium.

bacterium Member of the kingdom Prokaryotae, containing organisms without nuclear envelopes around the single, circular DNA genome; most are very small, unicellular or forming colonies of independent cells.

basal body Structure at the base of a cilium or flagellum, containing a circle of nine triple microtubules.

base (adj. basic) See alkali.

base pairing In nucleic acids, the attachment of one nucleotide base to another by hydrogen bonds, in unique pairs: adenine-thymine in DNA, adenine-uracil in RNA, or guanine-cytosine, in both).

bicarbonate ion HCO_3^-, a common buffer in biological systems.

binary fission Division (usually of a cell) into two equal parts.

biodiversity The number of different species of organisms found in an area.

biogeochemical cycle Description of the geological and biological processes that affect the movement of an element among different components of an ecosystem.

biological species A group containing all the members of one or more populations that interbreed or potentially interbreed with one another under natural conditions.

bioluminescence Production of light by a living organism (such as a firefly).

biomass Amount of material that is part of the bodies of living organisms.

biome Major type of terrestrial communities of organisms; e.g., tropical rain forest, desert.

biosphere Total of all areas on Earth where living organisms are found; includes deep ocean and part of the atmosphere.

biota The organisms living in an area.

biotic Having to do with living organisms.

biotic potential The rate at which a population can increase in size by reproduction under ideal conditions.

bipedalism Habit of walking on two legs.

bloom Generally means flower; ecologists also use it to mean a rapid increase in population of microorganisms, especially cyanobacteria or unicellular algae.

bone marrow Soft tissue in the center of the long bones in the limbs.

brachiation Mode of locomotion using the forelimbs to swing from branch to branch.

bromeliad Member of a large family of mostly epiphytic plants.

bryophyte A member of the plant division Bryophyta: embryophyte plants lacking vascular tissue; mosses and liverworts.

bug Member of the insect order Hemiptera.

-c-

C_3 cycle (= Calvin cycle) Metabolic pathway of photosynthesis in which NADPH and ATP are used to fix carbon dioxide into three-carbon organic molecules.

caecum A pouch leading off the digestive tract in some animals.

calcareous Composed of, or containing, calcium carbonate ($CaCO_3$).

calorie (adj. caloric) 1 Calorie = 1000 calories = 1 kilocalorie or kcal. A calorie is the amount of heat needed to raise one gram of water from 14.5°C to 15.5°C. The energy values of foods are expressed in Calories.

cambium (adj. cambial) Meristematic tissue that produces new cells which increase the diameter of a woody stem or root.

canine 1. (adj.) Of or pertaining to dogs. 2. (noun) The (usually) pointed tooth next to the incisors in mammals.
capillaries The smallest blood vessels in a circulatory system; exchange of substances between the blood and the extracellular fluid takes place across the thin walls of the capillaries.
capillarity Movement of fluid into a narrow tube or other space because of attraction of the walls of the space for the molecules of the fluid.
carbohydrates A class of compounds whose members have the general formula $(CH_2O)_n$ and contain at least one double-bonded oxygen.
carbon dioxide Gas (CO_2) produced during respiration by organisms, and by burning of organic matter or other carbon-containing substances.
carboxyl group A functional group (—COOH) consisting of a carbon atom double-bonded to an oxygen atom and single-bonded to another oxygen atom, which in turn is bonded to a hydrogen atom.
carcinogen Cancer-causing agent.
cardiac Of the heart.
cardiac muscle The type of muscle that makes up the vertebrate heart.
cardiovascular Of the heart and circulatory system.
carnivore 1. Animal that eats other animals. 2. Member of the mammalian order Carnivora; e.g., dogs, cats, weasels, bears, raccoons, skunks.
carotenoid Accessory photosynthetic pigment that usually appears yellow, orange, or brown.
carpals The bones in the wrist.
carpel A female flower part.
carrier In genetics, an individual heterozygous for a (usually damaging) recessive allele.
carrying capacity Number of individuals of a species that a particular environment can support indefinitely; represented in formulas by the letter K.
cartilage (adj. cartilaginous) l. Tissue composed of scattered cells surrounded by tough, flexible intercellular protein fibers. 2. A skeletal element composed of cartilage.
cation A positively charged ion.
caudal Of the tail.
cell wall Stiff, fibrous layer that lies outside the plasma membrane of a cell. Occurs in plants, fungi, bacteria, and some protists.
cellular respiration The stepwise release of energy from food molecules, accompanied by storage of the energy in short-lived energy intermediates such as ATP.
cellulose Polysaccharide that makes up fibers that form a large part of the cell walls of plant cells.
central nervous system The brain and spinal cord.
centrioles Pair of structures containing nine triple microtubules, found in most animal cells and in cells of lower plants and fungi; usually close to the nucleus outside the nuclear membrane.
centromere The constricted area where chromatids of a replicated chromosome remain joined until anaphase.
cephalization Concentration of nervous tissue and sense organs in the head.
Cephalopoda Class of molluscs containing octopuses, squids, *Nautilus.*
cereals Those grass species of the plant family Gramineae which make up the bulk of human food.
cerebral cortex Outer layer of the cerebral hemispheres.
cerebral hemispheres The two halves of the cerebrum in the forebrain.
cerebrospinal fluid A clear fluid, derived from blood, that bathes and cushions the brain and spinal cord.
cerebrum A large, rounded area of the forebrain divided by a fissure into right and left halves (= cerebral hemispheres).
Cetacea The order of mammals whose members are most highly adapted to aquatic life, including whales and porpoises.
chaparral Dry shrublands of temperate coastal regions such as California and the Mediterranean.
chelicerae Piercing and sucking mouthparts of spiders and their relatives; associated with poison glands in spiders.
Chelonia Order of reptiles including turtles and tortoises.
chemiosmosis (adj. chemiosmotic) A means of releasing useful energy by permitting hydrogen ions to pass through a membrane down their concentration gradient.
chemoreceptor A sense organ that responds to chemical stimuli.
chemosynthesis Production of organic molecules from inorganic molecules, using energy released by chemical reactions rather than light energy, which is used in the much more common process of photosynthesis. Also called chemoautotrophy or chemolithotrophy.
chitin (adj. chitinous) A stiff polysaccharide, a prominent component of the cuticle of arthropods (e.g., insects) and of the cell walls of many fungi.
chiton A member of the molluscan class Amphineura, having a body covered with eight crosswise skeletal plates.
chlorophyll Green pigment that traps light energy during photosynthesis.
Chlorophyta A division containing the green algae.
chloroplast Green plastid (containing chlorophyll) in which photosynthesis occurs.
Chondrichthyes Class of vertebrates containing cartilaginous fish with jaws; the sharks, skates, and rays (also called elasmobranchs).
chordate Member of the phylum Chordata; animal with a notochord and pharyngeal gill slits at some stage of its life.
chorion The embryonic membrane lying immediately under the egg shell in reptiles and birds; in mammals it forms part of the placenta.
chromatid One of the two replicated strands of a chromosome still held together at the centromere.
chromatin Combination of DNA and proteins, visible as a loosely arranged mass in the nucleus of an appropriately stained eukaryotic cell during interphase.
chromatography Separation of a mixture of substances into its components by differences in their electrical charges and particle sizes.
chromoplast Plastid that produces and stores yellow and orange pigments.
chromosome (adj. chromosomal) Thread-like structure in the nucleus of a eukaryotic cell, consisting of DNA and proteins and carrying genetic information.
Chrysophyta Phylum of protists: diatoms and their relatives. Most are photosynthetic.
ciliate Member of the protistan phylum Ciliophora.
Ciliophora Phylum of heterotrophic protists that move by means of cilia and have two types of nuclei in their cells.
cilium (pl. cilia; adj. ciliary, ciliated) Thread-like organelle containing microtubules, present on the surfaces of many eukaryotic cells and used in movement.
circadian rhythm Cycle of about 24 hours in the physiology and behavior of a eukaryotic organism.
clade The set of all species descended from a particular ancestral species, together with their common ancestor.
cleavage A series of cell divisions that convert the zygote into a multicellular blastula.
cleidoic egg (= amniotic egg) "Closed" egg of reptiles, birds, and mammals.
cline A series of interconnected populations extending from one area to another, with gradual, continuous changes from one population to the next.
cloaca The vestibule, found in most vertebrates, into which urine, feces, and sperm or eggs are discharged before they leave the body.
clone A population of cells or individuals descended from one original cell or individual by asexual propagation, and hence genetically identical.

Cnidaria Phylum of simple animals with only two well-developed layers of cells, only one opening into the gastrovascular cavity, tentacles, and stinging nematocysts (e.g., jellyfish, *Hydra,* corals, sea anemones).

coccus (pl. cocci) Sphere-shaped bacterium.

codominance Condition in which both alleles are expressed in the phenotype of a heterozygote.

codon Series of three messenger RNA nucleotides which, in the genetic code, specifies a particular amino acid.

Coelenterata Old name of a phylum whose members are now placed in the phyla Cnidaria and Ctenophora.

coelom (adj. coelomic) A body cavity lined with mesodermal tissue, in which the internal organs are suspended.

coelomates Animals with bodies containing coeloms.

coenocyte Structure in fungi or plants in which many nuclei occupy a mass of cytoplasm, without being separated into individual cells by plasma membranes.

coenzyme Organic molecule that must be present for an enzyme to function.

coevolution Evolution together of two or more species whose members exert selective pressures on one another.

cofactor Ion or molecule that must be associated with an enzyme for the enzyme's proper functioning.

cohesion The sticking together of molecules of the same substance.

coleoptile The sheath covering the embryonic shoot in a monocot seed.

collagen A structural protein that forms fibers; common intercellular component of connective tissue.

colony (adj. colonial) In animals, bacteria, or protists, a group of more or less independent individuals attached to one another.

commensalism Close association between members of different species in which one member benefits and the effect on the other is neutral or unknown.

community All the organisms living in a particular habitat.

competitive exclusion Name given to the idea that two species with identical niches cannot exist together in the same place and time.

concentration The proportion of one substance found in the total of a mixture of several substances; may be given in terms of weight, of proportion of molecules, and so on. Concentration is symbolized by square brackets, as in [sugar] = concentration of sugar.

concentration gradient The change in concentration of a substance over a distance.

condensation reaction Chemical reaction in which two molecules become covalently bonded by removing —H from one and —OH from the other, with the removed atoms joining to form a molecule of water.

conifer Member of the plant division Coniferophyta: cone-bearing gymnosperms, including pines and spruces (as well as junipers and yews, whose reproductive structures do not resemble cones).

conjugation Method of genetic recombination in which one cell passes DNA to another via a physical link formed between the two.

connective tissue Tissue composed of scattered cells and much intercellular material secreted by the cells.

conservative Of a character that has changed very little during the course of evolution.

consumer Organism that eats other organisms.

convergent evolution Evolution of similar features by unrelated organisms in response to similar selective pressures.

copulation (= mating, coitus) Linking of sexual organs of male and female that permits transfer of sperm (in semen, spermatophores, etc.) from male to female.

coronary Pertaining to the heart.

corpus luteum "Yellow body" that forms in an egg follicle of an ovary after the follicle has ruptured and released an egg.

cortex (adj. cortical) Layer lying just inside the outermost boundary (epidermal or epithelial) layer of a stem, root, kidney, brain, etc.

cotyledon A seed leaf of a plant embryo.

courtship Behavior that precedes mating.

Crocodilia Order of reptiles including crocodiles, alligators, etc.

cross section View of an organism from a cut made perpendicular to the long axis of the body.

cross-fertilization Fertilization of one plant by sperm nuclei from another plant (not to be confused with cross-pollination).

cross-pollination Transfer of pollen from male flower parts of one plant to female flower parts of another.

crossing over In genetics, the process by which homologous chromosomes exchange pieces of DNA and form new assortments of genes.

crustacean Member of the arthropod class Crustacea: lobsters, shrimp, crabs, barnacles, pill bugs, copepods, ostracods, etc.

cubic centimetre (abbr. cc, cm^3) A volume equal to one milliliter.

cutaneous Of the skin.

cuticle Layer of waxy waterproof substance secreted on the outer surface of an organism.

cyanobacteria Prokaryotic organisms that use chlorophyll *a* in their photosynthesis and produce O_2 as a byproduct (also called blue-green bacteria or cyanophytes).

cycad Palm-like gymnosperm; member of the division Cycadophyta.

cyst 1. A dormant organism within a resistant covering; stage in which some organisms pass through adverse conditions. 2. Sac-like nonmalignant tumor.

cytochrome An electron carrier molecule consisting of a protein and a porphyrin ring containing a metal ion.

cytokinesis Division of a eukaryotic cell in two, following nuclear division.

cytokinins Plant hormones that stimulate cell division.

cytoplasm The fluid that makes up all of a cell except its nucleus and organelles.

cytoplasmic streaming Flow of cytoplasm within a cell or between adjacent cells.

cytoskeleton The "skeleton" of a cell, composed of thin tubules and contractile filaments in the cytoplasm; much of the cytoskeleton can be assembled from subunits and disassembled as needed.

cytosol The soluble portion of cytoplasm.

cytotoxic T cell Lymphocyte that destroys a cell displaying both self and foreign antigens on its surface.

-d-

deamination Removal of an amino (—NH_2) group.

deciduous Of plants that lose their leaves during one season of the year, not evergreen.

decomposer In ecology, an organism that feeds on the dead bodies, body parts, or wastes of other organisms, thereby breaking down and recycling the nutrients they contain.

dendrite Extension of a neuron that receives stimuli or impulses from other neurons.

depolarization Decrease in the electrical potential difference across a membrane.

desiccation Drying out.

desmosome Attachment between two cells in which the cells' membranes are fastened together in areas of high stress.

detritus Molecules and larger particles of dead organic matter.

Deuteromycota Division of fungi whose members do not reproduce sexually.

diaphragm 1. In mammals, a muscular partition between the thorax and abdomen, used in breathing. 2. A disk-like rubber device inserted into the vagina to bar sperm from reaching an egg, used for birth control. 3. A disk with a central opening that controls the amount of light entering a microscope.

dichotomous Forking into two.
dicotyledon (abbr. dicot) Member of the group of flowering plants whose embryos have two cotyledons.
dikaryon Fungus or part of fungus in which each cell contains two haploid nuclei.
dinoflagellate One-celled organism with two flagella belonging to the phylum Pyrrophyta; most are photosynthetic.
diploid Containing twice the number of chromosomes found in a gamete; having paired homologous chromosomes.
disaccharide A molecule formed by a condensation reaction between two simple sugars (monosaccharides).
distal In a position or direction away from the point of an appendage's attachment to the body.
disulfide bond In protein structure, a covalent bond between sulfur atoms that are parts of two different cysteine monomers; the bond attaches parts of the same or different polypeptides to each other.
diurnal 1. Active during the daytime. 2. Daily.
division A taxon of plants or fungi equivalent to a phylum in the animal kingdom.
dizygotic twins Twins arising from two different zygotes (fertilized eggs); nonidentical twins.
DNA Deoxyribonucleic acid, the genetic material of organisms and of many viruses.
dominant allele Allele expressed in the heterozygote.
dominant generation Larger, more conspicuous stage in the life history of a species with alternation of generations.
dorsal Toward the back, or uppermost surface, of an animal.

-e-

echinoderm ("spiny-skinned") Member of the invertebrate phylum Echinodermata, e.g., sea stars, sea urchins, sand dollars, sea cucumbers.
ecology Study of the relationships of organisms with other organisms and with their physical environment.
ecosystem All of the organisms present in a particular area, together with their physical environment.
ecto- Prefix meaning external or outermost.
ectoderm Outermost of the three germ layers of the embryonic gastrula, giving rise to skin, nervous system, and associated structures.
ectothermy Regulation of body temperature by behavioral means (obtaining heat from outside the body).
EEG Electroencephalogram; recording of electrical activity in the brain.
effectors Structures that carry out an animal's response to a stimulus; e.g., muscles, glands.
EKG Electrocardiogram; recording of electrical activity in heart muscle.
elasmobranchs Cartilaginous fish with jaws; the sharks, skates, and rays. Another name for Chondrichthyes.
electrical potential Difference in concentration of electrically charged particles on two sides of a membrane.
electron Fundamental negatively charged particle occupying space outside the nucleus of an atom.
electron micrograph Photograph taken using an electron microscope.
electron transport chain (= electron transport system) A series of proteins found in certain biological membranes that splits hydrogen atoms into electrons and hydrogen ions, and removes the electrons.
electrophoresis A technique that separates substances, using the fact that they move at different rates (depending on their size and electrical charge) when subjected to an electric current.
embryo A multicellular developing plant or animal still enclosed inside the parent's body or in a seed or egg.
embryophyte A plant in which the zygote remains within the parent plant as it develops into an embryo; a land plant.
emigration Leaving an area of residence for some other place.
endemic (adj. sometimes used as a noun) Peculiar to a particular population or locality, where it originated.
endergonic reaction Reaction in which the energy of the products is greater than the energy of the reactants.
endo-, end- Prefix meaning inside or within.
endocrine gland A gland whose hormone secretion enters the body fluids directly rather than being transported to the site of action through a duct.
endocytosis Engulfing of a particle by a cell.
endoderm The innermost of the three germ layers in the early embryo of higher animals; gives rise to lining of the digestive tract.
endodermis Layer of cells between the cortex and the pericycle in a root.
endoplasmic reticulum System of membranous sacs and tunnels found in most eukaryotic cells.
endoskeleton Internal skeleton.
endosperm Nutritive tissue in an angiosperm seed.
endothermy Maintenance of a particular body temperature by physiological regulation of the loss of metabolically generated heat.
energy The capacity to do work.
entomology The study of insects.
environment An organism's physical and biological surroundings (often includes conditions within the organism's own body).
enzyme Protein that catalyzes a particular biochemical reaction involving specific substrate (reactant) molecules.
epi- Prefix meaning on, above, or around.
epidermis Layer of cells covering the outside of the body.
epinephrine A hormone released by the medulla of the adrenal glands; one of its effects is to bring about the physiological changes associated with stress. Also called adrenalin.
epiphyte Plant that grows on the surface of larger plants, which are not harmed by having the smaller plant growing there; e.g., most orchids.
epithelial tissue Animal tissue that forms a sheet covering an external or internal surface.
equator An imaginary circle round the Earth's surface formed by the intersection of a plane passing through the Earth's center perpendicular to its axis of rotation.
equilibrium 1. Point in a chemical reaction at which the rates of forward and reverse reactions are equal and energy change during the reaction is zero. 2. Sense of balance.
erythrocyte Red blood cell.
essential amino acid An amino acid that an animal must obtain in its diet because it cannot make enough of the molecule to survive.
estivation (also aestivation) Period of dormancy during the summer.
estrus (adj. estrous) Sexually receptive state of female mammals; usually called rut in males.
Eubacteria ("true bacteria") Prokaryote group containing most bacteria, including actinomycetes, mycoplasmas, spirochetes, and cyanobacteria. (Archaeobacteria are the only other prokaryotes.)
Euglenophyta Phylum of photosynthetic protists containing *Euglena* and its relatives.
eukaryotic Having a nuclear membrane surrounding the genetic material, and with other membrane-bounded organelles in the cytoplasm.
eutrophic Of a body of fresh water rich in nutrients and hence in living organisms.
eutrophication Process in which debris accumulates in a body of water, making it richer in nutrients and hence in organisms, until eventually it fills in and becomes dry land.
evaporative cooling Reduction of temperature by escape of the fastest-moving

(that is, warmest) water molecules as water vapor.

evolution 1. Descent of modern species of organisms from related, but different, species that lived in previous times. 2. Change in the gene pool of a population from generation to generation.

ex-, exo- Prefix meaning outside or proceeding from.

excitatory Of nerve or muscle, tending to cause depolarization of the postsynaptic membrane, enhancing the chances that the postsynaptic cell will fire an action potential.

exergonic reaction Reaction that releases energy, that is, one in which the energy of the products is less than the energy of the reactants.

exocytosis Method of expelling substances contained in a membrane-bounded vesicle to the exterior of the cell by fusion of the vesicle membrane with the plasma membrane.

exon The part of a structural gene that is translated into protein (as opposed to an intron, which is not).

exoskeleton External skeleton, outside the rest of the body.

exothermic Heat-releasing.

extinction (of a species) Disappearance of a species from Earth when its last surviving member dies.

extracellular Outside a cell.

extracellular fluid (ECF) Fluid surrounding cells.

-f-

FAD Flavin adenine dinucleotide, a hydrogen carrier molecule; a coenzyme in cellular respiration.

Fallopian tube Oviduct, in mammals.

fat A lipid made up of three fatty acids and glycerol; solid at room temperature.

fauna Animal life of an area.

fermentation Anaerobic breakdown of food molecules to release energy, in which the final electron acceptors (and end products) are organic molecules.

fertility 1. Of a female, the ability to reproduce. 2. Of soil, the ability to supply plants with the nutrients they need. 3. (rate) Of a population, number of offspring produced per female per unit time.

fertilization Union of an egg with a sperm.

filter feeders Animals that eat smaller organisms or particles of organic matter, which they strain out of the surrounding water.

filtrate Substance that has passed through a filter.

fission Division into (usually two) parts.

fitness In population genetics, a measure of the ability of an organism to produce surviving offspring.

fix 1. In chemistry, to incorporate into a less volatile compound. 2. In genetics, to establish one allele in place of all alternate alleles in a population.

flagellate 1. (noun) Protist that moves by means of one or more flagella. 2. (adj.) Bearing one or more flagella.

flagellum (pl. flagella) Long thin projection from a cell, which moves and propels the cell.

flatworm Member of the invertebrate phylum Platyhelminthes; e.g., planarians, flukes, tapeworms.

flavin adenine dinucleotide See FAD.

flora Plant life of an area (often includes bacteria and fungi, reflecting the old two-kingdom system).

fossil fuel A fuel created by decomposition and geological processes from the remains of dead organisms; includes oil, coal, natural gas, and peat.

fossils Remains of organisms, or other evidence of once-living organisms, preserved in rocks.

frameshift mutation Mutation that adds or deletes nucleotides, thereby altering the reading frame of the genetic message.

free energy Energy that is available or usable to do work.

frond A leaf, usually highly divided (usually applied to ferns or palms).

fruit Structure that develops from the ovary of a flower, surrounding one or more seeds.

fruiting body Rather large, prominent reproductive structure formed by some fungi.

fugitive species In ecology, species that occur in an area for only a short time, during succession.

Fungi Kingdom of organisms containing eukaryotic, unicellular (yeasts) or multicellular heterotrophs feeding by absorbing nutrients through their cell walls.

-g-

gamete Sexual reproductive cell: egg or sperm.

gametophyte Haploid plant that produces haploid gametes by mitosis.

ganglion (pl. ganglia) A group of neuron cell bodies.

gap junction An area where ions can move directly between the cytoplasm of two animal cells by way of many tiny tunnels through their plasma membranes.

gastrocoel The hollow in the embryonic gastrula that becomes the lumen of the gut.

gastropod Member of the molluscan class Gastropoda, including snails, slugs, nudibranchs, and limpets.

gastrovascular cavity A cavity in the body that serves for both digestion and distribution of food.

gated channel Channel through a membrane that can be opened or closed by chemical or electrical events.

gene A length of DNA that functions as a unit.

gene bank An institution where plant material is stored in a viable condition. Usually seeds are dried and frozen in a sealed container. Some plants have seeds that will not survive this treatment and they must be maintained as growing plants or in tissue culture.

gene flow Transfer of genes between one more-or-less isolated population and another.

gene pool All of the genes present in a population of organisms.

genera Plural of genus.

generation time Time elapsed from production (birth) of a new individual to production of its first offspring.

genetic drift Changes in the gene pool of a population caused by random events (as opposed to natural selection).

genetic engineering The isolation of useful genes from a donor organism or tissue and their incorporation into an organism that does not normally possess them.

genetic reassortment Production of new gene combinations in sexually reproducing organisms by crossing over and independent assortment during meiosis and by combination of gametes at fertilization.

genetic recombination Formation of new combinations of genes by joining DNA from two different molecules into one.

genital Of the reproductive system.

genome All the genetic material contained by an individual (or by a representative member of a population).

genotype The particular genes present in an individual, some of which may not be expressed in the phenotype; usually refers to only one or a few gene pairs.

genus (pl. genera; adj. generic) The taxon above species in the hierarchical classification of organisms. The genus name is the first word of the latin binomial for a species; e.g., *Ursus* is the generic name of the grizzly bear, *Ursus arctos*.

germ cells Cells that give rise to reproductive cells (eggs, sperm, spores).

germ layers The three layers of cells in the embryonic gastrula (ectoderm, mesoderm, endoderm).

gestation Period during which an embryo is carried within the mother's body before birth.

gibberellins Plant hormones that stimulate cell enlargement.

gills Thin extensions of the body surfaces of many aquatic animals, used for gas exchange and/or feeding.

Ginkgophyta Gymnosperm division containing *Ginkgo* trees.
gizzard Part of the stomach or gut modified as a heavy-walled grinding chamber.
glaciation (= Ice Age) One of four cold periods during the Pleistocene era when ice and glaciers extended farther south from the North Pole than they do now.
glucose A six-carbon monosaccharide.
glycogen A storage polysaccharide made from glucose monomers, commonly found in animals.
glycolipid Molecule made up of carbohydrate and lipid subunits.
glycolysis Biochemical pathway found in most organisms, in which glucose is broken down into pyruvate and hydrogen atoms, with the storage of energy in ATP.
glycoprotein Protein with carbohydrate attached.
Golgi complex Stack of membrane-enclosed sacs that modify and package molecules produced in a eukaryotic cell.
gonad In animals, an organ that produces gametes.
grafting A procedure in which a tissue or organ of one plant or animal is attached to and incorporated into the body of another.
grain 1. The fruit of members of the monocotyledonous plant family Gramineae from which most human food comes. 2. A cluster of molecules of starch.
gram molecular weight See mole.
gray matter Nervous tissue made up of unmyelinated neuron cell bodies and extensions.
green algae Members of the division Chlorophyta: plants with unicellular or multicellular body plans, using chlorophylls *a* and *b* for photosynthesis and storing food as starch.
green plants Photosynthetic organisms that give off oxygen during their photosynthesis, including prokaryotic cyanobacteria and all photosynthetic eukaryotes (protists and plants).
greenhouse effect Heating of the Earth caused by gases in the atmosphere that trap infrared radiation from the Earth and prevent it's escaping into space.
guard cells Two cells surrounding a stoma (pore) in the epidermis of a plant that are able to regulate the pore's opening and closing.
gustation Tasting.
guttation Exudation of water from tips of xylem veinlets in leaves, due to root pressure.
gymnosperm Nonflowering plant that produces seeds, e.g., pines, redwoods, cycads.

-h-

habitat The physical area where an organism lives.
haploid Containing the number of (unpaired) chromosomes found in a gamete; equal to half the number of chromosomes found in a body cell of most higher plants and animals.
haustorium (pl. haustoria) An extension of a fungus inside the cell wall of a living plant cell; the fungus lies against the plant cell's plasma membrane and absorbs nutrients from the plant cell.
hectare 10,000 square meters, or about 2.5 acres.
heme Iron-containing group in hemoglobin and some cytochromes.
hemocoel Body cavity containing a fluid that acts as the transport medium in many invertebrates.
hemoglobin Respiratory pigment that carries oxygen in the blood of vertebrates and various invertebrates.
hemolymph Fluid that fills the body cavity and acts as blood in animals with open circulatory systems.
hemolysis The bursting of blood cells.
hemophilia Condition in which blood fails to clot, resulting in excessive bleeding, caused by a defect in a gene for clotting factor.
hemorrhage Bleeding, e.g., from a ruptured blood vessel.
herbaceous Having nonwoody stems.
herbicide A chemical used to kill plants.
herbivore An animal that eats plants or parts of plants.
hermaphroditic Containing both male and female organs.
hetero- Prefix meaning "other" or "different."
heterospory In plants, production of spores of two distinct sizes, which give rise to gametophytes of the two sexes: microspores to male gametophytes, megaspores to female.
heterotroph (adj. heterotrophic) Organism dependent on other organisms to produce its organic (food) molecules.
heterozygote (adj. heterozygous) Individual with two different alleles in a given gene pair.
heterozygote advantage (= heterosis) Selective advantage of a heterozygote over either homozygote.
histone Type of protein with an alkaline pH, characteristic of eukaryotic chromosomes and playing a role in the packing of DNA and regulation of transcription.
homeostasis (adj. homeostatic) The maintenance of conditions inside the body within the narrow limits required for life.
homeothermy Maintenance of a constant, high body temperature and metabolic rate.
hominids Humans and their direct ancestors: members of the primate family Hominidae, with large brains, small teeth, and bipedal locomotion.
hominoids Humans and apes: large tail-less primates.
homologous 1. Of chromosomes, of the same origin and containing the same kinds of genetic information. 2. Of structures or organs, originating from the same structure in ancestral forms (e.g., a bird's wing and a seal's front flipper are homologous structures).
homozygote (adj. homozygous) Individual having same allele in both members of a given gene pair.
homozygous recessive Having two copies of a recessive allele in a given gene pair.
hormone Chemical messenger produced in one part of the body and specifically influencing certain activities of cells in another part of the body.
hunter-gatherers Members of human society in which food is obtained by collecting plants and hunting wild animals.
hybrid Offspring of a mating between genetically different individuals.
hydrogen bond Weak link between two molecules, or two parts of the same macromolecule, due to the attraction of a hydrogen atom with a partial positive charge to an oxygen or nitrogen atom with a partial negative charge.
hydrolysis The breaking apart of a molecule into its monomer subunits by addition of the components of a water molecule into each of the covalent bonds linking the monomers.
hydrophilic "Water-loving"; able to dissolve in water; polar or ionic.
hydrophobic "Water-hating"; unable to dissolve in water; nonpolar.
hydrostatic pressure Pressure exerted by confined fluid.
Hymenoptera The order of insects that includes bees, wasps, ants, etc.
hyperpolarization Increase in the membrane potential.
hypertonic Of a solution, tending to gain water from a solution to which it is being compared, when the two are separated by a membrane; usually this means having a higher concentration of dissolved particles (lower osmotic potential) than the other solution.
hypha (pl. hyphae) One of the thread-like structures that make up the body of a fungus.
hypothalamus A part of the brain, responsible for monitoring internal conditions in the body and initiating behaviors that tend to maintain physiological homeostasis.
hypothesis (pl. hypotheses) A proposed answer to a question.
hypotonic Of a solution, tending to lose water to a solution to which it is being compared, when the two are separated by a membrane; usually this means

having a lower concentration of dissolved particles (higher osmotic potential) than the other solution.

-i-

immigration Movement of new individuals into an area.

immunoglobulin See antibody.

immunology Study of the reactions that provide protection specifically against individual diseases.

inbreeding Mating of related individuals that share many of the same alleles, and therefore produce offspring with a high degree of homozygosity.

incisors Chisel-shaped cutting teeth found in the center of the lower and upper jaws in mammals.

incomplete dominance Condition in which neither of a pair of alleles hides the presence of the other in the phenotype.

independent assortment Creation of new haploid combinations of chromosomes due to random lining up of members of homologous chromosome pairs on either side of the spindle equator during meiosis.

inducer molecule Molecule (often a food molecule) that causes the genetic machinery of a cell to transcribe the genes for and to produce particular metabolic enzymes.

induction 1. The initiation of protein synthesis because a particular inducer substance is present. 2. In development, conversion of one type of tissue into another as a result of the tissue's contact with a particular stimulus.

infant mortality rate The death rate for humans in the first year of life.

ingest To take into the body through the mouth.

innervate To supply nervous connections.

insect Arthropod with three distinct body areas (head, thorax, abdomen); adults with three pairs of legs, and usually two pairs of wings, attached to the thorax.

insectivore (adj. insectivorous) Insect-eating.

insulin A small protein hormone, one of whose functions is to regulate cellular uptake of glucose from the blood; it is produced by the pancreas.

interferon Protein produced by mammalian cells that interferes with replication of viruses.

intermediary metabolism Total of all of a cell's enzyme-mediated reactions involved with extracting energy from food molecules and using it to synthesize the cell's own molecules.

interneuron Type of neuron in the central nervous system that is neither sensory nor motor, but transmits information between other neurons.

internode Portion of a stem between sites of leaf attachment.

intertidal Between tidemarks; covered by water at high tide and exposed to the air at low tide.

intervening sequence See intron.

intracellular Inside a cell.

intraspecific Within one species.

intron Part of a structural gene that is transcribed into messenger RNA but not translated into protein.

invertebrate An animal that lacks a backbone; e.g., earthworm, snail.

ion (adj. ionic) Particle carrying one or more positive or negative electrical charges.

iso- Prefix meaning "same."

isomers Molecules containing identical numbers and types of atoms, but with these atoms arranged differently.

isotonic Of a solution, having osmotic properties such that it neither gains nor loses water across a membrane separating it from a solution to which it is being compared.

isotopes Two or more forms of an element, differing by the numbers of neutrons in the nuclei of their atoms.

-k-

K 1. Chemical symbol for potassium, which usually occurs as an ion (K^+). 2. Symbol for carrying capacity.

kelps Large marine brown algae with considerable differentiation of tissues.

keratin A structural protein that makes up hair and fingernails.

killer cells Poorly understood lymphocyte-like immune cells, which are apparently specialized to destroy abnormal body cells.

kin selection Natural selection that causes an individual to act in such a way as to enhance the survival or reproduction of other, genetically related individuals (kin).

kinetochore Structure on a chromosome to which microtubules attach; usually located on or near the centromere.

krill Marine planktonic arthropods, up to about 15 centimeters in length.

-l-

labelled Prepared in such a way that it can be traced; e.g. containing a high proportion of a rare isotope allowing the fate of the atoms to be traced by methods that can distinguish between the isotopes of an element.

lactation Secretion of milk from mammary glands.

larva (pl. larvae) Immature stage of an animal with different appearance and way of life from the adult.

larynx Voice box ("Adam's apple") located at the top of the trachea (windpipe).

lateral 1. (adj.) Pertaining to the side. 2. (noun) A side branch or branch root of a plant.

lateral line organ Pressure-sensitive sense organ found in fish and larval amphibians, used to detect water currents, etc.

laterite A hard crust that may develop when vegetation is removed from the surface of soil containing metals such as aluminum and iron in tropical regions with wet and dry seasons. In the dry season, soil solution rises to the surface by capillarity, and aluminum and iron oxides accumulate and combine to form the crust. Laterization results in an infertile soil called latosol.

latitude Distance north or south of the equator.

leach To dissolve and carry out of; leached soil has lost mineral nutrients that have dissolved in rainwater running through the soil and have been carried away into streams.

leaf nodes Areas of stem at which leaves are attached.

legumes Members of the plant family Leguminosae, including beans, peas, peanuts, alfalfa, clover, *Acacia*.

Lepidoptera The order of insects that includes moths and butterflies.

lethal allele Allele whose expression causes premature death of the individual that carries it (usually, in the homozygous recessive condition).

leucoplast Colorless plastid.

leukocyte White blood cell.

life expectancy The number of years a particular person can expect to live, calculated from actuarial statistics.

linkage group All of the genes on the same chromosome, which are therefore all inherited together (except for crossing over).

lipid One of a large class of organic molecules not soluble in water; includes fats, waxes, oils, steroids, carotenes.

lipopolysaccharide Molecule composed of lipid joined to polysaccharide.

lipoprotein Molecule made up of lipid and polypeptide subunits.

locus (pl. loci) The position on a chromosome that is occupied by an allele for a particular gene; e.g., the hemoglobin beta chain locus may be occupied by an allele for normal or sickle hemoglobin. (This book uses the less technical term, gene location.)

lumen Space in the center of a tube.

Lycophyta A division of primitive vascular plants: club mosses and ground pines.

lymphocyte (= T cell or B cell) One of a group of white blood cells responsible for the specificity of immune responses: production of antibodies, recognition of antigens, and immunological memory.

lysis (adj. lytic) Bursting of a cell.

lysosome Membrane-bounded organelle filled with hydrolytic (digestive) enzymes.

-m-

macromolecule Large molecule. Usually refers to a polymer with a molecular weight of many thousands, made up of many (identical or different) monomers.

macronutrient Nutrient needed in relatively large amounts

macroscopic Visible to the unaided eye.

maize *Zea mays,* corn, important cereal crop.

major histocompatibility complex See MHC.

mammal Warm-blooded vertebrate with lower jaw consisting of only one bone (the mandible) on each side, with fur or hair, with young nourished by milk from the mammary glands of the female parent; e.g., humans, rabbits, cattle.

mantle Sheet of tissue covering the visceral mass of a mollusc and secreting the shell, if present.

map In genetics, the plan of a genome showing the relative positions of particular genes on the organism's chromosomes.

marine Of the sea.

marsupials Mammals whose young are born quite early in development and complete their development attached to a nipple in the mother's marsupium, or pouch.

mating types The equivalent of sexes in fungi, some bacteria, and protists.

medusa (pl. medusae) One of the two possible forms of the cnidarian body (the other is the polyp); often the reproductive stage in the life history; for example, the body of a jellyfish is a medusa.

megaspores Large spores produced by meiosis in some plants and giving rise to female gametophytes.

meiosis A series of nuclear divisions that produces four new nuclei, each with half the number of chromosomes contained in the original nucleus.

melanin Black pigment that gives dark color to organisms.

melanism Dark coloration.

membrane potential The difference in electrical charge across a membrane; the inside of a cell is negatively charged with respect to the outside of the cell so the membrane usually has a potential of about -40 millivolts (about -70 millivolts for a neuron).

meristem (adj. meristematic) Region of dividing cells in a growing area of a plant.

mesentery A sheet of mesodermal tissue that suspends the internal organs in the coelom.

mesoderm The middle one of the three embryonic germ layers of the gastrula; gives rise to most of the muscles, heart, kidneys, gonads, etc.

mesoglea The "middle glue" layer, containing few, scattered cells, between the outer and inner layers of a cnidarian.

messenger RNA The molecule that carries genetic information from DNA to ribosomes, where the information is used as a code to direct the order in which amino acids are joined to form a polypeptide.

metabolic heat Heat released by chemical reactions in the body.

metabolic rate The rate of the total of an organism's biochemical reactions; usually measured as the rate of oxygen consumption by respiration, since respiration produces the energy needed for the other biochemical processes.

metabolism All the chemical reactions taking place within an organism.

metamorphosis The radical change in shape, physiology, and behavior that occurs when a larva becomes a very different-looking adult.

methane Gas (CH_4) produced by the metabolic processes of anaerobic methanogen bacteria.

mg (milligram) A thousandth of a gram.

MHC (major histocompatibility complex) The group of genes that determines many mammalian cell surface antigens, including those that cause rejection of transplanted organs.

microfilament Filament assembled from actin subunits, found near the plasma membrane; part of the cytoskeleton; responsible for much of the movement within a eukaryotic cell.

micrograph Photograph taken using a microscope.

micrometer (mμ) 10^{-3} mm $= 10^{-6}$ m

micron Outmoded name for micrometer.

micronutrient Nutrient needed by an organism in relatively small amounts.

microorganisms Unicellular or simple many-celled organisms: bacteria, fungi, protists, or small algae.

microspores Small spores produced by meiosis in some plants and giving rise to male gametophytes.

microtubule Thickest type of filament in the cytoskeleton of a eukaryotic cell; present in eukaryotic centrioles and in cilia, flagella, and their basal bodies, and in mitotic and meiotic spindles; assembled from protein subunits.

microvillus Tiny, finger-like projection that increases the surface area of a cell.

middle lamella The shared partition between the cell walls of adjacent plant cells.

mimicry Resemblance of one organism to another, or to a nonliving object, providing an offensive or defensive advantage to the mimic.

mineral Any naturally occurring inorganic substance having a definite chemical composition and a particular crystalline structure, color, and hardness.

mites Small arthropods in the class Arachnida. Mites have eight legs, and the body is not divided into two parts as it is in other arachnids, such as spiders.

mitochondrion (pl. mitochondria) Large, self-replicating membrane-bounded organelle where most of a eukaryotic cell's ATP is produced.

mitosis (adj. mitotic) Series of events that results in the division of one cell nucleus into two nuclei containing sets of chromosomes identical to that in the original nucleus.

mole Gram molecular weight; for example, 1 mole of water = 18 grams because the molecular weight of H_2O is 18.

molecular biology Study of the molecular basis of inheritance.

molecular weight Sum of the atomic weights of the atoms in a compound.

Mollusca Phylum of soft-bodied invertebrate animals with a muscular head-foot and a mantle, which usually secretes a shell; e.g., snails, clams, squids.

molting Shedding of skin, exoskeleton, or feathers.

monocotyledon (abbr. monocot) Member of the group of flowering plants whose embryos have only one cotyledon.

monogamy The mating of one male with one female, either for life or for the duration of one breeding season.

monomers Small molecules that may become joined together to form large (macro) molecules; for example, amino acids are the monomers that make up polypeptides.

monophyletic Of a clade, i.e., a taxon that contains a common ancestor and all the species descended from it.

monosaccharide Simple sugar, with formula given by $(CH_2O)_n$; e.g., glucose, ribose.

monotremes Egg-laying mammals.

monozygotic twins Twins originating from the same fertilized egg; identical (genetically) twins.

morph Form, variety.

morphogenesis (adj. morphogenetic) The formation of shape or structures during development.

morphology (adj. morphological) Structure, anatomy.

mortality Death.

motile Able to move itself from place to place.

multicellular Composed of more than one cell.

mutagen Agent (e.g., chemicals, certain kinds of radiation) that causes mutation.
mutation Inheritable change in the genetic material (DNA).
mutualism Close association between members of different species that benefits both.
mycelium Tangled mass of fungal filaments that make the body of a fungus.
mycology Study of fungi.
mycorrhiza (pl. mycorrhizae) Mutualistic association between a fungus and the roots of a higher plant; the fungus takes up mineral nutrients from the soil and passes them to the plant, receiving some organic (food) molecules made by the plant in return.
myelin sheath Layers of fatty, insulating wrapping around the axons of some neurons.
myoglobin Oxygen-storing molecule in muscles.
myosin Protein that interacts with actin to produce movement in cells, such as contraction in muscle cells.
myotome A block of muscle that forms one of a series of such blocks along the back of an animal.
Myxomycota Acellular slime molds.

-n-

NAD (nicotinamide adenine dinucleotide) A hydrogen-carrying coenzyme in cellular respiration.
nanometer (abbr. nm) 10^{-9} meter.
natural selection Differential reproduction among the variety of genotypes in a population; leads to different genetic contributions to future generations.
negative feedback Mechanism whereby the change detected in some condition stimulates compensating physiological activity that brings the condition back toward its average value.
nematocyst Stinging structure characteristic of cnidarians.
nematode A roundworm, member of the phylum Nematoda.
neuron (adj. neuronal) Nerve cell.
neurotransmitter Chemical that transmits information in the nervous system; it travels across the synaptic cleft from a neuron that has just fired an action potential to another cell, whose electrical or chemical activity it affects.
niche The way of life of a species; includes the habitat, food, nest sites, etc., that it needs in order to survive.
nitrogen fixation Conversion of gaseous nitrogen (N_2) to ammonia (NH_3).
nocturnal Active at night.
nomadic Moving from place to place, with no single fixed home.
nondisjunction Failure of homologous chromosomes or of sister chromatids to separate; if this occurs during meiosis, one of the resulting nuclei will have an extra copy of the chromosome, while another will have one chromosome too few.
notochord Elastic rod dorsal to the gut in all chordate embryos; in most adult chordates (i.e., vertebrates), the notochord is replaced by vertebrae, which form around it.
nucleic acids Class of macromolecules, made up of nucleotide monomers, that contain the genetic information of organisms; DNA and RNA.
nucleolus (pl. nucleoli) Area of the cell nucleus where the nucleolar organizer (part of one or two particular chromosomes) lies and where ribosomes are made; often appears denser than the rest of the nucleus in micrographs.
nucleosome Particle-like structure formed by part of a chromosome in which DNA is wound around a cluster of histones.
nucleotide Monomer unit that makes up nucleic acids; consists of a single- or double-ring nitrogenous base, a pentose sugar (ribose or deoxyribose), and one to three phosphate groups.
nucleus (pl. nuclei) 1. That part of a eukaryotic cell surrounded by the nuclear envelope and containing the genetic material. 2. The more-or-less central part of an atom, consisting of one or more protons and (except in most hydrogen atoms) neutrons. 3. A cluster of neuron cell bodies in the brain.
nutrient Any chemical that an organism must take in from its environment because it cannot produce it (or cannot produce it as fast as it needs it).

-o-

olfactory Pertaining to the sense of smell.
oligotrophic Of a body of fresh water that contains few nutrients and few organisms.
omnivore Animal that eats both plants and animals.
oncogene A gene capable of taking part in the transformation of a normal cell into a cancerous cell.
oncogenic Of something that causes cancer.
oocyte Cell that undergoes meiosis during oogenesis.
oogamy Form of sexual reproduction involving an egg, which is large and nonmotile, and a sperm, which is small and motile.
oogenesis Formation of eggs or ova, the female gametes.
Oomycota Division of fungi that produce egg-like structures during sexual reproduction; "water molds."
operculum Common covering of all the gills on each side in bony fish.
operon A cluster of genes with related functions: usually one or more structural genes and the regulatory genes that control their activity.
organ Group of tissues assembled in such a way that the entire structure (organ) performs a particular function; e.g. liver, kidney, heart.
organelle Structure within a cell that takes part in carrying out the cell's life functions.
organism An individual living thing, made up of one or more cells.
osmoregulation Regulation of the body's salt and water content.
osmosis Movement of water through a membrane down a water potential gradient.
osmotic potential Tendency of a solution to gain water when separated from pure water by an ideal selectively permeable membrane; the negative of osmotic pressure.
osmotic pressure Pressure that must be exerted on a solution to keep it from gaining water when separated from pure water by an ideal selectively permeable membrane.
Osteichthyes Class of vertebrates containing the bony fishes.
oviduct Tube through which eggs pass from the ovary toward the exterior of the body.
oviparous Reproducing by laying eggs which develop outside the female's body.
oviposition Laying of eggs.
ovoviviparous Retaining eggs within the mother's body until hatching; e.g., most sharks and some reptiles.
ovule Structure inside the ovary of a flower enclosing a female gametophyte.
ovum (pl. ova) Female gamete, egg.
oxidation A chemical reaction involving removal of electrons or hydrogen atoms, or addition of oxygen; always paired with a reduction.
ozone O_3. A poisonous gas, a common pollutant in smog; also formed by the action of sunlight on oxygen in the ozone layer of the atmosphere.
ozone layer Layer in the upper stratosphere where solar radiation converts some O_2 atoms into ozone molecules. The ozone layer absorbs much ultraviolet radiation and prevents it from reaching the Earth.

-p-

pancreas In vertebrates, an organ near the stomach that produces hormones and digestive enzymes.
parasite Organism that feeds on another living organism without killing it.
parasympathetic nervous system Part of the autonomic nervous system; its activi-

ties promote digestion and elimination, slowing of heart rate, etc.

parthenogenesis Production of young from unfertilized eggs.

pathogenic Disease-causing.

pectoral Pertaining to the anterior fins or limbs of vertebrates.

pelvic Pertaining to the rear pair of appendages (limbs or fins) of vertebrates.

peptide bond Covalent bond joining the carboxyl carbon of one amino acid to the amino nitrogen of the next.

perennial Living for many years and surviving normal seasonal changes.

peripheral nervous system The part of the nervous system outside the brain and spinal cord; it consists of nerves to and from the muscles, sense organs, and internal organs.

permafrost Permanently frozen layer in the soil, found in arctic and antarctic regions.

pest Any organism that is undesirable at the time and in the place where it exists. Includes plant-eating insects, molluscs, and nematodes on crop plants, parasites on animals, weeds in fields of crops.

pesticide Substance used to kill undesirable organisms; includes insecticides, herbicides, nematocides, etc.

pH Logarithm to the base 10 of the hydrogen ion concentration of a solution; a measure of how acidic or basic a solution is, on a scale of 0 to 14 (0 = very acidic, 14 = very basic, 7 = neutral).

Phaeophyta Taxonomic division containing the brown algae.

phage See bacteriophage.

phagocyte Scavenger cell that ingests and destroys pathogens, dead body cells, and other debris.

pharynx Part of the gut just behind the mouth in many animals.

phenotype The characteristic produced by expression of an organism's genes.

pheromone Chemical released by one member of a species that influences the behavior of another member of the species.

phloem Tissue in plants that conducts food from sites of synthesis or storage to sites where food is used or stored.

phospholipids Group of structural lipids that are the main components of biological membranes; made up of fatty acid(s), phosphate, and (usually) nitrogen-containing choline.

phosphorylation Addition of a phosphate group.

photochemical reaction Chemical reaction powered by light energy.

photoperiodism Production of a physiological response (such as flowering or coming into breeding condition) in response to length of period of daylight each day.

photosynthesis Process whereby plants, and some bacteria and protists, capture solar energy and store it as chemical bonds in carbohydrate molecules, using CO_2 to build the carbohydrate.

photosystem Cluster of pigment molecules in which light-absorbing reactions of photosynthesis occur.

phototropism Growth toward or away from light.

phycobilins Accessory photosynthetic pigments found in Cyanobacteria and Rhodophyta; major ones are phycocyanin and phycoerythrin.

phycology Study of algae.

phylogeny (adj. phylogenetic) Line of evolutionary descent.

physiology The processes by which an organism carries out its various biological functions; how an organism works.

phytoplankton Plants floating in the upper layers of a body of water.

P_i Abbreviation for an inorganic phosphate group.

pigment Molecule that differentially absorbs particular wavelengths of visible light and so appears colored.

placenta Organ in mammals in which blood capillaries from mother and fetus lie close together and exchange substances via the extracellular fluid between the two bloodstreams.

placentals Mammals that undergo part of their development in the mother's uterus, where they receive food and oxygen via the placenta, until a fairly advanced stage of development.

plankton Organisms that drift around in water because they are not capable of swimming against currents in the water.

plasma membrane Phospholipid and protein membrane surrounding a cell. Also called plasmalemma, cell membrane.

plasmid Small, circular DNA molecule, found in some bacteria and fungi in addition to the organism's own genome.

plasmodesma (pl. plasmodesmata) Strand of cytoplasm that directly links the cytoplasms of two neighboring plant cells.

plastid Organelle found only in plant cells; depending on the cell's specialized function, its plastids may develop as chloroplasts, chromoplasts, amyloplasts, etc.

Platyhelminthes Phylum of animals containing the flatworms, e.g., planarians, tapeworms, flukes.

poikilothermy Condition of having a body temperature that changes with that of the environment (opposite of homeothermy).

polar 1. Electrically asymmetrical. 2. Pertaining to the poles (ends) of the mitotic or meiotic spindle.

pollen grain Immature male gametophyte of a gymnosperm or flowering plant; it will produce the sperm nuclei that fertilize the egg.

pollination Deposition of pollen on or near the female parts of a gymnosperm or angiosperm.

pollutant Substance that causes pollution.

pollution A change in the physical, chemical, or biological properties of air, water, or soil that can adversely affect the health, survival, or activities of humans and other living organisms.

poly- Prefix meaning many.

polyandry Mating system in which a female mates with more than one male.

polygamy Mating system in which an individual may have more than one mate.

polygenic character Trait whose phenotypic expression is governed by many pairs of alleles (e.g., human skin or hair color).

polygyny Mating system in which one male mates with more than one female.

polymer Large molecule made up of many subunits that are smaller molecules similar or identical to one another.

polymorphism Simultaneous presence in a population of two genetically different forms of a trait at frequencies higher than could be maintained by recurrent mutation.

polyp 1. One of the two alternate body forms found in members of the phylum Cnidaria. 2. A small tumor, often attached by a stalk.

polypeptide A polymer composed of amino acid monomers joined by peptide bonds.

polyphyletic Of a taxon containing members of several evolutionary lines but not including their common ancestor (if there is one).

polyploidy Possession of three or more haploid sets of chromosomes.

polysaccharide A macromolecule made up of many subunits which are simple sugars.

polytene chromosome A chromosome that is exceptionally large because it contains multiple copies of its DNA side by side.

population All members of a species living in a particular area and making up one breeding group.

portal system A group of blood vessels that carry blood from one capillary bed to another without passing through the heart.

posterior At or toward the rear end of an animal.

postsynaptic Receiving stimuli from a nerve cell across a small gap, or synapse.

prebiotic Before life arose.

Precambrian Before the Cambrian period, which began about 600 million years ago.

predator Animal that captures other organisms (usually animals) for food.

presynaptic Pertaining to part of the neuron that sends a signal across a synapse; the receiving cell is postsynaptic.

primary growth Growth in length and production of new stem and root branches in plants.

primary productivity The rate at which food is made from inorganic substances by photosynthetic and chemosynthetic organisms.

primary structure Of a protein, the order in which amino acids are joined together to form a polypeptide.

primary tissues Plant tissues that differentiate from cells laid down by the apical meristem.

Primates The order of mammals that contains monkeys, apes, humans, etc.

primitive Showing features believed to have arisen early in evolution.

producers Photosynthetic and chemosynthetic organisms.

productivity Amount of organic matter produced by members of a given trophic level during a given period of time.

progeny Offspring.

Prokaryotae Kingdom containing all prokaryotic organisms (bacteria and blue green algae).

prokaryotes Bacteria: organisms that lack both a nuclear envelope separating the DNA from the cytoplasm and membrane-bounded organelles.

promoter site A section of DNA to which RNA polymerase must bind before transcription can occur.

protein A functional unit made up of one or more polypeptides.

Protista Kingdom (under the five kingdom system of classification) of unicellular or colonial eukaryotes (e.g., *Amoeba, Paramecium, Euglena*).

protozoa Heterotrophic unicellular eukaryotes.

proximal Toward the center of the body or point of origin of a limb or other structure.

pseudopodium (pl. pseudopodia) Flowing extension of the plasma membrane and cytoplasm of a cell, used for locomotion in organisms such as amoebas.

Pterophyta Division of plants containing the ferns.

pulmonary Of the lungs.

punctuated equilibrium Idea that evolution sometimes proceeds in fits and starts, with new species forming or changing rapidly and then remaining unchanged for long periods of time.

pupa Stage between larva and adult in insects with complete metamorphosis, e.g., butterflies, flies, beetles.

pupation 1. Time of entry into the pupal stage. 2. State of being in the pupal stage.

-q-

quadrupeds Animals that walk on four legs.

-r-

receptor 1. A structure whose function is to recognize and bind a particular molecule; e.g., a T-cell receptor on the surface of a lymphocyte binds a specific antigen, and receptor proteins in the cytoplasm bind specific steroid hormones. 2. A cell or part of a cell in a sensory system that intercepts a stimulus and causes an effect such as generation of a nerve impulse.

recessive allele An allele not expressed in the heterozygote's phenotype.

recombinant DNA DNA produced by combining genes from more than one organism.

recombination In genetics, the formation of DNA molecules with new combinations of genes in an individual as a result of crossing over during sexual reproduction in eukaryotes or as a result of transduction, transformation, or conjugation in prokaryotes.

red blood cells Cells in the blood that contain hemoglobin, a red, oxygen-carrying pigment. Also called erythrocytes.

reduction A chemical reaction involving addition of electrons; often takes the form of adding entire hydrogen atoms.

reflex Unit of automatic action of the nervous system, controlled by a reflex arc, which consists of a sensory neuron, usually one or more interneurons, and one or more motor neurons.

regeneration The process by which an animal regrows an amputated organ.

renal Of the kidney.

replication (= duplication) The making of an exact copy.

repressor In genetics, a molecule that binds to part of an operon and prevents its transcription.

reproductive potential See biotic potential.

reptile Vertebrate with dry, scaly skin and eggs laid on land, e.g., snakes, lizards, alligators, turtles.

residue In molecular structure, what is left of a molecule when it has reacted with other molecules; e.g., a molecule of glucose and a molecule of fructose may condense together to form sucrose, which consists of a glucose residue and a fructose residue.

resorption Absorption back into the body.

respiration Series of oxidation-reduction reactions by which organisms break the chemical bonds in food molecules to release energy, using an inorganic substance (usually O_2) as the final electron receptor.

respiratory pigment Colored molecule that transports oxygen in an animal's blood.

restriction enzyme Bacterial enzyme that breaks DNA between specific nucleotides in a particular sequence; used in genetic engineering.

retina Layer of receptor cells in the eye that responds to light.

retrovirus A virus with an RNA genome that is replicated by the action of reverse transcriptase.

reverse transcriptase Viral enzyme that synthesizes DNA on an RNA template.

rhizoid Root-like structure that anchors bryophytes, some fungi, etc.

rhizome Underground stem of a vascular plant.

Rhodophyta Division containing the red algae.

rodents Members of the mammalian order Rodentia: rats, mice, and their relatives.

Rotifera Phylum of small freshwater animals with a mouth surrounded by cilia.

ruminants Mammalian herbivores in which the stomach contains fluids of an alkaline (basic) pH and is divided into fermentation chambers housing microorganisms that digest the food.

-s-

salinity Saltiness.

salt A substance composed of a positively charged ion other than H^+ and a negatively charged ion other than OH^-; an ionic compound whose cation comes from a base and whose anion comes from an acid.

sampling error Statistical error in scientific experiment resulting from sampling too few subjects.

sap Mixture of water, minerals, etc., conducted in xylem tissue of plants.

saprobe Organism using nonliving organic matter for food.

scavenger Animal that eats dead organisms or organic matter.

Schwann cell Type of cell forming the fatty myelin sheath that surrounds some peripheral neurons.

seaweed Multicellular algae in marine habitats; some members of the Rhodophyta, Phaeophyta, and Chlorophyta.

secondary growth Growth in girth in plants.

secretion 1. Expulsion of a product of a gland or cell. 2. Movement of substances from the blood into forming urine inside the nephron tube.

seed Dispersal unit of gymnosperms and angiosperms, consisting of a seed coat, embryonic plant, and food supply.

seed coat Outer covering of a seed, developed from the outer layers of the ovule.

selective permeability Property of allowing some substances to pass through more easily than others.

self-pollination The transfer of pollen from male reproductive structures such as cones or flower parts to female structures on the same plant.

semen Fluid containing sperm and attendant secretions.

sensory Having to do with detection of changes in the external or internal environment of an organism.

septum (pl. septa) Partition.

sessile (= sitting) Not moving from place to place.

sex chromosomes A pair of chromosomes that cause their carrier to develop as a member of one sex if homozygous for the chromosome pair and the other sex if heterozygous (e.g., X and Y chromosomes in most mammals).

sexual dimorphism Dimorphism = "two forms"; in sexual dimorphism, the two sexes differ in appearance or behavior.

sexual selection Type of natural selection in which females choose to mate with males with particular hereditary traits.

shoot system Part of a plant consisting of the stems, leaves, and any reproductive structures borne thereon.

siliceous Containing silica (SiO_2).

skeletal muscle (= striated muscle) The type of muscle that forms the muscles of the limbs, back, etc.; contains multinucleated fibers.

smooth muscle The type of muscle that lines the walls of internal organs, e.g., digestive tract, blood vessels; consists of single cells.

sociobiology The study of social behavior, the cooperative interactions of animals of the same species.

somatic cells Body cells (as opposed to germ cells, which give rise to gametes).

sorus (pl. sori) Cluster of sporangia in ferns.

speciation Formation of a new species.

species Group of organisms whose members share a common gene pool. See also biological species.

spermatocyte Cell that undergoes meiosis during spermatogenesis.

spermatogenesis Formation of spermatozoa (sperm), male gametes.

spermicidal Sperm-killing.

Sphenophyta A division of lower vascular plants: "horsetails."

sphincter A circular muscle whose contraction closes a tube.

spindle Structure made up of microtubules upon which chromosomes move during mitosis and meiosis.

spirillum (pl. spirilla) Spiral-shaped bacterium.

spontaneous generation The ancient belief that living things routinely arise from nonliving matter.

sporangium (pl. sporangia) Structure in which spores develop.

spore Reproductive cell that can grow into a new individual without fertilization; produced by meiosis in plants, by meiosis or mitosis in fungi. Bacterial spores form when an individual cell encases itself in a protective covering when conditions are unfavorable for growth.

sporophyte Plant that produces haploid spores following meiosis.

Sporozoa Old phylum of protists; most of its members are now placed in the phylum Apicomplexa, the rest in a small phylum, Microspora.

Squamata Order of reptiles containing lizards and snakes.

starch Storage polysaccharide made up of glucose monomers, commonly found in plants.

sterilization 1. Any more-or-less permanent change that prevents an animal from reproducing sexually. 2. Cleaning of an object by destroying all the organisms on it.

sternum Breastbone.

steroid A lipid containing four contiguous carbon rings; e.g., cholesterol, estrogen, testosterone.

stigma Tip of female flower part, usually sticky, allowing pollen to adhere to it easily.

stimulus (pl. stimuli) Energy (chemical, electrical, thermal, light, mechanical, etc.) in the external or internal environment of an organism, to which the organism may respond.

stoma (pl. stomata) In a plant, a pore between two guard cells through which gases are exchanged between the plant and the air.

striated muscle See skeletal muscle.

stroma In chloroplasts, the material surrounding the thylakoid membrane and containing the chloroplast's ribosomes, DNA, and enzymes of carbon fixation.

structural gene A section of chromosome that carries information determining the sequence of amino acids in a polypeptide or protein.

substrate 1. Reactant in an enzyme-mediated chemical reaction. 2. Underlying surface, e.g., a rock in the ocean floor.

subtropical (or semitropical) Lying near, but not within, the tropics.

succession In ecology, process in which the community of species inhabiting an area that has been disturbed changes with time in a regular sequence; succession finishes when the organisms of the climax community of the area have become established.

succulents Plants that store water in fleshy stems or leaves.

sucrose A disaccharide consisting of a glucose monomer and a fructose monomer; table sugar.

sustainable Capable of continuing indefinitely in approximately its present form.

symbiont Organism that lives in close association with a member of another species (often refers to a microorganism living in relationship with a larger host organism).

symbiosis (adj. symbiotic) Close association between members of two or more species (see mutualism, commensalism, parasite).

sympathetic nervous system The part of the nervous system that prepares the body to meet stressful or dangerous situations.

sympatric Living in the same area.

synapse Tiny gap across which information is transferred from one neuron to an adjacent one.

synapsis The lining up of sister chromatids with their homologues in the early stages of meiosis.

syncytium An animal "cell" that contains more than one nucleus within its plasma membrane.

-t-

tactile Of the sense of touch.

taxon (pl. taxa) Any one of the hierarchical categories into which organisms are classified; e.g., species, order, class.

taxonomy Study of the classification and identification of living organisms.

TCA cycle Abbreviation for tricarboxylic acid cycle, old name for the citric acid cycle.

template A pattern or mold.

temporal Of time.

terrestrial Of land (as opposed to water or air).

territory Area defended by one or more animals against intruders.

tetrad Foursome consisting of two sets of two (= 4 altogether) linked sister chromatids lined up at synapsis.

tetrapods "Four-footed" vertebrates; amphibians, reptiles, birds, and mammals.

thermochemical reaction Chemical reaction whose rate varies with the temperature.

thorax (adj. thoracic) Part of the body between the head and the abdomen.

thylakoid Flattened membranous sac in which the light-energy capturing reactions of photosynthesis take place.

tissues Groups of cells that perform a particular task in an organism; e.g., blood, cartilage, xylem.

tonoplast Membrane enclosing the central vacuole of a plant cell.

toxin (adj. toxic) Poison.

trachea 1. In tetrapod vertebrates, the windpipe: tube that conducts air from the pharynx to the lungs. 2. In insects, one of the tubes that conduct air from the outside throughout the interior of the body.

tracheophyte Vascular plant.

transcription Synthesis of RNA using a DNA template.

transduction 1. Conversion of the energy of a stimulus into electrical energy that can be transmitted by the nervous system. 2. Transfer of genetic material from one bacterium to another by a bacteriophage.

transfer RNA RNA molecule that transports amino acids to ribosomes during protein synthesis.

translation The assembly of amino acids to form a polypeptide, in a sequence specified by the order of nucleotides in a molecule of messenger RNA.

translocation 1. In genetics, the movement of a segment of nucleic acid from one part of a chromosome to another part of the same chromosome, or to a different chromosome. 2. In protein synthesis, the movement of messenger RNA and a transfer RNA molecule attached to a growing polypeptide, from the A site to the P site along a ribosome.

transpiration Loss of water by evaporation through pores (stomata) in the shoot system of a plant.

transposable element A length of DNA that can move from its position on one chromosome to another position on the same chromosome or to another chromosome.

trophic level The level in the food chain at which an organism functions; e.g., herbivores, members of the second trophic level, eat autotrophs, members of the first trophic level.

tropics That part of the Earth lying between the tropic of Cancer (at latitude 23 degrees 27 minutes north of the equator) and the tropic of Capricorn (at the same latitude south of the equator). These latitudes mark the limits of the sun's apparent movement north and south during the year.

tuber Underground storage stem, e.g., potato.

tumor Abnormal growth of tissue; sometimes malignant (cancerous), but may be benign.

turgid Swollen with fluid.

turgor Internal pressure that results from being filled with fluid.

type specimen Individual specimen preserved by a person who describes a new species, as the standard for comparison in determining whether other individuals are members of the same species.

-U-

ungulates Hoofed mammals.

unicellular With a body consisting of only one cell.

ureter Tube through which urine passes from the kidney to the urinary bladder.

urethra Tube from the urinary bladder to the exterior.

Urodela Order of amphibians including newts, salamanders, mud puppies, etc.

uterus (adj. uterine) The organ in females of most species of mammals where the young develop; situated between the oviducts and the vagina.

-V-

vacuole A membrane-enclosed sac filled with fluid.

variety A subdivision of the species in the hierarchy of taxonomic classification; the variety (sometimes called subspecies) name follows the species epithet.

vascular Of tissues that transport fluids around the body; e.g., veins, arteries, xylem, phloem.

vascular cambium Meristematic tissue that produces secondary xylem and phloem.

vascular tissue Tissue that conducts water, minerals, and food from one part of the plant to another.

vegetative Carrying out the basic life activities of photosynthesis, metabolism, etc., as opposed to reproduction.

vegetative reproduction Reproduction by growth of an individual's body or fragments of its body; reproduction without production of gametes or spores.

vein 1. Vessel carrying blood toward the heart. 2. Thickened ridge in the wing of an insect. 3. Bundle of vascular tissue in a leaf or flower part.

vena cava A large vein that delivers blood to the (right) atrium (of the heart).

venation The arrangement of veins in leaves or in the wings of insects.

venereal disease (VD) Traditional name for a disease, such as syphilis or AIDS, transmitted by sexual activity. These diseases are now generally known as sexually transmitted diseases (STDs).

ventral Pertaining to the undersurface of a bilaterally symmetrical animal; e.g., underside of a worm, belly of a dog.

ventricle (adj. ventricular) 1. A heart chamber that pumps blood into one or more arteries. 2. Fluid-filled cavity in the brain.

venule A small vein in the circulatory system.

vertebrate Animal with a backbone; e.g., fish, human.

vesicle Membrane-enclosed sac, holding secretory products, enzymes, etc., in a cell.

vestigial Small and nonfunctional.

virus Particle composed of nucleic acid and protein that can be reproduced only in a living cell.

vitamin An organic micronutrient.

viviparity (adj. viviparous) Condition of giving birth to young rather than laying eggs.

-W-

warbler Member of a group of small insect-eating birds.

wavelength Light and sound may be considered as travelling in wavy lines. The distance between adjacent peaks of the line is the wavelength, symbolized by λ (lambda).

white blood cells Cells in the blood that are involved in defending the body against foreign organisms and substances; also called leukocytes.

white matter Nervous tissue consisting mainly of myelinated axons.

wild type In genetics, showing the normal phenotype of wild members of the species for the trait in question.

wood Secondary xylem.

-X-

xylem Plant tissue that conducts sap from the roots to the leaves.

-Z-

Zoomastigina Phylum of heterotrophic protozoa that move by means of flagella.

zooplankton Animals and protozoa floating in the surface layers of a body of water.

Zygomycota Division of fungi whose members form zygospores during sexual reproduction.

zygote Fertilized egg.

Index

Note: **Boldface** page numbers indicate pages on which the index item is defined; *italicized* page numbers indicate pages containing illustrations. If both occur on a page, the number is ***boldfaced and italicized.***

Abdomen, of arthropods, 445
Abiotic factor, **749**
ABO blood groups, 254-255
Abortion, induced, 572
  spontaneous, 569
Abscisic acid, *713*, 715
Abscission, 721
Absorption, **362**
  by roots, 674
  of food, **483**
Acacia, *156-157*
Acetyl CoA, 130
  ATP yield, 134
Acetylcholine, 592
Acetylcholinesterase, 592
Achillea filipendulina, *306*
Acid hemoglobin, 502
Acid rain, 13, 34-35, *825-826*
Acorn, *295*
Acquired immunodeficiency syndrome. *see* AIDS
Acrasiomycota, 390-*391*
ACTH, 636
Actin, 615
Actinomycetes, 379
Action pattern, fixed, **650**
Action potential, 586-*588*
Activation energy, **30**, **114**
Active site of enzyme, 56
Active transport, by carrier proteins, *75-77*
  by membrane proteins, 74
Adaptation, 295-296
  in evolution, ***10-11***
Adaptive behavior, *646*
Adaptive radiation, 189
Addiction, effect on brain, 597
Adenine, *168*
Adenosine diphosphate, 129
Adenosine triphosphate. *see* ATP
Adhesion, 31
Adipose cells, *137*
Adipose tissue, 488
Adrenal gland, 516, 555
Adrenalin. *see* Epinephrine
Adrenocorticotropic hormone, 636
Adventitious roots, 673
Aerobes, **375**
Aerobic processes, **125**
Aerobic respiration, *126*
Affected individual (genetics), **251**
African sleeping sickness, 386
Age structure, of population, **789**
Aggregates, formation, 346
  prebiotic, 347
Aggression, among social animals, 661
Aging, **578**
  in plant, **721**
Agnatha, 453-454
Agricultural revolution, 813-815
Agriculture, **813**
  effects, 814-815
  efficiency, 816-817
  green revolution, 817-819
  irrigation, 817-819
  low-input, *819*
  tropical, 820-821
AIDS (Acquired immunodeficiency syndrome), 4
  elimination, 468
  sexually transmitted, 571
  understanding, 538-539
AIDS virus, 369
Air bladder, 408
Air pollution, 498-499, 822
Albinism, 252-*253*
  in humans, 230-*231*
  in peacock, *254*
  in snake, *246*
Alcohol, effect on brain, *596*-597
Alcohol groups, of molecules, *44*
Alcoholic fermentation, *136*
Aldehyde groups, *44*
Aldosterone, 555
Algae, 381
  brown, 408-*409*
  golden, 385-*386*
  green, *409*-411
  multicellular, 407
  red, 407-*408*
Alimentary canal, 480
Alkalinity, **33**
Allantois, 458, **573**
Allele(s), **229**
  dominant, 229-231
  lethal, 248-252
  multiple, 254-255
  recessive, 229-231
Allergen, **540**
Allergies, *540*-541
Alligator, *479*
Allografts, 664
Allopatric populations, 313
Allopatric speciation, **314**-316
Alpine newt, *360*
Altered vegetable states, 744-745
Altitude, effect on vegetation, *751*
Altruistic behavior, **333**, 829
Alveoli, 496
Amino acids, **51**-*52*, 474
Amino groups, *44*
Aminoacyl attachment site, 187
Ammocoete, 454
Ammonia, excretion, *547*
Amniocentesis, 266-*267*
Amnion, 458, **573**
Amniotic eggs, *459*
Amoeba, feeding, *78*
  plasma membrane, 2
Amoeboid motion, *383*
Amphibians, 457-*458*
  circulation, 510, 512
  extinction, 292
Amplification, of chemical messengers, 633
Amylase, 482
Amyloplasts, **97**-*98*
  and plant tropisms, 718
  of roots, 672
Anabolic steroids, 632
Anaerobes, **375**
Anaerobic processes, **125**
Anaerobic respiration, **375**
Analogous structures, **285**
Anamniotic eggs, *459*
Anaphase, 214
  meiosis, *216-217*
  mitosis, *213*
Anaphylactic shock, **541**
Anatomy, comparative, evidence for evolution, 285-288
Ancestral characters, 360
Anchoring, roots for, 674
Androgens, 566
Anemia, 518
  sickle cell, 250-252
Angiosperm(s), flowers, *426*
  phloem, *702*
  secondary xylem, *697*
Angiotensin, 557
Animal(s), digestion, *479*
  domestication, *814*, 816
  environments, 433-434
  evolutionary trends, *435*
  gas exchange, 490-505
  nutrition and digestion, 472-489
  structure, 434-436
Animal biology, 471-665
Animal groups, diversity of, *432*
Animal hormones, 628-643
Animal kingdom (Animalia), 364, 430-467
Animal tissues, 104-*105*
Annelida, 442-*443*
  transport in, 508, 510
Annual rhythms, 638
Annual rings, 695-*696*
Ant acacias, *156-157*
Antagonistic muscles, 618-620
Antenna pigments, 149
Anterior pituitary hormones, 636
Anthers, 728
Anthophyta, *425*-427
Anthropomorphism, 645
Antibiotics, *87*
  in AIDS, 538
  resistance to, 295-*296*
Antibodies, **531**
  classification, 531-*532*
  genetics, 532-533
Anticodon, 186-187
Antidiuretic hormone, 555
Antigens, **528**

histocompatibility, 533
recognition, 531-533
Anurans, 457-*458*
Aorta, 510, 514
Aphid, *325*, *485*
Apical dominance, **675**, ***716***-*717*
Apical meristem, 670
Apicomplexa, 387-388
Apnea, **503**
Appendicular skeleton, **620**
Apple blossom, *732-733*
Aquaspirillum sinosum, *374*
Aquatic communities, 758-762
Arachnids, 445-*446*
Archaeobacteria, **379**
Archaeopteryx, *285*, 460-*461*
Archenteron, 576
Arid regions, and soil erosion, 766
Arm, muscles of, *620*
Arteries, 510, 512-*513*
Arthropoda, 445-449
key to, *313*
Artificial selection, evidence for evolution, 283
in dogs, *284*
Ascomycota, *395*-396
Ascus, **396**
Asexual reproduction, *324*, 738
Asia, population growth and wealth in, 806
Association, by nervous system, 594
Associative learning, 653
Atherosclerosis, 476-*477*
Atmosphere, circulation of, 778
composition, 120
reducing, 343
Atom, ***25***
Atomic mass, **26**
Atomic number, **26**
ATP, **58**
chemiosmotic synthesis, *119*, **122**, 127-*128*, 132-134, *151*-152
energy storage, 112
in active transport, 76
in muscle contraction, 615
production, 96, *118*, 129
regeneration, 348
uses, *118*
ATP synthetase, 122, 134
ATPase, 118
Atrioventricular node, 612, *614*
Atrium, of heart, 510
Attack, sign stimulus for, *651*
Attitude change, with agriculture, 815
Aurelia, *508*
Autografts, 664
Autoimmunity, **537**, 540
Autonomic nervous system, 595, 602
Autosomes, 204, 259
Autotroph(s), **116**, **376**, **669**, **772**
heterotrophs and, **349**-**350**
Auxin, 712-*713*
and apical dominance, *716*-717
and plant tropisms, 718
response to, *714*
Aves, 460-462
Axial skeleton, **620**
Axil bud, 675
Axon, 584, *585*
AZT, 538

B allele, of human ABO blood group, *310*
B cell (lymphocyte), in immune system, 531, 535
Baboons, *333*, *644*
Bacteria, 374-381
and fungi, 399-400
antibiotic resistance, *296*
autotrophic, 376-*377*
blue-green, 379
cell shapes, 374
chemical composition, 51
chemosynthetic, ***349***
DNA, *164*
fossil, *342*
genetic recombination, *376*
Gram-stained, *379*
heterotrophic, 378
in soil, 690
metabolism, 375-378
nitrifying, 377
nitrogen-fixing, 377-378
nonvirulent, 166-*167*
oral, *377*
photosynthetic, *350*
protein synthesis, 190
Rhizobium, *378*
symbiotic, 380-381
taxonomy, 362
transformation, **166**-168
virulent, 166-*167*
Bacteriophage, 166-*167*, *367*
Badger, *796*
Bald eagle, *802*
Baldness, 262-*263*
Bananas, *317*, *716*
Barbary apes, *660*
Barbiturates, 597
Bark, **680**
Barnacles, *798*
Baroreceptors, 516
Basal body, *103*
Base, **33**
Basidiomycota, 396
Basidium, **396**
Basswood, *679*
Bat, *74*, 464
Beach life, chemical composition, 24
Beak variations, *487*
Bean seed, *670*
Bee(s), 658-*659*
as pollinators, 740
genetic relationships of, *337*
Bee orchid, *726*
Behavior, 644-663
adaptive, *646*, 649
causes, 646
conflict, *654*
development, 647-*648*
innate, **648**-649
learned, 649, *651*
neural basis, 649-652
selective pressures, 646
stereotyped, 650-*651*
territorial, 653-*654*
Benthos, 762
Bicarbonate ion, 36
Bighorn sheep, *331*
Bilateral symmetry, 434
Bile, 483
Bills, of birds, *487*
Binary fission, **374**-*375*
Binomial nomenclature, 358
Biochemistry, comparative, evidence for evolution, 288
Biodiversity, loss of, 292
Biogeochemical cycle, 778
Biogeography, evidence for evolution, 289-290
Biological chemistry, 40-63
Biological clocks, 638-640
Biological control, of Klamath weed, *799*
Biological information, beginnings of, 348-349
Biological membranes, 66-67
Biological polymers, 42
Biological rhythms, 638-641
Biological systems, energy in, *113*
Biology, **1**
complete picture, 2-4
developmental, evidence for evolution, 288-289
saving life and preserving earth, 4, 12
Biomass, **776**
Biome(s), **749**, *752*
temperate, 755-757
tropical, 752-754
Biosphere, **1**, **811**
Biotic factors, **749**
Biotic potential, **790**
Bipolar neuron, *585*
Birds, 460-462
as pollinators, 740
bills of, *487*
circulation, 512
ecological niches, *797*
feeding, 487
sex determination, **257**
Birth, *577*
Birth control, 570-572
Birth rate, 803
Bivalves, 444
Black-headed gull, 654
Blades, of brown algae, 408
Blastocoel, **574**
Blastocyst, **572**
Blastopore, **576**
Blastula, **572**
Blinding, in experimental method, 7
Blood, 518-519
components, 518
oxygen carrying by, *501*
production, 624
sodium content, *556*
Blood cells, *105*

in sickle cell anemia, *251*
Blood clotting, 518-*519*
Blood flow, 514
Blood groups, human, 254-255
Blood pressure, 514-516
Blood vessels, 512-*513*
Blue-ringed octopus, *280*
Body layers, *435*
Body surface, and ventilation, 494
Body symmetry, 434-*435*
Body temperature, *124*
variations, 640
Boiling point, 31
Bolus, food, 480
Bonds, chemical, 26-28
Bone, 622-*624*
Bone growth protein, 624
Bone marrow, and immunity, 529
Bony fishes, 455-*457*
Boreal forest, 757
Bougainvillea, *729*
Boy in the bubble, *199*
Brachydactyly, 249-*250*
Bracket fungi, *396*
Bract, *729*
Bradycardia, 521
Brain(s), **593**
effect of drugs on, 596-597
human, *599*
vertebrate, *595*, 598
Breakage mutation, **171**
Breast feeding, *569*
Breathing, 500
regulation, *503*
Breeding experiment, genetic, 228-233
Breeding programs, for seeds, 737-738
Brewer's yeast, 400
Bronchi, 496
Bronchitis, 498
Brown algae, 408-*409*
Bryophyta, 415-*416*
Bubonic plague, 529
Bud scales, *715*
Buffer, **36**
Bulbourethral gland, 563
Bull thistle, *425*
Butter, phase changes, *30*
Buttercup root tip, *671*
Butterflies, pollination by, 740
puddling behavior, *6-9*

C_4 photosynthesis, 154-*155*
Cactus, flowering, *425*
Caecilian, *458*
Caecum, 480
Caladium, *270*
Calcitonin, in calcium control, 624
Calcium, as second messenger, 633
in bone, 624
in plant nutrition, 688-689
plasma, *632*
Calcium pump, 76
Callus, bone, 623
Calvin cycle, *151*-152

CAM (crassulacean acid metabolism) photosynthesis, 154-*155*
Camouflage, 329, **804**-805
Cancer, 196-198
death, 198
elimination, 468
lung, *197*
macrophages attacking, *524*
Canthon pilularis, *447*
Capillarity, 688, *698*
Capillary, *506*
in vertebrates, 510
of mammals, 512-*513*
Capsule, bacterial, 86
Carbohydrates, **41**, 48-51
as macronutrients, 474
energy yield, 139
Carbon, 24-25
atoms, *43*
fixation, 152
structure, *25*
Carbon cycle, 778-781
Carbon dioxide, as greenhouse gas, 120
excretion, 546
transport, 502
Carbon monoxide, smoking and, 498
Carbon skeletons, 42-*43*
Carbonic anhydrase, 502
Carboxyl group, *44*, *46*
Carcinogen, **197**-198
Cardiac muscle, 612, 614
Cardinal fish, *457*
Cardiovascular disease, 516-518
diet and, 476-477
Carnivores, 464, **473**
cnidarians as, 436
in ecosystem, 772
mammalian, 486-487
Carolina parakeet, *16*
Carotenoids, 146
Carrier, **251**
Carrier proteins, *75-77*
Carrot, *706*
Carrying capacity, of earth, 834
of environment, 334, **793**-*794*
Cartilage, *105*, 622
Cartilaginous fishes, 454-*456*
Casparian strip, 692
Cat(s), coat color in, 261-*262*
Siamese, *58*
Cat-litter box, enzyme reactions in, *57*
Catalyst, 56, *114*
Cell(s), **12**
communication between, 79, 81
differences from viruses, 365
eukaryotic, **85**, 87-89
eukaryotic vs prokaryotic, 88-89
membrane attachments, 78-*80*
metabolism, *60*, *128*
nucleus, **65**
osmotic relations, 72-73
prokaryotic, **85**-87
structure, *90-91*
surface-to-volume ratio, *66*
Cell body, of neurons, 584

Cell cyle, **210**-211
Cell division, and surface area, *66*
cytokinesis, 210-*211*, *213-215*
meiosis, **204**, **215**-218
mitosis, **204**, 211-214
Cell membrane, 64-83. *see also* Plasma membrane
Cell plate, in cytokinesis, 215
Cell theory, 65
Cell walls, **65**
function, 97-99
in xylem, 694
plant, *99*
Cellular immunity, 533-*534*
Cellular respiration, 96, 117, **125**-141
Cellulose, 49-*50*
Centipede, *446*
Central nervous system, 594
Centrioles, *90*, 103-*104*
Centromere, 204
Cephalization, **434**, **593**
Cephalochordata, 451-*452*
Cephalopod, *444*
Cerebellum, 597, 600
Cerebral cortex, 598, 600
Cerebral hemispheres, 598
Cerebrospinal fluid, **545**, 595
Cerebrum, 598
Cervix, 563
Cetaceans, 464
CFCs (chlorofluorocarbons), and ozone depletion, 352-*353*, 826
as greenhouse gases, 121
Chameleon, *15*, *460*
Channel proteins, 75
Chaparral, **756**
Characters, genetic, 226
polygenic, **255**-256
Cheese, fungi and, 400
Cheetahs, genetic uniformity of, 308
Chemical bonds, 26-28
Chemical defenses, by leaves, 681
in Ascomycetes, *396*
of skin, *527*
Chemical elements, **24**
Chemical energy, **111**, *113*
Chemical fertilizers, effects of, 817
Chemical messenger(s), **629–630**, 633-*634*, 636-637
Chemical reactions, 30
and energy, 112-115
reversible, 30
Chemical regulation, hormones and, 628-643
Chemical selection, 347
Chemiosmotic ATP synthesis, **122**, 127
in chloroplast, *119*
in mitochondria, 132-134
in photosynthesis, *151*-152
Chemistry, basic, 22-39
biological, 40-63
Chemoreceptors, stimuli of, 606
Chemosynthesis, bacterial, ***349***, 377
Chiasma, 218
Chimera effect, 665

Chimpanzees, *430*, 660–661
Chitin, 49–*50*, 445
Chlamydia infections, 571
Chlorofluorocarbons (CFCs), and ozone depletion, 352–*353*, 826
  as greenhouse gases, 121
Chlorophyll, **145**
Chlorophyta, *409*–411
Chloroplast, *91*, **97**–*98*, 101, 145
Chlorosis, 688
Cholesterol, 47–*48*
  diet and, 476
  in biological membranes, 67
Chondrichthyes, 454–*456*, 549
Chordates, 449–*451*
Chorion, 458, **573**
Chorionic villus sampling, 266–*267*
Chromatids, sister, 204
Chromatin, **89**, *92*, **172**, 193
Chromoplasts, **97**
Chromosome(s), **89**, **164**, *206*
  abnormal movements, *265*
  arrangement, *220*
  DNA, *173*
  eukaryotic, 172, 204–209
  homologous, 229
  mapping, 209, *242*
  number in cells, 210
  polytene, *193*–194
  puffs, *193*–194
  sex, 204, 257, 259
  structural changes, 193–194
Chrysophyta, 385–*386*
Chylomicrons, 484
Chyme, 482
Cilia, 102–*103*
Ciliary motion, *383*
Ciliates, 388–*389*
Cinchona, *424*
Circadian rhythms, 638–*640*
Circulatory system, adjustment to exercise, 516–*517*
  blood pressure and, 514–516
  closed, 508–*509*
  diseases, 516–518
  double, 510
  earthworm, 508–*509*
  fish, **506**, 510
  human, *515*
  insect, *509*–510
  mammalian, 512–518
  open, *509*–510
  single, 510
Citric acid cycle, *126*, *128*, 130–*132*
  energy yield, 134
  preparation, 127
Clade vs grade, **361**
Class, **358**
Classification system, five-kingdom, 362–364
Clay bowl, Egyptian, *812*
Cleavage, 574
Cleavage furrow, 215
Climate, and organism location, 764
  and vegetation, 750–752
Climax community, **762**
Cline, of salamanders, *312*
Clitoris, 563
Cloaca, **547**
Clock, biological, 638–640
Clone, 196, **324**
  plant, 738
Club moss, *418*
Cnidarians, 436, *438*
  gastrovascular cavity, *508*
  nervous system, 593–*594*
  transport, 508
Coat color, in cats, 261–*262*
  temperature and, *264*
Coatimundis, *755*
Cocaine, effect on brain, 597
Cockroaches, *16*
Coconut palm trees, *668*
Codominance, **236**–237
Codons, **182**–183
Coelom, *435*–436, 508, 576
Coenocytic fungi, 393
Coenzyme(s), 56, 127
Coenzyme A, 127
Coevolution, flowers with pollinators, 740–741
  herbivores with plants, 156–157
  metabolism with reproduction, 349
Coexistence, *797*
Cofactor, 56
Cohesion, of water, 31
Colchicine, in cell division, 214
Coleoptile, 582, 718
Coleus shoot tip, *675*
Collagen, 622
Collar cell, 436
Collecting duct, 553
Colon, 482
Colorblindness, *260*
Combjelly, *438*
Commensalism, **380**
Commons, tragedy of, 828–829
Communication, among social animals, 658
  between cells, 79, 81
Community(ies), **749**, **771**
  aquatic, 758–762
  climax, **762**
  fire-maintained, 764–*765*
  productivity, 777
  species diversity, 783–785
  species turnover, 784–785
Compact bone, 623–*624*
Companion cell, 702
Comparative anatomy, evidence for evolution, 285–288
Compensation depth, 759
Competition, **795**–798
Competitive exclusion, **796**, *798*
Complement reactions, **526**
Complementary base pairing, 169
Compound, **28**–29
Compound leaves, **676**
Compound light microscope, 68–*69*
Concentration, gradient, 71
  of solution, **2**
Condensation, 172–*173*, 214
  of chromosomes, 204
Condensation reaction, **42**, *44*
Condenser microscope lens, 68–*69*
Conditioned reflexes, 652
Condom, 539, 570
Conflict behavior, *654*–*655*
Congenital insensitivity to androgen syndrome, 248
Coniferophyta, 423
Conjugation, **374**
Connective tissue, ***105***, *622*–*624*
Conservative characters, 360
Consumers, 772
Contact, plant responses to, *719*
Continental drift theory, *3*–*4*
Continental shelf, **760**
Contraception, **570**
Contractile proteins, 51
Contractile vacuoles, 382
Control treatments, 7
Convergent evolution, 360–**361**, *765*
Convoluted tubules, 553
Copulation, **562**
Coral, 436, *438*
Coral fungus, *396*
Coral reef, *761*
Cordgrass, 774
Cork, **680**
Cork cambium, 680
Corn, hybrid, *311*
  prop roots, *673*
  seed structure, *682*
  stem characteristics, *14*
Corpus callosum, 600
Corpus luteum, 567
Corpus striatum, 598
Cortex, kidney, 553
  root, 672
Cotyledon, 670, 734
Countershading, **804**
Courtship behavior, 330, **561**, 654–*656*
  of European jackdaw, *647*
  of peacock, *305*
Covalent bond, **26**–*27*
Crab, *282*
Cranial nerves, 602
Crassulacean acid metabolism (CAM), 154
Critical daylength/night length, 720
Crocodile, 458–460
Crop, and digestion, 480
Crop plants, genetics, 737
Cross-pollination, 730
Crossing over, 218, ***239***, 242
Crowning, in birth process, 577
Crustaceans, *446*–447
Cryptic coloration, **805**
Culture, and population, 835
Cut flowers, caring for, 705
Cuticle, 414, 674
  leaf, **147**–*148*
Cyanobacteria, *342*, 379
Cyanophora, *101*
Cycads, *423*
Cyclic AMP, as second messenger, 633
Cyclophosphamide, for immune suppression, 665
Cyclostomes, 453
Cysteine, *52*

Cystic fibrosis, 252
Cytokinesis, 214-*215*
  in cell cycle, 210-*211*
  in mitosis, *213*
  overview, 205
Cytokinins, *713-714*
Cytoplasm, **65**, **85**, *90*, *95*, *99*
Cytoplasmic bridge, *99*
Cytoplasmic streaming, **99**
Cytosine, *168*
Cytoskeleton, 87, 99, 102-104
Cytosol, 87

Dandelion, *326*
Darwin, and theory of evolution, 280-283
Darwin's finches, *290*
Day-neutral plants, 720
Deamination, 138, 546
Death, 578-579
Death rate, 308, 802
Deciduous trees, 753
Decomposer, 378, 772
Defenses, against disease, 524-543
  chemical, by leaves, 681
Defoliation, *448*
Deforestation, 13, 779, *820*-821, 826-*827*
Degeneracy of genetic code, 183
Deletion mutation, **171**
Demographic division, 803, 806
Demographic transition, 802-*803*
Demography, **802**
Denaturation of protein, **53**, 55
Dendrite, 584-*585*
Density, population, **789**
Density-dependent and -independent
    mortality factors, **794**
Deoxyribonucleic acid. *see* DNA
Depolarization, **584**, 586-*587*
Derived characters, 360
Dermal tissue, in plants, **669**
Desert, 754-*755*, 757
Desertification, 766
Desmid, *409*
Desmosome, 79-*80*
Detoxification, 546
Detritus, 759, 774
Deuteromycota, 397
Developmental biology, evidence for
    evolution, 288-289
Diaphragm, 500
  contraceptive, 570
Diastole, 514
Diatoms, 385-*386*
Dichotomous key, to species identification,
    313
Dicotyledons, 670
  comparison with monocotyledons,
    581-584
Didinium, *389*
Diencephalon, 597
Diet, and cardiovascular disease, 476-477
  and gene expression, 263-*264*
  effect on phenotypes, *264*
  simplification with agriculture, 815
Differentiation, 194
Diffusion, *72*
  facilitated, 76, 492
  through membranes, 71-72
Digestion, **473**, 478-480
  enzymes, 482-483
  human, 480-485
  in animals, 472-489
  in herbivores, *485*-486
Digger wasp, *650*
Dihybrid cross, **234**-236
  involving linkage, *238*
  Punnett square of, *236*
Dilation, in birth process, 577
Dimethyl ether, *29*
Dinoflagellates, 384
Dipeptide, **52**
Diploid chromosomes, 204, **206**
Dipylidium caninum, *439*
Directional selection, *304*
Disaccharide, 48-*49*
Disease, control, 399
  defenses against, 524-543
    external, 526
    internal, 526-528
    nonspecific, 526-528
  elimination, 468-469
  susceptibility in cheetahs, 308
Dispersal, and organism location, 764
Disruptive selection, *304*-305
Dissociation, **33**, 36-37
Distal convoluted tubule, 553
Diuretic drugs, 516
Diversity of life, evolution and, 275-469
Diving adaptations, of Emperor Penguins,
    520
DNA, **58**
  active segments, *193*
  as genetic material, 166-168
  function, *180*
  in chromatin, 89
  in chromosomes, *173*
  of bacteria, *164*
  recombinant, **174**
  repair, 171
  repetitive, 176
  replication, *170*-171
  satellite, 176
  structure, *59*, *168-170*, *180*
  transcription, *180*-181, 192
DNA polymerase, 170
Dog(s), artificial selection, *284*
  gas exchange, *492*
Dogwood, *729*
Dolphin, *464*
Dominance, genetic, 331
  hierarchy, **659**
  incomplete, **236**-237
Dominant alleles, 229-231
Dopamine, 592
Double fertilization, 427, 731
Double helix, in DNA, 169
Down syndrome, 265
Dragonfly, metamorphosis, *196*
Drive, 652
Drosophila, chromosome map of, *242*
Drugs, effect on brain, 596-597
Duck, *479*, *487*
Dugesia, *439*
Dung beetle, 293, *447*
Duodenum, 482

Eagle, bill of, *487*
Early fossils, 354
Earth, *18*
  biomes of, *750*
  maximum population of, 834-836
  primitive, *340*
Earthworm, 442-*443*
  digestion, *479*
  movement, *620*
  reproduction, *329*
Echinodermata, 449-*450*
Ecological succession, 762-764
Ecology, **1**, **749**-836
  and mating systems, 332-333
Economy, and population, 835
Ecosystem(s), **749**, **771**
  and communities, 770-787
  components, 772
  energy flow, *772-773*
  experimental, 782-*783*
Ectoderm, 435, 576
Ectoparasite, *440*
Ectothermic animals, **459**
Ectotourism, *723*
Education, and environment, 830-831
  in AIDS prevention, 539
Effectors, 583, 603, **611**
Egestion, **479**, **546**
Egg, 221, **561**
  amniotic, *459*
  fertilization, 561-*562*, 572-574
  production, *222*
  size relative to sperm, *328*
Ejaculation, **566**
Electrical energy, *113*
Electrical synapse, 592
Electrocardiogram, *614*
Electrolyte imbalance, 546
Electromagnetic radiation, *145*
Electron, **25**
  transport system, *119*, **122**, 132-*133*
    in chloroplast, 149
    in mitochondria, *128*
Electron microscopes, 68-*69*
Electron shells, **25**
Electronegative bond, 26
Electroreceptors, 606
Elements (chemical), *25*
Elephants, *463-464*
Elm wood, *696*
Embryo(s), **572**
  of flower, *731*
  of gymnosperm, 421
  of vertebrate, *288*
Embryo sac, in pollen maturation, 730
Embryonic development, 560-581
  cleavage, 574
  epidermis, 576
  fertilization, 324
  gastrulation, 574, 576

neurulation, 576
organogenesis, 576
stages, 574-576
Embryophyte, **414**
Emergents, 722
Emperor goose, *655*
Emperor Penguins, *520*-521
Emphysema, smoking and, 498
Endangered species, *5*, 800-*802*
Endemic species, 189
Endergonic reaction, 113-*114*
Endocrine glands, **629**-*630*
Endocytosis, 77-*78*
Endoderm, 435, 576
Endodermis, 672
Endogenous hypothesis, of mitochondrial origin, 101
Endogenous rhythms, 638
Endoparasite, *440*
Endoplasmic reticulum, *90-91*, **93**-*94*
Endorphins, 592
Endoskeleton, *619*
Endosperm, 427, 582, 731
Endosymbiont theory, *100*
Endothermic animals, **459**
Energy, **111**
and living cells, 110-123
capture in photosynthesis, 149-152
chemical reactions and, 112-115
flow in ecosystems, *772-773*
intermediates, **118**-119, 122
metabolism, 347-348
pyramids, 772, 776
renewable vs nonrenewable sources, 821-*822*
transformations, 112
Enhancer, 193
Enkephalins, 592
Ensatina salamander cline, *312*
Entropy, **112**
Environment, abuse of, *828*
and population, 835
animals and, 433-434
genes and, 263, 646-647
human impact, 810-833
plant response, 717-719
Environmental movement, 828-829
Environmental problems, 13
evaluation of, 827, 830-831
two views of, 823-827
Environmental resistance, 334-335
Enzyme, *56*-58
and protein denaturation, 55
factors affecting, *57*-58
function, 51, *56*
hydrolytic, **95**-*96*
Enzyme-substrate complexes, *56*
Epidermis, defense by, 526
in embryonic development, 576
plant, ***106***, ***147***-148, 414, *672*
Epididymis, 563
Epiglottis, 480
Epinephrine, 516
Epiphytes, 722, 752
Epithelial tissues, **104**-*105*
Equilibrium, Hardy-Weinberg, 302
metabolic, 115
punctuated, 318
Equisetum, *419*
Erection, **566**
Ergot, *396*
Erosion, soil, 766-767
Erythropoietin, 518
Escherichia coli, as experimental material, 166, 175, *185*
plasmid, *173*
Esophagus, 480
Essential hypertension, 516
Estrogen, 567
Ethics, lifeboat, 829
Ethyl ether, *29*
Ethylene, in plants, *713*, 715-716
Eubacteria, **379**-380
Euglena, 362, *385*
Euglenida, 384-385
Eukaryotic cell(s), **85**, 87-89
evolution, 100
nucleus, *92*
origin, 350-351, 381-400
prokaryotic vs, 88-89
relation to prokaryotes, 100-101
reproduction, 202-225
transcription control, *191*, 193-194
European jackdaw, *647*
Eutrophic lakes, 759
Evaporation, 31
from plants, 699
Evolution, **3**, **10–11**, **277**, *279*
and diversity of life, 275-469
and natural selection, 276-299
and reproduction, 322-339
causes, 302-309
convergent, *765*
evidence for, 283-290
of HeLa cells, 160-161
theory of, 278-283
viruses and, 370
Evolutionary characters, types of, 360
Excitability, of neurons, 584
Excitatory stimuli, 586-*587*
Excretion, 544-559
daily exchange and, *546*
substances, 546-548
systems, 550-*552*
Exergonic reaction, 113-*114*
Exocytosis, 77-*78*
Exons, 185
Exoskeleton, 445, *619*
Experimental method, 6-8
blind, 7
Exponential population growth, *790*
Expulsion, in birth process, 577
Extensor muscle, 619
External fertilization, 324, *562*
Extinction, 13, 16, 278, 292-293, **800**-801
Extracellular digestion, 479
Extracellular fluid, 492, **506**
Extra-embryonic membranes, 572
Eyes, of owl, *582*
Eyespots, **382**
Facilitated diffusion, 76
Facultative anaerobes, **375**
FAD, 127
Family, **358**
Farming, low-input, *819*
Fat, 47
as macronutrient, 474
energy yield, 139
Fatigue, in muscle contraction, 618
Fatty acids, *46*-47
Feedback, negative, in circulatory system, 516
of hormones, *632*
Feeding, birds, 487
carnivores, 486-487
herbivores, *485*-486
Female, evolutionary role, 328-333
Fermentation, **125**, 135-137
Fern, *419*-421
Fertility, low, in cheetahs, 308
world rates, 834
Fertilization, **561**-*562*, 572-574
double, 427
external vs internal, 324, *562*
in flowers, 427, 731
in gymnosperms, 421
in reproduction, 324
in vitro, 578
mechanisms that ensure, 328
Fertilizers, chemical, 817
organic vs inorganic, 690
Fetus, *560*
development, *575*, 577
testing, 266-267
Fever, 527
Fiber cells, of stems, 674, 695
Fibrin/fibrinogen, and blood clotting, 518
Fibrous root system, *673*
Fiddlehead ferns, *667*
Fight or flight reaction, 602, 634-*635*
Filaments, of skeletal muscle, 614
Filial generations, 228-229
Filter feeders, 434
Filtrate, from nephron, 553
Fin, evolution of, 454
Finches, Darwin's, *290*
Fire-maintained communities, 764
First foliage leaves, 670
First law of thermodynamics, 112
Fischer's lovebirds, *331*
Fishes, bony, 455-*456*
cartilaginous, 454-*456*
circulatory system, **506**, 510
classes, 453-455
embryos, *14*
evolution, 454
jawless, 453-454
Fitness, in natural selection, 304
Fixed action pattern, **650**
FK506, for immune suppression, 665
Flagella, 102-103
bacterial, 87
Flagellar motion, *383*

Flame anglerfish, *457*
Flame cell, 550-*551*
Flamingo, *788*
Flashlight fish, *380*
Flatworm, 438-*439*, *479*
Flavin adenine dinucleotide (FAD), 127
FLAVR SAVR tomatoes, *745*
Flexor muscle, 619
Flower(s), *728*-729
  anatomy, *728*
  coevolution with pollinators, 740-741
Flowering, *720*-724
Flowering plants, *425*-427
  reproduction, 726-743
Fluid mosaic model, 67
Flukes, 438
Fly, pollination by, *741*
Foliar feeding, 706
Follicle-stimulating hormone, 566-567
Food(s), 124-141
  alternative molecules, *137*-139
  bacteria and fungi, 399-400
  energy yield, 139, *778*
  irradiated, 744
  nutritional value and production costs, *816*
  plant storage, *138*, *706*
  production, 676, *815*, 818
  use by body, 488
Food chain, **772**
Food poisoning, 400
Food supply, limited, and population, 835
  of gymnosperm embryo, 422
Food webs, and energy relationships, 772, 776
Foraminiferan, *387*
Forebrain, 597-600
Foreign body, and immune response, 528
Forelimbs, homologous, *287*
Forest, boreal, 757
  death, 34
  deforestation, 13, 779, *820*-821, 826-*827*
  Hubbard Brook Experimental, 782-*783*
  rain forest, *404*, 755
    tropical, 722-723, 752-*753*, *827*
  temperate, 755-*756*
  tropical, agriculture in, 820-821
Fossil, *276*, **283**, 354
Fossil fuel, and acid rain, 35
Fossil record, evidence for evolution, 283-*285*
Founder effect, *307*-308
Frameshift mutation, 184
Freezing point, 32
Frog(s), 457-*458*
  as food, *17*
  breathing, 500
  mating, *14*
  metamorphosis, 195
  survival, *11*
Frond, 419
Fruit, development, 731, *734*
  dispersal, 734-737
  production, 714
  ripening, *735*
Fruiting bodies, 393, *396*
Fuel shortage, deforestation and, *820*
Fugitive plants, in secondary succession, 763
Functional groups, **42**, 44
Fungi, 364, 392-394
  bacteria and, 399-400
  body plan, 393
  classification, *394*-397
  in soil, 690
  overview, *392*
  reproduction, 393
  symbiotic relationsips, 397-399
Future generations, genetic contribution to, 294-295

GABA, 592
Galapagos Island species, *282*, *289*
Gall bladder, 482
Gamete, **204**, **430**, **561**
  formation, 219-223
Gametophyte, **411**, 729
Gamma-aminobutyric acid, 592
Ganglia, **593**
Gap junctions, **79**-*80*, 612
Gas, *29*
Gas exchange, 490-505
Gastrocoel, 576
Gastrointestinal tract, 480
Gastropods, 443-*444*
Gastrovascular cavity, 436, 479, *508*
Gastrula, 576
Gastrulation, 574, 576
Gated channels, 75
Gel electrophoresis, in gene mapping, 207
Gelada Baboon, *644*
Gene(s), **164**, **179**
  activity, 194-196
  and environment, 646-647
  independent assortment, **234**-236
  mapping, 207, 209
  modifier, 263
  regulation, in protein synthesis, 190-*191*
  sex-influenced, 262-263
  structural, 179
  transcription, *181*
Gene expression, **165**, 246-271
Gene flow, 305-307
Gene pairs, 229
Gene pool, *301*
Gene therapy, 199
Generations, alternation of, 411
Genetic code, **165**, 182-184
Genetic contribution, to future generations, 294-295
Genetic counseling, 266-267
Genetic cross, 233
Genetic diseases, elimination, 468
Genetic drift, *306*-309
Genetic engineering, 172, **174**-175
Genetic information, **164**
Genetic markers, 208
Genetic polymorphism, 309
Genetic reassortment, **218**-219
Genetic recombination, 218, *376*
Genetic uniformity, 308
Genetic variation, *270*, *306*, 309
Genetics, and evolution, 278-283, 322-339
  chromosome mapping, 209, *242*
  codominance, **236**-237
  crossing over, 218, ***239***, 242
  gene pairs, 229
  hybrid cross, **234**-*236*, *238*
  incomplete dominance, **236**-*237*
  independent assortment, **234**-236
  law of segregation, 231
  lethal alleles, 248-252
  linkage, 237-**239**
  mendelian, 226-245
  metabolic errors, 252-*253*
  monohybrid cross, *232*-233
  multiple alleles, 254-255
  phenotypic expression, **231**, *249*
  polygenic traits, **255**-*256*
  probability calculations, 343
  Punnett square, 233-*234*, *236*
  test cross, 233-*234*
  vocabulary, *230*
Genome, **172**
  human, 199
  organization, 172, 176
Genotype, **231**
Genus, **358**
Germ cells, **108**
Germ layers, 576
Germination, 737
Gestation, **569**
Giant kelp, *409*
Giardiasis, 387
Gibberellins, *713*-*714*
Gill, 442, 494-*496*
Gingko, *318*, *423*
Ginkgophyta, 423, 425
Giraffe, *281*
Girdling, **694**
Gizzard, 480
Glial cell, 584
Global warming, 13, 120-122
  in carbon cycle, *780*
  two views of, 824-825
Glomerulus, 553
Glucose, 48-*49*
  energy yield, 134-135
  oxidation, *127*
Glucose threshold, 553
Glyceraldehyde phosphate, 152
Glycogen, 48, *50*
Glycolipids, 67
Glycolysis, **127**-*130*
  energy yield, 134, *137*
Golden algae, 385-*386*
Golden toad, extinction, 292
Golgi complex, *90*-*91*, **94**-**95**
Gonads, **257**, **430**, **563**
Gondwanaland, *4*
Gonorrhea, 571
Goosegrass seeds, *736*
Gout, 263
Grade vs clade, **361**

Graded response, of intact muscle, 618
Gradients, growth response to, 718
Grafting, *739*
Gram-stained bacteria, *379*
Grana, *98*, **147**-*148*
Grass, armed, *680*
Grasshopper, *446*, *479*
Grassland, temperate, 756-757
Gravitropism, 718-*719*
Gray heron, *795*
Gray matter, 589
Great blue heron, *487*, *774*
Green algae, *409*-411
Green manure, 767
Green revolution, 817-819
Greenhouse gases, 120
Griffith's experiment, *167*
Grizzly bear, *358*
Grooming behavior, 660
Ground pines, 418
Ground tissue, ***106***, ***669***
Growth, secondary, 678-681
Growth factors, 210, 637
Growth rings, 695
Guanine, *168*
Guard cells, 678, *701*
Gut, 480
Guttation, *698*
Gymnosperms, *421*-425, *697*
Gypsy moth, *448*

Habitat, and species diversity, 783
  destruction, with agriculture, 814
  loss, and extinction, 292-293
Habituation, 652
Hairs, protective, on plant, *680*
Hammerhead shark, *456*
Hand, growth, *623*
Haploid chromosomes, 204, **206**
Hardy-Weinberg principle, 302-*303*
Hares, 464
Harlequin beetle, 471
Haustoria, 393
Haversian system, *624*
Hawaiian goose, *315*
Hazel bush, *728*
Head-foot, 442
Heart, conduction system, *614*
Heart attack, 476, 517
Heart cells, *102*
Heart cycle, in mammals, 514
Heartwood, 696
Heat, of respiration, 115
  production, 112
Height, genotype, *256*
HeLa cells, evolution, 160-161
Heliconia, flowering, *425*
Helix, DNA, 169
Hemoglobin, 469, 501
Hemophilia, **248**, 260-*261*
Hemorrhoids, 514
Henle's loop, 553
Hepatic portal vein, 488
Herb, 755
Herbivores, **473**, 772
  coevolution with plants, 156-157
  feeding and digestion, *485*-486
Heredity, principles, 240
Hermaphroditism, **328–329**
Herons, *487*, *774*, *795*
Herpes, 571
Hershey-Chase experiment, *167*-168
Heterotrophs, **117**, 772
  and autotrophs, **349**-350
Heterozygote advantage, 310-311
Heterozygous individuals, **229**
Hippocampus, 598
Histamine, 526, *540*
  in mast cell, *71*
Histones, **172**-*173*
HIV virus, 4
Holdfast, 408
Homeostasis, **64**, **545**
Homeothermic animals, **493**
Homing, 655-657
Homo sapiens, classification, 359
Homologous characters, 360
Homologous chromosomes, 229
Homologous structures, 204, **285**
Homozygous dominant/recessive genotype, **231**
Homozygous individuals, **229**
Honeybee societies, 658-*659*
Hormone(s), 47, 51, **630**-633
  and reproduction, 566-569
  and seasonal changes, 637-638
  control, 634-636
  in gene expression, 263
  male, *632*
  pituitary, 631
  plant, 414, **712**-*713*
  production, 263
  vertebrate, 631
  water-soluble, 633-634
Hornworm, feeding, *485*
Horse, digestion, *479*
  evolution, 285-*286*
Horse chestnut, *737*
Horsetails, 418-*419*
Hubbard Brook Experimental Forest, 782-*783*
Human beings, blood groups, 254-255, *310*
  brain, *599*
  digestion, 480-485
  early development, *575*
  hunter-gatherers, 812-813
  impact on environment, 810-833
  reproduction, 563-578
  sex determination, **258**
Human chorionic gonadotropin, 567
Human error, in experiment, 8
Human genome, improvement in, 199
Human genome project, 207-209
Human population, feeding, 815-819
  growth, *802*-803, 806
Hummingbird, *14*
  bill, *487*
Humoral immunity, 533-535
Hunter-gatherers, 812-813
Huntington's disease, 252
Hurricane, *122*
Hybrid(s), 228-*229*, **310**-*311*
  selection against, 317
Hybrid vigor, 310-311
Hydra, *14*
  asexual budding, *324*
  digestion, *479*
  gastrovascular cavity, *508*
Hydrocarbon chain, *44*, *46*
Hydrogen bond, *27*-**28**
  between water molecules, *32*
Hydrogen ion pump, 347-348
Hydrolysis reaction, **42**, *44*-*45*
Hydrotropism, 718
Hymen, 563
Hyperpolarization, *587*
Hypersensitivity reaction, 540
Hypertension, diet and, 476
Hypertonic fluids, **548**
Hypertonic solution, **73**-*74*
Hypha, 393
Hypothalamus, 598, **635**
Hypothalamus-pituitary connection, 635-*636*
Hypotheses, 6
Hypotonic fluids, **548**
Hypotonic solution, **73**-*74*
Hyracotherium, evolution, 285-*286*

Ice, *32*
"Iceman", *122*
Iguana, marine, *544*
  of Galapagos Islands, *282*, *289*
Imagination, and environment, 830-831
Immune response(s), 528-529, *534*
  main features, *529*
  medical aspects, 536-537
  primary, **535**
  types, 533
Immune system, **525**, 529-531
  cells, 530-531
  human, *530*
  malfunction, 537, 540-541
Immunity, **528**
  cellular, 533-*534*
  humoral, 533-535
  passive, 537
Immunodeficiency. *see also* AIDS
  severe combined, 199
Immunological memory, **528**, 535-*536*
Immunological proteins, 51
Immunosuppressant drugs, 537
Impala horns, *330*
Impermeability, 72
Implantation, 572-574
Imprinting, *647*
Inborn errors of metabolism, 252-*253*
Incisors, 480
Incomplete dominance, **236**-*237*
Independent assortment, **234**-236
Index of relatedness, 306
Indian corn, *173*
Individual distance, 654-*655*

Inducer, 190
Industrial melanism, 291
Industrial revolution, 821–822
Inflammation, 526
Inflammatory response, *528*
Ingestion, **362**
Inheritance patterns, 246–271
Inhibitors of enzyme action, 57
Inhibitory responses, 586–*587*
Initiation, of protein synthesis, 187
Innate behavior, **648**
Insect(s), 447–449
  circulatory system, *509*–510
  devastation by, 448
  learning by, *650*
  metamorphosis, *196*
  social, 336–*337*
Insectivora, 463
Insertion mutation, **171**
Insight learning, *653*
Instinct, learning vs, 648–649
Insulin structure, *53*
Integration, nervous system, 594
Interbreeding, barriers, 314
Intercalary meristem, 584
Intercellular junctions, *80*
Interfaces, in water, *33*
Interferon, 527
Interleukin, **535**
Internal fertilization, 324, *562*
Internal transport, 506–523
Interneuron, **583**
Interphase, 210, *212–213*
Intertidal zone, **760**
Intervening sequences, of DNA, 185
Intracellular digestion, 479
Intrauterine devices, 571
Introns, 185
Inversion mutation, **171**
Invertebrates, **430**, 433, 436–451
  excretory organs, 550–*551*
  nervous system, *594*
  transport, 508–510
Ion, **28**
  intracellular vs extracellular, *76*
Ionic bond, *27*–**28**
Iron, as plant nutrient, 688–689
  in soil, 691
Irrigation, 817–819
Irritability, 71
Isle Royal National Park, *830*
Isograft, 664
Isolation, as interbreeding barrier, 314
Isotonic fluids, **548**
Isotonic solution, **73**–*74*

Jackson's chameleon, *460*
Jaw, evolution, 454
Jawless fishes, 453–454
Jellyfish, 436, *438*, *508*
Jenner, Edward, 536
Jet lag, 641
Joint structure, *622*
Jumping genes, 172

Juvenile hormone, in metamorphosis, 196

Kalahari bushmen, *812*
Keystone species, **811**
Kidney, function, 555–557
  structure, *552*
Kin selection, **333**
Kinetic energy, 29, **111**, *113*
Kinetochore, 214
Kingdom(s), of organisms, **358**, 362–364
Kitchen chemistry, 55
Klamath weed, *799*
Klebsiella pneumoniae, *374*
Klinefelter syndrome, 265

Labia majora and minora, 563
Lactate fermentation, 135–*136*
Lactation, 569, *635*
Lakes, 759–760
Lamarckism, 280
Lampreys, *454*
Lancelet, 451–*452*
Land plants, 411–427
Large intestine, 482
Larva, **434**
Larynx, 496
Latency, in behavioral responses, 649
Latent learning, 653
Latent virus, 197
Lateral bud, 675
Lateral meristem, 678
Lateral root, 672–673
Latitude, effect on vegetation, *751*
Law of segregation, and meiosis, 231
Laws of thermodynamics, 112
Leaching, 690
Leaf, 676–678
  colors, *146–147*
  defense, 680–681
  primordia, 675
  senescence and abscission, *721*
  structure, 147–149, 676–678
Leaflets, **676**
Learning, instinct vs, 648–*649*, *651*
  types, 652–653
Lecithin, *48*
Leeches, 442–*443*
Legume, 377–378
Leishmaniasis, 386–387
Letdown response, *635*
Leukocytes, *84*
  polymorphonuclear, *531*
Lianas, 722
Lichens, *398*
Life, **12**, 15–19
  diversity, 275–469
  origin, 340–355
Life expectancy, 802
Life functions, basic, 392
Life table, of US population, 793
Lifeboat ethics, 829
Lifespan, **210**

Lifestyle, and consumption level, *824*
Ligaments, **621**
Light, 144
  absorption in photosynthesis, 149
  and gene expression, 263
  trapping energy of, 145–147
Lignin, 694
Limbic structures, 598
Limiting factor, 776
Linkage groups, 237–**239**
Linnaeus, and scientific names, 369
Lion, *270*, *464*, *658*
Lipids, **41**, **45**
  bilayers, 67
Liquid, *29*
Littoral zone, 759
Liver, functions, 488
  transplantation, 664–665
Liverworts, 415–*416*
Lizard, 458–460
Local potential, 586–587
Long-day plants, 720
Loop of Henle, 553
Low-input farming, *819*
Lumen, 483–*484*
Lung(s), 496, 500
  cancer, *197*, 499
Lungfish, *4*
Luteinizing hormone, 566–567, 632
Lycophyta, 418
Lycopods, 418
Lyell, and theory of evolution, 282
Lymph nodes, 529–*530*
Lymphatic system, **519**, 522, *530*
Lymphocyte(s), **528**, 531. *see also* B cells; T cells
Lysogenic cycle, 366
Lysosomes, *90–91*, **95**–*96*
Lysozyme, 526
Lytic cycle, 366

Macroevolution, **318**
Macromolecules, **41**, *43*
Macronucleus, 389
Macronutrients, 474
  plant, 688–689
Macrophages, *524*, 526, 531
Madagascar, crowding in, *807*
Madagascar periwinkle, *801*
Magnesium, as plant nutrient, 688–689
Major histocompatibility complex, **533**
Malaria, 387
  control, 424
  elimination, 468
  genetic protection from, 251
Male, evolutionary role, 328–333
Malignant tumor, **196**
Malpighian tubules, invertebrate, 550–*551*
Malthus's theory, of evolution, 282
  of population growth, 834
Mammal(s), 462–465
  adaptation to sodium-deficient environments, 554
  circulatory system, 512–518

diversity, *464*
evolutionary lines, *362*
sex determination, **257**
Mandible, 462
Mantle, 443
Manx cat, 249, *251*
Map sense, in migration, 657
Maple syrup, *686*
Marbled salamander, *458*
Marchantia, *416*
Marijuana, effect on brain, 597
Marine iguana, *544*
Marsupials, *463*
Mass flow, 703-*704*
Mast cell, *71*, 526, *540*
Mating, preferences, 305
systems, 330-333
Maturation, 578-579
Mechanical energy, *113*
Mechanoreceptor, 606
Medulla, of kidney, 553
Medusa form, of cnidarians, 436
Meerkat, *322*
Megaspores, 421
Meiosis, ***204, 215-218***
difference from mitosis, 219
law of segregation, 231
overview, 205
Melanin level, and albinism, 253
Melanism, industrial, 291
Membrane, attachments between cells, 78-*80*
uncharged molecules crossing, 71-74
Membrane potential, 77, *587*
Membrane proteins, 74-77, *75*
Membrane transport, 74-78
Memory, immunological, **528**, 535-*536*
Memory cells, 535
Mendel, Gregor, 226, *240*-241
Mendelian genetics, 226-245
Meninges, 595, *601*
Menstrual cycle, 567-*568*
Meristem, **669**
apical, 670
intercalary, 584
lateral, 678
Meristematic cell culture, 742
Mesoderm, 576
Mesophyll, **147**-*148*, 678
Mesosome, **86**
Messenger RNA, *180*-181
codons, 183
production, *185*
Metabolic pathways, **59**, *137*, *347*
Metabolic rate, ***493***
Metabolism, 58-61
beginnings, 347-348
by plasma membrane, 67
coevolution with reproduction, 349
inborn errors, 252-*253*
Metamorphosis, **194**-196
Metaphase, chromosome arrangement, *220*
in cell cycle, 214
in meiosis, *216-217*
in mitosis, *212*
Metastasis, 196
Methane, as greenhouse gas, 121
Methylation, **194**
Metric prefixes, **88**
Microbiota, **380**
Microevolution, **318**
Microfilaments, 103-104
Micrographs, **68**
Micronucleus, 389
Micronutrients, 474-478
in plant nutrition, 688-689
Microorganisms, **373**
Micropyle, 730
Microscopes, 68-*69*
Microspheres, *346*
Microspores, 421
Microtubules, *91*, *102*
Microvilli, *90*, 483-*484*
Midbrain, 597
Middle lamella, cell plate, 215
Migration, 655-*657*
Mildew, *393*
Miller's apparatus, *344*
Mimicry, for camouflage, **805**
Minerals, 475, 478
cycling, 778-782
plant absorption, 688-689
Minnows, *762*
Miscarriage, 569
Mitochondrion (pl., mitochondria), *90-91*, **96**-*97*, 129, *131*
as endosymbiont with chloroplasts, 101
matrix, **129**
Mitosis, ***204, 211-214***
difference from meiosis, 219
in cell cycle, *211*
overview, 205
Mitotic spindle, 214
Modifier genes, 263
Moisture, and vegetation, 750
Molars, 480
Mole, **37**
Molecular formula, 28
Molecule, **26**, 28-29
crossing membranes, 71-74
functional groups, **42**
movement, 29
Mollusca, 442-445
Molting, arthropod, 445
Monera. *see* Prokaryotes
Monocotyledons, **581**
comparison with dicotyledons, 581-584
Monoculture, 399
Monogamy, 332
Monohybrid cross, *232*-233
Monomers, **41**, *43*
organic, 344-345
Monophyletic taxon, 361
Monosaccharides, 48-*49*
Monotremes, *463*
Morel, *400*
Morpho butterfly, *723*
Morphological species, 312
Mortality factors, 794
Morula, **572**
Moss, 415-*418*
Moth, peppered, evolution, *291*, 294
pollination by, 740
Motility, **362**
Motivation, 652
Motor neuron, **583**, *585*
Motor pathways (brain), 600
Motor units, 618
Mountain lion, *464*
Mucous membranes, 526
Mucus, 526
Multicellular algae, 407-411
Multicellular organisms, 65
Multicellularity, 390-392
Multiple alleles, 254-255
Multiple births, 572
Multipolar neuron, *585*
Muscle(s), ***105***, *610-627*, *612-614*
antagonistic, 618-620
cardiac, 612, 614
contraction, 615-618
distribution and function, *613*
fibers, 614
graded response, 618
interaction with skeleton, 618-620
skeletal, 614
smooth, 612
Mushrooms, 396-*397*, 400
Mustard oils, 156-157
Mutagens, **171**, **197**
Mutation, **171**-172, **248**
frameshift, 184
phenotypic expression, 248
types, 171
Mutualism, **380**, *747*
Mycelia, *393*
Mycoplasma, 379
Mycorrhizae, 397-*398*, *692*
Myelin sheath, **589**
Myofibril, 614
Myoglobin, 521
Myosin, 103, 615
Myxomycota, 389-*390*

Natural selection, **1**, **3**, **10**-11, **278**, 290-294, **302**-305
Negative feedback, **59**
circulatory system, 516
hormones, *632*
Negative pressure breathing, 500
Nekton, 761-*762*
Nematocysts, 436
Nematoda, 439, *442*
Nephric capsule, 553
Nephridium, 550-*551*
Nephron, 551, 553
Nerve, **589**-*590*, 595
Nerve growth factor, 637
Nerve impulse, 584, 587-589
Nervous system, 582-609
autonomic, 595, 602
control, 634-636
neurons, 593-*594*
tissue, ***105***

vertebrate, 594–606
Net primary productivity, 776, 835
Neurectoderm, 576
Neuroglia, **584**
Neuromuscular junction, *591*–592, **612**
Neurons, **583**–593
electrical properties, 584–589
function, 584
organization, 593–*594*
structure, *585*
Neuropeptides, 592
Neuroregulators, 637
Neurotransmitters, **589**, 592–593
as local messengers, 637
Neurulation, 576
Neutron, **25**
Neutrophils, *531*
Newts, 457–*458*
Niche, **796**, 798
Nicotinamide adenine dinucleotide, 127
Nicotine, 498, 596
Nitrifying bacteria, 377
Nitrogen, as plant nutrient, 688–689
Nitrogen cycle, *781*
Nitrogen fixation, 377
Nitrogen oxides, in acid rain, 35
Nitrogenous wastes, 546–548
Nitrous oxide, as greenhouse gas, 121
Node of Ranvier, *585*, **589**
Nondisjunction, **265**
Nonpolar bond, 26–*27*
Norepinephrine, 592
Northern bear, *358*
Norway rat, *16*
Notochord, **450**, 576
Nuclear area, **86**
Nuclear envelope, ***92***
Nuclear genetics, 174
Nuclear membrane, *90–91*
Nuclear pore, *92*
Nucleic acids, **41**, 58
Nucleocapsids, *367*
Nucleolus, **89**–*90*, *92*, **186**
Nucleosomes, **172**–*173*
Nucleotide, 58
Nucleus, **25**, *90–91*
as genetic message center, 89–93
cell, **65**
Nudibranch, 443–*444*, *492*
Nutrients, 474–478
mineral, 778–782
Nutrition, animal, 472–489
human, 480–485
plant, 688–689

Objective microscope lens, 68–*69*
Obligate anaerobes, **375**
Observations, experimental, 6
Ocean communities, 760–762
Ocular microscope lens, 68–*69*
Oils, 47
Olfactory organs, 598
Oligosaccharins, 712
Oligotrophic lake, 759–*760*

Omnivore, **473**, 772
Oncogenes, **197**, 369
Onion, cells, *14*
food storage, *706*
Oocyte, oogenesis, oogonia, 221–*222*
Oomycota, 390
Operculum, **495**
Opiates, effect on brain, 597
Optic nerves, 597
Optic tectum, 597
Optical illusion, *603*
Opuntia cactus, *289*
Orangutan, *360*
Orchid, *425*
Order, **358**
Organ(s), **104**
transplantation, 537
Organ pipe cactus, *755*
Organic molecules, 344–345
structure, 42–45
Organism, **1**
characteristics, 359–361
classification, 356–371
distribution, 748–769
Organogenesis, 576
Orgasm, **566**
Origin of life, 340–355
Origin of Species by Means of Natural Selection, 283
Oscillatoria, *375*
Osmoregulation, **548**–550
Osmosis, 72–***73***, *549*
Osteichthyes, 455–*456*
Osteoporosis, *625*
Ostracoderms, 453
Ostrich, *482*
Ovary, flower, 427, 728
human, 563
Overfarming, 13
Overgrazing, *828*
Overlapping niches, 798
Overpopulation, 13, 334–335
two views of, 823–824
Ovulation, in human female, 563–*564*
Ovule, 728, 730–731
Ovum, 221
Owl, great grey, *461*
Oxaloacetic acid, 131
Oxidation-reduction reactions, *117*–118
Oxygen, and orgin of life, 342
blood carrying, *501*
in cellular respiration, 134
in soil, 690
partial pressure of, 501
production in photosynthesis, *15*
supply, 492–493
Oxygen debt, 135
Oxyhemoglobin, as respiratory pigment, 501
Oxytocin, 567, 569, 636
Ozone, as greenhouse gas, 121
depletion, 13, 352–*353*
two views of, 826

Paleontologists, 16

Palisade layer (mesophyll), **147**–*148*, 678
Pancreas, 482
Panther chameleon, *15*
Paramecium, *389*
aurelia carrying capacity, *794*
competition between, *797*
Parasites, 378, **440**
worms, 441
Parasitism, **380**, 440–441
Parasitoid, **441**
Parasympathetic nervous system, 602
Parathyroid hormone, 624, 632
Parenchyma, ***106***, ***669***
Parental behavior, 336
Parental generation, 228
Parrot beak, *487*
Parthenogenesis, **317**, 324
Passive immunity, 537
Passive transport, proteins in, 74–76
Pasteur, Louis, 536
Paternity testing, 255
Pathogen, **399**, **525**, 533, 535
Pea plants, genetics, *227*
Peacock, albinism, 252–*254*
courtship display, *305*
Peanuts, self-planting, *726*
Pecking order, 658–**659**
Pectin, *99*
Pectoral girdle, **620**
Pelican, *487*
Pellicle, **385**
Pelvic girdle, **620**
Penaeus duorarum, *775*
Penguins, *461*, *520*–521
Penicillin, 396
Penicillium, *395*
Penis, 563
Peppered moth, *291*, 294
Pepsin/pepsinogen, 482
Peptide bonds, **52**
Peptide chain formation, 187, *189*
Pericycle, 672
Peripheral nervous system, 595
Peristalsis, **482**
Periwinkle, Madagascar, *801*
Permafrost, 758
Persian calico cat, 261–*262*
Pest control, predation and, **798**–800
Pesticides, and green revolution, 817–819
for insects, 448
for plants, 707
resistance, 295
Petals, 728
pH, **33**, 36–37
and enzyme action, 57
Phaeophyta, 408–*409*
Phages, 166–*167*
Phagocyte, **525**, 530
Phagocytosis, 77–*78*, *535*
Pharyngeal gill slits, **450**
Pharynx, **450**, 479–480, 496
Phenotypes, **231**
and mutations, *249*
Phenylalanine, 253
Phenylketonuria, 252–*253*

Pheromones, **630**, **641**-*642*, 658
Phloem, 414, **669**
  fibers, 702
  functions, 694
  structure, 701-702
  transport, 702-705
Phosphate group, *44*
Phospholipids, 47-*48*
  in membranes, 67
Phosphorus, as plant nutrient, 688-689
  transport in plant, *703*
Phosphorus cycle, 781-*782*
Photochemical reactions, 149
Photolysis, *151*
Photoperiodism, 720
Photoreceptor, 606
Photosynthesis, **142**-161, 350
  and respiration, ***116***-117
  C_4, 154-*155*
  CAM (crassulacean acid metabolism), 154-*155*
  details, 149-152
  ecological aspects, 154
  equation, 142
  in bacteria, *350*, 376
  in protists, 383-386
  location, *148*
  pigments, 145-147
  rate control, 153-154
Photosystem, 149, *151*
Phototropism, 718-*719*
Phylogeny, 359
Phylum, **358**
Physarium polycephalum, *390*
Phytoplankton, **383**
Pigeons, homing, 657
Pigments, photosynthetic, 145-147
Pikas, 464
"Pill, the", 570
Pilus (pl., pili), **374**
  bacterial, 86
Pine, annual rings, *696*
  life history, *422*
  pollen, *305*
Pineal gland, 638
Pinocytosis, 77
Pioneer plants, 763
Pistils, 728
Pit, in tracheid, 694
Pith, 674
Pituitary gland, 635-*636*
Placenta, **463**, 567
Placental mammals, *463*
Plamodesmata, **79**, *80*-81
Planaria, 508-*509*
Plankton, 433-*434*, 761
Plant(s), biology, 667-745
  biotechnology, 744
  breeding, *738*
  coevolution with herbivores, 156-157
  cytokinesis, *215*
  domestication, *814*
  flowering, *425*-427
    reproduction, 726-743
  gene-engineered, 744
  growth, 717-719
  hormones, 712-716
  land, 411-427
    adaptations to terrestrial environment, 413
    economic importance, 426
    evolution, *415*
    water vs, 413
  life histories, 411
  major groups, 410
  nutritional requirements, 688-689
  regulation and response, 710-725
  reproduction on land, 414-415
  structure and growth, 668-685
  substances in, 705-707
  tissues, ***106***
  vascular, 416-417
    lower, 418-421
    nutrition and transport, 686-709
Plant kingdom (Plantae), 364, 404-429
Plasma, 518
Plasma cell, 535
Plasma membrane, **65**
  roles, 67, 71
  structure, *70*
Plasmids, **174**, 270
Plasmodesmata, 692
Plasmodium, 387-*388*
Plastids, 96-97
Platelets, 518
Platyhelminthes, 438-*439*
Pleistocene glaciations, 315-316
Poikilothermic animals, *494*
Polar body, 221
Polar bond, 26-*27*
Polar nuclei, 730
Polarization, **584**
Poles, in cell cycle, 214
Pollen, *729*
  gymnosperm, 421
  maturation, 730
Pollen tube, 730
Pollination, *728*, **730**
Pollinators, coevolution with flowers, 740-741
Pollution, air, 498-499, 822
  solid waste, 822
  water, 690, 817
Polyandry, **331**-332
Polychaetes, 442-*443*
Polygenic characters, **255**-*256*
Polygyny, **330**-331
Polymers, **41**, *43*
  biological, 42
  formation, *346*
Polymorphism, genetic, 309
Polymorphonuclear leukocytes, *531*
Polyp form, of cnidarians, 436
Polypeptide(s), *43*, **52**
  formation, 182
  production, *188*
Polyphyletic taxon, 361
Polyploidy, **171**, **316**
Polysaccharides, 48, *50*
Polytene chromosome, *193*-194
Population(s), **301**, **789**-809
  changes, *803*
  density, **789**, **795**
  growth, *790*-794
    human, *802*-803, 806
    with agriculture, 814
    zero, 834
  regulation, 794-795
  stable, *795*
  variablity, 309-311
Population explosion, 334-335
Population genetics, and speciation, 300-321
Porifera, 436-*437*
Positive pressure breathing, 500
Posterior pituitary hormones, 636
Postsynaptic membrane, 589
Postzygotic isolation, 314
Potassium, as plant nutrient, 688-689
Potato, *706*
Potential energy, **111**, *113*
Poverty, and birthrate, *824*
Powdery mildew, *393*
Prairie, 756-*757*
Prebiotic earth, 341, 343-344
Precipitation, and productivity, *777*
Predation, and pest control, **798**-800
  and species diversity, 783
Predator, 798
Prediction, in experimental method, 7
Prednisone, for immune suppression, 665
Pregnancy, 567, 569
  smoking in, 498
Pressure flow, 703
Presynaptic membrane, 589
Prey, 798
Prezygotic isolation, 314
Primary cell wall, 694
Primary consumers, 772
Primary growth, **669**
  roots, 671-673
  stems, 674-675
Primary immune response, **535**
Primary productivity, 776-777
Primary structure, of protein, 53-*54*
Primary succession, **762**-*763*
Primary tissue, roots, 672
  stems, *675*
Primates, 464, *660*
Probability, 343
Prochlorophytes, 380
Producers, 772
Productivity, 776-778
Progesterone, 567
Projector microscope lens, 68-*69*
Prokaryotes, 362, 374-381
  classification, 378-380
  protein synthesis control, 190-*191*
  reproduction and evolution, 374-375
  transcription control, *191*
Prokaryotic cells, **85**-87
  eukaryotic vs, 88-89, 100-101
Prolactin, 567, 569
Promotor, 181
Prop roots, *673*

Prophase, in cell cycle, 214
in meiosis, *216-217*
in mitosis, *212*
Prostaglandin(s), as local chemical messengers, 637
for immune suppression, 665
Prostate, 563
Protein(s), **41**, 51-53
as macronutrients, 474
energy yield, 139
functions, 51
structure, 53-*54*, 182
synthesis, 179, 187-190
control, 190, 193-194
overview, 184
RNA and, 178-201
Proteinoid microspheres, *346*
Prothrombin, 518
Protists, **1**, 362, 364, 381-400
colonies, 381
cysts of, 382
heterotrophic, 386-390
photosynthetic, 383-386
physiology, 381-383
sessile, 382
Protobionts, 347
Proton, **25**
Proto-oncogene, **197**
Protozoa, 381-*383*, 386-390
Proximal convoluted tubule, 553
Pseudomonas aeruginosa, *86*
Pseudopods, **382**
Pterophyta, *419*-421
Puddling, of butterflies, *6*, *9*
Pulmonary artery and veins, 514
Punctuated equilibrium, 318
Punnett square, 233-*234*, *236*
Purple carrion flower, *741*
Purple loosestrife, *16*
Pyrogens, **527**
Pyrrophyta, 384
Pyruvate, 127, 129-*130*

Quarternary structure, of protein, 53-*54*
Queen, among social insects, 336-*337*
Quinine, 424

Rabbits, 464
Radial symmetry, 434
Radiolarians, *14*
Rafting, *307*
Rain forest, *404*, 755
Rainbow, *142*
Ray (fish), 454-*456*
Ray (in wood), 695
Reaction, chemical, 30
Reaction center, 149
Receptacle, seed, 734
Receptor, 583, 603, 606
cell surface, 67
Receptor cell, 583, 603, 606
Receptor-mediated endocytosis, 77, *79*
Receptor potential, **606**
Recessive allele, 229-231
Reciprocal inhibition, 619
Recombinant, 239
Recombinant DNA, **174**
Rectum, 482
Red algae, 407-*408*
Red anemone, *438*
Red blood cells, 518-*519*
Red-eyed treefrog, *458*
Red footed Boobie, *282*
Red maple, *425*
Redox reactions, *117*-118
Reducing atmosphere, 343
Reef coral, *438*
Reflex, conditioned, 652
Reflex arc, 600-602
Reinforcement, 653
Rejection, in transplantation, 537
Relatedness index, 306
Releaser, 651
Releasing factor, **636**
Renal artery and vein, 551
Renin, 557
Repetitive DNA, 176
Reproduction, age of, and population growth, **791**
asexual vs sexual, 324
coevolution with metabolism, 349
environmental control, 637-638
evolution and, 322-339
hormones, 566-569
human, 563-578
parthenogenetic, **317**, 324
patterns, 562
sexual, 324, 560-581
strategies, and survivorship, **791**-793
Reproductive organs, human, 563-*565*
Reptiles, 458-*460*
circulation, 510, 512
Resin, 695
Resolving power, 68
Resorption by nephron, 553
Respiration, 126-135, 350
cellular, 117, **125**-141
mitochondria in, 96
coenzymes in, 127
photosynthesis and, *116*-**117**
preparation for, 129-130
Respiratory medium, 492
Respiratory pigment, 501-502
Respiratory surface, 494-501
Respiratory system, *497*
Resting potential, 584, *586*
Restriction enzyme, 207
Restriction fragment length polymorphism (RFLP), 208
Retroviruses, 369, 538
Reverse transcriptase, 368
RFLP, 208
Rhizobium bacteria, *378*
Rhizoid, **415**
Rhizome, 417-*418*
Rhizopus, *395*
Rhodophyta, 407-*408*
Rhynia, *418*
Rhythm method, 570
Ribonucleic acid. *see* RNA
Ribosomal RNA, *180*-181, 186
Ribosome, *86*, **89**, ***93***
Ribulose bisphosphate, 152
Rickettsiae, 379
Right lymph duct, 519
Rigor mortis, **615**
RNA, **58**, 180-181
and protein synthesis, 178-201
messenger, *180*-181, 183, *185*
ribosomal, *180*-181, 186
structure and function, *180*
transfer, *180*-181, *186*-187
types, *185*-187
Robin, *368*
Rodent, 464
Root(s), absorption, 692-694
functions, 674, *693*
primary growth, 671-673
primary structure, *672*
stimulation, 712
structure, *693*
Root cap, **671**
Root hairs, 672
Root nodules, *378*
Root pressure, *698*
Root system, 669, *671*-674
Roseate Spoonbill, *770*
Rough endoplasmic reticulum, *90-91*, **93**-*95*
Roundworm, 439, *442*
Rumen/ruminant, **486**
"Runner's high", 592

Sabellid worm, *485*
Salamander, 457-*458*, *490*
metamorphosis in, *195*
Salinization, **766**
Saliva, 480
Salt, **33**
dissolving, *33*
Salt gland, *549*
Salt marsh, food web, 774-775
Saltatory conduction, 589
Sampling error, 8
Sap, 694
Saprobes, **373**, 378
Sapwood, 695
Sarcodina, 387
Sarcolemma, 614
Sarcoplasmic reticulum, 615
Satellite DNA, 176
Saturated fats, 47
Savanna, tropical, 753-*754*
Scanning electron microscopes, 68-*69*
Scanning probe microscopes, 69
Scarlet kingsnake, *460*
Scent glands, in cat, *628*
Schistosoma mansoni, *439*
Scientific method, 2-**3**, 6-8
Scion, **739**
Scouring rushes, 418-*419*
Scrotum, 563

Sea anemone, 436, *438*
Sea cucumber, *450*
Sea fan, *472*
Sea shore, *761*
Sea squirt, 451-*452*
Sea turtle, *460*, *762*
Sea urchin, *450*
Seafood, salt marsh and, 774-775
Seasonal changes, hormones and, 637-638
Seaweed, 408-*409*
Second law of thermodynamics, 112
Second messengers, 633
Secondary cell walls, 694
Secondary consumers, 772
Secondary growth, **669**, 678-681
Secondary immune response, **535**
Secondary phloem, 678
Secondary productivity, 778
Secondary structure, of protein, 53-*54*
Secondary succession, **763**-*764*
Secondary vascular tissues, 678
Secondary xylem, 678
Secretion, by nephron, 553
  of hormones, *632*-633
Sedimentary nutrient cycle, 782
Seed(s), breeding programs, 737-738
  density, and survival, 795-*796*
  development, 731, 734
  dispersal, 734-737
  gymnosperm, 421
  predators, 735-736
  production, 714
  viability, 737
Seed coat, 422, 670, **731**
Seed masting, 736
Segmented worms, 442-*443*
Segregation (genetic), 231
Selection, chemical, 347
  kin, **333**
  natural, **302**-305. *see also* Natural selection
  types, *304*
Selection against, 302
Selection against young, *4*
Selection for, 303
Selective permeability, 72
Selective pressure, *4*, 11
Self-planting, *726*
Self-pollination, 730
Self-propagation, of action potential, 587, 589
Selfishness, and altruism, 333, 336-337
Semen, 563
Semidesert, *758*
Seminal vesicles, 563
Seminiferous tubules, *202*, 563
Senescence, **721**
Sense(s), *604-605*
Sense organs, 582-609
  function, 602-606
Sensitization, in allergy, 540
Sensory neurons, **583**, *585*
Sepal, 728
Septa, in fungi, 393
  in segmented worms, 442
Serine, *52*
Serotonin, and blood clotting, 518
  as neurotransmitter, 592
Serum, blood, 518
Sessile protists, 382
Severe combined immunodeficiency, 199
Sex chromosomes, 204, 257, 259
Sex determination, 256-259
Sex-influenced genes, 262-263
Sex linkage, 259-262
Sexual characteristics, secondary, 563
Sexual differences, 328-330
Sexual dimorphism, 329-*330*
Sexual intercourse, 563-566
Sexual reproduction, 324-325
  evolution of, 326-328
  in plants, **727**
Sexually transmitted diseases, 571
Shade plants, 154-*155*
Shark, 454-*456*
Shield bug, *446*
Shift rotations, biological rhythms and, 641
Shoot system, 669
Shoot tip, *675*
Short-day plants, 720
Shrimp, *446*, *775*
Shrubland, 756
Siamese cat, *58*
Siamese fighting fish, *330*
Sickle cell anemia, 250-252, 469
Sieve areas, 701
Sieve plates, 702
Sieve tube elements, 701
Sign stimuli, in behavior, 651-*652*
Silkworm, *14*
Sinoatrial node, 514, 612, *614*
Siphon, 444
Sister chromatid, 204
Skeletal muscle, 614, *616-617*
Skeleton, **618**-*619*
  human, *621*
  interaction with muscles, 618-620
  vertebrate, 620-622
Skin, cancer, 826
  defense by, 526
  fungal infection, *399*
  section through, *527*
Sliding filament mechanism, 615
Slime mold, 389-*391*
Sloth, three-toed, *723*
Small intestine, 482, *484*
Smallpox vaccination, 536
Smart genes, 192
Smoking, 498-499
Smooth endoplasmic reticulum, *90-91*, ***94***
Smooth muscle, 612
Snail, *300*
Snakes, 458-460
Snapdragon, *237*
Social animals, *658*
Social behavior (sociobiology), **657**-661
Social hierarchy, 661
Social insects, 336-*337*
Social status, *824*
Sodium-deficient environment, 554
Sodium-potassium pump, 76, 584
Soil, and plant nutrition, 688-692
  disappearing, 766-767
  erosion, 766-767, 814, 821
  fertility, 692
  improvement, 767
  particle size and water capacity, *689*
Soil water, 688, 690-691
Solar power, *822*
Solid, *29*
Solid waste pollution, 822
Solute/solution, **2**
Solvent, **2**
Somatic cell, **108**
Somatic nervous system, 595
Somatotropin, 637
Specialization, 783
Speciation, **314**-318
  population genetics and, 300-321
Species, **312**-314, **358**
  derived, *315*
  diversity, **783**-*785*
  formation, 318-319
  identification key, 313
  loss, 826-*827*
  naming, 368-369
  turnover in communities, 784-785
Specific heat, 31
Spectacled caiman, *460*
Spectrum, *142*
Sperm, *202*, **561**, *565*
  fern, *420*
  formation, 219
  nuclei, pollen, 730
  size relative to egg, *328*
Spermatid/spermatocyte, 220
Spermatogenesis, 220-*221*
Spermatogonium, 220
Sphagnum moss, 415
Sphenophyta, 418-*419*
Sphincter, 482
Spinal cord, 600-602
Spinal nerve, 602
Spinal reflexes, 600-602
Spindle, mitotic, 214
Spiracle, 500
Spirochete, 379
Sponges, 436-*437*
Spongy bone, 623-*624*
Spongy layer (mesophyll), **147**-*148*, 678
Sporangia, fungal, 393
  of ferns, *419*
Spore(s), **204**
  bacterial, 375
  flowers, 729
  fungal, 393
  plants, 411
Sporophyte, **411**
Squid, *496*
Stabilizing selection, *304*
Stamen, 728
Staphylococcus aureus, *374*
Starch, 48-*50*
Starfish, *450*
Starvation, **815**

Stem(s), 674-676
  dicot, primary tissues, *675*
  functions, 675-676
  primary growth, 674-675
  primary structure, 674-675
  specialized, *676*
Stem cells, elongation, 713
Stereotyped behavior, 650-*651*
Steroids, 47-*48*, 633
Stigma, 728
Stimulus(i), **583**, 651
  conditioned, 652
  of sense organs, 603
  reaction to, *646*
  sign, 651-652
Stingray, *456*
Stipe, of brown algae, 408
Stock, **739**
Stomach, 480, 482-*483*
Stomata, **147**-*148*, 414, 676, *701*
STOP codon, 183, *189*
Storage, roots, 674
  stems, 676
Storage proteins, 51
Stratification, *284*
Strawberry plant, guttation, *698*
  vegetative reproduction, *739*
Stroke, 517
Stroma, *98*, **147**-*148*
Stromatolites, *354*
Structural formula, 28
Structural genes, 179
Structural protein, 51
Style, 728
Substitution mutation, **171**
Substrate, of enzyme, **56**-57
Subtidal zone, **760**
Succession, ecological, 762-764
Succulents, 754
Sucrose sources/sinks, 704
Sulfur, as plant nutrient, 688-689
Sulfur butterflies, puddling, *6*
Sulfur dioxide, in acid rain, 35
Sulfur spring, *372*
Summation (in muscle), 618
Sun plants, 154-*155*
Sun star, *450*
Sunflower, *675*, *688*
Sunlight, plant responses, *719*
  vertical vs oblique rays, *751*
Supernormal stiumulus, 651-*652*
Support, stems for, 675
Surface tension, 31
Surface-to-volume ratio, of cell, *66*
Surrogate mothers, 578
Survival, *11*
Survival advantage, 278
Survivorship, **791**-793
Suspensor, 731
Sutures, joint, 620
Swans, *332*
Swimbladder, 455
Swiss starlings, 294-295
Symbiont, **380**-*381*
Symbiosis, **100**, *380*
  of fungi, 397-399
Symbiotic bacteria, 380-381
  in humans, 482
Sympathetic nervous system, 602
Sympatric populations, 313
Sympatric speciation, **316**-317
Synapse, 218, **589**, *591*
Synaptic cleft, 589
Synaptic transmission, 589, 592
Synovial fluid, 622
Synthesis, in cell cycle, 210
Syphilis, 571
Systemic pesticide, 707
Systole, 514

T cell (lymphocyte), 531
  killer, **533**
  receptors, **531**-533
Tadpole, *790*
Taiga, **757**-*758*
Tapeworm, 438-439
Taproot, ***673***
Tarantula, *446*
Target cells, 633
Tars, 498
Taxon, **358**
Taxonomy, **358**
Tay-Sachs disease, 252
Telencephalon, 598
Teleosts, 455
Telophase, in cell cycle, 214
  in meiosis, *216-217*
  in mitosis, *213*
Temperate biomes, 755-757
Temperate desert, 757
Temperate forest, 755-*756*
Temperate grassland, 756-757
Temperate semidesert, *758*
Temperate shrubland, 756
Temperate woodland, 756
Temperature, 29
  and enzyme action, 57
  and gene expression, 263-*264*
  and metabolic rate, *494*
  and vegetation, 750
  relationship to biome, *752*
Template, in DNA, 170
Tendon, **619**
Termination, in protein synthesis, 187, *189*-190
  signal, in transcription, 181
Termites, *381*
Terrestrial environment, 413
Territorial behavior, 653-*654*
  and mating, 332
Territory, **653**, **796**
Tertiary structure, of protein, 53-*54*
Test cross, 233-*234*
Test tube babies, 578
Testes, 563
Testosterone, *48*, 566
  feedback control, 632
  in human reproduction, 567
Tetanization, 615, 618
Tetrad, 218, **231**
Tetraploidy, **209**
Tetrapods, 455-465
Texas bluebonnet plants, *710*
Thalamus, 598
Thalassemias, 252
The Ethics of a Lifeboat, 829
The Origin of Species by Means of Natural Selection, 283
Theory, 8
Thermal conductivity, 31
Thermochemical reactions, 153
Thermodynamics, laws of, 112
Thermoreceptors, 606
Thick filaments, in muscle, 614
Thigmotropism, 718-*719*
Thin filaments, in muscle, 614
Thoracic duct, 519
Thorax, 445
Thornwood, tropical, 754
Threat displays, 654-*655*
Three-toed sloth, *723*
Threshold, of neuron, 589
Thrombin, 518
Thromboplastin, 518
Thylakoids, *98*, **147**-*148*
Thymine, in DNA, *168*
Thyroid hormones, 633
Thyrotropin, 636
Thyroxin, 195
Tidal flow (breathing), 496
Tiger swallowtail butterflies, puddling, *6*
Tight junctions, 78, *80*
Timber wolf, *464*
Time, and orgin of life, 342
Tissue(s), **104**
  transplantation, 537
Toads, 457-*458*
Tobacco plant, *174*
Tolerance, immunological, 529
Tongue, 480
Tooth (teeth), form and function, *486*
  specialization, *462*, *480*
Topical pesticide, 707
Topsoil, creation, *691*
  loss, 13
Tortoise, *282*
Toucan, *461*
Toxins, 51, **399**
Toxoplasma, 387
Trace minerals, 478
Trachea, 496
  in insects, 447
  in vertebrates, 500
Tracheal systems, 500-*501*
Tracheids, 694
Tragedy of the commons, 828-829
Trait, genetic, 226-*227*
Transcription, *181*-182, *184-185*
  control, *191*
Transduction, 366, **374**
Transfer RNA, *180*-181, *186*-187
Transformation, **374**
Transgenic crops, 175
Translation, *188*

genetic, 182, *184-185*
Translocation, **265**
in protein synthesis, 187
Transmission electron microscopes, 68-*69*
Transpiration, **699**-701
Transplantation, 664-665
Transport, protein, 51, 74-77
roots, 674
stems, 675-676
Transposable genes, 172
Tree shapes, *717*
Trial and error, 652-*653*
Trichocysts, **388**
Tridachna, *444*
Triglycerides, *46*-47
Triplet code, 182
Trophic level(s), limitation, 776
of food chain, 772
Trophoblast, **573**
Tropical biomes, 752-754
Tropical forest, agriculture in, 820-821
seasonal, 753
Tropical rain forest, 752-*753*
canopy, 722-723
settlement, *827*
Tropical savanna, 753-*754*
Tropical thornwood, 754
Tropisms, 718-*719*
Tropomyosin, 615
Troponin, 615
Trypanosomes, *386*
Trypsin/trypsinogen, 483
Tubal ligation, 572
Tubers, **738**
Tumor, **196**
Tumor suppressor gene, 197
Tundra, 398, **757**-*759*
Tunicates, 451-*452*
Turbellarians, 438
Turgor, **675**
Turkeys, *461*
Turner syndrome, 265
Turtles, 458-460
Twin births, 572
Type specimens, 312, **358**
Typhoid immunization, *803*

Ulothrix, 326-*327*
Ultraviolet radiation, 352-*353*
Ulva, *412*
Umbilicus, **573**
Understory, forest, 755
Ungulates, 465
Unicellular organisms, 65
Unity of life, 163-273
Unsaturated fats, 47
Urbanization, *822*
Urea, excretion, *547*
retention, *549*
Ureter, 551
Urethra, 551
Uric acid, *547*
Urinary bladder, 551
Urine, 553-555
Urochordata, 451-*452*
Urodeles, 457-*458*
Ursus arctos, *358*
Uterus, 563

Vaccination, 536-537
Vacuole, *91*, 99
in membrane transport, 77
Vagina, 563
Vagus nerve, 612
Valves, 512-*513*
Varicose veins, 514
Vas deferens, 563
Vascular bundle, of stems, 674, *703*
Vascular cambium, 678
Vascular plants, 416-417
lower, 418-421
nutrition and transport, 656-709
Vascular systems, **506**
Vascular tissue, ***106***
plant, 413, **669**
Vasectomy, 572
Vasodilatation, 516
Vasopressin, 555, 636
Vegetables, altered states, 744-745
Vegetation, and climate, 750-752
Vegetative reproduction, **727**, **738**-*739*, 742
Vein(s), 418
mammal, 512-*513*
plant, **147**-*148*, 678
vertebrate, 510
Venae cavae, 514
Venice, Italy, *825*
Ventilation, 492
regulation, 502-*503*
Ventricle of heart, 510
Vertebrate(s), **430**, 451-465
blood flow, *511*
brain, *595*, 598
circulation, 510-512
classes, 453
fossil record, *453*
gas exchange, *492*
homologous embryos, *288*
kidney, 551-555
nervous system, 594-606
osmostic adaptations, *549*
skeleton, 620-622
societies, 659-661
Vesicles, **94**
nervous system, 589
Vessel elements, **695**
Vestigial structures, 288
Viability of seeds, 737
Villi, 483-*484*
Vincristine, *801*
Viruses, **364**-365
and evolution, 370
as carcinogens, 197
differences from cells, 365
groups infecting animals, 366
problem of, 356-371
reproduction, 366-370
Vitamins, 475
Viviparous animals, **462**
Volvox, 21, *409*
Vultures, **361**
Vulva, 563

Wallace, and theory of evolution, 280-283
Warblers, coexistence, *797*
speciation, *316*
Wasp, digger, *650*
Water, *31*-33
and green revolution, 817-819
excretion, 546
pollution/loss, 13
Water cycle, *818*
Water hyacinth, *800*
root structure, *671*
Water lily, *425*
Water pollution, 13, 690, 817
Water-soluble hormones, 633-634
Watson-Crick model, of DNA, 169
Wavelength, **144**-*145*
Whales, *356*
Wheat, *256*
White blood cell, 518-*519*, *531*
AIDS infected, *365*
White-crowned sparrow, 647-*648*
White matter, 589
Wildebeest, *657*, *754*
Wind dispersal, of seeds, 734-*735*
Wind erosion, *767*
Wind power, *822*
Wings, analogous, *287*
Wood, 678-*679*
structure and function, *697*
Woodland, temperate, 756
Woodpecker, *487*
Workers, among social insects, 336-*337*
Worms, segmented, 442-*443*

X chromosomes, 204, 257, 259-260
Xenografts, 664
Xylem, 414, **669**, 695-696
functions, 694
secondary thickening, *694*
structure, 694-696
transport, 696-701

Y chromosomes, 204, 257, 259-260
Yarrow, *306*
Yeast artificial chromosome, 208
Yellowstone National Park fire, *748*
Yolk sac, 458

Z lines, 614
Zebra, *464*
Zidovudine, 538
Zone of elongation, 671
Zoomastigina (Zooflagellates), *386*-387
Zoospores, 327
Zygomycota, 395
Zygote, 326-327, **561**